There are a number of factors that affect heat dissipation, each of which affects the final conductor ampacity. For conductors in underground electrical ducts, there are several heat sources.

**Conductor loss** due to the load current $I^2R$. These losses vary with the load current, conductor material, and conductor cross-sectional area (conductor size).

**Skin-effect heating** if the current is alternating current. The heat developed by the skin effect is based on the shape and configuration of the conductors (i.e., whether they are solid, stranded, or compact).

**Hysteresis loss** if the duct is steel or other magnetic material. These losses are dependent on the shape and magnetic properties of the electrical duct.

**Heating from other conductors in the duct.** This heating is based on the number, location, and proximity of other conductors and on the losses in the other conductors. The more conductors in the raceway, the greater the heating effect. This factor replaces the adjustment factors to the ampacity tables in 310.15(B)(3)(a).

**Mutual heating from other ducts** or cables in the vicinity. The closer the other heat sources and the more they surround the duct for which calculations are being made, the greater the heating effect. For example, in the case of the symmetrical nine-duct bank shown below the center duct will receive the most heat as a result of mutual heating.

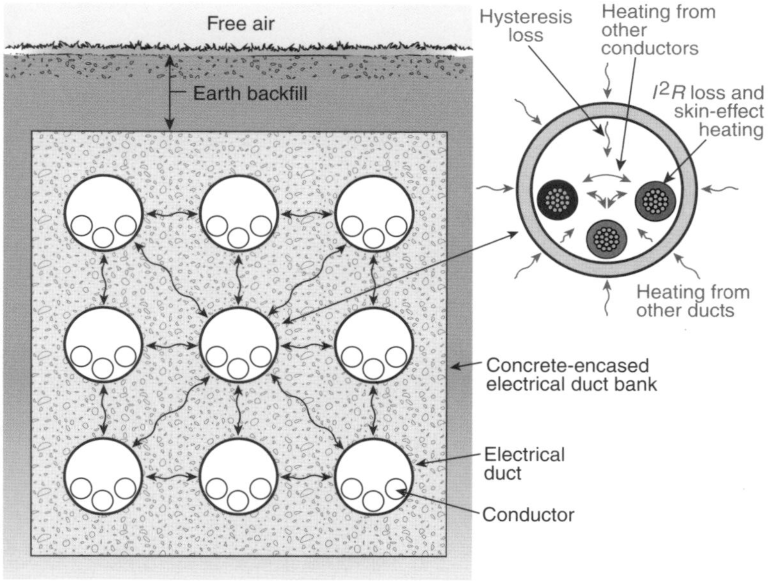


For more detailed information on the use of this method of calculation see the original paper at:

AIEE P-5-660, (*AIEE Transactions*, Part III (Power Apparatus and Systems), Vol. 76, October 1957, pp. 752–772.) *The Calculation of the Temperature Rise and Load Capability of Cable Systems*, by Neher and McGrath, [The AIEE (American Institute of Electrical Engineers) is now the Institute of Electrical and Electronics Engineers (IEEE).]

# How to Use this Book



Vertical rules indicate a change to the *Code*

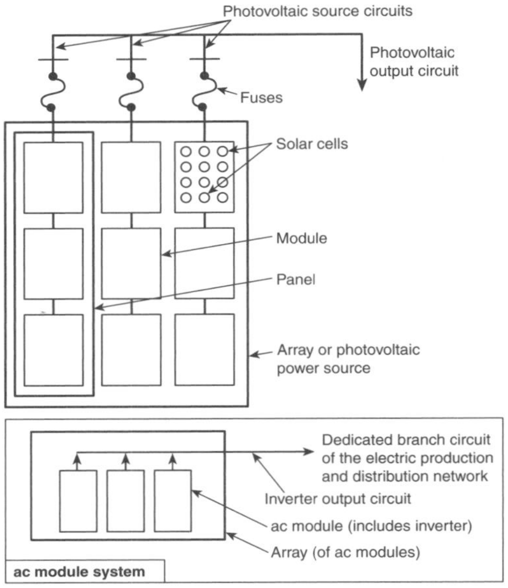


Notes:
1. These diagrams are intended to be a means of identification for photovoltaic system components, circuits, and connections.
2. Disconnecting means required by Article 690, Part III, are not shown.
3. System grounding and equipment grounding are not shown. See Article 690, Part V.

***FIGURE 690.1(a)*** *Identification of Solar Photovoltaic System Components.*

controller(s) for such systems. [See Figure 690.1(a) and Figure 690.1(b).] Solar PV systems covered by this article may be interactive with other electrical power production sources or stand-alone, with or without electrical energy storage such as batteries. These systems may have ac or dc output for utilization.

Commentary is printed in blue to distinguish it from *Code* text.

The use of photovoltaic (PV) systems as utility-interactive or stand-alone power-supply systems has steadily increased as the technology of PV equipment has evolved and its availability has improved. The requirements of Article 690 cover the use of stand-alone and utility-interactive PV systems. Utility-interactive photovoltaic systems are also subject to the requirements for interconnected electric power production sources contained in Article 705.

Exhibit 690.1 shows a typical installation of a PV array in a field.

## 690.2 Definitions

**Alternating-Current (ac) Module (Alternating-Current Photovoltaic Module).** A complete, environmentally protected unit consisting of solar cells, optics, inverter, and other components, exclusive of tracker, designed to generate ac power when exposed to sunlight.

***EXHIBIT 690.1*** *A PV array. (Courtesy of Solar Design Associates, LLC)*

An ac PV module consists of a single integrated mechanical unit. Because there is no accessible, field-installed dc wiring in this single unit, the dc PV source-circuit requirements in this *Code* are not applicable to the dc wiring in an ac PV module.

**Array.** A mechanically integrated assembly of modules or p[anels] with a support structure and foundation, tracker, and [other] components, as required, to form a direct-current power-pr[o]ducing unit.

Mandatory *Code* text is printed in black.

An array composed of multiple panels installed on a support structure is illustrated in Exhibit 690.2.

**Bipolar Photovoltaic Array.** A PV array that has two outputs, each having opposite polarity to a common reference point or center tap.

**Blocking Diode.** A diode used to block reverse flow of current into a PV source circuit.

Blocking diodes are not required by this *Code*, although the instructions or labels supplied with the PV module may require them. Blocking diodes are not overcurrent devices and may not be substituted for any overcurrent device required by the *NEC*.

**Building Integrated Photovoltaics.** Photovoltaic cells, devices, modules, or modular materials that are integrated into the ou[ter] surface or structure of a building and serve as the outer prote[ctive] surface of that building.

A bullet indicates where a portion of the *Code* has been removed.

•

**DC-to-DC Converter.** A device installed in the PV source [cir]cuit or PV output circuit that can provide an output dc volta[ge] and current at a higher or lower value than the input dc voltage and current.

**Direct-Current (dc) Combiner.** A device used in the PV source and PV output circuits to combine two or more dc circuit inputs and provide one dc circuit output.



Changes to *Code* text made in the 2014 edition are highlighted by yellow shading.

# National Electrical Code®
# Handbook

Thirteenth Edition

International Electrical Code® Series

**Mark W. Earley, P.E.**
Editor-in-Chief

**Christopher D. Coache**
Editor

**Mark Cloutier**
Editor

**Gil Moniz**
Editor

With the complete text of the 2014 edition of *NFPA 70®*, *National Electrical Code®*

NFPA®

**National Fire Protection Association®, Quincy, Massachusetts**

Product Management: Debra Rose
Development: Jennifer Harvey
Production: Jennifer Harvey, Jennifer Williams-Rapa
Permissions: Josiane Domenici
Copyediting: Cara Grady
Composition: Shepherd, Inc.
Art Coordination: Cheryl Langway
Original Illustrations: Rollin Graphics, George Nichols, J. Phillip Simmons
Revised Illustrations: George Nichols
Cover Design: Cameron Inc.
Manufacturing: Ellen Glisker
Printing and Binding: Courier/Kendallville



**Important Notices and Disclaimers:** Publication of this handbook is for the purpose of circulating information and opinion among those concerned for fire and electrical safety and related subjects. While every effort has been made to achieve a work of high quality, neither the NFPA® nor the contributors to this handbook guarantee the accuracy or completeness of or assume any liability in connection with the information and opinions contained in this handbook. The NFPA and the contributors shall in no event be liable for any personal injury, property, or other damages of any nature whatsoever, whether special, indirect, consequential, or compensatory, directly or indirectly resulting from the publication, use of, or reliance upon this handbook.

This handbook is published with the understanding that the NFPA and the contributors to this handbook are supplying information and opinion but are not attempting to render engineering or other professional services. If such services are required, the assistance of an appropriate professional should be sought.

NFPA codes, standards, recommended practices, and guides ("NFPA Documents"), including the NFPA Document that is the subject of this handbook, are made available for use subject to Important Notices and Legal Disclaimers, which appear at the end of this handbook and can also be viewed at *www.nfpa.org/disclaimers*.

**Notice Concerning Code Interpretations:** This thirteenth edition of the *National Electrical Code® Handbook* is based on the 2014 edition of *NFPA 70®*. All NFPA codes, standards, recommended practices, and guides ("NFPA Documents") are developed in accordance with the published procedures of the NFPA by technical committees comprised of volunteers drawn from a broad array of relevant interests. The handbook contains the complete text of *NFPA 70* and any applicable Formal Interpretations issued by the NFPA. This NFPA Document is accompanied by explanatory commentary and other supplementary materials.

The commentary and supplementary materials in this handbook are not a part of the NFPA Document and do not constitute Formal Interpretations of the NFPA (which can be obtained only through requests processed by the responsible technical committees in accordance with the published procedures of the NFPA). The commentary and supplementary materials, therefore, solely reflect the personal opinions of the editor or other contributors and do not necessarily represent the official position of the NFPA or its technical committees.

The following are registered trademarks of the National Fire Protection Association:

National Fire Protection Association®
NFPA®
National Electrical Code®, NEC®, and NFPA 70®
Standard for Electrical Safety in the Workplace® and NFPA 70E®
Building Construction and Safety Code® and NFPA 5000®
NFPA 72®
Life Safety Code® and 101®



NFPA No.: 70HB14
ISBN (book): 978-1-455-90544-7
ISBN (PDF): 978-1-455-90674-1
ISBN (Ebook) 978-1-455-90833-2
Library of Congress Card Control No.: 2013941415



Printed in the United States of America
13 14 15 16 17 5 4 3 2 1

# Dedication

This edition of the *National Electrical Code® Handbook* is dedicated to James T. Pauley, P.E., for his exceptional service to the *NEC®*. Jim has been a dedicated member of the National Electrical Code® Committee since the 1996 edition, representing the National Electrical Manufacturers Association (NEMA). He has been a member of Code-Making Panel 2 and the Technical Correlating Committee and has served on numerous task groups, including the Membership Task Group and the Usability Task Group, both of which are challenging and time-consuming commitments.

Jim's enthusiasm is contagious. It is said that if you want something done, give it to a busy person, and Jim is one of the busiest people we know. He has always been willing to help, and his service has benefited an entire generation of electrical inspectors, electricians, and contractors.

Jim's commitment to the *National Electrical Code®* was readily apparent when he first started attending meetings of the International Association of Electrical Inspectors (IAEI). His deep understanding of the *Code* was obvious, and he immediately became a code guru, attending chapter and section meetings all over the United States. Jim's knowledge of the NFPA Codes and Standards–making system earned him a nomination to the NFPA Standards Council, where he has served for 14 years, the last six as chair. He has a unique ability to see beyond the rhetoric to the fundamental issues at hand, an ability that has extended well beyond the electrical arena. Jim's knowledge of the process and his belief in the importance of fairness made him an ideal chair.

Before his *Code* work — in fact, before college — Jim worked as an electrician and an electrical contractor. Through his hands-on experience, he knows how products and codes and standards are used in the field — an insight that most engineers never get. He earned his B.S. in Electrical Engineering from the University of Kentucky in 1986 and began his engineering career at Square D (now Schneider Electric) prior to graduating from college. Jim has worked in a variety of roles, including product planning, codes and standards, and government relations. He is now Senior Vice President of External Affairs and Government Relations for Schneider Electric.

Jim has volunteered his expertise to a number of organizations. He currently serves as chair of the board of directors of the American National Standards Institute (ANSI), as the co-chair of the ANSI Electric Vehicle Standards Panel (EVSP), and as a member of the external advisory board of the University of Kentucky's Power and Energy Institute of Kentucky (PEIK). He also has served as a member of the Standards Board of the Institute of Electrical and Electronics Engineers (IEEE) and as Technical Advisor to the Technical Advisory Group for the International Electrotechnical Commission (IEC) Technical Committee 64 for the United States National Committee.

Jim has received several industry awards, including the Kite & Key Award from NEMA and the Meritorious Service Award from ANSI. He was one of the youngest inductees into the Gold Road Runner Club for his support of IAEI. In 2013, Jim was inducted into the University of Kentucky's College of Engineering Hall of Distinction.

The editors wish to thank Jim for his many contributions to the *NEC* and to the *NEC Handbook* over the years. We also thank his wife, Lisa, and their daughters, Taylor and Kira, for their patience when Jim's deep commitment to the quality of the *NEC* took his time and his attention. We also appreciate his family's participation in numerous industry events. Many of us in the electrical industry have known the Pauley family for years, and we've seen his daughters grow from toddlers to young women. They too are part of the NEC family. It is with deep appreciation and enduring respect that we dedicate this handbook to Jim Pauley.

# Contents

# Preface

This handbook contains the 53rd edition of the *National Electrical Code*®. Nearly 117 years have passed since those cold days in March of 1896 (a mere 17 years after the invention of the incandescent light bulb), when a group representing a variety of organizations met at the headquarters of the American Society of Mechanical Engineers in New York City to develop a national code of rules for electrical construction and operation. This was not the first attempt to establish consistent rules for electrical installations, but it was the first national effort. The need for standardization was becoming urgent; the number of electrical fires was increasing. By 1881, one insurer had reported electrical fires in 23 of the 65 insured textile mills in New England.

The major problem was the lack of an authoritative, nationwide electrical installation standard. As one of the early participants noted, "We were without standards and inspectors, while manufacturers were without experience and knowledge of real installation needs. The workmen frequently created the standards as they worked, and rarely did two men think and work alike." By 1895, five electrical installation codes were in use in the U.S. The manufacture of products that met the requirements of all five codes was difficult, so something had to be done to develop a single, national code. The committee that met in 1896 recognized that the five existing codes should be used collectively as the basis for the new code. In the first known instance of international harmonization, the group also referred to the German code, the code of the British Board of Trade, and the Phoenix Rules of England. The importance of industry consensus was immediately recognized; before the committee met again in 1897, the new code was reviewed by 1200 individuals in the U.S. and Europe. Shortly thereafter, the first standardized U.S. electrical code, the *National Electrical Code*, was published.

The *National Electrical Code* has become the most widely adopted code in the U.S. It is the installation code used in all 50 states and all U.S. territories and is now used in numerous other countries. Use of the *Code* continues to grow because it is a living document, constantly changing to reflect changes in technology. And it continues to offer an open-consensus process. Anyone can submit a proposal for change or a public comment, and all proposals and comments are subject to a rigorous public review process. The *NEC* provides the best technical information, ensuring the practical safeguarding of persons and property from the hazards arising from the use of electricity.

Throughout its history, the National Electrical Code Committee has been guided by giants in the electrical industry, too many to mention them all. The first chairman, William J. Hammer, provided the leadership necessary to get the *Code* started. More recently, the *Code* has been chaired by outstanding leaders such as Richard L. Loyd, Richard W. Osborne, Richard G. Biermann, D. Harold Ware, and James W. Carpenter. Each of these men has devoted many years to the National Electrical Code Committee. With this edition of the Code, Michael J. Johnston from the National Electrical Contractors Association began his tenure as chair of the NEC Correlating Committee. We are pleased to have in him another great leader at the helm of this the most important document in the electrical industry.

# Acknowledgments

This edition of this handbook has been more of a team effort than any previous edition. We got an early start so that we could give this book the most thorough review it has ever had. The team met frequently to discuss and debate the merits of the commentary, and we rewrote much of it in those meetings. A critical goal was to reduce "code speak" to make the book more readable. As a result, we hope you find this book easier and more helpful. These meetings also engendered some great ideas for future editions. Although the word "synergy" may be overused, the editors wish to thank Debra Rose for all she did to create the synergy of this team. When we added Jennifer Harvey from the editorial staff, the work became even better. Toward the end of the team editorial meetings, Jennifer Williams-Rapa, another very capable editor, joined the team. In addition to the editors, the technical team included Michael Fontaine. Michael's experience and his unique perspective were a valuable addition. The editors wish to thank Jeff Sargent for all of the quality work that he did on previous editions. Jeff's assignment at NFPA has changed recently, but he continues to make himself available for consultation. We value his ability to provide a "sanity check," and his ideas are always spot on.

One of the critical positions to the success of the *NEC*® is the recording secretary/project administrator. We have been blessed to have had two of the very best, Jean O'Connor and Kimberly Shea. Both are dedicated and both have amazing attention to detail. Jean was the longest serving recording secretary in the history of the *Code* when she retired in 2012. We wish Jean a long and happy retirement. Kim worked closely with Jean for several years, which ensured a smooth transition. We are pleased that Kim has taken on this new assignment, and we hope to be working together for many years.

The editors express special thanks to Jen Harvey and Jen Williams-Rapa for their long hours and extraordinary effort in attending to all of the editorial details that we technical types often overlook. We also thank Cheryl Langway and Josiane Domenici for their contributions to the design and artwork and for obtaining the necessary permissions. Special thanks are also due to Debra Rose, the project manager who kept this project on track and handled all the important logistical details. Without the efforts of Debra, Jen, Jen, Cheryl, and Josiane, this new and improved edition of the *NEC Handbook* would not have been possible. The editors also gratefully acknowledge the fine work of proofreader David March. David will have read every word in this book, including every equation and table entry. He finds the things that escape the attention of most people, and that makes us all look good.

We also wish to thank the electrical support staff, Carol Henderson and Mary Warren, for their support of this project. The editors express special appreciation to the International Association of Electrical Inspectors and IAEI CEO/executive director David Clements for their contribution of a significant number of outstanding photographs for this edition of the handbook. We have conferred closely with members of the National Electrical Code Committee in developing the revisions incorporated into the 2014 edition of the *Code*. The assistance and cooperation of code-making panel chairs and various committee members are gratefully acknowledged.

The editors also thank the manufacturers and their representatives who generously supplied photographs, drawings, and data upon request. Special thanks also to the editors of and contributors to past editions. Their work provided an excellent foundation on which to build.

**Notice to Users**

Throughout this handbook, the commentary text is printed in blue type to distinguish it from the *Code* text. Note that the commentary is not part of the *Code* and therefore is not enforceable.

# About the Editors

**Mark W. Earley, P.E.,** is Chief Electrical Engineer at NFPA. He has served as Secretary of the *NEC* since 1989, been co-author of NFPA's reference book, *Electrical Installations in Hazardous Locations*, and been editor for the *NFPA 70E® Handbook for Electrical Safety in the Workplace.* Prior to joining NFPA, he worked as an electrical engineer at Factory Mutual Research Corporation. Additionally, he has served on several of NFPA's electrical committees and *NEC* code-making panels (CMPs). Mr. Earley is a registered professional engineer (licensed in Rhode Island) and a member of International Association of Electrical Inspectors (IAEI), IEEE, the Society of Fire Protection Engineers (SFPE), the Automatic Fire Alarm Association (AFAA), the UL Electrical Council, the U.S. National Committee of the International Electrotechnical Commission, and the Canadian Electrical Code, Part 1 Committee. He is the recipient of the distinguished service award from the U.S. National Committee of the International Electrotechnical Commission and the meritorious service award from the American National Standards Institute (ANSI). He was recently awarded the honorary Artie's Apple award from the Southwestern Section of the International Association of Electrical Inspectors, in recognition of his dedication and exemplary efforts toward education.

**Christopher D. Coache** is Senior Electrical Engineer at NFPA, specializing in hazardous locations. Prior to joining NFPA, he was employed for more than 25 years as an electrical engineer and as a compliance engineer in the information technology industry. He has participated in the International Electrotechnical Commission (IEC), Underwriters Laboratories (UL), and the Instrument Society of America (ISA) standards development. He serves as the staff liaison for NFPA 73, *Standard for Electrical Inspections for Existing Dwellings,* NFPA 110, *Standard for Emergency and Standby Power Systems,* and NFPA 111, *Standard on Stored Electrical Energy Emergency and Standby Systems*. Mr. Coache is a member of International Association of Electrical Inspectors (IAEI) and IEEE.

**Mark Cloutier** is Senior Electrical Engineer at NFPA, where he supports a number of projects related to the *NEC®* along with other electrical standards. He serves as staff liaison for NFPA 79, *Electrical Standard for Industrial Machinery.* He has been involved in the electrical industry for over 30 years, working as an electrical engineer for the industrial consumer division of a major automatic test equipment company and as an electrical contractor and electrician. He holds electrician licenses in both Massachusetts and Maine and a B.S. degree in Electrical Engineering from Southeastern Massachusetts University, and is a member of IEEE and the International Association of Electrical Inspectors (IAEI). Mr. Cloutier also represents NFPA on several Underwriters Laboratories (UL) standards technical panels.

**Gil Moniz** is Senior Electrical Specialist at NFPA. Prior to joining NFPA in 2013, he served as the Northeast Field Representative for the National Electrical Manufacturers Association, an electrical inspector for the City of New Bedford, Massachusetts and a licensed Massachusetts master and journeyman electrician and Rhode Island journeyman electrician. He is a certified continuing education provider for electrical license renewal in Massachusetts, New Hampshire, and Rhode Island. He served as Chairman of Code Making Panel 1 for the 2011 and 2014 *NEC* and as a principal member of Code Making Panel 20 for the 2008 *NEC*. Mr. Moniz also served on the 2008 and 2012 New York State Residential Code Technical Subcommittees, Massachusetts Electrical Code Advisory Committee, New York City Electrical Advisory Board, New York City Electrical Code Revisions and Interpretations Committee, and as an advisor to the Rhode Island Electrical Code Subcommittee.

# 90 Introduction

## ARTICLE 90
## Introduction

### 90.1 Purpose

**(A) Practical Safeguarding.** The purpose of this *Code* is the practical safeguarding of persons and property from hazards arising from the use of electricity. This Code is not intended as a design specification or an instruction manual for untrained persons.

The *National Electrical Code®* *(NEC®)* is prepared by the National Electrical Code Committee, which consists of a Correlating Committee and 19 code-making panels. The code-making panels have specific subject responsibility within the *Code*. The scope of the National Electrical Code Committee is as follows:

> This committee shall have primary responsibility for documents on minimizing the risk of electricity as a source of electric shock and as a potential ignition source of fires and explosions. It shall also be responsible for text to minimize the propagation of fire and explosions due to electrical installations.

In addition to its overall responsibility for the *National Electrical Code*, the Correlating Committee is responsible for correlation of the following documents:

1. NFPA 70B, *Recommended Practice for Electrical Equipment Maintenance*
2. *NFPA 70E®*, *Standard for Electrical Safety in the Workplace®*
3. NFPA 73, *Standard for Electrical Inspections for Existing Dwellings*
4. NFPA 79, *Electrical Standard for Industrial Machinery*
5. NFPA 110, *Standard for Emergency and Standby Power Systems*
6. NFPA 111, *Standard on Stored Electrical Energy Emergency and Standby Power Systems*
7. NFPA 790, *Standard for Competency of Third-Party Field Evaluation Bodies*
8. NFPA 791, *Recommended Practice and Procedures for Unlabeled Electrical Equipment Evaluation*

**(B) Adequacy.** This *Code* contains provisions that are considered necessary for safety. Compliance therewith and proper maintenance results in an installation that is essentially free from hazard but not necessarily efficient, convenient, or adequate for good service or future expansion of electrical use.

> Informational Note: Hazards often occur because of overloading of wiring systems by methods or usage not in conformity with this *Code*. This occurs because initial wiring did not provide for increases in the use of electricity. An initial adequate installation and reasonable provisions for system changes provide for future increases in the use of electricity.

Consideration should always be given to future expansion of the electrical system. Future expansion might be unlikely in some occupancies, but for others it is wise to plan an initial installation – of service-entrance conductors and equipment, feeder conductors, and panelboards – that allows for future additions, alterations, or designs.

•

**(C) Relation to Other International Standards.** The requirements in this *Code* address the fundamental principles of protection for safety contained in Section 131 of International Electrotechnical Commission Standard 60364-1, *Electrical Installations of Buildings*.

> Informational Note: IEC 60364-1, Section 131, contains fundamental principles of protection for safety that encompass protection against electric shock, protection against thermal effects, protection against overcurrent, protection against fault currents, and protection against overvoltage. All of these potential hazards are addressed by the requirements in this *Code*.

In addition to being an essential part of the safety system of the Americas and the most widely adopted code for the built environment in the United States, the *NEC* is also adopted and used extensively in many other countries. The *NEC* is compatible with international safety principles, and installations meeting the requirements of the *NEC* are also in compliance with the fundamental principles outlined in IEC 60364-1, *Electrical Installations of Buildings*, Section 131. Countries that do not have formalized rules for electrical installations can adopt the *NEC* and be fully compatible with the safety principles of IEC 60364-1, Section 131.

## 90.2 Scope

**(A) Covered.** This *Code* covers the installation of electrical conductors, equipment, and raceways; signaling and communications conductors, equipment, and raceways; and optical fiber cables and raceways for the following:

(1) Public and private premises, including buildings, structures, mobile homes, recreational vehicles, and floating buildings
(2) Yards, lots, parking lots, carnivals, and industrial substations
(3) Installations of conductors and equipment that connect to the supply of electricity

Often, but not always, the source of supply of electricity is the serving electric utility. The point of connection from a premises wiring system to a serving electric utility system is, by definition, referred to as the *service point*. The conductors on the premises side of the service point are, by definition, referred to as *service conductors*. (These definitions are found in Article 100.) The requirements for service conductors as well as for service-related equipment are found in Article 230. Article 230 applies only where the source of supply of electricity is from a utility.

The source may be a stand-alone system, such as a generator, a battery system, a photovoltaic system, a fuel cell, a wind turbine, or a combination of those sources. Conductors from stand-alone systems are not service conductors, they are feeders. Service conductors are only supplied by a utility source. Where the source of supply includes a utility source(s) in combination with alternate energy sources, Article 705 also applies.

(4) Installations used by the electric utility, such as office buildings, warehouses, garages, machine shops, and recreational buildings, that are not an integral part of a generating plant, substation, or control center.

Exhibit 90.1 illustrates the distinction between electric utility facilities to which the *NEC* applies and those to which it does not apply. The electrical equipment in the generating plant is not governed by the rules of the *NEC*. The warehouse is a typical commercial facility in which the electrical installation would be governed by the rules of the *NEC*, regardless of its ownership. Office buildings and warehouses of electric utilities are functionally similar to like facilities owned by other commercial entities.

Industrial and multibuilding complexes and campus-style complexes often include substations and other installations that employ construction and wiring similar to those of electric utility installations. Because these installations are on the load side of the service point, they are within the purview of the *NEC*. At an increasing number of industrial, institutional, and other campus-style distribution systems, the service point is at an owner-maintained substation, and the conductors extending from that substation to the campus facilities are feeders (see definition in Article 100). *NEC* requirements cover these distribution systems in 225.60 and 225.61 and in Article 399. These overhead conductor and live parts clearance requirements in the *NEC* correlate with those in ANSI C2, *National Electrical Safety Code*® *(NESC)*, for overhead conductors under the control of an electric utility.

***EXHIBIT 90.1*** *The top photo is of a conventional generating plant, which is governed by the National Electrical Safety Code. The bottom photo is of a support facility that is subject to NEC requirements.*

**(B) Not Covered.** This *Code* does not cover the following:

(1) Installations in ships, watercraft other than floating buildings, railway rolling stock, aircraft, or automotive vehicles other than mobile homes and recreational vehicles

Informational Note: Although the scope of this *Code* indicates that the *Code* does not cover installations in ships, portions of this *Code* are incorporated by reference into Title 46, *Code of Federal Regulations*, Parts 110–113.

(2) Installations underground in mines and self-propelled mobile surface mining machinery and its attendant electrical trailing cable
(3) Installations of railways for generation, transformation, transmission, or distribution of power used exclusively for operation of rolling stock or installations used exclusively for signaling and communications purposes
(4) Installations of communications equipment under the exclusive control of communications utilities located

outdoors or in building spaces used exclusively for such installations

(5) Installations under the exclusive control of an electric utility where such installations

a. Consist of service drops or service laterals, and associated metering, or

b. Are on property owned or leased by the electric utility for the purpose of communications, metering, generation, control, transformation, transmission, or distribution of electric energy, or

c. Are located in legally established easements or rights-of-way, or

d. Are located by other written agreements either designated by or recognized by public service commissions, utility commissions, or other regulatory agencies having jurisdiction for such installations. These written agreements shall be limited to installations for the purpose of communications, metering, generation, control, transformation, transmission, or distribution of electric energy where legally established easements or rights-of-way cannot be obtained. These installations shall be limited to federal lands, Native American reservations through the U.S. Department of the Interior Bureau of Indian Affairs, military bases, lands controlled by port authorities and state agencies and departments, and lands owned by railroads.

Informational Note to (4) and (5): Examples of utilities may include those entities that are typically designated or recognized by governmental law or regulation by public service/utility commissions and that install, operate, and maintain electric supply (such as generation, transmission, or distribution systems) or communications systems (such as telephone, CATV, Internet, satellite, or data services). Utilities may be subject to compliance with codes and standards covering their regulated activities as adopted under governmental law or regulation. Additional information can be found through consultation with the appropriate governmental bodies, such as state regulatory commissions, the Federal Energy Regulatory Commission, and the Federal Communications Commission.

Section 90.2(B)(5) is not intended to prevent the *NEC* from being used as an installation regulatory document for these types of installations. The *NEC* is fully capable of being utilized for electrical installations in most cases. Rather, 90.2(B)(5) lists specific areas where the nature of the installation requires specialized rules or where other installation rules, standards, and guidelines have been developed for specific uses and industries. For example, the electric utility industry uses the *NESC* as its primary requirement in the generation, transmission, distribution, and metering of electric energy. See Exhibit 90.1 for examples of electric utility facilities covered or not covered by the *NEC*. In most cases, utility-owned installations are on legally established easements or rights of way. Easements or rights of way may not be available on federally owned lands, Native American reservations, military bases, lands controlled by port authorities or state agencies or departments, and lands owned by railroads. In these limited applications, a written agreement complying with this section may be used to establish the extent of the utility installation.

**(C) Special Permission.** The authority having jurisdiction for enforcing this *Code* may grant exception for the installation of conductors and equipment that are not under the exclusive control of the electric utilities and are used to connect the electric utility supply system to the service conductors of the premises served, provided such installations are outside a building or structure, or terminate inside at a readily accessible location nearest the point of entrance of the service conductors.

## 90.3 Code Arrangement

This *Code* is divided into the introduction and nine chapters, as shown in Figure 90.3. Chapters 1, 2, 3, and 4 apply generally; Chapters 5, 6, and 7 apply to special occupancies, special equipment, or other special conditions. These latter chapters supplement or modify the general rules. Chapters 1 through 4 apply except as amended by Chapters 5, 6, and 7 for the particular conditions.

Chapter 8 covers communications systems and is not subject to the requirements of Chapters 1 through 7 except where the requirements are specifically referenced in Chapter 8.

Chapter 9 consists of tables that are applicable as referenced.

Informative annexes are not part of the requirements of this *Code* but are included for informational purposes only.

An example of how the general rules of Chapter 3 are modified is 300.22, which is modified by 725.3(C) and 760.3(B) and is

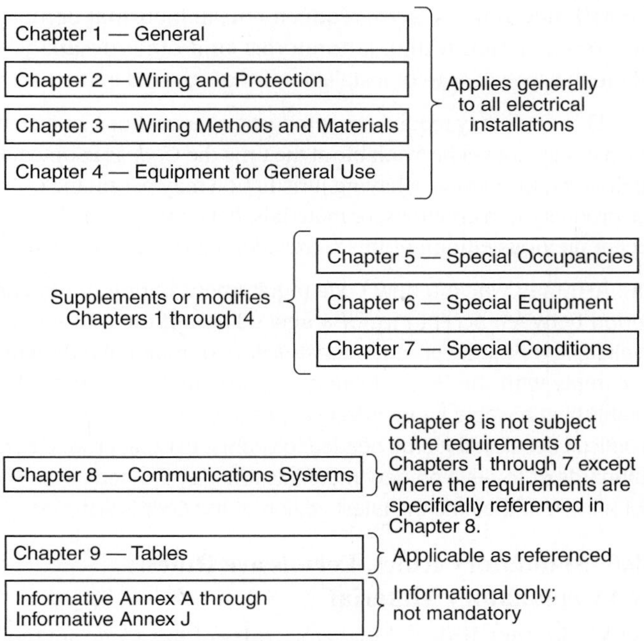


***FIGURE 90.3*** *Code Arrangement.*

specifically referenced in 800.3(B), 820.3(C), and 830.3(B). Figure 90.3 is a graphic explanation of the *NEC* arrangement.

## 90.4 Enforcement

This *Code* is intended to be suitable for mandatory application by governmental bodies that exercise legal jurisdiction over electrical installations, including signaling and communications systems, and for use by insurance inspectors. The authority having jurisdiction for enforcement of the *Code* has the responsibility for making interpretations of the rules, for deciding on the approval of equipment and materials, and for granting the special permission contemplated in a number of the rules.

Some localities do not adopt the *NEC*, but even in those localities, installations that comply with the current *Code* are prima facie evidence that the electrical installation is safe.

All materials and equipment used under the requirements of the *Code* are subject to the approval of the AHJ. Sections 90.7, 110.2, and 110.3, along with the definitions of the terms *approved*, *identified (as applied to equipment)*, *labeled*, and *listed*, are intended to provide a basis for the AHJ to make judgments about the approval of an installation.

The phrase "including signaling and communications systems" emphasizes that these systems are also subject to enforcement.

By special permission, the authority having jurisdiction may waive specific requirements in this *Code* or permit alternative methods where it is assured that equivalent objectives can be achieved by establishing and maintaining effective safety.

The AHJ is responsible for interpreting the *Code*. Using special permission (written consent), the AHJ may permit alternative methods where specific rules are not established in the *Code*. For example, the AHJ may waive specific requirements in industrial occupancies, research and testing laboratories, and other occupancies where the specific type of installation is not covered in the *Code*.

This *Code* may require new products, constructions, or materials that may not yet be available at the time the *Code* is adopted. In such event, the authority having jurisdiction may permit the use of the products, constructions, or materials that comply with the most recent previous edition of this *Code* adopted by the jurisdiction.

The AHJ may waive a new *Code* requirement during the interim period between acceptance of a new edition of the *NEC* and the availability of a new product, construction, or material redesigned to comply with the level of safety required by the latest edition. Establishing a viable future effective date in each section of the *NEC* is difficult because the time needed to change existing products and standards or to develop new materials and test methods usually is not known at the time the latest edition of the *Code* is adopted.

## 90.5 Mandatory Rules, Permissive Rules, and Explanatory Material

**(A) Mandatory Rules.** Mandatory rules of this *Code* are those that identify actions that are specifically required or prohibited and are characterized by the use of the terms *shall* or *shall not*.

**(B) Permissive Rules.** Permissive rules of this *Code* are those that identify actions that are allowed but not required, are normally used to describe options or alternative methods, and are characterized by the use of the terms *shall be permitted* or *shall not be required*.

Permissive rules are options or alternative methods of achieving equivalent safety – they are not requirements. A close reading of permissive terms is important because permissive rules are often misinterpreted. For example, the frequently used permissive term *shall be permitted* can be mistaken for a requirement. Substituting "the inspector must allow [item A or method A]" for "[item A or method A] shall be permitted" generally clarifies the interpretation.

**(C) Explanatory Material.** Explanatory material, such as references to other standards, references to related sections of this *Code*, or information related to a *Code* rule, is included in this *Code* in the form of informational notes. Such notes are informational only and are not enforceable as requirements of this *Code*.

Brackets containing section references to another NFPA document are for informational purposes only and are provided as a guide to indicate the source of the extracted text. These bracketed references immediately follow the extracted text.

A number of requirements in the *NEC* have been extracted from other NFPA codes and standards. Although *NEC* requirements based on extracted material are under the jurisdiction of the technical committee responsible for the particular document from which the material was extracted, Section 90.5(C) clarifies that the *NEC* requirements stand on their own as part of the *NEC*. The extracted material with bracketed references does not indicate that other NFPA documents are adopted through reference.

The *NEC* contains a number of informational notes. Prior to the 2011 *Code*, these notes were referred to as *fine print notes* or *FPNs*. They were renamed *informational notes* to clarify that they do not contain requirements, statements of intent, or recommendations. They present additional supplementary material that aids in the application of the requirement they follow. Because informational notes are not requirements of the *NEC*, they are not enforceable.

Footnotes to tables, although also in fine print, are not explanatory material unless they are identified as informational notes. Table footnotes are part of the tables and are necessary for proper use of the tables. Therefore, they are mandatory and enforceable *Code* text.

Additional explanatory material is located in the informative annexes. The term *informative annex* clarifies that these annexes contain additional information and do not contain recommendations or requirements.

Informational Note: The format and language used in this *Code* follows guidelines established by NFPA and published in the *NEC Style Manual*. Copies of this manual can be obtained from NFPA.

The *National Electrical Code® Style Manual* can be downloaded from the NFPA website at http://www.nfpa.org/assets/files/pdf/nec_stylemanual_2011.pdf.

**(D) Informative Annexes.** Nonmandatory information relative to the use of the *NEC* is provided in informative annexes. Informative annexes are not part of the enforceable requirements of the *NEC*, but are included for information purposes only.

## 90.6 Formal Interpretations

To promote uniformity of interpretation and application of the provisions of this *Code*, formal interpretation procedures have been established and are found in the NFPA Regulations Governing Committee Projects.

The procedures for Formal Interpretations of the provisions of the *NEC* are outlined in Section 6 of the NFPA Regulations Governing the Development of NFPA Standards (formerly the Regulations Governing Committee Projects). These regulations are included in the *NFPA Standards Directory*, which is published annually and can be downloaded from the NFPA website.

The National Electrical Code Committee cannot be responsible for subsequent actions of authorities enforcing the *NEC* that accept or reject its findings. The AHJ is responsible for interpreting *Code* rules and should attempt to resolve all disagreements at the local level. Two general forms of Formal Interpretations are recognized: (1) those that are interpretations of the literal text and (2) those that are interpretations of the intent of the Committee at the time the particular text was issued.

Interpretations of the *NEC* not subject to processing are those that involve (1) a determination of compliance of a design, installation, product, or equivalency of protection; (2) a review of plans or specifications or judgment or knowledge that can be acquired only as a result of on-site inspection; (3) text that clearly and decisively provides the requested information; or (4) subjects not previously considered by the Technical Committee or not addressed in the document. Formal Interpretations of *Code* rules are published in several venues, including *necplus*®, NFPA *News*, *NFPACodesOnline.org* (the *National Fire Codes*® subscription service), and various trade publications. They are also found on the NFPA website.

Most interpretations of the *NEC* are rendered as the personal opinions of NFPA electrical engineering staff, because most requests for interpretation do not qualify for processing as a Formal Interpretation in accordance with NFPA Regulations Governing the Development of NFPA Standards. Such opinions are rendered in writing only in response to written requests. The correspondence contains a disclaimer indicating that it is not a Formal Interpretation issued pursuant to NFPA Regulations and that any opinion expressed is the personal opinion of the author and does not necessarily represent the official position of NFPA or the National Electrical Code Committee.

## 90.7 Examination of Equipment for Safety

For specific items of equipment and materials referred to in this *Code*, examinations for safety made under standard conditions provide a basis for approval where the record is made generally available through promulgation by organizations properly equipped and qualified for experimental testing, inspections of the run of goods at factories, and service-value determination through field inspections. This avoids the necessity for repetition of examinations by different examiners, frequently with inadequate facilities for such work, and the confusion that would result from conflicting reports on the suitability of devices and materials examined for a given purpose.

It is the intent of this *Code* that factory-installed internal wiring or the construction of equipment need not be inspected at the time of installation of the equipment, except to detect alterations or damage, if the equipment has been listed by a qualified electrical testing laboratory that is recognized as having the facilities described in the preceding paragraph and that requires suitability for installation in accordance with this *Code*.

Informational Note No. 1: See requirements in 110.3.

Informational Note No. 2: *Listed* is defined in Article 100.

Informational Note No. 3: Informative Annex A contains an informative list of product safety standards for electrical equipment.

Testing laboratories, inspection agencies, and other organizations concerned with product evaluation publish lists of equipment and materials that have been tested and meet nationally recognized standards or that have been found suitable for use in a specified manner. The *Code* does not contain detailed information on equipment or materials but refers to products as *listed, labeled,* or *identified.* See Article 100 for definitions of these terms. Many certification agencies also perform field evaluations of specific installations in order to render an opinion on compliance with the *Code* requirements along with any applicable product standards. NFPA 790, *Standard for Competency of Third-Party Field Evaluation Bodies,* provides requirements for the qualification and competence of a body performing field evaluations on electrical products and assemblies with electrical components. NFPA 791, *Recommended Practice and Procedures for Unlabeled Electrical Equipment Evaluation,* covers recommended procedures for evaluating unlabeled electrical equipment for compliance with nationally recognized standards and with any requirements of the AHJ.

NFPA does not approve, inspect, or certify any installations, procedures, equipment, or materials, nor does it approve or evaluate testing laboratories. In determining the acceptability of installations or procedures, equipment, or materials, the AHJ may base acceptance on compliance with NFPA or other appropriate standards. The AHJ may also refer to the listing or labeling practices of an organization concerned with product evaluations in order to determine compliance with appropriate standards for the current production of listed items.

Informative Annex A contains a list of product safety standards used for product listing. The list includes only product safety standards for which a listing is required by the *Code*. For example, 344.6 requires that rigid metal conduit, Type RMC, be listed. By using Informative Annex A, the user finds that the listing standard for rigid metal conduit is UL 6, *Electrical Rigid Metal Conduit – Steel.* Because associated conduit fittings are required to be listed, UL 514B, *Conduit, Tubing, and Cable Fittings*, is found in Informative Annex A also.

## 90.8 Wiring Planning

**(A) Future Expansion and Convenience.** Plans and specifications that provide ample space in raceways, spare raceways, and additional spaces allow for future increases in electric power and communications circuits. Distribution centers located in readily accessible locations provide convenience and safety of operation.

The requirement for providing the exclusively dedicated equipment space mandated by 110.26(E) supports the intent of 90.8(A) regarding future increases in the use of electricity. Communications circuits are important to consider when planning for future needs. Electrical and communications distribution centers should contain additional space and capacity for future additions and should be conveniently located for easy accessibility.

If electrical and communications distribution equipment is installed where easy access cannot be achieved, a spare raceway(s) or pull line(s) should be run at the initial installation. In commercial and industrial facilities, a common practice is to purchase switchboards with additional capacity to accommodate future expansion. See Exhibit 90.2.

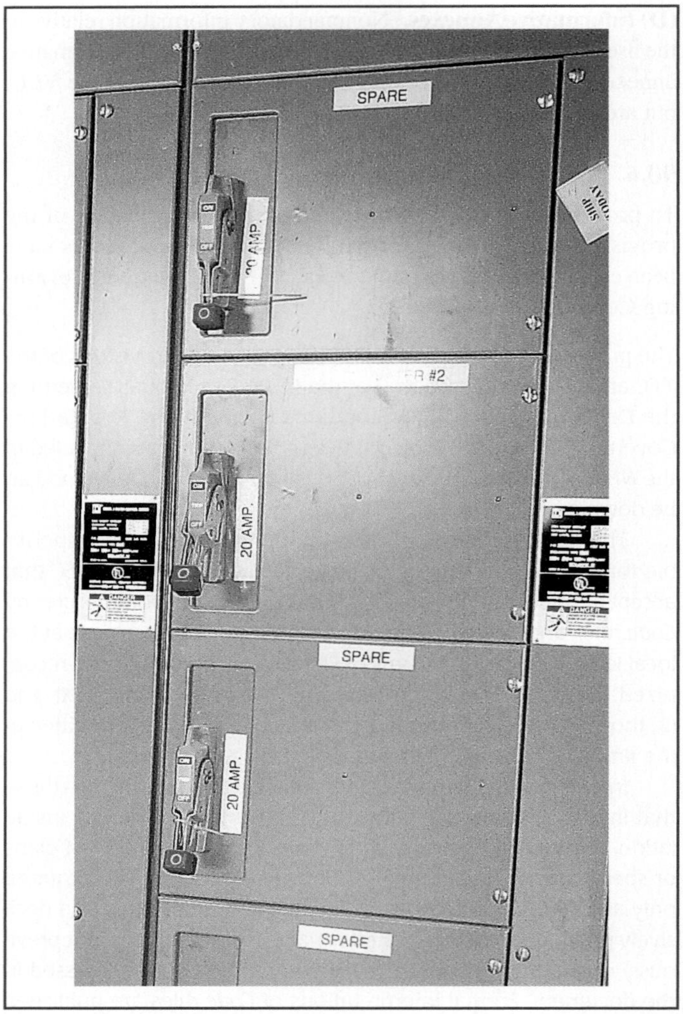


***EXHIBIT 90.2*** *A switchboard with spare circuit breakers installed to facilitate future growth. (Courtesy of the International Association of Electrical Inspectors)*

**(B) Number of Circuits in Enclosures.** It is elsewhere provided in this *Code* that the number of wires and circuits confined in a single enclosure be varyingly restricted. Limiting the number of circuits in a single enclosure minimizes the effects from a short circuit or ground fault.

## 90.9 Units of Measurement.

**(A) Measurement System of Preference.** For the purpose of this *Code*, metric units of measurement are in accordance with the modernized metric system known as the International System of Units (SI).

Most U.S. industries that do business abroad are predominantly metric already because of global sourcing of parts, service, components, and production. However, many domestic industries still use U.S. customary units (sometimes referred to as inch-pound units). Metric dimensions are beginning to appear in the domestic building construction industry because the national standards are being harmonized with international standards.

**(B) Dual System of Units.** SI units shall appear first, and inch-pound units shall immediately follow in parentheses. Conversion from inch-pound units to SI units shall be based on hard conversion except as provided in 90.9(C).

**(C) Permitted Uses of Soft Conversion.** The cases given in 90.9(C)(1) through (C)(4) shall not be required to use hard conversion and shall be permitted to use soft conversion.

**(1) Trade Sizes.** Where the actual measured size of a product is not the same as the nominal size, trade size designators shall be used rather than dimensions. Trade practices shall be followed in all cases.

In *NEC* raceway articles, metric designators (metric trade sizes) of conduits precede the trade size equivalents. For example, in 350.20(A), the size requirement is stated as follows: "LFMC smaller than metric designator 16 (trade size ½) shall not be used."

**(2) Extracted Material.** Where material is extracted from another standard, the context of the original material shall not be compromised or violated. Any editing of the extracted text shall be confined to making the style consistent with that of the *NEC*.

**(3) Industry Practice.** Where industry practice is to express units in inch-pound units, the inclusion of SI units shall not be required.

**(4) Safety.** Where a negative impact on safety would result, soft conversion shall be used.

**(D) Compliance.** Conversion from inch-pound units to SI units shall be permitted to be an approximate conversion. Compliance with the numbers shown in either the SI system or the inch-pound system shall constitute compliance with this *Code*.

Informational Note No. 1: Hard conversion is considered a change in dimensions or properties of an item into new sizes that might or might not be interchangeable with the sizes used in the original measurement. Soft conversion is considered a direct mathematical conversion and involves a change in the description of an existing measurement but not in the actual dimension.

Informational Note No. 2: SI conversions are based on IEEE/ASTM SI 10-1997, *Standard for the Use of the International System of Units (SI): The Modern Metric System.*

Soft conversions do not change the physical dimension. Hard conversions are rounded or approximate values of the original dimension. Soft conversions are required when an approximation would negatively impact safety. Commentary Table 90.1 offers some examples of the hard conversion process. The equivalent U.S. customary units show the small variance between customary units and the hard conversion units.

Warning signs that state specific clearances, such as required in 513.10(B), permit distance measurements in either inch-pound units or metric units.

**COMMENTARY TABLE 90.1** Conversions Using the Hard Conversion Method

| U.S. Customary Units | Soft Conversions to SI Units | Hard Conversions to SI Units | Equivalent U.S. Customary Units |
|---|---|---|---|
| ½ in. | 12.7 mm | 13 mm | 0.51 in. |
| ¾ in. | 19.05 mm | 19 mm | 0.75 in. |
| 1 in. | 25.4 mm | 25 mm | 0.98 in. |
| 4 in. | 102 mm | 100 mm | 3.94 in. |
| 12 in. | 305 mm | 300 mm | 11.81 in. |
| 2 ft | 610 mm | 600 mm | 1.97 ft |
| 3 ft | 914 mm | 900 mm | 2.95 ft |
| 6 ft | 1.83 m | 1.8 m | 5.91 ft |
| 15 ft | 4.57 m | 4.5 m | 14.76 ft |

# 1 General

## ARTICLE 100 Definitions

**Scope.** This article contains only those definitions essential to the proper application of this *Code*. It is not intended to include commonly defined general terms or commonly defined technical terms from related codes and standards. In general, only those terms that are used in two or more articles are defined in Article 100. Other definitions are included in the article in which they are used but may be referenced in Article 100.

Part I of this article contains definitions intended to apply wherever the terms are used throughout this *Code*. Part II contains definitions applicable only to articles and parts of articles specifically covering installations and equipment operating at over 600 volts, nominal.

Commonly defined general terms include those terms defined in general English language dictionaries and terms that are not used in a unique or restricted manner in the *NEC*. Commonly defined technical terms such as *volt* (abbreviated V) and *ampere* (abbreviated A) are found in the *Authoritative Dictionary of IEEE Standards Terms*.

Some definitions are not listed in Article 100 but are included in the *Code* article in which the term is used. For articles that follow the common format according to the *NEC Style Manual,* the section number is generally XXX.2 Definition(s). For example, the definition of *nonmetallic-sheathed cable* is found in 334.2.

## I. General

**Accessible (as applied to equipment).** Admitting close approach; not guarded by locked doors, elevation, or other effective means.

Exhibit 100.1 illustrates equipment considered accessible (as applied to equipment) in accordance with requirements elsewhere in the *Code*. The requirement for the accessibility of switches and circuit breakers used as switches is shown in (a) and is according to 404.8(A). In (b), the busway installation is according to

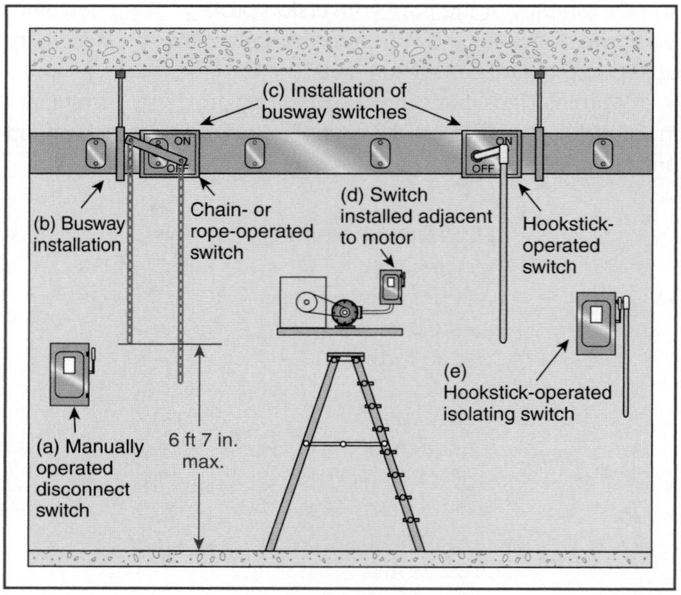


***EXHIBIT 100.1*** *An example of a busway and of switches considered accessible even if located above the 6 feet 7 inch maximum height of the switch handle specified in 404.8(A).*

368.17(C). The exceptions to 404.8(A) are illustrated as follows: in (c), busway switches installed according to Exception No. 1; in (d), a switch installed adjacent to a motor according to Exception No. 2; and in (e), a hookstick-operated isolating switch installed according to Exception No. 3.

**Accessible (as applied to wiring methods).** Capable of being removed or exposed without damaging the building structure or finish or not permanently closed in by the structure or finish of the building.

Wiring methods located behind removable panels designed to allow access are not considered permanently enclosed and are considered exposed (as applied to wiring methods). See 300.4(C) regarding cables located in spaces behind accessible panels. Exhibit 100.2 is an example of wiring methods and equipment that are considered accessible despite being located above a suspended ceiling.

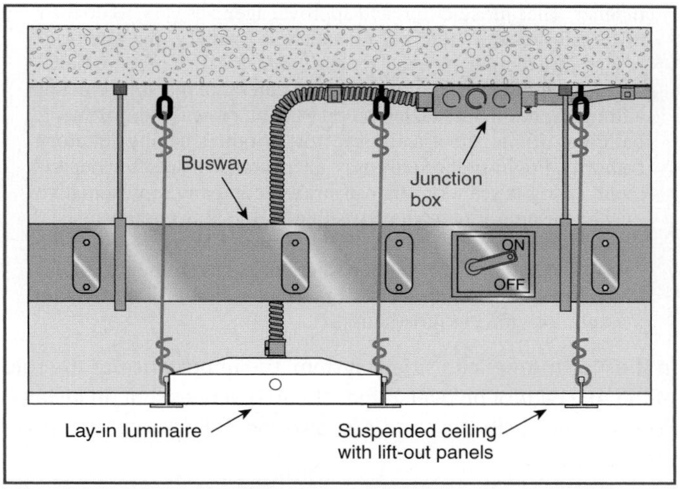


***EXHIBIT 100.2*** *An example of an accessible busway and junction box located above hung ceilings with lift-out panels.*

***EXHIBIT 100.3*** *A warning sign on a locked electrical equipment door.*

**Accessible, Readily (Readily Accessible).** Capable of being reached quickly for operation, renewal, or inspections without requiring those to whom ready access is requisite to actions such as to use tools, to climb over or remove obstacles, or to resort to portable ladders, and so forth.

The definition of *readily accessible* does not preclude using locks on service equipment doors or doors of rooms containing service equipment, provided a key or lock combination is available to those for whom ready access is necessary. If a tool is necessary to gain access, the equipment is not readily accessible. Sections 230.70(A)(1) and 230.205(A) require service-disconnecting means to be readily accessible. However, 230.205(A) permits overhead or underground primary distribution systems on services of over 1000 volts, nominal on private property to have a disconnecting means that is not readily accessible, provided there is a readily accessible means to operate the disconnecting means through a mechanical linkage or through an electronically actuated method.

Section 225.32 requires that feeder disconnecting means for separate buildings be readily accessible. A commonly used, permitted practice is to locate the disconnecting means in the electrical equipment room of an office building or large apartment building and to keep the door to that room locked to prevent access by unauthorized persons. Section 240.24(A) requires that overcurrent devices be located so as to be readily accessible. Exhibit 100.3 shows an example of a locked electrical equipment door with a warning sign that restricts access.

**Adjustable Speed Drive.** Power conversion equipment that provides a means of adjusting the speed of an electric motor.

The definitions for *adjustable speed drives* and *adjustable speed drive systems* were relocated from 430.2 for this edition.

Informational Note: A variable frequency drive is one type of electronic adjustable speed drive that controls the rotational speed of an ac electric motor by controlling the frequency and voltage of the electrical power supplied to the motor.

**Adjustable Speed Drive System.** A combination of an adjustable speed drive, its associated motor(s), and auxiliary equipment.

**Ampacity.** The maximum current, in amperes, that a conductor can carry continuously under the conditions of use without exceeding its temperature rating.

Ambient temperature is a condition of use. A conductor with insulation rated at 60°C and installed near a furnace where the ambient temperature is continuously maintained at 60°C has no current-carrying capacity. Any current flowing through the conductor will raise its temperature above the 60°C insulation rating. Therefore, the ampacity of this conductor, regardless of its size, is zero. See the ampacity correction factors for temperature in Tables 310.15(B)(2)(a) and 310.15(B)(2)(b), or see Informative Annex B. The temperature limitation of conductors is further explained and examples are given in the commentary following 310.15(B)(2). Another condition of use is the number of conductors in a raceway or cable, because the additional conductors raise the temperature in the raceway, decreasing the available ampacity. [See 310.15(B)(3)(a).]

**Appliance.** Utilization equipment, generally other than industrial, that is normally built in standardized sizes or types and is installed or connected as a unit to perform one or more functions such as clothes washing, air-conditioning, food mixing, deep frying, and so forth.

**Approved.** Acceptable to the authority having jurisdiction.

See the definition of *authority having jurisdiction* and 110.2 for a better understanding of the approval process. Typically, approval of listed equipment is more readily given by an AHJ where the authority accepts a laboratory's listing mark. Other options may be available for the jurisdiction to approve equipment, including evaluation by the inspection authority or field evaluation by a qualified laboratory or individual. Where an evaluation is conducted on site, industry standards such as NFPA 79, *Electrical Standard for Industrial Machinery,* if applicable, can be used. NFPA 790, *Standard for Competency of Third Party Field Evaluation Bodies,* can be used to qualify evaluation services. NFPA 791, *Recommended Practice and Procedures for Unlabeled Electrical Equipment Evaluation,* can be used to evaluate unlabeled equipment in accordance with nationally recognized standards and any requirements of the AHJ.

**Arc-Fault Circuit Interrupter (AFCI).** A device intended to provide protection from the effects of arc faults by recognizing characteristics unique to arcing and by functioning to de-energize the circuit when an arc fault is detected.

Arc-fault circuit interrupters are evaluated in accordance with UL 1699, *Standard for Arc-Fault Circuit-Interrupters,* using testing methods that create or simulate arcing conditions to determine a product's ability to detect and interrupt arcing faults. These devices are also tested to verify that arc detection is not inhibited by the presence of loads and circuit characteristics that mask the hazardous arcing condition. In addition, these devices are evaluated to determine resistance to unwanted tripping due to the presence of arcing that occurs in equipment under normal operating conditions or to a loading condition that closely mimics an arcing fault, such as a solid-state electronic ballast or a dimmed load.

**Askarel.** A generic term for a group of nonflammable synthetic chlorinated hydrocarbons used as electrical insulating media.

Informational Note: Askarels of various compositional types are used. Under arcing conditions, the gases produced, while consisting predominantly of noncombustible hydrogen chloride, can include varying amounts of combustible gases, depending on the askarel type.

**Attachment Plug (Plug Cap) (Plug).** A device that, by insertion in a receptacle, establishes a connection between the conductors of the attached flexible cord and the conductors connected permanently to the receptacle.

See 406.7 for requirements pertaining to attachment plugs and the configuration charts from NEMA WD 6, *Wiring Devices – Dimensional Requirements,* for general-purpose nonlocking and specific-purpose locking plugs and receptacles shown in Exhibit 406.3.

**Authority Having Jurisdiction (AHJ).** An organization, office, or individual responsible for enforcing the requirements of a code or standard, or for approving equipment, materials, an installation, or a procedure.

Informational Note: The phrase "authority having jurisdiction," or its acronym AHJ, is used in NFPA documents in a broad manner, since jurisdictions and approval agencies vary, as do their responsibilities. Where public safety is primary, the authority having jurisdiction may be a federal, state, local, or other regional department or individual such as a fire chief; fire marshal; chief of a fire prevention bureau, labor department, or health department; building official; electrical inspector; or others having statutory authority. For insurance purposes, an insurance inspection department, rating bureau, or other insurance company representative may be the authority having jurisdiction. In many circumstances, the property owner or his or her designated agent assumes the role of the authority having jurisdiction; at government installations, the commanding officer or departmental official may be the authority having jurisdiction.

In the North American safety system, the importance of the role of the AHJ cannot be overstated. The AHJ verifies that an installation complies with the *Code*. See also the definition of *approved,* 90.7, and 110.2.

**Automatic.** Performing a function without the necessity of human intervention.

**Bathroom.** An area including a basin with one or more of the following: a toilet, a urinal, a tub, a shower, a bidet, or similar plumbing fixtures.

**Battery System.** Interconnected battery subsystems consisting of one or more storage batteries and battery chargers, and can include inverters, converters, and associated electrical equipment.

This definition is suitable for use with all types of storage batteries. The system consists of all of the major components that are necessary to supply the desired output. A battery system can contain multiple batteries. Other equipment can include chargers, inverters, converters, and overcurrent protective devices.

**Bonded (Bonding).** Connected to establish electrical continuity and conductivity.

**Bonding Conductor or Jumper.** A reliable conductor to ensure the required electrical conductivity between metal parts required to be electrically connected.

Either of the two terms *bonding conductor* or *bonding jumper* may be used. The term *bonding jumper* is sometimes interpreted to mean a short conductor, although some bonding jumpers may be several feet in length. The primary purpose of a bonding conductor or jumper is to ensure electrical conductivity between two conductive bodies, such as between a box and a metal raceway. Bonding jumpers are particularly important where a box has either concentric- or eccentric-type knockouts. These knockouts can impair the electrical conductivity between metal parts and may actually introduce unnecessary impedance into the grounding path. Bonding jumpers may be found at service equipment [250.92(B)], equipment operating over 250 volts (250.97), and expansion fittings in metal raceways (250.98). Exhibit 100.4 shows the difference between concentric- and eccentric-type knockouts and illustrates one method of applying bonding jumpers at these types of knockouts.

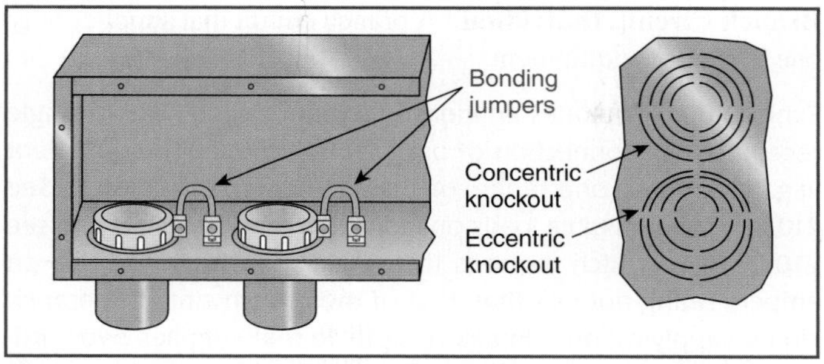


***EXHIBIT 100.4*** *Bonding jumpers installed around concentric or eccentric knockouts.*

**Bonding Jumper, Equipment.** The connection between two or more portions of the equipment grounding conductor.

Equipment bonding jumpers ensure that an effective ground-fault current path is not compromised by an interruption in mechanical or electrical continuity. For example, conduits entering an open-bottom switchboard are usually not mechanically connected to the switchboard. Bonding jumpers provide electrical continuity. An example of potential loss of both mechanical and electrical continuity would be an installation of an expansion fitting intended to allow for movement in a metal conduit system as illustrated in Exhibit 100.5. Expansion fittings consist of loosely joined raceways that allow expansion without deformation of the raceway. Some expansion fittings for metal conduit have an internal bonding jumper that is integral to the fitting, eliminating the need for the external bonding jumpers shown in Exhibit 100.5. Equipment bonding jumpers are also used to connect the grounding terminal of a receptacle to a metal box that in turn is grounded via an equipment grounding conductor (the raceway system).

***EXHIBIT 100.5*** *Equipment bonding jumpers installed to maintain electrical continuity around conduit expansion fittings. (Courtesy of the International Association of Electrical Inspectors)*

**Bonding Jumper, Main.** The connection between the grounded circuit conductor and the equipment grounding conductor at the service.

Exhibit 100.6 shows a main bonding jumper that provides the connection between the grounded service conductor and the equipment grounding conductor at the service. Bonding jumpers may be located throughout the electrical system, but a main bonding jumper is located only at the service. Main bonding jumper requirements are found in 250.28.

**Bonding Jumper, System.** The connection between the grounded circuit conductor and the supply-side bonding jumper, or the equipment grounding conductor, or both, at a separately derived system.

The system bonding jumper can be installed in several ways. For example, if a multi-barrel lug is connected to the XO terminal of a transformer, the system bonding jumper, grounding electrode conductor, grounded conductor, and bonding jumper can be connected at that connector. If a multi-barrel lug is connected to the transformer or generator enclosure, a common practice is to connect the system bonding jumper, the grounding electrode conductor, and the bonding jumper or conductor to that connector. The grounded conductor should always connect directly to the XO terminal. A system bonding jumper is used to connect the equipment grounding conductor(s) or the supply-side bonding jumper to the grounded conductor of a separately derived system

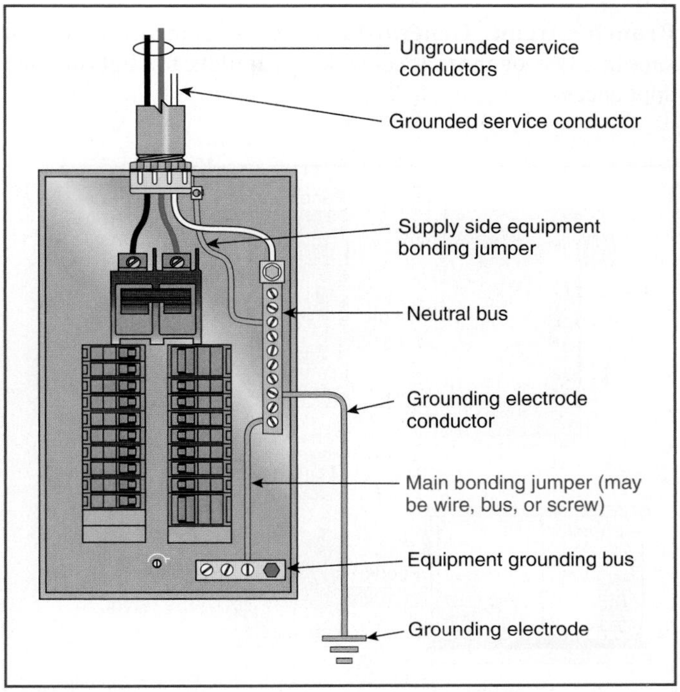


***EXHIBIT 100.6*** *A main bonding jumper installed at the service between the grounded service conductor and the equipment grounding conductor.*

either at the source (see Exhibit 250.13) or at the first system disconnecting means (see Exhibit 250.14). A system bonding jumper is used at the derived system if the derived system contains a grounded conductor.

Like the main bonding jumper at the service equipment, the system bonding jumper provides the necessary link between the equipment grounding conductors and the system grounded conductor in order to establish an effective path for ground-fault current to return to the source. The requirements for system bonding jumper(s) are found in 250.30(A)(1).

**Branch Circuit.** The circuit conductors between the final overcurrent device protecting the circuit and the outlet(s).

Exhibit 100.7 shows the difference between branch circuits and feeders. Conductors between the overcurrent devices in the panelboards and the duplex receptacles are branch-circuit conductors. Conductors between the service equipment or source of separately derived systems and the panelboards are feeders.

**Branch Circuit, Appliance.** A branch circuit that supplies energy to one or more outlets to which appliances are to be connected and that has no permanently connected luminaires that are not a part of an appliance.

Two or more 20-ampere small-appliance branch circuits are required by 210.11(C)(1) for dwelling units. Section 210.52(B)(1) requires that these circuits supply receptacle outlets located in rooms such as the kitchen and pantry. These small-appliance branch circuits are not permitted to supply other outlets or permanently connected luminaires. (See 210.52 for details.)

**Branch Circuit, General-Purpose.** A branch circuit that supplies two or more receptacles or outlets for lighting and appliances.

**Branch Circuit, Individual.** A branch circuit that supplies only one utilization equipment.

Exhibit 100.8 illustrates an individual branch circuit with a single receptacle for connection of one piece of utilization equipment (e.g., one dryer, one range, one space heater, one motor). See 210.23 for permissible loads on individual branch circuits and see 210.21(B)(1), which requires the single receptacle to have an ampere rating not less than that of the branch circuit. A branch circuit supplying one duplex receptacle that supplies two cord-and-plug-connected appliances or similar equipment is not an individual branch circuit.

**Branch Circuit, Multiwire.** A branch circuit that consists of two or more ungrounded conductors that have a voltage between them, and a grounded conductor that has equal voltage between it and each ungrounded conductor of the circuit and that is connected to the neutral or grounded conductor of the system.

See 210.4, 240.15(B)(1), and 300.13(B) for specific information about multiwire branch circuits.

**Building.** A structure that stands alone or that is cut off from adjoining structures by fire walls with all openings therein protected by approved fire doors.

A building is generally considered to be a roofed or walled structure that is intended for supporting or sheltering any use or occupancy. However, a separate structure such as a pole, billboard sign, or water tower may also be a building. Definitions of the terms *fire walls* and *fire doors* are the responsibility of building codes. Generically, a fire wall may be defined as a wall that separates buildings or subdivides a building to prevent the spread of fire and that has a fire resistance rating and structural stability. Fire doors (and fire windows) are used to protect openings in

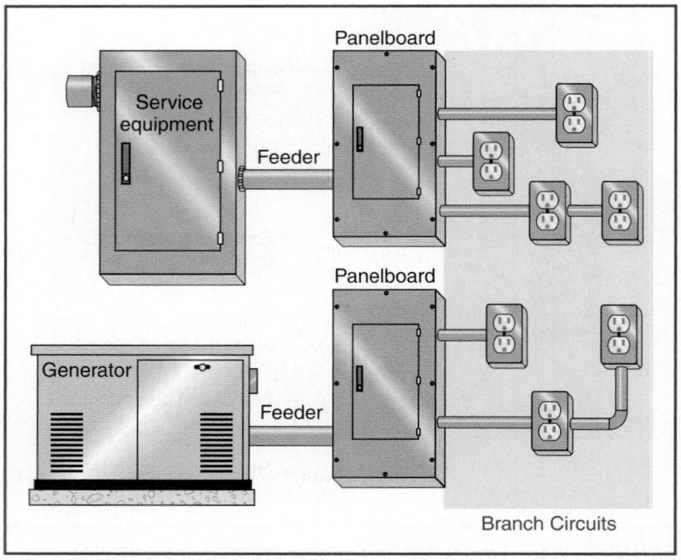


***EXHIBIT 100.7*** *Feeder (circuits) and branch circuits.*

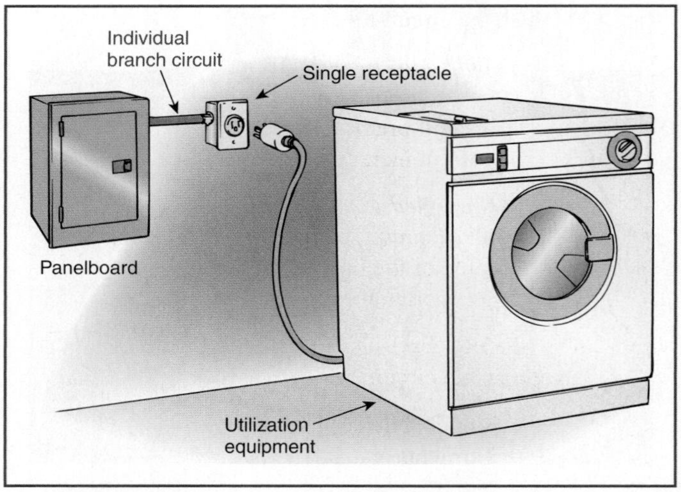


***EXHIBIT 100.8*** *An individual branch circuit, supplying only one piece of utilization equipment via a single receptacle.*

walls, floors, and ceilings against the spread of fire and smoke within, into, or out of buildings.

**Cabinet.** An enclosure that is designed for either surface mounting or flush mounting and is provided with a frame, mat, or trim in which a swinging door or doors are or can be hung.

Both cabinets and cutout boxes are covered in Article 312. Cabinets are designed for surface or flush mounting with a trim to which a swinging door(s) is hung. Cutout boxes are designed for surface mounting with a swinging door(s) secured directly to the box. Panelboards are electrical assemblies designed to be placed in a cabinet or cutout box. (See the definitions of *cutout box* and *panelboard.*)

**Cable Routing Assembly.** A single channel or connected multiple channels, as well as associated fittings, forming a structural system that is used to support and route communications wires and cables, optical fiber cables, data cables associated with information technology and communications equipment, Class 2 and Class 3 cables, and power-limited fire alarm cables.

**Charge Controller.** Equipment that controls dc voltage or dc current, or both, and that is used to charge a battery or other energy storage device.

**Circuit Breaker.** A device designed to open and close a circuit by nonautomatic means and to open the circuit automatically on a predetermined overcurrent without damage to itself when properly applied within its rating.

Informational Note: The automatic opening means can be integral, direct acting with the circuit breaker, or remote from the circuit breaker.

*Adjustable (as applied to circuit breakers).* A qualifying term indicating that the circuit breaker can be set to trip at various values of current, time, or both, within a predetermined range.

*Instantaneous Trip (as applied to circuit breakers).* A qualifying term indicating that no delay is purposely introduced in the tripping action of the circuit breaker.

*Inverse Time (as applied to circuit breakers).* A qualifying term indicating that there is purposely introduced a delay in the tripping action of the circuit breaker, which delay decreases as the magnitude of the current increases.

*Nonadjustable (as applied to circuit breakers).* A qualifying term indicating that the circuit breaker does not have any adjustment to alter the value of the current at which it will trip or the time required for its operation.

*Setting (of circuit breakers).* The value of current, time, or both, at which an adjustable circuit breaker is set to trip.

**Clothes Closet.** A nonhabitable room or space intended primarily for storage of garments and apparel.

This definition helps to determine whether the rules of 240.24(D), 410.16, and 550.11(A) apply to an installation. If the definition does not apply, the area may be classified as something other than a clothes closet, such as a bedroom. Other requirements may then be applied, such as 210.52 and 210.70.

**Communications Equipment.** The electronic equipment that performs the telecommunications operations for the transmission of audio, video, and data, and includes power equipment (e.g., dc converters, inverters, and batteries), technical support equipment (e.g., computers), and conductors dedicated solely to the operation of the equipment.

This definition indicates that communications equipment includes related power supplies and computers. These related items are subject to the same requirements that apply to communications equipment. This definition correlates with NFPA 76, *Standard for the Fire Protection of Telecommunications Facilities.*

**Communications Raceway.** An enclosed channel of nonmetallic materials designed expressly for holding communications wires and cables, typically communications wires and cables and optical fiber and data (Class 2 and Class 3) in plenum, riser, and general-purpose applications.

**Concealed.** Rendered inaccessible by the structure or finish of the building.

Raceways and cables supported or located within hollow frames or permanently closed in by the finish of buildings are considered concealed. Open-type work — such as raceways and cables in unfinished basements; in accessible underfloor areas or attics; or behind, above, or below panels designed to allow access; and that may be removed without damage to the building structure or finish — is not considered concealed. [See definition of *exposed (as applied to wiring methods).*]

Informational Note: Wires in concealed raceways are considered concealed, even though they may become accessible by withdrawing them.

**Conductor, Bare.** A conductor having no covering or electrical insulation whatsoever.

**Conductor, Covered.** A conductor encased within material of composition or thickness that is not recognized by this *Code* as electrical insulation.

The green-covered equipment grounding conductors contained within a nonmetallic-sheathed cable and the uninsulated grounded system conductors within the overall exterior jacket of a Type SE cable are examples of covered conductors. Covered conductors should always be treated as bare conductors for working clearances because the covering does not have an insulation rating, so the conductors are effectively uninsulated. See the definition of *insulated conductor* that follows and the requirements in Article 310.

**Conductor, Insulated.** A conductor encased within material of composition and thickness that is recognized by this *Code* as electrical insulation.

For the covering on a conductor to be considered insulation, it is generally required to pass minimum testing required by a product standard. One such product standard is UL 83, *Thermoplastic-Insulated Wires and Cables*. To meet the requirements of UL 83, specimens of finished single-conductor wires must pass specified tests that measure (1) resistance to flame propagation, (2) dielectric strength, even while immersed, and (3) resistance to abrasion, cracking, crushing, and impact. Only wires and cables that meet the minimum fire, electrical, and physical properties required by the applicable product standards are permitted to be marked with the letter designations found in Table 310.104(A). Unless a voltage rating is marked on the insulation, a conductor should be considered a covered conductor. See 310.104 for the requirements of insulated conductor construction and applications.

**Conduit Body.** A separate portion of a conduit or tubing system that provides access through a removable cover(s) to the interior of the system at a junction of two or more sections of the system or at a terminal point of the system.

Boxes such as FS and FD or larger cast or sheet metal boxes are not classified as conduit bodies.

Conduit bodies include the short-radius type as well as capped elbows and service-entrance elbows. Some conduit bodies are referred to in the trade as "condulets" and include the LB, LL, LR, C, T, and X designs. A typical conduit body is shown in Exhibit 100.9. See 300.15 and Article 314 for rules on the usage of conduit bodies. Type FS and Type FD boxes, which are not classified as conduit bodies, are listed with boxes in Table 314.16(A).

**Connector, Pressure (Solderless).** A device that establishes a connection between two or more conductors or between one or more conductors and a terminal by means of mechanical pressure and without the use of solder.

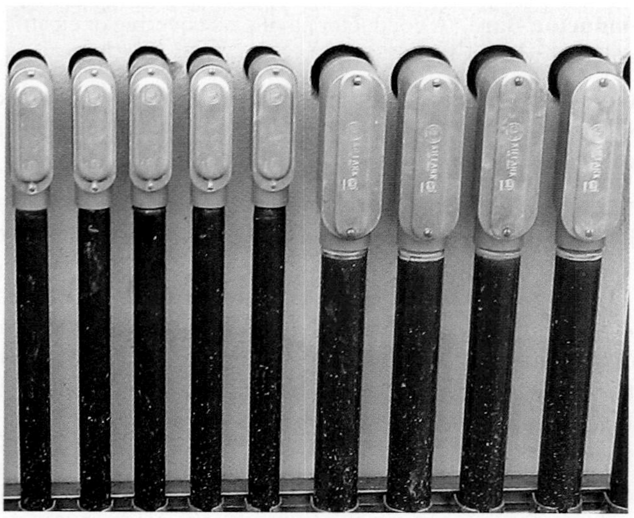

***EXHIBIT 100.9*** *Typical conduit bodies. (Courtesy of the National Electrical Contractors Association)*

**Continuous Load.** A load where the maximum current is expected to continue for 3 hours or more.

**Control Circuit.** The circuit of a control apparatus or system that carries the electric signals directing the performance of the controller but does not carry the main power current.

**Controller.** A device or group of devices that serves to govern, in some predetermined manner, the electric power delivered to the apparatus to which it is connected.

A controller may be a remote-controlled magnetic contactor, switch, circuit breaker, or other device that is normally used to start and stop motors and other apparatus and, in the case of motors, is required by 110.9 and 430.82 to be capable of interrupting the stalled rotor current of the motor. Stop-and-start stations and similar control circuit components that do not open the power conductors to the motor are not considered controllers.

**Cooking Unit, Counter-Mounted.** A cooking appliance designed for mounting in or on a counter and consisting of one or more heating elements, internal wiring, and built-in or mountable controls.

**Coordination (Selective).** Localization of an overcurrent condition to restrict outages to the circuit or equipment affected, accomplished by the selection and installation of overcurrent protective devices and their ratings or settings for the full range of available overcurrents, from overload to the maximum available fault current, and for the full range of overcurrent protective device opening times associated with those overcurrents.

Fuses and circuit breakers have time/current characteristics that determine the time it takes to clear the fault for a given value of fault current. Selectivity occurs when the device closest to the fault opens before the next device upstream operates. Any fault on a branch circuit should open the branch-circuit breaker rather than the feeder overcurrent protection. All faults on a feeder should open the feeder overcurrent protection rather than the service overcurrent protection.

With coordinated overcurrent protection, the faulted or overloaded circuit is isolated by the selective operation of only the overcurrent protective device closest to the overcurrent condition. The main goal of selective coordination is to isolate the faulted portion of the electrical circuit quickly while at the same time maintaining power to the remainder of the electrical system. The electrical system overcurrent protection must guard against short circuits and ground faults to ensure that the resulting damage is minimized while other parts of the system not directly involved with the fault are kept operational until other protective devices clear the fault. Where a series-rated system is used, an upstream device in the series will operate to protect a downstream device. For example, a current-limiting fuse will limit the available fault current to the downstream circuit breaker.

Selective coordination requirements include emergency systems, legally required standby systems, and critical operations power systems in 700.28, 701.27, and 708.54, respectively.

Requirements for selective coordination for elevator feeders are in 620.62. In 517.30(G), coordination is required only for faults that exceed 0.1 second in duration.

**Copper-Clad Aluminum Conductors.** Conductors drawn from a copper-clad aluminum rod, with the copper metallurgically bonded to an aluminum core, where the copper forms a minimum of 10 percent of the cross-sectional area of a solid conductor or each strand of a stranded conductor.

**Cutout Box.** An enclosure designed for surface mounting that has swinging doors or covers secured directly to and telescoping with the walls of the box proper.

**Dead Front.** Without live parts exposed to a person on the operating side of the equipment.

**Demand Factor.** The ratio of the maximum demand of a system, or part of a system, to the total connected load of a system or the part of the system under consideration.

**Device.** A unit of an electrical system, other than a conductor, that carries or controls electric energy as its principal function.

Switches, circuit breakers, fuseholders, receptacles, attachment plugs, and lampholders that distribute or control but do not consume electrical energy are considered devices. Devices that consume incidental amounts of electrical energy in the performance of carrying or controlling electricity — such as a switch with an internal pilot light, a GFCI receptacle, or a magnetic contactor — are also considered devices. Although conductors are units of the electrical system, they are not devices.

**Disconnecting Means.** A device, or group of devices, or other means by which the conductors of a circuit can be disconnected from their source of supply.

See also references for *disconnecting means* in the index.

**Dusttight.** Constructed so that dust will not enter the enclosing case under specified test conditions.

Requirements for enclosures are found in 110.28, and Table 110.28, Enclosure Selection, provides a basis for selecting enclosure types that are dusttight. (See also the commentary following the definition of *enclosure*.) Some dusttight constructions are only for wind-blown dust and may not be designed for combustible dusts found in Class II hazardous locations. The basic standard used to investigate dusttight enclosures for Class II, Division 2 locations is UL 1604, *Electrical Equipment for Use in Class I and II, Division 2 and Class III Hazardous (Classified) Locations*.

**Duty, Continuous.** Operation at a substantially constant load for an indefinitely long time.

**Duty, Intermittent.** Operation for alternate intervals of (1) load and no load; or (2) load and rest; or (3) load, no load, and rest.

**Duty, Periodic.** Intermittent operation in which the load conditions are regularly recurrent.

**Duty, Short-Time.** Operation at a substantially constant load for a short and definite, specified time.

**Duty, Varying.** Operation at loads, and for intervals of time, both of which may be subject to wide variation.

**Dwelling, One-Family.** A building that consists solely of one dwelling unit.

**Dwelling, Two-Family.** A building that consists solely of two dwelling units.

**Dwelling, Multifamily.** A building that contains three or more dwelling units.

**Dwelling Unit.** A single unit, providing complete and independent living facilities for one or more persons, including permanent provisions for living, sleeping, cooking, and sanitation.

Where dwelling units are referenced throughout the *Code*, it is important to note that some rooms in motels, hotels, and similar occupancies can be classified as dwelling units if they satisfy the requirements of the definition. Exhibit 100.10 illustrates a motel or hotel room that clearly meets the definition because it has permanent provisions for living, sleeping, cooking, and sanitation. See also the definition of *mobile home* in 550.2.

**Effective Ground-Fault Current Path.** An intentionally constructed, low-impedance electrically conductive path designed and intended to carry current under ground-fault conditions from the point of a ground fault on a wiring system to the electrical supply source and that facilitates the operation of the overcurrent protective device or ground-fault detectors.

**Electric Power Production and Distribution Network.** Power production, distribution, and utilization equipment and facilities, such as electric utility systems that deliver electric power to the connected loads, that are external to and not controlled by an interactive system.

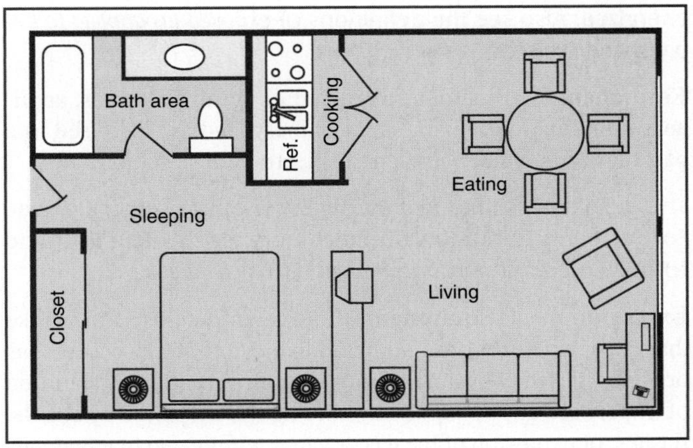


***EXHIBIT 100.10*** *An example of a motel or hotel room considered to be a dwelling unit.*

**Electric Sign.** A fixed, stationary, or portable self-contained, electrically illuminated utilization equipment with words or symbols designed to convey information or attract attention.

**Electric-Discharge Lighting.** Systems of illumination utilizing fluorescent lamps, high-intensity discharge (HID) lamps, or neon tubing.

**Electronically Actuated Fuse.** An overcurrent protective device that generally consists of a control module that provides current-sensing, electronically derived time–current characteristics, energy to initiate tripping, and an interrupting module that interrupts current when an overcurrent occurs. Such fuses may or may not operate in a current-limiting fashion, depending on the type of control selected.

**Enclosed.** Surrounded by a case, housing, fence, or wall(s) that prevents persons from accidentally contacting energized parts.

**Enclosure.** The case or housing of apparatus, or the fence or walls surrounding an installation to prevent personnel from accidentally contacting energized parts or to protect the equipment from physical damage.

Informational Note: See Table 110.28 for examples of enclosure types.

Enclosures are required by 110.28 to be marked with a number that identifies the environmental conditions in which that type of enclosure can be used. Enclosures that comply with the requirements for more than one type of enclosure are marked with multiple designations. See the commentary following 110.28 for details on enclosure markings and types as well as Table 110.28, which lists the types of enclosures required to be used in specific locations.

**Energized.** Electrically connected to, or is, a source of voltage.

The term *energized* is not limited to equipment that is "connected to a source of voltage." Equipment such as batteries, capacitors, and conductors with induced voltages must also be considered energized. Also see the definitions of *exposed (as applied to live parts)* and *live parts*.

**Equipment.** A general term, including fittings, devices, appliances, luminaires, apparatus, machinery, and the like used as a part of, or in connection with, an electrical installation.

This definition clarifies that machinery is considered equipment. For further information on machinery, see Article 670 and NFPA 79, *Electrical Standard for Industrial Machinery*.

**Explosionproof Equipment.** Equipment enclosed in a case that is capable of withstanding an explosion of a specified gas or vapor that may occur within it and of preventing the ignition of a specified gas or vapor surrounding the enclosure by sparks, flashes, or explosion of the gas or vapor within, and that operates at such an external temperature that a surrounding flammable atmosphere will not be ignited thereby.

Informational Note: For further information, see ANSI/UL 1203-2009, *Explosion-Proof and Dust-Ignition-Proof Electrical Equipment for Use in Hazardous (Classified) Locations.*

**Exposed (as applied to live parts).** Capable of being inadvertently touched or approached nearer than a safe distance by a person.

The installation instructions should be consulted to ensure that live parts are properly guarded. See 110.27 for the requirements for guarding live parts. Also see the definitions of *energized* and *live parts*.

Informational Note: This term applies to parts that are not suitably guarded, isolated, or insulated.

**Exposed (as applied to wiring methods).** On or attached to the surface or behind panels designed to allow access.

See Exhibit 100.2, which illustrates wiring methods that would be considered exposed because they are located above a suspended ceiling with lift-out panels.

**Externally Operable.** Capable of being operated without exposing the operator to contact with live parts.

**Feeder.** All circuit conductors between the service equipment, the source of a separately derived system, or other power supply source and the final branch-circuit overcurrent device.

See the commentary following the definition of *branch circuit*, including Exhibit 100.7, which illustrates the difference between branch circuits and feeders.

**Festoon Lighting.** A string of outdoor lights that is suspended between two points.

**Fitting.** An accessory such as a locknut, bushing, or other part of a wiring system that is intended primarily to perform a mechanical rather than an electrical function.

Condulets, conduit couplings, EMT connectors and couplings, and threadless connectors are considered fittings.

**Garage.** A building or portion of a building in which one or more self-propelled vehicles can be kept for use, sale, storage, rental, repair, exhibition, or demonstration purposes.

Informational Note: For commercial garages, repair and storage, see Article 511.

**Ground.** The earth.

**Ground Fault.** An unintentional, electrically conductive connection between an ungrounded conductor of an electrical circuit and the normally non–current-carrying conductors, metallic enclosures, metallic raceways, metallic equipment, or earth.

**Grounded (Grounding).** Connected (connecting) to ground or to a conductive body that extends the ground connection.

**Grounded, Solidly.** Connected to ground without inserting any resistor or impedance device.

**Grounded Conductor.** A system or circuit conductor that is intentionally grounded.

**Ground-Fault Circuit Interrupter (GFCI).** A device intended for the protection of personnel that functions to de-energize a circuit or portion thereof within an established period of time when a current to ground exceeds the values established for a Class A device.

Informational Note: Class A ground-fault circuit interrupters trip when the current to ground is 6 mA or higher and do not trip when the current to ground is less than 4 mA. For further information, see UL 943, *Standard for Ground-Fault Circuit Interrupters.*

The commentary following 210.8 contains a list of applicable cross-references for GFCIs. See Exhibits 210.6 through 210.8 for further information on GFCIs.

**Ground-Fault Current Path.** An electrically conductive path from the point of a ground fault on a wiring system through normally non–current-carrying conductors, equipment, or the earth to the electrical supply source.

Informational Note: Examples of ground-fault current paths are any combination of equipment grounding conductors, metallic raceways, metallic cable sheaths, electrical equipment, and any other electrically conductive material such as metal, water, and gas piping; steel framing members; stucco mesh; metal ducting; reinforcing steel; shields of communications cables; and the earth itself.

**Ground-Fault Protection of Equipment.** A system intended to provide protection of equipment from damaging line-to-ground fault currents by operating to cause a disconnecting means to open all ungrounded conductors of the faulted circuit. This protection is provided at current levels less than those required to protect conductors from damage through the operation of a supply circuit overcurrent device.

See the commentary following 230.95, 426.28, and 427.22.

**Grounding Conductor, Equipment (EGC).** The conductive path(s) that provides a ground-fault current path and connects normally non–current-carrying metal parts of equipment together and to the system grounded conductor or to the grounding electrode conductor, or both.

Informational Note No. 1: It is recognized that the equipment grounding conductor also performs bonding.

Informational Note No. 2: See 250.118 for a list of acceptable equipment grounding conductors.

**Grounding Electrode.** A conducting object through which a direct connection to earth is established.

**Grounding Electrode Conductor.** A conductor used to connect the system grounded conductor or the equipment to a grounding electrode or to a point on the grounding electrode system.

Grounding electrode conductors are covered extensively in Article 250, Part III. These conductors are required by 250.62 to be copper, aluminum, or copper-clad aluminum and are required to be sized according to 250.66 and Table 250.66. Exhibit 100.6 and Exhibit 250.1 show a grounding electrode conductor in a typical grounding system for a single-phase, 3-wire service.

**Guarded.** Covered, shielded, fenced, enclosed, or otherwise protected by means of suitable covers, casings, barriers, rails, screens, mats, or platforms to remove the likelihood of approach or contact by persons or objects to a point of danger.

**Guest Room.** An accommodation combining living, sleeping, sanitary, and storage facilities within a compartment.

**Guest Suite.** An accommodation with two or more contiguous rooms comprising a compartment, with or without doors between such rooms, that provides living, sleeping, sanitary, and storage facilities.

Some requirements for guest rooms in hotels, motels, and similar occupancies are found in 210.60. If a guest room or guest suite meets the definition of a dwelling unit, additional dwelling unit requirements may apply.

**Handhole Enclosure.** An enclosure for use in underground systems, provided with an open or closed bottom, and sized to allow personnel to reach into, but not enter, for the purpose of installing, operating, or maintaining equipment or wiring or both.

Exhibit 100.11 shows the installation of one type of handhole enclosure. Handhole enclosures are required by 314.30 to be "identified" for use in underground systems.

**Hermetic Refrigerant Motor-Compressor.** A combination consisting of a compressor and motor, both of which are enclosed in the same housing, with no external shaft or shaft seals, with the motor operating in the refrigerant.

***EXHIBIT 100.11*** *An example of a handhole enclosure. (Courtesy of Quazite/Hubbell Lenoir City, Inc.)*

**Hoistway.** Any shaftway, hatchway, well hole, or other vertical opening or space in which an elevator or dumbwaiter is designed to operate.

**Hybrid System.** A system comprised of multiple power sources. These power sources could include photovoltaic, wind, micro-hydro generators, engine-driven generators, and others, but do not include electric power production and distribution network systems. Energy storage systems such as batteries, flywheels, or superconducting magnetic storage equipment do not constitute a power source for the purpose of this definition. The energy regenerated by an overhauling (descending) elevator does not constitute a power source for the purpose of this definition.

**Identified (as applied to equipment).** Recognizable as suitable for the specific purpose, function, use, environment, application, and so forth, where described in a particular *Code* requirement.

> Informational Note: Some examples of ways to determine suitability of equipment for a specific purpose, environment, or application include investigations by a qualified testing laboratory (listing and labeling), an inspection agency, or other organizations concerned with product evaluation.

**In Sight From (Within Sight From, Within Sight).** Where this *Code* specifies that one equipment shall be "in sight from," "within sight from," or "within sight of," and so forth, another equipment, the specified equipment is to be visible and not more than 15 m (50 ft) distant from the other.

Several requirements are in the *Code* for a disconnecting means to be in sight from the equipment that it controls. For example, 430.102 requires the disconnecting means to be in sight from the controller. Of the three exceptions, two permit the disconnect to be located elsewhere. Both of these exceptions require that the disconnect be capable of being locked open. Exhibit 430.18 depicts requirements for the placement of a disconnecting means that is not in sight of a motor.

**Industrial Control Panel.** An assembly of two or more components consisting of one of the following: (1) power circuit components only, such as motor controllers, overload relays, fused disconnect switches, and circuit breakers; (2) control circuit components only, such as push buttons, pilot lights, selector switches, timers, switches, and control relays; (3) a combination of power and control circuit components. These components, with associated wiring and terminals, are mounted on, or contained within, an enclosure or mounted on a subpanel.

The industrial control panel does not include the controlled equipment.

**Interactive System.** An electric power production system that is operating in parallel with and capable of delivering energy to an electric primary source supply system.

**Interrupting Rating.** The highest current at rated voltage that a device is identified to interrupt under standard test conditions.

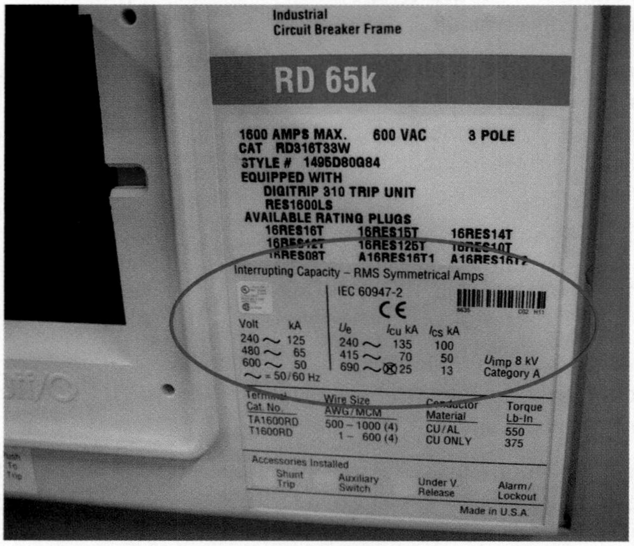


***EXHIBIT 100.12*** *Interrupting ratings information on the label of a 1600-ampere frame circuit breaker.*

> Informational Note: Equipment intended to interrupt current at other than fault levels may have its interrupting rating implied in other ratings, such as horsepower or locked rotor current.

Interrupting ratings are essential for the coordination of electrical systems. Sections specifically dealing with interrupting ratings are 110.9, 240.60(C), 240.83(C), and 240.86. Exhibit 100.12 depicts the label of a 1600-ampere frame circuit breaker, showing the interrupting capacity ratings. See also the definition of *short-circuit current rating*.

**Intersystem Bonding Termination.** A device that provides a means for connecting intersystem bonding conductors for communications systems to the grounding electrode system.

An intersystem bonding termination is a dedicated location for terminating the equipment grounding conductors required in Chapter 8 and 770.93 to the service, building, or structure's grounding electrode system. This important safety measure prevents voltage differences between the equipment grounding conductors of the communication system and the power system. Frequently in new construction, the grounding electrode, the raceway, and the grounding electrode conductor are hidden behind walls and are not accessible for bonding connection.

**Isolated (as applied to location).** Not readily accessible to persons unless special means for access are used.

**Kitchen.** An area with a sink and permanent provisions for food preparation and cooking.

**Labeled.** Equipment or materials to which has been attached a label, symbol, or other identifying mark of an organization that is acceptable to the authority having jurisdiction and concerned with product evaluation, that maintains periodic inspection of production of labeled equipment or materials, and by whose labeling the manufacturer indicates compliance with appropriate standards or performance in a specified manner.

Equipment and conductors required or permitted by this *Code* are acceptable only if they have been approved for a specific environment or application by the AHJ as stated in 110.2. See 90.7 regarding the examination of equipment for safety. Listing or labeling by a qualified testing laboratory provides a basis for approval.

**Lighting Outlet.** An outlet intended for the direct connection of a lampholder or luminaire.

**Lighting Track (Track Lighting).** A manufactured assembly designed to support and energize luminaires that are capable of being readily repositioned on the track. Its length can be altered by the addition or subtraction of sections of track.

**Listed.** Equipment, materials, or services included in a list published by an organization that is acceptable to the authority having jurisdiction and concerned with evaluation of products or services, that maintains periodic inspection of production of listed equipment or materials or periodic evaluation of services, and whose listing states that either the equipment, material, or service meets appropriate designated standards or has been tested and found suitable for a specified purpose.

Informational Note: The means for identifying listed equipment may vary for each organization concerned with product evaluation, some of which do not recognize equipment as listed unless it is also labeled. Use of the system employed by the listing organization allows the authority having jurisdiction to identify a listed product.

See also the definitions of *approved, authority having jurisdiction (AHJ), identified (as applied to equipment),* and *labeled,* which all have distinctly different meanings.

**Live Parts.** Energized conductive components.

The definition of *live parts* is associated with all voltage levels, not just voltage levels that present a shock hazard. See the definitions of *energized* and *exposed.*

**Location, Damp.** Locations protected from weather and not subject to saturation with water or other liquids but subject to moderate degrees of moisture.

Informational Note: Examples of such locations include partially protected locations under canopies, marquees, roofed open porches, and like locations, and interior locations subject to moderate degrees of moisture, such as some basements, some barns, and some cold-storage warehouses.

**Location, Dry.** A location not normally subject to dampness or wetness. A location classified as dry may be temporarily subject to dampness or wetness, as in the case of a building under construction.

**Location, Wet.** Installations underground or in concrete slabs or masonry in direct contact with the earth; in locations subject to saturation with water or other liquids, such as vehicle washing areas; and in unprotected locations exposed to weather.

The inside of a raceway in a wet location and a raceway installed underground are considered wet locations. Therefore, any conductors contained therein would be required to be suitable for wet locations.

A general requirement for protection against corrosion and deterioration is found in 300.6(D), which also identifies numerous examples of wet locations. The requirements for luminaires installed in wet locations are found in 410.10(A). See also *patient care area* in 517.2 for a definition of *wet procedure locations* in a patient care area. Suitable enclosures for wet locations can be found in Table 110.28.

**Luminaire.** A complete lighting unit consisting of a light source such as a lamp or lamps, together with the parts designed to position the light source and connect it to the power supply. It may also include parts to protect the light source or the ballast or to distribute the light. A lampholder itself is not a luminaire.

Lighting techniques such as light pipes and glass fiber optics are sometimes referred to as "lighting systems." The definition of *luminaire* includes such systems, because light pipes and fiber optics are actually "parts designed to distribute the light."

*Luminaire* is the term specified by IESNA, the ANSI/UL safety standards, and the ANSI/NEMA performance standards for lighting products previously referred to as "light fixtures" in the United States. *Luminaire* is also the term used in IEC standards and accepted globally.

**Motor Control Center.** An assembly of one or more enclosed sections having a common power bus and principally containing motor control units.

**Multioutlet Assembly.** A type of surface, flush, or freestanding raceway designed to hold conductors and receptacles, assembled in the field or at the factory.

The definition of *multioutlet assembly* includes a reference to a freestanding assembly with multiple outlets, commonly called a power pole. In dry locations, metallic and nonmetallic multioutlet assemblies are permitted; however, they are not permitted to be installed if concealed. See Article 380 for details on recessing multioutlet assemblies. Exhibit 100.13 shows a multioutlet assembly used for countertop appliances. Portable assemblies, often called "power strips" or "plug strips," are not multioutlet assemblies but are "relocatable power taps."

**Neutral Conductor.** The conductor connected to the neutral point of a system that is intended to carry current under normal conditions.

It is important to remember that the neutral conductor is a current-carrying conductor and is not safe to work on while energized.

**Neutral Point.** The common point on a wye-connection in a polyphase system or midpoint on a single-phase, 3-wire system,

***EXHIBIT 100.13*** *A multioutlet assembly installed to serve countertop appliances. (Courtesy of Legrand/Wiremold®)*

or midpoint of a single-phase portion of a 3-phase delta system, or a midpoint of a 3-wire, direct-current system.

Informational Note: At the neutral point of the system, the vectorial sum of the nominal voltages from all other phases within the system that utilize the neutral, with respect to the neutral point, is zero potential.

Exhibit 100.14 illustrates four examples of a neutral point in a system.

**Nonautomatic.** Requiring human intervention to perform a function.

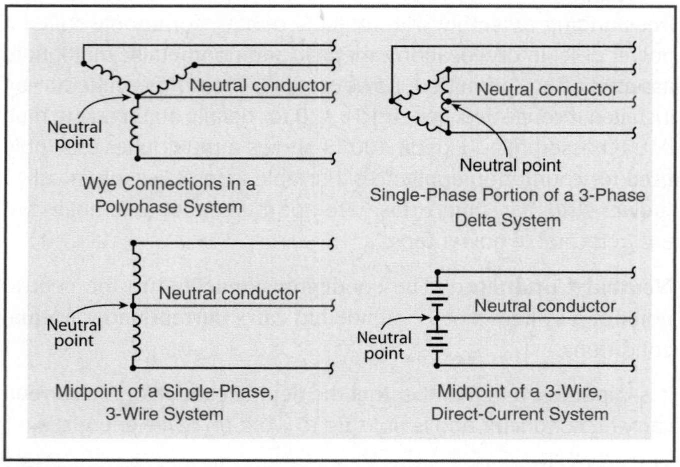


***EXHIBIT 100.14*** *Four examples of a neutral point.*

**Nonlinear Load.** A load where the wave shape of the steady-state current does not follow the wave shape of the applied voltage.

Informational Note: Electronic equipment, electronic/electric-discharge lighting, adjustable-speed drive systems, and similar equipment may be nonlinear loads.

Nonlinear loads are a major cause of harmonic currents in modern circuits. Additional conductor heating is just one of the undesirable operational effects often associated with harmonic currents. Informational Note No. 1 following 310.15(A)(3) addresses that harmonic current, as well as fundamental current, should be used in determining the heat generated internally in a conductor.

Actual circuit measurements of current for nonlinear loads should be made using only true rms-measuring ammeter instruments. Averaging ammeters produces inaccurate values if used to measure nonlinear loads. See the commentary following 310.15(B)(5)(c).

**Outlet.** A point on the wiring system at which current is taken to supply utilization equipment.

This term is frequently misused to refer to receptacles. Although receptacle outlets are outlets, not all outlets are receptacle outlets. Other common examples of outlets include lighting outlets and smoke alarm outlets.

**Outline Lighting.** An arrangement of incandescent lamps, electric-discharge lighting, or other electrically powered light sources to outline or call attention to certain features such as the shape of a building or the decoration of a window.

Outline lighting includes low-voltage, light-emitting diodes as well as other luminaires installed to form various shapes. See Article 600 for requirements for outline lighting.

**Overcurrent.** Any current in excess of the rated current of equipment or the ampacity of a conductor. It may result from overload, short circuit, or ground fault.

Informational Note: A current in excess of rating may be accommodated by certain equipment and conductors for a given set of conditions. Therefore, the rules for overcurrent protection are specific for particular situations.

**Overcurrent Protective Device, Branch-Circuit.** A device capable of providing protection for service, feeder, and branch circuits and equipment over the full range of overcurrents between its rated current and its interrupting rating. Such devices are provided with interrupting ratings appropriate for the intended use but no less than 5000 amperes.

The protection provided may be overload, short-circuit, or ground-fault or a combination, depending on the application.

**Overcurrent Protective Device, Supplementary.** A device intended to provide limited overcurrent protection for specific applications and utilization equipment such as luminaires and appliances. This limited protection is in addition to the protection provided in the required branch circuit by the branch-circuit overcurrent protective device.

There are two levels of overcurrent protection within branch circuits: branch-circuit overcurrent protection and supplementary overcurrent protection. The devices used to provide overcurrent protection are different, and the differences are found in the product standards UL 489, *Molded-Case Circuit Breakers, Molded-Case Switches and Circuit-Breaker Enclosures,* and UL 1077, *Supplementary Protectors for Use in Electrical Equipment.*

The *NEC* requires that all branch circuits use only branch-circuit-rated overcurrent protective devices to protect branch circuits, but it permits supplementary overcurrent protective devices for limited use downstream of the branch-circuit-rated overcurrent protective device.

The definition of *supplementary overcurrent protective device* makes two important distinctions between overcurrent protective devices. First, the use of a supplementary device is specifically limited to a few applications. Second, where it is used, the supplementary device must be in addition to and be protected by the more robust branch-circuit overcurrent protective device.

**Overload.** Operation of equipment in excess of normal, full-load rating, or of a conductor in excess of rated ampacity that, when it persists for a sufficient length of time, would cause damage or dangerous overheating. A fault, such as a short circuit or ground fault, is not an overload.

**Panelboard.** A single panel or group of panel units designed for assembly in the form of a single panel, including buses and automatic overcurrent devices, and equipped with or without switches for the control of light, heat, or power circuits; designed to be placed in a cabinet or cutout box placed in or against a wall, partition, or other support; and accessible only from the front.

**Photovoltaic (PV) System.** The total components and subsystem that, in combination, convert solar energy into electric energy suitable for connection to a utilization load.

**Plenum.** A compartment or chamber to which one or more air ducts are connected and that forms part of the air distribution system.

Because of concerns about the transfer of products of combustion through environmental air systems, the *NEC* provides specific requirements — in 300.22(B), (C), and (D) and in Articles 725, 760, 770, 800, 820, 830, and 840 — for the installation of wiring methods that are subject to the direct flow of environmental air. The *NEC* definition of the term *plenum* is similar to the definition of *plenum* contained in NFPA 90A, *Standard for the Installation of Air-Conditioning and Ventilating Systems.* The definition is used in conjunction with the requirements for the installation of wiring methods in spaces used for air transfer that are not specifically fabricated as ducts for environmental air. The air-handling space under a computer room floor has specific requirements as given in Article 645.

**Power Outlet.** An enclosed assembly that may include receptacles, circuit breakers, fuseholders, fused switches, buses, and watt-hour meter mounting means; intended to supply and control power to mobile homes, recreational vehicles, park trailers, or boats or to serve as a means for distributing power required to operate mobile or temporarily installed equipment.

**Premises Wiring (System).** Interior and exterior wiring, including power, lighting, control, and signal circuit wiring together with all their associated hardware, fittings, and wiring devices, both permanently and temporarily installed. This includes (a) wiring from the service point or power source to the outlets or (b) wiring from and including the power source to the outlets where there is no service point.

Such wiring does not include wiring internal to appliances, luminaires, motors, controllers, motor control centers, and similar equipment.

Informational Note: Power sources include, but are not limited to, interconnected or stand-alone batteries, solar photovoltaic systems, other distributed generation systems, or generators.

For this edition, a new informational note was added to clarify what a premises wiring system is. A premises wiring system does not have to be supplied by an electric utility. For example, portable generators and stand-alone photovoltaic systems can supply premises wiring systems. If there is no service point, there are no service conductors. The supply conductors are feeder conductors.

**Qualified Person.** One who has skills and knowledge related to the construction and operation of the electrical equipment and installations and has received safety training to recognize and avoid the hazards involved.

Informational Note: Refer to NFPA 70E-2012, *Standard for Electrical Safety in the Workplace*, for electrical safety training requirements.

This definition points out that safety training that qualifies a worker is training in hazard recognition and avoidance. See 110.6 in *NFPA 70E®*, *Standard for Electrical Safety in the Workplace®*, for the training requirements for qualified and unqualified persons who may be exposed to electrical hazards.

**Raceway.** An enclosed channel of metallic or nonmetallic materials designed expressly for holding wires, cables, or busbars, with additional functions as permitted in this *Code.*

Cable trays (see Article 392) are support systems for wiring methods and are not considered to be raceways.

Informational Note: A raceway is identified within specific article definitions.

**Rainproof.** Constructed, protected, or treated so as to prevent rain from interfering with the successful operation of the apparatus under specified test conditions.

See the commentary following 110.28.

**Raintight.** Constructed or protected so that exposure to a beating rain will not result in the entrance of water under specified test conditions.

Table 110.28 provides information on enclosure types that are considered to be raintight. Related requirements for raintight boxes and cabinets are found in 300.6(A)(2).

**Receptacle.** A receptacle is a contact device installed at the outlet for the connection of an attachment plug. A single receptacle is a single contact device with no other contact device on the same yoke. A multiple receptacle is two or more contact devices on the same yoke.

**Receptacle Outlet.** An outlet where one or more receptacles are installed.

**Remote-Control Circuit.** Any electrical circuit that controls any other circuit through a relay or an equivalent device.

**Retrofit Kit.** A general term for a complete subassembly of parts and devices for field conversion of utilization equipment.

**Sealable Equipment.** Equipment enclosed in a case or cabinet that is provided with a means of sealing or locking so that live parts cannot be made accessible without opening the enclosure.

Informational Note: The equipment may or may not be operable without opening the enclosure.

**Separately Derived System.** An electrical source, other than a service, having no direct connection(s) to circuit conductors of any other electrical source other than those established by grounding and bonding connections.

Examples of separately derived systems include generators, batteries, converter windings, transformers, and solar photovoltaic systems, provided they have no direct electrical connection to another source. The earth, metal enclosures, metal raceways, and equipment grounding conductors may provide incidental connection between systems. In addition, 250.30(A)(6) permits a common grounding electrode conductor to be installed for multiple separately derived systems. This definition clarifies that those systems can still be considered to be separately derived systems as long as the separately derived systems have no direct electrical connection to service-derived systems. The grounded circuit conductors are not intended to be directly connected.

**Service.** The conductors and equipment for delivering electric energy from the serving utility to the wiring system of the premises served.

A service can only be supplied by the serving utility. If electric energy is supplied by other than the serving utility, the supplied conductors and equipment are considered feeders and not a service.

**Service Cable.** Service conductors made up in the form of a cable.

**Service Conductors.** The conductors from the service point to the service disconnecting means.

The term *service conductors* is broad and may include overhead service conductors, underground service conductors, and service-entrance conductors. This term specifically excludes any wiring on the supply side (serving utility side) of the service point. The service conductors originate at the service point (where the serving utility ends) and end at the service disconnect. These service conductors may originate only from the serving utility. The definition no longer includes service drops and service laterals, which are now conductors under the control of the serving utility.

If the utility has specified that the service point is at the utility pole, the service conductors from an overhead distribution system originate at the utility pole and terminate at the service disconnecting means.

If the utility has specified that the service point is at the utility manhole, the service conductors from an underground distribution system originate at the utility manhole and terminate at the service disconnecting means. Where utility-owned primary conductors are extended to outdoor pad-mounted transformers on private property, the service conductors originate at the secondary connections of the transformers only if the utility has specified that the service point is at the secondary connections.

See Article 230, Part VIII, and the commentary following 230.200 for information on service conductors exceeding 1000 volts, nominal.

**Service Conductors, Overhead.** The overhead conductors between the service point and the first point of connection to the service-entrance conductors at the building or other structure.

**Service Conductors, Underground.** The underground conductors between the service point and the first point of connection to the service-entrance conductors in a terminal box, meter, or other enclosure, inside or outside the building wall.

Informational Note: Where there is no terminal box, meter, or other enclosure, the point of connection is considered to be the point of entrance of the service conductors into the building.

**Service Drop.** The overhead conductors between the utility electric supply system and the service point.

This definition correlates with the definition of the term *service lateral conductors*. Service-drop and service lateral conductors are conductors on the line side of the service point and are not subject to the *NEC*. Overhead conductors on the load side of the service point are overhead service conductors.

In Exhibit 100.15, the service-drop conductors run from the utility pole and connect to the service-entrance conductors at the service point. Conductors on the utility side of the service point are not covered by the *NEC*. The utility specifies the location of the

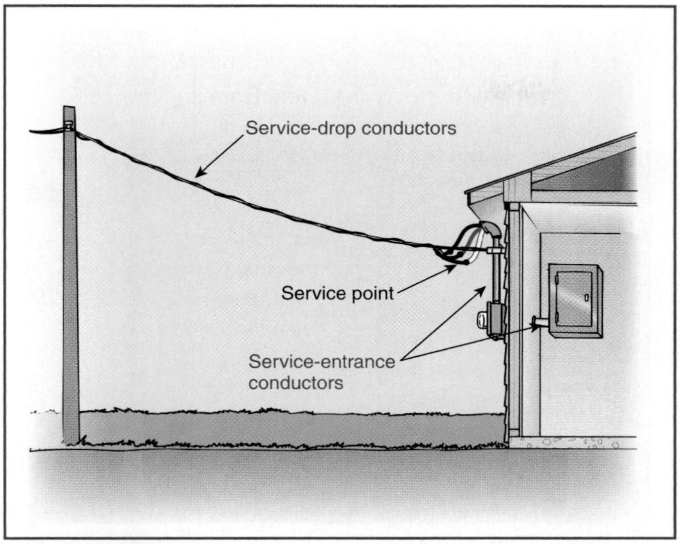


***EXHIBIT 100.15*** *Conductors in an overhead service.*

service point. Exact locations of the service point may vary from utility to utility as well as from occupancy to occupancy.

**Service-Entrance Conductors, Overhead System.** The service conductors between the terminals of the service equipment and a point usually outside the building, clear of building walls, where joined by tap or splice to the service drop or overhead service conductors.

See Exhibit 100.15 for an illustration of service-entrance conductors in an overhead system. The system shows a service drop from a utility pole to attachment on a house and service-entrance conductors from point of attachment (spliced to service-drop conductors), down the side of the house, through the meter socket, and terminating in the service equipment. In this instance, the service point is at the drip loop. The conductors on the line side of the service point are under the control of the utility.

**Service-Entrance Conductors, Underground System.** The service conductors between the terminals of the service equipment and the point of connection to the service lateral or underground service conductors.

See Exhibit 100.16 for an illustration of service-entrance conductors in an underground system. The illustration on the top shows underground service lateral conductors run from a pole to a service point underground. The conductors from the service point into the building are underground service conductors. The illustration on the bottom shows service lateral conductors run from a utility transformer.

Informational Note: Where service equipment is located outside the building walls, there may be no service-entrance conductors or they may be entirely outside the building.

**Service Equipment.** The necessary equipment, usually consisting of a circuit breaker(s) or switch(es) and fuse(s) and their

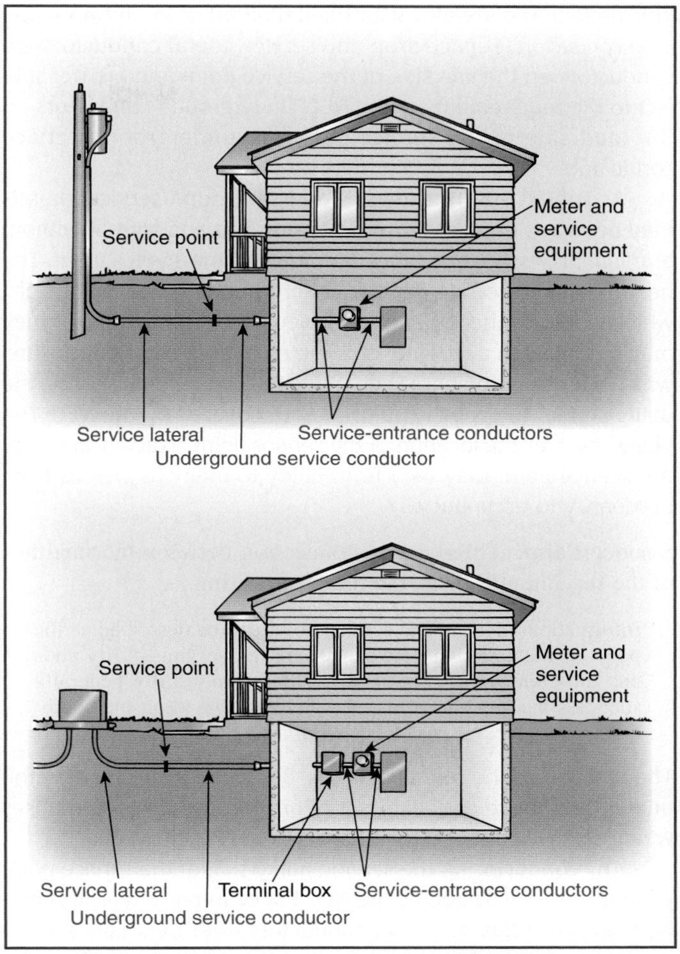


***EXHIBIT 100.16*** *Underground systems showing service laterals run from a pole (top) and from a transformer (bottom).*

accessories, connected to the load end of service conductors to a building or other structure, or an otherwise designated area, and intended to constitute the main control and cutoff of the supply.

Service equipment may consist of circuit breakers or fused switches that are provided to disconnect all ungrounded conductors in a building or other structure from the service-entrance conductors. Individual meter socket enclosures are not considered service equipment according to 230.66.

The disconnecting means at any one location in a building or other structure is not allowed to consist of more than six circuit breakers or six switches and is required to be readily accessible either outside or inside nearest the point of entrance of the service-entrance conductors. Requirements for service conductors outside the building are in 230.6, and those for disconnecting means are throughout Article 230, Part VI.

**Service Lateral.** The underground conductors between the utility electric supply system and the service point.

This definition correlates with the definition of the term *service-drop conductors.* Service-drop and service lateral conductors are conductors on the line side of the service point and are not subject to the requirements of the *NEC.* Underground conductors on the load side of the service point are underground service conductors.

As Exhibit 100.16 shows, the underground service laterals may be run from poles or from transformers and with or without terminal boxes, provided they terminate at the service point. The next transition would be to the underground service conductors, which would connect to the service-entrance conductors, or they may terminate in a terminal box, meter, or some other enclosure, which may be inside or outside of the building. Conductors on the utility side of the service point are not covered by the *NEC.* The utility specifies the location of the service point. Exact locations of the service point may vary from utility to utility as well as from occupancy to occupancy.

**Service Point.** The point of connection between the facilities of the serving utility and the premises wiring.

Informational Note: The service point can be described as the point of demarcation between where the serving utility ends and the premises wiring begins. The serving utility generally specifies the location of the service point based on the conditions of service.

The location of the service point is generally determined by the utility. Only those conductors that are located on the premises wiring side of the service point are covered by the *NEC.*

Any conductor on the serving utility side of the service point generally is not covered by the *NEC.* For example, a typical suburban residence has overhead conductors from the utility pole to the house. If the utility specifies that the service point is at the point of attachment of the overhead conductors to the house, the overhead conductors are service-drop conductors that are not covered by the *Code* because the conductors are not on the premises wiring side of the service point. Alternatively, if the utility specifies that the service point is at the pole, the overhead conductors are considered overhead service conductors, and the *NEC* would apply to these conductors.

Exact locations for a service point may vary from utility to utility as well as from occupancy to occupancy.

**Short-Circuit Current Rating.** The prospective symmetrical fault current at a nominal voltage to which an apparatus or system is able to be connected without sustaining damage exceeding defined acceptance criteria.

The short-circuit current rating is marked on the equipment nameplate as shown in Exhibit 100.17. This value must not be exceeded. Otherwise, the equipment can be damaged by short-circuit currents, posing a hazard to personnel and property. See also the definition of *interrupting rating.*

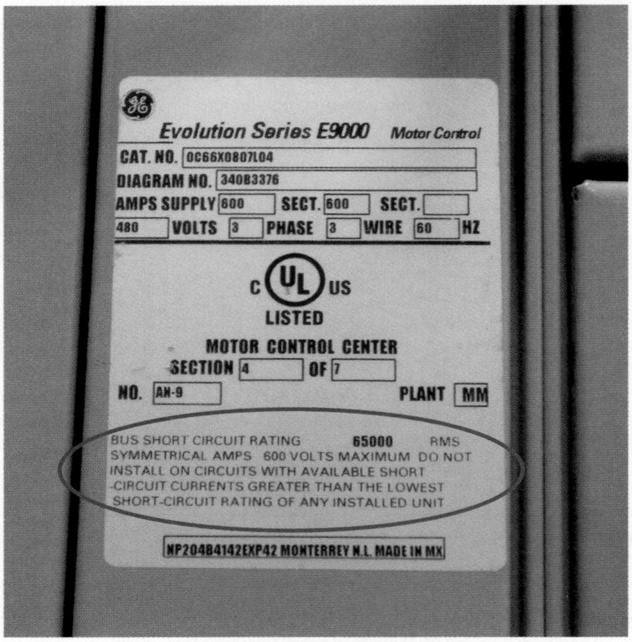


***EXHIBIT 100.17*** *Short-circuit current rating on equipment nameplate.*

**Show Window.** Any window used or designed to be used for the display of goods or advertising material, whether it is fully or partly enclosed or entirely open at the rear and whether or not it has a platform raised higher than the street floor level.

**Signaling Circuit.** Any electrical circuit that energizes signaling equipment.

•

**Special Permission.** The written consent of the authority having jurisdiction.

The AHJ for enforcement of the *Code* is responsible for making interpretations and granting special permission contemplated in a number of the rules as stated in 90.4. For requirements governing special permission situations, see 110.26(A)(1)(b), 230.2(B), and 426.14.

**Structure.** That which is built or constructed.

**Substation.** An enclosed assemblage of equipment (e.g., switches, interrupting devices, circuit breakers, buses, and transformers) through which electric energy is passed for the purpose of distribution, switching, or modifying its characteristics.

**Surge Arrester.** A protective device for limiting surge voltages by discharging or bypassing surge current; it also prevents continued flow of follow current while remaining capable of repeating these functions.

**Surge-Protective Device (SPD).** A protective device for limiting transient voltages by diverting or limiting surge current; it also prevents continued flow of follow current while remaining capable of repeating these functions and is designated as follows:

Type 1: Permanently connected SPDs intended for installation between the secondary of the service transformer and the line side of the service disconnect overcurrent device.

Type 2: Permanently connected SPDs intended for installation on the load side of the service disconnect overcurrent device, including SPDs located at the branch panel.

Type 3: Point of utilization SPDs.

Type 4: Component SPDs, including discrete components, as well as assemblies.

Informational Note: For further information on Type 1, Type 2, Type 3, and Type 4 SPDs, see UL 1449, *Standard for Surge Protective Devices*.

Type 1 and Type 2 surge-protective devices (SPDs) for permanently connected devices on circuits not exceeding 600 volts are included in UL 1449, *Transient Voltage Surge Suppressors*. Requirements related to SPDs are found in Article 280.

**Switch, Bypass Isolation.** A manually operated device used in conjunction with a transfer switch to provide a means of directly connecting load conductors to a power source and of disconnecting the transfer switch.

**Switch, General-Use.** A switch intended for use in general distribution and branch circuits. It is rated in amperes, and it is capable of interrupting its rated current at its rated voltage.

**Switch, General-Use Snap.** A form of general-use switch constructed so that it can be installed in device boxes or on box covers, or otherwise used in conjunction with wiring systems recognized by this *Code*.

**Switch, Isolating.** A switch intended for isolating an electrical circuit from the source of power. It has no interrupting rating, and it is intended to be operated only after the circuit has been opened by some other means.

**Switch, Motor-Circuit.** A switch rated in horsepower that is capable of interrupting the maximum operating overload current of a motor of the same horsepower rating as the switch at the rated voltage.

**Switch, Transfer.** An automatic or nonautomatic device for transferring one or more load conductor connections from one power source to another.

**Switchboard.** A large single panel, frame, or assembly of panels on which are mounted on the face, back, or both, switches, overcurrent and other protective devices, buses, and usually instruments. These assemblies are generally accessible from the rear as well as from the front and are not intended to be installed in cabinets.

Busbars are required to be arranged to avoid inductive overheating. Service busbars are required to be isolated by barriers from the remainder of the switchboard. Although not required, most modern switchboards are totally enclosed to minimize the probability of spreading fire to adjacent combustible materials and to guard live parts.

**Switchgear.** An assembly completely enclosed on all sides and top with sheet metal (except for ventilating openings and inspection windows) and containing primary power circuit switching, interrupting devices, or both, with buses and connections. The assembly may include control and auxiliary devices. Access to the interior of the enclosure is provided by doors, removable covers, or both.

This definition was revised for the 2014 *Code*. It formerly appeared as *metal-enclosed power switchgear*. The revised definition is inclusive of all types of switchgear governed by the *NEC*. The generic term can be used in most *Code* text where the term *switchboard* is already mentioned, and where the use of the term *switchgear* is appropriate. According to ANSI C37.20, *Standard for Metal-Enclosed Low-Voltage Power Circuit Breaker Switchgear*, the term *switchgear* includes "metal-enclosed low-voltage power circuit-breaker switchgear," "metal-clad switchgear," and "metal-enclosed interrupter switchgear."

Informational Note: All switchgear subject to *NEC* requirements is metal enclosed. Switchgear rated below 1000 V or less may be identified as "low-voltage power circuit breaker switchgear." Switchgear rated over 1000 V may be identified as "metal-enclosed switchgear" or "metal-clad switchgear." Switchgear is available in non–arc-resistant or arc-resistant constructions.

**Thermal Protector (as applied to motors).** A protective device for assembly as an integral part of a motor or motor-compressor that, when properly applied, protects the motor against dangerous overheating due to overload and failure to start.

Informational Note: The thermal protector may consist of one or more sensing elements integral with the motor or motor-compressor and an external control device.

**Thermally Protected (as applied to motors).** The words *Thermally Protected* appearing on the nameplate of a motor or motor-compressor indicate that the motor is provided with a thermal protector.

**Ungrounded.** Not connected to ground or to a conductive body that extends the ground connection.

**Uninterruptible Power Supply.** A power supply used to provide alternating current power to a load for some period of time in the event of a power failure.

Informational Note: In addition, it may provide a more constant voltage and frequency supply to the load, reducing the effects of voltage and frequency variations.

**Utility-Interactive Inverter.** An inverter intended for use in parallel with an electric utility to supply common loads that may deliver power to the utility.

**Utilization Equipment.** Equipment that utilizes electric energy for electronic, electromechanical, chemical, heating, lighting, or similar purposes.

**Ventilated.** Provided with a means to permit circulation of air sufficient to remove an excess of heat, fumes, or vapors.

See the commentary following 110.13(B).

**Volatile Flammable Liquid.** A flammable liquid having a flash point below 38°C (100°F), or a flammable liquid whose temperature is above its flash point, or a Class II combustible liquid that has a vapor pressure not exceeding 276 kPa (40 psia) at 38°C (100°F) and whose temperature is above its flash point.

The flash point of a liquid is defined as the minimum temperature at which it gives off sufficient vapor to form an ignitible mixture with the air near the surface of the liquid or within the vessel used to contain the liquid. An ignitible mixture is defined as a mixture within the explosive or flammable range (between upper and lower limits) that is capable of the propagation of flame away from the source of ignition when ignited. Some emission of vapors takes place below the flash point but not enough to form an ignitible mixture.

**Voltage (of a circuit).** The greatest root-mean-square (rms) (effective) difference of potential between any two conductors of the circuit concerned.

Informational Note: Some systems, such as 3-phase 4-wire, single-phase 3-wire, and 3-wire direct current, may have various circuits of various voltages.

Common 3-phase, 4-wire wye systems are 480/277 volts and 208/120 volts. The voltage of the circuit is the higher voltage between any two phase conductors (i.e., 480 volts or 208 volts). The voltage of the circuit of a 2-wire feeder or branch-circuit (single-phase and the grounded conductor) derived from these systems would be the lower voltage between two conductors (i.e., 277 volts or 120 volts). The same applies to direct current (dc) or single-phase, 3-wire systems where there are two voltages.

**Voltage, Nominal.** A nominal value assigned to a circuit or system for the purpose of conveniently designating its voltage class (e.g., 120/240 volts, 480Y/277 volts, 600 volts).

Informational Note No. 1: The actual voltage at which a circuit operates can vary from the nominal within a range that permits satisfactory operation of equipment.

Informational Note No. 2: See ANSI C84.1-2006, *Voltage Ratings for Electric Power Systems and Equipment (60 Hz).*

For a list of nominal voltages to use in computing branch-circuit and feeder loads, see 220.5(A).

**Voltage to Ground.** For grounded circuits, the voltage between the given conductor and that point or conductor of the circuit that is grounded; for ungrounded circuits, the greatest voltage between the given conductor and any other conductor of the circuit.

This definition can be best illustrated using examples.

| System | Voltage to Ground |
|---|---|
| 480/277-volt wye | 277 |
| 208/120-volt wye | 120 |
| 3-phase, 3-wire ungrounded 480-volt | 480 |

For a 3-phase, 4-wire delta system with the center of one leg grounded, there are three voltages to ground – that is, on a 240-volt system, two legs would each have 120 volts to ground, and the third, or high, leg would have 208 volts to ground. See 110.15, 230.56, and 408.3(E) for requirements pertaining to special markings and arrangements on such circuit conductors.

**Watertight.** Constructed so that moisture will not enter the enclosure under specified test conditions.

Unless an enclosure is hermetically sealed, it is possible for moisture to enter the enclosure. See the requirements related to watertight enclosures in 110.28 and Table 110.28 and the additional information given in the accompanying commentary.

**Weatherproof.** Constructed or protected so that exposure to the weather will not interfere with successful operation.

Informational Note: Rainproof, raintight, or watertight equipment can fulfill the requirements for weatherproof where varying weather conditions other than wetness, such as snow, ice, dust, or temperature extremes, are not a factor.

For information on enclosures considered to be weatherproof based on exposure to specific environmental conditions, see 110.28 and the accompanying commentary.

## II. Over 600 Volts, Nominal

Part II contains definitions applicable only to the articles and parts of articles specifically covering installations and equipment operating at over 600 volts, nominal.

The definitions in Part I are intended to apply wherever the terms are used throughout this *Code*. The definitions in Part II are applicable only to articles and parts of articles specifically covering installations and equipment operating at over 600 volts, nominal.

**Electronically Actuated Fuse.** An overcurrent protective device that generally consists of a control module that provides current sensing, electronically derived time–current characteristics, energy to initiate tripping, and an interrupting module that interrupts current when an overcurrent occurs. Electronically actuated fuses may or may not operate in a current-limiting fashion, depending on the type of control selected.

Although they are called fuses because they interrupt current by melting a fusible element, electronically actuated fuses respond to a signal from an electronic control, rather than from the heat generated by actual current passing through a fusible element. Electronically actuated fuses have controls similar to those of electronic circuit breakers.

**Fuse.** An overcurrent protective device with a circuit-opening fusible part that is heated and severed by the passage of overcurrent through it.

Informational Note: A fuse comprises all the parts that form a unit capable of performing the prescribed functions. It may or may not be the complete device necessary to connect it into an electrical circuit.

*Controlled Vented Power Fuse.* A fuse with provision for controlling discharge circuit interruption such that no solid material may be exhausted into the surrounding atmosphere.

Informational Note: The fuse is designed so that discharged gases will not ignite or damage insulation in the path of the discharge or propagate a flashover to or between grounded members or conduction members in the path of the discharge where the distance between the vent and such insulation or conduction members conforms to manufacturer's recommendations.

*Expulsion Fuse Unit (Expulsion Fuse).* A vented fuse unit in which the expulsion effect of gases produced by the arc and lining of the fuseholder, either alone or aided by a spring, extinguishes the arc.

*Nonvented Power Fuse.* A fuse without intentional provision for the escape of arc gases, liquids, or solid particles to the atmosphere during circuit interruption.

*Power Fuse Unit.* A vented, nonvented, or controlled vented fuse unit in which the arc is extinguished by being drawn through solid material, granular material, or liquid, either alone or aided by a spring.

*Vented Power Fuse.* A fuse with provision for the escape of arc gases, liquids, or solid particles to the surrounding atmosphere during circuit interruption.

**Multiple Fuse.** An assembly of two or more single-pole fuses.

**Switching Device.** A device designed to close, open, or both, one or more electrical circuits.

*Circuit Breaker.* A switching device capable of making, carrying, and interrupting currents under normal circuit conditions, and also of making, carrying for a specified time, and interrupting currents under specified abnormal circuit conditions, such as those of short circuit.

*Cutout.* An assembly of a fuse support with either a fuseholder, fuse carrier, or disconnecting blade. The fuseholder or fuse carrier may include a conducting element (fuse link) or may act as the disconnecting blade by the inclusion of a nonfusible member.

*Disconnecting Means.* A device, group of devices, or other means whereby the conductors of a circuit can be disconnected from their source of supply.

*Disconnecting (or Isolating) Switch (Disconnector, Isolator).* A mechanical switching device used for isolating a circuit or equipment from a source of power.

*Interrupter Switch.* A switch capable of making, carrying, and interrupting specified currents.

*Oil Cutout (Oil-Filled Cutout).* A cutout in which all or part of the fuse support and its fuse link or disconnecting blade is mounted in oil with complete immersion of the contacts and the fusible portion of the conducting element (fuse link) so that arc interruption by severing of the fuse link or by opening of the contacts will occur under oil.

*Oil Switch.* A switch having contacts that operate under oil (or askarel or other suitable liquid).

*Regulator Bypass Switch.* A specific device or combination of devices designed to bypass a regulator.

# ARTICLE 110 Requirements for Electrical Installations

## I. General

### 110.1 Scope

This article covers general requirements for the examination and approval, installation and use, access to and spaces about electrical conductors and equipment; enclosures intended for personnel entry; and tunnel installations.

Informational Note: See Informative Annex J for information regarding ADA accessibility design.

### 110.2 Approval

The conductors and equipment required or permitted by this *Code* shall be acceptable only if approved.

Informational Note: See 90.7, Examination of Equipment for Safety, and 110.3, Examination, Identification, Installation, and Use of Equipment. See definitions of *Approved, Identified, Labeled,* and *Listed.*

All electrical equipment is required to be *approved* and to be *acceptable to the authority having jurisdiction* (these terms are defined in Article 100). Approval of equipment is the responsibility of the electrical inspection authority, and many such approvals are based on tests and listings of testing laboratories. Unique equipment is often approved following a field evaluation by a qualified third-party laboratory or qualified individual.

### 110.3 Examination, Identification, Installation, and Use of Equipment

This section provides criteria and considerations for the evaluation of equipment, and it recognizes listing or labeling as a means of establishing suitability. In itself, 110.3 does not require listing or labeling of equipment. It does, however, require considerable evaluation of equipment. Section 110.2 requires that equipment be approved. Before issuing approval, the AHJ may require evidence of compliance with 110.3(A). The most common form of evidence considered acceptable by AHJ is a listing or labeling by a third party.

For wire-bending and connection space in cabinets and cutout boxes, see 312.6, Table 312.6(A), Table 312.6(B), 312.7, 312.9, and 312.11. For wire-bending and connection space in other equipment, see the appropriate *NEC* article and section. For example, see 314.16 and 314.28 for outlet, device, pull, and junction boxes as well as conduit bodies; 404.3 and 404.18 for switches; 408.3(F) for switchboards and panelboards; and 430.10 for motors and motor controllers.

**(A) Examination.** In judging equipment, considerations such as the following shall be evaluated:

(1) Suitability for installation and use in conformity with the provisions of this *Code*

Informational Note: Suitability of equipment use may be identified by a description marked on or provided with a product to identify the suitability of the product for a specific purpose, environment, or application. Special conditions of use or other limitations and other pertinent information may be marked on the equipment, included in the product instructions, or included in the appropriate listing and labeling information. Suitability of equipment may be evidenced by listing or labeling.

Examples of special conditions of use include elevated or reduced ambient temperatures, special environmental limitations, stringent power quality requirements, or specific types of overcurrent protective devices. The additional information needed for these special cases may be marked on the equipment, included as part of the listing information in a listing directory, or included in the information furnished with the equipment.

(2) Mechanical strength and durability, including, for parts designed to enclose and protect other equipment, the adequacy of the protection thus provided
(3) Wire-bending and connection space
(4) Electrical insulation
(5) Heating effects under normal conditions of use and also under abnormal conditions likely to arise in service
(6) Arcing effects
(7) Classification by type, size, voltage, current capacity, and specific use
(8) Other factors that contribute to the practical safeguarding of persons using or likely to come in contact with the equipment

**(B) Installation and Use.** Listed or labeled equipment shall be installed and used in accordance with any instructions included in the listing or labeling.

This section requires that manufacturers' listing and labeling installation instructions be followed even if the equipment itself is not required to be listed. For example, 210.52 permits permanently installed electric baseboard heaters to be equipped with receptacle outlets that meet the requirements for the wall space utilized by such heaters. The installation instructions for such permanent baseboard heaters indicate that the heaters should not be mounted beneath a receptacle. In dwelling units, the use of low-density heating units more than 12 feet in length is common. Therefore, to meet the requirements of 210.52(A) and also the installation instructions, a receptacle must either be part of the heating unit or be installed in the floor close to the wall but not above the heating unit. (See 210.52, Informational Note, and Exhibit 210.28 for more specific details.)

In itself, 110.3(B) does not require listing or labeling of equipment. It does, however, require considerable evaluation of equipment. Section 110.2 requires that equipment be acceptable only if approved. The term *approved* is defined in Article 100 as acceptable to the AHJ. Before issuing approval, the AHJ may require evidence of compliance with 110.3(A). The most common form of evidence considered acceptable by AHJ is a listing or labeling by a third party.

Some sections in the *Code* do require listed or labeled equipment. For example, 250.8 specifies "listed pressure connectors . . . pressure connectors listed as grounding and bonding equipment [or] . . . other listed means" as connection methods for grounding and bonding conductors.

## 110.4 Voltages

Throughout this *Code*, the voltage considered shall be that at which the circuit operates. The voltage rating of electrical equipment shall not be less than the nominal voltage of a circuit to which it is connected.

Voltages used for computing branch-circuit and feeder loads in accordance with Article 220 are nominal voltages as specified in 220.5. See the definitions of *voltage (of a circuit)*; *voltage, nominal*; and *voltage to ground* in Article 100. See also 300.2 and 300.3(C), which specify the voltage limitations of conductors of circuits rated 1000 V, nominal, or less, and over 1000 V, nominal.

## 110.5 Conductors

Conductors normally used to carry current shall be of copper unless otherwise provided in this *Code*. Where the conductor material is not specified, the material and the sizes given in this *Code* shall apply to copper conductors. Where other materials are used, the size shall be changed accordingly.

Informational Note: For aluminum and copper-clad aluminum conductors, see 310.15.

See 310.106(B), which specifies the alloy for aluminum conductors.

## 110.6 Conductor Sizes

Conductor sizes are expressed in American Wire Gage (AWG) or in circular mils.

For copper, aluminum, or copper-clad aluminum conductors up to size 4/0 AWG, the *Code* uses the American Wire Gage (AWG) for size identification, which is the same as the Brown and Sharpe (BS) Wire Gauge. Wire sizes up to size 4/0 AWG are expressed as XX AWG, with XX being the size wire. A wire size expressed as No. 12 in editions prior to 2002 is now expressed as 12 AWG.

Conductors larger than 4/0 AWG are sized in circular mils, beginning with 250,000 circular mils. Prior to the 1990 edition, a 250,000-circular-mil conductor was labeled 250 MCM. The term *MCM* was defined as 1000 circular mils (the first *M* being the roman numeral designation for 1000). Beginning in the 1990 edition, the notation was changed to 250 kcmil to recognize the accepted convention that *k* indicates 1000. UL standards and IEEE standards also use the notation kcmil rather than MCM.

**Calculation Example**

The circular mil area of a conductor is equal to its diameter in mils squared (1 in. = 1000 mils). What is the circular mil area of an 8 AWG solid conductor that has a 0.1285-inch diameter?

*Solution*

$$0.1285 \text{ in.} \times 1000 = 128.5 \text{ mils}$$
$$128.5 \times 128.5 = 16{,}512.25 \text{ circular mils}$$

or 16,510 circular mils (rounded off)

According to Table 8 in Chapter 9, this rounded value represents the circular mil area for one conductor. Where stranded conductors are used, the circular mil area of each strand must be multiplied by the number of strands to determine the circular mil area of the conductor.

## 110.7 Wiring Integrity

Completed wiring installations shall be free from short circuits, ground faults, or any connections to ground other than as required or permitted elsewhere in this *Code*.

Failure of the insulation system is one of the most common causes of problems in electrical installations, in both high-voltage and low-voltage systems. The principal causes of insulation failures are heat, moisture, dirt, and physical damage (abrasion or nicks) occurring during and after installation. Insulation can also fail due to chemical attack, sunlight, and excessive voltage stresses.

Overcurrent protective devices must be selected and coordinated using tables of insulation thermal-withstand ability to ensure that the damage point of an insulated conductor is never reached. These tables, entitled "Allowable Short-Circuit Currents for Insulated Copper (or Aluminum) Conductors," are contained in the Insulated Cable Engineers Association's publication ICEA P-32-382. See 110.10 for selection criteria for other circuit components.

Insulation tests are performed on new or existing installations to determine the quality or condition of the insulation of conductors and equipment. In an insulation resistance test, a voltage ranging from 100 to 5000 (usually 500 to 1000 V for systems of 1000 V or less), supplied from a source of constant potential, is applied across the insulation. A megohmmeter is usually the potential source, and it indicates the insulation resistance directly on a scale calibrated in megohms (MΩ). The quality of the insulation is evaluated based on the level of the insulation resistance.

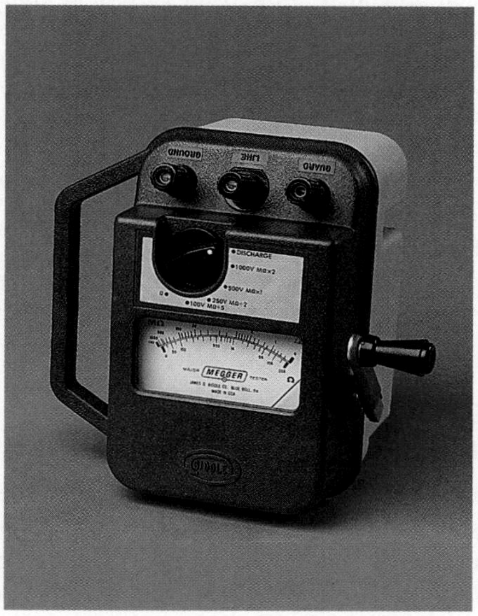


***EXHIBIT 110.1*** *A manual multivoltage, multirange insulation tester.*

The insulation resistance of many types of insulation varies with temperature, so the field data obtained should be corrected to the standard temperature for the class of equipment being tested. The megohm value of insulation resistance obtained is inversely proportional to the volume of insulation tested. For example, a cable 1000 feet long would be expected to have one-tenth the insulation resistance of a cable 100 feet long if all other conditions are identical.

NFPA 70B, *Recommended Practice for Electrical Equipment Maintenance*, provides useful information on test methods and on establishing a preventive maintenance program. Information on specific test methods is available from instrument manufacturers. Thorough knowledge in the use of insulation testers is essential if the test results are to be meaningful. Exhibit 110.1 shows a typical megohmmeter insulation tester.

## 110.8 Wiring Methods

Only wiring methods recognized as suitable are included in this *Code*. The recognized methods of wiring shall be permitted to be installed in any type of building or occupancy, except as otherwise provided in this *Code*.

Article 300 applies generally to all wiring methods, except as amended, modified, or supplemented by other *NEC* chapters. The application statement is found in 90.3.

## 110.9 Interrupting Rating

Equipment intended to interrupt current at fault levels shall have an interrupting rating at nominal circuit voltage sufficient for the current that is available at the line terminals of the equipment.

Equipment intended to interrupt current at other than fault levels shall have an interrupting rating at nominal circuit voltage sufficient for the current that must be interrupted.

Fuses or circuit breakers that do not have adequate interrupting ratings could rupture while attempting to clear a short circuit. The interrupting rating of an overcurrent protective device is determined under standard test conditions that should match the actual installation needs.

Interrupting ratings should not be confused with short-circuit current ratings. Short-circuit current ratings are further explained in the commentary following 110.10.

## 110.10 Circuit Impedance, Short-Circuit Current Ratings, and Other Characteristics

The overcurrent protective devices, the total impedance, the equipment short-circuit current ratings, and other characteristics of the circuit to be protected shall be selected and coordinated to permit the circuit protective devices used to clear a fault to do so without extensive damage to the electrical equipment of the circuit. This fault shall be assumed to be either between two or more of the circuit conductors or between any circuit conductor and the equipment grounding conductor(s) permitted in 250.118. Listed equipment applied in accordance with their listing shall be considered to meet the requirements of this section.

Short-circuit current ratings are marked on equipment such as panelboards, switchboards, switchgear, busways, contactors, and starters. Listed products are tested and are subjected to rigorous testing as part of their evaluation, which includes tests under fault conditions. Therefore, listed products used within their ratings are considered to have met the requirements of 110.10.

The purpose of overcurrent protection is to open the circuit before conductors or conductor insulation is damaged when an overcurrent condition occurs. An overcurrent condition can be the result of an overload, a ground fault, or a short circuit.

Overcurrent protective devices (such as fuses and circuit breakers) should be selected to ensure that the short-circuit current rating of the system components is not exceeded should a short circuit or high-level ground fault occur.

Wire, bus structures, switching, protection and disconnect devices, and distribution equipment all have limited short-circuit ratings and would be damaged or destroyed if those short-circuit ratings were exceeded. Merely providing overcurrent protective devices with sufficient interrupting ratings would not ensure adequate short-circuit protection for the system components. When the available short-circuit current exceeds the short-circuit current rating of an electrical component, the overcurrent protective device must limit the let-through energy to within the rating of that electrical component.

Utility companies usually determine and provide information on available short-circuit current levels at the service equipment. Literature on how to calculate short-circuit currents at each point in any distribution can generally be obtained by contacting the manufacturers of overcurrent protective devices or by referring to IEEE 141-1993 (R1999), *IEEE Recommended Practice for Electric Power Distribution for Industrial Plants* (Red Book).

Adequate short-circuit protection can be provided by fuses, molded-case circuit breakers, and low-voltage power circuit breakers, depending on specific circuit and installation requirements.

**Application Example**

For a typical one-family dwelling with a 100-ampere service using 2 AWG aluminum supplied by a 37½ kVA transformer with 1.72 percent impedance located at a distance of 25 feet, the available short-circuit current would be approximately 6000 amperes.

Available short-circuit current to multifamily structures, where pad-mounted transformers are located close to the multimetering location, can be relatively high. For example, the line-to-line fault current values close to a low-impedance transformer could exceed 22,000 amperes. At the secondary of a single-phase, center-tapped transformer, the line-to-neutral fault current is approximately one and one-half times that of the line-to-line fault current. The short-circuit current rating of utilization equipment located and connected near the service equipment should be known. For example, HVAC equipment is tested at 3500 amperes through a 40-ampere load rating and at 5000 amperes for loads rated more than 40 amperes.

## 110.11 Deteriorating Agents

Unless identified for use in the operating environment, no conductors or equipment shall be located in damp or wet locations; where exposed to gases, fumes, vapors, liquids, or other agents that have a deteriorating effect on the conductors or equipment; or where exposed to excessive temperatures.

Informational Note No. 1: See 300.6 for protection against corrosion.

Informational Note No. 2: Some cleaning and lubricating compounds can cause severe deterioration of many plastic materials used for insulating and structural applications in equipment.

Equipment not identified for outdoor use and equipment identified only for indoor use, such as "dry locations," "indoor use only," "damp locations," or enclosure Types 1, 2, 5, 12, 12K, and/or 13, shall be protected against damage from the weather during construction.

Informational Note No. 3: See Table 110.28 for appropriate enclosure-type designations.

## 110.12 Mechanical Execution of Work

Electrical equipment shall be installed in a neat and workmanlike manner.

Informational Note: Accepted industry practices are described in ANSI/NECA 1-2010, *Standard Practice of Good Workmanship in Electrical Construction,* and other ANSI-approved installation standards.

***EXHIBIT 110.2*** *ANSI/NECA 1-2010,* Standard for Good Workmanship in Electrical Construction, *is one example of the many ANSI standards that describe "neat and workmanlike" installations. (Courtesy of the National Electrical Contractors Association)*

The requirement for "neat and workmanlike" installations has appeared in the *NEC* as currently worded for more than a half-century. Such an installation represents pride in one's work and has been emphasized by persons involved in the training of apprentice electricians for many years. A neat and workmanlike installation is also easy to troubleshoot and is unlikely to have operational problems.

Many *Code* conflicts or violations cited by the AHJ have been based on the authority's interpretation of "neat and workmanlike manner." Many electrical inspection authorities use their own experience or precedents in their local areas as the basis for their judgments. The Informational Note directs the user to an industry-accepted ANSI standard that clearly describes and illustrates "neat and workmanlike" electrical installations. See Exhibit 110.2.

Installations that do not qualify as "neat and workmanlike" include exposed runs of cables or raceways that are improperly supported (that are sagging between supports, for instance, or supported by improper methods); field-bent and kinked, flattened, or poorly measured raceways; or cabinets, cutout boxes, and enclosures that are not plumb or not properly secured.

**(A) Unused Openings.** Unused openings, other than those intended for the operation of equipment, those intended for mounting purposes, or those permitted as part of the design for listed equipment, shall be closed to afford protection substantially equivalent to the wall of the equipment. Where metallic plugs or plates are used with nonmetallic enclosures, they shall be recessed at least 6 mm (¼ in.) from the outer surface of the enclosure.

This section requires all unused openings other than those openings used for mounting, cooling, or drainage to be closed up.

See 408.7 for requirements on unused openings in switchboard and panelboard enclosures.

**(B) Integrity of Electrical Equipment and Connections.** Internal parts of electrical equipment, including busbars, wiring terminals, insulators, and other surfaces, shall not be damaged or contaminated by foreign materials such as paint, plaster, cleaners, abrasives, or corrosive residues. There shall be no damaged parts that may adversely affect safe operation or mechanical strength of the equipment such as parts that are broken; bent; cut; or deteriorated by corrosion, chemical action, or overheating.

## 110.13 Mounting and Cooling of Equipment

**(A) Mounting.** Electrical equipment shall be firmly secured to the surface on which it is mounted. Wooden plugs driven into holes in masonry, concrete, plaster, or similar materials shall not be used.

**(B) Cooling.** Electrical equipment that depends on the natural circulation of air and convection principles for cooling of exposed surfaces shall be installed so that room airflow over such surfaces is not prevented by walls or by adjacent installed equipment. For equipment designed for floor mounting, clearance between top surfaces and adjacent surfaces shall be provided to dissipate rising warm air.

Electrical equipment provided with ventilating openings shall be installed so that walls or other obstructions do not prevent the free circulation of air through the equipment.

The term *ventilated* is defined in Article 100. Ventilating openings in equipment are provided to allow the circulation of room air around internal equipment components. Blocking these openings can cause dangerous overheating. For example, a ventilated busway must be located where there are no walls or other objects that might interfere with the natural circulation of air and convection principles for cooling. The surfaces of some enclosures, such as panelboards and transformers, may also require normal room air circulation to prevent overheating. Ventilation for motor locations is covered in 430.14(A) and 430.16. Ventilation for transformer locations is covered in 450.9 and 450.45. In addition to 110.13, proper placement of equipment requiring ventilation becomes enforceable using the requirements of 110.3(B). It is critical that ventilation openings not be blocked. Otherwise, overheating may result.

## 110.14 Electrical Connections

Because of different characteristics of dissimilar metals, devices such as pressure terminal or pressure splicing connectors and soldering lugs shall be identified for the material of the conductor and shall be properly installed and used. Conductors of dissimilar metals shall not be intermixed in a terminal or splicing connector where physical contact occurs between dissimilar conductors (such as copper and aluminum, copper and copper-clad aluminum, or aluminum and copper-clad aluminum), unless the device is identified for the purpose and conditions of use. Materials such as solder, fluxes, inhibitors, and compounds, where employed, shall be suitable for the use and shall be of a type that will not adversely affect the conductors, installation, or equipment.

Connectors and terminals for conductors more finely stranded than Class B and Class C stranding as shown in Chapter 9, Table 10, shall be identified for the specific conductor class or classes.

Informational Note: Many terminations and equipment are either marked with tightening torque or are identified as to tightening torque in the installation instructions provided.

Section 110.3(B) requires that the manufacturer's installation instructions be followed for listed equipment. This requirement applies where terminations and equipment are marked with tightening torques.

For the testing of wire connectors for which the manufacturer has not assigned a value appropriate for the design, Informative Annex I provides information on the tightening torques from UL 468A-B, *Wire Connectors*. These tables should be used for guidance only if no tightening information on a specific wire connector is available. They should not be used to replace the manufacturer's instructions, which should always be followed.

UL 486A-B refers to conductor stranding by class. Terminals and connectors for conductors that are more finely stranded than Class B and C stranding are now required to be identified for the class or classes of conductor stranding and the number of strands. This is consistent with 10.12 of UL 486A-B, which requires that connectors for other than Class B or C stranding be marked with the conductor class and the number of strands. Table 10 in Chapter 9 was added to provide information for the application of this information in the field.

**(A) Terminals.** Connection of conductors to terminal parts shall ensure a thoroughly good connection without damaging the conductors and shall be made by means of pressure connectors (including set-screw type), solder lugs, or splices to flexible leads. Connection by means of wire-binding screws or studs and nuts that have upturned lugs or the equivalent shall be permitted for 10 AWG or smaller conductors.

Terminals for more than one conductor and terminals used to connect aluminum shall be so identified.

**(B) Splices.** Conductors shall be spliced or joined with splicing devices identified for the use or by brazing, welding, or soldering with a fusible metal or alloy. Soldered splices shall first be spliced or joined so as to be mechanically and electrically secure without solder and then be soldered. All splices and joints and the free ends of conductors shall be covered with an insulation equivalent to that of the conductors or with an identified insulating device.

Wire connectors or splicing means installed on conductors for direct burial shall be listed for such use.

Field observations and trade magazine articles indicate that electrical connection failures are the cause of many equipment burnouts and fires. Many of these failures are attributable to improper terminations, poor workmanship, the differing characteristics of dissimilar metals, and improper binding screws or splicing devices.

UL's requirements for listing solid aluminum conductors in 12 AWG and 10 AWG and their requirements for listing snap switches and receptacles for use on 15- and 20-ampere branch circuits incorporate stringent tests. The tests are intended to detect many of the failures listed in the preceding paragraph. For further information regarding receptacles and switches using CO/ALR-rated terminals, refer to 404.14(C) and 406.3(C).

Screwless pressure terminal connectors of the conductor push-in type are for use only with solid copper and copper-clad aluminum conductors.

Instructions that describe proper installation techniques and emphasize the need to follow those techniques and practice good workmanship are required to be included with each coil of 12 AWG and 10 AWG insulated aluminum wire or cable. See also the commentary on tightening torque that follows 110.14, Informational Note.

The electrical industry has developed new product and material designs that provide increased levels of safety for aluminum wire terminations. To assist all concerned parties in the proper and safe use of solid aluminum wire in making connections to wiring devices used on 15- and 20-ampere branch circuits, the following information is presented.

## FOR NEW INSTALLATIONS

The following commentary is based on a report prepared by the Ad Hoc Committee on Aluminum Terminations prior to publication of the 1975 *Code*. This information is still pertinent today and is necessary for compliance with 110.14(A) when aluminum wire is used in new installations. New installation of aluminum conductors on 15- and 20-ampere branch circuits is not common, but many of these circuits continue to be in use.

### New Materials and Devices

For direct connection, only 15- and 20-ampere receptacles and switches marked "CO/ALR" and connected as follows should be used. The "CO/ALR" marking is on the device mounting yoke or strap. The "CO/ALR" marking means the devices have been tested to stringent heat-cycling requirements to determine their suitability for use with UL-labeled aluminum, copper, or copper-clad aluminum wire.

Listed solid aluminum wire, 12 AWG or 10 AWG, marked with the aluminum insulated wire label should be used. The installation instructions that are packaged with the wire should be followed.

### Installation Method

Exhibit 110.3 illustrates the following correct method of connection:

1. The freshly stripped end of the wire is wrapped two-thirds to three-quarters of the distance around the wire-binding screw post as shown in Step A. The loop is made so that rotation of the screw during tightening will tend to wrap the wire around the post rather than unwrap it.

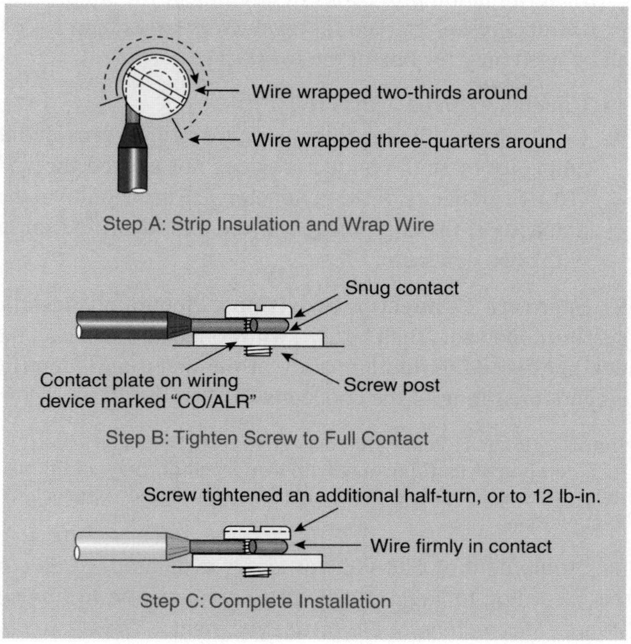


***EXHIBIT 110.3*** *Correct method of terminating aluminum wire at wire-binding screw terminals of receptacles and snap switches. (Courtesy of Underwriters Laboratories Inc.)*

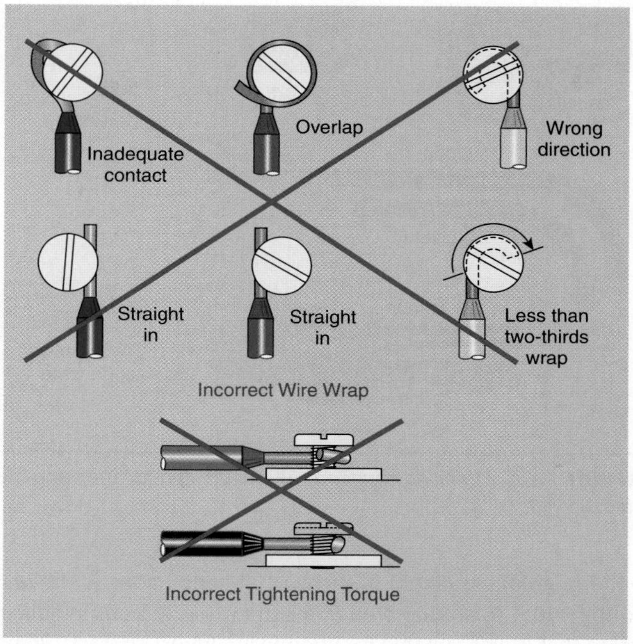


***EXHIBIT 110.4*** *Incorrect methods of terminating aluminum wire at wire-binding screw terminals of receptacles and snap switches. (Courtesy of Underwriters Laboratories Inc.)*

2. The screw is tightened until the wire is snugly in contact with the underside of the screw head and with the contact plate on the wiring device as shown in Step B.
3. The screw is tightened an additional half-turn, thereby providing a firm connection, as shown in Step C. When a torque screwdriver is used, the screw is tightened to 12 inch-pounds.
4. The wires should be positioned behind the wiring device to decrease the likelihood of the terminal screws loosening when the device is positioned into the outlet box.

Exhibit 110.4 illustrates incorrect methods of connection. These methods should *not* be used.

## Existing Inventory

Labeled 12 AWG or 10 AWG solid aluminum wire that does not bear the new aluminum wire label should be used with wiring devices marked "CO/ALR" and connected as described under Installation Method. This is the preferred and recommended method for using such wire.

For the following types of devices, the terminals should not be directly connected to aluminum conductors but may be used with labeled copper or copper-clad conductors:

1. Receptacles and snap switches marked "AL-CU"
2. Receptacles and snap switches having no conductor marking
3. Receptacles and snap switches that have back-wired terminals or screwless terminals of the push-in type

### FOR EXISTING INSTALLATIONS

If examination discloses overheating or loose connections, the recommendations described under Existing Inventory should be followed.

## Splicing Wire Connectors

Splicing wire connectors are required to be marked for the material of the conductor and for their suitability where intermixed. Splicing wire connectors, such as twist-on wire connectors, are not suitable for splicing aluminum conductors or copper-clad aluminum to copper conductors unless it is so stated and marked as such on the unit container or an information sheet supplied with the unit container. The required marking is "AL-CU (intermixed-dry locations)" where intermixing (direct contact) occurs. Other types of listed splicing wire connectors that are not rated for intermixing between the copper and the aluminum may also be used, as long as the conductors are not in direct physical contact. These connectors are just marked "AL-CU."

Underwriters Laboratories lists twist-on wire connectors that are suitable for use with aluminum-to-copper conductors, in accordance with UL 486C, *Splicing Wire Connectors*. The UL listing does *not* cover aluminum-to-aluminum combinations. However, more than one aluminum or copper conductor is allowed where used in combination. Suitable wire combinations are marked on the unit container or supplied on the information sheet with the unit container. These listed splicing wire-connecting devices are available for pigtailing short lengths of copper conductors directly

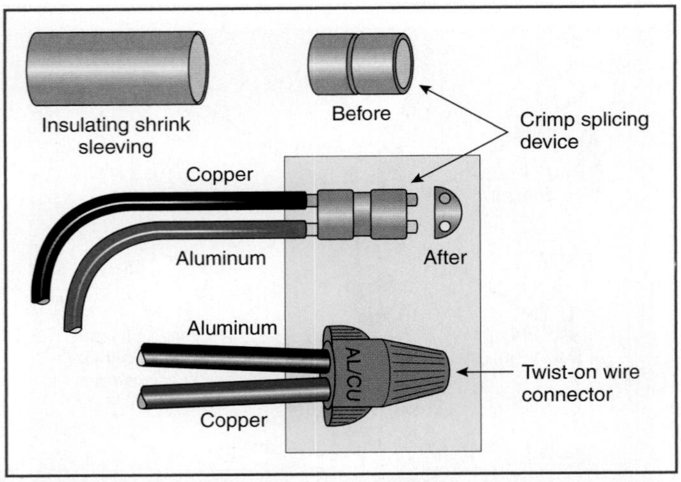


***EXHIBIT 110.5*** *Pigtailing copper-to-aluminum conductors using two listed devices.*

to the original aluminum branch-circuit conductors as shown in Exhibit 110.5. Also depicted is a similarly rated crimp splicing device that is also suitable for intermixing (direct contact). Primarily, these pigtailed conductors supply 15- and 20-ampere wiring devices. Pigtailing is permitted, provided suitable space is within the enclosure.

**(C) Temperature Limitations.** The temperature rating associated with the ampacity of a conductor shall be selected and coordinated so as not to exceed the lowest temperature rating of any connected termination, conductor, or device. Conductors with temperature ratings higher than specified for terminations shall be permitted to be used for ampacity adjustment, correction, or both.

**(1) Equipment Provisions.** The determination of termination provisions of equipment shall be based on 110.14(C)(1)(a) or (C)(1)(b). Unless the equipment is listed and marked otherwise, conductor ampacities used in determining equipment termination provisions shall be based on Table 310.15(B)(16) as appropriately modified by 310.15(B)(7).

(a) Termination provisions of equipment for circuits rated 100 amperes or less, or marked for 14 AWG through 1 AWG conductors, shall be used only for one of the following:

(1) Conductors rated 60°C (140°F).
(2) Conductors with higher temperature ratings, provided the ampacity of such conductors is determined based on the 60°C (140°F) ampacity of the conductor size used.
(3) Conductors with higher temperature ratings if the equipment is listed and identified for use with such conductors.
(4) For motors marked with design letters B, C, or D, conductors having an insulation rating of 75°C (167°F) or higher shall be permitted to be used, provided the ampacity of such conductors does not exceed the 75°C (167°F) ampacity.

(b) Termination provisions of equipment for circuits rated over 100 amperes, or marked for conductors larger than 1 AWG, shall be used only for one of the following:

(1) Conductors rated 75°C (167°F)
(2) Conductors with higher temperature ratings, provided the ampacity of such conductors does not exceed the 75°C (167°F) ampacity of the conductor size used, or up to their ampacity if the equipment is listed and identified for use with such conductors

**(2) Separate Connector Provisions.** Separately installed pressure connectors shall be used with conductors at the ampacities not exceeding the ampacity at the listed and identified temperature rating of the connector.

Informational Note: With respect to 110.14(C)(1) and (C)(2), equipment markings or listing information may additionally restrict the sizing and temperature ratings of connected conductors.

When equipment of 600 volts or less is evaluated, conductors sized according to Table 310.15(B)(16) are required to be used. UL *Guide Information for Electrical Equipment – The White Book* clearly indicates that the 60°C and 75°C termination temperature ratings for equipment have been determined using conductors from *NEC* Table 310.15(B)(16). However, installers or designers who are unaware of the UL guide information might attempt to select conductors based on a table other than Table 310.15(B)(16), especially if a wiring method is used that allows the use of ampacities such as those in Table 310.15(B)(17). That use can result in overheated terminations at the equipment. The ampacities shown in other tables [such as Table 310.15(B)(17)] could be used for various conditions to which the wiring method is subject (such as ambient or ampacity correction conditions), but the conductor size at the termination must be based on ampacities from Table 310.15(B)(16).

Conductor terminations, as well as conductors, must be rated for the operating temperature of the circuit. For example, the load on an 8 AWG THHN, 90°C copper conductor is limited to 40 amperes where connected to a disconnect switch with terminals rated at 60°C. The same conductor is limited to 50 amperes where connected to a fusible switch with terminals rated at 75°C. Not only do termination temperature ratings apply to conductor terminations, but the equipment enclosure marking must also permit terminations above 60°C. Exhibit 110.6 shows an example of termination temperature marking.

## 110.15 High-Leg Marking

On a 4-wire, delta-connected system where the midpoint of one phase winding is grounded, only the conductor or busbar having the higher phase voltage to ground shall be durably and permanently marked by an outer finish that is orange in color or by other effective means. Such identification shall be placed at each point on the system where a connection is made if the grounded conductor is also present.

***EXHIBIT 110.6*** *An example of termination temperature marking on a main circuit breaker. (Courtesy of International Association of Electrical Inspectors)*

The high leg is common on a 240/120-V 3-phase, 4-wire delta system. It is typically designated as "B phase." The high-leg marking, which is required to be the color orange or other similar effective means, is intended to prevent problems caused by the lack of standardization where metered and nonmetered equipment are installed in the same installation. See Exhibit 110.7. Electricians should always test each phase relative to ground with suitable equipment to determine exactly where the high leg is located in the system.

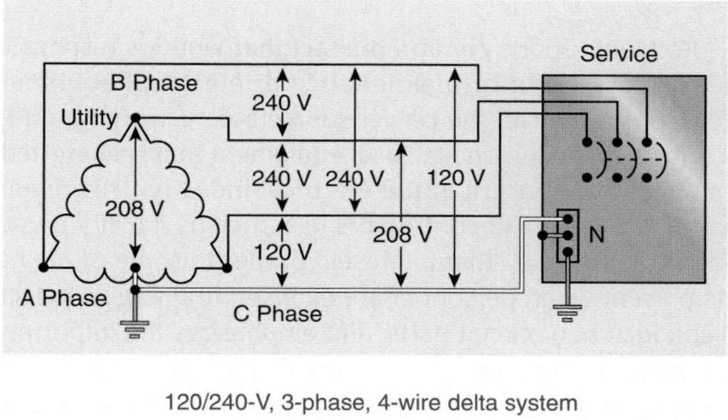


***EXHIBIT 110.7*** *A 240/120-volt 3-phase, 4-wire delta system.*

## 110.16 Arc-Flash Hazard Warning

Electrical equipment, such as switchboards, switchgear, panelboards, industrial control panels, meter socket enclosures, and motor control centers, that are in other than dwelling units, and are likely to require examination, adjustment, servicing, or maintenance while energized, shall be field or factory marked to warn qualified persons of potential electric arc flash hazards. The marking shall meet the requirements in 110.21(B) and shall be located so as to be clearly visible to qualified persons before examination, adjustment, servicing, or maintenance of the equipment.

This section requires switchboards, panelboards, motor control centers, and other equipment to be individually field or factory marked with proper warning labels to raise the level of awareness of electrical arc-flash hazards and to decrease the number of accidents that result when electricians do not wear the proper type of protective clothing when working on "hot" (energized) equipment. Exhibit 110.8 is one example of an equipment warning sign as required by this section.

Exhibit 110.9 shows an electrical employee wearing personal protective equipment (PPE) considered appropriate flash protection clothing for the flash hazard involved. Suitable PPE appropriate to a particular hazard is described in *NFPA 70E®, Standard for Electrical Safety in the Workplace®*.

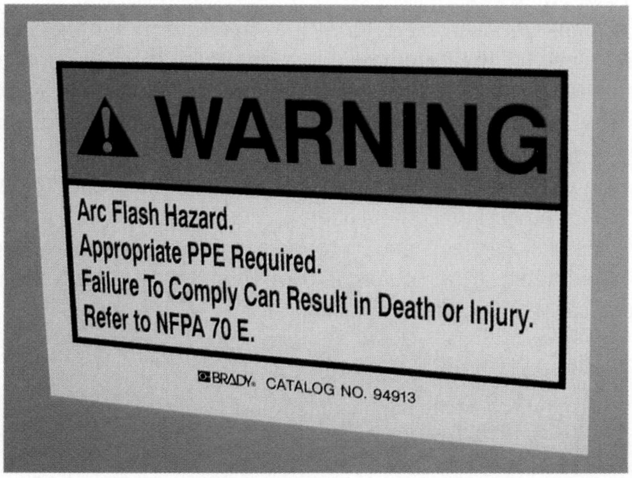


***EXHIBIT 110.8*** *One example of an arc-flash warning sign. (Courtesy of the International Association of Electrical Inspectors)*

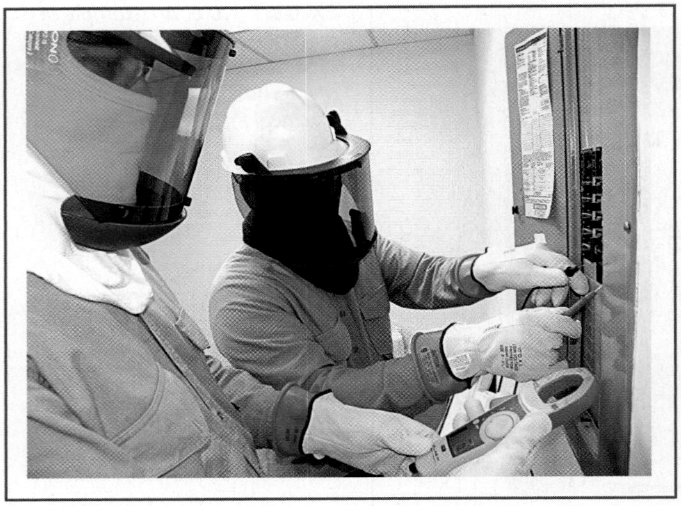

***EXHIBIT 110.9*** *An electrical worker clothed in personal protective equipment (PPE) appropriate for the hazard involved. (Courtesy of KTR Associates/ArcFlashPPE.com)*

Accident reports confirm the fact that workers responsible for the installation or maintenance of electrical equipment often do not turn off the power source before working on the equipment. Working on electrical equipment that is energized is a major safety concern in the electrical industry. This requirement alerts electrical contractors, electricians, facility owners and managers, and other interested parties to some of the hazards present when personnel are exposed to energized electrical conductors or circuit parts, and emphasizes the importance of turning off the power before working on electrical circuits. This section does not apply to equipment in dwelling units. However, dwelling occupancies include multifamily dwellings, which include multiple dwelling units and could have the same electric service as a commercial office building. The intent is to provide warnings to electricians working on these larger services.

Employers can be assured that they are providing a safe workplace for their employees if safety-related work practices required by *NFPA 70E* have been implemented and are being followed. (See also the commentary following the definition of *qualified person* in Article 100.)

In addition to the standards referenced in the informational notes and their individual bibliographies, additional information on electrical accidents can be found in the 1997 report "Hazards of Working Electrical Equipment Hot," published by the National Electrical Manufacturers Association.

Informational Note No. 1: NFPA 70E-2012, *Standard for Electrical Safety in the Workplace*, provides guidance, such as determining severity of potential exposure, planning safe work practices, arc flash labeling, and selecting personal protective equipment.

Informational Note No. 2: ANSI Z535.4-1998, *Product Safety Signs and Labels*, provides guidelines for the design of safety signs and labels for application to products.

## 110.18 Arcing Parts

Parts of electrical equipment that in ordinary operation produce arcs, sparks, flames, or molten metal shall be enclosed or separated and isolated from all combustible material.

Examples of electrical equipment that may produce sparks during ordinary operation include open motors having a centrifugal starting switch, open motors with commutators, and collector rings. Adequate separation from combustible material is essential if open motors with those features are used.

Informational Note: For hazardous (classified) locations, see Articles 500 through 517. For motors, see 430.14.

## 110.19 Light and Power from Railway Conductors

Circuits for lighting and power shall not be connected to any system that contains trolley wires with a ground return.

*Exception: Such circuit connections shall be permitted in car houses, power houses, or passenger and freight stations operated in connection with electric railways.*

## 110.21 Marking

**(A) Manufacturer's Markings.** The manufacturer's name, trademark, or other descriptive marking by which the organization responsible for the product can be identified shall be placed on all electrical equipment. Other markings that indicate voltage, current, wattage, or other ratings shall be provided as specified elsewhere in this *Code*. The marking or label shall be of sufficient durability to withstand the environment involved.

**(B) Field-Applied Hazard Markings.** Where caution, warning, or danger signs or labels are required by this *Code*, the labels shall meet the following requirements:

The *Code* requires that equipment be marked with the identification of the manufacturer by name or appropriate trademark, the equipment ratings, and hazard warnings. The markings must be located where they will be visible or easily accessible during or after installation. The marking must not be handwritten and must have adequate durability to survive the environment. Only information that is variable or subject to change is permitted to be handwritten. See Exhibit 110.10.

(1) The marking shall adequately warn of the hazard using effective words and/or colors and/or symbols.

Informational Note: ANSI Z535.4-2011, *Product Safety Signs and Labels*, provides guidelines for suitable font sizes, words, colors, symbols, and location requirements for labels.

(2) The label shall be permanently affixed to the equipment or wiring method and shall not be hand written.

*Exception to (2): Portions of labels or markings that are variable, or that could be subject to changes, shall be permitted to be hand written and shall be legible.*

(3) The label shall be of sufficient durability to withstand the environment involved.

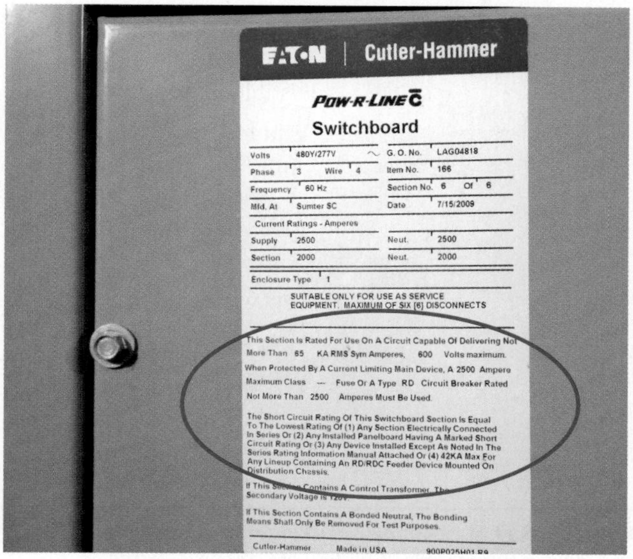


***EXHIBIT 110.10*** *A short-circuit current rating marking.*

Informational Note: ANSI Z535.4-2011, *Product Safety Signs and Labels*, provides guidelines for the design and durability of safety signs and labels for application to electrical equipment.

## 110.22 Identification of Disconnecting Means

**(A) General.** Each disconnecting means shall be legibly marked to indicate its purpose unless located and arranged so the purpose is evident. The marking shall be of sufficient durability to withstand the environment involved.

This section requires that markings of disconnecting means specifically identify the purpose of each piece of equipment – that is, the marking should not indicate simply "motor," but rather "motor, water pump," and not simply "lights," but rather "lights, front lobby." Consideration of the form of identification is also required to assure that the markings do not fade or wear off. Section 408.4 and its associated commentary provide additional requirements and information on circuit directories for switchboards and panelboards.

**(B) Engineered Series Combination Systems.** Equipment enclosures for circuit breakers or fuses applied in compliance with series combination ratings selected under engineering supervision in accordance with 240.86(A) shall be legibly marked in the field as directed by the engineer to indicate the equipment has been applied with a series combination rating. The marking shall meet the requirements in 110.21(B) and shall be readily visible and state the following:

CAUTION — ENGINEERED SERIES COMBINATION SYSTEM RATED _______ AMPERES. IDENTIFIED REPLACEMENT COMPONENTS REQUIRED.

If the ratings are determined under engineering supervision, the equipment must have a durable label, as specified in 110.21(B), indicating that the series combination rating has been used. It is important that the warnings on replacement components be heeded in order to maintain the level of protection provided by the design. Likewise, where components are replaced, new or updated warning labels with information based on the new component may be necessary.

**(C) Tested Series Combination Systems.** Equipment enclosures for circuit breakers or fuses applied in compliance with the series combination ratings marked on the equipment by the manufacturer in accordance with 240.86(B) shall be legibly marked in the field to indicate the equipment has been applied with a series combination rating. The marking shall meet the requirements in 110.21(B) and shall be readily visible and state the following:

CAUTION — SERIES COMBINATION SYSTEM RATED ____ AMPERES. IDENTIFIED REPLACEMENT COMPONENTS REQUIRED.

The equipment manufacturer can mark the equipment to be used with series combination ratings. If the equipment is installed in the field at its marked series combination rating, the equipment must have an additional durable label, as specified in 110.21(B), indicating that the series combination rating was used.

## 110.23 Current Transformers

Unused current transformers associated with potentially energized circuits shall be short-circuited.

Because Article 450 specifically exempts current transformers, the requirement in 110.23 provides the practical means of preventing damage to current transformers not connected to a load or for unused current transformers.

## 110.24 Available Fault Current

**(A) Field Marking.** Service equipment in other than dwelling units shall be legibly marked in the field with the maximum available fault current. The field marking(s) shall include the date the fault-current calculation was performed and be of sufficient durability to withstand the environment involved.

Informational Note: The available fault-current marking(s) addressed in 110.24 is related to required short-circuit current ratings of equipment. *NFPA 70E*-2012, *Standard for Electrical Safety in the Workplace*, provides assistance in determining the severity of potential exposure, planning safe work practices, and selecting personal protective equipment.

**(B) Modifications.** When modifications to the electrical installation occur that affect the maximum available fault current at the service, the maximum available fault current shall be verified or recalculated as necessary to ensure the service equipment ratings are sufficient for the maximum available fault current at the line terminals of the equipment. The required field marking(s) in

110.24(A) shall be adjusted to reflect the new level of maximum available fault current.

*Exception: The field marking requirements in 110.24(A) and 110.24(B) shall not be required in industrial installations where conditions of maintenance and supervision ensure that only qualified persons service the equipment.*

To be used safely, equipment must have an interrupting rating or short-circuit current rating equal to or greater than the available fault current. Any equipment operating with ratings less than the available fault current is potentially unsafe. Existing electrical distribution systems often experience change over the life of the system. As the system ages, the supply network to which it is connected is impacted by growth and is forced to increase capacity or increase efficiency by reducing transformer impedance. In some cases, alternative energy systems are added to existing installations. Such changes to the electrical distribution system can result in an increase of the available fault current. This increase in available fault can exceed the ratings of the originally installed equipment violating 110.9 and 110.10, creating an unsafe condition. This section requires an initial marking of maximum available fault current as well as the requirement to update the information when the system is modified.

In order to complete an arc-flash hazard analysis per Section 130.3 of *NFPA 70E, Standard for Electrical Safety in the Workplace,* the available fault current must be known. This analysis is used to determine the arc-flash protection boundary and required PPE in accordance with *NFPA 70E,* Section 130.3(A) and 130.3(B). The equipment must then be marked with the incident energy or required level of PPE per 130.3(C). In addition, per *NFPA 70E* 130.3, an arc-flash hazard analysis is also required to be updated when major modifications or renovations take place. The analysis must be reviewed periodically but not less than every 5 years, to account for changes in the electrical distribution system that could affect the original arc-flash analysis.

## 110.25 Lockable Disconnecting Means

Where a disconnecting means is required to be lockable open elsewhere in this Code, it shall be capable of being locked in the open position. The provisions for locking shall remain in place with or without the lock installed.

*Exception: Cord-and-plug connection locking provisions shall not be required to remain in place without the lock installed.*

This section, added for the 2014 *NEC,* consolidates and harmonizes with the lockable disconnecting means requirements throughout the *Code.* The means to lock the switch or circuit breaker in the open position must be an integral part of the enclosure or be an accessory that is not readily removed from the switch or circuit breaker. Portable locking mechanisms that are intended for temporary applications are not acceptable means of compliance. See Exhibit 110.11.

***EXHIBIT 110.11*** *An example of locking hardware that is not readily removable or transferable. (Courtesy of Schneider Electric)*

# II. 600 Volts, Nominal, or Less

## 110.26 Spaces About Electrical Equipment

Access and working space shall be provided and maintained about all electrical equipment to permit ready and safe operation and maintenance of such equipment.

Spaces about electrical equipment are divided into two separate and distinct categories: working space and dedicated equipment space. The term *working space* generally applies to the protection of the worker, and *dedicated equipment space* applies to the space reserved for future access to electrical equipment and to protection of the equipment from intrusion by nonelectrical equipment. The performance requirements for all spaces about electrical equipment are set forth in this section. Storage of material that blocks access or prevents safe work practices must be avoided at all times.

**(A) Working Space.** Working space for equipment operating at 600 volts, nominal, or less to ground and likely to require examination, adjustment, servicing, or maintenance while energized shall comply with the dimensions of 110.26(A)(1), (A)(2), and (A)(3) or as required or permitted elsewhere in this *Code.*

The intent is to provide enough space for personnel to perform any of the operations listed without jeopardizing worker safety. Examples of such equipment include panelboards, switches, circuit breakers, controllers, and controls on heating and air-conditioning equipment. Note that the word *examination,* as used in 110.26(A), includes tasks such as checking for the presence of voltage using a portable voltmeter.

Minimum working clearances are not required if the equipment is not likely to require examination, adjustment, servicing, or maintenance while energized. However, access and working space are still required by the opening paragraph of 110.26.

**(1) Depth of Working Space.** The depth of the working space in the direction of live parts shall not be less than that specified in Table 110.26(A)(1) unless the requirements of 110.26(A)(1)(a), (A)(1)(b), or (A)(1)(c) are met. Distances shall be measured from the exposed live parts or from the enclosure or opening if the live parts are enclosed.

***TABLE 110.26(A)(1)*** *Working Spaces*

| Nominal Voltage to Ground | Minimum Clear Distance | | |
|---|---|---|---|
| | Condition 1 | Condition 2 | Condition 3 |
| 0–150 | 914 mm (3 ft) | 914 mm (3 ft) | 914 mm (3 ft) |
| 151–600 | 914 mm (3 ft) | 1.07 m (3 ft 6 in.) | 1.22 m (4 ft) |

Note: Where the conditions are as follows:
**Condition 1** — Exposed live parts on one side of the working space and no live or grounded parts on the other side of the working space, or exposed live parts on both sides of the working space that are effectively guarded by insulating materials.
**Condition 2** — Exposed live parts on one side of the working space and grounded parts on the other side of the working space. Concrete, brick, or tile walls shall be considered as grounded.
**Condition 3** — Exposed live parts on both sides of the working space.

Included in the clearance requirements in Table 110.26(A)(1) is the step-back distance from the face of the equipment. This table provides requirements for clearances away from the equipment, based on the circuit voltage to ground and whether there are grounded or ungrounded objects in the step-back space or exposed live parts across from each other. The voltages to ground consist of two groups: 0 to 150, inclusive, and 151 to 600, inclusive.

Examples of common electrical supply systems covered in the 0 to 150 volts-to-ground group are 120/240-V, single-phase, 3-wire and 208Y/120-V, 3-phase, 4-wire systems. Examples of common electrical supply systems covered in the 151 to 600 volts-to-ground group are 240-V, 3-phase, 3-wire; 480Y/277-V, 3-phase, 4-wire; and 480-V, 3-phase, 3-wire (ungrounded and corner grounded) systems. Where an ungrounded system is utilized, the voltage to ground (by definition) is the greatest voltage between the given conductor and any other conductor of the circuit. For example, the voltage to ground for a 480-V ungrounded delta system is 480 V. See Exhibit 110.12 for the general working clearance requirements for each of the three conditions listed in Table 110.26(A)(1). For assemblies, such as switchboards, switchgear, or motor-control centers, that are accessible from the back and expose live parts, the working clearance dimensions are required at the rear of the equipment as illustrated. For Condition 3, where an enclosure is on opposite sides of the working space, the clearance for only one working space is required.

(a) *Dead-Front Assemblies.* Working space shall not be required in the back or sides of assemblies, such as dead-front switchboards, switchgear, or motor control centers, where all connections and all renewable or adjustable parts, such as fuses or switches, are accessible from locations other than the back or sides. Where rear access is required to work on nonelectrical parts on the back of enclosed equipment, a minimum horizontal working space of 762 mm (30 in.) shall be provided.

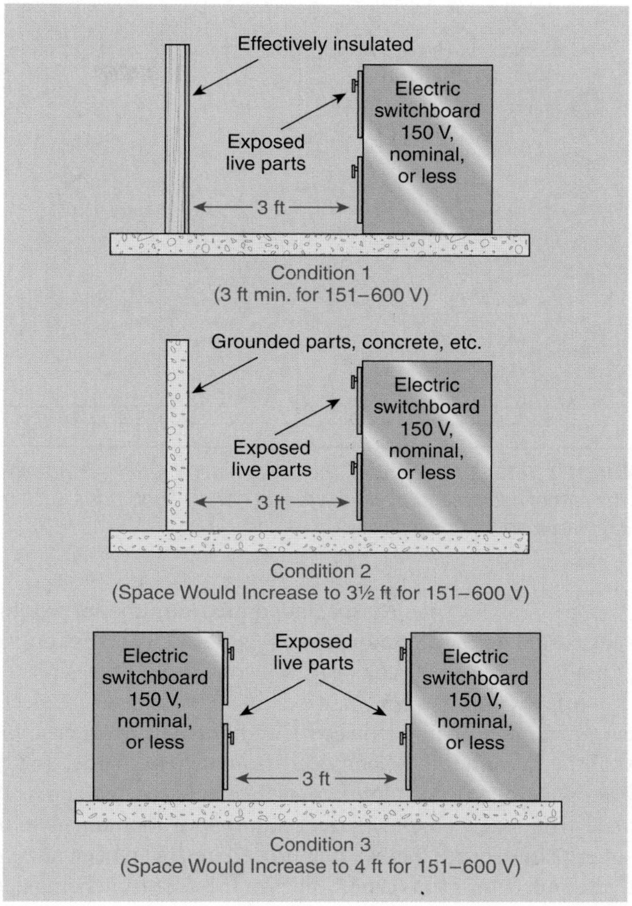


***EXHIBIT 110.12*** *Distances measured from the live parts if the live parts are exposed or from the enclosure front if the live parts are enclosed.*

The intent of this section is to point out that work space is required only from the side(s) of the enclosure requiring access. The general rule applies: Equipment that requires front, rear, or side access for the electrical activities described in 110.26(A) must meet the requirements of Table 110.26(A)(1). In many cases, equipment of "dead-front" assemblies requires only front access. For equipment that requires rear access for nonelectrical activity, however, a reduced working space of at least 30 inches must be provided. Exhibit 110.13 shows a reduced working space of 30 inches at the rear of equipment to allow work on nonelectrical parts.

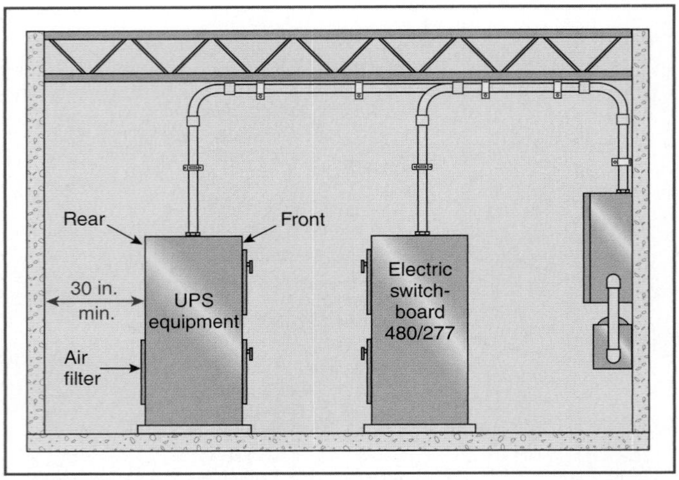


**EXHIBIT 110.13** *An example of the 30-inch minimum working space at the rear of equipment to allow work on nonelectrical parts, such as the replacement of an air filter.*

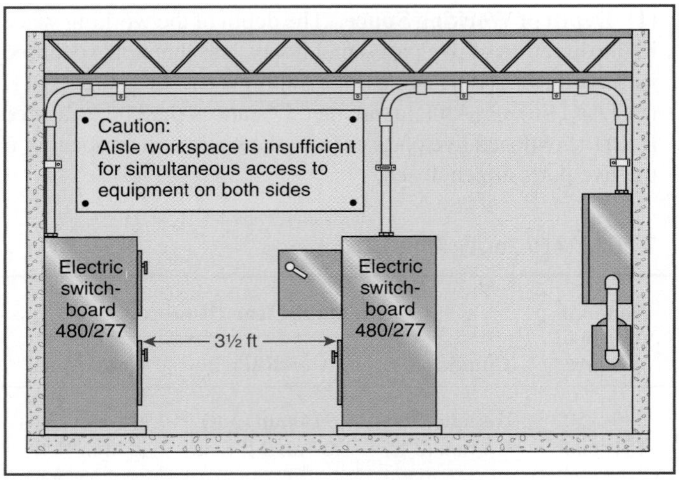


**EXHIBIT 110.14** *A permitted reduction from a Condition 3 to a Condition 2 clearance.*

(b) *Low Voltage.* By special permission, smaller working spaces shall be permitted where all exposed live parts operate at not greater than 30 volts rms, 42 volts peak, or 60 volts dc.

(c) *Existing Buildings.* In existing buildings where electrical equipment is being replaced, Condition 2 working clearance shall be permitted between dead-front switchboards, switchgear, panelboards, or motor control centers located across the aisle from each other where conditions of maintenance and supervision ensure that written procedures have been adopted to prohibit equipment on both sides of the aisle from being open at the same time and qualified persons who are authorized will service the installation.

This section permits some relief for installations that are being upgraded. When assemblies such as dead-front switchboards, panelboards, or motor-control centers are replaced in an existing building, the working clearance allowed is that required by Table 110.26(A)(1), Condition 2. The reduction from a Condition 3 to a Condition 2 clearance is allowed only where a written procedure prohibits facing doors of equipment from being open at the same time and where only authorized and qualified persons service the installation. Exhibit 110.14 illustrates this relief for existing buildings.

**(2) Width of Working Space.** The width of the working space in front of the electrical equipment shall be the width of the equipment or 762 mm (30 in.), whichever is greater. In all cases, the work space shall permit at least a 90 degree opening of equipment doors or hinged panels.

Regardless of the width of the electrical equipment, the working space cannot be less than 30 inches wide as required by Table 110.26(A)(1). This space allows an individual to have at least shoulder-width space in front of the equipment. The 30-inch measurement can be made from either the left or the right edge of the equipment and can overlap other electrical equipment, provided the other equipment does not extend beyond the clearance required by Table 110.26(A)(1). If the equipment is wider than 30 inches, the left-to-right space must be equal to the width of the equipment. Exhibit 110.15 illustrates the 30-inch width requirement.

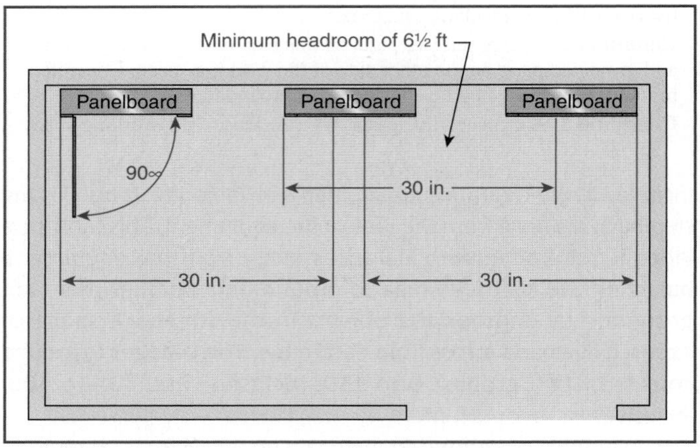


**EXHIBIT 110.15** *The 30-inch-wide front working space, which is not required to be directly centered on the electrical equipment and can overlap other electrical equipment.*

Sufficient depth in the working space is also required to allow a panel or a door to open at least 90 degrees. If doors or hinged panels are wider than 3 feet, more than a 3-foot deep working space must be provided to allow a full 90-degree opening. (See Exhibit 110.16.)

**(3) Height of Working Space.** The work space shall be clear and extend from the grade, floor, or platform to a height of 2.0 m (6½ ft) or the height of the equipment, whichever is greater. Within the height requirements of this section, other equipment that is associated with the electrical installation

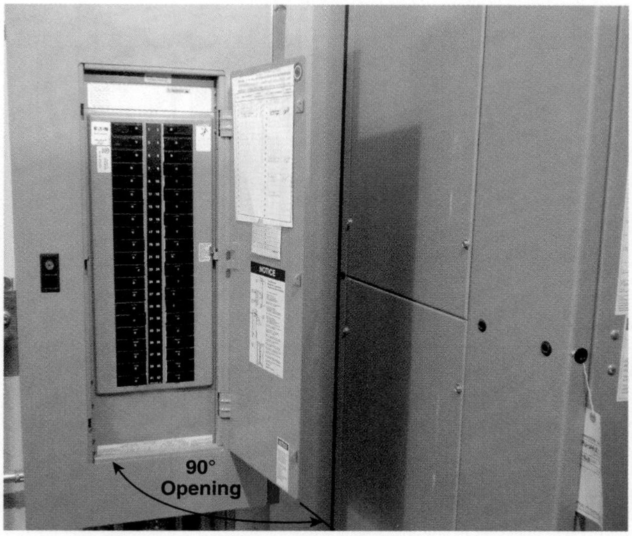


**EXHIBIT 110.16** *A full 90-degree opening of an equipment door in order to ensure a safe working approach.*

**EXHIBIT 110.17** *An equipment location that is free of storage to allow the equipment to be worked on safely. (Courtesy of the International Association of Electrical Inspectors)*

and is located above or below the electrical equipment shall be permitted to extend not more than 150 mm (6 in.) beyond the front of the electrical equipment.

In addition to requiring a working space to be clear from the floor to a height of 6½ feet or to the height of the equipment, whichever is greater, 110.26(A)(3) permits electrical equipment located above or below other electrical equipment to extend into the working space not more than 6 inches. This requirement allows the placement of a 12 inch × 12 inch wireway on the wall directly above or below a 6-inch-deep panelboard without impinging on the working space or compromising practical working clearances. The requirement prohibits large differences in depth of equipment below or above other equipment that specifically requires working space. To minimize the amount of space required for electrical equipment, large free-standing, dry-type transformers are commonly installed within the required work space for a wall-mounted panelboard, which compromises clear access to the panelboard and is clearly not permitted by this section. Electrical equipment that produces heat or that otherwise requires ventilation also must comply with 110.3(B) and 110.13.

*Exception No. 1: In existing dwelling units, service equipment or panelboards that do not exceed 200 amperes shall be permitted in spaces where the height of the working space is less than 2.0 m (6½ ft).*

*Exception No. 2: Meters that are installed in meter sockets shall be permitted to extend beyond the other equipment. The meter socket shall be required to follow the rules of this section.*

For meter socket installations, the penetration of the workspace is based on the meter socket itself, without the meter installed. The installation of the meter in the socket could extend beyond the 6 inches, which is permitted by 110.26(A)(3), Exception No. 2.

**(B) Clear Spaces.** Working space required by this section shall not be used for storage. When normally enclosed live parts are exposed for inspection or servicing, the working space, if in a passageway or general open space, shall be suitably guarded.

This section and the rest of 110.26 do not prohibit the placement of panelboards in corridors or passageways. For that reason, when the covers of corridor-mounted panelboards are removed for servicing or other work, access to the area around the panelboard should be guarded or limited to prevent injury to unqualified persons using the corridor.

Equipment that requires servicing while energized must be located in an area that is not used for storage as shown in Exhibit 110.17.

**(C) Entrance to and Egress from Working Space.**

**(1) Minimum Required.** At least one entrance of sufficient area shall be provided to give access to and egress from working space about electrical equipment.

The requirements in this section are intended to provide access to electrical equipment. However, the primary intent is to provide egress from the area so that workers can escape if an arc-flash incident occurs.

**(2) Large Equipment.** For equipment rated 1200 amperes or more and over 1.8 m (6 ft) wide that contains overcurrent devices, switching devices, or control devices, there shall be one entrance to and egress from the required working space not less than 610 mm (24 in.) wide and 2.0 m (6½ ft) high at each end of the working space.

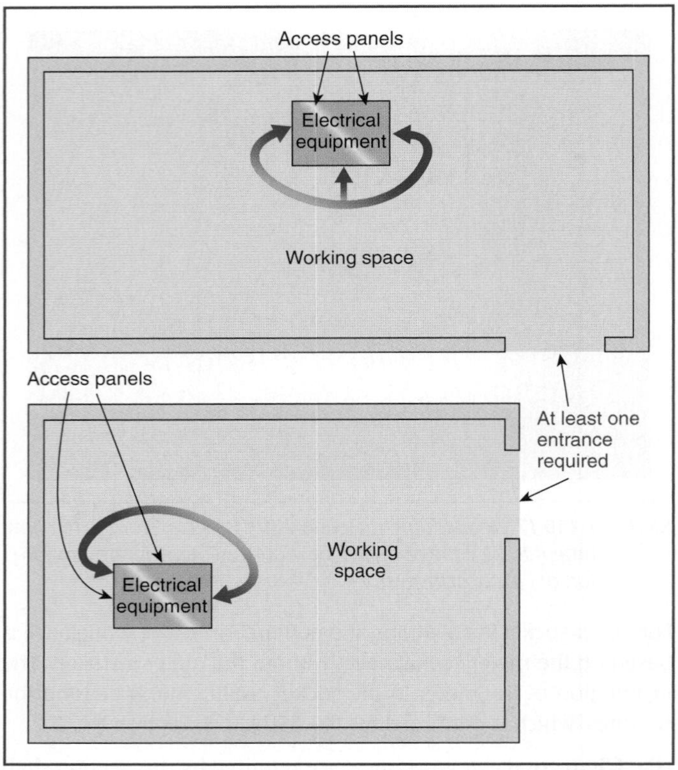


**EXHIBIT 110.18** *At least one entrance is required to provide access to the working space around electrical equipment.*

A single entrance to and egress from the required working space shall be permitted where either of the conditions in 110.26(C)(2)(a) or (C)(2)(b) is met.

Exhibits 110.18 and 110.19 provide a visual explanation of access and entrance requirements for working spaces. Exhibit 110.20 shows an unacceptable and hazardous work space arrangement. See Exhibits 110.21 and 110.22 for a representation of the single egress requirements for large equipment.

Where the entrance(s) to the working space is through a door, each door must comply with the requirements for swinging open in the direction of egress and have door opening hardware that does not require turning a door knob or similar action that may preclude quick exit from the area in the event of an emergency. This requirement affords safety for workers exposed to energized conductors by allowing an injured worker to safely and quickly exit an electrical room without having to turn knobs or pull doors open.

(a) *Unobstructed Egress.* Where the location permits a continuous and unobstructed way of egress travel, a single entrance to the working space shall be permitted.

(b) *Extra Working Space.* Where the depth of the working space is twice that required by 110.26(A)(1), a single entrance shall be permitted. It shall be located such that the distance from the equipment to the nearest edge of the entrance is not less than the minimum clear distance specified in Table 110.26(A)(1) for equipment operating at that voltage and in that condition.

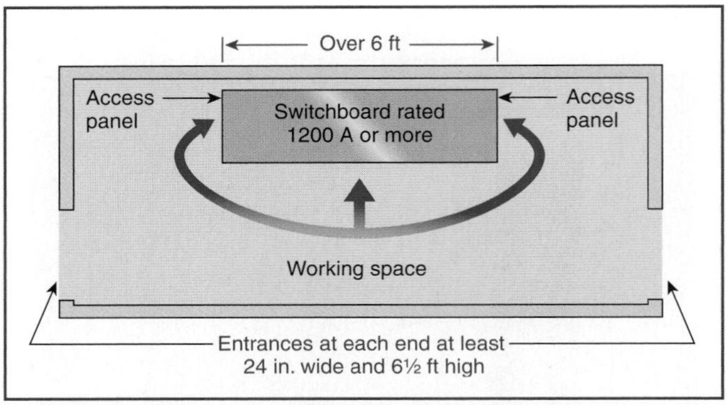


**EXHIBIT 110.19** *For equipment rated 1200 amperes or more and over 6 feet wide, one entrance not less than 24 inches wide and 6½ feet high is required at each end.*

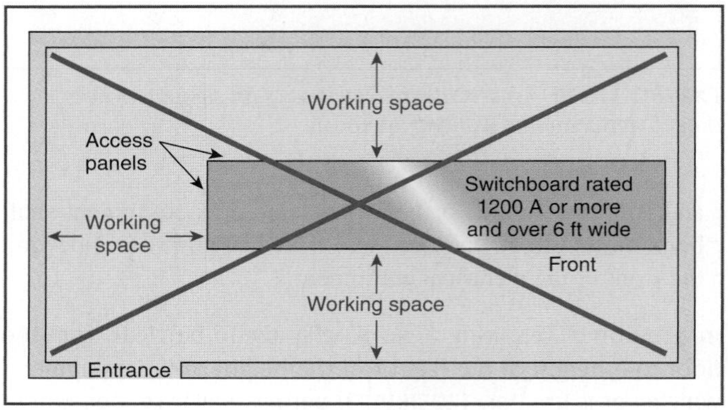


**EXHIBIT 110.20** *An unacceptable arrangement of a large switchboard in which a worker could be trapped behind arcing electrical equipment.*

For an explanation of paragraphs 110.26(C)(2)(a) and 110.26(C)(2)(b), see Exhibits 110.21 and 110.22.

**(3) Personnel Doors.** Where equipment rated 800 A or more that contains overcurrent devices, switching devices, or control devices is installed and there is a personnel door(s) intended for entrance to and egress from the working space less than 7.6 m (25 ft) from the nearest edge of the working space, the door(s) shall open in the direction of egress and be equipped with listed panic hardware.

The requirements in this section are based on equipment rated 800 amperes or more, not on the width of the equipment. The measurement for less than 25 feet for the personnel door(s) is made from the nearest edge of the working space, and the requirement for listed panic hardware and outward egress is independent of the need for two exits from the working space. Not every electrical installation is in an equipment room. This section requires

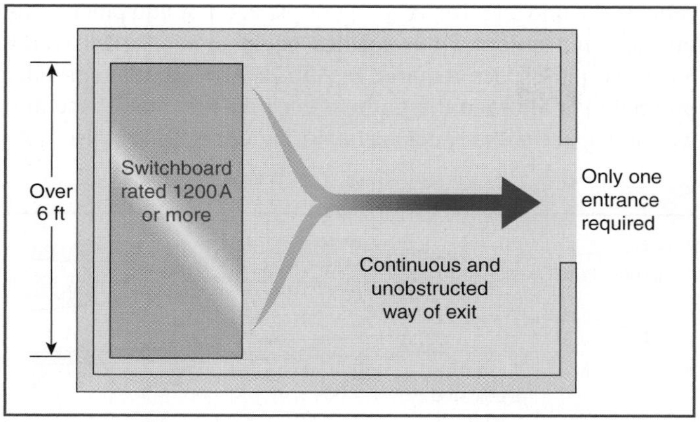


***EXHIBIT 110.21*** *An equipment location that allows a continuous and unobstructed way of exit travel.*

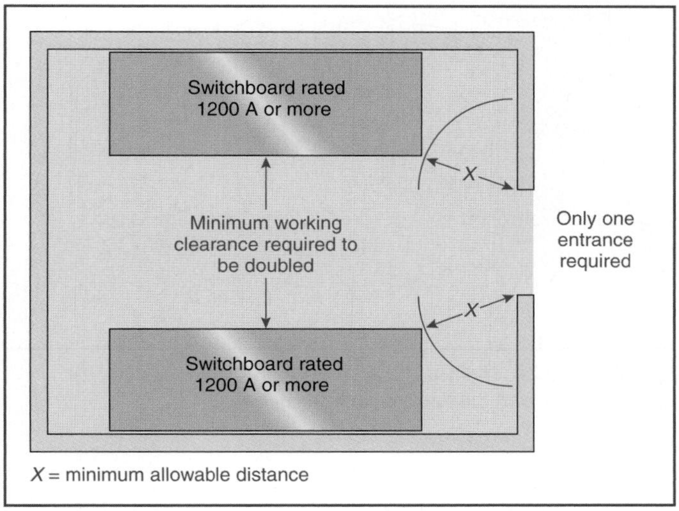


***EXHIBIT 110.22*** *A working space with one entrance, which is permitted if the working space required by 110.26(A) is doubled [see Table 110.26(A)(1) for permitted dimensions of X].*

personnel doors that are up to 25 feet from the working space to have listed panic hardware and to open in the direction of egress from the area. Exhibit 110.23 shows one of the required exits.

**(D) Illumination.** Illumination shall be provided for all working spaces about service equipment, switchboards, switchgear, panelboards, or motor control centers installed indoors and shall not be controlled by automatic means only. Additional lighting outlets shall not be required where the work space is illuminated by an adjacent light source or as permitted by 210.70(A)(1), Exception No. 1, for switched receptacles.

This section requires working spaces around service equipment, switchboards, switchgear, and motor control centers to have a nonautomatic means to control the lighting. Automatic lighting control through devices such as occupancy sensors and similar devices would not be prohibited, but a manual means to bypass the automatic control is required.

***EXHIBIT 110.23*** *An installation of large equipment showing one of the required exits. (Courtesy of the International Association of Electrical Inspectors)*

**(E) Dedicated Equipment Space.** All switchboards, switchgear, panelboards, and motor control centers shall be located in dedicated spaces and protected from damage.

*Exception: Control equipment that by its very nature or because of other rules of the Code must be adjacent to or within sight of its operating machinery shall be permitted in those locations.*

**(1) Indoor.** Indoor installations shall comply with 110.26(E)(1)(a) through (E)(1)(d).

(a) *Dedicated Electrical Space.* The space equal to the width and depth of the equipment and extending from the floor to a height of 1.8 m (6 ft) above the equipment or to the structural ceiling, whichever is lower, shall be dedicated to the electrical installation. No piping, ducts, leak protection apparatus, or other equipment foreign to the electrical installation shall be located in this zone.

*Exception: Suspended ceilings with removable panels shall be permitted within the 1.8-m (6-ft) zone.*

(b) *Foreign Systems.* The area above the dedicated space required by 110.26(E)(1)(a) shall be permitted to contain foreign systems, provided protection is installed to avoid damage to the electrical equipment from condensation, leaks, or breaks in such foreign systems.

(c) *Sprinkler Protection.* Sprinkler protection shall be permitted for the dedicated space where the piping complies with this section.

(d) *Suspended Ceilings.* A dropped, suspended, or similar ceiling that does not add strength to the building structure shall not be considered a structural ceiling.

The dedicated electrical space, which extends the footprint of the switchboard or panelboard from the floor to a height of 6 feet above the height of the equipment or to the structural ceiling (whichever is lower), is required to be clear of piping, ducts, leak protection apparatus, or equipment foreign to the electrical installation. Plumbing, heating, ventilation, and air-conditioning piping, ducts, and equipment must be installed outside the width and depth zone. Busways, conduits, raceways, and cables are permitted to enter equipment through this zone.

Foreign systems installed directly above the dedicated space reserved for electrical equipment are required to include protective equipment that ensures that occurrences such as leaks, condensation, and even breaks do not damage the electrical equipment located below.

Sprinkler protection is permitted for the dedicated spaces as long as the sprinkler or other suppression system piping complies with 110.26(E)(1)(c). A dropped, suspended, or similar ceiling is permitted to be located directly in the dedicated space, because they are not considered structural ceilings. Building structural members are also permitted in this space.

The electrical equipment also must be protected from physical damage. Damage can be caused by activities performed near the equipment, such as material handling by personnel or the operation of a forklift or other mobile equipment. See 110.27(B) for other requirements relating to the protection of electrical equipment.

Exhibits 110.24, 110.25, and 110.26 illustrate the two distinct indoor installation spaces required by 110.26(A) and 110.26(E), that is, the working space and the dedicated electrical space.

In Exhibit 110.24, the dedicated electrical space required by 110.26(E) is the space outlined by the width and the depth of the equipment (the footprint) and extending from the floor to 6 feet above the equipment or to the structural ceiling (whichever is lower). The dedicated electrical space is reserved for the installation of electrical equipment and for the installation of conduits, cable trays, and so forth, entering or exiting that equipment. The outlined area in front of the electrical equipment in Exhibit 110.24 is the working space required by 110.26(A). Note that sprinkler protection is afforded the entire dedicated electrical space and working space without actually entering either space. Also note

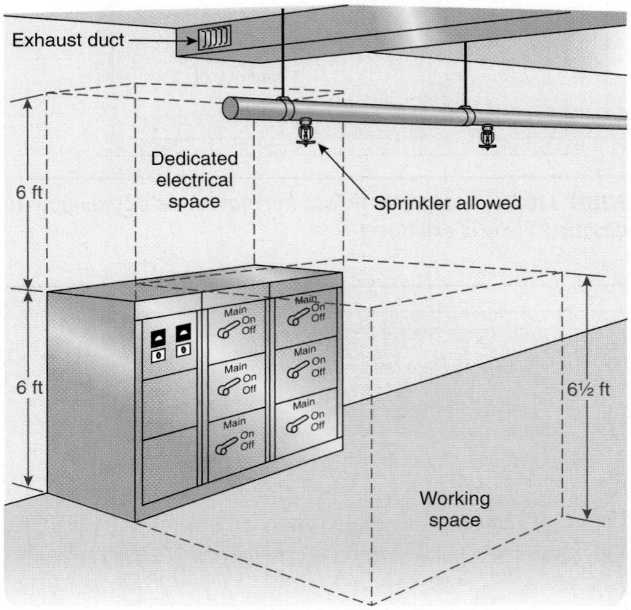


***EXHIBIT 110.24*** *The two distinct indoor installation spaces: the working space and the dedicated electrical space.*

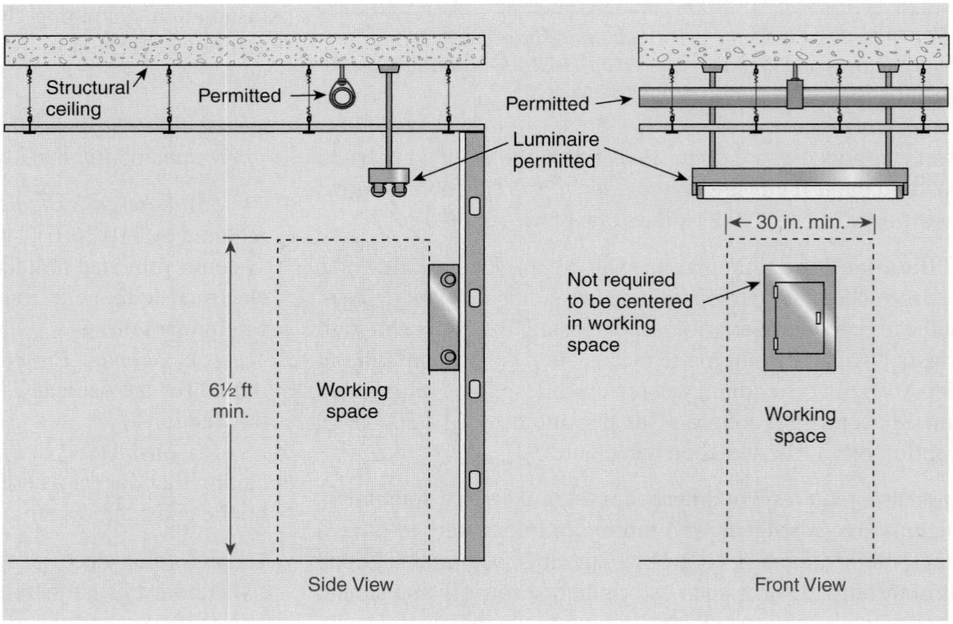


***EXHIBIT 110.25*** *The working space in front of a panelboard. (This illustration supplements the dedicated electrical space shown in Exhibit 110.24.)*

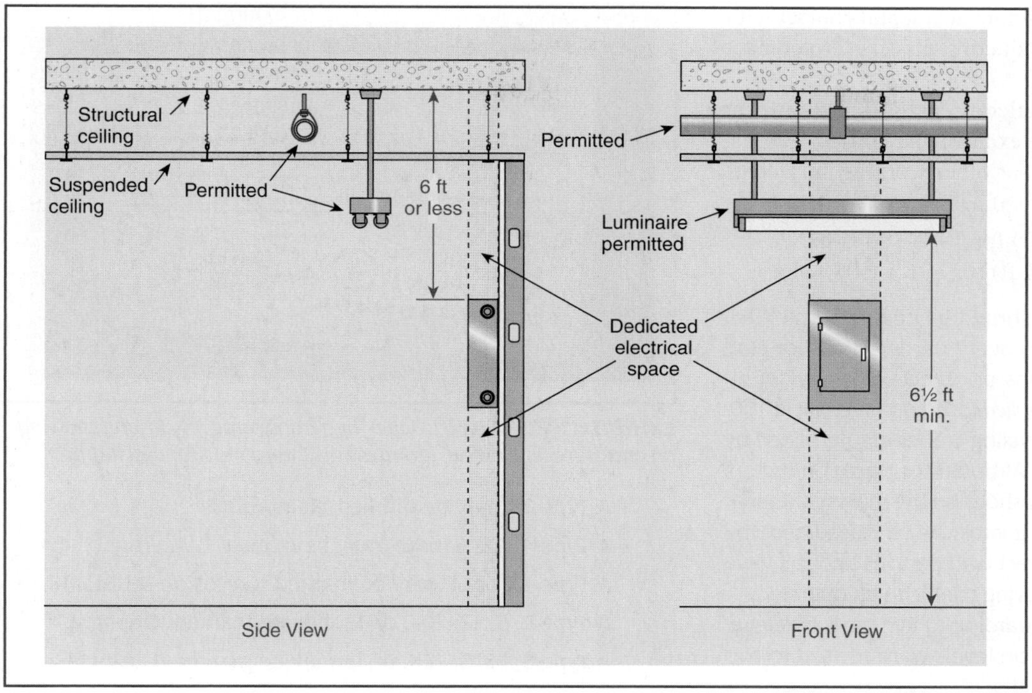


***EXHIBIT 110.26*** *The dedicated electrical space above and below a panelboard.*

that the exhaust duct is not located in or directly above the dedicated electrical space. Although not specifically required to be located here, this duct location may be a cost-effective solution that avoids the substantial physical protection requirements of 110.26(E)(1)(b).

Exhibit 110.25 illustrates the working space required in front of the panelboard by 110.26(A). No equipment, electrical or otherwise, is allowed in the working space. Exhibit 110.26 illustrates the dedicated electrical space above and below the panelboard required by 110.26(E)(1). This space is for the cables, raceways, and so on, that run to and from the panelboard.

**(2) Outdoor.** Outdoor installations shall comply with 110.26(E)(2)(a) and (b).

(a) *Installation Requirements.* Outdoor electrical equipment shall be installed in suitable enclosures and shall be protected from accidental contact by unauthorized personnel, or by vehicular traffic, or by accidental spillage or leakage from piping systems. The working clearance space shall include the zone described in 110.26(A). No architectural appurtenance or other equipment shall be located in this zone.

(b) *Dedicated Equipment Space.* The space equal to the width and depth of the equipment, and extending from grade to a height of 1.8 m (6 ft) above the equipment, shall be dedicated to the electrical installation. No piping or other equipment foreign to the electrical installation shall be located in this zone.

Extreme care should be taken where protection from unauthorized personnel or vehicular traffic is added to existing installations in order to comply with 110.26(E)(2). Any excavation or driving of steel into the ground for the placement of fencing, vehicle stops, or bollards should be done only after a thorough investigation of the belowgrade wiring.

The requirements for dedicated equipment space for outdoor electrical equipment are similar to the requirements for indoor equipment.

**(F) Locked Electrical Equipment Rooms or Enclosures.** Electrical equipment rooms or enclosures housing electrical apparatus that are controlled by a lock(s) shall be considered accessible to qualified persons.

This requirement allows equipment enclosures and rooms to be locked to prevent unauthorized access. Such rooms or enclosures are nonetheless considered accessible if qualified personnel have access.

## 110.27 Guarding of Live Parts

**(A) Live Parts Guarded Against Accidental Contact.** Except as elsewhere required or permitted by this *Code,* live parts of electrical equipment operating at 50 volts or more shall be guarded against accidental contact by approved enclosures or by any of the following means:

(1) By location in a room, vault, or similar enclosure that is accessible only to qualified persons.
(2) By suitable permanent, substantial partitions or screens arranged so that only qualified persons have access to the space within reach of the live parts. Any openings in such partitions or screens shall be sized and located so that

persons are not likely to come into accidental contact with the live parts or to bring conducting objects into contact with them.

(3) By location on a suitable balcony, gallery, or platform elevated and arranged so as to exclude unqualified persons.

(4) By elevation above the floor or other working surface as shown in 110.27(A)(4)(a) or (b) below:

a. A minimum of 2.5 m (8 ft) for 50 to 300 volts

b. A minimum of 2.6 m (8½ ft) for 301 to 600 volts

Live parts of electrical equipment should be covered, shielded, enclosed, or otherwise protected by covers, barriers, mats, or platforms to prevent contact by persons or objects. See the definitions of *dead front* and *isolated (as applied to location)* in Article 100.

Contact conductors used for traveling cranes are permitted by 610.13(B) and 610.21(A) to be bare. Although contact conductors obviously have to be bare for contact shoes on the moving member to make contact with the conductor, guards can be placed near the conductor to prevent accidental contact with persons and still have slots or spaces through which the moving contacts can operate.

The *Code* also recognizes the guarding of live parts by elevation. In the 2014 edition, the elevation levels were revised to correlate with similar requirements in ANSI/IEEE C2, *National Electrical Safety Code.*

**(B) Prevent Physical Damage.** In locations where electrical equipment is likely to be exposed to physical damage, enclosures or guards shall be so arranged and of such strength as to prevent such damage.

**(C) Warning Signs.** Entrances to rooms and other guarded locations that contain exposed live parts shall be marked with conspicuous warning signs forbidding unqualified persons to enter. The marking shall meet the requirements in 110.21(B).

Informational Note: For motors, see 430.232 and 430.233. For over 600 volts, see 110.34.

## 110.28 Enclosure Types

Enclosures (other than surrounding fences or walls) of switchboards, switchgear, panelboards, industrial control panels, motor control centers, meter sockets, enclosed switches, transfer switches, power outlets, circuit breakers, adjustable-speed drive systems, pullout switches, portable power distribution equipment, termination boxes, general-purpose transformers, fire pump controllers, fire pump motors, and motor controllers, rated not over 600 volts nominal and intended for such locations, shall be marked with an enclosure-type number as shown in Table 110.28.

Enclosures that comply with the requirements for more than one type of enclosure may be marked with multiple designations. Enclosures marked with a type may also be marked as follows:

A Type 1 may be marked "Indoor Use Only."

A Type 3, 3S, 4, 4X, 6, or 6P may be marked "Raintight."

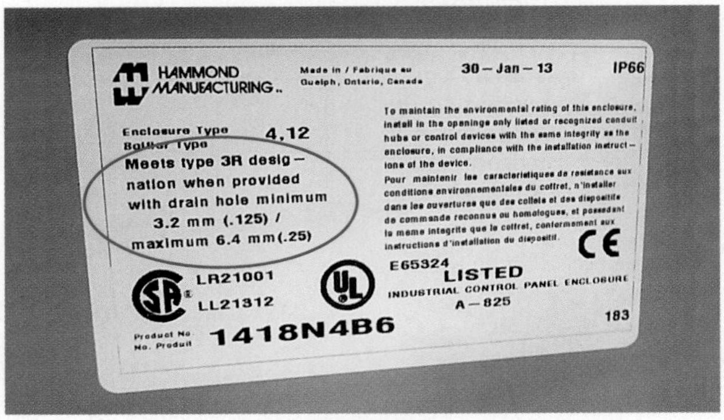


*EXHIBIT 110.27* *A typical label for a transformer enclosure, showing an enclosure type rating. (Courtesy of Hammond Manufacturing)*

A Type 3R may be marked "Rainproof."

A Type 4, 4X, 6, or 6P may be marked "Watertight."

A Type 4X or 6P may be marked "Corrosion Resistant."

A Type 2, 5, 12, 12K, or 13 may be marked "Driptight."

A Type 3, 3S, 5, 12K, or 13 may be marked "Dusttight."

For equipment designated "raintight," testing designed to simulate exposure to a beating rain will not result in entrance of water. For equipment designated "rainproof," testing designed to simulate exposure to a beating rain will not interfere with the operation of the apparatus or result in wetting of live parts and wiring within the enclosure. "Watertight" equipment is constructed so that water does not enter the enclosure when subjected to a stream of water. "Corrosion-resistant" equipment is constructed so that it provides a degree of protection against exposure to corrosive agents such as salt spray. "Driptight" equipment is constructed so that falling moisture or dirt does not enter the enclosure. "Dusttight" equipment is constructed so that circulating or airborne dust does not enter the enclosure.

Exhibit 110.27 illustrates a label for a transformer. The circled area on the label indicates the enclosure type, which is a Type 3R.

Table 110.28 shall be used for selecting these enclosures for use in specific locations other than hazardous (classified) locations. The enclosures are not intended to protect against conditions such as condensation, icing, corrosion, or contamination that may occur within the enclosure or enter via the conduit or unsealed openings.

# III. Over 600 Volts, Nominal

## 110.30 General

Conductors and equipment used on circuits over 600 volts, nominal, shall comply with Part I of this article and with 110.30 through 110.40, which supplement or modify Part I. In no case shall the provisions of this part apply to equipment on the supply side of the service point.

***TABLE 110.28*** *Enclosure Selection*

| Provides a Degree of Protection Against the Following Environmental Conditions | For Outdoor Use | | | | | | | | | |
|---|---|---|---|---|---|---|---|---|---|---|
| | Enclosure Type Number | | | | | | | | | |
| | **3** | **3R** | **3S** | **3X** | **3RX** | **3SX** | **4** | **4X** | **6** | **6P** |
| Incidental contact with the enclosed equipment | X | X | X | X | X | X | X | X | X | X |
| Rain, snow, and sleet | X | X | X | X | X | X | X | X | X | X |
| Sleet* | — | — | X | — | — | X | — | — | — | — |
| Windblown dust | X | — | X | X | — | X | X | X | X | X |
| Hosedown | — | — | — | — | — | — | X | X | X | X |
| Corrosive agents | — | — | — | X | X | X | — | X | — | X |
| Temporary submersion | — | — | — | — | — | — | — | — | X | X |
| Prolonged submersion | — | — | — | — | — | — | — | — | — | X |

| Provides a Degree of Protection Against the Following Environmental Conditions | For Indoor Use | | | | | | | | | |
|---|---|---|---|---|---|---|---|---|---|---|
| | Enclosure Type Number | | | | | | | | | |
| | **1** | **2** | **4** | **4X** | **5** | **6** | **6P** | **12** | **12K** | **13** |
| Incidental contact with the enclosed equipment | X | X | X | X | X | X | X | X | X | X |
| Falling dirt | X | X | X | X | X | X | X | X | X | X |
| Falling liquids and light splashing | — | X | X | X | X | X | X | X | X | X |
| Circulating dust, lint, fibers, and flyings | — | — | X | X | — | X | X | X | X | X |
| Settling airborne dust, lint, fibers, and flyings | — | — | X | X | X | X | X | X | X | X |
| Hosedown and splashing water | — | — | X | X | — | X | X | — | — | — |
| Oil and coolant seepage | — | — | — | — | — | — | — | X | X | X |
| Oil or coolant spraying and splashing | — | — | — | — | — | — | — | — | — | X |
| Corrosive agents | — | — | — | X | — | — | X | — | — | — |
| Temporary submersion | — | — | — | — | — | X | X | — | — | — |
| Prolonged submersion | — | — | — | — | — | — | X | — | — | — |

*Mechanism shall be operable when ice covered.

Informational Note No. 1: The term *raintight* is typically used in conjunction with Enclosure Types 3, 3S, 3SX, 3X, 4, 4X, 6, and 6P. The term *rainproof* is typically used in conjunction with Enclosure Types 3R, and 3RX. The term *watertight* is typically used in conjunction with Enclosure Types 4, 4X, 6, 6P. The term *driptight* is typically used in conjunction with Enclosure Types 2, 5, 12, 12K, and 13. The term *dusttight* is typically used in conjunction with Enclosure Types 3, 3S, 3SX, 3X, 5, 12, 12K, and 13.

Informational Note No. 2: Ingress protection (IP) ratings may be found in ANSI/NEMA 60529, *Degrees of Protection Provided by Enclosures*. IP ratings are not a substitute for Enclosure Type ratings.

Typical equipment covered by Part III of Article 110 is shown in Exhibit 110.28.

## 110.31 Enclosure for Electrical Installations

Electrical installations in a vault, room, or closet or in an area surrounded by a wall, screen, or fence, access to which is controlled by a lock(s) or other approved means, shall be considered to be accessible to qualified persons only. The type of enclosure used in a given case shall be designed and constructed according to the nature and degree of the hazard(s) associated with the installation.

For installations other than equipment as described in 110.31(D), a wall, screen, or fence shall be used to enclose an outdoor electrical installation to deter access by persons who are not qualified. A fence shall not be less than 2.1 m (7 ft) in height or a combination of 1.8 m (6 ft) or more of fence fabric and a 300 mm (1 ft) or more extension utilizing three or more strands of barbed wire or equivalent. The distance from the fence to live parts shall be not less than given in Table 110.31.

Informational Note: See Article 450 for construction requirements for transformer vaults.

**EXHIBIT 110.28** *Medium voltage metal-clad switchgear.*

***TABLE 110.31*** *Minimum Distance from Fence to Live Parts*

| | Minimum Distance to Live Parts | |
|---|---|---|
| **Nominal Voltage** | **m** | **ft** |
| 601 – 13,799 | 3.05 | 10 |
| 13,800– 230,000 | 4.57 | 15 |
| Over 230,000 | 5.49 | 18 |

Note: For clearances of conductors for specific system voltages and typical BIL ratings, see ANSI C2-2007, *National Electrical Safety Code.*

**(A) Electrical Vaults.** Where an electrical vault is required or specified for conductors and equipment operating at over 600 volts, nominal, the following shall apply.

**(1) Walls and Roof.** The walls and roof shall be constructed of materials that have adequate structural strength for the conditions, with a minimum fire rating of 3 hours. For the purpose of this section, studs and wallboard construction shall not be permitted.

**(2) Floors.** The floors of vaults in contact with the earth shall be of concrete that is not less than 102 mm (4 in.) thick, but where the vault is constructed with a vacant space or other stories below it, the floor shall have adequate structural strength for the load imposed on it and a minimum fire resistance of 3 hours.

**(3) Doors.** Each doorway leading into a vault from the building interior shall be provided with a tight-fitting door that has a minimum fire rating of 3 hours. The authority having jurisdiction shall be permitted to require such a door for an exterior wall opening where conditions warrant.

*Exception to (1), (2), and (3): Where the vault is protected with automatic sprinkler, water spray, carbon dioxide, or halon, construction with a 1-hour rating shall be permitted.*

**(4) Locks.** Doors shall be equipped with locks, and doors shall be kept locked, with access allowed only to qualified persons. Personnel doors shall swing out and be equipped with panic bars, pressure plates, or other devices that are normally latched but that open under simple pressure.

**(5) Transformers.** Where a transformer is installed in a vault as required by Article 450, the vault shall be constructed in accordance with the requirements of Part III of Article 450.

Informational Note No. 1: For additional information, see ANSI/ASTM E119-2011a, *Method for Fire Tests of Building Construction and Materials*, and NFPA 80-2013, *Standard for Fire Doors and Other Opening Protectives.*

Informational Note No. 2: A typical 3-hour construction is 150 mm (6 in.) thick reinforced concrete.

These requirements apply where a vault is required or specified. The requirements for doors and locks are based on the requirement in 450.43. Where a vault is required by Article 450, it must be constructed in accordance with Article 450, Part III. Although the language in Article 450 is similar to this section, some requirements for door sills and ventilation would not be applicable in an equipment/conductor vault. An exception to the vault construction requirements is added to allow for 1-hour construction when the vault is protected by a fire suppression system. This exception is based on 450.42 and 450.43.

**(B) Indoor Installations.**

**(1) In Places Accessible to Unqualified Persons.** Indoor electrical installations that are accessible to unqualified persons shall be made with metal-enclosed equipment. Switchgear, unit substations, transformers, pull boxes, connection boxes, and other similar associated equipment shall be marked with appropriate caution signs. Openings in ventilated dry-type transformers or similar openings in other equipment shall be designed so that foreign objects inserted through these openings are deflected from energized parts.

**(2) In Places Accessible to Qualified Persons Only.** Indoor electrical installations considered accessible only to qualified persons in accordance with this section shall comply with 110.34, 110.36, and 490.24.

**(C) Outdoor Installations.**

**(1) In Places Accessible to Unqualified Persons.** Outdoor electrical installations that are open to unqualified persons shall comply with Parts I, II, and III of Article 225.

**(2) In Places Accessible to Qualified Persons Only.** Outdoor electrical installations that have exposed live parts shall be accessible to qualified persons only in accordance with the first paragraph of this section and shall comply with 110.34, 110.36, and 490.24.

**(D) Enclosed Equipment Accessible to Unqualified Persons.** Ventilating or similar openings in equipment shall be designed such that foreign objects inserted through these openings are deflected from energized parts. Where exposed to physical damage from vehicular traffic, suitable guards shall be provided. Nonmetallic or metal-enclosed equipment located outdoors and accessible to the general public shall be designed such that exposed nuts or bolts cannot be readily removed, permitting access to live parts. Where nonmetallic or metal-enclosed equipment is accessible to the general public and the bottom of the enclosure is less than 2.5 m (8 ft) above the floor or grade level, the enclosure door or hinged cover shall be kept locked. Doors and covers of enclosures used solely as pull boxes, splice boxes, or junction boxes shall be locked, bolted, or screwed on. Underground box covers that weigh over 45.4 kg (100 lb) shall be considered as meeting this requirement.

## 110.32 Work Space About Equipment

Sufficient space shall be provided and maintained about electrical equipment to permit ready and safe operation and maintenance of such equipment. Where energized parts are exposed, the minimum clear work space shall be not less than 2.0 m (6½ ft) high (measured vertically from the floor or platform) or not less than 914 mm (3 ft) wide (measured parallel to the equipment). The depth shall be as required in 110.34(A). In all cases, the work space shall permit at least a 90 degree opening of doors or hinged panels.

## 110.33 Entrance to Enclosures and Access to Working Space

**(A) Entrance.** At least one entrance to enclosures for electrical installations as described in 110.31 not less than 610 mm (24 in.) wide and 2.0 m (6½ ft) high shall be provided to give access to the working space about electrical equipment.

**(1) Large Equipment.** On switchgear and control panels exceeding 1.8 m (6 ft) in width, there shall be one entrance at each end of the equipment. A single entrance to the required working space shall be permitted where either of the conditions in 110.33(A)(1)(a) or (A)(1)(b) is met.

(a) *Unobstructed Exit.* Where the location permits a continuous and unobstructed way of exit travel, a single entrance to the working space shall be permitted.

(b) *Extra Working Space.* Where the depth of the working space is twice that required by 110.34(A), a single entrance shall be permitted. It shall be located so that the distance from the equipment to the nearest edge of the entrance is not less than the minimum clear distance specified in Table 110.34(A) for equipment operating at that voltage and in that condition.

**(2) Guarding.** Where bare energized parts at any voltage or insulated energized parts above 600 volts, nominal, to ground are located adjacent to such entrance, they shall be suitably guarded.

Section 110.33(A) contains requirements very similar to those of 110.26(C). For further information, see the commentary following 110.26(C)(2), most of which also is valid for installations over 600 volts.

**(3) Personnel Doors.** Where there is a personnel door(s) intended for entrance to and egress from the working space less than 7.6 m (25 ft) from the nearest edge of the working space, the door(s) shall open in the direction of egress and be equipped with listed panic hardware.

The measurement for the personnel door is made from the nearest edge of the working space. Not every electrical installation is in an equipment room. This section requires personnel doors that are up to 25 feet from the working space to have panic hardware and to open in the direction of egress from the area.

**(B) Access.** Permanent ladders or stairways shall be provided to give safe access to the working space around electrical equipment installed on platforms, balconies, or mezzanine floors or in attic or roof rooms or spaces.

## 110.34 Work Space and Guarding

The requirements of 110.34 are conditional, just like the requirements in 110.26; that is, some of the requirements are applicable only where the equipment "is likely to require examination, adjustment, servicing, or maintenance while energized."

**(A) Working Space.** Except as elsewhere required or permitted in this *Code,* equipment likely to require examination, adjustment, servicing, or maintenance while energized shall have clear working space in the direction of access to live parts of the electrical equipment and shall be not less than specified in Table 110.34(A). Distances shall be measured from the live parts, if such are exposed, or from the enclosure front or opening if such are enclosed.

*Exception: Working space shall not be required in back of equipment such as switchgear or control assemblies where there are no renewable or adjustable parts (such as fuses or*

***TABLE 110.34(A)*** *Minimum Depth of Clear Working Space at Electrical Equipment*

| Nominal Voltage to Ground | Minimum Clear Distance | | |
|---|---|---|---|
| | **Condition 1** | **Condition 2** | **Condition 3** |
| 601–2500 V | 900 mm (3 ft) | 1.2 m (4 ft) | 1.5 m (5 ft) |
| 2501–9000 V | 1.2 m (4 ft) | 1.5 m (5 ft) | 1.8 m (6 ft) |
| 9001–25,000 V | 1.5 m (5 ft) | 1.8 m (6 ft) | 2.8 m (9 ft) |
| 25,001 V–75 kV | 1.8 m (6 ft) | 2.5 m (8 ft) | 3.0 m (10 ft) |
| Above 75 kV | 2.5 m (8 ft) | 3.0 m (10 ft) | 3.7 m (12 ft) |

Note: Where the conditions are as follows:

**Condition 1** — Exposed live parts on one side of the working space and no live or grounded parts on the other side of the working space, or exposed live parts on both sides of the working space that are effectively guarded by insulating materials.

**Condition 2** — Exposed live parts on one side of the working space and grounded parts on the other side of the working space. Concrete, brick, or tile walls shall be considered as grounded.

**Condition 3** — Exposed live parts on both sides of the working space.

*switches) on the back and where all connections are accessible from locations other than the back. Where rear access is required to work on nonelectrical parts on the back of enclosed equipment, a minimum working space of 762 mm (30 in.) horizontally shall be provided.*

**(B) Separation from Low-Voltage Equipment.** Where switches, cutouts, or other equipment operating at 600 volts, nominal, or less are installed in a vault, room, or enclosure where there are exposed live parts or exposed wiring operating at over 600 volts, nominal, the high-voltage equipment shall be effectively separated from the space occupied by the low-voltage equipment by a suitable partition, fence, or screen.

*Exception: Switches or other equipment operating at 600 volts, nominal, or less and serving only equipment within the high-voltage vault, room, or enclosure shall be permitted to be installed in the high-voltage vault, room, or enclosure without a partition, fence, or screen if accessible to qualified persons only.*

**(C) Locked Rooms or Enclosures.** The entrance to all buildings, vaults, rooms, or enclosures containing exposed live parts or exposed conductors operating at over 600 volts, nominal, shall be kept locked unless such entrances are under the observation of a qualified person at all times.

Permanent and conspicuous danger signs shall be provided. The danger sign shall meet the requirements in 110.21(B) and shall read as follows:

DANGER — HIGH VOLTAGE — KEEP OUT

***EXHIBIT 110.29*** *Enclosures of equipment operations at over 600 V required to have the warning label specified in 110.31(B)(1). (Courtesy of the International Association of Electrical Inspectors)*

Equipment used on circuits over 600 V, nominal, and containing exposed live parts or exposed conductors is required to be located in a locked room or in an enclosure. Locking is not required if the room or enclosure is under observation at all times, as is the case with some engine rooms. Where the room or enclosure is accessible to other than qualified persons, the entry to the room and equipment is required to be provided with warning labels. See Exhibit 110.29.

**(D) Illumination.** Illumination shall be provided for all working spaces about electrical equipment. The lighting outlets shall be arranged so that persons changing lamps or making repairs on the lighting system are not endangered by live parts or other equipment.

The points of control shall be located so that persons are not likely to come in contact with any live part or moving part of the equipment while turning on the lights.

**(E) Elevation of Unguarded Live Parts.** Unguarded live parts above working space shall be maintained at elevations not less than required by Table 110.34(E).

***TABLE 110.34(E)*** *Elevation of Unguarded Live Parts Above Working Space*

| Nominal Voltage Between Phases | Elevation | |
|---|---|---|
| | **m** | **ft** |
| 601–7500 V | 2.8 | 9 |
| 7501–35,000 V | 2.9 | 9 ft 6 in. |
| Over 35 kV | 2.9 m + 9.5 mm/kV above 35 | 9 ft 6 in. + 0.37 in./kV above 35 |

**(F) Protection of Service Equipment, Switchgear, and Industrial Control Assemblies.** Pipes or ducts foreign to the electrical installation and requiring periodic maintenance or whose malfunction would endanger the operation of the electrical system shall not be located in the vicinity of the service equipment, switchgear, or industrial control assemblies. Protection shall be provided where necessary to avoid damage from condensation leaks and breaks in such foreign systems. Piping and other facilities shall not be considered foreign if provided for fire protection of the electrical installation.

### 110.36 Circuit Conductors

Circuit conductors shall be permitted to be installed in raceways; in cable trays; as metal-clad cable Type MC; as bare wire, cable, and busbars; or as Type MV cables or conductors as provided in 300.37, 300.39, 300.40, and 300.50. Bare live conductors shall comply with 490.24.

Insulators, together with their mounting and conductor attachments, where used as supports for wires, single-conductor cables, or busbars, shall be capable of safely withstanding the maximum magnetic forces that would prevail if two or more conductors of a circuit were subjected to short-circuit current.

Exposed runs of insulated wires and cables that have a bare lead sheath or a braided outer covering shall be supported in a manner designed to prevent physical damage to the braid or sheath. Supports for lead-covered cables shall be designed to prevent electrolysis of the sheath.

### 110.40 Temperature Limitations at Terminations

Conductors shall be permitted to be terminated based on the 90°C (194°F) temperature rating and ampacity as given in Table 310.60(C)(67) through Table 310.60(C)(86), unless otherwise identified.

## IV. Tunnel Installations over 600 Volts, Nominal

### 110.51 General

**(A) Covered.** The provisions of this part shall apply to the installation and use of high-voltage power distribution and utilization equipment that is portable, mobile, or both, such as substations, trailers, cars, mobile shovels, draglines, hoists, drills, dredges, compressors, pumps, conveyors, underground excavators, and the like.

**(B) Other Articles.** The requirements of this part shall be additional to, or amendatory of, those prescribed in Articles 100 through 490 of this *Code*.

**(C) Protection Against Physical Damage.** Conductors and cables in tunnels shall be located above the tunnel floor and so placed or guarded to protect them from physical damage.

### 110.52 Overcurrent Protection

Motor-operated equipment shall be protected from overcurrent in accordance with Parts III, IV, and V of Article 430. Transformers shall be protected from overcurrent in accordance with 450.3.

### 110.53 Conductors

High-voltage conductors in tunnels shall be installed in metal conduit or other metal raceway, Type MC cable, or other approved multiconductor cable. Multiconductor portable cable shall be permitted to supply mobile equipment.

### 110.54 Bonding and Equipment Grounding Conductors

**(A) Grounded and Bonded.** All non–current-carrying metal parts of electrical equipment and all metal raceways and cable sheaths shall be solidly grounded and bonded to all metal pipes and rails at the portal and at intervals not exceeding 300 m (1000 ft) throughout the tunnel.

**(B) Equipment Grounding Conductors.** An equipment grounding conductor shall be run with circuit conductors inside the metal raceway or inside the multiconductor cable jacket. The equipment grounding conductor shall be permitted to be insulated or bare.

### 110.55 Transformers, Switches, and Electrical Equipment

All transformers, switches, motor controllers, motors, rectifiers, and other equipment installed belowground shall be protected from physical damage by location or guarding.

### 110.56 Energized Parts

Bare terminals of transformers, switches, motor controllers, and other equipment shall be enclosed to prevent accidental contact with energized parts.

### 110.57 Ventilation System Controls

Electrical controls for the ventilation system shall be arranged so that the airflow can be reversed.

### 110.58 Disconnecting Means

A switch or circuit breaker that simultaneously opens all ungrounded conductors of the circuit shall be installed within sight of each transformer or motor location for disconnecting the transformer or motor. The switch or circuit breaker for a transformer shall have an ampere rating not less than the ampacity of the transformer supply conductors. The switch or circuit breaker for a motor shall comply with the applicable requirements of Article 430.

### 110.59 Enclosures

Enclosures for use in tunnels shall be dripproof, weatherproof, or submersible as required by the environmental conditions. Switch or contactor enclosures shall not be used as junction boxes or as raceways for conductors feeding through or tapping off to other switches, unless the enclosures comply with 312.8.

## V. Manholes and Other Electrical Enclosures Intended for Personnel Entry, All Voltages

Manhole working space requirements and issues for cabling and other equipment here parallel those same working space issues elsewhere in Article 110. For handhole installations, see Article 314.

### 110.70 General

Electrical enclosures intended for personnel entry and specifically fabricated for this purpose shall be of sufficient size to provide safe work space about electrical equipment with live parts that is likely to require examination, adjustment, servicing, or maintenance while energized. Such enclosures shall have sufficient size to permit ready installation or withdrawal of the conductors employed without damage to the conductors or to their insulation. They shall comply with the provisions of this part.

*Exception: Where electrical enclosures covered by Part V of this article are part of an industrial wiring system operating under conditions of maintenance and supervision that ensure that only qualified persons monitor and supervise the system, they shall be permitted to be designed and installed in accordance with appropriate engineering practice. If required by the authority having jurisdiction, design documentation shall be provided.*

The requirements of Part V are conditional, just like the requirements in 110.26; that is, some of the requirements are applicable only where the equipment "is likely to require examination, adjustment, servicing, or maintenance while energized."

### 110.71 Strength

Manholes, vaults, and their means of access shall be designed under qualified engineering supervision and shall withstand all loads likely to be imposed on the structures.

Informational Note: See ANSI C2-2007, *National Electrical Safety Code,* for additional information on the loading that can be expected to bear on underground enclosures.

### 110.72 Cabling Work Space

A clear work space not less than 900 mm (3 ft) wide shall be provided where cables are located on both sides, and not less than 750 mm (2½ ft) where cables are only on one side. The vertical headroom shall be not less than 1.8 m (6 ft) unless the opening is within 300 mm (1 ft), measured horizontally, of the adjacent interior side wall of the enclosure.

*Exception: A manhole containing only one or more of the following shall be permitted to have one of the horizontal work space dimensions reduced to 600 mm (2 ft) where the other horizontal clear work space is increased so the sum of the two dimensions is not less than 1.8 m (6 ft):*

*(1) Optical fiber cables as covered in Article 770*
*(2) Power-limited fire alarm circuits supplied in accordance with 760.121*
*(3) Class 2 or Class 3 remote-control and signaling circuits, or both, supplied in accordance with 725.121*

### 110.73 Equipment Work Space

Where electrical equipment with live parts that is likely to require examination, adjustment, servicing, or maintenance while energized is installed in a manhole, vault, or other enclosure designed for personnel access, the work space and associated requirements in 110.26 shall be met for installations operating at 600 volts or less. Where the installation is over 600 volts, the work space and associated requirements in 110.34 shall be met. A manhole access cover that weighs over 45 kg (100 lb) shall be considered as meeting the requirements of 110.34(C).

### 110.74 Conductor Installation

Conductors installed in manholes and other enclosures intended for personnel entry shall be cabled, racked up, or arranged in an approved manner that provides ready and safe access for persons to enter for installation and maintenance. The installation shall comply with 110.74(A) or 110.74(B), as applicable.

**(A) 600 Volts, Nominal, or Less.** Wire bending space for conductors operating at 600 volts or less shall be provided in accordance with the requirements of 314.28.

**(B) Over 600 Volts, Nominal.** Conductors operating at over 600 volts shall be provided with bending space in accordance with 314.71(A) and (B), as applicable.

*Exception: Where 314.71(B) applies, each row or column of ducts on one wall of the enclosure shall be calculated individually, and the single row or column that provides the maximum distance shall be used.*

### 110.75 Access to Manholes

**(A) Dimensions.** Rectangular access openings shall not be less than 650 mm × 550 mm (26 in. × 22 in.). Round access openings in a manhole shall be not less than 650 mm (26 in.) in diameter.

*Exception: A manhole that has a fixed ladder that does not obstruct the opening or that contains only one or more of the following shall be permitted to reduce the minimum cover diameter to 600 mm (2 ft):*

*(1) Optical fiber cables as covered in Article 770*
*(2) Power-limited fire alarm circuits supplied in accordance with 760.121*
*(3) Class 2 or Class 3 remote-control and signaling circuits, or both, supplied in accordance with 725.121*

**(B) Obstructions.** Manhole openings shall be free of protrusions that could injure personnel or prevent ready egress.

**(C) Location.** Manhole openings for personnel shall be located where they are not directly above electrical equipment or conductors in the enclosure. Where this is not practicable, either a protective barrier or a fixed ladder shall be provided.

**(D) Covers.** Covers shall be over 45 kg (100 lb) or otherwise designed to require the use of tools to open. They shall be designed or restrained so they cannot fall into the manhole or protrude sufficiently to contact electrical conductors or equipment within the manhole.

**(E) Marking.** Manhole covers shall have an identifying mark or logo that prominently indicates their function, such as "electric."

## 110.76 Access to Vaults and Tunnels

**(A) Location.** Access openings for personnel shall be located where they are not directly above electrical equipment or conductors in the enclosure. Other openings shall be permitted over equipment to facilitate installation, maintenance, or replacement of equipment.

**(B) Locks.** In addition to compliance with the requirements of 110.34, if applicable, access openings for personnel shall be arranged such that a person on the inside can exit when the access door is locked from the outside, or in the case of normally locking by padlock, the locking arrangement shall be such that the padlock can be closed on the locking system to prevent locking from the outside.

## 110.77 Ventilation

Where manholes, tunnels, and vaults have communicating openings into enclosed areas used by the public, ventilation to open air shall be provided wherever practicable.

## 110.78 Guarding

Where conductors or equipment, or both, could be contacted by objects falling or being pushed through a ventilating grating, both conductors and live parts shall be protected in accordance with the requirements of 110.27(A)(2) or 110.31(B)(1), depending on the voltage.

## 110.79 Fixed Ladders

Fixed ladders shall be corrosion resistant.

# 2 Wiring and Protection

## ARTICLE 200 Use and Identification of Grounded Conductors

### 200.1 Scope

This article provides requirements for the following:

(1) Identification of terminals
(2) Grounded conductors in premises wiring systems
(3) Identification of grounded conductors

Informational Note: See Article 100 for definitions of *Grounded Conductor, Equipment Grounding Conductor,* and *Grounding Electrode Conductor*.

The grounded conductor is often, but not always, the neutral conductor. As defined in Article 100, a neutral conductor is one that is connected to the neutral point of an electrical system. For example, in a single-phase, 2-wire or in a 3-phase, corner-grounded delta system, the intentionally grounded conductor is not a neutral conductor, because it is not connected to a system neutral point. Because the conductor is connected to the same grounding electrode system as the non–current-carrying metal parts of the electrical equipment, generally no potential difference exists between the grounded conductor and those grounded metal parts. However, unlike an equipment grounding conductor, the grounded conductor is a circuit conductor and as such is a current-carrying conductor. Whether it is referred to as the grounded conductor or as the neutral conductor, the white or gray marking on a circuit conductor indicates that it is intentionally connected to the earth.

Electric shock injuries and electrocutions have occurred as a result of working on the grounded conductor while the circuit is energized. Extreme caution must be exercised where the grounded (neutral) conductor is part of a multiwire branch circuit [see 210.4(B) on disconnecting means for multiwire branch circuits]. Note that 300.13 does not permit the wiring terminals of a device, such as a receptacle, to be the means of maintaining the continuity of the grounded conductor in that type of branch circuit.

In addition to the requirements in this article, the use and installation of the grounded conductor are covered extensively in Article 250.

### 200.2 General

Grounded conductors shall comply with 200.2(A) and (B).

**(A) Insulation.** The grounded conductor, if insulated, shall have insulation that is (1) suitable, other than color, for any ungrounded conductor of the same circuit for systems of 1000 volts or less, or impedance grounded neutral systems of over 1000 volts, or (2) rated not less than 600 volts for solidly grounded neutral systems of over 1000 volts as described in 250.184(A).

**(B) Continuity.** The continuity of a grounded conductor shall not depend on a connection to a metallic enclosure, raceway, or cable armor.

Informational Note: See 300.13(B) for the continuity of grounded conductors used in multiwire branch circuits.

This section requires grounded conductors be connected to a terminal or busbar that is specifically intended and identified for connection of grounded or neutral conductors. Because grounded conductors are current carrying, connecting them to a separate equipment grounding terminal or bar (that is directly connected to a metal cabinet or enclosure) results in the enclosure becoming a neutral conductor between the equipment grounding terminal and the point of connection for the grounded conductor.

### 200.3 Connection to Grounded System

Premises wiring shall not be electrically connected to a supply system unless the latter contains, for any grounded conductor of the interior system, a corresponding conductor that is grounded. For the purpose of this section, *electrically connected* shall mean connected so as to be capable of carrying current, as distinguished from connection through electromagnetic induction.

*Exception: Listed utility-interactive inverters identified for use in distributed resource generation systems such as photovoltaic and fuel cell power systems shall be permitted to be connected to premises wiring without a grounded conductor where the connected premises wiring or utility system includes a grounded conductor.*

Grounded conductors of premises wiring are required to be connected to the supply system (service or applicable separately derived system) grounded conductor to ensure a common, continuous, grounded system.

## 200.4 Neutral Conductors

Neutral conductors shall be installed in accordance with 200.4(A) and (B).

Specific provisions in 215.4 and 225.7 permit multiple circuits to have a common or shared neutral conductor. A neutral conductor is not required to be installed with the ungrounded circuit conductors if line-to-neutral loads are not supplied by the branch circuit or feeder. However, each multiwire circuit (see Article 100 for definition) must have its own neutral conductor. Section 200.4(B) requires that where multiple circuits enter an enclosure, each circuit must have all of its conductors grouped together in at least one location to make it obvious which conductors are associated with the circuit. This grouping is not necessary if the conductors enter the enclosure as a single cable.

The concern with multiple circuits sharing a neutral conductor is the possibility of overloading the neutral conductor. Typically, an overcurrent protective device is not present, nor is it required to be installed in series with the neutral conductor. Therefore, an overload condition can exist without being detected or responded to, and insulation damage can occur. Because of the electrical relationship between the ungrounded conductors and the neutral conductor in multiwire single-phase and 3-phase circuits, the possibility of overload does not exist except where the load characteristics result in additive harmonic currents in the neutral conductor.

**(A) Installation.** Neutral conductors shall not be used for more than one branch circuit, for more than one multiwire branch circuit, or for more than one set of ungrounded feeder conductors unless specifically permitted elsewhere in this *Code*.

**(B) Multiple Circuits.** Where more than one neutral conductor associated with different circuits is in an enclosure, grounded circuit conductors of each circuit shall be identified or grouped to correspond with the ungrounded circuit conductor(s) by wire markers, cable ties, or similar means in at least one location within the enclosure.

*Exception No. 1: The requirement for grouping or identifying shall not apply if the branch-circuit or feeder conductors enter from a cable or a raceway unique to the circuit that makes the grouping obvious.*

*Exception No. 2: The requirement for grouping or identifying shall not apply where branch-circuit conductors pass though a box or conduit body without a loop as described in 314.16(B)(1) or without a splice or termination.*

## 200.6 Means of Identifying Grounded Conductors

**(A) Sizes 6 AWG or Smaller.** An insulated grounded conductor of 6 AWG or smaller shall be identified by one of the following means:

(1) A continuous white outer finish.
(2) A continuous gray outer finish.
(3) Three continuous white or gray stripes along the conductor's entire length on other than green insulation.
(4) Wires that have their outer covering finished to show a white or gray color but have colored tracer threads in the braid identifying the source of manufacture shall be considered as meeting the provisions of this section.

Using three white or gray stripes for identification is permitted for all conductor sizes. This method is the one most typically employed by a wire or cable manufacturer.

Other methods of identification are also permitted in 200.6(A). For example, the grounded conductor of mineral-insulated (MI) cable, due to its unique construction, is permitted to be identified at the time of installation. Aerial cable may have its grounded conductor identified by a ridge along its insulated surface, and fixture wires are permitted to have the grounded conductor identified by various methods, including colored insulation, stripes on the insulation, colored braid, colored separator, and tinned conductors. These identification methods are covered in 402.8 and explained in detail in 400.22(A) through (E).

For 6 AWG or smaller, identification of the grounded conductor solely through the use of white or gray marking tape or other distinctive white or gray marking applied at the time of installation is not permitted, except as described for flexible cords and multiconductor cables in 200.6(C) and (E), Exception No. 1, and for single conductors in outdoor photovoltaic power installations in accordance with 200.6(A)(2).

(5) The grounded conductor of a mineral-insulated, metal-sheathed cable (Type MI) shall be identified at the time of installation by distinctive marking at its terminations.
(6) A single-conductor, sunlight-resistant, outdoor-rated cable used as a grounded conductor in photovoltaic power systems, as permitted by 690.31, shall be identified at the time of installation by distinctive white marking at all terminations.
(7) Fixture wire shall comply with the requirements for grounded conductor identification as specified in 402.8.
(8) For aerial cable, the identification shall be as above, or by means of a ridge located on the exterior of the cable so as to identify it.

**(B) Sizes 4 AWG or Larger.** An insulated grounded conductor 4 AWG or larger shall be identified by one of the following means:

(1) A continuous white outer finish.
(2) A continuous gray outer finish.
(3) Three continuous white or gray stripes along the conductor's entire length on other than green insulation.

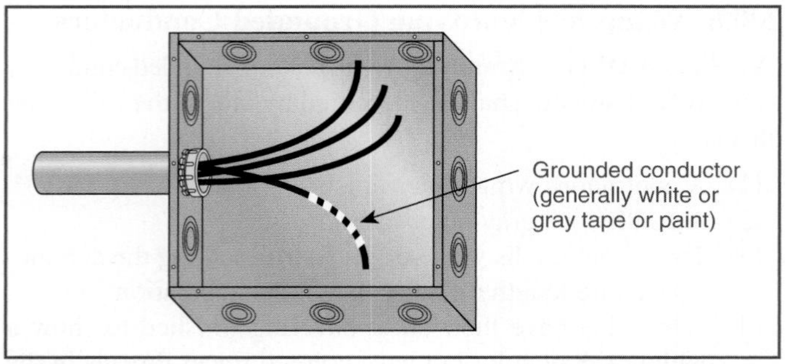


***EXHIBIT 200.1*** *Field-applied identification of a 4 AWG conductor to identify it as the grounded conductor.*

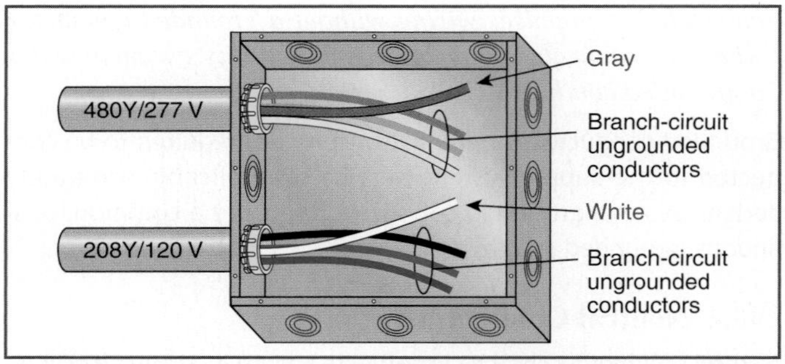


***EXHIBIT 200.2*** *Grounded conductors of different systems in the same enclosure.*

(4) At the time of installation, by a distinctive white or gray marking at its terminations. This marking shall encircle the conductor or insulation.

The most common method used by installers to identify a single conductor as a grounded conductor is the application of a white or gray marking to the insulation at all termination points at the time of installation. To be clearly visible, this field-applied white or gray marking must completely encircle the conductor insulation. This coloring can be applied by using marking tape or by painting the insulation. This method of identification is shown in Exhibit 200.1.

**(C) Flexible Cords.** An insulated conductor that is intended for use as a grounded conductor, where contained within a flexible cord, shall be identified by a white or gray outer finish or by methods permitted by 400.22.

**(D) Grounded Conductors of Different Systems.** Where grounded conductors of different systems are installed in the same raceway, cable, box, auxiliary gutter, or other type of enclosure, each grounded conductor shall be identified by system. Identification that distinguishes each system grounded conductor shall be permitted by one of the following means:

(1) One system grounded conductor shall have an outer covering conforming to 200.6(A) or (B).
(2) The grounded conductor(s) of other systems shall have a different outer covering conforming to 200.6(A) or 200.6(B) or by an outer covering of white or gray with a readily distinguishable colored stripe other than green running along the insulation.
(3) Other and different means of identification as allowed by 200.6(A) or (B) that will distinguish each system grounded conductor.

The means of identification shall be documented in a manner that is readily available or shall be permanently posted where the conductors of different systems originate.

Exhibit 200.2 illustrates an enclosure containing grounded conductors of two different systems that are distinguished from each other by color, which is one of the methods specified by 200.6(D). Gray and white colored insulation on conductors is recognized by the *Code* as two separate means of identifying grounded conductors.

The use of colored stripes (other than green) on white insulation for the entire conductor insulation length is also an acceptable method to distinguish one system grounded conductor from one with white insulation or white marking. This requirement applies only where grounded conductors of different systems are installed in a common enclosure, such as a junction or pull box or a wireway. Where grounded conductors of different supply systems are installed in a common enclosure, identifying the grounded conductors associated with one system from the grounded conductors of another system helps to ensure proper system connections. As more systems with grounded conductors are installed in a common enclosure, more means to distinguish each system grounded conductor from the other system grounded conductors in that enclosure become necessary. The only acceptable means of identifying grounded conductors are those specified in 200.6(A) and (B).

The *Code* requires the identification method or scheme used to distinguish the grounded conductors of different systems to be posted at the equipment (such as the panelboard, switchboard, or similar distribution equipment) where the feeders or branch circuits originate. Alternatively, the means of identification is permitted to be included in a facility's electrical system documentation as long as the documents can be readily obtained. The industry practice of using white for lower-voltage systems and gray for higher-voltage systems is permitted but not mandated by the *Code.* Exhibit 200.3 is an example of a permanent label posted at electrical distribution equipment, indicating the identification scheme for the grounded conductors of each nominal voltage system. Because this equipment contains two nominal voltage systems, the grounded (neutral) conductor of each system must be identified in a manner that distinguishes it from the grounded conductor(s) of other systems in that equipment.

**(E) Grounded Conductors of Multiconductor Cables.** The insulated grounded conductors in a multiconductor cable shall be identified by a continuous white or gray outer finish or by three continuous white or gray stripes on other than green insulation along its entire length. Multiconductor flat cable 4 AWG

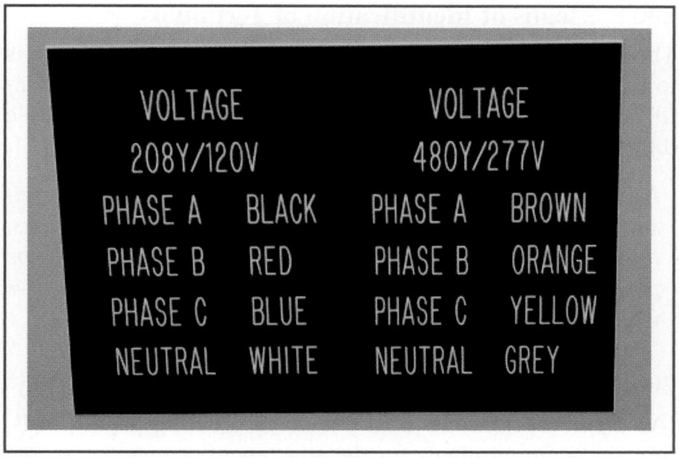

VOLTAGE
208Y/120V
PHASE A BLACK
PHASE B RED
PHASE C BLUE
NEUTRAL WHITE
VOLTAGE
480Y/277V
PHASE A BROWN
PHASE B ORANGE
PHASE C YELLOW
NEUTRAL GREY


***EXHIBIT 200.3*** *A label on a piece of equipment containing conductors of two nominal voltage systems that provides the required identification of the two systems.*

or larger shall be permitted to employ an external ridge on the grounded conductor.

*Exception No. 1: Where the conditions of maintenance and supervision ensure that only qualified persons service the installation, grounded conductors in multiconductor cables shall be permitted to be permanently identified at their terminations at the time of installation by a distinctive white marking or other equally effective means.*

This exception allows identification of the grounded conductor that is part of a multiconductor cable by use of a distinctive white marking as shown in Exhibit 200.4. Another effective means, such as numbering, lettering, or tagging, is also permitted. This exception applies only to installations in facilities that have a regulated system of maintenance and supervision performed by qualified persons.

*Exception No. 2: The grounded conductor of a multiconductor varnished-cloth-insulated cable shall be permitted to be identified at its terminations at the time of installation by a distinctive white marking or other equally effective means.*

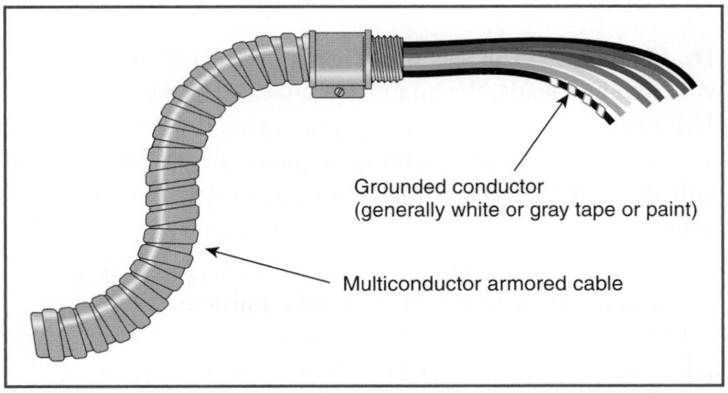


***EXHIBIT 200.4*** *Field-applied identification to the grounded conductor of a multiconductor armored cable.*

Informational Note: The color gray may have been used in the past as an ungrounded conductor. Care should be taken when working on existing systems.

All shades of gray insulation and marking are reserved for grounded conductors. The Informational Note following 200.6(E) warns the user to exercise caution when working on existing systems because gray may identify ungrounded conductors in those systems.

## 200.7 Use of Insulation of a White or Gray Color or with Three Continuous White or Gray Stripes

**(A) General.** The following shall be used only for the grounded circuit conductor, unless otherwise permitted in 200.7(B) and (C):

(1) A conductor with continuous white or gray covering
(2) A conductor with three continuous white or gray stripes on other than green insulation
(3) A marking of white or gray color at the termination

**(B) Circuits of Less Than 50 Volts.** A conductor with white or gray color insulation or three continuous white stripes or having a marking of white or gray at the termination for circuits of less than 50 volts shall be required to be grounded only as required by 250.20(A).

**(C) Circuits of 50 Volts or More.** The use of insulation that is white or gray or that has three continuous white or gray stripes for other than a grounded conductor for circuits of 50 volts or more shall be permitted only as in (1) and (2).

(1) If part of a cable assembly that has the insulation permanently reidentified to indicate its use as an ungrounded conductor by marking tape, painting, or other effective means at its termination and at each location where the conductor is visible and accessible. Identification shall encircle the insulation and shall be a color other than white, gray, or green. If used for single-pole, 3-way or 4-way switch loops, the reidentified conductor with white or gray insulation or three continuous white or gray stripes shall be used only for the supply to the switch, but not as a return conductor from the switch to the outlet.

Examples of applications where it might be necessary to re-identify and use a white or gray insulated conductor as an ungrounded conductor include water heaters, electric heat, motors, and switch loops. The insulation is re-identified to avoid confusing it with grounded conductors having white or gray insulation or neutral conductors at switch and outlet points and at other points in the wiring system where the conductors are accessible.

In previous editions, the *Code* permitted switch loops using a white insulated conductor to serve as an ungrounded conductor supplying the switch but not as the return ungrounded conductor to supply the lighting outlet. Prior to the 1999 *NEC*, re-identification of a white conductor used for this purpose was not required. However, electronic switching devices with small power supplies are available that can be installed at switch

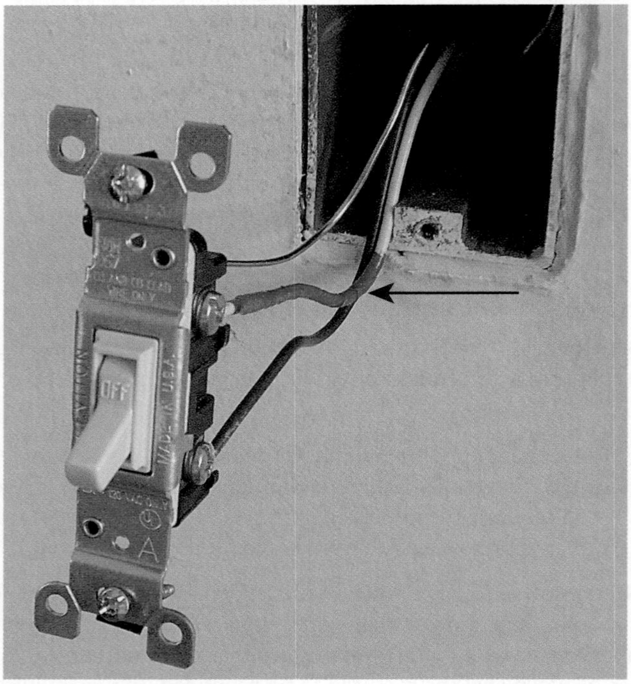

*EXHIBIT 200.5 The re-identified conductor at a switch location, limited to use only as the ungrounded conductor that supplies the switch.*

locations. These devices require a grounded conductor in order to power the internal components, and they help in clearly distinguishing a grounded conductor from ungrounded conductors, which aids in properly connecting such devices. Although in switch loops, the conductor with white or gray insulation is re-identified at all accessible locations to indicate that it is an ungrounded conductor, it is only permitted to supply the switch and cannot be the switched ungrounded conductor at the controlled outlet.

Exhibit 200.5 shows a switch location where the conductor with the white insulation is re-identified with red paint at the termination to indicate that it is the ungrounded supply conductor run to the switch.

(2) A flexible cord having one conductor identified by a white or gray outer finish or three continuous white or gray stripes, or by any other means permitted by 400.22, that is used for connecting an appliance or equipment permitted by 400.7. This shall apply to flexible cords connected to outlets whether or not the outlet is supplied by a circuit that has a grounded conductor.

Plating of all screws and terminals is commonly used to identify terminals, as well as to meet other requirements of specific applications, such as corrosion-resistant devices.

Informational Note: The color gray may have been used in the past as an ungrounded conductor. Care should be taken when working on existing systems.

## 200.9 Means of Identification of Terminals

The identification of terminals to which a grounded conductor is to be connected shall be substantially white in color. The identification of other terminals shall be of a readily distinguishable different color.

*Exception: Where the conditions of maintenance and supervision ensure that only qualified persons service the installations, terminals for grounded conductors shall be permitted to be permanently identified at the time of installation by a distinctive white marking or other equally effective means.*

## 200.10 Identification of Terminals

**(A) Device Terminals.** All devices, excluding panelboards, provided with terminals for the attachment of conductors and intended for connection to more than one side of the circuit shall have terminals properly marked for identification, unless the electrical connection of the terminal intended to be connected to the grounded conductor is clearly evident.

*Exception: Terminal identification shall not be required for devices that have a normal current rating of over 30 amperes, other than polarized attachment plugs and polarized receptacles for attachment plugs as required in 200.10(B).*

**(B) Receptacles, Plugs, and Connectors.** Receptacles, polarized attachment plugs, and cord connectors for plugs and polarized plugs shall have the terminal intended for connection to the grounded conductor identified as follows:

(1) Identification shall be by a metal or metal coating that is substantially white in color or by the word *white* or the letter *W* located adjacent to the identified terminal.
(2) If the terminal is not visible, the conductor entrance hole for the connection shall be colored white or marked with the word *white* or the letter *W*.

Informational Note: See 250.126 for identification of wiring device equipment grounding conductor terminals.

**(C) Screw Shells.** For devices with screw shells, the terminal for the grounded conductor shall be the one connected to the screw shell.

**(D) Screw Shell Devices with Leads.** For screw shell devices with attached leads, the conductor attached to the screw shell shall have a white or gray finish. The outer finish of the other conductor shall be of a solid color that will not be confused with the white or gray finish used to identify the grounded conductor.

Informational Note: The color gray may have been used in the past as an ungrounded conductor. Care should be taken when working on existing systems.

**(E) Appliances.** Appliances that have a single-pole switch or a single-pole overcurrent device in the line or any line-connected screw shell lampholders, and that are to be connected by (1) a permanent wiring method or (2) field-installed attachment plugs and cords with three or more wires (including the equipment grounding conductor), shall have means to identify the terminal for the grounded circuit conductor (if any).

### 200.11 Polarity of Connections

No grounded conductor shall be attached to any terminal or lead so as to reverse the designated polarity.

# ARTICLE 210 Branch Circuits

## I. General Provisions

### 210.1 Scope

This article covers branch circuits except for branch circuits that supply only motor loads, which are covered in Article 430. Provisions of this article and Article 430 apply to branch circuits with combination loads.

Although Article 210 provides the requirements for branch circuits, the requirements in Chapters 5, 6, and 7 may amend or modify these general requirements. For example, 668.3(C) covers electrolytic cell lines. This equipment is not covered by the general branch-circuit requirements of Article 210; rather, the requirements of Article 668 address the unique concerns associated with electrolytic cells.

### 210.2 Other Articles for Specific-Purpose Branch Circuits

Branch circuits shall comply with this article and also with the applicable provisions of other articles of this *Code*. The provisions for branch circuits supplying equipment listed in Table 210.2 amend or supplement the provisions in this article.

### 210.3 Rating

Branch circuits recognized by this article shall be rated in accordance with the maximum permitted ampere rating or setting of the overcurrent device. The rating for other than individual branch circuits shall be 15, 20, 30, 40, and 50 amperes. Where conductors of higher ampacity are used for any reason, the ampere rating or setting of the specified overcurrent device shall determine the circuit rating.

The rating of a branch circuit is determined by the rating of the overcurrent protective device, not by the ampacity of the circuit conductors. An increase in conductor ampacity from 20 amperes (12 AWG) to 30 amperes (10 AWG) for any reason does not change the rating of the circuit. If the rating of the branch-circuit overcurrent device is 20 amperes, then the rating of the circuit is 20 amperes.

***TABLE 210.2*** *Specific-Purpose Branch Circuits*

| Equipment | Article | Section |
|---|---|---|
| Air-conditioning and refrigerating equipment | | 440.6, 440.31, 440.32 |
| Audio signal processing, amplification, and reproduction equipment | | 640.8 |
| Busways | | 368.17 |
| Circuits and equipment operating at less than 50 volts | 720 | |
| Central heating equipment other than fixed electric space-heating equipment | | 422.12 |
| Class 1, Class 2, and Class 3 remote-control, signaling, and power-limited circuits | 725 | |
| Cranes and hoists | | 610.42 |
| Electric signs and outline lighting | | 600.6 |
| Electric welders | 630 | |
| Electrified truck parking space | 626 | |
| Elevators, dumbwaiters, escalators, moving walks, wheelchair lifts, and stairway chair lifts | | 620.61 |
| Fire alarm systems | 760 | |
| Fixed electric heating equipment for pipelines and vessels | | 427.4 |
| Fixed electric space-heating equipment | | 424.3 |
| Fixed outdoor electrical deicing and snow-melting equipment | | 426.4 |
| Information technology equipment | | 645.5 |
| Infrared lamp industrial heating equipment | | 422.48, 424.3 |
| Induction and dielectric heating equipment | 665 | |
| Marinas and boatyards | | 555.19 |
| Mobile homes, manufactured homes, and mobile home parks | 550 | |
| Motion picture and television studios and similar locations | 530 | |
| Motors, motor circuits, and controllers | 430 | |
| Pipe organs | | 650.7 |
| Recreational vehicles and recreational vehicle parks | 551 | |
| Switchboards and panelboards | | 408.52 |
| Theaters, audience areas of motion picture and television studios, and similar locations | | 520.41, 520.52, 520.62 |
| X-ray equipment | | 660.2, 517.73 |

*Exception: Multioutlet branch circuits greater than 50 amperes shall be permitted to supply nonlighting outlet loads on industrial premises where conditions of maintenance and supervision ensure that only qualified persons service the equipment.*

A common practice at industrial facilities is to provide several single receptacles with ratings of 50 amperes or higher on a single branch circuit to allow quick relocation of equipment (such as electric welders) for production or maintenance. Generally, only one piece of equipment at a time is supplied from this type of receptacle circuit. The type of receptacle used in this situation is generally a pin-and-sleeve receptacle. Pin-and-sleeve receptacles may not be horsepower rated.

## 210.4 Multiwire Branch Circuits

**(A) General.** Branch circuits recognized by this article shall be permitted as multiwire circuits. A multiwire circuit shall be permitted to be considered as multiple circuits. All conductors of a multiwire branch circuit shall originate from the same panelboard or similar distribution equipment.

Informational Note No. 1: A 3-phase, 4-wire, wye-connected power system used to supply power to nonlinear loads may necessitate that the power system design allow for the possibility of high harmonic currents on the neutral conductor.

Informational Note No. 2: See 300.13(B) for continuity of grounded conductors on multiwire circuits.

Power supplies for equipment such as computers, printers, and adjustable-speed motor drives can introduce harmonic currents in the system neutral conductor. The resulting total harmonic distortion current could exceed the load current of the device itself. See the commentary following 310.15(B)(5)(c) for a discussion of neutral conductor ampacity.

**(B) Disconnecting Means.** Each multiwire branch circuit shall be provided with a means that will simultaneously disconnect all ungrounded conductors at the point where the branch circuit originates.

This requirement for a simultaneous disconnecting means reduces the risk of shock to personnel working on equipment supplied by the multiwire branch circuit, and it takes the guesswork out of ensuring safe conditions for maintenance. In former editions of the *NEC*, this requirement applied only where the multiwire branch circuit supplied equipment mounted to a common yoke or strap.

For a single-phase installation, the disconnecting means could be two single-pole circuit breakers with an identified handle tie or a 2-pole circuit breaker, as shown in Exhibit 210.1 (top), or by a 2-pole switch, as shown in Exhibit 210.1 (bottom). For a 3-phase installation, a 3-pole circuit breaker, three single-pole circuit breakers with an identified handle tie, or a 3-pole switch provides the required simultaneous opening of the ungrounded conductors. The simultaneous opening of all "hot" conductors at the panelboard effectively protects personnel from inadvertent contact with an energized conductor or device terminal during servicing.

Informational Note: See 240.15(B) for information on the use of single-pole circuit breakers as the disconnecting means.

**(C) Line-to-Neutral Loads.** Multiwire branch circuits shall supply only line-to-neutral loads.

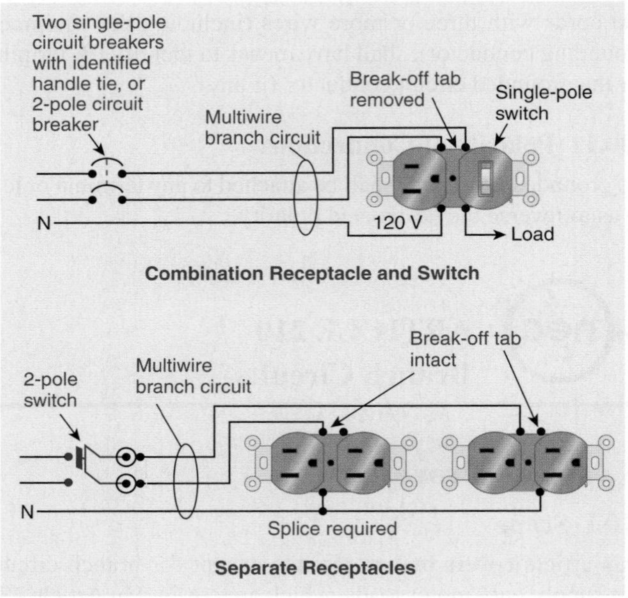


***EXHIBIT 210.1*** *Two examples of situations where 210.4(B) requires the simultaneous disconnection of all ungrounded conductors to multiwire branch circuits supplying more than one device or equipment.*

*Exception No. 1: A multiwire branch circuit that supplies only one utilization equipment.*

*Exception No. 2: Where all ungrounded conductors of the multiwire branch circuit are opened simultaneously by the branch-circuit overcurrent device.*

The term *multiwire branch circuit* is defined in Article 100 as "a branch circuit that consists of two or more ungrounded conductors that have a voltage between them, and a grounded conductor that has equal voltage between it and each ungrounded conductor of the circuit and that is connected to the neutral or grounded conductor of the system." Section 210.4(A) permits a multiwire branch circuit to be considered as multiple circuits. These circuits could be used to satisfy the requirement for providing two small-appliance branch circuits for countertop receptacle outlets in a dwelling unit kitchen.

The most commonly used multiwire branch circuit consists of two ungrounded conductors and one grounded conductor supplied from a 120/240-V, single-phase, 3-wire system. Such multiwire circuits supply appliances that have both line-to-line and line-to-neutral connected loads, such as electric ranges and clothes dryers, or supply loads that are line-to-neutral connected only, such as the split-wired combination device shown in Exhibit 210.1 (bottom). A multiwire branch circuit is also permitted to supply a device with a 250-V receptacle (line-to-line) and a 125-V receptacle (line-to-neutral), as shown in Exhibit 210.2, provided the branch-circuit overcurrent device simultaneously opens both of the ungrounded conductors.

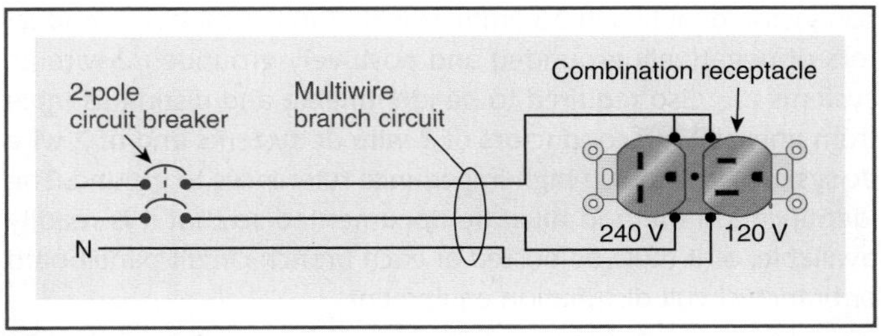


***EXHIBIT 210.2*** *A multiwire branch circuit supplying line-to-neutral and line-to-line connected loads, provided the ungrounded conductors can be opened simultaneously by the branch-circuit overcurrent device.*

Multiwire branch circuits have many advantages, including using three wires to do the work of four (in place of two 2-wire circuits), less raceway fill, easier balancing and phasing of a system, and less voltage drop. See the commentary following 215.2(A)(1), Informational Note No. 4, for further information on voltage drop for branch circuits.

Multiwire branch circuits may be derived from a 120/240-V, single-phase; a 208Y/120-V and 480Y/277-V, 3-phase, 4-wire; or a 240/120-V, 3-phase, 4-wire delta system. If two ungrounded conductors and a common neutral of a multiwire branch circuit are supplied from a 208Y/120-V, 3-phase, 4-wire system, the neutral carries the same current as the phase conductor with the highest current and, therefore, should be the same size. See the commentary following 210.4(A), Informational Note No. 2, for further information on 3-phase, 4-wire system neutral conductors.

If loads are connected line-to-line (i.e., utilization equipment connected between 2 or 3 phases), 2-pole or 3-pole circuit breakers are required to disconnect all ungrounded conductors simultaneously. With 240-V equipment, it is quite possible not to realize that the circuit is still energized with 120-V if one pole of the overcurrent device is open. See 210.10 and 240.15(B) for further information on circuit breaker overcurrent protection of ungrounded conductors. Other precautions concerning device removal on multiwire branch circuits are found in the commentary following 300.13(B).

•

**(D) Grouping.** The ungrounded and grounded circuit conductors of each multiwire branch circuit shall be grouped by cable ties or similar means in at least one location within the panelboard or other point of origination.

*Exception: The requirement for grouping shall not apply if the circuit enters from a cable or raceway unique to the circuit that makes the grouping obvious or if the conductors are identified at their terminations with numbered wire markers corresponding to the appropriate circuit number.*

The use of cables or raceways that are unique to the circuit makes it obvious which conductors are associated with the circuit. The use of tags bearing corresponding numbers at the termination point of a multiwire branch circuit also make it more apparent which conductors are associated with each circuit.

## 210.5 Identification for Branch Circuits

**(A) Grounded Conductor.** The grounded conductor of a branch circuit shall be identified in accordance with 200.6.

**(B) Equipment Grounding Conductor.** The equipment grounding conductor shall be identified in accordance with 250.119.

**(C) Identification of Ungrounded Conductors.** Ungrounded conductors shall be identified in accordance with 210.5(C)(1) or (2), as applicable.

**(1) Branch Circuits Supplied from More Than One Nominal Voltage System.** Where the premises wiring system has branch circuits supplied from more than one nominal voltage system, each ungrounded conductor of a branch circuit shall be identified by phase or line and system at all termination, connection, and splice points in compliance with 210.5(C)(1) (a) and (b).

The requirement for the identification of ungrounded branch-circuit conductors covers all branch-circuit configurations. The identification requirement applies only where more than one nominal voltage system supplies branch circuits (e.g., a 208Y/120-V system and a 480Y/277-V system) and requires that the conductors be identified by system and phase. Unlike the requirement of 200.6(D) for identifying the grounded conductors supplied from different voltage systems, application of this rule is not predicated on the different system conductors sharing a common raceway, cabinet, or enclosure.

The method of identification can be unique to the premises. Although color coding is a popular method, other types of marking or tagging are acceptable alternatives. Whatever identification method is used, it must be consistent throughout the premises. The identification legend must be posted at each piece of equipment supplying a branch circuit or be documented in an on-site manual or other form of record that is readily available to service personnel. If posted at electrical distribution equipment, the marking only has to describe the identification scheme for the ungrounded conductors supplied from that particular equipment. The basis for this requirement is to provide a higher level of safety for personnel working on premises electrical systems having ungrounded conductors supplied from multiple nominal voltage systems.

Exhibit 210.3 shows an example of two different nominal voltage systems in a building. Each ungrounded system conductor is identified by color-coded marking tape. A label indicating the means of the identification is permanently located at each panelboard. Although there are commonly employed color schemes, the *NEC* contains few specific color designations for ungrounded conductors, such as for heating cables, intrinsically safe circuits, isolated systems in health care facilities, or a 4-wire delta service.

For rules on identification of dc systems, see 210.5(C)(2) and associated commentary.

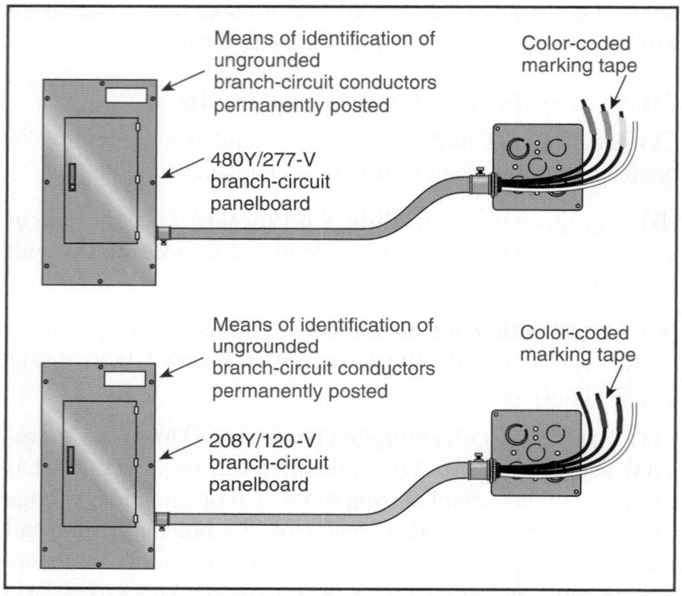


***EXHIBIT 210.3*** *Examples of accessible (ungrounded) phase conductors identified by marking tape at a junction or outlet location where the conductors will be spliced or terminated.*

(a) *Means of Identification.* The means of identification shall be permitted to be by separate color coding, marking tape, tagging, or other approved means.

(b) *Posting of Identification Means.* The method utilized for conductors originating within each branch-circuit panelboard or similar branch-circuit distribution equipment shall be documented in a manner that is readily available or shall be permanently posted at each branch-circuit panelboard or similar branch-circuit distribution equipment.

**(2) Branch Circuits Supplied From Direct-Current Systems.** Where a branch circuit is supplied from a dc system operating at more than 50 volts, each ungrounded conductor of 4 AWG or larger shall be identified by polarity at all termination, connection, and splice points by marking tape, tagging, or other approved means; each ungrounded conductor of 6 AWG or smaller shall be identified by polarity at all termination, connection, and splice points in compliance with 210.5(C)(2)(a) and (b). The identification methods utilized for conductors originating within each branch-circuit panelboard or similar branch-circuit distribution equipment shall be documented in a manner that is readily available or shall be permanently posted at each branch-circuit panelboard or similar branch-circuit distribution equipment.

This section was added to the 2014 *Code* to provide more detailed requirements for identification of conductors 4 AWG and larger for dc branch circuits. The requirements for the identification of ungrounded conductors are based on whether the system is negatively or positively grounded. Negatively grounded and positively grounded 2-wire dc systems must employ a grounded conductor identified in accordance with 200.6. Grounded conductors of negatively grounded and positively grounded 2-wire dc systems are also required to be identifiable and distinguishable from ungrounded conductors of 3-wire dc systems and of 2-wire dc systems employing high-impedance references to ground. The identification method must be documented so that it is readily available, or it must be posted at each branch-circuit panelboard or branch-circuit distribution equipment.

(a) *Positive Polarity, Sizes 6 AWG or Smaller.* Where the positive polarity of a dc system does not serve as the connection point for the grounded conductor, each positive ungrounded conductor shall be identified by one of the following means:

(1) A continuous red outer finish
(2) A continuous red stripe durably marked along the conductor's entire length on insulation of a color other than green, white, gray, or black
(3) Imprinted plus signs (+) or the word POSITIVE or POS durably marked on insulation of a color other than green, white, gray, or black, and repeated at intervals not exceeding 610 mm (24 in.) in accordance with 310.120(B)

(b) *Negative Polarity, Sizes 6 AWG or Smaller.* Where the negative polarity of a dc system does not serve as the connection point for the grounded conductor, each negative ungrounded conductor shall be identified by one of the following means:

(1) A continuous black outer finish
(2) A continuous black stripe durably marked along the conductor's entire length on insulation of a color other than green, white, gray, or red
(3) Imprinted minus signs (2) or the word NEGATIVE or NEG durably marked on insulation of a color other than green, white, gray, or red, and repeated at intervals not exceeding 610 mm (24 in.) in accordance with 310.120(B)

## 210.6 Branch-Circuit Voltage Limitations

The nominal voltage of branch circuits shall not exceed the values permitted by 210.6(A) through (E).

**(A) Occupancy Limitation.** In dwelling units and guest rooms or guest suites of hotels, motels, and similar occupancies, the voltage shall not exceed 120 volts, nominal, between conductors that supply the terminals of the following:

(1) Luminaires
(2) Cord-and-plug-connected loads 1440 volt-amperes, nominal, or less or less than ¼ hp

The term *similar occupancies* in 210.6(A) refers to sleeping rooms in dormitories, fraternities, sororities, nursing homes, and other such facilities. This requirement is intended to reduce the exposure of residents in these types of occupancies to electric shock hazards when using or servicing permanently installed luminaires and cord-and-plug-connected portable lamps and appliances.

Small loads, such as those of 1440 VA or less and motors of less than ¼ horsepower, are limited to 120-V circuits. High-wattage cord-and-plug-connected loads, such as electric ranges, clothes dryers, and some window air conditioners, could be connected to a 208-V or 240-V circuit.

**(B) 120 Volts Between Conductors.** Circuits not exceeding 120 volts, nominal, between conductors shall be permitted to supply the following:

(1) The terminals of lampholders applied within their voltage ratings
(2) Auxiliary equipment of electric-discharge lamps

*Auxiliary equipment* includes ballasts and starting devices for fluorescent and high-intensity-discharge (e.g., mercury vapor, metal halide, and sodium) lamps.

Informational Note: See 410.137 for auxiliary equipment limitations.

(3) Cord-and-plug-connected or permanently connected utilization equipment

**(C) 277 Volts to Ground.** Circuits exceeding 120 volts, nominal, between conductors and not exceeding 277 volts, nominal, to ground shall be permitted to supply the following:

(1) Listed electric-discharge or listed light-emitting diode-type luminaires
(2) Listed incandescent luminaires, where supplied at 120 volts or less from the output of a stepdown autotransformer that is an integral component of the luminaire and the outer shell terminal is electrically connected to a grounded conductor of the branch circuit

An incandescent luminaire is permitted on a 277-V circuit only if it is a listed luminaire with an integral autotransformer and an output to the lampholder that does not exceed 120-V. In this application, the autotransformer supplies 120 V to the lampholder, and the grounded conductor is connected to the screw shell of the lampholder. This application is similar to a branch circuit derived from an autotransformer, except that the 120-V circuit is the internal wiring of the luminaire.

(3) Luminaires equipped with mogul-base screw shell lampholders
(4) Lampholders, other than the screw shell type, applied within their voltage ratings
(5) Auxiliary equipment of electric-discharge lamps

Informational Note: See 410.137 for auxiliary equipment limitations.

(6) Cord-and-plug-connected or permanently connected utilization equipment

Exhibit 210.4 shows some examples of luminaires permitted to be connected to branch circuits. Medium-base screw shell lampholders cannot be directly connected to 277-V branch circuits. Other types of lampholders may be connected to 277-V circuits but only if the lampholders have a 277-V rating. A 277-V branch circuit may be connected to a listed electric-discharge fixture or to a listed autotransformer-type incandescent fixture with a medium-base screw shell lampholder.

Typical examples of cord-and-plug-connected equipment include through-the-wall heating and air-conditioning units and restaurant deep fat fryers that operate at 480-V, 3 phase, from a grounded wye system.

The requirements in 210.6 have to be read carefully because 210.6(D) describes the voltage as "volts, nominal, to ground," whereas 210.6(A), (B), (C), and (E) describe voltage as "volts, nominal, between conductors." Luminaires listed for and connected to a 480-V source may be used in applications permitted by 210.6(C), provided the 480-V system is in fact a grounded wye system that contains a grounded conductor (thus limiting the system "voltage to ground" to the 277-V level).

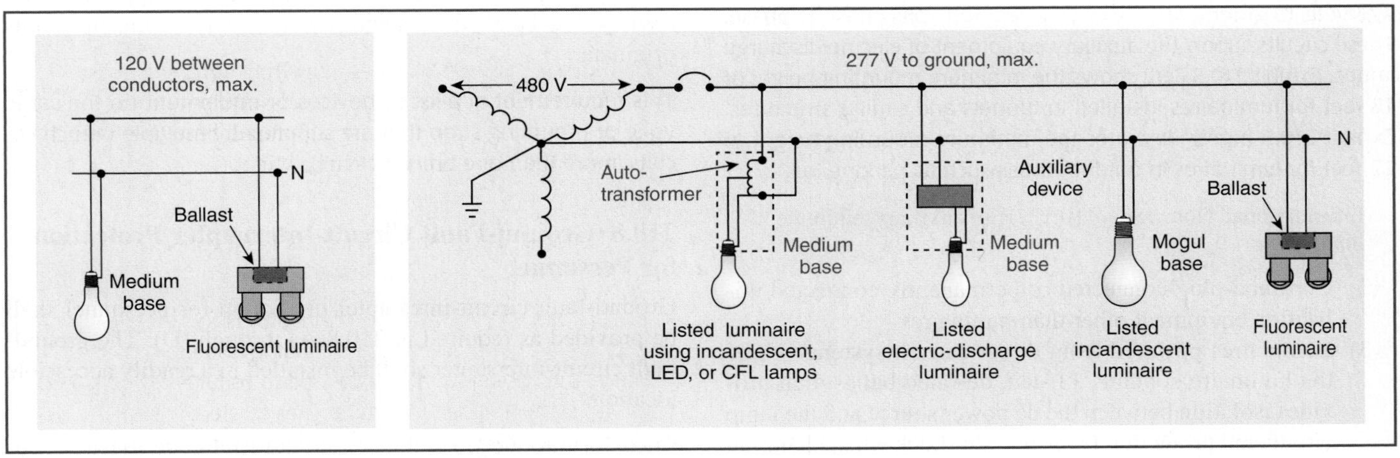


***EXHIBIT 210.4*** *Examples of luminaires permitted to be connected to branch circuits.*

*EXHIBIT 210.5 Minimum mounting heights for tunnel and parking lot lighting for circuits exceeding 277 volts to ground and not exceeding 600 volts between conductors supplying auxiliary equipment of electric-discharge lampholders. (Left: © Anese/ Dreamstime.com)*

Tunnel Lighting

Parking Lot Lighting

**(D) 600 Volts Between Conductors.** Circuits exceeding 277 volts, nominal, to ground and not exceeding 600 volts, nominal, between conductors shall be permitted to supply the following:

(1) The auxiliary equipment of electric-discharge lamps mounted in permanently installed luminaires where the luminaires are mounted in accordance with one of the following:

a. Not less than a height of 6.7 m (22 ft) on poles or similar structures for the illumination of outdoor areas such as highways, roads, bridges, athletic fields, or parking lots

b. Not less than a height of 5.5 m (18 ft) on other structures such as tunnels

The minimum mounting heights are for circuits that exceed 277 volts to ground and do not exceed 600 volts phase to phase. These circuits supply the auxiliary equipment of electric-discharge lamps. Exhibit 210.5 (left) shows the minimum mounting height of 18 feet for luminaires installed in tunnels and similar structures. Exhibit 210.5 (right) illustrates the minimum mounting height of 22 feet for luminaires in outdoor areas such as parking lots.

Informational Note: See 410.137 for auxiliary equipment limitations.

(2) Cord-and-plug-connected or permanently connected utilization equipment other than luminaires

(3) Luminaires powered from direct-current systems where the luminaire contains a listed, dc-rated ballast that provides isolation between the dc power source and the lamp circuit and protection from electric shock when changing lamps.

*Exception No. 1 to (B), (C), and (D): For lampholders of infrared industrial heating appliances as provided in 422.14.*

*Exception No. 2 to (B), (C), and (D): For railway properties as described in 110.19.*

**(E) Over 600 Volts Between Conductors.** Circuits exceeding 600 volts, nominal, between conductors shall be permitted to supply utilization equipment in installations where conditions of maintenance and supervision ensure that only qualified persons service the installation.

## 210.7 Multiple Branch Circuits

Where two or more branch circuits supply devices or equipment on the same yoke or mounting strap, a means to simultaneously disconnect the ungrounded conductors supplying those devices shall be provided at the point at which the branch circuits originate.

This requirement applies to devices or equipment on the same yoke or mounting strap that are supplied by multiple branch circuits (more than one branch circuit).

## 210.8 Ground-Fault Circuit-Interrupter Protection for Personnel

Ground-fault circuit-interrupter protection for personnel shall be provided as required in 210.8(A) through (D). The ground-fault circuit-interrupter shall be installed in a readily accessible location.

Since the introduction of the GFCI in the 1971 *Code*, these devices have proven their worth. Data show a decrease in the number of

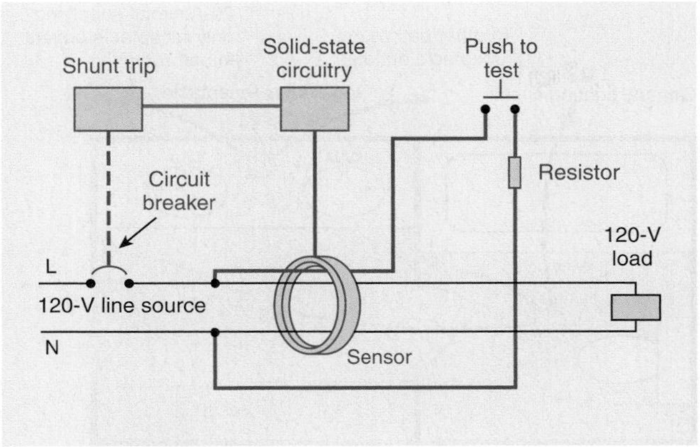


***EXHIBIT 210.6*** *The circuitry and components of a typical GFCI.*

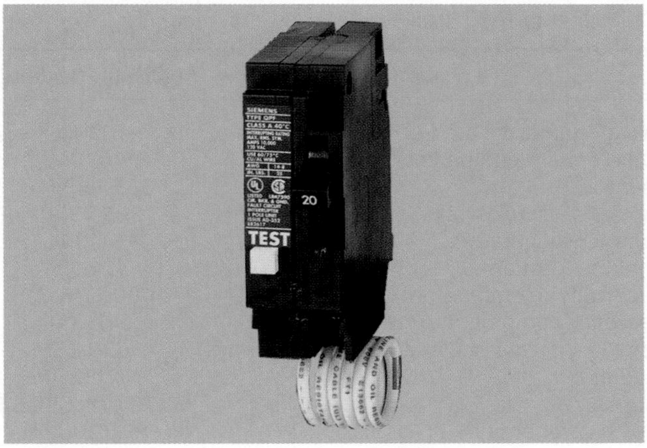


***EXHIBIT 210.7*** *A GFCI circuit breaker, which protects all outlets supplied by the branch circuit. (Courtesy of Siemens)*

electrocutions in the United States since the introduction of GFCI devices. Unfortunately, no statistics are available for the actual number of lives saved or injuries prevented by GFCI devices. However, most safety experts agree that GFCIs are directly responsible for saving numerous lives and preventing countless injuries.

To ensure proper operation of GFCI devices, the manufacturer's installation and use instructions specify periodic testing — typically every month. The specified test is to operate the test button on the device and verify that all receptacles protected through that GFCI device are de-energized. To facilitate this important ongoing safety check, GFCIs installed to protect the receptacles covered in 210.8(A) and (B) are required to be readily accessible (see Article 100 for the definition of *readily accessible*).

Exhibit 210.6 shows a typical circuit arrangement of a GFCI. The line conductors are passed through a sensor and are connected to a shunt-trip device. As long as the current in the conductors is equal, the device remains in a closed position. If one of the conductors comes in contact with a grounded object, either directly or through a person's body, some of the current returns by the alternative path, resulting in an unbalanced current. The toroidal coil senses the unbalanced current, and the shunt-trip mechanism reacts to open the circuit. The circuit design does not require the presence of an equipment grounding conductor, which is the reason 406.4(D)(2)(c) permits the use of GFCIs as replacements for receptacles where a grounding means does not exist.

GFCIs operate on fault currents of 4 to 6 mA. At trip levels of 5 mA (the instantaneous current could be much higher), a shock can be felt during the fault. The shock can lead to an involuntary reaction that may cause a secondary accident such as a fall. GFCIs do not protect persons from shock hazards where contact is between phase and neutral or between phase-to-phase conductors.

A variety of GFCIs are available, including portable and plug-in types and circuit-breaker types, types built into attachment plug caps, and receptacle types. Each type has a test switch so

***EXHIBIT 210.8*** *A 15-ampere duplex receptacle with integral GFCI that also protects downstream loads.*

that units can be checked periodically to ensure proper operation. See Exhibits 210.7 and 210.8.

Although 210.8 illustrates the main rule for GFCIs, other specific applications that require the use of GFCIs are listed in Commentary Table 210.1.

Informational Note: See 215.9 for ground-fault circuit-interrupter protection for personnel on feeders.

**(A) Dwelling Units.** All 125-volt, single-phase, 15- and 20-ampere receptacles installed in the locations specified

**COMMENTARY TABLE 210.1** Additional Requirements for the Application of GFCI Protection

| Location | Applicable Section(s) |
|---|---|
| Aircraft hangars | 513.12 |
| Audio system equipment | 640.10(A) |
| Boathouses | 555.19(B)(1) |
| Carnivals, circuses, fairs, and similar events | 525.23(A) |
| Commercial garages | 511.12 |
| Drinking fountains | 422.52 |
| Electrically operated pool covers | 680.27(B)(2) |
| Electronic equipment, sensitive | 647.7(A) |
| Elevators, escalators, and moving walkways | 620.85 |
| Feeders | 215.9 |
| Fountains | 680.51(A), 680.56 |
| Health care facilities | 517.20(A) |
| High-pressure spray washers | 422.49 |
| Hydromassage bathtubs | 680.71 |
| Marinas and boatyards | 555.19(B)(1) |
| Mobile and manufactured homes | 550.13(B), 550.13(E), 550.32(E) |
| Natural and artificially made bodies of water | 682.15 |
| Park trailers | 552.41(C), 552.42(D) |
| Pools, permanently installed | 680.22(A)(1), 680.22(A)(4), 680.22(B)(3), 680.22(B)(4), 680.23(A)(3) |
| Pools, storable | 680.32 |
| Sensitive electronic equipment | 647.7(A) |
| Space heating embedded in floor | 424.44(G) |
| Spas and hot tubs | 680.44 |
| Signs within fountains | 680.57(B) |
| Signs, mobile or portable | 600.10(C)(2) |
| Recreational vehicles | 551.40(C), 551.41(C) |
| Recreational vehicle parks | 551.71 |
| Replacement receptacles | 406.4(D)(3) |
| Temporary installations | 590.6 |
| Tubs, therapeutic | 680.62(A) |

in 210.8(A)(1) through (10) shall have ground-fault circuit-interrupter protection for personnel.

(1) Bathrooms

All 125-V, single-phase, 15- and 20-A receptacles in bathrooms must have GFCI protection, including receptacles that are integral with luminaires. There are no exceptions to the bathroom GFCI requirement. If a washing machine is located in the bathroom, the 15- or 20-A, 125-V receptacle required to be supplied from the laundry branch circuit must be GFCI protected.

A *bathroom* is defined in Article 100 as "an area including a basin with one or more of the following: a toilet, a urinal, a tub, a shower, a bidet, or similar plumbing fixtures." The term applies to

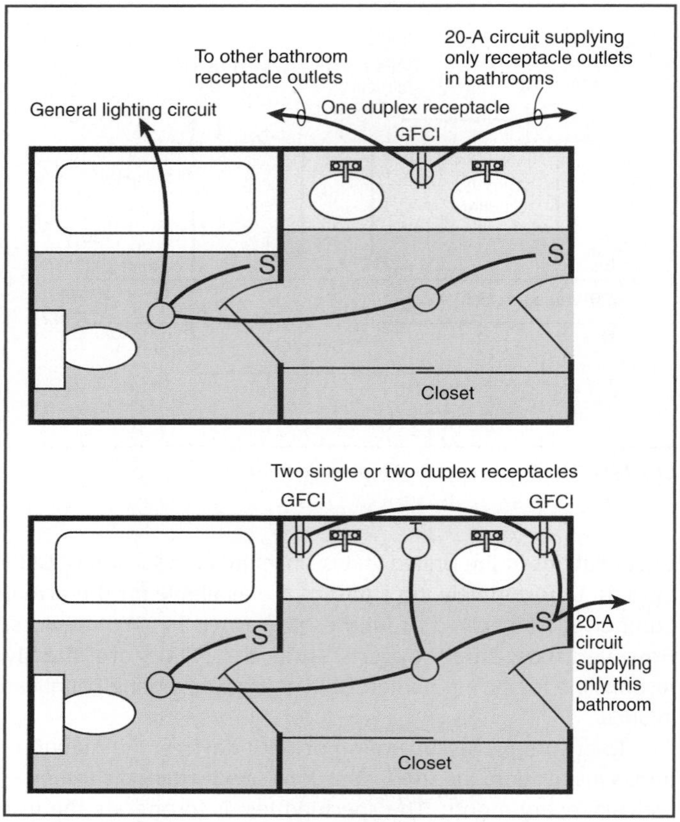


***EXHIBIT 210.9*** *GFCI-protected receptacles in bathrooms.*

the entire area, whether or not a separating door, as illustrated in Exhibit 210.9, is present. If the basins are adjacent and in close proximity, one receptacle outlet, meeting the proximity requirement of 210.52(D) for each basin, can be used to meet the receptacle outlet location requirement as shown in Exhibit 210.9 (top). Exhibit 210.9 also illustrates the requirements of 210.11(C)(3), which provides two acceptable supply circuit arrangements for the bathroom receptacle outlet(s).

(2) Garages, and also accessory buildings that have a floor located at or below grade level not intended as habitable rooms and limited to storage areas, work areas, and areas of similar use

The requirement for GFCI receptacles in garages and sheds, as illustrated in Exhibit 210.10, improves safety for persons using portable handheld tools, string trimmers, snow blowers, and similar tools that might be connected to these receptacles. GFCI protection is also required because auto repair work and general workshop electrical tools are often used.

There are no exceptions. All 125-V, single-phase, 15- and 20-A receptacles installed in garages must provide GFCI protection for the user of appliances or other equipment regardless of

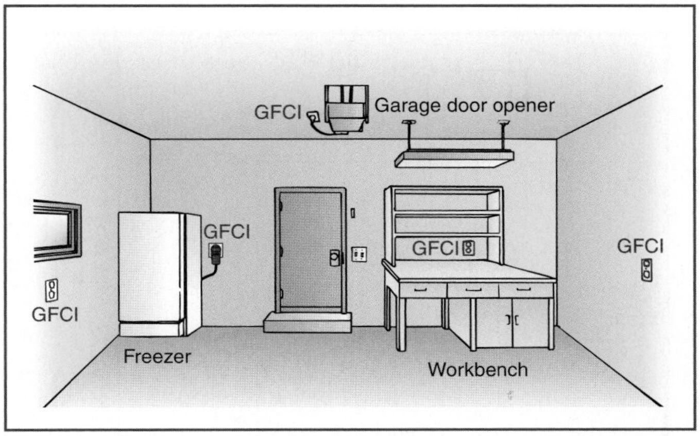


***EXHIBIT 210.10*** *Examples of receptacles in a garage required to have GFCI protection.*

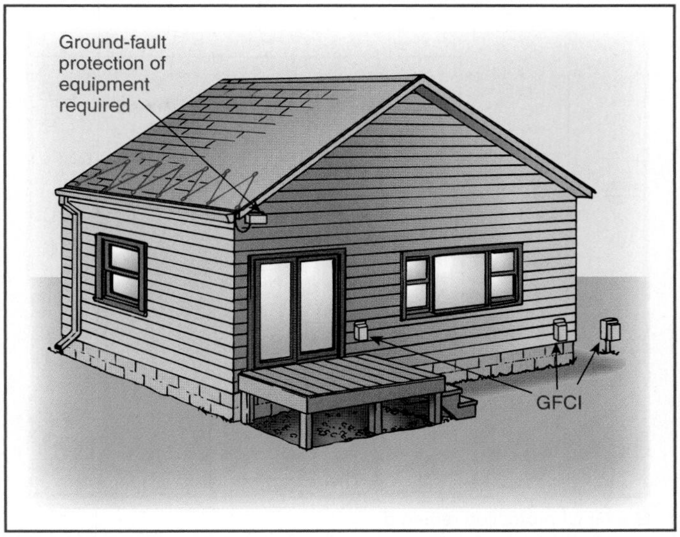


***EXHIBIT 210.11*** *A dwelling unit with three receptacles required to have GFCI protection and one that is not.*

where the receptacle is located in the garage. Appliance leakage currents permitted by today's product standards are far less than the operational threshold of a GFCI, so nuisance tripping is unlikely.

(3) Outdoors

*Exception to (3): Receptacles that are not readily accessible and are supplied by a branch circuit dedicated to electric snow-melting, deicing, or pipeline and vessel heating equipment shall be permitted to be installed in accordance with 426.28 or 427.22, as applicable.*

The dwelling unit shown in Exhibit 210.11, which has four outdoor receptacles, illustrates the requirement of 210.8(A)(3). Three receptacles must be provided with GFCI protection. The fourth receptacle, located adjacent to the gutter for the roof-mounted snow-melting cable, is not readily accessible and, therefore, is exempt from the GFCI requirements because of its dedicated function to supply the deicing equipment. This receptacle is, however, covered by the equipment ground-fault protection requirements of 426.28. See the commentary following 210.52(E) and 406.9(B)(1) regarding the installation of outdoor receptacles in wet and damp locations.

(4) Crawl spaces — at or below grade level
(5) Unfinished basements — for purposes of this section, unfinished basements are defined as portions or areas of the basement not intended as habitable rooms and limited to storage areas, work areas, and the like

*Exception to (5): A receptacle supplying only a permanently installed fire alarm or burglar alarm system shall not be required to have ground-fault circuit-interrupter protection.*

Informational Note: See 760.41(B) and 760.121(B) for power supply requirements for fire alarm systems.

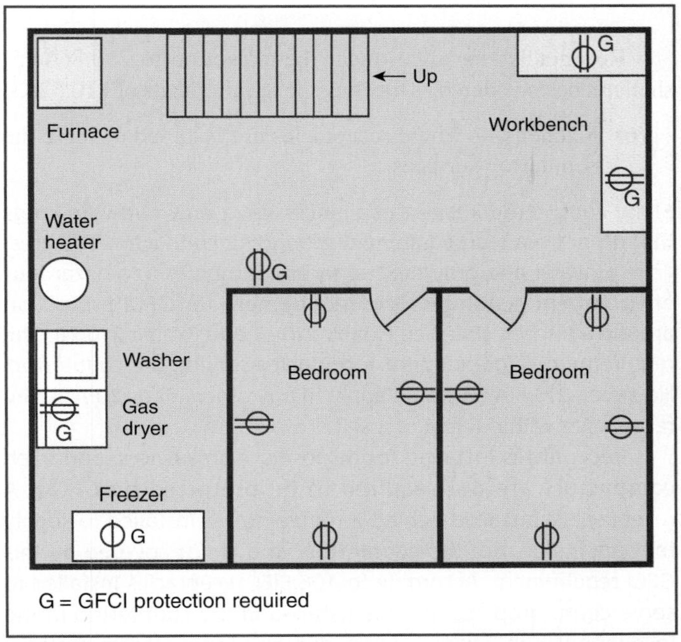


***EXHIBIT 210.12*** *A basement floor plan with GFCI-protected receptacles in the work area and non-GFCI receptacles in the finished areas.*

The receptacles in a work area of a basement, as shown in Exhibit 210.12, must have GFCI protection. This requirement does not apply to finished areas in basements, such as sleeping rooms or family rooms. The only exception to this requirement is for fire alarm and burglar systems. This correlates with the performance requirements covering fire alarm power supplies contained in *NFPA 72®*, *National Fire Alarm and Signaling Code.*

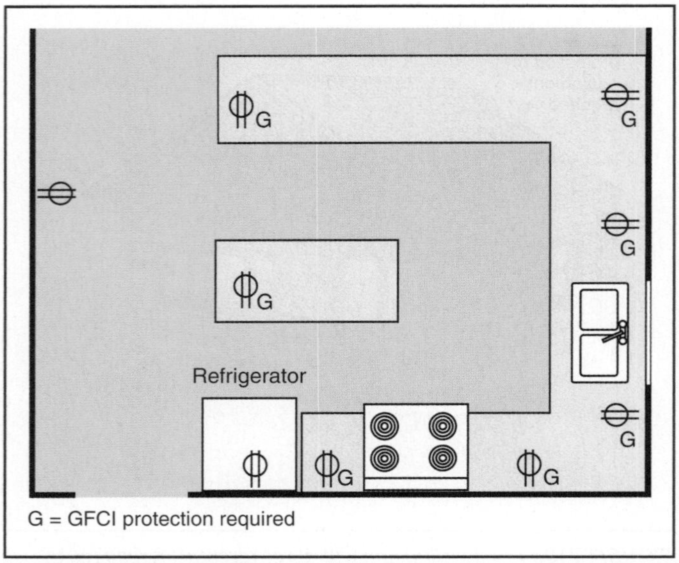


**EXHIBIT 210.13** *GFCI-protected receptacles serving countertop surfaces in dwelling unit kitchens.*

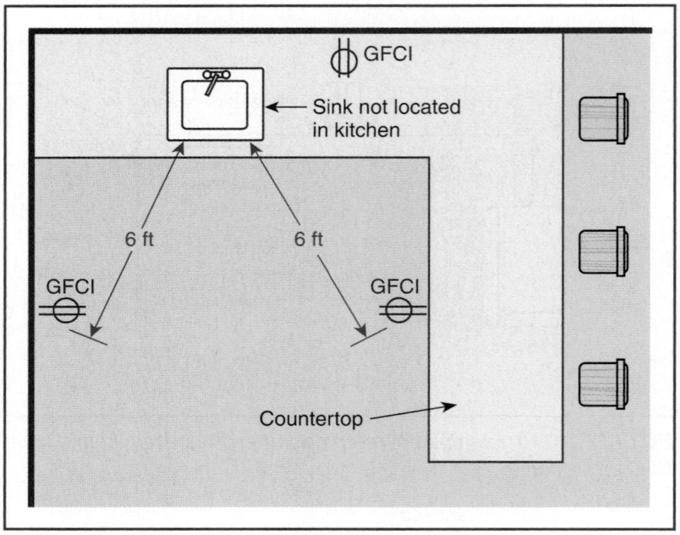


**EXHIBIT 210.14** *GFCI protection of receptacles located within 6 feet of a sink located in other than the kitchen.*

Receptacles installed under the exception to 210.8(A)(5) shall not be considered as meeting the requirements of 210.52(G)

(6) Kitchens — where receptacles are installed to serve the countertop surfaces

Many countertop kitchen appliances have only two-wire cords that do not have an equipment grounding conductor. The presence of water and grounded surfaces contributes to a hazardous environment, leading to the requirement for GFCI protection around a kitchen sink. See Exhibit 210.13 and Exhibit 210.28. The requirement is intended for receptacles serving the countertop. However, 210.8(A)(7) would apply to any other 15- or 20-A, 125-V receptacles within 6 feet of a sink.

Receptacles installed for disposals, dishwashers, and trash compactors are not required to be protected by GFCIs. A receptacle(s) installed behind a refrigerator is installed to supply that appliance, not the countertop, and is not covered by this GFCI requirement. According to 406.5(E), receptacles installed to serve countertops cannot be installed in the countertop in the face-up position because liquid, dirt, and other foreign material can enter the receptacle.

(7) Sinks — where receptacles are installed within 1.8 m (6 ft) of the outside edge of the sink

Sinks in kitchens are not the only sinks where a ground-fault shock hazard exists; therefore, this requirement covers all other sinks in a dwelling. This GFCI requirement is not limited to receptacles serving countertop surfaces; rather, it covers all 125-V, 15- and 20-A receptacles within 6 feet of any point along the outside edge of the sink. Many appliances used in these locations are ungrounded, and the presence of water and grounded surfaces contributes to a hazardous environment. As illustrated in Exhibit 210.14, any 125-V, 15- or 20-A receptacle installed within 6 feet of a sink located in other than a kitchen is also required to be GFCI protected.

(8) Boathouses

(9) Bathtubs or shower stalls — where receptacles are installed within 1.8 m (6 ft) of the outside edge of the bathtub or shower stall

(10) Laundry areas

Items (9) and (10) were added to 210.8(A) for the 2014 *Code*. In some instances, bathtubs and shower stalls are installed in areas that might not meet the *NEC* definition of a bathroom. Many of these areas may have tile or other conductive, and possibly grounded, floors. A shock hazard could be present. This change in *Code* language ensures that tubs and showers in dwelling units will have the same requirements for GFCI protection of receptacles, regardless of what the room is called.

The requirement for GFCI protection of receptacles in dwelling unit laundry areas is new for 2014. Wet clothes and puddles of water can pose a shock hazard to anyone using any appliance in laundry areas.

**(B) Other Than Dwelling Units.** All 125-volt, single-phase, 15- and 20-ampere receptacles installed in the locations specified in 210.8(B)(1) through (8) shall have ground-fault circuit-interrupter protection for personnel.

(1) Bathrooms

If 125-V, single-phase, 15- and 20-A receptacles are provided in the bathroom areas of occupancies other than dwelling units, they must be GFCI-protected. Some motel and hotel bathrooms, like the one shown in Exhibit 210.15, have the basin located outside the door to the room containing the tub, toilet, or shower. *Bathroom* as defined in Article 100 uses the word "area," so that

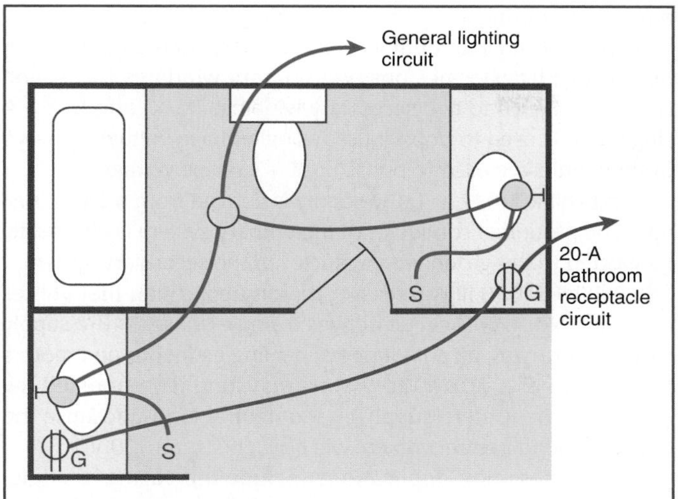


***EXHIBIT 210.15*** *GFCI protection of receptacles in a motel/hotel bathroom where one basin is located outside the bathroom door.*

the sink shown in Exhibit 210.15 is considered as being in the bathroom and is subject to the GFCI requirement. In Exhibit 210.15 the supply circuit requirement of 210.11(C)(3) for the bathroom receptacle outlets is depicted.

(2) Kitchens

All 15- and 20-A, 125-V receptacles in nondwelling-type kitchens must be GFCI protected. This requirement applies to all 15- and 20-A, 125-V kitchen receptacles, whether or not the receptacle serves countertop areas.

Electrical accident data indicate that there are many electrical hazards in nondwelling kitchens, including poorly maintained electrical equipment, damaged cords, wet floors, and employees without proper electrical safety training. Requiring GFCI protection in these kitchens protects personnel who might be exposed to conditions that are conducive to electric shock accidents. The definition of *kitchen* in Article 100 is "an area with a sink and permanent facilities for food preparation and cooking." A portable cooking appliance (e.g., cord-and-plug-connected microwave oven or hot plate) is not a permanent cooking facility. Kitchens in restaurants, hotels, schools, churches, dining halls, and similar facilities are examples of the types of kitchens covered by this requirement.

(3) Rooftops

For rooftops that also have heating, air-conditioning, and refrigeration equipment, see 210.63.

(4) Outdoors

*Exception No. 1 to (3): Receptacles on rooftops shall not be required to be readily accessible other than from the rooftop.*

Exception No. 1 was added for the 2014 *Code*. The main rule in 210.8 requires GFCI to be installed at a readily accessible location. By definition, a rooftop is not readily accessible. However, it must be readily accessible from the rooftop, which this exception clarifies.

*Exception No. 2 to (3) and (4): Receptacles that are not readily accessible and are supplied by a branch circuit dedicated to electric snow-melting, deicing, or pipeline and vessel heating equipment shall be permitted to be installed in accordance with 426.28 or 427.22, as applicable.*

*Exception No. 3 to (4): In industrial establishments only, where the conditions of maintenance and supervision ensure that only qualified personnel are involved, an assured equipment grounding conductor program as specified in 590.6(B)(2) shall be permitted for only those receptacle outlets used to supply equipment that would create a greater hazard if power is interrupted or having a design that is not compatible with GFCI protection.*

Although commercial, institutional, and industrial occupancies are not required to have outdoor receptacle outlets installed for general use, outdoor receptacle outlets may be installed to meet the requirement of 210.63 or at the discretion of the designer or owner. Except for the two limited exceptions [Exception No. 2 to (3) and (4) and Exception No. 3 to (4)], all outdoor 125-V, single-phase, 15- and 20-A receptacles at commercial, institutional, and industrial occupancies are required to be provided with GFCI protection.

(5) Sinks — where receptacles are installed within 1.8 m (6 ft) of the outside edge of the sink

This covers receptacles installed near sinks in lunchrooms, janitors' closets, classrooms, and all other areas not covered by the bathroom and kitchen requirements. Exhibit 210.16 shows two single-phase, 125-V, 15-A duplex receptacles installed within 6 feet of a sink that is covered by the GFCI requirement. The 240-V receptacle to the right of the duplex receptacles is not subject to the GFCI requirement.

***EXHIBIT 210.16*** *GFCI protection required for single-phase, 125-volt, 15-ampere duplex receptacles installed within 6 feet of a sink.*

*Exception No. 1 to (5): In industrial laboratories, receptacles used to supply equipment where removal of power would introduce a greater hazard shall be permitted to be installed without GFCI protection.*

*Exception No. 2 to (5): For receptacles located in patient bed locations of general care or critical care areas of health care facilities other than those covered under 210.8(B)(1), GFCI protection shall not be required.*

(6) Indoor wet locations
(7) Locker rooms with associated showering facilities
(8) Garages, service bays, and similar areas other than vehicle exhibition halls and showrooms

This requirement was expanded for the 2014 *Code* so that it now covers all garages, not just those in which vehicle maintenance is expected to take place. Many commercial garages have receptacles installed for purposes other than the use of hand tools. For instance, winter temperatures in some areas necessitate the use of engine block heaters. Cord-and-plug-connected engine block heaters may not be listed, and, therefore are not subject to the maximum leakage current requirements for appliances. If these receptacles are not GFCI protected, the frame of the vehicle can become energized, posing a shock hazard. This requirement does not apply to vehicle exhibit halls or showrooms. These areas are unlikely to be used for vehicle maintenance requiring the use of cord-and-plug-connected equipment.

**(C) Boat Hoists.** GFCI protection shall be provided for outlets not exceeding 240 volts that supply boat hoists installed in dwelling unit locations.

The proximity of this type of equipment to water and the wet or damp environment that is typical where boat hoists are used is the reason for this GFCI requirement. Documented cases of electrocutions associated with the use of boat hoists compiled by the U.S. Consumer Product Safety Commission substantiated the need for this requirement. The GFCI requirement applies only to dwelling unit locations and to boat hoists supplied by 15- or 20-A branch circuits rated 240 V or less. It is important to note that it applies to all outlets, not just to receptacle outlets. Therefore, both cord-and-plug-connected and hard-wired boat hoists are required to be GFCI protected.

**(D) Kitchen Dishwasher Branch Circuit.** GFCI protection shall be provided for outlets that supply dishwashers installed in dwelling unit locations.

## 210.9 Circuits Derived from Autotransformers

Branch circuits shall not be derived from autotransformers unless the circuit supplied has a grounded conductor that is electrically connected to a grounded conductor of the system supplying the autotransformer.

An autotransformer requires little physical space, is economical, and, above all, is efficient. A buck-boost autotransformer provides a means of raising (boosting) or lowering (bucking) a supply line voltage by a small amount (usually no more than 20 percent). A buck-boost transformer has two primary windings ($H_1$-$H_2$ and $H_3$-$H_4$) connected to two secondary windings ($X_1$-$X_2$ and $X_3$-$X_4$). A single unit is used to boost/buck a single-phase voltage, but two or three units are used to boost/buck a 3-phase voltage.

In Exhibit 210.17, a 120-V supply is derived from a 240-V system. The grounded conductor of the primary system is electrically connected to the grounded conductor of the secondary system.

Exhibit 210.18 illustrates a common application that utilizes Exception No. 1. This circuit derives a single-phase, 240-V supply system for ranges, air conditioners, heating elements, and motors from a 3-phase, 208Y/120-V source system. The boosted leg should not be used to supply line-to-neutral loads, because the boosted line-to-neutral voltage will be higher than 120 V.

Other common applications include increasing a single-phase, 240-V source to a single-phase, 277-V supply for lighting systems and transforming 240 V to 208 V for use with 208-V appliances. See Exhibit 210.19.

*Exception No. 1: An autotransformer shall be permitted without the connection to a grounded conductor where transforming from a nominal 208 volts to a nominal 240-volt supply or similarly from 240 volts to 208 volts.*

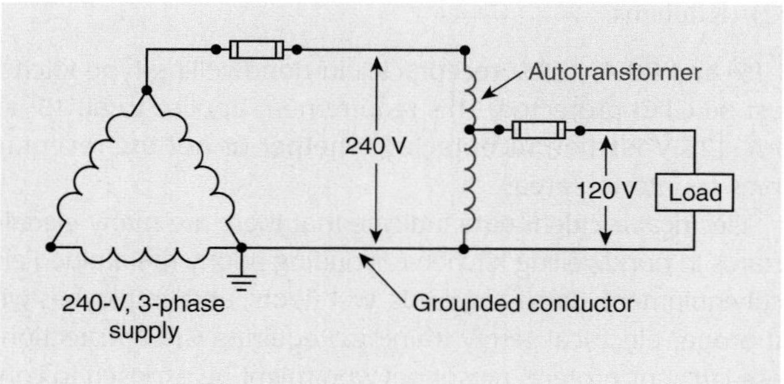


***EXHIBIT 210.17*** *Circuitry for an autotransformer used to derive a 2-wire, 120-volt system for lighting or convenience receptacles from a 240-V corner-grounded delta system.*

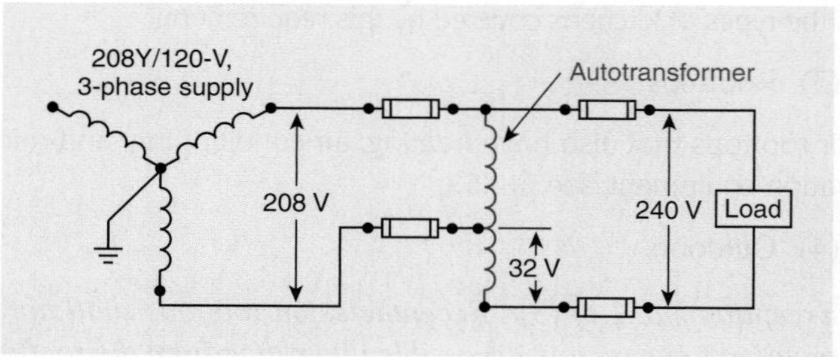


***EXHIBIT 210.18*** *Circuitry for an autotransformer used to derive a 240-volt system for appliances from a 208Y/120-V source.*

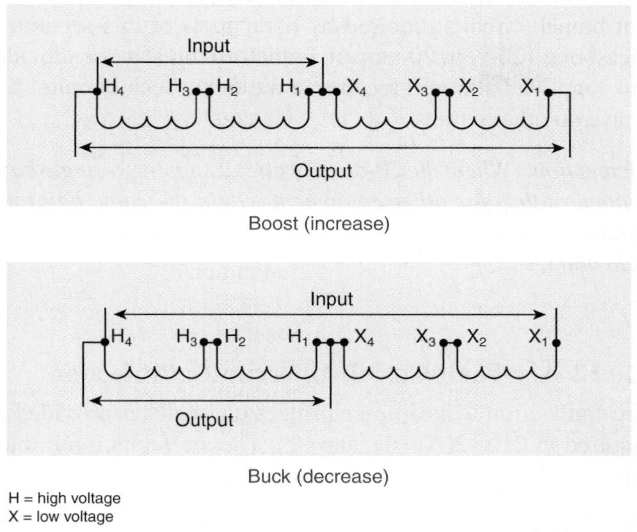


***EXHIBIT 210.19*** *Typical single-phase connection diagrams for buck-boost transformers connected as autotransformers to change 240 V single-phase to 208 V and vice versa.*

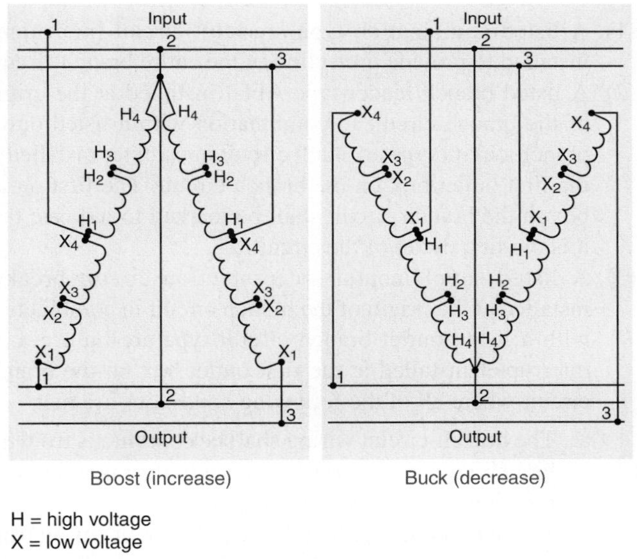


***EXHIBIT 210.20*** *Typical connection diagrams for buck-boost transformers connected in 3-phase open delta as autotransformers to change 240 V to 208 V and vice versa.*

An autotransformer (without an electrical connection to a grounded conductor) is permitted to extend or add an individual branch circuit in an existing installation where transforming (boosting) 208 V to 240 V as shown in Exhibit 210.18. Exhibit 210.20 illustrates typical 3-phase buck-boost transformers connected to change 240 V to 208 V and vice versa.

*Exception No. 2: In industrial occupancies, where conditions of maintenance and supervision ensure that only qualified persons service the installation, autotransformers shall be permitted to supply nominal 600-volt loads from nominal 480-volt systems, and 480-volt loads from nominal 600-volt systems, without the connection to a similar grounded conductor.*

## 210.10 Ungrounded Conductors Tapped from Grounded Systems

Two-wire dc circuits and ac circuits of two or more ungrounded conductors shall be permitted to be tapped from the ungrounded conductors of circuits that have a grounded neutral conductor. Switching devices in each tapped circuit shall have a pole in each ungrounded conductor. All poles of multipole switching devices shall manually switch together where such switching devices also serve as a disconnecting means as required by the following:

(1) 410.93 for double-pole switched lampholders
(2) 410.104(B) for electric-discharge lamp auxiliary equipment switching devices
(3) 422.31(B) for an appliance
(4) 424.20 for a fixed electric space-heating unit
(5) 426.51 for electric deicing and snow-melting equipment
(6) 430.85 for a motor controller
(7) 430.103 for a motor

Exhibit 210.21 (top) illustrates an ungrounded 2-wire branch circuit tapped from the ungrounded conductors of a dc or single-phase system to supply a small motor. Exhibit 210.21 (bottom) illustrates a 3-phase, 4-wire wye system. Based on the characteristics of the depicted loads (line-to-line connected) supplied by the tapped circuits, a neutral or grounded conductor is not required to be installed with the ungrounded conductors.

Circuit breakers or switches used as the disconnecting means for the tapped branch circuit must open all poles simultaneously using only the manual operation of the disconnecting means. If switches and fuses are used and one fuse blows, or if circuit breakers (two single-pole circuit breakers with a handle tie) are used and one breaker trips, one pole could possibly remain closed. The intention is not to require a common trip of fuses or circuit breakers but rather to disconnect the ungrounded conductors of the branch circuit with one manual operation. See 240.15(B) for information on the use of identified handle ties with single-pole circuit breakers.

## 210.11 Branch Circuits Required

Branch circuits for lighting and for appliances, including motor-operated appliances, shall be provided to supply the loads calculated in accordance with 220.10. In addition, branch circuits shall be provided for specific loads not covered by 220.10 where required elsewhere in this *Code* and for dwelling unit loads as specified in 210.11(C).

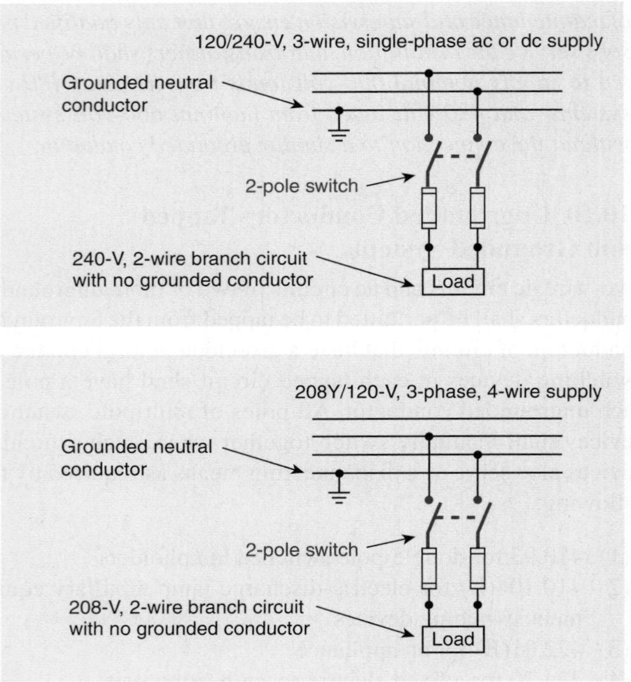


***EXHIBIT 210.21*** *Branch circuits tapped from ungrounded conductors of multiwire systems.*

**(A) Number of Branch Circuits.** The minimum number of branch circuits shall be determined from the total calculated load and the size or rating of the circuits used. In all installations, the number of circuits shall be sufficient to supply the load served. In no case shall the load on any circuit exceed the maximum specified by 220.18.

**(B) Load Evenly Proportioned Among Branch Circuits.** Where the load is calculated on the basis of volt-amperes per square meter or per square foot, the wiring system up to and including the branch-circuit panelboard(s) shall be provided to serve not less than the calculated load. This load shall be evenly proportioned among multioutlet branch circuits within the panelboard(s). Branch-circuit overcurrent devices and circuits shall be required to be installed only to serve the connected load.

**(C) Dwelling Units.**

**(1) Small-Appliance Branch Circuits.** In addition to the number of branch circuits required by other parts of this section, two or more 20-ampere small-appliance branch circuits shall be provided for all receptacle outlets specified by 210.52(B).

**(2) Laundry Branch Circuits.** In addition to the number of branch circuits required by other parts of this section, at least one additional 20-ampere branch circuit shall be provided to supply the laundry receptacle outlet(s) required by 210.52(F). This circuit shall have no other outlets.

**(3) Bathroom Branch Circuits.** In addition to the number of branch circuits required by other parts of this section, at least one 120-volt, 20-ampere branch circuit shall be provided to supply a bathroom receptacle outlet(s). Such circuits shall have no other outlets.

*Exception: Where the 20-ampere circuit supplies a single bathroom, outlets for other equipment within the same bathroom shall be permitted to be supplied in accordance with 210.23(A)(1) and (A)(2).*

•

## 210.12 Arc-Fault Circuit-Interrupter Protection

Arc-fault circuit-interrupter protection shall be provided as required in 210.12(A) (B), and (C). The arc-fault circuit interrupter shall be installed in a readily accessible location.

**(A) Dwelling Units.** All 120-volt, single-phase, 15- and 20-ampere branch circuits supplying outlets or devices installed in dwelling unit kitchens, family rooms, dining rooms, living rooms, parlors, libraries, dens, bedrooms, sunrooms, recreation rooms, closets, hallways, laundry areas, or similar rooms or areas shall be protected by any of the means described in 210.12(A)(1) through (6):

(1) A listed combination-type arc-fault circuit interrupter, installed to provide protection of the entire branch circuit
(2) A listed branch/feeder-type AFCI installed at the origin of the branch-circuit in combination with a listed outlet branch-circuit type arc-fault circuit interrupter installed at the first outlet box on the branch circuit. The first outlet box in the branch circuit shall be marked to indicate that it is the first outlet of the circuit.
(3) A listed supplemental arc protection circuit breaker installed at the origin of the branch circuit in combination with a listed outlet branch-circuit type arc-fault circuit interrupter installed at the first outlet box on the branch circuit where all of the following conditions are met:
   a. The branch-circuit wiring shall be continuous from the branch-circuit overcurrent device to the outlet branch-circuit arc-fault circuit interrupter.
   b. The maximum length of the branch-circuit wiring from the branch-circuit overcurrent device to the first outlet shall not exceed 15.2 m (50 ft) for a 14 AWG conductor or 21.3 m (70 ft) for a 12 AWG conductor.
   c. The first outlet box in the branch circuit shall be marked to indicate that it is the first outlet of the circuit.

AFCI devices are evaluated in accordance with UL 1699, *Standard for Arc-Fault Circuit-Interrupters.* Testing methods create or simulate arcing conditions to determine a product's ability to detect and interrupt arcing faults. These devices are also tested to verify that arc detection is not unduly inhibited by the presence of loads and circuit characteristics that may mask the hazardous arcing condition. In addition, these devices are evaluated to determine resistance to unwanted tripping due to the presence of arcing that occurs in control and utilization equipment under normal operating conditions or to a loading condition that closely mimics an arcing fault such as a solid-state electronic ballast or a dimmed load.

AFCI devices may also be capable of performing other functions such as overcurrent protection, ground-fault circuit interruption, and surge suppression. UL 1699 currently recognizes four types of AFCIs: branch/feeder, cord, outlet circuit, and portable. (See Exhibit 210.22 for an example of the required marking indicating the type of AFCI protection.) AFCI devices have a maximum rating of 20 A and are intended for use in 120 V ac, 60-Hz circuits. Cord AFCIs may be rated up to 30 A. Placement of the device in the circuit must be considered when complying with 210.12. Six possible configurations using listed equipment are permitted by the *Code*. The objective of the *NEC* is to provide protection of the entire branch circuit. The configurations may use a single combination-type AFCI at the origin of the circuit, a combination of devices, or a combination of physical protection for part of the circuit and a device-type AFCI located downstream of the origin of the circuit. Where the AFCI is not located at the origin of the circuit, a higher level of physical protection must be provided for branch-circuit conductors from the origin of the branch circuit to the device-type AFCI. Some configurations have length restrictions to the first outlet, based on the size of the conductors (50 feet for 14 AWG and 70 feet for 12 AWG). Commentary Table 210.2 summarizes the permitted methods of providing AFCI protection.

***EXHIBIT 210.22*** *A circuit breaker with the required marking indicating the type of AFCI protection.*

Branch-circuit/feeder-type AFCI devices provide arcing protection against parallel faults. An example of a parallel arcing fault is a cable stapled to a wooden stud where the staple has been driven deeply into the cable jacket, damaging the conductor installation. Combination-type AFCIs provide parallel arcing

**COMMENTARY TABLE 210.2** AFCI Protection Methods

| 210.12(A) Reference | AFCI Protection Method | Additional Installation Requirements |
|---|---|---|
| 210.12(A)(1) | • Combination type AFCI circuit breaker installed at origin of branch circuit. | • No additional requirements |
| 210.12(A)(2) | • Branch/feeder type AFCI circuit breaker installed at origin of branch circuit, plus<br>• Outlet branch circuit type AFCI device installed at first outlet in branch circuit. | • Marking of first outlet box in branch circuit |
| 210.12(A)(3) | • Supplemental arc protection type circuit breaker installed at origin of branch circuit, plus<br>• Outlet branch circuit type AFCI device installed at first outlet in branch circuit. | • Continuous branch circuit wiring;<br>• "Home run" conductor length restricted;<br>• Marking of first outlet box in branch circuit |
| 210.12(A)(4) | • *Branch circuit overcurrent protective device, plus<br>• *Outlet branch circuit type AFCI device installed at first outlet in branch circuit.<br>The combination of devices must be listed and identified to provide *system combination type* arc-fault protection for the "home run" conductors. | • Continuous branch circuit wiring;<br>• "Home run" conductor length restricted (14 AWG–50 ft., 12 AWG–70 ft.);<br>• Marking of first outlet box in branch circuit |
| 210.12(A)(5) | • Outlet branch circuit type AFCI device installed at first outlet in branch circuit. | • Branch circuit conductors installed in specific types of metal raceways or metal cables and metal boxes from origin of branch circuit to the first outlet |
| 210.12(A)(6) | • Outlet branch circuit type AFCI device installed at first outlet in branch circuit. | • Branch circuit conduit, tubing or cable encased in 2 in. of concrete from origin of branch circuit to the first outlet |

protection as well as providing protection against series arcing, such as could occur in a cord set.

AFCI protection is required for all 15- and 20-A, 120-V branch circuits that supply outlets (including receptacle, lighting, and other outlets; see definition of *outlet* in Article 100) located throughout a dwelling unit. For the 2014 *Code,* the requirement was expanded to include outlets installed in kitchens and laundry areas. The requirement does not include outlets in bathrooms, unfinished basements, garages, and outdoors. Because circuits are often shared between a bedroom and other areas such as closets and hallways, providing AFCI protection on the complete circuit would comply with 210.12. AFCI protection on other circuits or locations other than those specified in 210.12(A) is not prohibited.

(4) A listed outlet branch-circuit type arc-fault circuit interrupter installed at the first outlet on the branch circuit in combination with a listed branch-circuit overcurrent protective device where all of the following conditions are met:

a. The branch-circuit wiring shall be continuous from the branch-circuit overcurrent device to the outlet branch-circuit arc-fault circuit interrupter.

b. The maximum length of the branch-circuit wiring from the branch-circuit overcurrent device to the first outlet shall not exceed 15.2 m (50 ft) for a 14 AWG conductor or 21.3 m (70 ft) for a 12 AWG conductor.

c. The first outlet box in the branch circuit shall be marked to indicate that it is the first outlet of the circuit.

d. The combination of the branch-circuit overcurrent device and outlet branch-circuit AFCI shall be identified as meeting the requirements for a system combination–type AFCI and shall be listed as such.

(5) If RMC, IMC, EMT, Type MC, or steel-armored Type AC cables meeting the requirements of 250.118, metal wireways, metal auxiliary gutters, and metal outlet and junction boxes are installed for the portion of the branch circuit between the branch-circuit overcurrent device and the first outlet, it shall be permitted to install a listed outlet branch-circuit type AFCI at the first outlet to provide protection for the remaining portion of the branch circuit.

(6) Where a listed metal or nonmetallic conduit or tubing or Type MC cable is encased in not less than 50 mm (2 in.) of concrete for the portion of the branch circuit between the branch-circuit overcurrent device and the first outlet, it shall be permitted to install a listed outlet branch-circuit type AFCI at the first outlet to provide protection for the remaining portion of the branch circuit.

•

*Exception: Where an individual branch circuit to a fire alarm system installed in accordance with 760.41(B) or 760.121(B) is installed in RMC, IMC, EMT, or steel-sheathed cable, Type AC or Type MC, meeting the requirements of 250.118, with metal outlet and junction boxes, AFCI protection shall be permitted to be omitted.*

Informational Note No. 1: For information on combination-type and branch/feeder-type arc-fault circuit interrupters, see UL 1699-2011, *Standard for Arc-Fault Circuit Interrupters*. For information on outlet branch-circuit type arc-fault circuit interrupters, see UL Subject 1699A, *Outline of Investigation for Outlet Branch Circuit Arc-Fault Circuit-Interrupters*. For information on system combination AFCIs, see UL Subject 1699C, *Outline of Investigation for System Combination Arc-Fault Circuit Interrupters*.

Informational Note No. 2: See 29.6.3(5) of *NFPA 72-2013, National Fire Alarm and Signaling Code*, for information related to secondary power-supply requirements for smoke alarms installed in dwelling units.

Informational Note No. 3: See 760.41(B) and 760.121(B) for power-supply requirements for fire alarm systems.

**(B) Branch Circuit Extensions or Modifications — Dwelling Units.** In any of the areas specified in 210.12(A), where branch-circuit wiring is modified, replaced, or extended, the branch circuit shall be protected by one of the following:

(1) A listed combination-type AFCI located at the origin of the branch circuit

(2) A listed outlet branch-circuit type AFCI located at the first receptacle outlet of the existing branch circuit

This section details how to implement AFCI protection when performing work on existing branch-circuit wiring. To address potential existing wiring system obstacles to providing some level of AFCI protection, the *Code* provides the option of installing a combination-type device at the point where the branch circuit originates (as is required for new branch-circuit installations) or of installing an outlet branch-circuit-type AFCI at the first receptacle outlet in the branch circuit. Where the location of the first receptacle outlet in the existing branch circuit cannot be ascertained, installing a new receptacle outlet, and ensuring it is the first one in the branch circuit, is a means to implement the protection required by 210.12(B)(2).

*Exception: AFCI protection shall not be required where the extension of the existing conductors is not more than 1.8 m (6 ft) and does not include any additional outlets or devices.*

**(C) Dormitory Units.** All 120-volt, single-phase, 15- and 20-ampere branch circuits supplying outlets installed in dormitory unit bedrooms, living rooms, hallways, closets, and similar rooms shall be protected by a listed arc-fault circuit interrupter meeting the requirements of 210.12(A)(1) through (6) as appropriate.

## 210.13 Ground-Fault Protection of Equipment

Each branch-circuit disconnect rated 1000 A or more and installed on solidly grounded wye electrical systems of more than 150 volts to ground, but not exceeding 600 volts phase-to-phase, shall be provided with ground-fault protection of equipment in accordance with the provisions of 230.95.

This section is new in the 2014 *Code*. Feeders and services rated at 480/277 V where the disconnect is rated 1000 A or more are required to have ground-fault protection of equipment. A branch circuit supplying an industrial machine may be supplied by a feeder with a similar rating. This new section extends the ground-fault protection requirement to branch circuits. Exceptions are provided for continuous industrial processes where shutdown could introduce additional or increased hazards, and for installations where ground-fault protection is provided upstream of the branch circuit on the load side of a transformer supplying the branch circuit. For further information on ground-fault protection of equipment, see commentary following 230.95.

Informational Note: For buildings that contain health care occupancies, see the requirements of 517.17.

*Exception No. 1: The provisions of this section shall not apply to a disconnecting means for a continuous industrial process where a nonorderly shutdown will introduce additional or increased hazards.*

*Exception No. 2: The provisions of this section shall not apply if ground-fault protection of equipment is provided on the supply side of the branch circuit and on the load side of any transformer supplying the branch circuit.*

### 210.17 Electric Vehicle Branch Circuit

An outlet(s) installed for the purpose of charging electric vehicles shall be supplied by a separate branch circuit. This circuit shall have no other outlets.

This section is new for the 2014 *Code*. Electric vehicle supply equipment is a continuous load that usually operates for several hours. Most electric vehicle supply equipment for consumers is rated at 12 A at 120 V or 32 A at 240 V. On a shared branch circuit, the overcurrent protective device could trip due to overload. For these loads, the branch-circuit rating should be 125 percent of the load. In this example, the 120-V branch circuit would be supplied by a 15-A circuit, while the 240-V circuit would be supplied by a 40-A circuit.

Informational Note: See 625.2 for the definition of *Electric Vehicle.*

### 210.18 Guest Rooms and Guest Suites

Guest rooms and guest suites that are provided with permanent provisions for cooking shall have branch circuits installed to meet the rules for dwelling units.

Guest rooms and guest suites equipped with permanent provisions for cooking must meet all of the branch-circuit requirements for dwelling units contained in Article 210. The guest suite configuration shown in Exhibit 210.23 triggers the requirement to install the branch-circuit wiring in this unit using all of the branch-circuit provisions that apply to dwelling units.

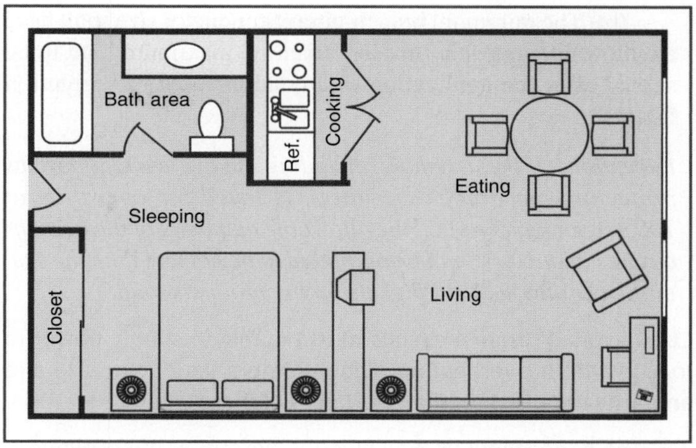


***EXHIBIT 210.23*** *Guest rooms or suites with permanent provisions for cooking in which the installation of branch circuit must follow all of the requirements in Article 210 covering dwelling units.*

## II. Branch-Circuit Ratings

### 210.19 Conductors — Minimum Ampacity and Size

**(A) Branch Circuits Not More Than 600 Volts.**

Informational Note No. 1: See 310.15 for ampacity ratings of conductors.

Informational Note No. 2: See Part II of Article 430 for minimum rating of motor branch-circuit conductors.

Informational Note No. 3: See 310.15(A)(3) for temperature limitation of conductors.

Informational Note No. 4: Conductors for branch circuits as defined in Article 100, sized to prevent a voltage drop exceeding 3 percent at the farthest outlet of power, heating, and lighting loads, or combinations of such loads, and where the maximum total voltage drop on both feeders and branch circuits to the farthest outlet does not exceed 5 percent, provide reasonable efficiency of operation. See Informational Note No. 2 of 215.2(A)(1) for voltage drop on feeder conductors.

Excessive voltage drop in supply conductors can cause trouble in and lead to inefficient operation of electrical equipment. Undervoltage conditions reduce the capability and reliability of motors, lighting sources, heaters, and solid-state equipment. It may be necessary to compensate for this with an installation beyond the minimum requirements. Sample voltage-drop calculations are found in the commentary following 215.2(A)(1), Informational Note No. 3, and following Table 9 in Chapter 9.

**(1) General.** Branch-circuit conductors shall have an ampacity not less than the maximum load to be served. Conductors shall be sized to carry not less than the larger of 210.19(A)(1)(a) or (b).

(a) Where a branch circuit supplies continuous loads or any combination of continuous and noncontinuous loads, the minimum branch-circuit conductor size shall have an allowable ampacity not less than the noncontinuous load plus 125 percent of the continuous load.

(b) The minimum branch-circuit conductor size shall have an allowable ampacity not less than the maximum load to be served after the application of any adjustment or correction factors.

*Exception: If the assembly, including the overcurrent devices protecting the branch circuit(s), is listed for operation at 100 percent of its rating, the allowable ampacity of the branch-circuit conductors shall be permitted to be not less than the sum of the continuous load plus the noncontinuous load.*

Conductors of branch circuits must be able to supply power to loads without overheating. The minimum conductor size and ampacity must be based on the larger of these criteria:

- The noncontinuous load plus 125 percent of the continuous load
- Not less than the maximum load to be served after the application of any adjustment or correction factors

The requirements for the minimum size of overcurrent protective devices are found in 210.20. An example showing these minimum-size calculations is found in the commentary following 210.20(A), Exception.

**(2) Branch Circuits with More than One Receptacle.** Conductors of branch circuits supplying more than one receptacle for cord-and-plug-connected portable loads shall have an ampacity of not less than the rating of the branch circuit.

Because the loading of branch-circuit conductors that supply receptacles for cord-and-plug-connected portable loads is unpredictable, the circuit conductors are required to have an ampacity that is not less than the rating of the branch circuit. The rating of the branch circuit is actually the rating of the overcurrent device according to 210.3.

**(3) Household Ranges and Cooking Appliances.** Branch-circuit conductors supplying household ranges, wall-mounted ovens, counter-mounted cooking units, and other household cooking appliances shall have an ampacity not less than the rating of the branch circuit and not less than the maximum load to be served. For ranges of 8¾ kW or more rating, the minimum branch-circuit rating shall be 40 amperes.

*Exception No. 1: Conductors tapped from a 50-ampere branch circuit supplying electric ranges, wall-mounted electric ovens, and counter-mounted electric cooking units shall have an ampacity of not less than 20 amperes and shall be sufficient for the load to be served. These tap conductors include any conductors that are a part of the leads supplied with the appliance that are smaller than the branch-circuit conductors. The taps shall not be longer than necessary for servicing the appliance.*

Both factory-installed pigtails and field-installed conductors are considered to be tap conductors in applying this exception. As

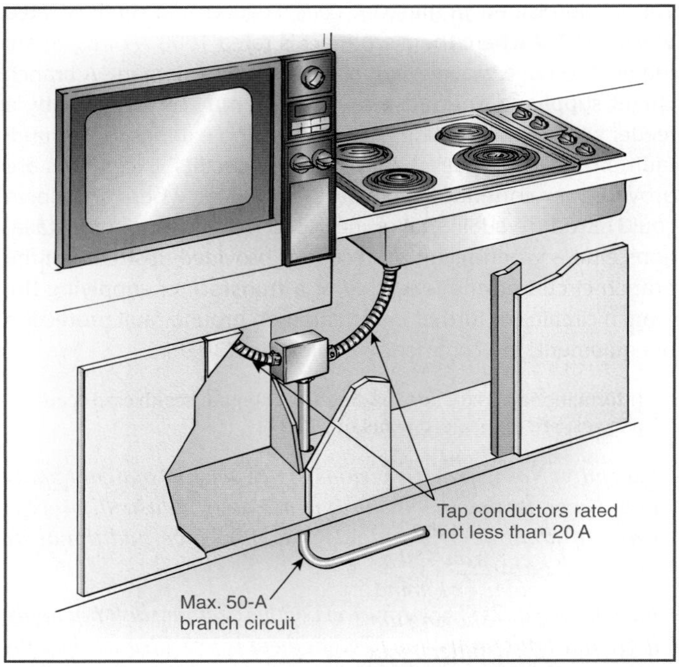


***EXHIBIT 210.24*** *Tap conductors sized smaller than the branch-circuit conductors and not longer than necessary for servicing the appliances.*

illustrated in Exhibit 210.24, this exception permits a 20-ampere tap conductor from a range, oven, or cooking unit to be connected to a 50-ampere branch circuit if the following four conditions are met:

1. The taps are not longer than necessary to service or permit access to the junction box.
2. The taps to each unit are properly spliced.
3. The junction box is adjacent to each unit.
4. The taps are of sufficient size for the load to be served.

*Exception No. 2: The neutral conductor of a 3-wire branch circuit supplying a household electric range, a wall-mounted oven, or a counter-mounted cooking unit shall be permitted to be smaller than the ungrounded conductors where the maximum demand of a range of 8¾-kW or more rating has been calculated according to Column C of Table 220.55, but such conductor shall have an ampacity of not less than 70 percent of the branch-circuit rating and shall not be smaller than 10 AWG.*

Column C of Table 220.55 indicates that the maximum demand for one range (not over 12 kW rating) is 8 kW (8000 VA; 8000 VA ÷ 240 V = 33.3 A). The allowable ampacity of an 8 AWG copper conductor from the 60°C column of Table 310.15(B)(16) is 40 amperes, and this conductor may be used for the range branch circuit. According to this exception, the neutral of this 3-wire circuit can be smaller than 8 AWG but not smaller than 10 AWG. A 10 AWG conductor has an allowable ampacity of 30 A (30 A is more than 70 percent of 40 A). The maximum demand for the neutral of an 8-kW range circuit seldom exceeds 25 A, because the only line-to-neutral connected loads are lights, clocks, timers, and the heating elements of some ranges when the control is adjusted to the low-heat setting.

**(4) Other Loads.** Branch-circuit conductors that supply loads other than those specified in 210.2 and other than cooking appliances as covered in 210.19(A)(3) shall have an ampacity sufficient for the loads served and shall not be smaller than 14 AWG.

*Exception No. 1: Tap conductors shall have an ampacity sufficient for the load served. In addition, they shall have an ampacity of not less than 15 for circuits rated less than 40 amperes and not less than 20 for circuits rated at 40 or 50 amperes and only where these tap conductors supply any of the following loads:*

*(a) Individual lampholders or luminaires with taps extending not longer than 450 mm (18 in.) beyond any portion of the lampholder or luminaire.*
*(b) A luminaire having tap conductors as provided in 410.117.*
*(c) Individual outlets, other than receptacle outlets, with taps not over 450 mm (18 in.) long.*
*(d) Infrared lamp industrial heating appliances.*
*(e) Nonheating leads of deicing and snow-melting cables and mats.*

Tap conductors are generally required to have the same ampacity as the branch-circuit overcurrent device. Exception No. 1 lists specific applications where the tap conductors are permitted to have reduced ampacities.

*Exception No. 2: Fixture wires and flexible cords shall be permitted to be smaller than 14 AWG as permitted by 240.5.*

**(B) Branch Circuits Over 600 Volts.** The ampacity of conductors shall be in accordance with 310.15 and 310.60, as applicable. Branch-circuit conductors over 600 volts shall be sized in accordance with 210.19(B)(1) or (B)(2).

**(1) General.** The ampacity of branch-circuit conductors shall not be less than 125 percent of the designed potential load of utilization equipment that will be operated simultaneously.

**(2) Supervised Installations.** For supervised installations, branch-circuit conductor sizing shall be permitted to be determined by qualified persons under engineering supervision. Supervised installations are defined as those portions of a facility where both of the following conditions are met:

(1) Conditions of design and installation are provided under engineering supervision.
(2) Qualified persons with documented training and experience in over 600-volt systems provide maintenance, monitoring, and servicing of the system.

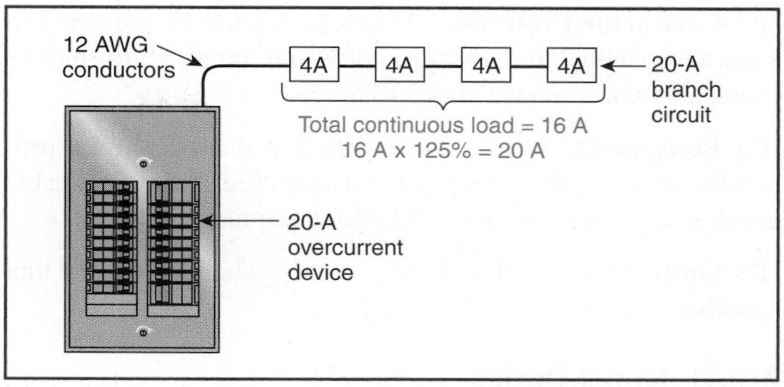


***EXHIBIT 210.25*** *A continuous load calculated at 125 percent.*

## 210.20 Overcurrent Protection

Branch-circuit conductors and equipment shall be protected by overcurrent protective devices that have a rating or setting that complies with 210.20(A) through (D).

**(A) Continuous and Noncontinuous Loads.** Where a branch circuit supplies continuous loads or any combination of continuous and noncontinuous loads, the rating of the overcurrent device shall not be less than the noncontinuous load plus 125 percent of the continuous load.

An example calculation to determine the conductor ampacity and the overcurrent device for a continuous load (store lighting) only is illustrated in Exhibit 210.25.

*Exception: Where the assembly, including the overcurrent devices protecting the branch circuit(s), is listed for operation at 100 percent of its rating, the ampere rating of the overcurrent device shall be permitted to be not less than the sum of the continuous load plus the noncontinuous load.*

An overcurrent device that supplies continuous and noncontinuous loads must have a rating that is not less than the sum of 100 percent of the noncontinuous load plus 125 percent of the continuous load. Because grounded/neutral conductors are generally not connected to the terminals of an overcurrent protective device, this requirement for sizing conductors subject to continuous loads does not apply. Grounded/neutral conductors are typically connected to a neutral bus or neutral terminal bar located in distribution equipment.

In addition, 210.19(A)(1) requires that the circuit conductors, chosen from the ampacity tables, have an ampacity based on the larger of these criteria:

- Not less than the sum of 100 percent of the noncontinuous load plus 125 percent of the continuous load
- Not less than the maximum load to be served after the application of any adjustment or correction factors

The rating of the overcurrent device cannot exceed whichever calculation yields the greater result. Visit the NEC Handbook page at www.nfpa.org/nech for calculation examples that demonstrate sizing of conductors and overcurrent protective devices for various circuits.

**(B) Conductor Protection.** Conductors shall be protected in accordance with 240.4. Flexible cords and fixture wires shall be protected in accordance with 240.5.

**(C) Equipment.** The rating or setting of the overcurrent protective device shall not exceed that specified in the applicable articles referenced in Table 240.3 for equipment.

**(D) Outlet Devices.** The rating or setting shall not exceed that specified in 210.21 for outlet devices.

## 210.21 Outlet Devices

Outlet devices shall have an ampere rating that is not less than the load to be served and shall comply with 210.21(A) and (B).

**(A) Lampholders.** Where connected to a branch circuit having a rating in excess of 20 amperes, lampholders shall be of the heavy-duty type. A heavy-duty lampholder shall have a rating of not less than 660 watts if of the admedium type, or not less than 750 watts if of any other type.

The intent is to restrict a fluorescent lighting branch-circuit rating to not more than 20 amperes, because most lampholders manufactured for use with fluorescent lights have a rating less than that required for heavy-duty lampholders (660 W for admedium type or 750 W for all other types).

Branch-circuit conductors for fluorescent electric-discharge lighting are usually connected to ballasts rather than to lampholders, and, by specifying a wattage rating for these lampholders, a limit of 20 amperes is applied to ballast circuits.

Only the admedium-base lampholder is recognized as heavy duty at the rating of 660 W. Other lampholders are required to have a rating of not less than 750 W to be recognized as heavy duty. The requirement of 210.21(A) prohibits the use of medium-base screw shell lampholders on branch circuits that are in excess of 20 amperes.

**(B) Receptacles.**

**(1) Single Receptacle on an Individual Branch Circuit.** A single receptacle installed on an individual branch circuit shall have an ampere rating not less than that of the branch circuit.

*Exception No. 1: A receptacle installed in accordance with 430.81(B).*

*Exception No. 2: A receptacle installed exclusively for the use of a cord-and-plug-connected arc welder shall be permitted to have an ampere rating not less than the minimum branch-circuit conductor ampacity determined by 630.11(A) for arc welders.*

A single receptacle installed on an individual branch circuit must have an ampere rating not less than that of the branch circuit. For example, a single receptacle on a 20-ampere individual branch circuit must be rated at 20 amperes in accordance with 210.21(B)(1); however, two or more 15-ampere single receptacles or a 15-ampere duplex receptacle are permitted on a 20-ampere branch circuit in accordance with 210.21(B)(3). This requirement does not apply to specific types of cord-and-plug-connected arc welders.

Informational Note: See the definition of *receptacle* in Article 100.

**(2) Total Cord-and-Plug-Connected Load.** Where connected to a branch circuit supplying two or more receptacles or outlets, a receptacle shall not supply a total cord-and-plug-connected load in excess of the maximum specified in Table 210.21(B)(2).

***TABLE 210.21(B)(2)*** *Maximum Cord-and-Plug-Connected Load to Receptacle*

| Circuit Rating (Amperes) | Receptacle Rating (Amperes) | Maximum Load (Amperes) |
|---|---|---|
| 15 or 20 | 15 | 12 |
| 20 | 20 | 16 |
| 30 | 30 | 24 |

**(3) Receptacle Ratings.** Where connected to a branch circuit supplying two or more receptacles or outlets, receptacle ratings shall conform to the values listed in Table 210.21(B)(3), or, where rated higher than 50 amperes, the receptacle rating shall not be less than the branch-circuit rating.

*Exception No. 1: Receptacles installed exclusively for the use of one or more cord-and plug-connected arc welders shall be permitted to have ampere ratings not less than the minimum branch-circuit conductor ampacity determined by 630.11(A) or (B) for arc welders.*

*Exception No. 2: The ampere rating of a receptacle installed for electric discharge lighting shall be permitted to be based on 410.62(C).*

***TABLE 210.21(B)(3)*** *Receptacle Ratings for Various Size Circuits*

| Circuit Rating (Amperes) | Receptacle Rating (Amperes) |
|---|---|
| 15 | Not over 15 |
| 20 | 15 or 20 |
| 30 | 30 |
| 40 | 40 or 50 |
| 50 | 50 |

**(4) Range Receptacle Rating.** The ampere rating of a range receptacle shall be permitted to be based on a single range demand load as specified in Table 220.55.

## 210.22 Permissible Loads, Individual Branch Circuits

An individual branch circuit shall be permitted to supply any load for which it is rated, but in no case shall the load exceed the branch-circuit ampere rating.

Electric vehicle supply equipment is an example of equipment that is supplied by an individual branch circuit. See 210.17.

## 210.23 Permissible Loads, Multiple-Outlet Branch Circuits

In no case shall the load exceed the branch-circuit ampere rating. A branch circuit supplying two or more outlets or receptacles shall supply only the loads specified according to its size as specified in 210.23(A) through (D) and as summarized in 210.24 and Table 210.24.

**(A) 15- and 20-Ampere Branch Circuits.** A 15- or 20-ampere branch circuit shall be permitted to supply lighting units or other utilization equipment, or a combination of both, and shall comply with 210.23(A)(1) and (A)(2).

*Exception: The small-appliance branch circuits, laundry branch circuits, and bathroom branch circuits required in a dwelling unit(s) by 210.11(C)(1), (C)(2), and (C)(3) shall supply only the receptacle outlets specified in that section.*

**(1) Cord-and-Plug-Connected Equipment Not Fastened in Place.** The rating of any one cord-and-plug-connected utilization equipment not fastened in place shall not exceed 80 percent of the branch-circuit ampere rating.

**(2) Utilization Equipment Fastened in Place.** The total rating of utilization equipment fastened in place, other than luminaires, shall not exceed 50 percent of the branch-circuit ampere rating where lighting units, cord-and-plug-connected utilization equipment not fastened in place, or both, are also supplied.

A 15- or 20-ampere branch circuit supplying lighting outlets may also supply utilization equipment fastened in place, such as a dishwasher or an air conditioner. The utilization equipment load, whether direct wired or cord and plug connected, must not exceed 50 percent of the branch-circuit ampere rating (7.5 amperes on a 15-ampere circuit and 10 amperes on a 20-ampere circuit). However, according to 210.52(B)(2), such fastened-in-place equipment is not permitted on the small-appliance branch circuits required in a kitchen, dining room, and so forth.

The requirement does not apply to a branch circuit that supplies only fastened-in-place utilization equipment. In that case, the entire rating of the branch circuit can be applied to the utilization equipment. For example, a 20-ampere branch circuit dedicated to supplying a waste disposer and a dishwasher is not restricted to either one of the appliances not exceeding 50 percent (10 amperes) of the branch-circuit rating although the combined load of the two appliances cannot exceed 20 amperes.

**(B) 30-Ampere Branch Circuits.** A 30-ampere branch circuit shall be permitted to supply fixed lighting units with heavy-duty lampholders in other than a dwelling unit(s) or utilization equipment in any occupancy. A rating of any one cord-and-plug-connected utilization equipment shall not exceed 80 percent of the branch-circuit ampere rating.

**(C) 40- and 50-Ampere Branch Circuits.** A 40- or 50-ampere branch circuit shall be permitted to supply cooking appliances that are fastened in place in any occupancy. In other than dwelling units, such circuits shall be permitted to supply fixed lighting units with heavy-duty lampholders, infrared heating units, or other utilization equipment.

**(D) Branch Circuits Larger Than 50 Amperes.** Branch circuits larger than 50 amperes shall supply only nonlighting outlet loads.

See the commentary following 210.3, Exception, regarding multioutlet branch circuits greater than 50 amperes that are permitted to supply nonlighting outlet loads at industrial premises.

## 210.24 Branch-Circuit Requirements — Summary

The requirements for circuits that have two or more outlets or receptacles, other than the receptacle circuits of 210.11(C)(1), (C)(2), and (C)(3), are summarized in Table 210.24. This table provides only a summary of minimum requirements. See 210.19, 210.20, and 210.21 for the specific requirements applying to branch circuits.

Table 210.24 summarizes the general branch-circuit requirements where two or more outlets are supplied. Small appliance, laundry, and bathroom circuits supplying receptacles are not included, and no allowance is made for conditions of use or adjustments.

The circuit rating is determined by the rating of the overcurrent device based on 210.3. The allowable tap conductor ampacity is governed by Exception No. 1 of 210.19(A)(4). The circuit conductor and tap conductor size are directly from the 60°C column of Table 310.15(B)(16). The receptacle ratings are from Table 210.21(B)(3), while the lampholder ratings are from 210.23.

## 210.25 Branch Circuits in Buildings with More Than One Occupancy

**(A) Dwelling Unit Branch Circuits.** Branch circuits in each dwelling unit shall supply only loads within that dwelling unit or loads associated only with that dwelling unit.

**(B) Common Area Branch Circuits.** Branch circuits installed for the purpose of lighting, central alarm, signal, communications, or other purposes for public or common areas of a two-family dwelling, a multifamily dwelling, or a multi-occupancy building shall not be supplied from equipment that supplies an individual dwelling unit or tenant space.

Not only does 210.25 prohibit branch circuits from feeding more than one dwelling unit, it also prohibits the sharing of systems, equipment, or common lighting if that equipment is fed from any of the dwelling units. In addition, common area circuits in occupancies other than dwelling units are subject to this requirement, and "house load" branch circuits must be supplied from equipment that does not directly supply branch circuits for an individual occupancy or tenant space. The systems, equipment, and lighting for public or common areas are required to be supplied

*TABLE 210.24* *Summary of Branch-Circuit Requirements*

| Circuit Rating | 15 A | 20 A | 30 A | 40 A | 50 A |
|---|---|---|---|---|---|
| Conductors (min. size): | | | | | |
| Circuit wires[1] | 14 | 12 | 10 | 8 | 6 |
| Taps | 14 | 14 | 14 | 12 | 12 |
| Fixture wires and cords — see 240.5 | | | | | |
| **Overcurrent Protection** | **15 A** | **20 A** | **30 A** | **40 A** | **50 A** |
| Outlet devices: | | | | | |
| Lampholders permitted | Any type | Any type | Heavy duty | Heavy duty | Heavy duty |
| Receptacle rating[2] | 15 max. A | 15 or 20 A | 30 A | 40 or 50 A | 50 A |
| **Maximum Load** | **15 A** | **20 A** | **30 A** | **40 A** | **50 A** |
| Permissible load | See 210.23(A) | See 210.23(A) | See 210.23(B) | See 210.23(C) | See 210.23(C) |

[1]These gauges are for copper conductors.

[2]For receptacle rating of cord-connected electric-discharge luminaires, see 410.62(C).

from a separate "house load" panelboard. This requirement permits access to the branch-circuit disconnecting means without the need to enter the space of any tenants. The requirement also prevents a tenant from turning off important circuits that may affect other tenants.

## III. Required Outlets

### 210.50 General

Receptacle outlets shall be installed as specified in 210.52 through 210.64.

Informational Note: See Informative Annex J for information regarding ADA accessibility design.

**(A) Cord Pendants.** A cord connector that is supplied by a permanently connected cord pendant shall be considered a receptacle outlet.

**(B) Cord Connections.** A receptacle outlet shall be installed wherever flexible cords with attachment plugs are used. Where flexible cords are permitted to be permanently connected, receptacles shall be permitted to be omitted for such cords.

Flexible cords are permitted to be permanently connected to boxes or fittings where specifically permitted by the *Code*. However, plugging a cord into a lampholder by inserting a screw-plug adapter is not permitted, because 410.90 requires lampholders of the screw shell type to be installed for use as lampholders only.

**(C) Appliance Receptacle Outlets.** Appliance receptacle outlets installed in a dwelling unit for specific appliances, such as laundry equipment, shall be installed within 1.8 m (6 ft) of the intended location of the appliance.

See 210.52(F) and 210.11(C)(2) for requirements regarding laundry receptacle outlets and branch circuits.

### 210.52 Dwelling Unit Receptacle Outlets

This section provides requirements for 125-volt, 15- and 20-ampere receptacle outlets. The receptacles required by this section shall be in addition to any receptacle that is:

(1) Part of a luminaire or appliance, or
(2) Controlled by a wall switch in accordance with 210.70(A)(1), Exception No. 1, or
(3) Located within cabinets or cupboards, or
(4) Located more than 1.7 m (5½ ft) above the floor

Permanently installed electric baseboard heaters equipped with factory-installed receptacle outlets or outlets provided as a separate assembly by the manufacturer shall be permitted as the required outlet or outlets for the wall space utilized by such permanently installed heaters. Such receptacle outlets shall not be connected to the heater circuits.

Informational Note: Listed baseboard heaters include instructions that may not permit their installation below receptacle outlets.

The requirements of 210.52 apply to dwelling unit receptacles that are rated 125 V and 15 or 20 A and that are not part of a luminaire or an appliance. These receptacles are normally used to supply lighting and general-purpose electrical equipment and are in addition to the ones that are 5½ feet above the floor and within cupboards and cabinets. An outlet containing a duplex receptacle and wired so that only one of the receptacles is controlled by a wall switch can be used to meet the receptacle outlet spacing requirement. However, if both halves are controlled by a wall switch(es), an additional unswitched receptacle has to be installed to meet the receptacle outlet spacing requirement. An outlet where both halves of the duplex receptacle are controlled by a wall switch may result in the occupant using an extension

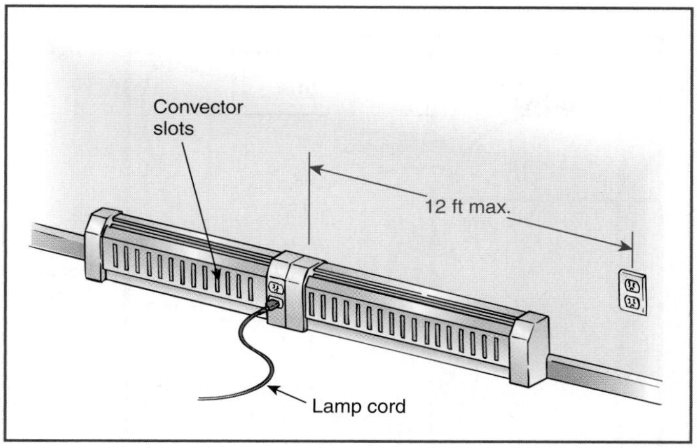


***EXHIBIT 210.26*** *Permanent electric baseboard heater equipped with a receptacle outlet that meets Code spacing requirements.*

cord, run from an outlet or device that is not controlled by a switch, to supply appliances or equipment that require continuous power, such as an electric clock.

According to listing requirements [see 110.3(B)], permanent electric baseboard heaters may not be located beneath wall receptacles. If the receptacle is part of the heater, appliance or lamp cords are less apt to be exposed to the heating elements, as might occur should the cords fall into convector slots. Many electric baseboard heaters are of the low-density type and are longer than 12 feet. To meet the spacing requirements of 210.52(A)(1), the required receptacle may be located as a part of the heater unit as shown in Exhibit 210.26.

**(A) General Provisions.** In every kitchen, family room, dining room, living room, parlor, library, den, sunroom, bedroom, recreation room, or similar room or area of dwelling units, receptacle outlets shall be installed in accordance with the general provisions specified in 210.52(A)(1) through (A)(4).

**(1) Spacing.** Receptacles shall be installed such that no point measured horizontally along the floor line of any wall space is more than 1.8 m (6 ft) from a receptacle outlet.

Receptacle outlets are to be installed so that an appliance or lamp with an attached flexible cord may be placed anywhere in the room near a wall and be within 6 feet of a receptacle, minimizing the need for occupants to use extension cords. The receptacle layout may be designed for intended utilization equipment or practical room use. For example, receptacles in a family room that are intended to serve home entertainment equipment or home office equipment may be placed in corners, may be grouped, or may be placed in a convenient location. Receptacles intended for window-type holiday lighting may be placed under windows. Even if more receptacles than the minimum required are installed in a room, no point in any wall space is permitted to be more than 6 feet from a receptacle.

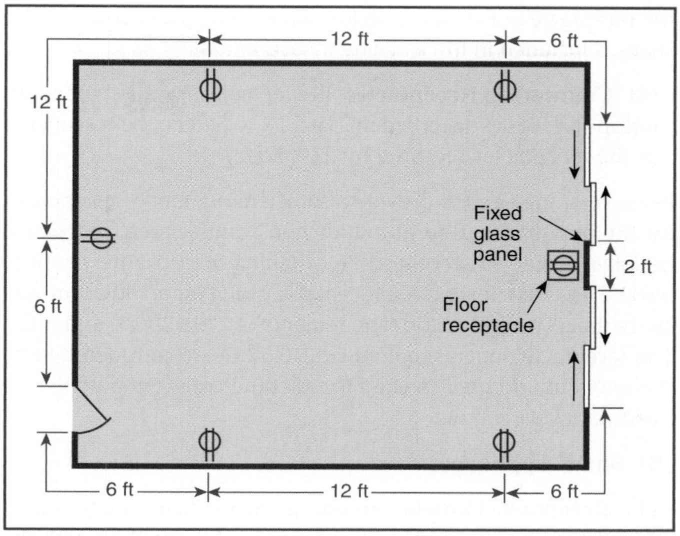


***EXHIBIT 210.27*** *A typical room plan view of the location of dwelling unit receptacles that meet Code requirements.*

**(2) Wall Space.** As used in this section, a wall space shall include the following:

(1) Any space 600 mm (2 ft) or more in width (including space measured around corners) and unbroken along the floor line by doorways and similar openings, fireplaces, and fixed cabinets
(2) The space occupied by fixed panels in exterior walls, excluding sliding panels
(3) The space afforded by fixed room dividers, such as freestanding bar-type counters or railings

**(3) Floor Receptacles.** Receptacle outlets in or on floors shall not be counted as part of the required number of receptacle outlets unless located within 450 mm (18 in.) of the wall.

Any wall space that is unbroken along the floor line by doorways, fireplaces, archways, and similar openings must be included in the measurement. The wall space may include two or more walls of a room (around corners) as illustrated in Exhibit 210.27.

Fixed room dividers, such as bar-type counters and railings, are required to be included in the 6-foot measurement. Fixed glass panels in exterior walls are counted as wall space, and a floor-type receptacle close to the wall can be used to meet the required spacing. Isolated, individual wall spaces 2 feet or more in width, which are often used for small pieces of furniture on which a lamp or an appliance may be placed, are required to have a receptacle outlet to preclude the use of an extension cord to supply equipment in such an isolated space.

The word *usable* does not appear at all in 210.52(A)(2) as a condition for determining compliance with the receptacle-spacing requirements. As an example, to correctly determine the dimension of the wall line in a room, the wall space behind the swing of a door is included in the measurement. This does not mean that the receptacle outlet has to be located in that space, only that the space is included in the wall-line measurement.

**(4) Countertop Receptacles.** Receptacles installed for countertop surfaces as specified in 210.52(C) shall not be considered as the receptacles required by 210.52(A).

Because of the need to provide a sufficient number of receptacles for the appliances used at the kitchen counter area, receptacle outlets installed to serve kitchen or dining area counters cannot also be used as the required receptacle outlet for an adjacent wall space that is subject to the requirements of 210.52(A)(1) and (A)(2). The receptacle outlets required by 210.52(C) are considered to be the minimum number needed to meet the countertop appliance needs in dwelling units.

**(B) Small Appliances.**

**(1) Receptacle Outlets Served.** In the kitchen, pantry, breakfast room, dining room, or similar area of a dwelling unit, the two or more 20-ampere small-appliance branch circuits required by 210.11(C)(1) shall serve all wall and floor receptacle outlets covered by 210.52(A), all countertop outlets covered by 210.52(C), and receptacle outlets for refrigeration equipment.

A minimum of two 20-ampere circuits is required for all receptacle outlets, including those for refrigeration equipment, in the kitchen/dining areas of a dwelling unit. The limited exceptions to 210.52(B)(2) keep loads for specific equipment to a minimum so that the majority of the circuit capacity is dedicated to supplying cord-and-plug-connected portable appliance loads. Connecting fastened-in-place appliances, such as waste disposers or dishwashers, to these circuits reduces the capacity to supply the typical higher wattage portable appliances used at a kitchen counter, such as toasters, coffee makers, skillets, and mixers. The *Code* restricts the loads supplied by these receptacle circuits because the number of cord-and-plug-connected portable appliances used by occupants is generally undetermined.

No restriction is placed on the number of outlets connected to a general-lighting or small-appliance branch circuit. The minimum number of receptacle outlets in a room is determined by 210.52(A) based on the room perimeter and on 210.52(C) for counter spaces. Installing more than the required minimum number of receptacle outlets can also help reduce the need for extension cords and cords lying across counters.

Exhibit 210.28 illustrates the application of the requirements of 210.52(B)(1), (B)(2), and (B)(3). The illustrated small-appliance branch circuits are not permitted to serve any other outlets, such as might be connected to exhaust hoods or fans, disposals, or dishwashers. The countertop receptacles are also required to be supplied by these two circuits if only the minimum of two circuits is provided for that dwelling. Only the counter area is required to be supplied by both of the small-appliance branch circuits. The wall receptacle outlets in the kitchen and dining room are permitted to be supplied by one or both of the circuits as shown in the two diagrams.

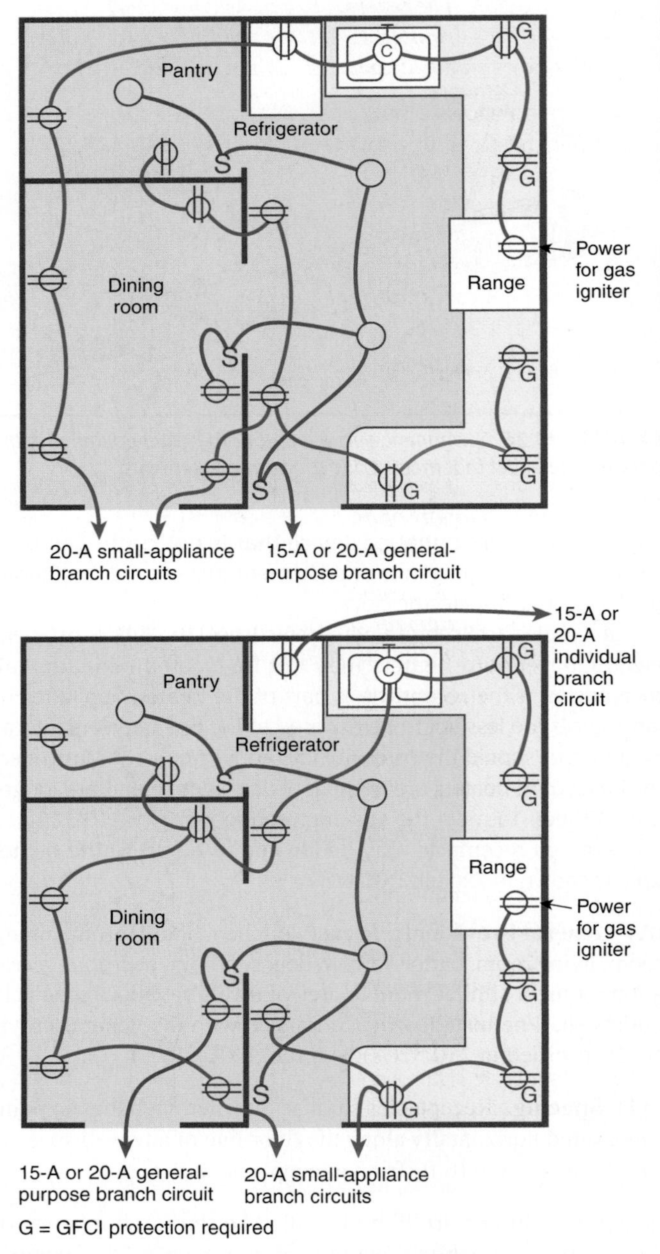


***EXHIBIT 210.28*** *Small-appliance branch circuits as required for all receptacle outlets in the kitchen (including refrigerator), pantry, and dining room.*

*Exception No. 1: In addition to the required receptacles specified by 210.52, switched receptacles supplied from a general-purpose branch circuit as defined in 210.70(A)(1), Exception No. 1, shall be permitted.*

Switched receptacles supplied from general-purpose 15-ampere branch circuits may be located in kitchens, pantries, breakfast rooms, and similar areas. See 210.70(A) and Exhibit 210.28 for details.

*Exception No. 2: The receptacle outlet for refrigeration equipment shall be permitted to be supplied from an individual branch circuit rated 15 amperes or greater.*

This exception allows a choice for refrigeration equipment receptacle outlets located in a kitchen or similar area. An individual 15-ampere or larger branch circuit may serve this equipment, or it may be included in the 20-ampere small-appliance branch circuit. Refrigeration equipment is also exempt from the GFCI requirements of 210.8 where the receptacle outlet for the refrigerator is located so that it cannot be used to serve countertop surfaces as shown in Exhibit 210.28.

**(2) No Other Outlets.** The two or more small-appliance branch circuits specified in 210.52(B)(1) shall have no other outlets.

*Exception No. 1: A receptacle installed solely for the electrical supply to and support of an electric clock in any of the rooms specified in 210.52(B)(1).*

*Exception No. 2: Receptacles installed to provide power for supplemental equipment and lighting on gas-fired ranges, ovens, or counter-mounted cooking units.*

Because of the comparatively small load associated with ignition controls and other electronics, gas-fired appliances are permitted to be supplied by a receptacle connected to one of the small-appliance branch circuits. See Exhibit 210.28 for an illustration.

**(3) Kitchen Receptacle Requirements.** Receptacles installed in a kitchen to serve countertop surfaces shall be supplied by not fewer than two small-appliance branch circuits, either or both of which shall also be permitted to supply receptacle outlets in the same kitchen and in other rooms specified in 210.52(B)(1). Additional small-appliance branch circuits shall be permitted to supply receptacle outlets in the kitchen and other rooms specified in 210.52(B)(1). No small-appliance branch circuit shall serve more than one kitchen.

In most dwellings, the countertop receptacle outlets supply more of the portable cooking appliances than the wall receptacles in the kitchen and dining areas, hence the requirement for the counter areas to be supplied by no fewer than two small-appliance branch circuits. The *Code* does not specify that both circuits be installed to serve the receptacle outlet(s) at each separate counter area in a kitchen, but rather that the total counter area of a kitchen must be supplied by no fewer than two circuits, and the arrangement of these circuits is determined by the designer or installer.

For example, a single receptacle outlet on a kitchen island is not required to be supplied by both of the small-appliance circuits serving the counter area. To provide efficient distribution of the small-appliance load, the number of receptacles connected to each small-appliance circuit should be carefully analyzed. The concept of evenly proportioning the load as specified in 210.11(B) (for loads calculated on the basis of volt-amperes per square foot) can be used as a best practice in distributing the number of receptacle outlets to be supplied by each of the small-appliance branch circuits. If additional small-appliance branch circuits are installed, they are subject to all the requirements that apply to the minimum two required circuits.

The two circuits that supply the countertop receptacle outlets may also supply receptacle outlets in the pantry, dining room, and breakfast room, as well as an electric clock receptacle and electric loads associated with gas-fired appliances, but these circuits are to supply no other outlets/loads. See 210.8(A)(6) for the GFCI requirements that apply to receptacles serving kitchen counters.

**(C) Countertops.** In kitchens, pantries, breakfast rooms, dining rooms, and similar areas of dwelling units, receptacle outlets for countertop spaces shall be installed in accordance with 210.52(C)(1) through (C)(5).

**(1) Wall Countertop Spaces.** A receptacle outlet shall be installed at each wall countertop space that is 300 mm (12 in.) or wider. Receptacle outlets shall be installed so that no point along the wall line is more than 600 mm (24 in.) measured horizontally from a receptacle outlet in that space.

*Exception: Receptacle outlets shall not be required on a wall directly behind a range, counter-mounted cooking unit, or sink in the installation described in Figure 210.52(C)(1).*

This exception and the associated figure [Figure 210.52(C)(1)] define the wall space behind a sink, range, or counter-mounted cooking unit that is not required to be provided with a receptacle outlet. Figure 210.52(C)(1) also shows where the edge of a sink, range, or counter-mounted cooking unit is considered to be on the wall behind it. Where the space behind a sink, range, or counter-mounted cooking unit is 12 inches or more or 18 inches or more (depending on the counter configuration), the space must be included in measuring the wall counter space. Receptacle outlets are not prohibited from being installed in this space.

**(2) Island Countertop Spaces.** At least one receptacle shall be installed at each island countertop space with a long dimension of 600 mm (24 in.) or greater and a short dimension of 300 mm (12 in.) or greater.

**(3) Peninsular Countertop Spaces.** At least one receptacle outlet shall be installed at each peninsular countertop space with a long dimension of 600 mm (24 in.) or greater and a short dimension of 300 mm (12 in.) or greater. A peninsular countertop is measured from the connecting edge.

**(4) Separate Spaces.** Countertop spaces separated by rangetops, refrigerators, or sinks shall be considered as separate countertop spaces in applying the requirements of 210.52(C)(1). If a range, counter-mounted cooking unit, or sink is installed in an island or peninsular countertop and the depth of the countertop behind the range, counter-mounted cooking unit, or sink is less than 300 mm (12 in.), the range, counter-mounted cooking unit, or sink shall be considered to divide the countertop space into two separate countertop spaces. Each separate

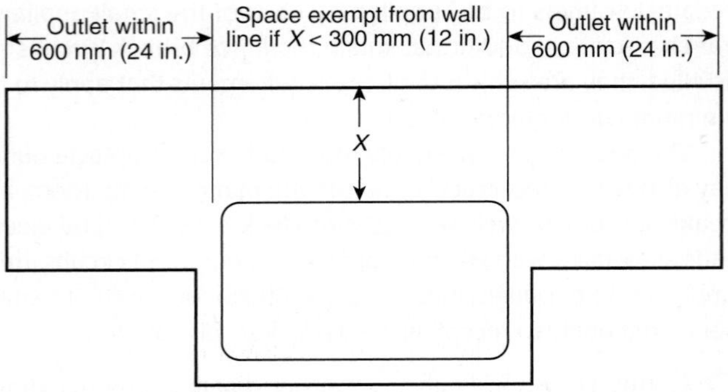


**Range, counter-mounted cooking unit extending from face of counter**

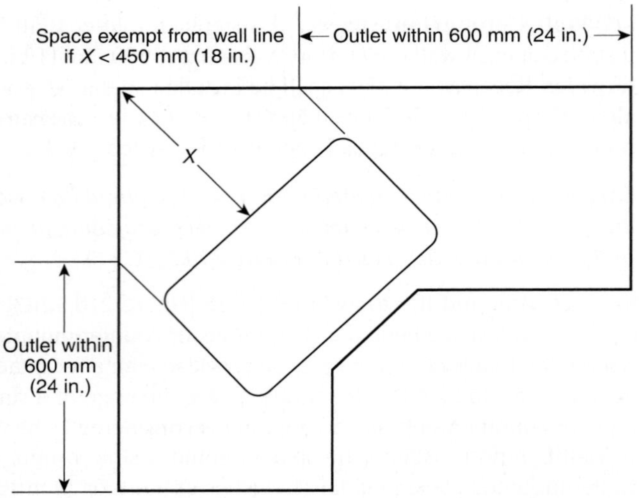


**Range, counter-mounted cooking unit mounted in corner**

*FIGURE 210.52(C)(1) Determination of Area Behind a Range, or Counter-Mounted Cooking Unit or Sink.*

countertop space shall comply with the applicable requirements in 210.52(C).

Wall, island, and peninsular countertops are subject to this requirement. The general receptacle outlet requirement for qualifying island and peninsula countertop spaces in 210.52(C)(2) and (C)(3) calls for one receptacle outlet regardless of how large the countertop space is. However, if the island or peninsula countertop is separated into two spaces according to this section and each space meets the minimum dimension criteria in 210.52(C)(2) and (C)(3), a minimum of two receptacle outlets is required for the island or peninsular countertop space.

**(5) Receptacle Outlet Location.** Receptacle outlets shall be located on or above, but not more than 500 mm (20 in.) above, the countertop. Receptacle outlet assemblies listed for the application shall be permitted to be installed in countertops. Receptacle outlets rendered not readily accessible by appliances fastened in place, appliance garages, sinks, or rangetops as covered in 210.52(C)(1), Exception, or appliances occupying dedicated space shall not be considered as these required outlets.

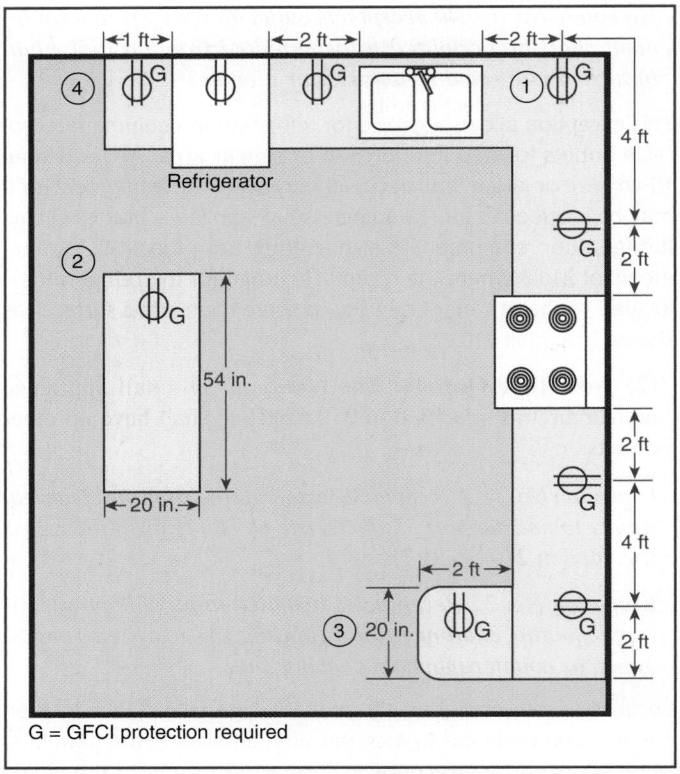


***EXHIBIT 210.29*** *Dwelling unit receptacles serving countertop spaces in a kitchen and arranged in accordance with 210.52(C).*

Informational Note: See 406.5(E) for requirements for installation of receptacles in countertops.

*Exception to (5): To comply with the conditions specified in (1) or (2), receptacle outlets shall be permitted to be mounted not more than 300 mm (12 in.) below the countertop. Receptacles mounted below a countertop in accordance with this exception shall not be located where the countertop extends more than 150 mm (6 in.) beyond its support base.*

*(1) Construction for the physically impaired*

*(2) On island and peninsular countertops where the countertop is flat across its entire surface (no backsplashes, dividers, etc.) and there are no means to mount a receptacle within 500 mm (20 in.) above the countertop, such as an overhead cabinet*

Dwelling unit receptacles that serve countertop spaces in kitchens, dining areas, and similar rooms, as illustrated in Exhibit 210.29, are required to be installed as follows:

1. In each wall space wider than 12 inches and spaced so that no point along the wall line is more than 24 inches from a receptacle

2. Not more than 20 inches above the countertop [According to 406.5(E), receptacles cannot be installed in a face-up position. Receptacles installed in a face-up position in a countertop could collect crumbs, liquids, and other debris, resulting in a potential fire or shock hazard.]
3. At each countertop island and peninsular countertop with a short dimension of at least 12 inches and a long dimension of at least 24 inches (The measurement of a peninsular-type countertop is from the edge connecting to the nonpeninsular counter.)
4. Accessible for use and not blocked by appliances occupying dedicated space or fastened in place
5. Fed from two or more of the required 20-ampere small-appliance branch circuits and GFCI protected according to 210.8(A)(6)

The maximum permitted height (20 inches) for a receptacle outlet serving a countertop is based on the standard dimension measured from the countertop to the bottom of the cabinets located above the countertop. This provision allows multioutlet assemblies installed on the bottom of the upper cabinets to be used as the required countertop receptacle outlet(s).

**(D) Bathrooms.** In dwelling units, at least one receptacle outlet shall be installed in bathrooms within 900 mm (3 ft) of the outside edge of each basin. The receptacle outlet shall be located on a wall or partition that is adjacent to the basin or basin countertop, located on the countertop, or installed on the side or face of the basin cabinet. In no case shall the receptacle be located more than 300 mm (12 in.) below the top of the basin. Receptacle outlet assemblies listed for the application shall be permitted to be installed in the countertop.

One wall receptacle must be installed adjacent to (within 36 inches of) the basin in each bathroom of a dwelling unit. An alternative location for the required receptacle outlet is on the side or face of the basin cabinet. Unlike kitchen counter receptacles, this permission to install a receptacle outlet on the basin cabinet is not contingent on the adjacent wall location being unfeasible or inaccessible to a handicapped person. Like the kitchen counter rule, the outlet must be located so that the receptacle(s) is not more than 12 inches below the basin countertop. This receptacle is also required to be GFCI protected in accordance with 210.8(A)(1).

This receptacle is required in addition to any receptacle that is part of any luminaire or medicine cabinet. If there is more than one basin, a receptacle outlet is required adjacent to each basin location. If the basins are in close proximity, one receptacle outlet can be used to satisfy this requirement. See 406.9(C), which prohibits installation of a receptacle over a bathtub or inside a shower stall. See Exhibit 210.9 for a sample electrical layout of a bathroom.

Section 210.11(C)(3) requires that receptacle outlets be supplied from a 20-ampere branch circuit with no other outlets. However, this circuit is permitted to supply the required receptacles in more than one bathroom. If the circuit supplies the required receptacle outlet in only one bathroom, it is also allowed to supply lighting and an exhaust fan in that bathroom, provided the lighting and fan load does not exceed that permitted by 210.23(A)(2).

Informational Note: See 406.5(E) for requirements for installation of receptacles in countertops.

**(E) Outdoor Outlets.** Outdoor receptacle outlets shall be installed in accordance with 210.52(E)(1) through (E)(3).

Outdoor receptacles must be installed so that the receptacle faceplate rests securely on the supporting surface to prevent moisture from entering the enclosure. On uneven surfaces such as brick, stone, or stucco, it may be necessary to close openings with caulking compound or mastic. See 406.9 for further information on receptacles installed in damp or wet locations.

Informational Note: See 210.8(A)(3).

**(1) One-Family and Two-Family Dwellings.** For a one-family dwelling and each unit of a two-family dwelling that is at grade level, at least one receptacle outlet readily accessible from grade and not more than 2.0 m (6½ ft) above grade level shall be installed at the front and back of the dwelling.

Two outdoor receptacle outlets are required for each dwelling unit. One receptacle outlet is required at the front and the back as shown in Exhibit 210.30. The two required receptacle outlets are to be available to a person standing on the ground (at grade level). Where outdoor heating, air-conditioning, or refrigeration (HACR) equipment is located at grade level, the receptacle outlets required by this section can be used to comply with the requirement of 210.63, provided that one of the outlets is located within 25 feet of the HACR equipment. Outdoor receptacle outlets on decks, porches, and similar structures can be used to meet 210.52(E) as long as the receptacle outlet is not more than 6½ feet above grade and can be accessed by a person standing at grade.

**(2) Multifamily Dwellings.** For each dwelling unit of a multifamily dwelling where the dwelling unit is located at grade level and provided with individual exterior entrance/egress, at least one receptacle outlet readily accessible from grade and not more than 2.0 m (6½ ft) above grade level shall be installed.

Multifamily dwellings (those with three or more dwelling units) are required to have at least one outdoor receptacle outlet accessible from grade level. This requirement applies to the dwelling units that are located at grade level and have a doorway that leads directly to the exterior of the structure. The unauthorized use of the outdoor receptacle outlet by other than the dwelling occupant(s) can be allayed by using a switch inside the dwelling unit to control the outlet. Due to the spatial constraints often associated with the construction of multifamily units, the required receptacle outlet is permitted to be accessible "from grade" rather than "while standing at grade," as is the case for one- and two-family dwellings. An outdoor receptacle outlet located 6½ feet or less above grade and accessible by walking up a set of deck or porch steps can be used to meet this requirement.

*EXHIBIT 210.30 Three one-family dwellings (separated from each other by fire walls) with GFCI-protected receptacles located at the front and the back of each dwelling.*

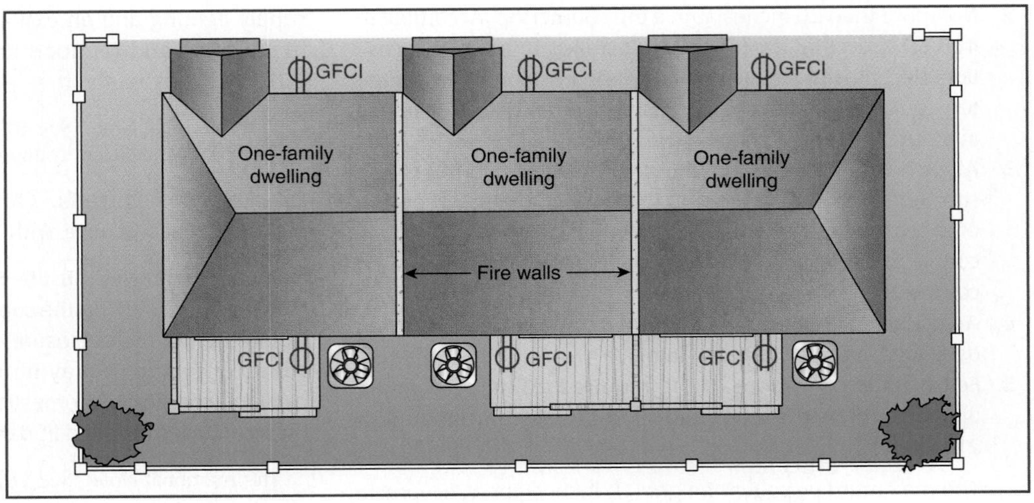


**(3) Balconies, Decks, and Porches.** Balconies, decks, and porches that are attached to the dwelling unit and are accessible from inside the dwelling unit shall have at least one receptacle outlet accessible from the balcony, deck, or porch. The receptacle outlet shall not be located more than 2.0 m (6½ ft) above the balcony, deck, or porch walking surface.

Regardless of area, a porch, balcony, or deck must have at least one receptacle outlet installed within its perimeter. This requirement only applies to porches, balconies, or decks that are accessible from inside the unit. Because it is an outdoor receptacle, GFCI protection is required. Depending on the location of the porch, balcony, or deck, the receptacle outlet can also be used to meet the receptacle requirements of 210.52(E)(1) and (E)(2). The receptacle must be not more that 6½ feet above the balcony, deck, or porch walking surface.

**(F) Laundry Areas.** In dwelling units, at least one receptacle outlet shall be installed in areas designated for the installation of laundry equipment.

*Exception No. 1: A receptacle for laundry equipment shall not be required in a dwelling unit of a multifamily building where laundry facilities are provided on the premises for use by all building occupants.*

*Exception No. 2: A receptacle for laundry equipment shall not be required in other than one-family dwellings where laundry facilities are not to be installed or permitted.*

A laundry receptacle outlet(s) is supplied by a 20-ampere branch circuit that can have no other outlets. See 210.11(C)(2) for further information.

**(G) Basements, Garages, and Accessory Buildings.** For a one-family dwelling, at least one receptacle outlet shall be installed in the areas specified in 210.52(G)(1) through (3). These receptacles shall be in addition to receptacles required for specific equipment.

**(1) Garages.** In each attached garage and in each detached garage with electric power. The branch circuit supplying this receptacle(s) shall not supply outlets outside of the garage. At least one receptacle outlet shall be installed for each car space.

**(2) Accessory Buildings.** In each accessory building with electric power.

A receptacle must be installed in the basement (in addition to the laundry receptacle), in each separate area of a basement, in each attached garage, and in each detached garage and accessory building with electric power. Receptacle outlets are only required in a detached garage if it is supplied with electricity.

GFCI protection is required by 210.8(A)(5) for receptacles in unfinished basements and by 210.8(A)(2) for receptacles installed in garages.

**(3) Basements.** In each separate unfinished portion of a basement.

**(H) Hallways.** In dwelling units, hallways of 3.0 m (10 ft) or more in length shall have at least one receptacle outlet.

As used in this subsection, the hallway length shall be considered the length along the centerline of the hallway without passing through a doorway.

This requirement is intended to minimize strain or damage to cords and receptacles for dwelling unit receptacles. The requirement does not apply to common hallways of hotels, motels, apartment buildings, condominiums, and similar occupancies.

**(I) Foyers.** Foyers that are not part of a hallway in accordance with 210.52(H) and that have an area that is greater than 5.6 m$^2$ (60 ft$^2$) shall have a receptacle(s) located in each wall space 900 mm (3 ft) or more in width. Doorways, door-side windows that extend to the floor, and similar openings shall not be considered wall space.

Foyers are not included in the living spaces covered by 210.52(A) even though they may have comparable dimensions in some cases. This requirement provides for receptacle outlets in these spaces to accommodate lamps or other utilization equipment. The primary objective of the requirements covering any of the dwelling areas included in 210.52 is to minimize the need to use extension cords to supply utilization equipment.

## 210.60 Guest Rooms, Guest Suites, Dormitories, and Similar Occupancies

**(A) General.** Guest rooms or guest suites in hotels, motels, sleeping rooms in dormitories, and similar occupancies shall have receptacle outlets installed in accordance with 210.52(A) and (D). Guest rooms or guest suites provided with permanent provisions for cooking shall have receptacle outlets installed in accordance with all of the applicable rules in 210.52.

**(B) Receptacle Placement.** In applying the provisions of 210.52(A), the total number of receptacle outlets shall not be less than the minimum number that would comply with the provisions of that section. These receptacle outlets shall be permitted to be located conveniently for permanent furniture layout. At least two receptacle outlets shall be readily accessible. Where receptacles are installed behind the bed, the receptacle shall be located to prevent the bed from contacting any attachment plug that may be installed or the receptacle shall be provided with a suitable guard.

The receptacles in guest rooms and guest suites of hotels and motels and in dormitories are permitted to be placed in accessible locations that are compatible with permanent furniture. However, the minimum number of receptacles required by 210.52 is not permitted to be reduced and should be determined by assuming furniture is not in the room. The practical locations of that minimum number of receptacles are then determined based on the permanent furniture layout.

Hotel and motel rooms and suites are commonly used as remote offices for businesspeople who use laptop computers and other plug-in devices. In dormitories, the use of electrical/electronic equipment can be significant, necessitating accessible receptacle outlets. The *Code* requires that two receptacle outlets be available without requiring the movement of furniture to access those receptacles. To reduce the risk of bedding material fires, receptacles located behind beds must include guards if attachment plugs could contact the bed.

Extended-stay hotels and motels are often equipped with permanent cooking equipment and countertop areas. All applicable receptacle spacing and supply requirements in 210.52 apply to guest rooms or suites that contain such provisions. A portable microwave oven is not considered to be a permanently installed cooking appliance. See 210.18 and its associated commentary for more information on hotel and motel guest rooms and guest suites that are equipped with permanent provisions for cooking.

Exhibit 210.31 shows a hotel guest room in which the receptacles are located based on the permanent furniture layout. Because of this rule, some spaces that are 2 feet or more in width have no receptacle outlets. However, the receptacle locations are compatible with the permanent furniture layout. In Exhibit 210.31, the receptacle outlet adjacent to the permanent dresser is needed because 210.60(B) applies only to the location of receptacle outlets, not to the minimum number of receptacle outlets.

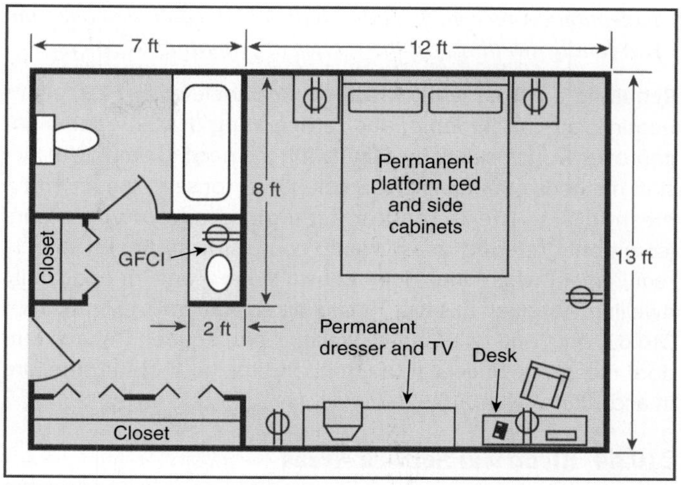


***EXHIBIT 210.31*** *Floor plan of a hotel guest room with receptacles located with respect to permanent furniture.*

## 210.62 Show Windows

At least one 125-volt, single-phase, 15- or 20-ampere-rated receptacle outlet shall be installed within 450 mm (18 in.) of the top of a show window for each 3.7 linear m (12 linear ft) or major fraction thereof of show window area measured horizontally at its maximum width.

Show windows usually extend from floor to ceiling for maximum display. To discourage floor receptacles and the use of extension cords, receptacles must be installed directly above a show window, and one receptacle is required for every 12 linear feet or "major fraction thereof" (6 feet or more). To further reduce the use of extension cords, the required receptacle outlet(s) must be installed within 18 inches of the top of the show window. See 220.14(G) and 220.43(A) for information regarding load calculations for show windows. This requirement requires the use of 125-V, 15- or 20-A receptacles. Such a receptacle would still be required if a circuit were provided for a 24 V lighting system.

## 210.63 Heating, Air-Conditioning, and Refrigeration Equipment Outlet

A 125-volt, single-phase, 15- or 20-ampere-rated receptacle outlet shall be installed at an accessible location for the servicing of heating, air-conditioning, and refrigeration equipment. The receptacle shall be located on the same level and within 7.5 m (25 ft) of the heating, air-conditioning, and refrigeration equipment. The receptacle outlet shall not be connected to the load side of the equipment disconnecting means.

Informational Note: See 210.8 for ground-fault circuit-interrupter requirements.

*Exception: A receptacle outlet shall not be required at one- and two-family dwellings for the service of evaporative coolers.*

Requiring a permanently installed receptacle within 25 feet of heating, air-conditioning, and refrigerating (HACR) equipment improves worker safety by eliminating the need to employ makeshift methods of obtaining 125-volt power for servicing and troubleshooting. The exception exempts evaporative coolers (commonly referred to as swamp coolers) from the receptacle requirement where the cooler is installed at a one- or two-family dwelling. Although this type of cooling equipment is exempt from 210.63, one- and two-family dwellings are required to have outdoor receptacle outlets at the front and the back of the structure in accordance with 210.52(E).

## 210.64 Electrical Service Areas

At least one 125-volt, single-phase, 15- or 20-ampere-rated receptacle outlet shall be installed within 15 m (50 ft) of the electrical service equipment.

This section is new in the 2014 *Code*. The required receptacle is intended to facilitate the use of portable test and diagnostic equipment that requires a 120 V power source. The receptacle is not required in one- or two-family dwellings.

*Exception: The receptacle outlet shall not be required to be installed in one-and two-family dwellings.*

## 210.70 Lighting Outlets Required

Lighting outlets shall be installed where specified in 210.70(A), (B), and (C).

**(A) Dwelling Units.** In dwelling units, lighting outlets shall be installed in accordance with 210.70(A)(1), (A)(2), and (A)(3).

**(1) Habitable Rooms.** At least one wall switch–controlled lighting outlet shall be installed in every habitable room and bathroom.

*Exception No. 1: In other than kitchens and bathrooms, one or more receptacles controlled by a wall switch shall be permitted in lieu of lighting outlets.*

*Exception No. 2: Lighting outlets shall be permitted to be controlled by occupancy sensors that are (1) in addition to wall switches or (2) located at a customary wall switch location and equipped with a manual override that will allow the sensor to function as a wall switch.*

A receptacle outlet controlled by a wall switch is not permitted to serve as the required lighting outlet in kitchens and bathrooms. Exhibit 210.28 shows a switched receptacle supplied by a 15-ampere general-purpose branch circuit in a dining room. A switched receptacle is not considered one of the receptacle outlets required by 210.52.

Occupancy sensors are permitted to be used for switching lighting outlets in habitable rooms, kitchens, and bathrooms, provided they are equipped with a manual override or are used in addition to regular switches.

**(2) Additional Locations.** Additional lighting outlets shall be installed in accordance with (A)(2)(a), (A)(2)(b), and (A)(2)(c).

(a) At least one wall switch–controlled lighting outlet shall be installed in hallways, stairways, attached garages, and detached garages with electric power.

(b) For dwelling units, attached garages, and detached garages with electric power, at least one wall switch–controlled lighting outlet shall be installed to provide illumination on the exterior side of outdoor entrances or exits with grade level access. A vehicle door in a garage shall not be considered as an outdoor entrance or exit.

(c) Where one or more lighting outlet(s) are installed for interior stairways, there shall be a wall switch at each floor level, and landing level that includes an entryway, to control the lighting outlet(s) where the stairway between floor levels has six risers or more.

*Exception to (A)(2)(a), (A)(2)(b), and (A)(2)(c): In hallways, in stairways, and at outdoor entrances, remote, central, or automatic control of lighting shall be permitted.*

Although 210.70(A)(2)(b) calls for a switched lighting outlet at outdoor entrances and exits, it does not prohibit a suitably located single lighting outlet from serving more than one door.

**(3) Storage or Equipment Spaces.** For attics, underfloor spaces, utility rooms, and basements, at least one lighting outlet containing a switch or controlled by a wall switch shall be installed where these spaces are used for storage or contain equipment requiring servicing. At least one point of control shall be at the usual point of entry to these spaces. The lighting outlet shall be provided at or near the equipment requiring servicing.

**(B) Guest Rooms or Guest Suites.** In hotels, motels, or similar occupancies, guest rooms or guest suites shall have at least one wall switch–controlled lighting outlet installed in every habitable room and bathroom.

A wall switch–controlled lighting outlet is required in every habitable room (the hotel room or rooms in a suite) and in bathrooms. If provided, a kitchen is required to have at least one lighting outlet controlled by a wall switch. Rooms other than bathrooms and kitchens may have switched receptacles to meet this lighting requirement. Exception No. 2 permits the use of occupancy sensors to control the lighting outlet, provided it is in the typical switch location and can be manually controlled.

*Exception No. 1: In other than bathrooms and kitchens where provided, one or more receptacles controlled by a wall switch shall be permitted in lieu of lighting outlets.*

*Exception No. 2: Lighting outlets shall be permitted to be controlled by occupancy sensors that are (1) in addition to wall switches or (2) located at a customary wall switch location and equipped with a manual override that allows the sensor to function as a wall switch.*

**(C) Other Than Dwelling Units.** For attics and underfloor spaces containing equipment requiring servicing, such as heating, air-conditioning, and refrigeration equipment, at least one lighting outlet containing a switch or controlled by a wall switch shall be installed in such spaces. At least one point of control shall be at the usual point of entry to these spaces. The lighting outlet shall be provided at or near the equipment requiring servicing.

# ARTICLE 215 Feeders

## 215.1 Scope

This article covers the installation requirements, overcurrent protection requirements, minimum size, and ampacity of conductors for feeders supplying branch-circuit loads.

*Exceptions: Feeders for electrolytic cells as covered in 668.8(C)(1) and (C)(4).*

## 215.2 Minimum Rating and Size

### (A) Feeders Not More Than 600 Volts.

**(1) General.** Feeder conductors shall have an ampacity not less than required to supply the load as calculated in Parts III, IV, and V of Article 220. Conductors shall be sized to carry not less than the larger of 215.2(A)(1)(a) or (b).

This section was revised for the 2014 *Code.* The ampacity of the feeder conductors is required to be based on the larger of the continuous load (at 125 percent) plus the continuous load (at 100 percent) or the maximum load to be served after any adjustments or correction factors.

(a) Where a feeder supplies continuous loads or any combination of continuous and noncontinuous loads, the minimum feeder conductor size shall have an allowable ampacity not less than the noncontinuous load plus 125 percent of the continuous load.

(b) The minimum feeder conductor size shall have an allowable ampacity not less than the maximum load to be served after the application of any adjustment or correction factors.

Informational Note No. 1: See Examples D1 through D11 in Informative Annex D.

Informational Note No. 2: Conductors for feeders, as defined in Article 100, sized to prevent a voltage drop exceeding 3 percent at the farthest outlet of power, heating, and lighting loads, or combinations of such loads, and where the maximum total voltage drop on both feeders and branch circuits to the farthest outlet does not exceed 5 percent, will provide reasonable efficiency of operation.

Informational Note No. 3: See 210.19(A), Informational Note No. 4, for voltage drop for branch circuits.

Reasonable operating efficiency is achieved if the voltage drop of a feeder or a branch circuit is limited to 3 percent. However, the total voltage drop of a branch circuit plus a feeder can reach 5 percent and still achieve reasonable operating efficiency. See Article 100 for the definitions of *feeder* and *branch circuit.*

The 5 percent voltage-drop value is explanatory material and, as such, appears as an informational note. The informational notes covering voltage drop are not mandatory (see 90.5). Where circuit conductors are increased due to voltage drop, 250.122(B) requires an increase in circular mil area for the associated equipment grounding conductors.

The resistance or impedance of conductors may cause a substantial difference between voltage at service equipment and voltage at the point-of-utilization equipment. Excessive voltage drop impairs the starting and the operation of electrical equipment. Undervoltage can result in inefficient operation of heating, lighting, and motor loads. An applied voltage of 10 percent below rating can result in a decrease in efficiency of substantially more than 10 percent – for example, fluorescent light output would be reduced by 15 percent, and incandescent light output would be reduced by 30 percent. Induction motors would run hotter and produce less torque. With an applied voltage of 10 percent below rating, the running current would increase 11 percent, and the operating temperature would increase 12 percent. At the same time, torque would be reduced 19 percent.

In addition to resistance or impedance, the type of raceway or cable enclosure, the type of circuit (ac, dc, single-phase, 3-phase), and the power factor should be considered to determine voltage drop.

This basic formula can be used to determine the voltage drop in a 2-wire dc circuit, a 2-wire ac circuit, or a 3-wire ac single-phase circuit, all with a balanced load at 100 percent power factor and where reactance can be neglected:

$$VD = \frac{2 \times L \times R \times I}{1000}$$

where:

$VD$ = voltage drop (based on conductor temperature of 75°C)
$L$ = one-way length of circuit (ft)
$R$ = conductor resistance in ohms (Ω) per 1000 ft (from Chapter 9, Table 8)
$I$ = load current (amperes)

For 3-phase circuits (at 100 percent power factor), the voltage drop between any two phase conductors is 0.866 times the voltage drop calculated by the preceding formula. See the commentary following Chapter 9, Table 9, for an example of voltage-drop calculation using ac reactance and resistance. Voltage-drop tables and calculations are also available from various manufacturers.

*Exception No. 1: If the assembly, including the overcurrent devices protecting the feeder(s), is listed for operation at 100 percent of its rating, the allowable ampacity of the feeder conductors shall be permitted to be not less than the sum of the continuous load plus the noncontinuous load.*

*Exception No. 2: Grounded conductors that are not connected to an overcurrent device shall be permitted to be sized at 100 percent of the continuous and noncontinuous load.*

Feeder grounded/neutral conductors that do not connect to the terminals of an overcurrent protective device are not required to be sized based on 125 percent of the continuous load. For example, if the maximum unbalanced load on a feeder neutral is calculated per 220.61 to be 200 amperes and the load is considered to be continuous, the use of a 3/0 AWG, Type THW conductor is permitted as long as the conductor terminates at a neutral bus or terminal bar within the electrical distribution equipment.

**(2) Grounded Conductor.** The size of the feeder circuit grounded conductor shall not be smaller than that required by 250.122, except that 250.122(F) shall not apply where grounded conductors are run in parallel.

Additional minimum sizes shall be as specified in 215.2(A)(2) and (A)(3) under the conditions stipulated.

Using 250.122 to establish the minimum size grounded conductor in a feeder circuit provides a relationship between the grounded conductor and the feeder circuit overcurrent protective device that is the same as is used for sizing equipment grounding conductors. It provides an adequate fault current path in the event of a fault between a line or phase conductor and the grounded conductor. For feeder circuits installed in parallel in separate raceways or cables, the requirements of 220.61 and 310.10(H) must be used to determine the minimum grounded conductor size. Sizing of the grounded feeder conductor is also covered by 215.2(B) for feeder circuits over 600 volts.

**(3) Ampacity Relative to Service Conductors.** The feeder conductor ampacity shall not be less than that of the service conductors where the feeder conductors carry the total load supplied by service conductors with an ampacity of 55 amperes or less.

According to Table 310.15(B)(16), a 3/0 AWG, Type THW copper wire has an ampacity of 200 amperes. However, for a 3-wire, single-phase dwelling unit service, as shown in Exhibit 215.1, 310.15(B)(7) permits a service conductor (or a main power feeder conductor) with an ampacity of 83 percent of the service rating. This permits a minimum of a 2/0 AWG, Type THW copper conductors or 4/0 AWG, Type THW aluminum conductors for services or a main power feeder rated at 200 amperes.

•

**(B) Feeders over 600 Volts.** The ampacity of conductors shall be in accordance with 310.15 and 310.60 as applicable. Where installed, the size of the feeder-circuit grounded conductor shall not be smaller than that required by 250.122, except that 250.122(F) shall not apply where grounded conductors are run in parallel. Feeder conductors over 600 volts shall be sized in accordance with 215.2(B)(1), (B)(2), or (B)(3).

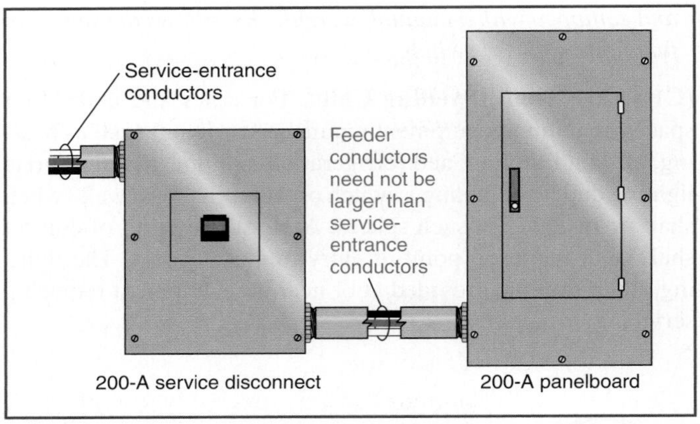


***EXHIBIT 215.1*** *A 3-wire, single-phase dwelling service with an ampacity of 200 amperes for 2/0 AWG copper or 4/0 AWG aluminum conductors used as service-entrance conductors and feeder conductors.*

**(1) Feeders Supplying Transformers.** The ampacity of feeder conductors shall not be less than the sum of the nameplate ratings of the transformers supplied when only transformers are supplied.

**(2) Feeders Supplying Transformers and Utilization Equipment.** The ampacity of feeders supplying a combination of transformers and utilization equipment shall not be less than the sum of the nameplate ratings of the transformers and 125 percent of the designed potential load of the utilization equipment that will be operated simultaneously.

**(3) Supervised Installations.** For supervised installations, feeder conductor sizing shall be permitted to be determined by qualified persons under engineering supervision. Supervised installations are defined as those portions of a facility where all of the following conditions are met:

(1) Conditions of design and installation are provided under engineering supervision.
(2) Qualified persons with documented training and experience in over 600-volt systems provide maintenance, monitoring, and servicing of the system.

## 215.3 Overcurrent Protection

Feeders shall be protected against overcurrent in accordance with the provisions of Part I of Article 240. Where a feeder supplies continuous loads or any combination of continuous and noncontinuous loads, the rating of the overcurrent device shall not be less than the noncontinuous load plus 125 percent of the continuous load.

*Exception No. 1: Where the assembly, including the overcurrent devices protecting the feeder(s), is listed for operation at 100 percent of its rating, the ampere rating of the overcurrent device shall be permitted to be not less than the sum of the continuous load plus the noncontinuous load.*

*Exception No. 2: Overcurrent protection for feeders between 600 to 1000 volts shall comply with Parts I through VII of Article 240. Feeders over 1000 volts, nominal, shall comply with Part IX of Article 240.*

## 215.4 Feeders with Common Neutral Conductor

**(A) Feeders with Common Neutral.** Up to three sets of 3-wire feeders or two sets of 4-wire or 5-wire feeders shall be permitted to utilize a common neutral.

**(B) In Metal Raceway or Enclosure.** Where installed in a metal raceway or other metal enclosure, all conductors of all feeders using a common neutral conductor shall be enclosed within the same raceway or other enclosure as required in 300.20.

If ac feeder conductors, including the neutral conductor, are installed in metal raceways, the conductors are required to be grouped together to avoid induction heating of the surrounding metal. If it is necessary to run parallel conductors through multiple metal raceways, conductors from each phase plus the neutral must be run in each raceway. See 250.102(E), 250.134(B), 300.3, 300.5(I), and 300.20 for requirements associated with conductor grouping of feeder circuits.

A 3-phase, 4-wire (208Y/120-V, 480Y/277-V) system is often used to supply both lighting and motor loads. The 3-phase motor loads are typically not connected to the neutral and, thus, will not cause current in the neutral conductor. The maximum current on the neutral, therefore, is due to lighting loads or circuits where the neutral is used. On this type of system (3-phase, 4-wire), a demand factor of 70 percent is permitted by 220.61 for that portion of the neutral load in excess of 200 A.

For example, if the maximum possible unbalanced load is 500 A, the neutral would have to be large enough to carry 410 A (200 A plus 70 percent of 300 A, or 410 A). No reduction of the neutral capacity for that portion of the load consisting of electric-discharge lighting is permitted.

Section 310.15(B)(5)(c) points out that a neutral conductor must be counted as a current-carrying conductor if the load it serves consists of harmonic currents. See 220.61 for other systems in which the 70-percent demand factor may be applied. The maximum unbalanced load for feeders supplying clothes dryers, household ranges, wall-mounted ovens, and counter-mounted cooking units is required to be considered 70 percent of the load on the ungrounded conductors. See Examples D1(a) through D5(b) of Informative Annex D.

## 215.5 Diagrams of Feeders

If required by the authority having jurisdiction, a diagram showing feeder details shall be provided prior to the installation of the feeders. Such a diagram shall show the area in square feet of the building or other structure supplied by each feeder, the total calculated load before applying demand factors, the demand factors used, the calculated load after applying demand factors, and the size and type of conductors to be used.

## 215.6 Feeder Equipment Grounding Conductor

Where a feeder supplies branch circuits in which equipment grounding conductors are required, the feeder shall include or provide an equipment grounding conductor in accordance with the provisions of 250.134, to which the equipment grounding conductors of the branch circuits shall be connected. Where the feeder supplies a separate building or structure, the requirements of 250.32(B) shall apply.

## 215.7 Ungrounded Conductors Tapped from Grounded Systems

Two-wire dc circuits and ac circuits of two or more ungrounded conductors shall be permitted to be tapped from the ungrounded conductors of circuits having a grounded neutral conductor. Switching devices in each tapped circuit shall have a pole in each ungrounded conductor.

A common trip or simultaneous opening of circuit breakers or fuses is not required, but rather a switching device is required to manually disconnect the ungrounded feeder conductors. See 210.10 for similar requirements related to the ungrounded conductors of the branch circuit.

## 215.9 Ground-Fault Circuit-Interrupter Protection for Personnel

Feeders supplying 15- and 20-ampere receptacle branch circuits shall be permitted to be protected by a ground-fault circuit interrupter in lieu of the provisions for such interrupters as specified in 210.8 and 590.6(A).

GFCI protection of the feeder circuit protects all branch-circuits supplied by that feeder. This type of GFCI installation is permitted in lieu of the requirements of 210.8(A) or (B). GFCI protection in the feeder can also be used to protect construction-site receptacles, as covered in 590.6(A), provided the feeder supplies no lighting branch circuits.

It may be more economical or convenient to install GFCIs for feeders. However, consideration should be given to the possibility that a GFCI may be monitoring several branch circuits and will de-energize all branch circuits in response to a line-to-ground fault from one branch circuit.

## 215.10 Ground-Fault Protection of Equipment

Each feeder disconnect rated 1000 amperes or more and installed on solidly grounded wye electrical systems of more than 150 volts to ground, but not exceeding 600 volts phase-to-phase, shall be provided with ground-fault protection of equipment in accordance with the provisions of 230.95.

Informational Note: For buildings that contain health care occupancies, see the requirements of 517.17.

*Exception No. 1: The provisions of this section shall not apply to a disconnecting means for a continuous industrial process where a nonorderly shutdown will introduce additional or increased hazards.*

A similar requirement for ground-fault protection of services is found in 230.95. The reason for the requirement was the unusually high number of burndowns reported on feeders and services operating in this voltage range. Solidly grounded systems operating at 480Y/277 V were the primary focus of this requirement when it was first introduced in the *Code*, but other solidly grounded, wye-connected systems operating over 150 volts to ground and not more than 600 volts phase-to-phase are covered by this requirement. Prior to being put into service, each ground-fault protection system must be performance tested and documented according to the requirements of 230.95(C).

Ground-fault protection of feeder equipment is not required if protection is provided on an upstream feeder or at the service. However, additional levels of ground-fault protection for feeders may be desired so that a single ground fault does not de-energize the whole electrical system. See 230.95 for further commentary on ground-fault protection of services. Also, see 517.17, which requires an additional level of ground-fault protection for health care facilities.

For emergency feeders covered within the scope of Article 700, the ground-fault protection requirements are different. See 700.26 for further details.

*Exception No. 2: The provisions of this section shall not apply if ground-fault protection of equipment is provided on the supply side of the feeder and on the load side of any transformer supplying the feeder.*

## 215.11 Circuits Derived from Autotransformers

Feeders shall not be derived from autotransformers unless the system supplied has a grounded conductor that is electrically connected to a grounded conductor of the system supplying the autotransformer.

*Exception No. 1: An autotransformer shall be permitted without the connection to a grounded conductor where transforming from a nominal 208 volts to a nominal 240-volt supply or similarly from 240 volts to 208 volts.*

*Exception No. 2: In industrial occupancies, where conditions of maintenance and supervision ensure that only qualified persons service the installation, autotransformers shall be permitted to supply nominal 600-volt loads from nominal 480-volt systems, and 480-volt loads from nominal 600-volt systems, without the connection to a similar grounded conductor.*

Ground-fault protection installed in equipment supplying the primary of a transformer will not function to protect equipment supplied by the secondary of the transformer. If the equipment supplied by the secondary of the transformer meets the parameters under which ground-fault protection of equipment (GFPE) is required by 215.10, protection must be installed to protect the equipment supplied by the secondary of the transformer.

## 215.12 Identification for Feeders

**(A) Grounded Conductor.** The grounded conductor of a feeder shall be identified in accordance with 200.6.

**(B) Equipment Grounding Conductor.** The equipment grounding conductor shall be identified in accordance with 250.119.

**(C) Identification of Ungrounded Conductors.** Ungrounded conductors shall be identified in accordance with 215.12(C)(1) or (C)(2), as applicable.

Parallel with the requirement for ungrounded branch circuit conductors in 210.5(C), 215.12(C) requires identification of ungrounded feeder conductors by system and phase where there is more than one nominal voltage supply system to a building, structure, or other premises. In addition, the 2014 *Code* was modified to provide separate identification requirements for ac and dc circuits.

For ac circuits, the identification scheme is not specified, but whatever is used is required to be consistent throughout the premises. A permanent legend or directory indicating the feeder identification system for the premises is required to be posted at each point in the distribution system from which feeder circuits are supplied, or the identification scheme is to be described in a facility log or other documentation and made readily available. Exhibit 215.2 is an example of the use of different colors to identify each ungrounded line or phase of a nominal voltage system.

For dc circuits, there are similar requirements for posting the method of identification at panelboards and similar distribution

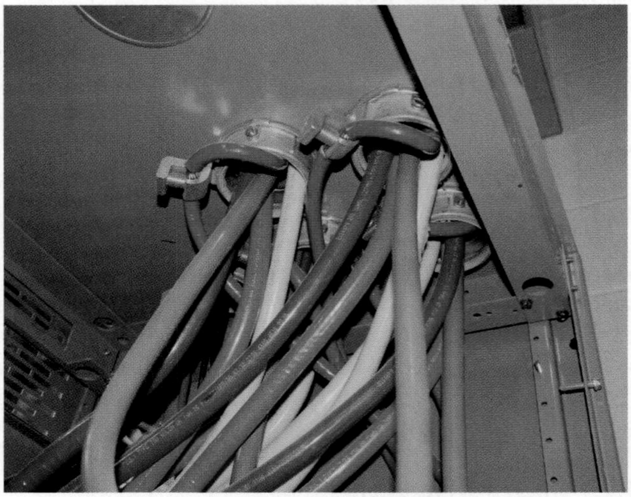

***EXHIBIT 215.2*** *Different colors being used to identify each ungrounded line or phase conductor of a nominal voltage system.*

points. For conductors 6 AWG and smaller, marking requirements for the ungrounded conductor are based on which conductor of the circuit is grounded. White, green, and gray are prohibited colors for ungrounded conductors to prevent confusion with grounded and equipment grounding conductors of ac circuits. In addition, where the positive terminal is not grounded, black is a prohibited color for conductors of positive polarity. Where the negative terminal is not grounded, red is a prohibited color for conductors of negative polarity.

**(1) Feeders Supplied from More Than One Nominal Voltage System.** Where the premises wiring system has feeders supplied from more than one nominal voltage system, each ungrounded conductor of a feeder shall be identified by phase or line and system at all termination, connection, and splice points in compliance with 215.12(C)(1)(a) and (b).

(a) *Means of Identification.* The means of identification shall be permitted to be by separate color coding, marking tape, tagging, or other approved means.

(b) *Posting of Identification Means.* The method utilized for conductors originating within each feeder panelboard or similar feeder distribution equipment shall be documented in a manner that is readily available or shall be permanently posted at each feeder panelboard or similar feeder distribution equipment.

**(2) Feeders Supplied from Direct-Current Systems.** Where a feeder is supplied from a dc system operating at more than 50 volts, each ungrounded conductor of 4 AWG or larger shall be identified by polarity at all termination, connection, and splice points by marking tape, tagging, or other approved means; each ungrounded conductor of 6 AWG or smaller shall be identified by polarity at all termination, connection, and splice points in compliance with 215.12(C)(2)(a) and (b). The identification methods utilized for conductors originating within each feeder panelboard or similar feeder distribution equipment shall be documented in a manner that is readily available or shall be permanently posted at each feeder panelboard or similar feeder distribution equipment.

(a) *Positive Polarity, Sizes 6 AWG or Smaller.* Where the positive polarity of a dc system does not serve as the connection for the grounded conductor, each positive ungrounded conductor shall be identified by one of the following means:

(1) A continuous red outer finish
(2) A continuous red stripe durably marked along the conductor's entire length on insulation of a color other than green, white, gray, or black
(3) Imprinted plus signs (+) or the word POSITIVE or POS durably marked on insulation of a color other than green, white, gray, or black, and repeated at intervals not exceeding 610 mm (24 in.) in accordance with 310.120(B)

(b) *Negative Polarity, Sizes 6 AWG or Smaller.* Where the negative polarity of a dc system does not serve as the connection for the grounded conductor, each negative ungrounded conductor shall be identified by one of the following means:

(1) A continuous black outer finish
(2) A continuous black stripe durably marked along the conductor's entire length on insulation of a color other than green, white, gray, or red
(3) Imprinted minus signs (−) or the word NEGATIVE or NEG durably marked on insulation of a color other than green, white, gray, or red, and repeated at intervals not exceeding 610 mm (24 in.) in accordance with 310.120(B)

# ARTICLE 220 Branch-Circuit, Feeder, and Service Calculations

## I. General

### 220.1 Scope

This article provides requirements for calculating branch-circuit, feeder, and service loads. Part I provides for general requirements for calculation methods. Part II provides calculation methods for branch-circuit loads. Parts III and IV provide calculation methods for feeders and services. Part V provides calculation methods for farms.

Informational Note No. 1: See examples in Informative Annex D.

Informational Note No. 2: See Figure 220.1 for information on the organization of Article 220.

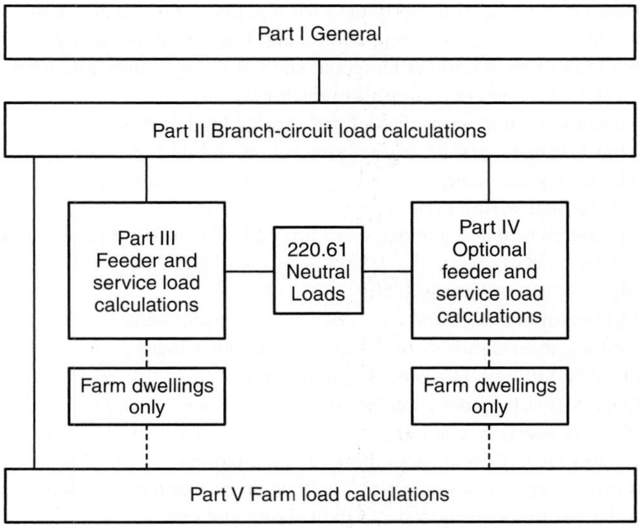


***FIGURE 220.1*** *Branch-Circuit, Feeder, and Service Load Calculation Methods.*

Although this article does not contain the requirements for determining the minimum number of branch circuits, the loads calculated in accordance with Article 220 are used in conjunction with the rules of 210.11 to determine how many branch circuits are needed at a premises. Table 220.3 identifies other articles and sections with load calculation requirements.

## 220.3 Application of Other Articles

In other articles applying to the calculation of loads in specialized applications, there are requirements provided in Table 220.3 that are in addition to, or modifications of, those within this article.

## 220.5 Calculations

**(A) Voltages.** Unless other voltages are specified, for purposes of calculating branch-circuit and feeder loads, nominal system voltages of 120, 120/240, 208Y/120, 240, 347, 480Y/277, 480, 600Y/347, and 600 volts shall be used.

**(B) Fractions of an Ampere.** Calculations shall be permitted to be rounded to the nearest whole ampere, with decimal fractions smaller than 0.5 dropped.

For uniform calculation of load, nominal voltages, as listed in 220.5(A), are required to be used in computing the ampere load on the conductors. To select conductor sizes, refer to 310.15(A) and (B).

Loads are calculated on the basis of volt-amperes (VA) or kilovolt-amperes (kVA), rather than watts or kilowatts (kW), to calculate the true ampere values. However, the rating of equipment is given in watts or kilowatts for noninductive loads. Such ratings are considered to be the equivalent of the same rating in volt-amperes or kilovolt-amperes. See, for example, 220.55. This concept recognizes that load calculations determine conductor and circuit sizes, that the power factor of the load is often unknown, and that the conductor "sees" the circuit volt-amperes only, not the circuit power (watts).

See Examples D1(a) through D5(b) in Informative Annex D. The results of these examples are generally expressed in amperes.

***TABLE 220.3*** *Additional Load Calculation References*

| Calculation | Article | Section (or Part) |
|---|---|---|
| Air-conditioning and refrigerating equipment, branch-circuit conductor sizing | 440 | Part IV |
| Cranes and hoists, rating and size of conductors | 610 | 610.14 |
| Electric vehicle charging system branch-circuit and feeder calculations | 625 | 625.41 |
| Electric welders, ampacity calculations | 630 | 630.11, 630.31 |
| Electrically driven or controlled irrigation machines | 675 | 675.7(A), 675.22(A) |
| Electrified truck parking space | 626 | |
| Electrolytic cell lines | 668 | 668.3(C) |
| Electroplating, branch-circuit conductor sizing | 669 | 669.5 |
| Elevator feeder demand factors | 620 | 620.14 |
| Fire pumps, voltage drop (mandatory calculation) | 695 | 695.7 |
| Fixed electric heating equipment for pipelines and vessels, branch-circuit sizing | 427 | 427.4 |
| Fixed electric space-heating equipment, branch-circuit sizing | 424 | 424.3 |
| Fixed outdoor electric deicing and snow-melting equipment, branch-circuit sizing | 426 | 426.4 |
| Industrial machinery, supply conductor sizing | 670 | 670.4(A) |
| Marinas and boatyards, feeder and service load calculations | 555 | 555.12 |
| Mobile homes, manufactured homes, and mobile home parks, total load for determining power supply | 550 | 550.18(B) |
| Mobile homes, manufactured homes, and mobile home parks, allowable demand factors for park electrical wiring systems | 550 | 550.31 |
| Motion picture and television studios and similar locations – sizing of feeder conductors for television studio sets | 530 | 530.19 |
| Motors, feeder demand factor | 430 | 430.26 |
| Motors, multimotor and combination-load equipment | 430 | 430.25 |
| Motors, several motors or a motor(s) and other load(s) | 430 | 430.24 |
| Over 600-volt branch-circuit calculations | 210 | 210.19(B) |
| Over 600-volt feeder calculations | 215 | 215.2(B) |
| Phase converters, conductors | 455 | 455.6 |
| Recreational vehicle parks, basis of calculations | 551 | 551.73(A) |
| Sensitive electrical equipment, voltage drop (mandatory calculation) | 647 | 647.4(D) |
| Solar photovoltaic systems, circuit sizing and current | 690 | 690.8 |
| Storage-type water heaters | 422 | 422.11(E) |
| Theaters, stage switchboard feeders | 520 | 520.27 |

In these examples, fractions of an ampere are rounded to the nearest whole ampere, with decimal fractions less than 0.5 amperes dropped per 220.5(B).

## II. Branch-Circuit Load Calculations

### 220.10 General

Branch-circuit loads shall be calculated as shown in 220.12, 220.14, and 220.16.

### 220.12 Lighting Load for Specified Occupancies

A unit load of not less than that specified in Table 220.12 for occupancies specified therein shall constitute the minimum lighting load. The floor area for each floor shall be calculated from the outside dimensions of the building, dwelling unit, or other area involved. For dwelling units, the calculated floor area shall not include open porches, garages, or unused or unfinished spaces not adaptable for future use.

Informational Note: The unit values herein are based on minimum load conditions and 100 percent power factor and may not provide sufficient capacity for the installation contemplated.

General lighting loads are in fact minimum lighting loads. A new exception permits the lighting load of a building to be calculated in accordance with the energy code adopted in the jurisdiction. The application has limitations, including the need for continuous monitoring of the lighting load and an alarm that is set to alert the building owner or manager if the load exceeds the values laid out in the energy code. The demand factors may not be applied on top of this reduced lighting load.

Examples of unused or unfinished spaces for dwelling units are some attics, cellars, and crawl spaces.

*Exception: Where the building is designed and constructed to comply with an energy code adopted by the local authority, the lighting load shall be permitted to be calculated at the values specified in the energy code where the following conditions are met:*

*(1) A power monitoring system is installed that will provide continuous information regarding the total general lighting load of the building.*
*(2) The power monitoring system will be set with alarm values to alert the building owner or manager if the lighting load exceeds the values set by the energy code.*
*(3) The demand factors specified in 220.42 are not applied to the general lighting load.*

**TABLE 220.12** *General Lighting Loads by Occupancy*

| Type of Occupancy | Unit Load: Volt-Amperes/ Square Meter | Unit Load: Volt-Amperes/ Square Foot |
|---|---|---|
| Armories and auditoriums | 11 | 1 |
| Banks | 39[b] | 3½[b] |
| Barber shops and beauty parlors | 33 | 3 |
| Churches | 11 | 1 |
| Clubs | 22 | 2 |
| Court rooms | 22 | 2 |
| Dwelling units[a] | 33 | 3 |
| Garages — commercial (storage) | 6 | ½ |
| Hospitals | 22 | 2 |
| Hotels and motels, including apartment houses without provision for cooking by tenants[a] | 22 | 2 |
| Industrial commercial (loft) buildings | 22 | 2 |
| Lodge rooms | 17 | 1½ |
| Office buildings | 39[b] | 3½[b] |
| Restaurants | 22 | 2 |
| Schools | 33 | 3 |
| Stores | 33 | 3 |
| Warehouses (storage) | 3 | ¼ |
| In any of the preceding occupancies except one-family dwellings and individual dwelling units of two-family and multifamily dwellings: | | |
| Assembly halls and auditoriums | 11 | 1 |
| Halls, corridors, closets, stairways | 6 | ½ |
| Storage spaces | 3 | ¼ |

[a]See 220.14(J).
[b]See 220.14(K).

### 220.14 Other Loads — All Occupancies

In all occupancies, the minimum load for each outlet for general-use receptacles and outlets not used for general illumination shall not be less than that calculated in 220.14(A) through (L), the loads shown being based on nominal branch-circuit voltages.

*Exception: The loads of outlets serving switchboards and switching frames in telephone exchanges shall be waived from the calculations.*

**(A) Specific Appliances or Loads.** An outlet for a specific appliance or other load not covered in 220.14(B) through (L) shall be calculated based on the ampere rating of the appliance or load served.

**(B) Electric Dryers and Electric Cooking Appliances in Dwellings and Household Cooking Appliances Used in Instructional Programs.** Load calculations shall be permitted as specified in 220.54 for electric dryers and in 220.55 for electric ranges and other cooking appliances.

Culinary school programs often use multiple household ranges rather than more costly commercial cooking equipment. These appliances tend to use less energy than those used in some commercial kitchens. The title change of this section permits these appliances to use lower household demand factors rather than those for a commercial kitchen.

**(C) Motor Outlets.** Loads for motor outlets shall be calculated in accordance with the requirements in 430.22, 430.24, and 440.6.

**(D) Luminaires.** An outlet supplying luminaire(s) shall be calculated based on the maximum volt-ampere rating of the equipment and lamps for which the luminaire(s) is rated.

In general, no additional calculation is required for luminaires (recessed and surface mounted) installed in or on a dwelling unit, because the load of such luminaires is covered in the 3 VA per square foot calculation specified by Table 220.12. Where the rating of the luminaires installed for general lighting exceeds the minimum load provided for in Table 220.12, the minimum general lighting load for that premises must be based on the installed luminaires. Distinguishing between the luminaires installed for general lighting versus those installed for accent, specialty, or display lighting is much easier to delineate in commercial (particularly mercantile) occupancies.

**(E) Heavy-Duty Lampholders.** Outlets for heavy-duty lampholders shall be calculated at a minimum of 600 volt-amperes.

**(F) Sign and Outline Lighting.** Sign and outline lighting outlets shall be calculated at a minimum of 1200 volt-amperes for each required branch circuit specified in 600.5(A).

If the specific load for signs and outline lighting outlets is known to be larger than that specified by 220.14(F), the actual load must be used for calculation purposes.

**(G) Show Windows.** Show windows shall be calculated in accordance with either of the following:

(1) The unit load per outlet as required in other provisions of this section
(2) At 200 volt-amperes per 300 mm (1 ft) of show window

As shown in Exhibit 220.1, the linear-foot calculation method is permitted in lieu of the specified unit load per outlet for branch circuits serving show windows.

**(H) Fixed Multioutlet Assemblies.** Fixed multioutlet assemblies used in other than dwelling units or the guest rooms or guest suites of hotels or motels shall be calculated in accordance with (H)(1) or (H)(2). For the purposes of this section, the calculation shall be permitted to be based on the portion that contains receptacle outlets.

(1) Where appliances are unlikely to be used simultaneously, each 1.5 m (5 ft) or fraction thereof of each separate and continuous length shall be considered as one outlet of not less than 180 volt-amperes.

***EXHIBIT 220.1*** *An example of the volt-ampere per linear-foot load calculation for branch circuits serving a show window. (© Deriufra/Dreamstime)*

(2) Where appliances are likely to be used simultaneously, each 300 mm (1 ft) or fraction thereof shall be considered as an outlet of not less than 180 volt-amperes.

In light-use commercial and industrial applications, not all of the cord-connected equipment is expected to be used at the same time. An example of light use is a workbench area where one worker uses one electrical tool at a time. In heavy-use applications, all of the cord-connected equipment is generally operating at the same time. An example of heavy use is a retail store displaying television sets, where most or all sets are operating simultaneously.

As shown in Exhibit 220.2, the requirement of 220.14(H)(1) states that each 5 feet of a fixed multioutlet assembly must be considered as one outlet rated 180 VA. The requirement of 220.14(H)(2) states that where appliances are likely to be used

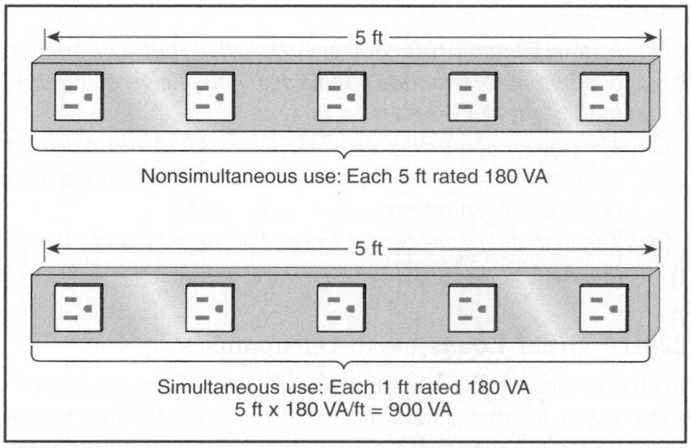


***EXHIBIT 220.2*** *The requirements of 220.14(H)(1) and (H)(2) as applied to fixed multioutlet assemblies.*

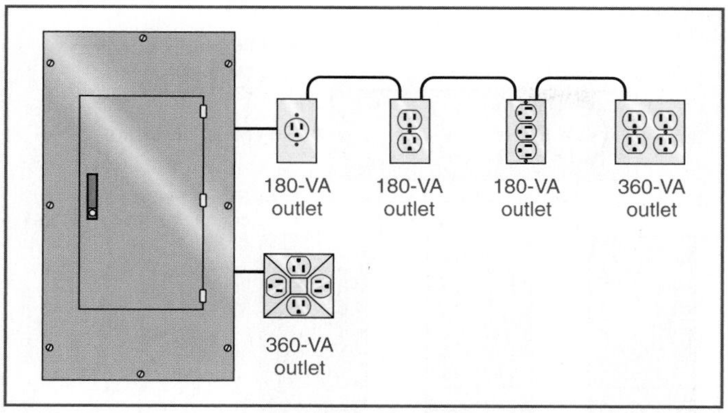


***EXHIBIT 220.3*** *The load requirement of 180 VA per 220.14(I) applied to single- and multiple-receptacle outlets on single straps and the load of 360 VA applied to a multiple device consisting of four receptacles.*

simultaneously, each foot of multioutlet assembly is to be considered as one outlet rated 180 VA.

**(I) Receptacle Outlets.** Except as covered in 220.14(J) and (K), receptacle outlets shall be calculated at not less than 180 volt-amperes for each single or for each multiple receptacle on one yoke. A single piece of equipment consisting of a multiple receptacle comprised of four or more receptacles shall be calculated at not less than 90 volt-amperes per receptacle. This provision shall not be applicable to the receptacle outlets specified in 210.11(C)(1) and (C)(2).

Exhibit 220.3 shows the load of 180 VA applied to single and multiple receptacles mounted on a single yoke or strap, and the load of 360 VA applied to the outlet containing two duplex receptacles and to the outlet containing the device with four receptacles. The receptacle outlets are not the lighting outlets installed for general illumination or the small-appliance branch circuits, as indicated in 220.14(J). The receptacle load for outlets for general illumination in one- and two-family and multifamily dwellings and in guest rooms of hotels and motels is included in the general lighting load value assigned by Table 220.12. The load requirement for the small-appliance branch circuits is 1500 VA per circuit, as described in 220.52(A).

Note in Exhibit 220.3 that the last outlet of the top circuit consists of two duplex receptacles on separate straps. That outlet is calculated at 360 VA because each duplex receptacle is on one yoke. The multiple receptacle supplied from the bottom circuit in the exhibit, which comprises four receptacles, is calculated at 90 VA per receptacle (4 × 90 VA = 360 VA). For example, single-strap and multiple-receptacle devices are calculated as follows:

| Device | Calculated Load |
|---|---|
| Duplex receptacle | 180 VA |
| Triplex receptacle | 180 VA |
| Double duplex receptacle | 360 VA (180 × 2) |
| Quad or four plex-type receptacle | 360 VA (90 × 4) |

A load of 180 VA is not required to be considered for outlets supplying recessed lighting fixtures, lighting outlets for general illumination, and small-appliance branch circuits. To apply the requirement of 180 VA in those cases would be unrealistic, because it would unnecessarily restrict the number of lighting or receptacle outlets on branch circuits in dwelling units. See the note below Table 220.12 that references 220.14(J). This note indicates that the requirement of 180 VA does not apply to most receptacle outlets in dwellings.

In Exhibit 220.4, the maximum number of outlets permitted on 15- and 20-A branch circuits is 10 and 13 outlets, respectively, based on the load assigned for each outlet by 220.14(I). This restriction does not apply to outlets connected to general lighting or small-appliance branch circuits in dwelling units.

**(J) Dwelling Occupancies.** In one-family, two-family, and multifamily dwellings and in guest rooms or guest suites of hotels and motels, the outlets specified in (J)(1), (J)(2), and (J)(3) are included in the general lighting load calculations of 220.12. No additional load calculations shall be required for such outlets.

(1) All general-use receptacle outlets of 20-ampere rating or less, including receptacles connected to the circuits in 210.11(C)(3)
(2) The receptacle outlets specified in 210.52(E) and (G)
(3) The lighting outlets specified in 210.70(A) and (B)

**(K) Banks and Office Buildings.** In banks or office buildings, the receptacle loads shall be calculated to be the larger of (1) or (2):

(1) The calculated load from 220.14(I)
(2) 11 volt-amperes/m$^2$ or 1 volt-ampere/ft$^2$

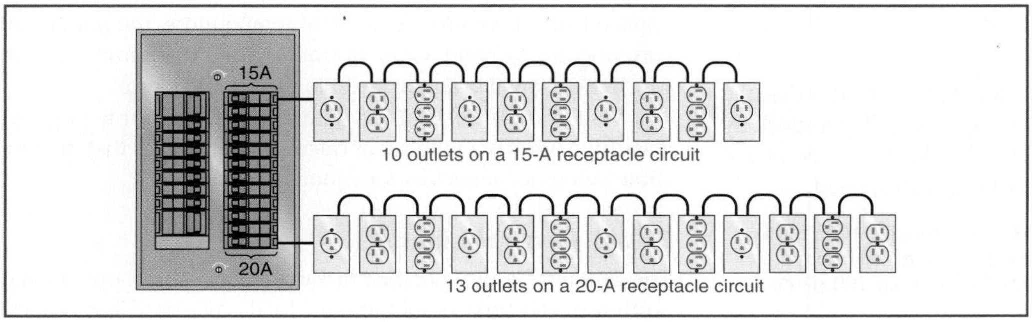


***EXHIBIT 220.4*** *Maximum number of outlets permitted on 15- and 20-ampere branch circuits.*

**(L) Other Outlets.** Other outlets not covered in 220.14(A) through (K) shall be calculated based on 180 volt-amperes per outlet.

### 220.16 Loads for Additions to Existing Installations

**(A) Dwelling Units.** Loads added to an existing dwelling unit(s) shall comply with the following as applicable:

(1) Loads for structural additions to an existing dwelling unit or for a previously unwired portion of an existing dwelling unit, either of which exceeds 46.5 $m^2$ (500 $ft^2$), shall be calculated in accordance with 220.12 and 220.14.

(2) Loads for new circuits or extended circuits in previously wired dwelling units shall be calculated in accordance with either 220.12 or 220.14, as applicable.

**(B) Other Than Dwelling Units.** Loads for new circuits or extended circuits in other than dwelling units shall be calculated in accordance with either 220.12 or 220.14, as applicable.

### 220.18 Maximum Loads

The total load shall not exceed the rating of the branch circuit, and it shall not exceed the maximum loads specified in 220.18(A) through (C) under the conditions specified therein.

**(A) Motor-Operated and Combination Loads.** Where a circuit supplies only motor-operated loads, Article 430 shall apply. Where a circuit supplies only air-conditioning equipment, refrigerating equipment, or both, Article 440 shall apply. For circuits supplying loads consisting of motor-operated utilization equipment that is fastened in place and has a motor larger than ⅛ hp in combination with other loads, the total calculated load shall be based on 125 percent of the largest motor load plus the sum of the other loads.

**(B) Inductive and LED Lighting Loads.** For circuits supplying lighting units that have ballasts, transformers, autotransformers, or LED drivers, the calculated load shall be based on the total ampere ratings of such units and not on the total watts of the lamps.

**(C) Range Loads.** It shall be permissible to apply demand factors for range loads in accordance with Table 220.55, including Note 4.

## III. Feeder and Service Load Calculations

### 220.40 General

The calculated load of a feeder or service shall not be less than the sum of the loads on the branch circuits supplied, as determined by Part II of this article, after any applicable demand factors permitted by Part III or IV or required by Part V have been applied.

Informational Note: See Examples D1(a) through D10 in Informative Annex D. See 220.18(B) for the maximum load in amperes permitted for lighting units operating at less than 100 percent power factor.

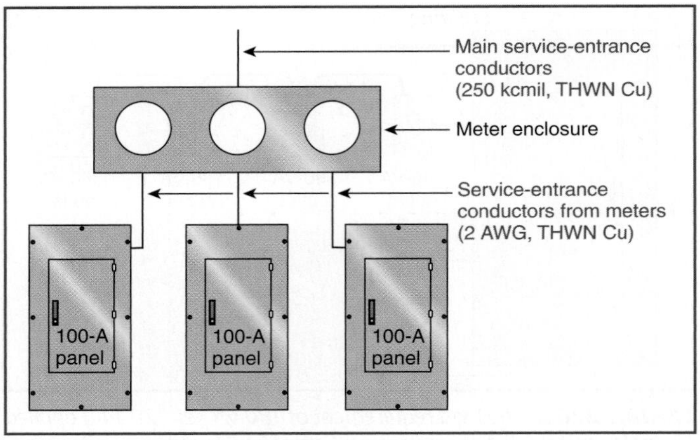


***EXHIBIT 220.5*** *Service conductors sized in accordance with 220.40.*

In the example shown in Exhibit 220.5, each panelboard supplies a calculated load of 80 A. The main set of service conductors is sized to carry the total calculated load of 240 amperes (3 × 80 A). The service conductors from the meter enclosure to each panelboard [2 AWG Cu = 95 A per 60°C column of Table 310.15(B)(16)] are sized to supply a calculated load of 80 amperes and to meet the requirement of 230.90 relative to overcurrent (overload) protection of service conductors terminating in a single service overcurrent protective device. The main set of service conductors [250 kcmil THWN Cu = 255 A per 75°C column of Table 310.15(B)(16)] is not required to be sized to carry 300 amperes based on the combined rating of the panelboards. The individual service-entrance conductors to each panelboard (2 AWG THWN) meet the requirement of 230.90.

See Exhibit 230.26 for a similar example. In that example, the ungrounded service conductors are not required to be sized for the sum of the main overcurrent device ratings of 350 amperes. Service conductors are required to have sufficient ampacity to carry the loads calculated in accordance with Article 220, with the appropriate demand factors applied. See 230.23, 230.31, and 230.42 for specifics on size and rating of service conductors.

Part III of Article 220 contains the requirements for calculating feeder and service loads. Part IV provides optional methods for calculating feeder and service loads in dwelling units and multifamily dwellings.

Except as permitted in 240.4 and 240.6, the rating of the overcurrent device cannot exceed the final ampacity of the circuit conductors after all the correction and adjustment factors are applied (such as where the ambient temperature, the number of current-carrying conductors, or both exceed the parameters on which the allowable ampacity table values are based).

Visit the National Electrical Code Handbook page at www.nfpa.org/nech for a set of calculation examples that demonstrate sizing of conductors for various circuits.

### 220.42 General Lighting

The demand factors specified in Table 220.42 shall apply to that portion of the total branch-circuit load calculated for general

***TABLE 220.42*** *Lighting Load Demand Factors*

| Type of Occupancy | Portion of Lighting Load to Which Demand Factor Applies (Volt-Amperes) | Demand Factor (%) |
|---|---|---|
| Dwelling units | First 3000 or less at | 100 |
| | From 3001 to 120,000 at | 35 |
| | Remainder over 120,000 at | 25 |
| Hospitals* | First 50,000 or less at | 40 |
| | Remainder over 50,000 at | 20 |
| Hotels and motels, including apartment houses without provision for cooking by tenants* | First 20,000 or less at | 50 |
| | From 20,001 to 100,000 at | 40 |
| | Remainder over 100,000 at | 30 |
| Warehouses (storage) | First 12,500 or less at | 100 |
| | Remainder over 12,500 at | 50 |
| All others | Total volt-amperes | 100 |

*The demand factors of this table shall not apply to the calculated load of feeders or services supplying areas in hospitals, hotels, and motels where the entire lighting is likely to be used at one time, as in operating rooms, ballrooms, or dining rooms.

illumination. They shall not be applied in determining the number of branch circuits for general illumination.

## 220.43 Show-Window and Track Lighting

**(A) Show Windows.** For show-window lighting, a load of not less than 660 volt-amperes/linear meter or 200 volt-amperes/linear foot shall be included for a show window, measured horizontally along its base.

Informational Note: See 220.14(G) for branch circuits supplying show windows.

**(B) Track Lighting.** For track lighting in other than dwelling units or guest rooms or guest suites of hotels or motels, an additional load of 150 volt-amperes shall be included for every 600 mm (2 ft) of lighting track or fraction thereof. Where multicircuit track is installed, the load shall be considered to be divided equally between the track circuits.

**Calculation Example**

A lighting plan shows 62.5 linear feet of single-circuit track lighting for a small clothing store. Because the actual track lighting fixtures are owner supplied, neither the quantity of track lighting fixtures nor the lamp size is specified. What is the minimum calculated load associated with the track lighting that must be added to the service or feeder supplying this store?

*Solution*

According to 220.43(B), the minimum calculated load to be added to the service or feeder supplying this track light installation is calculated as follows:

$$\frac{62.5\text{ ft}}{2\text{ ft}} = 31.25\text{, rounded up to 32}$$

$$32 \times 150\text{ VA} = 4800\text{ VA}$$

The minimum load for the lighting track that is added to the service and/or feeder calculation is 4800 VA.

It is important to note that the branch circuits supplying this installation are covered in 410.151(B). The maximum load on the track must not exceed the rating of the branch circuit supplying the track. Also, the track must be supplied by a branch circuit that has a rating not exceeding the rating of the track. The track length does not enter into the branch-circuit calculation.

Section 220.43(B) is not intended to limit the number of feet of track on a single branch circuit, nor is it intended to limit the number of fixtures on an individual track. It is meant to be used solely for load calculations of feeders and services.

*Exception: If the track lighting is supplied through a device that limits the current to the track, the load shall be permitted to be calculated based on the rating of the device used to limit the current.*

The rating of the branch-circuit overcurrent protective device can be used for this calculation, or the device used to limit current can be a supplementary overcurrent protective device. This exception enables load calculations to be more in line with the limitations placed on building lighting loads by local and federal energy codes.

**Calculation Example**

A lighting plan shows 62.5 linear feet of single-circuit track lighting for a small clothing store. Because the actual track lighting fixtures are owner supplied, neither the quantity of track lighting fixtures nor the lamp size is specified in the plan. Because the amount of track is to facilitate easy relocation of the luminaires to accommodate changes in the display of merchandise rather than to accommodate a large number of luminaires, the entire length will be supplied by a single 20-A, 120-V branch circuit. What is the minimum calculated load associated with the track lighting that must be added to the service or feeder supplying this store?

*Solution*

According to the exception to 220.43(B), the minimum calculated load to be added to the service or feeder supplying this track light installation is calculated as follows:

$$20\text{ A} \times 120\text{ V} = 2400\text{ VA}$$

The minimum load for the lighting track that is added to the service and/or feeder calculation is 2400 VA. This calculation could be further reduced where the track is supplied through a supplementary overcurrent protective device(s) having a current rating less than 20 A.

***TABLE 220.44** Demand Factors for Non-Dwelling Receptacle Loads*

| Portion of Receptacle Load to Which Demand Factor Applies (Volt-Amperes) | Demand Factor (%) |
|---|---|
| First 10 kVA or less at | 100 |
| Remainder over 10 kVA at | 50 |

## 220.44 Receptacle Loads — Other Than Dwelling Units

Receptacle loads calculated in accordance with 220.14(H) and (I) shall be permitted to be made subject to the demand factors given in Table 220.42 or Table 220.44.

Section 220.44 permits receptacle loads calculated at not more than 180 VA per strap to be computed by either of the following methods:

1. The receptacle loads are added to the lighting load. The demand factors (if applicable) in Table 220.12 are then applied to the combined load.
2. The receptacle loads are calculated (without the lighting load) with demand factors from Table 220.44 applied.

## 220.50 Motors

Motor loads shall be calculated in accordance with 430.24, 430.25, and 430.26 and with 440.6 for hermetic refrigerant motor compressors.

## 220.51 Fixed Electric Space Heating

Fixed electric space-heating loads shall be calculated at 100 percent of the total connected load. However, in no case shall a feeder or service load current rating be less than the rating of the largest branch circuit supplied.

*Exception: Where reduced loading of the conductors results from units operating on duty-cycle, intermittently, or from all units not operating at the same time, the authority having jurisdiction may grant permission for feeder and service conductors to have an ampacity less than 100 percent, provided the conductors have an ampacity for the load so determined.*

## 220.52 Small-Appliance and Laundry Loads — Dwelling Unit

**(A) Small-Appliance Circuit Load.** In each dwelling unit, the load shall be calculated at 1500 volt-amperes for each 2-wire small-appliance branch circuit as covered by 210.11(C)(1). Where the load is subdivided through two or more feeders, the calculated load for each shall include not less than 1500 volt-amperes for each 2-wire small-appliance branch circuit. These loads shall be permitted to be included with the general lighting load and subjected to the demand factors provided in Table 220.42.

*Exception: The individual branch circuit permitted by 210.52(B)(1), Exception No. 2, shall be permitted to be excluded from the calculation required by 220.52.*

See the commentary following 210.52(B) regarding required receptacle outlets for small-appliance branch circuits.

**(B) Laundry Circuit Load.** A load of not less than 1500 volt-amperes shall be included for each 2-wire laundry branch circuit installed as covered by 210.11(C)(2). This load shall be permitted to be included with the general lighting load and subjected to the demand factors provided in Table 220.42.

Where additional small-appliance and laundry branch circuits are provided, they also are calculated at 1500 VA per circuit. These loads are permitted to be totaled and then added to the general lighting load. The demand factors in Table 220.42 can then be applied to the combined total load of the small-appliance branch circuits, the laundry branch circuit, and the general lighting from Table 220.12.

## 220.53 Appliance Load — Dwelling Unit(s)

It shall be permissible to apply a demand factor of 75 percent to the nameplate rating load of four or more appliances fastened in place, other than electric ranges, clothes dryers, space-heating equipment, or air-conditioning equipment, that are served by the same feeder or service in a one-family, two-family, or multifamily dwelling.

For appliances fastened in place (other than ranges, clothes dryers, and space-heating and air-conditioning equipment), feeder capacity must be provided for the sum of these loads. For a total load of four or more such appliances, a demand factor of 75 percent is permitted by 220.53. See Table 430.248 for the full-load current, in amperes, for single-phase ac motors in accordance with 220.50. Visit the National Electrical Code Handbook page at www.nfpa.org/nech for a set of calculation examples that demonstrate sizing of conductors for various circuits.

## 220.54 Electric Clothes Dryers — Dwelling Unit(s)

The load for household electric clothes dryers in a dwelling unit(s) shall be either 5000 watts (volt-amperes) or the nameplate rating, whichever is larger, for each dryer served. The use of the demand factors in Table 220.54 shall be permitted. Where two or more single-phase dryers are supplied by a 3-phase, 4-wire feeder or service, the total load shall be calculated on the basis of twice the maximum number connected between any two phases. Kilovolt-amperes (kVA) shall be considered equivalent to kilowatts (kW) for loads calculated in this section.

The use of demand factors is permitted but not required, because the *NEC* does not prohibit applying the full load of all dryers to a service and/or feeder calculation. However, this method is not

***TABLE 220.54*** *Demand Factors for Household Electric Clothes Dryers*

| Number of Dryers | Demand Factor (%) |
|---|---|
| 1–4 | 100 |
| 5 | 85 |
| 6 | 75 |
| 7 | 65 |
| 8 | 60 |
| 9 | 55 |
| 10 | 50 |
| 11 | 47 |
| 12–23 | 47% minus 1% for each dryer exceeding 11 |
| 24–42 | 35% minus 0.5% for each dryer exceeding 23 |
| 43 and over | 25% |

necessary or practical, and experience has demonstrated that the use of the Table 220.54 demand factors provides sufficient capacity in the service for dryer loads. It is unlikely that all dryers will be in operation simultaneously, and Table 220.54 was developed based on utility demand that proves this point.

The minimum load to be used is the larger of either 5000 VA or the nameplate rating of the dryer. In the following calculation example, the nameplate rating of 5500 VA (watts) is used. In addition, because of the supply system characteristics, calculation is a little different from simply adding the total rating of all dryers and applying the appropriate demand factor. The load of 10 single-phase dryers will be distributed across a 3-phase supply system.

**Calculation Example**

Assuming the load of 10 single-phase dryers is connected as evenly as possible to the 3-phase system (3 dryers connected between phases A and B, 3 dryers connected between phases B and C, and 4 dryers connected between phases A and C), the maximum number of dryers connected between any two phases is 4.

*Solution*

STEP 1. Twice the maximum number of dryers connected between any two phases is used as the basis for calculating the demand load:

$$4 \text{ dryers} \times 2 = 8 \text{ dryers}$$

60% demand from Table 220.54

$$5500 \text{ VA} \times 8 \times 0.6 = 26{,}400 \text{ VA}$$

(connected between two phases of the 3-phase system)

$$26{,}400 \text{ VA} \div 2 = 13{,}200 \text{ VA per phase load}$$

3-phase dryer load on service:

$$13{,}200 \text{ VA} \times 3 = 39{,}600 \text{ VA}$$

Dryer load on each of the ungrounded service or feeder conductors:

$$39{,}600 \text{ VA} \div (208 \text{ V} \times 1.732) = 110 \text{ A}$$

STEP 2. For the grounded (neutral) service or feeder conductor, 220.61(B)(1) permits a 70-percent demand to be applied to the demand load of the ungrounded conductors:

$$39{,}600 \text{ VA} \times 0.70 = 27{,}720 \text{ VA}$$

Dryer load on the grounded (neutral) service or feeder conductor:

$$27{,}720 \text{ VA} \div (208 \text{ V} \times 1.732) = 77 \text{ A}$$

Dryer load without applying the 220.54 demand factor:
3-phase dryer load:

$$5500 \text{ VA} \times 10 = 55{,}000 \text{ VA}$$
$$55{,}000 \text{ VA} \div (208 \text{ V} \times 1.732) = 153 \text{ A}$$

If the service and feeder load for this multifamily dwelling is calculated using the optional calculation for multifamily dwellings from Part IV of Article 220, the dryer load is not subject to an individual demand factor as required by the Part III calculation. Instead, the calculation performed in accordance with 220.84 totals all of the loads in an individual dwelling unit to arrive at a total "connected load." The nameplate rating of the dryer and other appliance loads are part of a total load determined using 220.84(C)(1) through (C)(5). This connected load is then subjected to a single demand factor from Table 220.84 based on the total number of dwelling units served by a set of service or feeder conductors. For a set of conductors serving 10 dwelling units, the demand factor is 43 percent. The dryer load determined by the optional calculation is as follows:

$$5500 \text{ VA} \times 10 \times 0.43 = 23{,}650 \text{ VA of dryer load}$$
$$23{,}650 \div (208 \text{ V} \times 1.732) = 66 \text{ A of 3-phase dryer load}$$

The load to be applied to the grounded service or feeder conductor can be calculated using the demand permitted by 220.61(B)(1).

The optional calculation can be used, provided all of the conditions specified in 220.84(A)(1) through (A)(3) are met. Otherwise, the calculation for the multifamily dwelling is performed in accordance with the requirements of Part III of Article 220.

## 220.55 Electric Cooking Appliances in Dwelling Units and Household Cooking Appliances Used in Instructional Programs

The load for household electric ranges, wall-mounted ovens, counter-mounted cooking units, and other household cooking appliances individually rated in excess of 1¾ kW shall be permitted to be calculated in accordance with Table 220.55. Kilovolt-amperes (kVA) shall be considered equivalent to kilowatts (kW) for loads calculated under this section.

Where two or more single-phase ranges are supplied by a 3-phase, 4-wire feeder or service, the total load shall be calculated on the basis of twice the maximum number connected between any two phases.

Counter-mounted cooking appliances have a smaller load rating than that of a full-sized range with an oven. The load for a counter-mounted cooking appliance is typically covered by either Column A or Column B in Table 220.55. Where counter-mounted cooking appliances like the one pictured in Exhibit 220.6 are used with a separate wall oven, it is permissible to run a single branch circuit (sized according to Note 4 to Table 220.55) to the kitchen and supply each with branch-circuit tap conductors installed as specified in 210.19(A)(3), Exception No. 1.

Informational Note No. 1: See the examples in Informative Annex D.

•

Informational Note No. 2: See Table 220.56 for commercial cooking equipment.

***EXHIBIT 220.6*** *A counter-mounted cooking appliance installed in a dwelling-unit kitchen. (© Linea/Dreamstime.com)*

***TABLE 220.55*** *Demand Factors and Loads for Household Electric Ranges, Wall-Mounted Ovens, Counter-Mounted Cooking Units, and Other Household Cooking Appliances over 1¾ kW Rating (Column C to be used in all cases except as otherwise permitted in Note 3.)*

| | Demand Factor (%) (See Notes) | | |
|---|---|---|---|
| **Number of Appliances** | **Column A (Less than 3½ kW Rating)** | **Column B (3½ kW through 8¾ kW Rating)** | **Column C Maximum Demand (kW) (See Notes) (Not over 12 kW Rating)** |
| 1 | 80 | 80 | 8 |
| 2 | 75 | 65 | 11 |
| 3 | 70 | 55 | 14 |
| 4 | 66 | 50 | 17 |
| 5 | 62 | 45 | 20 |
| 6 | 59 | 43 | 21 |
| 7 | 56 | 40 | 22 |
| 8 | 53 | 36 | 23 |
| 9 | 51 | 35 | 24 |
| 10 | 49 | 34 | 25 |
| 11 | 47 | 32 | 26 |
| 12 | 45 | 32 | 27 |
| 13 | 43 | 32 | 28 |
| 14 | 41 | 32 | 29 |
| 15 | 40 | 32 | 30 |
| 16 | 39 | 28 | 31 |
| 17 | 38 | 28 | 32 |
| 18 | 37 | 28 | 33 |
| 19 | 36 | 28 | 34 |
| 20 | 35 | 28 | 35 |
| 21 | 34 | 26 | 36 |
| 22 | 33 | 26 | 37 |
| 23 | 32 | 26 | 38 |
| 24 | 31 | 26 | 39 |
| 25 | 30 | 26 | 40 |
| 26–30 | 30 | 24 | 15 kW + 1 kW for each range |
| 31–40 | 30 | 22 | |
| 41–50 | 30 | 20 | 25 kW + ¾ kW for each range |
| 51–60 | 30 | 18 | |
| 61 and over | 30 | 16 | |

*TABLE 220.55 Continued*

Notes:

1. Over 12 kW through 27 kW ranges all of same rating. For ranges individually rated more than 12 kW but not more than 27 kW, the maximum demand in Column C shall be increased 5 percent for each additional kilowatt of rating or major fraction thereof by which the rating of individual ranges exceeds 12 kW.

For household electric ranges and other cooking appliances, the size of the conductors must be determined by the rating of the range. According to Table 220.55, for one range rated 12 kW or less, the maximum demand load is 8 kW (8 kVA per 220.55), and 8 AWG copper conductors with 60°C insulation would suffice. Note that 210.19(A)(3) does not permit the branch-circuit rating of a circuit supplying household ranges with a nameplate rating of 8¾ kW to be less than 40 amperes.

2. Over 8¾ kW through 27 kW ranges of unequal ratings. For ranges individually rated more than 8¾ kW and of different ratings, but none exceeding 27 kW, an average value of rating shall be calculated by adding together the ratings of all ranges to obtain the total connected load (using 12 kW for any range rated less than 12 kW) and dividing by the total number of ranges. Then the maximum demand in Column C shall be increased 5 percent for each kilowatt or major fraction thereof by which this average value exceeds 12 kW.

Note 2 to Table 220.55 provides for ranges larger than 8¾ kW. Note 4 covers installations where the circuit supplies multiple cooking components, which are combined and treated as a single range.

3. Over 1¾ kW through 8¾ kW. In lieu of the method provided in Column C, it shall be permissible to add the nameplate ratings of all household cooking appliances rated more than 1¾ kW but not more than 8¾ kW and multiply the sum by the demand factors specified in Column A or Column B for the given number of appliances. Where the rating of cooking appliances falls under both Column A and Column B, the demand factors for each column shall be applied to the appliances for that column, and the results added together.

The branch-circuit load for one range is permitted to be computed by using either the nameplate rating of the appliance or Table 220.55. If a single branch circuit supplies a counter-mounted cooking unit and not more than two wall-mounted ovens, all of which are located in the same room, the nameplate ratings of these appliances can be added and the total treated as the equivalent of one range, according to Note 4 of Table 220.55.

### Calculation Example

Calculate the load for a single branch circuit that supplies the following cooking units:

- One counter-mounted cooking unit with rating of 8 kW
- One wall-mounted oven with rating of 7 kW
- A second wall-mounted oven with rating of 6 kW

*Solution*

STEP 1. The combined cooking appliances can be treated as one range, according to Note 4 of Table 220.55. In Table 220.55, find the maximum demand for one range not over 12 kW, which is 8 kW (from Column C).

STEP 2. According to Note 1 in Table 220.55, for ranges that are over 12 kW but not more than 27 kW, the maximum demand in Column C (8 kW) is increased 5 percent for each kW that exceeds 12 kW. Determine the additional kilowatts:

$$\text{Combined unit rating} = 8\text{ kW} + 7\text{ kW} + 6\text{ kW} = 21\text{ kW}$$

$$\text{Additional kW} = 21\text{ kW} - 12\text{ kW} = 9\text{ kW}$$

STEP 3. Calculate by how much the maximum load in Column C in Table 220.55 must be increased for the combined appliances:

$$\text{Increase} = 5\%\text{ per kW} \times 9\text{ kW} = 45\% = 0.45 \times 8\text{ kW} = 3.6\text{ kW}$$

STEP 4. Calculate the total load in amperes, as follows:

$$\text{Total load} = 8\text{ kW} + 3.6\text{ kW} = 11.6\text{ kW} = 11{,}600\text{ W} = 11{,}600\text{ VA} = \frac{11{,}600\text{ VA}}{240\text{ V}} = 48.3\text{ or }48\text{ A}$$

4. Branch-Circuit Load. It shall be permissible to calculate the branch-circuit load for one range in accordance with Table 220.55. The branch-circuit load for one wall-mounted oven or one counter-mounted cooking unit shall be the nameplate rating of the appliance. The branch-circuit load for a counter-mounted cooking unit and not more than two wall-mounted ovens, all supplied from a single branch circuit and located in the same room, shall be calculated by adding the nameplate rating of the individual appliances and treating this total as equivalent to one range.

5. This table shall also apply to household cooking appliances rated over 1¾ kW and used in instructional programs.

The nameplate ratings of all household cooking appliances rated more than 1¾ kW but not more than 8¾ kW may be added and the sum multiplied by the demand factor specified in Column A or B of Table 220.55 for the given number of appliances. For feeder demand factors for other than dwelling units, that is, commercial electric cooking equipment, dishwasher booster heaters, water heaters, and so forth, see Table 220.56.

The demand factors in the *Code* are based on the diversified use of household appliances, because it is unlikely that all appliances will be used simultaneously or that all cooking units and the oven of a range will be at maximum heat for any length of time.

*TABLE 220.56 Demand Factors for Kitchen Equipment — Other Than Dwelling Unit(s)*

| Number of Units of Equipment | Demand Factor (%) |
|---|---|
| 1 | 100 |
| 2 | 100 |
| 3 | 90 |
| 4 | 80 |
| 5 | 70 |
| 6 and over | 65 |

## 220.56 Kitchen Equipment — Other Than Dwelling Unit(s)

It shall be permissible to calculate the load for commercial electric cooking equipment, dishwasher booster heaters, water heaters, and other kitchen equipment in accordance with Table 220.56. These demand factors shall be applied to all equipment that has either thermostatic control or intermittent use as kitchen equipment. These demand factors shall not apply to space-heating, ventilating, or air-conditioning equipment.

However, in no case shall the feeder or service calculated load be less than the sum of the largest two kitchen equipment loads.

Because commercial electric cooking and other kitchen equipment is used more intensively and for longer periods of time than electric cooking equipment in a dwelling unit, the demand factors in Table 220.56 do not provide a comparable demand to that permitted for dwelling unit cooking appliances until there are six or more commercial appliances. The demand factors in Table 220.56 can be applied to a combination of cooking, dishwashing, water heating, and other type commercial kitchen appliances, and the combined total allows a deeper demand factor to be applied. Thermostatically controlled electric cooking appliances such as the one shown in Exhibit 220.7 are permitted, but not required, to have the demand applied to the appliance nameplate rating.

Commercial electric cooking appliances are sometimes used continuously from morning until night, thus the demand factors for commercial applications are not as generous as those permitted for dwelling units. However, household cooking equipment used in instructional programs, such as in culinary schools, is subject to the demands in 220.55. Such equipment is not subject to the continuous demands that would exist in a commercial restaurant. [See the commentary following 220.14(B) for more information.]

## 220.60 Noncoincident Loads

Where it is unlikely that two or more noncoincident loads will be in use simultaneously, it shall be permissible to use only the largest load(s) that will be used at one time for calculating the total load of a feeder or service.

***EXHIBIT 220.7*** *Electrically powered cooking/heating appliance with thermostatic control in a commercial kitchen. (Courtesy of the International Association of Electrical Inspectors)*

## 220.61 Feeder or Service Neutral Load

**(A) Basic Calculation.** The feeder or service neutral load shall be the maximum unbalance of the load determined by this article. The maximum unbalanced load shall be the maximum net calculated load between the neutral conductor and any one ungrounded conductor.

*Exception: For 3-wire, 2-phase or 5-wire, 2-phase systems, the maximum unbalanced load shall be the maximum net calculated load between the neutral conductor and any one ungrounded conductor multiplied by 140 percent.*

**(B) Permitted Reductions.** A service or feeder supplying the following loads shall be permitted to have an additional demand factor of 70 percent applied to the amount in 220.61(B)(1) or portion of the amount in 220.61(B)(2) determined by the basic calculation:

(1) A feeder or service supplying household electric ranges, wall-mounted ovens, counter-mounted cooking units, and electric dryers, where the maximum unbalanced load has been determined in accordance with Table 220.55 for ranges and Table 220.54 for dryers

(2) That portion of the unbalanced load in excess of 200 amperes where the feeder or service is supplied from a 3-wire dc or single-phase ac system; or a 4-wire, 3-phase, 3-wire, 2-phase system; or a 5-wire, 2-phase system

Informational Note: See Examples D1(a), D1(b), D2(b), D4(a), and D5(a) in Informative Annex D.

**(C) Prohibited Reductions.** There shall be no reduction of the neutral or grounded conductor capacity applied to the amount in 220.61(C)(1), or portion of the amount in (C)(2), from that determined by the basic calculation:

(1) Any portion of a 3-wire circuit consisting of 2 ungrounded conductors and the neutral conductor of a 4-wire, 3-phase, wye-connected system
(2) That portion consisting of nonlinear loads supplied from a 4-wire, wye-connected, 3-phase system

•

Informational Note: A 3-phase, 4-wire, wye-connected power system used to supply power to nonlinear loads may necessitate that the power system design allow for the possibility of high harmonic neutral conductor currents.

Section 220.61 describes the basis for calculating the neutral load of feeders or services as the maximum unbalanced load that can occur between the neutral and any other ungrounded conductor. For a household electric range or clothes dryer, the maximum unbalanced load may be assumed to be 70 percent, so the neutral can be sized on that basis. Section 220.61(B) permits the reduction of the feeder neutral conductor size under specific conditions of use, and 220.61(C)(1) and (C)(2) cite a circuit arrangement and a load characteristic as applications where the capacity of a neutral or grounded conductor of a feeder or service is not permitted to be reduced.

If the system also supplies nonlinear loads such as electric-discharge lighting (including fluorescent and HID) or data-processing or similar equipment, the neutral is considered a current-carrying conductor if the load of the electric-discharge lighting, data-processing, or similar equipment on the feeder neutral consists of more than half the total load, in accordance with 310.15(B)(5)(c). Electric-discharge lighting and data-processing equipment may have harmonic currents in the neutral that may exceed the load current in the ungrounded conductors. The Informational Note to item 2 cautions designers and installers to be aware of harmonic contribution and to design the electrical installation to accommodate the harmonic load imposed on neutral conductors. In some instances, the neutral current may exceed the current in the phase conductors. See the commentary following 310.15(B)(5)(c) regarding neutral conductor ampacity.

# IV. Optional Feeder and Service Load Calculations

## 220.80 General

Optional feeder and service load calculations shall be permitted in accordance with Part IV.

## 220.82 Dwelling Unit

**(A) Feeder and Service Load.** This section applies to a dwelling unit having the total connected load served by a single 120/240-volt or 208Y/120-volt set of 3-wire service or feeder conductors with an ampacity of 100 or greater. It shall be permissible to calculate the feeder and service loads in accordance with this section instead of the method specified in Part III of this article. The calculated load shall be the result of adding the loads from 220.82(B) and (C). Feeder and service-entrance conductors whose calculated load is determined by this optional calculation shall be permitted to have the neutral load determined by 220.61.

This optional calculation method applies to a single dwelling unit, whether it is a separate building or located in a multifamily dwelling. See Article 100 for the definition of *dwelling unit.* Examples of the optional calculation for a dwelling unit are given in Examples D2(a), D2(b), D2(c), and D4(b) of Informative Annex D.

**(B) General Loads.** The general calculated load shall be not less than 100 percent of the first 10 kVA plus 40 percent of the remainder of the following loads:

(1) 33 volt-amperes/m$^2$ or 3 volt-amperes/ft$^2$ for general lighting and general-use receptacles. The floor area for each floor shall be calculated from the outside dimensions of the dwelling unit. The calculated floor area shall not include open porches, garages, or unused or unfinished spaces not adaptable for future use.
(2) 1500 volt-amperes for each 2-wire, 20-ampere small-appliance branch circuit and each laundry branch circuit covered in 210.11(C)(1) and (C)(2).
(3) The nameplate rating of the following:
    a. All appliances that are fastened in place, permanently connected, or located to be on a specific circuit
    b. Ranges, wall-mounted ovens, counter-mounted cooking units
    c. Clothes dryers that are not connected to the laundry branch circuit specified in item (2)
    d. Water heaters
(4) The nameplate ampere or kVA rating of all permanently connected motors not included in item (3).

**(C) Heating and Air-Conditioning Load.** The largest of the following six selections (load in kVA) shall be included:

(1) 100 percent of the nameplate rating(s) of the air conditioning and cooling.
(2) 100 percent of the nameplate rating(s) of the heat pump when the heat pump is used without any supplemental electric heating.
(3) 100 percent of the nameplate rating(s) of the heat pump compressor and 65 percent of the supplemental electric heating for central electric space-heating systems. If the heat pump compressor is prevented from operating at the same time as the supplementary heat, it does not need to be added to the supplementary heat for the total central space heating load.

Where the heat pump compressor and supplemental heating operate at the same time, 100 percent of the compressor load plus 65 percent of the supplemental heating load is treated as the central space heating load. If the equipment operates such that the compressor cannot operate concurrently with the supplemental heating, the central space heating load is required to be based on only 65 percent of the supplemental heating load.

(4) 65 percent of the nameplate rating(s) of electric space heating if less than four separately controlled units.
(5) 40 percent of the nameplate rating(s) of electric space heating if four or more separately controlled units.
(6) 100 percent of the nameplate ratings of electric thermal storage and other heating systems where the usual load is expected to be continuous at the full nameplate value. Systems qualifying under this selection shall not be calculated under any other selection in 220.82(C).

For loads that do not operate simultaneously, 220.60 permits the total load to be calculated based on the largest load being considered. Similarly, 220.82(C) requires that only the largest of the six choices needs to be included in the feeder or service calculation. Examples of calculations using air conditioning and heating are found in Informative Annex D, Examples D2(a), (b), and (c).

## 220.83 Existing Dwelling Unit

This section shall be permitted to be used to determine if the existing service or feeder is of sufficient capacity to serve additional loads. Where the dwelling unit is served by a 120/240-volt or 208Y/120-volt, 3-wire service, it shall be permissible to calculate the total load in accordance with 220.83(A) or (B).

**(A) Where Additional Air-Conditioning Equipment or Electric Space-Heating Equipment Is Not to Be Installed.** The following percentages shall be used for existing and additional new loads.

| Load (kVA) | Percent of Load |
|---|---|
| First 8 kVA of load at | 100 |
| Remainder of load at | 40 |

Load calculations shall include the following:

(1) General lighting and general-use receptacles at 33 volt-amperes/m$^2$ or 3 volt-amperes/ft$^2$ as determined by 220.12
(2) 1500 volt-amperes for each 2-wire, 20-ampere small-appliance branch circuit and each laundry branch circuit covered in 210.11(C)(1) and (C)(2)
(3) The nameplate rating of the following:
  a. All appliances that are fastened in place, permanently connected, or located to be on a specific circuit
  b. Ranges, wall-mounted ovens, counter-mounted cooking units
  c. Clothes dryers that are not connected to the laundry branch circuit specified in item (2)
  d. Water heaters

**(B) Where Additional Air-Conditioning Equipment or Electric Space-Heating Equipment Is to Be Installed.** The following percentages shall be used for existing and additional new loads. The larger connected load of air-conditioning or space-heating, but not both, shall be used.

| Load | Percent of Load |
|---|---|
| Air-conditioning equipment | 100 |
| Central electric space heating | 100 |
| Less than four separately controlled space-heating units | 100 |
| First 8 kVA of all other loads | 100 |
| Remainder of all other loads | 40 |

Other loads shall include the following:

(1) General lighting and general-use receptacles at 33 volt-amperes/m$^2$ or 3 volt-amperes/ft$^2$ as determined by 220.12
(2) 1500 volt-amperes for each 2-wire, 20-ampere small-appliance branch circuit and each laundry branch circuit covered in 210.11(C)(1) and (C)(2)
(3) The nameplate rating of the following:
  a. All appliances that are fastened in place, permanently connected, or located to be on a specific circuit
  b. Ranges, wall-mounted ovens, counter-mounted cooking units
  c. Clothes dryers that are not connected to the laundry branch circuit specified in (2)
  d. Water heaters

The optional methods described in 220.83(A) and (B) allow an additional load to be supplied by an existing service.

### Calculation Example

An existing dwelling unit is served by a 100-A service. An additional load of a single 5-kVA, 240-V air-conditioning unit is to be installed. Because the existing load does not contain heating or air-conditioning equipment, the existing load is calculated according to 220.83(A). The load of the existing dwelling unit consists of the following:

| | |
|---|---|
| General lighting: | |
| 24 ft × 40 ft = 960 ft$^2$ × VA per ft$^2$ | 2,880 VA |
| Small-appliance circuits: 3 × 1500 VA | 4,500 VA |
| Laundry circuit at 1500 VA | 1,500 VA |
| Electric range rated 10.5 kW | 10,500 VA |
| Electric water heater rated 3.0 kW | 3,000 VA |
| Total existing load | 22,380 VA |

*Solution*

STEP 1. Following the requirements of 220.83(A), calculate the existing dwelling unit load before adding any equipment:

| | |
|---|---|
| First 8 kVA of load at 100% | 8,000 VA |
| Remainder of load at 40%: | |
| 22,380 VA − 8,000 VA = 14,380 VA × 40% | 5,752 VA |
| Total load | 13,752 VA |

(without air-conditioning equipment)

13,752 VA ÷ 240 V = 57.3 A

STEP 2. Add existing and new loads of the dwelling unit.

| | |
|---|---|
| Existing load | 22,380 VA |
| Added air-conditioning equipment | 5,000 VA |
| Total new load | 27,380 VA |

STEP 3. Following the requirements in 220.83(B), calculate the dwelling unit total load after adding any new equipment:

| | |
|---|---|
| First 8 kVA of other load at 100% | 8,000 VA |
| Remainder of other load at 40%: | |
| 22,380 VA − 8,000 VA = 14,380 VA × 40% | 5,752 VA |
| 100% of air-conditioning equipment | 5,000 VA |
| Total load | 18,752 VA |

(with added air-conditioning equipment)

18,752 VA ÷ 240 V = 78.13 or 78 A

The additional load contributed by the added 5-kVA air-conditioning equipment does not exceed the allowable load permitted on a 100-A service.

## 220.84 Multifamily Dwelling

**(A) Feeder or Service Load.** It shall be permissible to calculate the load of a feeder or service that supplies three or more dwelling units of a multifamily dwelling in accordance with Table 220.84 instead of Part III of this article if all the following conditions are met:

(1) No dwelling unit is supplied by more than one feeder.
(2) Each dwelling unit is equipped with electric cooking equipment.

*Exception: When the calculated load for multifamily dwellings without electric cooking in Part III of this article exceeds that calculated under Part IV for the identical load plus electric cooking (based on 8 kW per unit), the lesser of the two loads shall be permitted to be used.*

This calculation method is only permitted for dwelling units with electric cooking equipment. The exception to 220.84(A)(2) permits load calculation for dwelling units that do not have electric cooking equipment, by using a simulated electric cooking equipment load of 8 kW per unit and comparing this calculated load to the load for the dwellings without electric cooking equipment calculated in accordance with Part III of Article 220 (commonly referred to as the "standard calculation"). Whichever of these calculations yields the smallest load is permitted to be used.

(3) Each dwelling unit is equipped with either electric space heating or air conditioning, or both. Feeders and service conductors whose calculated load is determined by this optional calculation shall be permitted to have the neutral load determined by 220.61.

**(B) House Loads.** House loads shall be calculated in accordance with Part III of this article and shall be in addition to the dwelling unit loads calculated in accordance with Table 220.84.

***TABLE 220.84*** *Optional Calculations — Demand Factors for Three or More Multifamily Dwelling Units*

| Number of Dwelling Units | Demand Factor (%) |
|---|---|
| 3–5 | 45 |
| 6–7 | 44 |
| 8–10 | 43 |
| 11 | 42 |
| 12–13 | 41 |
| 14–15 | 40 |
| 16–17 | 39 |
| 18–20 | 38 |
| 21 | 37 |
| 22–23 | 36 |
| 24–25 | 35 |
| 26–27 | 34 |
| 28–30 | 33 |
| 31 | 32 |
| 32–33 | 31 |
| 34–36 | 30 |
| 37–38 | 29 |
| 39–42 | 28 |
| 43–45 | 27 |
| 46–50 | 26 |
| 51–55 | 25 |
| 56–61 | 24 |
| 62 and over | 23 |

The optional load calculation applies only where one service or feeder supplies the entire load of a dwelling unit. Where applicable, this method may be used instead of the standard calculation in Part III of Article 220.

**(C) Calculated Loads.** The calculated load to which the demand factors of Table 220.84 apply shall include the following:

(1) 33 volt-amperes/m$^2$ or 3 volt-amperes/ft$^2$ for general lighting and general-use receptacles
(2) 1500 volt-amperes for each 2-wire, 20-ampere small-appliance branch circuit and each laundry branch circuit covered in 210.11(C)(1) and (C)(2)
(3) The nameplate rating of the following:
   a. All appliances that are fastened in place, permanently connected, or located to be on a specific circuit
   b. Ranges, wall-mounted ovens, counter-mounted cooking units
   c. Clothes dryers that are not connected to the laundry branch circuit specified in item (2)
   d. Water heaters
(4) The nameplate ampere or kVA rating of all permanently connected motors not included in item (3)
(5) The larger of the air-conditioning load or the fixed electric space-heating load

## 220.85 Two Dwelling Units

Where two dwelling units are supplied by a single feeder and the calculated load under Part III of this article exceeds that for three identical units calculated under 220.84, the lesser of the two loads shall be permitted to be used.

## 220.86 Schools

The calculation of a feeder or service load for schools shall be permitted in accordance with Table 220.86 in lieu of Part III of this article where equipped with electric space heating, air conditioning, or both. The connected load to which the demand factors of Table 220.86 apply shall include all of the interior and exterior lighting, power, water heating, cooking, other loads, and the larger of the air-conditioning load or space-heating load within the building or structure.

Feeders and service conductors whose calculated load is determined by this optional calculation shall be permitted to have the neutral load determined by 220.61. Where the building or structure load is calculated by this optional method, feeders within the building or structure shall have ampacity as permitted in Part III of this article; however, the ampacity of an individual feeder shall not be required to be larger than the ampacity for the entire building.

This section shall not apply to portable classroom buildings.

Many schools add small, portable classroom buildings. The air-conditioning load in these portable classrooms must comply with Article 440, and the lighting load must be considered continuous. The demand factors in Table 220.86 do not apply to portable classrooms, because those demand factors would decrease the feeder or service size to below that required for the connected continuous load.

Table 220.86 provides a series of demand increments that are permitted to be applied to the initial calculated load for service or feeder conductors supplying the total load of a school building in lieu of the "standard calculation" covered in Part III of Article 220. The incremental steps of Table 220.86 are based on more significant reductions of the total load as the load per square foot increases. Any portion of the load exceeding 20 VA per square foot is permitted to have a 25-percent demand factor applied. This approach is similar in concept to that applied to dwellings in Table 220.42, except that the demands in Table 220.86 apply to the entire load of the building, not just the general lighting load.

To use Table 220.86, the school must have electric space heating or air conditioning, or both. The loads to which the demand factors can be applied are specified in 220.86. Feeder conductors that do not supply the entire load of the building or structure are to be calculated in accordance with Part III of Article 220. Feeder conductors supplying subdivided building loads are not required to be larger than the service or feeder conductors that supply the entire building or structure load.

To apply Table 220.86, the initial calculation based on all of the loads specified in 220.86 first is performed. Once this calculation is made, the demand factors are applied as indicated in the table.

***TABLE 220.86*** *Optional Method — Demand Factors for Feeders and Service Conductors for Schools*

| Connected Load | | Demand Factor (Percent) |
|---|---|---|
| First 33 VA/m$^2$ | (3 VA/ft$^2$) at | 100 |
| Plus, | | |
| Over 33 through 220 VA/m$^2$ | (3 through 20 VA/ft$^2$) at | 75 |
| Plus, | | |
| Remainder over 220 VA/m$^2$ | (20 VA/ft$^2$) at | 25 |

**Calculation Example 1**

Calculate the overall demand factor for a 100,000 ft$^2$ school building (based on outside dimensions and the number of stories) using Table 220.86.

*Solution*

The initial calculated load based on 220.86 is 1.5 MVA, or 1,500,000 VA.

$$1.5 \text{ MVA}/100{,}000 \text{ ft}^2 = 15 \text{ VA/ft}^2$$

First 3 VA/ft$^2$ at 100%:

$$3 \text{ VA} \times 100{,}000 = 300 \text{ kVA}$$

Remaining 8.5 VA at 75%:

$$8.5 \text{ VA} \times 100{,}000 \times 0.75 = 637.5 \text{ kVA}$$

Demand load for the school:

$$300 \text{ kVA} + 637.5 \text{ kVA} = 937.5 \text{ kVA}$$
$$937{,}500 \text{ VA}/1{,}500{,}000 \text{ VA} = 0.625$$

or 82 percent overall demand factor on entire building load obtained using Table 220.86 demand factors.

**Calculation Example 2**

Now calculate the overall demand factor for the same 100,000 ft$^2$ school building if the initial calculated load is 2.5 MVA (2,500,000 VA).

*Solution*

$$2.5 \text{ VA}/100{,}000 \text{ ft}^2 = 25 \text{ VA/ft}^2$$

First 3 VA/ft$^2$ at 100%:

$$3 \text{ VA} \times 100{,}000 = 300 \text{ kVA}$$

Next 17 VA/ft$^2$ at 75%:

$$17 \text{ VA} \times 100{,}000 \times 0.75 = 1{,}275 \text{ kVA}$$

Remaining 5 VA/ft$^2$ at 25%:

$$5 \text{ VA} \times 100{,}000 \times 0.25 = 125 \text{ kVA}$$

Demand load for the school:

$$300 + 1{,}275 + 125 = 1700 \text{ kVA } (1.7 \text{ MVA})$$
$$1.7 \text{ MVA}/2.5 \text{ MVA} = 0.68$$

or 68 percent overall demand factor on entire building load obtained using Table 220.86 demand factors.

## 220.87 Determining Existing Loads

The calculation of a feeder or service load for existing installations shall be permitted to use actual maximum demand to determine the existing load under all of the following conditions:

(1) The maximum demand data is available for a 1-year period.

*Exception: If the maximum demand data for a 1-year period is not available, the calculated load shall be permitted to be based on the maximum demand (measure of average power demand over a 15-minute period) continuously recorded over a minimum 30-day period using a recording ammeter or power meter connected to the highest loaded phase of the feeder or service, based on the initial loading at the start of the recording. The recording shall reflect the maximum demand of the feeder or service by being taken when the building or space is occupied and shall include by measurement or calculation the larger of the heating or cooling equipment load, and other loads that may be periodic in nature due to seasonal or similar conditions.*

(2) The maximum demand at 125 percent plus the new load does not exceed the ampacity of the feeder or rating of the service.

(3) The feeder has overcurrent protection in accordance with 240.4, and the service has overload protection in accordance with 230.90.

## 220.88 New Restaurants

Calculation of a service or feeder load, where the feeder serves the total load, for a new restaurant shall be permitted in accordance with Table 220.88 in lieu of Part III of this article.

The overload protection of the service conductors shall be in accordance with 230.90 and 240.4.

Feeder conductors shall not be required to be of greater ampacity than the service conductors.

Service or feeder conductors whose calculated load is determined by this optional calculation shall be permitted to have the neutral load determined by 220.61.

The demand factors in 220.88 recognize the effects of load diversity that are typical of restaurants. It also recognizes the amount of continuous loads as a percentage of the total connected load.

The National Restaurant Association, the Edison Electric Institute, and the Electric Power Research Institute based the data for 220.88 on load studies of 262 restaurants. These studies showed that the demand factors were lower for restaurants with larger connected loads. The service or feeder size is calculated by applying the appropriate demand factor from Table 220.88 to the total load kVA.

### Calculation Example 1

A new, all-electric restaurant has a total connected load of 348 kVA at 208Y/120 V. Using Table 220.88, calculate the demand load and determine the size of the service-entrance conductors and the maximum-size overcurrent device for the service.

*Solution*

STEP 1. Use the value in Table 220.88 for a connected load of 348 kVA (Row 3, Column 2) to calculate the demand load for an all-electric restaurant:

$$\begin{aligned}\text{Demand load} &= 50\% \text{ of amount over } 325 \text{ kVA} + 172.5 \text{ kVA}\\ &= (0.50 \times (348 \text{ kVA} - 325 \text{ kVA})) + 172.5 \text{ kVA}\\ &= 11.5 + 172.5\\ &= 184 \text{ kVA}\end{aligned}$$

STEP 2. Calculate the service size using the calculated demand load in Step 1:

$$\begin{aligned}\text{Service size} &= \frac{\text{kVA}_{\text{Demand load}} \times 1000}{\text{voltage}\sqrt{3}}\\ &= \frac{184 \text{ kVA} \times 1000}{208 \text{ V} \times \sqrt{3}}\\ &= 510.7 \text{ or } 511 \text{ A}\end{aligned}$$

STEP 3. Determine the size of the overcurrent device. The next larger standard-size overcurrent device is 600 A. The minimum size of the conductors must be adequate to handle the load, but 240.4(B) permits the next larger standard-rated overcurrent device to be used.

### Calculation Example 2

A new restaurant has gas cooking appliances plus a total connected electrical load of 348 kVA at 208Y/120 V. Calculate the demand load using Table 220.88 and the service size. Then determine the maximum-size overcurrent device for the service.

*Solution*

STEP 1. Calculate the demand load for a new restaurant using the value in Table 220.88 for a connected load of 348 kVA (Column 3, Row 3) as follows:

$$\begin{aligned}\text{Demand load} &= 45\% \text{ of amount over } 325 \text{ kVA} + 262.5 \text{ kVA}\\ &= (0.45 \times (348 - 325)) + 262.5\\ &= 10.35 + 262.5\\ &= 272.85 \text{ VA}\end{aligned}$$

***TABLE 220.88*** *Optional Method — Permitted Load Calculations for Service and Feeder Conductors for New Restaurants*

| Total Connected Load (kVA) | All Electric Restaurant Calculated Loads (kVA) | Not All Electric Restaurant Calculated Loads (kVA) |
|---|---|---|
| 0–200 | 80% | 100% |
| 201–325 | 10% (amount over 200) + 160.0 | 50% (amount over 200) + 200.0 |
| 326–800 | 50% (amount over 325) + 172.5 | 45% (amount over 325) + 262.5 |
| Over 800 | 50% (amount over 800) + 410.0 | 20% (amount over 800) + 476.3 |

Note: Add all electrical loads, including both heating and cooling loads, to calculate the total connected load. Select the one demand factor that applies from the table, then multiply the total connected load by this single demand factor.

STEP 2. Calculate the service size using the calculated demand load in Step 1.

$$\text{Service size} = \frac{\text{kVA}_{\text{Demand load}} \times 1000}{\text{voltage}\sqrt{3}}$$

$$= \frac{272.85\ \text{kVA} \times 1000}{208\ \text{V} \times \sqrt{3}}$$

$$= 757.36 \text{ or } 757\ \text{A}$$

STEP 3. Determine the size of the overcurrent device. The next higher standard rating for an overcurrent device, according to 240.6, is 800 A. Section 230.79 requires that the service disconnecting means have a rating that is not less than the calculated load (757 A). The minimum size of the conductors must be adequate to handle the load, but 240.4(B) permits the next larger standard-rated overcurrent device to be used.

# V. Farm Load Calculations

## 220.100 General

Farm loads shall be calculated in accordance with Part V.

## 220.102 Farm Loads — Buildings and Other Loads

**(A) Dwelling Unit.** The feeder or service load of a farm dwelling unit shall be calculated in accordance with the provisions for dwellings in Part III or IV of this article. Where the dwelling has electric heat and the farm has electric grain-drying systems, Part IV of this article shall not be used to calculate the dwelling load where the dwelling and farm loads are supplied by a common service.

**(B) Other Than Dwelling Unit.** Where a feeder or service supplies a farm building or other load having two or more separate branch circuits, the load for feeders, service conductors, and service equipment shall be calculated in accordance with demand factors not less than indicated in Table 220.102

***TABLE 220.102*** *Method for Calculating Farm Loads for Other Than Dwelling Unit*

| Ampere Load at 240 Volts Maximum | Demand Factor (%) |
|---|---|
| The greater of the following: | |
| All loads that are expected to operate simultaneously, or | 100 |
| 125 percent of the full load current of the largest motor, or | |
| First 60 amperes of the load | |
| Next 60 amperes of all other loads | 50 |
| Remainder of other loads | 25 |

***TABLE 220.103*** *Method for Calculating Total Farm Load*

| Individual Loads Calculated in Accordance with Table 220.102 | Demand Factor (%) |
|---|---|
| Largest load | 100 |
| Second largest load | 75 |
| Third largest load | 65 |
| Remaining loads | 50 |

Note: To this total load, add the load of the farm dwelling unit calculated in accordance with Part III or IV of this article. Where the dwelling has electric heat and the farm has electric grain-drying systems, Part IV of this article shall not be used to calculate the dwelling load.

## 220.103 Farm Loads — Total

Where supplied by a common service, the total load of the farm for service conductors and service equipment shall be calculated in accordance with the farm dwelling unit load and demand factors specified in Table 220.103. Where there is equipment in two or more farm equipment buildings or for loads having the same function, such loads shall be calculated in accordance with Table 220.102 and shall be permitted to be combined as a single load in Table 220.103 for calculating the total load.

# ARTICLE 225 Outside Branch Circuits and Feeders

## 225.1 Scope

This article covers requirements for outside branch circuits and feeders run on or between buildings, structures, or poles on the premises; and electrical equipment and wiring for the supply of utilization equipment that is located on or attached to the outside of buildings, structures, or poles.

Article 225 provides requirements unique to the installation of feeders and branch circuits on the outside (overhead and underground) of buildings and structures. These circuits may be supplying specific items of electrical equipment, or they may be the power supply to another building or structure. The power source for the outside feeders and/or branch circuits may one of the following:

- Supplied by a utility and subsequently distributed throughout a multibuilding industrial or institutional campus as feeder or branch circuits
- a separately derived system such as an emergency, standby system, or an alternative energy system
- a combination of the campus-style utility supplied system described above, along with the onsite power generation described above

These requirements are in addition to the general requirements for branch circuits and feeders in Articles 210 and 215.

Informational Note: For additional information on wiring over 1000 volts, see ANSI C2-2007, *National Electrical Safety Code*.

•

## 225.3 Other Articles

Application of other articles, including additional requirements to specific cases of equipment and conductors, is shown in Table 225.3.

***TABLE 225.3*** *Other Articles*

| Equipment/Conductors | Article |
|---|---|
| Branch circuits | 210 |
| Class 1, Class 2, and Class 3 remote-control, signaling, and power-limited circuits | 725 |
| Communications circuits | 800 |
| Community antenna television and radio distribution systems | 820 |
| Conductors for general wiring | 310 |
| Electrically driven or controlled irrigation machines | 675 |
| Electric signs and outline lighting | 600 |
| Feeders | 215 |
| Fire alarm systems | 760 |
| Fixed outdoor electric deicing and snow-melting equipment | 426 |
| Floating buildings | 553 |
| Grounding and bonding | 250 |
| Hazardous (classified) locations | 500 |
| Hazardous (classified) locations— specific | 510 |
| Marinas and boatyards | 555 |
| Messenger-supported wiring | 396 |
| Mobile homes, manufactured homes, and mobile home parks | 550 |
| Open wiring on insulators | 398 |
| Over 1000 volts, general | 490 |
| Overcurrent protection | 240 |
| Radio and television equipment | 810 |
| Services | 230 |
| Solar photovoltaic systems | 690 |
| Swimming pools, fountains, and similar installations | 680 |
| Use and identification of grounded conductors | 200 |

# I. General

## 225.4 Conductor Covering

Where within 3.0 m (10 ft) of any building or structure other than supporting poles or towers, open individual (aerial) overhead conductors shall be insulated for the nominal voltage. Conductors in cables or raceways, except Type MI cable, shall be of the rubber-covered type or thermoplastic type and, in wet locations, shall comply with 310.10(C). Conductors for festoon lighting shall be of the rubber-covered or thermoplastic type.

*Exception: Equipment grounding conductors and grounded circuit conductors shall be permitted to be bare or covered as specifically permitted elsewhere in this Code.*

This exception and 250.184(A)(1), Exception No. 2, correlate to permit the use of the bare messenger wire of an overhead cable assembly as the grounded (neutral) conductor of an outdoor feeder circuit. Section 250.32(B), Exception, permits a grounding electrode conductor connection to the grounded conductor at the load end of such outdoor overhead feeder circuits for existing installations. For new installations, an equipment grounding conductor is required to be included in the overhead feeder cable assembly and it is required by 250.32(B) to be connected to a grounding electrode system.

## 225.5 Size of Conductors 600 Volts, Nominal, or Less

The ampacity of outdoor branch-circuit and feeder conductors shall be in accordance with 310.15 based on loads as determined under 220.10 and Part III of Article 220.

## 225.6 Conductor Size and Support

**(A) Overhead Spans.** Open individual conductors shall not be smaller than the following:

(1) For 1000 volts, nominal, or less, 10 AWG copper or 8 AWG aluminum for spans up to 15 m (50 ft) in length, and 8 AWG

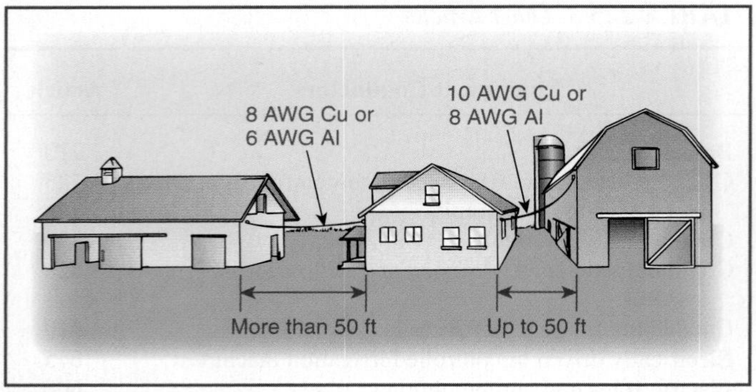


***EXHIBIT 225.1*** *Minimum sizes of conductors in overhead spans for 1000 volts, nominal, or less.*

copper or 6 AWG aluminum for a longer span unless supported by a messenger wire

(2) For over 1000 volts, nominal, 6 AWG copper or 4 AWG aluminum where open individual conductors, and 8 AWG copper or 6 AWG aluminum where in cable

The size limitation of copper and aluminum conductors for overhead spans is based on the need for adequate mechanical strength to support the weight of the conductors and to withstand wind, ice, and other similar conditions. Exhibit 225.1 illustrates overhead spans that are not messenger supported; that are run between buildings, structures, or poles; and that are 1000 volts or less. If the conductors are supported on a messenger, the messenger cable provides the necessary mechanical strength. See 396.10 for cable types permitted to be messenger supported.

**(B) Festoon Lighting.** Overhead conductors for festoon lighting shall not be smaller than 12 AWG unless the conductors are supported by messenger wires. In all spans exceeding 12 m (40 ft), the conductors shall be supported by messenger wire. The messenger wire shall be supported by strain insulators. Conductors or messenger wires shall not be attached to any fire escape, downspout, or plumbing equipment.

Article 100 defines *festoon lighting* as "a string of outdoor lights that is suspended between two points." Exhibit 225.2 illustrates an installation of festoon lighting that complies with the minimum clearance above-grade requirement specified in 225.18(1). The maximum distance between conductor supports is 40 feet unless a messenger wire is installed to support the current-carrying conductors.

Attachment of festoon lighting to fire escapes, plumbing equipment, or metal drain spouts is prohibited, because the attachment could provide a path to ground. Such methods of attachment could not be relied on for a permanent or secure means of support.

## 225.7 Lighting Equipment Installed Outdoors

**(A) General.** For the supply of lighting equipment installed outdoors, the branch circuits shall comply with Article 210 and 225.7(B) through (D).

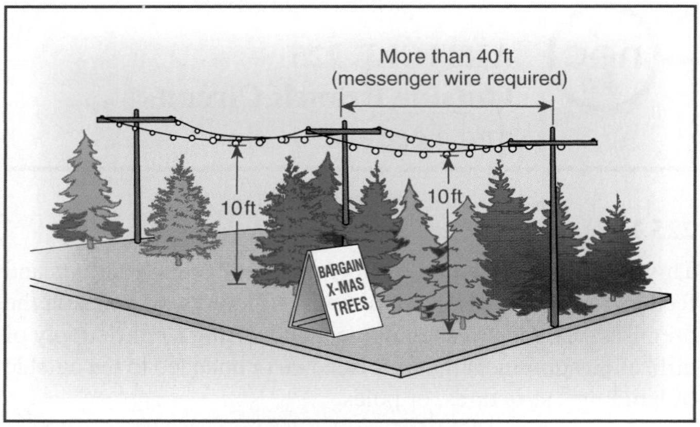


***EXHIBIT 225.2*** *Messenger wire required for festoon lighting conductors in a span exceeding 40 feet.*

**(B) Common Neutral.** The ampacity of the neutral conductor shall not be less than the maximum net calculated load current between the neutral conductor and all ungrounded conductors connected to any one phase of the circuit.

As indicated in 200.4, specific provisions in the *Code* permit the use of a single neutral conductor shared between multiple ungrounded branch-circuit conductors. For outdoor lighting, 225.7(B) permits the use of a common or shared neutral conductor for more than one branch circuit. The branch-circuit configuration can be 2-wire, line-to-neutral connected, in which multiple ungrounded conductors share a common neutral, or it can be configured in a single- or 3-phase multiwire arrangement. The neutral capacity is required to be not less than the total load of all ungrounded conductors connected to any one phase of the circuit.

Exhibit 225.3 illustrates a 120/240-V, single-phase, 3-wire system, and Exhibit 225.4 illustrates a 208Y/120-V, 3-phase, 4-wire system. In each of these illustrations, all branch circuits are rated at 20 A. The maximum unbalanced current that can occur on the

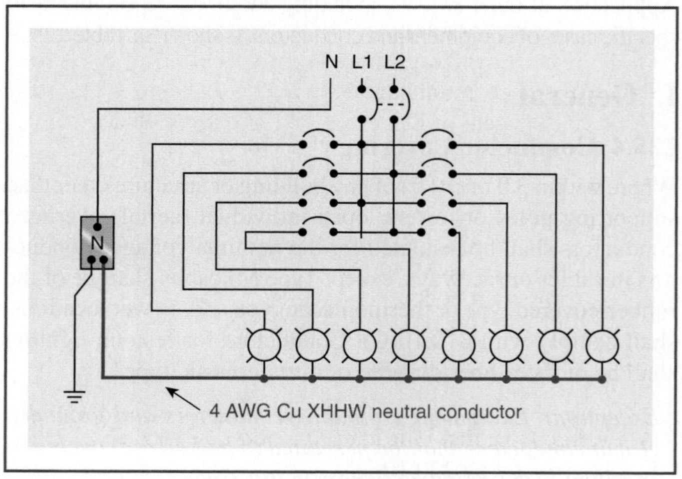


***EXHIBIT 225.3*** *A 120/240 V, single-phase, 3-wire system (maximum unbalanced current of 80 amperes is used to size the common neutral conductor).*

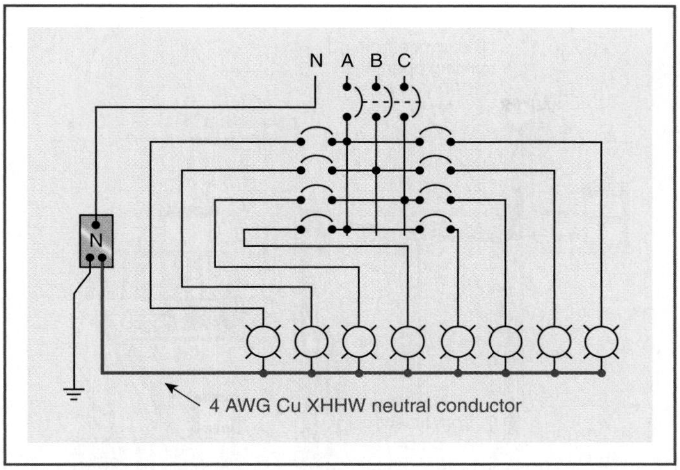


***EXHIBIT 225.4*** *A 208Y/120-V, 3-phase, 4-wire system (maximum unbalanced current is 80 amperes is used to size the common neutral conductor).*

4-wire system shown in Exhibit 225.3 is four times 20 amperes, or 80 amperes. The maximum unbalanced current that can occur on a 3-phase system with the load connected as shown in Exhibit 225.4 is 80 amperes, due to the load on phase A.

**(C) 277 Volts to Ground.** Circuits exceeding 120 volts, nominal, between conductors and not exceeding 277 volts, nominal, to ground shall be permitted to supply luminaires for illumination of outdoor areas of industrial establishments, office buildings, schools, stores, and other commercial or public buildings.

Branch circuits for the outdoor illumination of industrial, commercial, and institutional buildings are permitted to operate at a maximum voltage to ground of 277 volts as specified in 210.6. See 210.6(D) for tunnel and pole-mounted luminaires with voltages greater than 277 volts to ground. The restrictions here are in addition to those in 210.6(C).

**(D) 600 Volts Between Conductors.** Circuits exceeding 277 volts, nominal, to ground and not exceeding 600 volts, nominal, between conductors shall be permitted to supply the auxiliary equipment of electric-discharge lamps in accordance with 210.6(D)(1).

Section 210.6(D)(1) contains the minimum height requirements for circuits exceeding 277 V, nominal, to ground but not exceeding 600 V, nominal, between conductors for circuits that supply the auxiliary equipment of electric-discharge lamps.

## 225.8 Calculation of Loads 1000 Volts, Nominal, or Less

**(A) Branch Circuits.** The load on outdoor branch circuits shall be as determined by 220.10.

**(B) Feeders.** The load on outdoor feeders shall be as determined by Part III of Article 220.

## 225.10 Wiring on Buildings (or Other Structures)

The installation of outside wiring on surfaces of buildings (or other structures) shall be permitted for circuits of not over 1000 volts, nominal, as open wiring on insulators, as multiconductor cable, as Type MC cable, as Type UF cable, as Type MI cable, as messenger-supported wiring, in rigid metal conduit (RMC), in intermediate metal conduit (IMC), in rigid polyvinyl chloride (PVC) conduit, in reinforced thermosetting resin conduit (RTRC), in cable trays, as cablebus, in wireways, in auxiliary gutters, in electrical metallic tubing (EMT), in flexible metal conduit (FMC), in liquidtight flexible metal conduit (LFMC), in liquidtight flexible nonmetallic conduit (LFNC), and in busways. Circuits of over 1000 volts, nominal, shall be installed as provided in 300.37.

## 225.11 Feeder and Branch-Circuit Conductors Entering, Exiting, or Attached to Buildings or Structures

Feeder and branch-circuit conductors entering or exiting buildings or structures shall be in installed in accordance with the requirements of 230.52. Overhead branch circuits and feeders attached to buildings or structures shall be installed in accordance with the requirements of 230.54.

## 225.12 Open-Conductor Supports

Open conductors shall be supported on glass or porcelain knobs, racks, brackets, or strain insulators.

## 225.14 Open-Conductor Spacings

**(A) 1000 Volts, Nominal, or Less.** Conductors of 1000 volts, nominal, or less, shall comply with the spacings provided in Table 230.51(C).

**(B) Over 1000 Volts, Nominal.** Conductors of over 1000 volts, nominal, shall comply with the spacings provided in 110.36 and 490.24.

**(C) Separation from Other Circuits.** Open conductors shall be separated from open conductors of other circuits or systems by not less than 100 mm (4 in.).

**(D) Conductors on Poles.** Conductors on poles shall have a separation of not less than 300 mm (1 ft) where not placed on racks or brackets. Conductors supported on poles shall provide a horizontal climbing space not less than the following:

(1) Power conductors below communications conductors — 750 mm (30 in.)
(2) Power conductors alone or above communications conductors:
   a. 300 volts or less — 600 mm (24 in.)
   b. Over 300 volts — 750 mm (30 in.)
(3) Communications conductors below power conductors — same as power conductors

(4) Communications conductors alone — no requirement

Sufficient space is required for personnel to climb over or through conductors to safely work with conductors on the pole.

## 225.15 Supports over Buildings

Supports over a building shall be in accordance with 230.29.

## 225.16 Attachment to Buildings

**(A) Point of Attachment.** The point of attachment to a building shall be in accordance with 230.26.

**(B) Means of Attachment.** The means of attachment to a building shall be in accordance with 230.27.

## 225.17 Masts as Supports

Only feeder or branch-circuit conductors specified within this section shall be permitted to be attached to the feeder and/or branch-circuit mast. Masts used for the support of final spans of feeders or branch circuits shall be installed in accordance with 225.17(A) and (B).

The rules for masts supporting overhead branch circuits and feeders are similar to the requirements in 230.28 for masts for supporting service drops. The attachment requirements were revised for the 2014 *Code*. Conductors that are attached to the exterior of the mast are not permitted to be attached between the weatherhead and any coupling that is above a point of securement to the building. Attaching conductors in this space may put additional strain on the mast that could cause it to fail.

A mast supporting an overhead branch circuit or feeder span is not permitted to support conductors of other systems, such as overhead conductor spans for signaling, communications, or CATV systems.

**(A) Strength.** The mast shall be of adequate strength or be supported by braces or guys to withstand safely the strain imposed by the overhead feeder or branch-circuit conductors. Hubs intended for use with a conduit that serves as a mast for support of feeder or branch-circuit conductors shall be identified for use with a mast.

**(B) Attachment.** Feeder and/or branch-circuit conductors shall not be attached to a mast between a weatherhead or the end of the conduit and a coupling where the coupling is located above the last point of securement to the building or other structure or is located above the building or other structure.

## 225.18 Clearance for Overhead Conductors and Cables

Overhead spans of open conductors and open multiconductor cables of not over 1000 volts, nominal, shall have a clearance of not less than the following:

(1) 3.0 m (10 ft) — above finished grade, sidewalks, or from any platform or projection from which they might be reached where the voltage does not exceed 150 volts to ground and accessible to pedestrians only

(2) 3.7 m (12 ft) — over residential property and driveways, and those commercial areas not subject to truck traffic where the voltage does not exceed 300 volts to ground

(3) 4.5 m (15 ft) — for those areas listed in the 3.7-m (12-ft) classification where the voltage exceeds 300 volts to ground

(4) 5.5 m (18 ft) — over public streets, alleys, roads, parking areas subject to truck traffic, driveways on other than residential property, and other land traversed by vehicles, such as cultivated, grazing, forest, and orchard

(5) 7.5 m (24.5 ft) — over track rails of railroads

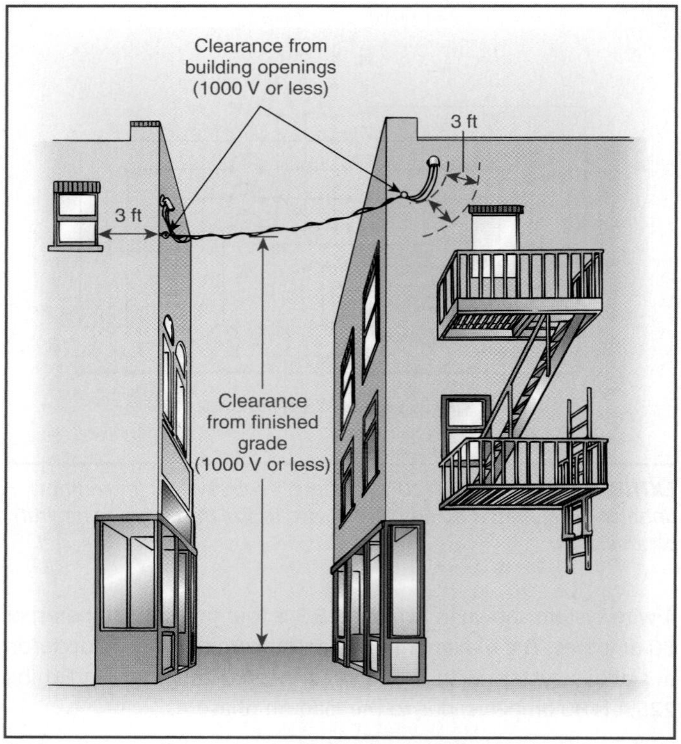


***EXHIBIT 225.5** Clearances from ground and buildings for conductors not over 1000 V.*

This section covers the requirements for clearances from ground, and 225.19 covers the requirements for clearances from buildings for conductors not over 1000 volts. Section 225.19(D) provides final span clearances from windows, doors, fire escapes, and similar areas where overhead conductors are subject to contact by persons. Exhibit 225.5 shows an example of where these clearance requirements apply.

## 225.19 Clearances from Buildings for Conductors of Not over 1000 Volts, Nominal

**(A) Above Roofs.** Overhead spans of open conductors and open multiconductor cables shall have a vertical clearance of not less than 2.5 m (8 ft) above the roof surface. The vertical clearance above the roof level shall be maintained for a distance not less than 900 mm (3 ft) in all directions from the edge of the roof.

*Exception No. 1: The area above a roof surface subject to pedestrian or vehicular traffic shall have a vertical clearance from the roof surface in accordance with the clearance requirements of 225.18.*

*Exception No. 2: Where the voltage between conductors does not exceed 300, and the roof has a slope of 100 mm in 300 mm (4 in. in 12 in.) or greater, a reduction in clearance to 900 mm (3 ft) shall be permitted.*

*Exception No. 3: Where the voltage between conductors does not exceed 300, a reduction in clearance above only the overhanging portion of the roof to not less than 450 mm (18 in.) shall be permitted if (1) not more than 1.8 m (6 ft) of the conductors, 1.2 m (4 ft) horizontally, pass above the roof overhang and (2) they are terminated at a through-the-roof raceway or approved support.*

*Exception No. 4: The requirement for maintaining the vertical clearance 900 mm (3 ft) from the edge of the roof shall not apply to the final conductor span where the conductors are attached to the side of a building.*

**(B) From Nonbuilding or Nonbridge Structures.** From signs, chimneys, radio and television antennas, tanks, and other nonbuilding or nonbridge structures, clearances — vertical, diagonal, and horizontal — shall not be less than 900 mm (3 ft).

**(C) Horizontal Clearances.** Clearances shall not be less than 900 mm (3 ft).

**(D) Final Spans.** Final spans of feeders or branch circuits shall comply with 225.19(D)(1), (D)(2), and (D)(3).

**(1) Clearance from Windows.** Final spans to the building they supply, or from which they are fed, shall be permitted to be attached to the building, but they shall be kept not less than 900 mm (3 ft) from windows that are designed to be opened, and from doors, porches, balconies, ladders, stairs, fire escapes, or similar locations.

*Exception: Conductors run above the top level of a window shall be permitted to be less than the 900-mm (3-ft) requirement.*

**(2) Vertical Clearance.** The vertical clearance of final spans above, or within 900 mm (3 ft) measured horizontally of, platforms, projections, or surfaces from which they might be reached shall be maintained in accordance with 225.18.

**(3) Building Openings.** The overhead branch-circuit and feeder conductors shall not be installed beneath openings through which materials may be moved, such as openings in farm and commercial buildings, and shall not be installed where they obstruct entrance to these buildings' openings.

**(E) Zone for Fire Ladders.** Where buildings exceed three stories or 15 m (50 ft) in height, overhead lines shall be arranged, where practicable, so that a clear space (or zone) at least 1.8 m (6 ft) wide will be left either adjacent to the buildings or beginning not over 2.5 m (8 ft) from them to facilitate the raising of ladders when necessary for fire fighting.

## 225.20 Mechanical Protection of Conductors

Mechanical protection of conductors on buildings, structures, or poles shall be as provided for services in 230.50.

## 225.21 Multiconductor Cables on Exterior Surfaces of Buildings (or Other Structures)

Supports for multiconductor cables on exterior surfaces of buildings (or other structures) shall be as provided in 230.51.

## 225.22 Raceways on Exterior Surfaces of Buildings or Other Structures

Raceways on exteriors of buildings or other structures shall be arranged to drain and shall be suitable for use in wet locations.

Raceways in exterior locations must be raintight; that is, "constructed or protected so that exposure to a beating rain will not result in the entrance of water under specified test conditions," as defined in Article 100.

As specified in 342.42(A) and 344.42(A), all conduit bodies, fittings, and boxes used in wet locations must be provided with threaded hubs or other approved means. Threadless couplings and connectors used with metal conduit or electrical metallic tubing installed on the exterior of a building must be of the raintight type.

If raceways are exposed to weather or rain through weatherhead openings, condensation is likely to occur, causing moisture to accumulate within raceways at low points of the installation and in junction boxes. Therefore, raceways are required by 225.22 to be installed so as to allow moisture to drain from the raceway through drain holes or other means provided at appropriate locations. The raintight requirement applies only to raceways installed in wet locations.

## 225.24 Outdoor Lampholders

Where outdoor lampholders are attached as pendants, the connections to the circuit wires shall be staggered. Where such lampholders have terminals of a type that puncture the insulation and make contact with the conductors, they shall be attached only to conductors of the stranded type.

Splices to branch-circuit conductors for outdoor lampholders of the Edison-base-type or "pigtail" sockets are required to be staggered so that splices will not be in close proximity to each other. Pin-type terminal sockets must be attached to stranded conductors only and are intended for installations for temporary lighting or decorations, signs, or specifically approved applications.

### 225.25 Location of Outdoor Lamps

Locations of lamps for outdoor lighting shall be below all energized conductors, transformers, or other electric utilization equipment, unless either of the following apply:

(1) Clearances or other safeguards are provided for relamping operations.
(2) Equipment is controlled by a disconnecting means that is lockable in accordance with 110.25.

The objective is to protect personnel while relamping outdoor luminaires. However, 225.18 requires a minimum clearance of 10 feet above grade or platforms for open conductors. Therefore, in some cases it may be difficult to keep all electrical equipment above the lamps. Section 225.25(1) allows other clearances or safeguards to permit safe relamping, while the reference to 110.25 provides another alternative for safe relamping through the use of a disconnecting means that can be locked in the open or off position.

### 225.26 Vegetation as Support

Vegetation such as trees shall not be used for support of overhead conductor spans.

Overhead conductor spans attached to a tree are subject to damage over the course of time as normal tree growth around the attachment device causes the mounting insulators to break. Normal growth can also cause tree bark to grow around the insulation. This requirement reduces the likelihood of chafing or degradation of the conductor insulation. These conditions can create a shock hazard for tree trimmers and tree climbers.

Outdoor luminaires and associated equipment are permitted by 410.36(G) to be supported by trees. To prevent the chafing damage, conductors are run up the tree from an underground wiring method. See 300.5(D) for requirements on the protection of direct-buried conductors emerging from below grade.

### 225.27 Raceway Seal

Where a raceway enters a building or structure from an underground distribution system, it shall be sealed in accordance with 300.5(G). Spare or unused raceways shall also be sealed. Sealants shall be identified for use with the cable insulation, conductor insulation, bare conductor, shield, or other components.

## II. Buildings or Other Structures Supplied by a Feeder(s) or Branch Circuit(s)

Part II of Article 225 covers outside branch circuits and feeders on single managed properties where outside branch circuits and feeders are the source of electrical supply for buildings and structures. Important in the application of the Part II requirements are the Article 100 definitions of *service point, service, service equipment, feeder,* and *branch circuit.* Determining what constitutes a set of feeder or branch-circuit conductors versus a set of service conductors depends on a clear understanding of where the service point is located and where the service and service equipment for a premises are located. In some cases, particularly with medium- and high-voltage distribution, the service location of a campus or multibuilding facility is a switchyard or substation. With the location of the service point and service equipment established, the requirements for outside branch circuits and feeders from Part II (and Part III if over 1000 volts) can be properly applied.

Included in Part II of Article 225 are the requirements for overhead and underground feeders that supply buildings or structures on college and other institutional campuses, multibuilding industrial facilities, multibuilding commercial facilities, and other facilities where the electrical supply is an outdoor feeder or branch circuit. The feeders and branch circuits covered in Part II of Article 225 may originate in one building or structure and supply another building or structure, or they may originate in outdoor equipment such as freestanding switchboards, switchgear, transformers, or generators and supply equipment located in buildings or structures. Such distribution is permitted under the condition that the entire premises is under a single management. Many of the requirements in Part II covering the number of feeders or branch circuits and the location and type of disconnecting means are similar to the requirements for services in Article 230.

### 225.30 Number of Supplies

A building or other structure that is served by a branch circuit or feeder on the load side of a service disconnecting means shall be supplied by only one feeder or branch circuit unless permitted in 225.30(A) through (E). For the purpose of this section, a multiwire branch circuit shall be considered a single circuit.

Where a branch circuit or feeder originates in these additional buildings or other structures, only one feeder or branch circuit shall be permitted to supply power back to the original building or structure, unless permitted in 225.30(A) through (E).

•

**(A) Special Conditions.** Additional feeders or branch circuits shall be permitted to supply the following:

(1) Fire pumps
(2) Emergency systems
(3) Legally required standby systems
(4) Optional standby systems
(5) Parallel power production systems
(6) Systems designed for connection to multiple sources of supply for the purpose of enhanced reliability

The fundamental requirement of 225.30 is that a building or structure be supplied by a single branch circuit or feeder (this is similar to the rule for a single service in 230.2). Sections 225.30(A) through (E) identify conditions under which a building or structure is permitted to be supplied by multiple sources. To address the need for increased reliability of the power source to a premises where critical operational loads (not classed as emergency or legally required standby) are supplied, 225.30(A)(6) provides a condition where multiple feeder or branch-circuit sources are permitted. Double-ended (main-tie-main) switchgear supplied by two feeders is an example of an installation covered by this provision.

**(B) Special Occupancies.** By special permission, additional feeders or branch circuits shall be permitted for either of the following:

(1) Multiple-occupancy buildings where there is no space available for supply equipment accessible to all occupants
(2) A single building or other structure sufficiently large to make two or more supplies necessary

**(C) Capacity Requirements.** Additional feeders or branch circuits shall be permitted where the capacity requirements are in excess of 2000 amperes at a supply voltage of 1000 volts or less.

**(D) Different Characteristics.** Additional feeders or branch circuits shall be permitted for different voltages, frequencies, or phases or for different uses, such as control of outside lighting from multiple locations.

**(E) Documented Switching Procedures.** Additional feeders or branch circuits shall be permitted to supply installations under single management where documented safe switching procedures are established and maintained for disconnection.

Buildings on college campuses, multibuilding industrial facilities, and multibuilding commercial facilities are permitted to be supplied by secondary loop supply (secondary selective) networks, provided that documented switching procedures are established. These switching procedures must establish a method to safely operate switches for the facility during maintenance and during alternative supply and emergency supply conditions. Keyed interlock systems are often used to reduce the likelihood of inappropriate switching procedures that could result in hazardous conditions.

## 225.31 Disconnecting Means

Means shall be provided for disconnecting all ungrounded conductors that supply or pass through the building or structure.

## 225.32 Location

The disconnecting means shall be installed either inside or outside of the building or structure served or where the conductors pass through the building or structure. The disconnecting means shall be at a readily accessible location nearest the point of entrance of the conductors. For the purposes of this section, the requirements in 230.6 shall be utilized.

*Exception No. 1: For installations under single management, where documented safe switching procedures are established and maintained for disconnection, and where the installation is monitored by qualified individuals, the disconnecting means shall be permitted to be located elsewhere on the premises.*

*Exception No. 2: For buildings or other structures qualifying under the provisions of Article 685, the disconnecting means shall be permitted to be located elsewhere on the premises.*

*Exception No. 3: For towers or poles used as lighting standards, the disconnecting means shall be permitted to be located elsewhere on the premises.*

*Exception No. 4: For poles or similar structures used only for support of signs installed in accordance with Article 600, the disconnecting means shall be permitted to be located elsewhere on the premises.*

The basic requirement for locating the disconnecting means for a feeder or branch circuit supplying a structure is essentially the same as that specified for services in 230.70(A) with an important difference. Unlike a premises supplied by a service – where a building or structure is supplied by a feeder or branch circuit – a feeder or branch-circuit disconnecting means must always be at the building or structure supplied, unless one of the conditions in Exception No. 1 through Exception No. 4 to 225.32 can be applied.

As is the case with many campus-style facilities supplied by a single utility service, the service disconnecting means may be remote from the buildings or structures supplied. In such installations, the supply conductors to the buildings or structures are feeders or branch circuits, and the main requirement of this section is that the feeder or branch-circuit disconnecting means is to be located inside or outside the building or structure supplied, at the point nearest to where the supply conductors enter the building or structure. This applies to conductors that supply a building and to conductors that pass through a building.

This requirement ensures that, where the service disconnecting means is remote from a building or structure, a disconnecting means for the feeder or branch circuit is located at the building or structure to facilitate ready disconnection of the power. Outside disconnecting means are not required to be physically attached to the building or structure supplied, as would be the case with feeder-supplied outside freestanding switchgear located at the building or structure. In addition, this disconnecting means requirement is modified by the rules in 700.12(B)(6), 701.12(B)(5), and 702.12 for emergency, legally required standby, and optional standby feeders supplied by an outdoor generator set.

Exhibit 225.6 shows an example of a disconnecting means for a generator feeder that can be used to meet the requirements of Article 225, Article 445, and Articles 700, 701, or 702 as applicable. The disconnecting means is required to be "within sight" of the building or structure supplied by the generator. A conditional exception to 700.12(B)(6) permits the disconnecting means for an outdoor generator supplying an emergency system to be located other than within sight of the building(s) or structure(s) supplied by the generator. This exception contains the same conditions that are used in 225.32, Exception No. 1.

***EXHIBIT 225.6*** *Disconnecting means for alternate source feeder supplying emergency, legally required standby, or optional standby loads or a combination of the three.*

## 225.33 Maximum Number of Disconnects

**(A) General.** The disconnecting means for each supply permitted by 225.30 shall consist of not more than six switches or six circuit breakers mounted in a single enclosure, in a group of separate enclosures, or in or on a switchboard or switchgear. There shall be no more than six disconnects per supply grouped in any one location.

*Exception: For the purposes of this section, disconnecting means used solely for the control circuit of the ground-fault protection system, or the control circuit of the power-operated supply disconnecting means, installed as part of the listed equipment, shall not be considered a supply disconnecting means.*

**(B) Single-Pole Units.** Two or three single-pole switches or breakers capable of individual operation shall be permitted on multiwire circuits, one pole for each ungrounded conductor, as one multipole disconnect, provided they are equipped with identified handle ties or a master handle to disconnect all ungrounded conductors with no more than six operations of the hand.

## 225.34 Grouping of Disconnects

**(A) General.** The two to six disconnects as permitted in 225.33 shall be grouped. Each disconnect shall be marked to indicate the load served.

*Exception: One of the two to six disconnecting means permitted in 225.33, where used only for a water pump also intended to provide fire protection, shall be permitted to be located remote from the other disconnecting means.*

**(B) Additional Disconnecting Means.** The one or more additional disconnecting means for fire pumps or for emergency, legally required standby or optional standby system permitted by 225.30 shall be installed sufficiently remote from the one to six disconnecting means for normal supply to minimize the possibility of simultaneous interruption of supply.

## 225.35 Access to Occupants

In a multiple-occupancy building, each occupant shall have access to the occupant's supply disconnecting means.

*Exception: In a multiple-occupancy building where electric supply and electrical maintenance are provided by the building management and where these are under continuous building management supervision, the supply disconnecting means supplying more than one occupancy shall be permitted to be accessible to authorized management personnel only.*

## 225.36 Type

The disconnecting means specified in 225.31 shall be comprised of a circuit breaker, molded case switch, general-use switch, snap switch, or other approved means. Where applied in accordance with 250.32(B), Exception, the disconnecting means shall be suitable for use as service equipment.

This section was revised for the 2014 *Code.* The feeder or branch-circuit disconnecting means is required to be suitable for use as service equipment only where the feeder grounded conductor is also used as the return path for ground-fault current per 250.32(B)(1), Exceptions No. 1 and No. 2. The exception permitting a three- or four-way snap switch to be used as a disconnecting means for an outside branch circuit or feeder was deleted because it would not provide a positive indication that the circuit was disconnected.

•

## 225.37 Identification

Where a building or structure has any combination of feeders, branch circuits, or services passing through it or supplying it, a permanent plaque or directory shall be installed at each feeder and branch-circuit disconnect location denoting all other services, feeders, or branch circuits supplying that building or structure or passing through that building or structure and the area served by each.

The requirement of 225.37 correlates with 230.2(E) in that where a building has multiple sources of supply, permanent identification at each supply (service, feeder, and branch circuit) disconnecting means is required. The term *permanent* indicates that this required identification must have long-term durability. This identification is an important safety feature in an emergency condition, because in many cases first responders are not familiar with the electrical distribution system of a facility. In addition to 225.37 and 230.2(E), identification of power sources is covered by the requirements of 700.7(A) for emergency sources, 701.7(A) for legally required standby sources, 702.7(A) for optional standby sources, and 705.10 for parallel power production sources.

*Exception No. 1: A plaque or directory shall not be required for large-capacity multibuilding industrial installations under single management, where it is ensured that disconnection can be accomplished by establishing and maintaining safe switching procedures.*

*Exception No. 2: This identification shall not be required for branch circuits installed from a dwelling unit to a second building or structure.*

### 225.38 Disconnect Construction

Disconnecting means shall meet the requirements of 225.38(A) through (D).

•

**(A) Manually or Power Operable.** The disconnecting means shall consist of either (1) a manually operable switch or a circuit breaker equipped with a handle or other suitable operating means or (2) a power-operable switch or circuit breaker, provided the switch or circuit breaker can be opened by hand in the event of a power failure.

**(B) Simultaneous Opening of Poles.** Each building or structure disconnecting means shall simultaneously disconnect all ungrounded supply conductors that it controls from the building or structure wiring system.

**(C) Disconnection of Grounded Conductor.** Where the building or structure disconnecting means does not disconnect the grounded conductor from the grounded conductors in the building or structure wiring, other means shall be provided for this purpose at the location of the disconnecting means. A terminal or bus to which all grounded conductors can be attached by means of pressure connectors shall be permitted for this purpose.

In a multisection switchboard or switchgear, disconnects for the grounded conductor shall be permitted to be in any section of the switchboard or switchgear, provided that any such switchboard section or switchgear section is marked.

**(D) Indicating.** The building or structure disconnecting means shall plainly indicate whether it is in the open or closed position.

### 225.39 Rating of Disconnect

The feeder or branch-circuit disconnecting means shall have a rating of not less than the calculated load to be supplied, determined in accordance with Parts I and II of Article 220 for branch circuits, Part III or IV of Article 220 for feeders, or Part V of Article 220 for farm loads. Where the branch circuit or feeder disconnecting means consists of more than one switch or circuit breaker, as permitted by 225.33, combining the ratings of all the switches or circuit breakers for determining the rating of the disconnecting means shall be permitted. In no case shall the rating be lower than specified in 225.39(A), (B), (C), or (D).

**(A) One-Circuit Installation.** For installations to supply only limited loads of a single branch circuit, the branch circuit disconnecting means shall have a rating of not less than 15 amperes.

**(B) Two-Circuit Installations.** For installations consisting of not more than two 2-wire branch circuits, the feeder or branch-circuit disconnecting means shall have a rating of not less than 30 amperes.

**(C) One-Family Dwelling.** For a one-family dwelling, the feeder disconnecting means shall have a rating of not less than 100 amperes, 3-wire.

**(D) All Others.** For all other installations, the feeder or branch-circuit disconnecting means shall have a rating of not less than 60 amperes.

### 225.40 Access to Overcurrent Protective Devices

Where a feeder overcurrent device is not readily accessible, branch-circuit overcurrent devices shall be installed on the load side, shall be mounted in a readily accessible location, and shall be of a lower ampere rating than the feeder overcurrent device.

## III. Over 1000 Volts

### 225.50 Sizing of Conductors

The sizing of conductors over 1000 volts shall be in accordance with 210.19(B) for branch circuits and 215.2(B) for feeders.

### 225.51 Isolating Switches

Where oil switches or air, oil, vacuum, or sulfur hexafluoride circuit breakers constitute a building disconnecting means, an isolating switch with visible break contacts and meeting the requirements of 230.204(B), (C), and (D) shall be installed on the supply side of the disconnecting means and all associated equipment.

*Exception: The isolating switch shall not be required where the disconnecting means is mounted on removable truck panels or switchgear units that cannot be opened unless the circuit is disconnected and that, when removed from the normal operating position, automatically disconnect the circuit breaker or switch from all energized parts.*

### 225.52 Disconnecting Means

**(A) Location.** A building or structure disconnecting means shall be located in accordance with 225.32, or, if not readily accessible, it shall be operable by mechanical linkage from a readily accessible point. For multibuilding industrial installations under single management, it shall be permitted to be electrically operated by a readily accessible, remote-control device in a separate building or structure.

**(B) Type.** Each building or structure disconnect shall simultaneously disconnect all ungrounded supply conductors it controls and shall have a fault-closing rating not less than the maximum available short-circuit current available at its supply terminals.

*Exception: Where the individual disconnecting means consists of fused cutouts, the simultaneous disconnection of all ungrounded supply conductors shall not be required if there is a means to disconnect the load before opening the cutouts.*

*A permanent legible sign shall be installed adjacent to the fused cutouts and shall read DISCONNECT LOAD BEFORE OPENING CUTOUTS.*

Where fused switches or separately mounted fuses are installed, the fuse characteristics shall be permitted to contribute to the fault closing rating of the disconnecting means.

The requirement for a disconnecting means for over-1000 V feeders to buildings or structures is similar to the requirements found in 230.205(B). This disconnect can be an air, oil, sulfur hexafluoride, or vacuum breaker or switch. Where a switch is used, fuses are permitted to help with the fault-closing capability of the switch. Using fused load-break cutouts to switch sections of overhead lines and using load-break elbows to switch sections of underground lines are permitted. However, the building disconnecting means must be gang-operated to simultaneously open and close all ungrounded supply conductors. Load-break elbows and fused cutouts cannot be used as the building disconnecting means.

**(C) Locking.** Disconnecting means shall be lockable in accordance with 110.25.

*Exception: Where an individual disconnecting means consists of fused cutouts, a suitable enclosure capable of being locked and sized to contain all cutout fuse holders shall be installed at a convenient location to the fused cutouts*

**(D) Indicating.** Disconnecting means shall clearly indicate whether they are in the open "off" or closed "on" position.

**(E) Uniform Position.** Where disconnecting means handles are operated vertically, the "up" position of the handle shall be the "on" position.

*Exception: A switching device having more than one "on" position, such as a double throw switch, shall not be required to comply with this requirement.*

**(F) Identification.** Where a building or structure has any combination of feeders, branch circuits, or services passing through or supplying it, a permanent plaque or directory shall be installed at each feeder and branch-circuit disconnect location that denotes all other services, feeders, or branch circuits supplying that building or structure or passing through that building or structure and the area served by each.

## 225.56 Inspections and Tests

**(A) Pre-Energization and Operating Tests.** The complete electrical system design, including settings for protective, switching, and control circuits, shall be prepared in advance and made available on request to the authority having jurisdiction and shall be performance tested when first installed on-site. Each protective, switching, and control circuit shall be adjusted in accordance with the system design and tested by actual operation using current injection or equivalent methods as necessary to ensure that each and every such circuit operates correctly to the satisfaction of the authority having jurisdiction.

To ensure proper operation of protective and switching devices and of control circuits, outdoor feeder and branch-circuit installations rated over 1000 volts are required to be performance tested prior to being energized. The AHJ must be satisfied that the tests demonstrate proper operation. Section 225.56(B) requires that a report of all tests performed in accordance with 225.56(A) be provided to the AHJ prior to energizing the system. In addition, revisions to the 2014 *Code* require more detailed information about the electrical system design that must be provided to the AHJ upon request. Adjustments of settings must be in accordance with the electrical system design.

**(1) Instrument Transformers.** All instrument transformers shall be tested to verify correct polarity and burden.

**(2) Protective Relays.** Each protective relay shall be demonstrated to operate by injecting current or voltage, or both, at the associated instrument transformer output terminal and observing that the associated switching and signaling functions occur correctly and in proper time and sequence to accomplish the protective function intended.

**(3) Switching Circuits.** Each switching circuit shall be observed to operate the associated equipment being switched.

**(4) Control and Signal Circuits.** Each control or signal circuit shall be observed to perform its proper control function or produce a correct signal output.

**(5) Metering Circuits.** All metering circuits shall be verified to operate correctly from voltage and current sources in a similar manner to protective relay circuits.

**(6) Acceptance Tests.** Complete acceptance tests shall be performed, after the substation installation is completed, on all assemblies, equipment, conductors, and control and protective systems, as applicable, to verify the integrity of all the systems.

**(7) Relays and Metering Utilizing Phase Differences.** All relays and metering that use phase differences for operation shall be verified by measuring phase angles at the relay under actual load conditions after operation commences.

**(B) Test Report.** A test report covering the results of the tests required in 225.56(A) shall be delivered to the authority having jurisdiction prior to energization.

Informational Note: For an example of acceptance specifications, see NETA ATS-2007, *Acceptance Testing Specifications for Electrical Power Distribution Equipment and Systems*, published by the InterNational Electrical Testing Association.

***TABLE 225.60*** *Clearances over Roadways, Walkways, Rail, Water, and Open Land*

| Location | Clearance m | Clearance ft |
|---|---|---|
| Open land subject to vehicles, cultivation, or grazing | 5.6 | 18.5 |
| Roadways, driveways, parking lots, and alleys | 5.6 | 18.5 |
| Walkways | 4.1 | 13.5 |
| Rails | 8.1 | 26.5 |
| Spaces and ways for pedestrians and restricted traffic | 4.4 | 14.5 |
| Water areas not suitable for boating | 5.2 | 17.0 |

## 225.60 Clearances over Roadways, Walkways, Rail, Water, and Open Land

**(A) 22 kV, Nominal, to Ground or Less.** The clearances over roadways, walkways, rail, water, and open land for conductors and live parts up to 22 kV, nominal, to ground or less shall be not less than the values shown in Table 225.60.

**(B) Over 22 kV Nominal to Ground.** Clearances for the categories shown in Table 225.60 shall be increased by 10 mm (0.4 in.) per kV above 22,000 volts.

**(C) Special Cases.** For special cases, such as where crossings will be made over lakes, rivers, or areas using large vehicles such as mining operations, specific designs shall be engineered considering the special circumstances and shall be approved by the authority having jurisdiction.

Informational Note: For additional information, see ANSI C2-2007, *National Electrical Safety Code.*

## 225.61 Clearances over Buildings and Other Structures

**(A) 22 kV Nominal to Ground or Less.** The clearances over buildings and other structures for conductors and live parts up to 22 kV, nominal, to ground or less shall be not less than the values shown in Table 225.61.

**(B) Over 22 kV Nominal to Ground.** Clearances for the categories shown in Table 225.61 shall be increased by 10 mm (0.4 in.) per kV above 22,000 volts.

Informational Note: For additional information, see ANSI C2-2007, *National Electrical Safety Code.*

Section 225.61 and its associated table provide clearance requirements and specific distances over buildings and structures that correlate with requirements in the *National Electrical Safety Code (NESC).*

***TABLE 225.61*** *Clearances over Buildings and Other Structures*

| Clearance from Conductors or Live Parts from: | Horizontal m | Horizontal ft | Vertical m | Vertical ft |
|---|---|---|---|---|
| Building walls, projections, and windows | 2.3 | 7.5 | — | — |
| Balconies, catwalks, and similar areas accessible to people | 2.3 | 7.5 | 4.1 | 13.5 |
| Over or under roofs or projections not readily accessible to people | — | — | 3.8 | 12.5 |
| Over roofs accessible to vehicles but not trucks | — | — | 4.1 | 13.5 |
| Over roofs accessible to trucks | — | — | 5.6 | 18.5 |
| Other structures | 2.3 | 7.5 | — | — |

# ARTICLE 230 Services

## 230.1 Scope

This article covers service conductors and equipment for control and protection of services and their installation requirements.

Informational Note: See Figure 230.1.

# I. General

## 230.2 Number of Services

A building or other structure served shall be supplied by only one service unless permitted in 230.2(A) through (D). For the purpose of 230.40, Exception No. 2 only, underground sets of conductors, 1/0 AWG and larger, running to the same location and connected together at their supply end but not connected together at their load end shall be considered to be supplying one service.

The general requirement is for a building or structure to be supplied by only one service. However, under some conditions, a single service may not be adequate. Therefore, the installation of additional services is permitted as specified in conditions covered by 230.2(A) through (D). Where more than one service (or combination of service, feeder, and branch circuit) is installed, 230.2(E) requires that a permanent plaque or directory with the pertinent information on the multiple sources of supply be located at each supply source disconnecting means. This information is very important to first responders or other persons who need to disconnect the building or structure from all of its supplies.

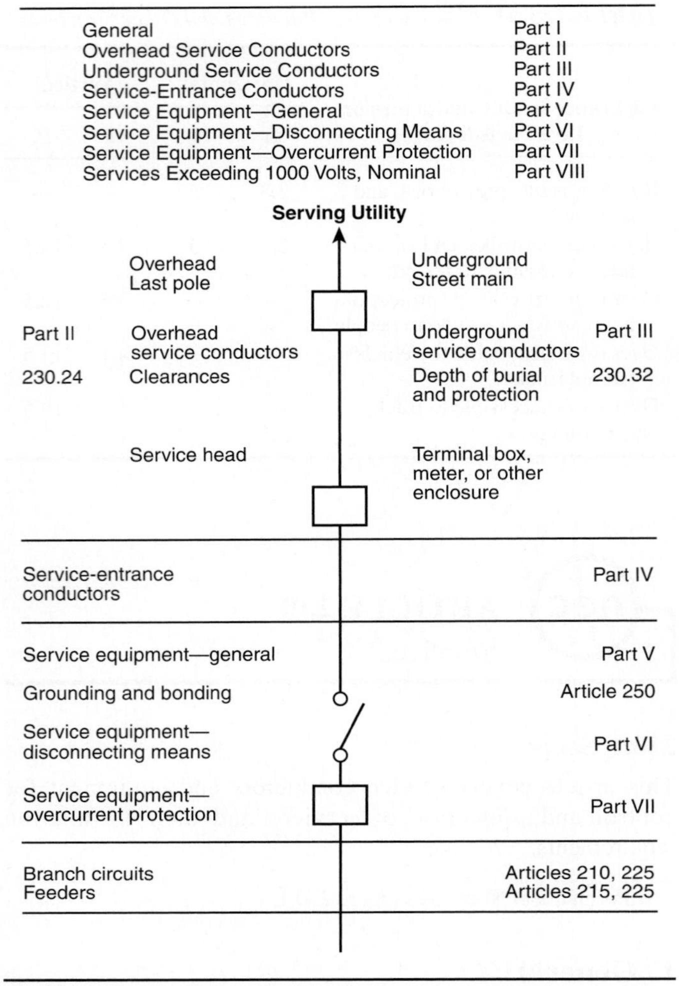


***FIGURE 230.1*** *Services.*

**(A) Special Conditions.** Additional services shall be permitted to supply the following:

(1) Fire pumps
(2) Emergency systems
(3) Legally required standby systems
(4) Optional standby systems
(5) Parallel power production systems
(6) Systems designed for connection to multiple sources of supply for the purpose of enhanced reliability

Allowing for completely separate services to the systems and equipment covered in 230.2(A) increases the overall reliability of the power supply to those systems by negating the impact of a disruption of the main building service. Unless the separate service is supplied from a different utility circuit, no protection is provided against an overall utility system outage. However, the effect of an outage resulting from a problem between the service point and the "normal," or main, building service equipment will be limited to the normal service. Continuity of power to the portion of the electrical system supplied by the additional service will be maintained.

**(B) Special Occupancies.** By special permission, additional services shall be permitted for either of the following:

(1) Multiple-occupancy buildings where there is no available space for service equipment accessible to all occupants
(2) A single building or other structure sufficiently large to make two or more services necessary

Additional services for certain occupancies may be allowed by special permission (written consent of the AHJ; see the definition of *authority having jurisdiction* in Article 100). The expansion of buildings, shopping centers, and industrial plants often necessitates the addition of one or more services. It may be impractical or impossible to install one service for an industrial plant with sufficient capacity for any and all future loads. It is also impractical to run extremely long feeders. In order to apply either of the requirements in 230.2(B), written consent from the AHJ is needed. Therefore, the AHJ should be consulted early in the planning stages to ascertain whether the special permission can be obtained.

**(C) Capacity Requirements.** Additional services shall be permitted under any of the following:

(1) Where the capacity requirements are in excess of 2000 amperes at a supply voltage of 1000 volts or less
(2) Where the load requirements of a single-phase installation are greater than the serving agency normally supplies through one service
(3) By special permission

Where two or more services are needed, 230.2(C) does not require that each service be rated 2000 amperes or that there be one service rated 2000 amperes and the additional service(s) be rated for the calculated load in excess of 2000 amperes. For example, a building with a calculated load of 2300 amperes could have two 1200-ampere services. Additional services for lesser loads are also allowed by special permission.

Many electric power companies have specifications for and have adopted special regulations covering certain types of electrical loads and service equipment. Before electrical services for large buildings and facilities are designed, the serving utility should be consulted to determine line and transformer capacities.

**(D) Different Characteristics.** Additional services shall be permitted for different voltages, frequencies, or phases, or for different uses, such as for different rate schedules.

An example of different service characteristics is a facility served by a 3-wire, 120/240-V, single-phase service and a 3-phase, 4-wire, 480Y/277-V service. Where rate schedules are different, a second service is permitted for supplying a second meter on a different rate. Curtailable loads, interruptible loads, electric heating, and electric water heating are examples of loads that may be on a different rate schedule.

**(E) Identification.** Where a building or structure is supplied by more than one service, or any combination of branch circuits, feeders, and services, a permanent plaque or directory shall be installed at each service disconnect location denoting all other services, feeders, and branch circuits supplying that building or structure and the area served by each. See 225.37.

The permanent plaque or directory must indicate where the other disconnects that feed the building are located, as illustrated in Exhibit 230.1. All the other services on or in the building or structure and the area served by each must also be noted on the plaques or directories. The plaques or directories must be of sufficient durability to withstand the environment in which the service disconnecting means is installed. It must be able to convey its information for as long as the service at which it has been installed is operational. See the commentary following 225.37 for further information on identification of multiple supply sources to a building or structure.

Exhibits 230.2 through 230.13 illustrate examples of service configurations permitted by 230.2(B) and (C); 230.40, Exceptions No. 1 and No. 2; 230.71; and 230.72. The exhibits depict a number of service arrangements where more than a single service is – or where multiple sets of service-entrance conductors are – used to supply buildings with one or more than one occupancy. Other service arrangements not depicted in these illustrations are acceptable under 230.2(A) through (D). The exhibits also illustrate the 230.71(A) requirement that not more than six service disconnecting means be grouped in one location; defining and determining this "one location" is a matter for the AHJ.

Additionally, a building may be supplied by a service and also by a feeder. This commonly occurs where the building is supplied with emergency, legally required standby, and/or optional standby power. Such installations are subject to the requirements of this article as well as Articles 225, 700, 701, and 702 as applicable.

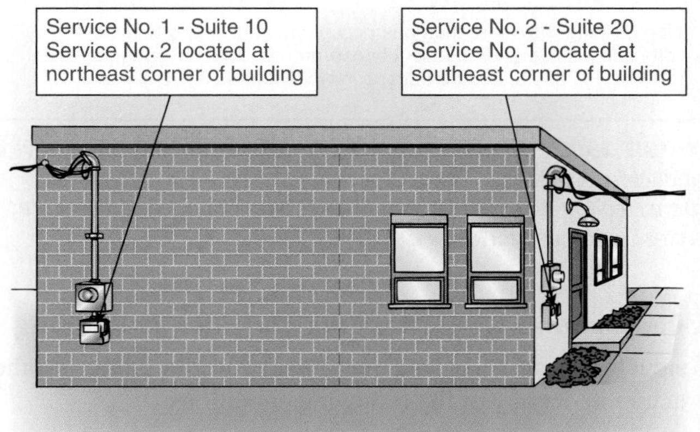


***EXHIBIT 230.1*** *An example of two separate services installed at one building with permanent plaques or directories at each service disconnecting means location containing information describing all other services and the area served by each.*

One building with more than one occupancy served by two separate services, per 230.2(B)(1) and 230.71(A)

***EXHIBIT 230.2*** *Two service drops or two sets of overhead service conductors supplying two services, installed at separate locations for a building where there is no available space for service equipment accessible to all occupants.*

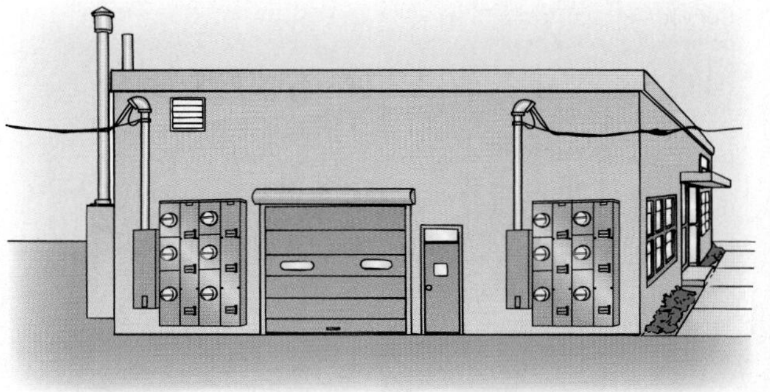

A single-occupancy building with more than one service supplied by two services, per 230.2(C)(1) and 230.71(A)

***EXHIBIT 230.3*** *Two service drops or two sets of overhead service conductors supplying two services, installed at separate locations for a building with capacity requirements exceeding 2000 amperes.*

## 230.3 One Building or Other Structure Not to Be Supplied Through Another

Service conductors supplying a building or other structure shall not pass through the interior of another building or other structure.

Although service conductors that supply one building or structure are prohibited from being run through the interior of another building or structure, service conductors are permitted to be installed along the exterior of one building to supply another building. Each building served in this manner is required to be provided with a disconnecting means for all ungrounded conductors, in accordance with Part VI.

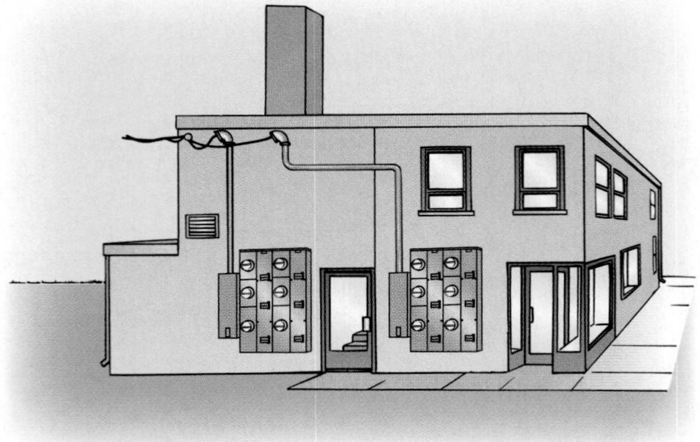

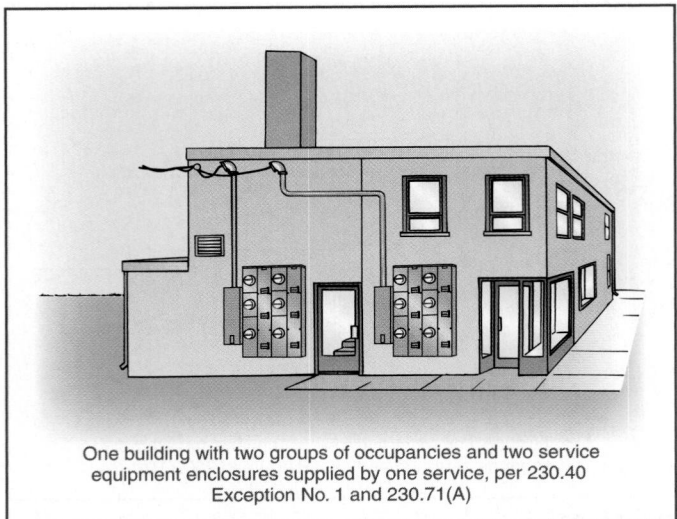

One building with two groups of occupancies and two service equipment enclosures supplied by one service, per 230.40 Exception No. 1 and 230.71(A)

***EXHIBIT 230.4*** *One service drop or one set of overhead service conductors supplying two service equipment enclosures installed at separate locations.*

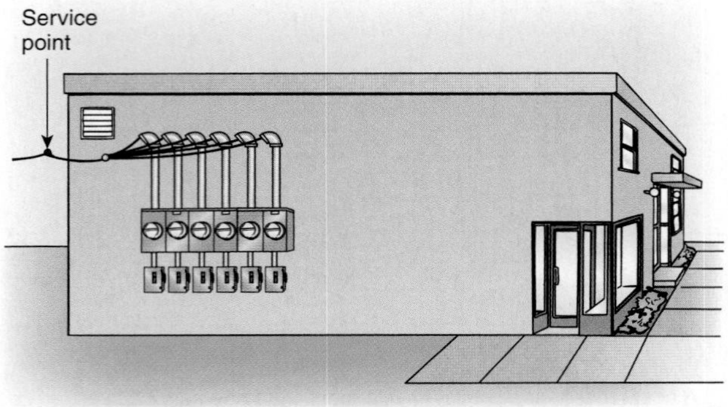


One building with more than one occupancy with separate loads fed by one service, per 230.40 Exception No. 2 and 230.71(A)

***EXHIBIT 230.5*** *One set of overhead service conductors supplying a maximum of six separate service disconnecting means enclosures.*

For example, in Exhibit 230.14, the service disconnecting means shown for Building No. 1 and Building No. 2 are located on the exterior walls. A disconnecting means suitable for use as service equipment is provided for each building. The prohibition against passing through one building en route to another applies only to service conductors. It does not prohibit feeders or branch-circuit conductors from running through the interior of a building, exiting that building, and continuing on to supply a separate building or structure. A significant difference between service conductors and feeder and branch-circuit conductors is that the feeders and branch circuits are generally provided with overcurrent protection at the point they receive their supply unless otherwise permitted by 240.21.

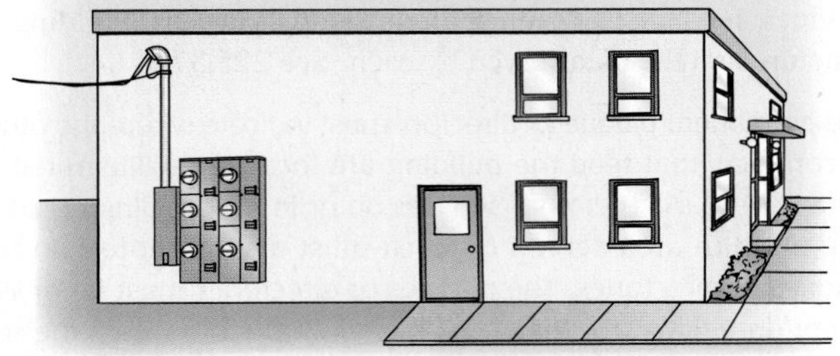

One building with more than one occupancy with separate loads fed by one service, per 230.2 and 230.71(A)

***EXHIBIT 230.6*** *One service drop or one set of overhead service conductors supplying a single service equipment enclosure. (Optional arrangement to Exhibit 230.5.)*

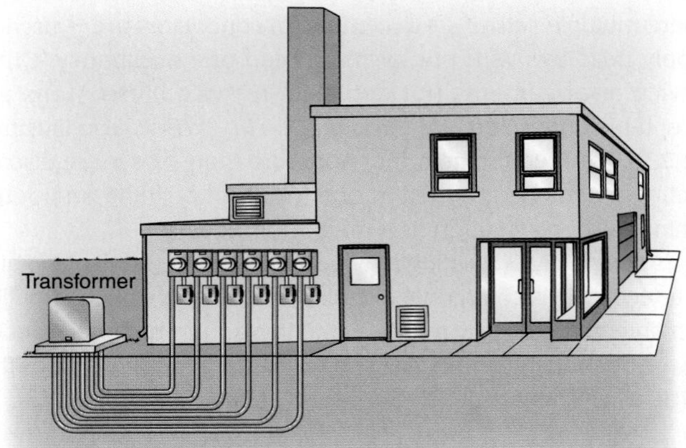


One building with more than one occupancy with separate loads fed by one service lateral or one set of underground service conductors, per 230.2, 230.40 Exception No. 2, and 230.71(A)

***EXHIBIT 230.7*** *One service lateral or one set of underground service conductors consisting of six sets of conductors 1/0 AWG or larger (connected together at their supply end), terminating in six separate service equipment enclosures.*

## 230.6 Conductors Considered Outside the Building

Conductors shall be considered outside of a building or other structure under any of the following conditions:

(1) Where installed under not less than 50 mm (2 in.) of concrete beneath a building or other structure
(2) Where installed within a building or other structure in a raceway that is encased in concrete or brick not less than 50 mm (2 in.) thick

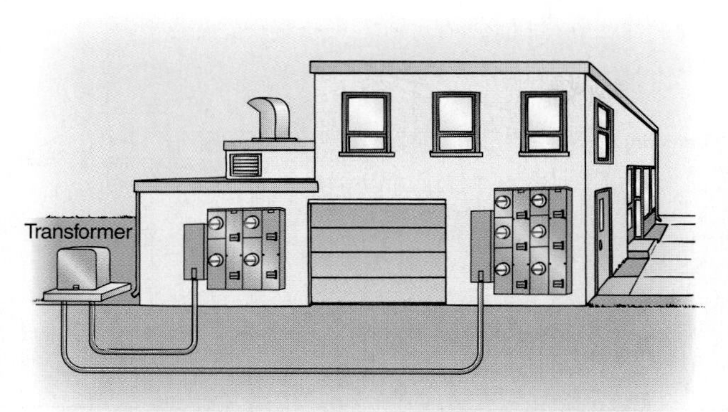


One building with more than one occupancy with separate loads each supplied by a separate service, per 230.2(B)(1) and 230.71(A)

***EXHIBIT 230.8*** *Two service laterals or two sets of underground service conductors, terminating in two service equipment enclosures installed at separate locations.*

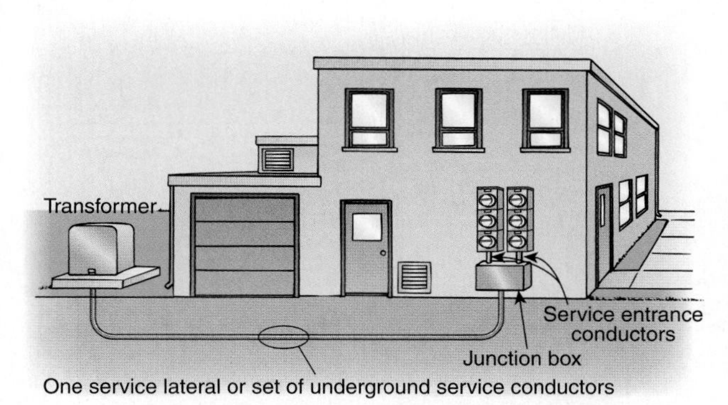


One building with more than one occupancy with separate loads supplied by one service, per 230.40 Exception No. 1 and 230.71(A)

***EXHIBIT 230.9*** *One service lateral or set of underground service conductors supplying two sets of service-entrance conductors, terminating in two service equipment enclosures grouped in one location in which the combined number of service disconnecting means in the two enclosures cannot exceed six.*

Exhibit 230.15 illustrates two of the conditions that permit service conductors to be considered "outside" a building.

Service conductors installed in an interior vault complying with the construction requirements of 450.41 through 450.48 are considered to be outside of a building. The vault does not have to contain a transformer but does have to meet the vault construction requirements contained in Part III of Article 450. Once the conductors leave the vault, the service disconnecting means has to be installed as required by 230.70(A)(1). Service conductors installed under 18 inches of earth beneath the building are also considered outside the building according to 230.6(4). An example of this provision is a building or structure built on piers with

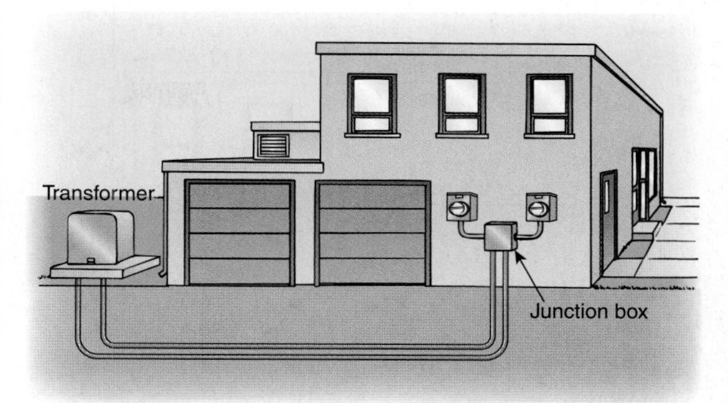


One building with more than one occupancy with separate loads each supplied by a separate service, per 230.2, 230.40 Exception No. 2, and 230.71(A)

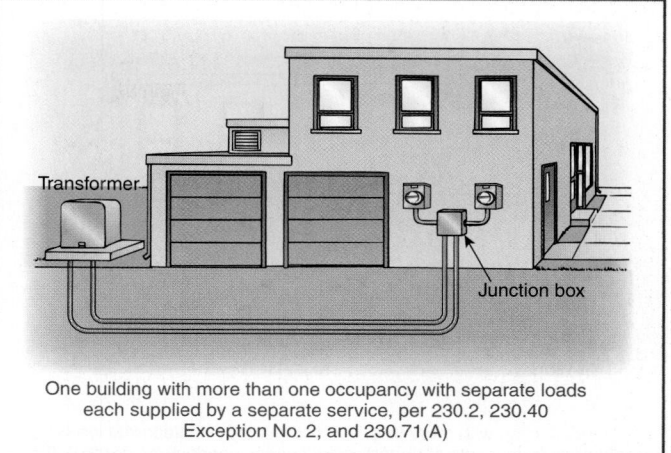

***EXHIBIT 230.10*** *Two service laterals or two sets of underground service conductors, each consisting of conductors 1/0 AWG or larger, supplying two sets of service-entrance conductors terminating in two service equipment enclosures grouped in one location in which the combined number of service disconnecting means in the two enclosures cannot exceed six.*

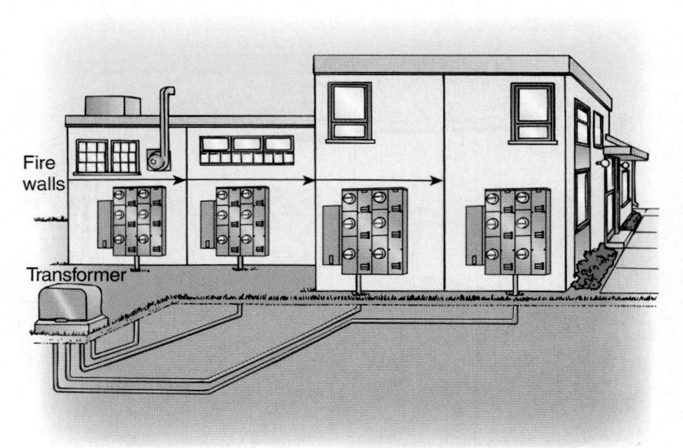


Four services to four separate buildings supplied by four underground service conductors, per 230.2 and the Article 100 definition of *Building*

***EXHIBIT 230.11*** *Four service laterals or four sets of underground service conductors supplying four service equipment enclosures installed at separate locations on a contiguous structure. Note presence of firewalls.*

service conductors buried beneath and running to a readily accessible service disconnecting means located within the interior of the building or structure. Service conductors passing through a roof overhang as covered in 230.24(A), Exception No. 3, and depicted in Exhibit 230.20 are considered to be outside the building per 230.6(5).

(3) Where installed in any vault that meets the construction requirements of Article 450, Part III

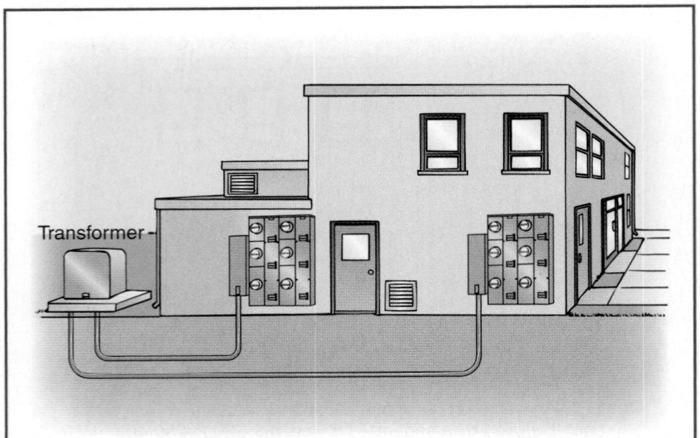


One building with more than one occupancy with separate loads supplied by two sets of underground service conductors, per 230.40 Exception No. 1 and 230.71(A)

***EXHIBIT 230.12*** *Two service laterals or two sets of underground service conductors supplying two service equipment enclosures installed at separate locations.*

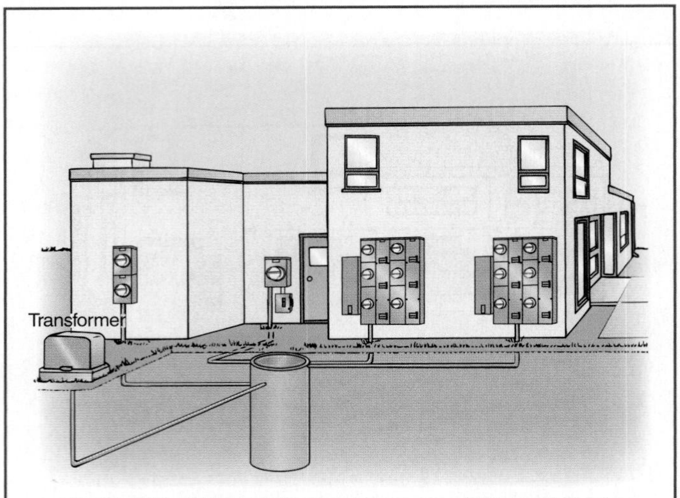


One building with more than one occupancy with separate loads supplied by one set of underground service conductors, per 230.40 Exception No.1 and 230.71(A)

***EXHIBIT 230.13*** *One service lateral or one set of underground service conductors supplying four service equipment enclosures installed at different locations.*

(4) Where installed in conduit and under not less than 450 mm (18 in.) of earth beneath a building or other structure

(5) Where installed within rigid metal conduit (Type RMC) or intermediate metal conduit (Type IMC) used to accommodate the clearance requirements in 230.24 and routed directly through an eave but not a wall of a building.

This section was revised for the 2014 *Code.* Where properly installed, rigid metal conduit and intermediate metal conduit can provide the strength and durability necessary to withstand the long-term exposure to wind and icing that a service mast could encounter.

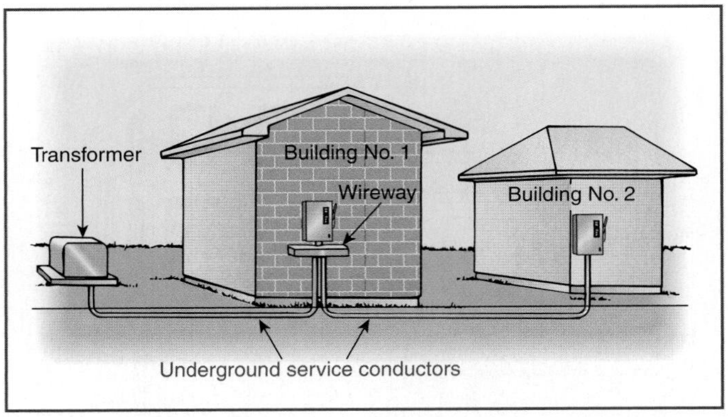


***EXHIBIT 230.14*** *Service conductors installed to not pass through the interior of Building No. 1 to supply Building No. 2.*

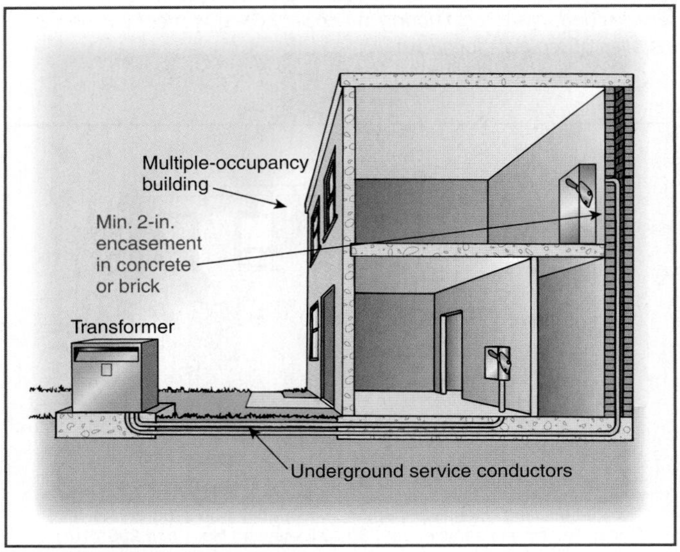


***EXHIBIT 230.15*** *Service conductors considered outside a building where installed under not less than 2 inches of concrete beneath the building or in a raceway encased by not less than 2 inches of concrete or brick within the building.*

## 230.7 Other Conductors in Raceway or Cable

Conductors other than service conductors shall not be installed in the same service raceway or service cable.

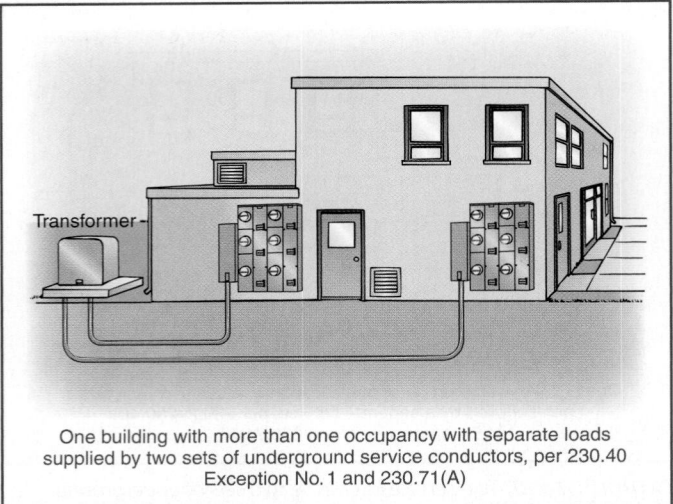

Service conductors are not provided with overcurrent protection where they receive their supply; they are protected against overload conditions at their load end by the service disconnect fuses or circuit breakers. The amount of current that could be imposed on feeder or branch-circuit conductors, should they be in the same raceway with service conductors during a fault, would be much higher than the ampacity of the feeder or branch-circuit

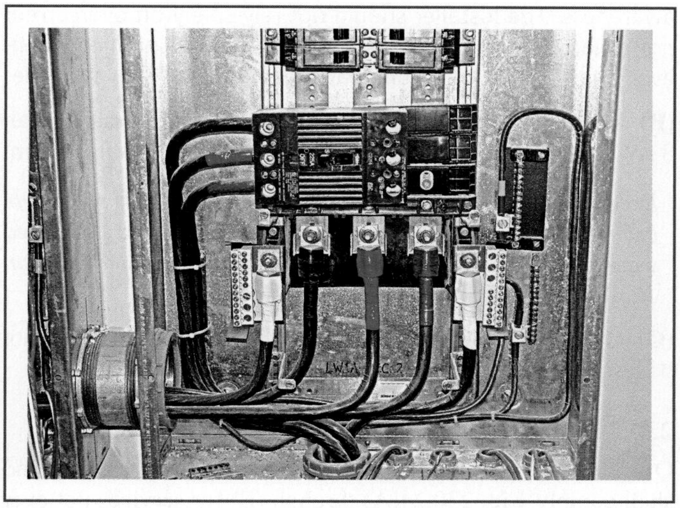

***EXHIBIT 230.16*** *A panelboard cabinet gutter space can be shared by service conductors, feeder conductors, and branch-circuit conductors. (Courtesy of the International Association of Electrical Inspectors)*

conductors. The gutter space of a panelboard cabinet or other electrical equipment enclosure is not a raceway (see definition of *raceway* in Article 100), and, therefore, is not subject to the requirement of 230.7. Service conductors, feeder conductors, and branch-circuit conductors can share the same gutter space, as shown in Exhibit 230.16. The panelboard cabinet gutter space accommodates a set of service conductors terminating on the 200-ampere main breaker, a set of feeder conductors supplying the adjacent panelboard, and several sets of branch-circuit conductors entering the bottom of the cabinet that will connect to OCPDs installed on the panelboard.

*Exception No. 1: Grounding electrode conductors and equipment bonding jumpers or conductors.*

*Exception No. 2: Load management control conductors having overcurrent protection.*

Load management control conductors, control circuits, and switch leg conductors for use with special rate meters are permitted to be installed in the service raceway or cable because they are usually short and are directly associated with control or operation of the service conductors.

## 230.8 Raceway Seal

Where a service raceway enters a building or structure from an underground distribution system, it shall be sealed in accordance with 300.5(G). Spare or unused raceways shall also be sealed. Sealants shall be identified for use with the cable insulation, shield, or other components.

Sealant, such as duct seal or a bushing incorporating the physical characteristics of a seal, is required by 230.8 to be used to seal the ends of service raceways. The intent is to prevent water — usually the result of condensation due to temperature differences — from entering the service equipment via the raceway. The sealant material should be compatible with the conductor insulation and should not cause deterioration of the insulation over time. For underground services over 1000 volts, nominal, refer to 300.50(F) for raceway seal requirements. See Exhibit 300.9 for an example.

## 230.9 Clearances on Buildings

Service conductors and final spans shall comply with 230.9(A), (B), and (C).

**(A) Clearances.** Service conductors installed as open conductors or multiconductor cable without an overall outer jacket shall have a clearance of not less than 900 mm (3 ft) from windows that are designed to be opened, doors, porches, balconies, ladders, stairs, fire escapes, or similar locations.

*Exception: Conductors run above the top level of a window shall be permitted to be less than the 900-mm (3-ft) requirement.*

The 3-foot clearance applies to open conductors, not to a raceway or to a cable assembly that has an overall outer jacket, such as Types SE, MC, and MI cables. The intent is to protect the conductors from physical damage and to protect persons from accidental contact with the conductors. The exception permits service conductors, including service-entrance conductors, overhead service conductors, and service-drop conductors, to be located just above window openings, because they are considered out of reach, as illustrated in Exhibit 230.17.

**(B) Vertical Clearance.** The vertical clearance of final spans above, or within 900 mm (3 ft) measured horizontally of, platforms, projections, or surfaces from which they might be reached shall be maintained in accordance with 230.24(B).

Service conductors must not be located where a person could reach and touch them. The amount of vertical clearance required is a function of the location and voltage of the overhead conductors.

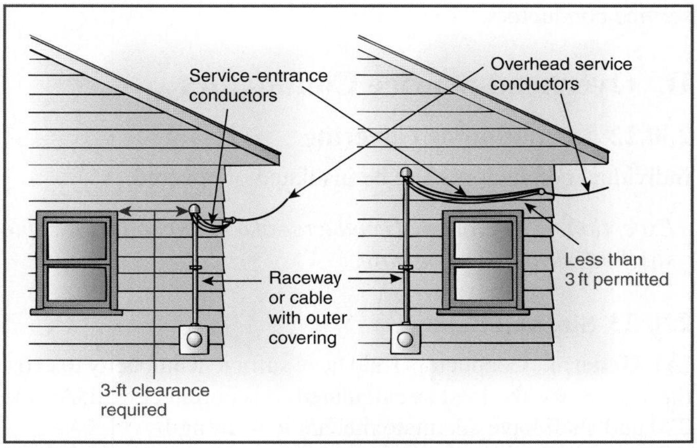


***EXHIBIT 230.17*** *Required dimensions for service conductors located alongside a window (left) and overhead service conductors above the top edge of a window designed to be opened (right).*

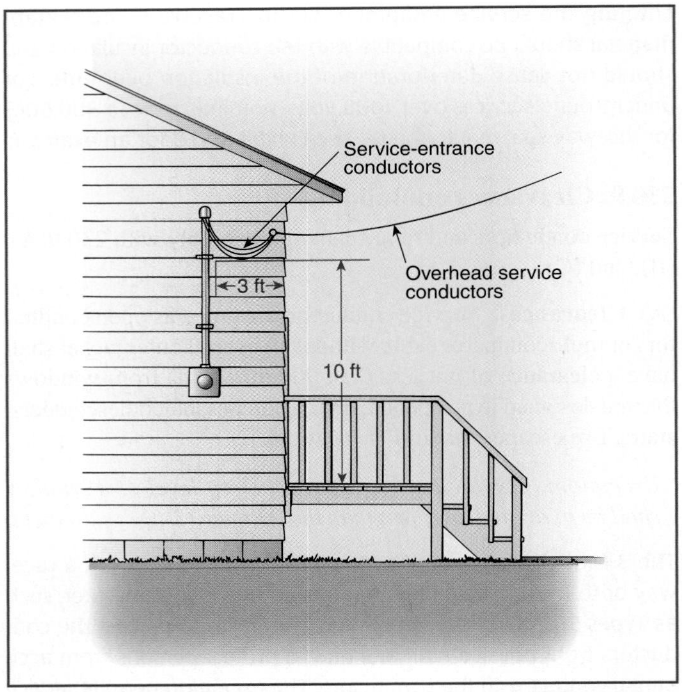


***EXHIBIT 230.18*** *Required dimensions for service conductors located above a stair landing.*

Exhibit 230.18 illustrates an installation where the 10-foot clearance is based on the conditions described in 230.24(B)(1).

**(C) Building Openings.** Overhead service conductors shall not be installed beneath openings through which materials may be moved, such as openings in farm and commercial buildings, and shall not be installed where they obstruct entrance to these building openings.

### 230.10 Vegetation as Support

Vegetation such as trees shall not be used for support of overhead service conductors.

## II. Overhead Service Conductors

### 230.22 Insulation or Covering

Individual conductors shall be insulated or covered.

*Exception: The grounded conductor of a multiconductor cable shall be permitted to be bare.*

### 230.23 Size and Rating

**(A) General.** Conductors shall have sufficient ampacity to carry the current for the load as calculated in accordance with Article 220 and shall have adequate mechanical strength.

When a load is added to any service, the installer must be aware of all existing loads. The potential for overloading the service conductors must be governed by installer responsibility and inspector awareness. The installer should not rely solely on overcurrent protection. The serving electric utility should be notified whenever load is added, to ensure that adequate capacity is available.

**(B) Minimum Size.** The conductors shall not be smaller than 8 AWG copper or 6 AWG aluminum or copper-clad aluminum.

*Exception: Conductors supplying only limited loads of a single branch circuit — such as small polyphase power, controlled water heaters, and similar loads — shall not be smaller than 12 AWG hard-drawn copper or equivalent.*

**(C) Grounded Conductors.** The grounded conductor shall not be less than the minimum size as required by 250.24(C).

### 230.24 Clearances

Overhead service conductors shall not be readily accessible and shall comply with 230.24(A) through (E) for services not over 1000 volts, nominal.

**(A) Above Roofs.** Conductors shall have a vertical clearance of not less than 2.5 m (8 ft) above the roof surface. The vertical clearance above the roof level shall be maintained for a distance of not less than 900 mm (3 ft) in all directions from the edge of the roof.

The 8-foot vertical clearance over the roof surface extends 3 feet in all directions from the edge. Exception No. 4 to 230.24(A), however, allows the final span of the overhead service conductors to enter this space in order to attach to the building or service mast.

*Exception No. 1: The area above a roof surface subject to pedestrian or vehicular traffic shall have a vertical clearance from the roof surface in accordance with the clearance requirements of 230.24(B).*

*Exception No. 2: Where the voltage between conductors does not exceed 300 and the roof has a slope of 100 mm in 300 mm (4 in. in 12 in.) or greater, a reduction in clearance to 900 mm (3 ft) shall be permitted.*

A reduction in overhead service conductor clearance above the roof from 8 feet to 3 feet is permitted, as illustrated in Exhibit 230.19, where the voltage between conductors does not exceed 300 V (for example, 120/240-V and 208Y/120-V services) and the roof is sloped not less than 4 inches vertically in 12 inches horizontally. Steeply sloped roofs are less likely to be walked on by other than those who have to work on the roof. The conductors' length over the roof is not restricted.

*Exception No. 3: Where the voltage between conductors does not exceed 300, a reduction in clearance above only the overhanging portion of the roof to not less than 450 mm (18 in.) shall be permitted if (1) not more than 1.8 m (6 ft) of overhead service conductors, 1.2 m (4 ft) horizontally, pass above the roof overhang, and (2) they are terminated at a through-the-roof raceway or approved support.*

Informational Note: See 230.28 for mast supports.

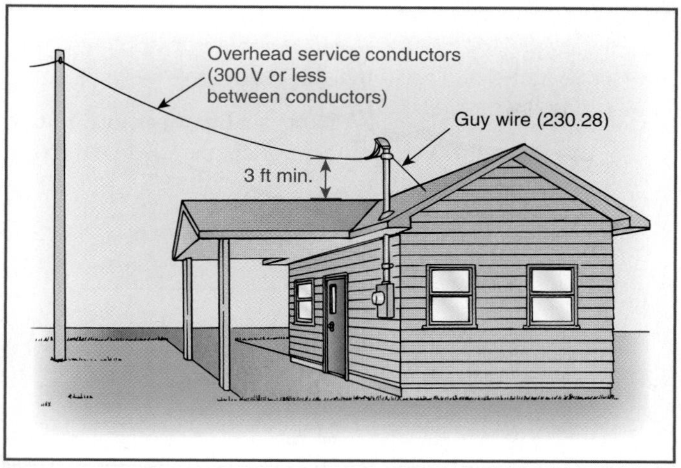


***EXHIBIT 230.19*** *Permitted reduction to 3-foot clearance above a sloped roof.*

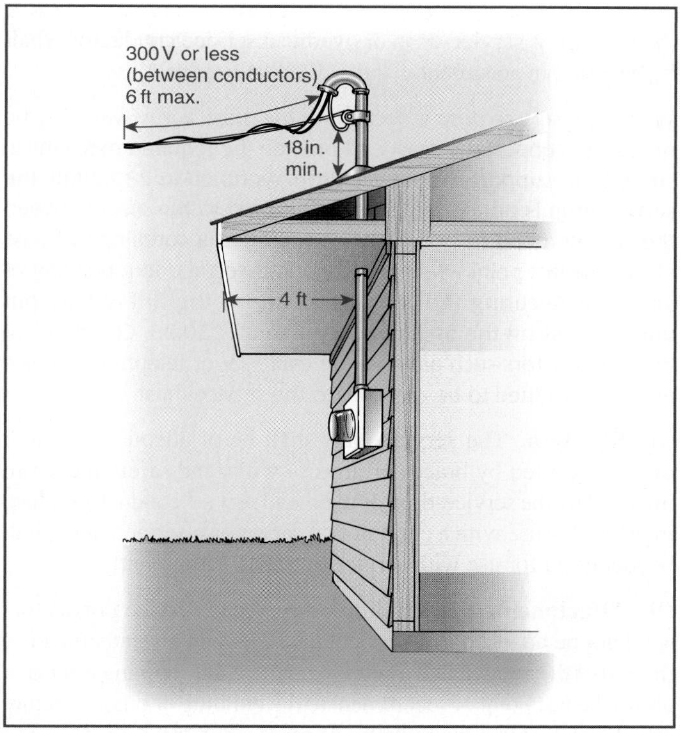


***EXHIBIT 230.20*** *Permitted reduction to 18-inch clearance above an overhang roof penetration.*

A reduction of overhead service conductor clearances to 18 inches above the roof is permitted as illustrated in Exhibit 230.20. This reduction is for service-mast (through-the-roof) installations where the voltage between conductors does not exceed 300 volts (for example, 120/240-V and 208Y/120-V services) and the mast is located within 4 feet of the edge of the roof, measured horizontally. Exception No. 3 applies to the overhanging portion of sloped and flat roofs. Not more than 6 feet of conductors is permitted to pass over the roof.

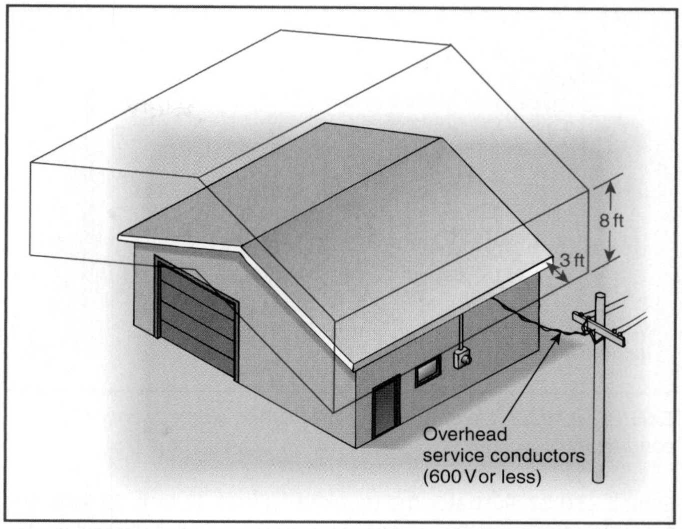


***EXHIBIT 230.21*** *Permitted clearance of the final span of overhead service conductors.*

*Exception No. 4: The requirement for maintaining the vertical clearance 900 mm (3 ft) from the edge of the roof shall not apply to the final conductor span where the service drop or overhead service conductors are attached to the side of a building.*

The final span of service drop or overhead service conductors attached to the side of a building is exempt from the 8-foot and 3-foot clearance requirements to allow the service conductors to be attached to the building as illustrated in Exhibit 230.21.

*Exception No. 5: Where the voltage between conductors does not exceed 300 and the roof area is guarded or isolated, a reduction in clearance to 900 mm (3 ft) shall be permitted.*

**(B) Vertical Clearance for Overhead Service Conductors.** Overhead service conductors, where not in excess of 600 volts, nominal, shall have the following minimum clearance from final grade:

(1) 3.0 m (10 ft) — at the electrical service entrance to buildings, also at the lowest point of the drip loop of the building electrical entrance, and above areas or sidewalks accessible only to pedestrians, measured from final grade or other accessible surface only for overhead service conductors supported on and cabled together with a grounded bare messenger where the voltage does not exceed 150 volts to ground

(2) 3.7 m (12 ft) — over residential property and driveways, and those commercial areas not subject to truck traffic where the voltage does not exceed 300 volts to ground

(3) 4.5 m (15 ft) — for those areas listed in the 3.7-m (12-ft) classification where the voltage exceeds 300 volts to ground

(4) 5.5 m (18 ft) — over public streets, alleys, roads, parking areas subject to truck traffic, driveways on other than residential property, and other land such as cultivated, grazing, forest, and orchard

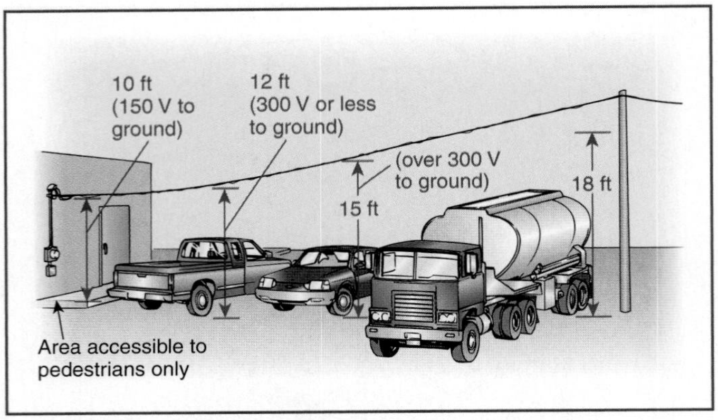


***EXHIBIT 230.22*** *Clearances from grade for overhead service conductors.*

Exhibit 230.22 illustrates the required 10-foot, 12-foot, 15-foot, and 18-foot vertical clearances from ground for overhead service conductors up to 600 volts. The voltages given are nominal voltages to ground, not the nominal voltage between circuit conductors specified in 230.24(A), Exceptions No. 2 and No. 3. A 480Y/277-V system (277 volts to ground) is covered by the 12-foot clearance requirement in 230.24(B)(2), but overhead service conductors supplied by an ungrounded 480-V system (considered to be 480 volts to ground) are required to have a 15-foot clearance over commercial areas not subject to truck traffic, in accordance with 230.24(B)(3).

**(C) Clearance from Building Openings.** See 230.9.

**(D) Clearance from Swimming Pools.** See 680.8.

**(E) Clearance from Communication Wires and Cables.** Clearance from communication wires and cables shall be in accordance with 800.44(A)(4).

## 230.26 Point of Attachment

The point of attachment of the overhead service conductors to a building or other structure shall provide the minimum clearances as specified in 230.9 and 230.24. In no case shall this point of attachment be less than 3.0 m (10 ft) above finished grade.

## 230.27 Means of Attachment

Multiconductor cables used for overhead service conductors shall be attached to buildings or other structures by fittings identified for use with service conductors. Open conductors shall be attached to fittings identified for use with service conductors or to noncombustible, nonabsorbent insulators securely attached to the building or other structure.

See 230.51 for mounting and supporting of service cables and individual open service conductors and 230.54 for connections at service heads.

## 230.28 Service Masts as Supports

Only power service-drop or overhead service conductors shall be permitted to be attached to a service mast. Service masts used for the support of service-drop or overhead service conductors shall be installed in accordance with 230.28(A) and (B).

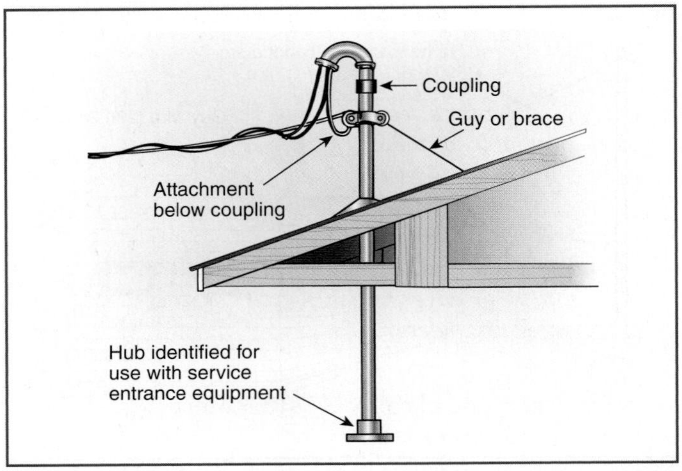


***EXHIBIT 230.23*** *The service conductors must not be secured above the fitting because the stress above the fitting could weaken the assembly.*

Where the service drop is secured to the mast, a guy wire may be needed to support the mast and provide the required mechanical strength to support the service drop. As noted in 230.28(B), the service drop is not permitted to be secured to the mast between the weatherhead or end of the conduit and a coupling installed above the last point where the conduit is secured to a building or structure. Securing the conductors above the fitting may put undue stress on the assembly. See Exhibit 230.23. Communications conductors such as those for cable TV or telephone service are not permitted to be attached to the service mast.

**(A) Strength.** The service mast shall be of adequate strength or be supported by braces or guys to withstand safely the strain imposed by the service-drop or overhead service conductors. Hubs intended for use with a conduit that serves as a service mast shall be identified for use with service-entrance equipment.

**(B) Attachment.** Service-drop or overhead service conductors shall not be attached to a service mast between a weatherhead or the end of the conduit and a coupling, where the coupling is located above the last point of securement to the building or other structure or is located above the building or other structure.

## 230.29 Supports over Buildings

Service conductors passing over a roof shall be securely supported by substantial structures. Where practicable, such supports shall be independent of the building.

# III. Underground Service Conductors

## 230.30 Installation

**(A) Insulation.** Underground service conductors shall be insulated for the applied voltage.

*Exception: A grounded conductor shall be permitted to be uninsulated as follows:*

*(1) Bare copper used in a raceway*
*(2) Bare copper for direct burial where bare copper is judged to be suitable for the soil conditions*
*(3) Bare copper for direct burial without regard to soil conditions where part of a cable assembly identified for underground use*
*(4) Aluminum or copper-clad aluminum without individual insulation or covering where part of a cable assembly identified for underground use in a raceway or for direct burial*

Due to the likelihood of corrosion, aluminum and copper-clad aluminum conductors must be insulated if they are run in a raceway or direct buried, unless they are part of a cable assembly identified for underground use.

**(B) Wiring Methods.** Underground service conductors shall be installed in accordance with the applicable requirements of this Code covering the type of wiring method used and shall be limited to the following methods:

(1) Type RMC conduit
(2) Type IMC conduit
(3) Type NUCC conduit
(4) Type HDPE conduit
(5) Type PVC conduit
(6) Type RTRC conduit
(7) Type IGS cable
(8) Type USE conductors or cables
(9) Type MV or Type MC cable identified for direct burial applications
(10) Type MI cable, where suitably protected against physical damage and corrosive conditions

### 230.31 Size and Rating

**(A) General.** Underground service conductors shall have sufficient ampacity to carry the current for the load as calculated in accordance with Article 220 and shall have adequate mechanical strength.

**(B) Minimum Size.** The conductors shall not be smaller than 8 AWG copper or 6 AWG aluminum or copper-clad aluminum.

*Exception: Conductors supplying only limited loads of a single branch circuit — such as small polyphase power, controlled water heaters, and similar loads — shall not be smaller than 12 AWG copper or 10 AWG aluminum or copper-clad aluminum.*

**(C) Grounded Conductors.** The grounded conductor shall not be less than the minimum size required by 250.24(C).

In addition to the minimum size requirement for the grounded conductor referenced in 230.31(C), see 310.15(B)(4) for the allowable ampacity of bare and covered conductors. For further information on sizing the grounded conductor for dwelling services, refer to 310.15(B)(7)(4); for more information on sizing the grounded service conductor for any occupancy, see the commentary following 230.42(C).

### 230.32 Protection Against Damage

Underground service conductors shall be protected against damage in accordance with 300.5. Service conductors entering a building or other structure shall be installed in accordance with 230.6 or protected by a raceway wiring method identified in 230.43.

### 230.33 Spliced Conductors

Service conductors shall be permitted to be spliced or tapped in accordance with 110.14, 300.5(E), 300.13, and 300.15.

## IV. Service-Entrance Conductors

### 230.40 Number of Service-Entrance Conductor Sets

Each service drop, set of overhead service conductors, set of underground service conductors, or service lateral shall supply only one set of service-entrance conductors.

*Exception No. 1: A building with more than one occupancy shall be permitted to have one set of service-entrance conductors for each service, as defined in 230.2, run to each occupancy or group of occupancies. If the number of service disconnect locations for any given classification of service does not exceed six, the requirements of 230.2(E) shall apply at each location. If the number of service disconnect locations exceeds six for any given supply classification, all service disconnect locations for all supply characteristics, together with any branch circuit or feeder supply sources, if applicable, shall be clearly described using suitable graphics or text, or both, on one or more plaques located in an approved, readily accessible location(s) on the building or structure served and as near as practicable to the point(s) of attachment or entry(ies) for each service drop or service lateral, and for each set of overhead or underground service conductors.*

If a building has more than one occupancy — such as multifamily dwellings, strip malls, and office buildings — each service drop, set of overhead service conductors, set of underground service conductors, or service lateral is allowed to supply more than one set of service-entrance conductors, provided they are run to each occupancy or group of occupancies. Although the building is supplied by a single service, this requirement allows for multiple service disconnecting means locations for each service that supplies the building or structure. Based on this exception, one service may be arranged similarly to a building or structure that is supplied by multiple services. This exception does not limit the number of disconnecting means locations supplied by each classification of service.

Where a building is supplied by multiple services, 230.2(E) requires a permanent plaque or directory at each service disconnecting means location with information about the multiple services or other circuits supplying the building and the areas served by each. Because this exception permits more than one service equipment location on a single building or structure, the information about the multiple equipment locations is required to be provided at each location.

Where the number of service equipment locations for any class of service exceeds six, a master plaque(s) or directory(s) is required near the point where service-drop, overhead service, underground service, or service-lateral conductors attach to or enter a building or structure. The information must describe the multiple service equipment locations using either text or a graphic or a combination. At the individual service equipment locations, the general marking required in 230.70(B) must be provided.

For example, if a mercantile building has eight storefronts and the building is supplied by a single 208Y/120-V service, eight sets of service-entrance conductors can be installed with one set run to each occupancy. The service equipment at each occupancy can have up to six service disconnecting means in accordance with 230.71(A). Because the number of service disconnecting means locations for this service exceeds six, a permanent plaque(s) with either a diagram of the supply equipment arrangement for the building or text or a combination of the two explaining the multiple supply equipment locations is required. The location of this plaque or display must be readily accessible and acceptable to the AHJ.

*Exception No. 2: Where two to six service disconnecting means in separate enclosures are grouped at one location and supply separate loads from one service drop, set of overhead service conductors, set of underground service conductors, or service lateral, one set of service-entrance conductors shall be permitted to supply each or several such service equipment enclosures.*

Exhibits 230.2 through 230.13 provide examples of service configurations permitted by Exception No. 2 to 230.40. In Exhibit 230.7 and Exhibit 230.10, the multiple sets of service-lateral conductors are treated as a single-service lateral in accordance with the first paragraph of 230.2.

*Exception No. 3: A single-family dwelling unit and its accessory structures shall be permitted to have one set of service-entrance conductors run to each from a single service drop, set of overhead service conductors, set of underground service conductors, or service lateral.*

A second set of service-entrance conductors supplied by a single service drop or lateral at a single-family dwelling unit is permitted to supply another building on the premises, such as a garage or storage shed. The utility meters may be grouped at one location, but in this application, the service disconnecting means are not required to be grouped at one location. A service disconnecting means at the dwelling and at the other building is acceptable where this exception is used.

*Exception No. 4: Two-family dwellings, multifamily dwellings, and multiple occupancy buildings shall be permitted to have one set of service-entrance conductors installed to supply the circuits covered in 210.25.*

*Exception No. 5: One set of service-entrance conductors connected to the supply side of the normal service disconnecting means shall be permitted to supply each or several systems covered by 230.82(5) or 230.82(6).*

## 230.41 Insulation of Service-Entrance Conductors

Service-entrance conductors entering or on the exterior of buildings or other structures shall be insulated.

*Exception: A grounded conductor shall be permitted to be uninsulated as follows:*

*(1) Bare copper used in a raceway or part of a service cable assembly*
*(2) Bare copper for direct burial where bare copper is judged to be suitable for the soil conditions*
*(3) Bare copper for direct burial without regard to soil conditions where part of a cable assembly identified for underground use*
*(4) Aluminum or copper-clad aluminum without individual insulation or covering where part of a cable assembly or identified for underground use in a raceway, or for direct burial*
*(5) Bare conductors used in an auxiliary gutter*

## 230.42 Minimum Size and Rating

**(A) General.** The ampacity of service-entrance conductors shall not be less than either 230.42(A)(1), (A)(2), or (A)(3). Loads shall be determined in accordance with Part III, IV, or V of Article 220, as applicable. Ampacity shall be determined from 310.15. The maximum allowable current of busways shall be that value for which the busway has been listed or labeled.

(1) The sum of the noncontinuous loads plus 125 percent of continuous loads

*Exception: Grounded conductors that are not connected to an overcurrent device shall be permitted to be sized at 100 percent of the continuous and noncontinuous load.*

(2) The sum of the noncontinuous load plus the continuous load after the application of any adjustment or correction factors.
(3) The sum of the noncontinuous load plus the continuous load if the service-entrance conductors terminate in an overcurrent device where both the overcurrent device and its assembly are listed for operation at 100 percent of their rating

**(B) Specific Installations.** In addition to the requirements of 230.42(A), the minimum ampacity for ungrounded conductors for specific installations shall not be less than the rating of the service disconnecting means specified in 230.79(A) through (D).

**(C) Grounded Conductors.** The grounded conductor shall not be smaller than the minimum size as required by 250.24(C).

The reference to 250.24(C) for determining the minimum size grounded service conductor reflects the fact that in addition to its role as a circuit conductor for loads supplied by the service, the grounded service conductor also provides the ground-fault current path from the load end of the service conductors in the service equipment to the transformer or other equipment at which the service conductors receive their supply. This is the same function the equipment grounding conductor provides in feeder and branch circuits. The normal circuit conductor function cannot be ignored in determining the minimum size of the ground service conductor, and the provision of 220.61 has to be applied. Using 220.61 and 250.24(C), the requirement that results in the largest conductor is what is used as the minimum size grounded service conductor. The *Code* does not prohibit sizing the grounded conductor simply by using the same size as the ungrounded (hot) service conductors.

In addition, the heating effect of harmonic currents due to nonlinear loads should be considered in the sizing of the neutral conductor of a 3-phase, 4-wire wye system.

## 230.43 Wiring Methods for 1000 Volts, Nominal, or Less

Service-entrance conductors shall be installed in accordance with the applicable requirements of this *Code* covering the type of wiring method used and shall be limited to the following methods:

(1) Open wiring on insulators
(2) Type IGS cable
(3) Rigid metal conduit (RMC)
(4) Intermediate metal conduit (IMC)
(5) Electrical metallic tubing (EMT)
(6) Electrical nonmetallic tubing
(7) Service-entrance cables
(8) Wireways
(9) Busways
(10) Auxiliary gutters
(11) Rigid polyvinyl chloride conduit (PVC)
(12) Cablebus
(13) Type MC cable
(14) Mineral-insulated, metal-sheathed cable, Type MI
(15) Flexible metal conduit (FMC) not over 1.8 m (6 ft) long or liquidtight flexible metal conduit (LFMC) not over 1.8 m (6 ft) long between a raceway, or between a raceway and service equipment, with a supply-side bonding jumper routed with the flexible metal conduit (FMC) or the liquidtight flexible metal conduit (LFMC) according to the provisions of 250.102(A), (B), (C), and (E)
(16) Liquidtight flexible nonmetallic conduit (LFNC)
(17) High density polyethylene conduit (HDPE)
(18) Nonmetallic underground conduit with conductors (NUCC)
(19) Reinforced thermosetting resin conduit (RTRC)

Section 230.43(15) permits no more than 6 feet of flexible metal conduit or liquidtight flexible metal conduit to be used as a service wiring method. Because of the high levels of fault energy available on the line side of the service disconnecting means, a bonding jumper must be installed where these raceway types are used for service conductors. The bonding jumper is allowed to be installed inside or outside the raceway, but it must follow the path of the raceway and cannot exceed 6 feet in length. In order to minimize the impedance of the ground fault current return path, the bonding jumper must not be wrapped or spiraled around the flexible conduit.

## 230.44 Cable Trays

Cable tray systems shall be permitted to support service-entrance conductors. Cable trays used to support service-entrance conductors shall contain only service-entrance conductors and shall be limited to the following methods:

(1) Type SE cable
(2) Type MC cable
(3) Type MI cable
(4) Type IGS cable
(5) Single conductors 1/0 and larger with CT rating

Such cable trays shall be identified with permanently affixed labels with the wording "Service-Entrance Conductors." The labels shall be located so as to be visible after installation with a spacing not to exceed 3 m (10 ft) so that the service-entrance conductors are able to be readily traced through the entire length of the cable tray.

*Exception: Conductors, other than service-entrance conductors, shall be permitted to be installed in a cable tray with service-entrance conductors, provided a solid fixed barrier of a material compatible with the cable tray is installed to separate the service-entrance conductors from other conductors installed in the cable tray.*

The solid barrier of material essentially divides the cable tray into two cable trays. The service conductors must be on the other side of the barrier from non-service conductors. The warning labels must be installed at intervals not greater than 10 feet.

## 230.46 Spliced Conductors

Service-entrance conductors shall be permitted to be spliced or tapped in accordance with 110.14, 300.5(E), 300.13, and 300.15.

Splices must be in an enclosure or be direct buried using a listed underground splice kit. The termination of an underground service lateral at a terminal box either inside or outside the building is common. At that point, service conductors may be spliced or run directly to the service equipment.

Splices are permitted where the cable enters a terminal box and a different wiring method, such as conduit, continues to the service equipment. Tapped sets of conductors are also recognized, and this method is commonly used to supply multiple service disconnecting means installed in separate enclosures as permitted by 230.71(A).

## 230.50 Protection Against Physical Damage

**(A) Underground Service-Entrance Conductors.** Underground service-entrance conductors shall be protected against physical damage in accordance with 300.5.

**(B) All Other Service-Entrance Conductors.** All other service-entrance conductors, other than underground service entrance conductors, shall be protected against physical damage as specified in 230.50(B)(1) or (B)(2).

**(1) Service-Entrance Cables.** Service-entrance cables, where subject to physical damage, shall be protected by any of the following:

(1) Rigid metal conduit (RMC)
(2) Intermediate metal conduit (IMC)
(3) Schedule 80 PVC conduit
(4) Electrical metallic tubing (EMT)
(5) Reinforced thermosetting resin conduit (RTRC)
(6) Other approved means

**(2) Other Than Service-Entrance Cables.** Individual open conductors and cables, other than service-entrance cables, shall not be installed within 3.0 m (10 ft) of grade level or where exposed to physical damage.

*Exception: Type MI and Type MC cable shall be permitted within 3.0 m (10 ft) of grade level where not exposed to physical damage or where protected in accordance with 300.5(D).*

## 230.51 Mounting Supports

Service-entrance cables or individual open service-entrance conductors shall be supported as specified in 230.51(A), (B), or (C).

**(A) Service-Entrance Cables.** Service-entrance cables shall be supported by straps or other approved means within 300 mm (12 in.) of every service head, gooseneck, or connection to a raceway or enclosure and at intervals not exceeding 750 mm (30 in.).

**(B) Other Cables.** Cables that are not approved for mounting in contact with a building or other structure shall be mounted on insulating supports installed at intervals not exceeding 4.5 m (15 ft) and in a manner that maintains a clearance of not less than 50 mm (2 in.) from the surface over which they pass.

**(C) Individual Open Conductors.** Individual open conductors shall be installed in accordance with Table 230.51(C). Where exposed to the weather, the conductors shall be mounted on insulators or on insulating supports attached to racks, brackets, or other approved means. Where not exposed to the weather, the conductors shall be mounted on glass or porcelain knobs.

## 230.52 Individual Conductors Entering Buildings or Other Structures

Where individual open conductors enter a building or other structure, they shall enter through roof bushings or through the wall in an upward slant through individual, noncombustible, nonabsorbent insulating tubes. Drip loops shall be formed on the conductors before they enter the tubes.

## 230.53 Raceways to Drain

Where exposed to the weather, raceways enclosing service-entrance conductors shall be suitable for use in wet locations and arranged to drain. Where embedded in masonry, raceways shall be arranged to drain.

The objective is to prevent water from entering electrical equipment through the raceway system. Service raceways exposed to the weather must have raintight fittings and drain holes or other means to preclude water from coming in contact with live parts of electrical equipment. Surface water, rain, or water from poured concrete are problematic for raceways embedded in masonry, and means must be provided for draining and diverting the water from these raceways.

## 230.54 Overhead Service Locations

**(A) Service Head.** Service raceways shall be equipped with a service head at the point of connection to service-drop or overhead service conductors. The service head shall be listed for use in wet locations.

**(B) Service-Entrance Cables Equipped with Service Head or Gooseneck** Service-entrance cables shall be equipped with

***TABLE 230.51(C)*** *Supports*

| Maximum Volts | Maximum Distance Between Supports | | Minimum Clearance | | | |
|---|---|---|---|---|---|---|
| | | | Between Conductors | | From Surface | |
| | m | ft | mm | in. | mm | in. |
| 1000 | 2.7 | 9 | 150 | 6 | 50 | 2 |
| 1000 | 4.5 | 15 | 300 | 12 | 50 | 2 |
| 300 | 1.4 | 4½ | 75 | 3 | 50 | 2 |
| 1000* | 1.4* | 4½* | 65* | 2½* | 25* | 1* |

*Where not exposed to weather.

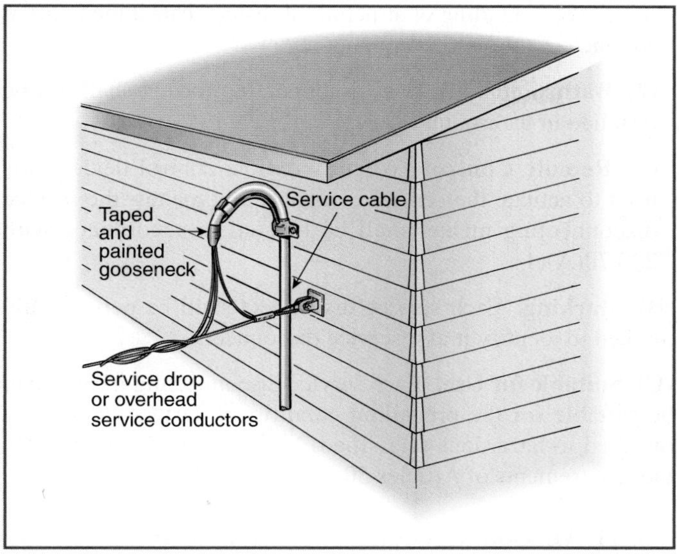


*EXHIBIT 230.24 A service-entrance cable that terminates in a gooseneck without a raintight service head (weatherhead).*

a service head. The service head shall be listed for use in wet locations.

Type SE service-entrance cables may be installed without a service head (weatherhead) if they are run continuously from a utility pole to metering or service equipment, or if they are shaped in a downward direction (forming a "gooseneck") and sealed by taping and painting, as shown in Exhibit 230.24.

*Exception: Type SE cable shall be permitted to be formed in a gooseneck and taped with a self-sealing weather-resistant thermoplastic.*

**(C) Service Heads and Goosenecks Above Service-Drop or Overhead Service Attachment.** Service heads and goosenecks in service-entrance cables shall be located above the point of attachment of the service-drop or overhead service conductors to the building or other structure.

Service heads and goosenecks are required to be located above the point of attachment to the building or structure for the service drop or overhead service conductors unless such location is not feasible, in which case the service head or gooseneck is permitted to be located not farther than 24 inches from the point of attachment, in accordance with the exception to 230.54(C). Individual conductors should extend in a downward direction, as shown in Exhibit 230.24, or drip loops should be formed so that any splices that are made are at the lowest point of the drip loop.

*Exception: Where it is impracticable to locate the service head or gooseneck above the point of attachment, the service head or gooseneck location shall be permitted not farther than 600 mm (24 in.) from the point of attachment.*

**(D) Secured.** Service-entrance cables shall be held securely in place.

**(E) Separately Bushed Openings.** Service heads shall have conductors of different potential brought out through separately bushed openings.

*Exception: For jacketed multiconductor service-entrance cable without splice.*

**(F) Drip Loops.** Drip loops shall be formed on individual conductors. To prevent the entrance of moisture, service-entrance conductors shall be connected to the service-drop or overhead service conductors either (1) below the level of the service head or (2) below the level of the termination of the service-entrance cable sheath.

**(G) Arranged That Water Will Not Enter Service Raceway or Equipment.** Service-entrance and overhead service conductors shall be arranged so that water will not enter service raceway or equipment.

### 230.56 Service Conductor with the Higher Voltage to Ground

On a 4-wire, delta-connected service where the midpoint of one phase winding is grounded, the service conductor having the higher phase voltage to ground shall be durably and permanently marked by an outer finish that is orange in color, or by other effective means, at each termination or junction point.

The special marking of the service conductor with the higher voltage to ground is intended to provide warning of a potential hazard; connection to this higher voltage conductor can damage equipment or injure personnel. The marking should be at both the point of connection to the service-drop or lateral conductors and the point of connection to the service disconnecting means. See 110.15 and 408.3(E) for similar "high-leg" marking and phase arrangement requirements.

## V. Service Equipment — General

### 230.62 Service Equipment — Enclosed or Guarded

Energized parts of service equipment shall be enclosed as specified in 230.62(A) or guarded as specified in 230.62(B).

**(A) Enclosed.** Energized parts shall be enclosed so that they will not be exposed to accidental contact or shall be guarded as in 230.62(B).

**(B) Guarded.** Energized parts that are not enclosed shall be installed on a switchboard, panelboard, or control board and guarded in accordance with 110.18 and 110.27. Where energized parts are guarded as provided in 110.27(1) and (A)(2), a means for locking or sealing doors providing access to energized parts shall be provided.

### 230.66 Marking

Service equipment rated at 1000 volts or less shall be marked to identify it as being suitable for use as service equipment. All service equipment shall be listed. Individual meter socket enclosures shall not be considered service equipment.

Service equipment is required to be listed. "Suitable for Use as Service Equipment" is a common marking found on equipment that can be used at the service location. The marking is not a field marking, but is applied by the equipment manufacturer and indicates the equipment meets requirements in the applicable product standard (i.e., panelboard, switchboard, enclosed switch, or other equipment product standard) that enables it to carry a marking indicating its suitability as service equipment. "Suitable Only for Use as Service Equipment" is a marking that indicates the grounded conductor or neutral terminal bus is not able to be electrically isolated from the metal equipment enclosure. This inability precludes most feeder applications for this equipment where the equipment grounding terminals and the grounded conductor terminals are required to be electrically isolated. A similar requirement is contained in 225.36 for outside feeder and branch-circuit disconnecting means.

# VI. Service Equipment — Disconnecting Means

## 230.70 General

Means shall be provided to disconnect all conductors in a building or other structure from the service-entrance conductors.

**(A) Location.** The service disconnecting means shall be installed in accordance with 230.70(A)(1), (A)(2), and (A)(3).

No maximum distance between the point of entrance of service conductors to a readily accessible location for the installation of a service disconnecting means is specified by 230.70(A). The authority enforcing this *Code* is responsible for the decision on how far inside the building the service-entrance conductors are allowed to travel to the service disconnecting means. The length of service-entrance conductors should be kept to a minimum inside buildings, because power utilities provide limited overcurrent protection. In the event of a fault, the service conductors could ignite nearby combustible materials.

Some local jurisdictions have adopted ordinances that allow service-entrance conductors to run within the building up to a specified length to terminate at the disconnecting means. The AHJ may permit service conductors to bypass fuel storage tanks or gas meters and the like, permitting the service disconnecting means to be located in a readily accessible location.

However, if the authority judges the distance to be excessive, the disconnecting means may be required to be located on the outside of the building or near the building at a readily accessible location that is not necessarily nearest the point of entrance of the conductors. See 230.6 and Exhibit 230.15 for conductors considered to be outside a building.

See 404.8(A) for mounting-height restrictions for switches and for circuit breakers used as switches.

**(1) Readily Accessible Location.** The service disconnecting means shall be installed at a readily accessible location either outside of a building or structure or inside nearest the point of entrance of the service conductors.

**(2) Bathrooms.** Service disconnecting means shall not be installed in bathrooms.

**(3) Remote Control.** Where a remote control device(s) is used to actuate the service disconnecting means, the service disconnecting means shall be located in accordance with 230.70(A)(1).

**(B) Marking.** Each service disconnect shall be permanently marked to identify it as a service disconnect.

**(C) Suitable for Use.** Each service disconnecting means shall be suitable for the prevailing conditions. Service equipment installed in hazardous (classified) locations shall comply with the requirements of Articles 500 through 517.

## 230.71 Maximum Number of Disconnects

**(A) General.** The service disconnecting means for each service permitted by 230.2, or for each set of service-entrance conductors permitted by 230.40, Exception No. 1, 3, 4, or 5, shall consist of not more than six switches or sets of circuit breakers, or a combination of not more than six switches and sets of circuit breakers, mounted in a single enclosure, in a group of separate enclosures, or in or on a switchboard or in switchgear. There shall be not more than six sets of disconnects per service grouped in any one location.

For the purpose of this section, disconnecting means installed as part of listed equipment and used solely for the following shall not be considered a service disconnecting means:

(1) Power monitoring equipment
(2) Surge-protective device(s)
(3) Control circuit of the ground-fault protection system
(4) Power-operable service disconnecting means

One set of service-entrance conductors, either overhead or underground, is permitted to supply two to six service disconnecting means in lieu of a single main disconnect. A single-occupancy building can have up to six disconnects for each set of service-entrance conductors. Multiple-occupancy buildings (residential or other than residential) can be provided with one main service disconnect or up to six main disconnects for each set of service-entrance conductors. Conductors from renewable energy sources, such as photovoltaic systems and wind generators, are feeder conductors, not service conductors. Where alternative energy systems are run in parallel with a utility source, Article 705 applies.

Multiple occupancy buildings may have service-entrance conductors run to each occupancy, and each such set of service-entrance conductors may have from one to six disconnects (see 230.40, Exception No. 1).

Exhibit 230.25 shows a single enclosure for grouping service equipment that consists of six circuit breakers or six fused switches. This arrangement does not require a single main service disconnecting means. Six separate enclosures also would be

***EXHIBIT 230.25*** *A service equipment enclosure that groups six service disconnecting means.*

permitted as the service equipment. Where factory-installed switches that disconnect power to surge protective devices and power monitoring equipment are included as part of listed equipment, the last sentence of 230.71(A) specifies that the disconnect switch for such equipment installed as part of the listed equipment does *not* count as one of the six service disconnecting means permitted by 230.71(A). The disconnecting means for the control circuit of ground-fault protection equipment or for a power-operable service disconnecting means are also not considered to be service disconnecting means where such disconnecting means are installed as a component of listed equipment.

**(B) Single-Pole Units.** Two or three single-pole switches or breakers, capable of individual operation, shall be permitted on multiwire circuits, one pole for each ungrounded conductor, as one multipole disconnect, provided they are equipped with identified handle ties or a master handle to disconnect all conductors of the service with no more than six operations of the hand.

Informational Note: See 408.36, Exception No. 1 and Exception No. 3, for service equipment in certain panelboards, and see 430.95 for service equipment in motor control centers.

## 230.72 Grouping of Disconnects

**(A) General.** The two to six disconnects as permitted in 230.71 shall be grouped. Each disconnect shall be marked to indicate the load served.

*Exception: One of the two to six service disconnecting means permitted in 230.71, where used only for a water pump also intended to provide fire protection, shall be permitted to be located remote from the other disconnecting means. If remotely installed in accordance with this exception, a plaque shall be posted at the location of the remaining grouped disconnects denoting its location.*

The water pump in the exception is not the fire pump covered by the requirements of Article 695; rather, it is a water pump used for normal water supply and also for fire protection. This application is used in agricultural settings and permits separation of the water pump disconnect so it can remain operational in the event of a problem at the location of the other service disconnecting means.

**(B) Additional Service Disconnecting Means.** The one or more additional service disconnecting means for fire pumps, emergency systems, legally required standby, or optional standby services permitted by 230.2 shall be installed remote from the one to six service disconnecting means for normal service to minimize the possibility of simultaneous interruption of supply.

Reliability of power to important safety equipment or systems, such as fire pumps and building emergency power systems, is increased by locating the disconnecting means for such equipment at a location that is remote from the normal service disconnecting means. Separating the service equipment in this way protects against failure of other building systems that could endanger the electrical equipment. It also protects against inadvertent operation of the disconnecting means. The AHJ is responsible for determining what is a suitably "remote" location for the additional service disconnecting means. Requirements in Articles 695 and 700 as well as in other standards such as NFPA 110, *Standard for Emergency and Standby Power Systems,* contain similar requirements that are focused on a high degree of reliability for power to safety equipment and systems. The disconnecting means allowed for the equipment or systems covered by 230.72(B) are in addition to the one or more disconnecting means allowed for the normal supply. The separate services are required to be installed in accordance with all the applicable requirements of Article 230, including the identification requirement specified in 230.2(E).

**(C) Access to Occupants.** In a multiple-occupancy building, each occupant shall have access to the occupant's service disconnecting means.

In multiple-occupancy buildings the different units are generally independent of each other, so access to the service disconnecting means for each unit may be difficult due to locked doors or other impediments. Unless electric service and maintenance are provided by and under continuous supervision of the building management, the occupants of a multiple-occupancy building must have ready access to their service disconnecting means, and this access has to be incorporated into the building service equipment layout or location. Section 240.24(B) contains a similar requirement for access to service, feeder, and branch-circuit OCPDs.

*Exception: In a multiple-occupancy building where electric service and electrical maintenance are provided by the building management and where these are under continuous building management supervision, the service disconnecting means supplying more than one occupancy shall be permitted to be accessible to authorized management personnel only.*

### 230.74 Simultaneous Opening of Poles

Each service disconnect shall simultaneously disconnect all ungrounded service conductors that it controls from the premises wiring system.

### 230.75 Disconnection of Grounded Conductor

Where the service disconnecting means does not disconnect the grounded conductor from the premises wiring, other means shall be provided for this purpose in the service equipment. A terminal or bus to which all grounded conductors can be attached by means of pressure connectors shall be permitted for this purpose. In a multisection switchboard or switchgear, disconnects for the grounded conductor shall be permitted to be in any section of the switchboard or switchgear, provided that any such switchboard or switchgear section is marked.

This disconnection does not have to be by operation of the service disconnecting means. Disconnection can be, and most commonly is, accomplished by manually removing the grounded conductor from the bus or terminal bar to which it is lugged or bolted. This location is often referred to as the neutral disconnect link.

Manufacturers design neutral terminal bars for service equipment so that grounded conductors must be cut to be attached; that is, the grounded conductor cannot be run straight through the service equipment without a means of disconnection from the premises wiring.

### 230.76 Manually or Power Operable

The service disconnecting means for ungrounded service conductors shall consist of one of the following:

(1) A manually operable switch or circuit breaker equipped with a handle or other suitable operating means
(2) A power-operated switch or circuit breaker, provided the switch or circuit breaker can be opened by hand in the event of a power supply failure

### 230.77 Indicating

The service disconnecting means shall plainly indicate whether it is in the open (off) or closed (on) position.

### 230.79 Rating of Service Disconnecting Means

The service disconnecting means shall have a rating not less than the calculated load to be carried, determined in accordance with Part III, IV, or V of Article 220, as applicable. In no case shall the rating be lower than specified in 230.79(A), (B), (C), or (D).

**(A) One-Circuit Installations.** For installations to supply only limited loads of a single branch circuit, the service disconnecting means shall have a rating of not less than 15 amperes.

**(B) Two-Circuit Installations.** For installations consisting of not more than two 2-wire branch circuits, the service disconnecting means shall have a rating of not less than 30 amperes.

**(C) One-Family Dwellings.** For a one-family dwelling, the service disconnecting means shall have a rating of not less than 100 amperes, 3-wire.

**(D) All Others.** For all other installations, the service disconnecting means shall have a rating of not less than 60 amperes.

### 230.80 Combined Rating of Disconnects

Where the service disconnecting means consists of more than one switch or circuit breaker, as permitted by 230.71, the combined ratings of all the switches or circuit breakers used shall not be less than the rating required by 230.79.

Where more than one switch or circuit breaker is used as the disconnecting means, the combined rating of all the switches or circuit breakers used cannot be less than the rating required for a single switch or circuit breaker.

Section 230.90 requires an overcurrent device to provide overload protection in each ungrounded service conductor. A single overcurrent device must have a rating or setting that is not higher than the allowable ampacity of the service conductors. However, Exception No. 3 to 230.90(A) allows not more than six circuit breakers or six sets of fuses to be considered the overcurrent device. None of these individual overcurrent devices can have a rating or setting higher than the ampacity of the service conductors.

In complying with these rules, it is possible for the total of the six overcurrent devices to be greater than the rating of the service-entrance conductors. However, the size of the service-entrance conductors is required to be adequate for the computed load only, and each individual service disconnecting means is required to be large enough for the individual load it supplies. See the commentary following 230.90(A), Exception No. 3.

### 230.81 Connection to Terminals

The service conductors shall be connected to the service disconnecting means by pressure connectors, clamps, or other approved means. Connections that depend on solder shall not be used.

### 230.82 Equipment Connected to the Supply Side of Service Disconnect

Only the following equipment shall be permitted to be connected to the supply side of the service disconnecting means:

(1) Cable limiters or other current-limiting devices.

Cable limiters or other current-limiting devices are permitted to be applied ahead of the service disconnecting means for the following reasons:

1. To individually isolate faulted cable(s) from the remainder of the circuit or paralleled set of conductors
2. To maintain continuity of service even though one or more cables are faulted

3. To reduce the possibility of severe equipment damage or burndown as a result of a fault on the service conductors
4. To provide protection against high short-circuit currents for services and to provide compliance with 110.10

(2) Meters and meter sockets nominally rated not in excess of 1000 volts, provided that all metal housings and service enclosures are grounded in accordance with Part VII and bonded in accordance with Part V of Article 250.

(3) Meter disconnect switches nominally rated not in excess of 1000 V that have a short-circuit current rating equal to or greater than the available short-circuit current, provided that all metal housings and service enclosures are grounded in accordance with Part VII and bonded in accordance with Part V of Article 250. A meter disconnect switch shall be capable of interrupting the load served. A meter disconnect shall be legibly field marked on its exterior in a manner suitable for the environment as follows:

METER DISCONNECT
NOT SERVICE EQUIPMENT

Meter sockets and meter disconnect switches are permitted by 230.82(2) and (3) to be connected on the supply side of the service disconnecting means. The meter disconnect is a load-break disconnect switch designed to interrupt the service load on 480Y/277-V services with self-contained meter sockets. The meter disconnect is not the service disconnecting means. The purpose of the meter disconnect switch is to facilitate meter change, maintenance, or disconnecting of the service. Section 230.82(3) requires meter disconnect switches to have a short-circuit current rating that is not less than the available short-circuit current at the line terminals of the meter disconnect switch.

Self-contained meters do not have external potential transformers or current transformers. The load current of the service travels through the meter itself. Neither the self-contained meter nor the meter bypass switch in the meter socket is designed to break the load current on a 480Y/277-V system.

Self-contained meters or internal meter bypass switches should not be used to break the load current of a service having a voltage of over 150 volts to ground, because a hazardous arc could be generated. Arcs generated at voltages greater than 150 volts are considered self-sustaining and can transfer from the energized portions of the equipment to the grounded portions of the equipment.

An arc created while breaking load current on a 480Y/277-V system (277 volts to ground) could transfer to the grounded equipment enclosure, creating a high-energy arcing ground fault and arc flash that could develop into a 3-phase short circuit. This hazardous arcing could burn down the meter socket and injure the person performing the work.

(4) Instrument transformers (current and voltage), impedance shunts, load management devices, surge arresters, and Type 1 surge-protective devices.

(5) Taps used only to supply load management devices, circuits for standby power systems, fire pump equipment, and fire and sprinkler alarms, if provided with service equipment and installed in accordance with requirements for service-entrance conductors.

Emergency lighting, fire alarm systems, fire pumps, standby power, and sprinkler alarms are permitted to be connected ahead of the normal service disconnecting means only if such systems are provided with a separate disconnecting means and overcurrent protection.

(6) Solar photovoltaic systems, fuel cell systems, or interconnected electric power production sources.

(7) Control circuits for power-operable service disconnecting means, if suitable overcurrent protection and disconnecting means are provided.

(8) Ground-fault protection systems or Type 2 surge-protective devices, where installed as part of listed equipment, if suitable overcurrent protection and disconnecting means are provided.

(9) Connections used only to supply listed communications equipment under the exclusive control of the serving electric utility, if suitable overcurrent protection and disconnecting means are provided. For installations of equipment by the serving electric utility, a disconnecting means is not required if the supply is installed as part of a meter socket, such that access can only be gained with the meter removed.

Listed communications equipment includes equipment associated with Smart Grid applications (two-way communications between the premises electrical system and the supplying utility) and equipment associated with premises-powered broadband communication systems. Where building safety functions such as automatic or manual transmission of fire, security, medical, or 911 messages are dependent on premises-powered communications equipment, this permission to connect ahead of the service disconnecting means enhances the reliability of the power supply. Opening (turning off) of the service disconnecting means may occur for any number of reasons including planned shutdowns, building emergencies, or vandalism. Regardless of the reason, power for the communications system remains uninterrupted.

## VII. Service Equipment — Overcurrent Protection

### 230.90 Where Required

Each ungrounded service conductor shall have overload protection.

Service equipment constitutes the main control and means of cutoff of the electrical supply to the premises wiring system. It usually consists of an overcurrent device, such as a circuit breaker or a fuse, which is installed in series with each ungrounded service conductor to provide overload protection only.

The service overcurrent device does not protect the service conductors under short-circuit or ground-fault conditions on the line side of the disconnect. Protection against ground faults and short circuits is provided by the special requirements for service conductor protection and the location of the conductors.

On multiwire circuits, two or three single-pole switches or circuit breakers that are capable of individual operation are permitted as one protective device, provided the switches or circuit breakers are equipped with handle ties or a master handle, so that all ungrounded conductors of a service can be disconnected with not more than six operations of the hand, per 230.71(B).

**(A) Ungrounded Conductor.** Such protection shall be provided by an overcurrent device in series with each ungrounded service conductor that has a rating or setting not higher than the allowable ampacity of the conductor. A set of fuses shall be considered all the fuses required to protect all the ungrounded conductors of a circuit. Single-pole circuit breakers, grouped in accordance with 230.71(B), shall be considered as one protective device.

*Exception No. 1: For motor-starting currents, ratings that comply with 430.52, 430.62, and 430.63 shall be permitted.*

The service OCPD for a load that includes motors as well as lighting or a lighting and appliance load is subject to the motor-starting currents for all of the motors within a building. This exception allows the service OCPD to be sized using the requirements of Article 430 to accommodate the motor-starting and running current plus the other loads within the building. All of the building loads are determined in accordance with Article 220, and the service conductors and OCPD must be sized to carry that load. OCPDs for motor loads are permitted to have a rating or setting that exceeds the allowable ampacity of the circuit conductors, and this exception extends that permission to the service OCPD. For an individual motor, the rating is specified by 430.52; for two or more motors, the rating is specified by 430.62; and for a motor(s) load plus lighting and appliance load, the rating is specified by 430.63.

*Exception No. 2: Fuses and circuit breakers with a rating or setting that complies with 240.4(B) or (C) and 240.6 shall be permitted.*

Where the conductor ampacity does not correspond to the standard ampere rating of a circuit breaker or fuse, this exception permits the next-larger-size circuit breaker or fuse to be installed. The permission to "round up" is limited by 240.4(B)(3) to ratings not exceeding 800 amperes. This provision only permits rounding up to the next standard size fuse or circuit breaker rating and does not permit the load to exceed the allowable ampacity of the service conductors. See 240.6 for standard ampere ratings of fuses and circuit breakers.

*Exception No. 3: Two to six circuit breakers or sets of fuses shall be permitted as the overcurrent device to provide the overload protection. The sum of the ratings of the circuit breakers or fuses shall be permitted to exceed the ampacity of the service conductors, provided the calculated load does not exceed the ampacity of the service conductors.*

If multiple switches or OCPDs are used as the disconnecting means, the ampacity of the service conductors must be equal to

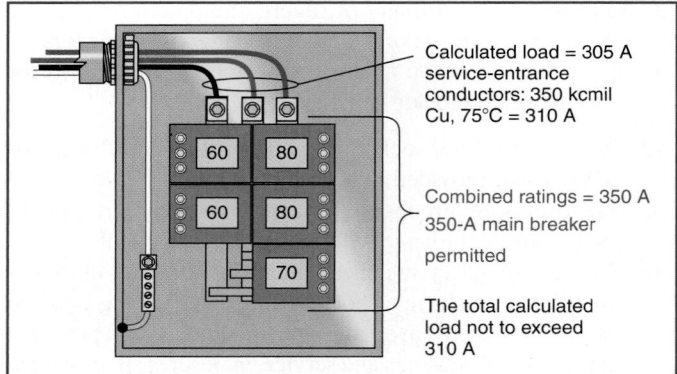

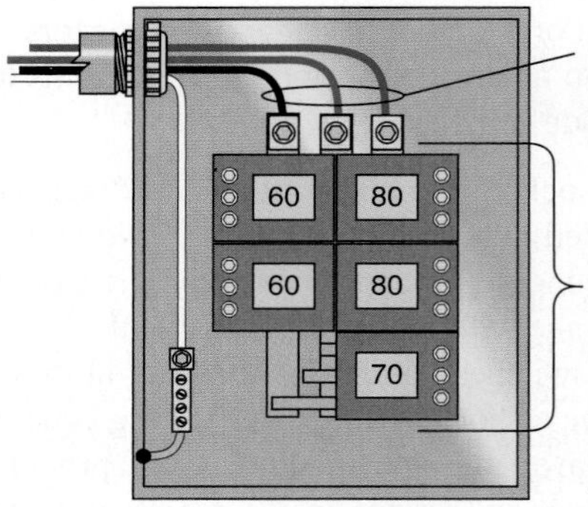


***EXHIBIT 230.26*** *An example in which the combined ratings of the five overcurrent devices are permitted to exceed the ampacity of the service conductors.*

or greater than the load calculated in accordance with Article 220; however, the conductor ampacity is not required to be equal to or greater than the combined rating of the multiple service OCPDs. The combined rating of service disconnecting means is covered in 230.80.

The combined ratings of the five OCPDs (350 amperes) shown in Exhibit 230.26 exceed the ampacity of the service-entrance conductors (310 amperes) that is permitted by this exception. As specified, the ampacity of the service-entrance conductors is sufficient to carry the calculated load. The combined rating of the five service disconnecting means also complies with 230.80, which requires that the combined rating (350 amperes) be not less than the calculated load (305 amperes), the minimum size required for the service OCPD specified by 240.4 and 230.80. In addition, the rating of the equipment (panelboard) in which the five OCPDs or service disconnecting means are installed cannot be less than the calculated load in accordance with the requirements of 408.30. This exception allows for some design flexibility by not requiring the sum of the ratings to be equal to or less than the ampacity of the service-entrance conductors.

For example, the calculated load supplied by the two 80-ampere devices shown in Exhibit 230.26 may be 73 amperes. Section 240.6 identifies 70 amperes and 80 amperes as standard OCPD sizes. A 70-ampere device is too small, but an 80-ampere device can be used in accordance with 240.4. The feeder conductors supplied by the 80-ampere device are required to have an ampacity not less than the calculated load, which in this example is 73 amperes.

*Exception No. 4: Overload protection for fire pump supply conductors shall comply with 695.4(B)(2)(a).*

*Exception No. 5: Overload protection for 120/240-volt, 3-wire, single-phase dwelling services shall be permitted in accordance with the requirements of 310.15(B)(7).*

**(B) Not in Grounded Conductor.** No overcurrent device shall be inserted in a grounded service conductor except a circuit breaker that simultaneously opens all conductors of the circuit.

## 230.91 Location

The service overcurrent device shall be an integral part of the service disconnecting means or shall be located immediately adjacent thereto.

## 230.92 Locked Service Overcurrent Devices

Where the service overcurrent devices are locked or sealed or are not readily accessible to the occupant, branch-circuit or feeder overcurrent devices shall be installed on the load side, shall be mounted in a readily accessible location, and shall be of lower ampere rating than the service overcurrent device.

## 230.93 Protection of Specific Circuits

Where necessary to prevent tampering, an automatic overcurrent device that protects service conductors supplying only a specific load, such as a water heater, shall be permitted to be locked or sealed where located so as to be accessible.

## 230.94 Relative Location of Overcurrent Device and Other Service Equipment

The overcurrent device shall protect all circuits and devices.

*Exception No. 1: The service switch shall be permitted on the supply side.*

*Exception No. 2: High-impedance shunt circuits, surge arresters, Type 1 surge-protective devices, surge-protective capacitors, and instrument transformers (current and voltage) shall be permitted to be connected and installed on the supply side of the service disconnecting means as permitted by 230.82.*

*Exception No. 3: Circuits for load management devices shall be permitted to be connected on the supply side of the service overcurrent device where separately provided with overcurrent protection.*

*Exception No. 4: Circuits used only for the operation of fire alarm, other protective signaling systems, or the supply to fire pump equipment shall be permitted to be connected on the supply side of the service overcurrent device where separately provided with overcurrent protection.*

*Exception No. 5: Meters nominally rated not in excess of 600 volts shall be permitted, provided all metal housings and service enclosures are grounded.*

*Exception No. 6: Where service equipment is power operable, the control circuit shall be permitted to be connected ahead of the service equipment if suitable overcurrent protection and disconnecting means are provided.*

## 230.95 Ground-Fault Protection of Equipment

Ground-fault protection of equipment shall be provided for solidly grounded wye electric services of more than 150 volts to ground but not exceeding 1000 volts phase-to-phase for each service disconnect rated 1000 amperes or more. The grounded conductor for the solidly grounded wye system shall be connected directly to ground through a grounding electrode system, as specified in 250.50, without inserting any resistor or impedance device.

The rating of the service disconnect shall be considered to be the rating of the largest fuse that can be installed or the highest continuous current trip setting for which the actual overcurrent device installed in a circuit breaker is rated or can be adjusted.

Ground-fault protection of equipment (GFPE) for service disconnecting means was first required in the 1971 *Code* due to the unusually high number of equipment burndowns reported on large capacity 480Y/277-V solidly grounded services. This section requires GFPE for each service disconnecting means rated 1000 amperes or more where the supply system is solidly grounded, is wye-connected, and operates at more than 150 volts to ground and not more than 1000 V phase-to-phase. See the definition of *ground-fault protection of equipment* in Article 100. Systems operating at 480Y/277 V were cited in the original substantiation for this requirement, but other solidly grounded, wye-connected systems (i.e., 600Y/347 V) are covered within the parameters specified by this section. This requirement does not apply to systems where the grounded conductor is not solidly grounded, as is the case with high-impedance grounded neutral systems covered in 250.36.

Ground-fault protection of services does not protect the conductors on the supply side of the service disconnecting means, but it is designed to provide protection from line-to-ground faults that occur on the load side of the service disconnecting means. An alternative to installing ground-fault protection is to provide multiple disconnects rated less than 1000 amperes. For instance, up to six 800-A disconnecting means may be used, and in that case ground-fault protection would not be required. Informational Note No. 2 to 230.95(C) recognizes that ground-fault protection may be desirable at lesser amperages on solidly grounded systems for voltages exceeding 150 volts to ground but not exceeding 1000 volts phase to phase.

Engineering studies are recommended to determine the circuit impedance and short-circuit currents that would be available at the supply terminals, so that equipment and overcurrent protection of the proper interrupting rating are used. See 110.9 and 110.10 for details on interrupting rating and circuit impedance.

The two basic types of ground-fault equipment protectors are illustrated in Exhibits 230.27 and 230.28. In Exhibit 230.27, the ground-fault sensor is installed around all the circuit conductors, and a stray current on a line-to-ground fault sets up an imbalance of the currents flowing in individual conductors installed through the ground-fault sensor. When this current exceeds the setting of the ground-fault sensor, the shunt trip operates and opens the circuit breakers.

The ground-fault sensor illustrated in Exhibit 230.28 is installed around the bonding jumper only. When an unbalanced current from a line-to-ground fault occurs, the current flows through the bonding jumper and the shunt trip causes the circuit breaker to operate, removing the load from the line. See also 250.24(A)(4), which permits a grounding electrode conductor connection to the equipment grounding terminal bar or bus.

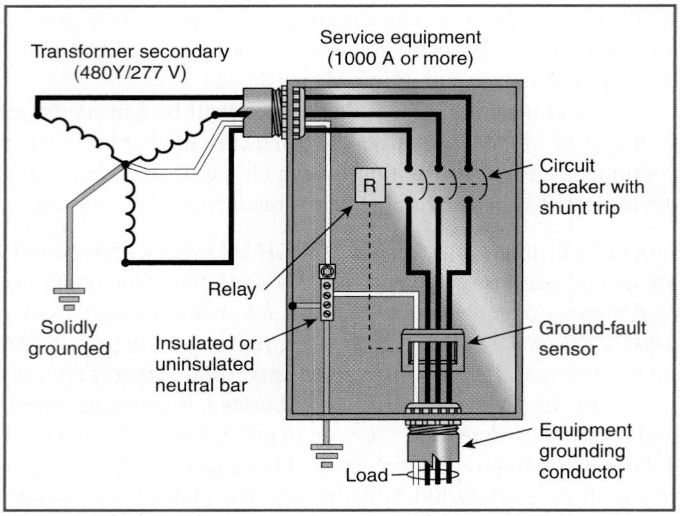


***EXHIBIT 230.27*** *A ground-fault sensor encircling all circuit conductors, including the neutral.*

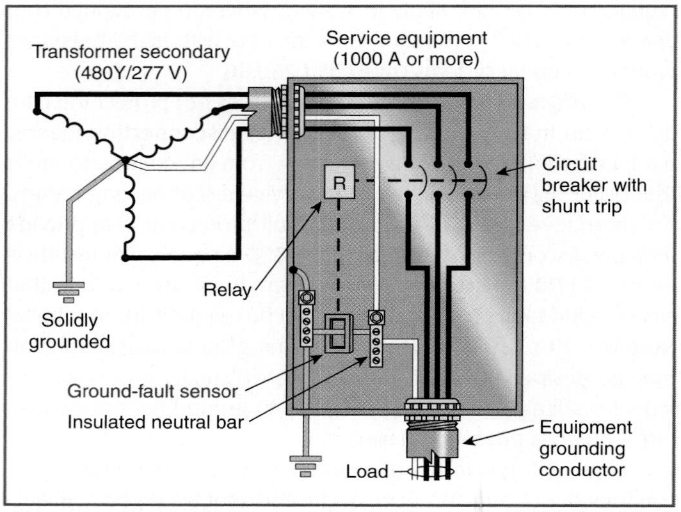


***EXHIBIT 230.28*** *A ground-fault sensor encircling only the bonding jumper conductor.*

*Exception: The ground-fault protection provisions of this section shall not apply to a service disconnect for a continuous industrial process where a nonorderly shutdown will introduce additional or increased hazards.*

Unplanned interruption of power to some industrial processes results in a life-safety or catastrophic-failure hazards more significant than the electrical hazard mitigated by this requirement. In addition, 695.6(G) prohibits fire pumps from being protected by GFPE-type devices. The use of high-impedance grounded neutral systems as covered in 250.36 is a means to maintain continuity of power where the supply system is 3-phase, 4-wire, wye-connected and no line-to-neutral loads are supplied.

**(A) Setting.** The ground-fault protection system shall operate to cause the service disconnect to open all ungrounded conductors of the faulted circuit. The maximum setting of the ground-fault protection shall be 1200 amperes, and the maximum time delay shall be one second for ground-fault currents equal to or greater than 3000 amperes.

Ground-fault sensors have a maximum setting of 1200 amperes and no minimum. Settings at low levels can increase the likelihood of unwanted shutdowns. The requirements of 230.95 place a restriction on fault currents greater than 3000 amperes and limit the duration of the fault to not more than 1 second. This restriction minimizes the amount of damage done by an arcing fault, which is directly proportional to the time the arcing fault is allowed to burn.

Care should be taken to ensure that interconnecting multiple supply systems does not interfere with proper sensing by the ground-fault protection equipment. A careful engineering study must be made to ensure that fault currents do not take parallel paths to the supply system, thereby bypassing the ground-fault detection device. See 215.10, 240.13, 517.17, and 705.32 for further requirements covering GFPE.

**(B) Fuses.** If a switch and fuse combination is used, the fuses employed shall be capable of interrupting any current higher than the interrupting capacity of the switch during a time that the ground-fault protective system will not cause the switch to open.

**(C) Performance Testing.** The ground-fault protection system shall be performance tested when first installed on site. The test shall be conducted in accordance with instructions that shall be provided with the equipment. A written record of this test shall be made and shall be available to the authority having jurisdiction.

The requirement for ground-fault protection system performance testing is a result of numerous reports of ground-fault protection systems that were improperly wired and could not or did not provide the intended protection. This section and third-party testing organizations that evaluate GFPE equipment require a set of performance testing instructions to be supplied with the equipment. Listed GFPE equipment that is not installed and tested as specified in the product's installation and use instructions does not comply with 110.3(B).

Informational Note No. 1: Ground-fault protection that functions to open the service disconnect affords no protection from faults on the line side of the protective element. It serves only to limit damage to conductors and equipment on the load side in the event of an arcing ground fault on the load side of the protective element.

Informational Note No. 2: This added protective equipment at the service equipment may make it necessary to review the overall wiring system for proper selective overcurrent protection coordination. Additional installations of ground-fault protective equipment may be needed on feeders and branch circuits where maximum continuity of electric service is necessary.

Informational Note No. 3: Where ground-fault protection is provided for the service disconnect and interconnection is made with another supply system by a transfer device, means or devices may be needed to ensure proper ground-fault sensing by the ground-fault protection equipment.

Informational Note No. 4: See 517.17(A) for information on where an additional step of ground-fault protection is required for hospitals and other buildings with critical areas or life support equipment.

## VIII. Services Exceeding 1000 Volts, Nominal

### 230.200 General

Service conductors and equipment used on circuits exceeding 1000 volts, nominal, shall comply with all the applicable provisions of the preceding sections of this article and with the following sections that supplement or modify the preceding sections. In no case shall the provisions of Part VIII apply to equipment on the supply side of the service point.

Informational Note: For clearances of conductors of over 1000 volts, nominal, see ANSI C2-2007, *National Electrical Safety Code*.

Where services rated over 1000 volts supply utility-owned and utility-maintained transformers, the underground conductors (including those rising up the pole) on the line side and load side of the transformer up to the underground enclosure at which the *service point* is established are *service-lateral* conductors. Only those conductors on the load side of the service-point connection are subject to the requirements of the *NEC*. The service point is a specific location where the supply conductors of the electric utility and the customer-owned (premises wiring) conductors connect.

Exhibit 230.29 depicts an installation where the transformer and service lateral to the service point are owned by the electric utility. The transformer secondary conductors between the service point (separate connection point outside of the transformer in this scenario) and the service disconnecting means at the building are *underground service conductors* until the point at which they enter the building. From that point to the termination in the service equipment they are *service-entrance conductors*.

In the installation depicted in Exhibit 230.30, the service disconnecting means is located at the customer-owned transformer primary. The conductors that connect the service point (at the top of the pole), the service disconnecting means, and the service/transformer OCPD are underground service conductors. The conductors between the transformer secondary and the line side of the building disconnecting means are *feeders* and are subject to the requirements for outside conductors connected to a transformer secondary contained in 240.21(C)(4). Conductors (not shown) on the load side of the building disconnecting means are also feeders. Each building or structure is required to have a disconnecting means, in accordance with 225.31.

### 230.202 Service-Entrance Conductors

Service-entrance conductors to buildings or enclosures shall be installed to conform to 230.202(A) and (B).

**(A) Conductor Size.** Service-entrance conductors shall not be smaller than 6 AWG unless in multiconductor cable. Multiconductor cable shall not be smaller than 8 AWG.

**(B) Wiring Methods.** Service-entrance conductors shall be installed by one of the wiring methods covered in 300.37 and 300.50.

### 230.204 Isolating Switches.

**(A) Where Required.** Where oil switches or air, oil, vacuum, or sulfur hexafluoride circuit breakers constitute the service disconnecting means, an isolating switch with visible break contacts shall be installed on the supply side of the disconnecting means and all associated service equipment.

*Exception: An isolating switch shall not be required where the circuit breaker or switch is mounted on removable truck*

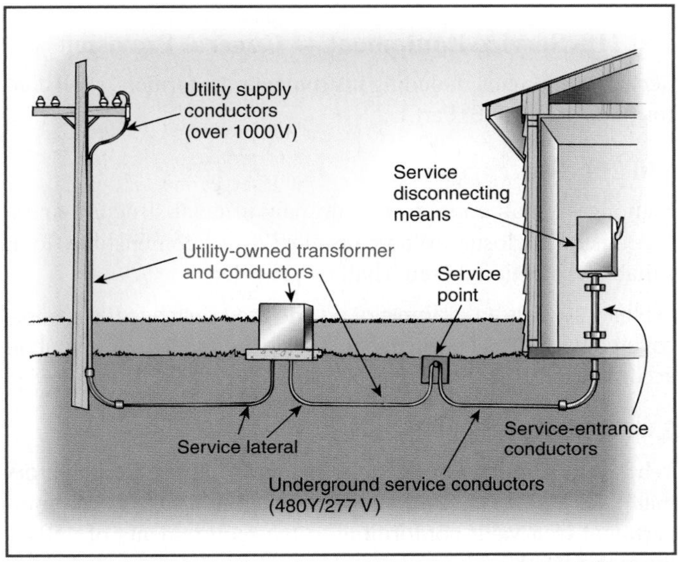


***EXHIBIT 230.29*** *Service rated over 1000 volts supplied by a utility-owned transformer.*

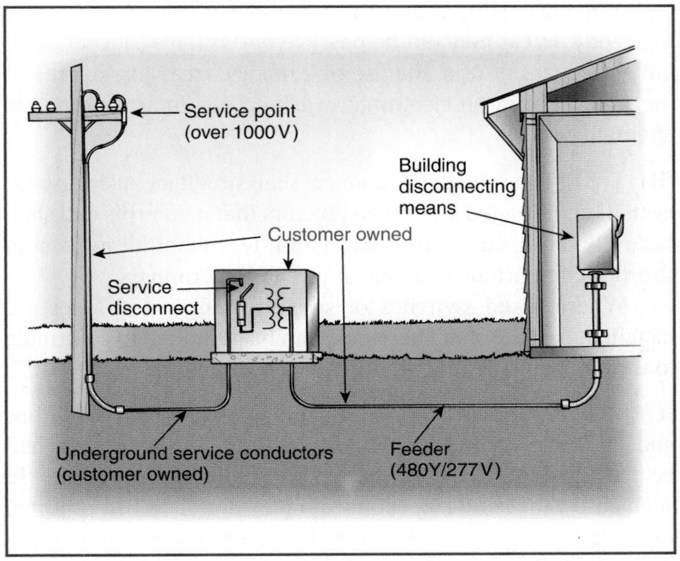


***EXHIBIT 230.30*** *Service rated over 1000 volts supplying a customer-owned transformer.*

*panels or switchgear units where both of the following conditions apply:*

*(1) Cannot be opened unless the circuit is disconnected*
*(2) Where all energized parts are automatically disconnected when the circuit breaker or switch is removed from the normal operating position*

**(B) Fuses as Isolating Switch.** Where fuses are of the type that can be operated as a disconnecting switch, a set of such fuses shall be permitted as the isolating switch.

**(C) Accessible to Qualified Persons Only.** The isolating switch shall be accessible to qualified persons only.

**(D) Connection to Ground.** Isolating switches shall be provided with a means for readily connecting the load side conductors to a grounding electrode system, equipment ground busbar, or grounded steel structure when disconnected from the source of supply.

A means for grounding the load side conductors to a grounding electrode system, equipment grounding busbar, or grounded structural steel shall not be required for any duplicate isolating switch installed and maintained by the electric supply company.

## 230.205 Disconnecting Means

**(A) Location.** The service disconnecting means shall be located in accordance with 230.70.

For either overhead or underground primary distribution systems on private property, the service disconnect shall be permitted to be located in a location that is not readily accessible, if the disconnecting means can be operated by mechanical linkage from a readily accessible point, or electronically in accordance with 230.205(C), where applicable.

As is the case for services operating at 1000 volts nominal or less, the general requirement for service disconnecting means in systems rated over 1000 volts is that it be readily accessible. Because this ready access may not be possible or desirable (since it could put employees at risk), the use of a readily accessible operating mechanism or of an electronic switching device are acceptable alternatives.

**(B) Type.** Each service disconnect shall simultaneously disconnect all ungrounded service conductors that it controls and shall have a fault-closing rating that is not less than the maximum short-circuit current available at its supply terminals.

Where fused switches or separately mounted fuses are installed, the fuse characteristics shall be permitted to contribute to the fault-closing rating of the disconnecting means.

**(C) Remote Control.** For multibuilding, industrial installations under single management, the service disconnecting means shall be permitted to be located at a separate building or structure. In such cases, the service disconnecting means shall be permitted to be electrically operated by a readily accessible, remote-control device.

## 230.206 Overcurrent Devices as Disconnecting Means

Where the circuit breaker or alternative for it, as specified in 230.208 for service overcurrent devices, meets the requirements specified in 230.205, they shall constitute the service disconnecting means.

## 230.208 Protection Requirements

A short-circuit protective device shall be provided on the load side of, or as an integral part of, the service disconnect, and shall protect all ungrounded conductors that it supplies. The protective device shall be capable of detecting and interrupting all values of current, in excess of its trip setting or melting point, that can occur at its location. A fuse rated in continuous amperes not to exceed three times the ampacity of the conductor, or a circuit breaker with a trip setting of not more than six times the ampacity of the conductors, shall be considered as providing the required short-circuit protection.

Informational Note: See Table 310.60(C)(67) through Table 310.60(C)(86) for ampacities of conductors rated 2001 volts and above.

Overcurrent devices shall conform to 230.208(A) and (B).

**(A) Equipment Type.** Equipment used to protect service-entrance conductors shall meet the requirements of Article 490, Part II.

**(B) Enclosed Overcurrent Devices.** The restriction to 80 percent of the rating for an enclosed overcurrent device for continuous loads shall not apply to overcurrent devices installed in systems operating at over 1000 volts.

## 230.209 Surge Arresters (Lightning Arresters)

Surge arresters installed in accordance with the requirements of Article 280 shall be permitted on each ungrounded overhead service conductor.

## 230.210 Service Equipment — General Provisions

Service equipment, including instrument transformers, shall conform to Article 490, Part I.

## 230.211 Switchgear

Switchgear shall consist of a substantial metal structure and a sheet metal enclosure. Where installed over a combustible floor, suitable protection thereto shall be provided.

Exhibit 230.31 is an example of an enclosed switchgear assembly covered by 230.211. Switchgear can be used in lieu of a vault in accordance with 230.212.

## 230.212 Over 35,000 Volts

Where the voltage exceeds 35,000 volts between conductors that enter a building, they shall terminate in a switchgear compartment or a vault conforming to the requirements of 450.41 through 450.48.

***EXHIBIT 230.31*** *Assembly of metal-enclosed switchgear. (Courtesy of Schneider Electric)*

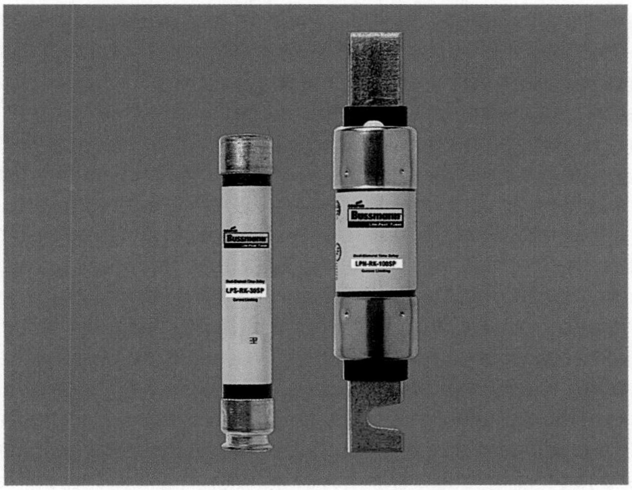

***EXHIBIT 240.1*** *Class R current-limiting fuses with rejection feature to prohibit the installation of non–current-limiting fuses. (Courtesy of Cooper Bussmann, a division of Cooper Industries PLC)*

# ARTICLE 240 Overcurrent Protection

## I. General

### 240.1 Scope

Parts I through VII of this article provide the general requirements for overcurrent protection and overcurrent protective devices not more than 1000 volts, nominal. Part VIII covers overcurrent protection for those portions of supervised industrial installations operating at voltages of not more than 1000 volts, nominal. Part IX covers overcurrent protection over 1000 volts, nominal.

Informational Note: Overcurrent protection for conductors and equipment is provided to open the circuit if the current reaches a value that will cause an excessive or dangerous temperature in conductors or conductor insulation. See also 110.9 for requirements for interrupting ratings and 110.10 for requirements for protection against fault currents.

### 240.2 Definitions

**Current-Limiting Overcurrent Protective Device.** A device that, when interrupting currents in its current-limiting range, reduces the current flowing in the faulted circuit to a magnitude substantially less than that obtainable in the same circuit if the device were replaced with a solid conductor having comparable impedance.

A current-limiting protective device is one that cuts off a fault current in less than one-half cycle, thus preventing short-circuit currents from building up to their full available values. Most electrical distribution systems can deliver high ground-fault or short-circuit currents to components such as conductors and service equipment. These components may be damaged or destroyed by high fault currents, resulting in serious burndowns and fires. Properly selected current-limiting OCPDs, such as the ones shown in Exhibit 240.1, limit the let-through energy to an amount that does not exceed the rating of the components in spite of high available short-circuit currents.

Proper selection of current-limiting devices depends on the type of device selected. For example, a Class RK5 fuse is not as current limiting as a Class RK1 fuse. Furthermore, a Class RK1 fuse is not as current limiting as a high-speed semiconductor fuse. See 110.9 and 110.10 for details on interrupting ratings, circuit impedance, and other characteristics.

**Supervised Industrial Installation.** For the purposes of Part VIII, the industrial portions of a facility where all of the following conditions are met:

(1) Conditions of maintenance and engineering supervision ensure that only qualified persons monitor and service the system.
(2) The premises wiring system has 2500 kVA or greater of load used in industrial process(es), manufacturing activities, or both, as calculated in accordance with Article 220.
(3) The premises has at least one service or feeder that is more than 150 volts to ground and more than 300 volts phase-to-phase.

This definition excludes installations in buildings used by the industrial facility for offices, warehouses, garages, machine shops, and recreational facilities that are not an integral part of the industrial plant, substation, or control center.

To qualify as an industrial establishment in accordance with this definition, the facility must have a combined process and manufacturing load greater than 2500 kVA. All process or manufacturing loads from each low-, medium-, and high-voltage system can be added together to satisfy the load requirements of Part VIII of Article 240. Loads not associated with manufacturing or processing cannot be used to meet the 2500 kVA minimum requirement. Loads are calculated in accordance with Article 220.

The requirements of Part VIII apply only to electrical systems operating at 1000 volts, nominal, or less that are used for process or manufacturing. Part VIII does not apply to electrical systems operating at over 1000 volts, nominal, or to electrical systems that serve separate facilities – such as offices, warehouses, garages, machine shops, or recreational buildings – that are not a part of the manufacturing or industrial process. However, if a part of the industrial process or manufacturing electrical system is used to serve an office, warehouse, garage, machine shop, or recreational facility that is an integral part of the industrial plant, control center, or substation, Part VIII can still apply to the total process or manufacturing electrical system.

**Tap Conductors.** As used in this article, a tap conductor is defined as a conductor, other than a service conductor, that has overcurrent protection ahead of its point of supply that exceeds the value permitted for similar conductors that are protected as described elsewhere in 240.4.

Tap conductors are branch-circuit and feeder conductors subject to the special overcurrent protection requirements specified in 240.21. Overcurrent protection includes short-circuit, ground-fault, and overload protection, and the general requirement in 240.21 is for conductor overcurrent protection to be provided at the point a conductor is supplied. Section 240.21 allows for the installation of branch-circuit and feeder conductors that are protected against overcurrent using an OCPD downstream of the point of supply.

## 240.3 Other Articles

Equipment shall be protected against overcurrent in accordance with the article in this *Code* that covers the type of equipment specified in Table 240.3.

***TABLE 240.3*** *Other Articles*

| Equipment | Article |
|---|---|
| Air-conditioning and refrigerating equipment | 440 |
| Appliances | 422 |
| Assembly occupancies | 518 |
| Audio signal processing, amplification, and reproduction equipment | 640 |
| Branch circuits | 210 |
| Busways | 368 |
| Capacitors | 460 |
| Class 1, Class 2, and Class 3 remote-control, signaling, and power-limited circuits | 725 |
| Cranes and hoists | 610 |
| Electric signs and outline lighting | 600 |
| Electric welders | 630 |
| Electrolytic cells | 668 |
| Elevators, dumbwaiters, escalators, moving walks, wheelchair lifts, and stairway chairlifts | 620 |
| Emergency systems | 700 |
| Fire alarm systems | 760 |
| Fire pumps | 695 |
| Fixed electric heating equipment for pipelines and vessels | 427 |
| Fixed electric space-heating equipment | 424 |
| Fixed outdoor electric deicing and snow-melting equipment | 426 |
| Generators | 445 |
| Health care facilities | 517 |
| Induction and dielectric heating equipment | 665 |
| Industrial machinery | 670 |
| Luminaires, lampholders, and lamps | 410 |
| Motion picture and television studios and similar locations | 530 |
| Motors, motor circuits, and controllers | 430 |
| Phase converters | 455 |
| Pipe organs | 650 |
| Receptacles | 406 |
| Services | 230 |
| Solar photovoltaic systems | 690 |
| Switchboards and panelboards | 408 |
| Theaters, audience areas of motion picture and television studios, and similar locations | 520 |
| Transformers and transformer vaults | 450 |
| X-ray equipment | 660 |

## 240.4 Protection of Conductors

Conductors, other than flexible cords, flexible cables, and fixture wires, shall be protected against overcurrent in accordance with their ampacities specified in 310.15, unless otherwise permitted or required in 240.4(A) through (G).

Informational Note: See ICEA P-32-382-2007 for information on allowable short-circuit currents for insulated copper and aluminum conductors.

**(A) Power Loss Hazard.** Conductor overload protection shall not be required where the interruption of the circuit would create a hazard, such as in a material-handling magnet circuit or fire pump circuit. Short-circuit protection shall be provided.

Informational Note: See NFPA 20-2013, *Standard for the Installation of Stationary Pumps for Fire Protection.*

A refinery is an example of a facility in which interruption of some processes could pose a significant hazard.

**(B) Overcurrent Devices Rated 800 Amperes or Less.** The next higher standard overcurrent device rating (above the ampacity of the conductors being protected) shall be permitted to be used, provided all of the following conditions are met:

(1) The conductors being protected are not part of a branch circuit supplying more than one receptacle for cord-and-plug-connected portable loads.

(2) The ampacity of the conductors does not correspond with the standard ampere rating of a fuse or a circuit breaker without overload trip adjustments above its rating (but that shall be permitted to have other trip or rating adjustments).
(3) The next higher standard rating selected does not exceed 800 amperes.

Article 310 contains conductor ampacity tables based on conductor types, voltage rating, and conditions of use and application. For premises electrical installations operating at 2000 volts and below, the most widely used ampacity table is Table 310.15(B)(16). Section 240.6 lists the standard ratings of overcurrent devices. Where the ampacity of the conductor specified in these tables does not match the rating of the standard overcurrent device, 240.4(B) permits the use of the next larger standard overcurrent device. All three conditions must be met for this permission to apply. However, if the ampacity of a conductor corresponds with a standard rating in 240.6, the conductor is required to be protected by the standard rated device or by one with a smaller rating. For example, in Table 310.15(B)(16), 3 AWG, 75°C copper, Type THWN, the ampacity is listed as 100 amperes. That conductor would have to be protected by an OCPD rated not more than 100 amperes unless otherwise permitted in 240.4(E), (F), or (G).

Section 240.4(B) does not modify or change the allowable ampacity of the conductor – it only serves to provide a reasonable increase in the permitted OCPD rating where the allowable ampacity and the standard OCPD ratings do not correspond. The allowable ampacity of branch circuits or feeders rated 600 volts or less must always be capable of supplying the calculated load in accordance with the requirements of 210.19(A)(1) and 215.2(A)(1).

For services rated 1000 volts or less, the allowable ampacity of the service conductors must always be capable of supplying the calculated load in accordance with the requirements of 230.42(A).

For example, a 500-kcmil THWN copper conductor has an allowable ampacity of 380 amperes, specified in Table 310.15(B)(16). This conductor can supply a load not exceeding 380 amperes and, in accordance with 240.4(B), can be protected by a 400-ampere OCPD.

Section 240.4 references 310.15; therefore, all of the requirements of that section are applicable. Section 310.15(B)(7) permits the conductor types and sizes specified in that section to supply calculated loads based on the ratings specified. The service and main power feeder loads permitted to be supplied by the conductor types and sizes covered in 310.15(B)(7) exceed the conductor ampacities for the same conductor types and sizes specified in 310.15(B)(16). The overcurrent protection for these residential supply conductors is also permitted to be based on the increased rating allowed by 310.15(B)(7). Application of 310.15(B)(7) is permitted only for single-phase, 120/240-V, residential services and main power feeders. The increased ratings given in 310.15(B)(7) are based on the significant diversity inherent to most dwelling unit loads and the fact that only the two ungrounded service or feeder conductors are considered to be current carrying.

**(C) Overcurrent Devices Rated over 800 Amperes.** Where the overcurrent device is rated over 800 amperes, the ampacity of the conductors it protects shall be equal to or greater than the rating of the overcurrent device defined in 240.6.

**(D) Small Conductors.** Unless specifically permitted in 240.4(E) or (G), the overcurrent protection shall not exceed that required by (D)(1) through (D)(7) after any correction factors for ambient temperature and number of conductors have been applied.

**(1) 18 AWG Copper.** 7 amperes, provided all the following conditions are met:

(1) Continuous loads do not exceed 5.6 amperes.
(2) Overcurrent protection is provided by one of the following:
  a. Branch-circuit-rated circuit breakers listed and marked for use with 18 AWG copper wire
  b. Branch-circuit-rated fuses listed and marked for use with 18 AWG copper wire
  c. Class CC, Class J, or Class T fuses

**(2) 16 AWG Copper.** 10 amperes, provided all the following conditions are met:

(1) Continuous loads do not exceed 8 amperes.
(2) Overcurrent protection is provided by one of the following:
  a. Branch-circuit-rated circuit breakers listed and marked for use with 16 AWG copper wire
  b. Branch-circuit-rated fuses listed and marked for use with 16 AWG copper wire
  c. Class CC, Class J, or Class T fuses

**(3) 14 AWG Copper.** 15 amperes

**(4) 12 AWG Aluminum and Copper-Clad Aluminum.** 15 amperes

**(5) 12 AWG Copper.** 20 amperes

**(6) 10 AWG Aluminum and Copper-Clad Aluminum.** 25 amperes

**(7) 10 AWG Copper.** 30 amperes

**(E) Tap Conductors.** Tap conductors shall be permitted to be protected against overcurrent in accordance with the following:

(1) 210.19(A)(3) and (A)(4), Household Ranges and Cooking Appliances and Other Loads
(2) 240.5(B)(2), Fixture Wire
(3) 240.21, Location in Circuit
(4) 368.17(B), Reduction in Ampacity Size of Busway
(5) 368.17(C), Feeder or Branch Circuits (busway taps)
(6) 430.53(D), Single Motor Taps

**(F) Transformer Secondary Conductors.** Single-phase (other than 2-wire) and multiphase (other than delta-delta, 3-wire) transformer secondary conductors shall not be considered to be protected by the primary overcurrent protective device. Conductors supplied by the secondary side of a single-phase transformer having a 2-wire (single-voltage) secondary, or a three-phase, delta-delta connected transformer having a 3-wire (single-voltage) secondary, shall be permitted to be protected by overcurrent protection provided on the primary (supply) side of the transformer, provided this protection is in accordance with 450.3 and does not exceed the value determined by multiplying the secondary conductor ampacity by the secondary-to-primary transformer voltage ratio.

Section 240.4 requires conductors to be protected against overcurrent in accordance with their ampacity, and 240.21 requires that the protection be provided at the point the conductor receives its supply. Section 240.4(F) permits the secondary circuit conductors from a transformer to be protected by overcurrent devices in the primary circuit conductors of the transformer only in the following two special cases:

1. A transformer with a 2-wire primary and a 2-wire secondary, provided the transformer primary is protected in accordance with 450.3
2. A 3-phase, delta-delta-connected transformer having a 3-wire, single-voltage secondary, provided its primary is protected in accordance with 450.3

Except for those two special cases, transformer secondary conductors must be protected by the use of overcurrent devices, because the primary overcurrent devices do not provide such protection.

**(G) Overcurrent Protection for Specific Conductor Applications.** Overcurrent protection for the specific conductors shall be permitted to be provided as referenced in Table 240.4(G).

***TABLE 240.4(G)*** *Specific Conductor Applications*

| Conductor | Article | Section |
|---|---|---|
| Air-conditioning and refrigeration equipment circuit conductors | 440, Parts III, VI | |
| Capacitor circuit conductors | 460 | 460.8(B) and 460.25(A)–(D) |
| Control and instrumentation circuit conductors (Type ITC) | 727 | 727.9 |
| Electric welder circuit conductors | 630 | 630.12 and 630.32 |
| Fire alarm system circuit conductors | 760 | 760.43, 760.45, 760.121, and Chapter 9, Tables 12(A) and 12(B) |
| Motor-operated appliance circuit conductors | 422, Part II | |
| Motor and motor-control circuit conductors | 430, Parts II, III, IV, V, VI, VII | |
| Phase converter supply conductors | 455 | 455.7 |
| Remote-control, signaling, and power-limited circuit conductors | 725 | 725.43, 725.45, 725.121, and Chapter 9, Tables 11(A) and 11(B) |
| Secondary tie conductors | 450 | 450.6 |

## 240.5 Protection of Flexible Cords, Flexible Cables, and Fixture Wires

Flexible cord and flexible cable, including tinsel cord and extension cords, and fixture wires shall be protected against overcurrent by either 240.5(A) or (B).

**(A) Ampacities.** Flexible cord and flexible cable shall be protected by an overcurrent device in accordance with their ampacity as specified in Table 400.5(A)(1) and Table 400.5(A)(2). Fixture wire shall be protected against overcurrent in accordance with its ampacity as specified in Table 402.5. Supplementary overcurrent protection, as covered in 240.10, shall be permitted to be an acceptable means for providing this protection.

**(B) Branch-Circuit Overcurrent Device.** Flexible cord shall be protected, where supplied by a branch circuit, in accordance with one of the methods described in 240.5(B)(1), (B)(3), or (B)(4). Fixture wire shall be protected, where supplied by a branch circuit, in accordance with 240.5(B)(2).

**(1) Supply Cord of Listed Appliance or Luminaire.** Where flexible cord or tinsel cord is approved for and used with a specific listed appliance or luminaire, it shall be considered to be protected when applied within the appliance or luminaire listing requirements. For the purposes of this section, a luminaire may be either portable or permanent.

A flexible cord connected to a listed appliance or portable lamp or used in a listed extension cord set is considered to be protected by the branch-circuit OCPD as long as the appliance, lamp, or extension cord is used in accordance with its listing requirements. These listing requirements are developed by the third-party testing and listing organizations with technical input from cord, appliance, and lamp manufacturers. For fixture wire, 240.5(B)(2) establishes a maximum protective device rating based on a minimum conductor size and a maximum conductor length.

**(2) Fixture Wire.** Fixture wire shall be permitted to be tapped to the branch-circuit conductor of a branch circuit in accordance with the following:

(1) 20-ampere circuits — 18 AWG, up to 15 m (50 ft) of run length
(2) 20-ampere circuits — 16 AWG, up to 30 m (100 ft) of run length

(3) 20-ampere circuits — 14 AWG and larger
(4) 30-ampere circuits — 14 AWG and larger
(5) 40-ampere circuits — 12 AWG and larger
(6) 50-ampere circuits — 12 AWG and larger

**(3) Extension Cord Sets.** Flexible cord used in listed extension cord sets shall be considered to be protected when applied within the extension cord listing requirements.

**(4) Field Assembled Extension Cord Sets.** Flexible cord used in extension cords made with separately listed and installed components shall be permitted to be supplied by a branch circuit in accordance with the following:

20-ampere circuits — 16 AWG and larger

Field-assembled extension cords are permitted, provided the conductors are 16 AWG or larger and the overcurrent protection for the branch circuit to which the cord is connected does not exceed 20 amperes. If a field-assembled extension cord is connected to a circuit rated higher than 20 amperes, the cord ampacity cannot be less than the rating of the circuit OCPD. The cord and the cord caps and connectors used for this type of assembly are required to be listed.

***EXHIBIT 240.2*** *An adjustable-trip circuit breaker with a transparent, removable, and sealable cover. (Courtesy of Schneider Electric)*

## 240.6 Standard Ampere Ratings

**(A) Fuses and Fixed-Trip Circuit Breakers.** The standard ampere ratings for fuses and inverse time circuit breakers shall be considered 15, 20, 25, 30, 35, 40, 45, 50, 60, 70, 80, 90, 100, 110, 125, 150, 175, 200, 225, 250, 300, 350, 400, 450, 500, 600, 700, 800, 1000, 1200, 1600, 2000, 2500, 3000, 4000, 5000, and 6000 amperes. Additional standard ampere ratings for fuses shall be 1, 3, 6, 10, and 601. The use of fuses and inverse time circuit breakers with nonstandard ampere ratings shall be permitted.

**(B) Adjustable-Trip Circuit Breakers.** The rating of adjustable-trip circuit breakers having external means for adjusting the current setting (long-time pickup setting), not meeting the requirements of 240.6(C), shall be the maximum setting possible.

**(C) Restricted Access Adjustable-Trip Circuit Breakers.** A circuit breaker(s) that has restricted access to the adjusting means shall be permitted to have an ampere rating(s) that is equal to the adjusted current setting (long-time pickup setting). Restricted access shall be defined as located behind one of the following:

(1) Removable and sealable covers over the adjusting means
(2) Bolted equipment enclosure doors
(3) Locked doors accessible only to qualified personnel

The set long-time pickup rating, as opposed to the instantaneous trip rating, of an adjustable-trip circuit breaker is permitted to be considered the circuit-breaker rating where access to the adjustment means is limited. This access limitation can be provided by locating the adjustment means behind sealable covers, as shown in Exhibit 240.2, behind bolted equipment enclosures, or behind locked equipment room doors with access available only to qualified personnel. The purpose of limiting access to the adjustment prevents tampering or readjustment by unqualified personnel.

## 240.8 Fuses or Circuit Breakers in Parallel

Fuses and circuit breakers shall be permitted to be connected in parallel where they are factory assembled in parallel and listed as a unit. Individual fuses, circuit breakers, or combinations thereof shall not otherwise be connected in parallel.

Parallel low-voltage circuit breakers or fuses are permitted if they are tested and factory assembled in parallel and listed as a unit. High-voltage fuses have long been used in parallel if they are assembled in an identified common mounting. Section 404.17 prohibits the use of fuses in parallel in fused switches except as permitted by 240.8 for listed assemblies.

## 240.9 Thermal Devices

Thermal relays and other devices not designed to open short circuits or ground faults shall not be used for the protection of conductors against overcurrent due to short circuits or ground faults, but the use of such devices shall be permitted to protect motor branch-circuit conductors from overload if protected in accordance with 430.40.

## 240.10 Supplementary Overcurrent Protection

Where supplementary overcurrent protection is used for luminaires, appliances, and other equipment or for internal circuits and components of equipment, it shall not be used as a substitute for required branch-circuit overcurrent devices or in place of the required branch-circuit protection. Supplementary overcurrent devices shall not be required to be readily accessible.

## 240.12 Electrical System Coordination

Where an orderly shutdown is required to minimize the hazard(s) to personnel and equipment, a system of coordination based on the following two conditions shall be permitted:

(1) Coordinated short-circuit protection

With coordinated overcurrent protection, the faulted or overloaded circuit is isolated by the selective operation of only the OCPD closest to the overcurrent condition. This selective operation prevents power loss to unaffected loads. Coordinated short-circuit protection will automatically open the circuit by localizing and de-energizing the faulted portion of the circuit, but an overload condition is not required to result in automatic opening of a protective device. Instead, an alarm can be used to warn of the overload condition, and remedial action can be taken. In some circumstances, an orderly shutdown of a system or process is more critical to personnel and equipment safety than is the automatic operation of the OCPD in response to an overload. Selective coordination requirements for specific systems or equipment are contained in 620.62, 645.27, 695.3(C)(3), 700.27, 701.27, and 708.54. Examples of overcurrent protection with and without coordinated protection are illustrated in Exhibit 240.3. The system with coordinated overcurrent protection (short-circuit in this case) localizes the fault and only the OCPD closest to the faulted portion of the circuit opens.

(2) Overload indication based on monitoring systems or devices

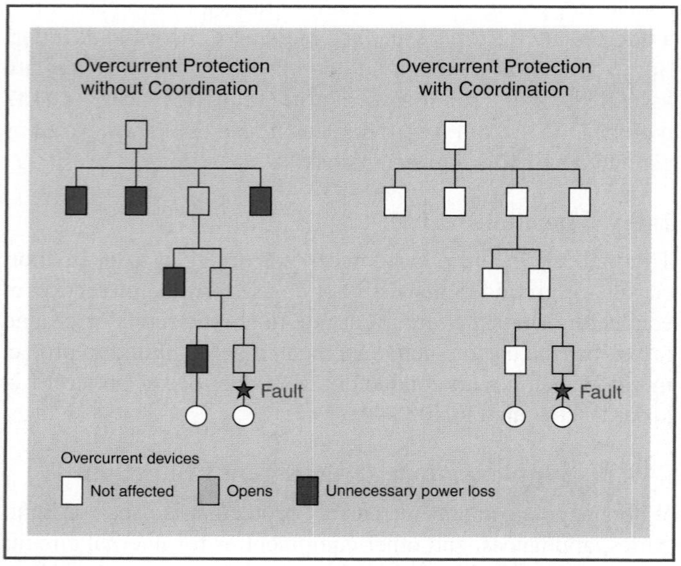


***EXHIBIT 240.3*** *Overcurrent protection schemes without system coordination and with system coordination.*

Informational Note: The monitoring system may cause the condition to go to alarm, allowing corrective action or an orderly shutdown, thereby minimizing personnel hazard and equipment damage.

## 240.13 Ground-Fault Protection of Equipment

Ground-fault protection of equipment shall be provided in accordance with the provisions of 230.95 for solidly grounded wye electrical systems of more than 150 volts to ground but not exceeding 1000 volts phase-to-phase for each individual device used as a building or structure main disconnecting means rated 1000 amperes or more.

The provisions of this section shall not apply to the disconnecting means for the following:

(1) Continuous industrial processes where a nonorderly shutdown will introduce additional or increased hazards
(2) Installations where ground-fault protection is provided by other requirements for services or feeders
(3) Fire pumps

Section 240.13 extends the requirement of 230.95 to building disconnects, regardless of how the disconnects are classified (whether they are service disconnects, or building disconnects for feeders or even branch circuits). See 215.10 and Article 225, Part II, for the requirements for building disconnects not on the utility service.

Section 240.13 requires each building or structure disconnect that is rated 1000 amperes or more, on a solidly grounded system of more than 150 volts to ground (e.g., a 480Y/277-V system), to be provided with ground-fault protection for equipment. Allowances are included for fire pumps and for continuous industrial processes in which nonorderly shutdowns would introduce additional hazards.

Where ground-fault protection for equipment is installed at the service equipment, and feeders or branch circuits are installed from that service to supply other buildings or structures, the disconnecting means at any subsequent building is not required to be provided with ground-fault protection if the service device provides the required protection. Installations performed prior to the 2008 *Code* permitted "re-grounding" of the grounded conductor at the separate building if an equipment grounding conductor was not included with the supply circuit. Where re-grounding of the neutral occurs downstream from the service, the re-grounding may nullify the ground-fault protection (or result in unwanted operation of the protection), because the neutral current has parallel paths on which to return to the source. Feeders or branch circuits supplying other buildings or structures may have to be isolated to allow for proper operation of the service ground-fault protection, and separate ground-fault protection installed at the building disconnecting means is then necessary to meet the requirements of 240.13.

## 240.15 Ungrounded Conductors

**(A) Overcurrent Device Required.** A fuse or an overcurrent trip unit of a circuit breaker shall be connected in series with each ungrounded conductor. A combination of a current transformer and overcurrent relay shall be considered equivalent to an overcurrent trip unit.

Informational Note: For motor circuits, see Parts III, IV, V, and XI of Article 430.

**(B) Circuit Breaker as Overcurrent Device.** Circuit breakers shall open all ungrounded conductors of the circuit both manually and automatically unless otherwise permitted in 240.15(B)(1), (B)(2), (B)(3), and (B)(4).

**(1) Multiwire Branch Circuits.** Individual single-pole circuit breakers, with identified handle ties, shall be permitted as the protection for each ungrounded conductor of multiwire branch circuits that serve only single-phase line-to-neutral loads.

**(2) Grounded Single-Phase Alternating-Current Circuits.** In grounded systems, individual single-pole circuit breakers rated 120/240 volts ac, with identified handle ties, shall be permitted as the protection for each ungrounded conductor for line-to-line connected loads for single-phase circuits.

**(3) 3-Phase and 2-Phase Systems.** For line-to-line loads in 4-wire, 3-phase systems or 5-wire, 2-phase systems, individual single-pole circuit breakers rated 120/240 volts ac with identified handle ties shall be permitted as the protection for each ungrounded conductor, if the systems have a grounded neutral point and the voltage to ground does not exceed 120 volts.

**(4) 3-Wire Direct-Current Circuits.** Individual single-pole circuit breakers rated 125/250 volts dc with identified handle ties shall be permitted as the protection for each ungrounded conductor for line-to-line connected loads for 3-wire, direct-current circuits supplied from a system with a grounded neutral where the voltage to ground does not exceed 125 volts.

Multiwire branch circuits are permitted to supply line-to-line connected loads where the loads are associated with a single piece of utilization equipment or where all of the ungrounded conductors are opened simultaneously by the branch-circuit overcurrent device (automatic opening in response to overcurrent). See the commentary following 210.4(C) for additional information.

The basic rule in 240.15 requires circuit breakers to open all ungrounded conductors of the circuit when they trip (automatic operation in response to overcurrent) or are manually operated as a disconnecting means. For 2-wire circuits with one conductor grounded, this rule is simple and needs no further explanation.

Two or more individual single-pole circuit breakers with their handles joined together using an identified handle tie can be used as the protection of line-to-line connected loads in single-phase and 3-phase alternating current systems, but this requirement is limited to circuit breakers rated 120/240 V. For multiwire branch circuits of 600 volts or less, however, two acceptable methods comply with this rule.

The first, and most widely used, method is to install a multipole circuit breaker with an internal common trip mechanism. The use of such multipole devices ensures compliance with all of the *Code* requirements for overcurrent protection and disconnecting of multiwire branch circuits. This breaker is operated by an external single lever internally attached to the two or three poles of the circuit breaker, or the external lever may be attached to multiple handles operated as one, provided the breaker is a factory-assembled unit in accordance with 240.8. Underwriters Laboratories refers to these devices as *multipole common trip circuit breakers*. This type of circuit breaker is required to be used for branch circuits that comprise multiple ungrounded conductors supplied by ungrounded 3-phase and single-phase systems. Where circuit breakers are used on ungrounded systems, verification of compliance with the application requirements in 240.85 is important. Of course, multipole common trip circuit breakers are permitted to be installed on any branch circuit supplied from a grounded system where used within their ratings.

The second method permitted for multiwire branch circuits is to use two or three single-pole circuit breakers and add an identified handle tie to function as a common operating handle. This multipole circuit breaker is field assembled by externally attaching an identified common lever (handle tie) onto the two or three individual circuit breakers. Handle ties do not cause the circuit breaker to function as a common trip device; rather, they only allow common operation as a disconnecting means. Handle tie mechanism circuit breakers are permitted as a substitute for internal common trip mechanism circuit breakers only for limited applications. Unless specifically prohibited elsewhere, circuit breakers with identified handle ties are permitted for multiwire branch circuits only where the circuit is supplied from grounded 3-phase or grounded single-phase systems. The single-pole circuit breakers used together in this fashion must be rated for the dual voltage encountered, such as 120/240 V. The term *identified* requires the use of hardware designed specifically to perform this common disconnecting means function. The use of homemade hardware such as nails or pieces of wire to perform this function is not an *identified* means.

Section 240.15(B)(4) covers protection of direct current circuits and permits the use of two single-pole circuit breakers (rated for dc application) with identified handle ties to be used for the overcurrent protection of line-to-line connected loads. The 3-wire circuits covered by this requirement are *multiwire branch circuits* per the definition of that term in Article 100, and as such are subject to all of the requirements specified in 210.4.

Exhibits 240.4 through 240.6 illustrate the application of 240.15(B). In Exhibit 240.4, where multipole common trip circuit breakers are required, handle ties are not permitted because the circuits are supplied from ungrounded systems. In Exhibit 240.5, where the supply systems are grounded, single-pole circuit breakers are permitted and handle ties or common trip operation is not required because the circuits supply line-to-neutral loads. In Exhibit 240.6, in which line-to-line loads are supplied from single-phase or 4-wire, 3-phase systems, identified handle ties or multipole common trip circuit breakers are permitted. The multiwire branch circuit shown in the top and middle diagrams supplies a single piece of utilization equipment in accordance with 210.4(C), Exception No. 1.

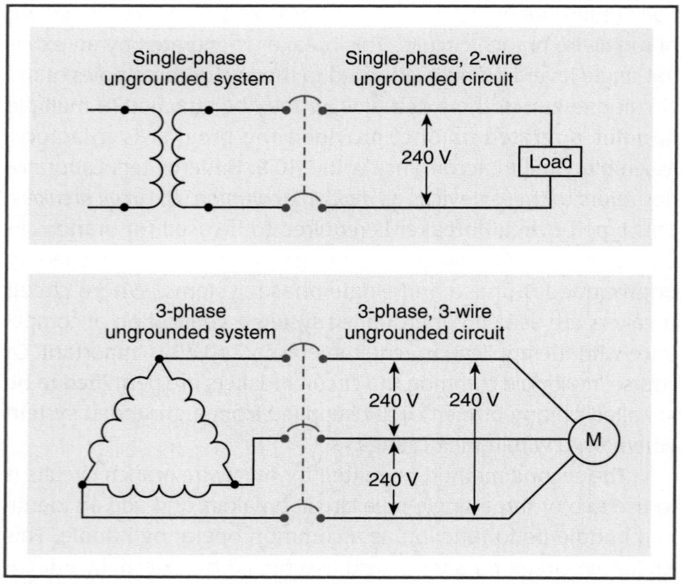


***EXHIBIT 240.4*** *Examples of circuits that require multipole common trip circuit breakers.*

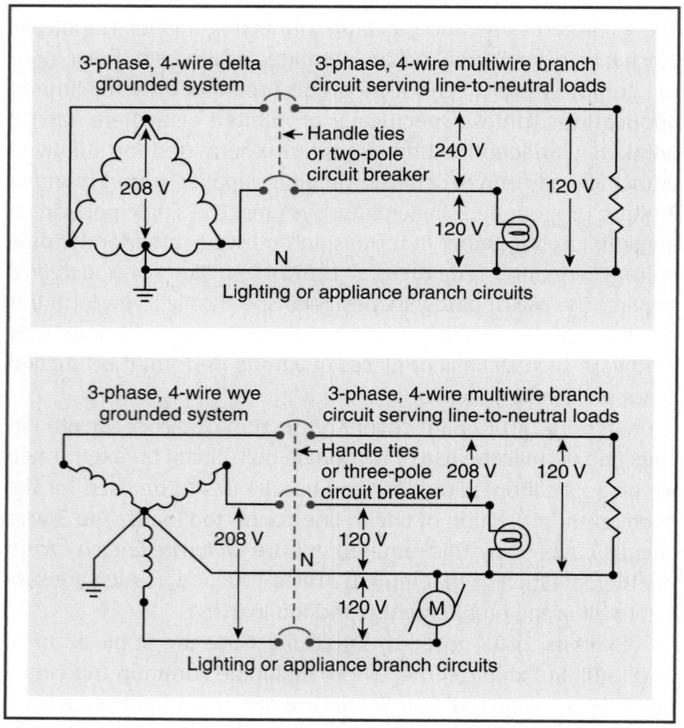


***EXHIBIT 240.5*** *Examples of circuits in which single-pole circuit breakers are permitted, because they open the ungrounded conductor of the circuit.*

## II. Location

### 240.21 Location in Circuit

Overcurrent protection shall be provided in each ungrounded circuit conductor and shall be located at the point where the

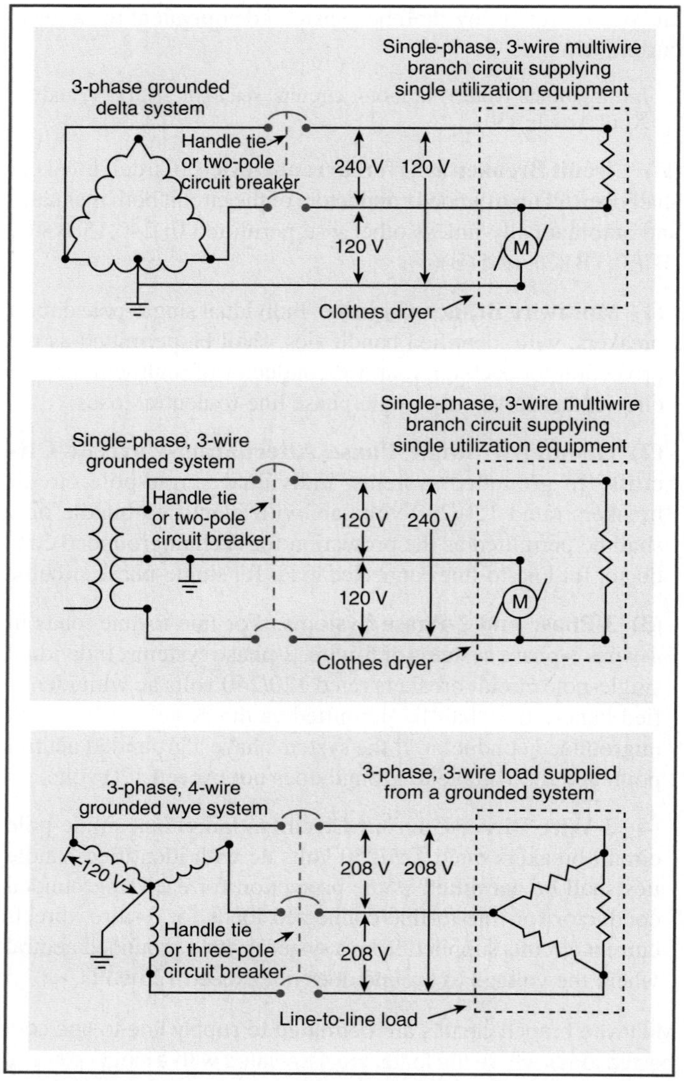


***EXHIBIT 240.6*** *Examples of circuits in which identified handle ties are permitted to provide the simultaneous disconnecting function.*

conductors receive their supply except as specified in 240.21(A) through (H). Conductors supplied under the provisions of 240.21(A) through (H) shall not supply another conductor except through an overcurrent protective device meeting the requirements of 240.4.

**(A) Branch-Circuit Conductors.** Branch-circuit tap conductors meeting the requirements specified in 210.19 shall be permitted to have overcurrent protection as specified in 210.20.

**(B) Feeder Taps.** Conductors shall be permitted to be tapped, without overcurrent protection at the tap, to a feeder as specified in 240.21(B)(1) through (B)(5). The provisions of 240.4(B) shall not be permitted for tap conductors.

The use of the next highest standard-sized device allowed by 240.4(B) is not permitted for feeder tap conductor applications.

For instance, the use of a 500-kcmil THWN copper conductor [380 A, per Table 310.15(B)(16)] as a tap conductor to supply a 400-A rated device is not permitted, nor is it permitted to use the requirement in 240.6 to establish the relationship between the size of the tap conductor and the rating or setting of the feeder OCPD. If the feeder OCPD is 1200 A, a conductor with an ampacity of 400 A is required.

Exhibit 240.7 illustrates a 1/0 AWG, Type THW copper conductor [150 A, from Table 310.15(B)(16)] connected to a 3/0 AWG, Type THW copper feeder conductor with an ampacity of 200 A (increased in size to compensate for voltage drop) that is protected by a 150-A OCPD. Because the ampacity of the 1/0 AWG conductor is not exceeded by the rating of the overcurrent device, the 1/0 AWG conductor is not considered to be a tap conductor based on the definition of *tap conductors* in 240.2. The overcurrent device protects both sets of conductors in accordance with the basic rule of 240.4, and additional overcurrent protection is not required at the supply or termination point of the 1/0 AWG conductors. Both the 1/0 AWG and the 3/0 AWG conductors are protected by the 150-ampere circuit breaker. In this application, the 1/0 AWG conductors are not tap conductors as defined in 240.2.

**(1) Taps Not over 3 m (10 ft) Long.** If the length of the tap conductors does not exceed 3 m (10 ft) and the tap conductors comply with all of the following:

(1) The ampacity of the tap conductors is

a. Not less than the combined calculated loads on the circuits supplied by the tap conductors, and

b. Not less than the rating of the equipment containing an overcurrent device(s) supplied by the tap conductors or not less than the rating of the overcurrent protective device at the termination of the tap conductors.

*Exception to b: Where listed equipment, such as a surge protective device(s) [SPD(s)], is provided with specific instructions on minimum conductor sizing, the ampacity of the tap conductors supplying that equipment shall be permitted to be determined based on the manufacturer's instructions.*

(2) The tap conductors do not extend beyond the switchboard, switchgear, panelboard, disconnecting means, or control devices they supply.

(3) Except at the point of connection to the feeder, the tap conductors are enclosed in a raceway, which extends from the tap to the enclosure of an enclosed switchboard, switchgear, a panelboard, or control devices, or to the back of an open switchboard.

Exhibit 240.8 illustrates the feeder tap requirements of 240.21(B)(2). In this example, three 3/0 AWG, Type THW copper tap conductors are protected from physical damage by installation in a raceway. The tap conductors are not more than 25 feet in length between terminations, and the conductors are tapped from 500 kcmil, Type THW copper feeders and terminate in a single circuit breaker with a rating not greater than the ampacity of the tap conductors. The ampacity of the 3/0 AWG, Type THW copper conductor (200 A) is more than one-third the rating of the overcurrent device (400 A) protecting the feeder circuit. See Table 310.15(B)(16) for the ampacity of copper conductors in conduit. Note that the lengths specified in 240.21(B) and (C) apply to the conductors, not to a raceway enclosing the conductors or to the distance between the enclosures in which the tap conductors originate and terminate.

(4) For field installations, if the tap conductors leave the enclosure or vault in which the tap is made, the ampacity of the tap conductors is not less than one-tenth of the rating of the overcurrent device protecting the feeder conductors.

Informational Note: For overcurrent protection requirements for panelboards, see 408.36.

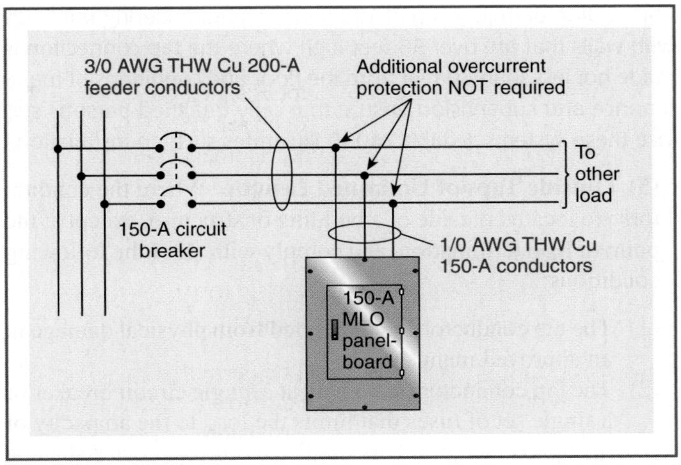


***EXHIBIT 240.7*** *An example in which the sets of 1/0 AWG and the 3/0 AWG conductors are both protected by the 150-ampere circuit breaker.*

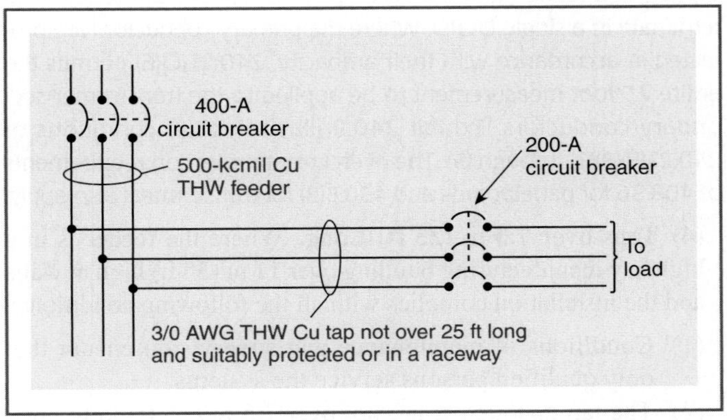


***EXHIBIT 240.8*** *An example in which the feeder taps terminate in a single circuit breaker.*

**(2) Taps Not over 7.5 m (25 ft) Long.** Where the length of the tap conductors does not exceed 7.5 m (25 ft) and the tap conductors comply with all the following:

(1) The ampacity of the tap conductors is not less than one-third of the rating of the overcurrent device protecting the feeder conductors.
(2) The tap conductors terminate in a single circuit breaker or a single set of fuses that limit the load to the ampacity of the tap conductors. This device shall be permitted to supply any number of additional overcurrent devices on its load side.
(3) The tap conductors are protected from physical damage by being enclosed in an approved raceway or by other approved means.

**(3) Taps Supplying a Transformer [Primary Plus Secondary Not over 7.5 m (25 ft) Long].** Where the tap conductors supply a transformer and comply with all the following conditions:

(1) The conductors supplying the primary of a transformer have an ampacity at least one-third the rating of the overcurrent device protecting the feeder conductors.
(2) The conductors supplied by the secondary of the transformer shall have an ampacity that is not less than the value of the primary-to-secondary voltage ratio multiplied by one-third of the rating of the overcurrent device protecting the feeder conductors.
(3) The total length of one primary plus one secondary conductor, excluding any portion of the primary conductor that is protected at its ampacity, is not over 7.5 m (25 ft).
(4) The primary and secondary conductors are protected from physical damage by being enclosed in an approved raceway or by other approved means.
(5) The secondary conductors terminate in a single circuit breaker or set of fuses that limit the load current to not more than the conductor ampacity that is permitted by 310.15.

This section covers applications where the conductor length of 25 feet is applied to the primary and secondary conductors (using the length of one primary conductor plus the length of one secondary conductor for the measurement). The primary conductors are tapped to a feeder, and the secondary conductors are required to terminate in a single OCPD. Where the primary conductors are protected in accordance with their ampacity, 240.21(C)(6) permits the entire 25-foot measurement to be applied to the transformer secondary conductors. Exhibit 240.9 illustrates the conditions of 240.21(B)(3)(1) through (5). The overcurrent protection requirements of 408.36 for panelboards and 450.3(B) for transformers also apply.

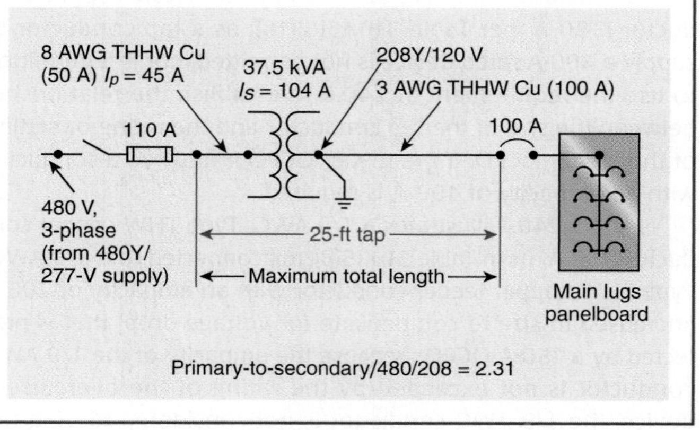


***EXHIBIT 240.9*** *An example in which the transformer feeder taps (primary plus secondary) are not over 25 feet long.*

**(4) Taps over 7.5 m (25 ft) Long.** Where the feeder is in a high bay manufacturing building over 11 m (35 ft) high at walls and the installation complies with all the following conditions:

(1) Conditions of maintenance and supervision ensure that only qualified persons service the systems.
(2) The tap conductors are not over 7.5 m (25 ft) long horizontally and not over 30 m (100 ft) total length.
(3) The ampacity of the tap conductors is not less than one-third the rating of the overcurrent device protecting the feeder conductors.
(4) The tap conductors terminate at a single circuit breaker or a single set of fuses that limit the load to the ampacity of the tap conductors. This single overcurrent device shall be permitted to supply any number of additional overcurrent devices on its load side.
(5) The tap conductors are protected from physical damage by being enclosed in an approved raceway or by other approved means.
(6) The tap conductors are continuous from end-to-end and contain no splices.
(7) The tap conductors are sized 6 AWG copper or 4 AWG aluminum or larger.
(8) The tap conductors do not penetrate walls, floors, or ceilings.
(9) The tap is made no less than 9 m (30 ft) from the floor.

This section permits a tap of 100 feet for manufacturing buildings with walls that are over 35 feet high where the tap connection is made not less than 30 feet from the floor and conditions of maintenance and supervision ensure that only qualified persons service these systems. Exhibit 240.10 illustrates such an installation.

**(5) Outside Taps of Unlimited Length.** Where the conductors are located outside of a building or structure, except at the point of load termination, and comply with all of the following conditions:

(1) The tap conductors are protected from physical damage in an approved manner.
(2) The tap conductors terminate at a single circuit breaker or a single set of fuses that limits the load to the ampacity of the tap conductors. This single overcurrent device shall be permitted to supply any number of additional overcurrent devices on its load side.

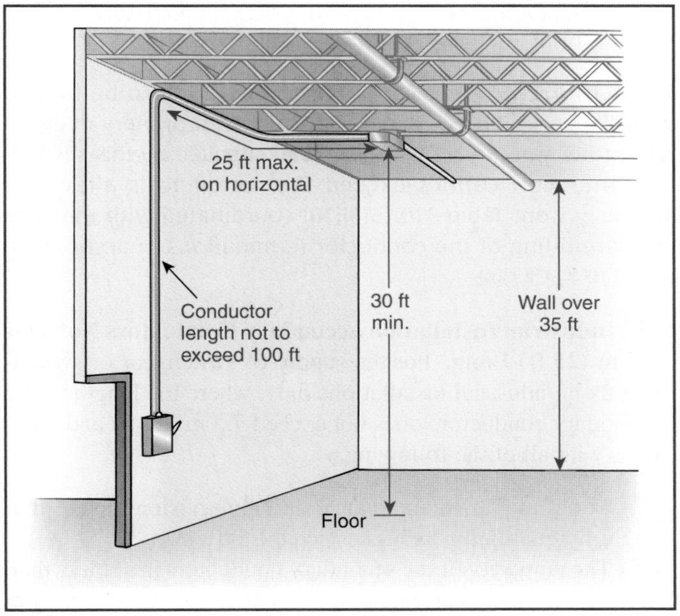


***EXHIBIT 240.10*** *An illustration of a feeder tap in a high bay building.*

(3) The overcurrent device for the tap conductors is an integral part of a disconnecting means or shall be located immediately adjacent thereto.

(4) The disconnecting means for the tap conductors is installed at a readily accessible location complying with one of the following:

a. Outside of a building or structure
b. Inside, nearest the point of entrance of the tap conductors
c. Where installed in accordance with 230.6, nearest the point of entrance of the tap conductors

Section 240.21(B)(5) is a tap conductor requirement that is similar in some respects to an installation of service conductors. The conductors are supplied from a feeder at an outdoor location and run to a building or structure without limitations on the tap conductor length. The tap conductors must be protected against physical damage and must terminate in a single, fused disconnect or a single circuit breaker with a rating that does not exceed the ampacity of the tap conductors. This OCPD provides overload protection for the tap conductors. The fused switch or circuit breaker is installed at a readily accessible location either inside or outside a building or structure and is subject to the applicable requirements covering feeder disconnecting means in Part II of Article 225. The fused switch or circuit breaker is required to be installed inside or outside of the building or structure at a point nearest to where the tap conductors enter the building or structure.

**(C) Transformer Secondary Conductors.** A set of conductors feeding a single load, or each set of conductors feeding separate loads, shall be permitted to be connected to a transformer secondary, without overcurrent protection at the secondary, as specified in 240.21(C)(1) through (C)(6). The provisions of 240.4(B) shall not be permitted for transformer secondary conductors.

Informational Note: For overcurrent protection requirements for transformers, see 450.3.

The six applications covered in 240.21(C) permit transformer secondary conductors without an OCPD at the point the secondary conductors receive their supply.

Like 240.21(B) for conductors tapped to a feeder, 240.21(C) specifically prohibits application of 240.4(B) with transformer secondary conductors covered by the requirements of 240.21(C)(1) through (C)(6). See the commentary for 240.21(B).

The secondary terminals of a transformer are permitted to supply one or more than one set of secondary conductors. The first sentence specifies that the requirements apply to "a set of conductors feeding a single load" or "to each set of conductors feeding separate loads." For example, the secondary terminals could supply two separate sets of secondary conductors that feed two panelboards. One set of conductors could be installed using the 25-foot secondary conductor rule of 240.21(C)(6), while the other set of conductors could be installed using the 10-foot secondary conductor rule of 240.21(C)(2). Each set is treated individually in applying the applicable secondary conductor requirement. It is necessary to coordinate the secondary conductor protection requirements of 240.21(C) and, where applicable, the transformer secondary protection requirements in 450.3(A) and (B).

**(1) Protection by Primary Overcurrent Device.** Conductors supplied by the secondary side of a single-phase transformer having a 2-wire (single-voltage) secondary, or a three-phase, delta-delta connected transformer having a 3-wire (single-voltage) secondary, shall be permitted to be protected by overcurrent protection provided on the primary (supply) side of the transformer, provided this protection is in accordance with 450.3 and does not exceed the value determined by multiplying the secondary conductor ampacity by the secondary-to-primary transformer voltage ratio.

Single-phase (other than 2-wire) and multiphase (other than delta-delta, 3-wire) transformer secondary conductors are not considered to be protected by the primary overcurrent protective device.

**(2) Transformer Secondary Conductors Not over 3 m (10 ft) Long.** If the length of secondary conductor does not exceed 3 m (10 ft) and complies with all of the following:

(1) The ampacity of the secondary conductors is

a. Not less than the combined calculated loads on the circuits supplied by the secondary conductors, and
b. Not less than the rating of the equipment containing an overcurrent device(s) supplied by the secondary conductors or not less than the rating of the overcurrent protective device at the termination of the secondary conductors.

*Exception: Where listed equipment, such as a surge protective device(s) [SPD(s)], is provided with specific instructions on minimum conductor sizing, the ampacity of the tap conductors supplying that equipment shall be permitted to be determined based on the manufacturer's instructions.*

(2) The secondary conductors do not extend beyond the switchboard, switchgear, panelboard, disconnecting means, or control devices they supply.
(3) The secondary conductors are enclosed in a raceway, which shall extend from the transformer to the enclosure of an enclosed switchboard, switchgear, a panelboard, or control devices or to the back of an open switchboard.
(4) For field installations where the secondary conductors leave the enclosure or vault in which the supply connection is made, the rating of the overcurrent device protecting the primary of the transformer, multiplied by the primary to secondary transformer voltage ratio, shall not exceed 10 times the ampacity of the secondary conductor.

Informational Note: For overcurrent protection requirements for panelboards, see 408.36.

The minimum size requirement in 240.21(C)(2) for 10-foot transformer secondary conductors establishes a relationship between the size of the ungrounded secondary conductors and the rating of the transformer primary OCPD similar to the sizing requirements applied in 240.21(B)(1) for 10-foot feeder tap conductors. This size–rating relationship is necessary because the transformer primary device also provides short-circuit ground-fault protection for the transformer secondary conductors. It is necessary to also ensure that the ampacity of the conductors is adequate for the calculated load and is not less than the rating of the device or OCPD in which the conductors terminate. The following example illustrates the application of this requirement.

### Calculation Example

Apply the 10-foot secondary conductor protection criteria of 240.21(C)(2) to a transformer rated 75 kVA, 3-phase, 480 V primary to 208Y/120 V secondary. The transformer primary OCPD is rated 125 A. Determine the minimum secondary conductor size for this installation.

*Solution*

STEP 1. Determine 1/10 of the primary OCPD rating using the following calculation:

$$125 \text{ A} \div 10 = 12.5 \text{ A}$$

STEP 2. Determine the line-to-line primary-to-secondary voltage ratio:

$$\left(\frac{480}{208}\right) = 2.31$$

STEP 3. Determine the minimum ampacity for ungrounded transformer secondary conductor:

$$12.5 \text{ A} \times 2.31 = 29 \text{ A} \rightarrow 10 \text{ AWG copper THWN}$$
[Table 310.15(B)(16)]: 30 A from 60°C column

A 10 AWG copper conductor is permitted to be tapped from the secondary of this transformer with primary overcurrent protection rated 125 A. The load supplied by this secondary conductor cannot exceed the conductor's allowable ampacity from Table 310.15(B)(16) coordinated with the temperature rating of the conductor terminations in accordance with 110.14(C)(1)(a).

**(3) Industrial Installation Secondary Conductors Not over 7.5 m (25 ft) Long.** For the supply of switchgear or switchboards in industrial installations only, where the length of the secondary conductors does not exceed 7.5 m (25 ft) and complies with all of the following:

(1) Conditions of maintenance and supervision ensure that only qualified persons service the systems.
(2) The ampacity of the secondary conductors is not less than the secondary current rating of the transformer, and the sum of the ratings of the overcurrent devices does not exceed the ampacity of the secondary conductors.
(3) All overcurrent devices are grouped.
(4) The secondary conductors are protected from physical damage by being enclosed in an approved raceway or by other approved means.

**(4) Outside Secondary Conductors.** Where the conductors are located outdoors of a building or structure, except at the point of load termination, and comply with all of the following conditions:

(1) The conductors are protected from physical damage in an approved manner.
(2) The conductors terminate at a single circuit breaker or a single set of fuses that limit the load to the ampacity of the conductors. This single overcurrent device shall be permitted to supply any number of additional overcurrent devices on its load side.
(3) The overcurrent device for the conductors is an integral part of a disconnecting means or shall be located immediately adjacent thereto.
(4) The disconnecting means for the conductors is installed at a readily accessible location complying with one of the following:
   a. Outside of a building or structure
   b. Inside, nearest the point of entrance of the conductors
   c. Where installed in accordance with 230.6, nearest the point of entrance of the conductors.

**(5) Secondary Conductors from a Feeder Tapped Transformer.** Transformer secondary conductors installed in accordance with 240.21(B)(3) shall be permitted to have overcurrent protection as specified in that section.

**(6) Secondary Conductors Not over 7.5 m (25 ft) Long.** Where the length of secondary conductor does not exceed 7.5 m (25 ft) and complies with all of the following:

(1) The secondary conductors shall have an ampacity that is not less than the value of the primary-to-secondary voltage ratio multiplied by one-third of the rating of the overcurrent device protecting the primary of the transformer.
(2) The secondary conductors terminate in a single circuit breaker or set of fuses that limit the load current to not more than the conductor ampacity that is permitted by 310.15.
(3) The secondary conductors are protected from physical damage by being enclosed in an approved raceway or by other approved means.

**(D) Service Conductors.** Service conductors shall be permitted to be protected by overcurrent devices in accordance with 230.91.

**(E) Busway Taps.** Busways and busway taps shall be permitted to be protected against overcurrent in accordance with 368.17.

**(F) Motor Circuit Taps.** Motor-feeder and branch-circuit conductors shall be permitted to be protected against overcurrent in accordance with 430.28 and 430.53, respectively.

**(G) Conductors from Generator Terminals.** Conductors from generator terminals that meet the size requirement in 445.13 shall be permitted to be protected against overload by the generator overload protective device(s) required by 445.12.

**(H) Battery Conductors.** Overcurrent protection shall be permitted to be installed as close as practicable to the storage battery terminals in an unclassified location. Installation of the overcurrent protection within a hazardous (classified) location shall also be permitted.

## 240.22 Grounded Conductor

No overcurrent device shall be connected in series with any conductor that is intentionally grounded, unless one of the following two conditions is met:

(1) The overcurrent device opens all conductors of the circuit, including the grounded conductor, and is designed so that no pole can operate independently.
(2) Where required by 430.36 or 430.37 for motor overload protection.

## 240.23 Change in Size of Grounded Conductor

Where a change occurs in the size of the ungrounded conductor, a similar change shall be permitted to be made in the size of the grounded conductor.

The size of the grounded circuit conductor may be increased (e.g., because of voltage-drop problems) or reduced to correspond to a reduction made in the size of the ungrounded circuit conductor(s), as in the case of feeder tap conductors, provided that the grounded and ungrounded conductors comprise the same circuit.

## 240.24 Location in or on Premises

**(A) Accessibility.** Overcurrent devices shall be readily accessible and shall be installed so that the center of the grip of the operating handle of the switch or circuit breaker, when in its highest position, is not more than 2.0 m (6 ft 7 in.) above the floor or working platform, unless one of the following applies:

(1) For busways, as provided in 368.17(C).
(2) For supplementary overcurrent protection, as described in 240.10.
(3) For overcurrent devices, as described in 225.40 and 230.92.
(4) For overcurrent devices adjacent to utilization equipment that they supply, access shall be permitted to be by portable means.

This section recognizes the need for overcurrent protection in locations that are not readily accessible, such as above suspended ceilings. Overcurrent devices are permitted to be located so that they are not readily accessible, as long as they are located next to the appliance, motor, or other equipment they supply and can be reached by using a ladder. For the purposes of this requirement, ready access to the operating handle of a fusible switch or circuit breaker is considered to be not more than 6 feet 7 inches above the finished floor or working platform.

The measurement is made from the center of the device operating handle where the handle is at its highest position. This text parallels the requirement of 404.8(A), which applies to all switches and circuit breakers used as switches. For information regarding the accessibility of supplementary overcurrent devices, refer to 240.10.

**(B) Occupancy.** Each occupant shall have ready access to all overcurrent devices protecting the conductors supplying that occupancy, unless otherwise permitted in 240.24(B)(1) and (B)(2).

**(1) Service and Feeder Overcurrent Devices.** Where electric service and electrical maintenance are provided by the building management and where these are under continuous building management supervision, the service overcurrent devices and feeder overcurrent devices supplying more than one occupancy shall be permitted to be accessible only to authorized management personnel in the following:

(1) Multiple-occupancy buildings
(2) Guest rooms or guest suites

**(2) Branch-Circuit Overcurrent Devices.** Where electric service and electrical maintenance are provided by the building management and where these are under continuous building management supervision, the branch-circuit overcurrent devices supplying any guest rooms or guest suites without permanent provisions for cooking shall be permitted to be accessible only to authorized management personnel.

**(C) Not Exposed to Physical Damage.** Overcurrent devices shall be located where they will not be exposed to physical damage.

Informational Note: See 110.11, Deteriorating Agents.

**(D) Not in Vicinity of Easily Ignitible Material.** Overcurrent devices shall not be located in the vicinity of easily ignitible material, such as in clothes closets.

**(E) Not Located in Bathrooms.** In dwelling units, dormitories, and guest rooms or guest suites, overcurrent devices, other than supplementary overcurrent protection, shall not be located in bathrooms.

**(F) Not Located over Steps.** Overcurrent devices shall not be located over steps of a stairway.

## III. Enclosures

### 240.30 General

**(A) Protection from Physical Damage.** Overcurrent devices shall be protected from physical damage by one of the following:

(1) Installation in enclosures, cabinets, cutout boxes, or equipment assemblies
(2) Mounting on open-type switchboards, panelboards, or control boards that are in rooms or enclosures free from dampness and easily ignitible material and are accessible only to qualified personnel

Physical damage includes exposure to deteriorating agents that adversely affect the operation of the OCPD. Enclosures for OCPDs provide a number of functions, including protecting the devices against physical damage. Guarding through controlled access is another means of affording protection against physical damage.

**(B) Operating Handle.** The operating handle of a circuit breaker shall be permitted to be accessible without opening a door or cover.

### 240.32 Damp or Wet Locations

Enclosures for overcurrent devices in damp or wet locations shall comply with 312.2.

### 240.33 Vertical Position

Enclosures for overcurrent devices shall be mounted in a vertical position unless that is shown to be impracticable. Circuit breaker enclosures shall be permitted to be installed horizontally where the circuit breaker is installed in accordance with 240.81. Listed busway plug-in units shall be permitted to be mounted in orientations corresponding to the busway mounting position.

A wall-mounted vertical position for enclosures for overcurrent devices affords easier access, natural hand operation, normal swinging or closing of doors or covers, and legibility of the manufacturer's markings. In addition, this section does not permit a panelboard or fusible switch enclosure to be installed in a horizontal position such that the back of the enclosure is mounted on the ceiling or the floor. Compliance with the requirement that the up position of the handle is on or closed, and the down position of the handle is off or open, in accordance with 240.81, limits the number of pole spaces available in a panelboard where its cabinet is mounted in a horizontal position on or in a wall.

## IV. Disconnecting and Guarding

### 240.40 Disconnecting Means for Fuses

Cartridge fuses in circuits of any voltage where accessible to other than qualified persons, and all fuses in circuits over 150 volts to ground, shall be provided with a disconnecting means on their supply side so that each circuit containing fuses can be independently disconnected from the source of power. A current-limiting device without a disconnecting means shall be permitted on the supply side of the service disconnecting means as permitted by 230.82. A single disconnecting means shall be permitted on the supply side of more than one set of fuses as permitted by 430.112, Exception, for group operation of motors and 424.22(C) for fixed electric space-heating equipment.

The installation of cable limiters or similar current-limiting devices on the supply side of the service disconnecting means is permitted by 230.82(1). A disconnecting means is not required on the supply side of such devices.

### 240.41 Arcing or Suddenly Moving Parts

Arcing or suddenly moving parts shall comply with 240.41(A) and (B).

**(A) Location.** Fuses and circuit breakers shall be located or shielded so that persons will not be burned or otherwise injured by their operation.

**(B) Suddenly Moving Parts.** Handles or levers of circuit breakers, and similar parts that may move suddenly in such a way that persons in the vicinity are likely to be injured by being struck by them, shall be guarded or isolated.

To comply with workplace safety requirements such as those contained in *NFPA 70E®, Standard for Electrical Safety in the Workplace®*, open switchboards and control boards must be under competent supervision and accessible only to persons qualified to work on or in the vicinity of such equipment. Fuses or circuit breakers must be located or shielded so that any arc across the opening device will not injure persons in the vicinity.

Guardrails are provided in the vicinity of disconnecting means as a means of protecting persons against injury caused by suddenly moving handles of OCPDs. Switchboards equipped with removable handles help mitigate this personal injury hazard. See Article 100 for the definition of *guarded*. See also 110.27 for the guarding of live parts (600 V, nominal, or less).

# V. Plug Fuses, Fuseholders, and Adapters

## 240.50 General

**(A) Maximum Voltage.** Plug fuses shall be permitted to be used in the following circuits:

(1) Circuits not exceeding 125 volts between conductors
(2) Circuits supplied by a system having a grounded neutral point where the line-to-neutral voltage does not exceed 150 volts

**(B) Marking.** Each fuse, fuseholder, and adapter shall be marked with its ampere rating.

**(C) Hexagonal Configuration.** Plug fuses of 15-ampere and lower rating shall be identified by a hexagonal configuration of the window, cap, or other prominent part to distinguish them from fuses of higher ampere ratings.

The 10- and 15-ampere plug fuses shown in Exhibit 240.11 are equipped with the hexagonal window required by 240.50(C). The 10-ampere fuse is a Type S plug fuse, which cannot be interchanged with a higher rated fuse where used with its corresponding fuseholder or adapter.

**(D) No Energized Parts.** Plug fuses, fuseholders, and adapters shall have no exposed energized parts after fuses or fuses and adapters have been installed.

**(E) Screw Shell.** The screw shell of a plug-type fuseholder shall be connected to the load side of the circuit.

Exhibit 240.12 shows a Type S nonrenewable plug fuse adapter designed to meet the requirements of 240.53 and 240.54. To prevent installation of a fuse exceeding the ampacity of the conductor being protected, the adapter will not accept Type S plug fuses having an ampere rating other than that for which it is specifically designed.

***EXHIBIT 240.11*** *Two plug fuses and a Type S fuse. (Courtesy of Cooper Bussmann, a division of Cooper Industries PLC)*

***EXHIBIT 240.12*** *Type S nonrenewable plug fuse adapter. (Courtesy of Cooper Bussmann, a division of Cooper Industries PLC)*

## 240.51 Edison-Base Fuses

**(A) Classification.** Plug fuses of the Edison-base type shall be classified at not over 125 volts and 30 amperes and below.

**(B) Replacement Only.** Plug fuses of the Edison-base type shall be used only for replacements in existing installations where there is no evidence of overfusing or tampering.

## 240.52 Edison-Base Fuseholders

Fuseholders of the Edison-base type shall be installed only where they are made to accept Type S fuses by the use of adapters.

## 240.53 Type S Fuses

Type S fuses shall be of the plug type and shall comply with 240.53(A) and (B).

**(A) Classification.** Type S fuses shall be classified at not over 125 volts and 0 to 15 amperes, 16 to 20 amperes, and 21 to 30 amperes.

**(B) Noninterchangeable.** Type S fuses of an ampere classification as specified in 240.53(A) shall not be interchangeable with a lower ampere classification. They shall be designed so that they cannot be used in any fuseholder other than a Type S fuseholder or a fuseholder with a Type S adapter inserted.

## 240.54 Type S Fuses, Adapters, and Fuseholders

**(A) To Fit Edison-Base Fuseholders.** Type S adapters shall fit Edison-base fuseholders.

**(B) To Fit Type S Fuses Only.** Type S fuseholders and adapters shall be designed so that either the fuseholder itself or the fuseholder with a Type S adapter inserted cannot be used for any fuse other than a Type S fuse.

**(C) Nonremovable.** Type S adapters shall be designed so that once inserted in a fuseholder, they cannot be removed.

**(D) Nontamperable.** Type S fuses, fuseholders, and adapters shall be designed so that tampering or shunting (bridging) would be difficult.

**(E) Interchangeability.** Dimensions of Type S fuses, fuseholders, and adapters shall be standardized to permit interchangeability regardless of the manufacturer.

## VI. Cartridge Fuses and Fuseholders

### 240.60 General

**(A) Maximum Voltage — 300-Volt Type.** Cartridge fuses and fuseholders of the 300-volt type shall be permitted to be used in the following circuits:

(1) Circuits not exceeding 300 volts between conductors
(2) Single-phase line-to-neutral circuits supplied from a 3-phase, 4-wire, solidly grounded neutral source where the line-to-neutral voltage does not exceed 300 volts

**(B) Noninterchangeable — 0–6000-Ampere Cartridge Fuseholders.** Fuseholders shall be designed so that it will be difficult to put a fuse of any given class into a fuseholder that is designed for a current lower, or voltage higher, than that of the class to which the fuse belongs. Fuseholders for current-limiting fuses shall not permit insertion of fuses that are not current-limiting.

**(C) Marking.** Fuses shall be plainly marked, either by printing on the fuse barrel or by a label attached to the barrel showing the following:

(1) Ampere rating
(2) Voltage rating
(3) Interrupting rating where other than 10,000 amperes
(4) Current limiting where applicable
(5) The name or trademark of the manufacturer

The interrupting rating shall not be required to be marked on fuses used for supplementary protection.

Exhibit 240.13 shows two examples of Class G fuses rated 300 volts that bear the markings required by 240.60(C). Class H-type cartridge fuses have an interrupting capacity (IC) rating of 10,000 amperes, which does not need to be marked on the fuse. However, Class CC, G, J, K, L, R, and T cartridge fuses exceed the 10,000-A IC rating and must be marked with the IC rating. Section 240.83(C) requires that an IC rating for circuit breakers other than 5000 A be indicated on the circuit breaker. Fuses or circuit breakers used for supplementary overcurrent protection

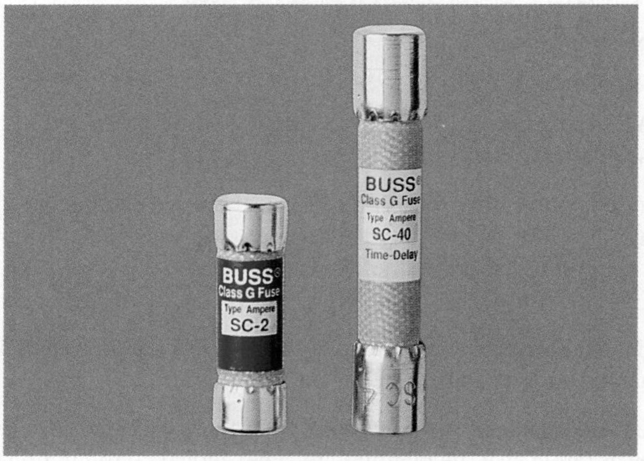


*EXHIBIT 240.13 Two fuses rated 300 volts with marking to indicate they are Class G. (Courtesy of Cooper Bussmann, a division of Cooper Industries PLC)*

of fluorescent fixtures, semiconductor rectifiers, motor-operated appliances, and so forth, do not need to be marked for IC.

**(D) Renewable Fuses.** Class H cartridge fuses of the renewable type shall be permitted to be used only for replacement in existing installations where there is no evidence of overfusing or tampering.

Where overfusing and/or tampering are detected with an existing installation, the use of nonrenewable fuses as a replacement is mandatory. An important consideration in the use of the traditional Class H renewable fuse is its 10,000-A interrupting rating. Caution must be exercised where the use of renewable fuses is contemplated, because the manufacturer's directions provided in some modern fusible switches do not permit the use of renewable fuses or strongly recommend against their use.

### 240.61 Classification

Cartridge fuses and fuseholders shall be classified according to voltage and amperage ranges. Fuses rated 1000 volts, nominal, or less shall be permitted to be used for voltages at or below their ratings.

See Part II of Article 100 for definitions covering fuses rated over 600 V. Section 490.21(B) covers applications of high-voltage fuses and fuseholders.

## VII. Circuit Breakers

### 240.80 Method of Operation

Circuit breakers shall be trip free and capable of being closed and opened by manual operation. Their normal method of operation by other than manual means, such as electrical or pneumatic, shall be permitted if means for manual operation are also provided.

## 240.81 Indicating

Circuit breakers shall clearly indicate whether they are in the open "off" or closed "on" position.

Where circuit breaker handles are operated vertically rather than rotationally or horizontally, the "up" position of the handle shall be the "on" position.

This section standardizes the switching operation of circuit breakers. Opening or turning off a circuit breaker or switch is generally the initial operation in de-energizing equipment or circuit conductors for servicing or maintenance. Although this standardized position requirement exists in the *Code*, *NFPA 70E, Standard for Electrical Safety in the Workplace,* contains additional steps that must be followed to ascertain that electrical conductors or circuit parts are de-energized (that is, placed in an "electrically safe work condition" per *NFPA 70E*). See 240.83(D), 404.11, and 410.141(A) for requirements for circuit breakers used as switches.

## 240.82 Nontamperable

A circuit breaker shall be of such design that any alteration of its trip point (calibration) or the time required for its operation requires dismantling of the device or breaking of a seal for other than intended adjustments.

## 240.83 Marking

**(A) Durable and Visible.** Circuit breakers shall be marked with their ampere rating in a manner that will be durable and visible after installation. Such marking shall be permitted to be made visible by removal of a trim or cover.

**(B) Location.** Circuit breakers rated at 100 amperes or less and 1000 volts or less shall have the ampere rating molded, stamped, etched, or similarly marked into their handles or escutcheon areas.

**(C) Interrupting Rating.** Every circuit breaker having an interrupting rating other than 5000 amperes shall have its interrupting rating shown on the circuit breaker. The interrupting rating shall not be required to be marked on circuit breakers used for supplementary protection.

**(D) Used as Switches.** Circuit breakers used as switches in 120-volt and 277-volt fluorescent lighting circuits shall be listed and shall be marked SWD or HID. Circuit breakers used as switches in high-intensity discharge lighting circuits shall be listed and shall be marked as HID.

Circuit breakers marked SWD are 15- or 20-A breakers that have been subjected to additional endurance and temperature testing to assess their ability for use as the regular control device for fluorescent lighting circuits. Circuit breakers marked HID are also acceptable for switching applications, and this marking must be on circuit breakers used as the regular switching device to control high-intensity discharge (HID) lighting such as mercury vapor, high-pressure or low-pressure sodium, or metal halide lighting. Circuit breakers marked HID can be used for switching both HID and fluorescent lighting loads; however, a circuit breaker marked SWD can be used only as a switching device for fluorescent lighting loads.

**(E) Voltage Marking.** Circuit breakers shall be marked with a voltage rating not less than the nominal system voltage that is indicative of their capability to interrupt fault currents between phases or phase to ground.

## 240.85 Applications

A circuit breaker with a straight voltage rating, such as 240V or 480V, shall be permitted to be applied in a circuit in which the nominal voltage between any two conductors does not exceed the circuit breaker's voltage rating. A two-pole circuit breaker shall not be used for protecting a 3-phase, corner-grounded delta circuit unless the circuit breaker is marked 1ϕ–3ϕ to indicate such suitability.

A circuit breaker with a slash rating, such as 120/240V or 480Y/277V, shall be permitted to be applied in a solidly grounded circuit where the nominal voltage of any conductor to ground does not exceed the lower of the two values of the circuit breaker's voltage rating and the nominal voltage between any two conductors does not exceed the higher value of the circuit breaker's voltage rating.

Informational Note: Proper application of molded case circuit breakers on 3-phase systems, other than solidly grounded wye, particularly on corner grounded delta systems, considers the circuit breakers' individual pole-interrupting capability.

A circuit breaker marked 480Y/277 V is not intended for use on a 480-V system with up to 480 volts to ground, such as a 480-V circuit derived from a corner-grounded, delta-connected system. A circuit breaker marked either 480 V or 600 V should be used on such a system. Similarly, a circuit breaker marked 120/240 V is not intended for use on a delta-connected 240-V circuit. A 240-V, 480-V, or 600-V circuit breaker should be used on such a circuit. The slash (/) between the lower and higher voltage ratings in the marking indicates that the circuit breaker has been tested for use on a circuit with the higher voltage between phases and with the lower voltage to ground.

## 240.86 Series Ratings

Where a circuit breaker is used on a circuit having an available fault current higher than the marked interrupting rating by being connected on the load side of an acceptable overcurrent protective device having a higher rating, the circuit breaker shall meet the requirements specified in (A) or (B), and (C).

A series-rated system is a combination of circuit breakers or fuses and circuit breakers that can be applied at available short-circuit levels above the interrupting rating of the load-side circuit breakers but not above that of the main or line-side device. Series-rated systems can consist of fuses that protect circuit breakers or of circuit breakers that protect circuit breakers. The arrangement of protective components in a series-rated system can be as specified in 240.86(A) for engineered systems applied to existing installations or in 240.86(B) for tested combinations that can be applied in any new or existing installation.

**(A) Selected Under Engineering Supervision in Existing Installations.** The series rated combination devices shall be selected by a licensed professional engineer engaged primarily in the design or maintenance of electrical installations. The selection shall be documented and stamped by the professional engineer. This documentation shall be available to those authorized to design, install, inspect, maintain, and operate the system. This series combination rating, including identification of the upstream device, shall be field marked on the end use equipment.

For calculated applications, the engineer shall ensure that the downstream circuit breaker(s) that are part of the series combination remain passive during the interruption period of the line side fully rated, current-limiting device.

Section 240.86(A) allows for an engineering solution at existing facilities where increases in transformer size, lowering of transformer impedances, and changes in utility distribution systems increase the available fault current beyond the interrupting rating of the existing circuit overcurrent protection equipment required by 110.9. An engineered system can be used to maintain compliance with 110.9 because it essentially redesigns the overcurrent protection scheme to accommodate the increase in available fault current. This eliminates the need for a wholesale replacement of electrical distribution equipment. Where the increase in fault current causes existing equipment to be "underrated," the engineering approach is to provide upstream protection that functions in concert with the existing protective devices to safely open the circuit under fault conditions. The requirement specifies that the design of such systems is to be performed only by licensed professional engineers whose credentials substantiate their ability to perform this type of engineering. Documentation in the form of stamped drawings and field marking of end-use equipment to indicate it is a component of a series-rated system is required.

Designing a series-rated system requires consideration of the fault-clearing characteristics of the existing protective devices, and of their ability to interact with the newly installed upstream protective device(s) under fault conditions. An engineered series-rated system can be applied to all existing installations. The operating parameters of the existing overcurrent protection equipment dictate what can be done in a field-engineered protection scheme.

The circuit breakers that are most likely to be compatible with series-rated systems are those that will remain closed during the interruption period of the fully rated OCPD installed on their line side, and those that have an interrupting rating not less than the let-through current of an upstream protective device (such as a current-limiting fuse). In cases where the opening of a circuit breaker (under any level of fault current) begins in less than one-half cycle, the use of a field-engineered series-rated system is likely to be contrary to acceptable application practices specified by the circuit breaker manufacturer. In an engineered system, the upstream device must operate before the downstream device for any fault that exceeds the interrupting rating of the downstream device in the series.

Where in doubt about the proper application of existing downstream circuit breakers with new upstream OCPDs, the manufacturers of the existing circuit breakers and the new upstream OCPDs must be consulted.

The safety objective of any overcurrent protection scheme is to ensure compliance with 110.9.

**(B) Tested Combinations.** The combination of line-side overcurrent device and load-side circuit breaker(s) is tested and marked on the end use equipment, such as switchboards and panelboards.

Informational Note to (A) and (B): See 110.22 for marking of series combination systems.

Section 240.86(B) requires that, where a series rating is used, the switchboards, panelboards, and load centers be marked for use with the series-rated combinations that may be used. Therefore, the enclosures must have a label affixed by the equipment manufacturer that provides the series rating of the combination(s). Because the equipment often does not have enough room to show all the legitimate series-rated combinations, UL 67, *Standard for Panelboards*, allows a bulletin to be referenced and supplied with the panelboard. These bulletins typically provide all the acceptable combinations. The installer of a series-rated system must also provide the additional labeling on equipment enclosures required by 110.22, indicating that the equipment has been applied in a series-rated system.

**(C) Motor Contribution.** Series ratings shall not be used where

(1) Motors are connected on the load side of the higher-rated overcurrent device and on the line side of the lower-rated overcurrent device, and
(2) The sum of the motor full-load currents exceeds 1 percent of the interrupting rating of the lower-rated circuit breaker.

The requirements in 240.86(C) limit the use of series-rated systems in which motors are connected between the line-side (protecting) device and the load-side (protected) circuit breaker. Section 240.86(C)(2) requires that series ratings developed under the parameters of either 240.86(A) or (B) are not to be used where the sum of motor full-load currents exceeds 1 percent of the interrupting rating of the load-side (protected) circuit breaker, as illustrated in Exhibit 240.14.

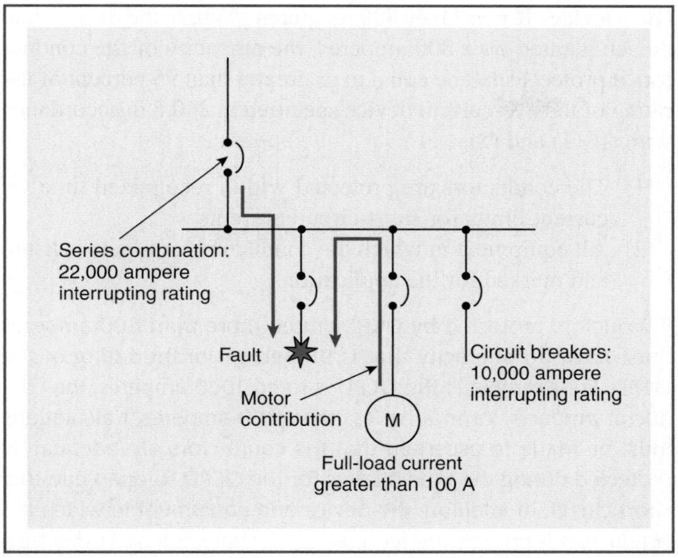


***EXHIBIT 240.14*** *Example of an installation where the level of motor contribution exceeds 1 percent of the interrupting rating for the lowest-rated circuit breaker in this series-rated system.*

## 240.87 Arc Energy Reduction

Where the highest continuous current trip setting for which the actual overcurrent device installed in a circuit breaker is rated or can be adjusted is 1200 A or higher, 240.87(A) and (B) shall apply.

**(A) Documentation.** Documentation shall be available to those authorized to design, install, operate, or inspect the installation as to the location of the circuit breaker(s).

**(B) Method to Reduce Clearing Time.** One of the following or approved equivalent means shall be provided:

(1) Zone-selective interlocking
(2) Differential relaying
(3) Energy-reducing maintenance switching with local status indicator
(4) Energy-reducing active arc flash mitigation system
(5) An approved equivalent means

Informational Note No. 1: An energy-reducing maintenance switch allows a worker to set a circuit breaker trip unit to "no intentional delay" to reduce the clearing time while the worker is working within an arc-flash boundary as defined in *NFPA 70E-2012, Standard for Electrical Safety in the Workplace*, and then to set the trip unit back to a normal setting after the potentially hazardous work is complete.

Informational Note No. 2: An energy-reducing active arc flash mitigation system helps in reducing arcing duration in the electrical distribution system. No change in the circuit breaker or the settings of other devices is required during maintenance when a worker is working within an arc flash boundary as defined in NFPA 70E-2012, *Standard for Electrical Safety in the Workplace*.

Where a circuit-breaker trip unit does not provide the capability for an instantaneous response to a short circuit, 240.87 requires that (1) the location of the device in the electrical system be documented, and (2) a means be provided in the system to limit the energy that personnel may be exposed to while working on energized equipment.

Circuit breakers without the capability of an instantaneous response can be used to selectively coordinate with other OCPDs where required by the *Code* or where incorporated as part of an electrical system design. However, a longer protective device opening time can expose the person(s) who may have initiated the fault while working on energized electrical equipment to more incident energy, thus increasing the severity of the hazard. This section identifies four specific methods that can be used to comply with this requirement and also permits use of an equivalent, which is acceptable to the AHJ. Zone-selective interlocking and differential relaying are methods that can be implemented as part of the system design and do not require any manual intervention.

An energy-reducing maintenance switch is a means by which an intentional delay in the opening of a circuit breaker can be overridden while maintenance, service, or diagnostic tasks are being performed. The circuit breaker can then be restored upon completion of the tasks to enable the system to be selectively coordinated or for other system design purposes.

The *Code* does not prohibit the use of these methods to limit fault current in systems where circuit breakers having an instantaneous capability are installed and adjusted to allow for a longer clearing time. This is a design consideration and is not required by this section.

Exhibit 240.15 is an example of an electronic trip unit for a circuit breaker employing an energy-reducing maintenance switch in addition to means for adjusting other settings.

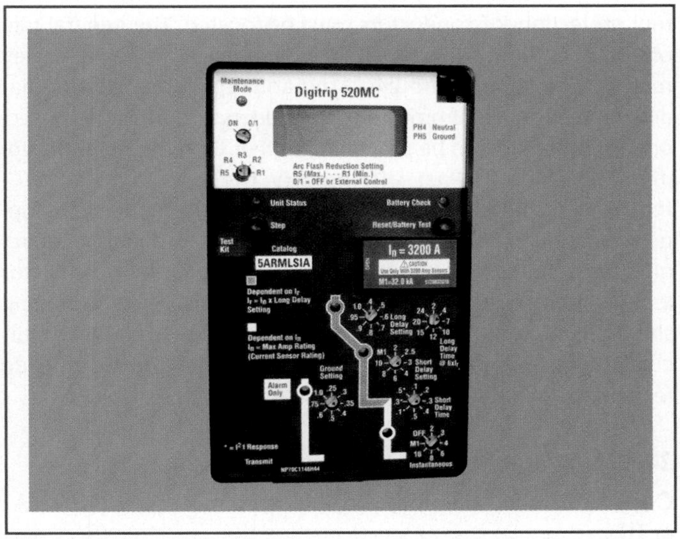


***EXHIBIT 240.15*** *An electronic trip unit for a circuit breaker that has its instantaneous setting set to "off" and that incorporates an energy-reducing maintenance setting. (Courtesy of Eaton Corporation)*

The function of the energy reduction maintenance switch is to override any settings that intentionally delay the opening time of the circuit breaker. A selectively coordinated electrical system is one example of where the response time adjustments available on some circuit breakers are set to allow for localization of a fault in the circuit. When these settings are overridden, the circuit breaker responds faster to a downstream fault in the circuit. The faster response time provides the important benefit of reducing the level of incident energy that personnel may be exposed to where tasks are being performed within the arc-flash boundary. An example of a local status indicator is an LED signifying "maintenance mode," which provides qualified persons with the indication that trip settings are adjusted to reduce incident energy should a fault occur anywhere electrically downstream of the device. (See *NFPA 70E* for definition of *arc-flash boundary,* and see the Article 100 definition of *qualified person*).

Once the task is completed, any settings that previously introduced an intentional delay in the circuit-breaker clearing time can be restored.

## VIII. Supervised Industrial Installations

### 240.90 General

Overcurrent protection in areas of supervised industrial installations shall comply with all of the other applicable provisions of this article, except as provided in Part VIII. The provisions of Part VIII shall be permitted only to apply to those portions of the electrical system in the supervised industrial installation used exclusively for manufacturing or process control activities.

For more information on supervised industrial installation, see the commentary following 240.2. Section 240.21 contains the requirements that specify the point in the circuit at which overcurrent protection for conductors must be located. The general rule of 240.21 is that conductors must be protected at the point they receive their supply, and 240.21(B) and (C) contain criteria that allow the overcurrent protection for feeders and transformer secondary conductors to be provided at other than the point of supply. The rules in Part VIII modify the 240.21(B) and (C) requirements based on the condition that only qualified personnel monitor and maintain the installation. Included in the requirements of 240.92(A) through (D) are criteria for longer conductor lengths, for use of differential relays as the means for providing short-circuit ground-fault protection, and for use of up to six circuit breakers or fuses as the overload protection for outside feeder taps and outside transformer secondary conductors.

### 240.91 Protection of Conductors

Conductors shall be protected in accordance with 240.91(A) or (B).

**(A) General.** Conductors shall be protected in accordance with 240.4.

**(B) Devices Rated Over 800 Amperes.** Where the overcurrent device is rated over 800 amperes, the ampacity of the conductors it protects shall be equal to or greater than 95 percent of the rating of the overcurrent device specified in 240.6 in accordance with (B)(1) and (2).

(1) The conductors are protected within recognized time vs. current limits for short-circuit currents
(2) All equipment in which the conductors terminate is listed and marked for the application

Conductors protected by OCPDs rated more than 800 amperes must have an ampacity that is 95 percent of the rating of the OCPD. For example, if the OCPD is rated 1000 amperes, the conductor ampacity cannot be less than 950 amperes. Calculations must be made to ascertain that the conductors are adequately protected during the time it takes for the OCPD to open due to a short circuit. In addition, the device and equipment in which the conductors terminate are required to be evaluated and listed by a qualified electrical testing laboratory for this specific application. This provision does not obviate the requirements in Articles 210, 215, and 230 that require branch circuit, feeder, and service conductors to have an ampacity not less than the calculated load being supplied.

### 240.92 Location in Circuit

An overcurrent device shall be connected in each ungrounded circuit conductor as required in 240.92(A) through (E).

**(A) Feeder and Branch-Circuit Conductors.** Feeder and branch-circuit conductors shall be protected at the point the conductors receive their supply as permitted in 240.21 or as otherwise permitted in 240.92(B), (C), (D), or (E).

**(B) Feeder Taps.** For feeder taps specified in 240.21(B)(2), (B)(3), and (B)(4), the tap conductors shall be permitted to be sized in accordance with Table 240.92(B).

**(C) Transformer Secondary Conductors of Separately Derived Systems.** Conductors shall be permitted to be connected to a transformer secondary of a separately derived system, without overcurrent protection at the connection, where the conditions of 240.92(C)(1), (C)(2), and (C)(3) are met.

**(1) Short-Circuit and Ground-Fault Protection.** The conductors shall be protected from short-circuit and ground-fault conditions by complying with one of the following conditions:

(1) The length of the secondary conductors does not exceed 30 m (100 ft) and the transformer primary overcurrent device has a rating or setting that does not exceed 150 percent of the value determined by multiplying the secondary conductor ampacity by the secondary-to-primary transformer voltage ratio.
(2) The conductors are protected by a differential relay with a trip setting equal to or less than the conductor ampacity.

*TABLE 240.92(B) Tap Conductor Short-Circuit Current Ratings*

Tap conductors are considered to be protected under short-circuit conditions when their short-circuit temperature limit is not exceeded. Conductor heating under short-circuit conditions is determined by (1) or (2):

(1) *Short-Circuit Formula for Copper Conductors*

$$(I^2/A^2)t = 0.0297 \log_{10} [(T_2 + 234)/(T_1 + 234)]$$

(2) *Short-Circuit Formula for Aluminum Conductors*

$$(I^2/A^2)t = 0.0125 \log_{10} [(T_2 + 228)/(T_1 + 228)]$$

where:

$I$ = short-circuit current in amperes

$A$ = conductor area in circular mils

$t$ = time of short circuit in seconds (for times less than or equal to 10 seconds)

$T_1$ = initial conductor temperature in degrees Celsius.

$T_2$ = final conductor temperature in degrees Celsius.

Copper conductor with paper, rubber, varnished cloth insulation, $T_2 = 200$

Copper conductor with thermoplastic insulation, $T_2 = 150$

Copper conductor with cross-linked polyethylene insulation, $T_2 = 250$

Copper conductor with ethylene propylene rubber insulation, $T_2 = 250$

Aluminum conductor with paper, rubber, varnished cloth insulation, $T_2 = 200$

Aluminum conductor with thermoplastic insulation, $T_2 = 150$

Aluminum conductor with cross-linked polyethylene insulation, $T_2 = 250$

Aluminum conductor with ethylene propylene rubber insulation, $T_2 = 250$

Informational Note: A differential relay is connected to be sensitive only to short-circuit or fault currents within the protected zone and is normally set much lower than the conductor ampacity. The differential relay is connected to trip protective devices that de-energize the protected conductors if a short-circuit condition occurs.

(3) The conductors shall be considered to be protected if calculations, made under engineering supervision, determine that the system overcurrent devices will protect the conductors within recognized time vs. current limits for all short-circuit and ground-fault conditions.

**(2) Overload Protection.** The conductors shall be protected against overload conditions by complying with one of the following:

(1) The conductors terminate in a single overcurrent device that will limit the load to the conductor ampacity.

(2) The sum of the overcurrent devices at the conductor termination limits the load to the conductor ampacity. The overcurrent devices shall consist of not more than six circuit breakers or sets of fuses mounted in a single enclosure, in a group of separate enclosures, or in or on a switchboard or switchgear. There shall be no more than six overcurrent devices grouped in any one location.

(3) Overcurrent relaying is connected [with a current transformer(s), if needed] to sense all of the secondary conductor current and limit the load to the conductor ampacity by opening upstream or downstream devices.

(4) Conductors shall be considered to be protected if calculations, made under engineering supervision, determine that the system overcurrent devices will protect the conductors from overload conditions.

**(3) Physical Protection.** The secondary conductors are protected from physical damage by being enclosed in an approved raceway or by other approved means.

**(D) Outside Feeder Taps.** Outside conductors shall be permitted to be tapped to a feeder or to be connected at a transformer secondary, without overcurrent protection at the tap or connection, where all the following conditions are met:

(1) The conductors are protected from physical damage in an approved manner.

(2) The sum of the overcurrent devices at the conductor termination limits the load to the conductor ampacity. The overcurrent devices shall consist of not more than six circuit breakers or sets of fuses mounted in a single enclosure, in a group of separate enclosures, or in or on a switchboard or switchgear. There shall be no more than six overcurrent devices grouped in any one location.

(3) The tap conductors are installed outdoors of a building or structure except at the point of load termination.

(4) The overcurrent device for the conductors is an integral part of a disconnecting means or shall be located immediately adjacent thereto.

(5) The disconnecting means for the conductors are installed at a readily accessible location complying with one of the following:

   a. Outside of a building or structure
   b. Inside, nearest the point of entrance of the conductors
   c. Where installed in accordance with 230.6, nearest the point of entrance of the conductors

**(E) Protection by Primary Overcurrent Device.** Conductors supplied by the secondary side of a transformer shall be permitted to be protected by overcurrent protection provided on the primary (supply) side of the transformer, provided the primary device time–current protection characteristic, multiplied by the maximum effective primary-to-secondary transformer voltage ratio, effectively protects the secondary conductors.

## IX. Overcurrent Protection over 1000 Volts, Nominal

### 240.100 Feeders and Branch Circuits

**(A) Location and Type of Protection.** Feeder and branch-circuit conductors shall have overcurrent protection in each ungrounded conductor located at the point where the conductor receives its supply or at an alternative location in the circuit when designed under engineering supervision that includes but is not limited to considering the appropriate fault studies and time–current coordination analysis of the protective devices and the conductor damage curves. The overcurrent protection shall be permitted to be provided by either 240.100(A)(1) or (A)(2).

**(1) Overcurrent Relays and Current Transformers.** Circuit breakers used for overcurrent protection of 3-phase circuits shall have a minimum of three overcurrent relay elements operated from three current transformers. The separate overcurrent relay elements (or protective functions) shall be permitted to be part of a single electronic protective relay unit.

On 3-phase, 3-wire circuits, an overcurrent relay element in the residual circuit of the current transformers shall be permitted to replace one of the phase relay elements.

An overcurrent relay element, operated from a current transformer that links all phases of a 3-phase, 3-wire circuit, shall be permitted to replace the residual relay element and one of the phase-conductor current transformers. Where the neutral conductor is not regrounded on the load side of the circuit as permitted in 250.184(B), the current transformer shall be permitted to link all 3-phase conductors and the grounded circuit conductor (neutral).

**(2) Fuses.** A fuse shall be connected in series with each ungrounded conductor.

**(B) Protective Devices.** The protective device(s) shall be capable of detecting and interrupting all values of current that can occur at their location in excess of their trip-setting or melting point.

**(C) Conductor Protection.** The operating time of the protective device, the available short-circuit current, and the conductor used shall be coordinated to prevent damaging or dangerous temperatures in conductors or conductor insulation under short-circuit conditions.

### 240.101 Additional Requirements for Feeders

**(A) Rating or Setting of Overcurrent Protective Devices.** The continuous ampere rating of a fuse shall not exceed three times the ampacity of the conductors. The long-time trip element setting of a breaker or the minimum trip setting of an electronically actuated fuse shall not exceed six times the ampacity of the conductor. For fire pumps, conductors shall be permitted to be protected for overcurrent in accordance with 695.4(B)(2).

**(B) Feeder Taps.** Conductors tapped to a feeder shall be permitted to be protected by the feeder overcurrent device where that overcurrent device also protects the tap conductor.

# ARTICLE 250
# Grounding and Bonding

## I. General

### 250.1 Scope

This article covers general requirements for grounding and bonding of electrical installations, and the specific requirements in (1) through (6).

(1) Systems, circuits, and equipment required, permitted, or not permitted to be grounded
(2) Circuit conductor to be grounded on grounded systems
(3) Location of grounding connections
(4) Types and sizes of grounding and bonding conductors and electrodes
(5) Methods of grounding and bonding
(6) Conditions under which guards, isolation, or insulation may be substituted for grounding

Informational Note: See Figure 250.1 for information on the organization of Article 250 covering grounding and bonding requirements.

The title of Article 250, *Grounding and Bonding,* conveys that grounding and bonding are two separate concepts. The two concepts are not mutually exclusive, and in many cases they are directly interrelated through the requirements of Article 250.

### 250.2 Definition

**Bonding Jumper, Supply-Side.** A conductor installed on the supply side of a service or within a service equipment enclosure(s), or for a separately derived system, that ensures the required electrical conductivity between metal parts required to be electrically connected.

Metal equipment enclosures, metal raceways, and metal cable trays are examples of equipment containing supply-side conductors that are required to be bonded. Where bonding jumpers are used, they are required to be installed and sized as specified in 250.102(A), (B), (C), and (E). Bonding jumpers installed on the load side of a service, feeder, or branch-circuit OCPD are *equipment bonding jumpers.*

•

### 250.3 Application of Other Articles

For other articles applying to particular cases of installation of conductors and equipment, grounding and bonding requirements are identified in Table 250.3 that are in addition to, or modifications of, those of this article.

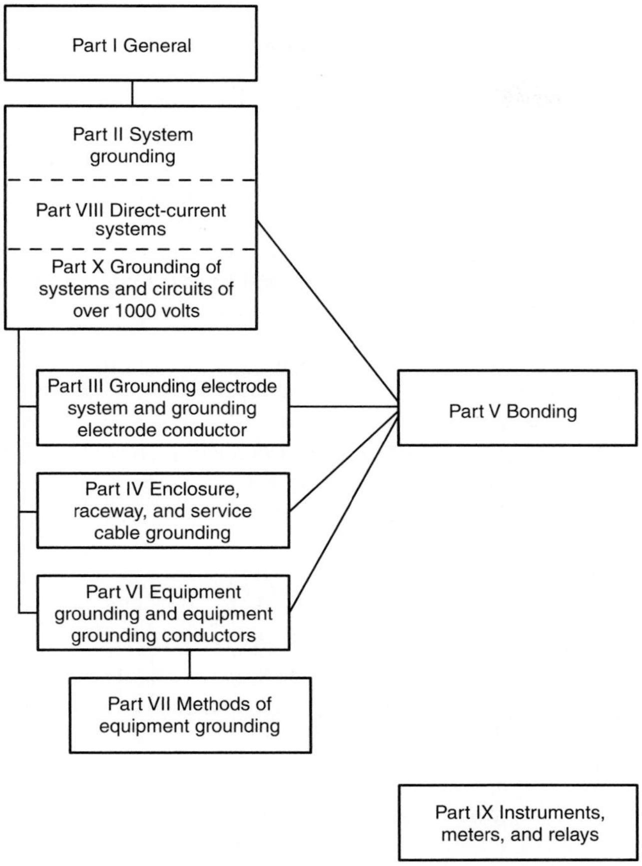


*FIGURE 250.1 Grounding and Bonding.*

## 250.4 General Requirements for Grounding and Bonding

The following general requirements identify what grounding and bonding of electrical systems are required to accomplish. The prescriptive methods contained in Article 250 shall be followed to comply with the performance requirements of this section.

Performance-based requirements provide an overall objective without mandating specifics for accomplishing that objective. The first paragraph of 250.4 indicates that the performance objectives stated in 250.4(A) for grounded systems and in 250.4(B) for ungrounded systems are accomplished by complying with the prescriptive requirements found in the rest of Article 250.

The requirements of 250.4 do not provide a specific rule for the sizing or connection of grounding conductors. Rather, the section outlines overall performance objectives for grounding conductors as applied to both grounded and ungrounded systems.

**(A) Grounded Systems.**

**(1) Electrical System Grounding.** Electrical systems that are grounded shall be connected to earth in a manner that will limit the voltage imposed by lightning, line surges, or unintentional contact with higher-voltage lines and that will stabilize the voltage to earth during normal operation.

Informational Note: An important consideration for limiting the imposed voltage is the routing of bonding and grounding electrode conductors so that they are not any longer than necessary to complete the connection without disturbing the permanent parts of the installation and so that unnecessary bends and loops are avoided.

**(2) Grounding of Electrical Equipment.** Normally non–current-carrying conductive materials enclosing electrical conductors or equipment, or forming part of such equipment, shall be connected to earth so as to limit the voltage to ground on these materials.

**(3) Bonding of Electrical Equipment.** Normally non–current-carrying conductive materials enclosing electrical conductors or equipment, or forming part of such equipment, shall be connected together and to the electrical supply source in a manner that establishes an effective ground-fault current path.

**(4) Bonding of Electrically Conductive Materials and Other Equipment.** Normally non–current-carrying electrically conductive materials that are likely to become energized shall be connected together and to the electrical supply source in a manner that establishes an effective ground-fault current path.

**(5) Effective Ground-Fault Current Path.** Electrical equipment and wiring and other electrically conductive material likely to become energized shall be installed in a manner that creates a low-impedance circuit facilitating the operation of the overcurrent device or ground detector for high-impedance grounded systems. It shall be capable of safely carrying the maximum ground-fault current likely to be imposed on it from any point on the wiring system where a ground fault may occur to the electrical supply source. The earth shall not be considered as an effective ground-fault current path.

The performance objective for the effective ground-fault current path is not always to facilitate operation of an OCPD. For high-impedance grounded systems, for example, the performance objective is to ensure operation of the required ground detector in order to activate some type of an alarm or other signal indicating the existence of a ground-fault condition.

**(B) Ungrounded Systems.**

**(1) Grounding Electrical Equipment.** Non–current-carrying conductive materials enclosing electrical conductors or equipment, or forming part of such equipment, shall be connected to earth in a manner that will limit the voltage imposed by lightning or unintentional contact with higher-voltage lines and limit the voltage to ground on these materials.

*TABLE 250.3 Additional Grounding and Bonding Requirements*

| Conductor/Equipment | Article | Section |
|---|---|---|
| Agricultural buildings | | 547.9 and 547.10 |
| Audio signal processing, amplification, and reproduction equipment | | 640.7 |
| Branch circuits | | 210.5, 210.6, 406.3 |
| Cablebus | | 370.9 |
| Cable trays | 392 | 392.60 |
| Capacitors | | 460.10, 460.27 |
| Circuits and equipment operating at less than 50 volts | 720 | |
| Communications circuits | 800 | |
| Community antenna television and radio distribution systems | | 820.93, 820.100, 820.103 |
| Conductors for general wiring | 310 | |
| Cranes and hoists | 610 | |
| Electrically driven or controlled irrigation machines | | 675.11(C), 675.12, 675.13, 675.14, 675.15 |
| Electric signs and outline lighting | 600 | |
| Electrolytic cells | 668 | |
| Elevators, dumbwaiters, escalators, moving walks, wheelchair lifts, and stairway chairlifts | 620 | |
| Fixed electric heating equipment for pipelines and vessels | | 427.29, 427.48 |
| Fixed outdoor electric deicing and snow-melting equipment | | 426.27 |
| Flexible cords and cables | | 400.22, 400.23 |
| Floating buildings | | 553.8, 553.10, 553.11 |
| Grounding-type receptacles, adapters, cord connectors, and attachment plugs | | 406.9 |
| Hazardous (classified) locations | 500–517 | |
| Health care facilities | 517 | |
| Induction and dielectric heating equipment | 665 | |
| Industrial machinery | 670 | |
| Information technology equipment | | 645.15 |
| Intrinsically safe systems | | 504.50 |
| Luminaires and lighting equipment | | 410.40, 410.42, 410.46, 410.155(B) |
| Luminaires, lampholders, and lamps | 410 | |
| Marinas and boatyards | | 555.15 |
| Mobile homes and mobile home park | 550 | |
| Motion picture and television studios and similar locations | | 530.20, 530.64(B) |
| Motors, motor circuits, and controllers | 430 | |
| Natural and artificially made bodies of water | 682 | 682.30, 682.31, 682.32, 682.33 |
| Outlet, device, pull, and junction boxes; conduit bodies; and fittings | | 314.4, 314.25 |
| Over 600 volts, nominal, underground wiring methods | | 300.50(C) |
| Panelboards | | 408.40 |
| Pipe organs | 650 | |
| Radio and television equipment | 810 | |
| Receptacles and cord connectors | | 406.3 |
| Recreational vehicles and recreational vehicle parks | 551 | |
| Services | 230 | |
| Solar photovoltaic systems | | 690.41, 690.42, 690.43, 690.45, 690.47 |
| Swimming pools, fountains, and similar installations | 680 | |
| Switchboards and panelboards | | 408.3(D) |
| Switches | | 404.12 |
| Theaters, audience areas of motion picture and television studios, and similar locations | | 520.81 |
| Transformers and transformer vaults | | 450.10 |
| Use and identification of grounded conductors | 200 | |
| X-ray equipment | 660 | 517.78 |

**(2) Bonding of Electrical Equipment.** Non–current-carrying conductive materials enclosing electrical conductors or equipment, or forming part of such equipment, shall be connected together and to the supply system grounded equipment in a manner that creates a low-impedance path for ground-fault current that is capable of carrying the maximum fault current likely to be imposed on it.

**(3) Bonding of Electrically Conductive Materials and Other Equipment.** Electrically conductive materials that are likely to become energized shall be connected together and to the supply system grounded equipment in a manner that creates a low-impedance path for ground-fault current that is capable of carrying the maximum fault current likely to be imposed on it.

**(4) Path for Fault Current.** Electrical equipment, wiring, and other electrically conductive material likely to become energized shall be installed in a manner that creates a low-impedance circuit from any point on the wiring system to the electrical supply source to facilitate the operation of overcurrent devices should a second ground fault from a different phase occur on the wiring system. The earth shall not be considered as an effective fault-current path.

The performance requirements for grounding in both grounded and ungrounded systems can be categorized into two functions: system grounding and equipment grounding. These two functions are kept separate except at the point of supply, such as at the service equipment or at a separately derived system.

Grounding is the intentional connection of a current-carrying conductor to ground or to something that serves in place of ground. In most instances, this connection is made at the supply source, such as a transformer, and at the main service disconnecting means of the premises using the energy. Where a system operates "ungrounded," it does not have an intentionally grounded circuit conductor (i.e., grounded conductor), but equipment grounding through the use of an equipment grounding conductor (EGC) is required.

The two reasons for grounding are:

1. To limit the voltages caused by lightning or by accidental contact of the supply conductors with conductors of higher voltage
2. To stabilize the voltage under normal operating conditions (which maintains the voltage at one level relative to ground, so that any equipment connected to the system will be subject only to that potential difference)

Exhibit 250.1 shows a grounded single-phase, 3-wire service supplied from a utility transformer. Inside the service disconnecting means enclosure, the grounded conductor of the system is intentionally connected to a grounding electrode via the grounding electrode conductor. Bonding the equipment grounding bus to the grounded or neutral bus via the main bonding jumper within this enclosure provides a ground reference for exposed

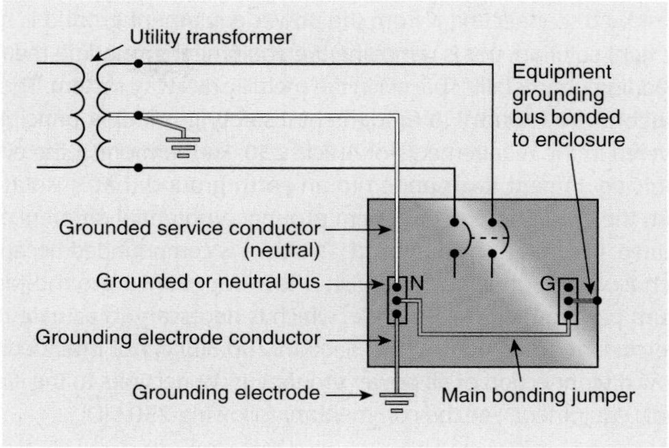


***EXHIBIT 250.1*** *Grounding and bonding arrangement for a single-phase, 3-wire service.*

non–current-carrying parts of the electrical system. It also provides a circuit for ground-fault current through the grounded service conductor back to the utility transformer (source of supply). At the utility transformer, often an additional connection is made from the grounded conductor to a separate grounding electrode. This bonding of the EGC bus to the neutral bus facilitates the operation of OCPDs or relays under ground-fault conditions, not the connection to earth.

## 250.6 Objectionable Current

**(A) Arrangement to Prevent Objectionable Current.** The grounding of electrical systems, circuit conductors, surge arresters, surge-protective devices, and conductive normally non–current-carrying metal parts of equipment shall be installed and arranged in a manner that will prevent objectionable current.

**(B) Alterations to Stop Objectionable Current.** If the use of multiple grounding connections results in objectionable current, one or more of the following alterations shall be permitted to be made, provided that the requirements of 250.4(A)(5) or (B)(4) are met:

(1) Discontinue one or more but not all of such grounding connections.
(2) Change the locations of the grounding connections.
(3) Interrupt the continuity of the conductor or conductive path causing the objectionable current.
(4) Take other suitable remedial and approved action.

Many electronic controls and computer equipment are sensitive to stray currents. Circulating currents on EGCs, metal raceways, and building steel develop potential differences between ground and the neutral of electronic equipment. Installation designers must look for ways to isolate electronic equipment from the effects of such stray circulating currents.

Isolating the electronic equipment from all other power equipment by disconnecting it from the power equipment ground is not the right solution, nor is removing the equipment grounding means or adding nonmetallic spacers in the metallic raceway system. These solutions are contrary to fundamental safety grounding principles covered in the requirements of Article 250. Furthermore, if the electronic equipment is grounded to an earth ground that is isolated from the common power system ground, a potential difference is created, which is a shock hazard. The error is compounded because such isolation does not establish a low-impedance ground-fault return path to the power source, which is necessary to actuate the overcurrent protection device. Section 250.6(B) is not intended to allow disconnection of all power grounding connections to the electronic equipment. See the commentary following 250.6(D).

**(C) Temporary Currents Not Classified as Objectionable Currents.** Temporary currents resulting from abnormal conditions, such as ground faults, shall not be classified as objectionable current for the purposes specified in 250.6(A) and (B).

**(D) Limitations to Permissible Alterations.** The provisions of this section shall not be considered as permitting electronic equipment from being operated on ac systems or branch circuits that are not connected to an equipment grounding conductor as required by this article. Currents that introduce noise or data errors in electronic equipment shall not be considered the objectionable currents addressed in this section.

Section 250.6(D) indicates that currents that result in noise or data errors in electronic equipment are not considered to be the objectionable currents referred to in 250.6, which limits the alterations permitted by 250.6(C). See 250.96(B) and 250.146(D) for requirements that provide safe bonding and grounding methods to minimize noise and data errors.

**(E) Isolation of Objectionable Direct-Current Ground Currents.** Where isolation of objectionable dc ground currents from cathodic protection systems is required, a listed ac coupling/dc isolating device shall be permitted in the equipment grounding conductor path to provide an effective return path for ac ground-fault current while blocking dc current.

The listed ac coupling/dc isolating device allowed by this section blocks the dc current on grounding and bonding conductors and allows the ground-fault return path to function properly. These devices are evaluated by a product testing organization for proper performance under ground-fault conditions.

Where cathodic protection for the piping system is provided, the required grounding and bonding connections associated with metal piping systems allow dc current to be imposed on grounding and bonding conductors.

## 250.8 Connection of Grounding and Bonding Equipment

**(A) Permitted Methods.** Equipment grounding conductors, grounding electrode conductors, and bonding jumpers shall be connected by one or more of the following means:

(1) Listed pressure connectors
(2) Terminal bars
(3) Pressure connectors listed as grounding and bonding equipment
(4) Exothermic welding process
(5) Machine screw-type fasteners that engage not less than two threads or are secured with a nut
(6) Thread-forming machine screws that engage not less than two threads in the enclosure
(7) Connections that are part of a listed assembly
(8) Other listed means

By specifically identifying machine screws and thread-forming machine screws as acceptable connection methods, this section indicates that no other type of screw (not a sheet metal screw or a wood screw, for instance) can be used for the connection of grounding and bonding conductors or terminals.

Listed pressure connectors, such as twist-on wire connectors that are not specifically listed for grounding or those that are specifically listed as grounding and bonding equipment, can be used for connection of grounding and bonding conductors. The use of listed pressure connectors other than those that are green in color is permitted for the connection of grounding and bonding conductors. Exhibits 250.2 and 250.3 illustrate two acceptable methods of attaching an equipment bonding jumper to a grounded metal box.

**(B) Methods Not Permitted.** Connection devices or fittings that depend solely on solder shall not be used.

## 250.10 Protection of Ground Clamps and Fittings

Ground clamps or other fittings exposed to physical damage shall be enclosed in metal, wood, or equivalent protective covering.

## 250.12 Clean Surfaces

Nonconductive coatings (such as paint, lacquer, and enamel) on equipment to be grounded shall be removed from threads and

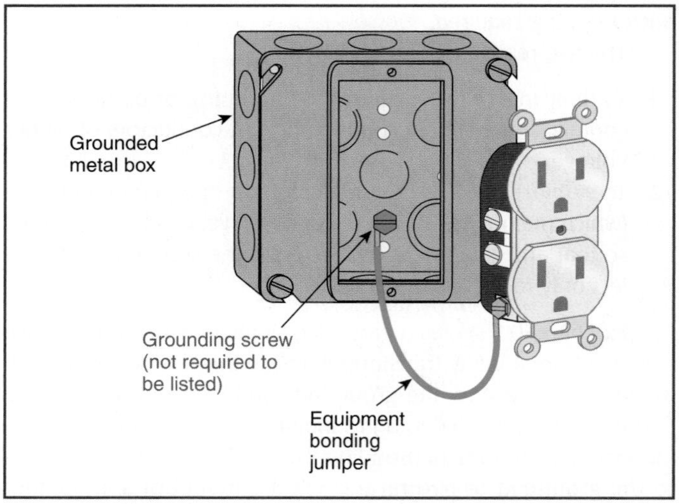


***EXHIBIT 250.2*** *Use of a grounding screw to attach equipment bonding jumper to a metal box.*

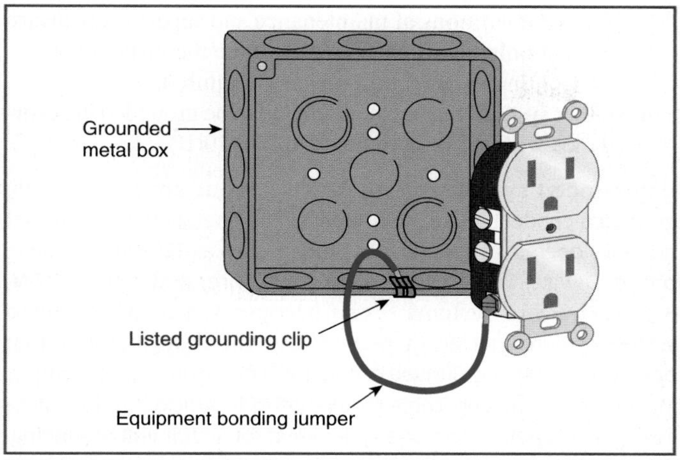


***EXHIBIT 250.3*** *Use of a listed grounding clip to attach a grounding conductor to a metal box.*

other contact surfaces to ensure good electrical continuity or be connected by means of fittings designed so as to make such removal unnecessary.

## II. System Grounding

### 250.20 Alternating-Current Systems to Be Grounded

Alternating-current systems shall be grounded as provided for in 250.20(A), (B), (C), or (D). Other systems shall be permitted to be grounded. If such systems are grounded, they shall comply with the applicable provisions of this article.

Informational Note: An example of a system permitted to be grounded is a corner-grounded delta transformer connection. See 250.26(4) for conductor to be grounded.

**(A) Alternating-Current Systems of Less Than 50 Volts.** Alternating-current systems of less than 50 volts shall be grounded under any of the following conditions:

(1) Where supplied by transformers, if the transformer supply system exceeds 150 volts to ground
(2) Where supplied by transformers, if the transformer supply system is ungrounded
(3) Where installed outside as overhead conductors

**(B) Alternating-Current Systems of 50 Volts to 1000 Volts.** Alternating-current systems of 50 volts to 1000 volts that supply premises wiring and premises wiring systems shall be grounded under any of the following conditions:

(1) Where the system can be grounded so that the maximum voltage to ground on the ungrounded conductors does not exceed 150 volts

Exhibit 250.4 illustrates the grounding requirements of 250.20(B)(1) for a 120-volt, single-phase, 2-wire system and for a 120/240-volt, single-phase, 3-wire system. The selection of which conductor to be grounded is covered in 250.26.

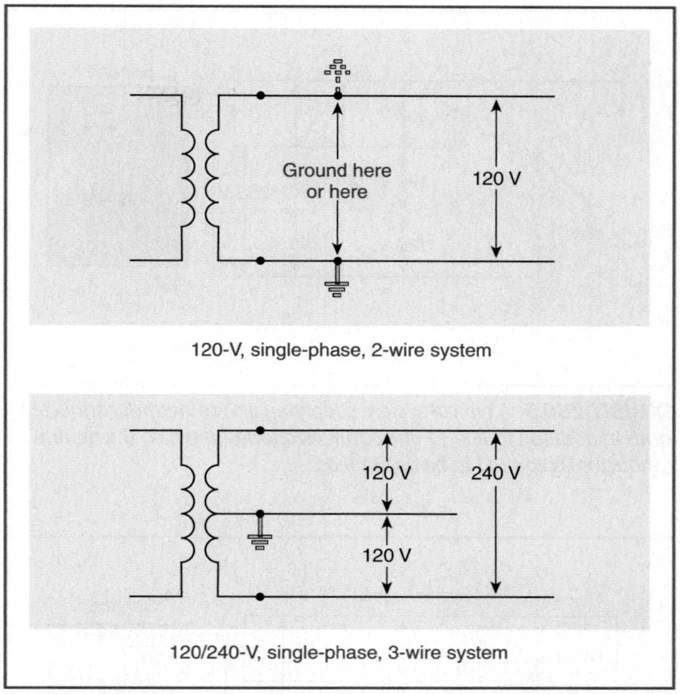


***EXHIBIT 250.4*** *Typical systems and conductor required to be grounded so that the maximum voltage to ground on the ungrounded conductors does not exceed 150 volts.*

(2) Where the system is 3-phase, 4-wire, wye connected in which the neutral conductor is used as a circuit conductor
(3) Where the system is 3-phase, 4-wire, delta connected in which the midpoint of one phase winding is used as a circuit conductor

Exhibit 250.5 illustrates a 3-phase, 4-wire, wye-connected system covered by 250.20(B)(2). Because this system supplies line-to-neutral connected loads in addition to line-to-line connected loads, the neutral conductor is required to be grounded.

Exhibit 250.6 shows a 3-phase, 4-wire, delta-connected system covered by 250.20(B)(3). A connection is made at the midpoint of one phase to enable line-to-neutral loads to be supplied. The same voltage is developed between this grounded conductor and the two phase conductors connected at either end of the winding that is midpoint grounded. The voltage between the third phase conductor and the grounded conductor is higher, and other requirements in the *Code* cover the arrangement and identification of the "high leg" in this system. For systems that are required by 250.20(B)(1), (2), or (3) to be grounded, 250.26 contains the requirements covering which conductor is to be grounded.

**(C) Alternating-Current Systems of over 1000 Volts.** Alternating-current systems supplying mobile or portable equipment shall be grounded as specified in 250.188. Where supplying other than mobile or portable equipment, such systems shall be permitted to be grounded.

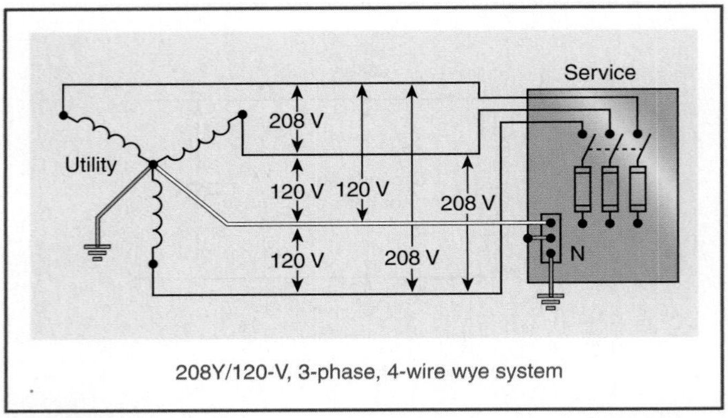


***EXHIBIT 250.5*** *A typical system supplying line-to-neutral connected loads in addition to line-to-line connected loads in which the neutral conductor is required to be grounded.*

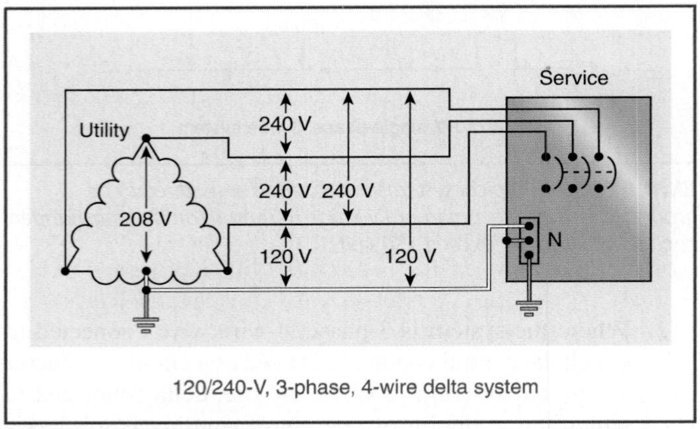


***EXHIBIT 250.6*** *A typical system in which the midpoint of one phase winding is used as a circuit conductor.*

**(D) Impedance Grounded Neutral Systems.** Impedance grounded neutral systems shall be grounded in accordance with 250.36 or 250.187.

## 250.21 Alternating-Current Systems of 50 Volts to 1000 Volts Not Required to Be Grounded

**(A) General.** The following ac systems of 50 volts to 1000 volts shall be permitted to be grounded but shall not be required to be grounded:

(1) Electrical systems used exclusively to supply industrial electric furnaces for melting, refining, tempering, and the like
(2) Separately derived systems used exclusively for rectifiers that supply only adjustable-speed industrial drives
(3) Separately derived systems supplied by transformers that have a primary voltage rating of 1000 volts or less, provided that all the following conditions are met:
   a. The system is used exclusively for control circuits.
   b. The conditions of maintenance and supervision ensure that only qualified persons service the installation.
   c. Continuity of control power is required.
(4) Other systems that are not required to be grounded in accordance with the requirements of 250.20(B)

Ungrounded systems are systems without an intentionally grounded conductor used in normal circuit operation. Systems that normally operate as grounded systems — such as 120/240-V, single-phase, 3-wire; 208Y/120-V, 3-phase, 4-wire; and 480Y/277-V, 3-phase, 4-wire systems — can be operated as ungrounded systems where specifically permitted by the *Code.* A system that operates without a grounded conductor is not exempt from complying with all of the applicable requirements in Article 250 for establishing a grounding electrode system and for equipment grounding. These protective features are required for grounded and ungrounded electrical distribution systems.

Ungrounded electrical systems are permitted for the specific functions described in 250.21(A)(1), (2), and (3) and for general power distribution systems in accordance with 250.21(A)(4). Delta-connected, 3-phase, 3-wire, 240-V and 480-V systems are examples of common electrical distribution systems that are permitted but are not required to have a circuit conductor that is intentionally grounded. The operational advantage of using an ungrounded electrical system is continuity of operation, which in some processes might be a safer condition than that created by the automatic and unplanned opening of the supply circuit. The disadvantage of operating systems ungrounded is increased susceptibility to high transient voltages that can hasten insulation deterioration. As stated in 250.4(A)(1), limiting voltage impressed on the system due to lightning or line surges is a primary function of system grounding. In some limited applications, it may be desirable to not establish a voltage to ground. However, the consensus has been that for general applications, grounded systems provide a higher level of safety.

Unlike solidly grounded systems, in which the first line-to-ground fault causes the OCPD to automatically open the circuit, the same line-to-ground fault in an ungrounded system instead results in the faulted circuit conductor becoming a grounded conductor until the damaged conductor insulation can be repaired. However, this latent ground-fault condition will remain undetected unless ground detectors are installed in the ungrounded system or until another insulation failure on a different ungrounded conductor results in a line-to-line-to-ground fault, with the potential for more extensive damage to electrical equipment.

**(B) Ground Detectors.** Ground detectors shall be installed in accordance with 250.21(B)(1) and (B)(2).

(1) Ungrounded ac systems as permitted in 250.21(A)(1) through (A)(4) operating at not less than 120 volts and at 1000 volts or less shall have ground detectors installed on the system.
(2) The ground detection sensing equipment shall be connected as close as practicable to where the system receives its supply.

Ground detectors provide a visual indication, an audible signal, or both, to alert system operators and maintainers of a ground-fault condition in the electrical system. The notification of the ground-fault condition, rather than automatic interruption of the circuit, allows the operators of the process to then take the necessary steps to effect an orderly shutdown, determine where the ground fault is located in the system, and to safely perform the necessary repair.

**(C) Marking.** Ungrounded systems shall be legibly marked "Caution: Ungrounded System Operating — _____Volts Between Conductors" at the source or first disconnecting means of the system. The marking shall be of sufficient durability to withstand the environment involved.

Section 250.21(C) incorporates the marking requirements from 408.3(F)(2) that specify marking to provide a safety warning. The permanency of the marking is critical to ensure that the safety message of the marking is conveyed to those who may install or work on equipment supplied by this system after its initial installation. In the 2002 and previous editions of the *Code*, the installation of ground detectors was required only for some very specific applications of ungrounded systems (and in impedance grounded neutral systems). For further information on what is considered to be the voltage to ground in an ungrounded system, see the definition of *voltage to ground* in Article 100.

## 250.22 Circuits Not to Be Grounded

The following circuits shall not be grounded:

(1) Circuits for electric cranes operating over combustible fibers in Class III locations, as provided in 503.155
(2) Circuits in health care facilities as provided in 517.61 and 517.160
(3) Circuits for equipment within electrolytic cell working zone as provided in Article 668
(4) Secondary circuits of lighting systems as provided in 411.6(A)
(5) Secondary circuits of lighting systems as provided in 680.23(A)(2).

## 250.24 Grounding Service-Supplied Alternating-Current Systems

**(A) System Grounding Connections.** A premises wiring system supplied by a grounded ac service shall have a grounding electrode conductor connected to the grounded service conductor, at each service, in accordance with 250.24(A)(1) through (A)(5).

The power for ac premises wiring systems is either separately derived, in accordance with 250.30, or supplied by the service. See the definition of *service* in Article 100. Section 250.30 covers grounding requirements for separately derived ac systems, and 250.24(A)(1) through (5) covers system grounding requirements for premises supplied by an ac service. The grounded conductor of an ac service is connected to a grounding electrode system to limit the voltage to ground imposed on the system by lightning, line surges, and (unintentional) high-voltage crossovers. Another reason for requiring this connection is to stabilize the voltage to ground during normal operation, including short circuits. These performance requirements are stated in 250.4(A) and 250.4(B).

**(1) General.** The grounding electrode conductor connection shall be made at any accessible point from the load end of the overhead service conductors, service drop, underground service conductors, or service lateral to, including the terminal or bus to which the grounded service conductor is connected at the service disconnecting means.

Informational Note: See definitions of *Service Conductors, Overhead; Service Conductors, Underground; Service Drop;* and *Service Lateral* in Article 100.

Allowing various locations for the connection to be made continues to meet the overall objectives for grounding while allowing the installer a variety of practical solutions. Exhibit 250.7 illustrates three possible connection point locations where the grounding electrode conductor is permitted to be connected to the grounded service conductor. Determination of whether a point of connection is *accessible* has to be determined by the AHJ based on local conditions such as locked meter socket enclosures.

**(2) Outdoor Transformer.** Where the transformer supplying the service is located outside the building, at least one additional grounding connection shall be made from the grounded service conductor to a grounding electrode, either at the transformer or elsewhere outside the building.

*Exception: The additional grounding electrode conductor connection shall not be made on high-impedance grounded neutral systems. The system shall meet the requirements of 250.36.*

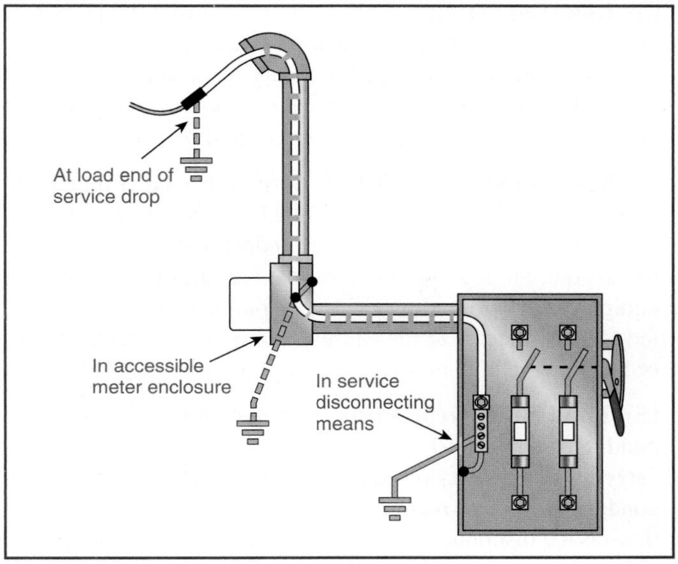


***EXHIBIT 250.7*** *Three locations where the grounding electrode conductor is permitted to be connected to the grounded service conductor.*

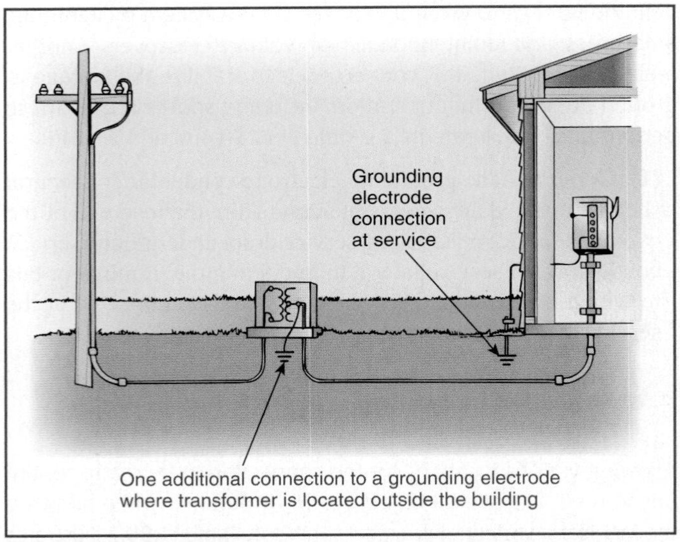


***EXHIBIT 250.8*** *Grounding connection to the grounded conductor at the transformer and at the service.*

A grounding electrode conductor connected to the grounded service conductor at an outdoor transformer is one example of how the requirement of 250.24(A)(2) can be met. Outdoor installations are susceptible to lightning as well as accidental primary-to-secondary crossovers. This requirement for a connection outside of a building helps mitigate the effects of these influences on the interior portion of the premises wiring system. Exhibit 250.8 illustrates two grounding electrode connections, one at the service equipment installed inside the building and one installed at the transformer, located outside of the building as required by 250.24(A)(2).

**(3) Dual-Fed Services.** For services that are dual fed (double ended) in a common enclosure or grouped together in separate enclosures and employing a secondary tie, a single grounding electrode conductor connection to the tie point of the grounded conductor(s) from each power source shall be permitted.

**(4) Main Bonding Jumper as Wire or Busbar.** Where the main bonding jumper specified in 250.28 is a wire or busbar and is installed from the grounded conductor terminal bar or bus to the equipment grounding terminal bar or bus in the service equipment, the grounding electrode conductor shall be permitted to be connected to the equipment grounding terminal, bar, or bus to which the main bonding jumper is connected.

**(5) Load-Side Grounding Connections.** A grounded conductor shall not be connected to normally non–current-carrying metal parts of equipment, to equipment grounding conductor(s), or be reconnected to ground on the load side of the service disconnecting means except as otherwise permitted in this article.

Informational Note: See 250.30 for separately derived systems, 250.32 for connections at separate buildings or structures, and 250.142 for use of the grounded circuit conductor for grounding equipment.

Section 250.24(A)(5) prohibits re-grounding of the grounded conductor on the load side of the service disconnecting means. This requirement correlates with the requirement of 250.142(B), which is a general prohibition on the use of the grounded conductor for grounding equipment. This prevents parallel paths for neutral current on the load side of the service disconnecting means. Parallel paths could include metal raceways, metal piping systems, metal ductwork, structural steel, and other continuous metal paths that are not intended to be normal current-carrying conductors.

**(B) Main Bonding Jumper.** For a grounded system, an unspliced main bonding jumper shall be used to connect the equipment grounding conductor(s) and the service-disconnect enclosure to the grounded conductor within the enclosure for each service disconnect in accordance with 250.28.

Where the service equipment of a grounded system consists of multiple disconnecting means, a main bonding jumper for each separate service disconnecting means is required to connect the grounded service conductor, the EGC, and the service equipment enclosure. The size of the main bonding jumper in each enclosure is selected from Table 250.102(C)(1), based on the size of the ungrounded service conductors, or it is calculated in accordance with 250.28(D)(1). See Exhibits 250.9 and 250.10.

*Exception No. 1: Where more than one service disconnecting means is located in an assembly listed for use as service*

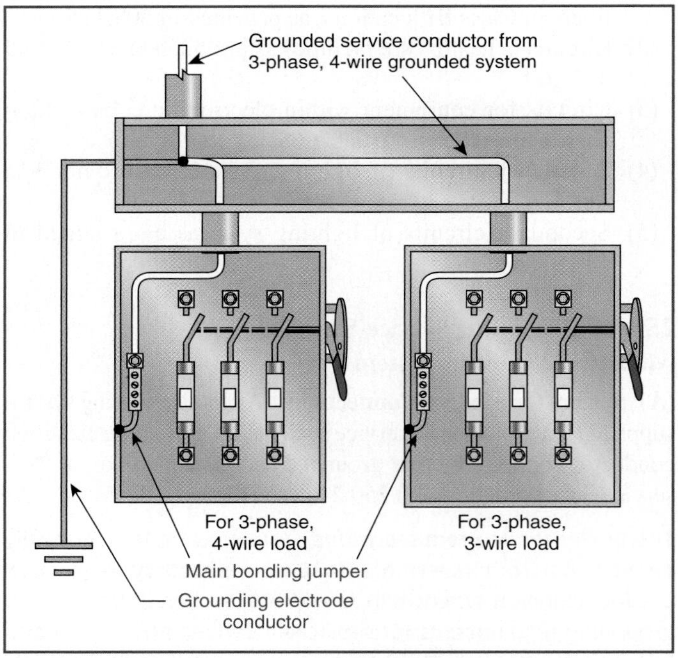


***EXHIBIT 250.9*** *A grounded system in which the grounded service conductor is bonded to the enclosure supplying 3-phase, 4-wire service loads and to the enclosure supplying 3-phase, 3-wire loads.*

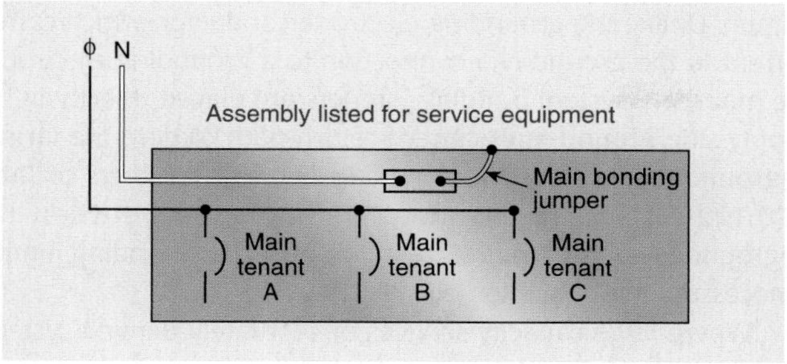


***EXHIBIT 250.10*** *One connection of the grounded service conductor to a listed service assembly containing multiple service disconnecting means.*

*equipment, an unspliced main bonding jumper shall bond the grounded conductor(s) to the assembly enclosure.*

Where multiple service disconnecting means are part of an assembly listed as service equipment, all grounded service conductors are required to be run to and bonded to the assembly. However, only one section of the assembly is required to have the main bonding jumper connection. See Exhibit 250.11, which accompanies the commentary following 250.28(D).

*Exception No. 2: Impedance grounded neutral systems shall be permitted to be connected as provided in 250.36 and 250.187.*

**(C) Grounded Conductor Brought to Service Equipment.** Where an ac system operating at 1000 volts or less is grounded at any point, the grounded conductor(s) shall be routed with the ungrounded conductors to each service disconnecting means and shall be connected to each disconnecting means grounded conductor(s) terminal or bus. A main bonding jumper shall connect the grounded conductor(s) to each service disconnecting means enclosure. The grounded conductor(s) shall be installed in accordance with 250.24(C)(1) through (C)(4).

*Exception: Where two or more service disconnecting means are located in a single assembly listed for use as service equipment, it shall be permitted to connect the grounded conductor(s) to the assembly common grounded conductor(s) terminal or bus. The assembly shall include a main bonding jumper for connecting the grounded conductor(s) to the assembly enclosure.*

If the utility service supplying the premises wiring system is grounded, the grounded conductor, whether or not it is used to supply a load, must be run to the service equipment. The conductor must be bonded to the equipment and be connected to a grounding electrode system. Exhibit 250.9 shows an example of the main rule in 250.24(C), which requires the grounded service conductor to be installed and bonded to each service disconnecting means enclosure. On the line side of the service disconnecting means, the grounded conductor is used to complete the ground-fault current path between the service equipment and the utility source. The grounded service conductor's other function, as a circuit conductor for line-to-neutral connected loads, is covered in 200.3 and 220.61. The exception to 250.24(C) permits a single connection of the grounded service conductor to a listed service assembly (such as a switchboard) that contains more than one service disconnecting means, as shown in Exhibit 250.10.

**(1) Sizing for a Single Raceway.** The grounded conductor shall not be smaller than specified in Table 250.102(C)(1).

**(2) Parallel Conductors in Two or More Raceways.** If the ungrounded service-entrance conductors are installed in parallel in two or more raceways, the grounded conductor shall also be installed in parallel. The size of the grounded conductor in each raceway shall be based on the total circular mil area of the parallel ungrounded conductors in the raceway, as indicated in 250.24(C)(1), but not smaller than 1/0 AWG.

Informational Note: See 310.10(H) for grounded conductors connected in parallel.

For a multiple raceway or cable service installation, the minimum size for the grounded conductor in each raceway or cable where conductors are in parallel cannot be less than 1/0 AWG. Although the cumulative size of the parallel grounded conductors may be larger than is required by 250.24(C)(1), the minimum 1/0 AWG per raceway or cable correlates with the requirements for parallel conductors contained in 310.10(H).

**(3) Delta-Connected Service.** The grounded conductor of a 3-phase, 3-wire delta service shall have an ampacity not less than that of the ungrounded conductors.

**(4) High Impedance.** The grounded conductor on a high-impedance grounded neutral system shall be grounded in accordance with 250.36.

**(D) Grounding Electrode Conductor.** A grounding electrode conductor shall be used to connect the equipment grounding conductors, the service-equipment enclosures, and, where the system is grounded, the grounded service conductor to the grounding electrode(s) required by Part III of this article. This conductor shall be sized in accordance with 250.66.

High-impedance grounded neutral system connections shall be made as covered in 250.36.

•

**(E) Ungrounded System Grounding Connections.** A premises wiring system that is supplied by an ac service that is ungrounded shall have, at each service, a grounding electrode conductor connected to the grounding electrode(s) required by Part III of this article. The grounding electrode conductor shall be connected to a metal enclosure of the service conductors at any accessible point from the load end of the overhead service conductors, service drop, underground service conductors, or service lateral to the service disconnecting means.

## 250.26 Conductor to Be Grounded — Alternating-Current Systems

For ac premises wiring systems, the conductor to be grounded shall be as specified in the following:

(1) Single-phase, 2-wire — one conductor
(2) Single-phase, 3-wire — the neutral conductor
(3) Multiphase systems having one wire common to all phases — the neutral conductor
(4) Multiphase systems where one phase is grounded — one phase conductor
(5) Multiphase systems in which one phase is used as in (2) — the neutral conductor

## 250.28 Main Bonding Jumper and System Bonding Jumper

For a grounded system, main bonding jumpers and system bonding jumpers shall be installed as follows:

The system bonding jumper performs the same electrical function as the main bonding jumper in a grounded ac system by connecting the EGC(s) to the grounded circuit conductor either at the source of a separately derived system or at the first disconnecting means supplied by the source. The term *system bonding jumper* is used to distinguish it from the main bonding jumper, which is installed in service equipment. See the commentary following the definition of *bonding jumper, system* in Article 100.

**(A) Material.** Main bonding jumpers and system bonding jumpers shall be of copper or other corrosion-resistant material. A main bonding jumper and a system bonding jumper shall be a wire, bus, screw, or similar suitable conductor.

**(B) Construction.** Where a main bonding jumper or a system bonding jumper is a screw only, the screw shall be identified with a green finish that shall be visible with the screw installed.

This identification requirement makes it possible to readily distinguish the bonding jumper screw from other screws in the grounded conductor terminal bar, to ensure that the required bonding connection has been made.

**(C) Attachment.** Main bonding jumpers and system bonding jumpers shall be connected in the manner specified by the applicable provisions of 250.8.

**(D) Size.** Main bonding jumpers and system bonding jumpers shall be sized in accordance with 250.28(D)(1) through (D)(3).

**(1) General.** Main bonding jumpers and system bonding jumpers shall not be smaller than specified in Table 250.102(C)(1).

In a grounded system, the primary function of the main bonding jumper and of the system bonding jumper is to create the link for ground-fault current between the EGCs and the grounded conductor. Table 250.102(C)(1) is used to establish the minimum size of main and system bonding jumpers where the ungrounded conductors do not exceed 1100 kcmil copper or 1750 kcmil aluminum. Unlike the grounding electrode conductor, which carries current to the ground (via connection to a grounding electrode), the main and system bonding jumpers are placed directly in the supply-side ground-fault current return path. Where the largest ungrounded supply conductor exceeds the parameters of Table 250.102(C)(1), a proportional relationship between the ungrounded conductor and the main or system bonding jumper is necessary to be maintained.

Where large capacity services or separately derived systems are installed, the main or system bonding jumper is likely to be larger than the grounding electrode conductor. Section 250.28(D)(1) requires that where the service-entrance conductors are larger than 1100 kcmil copper or 1750 kcmil aluminum, the bonding jumper must have a cross-sectional area of not less than 12½ percent of the cross-sectional area of the largest phase conductor or largest phase conductor set. In equipment such as panelboards or switchboards that are listed for use as service equipment, the manufacturer provides a bonding jumper that can be installed as the main or system bonding jumper. Thus, it is not necessary to duplicate this bonding jumper with another one sized in accordance with 250.28(D)(1).

**(2) Main Bonding Jumper for Service with More Than One Enclosure.** Where a service consists of more than a single enclosure as permitted in 230.71(A), the main bonding jumper for each enclosure shall be sized in accordance with 250.28(D)(1) based on the largest ungrounded service conductor serving that enclosure.

Where a service consists of more than one disconnecting means in separate enclosures, each line-side service equipment enclosure is treated separately, as depicted in Exhibit 250.11. The main bonding jumper in the left enclosure is a 4 AWG copper conductor. Based on the 3/0 AWG ungrounded service conductors supplying the 200 ampere circuit breaker and Table 250.102(C), the minimum-size main bonding jumper for this service equipment enclosure is 4 AWG copper. Similarly, the 1/0 AWG main bonding jumper for the enclosure on the right is derived from Table 250.102(C)(1) using the 500 kcmil ungrounded service conductors.

The main bonding jumpers and the grounded conductors for the two disconnecting means enclosures shown in Exhibit 250.11 are sized using Table 250.102(C)(1) and the grounding electrode conductor is sized using Table 250.66. First, the grounding electrode conductor at 2/0 AWG is full-sized based on Table 250.66 using the 750 kcmil ungrounded service conductor as the basis for selection. In some conditions, the grounding electrode conductor is permitted to be sized smaller than what is required in the table, which depends on the type of electrode that is used.

Next, the grounded conductor run to each enclosure in accordance with 250.24(B) is sized using Table 250.102(C)(1) as the minimum size permitted. For each enclosure, the minimum size grounded conductor is established based on the largest ungrounded conductor serving that enclosure. The grounded

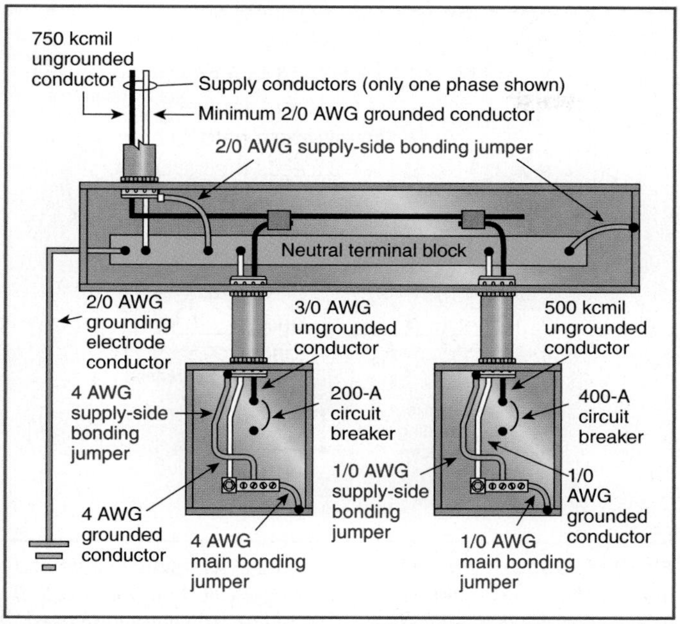


***EXHIBIT 250.11*** *An example of the bonding requirements for service equipment.*

conductor is also subject to the requirements of 220.61 covering the conductor's capacity for unbalanced load, which could result in having to increase the size to larger than what was determined from Table 250.102(C)(1).

Finally, supply-side bonding jumpers are used to bond the three metal conduits containing service conductors and the metal wireway located above the two service equipment enclosures. These bonding jumpers are also sized from Table 250.102(C)(1). The bonding jumpers for the raceways are sized based on the ungrounded conductors contained in each metal service raceway.

For the metal conduit entering the top of the wireway and for the wireway itself shown in Exhibit 250.11, the bonding jumper is sized from Table 250.102(C)(1) based on the 750 kcmil main service-entrance conductors and cannot be less than a 2/0 AWG copper conductor. The service-entrance conductors to the enclosures are 3/0 AWG and 500 kcmil copper, based on the loads supplied from each enclosure. The supply-side bonding jumpers for the short nipples are sized based on the size of the phase conductors supplying each disconnecting means. In this case, the metal raceway nipples containing the 3/0 AWG and 500 kcmil ungrounded service conductors require minimum 4 AWG and 1/0 AWG copper supply-side bonding jumpers, respectively.

### Application Example

A service is supplied by four 500 kcmil conductors in parallel for each phase; the minimum cross-sectional area of the bonding jumper is calculated as follows:

$$4 \times 500 \text{ kcmil} = 2000 \text{ kcmil}$$

Therefore, the main or system bonding jumper cannot be less than 12½ percent of 2000 kcmil, which results in a 250 kcmil copper conductor. The copper grounding electrode conductor for this set of conductors, based on Table 250.66, is not required to be larger than 3/0 AWG.

**(3) Separately Derived System with More Than One Enclosure.** Where a separately derived system supplies more than a single enclosure, the system bonding jumper for each enclosure shall be sized in accordance with 250.28(D)(1) based on the largest ungrounded feeder conductor serving that enclosure, or a single system bonding jumper shall be installed at the source and sized in accordance with 250.28(D)(1) based on the equivalent size of the largest supply conductor determined by the largest sum of the areas of the corresponding conductors of each set.

To determine the size of the system bonding jumper, all of the ungrounded conductors serving the multiple enclosures must be evaluated and the phase that yields the largest collective circular mil area is the basis for sizing the system bonding jumper.

### Calculation Example 1

A 225-kVA transformer supplies three, 3-phase, 208Y/120-volt secondary feeders that terminate in panelboards with 200-ampere main breakers. The ungrounded conductors for all three phases are 3/0 AWG copper with THHW insulation. A system bonding jumper is installed at each of the panelboards. Determine the size of the system bonding jumpers at each enclosure using Option 1.

*Solution*

Size of largest ungrounded conductor supplying individual panelboard:

3/0 AWG copper

System bonding jumper [from Table 250.102(C)(1)]:

3/0 AWG copper ungrounded conductors → 4 AWG copper

or 2 AWG aluminum system bonding jumper

Depending on how the transformer and the panelboards are arranged to prevent parallel neutral current paths on raceways and enclosures, the system bonding jumper may be internal to the panelboard using the manufacturer-supplied equipment or it may be installed in the transformer enclosure so that it connects any supply-side bonding jumpers to the neutral terminal (XO) of the transformer.

### Calculation Example 2

The same electrical equipment arrangement as in the previous example applies. In this case, however, the system bonding jumper is installed at the transformer and connects the transformer neutral terminal (XO) to individual supply-side bonding jumpers that are installed between the panelboard EGC terminals and a terminal bus attached to the transformer enclosure. Determine the size of the system bonding jumper using Option 2.

*Solution*

STEP 1. Size the individual supply-side bonding jumpers from each of the panelboard equipment grounding terminals to the terminal bus in the transformer.

From Table 250.102(C)(1):

3/0 AWG copper ungrounded conductors → 4 AWG copper

or 2 AWG aluminum system bonding jumper

STEP 2. Size the system bonding jumper from the terminal bus in the transformer to the transformer neutral terminal (XO). There are three secondary feeder circuits with 3/0 AWG ungrounded conductors for all phases. Find the cumulative circular mil area of one phase.

From Chapter 9, Table 8:

3/0 AWG = 167,800 circular mils

167,800 circular mils × 3 (number of sets of secondary conductors) = 503,400 circular mils

From Table 250.102(C):

503,400 circular mils copper ungrounded conductors → 1/0 AWG copper

or 3/0 AWG aluminum system bonding jumper

## 250.30 Grounding Separately Derived Alternating-Current Systems

In addition to complying with 250.30(A) for grounded systems, or as provided in 250.30(B) for ungrounded systems, separately derived systems shall comply with 250.20, 250.21, 250.22, or 250.26, as applicable. Multiple separately derived systems that are connected in parallel shall be installed in accordance with 250.30.

Informational Note No. 1: An alternate ac power source, such as an on-site generator, is not a separately derived system if the grounded conductor is solidly interconnected to a service-supplied system grounded conductor. An example of such a situation is where alternate source transfer equipment does not include a switching action in the grounded conductor and allows it to remain solidly connected to the service-supplied grounded conductor when the alternate source is operational and supplying the load served.

Informational Note No. 2: See 445.13 for the minimum size of conductors that carry fault current.

Exhibits 250.12 and 250.13 depict a 208Y/120-volt, 3-phase, 4-wire electrical service supplying a service disconnecting means to a building. The building also has an alternate or emergency electrical system. A feeder is installed from the service equipment to the normal power terminals of a transfer switch. The emergency or alternate power terminals of the transfer switch are supplied by a feeder that is supplied by a generator with a 208Y/120-volt, 3-phase output. Emergency, legally required standby, and/or optional standby loads can be supplied from the load terminals of the transfer switch in accordance with the applicable requirements of Articles 700, 701, and 702.

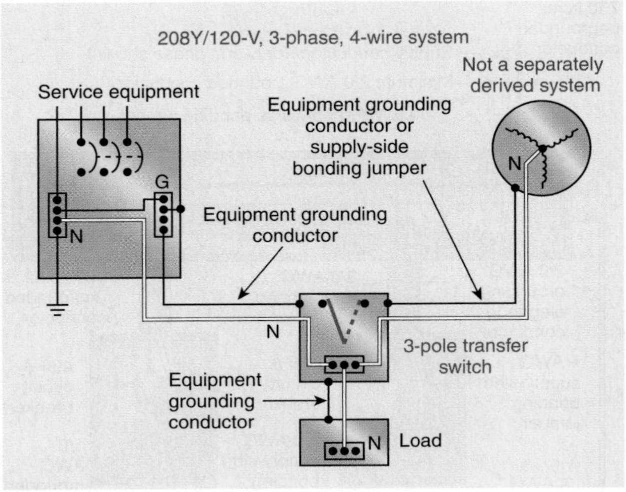


***EXHIBIT 250.12*** *A 208Y/120-volt, 3-phase, 4-wire system that has a direct electrical connection of the grounded circuit conductor (neutral) to the generator and is therefore not considered a separately derived system.*

In Exhibit 250.12, the neutral conductor from the generator to the load is not disconnected by the transfer switch. The system has a direct electrical connection between the normal grounded system conductor (neutral) and the generator neutral through the neutral bus in the transfer switch, thereby grounding the generator neutral. Because the generator is grounded by connection to the normal system ground, it is not a separately derived system and there are no requirements for grounding the neutral at the generator (see Informational Note No. 1 to 250.30). The conductor installed between the equipment grounding terminal of the transfer switch and the generator frame/equipment grounding terminal is either an EGC or a supply-side bonding jumper, depending on where the first overcurrent device in the generator feeder circuit is located. See 250.35(B) for the requirement covering generators supplying systems that are not separately derived.

In Exhibit 250.13, the grounded conductor (neutral) is connected to the switching contacts of a 4-pole transfer switch. Therefore, the generator system does not have a direct electrical connection to the other supply system grounded conductor (neutral), and the system supplied by the generator is considered separately derived. This separately derived system (3-phase, 4-wire, wye-connected system that supplies line-to-neutral loads) is required to be grounded in accordance with 250.20(B). The methods for grounding the system are specified in 250.30(A).

Section 250.30(A)(1) requires separately derived systems to have a system bonding jumper connected between the generator frame and the grounded circuit conductor (neutral). The grounding electrode conductor from the generator is required to be connected to a grounding electrode. This conductor and the grounding electrode are required to be located as close to the generator as practicable, according to 250.30(A)(4). However,

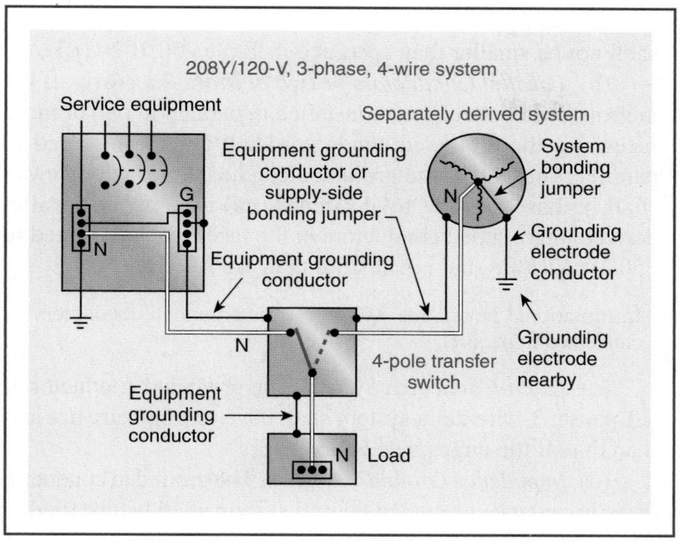


***EXHIBIT 250.13*** *A 208Y/120-volt, 3-phase, 4-wire system that does not have a direct electrical connection of the grounded circuit conductor (neutral) to the generator and is therefore considered a separately derived system.*

because the generator feeder is supplying a building that is also supplied by an ac service, 250.58 requires both supply systems to be connected to the same grounding electrode system. As is the case with the generator supplying the system in Exhibit 250.12 that is not separately derived, the conductor installed between the equipment grounding terminal of the transfer switch and the generator frame/equipment grounding terminal is either an EGC or a supply-side bonding jumper, depending on where the first overcurrent device in the generator feeder circuit is located.

**(A) Grounded Systems.** A separately derived ac system that is grounded shall comply with 250.30(A)(1) through (A)(8). Except as otherwise permitted in this article, a grounded conductor shall not be connected to normally non–current-carrying metal parts of equipment, be connected to equipment grounding conductors, or be reconnected to ground on the load side of the system bonding jumper.

Informational Note: See 250.32 for connections at separate buildings or structures and 250.142 for use of the grounded circuit conductor for grounding equipment.

*Exception: Impedance grounded neutral system grounding connections shall be made as specified in 250.36 or 250.187, as applicable.*

Because separately derived systems have no direct electrical connection, including a solidly connected grounded circuit conductor, to supply conductors originating in another system, grounding and bonding connections for each separately derived system are necessary to establish. A common installation where the requirements in 250.30(A) are applied is a transformer installed as part of the premises wiring system. For example, if the service voltage is 480Y/277 V and loads in the building or structure operate at 120 volts, a transformer is commonly used to develop a separate 208Y/120-V system for lighting and appliance loads.

**(1) System Bonding Jumper.** An unspliced system bonding jumper shall comply with 250.28(A) through (D). This connection shall be made at any single point on the separately derived system from the source to the first system disconnecting means or overcurrent device, or it shall be made at the source of a separately derived system that has no disconnecting means or overcurrent devices, in accordance with 250.30(A)(1)(a) or (b). The system bonding jumper shall remain within the enclosure where it originates. If the source is located outside the building or structure supplied, a system bonding jumper shall be installed at the grounding electrode connection in compliance with 250.30(C).

See the commentary following 250.28(D) for further information on sizing the system bonding jumper.

*Exception No. 1: For systems installed in accordance with 450.6, a single system bonding jumper connection to the tie point of the grounded circuit conductors from each power source shall be permitted.*

*Exception No. 2: If a building or structure is supplied by a feeder from an outdoor transformer, a system bonding jumper at both the source and the first disconnecting means shall be permitted if doing so does not establish a parallel path for the grounded conductor. If a grounded conductor is used in this manner, it shall not be smaller than the size specified for the system bonding jumper but shall not be required to be larger than the ungrounded conductor(s). For the purposes of this exception, connection through the earth shall not be considered as providing a parallel path.*

*Exception No. 3: The size of the system bonding jumper for a system that supplies a Class 1, Class 2, or Class 3 circuit, and is derived from a transformer rated not more than 1000 volt-amperes, shall not be smaller than the derived ungrounded conductors and shall not be smaller than 14 AWG copper or 12 AWG aluminum.*

(a) *Installed at the Source.* The system bonding jumper shall connect the grounded conductor to the supply-side bonding jumper and the normally non–current-carrying metal enclosure.

(b) *Installed at the First Disconnecting Means.* The system bonding jumper shall connect the grounded conductor to the supply-side bonding jumper, the disconnecting means enclosure, and the equipment grounding conductor(s).

The supply-side bonding jumper installed between the separately derived system enclosure and the enclosure of the first system disconnecting means provides the circuit ground-fault current, whether the system bonding jumper is installed at the source enclosure or at the disconnecting means enclosure. The supply-side bonding jumper can be a wire that is sized per 250.102(C), or it can be rigid metal conduit, intermediate metal conduit, or electrical metallic tubing installed between the two enclosures. Where the system operates at over 250 volts to ground, the raceway connections to enclosures must be made as specified in 250.97.

### Application Example

The source of a separately derived system is a 75-kVA dry-type transformer. Liquidtight flexible metal conduit is used as the wiring method between the transformer and the 200-ampere fusible safety switch that is the first system disconnecting means. The system bonding jumper is installed in the safety switch enclosure. The ungrounded conductors are 3/0 AWG copper. The wiring method necessitates the installation of a wire-type supply-side bonding jumper. What is the minimum size for this bonding jumper?

*Solution*

The requirement covering the minimum size for the supply-side bonding jumper is 250.102(C). This section refers to Table 250.102(C)(1) for sizing supply-side bonding jumpers where the ungrounded supply conductors are not greater than 1100 kcmil copper or 1750 kcmil aluminum. In this example, the largest ungrounded conductor size (3/0 AWG copper) is not greater than 1100 kcmil copper, which is the point at which 250.102(C) requires the bonding jumper size be calculated. Therefore, the supply-side bonding jumper size can be selected directly from Table 250.102(C)(1).

From Table 250.102(C)(1), 3/0 AWG ungrounded conductors = 4 AWG copper or 2 AWG aluminum supply-side bonding jumper. The bonding jumper installed can be a bare, covered, or insulated conductor and can be installed inside the raceway or, where the length of the bonding jumper does not exceed 6 feet, it can be installed outside and routed with the raceway.

**(2) Supply-Side Bonding Jumper.** If the source of a separately derived system and the first disconnecting means are located in separate enclosures, a supply-side bonding jumper shall be installed with the circuit conductors from the source enclosure to the first disconnecting means. A supply-side bonding jumper shall not be required to be larger than the derived ungrounded conductors. The supply-side bonding jumper shall be permitted to be of nonflexible metal raceway type or of the wire or bus type as follows:

(a) A supply-side bonding jumper of the wire type shall comply with 250.102(C), based on the size of the derived ungrounded conductors.

(b) A supply-side bonding jumper of the bus type shall have a cross-sectional area not smaller than a supply-side bonding jumper of the wire type as determined in 250.102(C).

*Exception: A supply-side bonding jumper shall not be required between enclosures for installations made in compliance with 250.30(A)(1), Exception No. 2.*

**(3) Grounded Conductor.** If a grounded conductor is installed and the system bonding jumper connection is not located at the source, 250.30(A)(3)(a) through (A)(3)(d) shall apply.

(a) *Sizing for a Single Raceway.* The grounded conductor shall not be smaller than specified in Table 250.102(C)(1).

(b) *Parallel Conductors in Two or More Raceways.* If the ungrounded conductors are installed in parallel in two or more raceways, the grounded conductor shall also be installed in parallel. The size of the grounded conductor in each raceway shall be based on the total circular mil area of the parallel derived ungrounded conductors in the raceway as indicated in 250.30(A)(3)(a), but not smaller than 1/0 AWG.

Informational Note: See 310.10(H) for grounded conductors connected in parallel.

(c) *Delta-Connected System.* The grounded conductor of a 3-phase, 3-wire delta system shall have an ampacity not less than that of the ungrounded conductors.

(d) *Impedance Grounded System.* The grounded conductor of an impedance grounded neutral system shall be installed in accordance with 250.36 or 250.187, as applicable.

The requirements of 250.30(A)(3)(a) and (b) are similar to those in 250.24(C)(1) and (C)(2) for the grounded service conductor. In addition, grounded and neutral conductors that carry current as part of normal circuit operation are required to be sized in accordance with 220.61.

**(4) Grounding Electrode.** The grounding electrode shall be as near as practicable to, and preferably in the same area as, the grounding electrode conductor connection to the system. The grounding electrode shall be the nearest of one of the following:

(1) Metal water pipe grounding electrode as specified in 250.52(A)(1)
(2) Structural metal grounding electrode as specified in 250.52(A)(2)

*Exception No. 1: Any of the other electrodes identified in 250.52(A) shall be used if the electrodes specified by 250.30(A)(4) are not available.*

*Exception No. 2 to (1) and (2): If a separately derived system originates in listed equipment suitable for use as service equipment, the grounding electrode used for the service or feeder equipment shall be permitted as the grounding electrode for the separately derived system.*

Informational Note No. 1: See 250.104(D) for bonding requirements for interior metal water piping in the area served by separately derived systems.

Informational Note No. 2: See 250.50 and 250.58 for requirements for bonding all electrodes together if located at the same building or structure.

The impedance of the connection to the grounding electrode is important to minimize, which is why this section specifies connection to the nearest of the specified electrodes. Where an effectively grounded metal water pipe is used as the grounding electrode for a separately derived system, 250.68(C)(1) specifies that only the first 5 feet of water piping entering the building can be used as the point to make grounding electrode conductor connections or as a conductor to interconnect grounding electrodes.

The exception to 250.68(C)(1) permits metal water piping beyond the first 5 feet of where it enters the building to be used as a bonding conductor to interconnect other electrodes or as a grounding electrode conductor. This enables grounding electrode conductor connections to be made at other locations in the building or structure. However, the piping has to meet all of the conditions specified in this exception from the point where the piping enters the building to the point where the grounding electrode conductor connection is made. The other grounding electrodes covered in 250.52(A) can be used, but only where a structural metal or metal water pipe–type grounding electrode is not available.

The practice of grounding the secondary of an isolating transformer to a ground rod or running the grounding electrode conductor back to the service ground (usually to reduce electrical noise on data-processing systems) is not permitted where either of the electrodes covered in item (1) or item (2) of 250.30(A)(4) is available. An isolation transformer that is part of a listed power supply for a data-processing room is not required to be grounded in accordance with 250.30(A)(4), but it must be grounded in accordance with the manufacturer's instructions.

**(5) Grounding Electrode Conductor, Single Separately Derived System.** A grounding electrode conductor for a single separately derived system shall be sized in accordance with 250.66 for the derived ungrounded conductors. It shall be used to connect the grounded conductor of the derived system to the grounding electrode as specified in 250.30(A)(4). This connection shall be made at the same point on the separately derived system where the system bonding jumper is connected.

*Exception No. 1: If the system bonding jumper specified in 250.30(A)(1) is a wire or busbar, it shall be permitted to connect the grounding electrode conductor to the equipment grounding terminal, bar, or bus, provided the equipment grounding terminal, bar, or bus is of sufficient size for the separately derived system.*

*Exception No. 2: If the source of a separately derived system is located within equipment listed and identified as suitable for use as service equipment, the grounding electrode conductor from the service or feeder equipment to the grounding electrode shall be permitted as the grounding electrode conductor for the separately derived system, provided that the grounding electrode conductor is of sufficient size for the separately derived system. If the equipment grounding bus internal to the equipment is not smaller than the required grounding electrode conductor for the separately derived system, the grounding electrode connection for the separately derived system shall be permitted to be made to the bus.*

*Exception No. 3: A grounding electrode conductor shall not be required for a system that supplies a Class 1, Class 2, or Class 3 circuit and is derived from a transformer rated not more than 1000 volt-amperes, provided the grounded conductor is bonded to the transformer frame or enclosure by a jumper sized in accordance with 250.30(A)(1), Exception No. 3, and the transformer frame or enclosure is grounded by one of the means specified in 250.134.*

If a separately derived system is required to be grounded, the conductor to be grounded is allowed to be connected to the grounding electrode system at any location between the source terminals (such as the transformer or generator) and the first disconnecting means or overcurrent device. The location of the grounding electrode conductor connection to the grounded conductor must be at the point at which the system bonding jumper is connected to the grounded conductor, so that normal neutral current will be carried only on the system grounded conductor and will not be imposed on parallel paths such as metal raceways, piping systems, and structural steel. Exhibits 250.14 and 250.15 illustrate two acceptable locations for connecting the grounding electrode conductor to the grounded conductor of a separately derived system. These exhibits depict typical wiring arrangements for dry-type transformers supplied from a 480-V, 3-phase feeder to derive a 208Y/120-V or 480Y/277-V secondary.

In Exhibit 250.14, the grounding electrode conductor connection is made at the source of the separately derived system (in the

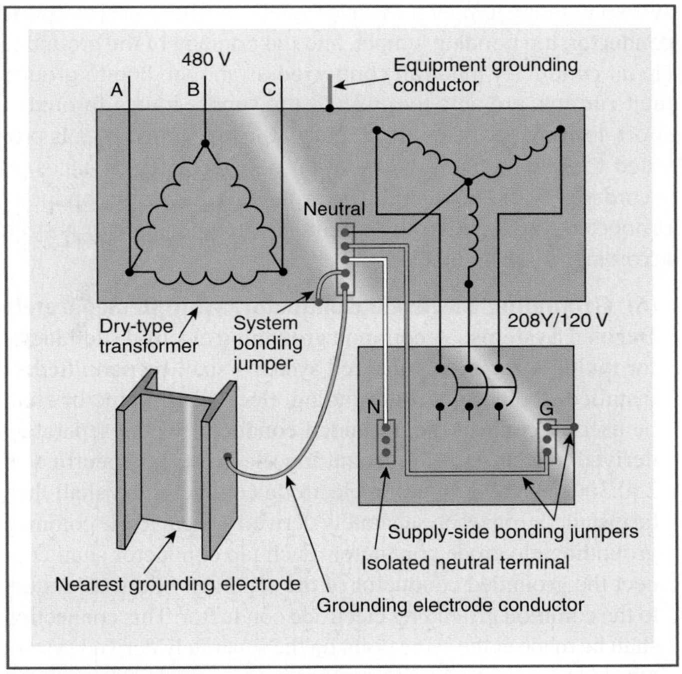


***EXHIBIT 250.14*** *A grounding arrangement for a separately derived system in which the grounding electrode conductor connection is made at the source of the separately derived system (transformer).*

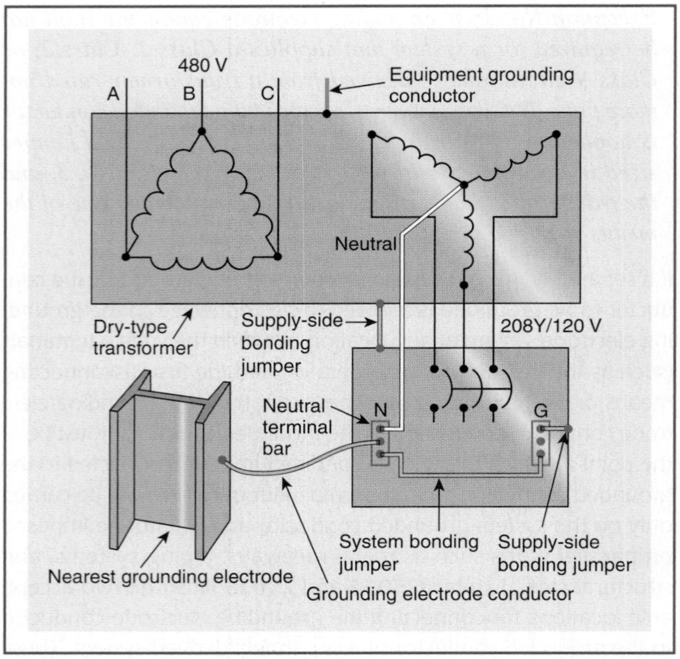


***EXHIBIT 250.15*** *A grounding arrangement for a separately derived system in which the grounding electrode conductor connection is made at the first system disconnecting means.*

***EXHIBIT 250.16*** *Listed connectors used to connect the common grounding electrode conductor and individual taps to a centrally located copper busbar with a minimum dimension of ¼ inch thick by 2 inches wide.*

transformer enclosure where the system bonding jumper is also installed). In Exhibit 250.15, the grounding electrode conductor connection is made at the first disconnecting means where the system bonding jumper is installed. With the grounding electrode conductor, the bonding jumper, and the bonding of the grounded circuit conductor (neutral) connected as shown, line-to-ground fault currents are able to return to the supply source through a short, low-impedance path. A path of lower impedance is provided that facilitates the operation of overcurrent devices, in accordance with 250.4(A)(5). The grounding electrode conductor connected to the secondary grounded circuit conductor is sized according to Table 250.66.

**(6) Grounding Electrode Conductor, Multiple Separately Derived Systems.** A common grounding electrode conductor for multiple separately derived systems shall be permitted. If installed, the common grounding electrode conductor shall be used to connect the grounded conductor of the separately derived systems to the grounding electrode as specified in 250.30(A)(4). A grounding electrode conductor tap shall then be installed from each separately derived system to the common grounding electrode conductor. Each tap conductor shall connect the grounded conductor of the separately derived system to the common grounding electrode conductor. This connection shall be made at the same point on the separately derived system where the system bonding jumper is connected.

A common grounding electrode conductor serving several separately derived systems is an alternative to installing individual grounding electrode conductors from each separately derived system to the grounding electrode system. In such an arrangement, a tapped grounding electrode conductor is installed from the common grounding electrode conductor to the point of connection to the individual separately derived system grounded conductor. This tap is sized from Table 250.66 based on the size of the ungrounded conductors for that individual separately derived system.

The minimum size for this conductor is 3/0 AWG copper or 250 kcmil aluminum so that the grounding electrode conductor always is of sufficient size to accommodate the multiple separately derived systems it serves. This minimum size for the common grounding electrode conductor correlates with the maximum size grounding electrode conductor required by Table 250.66. Therefore, the 3/0 AWG copper or 250 kcmil aluminum becomes the maximum size required for the common grounding electrode conductor. The sizing requirement for the common grounding electrode conductor is specified in 250.30(A)(6)(a), and the sizing requirement for the individual taps to the common grounding electrode conductor is specified in 250.30(A)(6)(b).

The methods of connecting the individual tap conductor(s) to the common grounding electrode conductor are specified in 250.30(A)(6)(c). The permitted methods include the use of busbar as a point of connection between the taps and the common grounding electrode conductor. The connections to the busbar must be made using a listed means. Exhibit 250.16 shows a copper busbar used as a connection point for the individual taps from multiple separately derived systems to be connected to the common grounding electrode.

*Exception No. 1: If the system bonding jumper specified in 250.30(A)(1) is a wire or busbar, it shall be permitted to connect the grounding electrode conductor tap to the equipment grounding terminal, bar, or bus, provided the equipment grounding terminal, bar, or bus is of sufficient size for the separately derived system.*

*Exception No. 2: A grounding electrode conductor shall not be required for a system that supplies a Class 1, Class 2, or Class 3 circuit and is derived from a transformer rated not more than 1000 volt-amperes, provided the system grounded conductor is bonded to the transformer frame or enclosure by a jumper sized in accordance with 250.30(A)(1), Exception No. 3, and the transformer frame or enclosure is grounded by one of the means specified in 250.134.*

(a) *Common Grounding Electrode Conductor.* The common grounding electrode conductor shall be permitted to be one of the following:

(1) A conductor of the wire type not smaller than 3/0 AWG copper or 250 kcmil aluminum
(2) The metal frame of the building or structure that complies with 250.52(A)(2) or is connected to the grounding electrode system by a conductor that shall not be smaller than 3/0 AWG copper or 250 kcmil aluminum

(b) *Tap Conductor Size.* Each tap conductor shall be sized in accordance with 250.66 based on the derived ungrounded conductors of the separately derived system it serves.

*Exception: If the source of a separately derived system is located within equipment listed and identified as suitable for use as service equipment, the grounding electrode conductor from the service or feeder equipment to the grounding electrode shall be permitted as the grounding electrode conductor for the separately derived system, provided that the grounding electrode conductor is of sufficient size for the separately derived system. If the equipment grounding bus internal to the equipment is not smaller than the required grounding electrode conductor for the separately derived system, the grounding electrode connection for the separately derived system shall be permitted to be made to the bus.*

(c) *Connections.* All tap connections to the common grounding electrode conductor shall be made at an accessible location by one of the following methods:

(1) A connector listed as grounding and bonding equipment.
(2) Listed connections to aluminum or copper busbars not smaller than 6 mm × 50 mm (¼ in. × 2 in.). If aluminum busbars are used, the installation shall comply with 250.64(A).
(3) The exothermic welding process.

Tap conductors shall be connected to the common grounding electrode conductor in such a manner that the common grounding electrode conductor remains without a splice or joint.

**(7) Installation.** The installation of all grounding electrode conductors shall comply with 250.64(A), (B), (C), and (E).

**(8) Bonding.** Structural steel and metal piping shall be connected to the grounded conductor of a separately derived system in accordance with 250.104(D).

**(B) Ungrounded Systems.** The equipment of an ungrounded separately derived system shall be grounded and bonded as specified in 250.30(B)(1) through (B)(3).

**(1) Grounding Electrode Conductor.** A grounding electrode conductor, sized in accordance with 250.66 for the largest derived ungrounded conductor (s) or set of derived ungrounded conductors, shall be used to connect the metal enclosures of the derived system to the grounding electrode as specified in 250.30(A)(5) or (6), as applicable. This connection shall be made at any point on the separately derived system from the source to the first system disconnecting means. If the source is located outside the building or structure supplied, a grounding electrode connection shall be made in compliance with 250.30(C).

For ungrounded separately derived systems, a grounding electrode conductor is required to be connected to the metal enclosure of the system disconnecting means. The grounding electrode conductor is sized from Table 250.66 based on the largest ungrounded supply conductor. This connection establishes a reference to ground for all exposed non–current-carrying metal equipment supplied from the ungrounded system. The EGCs of circuits supplied from the ungrounded system are connected to ground via this grounding electrode conductor connection.

**(2) Grounding Electrode.** Except as permitted by 250.34 for portable and vehicle-mounted generators, the grounding electrode shall comply with 250.30(A)(4).

**(3) Bonding Path and Conductor.** A supply-side bonding jumper shall be installed from the source of a separately derived system to the first disconnecting means in compliance with 250.30(A)(2).

**(C) Outdoor Source.** If the source of the separately derived system is located outside the building or structure supplied, a grounding electrode connection shall be made at the source location to one or more grounding electrodes in compliance with 250.50. In addition, the installation shall comply with 250.30(A) for grounded systems or with 250.30(B) for ungrounded systems.

*Exception: The grounding electrode conductor connection for impedance grounded neutral systems shall comply with 250.36 or 250.187, as applicable.*

This exception is similar in function to the requirement of 250.24(A)(2) in that it allows an outdoor grounding connection at the source of a separately derived system. This connection provides a first line of defense for the system against the effects of overvoltages due to lightning, transients, or accidental contact between conductors of systems operating at different voltages.

## 250.32 Buildings or Structures Supplied by a Feeder(s) or Branch Circuit(s)

**(A) Grounding Electrode.** Building(s) or structure(s) supplied by feeder(s) or branch circuit(s) shall have a grounding electrode

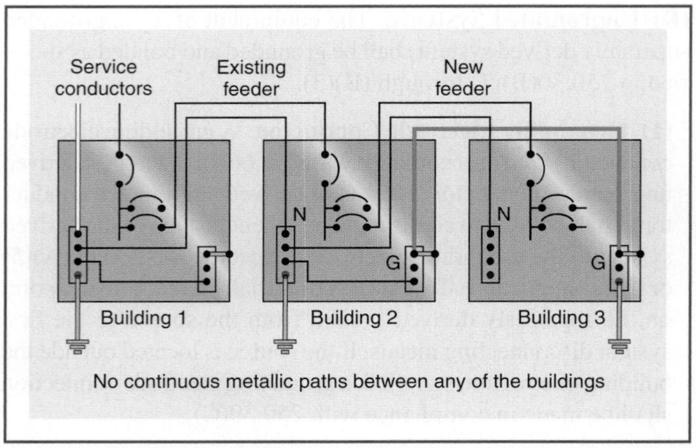


***EXHIBIT 250.17*** *Example of grounding electrode systems required at feeder-supplied Building 2 and Building 3.*

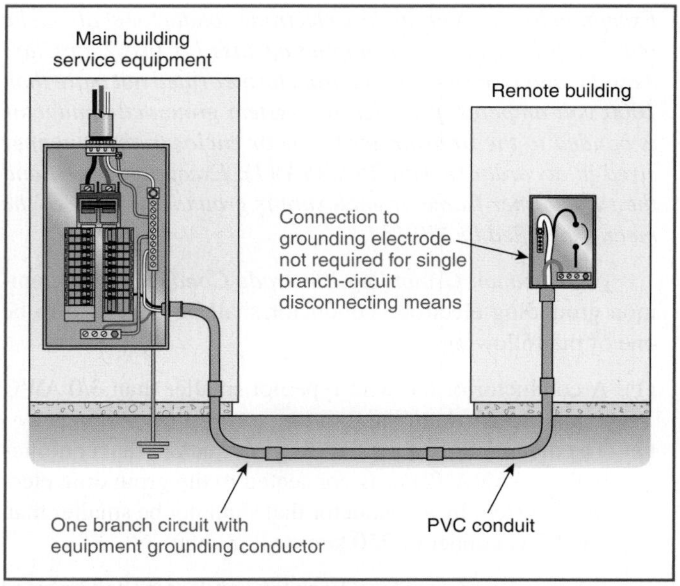


***EXHIBIT 250.18*** *An installation where a connection from the single branch-circuit disconnecting means enclosure to a grounding electrode system is not required at the remote building because an EGC is installed with the circuit conductors.*

or grounding electrode system installed in accordance with Part III of Article 250. The grounding electrode conductor(s) shall be connected in accordance with 250.32(B) or (C). Where there is no existing grounding electrode, the grounding electrode(s) required in 250.50 shall be installed.

The equipment grounding bus must be bonded to the grounding electrode system as is shown for Buildings 2 and 3 in Exhibit 250.17. Building 1 is supplied by a service and is grounded in accordance with 250.24(A) through (D), and the disconnecting means enclosure, building steel, and interior metal water piping are also required to be bonded to the grounding electrode system. All exposed non–current-carrying metal parts of electrical equipment are required to be grounded through EGC connections to the equipment grounding bus at the building disconnecting means. The grounded conductor of the feeder supplying Building 2 is permitted to be re-grounded per 250.32(B)(1), Exception. An EGC is run with the feeder to Building 3 as specified in the general requirement of 250.32(B).

*Exception: A grounding electrode shall not be required where only a single branch circuit, including a multiwire branch circuit, supplies the building or structure and the branch circuit includes an equipment grounding conductor for grounding the normally non–current-carrying metal parts of equipment.*

Detached garages, sheds, and similar structures supplied by a single branch circuit are examples of where this exception can be applied as illustrated in Exhibit 250.18.

**(B) Grounded Systems.**

**(1) Supplied by a Feeder or Branch Circuit.** An equipment grounding conductor, as described in 250.118, shall be run with the supply conductors and be connected to the building or structure disconnecting means and to the grounding electrode(s). The equipment grounding conductor shall be used for grounding or bonding of equipment, structures, or frames required to be grounded or bonded. The equipment grounding conductor shall be sized in accordance with 250.122. Any installed grounded conductor shall not be connected to the equipment grounding conductor or to the grounding electrode(s).

*Exception No. 1: For installations made in compliance with previous editions of this Code that permitted such connection, the grounded conductor run with the supply to the building or structure shall be permitted to serve as the ground-fault return path if all of the following requirements continue to be met:*

*(1) An equipment grounding conductor is not run with the supply to the building or structure.*
*(2) There are no continuous metallic paths bonded to the grounding system in each building or structure involved.*
*(3) Ground-fault protection of equipment has not been installed on the supply side of the feeder(s).*

*If the grounded conductor is used for grounding in accordance with the provision of this exception, the size of the grounded conductor shall not be smaller than the larger of either of the following:*

*(1) That required by 220.61*
*(2) That required by 250.122*

*Exception No. 2: If system bonding jumpers are installed in accordance with 250.30(A)(1), Exception No. 2, the feeder grounded circuit conductor at the building or structure served shall be connected to the equipment grounding conductors,*

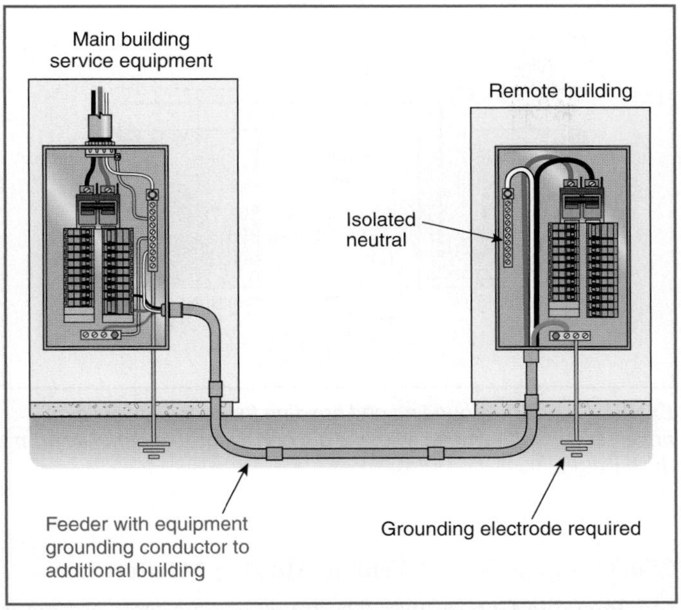


***EXHIBIT 250.19*** *An installation in which a connection between the grounded conductor (neutral) and equipment grounding terminal bar is not permitted. A connection from the equipment grounding terminal bus to the grounding electrode is required.*

*grounding electrode conductor, and the enclosure for the first disconnecting means.*

Using the grounded conductor to ground equipment, in lieu of installing a separate EGC, creates parallel paths for normal neutral current along metal raceways, metal piping, metal cable sheaths or shields, and other metal structures such as ductwork. Installing an EGC with the supply circuit conductors helps ensure that normal circuit current is not imposed on continuous metal paths other than the insulated grounded or neutral conductor. At a building or structure supplied by a feeder or branch circuit, the EGC is connected to the grounding electrode system [unless the installation complies with 250.32(A), Exception] in the equipment supplied by the feeder or branch circuit. Where installed, the grounded or neutral conductor is electrically isolated from the EGC and any grounding electrodes at the building or structure supplied by the feeder or branch circuit, which is illustrated in Exhibit 250.19.

Because "re-grounding" the neutral or grounded conductor was permitted in previous editions of the *Code*, the exception to 250.32(B) permits limited applications where the grounded conductor is used for grounding and bonding of equipment and systems, but only for circuits that were installed in compliance with the *Code* prior to the 2008 edition and where the three conditions specified in the current exception are met.

A connection from the equipment grounding terminal bus to the grounding electrode is required, but a connection to grounding electrode is not required for single branch-circuit disconnecting means.

**(2) Supplied by Separately Derived System.**

(a) *With Overcurrent Protection.* If overcurrent protection is provided where the conductors originate, the installation shall comply with 250.32(B)(1).

(b) *Without Overcurrent Protection.* If overcurrent protection is not provided where the conductors originate, the installation shall comply with 250.30(A). If installed, the supply-side bonding jumper shall be connected to the building or structure disconnecting means and to the grounding electrode(s).

Sections 250.32(B)(2)(a) and (b) correlate the grounding and bonding requirements for separately derived systems with the requirements for grounding and bonding at buildings or structures supplied by a feeder or branch circuit that originates in outdoor equipment such as a transformer or generator. The grounding and bonding requirements are dependent on the location of the OCPD. For example, if a separately derived system originates in a generator and an OCPD is installed at the generator to protect the feeder, an EGC is required to be installed with the feeder conductors. The separately derived system is grounded as specified in 250.30(C) for outdoor sources. Where the building has a disconnecting means [700.12(B)(6), 701.12(B)(5), and 702.12 amend the general disconnecting means requirement in 225.31 and 225.32], the EGC is connected to the disconnecting means enclosure and to the building's grounding electrode system. This installation is no different from what is required for feeders originating in another building. Where an OCPD is not located at the source (as is the case with many outdoor transformer installations), the grounding and bonding provisions of 250.30(A) apply and a supply-side bonding jumper is used to complete the ground-fault current path between the source and the building or structure supplied.

**(C) Ungrounded Systems.**

**(1) Supplied by a Feeder or Branch Circuit.** An equipment grounding conductor, as described in 250.118, shall be installed with the supply conductors and be connected to the building or structure disconnecting means and to the grounding electrode(s). The grounding electrode(s) shall also be connected to the building or structure disconnecting means.

**(2) Supplied by a Separately Derived System.**

(a) *With Overcurrent Protection.* If overcurrent protection is provided where the conductors originate, the installation shall comply with (C)(1).

(b) *Without Overcurrent Protection.* If overcurrent protection is not provided where the conductors originate, the installation shall comply with 250.30(B). If installed, the supply-side bonding jumper shall be connected to the building or structure disconnecting means and to the grounding electrode(s).

**(D) Disconnecting Means Located in Separate Building or Structure on the Same Premises.** Where one or more disconnecting means supply one or more additional buildings or structures under single management, and where these disconnecting means are located remote from those buildings or structures in accordance with the provisions of 225.32, Exception No. 1 and No. 2, 700.12(B)(6), 701.12(B)(5), or 702.12, all of the following conditions shall be met:

(1) The connection of the grounded conductor to the grounding electrode, to normally non–current-carrying metal parts of equipment, or to the equipment grounding conductor at a separate building or structure shall not be made.
(2) An equipment grounding conductor for grounding and bonding any normally non–current-carrying metal parts of equipment, interior metal piping systems, and building or structural metal frames is run with the circuit conductors to a separate building or structure and connected to existing grounding electrode(s) required in Part III of this article, or, where there are no existing electrodes, the grounding electrode(s) required in Part III of this article shall be installed where a separate building or structure is supplied by more than one branch circuit.
(3) The connection between the equipment grounding conductor and the grounding electrode at a separate building or structure shall be made in a junction box, panelboard, or similar enclosure located immediately inside or outside the separate building or structure.

Exceptions to 225.32 and provisions within 700.12(B)(6), 701.12(B)(5), and 702.12 permit the disconnecting means to be located elsewhere from the building or structure being supplied. The requirement to make a connection to a grounding electrode system at these buildings or structures still applies, but no disconnecting means is available in which the connection can be made. Sections 250.32(D)(1) through (3) allow for the required connection to be made in a panelboard, junction box, or similar enclosure that is located either inside or outside of the building or structure being supplied. This enclosure must be located at a point nearest to where the supply conductors enter the building or structure. An EGC must be run with the supply conductors. The grounded conductor (where installed) must not be bonded to the enclosure or equipment grounding bus. The equipment grounding bus must be connected to a new or existing grounding electrode system at the second building. All non–current-carrying metal parts of equipment, building steel, and interior metal piping systems must be connected to the grounding electrode system. Exhibit 250.20 illustrates an installation in which Building 1 houses the disconnecting means for Building 2.

**(E) Grounding Electrode Conductor.** The size of the grounding electrode conductor to the grounding electrode(s) shall not be smaller than given in 250.66, based on the largest ungrounded supply conductor. The installation shall comply with Part III of this article.

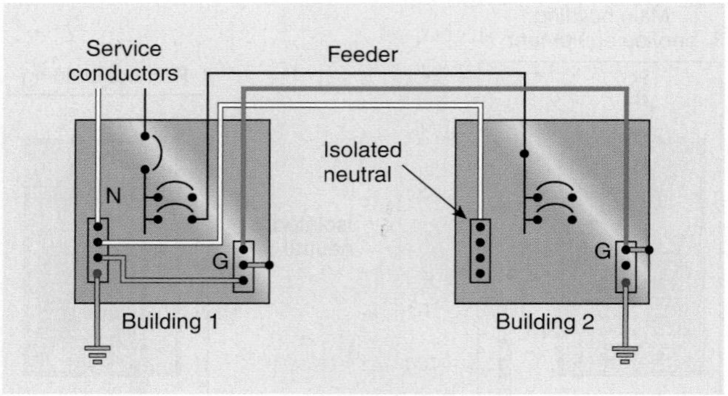


***EXHIBIT 250.20*** *Grounding and bonding for a separate building under single management, with the disconnect located remotely from the building.*

## 250.34 Portable and Vehicle-Mounted Generators

**(A) Portable Generators.** The frame of a portable generator shall not be required to be connected to a grounding electrode as defined in 250.52 for a system supplied by the generator under the following conditions:

(1) The generator supplies only equipment mounted on the generator, cord-and-plug-connected equipment through receptacles mounted on the generator, or both, and
(2) The normally non–current-carrying metal parts of equipment and the equipment grounding conductor terminals of the receptacles are connected to the generator frame.

The frame of a portable generator is not required to be connected to earth (as through a ground rod or water pipe) if the generator has receptacles mounted on the generator panel and the receptacles have equipment grounding terminals bonded to the generator frame. "Portable" describes equipment that is easily carried by personnel from one location to another. "Mobile" describes equipment, such as vehicle-mounted generators, that is capable of being moved on wheels or rollers.

**(B) Vehicle-Mounted Generators.** The frame of a vehicle shall not be required to be connected to a grounding electrode as defined in 250.52 for a system supplied by a generator located on this vehicle under the following conditions:

(1) The frame of the generator is bonded to the vehicle frame, and
(2) The generator supplies only equipment located on the vehicle or cord-and-plug-connected equipment through receptacles mounted on the vehicle, or both equipment located on the vehicle and cord-and-plug-connected equipment through receptacles mounted on the vehicle or on the generator, and

(3) The normally non–current-carrying metal parts of equipment and the equipment grounding conductor terminals of the receptacles are connected to the generator frame.

Vehicle-mounted generators that provide a neutral conductor and are installed as separately derived systems supplying equipment and receptacles on the vehicle are required to have the neutral conductor bonded to the generator frame and to the vehicle frame. The non–current-carrying parts of the equipment must be bonded to the generator frame.

**(C) Grounded Conductor Bonding.** A system conductor that is required to be grounded by 250.26 shall be connected to the generator frame where the generator is a component of a separately derived system.

Informational Note: For grounding portable generators supplying fixed wiring systems, see 250.30.

Portable and vehicle-mounted generators that are installed as separately derived systems and that provide a neutral conductor (such as 3-phase, 4-wire, wye-connected; single-phase 240/120-V; or 3-phase, 4-wire, delta-connected) are required to have the neutral conductor bonded to the generator frame.

## 250.35 Permanently Installed Generators

A conductor that provides an effective ground-fault current path shall be installed with the supply conductors from a permanently installed generator(s) to the first disconnecting mean(s) in accordance with (A) or (B).

**(A) Separately Derived System.** If the generator is installed as a separately derived system, the requirements in 250.30 shall apply.

**(B) Nonseparately Derived System.** If the generator is installed as a nonseparately derived system, and overcurrent protection is not integral with the generator assembly, a supply-side bonding jumper shall be installed between the generator equipment grounding terminal and the equipment grounding terminal, bar, or bus of the disconnecting mean(s). It shall be sized in accordance with 250.102(C) based on the size of the conductors supplied by the generator.

The requirements of 250.35(B) create a return path for ground-fault current for permanently installed generators supplying a system that is not separately derived and in which the first system OCPD is not installed at the generator. The conductor used to conduct ground-fault current between the first system disconnecting means and the generator is a supply-side bonding jumper, which is permitted to be a nonflexible metal raceway or a wire. Wire-type supply-side bonding jumpers are sized using 250.102(C). For many installations, the size can be selected directly from Table 250.102(C)(1).

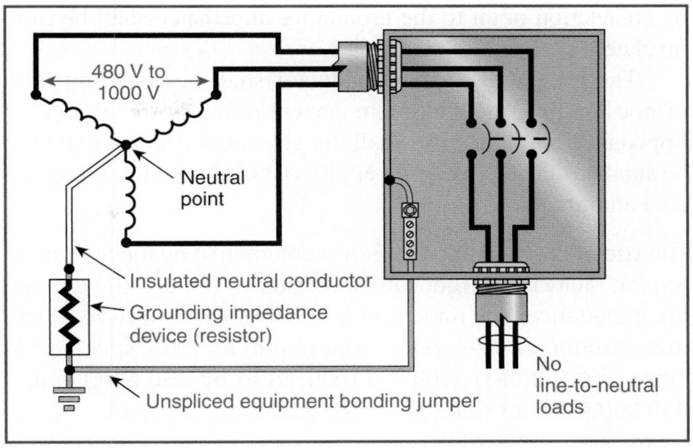


***EXHIBIT 250.21*** *Diagram of a high-impedance grounded neutral system.*

## 250.36 High-Impedance Grounded Neutral Systems

High-impedance grounded neutral systems in which a grounding impedance, usually a resistor, limits the ground-fault current to a low value shall be permitted for 3-phase ac systems of 480 volts to 1000 volts if all the following conditions are met:

(1) The conditions of maintenance and supervision ensure that only qualified persons service the installation.
(2) Ground detectors are installed on the system.
(3) Line-to-neutral loads are not served.

High-impedance grounded neutral systems shall comply with the provisions of 250.36(A) through (G).

Exhibit 250.21 shows the location of the grounding impedance device (resistor used in this example) in a high-impedance grounded neutral system, which limits the amount of ground-fault current in the neutral conductor when a line-to-ground fault occurs. The grounding impedance is selected to limit fault current to a value that is slightly greater than or equal to the capacitive charging current. This system is used where continuity of power is required. Therefore, a ground fault results in an alarm condition rather than in the tripping of a circuit breaker. This alarm allows for the safe and orderly shutdown of a process for which a non-orderly shutdown could introduce additional or increased hazards. Systems rated over 1000 volts are covered in 250.186.

**(A) Grounding Impedance Location.** The grounding impedance shall be installed between the grounding electrode conductor and the system neutral point. If a neutral point is not available, the grounding impedance shall be installed between the grounding electrode conductor and the neutral point derived from a grounding transformer.

**(B) Grounded System Conductor.** The grounded system conductor from the neutral point of the transformer or generator to its connection point to the grounding impedance shall be fully insulated.

The grounded system conductor shall have an ampacity of not less than the maximum current rating of the grounding impedance but in no case shall the grounded system conductor be smaller than 8 AWG copper or 6 AWG aluminum or copper-clad aluminum.

The size of the neutral conductor is determined by the amount of ground-fault current that can be conducted through the grounding impedance. The minimum 8 AWG copper or 6 AWG aluminum conductor sizes reflect the minimum sizes specified in Table 250.102(C)(1), which is referred to by 250.24(C)(1) and 250.30(A)(3)(a) for sizing grounded (neutral) conductors.

**(C) System Grounding Connection.** The system shall not be connected to ground except through the grounding impedance.

Informational Note: The impedance is normally selected to limit the ground-fault current to a value slightly greater than or equal to the capacitive charging current of the system. This value of impedance will also limit transient overvoltages to safe values. For guidance, refer to criteria for limiting transient overvoltages in ANSI/IEEE 142-2007, *Recommended Practice for Grounding of Industrial and Commercial Power Systems.*

**(D) Neutral Point to Grounding Impedance Conductor Routing.** The conductor connecting the neutral point of the transformer or generator to the grounding impedance shall be permitted to be installed in a separate raceway from the ungrounded conductors. It shall not be required to run this conductor with the phase conductors to the first system disconnecting means or overcurrent device.

**(E) Equipment Bonding Jumper.** The equipment bonding jumper (the connection between the equipment grounding conductors and the grounding impedance) shall be an unspliced conductor run from the first system disconnecting means or overcurrent device to the grounded side of the grounding impedance.

**(F) Grounding Electrode Conductor Connection Location.** For services or separately derived systems, the grounding electrode conductor shall be connected at any point from the grounded side of the grounding impedance to the equipment grounding connection at the service equipment or the first system disconnecting means of a separately derived system.

**(G) Equipment Bonding Jumper Size.** The equipment bonding jumper shall be sized in accordance with (1) or (2) as follows:

(1) If the grounding electrode conductor connection is made at the grounding impedance, the equipment bonding jumper shall be sized in accordance with 250.66, based on the size of the service entrance conductors for a service or the derived phase conductors for a separately derived system.

(2) If the grounding electrode conductor is connected at the first system disconnecting means or overcurrent device, the equipment bonding jumper shall be sized the same as the neutral conductor in 250.36(B).

## III. Grounding Electrode System and Grounding Electrode Conductor

### 250.50 Grounding Electrode System

All grounding electrodes as described in 250.52(A)(1) through (A)(7) that are present at each building or structure served shall be bonded together to form the grounding electrode system. Where none of these grounding electrodes exist, one or more of the grounding electrodes specified in 250.52(A)(4) through (A)(8) shall be installed and used.

Section 250.50 introduces the important concept of a *grounding electrode system,* in which all electrodes that are present at a building or structure are bonded together, as illustrated in Exhibit 250.22. Rather than total reliance on a single grounding electrode to perform its function over the life of the electrical installation, the *NEC* requires the formation of a system of electrodes where multiple grounding electrodes are at the building being served. Metal structural members, metal water pipe, and concrete footings or foundations are found in many buildings and structures and are required to be integrated into the grounding electrode system if they qualify under the conditions specified in 250.52(A). The *Code* does not specify that metal water pipe, structural metal frame, or concrete-encased–type electrodes have to be

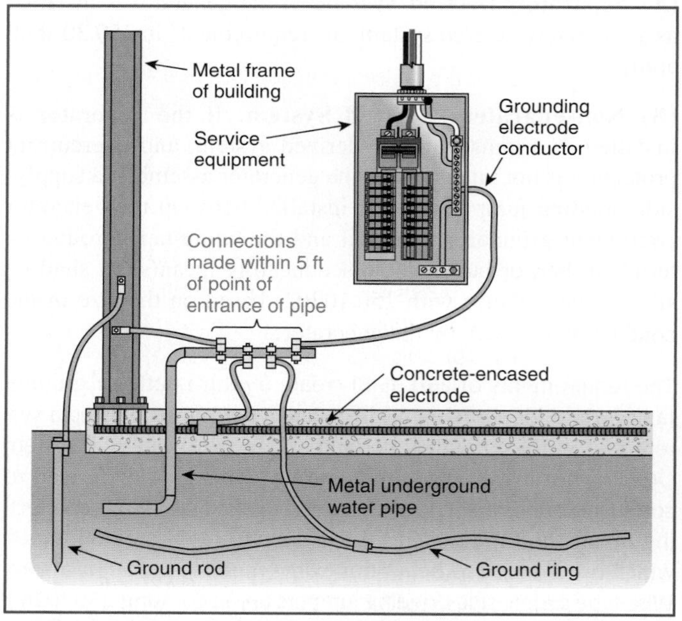


***EXHIBIT 250.22** A grounding electrode system that uses the metal frame of a building, a ground ring, a concrete-encased electrode, a metal underground water pipe, and a ground rod.*

installed, only that where they have been installed as part of the building construction they are to be used as components of the grounding electrode system. A system of electrodes adds a level of reliability and helps ensure system performance over a long period of time.

*Exception: Concrete-encased electrodes of existing buildings or structures shall not be required to be part of the grounding electrode system where the steel reinforcing bars or rods are not accessible for use without disturbing the concrete.*

The exception exempts existing buildings and structures where access to the concrete-encased electrode would involve some type of demolition or similar activity that would damage the existing construction. Because the installation of the footings and foundation is one of the first elements of a construction project, one that in most cases has long been completed by the time the electric service is installed, this rule necessitates an awareness and coordinated effort on the part of designers and the construction trades to make sure that the concrete-encased electrode is incorporated into the grounding electrode system.

## 250.52 Grounding Electrodes

### (A) Electrodes Permitted for Grounding.

**(1) Metal Underground Water Pipe.** A metal underground water pipe in direct contact with the earth for 3.0 m (10 ft) or more (including any metal well casing bonded to the pipe) and electrically continuous (or made electrically continuous by bonding around insulating joints or insulating pipe) to the points of connection of the grounding electrode conductor and the bonding conductor(s) or jumper(s), if installed.

In the early years of the *NEC*, concerns over the effect of electric current on metal water piping created some uncertainty as to whether metal water piping systems should be used as grounding electrodes. To address those concerns, the electrical industry and the waterworks industry formed a committee to evaluate the use of metal underground water piping systems as grounding electrodes. Based on its findings, the committee issued an authoritative report on the subject. The International Association of Electrical Inspectors published the report, *Interim Report of the American Research Committee on Grounding*, in January 1944 (reprinted March 1949). This report serves as part of the basis for the continuation of the practice of using metal water pipes as part of the grounding electrode system.

**(2) Metal Frame of the Building or Structure.** The metal frame of the building or structure that is connected to the earth by one or more of the following methods:

(1) At least one structural metal member that is in direct contact with the earth for 3.0 m (10 ft) or more, with or without concrete encasement.
(2) Hold-down bolts securing the structural steel column that are connected to a concrete-encased electrode that complies with 250.52(A)(3) and is located in the support footing or foundation. The hold-down bolts shall be connected to the concrete-encased electrode by welding, exothermic welding, the usual steel tie wires, or other approved means.

Connection to more than one building column or structural member is not necessary. As long as electrical continuity is through the entire metal frame of the building, the single connection to earth allows the entire structural metal building frame to be used as a grounding electrode. Building steel connected to earth using means other than those specified in 250.52(A)(2) is not permitted to be used as a grounding electrode. Such steel can, however, be used as a bonding conductor or grounding electrode conductor in accordance with the requirements of 250.68(C)(2).

**(3) Concrete-Encased Electrode.** A concrete-encased electrode shall consist of at least 6.0 m (20 ft) of either (1) or (2):

(1) One or more bare or zinc galvanized or other electrically conductive coated steel reinforcing bars or rods of not less than 13 mm (½ in.) in diameter, installed in one continuous 6.0 m (20 ft) length, or if in multiple pieces connected together by the usual steel tie wires, exothermic welding, welding, or other effective means to create a 6.0 m (20 ft) or greater length; or
(2) Bare copper conductor not smaller than 4 AWG

Metallic components shall be encased by at least 50 mm (2 in.) of concrete and shall be located horizontally within that portion of a concrete foundation or footing that is in direct contact with the earth or within vertical foundations or structural components or members that are in direct contact with the earth. If multiple concrete-encased electrodes are present at a building or structure, it shall be permissible to bond only one into the grounding electrode system.

Informational Note: Concrete installed with insulation, vapor barriers, films or similar items separating the concrete from the earth is not considered to be in "direct contact" with the earth.

To qualify as a grounding electrode, the horizontal or vertical installation of the steel reinforcing rod or the 4 AWG bare copper conductor within the concrete encasement is required to be in one continuous 20-foot length so that a 20-foot-long electrode is in contact with the earth. Shorter lengths of reinforcing rod can be connected together to form an electrode 20 feet or longer using the connection methods identified in this requirement. Section 250.52(A)(3) requires that only a single concrete-encased electrode be incorporated into the grounding electrode system. Some buildings or structures may have discontinuous segments of a footing or foundation that individually qualify as grounding electrodes per this section, and once one has been bonded to the grounding electrode system, the remaining ones are exempt from any bonding or grounding requirements. Exhibit 250.23 shows an example of a concrete-encased electrode embedded horizontally. As indicated in the informational note, direct contact with

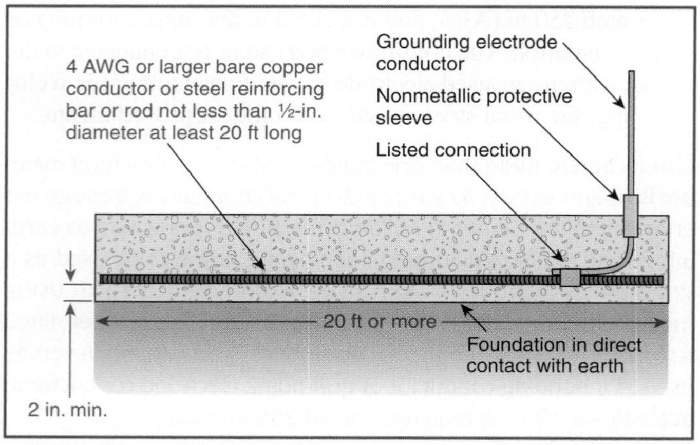


***EXHIBIT 250.23*** *An example of a concrete-encased electrode that is required to be incorporated into the grounding electrode system.*

the earth means that no medium is between the concrete and the earth that impedes the grounding connection or insulates the concrete from being in direct contact with the earth.

**(4) Ground Ring.** A ground ring encircling the building or structure, in direct contact with the earth, consisting of at least 6.0 m (20 ft) of bare copper conductor not smaller than 2 AWG.

**(5) Rod and Pipe Electrodes.** Rod and pipe electrodes shall not be less than 2.44 m (8 ft) in length and shall consist of the following materials.

(a) Grounding electrodes of pipe or conduit shall not be smaller than metric designator 21 (trade size ¾) and, where of steel, shall have the outer surface galvanized or otherwise metal-coated for corrosion protection.

(b) Rod-type grounding electrodes of stainless steel and copper or zinc coated steel shall be at least 15.87 mm (⅝ in.) in diameter, unless listed.

**(6) Other Listed Electrodes.** Other listed grounding electrodes shall be permitted.

**(7) Plate Electrodes.** Each plate electrode shall expose not less than 0.186 $m^2$ (2 $ft^2$) of surface to exterior soil. Electrodes of bare or conductively coated iron or steel plates shall be at least 6.4 mm (¼ in.) in thickness. Solid, uncoated electrodes of nonferrous metal shall be at least 1.5 mm (0.06 in.) in thickness.

**(8) Other Local Metal Underground Systems or Structures.** Other local metal underground systems or structures such as piping systems, underground tanks, and underground metal well casings that are not bonded to a metal water pipe.

**(B) Not Permitted for Use as Grounding Electrodes.** The following systems and materials shall not be used as grounding electrodes:

(1) Metal underground gas piping systems
(2) Aluminum

Informational Note: See 250.104(B) for bonding requirements of gas piping.

## 250.53 Grounding Electrode System Installation

•

**(A) Rod, Pipe, and Plate Electrodes.** Rod, pipe, and plate electrodes shall meet the requirements of 250.53(A)(1) through (A)(3).

**(1) Below Permanent Moisture Level.** If practicable, rod, pipe, and plate electrodes shall be embedded below permanent moisture level. Rod, pipe, and plate electrodes shall be free from nonconductive coatings such as paint or enamel.

**(2) Supplemental Electrode Required.** A single rod, pipe, or plate electrode shall be supplemented by an additional electrode of a type specified in 250.52(A)(2) through (A)(8). The supplemental electrode shall be permitted to be bonded to one of the following:

(1) Rod, pipe, or plate electrode
(2) Grounding electrode conductor
(3) Grounded service-entrance conductor
(4) Nonflexible grounded service raceway
(5) Any grounded service enclosure

*Exception: If a single rod, pipe, or plate grounding electrode has a resistance to earth of 25 ohms or less, the supplemental electrode shall not be required.*

**(3) Supplemental Electrode.** If multiple rod, pipe, or plate electrodes are installed to meet the requirements of this section, they shall not be less than 1.8 m (6 ft) apart.

Informational Note: The paralleling efficiency of rods is increased by spacing them twice the length of the longest rod.

The spacing requirement in 250.53(A)(3) for multiple auxiliary rod, pipe, or plate electrodes is illustrated in Exhibit 250.24. The direct-buried clamps shown are required to be listed for underground installation per 250.70.

Several methods measure the resistance to ground of a rod, pipe, or plate electrode. Exhibit 250.25 illustrates one method of determining the ground resistance of a rod-type electrode in which a ground tester is used to measure the "fall of potential" between the rod being tested and the reference rod (stake) connected to the "$P_1$" or "$P_2$" terminal of the tester. The clamp-on ground resistance tester is a simpler method of testing the earth resistance of a grounding electrode, as it does not involve the use of reference electrodes to perform the earth resistance test.

**(B) Electrode Spacing.** Where more than one of the electrodes of the type specified in 250.52(A)(5) or (A)(7) are used, each electrode of one grounding system (including that used for strike

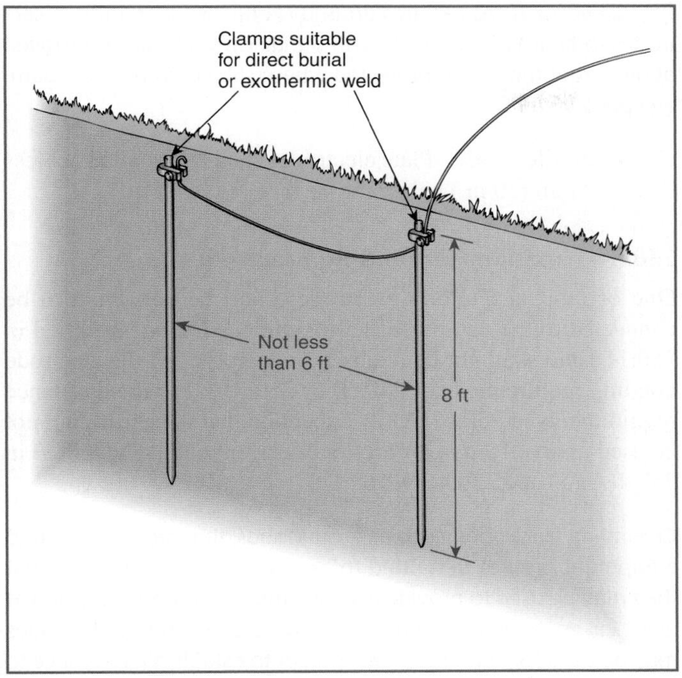


*EXHIBIT 250.24* *The 6-foot spacing required between electrodes.*

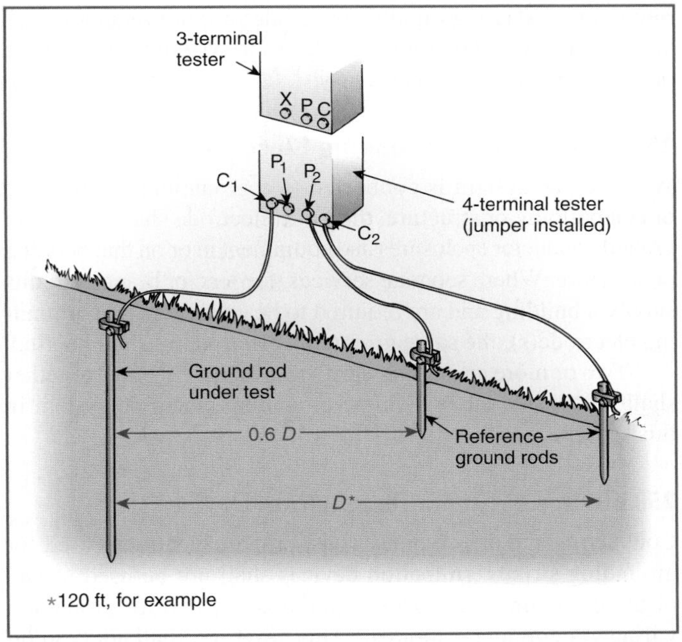


*EXHIBIT 250.25* *The resistance to ground of a ground rod being measured using a fall of potential ground tester. The clamp-on tester is another method of determining the earth resistance of a grounding electrode.*

termination devices) shall not be less than 1.83 m (6 ft) from any other electrode of another grounding system. Two or more grounding electrodes that are bonded together shall be considered a single grounding electrode system.

**(C) Bonding Jumper.** The bonding jumper(s) used to connect the grounding electrodes together to form the grounding electrode system shall be installed in accordance with 250.64(A), (B), and (E), shall be sized in accordance with 250.66, and shall be connected in the manner specified in 250.70.

**(D) Metal Underground Water Pipe.** If used as a grounding electrode, metal underground water pipe shall meet the requirements of 250.53(D)(1) and (D)(2).

**(1) Continuity.** Continuity of the grounding path or the bonding connection to interior piping shall not rely on water meters or filtering devices and similar equipment.

**(2) Supplemental Electrode Required.** A metal underground water pipe shall be supplemented by an additional electrode of a type specified in 250.52(A)(2) through (A)(8). If the supplemental electrode is of the rod, pipe, or plate type, it shall comply with 250.53(A). The supplemental electrode shall be bonded to one of the following:

(1) Grounding electrode conductor
(2) Grounded service-entrance conductor
(3) Nonflexible grounded service raceway
(4) Any grounded service enclosure
(5) As provided by 250.32(B)

*Exception: The supplemental electrode shall be permitted to be bonded to the interior metal water piping at any convenient point as specified in 250.68(C)(1), Exception.*

This requirement clarifies that the supplemental electrode system must be installed as if it were the sole grounding electrode for the system. As specified in the exception to 250.53(A)(2), if a single rod, pipe, or plate electrode has a resistance to earth of 25 ohms or less, it is not necessary to supplement that electrode with one of the types from 250.52(A)(2) through (A)(8). In other words, a single rod, pipe, or plate electrode being used to supplement a metal underground water pipe–type electrode is itself required to be provided with a supplemental electrode unless the condition of 250.53(A)(2), Exception, can be met. One of the permitted methods of bonding a supplemental grounding electrode conductor to the grounding electrode system is to connect it to the grounded service enclosure.

The need for supplemental electrodes for metal water pipe is due to the common practice of using a plastic pipe for replacement when the original metal water pipe fails. Plastic replacement pipe leaves the system without a grounding electrode unless a supplemental electrode is provided.

**(E) Supplemental Electrode Bonding Connection Size.** Where the supplemental electrode is a rod, pipe, or plate electrode, that portion of the bonding jumper that is the sole connection to the supplemental grounding electrode shall not be required to be larger than 6 AWG copper wire or 4 AWG aluminum wire.

Section 250.53(E) correlates with 250.52(A)(5) or (A)(7) and with 250.66(A). For example, if a metal underground water pipe or the metal frame of the building or structure is used as the grounding electrode or as part of the grounding electrode system, Table 250.66 must be used for sizing the grounding electrode conductor. The size of the grounding electrode conductor or bonding jumper that is the sole connection to the supplemental electrode is not required to be larger than 6 AWG copper or 4 AWG aluminum.

**(F) Ground Ring.** The ground ring shall be buried at a depth below the earth's surface of not less than 750 mm (30 in.).

**(G) Rod and Pipe Electrodes.** The electrode shall be installed such that at least 2.44 m (8 ft) of length is in contact with the soil. It shall be driven to a depth of not less than 2.44 m (8 ft) except that, where rock bottom is encountered, the electrode shall be driven at an oblique angle not to exceed 45 degrees from the vertical or, where rock bottom is encountered at an angle up to 45 degrees, the electrode shall be permitted to be buried in a trench that is at least 750 mm (30 in.) deep. The upper end of the electrode shall be flush with or below ground level unless the aboveground end and the grounding electrode conductor attachment are protected against physical damage as specified in 250.10.

Where rock bottom is encountered, the electrodes must be either driven at not more than a 45-degree angle or buried in a 2½-foot-deep trench. Driving the rod at an angle is permitted only if it is not possible to drive the rod vertically to obtain at least 8 feet of earth contact. Burying the ground rod is permitted only if driving the rod vertically or at an angle is not possible. Exhibit 250.26 illustrates these requirements.

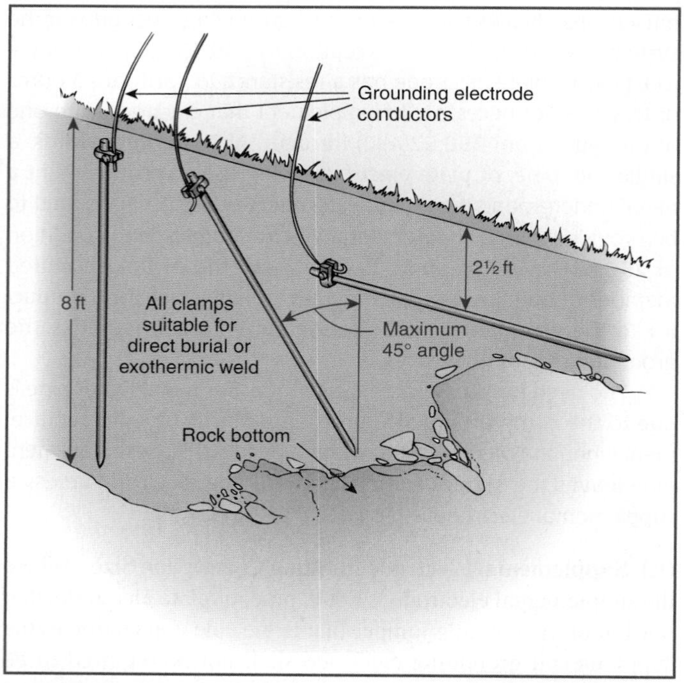


***EXHIBIT 250.26*** *Installation requirements for rod and pipe electrodes.*

Section 250.70 requires ground clamps used on buried electrodes to be listed for direct earth burial. Ground clamps installed above ground must be protected where subject to physical damage per 250.53(G).

**(H) Plate Electrode.** Plate electrodes shall be installed not less than 750 mm (30 in.) below the surface of the earth.

## 250.54 Auxiliary Grounding Electrodes

One or more grounding electrodes shall be permitted to be connected to the equipment grounding conductors specified in 250.118 and shall not be required to comply with the electrode bonding requirements of 250.50 or 250.53(C) or the resistance requirements of 250.53(A)(2) Exception, but the earth shall not be used as an effective ground-fault current path as specified in 250.4(A)(5) and 250.4(B)(4).

Grounding electrodes, such as ground rods, that are connected to equipment are not permitted to be used in lieu of the EGC, but they may be used to provide a local earth reference connection at electrical equipment locations. For example, grounding electrodes may be used for lightning protection or to establish a reference to ground in the area of electrically operated equipment. The earth may not be used as the sole EGC or effective (ground) fault current path. Auxiliary grounding electrodes are not required to be incorporated into the grounding electrode system for the service or other source of electrical supply.

## 250.58 Common Grounding Electrode

Where an ac system is connected to a grounding electrode in or at a building or structure, the same electrode shall be used to ground conductor enclosures and equipment in or on that building or structure. Where separate services, feeders, or branch circuits supply a building and are required to be connected to a grounding electrode(s), the same grounding electrode(s) shall be used.

Two or more grounding electrodes that are bonded together shall be considered as a single grounding electrode system in this sense.

## 250.60 Use of Strike Termination Devices

Conductors and driven pipes, rods, or plate electrodes used for grounding strike termination devices shall not be used in lieu of the grounding electrodes required by 250.50 for grounding wiring systems and equipment. This provision shall not prohibit the required bonding together of grounding electrodes of different systems.

Informational Note No. 1: See 250.106 for spacing from strike termination devices. See 800.100(D), 810.21(J), and 820.100(D) for bonding of electrodes.

Informational Note No. 2: Bonding together of all separate grounding electrodes will limit potential differences between them and between their associated wiring systems.

## 250.62 Grounding Electrode Conductor Material

The grounding electrode conductor shall be of copper, aluminum, copper-clad aluminum, or the items as permitted in 250.68(C). The material selected shall be resistant to any corrosive condition existing at the installation or shall be protected against corrosion. Conductors of the wire type shall be solid or stranded, insulated, covered, or bare.

## 250.64 Grounding Electrode Conductor Installation

Grounding electrode conductors at the service, at each building or structure where supplied by a feeder(s) or branch circuit(s), or at a separately derived system shall be installed as specified in 250.64(A) through (F).

**(A) Aluminum or Copper-Clad Aluminum Conductors.** Bare aluminum or copper-clad aluminum grounding electrode conductors shall not be used where in direct contact with masonry or the earth or where subject to corrosive conditions. Where used outside, aluminum or copper-clad aluminum grounding electrode conductors shall not be terminated within 450 mm (18 in.) of the earth.

**(B) Securing and Protection Against Physical Damage.** Where exposed, a grounding electrode conductor or its enclosure shall be securely fastened to the surface on which it is carried. Grounding electrode conductors shall be permitted to be installed on or through framing members. A 4 AWG or larger copper or aluminum grounding electrode conductor shall be protected if exposed to physical damage. A 6 AWG grounding electrode conductor that is free from exposure to physical damage shall be permitted to be run along the surface of the building construction without metal covering or protection if it is securely fastened to the construction; otherwise, it shall be protected in rigid metal conduit RMC, intermediate metal conduit (IMC), rigid polyvinyl chloride conduit (PVC), reinforced thermosetting resin conduit (RTRC), electrical metallic tubing EMT, or cable armor. Grounding electrode conductors smaller than 6 AWG shall be protected in (RMC), IMC, PVC, RTRC, (EMT), or cable armor. Grounding electrode conductors and grounding electrode bonding jumpers shall not be required to comply with 300.5.

See 250.64(E) for additional information on situations in which metal raceways enclose the grounding electrode conductor. Also see the commentary following 250.64(E) and Exhibit 250.28 for information on installation requirements for metal raceways used to install and physically protect the grounding electrode conductor(s).

**(C) Continuous.** Except as provided in 250.30(A)(5) and (A)(6), 250.30(B)(1), and 250.68(C), grounding electrode conductor(s) shall be installed in one continuous length without a splice or joint. If necessary, splices or connections shall be made as permitted in (1) through (4):

(1) Splicing of the wire-type grounding electrode conductor shall be permitted only by irreversible compression-type connectors listed as grounding and bonding equipment or by the exothermic welding process.
(2) Sections of busbars shall be permitted to be connected together to form a grounding electrode conductor.
(3) Bolted, riveted, or welded connections of structural metal frames of buildings or structures.
(4) Threaded, welded, brazed, soldered or bolted-flange connections of metal water piping.

A building remodeling project or the replacement of existing electrical equipment may necessitate splicing the grounding electrode conductor. Section 250.64(C)(1) contains the permitted means by which the splice can be considered to be a permanent connection and to provide the intended grounding electrode conductor continuity. Section 250.64(C)(2) recognizes the normal bolted connections between sections of busbar that are joined to form the grounding electrode conductor. The connections specified in (3) and (4) correlate with the permission in 250.68(C) to use metal water piping and the structural metal frame of a building as a grounding electrode conductor or as a conductor used to interconnect other grounding electrodes.

**(D) Building or Structure with Multiple Disconnecting Means in Separate Enclosures.** For a service or feeder with two or more disconnecting means in separate enclosures supplying a building or structure, the grounding electrode connections shall be made in accordance with 250.64(D)(1), (D)(2), or (D)(3).

**(1) Common Grounding Electrode Conductor and Taps.** A common grounding electrode conductor and grounding electrode conductor taps shall be installed. The common grounding electrode conductor shall be sized in accordance with 250.66, based on the sum of the circular mil area of the largest ungrounded conductor(s) of each set of conductors that supplies the disconnecting means. If the service-entrance conductors connect directly to the overhead service conductors, service drop, underground service conductors, or service lateral, the common grounding electrode conductor shall be sized in accordance with Table 250.66, note 1.

A grounding electrode conductor tap shall extend to the inside of each disconnecting means enclosure. The grounding electrode conductor taps shall be sized in accordance with 250.66 for the largest service-entrance or feeder conductor serving the individual enclosure. The tap conductors shall be connected to the common grounding electrode conductor by one of the following methods in such a manner that the common grounding electrode conductor remains without a splice or joint:

(1) Exothermic welding.
(2) Connectors listed as grounding and bonding equipment.
(3) Connections to an aluminum or copper busbar not less than 6 mm thick × 50 mm wide (¼ in. thick × 2 in. wide) and of sufficient length to accommodate the number of

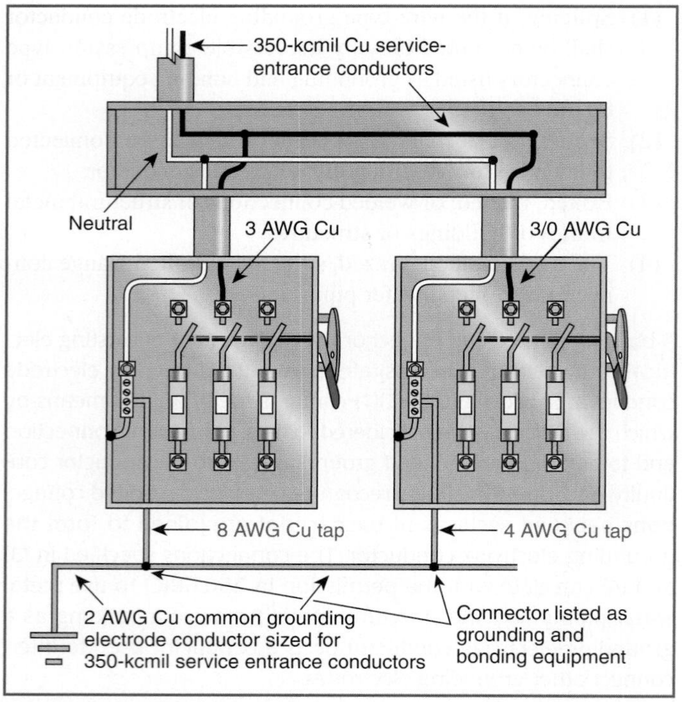


***EXHIBIT 250.27*** *The tap method of connecting the grounding electrode conductor to multiple service disconnecting means enclosures.*

terminations necessary for the installation. The busbar shall be securely fastened and shall be installed in an accessible location. Connections shall be made by a listed connector or by the exothermic welding process. If aluminum busbars are used, the installation shall comply with 250.64(A).

The common grounding electrode conductor from which the taps are made must be sized from Table 250.66 based on the sum of the circular mil area of the largest ungrounded conductor(s) of each set of conductors that supply the disconnecting means or the sum of the equivalent cross-sectional area for parallel conductors.

As illustrated in Exhibit 250.27, the tap method permitted by this section eliminates the difficulties found in looping grounding electrode conductors from one enclosure to another. The 2 AWG grounding electrode conductor (based on the 350 kcmil ungrounded conductor) shown in Exhibit 250.27 is required to be installed without a splice or joint, except as permitted in 250.64(C), and the 8 AWG and 4 AWG taps are sized from Table 250.66 based on the size of the ungrounded conductor serving the respective service disconnecting means. This section recognizes exothermic welding, connectors specifically listed for grounding and bonding, and connections to an accessible copper or aluminum busbar with minimum dimensions of ¼ inch thick by 2 inches wide as acceptable methods to connect the taps to the common grounding electrode conductor.

**(2) Individual Grounding Electrode Conductors.** A grounding electrode conductor shall be connected between the grounding electrode system and one or more of the following, as applicable:

(1) Grounded conductor in each service equipment disconnecting means enclosure
(2) Equipment grounding conductor installed with the feeder
(3) Supply-side bonding jumper

Each grounding electrode conductor shall be sized in accordance with 250.66 based on the service-entrance or feeder conductor(s) supplying the individual disconnecting means.

**(3) Common Location.** A grounding electrode conductor shall be connected in a wireway or other accessible enclosure on the supply side of the disconnecting means to one or more of the following, as applicable:

(1) Grounded service conductor(s)
(2) Equipment grounding conductor installed with the feeder
(3) Supply-side bonding jumper

The connection shall be made with exothermic welding or a connector listed as grounding and bonding equipment. The grounding electrode conductor shall be sized in accordance with 250.66 based on the service-entrance or feeder conductor(s) at the common location where the connection is made.

**(E) Raceways and Enclosures for Grounding Electrode Conductors.**

Bonding jumpers installed to ensure the electrical continuity of ferrous metal enclosures must be sized in accordance with 250.102(C). Exhibit 250.28 shows the required bonding of a ferrous metal raceway to the grounding electrode conductor at both ends of the raceway to ensure that the raceway and conductor are in parallel. These bonding connections are necessary so that the ferrous raceway does not create an inductive choke on the grounding electrode conductor.

**(1) General.** Ferrous metal raceways and enclosures for grounding electrode conductors shall be electrically continuous from the point of attachment to cabinets or equipment to the grounding electrode and shall be securely fastened to the ground clamp or fitting. Ferrous metal raceways and enclosures shall be bonded at each end of the raceway or enclosure to the grounding electrode or grounding electrode conductor. Nonferrous metal raceways and enclosures shall not be required to be electrically continuous.

**(2) Methods.** Bonding shall be in compliance with 250.92(B) and ensured by one of the methods in 250.92(B)(2) through (B)(4).

**(3) Size.** The bonding jumper for a grounding electrode conductor raceway or cable armor shall be the same size as, or larger than, the enclosed grounding electrode conductor.

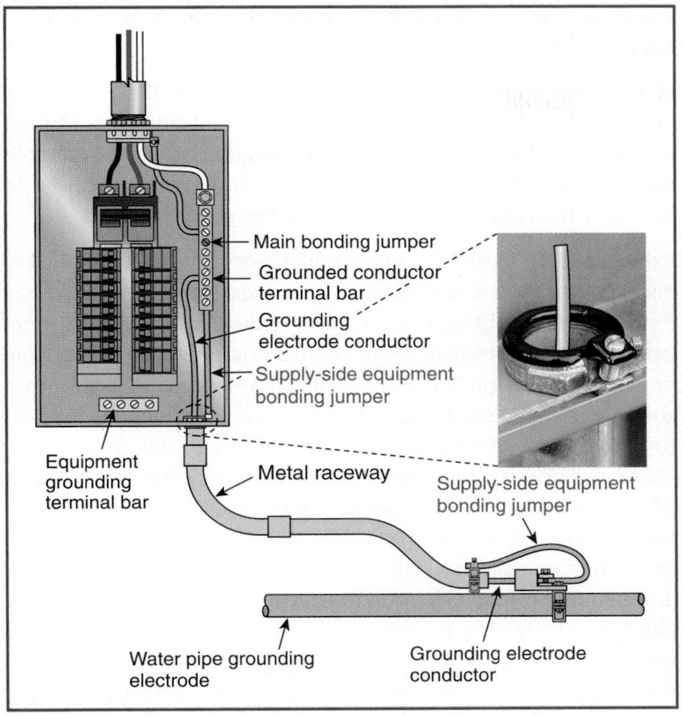


***EXHIBIT 250.28*** *Bonding of a metal raceway that contains a grounding electrode conductor to the conductor at both ends.*

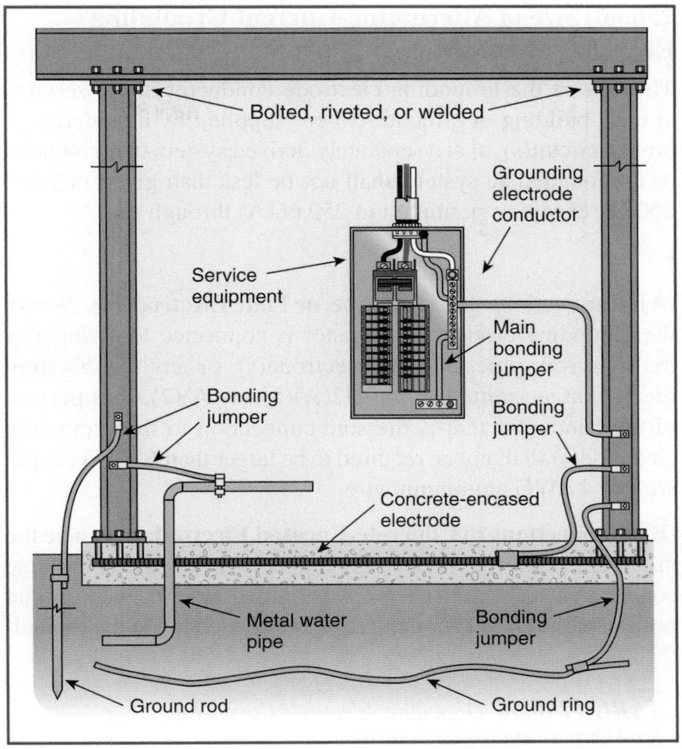


***EXHIBIT 250.29*** *An example of running the grounding electrode conductor to any convenient electrode available as well as bonding all electrodes that are present at a building or structure together to form the required grounding electrode system.*

**(4) Wiring Methods.** If a raceway is used as protection for a grounding electrode conductor, the installation shall comply with the requirements of the appropriate raceway article.

**(F) Installation to Electrode(s).** Grounding electrode conductor(s) and bonding jumpers interconnecting grounding electrodes shall be installed in accordance with (1), (2), or (3). The grounding electrode conductor shall be sized for the largest grounding electrode conductor required among all the electrodes connected to it.

(1) The grounding electrode conductor shall be permitted to be run to any convenient grounding electrode available in the grounding electrode system where the other electrode(s), if any, is connected by bonding jumpers that are installed in accordance with 250.53(C).
(2) Grounding electrode conductor(s) shall be permitted to be run to one or more grounding electrode(s) individually.
(3) Bonding jumper(s) from grounding electrode(s) shall be permitted to be connected to an aluminum or copper busbar not less than 6 mm × 50 mm (¼ in. × 2 in.). The busbar shall be securely fastened and shall be installed in an accessible location. Connections shall be made by a listed connector or by the exothermic welding process. The grounding electrode conductor shall be permitted to be run to the busbar. Where aluminum busbars are used, the installation shall comply with 250.64(A).

Exhibit 250.29 illustrates a grounding electrode system. The single grounding electrode conductor is permitted to run "to any convenient grounding electrode available," per 250.64(F)(1). The other electrodes are connected together using bonding jumpers sized in accordance with 250.53(C), which references 250.66. A permitted alternative is to run the grounding electrode conductor to a busbar that can be used as a connection point for bonding jumpers from multiple electrodes that form the grounding electrode system.

A securely fastened section of copper or aluminum busbar, not less than ¼ inch thick by 2 inches wide (the length can be whatever is necessary to make the connections), is permitted as a connection point for multiple grounding electrode conductors or for bonding jumpers used to connect multiple grounding electrodes together. The connection of the wire to the busbar must be via an exothermic weld or by a listed connector that is attached to the busbar using the typical bolted connection. A fully sized grounding electrode conductor, sized in accordance with 250.66 for the largest electrode used, is permitted to be run to the busbar from the point at which it is connected to the grounded or ungrounded system equipment.

## 250.66 Size of Alternating-Current Grounding Electrode Conductor

The size of the grounding electrode conductor at the service, at each building or structure where supplied by a feeder(s) or branch circuit(s), or at a separately derived system of a grounded or ungrounded ac system shall not be less than given in Table 250.66, except as permitted in 250.66(A) through (C).

•

**(A) Connections to a Rod, Pipe, or Plate Electrode(s).** Where the grounding electrode conductor is connected to a single or multiple rod, pipe, or plate electrode(s), or any combination thereof, as permitted in 250.52(A)(5) or (A)(7), that portion of the conductor that is the sole connection to the grounding electrode(s) shall not be required to be larger than 6 AWG copper wire or 4 AWG aluminum wire.

**(B) Connections to Concrete-Encased Electrodes.** Where the grounding electrode conductor is connected to a single or multiple concrete-encased electrode(s) as permitted in 250.52(A)(3), that portion of the conductor that is the sole connection to the grounding electrode(s) shall not be required to be larger than 4 AWG copper wire.

**(C) Connections to Ground Rings.** Where the grounding electrode conductor is connected to a ground ring as permitted in 250.52(A)(4), that portion of the conductor that is the sole connection to the grounding electrode shall not be required to be larger than the conductor used for the ground ring.

Exhibit 250.30 illustrates a grounding electrode conductor (GEC) installed from service equipment or from a separately derived system to a water pipe grounding electrode. The grounding electrode conductor between the service equipment or separately derived system and the water pipe is required to be a full-sized conductor based on the size of the ungrounded supply conductors, per Table 250.66. The bonding jumpers that connect the other grounding electrodes together are also sized using 250.53(C), which refers to Table 250.66, but they are not necessarily required to be full sized if they are covered by 250.66(A), (B), or (C). In this illustration, the size of the GEC and bonding jumpers is dependent on the electrode to which the GEC is connected. For example, if the GEC from the service equipment is run to the ground rod first and then to the water pipe, the GEC to the ground rod is required to be full sized per Table 250.66. In this illustration, it can be no smaller than a 4 AWG copper conductor. Exhibit 250.30 is not intended to show a mandatory physical routing and connection order of the bonding jumpers and the grounding electrode conductor, since the *Code* does not specify an order or hierarchy for these connections. The sizes for the bonding jumpers to the ground rod and the concrete-encased electrode shown in Exhibit 250.30 are the maximum sizes

***TABLE 250.66*** *Grounding Electrode Conductor for Alternating-Current Systems*

| Size of Largest Ungrounded Service-Entrance Conductor or Equivalent Area for Parallel Conductors[a] (AWG/kcmil) | | Size of Grounding Electrode Conductor (AWG/kcmil) | |
|---|---|---|---|
| **Copper** | **Aluminum or Copper-Clad Aluminum** | **Copper** | **Aluminum or Copper-Clad Aluminum[b]** |
| 2 or smaller | 1/0 or smaller | 8 | 6 |
| 1 or 1/0 | 2/0 or 3/0 | 6 | 4 |
| 2/0 or 3/0 | 4/0 or 250 | 4 | 2 |
| Over 3/0 through 350 | Over 250 through 500 | 2 | 1/0 |
| Over 350 through 600 | Over 500 through 900 | 1/0 | 3/0 |
| Over 600 through 1100 | Over 900 through 1750 | 2/0 | 4/0 |
| Over 1100 | Over 1750 | 3/0 | 250 |

Notes:

1. If multiple sets of service-entrance conductors connect directly to a service drop, set of overhead service conductors, set of underground service conductors, or service lateral, the equivalent size of the largest service-entrance conductor shall be determined by the largest sum of the areas of the corresponding conductors of each set.

2. Where there are no service-entrance conductors, the grounding electrode conductor size shall be determined by the equivalent size of the largest service-entrance conductor required for the load to be served.

[a]This table also applies to the derived conductors of separately derived ac systems.

[b]See installation restrictions in 250.64(A).

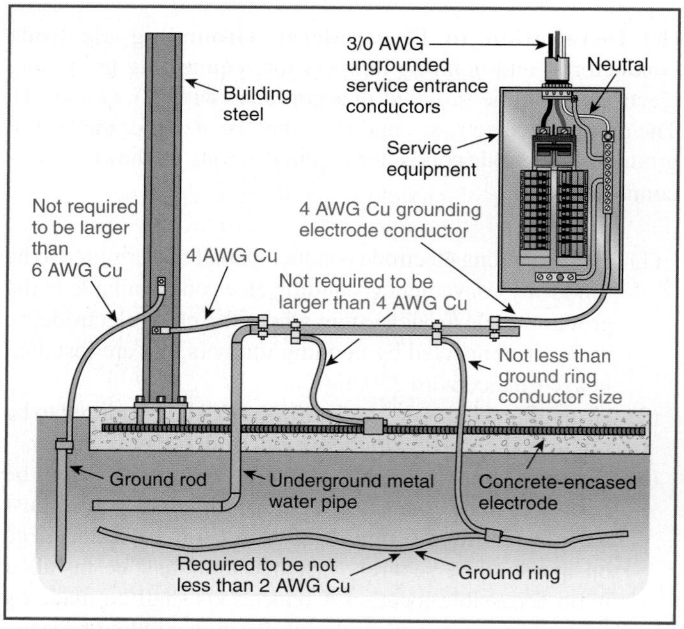


***EXHIBIT 250.30*** *Grounding electrode conductor and bonding jumpers sized in accordance with 250.66 for a service supplied by 3/0 AWG copper ungrounded conductors.*

required by the *Code* based on 250.66. The use of bonding jumpers or grounding electrode conductors larger than required by 250.66 is not prohibited.

## 250.68 Grounding Electrode Conductor and Bonding Jumper Connection to Grounding Electrodes

The connection of a grounding electrode conductor at the service, at each building or structure where supplied by a feeder(s) or branch circuit(s), or at a separately derived system and associated bonding jumper(s) shall be made as specified 250.68(A) through (C).

**(A) Accessibility.** All mechanical elements used to terminate a grounding electrode conductor or bonding jumper to a grounding electrode shall be accessible.

*Exception No. 1: An encased or buried connection to a concrete-encased, driven, or buried grounding electrode shall not be required to be accessible.*

*Exception No. 2: Exothermic or irreversible compression connections used at terminations, together with the mechanical means used to attach such terminations to fireproofed structural metal whether or not the mechanical means is reversible, shall not be required to be accessible.*

Where the exposed portion of an encased, driven, or buried electrode is used for the termination of a grounding electrode conductor, the terminations must be accessible. However, if the connection is buried or encased, Exception No. 1 does not require the terminations to be accessible. Ground clamps and other connectors suitable for use where buried in earth or embedded in concrete must be listed for such use per 250.70. Indication of this listing is either by a marking on the connector or by a tag attached to the connector. See Exhibits 250.22 and 250.24 for illustrations of encased and buried electrodes. Connections, including the mechanical attachment of a compression lug to structural steel, are permitted to be encapsulated by fireproofing material and are not required to be accessible for inspection. This recognizes the importance of maintaining the integrity of the structural fireproofing.

**(B) Effective Grounding Path.** The connection of a grounding electrode conductor or bonding jumper to a grounding electrode shall be made in a manner that will ensure an effective grounding path. Where necessary to ensure the grounding path for a metal piping system used as a grounding electrode, bonding shall be provided around insulated joints and around any equipment likely to be disconnected for repairs or replacement. Bonding jumpers shall be of sufficient length to permit removal of such equipment while retaining the integrity of the grounding path.

Water meters and water filter systems are examples of equipment likely to be disconnected for repairs or replacement. Shorter bonding jumpers are more likely to be disconnected to facilitate removal or reinstallation of a water meter or filter cartridge. This section specifies that the bonding jumper must be long enough to permit removal of such equipment without disconnecting or otherwise interrupting the bonding jumper.

**(C) Grounding Electrode Connections.** Grounding electrode conductors and bonding jumpers shall be permitted to be connected at the following locations and used to extend the connection to an electrode(s):

(1) Interior metal water piping located not more than 1.52 m (5 ft) from the point of entrance to the building shall be permitted to be used as a conductor to interconnect electrodes that are part of the grounding electrode system.

*Exception: In industrial, commercial, and institutional buildings or structures, if conditions of maintenance and supervision ensure that only qualified persons service the installation, interior metal water piping located more than 1.52 m (5 ft) from the point of entrance to the building shall be permitted as a bonding conductor to interconnect electrodes that are part of the grounding electrode system, or as a grounding electrode conductor, if the entire length, other than short sections passing perpendicularly through walls, floors, or ceilings, of the interior metal water pipe that is being used for the conductor is exposed.*

The piping at this point is not a grounding electrode [only the underground portion is an electrode per 250.52(A)(1)]. Rather, it is used to extend grounding and bonding conductor connections to the grounding electrode. The exception permits connections beyond the first 5 feet, and at that point the water piping is considered a conductor used for bonding grounding electrodes together or is considered the actual grounding electrode conductor. All of the conditions of the exception, including the use of qualified persons to service the water piping system, must be met in order to extend the permitted point of connection beyond the first 5 feet of where the piping enters the building.

(2) The metal structural frame of a building shall be permitted to be used as a conductor to interconnect electrodes that are part of the grounding electrode system, or as a grounding electrode conductor.

The structural metal frame of a building is not a grounding electrode unless it is connected to earth by one or more of the means specified in 250.52(A)(2). The required bonding of metal water piping and metal structural frames is covered in 250.104(A) and (C).

•

(3) A concrete-encased electrode of either the conductor type, reinforcing rod or bar installed in accordance with 250.52(A)(3) extended from its location within the concrete to an accessible location above the concrete shall be permitted.

## 250.70 Methods of Grounding and Bonding Conductor Connection to Electrodes

The grounding or bonding conductor shall be connected to the grounding electrode by exothermic welding, listed lugs, listed pressure connectors, listed clamps, or other listed means. Connections depending on solder shall not be used. Ground clamps shall be listed for the materials of the grounding electrode and the grounding electrode conductor and, where used on pipe, rod, or other buried electrodes, shall also be listed for direct soil burial or concrete encasement. Not more than one conductor shall be connected to the grounding electrode by a single clamp or fitting unless the clamp or fitting is listed for multiple conductors. One of the following methods shall be used:

(1) A pipe fitting, pipe plug, or other approved device screwed into a pipe or pipe fitting
(2) A listed bolted clamp of cast bronze or brass, or plain or malleable iron
(3) For indoor communications purposes only, a listed sheet metal strap-type ground clamp having a rigid metal base that seats on the electrode and having a strap of such material and dimensions that it is not likely to stretch during or after installation
(4) An equally substantial approved means

Where a ground clamp terminates on a galvanized water pipe, the clamp must be of a material compatible with steel to prevent galvanic corrosion. The same requirement applies to ground clamps used with grounding electrodes made of other materials such as copper or steel reinforcing rods or bars. The clamp must also be compatible with the material for the grounding electrode conductor and, if buried, listed for that use.

Exhibit 250.31 shows a listed water pipe ground clamp generally used with 8 AWG through 4 AWG grounding electrode conductors. Exothermic weld kits acceptable for this purpose are commercially available.

***EXHIBIT 250.31*** *An application of a listed ground clamp.*

# IV. Enclosure, Raceway, and Service Cable Connections

## 250.80 Service Raceways and Enclosures

Metal enclosures and raceways for service conductors and equipment shall be connected to the grounded system conductor if the electrical system is grounded or to the grounding electrode conductor for electrical systems that are not grounded.

*Exception: A metal elbow that is installed in an underground nonmetallic raceway and is isolated from possible contact by a minimum cover of 450 mm (18 in.) to any part of the elbow shall not be required to be connected to the grounded system conductor or grounding electrode conductor.*

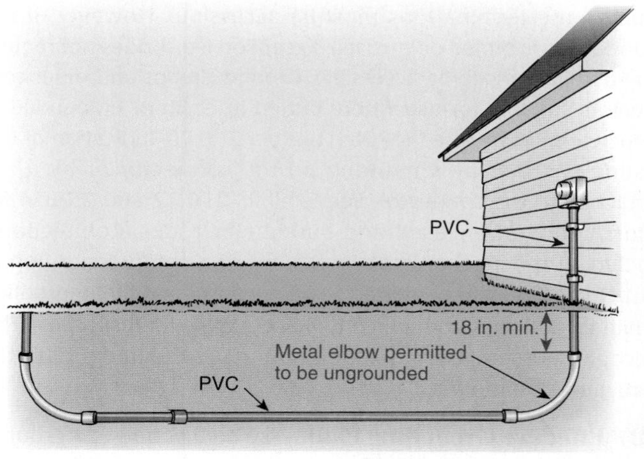


***EXHIBIT 250.32*** *Metal elbows permitted to be ungrounded, provided they are isolated from contact by a minimum cover of 18 inches to any part of the elbow.*

Metal elbows are installed because nonmetallic elbows can be damaged from the friction caused by taut conductor pull lines or ropes rubbing against the interior of the elbow throat. The elbows are isolated from physical contact by burying the entire elbow at a depth not less than 18 inches below grade. This exception applies where isolated metal elbows are used in a service raceway installation. Installations of raceways containing other than service conductors can be found in 250.86, Exception No. 3. See Exhibit 250.32 for an example of this application.

### 250.84 Underground Service Cable or Raceway

**(A) Underground Service Cable.** The sheath or armor of a continuous underground metal-sheathed or armored service cable system that is connected to the grounded system conductor on the supply side shall not be required to be connected to the grounded system conductor at the building or structure. The sheath or armor shall be permitted to be insulated from the interior metal raceway or piping.

**(B) Underground Service Raceway Containing Cable.** An underground metal service raceway that contains a metal-sheathed or armored cable connected to the grounded system conductor shall not be required to be connected to the grounded system conductor at the building or structure. The sheath or armor shall be permitted to be insulated from the interior metal raceway or piping.

### 250.86 Other Conductor Enclosures and Raceways

Except as permitted by 250.112(I), metal enclosures and raceways for other than service conductors shall be connected to the equipment grounding conductor.

*Exception No. 1: Metal enclosures and raceways for conductors added to existing installations of open wire, knob-and-tube wiring, and nonmetallic-sheathed cable shall not be required to be connected to the equipment grounding conductor where these enclosures or wiring methods comply with (1) through (4) as follows:*

*(1) Do not provide an equipment ground*
*(2) Are in runs of less than 7.5 m (25 ft)*
*(3) Are free from probable contact with ground, grounded metal, metal lath, or other conductive material*
*(4) Are guarded against contact by persons*

*Exception No. 2: Short sections of metal enclosures or raceways used to provide support or protection of cable assemblies from physical damage shall not be required to be connected to the equipment grounding conductor.*

*Exception No. 3: A metal elbow shall not be required to be connected to the equipment grounding conductor where it is installed in a run of nonmetallic raceway and is isolated from possible contact by a minimum cover of 450 mm (18 in.) to any part of the elbow or is encased in not less than 50 mm (2 in.) of concrete.*

Connectors, couplings, or other similar fittings that perform mechanical and electrical functions must ensure bonding and grounding continuity between the fitting, the metal raceway, and the enclosure. Metal enclosures must be grounded so that when a fault occurs between an ungrounded (hot) conductor and ground, the potential difference between the non–current-carrying parts of the electrical installation is minimized, thereby reducing the risk of shock.

## V. Bonding

### 250.90 General

Bonding shall be provided where necessary to ensure electrical continuity and the capacity to conduct safely any fault current likely to be imposed.

### 250.92 Services

**(A) Bonding of Equipment for Services.** The normally non–current-carrying metal parts of equipment indicated in 250.92(A)(1) and (A)(2) shall be bonded together.

(1) All raceways, cable trays, cablebus framework, auxiliary gutters, or service cable armor or sheath that enclose, contain, or support service conductors, except as permitted in 250.80
(2) All enclosures containing service conductors, including meter fittings, boxes, or the like, interposed in the service raceway or armor

**(B) Method of Bonding at the Service.** Bonding jumpers meeting the requirements of this article shall be used around impaired connections, such as reducing washers or oversized, concentric, or eccentric knockouts. Standard locknuts or bushings shall not be the only means for the bonding required by this section but shall be permitted to be installed to make a mechanical connection of the raceway(s).

Standard locknuts, sealing locknuts, and metal bushings are not acceptable as the sole means for bonding a raceway or cable to an enclosure on the line side of the service disconnecting means, regardless of the type of or condition of the knockout. For knockouts that are concentric, eccentric, or oversized, electrical continuity must be ensured through the use of a supply-side bonding jumper that connects the raceway to the enclosure. Oversized, concentric, and eccentric knockouts in service enclosures impede the bonding connections. Bonding jumpers are required in these situations and also where reducing washers are used to ensure that any bonding connection provides a suitable path for the high level of ground-fault current that is available on the line side of the service disconnecting means and OCPD. For the same reason, standard locknuts and metal bushings – which may be a suitable bonding connection at some locations on the load side of the service equipment – are not considered to be a reliable bonding connection at this point in the electrical system and cannot be relied upon as the sole bonding connection. For further information on concentric and eccentric knockouts, see the commentary following the definition of *bonding jumper* in Article 100 and the example shown in Exhibit 100.4.

Electrical continuity at service equipment, service raceways, and service conductor enclosures shall be ensured by one of the following methods:

(1) Bonding equipment to the grounded service conductor in a manner provided in 250.8

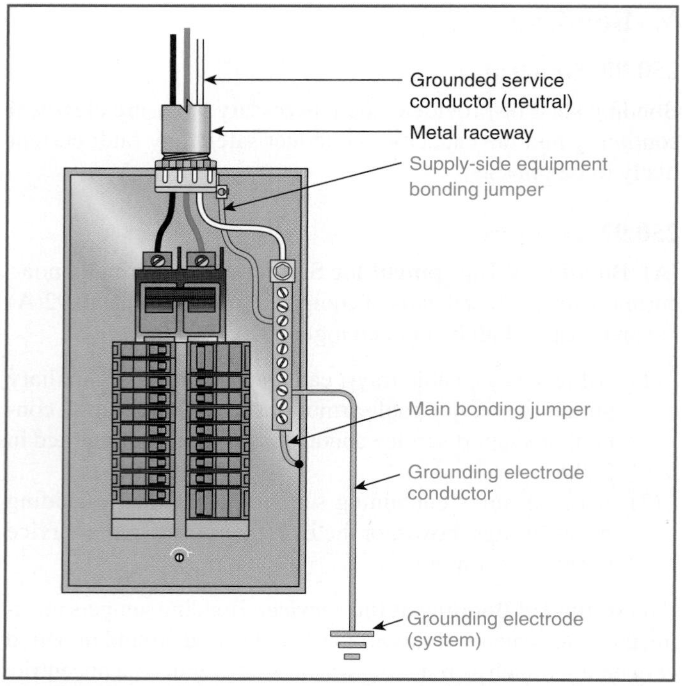


**EXHIBIT 250.33** *Grounding and bonding arrangement for a service with one disconnecting means.*

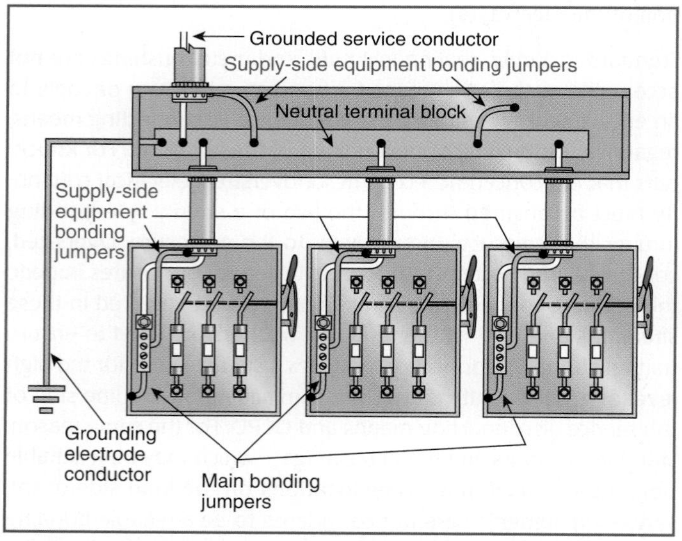


**EXHIBIT 250.34** *Grounding and bonding arrangement for a service with three disconnecting means.*

Exhibit 250.33 illustrates one acceptable grounding and bonding arrangement at a service that has one disconnecting means. Exhibit 250.34 illustrates one acceptable grounding and bonding arrangement for a service that has three disconnecting means as permitted by 230.71(A). Section 250.24(C) specifies that the grounded service conductor must be run to each service disconnecting means and be bonded to the disconnecting means enclosure, as illustrated in Exhibit 250.34. Section 250.92(B)(1) permits

**EXHIBIT 250.35** *Grounding bushings used to connect a copper bonding or grounding wire to conduits. (Courtesy of Thomas & Betts Corp.)*

the bonding of service equipment enclosures to be accomplished by bonding the grounded service conductor to the enclosure.

(2) Connections utilizing threaded couplings or threaded hubs on enclosures if made up wrenchtight

(3) Threadless couplings and connectors if made up tight for metal raceways and metal-clad cables

(4) Other listed devices, such as bonding-type locknuts, bushings, or bushings with bonding jumpers

Bonding-type locknuts and grounding and bonding bushings for use with rigid or intermediate metal conduit are provided with means (usually one or more set screws that make positive contact with the conduit) for reliably bonding the bushing and the conduit on which it is threaded to the metal equipment enclosure or box.

Grounding bushings used with fittings for rigid or intermediate metal conduit or with EMT have means for connecting a bonding jumper or have means provided by the manufacturer for use in mounting a wire connector. (See Exhibit 250.35.) This type of bushing may also have one or more set screws to reliably bond the bushing to the conduit. Exhibit 250.36 shows a listed bonding-type wedge lug used to connect a conduit to a box and provide the bonding connection required by 250.92(B).

## 250.94 Bonding for Other Systems

An intersystem bonding termination for connecting intersystem bonding conductors required for other systems shall be provided external to enclosures at the service equipment or metering equipment enclosure and at the disconnecting means for any additional buildings or structures. The intersystem bonding termination shall comply with the following:

(1) Be accessible for connection and inspection.

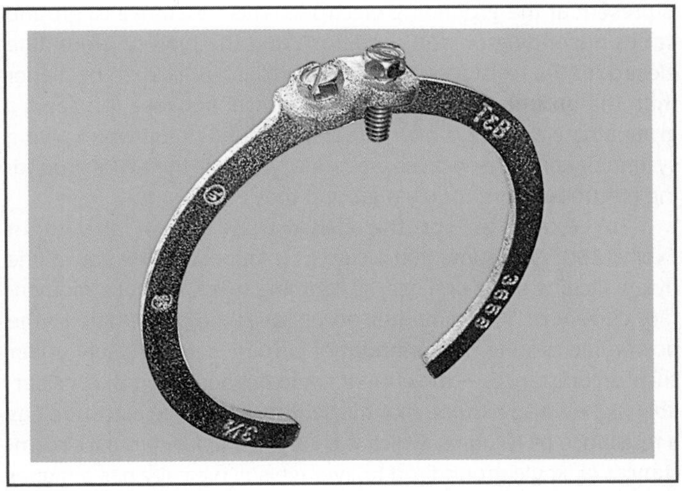

**EXHIBIT 250.36** *A grounding wedge lug used to provide an electrical connection between a conduit and a box. (Courtesy of Thomas & Betts Corp.)*

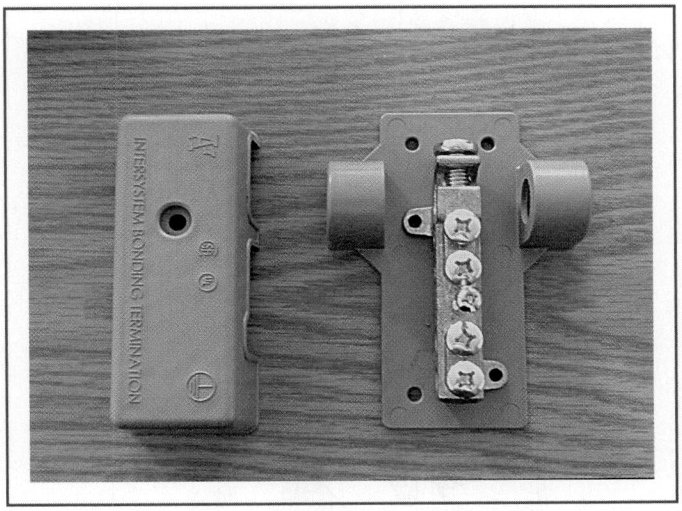


**EXHIBIT 250.37** *A listed intersystem bonding termination providing the required number of terminals (minimum of three) for connecting other building systems to the grounding system of the electrical power supply.*

(2) Consist of a set of terminals with the capacity for connection of not less than three intersystem bonding conductors.
(3) Not interfere with opening the enclosure for a service, building or structure disconnecting means, or metering equipment.

*Intersystem* means that the electrical system and other systems such as optical fiber (Article 770), communications (Article 800), CATV (Article 820), and broadband systems (Articles 830 and 840) are bonded together to minimize the occurrence of potential differences between equipment of different systems. Exhibit 250.37 is an example of an intersystem bonding termination that is to be installed and connected as specified in 250.94(1) through (6).

(4) At the service equipment, be securely mounted and electrically connected to an enclosure for the service equipment, to the meter enclosure, or to an exposed nonflexible metallic service raceway, or be mounted at one of these enclosures and be connected to the enclosure or to the grounding electrode conductor with a minimum 6 AWG copper conductor
(5) At the disconnecting means for a building or structure, be securely mounted and electrically connected to the metallic enclosure for the building or structure disconnecting means, or be mounted at the disconnecting means and be connected to the metallic enclosure or to the grounding electrode conductor with a minimum 6 AWG copper conductor.
(6) The terminals shall be listed as grounding and bonding equipment.

*Exception: In existing buildings or structures where any of the intersystem bonding and grounding electrode conductors required by 770.100(B)(2), 800.100(B)(2), 810.21(F)(2), 820.100(B)(2), and 830.100(B)(2) exist, installation of the intersystem bonding termination is not required. An accessible means external to enclosures for connecting intersystem bonding and grounding electrode conductors shall be permitted at the service equipment and at the disconnecting means for any additional buildings or structures by at least one of the following means:*

*(1) Exposed nonflexible metallic raceways*
*(2) An exposed grounding electrode conductor*
*(3) Approved means for the external connection of a copper or other corrosion-resistant bonding or grounding electrode conductor to the grounded raceway or equipment*

The intersystem bonding points at an existing building or structure can be installed with the connection points as shown in Exhibit 250.38. An intersystem bonding termination of the type shown in Exhibit 250.37 is not required to be installed under the conditions of this exception.

Informational Note No. 1: A 6 AWG copper conductor with one end bonded to the grounded nonflexible metallic raceway or equipment and with 150 mm (6 in.) or more of the other end made accessible on the outside wall is an example of the approved means covered in 250.94, Exception item (3).

Informational Note No. 2: See 770.100, 800.100, 810.21, 820.100, and 830.100 for intersystem bonding and grounding requirements for conductive optical fiber cables, communications circuits, radio and television equipment, CATV circuits and network-powered broadband communications systems, respectively.

Lightning protection systems, communications, radio and TV, and CATV systems are required to be bonded together to minimize the potential differences between the systems that can occur where there is a common interface. Lack of interconnection can

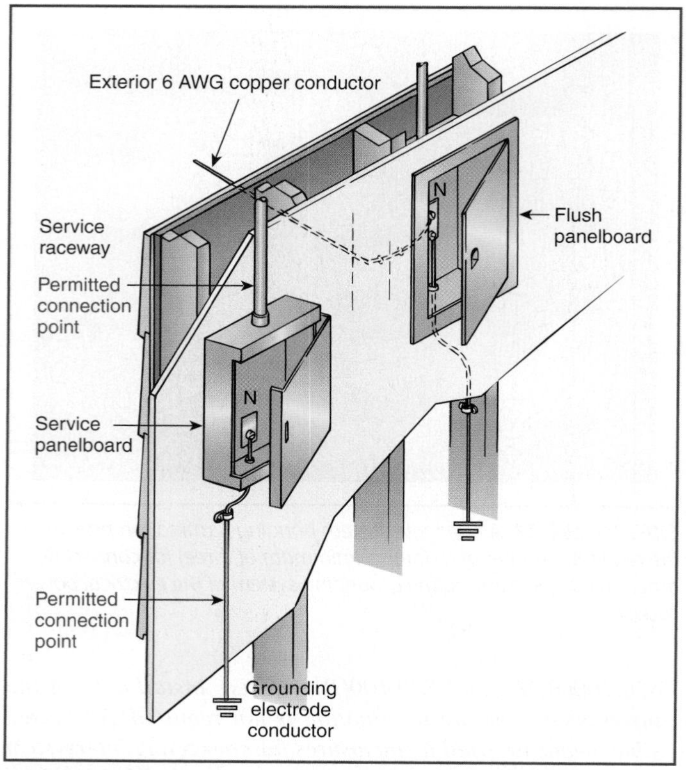


***EXHIBIT 250.38*** *Methods of providing intersystem bonding at an existing building or structure.*

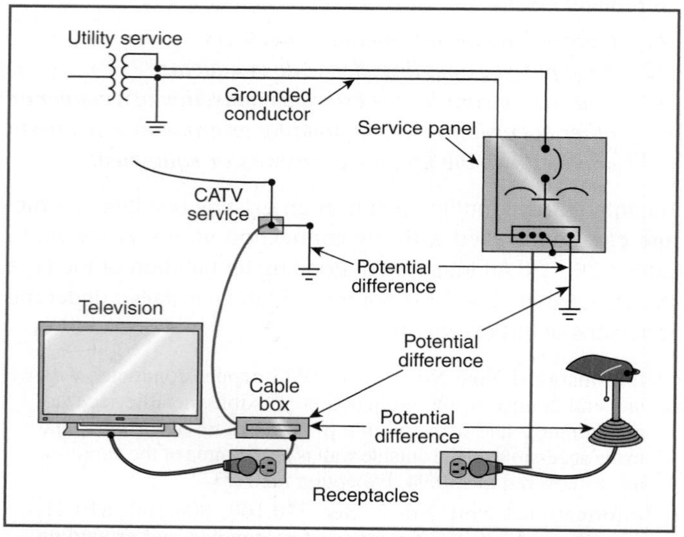


***EXHIBIT 250.39*** *A CATV installation that does not comply with the* Code, *illustrating why bonding between different systems is necessary.*

result in a severe shock and fire hazard due to these differences in potential. The hazard is illustrated in Exhibit 250.39. In this exhibit, the CATV cable is connected to the cable decoder and the tuner of a television set. Also connected to the decoder and the television is the 120-volt supply, with one conductor grounded at the service (the power ground). In each case, resistance to ground is present at the grounding electrode. This resistance to ground varies depending on soil conditions and the type of grounding electrode. The resistance at the CATV ground is likely to be higher than the ground resistance of the service, because the service grounding electrode is often an underground metal water piping system or concrete-encased electrode, whereas the CATV grounding electrode is commonly a ground rod.

For example, for the CATV installation shown in Exhibit 250.39, assume that a current is induced in the power line by a switching surge or a nearby lightning strike, so that a momentary current of 1000 amperes occurs over the power line to the power line ground. This amount of current is not unusual under such circumstances – the current could be, and often is, considerably higher. Also assume that the service grounding electrode has a resistance of 10 ohms, which is a very low value in most circumstances (a single ground rod in average soil typically has a higher resistance to ground).

Using Ohm's law, we can calculate that the current through the equipment connected to the electrical system will be raised momentarily to a potential of 10,000 volts (1000 amperes × 10 ohms). This potential of 10,000 volts would exist between the CATV system and the electrical system and between the grounded conductor within the CATV cable and the grounded surfaces in the walls of the home, such as water pipes (which are connected to the power ground), over which the cable runs. This potential could also appear across a person with one hand on the CATV cable and the other hand on a metal surface connected to the power ground (such as a radiator or a refrigerator).

Actual voltage is likely to be many times the 10,000 volts calculated, because extremely low (below normal) values were assumed for both resistance to ground and current. Most insulation systems, however, are not designed to withstand even 10,000 volts. Even if the insulation system does withstand a 10,000-volt surge, it is likely to be damaged, and breakdown of the insulation system will result in sparking.

The same situation would exist if the current surge were on the CATV cable or a telephone line. The only difference would be the voltage involved, which would depend on the individual resistance to ground of the grounding electrodes.

The solution required by the *Code* is to bond the two grounding electrode systems together, as shown in Exhibit 250.40, or to connect the CATV cable jacket to the power ground. When one system is raised above ground potential, the second system rises to the same potential, and no voltage exists between the two grounding systems.

The bonding requirement of 250.94 addresses the difficulties sometimes encountered by communications and CATV installers trying to properly bond their respective systems together and to the electrical supply system. In the past, bonding between communications, CATV, and power systems was usually achieved by connecting the communications protector grounds or cable shield to an interior metal water pipe, because the pipe was often used as the power grounding electrode. Thus, the requirement that the power, communications, CATV cable shield, and metal water piping systems be bonded together was easily satisfied. If the power

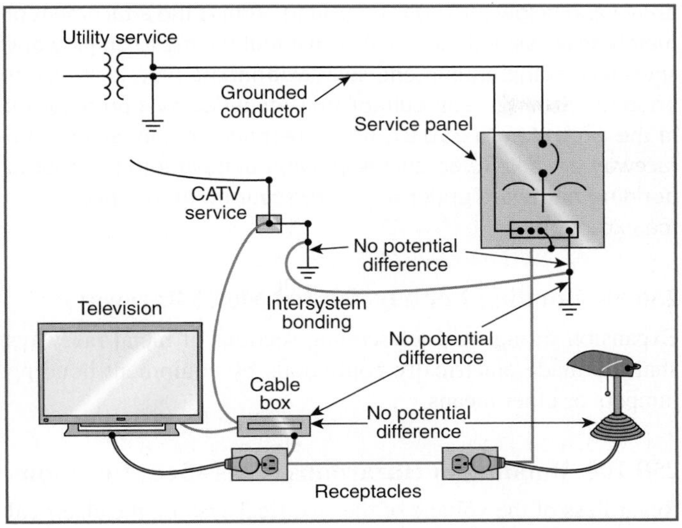


***EXHIBIT 250.40*** *A cable TV installation that complies with 250.94.*

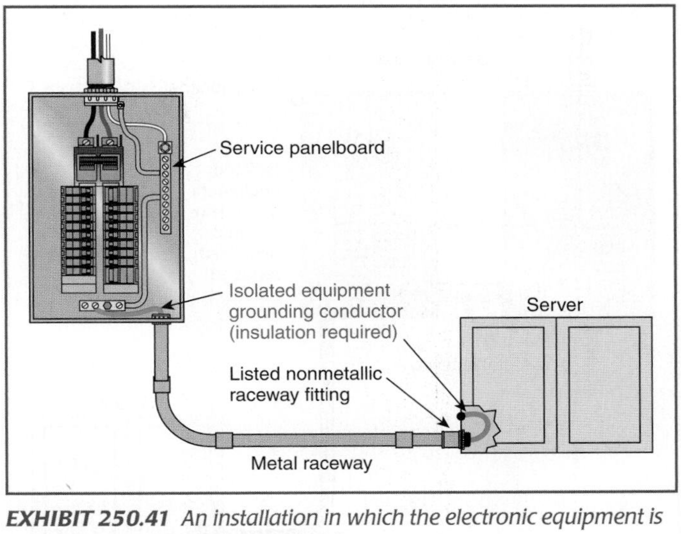


***EXHIBIT 250.41*** *An installation in which the electronic equipment is grounded through the isolated EGC.*

system was grounded to one of the other electrodes permitted by the *Code* (usually by a made electrode such as a ground rod), the bond was connected to the power grounding electrode conductor or to a metal service raceway, since at least one of these was usually accessible.

With the proliferation of plastic water pipe and the service equipment sometimes being installed in finished areas (often flush-mounted), where the grounding electrode conductor is typically concealed, as well as with the increased use of nonmetallic service-entrance conduit, communications and CATV installers often do not have access to a suitable point for connecting bonding jumpers or grounding electrode conductors. For further information, see the commentary following 820.100(D), Informational Note No. 2.

## 250.96 Bonding Other Enclosures

**(A) General.** Metal raceways, cable trays, cable armor, cable sheath, enclosures, frames, fittings, and other metal non–current-carrying parts that are to serve as equipment grounding conductors, with or without the use of supplementary equipment grounding conductors, shall be bonded where necessary to ensure electrical continuity and the capacity to conduct safely any fault current likely to be imposed on them. Any nonconductive paint, enamel, or similar coating shall be removed at threads, contact points, and contact surfaces or be connected by means of fittings designed so as to make such removal unnecessary.

**(B) Isolated Grounding Circuits.** Where installed for the reduction of electrical noise (electromagnetic interference) on the grounding circuit, an equipment enclosure supplied by a branch circuit shall be permitted to be isolated from a raceway containing circuits supplying only that equipment by one or more listed nonmetallic raceway fittings located at the point of attachment of the raceway to the equipment enclosure. The metal raceway shall comply with provisions of this article and shall be supplemented by an internal insulated equipment grounding conductor installed in accordance with 250.146(D) to ground the equipment enclosure.

Informational Note: Use of an isolated equipment grounding conductor does not relieve the requirement for grounding the raceway system.

To reduce electromagnetic interference, electronic equipment is permitted to be isolated from the raceway in a manner similar to that for cord-and-plug-connected equipment. A metal equipment enclosure supplied by a branch circuit is covered by this requirement. Additional wiring, raceways, or other equipment beyond the insulating fitting is not permitted.

Exhibits 250.41 and 250.42 illustrate installations with isolated EGCs. In Exhibit 250.41, the metal raceway is bonded through its attachment to the grounded service equipment enclosure, which provides equipment grounding for the raceway. In Exhibit 250.42, the isolated EGC (which is required to be insulated in order to prevent inadvertent contact with grounded enclosures and raceways) passes through the downstream feeder panelboard and terminates in the service equipment, as permitted by 408.40, Exception. In order to meet the performance objectives for the grounding and bonding of electrical equipment specified in 250.4 and more specifically 250.4(A)(5), the insulated EGC, regardless of where it terminates in the distribution system, must be connected in a manner that creates an effective path for ground-fault current.

## 250.97 Bonding for Over 250 Volts

For circuits of over 250 volts to ground, the electrical continuity of metal raceways and cables with metal sheaths that contain any conductor other than service conductors shall be ensured by one or more of the methods specified for services in 250.92(B), except for (B)(1).

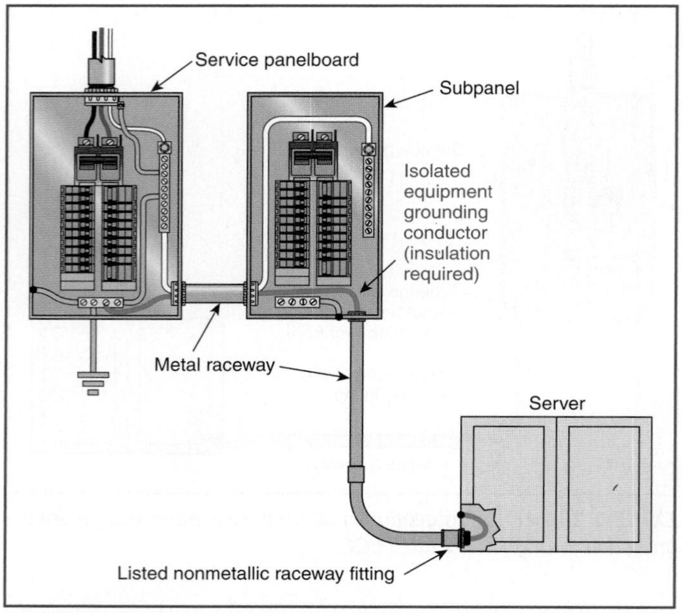


***EXHIBIT 250.42*** *An installation in which the isolated EGC is allowed to pass through the subpanel without connecting to the grounding bus to terminate at the service grounding bus.*

*Exception: Where oversized, concentric, or eccentric knockouts are not encountered, or where a box or enclosure with concentric or eccentric knockouts is listed to provide a reliable bonding connection, the following methods shall be permitted:*

*(1) Threadless couplings and connectors for cables with metal sheaths*
*(2) Two locknuts, on rigid metal conduit or intermediate metal conduit, one inside and one outside of boxes and cabinets*
*(3) Fittings with shoulders that seat firmly against the box or cabinet, such as electrical metallic tubing connectors, flexible metal conduit connectors, and cable connectors, with one locknut on the inside of boxes and cabinets*
*(4) Listed fittings*

Bonding around prepunched concentric or eccentric knockouts is not required if the enclosure containing the knockouts is listed as suitable for bonding. Fittings, such as EMT connectors, cable connectors, and similar fittings with shoulders that seat firmly against the metal of a box or cabinet, are permitted to be installed with only one locknut on the inside of the box.

Guide card information from the UL *Guide Information for Electrical Equipment – The White Book* indicates that concentric and eccentric knockouts of all metal outlet boxes evaluated in accordance with UL 514A, *Metallic Outlet Boxes,* are suitable for bonding in circuits of above or below 250 volts to ground without the use of additional bonding equipment. The guide card information further indicates that metal outlet boxes may be marked to indicate this condition of use. The characteristic of having concentric or eccentric knockouts suitable for bonding above and below 250 volts to ground permits the attachment of metal raceways and metal cables without having to employ any special bonding equipment, such as bonding-type locknuts or bonding bushings. This suitability is identified as a prerequisite in the exception to 250.97, which permits the connection of a raceway or cable to an enclosure without having to use special bonding hardware under any of the four methods specified in the exception.

## 250.98 Bonding Loosely Jointed Metal Raceways

Expansion fittings and telescoping sections of metal raceways shall be made electrically continuous by equipment bonding jumpers or other means.

## 250.100 Bonding in Hazardous (Classified) Locations

Regardless of the voltage of the electrical system, the electrical continuity of non–current-carrying metal parts of equipment, raceways, and other enclosures in any hazardous (classified) location, as defined in 500.5, 505.5, and 506.5, shall be ensured by any of the bonding methods specified in 250.92(B)(2) through (B)(4). One or more of these bonding methods shall be used whether or not equipment grounding conductors of the wire type are installed.

Informational Note: See 501.30, 502.30, 503.30, 505.25, or 506.25 for specific bonding requirements.

## 250.102 Bonding Conductors and Jumpers

**(A) Material.** Bonding jumpers shall be of copper or other corrosion-resistant material. A bonding jumper shall be a wire, bus, screw, or similar suitable conductor.

**(B) Attachment.** Bonding jumpers shall be attached in the manner specified by the applicable provisions of 250.8 for circuits and equipment and by 250.70 for grounding electrodes.

**(C) Size — Supply-Side Bonding Jumper.**

**(1) Size for Supply Conductors in a Single Raceway or Cable.** The supply-side bonding jumper shall not be smaller than specified in Table 250.102(C)(1).

**(2) Size for Parallel Conductor Installations in Two or More Raceways.** Where the ungrounded supply conductors are paralleled in two or more raceways or cables, and an individual supply-side bonding jumper is used for bonding these raceways or cables, the size of the supply-side bonding jumper for each raceway or cable shall be selected from Table 250.102(C)(1) based on the size of the ungrounded supply conductors in each raceway or cable. A single supply-side bonding jumper installed for bonding two or more raceways or cables shall be sized in accordance with 250.102(C)(1).

***TABLE 250.102(C)(1)*** *Grounded Conductor, Main Bonding Jumper, System Bonding Jumper, and Supply-Side Bonding Jumper for Alternating-Current Systems*

| Size of Largest Ungrounded Conductor or Equivalent Area for Parallel Conductors (AWG/kcmil) | | Size of Grounded Conductor or Bonding Jumper* (AWG/kcmil) | |
|---|---|---|---|
| **Copper** | **Aluminum or Copper-Clad Aluminum** | **Copper** | **Aluminum or Copper-Clad Aluminum** |
| 2 or smaller | 1/0 or smaller | 8 | 6 |
| 1 or 1/0 | 2/0 or 3/0 | 6 | 4 |
| 2/0 or 3/0 | 4/0 or 250 | 4 | 2 |
| Over 3/0 through 350 | Over 250 through 500 | 2 | 1/0 |
| Over 350 through 600 | Over 500 through 900 | 1/0 | 3/0 |
| Over 600 through 1100 | Over 900 through 1750 | 2/0 | 4/0 |
| Over 1100 | Over 1750 | See Notes | |

Notes:

1. If the ungrounded supply conductors are larger than 1100 kcmil copper or 1750 kcmil aluminum, the grounded conductor or bonding jumper shall have an area not less than 12½ percent of the area of the largest ungrounded supply conductor or equivalent area for parallel supply conductors. The grounded conductor or bonding jumper shall not be required to be larger than the largest ungrounded conductor or set of ungrounded conductors.

2. If the ungrounded supply conductors and the bonding jumper are of different materials (copper, aluminum, or copper-clad aluminum), the minimum size of the grounded conductor or bonding jumper shall be based on the assumed use of ungrounded supply conductors of the same material as the grounded conductor or bonding jumper and will have an ampacity equivalent to that of the installed ungrounded supply conductors.

3. If multiple sets of service-entrance conductors are used as permitted in 230.40, Exception No. 2, or if multiple sets of ungrounded supply conductors are installed for a separately derived system, the equivalent size of the largest ungrounded supply conductor(s) shall be determined by the largest sum of the areas of the corresponding conductors of each set.

4. If there are no service-entrance conductors, the supply conductor size shall be determined by the equivalent size of the largest service-entrance conductor required for the load to be served.

*For the purposes of this table, the term *bonding jumper* refers to main bonding jumpers, system bonding jumpers, and supply-side bonding jumpers.

Similar to the way that the main bonding jumper is sized per 250.28(D), supply-side bonding jumpers are sized based on the size of the ungrounded conductors of which they are associated. For small capacity installations, the bonding jumper size is based on the size of the largest ungrounded conductor in the supply circuit. The supply-side bonding jumper is selected from Table 250.102(C)(1). If the ungrounded conductors are larger than 1100 kcmil copper or 1750 kcmil aluminum, the size of the supply-side bonding jumper(s) is calculated based on a 12.5-percent sizing relationship between it and the ungrounded conductor. Where an installation consists of multiple raceways for parallel conductors, an individual supply-side bonding jumper can be installed for each raceway, and this jumper is sized based on the size of the ungrounded conductors in that raceway.

### Application Example

A 3-phase, 1600-ampere service is supplied using five 350 kcmil THWN conductors per phase. The parallel conductors are installed in five separate runs of rigid metal conduit. In accordance with 300.12, Exception No. 2, a supply-side bonding jumper is needed for each raceway at the point it enters the open-bottom switchboard. Using 250.102(C)(2), determine the size of the supply-side bonding jumper for each raceway.

Step 1. Determine the size of the largest ungrounded conductor in each raceway – 350 kcmil.

Step 2. Determine the size of the supply-side bonding jumper for each raceway using the "over 3/0 through 350" row in Table 250.102(C). This results in a 2 AWG copper or 1/0 AWG aluminum supply-side bonding jumper.

*Solution*

If a single supply-side bonding jumper is used to bond all five raceways, the size is determined in accordance with 250.102(C)(1) using the following calculation:

5 parallel 350 kcmil conductors:

$$5 \times 350 = 1750 \text{ kcmil}$$

This total exceeds 1100 kcmil; therefore, the total is multiplied by 0.125 (12.5 percent)

$$1750 \text{ kcmil} \times 0.125 = 218.75 \text{ kcmil}$$

From Table 8 of Chapter 9, the next standard conductor size is 250 kcmil copper. If an aluminum supply-side bonding jumper is used, the requirements of 250.102(C)(3) must be applied.

From a practical standpoint, installing the five smaller supply-side bonding jumpers (one to each conduit) is likely to facilitate proper bonding of the raceways when compared to installing a single, 250 kcmil conductor through the five fittings that are used to connect the bonding jumper to the rigid metal conduits.

Informational Note: The term supply conductors includes ungrounded conductors that do not have overcurrent protection on their supply side and terminate at service equipment or the first disconnecting means of a separately derived system.

Informational Note: See Chapter 9, Table 8, for the circular mil area of conductors 18 AWG through 4/0 AWG.

•

**(D) Size — Equipment Bonding Jumper on Load Side of an Overcurrent Device.** The equipment bonding jumper on the load side of an overcurrent device(s) shall be sized in accordance with 250.122.

A single common continuous equipment bonding jumper shall be permitted to connect two or more raceways or cables if the bonding jumper is sized in accordance with 250.122 for the largest overcurrent device supplying circuits therein.

**(E) Installation.** Bonding jumpers or conductors and equipment bonding jumpers shall be permitted to be installed inside or outside of a raceway or an enclosure.

**(1) Inside a Raceway or an Enclosure.** If installed inside a raceway, equipment bonding jumpers and bonding jumpers or conductors shall comply with the requirements of 250.119 and 250.148.

**(2) Outside a Raceway or an Enclosure.** If installed on the outside, the length of the bonding jumper or conductor or equipment bonding jumper shall not exceed 1.8 m (6 ft) and shall be routed with the raceway or enclosure.

*Exception: An equipment bonding jumper or supply-side bonding jumper longer than 1.8 m (6 ft) shall be permitted at outside pole locations for the purpose of bonding or grounding isolated sections of metal raceways or elbows installed in exposed risers of metal conduit or other metal raceway, and for bonding grounding electrodes, and shall not be required to be routed with a raceway or enclosure.*

The installation of the bonding jumper for a conduit expansion joint on the inside of the conduit is often impractical. For some metal raceway and rigid conduit systems and conduit systems in hazardous (classified) locations, installing the bonding jumper where it is visible and accessible for inspection and maintenance is desirable. An external bonding jumper has a higher impedance than an internal bonding jumper, but by limiting the length of the bonding jumper to 6 feet and routing it with the raceway, the

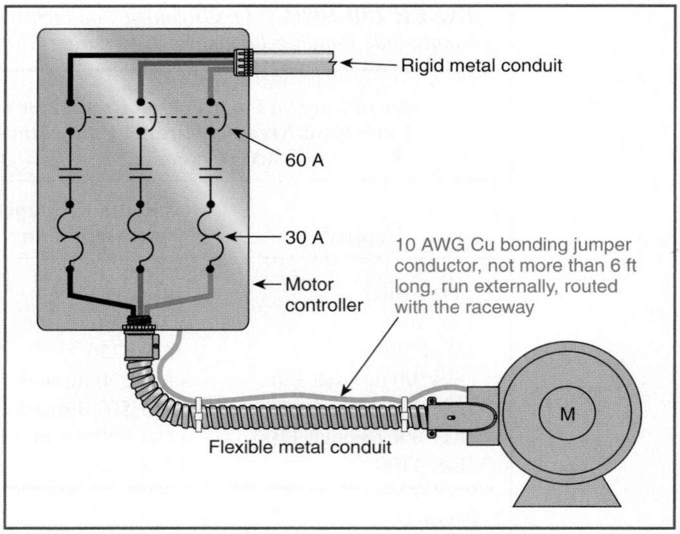


***EXHIBIT 250.43*** *A bonding jumper along the outside of a flexible metal conduit.*

increase in the impedance of the equipment grounding circuit is insignificant. Exhibit 250.43 illustrates a bonding jumper run outside a length of flexible metal conduit. Because the function of a bonding jumper is readily apparent, it does not have to be identified using one of the means specified in 250.119 for EGCs. Using green marking or insulation on an equipment bonding jumper is neither required nor prohibited.

The exception provides a practical solution to bonding sections of metal raceways installed on exterior poles. The 6-foot length restriction in the main requirement is relaxed in the exception so that a bonding jumper sized in accordance with either 250.102(C) or (D), as applicable, can be run up the pole to a point where it is practical to connect it to either the grounded conductor or the EGC, depending on whether the conductors are service conductors or conductors installed on the load side of the service point.

**(3) Protection.** Bonding jumpers or conductors and equipment bonding jumpers shall be installed in accordance with 250.64(A) and (B).

## 250.104 Bonding of Piping Systems and Exposed Structural Metal

**(A) Metal Water Piping.** The metal water piping system shall be bonded as required in (A)(1), (A)(2), or (A)(3) of this section. The bonding jumper(s) shall be installed in accordance with 250.64(A), (B), and (E). The points of attachment of the bonding jumper(s) shall be accessible.

**(1) General.** Metal water piping system(s) installed in or attached to a building or structure shall be bonded to the service equipment enclosure, the grounded conductor at the service, the grounding electrode conductor where of sufficient size, or to the one or more grounding electrodes used. The bonding jumper(s) shall be sized in accordance with Table 250.66 except as permitted in 250.104(A)(2) and (A)(3).

Bonding a metal water piping system is not the same as using the metal water piping system as a grounding electrode. Bonding to the grounding electrode system places the bonded components at the same voltage level. For example, a current of 2000 amperes through 25 feet of 6 AWG copper conductor produces a voltage differential of approximately 26 volts.

Where it is not certain that the hot and cold water pipes are reliably bonded through mechanical connections, an electrical bonding jumper is required to ensure that this connection is made. Judgment must be exercised for each installation. Isolated sections of metal water piping (such as may be used for a plumbing fixture connection) that are connected to an overall nonmetallic water piping system are not subject to the requirements of 250.104(A). The isolated sections are not a metal water piping system. The special installation requirements provided in 250.64(A), (B), and (E) also apply to the water piping bonding jumper.

**(2) Buildings of Multiple Occupancy.** In buildings of multiple occupancy where the metal water piping system(s) installed in or attached to a building or structure for the individual occupancies is metallically isolated from all other occupancies by use of nonmetallic water piping, the metal water piping system(s) for each occupancy shall be permitted to be bonded to the equipment grounding terminal of the switchgear, switchboard, or panelboard enclosure (other than service equipment) supplying that occupancy. The bonding jumper shall be sized in accordance with Table 250.122, based on the rating of the overcurrent protective device for the circuit supplying the occupancy.

Due to increasing use of nonmetallic piping, such as plastic, metal water piping in multiple-occupancy buildings may be isolated by plastic fittings and lengths of plastic pipe. Therefore, the metal water pipe of an occupancy is permitted to be bonded to the panelboard or switchboard that serves only that particular occupancy. The bonding jumper, in this case, is permitted to be sized according to Table 250.122, based on the size of the main overcurrent device supplying the occupancy.

**(3) Multiple Buildings or Structures Supplied by a Feeder(s) or Branch Circuit(s).** The metal water piping system(s) installed in or attached to a building or structure shall be bonded to the building or structure disconnecting means enclosure where located at the building or structure, to the equipment grounding conductor run with the supply conductors, or to the one or more grounding electrodes used. The bonding jumper(s) shall be sized in accordance with 250.66, based on the size of the feeder or branch-circuit conductors that supply the building or structure. The bonding jumper shall not be required to be larger than the largest ungrounded feeder or branch-circuit conductor supplying the building or structure.

**(B) Other Metal Piping.** If installed in, or attached to, a building or structure, a metal piping system(s), including gas piping, that is likely to become energized shall be bonded to any of the following:

(1) Equipment grounding conductor for the circuit that is likely to energize the piping system
(2) Service equipment enclosure
(3) Grounded conductor at the service
(4) Grounding electrode conductor, if of sufficient size
(5) One or more grounding electrodes used

The bonding conductor(s) or jumper(s) shall be sized in accordance with 250.122, using the rating of the circuit that is likely to energize the piping system(s). The points of attachment of the bonding jumper(s) shall be accessible.

Informational Note No. 1: Bonding all piping and metal air ducts within the premises will provide additional safety.

Informational Note No. 2: Additional information for gas piping systems can be found in Section 7.13 of NFPA 54-2012, *National Fuel Gas Code*.

Where the phrase "likely to become energized" is used in the *Code*, it means that the failure of electrical insulation on a conductor can cause normally non–current-carrying metal parts to become energized. Where metal piping systems and electrical circuits interface through mechanical and electrical connections within equipment, a failure of electrical insulation can result in the connected piping system(s) becoming energized. A gas range is an example of equipment in which an insulation failure in an electrical circuit could energize metal gas piping.

Typically, the use of an additional bonding jumper is not necessary to comply with 250.104(B), because the equipment grounding connection to the non–current-carrying metal parts of the appliance also provides a bonding connection to the metal piping attached to the appliance. Section 250.104(B) is a requirement to bond the piping system, and not to make the piping a component of the grounding electrode system. Therefore, this requirement does not conflict with 250.52(B)(1), which prohibits the use of metal underground gas piping as a grounding electrode for electrical services or other sources of supply. To prevent the underground gas piping from inadvertently becoming a grounding electrode, gas utilities usually provide an isolating fitting. This fitting provides compliance with both 250.52(B)(1) and 250.104(B).

**(C) Structural Metal.** Exposed structural metal that is interconnected to form a metal building frame and is not intentionally grounded or bonded and is likely to become energized shall be bonded to the service equipment enclosure; the grounded conductor at the service; the disconnecting means for buildings or structures supplied by a feeder or branch circuit; the grounding electrode conductor, if of sufficient size; or to one or more grounding electrodes used. The bonding jumper(s) shall be sized in accordance with Table 250.66 and installed in accordance with 250.64(A), (B), and (E). The points of attachment of the bonding jumper(s) shall be accessible unless installed in compliance with 250.68(A), Exception No. 2.

**(D) Separately Derived Systems.** Metal water piping systems and structural metal that is interconnected to form a building frame shall be bonded to separately derived systems in accordance with (D)(1) through (D)(3).

**(1) Metal Water Piping System(s).** The grounded conductor of each separately derived system shall be bonded to the nearest available point of the metal water piping system(s) in the area served by each separately derived system. This connection shall be made at the same point on the separately derived system where the grounding electrode conductor is connected. Each bonding jumper shall be sized in accordance with Table 250.66 based on the largest ungrounded conductor of the separately derived system.

*Exception No. 1: A separate bonding jumper to the metal water piping system shall not be required where the metal water piping system is used as the grounding electrode for the separately derived system and the water piping system is in the area served.*

The grounded conductor of a separately derived system must be bonded to the nearest point of the metal piping system in the area. It must also be bonded to any exposed structural metal in the area supplied by the separately derived system. The connection must be made at the point nearest the derived system, and it must be accessible. Where either the piping system or the structural steel is used as the grounding electrode for the separately derived system, it is not necessary to provide an additional bonding jumper to that electrode, per Exception No. 1 to 250.104(D)(1) and (2).

*Exception No. 2: A separate water piping bonding jumper shall not be required where the metal frame of a building or structure is used as the grounding electrode for a separately derived system and is bonded to the metal water piping in the area served by the separately derived system.*

**(2) Structural Metal.** Where exposed structural metal that is interconnected to form the building frame exists in the area served by the separately derived system, it shall be bonded to the grounded conductor of each separately derived system. This connection shall be made at the same point on the separately derived system where the grounding electrode conductor is connected. Each bonding jumper shall be sized in accordance with Table 250.66 based on the largest ungrounded conductor of the separately derived system.

*Exception No. 1: A separate bonding jumper to the building structural metal shall not be required where the metal frame of a building or structure is used as the grounding electrode for the separately derived system.*

*Exception No. 2: A separate bonding jumper to the building structural metal shall not be required where the water piping of a building or structure is used as the grounding electrode for a separately derived system and is bonded to the building structural metal in the area served by the separately derived system.*

**(3) Common Grounding Electrode Conductor.** Where a common grounding electrode conductor is installed for multiple separately derived systems as permitted by 250.30(A)(6), and exposed structural metal that is interconnected to form the building frame or interior metal piping exists in the area served by the separately derived system, the metal piping and the structural metal member shall be bonded to the common grounding electrode conductor in the area served by the separately derived system.

*Exception: A separate bonding jumper from each derived system to metal water piping and to structural metal members shall not be required where the metal water piping and the structural metal members in the area served by the separately derived system are bonded to the common grounding electrode conductor.*

## 250.106 Lightning Protection Systems

The lightning protection system ground terminals shall be bonded to the building or structure grounding electrode system.

Informational Note No. 1: See 250.60 for use of strike termination devices. For further information, see NFPA 780-2014, *Standard for the Installation of Lightning Protection Systems*, which contains detailed information on grounding, bonding, and sideflash distance from lightning protection systems.

Informational Note No. 2: Metal raceways, enclosures, frames, and other non–current-carrying metal parts of electrical equipment installed on a building equipped with a lightning protection system may require bonding or spacing from the lightning protection conductors in accordance with NFPA 780-2014, *Standard for the Installation of Lightning Protection Systems*.

Exhibit 250.44 illustrates the bonding of the lightning protection system to the electrical service grounding electrode system. A similar requirement is found in Section 4.14 of NFPA 780, *Standard for the Installation of Lightning Protection Systems*. Additional bonding between the lightning protection system and the electrical system may be necessary based on proximity and whether separation between the systems is through air or building materials. Subsection 4.21.2 of NFPA 780 includes a method for calculating flashover distances.

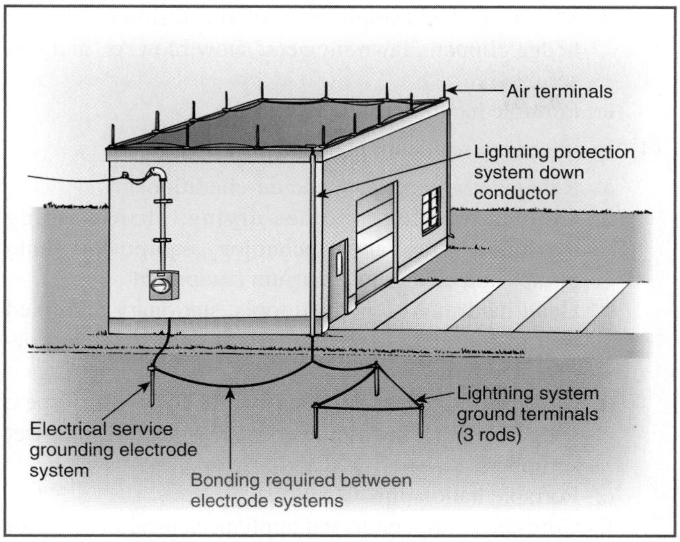


***EXHIBIT 250.44*** *Bonding between the lightning system ground terminals and the electrical service grounding electrode system.*

# VI. Equipment Grounding and Equipment Grounding Conductors

## 250.110 Equipment Fastened in Place (Fixed) or Connected by Permanent Wiring Methods

Exposed, normally non–current-carrying metal parts of fixed equipment supplied by or enclosing conductors or components that are likely to become energized shall be connected to an equipment grounding conductor under any of the following conditions:

(1) Where within 2.5 m (8 ft) vertically or 1.5 m (5 ft) horizontally of ground or grounded metal objects and subject to contact by persons
(2) Where located in a wet or damp location and not isolated
(3) Where in electrical contact with metal
(4) Where in a hazardous (classified) location as covered by Articles 500 through 517
(5) Where supplied by a wiring method that provides an equipment grounding conductor, except as permitted by 250.86 Exception No. 2 for short sections of metal enclosures
(6) Where equipment operates with any terminal at over 150 volts to ground

Exposed, non–current-carrying metal parts of fixed equipment that are not likely to become energized are not required to be grounded. These parts include some metal nameplates on nonmetallic enclosures and small parts such as bolts and screws.

*Exception No. 1: If exempted by special permission, the metal frame of electrically heated appliances that have the frame permanently and effectively insulated from ground shall not be required to be grounded.*

*Exception No. 2: Distribution apparatus, such as transformer and capacitor cases, mounted on wooden poles at a height exceeding 2.5 m (8 ft) above ground or grade level shall not be required to be grounded.*

*Exception No. 3: Listed equipment protected by a system of double insulation, or its equivalent, shall not be required to be connected to the equipment grounding conductor. Where such a system is employed, the equipment shall be distinctively marked.*

## 250.112 Specific Equipment Fastened in Place (Fixed) or Connected by Permanent Wiring Methods

Except as permitted in 250.112(F) and (I), exposed, normally non–current-carrying metal parts of equipment described in 250.112(A) through (K), and normally non–current-carrying metal parts of equipment and enclosures described in 250.112(L) and (M), shall be connected to an equipment grounding conductor, regardless of voltage.

**(A) Switchgear and Switchboard Frames and Structures.** Switchgear or switchboard frames and structures supporting switching equipment, except frames of 2-wire dc switchgear or switchboards where effectively insulated from ground.

**(B) Pipe Organs.** Generator and motor frames in an electrically operated pipe organ, unless effectively insulated from ground and the motor driving it.

**(C) Motor Frames.** Motor frames, as provided by 430.242.

**(D) Enclosures for Motor Controllers.** Enclosures for motor controllers unless attached to ungrounded portable equipment.

**(E) Elevators and Cranes.** Electrical equipment for elevators and cranes.

**(F) Garages, Theaters, and Motion Picture Studios.** Electrical equipment in commercial garages, theaters, and motion picture studios, except pendant lampholders supplied by circuits not over 150 volts to ground.

**(G) Electric Signs.** Electric signs, outline lighting, and associated equipment as provided in 600.7.

**(H) Motion Picture Projection Equipment.** Motion picture projection equipment.

**(I) Remote-Control, Signaling, and Fire Alarm Circuits.** Equipment supplied by Class 1 circuits shall be grounded unless operating at less than 50 volts. Equipment supplied by Class 1 power-limited circuits, by Class 2 and Class 3 remote-control and signaling circuits, and by fire alarm circuits shall be grounded where system grounding is required by Part II or Part VIII of this article.

**(J) Luminaires.** Luminaires as provided in Part V of Article 410.

**(K) Skid-Mounted Equipment.** Permanently mounted electrical equipment and skids shall be connected to the equipment grounding conductor sized as required by 250.122.

**(L) Motor-Operated Water Pumps.** Motor-operated water pumps, including the submersible type.

During normal operation or maintenance, a ground fault to the casing of a water pump can be a shock hazard if exposed or it could energize a metal well casing if the pump is a submersible type.

**(M) Metal Well Casings.** Where a submersible pump is used in a metal well casing, the well casing shall be connected to the pump circuit equipment grounding conductor.

This requirement is intended to prevent a shock hazard that could exist due to a potential difference between the pump (which is grounded to the system ground) and the metal well casing. This grounding and bonding connection also provides a return path for ground-fault current.

## 250.114 Equipment Connected by Cord and Plug

Under any of the conditions described in 250.114(1) through (4), exposed, normally non–current-carrying metal parts of cord-and-plug-connected equipment shall be connected to the equipment grounding conductor.

*Exception: Listed tools, listed appliances, and listed equipment covered in 250.114(2) through (4) shall not be required to be connected to an equipment grounding conductor where protected by a system of double insulation or its equivalent. Double insulated equipment shall be distinctively marked.*

(1) In hazardous (classified) locations (see Articles 500 through 517)
(2) Where operated at over 150 volts to ground

*Exception No. 1: Motors, where guarded, shall not be required to be connected to an equipment grounding conductor.*

*Exception No. 2: Metal frames of electrically heated appliances, exempted by special permission, shall not be required to be connected to an equipment grounding conductor, in which case the frames shall be permanently and effectively insulated from ground.*

(3) In residential occupancies:
   a. Refrigerators, freezers, and air conditioners
   b. Clothes-washing, clothes-drying, dish-washing machines; ranges; kitchen waste disposers; information technology equipment; sump pumps and electrical aquarium equipment
   c. Hand-held motor-operated tools, stationary and fixed motor-operated tools, and light industrial motor-operated tools
   d. Motor-operated appliances of the following types: hedge clippers, lawn mowers, snow blowers, and wet scrubbers
   e. Portable handlamps
(4) In other than residential occupancies:
   a. Refrigerators, freezers, and air conditioners
   b. Clothes-washing, clothes-drying, dish-washing machines; information technology equipment; sump pumps and electrical aquarium equipment
   c. Hand-held motor-operated tools, stationary and fixed motor-operated tools, and light industrial motor-operated tools
   d. Motor-operated appliances of the following types: hedge clippers, lawn mowers, snow blowers, and wet scrubbers
   e. Portable handlamps
   f. Cord-and-plug-connected appliances used in damp or wet locations or by persons standing on the ground or on metal floors or working inside of metal tanks or boilers
   g. Tools likely to be used in wet or conductive locations

*Exception: Tools and portable handlamps likely to be used in wet or conductive locations shall not be required to be connected to an equipment grounding conductor where supplied through an isolating transformer with an ungrounded secondary of not over 50 volts.*

Double insulated portable tools and appliances must be listed by a qualified electrical testing laboratory, and they must be distinctively marked as double insulated.

Cord-connected portable tools or appliances are not intended to be used in damp, wet, or conductive locations unless they are grounded, supplied by an isolation transformer with a secondary of not more than 50 volts, or double insulated.

## 250.116 Nonelectrical Equipment

The metal parts of the following nonelectrical equipment described in this section shall be connected to the equipment grounding conductor:

(1) Frames and tracks of electrically operated cranes and hoists
(2) Frames of nonelectrically driven elevator cars to which electrical conductors are attached
(3) Hand-operated metal shifting ropes or cables of electric elevators

Informational Note: Where extensive metal in or on buildings or structures may become energized and is subject to personal contact, adequate bonding and grounding will provide additional safety.

Because metal siding on buildings is not electrical equipment, it is outside the scope of the *Code* [see 90.2(A)] and, therefore, is not subject to the *Code* requirements for grounding and bonding.

However, luminaires, signs, or receptacles installed on buildings with metal siding could cause the siding to become energized. Grounding of metal siding, while not required, does reduce the shock risk from contacting the siding.

## 250.118 Types of Equipment Grounding Conductors

The equipment grounding conductor run with or enclosing the circuit conductors shall be one or more or a combination of the following:

(1) A copper, aluminum, or copper-clad aluminum conductor. This conductor shall be solid or stranded; insulated, covered, or bare; and in the form of a wire or a busbar of any shape.
(2) Rigid metal conduit.

Rigid metal conduit (RMC) is an EGC , and, like other metal raceways such as intermediate metal conduit (IMC) and electrical metallic tubing (EMT), it can be used as the sole EGC without a wire-type EGC having to be installed in the raceway. As a general rule, the *Code* only requires one EGC, and it can be any of the types specified in this section. Specific requirements in the *Code*, such as 517.13(B), require a wire-type EGC regardless of the wiring method. However, if a wire-type EGC is installed in a metal raceway and it is not isolated per 250.96(B), the raceway is in parallel with the wire, and the combination of the metal raceway and the wire is the EGC for enclosed circuit(s). See 250.122 for sizing of wire-type EGCs.

(3) Intermediate metal conduit.
(4) Electrical metallic tubing.
(5) Listed flexible metal conduit meeting all the following conditions:
   a. The conduit is terminated in listed fittings.
   b. The circuit conductors contained in the conduit are protected by overcurrent devices rated at 20 amperes or less.
   c. The combined length of flexible metal conduit and flexible metallic tubing and liquidtight flexible metal conduit in the same ground-fault current path does not exceed 1.8 m (6 ft).
   d. If used to connect equipment where flexibility is necessary to minimize the transmission of vibration from equipment or to provide flexibility for equipment that requires movement after installation, an equipment grounding conductor shall be installed.
(6) Listed liquidtight flexible metal conduit meeting all the following conditions:
   a. The conduit is terminated in listed fittings.
   b. For metric designators 12 through 16 (trade sizes ⅜ through ½), the circuit conductors contained in the conduit are protected by overcurrent devices rated at 20 amperes or less.
   c. For metric designators 21 through 35 (trade sizes ¾ through 1¼), the circuit conductors contained in the conduit are protected by overcurrent devices rated not more than 60 amperes and there is no flexible metal conduit, flexible metallic tubing, or liquidtight flexible metal conduit in trade sizes metric designators 12 through 16 (trade sizes ⅜ through ½) in the ground-fault current path.
   d. The combined length of flexible metal conduit and flexible metallic tubing and liquidtight flexible metal conduit in the same ground-fault current path does not exceed 1.8 m (6 ft).
   e. If used to connect equipment where flexibility is necessary to minimize the transmission of vibration from equipment or to provide flexibility for equipment that requires movement after installation, an equipment grounding conductor shall be installed.

Flexible raceways that are used to facilitate connection to equipment but that remain stationary after the connection is made are not covered by 250.118(5)(d) or 250.118(6)(e).

(7) Flexible metallic tubing where the tubing is terminated in listed fittings and meeting the following conditions:
   a. The circuit conductors contained in the tubing are protected by overcurrent devices rated at 20 amperes or less.
   b. The combined length of flexible metal conduit and flexible metallic tubing and liquidtight flexible metal conduit in the same ground-fault current path does not exceed 1.8 m (6 ft).
(8) Armor of Type AC cable as provided in 320.108.
(9) The copper sheath of mineral-insulated, metal-sheathed cable Type MI.
(10) Type MC cable that provides an effective ground-fault current path in accordance with one or more of the following:
   a. It contains an insulated or uninsulated equipment grounding conductor in compliance with 250.118(1)
   b. The combined metallic sheath and uninsulated equipment grounding/bonding conductor of interlocked metal tape–type MC cable that is listed and identified as an equipment grounding conductor
   c. The metallic sheath or the combined metallic sheath and equipment grounding conductors of the smooth or corrugated tube-type MC cable that is listed and identified as an equipment grounding conductor
(11) Cable trays as permitted in 392.10 and 392.60.
(12) Cablebus framework as permitted in 370.60(1).
(13) Other listed electrically continuous metal raceways and listed auxiliary gutters.
(14) Surface metal raceways listed for grounding.

Informational Note: For a definition of *Effective Ground-Fault Current Path*, see Article 100.

## 250.119 Identification of Equipment Grounding Conductors

Unless required elsewhere in this *Code,* equipment grounding conductors shall be permitted to be bare, covered, or insulated. Individually covered or insulated equipment grounding conductors shall have a continuous outer finish that is either green or green with one or more yellow stripes except as permitted in this section. Conductors with insulation or individual covering that is green, green with one or more yellow stripes, or otherwise identified as permitted by this section shall not be used for ungrounded or grounded circuit conductors.

*Exception No. 1: Power-limited Class 2 or Class 3 cables, power-limited fire alarm cables, or communications cables containing only circuits operating at less than 50 volts where connected to equipment not required to be grounded in accordance with 250.112(I) shall be permitted to use a conductor with green insulation or green with one or more yellow stripes for other than equipment grounding purposes.*

Exception No. 1 permits limited use of conductors with green-colored insulation for circuit conductors in signaling or other Class 2 or Class 3 circuits operating at less than 50 volts to ground. Many limited-energy ac systems are not required to be grounded per 250.20(B); thus, per 250.112(I), equipment supplied by these circuits does not have grounding requirements.

*Exception No. 2: Flexible cords having an integral insulation and jacket without an equipment grounding conductor shall be permitted to have a continuous outer finish that is green.*

Informational Note: An example of a flexible cord with integral-type insulation is Type SPT-2, 2 conductor.

*Exception No. 3: Conductors with green insulation shall be permitted to be used as ungrounded signal conductors where installed between the output terminations of traffic signal control and traffic signal indicating heads. Signaling circuits installed in accordance with this exception shall include an equipment grounding conductor in accordance with 250.118. Wire-type equipment grounding conductors shall be bare or have insulation or covering that is green with one or more yellow stripes.*

The proper operation of traffic signal equipment is an important part of public safety. A long-standing practice is to use a conductor with green insulation as the signal conductor installed between the controller and the green indicating light located in the traffic signal head.

**(A) Conductors 4 AWG and Larger.** Equipment grounding conductors 4 AWG and larger shall comply with 250.119(A)(1) and (A)(2).

(1) An insulated or covered conductor 4 AWG and larger shall be permitted, at the time of installation, to be permanently identified as an equipment grounding conductor at each end and at every point where the conductor is accessible.

*Exception: Conductors 4 AWG and larger shall not be required to be marked in conduit bodies that contain no splices or unused hubs.*

(2) Identification shall encircle the conductor and shall be accomplished by one of the following:
   a. Stripping the insulation or covering from the entire exposed length
   b. Coloring the insulation or covering green at the termination
   c. Marking the insulation or covering with green tape or green adhesive labels at the termination

**(B) Multiconductor Cable.** Where the conditions of maintenance and supervision ensure that only qualified persons service the installation, one or more insulated conductors in a multiconductor cable, at the time of installation, shall be permitted to be permanently identified as equipment grounding conductors at each end and at every point where the conductors are accessible by one of the following means:

(1) Stripping the insulation from the entire exposed length
(2) Coloring the exposed insulation green
(3) Marking the exposed insulation with green tape or green adhesive labels

**(C) Flexible Cord.** An uninsulated equipment grounding conductor shall be permitted, but, if individually covered, the covering shall have a continuous outer finish that is either green or green with one or more yellow stripes.

## 250.120 Equipment Grounding Conductor Installation

An equipment grounding conductor shall be installed in accordance with 250.120(A), (B), and (C).

**(A) Raceway, Cable Trays, Cable Armor, Cablebus, or Cable Sheaths.** Where it consists of a raceway, cable tray, cable armor, cablebus framework, or cable sheath or where it is a wire within a raceway or cable, it shall be installed in accordance with the applicable provisions in this *Code* using fittings for joints and terminations approved for use with the type raceway or cable used. All connections, joints, and fittings shall be made tight using suitable tools.

Informational Note: See the UL guide information on FHIT systems for equipment grounding conductors installed in a raceway that are part of an electrical circuit protective system or a fire-rated cable listed to maintain circuit integrity.

**(B) Aluminum and Copper-Clad Aluminum Conductors.** Equipment grounding conductors of bare or insulated aluminum or copper-clad aluminum shall be permitted. Bare conductors shall not come in direct contact with masonry or the earth or where subject to corrosive conditions. Aluminum or copper-clad aluminum conductors shall not be terminated within 450 mm (18 in.) of the earth.

**(C) Equipment Grounding Conductors Smaller Than 6 AWG.** Where not routed with circuit conductors as permitted in 250.130(C) and 250.134(B) Exception No. 2, equipment grounding conductors smaller than 6 AWG shall be protected from physical damage by an identified raceway or cable armor unless installed within hollow spaces of the framing members of buildings or structures and where not subject to physical damage.

## 250.121 Use of Equipment Grounding Conductors

An equipment grounding conductor shall not be used as a grounding electrode conductor.

*Exception: A wire-type equipment grounding conductor installed in compliance with 250.6(A) and the applicable requirements for both the equipment grounding conductor and the grounding electrode conductor in Parts II, III, and VI of this article shall be permitted to serve as both an equipment grounding conductor and a grounding electrode conductor.*

EGCs and grounding electrode conductors have specific functions. Grounding electrode conductors are required to be a wire- or busbar-type conductor in accordance with 250.62, whereas EGCs are permitted to be any of the types listed in 250.118. This exception permits a wire-type conductor to be used for both purposes if it satisfies all applicable requirements for both the EGC and the grounding electrode conductor and it does not carry current during normal operating conditions.

## 250.122 Size of Equipment Grounding Conductors

**(A) General.** Copper, aluminum, or copper-clad aluminum equipment grounding conductors of the wire type shall not be smaller than shown in Table 250.122, but in no case shall they be required to be larger than the circuit conductors supplying the equipment. Where a cable tray, a raceway, or a cable armor or sheath is used as the equipment grounding conductor, as provided in 250.118 and 250.134(A), it shall comply with 250.4(A)(5) or (B)(4).

Equipment grounding conductors shall be permitted to be sectioned within a multiconductor cable, provided the combined circular mil area complies with Table 250.122.

An installation condition (such as a long distance between the power source and utilization equipment) may necessitate additional means or increased EGC sizing to lower the overall impedance of the ground-fault current return path in order to facilitate quick operation of the OCPD in the event of a line-to-ground fault.

**(B) Increased in Size.** Where ungrounded conductors are increased in size from the minimum size that has sufficient ampacity for the intended installation, wire-type equipment grounding conductors, where installed, shall be increased in size proportionately according to the circular mil area of the ungrounded conductors.

Generally, the minimum-sized EGC is selected from Table 250.122 based on the rating or setting of the feeder or branch-circuit OCPD(s). Where the ungrounded circuit conductors are increased in size to compensate for voltage drop or for any other reason related to proper circuit operation, the EGCs must be increased proportionately. In some cases, use of a conductor with a higher insulation temperature rating allows for compliance with ampacity adjustment and correction requirements without having to increase the circular mil area of the conductor.

### Calculation Example

A 240-volt, single-phase, 250-ampere load is supplied from a 300-ampere breaker located in a panelboard 500 feet away. The conductors are 250 kcmil copper, installed in rigid nonmetallic conduit, with a 4 AWG copper EGC. If the conductors are increased to 350 kcmil, what is the minimum size for the EGC based on the proportional-increase requirement?

*Solution*

STEP 1. Calculate the size ratio of the new conductors to the existing conductors:

$$\text{Size ratio} = \frac{350{,}000 \text{ circular mils}}{250{,}000 \text{ circular mils}} = 1.4$$

STEP 2. Calculate the cross-sectional area of the new EGC:

$$41{,}740 \text{ circular mils} \times 1.4 = 58{,}436 \text{ circular mils}$$

According to Chapter 9, Table 8, 4 AWG, the size of the existing grounding conductor, has a cross-sectional area of 41,740 circular mils.

STEP 3. Determine the size of the new EGC. Again, referring to Chapter 9, Table 8, we find that 58,436 circular mils is larger than 3 AWG. The next larger size is 66,360 circular mils, which converts to a 2 AWG copper EGC.

**(C) Multiple Circuits.** Where a single equipment grounding conductor is run with multiple circuits in the same raceway, cable, or cable tray, it shall be sized for the largest overcurrent device protecting conductors in the raceway, cable, or cable tray. Equipment grounding conductors installed in cable trays shall meet the minimum requirements of 392.10(B)(1)(c).

A single EGC serving multiple circuits is required to be sized based on the size of the largest circuit supplied; it is not likely that all circuits will develop faults at the same time. For example, three 3-phase circuits in the same raceway, protected by overcurrent devices rated 30, 60, and 100 amperes, would require only one EGC, sized according to the largest overcurrent device (in this case, 100 amperes). Therefore, an 8 AWG copper or 6 AWG aluminum conductor or copper-clad aluminum conductor is required, according to Table 250.122.

**(D) Motor Circuits.** Equipment grounding conductors for motor circuits shall be sized in accordance with (D)(1) or (D)(2).

*TABLE 250.122 Minimum Size Equipment Grounding Conductors for Grounding Raceway and Equipment*

| Rating or Setting of Automatic Overcurrent Device in Circuit Ahead of Equipment, Conduit, etc., Not Exceeding (Amperes) | Size (AWG or kcmil) | |
|---|---|---|
| | **Copper** | **Aluminum or Copper-Clad Aluminum*** |
| 15 | 14 | 12 |
| 20 | 12 | 10 |
| 60 | 10 | 8 |
| 100 | 8 | 6 |
| 200 | 6 | 4 |
| 300 | 4 | 2 |
| 400 | 3 | 1 |
| 500 | 2 | 1/0 |
| 600 | 1 | 2/0 |
| 800 | 1/0 | 3/0 |
| 1000 | 2/0 | 4/0 |
| 1200 | 3/0 | 250 |
| 1600 | 4/0 | 350 |
| 2000 | 250 | 400 |
| 2500 | 350 | 600 |
| 3000 | 400 | 600 |
| 4000 | 500 | 750 |
| 5000 | 700 | 1200 |
| 6000 | 800 | 1200 |

Note: Where necessary to comply with 250.4(A)(5) or (B)(4), the equipment grounding conductor shall be sized larger than given in this table.
*See installation restrictions in 250.120.

**(1) General.** The equipment grounding conductor size shall not be smaller than determined by 250.122(A) based on the rating of the branch-circuit short-circuit and ground-fault protective device.

**(2) Instantaneous-Trip Circuit Breaker and Motor Short-Circuit Protector.** Where the overcurrent device is an instantaneous-trip circuit breaker or a motor short-circuit protector, the equipment grounding conductor shall be sized not smaller than that given by 250.122(A) using the maximum permitted rating of a dual element time-delay fuse selected for branch-circuit short-circuit and ground-fault protection in accordance with 430.52(C)(1), Exception No. 1.

**(E) Flexible Cord and Fixture Wire.** The equipment grounding conductor in a flexible cord with the largest circuit conductor 10 AWG or smaller, and the equipment grounding conductor used with fixture wires of any size in accordance with 240.5, shall not be smaller than 18 AWG copper and shall not be smaller than the circuit conductors. The equipment grounding conductor in a flexible cord with a circuit conductor larger than 10 AWG shall be sized in accordance with Table 250.122.

**(F) Conductors in Parallel.** Where conductors are installed in parallel in multiple raceways or cables as permitted in 310.10(H), the equipment grounding conductors, where used, shall be installed in parallel in each raceway or cable. Where conductors are installed in parallel in the same raceway, cable, or cable tray as permitted in 310.10(H), a single equipment grounding conductor shall be permitted. Equipment grounding conductors installed in cable tray shall meet the minimum requirements of 392.10(B)(1)(c).

Each equipment grounding conductor shall be sized in compliance with 250.122.

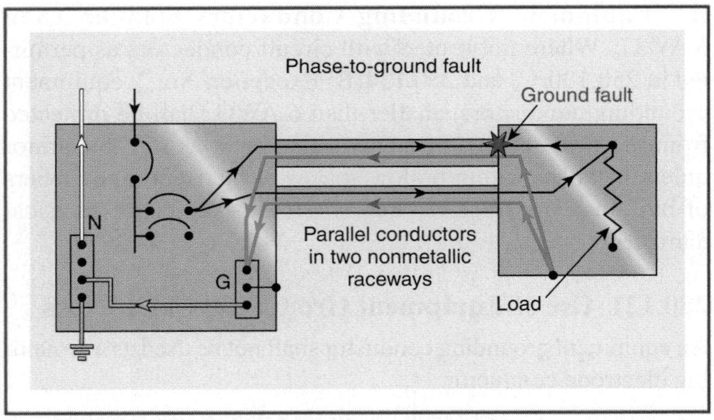


***EXHIBIT 250.45*** *Grounding paths for ground fault at the load supplied by parallel conductors in two nonmetallic raceways, illustrating the reason for the requirement of 250.122(F).*

A full-sized EGC is required to prevent overloading and possible burnout of the conductor should a ground fault occur along one of the parallel branches. The installation conditions for paralleled conductors prescribed in 310.10(H) result in proportional distribution of the current-time duty among the several paralleled grounding conductors only for overcurrent conditions downstream of the paralleled set of circuit conductors.

Exhibit 250.45 shows a parallel arrangement with two nonmetallic conduits installed underground. A ground fault at the enclosure will cause the EGC in the top conduit to carry more than its proportionate share of fault current. The fault is fed by two different conductors of the same phase, one from the left and one from the right. The shortest and lowest-impedance path to ground from the fault to the supply panelboard is through the EGC in the top conduit. The grounding path from the fault through the bottom conduit is longer and of higher impedance. Therefore, the EGC in each raceway must be capable of carrying a major portion of the fault current without burning open.

Where an installation consists of raceways, compliance with this requirement is accomplished by installing a wire-type EGC selected from Table 250.122 in the raceway. Cables, on the other hand, are manufactured with the conductors already installed, including the EGC. Therefore, a cable in which the EGC meets the minimum size per Table 250.122 must be selected.

Where cables are used in parallel, the EGC in each cable is required to be sized in accordance with 250.122 based on the rating or setting of the OCPD that is protecting the conductors connected in parallel. In large capacity circuits the EGC may have to be larger than the ungrounded conductor in an individual cable that is used as part of a larger parallel set. EGCs are permitted to be connected in parallel per 310.10(H), but each parallel EGC is required to be fully sized per 250.122. The cross-sectional area of sectioned EGCs in an individual cable are permitted to be added together to meet the requirement of a fully sized EGC in each cable.

Where parallel conductors are installed in a single raceway or cable tray, a single EGC is permitted to be installed in the raceway or cable tray that is sized based on the rating of the OCPD protecting the parallel circuit conductors.

**(G) Feeder Taps.** Equipment grounding conductors run with feeder taps shall not be smaller than shown in Table 250.122 based on the rating of the overcurrent device ahead of the feeder but shall not be required to be larger than the tap conductor.

For a circuit tapped from a feeder, the OCPD on the supply side of the tap conductors will respond to a ground-fault condition between the point at which the tap conductors are supplied and the point at which they terminate. In accordance with this paragraph and 250.122(A), the EGC is not required to be larger than the ungrounded conductors. This applies only where a wire-type EGC is run with the feeder tap conductors. Other EGCs permitted in 250.118 can be used where they meet the requirements for tap conductor wiring methods specified in 240.21(B)(1) through (B)(5).

### 250.124 Equipment Grounding Conductor Continuity

**(A) Separable Connections.** Separable connections such as those provided in drawout equipment or attachment plugs and mating connectors and receptacles shall provide for first-make, last-break of the equipment grounding conductor. First-make, last-break shall not be required where interlocked equipment, plugs, receptacles, and connectors preclude energization without grounding continuity.

**(B) Switches.** No automatic cutout or switch shall be placed in the equipment grounding conductor of a premises wiring system unless the opening of the cutout or switch disconnects all sources of energy.

### 250.126 Identification of Wiring Device Terminals

The terminal for the connection of the equipment grounding conductor shall be identified by one of the following:

(1) A green, not readily removable terminal screw with a hexagonal head.
(2) A green, hexagonal, not readily removable terminal nut.
(3) A green pressure wire connector. If the terminal for the equipment grounding conductor is not visible, the conductor entrance hole shall be marked with the word *green* or *ground*, the letters *G* or *GR*, a grounding symbol, or otherwise identified by a distinctive green color. If the terminal for the equipment grounding conductor is readily removable, the area adjacent to the terminal shall be similarly marked.

Informational Note: See Informational Note Figure 250.126.

***INFORMATIONAL NOTE FIGURE 250.126*** *One Example of a Symbol Used to Identify the Grounding Termination Point for an Equipment Grounding Conductor.*

## VII. Methods of Equipment Grounding

### 250.130 Equipment Grounding Conductor Connections

Equipment grounding conductor connections at the source of separately derived systems shall be made in accordance with 250.30(A)(1). Equipment grounding conductor connections at service equipment shall be made as indicated in 250.130(A) or (B). For replacement of non–grounding-type receptacles with grounding-type receptacles and for branch-circuit extensions only in existing installations that do not have an equipment grounding conductor in the branch circuit, connections shall be permitted as indicated in 250.130(C).

**(A) For Grounded Systems.** The connection shall be made by bonding the equipment grounding conductor to the grounded service conductor and the grounding electrode conductor.

The grounding and bonding arrangement required by 250.130(A) for a grounded system is illustrated in Exhibit 250.46.

**(B) For Ungrounded Systems.** The connection shall be made by bonding the equipment grounding conductor to the grounding electrode conductor.

**(C) Nongrounding Receptacle Replacement or Branch Circuit Extensions.** The equipment grounding conductor of a grounding-type receptacle or a branch-circuit extension shall be permitted to be connected to any of the following:

(1) Any accessible point on the grounding electrode system as described in 250.50
(2) Any accessible point on the grounding electrode conductor

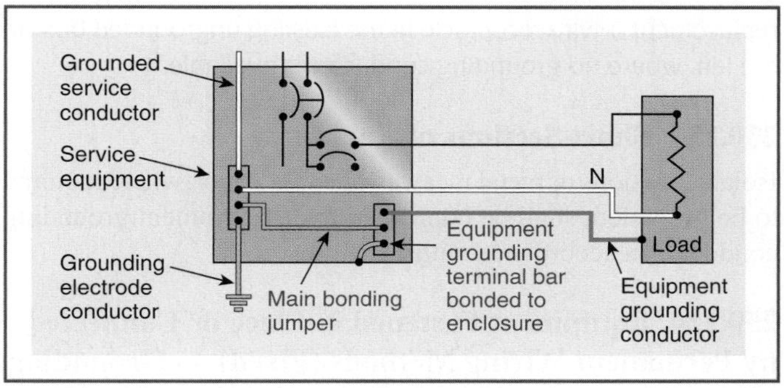


***EXHIBIT 250.46*** *Grounding and bonding arrangement for grounded systems, per 250.130(A), illustrating connection of the EGC (bus) to the enclosures and the grounded service conductor.*

(3) The equipment grounding terminal bar within the enclosure where the branch circuit for the receptacle or branch circuit originates

(4) An equipment grounding conductor that is part of another branch circuit that originates from the enclosure where the branch circuit for the receptacle or branch circuit originates

(5) For grounded systems, the grounded service conductor within the service equipment enclosure

(6) For ungrounded systems, the grounding terminal bar within the service equipment enclosure

Informational Note: See 406.4(D) for the use of a ground-fault circuit-interrupting type of receptacle.

Section 250.130(C) applies to both ungrounded and grounded systems, but its most common application is for receptacle replacement of branch-circuit extensions in single-phase, 120-volt, 15- and 20-ampere branch circuits, which are required to be supplied by a grounded system per 250.20(B). This section permits a nongrounding-type receptacle to be replaced with a grounding-type receptacle under the following conditions:

1. The branch circuit does not contain an equipment ground.
2. An existing branch circuit is being extended for additional receptacle outlets.
3. An EGC is connected from the receptacle grounding terminal to any accessible point on the grounding electrode system, to any accessible point on the grounding electrode conductor, to the grounded service conductor within the service equipment enclosure, to the equipment grounding terminal bar in the enclosure from which the circuit is supplied, or to an EGC that is part of another branch circuit that originates from the same enclosure where the branch circuit for the receptacle originates.

The requirement in 250.68(C)(1) does not permit this separate EGC to be connected to the metal water piping of a building or structure beyond the first 5 feet of where the piping enters the building or structure unless the conditions of the exception to 250.68(C)(1) can be met.

Exhibit 250.47 shows a branch-circuit extension made from an existing installation. This method is also permitted to ground a replacement 3-wire receptacle in the existing ungrounded box on the left, where no grounding conductor is available.

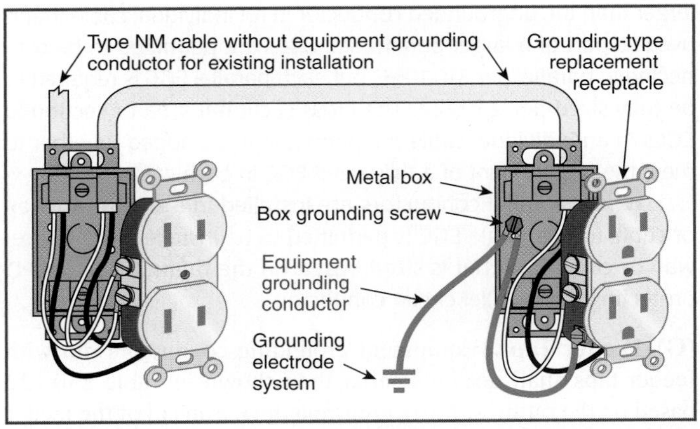


***EXHIBIT 250.47*** *Branch-circuit extension to an existing installation, per 250.130(C), illustrating a separate EGC connected to the grounding electrode system.*

## 250.132 Short Sections of Raceway

Isolated sections of metal raceway or cable armor, where required to be grounded, shall be connected to an equipment grounding conductor in accordance with 250.134.

## 250.134 Equipment Fastened in Place or Connected by Permanent Wiring Methods (Fixed) — Grounding

Unless grounded by connection to the grounded circuit conductor as permitted by 250.32, 250.140, and 250.142, non–current-carrying metal parts of equipment, raceways, and other enclosures, if grounded, shall be connected to an equipment grounding conductor by one of the methods specified in 250.134(A) or (B).

**(A) Equipment Grounding Conductor Types.** By connecting to any of the equipment grounding conductors permitted by 250.118.

**(B) With Circuit Conductors.** By connecting to an equipment grounding conductor contained within the same raceway, cable, or otherwise run with the circuit conductors.

The EGC run in the same raceway or cable as the circuit conductor(s) allows the magnetic field developed by the circuit conductor and the EGC to cancel, reducing their impedance.

Magnetic flux strength is inversely proportional to the square of the distance between the two conductors. By placing an EGC away from the conductor delivering the fault current, the magnetic flux cancellation decreases. This increases the impedance of the fault path and delays operation of the protective device contrary to the performance requirements specified in 250.4(A)(5) and (B)(4) for the ground-fault current return path.

*Exception No. 1: As provided in 250.130(C), the equipment grounding conductor shall be permitted to be run separately from the circuit conductors.*

This practice applies only where a grounding-type receptacle is used on a circuit that does not include an EGC. See the commentary following 250.130(C) for further explanation.

*Exception No. 2: For dc circuits, the equipment grounding conductor shall be permitted to be run separately from the circuit conductors.*

Informational Note No. 1: See 250.102 and 250.168 for equipment bonding jumper requirements.

Informational Note No. 2: See 400.7 for use of cords for fixed equipment.

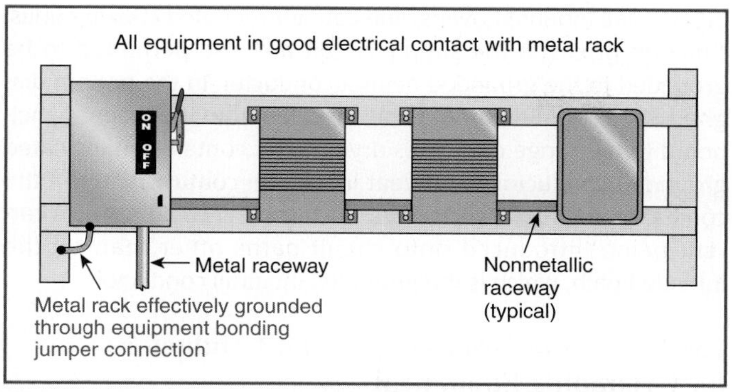


***EXHIBIT 250.48*** *Mounting equipment to a grounded support structure or rack is an acceptable means of providing the required equipment grounding connection.*

## 250.136 Equipment Considered Grounded

Under the conditions specified in 250.136(A) and (B), the normally non–current-carrying metal parts of the equipment shall be considered grounded.

**(A) Equipment Secured to Grounded Metal Supports.** Electrical equipment secured to and in electrical contact with a metal rack or structure provided for its support and connected to an equipment grounding conductor by one of the means indicated in 250.134. The structural metal frame of a building shall not be used as the required equipment grounding conductor for ac equipment.

Equipment bolted or securely clamped to the rack will typically provide the necessary electrical contact to ensure a low impedance connection between the rack and the equipment. If the rack has been painted, 250.12 requires the paint to be removed to ensure that the connection is not impeded.

Exhibit 250.48 is an example of electrical equipment secured to and in electrical contact with a metal rack that is effectively grounded by the equipment bonding jumper installed between the safety switch and the rack. The permission to use this method does not include installations where equipment is supported by the structural metal of a building. The physical separation between the circuit conductors and the building steel raises the impedance of the equipment grounding and bonding path and is not permitted by 250.134(B) and 300.3(B).

**(B) Metal Car Frames.** Metal car frames supported by metal hoisting cables attached to or running over metal sheaves or drums of elevator machines that are connected to an equipment grounding conductor by one of the methods indicated in 250.134.

## 250.138 Cord-and-Plug-Connected Equipment

Non–current-carrying metal parts of cord-and-plug-connected equipment, if grounded, shall be connected to an equipment grounding conductor by one of the methods in 250.138(A) or (B).

**(A) By Means of an Equipment Grounding Conductor.** By means of an equipment grounding conductor run with the power supply conductors in a cable assembly or flexible cord properly terminated in a grounding-type attachment plug with one fixed grounding contact.

*Exception: The grounding contacting pole of grounding-type plug-in ground-fault circuit interrupters shall be permitted to be of the movable, self-restoring type on circuits operating at not over 150 volts between any two conductors or over 150 volts between any conductor and ground.*

**(B) By Means of a Separate Flexible Wire or Strap.** By means of a separate flexible wire or strap, insulated or bare, connected to an equipment grounding conductor, and protected as well as practicable against physical damage, where part of equipment.

## 250.140 Frames of Ranges and Clothes Dryers

Frames of electric ranges, wall-mounted ovens, counter-mounted cooking units, clothes dryers, and outlet or junction boxes that are part of the circuit for these appliances shall be connected to the equipment grounding conductor in the manner specified by 250.134 or 250.138.

*Exception: For existing branch-circuit installations only where an equipment grounding conductor is not present in the outlet or junction box, the frames of electric ranges, wall-mounted ovens, counter-mounted cooking units, clothes dryers, and outlet or junction boxes that are part of the circuit for these appliances shall be permitted to be connected to the grounded circuit conductor if all the following conditions are met.*

*(1) The supply circuit is 120/240-volt, single-phase, 3-wire; or 208Y/120-volt derived from a 3-phase, 4-wire, wye-connected system.*
*(2) The grounded conductor is not smaller than 10 AWG copper or 8 AWG aluminum.*
*(3) The grounded conductor is insulated, or the grounded conductor is uninsulated and part of a Type SE service-entrance cable and the branch circuit originates at the service equipment.*
*(4) Grounding contacts of receptacles furnished as part of the equipment are bonded to the equipment.*

Prior to the 1996 *Code,* use of the grounded circuit conductor as a grounding conductor was permitted for all installations. In many instances, the wiring method was service-entrance cable with an uninsulated neutral conductor covered by the cable jacket. Where Type SE cable was used to supply ranges and dryers, the branch circuit was required to originate at the service equipment to avoid neutral current from downstream panelboards being imposed on metal objects, such as pipes or ducts.The grounded conductor (neutral) of newly installed branch circuits supplying ranges and clothes dryers is not permitted to be used for grounding

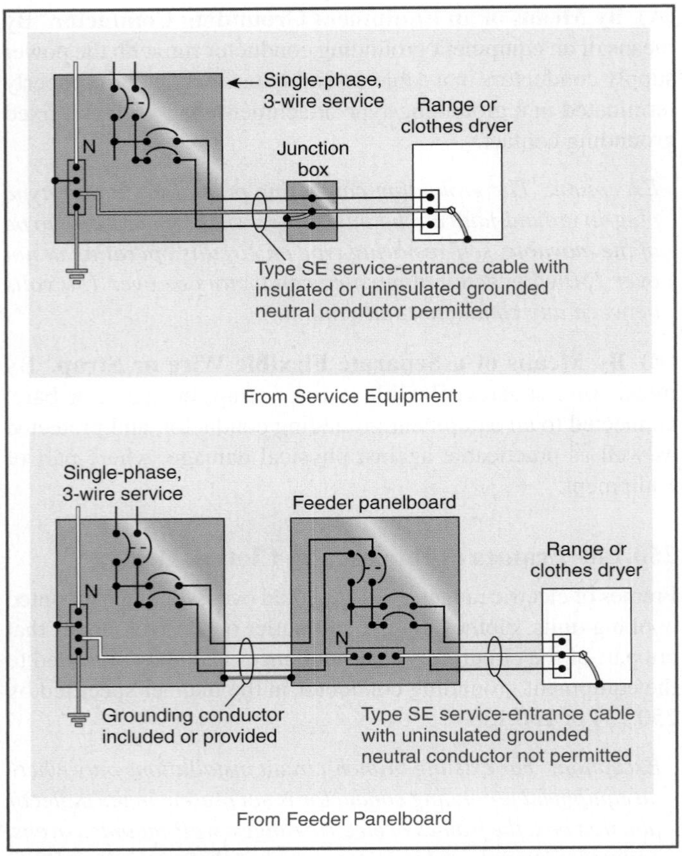


***EXHIBIT 250.49*** *Two existing installations in which the grounded conductor in Type SE service-entrance cable is used for grounding the frames of ranges and clothes dryers, plus associated metal junction boxes, in accordance with 250.140.*

the non–current-carrying metal parts of the appliances. Branch circuits for new appliance installations are required to provide an EGC sized in accordance with 250.122 for grounding the non–current-carrying metal parts.

An older appliance connected to a new branch circuit must have its 3-wire cord and plug replaced with a 4-conductor cord, with one of those conductors being an EGC. The appliance bonding jumper between the neutral and the frame of the appliance must be removed. Where a new range or clothes dryer is connected to an existing branch circuit without an EGC, an appliance bonding jumper must be connected between the neutral terminal and the frame of the appliance.

The grounded circuit conductor of an existing branch circuit is permitted to be used to ground the frame of an electric range, wall-mounted oven, or counter-mounted cooking unit, provided all four conditions of 250.140, Exception, are met. In addition, the grounded circuit conductor of these existing branch circuits is also permitted to be used to ground any junction boxes in the circuit supplying the appliance, and a 3-wire pigtail and range receptacle are permitted to be used.

Exhibit 250.49 shows two examples of existing installations in which Type SE service-entrance cable is used to supply ranges, dryers, wall-mounted ovens, and counter-mounted cooking units. Junction boxes in the supply circuit are also permitted to be grounded to the grounded neutral conductor. In the bottom diagram, the service-entrance cable installed from the feeder panelboard to the range or clothes dryer outlet contains an insulated grounded conductor to prevent incidental contact between the conductor and metal enclosures. Such contact could result in current being introduced onto circuit paths other than on the intended path, which is the grounded (neutral) conductor.

## 250.142 Use of Grounded Circuit Conductor for Grounding Equipment

**(A) Supply-Side Equipment.** A grounded circuit conductor shall be permitted to ground non–current-carrying metal parts of equipment, raceways, and other enclosures at any of the following locations:

(1) On the supply side or within the enclosure of the ac service-disconnecting means
(2) On the supply side or within the enclosure of the main disconnecting means for separate buildings as provided in 250.32(B)
(3) On the supply side or within the enclosure of the main disconnecting means or overcurrent devices of a separately derived system where permitted by 250.30(A)(1)

**(B) Load-Side Equipment.** Except as permitted in 250.30(A)(1) and 250.32(B) Exception, a grounded circuit conductor shall not be used for grounding non–current-carrying metal parts of equipment on the load side of the service disconnecting means or on the load side of a separately derived system disconnecting means or the overcurrent devices for a separately derived system not having a main disconnecting means.

*Exception No. 1: The frames of ranges, wall-mounted ovens, counter-mounted cooking units, and clothes dryers under the conditions permitted for existing installations by 250.140 shall be permitted to be connected to the grounded circuit conductor.*

*Exception No. 2: It shall be permissible to ground meter enclosures by connection to the grounded circuit conductor on the load side of the service disconnect where all of the following conditions apply:*

*(1) No service ground-fault protection is installed.*
*(2) All meter enclosures are located immediately adjacent to the service disconnecting means.*
*(3) The size of the grounded circuit conductor is not smaller than the size specified in Table 250.122 for equipment grounding conductors.*

*Exception No. 3: Direct-current systems shall be permitted to be grounded on the load side of the disconnecting means or overcurrent device in accordance with 250.164.*

*Exception No. 4: Electrode-type boilers operating at over 1000 volts shall be grounded as required in 490.72(E)(1) and 490.74.*

If the grounded circuit conductor was re-grounded on the load side of the service and the grounded conductor became disconnected at any point on the line side of the service, the EGC and all conductive parts connected to it would carry the neutral current. Under this condition, the potential to ground of exposed metal parts not normally intended to carry current is raised. This rise in potential on non–current-carrying conductive parts could result in arcing in concealed spaces and could pose a severe shock hazard, particularly if the path is inadvertently opened by a person servicing or repairing piping or ductwork.

Even without an open grounded conductor (usually referred to as an open neutral), a connection between the grounded conductor and the EGC on the load side of the service places the EGC in a parallel circuit path with the grounded conductor. There could be some potential drop on exposed and concealed non–current-carrying metal parts. The magnitude of this potential difference would be determined by the relative impedances of the equipment grounding path and the grounded conductor circuits. Not only would the EGC path be affected, but all parallel paths not intended as EGCs would be affected as well. The parallel current paths could be metal building structures, metal piping, and metal ducts. The requirements of 250.30 and 250.32(B) were revised in recent editions of the *Code* to prohibit the creation of parallel paths for normal neutral current.

## 250.144 Multiple Circuit Connections

Where equipment is grounded and is supplied by separate connection to more than one circuit or grounded premises wiring system, an equipment grounding conductor termination shall be provided for each such connection as specified in 250.134 and 250.138.

## 250.146 Connecting Receptacle Grounding Terminal to Box

An equipment bonding jumper shall be used to connect the grounding terminal of a grounding-type receptacle to a grounded box unless grounded as in 250.146(A) through (D). The equipment bonding jumper shall be sized in accordance with Table 250.122 based on the rating of the overcurrent device protecting the circuit conductors.

**(A) Surface-Mounted Box.** Where the box is mounted on the surface, direct metal-to-metal contact between the device yoke and the box or a contact yoke or device that complies with 250.146(B) shall be permitted to ground the receptacle to the box. At least one of the insulating washers shall be removed from receptacles that do not have a contact yoke or device that complies with 250.146(B) to ensure direct metal-to-metal contact. This provision shall not apply to cover-mounted receptacles unless the box and cover combination are listed as providing satisfactory ground continuity between the box and the receptacle. A listed exposed work cover shall be permitted to be the grounding and bonding means when (1) the device is attached to the cover with at least two fasteners that are permanent (such as a rivet) or have a thread locking or screw or nut locking means and (2) when the cover mounting holes are located on a flat non-raised portion of the cover.

Section 250.146(A) permits the equipment bonding jumper to be omitted where the metal yoke of the device is in direct metal-to-metal contact with the metal device box and at least one of the fiber retention washers for the receptacle mounting screws is removed, as illustrated in Exhibit 250.50.

Cover-mounted wiring devices, such as on 4-inch-square covers, are not considered grounded. Section 250.146(A) does not apply to cover-mounted receptacles, such as the one illustrated in Exhibit 250.51. Box-cover and device combinations listed as providing grounding continuity are permitted. The mounting holes for the cover must be located on a *flat, non-raised* portion of the cover to provide the best possible surface-to-surface contact, and the receptacle must be secured to the cover using not less than two rivets or locking means for threaded attachment means.

**(B) Contact Devices or Yokes.** Contact devices or yokes designed and listed as self-grounding shall be permitted in conjunction with the supporting screws to establish equipment bonding between the device yoke and flush-type boxes.

An example of the contact devices permitted by this section is illustrated in Exhibit 250.52, which shows a receptacle designed with a spring-type grounding strap for holding the mounting screw and establishing the grounding circuit so that an equipment bonding jumper is not required. Such devices are listed as "self-grounding" and are permitted to be used for bonding the flush-mounted receptacle to a grounded metal box.

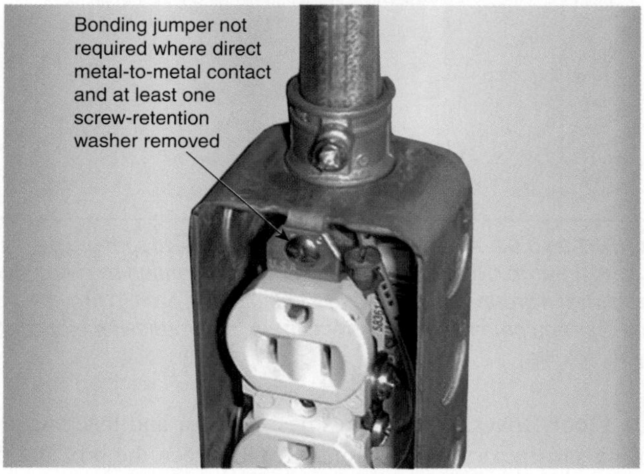


***EXHIBIT 250.50*** *An example of a box-mounted receptacle attached to a surface box where a bonding jumper from the grounded metal box to the receptacle is not required provided at least one of the insulating washers is removed.*

***EXHIBIT 250.51*** *An example of a cover-mounted receptacle attached to a surface box where a bonding jumper from the grounded metal box to the receptacle is not required.*

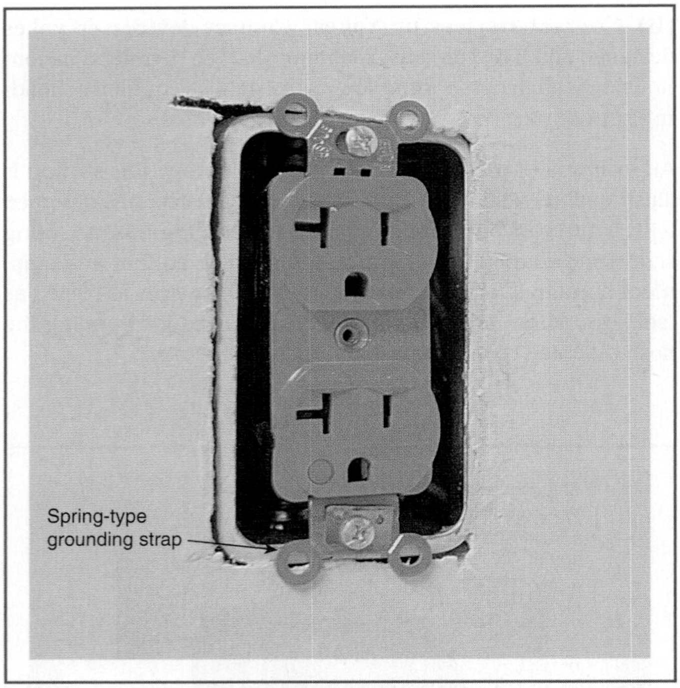


***EXHIBIT 250.52*** *A receptacle designed with a listed spring-type grounding strap. The clip holding the bottom mounting screw captive establishes a grounding circuit and eliminates the need to provide a wire-type equipment bonding jumper from the grounded metal box to the receptacle.*

**(C) Floor Boxes.** Floor boxes designed for and listed as providing satisfactory ground continuity between the box and the device shall be permitted.

**(D) Isolated Ground Receptacles.** Where installed for the reduction of electrical noise (electromagnetic interference) on

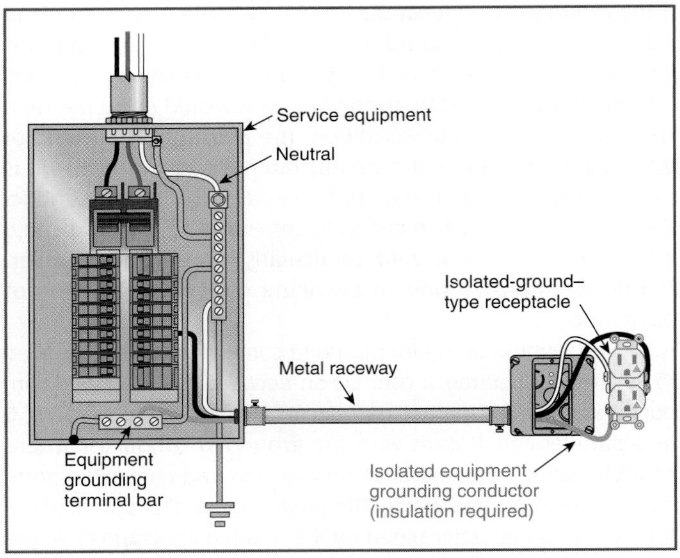


***EXHIBIT 250.53*** *An isolated-ground–type receptacle with an insulated EGC and the device box grounded through the metal raceway.*

the grounding circuit, a receptacle in which the grounding terminal is purposely insulated from the receptacle mounting means shall be permitted. The receptacle grounding terminal shall be connected to an insulated equipment grounding conductor run with the circuit conductors. This equipment grounding conductor shall be permitted to pass through one or more panelboards without a connection to the panelboard grounding terminal bar as permitted in 408.40, Exception, so as to terminate within the same building or structure directly at an equipment grounding conductor terminal of the applicable derived system or service. Where installed in accordance with the provisions of this section, this equipment grounding conductor shall also be permitted to pass through boxes, wireways, or other enclosures without being connected to such enclosures.

Informational Note: Use of an isolated equipment grounding conductor does not relieve the requirement for grounding the raceway system and outlet box.

An isolated-ground–type receptacle is permitted to be installed without a bonding jumper between the metal device box and the receptacle grounding terminal. However, the isolated EGC must provide an effective path for ground-fault current between the receptacle grounding terminal and the source of the branch circuit supplying the receptacle. An insulated EGC, as shown in Exhibit 250.53, is installed with the branch-circuit conductors. This conductor may originate in the service panel, pass through any number of subpanels without being connected to the equipment grounding bus, and terminate at the isolated-ground–type receptacle ground terminal. Termination of the isolated EGC at the service is not necessary. It may be terminated at any of the intervening panelboards. The objective is to terminate it where the noise is eliminated.

This isolated EGC arrangement does not exempt the metal device box from being grounded. The metal device box must be grounded either by an EGC run with the circuit conductors or by a wiring method that serves as an EGC.

According to 250.146(D), where isolated-ground–type receptacles are used, the isolated EGC can terminate at an equipment grounding terminal of the applicable service or derived system in the same building as the receptacle. If the isolated EGC terminates at a separate building, a large voltage difference may exist between buildings during lightning transients. Such transients could cause damage to equipment connected to an isolated-ground–type receptacle and present a shock hazard between the isolated equipment frame and other grounded surfaces.

The informational note to 250.146(D) is a reminder that metal raceways and boxes are required to be grounded by one of the EGC types specified in 250.118. A wire-type EGC, in addition to the isolated EGC, could be installed where a nonmetallic raceway is used and the isolated ground receptacle is installed in a metal outlet box that is required to be grounded per 250.148. Similarly, an additional wire-type EGC could be installed where metal outlet boxes are used and flexible metal conduit installed in lengths exceeding 6 feet is the wiring method. Where an ordinary grounding-type receptacle is being replaced with an isolated-ground–type receptacle, use of an existing insulated EGC as the isolated EGC could effectively defeat or seriously compromise the required grounding of the box and raceway if the wiring method does not also qualify as an EGC per 250.118.

### 250.148 Continuity and Attachment of Equipment Grounding Conductors to Boxes

Where circuit conductors are spliced within a box, or terminated on equipment within or supported by a box, any equipment grounding conductor(s) associated with those circuit conductors shall be connected within the box or to the box with devices suitable for the use in accordance with 250.148(A) through (E).

*Exception: The equipment grounding conductor permitted in 250.146(D) shall not be required to be connected to the other equipment grounding conductors or to the box.*

**(A) Connections.** Connections and splices shall be made in accordance with 110.14(B) except that insulation shall not be required.

**(B) Grounding Continuity.** The arrangement of grounding connections shall be such that the disconnection or the removal of a receptacle, luminaire, or other device fed from the box does not interfere with or interrupt the grounding continuity.

**(C) Metal Boxes.** A connection shall be made between the one or more equipment grounding conductors and a metal box by means of a grounding screw that shall be used for no other purpose, equipment listed for grounding, or a listed grounding device.

Where a metal box is used in a metal raceway system and a wire-type EGC is installed in the raceway, 250.148 does not require that the wire-type EGC be connected to the box if the box is grounded by the metal raceway and the circuit conductors are not spliced or terminated to equipment in the metal box. A metal box used as a point at which to pull conductors into the raceway system is covered by this section, as long as the conductors are not spliced or otherwise terminated in the box. In this respect, the box is treated the same as a metal conduit body such as an "L" or "T" type installed to provide a conductor pull point in a conduit or tubing system.

**(D) Nonmetallic Boxes.** One or more equipment grounding conductors brought into a nonmetallic outlet box shall be arranged such that a connection can be made to any fitting or device in that box requiring grounding.

**(E) Solder.** Connections depending solely on solder shall not be used.

## VIII. Direct-Current Systems

### 250.160 General

Direct-current systems shall comply with Part VIII and other sections of Article 250 not specifically intended for ac systems.

### 250.162 Direct-Current Circuits and Systems to Be Grounded

Direct-current circuits and systems shall be grounded as provided for in 250.162(A) and (B).

**(A) Two-Wire, Direct-Current Systems.** A 2-wire, dc system supplying premises wiring and operating at greater than 60 volts but not greater than 300 volts shall be grounded.

*Exception No. 1: A system equipped with a ground detector and supplying only industrial equipment in limited areas shall not be required to be grounded where installed adjacent to or integral with the source of supply.*

*Exception No. 2: A rectifier-derived dc system supplied from an ac system complying with 250.20 shall not be required to be grounded.*

*Exception No. 3: Direct-current fire alarm circuits having a maximum current of 0.030 ampere as specified in Article 760, Part III, shall not be required to be grounded.*

**(B) Three-Wire, Direct-Current Systems.** The neutral conductor of all 3-wire, dc systems supplying premises wiring shall be grounded.

### 250.164 Point of Connection for Direct-Current Systems

**(A) Off-Premises Source.** Direct-current systems to be grounded and supplied from an off-premises source shall have the grounding connection made at one or more supply stations.

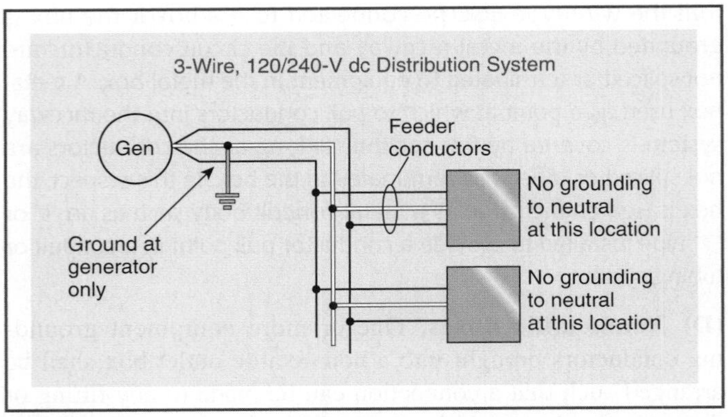


*EXHIBIT 250.54 A 3-wire, 120/240-volt dc distribution system with the neutral grounded at the off-premises generator site.*

A grounding connection shall not be made at individual services or at any point on the premises wiring.

As shown in Exhibit 250.54, the neutral of the 3-wire dc distribution system is grounded at the off-premises generator site. Grounding of a 2-wire dc system would be accomplished in the same manner. For an on-premises generator, a grounding connection is required to be located at the source of the first system disconnecting means or OCPD. Other equivalent means that use equipment listed and identified for such use are permitted.

**(B) On-Premises Source.** Where the dc system source is located on the premises, a grounding connection shall be made at one of the following:

(1) The source
(2) The first system disconnection means or overcurrent device
(3) By other means that accomplish equivalent system protection and that utilize equipment listed and identified for the use

## 250.166 Size of the Direct-Current Grounding Electrode Conductor

The size of the grounding electrode conductor for a dc system shall be as specified in 250.166(A) and (B), except as permitted by 250.166(C) through (E). The grounding electrode conductor for a dc system shall meet the sizing requirements in this section but shall not be required to be larger than 3/0 copper or 250 kcmil aluminum.

**(A) Not Smaller Than the Neutral Conductor.** Where the dc system consists of a 3-wire balancer set or a balancer winding with overcurrent protection as provided in 445.12(D), the grounding electrode conductor shall not be smaller than the neutral conductor and not smaller than 8 AWG copper or 6 AWG aluminum.

**(B) Not Smaller Than the Largest Conductor.** Where the dc system is other than as in 250.166(A), the grounding electrode conductor shall not be smaller than the largest conductor supplied by the system, and not smaller than 8 AWG copper or 6 AWG aluminum.

**(C) Connected to Rod, Pipe, or Plate Electrodes.** Where connected to rod, pipe, or plate electrodes as in 250.52(A)(5) or (A)(7), that portion of the grounding electrode conductor that is the sole connection to the grounding electrode shall not be required to be larger than 6 AWG copper wire or 4 AWG aluminum wire.

**(D) Connected to a Concrete-Encased Electrode.** Where connected to a concrete-encased electrode as in 250.52(A)(3), that portion of the grounding electrode conductor that is the sole connection to the grounding electrode shall not be required to be larger than 4 AWG copper wire.

**(E) Connected to a Ground Ring.** Where connected to a ground ring as in 250.52(A)(4), that portion of the grounding electrode conductor that is the sole connection to the grounding electrode shall not be required to be larger than the conductor used for the ground ring.

## 250.167 Direct-Current Ground-Fault Detection

**(A) Ungrounded Systems.** Ground-fault detection systems shall be required for ungrounded systems.

**(B) Grounded Systems.** Ground-fault detection shall be permitted for grounded systems.

Some dc applications cannot utilize a grounded system. In these cases , a "floating" power is necessary. An unintentional ground can result in fires or shock hazard. Grounded systems are not all grounded in the same manner, and ground fault detection would not be appropriate for all types of systems.

**(C) Marking.** Direct-current systems shall be legibly marked to indicate the grounding type at the dc source or the first disconnecting means of the system. The marking shall be of sufficient durability to withstand the environment involved.

Informational Note: *NFPA 70E-2012* identifies four dc grounding types in detail.

## 250.168 Direct-Current System Bonding Jumper

For direct-current systems that are to be grounded, an unspliced bonding jumper shall be used to connect the equipment grounding conductor(s) to the grounded conductor at the source or the first system disconnecting means where the system is grounded. The size of the bonding jumper shall not be smaller than the system grounding electrode conductor specified in 250.166 and shall comply with the provisions of 250.28(A), (B), and (C).

### 250.169 Ungrounded Direct-Current Separately Derived Systems

Except as otherwise permitted in 250.34 for portable and vehicle-mounted generators, an ungrounded dc separately derived system supplied from a stand-alone power source (such as an engine–generator set) shall have a grounding electrode conductor connected to an electrode that complies with Part III of this article to provide for grounding of metal enclosures, raceways, cables, and exposed non–current-carrying metal parts of equipment. The grounding electrode conductor connection shall be to the metal enclosure at any point on the separately derived system from the source to the first system disconnecting means or overcurrent device, or it shall be made at the source of a separately derived system that has no disconnecting means or overcurrent devices.

The size of the grounding electrode conductor shall be in accordance with 250.166.

## IX. Instruments, Meters, and Relays

### 250.170 Instrument Transformer Circuits

Secondary circuits of current and potential instrument transformers shall be grounded where the primary windings are connected to circuits of 300 volts or more to ground and, where installed on or in switchgear and on switchboards, shall be grounded irrespective of voltage.

*Exception No. 1: Circuits where the primary windings are connected to circuits of 1000 volts or less with no live parts or wiring exposed or accessible to other than qualified persons.*

*Exception No. 2: Current transformer secondaries connected in a three-phase delta configuration shall not be required to be grounded.*

### 250.172 Instrument Transformer Cases

Cases or frames of instrument transformers shall be connected to the equipment grounding conductor where accessible to other than qualified persons.

*Exception: Cases or frames of current transformers, the primaries of which are not over 150 volts to ground and that are used exclusively to supply current to meters.*

### 250.174 Cases of Instruments, Meters, and Relays Operating at 1000 Volts or Less

Instruments, meters, and relays operating with windings or working parts at 1000 volts or less shall be connected to the equipment grounding conductor as specified in 250.174(A), (B), or (C).

**(A) Not on Switchgear or Switchboards.** Instruments, meters, and relays not located on switchgear or switchboards operating with windings or working parts at 300 volts or more to ground, and accessible to other than qualified persons, shall have the cases and other exposed metal parts connected to the equipment grounding conductor.

**(B) On Switchgear or Dead-Front Switchboards.** Instruments, meters, and relays (whether operated from current and potential transformers or connected directly in the circuit) on switchgear or switchboards having no live parts on the front of the panels shall have the cases connected to the equipment grounding conductor.

**(C) On Live-Front Switchboards.** Instruments, meters, and relays (whether operated from current and potential transformers or connected directly in the circuit) on switchboards having exposed live parts on the front of panels shall not have their cases connected to the equipment grounding conductor. Mats of insulating rubber or other suitable floor insulation shall be provided for the operator where the voltage to ground exceeds 150.

### 250.176 Cases of Instruments, Meters, and Relays — Operating at 1000 Volts and Over

Where instruments, meters, and relays have current-carrying parts of 1000 volts and over to ground, they shall be isolated by elevation or protected by suitable barriers, grounded metal, or insulating covers or guards. Their cases shall not be connected to the equipment grounding conductor.

*Exception: Cases of electrostatic ground detectors where the internal ground segments of the instrument are connected to the instrument case and grounded and the ground detector is isolated by elevation.*

### 250.178 Instrument Equipment Grounding Conductor

The equipment grounding conductor for secondary circuits of instrument transformers and for instrument cases shall not be smaller than 12 AWG copper or 10 AWG aluminum. Cases of instrument transformers, instruments, meters, and relays that are mounted directly on grounded metal surfaces of enclosures or grounded metal of switchgear or switchboard panels shall be considered to be grounded, and no additional equipment grounding conductor shall be required.

## X. Grounding of Systems and Circuits of over 1000 Volts

### 250.180 General

Where systems over 1000 volts are grounded, they shall comply with all applicable provisions of the preceding sections of this article and with 250.182 through 250.194, which supplement and modify the preceding sections.

The general requirements in Parts I through IX for systems operating under 1 kV also apply to systems 1 kV and over, except as modified in Part X. As a general rule, Tables 250.66 (for sizing grounding electrode conductors) and 250.122 (for sizing EGCs) apply to systems operating over 1 kV; however, special requirements are in Part X for EGCs for electrical systems utilizing shielded solid dielectric insulated cables rated 2001 to 35,000 volts. These electrical systems are commonly referred to as medium- and high-voltage systems.

## 250.182 Derived Neutral Systems

A system neutral point derived from a grounding transformer shall be permitted to be used for grounding systems over 1 kV.

## 250.184 Solidly Grounded Neutral Systems

Solidly grounded neutral systems shall be permitted to be either single point grounded or multigrounded neutral.

**(A) Neutral Conductor.**

**(1) Insulation Level.** The minimum insulation level for neutral conductors of solidly grounded systems shall be 600 volts.

*Exception No. 1: Bare copper conductors shall be permitted to be used for the neutral conductor of the following:*

*(1) Service-entrance conductors*
*(2) Service laterals or underground service conductors*
*(3) Direct-buried portions of feeders*

*Exception No. 2: Bare conductors shall be permitted for the neutral conductor of overhead portions installed outdoors.*

*Exception No. 3: The grounded neutral conductor shall be permitted to be a bare conductor if isolated from phase conductors and protected from physical damage.*

Informational Note: See 225.4 for conductor covering where within 3.0 m (10 ft) of any building or other structure.

**(2) Ampacity.** The neutral conductor shall be of sufficient ampacity for the load imposed on the conductor but not less than 33½ percent of the ampacity of the phase conductors.

*Exception: In industrial and commercial premises under engineering supervision, it shall be permissible to size the ampacity of the neutral conductor to not less than 20 percent of the ampacity of the phase conductor.*

**(B) Single-Point Grounded Neutral System.** Where a single-point grounded neutral system is used, the following shall apply:

(1) A single-point grounded neutral system shall be permitted to be supplied from (a) or (b):
  a. A separately derived system
  b. A multigrounded neutral system with an equipment grounding conductor connected to the multigrounded neutral conductor at the source of the single-point grounded neutral system
(2) A grounding electrode shall be provided for the system.
(3) A grounding electrode conductor shall connect the grounding electrode to the system neutral conductor.
(4) A bonding jumper shall connect the equipment grounding conductor to the grounding electrode conductor.
(5) An equipment grounding conductor shall be provided to each building, structure, and equipment enclosure.
(6) A neutral conductor shall only be required where phase-to-neutral loads are supplied.
(7) The neutral conductor, where provided, shall be insulated and isolated from earth except at one location.
(8) An equipment grounding conductor shall be run with the phase conductors and shall comply with (a), (b), and (c):
  a. Shall not carry continuous load
  b. May be bare or insulated
  c. hall have sufficient ampacity for fault current duty

Circuits supplied from a single-point grounded system are required to have an EGC run with the circuit conductors. This conductor is not to be used as a conductor for continuous line-to-neutral loads.

**(C) Multigrounded Neutral Systems.** Where a multigrounded neutral system is used, the following shall apply:

(1) The neutral conductor of a solidly grounded neutral system shall be permitted to be grounded at more than one point. Grounding shall be permitted at one or more of the following locations:
  a. Transformers supplying conductors to a building or other structure
  b. Underground circuits where the neutral conductor is exposed
  c. Overhead circuits installed outdoors
(2) The multigrounded neutral conductor shall be grounded at each transformer and at other additional locations by connection to a grounding electrode.
(3) At least one grounding electrode shall be installed and connected to the multigrounded neutral conductor every 400 m (1300 ft).
(4) The maximum distance between any two adjacent electrodes shall not be more than 400 m (1300 ft).
(5) In a multigrounded shielded cable system, the shielding shall be grounded at each cable joint that is exposed to personnel contact.

## 250.186 Ground-Fault Circuit Conductor Brought to Service Equipment

**(A) Systems with a Grounded Conductor at the Service Point.** Where an ac system operating at over 1000 volts is grounded at any point and is provided with a grounded conductor at the service point, a grounded conductor(s) shall be installed and routed with the ungrounded conductors to each service disconnecting means and shall be connected to each disconnecting means grounded conductor(s) terminal or bus. A main bonding jumper shall connect the grounded conductor(s) to each service disconnecting means enclosure. The grounded conductor(s) shall be installed in accordance with 250.186(A)(1) through (A)(4). The size of the solidly grounded circuit conductor(s) shall be the larger of that determined by 250.184 or 250.186(A)(1) or (A)(2).

*Exception: Where two or more service disconnecting means are located in a single assembly listed for use as service equipment, it shall be permitted to connect the grounded conductor(s) to the assembly common grounded conductor(s) terminal or bus. The assembly shall include a main bonding jumper for connecting the grounded conductor(s) to the assembly enclosure.*

**(1) Sizing for a Single Raceway or Overhead Conductor.** The grounded conductor shall not be smaller than the required grounding electrode conductor specified in Table 250.66 but shall not be required to be larger than the largest ungrounded service-entrance conductor(s). In addition, for sets of ungrounded service-entrance conductors larger than 1100 kcmil copper or 1750 kcmil aluminum, the grounded conductor shall not be smaller than 12½ percent of the circular mil area of the largest set of service-entrance ungrounded conductor(s).

**(2) Parallel Conductors in Two or More Raceways or Overhead Conductors.** If the ungrounded service-entrance conductors are installed in parallel in two or more raceways or as overhead parallel conductors, the grounded conductors shall also be installed in parallel. The size of the grounded conductor in each raceway or overhead shall be based on the total circular mil area of the parallel ungrounded conductors in the raceway or overhead, as indicated in 250.186(A)(1), but not smaller than 1/0 AWG.

Informational Note: See 310.10(H) for grounded conductors connected in parallel.

**(3) Delta-Connected Service.** The grounded conductor of a 3-phase, 3-wire delta service shall have an ampacity not less than that of the ungrounded conductors.

**(4) Impedance Grounded Neutral Systems.** Impedance grounded neutral systems shall be installed in accordance with 250.187.

**(B) Systems Without a Grounded Conductor at the Service Point.** Where an ac system operating at greater than 1000 volts is grounded at any point and is not provided with a grounded conductor at the service point, a supply-side bonding jumper shall be installed and routed with the ungrounded conductors to each service disconnecting means and shall be connected to each disconnecting means equipment grounding conductor terminal or bus. The supply-side bonding jumper shall be installed in accordance with 250.186(B)(1) through (B)(3).

*Exception: Where two or more service disconnecting means are located in a single assembly listed for use as service equipment, it shall be permitted to connect the supply-side bonding jumper to the assembly common equipment grounding terminal or bus.*

**(1) Sizing for a Single Raceway or Overhead Conductor.** The supply-side bonding jumper shall not be smaller than the required grounding electrode conductor specified in Table 250.66 but shall not be required to be larger than the largest ungrounded service-entrance conductor(s). In addition, for sets of ungrounded service-entrance conductors larger than 1100 kcmil copper or 1750 kcmil aluminum, the supply-side bonding jumper shall not be smaller than 12½ percent of the circular mil area of the largest set of service-entrance ungrounded conductor(s).

**(2) Parallel Conductors in Two or More Raceways or Overhead Conductors.** If the ungrounded service-entrance conductors are installed in parallel in two or more raceways or overhead conductors, the supply-side bonding jumper shall also be installed in parallel. The size of the supply-side bonding jumper in each raceway or overhead shall be based on the total circular mil area of the parallel ungrounded conductors in the raceway or overhead, as indicated in 250.186(A)(1), but not smaller than 1/0 AWG.

**(3) Impedance Grounded Neutral Systems.** Impedance grounded neutral systems shall be installed in accordance with 250.187.

### 250.187 Impedance Grounded Neutral Systems

Impedance grounded neutral systems in which a grounding impedance, usually a resistor, limits the ground-fault current shall be permitted where all of the following conditions are met:

(1) The conditions of maintenance and supervision ensure that only qualified persons service the installation.
(2) Ground detectors are installed on the system.
(3) Line-to-neutral loads are not served.

Impedance grounded neutral systems shall comply with the provisions of 250.187(A) through (D).

**(A) Location.** The grounding impedance shall be inserted in the grounding electrode conductor between the grounding electrode of the supply system and the neutral point of the supply transformer or generator.

**(B) Identified and Insulated.** The neutral conductor of an impedance grounded neutral system shall be identified, as well as fully insulated with the same insulation as the phase conductors.

**(C) System Neutral Conductor Connection.** The system neutral conductor shall not be connected to ground, except through the neutral grounding impedance.

**(D) Equipment Grounding Conductors.** Equipment grounding conductors shall be permitted to be bare and shall be electrically connected to the ground bus and grounding electrode conductor.

## 250.188 Grounding of Systems Supplying Portable or Mobile Equipment

Systems supplying portable or mobile equipment over 1000 volts, other than substations installed on a temporary basis, shall comply with 250.188(A) through (F).

**(A) Portable or Mobile Equipment.** Portable or mobile equipment over 1000 volts shall be supplied from a system having its neutral conductor grounded through an impedance. Where a delta-connected system over 1000 volts is used to supply portable or mobile equipment, a system neutral point and associated neutral conductor shall be derived.

"Portable" describes equipment that is easily carried from one location to another. "Mobile" describes equipment that is easily moved on wheels, treads, skids, or similar means.

**(B) Exposed Non–Current-Carrying Metal Parts.** Exposed non–current-carrying metal parts of portable or mobile equipment shall be connected by an equipment grounding conductor to the point at which the system neutral impedance is grounded.

**(C) Ground-Fault Current.** The voltage developed between the portable or mobile equipment frame and ground by the flow of maximum ground-fault current shall not exceed 100 volts.

**(D) Ground-Fault Detection and Relaying.** Ground-fault detection and relaying shall be provided to automatically de-energize any component of a system over 1000 volts that has developed a ground fault. The continuity of the equipment grounding conductor shall be continuously monitored so as to automatically de-energize the circuit of the system over 1000 volts to the portable or mobile equipment upon loss of continuity of the equipment grounding conductor.

**(E) Isolation.** The grounding electrode to which the portable or mobile equipment system neutral impedance is connected shall be isolated from and separated in the ground by at least 6.0 m (20 ft) from any other system or equipment grounding electrode, and there shall be no direct connection between the grounding electrodes, such as buried pipe and fence, and so forth.

**(F) Trailing Cable and Couplers.** Trailing cable and couplers of systems over 1000 volts for interconnection of portable or mobile equipment shall meet the requirements of Part III of Article 400 for cables and 490.55 for couplers.

## 250.190 Grounding of Equipment

**(A) Equipment Grounding.** All non–current-carrying metal parts of fixed, portable, and mobile equipment and associated fences, housings, enclosures, and supporting structures shall be grounded.

*Exception: Where isolated from ground and located such that any person in contact with ground cannot contact such metal parts when the equipment is energized, the metal parts shall not be required to be grounded.*

Informational Note: See 250.110, Exception No. 2, for pole-mounted distribution apparatus.

**(B) Grounding Electrode Conductor.** If a grounding electrode conductor connects non–current-carrying metal parts to ground, the grounding electrode conductor shall be sized in accordance with Table 250.66, based on the size of the largest ungrounded service, feeder, or branch-circuit conductors supplying the equipment. The grounding electrode conductor shall not be smaller than 6 AWG copper or 4 AWG aluminum.

**(C) Equipment Grounding Conductor.** Equipment grounding conductors shall comply with 250.190(C)(1) through (C)(3).

**(1) General.** Equipment grounding conductors that are not an integral part of a cable assembly shall not be smaller than 6 AWG copper or 4 AWG aluminum.

**(2) Shielded Cables.** The metallic insulation shield encircling the current carrying conductors shall be permitted to be used as an equipment grounding conductor, if it is rated for clearing time of ground fault current protective device operation without damaging the metallic shield. The metallic tape insulation shield and drain wire insulation shield shall not be used as an equipment grounding conductor for solidly grounded systems.

Shields comprised of copper tape and drain wires cannot be used as the EGC in solidly grounded systems. The use of shielded cables is specified in 310.10(E). These requirements are based on the system voltage, installation conditions, and cable construction. As explained in the informational note to 310.10(E), the cable shield provides several different functions related to the safe operation of insulated cables used in medium- and high-voltage systems. Grounding and bonding are included in those functions.

Exhibit 250.55 shows three different types of single-conductor shielded cable construction. Shielded cables are also available in multiconductor configurations such as Type MV (Article 328) or Type MC (Article 330) cables. Where the ground-fault current is relatively low (as in impedance grounded neutral systems), the metallic shield of any of the cable types pictured in Exhibit 250.55 is permitted to serve as the EGC if it is rated for clearing time of ground-fault current without being damaged. Cable manufacturers can provide permissible short-circuit currents for a metallic shield based on fault clearing time of the OCPD.

Where the system is solidly grounded, neither the metallic tape insulation shield (top cable shown in Exhibit 250.55) nor the drain wire insulation shield (middle cable shown) can be used as the EGC, because they do not have sufficient circular mil area to provide the effective ground-fault return path required by 250.4(A)(5). A metallic insulation shield encircling the conductor (bottom cable shown),

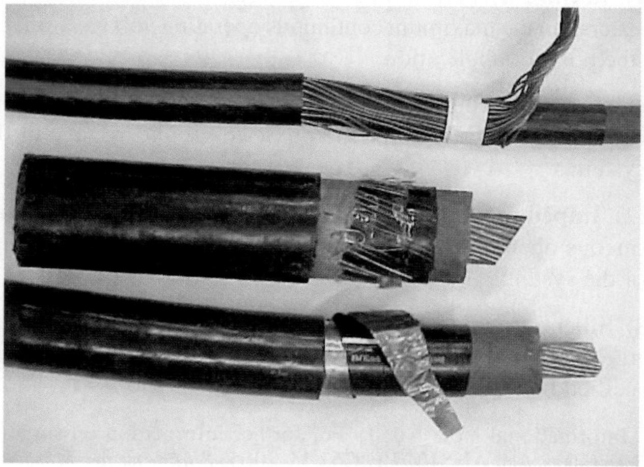

*EXHIBIT 250.55 Three examples of single-conductor shielded cable, each having a different type of shielding. (Courtesy of Chuck Mello)*

commonly called concentric neutral cable, has larger conductor strands in the concentric wrap. Because of its larger overall circular mil area, this concentric neutral cable is permitted to be used as an EGC in a solidly grounded system if the metallic shield will not be damaged during the time it takes to open the circuit OCPD. The copper metallic tape is typically 5 mils thick and is helically applied with a 12.5 percent or larger overlap over the insulation shield. Drain wires are typically 24 AWG bare copper wires. The concentric neutral wires are normally bare or tinned wires with the conductance equivalent of neutral-to-phase conductor for single-conductor cables and typically one-third of the phase conductor for 3-conductor cables.

Where the cable shield cannot carry ground-fault current without damage, a separate EGC must be installed. A separate EGC would also be required where tape or drain-wire–type shields are not permitted to carry fault current. The EGC can be integral to a cable assembly, can be run as a separate conductor in a raceway or cable tray, or can be a wiring method–type EGC of one of the types specified in 250.118, such as rigid metal conduit or intermediate metal conduit. Wire-type EGCs are sized in accordance with 250.122. See the informational note to 250.190(C)(3) for more information on how the rating of OCPDs used in systems operating over 1000 volts is determined.

**(3) Sizing.** Equipment grounding conductors shall be sized in accordance with Table 250.122 based on the current rating of the fuse or the overcurrent setting of the protective relay.

Informational Note: The overcurrent rating for a circuit breaker is the combination of the current transformer ratio and the current pickup setting of the protective relay.

This requirement applies to EGCs that are separately installed as specified in 250.190(C)(1). It also applies to a conductor installed in a cable assembly, other than the cable shield, that is used as an EGC. An EGC contained within a cable assembly may be a single conductor, or it may be sectioned (comprised of multiple conductors within the cable jacket or sheath to form a single EGC) as permitted by 310.10(H)(5).

## 250.191 Grounding System at Alternating-Current Substations

For ac substations, the grounding system shall be in accordance with Part III of Article 250.

Informational Note: For further information on outdoor ac substation grounding, see ANSI/IEEE 80-2000, *IEEE Guide for Safety in AC Substation Grounding*.

## 250.194 Grounding and Bonding of Fences and Other Metal Structures

Metallic fences enclosing, and other metal structures in or surrounding, a substation with exposed electrical conductors and equipment shall be grounded and bonded to limit step, touch, and transfer voltages.

**(A) Metal Fences.** Where metal fences are located within 5 m (16 ft) of the exposed electrical conductors or equipment, the fence shall be bonded to the grounding electrode system with wire-type bonding jumpers as follows:

(1) Bonding jumpers shall be installed at each fence corner and at maximum 50 m (160 ft) intervals along the fence.
(2) Where bare overhead conductors cross the fence, bonding jumpers shall be installed on each side of the crossing.
(3) Gates shall be bonded to the gate support post, and each gate support post shall be bonded to the grounding electrode system.
(4) Any gate or other opening in the fence shall be bonded across the opening by a buried bonding jumper.
(5) The grounding grid or grounding electrode systems shall be extended to cover the swing of all gates.
(6) The barbed wire strands above the fence shall be bonded to the grounding electrode system.

Alternate designs performed under engineering supervision shall be permitted for grounding or bonding of metal fences.

Informational Note No. 1: A nonconducting fence or section may provide isolation for transfer of voltage to other areas.

Informational Note No. 2: See IEEE 80-2000, IEEE Guide for Safety In AC Substation Grounding, for design and installation of fence grounding.

**(B) Metal Structures.** All exposed conductive metal structures, including guy wires within 2.5 m (8 ft) vertically or 5 m (16 ft) horizontally of exposed conductors or equipment and subject to contact by persons, shall be bonded to the grounding electrode systems in the area.

For reasons of security and economics, metal fences are often built around substations. These fences must be grounded to limit the rise of hazardous potential on the fence. This section establishes basic prescriptive requirements for grounding and bonding of metal fences built in and around substations. For situations where step and touch potential considerations indicate that additional grounding and bonding design is required, alternate designs performed under engineering supervision are permitted.

# ARTICLE 280
# Surge Arresters, Over 1000 Volts

## I. General

### 280.1 Scope

This article covers general requirements, installation requirements, and connection requirements for surge arresters installed on premises wiring systems over 1000 volts.

Voltage surges with peaks of several thousand volts, even on 120 V circuits, are not uncommon. These surges occur because of induced voltages in power and transmission lines resulting from lightning strikes in the vicinity of the line. Surges also occur as a result of switching inductive circuits on the premises. Surge arresters for installation as part of a premises wiring system are commercially available. The basic standard on surge arresters is ANSI/IEEE C62.11, *Standard for Metal-Oxide Surge Arresters for Alternating-Current Power Systems (> 1 kV).*

### 280.2 Uses Not Permitted

A surge arrester shall not be installed where the rating of the surge arrester is less than the maximum continuous phase-to-ground power frequency voltage available at the point of application.

### 280.3 Number Required

Where used at a point on a circuit, a surge arrester shall be connected to each ungrounded conductor. A single installation of such surge arresters shall be permitted to protect a number of interconnected circuits, provided that no circuit is exposed to surges while disconnected from the surge arresters.

Means must be provided for protection of circuits that may be disconnected from the generating station bus, because the circuits could still be exposed to surges from lightning. A switch with double-throw action used to disconnect the outside circuits from the station generator and alternatively connect those circuits to ground would satisfy the condition of a single set of arresters protecting more than one circuit.

### 280.4 Surge Arrester Selection

The surge arresters shall comply with 280.4(A) and (B).

**(A) Rating.** The rating of a surge arrester shall be equal to or greater than the maximum continuous operating voltage available at the point of application.

**(1) Solidly Grounded Systems.** The maximum continuous operating voltage shall be the phase-to-ground voltage of the system.

**(2) Impedance or Ungrounded System.** The maximum continuous operating voltage shall be the phase-to-phase voltage of the system.

**(B) Silicon Carbide Types.** The rating of a silicon carbide-type surge arrester shall be not less than 125 percent of the rating specified in 280.4(A).

Informational Note No. 1: For further information on surge arresters, see ANSI/IEEE C62.11-2005, *Standard for Metal-Oxide Surge Arresters for Alternating-Current Power Circuits (>1 kV)*; and ANSI/IEEE C62.22-2009, *Guide for the Application of Metal-Oxide Surge Arresters for Alternating-Current Systems.*

Informational Note No. 2: The selection of a properly rated metal oxide arrester is based on considerations of maximum continuous operating voltage and the magnitude and duration of overvoltages at the arrester location as affected by phase-to-ground faults, system grounding techniques, switching surges, and other causes. See the manufacturer's application rules for selection of the specific arrester to be used at a particular location.

## II. Installation

### 280.11 Location

Surge arresters shall be permitted to be located indoors or outdoors. Surge arresters shall be made inaccessible to unqualified persons, unless listed for installation in accessible locations.

Maximum protection is achieved where the surge protective device is located as close as practicable to the equipment being protected. When a surge passes through an arrester, a wave is reflected in both directions on the conductors connected to the surge arrester. The magnitude of the reflected wave increases as the distance from the arrester increases. If the length of the conductor between the protected equipment and the surge arrester is short, the magnitude of the wave reflected through the equipment is minimized.

### 280.12 Routing of Surge Arrester Grounding Conductors

The conductor used to connect the surge arrester to line, bus, or equipment and to a grounding conductor connection point as provided in 280.21 shall not be any longer than necessary and shall avoid unnecessary bends.

## III. Connecting Surge Arresters

### 280.21 Connection

The arrester shall be connected to one of the following:

(1) Grounded service conductor
(2) Grounding electrode conductor

(3) Grounding electrode for the service
(4) Equipment grounding terminal in the service equipment

### 280.23 Surge-Arrester Conductors

The conductor between the surge arrester and the line and the surge arrester and the grounding connection shall not be smaller than 6 AWG copper or aluminum.

### 280.24 Interconnections

The surge arrester protecting a transformer that supplies a secondary distribution system shall be interconnected as specified in 280.24(A), (B), or (C).

**(A) Metallic Interconnections.** A metallic interconnection shall be made to the secondary grounded circuit conductor or the secondary circuit grounding electrode conductor provided that, in addition to the direct grounding connection at the surge arrester, the following occurs:

**(1) Additional Grounding Connection.** The grounded conductor of the secondary has elsewhere a grounding connection to a continuous metal underground water piping system. In urban water-pipe areas where there are at least four water-pipe connections on the neutral conductor and not fewer than four such connections in each mile of neutral conductor, the metallic interconnection shall be permitted to be made to the secondary neutral conductor with omission of the direct grounding connection at the surge arrester.

**(2) Multigrounded Neutral System Connection.** The grounded conductor of the secondary system is a part of a multigrounded neutral system or static wire of which the primary neutral conductor or static wire has at least four grounding connections in each mile of line in addition to a grounding connection at each service.

**(B) Through Spark Gap or Device.** Where the surge arrester grounding electrode conductor is not connected as in 280.24(A), or where the secondary is not grounded as in 280.24(A) but is otherwise grounded as in 250.52, an interconnection shall be made through a spark gap or listed device as required by 280.24(B)(1) or (B)(2).

**(1) Ungrounded or Unigrounded Primary System.** For ungrounded or unigrounded primary systems, the spark gap or listed device shall have a 60-Hz breakdown voltage of at least twice the primary circuit voltage but not necessarily more than 10 kV, and there shall be at least one other ground on the grounded conductor of the secondary that is not less than 6.0 m (20 ft) distant from the surge-arrester grounding electrode.

**(2) Multigrounded Neutral Primary System.** For multigrounded neutral primary systems, the spark gap or listed device shall have a 60-Hz breakdown of not more than 3 kV, and there shall be at least one other ground on the grounded conductor of the secondary that is not less than 6.0 m (20 ft) distant from the surge-arrester grounding electrode.

**(C) By Special Permission.** An interconnection of the surge-arrester ground and the secondary neutral conductor, other than as provided in 280.24(A) or (B), shall be permitted to be made only by special permission.

### 280.25 Grounding Electrode Conductor Connections and Enclosures

Except as indicated in this article, surge-arrester grounding electrode conductor connections shall be made as specified in Article 250, Parts III and X. Grounding electrode conductors installed in metal enclosures shall comply with 250.64(E).

# ARTICLE 285 Surge-Protective Devices (SPDs), 1000 Volts or Less

## I. General

### 285.1 Scope

This article covers general requirements, installation requirements, and connection requirements for surge-protective devices (SPDs) permanently installed on premises wiring systems of 1000 volts or less.

Informational Note: Surge arresters 1000 volts or less are also known as Type 1 SPDs.

•

Surge-protective devices (SPDs) protect electrical systems operating at 1 kV or less against the effects of surges. Where the SPD will be installed in the premises wiring system determines the selection of appropriate SPD. Historically, surge arresters were installed on the line side of service or other supply system disconnecting means (and were covered by Article 280), while transient voltage surge suppressors (TVSSs) were installed on the load side or downstream of the main disconnecting means (and were covered by Article 285).

Now, the delineation between SPDs covered by Article 280 and those covered by Article 285 is the voltage rating of the supply system. The designations of SPDs are varied, depending on their location in the premises wiring system. For instance, a Type 1 SPD is permitted to be connected on the supply side of the service or building disconnecting means. Type 2 and Type 3 SPDs must be installed on the load side of overcurrent protective devices and are the devices formerly referred to by this article as TVSSs. Two examples of SPDs are shown in Exhibit 285.1.

### 285.3 Uses Not Permitted

An SPD device shall not be installed in the following:

(1) Circuits over 1000 volts

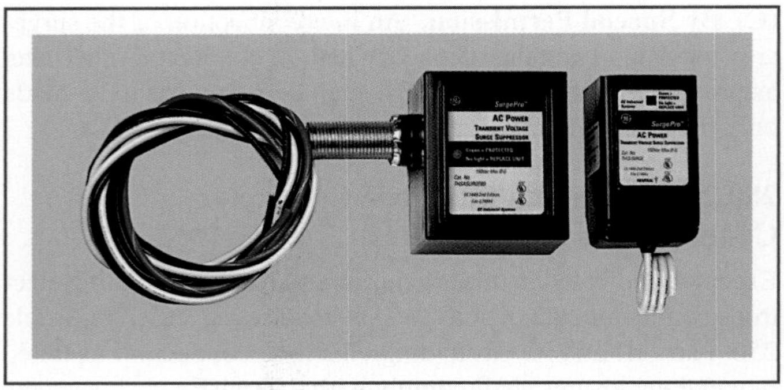

***EXHIBIT 285.1*** *Two SPDs suitable for service-entrance installation, one for direct connection to panelboard busbars and one for mounting in a cabinet or enclosure knockout. (Courtesy of General Electric)*

(2) On ungrounded systems, impedance grounded systems, or corner grounded delta systems unless listed specifically for use on these systems

(3) Where the rating of the SPD is less than the maximum continuous phase-to-ground power frequency voltage available at the point of application

### 285.4 Number Required

Where used at a point on a circuit, the SPD shall be connected to each ungrounded conductor.

### 285.5 Listing

An SPD shall be a listed device.

UL 1449, *Standard for Surge Protective Devices*, covers Types 1, 2, 3, and 4 devices. SPDs are permitted to be installed on ungrounded systems, impedance grounded systems, and corner grounded systems where the device is listed for the specific characteristic of the system per 285.3(2).

### 285.6 Short-Circuit Current Rating

The SPD shall be marked with a short-circuit current rating and shall not be installed at a point on the system where the available fault current is in excess of that rating. This marking requirement shall not apply to receptacles.

In residential and small commercial electrical systems, the first SPD is commonly installed either as an integral component of or near to the service-entrance equipment.

Depending on the system voltage, surge protection in larger commercial and industrial electrical systems can be provided by installing a Type 1 SPD (a surge arrester for systems 1 kV and less) or surge arrester (the devices covered in Article 280 for systems over 1 kV) on the line side of the service equipment. Subsequent levels of SPDs are then provided at intermediate points in the distribution system (such as at panelboards that serve loads

***EXHIBIT 285.2*** *An SPD as an integral component of a receptacle, providing local point-of-use protection of equipment when transient events occur within the facility. (Courtesy of Legrand/Pass & Seymour®)*

susceptible to transients) and at the point where utilization equipment connects to the electrical system.

Point-of-use SPDs such as receptacles and permanently installed power strips may be installed at the equipment (such as computers or equipment with electronic controls). The function of a point-of-use SPD is to remove small transients that pass through the more robust surge devices located at the service. Point-of-use SPD devices are also useful in removing small transients that have been generated within the building. See Exhibit 285.2 for a point-of-use or Type 3 SPD.

## II. Installation

### 285.11 Location

SPDs shall be permitted to be located indoors or outdoors and shall be made inaccessible to unqualified persons, unless listed for installation in accessible locations.

### 285.12 Routing of Connections

The conductors used to connect the SPD to the line or bus and to ground shall not be any longer than necessary and shall avoid unnecessary bends.

In order to optimize performance of SPDs, the length and routing of the conductor that connects the device to ground is limited. High-frequency currents, such as those common to lightning discharges, tend to reduce the effectiveness of a conductor that

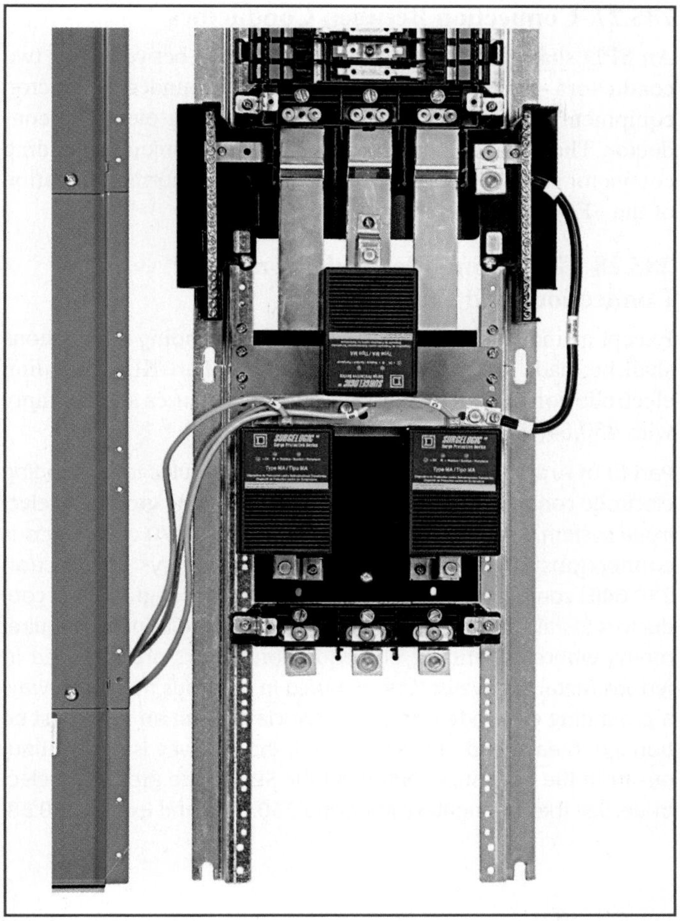

***EXHIBIT 285.3*** *A Type 2 SPD mounted as an integral component of a panelboard, which minimizes conductor length between the electrical system and the SPD. (Courtesy of Schneider Electric)*

connects the device to ground. Short conductors with few bends will have a lower impedance to surge current. Higher impedance drives the clamping voltage higher and reduces the protection provided by the SPD unit. Maximum protection is achieved where the SPD is located as close as practicable to the equipment being protected as shown in Exhibit 285.3.

### 285.13 Type 4 and Other Component Type SPDs

Type 4 component assemblies and other component type SPDs shall only be installed by the equipment manufacturer.

Type 4 and other component-type SPDs are incomplete devices that are only acceptable when installed as part of listed equipment.

## III. Connecting SPDs

### 285.21 Connection

Where an SPD device is installed, it shall comply with 285.23 through 285.28.

The point in the electrical system where SPDs are connected is dependent on the type of SPD. UL 1449, *Standard for Surge Protection Devices,* is the product standard used to evaluate safe performance of SPDs. In accordance with the scope of UL 1449, a Type 2 SPD must be installed on the load side of the service disconnect overcurrent protection, and a Type 3 SPD must be installed on the load side of a branch-circuit overcurrent protective device. The requirement for Type 2 SPDs to be connected on the load side of the first overcurrent protection device in a feeder-supplied structure is necessary due to the exposure of external feeder conductors to a more hostile surge environment such as lightning.

Two requirements in Article 230 permit the installation of SPDs on the line side of the service disconnecting means. First, 230.71(A) permits an additional disconnecting means at the service equipment for SPDs installed as part of listed equipment. The disconnecting means for the SPD does not count as one of the six permitted by 230.71(A) where the SPD and its disconnecting means are provided in the listed equipment by the manufacturer. The second is in 230.82(8) in which Type 2 SPDs installed in listed equipment are permitted to be connected on the line side of the service disconnecting means where the SPD is provided with a disconnecting means and overcurrent protection.

### 285.23 Type 1 SPDs

Type 1 SPDs shall be installed in accordance with 285.23(A) and (B).

**(A) Installation.** Type 1 SPDs shall be installed as follows:

(1) Type 1 SPDs shall be permitted to be connected to the supply side of the service disconnect as permitted in 230.82(4), or

(2) Type 1 SPDs shall be permitted to be connected as specified in 285.24.

**(B) At the Service.** When installed at services, Type 1 SPDs shall be connected to one of the following:

(1) Grounded service conductor
(2) Grounding electrode conductor
(3) Grounding electrode for the service
(4) Equipment grounding terminal in the service equipment

Although four locations for connecting the SPD grounding lead are acceptable, the requirement in 285.12 covering the length and physical routing of the conductor must be followed. See the commentary following 285.12 for more information on the importance of controlling the length and physical routing of the conductor.

### 285.24 Type 2 SPDs

Type 2 SPDs shall be installed in accordance with 285.24(A) through (C).

**(A) Service-Supplied Building or Structure.** Type 2 SPDs shall be connected anywhere on the load side of a service disconnect overcurrent device required in 230.91, unless installed in accordance with 230.82(8).

**(B) Feeder-Supplied Building or Structure.** Type 2 SPDs shall be connected at the building or structure anywhere on the load side of the first overcurrent device at the building or structure.

**(C) Separately Derived System.** The SPD shall be connected on the load side of the first overcurrent device in a separately derived system.

### 285.25 Type 3 SPDs

Type 3 SPDs shall be permitted to be installed on the load side of branch-circuit overcurrent protection up to the equipment served. If included in the manufacturer's instructions, the Type 3 SPD connection shall be a minimum 10 m (30 ft) of conductor distance from the service or separately derived system disconnect.

### 285.26 Conductor Size

Line and grounding conductors shall not be smaller than 14 AWG copper or 12 AWG aluminum.

### 285.27 Connection Between Conductors

An SPD shall be permitted to be connected between any two conductors — ungrounded conductor(s), grounded conductor, equipment grounding conductor, or grounding electrode conductor. The grounded conductor and the equipment grounding conductor shall be interconnected only by the normal operation of the SPD during a surge.

### 285.28 Grounding Electrode Conductor Connections and Enclosures

Except as indicated in this article, SPD grounding connections shall be made as specified in Article 250, Part III. Grounding electrode conductors installed in metal enclosures shall comply with 250.64(E).

Part III of Article 250 contains the connection rules for grounding electrode conductors and other components of the grounding electrode system. Sections 250.64, 250.68, and 250.70 cover various connections within the grounding electrode system. Section 250.64(E) contains the installation requirements for grounding conductors installed in metal raceway, including the bonding requirements where grounding electrode conductors are installed in ferrous metal raceways. Where routed in a ferrous metal raceway, a grounding electrode conductor associated with an SPD must be bonded to each end of the raceway if the raceway is not continuous from the enclosure containing the SPD to the grounding electrode. See the commentary following 250.64(E) and Exhibit 250.28.

# 3 Wiring Methods and Materials

## ARTICLE 300 General Requirements for Wiring Methods and Materials

## I. General Requirements

### 300.1 Scope

**(A) All Wiring Installations.** This article covers general requirements for wiring methods and materials for all wiring installations unless modified by other articles in Chapter 3.

**(B) Integral Parts of Equipment.** The provisions of this article are not intended to apply to the conductors that form an integral part of equipment, such as motors, controllers, motor control centers, or factory assembled control equipment or listed utilization equipment.

Requirements for specific wiring methods can be found in the Chapter 3 article governing that particular wiring method, but the overarching requirements for wiring methods are covered in this first article of Chapter 3. Chapters 5 through 7 modify some of the requirements of Article 300. Chapter 8 is not subject to the requirements of Article 300, except where specifically referenced. Article 300 also covers wiring requirements within boxes, conduit bodies, and fittings. Additional wiring requirements are found in Articles 312 and 314.

Wiring within equipment is not within the scope of this article. Integral wiring of equipment is generally covered by product standards. For example, integral wiring of motors is covered by NEMA MG 1, *Motors and Generators*; of industrial control panels by UL 508A, *Standard for Industrial Control Panels*; and of industrial machinery by NFPA 79, *Electrical Standard for Industrial Machinery*.

**(C) Metric Designators and Trade Sizes.** Metric designators and trade sizes for conduit, tubing, and associated fittings and accessories shall be as designated in Table 300.1(C).

***TABLE 300.1(C)*** *Metric Designators and Trade Sizes*

| Metric Designator | Trade Size |
|---|---|
| 12 | 3/8 |
| 16 | 1/2 |
| 21 | 3/4 |
| 27 | 1 |
| 35 | 1 1/4 |
| 41 | 1 1/2 |
| 53 | 2 |
| 63 | 2 1/2 |
| 78 | 3 |
| 91 | 3 1/2 |
| 103 | 4 |
| 129 | 5 |
| 155 | 6 |

Note: The metric designators and trade sizes are for identification purposes only and are not actual dimensions.

Metric designators are used for traditional trade size threaded conduit. They do not change the physical dimensions or the traditional "NPT-type" threads of the conduit. Metric designators are simply another method of identifying the size of a circular raceway. Table 300.1(C) identifies a distinct metric designator for each circular raceway trade size. The unit of measure has not been included because it reflects a "modular" or "relative" measure rather than an exact dimension. As stated in the table footnote, the metric designators and trade sizes are not actual dimensions.

Each metric designator–sized circular raceway is identical in dimension (including manufacturing tolerances) to its trade size counterpart in Table 4 of Chapter 9. Therefore, the Informative Annex C wire fill tables are applicable to both metric designator and trade size circular raceways.

Threaded joints on circular raceways are a concern. For example, 344.6 requires RMC to be listed and the appropriate product standard is ANSI/UL 6, *Electrical Rigid Metal Conduit – Steel*. Listed conduit must be threaded in accordance with ANSI/ASME B.1.20.1-1983, *Pipe Threads, General Purpose (Inch)*. Therefore, only conduit threaded to the traditional dimension of 3/4-inch taper per foot is acceptable. Simply stated, although conduit with a metric designator is permitted, metric threaded conduit is not permitted by the *NEC*. This aligns with 500.8(E)(2) for example, which states that although metric threads are permitted on equipment, an adapter must be used for connection to conduit.

## 300.2 Limitations

**(A) Voltage.** Wiring methods specified in Chapter 3 shall be used for 1000 volts, nominal, or less where not specifically limited in some section of Chapter 3. They shall be permitted for over 1000 volts, nominal, where specifically permitted elsewhere in this *Code*.

For the 2014 edition, many of the voltage ranges used in the *Code* have changed. Prior to this edition, the wiring methods in Chapter 3 were permitted for installations of 600 volts and less. A few requirements in the *Code* continue to be linked to installations limited to 600 volts. Where an installation involves requirements using different voltage range limits, the limits within each applicable article must be observed.

**(B) Temperature.** Temperature limitation of conductors shall be in accordance with 310.15(A)(3).

Maintaining the temperature of the conductor below its temperature rating depends on environmental factors, as well as not exceeding the termination temperature ratings. See 110.14(C) and its commentary for information on temperature limitations of conductor terminations.

## 300.3 Conductors

**(A) Single Conductors.** Single conductors specified in Table 310.104(A) shall only be installed where part of a recognized wiring method of Chapter 3.

*Exception: Individual conductors shall be permitted where installed as separate overhead conductors in accordance with 225.6.*

Individual insulated conductors, such as THHN, are prohibited from use in other than recognized wiring methods. The exception points out two long-time permissions: allowing individual conductors as overhead spans and as festoon lighting.

**(B) Conductors of the Same Circuit.** All conductors of the same circuit and, where used, the grounded conductor and all equipment grounding conductors and bonding conductors shall be contained within the same raceway, auxiliary gutter, cable tray, cablebus assembly, trench, cable, or cord, unless otherwise permitted in accordance with 300.3(B)(1) through (B)(4).

All circuit conductors of an individual circuit must be grouped in order to reduce inductive heating and to avoid increases in overall circuit impedance. A similar rule is found in 300.5(I).

**(1) Paralleled Installations.** Conductors shall be permitted to be run in parallel in accordance with the provisions of 310.10(H). The requirement to run all circuit conductors within the same raceway, auxiliary gutter, cable tray, trench, cable, or cord shall apply separately to each portion of the paralleled installation, and the equipment grounding conductors shall comply with the provisions of 250.122. Parallel runs in cable tray shall comply with the provisions of 392.20(C).

*Exception: Conductors installed in nonmetallic raceways run underground shall be permitted to be arranged as isolated phase installations. The raceways shall be installed in close proximity, and the conductors shall comply with the provisions of 300.20(B).*

**(2) Grounding and Bonding Conductors.** Equipment grounding conductors shall be permitted to be installed outside a raceway or cable assembly where in accordance with the provisions of 250.130(C) for certain existing installations or in accordance with 250.134(B), Exception No. 2, for dc circuits. Equipment bonding conductors shall be permitted to be installed on the outside of raceways in accordance with 250.102(E).

**(3) Nonferrous Wiring Methods.** Conductors in wiring methods with a nonmetallic or other nonmagnetic sheath, where run in different raceways, auxiliary gutters, cable trays, trenches, cables, or cords, shall comply with the provisions of 300.20(B). Conductors in single-conductor Type MI cable with a nonmagnetic sheath shall comply with the provisions of 332.31. Conductors of single-conductor Type MC cable with a nonmagnetic sheath shall comply with the provisions of 330.31, 330.116, and 300.20(B).

**(4) Enclosures.** Where an auxiliary gutter runs between a column-width panelboard and a pull box, and the pull box includes neutral terminations, the neutral conductors of circuits supplied from the panelboard shall be permitted to originate in the pull box.

**(C) Conductors of Different Systems.**

**(1) 1000 Volts, Nominal, or Less.** Conductors of ac and dc circuits, rated 1000 volts, nominal, or less, shall be permitted to occupy the same equipment wiring enclosure, cable, or raceway. All conductors shall have an insulation rating equal to at least the maximum circuit voltage applied to any conductor within the enclosure, cable, or raceway.

Secondary wiring to electric-discharge lamps of 1000 volts or less, if insulated for the secondary voltage involved, shall be permitted to occupy the same luminaire, sign, or outline lighting enclosure as the branch-circuit conductors.

Informational Note No. 1: See 725.136(A) for Class 2 and Class 3 circuit conductors.

Informational Note No. 2: See 690.4(B) for photovoltaic source and output circuits.

For systems of 1000 volts or less, the maximum circuit voltage in the raceway is what determines the minimum voltage rating required for the insulation of conductors, not the maximum insulation voltage rating of the conductors in the raceway.

Informational Note No. 2 references 690.4(B), which prohibits the location of solar photovoltaic circuits within the same enclosure as conductors of other systems unless separated by a partition. Additionally, 700.10(B) requires that circuit wiring for emergency systems be kept entirely independent of all other wiring and equipment.

**(2) Over 1000 Volts, Nominal.** Conductors of circuits rated over 1000 volts, nominal, shall not occupy the same equipment wiring enclosure, cable, or raceway with conductors of circuits rated 1000 volts, nominal, or less unless otherwise permitted in 300.3(C)(2)(a) through (C)(2)(d).

•

(a) Primary leads of electric-discharge lamp ballasts insulated for the primary voltage of the ballast, where contained within the individual wiring enclosure, shall be permitted to occupy the same luminaire, sign, or outline lighting enclosure as the branch-circuit conductors.

(b) Excitation, control, relay, and ammeter conductors used in connection with any individual motor or starter shall be permitted to occupy the same enclosure as the motor-circuit conductors.

(c) In motors, transformers, switchgear, switchboards, control assemblies, and similar equipment, conductors of different voltage ratings shall be permitted.

(d) In manholes, if the conductors of each system are permanently and effectively separated from the conductors of the other systems and securely fastened to racks, insulators, or other approved supports, conductors of different voltage ratings shall be permitted.

Conductors having nonshielded insulation and operating at different voltage levels shall not occupy the same enclosure, cable, or raceway.

## 300.4 Protection Against Physical Damage

Where subject to physical damage, conductors, raceways, and cables shall be protected.

### (A) Cables and Raceways Through Wood Members.

**(1) Bored Holes.** In both exposed and concealed locations, where a cable- or raceway-type wiring method is installed through bored holes in joists, rafters, or wood members, holes shall be bored so that the edge of the hole is not less than 32 mm (1¼ in.) from the nearest edge of the wood member. Where this distance cannot be maintained, the cable or raceway shall be protected from penetration by screws or nails by a steel plate(s) or bushing(s), at least 1.6 mm (¹⁄₁₆ in.) thick, and of appropriate length and width installed to cover the area of the wiring.

*Exception No. 1: Steel plates shall not be required to protect rigid metal conduit, intermediate metal conduit, rigid nonmetallic conduit, or electrical metallic tubing.*

*Exception No. 2: A listed and marked steel plate less than 1.6 mm (¹⁄₁₆ in.) thick that provides equal or better protection against nail or screw penetration shall be permitted.*

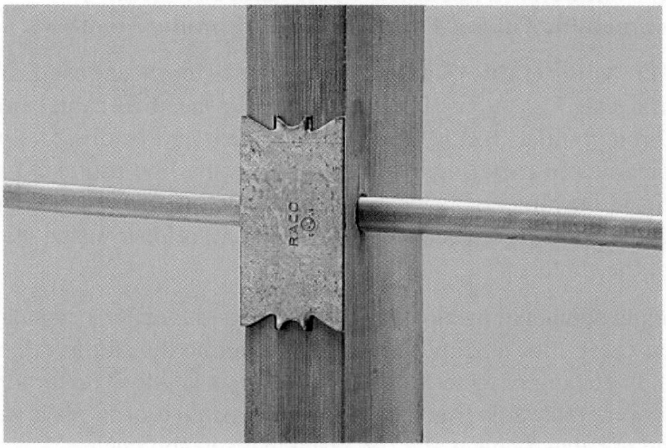


***EXHIBIT 300.1*** *A listed and marked steel plate, permitted to be less than ¹⁄₁₆ inch only if it is listed and marked, used to protect a nonmetallic-sheathed cable less than 1¼ inches from the edge of a wood stud. (Courtesy of Hubbell RACO)*

The intent is to prevent nails and screws from being driven into conductors, cables, and raceways. As shown in Exhibit 300.1, if the edge of a drilled hole is less than 1¼ inch from the nearest edge of a wood stud, a plate is required to prevent screws or nails from penetrating the stud far enough to injure a cable. Where a single metal plate does not provide for the protection of a cable, multiple steel plates are required. Building codes limit the maximum size of bored or notched holes in studs, and 300.4(A)(2) indicates that consideration should be given to the size of notches in studs, so as not to affect the strength of the structure.

Exception No. 1 to 300.4(A)(1) and Exception No. 1 to 300.4(A)(2) permit IMC, RMC, PVC conduit, and EMT to be installed less than 1¼ inches from the nearest edge of the stud without using a steel plate or bushing. Exception No. 2 to 300.4(A)(1) and Exception No. 2 to 300.4(A)(2) permit steel plates thinner than ¹⁄₁₆ inch to protect cables and raceways, but only if the plates are specifically listed and marked.

**(2) Notches in Wood.** Where there is no objection because of weakening the building structure, in both exposed and concealed locations, cables or raceways shall be permitted to be laid in notches in wood studs, joists, rafters, or other wood members where the cable or raceway at those points is protected against nails or screws by a steel plate at least 1.6 mm (¹⁄₁₆ in.) thick, and of appropriate length and width, installed to cover the area of the wiring. The steel plate shall be installed before the building finish is applied.

*Exception No. 1: Steel plates shall not be required to protect rigid metal conduit, intermediate metal conduit, rigid nonmetallic conduit, or electrical metallic tubing.*

*Exception No. 2: A listed and marked steel plate less than 1.6 mm (¹⁄₁₆ in.) thick that provides equal or better protection against nail or screw penetration shall be permitted.*

**(B) Nonmetallic-Sheathed Cables and Electrical Nonmetallic Tubing Through Metal Framing Members.**

**(1) Nonmetallic-Sheathed Cable.** In both exposed and concealed locations where nonmetallic-sheathed cables pass through either factory- or field-punched, cut, or drilled slots or holes in metal members, the cable shall be protected by listed bushings or listed grommets covering all metal edges that are securely fastened in the opening prior to installation of the cable.

Should additional metal studs become necessary after installation of a cable, the cable must be removed before the stud is added. Field notching of metal studs and then installing the stud around a nonmetallic sheathed cable already installed or in place can lead to cable damage and can result in insulation failure.

**(2) Nonmetallic-Sheathed Cable and Electrical Nonmetallic Tubing.** Where nails or screws are likely to penetrate nonmetallic-sheathed cable or electrical nonmetallic tubing, a steel sleeve, steel plate, or steel clip not less than 1.6 mm (1/16 in.) in thickness shall be used to protect the cable or tubing.

*Exception: A listed and marked steel plate less than 1.6 mm (1/16 in.) thick that provides equal or better protection against nail or screw penetration shall be permitted.*

**(C) Cables Through Spaces Behind Panels Designed to Allow Access.** Cables or raceway-type wiring methods, installed behind panels designed to allow access, shall be supported according to their applicable articles.

Cable- or raceway-type wiring installed above suspended ceilings with lift-up panels must not be laid on the suspended ceiling; this would inhibit access. Such wiring is required to be supported according to 300.11(A), 300.23, and the requirements of the Chapter 3 article applicable to the particular wiring method.

Similarly, low-voltage, optical fiber, broadband, and communications cables are not permitted to block access to equipment above the suspended ceiling. Examples of this requirement are found in 725.21, 760.21, 770.21, 800.21, 820.21, and 830.21. For support of low-voltage cables, optical fiber, broadband, and communications cables, see 720.11, 725.24, 760.24, 770.24, 800.24, 820.24, and 830.24.

**(D) Cables and Raceways Parallel to Framing Members and Furring Strips.** In both exposed and concealed locations, where a cable- or raceway-type wiring method is installed parallel to framing members, such as joists, rafters, or studs, or is installed parallel to furring strips, the cable or raceway shall be installed and supported so that the nearest outside surface of the cable or raceway is not less than 32 mm (1¼ in.) from the nearest edge of the framing member or furring strips where nails or screws are likely to penetrate. Where this distance cannot be maintained, the cable or raceway shall be protected from penetration by nails or screws by a steel plate, sleeve, or equivalent at least 1.6 mm (1/16 in.) thick.

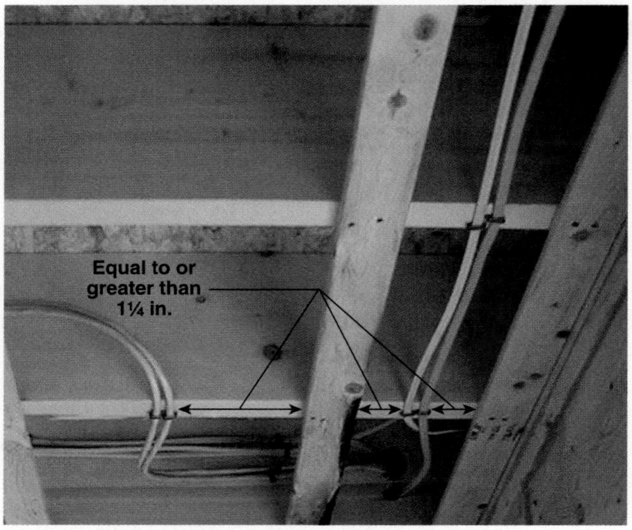


***EXHIBIT 300.2*** *Nonmetallic sheathed cables adjacent to furring strips in a wood frame structure.*

To prevent mechanical damage to cables and raceways from nails and screws, the raceways need the same level of physical protection at furring strips as they do at framing members. Two means of protection that generally apply to exposed or concealed work are described. The first method, shown in Exhibit 300.2, is to position the NM cables to equal or exceed the minimum clear distance of 1¼ inches that is required between the furring strip (wood strapping in this case) and the nearest edge of the NM cable. Exhibit 300.3 illustrates the NM cable installed parallel to framing members and fastened to maintain at least 1¼ inches clear space from the edge. The second method permits the cable or raceway to be installed closer than 1¼ inches from the edge of the framing member if physical protection, such as a steel plate, its equivalent, or a sleeve, is provided. (A steel plate is illustrated in Exhibit 300.1.)

*Exception No. 1: Steel plates, sleeves, or the equivalent shall not be required to protect rigid metal conduit, intermediate metal conduit, rigid nonmetallic conduit, or electrical metallic tubing.*

*Exception No. 2: For concealed work in finished buildings, or finished panels for prefabricated buildings where such supporting is impracticable, it shall be permissible to fish the cables between access points.*

*Exception No. 3: A listed and marked steel plate less than 1.6 mm (1/16 in.) thick that provides equal or better protection against nail or screw penetration shall be permitted.*

**(E) Cables, Raceways, or Boxes Installed in or Under Roof Decking.** A cable, raceway, or box, installed in exposed or concealed locations under metal-corrugated sheet roof decking, shall be installed and supported so there is not less than 38 mm (1½ in.) measured from the lowest surface of the roof decking to the top

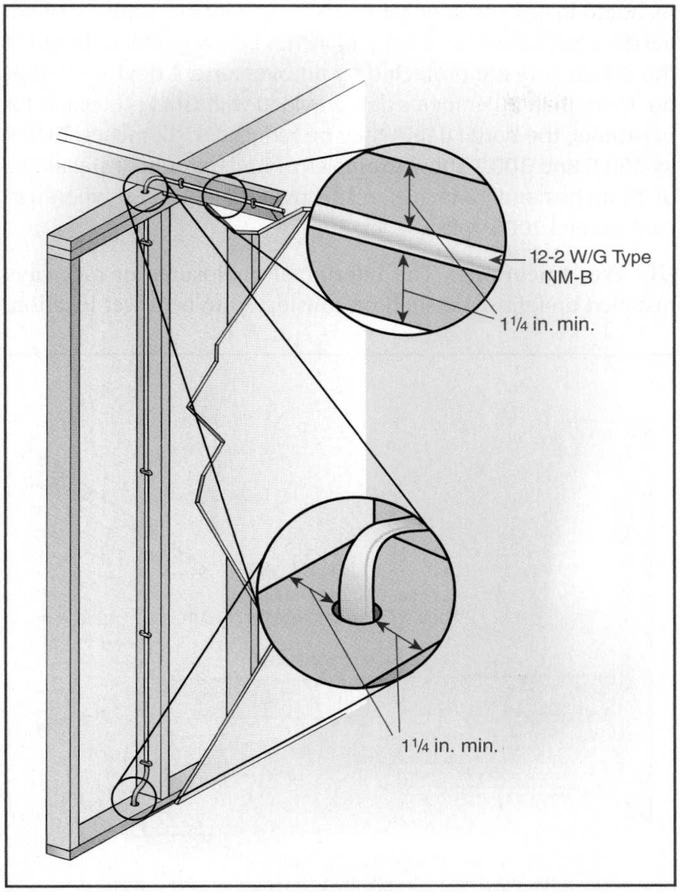


**EXHIBIT 300.3** *Cables and raceways installed parallel to framing members.*

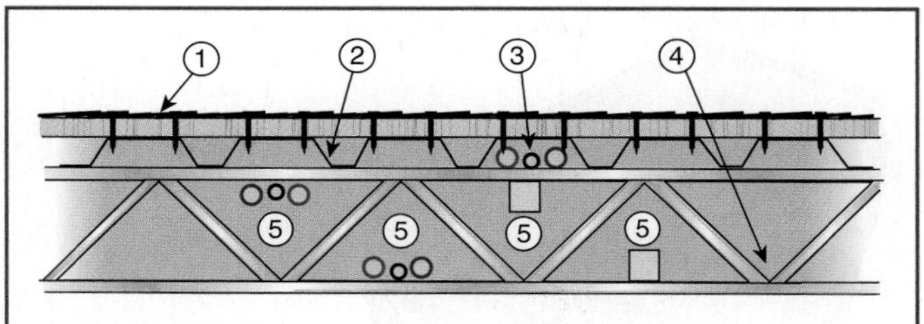


1. Roofing material
2. Metal-corrugated sheet roof decking
3. Only RMC and IMC permitted within 1½ in. of roof decking
4. Steel or wood joists
5. Any wiring method or box permitted within or below the roof framing where installed not less than 1½ in. below the nearest point of the metal corrugated roof decking surface

**EXHIBIT 300.4** *Placement of cables and raceways under metal-corrugated sheet roof decking to avoid physical damage.*

of the cable, raceway, or box. A cable, raceway, or box shall not be installed in concealed locations in metal-corrugated, sheet decking–type roof.

Informational Note: Roof decking material is often repaired or replaced after the initial raceway or cabling and roofing installation and may be penetrated by the screws or other mechanical devices designed to provide "hold down" strength of the waterproof membrane or roof insulating material.

*Exception: Rigid metal conduit and intermediate metal conduit shall not be required to comply with 300.4(E).*

For cables, raceways, and boxes installed below a metal-corrugated sheet roof decking installation, 300.4(E) requires at least 1½ inches of separation from any of the roof decking surface, as illustrated in Exhibit 300.4. The 1½-inch dimension is measured from the lowest point of the roof deck to the top surface of the cable, raceway, or box. This section prohibits cables, raceways, or boxes from being installed in the space between the metal deck and the roofing material.

This spacing is required to avoid future damage when the roofing is repaired or replaced. Roof replacement materials are fastened in place with long screws from atop the roof. These fasteners usually penetrate the roof decking installation and could continue into cables and raceways installed below. Adequate space between the corrugated sheet metal roof deck installation and the cables and raceways below the deck will prevent future damage to the cable or raceway installation. The exception clarifies that this requirement does not apply to more robust metal raceways such as RMC and IMC.

**(F) Cables and Raceways Installed in Shallow Grooves.** Cable- or raceway-type wiring methods installed in a groove, to be covered by wallboard, siding, paneling, carpeting, or similar finish, shall be protected by 1.6 mm (1⁄16 in.) thick steel plate, sleeve, or equivalent or by not less than 32-mm (1¼-in.) free space for the full length of the groove in which the cable or raceway is installed.

*Exception No. 1: Steel plates, sleeves, or the equivalent shall not be required to protect rigid metal conduit, intermediate metal conduit, rigid nonmetallic conduit, or electrical metallic tubing.*

*Exception No. 2: A listed and marked steel plate less than 1.6 mm (1⁄16 in.) thick that provides equal or better protection against nail or screw penetration shall be permitted.*

**(G) Insulated Fittings.** Where raceways contain 4 AWG or larger insulated circuit conductors, and these conductors enter a cabinet, a box, an enclosure, or a raceway, the conductors shall be protected by an identified fitting providing a smoothly rounded insulating surface, unless the conductors are separated from the fitting or raceway by identified insulating material that is securely fastened in place.

*Exception: Where threaded hubs or bosses that are an integral part of a cabinet, box, enclosure, or raceway provide a smoothly rounded or flared entry for conductors.*

Heavy conductors and cables tend to stress the conductor insulation at raceway terminating points. Insulated bushing or smooth rounded entries are required at raceway and cable terminations to reduce the risk of insulation failure at conductor insulation

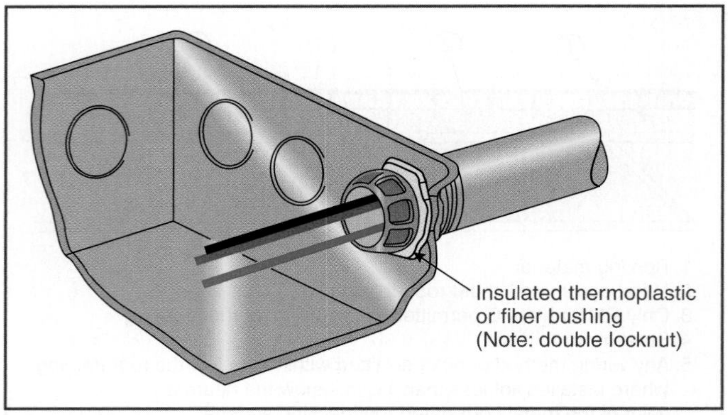


***EXHIBIT 300.5*** *An insulated bushing used to protect conductors from chafing against a metal conduit fitting.*

stress points. The temperature ratings of the insulating bushing must coordinate with the insulation of the conductor to ensure that the conductor is protected for its entire life cycle.

Whether provided separately or as part of a fitting, listed insulating bushings are colored black or brown to indicate a temperature rating of 150°C and any other color to indicate a rating of 90°C unless specifically marked for a higher temperature. Exhibit 300.5 shows an insulated thermoplastic or fiber bushing used to protect the conductors from chafing against a metal conduit fitting. Note the use of a double locknut, with one on the inside and one on the outside of the enclosure. The locknuts are necessary because the raceway cannot be secured by the fiber or plastic bushing.

The main rule in 300.4(G) applies to all wiring methods and all enclosures. See also 342.46, 344.46, and 352.46 for further information relating to bushings.

Conduit bushings constructed wholly of insulating material shall not be used to secure a fitting or raceway. The insulating fitting or insulating material shall have a temperature rating not less than the insulation temperature rating of the installed conductors.

**(H) Structural Joints.** A listed expansion/deflection fitting or other approved means shall be used where a raceway crosses a structural joint intended for expansion, contraction or deflection, used in buildings, bridges, parking garages, or other structures.

Raceways can be damaged when improperly installed in structural construction joints, leaving conductors or cables exposed. Structural construction joints can experience shear and lateral loads due to gravity, expansion and contraction, and movement of the structure. Listed expansion/deflection fittings are available for use at structural joints. However, other approved means of protecting the integrity of the raceways system are also permitted.

## 300.5 Underground Installations

**(A) Minimum Cover Requirements.** Direct-buried cable or conduit or other raceways shall be installed to meet the minimum cover requirements of Table 300.5.

As noted in Column 4 of Table 300.5, conductors under residential driveways must be at least 18 inches below grade. However, if the conductors are protected by an overcurrent device rated at not more than 20 amperes and provided with GFCI protection for personnel, the burial depth may be reduced to 12 inches. Exhibits 300.6 and 300.7 show examples of underground installations of 18 inches and 12 inches, respectively. See 300.50 where circuits exceed 1000 volts.

**(B) Wet Locations.** The interior of enclosures or raceways installed underground shall be considered to be a wet location.

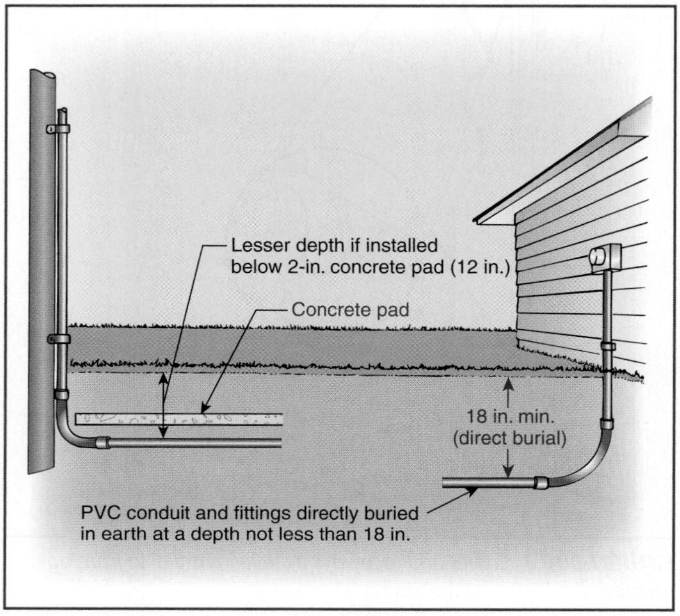


***EXHIBIT 300.6*** *Type PVC conduit buried 18 inches below grade.*

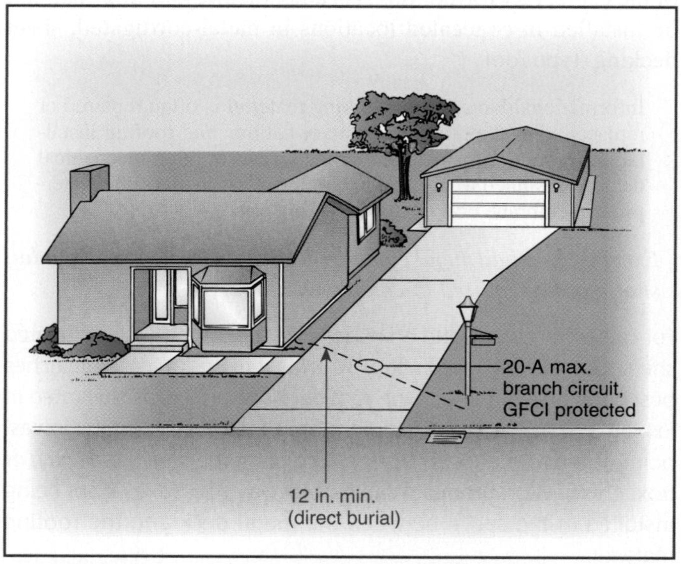


***EXHIBIT 300.7*** *A 20-ampere, GFCI-protected residential branch circuit installed with a minimum burial depth of 12 inches beneath a residential driveway.*

***TABLE 300.5*** *Minimum Cover Requirements, 0 to 1000 Volts, Nominal, Burial in Millimeters (Inches)*

| Location of Wiring Method or Circuit | Type of Wiring Method or Circuit: Column 1 Direct Burial Cables or Conductors | | Column 2 Rigid Metal Conduit or Intermediate Metal Conduit | | Column 3 Nonmetallic Raceways Listed for Direct Burial Without Concrete Encasement or Other Approved Raceways | | Column 4 Residential Branch Circuits Rated 120 Volts or Less with GFCI Protection and Maximum Overcurrent Protection of 20 Amperes | | Column 5 Circuits for Control of Irrigation and Landscape Lighting Limited to Not More Than 30 Volts and Installed with Type UF or in Other Identified Cable or Raceway | |
|---|---|---|---|---|---|---|---|---|---|---|
| | **mm** | **in.** | **mm** | **in.** | **mm** | **in.** | **mm** | **in.** | **mm** | **in.** |
| All locations not specified below | 600 | 24 | 150 | 6 | 450 | 18 | 300 | 12 | 150 | 6 |
| In trench below 50 mm (2 in.) thick concrete or equivalent | 450 | 18 | 150 | 6 | 300 | 12 | 150 | 6 | 150 | 6 |
| Under a building | 0 (in raceway or Type MC or Type MI cable identified for direct burial) | 0 | 0 | 0 | 0 | 0 | 0 (in raceway or Type MC or Type MI cable identified for direct burial) | 0 | 0 (in raceway or Type MC or Type MI cable identified for direct burial) | 0 |
| Under minimum of 102 mm (4 in.) thick concrete exterior slab with no vehicular traffic and the slab extending not less than 152 mm (6 in.) beyond the underground installation | 450 | 18 | 100 | 4 | 100 | 4 | 150 (direct burial) 100 (in raceway) | 6 4 | 150 (direct burial) 100 (in raceway) | 6 4 |
| Under streets, highways, roads, alleys, driveways, and parking lots | 600 | 24 | 600 | 24 | 600 | 24 | 600 | 24 | 600 | 24 |
| One- and two-family dwelling driveways and outdoor parking areas, and used only for dwelling-related purposes | 450 | 18 | 450 | 18 | 450 | 18 | 300 | 12 | 450 | 18 |
| In or under airport runways, including adjacent areas where trespassing prohibited | 450 | 18 | 450 | 18 | 450 | 18 | 450 | 18 | 450 | 18 |

Notes:

1. Cover is defined as the shortest distance in millimeters (inches) measured between a point on the top surface of any direct-buried conductor, cable, conduit, or other raceway and the top surface of finished grade, concrete, or similar cover.

2. Raceways approved for burial only where concrete encased shall require concrete envelope not less than 50 mm (2 in.) thick.

3. Lesser depths shall be permitted where cables and conductors rise for terminations or splices or where access is otherwise required.

4. Where one of the wiring method types listed in Columns 1 through 3 is used for one of the circuit types in Columns 4 and 5, the shallowest depth of burial shall be permitted.

5. Where solid rock prevents compliance with the cover depths specified in this table, the wiring shall be installed in metal or nonmetallic raceway permitted for direct burial. The raceways shall be covered by a minimum of 50 mm (2 in.) of concrete extending down to rock.

Insulated conductors and cables installed in these enclosures or raceways in underground installations shall comply with 310.10(C). Any connections or splices in an underground installation shall be approved for wet locations.

The inside of all raceways and enclosures installed underground is classified as a wet location. Conductors installed in such underground locations must meet one of the following criteria:

1. Be moisture-impervious metal-sheathed
2. Be Types MTW, RHW, RHW-2, TW, THW, THW-2, THHW, THWN, THWN-2, XHHW, XHHW-2, ZW
3. Be of a type listed for use in wet locations

**(C) Underground Cables and Conductors Under Buildings.** Underground cable and conductors installed under a building shall be in a raceway.

*Exception No. 1: Type MI cable shall be permitted under a building without installation in a raceway where embedded in concrete, fill, or other masonry in accordance with 332.10(6) or in underground runs where suitably protected against physical damage and corrosive conditions in accordance with 332.10(10).*

*Exception No. 2: Type MC cable listed for direct burial or concrete encasement shall be permitted under a building without installation in a raceway in accordance with 330.10(A)(5) and in wet locations in accordance with 330.10(A)(11).*

**(D) Protection from Damage.** Direct-buried conductors and cables shall be protected from damage in accordance with 300.5(D)(1) through (D)(4).

**(1) Emerging from Grade.** Direct-buried conductors and cables emerging from grade and specified in columns 1 and 4 of Table 300.5 shall be protected by enclosures or raceways extending from the minimum cover distance below grade required by 300.5(A) to a point at least 2.5 m (8 ft) above finished grade. In no case shall the protection be required to exceed 450 mm (18 in.) below finished grade.

**(2) Conductors Entering Buildings.** Conductors entering a building shall be protected to the point of entrance.

**(3) Service Conductors.** Underground service conductors that are not encased in concrete and that are buried 450 mm (18 in.) or more below grade shall have their location identified by a warning ribbon that is placed in the trench at least 300 mm (12 in.) above the underground installation.

The warning ribbon reduces the risk of an accident, an electrocution, or an arc-flash incident during excavation near underground service conductors that are not encased in concrete. This requirement does not extend to feeders and branch circuits, because these circuits contain short-circuit and overload protection.

**(4) Enclosure or Raceway Damage.** Where the enclosure or raceway is subject to physical damage, the conductors shall be installed in rigid metal conduit, intermediate metal conduit, RTRC-XW, Schedule 80 PVC conduit, or equivalent.

**(E) Splices and Taps.** Direct-buried conductors or cables shall be permitted to be spliced or tapped without the use of splice boxes. The splices or taps shall be made in accordance with 110.14(B).

Underground splices are not required to be in a box where they are made in accordance with 110.14(B), which requires the splicing means to be listed for underground use. Sealed wire connector systems are listed for underground splicing. They restore the insulation integrity of the spliced conductors in a permanent joint. Sealed wire connectors are used where future access to the splices will not be necessary.

A difference exists between multiconductor cables labeled for direct burial and single conductors labeled for direct burial. Because direct-burial multiconductor cables may or may not contain individual conductors labeled for direct burial, the overall cable jacket may be the only underground protection technique for the contained conductors. Although the direct-burial splicing techniques used on multiconductor cables can differ widely from the techniques used on direct-burial single-conductor cables, the *Code* requirements are generally the same. The splicing technique must be listed for the cable type and listed for direct burial. See 250.8 for spliced grounding and bonding conductors. An example of a sealed wire connector system listed for direct burial used to splice single-conductor installed underground cables is shown in Exhibit 300.8.

**(F) Backfill.** Backfill that contains large rocks, paving materials, cinders, large or sharply angular substances, or corrosive material shall not be placed in an excavation where materials may damage raceways, cables, or other substructures or prevent adequate compaction of fill or contribute to corrosion of raceways, cables, or other substructures.

Where necessary to prevent physical damage to the raceway or cable, protection shall be provided in the form of granular or

***EXHIBIT 300.8*** *One underground splicing method. (Courtesy of 3M Co., Electrical Markets Division)*

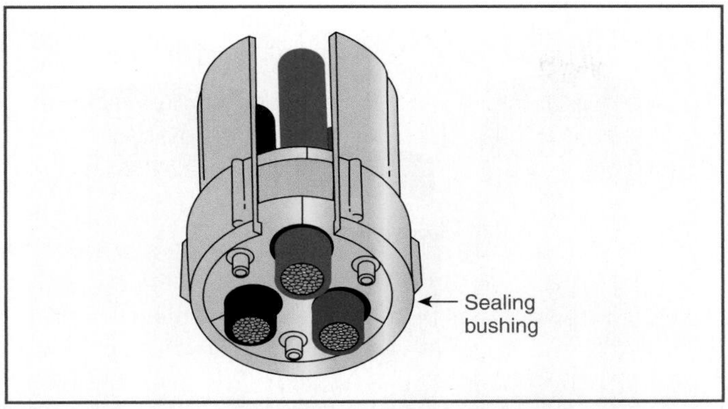


***EXHIBIT 300.9*** *A conduit sealing bushing used to prevent the entrance of gas or moisture. (Redrawn courtesy of O-Z/Gedney, a division of EGS Electrical Group)*

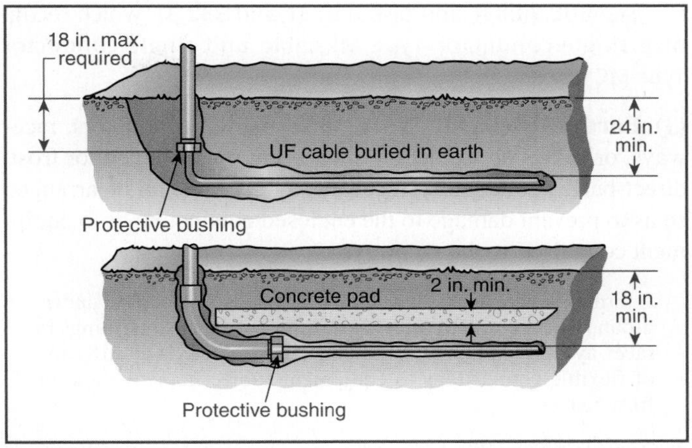


***EXHIBIT 300.10*** *Type UF cable buried in compliance with Table 300.5.*

selected material, suitable running boards, suitable sleeves, or other approved means.

**(G) Raceway Seals.** Conduits or raceways through which moisture may contact live parts shall be sealed or plugged at either or both ends.

Informational Note: Presence of hazardous gases or vapors may also necessitate sealing of underground conduits or raceways entering buildings.

A conduit sealing bushing, shown in Exhibit 300.9, is one method that prevents the entrance of gas or moisture. See 230.8 for sealing service raceways.

**(H) Bushing.** A bushing, or terminal fitting, with an integral bushed opening shall be used at the end of a conduit or other raceway that terminates underground where the conductors or cables emerge as a direct burial wiring method. A seal incorporating the physical protection characteristics of a bushing shall be permitted to be used in lieu of a bushing.

A raceway that terminates underground and emerges as a direct-burial cable requires a protective fitting. Exhibit 300.10 shows such a fitting with a Type UF cable buried in compliance with Table 300.5. The bushing will protect the cable from damage by the conduit. A metal conduit fitting is usually used with nonmetallic conduit, because a nonmetallic fitting can be damaged when pulling conductors in an underground installation. See 300.10, Exception No. 1 as well as the commentary following 300.4(G).

**(I) Conductors of the Same Circuit.** All conductors of the same circuit and, where used, the grounded conductor and all equipment grounding conductors shall be installed in the same raceway or cable or shall be installed in close proximity in the same trench.

*Exception No. 1: Conductors shall be permitted to be installed in parallel in raceways, multiconductor cables, or direct-buried single conductor cables. Each raceway or multiconductor cable shall contain all conductors of the same circuit, including equipment grounding conductors. Each direct-buried single conductor cable shall be located in close proximity in the trench to the other single conductor cables in the same parallel set of conductors in the circuit, including equipment grounding conductors.*

Keeping all circuit conductors together reduces inductive heating and reduces circuit impedance. Section 300.5(I), Exception No. 1, permits the installation of paralleled conductors in different raceways provided all circuit conductors, including equipment grounding conductors, are installed in each of the parallel raceways. Conductors of the same circuit are addressed in 300.3(B). Also see 310.10(H) for requirements for conductors in parallel.

*Exception No. 2: Isolated phase, polarity, grounded conductor, and equipment grounding and bonding conductor installations shall be permitted in nonmetallic raceways or cables with a nonmetallic covering or nonmagnetic sheath in close proximity where conductors are paralleled as permitted in 310.10(H), and where the conditions of 300.20(B) are met.*

Isolated phase installations contain only one phase per raceway or cable. In an ac circuit installation, the spacing between isolated phase raceways and cables should be as small as possible and the length of the run limited to avoid increased circuit impedance and the resulting increase in voltage drop. Isolated phase installations may be used in ac circuits to limit available fault current at downstream equipment.

Isolated phase installations present an inherent hazard of overheating, which is a risk that must be understood and carefully controlled. This hazard results from induced currents in metal surrounding a raceway that contains only one phase conductor. [See 300.20(A) and (B) for more information on induced currents in raceways.] The surrounding metal acts as a shorted transformer turn. In underground installations, a single conductor is unlikely to be installed in a metal raceway or, if it were, is unlikely to present a fire hazard. This is not true, however, for aboveground raceways, which is the reason isolated phase installations have limited application for aboveground installations.

See 300.3(B)(3), and also 330.31 and 332.31, which recognize single-conductor Type MI cable and single-conductor Type MC cable.

**(J) Earth Movement.** Where direct-buried conductors, raceways, or cables are subject to movement by settlement or frost, direct-buried conductors, raceways, or cables shall be arranged so as to prevent damage to the enclosed conductors or to equipment connected to the raceways.

Informational Note: This section recognizes "S" loops in underground direct burial to raceway transitions, expansion fittings in raceway risers to fixed equipment, and, generally, the provision of flexible connections to equipment subject to settlement or frost heaves.

Slack must be provided in cables, expansion joints must be used for raceways, or other measures must be taken, if earth movement due to frost or settlement is anticipated.

**(K) Directional Boring.** Cables or raceways installed using directional boring equipment shall be approved for the purpose.

A number of metal and nonmetallic raceways are suitable for boring installations, such as HDPE conduit. See Article 353 for more information.

## 300.6 Protection Against Corrosion and Deterioration

Raceways, cable trays, cablebus, auxiliary gutters, cable armor, boxes, cable sheathing, cabinets, elbows, couplings, fittings, supports, and support hardware shall be of materials suitable for the environment in which they are to be installed.

This correlates with the requirement in 110.11 that equipment be suitable for the environment in which it is installed. Section 300.6 applies generally, and for specific applications the manufacturers' information must be reviewed.

**(A) Ferrous Metal Equipment.** Ferrous metal raceways, cable trays, cablebus, auxiliary gutters, cable armor, boxes, cable sheathing, cabinets, metal elbows, couplings, nipples, fittings, supports, and support hardware shall be suitably protected against corrosion inside and outside (except threads at joints) by a coating of approved corrosion-resistant material. Where corrosion protection is necessary and the conduit is threaded in the field, the threads shall be coated with an approved electrically conductive, corrosion-resistant compound.

Ferrous metal equipment must be protected from corrosion with an approved anti-corrosion compound, such as that shown in Exhibit 300.11.

Informational Note: Field-cut threads are those threads that are cut in conduit, elbows, or nipples anywhere other than at the factory where the product is listed.

*Exception: Stainless steel shall not be required to have protective coatings.*

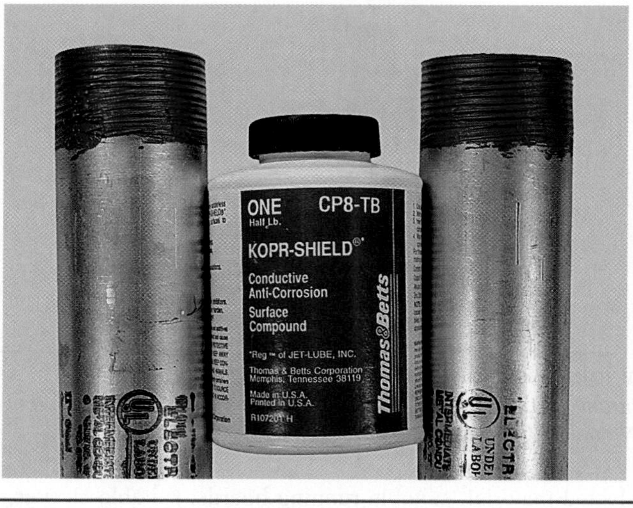


***EXHIBIT 300.11*** *KOPR-Shield® (a registered trademark of Jet Lube) is a conductive anti-corrosion surface compound suitable for application on field-cut conduit threads where protection from corrosion is necessary. (Courtesy of Thomas & Betts Corp.)*

**(1) Protected from Corrosion Solely by Enamel.** Where protected from corrosion solely by enamel, ferrous metal raceways, cable trays, cablebus, auxiliary gutters, cable armor, boxes, cable sheathing, cabinets, metal elbows, couplings, nipples, fittings, supports, and support hardware shall not be used outdoors or in wet locations as described in 300.6(D).

**(2) Organic Coatings on Boxes or Cabinets.** Where boxes or cabinets have an approved system of organic coatings and are marked "Raintight," "Rainproof," or "Outdoor Type," they shall be permitted outdoors.

**(3) In Concrete or in Direct Contact with the Earth.** Ferrous metal raceways, cable armor, boxes, cable sheathing, cabinets, elbows, couplings, nipples, fittings, supports, and support hardware shall be permitted to be installed in concrete or in direct contact with the earth, or in areas subject to severe corrosive influences where made of material approved for the condition, or where provided with corrosion protection approved for the condition.

Metal raceways installed in the earth can be coated with an asphalt compound, plastic sheath, or other equivalent protection to help prevent deterioration. Also, metal raceways are available with a bonded PVC coating. Special precautions are normally necessary for installing aluminum conduits in concrete to ensure adequate corrosion protection.

Galvanized steel RMC and steel IMC do not generally require supplementary corrosion protection.

**(B) Aluminum Metal Equipment.** Aluminum raceways, cable trays, cablebus, auxiliary gutters, cable armor, boxes, cable sheathing, cabinets, elbows, couplings, nipples, fittings, supports, and support hardware embedded or encased in concrete or in direct contact with the earth shall be provided with supplementary corrosion protection.

**(C) Nonmetallic Equipment.** Nonmetallic raceways, cable trays, cablebus, auxiliary gutters, boxes, cables with a nonmetallic outer jacket and internal metal armor or jacket, cable sheathing, cabinets, elbows, couplings, nipples, fittings, supports, and support hardware shall be made of material approved for the condition and shall comply with (C)(1) and (C)(2) as applicable to the specific installation.

**(1) Exposed to Sunlight.** Where exposed to sunlight, the materials shall be listed as sunlight resistant or shall be identified as sunlight resistant.

**(2) Chemical Exposure.** Where subject to exposure to chemical solvents, vapors, splashing, or immersion, materials or coatings shall either be inherently resistant to chemicals based on their listing or be identified for the specific chemical reagent.

**(D) Indoor Wet Locations.** In portions of dairy processing facilities, laundries, canneries, and other indoor wet locations, and in locations where walls are frequently washed or where there are surfaces of absorbent materials, such as damp paper or wood, the entire wiring system, where installed exposed, including all boxes, fittings, raceways, and cable used therewith, shall be mounted so that there is at least a 6-mm (¼-in.) airspace between it and the wall or supporting surface.

*Exception: Nonmetallic raceways, boxes, and fittings shall be permitted to be installed without the airspace on a concrete, masonry, tile, or similar surface.*

Informational Note: In general, areas where acids and alkali chemicals are handled and stored may present such corrosive conditions, particularly when wet or damp. Severe corrosive conditions may also be present in portions of meatpacking plants, tanneries, glue houses, and some stables; in installations immediately adjacent to a seashore and swimming pool areas; in areas where chemical deicers are used; and in storage cellars or rooms for hides, casings, fertilizer, salt, and bulk chemicals.

The exception coincides with 547.5(B), which permits nonmetallic boxes, fittings, conduit, and cables to be installed without the airspace in corrosive locations of agricultural buildings.

## 300.7 Raceways Exposed to Different Temperatures

**(A) Sealing.** Where portions of a raceway or sleeve are known to be subjected to different temperatures, and where condensation is known to be a problem, as in cold storage areas of buildings or where passing from the interior to the exterior of a building, the raceway or sleeve shall be filled with an approved material to prevent the circulation of warm air to a colder section of the raceway or sleeve. An explosionproof seal shall not be required for this purpose.

Condensation can form in raceways or sleeves that are subjected to temperature differences as a result of air circulating through the raceway from a warmer to a colder section. Condensation could accumulate, for example, in a raceway used to supply lighting or branch-circuit conductors within a walk-in refrigerator or freezer. Circulation of air can be prevented by sealing the raceway with a suitable pliable compound at a conduit body or junction box, usually installed in the raceway before it enters the colder section. Special sealing fittings such as those used in hazardous (classified) locations are not necessary.

**(B) Expansion Fittings.** Raceways shall be provided with expansion fittings where necessary to compensate for thermal expansion and contraction.

Informational Note: Table 352.44 and Table 355.44 provide the expansion information for polyvinyl chloride (PVC) and for reinforced thermosetting resin conduit (RTRC), respectively. A nominal number for steel conduit can be determined by multiplying the expansion length in Table 352.44 by 0.20. The coefficient of expansion for steel electrical metallic tubing, intermediate metal conduit, and rigid metal conduit is $1.170 \times 10^{-5}$ (0.0000117 mm per mm of conduit for each °C in temperature change) [$0.650 \times 10^{-5}$ (0.0000065 in. per inch of conduit for each °F in temperature change)].

A nominal number for aluminum conduit and aluminum electrical metallic tubing can be determined by multiplying the expansion length in Table 352.44 by 0.40. The coefficient of expansion for aluminum electrical metallic tubing and aluminum rigid metal conduit is $2.34 \times 10^{-5}$ (0.0000234 mm per mm of conduit for each °C in temperature change) [$1.30 \times 10^{-5}$ (0.000013) in. per inch of conduit for each °F in temperature change].

Substantial changes in temperature cause destructive amounts of expansion and contraction in a raceway system. Properly designed and installed expansion fittings allow expansion and contraction without damage to enclosures, raceways, and their conductors.

The informational note addresses expansion of aluminum RMC and RTRC. The note also provides a few simple relationships (or ratios) of linear expansion in length of PVC conduit to other types of conduit.

Knowing the coefficient of expansion of the conduit material (e.g., steel, aluminum, PVC, or fiberglass) and understanding that the change in length for a conduit is equal to the coefficient of expansion times the change in temperature times the initial length of the conduit is key to determining any temperature-related change in conduit length. A cross reference to the coefficient of expansion for various conduit and tubing types is provided in Commentary Table 300.1.

The approximate linear expansion for 40 feet of PVC conduit is ⅞ inch (actual 0.892), of steel RMC is ⅜ inch (actual 0.396), and of RTRC is 3/16 inch (actual 0.178).

## 300.8 Installation of Conductors with Other Systems

Raceways or cable trays containing electrical conductors shall not contain any pipe, tube, or equal for steam, water, air, gas, drainage, or any service other than electrical.

**COMMENTARY TABLE 300.1** Expansion Coefficients with a 55°F Change

| Raceway Type | Reference Table | Expansion Coefficient in./100 feet | Expansion Length (inches) 40 feet |
|---|---|---|---|
| PVC | 352.44 | 2.23 | 0.892 |
| RTRC | 355.44 | 0.99 | 0.396 |
| Steel RMC, IMC, or EMT | 352.44 (from 300.7(B) Informational Note) | 0.446 | 0.178 |

## 300.9 Raceways in Wet Locations Abovegrade

Where raceways are installed in wet locations abovegrade, the interior of these raceways shall be considered to be a wet location. Insulated conductors and cables installed in raceways in wet locations abovegrade shall comply with 310.10(C).

The insides of these raceways are considered wet locations requiring conductors and cables that are listed for use in wet locations.

## 300.10 Electrical Continuity of Metal Raceways and Enclosures

Metal raceways, cable armor, and other metal enclosures for conductors shall be metallically joined together into a continuous electrical conductor and shall be connected to all boxes, fittings, and cabinets so as to provide effective electrical continuity. Unless specifically permitted elsewhere in this *Code*, raceways and cable assemblies shall be mechanically secured to boxes, fittings, cabinets, and other enclosures.

Sections 250.4(A) and (B) detail what must be accomplished by the grounding and bonding of metal parts of the electrical system. The metal parts must form an effective low-impedance path to ground in order to safely conduct any fault current and facilitate the operation of overcurrent devices protecting the enclosed circuit conductors.

*Exception No. 1: Short sections of raceways used to provide support or protection of cable assemblies from physical damage shall not be required to be made electrically continuous.*

*Exception No. 2: Equipment enclosures to be isolated, as permitted by 250.96(B), shall not be required to be metallically joined to the metal raceway.*

## 300.11 Securing and Supporting

**(A) Secured in Place.** Raceways, cable assemblies, boxes, cabinets, and fittings shall be securely fastened in place. Support wires that do not provide secure support shall not be permitted as the sole support. Support wires and associated fittings that provide secure support and that are installed in addition to the ceiling grid support wires shall be permitted as the sole support. Where independent support wires are used, they shall be secured at both ends. Cables and raceways shall not be supported by ceiling grids.

**(1) Fire-Rated Assemblies.** Wiring located within the cavity of a fire-rated floor–ceiling or roof–ceiling assembly shall not be secured to, or supported by, the ceiling assembly, including the ceiling support wires. An independent means of secure support shall be provided and shall be permitted to be attached to the assembly. Where independent support wires are used, they shall be distinguishable by color, tagging, or other effective means from those that are part of the fire-rated design.

*Exception: The ceiling support system shall be permitted to support wiring and equipment that have been tested as part of the fire-rated assembly.*

Wiring methods for any type and all luminaires are not allowed to be supported or secured to the support wires or T-bars of a fire-rated ceiling assembly unless the assembly has been tested and listed for that use. If wire is selected as the supporting means for the electrical system within a fire-rated ceiling cavity, it must be distinguishable from the ceiling support wires by color, tagging, or other effective means and must be secured at both ends. Independent support, and securing both ends of the support wire, does preclude a connection to the ceiling grid on one end.

Generally, the rule for supporting electrical equipment is that the equipment must be "securely fastened in place." This phrase means not only that vertical support for the weight of the equipment must be provided but also that the equipment must be secured to prevent horizontal movement or sway. The intention is to prevent the loss of grounding continuity provided by the raceway that could result from horizontal movement.

Sections 300.11(A)(1) and (A)(2) are quite similar. Unless the exceptions apply, these sections prohibit all types of wiring from being attached in any way to the support wires of a ceiling assembly or ceiling grid not part of the building structure.

Refer to the appropriate wiring method article in Chapter 3 for cable and raceway supporting requirements. See Article 410, Part IV, for the proper support of luminaires; 314.23 for the support of outlet boxes; and 725.24, 760.24, and 770.24 for various low-voltage fire alarm and optical fiber cable supports. See Chapter 8 for requirements on communications cable supports.

Informational Note: One method of determining fire rating is testing in accordance with ANSI/ASTM E119-2012a, *Method for Fire Tests of Building Construction and Materials*.

**(2) Non–Fire-Rated Assemblies.** Wiring located within the cavity of a non–fire-rated floor–ceiling or roof–ceiling assembly shall not be secured to, or supported by, the ceiling assembly, including the ceiling support wires. An independent means of secure support shall be provided and shall be permitted to be attached to the assembly. Where independent support wires are used, they shall be distinguishable by color, tagging, or other effective means.

*Exception: The ceiling support system shall be permitted to support branch-circuit wiring and associated equipment where installed in accordance with the ceiling system manufacturer's instructions.*

**(B) Raceways Used as Means of Support.** Raceways shall be used only as a means of support for other raceways, cables, or nonelectrical equipment under any of the following conditions:

(1) Where the raceway or means of support is identified as a means of support
(2) Where the raceway contains power supply conductors for electrically controlled equipment and is used to support Class 2 circuit conductors or cables that are solely for the purpose of connection to the equipment control circuits
(3) Where the raceway is used to support boxes or conduit bodies in accordance with 314.23 or to support luminaires in accordance with 410.36(E)

As a general rule, this section prohibits supporting cables by securing them to the exterior of a raceway. Electrical, telephone, and data cables wrapped around a raceway impede dissipation of heat from the raceway, thus affecting the temperature of the conductors contained in the raceway. The weight from large bundles of cables can compromise the mechanical integrity of the raceway system. For these reasons, this section also prohibits the use of a raceway as a means of support for nonelectrical equipment, such as suspended ceilings, water pipes, and nonelectric signs.

Class 2 conductors or cables are allowed to be supported by a raceway as long as the power supply conductors are inside the raceway or functionally associated with the attached Class 2 circuit conductors. For example, the thermostat conductors for heating or air-conditioner units are permitted to be supported by the conduit supplying power to the unit. Exhibit 300.12 shows an example of this application, with the Class 2 circuit cables for control of the heating equipment being secured to and supported by the electrical metallic tubing containing the power conductors for the equipment. These Class 2 circuits are functionally associated with the branch-circuit wiring method.

**(C) Cables Not Used as Means of Support.** Cable wiring methods shall not be used as a means of support for other cables, raceways, or nonelectrical equipment.

## 300.12 Mechanical Continuity — Raceways and Cables

Metal or nonmetallic raceways, cable armors, and cable sheaths shall be continuous between cabinets, boxes, fittings, or other enclosures or outlets.

***EXHIBIT 300.12*** *Raceway used to support Class 2 thermostat cables.*

*Exception No. 1: Short sections of raceways used to provide support or protection of cable assemblies from physical damage shall not be required to be mechanically continuous.*

*Exception No. 2: Raceways and cables installed into the bottom of open bottom equipment, such as switchboards, motor control centers, and floor or pad-mounted transformers, shall not be required to be mechanically secured to the equipment.*

Exception No. 2 permits raceways and cables to enter equipment from below without actually being attached to the underside of the equipment, if it is constructed with an open-bottom feature. Where metal raceways or cables enter open-bottom–type equipment, electrical continuity between the wiring method and the equipment must be maintained as required by 300.10.

## 300.13 Mechanical and Electrical Continuity — Conductors

**(A) General.** Conductors in raceways shall be continuous between outlets, boxes, devices, and so forth. There shall be no splice or tap within a raceway unless permitted by 300.15; 368.56(A); 376.56; 378.56; 384.56; 386.56; 388.56; or 390.7.

Splices or taps are prohibited within raceways unless the raceways are equipped with hinged or removable covers. Busway conductors are exempt from this requirement. Splices and taps must be accessible according to 300.15.

**(B) Device Removal.** In multiwire branch circuits, the continuity of a grounded conductor shall not depend on device connections such as lampholders, receptacles, and so forth, where the removal of such devices would interrupt the continuity.

Grounded conductors (neutrals) of multiwire branch circuits supplying receptacles, lampholders, or other devices are not permitted to depend on terminal connections for continuity between devices. For such installations (3- or 4-wire circuits), a splice may be made with a jumper connected to the terminal or the neutral may be looped. This allows a receptacle or device to be replaced without interrupting the continuity of energized downstream line-to-neutral loads (see the commentary following 300.14). Opening the neutral could cause unbalanced voltages, and a considerably higher voltage would be impressed on one part of a multiwire branch circuit, especially if the downstream line-to-neutral loads were appreciably unbalanced. Section 210.4(B) requires simultaneous disconnection of all ungrounded conductors for each multiwire branch circuit. See the associated commentary following 210.4(B). This requirement does not apply to other circuits that do not contain a grounded (neutral) conductor.

## 300.14 Length of Free Conductors at Outlets, Junctions, and Switch Points

At least 150 mm (6 in.) of free conductor, measured from the point in the box where it emerges from its raceway or cable sheath, shall be left at each outlet, junction, and switch point for splices or the connection of luminaires or devices. Where the opening to an outlet, junction, or switch point is less than 200 mm (8 in.) in any dimension, each conductor shall be long enough to extend at least 75 mm (3 in.) outside the opening.

*Exception: Conductors that are not spliced or terminated at the outlet, junction, or switch point shall not be required to comply with 300.14.*

A conductor intended for connection to a receptacle, switch, or other device in an outlet box must have enough slack for the terminal connections to be made easily. The length of slack (free conductor) required for the box size is shown in Exhibit 300.13. The exception excludes conductors running through a box, which should have sufficient slack to prevent physical damage from the insertion of devices or from the use of luminaire studs, hickeys, or other luminaire supports within the box.

## 300.15 Boxes, Conduit Bodies, or Fittings — Where Required

A box shall be installed at each outlet and switch point for concealed knob-and-tube wiring.

Fittings and connectors shall be used only with the specific wiring methods for which they are designed and listed.

Where the wiring method is conduit, tubing, Type AC cable, Type MC cable, Type MI cable, nonmetallic-sheathed cable, or other cables, a box or conduit body shall be installed at each conductor splice point, outlet point, switch point, junction point, termination point, or pull point, unless otherwise permitted in 300.15(A) through (L).

**(A) Wiring Methods with Interior Access.** A box or conduit body shall not be required for each splice, junction, switch, pull, termination, or outlet points in wiring methods with removable covers, such as wireways, multioutlet assemblies, auxiliary gutters, and surface raceways. The covers shall be accessible after installation.

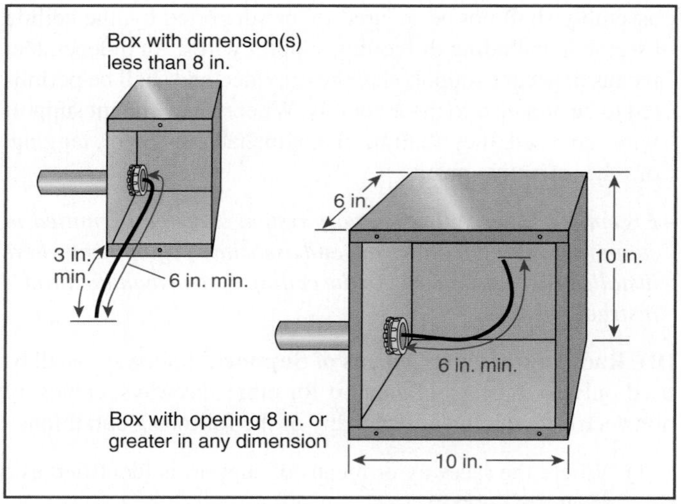


***EXHIBIT 300.13*** *Two different boxes with free conductor lengths.*

**(B) Equipment.** An integral junction box or wiring compartment as part of approved equipment shall be permitted in lieu of a box.

**(C) Protection.** A box or conduit body shall not be required where cables enter or exit from conduit or tubing that is used to provide cable support or protection against physical damage. A fitting shall be provided on the end(s) of the conduit or tubing to protect the cable from abrasion.

Conduit or tubing used as a sleeve to provide support and protection against physical damage for the wires or cables is permitted without terminating in a box. Conduit or tubing used as physical protection for underground cables that exit from buildings or that are located outdoors on poles is also permitted without a box on the end of the conduit. A fitting to protect the wires or cables against physical damage is required on the ends of the conduit or tubing.

**(D) Type MI Cable.** A box or conduit body shall not be required where accessible fittings are used for straight-through splices in mineral-insulated metal-sheathed cable.

**(E) Integral Enclosure.** A wiring device with integral enclosure identified for the use, having brackets that securely fasten the device to walls or ceilings of conventional on-site frame construction, for use with nonmetallic-sheathed cable, shall be permitted in lieu of a box or conduit body.

Informational Note: See 334.30(C); 545.10; 550.15(I); 551.47(E), Exception No. 1; and 552.48(E), Exception No. 1.

This requirement applies to a device with an integral enclosure (boxless device) such as the one shown in Exhibit 300.14.

**(F) Fitting.** A fitting identified for the use shall be permitted in lieu of a box or conduit body where conductors are not spliced

***EXHIBIT 300.14*** *A self-contained device (SCD) receptacle, which is one example of a boxless device. (Courtesy of Legrand/Pass & Seymour®)*

or terminated within the fitting. The fitting shall be accessible after installation.

A fitting is permitted instead of a box where a cable system transitions to a raceway to protect it against physical damage. For example, when a nonmetallic-sheathed cable runs overhead on floor joists then needs protection where it drops down on a masonry wall to supply a receptacle, a short length of raceway is installed to the outlet device box. The cable is then inserted in the raceway and secured by a combination fitting that is fastened to the end of the raceway.

**(G) Direct-Buried Conductors.** As permitted in 300.5(E), a box or conduit body shall not be required for splices and taps in direct-buried conductors and cables.

**(H) Insulated Devices.** As permitted in 334.40(B), a box or conduit body shall not be required for insulated devices supplied by nonmetallic-sheathed cable.

**(I) Enclosures.** A box or conduit body shall not be required where a splice, switch, terminal, or pull point is in a cabinet or cutout box, in an enclosure for a switch or overcurrent device as permitted in 312.8, in a motor controller as permitted in 430.10(A), or in a motor control center.

**(J) Luminaires.** A box or conduit body shall not be required where a luminaire is used as a raceway as permitted in 410.64.

**(K) Embedded.** A box or conduit body shall not be required for splices where conductors are embedded as permitted in 424.40, 424.41(D), 426.22(B), 426.24(A), and 427.19(A).

**(L) Manholes and Handhole Enclosures.** A box or conduit body shall not be required for conductors in manholes or handhole enclosures, except where connecting to electrical equipment. The installation shall comply with the provisions of Part V of Article 110 for manholes, and 314.30 for handhole enclosures.

## 300.16 Raceway or Cable to Open or Concealed Wiring

**(A) Box, Conduit Body, or Fitting.** A box, conduit body, or terminal fitting having a separately bushed hole for each conductor shall be used wherever a change is made from conduit, electrical metallic tubing, electrical nonmetallic tubing, nonmetallic-sheathed cable, Type AC cable, Type MC cable, or mineral-insulated, metal-sheathed cable and surface raceway wiring to open wiring or to concealed knob-and-tube wiring. A fitting used for this purpose shall contain no taps or splices and shall not be used at luminaire outlets. A conduit body used for this purpose shall contain no taps or splices, unless it complies with 314.16(C)(2).

**(B) Bushing.** A bushing shall be permitted in lieu of a box or terminal where the conductors emerge from a raceway and enter or terminate at equipment, such as open switchboards, unenclosed control equipment, or similar equipment. The bushing shall be of the insulating type for other than lead-sheathed conductors.

## 300.17 Number and Size of Conductors in Raceway

The number and size of conductors in any raceway shall not be more than will permit dissipation of the heat and ready installation or withdrawal of the conductors without damage to the conductors or to their insulation.

Informational Note: See the following sections of this *Code*: intermediate metal conduit, 342.22; rigid metal conduit, 344.22; flexible metal conduit, 348.22; liquidtight flexible metal conduit, 350.22; PVC conduit, 352.22; HDPE conduit, 353.22; RTRC, 355.22; liquidtight nonmetallic flexible conduit, 356.22; electrical metallic tubing, 358.22; flexible metallic tubing, 360.22; electrical nonmetallic tubing, 362.22; cellular concrete floor raceways, 372.11; cellular metal floor raceways, 374.5; metal wireways, 376.22; nonmetallic wireways, 378.22; surface metal raceways, 386.22; surface nonmetallic raceways, 388.22; underfloor raceways, 390.6; fixture wire, 402.7; theaters, 520.6; signs, 600.31(C); elevators, 620.33; audio signal processing, amplification, and reproduction equipment, 640.23(A) and 640.24; Class 1, Class 2, and Class 3 circuits, Article 725; fire alarm circuits, Article 760; and optical fiber cables and raceways, Article 770.

As covered in 310.10(G), wire-pulling compounds must not be used if they have a harmful effect on either the conductor or the conductor insulation.

## 300.18 Raceway Installations

**(A) Complete Runs.** Raceways, other than busways or exposed raceways having hinged or removable covers, shall be installed complete between outlet, junction, or splicing points prior to the installation of conductors. Where required to facilitate the installation of utilization equipment, the raceway shall be permitted to be initially installed without a terminating connection at the equipment. Prewired raceway assemblies shall be permitted only where specifically permitted in this *Code* for the applicable wiring method.

*Exception: Short sections of raceways used to contain conductors or cable assemblies for protection from physical damage shall not be required to be installed complete between outlet, junction, or splicing points.*

One of the primary functions of a raceway is to provide physical protection for conductors. If raceways are incomplete at the time of conductor installation, a greater possibility of damage to the conductors exists.

Section 300.18(A) permits the installation of conductors in a raceway prior to the complete installation of the raceway up to the point of utilization. The motor installation shown in Exhibit 300.15 is a typical example, where the motor will be supplied through liquidtight flexible metal conduit that terminates in the motor terminal box through a 90-degree angle connector. Wiring a luminaire whip prior to connecting a luminaire is also permitted by this section.

**(B) Welding.** Metal raceways shall not be supported, terminated, or connected by welding to the raceway unless specifically designed to be or otherwise specifically permitted to be in this *Code*.

## 300.19 Supporting Conductors in Vertical Raceways

**(A) Spacing Intervals — Maximum.** Conductors in vertical raceways shall be supported if the vertical rise exceeds the values in Table 300.19(A). One cable support shall be provided at the top of the vertical raceway or as close to the top as practical. Intermediate supports shall be provided as necessary to limit supported conductor lengths to not greater than those values specified in Table 300.19(A).

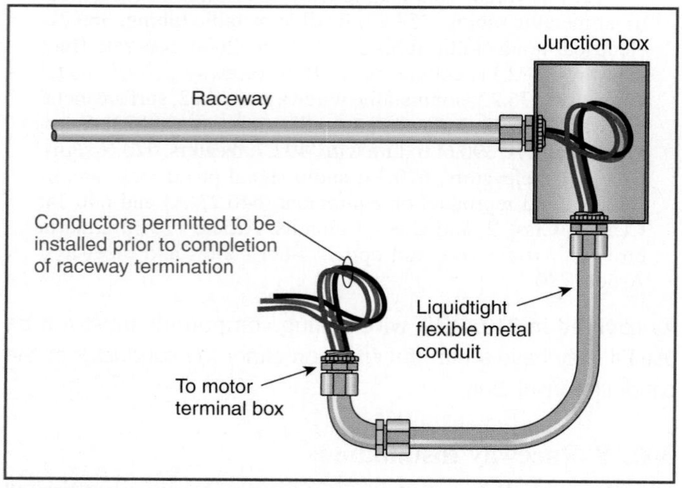


***EXHIBIT 300.15*** *The conductors supplying a motor through LFMC are permitted to be installed prior to the connection of the raceway to the motor terminal box.*

This requirement prevents the weight of the conductors from damaging the insulation where they leave the conduit and prevents the conductors from being pulled out of the terminals. Support bushings or cleats such as those shown in Exhibits 300.16 and 300.17 may be used, in addition to many other types of grips manufactured for this purpose.

### Application Example

A vertical raceway contains 1/0 AWG copper conductors. Using Table 300.19(A), if the vertical run is not greater than 100 feet,

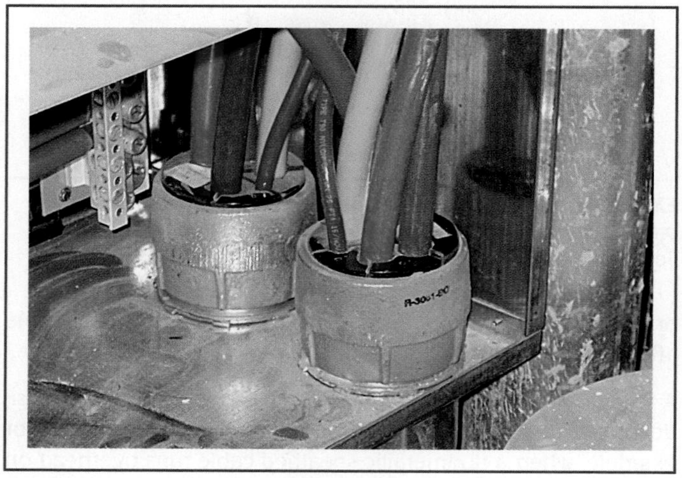

***EXHIBIT 300.16*** *A support bushing, located at the top of a vertical conduit at a cabinet or pull box, used to prevent the weight of the conductors from damaging insulation or placing strain on termination points.*

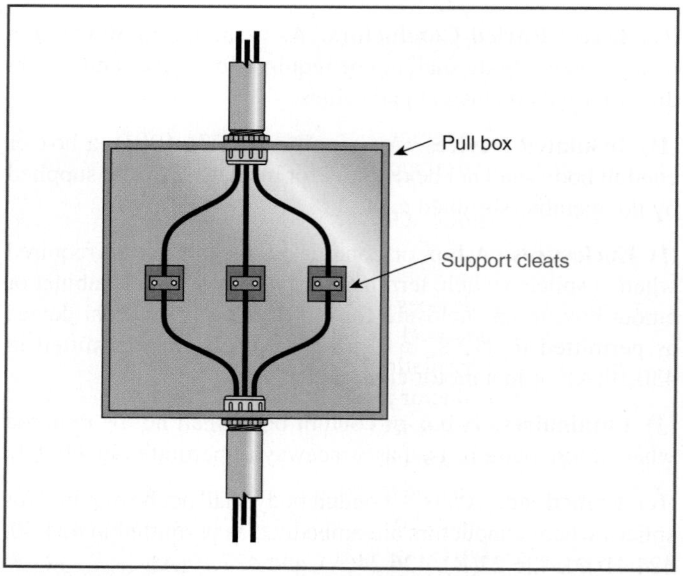


***EXHIBIT 300.17*** *Support cleats used to prevent the weight of vertical conductors from damaging insulation or placing strain on termination points.*

*TABLE 300.19(A) Spacings for Conductor Supports*

| Conductor Size | Support of Conductors in Vertical Raceways | Conductors: Aluminum or Copper-Clad Aluminum | | Conductors: Copper | |
|---|---|---|---|---|---|
| | | m | ft | m | ft |
| 18 AWG through 8 AWG | Not greater than | 30 | 100 | 30 | 100 |
| 6 AWG through 1/0 AWG | Not greater than | 60 | 200 | 30 | 100 |
| 2/0 AWG through 4/0 AWG | Not greater than | 55 | 180 | 25 | 80 |
| Over 4/0 AWG through 350 kcmil | Not greater than | 41 | 135 | 18 | 60 |
| Over 350 kcmil through 500 kcmil | Not greater than | 36 | 120 | 15 | 50 |
| Over 500 kcmil through 750 kcmil | Not greater than | 28 | 95 | 12 | 40 |
| Over 750 kcmil | Not greater than | 26 | 85 | 11 | 35 |

300.19(A) requires one cable support near the top of the run. For longer vertical runs, 300.19(A) requires intermediate supports to limit the supported length to the table values. If the vertical run is less than 100 feet, cable supports are not required.

*Exception: Steel wire armor cable shall be supported at the top of the riser with a cable support that clamps the steel wire armor. A safety device shall be permitted at the lower end of the riser to hold the cable in the event there is slippage of the cable in the wire-armored cable support. Additional wedge-type supports shall be permitted to relieve the strain on the equipment terminals caused by expansion of the cable under load.*

**(B) Fire-Rated Cables and Conductors.** Support methods and spacing intervals for fire-rated cables and conductors shall comply with any restrictions provided in the listing of the electrical circuit protective system used and in no case shall exceed the values in Table 300.19(A).

Electrical circuit protective systems using fire-rated cable assemblies require robust support systems. For example, one manufacturer of Type MI fire-rated cable requires that vertical runs of a two-hour fire-rated assembly be supported every 4 feet, varying substantially from the support requirements of Table 300.19(A).

**(C) Support Methods.** One of the following methods of support shall be used:

(1) By clamping devices constructed of or employing insulating wedges inserted in the ends of the raceways. Where clamping of insulation does not adequately support the cable, the conductor also shall be clamped.
(2) By inserting boxes at the required intervals in which insulating supports are installed and secured in a satisfactory manner to withstand the weight of the conductors attached thereto, the boxes being provided with covers.
(3) In junction boxes, by deflecting the cables not less than 90 degrees and carrying them horizontally to a distance not less than twice the diameter of the cable, the cables being carried on two or more insulating supports and additionally secured thereto by tie wires if desired. Where this method is used, cables shall be supported at intervals not greater than 20 percent of those mentioned in the preceding tabulation.
(4) By a method of equal effectiveness.

## 300.20 Induced Currents in Ferrous Metal Enclosures or Ferrous Metal Raceways

**(A) Conductors Grouped Together.** Where conductors carrying alternating current are installed in ferrous metal enclosures or ferrous metal raceways, they shall be arranged so as to avoid heating the surrounding ferrous metal by induction. To accomplish this, all phase conductors and, where used, the grounded conductor and all equipment grounding conductors shall be grouped together.

Nonferrous metals are defined as those with little or no iron in their composition. Some of the more common nonferrous (nonmagnetic) metals include aluminum, brass, bronze, copper, lead, tin, and zinc. Section 300.20(A) addresses the problem of induction from ac conductors into ferrous (magnetic) metal enclosures and ferrous raceways. Induction into raceways and enclosures can lead to overheating and is also a shock hazard.

*Exception No. 1: Equipment grounding conductors for certain existing installations shall be permitted to be installed separate from their associated circuit conductors where run in accordance with the provisions of 250.130(C).*

*Exception No. 2: A single conductor shall be permitted to be installed in a ferromagnetic enclosure and used for skin-effect heating in accordance with the provisions of 426.42 and 427.47.*

**(B) Individual Conductors.** Where a single conductor carrying alternating current passes through metal with magnetic properties, the inductive effect shall be minimized by (1) cutting slots in the metal between the individual holes through which the individual conductors pass or (2) passing all the conductors in the circuit through an insulating wall sufficiently large for all of the conductors of the circuit.

*Exception: In the case of circuits supplying vacuum or electric-discharge lighting systems or signs or X-ray apparatus, the currents carried by the conductors are so small that the inductive heating effect can be ignored where these conductors are placed in metal enclosures or pass through metal.*

Informational Note: Because aluminum is not a magnetic metal, there will be no heating due to hysteresis; however, induced currents will be present. They will not be of sufficient magnitude to require grouping of conductors or special treatment in passing conductors through aluminum wall sections.

## 300.21 Spread of Fire or Products of Combustion

Electrical installations in hollow spaces, vertical shafts, and ventilation or air-handling ducts shall be made so that the possible spread of fire or products of combustion will not be substantially increased. Openings around electrical penetrations into or through fire-resistant-rated walls, partitions, floors, or ceilings shall be firestopped using approved methods to maintain the fire resistance rating.

Informational Note: Directories of electrical construction materials published by qualified testing laboratories contain many listing installation restrictions necessary to maintain the fire-resistive rating of assemblies where penetrations or openings are made. Building codes also contain restrictions on membrane penetrations on opposite sides of a fire-resistance-rated wall assembly. An example is the 600-mm (24-in.) minimum horizontal separation that usually applies between boxes installed on opposite sides of the wall. Assistance in complying with 300.21 can be found in building codes, fire resistance directories, and product listings.

Cables, cable trays, and raceways must be installed through fire-rated walls, floors, and ceiling assemblies using an approved firestop method so that they do not contribute to the spread of fire or the products of combustion. In the UL *Guide Information for Electrical Equipment – The White Book*, Category XHEZ covers through-penetration firestop systems and Category XHLI covers firestop devices. These two category sections provide valuable information concerning application, installation, and use of firestop systems and firestop devices. A firestop system, the seals for which are shown in Exhibit 300.18, meets the requirements of 300.21.

Using the proper protection techniques is crucial in limiting or stopping flame, excessive temperature, and smoke from passing through fire-rated construction. The structural integrity of the floor or wall assembly needs to be evaluated when providing openings for the penetrating items.

First, the rating of the building assembly being penetrated by electrical cables or conduits must be determined. This information is available from construction documents or building codes. Once the fire resistance rating of the penetrated wall, floor, or ceiling–floor assembly has been determined, the properties and types of electrical penetration required [whether metal or nonmetallic cable(s) or conduit(s) or even cable trays] must be determined. Next, the method of firestopping must be selected. According to 300.21, the selected method must first be approved and

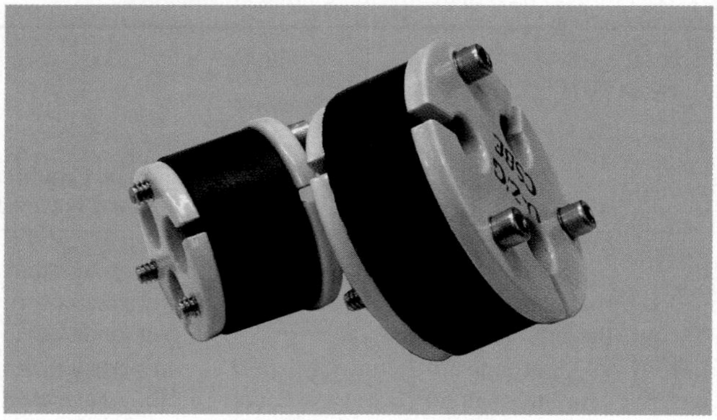

***EXHIBIT 300.18*** *Fire seals used in a through-penetration firestop system to maintain the fire resistance rating of the wall. (Courtesy of O-Z/Gedney, a division of EGS Electrical Group)*

satisfactory to the AHJ. The building codes generally require that a firestop system or device be tested in accordance with ASTM E 814-2006, *Standard Test Method for Fire Tests of Through-Penetration Fire Stops*, or ANSI/UL 1479-2006, *Fire Tests of Through-Penetration Firestops*. But, as an alternative to these tested systems or devices, NFPA 221, *Standard for High Challenge Fire Walls, Fire Walls, and Fire Barrier Walls* (as well as some building codes), contains a statement somewhat similar to the following:

> Where concrete, grout, or mortar has been used to fill the annular spaces around . . . steel conduit or tubing that penetrates one or more concrete or masonry walls, the nominal diameter of each penetrating item shall not exceed 6 in., the opening size shall not exceed 144 in.$^2$ and the thickness of the concrete, grout, or mortar shall be the full thickness of the assembly.

## 300.22 Wiring in Ducts Not Used for Air Handling, Fabricated Ducts for Environmental Air, and Other Spaces for Environmental Air (Plenums)

The provisions of this section shall apply to the installation and uses of electrical wiring and equipment in ducts used for dust, loose stock, or vapor removal; ducts specifically fabricated for environmental air; and other spaces used for environmental air (plenums).

Informational Note: See Article 424, Part VI, for duct heaters.

Other codes and standards refer to all spaces that move air as plenums, including spaces that the *NEC* identifies as "other spaces." Some of the terms used to describe this space include plenum, ceiling cavity plenum, and raised floor plenum. Products are often required to be marked as "suitable for use in other spaces for environmental air" or equivalent language to comply with requirements in the *NEC*.

Section 300.22(B) addresses ducts and spaces used solely for the movement of environmental air. The term *plenum* appears with *other spaces used for environmental air* to make it clear that it applies to structures that are not fabricated specifically to handle environmental air as the primary purpose, but handle the air nonetheless.

**(A) Ducts for Dust, Loose Stock, or Vapor Removal.** No wiring systems of any type shall be installed in ducts used to transport dust, loose stock, or flammable vapors. No wiring system of any type shall be installed in any duct, or shaft containing only such ducts, used for vapor removal or for ventilation of commercial-type cooking equipment.

**(B) Ducts Specifically Fabricated for Environmental Air.** Equipment, devices, and the wiring methods specified in this section shall be permitted within such ducts only if necessary for the direct action upon, or sensing of, the contained air. Where equipment or devices are installed and illumination is necessary to facilitate maintenance and repair, enclosed gasketed-type luminaires shall be permitted.

Only wiring methods consisting of Type MI cable without an overall nonmetallic covering, Type MC cable employing a smooth or corrugated impervious metal sheath without an overall nonmetallic covering, electrical metallic tubing, flexible metallic tubing, intermediate metal conduit, or rigid metal conduit without an overall nonmetallic covering shall be installed in ducts specifically fabricated to transport environmental air. Flexible metal conduit shall be permitted, in lengths not to exceed 1.2 m (4 ft), to connect physically adjustable equipment and devices permitted to be in these fabricated ducts. The connectors used with flexible metal conduit shall effectively close any openings in the connection.

The use of materials that would contribute smoke and products of combustion during a fire in an area that handles environmental air is limited. Section 300.22(B) provides for an effective barrier against the spread of products of combustion through sheet metal ducts and other ducts specifically fabricated to transport environmental air.

Because equipment and devices such as luminaires and motors are not normally permitted in ducts or plenums, the wiring methods in 300.22(B) differ from those permitted in 300.22(C).

**(C) Other Spaces Used for Environmental Air (Plenums).** This section shall apply to spaces not specifically fabricated for environmental air-handling purposes but used for air-handling purposes as a plenum. This section shall not apply to habitable rooms or areas of buildings, the prime purpose of which is not air handling.

Informational Note No. 1: The space over a hung ceiling used for environmental air-handling purposes is an example of the type of other space to which this section applies.

Informational Note No. 2: The phrase "Other Spaces Used for Environmental Air (Plenum)" as used in this section correlates with the use of the term "plenum" in NFPA 90A-2012, *Standard for the Installation of Air-Conditioning and Ventilating Systems*, and other mechanical codes where the plenum is used for return air purposes, as well as some other air-handling spaces.

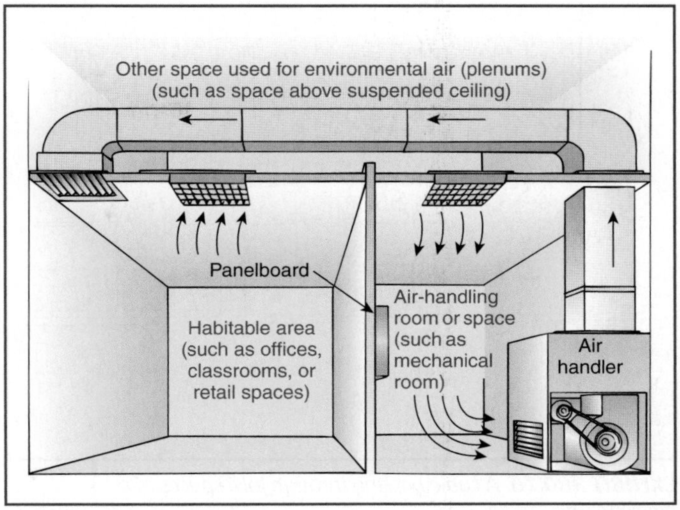


***EXHIBIT 300.19*** *An example of other spaces used for environmental air.*

Other spaces or plenums — such as the space or cavity between a structural floor or roof and a suspended (hung) ceiling — are used to transport environmental air and are not specifically manufactured as ducts. Many spaces above suspended ceilings are intended to transport return air. Informational Note No. 2 correlates the term *other spaces used with environmental air* with the term *plenum* as used in NFPA 90A, *Standard for the Installation of Air-Conditioning and Ventilating Systems*

This section does not apply to habitable rooms and other areas whose prime purpose is other than air handling and ordinary wiring methods are permitted in these areas. If the prime purpose of the room or space is air handling as depicted in Exhibit 300.19, the restrictions in 300.22(C) apply, whether or not electrical equipment is located in the room. Only wiring methods described in 300.22(C)(1) are permitted in such rooms or spaces.

*Exception: This section shall not apply to the joist or stud spaces of dwelling units where the wiring passes through such spaces perpendicular to the long dimension of such spaces.*

This exception permits cable to pass through joist or stud spaces of a dwelling unit, where the joist space is used as a return for a forced-air central heating or air-conditioning system. As shown in Exhibit 300.20, the joist space is covered with appropriate material, and the cable passes through the space perpendicular to the vertical run. The exception does not permit equipment such as junction boxes or device enclosures within this space unless the wiring method employed and enclosure type used meet the requirements of both 300.22(C)(1) and (C)(2).

**(1) Wiring Methods.** The wiring methods for such other space shall be limited to totally enclosed, nonventilated, insulated busway having no provisions for plug-in connections, Type MI cable without an overall nonmetallic covering, Type MC cable

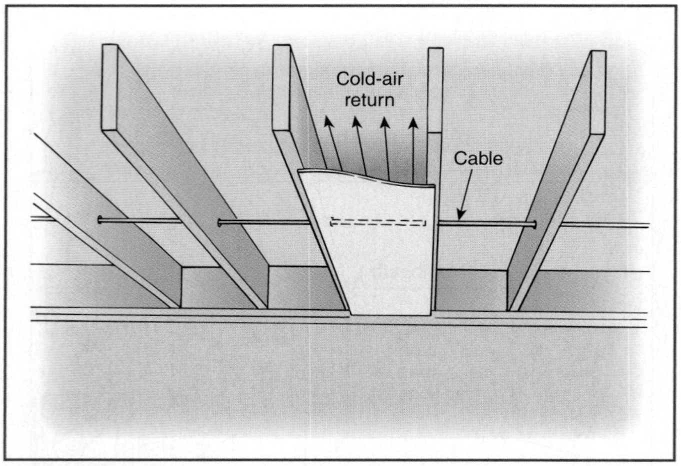


**EXHIBIT 300.20** *A cable passing through joist spaces of a dwelling unit.*

without an overall nonmetallic covering, Type AC cable, or other factory-assembled multiconductor control or power cable that is specifically listed for use within an air-handling space, or listed prefabricated cable assemblies of metallic manufactured wiring systems without nonmetallic sheath. Other types of cables, conductors, and raceways shall be permitted to be installed in electrical metallic tubing, flexible metallic tubing, intermediate metal conduit, rigid metal conduit without an overall nonmetallic covering, flexible metal conduit, or, where accessible, surface metal raceway or metal wireway with metal covers.

Nonmetallic cable ties and other nonmetallic cable accessories used to secure and support cables shall be listed as having low smoke and heat release properties.

Informational Note: One method to determine low smoke and heat release properties is that the nonmetallic cable ties and other nonmetallic cable accessories exhibit a maximum peak optical density of 0.50 or less, an average optical density of 0.15 or less, and a peak heat release rate of 100 kW or less when tested in accordance with ANSI/UL 2043-2008, *Fire Test for Heat and Visible Smoke Release for Discrete Products and Their Accessories Installed in Air-Handling Spaces.*

**(2) Cable Tray Systems.** The provisions in (a) or (b) shall apply to the use of metallic cable tray systems in other spaces used for environmental air (plenums), where accessible, as follows:

(a) *Metal Cable Tray Systems.* Metal cable tray systems shall be permitted to support the wiring methods in 300.22(C)(1).

(b) *Solid Side and Bottom Metal Cable Tray Systems.* Solid side and bottom metal cable tray systems with solid metal covers shall be permitted to enclose wiring methods and cables, not already covered in 300.22(C)(1), in accordance with 392.10(A) and (B).

Cable trays are a support system. They are permitted to support any wiring permitted to be installed in plenums per 300.22(C)(1), as long as the cable tray is suitable for use in a plenum. Solid-bottom cable trays with solid metal covers and solid sides can support any system allowed in 392.10 because they would completely enclose the wiring method.

**(3) Equipment.** Electrical equipment with a metal enclosure, or electrical equipment with a nonmetallic enclosure listed for use within an air-handling space and having adequate fire-resistant and low-smoke-producing characteristics, and associated wiring material suitable for the ambient temperature shall be permitted to be installed in such other space unless prohibited elsewhere in this *Code*.

Informational Note: One method of defining adequate fire-resistant and low-smoke-producing characteristics for electrical equipment with a nonmetallic enclosure is in ANSI/UL 2043-2008, *Fire Test for Heat and Visible Smoke Release for Discrete Products and Their Accessories Installed in Air-Handling Spaces.*

*Exception: Integral fan systems shall be permitted where specifically identified for use within an air-handling space.*

Electrical equipment with metal enclosures is allowed within spaces used for environmental air. Nonmetallic enclosures are only permitted if they are specifically listed for use within air-handling spaces.

**(D) Information Technology Equipment.** Electrical wiring in air-handling areas beneath raised floors for information technology equipment shall be permitted in accordance with Article 645.

The requirements of 300.22(B) or (C) are not intended to apply to air-handling areas beneath raised floors in information technology equipment rooms that meet the requirements specified in Article 645. See Article 645 for more information on information technology equipment.

### 300.23 Panels Designed to Allow Access

Cables, raceways, and equipment installed behind panels designed to allow access, including suspended ceiling panels, shall be arranged and secured so as to allow the removal of panels and access to the equipment.

## II. Requirements for over 1000 Volts, Nominal

### 300.31 Covers Required

Suitable covers shall be installed on all boxes, fittings, and similar enclosures to prevent accidental contact with energized parts or physical damage to parts or insulation.

### 300.32 Conductors of Different Systems

See 300.3(C)(2).

### 300.34 Conductor Bending Radius

The conductor shall not be bent to a radius less than 8 times the overall diameter for nonshielded conductors or 12 times the overall diameter for shielded or lead-covered conductors during or after installation. For multiconductor or multiplexed single-conductor cables having individually shielded conductors, the minimum bending radius is 12 times the diameter of the individually shielded conductors or 7 times the overall diameter, whichever is greater.

### 300.35 Protection Against Induction Heating

Metallic raceways and associated conductors shall be arranged so as to avoid heating of the raceway in accordance with the applicable provisions of 300.20.

### 300.37 Aboveground Wiring Methods

Aboveground conductors shall be installed in rigid metal conduit, in intermediate metal conduit, in electrical metallic tubing, in RTRC and PVC conduit, in cable trays, in auxiliary gutters, as busways, as cablebus, in other identified raceways, or as exposed runs of metal-clad cable suitable for the use and purpose. In locations accessible to qualified persons only, exposed runs of Type MV cables, bare conductors, and bare busbars shall also be permitted. Busbars shall be permitted to be either copper or aluminum.

Any suitable wiring method may be used in transformer vaults, switch rooms, and similar areas restricted to qualified personnel. Exposed wiring using bare or insulated conductors on insulators is commonly employed, as is RMC, rigid PVC conduit, and RTRC.

### 300.38 Raceways in Wet Locations Above Grade

Where raceways are installed in wet locations above grade, the interior of these raceways shall be considered to be a wet location. Insulated conductors and cables installed in raceways in wet locations above grade shall comply with 310.10(C).

### 300.39 Braid-Covered Insulated Conductors — Exposed Installation

Exposed runs of braid-covered insulated conductors shall have a flame-retardant braid. If the conductors used do not have this protection, a flame-retardant saturant shall be applied to the braid covering after installation. This treated braid covering shall be stripped back a safe distance at conductor terminals, according to the operating voltage. Where practicable, this distance shall not be less than 25 mm (1 in.) for each kilovolt of the conductor-to-ground voltage of the circuit.

### 300.40 Insulation Shielding

Metallic and semiconducting insulation shielding components of shielded cables shall be removed for a distance dependent on the circuit voltage and insulation. Stress reduction means shall be provided at all terminations of factory-applied shielding.

Metallic shielding components such as tapes, wires, or braids, or combinations thereof, shall be connected to a grounding conductor, grounding busbar, or a grounding electrode.

### 300.42 Moisture or Mechanical Protection for Metal-Sheathed Cables

Where cable conductors emerge from a metal sheath and where protection against moisture or physical damage is necessary, the insulation of the conductors shall be protected by a cable sheath terminating device.

### 300.45 Warning Signs

Warning signs shall be conspicuously posted at points of access to conductors in all conduit systems and cable systems. The warning sign(s) shall be legible and permanent and shall carry the following wording:

DANGER—HIGH VOLTAGE—KEEP OUT

### 300.50 Underground Installations

**(A) General.** Underground conductors shall be identified for the voltage and conditions under which they are installed. Direct-burial cables shall comply with the provisions of 310.10(F). Underground cables shall be installed in accordance with 300.50(A)(1), (A)(2), or (A)(3), and the installation shall meet the depth requirements of Table 300.50.

Prior to backfilling a ditch or trench, a warning ribbon must be placed near underground direct-buried conductors over 1000 volts, nominal, that are not encased in concrete. This requirement is intended to reduce the risk of an accident, electrocution, or arc-flash incident during excavation.

**(1) Shielded Cables and Nonshielded Cables in Metal-Sheathed Cable Assemblies.** Underground cables, including nonshielded, Type MC and moisture-impervious metal sheath cables, shall have those sheaths grounded through an effective grounding path meeting the requirements of 250.4(A)(5) or (B)(4). They shall be direct buried or installed in raceways identified for the use.

**(2) Industrial Establishments.** In industrial establishments, where conditions of maintenance and supervision ensure that only qualified persons service the installed cable, nonshielded single-conductor cables with insulation types up to 2000 volts that are listed for direct burial shall be permitted to be directly buried.

**(3) Other Nonshielded Cables.** Other nonshielded cables not covered in 300.50(A)(1) or (A)(2) shall be installed in rigid metal conduit, intermediate metal conduit, or rigid nonmetallic conduit encased in not less than 75 mm (3 in.) of concrete.

**(B) Wet Locations.** The interior of enclosures or raceways installed underground shall be considered to be a wet location.

*TABLE 300.50 Minimum Cover[a] Requirements*

| | General Conditions (not otherwise specified) | | | | | | Special Conditions (use if applicable) | | | | | |
|---|---|---|---|---|---|---|---|---|---|---|---|---|
| | Column 1 | | Column 2 | | Column 3 | | Column 4 | | Column 5 | | Column 6 | |
| | Direct-Buried Cables[b] | | RTRC, PVC, and HDPE Conduit[c] | | Rigid Metal Conduit and Intermediate Metal Conduit | | Raceways Under Buildings or Exterior Concrete Slabs, 100 mm (4 in.) Minimum Thickness[d] | | Cables in Airport Runways or Adjacent Areas Where Trespass Is Prohibited | | Areas Subject to Vehicular Traffic, Such as Thoroughfares and Commercial Parking Areas | |
| **Circuit Voltage** | **mm** | **in.** | **mm** | **in.** | **mm** | **in.** | **mm** | **in.** | **mm** | **in.** | **mm** | **in.** |
| Over 1000 V through 22 kV | 750 | 30 | 450 | 18 | 150 | 6 | 100 | 4 | 450 | 18 | 600 | 24 |
| Over 22 kV through 40 kV | 900 | 36 | 600 | 24 | 150 | 6 | 100 | 4 | 450 | 18 | 600 | 24 |
| Over 40 kV | 1000 | 42 | 750 | 30 | 150 | 6 | 100 | 4 | 450 | 18 | 600 | 24 |

General Notes:

1. Lesser depths shall be permitted where cables and conductors rise for terminations or splices or where access is otherwise required.
2. Where solid rock prevents compliance with the cover depths specified in this table, the wiring shall be installed in a metal or nonmetallic raceway permitted for direct burial. The raceways shall be covered by a minimum of 50 mm (2 in.) of concrete extending down to rock.
3. In industrial establishments, where conditions of maintenance and supervision ensure that qualified persons will service the installation, the minimum cover requirements, for other than rigid metal conduit and intermediate metal conduit, shall be permitted to be reduced 150 mm (6 in.) for each 50 mm (2 in.) of concrete or equivalent placed entirely within the trench over the underground installation.

Specific Footnotes:

[a]Cover is defined as the shortest distance in millimeters (inches) measured between a point on the top surface of any direct-buried conductor, cable, conduit, or other raceway and the top surface of finished grade, concrete, or similar cover.

[b]Underground direct-buried cables that are not encased or protected by concrete and are buried 750 mm (30 in.) or more below grade shall have their location identified by a warning ribbon that is placed in the trench at least 300 mm (12 in.) above the cables.

[c]Listed by a qualified testing agency as suitable for direct burial without encasement. All other nonmetallic systems shall require 50 mm (2 in.) of concrete or equivalent above conduit in addition to the table depth.

[d]The slab shall extend a minimum of 150 mm (6 in.) beyond the underground installation, and a warning ribbon or other effective means suitable for the conditions shall be placed above the underground installation.

Insulated conductors and cables installed in these enclosures or raceways in underground installations shall be listed for use in wet locations and shall comply with 310.10(C). Any connections or splices in an underground installation shall be approved for wet locations.

The inside of all raceways and enclosures installed underground is classified as a wet location. Conductors installed in such underground locations must be listed for use in wet locations and comply with 310.10(C).

**(C) Protection from Damage.** Conductors emerging from the ground shall be enclosed in listed raceways. Raceways installed on poles shall be of rigid metal conduit, intermediate metal conduit, RTRC-XW, Schedule 80 PVC conduit, or equivalent, extending from the minimum cover depth specified in Table 300.50 to a point 2.5 m (8 ft) above finished grade. Conductors entering a building shall be protected by an approved enclosure or raceway from the minimum cover depth to the point of entrance. Where direct-buried conductors, raceways, or cables are subject to movement by settlement or frost, they shall be installed to prevent damage to the enclosed conductors or to the equipment connected to the raceways. Metallic enclosures shall be grounded.

**(D) Splices.** Direct burial cables shall be permitted to be spliced or tapped without the use of splice boxes, provided they are installed using materials suitable for the application. The taps and splices shall be watertight and protected from mechanical damage. Where cables are shielded, the shielding shall be continuous across the splice or tap.

*Exception: At splices of an engineered cabling system, metallic shields of direct-buried single-conductor cables with maintained spacing between phases shall be permitted to be interrupted and overlapped. Where shields are interrupted and overlapped, each shield section shall be grounded at one point.*

**(E) Backfill.** Backfill containing large rocks, paving materials, cinders, large or sharply angular substances, or corrosive materials shall not be placed in an excavation where materials can damage or contribute to the corrosion of raceways, cables, or other substructures or where it may prevent adequate compaction of fill.

Protection in the form of granular or selected material or suitable sleeves shall be provided to prevent physical damage to the raceway or cable.

**(F) Raceway Seal.** Where a raceway enters from an underground system, the end within the building shall be sealed with an identified compound so as to prevent the entrance of moisture or gases, or it shall be so arranged to prevent moisture from contacting live parts.

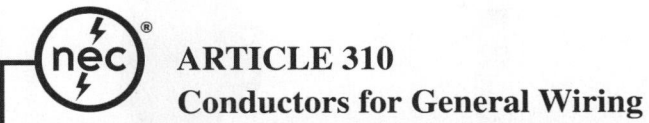

# ARTICLE 310
# Conductors for General Wiring

## I. General

### 310.1 Scope

This article covers general requirements for conductors and their type designations, insulations, markings, mechanical strengths, ampacity ratings, and uses. These requirements do not apply to conductors that form an integral part of equipment, such as motors, motor controllers, and similar equipment, or to conductors specifically provided for elsewhere in this *Code*.

Informational Note: For flexible cords and cables, see Article 400. For fixture wires, see Article 402.

### 310.2 Definitions

**Electrical Ducts.** Electrical conduits, or other raceways round in cross section, that are suitable for use underground or embedded in concrete.

**Thermal Resistivity.** As used in this *Code*, the heat transfer capability through a substance by conduction.

Informational Note: Thermal resistivity is the reciprocal of thermal conductivity and is designated Rho, which is expressed in the units °C-cm/W.

## II. Installation

### 310.10 Uses Permitted

The conductors described in 310.104 shall be permitted for use in any of the wiring methods covered in Chapter 3 and as specified in their respective tables or as permitted elsewhere in this *Code*.

•

**(A) Dry Locations.** Insulated conductors and cables used in dry locations shall be any of the types identified in this *Code*.

**(B) Dry and Damp Locations.** Insulated conductors and cables used in dry and damp locations shall be Types FEP, FEPB, MTW, PFA, RHH, RHW, RHW-2, SA, THHN, THW, THW-2, THHW, THWN, THWN-2, TW, XHH, XHHW, XHHW-2, Z, or ZW.

**(C) Wet Locations.** Insulated conductors and cables used in wet locations shall comply with one of the following:

(1) Be moisture-impervious metal-sheathed
(2) Be types MTW, RHW, RHW-2, TW, THW, THW-2, THHW, THWN, THWN-2, XHHW, XHHW-2, ZW
(3) Be of a type listed for use in wet locations

**(D) Locations Exposed to Direct Sunlight.** Insulated conductors or cables used where exposed to direct rays of the sun shall comply with (D)(1) or (D)(2):

(1) Conductors and cables shall be listed, or listed and marked, as being sunlight resistant
(2) Conductors and cables shall be covered with insulating material, such as tape or sleeving, that is listed, or listed and marked, as being sunlight resistant

**(E) Shielding.** Shielded, ozone-resistant insulated conductors with a maximum phase-to-phase voltage of 5000 volts shall be permitted in Type MC cables in industrial establishments where the conditions of maintenance and supervision ensure that only qualified persons service the installation. For other establishments, solid dielectric insulated conductors operated above 2000 volts in permanent installations shall have ozone-resistant insulation and shall be shielded. All metallic insulation shields shall be connected to a grounding electrode conductor, a grounding busbar, an equipment grounding conductor, or a grounding electrode.

The construction of metal-armored cable provides enhanced reliability because the conductors have a concentric lay-orientation and their insulation is protected from damage during installation. Nonshielded conductors within metal raceways do not provide the same level of reliability. Conductors-into-conduit installation is inconsistent and cannot guarantee insulation will not be damaged nor that cables will be in concentric lay-orientation.

Solid dielectric insulated conductors that are permanently installed and that operate at greater than 2000 volts are required to have ozone-resistant insulation and must be shielded with a grounded metallic shield. Shielding is accomplished by applying a metal tape or nonmetallic semiconducting tape around the conductor surface to prevent corona from forming and to reduce high-voltage stresses.

Corona is a faint glow adjacent to the surface of the electrical conductor at high voltage. If high-voltage stresses and a charging current are flowing between the conductor and ground (usually due to moisture), the surrounding atmosphere is ionized, and ozone – generated by an electric discharge in ordinary oxygen or air – is formed and will attack the conductor jacket and insulation, eventually breaking them down. The shield is at ground potential; therefore, no voltage above ground is present on the jacket outside the shield, thus preventing a discharge from the jacket and the subsequent formation of ozone.

Exhibits 310.1 and 310.2 illustrate shielded cable installations: a three-conductor cable of the shielded type, a stress-relief cone for an indoor cable terminator, and a stress cone on a single-conductor shielded cable terminating inside a pothead. In Exhibit 310.2 (right), a clamping ring provides a grounding connection between the copper shielding tape and the shield to the metallic base of the pothead.

Informational Note: The primary purposes of shielding are to confine the voltage stresses to the insulation, dissipate insulation leakage current, drain off the capacitive charging current, and carry ground-fault current to facilitate operation of ground-fault protective devices in the event of an electrical cable fault.

*Exception No. 1: Nonshielded insulated conductors listed by a qualified testing laboratory shall be permitted for use up to 2400 volts under the following conditions:*

*(a) Conductors shall have insulation resistant to electric discharge and surface tracking, or the insulated conductor(s) shall be covered with a material resistant to ozone, electric discharge, and surface tracking.*

*(b) Where used in wet locations, the insulated conductor(s) shall have an overall nonmetallic jacket or a continuous metallic sheath.*

*(c) Insulation and jacket thicknesses shall be in accordance with Table 310.104(D).*

*Exception No. 2: Nonshielded insulated conductors listed by a qualified testing laboratory shall be permitted for use up to 5000 volts to replace existing nonshielded conductors, on existing equipment in industrial establishments only, under the following conditions:*

*(a) Where the condition of maintenance and supervision ensures that only qualified personnel install and service the installation.*

*(b) Conductors shall have insulation resistant to electric discharge and surface tracking, or the insulated conductor(s) shall be covered with a material resistant to ozone, electric discharge, and surface tracking.*

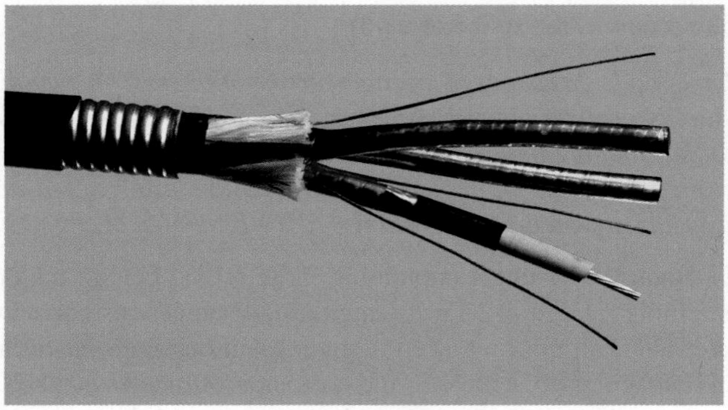

***EXHIBIT 310.1*** *A three-conductor cable of the shielded type.*

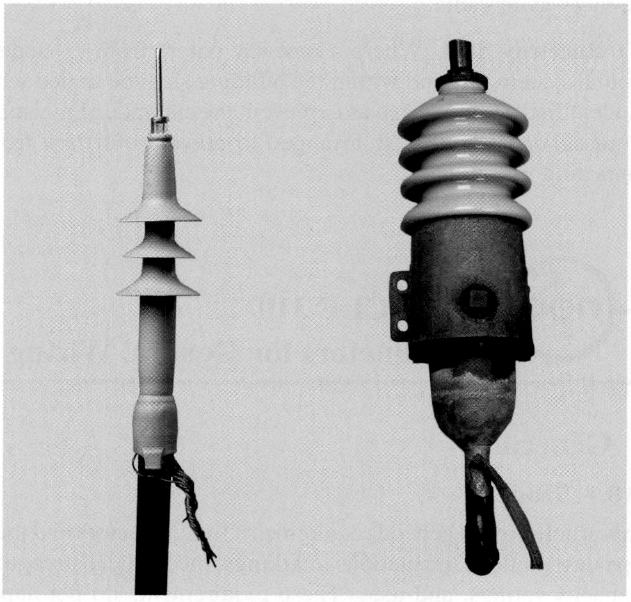

***EXHIBIT 310.2*** *(Left) A one-piece, premolded stress-relief cone for indoor cable terminations of up to 35 kV phase-to-phase. (Right) A stress cone on a single-conductor shielded cable terminating inside a pothead.*

*(c) Where used in wet locations, the insulated conductor(s) shall have an overall nonmetallic jacket or a continuous metallic sheath.*

*(d) Insulation and jacket thicknesses shall be in accordance with Table 310.104(D).*

Informational Note: Relocation or replacement of equipment may not comply with the term existing as related to this exception.

Exception No. 1 limits the omission of shielding up to the 2.4-kV level only, which correlates with the requirements in 310.10(F) for direct-burial conductors. Exception No. 2 recognizes that nonshielded cables with insulation ratings of up to 5000 volts were previously permitted. Where cable in an existing installation requires replacement, it may be preferable to replace a nonshielded cable with another nonshielded cable where existing raceways and termination enclosures do not provide adequate space for shielded conductors and their associated terminations.

Federal Aviation Administration Advisory Circulars for airfield lighting cable permit certain circuits up to 5 kV to be unshielded. See 310.10(F), Exception No. 2, for permitted uses of nonshielded cable in airfield lighting.

Specialized training and close adherence to manufacturers' instructions are absolutely essential for high-voltage cable installations.

*Exception No. 3: Where permitted in 310.10(F), Exception No. 2.*

**(F) Direct-Burial Conductors.** Conductors used for direct-burial applications shall be of a type identified for such use.

Cables rated above 2000 volts shall be shielded.

*Exception No. 1: Nonshielded multiconductor cables rated 2001–2400 volts shall be permitted if the cable has an overall metallic sheath or armor.*

The metallic shield, sheath, or armor shall be connected to a grounding electrode conductor, grounding busbar, or a grounding electrode.

*Exception No. 2: Airfield lighting cable used in series circuits that are rated up to 5000 volts and are powered by regulators shall be permitted to be nonshielded.*

Informational Note to Exception No. 2: Federal Aviation Administration (FAA) Advisory Circulars (ACs) provide additional practices and methods for airport lighting.

Informational Note No. 1: See 300.5 for installation requirements for conductors rated 1000 volts or less.

Informational Note No. 2: See 300.50 for installation requirements for conductors rated over 1000 volts.

**(G) Corrosive Conditions.** Conductors exposed to oils, greases, vapors, gases, fumes, liquids, or other substances having a deleterious effect on the conductor or insulation shall be of a type suitable for the application.

Nylon-jacketed conductors, such as Type THWN, are suitable for use where exposed to gasoline. The UL guide *Information for Electrical Equipment – The White Book* states in the category for Thermoplastic-Insulated Wire (ZLGR) in part:

THWN – wire that is suitable for exposure to mineral oil, and to liquid gasoline and gasoline vapors at ordinary ambient temperature, is marked "Gasoline and Oil Resistant I" if suitable for exposure to mineral oil at 60°C, or "Gasoline and Oil Resistant II" if the compound is suitable for exposure to mineral oil at 75°C. Gasoline-resistant wire has been tested at 23°C when immersed in gasoline and is considered inherently resistant to gasoline vapors within the limits of the temperature rating of the wire type.

Before using a wire-pulling compound, it should first be investigated to determine compliance with 310.10(G).

**(H) Conductors in Parallel.**

**(1) General.** Aluminum, copper-clad aluminum, or copper conductors, for each phase, polarity, neutral, or grounded circuit shall be permitted to be connected in parallel (electrically joined at both ends) only in sizes 1/0 AWG and larger where installed in accordance with 310.10(H)(2) through (H)(6).

Conductors connected in parallel are treated by the *Code* as a single conductor with a total cross-sectional area of all conductors in parallel. The use of parallel conductors is a practical and cost-effective means of installing large-capacity feeders or services. Using conductors larger than 1000 kcmil in raceways is neither economical nor practical unless the conductor size is governed by voltage drop. The ampacity of larger sizes of conductors would increase very little in proportion to the increase in the size of the conductor. Where the cross-sectional area of a conductor increases 50 percent (e.g., from 1000 to 1500 kcmil), a Type THW conductor ampacity increases only 80 amperes (less than 15 percent). A 100-percent increase (from 1000 to 2000 kcmil) causes an increase of only 120 amperes (approximately 22 percent). Generally, where cost is a factor, installation of two (or more) paralleled conductors per phase may be beneficial.

The parallel connection of two or more conductors in place of using one large conductor depends on compliance with 310.10(H)(2) to ensure equal current division in order to prevent overloading any of the individual paralleled conductors.

Where individual conductors are tapped from conductors in parallel, the tap connection must include all the conductors in parallel for that particular phase. Tapping into only one of the parallel conductors would result in unbalanced distribution of tap load current between parallel conductors, resulting in one of the conductors carrying more than its share of the load, which could cause overheating and conductor insulation failure. For example, if a 250-kcmil conductor is tapped from a set of two 500-kcmil conductors in parallel, the splicing device must include both 500-kcmil conductors and the single 250-kcmil tap conductor.

*Exception No. 1: Conductors in sizes smaller than 1/0 AWG shall be permitted to be run in parallel to supply control power to indicating instruments, contactors, relays, solenoids, and similar control devices, or for frequencies of 360 Hz and higher, provided all of the following apply:*

*(a) They are contained within the same raceway or cable.*
*(b) The ampacity of each individual conductor is sufficient to carry the entire load current shared by the parallel conductors.*
*(c) The overcurrent protection is such that the ampacity of each individual conductor will not be exceeded if one or*

*more of the parallel conductors become inadvertently disconnected.*

In control wiring and circuits that operate at frequencies greater than 360 Hz, a reduction of cable capacitance or voltage drop in long lengths of wire may be necessary. A 14 AWG conductor might have more than sufficient capacity to carry the load, but by installing two conductors in parallel, the voltage drop can be reduced to acceptable limits. The presence of the word *polarity* in 310.10(H)(1) specifically allows the inclusion of dc circuits.

*Exception No. 2: Under engineering supervision, 2 AWG and 1 AWG grounded neutral conductors shall be permitted to be installed in parallel for existing installations.*

Informational Note to Exception No. 2: Exception No. 2 can be used to alleviate overheating of neutral conductors in existing installations due to high content of triplen harmonic currents.

The word *triplen* refers to a third-order harmonic current, such as the third, sixth, ninth, and so on. The concern is limited to odd-number triplen harmonic currents, such as the third, ninth, and fifteenth, since these are additive currents in the neutral conductor and do not cancel. See Chapter 10 of NFPA 70B, *Recommended Practice for Electrical Equipment Maintenance*, for additional information on power quality and harmonics.

**(2) Conductor and Installation Characteristics.** The paralleled conductors in each phase, polarity, neutral, grounded circuit conductor, equipment grounding conductor, or equipment bonding jumper shall comply with all of the following:

(1) Be the same length.
(2) Consist of the same conductor material.
(3) Be the same size in circular mil area.
(4) Have the same insulation type.
(5) Be terminated in the same manner.

In order to avoid excessive voltage drop and also to ensure equal division of current, different phase conductors must be located close together and each phase conductor, grounded conductor, and the grounding conductor (if used) must be grouped together in each raceway or cable. However, isolated phase installations are permitted underground where the phase conductors are run in nonmetallic raceways that are in close proximity.

The impedance of a circuit in an aluminum raceway or aluminum-sheathed cable differs from the impedance of the same circuit in a steel raceway or steel-sheathed cable; therefore, separate raceways and cables must have the same physical characteristics. Also, the same number of conductors must be used in each raceway or cable. See 300.20 regarding induced currents in metal enclosures or raceways.

All conductors of the same phase or neutral are required by 310.10(H)(2) to be of the same conductor material. For example, if 12 conductors are paralleled for a 3-phase, 4-wire, 480Y/277 V ac circuit, 4 conductors could be installed in each of three raceways. The *Code* does not intend that all 12 conductors be copper or aluminum but does intend that the individual conductors in parallel for each phase, grounded conductor, and neutral be the same material, insulation type, length, and so forth. Also, the three raceways are intended to have the same physical characteristics (e.g., three rigid aluminum conduits, three steel IMC conduits, three EMTs, or three nonmetallic conduits), not a mixture (e.g., two rigid aluminum conduits and one rigid steel conduit).

The presence of the word *polarity* throughout the section specifically allows the inclusion of dc circuits.

**(3) Separate Cables or Raceways.** Where run in separate cables or raceways, the cables or raceways with conductors shall have the same number of conductors and shall have the same electrical characteristics. Conductors of one phase, polarity, neutral, grounded circuit conductor, or equipment grounding conductor shall not be required to have the same physical characteristics as those of another phase, polarity, neutral, grounded circuit conductor, or equipment grounding conductor.

All parallel raceways or cables for a circuit are required to be of the same size, material, and length. "Cables," in this case, means wiring method–type cables such as Type MC. For example, the conductors in phases A and B may be copper, and those in phase C may be aluminum.

**(4) Ampacity Adjustment.** Conductors installed in parallel shall comply with the provisions of 310.15(B)(3)(a).

**(5) Equipment Bonding Conductors.** Where parallel equipment bonding conductors are used, they shall be sized in accordance with 250.122. Sectioned equipment bonding conductors smaller than 1/0 AWG shall be permitted in multiconductor cables, provided that the combined circular mil area of the sectioned equipment bonding conductors in each cable complies with 250.122.

**(6) Bonding Jumpers.** Where parallel equipment bonding jumpers or supply-side bonding jumpers are installed in raceways, they shall be sized and installed in accordance with 250.102.

The equipment bonding jumper size requirements may be different from the requirements for equipment grounding conductors. On the supply side of the service, the size of the bonding jumper is based on 250.102(C)(1), which is the same as the requirement in 250.66 for grounding electrode conductors. On the load side of the service, the size is based on 250.122, which is also the size requirement for equipment grounding conductors. The 1/0 AWG minimum size limitation on paralleled conductors does not apply to the equipment bonding jumper.

## 310.15 Ampacities for Conductors Rated 0–2000 Volts

### (A) General.

**(1) Tables or Engineering Supervision.** Ampacities for conductors shall be permitted to be determined by tables as provided in 310.15(B) or under engineering supervision, as provided in 310.15(C).

Informational Note No. 1: Ampacities provided by this section do not take voltage drop into consideration. See 210.19(A), Informational Note No. 4, for branch circuits and 215.2(A), Informational Note No. 2, for feeders.

Two methods are permitted for determining conductor ampacity for conductors rated 0 through 2000 volts: selecting the ampacity from a table, using correction factors in the table or notes where necessary, or calculating the ampacity. The latter method can be complex and time consuming and requires engineering supervision. It can, however, result in lower installation costs in some cases, and if calculated properly, it provides a mathematically exact ampacity. See the commentary following 310.15(C) and Informative Annex B for further explanation.

Informational Note No. 2: For the allowable ampacities of Type MTW wire, see Table 13.5.1 in NFPA 79-2012, *Electrical Standard for Industrial Machinery*.

**(2) Selection of Ampacity.** Where more than one ampacity applies for a given circuit length, the lowest value shall be used.

*Exception: Where two different ampacities apply to adjacent portions of a circuit, the higher ampacity shall be permitted to be used beyond the point of transition, a distance equal to 3.0 m (10 ft) or 10 percent of the circuit length figured at the higher ampacity, whichever is less.*

Informational Note: See 110.14(C) for conductor temperature limitations due to termination provisions.

#### Calculation Example

Three 500-kcmil THW conductors in a RMC are run from a motor control center for 12 feet past a heat-treating furnace to a pump motor located 150 feet from the motor control center. Where run in a 78°F to 86°F ambient temperature, the conductors have an ampacity of 380 amperes, per Table 310.15(B)(16). The ambient temperature near the furnace, where the conduit is run, is found to be 113°F, and the length of this particular part of the run is greater than 10 feet and more than 10 percent of the total length of the run at the 78°F to 86°F ambient. Determine the ampacity of the total run in accordance with 310.15(A)(2).

*Solution*

Using the ambient temperature correction factors in Table 310.15(B)(2)(a) for 113°F, the ampacity is calculated:

$$0.82 \times 380 \text{ A} = 311.6 \text{ A}$$

which is the ampacity of the total run, in accordance with 310.15(A)(2).

If the run near the furnace at the 113°F ambient temperature was 10 feet or less in length, then the ampacity of the entire run would have been 380 A, according to the exception to 310.15(A)(2). The heat-sinking effect of the run at the lower ambient temperature is sufficient to reduce the conductor temperature near the furnace.

**(3) Temperature Limitation of Conductors.** No conductor shall be used in such a manner that its operating temperature exceeds that designated for the type of insulated conductor involved. In no case shall conductors be associated together in such a way, with respect to type of circuit, the wiring method employed, or the number of conductors, that the limiting temperature of any conductor is exceeded.

Most terminations are normally designed for 60°C and/or 75°C maximum temperatures. The higher-rated ampacities for conductors of 90°C, 105°C, and so forth cannot be used unless the terminals at which the conductors terminate have comparable ratings.

Ambient temperature must also be considered in determining the allowable ampacity of conductors. Tables 310.15(B)(16) through 310.15(B)(20) have ampacities based on a 30°C or 40°C ambient, as indicated in the table heading. Where the ambient temperature is different, Table 310.15(B)(2)(a) or Table 310.15(B)(2)(b) is used to correct the ampacity. This correction factor is applied in addition to any adjustment factor, such as in 310.15(B)(3)(a).

Chosen conductors should have a rating above the anticipated maximum ambient temperature. The operating temperature of conductors should be controlled at or below the conductor rating by coordinating conductor size, number of associated conductors, and ampacity for the particular conductor rating and ambient temperature. All tabulations should be corrected for the anticipated ambient temperature, using the correction factors at the bottom of the ampacity tables. If more than three conductors are associated together, the additional adjustment shown in 310.15(B)(3)(a) must be applied.

#### Calculation Example

Determine the ampacity of 2 AWG THHN copper conductors to be installed in a raceway in an ambient temperature of 50°C (122°F).

*Solution*

Table 310.15(B)(16) shows that the allowable ampacity of the conductor at 30°C is 130 A, which is multiplied by 0.82 [the ambient temperature correction factor in Table 310.15(B)(2)(a)].

$$130 \text{ A} \times 0.82 = 106.6 \text{ A}$$

Thus, the allowable ampacity of the 2 AWG conductor at 50°C is reduced to 106.6 amperes. For six of these conductors run in the raceway, 310.15(B)(3)(a) requires the allowable ampacity to be further reduced to 80 percent:

$$106.6 \text{ A} \times 0.8 = 85.28 \text{ A}$$

Under these conditions, the 2 AWG conductors would be suitable for an 80-A circuit, based on the standard ampere ratings of circuit breaker and fuses in 240.6(A).

The basis for determining the ampacities of conductors for Tables 310.15(B)(16) and 310.15(B)(17) was the NEMA *Report of Determination of Maximum Permissible Current-Carrying Capacity of Code Insulated Wires and Cables for Building Purposes*, dated June 27, 1938. The basis for determining the ampacities of conductors for Tables 310.15(B)(18) and 310.15(B)(19) and the ampacity tables in Informative Annex B was the Neher–McGrath method. See the commentary following 310.15(C) and the inside cover of this book for further explanation.

Informational Note No. 1: The temperature rating of a conductor [see Table 310.104(A) and Table 310.104(C)] is the maximum temperature, at any location along its length, that the conductor can withstand over a prolonged time period without serious degradation. The allowable ampacity tables, the ampacity tables of Article 310 and the ampacity tables of Informative Annex B, the ambient temperature correction factors in 310.15(B)(2), and the notes to the tables provide guidance for coordinating conductor sizes, types, allowable ampacities, ampacities, ambient temperatures, and number of associated conductors. The principal determinants of operating temperature are as follows:

(1) Ambient temperature — ambient temperature may vary along the conductor length as well as from time to time.
(2) Heat generated internally in the conductor as the result of load current flow, including fundamental and harmonic currents.
(3) The rate at which generated heat dissipates into the ambient medium. Thermal insulation that covers or surrounds conductors affects the rate of heat dissipation.
(4) Adjacent load-carrying conductors — adjacent conductors have the dual effect of raising the ambient temperature and impeding heat dissipation.

Informational Note No. 1 focuses attention on the necessity for derating conductors where high ambient temperatures are encountered. It also provides helpful information in coordinating ampacities, ambient temperatures, conductor size and number, and so forth to ensure operation at or below the temperature rating.

Item (2) of the informational note explains that heating due to harmonic current should also be considered in determinations of operating temperature. In certain cases, larger-sized conductors may be required. For existing installations, see 310.10(H)(1), Exception No. 2, and the informational note and the associated commentary that follows.

Informational Note No. 2: Refer to 110.14(C) for the temperature limitation of terminations.

**(B) Tables.** Ampacities for conductors rated 0 to 2000 volts shall be as specified in the Allowable Ampacity Table 310.15(B)(16) through Table 310.15(B)(19), and Ampacity Table 310.15(B)(20) and Table 310.15(B)(21) as modified by 310.15(B)(1) through (B)(7).

The temperature correction and adjustment factors shall be permitted to be applied to the ampacity for the temperature rating of the conductor, if the corrected and adjusted ampacity does not exceed the ampacity for the temperature rating of the termination in accordance with the provisions of 110.14(C).

Each of the table numbers corresponds with the applicable *Code* section. The ampacity table numbers also include a reference to the former table numbers. For example, former Table 310.16 is Table 310.15(B)(16). In addition, Table 310.15(B)(2)(a) is utilized for ambient temperature correction based on 30°C for use with Tables 310.15(B)(16) and 310.15(B)(17). Table 310.15(B)(2)(b) is employed for ambient temperature correction based on 40°C for use with Tables 310.15(B)(18) and 310.15(B)(20).

Informational Note: Table 310.15(B)(16) through Table 310.15(B)(19) are application tables for use in determining conductor sizes on loads calculated in accordance with Article 220. Allowable ampacities result from consideration of one or more of the following:

(1) Temperature compatibility with connected equipment, especially the connection points.
(2) Coordination with circuit and system overcurrent protection.
(3) Compliance with the requirements of product listings or certifications. See 110.3(B).
(4) Preservation of the safety benefits of established industry practices and standardized procedures.

Ampacity tables, particularly Table 310.15(B)(16), do not take into account all the many factors affecting ampacity. However, experience has proven the table values to be adequate for loads calculated in accordance with Article 220, because not all of the load diversity occurring in most actual installations is specifically provided for in Article 220. If loads are not calculated in accordance with the requirements of Article 220, the table ampacities, even when corrected in accordance with ambient correction factors and the notes to the tables, might be too high. This result can be particularly true where many cables or raceways are routed close to one another underground. However, load diversity and thermal conductance fill around buried cable could result in increased ampacity. For further information, see the commentary following 310.15(C) and Informative Annex B.

The factors in the second column of Table 310.15(B)(3)(a) are based on no diversity, meaning that all conductors in the raceway or cable are loaded to their maximum rated load. The table values can be used even if there is diversity, because the table values would represent a worst-case scenario. If there is diversity and a more favorable adjustment factor is desired, a calculation can be performed in accordance with Informative Annex B.

Specific cross references for raceway fill and adjustment factors of 310.15(B)(2) can be found in the informational note following 300.17. For fill requirements and adjustment factors for Class 1 conductors, see 725.51; for fire alarm systems, 760.51 and 760.130; for optical fiber cables and raceways, 770.110; and for communications wires and cables within buildings, 800.110.

**(1) General.** For explanation of type letters used in tables and for recognized sizes of conductors for the various conductor insulations, see Table 310.104(A) and Table 310.104(B). For installation requirements, see 310.1 through 310.15(A)(3) and the various articles of this *Code*. For flexible cords, see Table 400.4, Table 400.5(A)(1), and Table 400.5(A)(2).

**(2) Ambient Temperature Correction Factors.** Ampacities for ambient temperatures other than those shown in the ampacity tables shall be corrected in accordance with Table 310.15(B)(2)(a) or Table 310.15(B)(2)(b), or shall be permitted to be calculated using the following equation:

$$I' = I\sqrt{\frac{T_c - T_a'}{T_c - T_a}}$$

where:

$I'$ = ampacity corrected for ambient temperature
$I$ = ampacity shown in the tables
$T_c$ = temperature rating of conductor (°C)
$T_a'$ = new ambient temperature (°C)
$T_a$ = ambient temperature used in the table (°C)

**(3) Adjustment Factors.**

(a) *More Than Three Current-Carrying Conductors.* Where the number of current-carrying conductors in a raceway or cable exceeds three, or where single conductors or multiconductor cables are installed without maintaining spacing for a continuous length longer than 600 mm (24 in.) and are not installed in raceways, the allowable ampacity of each conductor shall be reduced as shown in Table 310.15(B)(3)(a). Each current-carrying conductor of a paralleled set of conductors shall be counted as a current-carrying conductor.

Where conductors of different systems, as provided in 300.3, are installed in a common raceway or cable, the adjustment factors shown in Table 310.15(B)(3)(a) shall apply only to the number of power and lighting conductors (Articles 210, 215, 220, and 230).

The basis for the last paragraph of 310.15(B)(3)(a) is the assumption that the watt loss (heating) from any control and signal conductors in the same raceway or cable will not be enough to significantly increase the temperature of the power and lighting conductors. See 725.48 and 725.133 for limitations on the installation of control and signal conductors in the same raceway or cable as power and lighting conductors.

Informational Note No. 1: See Annex B for adjustment factors for more than three current-carrying conductors in a raceway or cable with load diversity.

Informational Note No. 2: See 366.23(A) for adjustment factors for conductors and ampacity for bare copper and aluminum bars in sheet metal auxiliary gutters and 376.22(B) for adjustment factors for conductors in metal wireways.

(1) Where conductors are installed in cable trays, the provisions of 392.80 shall apply.
(2) Adjustment factors shall not apply to conductors in raceways having a length not exceeding 600 mm (24 in.).
(3) Adjustment factors shall not apply to underground conductors entering or leaving an outdoor trench if those conductors have physical protection in the form of rigid metal conduit, intermediate metal conduit, rigid polyvinyl chloride conduit (PVC), or reinforced thermosetting resin conduit (RTRC) having a length not exceeding 3.05 m (10 ft), and if the number of conductors does not exceed four.

***TABLE 310.15(B)(2)(a)*** *Ambient Temperature Correction Factors Based on 30°C (86°F)*

**For ambient temperatures other than 30°C (86°F), multiply the allowable ampacities specified in the ampacity tables by the appropriate correction factor shown below.**

| Ambient Temperature (°C) | Temperature Rating of Conductor | | | Ambient Temperature (°F) |
|---|---|---|---|---|
| | 60°C | 75°C | 90°C | |
| 10 or less | 1.29 | 1.20 | 1.15 | 50 or less |
| 11–15 | 1.22 | 1.15 | 1.12 | 51–59 |
| 16–20 | 1.15 | 1.11 | 1.08 | 60–68 |
| 21–25 | 1.08 | 1.05 | 1.04 | 69–77 |
| 26–30 | 1.00 | 1.00 | 1.00 | 78–86 |
| 31–35 | 0.91 | 0.94 | 0.96 | 87–95 |
| 36–40 | 0.82 | 0.88 | 0.91 | 96–104 |
| 41–45 | 0.71 | 0.82 | 0.87 | 105–113 |
| 46–50 | 0.58 | 0.75 | 0.82 | 114–122 |
| 51–55 | 0.41 | 0.67 | 0.76 | 123–131 |
| 56–60 | — | 0.58 | 0.71 | 132–140 |
| 61–65 | — | 0.47 | 0.65 | 141–149 |
| 66–70 | — | 0.33 | 0.58 | 150–158 |
| 71–75 | — | — | 0.50 | 159–167 |
| 76–80 | — | — | 0.41 | 168–176 |
| 81–85 | — | — | 0.29 | 177–185 |

*TABLE 310.15(B)(2)(b) Ambient Temperature Correction Factors Based on 40°C (104°F)*

**For ambient temperatures other than 40°C (104°F), multiply the allowable ampacities specified in the ampacity tables by the appropriate correction factor shown below.**

| Ambient Temperature (°C) | Temperature Rating of Conductor | | | | | | Ambient Temperature (°F) |
|---|---|---|---|---|---|---|---|
| | 60°C | 75°C | 90°C | 150°C | 200°C | 250°C | |
| 10 or less | 1.58 | 1.36 | 1.26 | 1.13 | 1.09 | 1.07 | 50 or less |
| 11–15 | 1.50 | 1.31 | 1.22 | 1.11 | 1.08 | 1.06 | 51–59 |
| 16–20 | 1.41 | 1.25 | 1.18 | 1.09 | 1.06 | 1.05 | 60–68 |
| 21–25 | 1.32 | 1.2 | 1.14 | 1.07 | 1.05 | 1.04 | 69–77 |
| 26–30 | 1.22 | 1.13 | 1.10 | 1.04 | 1.03 | 1.02 | 78–86 |
| 31–35 | 1.12 | 1.07 | 1.05 | 1.02 | 1.02 | 1.01 | 87–95 |
| 36–40 | 1.00 | 1.00 | 1.00 | 1.00 | 1.00 | 1.00 | 96–104 |
| 41–45 | 0.87 | 0.93 | 0.95 | 0.98 | 0.98 | 0.99 | 105–113 |
| 46–50 | 0.71 | 0.85 | 0.89 | 0.95 | 0.97 | 0.98 | 114–122 |
| 51–55 | 0.50 | 0.76 | 0.84 | 0.93 | 0.95 | 0.96 | 123–131 |
| 56–60 | — | 0.65 | 0.77 | 0.90 | 0.94 | 0.95 | 132–140 |
| 61–65 | — | 0.53 | 0.71 | 0.88 | 0.92 | 0.94 | 141–149 |
| 66–70 | — | 0.38 | 0.63 | 0.85 | 0.90 | 0.93 | 150–158 |
| 71–75 | — | — | 0.55 | 0.83 | 0.88 | 0.91 | 159–167 |
| 76–80 | — | — | 0.45 | 0.80 | 0.87 | 0.90 | 168–176 |
| 81–90 | — | — | — | 0.74 | 0.83 | 0.87 | 177–194 |
| 91–100 | — | — | — | 0.67 | 0.79 | 0.85 | 195–212 |
| 101–110 | — | — | — | 0.60 | 0.75 | 0.82 | 213–230 |
| 111–120 | — | — | — | 0.52 | 0.71 | 0.79 | 231–248 |
| 121–130 | — | — | — | 0.43 | 0.66 | 0.76 | 249–266 |
| 131–140 | — | — | — | 0.30 | 0.61 | 0.72 | 267–284 |
| 141–160 | — | — | — | — | 0.50 | 0.65 | 285–320 |
| 161–180 | — | — | — | — | 0.35 | 0.58 | 321–356 |
| 181–200 | — | — | — | — | — | 0.49 | 357–392 |
| 201–225 | — | — | — | — | — | 0.35 | 393–437 |

The conditions under which derating factors do not apply are defined in this section. Exhibit 310.3 illustrates the conditions specified in 310.15(B)(3)(a)(3),

(4) Adjustment factors shall not apply to Type AC cable or to Type MC cable under the following conditions:

a. The cables do not have an overall outer jacket.
b. Each cable has not more than three current-carrying conductors.
c. The conductors are 12 AWG copper.
d. Not more than 20 current-carrying conductors are installed without maintaining spacing, are stacked, or are supported on "bridle rings."

### Calculation Example

A commercial office space requires fourteen 277-V fluorescent lighting circuits to serve the office area. The office area lighting is assumed to be a continuous load, and the office ambient temperature does not exceed 30°C (86°F). Each circuit is arranged so that it has a calculated load not exceeding 16 amperes. The wiring method is Type MC cable, 3-conductor (with an additional equipment grounding conductor), 12 AWG THHN copper. Each individual MC cable contains a 3-wire multiwire branch circuit. To serve the entire area, this arrangement requires a total of seven cables bundled for a distance of 25 feet, without maintaining spacing between them where they leave the electrical room and enter the office area.

Determine the ampacity of each circuit conductor in accordance with 310.15, applying 310.15(B)(3)(a)(4) to account for the bundled cables. Then determine the maximum permitted branch-circuit overcurrent protection.

*Solution*

Step 1. Determine the quantity of current-carrying conductors. According to 310.15(B)(5), equipment grounding conductors are not counted as current-carrying conductors. According to 310.15(B)(4)(c), fluorescent lighting is considered a nonlinear load, so the grounded conductor of each Type MC cable must be counted as a current-carrying conductor:

$$7 \text{ cables} \times 3 \text{ conductors each} = 21 \text{ current-carrying conductors}$$

Step 2. Determine the ampacity of each current-carrying conductor. Because the quantity of current-carrying conductors being

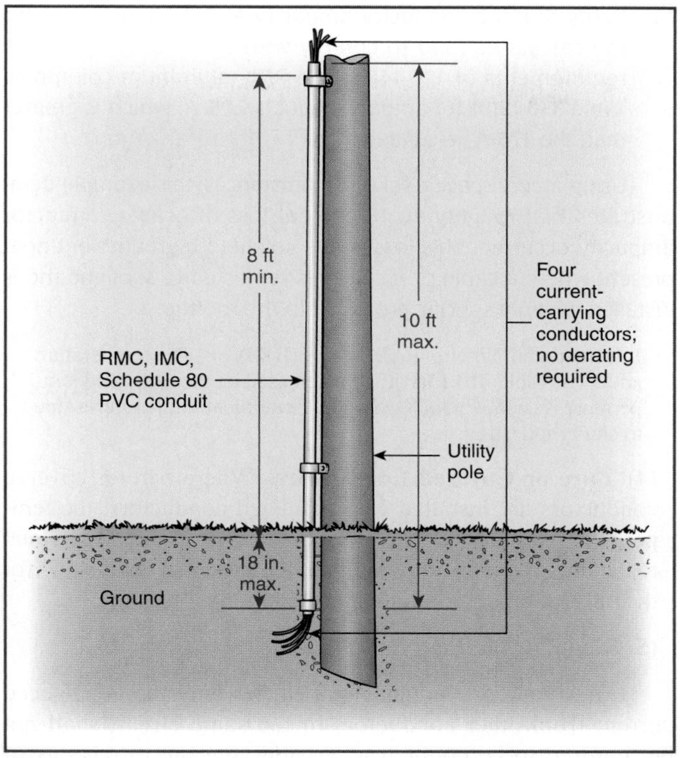


***EXHIBIT 310.3*** *Conditions under which derating factors do not apply.*

bundled exceeds 20, a 60-percent adjustment factor is required by 310.15(B)(3)(a)(5). From Table 310.15(B)(16):

$$12 \text{ AWG THHN} = 30 \text{ A}$$
$$30 \text{ A} \times 0.60 = 18 \text{ A}$$

Because the actual calculated load is 16 amperes of continuous load, 210.19(A)(1) is applicable. The conductors must have an ampacity equal to or greater than the load before the adjustment factor is applied. Because the ampacity of the conductors after the adjustment factor is applied is 18 amperes, the conductors are suitable for this installation.

STEP 3. Finally, determine the maximum size of the OCPD permitted for these bundled cables.. Section 240.4(B) permits the use of the next higher standard rating of OCPD. Therefore, although the conductors have a calculated ampacity of 18 A, a 20-A OCPD is permitted. In addition, and of significance, the 20-A OCPD is in compliance with 210.20(A), given that the actual 16-A continuous load would require a 20-A OCPD.

(5) An adjustment factor of 60 percent shall be applied to Type AC cable or Type MC cable under the following conditions:

a. The cables do not have an overall outer jacket.
b. The number of current carrying conductors exceeds 20.
c. The cables are stacked or bundled longer that 600 mm (24 in) without spacing being maintained.

***TABLE 310.15(B)(3)(a)*** *Adjustment Factors for More Than Three Current-Carrying* **Conductors**

| Number of Conductors[1] | Percent of Values in Table 310.15(B)(16) through Table 310.15(B)(19) as Adjusted for Ambient Temperature if Necessary |
|---|---|
| 4–6 | 80 |
| 7–9 | 70 |
| 10–20 | 50 |
| 21–30 | 45 |
| 31–40 | 40 |
| 41 and above | 35 |

[1]Number of conductors is the total number of conductors in the raceway or cable, including spare conductors. The count shall be adjusted in accordance with 310.15(B)(5) and (6). The count shall not include conductors that are connected to electrical components but that cannot be simultaneously energized.

(b) *Raceway Spacing.* Spacing between raceways shall be maintained.

Spacing is normally maintained between individual conduits in groups of conduit runs from junction box to junction box because the conduits need to be separated where they enter the junction box, to allow room for locknuts and bushings.

(c) *Raceways and Cables Exposed to Sunlight on Rooftops.* Where raceways or cables are exposed to direct sunlight on or above rooftops, the adjustments shown in Table 310.15(B)(3)(c) shall be added to the outdoor temperature to determine the applicable ambient temperature for application of the correction factors in Table 310.15(B)(2)(a) or Table 310.15(B)(2)(b).

*Exception: Type XHHW-2 insulated conductors shall not be subject to this ampacity adjustment.*

Informational Note: One source for the ambient temperatures in various locations is the ASHRAE *Handbook — Fundamentals.*

The conductors in outdoor conduits installed on or near the surface of the roof are subject to a significant increase in temperature when the roof is exposed to direct sunlight. The closer the conduit is to the roof, the greater the ambient temperature adjustment. See the following example and Exhibit 310.4.

### Calculation Example

A feeder installed in IMC runs across the top of a commercial building in St. Louis, MO, as shown in Exhibit 310.4. The calculated load on the feeder is 175 A. The lateral portion of the raceway is secured to supports elevated not less than 15 inches above the finished rooftop surface and is exposed to sunlight. Determine the minimum size circuit conductor using aluminum 90°C XHHW-2

***EXHIBIT 310.4*** *Type IMC crossing a rooftop and exposed to sunlight. (Courtesy of the International Association of Electrical Inspectors)*

insulation, taking into consideration only the exposure to sunlight. None of the loads are continuous, and the neutral is not considered a current-carrying conductor. The design temperature is based on the averaged June, July, and August 2-percent design temperature from the 2009 ASHRAE Handbook.

*Solution*

STEP 1. Determine the ambient temperature (compensated for proximity of conduit to the rooftop exposure to sunlight):

a. Compensated ambient temperature = design temperature + value from Table 310.15(B)(3)(c)
b. Design temperature for St. Louis area = 95°F
c. Temperature adjustment from Table 310.15(B)(3)(c) for a raceway elevated 15 in. above rooftop = 25°F
d. Compensated ambient temperature: 95°F + 25°F = 120°F

STEP 2. Determine the temperature correction factor for this application from Table 310.15(B)(2)(a). Using the 90°C column and the temperature correction factor row for 120°F, the temperature correction factor is 0.82.

STEP 3. Determine the proper conductor size to supply the 175-A load.

a. Because the load is calculated at 175 A noncontinuous, and the neutral conductor is not considered to be a current-carrying conductor, the conductor ampacity is calculated as follows:

$$175 \text{ A} \div 0.82 = 213 \text{ A}$$

b. Select a conductor not less than 213 amperes from the aluminum 90°C column of Table 310.15(B)(16):

250 kcmil aluminum XHHW-2

c. Verify that the conductor ampacity at 75°C is sufficient for the calculated load to comply with terminal temperature requirements of 110.14(C): The 75°C aluminum column of Table 310.15(B)(16) ampacity equals 205 A, which is greater than the 175-A calculated load.

Using a very specific set of circumstances, this example demonstrates that roughly an 18-percent loss of usable conductor ampacity occurred. This loss is due solely to high ambient heat present where a cable or raceway is subjected to sunlight and is installed within a specific proximity to the rooftop.

Informational Note to Table 310.15(B)(3)(c): The temperature adders in Table 310.15(B)(3)(c) are based on the measured temperature rise above the local climatic ambient temperatures due to sunlight heating.

**(4) Bare or Covered Conductors.** Where bare or covered conductors are installed with insulated conductors, the temperature rating of the bare or covered conductor shall be equal to the lowest temperature rating of the insulated conductors for the purpose of determining ampacity.

**(5) Neutral Conductor.**

(a) A neutral conductor that carries only the unbalanced current from other conductors of the same circuit shall not be required to be counted when applying the provisions of 310.15(B)(3)(a).

(b) In a 3-wire circuit consisting of two phase conductors and the neutral conductor of a 4-wire, 3-phase, wye-connected system, a common conductor carries approximately the same current as the line-to-neutral load currents of the other conductors and shall be counted when applying the provisions of 310.15(B)(3)(a).

(c) On a 4-wire, 3-phase wye circuit where the major portion of the load consists of nonlinear loads, harmonic currents are present in the neutral conductor; the neutral conductor shall therefore be considered a current-carrying conductor.

Nonlinear loads on 3-phase circuits can cause an increase in neutral conductor current. See the discussion of nonlinear loads on the inside front cover of this handbook.

***TABLE 310.15(B)(3)(C)*** *Ambient Temperature Adjustment for* **Raceways or Cables** *Exposed to Sunlight on or Above Rooftops*

| Distance Above Roof to Bottom of Raceway or Cable | Temperature Adder °C | Temperature Adder °F |
|---|---|---|
| On roof 0 – 13 mm (0–½ in.) | 33 | 60 |
| Above roof 13 mm (½ in.) | 22 | 40 |
| Above 90 mm – 300 mm (3½ in. – 12 in.) | 17 | 30 |
| Above 300 mm – 900 mm (12 in. – 36 in.) | 14 | 25 |

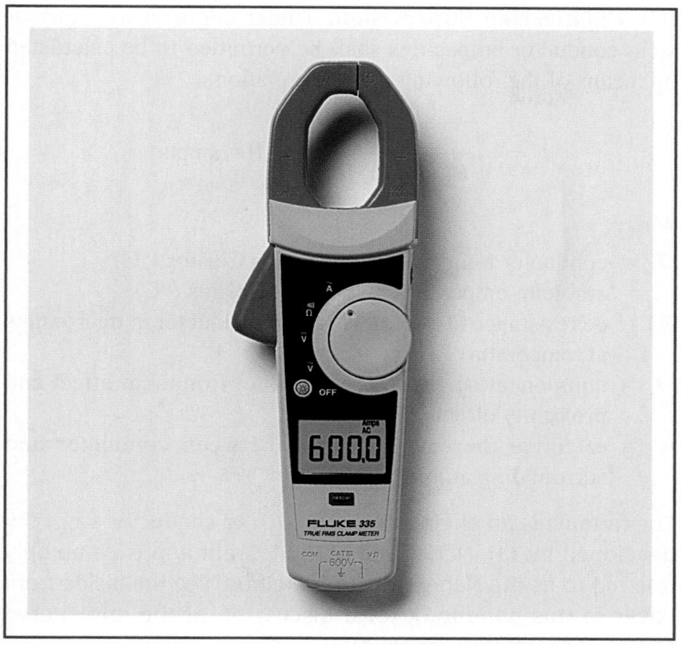


***EXHIBIT 310.5*** *A clamp-on ammeter that uses true rms measurements, where accurate measurement of neutral current becomes necessary. (Courtesy of Fluke Corp.)*

***EXHIBIT 310.6*** *A portable tool for such tasks as diagnostic power analysis, including harmonic distortion, where accurate measurement and thorough circuit analysis is desirable. (Courtesy of Dranetz-BMI)*

Exhibit 310.5 illustrates a clamp-on ammeter that uses true rms measurements. Exhibit 310.6 is an example of a portable diagnostic analyzer used for more sophisticated power measurements, including measuring harmonic distortion.

**(6) Grounding or Bonding Conductor.** A grounding or bonding conductor shall not be counted when applying the provisions of 310.15(B)(3)(a).

**(7) 120/240-Volt, Single-Phase Dwelling Services and Feeders.** For one-family dwellings and the individual dwelling units of two-family and multifamily dwellings, service and feeder conductors supplied by a single-phase, 120/240-volt system shall be permitted be sized in accordance with 310.15(B)(7)(1) through (4).

The main service or feeder to a dwelling unit is permitted to be sized at 83 percent of the disconnect rating. The calculation is not based on the rating of the overcurrent device protecting the main feeder. The minimum disconnect rating for a dwelling unit is 100 amperes according to 225.39 and 230.79. This calculation applies only to conductors carrying 100 percent of the dwelling unit's diversified load.

Provided a single set of 120/240-volt single-phase, service-entrance conductors supplies a one-family dwelling, or an individual unit of two-family or multifamily dwelling, the reduced conductor size is applicable to the service-entrance conductors, service-lateral conductors, or feeder conductors that supply a dwelling unit. The feeder conductors to a dwelling unit are not required to be larger than its service-entrance conductors.

Exhibits 310.7 and 310.8 illustrate the application of 310.15(B)(7). In Exhibit 310.7, the reduced conductor size permitted is applicable only to the service-entrance conductors run to each apartment from the meters. In Exhibit 310.8, the reduced conductor size permitted is also applicable to the feeder conductors run to each apartment from the service disconnecting means, because these feeders carry the entire load to each apartment.

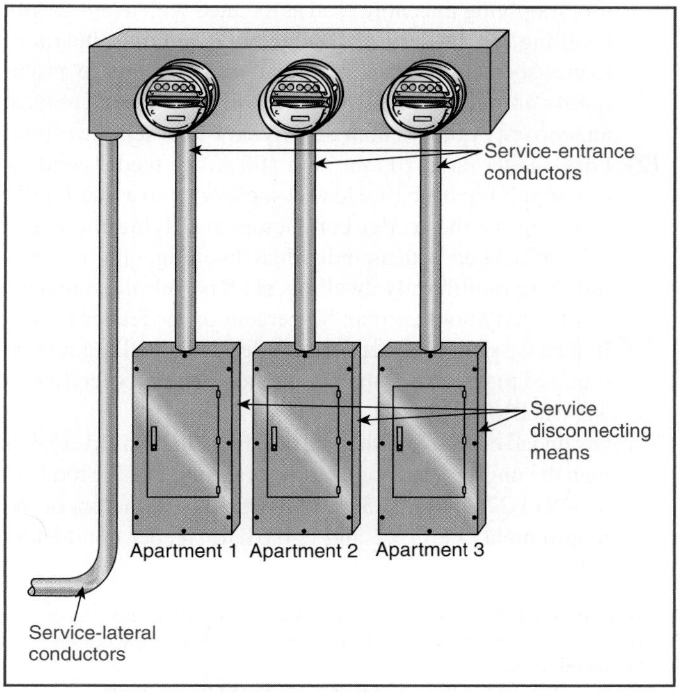


***EXHIBIT 310.7*** *One application where the reduced conductor size is applicable to the service-entrance conductors.*

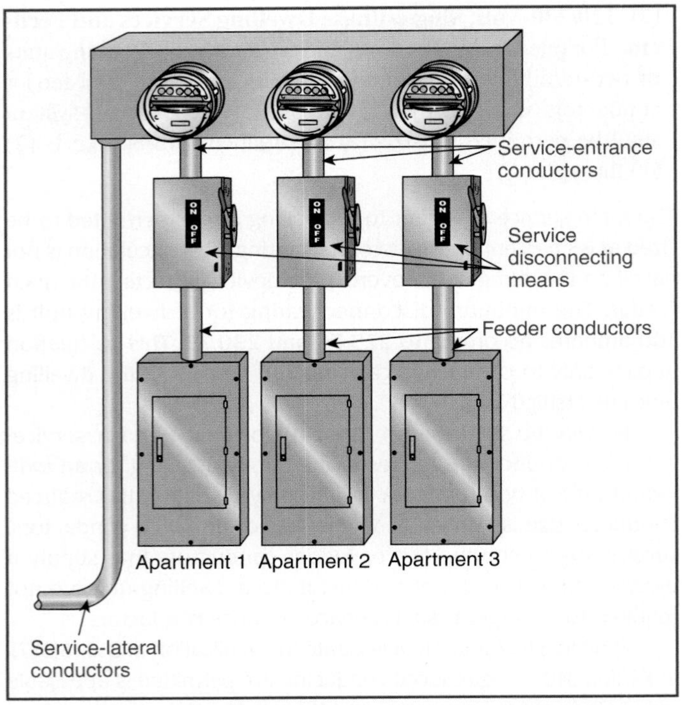


***EXHIBIT 310.8*** *A second application where the reduced conductor size is applicable to the service-entrance conductors and to the feeder conductors.*

(1) For a service rated 100 through 400 A, the service conductors supplying the entire load associated with a one-family dwelling, or the service conductors supplying the entire load associated with an individual dwelling unit in a two-family or multifamily dwelling, shall be permitted to have an ampacity not less than 83 percent of the service rating.

(2) For a feeder rated 100 through 400 A, the feeder conductors supplying the entire load associated with a one-family dwelling, or the feeder conductors supplying the entire load associated with an individual dwelling, unit in a two-family or multifamily dwelling, shall be permitted to have an ampacity not less than 83 percent of the feeder rating.

(3) In no case shall a feeder for an individual dwelling unit be required to have an ampacity greater than that specified in 310.15(B)(7)(1) or (2).

(4) Grounded conductors shall be permitted to be sized smaller than the ungrounded conductors, provided that the requirements of 220.61 and 230.42 for service conductors or the requirements of 215.2 and 220.61 for feeder conductors are met.

Informational Note No. 1: The conductor ampacity may require other correction or adjustment factors applicable to the conductor installation.

Informational Note No. 2: See Example D7 in Annex D.

•

**(C) Engineering Supervision.** Under engineering supervision, conductor ampacities shall be permitted to be calculated by means of the following general equation:

$$I = \sqrt{\frac{T_c - T_a}{R_{dc}(1 + Y_c)R_{ca}}} \times 10^3 \text{ amperes}$$

where:

$T_c$ = conductor temperature in degrees Celsius (°C)
$T_a$ = ambient temperature in degrees Celsius (°C)
$R_{dc}$ = dc resistance of 305 mm (1 ft) of conductor in microohms at temperature, $T_c$
$Y_c$ = component ac resistance resulting from skin effect and proximity effect
$R_{ca}$ = effective thermal resistance between conductor and surrounding ambient

The formula is an engineered approach to conductor ampacity developed by J.H. Neher and M.H. McGrath and is commonly referred to as the Neher–McGrath Method. See the inside front cover of this handbook for a discussion of the engineered approach to conductor ampacity.

## 310.60 Conductors Rated 2001 to 35,000 Volts

•

**(A) Ampacities of Conductors Rated 2001 to 35,000 Volts.** Ampacities for solid dielectric-insulated conductors shall be permitted to be determined by tables or under engineering supervision, as provided in 310.60(B) and (C).

**(1) Selection of Ampacity.** Where more than one calculated or tabulated ampacity could apply for a given circuit length, the lowest value shall be used.

*Exception: Where two different ampacities apply to adjacent portions of a circuit, the higher ampacity shall be permitted to be used beyond the point of transition, a distance equal to 3.0 m (10 ft) or 10 percent of the circuit length calculated at the higher ampacity, whichever is less.*

Informational Note: See 110.40 for conductor temperature limitations due to termination provisions.

**(B) Tables.** Ampacities for conductors rated 2001 to 35,000 volts shall be as specified in Table 310.60(C)(67) through Table 310.60(C)(86). Ampacities for ambient temperatures other than those specified in the ampacity tables shall be corrected in accordance with 310.60(B)(4).

Informational Note No. 1: For ampacities calculated in accordance with 310.60(A), reference IEEE 835-1994 (IPCEA Pub. No. P-46-426), *Standard Power Cable Ampacity Tables,* and the references therein for availability of all factors and constants.

Informational Note No. 2: Ampacities provided by this section do not take voltage drop into consideration. See 210.19(A),

***TABLE 310.15(B)(16)*** *(formerly Table 310.16) Allowable Ampacities of Insulated Conductors Rated Up to and Including 2000 Volts, 60°C Through 90°C (140°F Through 194°F), Not More Than Three Current-Carrying Conductors in Raceway, Cable, or Earth (Directly Buried), Based on Ambient Temperature of 30°C (86°F)**

| Size AWG or kcmil | Temperature Rating of Conductor [See Table 310.104(A).] | | | | | | Size AWG or kcmil |
|---|---|---|---|---|---|---|---|
| | 60°C (140°F) | 75°C (167°F) | 90°C (194°F) | 60°C (140°F) | 75°C (167°F) | 90°C (194°F) | |
| | Types TW, UF | Types RHW, THHW, THW, THWN, XHHW, USE, ZW | Types TBS, SA, SIS, FEP, FEPB, MI, RHH, RHW-2, THHN, THHW, THW-2, THWN-2, USE-2, XHH, XHHW, XHHW-2, ZW-2 | Types TW, UF | Types RHW, THHW, THW, THWN, XHHW, USE | Types TBS, SA, SIS, THHN, THHW, THW-2, THWN-2, RHH, RHW-2, USE-2, XHH, XHHW, XHHW-2, ZW-2 | |
| | COPPER | | | ALUMINUM OR COPPER-CLAD ALUMINUM | | | |
| 18** | — | — | 14 | — | — | — | — |
| 16** | — | — | 18 | — | — | — | — |
| 14** | 15 | 20 | 25 | — | — | — | — |
| 12** | 20 | 25 | 30 | 15 | 20 | 25 | 12** |
| 10** | 30 | 35 | 40 | 25 | 30 | 35 | 10** |
| 8 | 40 | 50 | 55 | 35 | 40 | 45 | 8 |
| 6 | 55 | 65 | 75 | 40 | 50 | 55 | 6 |
| 4 | 70 | 85 | 95 | 55 | 65 | 75 | 4 |
| 3 | 85 | 100 | 115 | 65 | 75 | 85 | 3 |
| 2 | 95 | 115 | 130 | 75 | 90 | 100 | 2 |
| 1 | 110 | 130 | 145 | 85 | 100 | 115 | 1 |
| 1/0 | 125 | 150 | 170 | 100 | 120 | 135 | 1/0 |
| 2/0 | 145 | 175 | 195 | 115 | 135 | 150 | 2/0 |
| 3/0 | 165 | 200 | 225 | 130 | 155 | 175 | 3/0 |
| 4/0 | 195 | 230 | 260 | 150 | 180 | 205 | 4/0 |
| 250 | 215 | 255 | 290 | 170 | 205 | 230 | 250 |
| 300 | 240 | 285 | 320 | 195 | 230 | 260 | 300 |
| 350 | 260 | 310 | 350 | 210 | 250 | 280 | 350 |
| 400 | 280 | 335 | 380 | 225 | 270 | 305 | 400 |
| 500 | 320 | 380 | 430 | 260 | 310 | 350 | 500 |
| 600 | 350 | 420 | 475 | 285 | 340 | 385 | 600 |
| 700 | 385 | 460 | 520 | 315 | 375 | 425 | 700 |
| 750 | 400 | 475 | 535 | 320 | 385 | 435 | 750 |
| 800 | 410 | 490 | 555 | 330 | 395 | 445 | 800 |
| 900 | 435 | 520 | 585 | 355 | 425 | 480 | 900 |
| 1000 | 455 | 545 | 615 | 375 | 445 | 500 | 1000 |
| 1250 | 495 | 590 | 665 | 405 | 485 | 545 | 1250 |
| 1500 | 525 | 625 | 705 | 435 | 520 | 585 | 1500 |
| 1750 | 545 | 650 | 735 | 455 | 545 | 615 | 1750 |
| 2000 | 555 | 665 | 750 | 470 | 560 | 630 | 2000 |

*Refer to 310.15(B)(2) for the ampacity correction factors where the ambient temperature is other than 30°C (86°F).
**Refer to 240.4(D) for conductor overcurrent protection limitations.

*TABLE 310.15(B)(17)* *(formerly Table 310.17) Allowable Ampacities of Single-Insulated Conductors Rated Up to and Including 2000 Volts in Free Air, Based on Ambient Temperature of 30°C (86°F)**

| | Temperature Rating of Conductor [See Table 310.104(A).] | | | | | | |
|---|---|---|---|---|---|---|---|
| | **60°C (140°F)** | **75°C (167°F)** | **90°C (194°F)** | **60°C (140°F)** | **75°C (167°F)** | **90°C (194°F)** | |
| **Size AWG or kcmil** | **Types TW, UF** | **Types RHW, THHW, THW, THWN, XHHW, ZW** | **Types TBS, SA, SIS, FEP, FEPB, MI, RHH, RHW-2, THHN, THHW, THW-2, THWN-2, USE-2, XHH, XHHW, XHHW-2, ZW-2** | **Types TW, UF** | **Types RHW, THHW, THW, THWN, XHHW** | **Types TBS, SA, SIS, THHN, THHW, THW-2, THWN-2, RHH, RHW-2, USE-2, XHH, XHHW, XHHW-2, ZW-2** | **Size AWG or kcmil** |
| | **COPPER** | | | **ALUMINUM OR COPPER-CLAD ALUMINUM** | | | |
| 18 | — | — | 18 | — | — | — | — |
| 16 | — | — | 24 | — | — | — | — |
| 14** | 25 | 30 | 35 | — | — | — | — |
| 12** | 30 | 35 | 40 | 25 | 30 | 35 | 12** |
| 10** | 40 | 50 | 55 | 35 | 40 | 45 | 10** |
| 8 | 60 | 70 | 80 | 45 | 55 | 60 | 8 |
| 6 | 80 | 95 | 105 | 60 | 75 | 85 | 6 |
| 4 | 105 | 125 | 140 | 80 | 100 | 115 | 4 |
| 3 | 120 | 145 | 165 | 95 | 115 | 130 | 3 |
| 2 | 140 | 170 | 190 | 110 | 135 | 150 | 2 |
| 1 | 165 | 195 | 220 | 130 | 155 | 175 | 1 |
| 1/0 | 195 | 230 | 260 | 150 | 180 | 205 | 1/0 |
| 2/0 | 225 | 265 | 300 | 175 | 210 | 235 | 2/0 |
| 3/0 | 260 | 310 | 350 | 200 | 240 | 270 | 3/0 |
| 4/0 | 300 | 360 | 405 | 235 | 280 | 315 | 4/0 |
| 250 | 340 | 405 | 455 | 265 | 315 | 355 | 250 |
| 300 | 375 | 445 | 500 | 290 | 350 | 395 | 300 |
| 350 | 420 | 505 | 570 | 330 | 395 | 445 | 350 |
| 400 | 455 | 545 | 615 | 355 | 425 | 480 | 400 |
| 500 | 515 | 620 | 700 | 405 | 485 | 545 | 500 |
| 600 | 575 | 690 | 780 | 455 | 545 | 615 | 600 |
| 700 | 630 | 755 | 850 | 500 | 595 | 670 | 700 |
| 750 | 655 | 785 | 885 | 515 | 620 | 700 | 750 |
| 800 | 680 | 815 | 920 | 535 | 645 | 725 | 800 |
| 900 | 730 | 870 | 980 | 580 | 700 | 790 | 900 |
| 1000 | 780 | 935 | 1055 | 625 | 750 | 845 | 1000 |
| 1250 | 890 | 1065 | 1200 | 710 | 855 | 965 | 1250 |
| 1500 | 980 | 1175 | 1325 | 795 | 950 | 1070 | 1500 |
| 1750 | 1070 | 1280 | 1445 | 875 | 1050 | 1185 | 1750 |
| 2000 | 1155 | 1385 | 1560 | 960 | 1150 | 1295 | 2000 |

*Refer to 310.15(B)(2) for the ampacity correction factors where the ambient temperature is other than 30°C (86°F).

**Refer to 240.4(D) for conductor overcurrent protection limitations.

***TABLE 310.15(B)(18)*** *(formerly Table 310.18) Allowable Ampacities of Insulated Conductors Rated Up to and Including 2000 Volts, 150°C Through 250°C (302°F Through 482°F). Not More Than Three Current-Carrying Conductors in Raceway or Cable, Based on Ambient Air Temperature of 40°C (104°F)**

| | Temperature Rating of Conductor [See Table 310.104(A).] | | | | |
|---|---|---|---|---|---|
| | 150°C (302°F) | 200°C (392°F) | 250°C (482°F) | 150°C (302°F) | |
| | Type Z | Types FEP, FEPB, PFA, SA | Types PFAH, TFE | Type Z | |
| Size AWG or kcmil | COPPER | | NICKEL OR NICKEL-COATED COPPER | ALUMINUM OR COPPER-CLAD ALUMINUM | Size AWG or kcmil |
| 14 | 34 | 36 | 39 | — | 14 |
| 12 | 43 | 45 | 54 | 30 | 12 |
| 10 | 55 | 60 | 73 | 44 | 10 |
| 8 | 76 | 83 | 93 | 57 | 8 |
| 6 | 96 | 110 | 117 | 75 | 6 |
| 4 | 120 | 125 | 148 | 94 | 4 |
| 3 | 143 | 152 | 166 | 109 | 3 |
| 2 | 160 | 171 | 191 | 124 | 2 |
| 1 | 186 | 197 | 215 | 145 | 1 |
| 1/0 | 215 | 229 | 244 | 169 | 1/0 |
| 2/0 | 251 | 260 | 273 | 198 | 2/0 |
| 3/0 | 288 | 297 | 308 | 227 | 3/0 |
| 4/0 | 332 | 346 | 361 | 260 | 4/0 |

*Refer to 310.15(B)(2) for the ampacity correction factors where the ambient temperature is other than 40°C (104°F).

***TABLE 310.15(B)(19)*** *(formerly Table 310.19) Allowable Ampacities of Single-Insulated Conductors, Rated Up to and Including 2000 Volts, 150°C Through 250°C (302°F Through 482°F), in Free Air, Based on Ambient Air Temperature of 40°C (104°F)**

| | Temperature Rating of Conductor [See Table 310.104(A).] | | | | |
|---|---|---|---|---|---|
| | 150°C (302°F) | 200°C (392°F) | 250°C (482°F) | 150°C (302°F) | |
| | Type Z | Types FEP, FEPB, PFA, SA | Types PFAH, TFE | Type Z | |
| Size AWG or kcmil | COPPER | | NICKEL, OR NICKEL-COATED COPPER | ALUMINUM OR COPPER-CLAD ALUMINUM | Size AWG or kcmil |
| 14 | 46 | 54 | 59 | — | 14 |
| 12 | 60 | 68 | 78 | 47 | 12 |
| 10 | 80 | 90 | 107 | 63 | 10 |
| 8 | 106 | 124 | 142 | 83 | 8 |
| 6 | 155 | 165 | 205 | 112 | 6 |
| 4 | 190 | 220 | 278 | 148 | 4 |
| 3 | 214 | 252 | 327 | 170 | 3 |
| 2 | 255 | 293 | 381 | 198 | 2 |
| 1 | 293 | 344 | 440 | 228 | 1 |
| 1/0 | 339 | 399 | 532 | 263 | 1/0 |
| 2/0 | 390 | 467 | 591 | 305 | 2/0 |
| 3/0 | 451 | 546 | 708 | 351 | 3/0 |
| 4/0 | 529 | 629 | 830 | 411 | 4/0 |

*Refer to 310.15(B)(2) for the ampacity correction factors where the ambient temperature is other than 40°C (104°F).

*TABLE 310.15(B)(20) (formerly Table 310.20) Ampacities of Not More Than Three Single Insulated Conductors, Rated Up to and Including 2000 Volts, Supported on a Messenger, Based on Ambient Air Temperature of 40°C (104°F)**

| | Temperature Rating of Conductor [See Table 310.104(A).] | | | | |
|---|---|---|---|---|---|
| | 75°C (167°F) | 90°C (194°F) | 75°C (167°F) | 90°C (194°F) | |
| | Types RHW, THHW, THW, THWN, XHHW, ZW | Types MI, THHN, THHW, THW-2, THWN-2, RHH, RHW-2, USE-2, XHHW, XHHW-2, ZW-2 | Types RHW, THW, THWN, THHW, XHHW | Types THHN, THHW, RHH, XHHW, RHW-2, XHHW-2, THW-2, THWN-2, USE-2, ZW-2 | |
| Size AWG or kcmil | COPPER | | ALUMINUM OR COPPER-CLAD ALUMINUM | | Size AWG or kcmil |
| 8 | 57 | 66 | 44 | 51 | 8 |
| 6 | 76 | 89 | 59 | 69 | 6 |
| 4 | 101 | 117 | 78 | 91 | 4 |
| 3 | 118 | 138 | 92 | 107 | 3 |
| 2 | 135 | 158 | 106 | 123 | 2 |
| 1 | 158 | 185 | 123 | 144 | 1 |
| 1/0 | 183 | 214 | 143 | 167 | 1/0 |
| 2/0 | 212 | 247 | 165 | 193 | 2/0 |
| 3/0 | 245 | 287 | 192 | 224 | 3/0 |
| 4/0 | 287 | 335 | 224 | 262 | 4/0 |
| 250 | 320 | 374 | 251 | 292 | 250 |
| 300 | 359 | 419 | 282 | 328 | 300 |
| 350 | 397 | 464 | 312 | 364 | 350 |
| 400 | 430 | 503 | 339 | 395 | 400 |
| 500 | 496 | 580 | 392 | 458 | 500 |
| 600 | 553 | 647 | 440 | 514 | 600 |
| 700 | 610 | 714 | 488 | 570 | 700 |
| 750 | 638 | 747 | 512 | 598 | 750 |
| 800 | 660 | 773 | 532 | 622 | 800 |
| 900 | 704 | 826 | 572 | 669 | 900 |
| 1000 | 748 | 879 | 612 | 716 | 1000 |

*Refer to 310.15(B)(2) for the ampacity correction factors where the ambient temperature is other than 40°C (104°F).

Informational Note No. 4, for branch circuits and 215.2(A), Informational Note No. 2, for feeders.

**(1) Grounded Shields.** Ampacities shown in Table 310.60(C)(69), Table 310.60(C)(70), Table 310.60(C)(81), and Table 310.60(C)(82) shall apply for cables with shields grounded at one point only. Where shields for these cables are grounded at more than one point, ampacities shall be adjusted to take into consideration the heating due to shield currents.

Informational Note: Tables other than those listed contain the ampacity of cables with shields grounded at multiple points.

**(2) Burial Depth of Underground Circuits.** Where the burial depth of direct burial or electrical duct bank circuits is modified from the values shown in a figure or table, ampacities shall be permitted to be modified as indicated in (C)(2)(a) and (C)(2)(b).

(a) Where burial depths are increased in part(s) of an electrical duct run, no decrease in ampacity of the conductors is needed, provided the total length of parts of the duct run increased in depth is less than 25 percent of the total run length.

(b) Where burial depths are deeper than shown in a specific underground ampacity table or figure, an ampacity derating factor of 6 percent per 300-mm (1-ft) increase in depth for all values of rho shall be permitted.

No rating change is needed where the burial depth is decreased.

***TABLE 310.15(B)(21)*** *(formerly Table 310.21) Ampacities of Bare or Covered Conductors in Free Air, Based on 40°C (104°F) Ambient, 80°C (176°F) Total Conductor Temperature, 610 mm/sec (2 ft/sec) Wind Velocity*

| **Copper Conductors** | | | | **AAC Aluminum Conductors** | | | |
|---|---|---|---|---|---|---|---|
| **Bare** | | **Covered** | | **Bare** | | **Covered** | |
| **AWG or kcmil** | **Amperes** | **AWG or kcmil** | **Amperes** | **AWG or kcmil** | **Amperes** | **AWG or kcmil** | **Amperes** |
| 8 | 98 | 8 | 103 | 8 | 76 | 8 | 80 |
| 6 | 124 | 6 | 130 | 6 | 96 | 6 | 101 |
| 4 | 155 | 4 | 163 | 4 | 121 | 4 | 127 |
| 2 | 209 | 2 | 219 | 2 | 163 | 2 | 171 |
| 1/0 | 282 | 1/0 | 297 | 1/0 | 220 | 1/0 | 231 |
| 2/0 | 329 | 2/0 | 344 | 2/0 | 255 | 2/0 | 268 |
| 3/0 | 382 | 3/0 | 401 | 3/0 | 297 | 3/0 | 312 |
| 4/0 | 444 | 4/0 | 466 | 4/0 | 346 | 4/0 | 364 |
| 250 | 494 | 250 | 519 | 266.8 | 403 | 266.8 | 423 |
| 300 | 556 | 300 | 584 | 336.4 | 468 | 336.4 | 492 |
| 500 | 773 | 500 | 812 | 397.5 | 522 | 397.5 | 548 |
| 750 | 1000 | 750 | 1050 | 477.0 | 588 | 477.0 | 617 |
| 1000 | 1193 | 1000 | 1253 | 556.5 | 650 | 556.5 | 682 |
| — | — | — | — | 636.0 | 709 | 636.0 | 744 |
| — | — | — | — | 795.0 | 819 | 795.0 | 860 |
| — | — | — | — | 954.0 | 920 | — | — |
| — | — | — | — | 1033.5 | 968 | 1033.5 | 1017 |
| — | — | — | — | 1272 | 1103 | 1272 | 1201 |
| — | — | — | — | 1590 | 1267 | 1590 | 1381 |
| — | — | — | — | 2000 | 1454 | 2000 | 1527 |

**(3) Electrical Ducts in Figure 310.60.** At locations where electrical ducts enter equipment enclosures from under ground, spacing between such ducts, as shown in Figure 310.60, shall be permitted to be reduced without requiring the ampacity of conductors therein to be reduced.

The term *electrical ducts* is used to differentiate these from other ducts, such as those used for air handling. The term is intended to include nonmetallic electrical ducts commonly used for underground wiring, as well as other raceways (such as RMC, IMC, PVC conduit, and HDPE conduit) listed for use underground in earth or concrete.

**(4) Ambient Temperature Correction.** Ampacities for ambient temperatures other than those specified in the ampacity tables shall be corrected in accordance with Table 310.60(C)(4) or shall be permitted to be calculated using the following equation:

$$I' = I\sqrt{\frac{T_c - T_a'}{T_c - T_a}}$$

where:

$I'$ = ampacity corrected for ambient temperature
$I$ = ampacity shown in the table for $T_c$ and $T_a$
$T_c$ = temperature rating of conductor (°C)
$T_a'$ = new ambient temperature (°C)
$T_a$ = ambient temperature used in the table (°C)

**(C) Engineering Supervision.** Under engineering supervision, conductor ampacities shall be permitted to be calculated by using the following general equation:

$$I = \sqrt{\frac{T_c - (T_a + \Delta T_d)}{R_{dc}(1 + Y_c)R_{ca}}} \times 10^3 \text{ amperes}$$

where:

$T_c$ = conductor temperature (°C)
$T_a$ = ambient temperature (°C)
$\Delta T_d$ = dielectric loss temperature rise
$R_{dc}$ = dc resistance of conductor at temperature $T_c$
$Y_c$ = component ac resistance resulting from skin effect and proximity effect
$R_{ca}$ = effective thermal resistance between conductor and surrounding ambient

Informational Note: The dielectric loss temperature rise ($\Delta Td$) is negligible for single circuit extruded dielectric cables rated below 46 kV.

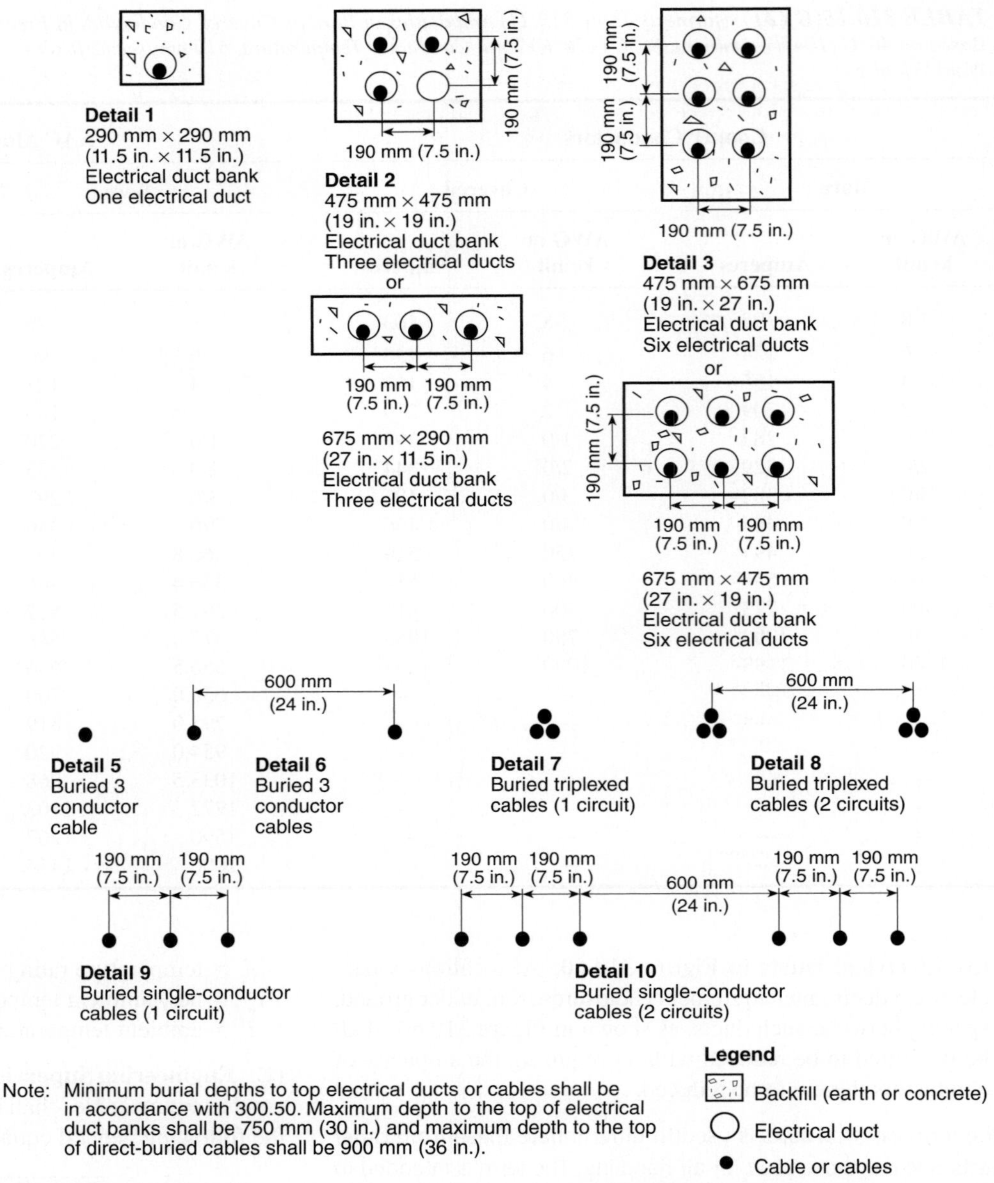
Detail 1
290 mm × 290 mm
(11.5 in. × 11.5 in.)
Electrical duct bank
One electrical duct
190 mm (7.5 in.)
190 mm (7.5 in.)
Detail 2
475 mm × 475 mm
(19 in. × 19 in.)
Electrical duct bank
Three electrical ducts
or
190 mm (7.5 in.) 190 mm (7.5 in.)
675 mm × 290 mm
(27 in. × 11.5 in.)
Electrical duct bank
Three electrical ducts
190 mm (7.5 in.) 190 mm (7.5 in.)
190 mm (7.5 in.)
Detail 3
475 mm × 675 mm
(19 in. × 27 in.)
Electrical duct bank
Six electrical ducts
or
190 mm (7.5 in.)
190 mm (7.5 in.) 190 mm (7.5 in.)
675 mm × 475 mm
(27 in. × 19 in.)
Electrical duct bank
Six electrical ducts
600 mm (24 in.)
Detail 5
Buried 3 conductor cable
Detail 6
Buried 3 conductor cables
Detail 7
Buried triplexed cables (1 circuit)
600 mm (24 in.)
Detail 8
Buried triplexed cables (2 circuits)
190 mm (7.5 in.) 190 mm (7.5 in.)
Detail 9
Buried single-conductor cables (1 circuit)
190 mm (7.5 in.) 190 mm (7.5 in.) 600 mm (24 in.) 190 mm (7.5 in.) 190 mm (7.5 in.)
Detail 10
Buried single-conductor cables (2 circuits)
Legend
Backfill (earth or concrete)
Electrical duct
Cable or cables
Note: Minimum burial depths to top electrical ducts or cables shall be in accordance with 300.50. Maximum depth to the top of electrical duct banks shall be 750 mm (30 in.) and maximum depth to the top of direct-buried cables shall be 900 mm (36 in.).

***FIGURE 310.60*** *Cable Installation Dimensions for Use with Table 310.60(C)(77) Through Table 310.60(C)(86).*

***TABLE 310.60(C)(4)*** *Ambient Temperature Correction Factors*

**For ambient temperatures other than 40°C (104°F), multiply the allowable ampacities specified in the ampacity tables by the appropriate factor shown below.**

| Ambient Temperature (°C) | Temperature Rating of Conductor | | Ambient Temperature (°F) |
|---|---|---|---|
| | 90°C | 105°C | |
| 10 or less | 1.26 | 1.21 | 50 or less |
| 11–15 | 1.22 | 1.18 | 51–59 |
| 16–20 | 1.18 | 1.14 | 60–68 |
| 21–25 | 1.14 | 1.11 | 69–77 |
| 26–30 | 1.10 | 1.07 | 78–86 |
| 31–35 | 1.05 | 1.04 | 87–95 |
| 36–40 | 1.00 | 1.00 | 96–104 |
| 41–45 | 0.95 | 0.96 | 105–113 |
| 46–50 | 0.89 | 0.92 | 114–122 |
| 51–55 | 0.84 | 0.88 | 123–131 |
| 56–60 | 0.77 | 0.83 | 132–140 |
| 61–65 | 0.71 | 0.78 | 141–149 |
| 66–70 | 0.63 | 0.73 | 150–158 |
| 71–75 | 0.55 | 0.68 | 159–167 |
| 76–80 | 0.45 | 0.62 | 168–176 |
| 81–85 | 0.32 | 0.55 | 177–185 |
| 86–90 | — | 0.48 | 186–194 |
| 91–95 | — | 0.39 | 195–203 |
| 96–100 | — | 0.28 | 204–212 |

***TABLE 310.60(C)(67)*** *Ampacities of Insulated Single Copper Conductor Cables Triplexed in Air Based on Conductor Temperatures of 90°C (194°F) and 105°C (221°F) and Ambient Air Temperature of 40°C (104°F)**

| Conductor Size (AWG or kcmil) | Temperature Rating of Conductor [See Table 310.104(C).] | | | |
|---|---|---|---|---|
| | 2001–5000 Volts Ampacity | | 5001–35,000 Volts Ampacity | |
| | 90°C (194°F) Type MV-90 | 105°C (221°F) Type MV-105 | 90°C (194°F) Type MV-90 | 105°C (221°F) Type MV-105 |
| 8 | 65 | 74 | — | — |
| 6 | 90 | 99 | 100 | 110 |
| 4 | 120 | 130 | 130 | 140 |
| 2 | 160 | 175 | 170 | 195 |
| 1 | 185 | 205 | 195 | 225 |
| 1/0 | 215 | 240 | 225 | 255 |
| 2/0 | 250 | 275 | 260 | 295 |
| 3/0 | 290 | 320 | 300 | 340 |
| 4/0 | 335 | 375 | 345 | 390 |
| 250 | 375 | 415 | 380 | 430 |
| 350 | 465 | 515 | 470 | 525 |
| 500 | 580 | 645 | 580 | 650 |
| 750 | 750 | 835 | 730 | 820 |
| 1000 | 880 | 980 | 850 | 950 |

*Refer to 310.60(C)(4) for the ampacity correction factors where the ambient air temperature is other than 40°C (104°F).

***TABLE 310.60(C)(68)*** *Ampacities of Insulated Single Aluminum Conductor Cables Triplexed in Air Based on Conductor Temperatures of 90°C (194°F) and 105°C (221°F) and Ambient Air Temperature of 40°C (104°F)**

| Conductor Size (AWG or kcmil) | Temperature Rating of Conductor [See Table 310.104(C).] | | | |
|---|---|---|---|---|
| | 2001–5000 Volts Ampacity | | 5001–35,000 Volts Ampacity | |
| | 90°C (194°F) Type MV-90 | 105°C (221°F) Type MV-105 | 90°C (194°F) Type MV-90 | 105°C (221°F) Type MV-105 |
| 8 | 50 | 57 | — | — |
| 6 | 70 | 77 | 75 | 84 |
| 4 | 90 | 100 | 100 | 110 |
| 2 | 125 | 135 | 130 | 150 |
| 1 | 145 | 160 | 150 | 175 |
| 1/0 | 170 | 185 | 175 | 200 |
| 2/0 | 195 | 215 | 200 | 230 |
| 3/0 | 225 | 250 | 230 | 265 |
| 4/0 | 265 | 290 | 270 | 305 |
| 250 | 295 | 325 | 300 | 335 |
| 350 | 365 | 405 | 370 | 415 |
| 500 | 460 | 510 | 460 | 515 |
| 750 | 600 | 665 | 590 | 660 |
| 1000 | 715 | 800 | 700 | 780 |

*Refer to 310.60(C)(4) for the ampacity correction factors where the ambient air temperature is other than 40°C (104°F).

*TABLE 310.60(C)(69) Ampacities of Insulated Single Copper Conductor Isolated in Air Based on Conductor Temperatures of 90°C (194°F) and 105°C (221°F) and Ambient Air Temperature of 40°C (104°F)**

| | Temperature Rating of Conductor [See Table 310.104(C).] | | | | | |
|---|---|---|---|---|---|---|
| | 2001–5000 Volts Ampacity | | 5001–15,000 Volts Ampacity | | 15,001–35,000 Volts Ampacity | |
| **Conductor Size (AWG or kcmil)** | **90°C (194°F) Type MV-90** | **105°C (221°F) Type MV-105** | **90°C (194°F) Type MV-90** | **105°C (221°F) Type MV-105** | **90°C (194°F) Type MV-90** | **105°C (221°F) Type MV-105** |
| 8 | 83 | 93 | — | — | — | — |
| 6 | 110 | 120 | 110 | 125 | — | — |
| 4 | 145 | 160 | 150 | 165 | — | — |
| 2 | 190 | 215 | 195 | 215 | — | — |
| 1 | 225 | 250 | 225 | 250 | 225 | 250 |
| 1/0 | 260 | 290 | 260 | 290 | 260 | 290 |
| 2/0 | 300 | 330 | 300 | 335 | 300 | 330 |
| 3/0 | 345 | 385 | 345 | 385 | 345 | 380 |
| 4/0 | 400 | 445 | 400 | 445 | 395 | 445 |
| 250 | 445 | 495 | 445 | 495 | 440 | 490 |
| 350 | 550 | 615 | 550 | 610 | 545 | 605 |
| 500 | 695 | 775 | 685 | 765 | 680 | 755 |
| 750 | 900 | 1000 | 885 | 990 | 870 | 970 |
| 1000 | 1075 | 1200 | 1060 | 1185 | 1040 | 1160 |
| 1250 | 1230 | 1370 | 1210 | 1350 | 1185 | 1320 |
| 1500 | 1365 | 1525 | 1345 | 1500 | 1315 | 1465 |
| 1750 | 1495 | 1665 | 1470 | 1640 | 1430 | 1595 |
| 2000 | 1605 | 1790 | 1575 | 1755 | 1535 | 1710 |

*Refer to 310.60(C)(4) for the ampacity correction factors where the ambient air temperature is other than 40°C (104°F).

*TABLE 310.60(C)(70) Ampacities of Insulated Single Aluminum Conductor Isolated in Air Based on Conductor Temperatures of 90°C (194°F) and 105°C (221°F) and Ambient Air Temperature of 40°C (104°F)**

| | Temperature Rating of Conductor [See Table 310.104(C).] | | | | | |
|---|---|---|---|---|---|---|
| | 2001–5000 Volts Ampacity | | 5001–15,000 Volts Ampacity | | 15,001–35,000 Volts Ampacity | |
| **Conductor Size (AWG or kcmil)** | **90°C (194°F) Type MV-90** | **105°C (221°F) Type MV-105** | **90°C (194°F) Type MV-90** | **105°C (221°F) Type MV-105** | **90°C (194°F) Type MV-90** | **105°C (221°F) Type MV-105** |
| 8 | 64 | 71 | — | — | — | — |
| 6 | 85 | 95 | 87 | 97 | — | — |
| 4 | 115 | 125 | 115 | 130 | — | — |
| 2 | 150 | 165 | 150 | 170 | — | — |
| 1 | 175 | 195 | 175 | 195 | 175 | 195 |
| 1/0 | 200 | 225 | 200 | 225 | 200 | 225 |
| 2/0 | 230 | 260 | 235 | 260 | 230 | 260 |
| 3/0 | 270 | 300 | 270 | 300 | 270 | 300 |
| 4/0 | 310 | 350 | 310 | 350 | 310 | 345 |
| 250 | 345 | 385 | 345 | 385 | 345 | 380 |
| 350 | 430 | 480 | 430 | 480 | 430 | 475 |
| 500 | 545 | 605 | 535 | 600 | 530 | 590 |
| 750 | 710 | 790 | 700 | 780 | 685 | 765 |
| 1000 | 855 | 950 | 840 | 940 | 825 | 920 |
| 1250 | 980 | 1095 | 970 | 1080 | 950 | 1055 |
| 1500 | 1105 | 1230 | 1085 | 1215 | 1060 | 1180 |
| 1750 | 1215 | 1355 | 1195 | 1335 | 1165 | 1300 |
| 2000 | 1320 | 1475 | 1295 | 1445 | 1265 | 1410 |

*Refer to 310.60(C)(4) for the ampacity correction factors where the ambient air temperature is other than 40°C (104°F).

***TABLE 310.60(C)(71)*** *Ampacities of an Insulated Three-Conductor Copper Cable Isolated in Air Based on Conductor Temperatures of 90°C (194°F) and 105°C (221°F) and Ambient Air Temperature of 40°C (104°F)**

| | **Temperature Rating of Conductor [See Table 310.104(C).]** | | | |
|---|---|---|---|---|
| | **2001–5000 Volts Ampacity** | | **5001–35,000 Volts Ampacity** | |
| **Conductor Size (AWG or kcmil)** | **90°C (194°F) Type MV-90** | **105°C (221°F) Type MV-105** | **90°C (194°F) Type MV-90** | **105°C (221°F) Type MV-105** |
| 8 | 59 | 66 | — | — |
| 6 | 79 | 88 | 93 | 105 |
| 4 | 105 | 115 | 120 | 135 |
| 2 | 140 | 154 | 165 | 185 |
| 1 | 160 | 180 | 185 | 210 |
| 1/0 | 185 | 205 | 215 | 240 |
| 2/0 | 215 | 240 | 245 | 275 |
| 3/0 | 250 | 280 | 285 | 315 |
| 4/0 | 285 | 320 | 325 | 360 |
| 250 | 320 | 355 | 360 | 400 |
| 350 | 395 | 440 | 435 | 490 |
| 500 | 485 | 545 | 535 | 600 |
| 750 | 615 | 685 | 670 | 745 |
| 1000 | 705 | 790 | 770 | 860 |

*Refer to 310.60(C)(4) for the ampacity correction factors where the ambient air temperature is other than 40°C (104°F).

***TABLE 310.60(C)(72)*** *Ampacities of an Insulated Three-Conductor Aluminum Cable Isolated in Air Based on Conductor Temperatures of 90°C (194°F) and 105°C (221°F) and Ambient Air Temperature of 40°C (104°F)**

| | **Temperature Rating of Conductor [See Table 310.104(C).]** | | | |
|---|---|---|---|---|
| | **2001–5000 Volts Ampacity** | | **5001–35,000 Volts Ampacity** | |
| **Conductor Size (AWG or kcmil)** | **90°C (194°F) Type MV-90** | **105°C (221°F) Type MV-105** | **90°C (194°F) Type MV-90** | **105°C (221°F) Type MV-105** |
| 8 | 46 | 51 | — | — |
| 6 | 61 | 68 | 72 | 80 |
| 4 | 81 | 90 | 95 | 105 |
| 2 | 110 | 120 | 125 | 145 |
| 1 | 125 | 140 | 145 | 165 |
| 1/0 | 145 | 160 | 170 | 185 |
| 2/0 | 170 | 185 | 190 | 215 |
| 3/0 | 195 | 215 | 220 | 245 |
| 4/0 | 225 | 250 | 255 | 285 |
| 250 | 250 | 280 | 280 | 315 |
| 350 | 310 | 345 | 345 | 385 |
| 500 | 385 | 430 | 425 | 475 |
| 750 | 495 | 550 | 540 | 600 |
| 1000 | 585 | 650 | 635 | 705 |

*Refer to 310.60(C)(4) for the ampacity correction factors where the ambient air temperature is other than 40°C (104°F).

***TABLE 310.60(C)(73)*** *Ampacities of an Insulated Triplexed or Three Single-Conductor Copper Cables in Isolated Conduit in Air Based on Conductor Temperatures of 90°C (194°F) and 105°C (221°F) and Ambient Air Temperature of 40°C (104°F)**

| | **Temperature Rating of Conductor [See Table 310.104(C).]** | | | |
|---|---|---|---|---|
| | **2001–5000 Volts Ampacity** | | **5001–35,000 Volts Ampacity** | |
| **Conductor Size (AWG or kcmil)** | **90°C (194°F) Type MV-90** | **105°C (221°F) Type MV-105** | **90°C (194°F) Type MV-90** | **105°C (221°F) Type MV-105** |
| 8 | 55 | 61 | — | — |
| 6 | 75 | 84 | 83 | 93 |
| 4 | 97 | 110 | 110 | 120 |
| 2 | 130 | 145 | 150 | 165 |
| 1 | 155 | 175 | 170 | 190 |
| 1/0 | 180 | 200 | 195 | 215 |
| 2/0 | 205 | 225 | 225 | 255 |
| 3/0 | 240 | 270 | 260 | 290 |
| 4/0 | 280 | 305 | 295 | 330 |
| 250 | 315 | 355 | 330 | 365 |
| 350 | 385 | 430 | 395 | 440 |
| 500 | 475 | 530 | 480 | 535 |
| 750 | 600 | 665 | 585 | 655 |
| 1000 | 690 | 770 | 675 | 755 |

*Refer to 310.60(C)(4) for the ampacity correction factors where the ambient air temperature is other than 40°C (104°F).

***TABLE 310.60(C)(74)*** *Ampacities of an Insulated Triplexed or Three Single-Conductor Aluminum Cables in Isolated Conduit in Air Based on Conductor Temperatures of 90°C (194°F) and 105°C (221°F) and Ambient Air Temperature of 40°C (104°F)**

| | **Temperature Rating of Conductor [See Table 310.104(C).]** | | | |
|---|---|---|---|---|
| | **2001–5000 Volts Ampacity** | | **5001–35,000 Volts Ampacity** | |
| **Conductor Size (AWG or kcmil)** | **90°C (194°F) Type MV-90** | **105°C (221°F) Type MV-105** | **90°C (194°F) Type MV-90** | **105°C (221°F) Type MV-105** |
| 8 | 43 | 48 | — | — |
| 6 | 58 | 65 | 65 | 72 |
| 4 | 76 | 85 | 84 | 94 |
| 2 | 100 | 115 | 115 | 130 |
| 1 | 120 | 135 | 130 | 150 |
| 1/0 | 140 | 155 | 150 | 170 |
| 2/0 | 160 | 175 | 175 | 200 |
| 3/0 | 190 | 210 | 200 | 225 |
| 4/0 | 215 | 240 | 230 | 260 |
| 250 | 250 | 280 | 255 | 290 |
| 350 | 305 | 340 | 310 | 350 |
| 500 | 380 | 425 | 385 | 430 |
| 750 | 490 | 545 | 485 | 540 |
| 1000 | 580 | 645 | 565 | 640 |

*Refer to 310.60(C)(4) for the ampacity correction factors where the ambient air temperature is other than 40°C (104°F).

*TABLE 310.60(C)(75) Ampacities of an Insulated Three-Conductor Copper Cable in Isolated Conduit in Air Based on Conductor Temperatures of 90°C (194°F) and 105°C (221°F) and Ambient Air Temperature of 40°C (104°F)**

| | Temperature Rating of Conductor [See Table 310.104(C).] | | | |
|---|---|---|---|---|
| | 2001–5000 Volts Ampacity | | 5001–35,000 Volts Ampacity | |
| Conductor Size (AWG or kcmil) | 90°C (194°F) Type MV-90 | 105°C (221°F) Type MV-105 | 90°C (194°F) Type MV-90 | 105°C (221°F) Type MV-105 |
| 8 | 52 | 58 | — | — |
| 6 | 69 | 77 | 83 | 92 |
| 4 | 91 | 100 | 105 | 120 |
| 2 | 125 | 135 | 145 | 165 |
| 1 | 140 | 155 | 165 | 185 |
| 1/0 | 165 | 185 | 195 | 215 |
| 2/0 | 190 | 210 | 220 | 245 |
| 3/0 | 220 | 245 | 250 | 280 |
| 4/0 | 255 | 285 | 290 | 320 |
| 250 | 280 | 315 | 315 | 350 |
| 350 | 350 | 390 | 385 | 430 |
| 500 | 425 | 475 | 470 | 525 |
| 750 | 525 | 585 | 570 | 635 |
| 1000 | 590 | 660 | 650 | 725 |

*Refer to 310.60(C)(4) for the ampacity correction factors where the ambient air temperature is other than 40°C (104°F).

*TABLE 310.60(C)(76) Ampacities of an Insulated Three-Conductor Aluminum Cable in Isolated Conduit in Air Based on Conductor Temperatures of 90°C (194°F) and 105°C (221°F) and Ambient Air Temperature of 40°C (104°F)**

| | Temperature Rating of Conductor [See Table 310.104(C).] | | | |
|---|---|---|---|---|
| | 2001–5000 Volts Ampacity | | 5001–35,000 Volts Ampacity | |
| Conductor Size (AWG or kcmil) | 90°C (194°F) Type MV-90 | 105°C (221°F) Type MV-105 | 90°C (194°F) Type MV-90 | 105°C (221°F) Type MV-105 |
| 8 | 41 | 46 | — | — |
| 6 | 53 | 59 | 64 | 71 |
| 4 | 71 | 79 | 84 | 94 |
| 2 | 96 | 105 | 115 | 125 |
| 1 | 110 | 125 | 130 | 145 |
| 1/0 | 130 | 145 | 150 | 170 |
| 2/0 | 150 | 165 | 170 | 190 |
| 3/0 | 170 | 190 | 195 | 220 |
| 4/0 | 200 | 225 | 225 | 255 |
| 250 | 220 | 245 | 250 | 280 |
| 350 | 275 | 305 | 305 | 340 |
| 500 | 340 | 380 | 380 | 425 |
| 750 | 430 | 480 | 470 | 520 |
| 1000 | 505 | 560 | 550 | 615 |

*Refer to 310.60(C)(4) for the ampacity correction factors where the ambient air temperature is other than 40°C (104°F).

*TABLE 310.60(C)(77) Ampacities of Three Single-Insulated Copper Conductors in Underground Electrical Ducts (Three Conductors per Electrical Duct) Based on Ambient Earth Temperature of 20°C (68°F), Electrical Duct Arrangement in Accordance with Figure 310.60, 100 Percent Load Factor, Thermal Resistance (RHO) of 90, Conductor Temperatures of 90°C (194°F) and 105°C (221°F)*

| | Temperature Rating of Conductor [See Table 310.104(C).] | | | |
|---|---|---|---|---|
| | 2001–5000 Volts Ampacity | | 5001–35,000 Volts Ampacity | |
| Conductor Size (AWG or kcmil) | 90°C (194°F) Type MV-90 | 105°C (221°F) Type MV-105 | 90°C (194°F) Type MV-90 | 105°C (221°F) Type MV-105 |
| **One Circuit (See Figure 310.60, Detail 1.)** | | | | |
| 8 | 64 | 69 | — | — |
| 6 | 85 | 92 | 90 | 97 |
| 4 | 110 | 120 | 115 | 125 |
| 2 | 145 | 155 | 155 | 165 |
| 1 | 170 | 180 | 175 | 185 |
| 1/0 | 195 | 210 | 200 | 215 |
| 2/0 | 220 | 235 | 230 | 245 |
| 3/0 | 250 | 270 | 260 | 275 |
| 4/0 | 290 | 310 | 295 | 315 |
| 250 | 320 | 345 | 325 | 345 |
| 350 | 385 | 415 | 390 | 415 |
| 500 | 470 | 505 | 465 | 500 |
| 750 | 585 | 630 | 565 | 610 |
| 1000 | 670 | 720 | 640 | 690 |
| **Three Circuits (See Figure 310.60, Detail 2.)** | | | | |
| 8 | 56 | 60 | — | — |
| 6 | 73 | 79 | 77 | 83 |
| 4 | 95 | 100 | 99 | 105 |
| 2 | 125 | 130 | 130 | 135 |
| 1 | 140 | 150 | 145 | 155 |
| 1/0 | 160 | 175 | 165 | 175 |
| 2/0 | 185 | 195 | 185 | 200 |
| 3/0 | 210 | 225 | 210 | 225 |
| 4/0 | 235 | 255 | 240 | 255 |
| 250 | 260 | 280 | 260 | 280 |
| 350 | 315 | 335 | 310 | 330 |
| 500 | 375 | 405 | 370 | 395 |
| 750 | 460 | 495 | 440 | 475 |
| 1000 | 525 | 565 | 495 | 535 |
| **Six Circuits (See Figure 310.60, Detail 3.)** | | | | |
| 8 | 48 | 52 | — | — |
| 6 | 62 | 67 | 64 | 68 |
| 4 | 80 | 86 | 82 | 88 |
| 2 | 105 | 110 | 105 | 115 |
| 1 | 115 | 125 | 120 | 125 |
| 1/0 | 135 | 145 | 135 | 145 |
| 2/0 | 150 | 160 | 150 | 165 |
| 3/0 | 170 | 185 | 170 | 185 |
| 4/0 | 195 | 210 | 190 | 205 |
| 250 | 210 | 225 | 210 | 225 |
| 350 | 250 | 270 | 245 | 265 |
| 500 | 300 | 325 | 290 | 310 |
| 750 | 365 | 395 | 350 | 375 |
| 1000 | 410 | 445 | 390 | 415 |

*TABLE 310.60(C)(78) Ampacities of Three Single-Insulated Aluminum Conductors in Underground Electrical Ducts (Three Conductors per Electrical Duct) Based on Ambient Earth Temperature of 20°C (68°F), Electrical Duct Arrangement in Accordance with Figure 310.60, 100 Percent Load Factor, Thermal Resistance (RHO) of 90, Conductor Temperatures of 90°C (194°F) and 105°C (221°F)*

| | Temperature Rating of Conductor [See Table 310.104(C).] | | | |
|---|---|---|---|---|
| | 2001–5000 Volts Ampacity | | 5001–35,000 Volts Ampacity | |
| Conductor Size (AWG or kcmil) | 90°C (194°F) Type MV-90 | 105°C (221°F) Type MV-105 | 90°C (194°F) Type MV-90 | 105°C (221°F) Type MV-105 |
| **One Circuit (See Figure 310.60, Detail 1.)** | | | | |
| 8 | 50 | 54 | — | — |
| 6 | 66 | 71 | 70 | 75 |
| 4 | 86 | 93 | 91 | 98 |
| 2 | 115 | 125 | 120 | 130 |
| 1 | 130 | 140 | 135 | 145 |
| 1/0 | 150 | 160 | 155 | 165 |
| 2/0 | 170 | 185 | 175 | 190 |
| 3/0 | 195 | 210 | 200 | 215 |
| 4/0 | 225 | 245 | 230 | 245 |
| 250 | 250 | 270 | 250 | 270 |
| 350 | 305 | 325 | 305 | 330 |
| 500 | 370 | 400 | 370 | 400 |
| 750 | 470 | 505 | 455 | 490 |
| 1000 | 545 | 590 | 525 | 565 |
| **Three Circuits (See Figure 310.60, Detail 2.)** | | | | |
| 8 | 44 | 47 | — | — |
| 6 | 57 | 61 | 60 | 65 |
| 4 | 74 | 80 | 77 | 83 |
| 2 | 96 | 105 | 100 | 105 |
| 1 | 110 | 120 | 110 | 120 |
| 1/0 | 125 | 135 | 125 | 140 |
| 2/0 | 145 | 155 | 145 | 155 |
| 3/0 | 160 | 175 | 165 | 175 |
| 4/0 | 185 | 200 | 185 | 200 |
| 250 | 205 | 220 | 200 | 220 |
| 350 | 245 | 265 | 245 | 260 |
| 500 | 295 | 320 | 290 | 315 |
| 750 | 370 | 395 | 355 | 385 |
| 1000 | 425 | 460 | 405 | 440 |
| **Six Circuits (See Figure 310.60, Detail 3.)** | | | | |
| 8 | 38 | 41 | — | — |
| 6 | 48 | 52 | 50 | 54 |
| 4 | 62 | 67 | 64 | 69 |
| 2 | 80 | 86 | 80 | 88 |
| 1 | 91 | 98 | 90 | 99 |
| 1/0 | 105 | 110 | 105 | 110 |
| 2/0 | 115 | 125 | 115 | 125 |
| 3/0 | 135 | 145 | 130 | 145 |
| 4/0 | 150 | 165 | 150 | 160 |
| 250 | 165 | 180 | 165 | 175 |
| 350 | 195 | 210 | 195 | 210 |
| 500 | 240 | 255 | 230 | 250 |
| 750 | 290 | 315 | 280 | 305 |
| 1000 | 335 | 360 | 320 | 345 |

*TABLE 310.60(C)(79) Ampacities of Three Insulated Copper Conductors Cabled Within an Overall Covering (Three-Conductor Cable) in Underground Electrical Ducts (One Cable per Electrical Duct) Based on Ambient Earth Temperature of 20°C (68°F), Electrical Duct Arrangement in Accordance with Figure 310.60, 100 Percent Load Factor, Thermal Resistance (RHO) of 90, Conductor Temperatures of 90°C (194°F) and 105°C (221°C)*

| | Temperature Rating of Conductor [See Table 310.104(C).] | | | |
|---|---|---|---|---|
| | 2001–5000 Volts Ampacity | | 5001–35,000 Volts Ampacity | |
| Conductor Size (AWG or kcmil) | 90°C (194°F) Type MV-90 | 105°C (221°F) Type MV-105 | 90°C (194°F) Type MV-90 | 105°C (221°F) Type MV-105 |
| **One Circuit (See Figure 310.60, Detail 1.)** | | | | |
| 8 | 59 | 64 | — | — |
| 6 | 78 | 84 | 88 | 95 |
| 4 | 100 | 110 | 115 | 125 |
| 2 | 135 | 145 | 150 | 160 |
| 1 | 155 | 165 | 170 | 185 |
| 1/0 | 175 | 190 | 195 | 210 |
| 2/0 | 200 | 220 | 220 | 235 |
| 3/0 | 230 | 250 | 250 | 270 |
| 4/0 | 265 | 285 | 285 | 305 |
| 250 | 290 | 315 | 310 | 335 |
| 350 | 355 | 380 | 375 | 400 |
| 500 | 430 | 460 | 450 | 485 |
| 750 | 530 | 570 | 545 | 585 |
| 1000 | 600 | 645 | 615 | 660 |
| **Three Circuits (See Figure 310.60, Detail 2.)** | | | | |
| 8 | 53 | 57 | — | — |
| 6 | 69 | 74 | 75 | 81 |
| 4 | 89 | 96 | 97 | 105 |
| 2 | 115 | 125 | 125 | 135 |
| 1 | 135 | 145 | 140 | 155 |
| 1/0 | 150 | 165 | 160 | 175 |
| 2/0 | 170 | 185 | 185 | 195 |
| 3/0 | 195 | 210 | 205 | 220 |
| 4/0 | 225 | 240 | 230 | 250 |
| 250 | 245 | 265 | 255 | 270 |
| 350 | 295 | 315 | 305 | 325 |
| 500 | 355 | 380 | 360 | 385 |
| 750 | 430 | 465 | 430 | 465 |
| 1000 | 485 | 520 | 485 | 515 |
| **Six Circuits (See Figure 310.60, Detail 3.)** | | | | |
| 8 | 46 | 50 | — | — |
| 6 | 60 | 65 | 63 | 68 |
| 4 | 77 | 83 | 81 | 87 |
| 2 | 98 | 105 | 105 | 110 |
| 1 | 110 | 120 | 115 | 125 |
| 1/0 | 125 | 135 | 130 | 145 |
| 2/0 | 145 | 155 | 150 | 160 |
| 3/0 | 165 | 175 | 170 | 180 |
| 4/0 | 185 | 200 | 190 | 200 |
| 250 | 200 | 220 | 205 | 220 |
| 350 | 240 | 270 | 245 | 275 |
| 500 | 290 | 310 | 290 | 305 |
| 750 | 350 | 375 | 340 | 365 |
| 1000 | 390 | 420 | 380 | 405 |

***TABLE 310.60(C)(80)*** *Ampacities of Three Insulated Aluminum Conductors Cabled Within an Overall Covering (Three-Conductor Cable) in Underground Electrical Ducts (One Cable per Electrical Duct) Based on Ambient Earth Temperature of 20°C (68°F), Electrical Duct Arrangement in Accordance with Figure 310.60, 100 Percent Load Factor, Thermal Resistance (RHO) of 90, Conductor Temperatures of 90°C (194°F) and 105°C (221°C)*

| | Temperature Rating of Conductor [See Table 310.104(C).] | | | |
|---|---|---|---|---|
| | 2001–5000 Volts Ampacity | | 5001–35,000 Volts Ampacity | |
| Conductor Size (AWG or kcmil) | 90°C (194°F) Type MV-90 | 105°C (221°F) Type MV-105 | 90°C (194°F) Type MV-90 | 105°C (221°F) Type MV-105 |
| **One Circuit (See Figure 310.60, Detail 1.)** | | | | |
| 8 | 46 | 50 | — | — |
| 6 | 61 | 66 | 69 | 74 |
| 4 | 80 | 86 | 89 | 96 |
| 2 | 105 | 110 | 115 | 125 |
| 1 | 120 | 130 | 135 | 145 |
| 1/0 | 140 | 150 | 150 | 165 |
| 2/0 | 160 | 170 | 170 | 185 |
| 3/0 | 180 | 195 | 195 | 210 |
| 4/0 | 205 | 220 | 220 | 240 |
| 250 | 230 | 245 | 245 | 265 |
| 350 | 280 | 310 | 295 | 315 |
| 500 | 340 | 365 | 355 | 385 |
| 750 | 425 | 460 | 440 | 475 |
| 1000 | 495 | 535 | 510 | 545 |
| **Three Circuits (See Figure 310.60, Detail 2.)** | | | | |
| 8 | 41 | 44 | — | — |
| 6 | 54 | 58 | 59 | 64 |
| 4 | 70 | 75 | 75 | 81 |
| 2 | 90 | 97 | 100 | 105 |
| 1 | 105 | 110 | 110 | 120 |
| 1/0 | 120 | 125 | 125 | 135 |
| 2/0 | 135 | 145 | 140 | 155 |
| 3/0 | 155 | 165 | 160 | 175 |
| 4/0 | 175 | 185 | 180 | 195 |
| 250 | 190 | 205 | 200 | 215 |
| 350 | 230 | 250 | 240 | 255 |
| 500 | 280 | 300 | 285 | 305 |
| 750 | 345 | 375 | 350 | 375 |
| 1000 | 400 | 430 | 400 | 430 |
| **Six Circuits (See Figure 310.60, Detail 3.)** | | | | |
| 8 | 36 | 39 | — | — |
| 6 | 46 | 50 | 49 | 53 |
| 4 | 60 | 65 | 63 | 68 |
| 2 | 77 | 83 | 80 | 86 |
| 1 | 87 | 94 | 90 | 98 |
| 1/0 | 99 | 105 | 105 | 110 |
| 2/0 | 110 | 120 | 115 | 125 |
| 3/0 | 130 | 140 | 130 | 140 |
| 4/0 | 145 | 155 | 150 | 160 |
| 250 | 160 | 170 | 160 | 170 |
| 350 | 190 | 205 | 190 | 205 |
| 500 | 230 | 245 | 230 | 245 |
| 750 | 280 | 305 | 275 | 295 |
| 1000 | 320 | 345 | 315 | 335 |

***TABLE 310.60(C)(81)*** *Ampacities of Single Insulated Copper Conductors Directly Buried in Earth Based on Ambient Earth Temperature of 20°C (68°F), Arrangement per Figure 310.60, 100 Percent Load Factor, Thermal Resistance (RHO) of 90, Conductor Temperatures of 90°C (194°F) and 105°C (221°C)*

| | Temperature Rating of Conductor [See Table 310.104(C).] | | | |
|---|---|---|---|---|
| | 2001–5000 Volts Ampacity | | 5001–35,000 Volts Ampacity | |
| Conductor Size (AWG or kcmil) | 90°C (194°F) Type MV-90 | 105°C (221°F) Type MV-105 | 90°C (194°F) Type MV-90 | 105°C (221°F) Type MV-105 |
| **One Circuit, Three Conductors (See Figure 310.60, Detail 9.)** | | | | |
| 8 | 110 | 115 | — | — |
| 6 | 140 | 150 | 130 | 140 |
| 4 | 180 | 195 | 170 | 180 |
| 2 | 230 | 250 | 210 | 225 |
| 1 | 260 | 280 | 240 | 260 |
| 1/0 | 295 | 320 | 275 | 295 |
| 2/0 | 335 | 365 | 310 | 335 |
| 3/0 | 385 | 415 | 355 | 380 |
| 4/0 | 435 | 465 | 405 | 435 |
| 250 | 470 | 510 | 440 | 475 |
| 350 | 570 | 615 | 535 | 575 |
| 500 | 690 | 745 | 650 | 700 |
| 750 | 845 | 910 | 805 | 865 |
| 1000 | 980 | 1055 | 930 | 1005 |
| **Two Circuits, Six Conductors (See Figure 310.60, Detail 10.)** | | | | |
| 8 | 100 | 110 | — | — |
| 6 | 130 | 140 | 120 | 130 |
| 4 | 165 | 180 | 160 | 170 |
| 2 | 215 | 230 | 195 | 210 |
| 1 | 240 | 260 | 225 | 240 |
| 1/0 | 275 | 295 | 255 | 275 |
| 2/0 | 310 | 335 | 290 | 315 |
| 3/0 | 355 | 380 | 330 | 355 |
| 4/0 | 400 | 430 | 375 | 405 |
| 250 | 435 | 470 | 410 | 440 |
| 350 | 520 | 560 | 495 | 530 |
| 500 | 630 | 680 | 600 | 645 |
| 750 | 775 | 835 | 740 | 795 |
| 1000 | 890 | 960 | 855 | 920 |

***TABLE 310.60(C)(82)*** *Ampacities of Single Insulated Aluminum Conductors Directly Buried in Earth Based on Ambient Earth Temperature of 20°C (68°F), Arrangement per Figure 310.60, 100 Percent Load Factor, Thermal Resistance (RHO) of 90, Conductor Temperatures of 90°C (194°F) and 105°C (221°F)*

| | Temperature Rating of Conductor [See Table 310.104(C).] | | | |
|---|---|---|---|---|
| | 2001–5000 Volts Ampacity | | 5001–35,000 Volts Ampacity | |
| **Conductor Size (AWG or kcmil)** | **90°C (194°F) Type MV-90** | **105°C (221°F) Type MV-105** | **90°C (194°F) Type MV-90** | **105°C (221°F) Type MV-105** |
| **One Circuit, Three Conductors (See Figure 310.60, Detail 9.)** | | | | |
| 8 | 85 | 90 | — | — |
| 6 | 110 | 115 | 100 | 110 |
| 4 | 140 | 150 | 130 | 140 |
| 2 | 180 | 195 | 165 | 175 |
| 1 | 205 | 220 | 185 | 200 |
| 1/0 | 230 | 250 | 215 | 230 |
| 2/0 | 265 | 285 | 245 | 260 |
| 3/0 | 300 | 320 | 275 | 295 |
| 4/0 | 340 | 365 | 315 | 340 |
| 250 | 370 | 395 | 345 | 370 |
| 350 | 445 | 480 | 415 | 450 |
| 500 | 540 | 580 | 510 | 545 |
| 750 | 665 | 720 | 635 | 680 |
| 1000 | 780 | 840 | 740 | 795 |
| **Two Circuits, Six Conductors (See Figure 310.60, Detail 10.)** | | | | |
| 8 | 80 | 85 | — | — |
| 6 | 100 | 110 | 95 | 100 |
| 4 | 130 | 140 | 125 | 130 |
| 2 | 165 | 180 | 155 | 165 |
| 1 | 190 | 200 | 175 | 190 |
| 1/0 | 215 | 230 | 200 | 215 |
| 2/0 | 245 | 260 | 225 | 245 |
| 3/0 | 275 | 295 | 255 | 275 |
| 4/0 | 310 | 335 | 290 | 315 |
| 250 | 340 | 365 | 320 | 345 |
| 350 | 410 | 440 | 385 | 415 |
| 500 | 495 | 530 | 470 | 505 |
| 750 | 610 | 655 | 580 | 625 |
| 1000 | 710 | 765 | 680 | 730 |

***TABLE 310.60(C)(83)*** *Ampacities of Three Insulated Copper Conductors Cabled Within an Overall Covering (Three-Conductor Cable), Directly Buried in Earth Based on Ambient Earth Temperature of 20°C (68°F), Arrangement per Figure 310.60, 100 Percent Load Factor, Thermal Resistance (RHO) of 90, Conductor Temperatures of 90°C (194°F) and 105°C (221°F)*

| | Temperature Rating of Conductor [See Table 310.104(C).] | | | |
|---|---|---|---|---|
| | 2001–5000 Volts Ampacity | | 5001–35,000 Volts Ampacity | |
| **Conductor Size (AWG or kcmil)** | **90°C (194°F) Type MV-90** | **105°C (221°F) Type MV-105** | **90°C (194°F) Type MV-90** | **105°C (221°F) Type MV-105** |
| **One Circuit (See Figure 310.60, Detail 5.)** | | | | |
| 8 | 85 | 89 | — | — |
| 6 | 105 | 115 | 115 | 120 |
| 4 | 135 | 150 | 145 | 155 |
| 2 | 180 | 190 | 185 | 200 |
| 1 | 200 | 215 | 210 | 225 |
| 1/0 | 230 | 245 | 240 | 255 |
| 2/0 | 260 | 280 | 270 | 290 |
| 3/0 | 295 | 320 | 305 | 330 |
| 4/0 | 335 | 360 | 350 | 375 |
| 250 | 365 | 395 | 380 | 410 |
| 350 | 440 | 475 | 460 | 495 |
| 500 | 530 | 570 | 550 | 590 |
| 750 | 650 | 700 | 665 | 720 |
| 1000 | 730 | 785 | 750 | 810 |
| **Two Circuits (See Figure 310.60, Detail 6.)** | | | | |
| 8 | 80 | 84 | — | — |
| 6 | 100 | 105 | 105 | 115 |
| 4 | 130 | 140 | 135 | 145 |
| 2 | 165 | 180 | 170 | 185 |
| 1 | 185 | 200 | 195 | 210 |
| 1/0 | 215 | 230 | 220 | 235 |
| 2/0 | 240 | 260 | 250 | 270 |
| 3/0 | 275 | 295 | 280 | 305 |
| 4/0 | 310 | 335 | 320 | 345 |
| 250 | 340 | 365 | 350 | 375 |
| 350 | 410 | 440 | 420 | 450 |
| 500 | 490 | 525 | 500 | 535 |
| 750 | 595 | 640 | 605 | 650 |
| 1000 | 665 | 715 | 675 | 730 |

***TABLE 310.60(C)(84)*** *Ampacities of Three Insulated Aluminum Conductors Cabled Within an Overall Covering (Three-Conductor Cable), Directly Buried in Earth Based on Ambient Earth Temperature of 20°C (68°F), Arrangement per Figure 310.60, 100 Percent Load Factor, Thermal Resistance (RHO) of 90, Conductor Temperatures of 90°C (194°F) and 105°C (221°F)*

| | Temperature Rating of Conductor [See Table 310.104(C).] | | | |
|---|---|---|---|---|
| | 2001–5000 Volts Ampacity | | 5001–35,000 Volts Ampacity | |
| Conductor Size (AWG or kcmil) | 90°C (194°F) Type MV-90 | 105°C (221°F) Type MV-105 | 90°C (194°F) Type MV-90 | 105°C (221°F) Type MV-105 |
| **One Circuit (See Figure 310.60, Detail 5.)** | | | | |
| 8 | 65 | 70 | — | — |
| 6 | 80 | 88 | 90 | 95 |
| 4 | 105 | 115 | 115 | 125 |
| 2 | 140 | 150 | 145 | 155 |
| 1 | 155 | 170 | 165 | 175 |
| 1/0 | 180 | 190 | 185 | 200 |
| 2/0 | 205 | 220 | 210 | 225 |
| 3/0 | 230 | 250 | 240 | 260 |
| 4/0 | 260 | 280 | 270 | 295 |
| 250 | 285 | 310 | 300 | 320 |
| 350 | 345 | 375 | 360 | 390 |
| 500 | 420 | 450 | 435 | 470 |
| 750 | 520 | 560 | 540 | 580 |
| 1000 | 600 | 650 | 620 | 665 |
| **Two Circuits (See Figure 310.60, Detail 6.)** | | | | |
| 8 | 60 | 66 | — | — |
| 6 | 75 | 83 | 80 | 95 |
| 4 | 100 | 110 | 105 | 115 |
| 2 | 130 | 140 | 135 | 145 |
| 1 | 145 | 155 | 150 | 165 |
| 1/0 | 165 | 180 | 170 | 185 |
| 2/0 | 190 | 205 | 195 | 210 |
| 3/0 | 215 | 230 | 220 | 240 |
| 4/0 | 245 | 260 | 250 | 270 |
| 250 | 265 | 285 | 275 | 295 |
| 350 | 320 | 345 | 330 | 355 |
| 500 | 385 | 415 | 395 | 425 |
| 750 | 480 | 515 | 485 | 525 |
| 1000 | 550 | 590 | 560 | 600 |

***TABLE 310.60(C)(85)*** *Ampacities of Three Triplexed Single Insulated Copper Conductors Directly Buried in Earth Based on Ambient Earth Temperature of 20°C (68°F), Arrangement per Figure 310.60, 100 Percent Load Factor, Thermal Resistance (RHO) of 90, Conductor Temperatures 90°C (194°F) and 105°C (221°F)*

| | Temperature Rating of Conductor [See Table 310.104(C).] | | | |
|---|---|---|---|---|
| | 2001–5000 Volts Ampacity | | 5001–35,000 Volts Ampacity | |
| Conductor Size (AWG or kcmil) | 90°C (194°F) Type MV-90 | 105°C (221°F) Type MV-105 | 90°C (194°F) Type MV-90 | 105°C (221°F) Type MV-105 |
| **One Circuit, Three Conductors (See Figure 310.60, Detail 7.)** | | | | |
| 8 | 90 | 95 | — | — |
| 6 | 120 | 130 | 115 | 120 |
| 4 | 150 | 165 | 150 | 160 |
| 2 | 195 | 205 | 190 | 205 |
| 1 | 225 | 240 | 215 | 230 |
| 1/0 | 255 | 270 | 245 | 260 |
| 2/0 | 290 | 310 | 275 | 295 |
| 3/0 | 330 | 360 | 315 | 340 |
| 4/0 | 375 | 405 | 360 | 385 |
| 250 | 410 | 445 | 390 | 410 |
| 350 | 490 | 580 | 470 | 505 |
| 500 | 590 | 635 | 565 | 605 |
| 750 | 725 | 780 | 685 | 740 |
| 1000 | 825 | 885 | 770 | 830 |
| **Two Circuits, Six Conductors (See Figure 310.60, Detail 8.)** | | | | |
| 8 | 85 | 90 | — | — |
| 6 | 110 | 115 | 105 | 115 |
| 4 | 140 | 150 | 140 | 150 |
| 2 | 180 | 195 | 175 | 190 |
| 1 | 205 | 220 | 200 | 215 |
| 1/0 | 235 | 250 | 225 | 240 |
| 2/0 | 265 | 285 | 255 | 275 |
| 3/0 | 300 | 320 | 290 | 315 |
| 4/0 | 340 | 365 | 325 | 350 |
| 250 | 370 | 395 | 355 | 380 |
| 350 | 445 | 480 | 425 | 455 |
| 500 | 535 | 575 | 510 | 545 |
| 750 | 650 | 700 | 615 | 660 |
| 1000 | 740 | 795 | 690 | 745 |

***TABLE 310.60(C)(86)*** *Ampacities of Three Triplexed Single Insulated Aluminum Conductors Directly Buried in Earth Based on Ambient Earth Temperature of 20°C (68°F), Arrangement per Figure 310.60, 100 Percent Load Factor, Thermal Resistance (RHO) of 90, Conductor Temperatures 90°C (194°F) and 105°C (221°F)*

| | Temperature Rating of Conductor [See Table 310.104(C).] | | | |
|---|---|---|---|---|
| | 2001–5000 Volts Ampacity | | 5001–35,000 Volts Ampacity | |
| Conductor Size (AWG or kcmil) | 90°C (194°F) Type MV-90 | 105°C (221°F) Type MV-105 | 90°C (194°F) Type MV-90 | 105°C (221°F) Type MV-105 |
| **One Circuit, Three Conductors (See Figure 310.60, Detail 7.)** | | | | |
| 8 | 70 | 75 | — | — |
| 6 | 90 | 100 | 90 | 95 |
| 4 | 120 | 130 | 115 | 125 |
| 2 | 155 | 165 | 145 | 155 |
| 1 | 175 | 190 | 165 | 175 |
| 1/0 | 200 | 210 | 190 | 205 |
| 2/0 | 225 | 240 | 215 | 230 |
| 3/0 | 255 | 275 | 245 | 265 |
| 4/0 | 290 | 310 | 280 | 305 |
| 250 | 320 | 350 | 305 | 325 |
| 350 | 385 | 420 | 370 | 400 |
| 500 | 465 | 500 | 445 | 480 |
| 750 | 580 | 625 | 550 | 590 |
| 1000 | 670 | 725 | 635 | 680 |
| **Two Circuits, Six Conductors (See Figure 310.60, Detail 8.)** | | | | |
| 8 | 65 | 70 | — | — |
| 6 | 85 | 95 | 85 | 90 |
| 4 | 110 | 120 | 105 | 115 |
| 2 | 140 | 150 | 135 | 145 |
| 1 | 160 | 170 | 155 | 170 |
| 1/0 | 180 | 195 | 175 | 190 |
| 2/0 | 205 | 220 | 200 | 215 |
| 3/0 | 235 | 250 | 225 | 245 |
| 4/0 | 265 | 285 | 255 | 275 |
| 250 | 290 | 310 | 280 | 300 |
| 350 | 350 | 375 | 335 | 360 |
| 500 | 420 | 455 | 405 | 435 |
| 750 | 520 | 560 | 485 | 525 |
| 1000 | 600 | 645 | 565 | 605 |

# III. Construction Specifications

## 310.104 Conductor Constructions and Applications

Insulated conductors shall comply with the applicable provisions of Table 310.104(A) through Table 310.104(E).

Informational Note: Thermoplastic insulation may stiffen at temperatures lower than −10°C (+14°F). Thermoplastic insulation may also be deformed at normal temperatures where subjected to pressure, such as at points of support.

The primary information in Tables 310.104(A) through Table 310.104(E) is the minimum required insulation thickness for each insulation type and voltage level. Additional detailed wire classification information for sizes 14 AWG through 2000 kcmil is available in standards and directories such as those published by Underwriters Laboratories Inc.

Table 310.104(A) includes conductor applications and maximum operating temperatures. Some conductors with dual ratings are listed, such as Type XHHW, which is rated 90°C for dry and damp locations and 75°C for wet locations, and Type THW, which is rated 75°C for dry and wet locations and 90°C for special applications within electric-discharge lighting equipment.

Types RHW-2, XHHW-2, and other types identified by the suffix "2" are rated 90°C for wet locations as well as dry and damp locations.

Copper-clad aluminum conductors are drawn from a copper-clad aluminum rod with the copper metallurgically bonded to an aluminum core. The copper forms a minimum of 10 percent of the cross-sectional area of a solid conductor or of each strand of a stranded conductor. See the commentary following 310.106(B) for making electrical connections with different types of conductor material.

## 310.106 Conductors

Only wires and cables that meet the minimum fire, electrical, and physical properties required by the applicable standards are permitted to be marked with the letter designations found in Table 310.104(A) and (C). See 310.104 for the requirements of insulated conductor construction and applications.

**(A) Minimum Size of Conductors.** The minimum size of conductors shall be as shown in Table 310.106(A), except as permitted elsewhere in this *Code*.

Section 310.106 requires the minimum conductor sizes listed in Table 310.106(A), except as permitted in other sections, such as the following:

1. Small conductor sizes 18 and 16 AWG as permitted by 240.4(D)(1) and (2)
2. Flexible cords as permitted by Table 400.4

***TABLE 310.104(A)** Conductor Applications and Insulations Rated 600 Volts*[1]

| Trade Name | Type Letter | Maximum Operating Temperature | Application Provisions | Insulation | Thickness of Insulation: AWG or kcmil | Thickness of Insulation: mm | Thickness of Insulation: mils | Outer Covering[2] |
|---|---|---|---|---|---|---|---|---|
| Fluorinated ethylene propylene | FEP or FEPB | 90°C<br>194°F | Dry and damp locations | Fluorinated ethylene propylene | 14–10<br>8–2 | 0.51<br>0.76 | 20<br>30 | None |
| | | 200°C<br>392°F | Dry locations — special applications[3] | Fluorinated ethylene propylene | 14–8 | 0.36 | 14 | Glass braid |
| | | | | | 6–2 | 0.36 | 14 | Glass or other suitable braid material |
| Mineral insulation (metal sheathed) | MI | 90°C<br>194°F<br>250°C<br>482°F | Dry and wet locations<br>For special applications[3] | Magnesium oxide | 18–16[4]<br>16–10<br>9–4<br>3–500 | 0.58<br>0.91<br>1.27<br>1.40 | 23<br>36<br>50<br>55 | Copper or alloy steel |
| Moisture-, heat-, and oil-resistant thermoplastic | MTW | 60°C<br>140°F<br>90°C<br>194°F | Machine tool wiring in wet locations<br>Machine tool wiring in dry locations.<br>Informational Note: See NFPA 79. | Flame-retardant, moisture-, heat-, and oil-resistant thermoplastic | <br>22–12<br>10<br>8<br>6<br>4–2<br>1–4/0<br>213–500<br>501–1000 | (A) (B)<br>0.76 0.38<br>0.76 0.51<br>1.14 0.76<br>1.52 0.76<br>1.52 1.02<br>2.03 1.27<br>2.41 1.52<br>2.79 1.78 | (A) (B)<br>30 15<br>30 20<br>45 30<br>60 30<br>60 40<br>80 50<br>95 60<br>110 70 | (A) None<br>(B) Nylon jacket or equivalent |
| Paper | | 85°C<br>185°F | For underground service conductors, or by special permission | Paper | | | | Lead sheath |
| Perfluoro-alkoxy | PFA | 90°C<br>194°F<br>200°C<br>392°F | Dry and damp locations<br>Dry locations — special applications[3] | Perfluoro-alkoxy | 14–10<br>8–2<br>1–4/0 | 0.51<br>0.76<br>1.14 | 20<br>30<br>45 | None |
| Perfluoro-alkoxy | PFAH | 250°C<br>482°F | Dry locations only. Only for leads within apparatus or within raceways connected to apparatus (nickel or nickel-coated copper only) | Perfluoro-alkoxy | 14–10<br>8–2<br>1–4/0 | 0.51<br>0.76<br>1.14 | 20<br>30<br>45 | None |
| Thermoset | RHH | 90°C<br>194°F | Dry and damp locations | | 14–10<br>8–2<br>1–4/0<br>213–500<br>501–1000<br>1001–2000 | 1.14<br>1.52<br>2.03<br>2.41<br>2.79<br>3.18 | 45<br>60<br>80<br>95<br>110<br>125 | Moisture-resistant, flame-retardant, nonmetallic covering[2] |
| Moisture-resistant thermoset | RHW<br>RHW-2 | 75°C<br>167°F<br>90°C<br>194°F | Dry and wet locations | Flame-retardant, moisture-resistant thermoset | 14–10<br>8–2<br>1–4/0<br>213–500<br>501–1000<br>1001–2000 | 1.14<br>1.52<br>2.03<br>2.41<br>2.79<br>3.18 | 45<br>60<br>80<br>95<br>110<br>125 | Moisture-resistant, flame-retardant, nonmetallic covering |
| Silicone | SA | 90°C<br>194°F<br>200°C<br>392°F | Dry and damp locations<br>For special application[3] | Silicone rubber | 14–10<br>8–2<br>1–4/0<br>213–500<br>501–1000<br>1001–2000 | 1.14<br>1.52<br>2.03<br>2.41<br>2.79<br>3.18 | 45<br>60<br>80<br>95<br>110<br>125 | Glass or other suitable braid material |
| Thermoset | SIS | 90°C<br>194°F | Switchboard and switchgear wiring only | Flame-retardant thermoset | 14–10<br>8–2<br>1–4/0 | 0.76<br>1.14<br>2.41 | 30<br>45<br>55 | None |
| Thermoplastic and fibrous outer braid | TBS | 90°C<br>194°F | Switchboard and switchgear wiring only | Thermoplastic | 14–10<br>8<br>6–2<br>1–4/0 | 0.76<br>1.14<br>1.52<br>2.03 | 30<br>45<br>60<br>80 | Flame-retardant, nonmetallic covering |

***TABLE 310.104(A)*** *Continued*

| Trade Name | Type Letter | Maximum Operating Temperature | Application Provisions | Insulation | Thickness of Insulation | | | Outer Covering[2] |
|---|---|---|---|---|---|---|---|---|
| | | | | | AWG or kcmil | mm | mils | |
| Extended polytetra-fluoro-ethylene | TFE | 250°C<br>482°F | Dry locations only. Only for leads within apparatus or within raceways connected to apparatus, or as open wiring (nickel or nickel-coated copper only) | Extruded polytetra-fluoroethylene | 14–10<br>8–2<br>1–4/0 | 0.51<br>0.76<br>1.14 | 20<br>30<br>45 | None |
| Heat-resistant thermoplastic | THHN | 90°C<br>194°F | Dry and damp locations | Flame-retardant, heat-resistant thermoplastic | 14–12<br>10<br>8–6<br>4–2 1–4/0<br>250–500<br>501–1000 | 0.38<br>0.51<br>0.76<br>1.02<br>1.27<br>1.52<br>1.78 | 15<br>20<br>30<br>40<br>50<br>60<br>70 | Nylon jacket or equivalent |
| Moisture- and heat-resistant thermoplastic | THHW | 75°C<br>167°F<br><br><br>90°C<br>194°F | Wet location<br><br><br><br>Dry location | Flame-retardant, moisture- and heat-resistant thermoplastic | 14–10<br>8<br>6–2<br>1–4/0<br>213–500<br>501–1000<br>1001–2000 | 0.76<br>1.14<br>1.52<br>2.03<br>2.41<br>2.79<br>3.18 | 30<br>45<br>60<br>80<br>95<br>110<br>125 | None |
| Moisture- and heat-resistant thermoplastic | THW | 75°C<br>167°F<br>90°C<br>194°F | Dry and wet locations<br><br>Special applications within electric discharge lighting equipment. Limited to 1000 open-circuit volts or less. (size 14-8 only as permitted in 410.68) | Flame-retardant, moisture- and heat-resistant thermoplastic | 14–10<br>8<br>6–2<br>1–4/0<br>213–500<br>501–1000<br>1001–2000 | 0.76<br>1.14<br>1.52<br>2.03<br>2.41<br>2.79<br>3.18 | 30<br>45<br>60<br>80<br>95<br>110<br>125 | None |
| | THW-2 | 90°C<br>194°F | Dry and wet locations | | | | | |
| Moisture- and heat-resistant thermoplastic | THWN | 75°C<br>167°F | Dry and wet locations | Flame-retardant, moisture- and heat-resistant thermoplastic | 14–12<br>10<br>8–6<br>4–2<br>1–4/0<br>250–500<br>501–1000 | 0.38<br>0.51<br>0.76<br>1.02<br>1.27<br>1.52<br>1.78 | 15<br>20<br>30<br>40<br>50<br>60<br>70 | Nylon jacket or equivalent |
| | THWN-2 | 90°C<br>194°F | | | | | | |
| Moisture-resistant thermoplastic | TW | 60°C<br>140°F | Dry and wet locations | Flame-retardant, moisture-resistant thermoplastic | 14–10<br>8<br>6–2<br>1–4/0<br>213–500<br>0 | 0.76<br>1.14<br>1.52<br>2.03<br>2.41<br>2.79<br>3.18 | 30<br>45<br>60<br>80<br>95<br>110<br>125 | None |
| Underground feeder and branch-circuit cable — single conductor (for Type UF cable employing more than one conductor, see Article 340.) | UF | 60°C<br>140°F<br><br><br>75°C<br>167°F[5] | See Article 340. | Moisture-resistant<br><br><br><br>Moisture- and heat-resistant | 14–10<br>8–2<br>1–4/0 | 1.52<br>2.03<br>2.41 | 60[6]<br>80[6]<br>95[6] | Integral with insulation |

*(continues)*

*TABLE 310.104(A) Continued*

| Trade Name | Type Letter | Maximum Operating Temperature | Application Provisions | Insulation | Thickness of Insulation | | | Outer Covering[2] |
|---|---|---|---|---|---|---|---|---|
| | | | | | AWG or kcmil | mm | mils | |
| Underground service-entrance cable — single conductor (for Type USE cable employing more than one conductor, see Article 338.) | USE | 75°C<br>167°F[5] | See Article 338. | Heat- and moisture-resistant | 14–10 | 1.14 | 45 | Moisture-resistant nonmetallic covering (See 338.2.) |
| | | | | | 8–2 | 1.52 | 60 | |
| | | | | | 1–4/0 | 2.03 | 80 | |
| | | | | | 213–500 | 2.41 | 95[7] | |
| | | | | | 501–1000 | 2.79 | 110 | |
| | | | | | 1001–2000 | 3.18 | 125 | |
| | USE-2 | 90°C<br>194°F | Dry and wet locations | | | | | |
| Thermoset | XHH | 90°C<br>194°F | Dry and damp locations | Flame-retardant thermoset | 14–10 | 0.76 | 30 | None |
| | | | | | 8–2 | 1.14 | 45 | |
| | | | | | 1–4/0 | 1.40 | 55 | |
| | | | | | 213–500 | 1.65 | 65 | |
| | | | | | 501–1000 | 2.03 | 80 | |
| | | | | | 1001–2000 | 2.41 | 95 | |
| Moisture-resistant thermoset | XHHW | 90°C<br>194°F<br>75°C<br>167°F | Dry and damp locations<br>Wet locations | Flame-retardant, moisture-resistant thermoset | 14–10 | 0.76 | 30 | None |
| | | | | | 8–2 | 1.14 | 45 | |
| | | | | | 1–4/0 | 1.40 | 55 | |
| | | | | | 213–500 | 1.65 | 65 | |
| | | | | | 501–1000 | 2.03 | 80 95 | |
| | | | | | 1001–2000 | 2.41 | | |
| Moisture-resistant thermoset | XHHW-2 | 90°C<br>194°F | Dry and wet locations | Flame-retardant, moisture-resistant thermoset | 14–10 | 0.76 | 30 | None |
| | | | | | 8–2 | 1.14 | 45 | |
| | | | | | 1–4/0 | 1.40 | 55 | |
| | | | | | 213–500 | 1.65 | 65 | |
| | | | | | 501–1000 | 2.03 | 80 | |
| | | | | | 1001–2000 | 2.41 | 95 | |
| Modified ethylene tetrafluoro-ethylene | Z | 90°C<br>194°F<br>150°C<br>302°F | Dry and damp locations<br>Dry locations — special applications[3] | Modified ethylene tetrafluoro-ethylene | 14–12 | 0.38 | 15 | None |
| | | | | | 10 | 0.51 | 20 | |
| | | | | | 8–4 | 0.64 | 25 | |
| | | | | | 3–1 | 0.89 | 35 | |
| | | | | | 1/0–4/0 | 1.14 | 45 | |
| Modified ethylene tetrafluoro-ethylene | ZW | 75°C<br>167°F<br>90°C<br>194°F<br>150°C<br>302°F | Wet locations<br>Dry and damp locations<br>Dry locations — special applications[3] | Modified ethylene tetrafluoro-ethylene | 14–10 | 0.76 | 30 | None |
| | | | | | 8–2 | 1.14 | 45 | |
| | ZW-2 | 90°C<br>194°F | Dry and wet locations | | | | | |

[1]Conductors can be rated up to 1000 V if listed and marked.

[2]Some insulations do not require an outer covering.

[3]Where design conditions require maximum conductor operating temperatures above 90°C (194°F).

[4]For signaling circuits permitting 300-volt insulation.

[5]For ampacity limitation, see 340.80.

[6]Includes integral jacket.

[7]Insulation thickness shall be permitted to be 2.03 mm (80 mils) for listed Type USE conductors that have been subjected to special investigations. The nonmetallic covering over individual rubber-covered conductors of aluminum-sheathed cable and of lead-sheathed or multiconductor cable shall not be required to be flame retardant. For Type MC cable, see 330.104. For nonmetallic-sheathed cable, see Article 334, Part III. For Type UF cable, see Article 340, Part III.

***TABLE 310.104(B)*** *Thickness of Insulation for Nonshielded Types RHH and RHW Solid Dielectric Insulated Conductors Rated 2000 Volts*

| Conductor Size (AWG or kcmil) | Column A[1] | | Column B[2] | |
|---|---|---|---|---|
| | mm | mils | mm | mils |
| 14–10 | 2.03 | 80 | 1.52 | 60 |
| 8 | 2.03 | 80 | 1.78 | 70 |
| 6–2 | 2.41 | 95 | 1.78 | 70 |
| 1–2/0 | 2.79 | 110 | 2.29 | 90 |
| 3/0–4/0 | 2.79 | 110 | 2.29 | 90 |
| 213–500 | 3.18 | 125 | 2.67 | 105 |
| 501–1000 | 3.56 | 140 | 3.05 | 120 |
| 1001–2000 | 3.56 | 140 | 3.56 | 140 |

[1]Column A insulations are limited to natural, SBR, and butyl rubbers.
[2]Column B insulations are materials such as cross-linked polyethylene, ethylene propylene rubber, and composites thereof.

***TABLE 310.104(C)*** *Conductor Application and Insulation Rated 2001 Volts and Higher*

| Trade Name | Type Letter | Maximum Operating Temperature | Application Provision | Insulation | Outer Covering |
|---|---|---|---|---|---|
| Medium voltage solid dielectric | MV-90<br>MV-105* | 90°C<br>105°C | Dry or wet locations | Thermo-plastic or thermo-setting | Jacket, sheath, or armor |

*Where design conditions require maximum conductor temperatures above 90°C.

***TABLE 310.104(D)*** *Thickness of Insulation and Jacket for Nonshielded Solid Dielectric Insulated Conductors Rated 2001 to 5000 Volts*

| Conductor Size (AWG or kcmil) | Dry Locations, Single Conductor | | | | | | Wet or Dry Locations | | | | | |
|---|---|---|---|---|---|---|---|---|---|---|---|---|
| | Without Jacket Insulation | | With Jacket | | | | Single Conductor | | | | Multiconductor Insulation* | |
| | | | Insulation | | Jacket | | Insulation | | Jacket | | | |
| | mm | mils | mm | mils | mm | mils | mm | mils | mm | mils | mm | mils |
| 8 | 2.79 | 110 | 2.29 | 90 | 0.76 | 30 | 3.18 | 125 | 2.03 | 80 | 2.29 | 90 |
| 6 | 2.79 | 110 | 2.29 | 90 | 0.76 | 30 | 3.18 | 125 | 2.03 | 80 | 2.29 | 90 |
| 4–2 | 2.79 | 110 | 2.29 | 90 | 1.14 | 45 | 3.18 | 125 | 2.03 | 80 | 2.29 | 90 |
| 1–2/0 | 2.79 | 110 | 2.29 | 90 | 1.14 | 45 | 3.18 | 125 | 2.03 | 80 | 2.29 | 90 |
| 3/0–4/0 | 2.79 | 110 | 2.29 | 90 | 1.65 | 65 | 3.18 | 125 | 2.41 | 95 | 2.29 | 90 |
| 213–500 | 3.05 | 120 | 2.29 | 90 | 1.65 | 65 | 3.56 | 140 | 2.79 | 110 | 2.29 | 90 |
| 501–750 | 3.30 | 130 | 2.29 | 90 | 1.65 | 65 | 3.94 | 155 | 3.18 | 125 | 2.29 | 90 |
| 751–1000 | 3.30 | 130 | 2.29 | 90 | 1.65 | 65 | 3.94 | 155 | 3.18 | 125 | 2.29 | 90 |
| 1001–1250 | 3.56 | 140 | 2.92 | 115 | 1.65 | 65 | 4.32 | 170 | 3.56 | 140 | 2.92 | 115 |
| 1251–1500 | 3.56 | 140 | 2.92 | 115 | 2.03 | 80 | 4.32 | 170 | 3.56 | 140 | 2.92 | 115 |
| 1501–2000 | 3.56 | 140 | 2.92 | 115 | 2.03 | 80 | 4.32 | 170 | 3.94 | 155 | 3.56 | 140 |

*Under a common overall covering such as a jacket, sheath, or armor.

***TABLE 310.104(E)*** *Thickness of Insulation for Shielded Solid Dielectric Insulated Conductors Rated 2001 to 35,000 Volts*

| Conductor Size (AWG or kcmil) | 2001–5000 Volts | | 5001–8000 Volts | | | | | | 8001–15,000 Volts | | | | | | 15,001–25,000 Volts | | | | | |
|---|---|---|---|---|---|---|---|---|---|---|---|---|---|---|---|---|---|---|---|---|
| | 100 Percent Insulation Level[1] | | 100 Percent Insulation Level[1] | | 133 Percent Insulation Level[2] | | 173 Percent Insulation Level[3] | | 100 Percent Insulation Level[1] | | 133 Percent Insulation Level[2] | | 173 Percent Insulation Level[3] | | 100 Percent Insulation Level[1] | | 133 Percent Insulation Level[2] | | 173 Percent Insulation Level[3] | |
| | mm | mils | mm | mils | mm | mils | mm | mils | mm | mils | mm | mils | mm | mils | mm | mils | mm | mils | mm | mils |
| 8 | 2.29 | 90 | — | — | — | — | — | — | — | — | — | — | — | — | — | — | — | — | — | — |
| 6–4 | 2.29 | 90 | 2.92 | 115 | 3.56 | 140 | 4.45 | 175 | — | — | — | — | — | — | — | — | — | — | — | — |
| 2 | 2.29 | 90 | 2.92 | 115 | 3.56 | 140 | 4.45 | 175 | 4.45 | 175 | 5.59 | 220 | 6.60 | 260 | — | — | — | — | — | — |
| 1 | 2.29 | 90 | 2.92 | 115 | 3.56 | 140 | 4.45 | 175 | 4.45 | 175 | 5.59 | 220 | 6.60 | 260 | 6.60 | 260 | 8.13 | 320 | 10.67 | 420 |
| 1/0–2000 | 2.29 | 90 | 2.92 | 115 | 3.56 | 140 | 4.45 | 175 | 4.45 | 175 | 5.59 | 220 | 6.60 | 260 | 6.60 | 260 | 8.13 | 320 | 10.67 | 420 |

| Conductor Size (AWG or kcmil) | 25,001–28,000 volts | | | | | | 28,001–35,000 volts | | | | | |
|---|---|---|---|---|---|---|---|---|---|---|---|---|
| | 100 Percent Insulation Level[1] | | 133 Percent Insulation Level[2] | | 173 Percent Insulation Level[3] | | 100 Percent Insulation Level[1] | | 133 Percent Insulation Level[2] | | 173 Percent Insulation Level[3] | |
| | mm | mils | mm | mils | mm | mils | mm | mils | mm | mils | mm | mils |
| 1 | 7.11 | 280 | 8.76 | 345 | 11.30 | 445 | — | — | — | — | — | — |
| 1/0–2000 | 7.11 | 280 | 8.76 | 345 | 11.30 | 445 | 8.76 | 345 | 10.67 | 420 | 14.73 | 580 |

[1]**100 Percent Insulation Level.** Cables in this category shall be permitted to be applied where the system is provided with relay protection such that ground faults will be cleared as rapidly as possible but, in any case, within 1 minute. While these cables are applicable to the great majority of cable installations that are on grounded systems, they shall be permitted to be used also on other systems for which the application of cables is acceptable, provided the above clearing requirements are met in completely de-energizing the faulted section.

[2]**133 Percent Insulation Level.** This insulation level corresponds to that formerly designated for ungrounded systems. Cables in this category shall be permitted to be applied in situations where the clearing time requirements of the 100 percent level category cannot be met and yet there is adequate assurance that the faulted section will be de-energized in a time not exceeding 1 hour. Also, they shall be permitted to be used in 100 percent insulation level applications where additional insulation is desirable.

[3]**173 Percent Insulation Level.** Cables in this category shall be permitted to be applied under all of the following conditions:

(1) In industrial establishments where the conditions of maintenance and supervision ensure that only qualified persons service the installation

(2) Where the fault clearing time requirements of the 133 percent level category cannot be met

(3) Where an orderly shutdown is essential to protect equipment and personnel

(4) There is adequate assurance that the faulted section will be de-energized in an orderly shutdown

Also, cables with this insulation thickness shall be permitted to be used in 100 or 133 percent insulation level applications where additional insulation strength is desirable.

3. Fixture wire as permitted by 402.6
4. Motors rated 1 hp or less as permitted by 430.22(F)
5. Cranes and hoists as permitted by 610.14
6. Elevator control and signaling circuits as permitted by 620.12
7. Class 1, Class 2, and Class 3 circuits as permitted by 725.49(A) and 725.127, Exception
8. Fire alarm circuits as permitted by 760.49(A); 760.127, Exception; and 760.179(B)
9. Motor-control circuits as permitted by 430.72
10. Control and instrumentation circuits as permitted by 727.6
11. Electric signs and outline lighting as permitted in 600.31(B) and 600.32(B)

***TABLE 310.106(A)*** *Minimum Size of Conductors*

| Conductor Voltage Rating (Volts) | Minimum Conductor Size (AWG) | |
|---|---|---|
| | Copper | Aluminum or Copper-Clad Aluminum |
| 0–2000 | 14 | 12 |
| 2001–5000 | 8 | 8 |
| 5001–8000 | 6 | 6 |
| 8001–15,000 | 2 | 2 |
| 15,001–28,000 | 1 | 1 |
| 28,001–35,000 | 1/0 | 1/0 |

**(B) Conductor Material.** Conductors in this article shall be of aluminum, copper-clad aluminum, or copper unless otherwise specified.

Solid aluminum conductors 8, 10, and 12 AWG shall be made of an AA-8000 series electrical grade aluminum alloy conductor material. Stranded aluminum conductors 8 AWG through 1000 kcmil marked as Type RHH, RHW, XHHW, THW, THHW, THWN, THHN, service-entrance Type SE Style U and SE Style R shall be made of an AA-8000 series electrical grade aluminum alloy conductor material.

This coordinates with the UL listing requirements for testing terminations – such as CO/ALR devices and other connectors – suitable for use with aluminum conductors. The electrical industry has developed AA-8000 series aluminum alloy materials and the connectors suitable for use with aluminum conductors to provide for safe and stable connections. Connections suitable for use with aluminum conductors are also generally listed as suitable for use with copper conductors and are marked accordingly, such as AL7CU or AL9CU. Numbers 7 and 9 identify the temperature ratings of 75°C and 90°C, respectively, for these connectors.

**(C) Stranded Conductors.** Where installed in raceways, conductors 8 AWG and larger, not specifically permitted or required elsewhere in this *Code* to be solid, shall be stranded.

Large-size conductors are required to be stranded for greater flexibility. This requirement does not apply to conductors outside of raceways, such as busbars and the conductors of Type MI metal-sheathed cable. Special applications elsewhere in the *Code* may require or permit different requirements for stranded conductors. For example, the bonding conductors of a permanently installed swimming pool are required to be solid copper conductors of 8 AWG or larger, according to 680.26(B).

**(D) Insulated.** Conductors, not specifically permitted elsewhere in this *Code* to be covered or bare, shall be insulated.

Informational Note: See 250.184 for insulation of neutral conductors of a solidly grounded high-voltage system.

## 310.110 Conductor Identification

**(A) Grounded Conductors.** Insulated or covered grounded conductors shall be identified in accordance with 200.6.

**(B) Equipment Grounding Conductors.** Equipment grounding conductors shall be in accordance with 250.119.

**(C) Ungrounded Conductors.** Conductors that are intended for use as ungrounded conductors, whether used as a single conductor or in multiconductor cables, shall be finished to be clearly distinguishable from grounded and grounding conductors. Distinguishing markings shall not conflict in any manner with the surface markings required by 310.120(B)(1). Branch-circuit ungrounded conductors shall be identified in accordance with 210.5(C). Feeders shall be identified in accordance with 215.12.

The identification requirements apply only where a premises wiring system has circuits from more than one nominal voltage system. See the commentary following 210.5(C) and 215.12 for additional information.

*Exception: Conductor identification shall be permitted in accordance with 200.7.*

Ungrounded conductors with white or gray insulation are permitted if the conductors are permanently re-identified at termination points and are visible and accessible. The normal methods of re-identification include colored tape, tagging, or paint. Other applications where white conductors are permitted include flexible cords and circuits less than 50 volts. A white conductor used in single-pole, 3-way and 4-way switch loops also requires re-identification (a color other than white, gray, or green) if it is used as an ungrounded conductor. See 200.7(C)(2) for further information about re-identification of conductors.

## 310.120 Marking

**(A) Required Information.** All conductors and cables shall be marked to indicate the following information, using the applicable method described in 310.120(B):

(1) The maximum rated voltage.
(2) The proper type letter or letters for the type of wire or cable as specified elsewhere in this *Code*.
(3) The manufacturer's name, trademark, or other distinctive marking by which the organization responsible for the product can be readily identified.
(4) The AWG size or circular mil area.

Informational Note: See Conductor Properties, Table 8 of Chapter 9, for conductor area expressed in SI units for conductor sizes specified in AWG or circular mil area.

(5) Cable assemblies where the neutral conductor is smaller than the ungrounded conductors shall be so marked.

**(B) Method of Marking.**

**(1) Surface Marking.** The following conductors and cables shall be durably marked on the surface. The AWG size or circular mil area shall be repeated at intervals not exceeding 610 mm (24 in.). All other markings shall be repeated at intervals not exceeding 1.0 m (40 in.).

(1) Single-conductor and multiconductor rubber- and thermoplastic-insulated wire and cable
(2) Nonmetallic-sheathed cable
(3) Service-entrance cable
(4) Underground feeder and branch-circuit cable
(5) Tray cable
(6) Irrigation cable
(7) Power-limited tray cable
(8) Instrumentation tray cable

**(2) Marker Tape.** Metal-covered multiconductor cables shall employ a marker tape located within the cable and running for its complete length.

*Exception No. 1: Type MI cable.*

*Exception No. 2: Type AC cable.*

*Exception No. 3: The information required in 310.120(A) shall be permitted to be durably marked on the outer nonmetallic covering of Type MC, Type ITC, or Type PLTC cables at intervals not exceeding 1.0 m (40 in.).*

*Exception No. 4: The information required in 310.120(A) shall be permitted to be durably marked on a nonmetallic covering under the metallic sheath of Type ITC or Type PLTC cable at intervals not exceeding 1.0 m (40 in.).*

Informational Note: Included in the group of metal-covered cables are Type AC cable (Article 320), Type MC cable (Article 330), and lead-sheathed cable.

Type PLTC cable may have a metallic sheath or armor over a nonmetallic jacketed cable. A second nonmetallic jacket covering the metallic sheath is optional. Exceptions No. 3 and No. 4 define the marking requirements for either case.

**(3) Tag Marking.** The following conductors and cables shall be marked by means of a printed tag attached to the coil, reel, or carton:

(1) Type MI cable
(2) Switchboard wires
(3) Metal-covered, single-conductor cables
(4) Type AC cable

**(4) Optional Marking of Wire Size.** The information required in 310.120(A)(4) shall be permitted to be marked on the surface of the individual insulated conductors for the following multiconductor cables:

(1) Type MC cable
(2) Tray cable
(3) Irrigation cable
(4) Power-limited tray cable
(5) Power-limited fire alarm cable
(6) Instrumentation tray cable

**(C) Suffixes to Designate Number of Conductors.** A type letter or letters used alone shall indicate a single insulated conductor. The letter suffixes shall be indicated as follows:

(1) D — For two insulated conductors laid parallel within an outer nonmetallic covering
(2) M — For an assembly of two or more insulated conductors twisted spirally within an outer nonmetallic covering

**(D) Optional Markings.** All conductors and cables contained in Chapter 3 shall be permitted to be surface marked to indicate special characteristics of the cable materials. These markings include, but are not limited to, markings for limited smoke, sunlight resistant, and so forth.

Cable insulations that have special characteristics are permitted to carry surface markings that indicate their characteristics. For example, the limited-smoke cables are permitted to be marked "LS" or "ST1." Other characteristics permitted to be marked include sunlight resistance and low corrosiveness. For a detailed list of optional wire and cable marking, see the UL *Wire Marking Guide*, available from Underwriters Laboratories Inc.

# ARTICLE 312 Cabinets, Cutout Boxes, and Meter Socket Enclosures

## I. Scope and Installation

### 312.1 Scope

This article covers the installation and construction specifications of cabinets, cutout boxes, and meter socket enclosures.

Cabinets and cutout boxes are designed with a swinging door(s) to enclose potential transformers, current transformers, switches, overcurrent devices, meters, or control equipment. (See the definitions of these terms in Article 100.) Cabinets and cutout boxes are required to be of sufficient size to accommodate all devices and conductors without overcrowding or jamming. Additional space is often provided through auxiliary gutters (Article 366).

The serving electric utility often has equipment specifications or service requirements beyond the *Code* for meter sockets, metering cabinets, and metering compartments within switchgear, switchboards, and panelboards. Consulting with the local electric utility on these requirements helps identify suitable equipment for an installation. One organization, the Electric Utility Service Equipment Requirements Committee (EUSERC), promotes uniform electric utility metering service requirements for these enclosures that meet the requirements of the *Code* and the serving utility. Many electrical equipment manufacturers identify their equipment as meeting the EUSERC metering space requirements.

### 312.2 Damp and Wet Locations

In damp or wet locations, surface-type enclosures within the scope of this article shall be placed or equipped so as to prevent moisture or water from entering and accumulating within the cabinet or cutout box, and shall be mounted so there is at least 6-mm (¼-in.) airspace between the enclosure and the wall or other supporting surface. Enclosures installed in wet locations shall be weatherproof. For enclosures in wet locations, raceways or cables entering above the level of uninsulated live parts shall use fittings listed for wet locations.

*Exception: Nonmetallic enclosures shall be permitted to be installed without the airspace on a concrete, masonry, tile, or similar surface.*

Informational Note: For protection against corrosion, see 300.6.

## 312.3 Position in Wall

In walls of concrete, tile, or other noncombustible material, cabinets shall be installed so that the front edge of the cabinet is not set back of the finished surface more than 6 mm (¼ in.). In walls constructed of wood or other combustible material, cabinets shall be flush with the finished surface or project therefrom.

## 312.4 Repairing Noncombustible Surfaces

Noncombustible surfaces that are broken or incomplete shall be repaired so there will be no gaps or open spaces greater than 3 mm (⅛ in.) at the edge of the cabinet or cutout box employing a flush-type cover.

Section 312.4 applies to all noncombustible surfaces and is not limited to plaster or drywall types of construction. It provides requirements for cabinets and cutout boxes similar to the outlet box requirements found in 314.21.

## 312.5 Cabinets, Cutout Boxes, and Meter Socket Enclosures

Conductors entering enclosures within the scope of this article shall be protected from abrasion and shall comply with 312.5(A) through (C).

**(A) Openings to Be Closed.** Openings through which conductors enter shall be closed in an approved manner.

**(B) Metal Cabinets, Cutout Boxes, and Meter Socket Enclosures.** Where metal enclosures within the scope of this article are installed with messenger-supported wiring, open wiring on insulators, or concealed knob-and-tube wiring, conductors shall enter through insulating bushings or, in dry locations, through flexible tubing extending from the last insulating support and firmly secured to the enclosure.

**(C) Cables.** Where cable is used, each cable shall be secured to the cabinet, cutout box, or meter socket enclosure.

The installation of several cables bunched together and run through a knockout or chase nipple is prohibited. Individual cable clamps or connectors are required to be used with only one cable per clamp or connector, unless the clamp or connector is identified for more than a single cable.

*Exception: Cables with entirely nonmetallic sheaths shall be permitted to enter the top of a surface-mounted enclosure through one or more nonflexible raceways not less than 450 mm (18 in.) and not more than 3.0 m (10 ft) in length, provided all of the following conditions are met:*

*(a) Each cable is fastened within 300 mm (12 in.), measured along the sheath, of the outer end of the raceway.*
*(b) The raceway extends directly above the enclosure and does not penetrate a structural ceiling.*
*(c) A fitting is provided on each end of the raceway to protect the cable(s) from abrasion and the fittings remain accessible after installation.*
*(d) The raceway is sealed or plugged at the outer end using approved means so as to prevent access to the enclosure through the raceway.*
*(e) The cable sheath is continuous through the raceway and extends into the enclosure beyond the fitting not less than 6 mm (¼ in.).*
*(f) The raceway is fastened at its outer end and at other points in accordance with the applicable article.*
*(g) Where installed as conduit or tubing, the cable fill does not exceed the amount that would be permitted for complete conduit or tubing systems by Table 1 of Chapter 9 of this Code and all applicable notes thereto.*

If the nipple length exceeds 24 inches, the ampacity adjustment factors of 310.15(B)(3) apply.

Informational Note: See Table 1 in Chapter 9, including Note 9, for allowable cable fill in circular raceways. See 310.15(B)(3)(a) for required ampacity reductions for multiple cables installed in a common raceway.

## 312.6 Deflection of Conductors

Conductors at terminals or conductors entering or leaving cabinets or cutout boxes and the like shall comply with 312.6(A) through (C).

*Exception: Wire-bending space in enclosures for motor controllers with provisions for one or two wires per terminal shall comply with 430.10(B).*

**(A) Width of Wiring Gutters.** Conductors shall not be deflected within a cabinet or cutout box unless a gutter having a width in accordance with Table 312.6(A) is provided. Conductors in parallel in accordance with 310.10(H) shall be judged on the basis of the number of conductors in parallel.

**(B) Wire-Bending Space at Terminals.** Wire-bending space at each terminal shall be provided in accordance with 312.6(B)(1) or (B)(2).

**(1) Conductors Not Entering or Leaving Opposite Wall.** Table 312.6(A) shall apply where the conductor does not enter or leave the enclosure through the wall opposite its terminal.

**(2) Conductors Entering or Leaving Opposite Wall.** Table 312.6(B) shall apply where the conductor does enter or leave the enclosure through the wall opposite its terminal.

*Exception No. 1: Where the distance between the wall and its terminal is in accordance with Table 312.6(A), a conductor*

***TABLE 312.6(A)*** *Minimum Wire-Bending Space at Terminals and Minimum Width of Wiring Gutters*

| | Wires per Terminal | | | | | | | | | |
|---|---|---|---|---|---|---|---|---|---|---|
| | 1 | | 2 | | 3 | | 4 | | 5 | |
| **Wire Size (AWG or kcmil)** | **mm** | **in.** | **mm** | **in.** | **mm** | **in.** | **mm** | **in.** | **mm** | **in.** |
| 14–10 | Not specified | | — | — | — | — | — | — | — | — |
| 8–6 | 38.1 | 1½ | — | — | — | — | — | — | — | — |
| 4–3 | 50.8 | 2 | — | — | — | — | — | — | — | — |
| 2 | 63.5 | 2½ | — | — | — | — | — | — | — | — |
| 1 | 76.2 | 3 | — | — | — | — | — | — | — | — |
| 1/0–2/0 | 88.9 | 3½ | 127 | 5 | 178 | 7 | — | — | — | — |
| 3/0–4/0 | 102 | 4 | 152 | 6 | 203 | 8 | — | — | — | — |
| 250 | 114 | 4½ | 152 | 6 | 203 | 8 | 254 | 10 | — | — |
| 300–350 | 127 | 5 | 203 | 8 | 254 | 10 | 305 | 12 | — | — |
| 400–500 | 152 | 6 | 203 | 8 | 254 | 10 | 305 | 12 | 356 | 14 |
| 600–700 | 203 | 8 | 254 | 10 | 305 | 12 | 356 | 14 | 406 | 16 |
| 750–900 | 203 | 8 | 305 | 12 | 356 | 14 | 406 | 16 | 457 | 18 |
| 1000–1250 | 254 | 10 | — | — | — | — | — | — | — | — |
| 1500–2000 | 305 | 12 | — | — | — | — | — | — | — | — |

Note: Bending space at terminals shall be measured in a straight line from the end of the lug or wire connector (in the direction that the wire leaves the terminal) to the wall, barrier, or obstruction.

*shall be permitted to enter or leave an enclosure through the wall opposite its terminal, provided the conductor enters or leaves the enclosure where the gutter joins an adjacent gutter that has a width that conforms to Table 312.6(B) for the conductor.*

*Exception No. 2: A conductor not larger than 350 kcmil shall be permitted to enter or leave an enclosure containing only a meter socket(s) through the wall opposite its terminal, provided the distance between the terminal and the opposite wall is not less than that specified in Table 312.6(A) and the terminal is a lay-in type, where the terminal is either of the following:*

*(a) Directed toward the opening in the enclosure and within a 45 degree angle of directly facing the enclosure wall*

*(b) Directly facing the enclosure wall and offset not greater than 50 percent of the bending space specified in Table 312.6(A)*

Informational Note: *Offset* is the distance measured along the enclosure wall from the axis of the centerline of the terminal to a line passing through the center of the opening in the enclosure.

Section 312.6(B)(2) and Table 312.6(B) provide the requirements for wire-bending space where straight-in wiring or offset (double bends) is employed at terminals. Section 312.6(B)(1) applies only to 90-degree bends.

The notes to Table 312.6(B) permit a reduction in required bending space for removable and lay-in wire terminals. The removable terminal wire connectors can be either the compression type or the setscrew type. However, connectors are required to be of the type intended for a single conductor (single barrel). Removable connectors designed for multiple wires are not permitted to have a reduction in bending space.

To facilitate wiring, a terminal may be placed on the stripped end of the conductor cut to the proper length. The terminal is crimped or lightly torqued on the wire as intended. The wire is bent and routed to facilitate mounting onto the stud or landing pad for proper connection. All mechanical screws, bolts, and nuts involved should then be torqued to the proper value. See the commentary following the informational note to 110.14 regarding tightening torques.

In accordance with the notes to Tables 312.6(A) and 312.6(B), when Table 312.6(A) is used, bending space is measured in the direction in which the wire leaves the terminal, and when Table 312.6(B) is used, it is measured in a direction perpendicular to the enclosure wall.

A lay-in–type terminal is a pressure wire connector in which part of the connector is removable or swings away so that the stripped end of the conductor can be laid into the fixed portion of the connector. The removable or swing-away portion is then put back in place and the connector tightened down on the conductor.

***TABLE 312.6(B)*** *Minimum Wire-Bending Space at Terminals*

| **Wire Size (AWG or kcmil)** | | **Wires per Terminal** | | | | | | | |
|---|---|---|---|---|---|---|---|---|---|
| | | **1** | | **2** | | **3** | | **4 or More** | |
| **All Other Conductors** | **Compact Stranded AA-8000 Aluminum Alloy Conductors (See Note 3.)** | **mm** | **in.** | **mm** | **in.** | **mm** | **in.** | **mm** | **in.** |
| 14–10 | 12–8 | Not specified | | — | — | — | | — | — |
| 8 | 6 | 38.1 | 1½ | — | — | — | | — | — |
| 6 | 4 | 50.8 | 2 | — | — | — | | — | — |
| 4 | 2 | 76.2 | 3 | — | — | — | | — | — |
| 3 | 1 | 76.2 | 3 | — | — | — | | — | — |
| 2 | 1/0 | 88.9 | 3½ | — | — | — | | — | — |
| 1 | 2/0 | 114 | 4½ | — | — | — | | — | — |
| 1/0 | 3/0 | 140 | 5½ | 140 | 5½ | 178 | 7 | — | — |
| 2/0 | 4/0 | 152 | 6 | 152 | 6 | 190 | 7½ | — | — |
| 3/0 | 250 | 165[a] | 6½[a] | 165[a] | 6½[a] | 203 | 8 | — | — |
| 4/0 | 300 | 178[b] | 7[b] | 190[c] | 7½[c] | 216[a] | 8½[a] | — | — |
| 250 | 350 | 216[d] | 8½[d] | 229[d] | 8½[d] | 254[b] | 9[b] | 254 | 10 |
| 300 | 400 | 254[e] | 10[e] | 254[d] | 10[d] | 279[b] | 11[b] | 305 | 12 |
| 350 | 500 | 305[e] | 12[e] | 305[e] | 12[e] | 330[e] | 13[e] | 356[d] | 14[d] |
| 400 | 600 | 330[e] | 13[e] | 330[e] | 13[e] | 356[e] | 14[e] | 381[e] | 15[e] |
| 500 | 700–750 | 356[e] | 14[e] | 356[e] | 14[e] | 381[e] | 15[e] | 406[e] | 16[e] |
| 600 | 800–900 | 381[e] | 15[e] | 406[e] | 16[e] | 457[e] | 18[e] | 483[e] | 19[e] |
| 700 | 1000 | 406[e] | 16[e] | 457[e] | 18[e] | 508[e] | 20[e] | 559[e] | 22[e] |
| 750 | — | 432[e] | 17[e] | 483[e] | 19[e] | 559[e] | 22[e] | 610[e] | 24[e] |
| 800 | — | 457 | 18 | 508 | 20 | 559 | 22 | 610 | 24 |
| 900 | — | 483 | 19 | 559 | 22 | 610 | 24 | 610 | 24 |
| 1000 | — | 508 | 20 | — | — | — | | — | |
| 1250 | — | 559 | 22 | — | — | — | | — | |
| 1500 | — | 610 | 24 | — | — | — | | — | |
| 1750 | — | 610 | 24 | — | — | — | | — | |
| 2000 | — | 610 | 24 | — | — | — | | — | |

Notes:

1. Bending space at terminals shall be measured in a straight line from the end of the lug or wire connector in a direction perpendicular to the enclosure wall.

2. For removable and lay-in wire terminals intended for only one wire, bending space shall be permitted to be reduced by the following number of millimeters (inches):
[a]12.7 mm (½ in.)
[b]25.4 mm (1 in.)
[c]38.1 mm (1½ in.)
[d]50.8 mm (2 in.)
[e]76.2 mm (3 in.)

3. This column shall be permitted to determine the required wire-bending space for compact stranded aluminum conductors in sizes up to 1000 kcmil and manufactured using AA-8000 series electrical grade aluminum alloy conductor material in accordance with 310.106(B).

Exhibit 312.1 applies the rules of 312.6(B)(1), 312.6(B)(2), and Tables 312.6(A) and 312.6(B) to the wiring for a lay-in–type terminal. The applicable table for determining the required bending space, where *T* stands for the dimension to the terminal and *G* stands for dimension of the gutter, is as follows.

$T_1$, 312.6(B)(2): Table 312.6(B) applies for conductors *M*.

$T_2$, 312.6(B)(2): Table 312.6(B) applies for conductors $BR_2$ unless Exception No. 2 to 312.6(B)(2) applies. This exception allows Table 312.6(A) to apply to $T_2$ as long as $BR_2$ enters or leaves the enclosure where gutter $G_2$ joins gutter $G_1$ and gutter $G_1$ has a width conforming to Table 312.6(B) for $BR_2$.

$T_3$, 312.6(B)(2): Table 312.6(B) applies for conductors $BR_3$.

$T_4$, 312.6(B)(1): Table 312.6(A) applies for conductor *N*.

$G_1$, 312.6(A): Table 312.6(A) applies for conductors *M*. Table 312.6(B) applies for conductors $BR_2$ where $T_2$ does not comply with Table 312.6(B).

$G_2$, 312.6(A): Table 312.6(A) applies for conductors $BR_2$.

$G_3$, 312.6(A): Table 312.6(A) applies for conductors $BR_3$.

$G_4$, 312.6(A): Table 312.6(A) applies for conductor *N*.

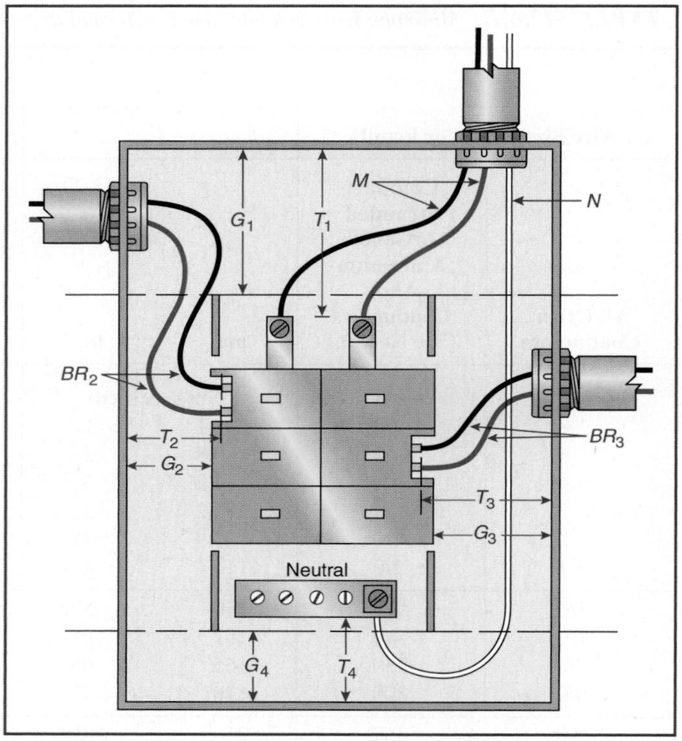


***EXHIBIT 312.1*** *Wiring with a 50 percent offset for meter socket lay-in–type terminals.*

Exhibit 312.2 illustrates the conditions under which 312.6(B)(2), Exception No. 2, is applicable. The terminal on the left has an offset not greater than 50 percent of bending space, per condition (b) of Exception No. 2. The terminal on the right is within a 45-degree angle of the enclosure, per condition (a) of Exception No. 2. See also Table 430.10(B) and the associated Exhibit 430.1, which shows an example of wire-bending space in enclosures for motor controllers.

**(C) Conductors 4 AWG or Larger.** Installation shall comply with 300.4(G).

## 312.7 Space in Enclosures

Cabinets and cutout boxes shall have approved space to accommodate all conductors installed in them without crowding.

## 312.8 Switch and Overcurrent Device Enclosures with Splices, Taps, and Feed-Through Conductors

The wiring space of enclosures for switches or overcurrent devices shall be permitted for conductors feeding through, spliced, or tapping off to other enclosures, switches, or overcurrent devices where all of the following conditions are met:

(1) The total of all conductors installed at any cross section of the wiring space does not exceed 40 percent of the cross-sectional area of that space.

(2) The total area of all conductors, splices, and taps installed at any cross section of the wiring space does not exceed 75 percent of the cross-sectional area of that space.

(3) A warning label complying with 110.21(B) is applied to the enclosure that identifies the closest disconnecting means for any feed-through conductors.

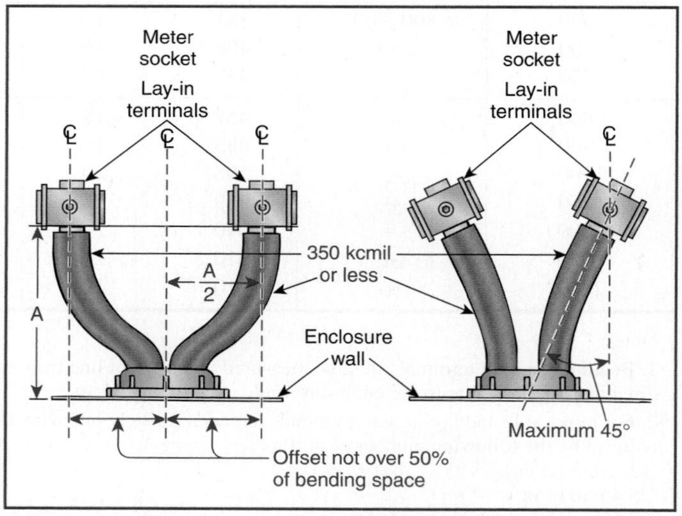


***EXHIBIT 312.2*** *Wiring with a 45 degree angle for meter socket lay-in–type terminals.*

Feed-through conductors, splices, and taps are permitted in the wiring space where conditions (1) through (3) are met. The total conductor fill in the enclosure must not exceed 40 percent of the cross section of the enclosure wiring space and no more than 75 percent if splices or taps are necessary.

**Application Example**

An enclosure having a wiring space of 4 inch wide by 3 inch deep has a cross-sectional area of 12 inch.$^2$ Thus, the total conductor fill (see Chapter 9, Table 5, for dimensions of conductors) at any cross section cannot exceed 4.8 in.$^2$ (40 percent of 12 in.$^2$), and the maximum space for conductors and splices or taps at any cross section cannot exceed 9 in.$^2$ (75 percent of 12 inch$_2$).

In general, the best way to avoid overcrowding enclosures is to properly plan and lay out work before installation and use properly sized auxiliary gutters (366.22, 366.56, and 366.58) or junction boxes (314.16 and 314.28). For wiring space in enclosures for motor controllers and disconnecting means, see 430.10 and the associated commentary. See also 110.59 for tunnel installations over 600 volts.

### 312.9 Side or Back Wiring Spaces or Gutters

Cabinets and cutout boxes shall be provided with back-wiring spaces, gutters, or wiring compartments as required by 312.11(C) and (D).

## II. Construction Specifications

### 312.10 Material

Cabinets, cutout boxes, and meter socket enclosures shall comply with 312.10(A) through (C).

**(A) Metal Cabinets and Cutout Boxes.** Metal enclosures within the scope of this article shall be protected both inside and outside against corrosion.

•

**(B) Strength.** The design and construction of enclosures within the scope of this article shall be such as to secure ample strength and rigidity. If constructed of sheet steel, the metal thickness shall not be less than 1.35 mm (0.053 in.) uncoated.

**(C) Nonmetallic Cabinets.** Nonmetallic cabinets shall be listed, or they shall be submitted for approval prior to installation.

### 312.11 Spacing

The spacing within cabinets and cutout boxes shall comply with 312.11(A) through (D).

**(A) General.** Spacing within cabinets and cutout boxes shall provide approved spacing for the distribution of wires and cables placed in them and for a separation between metal parts of devices and apparatus mounted within them in accordance with 312.11(A)(1), (A)(2), and (A)(3).

**(1) Base.** Other than at points of support, there shall be an airspace of at least 1.59 mm (0.0625 in.) between the base of the device and the wall of any metal cabinet or cutout box in which the device is mounted.

**(2) Doors.** There shall be an airspace of at least 25.4 mm (1.00 in.) between any live metal part, including live metal parts of enclosed fuses, and the door.

*Exception: Where the door is lined with an approved insulating material or is of a thickness of metal not less than 2.36 mm (0.093 in.) uncoated, the airspace shall not be less than 12.7 mm (0.500 in.).*

**(3) Live Parts.** There shall be an airspace of at least 12.7 mm (0.500 in.) between the walls, back, gutter partition, if of metal, or door of any cabinet or cutout box and the nearest exposed current-carrying part of devices mounted within the cabinet where the voltage does not exceed 250. This spacing shall be increased to at least 25.4 mm (1.00 in.) for voltages of 251 to 1000, nominal.

*Exception: Where the conditions in 312.11(A)(2), Exception, are met, the airspace for nominal voltages from 251 to 600 shall be permitted to be not less than 12.7 mm (0.500 in.).*

**(B) Switch Clearance.** Cabinets and cutout boxes shall be deep enough to allow the closing of the doors when 30-ampere branch-circuit panelboard switches are in any position, when combination cutout switches are in any position, or when other single-throw switches are opened as far as their construction permits.

**(C) Wiring Space.** Cabinets and cutout boxes that contain devices or apparatus connected within the cabinet or box to more than eight conductors, including those of branch circuits, meter loops, feeder circuits, power circuits, and similar circuits, but not including the supply circuit or a continuation thereof, shall have back-wiring spaces or one or more side-wiring spaces, side gutters, or wiring compartments.

**(D) Wiring Space — Enclosure.** Side-wiring spaces, side gutters, or side-wiring compartments of cabinets and cutout boxes shall be made tight enclosures by means of covers, barriers, or partitions extending from the bases of the devices contained in the cabinet, to the door, frame, or sides of the cabinet.

*Exception: Side-wiring spaces, side gutters, and side-wiring compartments of cabinets shall not be required to be made tight enclosures where those side spaces contain only conductors that enter the cabinet directly opposite to the devices where they terminate.*

Partially enclosed back-wiring spaces shall be provided with covers to complete the enclosure. Wiring spaces that are required by 312.11(C) and are exposed when doors are open shall be provided with covers to complete the enclosure. Where space is provided for feed-through conductors and for splices as required in 312.8, additional barriers shall not be required.

# ARTICLE 314 Outlet, Device, Pull, and Junction Boxes; Conduit Bodies; Fittings; and Handhole Enclosures

## I. Scope and General

### 314.1 Scope

This article covers the installation and use of all boxes and conduit bodies used as outlet, device, junction, or pull boxes, depending on their use, and handhole enclosures. Cast, sheet metal, nonmetallic, and other boxes such as FS, FD, and larger boxes are not classified as conduit bodies. This article also includes installation requirements for fittings used to join raceways and to connect raceways and cables to boxes and conduit bodies.

### 314.2 Round Boxes

Round boxes shall not be used where conduits or connectors requiring the use of locknuts or bushings are to be connected to the side of the box.

Rectangular or octagonal boxes must have a flat bearing surface at each knockout for locknuts and bushings to ensure effective grounding continuity. Round boxes, however, can be used if the conduit or cable is secured by clamps within the box or if the cable does not need attachment to the box, as permitted by 314.17(C), Exception.

### 314.3 Nonmetallic Boxes

Nonmetallic boxes shall be permitted only with open wiring on insulators, concealed knob-and-tube wiring, cabled wiring methods with entirely nonmetallic sheaths, flexible cords, and nonmetallic raceways.

*Exception No. 1: Where internal bonding means are provided between all entries, nonmetallic boxes shall be permitted to be used with metal raceways or metal-armored cables.*

This exception applies to nonmetallic boxes without threaded entries and permits the use of metal raceways and metal-armored cables with nonmetallic boxes. Internal bonding means must be installed to ensure ground continuity between the metal raceways or metal-armored cables. For the purposes of this exception, the term *metal-armored cable* includes other cables with a metal covering such as mineral-insulated, metal-sheathed cable (Type MI), metal-clad cable (Type MC), and armored cable (Type AC).

*Exception No. 2: Where integral bonding means with a provision for attaching an equipment bonding jumper inside the box are provided between all threaded entries in nonmetallic boxes listed for the purpose, nonmetallic boxes shall be permitted to be used with metal raceways or metal-armored cables.*

An integral bonding means is required to ensure ground continuity between the threaded entries. The requirement for means to attach a bonding jumper accommodates devices or equipment attached to the box. For the purposes of this exception, the term *metal-armored cable* includes other cables with a metal covering such as mineral-insulated, metal-sheathed cable (Type MI), metal-clad cable (Type MC), and armored cable (Type AC).

### 314.4 Metal Boxes

Metal boxes shall be grounded and bonded in accordance with Parts I, IV, V, VI, VII, and X of Article 250 as applicable, except as permitted in 250.112(I).

## II. Installation

### 314.15 Damp or Wet Locations

In damp or wet locations, boxes, conduit bodies, and fittings shall be placed or equipped so as to prevent moisture from entering or accumulating within the box, conduit body, or fitting. Boxes, conduit bodies, and fittings installed in wet locations shall be listed for use in wet locations. Approved drainage openings not larger than 6 mm (¼ in.) shall be permitted to be installed in the field in boxes or conduit bodies listed for use in damp or wet locations. For installation of listed drain fittings, larger openings are permitted to be installed in the field in accordance with manufacturer's instructions.

Informational Note No. 1: For boxes in floors, see 314.27(B).

Informational Note No. 2: For protection against corrosion, see 300.6.

Wet locations include those where the boxes, fittings, or conduit bodies are exposed to weather. Article 100 defines the term *weatherproof* as "constructed or protected so that exposure to the weather will not interfere with successful operation." Rainproof, raintight, or watertight equipment can fulfill the requirements of this definition where varying weather conditions other than wetness, such as snow, ice, dust, or temperature extremes, are not a factor. A weatherhead fitting is considered to be weatherproof because the openings for the conductors are placed in a downward position so that rain or snow cannot enter the fitting.

See the definitions of *location, damp* and *location, wet* in Article 100, as well as the commentary following the definition of *enclosure*, for further explanation. See also 110.28 and Table 110.28 for enclosure-type number selections (e.g., Type 1, Type 3R, and Type 4X).

### 314.16 Number of Conductors in Outlet, Device, and Junction Boxes, and Conduit Bodies

Boxes and conduit bodies shall be of an approved size to provide free space for all enclosed conductors. In no case shall the volume of the box, as calculated in 314.16(A), be less than the fill calculation as calculated in 314.16(B). The minimum volume for conduit bodies shall be as calculated in 314.16(C).

The provisions of this section shall not apply to terminal housings supplied with motors or generators.

Informational Note: For volume requirements of motor or generator terminal housings, see 430.12.

Boxes and conduit bodies enclosing conductors 4 AWG or larger shall also comply with the provisions of 314.28.

**(A) Box Volume Calculations.** The volume of a wiring enclosure (box) shall be the total volume of the assembled sections and, where used, the space provided by plaster rings, domed covers, extension rings, and so forth, that are marked with their volume or are made from boxes the dimensions of which are listed in Table 314.16(A).

**(1) Standard Boxes.** The volumes of standard boxes that are not marked with their volume shall be as given in Table 314.16(A).

**(2) Other Boxes.** Boxes 1650 cm$^3$ (100 in.$^3$) or less, other than those described in Table 314.16(A), and nonmetallic boxes shall be durably and legibly marked by the manufacturer with their volume. Boxes described in Table 314.16(A) that have a volume larger than is designated in the table shall be permitted to have their volume marked as required by this section.

**(B) Box Fill Calculations.** The volumes in paragraphs 314.16(B)(1) through (B)(5), as applicable, shall be added together. No allowance shall be required for small fittings such as locknuts and bushings.

**(1) Conductor Fill.** Each conductor that originates outside the box and terminates or is spliced within the box shall be counted once, and each conductor that passes through the box without splice or termination shall be counted once. Each loop or coil of unbroken conductor not less than twice the minimum length required for free conductors in 300.14 shall be counted twice. The conductor fill shall be calculated using Table 314.16(B). A conductor, no part of which leaves the box, shall not be counted.

***TABLE 314.16(A)*** *Metal Boxes*

| Box Trade Size | | | Minimum Volume | | Maximum Number of Conductors* (arranged by AWG size) | | | | | | |
|---|---|---|---|---|---|---|---|---|---|---|---|
| mm | in. | | cm$^3$ | in.$^3$ | 18 | 16 | 14 | 12 | 10 | 8 | 6 |
| 100 × 32 | (4 × 1¼) | round/octagonal | 205 | 12.5 | 8 | 7 | 6 | 5 | 5 | 5 | 2 |
| 100 × 38 | (4 × 1½) | round/octagonal | 254 | 15.5 | 10 | 8 | 7 | 6 | 6 | 5 | 3 |
| 100 × 54 | (4 × 2⅛) | round/octagonal | 353 | 21.5 | 14 | 12 | 10 | 9 | 8 | 7 | 4 |
| 100 × 32 | (4× 1¼) | square | 295 | 18.0 | 12 | 10 | 9 | 8 | 7 | 6 | 3 |
| 100 × 38 | (4 × 1½) | square | 344 | 21.0 | 14 | 12 | 10 | 9 | 8 | 7 | 4 |
| 100 × 54 | (4 × 2⅛) | square | 497 | 30.3 | 20 | 17 | 15 | 13 | 12 | 10 | 6 |
| 120 × 32 | (4¹¹⁄₁₆ × 1¼) | square | 418 | 25.5 | 17 | 14 | 12 | 11 | 10 | 8 | 5 |
| 120 × 38 | (4¹¹⁄₁₆ × 1½) | square | 484 | 29.5 | 19 | 16 | 14 | 13 | 11 | 9 | 5 |
| 120 × 54 | (4¹¹⁄₁₆ × 2⅛) | square | 689 | 42.0 | 28 | 24 | 21 | 18 | 16 | 14 | 8 |
| 75 × 50 × 38 | (3 × 2 × 1½) | device | 123 | 7.5 | 5 | 4 | 3 | 3 | 3 | 2 | 1 |
| 75 × 50 × 50 | (3 × 2 × 2) | device | 164 | 10.0 | 6 | 5 | 5 | 4 | 4 | 3 | 2 |
| 75× 50 × 57 | (3× 2 × 2¼) | device | 172 | 10.5 | 7 | 6 | 5 | 4 | 4 | 3 | 2 |
| 75 × 50 × 65 | (3 × 2 × 2½) | device | 205 | 12.5 | 8 | 7 | 6 | 5 | 5 | 4 | 2 |
| 75 × 50 × 70 | (3 × 2 × 2¾) | device | 230 | 14.0 | 9 | 8 | 7 | 6 | 5 | 4 | 2 |
| 75 × 50 × 90 | (3 × 2 × 3½) | device | 295 | 18.0 | 12 | 10 | 9 | 8 | 7 | 6 | 3 |
| 100 × 54 × 38 | (4 × 2⅛ × 1½) | device | 169 | 10.3 | 6 | 5 | 5 | 4 | 4 | 3 | 2 |
| 100 × 54 × 48 | (4 × 2⅛ × 1⅞) | device | 213 | 13.0 | 8 | 7 | 6 | 5 | 5 | 4 | 2 |
| 100 × 54 × 54 | (4 × 2⅛ × 2⅛) | device | 238 | 14.5 | 9 | 8 | 7 | 6 | 5 | 4 | 2 |
| 95 × 50 × 65 | (3¾ × 2 × 2½) | masonry box/gang | 230 | 14.0 | 9 | 8 | 7 | 6 | 5 | 4 | 2 |
| 95 × 50 × 90 | (3¾ × 2 × 3½) | masonry box/gang | 344 | 21.0 | 14 | 12 | 10 | 9 | 8 | 7 | 4 |
| min. 44.5 depth | FS — single cover/gang (1¾) | | 221 | 13.5 | 9 | 7 | 6 | 6 | 5 | 4 | 2 |
| min. 60.3 depth | FD — single cover/gang (2⅜) | | 295 | 18.0 | 12 | 10 | 9 | 8 | 7 | 6 | 3 |
| min. 44.5 depth | FS — multiple cover/gang (1¾) | | 295 | 18.0 | 12 | 10 | 9 | 8 | 7 | 6 | 3 |
| min. 60.3 depth | FD — multiple cover/gang (2⅜) | | 395 | 24.0 | 16 | 13 | 12 | 10 | 9 | 8 | 4 |

*Where no volume allowances are required by 314.16(B)(2) through (B)(5).

*Exception: An equipment grounding conductor or conductors or not over four fixture wires smaller than 14 AWG, or both, shall be permitted to be omitted from the calculations where they enter a box from a domed luminaire or similar canopy and terminate within that box.*

**(2) Clamp Fill.** Where one or more internal cable clamps, whether factory or field supplied, are present in the box, a single volume allowance in accordance with Table 314.16(B) shall be made based on the largest conductor present in the box. No allowance shall be required for a cable connector with its clamping mechanism outside the box.

A clamp assembly that incorporates a cable termination for the cable conductors shall be listed and marked for use with specific nonmetallic boxes. Conductors that originate within the clamp assembly shall be included in conductor fill calculations covered in 314.16(B)(1) as though they entered from outside the box. The clamp assembly shall not require a fill allowance, but the volume of the portion of the assembly that remains within the box after installation shall be excluded from the box volume as marked in 314.16(A)(2).

**(3) Support Fittings Fill.** Where one or more luminaire studs or hickeys are present in the box, a single volume allowance in accordance with Table 314.16(B) shall be made for each type of fitting based on the largest conductor present in the box.

**(4) Device or Equipment Fill.** For each yoke or strap containing one or more devices or equipment, a double volume allowance in accordance with Table 314.16(B) shall be made for each yoke or strap based on the largest conductor connected to a device(s) or equipment supported by that yoke or strap. A device or utilization equipment wider than a single 50 mm (2 in.) device box as described in Table 314.16(A) shall have double volume allowances provided for each gang required for mounting.

**(5) Equipment Grounding Conductor Fill.** Where one or more equipment grounding conductors or equipment bonding jumpers enter a box, a single volume allowance in accordance with Table 314.16(B) shall be made based on the largest equipment grounding conductor or equipment bonding jumper present in the box. Where an additional set of equipment grounding conductors, as permitted by 250.146(D), is present in the box, an additional volume allowance shall be made based on the largest equipment grounding conductor in the additional set.

This section requires that the total box "volume" be equal to or greater than the total box "fill." The total box volume is determined by adding the individual volumes of the box components. The components include the box itself plus any attachments to it, such as a plaster ring, an extension ring, or a dome cover. The volume of each component is determined either from the volume marking on the component itself or from the standard volumes listed in Table 314.16(A). If a box is marked with a volume larger than listed in Table 314.16(A), the larger volume can be used instead of the table value.

Adding all the volume allowances for all items contributing to box fill determines the total box fill. The volume allowance for each fill item is based on the volume listed in Table 314.16(B) for the conductor size indicated. Commentary Table 314.1 summarizes the components contributing to box fill.

**TABLE 314.16(B)** *Volume Allowance Required per Conductor*

| Size of Conductor (AWG) | Free Space Within Box for Each Conductor | |
|---|---|---|
| | $cm^3$ | $in.^3$ |
| 18 | 24.6 | 1.50 |
| 16 | 28.7 | 1.75 |
| 14 | 32.8 | 2.00 |
| 12 | 36.9 | 2.25 |
| 10 | 41.0 | 2.50 |
| 8 | 49.2 | 3.00 |
| 6 | 81.9 | 5.00 |

### Calculation Example 1

Select a standard-sized box for use where all the conductors are the same size and, as shown in Exhibit 314.1, the box does not contain any cable clamps, support fittings, devices, or equipment grounding conductors.

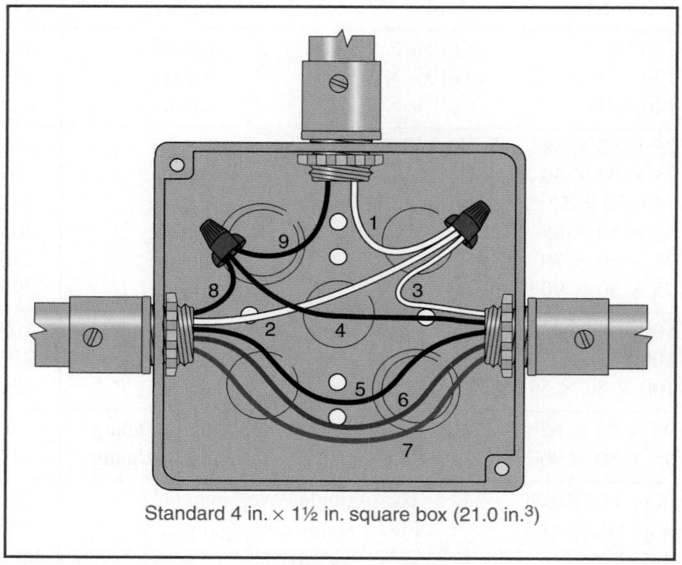


***EXHIBIT 314.1*** *A standard-sized square box containing no fittings or devices, such as luminaire studs, cable clamps, switches, receptacles, or EGCs.*

**COMMENTARY TABLE 314.1** Summary of Items Contributing to Box Fill

| Items Contained Within Box | Volume Allowance | Based on [see Table 314.16(B)] |
|---|---|---|
| Conductors that originate outside box | One for each conductor | Actual conductor size |
| Conductors that pass through box without splice or connection (less than 12 in. in total length) | One for each conductor | Actual conductor size |
| Conductors 12 in. or greater that are looped (or coiled) and unbroken (see 300.14 for exact measurement) | Two for a single (entire) unbroken conductor | Actual conductor size |
| Conductors that originate within box and do not leave box | None (these conductors not counted) | n.a. |
| Fixture wires [per 314.16(B)(1), Exception] | None (these conductors not counted) | n.a. |
| Internal cable clamps (one or more) | One only | Largest-sized conductor present |
| Support fittings (such as luminaire studs or hickeys) | One for each type of support fitting | Largest-sized conductor present |
| Devices (such as receptacles, switches) or utilization equipment (such as timers, dimmers, AFCI receptacles, GFCI receptacles, TVSS receptacles) | Two for each yoke or mounting strap | Largest-sized conductor connected to device or utilization equipment |
| Equipment grounding conductor (one or more) | One only | Largest equipment grounding conductor present |
| Isolated equipment grounding conductor (one or more) [see 250.146(D)] | One only | Largest isolated and insulated equipment grounding conductor present |

n.a.= not applicable.

*Solution*

To determine the minimum standard-sized square box for the number of 12 AWG conductors being installed, count the conductors and compare the total to the maximum number of conductors permitted by Table 314.16(A). Each unspliced conductor running through the box is counted as one conductor, and each other conductor is counted as one conductor. Therefore, the total conductor count is nine. Table 314.16(A) indicates that the maximum fill for standard 4 inch × 1½ inch square box is nine 12 AWG conductors, so this is the minimum square box size.

### Calculation Example 2

The standard method for determining adequate box size calculates the total box volume first and then subtracts the total box fill to ensure compliance. Using this method, determine whether the box in Exhibit 314.2 is adequately sized.

*Solution*

Table 314.16(A) shows the minimum volume of a standard 3 in. × 2 in. × 3½ in. device box to be 18 in.$^3$. The box fill for this situation as given in Commentary Table 314.2 is 16 in.$^3$. Because the total box fill of 16 in.$^3$ is less than the 18 in.$^3$ total box volume, the box is adequately sized.

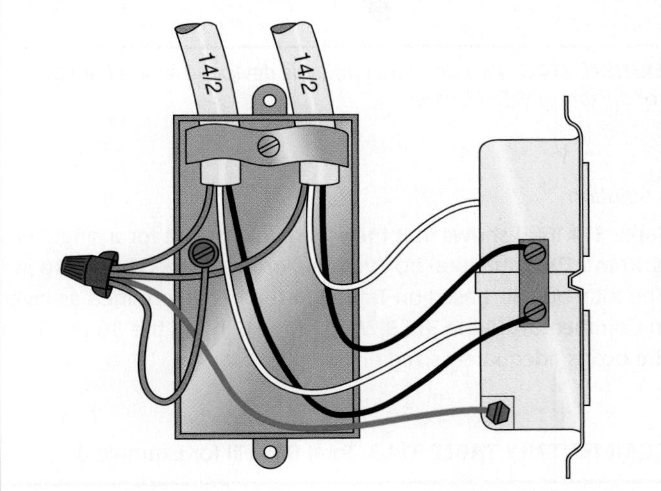


***EXHIBIT 314.2*** *A standard-sized device box containing a device and conductors requiring deductions.*

### Calculation Example 3

Using the standard method, determine the adequacy of the device box illustrated in Exhibit 314.3, where two standard-sized 3 in. × 2 in. × 3½ in. device boxes are ganged to form a single box.

**COMMENTARY TABLE 314.2** Total Box Fill for Example 2

| Items Contained Within Box | Volume Allowance | Unit Volume Based on Table 314.16(B) (in.$^3$) | Total Box Fill (in.$^3$) |
|---|---|---|---|
| 4 conductors | 4 volume allowances for 14 AWG conductors | 2.00 | 8.00 |
| 1 clamp | 1 volume allowance (based on 14 AWG conductors) | 2.00 | 2.00 |
| 1 device | 2 volume allowances (based on 14 AWG conductors) | 2.00 | 4.00 |
| Equipment grounding conductors (all) | 1 volume allowance (based on 14 AWG conductors) | 2.00 | 2.00 |
| Total | | | 16.00 |

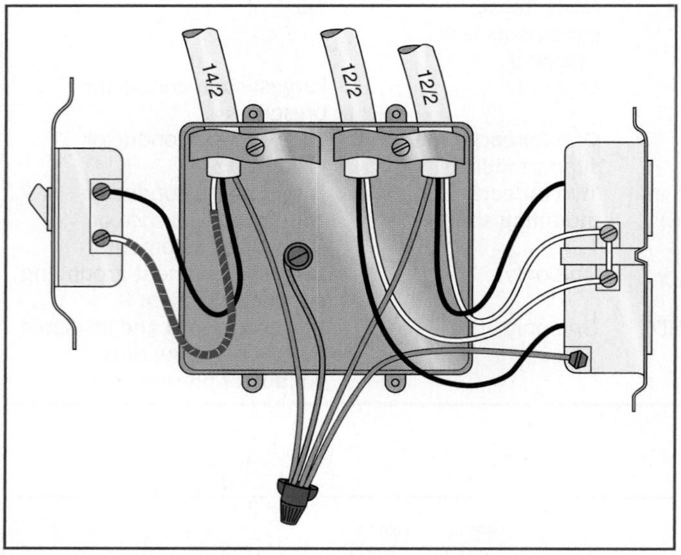


***EXHIBIT 314.3*** *Two standard gangable device boxes containing conductors of different sizes.*

*Solution*

Table 314.16(A) shows that the minimum volume for a single box is 18 in$^3$. Thus, the total box volume for the ganged box is 36 in.$^3$. The total box fill, based on Table 314.16(B), is determined as given in Commentary Table 314.3. With only 26 in.$^3$ of the 36 in.$^3$ filled, the box is adequately sized.

**(C) Conduit Bodies.**

**(1) General.** Conduit bodies enclosing 6 AWG conductors or smaller, other than short-radius conduit bodies as described in 314.16(C)(3), shall have a cross-sectional area not less than twice the cross-sectional area of the largest conduit or tubing to which they can be attached. The maximum number of conductors permitted shall be the maximum number permitted by Table 1 of Chapter 9 for the conduit or tubing to which it is attached.

**(2) With Splices, Taps, or Devices.** Only those conduit bodies that are durably and legibly marked by the manufacturer with their volume shall be permitted to contain splices, taps, or devices. The maximum number of conductors shall be calculated in accordance with 314.16(B). Conduit bodies shall be supported in a rigid and secure manner.

Conduit bodies other than the short-radius type are permitted to contain splices or taps, provided the conduit bodies are marked with their cubic inch capacity as illustrated in Exhibit 314.4. Such conduit bodies are required to have a cross-sectional area not less than twice that of the conduit to which they are attached and are not permitted to contain more conductors than the attached raceway. The volume requirements for splicing or tapping are provided in 314.16(C)(2).

Conduit bodies must be rigidly supported. See the Exception to 314.23(E), and Exception No. 1 to 314.23(F), which permits the raceway to support the conduit body, provided the conduit body is not larger than the attached raceway. See 314.28 for requirements that apply to conduit bodies used as pull and junction boxes.

**COMMENTARY TABLE 314.3** Total Box Fill for Example 3

| Items Contained Within Box | Volume Allowance | Unit Volume Based on Table 314.16(B) (in.$^3$) | Total Box Fill (in.$^3$) |
|---|---|---|---|
| 6 conductors | 2 volume allowances for 14 AWG conductors | 2.00 | 4.00 |
| | 4 volume allowances for 12 AWG conductors | 2.25 | 9.00 |
| 2 clamps | 1 volume allowance (based on 12 AWG conductors) | 2.25 | 2.25 |
| 2 devices | 2 volume allowances (based on 14 AWG conductors) | 2.00 | 4.00 |
| | 2 volume allowances (based on 12 AWG conductors) | 2.25 | 4.50 |
| Equipment grounding conductors (all) | 1 volume allowance (based on 12 AWG conductors) | 2.25 | 2.25 |
| Total | | | 26.00 |

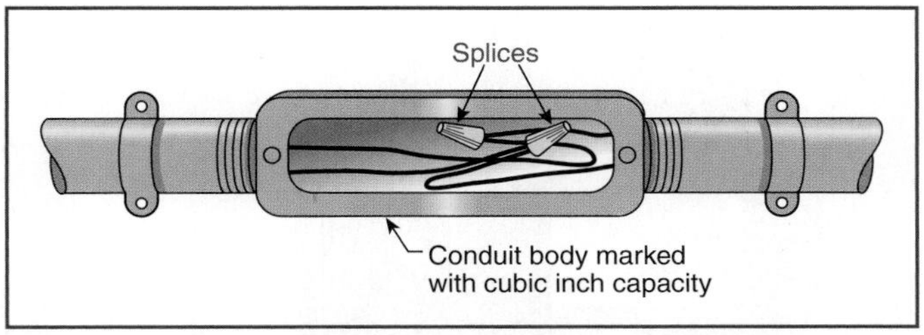


***EXHIBIT 314.4*** *An example of permitted splices in a raceway-supported conduit body.*

**(3) Short Radius Conduit Bodies.** Conduit bodies such as capped elbows and service-entrance elbows that enclose conductors 6 AWG or smaller, and are only intended to enable the installation of the raceway and the contained conductors, shall not contain splices, taps, or devices and shall be of an approved size to provide free space for all conductors enclosed in the conduit body.

## 314.17 Conductors Entering Boxes, Conduit Bodies, or Fittings

Conductors entering boxes, conduit bodies, or fittings shall be protected from abrasion and shall comply with 314.17(A) through (D).

**(A) Openings to Be Closed.** Openings through which conductors enter shall be closed in an approved manner.

**(B) Metal Boxes and Conduit Bodies.** Where metal boxes or conduit bodies are installed with messenger-supported wiring, open wiring on insulators, or concealed knob-and-tube wiring, conductors shall enter through insulating bushings or, in dry locations, through flexible tubing extending from the last insulating support to not less than 6 mm (¼ in.) inside the box and beyond any cable clamps. Except as provided in 300.15(C), the wiring shall be firmly secured to the box or conduit body. Where raceway or cable is installed with metal boxes or conduit bodies, the raceway or cable shall be secured to such boxes and conduit bodies.

**(C) Nonmetallic Boxes and Conduit Bodies.** Nonmetallic boxes and conduit bodies shall be suitable for the lowest temperature-rated conductor entering the box. Where nonmetallic boxes and conduit bodies are used with messenger-supported wiring, open wiring on insulators, or concealed knob-and-tube wiring, the conductors shall enter the box through individual holes. Where flexible tubing is used to enclose the conductors, the tubing shall extend from the last insulating support to not less than 6 mm (¼ in.) inside the box and beyond any cable clamp. Where nonmetallic-sheathed cable or multiconductor Type UF cable is used, the sheath shall extend not less than 6 mm (¼ in.) inside the box and beyond any cable clamp. In all instances, all permitted wiring methods shall be secured to the boxes.

Standard nonmetallic boxes are permitted for use with 90°C insulated conductors. A nonmetallic box used for splicing a conductor of a higher temperature rating to a conductor of a lower temperature rating is required to be identified as suitable for the temperature rating of the lower-rated conductor because the maximum temperature permitted will be that of the lower-rated conductor. The intent is to avoid the necessity of giving a high temperature rating to boxes in a normal temperature location simply because high-temperature conductors enter the box from or exit to a high-temperature location. However, where insulated conductors rated at higher temperatures are necessary in a high-temperature environment, the box is required to be identified by a marking on the box or in the listing of the box to comply with 110.3(B).

*Exception: Where nonmetallic-sheathed cable or multiconductor Type UF cable is used with single gang boxes not larger than a nominal size 57 mm × 100 mm (2¼ in. × 4 in.) mounted in walls or ceilings, and where the cable is fastened within 200 mm (8 in.) of the box measured along the sheath and where the sheath extends through a cable knockout not less than 6 mm (¼ in.), securing the cable to the box shall not be required. Multiple cable entries shall be permitted in a single cable knockout opening.*

Exhibit 314.5 is an example of an installation that complies with 314.17(C), Exception.

**(D) Conductors 4 AWG or Larger.** Installation shall comply with 300.4(G).

Informational Note: See 110.12(A) for requirements on closing unused cable and raceway knockout openings.

***EXHIBIT 314.5*** *Installation of a single gang outlet box provided the cables are securely fastened within 8 inches of the box. (Courtesy of the International Association of Electrical Inspectors)*

## 314.19 Boxes Enclosing Flush Devices

Boxes used to enclose flush devices shall be of such design that the devices will be completely enclosed on back and sides and substantial support for the devices will be provided. Screws for supporting the box shall not be used in attachment of the device contained therein.

## 314.20 In Wall or Ceiling

In walls or ceilings with a surface of concrete, tile, gypsum, plaster, or other noncombustible material, boxes employing a flush-type cover or faceplate shall be installed so that the front edge of the box, plaster ring, extension ring, or listed extender will not be set back of the finished surface more than 6 mm (¼ in.).

In walls and ceilings constructed of wood or other combustible surface material, boxes, plaster rings, extension rings, or listed extenders shall be flush with the finished surface or project therefrom.

Section 314.20 applies only to the construction of the surface of the wall or ceiling and not to its structure or subsurface. Therefore, a wall constructed of wood but sheathed with an outer layer of gypsum board is permitted to contain boxes set back or recessed not more than ¼ inch. Using an opposite example, a wall constructed of metal studs but finished with wood panels requires that contained outlet boxes be mounted flush with the combustible finish.

## 314.21 Repairing Noncombustible Surfaces

Noncombustible surfaces that are broken or incomplete around boxes employing a flush-type cover or faceplate shall be repaired so there will be no gaps or open spaces greater than 3 mm (⅛ in.) at the edge of the box.

## 314.22 Surface Extensions

Surface extensions shall be made by mounting and mechanically securing an extension ring over the box. Equipment grounding shall be in accordance with Part VI of Article 250.

*Exception: A surface extension shall be permitted to be made from the cover of a box where the cover is designed so it is unlikely to fall off or be removed if its securing means becomes loose. The wiring method shall be flexible for an approved length that permits removal of the cover and provides access to the box interior and shall be arranged so that any grounding continuity is independent of the connection between the box and cover.*

A flexible surface extension is permitted to be made from a flush-mounted outlet box as illustrated in Exhibit 314.6. This technique requires the use of a cover that will not fall off if the securing screws become loose. Grounding continuity must not rely on the means to secure the box to the cover.

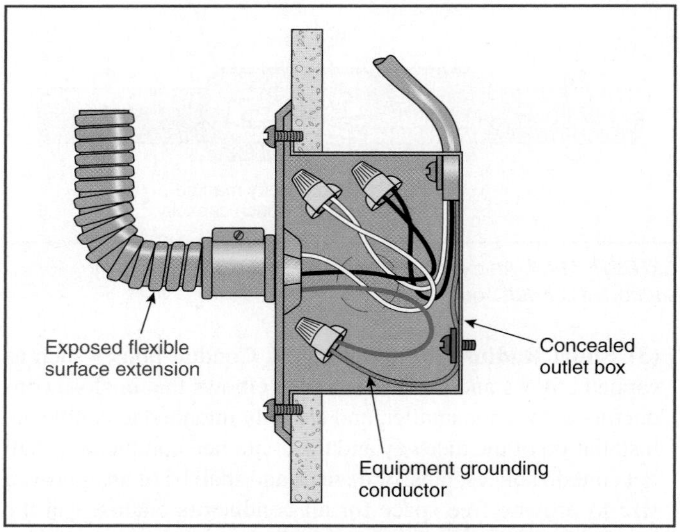


***EXHIBIT 314.6*** *An example of a flexible surface extension made from the cover of a flush-mounted outlet box.*

## 314.23 Supports

Enclosures within the scope of this article shall be supported in accordance with one or more of the provisions in 314.23(A) through (H).

**(A) Surface Mounting.** An enclosure mounted on a building or other surface shall be rigidly and securely fastened in place. If the surface does not provide rigid and secure support, additional support in accordance with other provisions of this section shall be provided.

**(B) Structural Mounting.** An enclosure supported from a structural member or from grade shall be rigidly supported either directly or by using a metal, polymeric, or wood brace.

**(1) Nails and Screws.** Nails and screws, where used as a fastening means, shall be attached by using brackets on the outside of the enclosure, or they shall pass through the interior within 6 mm (¼ in.) of the back or ends of the enclosure. Screws shall not be permitted to pass through the box unless exposed threads in the box are protected using approved means to avoid abrasion of conductor insulation.

The intent of these requirements is to prevent nails and screws from interfering with the installation of devices. Permitting nails inside the box within ¼ inch of the ends reduces splitting of the smaller wooden studs used in some frame-type construction. However, splitting sometimes occurs where nails are within ¼ inch of the back of the box.

**(2) Braces.** Metal braces shall be protected against corrosion and formed from metal that is not less than 0.51 mm (0.020 in.) thick uncoated. Wood braces shall have a cross section not less than nominal 25 mm × 50 mm (1 in. × 2 in.). Wood braces

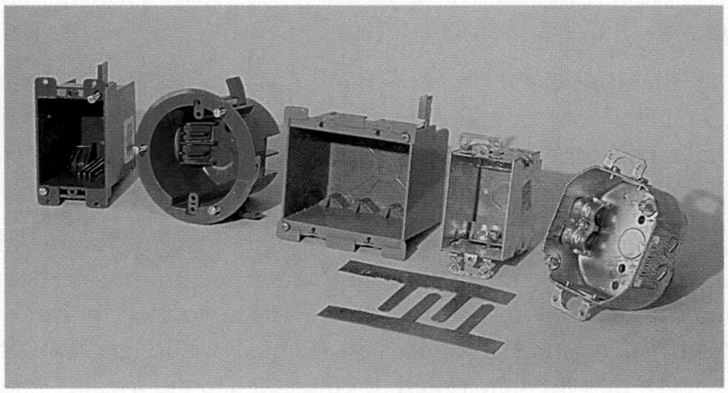

***EXHIBIT 314.7*** *Metal and nonmetallic outlet boxes designed for mounting in finished wall and ceiling surfaces.*

in wet locations shall be treated for the conditions. Polymeric braces shall be identified as being suitable for the use.

**(C) Mounting in Finished Surfaces.** An enclosure mounted in a finished surface shall be rigidly secured thereto by clamps, anchors, or fittings identified for the application.

Where structural members are lacking or where boxes are cut into existing walls, boxes are permitted to be secured by clamps or anchors. Exhibit 314.7 shows several examples of "old work" boxes that are designed to be installed into a finished wall or ceiling. The three nonmetallic boxes have mounting tabs that secure to the back of the finished surface as the screws on the front of the box are tightened. The metal boxes have mounting tabs that are designed to accept screws that secure the box to a surface such as wood paneling or wainscoting. The metal brackets shown can be used to secure a box to a wall surface such as gypsum wall board.

**(D) Suspended Ceilings.** An enclosure mounted to structural or supporting elements of a suspended ceiling shall be not more than 1650 cm$^3$ (100 in.$^3$) in size and shall be securely fastened in place in accordance with either (D)(1) or (D)(2).

**(1) Framing Members.** An enclosure shall be fastened to the framing members by mechanical means such as bolts, screws, or rivets, or by the use of clips or other securing means identified for use with the type of ceiling framing member(s) and enclosure(s) employed. The framing members shall be supported in an approved manner and securely fastened to each other and to the building structure.

**(2) Support Wires.** The installation shall comply with the provisions of 300.11(A). The enclosure shall be secured, using identified methods, to ceiling support wire(s), including any additional support wire(s) installed for ceiling support. Support wire(s) used for enclosure support shall be fastened at each end so as to be taut within the ceiling cavity.

**(E) Raceway-Supported Enclosure, Without Devices, Luminaires, or Lampholders.** An enclosure that does not contain a device(s), other than splicing devices, or supports a luminaire(s), a lampholder, or other equipment and is supported by entering raceways shall not exceed 1650 cm$^3$ (100 in.$^3$) in size. It shall have threaded entries or identified hubs. It shall be supported by two or more conduits threaded wrenchtight into the enclosure or hubs. Each conduit shall be secured within 900 mm (3 ft) of the enclosure, or within 450 mm (18 in.) of the enclosure if all conduit entries are on the same side.

Boxes are not permitted to be supported by rigid raceways using locknuts and bushings. Enclosures without devices or luminaires are considered to be adequately supported provided the conduit is connected to the enclosure by threaded hubs and the threaded conduits enter the box on two or more sides and are supported within 3 feet of the enclosure. A box is not permitted to be supported by a single raceway.

*Exception: The following wiring methods shall be permitted to support a conduit body of any size, including a conduit body constructed with only one conduit entry, provided that the trade size of the conduit body is not larger than the largest trade size of the conduit or tubing:*

*(1) Intermediate metal conduit, Type IMC*
*(2) Rigid metal conduit, Type RMC*
*(3) Rigid polyvinyl chloride conduit, Type PVC*
*(4) Reinforced thermosetting resin conduit, Type RTRC*
*(5) Electrical metallic tubing, Type EMT*

**(F) Raceway-Supported Enclosures, with Devices, Luminaires, or Lampholders.** An enclosure that contains a device(s), other than splicing devices, or supports a luminaire(s), a lampholder, or other equipment and is supported by entering raceways shall not exceed 1650 cm$^3$ (100 in.$^3$) in size. It shall have threaded entries or identified hubs. It shall be supported by two or more conduits threaded wrenchtight into the enclosure or hubs. Each conduit shall be secured within 450 mm (18 in.) of the enclosure.

*Exception No. 1: Rigid metal or intermediate metal conduit shall be permitted to support a conduit body of any size, including a conduit body constructed with only one conduit entry, provided the trade size of the conduit body is not larger than the largest trade size of the conduit.*

*Exception No. 2: An unbroken length(s) of rigid or intermediate metal conduit shall be permitted to support a box used for luminaire or lampholder support, or to support a wiring enclosure that is an integral part of a luminaire and used in lieu of a box in accordance with 300.15(B), where all of the following conditions are met:*

*(a) The conduit is securely fastened at a point so that the length of conduit beyond the last point of conduit support does not exceed 900 mm (3 ft).*

*(b) The unbroken conduit length before the last point of conduit support is 300 mm (12 in.) or greater, and that portion of the conduit is securely fastened at some point not less than 300 mm (12 in.) from its last point of support.*

*(c) Where accessible to unqualified persons, the luminaire or lampholder, measured to its lowest point, is at least 2.5 m (8 ft) above grade or standing area and at least 900 mm (3 ft) measured horizontally to the 2.5 m (8 ft) elevation from windows, doors, porches, fire escapes, or similar locations.*

*(d) A luminaire supported by a single conduit does not exceed 300 mm (12 in.) in any direction from the point of conduit entry.*

*(e) The weight supported by any single conduit does not exceed 9 kg (20 lb).*

*(f) At the luminaire or lampholder end, the conduit(s) is threaded wrenchtight into the box, conduit body, integral wiring enclosure, or identified hubs. Where a box or conduit body is used for support, the luminaire shall be secured directly to the box or conduit body, or through a threaded conduit nipple not over 75 mm (3 in.) long.*

**(G) Enclosures in Concrete or Masonry.** An enclosure supported by embedment shall be identified as suitably protected from corrosion and securely embedded in concrete or masonry.

Boxes are permitted to be embedded in masonry or concrete, provided they are rigid and secure. Corrosion protection is required due to the corrosive effects of concrete. Exhibit 314.8 shows a mud box installed in a brick wall. Additional support is not required.

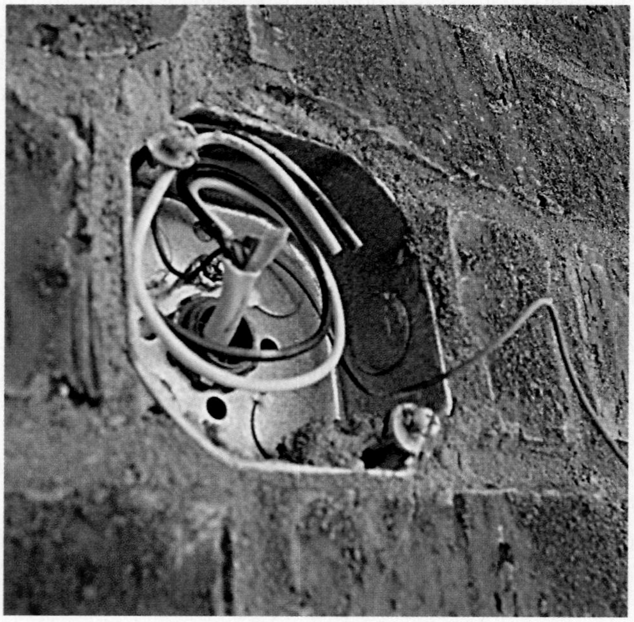

***EXHIBIT 314.8** A mud box installed in a brick wall.*

**(H) Pendant Boxes.** An enclosure supported by a pendant shall comply with 314.23(H)(1) or (H)(2).

**(1) Flexible Cord.** A box shall be supported from a multiconductor cord or cable in an approved manner that protects the conductors against strain, such as a strain-relief connector threaded into a box with a hub.

**(2) Conduit.** A box supporting lampholders or luminaires, or wiring enclosures within luminaires used in lieu of boxes in accordance with 300.15(B), shall be supported by rigid or intermediate metal conduit stems. For stems longer than 450 mm (18 in.), the stems shall be connected to the wiring system with flexible fittings suitable for the location. At the luminaire end, the conduit(s) shall be threaded wrenchtight into the box, wiring enclosure, or identified hubs.

Where supported by only a single conduit, the threaded joints shall be prevented from loosening by the use of setscrews or other effective means, or the luminaire, at any point, shall be at least 2.5 m (8 ft) above grade or standing area and at least 900 mm (3 ft) measured horizontally to the 2.5 m (8 ft) elevation from windows, doors, porches, fire escapes, or similar locations. A luminaire supported by a single conduit shall not exceed 300 mm (12 in.) in any horizontal direction from the point of conduit entry.

## 314.24 Depth of Boxes

Outlet and device boxes shall have an approved depth to allow equipment installed within them to be mounted properly and without likelihood of damage to conductors within the box.

**(A) Outlet Boxes Without Enclosed Devices or Utilization Equipment.** Outlet boxes that do not enclose devices or utilization equipment shall have a minimum internal depth of 12.7 mm (½ in.).

**(B) Outlet and Device Boxes with Enclosed Devices or Utilization Equipment.** Outlet and device boxes that enclose devices or utilization equipment shall have a minimum internal depth that accommodates the rearward projection of the equipment and the size of the conductors that supply the equipment. The internal depth shall include, where used, that of any extension boxes, plaster rings, or raised covers. The internal depth shall comply with all applicable provisions of (B)(1) through (B)(5).

The use of a shallow box might become necessary because of old work or existing construction where, for example, very shallow partitions, structural members very close to finish surfaces, plumbing pipes, or ductwork is encountered within the partition. In addition to meeting minimum depth requirements, the box selection must be based on its having sufficient cubic-inch capacity.

**(1) Large Equipment.** Boxes that enclose devices or utilization equipment that projects more than 48 mm (1⅞ in.) rearward from the mounting plane of the box shall have a depth that is not less than the depth of the equipment plus 6 mm (¼ in.).

**(2) Conductors Larger Than 4 AWG.** Boxes that enclose devices or utilization equipment supplied by conductors larger than 4 AWG shall be identified for their specific function.

*Exception to (2): Devices or utilization equipment supplied by conductors larger than 4 AWG shall be permitted to be mounted on or in junction and pull boxes larger than 1650 cm³ (100 in.³) if the spacing at the terminals meets the requirements of 312.6.*

**(3) Conductors 8, 6, or 4 AWG.** Boxes that enclose devices or utilization equipment supplied by 8, 6, or 4 AWG conductors shall have an internal depth that is not less than 52.4 mm (2¹⁄₁₆ in.).

**(4) Conductors 12 or 10 AWG.** Boxes that enclose devices or utilization equipment supplied by 12 or 10 AWG conductors shall have an internal depth that is not less than 30.2 mm (1³⁄₁₆ in.). Where the equipment projects rearward from the mounting plane of the box by more than 25 mm (1 in.), the box shall have a depth not less than that of the equipment plus 6 mm (¼ in.).

**(5) Conductors 14 AWG and Smaller.** Boxes that enclose devices or utilization equipment supplied by 14 AWG or smaller conductors shall have a depth that is not less than 23.8 mm (¹⁵⁄₁₆ in.).

The intent of this requirement is to prevent damage to conductors that can occur where clearance from the device or utilization equipment to the back of the box is insufficient. Where clearance is insufficient, conductors are often pinched or the insulation is damaged as the device or utilization equipment is pushed into the box.

Examples of utilization equipment often located within an outlet or device box include speakers, timers, motion detectors, alarms, and video and audio surveillance equipment. Typically, minimum box dimensions are included in product installation instructions for listed utilization equipment.

*Exception to (1) through (5): Devices or utilization equipment that is listed to be installed with specified boxes shall be permitted.*

## 314.25 Covers and Canopies

In completed installations, each box shall have a cover, faceplate, lampholder, or luminaire canopy, except where the installation complies with 410.24(B). Screws used for the purpose of attaching covers, or other equipment, to the box shall be either machine screws matching the thread gauge or size that is integral to the box or shall be in accordance with the manufacturer's instructions.

**(A) Nonmetallic or Metal Covers and Plates.** Nonmetallic or metal covers and plates shall be permitted. Where metal covers or plates are used, they shall comply with the grounding requirements of 250.110.

Informational Note: For additional grounding requirements, see 410.42 for metal luminaire canopies, and 404.12 and 406.6(B) for metal faceplates.

**(B) Exposed Combustible Wall or Ceiling Finish.** Where a luminaire canopy or pan is used, any combustible wall or ceiling finish exposed between the edge of the canopy or pan and the outlet box shall be covered with noncombustible material if required by 410.23.

Because heat from a short circuit, from a ground fault, or due to overlamping could create a fire hazard within a luminaire canopy or pan, any exposed combustible wall or ceiling space between the edge of the outlet box and the perimeter of the luminaire is required by 314.25(B) to be covered with noncombustible material. The noncombustible material need not be metal. Glass fiber pads, commonly provided as thermal barriers within the ceiling pan of luminaires, can be used to meet this requirement. Where the wall or ceiling finish is concrete, tile, gypsum, plaster, or other noncombustible material, the requirements of this section do not apply. Section 314.20 contains requirements for flush-mounted boxes and boxes "set back" from finish surfaces.

**(C) Flexible Cord Pendants.** Covers of outlet boxes and conduit bodies having holes through which flexible cord pendants pass shall be provided with identified bushings or shall have smooth, well-rounded surfaces on which the cords may bear. So-called hard rubber or composition bushings shall not be used.

## 314.27 Outlet Boxes

**(A) Boxes at Luminaire or Lampholder Outlets.** Outlet boxes or fittings designed for the support of luminaires and lampholders, and installed as required by 314.23, shall be permitted to support a luminaire or lampholder.

An outlet box "designed for the support of luminaires" will adequately support a luminaire weighing up to 50 pounds. If the box is designed to support other than 50 pounds, the maximum design weight must be marked on the inside of the box.

**(1) Vertical Surface Outlets.** Boxes used at luminaire or lampholder outlets in or on a vertical surface shall be identified and marked on the interior of the box to indicate the maximum weight of the luminaire that is permitted to be supported by the box if other than 23 kg (50 lb).

*Exception: A vertically mounted luminaire or lampholder weighing not more than 3 kg (6 lb) shall be permitted to be supported on other boxes or plaster rings that are secured to other boxes, provided that the luminaire or its supporting yoke, or the lampholder, is secured to the box with no fewer than two No. 6 or larger screws.*

Device boxes designed for the mounting of snap switches, receptacles, and other devices are usually provided with 6-32 screws. They are not usually suitable for supporting other than lightweight wall-mounted luminaires.

**(2) Ceiling Outlets.** At every outlet used exclusively for lighting, the box shall be designed or installed so that a luminaire or lampholder may be attached. Boxes shall be required to support a luminaire weighing a minimum of 23 kg (50 lb). A luminaire that weighs more than 23 kg (50 lb) shall be supported independently of the outlet box, unless the outlet box is listed and marked on the interior of the box to indicate the maximum weight the box shall be permitted to support.

Whether a luminaire is attached to an outlet box or is supported independently of the outlet box, the supporting means of the luminaire should be securely fastened. Boxes designed to support luminaires weighing more than 50 pounds must be listed and marked with the maximum weight.

**(B) Floor Boxes.** Boxes listed specifically for this application shall be used for receptacles located in the floor.

*Exception: Where the authority having jurisdiction judges them free from likely exposure to physical damage, moisture, and dirt, boxes located in elevated floors of show windows and similar locations shall be permitted to be other than those listed for floor applications. Receptacles and covers shall be listed as an assembly for this type of location.*

**(C) Boxes at Ceiling-Suspended (Paddle) Fan Outlets.** Outlet boxes or outlet box systems used as the sole support of a ceiling-suspended (paddle) fan shall be listed, shall be marked by their manufacturer as suitable for this purpose, and shall not support ceiling-suspended (paddle) fans that weigh more than 32 kg (70 lb). For outlet boxes or outlet box systems designed to support ceiling-suspended (paddle) fans that weigh more than 16 kg (35 lb), the required marking shall include the maximum weight to be supported.

Where spare, separately switched, ungrounded conductors are provided to a ceiling-mounted outlet box, in a location acceptable for a ceiling-suspended (paddle) fan in single-family, two-family, or multi-family dwellings, the outlet box or outlet box system shall be listed for sole support of a ceiling-suspended (paddle) fan.

Outlet boxes specifically listed to support ceiling-mounted paddle fans are available, as are several alternative and retrofit methods that can provide suitable support for a paddle fan. Exhibit 314.9 illustrates two methods of supporting a fan from an outlet box listed for fan support. For a detailed view of an outlet box for a fan where the box does not serve as the sole support, see Exhibit 422.2. In new residential construction, it is common to provide a wall-mounted switch with wiring for paddle fans to ceiling-mounted outlet boxes, even where there is not yet a paddle fan, to allow a fan to be installed in the future. Such installations are required to have an outlet box or outlet box system that is listed for the sole support of a fan.

**(D) Utilization Equipment.** Boxes used for the support of utilization equipment other than ceiling-suspended (paddle) fans shall meet the requirements of 314.27(A) for the support of a luminaire that is the same size and weight.

***EXHIBIT 314.9*** *Two methods of supporting a ceiling fan from a listed outlet box. (Top) A listed outlet box used as the sole support of a ceiling-suspended (paddle) fan. Note the two clamps on the perimeter of the box that will support a fixture. (Courtesy of the International Association of Electrical Inspectors). (Bottom) Supporting a ceiling-suspended (paddle) fan (35 lb or less) with a box identified for such use. (Courtesy of Hubbell/RACO)*

*Exception: Utilization equipment weighing not more than 3 kg (6 lb) shall be permitted to be supported on other boxes or plaster rings that are secured to other boxes, provided the equipment or its supporting yoke is secured to the box with no fewer than two No. 6 or larger screws.*

## 314.28 Pull and Junction Boxes and Conduit Bodies

Boxes and conduit bodies used as pull or junction boxes shall comply with 314.28(A) through (E).

*Exception: Terminal housings supplied with motors shall comply with the provisions of 430.12.*

**(A) Minimum Size.** For raceways containing conductors of 4 AWG or larger that are required to be insulated, and for cables containing conductors of 4 AWG or larger, the minimum dimensions of pull or junction boxes installed in a raceway or cable run shall comply with (A)(1) through (A)(3). Where an enclosure dimension is to be calculated based on the diameter of entering raceways, the diameter shall be the metric designator (trade size) expressed in the units of measurement employed.

**(1) Straight Pulls.** In straight pulls, the length of the box or conduit body shall not be less than eight times the metric designator (trade size) of the largest raceway.

Straight pulls — such as trade size 2 conduit containing four 4/0 AWG, Type THHN conductors (see Informative Annex C, Table C.8) — require a 16-inch-long pull box (8 × 2 in. = 16 in.). However, although 16 inches is the required minimum length, a longer pull box may be desired for maximum ease in handling this size conductor.

The six times rule of 314.28(A)(2) applies to straight-through conduit entries if the conductors are spliced as part of the straight-through wiring. Adjusting the previous example of trade size 2 conduit containing four 4/0 AWG, Type THHN conductors, provided the conductors were spliced within the enclosure, the required pull box dimension could be reduced to a 12-inch-long pull box (6 × 2 in. = 12 in.).

Section 314.28(A) addresses raceways or cables that contain conductors 4 AWG or larger and are required to be insulated. One conductor that is not required to be insulated is a grounding electrode conductor. So, where a grounding electrode conductor is installed as the sole conductor within a raceway, conduit bodies used as part of that raceway system are not required to comply with enclosure dimensions in 314.28(A).

**(2) Angle or U Pulls, or Splices.** Where splices or where angle or U pulls are made, the distance between each raceway entry inside the box or conduit body and the opposite wall of the box or conduit body shall not be less than six times the metric designator (trade size) of the largest raceway in a row. This distance shall be increased for additional entries by the amount of the sum of the diameters of all other raceway entries in the same row on the same wall of the box. Each row shall be calculated individually, and the single row that provides the maximum distance shall be used.

*Exception: Where a raceway or cable entry is in the wall of a box or conduit body opposite a removable cover, the distance from that wall to the cover shall be permitted to comply with the distance required for one wire per terminal in Table 312.6(A).*

The distance between raceway entries enclosing the same conductor shall not be less than six times the metric designator (trade size) of the larger raceway.

When transposing cable size into raceway size in 314.28(A)(1) and (A)(2), the minimum metric designator (trade size) raceway required for the number and size of conductors in the cable shall be used.

Where splices, angle pulls, or U pulls are made, the distance between each raceway entry inside the box and the opposite wall of the box must not be less than six times the trade diameter of the largest raceway, plus the distance for additional raceway entries (see Exhibit 314.10). This additional distance is calculated by adding the diameters of the other raceway entries in one row on the same side of the box. Raceway entries enclosing the same conductor are required to have a minimum separation between them (see Exhibit 314.11). The intent is to provide adequate space for the conductor to make the bend.

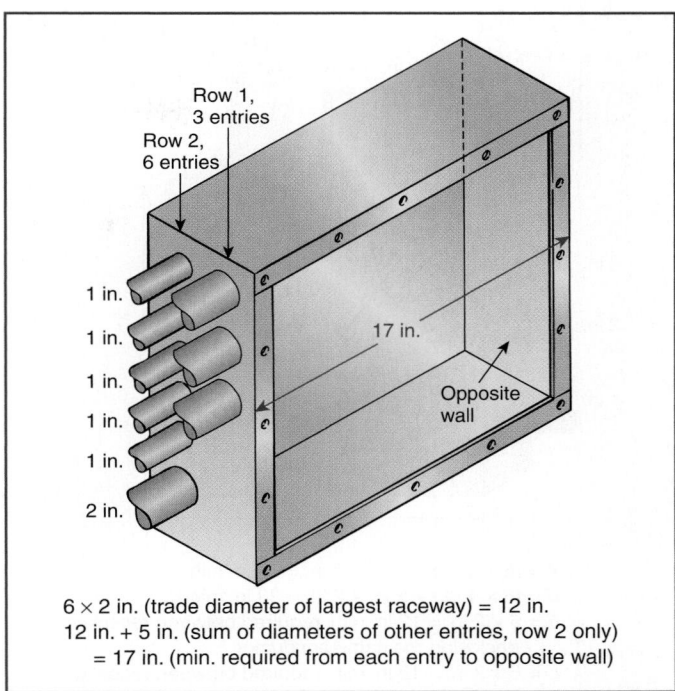

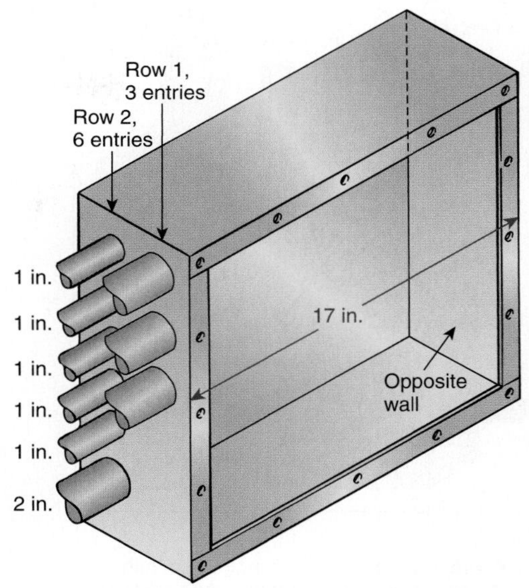


***EXHIBIT 314.10*** *An example showing calculations for splices, angle pulls, or U pulls.*

**(3) Smaller Dimensions.** Listed boxes or listed conduit bodies of dimensions less than those required in 314.28(A)(1) and (A)(2) shall be permitted for installations of combinations of conductors that are less than the maximum conduit or tubing fill (of conduits or tubing being used) permitted by Table 1 of Chapter 9.

Listed conduit bodies of dimensions less than those required in 314.28(A)(2), and having a radius of the curve to the centerline not less than that indicated in Table 2 of Chapter 9 for one-shot and full-shoe benders, shall be permitted for installations of combinations of conductors permitted by Table 1 of Chapter 9. These conduit bodies shall be marked to show they have been specifically evaluated in accordance with this provision.

Where the permitted combinations of conductors for which the box or conduit body has been listed are less than the maximum conduit or tubing fill permitted by Table 1 of Chapter 9,

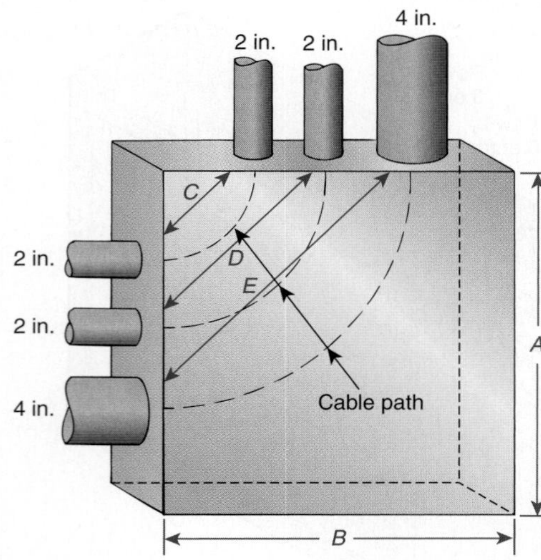


A = (6 × 4 in.) + 2 in. + 2 in. = 28 in. min.
B = (6 × 4 in.) + 2 in. + 2 in. = 28 in. min.
C = 6 × 2 in. = 12 in. min. required between raceways enclosing the same conductor
D = 6 × 2 in. = 12 in. min. required between raceways enclosing the same conductor
E = 6 × 4 in. = 24 in. min. required between raceways enclosing the same conductor

***EXHIBIT 314.11*** *An example showing calculations for raceways enclosing the same conductor.*

the box or conduit body shall be permanently marked with the maximum number and maximum size of conductors permitted.

**(B) Conductors in Pull or Junction Boxes.** In pull boxes or junction boxes having any dimension over 1.8 m (6 ft), all conductors shall be cabled or racked up in an approved manner.

**(C) Covers.** All pull boxes, junction boxes, and conduit bodies shall be provided with covers compatible with the box or conduit body construction and suitable for the conditions of use. Where used, metal covers shall comply with the grounding requirements of 250.110.

**(D) Permanent Barriers.** Where permanent barriers are installed in a box, each section shall be considered as a separate box.

**(E) Power Distribution Blocks.** Power distribution blocks shall be permitted in pull and junction boxes over 1650 $cm^3$ (100 $in.^3$) for connections of conductors where installed in boxes and where the installation complies with (1) through (5).

*Exception: Equipment grounding terminal bars shall be permitted in smaller enclosures.*

**(1) Installation.** Power distribution blocks installed in boxes shall be listed.

**(2) Size.** In addition to the overall size requirement in the first sentence of 314.28(A)(2), the power distribution block shall be installed in a box with dimensions not smaller than specified in the installation instructions of the power distribution block.

**(3) Wire Bending Space.** Wire bending space at the terminals of power distribution blocks shall comply with 312.6.

**(4) Live Parts.** Power distribution blocks shall not have uninsulated live parts exposed within a box, whether or not the box cover is installed.

**(5) Through Conductors.** Where the pull or junction boxes are used for conductors that do not terminate on the power distribution block(s), the through conductors shall be arranged so the power distribution block terminals are unobstructed following installation.

Similar provisions exist in Article 376 for power distribution blocks in wireways. The use in junction or pull boxes is limited to enclosures over 100 $in.^3$ However, equipment grounding terminal bars are permitted in smaller enclosures.

## 314.29 Boxes, Conduit Bodies, and Handhole Enclosures to Be Accessible

Boxes, conduit bodies, and handhole enclosures shall be installed so that the wiring contained in them can be rendered accessible without removing any part of the building or structure or, in underground circuits, without excavating sidewalks, paving, earth, or other substance that is to be used to establish the finished grade.

*Exception: Listed boxes and handhole enclosures shall be permitted where covered by gravel, light aggregate, or noncohesive granulated soil if their location is effectively identified and accessible for excavation.*

A junction box installed on a structural ceiling above a suspended ceiling is permitted to be used at any point for the connection of conduit, tubing, or cable, provided it remains accessible. See Article 100 for the definition of *accessible (as applied to wiring methods)*. See 300.15 for other requirements for boxes, conduit bodies, fittings, or handholes.

## 314.30 Handhole Enclosures

Handhole enclosures shall be designed and installed to withstand all loads likely to be imposed on them. They shall be identified for use in underground systems.

Informational Note: See ANSI/SCTE 77-2002, *Specification for Underground Enclosure Integrity*, for additional information on deliberate and nondeliberate traffic loading that can be expected to bear on underground enclosures.

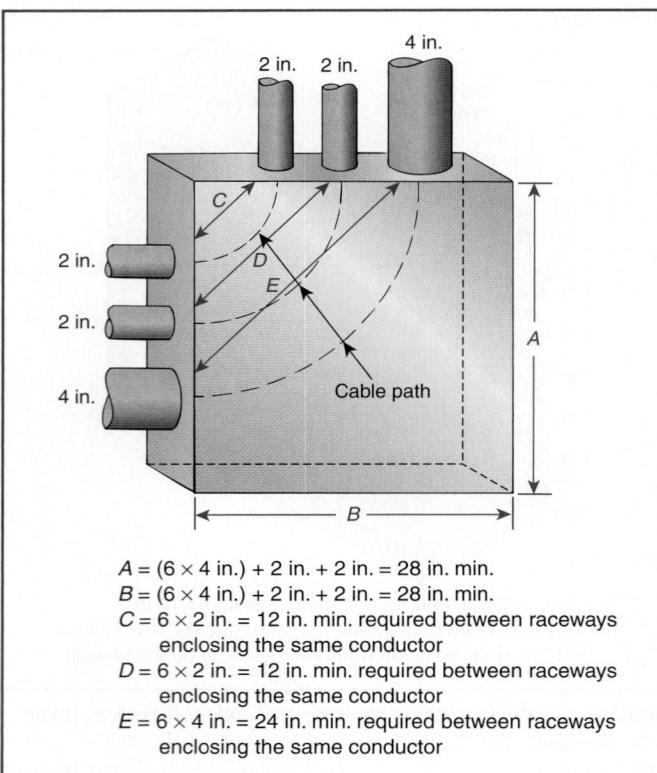

The load referred to in this requirement is the weight or force of traffic loads on handhole enclosures, not electrical loads. The definition of *handhole enclosure*, found in Article 100, states that it is "an enclosure for use in underground systems, provided with an open or closed bottom, and sized to allow personnel to reach into, but not enter, for the purpose of installing, operating, or maintaining equipment or wiring or both." The term *identified (as applied to equipment)* is defined in Article 100 as recognizably suitable for the intended purpose.

**(A) Size.** Handhole enclosures shall be sized in accordance with 314.28(A) for conductors operating at 1000 volts or below, and in accordance with 314.71 for conductors operating at over 1000 volts. For handhole enclosures without bottoms where the provisions of 314.28(A)(2), Exception, or 314.71(B)(1), Exception No. 1, apply, the measurement to the removable cover shall be taken from the end of the conduit or cable assembly.

**(B) Wiring Entries.** Underground raceways and cable assemblies entering a handhole enclosure shall extend into the enclosure, but they shall not be required to be mechanically connected to the enclosure.

**(C) Enclosed Wiring.** All enclosed conductors and any splices or terminations, if present, shall be listed as suitable for wet locations.

**(D) Covers.** Handhole enclosure covers shall have an identifying mark or logo that prominently identifies the function of the enclosure, such as "electric." Handhole enclosure covers shall require the use of tools to open, or they shall weigh over 45 kg (100 lb). Metal covers and other exposed conductive surfaces shall be bonded in accordance with 250.92 if the conductors in the handhole are service conductors, or in accordance with 250.96(A) if the conductors in the handhole are feeder or branch-circuit conductors.

Other exposed conductive surfaces often include metal frames used to secure metal covers in place. No exceptions to the requirement are given for bonding the metal covers and other exposed conductive surfaces.

## III. Construction Specifications

### 314.40 Metal Boxes, Conduit Bodies, and Fittings

**(A) Corrosion Resistant.** Metal boxes, conduit bodies, and fittings shall be corrosion resistant or shall be well-galvanized, enameled, or otherwise properly coated inside and out to prevent corrosion.

Informational Note: See 300.6 for limitation in the use of boxes and fittings protected from corrosion solely by enamel.

**(B) Thickness of Metal.** Sheet steel boxes not over 1650 $cm^3$ (100 $in.^3$) in size shall be made from steel not less than 1.59 mm (0.0625 in.) thick. The wall of a malleable iron box or conduit body and a die-cast or permanent-mold cast aluminum, brass, bronze, or zinc box or conduit body shall not be less than 2.38 mm (3⁄32 in.) thick. Other cast metal boxes or conduit bodies shall have a wall thickness not less than 3.17 mm (1⁄8 in.).

*Exception No. 1: Listed boxes and conduit bodies shown to have equivalent strength and characteristics shall be permitted to be made of thinner or other metals.*

*Exception No. 2: The walls of listed short radius conduit bodies, as covered in 314.16(C)(2), shall be permitted to be made of thinner metal.*

**(C) Metal Boxes Over 1650 $cm^3$ (100 $in.^3$).** Metal boxes over 1650 $cm^3$ (100 $in.^3$) in size shall be constructed so as to be of ample strength and rigidity. If of sheet steel, the metal thickness shall not be less than 1.35 mm (0.053 in.) uncoated.

**(D) Grounding Provisions.** A means shall be provided in each metal box for the connection of an equipment grounding conductor. The means shall be permitted to be a tapped hole or equivalent.

For device boxes and other standard outlet boxes, the means for connecting the equipment grounding conductor is usually provided by the box manufacturer in the form of a 10-32 tapped hole marked "GR" or "GRD," or the equivalent, next to the hole. However, the means provided may not necessarily be used.

### 314.41 Covers

Metal covers shall be of the same material as the box or conduit body with which they are used, or they shall be lined with firmly attached insulating material that is not less than 0.79 mm (1⁄32 in.) thick, or they shall be listed for the purpose. Metal covers shall be the same thickness as the boxes or conduit body for which they are used, or they shall be listed for the purpose. Covers of porcelain or other approved insulating materials shall be permitted if of such form and thickness as to afford the required protection and strength.

### 314.42 Bushings

Covers of outlet boxes and conduit bodies having holes through which flexible cord pendants may pass shall be provided with approved bushings or shall have smooth, well-rounded surfaces on which the cord may bear. Where individual conductors pass through a metal cover, a separate hole equipped with a bushing of suitable insulating material shall be provided for each conductor. Such separate holes shall be connected by a slot as required by 300.20.

### 314.43 Nonmetallic Boxes

Provisions for supports or other mounting means for nonmetallic boxes shall be outside of the box, or the box shall be constructed so as to prevent contact between the conductors in the box and the supporting screws.

### 314.44 Marking

All boxes and conduit bodies, covers, extension rings, plaster rings, and the like shall be durably and legibly marked with the manufacturer's name or trademark.

## IV. Pull and Junction Boxes, Conduit Bodies, and Handhole Enclosures for Use on Systems over 1000 Volts, Nominal

### 314.70 General

**(A) Pull and Junction Boxes.** Where pull and junction boxes are used on systems over 1000 volts, the installation shall comply with the provisions of Part IV and with the following general provisions of this article:

(1) Part I, 314.2; 314.3; and 314.4
(2) Part II, 314.15; 314.17; 314.20; 314.23(A), (B), or (G); 314.28(B); and 314.29
(3) Part III, 314.40(A) and (C); and 314.41

**(B) Conduit Bodies.** Where conduit bodies are used on systems over 1000 volts, the installation shall comply with the provisions of Part IV and with the following general provisions of this article:

(1) Part I, 314.4
(2) Part II, 314.15; 314.17; 314.23(A), (E), or (G); and 314.29
(3) Part III, 314.40(A); and 314.41

**(C) Handhole Enclosures.** Where handhole enclosures are used on systems over 1000 volts, the installation shall comply with the provisions of Part IV and with the following general provisions of this article:

(1) Part I, 314.3; and 314.4
(2) Part II, 314.15; 314.17; 314.23(G); 314.28(B); 314.29; and 314.30

### 314.71 Size of Pull and Junction Boxes, Conduit Bodies, and Handhole Enclosures

Pull and junction boxes and handhole enclosures shall provide approved space and dimensions for the installation of conductors, and they shall comply with the specific requirements of this section. Conduit bodies shall be permitted if they meet the dimensional requirements for boxes.

*Exception: Terminal housings supplied with motors shall comply with the provisions of 430.12.*

**(A) For Straight Pulls.** The length of the box shall not be less than 48 times the outside diameter, over sheath, of the largest shielded or lead-covered conductor or cable entering the box. The length shall not be less than 32 times the outside diameter of the largest nonshielded conductor or cable.

**(B) For Angle or U Pulls.**

**(1) Distance to Opposite Wall.** The distance between each cable or conductor entry inside the box and the opposite wall of the box shall not be less than 36 times the outside diameter, over sheath, of the largest cable or conductor. This distance shall be increased for additional entries by the amount of the sum of the outside diameters, over sheath, of all other cables or conductor entries through the same wall of the box.

*Exception No. 1: Where a conductor or cable entry is in the wall of a box opposite a removable cover, the distance from that wall to the cover shall be permitted to be not less than the bending radius for the conductors as provided in 300.34.*

*Exception No. 2: Where cables are nonshielded and not lead covered, the distance of 36 times the outside diameter shall be permitted to be reduced to 24 times the outside diameter.*

**(2) Distance Between Entry and Exit.** The distance between a cable or conductor entry and its exit from the box shall not be less than 36 times the outside diameter, over sheath, of that cable or conductor.

*Exception: Where cables are nonshielded and not lead covered, the distance of 36 times the outside diameter shall be permitted to be reduced to 24 times the outside diameter.*

**(C) Removable Sides.** One or more sides of any pull box shall be removable.

### 314.72 Construction and Installation Requirements

**(A) Corrosion Protection.** Boxes shall be made of material inherently resistant to corrosion or shall be suitably protected, both internally and externally, by enameling, galvanizing, plating, or other means.

**(B) Passing Through Partitions.** Suitable bushings, shields, or fittings having smooth, rounded edges shall be provided where conductors or cables pass through partitions and at other locations where necessary.

**(C) Complete Enclosure.** Boxes shall provide a complete enclosure for the contained conductors or cables.

**(D) Wiring Is Accessible.** Boxes and conduit bodies shall be installed so that the conductors are accessible without removing any fixed part of the building or structure. Working space shall be provided in accordance with 110.34.

**(E) Suitable Covers.** Boxes shall be closed by suitable covers securely fastened in place. Underground box covers that weigh over 45 kg (100 lb) shall be considered meeting this requirement. Covers for boxes shall be permanently marked "DANGER — HIGH VOLTAGE — KEEP OUT." The marking shall be on the outside of the box cover and shall be readily visible. Letters shall be block type and at least 13 mm (½ in.) in height.

**(F) Suitable for Expected Handling.** Boxes and their covers shall be capable of withstanding the handling to which they are likely to be subjected.

# ARTICLE 320
# Armored Cable: Type AC

## I. General

### 320.1 Scope

This article covers the use, installation, and construction specifications for armored cable, Type AC.

Type AC cable is listed in sizes 14 AWG through 1 AWG copper and 12 AWG through 1 AWG aluminum or copper-clad aluminum and is rated at 600 volts or less. However, the *NEC* does not specifically require that AC cable be listed. Exhibit 320.1 shows an example of AC cable.

### 320.2 Definition

**Armored Cable, Type AC.** A fabricated assembly of insulated conductors in a flexible interlocked metallic armor. See 320.100.

## II. Installation

### 320.10 Uses Permitted

Type AC cable shall be permitted as follows:

(1) For feeders and branch circuits in both exposed and concealed installations
(2) In cable trays
(3) In dry locations
(4) Embedded in plaster finish on brick or other masonry, except in damp or wet locations
(5) To be run or fished in the air voids of masonry block or tile walls where such walls are not exposed or subject to excessive moisture or dampness

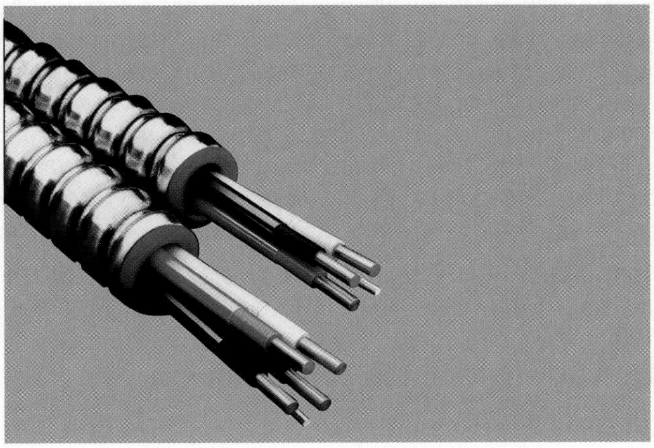

***EXHIBIT 320.1*** *An example of Type AC cable. (Courtesy of AFC Cable Systems, Inc.)*

Informational Note: The "Uses Permitted" is not an all-inclusive list.

### 320.12 Uses Not Permitted

Type AC cable shall not be used as follows:

(1) Where subject to physical damage
(2) In damp or wet locations
(3) In air voids of masonry block or tile walls where such walls are exposed or subject to excessive moisture or dampness
(4) Where exposed to corrosive conditions
(5) Embedded in plaster finish on brick or other masonry in damp or wet locations

### 320.15 Exposed Work

Exposed runs of cable, except as provided in 300.11(A), shall closely follow the surface of the building finish or of running boards. Exposed runs shall also be permitted to be installed on the underside of joists where supported at each joist and located so as not to be subject to physical damage.

### 320.17 Through or Parallel to Framing Members

Type AC cable shall be protected in accordance with 300.4(A), (C), and (D) where installed through or parallel to framing members.

### 320.23 In Accessible Attics

Type AC cables in accessible attics or roof spaces shall be installed as specified in 320.23(A) and (B).

**(A) Cables Run Across the Top of Floor Joists.** Where run across the top of floor joists, or within 2.1 m (7 ft) of the floor or floor joists across the face of rafters or studding, the cable shall be protected by guard strips that are at least as high as the cable. Where this space is not accessible by permanent stairs or ladders, protection shall only be required within 1.8 m (6 ft) of the nearest edge of the scuttle hole or attic entrance.

Type AC cable installed in accessible attics across the top of floor joists or within 7 feet of the floor or floor joists across the face of rafters or studs must be protected by guard strips. Where the attic is not accessible by a permanent ladder or stairs, guard strips are required only within 6 feet of the scuttle hole or opening.

**(B) Cable Installed Parallel to Framing Members.** Where the cable is installed parallel to the sides of rafters, studs, or ceiling or floor joists, neither guard strips nor running boards shall be required, and the installation shall also comply with 300.4(D).

### 320.24 Bending Radius

Bends in Type AC cable shall be made such that the cable is not damaged. The radius of the curve of the inner edge of any bend shall not be less than five times the diameter of the Type AC cable.

### 320.30 Securing and Supporting

**(A) General.** Type AC cable shall be supported and secured by staples, cable ties, straps, hangers, or similar fittings, designed and installed so as not to damage the cable.

Simply draping the cable over air ducts or lower members of bar joists, pipes, and ceiling grid members is not permitted.

**(B) Securing.** Unless otherwise permitted, Type AC cable shall be secured within 300 mm (12 in.) of every outlet box, junction box, cabinet, or fitting and at intervals not exceeding 1.4 m (4½ ft) where installed on or across framing members.

**(C) Supporting.** Unless otherwise permitted, Type AC cable shall be supported at intervals not exceeding 1.4 m (4½ ft).

Horizontal runs of Type AC cable installed in wooden or metal framing members or similar supporting means shall be considered supported where such support does not exceed 1.4-m (4½-ft) intervals.

Type AC cable, where run horizontally through framing members, is permitted to be passed through bored or punched holes in framing members without additional securing, provided the cable is secured within 12 inches of the outlet and the framing members are less than 54 inches apart.

**(D) Unsupported Cables.** Type AC cable shall be permitted to be unsupported where the cable complies with any of the following:

(1) Is fished between access points through concealed spaces in finished buildings or structures and supporting is impracticable
(2) Is not more than 600 mm (2 ft) in length at terminals where flexibility is necessary
(3) Is not more than 1.8 m (6 ft) in length from the last point of cable support to the point of connection to a luminaire(s) or other electrical equipment and the cable and point of connection are within an accessible ceiling. For the purposes of this section, Type AC cable fittings shall be permitted as a means of cable support.

### 320.40 Boxes and Fittings

At all points where the armor of AC cable terminates, a fitting shall be provided to protect wires from abrasion, unless the design of the outlet boxes or fittings is such as to afford equivalent protection, and, in addition, an insulating bushing or its equivalent protection shall be provided between the conductors and the armor. The connector or clamp by which the Type AC cable is fastened to boxes or cabinets shall be of such design that the insulating bushing or its equivalent will be visible for inspection. Where change is made from Type AC cable to other cable or raceway wiring methods, a box, fitting, or conduit body shall be installed at junction points as required in 300.15.

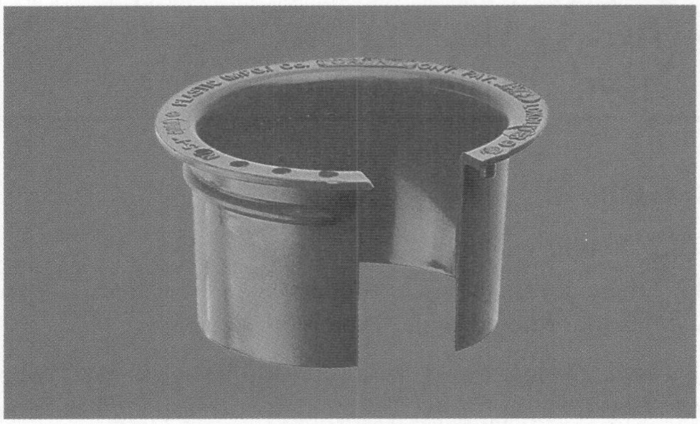

***EXHIBIT 320.2*** *An anti-short bushing designed to protect insulated conductors from abrasion. (Courtesy of Cooper Crouse-Hinds)*

An anti-short bushing (sometimes referred to as a "red head") is shown in Exhibit 320.2. It is a plastic insert placed between the metal jacket of an AC cable and the insulated conductors at the point where the conductors emerge from the metal jacket of an AC cable. The anti-short bushing provides the insulated conductors with an additional level of short-circuit and ground-fault protection where they are most vulnerable.

Armored cable connectors are considered suitable for equipment grounding if installed in accordance with 300.10.

### 320.80 Ampacity

The ampacity shall be determined in accordance with 310.15.

**(A) Thermal Insulation.** Armored cable installed in thermal insulation shall have conductors rated at 90°C (194°F). The ampacity of cable installed in these applications shall not exceed that of a 60°C (140°F) rated conductor. The 90°C (194°F) rating shall be permitted to be used for ampacity adjustment and correction calculations; however, the ampacity shall not exceed that of a 60°C (140°F) rated conductor.

Armored cable installed in thermal insulation has a decreased heat dissipation capacity. Cable marked "ACTH" indicates an armored cable rated 75°C and employing conductors having thermoplastic insulation. Cable marked "ACTHH" indicates an armored cable rated 90°C and employing conductors having thermoplastic insulation. Cable marked "ACHH" indicates armored cable rated 90°C and employing conductors having thermosetting insulation. Where conductors are rated 90°C, the 90°C ampacity can be used for derating.

**(B) Cable Tray.** The ampacity of Type AC cable installed in cable tray shall be determined in accordance with 392.80(A).

## III. Construction Specifications

### 320.100 Construction

Type AC cable shall have an armor of flexible metal tape and shall have an internal bonding strip of copper or aluminum in intimate contact with the armor for its entire length.

The armor of AC cable is recognized as an equipment grounding conductor by 250.118, and the internal bonding strip required by 320.100 can simply be cut off at the termination of the armored cable, or it can be bent back on the armor. This bonding strip is not required to be connected to an equipment grounding terminal. It reduces the inductive reactance of the spiral armor and increases the armor's effectiveness as an equipment ground. Many installers use this strip to help prevent the insulating (anti-short) bushing required by 320.40 (the "red head") from falling out during rough wiring.

### 320.104 Conductors

Insulated conductors shall be of a type listed in Table 310.104(A) or those identified for use in this cable. In addition, the conductors shall have an overall moisture-resistant and fire-retardant fibrous covering. For Type ACT, a moisture-resistant fibrous covering shall be required only on the individual conductors.

### 320.108 Equipment Grounding Conductor

Type AC cable shall provide an adequate path for fault current as required by 250.4(A)(5) or (B)(4) to act as an equipment grounding conductor.

### 320.120 Marking

The cable shall be marked in accordance with 310.120, except that Type AC shall have ready identification of the manufacturer by distinctive external markings on the cable armor throughout its entire length.

# ARTICLE 322 Flat Cable Assemblies: Type FC

## I. General

### 322.1 Scope

This article covers the use, installation, and construction specifications for flat cable assemblies, Type FC.

Type FC cable is an assembly of three or four parallel 10 AWG special stranded copper wires formed integrally with an insulating material web. The cable is marked with the size of the maximum branch circuit to which it can be connected, the cable type designation, manufacturer's identification, maximum working voltage, conductor size, and temperature rating. A marking accompanying the cable on a tag or reel indicates the special metal raceways and specific FC cable fittings with which the cable is intended to be used. Exhibits 322.1 and 322.2 show the basic components of this wiring method.

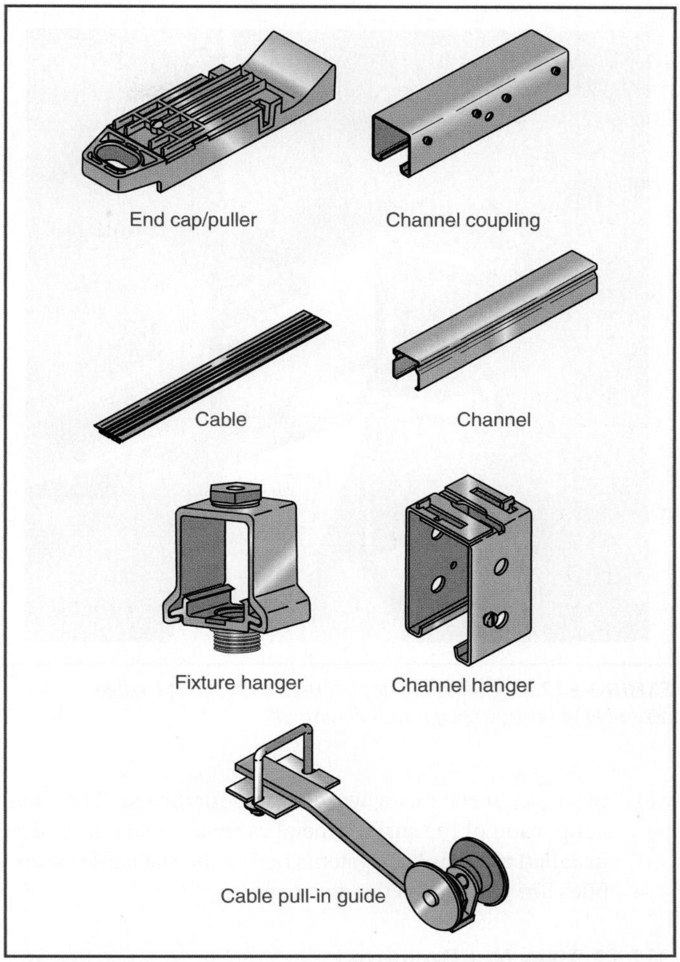


***EXHIBIT 322.1*** *Basic components and accessories used for an installation of Type FC cable assembly. (Courtesy of Legrand/ Wiremold®)*

### 322.2 Definition

**Flat Cable Assembly, Type FC.** An assembly of parallel conductors formed integrally with an insulating material web specifically designed for field installation in surface metal raceway.

## II. Installation

### 322.10 Uses Permitted

Flat cable assemblies shall be permitted only as follows:

(1) As branch circuits to supply suitable tap devices for lighting, small appliances, or small power loads. The rating of the branch circuit shall not exceed 30 amperes.
(2) Where installed for exposed work.
(3) In locations where they will not be subjected to physical damage. Where a flat cable assembly is installed less than 2.5 m (8 ft) above the floor or fixed working platform, it shall be protected by a cover identified for the use.

**EXHIBIT 322.2** *A luminaire hanger used with Type FC cable assembly. (Courtesy of Legrand/Wiremold®)*

(4) In surface metal raceways identified for the use. The channel portion of the surface metal raceway systems shall be installed as complete systems before the flat cable assemblies are pulled into the raceways.

### 322.12 Uses Not Permitted

Flat cable assemblies shall not be used as follows:

(1) Where exposed to corrosive conditions, unless suitable for the application
(2) In hoistways or on elevators or escalators
(3) In any hazardous (classified) location, except as specifically permitted by other articles in this *Code*
(4) Outdoors or in wet or damp locations unless identified for the use

### 322.30 Securing and Supporting

The flat cable assemblies shall be supported by means of their special design features, within the surface metal raceways.

The surface metal raceways shall be supported as required for the specific raceway to be installed.

### 322.40 Boxes and Fittings

**(A) Dead Ends.** Each flat cable assembly dead end shall be terminated in an end-cap device identified for the use.

The dead-end fitting for the enclosing surface metal raceway shall be identified for the use.

**(B) Luminaire Hangers.** Luminaire hangers installed with the flat cable assemblies shall be identified for the use.

**(C) Fittings.** Fittings to be installed with flat cable assemblies shall be designed and installed to prevent physical damage to the cable assemblies.

**(D) Extensions.** All extensions from flat cable assemblies shall be made by approved wiring methods, within the junction boxes, installed at either end of the flat cable assembly runs.

### 322.56 Splices and Taps

**(A) Splices.** Splices shall be made in listed junction boxes.

**(B) Taps.** Taps shall be made between any phase conductor and the grounded conductor or any other phase conductor by means of devices and fittings identified for the use. Tap devices shall be rated at not less than 15 amperes, or more than 300 volts to ground, and shall be color-coded in accordance with the requirements of 322.120(C).

## III. Construction

### 322.100 Construction

Flat cable assemblies shall consist of two, three, four, or five conductors.

### 322.104 Conductors

Flat cable assemblies shall have conductors of 10 AWG special stranded copper wires.

### 322.112 Insulation

The entire flat cable assembly shall be formed to provide a suitable insulation covering all the conductors and using one of the materials recognized in Table 310.104(A) for general branch-circuit wiring.

### 322.120 Marking

**(A) Temperature Rating.** In addition to the provisions of 310.120, Type FC cable shall have the temperature rating durably marked on the surface at intervals not exceeding 600 mm (24 in.).

**(B) Identification of Grounded Conductor.** The grounded conductor shall be identified throughout its length by means of a distinctive and durable white or gray marking.

Informational Note: The color gray may have been used in the past as an ungrounded conductor. Care should be taken when working on existing systems.

**(C) Terminal Block Identification.** Terminal blocks identified for the use shall have distinctive and durable markings for color or word coding. The grounded conductor section shall have a white marking or other suitable designation. The next adjacent section of the terminal block shall have a black marking or other suitable designation. The next section shall have a red marking or other suitable designation. The final or outer section, opposite the grounded conductor section of the terminal block, shall have a blue marking or other suitable designation.

# ARTICLE 324 Flat Conductor Cable: Type FCC

## I. General

### 324.1 Scope

This article covers a field-installed wiring system for branch circuits incorporating Type FCC cable and associated accessories as defined by the article. The wiring system is designed for installation under carpet squares.

The FCC system is designed to provide a completely accessible, flexible power system. As shown in Exhibit 324.1, it also provides an easy method for reworking obsolete wiring systems currently in use in many office facilities. The carpet squares are not permitted to be larger than 1.0 meter by 1.0 meter to comply with 324.41. This limitation provides ready access to the cable by lifting a carpet square. It also reduces the likelihood of an individual cutting through the carpet above the cable with a knife or razor blade and possibly penetrating the top shield of the cable.

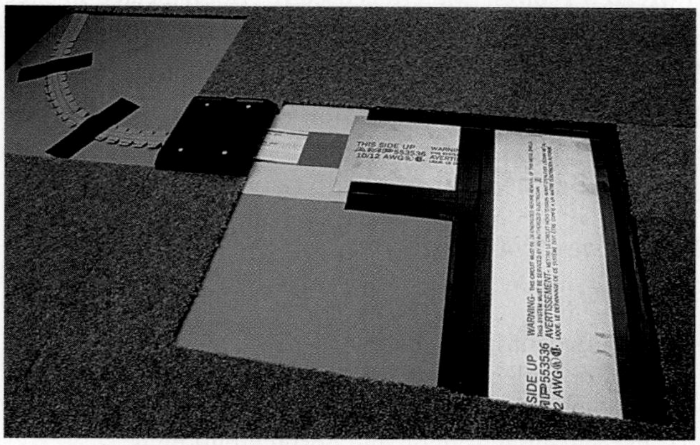

***EXHIBIT 324.1*** *Type FCC cable installed beneath carpet squares. (Courtesy of Tyco Electronics)*

### 324.2 Definitions

**Bottom Shield.** A protective layer that is installed between the floor and Type FCC flat conductor cable to protect the cable from physical damage and may or may not be incorporated as an integral part of the cable.

**Cable Connector.** A connector designed to join Type FCC cables without using a junction box.

**FCC System.** A complete wiring system for branch circuits that is designed for installation under carpet squares.

Informational Note: The FCC system includes Type FCC cable and associated shielding, connectors, terminators, adapters, boxes, and receptacles.

**Insulating End.** An insulator designed to electrically insulate the end of a Type FCC cable.

**Metal Shield Connections.** Means of connection designed to electrically and mechanically connect a metal shield to another metal shield, to a receptacle housing or self-contained device, or to a transition assembly.

**Top Shield.** A grounded metal shield covering under-carpet components of the FCC system for the purposes of providing protection against physical damage.

**Transition Assembly.** An assembly to facilitate connection of the FCC system to other wiring systems, incorporating (1) a means of electrical interconnection and (2) a suitable box or covering for providing electrical safety and protection against physical damage.

**Type FCC Cable.** Three or more flat copper conductors placed edge-to-edge and separated and enclosed within an insulating assembly.

### 324.6 Listing Requirements

Type FCC cable and associated fittings shall be listed.

## II. Installation

### 324.10 Uses Permitted

**(A) Branch Circuits.** Use of FCC systems shall be permitted both for general-purpose and appliance branch circuits and for individual branch circuits.

**(B) Branch-Circuit Ratings.**

**(1) Voltage.** Voltage between ungrounded conductors shall not exceed 300 volts. Voltage between ungrounded conductors and the grounded conductor shall not exceed 150 volts.

**(2) Current.** General-purpose and appliance branch circuits shall have ratings not exceeding 20 amperes. Individual branch circuits shall have ratings not exceeding 30 amperes.

**(C) Floors.** Use of FCC systems shall be permitted on hard, sound, smooth, continuous floor surfaces made of concrete, ceramic, or composition flooring, wood, and similar materials.

**(D) Walls.** Use of FCC systems shall be permitted on wall surfaces in surface metal raceways.

**(E) Damp Locations.** Use of FCC systems in damp locations shall be permitted.

**(F) Heated Floors.** Materials used for floors heated in excess of 30°C (86°F) shall be identified as suitable for use at these temperatures.

**(G) System Height.** Any portion of an FCC system with a height above floor level exceeding 2.3 mm (0.090 in.) shall be tapered or feathered at the edges to floor level.

## 324.12 Uses Not Permitted

FCC systems shall not be used in the following locations:

(1) Outdoors or in wet locations
(2) Where subject to corrosive vapors
(3) In any hazardous (classified) location
(4) In residential, school, and hospital buildings

Type FCC wiring systems are not permitted throughout school and hospital buildings even though parts of these buildings may be office or administrative spaces.

## 324.18 Crossings

Crossings of more than two Type FCC cable runs shall not be permitted at any one point. Crossings of a Type FCC cable over or under a flat communications or signal cable shall be permitted. In each case, a grounded layer of metal shielding shall separate the two cables, and crossings of more than two flat cables shall not be permitted at any one point.

## 324.30 Securing and Supporting

All FCC system components shall be firmly anchored to the floor or wall using an adhesive or mechanical anchoring system identified for this use. Floors shall be prepared to ensure adherence of the FCC system to the floor until the carpet squares are placed.

## 324.40 Boxes and Fittings

**(A) Cable Connections and Insulating Ends.** All Type FCC cable connections shall use connectors identified for their use, installed such that electrical continuity, insulation, and sealing against dampness and liquid spillage are provided. All bare cable ends shall be insulated and sealed against dampness and liquid spillage using listed insulating ends.

**(B) Polarization of Connections.** All receptacles and connections shall be constructed and installed so as to maintain proper polarization of the system.

**(C) Shields.**

**(1) Top Shield.** A metal top shield shall be installed over all floor-mounted Type FCC cable, connectors, and insulating ends. The top shield shall completely cover all cable runs, corners, connectors, and ends.

**(2) Bottom Shield.** A bottom shield shall be installed beneath all Type FCC cable, connectors, and insulating ends.

**(D) Connection to Other Systems.** Power feed, grounding connection, and shield system connection between the FCC system and other wiring systems shall be accomplished in a transition assembly identified for this use.

**(E) Metal-Shield Connectors.** Metal shields shall be connected to each other and to boxes, receptacle housings, self-contained devices, and transition assemblies using metal-shield connectors.

## 324.41 Floor Coverings

Floor-mounted Type FCC cable, cable connectors, and insulating ends shall be covered with carpet squares not larger than 1.0 m (39.37 in.) square. Carpet squares that are adhered to the floor shall be attached with release-type adhesives.

## 324.42 Devices

**(A) Receptacles.** All receptacles, receptacle housings, and self-contained devices used with the FCC system shall be identified for this use and shall be connected to the Type FCC cable and metal shields. Connection from any grounding conductor of the Type FCC cable shall be made to the shield system at each receptacle.

**(B) Receptacles and Housings.** Receptacle housings and self-contained devices designed either for floor mounting or for in-wall or on-wall mounting shall be permitted for use with the FCC system. Receptacle housings and self-contained devices shall incorporate means for facilitating entry and termination of Type FCC cable and for electrically connecting the housing or device with the metal shield. Receptacles and self-contained devices shall comply with 406.4. Power and communications outlets installed together in common housing shall be permitted in accordance with 800.133(A)(1)(d), Exception No. 2.

## 324.56 Splices and Taps

**(A) FCC Systems Alterations.** Alterations to FCC systems shall be permitted. New cable connectors shall be used at new connection points to make alterations. It shall be permitted to leave unused cable runs and associated cable connectors in place and energized. All cable ends shall be covered with insulating ends.

**(B) Transition Assemblies.** All transition assemblies shall be identified for their use. Each assembly shall incorporate means for facilitating entry of the Type FCC cable into the assembly, for connecting the Type FCC cable to grounded conductors, and for electrically connecting the assembly to the metal cable shields and to equipment grounding conductors.

## 324.60 Grounding

All metal shields, boxes, receptacle housings, and self-contained devices shall be electrically continuous to the equipment grounding conductor of the supplying branch circuit. All such electrical connections shall be made with connectors identified for this use. The electrical resistivity of such shield system shall not be more than that of one conductor of the Type FCC cable used in the installation.

## III. Construction

### 324.100 Construction

**(A) Type FCC Cable.** Type FCC cable shall be listed for use with the FCC system and shall consist of three, four, or five flat copper conductors, one of which shall be an equipment grounding conductor.

**(B) Shields.**

**(1) Materials and Dimensions.** All top and bottom shields shall be of designs and materials identified for their use. Top shields shall be metal. Both metallic and nonmetallic materials shall be permitted for bottom shields.

**(2) Resistivity.** Metal shields shall have cross-sectional areas that provide for electrical resistivity of not more than that of one conductor of the Type FCC cable used in the installation.

### 324.101 Corrosion Resistance

Metal components of the system shall be either corrosion resistant, coated with corrosion-resistant materials, or insulated from contact with corrosive substances.

### 324.112 Insulation

The insulating material of the cable shall be moisture resistant and flame retardant. All insulating materials in the FCC systems shall be identified for their use.

### 324.120 Markings

**(A) Cable Marking.** Type FCC cable shall be clearly and durably marked on both sides at intervals of not more than 610 mm (24 in.) with the information required by 310.120(A) and with the following additional information:

(1) Material of conductors
(2) Maximum temperature rating
(3) Ampacity

**(B) Conductor Identification.** Conductors shall be clearly and durably identified on both sides throughout their length as specified in 310.110.

# ARTICLE 326 Integrated Gas Spacer Cable: Type IGS

## I. General

### 326.1 Scope

This article covers the use, installation, and construction specifications for integrated gas spacer cable, Type IGS.

As illustrated in Exhibit 326.1, IGS cable consists of solid aluminum rod conductors, 250 kcmil minimum size. These conductors are insulated with dry kraft paper and are installed in a medium-density polyethylene gas pipe, minimum trade size 2, which is then filled with sulfur hexafluoride ($SF_6$) gas at a pressure of approximately 20 psi.

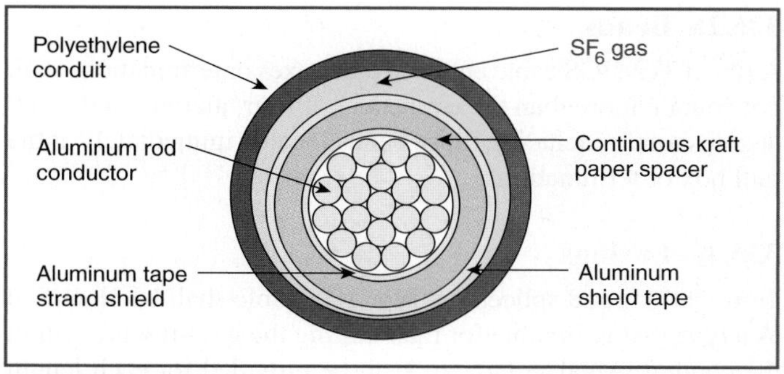


***EXHIBIT 326.1*** *Cross section of single-conductor, 4750-kcmil Type IGS cable.*

### 326.2 Definition

**Integrated Gas Spacer Cable, Type IGS.** A factory assembly of one or more conductors, each individually insulated and enclosed in a loose fit, nonmetallic flexible conduit as an integrated gas spacer cable rated 0 through 600 volts.

## II. Installation

### 326.10 Uses Permitted

Type IGS cable shall be permitted for use underground, including direct burial in the earth, as the following:

(1) Service-entrance conductors
(2) Feeder or branch-circuit conductors
(3) Service conductors, underground

### 326.12 Uses Not Permitted

Type IGS cable shall not be used as interior wiring or be exposed in contact with buildings.

### 326.24 Bending Radius

Where the coilable nonmetallic conduit and cable is bent for installation purposes or is flexed or bent during shipment or installation, the radii of bends measured to the inside of the bend shall not be less than specified in Table 326.24.

***TABLE 326.24*** *Minimum Radii of Bends*

| Conduit Size | | Minimum Radii | |
|---|---|---|---|
| **Metric Designator** | **Trade Size** | **mm** | **in.** |
| 53 | 2 | 600 | 24 |
| 78 | 3 | 900 | 35 |
| 103 | 4 | 1150 | 45 |

### 326.26 Bends

A run of Type IGS cable between pull boxes or terminations shall not contain more than the equivalent of four quarter bends (360 degrees total), including those bends located immediately at the pull box or terminations.

### 326.40 Fittings

Terminations and splices for Type IGS cable shall be identified as a type that is suitable for maintaining the gas pressure within the conduit. A valve and cap shall be provided for each length of the cable and conduit to check the gas pressure or to inject gas into the conduit.

### 326.80 Ampacity

The ampacity of Type IGS cable shall not exceed the values shown in Table 326.80.

## III. Construction Specifications

### 326.104 Conductors

The conductors shall be solid aluminum rods, laid parallel, consisting of one to nineteen 12.7 mm (½ in.) diameter rods. The minimum conductor size shall be 250 kcmil, and the maximum size shall be 4750 kcmil.

### 326.112 Insulation

The insulation shall be dry kraft paper tapes and a pressurized sulfur hexafluoride gas ($SF_6$), both approved for electrical use. The nominal gas pressure shall be 138 kPa gauge (20 lb/in.$^2$ gauge). The thickness of the paper spacer shall be as specified in Table 326.112.

### 326.116 Conduit

The conduit shall be a medium density polyethylene identified as suitable for use with natural gas rated pipe in metric designator 53, 78, or 103 (trade size 2, 3, or 4). The percent fill dimensions for the conduit are shown in Table 326.116.

***TABLE 326.80*** *Ampacity of Type IGS Cable*

| Size (kcmil) | Amperes | Size (kcmil) | Amperes |
|---|---|---|---|
| 250 | 119 | 2500 | 376 |
| 500 | 168 | 3000 | 412 |
| 750 | 206 | 3250 | 429 |
| 1000 | 238 | 3500 | 445 |
| 1250 | 266 | 3750 | 461 |
| 1500 | 292 | 4000 | 476 |
| 1750 | 315 | 4250 | 491 |
| 2000 | 336 | 4500 | 505 |
| 2250 | 357 | 4750 | 519 |

***TABLE 326.112*** *Paper Spacer Thickness*

| Size (kcmil) | Thickness | |
|---|---|---|
| | mm | in. |
| 250–1000 | 1.02 | 0.040 |
| 1250–4750 | 1.52 | 0.060 |

***TABLE 326.116*** *Conduit Dimensions*

| Conduit Size | | Actual Outside Diameter | | Actual Inside Diameter | |
|---|---|---|---|---|---|
| Metric Designator | Trade Size | mm | in. | mm | in. |
| 53 | 2 | 60 | 2.375 | 49.46 | 1.947 |
| 78 | 3 | 89 | 3.500 | 73.30 | 2.886 |
| 103 | 4 | 114 | 4.500 | 94.23 | 3.710 |

The size of the conduit permitted for each conductor size shall be calculated for a percent fill not to exceed those found in Table 1, Chapter 9.

### 326.120 Marking

The cable shall be marked in accordance with 310.120(A), 310.120(B)(1), and 310.120(D).

# ARTICLE 328 Medium Voltage Cable: Type MV

## I. General

### 328.1 Scope

This article covers the use, installation, and construction specifications for medium voltage cable, Type MV.

Type MV cables are rated 2001 to 35,000 volts. Requirements for shielding are found in 310.10(E) and 310.10(F). If MV cables are installed in underground installations, they must comply with 300.50.

### 328.2 Definition

**Medium Voltage Cable, Type MV.** A single or multiconductor solid dielectric insulated cable rated 2001 volts or higher.

## II. Installation

### 328.10 Uses Permitted

Type MV cable shall be permitted for use on power systems rated up to and including 35,000 volts, nominal, as follows:

(1) In wet or dry locations.
(2) In raceways.
(3) In cable trays, where identified for the use, in accordance with 392.10, 392.20(B), (C), and (D), 392.22(C), 392.30(B)(1), 392.46, 392.56, and 392.60. Type MV cable that has an overall metallic sheath or armor, complies with the requirements for Type MC cable, and is identified as "MV or MC" shall be permitted to be installed in cable trays in accordance with 392.10(B)(2).
(4) Direct buried in accordance with 300.50.
(5) In messenger-supported wiring in accordance with Part II of Article 396.
(6) As exposed runs in accordance with 300.37. Type MV cable that has an overall metallic sheath or armor, complies with the requirements for Type MC cable, and is identified as "MV or MC" shall be permitted to be installed as exposed runs of metal-clad cable in accordance with 300.37.

Informational Note: The "Uses Permitted" is not an all-inclusive list.

Type MV cables intended for installation in cable trays in accordance with Article 392 are marked "For CT Use" or "For Use in Cable Trays." Where marked "MV or MC," the cable complies with the crush and impact rating associated with MC cable. Cable marked "MV or MC" is permitted to be installed in accordance with Article 330 as well as Article 392.

### 328.12 Uses Not Permitted

Type MV cable shall not be used where exposed to direct sunlight, unless identified for the use.

### 328.14 Installation

Type MV cable shall be installed, terminated, and tested by qualified persons.

Informational Note: IEEE 576-2000, *Recommended Practice for Installation, Termination, and Testing of Insulated Power Cables as Used in Industrial and Commercial Applications*, includes installation information and testing criteria for MV cable.

### 328.80 Ampacity

The ampacity of Type MV cable shall be determined in accordance with 310.60. The ampacity of Type MV cable installed in cable tray shall be determined in accordance with 392.80(B).

## III. Construction Specifications

### 328.100 Construction

Type MV cables shall have copper, aluminum, or copper-clad aluminum conductors and shall comply with Table 310.104(C) and Table 310.104(D) or Table 310.104(E).

### 328.120 Marking

Medium voltage cable shall be marked as required by 310.120.

In addition to the marking requirements of 310.120, insulation-level marking requirements include 100 percent, 133 percent, and 173 percent. These insulation levels are explained in the notes to Table 310.104(E). Shielded cable is marked "MV-90" and "MV-105" and is suitable for use in both dry and wet locations at both 90°C and 105°C. Additionally, the UL *Guide Information for Electrical Equipment – The White Book*, in category PITY, gives examples of required and optional marking.

# ARTICLE 330
# Metal-Clad Cable: Type MC

## I. General

### 330.1 Scope

This article covers the use, installation, and construction specifications of metal-clad cable, Type MC.

Type MC cable is rated up to 2000 volts in sizes 18 AWG and larger for copper or nickel-coated copper and 12 AWG and larger for aluminum or copper-clad aluminum. Type MC cable rated 2400 to 35,000 volts is classified as medium-voltage cable, is marked "Type MV or MC," and is covered by Article 328. Type MC-HL cable rated up to 35,000 volts is suitable for hazardous locations and is provided with a gas/vaportight continuous sheath. Composite electrical MC and optical fiber cables are classified as Type MC cable and are marked "MC-OF." See Exhibit 770.2 for an example.

Type MC cable is available in three designs: interlocked metal tape, corrugated metal tube, and smooth metal tube. A nonmetallic jacket may be provided over the metal sheath. Cable construction must comply with 250.118(10) in order for the cable to be used as an equipment grounding conductor. It must be marked with the maximum rated voltage, the proper insulation type letter or letters, and the AWG size or circular mil area. See 310.120 for marking requirements. Exhibit 330.1 shows some examples of MC cable. The basic standard to investigate cable in this category is ANSI/UL 1569, *Standard for Metal-Clad Cables*. Summary information regarding listed metal-clad cable may be found in the UL *Guide Information for Electrical Equipment – The White Book*, under category PJAZ.

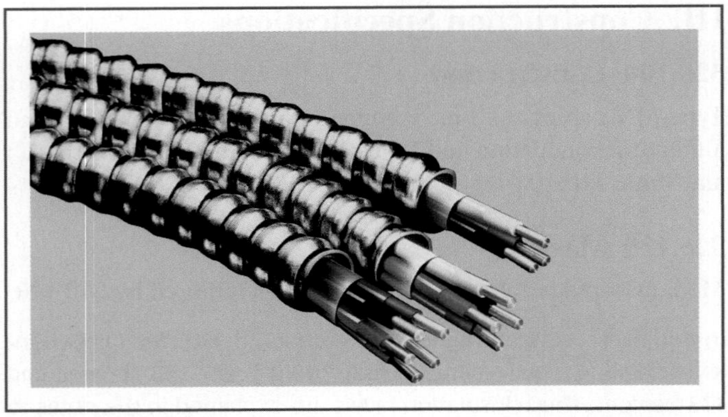

*EXHIBIT 330.1 Examples of Type MC cable. (Courtesy of AFC Cable Systems, Inc.)*

### 330.2 Definition

**Metal Clad Cable, Type MC.** A factory assembly of one or more insulated circuit conductors with or without optical fiber members enclosed in an armor of interlocking metal tape, or a smooth or corrugated metallic sheath.

## II. Installation

### 330.10 Uses Permitted

**(A) General Uses.** Type MC cable shall be permitted as follows:

(1) For services, feeders, and branch circuits.
(2) For power, lighting, control, and signal circuits.
(3) Indoors or outdoors.
(4) Exposed or concealed.
(5) To be direct buried where identified for such use.
(6) In cable tray where identified for such use.
(7) In any raceway.
(8) As aerial cable on a messenger.
(9) In hazardous (classified) locations where specifically permitted by other articles in this *Code*.
(10) In dry locations and embedded in plaster finish on brick or other masonry except in damp or wet locations.
(11) In wet locations where a corrosion-resistant jacket is provided over the metallic covering and any of the following conditions are met:

a. The metallic covering is impervious to moisture.
b. A jacket resistant to moisture is provided under the metal covering.
c. The insulated conductors under the metallic covering are listed for use in wet locations.

(12) Where single-conductor cables are used, all phase conductors and, where used, the grounded conductor shall be grouped together to minimize induced voltage on the sheath.

Installation practice for single-conductor wiring methods dictates close circuit conductor spacing not only to minimize induced voltage on the sheath but also to minimize overall circuit impedance.

**(B) Specific Uses.** Type MC cable shall be permitted to be installed in compliance with Parts II and III of Article 725 and 770.133 as applicable and in accordance with 330.10(B)(1) through (B)(4).

**(1) Cable Tray.** 'Type MC cable installed in cable tray shall comply with 392.10, 392.12, 392.18, 392.20, 392.22, 392.30, 392.46, 392.56, 392.60(C), and 392.80.

**(2) Direct Buried.** Direct-buried cable shall comply with 300.5 or 300.50, as appropriate.

**(3) Installed as Service-Entrance Cable.** Type MC cable installed as service-entrance cable shall be permitted in accordance with 230.43.

**(4) Installed Outside of Buildings or Structures or as Aerial Cable.** Type MC cable installed outside of buildings or structures or as aerial cable shall comply with 225.10, 396.10, and 396.12.

Informational Note: The "Uses Permitted" is not an all-inclusive list.

### 330.12 Uses Not Permitted

Type MC cable shall not be used under either of the following conditions:

(1) Where subject to physical damage
(2) Where exposed to any of the destructive corrosive conditions in (a) or (b), unless the metallic sheath or armor is resistant to the conditions or is protected by material resistant to the conditions:

a. Direct buried in the earth or embedded in concrete unless identified for direct burial
b. Exposed to cinder fills, strong chlorides, caustic alkalis, or vapors of chlorine or of hydrochloric acids

### 330.17 Through or Parallel to Framing Members

Type MC cable shall be protected in accordance with 300.4(A), (C), and (D) where installed through or parallel to framing members.

### 330.23 In Accessible Attics

The installation of Type MC cable in accessible attics or roof spaces shall also comply with 320.23.

In accessible attics, cable installed across the top of floor joists or within 7 feet of the floor or floor joists across the face of rafters or studs must be protected by guard strips. Where the attic is not accessible by a permanent ladder or stairs, guard strips are required only within 6 feet of the scuttle hole or opening.

### 330.24 Bending Radius

Bends in Type MC cable shall be so made that the cable will not be damaged. The radius of the curve of the inner edge of any bend shall not be less than required in 330.24(A) through (C).

**(A) Smooth Sheath.**

(1) Ten times the external diameter of the metallic sheath for cable not more than 19 mm (¾ in.) in external diameter
(2) Twelve times the external diameter of the metallic sheath for cable more than 19 mm (¾ in.) but not more than 38 mm (1½ in.) in external diameter
(3) Fifteen times the external diameter of the metallic sheath for cable more than 38 mm (1½ in.) in external diameter

**(B) Interlocked-Type Armor or Corrugated Sheath.** Seven times the external diameter of the metallic sheath.

**(C) Shielded Conductors.** Twelve times the overall diameter of one of the individual conductors or seven times the overall diameter of the multiconductor cable, whichever is greater.

### 330.30 Securing and Supporting

**(A) General.** Type MC cable shall be supported and secured by staples, cable ties, straps, hangers, or similar fittings or other approved means designed and installed so as not to damage the cable.

A difference exists between securing and supporting. Cable that runs horizontally through or on framing members or racks (spaced less than 6 feet apart) without additional securing is considered supported. Staples or cable ties are not required as the cable passes through or on these members. However, the cable must be secured (fastened in place) within 12 inches of the outlet box. Staples, cable ties, or clamps would be necessary to secure a cable. Both requirements are illustrated in Exhibit 330.2.

**(B) Securing.** Unless otherwise provided, cables shall be secured at intervals not exceeding 1.8 m (6 ft). Cables containing four or fewer conductors sized no larger than 10 AWG shall be secured within 300 mm (12 in.) of every box, cabinet, fitting, or other cable termination. In vertical installations, listed cables with ungrounded conductors 250 kcmil and larger shall be permitted to be secured at intervals not exceeding 3 m (10 ft).

**(C) Supporting.** Unless otherwise provided, cables shall be supported at intervals not exceeding 1.8 m (6 ft).

Horizontal runs of Type MC cable installed in wooden or metal framing members or similar supporting means shall be considered supported and secured where such support does not exceed 1.8-m (6-ft) intervals.

**(D) Unsupported Cables.** Type MC cable shall be permitted to be unsupported where the cable:

(1) Is fished between access points through concealed spaces in finished buildings or structures and supporting is impractical.
(2) Is not more than 1.8 m (6 ft) in length from the last point of cable support to the point of connection to luminaires or other electrical equipment and the cable and point of connection are within an accessible ceiling. For the purpose of this section, Type MC cable fittings shall be permitted as a means of cable support.

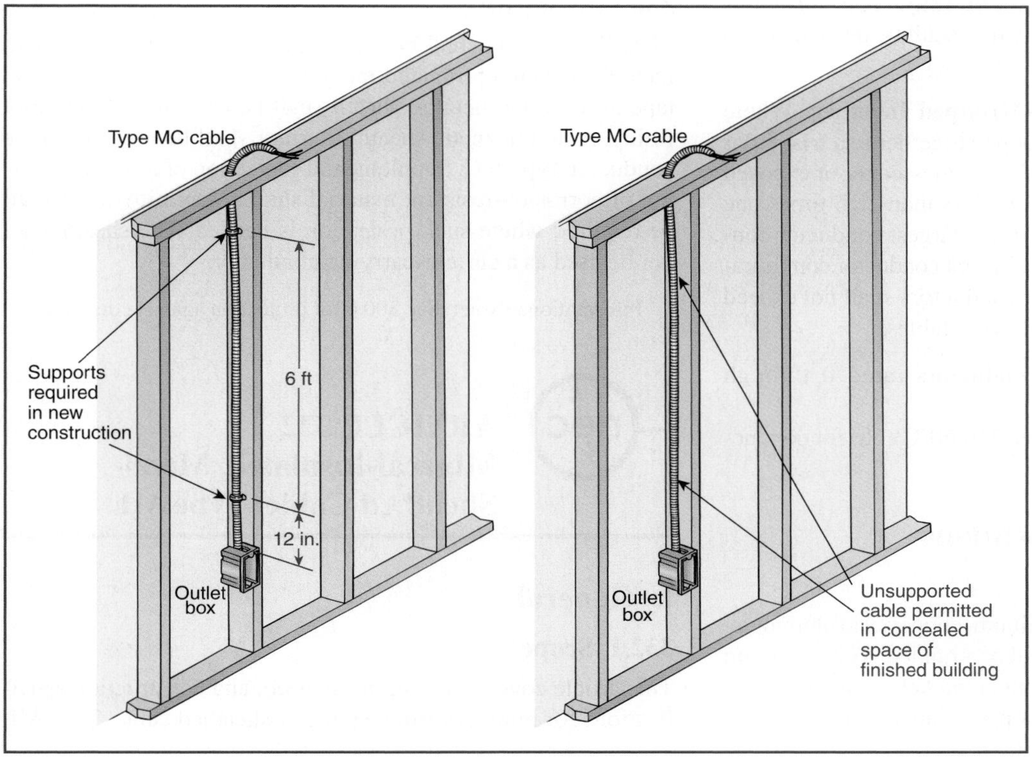


***EXHIBIT 330.2*** *Type MC cable supported and secured at intervals not exceeding 6 feet and within 12 inches of the box [per 330.30(B)] and Type MC cable to be fished in walls, floors, or ceilings [per 330.30(D)(1)].*

(3) Is Type MC of the interlocked armor type in lengths not exceeding 900 mm (3 ft) from the last point where it is securely fastened and is used to connect equipment where flexibility is necessary to minimize the transmission of vibration from equipment or to provide flexibility for equipment that requires movement after installation.

### 330.31 Single Conductors

Where single-conductor cables with a nonferrous armor or sheath are used, the installation shall comply with 300.20.

### 330.40 Boxes and Fittings

Fittings used for connecting Type MC cable to boxes, cabinets, or other equipment shall be listed and identified for such use.

Connectors should be selected in accordance with the size and type of cable for which they are designated. Bronze connectors are intended for use only with cable employing corrugated copper armor. Some Type AC cable connectors are also acceptable and listed for use with MC cable when specifically indicated on the fitting or the shipping carton.

### 330.80 Ampacity

The ampacity of Type MC cable shall be determined in accordance with 310.15 or 310.60 for 14 AWG and larger conductors and in accordance with Table 402.5 for 18 AWG and 16 AWG conductors. The installation shall not exceed the temperature ratings of terminations and equipment.

**(A) Type MC Cable Installed in Cable Tray.** The ampacities for Type MC cable installed in cable tray shall be determined in accordance with 392.80.

**(B) Single Type MC Conductors Grouped Together.** Where single Type MC conductors are grouped together in a triangular or square configuration and installed on a messenger or exposed with a maintained free airspace of not less than 2.15 times one conductor diameter (2.15 × O.D.) of the largest conductor contained within the configuration and adjacent conductor configurations or cables, the ampacity of the conductors shall not exceed the allowable ampacities in the following tables:

(1) Table 310.15(B)(20) for conductors rated 0 through 2000 volts
(2) Table 310.60(C)(67) and Table 310.60(C)(68) for conductors rated over 2000 volts

## III. Construction Specifications

### 330.104 Conductors

Conductors shall be of copper, aluminum, copper-clad aluminum, nickel or nickel-coated copper, solid or stranded. The minimum conductor size shall be 18 AWG copper, nickel or nickel-coated copper, or 12 AWG aluminum or copper-clad aluminum.

Nickel and nickel-coated copper are used as conductors in some fire-rated MC cables. The use of conductor material of nickel and nickel-coated copper is already permitted by 310.106(B) and is listed in Table 310.104(A) for insulation types PFAH and TFE (and used in high-temperature applications).

### 330.108 Equipment Grounding Conductor

Where Type MC cable is used to provide an equipment grounding conductor, it shall comply with 250.118(10) and 250.122.

The interlocked sheath cable construction is not recognized as an EGC. A discrete conductor must be integral to this type of MC cable to qualify as an EGC. The EGC within a Type MC cable may be either single or segmented. Where segmented, the individual segments must have a total cross-sectional area of not less than that required by 250.122 for a single conductor.

### 330.112 Insulation

Insulated conductors shall comply with 330.112(A) or (B).

**(A) 1000 Volts or Less.** Insulated conductors in sizes 18 AWG and 16 AWG shall be of a type listed in Table 402.3, with a maximum operating temperature not less than 90°C (194°F) and as permitted by 725.49. Conductors larger than 16 AWG shall be of a type listed in Table 310.104(A) or of a type identified for use in Type MC cable.

**(B) Over 1000 Volts.** Insulated conductors shall be of a type listed in Table 310.104(B) and Table 310.104(C).

### 330.116 Sheath

Metallic covering shall be one of the following types: smooth metallic sheath, corrugated metallic sheath, interlocking metal tape armor. The metallic sheath shall be continuous and close fitting. A nonmagnetic sheath or armor shall be used on single conductor Type MC. Supplemental protection of an outer covering of corrosion-resistant material shall be permitted and shall be required where such protection is needed. The sheath shall not be used as a current-carrying conductor.

Informational Note: See 300.6 for protection against corrosion.

# ARTICLE 332 Mineral-Insulated, Metal-Sheathed Cable: Type MI

## I. General

### 332.1 Scope

This article covers the use, installation, and construction specifications for mineral-insulated, metal-sheathed cable, Type MI.

Type MI cable consists of one or more solid copper conductors insulated with highly compressed magnesium oxide and enclosed in a continuous copper or alloy steel (e.g., stainless steel) sheath with or without a nonmetallic jacket. It is manufactured in size 16 AWG to 500 kcmil, single conductor; 16 AWG to 4 AWG, 2 and 3 conductor; 16 AWG to 6 AWG, four conductor; and 16 AWG to 10 AWG, 7 conductor. The cable is rated 600 volts. Two, three, four, and seven conductor MI cable rated 300 volts is available in 18 to 16 AWG for signaling circuits.

### 332.2 Definition

**Mineral-Insulated, Metal-Sheathed Cable, Type MI.** A factory assembly of one or more conductors insulated with a highly compressed refractory mineral insulation and enclosed in a liquidtight and gastight continuous copper or alloy steel sheath.

## II. Installation

### 332.10 Uses Permitted

Type MI cable shall be permitted as follows:

(1) For services, feeders, and branch circuits
(2) For power, lighting, control, and signal circuits
(3) In dry, wet, or continuously moist locations
(4) Indoors or outdoors
(5) Where exposed or concealed
(6) Where embedded in plaster, concrete, fill, or other masonry, whether above or below grade
(7) In hazardous (classified) locations where specifically permitted by other articles in this *Code*
(8) Where exposed to oil and gasoline
(9) Where exposed to corrosive conditions not deteriorating to its sheath
(10) In underground runs where suitably protected against physical damage and corrosive conditions
(11) In or attached to cable tray

Informational Note: The "Uses Permitted" is not an all-inclusive list.

### 332.12 Uses Not Permitted

Type MI cable shall not be used under the following conditions or in the following locations:

(1) In underground runs unless protected from physical damage, where necessary
(2) Where exposed to conditions that are destructive and corrosive to the metallic sheath, unless additional protection is provided

### 332.17 Through or Parallel to Framing Members

Type MI cable shall be protected in accordance with 300.4 where installed through or parallel to framing members.

### 332.24 Bending Radius

Bends in Type MI cable shall be so made that the cable will not be damaged. The radius of the inner edge of any bend shall not be less than required as follows:

(1) Five times the external diameter of the metallic sheath for cable not more than 19 mm (¾ in.) in external diameter
(2) Ten times the external diameter of the metallic sheath for cable greater than 19 mm (¾ in.) but not more than 25 mm (1 in.) in external diameter

The minimum bending radius is intended to prevent mechanical damage to the conductor insulation or the sheath that could result in cracking, a hot spot at the point of damage, or both. As illustrated in Exhibit 332.1, for cables with an external or outside diameter (OD) not greater than ¾ inch, the minimum bending radius ($R_{min.}$) is 5 times the cable OD. For cables greater than ¾ inch but not greater than 1 inch in diameter, the minimum bending radius is 10 times the cable OD.

### 332.30 Securing and Supporting

Type MI cable shall be supported and secured by staples, straps, hangers, or similar fittings, designed and installed so as not to damage the cable, at intervals not exceeding 1.8 m (6 ft).

For support requirements of vertical runs of fire-rated cable assemblies, see 300.19(B), including the associated commentary. Generally, the manufacturer's instructions (to achieve the desired fire rating) require more supports than for non–fire-rated installations.

**(A) Horizontal Runs Through Holes and Notches.** In other than vertical runs, cables installed in accordance with 300.4 shall be considered supported and secured where such support does not exceed 1.8 m (6 ft) intervals.

**(B) Unsupported Cable.** Type MI cable shall be permitted to be unsupported where the cable is fished between access points through concealed spaces in finished buildings or structures and supporting is impracticable.

**(C) Cable Trays.** All MI cable installed in cable trays shall comply with 392.30(A).

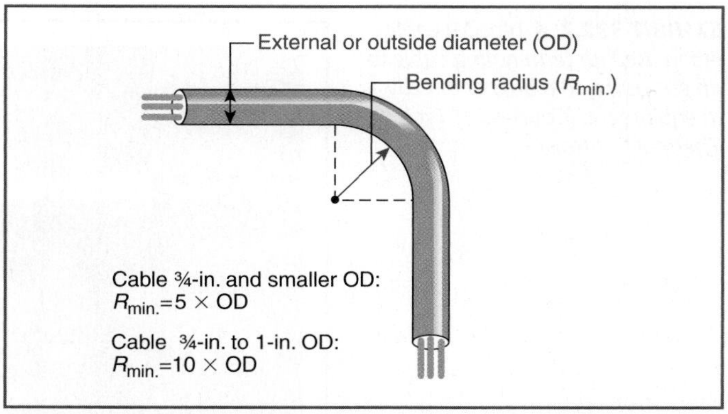


***EXHIBIT 332.1** An illustration of the bending radius in Type MI cable.*

## 332.31 Single Conductors

Where single-conductor cables are used, all phase conductors and, where used, the neutral conductor shall be grouped together to minimize induced voltage on the sheath.

The larger sizes of MI cable are available only as single-conductor cables. Because single conductors in a metal sheath can result in induced voltage on the sheath, this section requires all conductors of the circuit to be grouped together to minimize the voltage on the sheath.

Where single conductors enter a ferrous metal enclosure, inductive heating can occur due to hysteresis loss caused by the magnetic flux occurring in ferrous metals and $I^2R$ losses from the currents induced by the conductor. To minimize this magnetic heating of enclosures, 300.20 requires additional measures, including cutting slots in the metal between the individual holes for each conductor connector. Cable manufacturers offer nonferrous connecting plates that accept individual threaded connections of all circuit conductors, thereby eliminating circulating currents and fully complying with 300.20.

## 332.40 Boxes and Fittings

**(A) Fittings.** Fittings used for connecting Type MI cable to boxes, cabinets, or other equipment shall be identified for such use.

Terminations specifically investigated for use with this cable are listed as "mineral-insulated cable fittings." Fittings for use on MI cable are suitable for use at a maximum operating temperature of 90°C in dry locations and 60°C in wet locations. As shown in Exhibit 332.2, a complete box connector consists of a connector body and a screw-on potting fitting that may be used separately as an end fitting for change to open wiring. The screw-on potting fitting is assembled with a special tool and consists of a screw-on pot, insulating cap, insulating sleeving, anchoring bead, and sealing compound.

**(B) Terminal Seals.** Where Type MI cable terminates, an end seal fitting shall be installed immediately after stripping to prevent the entrance of moisture into the insulation. The conductors extending beyond the sheath shall be individually provided with an insulating material.

## 332.80 Ampacity

The ampacity of Type MI cable shall be determined in accordance with 310.15. The conductor temperature at the end seal fitting shall not exceed the temperature rating of the listed end seal fitting, and the installation shall not exceed the temperature ratings of terminations or equipment.

**(A) Type MI Cable Installed in Cable Tray.** The ampacities for Type MI cable installed in cable tray shall be determined in accordance with 392.80(A).

**(B) Single Type MI Conductors Grouped Together.** Where single Type MI conductors are grouped together in a triangular or square configuration, as required by 332.31, and installed on a messenger or exposed with a maintained free air space of not less than 2.15 times one conductor diameter (2.15 × O.D.) of the largest conductor contained within the configuration and adjacent conductor configurations or cables, the ampacity of the conductors shall not exceed the allowable ampacities of Table 310.15(B)(17).

Using Table 310.15(B)(16) or Table 310.15(B)(17) is the most common approach to determining the ampacities specified in 332.80(B). However, where MI cables are terminated at distribution and control equipment, such as circuit breakers, switchgear, and transfer switches, the temperature limitations of electrical equipment terminals must be coordinated with the ampacity of the cables. As stated in both the UL *Guide Information for Electrical Equipment – The White Book* and in 110.14(C)(1), unless equipment is listed and marked otherwise, conductor ampacities used in determining equipment terminations must be based on Table 310.15(B)(16) as appropriately modified by 310.15(B)(7). Where MI cable is constructed using nickel conductors or nickel-coated copper conductors, the allowable ampacities of Tables 310.15(B)(16) through 310.15(B)(18) do not apply, and the product manufacturer must be consulted for proper ampacities.

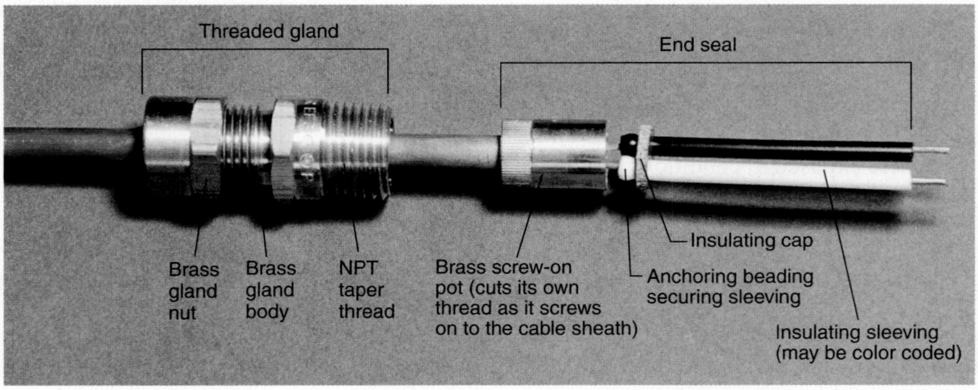


***EXHIBIT 332.2*** *A Type MI cable fitting used for terminating cable to an enclosure, to a box, or directly to equipment. (Courtesy of Tyco Thermal Controls)*

# III. Construction Specifications

## 332.104 Conductors

Type MI cable conductors shall be of solid copper, nickel, or nickel-coated copper with a resistance corresponding to standard AWG and kcmil sizes.

## 332.108 Equipment Grounding Conductor

Where the outer sheath is made of copper, it shall provide an adequate path to serve as an equipment grounding conductor. Where the outer sheath is made of steel, a separate equipment grounding conductor shall be provided.

The copper sheath of Type MI cable must be constructed as an EGC and is permitted to be used as an EGC according to 250.118(9), but an alloy steel outer sheath is not. For Type MI cables with an alloy steel outer sheath, one of the conductors (within the cable assembly) is required to be used for equipment grounding.

## 332.112 Insulation

The conductor insulation in Type MI cable shall be a highly compressed refractory mineral that provides proper spacing for all conductors.

## 332.116 Sheath

The outer sheath shall be of a continuous construction to provide mechanical protection and moisture seal.

# ARTICLE 334 Nonmetallic-Sheathed Cable: Types NM, NMC, and NMS

# I. General

## 334.1 Scope

This article covers the use, installation, and construction specifications of nonmetallic-sheathed cable.

Nonmetallic-sheathed cable was first recognized in the 1928 *NEC* as a substitute for concealed knob-and-tube wiring (Article 394) and open wiring on insulators (Article 398). The original advantages of nonmetallic-sheathed cable over knob-and-tube wiring were that the outer sheath provided continuous protection in addition to the insulation applied to the conductors; the cable was easily fished in partitions of finished buildings; no insulating supports were required; and only one hole needed to be bored, and that hole could accommodate more than one cable passing through a wood cross member.

## 334.2 Definitions

**Nonmetallic-Sheathed Cable.** A factory assembly of two or more insulated conductors enclosed within an overall nonmetallic jacket.

**Type NM.** Insulated conductors enclosed within an overall nonmetallic jacket.

**Type NMC.** Insulated conductors enclosed within an overall, corrosion resistant, nonmetallic jacket.

**Type NMS.** Insulated power or control conductors with signaling, data, and communications conductors within an overall nonmetallic jacket.

## 334.6 Listed

Type NM, Type NMC, and Type NMS cables shall be listed.

ANSI/UL 719, *Standard for Nonmetallic-Sheathed Cables*, requires a construction and performance evaluation, including testing related to flammability, dielectric voltage-withstand, unwinding at low temperatures, pulling through joists, conductor pullout, crushing, and abrasion.

# II. Installation

## 334.10 Uses Permitted

Type NM, Type NMC, and Type NMS cables shall be permitted to be used in the following, except as prohibited in 334.12:

(1) One- and two-family dwellings and their attached or detached garages, and their storage buildings.

The allowance to use of NM cable in garages of one- and two-family dwellings applies regardless of whether the garage is attached or detached. The allowance also applies to detached sheds or storage buildings associated with one- and two-family dwellings.

(2) Multi-family dwellings permitted to be of Types III, IV, and V construction.
(3) Other structures permitted to be of Types III, IV, and V construction. Cables shall be concealed within walls, floors, or ceilings that provide a thermal barrier of material that has at least a 15-minute finish rating as identified in listings of fire-rated assemblies.

Informational Note No. 1: Types of building construction and occupancy classifications are defined in NFPA 220-2012, *Standard on Types of Building Construction*, or the applicable building code, or both.

Informational Note No. 2: See Informative Annex E for determination of building types [NFPA 220, Table 3-1].

(4) Cable trays in structures permitted to be Types III, IV, or V where the cables are identified for the use.

Informational Note: See 310.15(A)(3) for temperature limitation of conductors.

(5) Types I and II construction where installed within raceways permitted to be installed in Types I and II construction.

A well-established means of codifying fire protection and fire safety requirements is to classify buildings by types of construction, based on materials used for the structural elements and the degree of fire resistance afforded by each element. The five fundamental construction types used by the model building codes are Type I (fire resistive), Type II (noncombustible), Type III (combination of combustible and noncombustible), Type IV (heavy timber), and Type V (wood frame). Types I and II basically require all structural elements to be noncombustible, whereas Types III, IV, and V allow some or all of the structural elements to be combustible (wood).

The selection of building construction types is regulated by the local building code, based on the occupancy, height, and area of the building. When a building of a selected height (in feet or stories above grade) and area is permitted to be built of combustible construction (i.e., Types III, IV, or V), the installation of nonmetallic-sheathed cable is permitted. The common areas (corridors) and incidental and subordinate uses (such as laundry rooms or lounge rooms) that serve a multifamily dwelling occupancy are also considered part of the multifamily occupancy, so NM cable is allowed in those areas.

If a building is to be of noncombustible construction (Type I or II) by the owner's choice, even though the building code would permit combustible construction, the building is allowed to be wired with NM cable. In such an instance, NM cable may be installed in the noncombustible building because the *Code* would have permitted the building to be of combustible construction.

Informative Annex E provides information on the types of construction as well as a table that cross references the five construction types to the types described in the model building codes.

Section 334.10(5) permits NM cables to be installed in Type I and Type II construction if the cables are within permitted raceways. A raceway is only permitted to be used if it complies with the article for the raceway and its use does not violate another article in the *Code*.

**(A) Type NM.** Type NM cable shall be permitted as follows:

(1) For both exposed and concealed work in normally dry locations except as prohibited in 334.10(3)
(2) To be installed or fished in air voids in masonry block or tile walls

For concealed work, cable should be installed where it is protected from physical damage often caused by nails or screws. Where practical, care should be taken to avoid areas where trim, door and window casings, baseboards, moldings, and so forth are likely to be nailed. See 300.4 for details on protection against physical damage.

**(B) Type NMC.** Type NMC cable shall be permitted as follows:

(1) For both exposed and concealed work in dry, moist, damp, or corrosive locations, except as prohibited by 334.10(3)
(2) In outside and inside walls of masonry block or tile
(3) In a shallow chase in masonry, concrete, or adobe protected against nails or screws by a steel plate at least 1.59 mm (1⁄16 in.) thick and covered with plaster, adobe, or similar finish

If NM cable is used in dairy barns and similar farm buildings (see Article 547), it must be Type NMC (corrosion resistant). The cable will be exposed to fumes, vapors, or liquids such as ammonia and barnyard acids. Under such circumstances, ordinary types of NM cable can deteriorate rapidly due to ammonia fumes or the growth of fungus or mold.

**(C) Type NMS.** Type NMS cable shall be permitted as follows:

(1) For both exposed and concealed work in normally dry locations except as prohibited by 334.10(3)
(2) To be installed or fished in air voids in masonry block or tile walls

## 334.12 Uses Not Permitted

The list of uses not permitted for NM cable is not complete. Restrictions exist elsewhere in the *Code*. For example, NM cables are not permitted to be installed in ducts, plenums, and other air-handling spaces. See 300.22 limiting the use of materials in ducts, plenums, and other air-handling spaces that may contribute smoke and products of combustion during a fire.

**(A) Types NM, NMC, and NMS.** Types NM, NMC, and NMS cables shall not be permitted as follows:

(1) In any dwelling or structure not specifically permitted in 334.10(1), (2), (3), and (5)
(2) Exposed in dropped or suspended ceilings in other than one- and two-family and multifamily dwellings

Nonmetallic-sheathed cables are prohibited in the space above hung ceilings that allow access. This requirement does not affect dwelling-type occupancies. The term *exposed*, as used in this requirement, closely follows the definition of *exposed (as applied to wiring methods)* found in Article 100, which states "on or attached to the surface or behind panels designed to allow access."

For example, cables installed above a dropped gypsum board ceiling or dropped gypsum board soffit would not be considered exposed cable, provided the area above the ceiling is not accessible (does not have removable tiles or does not contain an access panel). Often hung or dropped ceilings are accessible; therefore, cables installed above these types of ceilings would be considered exposed cables if the cables do not have additional physical protection.

A simple change to an architectural finish schedule during construction could change the acceptability of the wiring method. For example, if a corridor ceiling in an occupancy (other than a dwelling type) called for a painted gypsum board ceiling and the finish schedule changed the ceiling construction to a 2 foot by 2 foot accessible tile ceiling, the wiring method would no longer be permitted to be NM cable unless the cable was installed using additional protection. Examples of additional protection are found in 334.15(B).

(3) As service-entrance cable
(4) In commercial garages having hazardous (classified) locations as defined in 511.3
(5) In theaters and similar locations, except where permitted in 518.4(B)
(6) In motion picture studios
(7) In storage battery rooms
(8) In hoistways or on elevators or escalators
(9) Embedded in poured cement, concrete, or aggregate
(10) In hazardous (classified) locations, except where specifically permitted by other articles in this *Code*.

**(B) Types NM and NMS.** Types NM and NMS cables shall not be used under the following conditions or in the following locations:

(1) Where exposed to corrosive fumes or vapors
(2) Where embedded in masonry, concrete, adobe, fill, or plaster
(3) In a shallow chase in masonry, concrete, or adobe and covered with plaster, adobe, or similar finish
(4) In wet or damp locations

## 334.15 Exposed Work

In exposed work, except as provided in 300.11(A), cable shall be installed as specified in 334.15(A) through (C).

**(A) To Follow Surface.** Cable shall closely follow the surface of the building finish or of running boards.

**(B) Protection from Physical Damage.** Cable shall be protected from physical damage where necessary by rigid metal conduit, intermediate metal conduit, electrical metallic tubing, Schedule 80 PVC conduit, Type RTRC marked with the suffix -XW, or other approved means. Where passing through a floor, the cable shall be enclosed in rigid metal conduit, intermediate metal conduit, electrical metallic tubing, Schedule 80 PVC conduit, Type RTRC marked with the suffix -XW, or other approved means extending at least 150 mm (6 in.) above the floor.

Type NMC cable installed in shallow chases or grooves in masonry, concrete, or adobe shall be protected in accordance with the requirements in 300.4(F) and covered with plaster, adobe, or similar finish.

**(C) In Unfinished Basements and Crawl Spaces.** Where cable is run at angles with joists in unfinished basements and crawl spaces, it shall be permissible to secure cables not smaller than two 6 AWG or three 8 AWG conductors directly to the lower edges of the joists. Smaller cables shall be run either through bored holes in joists or on running boards. Nonmetallic-sheathed cable installed on the wall of an unfinished basement shall be permitted to be installed in a listed conduit or tubing or shall be protected in accordance with 300.4. Conduit or tubing shall be provided with a suitable insulating bushing or adapter at the point the cable enters the raceway. The sheath of the nonmetallic-sheathed cable shall extend through the conduit or tubing and into the outlet or device box not less than 6 mm (¼ in.). The cable shall be secured within 300 mm (12 in.) of the point where the cable enters the conduit or tubing. Metal conduit, tubing, and metal outlet boxes shall be connected to an equipment grounding conductor complying with the provisions of 250.86 and 250.148.

Crawl spaces pose dangers similar to those of unfinished basements and in some case are more dangerous due to limited height. The means of providing physical protection in crawl spaces and unfinished basements includes specific protection techniques. Where NMC is installed close to the surface in masonry, concrete, or adobe-type construction, physical protection must be afforded to the cable by using steel plate–type protectors as described in 300.4(F).

Nonmetallic-sheathed cables installed in an unfinished basement or crawl space can be run through holes in joists, attached to the side of joists or beams, and installed on running boards as shown on the left side in Exhibit 334.1. Section 300.4(D) requires

***EXHIBIT 334.1*** *Nonmetallic-sheathed cables and Type SE cables installed in an unfinished basement or crawl space.*

cables that run parallel to framing members be installed at least 1¼ inches from the nearest edge of studs, joists, or rafters.

## 334.17 Through or Parallel to Framing Members

Types NM, NMC, or NMS cable shall be protected in accordance with 300.4 where installed through or parallel to framing members. Grommets used as required in 300.4(B)(1) shall remain in place and be listed for the purpose of cable protection.

Cable that passes through factory- or field-punched holes in metal studs or similar members is required to be protected in accordance with 300.4(B)(1). Listed bushings or listed grommets covering all metal edges must be securely fastened in the opening before the cable is installed. See the commentary following 300.4(B)(1) for further information regarding physical protection of NM cables.

## 334.23 In Accessible Attics

The installation of cable in accessible attics or roof spaces shall also comply with 320.23.

## 334.24 Bending Radius

Bends in Types NM, NMC, and NMS cable shall be so made that the cable will not be damaged. The radius of the curve of the inner edge of any bend during or after installation shall not be less than five times the diameter of the cable.

## 334.30 Securing and Supporting

Nonmetallic-sheathed cable shall be supported and secured by staples, cable ties, straps, hangers, or similar fittings designed and installed so as not to damage the cable, at intervals not exceeding 1.4 m (4½ ft) and within 300 mm (12 in.) of every outlet box, junction box, cabinet, or fitting. Flat cables shall not be stapled on edge.

Sections of cable protected from physical damage by raceway shall not be required to be secured within the raceway.

Simply draping the cable over air ducts, rafters, timbers, joists, pipes, and ceiling grid members is not permitted, except where fished as allowed in 334.30(B)(1).

Two-conductor NM cable (or other flat configurations) is prohibited from being stapled on edge (that is, with its short dimension against a wood joist). When stapled in this manner, two cables are usually placed side by side under the staple. If the staple is driven too far into the stud, damage to the insulation and conductors could occur. See 300.4(C) for support requirements of cables through spaces behind panels designed to allow access.

**(A) Horizontal Runs Through Holes and Notches.** In other than vertical runs, cables installed in accordance with 300.4 shall be considered to be supported and secured where such support does not exceed 1.4-m (4½-ft) intervals and the nonmetallic-sheathed cable is securely fastened in place by an approved means within 300 mm (12 in.) of each box, cabinet, conduit body, or other nonmetallic-sheathed cable termination.

Informational Note: See 314.17(C) for support where nonmetallic boxes are used.

Cable running horizontally through framing members (spaced less than 54 inches apart) and passing through bored or punched holes in the framing members is considered to be supported by the framing members and does not require additional securing or fastening with cable ties. However, the cable must be secured within 12 inches of the outlet box. Where the cable terminates at a nonmetallic outlet box that does not contain a cable clamping device, the cable may be secured (fastened in place) within 8 inches of the outlet box, according to 314.17(C), Exception.

**(B) Unsupported Cables.** Nonmetallic-sheathed cable shall be permitted to be unsupported where the cable:

(1) Is fished between access points through concealed spaces in finished buildings or structures and supporting is impracticable.
(2) Is not more than 1.4 m (4½ ft) from the last point of cable support to the point of connection to a luminaire or other piece of electrical equipment and the cable and point of connection are within an accessible ceiling.

**(C) Wiring Device Without a Separate Outlet Box.** A wiring device identified for the use, without a separate outlet box, and incorporating an integral cable clamp shall be permitted where the cable is secured in place at intervals not exceeding 1.4 m (4½ ft) and within 300 mm (12 in.) from the wiring device wall opening, and there shall be at least a 300 mm (12 in.) loop of unbroken cable or 150 mm (6 in.) of a cable end available on the interior side of the finished wall to permit replacement.

## 334.40 Boxes and Fittings

**(A) Boxes of Insulating Material.** Nonmetallic outlet boxes shall be permitted as provided by 314.3.

Nonmetallic boxes and nonmetallic wiring systems are recommended for use in some corrosive atmospheres, such as farm buildings. See Article 547 for details.

**(B) Devices of Insulating Material.** Self-contained switches, self-contained receptacles, and nonmetallic-sheathed cable interconnector devices of insulating material that are listed shall be permitted to be used without boxes in exposed cable wiring and for repair wiring in existing buildings where the cable is concealed. Openings in such devices shall form a close fit around the outer covering of the cable, and the device shall fully enclose the part of the cable from which any part of the covering has been removed. Where connections to conductors are by binding-screw terminals, there shall be available as many terminals as conductors.

**(C) Devices with Integral Enclosures.** Wiring devices with integral enclosures identified for such use shall be permitted as provided by 300.15(E).

### 334.80 Ampacity

The ampacity of Types NM, NMC, and NMS cable shall be determined in accordance with 310.15. The allowable ampacity shall not exceed that of a 60°C (140°F) rated conductor. The 90°C (194°F) rating shall be permitted to be used for ampacity adjustment and correction calculations, provided the final derated ampacity does not exceed that of a 60°C (140°F) rated conductor. The ampacity of Types NM, NMC, and NMS cable installed in cable tray shall be determined in accordance with 392.80(A).

Where more than two NM cables containing two or more current-carrying conductors are installed, without maintaining spacing between the cables, through the same opening in wood framing that is to be sealed with thermal insulation, caulk, or sealing foam, the allowable ampacity of each conductor shall be adjusted in accordance with Table 310.15(B)(3)(a) and the provisions of 310.15(A)(2), Exception, shall not apply.

Where more than two NM cables containing two or more current-carrying conductors are installed in contact with thermal insulation without maintaining spacing between cables, the allowable ampacity of each conductor shall be adjusted in accordance with Table 310.15(B)(3)(a).

Section 310.15 is referenced for determining the ampacities of Types NM, NMC, and NMS cable. As stated in 310.15(B)(3)(a): ". . . where single conductors or multiconductor cables are installed without maintaining spacing for a continuous length longer than 600 millimeters (24 inches) and are not installed in raceways, the allowable ampacity of each conductor shall be reduced as shown in Table 310.15(B)(3)(a)." Failure to apply the appropriate ampacity adjustment factor called for by this table, where NM cables are stacked or bundled without maintaining spacing, can lead to overheating of conductors. The ampacity adjustment requirements prevent overheating of the conductors where passing though wood-framed draft- and fire-stopping material. The ampacity adjustment requirement for installations of more than two cables that are in direct contact with thermal insulation is intended to prevent overheating of the conductors. Not only is thermal insulation provided within structures to reduce heat loss or heat gain, but the same thermal insulation material may be used to control sound within structures as well.

**Calculation Example**

Four 2-conductor, size 12 AWG, copper with ground, Type NM cables are installed in direct contact with thermal insulation without maintaining spacing. Calculate the ampacity of the conductors according to the requirements of 334.80, and determine the maximum overcurrent protection permitted for the four circuits.

*Solution*

Step 1. Determine the number of current-carrying conductors.

$$4 \text{ cables} \times 2 \text{ conductors per cable} = 8 \text{ current-carrying conductors}$$

Step 2. Determine the initial conductor ampacity. Using the 90°C copper ampacity from Table 310.15(B)(16) for derating purposes, the initial ampacity of 12 AWG is 30 amperes.

Step 3. Determine the adjusted conductor ampacity. Due to the direct contact with thermal insulation, use Table 310.15(B)(3)(a). Eight current-carrying conductors require an adjustment factor of 70 percent.

$$30 \text{ A} \times 0.7 = 21 \text{ adjusted A}$$

Step 4. Determine the maximum permitted overcurrent device for each circuit. Section 334.80 does not allow an ampacity greater than given in the 60°C column of Table 310.15(B)(16). And, according to the footnote of Table 310.15(B)(16), conductor sizes of 14 AWG through 10 AWG must also comply with 240.4(D). Section 240.4(D) limits a 12 AWG copper conductor to a maximum of 20 amperes. Therefore, the 21 amperes of adjusted ampacity must be further reduced or limited by being protected by an overcurrent device not to exceed 20 amperes.

Conclusion: The final ampacity for each current-carrying conductor is 20 amperes, and the maximum overcurrent device permitted for each of the four circuits is 20 amperes.

This example points out that the 14 AWG to 10 AWG NM cable typically used for branch circuits can be installed without spacing and placed within thermal insulation with little impact on most installations. For similar installations, as long as the bundle is limited to not more than nine current-carrying conductors, the adjusted ampacity will not be below the small conductor limit set in 240.4(D).

## III. Construction Specifications

### 334.100 Construction

The outer cable sheath of nonmetallic-sheathed cable shall be a nonmetallic material.

### 334.104 Conductors

The 600-volt insulated conductors shall be sizes 14 AWG through 2 AWG copper conductors or sizes 12 AWG through 2 AWG aluminum or copper-clad aluminum conductors. The communications conductors shall comply with Part V of Article 800.

### 334.108 Equipment Grounding Conductor

In addition to the insulated conductors, the cable shall have an insulated, covered, or bare equipment grounding conductor.

### 334.112 Insulation

The insulated power conductors shall be one of the types listed in Table 310.104(A) that are suitable for branch-circuit wiring or one that is identified for use in these cables. Conductor insulation shall be rated at 90°C (194°F).

Informational Note: Types NM, NMC, and NMS cable identified by the markings NM-B, NMC-B, and NMS-B meet this requirement.

### 334.116 Sheath

The outer sheath of nonmetallic-sheathed cable shall comply with 334.116(A), (B), and (C).

**(A) Type NM.** The overall covering shall be flame retardant and moisture resistant.

**(B) Type NMC.** The overall covering shall be flame retardant, moisture resistant, fungus resistant, and corrosion resistant.

**(C) Type NMS.** The overall covering shall be flame retardant and moisture resistant. The sheath shall be applied so as to separate the power conductors from the communications conductors.

# ARTICLE 336 Power and Control Tray Cable: Type TC

## I. General

### 336.1 Scope

This article covers the use, installation, and construction specifications for power and control tray cable, Type TC.

The basic standard to investigate tray cable is ANSI/UL 1277, *Electrical Power and Control Tray Cables with Optional Optical-Fiber Members.* Summary information regarding listed power and control tray cable may be found in the UL *Guide Information for Electrical Equipment – The White Book,* under category QPOR.

### 336.2 Definition

**Power and Control Tray Cable, Type TC.** A factory assembly of two or more insulated conductors, with or without associated bare or covered grounding conductors, under a nonmetallic jacket.

## II. Installation

### 336.10 Uses Permitted

Type TC cable shall be permitted to be used as follows:

(1) For power, lighting, control, and signal circuits.
(2) In cable trays.
(3) In raceways.
(4) In outdoor locations supported by a messenger wire.
(5) For Class 1 circuits as permitted in Parts II and III of Article 725.
(6) For non–power-limited fire alarm circuits if conductors comply with the requirements of 760.49.

Type TC cable is permitted to be used for non–power-limited fire alarm circuits. According to 760.49, the cable must be listed and the conductor material must be copper. Aluminum and copper-clad aluminum conductors are not permitted for fire alarm circuits.

(7) In industrial establishments where the conditions of maintenance and supervision ensure that only qualified persons service the installation, and where the cable is continuously supported and protected against physical damage using mechanical protection, such as struts, angles, or channels, Type TC tray cable that complies with the crush and impact requirements of Type MC cable and is identified for such use with the marking Type TC–ER shall be permitted between a cable tray and the utilization equipment or device. The cable shall be secured at intervals not exceeding 1.8 m (6 ft). Equipment grounding for the utilization equipment shall be provided by an equipment grounding conductor within the cable. In cables containing conductors sized 6 AWG or smaller, the equipment grounding conductor shall be provided within the cable or, at the time of installation, one or more insulated conductors shall be permanently identified as an equipment grounding conductor in accordance with 250.119(B).

*Exception: Where not subject to physical damage, Type TC-ER shall be permitted to transition between cable trays and between cable trays and utilization equipment or devices for a distance not to exceed 1.8 m (6 ft) without continuous support. The cable shall be mechanically supported where exiting the cable tray to ensure that the minimum bending radius is not exceeded.*

Specific types of tray cable used in qualifying occupancies are permitted to extend from a cable tray to a piece of equipment without the use of conduit. According to UL 1277, *Electrical Power and Control Tray Cables with Optional Optical-Fiber Members,* cables suitable for use as exposed wiring between cable tray and the utilization equipment according to this requirement are surface-marked "Type TC-ER" (tray cable for exposed runs).

The exception permits TC-ER cable to exit a cable tray without continuous support for a maximum of 6 feet. However, extension from cable tray to cable tray or from cable tray to equipment is not allowed beyond 6 feet without continuous support.

(8) Where installed in wet locations, Type TC cable shall also be resistant to moisture and corrosive agents.

Informational Note: See 310.15(A)(3) for temperature limitation of conductors.

### 336.12 Uses Not Permitted

Type TC tray cable shall not be installed or used as follows:

(1) Installed where it will be exposed to physical damage
(2) Installed outside a raceway or cable tray system, except as permitted in 336.10(4) and 336.10(7)
(3) Used where exposed to direct rays of the sun, unless identified as sunlight resistant
(4) Direct buried, unless identified for such use

### 336.24 Bending Radius

Bends in Type TC cable shall be made so as not to damage the cable. For Type TC cable without metal shielding, the minimum bending radius shall be as follows:

(1) Four times the overall diameter for cables 25 mm (1 in.) or less in diameter
(2) Five times the overall diameter for cables larger than 25 mm (1 in.) but not more than 50 mm (2 in.) in diameter
(3) Six times the overall diameter for cables larger than 50 mm (2 in.) in diameter

Type TC cables with metallic shielding shall have a minimum bending radius of not less than 12 times the cable overall diameter.

### 336.80 Ampacity

The ampacity of Type TC tray cable shall be determined in accordance with 392.80(A) for 14 AWG and larger conductors, in accordance with 402.5 for 18 AWG through 16 AWG conductors where installed in cable tray, and in accordance with 310.15 where installed in a raceway or as messenger-supported wiring.

## III. Construction Specifications

### 336.100 Construction

A metallic sheath or armor as defined in 330.116 shall not be permitted either under or over the nonmetallic jacket. Metallic shield(s) shall be permitted over groups of conductors, under the outer jacket, or both.

Type TC cables are permitted to have one or more metal shields but not a metal sheath or armor. Electrical cables with a metal sheath or armor are covered in either Article 320 as armored cable, Type AC, or Article 330 as metal-clad cable, Type MC.

### 336.104 Conductors

The insulated conductors of Type TC cables shall be in sizes 18 AWG to 1000 kcmil copper, nickel, or nickel-coated copper, and sizes 12 AWG through 1000 kcmil aluminum or copper-clad aluminum. Insulated conductors of sizes 14 AWG, and larger copper, nickel, or nickel-coated copper, and sizes 12 AWG through 1000 kcmil aluminum or copper-clad aluminum shall be one of the types listed in Table 310.104(A) or Table 310.104(B) that is suitable for branch circuit and feeder circuits or one that is identified for such use.

The use of nickel and nickel-coated copper conductor material is generally permitted by 310.106(B) and shown in Table 310.104(A) for insulation types PFAH and TFE (and used in high-temperature applications).

**(A) Fire Alarm Systems.** Where used for fire alarm systems, conductors shall also be in accordance with 760.49.

**(B) Thermocouple Circuits.** Conductors in Type TC cable used for thermocouple circuits in accordance with Part III of Article 725 shall also be permitted to be any of the materials used for thermocouple extension wire.

**(C) Class 1 Circuit Conductors.** Insulated conductors of 18 AWG and 16 AWG copper shall also be in accordance with 725.49.

### 336.116 Jacket

The outer jacket shall be a flame-retardant, nonmetallic material.

### 336.120 Marking

There shall be no voltage marking on a Type TC cable employing thermocouple extension wire.

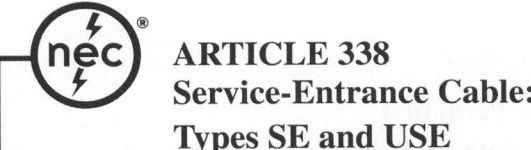

# ARTICLE 338 Service-Entrance Cable: Types SE and USE

## I. General

### 338.1 Scope

This article covers the use, installation, and construction specifications of service-entrance cable.

According to the UL *Guide Information for Electrical Equipment – The White Book*, category TYLZ cable (service-entrance cable rated 600 volts) is listed in sizes 14 AWG and larger for copper and 12 AWG and larger for aluminum or copper-clad aluminum. Type SE cable contains Types RHW, RHW-2, XHHW, XHHW-2, THWN, or THWN-2 conductors. Type USE cable contains conductors with insulation equivalent to RHW or XHHW. Type USE-2 contains insulation equivalent to RHW-2 or XHHW-2 and is rated 90°C, wet or dry.

The type designation of the conductors may be marked on the cable surface. When used, this marking indicates the temperature rating for the cable corresponding to the temperature rating of the conductors. When this marking does not appear, the temperature rating of the cable is 75°C. The cables are designated as Type SE, Type USE or USE-2, and submersible water pump cable.

*Type SE* — Cable suitable for aboveground installations. Both the insulated conductors and the outer jacket are suitable for use where exposed to sun.

*Type USE or USE-2* — Cable suitable for underground installations, including direct burial. Although both the conductor insulation and outer covering are suitable for use where exposed to sun, the cables are not suitable inside the premises nor aboveground other than to terminate at service or metering equipment.

*Submersible water pump cable* — A multiconductor cable containing two, three, or four single-conductor, Type USE or USE-2 cables in a flat or twisted assembly. The cable is tag-marked "For use within the well casing for wiring deep-well water pumps where the cable is not subject to repetitive handling caused by frequent servicing of the pump units." The cable may be directly buried.

### 338.2 Definitions

**Service-Entrance Cable.** A single conductor or multiconductor assembly provided with or without an overall covering, primarily used for services, and of the following types:

*Type SE.* Service-entrance cable having a flame-retardant, moisture-resistant covering.

*Type USE.* Service-entrance cable, identified for underground use, having a moisture-resistant covering, but not required to have a flame-retardant covering.

## II. Installation

### 338.10 Uses Permitted

**(A) Service-Entrance Conductors.** Service-entrance cable shall be permitted to be used as service-entrance conductors and shall be installed in accordance with 230.6, 230.7, and Parts II, III, and IV of Article 230.

**(B) Branch Circuits or Feeders.**

**(1) Grounded Conductor Insulated.** Type SE service-entrance cables shall be permitted in wiring systems where all of the circuit conductors of the cable are of the thermoset or thermoplastic type.

Branch circuits using Type SE cable as a wiring method are permitted only if all circuit conductors within the cable are fully insulated according to 310.104. The EGC is the only conductor permitted to be bare or covered within Type SE cable used for branch circuits.

**(2) Use of Uninsulated Conductor.** Type SE service-entrance cable shall be permitted for use where the insulated conductors are used for circuit wiring and the uninsulated conductor is used only for equipment grounding purposes.

*Exception: In existing installations, uninsulated conductors shall be permitted as a grounded conductor in accordance with 250.32 and 250.140, where the uninsulated grounded conductor of the cable originates in service equipment, and with 225.30 through 225.40.*

Service-entrance cable containing a bare grounded (neutral) conductor is not permitted for new installations where it is used as a branch circuit to supply appliances such as ranges, wall-mounted ovens, counter-mounted cooking units, or clothes dryers. The exception permits a bare neutral service-entrance cable for existing installations only and is coordinated with the sections listed.

**(3) Temperature Limitations.** Type SE service-entrance cable used to supply appliances shall not be subject to conductor temperatures in excess of the temperature specified for the type of insulation involved.

**(4) Installation Methods for Branch Circuits and Feeders.**

(a) *Interior Installations.* In addition to the provisions of this article, Type SE service-entrance cable used for interior wiring shall comply with the installation requirements of Part II of Article 334, excluding 334.80.

Where installed in thermal insulation the ampacity shall be in accordance with the 60°C (140°F) conductor temperature rating. The maximum conductor temperature rating shall be permitted to be used for ampacity adjustment and correction purposes, if the final derated ampacity does not exceed that for a 60°C (140°F) rated conductor.

Informational Note No. 1: See 310.15(A)(3) for temperature limitation of conductors.

Informational Note No. 2: For the installation of main power feeder conductors in dwelling units refer to 310.15(B)(7).

Type SE cable is permitted as an interior branch circuit or feeder wiring method and where used in this manner, the installation is to comply with the requirements in Part II of Article 334 (Type NM cable) except for the ampacity rules in 334.80. The different configurations of Type SE cable, including SEU and SER, are used for a variety of interior circuits including ranges, clothes dryers, heating, and air-conditioning equipment and as feeders to supply panelboards that are not the service equipment.

While all conductors in nonmetallic-sheathed cable are required to have 90°C-rated insulation, 334.80 limits the operating (calculated load) ampacity to those contained in the 60°C column of Table 310.15(B)(16). This limitation applies to all uses of NM cable. In contrast to this restriction, Type SE cable is only limited to operating at a 60°C ampacity when it is installed in thermal insulation. Type SE cable is permitted to have conductors with either 75°C or 90°C insulation. If the cable surface has no temperature marking, the conductors have a 75°C insulation temperature rating, and ampacity adjustment or correction is based on that rating. If the cable is marked with a conductor insulation temperature rating, ampacity adjustment or correction of the conductor can be made based on the conductor temperature marked on the cable.

Where the cable is installed in thermal insulation, the adjusted and/or corrected operating ampacity of the conductors cannot exceed those contained in the 60°C column of Table 310.15(B)(16). If the cable is not installed in thermal insulation, the limiting factor on conductor ampacity is the requirement for coordinating the operating ampacity of the conductor with the terminal temperature ratings of electrical equipment contained in 110.14(C).

(b) *Exterior Installations.* In addition to the provisions of this article, service-entrance cable used for feeders or branch circuits, where installed as exterior wiring, shall be installed in accordance with Part I of Article 225. The cable shall be supported in accordance with 334.30. Type USE cable installed as underground feeder and branch circuit cable shall comply with Part II of Article 340.

*Exception: Single-conductor Type USE and multi-rated USE conductors shall not be subject to the ampacity limitations of Part II of Article 340.*

### 338.12 Uses Not Permitted

**(A) Service-Entrance Cable.** Service-entrance cable (SE) shall not be used under the following conditions or in the following locations:

(1) Where subject to physical damage unless protected in accordance with 230.50(B)
(2) Underground with or without a raceway
(3) For exterior branch circuits and feeder wiring unless the installation complies with the provisions of Part I of Article 225 and is supported in accordance with 334.30 or is used as messenger-supported wiring as permitted in Part II of Article 396

**(B) Underground Service-Entrance Cable.** Underground service-entrance cable (USE) shall not be used under the following conditions or in the following locations:

(1) For interior wiring
(2) For aboveground installations except where USE cable emerges from the ground and is terminated in an enclosure at an outdoor location and the cable is protected in accordance with 300.5(D)
(3) As aerial cable unless it is a multiconductor cable identified for use aboveground and installed as messenger-supported wiring in accordance with 225.10 and Part II of Article 396

Cables marked only as "Type USE service-entrance cable" are not required to have a flame-retardant covering, according to 338.2.

### 338.24 Bending Radius

Bends in Types USE and SE cable shall be so made that the cable will not be damaged. The radius of the curve of the inner edge of any bend, during or after installation, shall not be less than five times the diameter of the cable.

## III. Construction

### 338.100 Construction

Cabled, single-conductor, Type USE constructions recognized for underground use shall be permitted to have a bare copper conductor cabled with the assembly. Type USE single, parallel, or cabled conductor assemblies recognized for underground use shall be permitted to have a bare copper concentric conductor applied. These constructions shall not require an outer overall covering.

Informational Note: See 230.41, Exception, item (2), for directly buried, uninsulated service-entrance conductors.

Type SE or USE cable containing two or more conductors shall be permitted to have one conductor uninsulated.

### 338.120 Marking

Service-entrance cable shall be marked as required in 310.120. Cable with the neutral conductor smaller than the ungrounded conductors shall be so marked.

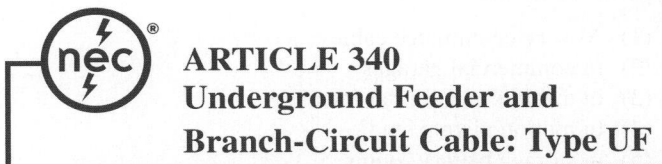

# ARTICLE 340 Underground Feeder and Branch-Circuit Cable: Type UF

## I. General

### 340.1 Scope

This article covers the use, installation, and construction specifications for underground feeder and branch-circuit cable, Type UF.

### 340.2 Definition

**Underground Feeder and Branch-Circuit Cable, Type UF.** A factory assembly of one or more insulated conductors with an integral or an overall covering of nonmetallic material suitable for direct burial in the earth.

### 340.6 Listing Requirements

Type UF cable shall be listed.

## II. Installation

### 340.10 Uses Permitted

Type UF cable shall be permitted as follows:

(1) For use underground, including direct burial in the earth. For underground requirements, see 300.5.
(2) As single-conductor cables. Where installed as single-conductor cables, all conductors of the feeder grounded conductor or branch circuit, including the grounded

conductor and equipment grounding conductor, if any, shall be installed in accordance with 300.3.

(3) For wiring in wet, dry, or corrosive locations under the recognized wiring methods of this *Code*.

(4) Installed as nonmetallic-sheathed cable. Where so installed, the installation and conductor requirements shall comply with Parts II and III of Article 334 and shall be of the multiconductor type.

If Type UF cable is installed as nonmetallic-sheathed cable, the installation and construction requirements of Article 334 for Type NM cable apply, including the ampacity adjustment in 334.80.

(5) For solar photovoltaic systems in accordance with 690.31.

(6) As single-conductor cables as the nonheating leads for heating cables as provided in 424.43.

(7) Supported by cable trays. Type UF cable supported by cable trays shall be of the multiconductor type.

Informational Note: See 310.15(A)(3) for temperature limitation of conductors.

### 340.12 Uses Not Permitted

Type UF cable shall not be used as follows:

(1) As service-entrance cable
(2) In commercial garages
(3) In theaters and similar locations
(4) In motion picture studios
(5) In storage battery rooms
(6) In hoistways or on elevators or escalators
(7) In hazardous (classified) locations, except as specifically permitted by other articles in this *Code*
(8) Embedded in poured cement, concrete, or aggregate, except where embedded in plaster as nonheating leads where permitted in 424.43
(9) Where exposed to direct rays of the sun, unless identified as sunlight resistant
(10) Where subject to physical damage
(11) As overhead cable, except where installed as messenger-supported wiring in accordance with Part II of Article 396

Type UF cable suitable for exposure to the direct rays of the sun is tagged and marked with the designation "Sunlight Resistant." This physical protection requirement ensures that Type UF cable, as it emerges from underground, is protected from UV damage.

### 340.24 Bending Radius

Bends in Type UF cable shall be so made that the cable is not damaged. The radius of the curve of the inner edge of any bend shall not be less than five times the diameter of the cable.

### 340.80 Ampacity

The ampacity of Type UF cable shall be that of 60°C (140°F) conductors in accordance with 310.15.

If Type UF cable is installed as nonmetallic-sheathed cable, the ampacity of the cable is determined according to rules for Type NM cable in 334.80. See 340.10(4).

## III. Construction Specifications

### 340.104 Conductors

The conductors shall be sizes 14 AWG copper or 12 AWG aluminum or copper-clad aluminum through 4/0 AWG.

### 340.108 Equipment Grounding Conductor

In addition to the insulated conductors, the cable shall be permitted to have an insulated or bare equipment grounding conductor.

### 340.112 Insulation

The conductors of Type UF shall be one of the moisture-resistant types listed in Table 310.104(A) that is suitable for branch-circuit wiring or one that is identified for such use. Where installed as a substitute wiring method for NM cable, the conductor insulation shall be rated 90°C (194°F).

### 340.116 Sheath

The overall covering shall be flame retardant; moisture, fungus, and corrosion resistant; and suitable for direct burial in the earth.

# ARTICLE 342 Intermediate Metal Conduit: Type IMC

## I. General

### 342.1 Scope

This article covers the use, installation, and construction specifications for intermediate metal conduit (IMC) and associated fittings.

IMC is thinner-walled and lighter in weight than RMC and is satisfactory for uses in all locations where RMC is permitted to be used. Threaded fittings, couplings, connectors, and so forth are interchangeable between IMC and RMC. Threadless fittings for IMC are suitable only for the type of conduit indicated by the carton marking.

### 342.2 Definition

**Intermediate Metal Conduit (IMC).** A steel threadable raceway of circular cross section designed for the physical protection and routing of conductors and cables and for use as an equipment grounding conductor when installed with its integral or associated coupling and appropriate fittings.

### 342.6 Listing Requirements

IMC, factory elbows and couplings, and associated fittings shall be listed.

## II. Installation

### 342.10 Uses Permitted

**(A) All Atmospheric Conditions and Occupancies.** Use of IMC shall be permitted under all atmospheric conditions and occupancies.

**(B) Corrosion Environments.** IMC, elbows, couplings, and fittings shall be permitted to be installed in concrete, in direct contact with the earth, or in areas subject to severe corrosive influences where protected by corrosion protection and judged suitable for the condition.

**(C) Cinder Fill.** IMC shall be permitted to be installed in or under cinder fill where subject to permanent moisture where protected on all sides by a layer of noncinder concrete not less than 50 mm (2 in.) thick; where the conduit is not less than 450 mm (18 in.) under the fill; or where protected by corrosion protection and judged suitable for the condition.

**(D) Wet Locations.** All supports, bolts, straps, screws, and so forth, shall be of corrosion-resistant materials or protected against corrosion by corrosion-resistant materials.

Informational Note: See 300.6 for protection against corrosion.

Galvanized IMC installed in concrete does not require supplementary corrosion protection. Similarly, galvanized IMC installed in contact with soil does not generally require supplementary corrosion protection. As a guide in the absence of experience with the corrosive effects of soil in a specific location, soils producing severe corrosive effects are generally characterized by low resistivity of less than 2000 ohm-cm.

Wherever ferrous metal conduit runs directly from concrete encasement to soil burial, the metal in contact with the soil can be severely corroded.

### 342.14 Dissimilar Metals

Where practicable, dissimilar metals in contact anywhere in the system shall be avoided to eliminate the possibility of galvanic action.

Aluminum fittings and enclosures shall be permitted to be used with IMC.

### 342.20 Size

**(A) Minimum.** IMC smaller than metric designator 16 (trade size ½) shall not be used.

**(B) Maximum.** IMC larger than metric designator 103 (trade size 4) shall not be used.

Informational Note: See 300.1(C) for the metric designators and trade sizes. These are for identification purposes only and do not relate to actual dimensions.

For further explanation of metric designators, see 90.9 and the commentary following Table 300.1(C).

### 342.22 Number of Conductors

The number of conductors shall not exceed that permitted by the percentage fill specified in Table 1, Chapter 9.

Cables shall be permitted to be installed where such use is not prohibited by the respective cable articles. The number of cables shall not exceed the allowable percentage fill specified in Table 1, Chapter 9.

Table 4 of Chapter 9 provides the usable area within the selected conduit or tubing, and Table 5 provides the required area for each conductor. Examples using these tables to calculate a conduit or tubing size are provided in the commentary following Chapter 9, Notes to Tables, Note 6.

If the conductors are of the same wire size, Informative Annex C can be used instead of performing calculations. This annex, through 12 sets of tables, accurately indicates the maximum number of conductors permitted in conduit or tubing. Examples using Informative Annex C to select a conduit or tubing size are provided in the commentary following the annex introduction.

To select the proper trade size IMC, the section entitled "Article 342 – Intermediate Metal Conduit (IMC)" in Table 4 of Chapter 9 should be followed. Use of Tables C.4 and C.4(A) for IMC in Informative Annex C is also permissible.

### 342.24 Bends — How Made

Bends of IMC shall be so made that the conduit will not be damaged and the internal diameter of the conduit will not be effectively reduced. The radius of the curve of any field bend to the centerline of the conduit shall not be less than indicated in Table 2, Chapter 9.

### 342.26 Bends — Number in One Run

There shall not be more than the equivalent of four quarter bends (360 degrees total) between pull points, for example, conduit bodies and boxes.

The number of bends in one run is limited to reduce pulling tension on conductors and helps ensure easy insertion or removal of conductors during later phases of construction, when the conduit may be permanently enclosed by the building's finish.

### 342.28 Reaming and Threading

All cut ends shall be reamed or otherwise finished to remove rough edges. Where conduit is threaded in the field, a standard cutting die with a taper of 1 in 16 (¾ in. taper per foot) shall be used.

Informational Note: See ANSI/ASME B.1.20.1-1983, *Standard for Pipe Threads, General Purpose (Inch)*.

Conduit is cut using a saw or a roll cutter (pipe cutter). Care should be taken to ensure a straight cut, given that crooked threads

*EXHIBIT 342.1 Minimum IMC fastening requirements.*

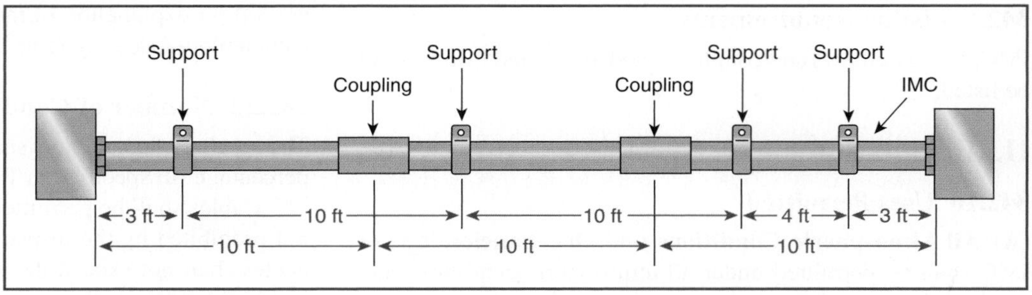


result from a die not started on the pipe squarely. After the cut is made, the conduit must be reamed. Proper reaming removes burrs from the interior of the cut conduit so that as wires and cables are pulled through the conduit, no chafing of the insulation or cable jacket occurs. Finally, the conduit is threaded. The number of threads is important, because cutting too many threads prevents a conduit from being made up properly. If a threaded ring gauge is not available, the same number of threads should be cut on the conduit as are present on the factory (threaded) end of the conduit.

## 342.30 Securing and Supporting

IMC shall be installed as a complete system in accordance with 300.18 and shall be securely fastened in place and supported in accordance with 342.30(A) and (B).

**(A) Securely Fastened.** IMC shall be secured in accordance with one of the following:

(1) IMC shall be securely fastened within 900 mm (3 ft) of each outlet box, junction box, device box, cabinet, conduit body, or other conduit termination.
(2) Where structural members do not readily permit fastening within 900 mm (3 ft), fastening shall be permitted to be increased to a distance of 1.5 m (5 ft).
(3) Where approved, conduit shall not be required to be securely fastened within 900 mm (3 ft) of the service head for above-the-roof termination of a mast.

As illustrated in Exhibit 342.1, IMC is required to be securely fastened within 3 feet of outlet boxes, junction boxes, cabinets, conduit bodies, or other conduit terminations. Couplings are not considered conduit terminations. However, where structural support members do not permit fastening within 3 feet, the support may be located up to 5 feet away. In addition, IMC is required to be supported at least every 10 feet unless permitted otherwise by 342.30(B).

**(B) Supports.** IMC shall be supported in accordance with one of the following:

(1) Conduit shall be supported at intervals not exceeding 3 m (10 ft).
(2) The distance between supports for straight runs of conduit shall be permitted in accordance with Table 344.30(B)(2), provided the conduit is made up with threaded couplings and such supports prevent transmission of stresses to termination where conduit is deflected between supports.
(3) Exposed vertical risers from industrial machinery or fixed equipment shall be permitted to be supported at intervals not exceeding 6 m (20 ft) if the conduit is made up with threaded couplings, the conduit is supported and securely fastened at the top and bottom of the riser, and no other means of intermediate support is readily available.
(4) Horizontal runs of IMC supported by openings through framing members at intervals not exceeding 3 m (10 ft) and securely fastened within 900 mm (3 ft) of termination points shall be permitted.

Lengths of IMC are permitted to be supported (but not necessarily secured) by framing members at 10-foot intervals, provided the IMC is secured and supported at least 3 feet from the box or enclosure. Exhibit 342.2 illustrates an installation where the IMC is installed through the bar joists.

## 342.42 Couplings and Connectors

**(A) Threadless.** Threadless couplings and connectors used with conduit shall be made tight. Where buried in masonry or concrete, they shall be the concretetight type. Where installed in wet locations, they shall comply with 314.15. Threadless couplings and

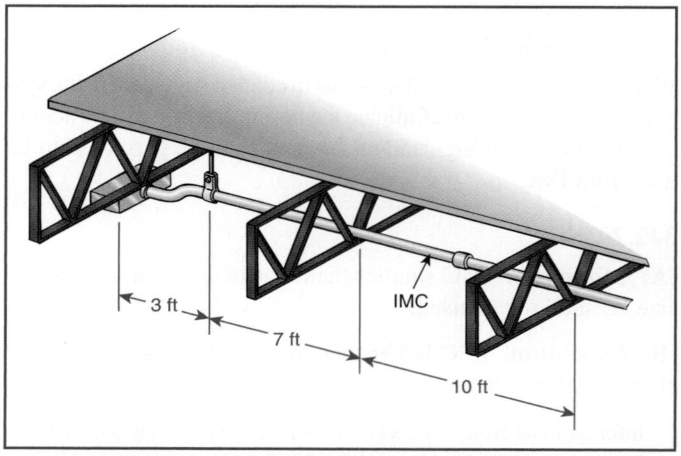


*EXHIBIT 342.2 An example of IMC supported by framing members and securely fastened at the 3-foot distance from the box.*

connectors shall not be used on threaded conduit ends unless listed for the purpose.

See the commentary following 344.42(A) for examples of threadless fittings.

**(B) Running Threads.** Running threads shall not be used on conduit for connection at couplings.

### 342.46 Bushings

Where a conduit enters a box, fitting, or other enclosure, a bushing shall be provided to protect the wires from abrasion unless the box, fitting, or enclosure is designed to provide such protection.

Informational Note: See 300.4(G) for the protection of conductors 4 AWG and larger at bushings.

### 342.56 Splices and Taps

Splices and taps shall be made in accordance with 300.15.

### 342.60 Grounding

IMC shall be permitted as an equipment grounding conductor.

## III. Construction Specifications

### 342.120 Marking

Each length shall be clearly and durably marked at least every 1.5 m (5 ft) with the letters IMC. Each length shall be marked as required in 110.21.

### 342.130 Standard Lengths

The standard length of IMC shall be 3.05 m (10 ft), including an attached coupling, and each end shall be threaded. Longer or shorter lengths with or without coupling and threaded or unthreaded shall be permitted.

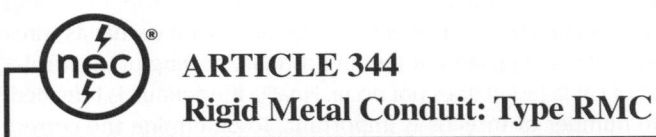

# ARTICLE 344 Rigid Metal Conduit: Type RMC

## I. General

### 344.1 Scope

This article covers the use, installation, and construction specifications for rigid metal conduit (RMC) and associated fittings.

### 344.2 Definition

**Rigid Metal Conduit (RMC).** A threadable raceway of circular cross section designed for the physical protection and routing of conductors and cables and for use as an equipment grounding conductor when installed with its integral or associated coupling and appropriate fittings.

### 344.6 Listing Requirements

RMC, factory elbows and couplings, and associated fittings shall be listed.

## II. Installation

### 344.10 Uses Permitted

**(A) Atmospheric Conditions and Occupancies.**

**(1) Galvanized Steel and Stainless Steel RMC.** Galvanized steel and stainless steel RMC shall be permitted under all atmospheric conditions and occupancies.

**(2) Red Brass RMC.** Red brass RMC shall be permitted to be installed for direct burial and swimming pool applications.

**(3) Aluminum RMC.** Aluminum RMC shall be permitted to be installed where judged suitable for the environment. Rigid aluminum conduit encased in concrete or in direct contact with the earth shall be provided with approved supplementary corrosion protection.

**(4) Ferrous Raceways and Fittings.** Ferrous raceways and fittings protected from corrosion solely by enamel shall be permitted only indoors and in occupancies not subject to severe corrosive influences.

**(B) Corrosive Environments.**

The AHJ for enforcing the *Code* should be consulted for approval of corrosion-resistant materials or for requirements prior to the installation of nonferrous metal (aluminum) conduit in concrete, since chloride additives in the concrete mix may cause corrosion.

**(1) Galvanized Steel, Stainless Steel, and Red Brass RMC, Elbows, Couplings, and Fittings.** Galvanized steel, stainless steel, and red brass RMC elbows, couplings, and fittings shall be permitted to be installed in concrete, in direct contact with the earth, or in areas subject to severe corrosive influences where protected by corrosion protection and judged suitable for the condition.

**(2) Supplementary Protection of Aluminum RMC.** Aluminum RMC shall be provided with approved supplementary corrosion protection where encased in concrete or in direct contact with the earth.

**(C) Cinder Fill.** Galvanized steel, stainless steel, and red brass RMC shall be permitted to be installed in or under cinder fill where subject to permanent moisture where protected on all sides by a layer of noncinder concrete not less than 50 mm (2 in.) thick; where the conduit is not less than 450 mm (18 in.) under the fill; or where protected by corrosion protection and judged suitable for the condition.

Although cinder fill is not commonly used in modern construction, it is still encountered at older building sites. Cinders used as fill may contain sulfur, and when they combine with moisture, sulfuric acid is formed, which can corrode metal raceways.

**(D) Wet Locations.** All supports, bolts, straps, screws, and so forth, shall be of corrosion-resistant materials or protected against corrosion by corrosion-resistant materials.

Informational Note: See 300.6 for protection against corrosion.

### 344.14 Dissimilar Metals

Where practicable, dissimilar metals in contact anywhere in the system shall be avoided to eliminate the possibility of galvanic action. Aluminum fittings and enclosures shall be permitted to be used with steel RMC, and steel fittings and enclosures shall be permitted to be used with aluminum RMC where not subject to severe corrosive influences.

Aluminum rigid conduit can be used with steel fittings and enclosures, as can aluminum fittings and enclosures with steel rigid conduit. Tests show that the galvanic corrosion at steel and aluminum interfaces is minor compared to the natural corrosion on the combination of steel and steel or of aluminum and aluminum.

### 344.20 Size

**(A) Minimum.** RMC smaller than metric designator 16 (trade size ½) shall not be used.

*Exception: For enclosing the leads of motors as permitted in 430.245(B).*

**(B) Maximum.** RMC larger than metric designator 155 (trade size 6) shall not be used.

Informational Note: See 300.1(C) for the metric designators and trade sizes. These are for identification purposes only and do not relate to actual dimensions.

For further explanation of metric designators, see 90.9 and the commentary following Table 300.1(C).

### 344.22 Number of Conductors

The number of conductors shall not exceed that permitted by the percentage fill specified in Table 1, Chapter 9.

Cables shall be permitted to be installed where such use is not prohibited by the respective cable articles. The number of cables shall not exceed the allowable percentage fill specified in Table 1, Chapter 9.

Table 4 of Chapter 9 provides the usable area within the selected conduit or tubing, and Table 5 provides the required area for each conductor. Examples using these tables to calculate a conduit or tubing size are provided in the commentary following Chapter 9, Notes to Tables, Note 6.

If the conductors are of the same wire size, Informative Annex C can be used instead of performing the calculations. This annex, through 12 sets of tables, accurately indicates the maximum number of conductors permitted in conduit or tubing. Examples using Informative Annex C to select a conduit or tubing size are provided in the commentary following the annex introduction.

To select the proper trade size of RMC, see "Article 344 – Rigid Metal Conduit (RMC)" in Table 4 of Chapter 9. Use of Tables C.8 and C.8(A) for RMC in Informative Annex C is also permissible.

### 344.24 Bends — How Made

Bends of RMC shall be so made that the conduit will not be damaged and so that the internal diameter of the conduit will not be effectively reduced. The radius of the curve of any field bend to the centerline of the conduit shall not be less than indicated in Table 2, Chapter 9.

### 344.26 Bends — Number in One Run

There shall not be more than the equivalent of four quarter bends (360 degrees total) between pull points, for example, conduit bodies and boxes.

Limiting the number of bends in a conduit run reduces pulling tension on conductors. It also helps ensure easy insertion or removal of conductors during later phases of construction when the conduit may be permanently enclosed by the finish of the building.

### 344.28 Reaming and Threading

All cut ends shall be reamed or otherwise finished to remove rough edges. Where conduit is threaded in the field, a standard cutting die with a 1 in 16 taper (¾ in. taper per foot) shall be used.

Inforzmational Note: See ANSI/ASME B.1.20.1-1983, *Standard for Pipe Threads, General Purpose (Inch).*

Conduit is cut using a saw or a roll cutter (pipe cutter). Crooked threads result from a die not started on the pipe squarely. After the cut is made, the conduit must be reamed. Proper reaming removes burrs from the interior of the cut conduit so that as wires and cables are pulled through the conduit, chafing of the insulation or cable jacket does not occur. Finally, the conduit is threaded. The number of threads is important. To determine the correct number of threads for a conduit end, the same number of threads should be cut on the conduit as are present on the factory (threaded) end of the conduit. Where excessive threads are cut on the conduit and threaded couplings are installed, the conduits within the coupling will butt, resulting in a weak mechanical joint and poor grounding continuity.

### 344.30 Securing and Supporting

RMC shall be installed as a complete system in accordance with 300.18 and shall be securely fastened in place and supported in accordance with 344.30(A) and (B).

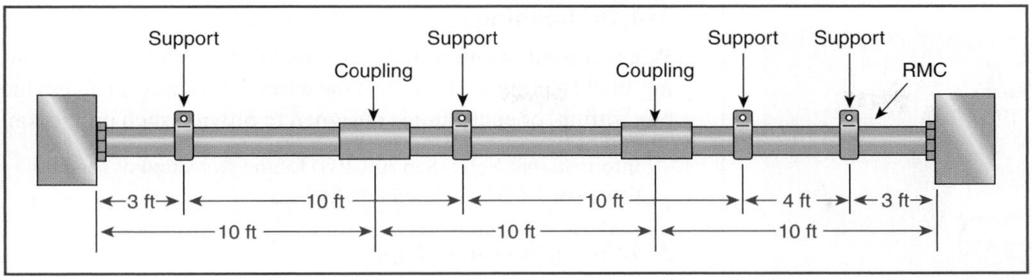


***EXHIBIT 344.1*** *Minimum support required for RMC.*

**(A) Securely Fastened.** RMC shall be secured in accordance with one of the following:

(1) RMC shall be securely fastened within 900 mm (3 ft) of each outlet box, junction box, device box, cabinet, conduit body, or other conduit termination.
(2) Fastening shall be permitted to be increased to a distance of 1.5 m (5 ft) where structural members do not readily permit fastening within 900 mm (3 ft).
(3) Where approved, conduit shall not be required to be securely fastened within 900 mm (3 ft) of the service head for above-the-roof termination of a mast.

As illustrated in Exhibit 344.1, RMC is required to be securely fastened within 3 feet of outlet boxes, junction boxes, cabinets, and conduit bodies, or other conduit terminations. Couplings are not considered conduit terminations. However, where structural support members do not permit fastening within 3 feet, secure fastening may be located up to 5 feet away. In addition, RMC is required to be supported at least every 10 feet unless permitted otherwise by 344.30(B).

**(B) Supports.** RMC shall be supported in accordance with one of the following:

(1) Conduit shall be supported at intervals not exceeding 3 m (10 ft).
(2) The distance between supports for straight runs of conduit shall be permitted in accordance with Table 344.30(B)(2), provided the conduit is made up with threaded couplings and such supports prevent transmission of stresses to termination where conduit is deflected between supports.
(3) Exposed vertical risers from industrial machinery or fixed equipment shall be permitted to be supported at intervals not exceeding 6 m (20 ft) if the conduit is made up with threaded couplings, the conduit is supported and securely fastened at the top and bottom of the riser, and no other means of intermediate support is readily available.
(4) Horizontal runs of RMC supported by openings through framing members at intervals not exceeding 3 m (10 ft) and securely fastened within 900 mm (3 ft) of terminatino points shall be permitted.

Lengths of RMC are permitted to be supported (but not necessarily secured) by framing members at 10-foot intervals, provided the

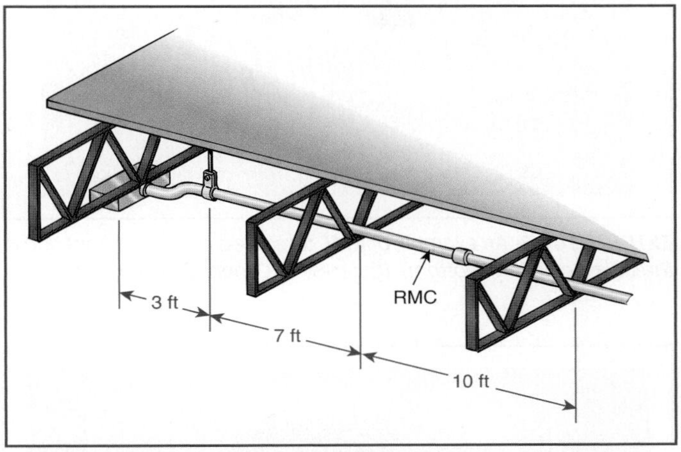


***EXHIBIT 344.2*** *An example of RMC supported by framing members and securely fastened at the 3-foot distance from the box.*

***TABLE 344.30(B)(2)*** *Supports for Rigid Metal Conduit*

| Conduit Size | | Maximum Distance Between Rigid Metal Conduit Supports | |
|---|---|---|---|
| **Metric Designator** | **Trade Size** | **m** | **ft** |
| 16–21 | ½–¾ | 3.0 | 10 |
| 27 | 1 | 3.7 | 12 |
| 35–41 | 1¼–1½ | 4.3 | 14 |
| 53–63 | 2–2½ | 4.9 | 16 |
| 78 and larger | 3 and larger | 6.1 | 20 |

RMC is secured and supported at least 3 feet from the box or enclosure. Installations where the RMC is installed through bar joists are just one example, as illustrated in Exhibit 344.2.

## 344.42 Couplings and Connectors

**(A) Threadless.** Threadless couplings and connectors used with conduit shall be made tight. Where buried in masonry or concrete, they shall be the concrete tight type. Where installed in wet locations, they shall comply with 314.15. Threadless couplings and connectors shall not be used on threaded conduit ends unless listed for the purpose.

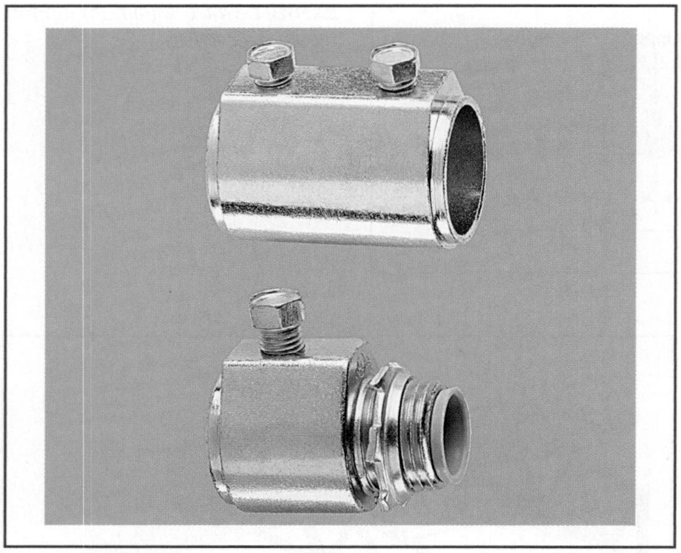

***EXHIBIT 344.3*** *An example of RMC threadless connector (top) and a threadless coupling (bottom). (Courtesy of Cooper Crouse-Hinds)*

***EXHIBIT 344.4*** *A three-piece (union-type) coupling. (Courtesy of Appleton Electric Co., EGS Electrical Group)*

Exhibit 344.3 illustrates two different conduit fittings: a threadless conduit coupling and a threadless conduit connector. Threadless fittings may be suitable for other applications such as in raintight or concretetight locations, or even wet locations, provided the product itself or the product carton is marked as such.

In general, threadless fittings are not intended for use over threads, because the fitting will not seat properly. The threaded end of the conduit should be cut off and reamed before installation.

Exhibit 344.4 illustrates a three-piece threaded coupling (the electrical equivalent of a pipe union), which is used to join two lengths of conduit where turning either length is impossible, such as in underground or concrete slab construction. Another fitting for joining conduit is a bolted split coupling.

**(B) Running Threads.** Running threads shall not be used on conduit for connection at couplings.

### 344.46 Bushings

Where a conduit enters a box, fitting, or other enclosure, a bushing shall be provided to protect the wires from abrasion unless the box, fitting, or enclosure is designed to provide such protection.

Informational Note: See 300.4(G) for the protection of conductors sizes 4 AWG and larger at bushings.

### 344.56 Splices and Taps

Splices and taps shall be made in accordance with 300.15.

### 344.60 Grounding

RMC shall be permitted as an equipment grounding conductor.

## III. Construction Specifications

### 344.100 Construction

RMC shall be made of one of the following:

(1) Steel (ferrous), with or without protective coatings
(2) Aluminum (nonferrous)
(3) Red brass
(4) Stainless steel

### 344.120 Marking

Each length shall be clearly and durably identified in every 3 m (10 ft) as required in the first sentence of 110.21(A). Nonferrous conduit of corrosion-resistant material shall have suitable markings.

### 344.130 Standard Lengths

The standard length of RMC shall be 3.05 m (10 ft), including an attached coupling, and each end shall be threaded. Longer or shorter lengths with or without coupling and threaded or unthreaded shall be permitted.

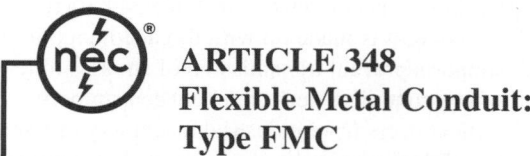

# ARTICLE 348 Flexible Metal Conduit: Type FMC

## I. General

### 348.1 Scope

This article covers the use, installation, and construction specifications for flexible metal conduit (FMC) and associated fittings.

### 348.2 Definition

**Flexible Metal Conduit (FMC).** A raceway of circular cross section made of helically wound, formed, interlocked metal strip.

### 348.6 Listing Requirements

FMC and associated fittings shall be listed.

## II. Installation

### 348.10 Uses Permitted

FMC shall be permitted to be used in exposed and concealed locations.

FMC ½ inch and larger may be installed in unlimited lengths, provided an EGC is installed with the circuit conductors. See 250.118(5) as well as 348.60 for specific requirements related to the use of FMC as an EGC.

### 348.12 Uses Not Permitted

FMC shall not be used in the following:

(1) In wet locations
(2) In hoistways, other than as permitted in 620.21(A)(1)
(3) In storage battery rooms
(4) In any hazardous (classified) location except as permitted by other articles in this *Code*
(5) Where exposed to materials having a deteriorating effect on the installed conductors, such as oil or gasoline
(6) Underground or embedded in poured concrete or aggregate
(7) Where subject to physical damage

### 348.20 Size

**(A) Minimum.** FMC less than metric designator 16 (trade size ½) shall not be used unless permitted in 348.20(A)(1) through (A)(5) for metric designator 12 (trade size ⅜).

(1) For enclosing the leads of motors as permitted in 430.245(B)
(2) In lengths not in excess of 1.8 m (6 ft) for any of the following uses:
   a. For utilization equipment
   b. As part of a listed assembly
   c. For tap connections to luminaires as permitted in 410.117(C)

Trade size ⅜ FMC is permitted to be used as the manufactured or field-installed metal raceway (1½ feet to 6 feet in length) to enclose tap conductors between the outlet box and the terminal housing of recessed luminaires. FMC is also permitted to be used as a 6-foot luminaire whip from an outlet box to a luminaire.

A smaller minimum size for manufactured wiring systems is permitted [see Section 604.6(A)] because the conductors are not as prone to physical damage where assembled under factory-controlled conditions.

(3) For manufactured wiring systems as permitted in 604.6(A)
(4) In hoistways as permitted in 620.21(A)(1)
(5) As part of a listed assembly to connect wired luminaire sections as permitted in 410.137(C)

**(B) Maximum.** FMC larger than metric designator 103 (trade size 4) shall not be used.

Informational Note: See 300.1(C) for the metric designators and trade sizes. These are for identification purposes only and do not relate to actual dimensions.

### 348.22 Number of Conductors

The number of conductors shall not exceed that permitted by the percentage fill specified in Table 1, Chapter 9, or as permitted in Table 348.22, or for metric designator 12 (trade size ⅜).

Cables shall be permitted to be installed where such use is not prohibited by the respective cable articles. The number of cables shall not exceed the allowable percentage fill specified in Table 1, Chapter 9.

Table 4 of Chapter 9 provides the usable area within the selected conduit or tubing, and Table 5 provides the required area for each conductor. Examples using these tables to calculate a conduit or tubing size are provided in the commentary following Chapter 9, Notes to Tables, Note 6.

***TABLE 348.22*** *Maximum Number of Insulated Conductors in Metric Designator 12 (Trade Size ⅜) Flexible Metal Conduit (FMC)**

| | Types RFH-2, SF-2 | | Types TF, XHHW, TW | | Types TFN, THHN, THWN | | Types FEP, FEBP, PF, PGF | |
|---|---|---|---|---|---|---|---|---|
| Size (AWG) | Fittings Inside Conduit | Fittings Outside Conduit | Fittings Inside Conduit | Fittings Outside Conduit | Fittings Inside Conduit | Fittings Outside Conduit | Fittings Inside Conduit | Fittings Outside Conduit |
| 18 | 2 | 3 | 3 | 5 | 5 | 8 | 5 | 8 |
| 16 | 1 | 2 | 3 | 4 | 4 | 6 | 4 | 6 |
| 14 | 1 | 2 | 2 | 3 | 3 | 4 | 3 | 4 |
| 12 | — | — | 1 | 2 | 2 | 3 | 2 | 3 |
| 10 | — | — | 1 | 1 | 1 | 1 | 1 | 2 |

*In addition, one insulated, covered, or bare equipment grounding conductor of the same size shall be permitted.

If the conductors are the same wire size, Informative Annex C can be used instead of performing the calculations. This annex, through 12 sets of tables, accurately indicates the maximum number of conductors permitted in conduit or tubing. Examples using Informative Annex C to select a conduit or tubing size are provided in the commentary following the introduction in the annex.

To select the proper trade size of FMC, see "Article 348 — Flexible Metal Conduit (FMC)" in Table 4 of Chapter 9. Use of Tables C.3 and C.3(A) for FMC is also permissible.

## 348.24 Bends — How Made

Bends in conduit shall be made so that the conduit is not damaged and the internal diameter of the conduit is not effectively reduced. Bends shall be permitted to be made manually without auxiliary equipment. The radius of the curve to the centerline of any bend shall not be less than shown in Table 2, Chapter 9 using the column "Other Bends."

## 348.26 Bends — Number in One Run

There shall not be more than the equivalent of four quarter bends (360 degrees total) between pull points, for example, conduit bodies and boxes.

Proper shaping and support of this flexible wiring method ensures that conductors can be easily installed or withdrawn at any time.

## 348.28 Trimming

All cut ends shall be trimmed or otherwise finished to remove rough edges, except where fittings that thread into the convolutions are used.

## 348.30 Securing and Supporting

FMC shall be securely fastened in place and supported in accordance with 348.30(A) and (B).

**(A) Securely Fastened.** FMC shall be securely fastened in place by an approved means within 300 mm (12 in.) of each box, cabinet, conduit body, or other conduit termination and shall be supported and secured at intervals not to exceed 1.4 m (4½ ft).

*Exception No. 1: Where FMC is fished between access points through concealed spaces in finished buildings or structures and supporting is impracticable.*

*Exception No. 2: Where flexibility is necessary after installation, lengths from the last point where the raceway is securely fastened shall not exceed the following:*

*(1) 900 mm (3 ft) for metric designators 16 through 35 (trade sizes ½ through 1¼)*
*(2) 1200 mm (4 ft) for metric designators 41 through 53 (trade sizes 1½ through 2)*
*(3) 1500 mm (5 ft) for metric designators 63 (trade size 2½) and larger*

An example of the phrase "where flexibility is necessary after installation" is an installation of FMC to a motor mounted on an adjustable or sliding frame, where the frame is required to be movable to perform drive belt maintenance. The length that the exception addresses is the length from the last point where the FMC is securely fastened.

*Exception No. 3: Lengths not exceeding 1.8 m (6 ft) from a luminaire terminal connection for tap connections to luminaires as permitted in 410.117(C).*

*Exception No. 4: Lengths not exceeding 1.8 m (6 ft) from the last point where the raceway is securely fastened for connections within an accessible ceiling to a luminaire(s) or other equipment. For the purposes of this exception, listed flexible metal conduit fittings shall be permitted as a means of support.*

**(B) Supports.** Horizontal runs of FMC supported by openings through framing members at intervals not greater than 1.4 m (4½ ft) and securely fastened within 300 mm (12 in.) of termination points shall be permitted.

## 348.42 Couplings and Connectors

Angle connectors shall not be concealed.

## 348.56 Splices and Taps

Splices and taps shall be made in accordance with 300.15.

## 348.60 Grounding and Bonding

If used to connect equipment where flexibility is necessary to minimize the transmission of vibration from equipment or to provide flexibility for equipment that requires movement after installation, an equipment grounding conductor shall be installed.

Where flexibility is not required after installation, FMC shall be permitted to be used as an equipment grounding conductor when installed in accordance with 250.118(5).

Where required or installed, equipment grounding conductors shall be installed in accordance with 250.134(B).

Where required or installed, equipment bonding jumpers shall be installed in accordance with 250.102.

An additional EGC is always required where FMC is used for flexibility. Examples of such installations include using FMC to minimize the transmission of equipment vibration such as motors or to provide flexibility for floodlights, spotlights, or other equipment that requires adjustment after installation.

According to ANSI/UL 1, *Standard for Flexible Metal Conduit*, FMC longer than 6 feet has not been judged to be suitable for grounding purposes. If the length of the total ground-fault return path exceeds 6 feet or the circuit overcurrent protection exceeds 20 amperes, a separate EGC must be installed with the circuit conductors according to 250.118(5). The upper sketch in Exhibit 348.1 shows an acceptable application of FMC where the total length of any ground return path is limited to 6 feet. The lower sketch shows an application that is unacceptable because the grounding return path for Luminaire 2 exceeds the permitted maximum of 6 feet to the box.

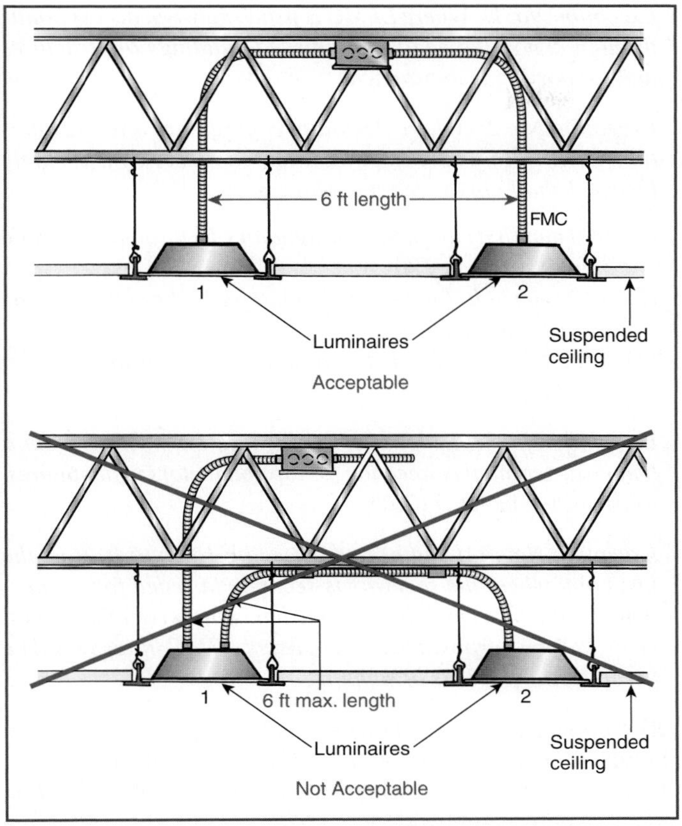


***EXHIBIT 348.1*** *An example of acceptable and unacceptable applications of FMC without separate EGCs used as a luminaire whip, in accordance with 250.118(5)(c).*

Where FMC is used in hazardous (classified) locations, a bonding jumper is required. See 501.30(B), 502.30(B), and 503.30(B) for details on types of EGCs. Section 250.102(E) permits the routing of equipment bonding jumpers on the outside of the raceway in lengths that are no longer than 6 feet and bonded at each end.

# ARTICLE 350 Liquidtight Flexible Metal Conduit: Type LFMC

## I. General

### 350.1 Scope

This article covers the use, installation, and construction specifications for liquidtight flexible metal conduit (LFMC) and associated fittings.

Liquidtight flexible metal conduit (LFMC) is intended for use in wet locations or where exposed to oil or coolants, at a maximum temperature of 140°F. LFMC is not intended for use where exposed to gasoline or similar light petroleum solvents unless so marked on the product. If properly marked for the application, LFMC is permitted for direct burial in the earth. LFMC is on the permitted list of wiring methods for services (230.43), provided the length does not exceed 6 feet and an equipment bonding jumper is installed in accordance with 250.102. LFMC may be installed in unlimited lengths, provided it meets the other requirements of Article 350 and a separate EGC is installed with the circuit conductors.

### 350.2 Definition

**Liquidtight Flexible Metal Conduit (LFMC).** A raceway of circular cross section having an outer liquidtight, nonmetallic, sunlight-resistant jacket over an inner flexible metal core with associated couplings, connectors, and fittings for the installation of electric conductors.

### 350.6 Listing Requirements

LFMC and associated fittings shall be listed.

## II. Installation

### 350.10 Uses Permitted

LFMC shall be permitted to be used in exposed or concealed locations as follows:

(1) Where conditions of installation, operation, or maintenance require flexibility or protection from liquids, vapors, or solids
(2) As permitted by 501.10(B), 502.10, 503.10, and 504.20 and in other hazardous (classified) locations where specifically approved, and by 553.7(B)
(3) For direct burial where listed and marked for the purpose

### 350.12 Uses Not Permitted

LFMC shall not be used as follows:

(1) Where subject to physical damage
(2) Where any combination of ambient and conductor temperature produces an operating temperature in excess of that for which the material is approved

### 350.20 Size

**(A) Minimum.** LFMC smaller than metric designator 16 (trade size ½) shall not be used.

*Exception: LFMC of metric designator 12 (trade size ⅜) shall be permitted as covered in 348.20(A).*

**(B) Maximum.** The maximum size of LFMC shall be metric designator 103 (trade size 4).

Informational Note: See 300.1(C) for the metric designators and trade sizes. These are for identification purposes only and do not relate to actual dimensions.

## 350.22 Number of Conductors or Cables

**(A) Metric Designators 16 through 103 (Trade Sizes ½ through 4).** The number of conductors shall not exceed that permitted by the percentage fill specified in Table 1, Chapter 9.

Cables shall be permitted to be installed where such use is not prohibited by the respective cable articles. The number of cables shall not exceed the allowable percentage fill specified in Table 1, Chapter 9.

**(B) Metric Designator 12 (Trade Size ⅜).** The number of conductors shall not exceed that permitted in Table 348.22, "Fittings Outside Conduit" columns.

Table 4 of Chapter 9 provides the usable area within the selected conduit or tubing, and Table 5 provides the required area for each conductor. Examples using these tables to calculate a conduit or tubing size are provided in the commentary following Chapter 9, Notes to Tables, Note 6.

If the conductors are of the same wire size, Informative Annex C can be used instead of performing the calculations. This annex, through 12 sets of tables, accurately indicates the maximum number of conductors permitted in conduit or tubing. Examples using Informative Annex C to select a conduit or tubing size are provided in the commentary following the annex introduction.

To select the proper trade size of LFMC, see "Article 350 — Liquidtight Flexible Metal Conduit (LFMC)" in Table 4 of Chapter 9. Use of Tables C.7 and C.7(A) for LFMC in Informative Annex C is also permissible.

The exception to 350.20(A) permits the use of trade size ⅜ LFMC under the limited conditions specified for flexible metal conduit (FMC) in 348.20(A). The number of conductors permitted in the LFMC is also limited by Table 348.22 for FMC.

## 350.24 Bends — How Made

Bends in conduit shall be so made that the conduit will not be damaged and the internal diameter of the conduit will not be effectively reduced. Bends shall be permitted to be made manually without auxiliary equipment. The radius of the curve to the centerline of any bend shall not be less than required in Table 2, Chapter 9 using the column "Other Bends."

## 350.26 Bends — Number in One Run

There shall not be more than the equivalent of four quarter bends (360 degrees total) between pull points, for example, conduit bodies and boxes.

## 350.30 Securing and Supporting

LFMC shall be securely fastened in place and supported in accordance with 350.30(A) and (B).

**(A) Securely Fastened.** LFMC shall be securely fastened in place by an approved means within 300 mm (12 in.) of each box, cabinet, conduit body, or other conduit termination and shall be supported and secured at intervals not to exceed 1.4 m (4½ ft).

*Exception No. 1: Where LFMC is fished between access points through concealed spaces in finished buildings or structures and supporting is impractical.*

*Exception No. 2: Where flexibility is necessary after installation, lengths from the last point where the raceway is securely fastened shall not exceed the following:*

*(1) 900 mm (3 ft) for metric designators 16 through 35 (trade sizes ½ through 1¼)*
*(2) 1200 mm (4 ft) for metric designators 41 through 53 (trade sizes 1½ through 2)*
*(3) 1500 mm (5 ft) for metric designators 63 (trade size 2½) and larger*

*Exception No. 3: Lengths not exceeding 1.8 m (6 ft) from a luminaire terminal connection for tap conductors to luminaires, as permitted in 410.117(C).*

*Exception No. 4: Lengths not exceeding 1.8 m (6 ft) from the last point where the raceway is securely fastened for connections within an accessible ceiling to luminaire(s) or other equipment. For the purposes of 350.30, listed LFMC fittings shall be permitted as a means of support.*

**(B) Supports.** Horizontal runs of LFMC supported by openings through framing members at intervals not greater than 1.4 m (4½ ft) and securely fastened within 300 mm (12 in.) of termination points shall be permitted.

## 350.42 Couplings and Connectors

Only fittings listed for use with LFMC shall be used. Angle connectors shall not be concealed. Straight LFMC fittings shall be permitted for direct burial where marked.

## 350.56 Splices and Taps

Splices and taps shall be made in accordance with 300.15.

## 350.60 Grounding and Bonding

If used to connect equipment where flexibility is necessary to minimize the transmission of vibration from equipment or to provide flexibility for equipment that requires movement after installation, an equipment grounding conductor shall be installed.

Where flexibility is not required after installation, LFMC shall be permitted to be used as an equipment grounding conductor when installed in accordance with 250.118(6).

Where required or installed, equipment grounding conductors shall be installed in accordance with 250.134(B).

Where required or installed, equipment bonding jumpers shall be installed in accordance with 250.102.

Informational Note: See 501.30(B), 502.30(B), 503.30(B), 505.25(B), and 506.25(B) for types of equipment grounding conductors.

## III. Construction Specifications

### 350.120 Marking

LFMC shall be marked according to 110.21. The trade size and other information required by the listing shall also be marked on the conduit. Conduit suitable for direct burial shall be so marked.

# ARTICLE 352 Rigid Polyvinyl Chloride Conduit: Type PVC

## I. General

### 352.1 Scope

This article covers the use, installation, and construction specifications for rigid polyvinyl chloride conduit (PVC) and associated fittings.

Informational Note: Refer to Article 353 for High Density Polyethylene Conduit: Type HDPE, and Article 355 for Reinforced Thermosetting Resin Conduit: Type RTRC.

Commentary Table 352.1 includes the following Type PVC conduits referenced with the appropriate UL category and product standard.

The UL *Guide Information for Electrical Equipment – The White Book* describes rigid PVC conduit, Type PVC, for use in accordance with Article 352. Schedule 40 is suitable where not subject to physical damage for underground, aboveground, indoor, and outdoor locations. Schedule 80 is suitable for the locations where the conduit will be subject to damage. Types A and EB are intended for underground installations.

Unless marked for higher temperature, PVC conduit is intended for use with wire rated 75°C or less, including where encased in concrete within buildings, and PVC is suitable for use with wires rated 90°C or less where encased in concrete outside of buildings.

**COMMENTARY TABLE 352.1** UL Category and Product Standards for Type PVC Conduits

| Type PVC Conduits | UL Product Category Code | UL Standard |
|---|---|---|
| Rigid PVC Conduit, Type A | EAZX | ANSI/UL 651A |
| Rigid PVC Conduit, Type EB | EAZX | ANSI/UL 651A |
| Rigid PVC Conduit, Schedule 40 | DZYR | ANSI/UL 651 |
| Rigid PVC Conduit, Schedule 80 | DZYR | ANSI/UL 651 |

### 352.2 Definition

**Rigid Polyvinyl Chloride Conduit (PVC).** A rigid nonmetallic raceway of circular cross section, with integral or associated couplings, connectors, and fittings for the installation of electrical conductors and cables.

### 352.6 Listing Requirements

PVC conduit, factory elbows, and associated fittings shall be listed.

## II. Installation

### 352.10 Uses Permitted

The use of PVC conduit shall be permitted in accordance with 352.10(A) through (I).

Informational Note: Extreme cold may cause some nonmetallic conduits to become brittle and, therefore, more susceptible to damage from physical contact.

**(A) Concealed.** PVC conduit shall be permitted in walls, floors, and ceilings.

**(B) Corrosive Influences.** PVC conduit shall be permitted in locations subject to severe corrosive influences as covered in 300.6 and where subject to chemicals for which the materials are specifically approved.

**(C) Cinders.** PVC conduit shall be permitted in cinder fill.

**(D) Wet Locations.** PVC conduit shall be permitted in portions of dairies, laundries, canneries, or other wet locations, and in locations where walls are frequently washed, the entire conduit system, including boxes and fittings used therewith, shall be installed and equipped so as to prevent water from entering the conduit. All supports, bolts, straps, screws, and so forth, shall be of corrosion-resistant materials or be protected against corrosion by approved corrosion-resistant materials.

**(E) Dry and Damp Locations.** PVC conduit shall be permitted for use in dry and damp locations not prohibited by 352.12.

**(F) Exposed.** PVC conduit shall be permitted for exposed work. PVC conduit used exposed in areas of physical damage shall be identified for the use.

Informational Note: PVC Conduit, Type Schedule 80, is identified for areas of physical damage.

**(G) Underground Installations.** For underground installations, PVC shall be permitted for direct burial and underground encased in concrete. See 300.5 and 300.50.

**(H) Support of Conduit Bodies.** PVC conduit shall be permitted to support nonmetallic conduit bodies not larger than the largest trade size of an entering raceway. These conduit bodies shall not support luminaires or other equipment and shall not contain devices other than splicing devices as permitted by 110.14(B) and 314.16(C)(2).

**(I) Insulation Temperature Limitations.** Conductors or cables rated at a temperature higher than the listed temperature rating of PVC conduit shall be permitted to be installed in PVC conduit, provided the conductors or cables are not operated at a temperature higher than the listed temperature rating of the PVC conduit.

Conductors marked with a rated temperature higher than that of the raceway can be used when the conductors are operated within the raceway temperature rating. One application is the use of 105°C-rated medium voltage cables, Type MV, where the cable ampacity at the 105°C rating is reduced to the cable ampacity at 75°C or 90°C to match the listed operating temperature rating of the PVC conduit (75°C or 90°C).

## 352.12 Uses Not Permitted

PVC conduit shall not be used under the conditions specified in 352.12(A) through (E).

**(A) Hazardous (Classified) Locations.** In any hazardous (classified) location, except as permitted by other articles of this *Code*.

**(B) Support of Luminaires.** For the support of luminaires or other equipment not described in 352.10(H).

**(C) Physical Damage.** Where subject to physical damage unless identified for such use.

**(D) Ambient Temperatures.** Where subject to ambient temperatures in excess of 50°C (122°F) unless listed otherwise.

**(E) Theaters and Similar Locations.** In theaters and similar locations, except as provided in 518.4 and 520.5.

In addition to the conditions in 352.12(A) through (E), PVC conduits are not permitted to be installed in ducts, plenums, and other air-handling spaces. See 300.22, which limits the use of materials in ducts, plenums, and other air-handling spaces that may contribute smoke and products of combustion during a fire.

## 352.20 Size

**(A) Minimum.** PVC conduit smaller than metric designator 16 (trade size ½) shall not be used.

**(B) Maximum.** PVC conduit larger than metric designator 155 (trade size 6) shall not be used.

Informational Note: The trade sizes and metric designators are for identification purposes only and do not relate to actual dimensions. See 300.1(C).

## 352.22 Number of Conductors

The number of conductors shall not exceed that permitted by the percentage fill specified in Table 1, Chapter 9.

Cables shall be permitted to be installed where such use is not prohibited by the respective cable articles. The number of cables shall not exceed the allowable percentage fill specified in Table 1, Chapter 9.

Table 4 of Chapter 9 provides the usable area within the selected conduit or tubing, and Table 5 provides the required area for each conductor. Examples using these tables to calculate a conduit or tubing size are provided in the commentary following Chapter 9, Notes to Tables, Note 6.

If the conductors are of the same wire size, Informative Annex C can be used instead of performing the calculations. This annex, through 12 sets of tables, accurately indicates the maximum number of conductors permitted in conduit or tubing. Examples using Informative Annex C to select a conduit or tubing size are provided in the commentary following the annex introduction.

To select the proper trade size of PVC conduit, four separate sub-tables in Table 4 of Chapter 9 address each type. The appropriate sub-table for Article 352, Rigid PVC, should be followed. Use of Tables C.9 and C.9(A) through C.12 and C.12(A) is also permissible, provided the appropriate table for the given type of PVC conduit is used.

## 352.24 Bends — How Made

Bends shall be so made that the conduit will not be damaged and the internal diameter of the conduit will not be effectively reduced. Field bends shall be made only with identified bending equipment. The radius of the curve to the centerline of such bends shall not be less than shown in Table 2, Chapter 9.

Pulling conductors in underground conduit can damage nonmetallic elbows. Metal elbows are often used to ensure the raceway's integrity. Metal elbows in runs of PVC conduit that are buried at least 18 inches are not required to be bonded to the grounded system conductor or the grounding electrode conductor. See 250.80 and the associated commentary. Table 2 of Chapter 9 is the common table for raceway field bend measurements.

## 352.26 Bends — Number in One Run

There shall not be more than the equivalent of four quarter bends (360 degrees total) between pull points, for example, conduit bodies and boxes.

The number of bends in a conduit run is limited to reduce pulling tension on the conductors and to help ensure easy insertion or removal of conductors during later phases of construction, when the conduit may be permanently enclosed by the building finish. The *Code* does not limit the pull points to conduit bodies and boxes, which are only examples of pull points.

## 352.28 Trimming

All cut ends shall be trimmed inside and outside to remove rough edges.

## 352.30 Securing and Supporting

PVC conduit shall be installed as a complete system as provided in 300.18 and shall be fastened so that movement from thermal expansion or contraction is permitted. PVC conduit shall be securely fastened and supported in accordance with 352.30(A) and (B).

Expansion fittings and supports, installed as prescribed, allow for expansion/contraction cycles without damage. See the commentary on expansion fittings following 352.44 for details.

**(A) Securely Fastened.** PVC conduit shall be securely fastened within 900 mm (3 ft) of each outlet box, junction box, device box, conduit body, or other conduit termination. Conduit listed for securing at other than 900 mm (3 ft) shall be permitted to be installed in accordance with the listing.

**(B) Supports.** PVC conduit shall be supported as required in Table 352.30. Conduit listed for support at spacings other than as shown in Table 352.30 shall be permitted to be installed in accordance with the listing. Horizontal runs of PVC conduit supported by openings through framing members at intervals not exceeding those in Table 352.30 and securely fastened within 900 mm (3 ft) of termination points shall be permitted.

***TABLE 352.30*** *Support of Rigid Polyvinyl Chloride Conduit (PVC)*

| Conduit Size | | Maximum Spacing Between Supports | |
|---|---|---|---|
| Metric Designator | Trade Size | mm or m | ft |
| 16–27 | ½–1 | 900 mm | 3 |
| 35–53 | 1¼–2 | 1.5 m | 5 |
| 63–78 | 2½–3 | 1.8 m | 6 |
| 91–129 | 3½–5 | 2.1 m | 7 |
| 155 | 6 | 2.5 m | 8 |

## 352.44 Expansion Fittings

Expansion fittings for PVC conduit shall be provided to compensate for thermal expansion and contraction where the length change, in accordance with Table 352.44, is expected to be 6 mm (¼ in.) or greater in a straight run between securely mounted items such as boxes, cabinets, elbows, or other conduit terminations.

Since PVC conduit exhibits a considerably greater change in length per degree change in temperature than do metal raceway systems, expansion fittings are required for specific temperature variations. According to Table 352.44, a 100-foot run of PVC conduit will change 4.06 inches of length if the temperature change is 100°F. See the commentary example following 300.7(B).

The allowable range of expansion for many PVC conduit expansion couplings is generally 6 inches. Information concerning installation and application of this type of coupling may be obtained from manufacturers' instructions.

Expansion fittings are seldom used underground, where temperatures are relatively constant. If PVC conduit is buried or covered immediately, expansion and contraction are not considered a problem.

***TABLE 352.44*** *Expansion Characteristics of PVC Rigid Nonmetallic Conduit Coefficient of Thermal Expansion = 6.084 × $10^{-5}$ mm/mm/°C (3.38 × $10^{-5}$ in./in./°F)*

| Temperature Change (°C) | Length Change of PVC Conduit (mm/m) | Temperature Change (°F) | Length Change of PVC Conduit (in./100 ft) | Temperature Change (°F) | Length Change of PVC Conduit (in./100 ft) |
|---|---|---|---|---|---|
| 5 | 0.30 | 5 | 0.20 | 105 | 4.26 |
| 10 | 0.61 | 10 | 0.41 | 110 | 4.46 |
| 15 | 0.91 | 15 | 0.61 | 115 | 4.66 |
| 20 | 1.22 | 20 | 0.81 | 120 | 4.87 |
| 25 | 1.52 | 25 | 1.01 | 125 | 5.07 |
| 30 | 1.83 | 30 | 1.22 | 130 | 5.27 |
| 35 | 2.13 | 35 | 1.42 | 135 | 5.48 |
| 40 | 2.43 | 40 | 1.62 | 140 | 5.68 |
| 45 | 2.74 | 45 | 1.83 | 145 | 5.88 |
| 50 | 3.04 | 50 | 2.03 | 150 | 6.08 |
| 55 | 3.35 | 55 | 2.23 | 155 | 6.29 |
| 60 | 3.65 | 60 | 2.43 | 160 | 6.49 |
| 65 | 3.95 | 65 | 2.64 | 165 | 6.69 |
| 70 | 4.26 | 70 | 2.84 | 170 | 6.90 |
| 75 | 4.56 | 75 | 3.04 | 175 | 7.10 |
| 80 | 4.87 | 80 | 3.24 | 180 | 7.30 |
| 85 | 5.17 | 85 | 3.45 | 185 | 7.50 |
| 90 | 5.48 | 90 | 3.65 | 190 | 7.71 |
| 95 | 5.78 | 95 | 3.85 | 195 | 7.91 |
| 100 | 6.08 | 100 | 4.06 | 200 | 8.11 |

### 352.46 Bushings

Where a conduit enters a box, fitting, or other enclosure, a bushing or adapter shall be provided to protect the wire from abrasion unless the box, fitting, or enclosure design provides equivalent protection.

Informational Note: See 300.4(G) for the protection of conductors 4 AWG and larger at bushings.

### 352.48 Joints

All joints between lengths of conduit, and between conduit and couplings, fittings, and boxes, shall be made by an approved method.

### 352.56 Splices and Taps

Splices and taps shall be made in accordance with 300.15.

### 352.60 Grounding

Where equipment grounding is required, a separate equipment grounding conductor shall be installed in the conduit.

*Exception No. 1: As permitted in 250.134(B), Exception No. 2, for dc circuits and 250.134(B), Exception No. 1, for separately run equipment grounding conductors.*

*Exception No. 2: Where the grounded conductor is used to ground equipment as permitted in 250.142.*

## III. Construction Specifications

### 352.100 Construction

PVC conduit shall be made of rigid (nonplasticized) polyvinyl chloride (PVC). PVC conduit and fittings shall be composed of suitable nonmetallic material that is resistant to moisture and chemical atmospheres. For use aboveground, it shall also be flame retardant, resistant to impact and crushing, resistant to distortion from heat under conditions likely to be encountered in service, and resistant to low temperature and sunlight effects. For use underground, the material shall be acceptably resistant to moisture and corrosive agents and shall be of sufficient strength to withstand abuse, such as by impact and crushing, in handling and during installation. Where intended for direct burial, without encasement in concrete, the material shall also be capable of withstanding continued loading that is likely to be encountered after installation.

### 352.120 Marking

Each length of PVC conduit shall be clearly and durably marked at least every 3 m (10 ft) as required in the first sentence of 110.21(A). The type of material shall also be included in the marking unless it is visually identifiable. For conduit recognized for use aboveground, these markings shall be permanent. For conduit limited to underground use only, these markings shall be sufficiently durable to remain legible until the material is installed. Conduit shall be permitted to be surface marked to indicate special characteristics of the material.

Informational Note: Examples of these markings include but are not limited to "limited smoke" and "sunlight resistant."

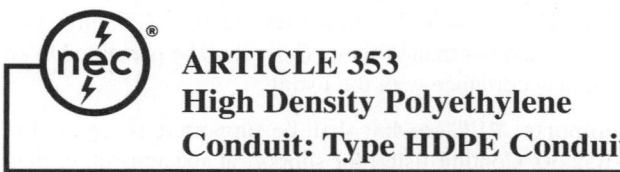

# ARTICLE 353
# High Density Polyethylene Conduit: Type HDPE Conduit

## I. General

### 353.1 Scope

This article covers the use, installation, and construction specifications for high density polyethylene (HDPE) conduit and associated fittings.

Commentary Table 353.1 includes all of the following types of HDPE raceways referenced with the appropriate UL category and product standard.

Informational Note: Refer to Article 352 for Rigid Polyvinyl Chloride Conduit: Type PVC and Article 355 for Reinforced Thermosetting Resin Conduit: Type RTRC.

### 353.2 Definition

**High Density Polyethylene (HDPE) Conduit.** A nonmetallic raceway of circular cross section, with associated couplings, connectors, and fittings for the installation of electrical conductors.

### 353.6 Listing Requirements

HDPE conduit and associated fittings shall be listed.

## II. Installation

### 353.10 Uses Permitted

The use of HDPE conduit shall be permitted under the following conditions:

**COMMENTARY TABLE 353.1** UL Category and Product Standard for Type HDPE Conduit

| Type HDPE Conduit | UL Product Category Code | UL Standard |
|---|---|---|
| HDPE Conduit, EPEC A | EAZX | ANSI/UL 651A |
| HDPE Conduit, EPEC B | EAZX | ANSI/UL 651A |
| HDPE Conduit, Schedule 40 | EAZX | ANSI/UL 651A |
| HDPE Conduit, Schedule 80 | EAZX | ANSI/UL 651A |
| Continuous Length HDPE Conduit | EAZX | ANSI/UL 651B |

(1) In discrete lengths or in continuous lengths from a reel
(2) In locations subject to severe corrosive influences as covered in 300.6 and where subject to chemicals for which the conduit is listed
(3) In cinder fill
(4) In direct burial installations in earth or concrete

Informational Note to (4): Refer to 300.5 and 300.50 for underground installations.

(5) Above ground, except as prohibited in 353.12, where encased in not less than 50 mm (2 in.) of concrete.
(6) Conductors or cables rated at a temperature higher than the listed temperature rating of HDPE conduit shall be permitted to be installed in HDPE conduit, provided the conductors or cables are not operated at a temperature higher than the listed temperature rating of the HDPE conduit.

Conductors marked with a rated temperature higher than that of the raceway can be used when the conductors are operated within the raceway temperature rating.

One application of 353.10(6) is the use of 105°C-rated medium voltage cables, Type MV, where the cable ampacity at the 105°C rating is reduced to the cable ampacity at 75°C or 90°C to match the listed operating temperature rating of HDPE (75°C or 90°C).

### 353.12 Uses Not Permitted

HDPE conduit shall not be used under the following conditions:

(1) Where exposed
(2) Within a building
(3) In any hazardous (classified) location, except as permitted by other articles in this *Code*
(4) Where subject to ambient temperatures in excess of 50°C (122°F) unless listed otherwise

### 353.20 Size

**(A) Minimum.** HDPE conduit smaller than metric designator 16 (trade size ½) shall not be used.

**(B) Maximum.** HDPE conduit larger than metric designator 155 (trade size 6) shall not be used.

Informational Note: The trade sizes and metric designators are for identification purposes only and do not relate to actual dimensions. See 300.1(C).

### 353.22 Number of Conductors

The number of conductors shall not exceed that permitted by the percentage fill specified in Table 1, Chapter 9.

Cables shall be permitted to be installed where such use is not prohibited by the respective cable articles. The number of cables shall not exceed the allowable percentage fill specified in Table 1, Chapter 9.

Table 4 of Chapter 9 provides the usable area within the selected conduit or tubing, and Table 5 provides the required area for each conductor. Examples using these tables to calculate a conduit or tubing size are provided in the commentary following Chapter 9, Notes to Tables, Note 6.

If the conductors are of the same wire size, Informative Annex C can be used instead of performing the calculations. This annex, through 12 sets of tables, accurately indicates the maximum number of conductors permitted in conduit or tubing. Examples using Informative Annex C to select a conduit or tubing size are provided in the commentary following the annex introduction.

To select the proper trade size of HDPE, see "Articles 352 and 353 – Rigid PVC Conduit (PVC), Schedule 40, and HDPE Conduit" in Table 4 of Chapter 9. Use of Tables C.10 and C.10(A) is also permissible.

### 353.24 Bends — How Made

Bends shall be so made that the conduit will not be damaged and the internal diameter of the conduit will not be effectively reduced. Bends shall be permitted to be made manually without auxiliary equipment, and the radius of the curve to the centerline of such bends shall not be less than shown in Table 354.24. For conduits of metric designators 129 and 155 (trade sizes 5 and 6) the allowable radii of bends shall be in accordance with specifications provided by the manufacturer.

### 353.26 Bends — Number in One Run

There shall not be more than the equivalent of four quarter bends (360 degrees total) between pull points, for example, conduit bodies and boxes.

### 353.28 Trimming

All cut ends shall be trimmed inside and outside to remove rough edges.

### 353.46 Bushings

Where a conduit enters a box, fitting, or other enclosure, a bushing or adapter shall be provided to protect the wire from abrasion unless the box, fitting, or enclosure design provides equivalent protection.

Informational Note: See 300.4(G) for the protection of conductors 4 AWG and larger at bushings.

### 353.48 Joints

All joints between lengths of conduit, and between conduit and couplings, fittings, and boxes, shall be made by an approved method.

Informational Note: HDPE conduit can be joined using either heat fusion, electrofusion, or mechanical fittings.

### 353.56 Splices and Taps

Splices and taps shall be made in accordance with 300.15.

### 353.60 Grounding

Where equipment grounding is required, a separate equipment grounding conductor shall be installed in the conduit.

*Exception No. 1: The equipment grounding conductor shall be permitted to be run separately from the conduit where used for grounding dc circuits as permitted in 250.134, Exception No. 2.*

*Exception No. 2: The equipment grounding conductor shall not be required where the grounded conductor is used to ground equipment as permitted in 250.142.*

## III. Construction Specifications

### 353.100 Construction

HDPE conduit shall be composed of high density polyethylene that is resistant to moisture and chemical atmospheres. The material shall be resistant to moisture and corrosive agents and shall be of sufficient strength to withstand abuse, such as by impact and crushing, in handling and during installation. Where intended for direct burial, without encasement in concrete, the material shall also be capable of withstanding continued loading that is likely to be encountered after installation.

### 353.120 Marking

Each length of HDPE shall be clearly and durably marked at least every 3 m (10 ft) as required in 110.21. The type of material shall also be included in the marking.

# ARTICLE 354 Nonmetallic Underground Conduit with Conductors: Type NUCC

## I. General

### 354.1 Scope

This article covers the use, installation, and construction specifications for nonmetallic underground conduit with conductors (NUCC).

NUCC (preassembled conductors in conduit) has been used by electric utilities for outdoor lighting for several years. It is supplied in continuous lengths on coils or reels or in cartons. NUCC consists of nonmetallic conduit with the conductors pre-installed by the manufacturer. The product is designed to allow conductors to be removed and reinserted.

### 354.2 Definition

**Nonmetallic Underground Conduit with Conductors (NUCC).** A factory assembly of conductors or cables inside a nonmetallic, smooth wall raceway with a circular cross section.

### 354.6 Listing Requirements

NUCC and associated fittings shall be listed.

## II. Installation

### 354.10 Uses Permitted

The use of NUCC and fittings shall be permitted in the following:

(1) For direct burial underground installation (For minimum cover requirements, see Table 300.5 and Table 300.50 under Rigid Nonmetallic Conduit.)
(2) Encased or embedded in concrete
(3) In cinder fill
(4) In underground locations subject to severe corrosive influences as covered in 300.6 and where subject to chemicals for which the assembly is specifically approved
(5) Aboveground, except as prohibited in 354.12, where encased in not less than 50 mm (2 in.) of concrete

### 354.12 Uses Not Permitted

NUCC shall not be used in the following:

(1) In exposed locations
(2) Inside buildings

*Exception: The conductor or the cable portion of the assembly, where suitable, shall be permitted to extend within the building for termination purposes in accordance with 300.3.*

(3) In any hazardous (classified) location, except as permitted by other articles of this *Code*

### 354.20 Size

**(A) Minimum.** NUCC smaller than metric designator 16 (trade size ½) shall not be used.

**(B) Maximum.** NUCC larger than metric designator 103 (trade size 4) shall not be used.

Informational Note: See 300.1(C) for the metric designators and trade sizes. These are for identification purposes only and do not relate to actual dimensions.

### 354.22 Number of Conductors

The number of conductors or cables shall not exceed that permitted by the percentage fill in Table 1, Chapter 9.

### 354.24 Bends — How Made

Bends shall be manually made so that the conduit will not be damaged and the internal diameter of the conduit will not be effectively reduced. The radius of the curve of the centerline of such bends shall not be less than shown in Table 354.24.

***TABLE 354.24** Minimum Bending Radius for Nonmetallic Underground Conduit with Conductors (NUCC)*

| Conduit Size | | Minimum Bending Radius | |
|---|---|---|---|
| **Metric Designator** | **Trade Size** | **mm** | **in.** |
| 16 | ½ | 250 | 10 |
| 21 | ¾ | 300 | 12 |
| 27 | 1 | 350 | 14 |
| 35 | 1¼ | 450 | 18 |
| 41 | 1½ | 500 | 20 |
| 53 | 2 | 650 | 26 |
| 63 | 2½ | 900 | 36 |
| 78 | 3 | 1200 | 48 |
| 103 | 4 | 1500 | 60 |

The bending radius for NUCC does not follow Chapter 9, Table 2, but rather must conform to Table 354.24.

### 354.26 Bends — Number in One Run

There shall not be more than the equivalent of four quarter bends (360 degrees total) between termination points.

### 354.28 Trimming

For termination, the conduit shall be trimmed away from the conductors or cables using an approved method that will not damage the conductor or cable insulation or jacket. All conduit ends shall be trimmed inside and out to remove rough edges.

### 354.46 Bushings

Where the NUCC enters a box, fitting, or other enclosure, a bushing or adapter shall be provided to protect the conductor or cable from abrasion unless the design of the box, fitting, or enclosure provides equivalent protection.

Informational Note: See 300.4(G) for the protection of conductors size 4 AWG or larger.

### 354.48 Joints

All joints between conduit, fittings, and boxes shall be made by an approved method.

### 354.50 Conductor Terminations

All terminations between the conductors or cables and equipment shall be made by an approved method for that type of conductor or cable.

### 354.56 Splices and Taps

Splices and taps shall be made in junction boxes or other enclosures.

### 354.60 Grounding

Where equipment grounding is required, an assembly containing a separate equipment grounding conductor shall be used.

## III. Construction Specifications

### 354.100 Construction

**(A) General.** NUCC is an assembly that is provided in continuous lengths shipped in a coil, reel, or carton.

**(B) Nonmetallic Underground Conduit.** The nonmetallic underground conduit shall be listed and composed of a material that is resistant to moisture and corrosive agents. It shall also be capable of being supplied on reels without damage or distortion and shall be of sufficient strength to withstand abuse, such as impact or crushing, in handling and during installation without damage to conduit or conductors.

**(C) Conductors and Cables.** Conductors and cables used in NUCC shall be listed and shall comply with 310.10(C). Conductors of different systems shall be installed in accordance with 300.3(C).

**(D) Conductor Fill.** The maximum number of conductors or cables in NUCC shall not exceed that permitted by the percentage fill in Table 1, Chapter 9.

### 354.120 Marking

NUCC shall be clearly and durably marked at least every 3.05 m (10 ft) as required by 110.21. The type of conduit material shall also be included in the marking.

Identification of conductors or cables used in the assembly shall be provided on a tag attached to each end of the assembly or to the side of a reel. Enclosed conductors or cables shall be marked in accordance with 310.120.

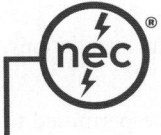

# ARTICLE 355 Reinforced Thermosetting Resin Conduit: Type RTRC

## I. General

### 355.1 Scope

This article covers the use, installation, and construction specification for reinforced thermosetting resin conduit (RTRC) and associated fittings.

Informational Note: Refer to Article 352 for Rigid Polyvinyl Chloride Conduit: Type PVC, and Article 353 for High Density Polyethylene Conduit: Type HDPE.

### 355.2 Definition

**Reinforced Thermosetting Resin Conduit (RTRC).** A rigid nonmetallic raceway of circular cross section, with integral or associated couplings, connectors, and fittings for the installation of electrical conductors and cables.

### 355.6 Listing Requirements

RTRC, factory elbows, and associated fittings shall be listed.

## II. Installation

### 355.10 Uses Permitted

The use of RTRC shall be permitted in accordance with 355.10(A) through (I).

**(A) Concealed.** RTRC shall be permitted in walls, floors, and ceilings.

**(B) Corrosive Influences.** RTRC shall be permitted in locations subject to severe corrosive influences as covered in 300.6 and where subject to chemicals for which the materials are specifically approved.

**(C) Cinders.** RTRC shall be permitted in cinder fill.

**(D) Wet Locations.** RTRC shall be permitted in portions of dairies, laundries, canneries, or other wet locations, and in locations where walls are frequently washed, the entire conduit system, including boxes and fittings used therewith, shall be installed and equipped so as to prevent water from entering the conduit. All supports, bolts, straps, screws, and so forth, shall be of corrosion-resistant materials or be protected against corrosion by approved corrosion-resistant materials.

**(E) Dry and Damp Locations.** RTRC shall be permitted for use in dry and damp locations not prohibited by 355.12.

**(F) Exposed.** RTRC shall be permitted for exposed work if identified for such use.

Informational Note: RTRC, Type XW, is identified for areas of physical damage.

**(G) Underground Installations.** For underground installations, see 300.5 and 300.50.

**(H) Support of Conduit Bodies.** RTRC shall be permitted to support nonmetallic conduit bodies not larger than the largest trade size of an entering raceway. These conduit bodies shall not support luminaires or other equipment and shall not contain devices other than splicing devices as permitted by 110.14(B) and 314.16(C)(2).

**(I) Insulation Temperature Limitations.** Conductors or cables rated at a temperature higher than the listed temperature rating of RTRC conduit shall be permitted to be installed in RTRC conduit, if the conductors or cables are not operated at a temperature higher than the listed temperature rating of the RTRC conduit.

### 355.12 Uses Not Permitted

RTRC shall not be used under the following conditions.

**(A) Hazardous (Classified) Locations.**

(1) In any hazardous (classified) location, except as permitted by other articles in this *Code*
(2) In Class I, Division 2 locations, except as permitted in 501.10(B)(3)

**(B) Support of Luminaires.** For the support of luminaires or other equipment not described in 355.10(H).

**(C) Physical Damage.** Where subject to physical damage unless identified for such use.

RTRC installed in a location where the raceway is subject to physical damage must be marked "XW." If the location is above ground and exposed to physical damage, the conduit must be marked "AG XW RTRC." For examples of requirements specifying the use of RTRC-XW, see 300.50(C), 334.15(B), 501.10(B)(6), and 551.80(B).

**(D) Ambient Temperatures.** Where subject to ambient temperatures in excess of 50°C (122°F) unless listed otherwise.

**(E) Theaters and Similar Locations.** In theaters and similar locations, except as provided in 518.4 and 520.5.

In addition to the conditions in 355.12(A) through (E), nonmetallic conduits are not permitted to be installed in ducts, plenums, and other air-handling spaces. The use of materials that may contribute smoke and products of combustion during a fire is limited in ducts, plenums, and other air-handling spaces in accordance with 300.22.

### 355.20 Size

**(A) Minimum.** RTRC smaller than metric designator 16 (trade size ½) shall not be used.

**(B) Maximum.** RTRC larger than metric designator 155 (trade size 6) shall not be used.

Informational Note: The trade sizes and metric designators are for identification purposes only and do not relate to actual dimensions. See 300.1(C).

### 355.22 Number of Conductors

The number of conductors shall not exceed that permitted by the percentage fill specified in Table 1, Chapter 9. Cables shall be permitted to be installed where such use is not prohibited by the respective cable articles. The number of cables shall not exceed the allowable percentage fill specified in Table 1, Chapter 9.

Table 1 of Chapter 9 specifies the maximum percent fill of conduit or tubing. No internal dimensions for Type RTRC are given in the *Code* for calculating the allowable number of conductors. Conductor fill calculation may be in accordance with provided dimensions marked on the conduit. For the exact dimensions for fill calculations of RTRC, refer to the product standard or to the manufacturer's product information.

### 355.24 Bends — How Made

Bends shall be so made that the conduit will not be damaged and the internal diameter of the conduit will not be effectively reduced. Field bends shall be made only with identified bending equipment. The radius of the curve to the centerline of such bends shall not be less than shown in Table 2, Chapter 9.

*TABLE 355.30 Support of Reinforced Thermosetting Resin Conduit (RTRC)*

| Conduit Size | | Maximum Spacing Between Supports | |
|---|---|---|---|
| Metric Designator | Trade Size | mm or m | ft |
| 16–27 | ½–1 | 900 mm | 3 |
| 35–53 | 1¼–2 | 1.5 m | 5 |
| 63–78 | 2½–3 | 1.8 m | 6 |
| 91–129 | 3½–5 | 2.1 m | 7 |
| 155 | 6 | 2.5 m | 8 |

### 355.26 Bends — Number in One Run

There shall not be more than the equivalent of four quarter bends (360 degrees total) between pull points, for example, conduit bodies and boxes.

The number of bends in a conduit run is limited to reduce pulling tension on conductors and to help ensure easy insertion or removal of conductors during later phases of construction, when the conduit may be permanently enclosed by the building finish. The *Code* does not limit the pull points to conduit bodies and boxes, which are only examples of pull points.

### 355.28 Trimming

All cut ends shall be trimmed inside and outside to remove rough edges.

### 355.30 Securing and Supporting

RTRC shall be installed as a complete system in accordance with 300.18 and shall be securely fastened in place and supported in accordance with 355.30(A) and (B).

**(A) Securely Fastened.** RTRC shall be securely fastened within 900 mm (3 ft) of each outlet box, junction box, device box, conduit body, or other conduit termination. Conduit listed for securing at other than 900 mm (3 ft) shall be permitted to be installed in accordance with the listing.

**(B) Supports.** RTRC shall be supported as required in Table 355.30. Conduit listed for support at spacing other than as shown in Table 355.30 shall be permitted to be installed in accordance with the listing. Horizontal runs of RTRC supported by openings through framing members at intervals not exceeding those in Table 355.30 and securely fastened within 900 mm (3 ft) of termination points shall be permitted.

### 355.44 Expansion Fittings

Expansion fittings for RTRC shall be provided to compensate for thermal expansion and contraction where the length change, in accordance with Table 355.44, is expected to be 6 mm (¼ in.) or

*TABLE 355.44 Expansion Characteristics of Reinforced Thermosetting Resin Conduit (RTRC) Coefficient of Thermal Expansion = $2.7 \times 10^{-5}$ mm/mm/°C ($1.5 \times 10^{-5}$ in./in./°F)*

| Temperature Change (°C) | Length Change of RTRC Conduit (mm/m) | Temperature Change (°F) | Length Change of RTRC Conduit (in./100 ft) | Temperature Change (°F) | Length Change of RTRC Conduit (in./100 ft) |
|---|---|---|---|---|---|
| 5 | 0.14 | 5 | 0.09 | 105 | 1.89 |
| 10 | 0.27 | 10 | 0.18 | 110 | 1.98 |
| 15 | 0.41 | 15 | 0.27 | 115 | 2.07 |
| 20 | 0.54 | 20 | 0.36 | 120 | 2.16 |
| 25 | 0.68 | 25 | 0.45 | 125 | 2.25 |
| 30 | 0.81 | 30 | 0.54 | 130 | 2.34 |
| 35 | 0.95 | 35 | 0.63 | 135 | 2.43 |
| 40 | 1.08 | 40 | 0.72 | 140 | 2.52 |
| 45 | 1.22 | 45 | 0.81 | 145 | 2.61 |
| 50 | 1.35 | 50 | 0.90 | 150 | 2.70 |
| 55 | 1.49 | 55 | 0.99 | 155 | 2.79 |
| 60 | 1.62 | 60 | 1.08 | 160 | 2.88 |
| 65 | 1.76 | 65 | 1.17 | 165 | 2.97 |
| 70 | 1.89 | 70 | 1.26 | 170 | 3.06 |
| 75 | 2.03 | 75 | 1.35 | 175 | 3.15 |
| 80 | 2.16 | 80 | 1.44 | 180 | 3.24 |
| 85 | 2.30 | 85 | 1.53 | 185 | 3.33 |
| 90 | 2.43 | 90 | 1.62 | 190 | 3.42 |
| 95 | 2.57 | 95 | 1.71 | 195 | 3.51 |
| 100 | 2.70 | 100 | 1.80 | 200 | 3.60 |

greater in a straight run between securely mounted items such as boxes, cabinets, elbows, or other conduit terminations.

Since RTRC exhibits a considerably greater change in length per degree change in temperature than do metal raceway systems, expansion fittings are required for specific variations in temperature. In some areas, outdoor temperature variations of over 100°F are common. According to Table 355.44, a 100-foot run of RTRC will change 1.80 inches in length if the temperature change is 100°F. See the commentary example following 300.7(B).

The allowable range of expansion for many conduit expansion couplings is generally 6 inches. Information concerning installation and application of this type of coupling may be obtained from manufacturers' instructions. Expansion fittings are seldom used underground, where temperatures are relatively constant.

### 355.46 Bushings

Where a conduit enters a box, fitting, or other enclosure, a bushing or adapter shall be provided to protect the wire from abrasion unless the box, fitting, or enclosure design provides equivalent protection.

Informational Note: See 300.4(G) for the protection of conductors 4 AWG and larger at bushings.

### 355.48 Joints

All joints between lengths of conduit, and between conduit and couplings, fitting, and boxes, shall be made by an approved method.

### 355.56 Splices and Taps

Splices and taps shall be made in accordance with 300.15.

### 355.60 Grounding

Where equipment grounding is required, a separate equipment grounding conductor shall be installed in the conduit.

*Exception No. 1: As permitted in 250.134(B), Exception No. 2, for dc circuits and 250.134(B), Exception No. 1, for separately run equipment grounding conductors.*

*Exception No. 2: Where the grounded conductor is used to ground equipment as permitted in 250.142.*

## III. Construction Specifications

### 355.100 Construction

RTRC and fittings shall be composed of suitable nonmetallic material that is resistant to moisture and chemical atmospheres. For use aboveground, it shall also be flame retardant, resistant to impact and crushing, resistant to distortion from heat under conditions likely to be encountered in service, and resistant to low temperature and sunlight effects. For use underground, the material shall be acceptably resistant to moisture and corrosive agents and shall be of sufficient strength to withstand abuse, such as by impact and crushing, in handling and during installation. Where intended for direct burial, without encasement in concrete, the material shall also be capable of withstanding continued loading that is likely to be encountered after installation.

### 355.120 Marking

Each length of RTRC shall be clearly and durably marked at least every 3 m (10 ft) as required in the first sentence of 110.21(A). The type of material shall also be included in the marking unless it is visually identifiable. For conduit recognized for use aboveground, these markings shall be permanent. For conduit limited to underground use only, these markings shall be sufficiently durable to remain legible until the material is installed. Conduit shall be permitted to be surface marked to indicate special characteristics of the material.

Informational Note: Examples of these markings include but are not limited to "limited smoke" and "sunlight resistant."

# ARTICLE 356 Liquidtight Flexible Nonmetallic Conduit: Type LFNC

## I. General

### 356.1 Scope

This article covers the use, installation, and construction specifications for liquidtight flexible nonmetallic conduit (LFNC) and associated fittings.

LFNC may be prewired as a listed assembly where the conductors are installed at the manufacturing facility and controlled conditions prevent damage to the conductor insulation. Special cutting tools are required to be used when cutting prewired Type LFNC to prevent nicking the conductor installation. This prewired assembly is shown in Exhibit 356.1. LFNC is also used extensively in the machine tool and related industries. See 13.5.5 in NFPA 79-2012,

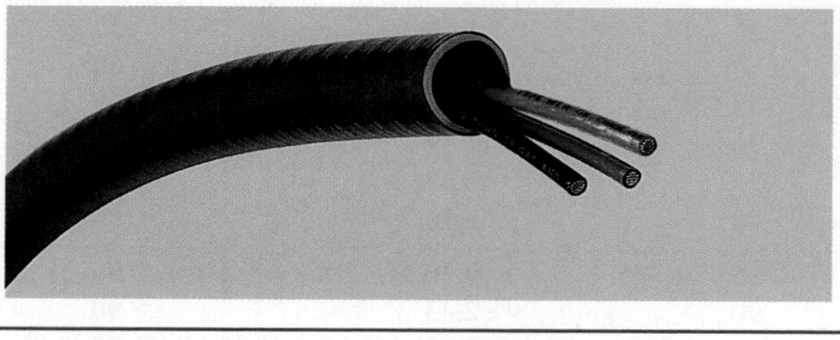

***EXHIBIT 356.1*** *Listed manufactured prewired assembly of LFNC, Type B. (Courtesy of Thomas & Betts Corp.)*

*Electrical Standard for Industrial Machinery,* for the uses permitted on an industrial machine.

### 356.2 Definition

**Liquidtight Flexible Nonmetallic Conduit (LFNC).** A raceway of circular cross section of various types as follows:

(1) A smooth seamless inner core and cover bonded together and having one or more reinforcement layers between the core and covers, designated as Type LFNC-A
(2) A smooth inner surface with integral reinforcement within the raceway wall, designated as Type LFNC-B
(3) A corrugated internal and external surface without integral reinforcement within the raceway wall, designated as LFNC-C

•

Informational Note: FNMC is an alternative designation for LFNC.

### 356.6 Listing Requirements

LFNC and associated fittings shall be listed.

## II. Installation

### 356.10 Uses Permitted

LFNC shall be permitted to be used in exposed or concealed locations for the following purposes:

Informational Note: Extreme cold may cause some types of nonmetallic conduits to become brittle and therefore more susceptible to damage from physical contact.

(1) Where flexibility is required for installation, operation, or maintenance.
(2) Where protection of the contained conductors is required from vapors, liquids, or solids.
(3) For outdoor locations where listed and marked as suitable for the purpose.
(4) For direct burial where listed and marked for the purpose.
(5) Type LFNC-B shall be permitted to be installed in lengths longer than 1.8 m (6 ft) where secured in accordance with 356.30.
(6) Type LFNC-B as a listed manufactured prewired assembly, metric designator 16 through 27 (trade size ½ through 1) conduit.
(7) For encasement in concrete where listed for direct burial and installed in compliance with 356.42.

### 356.12 Uses Not Permitted

LFNC shall not be used as follows:

(1) Where subject to physical damage
(2) Where any combination of ambient and conductor temperatures is in excess of that for which the LFNC is approved
(3) In lengths longer than 1.8 m (6 ft), except as permitted by 356.10(5) or where a longer length is approved as essential for a required degree of flexibility

•

(4) In any hazardous (classified) location, except as permitzted by other articles in this *Code*

### 356.20 Size

**(A) Minimum.** LFNC smaller than metric designator 16 (trade size ½) shall not be used unless permitted in 356.20(A)(1) or (A)(2) for metric designator 12 (trade size ⅜).

(1) For enclosing the leads of motors as permitted in 430.245(B)
(2) In lengths not exceeding 1.8 m (6 ft) as part of a listed assembly for tap connections to luminaires as required in 410.117(C), or for utilization equipment

**(B) Maximum.** LFNC larger than metric designator 103 (trade size 4) shall not be used.

Informational Note: See 300.1(C) for the metric designators and trade sizes. These are for identification purposes only and do not relate to actual dimensions.

### 356.22 Number of Conductors

The number of conductors shall not exceed that permitted by the percentage fill specified in Table 1, Chapter 9.

Cables shall be permitted to be installed where such use is not prohibited by the respective cable articles. The number of cables shall not exceed the allowable percentage fill specified in Table 1, Chapter 9.

Table 4 of Chapter 9 provides the usable area within the selected conduit or tubing, and Table 5 provides the required area for each conductor. Examples using these tables to calculate a conduit or tubing size are provided in the commentary following Chapter 9, Notes to Tables, Note 6.

If the conductors are of the same wire size, Informative Annex C can be used instead of performing the calculations. This annex, through 12 sets of tables, accurately indicates the maximum number of conductors permitted in conduit or tubing. Examples using Informative Annex C to select a conduit or tubing size are provided in the commentary following the annex introduction.

To select the proper trade size of LFNC, see "Article 356 — Liquidtight Flexible Nonmetallic Conduit (LFNC-B)" or "Article 356 — Liquidtight Flexible Nonmetallic Conduit (LFNC-A)" in Table 4 of Chapter 9. The use of Tables C.5 and C.5(A), or C.6 and C.6(A), for liquidtight flexible nonmetallic conduit (LFNC-B or LFNC-A) is also permissible.

### 356.24 Bends — How Made

Bends in conduit shall be so made that the conduit is not damaged and the internal diameter of the conduit is not effectively reduced. Bends shall be permitted to be made manually without auxiliary equipment. The radius of the curve to the centerline of any bend shall not be less than shown in Table 2, Chapter 9 using the column "Other Bends."

### 356.26 Bends — Number in One Run

There shall not be more than the equivalent of four quarter bends (360 degrees total) between pull points, for example, conduit bodies and boxes.

### 356.28 Trimming

All cut ends of conduit shall be trimmed inside and outside to remove rough edges.

### 356.30 Securing and Supporting

Type LFNC-B shall be securely fastened and supported in accordance with one of the following:

(1) Where installed in lengths exceeding 1.8 m (6 ft), the conduit shall be securely fastened at intervals not exceeding 900 mm (3 ft) and within 300 mm (12 in.) on each side of every outlet box, junction box, cabinet, or fitting.
(2) Securing or supporting of the conduit shall not be required where it is fished, installed in lengths not exceeding 900 mm (3 ft) at terminals where flexibility is required, or installed in lengths not exceeding 1.8 m (6 ft) from a luminaire terminal connection for tap conductors to luminaires permitted in 410.117(C).
(3) Horizontal runs of LFNC supported by openings through framing members at intervals not exceeding 900 mm (3 ft) and securely fastened within 300 mm (12 in.) of termination points shall be permitted.
(4) Securing or supporting of LFNC-B shall not be required where installed in lengths not exceeding 1.8 m (6 ft) from the last point where the raceway is securely fastened for connections within an accessible ceiling to a luminaire(s) or other equipment. For the purpose of 356.30, listed liquidtight flexible nonmetallic conduit fittings shall be permitted as a means of support.

### 356.42 Couplings and Connectors

Only fittings listed for use with LFNC shall be used. Angle connectors shall not be used for concealed raceway installations. Straight LFNC fittings are permitted for direct burial or encasement in concrete.

### 356.56 Splices and Taps

Splices and taps shall be made in accordance with 300.15.

### 356.60 Grounding

Where equipment grounding is required, a separate equipment grounding conductor shall be installed in the conduit.

*Exception No. 1: As permitted in 250.134(B), Exception No. 2, for dc circuits and 250.134(B), Exception No. 1, for separately run equipment grounding conductors.*

*Exception No. 2: Where the grounded conductor is used to ground equipment as permitted in 250.142.*

## III. Construction Specifications

### 356.100 Construction

LFNC-B as a prewired manufactured assembly shall be provided in continuous lengths capable of being shipped in a coil, reel, or carton without damage.

### 356.120 Marking

LFNC shall be marked at least every 600 mm (2 ft) in accordance with 110.21. The marking shall include a type designation in accordance with 356.2 and the trade size. Conduit that is intended for outdoor use or direct burial shall be marked.

The type, size, and quantity of conductors used in prewired manufactured assemblies shall be identified by means of a printed tag or label attached to each end of the manufactured assembly and either the carton, coil, or reel. The enclosed conductors shall be marked in accordance with 310.120.

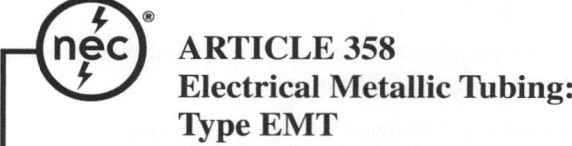

# ARTICLE 358 Electrical Metallic Tubing: Type EMT

## I. General

### 358.1 Scope

This article covers the use, installation, and construction specifications for electrical metallic tubing (EMT) and associated fittings.

### 358.2 Definition

**Electrical Metallic Tubing (EMT).** An unthreaded thinwall raceway of circular cross section designed for the physical protection and routing of conductors and cables and for use as an equipment grounding conductor when installed utilizing appropriate fittings. EMT is generally made of steel (ferrous) with protective coatings or aluminum (nonferrous).

### 358.6 Listing Requirements

EMT, factory elbows, and associated fittings shall be listed.

## II. Installation

### 358.10 Uses Permitted

**(A) Exposed and Concealed.** The use of EMT shall be permitted for both exposed and concealed work.

**(B) Corrosion Protection.** Ferrous or nonferrous EMT, elbows, couplings, and fittings shall be permitted to be installed in concrete, in direct contact with the earth, or in areas subject to severe corrosive influences where protected by corrosion protection and approved as suitable for the condition.

According to the UL *Guide Information for Electrical Equipment – The White Book*, category FJMX, galvanized steel EMT installed in concrete, on grade or above, generally requires no supplementary corrosion protection. Galvanized steel EMT in concrete slab below grade level may require supplementary corrosion protection. In general, galvanized steel EMT in contact with soil requires supplementary corrosion protection. Where galvanized steel EMT without supplementary corrosion protection extends directly from concrete encasement to soil burial, severe corrosive effects are likely to occur on the metal in contact with the soil.

**(C) Wet Locations.** All supports, bolts, straps, screws, and so forth shall be of corrosion-resistant materials or protected against corrosion by corrosion-resistant materials.

Informational Note: See 300.6 for protection against corrosion.

### 358.12 Uses Not Permitted

EMT shall not be used under the following conditions:

(1) Where, during installation or afterward, it will be subject to severe physical damage.
(2) Where protected from corrosion solely by enamel.
(3) In cinder concrete or cinder fill where subject to permanent moisture unless protected on all sides by a layer of noncinder concrete at least 50 mm (2 in.) thick or unless the tubing is at least 450 mm (18 in.) under the fill.
(4) In any hazardous (classified) location except as permitted by other articles in this *Code*.
(5) For the support of luminaires or other equipment except conduit bodies no larger than the largest trade size of the tubing.
(6) Where practicable, dissimilar metals in contact anywhere in the system shall be avoided to eliminate the possibility of galvanic action.

*Exception: Aluminum fittings and enclosures shall be permitted to be used with steel EMT where not subject to severe corrosive influences.*

### 358.20 Size

**(A) Minimum.** EMT smaller than metric designator 16 (trade size ½) shall not be used.

*Exception: For enclosing the leads of motors as permitted in 430.245(B).*

**(B) Maximum.** The maximum size of EMT shall be metric designator 103 (trade size 4).

Informational Note: See 300.1(C) for the metric designators and trade sizes. These are for identification purposes only and do not relate to actual dimensions.

### 358.22 Number of Conductors

The number of conductors shall not exceed that permitted by the percentage fill specified in Table 1, Chapter 9.

Cables shall be permitted to be installed where such use is not prohibited by the respective cable articles. The number of cables shall not exceed the allowable percentage fill specified in Table 1, Chapter 9.

Table 4 of Chapter 9 provides the usable area within the selected conduit or tubing, and Table 5 provides the required area for each conductor. Examples using these tables to calculate a conduit or tubing size are provided in the commentary following Chapter 9, Notes to Tables, Note 6.

If the conductors are of the same wire size, Informative Annex C can be used instead of performing the calculations. This annex, through 12 sets of tables, accurately indicates the maximum number of conductors permitted in conduit or tubing. Examples using Informative Annex C to select a conduit or tubing size are provided in the commentary following the annex introduction.

To select the proper trade size of EMT, see "Article 358 – Electrical Metallic Tubing (EMT)" in Table 4 of Chapter 9. Use of Tables C.1 and C.1(A) for EMT in Informative Annex C is also permissible.

### 358.24 Bends — How Made

Bends shall be made so that the tubing is not damaged and the internal diameter of the tubing is not effectively reduced. The radius of the curve of any field bend to the centerline of the tubing shall not be less than shown in Table 2, Chapter 9 for one-shot and full shoe benders.

### 358.26 Bends — Number in One Run

There shall not be more than the equivalent of four quarter bends (360 degrees total) between pull points, for example, conduit bodies and boxes.

### 358.28 Reaming and Threading

**(A) Reaming.** All cut ends of EMT shall be reamed or otherwise finished to remove rough edges.

In addition to a reamer, a half-round file has proved practical for removing rough edges. The steel handle of a pair of pump pliers, the nose of side-cutting pliers, or an electrician's knife can be effective on the smaller sizes of EMT as well.

**(B) Threading.** EMT shall not be threaded.

*Exception: EMT with factory threaded integral couplings complying with 358.100.*

## 358.30 Securing and Supporting

EMT shall be installed as a complete system in accordance with 300.18 and shall be securely fastened in place and supported in accordance with 358.30(A) and (B).

**(A) Securely Fastened.** EMT shall be securely fastened in place at least every 3 m (10 ft). In addition, each EMT run between termination points shall be securely fastened within 900 mm (3 ft) of each outlet box, junction box, device box, cabinet, conduit body, or other tubing termination.

Type EMT is required to be "securely fastened" at the prescribed intervals as illustrated in Exhibit 358.1. See the commentary following 344.30(A).

*Exception No. 1: Fastening of unbroken lengths shall be permitted to be increased to a distance of 1.5 m (5 ft) where structural members do not readily permit fastening within 900 mm (3 ft).*

As illustrated in Exhibit 358.2, boxes are permitted to be secured to ceiling or roof support structural members that are spaced not more than 5 feet apart to serve as support for runs of EMT perpendicular to the axis of the ceiling or roof support members.

*Exception No. 2: For concealed work in finished buildings or prefinished wall panels where such securing is impracticable, unbroken lengths (without coupling) of EMT shall be permitted to be fished.*

**(B) Supports.** Horizontal runs of EMT supported by openings through framing members at intervals not greater than 3 m (10 ft) and securely fastened within 900 mm (3 ft) of termination points shall be permitted.

Horizontal runs of EMT are permitted to be supported (but not necessarily secured) by framing members at 10-foot intervals, provided the EMT is secured at least 3 feet from the box or enclosure. See Exhibit 342.2 in the commentary following 342.30(B)(4) for an example.

## 358.42 Couplings and Connectors

Couplings and connectors used with EMT shall be made up tight. Where buried in masonry or concrete, they shall be concretetight type. Where installed in wet locations, they shall comply with 314.15.

Only listed fittings are permitted per 358.6, and 314.15 specifically requires that fittings for use in wet locations be listed for such use.

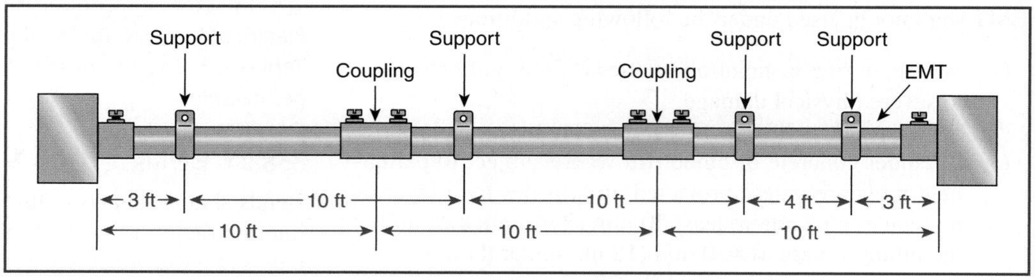


***EXHIBIT 358.1*** *Minimum requirements for securely fastening EMT unless an exception applies.*

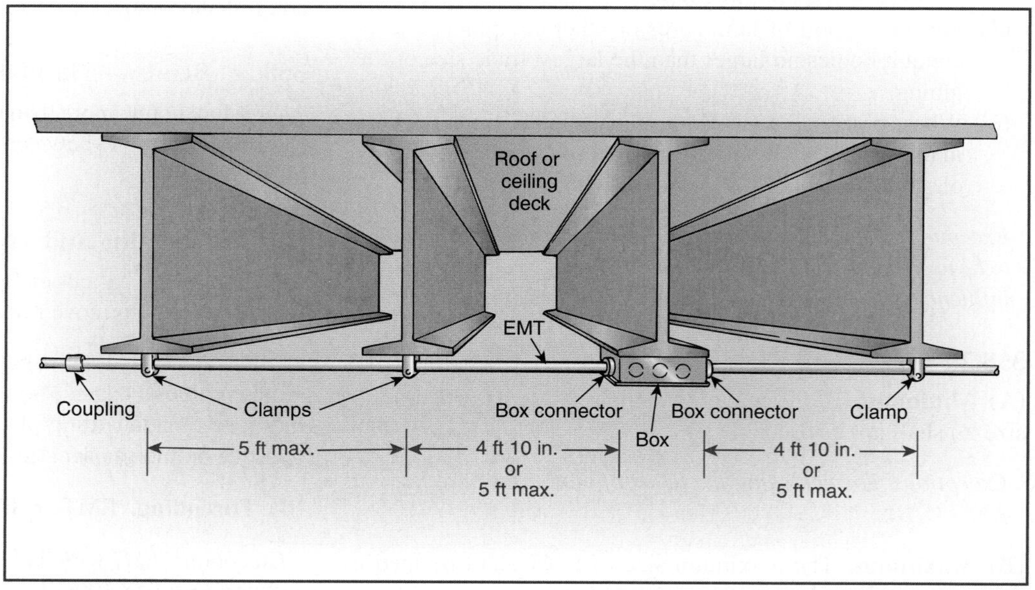


***EXHIBIT 358.2*** *An EMT installation in which the fastening spacing is increased to a maximum of 5 feet.*

According to ANSI/UL 797, *Electrical Metallic Tubing – Steel*, listed fittings suitable for use in poured concrete or where exposed to rain are indicated on the fitting or carton. The term *concretetight* or equivalent on the carton indicates suitability for use in poured concrete. The term *raintight* or the equivalent on the carton indicates suitability for use where directly exposed to rain. See 225.22 and 230.54(A) for raintight requirements as applied to raceways on exterior surfaces of buildings and to service raceways.

Indentor-type fittings, utilized only with metallic-coated EMT, require a special tool supplied by the manufacturer for proper installation. Fittings are tested for use only with steel EMT, unless specific marking on the device or carton indicates the fittings are suitable for use with aluminum or other material.

### 358.56 Splices and Taps

Splices and taps shall be made in accordance with 300.15.

### 358.60 Grounding

EMT shall be permitted as an equipment grounding conductor.

## III. Construction Specifications

### 358.100 Construction

Factory-threaded integral couplings shall be permitted. Where EMT with a threaded integral coupling is used, threads for both the tubing and coupling shall be factory-made. The coupling and EMT threads shall be designed so as to prevent bending of the tubing at any part of the thread.

### 358.120 Marking

EMT shall be clearly and durably marked at least every 3 m (10 ft) as required in the first sentence of 110.21(A).

# ARTICLE 360 Flexible Metallic Tubing: Type FMT

## I. General

### 360.1 Scope

This article covers the use, installation, and construction specifications for flexible metallic tubing (FMT) and associated fittings.

FMT is a type of raceway used for certain specific applications, particularly under the requirements of 300.22(B) and (C) for wiring in ducts and other air-handling spaces. Initially intended for use in these locations, FMT is an effective barrier to the gases and products of combustion. It is very flexible and rarely affected by vibration or other movement.

### 360.2 Definition

**Flexible Metallic Tubing (FMT).** A raceway that is circular in cross section, flexible, metallic, and liquidtight without a nonmetallic jacket.

### 360.6 Listing Requirements

FMT and associated fittings shall be listed.

## II. Installation

### 360.10 Uses Permitted

FMT shall be permitted to be used for branch circuits as follows:

(1) In dry locations
(2) Where concealed
(3) In accessible locations
(4) For system voltages of 1000 volts maximum

The 1000-volt limitation prohibits the use of FMT for the secondary circuits of sign ballasts, sign transformers, electronic sign power supplies, or oil burner ignition transformers unless these circuits are less than 1000 volts.

### 360.12 Uses Not Permitted

FMT shall not be used as follows:

(1) In hoistways
(2) In storage battery rooms
(3) In hazardous (classified) locations unless otherwise permitted under other articles in this *Code*
(4) Underground for direct earth burial, or embedded in poured concrete or aggregate
(5) Where subject to physical damage
(6) In lengths over 1.8 m (6 ft)

### 360.20 Size

**(A) Minimum.** FMT smaller than metric designator 16 (trade size ½) shall not be used.

*Exception No. 1: FMT of metric designator 12 (trade size ⅜) shall be permitted to be installed in accordance with 300.22(B) and (C).*

*Exception No. 2: FMT of metric designator 12 (trade size ⅜) shall be permitted in lengths not in excess of 1.8 m (6 ft) as part of a listed assembly or for luminaires. See 410.117(C).*

**(B) Maximum.** The maximum size of FMT shall be metric designator 21 (trade size ¾).

Informational Note: See 300.1(C) for the metric designators and trade sizes. These are for identification purposes only and do not relate to actual dimensions.

### 360.22 Number of Conductors

**(A) FMT — Metric Designators 16 and 21 (Trade Sizes ½ and ¾).** The number of conductors in metric designators 16 (trade size ½) and 21 (trade size ¾) shall not exceed that permitted by the percentage fill specified in Table 1, Chapter 9.

Cables shall be permitted to be installed where such use is not prohibited by the respective cable articles. The number of cables shall not exceed the allowable percentage fill specified in Table 1, Chapter 9.

Table 4 of Chapter 9 provides the usable area within the selected conduit or tubing, and Table 5 provides the required area for each conductor. Examples using these tables to calculate a conduit or tubing size are provided in the commentary following Chapter 9, Notes to Tables, Note 6.

If the conductors are of the same wire size, Informative Annex C can be used instead of performing the calculations. This annex, through 12 sets of tables, accurately indicates the maximum number of conductors permitted in conduit or tubing. Examples using Informative Annex C to select a conduit or tubing size are provided in the commentary following the annex introduction.

To select the proper trade size of FMT, see "Article 348 – Flexible Metal Conduit (FMC)" in Table 4 of Chapter 9. Use of Tables C.3 and C.3(A) for FMT sizes ½ inch and ¾ inch in Informative Annex C is also permissible.

**(B) FMT — Metric Designator 12 (Trade Size ⅜).** The number of conductors in metric designator 12 (trade size ⅜) shall not exceed that permitted in Table 348.22.

### 360.24 Bends

**(A) Infrequent Flexing Use.** When FMT is infrequently flexed in service after installation, the radii of bends measured to the inside of the bend shall not be less than specified in Table 360.24(A).

**(B) Fixed Bends.** Where FMT is bent for installation purposes and is not flexed or bent as required by use after installation, the radii of bends measured to the inside of the bend shall not be less than specified in Table 360.24(B).

### 360.56 Splices and Taps

Splices and taps shall be made in accordance with 300.15.

***TABLE 360.24(A)*** *Minimum Radii for Flexing Use*

| Metric Designator | Trade Size | Minimum Radii for Flexing Use | |
|---|---|---|---|
| | | mm | in. |
| 12 | ⅜ | 254.0 | 10 |
| 16 | ½ | 317.5 | 12½ |
| 21 | ¾ | 444.5 | 17½ |

***TABLE 360.24(B)*** *Minimum Radii for Fixed Bends*

| Metric Designator | Trade Size | Minimum Radii for Fixed Bends | |
|---|---|---|---|
| | | mm | in. |
| 12 | ⅜ | 88.9 | 3½ |
| 16 | ½ | 101.6 | 4 |
| 21 | ¾ | 127.0 | 5 |

### 360.60 Grounding

FMT shall be permitted as an equipment grounding conductor where installed in accordance with 250.118(7).

## III. Construction Specifications

### 360.120 Marking

FMT shall be marked according to 110.21.

# ARTICLE 362 Electrical Nonmetallic Tubing: Type ENT

## I. General

### 362.1 Scope

This article covers the use, installation, and construction specifications for electrical nonmetallic tubing (ENT) and associated fittings.

ENT is made of the same material used for PVC conduit. The outside diameters of ENT (½-inch through 2-inch trade sizes only) are such that standard couplings and other fittings for rigid PVC conduit can be used.

Because of the corrugations, the raceway can be bent by hand and has some degree of flexibility. ENT is not intended for use where flexibility is necessary, such as at motor terminations to prevent transmission of noise and vibration, or for connection of adjustable luminaires or moving parts. ENT is suitable for the installation of conductors having a temperature rating as indicated on the product. The maximum allowable ambient temperature is 122°F. Exhibit 362.1 shows an example of ENT.

### 362.2 Definition

**Electrical Nonmetallic Tubing (ENT).** A nonmetallic, pliable, corrugated raceway of circular cross section with integral or associated couplings, connectors, and fittings for the installation of electrical conductors. ENT is composed of a material that is resistant to moisture and chemical atmospheres and is flame retardant.

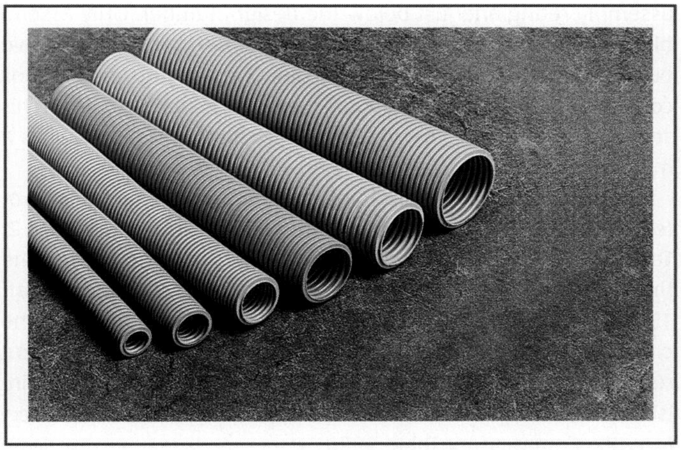

**EXHIBIT 362.1** *Various sizes of ENT suitable for use. (Courtesy of Thomas & Betts Corp.)*

A pliable raceway is a raceway that can be bent by hand with a reasonable force but without other assistance.

### 362.6 Listing Requirements

ENT and associated fittings shall be listed.

## II. Installation

### 362.10 Uses Permitted

For the purpose of this article, the first floor of a building shall be that floor that has 50 percent or more of the exterior wall surface area level with or above finished grade. One additional level that is the first level and not designed for human habitation and used only for vehicle parking, storage, or similar use shall be permitted. The use of ENT and fittings shall be permitted in the following:

(1) In any building not exceeding three floors above grade as follows:

a. For exposed work, where not prohibited by 362.12

b. Concealed within walls, floors, and ceilings

Where exposed and subject to physical damage, ENT is required to be protected and is limited to use in buildings not exceeding three floors above grade. Where concealed or above a suspended ceiling (exposed), ENT is permitted to be installed within walls, floors, or ceilings in buildings of three floors or less without the need for fire-rated construction. The three-floor limitation is based on the likelihood that only a small quantity of ENT would be exposed to fire and that the occupants would have adequate time to exit the building before the products of combustion make the building untenable. Exhibit 362.2 illustrates permitted uses of ENT in a building of three floors or less.

(2) In any building exceeding three floors above grade, ENT shall be concealed within walls, floors, and ceilings where the walls, floors, and ceilings provide a thermal barrier

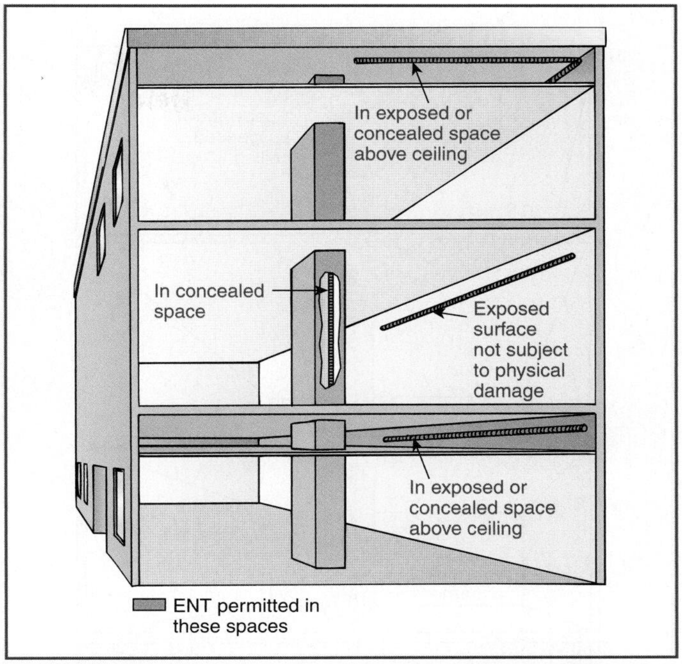


**EXHIBIT 362.2** *Examples of permitted uses of ENT in a building not exceeding three floors.*

of material that has at least a 15-minute finish rating as identified in listings of fire-rated assemblies. The 15-minute-finish-rated thermal barrier shall be permitted to be used for combustible or noncombustible walls, floors, and ceilings.

*Exception to (2): Where a fire sprinkler system(s) is installed in accordance with NFPA 13-2013, Standard for the Installation of Sprinkler Systems, on all floors, ENT shall be permitted to be used within walls, floors, and ceilings, exposed or concealed, in buildings exceeding three floors abovegrade.*

Informational Note: A finish rating is established for assemblies containing combustible (wood) supports. The finish rating is defined as the time at which the wood stud or wood joist reaches an average temperature rise of 121°C (250°F) or an individual temperature of 163°C (325°F) as measured on the plane of the wood nearest the fire. A finish rating is not intended to represent a rating for a membrane ceiling.

ENT is permitted to be installed within the walls, floors, or ceilings of a building of any height where the walls, floors, or ceilings provide a thermal barrier of material that has at least a 15-minute finish rating. Exposed ENT in the first three floors of a building that exceeds three floors is not permitted or intended except as permitted in 362.10(5). Where installed in a building over three floors, ENT must be installed behind the 15-minute thermal barrier on all floors. Exhibit 362.3 illustrates permitted uses of ENT in a building exceeding three floors. In accordance with the exception, fire sprinkler systems can also be used as a construction condition under which an expanded use of ENT is allowed.

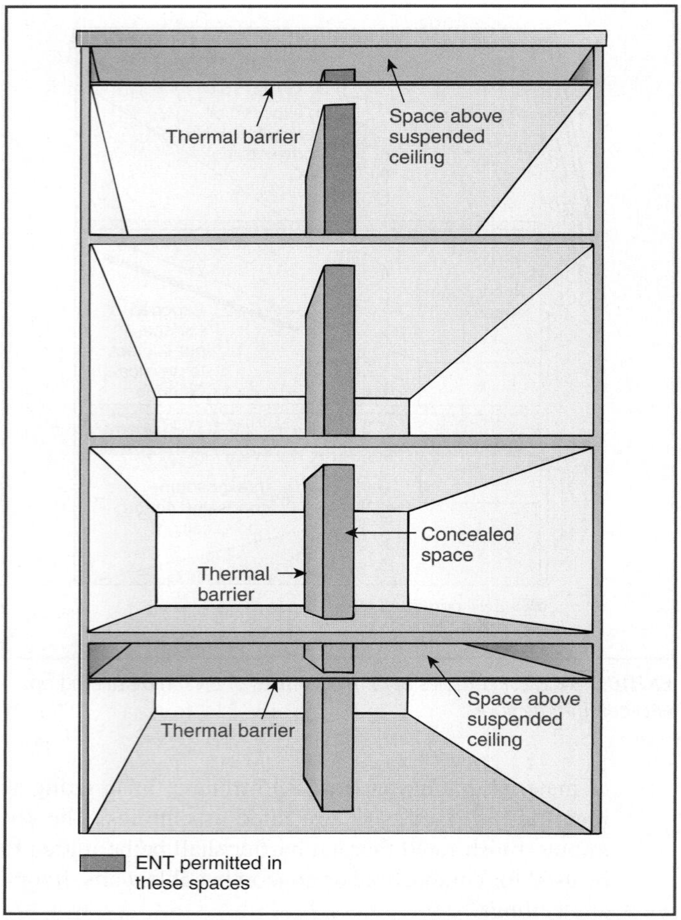


***EXHIBIT 362.3*** *Examples of permitted uses of ENT in a building exceeding three floors.*

Interior finish is generally considered to consist of those materials or combinations of materials that form the exposed interior surface of walls and ceilings in a building. Common interior finish materials include plaster, gypsum wallboard, wood, plywood paneling, fibrous ceiling tiles, and a variety of wall coverings. Ordinary paint, wallpaper, or other similar wall coverings not exceeding 1⁄28 inch in thickness are generally considered incidental to interior finish, except where the AHJ deems them a hazard. For more information regarding classification of interior finish material, refer to 10.2.1 of NFPA *101®*, *Life Safety Code®*.

The finish rating of a wall or ceiling finish material is the time required for the unexposed surface of the finish membrane to reach an average temperature rise of 250°F above ambient or an individual temperature rise at any one point not exceeding 325°F when the assembly is tested in accordance with ANSI/UL 263, *Standard for Fire Tests of Building Construction and Materials*, or ASTM E119, *Standard Test Methods for Fire Tests of Building Construction and Materials*.

The finish rating of wall and ceiling finish materials tested and rated by UL as part of wall and ceiling assemblies can be found in the UL *Fire Resistance Directory*, immediately following the assembly rating and just below the design number. Only assemblies containing combustible support members, however, have published finish ratings. Obviously, limiting ENT to constructions consisting of combustible support members is not the intent. This section is intended to provide a 15-minute thermal barrier as a minimum threshold of acceptability.

Commentary Table 362.1, reproduced from the NFPA *Fire Protection Handbook*, 20th edition (Volume 2, Section 19, Chapter 2, Table 19.2.13), provides ratings for common finish materials. If the finish rating concealing the ENT is unknown or is less than 15 minutes, the ENT can still be used if the installation meets the criteria in 362.10, including the three-floor limitation, where required, and the installation is not prohibited by 362.12. For finish materials not tested and rated in the UL *Fire Resistance Directory*, use Commentary Table 362.1.

(3) In locations subject to severe corrosive influences as covered in 300.6 and where subject to chemicals for which the materials are specifically approved.
(4) In concealed, dry, and damp locations not prohibited by 362.12.
(5) Above suspended ceilings where the suspended ceilings provide a thermal barrier of material that has at least a 15-minute finish rating as identified in listings of fire-rated assemblies, except as permitted in 362.10(1)(a).

*Exception to (5): ENT shall be permitted to be used above suspended ceilings in buildings exceeding three floors above grade where the building is protected throughout by a fire sprinkler system installed in accordance with NFPA 13-2013, Standard for the Installation of Sprinkler Systems.*

(6) Encased in poured concrete, or embedded in a concrete slab on grade where ENT is placed on sand or approved screenings, provided fittings identified for this purpose are used for connections.
(7) For wet locations indoors as permitted in this section or in a concrete slab on or belowgrade, with fittings listed for the purpose.
(8) Metric designator 16 through 27 (trade size ½ through 1) as listed manufactured prewired assembly.

Informational Note: Extreme cold may cause some types of nonmetallic conduits to become brittle and therefore more susceptible to damage from physical contact.

Prewired ENT is a listed assembly whose conductors must be installed at the manufacturing facility, where controlled conditions prevent damage to the conductor insulation. Special tools are required when cutting prewired ENT to prevent nicking of the conductor insulation.

(9) Conductors or cables rated at a temperature higher than the listed temperature rating of ENT shall be permitted to be installed in ENT, if the conductors or cables are not operated at a temperature higher than the listed temperature rating of the ENT.

**COMMENTARY TABLE 362.1** Various Finishes over Wood Framing, One Side (Combustible) with Exposure on Finish Side

| Material | Fire Resistance Rating[a] (min.) |
|---|---|
| Fiberboard, ½ in. thick | 5 |
| Fiberboard, flameproofed, ½ in. thick | 10 |
| Fiberboard, ½ in. thick, with ½ in.-1:2, 1:2 gypsum-sand plaster | 15 |
| Gypsum wallboard, ⅜ in. thick | 10 |
| Gypsum wallboard, ½ in. thick | 15 |
| Gypsum wallboard, ⅝ in. thick | 20 |
| Gypsum wallboard, laminated, two ⅜ in. | 28 |
| Gypsum wallboard, laminated, one ⅜ in. plus one ½ in. thick | 37 |
| Gypsum wallboard, laminated, two ½ in. thick | 47 |
| Gypsum wallboard, laminated, two ⅝ in. thick | 60 |
| Gypsum lath, plain or indented, ⅜ in. thick, with ½ in.-1:2, 1:2 gypsum-sand plaster | 20 |
| Gypsum lath, perforated, ⅜ in. thick, with ½ in.-1:2, 1:2 gypsum-sand plaster | 30 |
| Gypsum-sand plaster, 1:2, 1:3, ½ in. thick, on wood lath | 15 |
| Lime-sand plaster, 1:5, 1:7.5, ½ in. thick, on wood lath | 15 |
| Gypsum-sand plaster, 1:2, 1:2, ¾ in. thick, on metal lath (no paper backing) | 15 |
| Neat gypsum plaster, ¾ in. thick, on metal lath (no paper backing)[b] | 30 |
| Neat gypsum plaster, 1 in. thick, on metal lath (no paper backing)[b] | 35 |
| Lime-sand plaster, 1:5, 1:7.5, ¾ in. thick, on metal lath (no paper backing) | 10 |
| Portland cement plaster, ¾ in. thick, on metal lath (no paper backing) | 10 |
| Gypsum-sand plaster, 1:2, 1:3, ¾ in. thick, on paper-backed metal lath | 20 |

Note: For SI units, 1 in. = 25.4 mm.
[a]From National Institute for Standards and Technology, BMS-92.
[b]Unsanded wood-fiber plaster.

Conductors marked with a rated temperature higher than that of the raceway can be used when the conductors are operated within the raceway temperature rating.

## 362.12 Uses Not Permitted

ENT shall not be used in the following:

(1) In any hazardous (classified) location, except as permitted by other articles in this *Code*
(2) For the support of luminaires and other equipment
(3) Where subject to ambient temperatures in excess of 50°C (122°F) unless listed otherwise
(4) For direct earth burial
(5) Where the voltage is over 600 volts
(6) In exposed locations, except as permitted by 362.10(1), 362.10(5), and 362.10(7)
(7) In theaters and similar locations, except as provided in 518.4 and 520.5
(8) Where exposed to the direct rays of the sun, unless identified as sunlight resistant
(9) Where subject to physical damage

## 362.20 Size

**(A) Minimum.** ENT smaller than metric designator 16 (trade size ½) shall not be used.

**(B) Maximum.** ENT larger than metric designator 53 (trade size 2) shall not be used.

Informational Note: See 300.1(C) for the metric designators and trade sizes. These are for identification purposes only and do not relate to actual dimensions.

## 362.22 Number of Conductors

The number of conductors shall not exceed that permitted by the percentage fill in Table 1, Chapter 9.

Cables shall be permitted to be installed where such use is not prohibited by the respective cable articles. The number of cables shall not exceed the allowable percentage fill specified in Table 1, Chapter 9.

Table 4 of Chapter 9 provides the usable area within the selected conduit or tubing, and Table 5 provides the required area for each conductor. Examples using these tables to calculate a conduit or tubing size are provided in the commentary following Chapter 9, Notes to Tables, Note 6.

If the conductors are of the same wire size, Informative Annex C can be used instead of performing the calculations. This annex, through 12 sets of tables, accurately indicates the maximum number of conductors permitted in conduit or tubing. Examples using Informative Annex C to select a conduit or tubing size are provided in the commentary following the annex introduction.

To select the proper trade size of ENT, see "Article 362 — Electrical Nonmetallic Tubing (ENT)" in Table 4 of Chapter 9. Using Tables C.2 and C.2(A) in Informative Annex C for ENT is also permissible.

## 362.24 Bends — How Made

Bends shall be so made that the tubing will not be damaged and the internal diameter of the tubing will not be effectively reduced. Bends shall be permitted to be made manually without auxiliary equipment, and the radius of the curve to the centerline of such bends shall not be less than shown in Table 2, Chapter 9 using the column "Other Bends."

## 362.26 Bends — Number in One Run

There shall not be more than the equivalent of four quarter bends (360 degrees total) between pull points, for example, conduit bodies and boxes.

### 362.28 Trimming

All cut ends shall be trimmed inside and outside to remove rough edges.

### 362.30 Securing and Supporting

ENT shall be installed as a complete system in accordance with 300.18 and shall be securely fastened in place and supported in accordance with 362.30(A) and (B).

**(A) Securely Fastened.** ENT shall be securely fastened at intervals not exceeding 900 mm (3 ft). In addition, ENT shall be securely fastened in place within 900 mm (3 ft) of each outlet box, device box, junction box, cabinet, or fitting where it terminates.

*Exception No. 1: Lengths not exceeding a distance of 1.8 m (6 ft) from a luminaire terminal connection for tap connections to lighting luminaires shall be permitted without being secured.*

*Exception No. 2: Lengths not exceeding 1.8 m (6 ft) from the last point where the raceway is securely fastened for connections within an accessible ceiling to luminaire(s) or other equipment.*

As illustrated in Exhibit 362.4, where run on the surface of framing members, ENT is required to be fastened to the framing member every 3 feet and within 3 feet of every box. See 300.4(D) for requirements for protection against physical damage.

As illustrated in Exhibit 362.5, ENT is permitted by Exception No. 1 to be used as luminaire whip without support for lengths not exceeding 6 feet. See 410.117(C) for details on tap conductor wiring.

*Exception No. 3: For concealed work in finished buildings or prefinished wall panels where such securing is impracticable, unbroken lengths (without coupling) of ENT shall be permitted to be fished.*

**(B) Supports.** Horizontal runs of ENT supported by openings in framing members at intervals not exceeding 900 mm (3 ft) and securely fastened within 900 mm (3 ft) of termination points shall be permitted.

### 362.46 Bushings

Where a tubing enters a box, fitting, or other enclosure, a bushing or adapter shall be provided to protect the wire from abrasion unless the box, fitting, or enclosure design provides equivalent protection.

Informational Note: See 300.4(G) for the protection of conductors size 4 AWG or larger.

### 362.48 Joints

All joints between lengths of tubing and between tubing and couplings, fittings, and boxes shall be by an approved method.

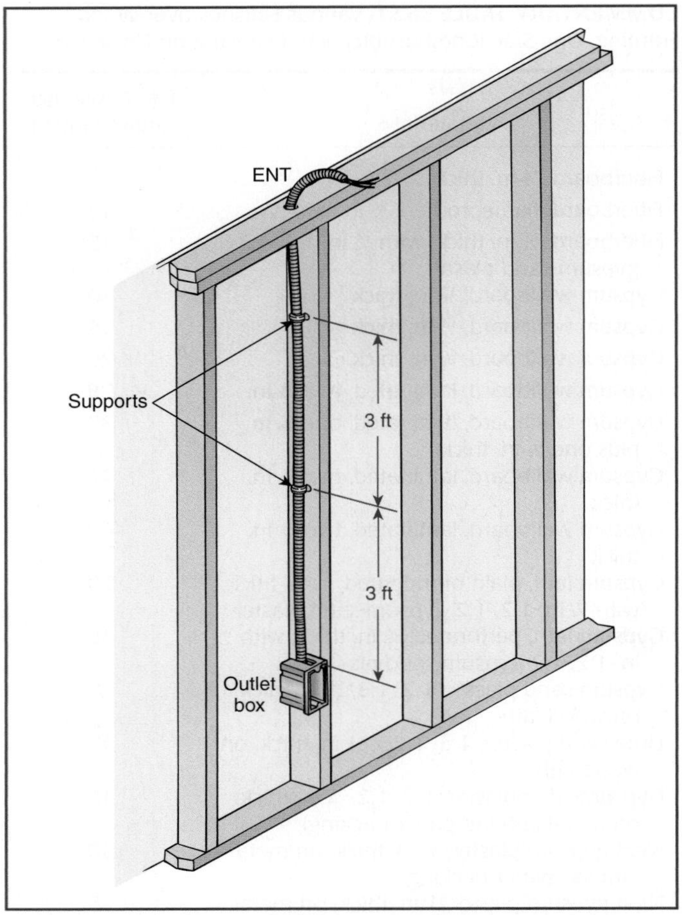


***EXHIBIT 362.4*** *An example showing ENT supported every 3 feet and within 3 feet of the outlet box.*

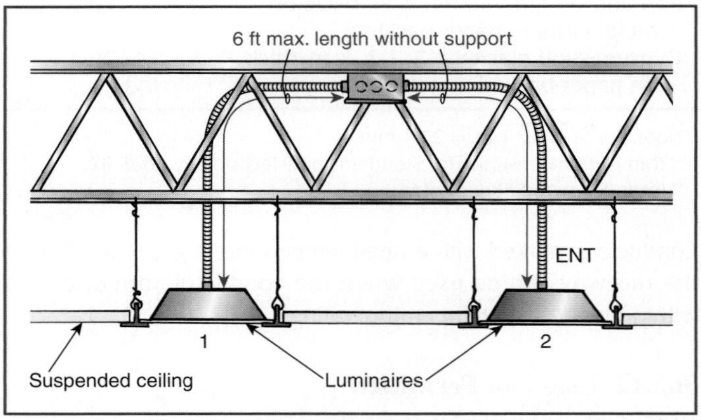


***EXHIBIT 362.5*** *An example showing two unsupported lengths of ENT, each permitted to be installed in a length not to exceed 6 feet.*

### 362.56 Splices and Taps

Splices and taps shall be made only in accordance with 300.15.

Informational Note: See Article 314 for rules on the installation and use of boxes and conduit bodies.

### 362.60 Grounding

Where equipment grounding is required, a separate equipment grounding conductor shall be installed in the raceway in compliance with Article 250, Part VI.

## III. Construction Specifications

### 362.100 Construction

ENT shall be made of material that does not exceed the ignitibility, flammability, smoke generation, and toxicity characteristics of rigid (nonplasticized) polyvinyl chloride.

ENT, as a prewired manufactured assembly, shall be provided in continuous lengths capable of being shipped in a coil, reel, or carton without damage.

### 362.120 Marking

ENT shall be clearly and durably marked at least every 3 m (10 ft) as required in the first sentence of 110.21(A). The type of material shall also be included in the marking. Marking for limited smoke shall be permitted on the tubing that has limited smoke-producing characteristics.

The type, size, and quantity of conductors used in prewired manufactured assemblies shall be identified by means of a printed tag or label attached to each end of the manufactured assembly and either the carton, coil, or reel. The enclosed conductors shall be marked in accordance with 310.120.

# ARTICLE 366 Auxiliary Gutters

## I. General

### 366.1 Scope

This article covers the use, installation, and construction requirements of metallic auxiliary gutters and nonmetallic auxiliary gutters and associated fittings.

An auxiliary gutter provides additional gutter space for wiring in various types of electrical enclosures and equipment. This additional gutter space may be necessary to provide sufficient room for the number of conductors in an enclosure or to provide adequate wiring bending/deflection space where conductors connect to a terminal. Although the construction of an auxiliary gutter is no different from that of a wireway, the field application of this equipment differentiates an auxiliary gutter from a wireway.

A wireway is a raceway in accordance with the definition of *raceway* in Article 100. Auxiliary gutters supplement enclosure wiring spaces and are not encompassed by the definition of raceway. Therefore, *Code* requirements that apply only to raceways do not apply to auxiliary gutters. An example of such a requirement is 230.7, which prohibits service conductors from being installed in a raceway with conductors that are not service conductors. This rule applies to wireways installed in accordance with Article 376 and 378. However, an auxiliary gutter installed to supplement the wiring space of a service equipment enclosure is not a wireway and, therefore, is not subject to 230.7.

### 366.2 Definitions

**Metallic Auxiliary Gutter.** A sheet metal enclosure used to supplement wiring spaces at meter centers, distribution centers, switchgear, switchboards, and similar points of wiring systems. The enclosure has hinged or removable covers for housing and protecting electrical wires, cable, and busbars. The enclosure is designed for conductors to be laid or set in place after the enclosures have been installed as a complete system.

**Nonmetallic Auxiliary Gutter.** A flame-retardant, nonmetallic enclosure used to supplement wiring spaces at meter centers, distribution centers, switchgear, switchboards, and similar points of wiring systems. The enclosure has hinged or removable covers for housing and protecting electrical wires, cable, and busbars. The enclosure is designed for conductors to be laid or set in place after the enclosures have been installed as a complete system.

### 366.6 Listing Requirements

**(A) Outdoors.** Nonmetallic auxiliary gutters installed outdoors shall comply with the following:

(1) Be listed as suitable for exposure to sunlight
(2) Be listed as suitable for use in wet locations
(3) Be listed for maximum ambient temperature of the installation

**(B) Indoors.** Nonmetallic auxiliary gutters installed indoors shall be listed for the maximum ambient temperature of the installation.

The sections and associated fittings of auxiliary gutters are identical to those of wireways. They differ only in their intended use. If listed, these may be marked as "Wireway," "Auxiliary Gutter," or "Wireway or Auxiliary Gutter" depending on the intended application. See 366.1 and associated commentary for a further understanding of auxiliary gutters. See the commentary following 376.1 for a comparative discussion.

## II. Installation

### 366.10 Uses Permitted

**(A) Sheet Metallic Auxiliary Gutters.**

**(1) Indoor and Outdoor Use.** Sheet metallic auxiliary gutters shall be permitted for indoor and outdoor use.

**(2) Wet Locations.** Sheet metallic auxiliary gutters installed in wet locations shall be suitable for such locations.

**(B) Nonmetallic Auxiliary Gutters.** Nonmetallic auxiliary gutters shall be listed for the maximum ambient temperature of the installation and marked for the installed conductor insulation temperature rating.

**(1) Outdoors.** Nonmetallic auxiliary gutters shall be permitted to be installed outdoors where listed and marked as suitable for the purpose.

Informational Note: Extreme cold may cause nonmetallic auxiliary gutters to become brittle and therefore more susceptible to damage from physical contact.

**(2) Indoors.** Nonmetallic auxiliary gutters shall be permitted to be installed indoors.

Both metal and nonmetallic gutters must have expansion fittings where temperature fluctuations are expected to change gutter length more than ¼ inch, according to 366.44. See the informational note following 378.44 regarding expansion characteristics of PVC conduit and PVC nonmetallic wireway. Also see the commentary example following 300.7(B).

### 366.12 Uses Not Permitted

Auxiliary gutters shall not be used:

(1) To enclose switches, overcurrent devices, appliances, or other similar equipment
(2) To extend a greater distance than 9 m (30 ft) beyond the equipment that it supplements

*Exception: As permitted in 620.35 for elevators, an auxiliary gutter shall be permitted to extend a distance greater than 9 m (30 ft) beyond the equipment it supplements.*

Informational Note: For wireways, see Articles 376 and 378. For busways, see Article 368.

### 366.22 Number of Conductors

**(A) Sheet Metallic Auxiliary Gutters.** The sum of the cross-sectional areas of all contained conductors at any cross section of a sheet metallic auxiliary gutter shall not exceed 20 percent of the interior cross-sectional area of the sheet metallic auxiliary gutter. The adjustment factors in 310.15(B)(3)(a) shall be applied only where the number of current-carrying conductors, including neutral conductors classified as current-carrying under the provisions of 310.15(B)(5), exceeds 30. Conductors for signaling circuits or controller conductors between a motor and its starter and used only for starting duty shall not be considered as current-carrying conductors.

**(B) Nonmetallic Auxiliary Gutters.** The sum of cross-sectional areas of all contained conductors at any cross section of the nonmetallic auxiliary gutter shall not exceed 20 percent of the interior cross-sectional area of the nonmetallic auxiliary gutter.

The dimensions of insulated conductors, found in Tables 5 and 5A of Chapter 9, can be used to calculate the size of auxiliary gutters.

Where sheet metal auxiliary gutters contain 30 or fewer current-carrying conductors, the correction factors in 310.15(B)(2) do not apply. However, if more than 30 conductors are installed in a sheet metal auxiliary gutter, the ampacity adjustment factors of 310.15(B)(3)(a) apply, and the number of current-carrying conductors is not limited up to the 20-percent fill.

The requirements for nonmetallic auxiliary gutters limit the cross-sectional area of all conductors to 20 percent. There is no 30-conductor allowance. The derating factors specified in 310.15(B)(3)(a) must be applied as stated in 366.23(B).

See the example for calculating the size of a wireway in the commentary following 376.22. This calculation method is also applicable to auxiliary gutters.

No limit is placed on the size of conductors that may be installed in an auxiliary gutter; however, see 366.23(A) for ampacity limitations of bare copper or aluminum busbars enclosed in gutters.

### 366.23 Ampacity of Conductors

**(A) Sheet Metallic Auxiliary Gutters.** Where the number of current-carrying conductors contained in the sheet metallic auxiliary gutter is 30 or less, the adjustment factors specified in 310.15(B)(3)(a) shall not apply. The current carried continuously in bare copper bars in sheet metallic auxiliary gutters shall not exceed 1.55 amperes/mm$^2$ (1000 amperes/in.$^2$) of cross section of the conductor. For aluminum bars, the current carried continuously shall not exceed 1.09 amperes/mm$^2$ (700 amperes/in.$^2$) of cross section of the conductor.

**(B) Nonmetallic Auxiliary Gutters.** The adjustment factors specified in 310.15(B)(3)(a) shall be applicable to the current-carrying conductors in the nonmetallic auxiliary gutter.

### 366.30 Securing and Supporting

**(A) Sheet Metallic Auxiliary Gutters.** Sheet metallic auxiliary gutters shall be supported and secured throughout their entire length at intervals not exceeding 1.5 m (5 ft).

**(B) Nonmetallic Auxiliary Gutters.** Nonmetallic auxiliary gutters shall be supported and secured at intervals not to exceed 900 mm (3 ft) and at each end or joint, unless listed for other support intervals. In no case shall the distance between supports exceed 3 m (10 ft).

### 366.44 Expansion Fittings

Expansion fittings shall be installed where expected length change, due to expansion and contraction due to temperature change, is more than 6 mm (0.25 in.).

### 366.56 Splices and Taps

Splices and taps shall comply with 366.56(A) through (D).

**(A) Within Gutters.** Splices or taps shall be permitted within gutters where they are accessible by means of removable covers or doors. The conductors, including splices and taps, shall not fill the gutter to more than 75 percent of its area.

**(B) Bare Conductors.** Taps from bare conductors shall leave the gutter opposite their terminal connections, and conductors shall not be brought in contact with uninsulated current-carrying parts of different potential.

**(C) Suitably Identified.** All taps shall be suitably identified at the gutter as to the circuit or equipment that they supply.

**(D) Overcurrent Protection.** Tap connections from conductors in auxiliary gutters shall be provided with overcurrent protection as required in 240.21.

### 366.58 Insulated Conductors

**(A) Deflected Insulated Conductors.** Where insulated conductors are deflected within an auxiliary gutter, either at the ends or where conduits, fittings, or other raceways or cables enter or leave the gutter, or where the direction of the gutter is deflected greater than 30 degrees, dimensions corresponding to one wire per terminal in Table 312.6(A) shall apply.

**(B) Auxiliary Gutters Used as Pull Boxes.** Where insulated conductors 4 AWG or larger are pulled through an auxiliary gutter, the distance between raceway and cable entries enclosing the same conductor shall not be less than that required in 314.28(A)(1) for straight pulls and 314.28(A)(2) for angle pulls.

### 366.60 Grounding

Metallic auxiliary gutters shall be connected to an equipment grounding conductor(s), to an equipment bonding jumper, or to the grounded conductor where permitted or required by 250.92(B)(1) or 250.142.

## III. Construction Specifications

### 366.100 Construction

**(A) Electrical and Mechanical Continuity.** Gutters shall be constructed and installed so that adequate electrical and mechanical continuity of the complete system is secured.

**(B) Substantial Construction.** Gutters shall be of substantial construction and shall provide a complete enclosure for the contained conductors. All surfaces, both interior and exterior, shall be suitably protected from corrosion. Corner joints shall be made tight, and where the assembly is held together by rivets, bolts, or screws, such fasteners shall be spaced not more than 300 mm (12 in.) apart.

**(C) Smooth Rounded Edges.** Suitable bushings, shields, or fittings having smooth, rounded edges shall be provided where conductors pass between gutters, through partitions, around bends, between gutters and cabinets or junction boxes, and at other locations where necessary to prevent abrasion of the insulation of the conductors.

**(D) Covers.** Covers shall be securely fastened to the gutter.

**(E) Clearance of Bare Live Parts.** Bare conductors shall be securely and rigidly supported so that the minimum clearance between bare current-carrying metal parts of different potential mounted on the same surface will not be less than 50 mm (2 in.), nor less than 25 mm (1 in.) for parts that are held free in the air. A clearance not less than 25 mm (1 in.) shall be secured between bare current-carrying metal parts and any metal surface. Adequate provisions shall be made for the expansion and contraction of busbars.

### 366.120 Marking

**(A) Outdoors.** Nonmetallic auxiliary gutters installed outdoors shall have the following markings:

(1) Suitable for exposure to sunlight
(2) Suitable for use in wet locations
(3) Installed conductor insulation temperature rating

**(B) Indoors.** Nonmetallic auxiliary gutters installed indoors shall be marked with the installed conductor insulation temperature rating.

# ARTICLE 368
# Busways

## I. General Requirements

### 368.1 Scope

This article covers service-entrance, feeder, and branch-circuit busways and associated fittings.

The metal enclosure of a listed busway is intended for use as an EGC. In some cases, an additional grounding bus may also act as an EGC. Busways not intended for use ahead of service equipment are marked with the maximum rating of overcurrent protection required on the supply side.

Four busway designs are available:

1. A lighting busway, with a maximum current rating of 50 amperes, supplies and supports luminaires.
2. A trolley busway allows continuous contact with a trolley through a slot in the enclosure and may also be marked as lighting busway if intended for use with luminaires.
3. A continuous plug-in busway allows for the insertion of plug-in devices at any point along its length. This busway, limited to a maximum current rating of 225 amperes, is intended for general use and may be installed within reach of persons.
4. A short-run busway is intended primarily to feed switchboards and is limited to a run of 30 feet horizontal or 10 feet vertical.

Busways are marked when suitable for installation in a specified position, for use in vertical runs, for support at intervals

*EXHIBIT 368.1* *A 10-foot section of feeder busway, which is one of four types of busways covered.*

greater than 5 feet, or for use outdoors. A busway or fitting containing a vapor seal has not been investigated for passage through a fire-rated wall unless marked otherwise.

Exhibit 368.1 illustrates busways as covered in this article. See also Exhibit 100.1.

### 368.2 Definition

**Busway.** A raceway consisting of a grounded metal enclosure containing factory-mounted, bare or insulated conductors, which are usually copper aluminum bars, rods, or tubes.

Informational Note: For cablebus, refer to Article 370.

## II. Installation

### 368.10 Uses Permitted

Busways shall be permitted to be installed where they are located in accordance with 368.10(A) through (C).

**(A) Exposed.** Busways shall be permitted to be located in the open where visible, except as permitted in 368.10(C).

**(B) Behind Access Panels.** Busways shall be permitted to be installed behind access panels, provided the busways are totally enclosed, of nonventilating-type construction, and installed so that the joints between sections and at fittings are accessible for maintenance purposes. Where installed behind access panels, means of access shall be provided, and either of the following conditions shall be met:

(1) The space behind the access panels shall not be used for air-handling purposes.
(2) Where the space behind the access panels is used for environmental air, other than ducts and plenums, there shall be no provisions for plug-in connections, and the conductors shall be insulated.

Busways are commonly used as feeders and are mounted horizontally in industrial buildings or mounted vertically in high-rise buildings. See Exhibit 368.2 for an illustration of a busway installed in the space above a dropped or hung ceiling. The space is not being used as an "other space for environmental air" and

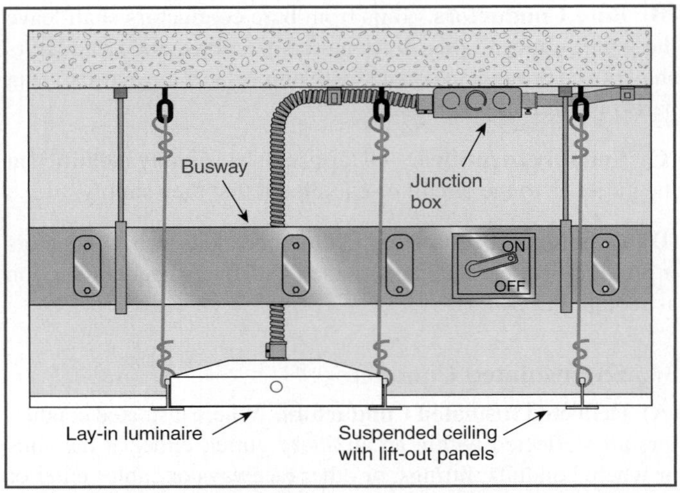


*EXHIBIT 368.2* *An example of a busway mounted in the space above a hung ceiling.*

can contain the plug-in devices for supplying luminaires or other electrical equipment.

**(C) Through Walls and Floors.** Busways shall be permitted to be installed through walls or floors in accordance with (C)(1) and (C)(2).

**(1) Walls.** Unbroken lengths of busway shall be permitted to be extended through dry walls.

**(2) Floors.** Floor penetrations shall comply with (a) and (b):

(a) Busways shall be permitted to be extended vertically through dry floors if totally enclosed (unventilated) where passing through and for a minimum distance of 1.8 m (6 ft) above the floor to provide adequate protection from physical damage.

(b) In other than industrial establishments, where a vertical riser penetrates two or more dry floors, a minimum 100-mm (4-in.) high curb shall be installed around all floor openings for riser busways to prevent liquids from entering the opening. The curb shall be installed within 300 mm (12 in.) of the floor opening. Electrical equipment shall be located so that it will not be damaged by liquids that are retained by the curb.

Informational Note: See 300.21 for information concerning the spread of fire or products of combustion.

A busway or fitting containing a vapor seal has not been investigated for passage through a fire-rated wall unless marked otherwise. The requirements of 300.21 are extremely important in order to confine a fire and the products of combustion at their origin.

The addition of a 4-inch curb encircling the busway can help eliminate the possibility that liquid spilled on an upper floor will trickle down the vertical rise of the busway, causing damage to the electrical system.

### 368.12 Uses Not Permitted

**(A) Physical Damage.** Busways shall not be installed where subject to severe physical damage or corrosive vapors.

**(B) Hoistways.** Busways shall not be installed in hoistways.

**(C) Hazardous Locations.** Busways shall not be installed in any hazardous (classified) location, unless specifically approved for such use.

Informational Note: See 501.10(B).

**(D) Wet Locations.** Busways shall not be installed outdoors or in wet or damp locations unless identified for such use.

**(E) Working Platform.** Lighting busway and trolley busway shall not be installed less than 2.5 m (8 ft) above the floor or working platform unless provided with an identified cover.

### 368.17 Overcurrent Protection

Overcurrent protection shall be provided in accordance with 368.17(A) through (D).

**(A) Rating of Overcurrent Protection — Feeders.** A busway shall be protected against overcurrent in accordance with the allowable current rating of the busway.

*Exception No. 1: The applicable provisions of 240.4 shall be permitted.*

*Exception No. 2: Where used as transformer secondary ties, the provisions of 450.6(A)(3) shall be permitted.*

The rated ampacity of a busway is based on the allowable temperature rise of the conductors and can be determined in the field only by reference to the nameplate data. The applicable sections of 240.4 referenced in Exception No. 1 are 240.4(B) and 240.4(C).

**(B) Reduction in Ampacity Size of Busway.** Overcurrent protection shall be required where busways are reduced in ampacity.

*Exception: For industrial establishments only, omission of overcurrent protection shall be permitted at points where busways are reduced in ampacity, provided that the length of the busway having the smaller ampacity does not exceed 15 m (50 ft) and has an ampacity at least equal to one-third the rating or setting of the overcurrent device next back on the line, and provided that such busway is free from contact with combustible material.*

In industrial establishments, where the size of a smaller busway is kept within the specified limits, the additional cost of providing overcurrent protection at the point where the size is changed is not warranted. For example, a busway protected by a 1200-ampere overcurrent device may be reduced in size, provided the smaller busway has a current rating of 400 amperes (⅓ of 1200 amperes) and does not extend more than 50 feet. In this case, overcurrent protection would be required if the smaller busway were rated less than 400 amperes (e.g., 200 amperes, 300 amperes).

**(C) Feeder or Branch Circuits.** Where a busway is used as a feeder, devices or plug-in connections for tapping off feeder or branch circuits from the busway shall contain the overcurrent devices required for the protection of the feeder or branch circuits. The plug-in device shall consist of an externally operable circuit breaker or an externally operable fusible switch. Where such devices are mounted out of reach and contain disconnecting means, suitable means such as ropes, chains, or sticks shall be provided for operating the disconnecting means from the floor.

*Exception No. 1: As permitted in 240.21.*

*Exception No. 2: For fixed or semifixed luminaires, where the branch-circuit overcurrent device is part of the luminaire cord plug on cord-connected luminaires.*

*Exception No. 3: Where luminaires without cords are plugged directly into the busway and the overcurrent device is mounted on the luminaire.*

**(D) Rating of Overcurrent Protection — Branch Circuits.** A busway used as a branch circuit shall be protected against overcurrent in accordance with 210.20.

### 368.30 Support

Busways shall be securely supported at intervals not exceeding 1.5 m (5 ft) unless otherwise designed and marked.

### 368.56 Branches from Busways

Branches from busways shall be permitted to be made in accordance with 368.56(A), (B), and (C).

**(A) General.** Branches from busways shall be permitted to use any of the following wiring methods:

(1) Type AC armored cable
(2) Type MC metal-clad cable
(3) Type MI mineral-insulated, metal-sheathed cable
(4) Type IMC intermediate metal conduit
(5) Type RMC rigid metal conduit
(6) Type FMC flexible metal conduit
(7) Type LFMC liquidtight flexible metal conduit
(8) Type PVC rigid polyvinyl chloride conduit
(9) Type RTRC reinforced thermosetting resin conduit
(10) Type LFNC liquidtight flexible nonmetallic conduit
(11) Type EMT electrical metallic tubing
(12) Type ENT electrical nonmetallic tubing
(13) Busways
(14) Strut-type channel raceway
(15) Surface metal raceway
(16) Surface nonmetallic raceway

Where a separate equipment grounding conductor is used, connection of the equipment grounding conductor to the busway shall comply with 250.8 and 250.12.

**(B) Cord and Cable Assemblies.** Suitable cord and cable assemblies approved for extra-hard usage or hard usage and listed bus drop cable shall be permitted as branches from busways for the connection of portable equipment or the connection of stationary equipment to facilitate their interchange in accordance with 400.7 and 400.8 and the following conditions:

(1) The cord or cable shall be attached to the building by an approved means.
(2) The length of the cord or cable from a busway plug-in device to a suitable tension take-up support device shall not exceed 1.8 m (6 ft).
(3) The cord and cable shall be installed as a vertical riser from the tension take-up support device to the equipment served.
(4) Strain relief cable grips shall be provided for the cord or cable at the busway plug-in device and equipment terminations.

*Exception to (B)(2): In industrial establishments only, where the conditions of maintenance and supervision ensure that only qualified persons service the installation, lengths exceeding 1.8 m (6 ft) shall be permitted between the busway plug-in device and the tension take-up support device where the cord or cable is supported at intervals not exceeding 2.5 m (8 ft).*

Exhibit 368.3 illustrates a cable or cord branch from a busway installed according to the requirements of 368.56(B)(2). Section 400.9 specifically prohibits the installation of spliced cords.

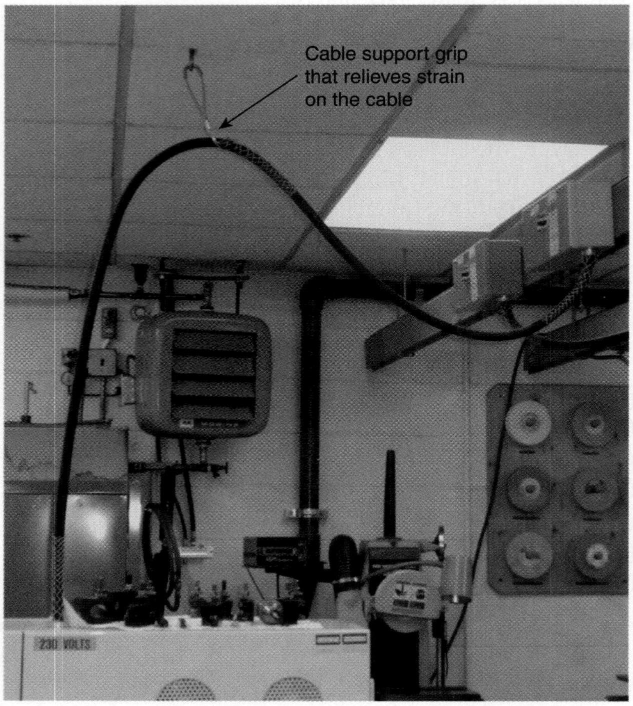


***EXHIBIT 368.3*** *An example of a cable or cord branch from a busway installed according to 368.56(B).*

Exhibit 368.4 illustrates a cable or cord branch from a busway installed according to 368.56(B)(2), Exception.

**(C) Branches from Trolley-Type Busways.** Suitable cord and cable assemblies approved for extra-hard usage or hard usage and listed bus drop cable shall be permitted as branches from trolley-type busways for the connection of movable equipment in accordance with 400.7 and 400.8.

### 368.58 Dead Ends

A dead end of a busway shall be closed.

### 368.60 Grounding

Busway shall be connected to an equipment grounding conductor(s), to an equipment bonding jumper, or to the grounded conductor where permitted or required by 250.92(B)(1) or 250.142.

## III. Construction

### 368.120 Marking

Busways shall be marked with the voltage and current rating for which they are designed, and with the manufacturer's name or trademark in such a manner as to be visible after installation.

## IV. Requirements for Over 600 Volts, Nominal

### 368.214 Adjacent and Supporting Structures

Metal-enclosed busways shall be installed so that temperature rise from induced circulating currents in any adjacent metallic parts will not be hazardous to personnel or constitute a fire hazard.

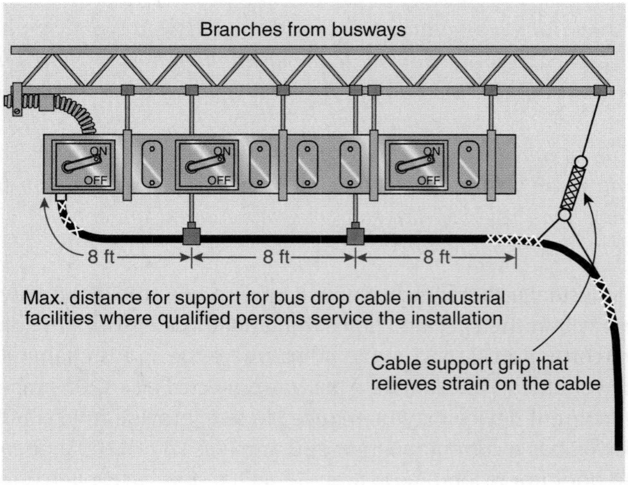


***EXHIBIT 368.4*** *An example of an installation permitted only in industrial occupancies with other restrictions.*

### 368.234 Barriers and Seals

**(A) Vapor Seals.** Busway runs that have sections located both inside and outside of buildings shall have a vapor seal at the building wall to prevent interchange of air between indoor and outdoor sections.

*Exception: Vapor seals shall not be required in forced-cooled bus.*

**(B) Fire Barriers.** Fire barriers shall be provided where fire walls, floors, or ceilings are penetrated.

Informational Note: See 300.21 for information concerning the spread of fire or products of combustion.

### 368.236 Drain Facilities

Drain plugs, filter drains, or similar methods shall be provided to remove condensed moisture from low points in busway run.

### 368.237 Ventilated Bus Enclosures

Ventilated busway enclosures shall be installed in accordance with Article 110, Part III, and 490.24.

### 368.238 Terminations and Connections

Where bus enclosures terminate at machines cooled by flammable gas, seal-off bushings, baffles, or other means shall be provided to prevent accumulation of flammable gas in the busway enclosures.

All conductor termination and connection hardware shall be accessible for installation, connection, and maintenance.

### 368.239 Switches

Switching devices or disconnecting links provided in the busway run shall have the same momentary rating as the busway. Disconnecting links shall be plainly marked to be removable only when bus is de-energized. Switching devices that are not load-break shall be interlocked to prevent operation under load, and disconnecting link enclosures shall be interlocked to prevent access to energized parts.

### 368.240 Wiring 600 Volts or Less, Nominal

Secondary control devices and wiring that are provided as part of the metal-enclosed bus run shall be insulated by fire-retardant barriers from all primary circuit elements with the exception of short lengths of wire, such as at instrument transformer terminals.

### 368.244 Expansion Fittings

Flexible or expansion connections shall be provided in long, straight runs of bus to allow for temperature expansion or contraction, or where the busway run crosses building vibration insulation joints.

### 368.258 Neutral Conductor

Neutral bus, where required, shall be sized to carry all neutral load current, including harmonic currents, and shall have adequate momentary and short-circuit rating consistent with system requirements.

### 368.260 Grounding

Metal-enclosed busway shall be grounded.

### 368.320 Marking

Each busway run shall be provided with a permanent nameplate on which the following information shall be provided:

(1) Rated voltage.
(2) Rated continuous current; if bus is forced-cooled, both the normal forced-cooled rating and the self-cooled (not forced-cooled) rating for the same temperature rise shall be given.
(3) Rated frequency.
(4) Rated impulse withstand voltage.
(5) Rated 60-Hz withstand voltage (dry).
(6) Rated momentary current.
(7) Manufacturer's name or trademark.

Informational Note: See ANSI C37.23-1987 (R1991), *Guide for Metal-Enclosed Bus and Calculating Losses in Isolated-Phase Bus*, for construction and testing requirements for metal-enclosed buses.

# ARTICLE 370
# Cablebus

## I. General

### 370.1 Scope

This article covers the use and installation requirements of cablebus and associated fittings.

Cablebus consists of a metal structure or framework installed in a manner similar to that of a cable tray support system. As illustrated in Exhibit 370.1, insulated conductors of 1/0 AWG or larger are field installed within the framework on special insulating blocks at specified intervals to provide controlled spacing between conductors. To completely enclose the conductors, a ventilated top cover is attached to the framework.

### 370.2 Definition

**Cablebus.** An assembly of units or sections with insulated conductors having associated fittings forming a structural system used to securely fasten or support conductors and conductor terminations in a completely enclosed, ventilated, protective metal housing. This assembly is designed to carry fault current and to withstand the magnetic forces of such current.

Informational Note: Cablebus is ordinarily assembled at the point of installation from the components furnished or specified by the manufacturer in accordance with instructions for the specific job.

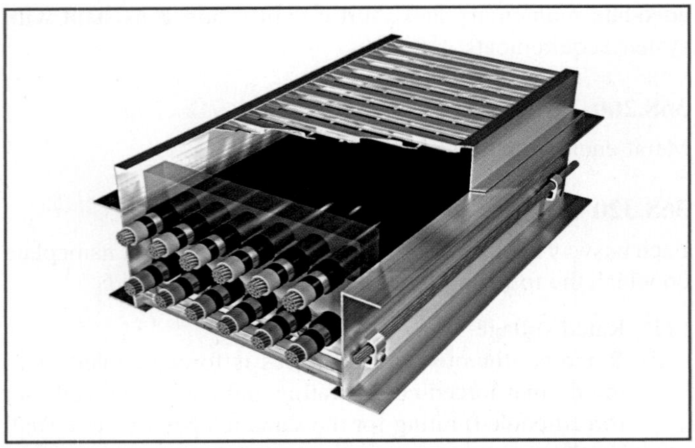

*EXHIBIT 370.1 A section of cablebus with conductors in place and the ventilated top cover ready to be attached to the busway frame. (Courtesy of MP Husky Cable Bus & Cable Tray)*

# II. Installation

## 370.10 Uses Permitted

Approved cablebus shall be permitted:

(1) At any voltage or current for which spaced conductors are rated and where installed only for exposed work, except as permitted in 370.18
(2) For branch circuits, feeders, and services
(3) To be installed outdoors or in corrosive, wet, or damp locations where identified for the use

## 370.12 Uses Not Permitted

Cablebus shall not be permitted to be installed:

(1) In hoistways
(2) In hazardous (classified) locations, unless specifically approved for the use

•

## 370.18 Cablebus Installation

•

**(A) Transversely Routed.** Cablebus shall be permitted to extend transversely through partitions or walls, other than fire walls, provided that the section within the wall is continuous, protected against physical damage, and unventilated.

**(B) Through Dry Floors and Platforms.** Except where firestops are required, cablebus shall be permitted to extend vertically through dry floors and platforms, provided that the cablebus is totally enclosed at the point where it passes through the floor or platform and for a distance of 1.8 m (6 ft) above the floor or platform.

**(C) Through Floors and Platforms in Wet Locations.** Except where firestops are required, cablebus shall be permitted to extend vertically through floors and platforms in wet locations where:

(1) There are curbs or other suitable means to prevent waterflow through the floor or platform opening, and
(2) Where the cablebus is totally enclosed at the point where it passes through the floor or platform and for a distance of 1.8 m (6 ft) above the floor or platform.

## 370.20 Conductor Size and Termination

**(A) Conductors.** The current-carrying conductors in cablebus shall:

(1) Have an insulation rating of 75°C (167°F) or higher and be of an approved type suitable for the application.
(2) Be sized in accordance with the design of the cablebus but in no case be smaller than 1/0.

**(B) Termination.** Approved terminating means shall be used for connections to cablebus conductors.

## 370.22 Number of Conductors

The number of conductors shall be that for which the cablebus is designed.

## 370.23 Overcurrent Protection

Cablebus shall be protected against overcurrent in accordance with the allowable ampacity of the cablebus conductors in accordance with 240.4.

*Exception: Overcurrent protection shall be permitted in accordance with 240.100 and 240.101 for over 1000 volts, nominal.*

## 370.30 Securing and Supporting

**(A) Cablebus Supports.** Cablebus shall be securely supported at intervals not exceeding 3.7 m (12 ft). Where spans longer than 3.7 m (12 ft) are required, the structure shall be specifically designed for the required span length.

**(B) Conductor Supports.** The insulated conductors shall be supported on blocks or other identified mounting means.

The individual conductors in a cablebus shall be supported at intervals not greater than 900 mm (3 ft) for horizontal runs and 450 mm (1½ ft) for vertical runs. Vertical and horizontal spacing between supported conductors shall be not less than one conductor diameter at the points of support.

## 370.42 Fittings

A cablebus system shall include approved fittings for the following:

(1) Changes in horizontal or vertical direction of the run
(2) Dead ends
(3) Terminations in or on connected apparatus or equipment or the enclosures for such equipment

(4) Additional physical protection where required, such as guards where subject to severe physical damage

### 370.60 Grounding

A cablebus system shall be grounded and/or bonded as applicable:

(1) Cablebus framework, where bonded, shall be permitted to be used as the equipment grounding conductor for branch circuits and feeders.
(2) A cablebus installation shall be grounded and bonded in accordance with Article 250, excluding 250.86, Exception No. 2.

### 370.80 Ampacity of Conductors

The ampacity of conductors in cablebus shall be in accordance with Table 310.15(B)(17) and Table 310.15(B)(19) for installations up to and including 2000 volts, or with Table 310.60(C)(69) and Table 310.60(C)(70) for installations 2001 to 35,000 volts.

## III. Construction Specifications

### 370.120 Marking

Each section of cablebus shall be marked with the manufacturer's name or trade designation and the maximum diameter, number, voltage rating, and ampacity of the conductors to be installed. Markings shall be located so as to be visible after installation.

# ARTICLE 372 Cellular Concrete Floor Raceways

### 372.1 Scope

This article covers cellular concrete floor raceways, the hollow spaces in floors constructed of precast cellular concrete slabs, together with suitable metal fittings designed to provide access to the floor cells.

Cellular concrete floor raceways are a form of floor deck construction commonly used in high-rise office buildings. This construction method is very similar in design, application, and adaptation to cellular metal floor raceways. Basically, this wiring method consists of floor cells (that are part of the structural floor system), header ducts laid at right angles to the cells (used to carry conductors from cabinets to cells), and junction boxes.

### 372.2 Definitions

**Cell.** A single, enclosed tubular space in a floor made of precast cellular concrete slabs, the direction of the cell being parallel to the direction of the floor member.

**Header.** Transverse metal raceways for electrical conductors, providing access to predetermined cells of a precast cellular concrete floor, thereby permitting the installation of electrical conductors from a distribution center to the floor cells.

### 372.4 Uses Not Permitted

Conductors shall not be installed in precast cellular concrete floor raceways as follows:

(1) Where subject to corrosive vapor
(2) In any hazardous (classified) location, except as permitted by other articles in this Code
(3) In commercial garages, other than for supplying ceiling outlets or extensions to the area below the floor but not above

Informational Note: See 300.8 for installation of conductors with other systems.

The installation of electrical conductors is prohibited in raceways or cable trays containing any pipes, tubes, or other means for carrying steam, water, air, gas, or drainage or for any service other than electrical.

### 372.5 Header

The header shall be installed in a straight line at right angles to the cells. The header shall be mechanically secured to the top of the precast cellular concrete floor. The end joints shall be closed by a metal closure fitting and sealed against the entrance of concrete. The header shall be electrically continuous throughout its entire length and shall be electrically bonded to the enclosure of the distribution center.

### 372.6 Connection to Cabinets and Other Enclosures

Connections from headers to cabinets and other enclosures shall be made by means of listed metal raceways and listed fittings.

### 372.7 Junction Boxes

Junction boxes shall be leveled to the floor grade and sealed against the free entrance of water or concrete. Junction boxes shall be of metal and shall be mechanically and electrically continuous with the header.

### 372.8 Markers

A suitable number of markers shall be installed for the future location of cells.

### 372.9 Inserts

Inserts shall be leveled and sealed against the entrance of concrete. Inserts shall be of metal and shall be fitted with grounded-type receptacles. A grounding conductor shall connect the insert receptacles to a positive ground connection provided on the header. Where cutting through the cell wall for setting inserts or other purposes (such as providing access openings between header and cells), chips and other dirt shall not be allowed to remain in the raceway, and the tool used shall be designed so as to prevent the tool from entering the cell and damaging the conductors.

### 372.10 Size of Conductors

No conductor larger than 1/0 AWG shall be installed, except by special permission.

### 372.11 Maximum Number of Conductors

The combined cross-sectional area of all conductors or cables shall not exceed 40 percent of the cross-sectional area of the cell or header.

### 372.12 Splices and Taps

Splices and taps shall be made only in header access units or junction boxes. A continuous unbroken conductor connecting the individual outlets is not a splice or tap.

### 372.13 Discontinued Outlets

When an outlet is abandoned, discontinued, or removed, the sections of circuit conductors supplying the outlet shall be removed from the raceway. No splices or reinsulated conductors, such as would be the case of abandoned outlets on loop wiring, shall be allowed in raceways.

### 372.17 Ampacity of Conductors

The ampacity adjustment factors, provided in 310.15(B)(3), shall apply to conductors installed in cellular concrete floor raceways.

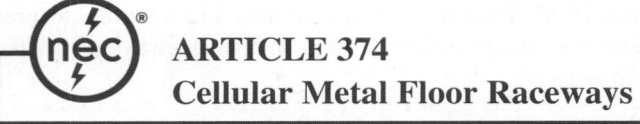

## ARTICLE 374 Cellular Metal Floor Raceways

### 374.1 Scope

This article covers the use and installation requirements for cellular metal floor raceways.

Cellular metal floor raceways, as shown in Exhibit 374.1, are a form of metal floor deck construction designed for use in steel-frame buildings and consist of sheet metal formed into shapes that are combined to form cells or raceways. The cells extend across the building and, depending on the structural strength required, can have various shapes and sizes.

Connections to the cells are made by means of headers extending across the cells and connecting only to those cells to be used as raceways for the conductors. Two or three separate headers, connecting to different sets of cells, may be used for different systems, such as light and power, signaling, and communications systems.

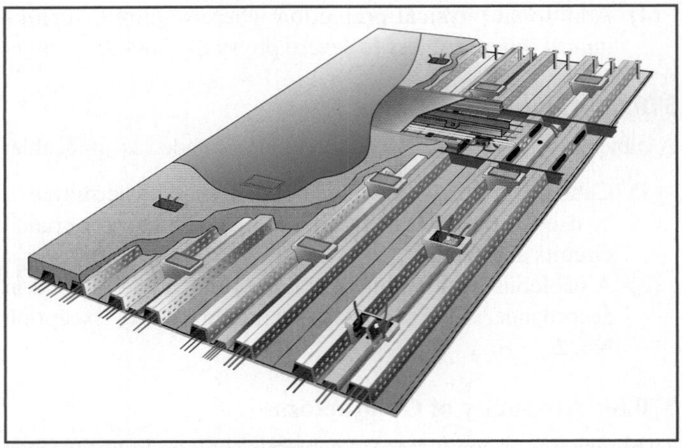

***EXHIBIT 374.1*** *An illustration of a cross section of cellular metal floor raceway system. (Courtesy of H.H. Robertson Floor Systems)*

### 374.2 Definitions

**Cellular Metal Floor Raceway.** The hollow spaces of cellular metal floors, together with suitable fittings, that may be approved as enclosed channel for electrical conductors.

*Cell.* A single enclosed tubular space in a cellular metal floor member, the axis of the cell being parallel to the axis of the metal floor member.

*Header.* A transverse raceway for electrical conductors, providing access to predetermined cells of a cellular metal floor, thereby permitting the installation of electrical conductors from a distribution center to the cells.

### 374.3 Uses Not Permitted

Conductors shall not be installed in cellular metal floor raceways as follows:

(1) Where subject to corrosive vapor
(2) In any hazardous (classified) location, except as permitted by other articles in this *Code*
(3) In commercial garages, other than for supplying ceiling outlets or extensions to the area below the floor but not above

Informational Note: See 300.8 for installation of conductors with other systems.

The installation of electrical conductors is prohibited in raceways or cable trays containing any pipes, tubes, or other carriers of steam, water, air, gas, or drainage, or for any service other than electrical.

## I. Installation

### 374.4 Size of Conductors

No conductor larger than 1/0 AWG shall be installed, except by special permission.

### 374.5 Maximum Number of Conductors in Raceway

The combined cross-sectional area of all conductors or cables shall not exceed 40 percent of the interior cross-sectional area of the cell or header.

### 374.6 Splices and Taps

Splices and taps shall be made only in header access units or junction boxes.

For the purposes of this section, so-called loop wiring (continuous unbroken conductor connecting the individual outlets) shall not be considered to be a splice or tap.

### 374.7 Discontinued Outlets

When an outlet is abandoned, discontinued, or removed, the sections of circuit conductors supplying the outlet shall be removed from the raceway. No splices or reinsulated conductors, such as would be the case with abandoned outlets on loop wiring, shall be allowed in raceways.

### 374.8 Markers

A suitable number of markers shall be installed for locating cells in the future.

Markers are typically brass flat-head screws set into the top side of the cells and adjusted so that their heads are flush with the floor finish. They are exposed in order to aid in the location of cells for future installations.

### 374.9 Junction Boxes

Junction boxes shall be leveled to the floor grade and sealed against the free entrance of water or concrete. Junction boxes used with these raceways shall be of metal and shall be electrically continuous with the raceway.

### 374.10 Inserts

Inserts shall be leveled to the floor grade and sealed against the entrance of concrete. Inserts shall be of metal and shall be electrically continuous with the raceway. In cutting through the cell wall and setting inserts, chips and other dirt shall not be allowed to remain in the raceway, and tools shall be used that are designed to prevent the tool from entering the cell and damaging the conductors.

### 374.11 Connection to Cabinets and Extensions from Cells

Connections between raceways and distribution centers and wall outlets shall be made by means of liquidtight flexible metal conduit, flexible metal conduit where not installed in concrete, rigid metal conduit, intermediate metal conduit, electrical metallic tubing, or approved fittings. Where there are provisions for the termination of an equipment grounding conductor, rigid polyvinyl chloride conduit, reinforced thermosetting resin conduit, electrical nonmetallic tubing, or liquidtight flexible nonmetallic conduit shall be permitted. Where installed in concrete, liquidtight flexible metal conduit and liquidtight flexible nonmetallic conduit shall be listed and marked for direct burial.

•

### 374.17 Ampacity of Conductors

The ampacity adjustment factors in 310.15(B)(3) shall apply to conductors installed in cellular metal floor raceways.

## II. Construction Specifications

### 374.100 General

Cellular metal floor raceways shall be constructed so that adequate electrical and mechanical continuity of the complete system will be secured. They shall provide a complete enclosure for the conductors. The interior surfaces shall be free from burrs and sharp edges, and surfaces over which conductors are drawn shall be smooth. Suitable bushings or fittings having smooth rounded edges shall be provided where conductors pass.

# ARTICLE 376 Metal Wireways

## I. General

### 376.1 Scope

This article covers the use, installation, and construction specifications for metal wireways and associated fittings.

Wireways are sheet-metal enclosures equipped with hinged or removable covers and are manufactured in 1-foot to 10-foot lengths and various widths and depths. Couplings, elbows, end plates, and accessories such as T and X fittings are available. Unlike auxiliary gutters, which are not permitted to extend more than 30 feet from the equipment they supplement, wireways may be run throughout an entire area. See the commentary following 366.2 for more information on the differences between wireways and auxiliary gutters.

### 376.2 Definition

**Metal Wireways.** Sheet metal troughs with hinged or removable covers for housing and protecting electrical wires and cable and in which conductors are laid in place after the raceway has been installed as a complete system.

## II. Installation

### 376.10 Uses Permitted

The use of metal wireways shall be permitted as follows:

(1) For exposed work.
(2) In any hazardous (classified) location, as permitted by other articles in this *Code*.

(3) In wet locations where wireways are listed for the purpose.
(4) In concealed spaces as an extension that passes transversely through walls, if the length passing through the wall is unbroken. Access to the conductors shall be maintained on both sides of the wall.

### 376.12 Uses Not Permitted

Metal wireways shall not be used in the following:

(1) Where subject to severe physical damage
(2) Where subject to severe corrosive environments

### 376.21 Size of Conductors

No conductor larger than that for which the wireway is designed shall be installed in any wireway.

### 376.22 Number of Conductors and Ampacity

The number of conductors and their ampacity shall comply with 376.22(A) and (B).

The requirements for the number of conductors for metal wireways are as follows:

- All conductors count in determining wireway fill.
- Total conductor fill must not exceed 20 percent of the wireway.

The conductor ampacity requirements for metal wireways are as follows:

- Conductor ampacity is determined according to 310.15.
- The adjustment factors of 310.15(B)(3)(a) only apply when the number of current-carrying conductors, including any current-carrying neutrals, exceeds 30.
- Signaling circuit conductors are not considered current-carrying conductors.
- Conductors used only for motor-starting duty as limited by 376.22(B) are not considered current-carrying conductors.

**(A) Cross-Sectional Areas of Wireway.** The sum of the cross-sectional areas of all contained conductors at any cross section of a wireway shall not exceed 20 percent of the interior cross-sectional area of the wireway.

**(B) Adjustment Factors.** The adjustment factors in 310.15(B)(3)(a) shall be applied only where the number of current-carrying conductors, including neutral conductors classified as current-carrying under the provisions of 310.15(B)(5), exceeds 30 at any cross section of the wireway. Conductors for signaling circuits or controller conductors between a motor and its starter and used only for starting duty shall not be considered as current-carrying conductors.

### 376.23 Insulated Conductors

Insulated conductors installed in a metallic wireway shall comply with 376.23(A) and (B).

**(A) Deflected Insulated Conductors.** Where insulated conductors are deflected within a metallic wireway, either at the ends or where conduits, fittings, or other raceways or cables enter or leave the metallic wireway, or where the direction of the metallic wireway is deflected greater than 30 degrees, dimensions corresponding to one wire per terminal in Table 312.6(A) shall apply.

**(B) Metallic Wireways Used as Pull Boxes.** Where insulated conductors 4 AWG or larger are pulled through a wireway, the distance between raceway and cable entries enclosing the same conductor shall not be less than that required by 314.28(A)(1) for straight pulls and 314.28(A)(2) for angle pulls. When transposing cable size into raceway size, the minimum metric designator (trade size) raceway required for the number and size of conductors in the cable shall be used.

The same minimum dimension requirements associated with raceway entries of pull boxes apply to metal wireways.

### 376.30 Securing and Supporting

Metal wireways shall be supported in accordance with 376.30(A) and (B).

**(A) Horizontal Support.** Wireways shall be supported where run horizontally at each end and at intervals not to exceed 1.5 m (5 ft) or for individual lengths longer than 1.5 m (5 ft) at each end or joint, unless listed for other support intervals. The distance between supports shall not exceed 3 m (10 ft).

**(B) Vertical Support.** Vertical runs of wireways shall be securely supported at intervals not exceeding 4.5 m (15 ft) and shall not have more than one joint between supports. Adjoining wireway sections shall be securely fastened together to provide a rigid joint.

### 376.56 Splices, Taps, and Power Distribution Blocks

**(A) Splices and Taps.** Splices and taps shall be permitted within a wireway, provided they are accessible. The conductors, including splices and taps, shall not fill the wireway to more than 75 percent of its area at that point.

Conductors in wireways can be accessed through hinged or removable covers. See 376.22 and the associated commentary example regarding the number of conductors permitted.

**(B) Power Distribution Blocks.**

**(1) Installation.** Power distribution blocks installed in metal wireways shall be listed. Power distribution blocks installed on the line side of the service equipment shall be listed for the purpose.

**(2) Size of Enclosure.** In addition to the wiring space requirement in 376.56(A), the power distribution block shall be installed in a wireway with dimensions not smaller than specified in the installation instructions of the power distribution block.

**(3) Wire Bending Space.** Wire bending space at the terminals of power distribution blocks shall comply with 312.6(B).

**(4) Live Parts.** Power distribution blocks shall not have uninsulated live parts exposed within a wireway, whether or not the wireway cover is installed.

**(5) Conductors.** Conductors shall be arranged so the power distribution block terminals are unobstructed following installation.

### 376.58 Dead Ends

Dead ends of metal wireways shall be closed.

### 376.70 Extensions from Metal Wireways

Extensions from wireways shall be made with cord pendants installed in accordance with 400.10 or with any wiring method in Chapter 3 that includes a means for equipment grounding. Where a separate equipment grounding conductor is employed, connection of the equipment grounding conductors in the wiring method to the wireway shall comply with 250.8 and 250.12.

Extensions from wireways are made through knockouts provided on the wireway or field punched. The extension wiring method must provide for equipment grounding. Cables and nonmetallic raceways as well as the wireway must include a wire-type EGC to ensure effective continuation of the ground path.

## III. Construction Specifications

### 376.100 Construction

**(A) Electrical and Mechanical Continuity.** Wireways shall be constructed and installed so that electrical and mechanical continuity of the complete system are assured.

**(B) Substantial Construction.** Wireways shall be of substantial construction and shall provide a complete enclosure for the contained conductors. All surfaces, both interior and exterior, shall be suitably protected from corrosion. Corner joints shall be made tight, and where the assembly is held together by rivets, bolts, or screws, such fasteners shall be spaced not more than 300 mm (12 in.) apart.

**(C) Smooth Rounded Edges.** Suitable bushings, shields, or fittings having smooth, rounded edges shall be provided where conductors pass between wireways, through partitions, around bends, between wireways and cabinets or junction boxes, and at other locations where necessary to prevent abrasion of the insulation of the conductors.

**(D) Covers.** Covers shall be securely fastened to the wireway.

The intent of this requirement is to aid in construction of one of a kind or custom-made wireways and fittings.

### 376.120 Marking

Metal wireways shall be so marked that their manufacturer's name or trademark will be visible after installation.

# ARTICLE 378 Nonmetallic Wireways

## I. General

### 378.1 Scope

This article covers the use, installation, and construction specifications for nonmetallic wireways and associated fittings.

Nonmetallic wireways are troughs with removable covers in which conductors are laid after the wireway has been installed as a complete system. The wireway must be installed on the surface of the structure, not concealed, and it is allowed to pass through walls, provided that an unbroken length passes through the wall.

### 378.2 Definition

**Nonmetallic Wireways.** Flame-retardant, nonmetallic troughs with removable covers for housing and protecting electrical wires and cables in which conductors are laid in place after the raceway has been installed as a complete system.

### 378.6 Listing Requirements

Nonmetallic wireways and associated fittings shall be listed.

## II. Installation

### 378.10 Uses Permitted

The use of nonmetallic wireways shall be permitted in the following:

(1) Only for exposed work, except as permitted in 378.10(4).
(2) Where subject to corrosive environments where identified for the use.
(3) In wet locations where listed for the purpose.

Informational Note: Extreme cold may cause nonmetallic wireways to become brittle and therefore more susceptible to damage from physical contact.

(4) As extensions to pass transversely through walls if the length passing through the wall is unbroken. Access to the conductors shall be maintained on both sides of the wall.

### 378.12 Uses Not Permitted

Nonmetallic wireways shall not be used in the following:

(1) Where subject to physical damage
(2) In any hazardous (classified) location, except as permitted by other articles in this *Code*
(3) Where exposed to sunlight unless listed and marked as suitable for the purpose

(4) Where subject to ambient temperatures other than those for which nonmetallic wireway is listed
(5) For conductors whose insulation temperature limitations would exceed those for which the nonmetallic wireway is listed

### 378.21 Size of Conductors

No conductor larger than that for which the nonmetallic wireway is designed shall be installed in any nonmetallic wireway.

### 378.22 Number of Conductors

The sum of cross-sectional areas of all contained conductors at any cross section of the nonmetallic wireway shall not exceed 20 percent of the interior cross-sectional area of the nonmetallic wireway. Conductors for signaling circuits or controller conductors between a motor and its starter and used only for starting duty shall not be considered as current-carrying conductors.

The adjustment factors specified in 310.15(B)(3)(a) shall be applicable to the current-carrying conductors up to and including the 20 percent fill specified above.

### 378.23 Insulated Conductors

Insulated conductors installed in a nonmetallic wireway shall comply with 378.23(A) and (B).

**(A) Deflected Insulated Conductors.** Where insulated conductors are deflected within a nonmetallic wireway, either at the ends or where conduits, fittings, or other raceways or cables enter or leave the nonmetallic wireway, or where the direction of the nonmetallic wireway is deflected greater than 30 degrees, dimensions corresponding to one wire per terminal in Table 312.6(A) shall apply.

**(B) Nonmetallic Wireways Used as Pull Boxes.** Where insulated conductors 4 AWG or larger are pulled through a wireway, the distance between raceway and cable entries enclosing the same conductor shall not be less than that required in 314.28(A)(1) for straight pulls and in 314.28(A)(2) for angle pulls. When transposing cable size into raceway size, the minimum metric designator (trade size) raceway required for the number and size of conductors in the cable shall be used.

These requirements provide adequate space for installing and bending conductors without damaging the conductor insulation. Section 378.23(A) requires that the Table 312.6(A) column of one wire per terminal be used. Section 378.23(B) requires that the same adequate space requirements that apply to conduits entering boxes also apply to cables as they enter a wireway. Where wireways are used as pull boxes, cable entries are converted to a minimum raceway trade size to determine wireway dimensions.

### 378.30 Securing and Supporting

Nonmetallic wireway shall be supported in accordance with 378.30(A) and (B).

**(A) Horizontal Support.** Nonmetallic wireways shall be supported where run horizontally at intervals not to exceed 900 mm (3 ft), and at each end or joint, unless listed for other support intervals. In no case shall the distance between supports exceed 3 m (10 ft).

**(B) Vertical Support.** Vertical runs of nonmetallic wireway shall be securely supported at intervals not exceeding 1.2 m (4 ft), unless listed for other support intervals, and shall not have more than one joint between supports. Adjoining nonmetallic wireway sections shall be securely fastened together to provide a rigid joint.

### 378.44 Expansion Fittings

Expansion fittings for nonmetallic wireway shall be provided to compensate for thermal expansion and contraction where the length change is expected to be 6 mm (0.25 in.) or greater in a straight run.

Informational Note: See Table 352.44 for expansion characteristics of PVC conduit. The expansion characteristics of PVC nonmetallic wireway are identical.

### 378.56 Splices and Taps

Splices and taps shall be permitted within a nonmetallic wireway, provided they are accessible. The conductors, including splices and taps, shall not fill the nonmetallic wireway to more than 75 percent of its area at that point.

### 378.58 Dead Ends

Dead ends of nonmetallic wireway shall be closed using listed fittings.

### 378.60 Grounding

Where equipment grounding is required, a separate equipment grounding conductor shall be installed in the nonmetallic wireway. A separate equipment grounding conductor shall not be required where the grounded conductor is used to ground equipment as permitted in 250.142.

### 378.70 Extensions from Nonmetallic Wireways

Extensions from nonmetallic wireway shall be made with cord pendants or any wiring method of Chapter 3. A separate equipment grounding conductor shall be installed in, or an equipment grounding connection shall be made to, any of the wiring methods used for the extension.

## III. Construction Specifications

### 378.120 Marking

Nonmetallic wireways shall be marked so that the manufacturer's name or trademark and interior cross-sectional area in square inches shall be visible after installation. Marking for limited smoke shall be permitted on the nonmetallic wireways that have limited smoke-producing characteristics.

# ARTICLE 380
# Multioutlet Assembly

## I. General

### 380.1 Scope

This article covers the use and installation requirements for multioutlet assemblies.

Informational Note: See the definition of multioutlet assembly in Article 100.

Multioutlet assemblies are metal or nonmetallic raceways that are usually surface mounted and designed to contain branch-circuit conductors and receptacles. Exhibit 380.1 provides an illustration of a multioutlet assembly. Receptables may be spaced at desired intervals and may be assembled at the factory or in the field. See 220.14(H) and Exhibit 220.4 for load calculations.

## II. Installation

### 380.10 Uses Permitted

The use of a multioutlet assembly shall be permitted in dry locations.

### 380.12 Uses Not Permitted

A multioutlet assembly shall not be installed as follows:

(1) Where concealed, except that it shall be permissible to surround the back and sides of a metal multioutlet assembly by the building finish or recess a nonmetallic multioutlet assembly in a baseboard
(2) Where subject to severe physical damage
(3) Where the voltage is 300 volts or more between conductors unless the assembly is of metal having a thickness of not less than 1.02 mm (0.040 in.)
(4) Where subject to corrosive vapors
(5) In hoistways
(6) In any hazardous (classified) location, except as permitted by other articles in this *Code*

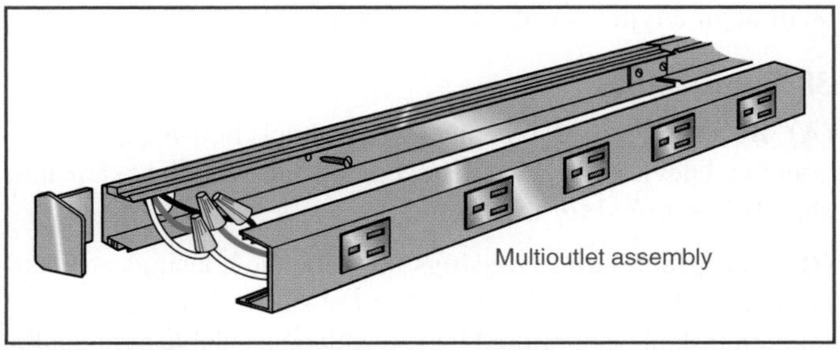


***EXHIBIT 380.1*** *A typical multioutlet assembly shown in an exploded view.*

### 380.23 Insulated Conductors

For field-assembled multioutlet assemblies, insulated conductors shall comply with 380.23(A) and (B), as applicable.

**(A) Deflected Insulated Conductors.** Where insulated conductors are deflected within a multioutlet assembly, either at the ends or where conduits, fittings, or other raceways or cables enter or leave the multioutlet assembly, or where the direction of the multioutlet assembly is deflected greater than 30 degrees, dimensions corresponding to one wire per terminal in Table 312.6(A) shall apply.

**(B) Multioutlet Assemblies Used as Pull Boxes.** Where insulated conductors 4 AWG or larger are pulled through a multioutlet assembly, the distance between raceway and cable entries enclosing the same conductor shall not be less than that required by 314.28(A)(1) for straight pulls and 314.28(A)(2) for angle pulls. When transposing cable size into raceway size, the minimum metric designator (trade size) raceway required for the number and size of conductors in the cable shall be used.

Safeguards to prevent overfill are provided by limiting the number of conductors that can be installed in multioutlet assemblies. For deflected insulated conductors, dimensions corresponding to the minimum width of wiring gutters must be maintained.

Where a multioutlet assembly is used as a pull box for insulated conductors of 4 AWG or larger, the distance between the raceway and the cable entries enclosing the conductor must not be less than eight times the trade size or metric designator of the raceway for straight pulls. For angle pulls, the distance must be six times the trade size or metric designator of the raceway.

### 380.76 Metal Multioutlet Assembly Through Dry Partitions

It shall be permissible to extend a metal multioutlet assembly through (not run within) dry partitions if arrangements are made for removing the cap or cover on all exposed portions and no outlet is located within the partitions.

# ARTICLE 382
# Nonmetallic Extensions

## I. General

### 382.1 Scope

This article covers the use, installation, and construction specifications for nonmetallic extensions.

### 382.2 Definitions

**Concealable Nonmetallic Extension.** A listed assembly of two, three, or four insulated circuit conductors within a nonmetallic jacket, an extruded thermoplastic covering, or a sealed nonmetallic covering. The classification includes surface extensions intended for mounting directly on the surface of walls or ceilings, and concealed with paint, texture, joint compound, plaster, wallpaper, tile, wall paneling, or other similar materials.

**Nonmetallic Extension.** An assembly of two insulated conductors within a nonmetallic jacket or an extruded thermoplastic covering. The classification includes surface extensions intended for mounting directly on the surface of walls or ceilings.

### 382.6 Listing Requirements

Concealable nonmetallic extensions and associated fittings and devices shall be listed. The starting/source tap device for the extension shall contain and provide the following protection for all load-side extensions and devices.

(1) Supplementary overcurrent protection
(2) Level of protection equivalent to a Class A GFCI
(3) Level of protection equivalent to a portable GFCI
(4) Line and load-side miswire protection
(5) Provide protection from the effects of arc faults

## II. Installation

### 382.10 Uses Permitted

Nonmetallic extensions shall be permitted only in accordance with 382.10(A), (B), and (C).

**(A) From an Existing Outlet.** The extension shall be from an existing outlet on a 15- or 20-ampere branch circuit. Where a concealable nonmetallic extension originates from a non–grounding-type receptacle, the installation shall comply with 250.130(C), 406.4(D)(2)(b), or 406.4(D)(2)(c).

**(B) Exposed and in a Dry Location.** The extension shall be run exposed, or concealed as permitted in 382.15, and in a dry location.

**(C) Residential or Offices.** For nonmetallic surface extensions mounted directly on the surface of walls or ceilings, the building shall be occupied for residential or office purposes and shall not exceed three floors abovegrade. Where identified for the use, concealable nonmetallic extensions shall be permitted more than three floors abovegrade.

Informational Note No. 1: See 310.15(A)(3) for temperature limitation of conductors.

Informational Note No. 2: See 362.10 for definition of *First Floor*.

### 382.12 Uses Not Permitted

Nonmetallic extensions shall not be used as follows:

(1) In unfinished basements, attics, or roof spaces
(2) Where the voltage between conductors exceeds 150 volts for nonmetallic surface extensions and 300 volts for aerial cable
(3) Where subject to corrosive vapors
(4) Where run through a floor or partition, or outside the room in which it originates

### 382.15 Exposed

**(A) Nonmetallic Extensions.** One or more extensions shall be permitted to be run in any direction from an existing outlet, but not on the floor or within 50 mm (2 in.) from the floor.

**(B) Concealable Nonmetallic Extensions.** Where identified for the use, nonmetallic extensions shall be permitted to be concealed with paint, texture, concealing compound, plaster, wallpaper, tile, wall paneling, or other similar materials and installed in accordance with 382.15(A).

### 382.26 Bends

**(A) Nonmetallic Extensions.** A bend that reduces the normal spacing between the conductors shall be covered with a cap to protect the assembly from physical damage.

**(B) Concealable Nonmetallic Extensions.** Concealable extensions shall be permitted to be folded back over themselves and flattened as required for installation.

### 382.30 Securing and Supporting

**(A) Nonmetallic Extensions.** Nonmetallic surface extensions shall be secured in place by approved means at intervals not exceeding 200 mm (8 in.), with an allowance for 300 mm (12 in.) to the first fastening where the connection to the supplying outlet is by means of an attachment plug. There shall be at least one fastening between each two adjacent outlets supplied. An extension shall be attached to only woodwork or plaster finish and shall not be in contact with any metal work or other conductive material other than with metal plates on receptacles.

**(B) Concealable Nonmetallic Extensions.** All surface-mounted concealable nonmetallic extension components shall be firmly anchored to the 382.40 Boxes and Fittings. Each run shall terminate in a fitting, connector, or box that covers the end of the assembly. All fittings, connectors, and devices shall be of a type identified for the use.

### 382.40 Boxes and Fittings

Each run shall terminate in a fitting, connector, or box that covers the end of the assembly. All fittings, connectors, and devices shall be of a type identified for the use.

### 382.42 Devices

**(A) Receptacles.** All receptacles, receptacle housings, and self-contained devices used with concealable nonmetallic extensions shall be identified for this use.

**(B) Receptacles and Housings.** Receptacle housings and self-contained devices designed either for surface or for recessed mounting shall be permitted for use with concealable nonmetallic extensions. Receptacle housings and self-contained devices shall incorporate means for facilitating entry and termination of concealable nonmetallic extensions and for electrically connecting the housing or device. Receptacle and self-contained devices shall comply with 406.4. Power and communications outlets installed together in common housing shall be permitted in accordance with 800.133(A)(1)(d), Exception No. 2.

### 382.56 Splices and Taps

Extensions shall consist of a continuous unbroken length of the assembly, without splices, and without exposed conductors between fittings, connectors, or devices. Taps shall be permitted where approved fittings completely covering the tap connections are used. Aerial cable and its tap connectors shall be provided with an approved means for polarization. Receptacle-type tap connectors shall be of the locking type.

## III. Construction Specifications (Concealable Nonmetallic Extensions Only)

### 382.100 Construction

Concealable nonmetallic extensions shall be a multilayer flat conductor design consisting of a center ungrounded conductor enclosed by a sectioned grounded conductor, and an overall sectioned grounding conductor.

### 382.104 Flat Conductors

Concealable nonmetallic extensions shall be constructed, using flat copper conductors equivalent to 14 AWG or 12 AWG conductor sizes, and constructed per 382.104(A), (B), and (C).

**(A) Ungrounded Conductor (Center Layer).** The ungrounded conductor shall consist of one or more ungrounded flat conductor(s) enclosed in accordance with 382.104(B) and (C) and identified in accordance with 310.110(C).

**(B) Grounded Conductor (Inner Sectioned Layers).** The grounded conductor shall consist of two sectioned inner flat conductors that enclose the center ungrounded conductor(s). The sectioned grounded conductor shall be enclosed by the sectioned grounding conductor and identified in accordance with 200.6.

**(C) Grounding Conductor (Outer Sectioned Layers).** The grounding conductor shall consist of two overall sectioned conductors that enclose the grounded conductor and ungrounded conductor(s) and shall comply with 250.4(A)(5). The grounding conductor layers shall be identified by any one of the following methods:

(1) As permitted in 250.119
(2) A clear covering

***EXHIBIT 382.1*** *One example of a flat conductor cable assembly used as a concealable nonmetallic extension. (Courtesy of FlatWire Technologies, a division of Southwire®)*

(3) One or more continuous green stripes or hash marks
(4) The term "Equipment Ground" printed at regular intervals throughout the cable

A multilayer flat conductor is a complete assembly of branch-circuit conductors, thinner than a business card and yet flexible enough to bend to any angle required for a customized installation. Exhibit 382.1 is one example of a concealable nonmetallic extension circuit conductor assembly of flat conductors meeting the requirement of Article 382.

A concealable extension is a five-layer design for circuit integrity and protection. Flat sectioned EGC layers fully encase the ungrounded conductor and grounded conductor layers of the cable. By design, any penetration in the flat wire cable assembly will penetrate the EGC layer first, then the grounded conductor layer, then the ungrounded conductor layer. A penetration will result in a short circuit between the three conductors described that will immediately trip the OCPD, causing automatic disconnection of the circuit.

### 382.112 Insulation

The ungrounded and grounded flat conductor layers shall be individually insulated and comply with 310.15(A)(3). The grounding conductor shall be covered or insulated.

### 382.120 Marking

**(A) Cable.** Concealable nonmetallic extensions shall be clearly and durably marked on both sides at intervals of not more than 610 mm (24 in.) with the information required by 310.120(A) and with the following additional information:

(1) Material of conductors
(2) Maximum temperature rating
(3) Ampacity

**(B) Conductor Identification.** Conductors shall be clearly and durably identified on both sides throughout their length as specified in 382.104.

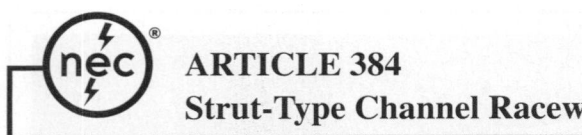

# ARTICLE 384
# Strut-Type Channel Raceway

## I. General

### 384.1 Scope

This article covers the use, installation, and construction specifications of strut-type channel raceway.

### 384.2 Definition

**Strut-Type Channel Raceway.** A metallic raceway that is intended to be mounted to the surface of or suspended from a structure, with associated accessories for the installation of electrical conductors and cables.

### 384.6 Listing Requirements

Strut-type channel raceways, closure strips, and accessories shall be listed and identified for such use.

## II. Installation

### 384.10 Uses Permitted

The use of strut-type channel raceways shall be permitted in the following:

(1) Where exposed.
(2) In dry locations.
(3) In locations subject to corrosive vapors where protected by finishes judged suitable for the condition.
(4) Where the voltage is 600 volts or less.
(5) As power poles.
(6) In Class I, Division 2 hazardous (classified) locations as permitted in 501.10(B)(3).
(7) As extensions of unbroken lengths through walls, partitions, and floors where closure strips are removable from either side and the portion within the wall, partition, or floor remains covered.
(8) Ferrous channel raceways and fittings protected from corrosion solely by enamel shall be permitted only indoors.

The installation shown in Exhibit 384.1 is typical of how a strut-type channel raceway can be used.

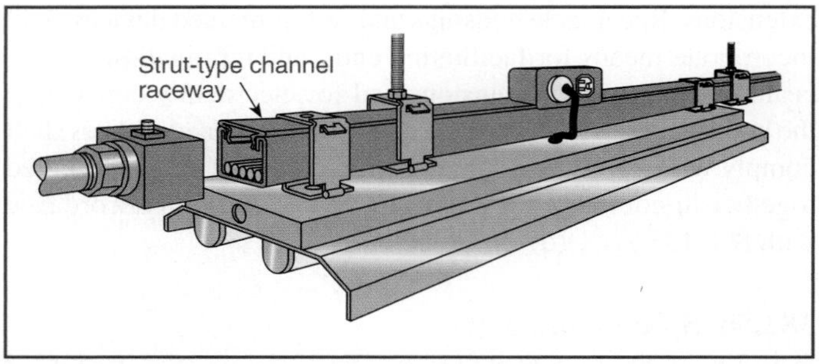


***EXHIBIT 384.1*** *An example of a strut-type channel raceway using accessories to support and supply power to luminaires.*

### 384.12 Uses Not Permitted

Strut-type channel raceways shall not be used as follows:

(1) Where concealed.
(2) Ferrous channel raceways and fittings protected from corrosion solely by enamel shall not be permitted where subject to severe corrosive influences.

### 384.21 Size of Conductors

No conductor larger than that for which the raceway is listed shall be installed in strut-type channel raceways.

### 384.22 Number of Conductors

The number of conductors permitted in strut-type channel raceways shall not exceed the percentage fill using Table 384.22 and applicable cross-sectional area of specific types and sizes of wire given in the tables in Chapter 9.

The adjustment factors of 310.15(B)(3)(a) shall not apply to conductors installed in strut-type channel raceways where all of the following conditions are met:

(1) The cross-sectional area of the raceway exceeds 2500 mm$^2$ (4 in.$^2$).
(2) The current-carrying conductors do not exceed 30 in number.

***TABLE 384.22*** *Channel Size and Inside Cross-Sectional Area*

| Size Channel | Area | | 40% Area* | | 25% Area† | |
|---|---|---|---|---|---|---|
| | in.$^2$ | mm$^2$ | in.$^2$ | mm$^2$ | in.$^2$ | mm$^2$ |
| 1⅝ × 13/16 | 0.887 | 572 | 0.355 | 229 | 0.222 | 143 |
| 1⅝ × 1 | 1.151 | 743 | 0.460 | 297 | 0.288 | 186 |
| 1⅝ × 1⅜ | 1.677 | 1076 | 0.671 | 433 | 0.419 | 270 |
| 1⅝ × 1⅝ | 2.028 | 1308 | 0.811 | 523 | 0.507 | 327 |
| 1⅝ × 2 7/16 | 3.169 | 2045 | 1.267 | 817 | 0.792 | 511 |
| 1⅝ × 3¼ | 4.308 | 2780 | 1.723 | 1112 | 1.077 | 695 |
| 1½ × ¾ | 0.849 | 548 | 0.340 | 219 | 0.212 | 137 |
| 1½ × 1½ | 1.828 | 1179 | 0.731 | 472 | 0.457 | 295 |
| 1½ × 1⅞ | 2.301 | 1485 | 0.920 | 594 | 0.575 | 371 |
| 1½ × 3 | 3.854 | 2487 | 1.542 | 995 | 0.964 | 622 |

*Raceways with external joiners shall use a 40 percent wire fill calculation to determine the number of conductors permitted.
†Raceways with internal joiners shall use a 25 percent wire fill calculation to determine the number of conductors permitted.

(3) The sum of the cross-sectional areas of all contained conductors does not exceed 20 percent of the interior cross-sectional area of the strut

### 384.30 Securing and Supporting.

**(A) Surface Mount.** A surface mount strut-type channel raceway shall be secured to the mounting surface with retention straps external to the channel at intervals not exceeding 3 m (10 ft) and within 900 mm (3 ft) of each outlet box, cabinet, junction box, or other channel raceway termination.

**(B) Suspension Mount.** Strut-type channel raceways shall be permitted to be suspension mounted in air with identified methods at intervals not to exceed 3 m (10 ft) and within 900 mm (3 ft) of channel raceway terminations and ends.

### 384.56 Splices and Taps

Splices and taps shall be permitted in raceways that are accessible after installation by having a removable cover. The conductors, including splices and taps, shall not fill the raceway to more than 75 percent of its area at that point. All splices and taps shall be made by approved methods.

### 384.60 Grounding

Strut-type channel raceway enclosures providing a transition to or from other wiring methods shall have a means for connecting an equipment grounding conductor. Strut-type channel raceways shall be permitted as an equipment grounding conductor in accordance with 250.118(13). Where a snap-fit metal cover for strut-type channel raceways is used to achieve electrical continuity in accordance with the listing, this cover shall not be permitted as the means for providing electrical continuity for a receptacle mounted in the cover.

## III. Construction Specifications

### 384.100 Construction

Strut-type channel raceways and their accessories shall be of a construction that distinguishes them from other raceways. Raceways and their elbows, couplings, and other fittings shall be designed such that the sections can be electrically and mechanically coupled together and installed without subjecting the wires to abrasion. They shall comply with 384.100(A), (B), and (C).

**(A) Material.** Raceways and accessories shall be formed of steel, stainless steel, or aluminum.

**(B) Corrosion Protection.** Steel raceways and accessories shall be protected against corrosion by galvanizing or by an organic coating.

Informational Note: Enamel and PVC coatings are examples of organic coatings that provide corrosion protection.

**(C) Cover.** Covers of strut-type channel raceways shall be either metallic or nonmetallic.

### 384.120 Marking

Each length of strut-type channel raceway shall be clearly and durably identified as required in the first sentence of 110.21(A).

# ARTICLE 386 Surface Metal Raceways

## I. General

### 386.1 Scope

This article covers the use, installation, and construction specifications for surface metal raceways and associated fittings.

The installation shown in Exhibit 386.1 is an example of how surface metal raceway can be used to extend branch-circuit wiring to new switch-controlled outlets.

### 386.2 Definition

**Surface Metal Raceway.** A metallic raceway that is intended to be mounted to the surface of a structure, with associated couplings, connectors, boxes, and fittings for the installation of electrical conductors.

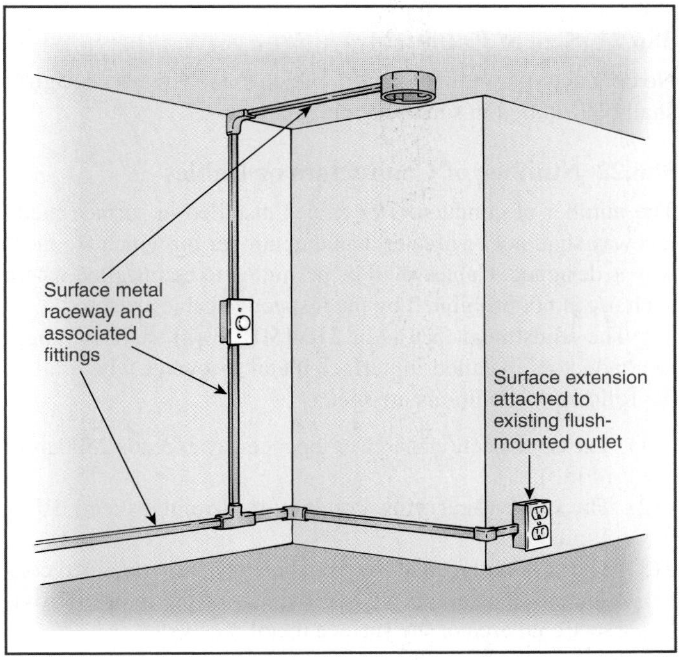


***EXHIBIT 386.1*** *An example of a surface metal raceway extending from an existing receptacle outlet.*

### 386.6 Listing Requirements

Surface metal raceway and associated fittings shall be listed.

## II. Installation

### 386.10 Uses Permitted

The use of surface metal raceways shall be permitted in the following:

(1) In dry locations.
(2) In Class I, Division 2 hazardous (classified) locations as permitted in 501.10(B)(3).
(3) Under raised floors, as permitted in 645.5(E)(2).
(4) Extension through walls and floors. Surface metal raceway shall be permitted to pass transversely through dry walls, dry partitions, and dry floors if the length passing through is unbroken. Access to the conductors shall be maintained on both sides of the wall, partition, or floor.

### 386.12 Uses Not Permitted

Surface metal raceways shall not be used in the following:

(1) Where subject to severe physical damage, unless otherwise approved
(2) Where the voltage is 300 volts or more between conductors, unless the metal has a thickness of not less than 1.02 mm (0.040 in.) nominal
(3) Where subject to corrosive vapors
(4) In hoistways
(5) Where concealed, except as permitted in 386.10

### 386.21 Size of Conductors

No conductor larger than that for which the raceway is designed shall be installed in surface metal raceway.

### 386.22 Number of Conductors or Cables

The number of conductors or cables installed in surface metal raceway shall not be greater than the number for which the raceway is designed. Cables shall be permitted to be installed where such use is not prohibited by the respective cable articles.

The adjustment factors of 310.15(B)(3)(a) shall not apply to conductors installed in surface metal raceways where all of the following conditions are met:

(1) The cross-sectional area of the raceway exceeds 2500 $mm^2$ (4 $in.^2$).
(2) The current-carrying conductors do not exceed 30 in number.
(3) The sum of the cross-sectional areas of all contained conductors does not exceed 20 percent of the interior cross-sectional area of the surface metal raceway.

The number, type, and sizes of conductors permitted to be installed in a listed surface metal raceway are marked on the raceway or on the package in which it is shipped.

### 386.30 Securing and Supporting

Surface metal raceways and associated fittings shall be supported in accordance with the manufacturer's installation instructions.

### 386.56 Splices and Taps

Splices and taps shall be permitted in surface metal raceways having a removable cover that is accessible after installation. The conductors, including splices and taps, shall not fill the raceway to more than 75 percent of its area at that point. Splices and taps in surface metal raceways without removable covers shall be made only in boxes. All splices and taps shall be made by approved methods.

Taps of Type FC cable installed in surface metal raceway shall be made in accordance with 322.56(B).

### 386.60 Grounding

Surface metal raceway enclosures providing a transition from other wiring methods shall have a means for connecting an equipment grounding conductor.

As the example in Exhibit 386.2 shows, where a surface metal raceway is supplied by Type MC or NM cable, a means (e.g., grounding terminal screw or lug) for terminating the EGC must be available at the surface metal raceway.

### 386.70 Combination Raceways

When combination surface metallic raceways are used for both signaling and for lighting and power circuits, the different systems shall be run in separate compartments identified by stamping, imprinting, or color coding of the interior finish.

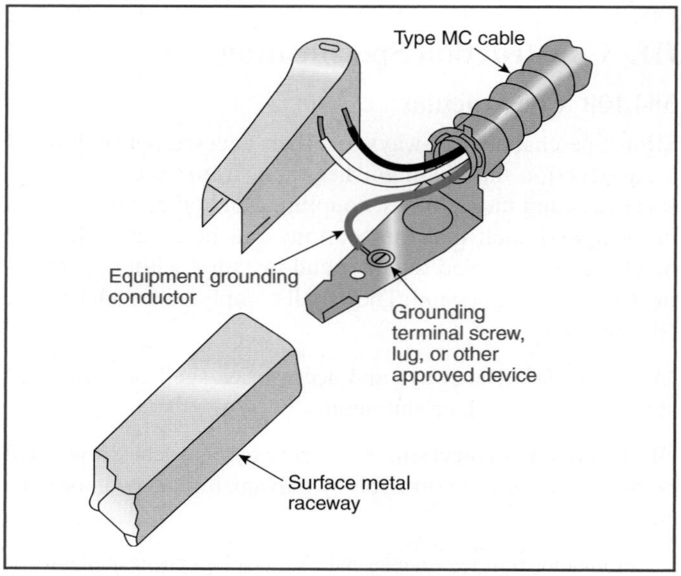


***EXHIBIT 386.2*** *One means for terminating an EGC at a surface metal raceway.*

## III. Construction Specifications

### 386.100 Construction

Surface metal raceways shall be of such construction as will distinguish them from other raceways. Surface metal raceways and their elbows, couplings, and similar fittings shall be designed so that the sections can be electrically and mechanically coupled together and installed without subjecting the wires to abrasion.

Where covers and accessories of nonmetallic materials are used on surface metal raceways, they shall be identified for such use.

### 386.120 Marking

Each length of surface metal raceway shall be clearly and durably identified as required in the first sentence of 110.21(A).

# ARTICLE 388
# Surface Nonmetallic Raceways

## I. General

### 388.1 Scope

This article covers the use, installation, and construction specifications for surface nonmetallic raceways and associated fittings.

Surface nonmetallic raceways may resemble base or chair rail molding and allow for circuit conductors to be installed without the need for wall penetration. The installation shown in Exhibit 388.1 is typical of how a surface nonmetallic raceway can be used to supply power, CATV, and communications outlets.

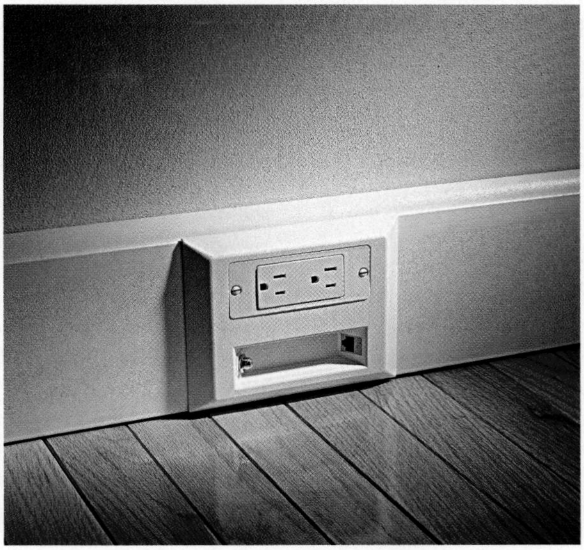

*EXHIBIT 388.1 An example of a surface nonmetallic raceway, resembling base molding, that supplies a receptacle outlet on the top and CATV and communications in the bottom compartment. (Courtesy of Legrand/Wiremold®)*

### 388.2 Definition

**Surface Nonmetallic Raceway.** A nonmetallic raceway that is intended to be mounted to the surface of a structure, with associated couplings, connectors, boxes, and fittings for the installation of electrical conductors.

### 388.6 Listing Requirements

Surface nonmetallic raceway and associated fittings shall be listed.

## II. Installation

### 388.10 Uses Permitted

Surface nonmetallic raceways shall be permitted as follows:

(1) The use of surface nonmetallic raceways shall be permitted in dry locations.

(2) Extension through walls and floors shall be permitted. Surface nonmetallic raceway shall be permitted to pass transversely through dry walls, dry partitions, and dry floors if the length passing through is unbroken. Access to the conductors shall be maintained on both sides of the wall, partition, or floor.

### 388.12 Uses Not Permitted

Surface nonmetallic raceways shall not be used in the following:

(1) Where concealed, except as permitted in 388.10(2)
(2) Where subject to severe physical damage
(3) Where the voltage is 300 volts or more between conductors, unless listed for higher voltage
(4) In hoistways
(5) In any hazardous (classified) location, except as permitted by other articles in this *Code*
(6) Where subject to ambient temperatures exceeding those for which the nonmetallic raceway is listed
(7) For conductors whose insulation temperature limitations would exceed those for which the nonmetallic raceway is listed

### 388.21 Size of Conductors

No conductor larger than that for which the raceway is designed shall be installed in surface nonmetallic raceway.

### 388.22 Number of Conductors or Cables

The number of conductors or cables installed in surface nonmetallic raceway shall not be greater than the number for which the raceway is designed. Cables shall be permitted to be installed where such use is not prohibited by the respective cable articles.

### 388.30 Securing and Supporting

Surface nonmetallic raceways and associated fittings shall be supported in accordance with the manufacturer's installation instructions.

### 388.56 Splices and Taps

Splices and taps shall be permitted in surface nonmetallic raceways having a cover capable of being opened in place that is accessible after installation. The conductors, including splices and taps, shall not fill the raceway to more than 75 percent of its area at that point. Splices and taps in surface nonmetallic raceways without covers capable of being opened in place shall be made only in boxes. All splices and taps shall be made by approved methods.

### 388.60 Grounding

Where equipment grounding is required, a separate equipment grounding conductor shall be installed in the raceway.

### 388.70 Combination Raceways

When combination surface nonmetallic raceways are used both for signaling and for lighting and power circuits, the different systems shall be run in separate compartments identified by stamping, imprinting, or color coding of the interior finish.

## III. Construction Specifications

### 388.100 Construction

Surface nonmetallic raceways shall be of such construction as will distinguish them from other raceways. Surface nonmetallic raceways and their elbows, couplings, and similar fittings shall be designed so that the sections can be mechanically coupled together and installed without subjecting the wires to abrasion.

Surface nonmetallic raceways and fittings are made of suitable nonmetallic material that is resistant to moisture and chemical atmospheres. It shall also be flame retardant, resistant to impact and crushing, resistant to distortion from heat under conditions likely to be encountered in service, and resistant to low-temperature effects.

### 388.120 Marking

Surface nonmetallic raceways that have limited smoke-producing characteristics shall be permitted to be so identified. Each length of surface nonmetallic raceway shall be clearly and durably identified as required in the first sentence of 110.21(A).

## ARTICLE 390 Underfloor Raceways

### 390.1 Scope

This article covers the use and installation requirements for underfloor raceways.

An underfloor raceway is a practical means of bringing light, power, and signal and communications systems to desks, work benches, or tables that are not located adjacent to wall space. This wiring method offers flexibility in layout where used with movable partitions and is commonly used in large retail stores and office buildings to supply power at any desired location.

Underfloor raceways are permitted beneath the surface of concrete, wood, or other flooring material. The wiring method between raceway junction boxes and cabinets or outlet boxes may be any appropriate Chapter 3 wiring method.

### 390.2 Definition

**Underfloor Raceway.** A raceway and associated components designed and intended for installation beneath or flush with the surface of a floor for the installation of cables and electrical conductors.

### 390.3 Use

**(A) Permitted.** The installation of underfloor raceways shall be permitted beneath the surface of concrete or other flooring material or in office occupancies where laid flush with the concrete floor and covered with linoleum or equivalent floor covering.

**(B) Not Permitted.** Underfloor raceways shall not be installed (1) where subject to corrosive vapors or (2) in any hazardous (classified) locations, except as permitted by 504.20 and in Class I, Division 2 locations as permitted in 501.10(B)(3). Unless made of a material judged suitable for the condition or unless corrosion protection approved for the condition is provided, ferrous or nonferrous metal underfloor raceways, junction boxes, and fittings shall not be installed in concrete or in areas subject to severe corrosive influences.

### 390.4 Covering

Raceway coverings shall comply with 390.4(A) through (D).

**(A) Raceways Not over 100 mm (4 in.) Wide.** Half-round and flat-top raceways not over 100 mm (4 in.) in width shall have not less than 20 mm (¾ in.) of concrete or wood above the raceway.

*Exception: As permitted in 390.4(C) and (D) for flat-top raceways.*

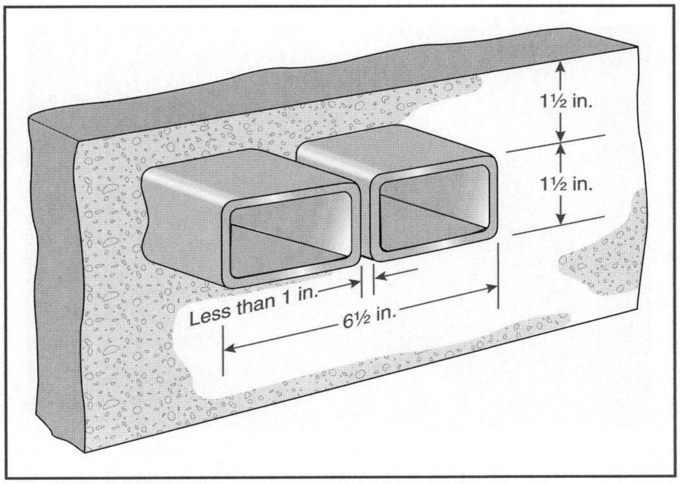


**EXHIBIT 390.1** *Two side-by-side underfloor raceways over 4 inches installed with the required covering. (Courtesy of Legrand/Wiremold®)*

**(B) Raceways over 100 mm (4 in.) Wide But Not over 200 mm (8 in.) Wide.** Flat-top raceways over 100 mm (4 in.) but not over 200 mm (8 in.) wide with a minimum of 25 mm (1 in.) spacing between raceways shall be covered with concrete to a depth of not less than 25 mm (1 in.). Raceways spaced less than 25 mm (1 in.) apart shall be covered with concrete to a depth of 38 mm (1½ in.).

As Exhibit 390.1 illustrates, flat-top underfloor raceways over 4 inches wide and spaced less than 1 inch apart must be covered with at least 1½ inches of concrete.

Approved trench-type underfloor raceways may be installed flush with the floor surface, provided they have covers that provide protection at least equal to those of junction box covers. Approved metal flat-top underfloor raceways, if not over 4 inches wide, may be installed flush with a concrete floor, provided they are equipped with covers that afford mechanical protection and rigidity equal to or exceeding that of junction box covers.

**(C) Trench-Type Raceways Flush with Concrete.** Trench-type flush raceways with removable covers shall be permitted to be laid flush with the floor surface. Such approved raceways shall be designed so that the cover plates provide adequate mechanical protection and rigidity equivalent to junction box covers.

**(D) Other Raceways Flush with Concrete.** In office occupancies, approved metal flat-top raceways, if not over 100 mm (4 in.) in width, shall be permitted to be laid flush with the concrete floor surface, provided they are covered with substantial linoleum that is not less than 1.6 mm (1/16 in.) thick or with equivalent floor covering. Where more than one and not more than three single raceways are each installed flush with the concrete, they shall be contiguous with each other and joined to form a rigid assembly.

## 390.5 Size of Conductors

No conductor larger than that for which the raceway is designed shall be installed in underfloor raceways.

## 390.6 Maximum Number of Conductors in Raceway

The combined cross-sectional area of all conductors or cables shall not exceed 40 percent of the interior cross-sectional area of the raceway.

## 390.7 Splices and Taps

Splices and taps shall be made only in junction boxes.

For the purposes of this section, so-called loop wiring (continuous, unbroken conductor connecting the individual outlets) shall not be considered to be a splice or tap.

*Exception: Splices and taps shall be permitted in trench-type flush raceway having a removable cover that is accessible after installation. The conductors, including splices and taps, shall not fill more than 75 percent of the raceway area at that point.*

## 390.8 Discontinued Outlets

When an outlet is abandoned, discontinued, or removed, the sections of circuit conductors supplying the outlet shall be removed from the raceway. No splices or reinsulated conductors, such as would be the case with abandoned outlets on loop wiring, shall be allowed in raceways.

## 390.9 Laid in Straight Lines

Underfloor raceways shall be laid so that a straight line from the center of one junction box to the center of the next junction box coincides with the centerline of the raceway system. Raceways shall be firmly held in place to prevent disturbing this alignment during construction.

## 390.10 Markers at Ends

A suitable marker shall be installed at or near each end of each straight run of raceways to locate the last insert.

## 390.11 Dead Ends

Dead ends of raceways shall be closed.

## 390.13 Junction Boxes

Junction boxes shall be leveled to the floor grade and sealed to prevent the free entrance of water or concrete. Junction boxes used with metal raceways shall be metal and shall be electrically continuous with the raceways.

## 390.14 Inserts

Inserts shall be leveled and sealed to prevent the entrance of concrete. Inserts used with metal raceways shall be metal and shall be electrically continuous with the raceway. Inserts set in or on fiber raceways before the floor is laid shall be mechanically secured to the raceway. Inserts set in fiber raceways after the floor is laid shall be screwed into the raceway. When cutting through the raceway wall and setting inserts, chips and other dirt shall not be allowed to remain in the raceway, and tools shall be used that are designed so as to prevent the tool from entering the raceway and damaging conductors that may be in place.

### 390.15 Connections to Cabinets and Wall Outlets

Connections from underfloor raceways to distribution centers and wall outlets shall be made by approved fittings or by any of the wiring methods in Chapter 3, where installed in accordance with the provisions of the respective articles.

### 390.17 Ampacity of Conductors

The ampacity adjustment factors, in 310.15(B)(3), shall apply to conductors installed in underfloor raceways.

# ARTICLE 392 Cable Trays

## I. General

### 392.1 Scope

This article covers cable tray systems, including ladder, ventilated trough, ventilated channel, solid bottom, and other similar structures.

Cable trays are mechanical support systems and not raceways. See the definition of *raceway* in Article 100. Cable tray installations are typically an industrial-type wiring method. However, they are sometimes installed in commercial facilities as a wire-and-cable management system for telecommunications/data installations and for feeder and branch-circuit wiring. Exhibit 392.1 is one example of MC cable installed in a wire mesh–type cable tray.

Informational Note: For further information on cable trays, see ANSI/NEMA–VE 1-2002, *Metal Cable Tray Systems*; NECA/NEMA 105-2007, *Standard for Installing Metal Cable Tray Systems*; and NEMA–FG 1-1998, *Nonmetallic Cable Tray Systems*.

### 392.2 Definition

**Cable Tray System.** A unit or assembly of units or sections and associated fittings forming a structural system used to securely fasten or support cables and raceways.

***EXHIBIT 392.1*** *MC cable installed in a cable tray. (Courtesy of Legrand®, Cablofil Cable Management Solutions)*

## II. Installation

### 392.10 Uses Permitted

Cable tray shall be permitted to be used as a support system for service conductors, feeders, branch circuits, communications circuits, control circuits, and signaling circuits. Cable tray installations shall not be limited to industrial establishments. Where exposed to direct rays of the sun, insulated conductors and jacketed cables shall be identified as being sunlight resistant. Cable trays and their associated fittings shall be identified for the intended use.

**(A) Wiring Methods.** The wiring methods in Table 392.10(A) shall be permitted to be installed in cable tray systems under the conditions described in their respective articles and sections.

Cable tray is rarely used as a major raceway support system. For raceway support systems, the versatility of strut systems exceeds that of cable tray support systems.

**(B) In Industrial Establishments.** The wiring methods in Table 392.10(A) shall be permitted to be used in any industrial establishment under the conditions described in their respective articles. In industrial establishments only, where conditions of maintenance and supervision ensure that only qualified persons service the installed cable tray system, any of the cables in 392.10(B)(1) and (B)(2) shall be permitted to be installed in ladder, ventilated trough, solid bottom, or ventilated channel cable trays.

**(1)** Single-conductor cables shall be permitted to be installed in accordance with (B)(1)(a) through (B)(1)(c).

(a) Single-conductor cable shall be 1/0 AWG or larger and shall be of a type listed and marked on the surface for use in cable trays. Where 1/0 AWG through 4/0 AWG single-conductor cables are installed in ladder cable tray, the maximum allowable rung spacing for the ladder cable tray shall be 225 mm (9 in.).

*TABLE 392.10(A) Wiring Methods*

| Wiring Method | Article |
|---|---|
| Armored cable: Type AC | 320 |
| CATV cables | 820 |
| Class 2 and Class 3 cables | 725 |
| Communications cables | 800 |
| Communications raceways | 725, 770, and 800 |
| Electrical metallic tubing: Type EMT | 358 |
| Electrical nonmetallic tubing: Type ENT | 362 |
| Fire alarm cables | 760 |
| Flexible metal conduit: Type FMC | 348 |
| Flexible metallic tubing: Type FMT | 360 |
| Instrumentation tray cable: Type ITC | 727 |
| Intermediate metal conduit: Type IMC | 342 |
| Liquidtight flexible metal conduit: Type LFMC | 350 |
| Liquidtight flexible nonmetallic conduit: Type LFNC | 356 |
| Metal-clad cable: Type MC | 330 |
| Mineral-insulated, metal-sheathed cable: Type MI | 332 |
| • | |
| Network-powered broadband communications cables | 830 |
| Nonmetallic-sheathed cable: Types NM, NMC, and NMS | 334 |
| Non–power-limited fire alarm cable | 760 |
| Optical fiber cables | 770 |
| • | |
| Other factory-assembled, multiconductor control, signal, or power cables that are specifically approved for installation in cable trays | |
| • | |
| Power and control tray cable: Type TC | 336 |
| Power-limited fire alarm cable | 760 |
| Power-limited tray cable | 725 |
| Rigid metal conduit: Type RMC | 344 |
| • | |
| Rigid polyvinyl chloride conduit: Type PVC | 352 |
| Reinforced thermosetting resin conduit: Type RTRC | 355 |
| Service-entrance cable: Types SE and USE | 338 |
| • | |
| Underground feeder and branch-circuit cable: Type UF | 340 |

(b) Welding cables shall comply with the provisions of Article 630, Part IV.

(c) Single conductors used as equipment grounding conductors shall be insulated, covered, or bare, and they shall be 4 AWG or larger.

Cable trays used to support welding cables are required by 630.42 to be dedicated for welding cable installation.

**(2)** Single- and multiconductor medium voltage cables shall be Type MV cable. Single conductors shall be installed in accordance with 392.10(B)(1).

**(C) Hazardous (Classified) Locations.** Cable trays in hazardous (classified) locations shall contain only the cable types and raceways permitted by other articles in this *Code*.

**(D) Nonmetallic Cable Tray.** In addition to the uses permitted elsewhere in 392.10, nonmetallic cable tray shall be permitted in corrosive areas and in areas requiring voltage isolation.

Fiberglass cable trays are often used to support cables in corrosive environments or in electrolytic cell rooms where voltage isolation is required. See Article 668 for more information on electrolytic cells.

## 392.12 Uses Not Permitted

Cable tray systems shall not be used in hoistways or where subject to severe physical damage.

Metal cable trays — ladder, ventilated trough, ventilated channel, or solid bottom type — can be used in other spaces used for environmental air (plenums) to support only the recognized wiring methods permitted in these spaces. Metal cable trays are not the limiting factor; rather, the cable or wiring method is the limiting factor. In addition, solid side and bottom metal cable tray systems with solid metal covers are permitted in accordance with 392.10(A) and (B).

## 392.18 Cable Tray Installation

**(A) Complete System.** Cable trays shall be installed as a complete system. Field bends or modifications shall be so made that the electrical continuity of the cable tray system and support for the cables is maintained. Cable tray systems shall be permitted to have mechanically discontinuous segments between cable tray runs or between cable tray runs and equipment.

Runs of cable tray are not required to be totally mechanically continuous from the equipment source to the equipment termination. Breaks in the mechanical continuity of cable tray systems are permitted and often occur at tees, crossovers, elevation changes, or firestops, or for thermal contraction and expansion. Also, cable tray systems are not required to be mechanically connected to the equipment they serve.

**(B) Completed Before Installation.** Each run of cable tray shall be completed before the installation of cables.

**(C) Covers.** In portions of runs where additional protection is required, covers or enclosures providing the required protection shall be of a material that is compatible with the cable tray.

**(D) Through Partitions and Walls.** Cable trays shall be permitted to extend transversely through partitions and walls or vertically through platforms and floors in wet or dry locations where the installations, complete with installed cables, are made in accordance with the requirements of 300.21.

**(E) Exposed and Accessible.** Cable trays shall be exposed and accessible, except as permitted by 392.18(D).

**(F) Adequate Access.** Sufficient space shall be provided and maintained about cable trays to permit adequate access for installing and maintaining the cables.

**(G) Raceways, Cables, Boxes, and Conduit Bodies Supported from Cable Tray Systems.** In industrial facilities where conditions of maintenance and supervision ensure that only qualified persons service the installation and where the cable tray systems are designed and installed to support the load, such systems shall be permitted to support raceways and cables, and boxes and conduit bodies covered in 314.1. For raceways terminating at the tray, a listed cable tray clamp or adapter shall be used to securely fasten the raceway to the cable tray system. Additional supporting and securing of the raceway shall be in accordance with the requirements of the appropriate raceway article. For raceways or cables running parallel to and attached to the bottom or side of a cable tray system, fastening and supporting shall be in accordance with the requirements of the appropriate raceway or cable article.

For boxes and conduit bodies attached to the bottom or side of a cable tray system, fastening and supporting shall be in accordance with the requirements of 314.23.

Conduit and cable termination supports as well as outlet boxes are permitted to be supported solely by the cable tray in qualifying industrial facilities only. These items are not permitted to be supported solely by the cable tray in commercial installations.

For commercial installations (and nonqualifying industrial facilities), conduits must be supported within 3 feet of the cable tray or within 5 feet if structural members do not permit fastening within 3 feet of the cable tray. Cables connecting to equipment outside the cable tray system must be supported according to their respective article. For example, Type MC cable in the larger sizes is required to be supported outside a cable tray system at intervals not exceeding 6 feet, according to 330.30.

**(H) Marking.** Cable trays containing conductors rated over 600 volts shall have a permanent, legible warning notice carrying the wording "DANGER — HIGH VOLTAGE — KEEP AWAY" placed in a readily visible position on all cable trays, with the spacing of warning notices not to exceed 3 m (10 ft). The danger marking(s) or labels shall comply with 110.21(B).

*Exception: Where not accessible (as applied to equipment), in industrial establishments where the conditions of maintenance and supervision ensure that only qualified persons service the installation, cable tray system warning notices shall be located where necessary for the installation to ensure safe maintenance and operation.*

## 392.20 Cable and Conductor Installation

**(A) Multiconductor Cables Operating at 600 Volts or Less.** Multiconductor cables operating at 600 volts or less shall be permitted to be installed in the same tray.

**(B) Cables Operating at Over 600 Volts.** Cables operating at over 600 volts and those operating at 600 volts or less installed in the same cable tray shall comply with either of the following:

(1) The cables operating at over 600 volts are Type MC.
(2) The cables operating at over 600 volts are separated from the cables operating at 600 volts or less by a solid fixed barrier of a material compatible with the cable tray.

**(C) Connected in Parallel.** Where single conductor cables comprising each phase, neutral, or grounded conductor of an alternating-current circuit are connected in parallel as permitted in 310.10(H), the conductors shall be installed in groups consisting of not more than one conductor per phase, neutral, or grounded conductor to prevent current imbalance in the paralleled conductors due to inductive reactance.

Single conductors shall be securely bound in circuit groups to prevent excessive movement due to fault-current magnetic forces unless single conductors are cabled together, such as triplexed assemblies.

The binding or otherwise grouping of 3-phase circuits results in the phase reactance of the conductors being balanced, which reduces the voltage imbalance between the phases of the 3-phase circuit.

**(D) Single Conductors.** Where any of the single conductors installed in ladder or ventilated trough cable trays are 1/0 through 4/0 AWG, all single conductors shall be installed in a single layer. Conductors that are bound together to comprise each circuit group shall be permitted to be installed in other than a single layer.

## 392.22 Number of Conductors or Cables

**(A) Number of Multiconductor Cables, Rated 2000 Volts or Less, in Cable Trays.** The number of multiconductor cables, rated 2000 volts or less, permitted in a single cable tray shall not exceed the requirements of this section. The conductor sizes apply to both aluminum and copper conductors.

**(1) Ladder or Ventilated Trough Cable Trays Containing Any Mixture of Cables.** Where ladder or ventilated trough cable trays contain multiconductor power or lighting cables, or any mixture of multiconductor power, lighting, control, and signal cables, the maximum number of cables shall conform to the following:

(a) Where all of the cables are 4/0 AWG or larger, the sum of the diameters of all cables shall not exceed the cable tray width, and the cables shall be installed in a single layer. Where the cable ampacity is determined according to 392.80(A)(1)(c), the cable tray width shall not be less than the sum of the diameters of the cables and the sum of the required spacing widths between the cables.

(b) Where all of the cables are smaller than 4/0 AWG, the sum of the cross-sectional areas of all cables shall not exceed

*TABLE 392.22(A) Allowable Cable Fill Area for Multiconductor Cables in Ladder, Ventilated Trough, or Solid Bottom Cable Trays for Cables Rated 2000 Volts or Less*

| | | Maximum Allowable Fill Area for Multiconductor Cables | | | | | | | |
|---|---|---|---|---|---|---|---|---|---|
| | | Ladder or Ventilated Trough or Wire Mesh Cable Trays, 392.22(A)(1) | | | | Solid Bottom Cable Trays, 392.22(A)(3) | | | |
| Inside Width of Cable Tray | | Column 1 Applicable for 392.22(A)(1)(b) Only | | Column 2[a] Applicable for 392.22(A)(1)(c) Only | | Column 3 Applicable for 392.22(A)(3)(b) Only | | Column 4[a] Applicable for 392.22(A)(3)(c) Only | |
| mm | in. | mm$^2$ | in.$^2$ | mm$^2$ | in.$^2$ | mm$^2$ | in.$^2$ | mm$^2$ | in.$^2$ |
| 50 | 2.0 | 1,500 | 2.5 | 1,500 – (30 Sd)[b] | 2.5 – (1.2 Sd)[b] | 1,200 | 2.0 | 1,200 – (25 Sd)[b] | 2.0 – Sd[b] |
| 100 | 4.0 | 3,000 | 4.5 | 3,000 – (30 Sd)[b] | 4.5 – (1.2 Sd) | 2,300 | 3.5 | 2,300 – (25 Sd) | 3.5 – Sd |
| 150 | 6.0 | 4,500 | 7.0 | 4,500 – (30 Sd)[b] | 7 – (1.2 Sd) | 3,500 | 5.5 | 3,500 – (25 Sd)[b] | 5.5–Sd |
| 200 | 8.0 | 6,000 | 9.5 | 6,000 – (30 Sd)[b] | 9.5 – (1.2 Sd) | 4,500 | 7.0 | 4,500 – (25 Sd) | 7.0 – Sd |
| 225 | 9.0 | 6,800 | 10.5 | 6,800 – (30 Sd) | 10.5 – (1.2 Sd) | 5,100 | 8.0 | 5,100 – (25 Sd) | 8.0 – Sd |
| 300 | 12.0 | 9,000 | 14.0 | 9,000 – (30 Sd) | 14 – (1.2 Sd) | 7,100 | 11.0 | 7,100 – (25 Sd) | 11.0 – Sd |
| 400 | 16.0 | 12,000 | 18.5 | 12,000 – (30 Sd) | 18.5 – (1.2 Sd) | 9,400 | 14.5 | 9,400 – (25 Sd) | 14.5 – Sd |
| 450 | 18.0 | 13,500 | 21.0 | 13,500 – (30 Sd) | 21 – (1.2 Sd) | 10,600 | 16.5 | 10,600 – (25 Sd) | 16.5 – Sd |
| 500 | 20.0 | 15,000 | 23.5 | 15,000 – (30 Sd) | 23.5 – (1.2 Sd) | 11,800 | 18.5 | 11,800 – (25 Sd) | 18.5 – Sd |
| 600 | 24.0 | 18,000 | 28.0 | 18,000 – (30 Sd) | 28 – (1.2 Sd) | 14,200 | 22.0 | 14,200 – (25 Sd) | 22.0 – Sd |
| 750 | 30.0 | 22,500 | 35.0 | 22,500 – (30 Sd) | 35 – (1.2 Sd) | 17,700 | 27.5 | 17,700 – (25 Sd) | 27.5 – Sd |
| 900 | 36.0 | 27,000 | 42.0 | 27,000 – (30 Sd) | 42 – (1.2 Sd) | 21,300 | 33.0 | 21,300 – (25 Sd) | 33.0 – Sd |

[a]The maximum allowable fill areas in Columns 2 and 4 shall be calculated. For example, the maximum allowable fill in mm$^2$ for a 150-mm wide cable tray in Column 2 shall be 4500 minus (30 multiplied by Sd) [the maximum allowable fill, in square inches, for a 6-in. wide cable tray in Column 2 shall be 7 minus (1.2 multiplied by Sd)].

[b]The term *Sd* in Columns 2 and 4 is equal to the sum of the diameters, in mm, of all cables 107.2 mm (in inches, of all 4/0 AWG) and larger multiconductor cables in the same cable tray with smaller cables.

the maximum allowable cable fill area in Column 1 of Table 392.22(A) for the appropriate cable tray width.

(c) Where 4/0 AWG or larger cables are installed in the same cable tray with cables smaller than 4/0 AWG, the sum of the cross-sectional areas of all cables smaller than 4/0 AWG shall not exceed the maximum allowable fill area resulting from the calculation in Column 2 of Table 392.22(A) for the appropriate cable tray width. The 4/0 AWG and larger cables shall be installed in a single layer, and no other cables shall be placed on them.

**(2) Ladder or Ventilated Trough Cable Trays Containing Multiconductor Control and/or Signal Cables Only.** Where a ladder or ventilated trough cable tray having a usable inside depth of 150 mm (6 in.) or less contains multiconductor control and/or signal cables only, the sum of the cross-sectional areas of all cables at any cross section shall not exceed 50 percent of the interior cross-sectional area of the cable tray. A depth of 150 mm (6 in.) shall be used to calculate the allowable interior cross-sectional area of any cable tray that has a usable inside depth of more than 150 mm (6 in.).

**(3) Solid Bottom Cable Trays Containing Any Mixture of Cables.** Where solid bottom cable trays contain multiconductor power or lighting cables, or any mixture of multiconductor power, lighting, control, and signal cables, the maximum number of cables shall conform to the following:

(a) Where all of the cables are 4/0 AWG or larger, the sum of the diameters of all cables shall not exceed 90 percent of the cable tray width, and the cables shall be installed in a single layer.

(b) Where all of the cables are smaller than 4/0 AWG, the sum of the cross- sectional areas of all cables shall not exceed the maximum allowable cable fill area in Column 3 of Table 392.22(A) for the appropriate cable tray width.

(c) Where 4/0 AWG or larger cables are installed in the same cable tray with cables smaller than 4/0 AWG, the sum of the cross-sectional areas of all cables smaller than 4/0 AWG shall not exceed the maximum allowable fill area resulting from the computation in Column 4 of Table 392.22(A) for the appropriate cable tray width. The 4/0 AWG and larger cables shall be installed in a single layer, and no other cables shall be placed on them.

**(4) Solid Bottom Cable Tray Containing Multiconductor Control and/or Signal Cables Only.** Where a solid bottom cable tray having a usable inside depth of 150 mm (6 in.)

or less contains multiconductor control and/or signal cables only, the sum of the cross sectional areas of all cables at any cross section shall not exceed 40 percent of the interior cross-sectional area of the cable tray. A depth of 150 mm (6 in.) shall be used to calculate the allowable interior cross-sectional area of any cable tray that has a usable inside depth of more than 150 mm (6 in.).

**(5) Ventilated Channel Cable Trays Containing Multiconductor Cables of Any Type.** Where ventilated channel cable trays contain multiconductor cables of any type, the following shall apply:

(a) Where only one multiconductor cable is installed, the cross-sectional area shall not exceed the value specified in Column 1 of Table 392.22(A)(5).

(b) Where more than one multiconductor cable is installed, the sum of the cross-sectional area of all cables shall not exceed the value specified in Column 2 of Table 392.22(A)(5).

**(6) Solid Channel Cable Trays Containing Multiconductor Cables of Any Type.** Where solid channel cable trays contain multiconductor cables of any type, the following shall apply:

(a) Where only one multiconductor cable is installed, the cross-sectional area of the cable shall not exceed the value specified in Column 1 of Table 392.22(A)(6).

(b) Where more than one multiconductor cable is installed, the sum of the cross-sectional area of all cable shall not exceed the value specified in Column 2 of Table 392.22(A)(6).

**(B) Number of Single-Conductor Cables, Rated 2000 Volts or Less, in Cable Trays.** The number of single conductor cables, rated 2000 volts or less, permitted in a single cable tray section shall not exceed the requirements of this section. The single conductors, or conductor assemblies, shall be evenly distributed across the cable tray. The conductor sizes apply to both aluminum and copper conductors.

***TABLE 392.22(A)(5)*** *Allowable Cable Fill Area for Multiconductor Cables in Ventilated Channel Cable Trays for Cables Rated 2000 Volts or Less*

| Inside Width of Cable Tray | | Maximum Allowable Fill Area for Multiconductor Cables: Column 1 One Cable | | Column 2 More Than One Cable | |
|---|---|---|---|---|---|
| mm | in. | $mm^2$ | $in.^2$ | $mm^2$ | $in.^2$ |
| 75 | 3 | 1500 | 2.3 | 850 | 1.3 |
| 100 | 4 | 2900 | 4.5 | 1600 | 2.5 |
| 150 | 6 | 4500 | 7.0 | 2450 | 3.8 |

***TABLE 392.22(A)(6)*** *Allowable Cable Fill Area for Multiconductor Cables in Solid Channel Cable Trays for Cables Rated 2000 Volts or Less*

| Inside Width of Cable Tray | | Column 1 One Cable | | Column 2 More Than One Cable | |
|---|---|---|---|---|---|
| mm | in. | $mm^2$ | $in.^2$ | $mm^2$ | $in.^2$ |
| 50 | 2 | 850 | 1.3 | 500 | 0.8 |
| 75 | 3 | 1300 | 2.0 | 700 | 1.1 |
| 100 | 4 | 2400 | 3.7 | 1400 | 2.1 |
| 150 | 6 | 3600 | 5.5 | 2100 | 3.2 |

**(1) Ladder or Ventilated Trough Cable Trays.** Where ladder or ventilated trough cable trays contain single-conductor cables, the maximum number of single conductors shall conform to the following:

(a) Where all of the cables are 1000 kcmil or larger, the sum of the diameters of all single-conductor cables shall not exceed the cable tray width, and the cables shall be installed in a single layer. Conductors that are bound together to comprise each circuit group shall be permitted to be installed in other than a single layer.

(b) Where all of the cables are from 250 kcmil through 900 kcmil, the sum of the cross-sectional areas of all single-conductor cables shall not exceed the maximum allowable cable fill area in Column 1 of Table 392.22(B)(1) for the appropriate cable tray width.

(c) Where 1000 kcmil or larger single-conductor cables are installed in the same cable tray with single-conductor cables smaller than 1000 kcmil, the sum of the cross sectional areas of all cables smaller than 1000 kcmil shall not exceed the maximum allowable fill area resulting from the computation in Column 2 of Table 392.22(B)(1) for the appropriate cable tray width.

(d) Where any of the single conductor cables are 1/0 through 4/0 AWG, the sum of the diameters of all single conductor cables shall not exceed the cable tray width.

**(2) Ventilated Channel Cable Trays.** Where 50 mm (2 in.), 75 mm (3 in.), 100 mm (4 in.), or 150 mm (6 in.) wide ventilated channel cable trays contain single-conductor cables, the sum of the diameters of all single conductors shall not exceed the inside width of the channel.

**(C) Number of Type MV and Type MC Cables (2001 Volts or Over) in Cable Trays.** The number of cables rated 2001 volts or over permitted in a single cable tray shall not exceed the requirements of this section.

The sum of the diameters of single-conductor and multiconductor cables shall not exceed the cable tray width, and the cables shall be installed in a single layer. Where single conductor cables are triplexed, quadruplexed, or bound together in circuit groups,

*TABLE 392.22(B)(1) Allowable Cable Fill Area for Single-Conductor Cables in Ladder, Ventilated Trough, or Wire Mesh Cable Trays for Cables Rated 2000 Volts or Less*

| Inside Width of Cable Tray | | Maximum Allowable Fill Area for Single-Conductor Cables in Ladder, Ventilated Trough, or Wire Mesh Cable Trays | | | |
|---|---|---|---|---|---|
| | | Column 1 Applicable for 392.22(B)(1)(b) Only | | Column 2[a] Applicable for 392.22(B)(1)(c) Only | |
| mm | in. | $mm^2$ | $in.^2$ | $mm^2$ | $in.^2$ |
| 50 | 2 | 1,400 | 2.0 | 1,400 – (28 Sd)[b] | 2.0 – (1.1 Sd)[b] |
| 100 | 4 | 2,800 | 4.5 | 2,800 – (28 Sd) | 4.5 – (1.1 Sd) |
| 150 | 6 | 4,200 | 6.5 | 4,200 – (28 Sd)[b] | 6.5 – (1.1 Sd)[b] |
| 200 | 8 | 5,600 | 8.5 | 5,600 – (28 Sd) | 8.5 – (1.1 Sd) |
| 225 | 9 | 6,100 | 9.5 | 6,100 – (28 Sd) | 9.5 – (1.1 Sd) |
| 300 | 12 | 8,400 | 13.0 | 8,400 – (28 Sd) | 13.0 – (1.1 Sd) |
| 400 | 16 | 11,200 | 17.5 | 11,200 – (28 Sd) | 17.5 – (1.1 Sd) |
| 450 | 18 | 12,600 | 19.5 | 12,600 – (28 Sd) | 19.5 – (1.1 Sd) |
| 500 | 20 | 14,000 | 21.5 | 14,000 – (28 Sd) | 21.5 – (1.1 Sd) |
| 600 | 24 | 16,800 | 26.0 | 16,800 – (28 Sd) | 26.0 – (1.1 Sd) |
| 750 | 30 | 21,000 | 32.5 | 21,000 – (28 Sd) | 32.5 – (1.1 Sd) |
| 900 | 36 | 25,200 | 39.0 | 25,200 – (28 Sd) | 39.0 – (1.1 Sd) |

[a]The maximum allowable fill areas in Column 2 shall be calculated. For example, the maximum allowable fill, in $mm^2$, for a 150 mm wide cable tray in Column 2 shall be 4200 minus (28 multiplied by Sd) [the maximum allowable fill, in square inches, for a 6-in. wide cable tray in Column 2 shall be 6.5 minus (1.1 multiplied by Sd)].

[b]The term *Sd* in Column 2 is equal to the sum of the diameters, in mm, of all cables 507 $mm^2$ (in inches, of all 1000 kcmil) and larger single-conductor cables in the same cable tray with small cables.

the sum of the diameters of the single conductors shall not exceed the cable tray width, and these groups shall be installed in single layer arrangement.

## 392.30 Securing and Supporting

**(A) Cable Trays.** Cable trays shall be supported at intervals in accordance with the installation instructions.

**(B) Cables and Conductors.** Cables and conductors shall be secured to and supported by the cable tray system in accordance with (1), (2) and (3) as applicable:

(1) In other than horizontal runs, the cables shall be fastened securely to transverse members of the cable runs.
(2) Supports shall be provided to prevent stress on cables where they enter raceways from cable tray systems.
(3) The system shall provide for the support of cables and raceway wiring methods in accordance with their corresponding articles. Where cable trays support individual conductors and where the conductors pass from one cable tray to another, or from a cable tray to raceway(s) or from a cable tray to equipment where the conductors are terminated, the distance between the cable trays or between the cable tray and the raceway(s) or the equipment shall not exceed 1.8 m (6 ft). The conductors shall be secured to the cable tray(s) at the transition, and they shall be protected, by guarding or by location, from physical damage.

The 6-foot distance limit applies to mechanically discontinuous cable tray segments for individual conductors but not to trays containing multiconductor cables. For further information regarding multiconductor Type TC tray cable used with discontinuous cable tray, refer to 336.10(7).

## 392.46 Bushed Conduit and Tubing

A box shall not be required where cables or conductors are installed in bushed conduit and tubing used for support or for protection against physical damage.

## 392.56 Cable Splices

Cable splices made and insulated by approved methods shall be permitted to be located within a cable tray, provided they are accessible. Splices shall be permitted to project above the side rails where not subject to physical damage.

## 392.60 Grounding and Bonding

**(A) Metallic Cable Trays.** Metallic cable trays shall be permitted to be used as equipment grounding conductors where continuous maintenance and supervision ensure that qualified persons service the installed cable tray system and the cable tray complies with provisions of this section. Metallic cable trays that support electrical conductors shall be grounded as required for conductor enclosures in accordance with 250.96 and Part IV of Article 250. Metal cable trays containing only non-power conductors shall be electrically continuous through approved connections or the use of a bonding jumper.

Informational Note: Examples of non-power conductors include nonconductive optical fiber cables and Class 2 and Class 3 Remote Control Signaling and Power Limiting Circuits.

Section 392.60(A), together with 250.96, requires all cable tray systems that support electrical conductors (whether mechanically continuous or with isolated segments) to be electrically continuous and effectively bonded and grounded. This requirement applies whether the cable tray is used as an equipment grounding conductor or is used for service conductors and connected to the grounded system conductor (or the grounding electrode conductor for ungrounded systems). Where a metal cable tray contains only non-power conductors, such as fire alarm, communications, CATV, or broadband conductors, the tray must be maintained electrically continuous through the use of approved connection or the use of a bonding jumper that is not smaller than 10 AWG. Where a bonding jumper is used, it is permitted to be stranded or solid.

**(B) Steel or Aluminum Cable Tray Systems.** Steel or aluminum cable tray systems shall be permitted to be used as equipment grounding conductors, provided all the following requirements are met:

(1) The cable tray sections and fittings are identified as an equipment grounding conductor.

*TABLE 392.60(A) Metal Area Requirements for Cable Trays Used as Equipment Grounding Conductor*

| Maximum Fuse Ampere Rating, Circuit Breaker Ampere Trip Setting, or Circuit Breaker Protective Relay Ampere Trip Setting for Ground-Fault Protection of Any Cable Circuit in the Cable Tray System | Minimum Cross-Sectional Area of Metal[a] | | | |
|---|---|---|---|---|
| | Steel Cable Trays | | Aluminum Cable Trays | |
| | mm$^2$ | in.$^2$ | mm$^2$ | in.$^2$ |
| 60 | 129 | 0.20 | 129 | 0.20 |
| 100 | 258 | 0.40 | 129 | 0.20 |
| 200 | 451.5 | 0.70 | 129 | 0.20 |
| 400 | 645 | 1.00 | 258 | 0.40 |
| 600 | 967.5 | 1.50[b] | 258 | 0.40 |
| 1000 | — | — | 387 | 0.60 |
| 1200 | — | — | 645 | 1.00 |
| 1600 | — | — | 967.5 | 1.50 |
| 2000 | — | — | 1290 | 2.00[b] |

[a]Total cross-sectional area of both side rails for ladder or trough cable trays; or the minimum cross-sectional area of metal in channel cable trays or cable trays of one-piece construction.

[b]Steel cable trays shall not be used as equipment grounding conductors for circuits with ground-fault protection above 600 amperes. Aluminum cable trays shall not be used as equipment grounding conductors for circuits with ground-fault protection above 2000 amperes.

(2) The minimum cross-sectional area of cable trays conform to the requirements in Table 392.60(A).

(3) All cable tray sections and fittings are legibly and durably marked to show the cross-sectional area of metal in channel cable trays, or cable trays of one-piece construction, and the total cross-sectional area of both side rails for ladder or trough cable trays.

(4) Cable tray sections, fittings, and connected raceways are bonded in accordance with 250.96, using bolted mechanical connectors or bonding jumpers sized and installed in accordance with 250.102.

Designers of cable tray systems, for use in establishments that qualify, are afforded the option to specify cables without EGCs and to use the cable tray system as the required EGC, provided the cable tray system meets the requirements of 392.60(A) and (B). Exhibit 392.2 illustrates an example of the grounding and bonding of multiconductor cables in cable trays with conduit runs to power equipment.

**(C) Transitions.** Where metallic cable tray systems are mechanically discontinuous, as permitted in 392.18(A), a bonding jumper sized in accordance with 250.102 shall connect the two sections of the cable tray, or the cable tray and the raceway or equipment. Bonding shall be in accordance with 250.96.

The bonding of the entire cable tray system is important, especially for discontinuous cable tray segments. According to 250.96(A), properly sized and installed bonding conductors must be installed across any mechanical discontinuities in the cable tray system and across any space between the cable tray and the conductor termination equipment enclosure or its equipment ground bus.

Cables installed within cable tray systems must meet or exceed the support requirements of the applicable article that covers the cables. This requirement either limits the gap distance in cable tray runs and between the cable tray and the equipment enclosures or requires intermediate cable supports at the appropriate distances in place of the cable tray.

## 392.80 Ampacity of Conductors

### (A) Ampacity of Cables, Rated 2000 Volts or Less, in Cable Trays.

**(1) Multiconductor Cables.** The allowable ampacity of multiconductor cables, nominally rated 2000 volts or less, installed according to the requirements of 392.22(A) shall be as given in Table 310.15(B)(16) and Table 310.15(B)(18), subject to the provisions of (A)(1)(a), (b), (c), and 310.15(A)(2).

(a) The adjustment factors of 310.15(B)(3)(a) shall apply only to multiconductor cables with more than three current-carrying conductors. Adjustment factors shall be limited to the number of current-carrying conductors in the cable and not to the number of conductors in the cable tray.

(b) Where cable trays are continuously covered for more than 1.8 m (6 ft) with solid unventilated covers, not over 95

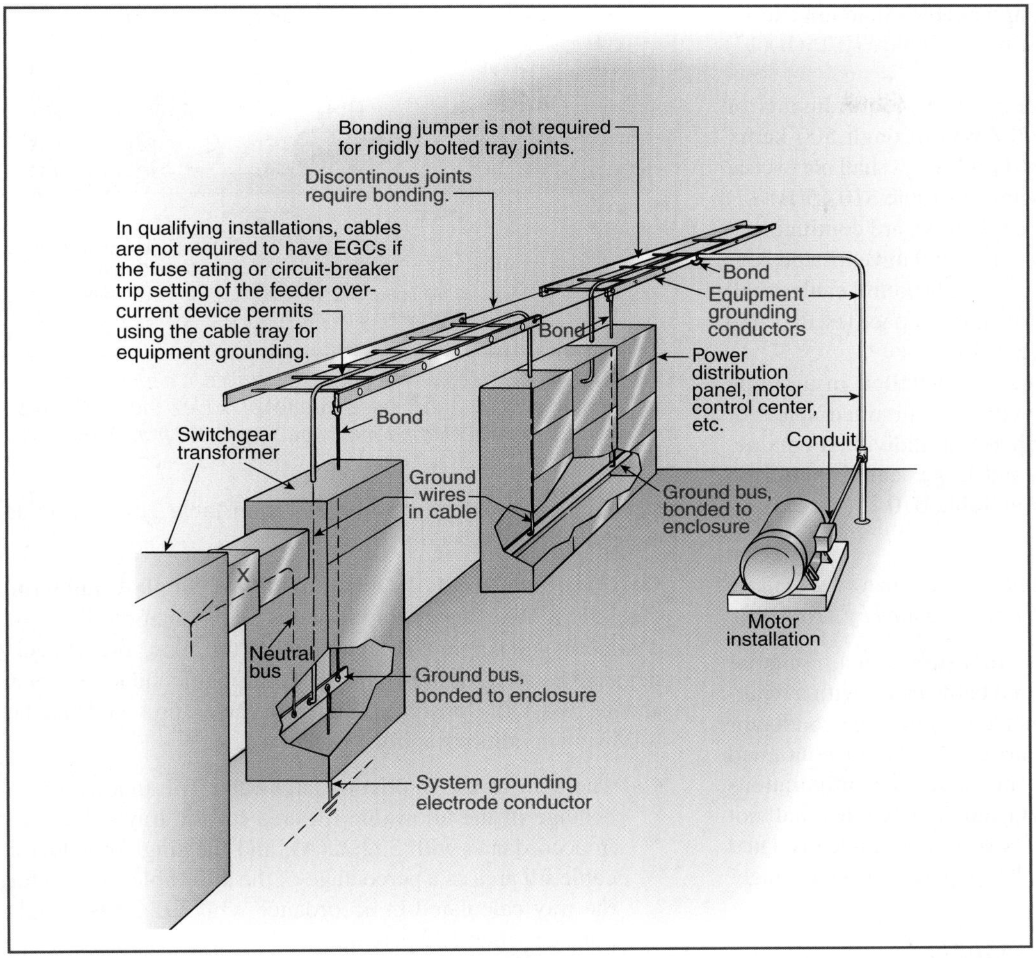


*EXHIBIT 392.2 An example of multiconductor cables in cable trays with conduit runs to power equipment where bonding is provided.*

percent of the allowable ampacities of Table 310.15(B)(16) and Table 310.15(B)(18) shall be permitted for multiconductor cables.

(c) Where multiconductor cables are installed in a single layer in uncovered trays, with a maintained spacing of not less than one cable diameter between cables, the ampacity shall not exceed the allowable ambient temperature-corrected ampacities of multiconductor cables, with not more than three insulated conductors rated 0 through 2000 volts in free air, in accordance with 310.15(C).

Informational Note: See Table B.310.15(B)(2)(3).

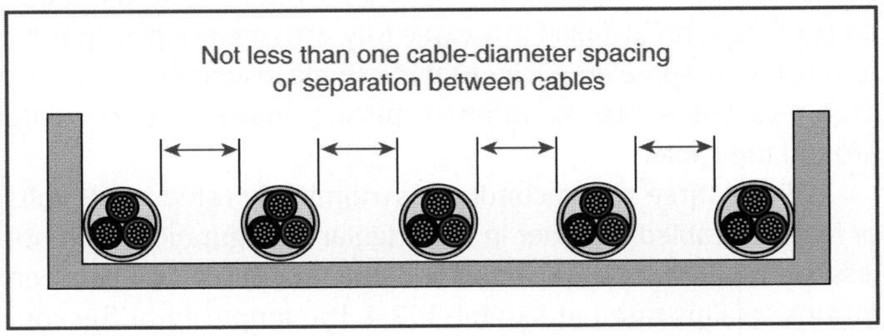


*EXHIBIT 392.3 Multiconductor cables, 2000 volts or less, with not more than three conductors per cable [ampacity to be determined from Table B.310.15(B)(2)(3) in Informative Annex B].*

The cables in Exhibit 392.3, rated 2000 volts or less, are installed in a single layer in an uncovered tray, with not less than one cable diameter between cables and not more than three conductors per cable. Refer to Table B.310.15(B)(2)(3) in Informative Annex B for the ampacity of the conductors in this configuration.

**(2) Single-Conductor Cables.** The allowable ampacity of single-conductor cables shall be as permitted by 310.15(A)(2). The adjustment factors of 310.15(B)(3)(a) shall not apply to the ampacity of cables in cable trays. The ampacity of single-conductor cables, or single conductors cabled together (triplexed, quadruplexed, etc.), nominally rated 2000 volts or less, shall comply with the following:

(a) Where installed according to the requirements of 392.22(B), the ampacities for 600 kcmil and larger single-conductor cables in uncovered cable trays shall not exceed 75 percent of the allowable ampacities in Table 310.15(B)(17) and Table 310.15(B)(19). Where cable trays are continuously covered for more than 1.8 m (6 ft) with solid unventilated covers, the ampacities for 600 kcmil and larger cables shall not exceed 70 percent of the allowable ampacities in Table 310.15(B)(17) and Table 310.15(B)(19).

(b) Where installed according to the requirements of 392.22(B), the ampacities for 1/0 AWG through 500 kcmil single-conductor cables in uncovered cable trays shall not exceed 65 percent of the allowable ampacities in Table 310.15(B)(17) and Table 310.15(B)(19). Where cable trays are continuously covered for more than 1.8 m (6 ft) with solid unventilated covers, the ampacities for 1/0 AWG through 500 kcmil cables shall not exceed 60 percent of the allowable ampacities in Table 310.15(B)(17) and Table 310.15(B)(19).

(c) Where single conductors are installed in a single layer in uncovered cable trays, with a maintained space of not less than one cable diameter between individual conductors, the ampacity of 1/0 AWG and larger cables shall not exceed the allowable ampacities in Table 310.15(B)(17) and Table 310.15(B)(19).

*Exception to (2)(3)(c): For solid bottom cable trays the ampacity of single conductor cables shall be determined by 310.15(C).*

(d) Where single conductors are installed in a triangular or square configuration in uncovered cable trays, with a maintained free airspace of not less than 2.15 times one conductor diameter (2.15 × O.D.) of the largest conductor contained within the configuration and adjacent conductor configurations or cables, the ampacity of 1/0 AWG and larger cables shall not exceed the allowable ampacities of two or three single insulated conductors rated 0 through 2000 volts supported on a messenger in accordance with 310.15(B).

Informational Note: See Table 310.15(B)(20).

Section 392.80(A)(2)(d) recognizes single conductors in a triangular configuration installed in a cable tray with maintained spacing as having the same ampacity as three single insulated conductors on a messenger. The maintained spacing allows air to circulate around the cable.

Where three single conductors, nominally rated 2000 volts or less, are cabled together in a triangular configuration, with not less than 2.15 times the conductor diameter (2.15 × OD) between groups, as illustrated in Exhibit 392.4, the ampacity of the conductors is determined in accordance with Table 310.15(B)(20).

Where single conductors are installed in cable trays, their ampacities are permitted to be calculated using many variations of Tables 310.15(B)(16), 310.15(B)(17), and 310.15(B)(19). Where these single-conductor cables emerge from a cable tray installation and are terminated at circuit breakers, distribution switchgear, and similar electrical equipment, the temperature limitations of the electrical equipment terminals should be coordinated with the ampacity of the single-conductor cables. As stated in both the UL *Guide Information for Electrical Equipment Directory – The White Book* and in 110.14(C)(1), unless the equipment is listed and marked otherwise, conductor ampacities used in determining

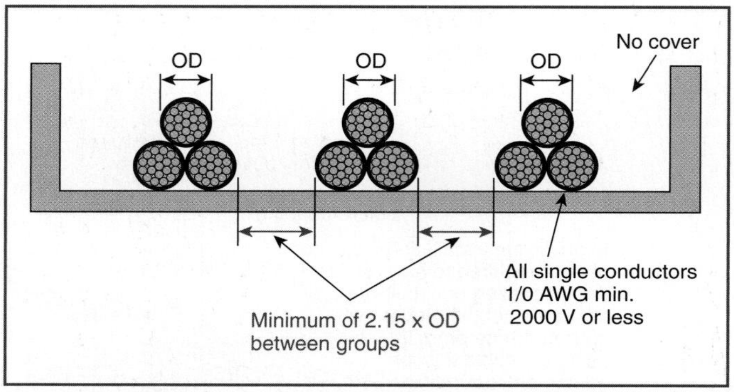


***EXHIBIT 392.4*** *Three single conductors installed in a triangular configuration with spacing between groups of not less than 2.15 times the conductor diameter [ampacities to be determined from Table 310.15(B)(20)].*

equipment terminations must be based on Table 310.15(B)(16) as modified by 310.15(B)(1) through (B)(7).

**(3) Combinations of Multiconductor and Single-Conductor Cables.** Where a cable tray contains a combination of multiconductor and single-conductor cables, the allowable ampacities shall be as given in 392.80(A)(1) for multiconductor cables and 392.80(A)(2) for single-conductor cables, provided that the following conditions apply:

(1) The sum of the multiconductor cable fill area as a percentage of the allowable fill area for the tray calculated in accordance with 392.22(A), and the single-conductor cable fill area as a percentage of the allowable fill area for the tray calculated in accordance with 392.22(B), totals not more than 100 percent.
(2) Multiconductor cables are installed according to 392.22(A) and single-conductor cables are installed according to 392.22(B) and 392.22(C).

Multiconductor and single-conductor cable installed in the same tray must not exceed the fill requirements for each type of cable as if it were installed in its own tray and all installation requirements and ampacity limits for each cable type apply.

**(B) Ampacity of Type MV and Type MC Cables (2001 Volts or Over) in Cable Trays.** The ampacity of cables, rated 2001 volts, nominal, or over, installed according to 392.22(C) shall not exceed the requirements of this section.

**(1) Multiconductor Cables (2001 Volts or Over).** The allowable ampacity of multiconductor cables shall be as given in Table 310.60(C)(75) and Table 310.60(C)(76), subject to the following provisions:

(a) Where cable trays are continuously covered for more than 1.8 m (6 ft) with solid unventilated covers, not more than 95 percent of the allowable ampacities of Table 310.60(C)(75) and Table 310.60(C)(76) shall be permitted for multiconductor cables.

(b) Where multiconductor cables are installed in a single layer in uncovered cable trays, with maintained spacing of not less than one cable diameter between cables, the ampacity shall not exceed the allowable ampacities of Table 310.60(C)(71) and Table 310.60(C)(72).

**(2) Single-Conductor Cables (2001 Volts or Over).** The ampacity of single-conductor cables, or single conductors cabled together (triplexed, quadruplexed, etc.), shall comply with the following:

(a) The ampacities for 1/0 AWG and larger single-conductor cables in uncovered cable trays shall not exceed 75 percent of the allowable ampacities in Table 310.60(C)(69) and Table 310.60(C)(70). Where the cable trays are covered for more than 1.8 m (6 ft) with solid unventilated covers, the ampacities for 1/0 AWG and larger single-conductor cables shall not exceed 70 percent of the allowable ampacities in Table 310.60(C)(69)and Table 310.60(C)(70).

(b) Where single-conductor cables are installed in a single layer in uncovered cable trays, with a maintained space of not less than one cable diameter between individual conductors, the ampacity of 1/0 AWG and larger cables shall not exceed the allowable ampacities in Table 310.60(C)(69) and Table 310.60(C)(70).

(c) Where single conductors are installed in a triangular or square configuration in uncovered cable trays, with a maintained free air space of not less than 2.15 times the diameter (2.15 × O.D.) of the largest conductor contained within the configuration and adjacent conductor configurations or cables, the ampacity of 1/0 AWG and larger cables shall not exceed the allowable ampacities in Table 310.60(C)(67) and Table 310.60(C)(68).

## III. Construction Specifications

### 392.100 Construction

**(A) Strength and Rigidity.** Cable trays shall have suitable strength and rigidity to provide adequate support for all contained wiring.

**(B) Smooth Edges.** Cable trays shall not have sharp edges, burrs, or projections that could damage the insulation or jackets of the wiring.

**(C) Corrosion Protection.** Cable tray systems shall be corrosion resistant. If made of ferrous material, the system shall be protected from corrosion as required by 300.6.

**(D) Side Rails.** Cable trays shall have side rails or equivalent structural members.

**(E) Fittings.** Cable trays shall include fittings or other suitable means for changes in direction and elevation of runs.

**(F) Nonmetallic Cable Tray.** Nonmetallic cable trays shall be made of flame-retardant material.

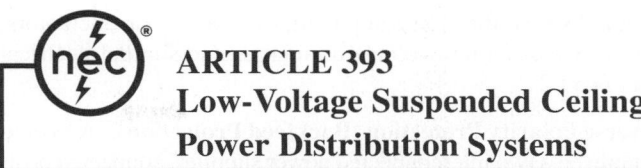

# ARTICLE 393
# Low-Voltage Suspended Ceiling Power Distribution Systems

## I. General

### 393.1 Scope

This article covers the installation of low-voltage suspended ceiling power distribution systems.

### 393.2 Definitions

**Busbar.** A noninsulated conductor electrically connected to the source of supply and physically supported on an insulator providing a power rail for connection to utilization equipment, such as sensors, actuators, A/V devices, low-voltage luminaire assemblies, and similar electrical equipment.

**Busbar Support.** An insulator that runs the length of a section of suspended ceiling bus rail that serves to support and isolate the busbars from the suspended grid rail.

**Connector.** A term used to refer to an electromechanical fitting.

**Connector, Load.** An electromechanical connector used for power from the busbar to utilization equipment.

**Connector, Pendant.** An electromechanical or mechanical connector used to suspend low-voltage luminaire or utilization equipment below the grid rail and to supply power to connect from the busbar to utilization equipment.

**Connector, Power Feed.** An electromechanical connector used to connect the power supply to a power distribution cable, to connect directly to the busbar, or to connect from a power distribution cable to the busbar.

**Connector, Rail to Rail.** An electromechanical connector used to interconnect busbars from one ceiling grid rail to another grid rail.

**Grid Bus Rail.** A combination of the busbar, the busbar support, and the structural suspended ceiling grid system.

**Low-Voltage Suspended Ceiling Power Distribution System.** A system that serves as a support for a finished ceiling surface and consists of a busbar and busbar support system to distribute power to utilization equipment supplied by a Class 2 power supply.

**Power Supply.** A Class 2 power supply connected between the branch-circuit power distribution system and the busbar low-voltage suspended ceiling power distribution system.

**Rail.** The structural support for the suspended ceiling system typically forming the ceiling grid supporting the ceiling tile and listed utilization equipment, such as sensors, actuators, A/V devices, and low-voltage luminaires and similar electrical equipment.

**Reverse Polarity Protection (Backfeed Protection).** A system that prevents two interconnected power supplies, connected positive to negative, from passing current from one power source into a second power source.

**Suspended Ceiling Grid.** A system that serves as a support for a finished ceiling surface and other utilization equipment.

### 393.6 Listing Requirements

Suspended ceiling power distribution systems and associated fittings shall be listed as in 393.6(A) or (B).

**(A) Listed System.** Low-voltage suspended ceiling distribution systems operating at 30 volts ac or less or 60 volts dc or less shall be listed as a complete system, with the utilization equipment, power supply, and fittings as part of the same identified system.

**(B) Assembly of Listed Parts.** A low-voltage suspended ceiling power distribution system assembled from the following parts, listed according to the appropriate function, shall be permitted:

(1) Listed low-voltage utilization equipment
(2) Listed Class 2 power supply
(3) Listed or identified fittings, including connectors and grid rails with bare conductors
(4) Listed low voltage cables in accordance with 725.179, conductors in raceways, or other fixed wiring methods for the secondary circuit

## II. Installation

### 393.10 Uses Permitted

Low-voltage suspended ceiling power distribution systems shall be permanently connected and shall be permitted as follows:

(1) For listed utilization equipment capable of operation at a maximum of 30 volts ac (42.4 volts peak) or 60 volts dc (24.8 volts peak for dc interrupted at a rate of 10 Hz to 200 Hz) and limited to Class 2 power levels in Chapter 9, Table 11(A) and Table 11(B) for lighting, control, and signaling circuits.
(2) In indoor dry locations.
(3) For residential, commercial, and industrial installations.
(4) In other spaces used for environmental air in accordance with 300.22(C), electrical equipment having a metal enclosure, or with a nonmetallic enclosure and fittings, shall be listed for use within an air-handling space and shall have adequate fire-resistant and low-smoke-producing characteristics and associated wiring material suitable for the ambient temperature.

Informational Note: One method of defining adequate fire-resistant and low-smoke producing characteristics for electrical equipment with a nonmetallic enclosure is in ANSI/ UL 2043-2008, *Fire Test for Heat and Visible Smoke Release for Discrete Products and Their Accessories Installed in Air-Handling Spaces*.

### 393.12 Uses Not Permitted

Suspended ceiling power distribution systems shall not be installed in the following:

(1) In damp or wet locations
(2) Where subject to corrosive fumes or vapors, such as storage battery rooms
(3) Where subject to physical damage
(4) In concealed locations
(5) In hazardous (classified) locations
(6) As part of a fire-rated floor-ceiling or roof-ceiling assembly, unless specifically listed as part of the assembly
(7) For lighting in general or critical patient care areas

### 393.14 Installation

**(A) General Requirements.** Support wiring shall be installed in a neat and workmanlike manner. Cables and conductors installed exposed on the surface of ceilings and sidewalls shall be supported by the building structure in such a manner that the cable is not damaged by normal building use. Such cables shall be supported by straps, staples, hangers, cable ties, or similar fittings designed and installed so as not to damage the cable.

Informational Note: Suspended ceiling low-voltage power grid distribution systems should be installed by qualified persons in accordance with the manufacturer's installation instructions.

**(B) Insulated Conductors.** Exposed insulated secondary circuit conductors shall be listed, of the type, and installed as described as follows:

(1) Class 2 cable supplied by a listed Class 2 power source and installed in accordance with Parts I and III of Article 725
(2) Wiring methods described in Chapter 3

### 393.21 Disconnecting Means

**(A) Location.** A disconnecting means for the Class 2 supply to the power grid system shall be located so as to be accessible and within sight of the Class 2 power source for servicing or maintenance of the grid system.

**(B) Multiwire Branch Circuits.** Where connected to a multiwire branch circuit, the disconnecting means shall simultaneously disconnect all the supply conductors, including the grounded conductors.

### 393.30 Securing and Supporting

**(A) Attached to Building Structure.** A suspended ceiling low-voltage power distribution system shall be secured to the mounting surface of the building structure by hanging wires, screws, or bolts in accordance with the installation and operation instructions. Mounting hardware, such as screws or bolts, shall be either packaged with the suspended ceiling low-voltage lighting power distribution system, or the installation instructions shall specify the types of mounting fasteners to be used.

**(B) Attachment of Power Grid Rails.** The individual power grid rails shall be mechanically secured to the overall ceiling grid assembly.

### 393.40 Connectors and Enclosures

**(A) Connectors.** Connections to busbar grid rails, cables, and conductors shall be made with listed insulating devices, and these connections shall be accessible after installation. A soldered connection shall be made mechanically secure before being soldered. Other means of securing leads, such as push-on terminals and spade-type connectors, shall provide a secure mechanical connection. The following connectors shall be permitted to be used as connection or interconnection devices:

(1) Load connectors shall be used for power from the busbar to listed utilization equipment.
(2) A pendant connector shall be permitted to suspend low-voltage luminaires or utilization equipment below the grid rail and to supply power from the busbar to the utilization equipment.
(3) A power feed connector shall be permitted to connect the power supply directly to a power distribution cable and to the busbar.
(4) Rail-to-rail connectors shall be permitted to interconnect busbars from one ceiling grid rail to another grid rail.

Informational Note: For quick-connect terminals, see UL 310, *Standard for Electrical Quick-Connect*, and for mechanical splicing devices, see UL 486A, *Standard for Wire Connectors and Soldering Lugs for Use with Copper Conductors*, and 486B, *Standard for Wire Connectors*.

**(B) Enclosures.** Where made in a wall, connections shall be installed in an enclosure in accordance with Parts I, II, and III of Article 314.

### 393.45 Overcurrent and Reverse Polarity (Backfeed) Protection

**(A) Overcurrent Protection.** The listed Class 2 power supply or transformer primary shall be protected at not greater than 20 amperes.

**(B) Interconnection of Power Sources.** Listed Class 2 sources shall not have the output connections paralleled or otherwise interconnected unless listed for such interconnection.

**(C) Reverse Polarity (Backfeed) Protection of Direct-Current Systems.** A suspended ceiling low-voltage power distribution system shall be permitted to have reverse polarity (backfeed) protection of dc circuits by one of the following means:

(1) If the power supply is provided as part of the system, the power supply is provided with reverse polarity (backfeed) protection; or
(2) If the power supply is not provided as part of the system, reverse polarity or backfeed protection can be provided as part of the grid rail busbar or as a part of the power feed connector.

### 393.56 Splices

A busbar splice shall be provided with insulation and mechanical protection equivalent to that of the grid rail busbars involved.

### 393.57 Connections

Connections in busbar grid rails, cables, and conductors shall be made with listed insulating devices and be accessible after installation. Where made in a wall, connections shall be installed in an enclosure in accordance with Parts I, II, and III of Article 314, as applicable.

### 393.60 Grounding

**(A) Grounding of Supply Side of Class 2 Power Source.** The supply side of the Class 2 power source shall be connected to an equipment grounding conductor in accordance with the applicable requirements in Part IV of Article 250.

**(B) Grounding of Load Side of Class 2 Power Source.** Class 2 load side circuits for suspended ceiling low-voltage power grid distribution systems shall not be grounded.

## III. Construction Specifications

### 393.104 Sizes and Types of Conductors

**(A) Load Side Utilization Conductor Size.** Current-carrying conductors for load side utilization equipment shall be copper and shall be 18 AWG minimum.

*Exception: Conductors of a size smaller than 18 AWG, but not smaller than 24 AWG, shall be permitted to be used for Class 2 circuits. Where used, these conductors shall be installed using a Chapter 3 wiring method, shall be totally enclosed, shall not be subject to movement or strain, and shall comply with the ampacity requirements in Table 522.22.*

**(B) Power Feed Bus Rail Conductor Size.** The power feed bus rail shall be 16 AWG minimum or equivalent. For a busbar with a circular cross section, the diameter shall be 1.29 mm (0.051 n.) minimum, and, for other than circular busbars, the area shall be 1.32 mm$^2$ (0.002 in.$^2$) minimum.

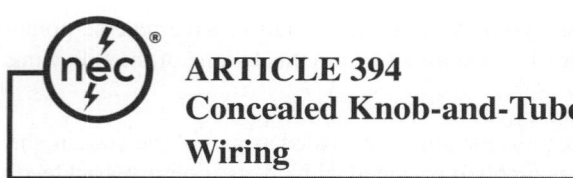

# ARTICLE 394 Concealed Knob-and-Tube Wiring

## I. General

### 394.1 Scope

This article covers the use, installation, and construction specifications of concealed knob-and-tube wiring.

### 394.2 Definition

**Concealed Knob-and-Tube Wiring.** A wiring method using knobs, tubes, and flexible nonmetallic tubing for the protection and support of single insulated conductors.

Knob-and-tube wiring is allowed to be concealed, while open wiring on insulators (Article 398) is required to be exposed. Concealed knob-and-tube wiring is designed for use in hollow spaces of walls, ceilings, and attics and utilizes the free air in such spaces for heat dissipation.

## II. Installation

### 394.10 Uses Permitted

Concealed knob-and-tube wiring shall be permitted to be installed in the hollow spaces of walls and ceilings, or in unfinished attics and roof spaces as provided by 394.23, only as follows:

(1) For extensions of existing installations
(2) Elsewhere by special permission

Concealed knob-and-tube wiring is permitted to be installed only for extensions of existing installations or where special permission is granted by the AHJ. See definition of *special permission* in Article 100.

### 394.12 Uses Not Permitted

Concealed knob-and-tube wiring shall not be used in the following:

(1) Commercial garages
(2) Theaters and similar locations
(3) Motion picture studios
(4) Hazardous (classified) locations
(5) Hollow spaces of walls, ceilings, and attics where such spaces are insulated by loose, rolled, or foamed-in-place insulating material that envelops the conductors

Blown-in, foamed-in, or rolled insulation prevents the dissipation of heat into the free air space, resulting in higher conductor temperature, which could cause insulation breakdown and possible insulation ignition. Section 394.12 prohibits installation of knob-and-tube wiring in hollow spaces that have been weatherized.

### 394.17 Through or Parallel to Framing Members

Conductors shall comply with 398.17 where passing through holes in structural members. Where passing through wood cross members in plastered partitions, conductors shall be protected by noncombustible, nonabsorbent, insulating tubes extending not less than 75 mm (3 in.) beyond the wood member.

### 394.19 Clearances

**(A) General.** A clearance of not less than 75 mm (3 in.) shall be maintained between conductors and a clearance of not less than 25 mm (1 in.) between the conductor and the surface over which it passes.

**(B) Limited Conductor Space.** Where space is too limited to provide these minimum clearances, such as at meters, panelboards, outlets, and switch points, the individual conductors shall be enclosed in flexible nonmetallic tubing, which shall be continuous in length between the last support and the enclosure or terminal point.

**(C) Clearance from Piping, Exposed Conductors, and So Forth.** Conductors shall comply with 398.19 for clearances from other exposed conductors, piping, and so forth.

### 394.23 In Accessible Attics

Conductors in unfinished attics and roof spaces shall comply with 394.23(A) or (B).

Informational Note: See 310.15(A)(3) for temperature limitation of conductors.

**(A) Accessible by Stairway or Permanent Ladder.** Conductors shall be installed along the side of or through bored holes in floor joists, studs, or rafters. Where run through bored holes, conductors in the joists and in studs or rafters to a height of not less than 2.1 m (7 ft) above the floor or floor joists shall be protected by substantial running boards extending not less than 25 mm (1 in.) on each side of the conductors. Running boards shall be securely fastened in place. Running boards and guard strips shall not be required where conductors are installed along the sides of joists, studs, or rafters.

**(B) Not Accessible by Stairway or Permanent Ladder.** Conductors shall be installed along the sides of or through bored holes in floor joists, studs, or rafters.

*Exception: In buildings completed before the wiring is installed, attic and roof spaces that are not accessible by stairway or permanent ladder and have headroom at all points less than 900 mm (3 ft), the wiring shall be permitted to be installed on the edges of rafters or joists facing the attic or roof space.*

### 394.30 Securing and Supporting

**(A) Supporting.** Conductors shall be rigidly supported on noncombustible, nonabsorbent insulating materials and shall not contact any other objects. Supports shall be installed as follows:

(1) Within 150 mm (6 in.) of each side of each tap or splice, and
(2) At intervals not exceeding 1.4 m (4½ ft).

Where it is impracticable to provide supports, conductors shall be permitted to be fished through hollow spaces in dry locations, provided each conductor is individually enclosed in flexible nonmetallic tubing that is in continuous lengths between supports, between boxes, or between a support and a box.

**(B) Securing.** Where solid knobs are used, conductors shall be securely tied thereto by tie wires having insulation equivalent to that of the conductor.

### 394.42 Devices

Switches shall comply with 404.4 and 404.10(B).

### 394.56 Splices and Taps

Splices shall be soldered unless approved splicing devices are used. In-line or strain splices shall not be used.

## III. Construction Specifications

### 394.104 Conductors

Conductors shall be of a type specified by Article 310.

# ARTICLE 396 Messenger-Supported Wiring

## I. General

### 396.1 Scope

This article covers the use, installation, and construction specifications for messenger-supported wiring.

For many years, messenger-supported wiring systems have been used in industrial installations as well as to supply services for commercial and residential installations. See references to messenger-supported wiring in 225.6(A)(1) and (B).

### 396.2 Definition

**Messenger-Supported Wiring.** An exposed wiring support system using a messenger wire to support insulated conductors by any one of the following:

(1) A messenger with rings and saddles for conductor support
(2) A messenger with a field-installed lashing material for conductor support
(3) Factory-assembled aerial cable
(4) Multiplex cables utilizing a bare conductor, factory assembled and twisted with one or more insulated conductors, such as duplex, triplex, or quadruplex type of construction

## II. Installation

### 396.10 Uses Permitted

**(A) Cable Types.** The cable types in Table 396.10(A) shall be permitted to be installed in messenger-supported wiring under the conditions described in the article or section referenced for each.

**(B) In Industrial Establishments.** In industrial establishments only, where conditions of maintenance and supervision ensure that only qualified persons service the installed messenger-supported wiring, the following shall be permitted:

(1) Any of the conductor types shown in Table 310.104(A) or Table 310.104(B)
(2) MV cable

Where exposed to weather, conductors shall be listed for use in wet locations. Where exposed to direct rays of the sun, conductors or cables shall be sunlight resistant.

Some of the triplex and quadruplex cable used by utilities as service-drop cable do not use conductors recognized in Table 310.104(A) and do not meet the requirements of Article 310.

See 310.15(B) and Table 310.15(B)(20) for two or three single-insulated conductors supported on a messenger. See 310.15(C) and Informative Annex B, Table B.310.15(B)(20)(3), for ampacities of conductors for other cable types.

***TABLE 396.10(A)*** *Cable Types*

| Cable Type | Section | Article |
|---|---|---|
| Medium-voltage cable | | 328 |
| Metal-clad cable | | 330 |
| Mineral-insulated, metal-sheathed cable | | 332 |
| Multiconductor service-entrance cable | | 338 |
| Multiconductor underground feeder and branch-circuit cable | | 340 |
| Other factory-assembled, multiconductor control, signal, or power cables that are identified for the use | | |
| Power and control tray cable | | 336 |
| Power-limited tray cable | 725.154(C) and 725.179(E) | |

**(C) Hazardous (Classified) Locations.** Messenger-supported wiring shall be permitted to be used in hazardous (classified) locations where the contained cables and messenger-supported wiring are specifically permitted by other articles in this *Code*.

### 396.12 Uses Not Permitted

Messenger-supported wiring shall not be used in hoistways or where subject to physical damage.

### 396.30 Messenger

**(A) Support.** The messenger shall be supported at dead ends and at intermediate locations so as to eliminate tension on the conductors. The conductors shall not be permitted to come into contact with the messenger supports or any structural members, walls, or pipes.

**(B) Neutral Conductor.** Where the messenger is used as a neutral conductor, it shall comply with the requirements of 225.4, 250.184(A), 250.184(B)(7), and 250.187(B).

**(C) Equipment Grounding Conductor.** Where the messenger is used as an equipment grounding conductor, it shall comply with the requirements of 250.32(B), 250.118, 250.184(B)(8), and 250.187(D).

### 396.56 Conductor Splices and Taps

Conductor splices and taps made and insulated by approved methods shall be permitted in messenger-supported wiring.

### 396.60 Grounding

The messenger shall be grounded as required by 250.80 and 250.86 for enclosure grounding.

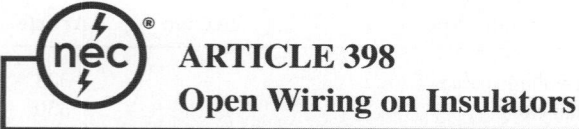

# ARTICLE 398 Open Wiring on Insulators

## I. General

### 398.1 Scope

This article covers the use, installation, and construction specifications of open wiring on insulators.

### 398.2 Definition

**Open Wiring on Insulators.** An exposed wiring method using cleats, knobs, tubes, and flexible tubing for the protection and support of single insulated conductors run in or on buildings.

## II. Installation

### 398.10 Uses Permitted

Open wiring on insulators shall be permitted only for industrial or agricultural establishments on systems of 600 volts, nominal, or less, as follows:

(1) Indoors or outdoors
(2) In wet or dry locations
(3) Where subject to corrosive vapors
(4) For services

Open wiring on insulators is an exposed wiring method that is not permitted to be concealed by the building structure or finish. It is permitted indoors or outdoors, in dry or wet locations, and where subject to corrosive vapors, provided the insulation choice from Table 310.104(A) is suitable for use in a corrosive environment.

This wiring method is not permitted for temporary lighting and power circuits on construction sites but is permitted for lighting and power circuits in agricultural buildings [see 547.5(A)]. It may also be used for services (see 230.43).

See Tables 310.15(B)(17) and 310.15(B)(19) for ampacities of conductors.

### 398.12 Uses Not Permitted

Open wiring on insulators shall not be installed where concealed by the building structure.

### 398.15 Exposed Work

**(A) Dry Locations.** In dry locations, where not exposed to physical damage, conductors shall be permitted to be separately enclosed in flexible nonmetallic tubing. The tubing shall be in continuous lengths not exceeding 4.5 m (15 ft) and secured to the surface by straps at intervals not exceeding 1.4 m (4½ ft).

**(B) Entering Spaces Subject to Dampness, Wetness, or Corrosive Vapors.** Conductors entering or leaving locations subject to dampness, wetness, or corrosive vapors shall have drip loops formed on them and shall then pass upward and inward from the outside of the buildings, or from the damp, wet, or corrosive location, through noncombustible, nonabsorbent insulating tubes.

Informational Note: See 230.52 for individual conductors entering buildings or other structures.

**(C) Exposed to Physical Damage.** Conductors within 2.1 m (7 ft) from the floor shall be considered exposed to physical damage. Where open conductors cross ceiling joists and wall studs and are exposed to physical damage, they shall be protected by one of the following methods:

(1) Guard strips not less than 25 mm (1 in.) nominal in thickness and at least as high as the insulating supports, placed on each side of and close to the wiring.

(2) A substantial running board at least 13 mm (½ in.) thick in back of the conductors with side protections. Running boards shall extend at least 25 mm (1 in.) outside the conductors, but not more than 50 mm (2 in.), and the protecting sides shall be at least 50 mm (2 in.) high and at least 25 mm (1 in.), nominal, in thickness.
(3) Boxing made in accordance with 398.15(C)(1) or (C)(2) and furnished with a cover kept at least 25 mm (1 in.) away from the conductors within. Where protecting vertical conductors on side walls, the boxing shall be closed at the top and the holes through which the conductors pass shall be bushed.
(4) Rigid metal conduit, intermediate metal conduit, rigid nonmetallic conduit, or electrical metallic tubing. When installed in metal piping, the conductors shall be encased in continuous lengths of approved flexible tubing.

## 398.17 Through or Parallel to Framing Members

Open conductors shall be separated from contact with walls, floors, wood cross members, or partitions through which they pass by tubes or bushings of noncombustible, nonabsorbent insulating material. Where the bushing is shorter than the hole, a waterproof sleeve of noninductive material shall be inserted in the hole and an insulating bushing slipped into the sleeve at each end in such a manner as to keep the conductors absolutely out of contact with the sleeve. Each conductor shall be carried through a separate tube or sleeve.

Informational Note: See 310.15(A)(3) for temperature limitation of conductors.

## 398.19 Clearances

Open conductors shall be separated at least 50 mm (2 in.) from metal raceways, piping, or other conducting material, and from any exposed lighting, power, or signaling conductor, or shall be separated therefrom by a continuous and firmly fixed nonconductor in addition to the insulation of the conductor. Where any insulating tube is used, it shall be secured at the ends. Where practicable, conductors shall pass over rather than under any piping subject to leakage or accumulations of moisture.

The requirement for additional protective insulation on open wiring is to prevent contact with metal piping, metal objects, or exposed conductors of other circuits.

## 398.23 In Accessible Attics

Conductors in unfinished attics and roof spaces shall comply with 398.23(A) or (B).

**(A) Accessible by Stairway or Permanent Ladder.** Conductors shall be installed along the side of or through bored holes in floor joists, studs, or rafters. Where run through bored holes, conductors in the joists and in studs or rafters to a height of not less than 2.1 m (7 ft) above the floor or floor joists shall be protected by substantial running boards extending not less than 25 mm (1 in.) on each side of the conductors. Running boards shall be securely fastened in place. Running boards and guard strips shall not be required for conductors installed along the sides of joists, studs, or rafters.

**(B) Not Accessible by Stairway or Permanent Ladder.** Conductors shall be installed along the sides of or through bored holes in floor joists, studs, or rafters.

*Exception: In buildings completed before the wiring is installed, in attic and roof spaces that are not accessible by stairway or permanent ladder and have headroom at all points less than 900 mm (3 ft), the wiring shall be permitted to be installed on the edges of rafters or joists facing the attic or roof space.*

## 398.30 Securing and Supporting

**(A) Conductor Sizes Smaller Than 8 AWG.** Conductors smaller than 8 AWG shall be rigidly supported on noncombustible, nonabsorbent insulating materials and shall not contact any other objects. Supports shall be installed as follows:

(1) Within 150 mm (6 in.) from a tap or splice
(2) Within 300 mm (12 in.) of a dead-end connection to a lampholder or receptacle
(3) At intervals not exceeding 1.4 m (4½ ft) and at closer intervals sufficient to provide adequate support where likely to be disturbed

**(B) Conductor Sizes 8 AWG and Larger.** Supports for conductors 8 AWG or larger installed across open spaces shall be permitted up to 4.5 m (15 ft) apart if noncombustible, nonabsorbent insulating spacers are used at least every 1.4 m (4½ ft) to maintain at least 65 mm (2½ in.) between conductors.

Where not likely to be disturbed in buildings of mill construction, 8 AWG and larger conductors shall be permitted to be run across open spaces if supported from each wood cross member on approved insulators maintaining 150 mm (6 in.) between conductors.

Mill construction is generally considered to be a building in which the floors and ceilings are supported by wood timbers or beams or wood cross members spaced approximately 15 feet apart. This type of construction is sometimes referred to as plank-on-timber construction. Conductors 8 AWG and larger are permitted to span the 15-foot distance where the ceilings are high and free of obstructions and the conductors are unlikely to come into contact with other objects.

**(C) Industrial Establishments.** In industrial establishments only, where conditions of maintenance and supervision ensure that only qualified persons service the system, conductors of sizes 250 kcmil and larger shall be permitted to be run across open spaces where supported at intervals up to 9.0 m (30 ft) apart.

The installation of open feeders on insulators mounted on the bottom of roof trusses at every bay location was once common in industrial buildings. Many bays are more than 15 feet wide. Therefore, size 250 kcmil and larger conductors are permitted to be supported at 30-foot intervals in industrial buildings where qualified persons must service the system.

In addition to the ease and economy of installation or alteration of open wiring, the close spacing of conductors reduces the reactance of a circuit and, hence, reduces the voltage drop.

**(D) Mounting of Conductor Supports.** Where nails are used to mount knobs, they shall not be smaller than tenpenny. Where screws are used to mount knobs, or where nails or screws are used to mount cleats, they shall be of a length sufficient to penetrate the wood to a depth equal to at least one-half the height of the knob and the full thickness of the cleat. Cushion washers shall be used with nails.

**(E) Tie Wires.** Conductors 8 AWG or larger and supported on solid knobs shall be securely tied thereto by tie wires having an insulation equivalent to that of the conductor.

### 398.42 Devices

Surface-type snap switches shall be mounted in accordance with 404.10(A), and boxes shall not be required. Other type switches shall be installed in accordance with 404.4.

## III. Construction Specifications

### 398.104 Conductors

Conductors shall be of a type specified by Article 310.

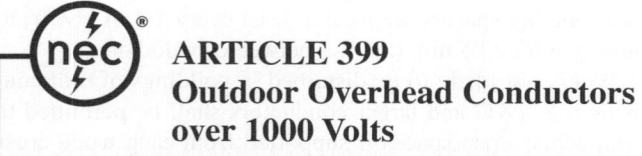

# ARTICLE 399 Outdoor Overhead Conductors over 1000 Volts

### 399.1 Scope

This article covers the use and installation for outdoor overhead conductors over 1000 volts, nominal.

### 399.2 Definition

**Outdoor Overhead Conductors.** Single conductors, insulated, covered, or bare, installed outdoors on support structures in free air.

### 399.10 Uses Permitted

Outdoor overhead conductors over 1000 volts, nominal, shall be permitted only for systems rated over 1000 volts, nominal, as follows:

(1) Outdoors in free air
(2) For service conductors, feeders, or branch circuits

Informational Note: For additional information on outdoor overhead conductors over 1000 volts, see ANSI/IEEE C2-2007, *National Electrical Safety Code*.

•

### 399.30 Support

**(A) Conductors.** Documentation of the engineered design by a licensed professional engineer engaged primarily in the design of such systems for the spacing between conductors shall be available upon request of the authority having jurisdiction and shall include consideration of the following:

(1) Applied voltage
(2) Conductor size
(3) Distance between support structures
(4) Type of structure
(5) Wind/ice loading
(6) Surge protection

**(B) Structures.** Structures of wood, metal, concrete, or combinations of those materials, shall be provided for support of overhead conductors over 1000 volts, nominal. Documentation of the engineered design by a licensed professional engineer engaged primarily in the design of such systems and the installation of each support structure shall be available upon request of the authority having jurisdiction and shall include consideration of the following:

(1) Soil conditions
(2) Foundations and structure settings
(3) Weight of all supported conductors and equipment
(4) Weather loading and other conditions such as, but not limited to, ice, wind, temperature, and lightning
(5) Angle where change of direction occurs
(6) Spans between adjacent structures
(7) Effect of dead-end structures
(8) Strength of guys and guy anchors
(9) Structure size and material(s)
(10) Hardware

**(C) Insulators.** Insulators used to support conductors shall be rated for all of the following:

(1) Applied phase-to-phase voltage
(2) Mechanical strength required for each individual installation
(3) Impulse withstand BIL in accordance with Table 490.24

Informational Note: 399.30(A), (B), and (C) are not all-inclusive lists.

# 4 Equipment for General Use

## ARTICLE 400 Flexible Cords and Cables

### I. General

#### 400.1 Scope

This article covers general requirements, applications, and construction specifications for flexible cords and flexible cables.

Flexible cords and cables, because of the nature of their use, are not considered to be wiring methods, which are covered in Chapter 3. A review of uses permitted in 400.7 and uses not permitted in 400.8 is necessary before choosing flexible cords or cables for a specific application. The flexible cords and cables referred to in Article 400 are not limited to use with portable equipment.

#### 400.2 Other Articles

Flexible cords and flexible cables shall comply with this article and with the applicable provisions of other articles of this *Code*.

#### 400.3 Suitability

Flexible cords and cables and their associated fittings shall be suitable for the conditions of use and location.

#### 400.4 Types

Flexible cords and flexible cables shall conform to the description in Table 400.4. The use of flexible cords and flexible cables other than those in Table 400.4 shall require permission by the authority having jurisdiction.

***TABLE 400.4*** *Flexible Cords and Cables (See 400.4.)*

| Trade Name | Type Letter | Voltage | AWG or kcmil | Number of Conductors | Insulation | AWG or kcmil | Nominal Insulation Thickness: mm | Nominal Insulation Thickness: mils | Braid on Each Conductor | Outer Covering | Use | | |
|---|---|---|---|---|---|---|---|---|---|---|---|---|---|
| Lamp cord | C | 300<br>600 | 18–16<br>15–10 | 2 or more | Thermoset or thermoplastic | 18–16<br>15–10 | 0.76<br>1.14 | 30<br>45 | Cotton | None | Pendant or portable | Dry locations | Not hard usage |
| Elevator cable | E[1,2,3,4] | 300 or 600 | 20–2 | 2 or more | Thermoset | 20–16<br>15–12<br>12–10<br>8–2 | 0.51<br>0.76<br>1.14<br>1.52 | 20<br>30<br>45<br>60 | Cotton | Three cotton; outer one flame-retardant & moisture-resistant | Elevator lighting and control | Unclassified locations | |
| | | | | | | 20–16<br>15–12<br>12–10<br>8–2 | 0.51<br>0.76<br>1.14<br>1.52 | 20<br>30<br>45<br>60 | Flexible nylon jacket | | | | |
| Elevator cable | EO[1,2,4] | 300 or 600 | 20–2 | 2 or more | Thermoset | 20–16<br>15–12<br>12–10<br>8–2 | 0.51<br>0.76<br>1.14<br>1.52 | 20<br>30<br>45<br>60 | Cotton | Three cotton; outer one flame-retardant & moisture-resistant | Elevator lighting and control | Unclassified locations | |
| | | | | | | | | | | One cotton and a neoprene jacket | | Hazardous (classified) locations | |

*(continues)*

*TABLE 400.4* Continued

| Trade Name | Type Letter | Voltage | AWG or kcmil | Number of Conductors | Insulation | AWG or kcmil | Nominal Insulation Thickness mm | Nominal Insulation Thickness mils | Braid on Each Conductor | Outer Covering | Use | | |
|---|---|---|---|---|---|---|---|---|---|---|---|---|---|
| Elevator cable | ETP[2,4] | 300 or 600 | | | | | | | Rayon | Thermoplastic | Hazardous (classified) location | | |
| | ETT[2,4] | 300 or 600 | | | | | | | None | One cotton or equivalent and a thermoplastic jacket | | | |
| Electric vehicle cable | EV[5,6] | 600 | 18–500 | 2 or more plus grounding conductor(s), plus optional hybrid data, signal communications, and optical fiber cables | Thermoset with optional nylon | 18–15<br>14–10<br>8–2<br>1–4/0<br>250–500 | 0.76<br>(0.51)<br>1.14<br>(0.76)<br>1.52<br>(1.14)<br>2.03<br>(1.52)<br>2.41<br>(1.90) | 30<br>(20)<br>45<br>(30)<br>60<br>(45)<br>80<br>(60)<br>95<br>(75) | Optional | Oil-resistant thermoset | Electric vehicle charging | Wet locations | Extra-hard usage |
| | EVJ[5,6] | 300 | 18–12 | | | 18–12 | 0.76<br>(0.51)] | 30<br>(20) | | | | | Hard usage |
| | EVE[5,6] | 600 | 18–500 | 2 or more plus grounding conductor(s), plus optional hybrid data, signal communications, and optical fiber cables | Thermoplastic elastomer with optional nylon | 18–15<br>14–10<br>8–2<br>1–4/0<br>250–500 | 0.76<br>(0.51)<br>1.14<br>(0.76)<br>1.52<br>(1.14)<br>2.03<br>(1.52)<br>2.41<br>(1.90) | 30<br>(20)<br>45<br>(30)<br>60<br>(45)<br>80<br>(60)<br>95<br>(75) | | Oil-resistant thermoplastic elastomer | | | Extra-hard usage |
| | EVJE[5,6] | 300 | 18–12 | | | 18–12 | 0.76<br>(0.51) | 30<br>(20) | | | | | Hard usage |
| | EVT[5,6] | 600 | 18–500 | 2 or more plus grounding conductor(s), plus optional hybrid data, signal communications, and optical fiber cables | Thermoplastic with optional nylon | 18–15<br>14–10<br>8–2<br>1–4/0<br>250–500 | 0.76<br>(0.51)<br>1.14<br>(0.76)<br>1.52<br>(1.14)<br>2.03<br>(1.52)<br>2.41<br>(1.90) | 30<br>(20)<br>45<br>(30)<br>60<br>(45)<br>80<br>(60)<br>95<br>(75) | Optional | Oil-resistant thermoplastic | Electric vehicle charging | Wet Locations | Extra-hard usage |
| | EVJT[5,6] | 300 | 18–12 | | | 18–12 | 0.76<br>(0.51) | 30<br>(20) | | | | | Hard usage |
| Portable power cable | G | 2000 | 12–500 | 2–6 plus grounding conductor(s) | Thermoset | 12–2<br>1–4/0<br>250–500 | 1.52<br>2.03<br>2.41 | 60<br>80<br>95 | | Oil-resistant thermoset | Portable and extra-hard usage | | |
| | G-GC[7] | 2000 | 12–500 | 3–6 plus grounding conductors and 1 ground check conductor | Thermoset | 12–2<br>1–4/0<br>250–500 | 1.52<br>2.03<br>2.41 | 60<br>80<br>95 | | Oil-resistant thermoset | | | |
| Heater cord | HPD | 300 | 18–12 | 2, 3, or 4 | Thermoset | 18–16<br>15–12 | 0.38<br>0.76 | 15<br>30 | None | Cotton or rayon | Portable heaters | Dry locations | Not hard usage |
| Parallel heater cord | HPN[8] | 300 | 18–12 | 2 or 3 | Oil-resistant thermoset | 18–16<br>15<br>14<br>12 | 1.14<br>1.52<br>2.41 | 45<br>60<br>95 | None | Oil-resistant thermoset | Portable | Damp locations | Not hard usage |

***TABLE 400.4*** *Continued*

| Trade Name | Type Letter | Voltage | AWG or kcmil | Number of Conductors | Insulation | AWG or kcmil | Nominal Insulation Thickness mm | Nominal Insulation Thickness mils | Braid on Each Conductor | Outer Covering | Use | | |
|---|---|---|---|---|---|---|---|---|---|---|---|---|---|
| Thermoset jacketed heater cords | HSJ | 300 | 18–12 | 2, 3, or 4 | Thermoset | 18–16<br>15–12 | 0.76<br>1.14 | 30<br>45 | None | Cotton and thermoset | Portable or portable heater | Damp locations | Hard usage |
| | HSJO | 300 | 18–12 | | | | | | | Cotton and oil-resistant thermoset | | Damp and wet locations | |
| | HSJOW[9] | 300 | 18–12 | | | | | | | | | Damp locations | |
| | HSJOO | 300 | 18–12 | | Oil-resistant thermoset | | | | | | | | |
| | HSJOOW[9] | 300 | 18–12 | | | | | | | | | Damp and wet locations | |
| Non-integral parallel cords | NISP-1 | 300 | 20–18 | 2 or 3 | Thermoset | 20–18 | 0.38 | 15 | None | Thermoset | Pendant or portable | Damp locations | Not hard usage |
| | NISP-2 | 300 | 18–16 | | | 18–16 | 0.76 | 30 | | | | | |
| | NISPE-1[8] | 300 | 20–18 | | Thermoplastic elastomer | 20–18 | 0.38 | 15 | | Thermoplastic elastomer | | | |
| | NISPE-2[8] | 300 | 18–16 | | | 18–16 | 0.76 | 30 | | | | | |
| | NISPT-1[8] | 300 | 20–18 | | Thermoplastic | 20–18 | 0.38 | 15 | | Thermoplastic | | | |
| | NISPT-2[8] | 300 | 18–16 | | | 18–16 | 0.76 | 30 | | | | | |
| Twisted portable cord | PD | 300<br>600 | 18–16<br>14–10 | 2 or more | Thermoset or thermoplastic | 18–16<br>15–10 | 0.76<br>1.14 | 30<br>45 | Cotton | Cotton or rayon | Pendant or portable | Dry locations | Not hard usage |
| Portable power cable | PPE[7] | 2000 | 12–500 | 1–6 plus optional grounding conductor(s) | Thermoplastic elastomer | 12–2<br>1–4/0<br>250–500 | 1.52<br>2.03<br>2.41 | 60<br>80<br>95 | | Oil-resistant thermoplastic elastomer | Portable, extra-hard usage | | |
| Hard service cord | S[7] | 600 | 18–2 | 2 or more | Thermoset | 18–15<br>14–10<br>8–2 | 0.76<br>1.14<br>1.52 | 30<br>45<br>60 | None | Thermoset | Pendant or portable | Damp locations | Extra-hard usage |
| Flexible stage and lighting power cable | SC[7,10] | 600 | 8–250 | 1 or more | Thermoset | 8–2<br>1–4/0<br>250 | 1.52<br>2.03<br>2.41 | 60<br>80<br>95 | | Thermoset | Portable, extra-hard usage | | |
| | SCE[7,10] | 600 | | | Thermoplastic elastomer | | | | | Thermoplastic elastomer | | | |
| | SCT[7,10] | 600 | | | Thermoplastic | | | | | Thermoplastic | | | |
| Hard service cord | SE[7] | 600 | 18–2 | 2 or more | Thermoplastic elastomer | 18–15<br>14–9<br>8–2 | 0.76<br>1.14<br>1.52 | 30<br>45<br>60 | None | Thermoplastic elastomer | Pendant or portable | Damp locations | Extra-hard usage |
| | SEW[7,9] | 600 | | | | | | | | | | Damp and wet locations | |
| | SEO[7] | 600 | | | | | | | | Oil-resistant thermoplastic elastomer | | Damp locations | |
| | SEOW[7,9] | 600 | | | | | | | | | | Damp and wet locations | |
| | SEOO[7] | 600 | | | Oil-resistant thermoplastic elastomer | | | | | | | Damp locations | |
| | SEOOW[7,9] | 600 | | | | | | | | | | Damp and wet locations | |
| Junior hard service cord | SJ | 300 | 18–10 | 2–6 | Thermoset | 18–11<br>10 | 0.76<br>1.14 | 30<br>45 | None | Thermoset | Pendant or portable | Damp locations | Hard usage |
| | SJE | 300 | | | Thermoplastic elastomer | | | | | Thermoplastic elastomer | | | |
| | SJEW[9] | 300 | | | | | | | | | | Damp and wet locations | |
| | SJEO | 300 | | | | | | | | Oil-resistant thermoplastic elastomer | | Damp locations | |

*(continues)*

*TABLE 400.4 Continued*

| Trade Name | Type Letter | Voltage | AWG or kcmil | Number of Conductors | Insulation | AWG or kcmil | Nominal Insulation Thickness mm | Nominal Insulation Thickness mils | Braid on Each Conductor | Outer Covering | Use | | |
|---|---|---|---|---|---|---|---|---|---|---|---|---|---|
| | SJEOW[9] | 300 | | | | | | | | | | Damp and wet locations | |
| | SJEOO | 300 | | | Oil-resistant thermoplastic elastomer | | | | | | | Damp locations | |
| | SJEOOW[9] | 300 | | | | | | | | | | Damp and wet locations | |
| | SJO | 300 | | | Thermoset | | | | | Oil-resistant thermoset | | Damp locations | |
| | SJOW[9] | 300 | | | | | | | | | | Damp and wet locations | |
| | SJOO | 300 | | | Oil-resistant thermoset | | | | | | | Damp locations | |
| | SJOOW[9] | 300 | | | | | | | | | | Damp and wet locations | |
| | SJT | 300 | | | Thermoplastic | | | | | Thermoplastic | | Damp locations | |
| | SJTW[9] | 300 | | | | | | | | | | Damp and wet locations | |
| | SJTO | 300 | | | | 18–12<br>10 | 0.76<br>1.14 | 30<br>45 | | Oil-resistant thermoplastic | | Damp locations | |
| | SJTOW[9] | 300 | | | | | | | | | | Damp and wet locations | |
| | SJTOO | 300 | | | Oil-resistant thermoplastic | | | | | | | Damp locations | |
| | SJTOOW[9] | 300 | | | | | | | | | | Damp and wet locations | |
| Hard service cord | SO[7] | 600 | 18–2 | 2 or more | Thermoset | 18–15 | 0.76 | 30 | None | Oil-resistant thermoset | Pendant or portable | Damp locations | Extra-hard usage |
| | SOW[7,9] | 600 | | | | | | | | | | Damp and wet locations | |
| | SOO[7] | 600 | | | Oil-resistant thermoset | 14–9<br>8–2 | 1.14<br>1.52 | 45<br>60 | | | | Damp locations | |
| | SOOW[7,9] | 600 | | | | | | | | | | Damp and wet locations | |
| All thermoset parallel cord | SP-1 | 300 | 20–18 | 2 or 3 | Thermoset | 20–18 | 0.76 | 30 | None | None | Pendant or portable | Damp locations | Not hard usage |
| | SP-2 | 300 | 18–16 | | | 18–16 | 1.14 | 45 | | | | | |
| | SP-3 | 300 | 18–10 | | | 18–16<br>15, 14<br>12<br>10 | 1.52<br>2.03<br>2.41<br>2.80 | 60<br>80<br>95<br>110 | | | Refrigerators, room air conditioners, and as permitted in 422.16(B) | | |

*TABLE 400.4 Continued*

| Trade Name | Type Letter | Voltage | AWG or kcmil | Number of Conductors | Insulation | AWG or kcmil | Nominal Insulation Thickness mm | Nominal Insulation Thickness mils | Braid on Each Conductor | Outer Covering | Use | | |
|---|---|---|---|---|---|---|---|---|---|---|---|---|---|
| All elastomer (thermoplastic) parallel cord | SPE-1[8] | 300 | 20–18 | 2 or 3 | Thermoplastic elastome | 20–18 | 0.76 | 30 | None | Non | Pendant or portable | Damp locations | Not hard usag |
| | SPE-2[8] | 300 | 18–16 | | | 18–16 | 1.14 | 45 | | | | | |
| | SPE-3[8] | 300 | 18–10 | | | 18–16<br>15<br>14<br>12<br>10 | 1.52<br>2.03<br>2.41<br>2.80 | 60<br>80<br>95<br>110 | | | Refrigerators, room air conditioners, and as permitted in 422.16(B) | | |
| All thermoplastic parallel cord | SPT-1 | 300 | 20–18 | 2 or 3 | Thermoplastic | 20–18 | 0.76 | 30 | None | None | Pendant or portable | Damp locations | Not hard usage |
| | SPT-1W[9] | 300 | | 2 | | | | | | | | Damp and wet locations | |
| | SPT-2 | 300 | 18–16 | 2 or 3 | | 18–16 | 1.14 | 45 | | | | Damp locations | |
| | SPT-2W[9] | 300 | | 2 | | | | | | | | Damp and wet locations | |
| | SPT-3 | 300 | 18–10 | 2 or 3 | | 18–16<br>15<br>14<br>12<br>10 | 1.52<br>2.03<br>2.41<br>2.80 | 60<br>80<br>95<br>110 | | | Refriger ators, room air conditioners, and as permitted in 422.16(B) | Damp locations | Not hard usage |
| Range, dryer cable | SRD | 300 | 10–4 | 3 or 4 | Thermoset | 10–4 | 1.14 | 45 | None | Thermoset | Portable | Damp locations | Ranges, dryers |
| | SRDE | 300 | 10–4 | 3 or 4 | Thermoplastic elastomer | | | | None | Thermoplastic elastomer | | | |
| | SRDT | 300 | 10–4 | 3 or 4 | Thermoplastic | | | | None | Thermoplastic | | | |
| Hard service cord | ST[7] | 600 | 18–2 | 2 or more | Thermoplastic | 18–15<br>14–9<br>8–2 | 0.76<br>1.14<br>1.52 | 30<br>45<br>60 | None | Thermoplastic | Pendant or portable | Damp locations | Extra-hard usage |
| | STW[7,9] | 600 | | | | | | | | | | Damp and wet locations | |
| | STO[7] | 600 | | | | | | | | Oil-resistant thermoplastic | | Damp locations | |
| | STOW[7,9] | 600 | | | | | | | | | | Damp and wet locations | |
| | STOO[7] | 600 | | | Oil-resistant thermoplastic | | | | | | | Damp locations | |
| | STOOW[7] | 600 | | | | | | | | | | Damp and wet locations | |
| Vacuum cleaner cord | SV | 300 | 18–16 | 2 or 3 | Thermoset | 18–16 | 0.38 | 15 | None | Thermoset | Pendant or portable | Damp locations | Not hard usage |
| | SVE | 300 | | | Thermoplastic elastomer | | | | | Thermoplastic elastomer | | | |
| | SVEO | 300 | | | | | | | | Oil-resistant thermoplastic elastomer | | | |
| | SVEOO | 300 | | | Oil-resistant thermoplastic elastomer | | | | | | | | |
| | SVO | 300 | | | Thermoset | | | | | Oil-resistant thermoset | | | |

*(continues)*

*TABLE 400.4 Continued*

| Trade Name | Type Letter | Voltage | AWG or kcmil | Number of Conductors | Insulation | AWG or kcmil | Nominal Insulation Thickness mm | Nominal Insulation Thickness mils | Braid on Each Conductor | Outer Covering | Use | | |
|---|---|---|---|---|---|---|---|---|---|---|---|---|---|
| | SVOO | 300 | | | Oil-resistant thermoset | | | | | Oil-resistant thermoset | | | |
| | SVT | 300 | | | Thermoplastic | | | | | Thermoplastic | | | |
| | SVTO | 300 | | | Thermoplastic | | | | | Oil-resistant thermoplastic | | | |
| | SVTOO | 300 | | | Oil-resistant thermoplastic | | | | | | | | |
| Parallel tinsel cord | TPT[11] | 300 | 27 | 2 | Thermoplastic | 27 | 0.76 | 30 | None | Thermoplastic | Attached to an appliance | Damp locations | Not hard usage |
| Jacketed tinsel cord | TST[11] | 300 | 27 | 2 | Thermoplastic | 27 | 0.38 | 15 | None | Thermoplastic | Attached to an appliance | Damp locations | Not hard usage |
| Portable power cable | W[7] | 2000 | 12–500<br>501–1000 | 1–6<br>1 | Thermoset | 12–2<br>1–4/0<br>250–500<br>501–1000 | 1.52<br>2.03<br>2.41<br>2.80 | 60<br>80<br>95<br>110 | | Oil-resistant thermoset | Portable, extra-hard usage | | |

Notes:

•

All types listed in Table 400.4 shall have individual conductors twisted together, except for Types HPN, SP-1, SP-2, SP-3, SPE-1, SPE-2, SPE-3, SPT-1, SPT-2, SPT-3, SPT-1W, SPT-2W, TPT, NISP-1, NISP-2, NISPT-1, NISPT-2, NISPE-1, NISPE-2, and three-conductor parallel versions of SRD, SRDE, and SRDT.

The individual conductors of all cords, except those of heat-resistant cords, shall have a thermoset or thermoplastic insulation, except that the equipment grounding conductor, where used, shall be in accordance with 400.23(B).

•

[1]Rubber-filled or varnished cambric tapes shall be permitted as a substitute for the inner braids.

[2]Elevator traveling cables for operating control and signal circuits shall contain nonmetallic fillers as necessary to maintain concentricity. Cables shall have steel supporting members as required for suspension by 620.41. In locations subject to excessive moisture or corrosive vapors or gases, supporting members of other materials shall be permitted. Where steel supporting members are used, they shall run straight through the center of the cable assembly and shall not be cabled with the copper strands of any conductor.

In addition to conductors used for control and signaling circuits, Types E, EO, ETP, and ETT elevator cables shall be permitted to incorporate in the construction one or more 20 AWG telephone conductor pairs, one or more coaxial cables, or one or more optical fibers. The 20 AWG conductor pairs shall be permitted to be covered with suitable shielding for telephone, audio, or higher frequency communications circuits; the coaxial cables consist of a center conductor, insulation, and a shield for use in video or other radio frequency communications circuits. The optical fiber shall be suitably covered with flame-retardant thermoplastic. The insulation of the conductors shall be rubber or thermoplastic of a thickness not less than specified for the other conductors of the particular type of cable. Metallic shields shall have their own protective covering. Where used, these components shall be permitted to be incorporated in any layer of the cable assembly but shall not run straight through the center.

[3]Insulations and outer coverings that meet the requirements as flame retardant, limited smoke, and are so listed, shall be permitted to be marked for limited smoke after the *Code* type designation.

[4]Elevator cables in sizes 20 AWG through 14 AWG are rated 300 volts, and sizes 10 AWG through 2 AWG are rated 600 volts. 12 AWG is rated 300 volts with a 0.76 mm (30 mil) insulation thickness and 600 volts with a 1.14 mm (45 mil) insulation thickness.

[5]Conductor size for Types EV, EVJ, EVE, EVJE, EVT, and EVJT cables apply to nonpower-limited circuits only. Conductors of power-limited (data, signal, or communications) circuits may extend beyond the stated AWG size range. All conductors shall be insulated for the same cable voltage rating.

[6]Insulation thickness for Types EV, EVJ, EVEJE, EVT, and EVJT cables of nylon construction is indicated in parentheses.

[7]Types G, G-GC, S, SC, SCE, SCT, SE, SEO, SEOO, SEW, SEOW, SEOOW, SO, SOO, SOW, SOOW, ST, STO, STOO, STW, STOW, STOOW, PPE, and W shall be permitted for use on theater stages, in garages, and elsewhere where flexible cords are permitted by this *Code*.

[8]The third conductor in Type HPN shall be used as an equipment grounding conductor only. The insulation of the equipment grounding conductor for Types SPE-1, SPE-2, SPE-3, SPT-1, SPT-2, SPT-3, NISPT-1, NISPT-2, NISPE-1, and NISPE-2 shall be permitted to be thermoset polymer.

[9]Cords that comply with the requirements for outdoor cords and are so listed shall be permitted to be designated as weather and water resistant with the suffix "W" after the *Code* type designation. Cords with the "W" suffix are suitable for use in wet locations and are sunlight resistant.

[10]The required outer covering on some single-conductor cables may be integral with the insulation.

[11]Types TPT and TST shall be permitted in lengths not exceeding 2.5 m (8 ft) where attached directly, or by means of a special type of plug, to a portable appliance rated at 50 watts or less and of such nature that extreme flexibility of the cord is essential.

## 400.5 Ampacities for Flexible Cords and Cables

**(A) Ampacity Tables.** Table 400.5(A)(1) provides the allowable ampacities, and Table 400.5(A)(2) provides the ampacities for flexible cords and cables with not more than three current-carrying conductors. These tables shall be used in conjunction with applicable end-use product standards to ensure selection of the proper size and type. Where cords and cables are used in ambient temperatures other than 30°C (86°F), the temperature correction factors from Table 310.15(B)(2)(a) that correspond to the temperature rating of the cord or cable shall be applied to the ampacity in Table 400.5(A)(1) and Table 400.5(A)(2). Cords and cables rated 105°C shall use correction factors in the 90°C column of Table 310.15(B)(2)(a) for temperature correction. Where the number of current-carrying conductors exceeds three, the allowable ampacity or the ampacity of each conductor shall be reduced from the three-conductor rating as shown in Table 400.5(A)(3).

Where power cables are used in an ambient temperature exceeding 30°C (86°F), "correction factors" are required to be applied to the ampacities in Table 400.5(A)(2). This parallels the Article 310 requirements for ampacity correction of conductors used in elevated ambient temperatures. In fact, the ambient correction factors that are to be used for power cables are those specified in Table 310.15(B)(16). The specific correction factor to be applied is predicated on the temperature rating of the power cable.

Informational Note: See Informative Annex B, Table B.310.15(B)(2)(11), for adjustment factors for more than three current-carrying conductors in a raceway or cable with load diversity.

***TABLE 400.5(A)(1)*** *Allowable Ampacity for Flexible Cords and Cables [Based on Ambient Temperature of 30°C (86°F). See 400.13 and Table 400.4.*

| Copper Conductor Size (AWG) | Thermoplastic Types TPT, TST | Thermoset Types C, E, EO, PD, S, SJ, SJO, SJOW, SJOO, SJOOW, SO, SOW, SOO, SOOW, SP-1, SP-2, SP-3, SRD, SV, SVO, SVOO, NISP-1, NISP-2 / Thermoplastic Types ETP, ETT, NISPE-1, NISPE-2, NISPT-1, NISPT-2, SE, SEW, SEO, SEOO, SEOW, SEOOW, SJE, SJEW, SJEO, SJEOO, SJEOW, SJEOOW, SJT, SJTW, SJTO, SJTOW, SJTOO, SJTOOW, SPE-1, SPE-2, SPE-3, SPT-1, SPT-1W, SPT-2, SPT-2W, SPT-3, ST, STW, SRDE, SRDT, STO, STOW, STOO, STOOW, SVE, SVEO, SVEOO, SVT, SVTO, SVTOO | | Types HPD, HPN, HSJ, HSJO, HSJOW, HSJOO, HSJOOW |
|---|---|---|---|---|
| | | **Column A**[a] | **Column B**[b] | |
| 27[c] | 0.5 | — | — | — |
| 20 | — | 5[d] | [e] | — |
| 18 | — | 7 | 10 | 10 |
| 17 | — | 9 | 12 | 13 |
| 16 | — | 10 | 13 | 15 |
| 15 | — | 12 | 16 | 17 |
| 14 | — | 15 | 18 | 20 |
| 13 | — | 17 | 21 | — |
| 12 | — | 20 | 25 | 30 |
| 11 | — | 23 | 27 | — |
| 10 | — | 25 | 30 | 35 |
| 9 | — | 29 | 34 | — |
| 8 | — | 35 | 40 | — |
| 6 | — | 45 | 55 | — |
| 4 | — | 60 | 70 | — |
| 2 | — | 80 | 95 | — |

[a]The allowable currents under Column A apply to three-conductor cords and other multiconductor cords connected to utilization equipment so that only three-conductors are current-carrying.

[b]The allowable currents under Column B apply to two-conductor cords and other multiconductor cords connected to utilization equipment so that only two conductors are current-carrying.

[c]Tinsel cord.

[d]Elevator cables only.

[e]7 amperes for elevator cables only; 2 amperes for other types.

*TABLE 400.5(A)(2) Ampacity of Cable Types SC, SCE, SCT, PPE, G, G-GC, and W. [Based on Ambient Temperature of 30°C (86°F). See Table 400.4.]*

| Copper Conductor Size (AWG or kmil) | Temperature Rating of Cable | | | | | | | | |
|---|---|---|---|---|---|---|---|---|---|
| | 60°C (140°F) | | | 75°C (167°F) | | | 90°C (194°F) | | |
| | D[1] | E[2] | F[3] | D[1] | E[2] | F[3] | D[1] | E[2] | F[3] |
| 12 | — | 31 | 26 | — | 37 | 31 | — | 42 | 35 |
| 10 | — | 44 | 37 | — | 52 | 43 | — | 59 | 49 |
| 8 | 60 | 55 | 48 | 70 | 65 | 57 | 80 | 74 | 65 |
| 6 | 80 | 72 | 63 | 95 | 88 | 77 | 105 | 99 | 87 |
| 4 | 105 | 96 | 84 | 125 | 115 | 101 | 140 | 130 | 114 |
| 3 | 120 | 113 | 99 | 145 | 135 | 118 | 165 | 152 | 133 |
| 2 | 140 | 128 | 112 | 170 | 152 | 133 | 190 | 174 | 152 |
| 1 | 165 | 150 | 131 | 195 | 178 | 156 | 220 | 202 | 177 |
| 1/0 | 195 | 173 | 151 | 230 | 207 | 181 | 260 | 234 | 205 |
| 2/0 | 225 | 199 | 174 | 265 | 238 | 208 | 300 | 271 | 237 |
| 3/0 | 260 | 230 | 201 | 310 | 275 | 241 | 350 | 313 | 274 |
| 4/0 | 300 | 265 | 232 | 360 | 317 | 277 | 405 | 361 | 316 |
| 250 | 340 | 296 | 259 | 405 | 354 | 310 | 455 | 402 | 352 |
| 300 | 375 | 330 | 289 | 445 | 395 | 346 | 505 | 449 | 393 |
| 350 | 420 | 363 | 318 | 505 | 435 | 381 | 570 | 495 | 433 |
| 400 | 455 | 392 | 343 | 545 | 469 | 410 | 615 | 535 | 468 |
| 500 | 515 | 448 | 392 | 620 | 537 | 470 | 700 | 613 | 536 |
| 600 | 575 | — | — | 690 | — | — | 780 | — | — |
| 700 | 630 | — | — | 755 | — | — | 855 | — | — |
| 750 | 655 | — | — | 785 | — | — | 885 | — | — |
| 800 | 680 | — | — | 815 | — | — | 920 | — | — |
| 900 | 730 | — | — | 870 | — | — | 985 | — | — |
| 1000 | 780 | — | — | 935 | — | — | 1055 | — | — |

[1]The ampacities under subheading D shall be permitted for single-conductor Types SC, SCE, SCT, PPE, and W cable only where the individual conductors are not installed in raceways and are not in physical contact with each other except in lengths not to exceed 600 mm (24 in.) where passing through the wall of an enclosure.

[2]The ampacities under subheading E apply to two-conductor cables and other multiconductor cables connected to utilization equipment so that only two conductors are current-carrying.

[3]The ampacities under subheading F apply to three-conductor cables and other multiconductor cables connected to utilization equipment so that only three conductors are current-carrying.

*TABLE 400.5(A)(3) Adjustment Factors for More Than Three Current-Carrying Conductors in a Flexible Cord or Cable*

| Number of Conductors | Percent of Value in Table 400.5(A)(1) and Table 400.5(A)(2) |
|---|---|
| 4–6 | 80 |
| 7–9 | 70 |
| 10–20 | 50 |
| 21–30 | 45 |
| 31–40 | 40 |
| 41 and above | 35 |

A neutral conductor that carries only the unbalanced current from other conductors of the same circuit shall not be required to meet the requirements of a current-carrying conductor.

In a 3-wire circuit consisting of two phase conductors and the neutral conductor of a 4-wire, 3-phase, wye-connected system, a common conductor carries approximately the same current as the line-to-neutral currents of the other conductors and shall be considered to be a current-carrying conductor.

On a 4-wire, 3-phase, wye circuit where more than 50 percent of the load consists of nonlinear loads, there are harmonic currents present in the neutral conductor and the neutral conductor shall be considered to be a current-carrying conductor.

An equipment grounding conductor shall not be considered a current-carrying conductor.

Where a single conductor is used for both equipment grounding and to carry unbalanced current from other conductors, as provided for in 250.140 for electric ranges and electric clothes dryers, it shall not be considered as a current-carrying conductor.

**(B) Ultimate Insulation Temperature.** In no case shall conductors be associated together in such a way with respect to the kind of circuit, the wiring method used, or the number of conductors such that the limiting temperature of the conductors is exceeded.

**(C) Engineering Supervision.** Under engineering supervision, conductor ampacities shall be permitted to be calculated in accordance with 310.15(C).

The ampacity of flexible cords and cables may be determined in accordance with the Neher–McGrath formula found in 310.15(C). See the inside cover of this book for more information about the Neher–McGrath formula.

## 400.6 Markings

**(A) Standard Markings.** Flexible cords and cables shall be marked by means of a printed tag attached to the coil reel or carton. The tag shall contain the information required in 310.120(A). Types S, SC, SCE, SCT, SE, SEO, SEOO, SJ, SJE, SJEO, SJEOO, SJO, SJT, SJTO, SJTOO, SO, SOO, ST, STO, STOO, SEW, SEOW, SEOOW, SJEW, SJEOW, SJEOOW, SJOW, SJTW, SJTOW, SJTOOW, SOW, SOOW, STW, STOW, and STOOW flexible cords and G, G-GC, PPE, and W flexible cables shall be durably marked on the surface at intervals not exceeding 610 mm (24 in.) with the type designation, size, and number of conductors. Required markings on tags, cords, and cables shall also include the maximum operating temperature of the flexible cord or cable.

**(B) Optional Markings.** Flexible cords and cable types listed in Table 400.4 shall be permitted to be surface marked to indicate special characteristics of the cable materials. These markings include, but are not limited to, markings for limited smoke, sunlight resistance, and so forth.

In addition to the markings identified in 400.6(A) and (B), the UL *Guide Information for Electrical Equipment – The White Book,* under the category Flexible Cord (ZJCZ), lists the following markings:

- "For Mobile Home Use," "For Recreational Vehicle Use," or "For Mobile Home and Recreational Vehicle Use" followed by the current rating in amperes indicates suitability for use in mobile homes or recreational vehicles.
- "W" indicates suitability for use outdoors and for immersion in water. The low temperature rating for these cords is −40°C, unless otherwise marked on the cord with optional ratings of −50°C, −60°C, or −70°C. The low temperature ratings are determined by means of a bend test (not a suppleness test) at the given temperature.
- "VW-1" indicates that the cord complies with a vertical flame test. Cord that has been evaluated for leakage currents between the circuit conductor and the grounding conductor and between the circuit conductor and the outer surface of the jacket may have the leakage current values marked on the cable jacket.

## 400.7 Uses Permitted

**(A) Uses.** Flexible cords and cables shall be used only for the following:

(1) Pendants.
(2) Wiring of luminaires.
(3) Connection of portable luminaires, portable and mobile signs, or appliances.
(4) Elevator cables.
(5) Wiring of cranes and hoists.
(6) Connection of utilization equipment to facilitate frequent interchange.
(7) Prevention of the transmission of noise or vibration.
(8) Appliances where the fastening means and mechanical connections are specifically designed to permit ready removal for maintenance and repair, and the appliance is intended or identified for flexible cord connection.
(9) Connection of moving parts.
(10) Where specifically permitted elsewhere in this *Code*.
(11) Between an existing receptacle outlet and an inlet, where the inlet provides power to an additional single receptacle outlet. The wiring interconnecting the inlet to the single receptacle outlet shall be a Chapter 3 wiring method. The inlet, receptacle outlet, and Chapter 3 wiring method, including the flexible cord and fittings, shall be a listed assembly specific for this application.

**(B) Attachment Plugs.** Where used as permitted in 400.7(A)(3), (A)(6), and (A)(8), each flexible cord shall be equipped with an attachment plug and shall be energized from a receptacle outlet or cord connector body.

*Exception: As permitted in 368.56.*

## 400.8 Uses Not Permitted

Unless specifically permitted in 400.7, flexible cords and cables shall not be used for the following:

(1) As a substitute for the fixed wiring of a structure
(2) Where run through holes in walls, structural ceilings, suspended ceilings, dropped ceilings, or floors
(3) Where run through doorways, windows, or similar openings
(4) Where attached to building surfaces

*Exception to (4): Flexible cord and cable shall be permitted to be attached to building surfaces in accordance with the provisions of 368.56(B).*

Section 368.56(B) provides the requirements for the installation of flexible cords installed as branches from busways.

(5) Where concealed by walls, floors, or ceilings or located above suspended or dropped ceilings
(6) Where installed in raceways, except as otherwise permitted in this *Code*

Flexible cords and cables are not limited to use with portable equipment. However, 400.8 prohibits the use of flexible cords and cables as a substitute for the fixed wiring of a structure or where concealed behind building walls, floors, or ceilings (including structural, suspended, or dropped-type ceilings). See 240.5, 590.4(B), and 590.4(C) for the uses of multiconductor flexible cords for feeder and branch-circuit installations and for overcurrent protection requirements for flexible cord. See 410.62 for cord-connected luminaires.

(7) Where subject to physical damage

## 400.9 Splices

Flexible cord shall be used only in continuous lengths without splice or tap where initially installed in applications permitted by 400.7(A). The repair of hard-service cord and junior hard-service cord (see Trade Name column in Table 400.4) 14 AWG and larger shall be permitted if conductors are spliced in accordance with 110.14(B) and the completed splice retains the insulation, outer sheath properties, and usage characteristics of the cord being spliced.

This section permits repair of a damaged cord in such a manner that the cord will retain its original operating and use integrity. However, if the repaired cord is reused or reinstalled at a new location, the in-line repair is no longer permitted, and the cord can be used only in lengths that do not contain a splice.

## 400.10 Pull at Joints and Terminals

Flexible cords and cables shall be connected to devices and to fittings so that tension is not transmitted to joints or terminals.

*Exception: Listed portable single-pole devices that are intended to accommodate such tension at their terminals shall be permitted to be used with single-conductor flexible cable.*

Informational Note: Some methods of preventing pull on a cord from being transmitted to joints or terminals include knotting the cord, winding with tape, and using support or strain-relief fittings.

## 400.11 In Show Windows and Showcases

Flexible cords used in show windows and showcases shall be Types S, SE, SEO, SEOO, SJ, SJE, SJEO, SJEOO, SJO, SJOO, SJT, SJTO, SJTOO, SO, SOO, ST, STO, STOO, SEW, SEOW, SEOOW, SJEW, SJEOW, SJEOOW, SJOW, SJOOW, SJTW, SJTOW, SJTOOW, SOW, SOOW, STW, STOW, or STOOW.

*Exception No. 1: For the wiring of chain-supported luminaires.*

*Exception No. 2: As supply cords for portable luminaires and other merchandise being displayed or exhibited.*

Flexible cords identified for hard usage or extra-hard usage should be used in show windows and showcases, because cords in these locations can come in contact with combustible materials such as fabrics or paper products, and they are exposed to wear and tear from continual housekeeping and display changes.

## 400.13 Overcurrent Protection

Flexible cords not smaller than 18 AWG, and tinsel cords or cords having equivalent characteristics of smaller size approved for use with specific appliances, shall be considered as protected against overcurrent in accordance with 240.5.

## 400.14 Protection from Damage

Flexible cords and cables shall be protected by bushings or fittings where passing through holes in covers, outlet boxes, or similar enclosures.

In industrial establishments where the conditions of maintenance and supervision ensure that only qualified persons service the installation, flexible cords and cables shall be permitted to be installed in aboveground raceways that are no longer than 15 m (50 ft) to protect the flexible cord or cable from physical damage. Where more than three current-carrying conductors are installed within the raceway, the allowable ampacity shall be reduced in accordance with Table 400.5(A)(3).

A variety of bushings and fittings, both insulated and noninsulated, are available for protecting flexible cords and cables. Some bushings or fittings include strain-relief fittings as required in 400.10. Many insulating bushings are listed in the following product categories:

1. Conduit fittings (bushings and fittings for use on the ends of conduit in boxes and gutters)
2. Insulating devices and materials
3. Outlet bushings and fittings (for use on the ends of conduit, electrical metallic tubing, or armored cable where a change to open wiring is made)

# II. Construction Specifications

## 400.20 Labels

Flexible cords shall be examined and tested at the factory and labeled before shipment.

See the definition of *labeled* in Article 100.

## 400.21 Construction

**(A) Conductors.** The individual conductors of a flexible cord or cable shall have flexible stranding and shall not be smaller than the sizes specified in Table 400.4.

Stranding is essential to ensure that the conductors are flexible. See the requirements in 110.14 regarding connectors and terminals used with conductors more finely stranded than Class B and Class C stranding.

**(B) Nominal Insulation Thickness.** The nominal thickness of insulation for conductors of flexible cords and cables shall not be less than specified in Table 400.4.

## 400.22 Grounded-Conductor Identification

One conductor of flexible cords that is intended to be used as a grounded circuit conductor shall have a continuous marker that readily distinguishes it from the other conductor or conductors. The identification shall consist of one of the methods indicated in 400.22(A) through (F).

**(A) Colored Braid.** A braid finished to show a white or gray color and the braid on the other conductor or conductors finished to show a readily distinguishable solid color or colors.

**(B) Tracer in Braid.** A tracer in a braid of any color contrasting with that of the braid and no tracer in the braid of the other conductor or conductors. No tracer shall be used in the braid of any conductor of a flexible cord that contains a conductor having a braid finished to show white or gray.

*Exception: In the case of Types C and PD and cords having the braids on the individual conductors finished to show white or gray. In such cords, the identifying marker shall be permitted to consist of the solid white or gray finish on one conductor, provided there is a colored tracer in the braid of each other conductor.*

**(C) Colored Insulation.** A white or gray insulation on one conductor and insulation of a readily distinguishable color or colors on the other conductor or conductors for cords having no braids on the individual conductors.

For jacketed cords furnished with appliances, one conductor having its insulation colored light blue, with the other conductors having their insulation of a readily distinguishable color other than white or gray.

*Exception: Cords that have insulation on the individual conductors integral with the jacket.*

The insulation shall be permitted to be covered with an outer finish to provide the desired color.

**(D) Colored Separator.** A white or gray separator on one conductor and a separator of a readily distinguishable solid color on the other conductor or conductors of cords having insulation on the individual conductors integral with the jacket.

Grounded conductors in flexible cords and cables are identified through the use of a white- or gray-colored braid, a white- or gray-colored tracer in the braid, white- or gray-colored insulation, or a white- or gray-colored separator. In existing installations where a gray-colored braid, tracer, or conductor insulation is encountered, caution should be exercised because gray could also signify ungrounded conductors.

**(E) Tinned Conductors.** One conductor having the individual strands tinned and the other conductor or conductors having the individual strands untinned for cords having insulation on the individual conductors integral with the jacket.

**(F) Surface Marking.** One or more ridges, grooves, or white stripes located on the exterior of the cord so as to identify one conductor for cords having insulation on the individual conductors integral with the jacket.

One method of surface marking is to identify the grounded conductor in a cord where the conductor insulation is part of the molded jacket and is not separable. A white stripe on the cord exterior serves to identify the segment of the cord that contains the grounded conductor. An example of a type of cord with these characteristics is zip cord, which is commonly used for floor lamps and table lamps. It is important that the grounded conductor be easily identified where zip cord is used for wiring lampholders, because the *Code* requires the grounded conductor to be connected to the device terminal that connects to the screw shell of the lampholder.

## 400.23 Equipment Grounding Conductor Identification

A conductor intended to be used as an equipment grounding conductor shall have a continuous identifying marker readily distinguishing it from the other conductor or conductors. Conductors having a continuous green color or a continuous green color with one or more yellow stripes shall not be used for other than equipment grounding conductors. Cords or cables consisting of integral insulation and a jacket without a nonintegral grounding conductor shall be permitted to be green. The identifying marker shall consist of one of the methods in 400.23(A) or (B).

**(A) Colored Braid.** A braid finished to show a continuous green color or a continuous green color with one or more yellow stripes.

**(B) Colored Insulation or Covering.** For cords having no braids on the individual conductors, an insulation of a continuous green color or a continuous green color with one or more yellow stripes.

## 400.24 Attachment Plugs

Where a flexible cord is provided with an equipment grounding conductor and equipped with an attachment plug, the attachment plug shall comply with 250.138(A) and (B).

# III. Portable Cables Over 600 Volts, Nominal

## 400.30 Scope

Part III applies to single and multiconductor portable cables used to connect mobile equipment and machinery.

## 400.31 Construction

**(A) Conductors.** The conductors shall be 12 AWG copper or larger and shall employ flexible stranding.

**(B) Equipment Grounding Conductor(s).** An equipment grounding conductor(s) shall be provided in cables with three or more conductors. The total area shall not be less than that of the size of the equipment grounding conductor required in 250.122.

### 400.32 Shielding

All shields shall be connected to an equipment grounding conductor.

### 400.33 Equipment Grounding Conductors

Equipment grounding conductors shall be connected in accordance with Parts VI and VII of Article 250.

### 400.34 Minimum Bending Radii

The minimum bending radii for portable cables during installation and handling in service shall be adequate to prevent damage to the cable.

### 400.35 Fittings

Connectors used to connect lengths of cable in a run shall be of a type that locks firmly together. Provisions shall be made to prevent opening or closing these connectors while energized. Suitable means shall be used to eliminate tension at connectors and terminations.

### 400.36 Splices and Terminations

Portable cables shall not contain splices unless the splices are of the permanent molded, vulcanized types in accordance with 110.14(B). Terminations on portable cables rated over 600 volts, nominal, shall be accessible only to authorized and qualified personnel.

## ARTICLE 402 Fixture Wires

### 402.1 Scope

This article covers general requirements and construction specifications for fixture wires.

### 402.2 Other Articles

Fixture wires shall comply with this article and also with the applicable provisions of other articles of this *Code*.

Informational Note: For application in luminaires, see Article 410.

### 402.3 Types

Fixture wires shall be of a type listed in Table 402.3, and they shall comply with all requirements of that table. The fixture wires listed in Table 402.3 are all suitable for service at 600 volts, nominal, unless otherwise specified.

Informational Note: Thermoplastic insulation may stiffen at temperatures colder than −10°C (+14°F), requiring that care be exercised during installation at such temperatures. Thermoplastic insulation may also be deformed at normal temperatures where subjected to pressure, requiring that care be exercised during installation and at points of support.

### 402.5 Allowable Ampacities for Fixture Wires

The allowable ampacity of fixture wire shall be as specified in Table 402.5.

No conductor shall be used under such conditions that its operating temperature exceeds the temperature specified in Table 402.3 for the type of insulation involved.

Informational Note: See 310.15(A)(3) for temperature limitation of conductors.

### 402.6 Minimum Size

Fixture wires shall not be smaller than 18 AWG.

### 402.7 Number of Conductors in Conduit or Tubing

The number of fixture wires permitted in a single conduit or tubing shall not exceed the percentage fill specified in Table 1, Chapter 9.

In Chapter 9, Table 4 provides the usable area within the selected conduit or tubing, and Table 5 provides the required area for each of the conductors.

The following examples show how to determine the minimum size conduit where the conductors are different sizes and how to select the minimum size conduit directly from the tables in Informative Annex C where the conductors are all the same size.

**Calculation Example 1**

A remote ballast installation requires a single FMC to contain fourteen 16 AWG TFFN fixture wires and three 12 AWG THHN conductors. What size FMC is required?

*Solution*

Step 1. Using Table 1, look up the maximum percent of cross section of conduit permitted for conductors. Table 1 sets the limit for over two conductors at 40 percent of the total cross-sectional area of the raceway. Note 6 to the table refers to Table 5 for conductor dimensions and Table 4 for the raceway dimensions.

Step 2. Find the individual conductor cross-sectional areas in Chapter 9, Table 5:

16 AWG TFFN = 0.0072 in.$^2$
12 AWG THHN = 0.0133 in.$^2$

Step 3. Calculate the total area occupied by the wires as follows:

Fourteen 16 AWG TFFN × 0.0072 = 0.1008 in.$^2$
Three 12 AWG THHN × 0.0133 = 0.0399 in.$^2$
Total area = 0.1407 in.$^2$

*TABLE 402.3* *Fixture Wires*

| Name | Type Letter | Insulation | AWG | Thickness of Insulation | | Outer Covering | Maximum Operating Temperature | Application Provisions |
|---|---|---|---|---|---|---|---|---|
| | | | | mm | mils | | | |
| Heat-resistant rubber-covered fixture wire — flexible stranding | FFH-2 | Heat-resistant rubber | 18–16 | 0.76 | 30 | Nonmetallic covering | 75°C 167°F | Fixture wiring |
| | | Cross-linked synthetic polymer | 18–16 | 0.76 | 30 | | | |
| ECTFE — solid or 7-strand | HF | Ethylene chlorotri- fluoroethylene | 18–14 | 0.38 | 15 | None | 150°C 302°F | Fixture wiring |
| ECTFE — flexible stranding | HFF | Ethylene chlorotriflu- oroethylene | 18–14 | 0.38 | 15 | None | 150°C 302°F | Fixture wiring |
| Tape insulated fixture wire — solid or 7-strand | KF-1 | Aromatic polyimide tape | 18–10 | 0.14 | 5.5 | None | 200°C 392°F | Fixture wiring — limited to 300 volts |
| | KF-2 | Aromatic polyimide tape | 18–10 | 0.21 | 8.4 | None | 200°C 392°F | Fixture wiring |
| Tape insulated fixture wire — flexible stranding | KFF-1 | Aromatic polyimide tape | 18–10 | 0.14 | 5.5 | None | 200°C 392°F | Fixture wiring — limited to 300 volts |
| | KFF-2 | Aromatic polyimide tape | 18–10 | 0.21 | 8.4 | None | 200°C 392°F | Fixture wiring |
| Perfluoro- alkoxy — solid or 7-strand (nickel or nickel-coated copper) | PAF | Perfluoroalkoxy | 18–14 | 0.51 | 20 | None | 250°C 482°F | Fixture wiring (nickel or nickel-coated copper) |
| Perfluoro- alkoxy — flexible stranding | PAFF | Perfluoroalkoxy | 18–14 | 0.51 | 20 | None | 150°C 302°F | Fixture wiring |
| Fluorinated ethylene propylene fixture wire — solid or 7-strand | PF | Fluorinated ethylene propylene | 18–14 | 0.51 | 20 | None | 200°C 392°F | Fixture wiring |
| Fluorinated ethylene propylene fixture wire — flexible stranding | PFF | Fluorinated ethylene propylene | 18–14 | 0.51 | 20 | None | 150°C 302°F | Fixture wiring |
| Fluorinated ethylene propylene fixture wire — solid or 7-strand | PGF | Fluorinated ethylene propylene | 18–14 | 0.36 | 14 | Glass braid | 200°C 392°F | Fixture wiring |

*(continues)*

*TABLE 402.3 Continued*

| Name | Type Letter | Insulation | AWG | Thickness of Insulation | | Outer Covering | Maximum Operating Temperature | Application Provisions |
|---|---|---|---|---|---|---|---|---|
| | | | | mm | mils | | | |
| Fluorinated ethylene propylene fixture wire — flexible stranding | PGFF | Fluorinated ethylene propylene | 18–14 | 0.36 | 14 | Glass braid | 150°C 302°F | Fixture wiring |
| Extruded polytetraflu-oroethylene — solid or 7-strand (nickel or nickel-coated copper) | PTF | Extruded polytetraflu-oroethylene | 18–14 | 0.51 | 20 | None | 250°C 482°F | Fixture wiring (nickel or nickel-coated copper) |
| Extruded polytetraflu-oroethylene — flexible stranding 26-36 (AWG silver or nickel-coated copper) | PTFF | Extruded polytetraflu-oroethylene | 18–14 | 0.51 | 20 | None | 150°C 302°F | Fixture wiring (silver or nickel-coated copper) |
| Heat-resistant rubber-covered fixture wire — solid or 7-strand | RFH-1 | Heat-resistant rubber | 18 | 0.38 | 15 | Nonmetallic covering | 75°C 167°F | Fixture wiring — limited to 300 volts |
| | RFH-2 | Heat-resistant rubber Cross-linked synthetic polymer | 18–16 | 0.76 | 30 | None or nonmetallic covering | 75°C 167°F | Fixture wiring |
| Heat-resistant cross-linked synthetic polymer-insulated fixture wire — solid or 7-strand | RFHH-2* | Cross-linked synthetic polymer | 18–16 | 0.76 | 30 | None or nonmetallic covering | 90°C 194°F | Fixture wiring — |
| | RFHH-3* | | 18–16 | 1.14 | 45 | | | |
| Silicone insulated fixture wire — solid or 7-strand | SF-1 | Silicone rubber | 18 | 0.38 | 15 | Nonmetallic covering | 200°C 392°F | Fixture wiring — limited to 300 volts |
| | SF-2 | Silicone rubber | 18–12<br>10 | 0.76<br>1.14 | 30<br>45 | Nonmetallic covering | 200°C 392°F | Fixture wiring |
| Silicone insulated fixture wire — flexible stranding | SFF-1 | Silicone rubber | 18 | 0.38 | 15 | Nonmetallic covering | 150°C 302°F | Fixture wiring — limited to 300 volts |
| | SFF-2 | Silicone rubber | 18–12<br>10 | 0.76<br>1.14 | 30<br>45 | Nonmetallic covering | 150°C 302°F | Fixture wiring |
| Thermoplastic covered fixture wire — solid or 7-strand | TF* | Thermoplastic | 18–16 | 0.76 | 30 | None | 60°C 140°F | Fixture wiring |
| Thermoplastic covered fixture wire — flexible stranding | TFF* | Thermoplastic | 18–16 | 0.76 | 30 | None | 60°C 140°F | Fixture wiring |

*TABLE 402.3 Continued*

| Name | Type Letter | Insulation | AWG | Thickness of Insulation | | Outer Covering | Maximum Operating Temperature | Application Provisions |
|---|---|---|---|---|---|---|---|---|
| | | | | mm | mils | | | |
| Heat-resistant thermoplastic covered fixture wire — solid or 7-strand | TFN* | Thermoplastic | 18–16 | 0.38 | 15 | Nylon-jacketed or equivalent | 90°C 194°F | Fixture wiring |
| Heat-resistant thermoplastic covered fixture wire — flexible stranded | TFFN* | Thermoplastic | 18–16 | 0.38 | 15 | Nylon-jacketed or equivalent | 90°C 194°F | Fixture wiring |
| Cross-linked polyolefin insulated fixture wire — solid or 7-strand | XF* | Cross-linked polyolefin | 18–14<br>12-10 | 0.76<br>1.14 | 30<br>45 | None | 150°C 302°F | Fixture wiring — limited to 300 volts |
| Cross-linked polyolefin insulated fixture wire — flexible stranded | XFF* | Cross-linked polyolefin | 18–14<br>12–10 | 0.76<br>1.14 | 30<br>45 | None | 150°C 302°F | Fixture wiring — limited to 300 volts |
| Modified ETFE — solid or 7-strand | ZF | Modified ethylene tetrafluoro-ethylene | 18–14 | 0.38 | 15 | None | 150°C 302°F | Fixture wiring |
| Flexible stranding | ZFF | Modified ethylene tetrafluoro-ethylene | 18–14 | 0.38 | 15 | None | 150°C 302°F | Fixture wiring |
| High temp. modified ETFE — solid or 7-strand | ZHF | Modified ethylene tetrafluoro-ethylene | 18–14 | 0.38 | 15 | None | 200°C 392°F | Fixture wiring |

*Insulations and outer coverings that meet the requirements of flame retardant, limited smoke, and are so listed, shall be permitted to be marked for limited smoke after the Code type designation.

*TABLE 402.5 Allowable Ampacity for Fixture Wires*

| Size (AWG) | Allowable Ampacity |
|---|---|
| 18 | 6 |
| 16 | 8 |
| 14 | 17 |
| 12 | 23 |
| 10 | 28 |

Step 4. Using the 40-percent column of Table 4, in the section entitled "Article 348, Flexible Metal Conduit (FMC)," find the appropriate FMC size based on 40-percent fill and a total conductor area fill of 0.1407 in.$^2$. Because 0.1407 in.$^2$ is greater than 0.127 and less than 0.213, select trade size ¾.

### Calculation Example 2

If the conductors in the FMC are all of the same wire size (16 AWG), Informative Annex C tables may be used instead of doing the calculations. This example uses Informative Annex C tables to determine FMC size.

What size FMC is required for seventeen 16 AWG TFFN conductors?

*Solution*

Step 1. In Informative Annex C, Table C.3, find TFFN insulation in the first column.

Step 2. Find 16 AWG in the second column. Proceed across the table until the desired number of conductors is equal to or less than the number shown in the table for the respective conduit sizes.

*Conclusion:* Trade size ½ is required.

## 402.8 Grounded Conductor Identification

Fixture wires that are intended to be used as grounded conductors shall be identified by one or more continuous white stripes on other than green insulation or by the means described in 400.22(A) through (E).

This requirement for fixture wires is similar to that required for flexible cords and cable to ensure that the grounded conductor in fixture wires is easily recognized. Because connection of the grounded conductor to the screw shell of lampholders is necessary, it must be easily recognized.

## 402.9 Marking

**(A) Method of Marking.** Thermoplastic insulated fixture wire shall be durably marked on the surface at intervals not exceeding 610 mm (24 in.). All other fixture wire shall be marked by means of a printed tag attached to the coil, reel, or carton.

**(B) Optional Marking.** Fixture wire types listed in Table 402.3 shall be permitted to be surface marked to indicate special characteristics of the cable materials. These markings include, but are not limited to, markings for limited smoke, sunlight resistance, and so forth.

## 402.10 Uses Permitted

Fixture wires shall be permitted (1) for installation in luminaires and in similar equipment where enclosed or protected and not subject to bending or twisting in use, or (2) for connecting luminaires to the branch-circuit conductors supplying the luminaires.

Fixture wire is permitted to be used as a tap conductor to connect a luminaire(s) to the branch-circuit conductors. The transition from the branch-circuit wiring method to the fixture wire tap conductors can be accomplished via a junction box or other fitting that is allowed to contain splices.

## 402.11 Uses Not Permitted

Fixture wires shall not be used as branch-circuit conductors except as permitted elsewhere in this *Code*.

The phrase "except as permitted elsewhere in this *Code*" correlates with limited application of fixture wire as branch-circuit conductors found in other sections of the *Code*. One example of such a use is in 725.49(B), where the insulation types required for Class 1 circuits wired with 18 AWG and 16 AWG conductors are the types specified in Table 402.3 for fixture wire.

## 402.12 Overcurrent Protection

Overcurrent protection for fixture wires shall be as specified in 240.5.

# ARTICLE 404 Switches

## I. Installation

### 404.1 Scope

The provisions of this article apply to all switches, switching devices, and circuit breakers used as switches operating at 1000 volts and below, unless specifically referenced elsewhere in this *Code* for higher voltages.

### 404.2 Switch Connections

**(A) Three-Way and Four-Way Switches.** Three-way and four-way switches shall be wired so that all switching is done only in the ungrounded circuit conductor. Where in metal raceways or metal-armored cables, wiring between switches and outlets shall be in accordance with 300.20(A).

*Exception: Switch loops shall not require a grounded conductor.*

The use of two 2-conductor nonmetallic-sheathed cables instead of a single 3-conductor cable for wiring 3-way and 4-way switches is not specifically prohibited by the *Code*. However, using two 2-conductor cables could result in a violation of 300.20 if metal boxes are used and the cables enter the box through separate knockouts. Also, use of the same clamp for both cables would, in most cases, not comply with 110.3(B), because clamps have been tested for only one cable per clamp.

The exception does not require a grounded conductor in a switch loop [see 300.20(A)] because the ungrounded conductor both enters and leaves the enclosure in the same cable or raceway, thus avoiding inductive heating.

**(B) Grounded Conductors.** Switches or circuit breakers shall not disconnect the grounded conductor of a circuit.

*Exception: A switch or circuit breaker shall be permitted to disconnect a grounded circuit conductor where all circuit conductors are disconnected simultaneously, or where the device is arranged so that the grounded conductor cannot be disconnected until all the ungrounded conductors of the circuit have been disconnected.*

**(C) Switches Controlling Lighting Loads.** The grounded circuit conductor for the controlled lighting circuit shall be provided at the location where switches control lighting loads that are supplied by a grounded general-purpose branch circuit for other than the following:

(1) Where conductors enter the box enclosing the switch through a raceway, provided that the raceway is large enough for all contained conductors, including a grounded conductor

(2) Where the box enclosing the switch is accessible for the installation of an additional or replacement cable without removing finish materials
(3) Where snap switches with integral enclosures comply with 300.15(E)
(4) Where a switch does not serve a habitable room or bathroom
(5) Where multiple switch locations control the same lighting load such that the entire floor area of the room or space is visible from the single or combined switch locations
(6) Where lighting in the area is controlled by automatic means
(7) Where a switch controls a receptacle load

Informational Note: The provision for a (future) grounded conductor is to complete a circuit path for electronic lighting control devices.

Many electronic lighting control devices require a standby current to maintain the ready state and detection capability of the device. This allows immediate switching of the load to the "on" condition. These devices require standby current when they are in the "off" state, that is, when there is no load current. In existing installations, many of these devices utilize the EGC for the standby current flow because a grounded conductor was not usually provided in the switch box.

Section 404.2(C) does not require a grounded circuit conductor in every installation. For example, a grounded conductor is not required at initial installation if a conductor can be readily added in the future, such as in raceway installations or where the construction of the framing cavity in which the switch box is located permits access. The grounded conductor is also not required where the area served is not a habitable room or bathroom, or the load is a switched receptacle.

## 404.3 Enclosure

**(A) General.** Switches and circuit breakers shall be of the externally operable type mounted in an enclosure listed for the intended use. The minimum wire-bending space at terminals and minimum gutter space provided in switch enclosures shall be as required in 312.6.

*Exception No. 1: Pendant- and surface-type snap switches and knife switches mounted on an open-face switchboard or panelboard shall be permitted without enclosures.*

*Exception No. 2: Switches and circuit breakers installed in accordance with 110.27(1), (A)(2), (A)(3), or (A)(4) shall be permitted without enclosures.*

**(B) Used as a Raceway.** Enclosures shall not be used as junction boxes, auxiliary gutters, or raceways for conductors feeding through or tapping off to other switches or overcurrent devices, unless the enclosure complies with 312.8.

## 404.4 Damp or Wet Locations

**(A) Surface-Mounted Switch or Circuit Breaker.** A surface-mounted switch or circuit breaker shall be enclosed in a weatherproof enclosure or cabinet that shall comply with 312.2.

**(B) Flush-Mounted Switch or Circuit Breaker.** A flush-mounted switch or circuit breaker shall be equipped with a weatherproof cover.

**(C) Switches in Tub or Shower Spaces.** Switches shall not be installed within tubs or shower spaces unless installed as part of a listed tub or shower assembly.

## 404.5 Time Switches, Flashers, and Similar Devices

Time switches, flashers, and similar devices shall be of the enclosed type or shall be mounted in cabinets or boxes or equipment enclosures. Energized parts shall be barriered to prevent operator exposure when making manual adjustments or switching.

*Exception: Devices mounted so they are accessible only to qualified persons shall be permitted without barriers, provided they are located within an enclosure such that any energized parts within 152 mm (6.0 in.) of the manual adjustment or switch are covered by suitable barriers.*

## 404.6 Position and Connection of Switches

**(A) Single-Throw Knife Switches.** Single-throw knife switches shall be placed so that gravity will not tend to close them. Single-throw knife switches, approved for use in the inverted position, shall be provided with an integral mechanical means that ensures that the blades remain in the open position when so set.

**(B) Double-Throw Knife Switches.** Double-throw knife switches shall be permitted to be mounted so that the throw is either vertical or horizontal. Where the throw is vertical, integral mechanical means shall be provided to hold the blades in the open position when so set.

An integral "mechanical means" that does not necessarily have to be a "locking device" is required for single-throw and double-throw switches to ensure the switch blades remain disengaged regardless of their orientation when the switch is in the "off" (open) position. New switch designs incorporate mechanical means other than a catch or a latch to ensure that the blades cannot accidentally close from the "off" position.

**(C) Connection of Switches.** Single-throw knife switches and switches with butt contacts shall be connected such that their blades are de-energized when the switch is in the open position. Bolted pressure contact switches shall have barriers that prevent inadvertent contact with energized blades. Single-throw knife switches, bolted pressure contact switches, molded case switches, switches with butt contacts, and circuit breakers used as switches shall be connected so that the terminals supplying the load are de-energized when the switch is in the open position.

Bolted pressure switches that have energized blades when open, such as bottom-feed designs, must be provided with barriers or other means to guard against inadvertent contact with the energized blades. This requirement is intended to provide protection against accidental contact with live parts when personnel are working on energized equipment.

*Exception: The blades and terminals supplying the load of a switch shall be permitted to be energized when the switch is in the open position where the switch is connected to circuits or equipment inherently capable of providing a backfeed source of power. For such installations, a permanent sign shall be installed on the switch enclosure or immediately adjacent to open switches with the following words or equivalent:*

*WARNING — LOAD SIDE TERMINALS MAY BE ENERGIZED BY BACKFEED. The warning sign or label shall comply with 110.21(B).*

Batteries, generators, PV systems, and double-ended switchboard ties are typical backfeed sources. These sources can cause the load side of the switch or circuit breaker to be energized when it is in the open position, which is a condition inherent to the circuitry.

## 404.7 Indicating

General-use and motor-circuit switches, circuit breakers, and molded case switches, where mounted in an enclosure as described in 404.3, shall clearly indicate whether they are in the open (off) or closed (on) position.

Where these switch or circuit breaker handles are operated vertically rather than rotationally or horizontally, the up position of the handle shall be the closed (on) position.

*Exception No. 1: Vertically operated double-throw switches shall be permitted to be in the closed (on) position with the handle in either the up or down position.*

*Exception No. 2: On busway installations, tap switches employing a center-pivoting handle shall be permitted to be open or closed with either end of the handle in the up or down position. The switch position shall be clearly indicating and shall be visible from the floor or from the usual point of operation.*

Exception No. 2 clarifies the operation of busway switches that are designed with a center pivot handle such that, at any time, one end of the handle is in the up position and the other is down, which means the handle is pulled down to turn the switch off and also pulled down to turn it on. While this configuration is not in accordance with the requirement in the main rule, the exception permits this time-proven method of operating busway switches.

## 404.8 Accessibility and Grouping

**(A) Location.** All switches and circuit breakers used as switches shall be located so that they may be operated from a readily accessible place. They shall be installed such that the center of the grip of the operating handle of the switch or circuit breaker, when in its highest position, is not more than 2.0 m (6 ft 7 in.) above the floor or working platform.

*Exception No. 1: On busway installations, fused switches and circuit breakers shall be permitted to be located at the same level as the busway. Suitable means shall be provided to operate the handle of the device from the floor.*

*Exception No. 2: Switches and circuit breakers installed adjacent to motors, appliances, or other equipment that they supply shall be permitted to be located higher than 2.0 m (6 ft 7 in.) and to be accessible by portable means.*

*Exception No. 3: Hookstick operable isolating switches shall be permitted at greater heights.*

**(B) Voltage Between Adjacent Devices.** A snap switch shall not be grouped or ganged in enclosures with other snap switches, receptacles, or similar devices, unless they are arranged so that the voltage between adjacent devices does not exceed 300 volts, or unless they are installed in enclosures equipped with identified, securely installed barriers between adjacent devices.

Barriers are required between switches that are ganged in a box and used to control 277-volt lighting on 480Y/277-V systems where two or more phase conductors enter the box. The barriers would be required between devices fed from two different phases of this system, because the voltage between the phase conductors would be 480 volts, nominal, and would exceed the 300-volt limit. Barriers are required even if one device space is left empty, because the two remaining devices fed from different phase conductors still would be adjacent to each other.

This requirement also applies to switches ganged together with any wiring device where the voltage between adjacent conductors exceeds 300 volts. Where switches or devices are on different systems and are in the same box, such as a switch on a 277-volt system and a receptacle on a 120-volt system, a careful analysis of the voltage between adjacent device terminals is required to ensure that the 300-volt limitation is not exceeded.

**(C) Multipole Snap Switches.** A multipole, general-use snap switch shall not be permitted to be fed from more than a single circuit unless it is listed and marked as a two-circuit or three-circuit switch.

Multiple-pole snap switches used to switch multiple circuits must be listed for the use. Where a switch is supplied by more than one circuit, 210.7 requires a means to simultaneously disconnect the ungrounded conductors supplying the switch.

Informational Note: See 210.7 for disconnect requirements where more than one circuit supplies a switch.

## 404.9 Provisions for General-Use Snap Switches

**(A) Faceplates.** Faceplates provided for snap switches mounted in boxes and other enclosures shall be installed so as to completely cover the opening and, where the switch is flush mounted, seat against the finished surface.

**(B) Grounding.** Snap switches, including dimmer and similar control switches, shall be connected to an equipment grounding conductor and shall provide a means to connect metal faceplates to the equipment grounding conductor, whether or not a metal faceplate is installed. Snap switches shall be considered to be part of an effective ground-fault current path if either of the following conditions is met:

(1) The switch is mounted with metal screws to a metal box or metal cover that is connected to an equipment grounding conductor or to a nonmetallic box with integral means for connecting to an equipment grounding conductor.
(2) An equipment grounding conductor or equipment bonding jumper is connected to an equipment grounding termination of the snap switch.

*Exception No. 1 to (B): Where no means exists within the snap-switch enclosure for connecting to the equipment grounding conductor, or where the wiring method does not include or provide an equipment grounding conductor, a snap switch without a connection to an equipment grounding conductor shall be permitted for replacement purposes only. A snap switch wired under the provisions of this exception and located within 2.5 m (8 ft) vertically, or 1.5 m (5 ft) horizontally, of ground or exposed grounded metal objects shall be provided with a faceplate of nonconducting noncombustible material with nonmetallic attachment screws, unless the switch mounting strap or yoke is nonmetallic or the circuit is protected by a ground-fault circuit interrupter.*

*Exception No. 2 to (B): Listed kits or listed assemblies shall not be required to be connected to an equipment grounding conductor if all of the following conditions are met:*

*(1) The device is provided with a nonmetallic faceplate that cannot be installed on any other type of device,*
*(2) The device does not have mounting means to accept other configurations of faceplates,*
*(3) The device is equipped with a nonmetallic yoke, and*
*(4) All parts of the device that are accessible after installation of the faceplate are manufactured of nonmetallic materials.*

*Exception No. 3 to (B): A snap switch with integral nonmetallic enclosure complying with 300.15(E) shall be permitted without a connection to an equipment grounding conductor.*

Although the non–current-carrying metal parts of switches typically are not subject to contact by personnel, metal faceplates would pose a shock hazard if they became energized. Sections 404.9(B)(1) and (B)(2) describe conditions under which the switch provides an effective ground fault current path to a metal cover plate. See Exhibit 404.1 for an example of the typical method by which a metal faceplate is grounded.

Exception No. 1 covers switch replacement where an EGC is not available. This requires either a switch plate made of insulating material or GFCI protection for the circuit. This exception provides additional safety to persons where older electrical installations did not provide an EGC in the wiring method.

Exception No. 2 permits listed kits or assemblies of snap switches – or other devices with nonmetallic yokes and faceplates – to be installed without an EGC connection. These assemblies are evaluated for dielectric breakdown to ensure there will not be a shock hazard. A faceplate that is not interchangeable with other types of faceplates ensures that a conductive faceplate will not be used on switches that do not provide an equipment grounding means through the yoke. Making sure that all parts accessible after installation are manufactured of nonmetallic materials provides further protection from shock hazards.

Exception No. 3 recognizes that listed switches of the box-less type incorporate integral nonconductive (nonmetallic) faceplates that do not permit the interchange of faceplates, which precludes the substitution of a metal faceplace. No equipment in the assembly requires grounding; therefore, an EGC is not necessary.

**(C) Construction.** Metal faceplates shall be of ferrous metal not less than 0.76 mm (0.030 in.) in thickness or of nonferrous metal not less than 1.02 mm (0.040 in.) in thickness. Faceplates of insulating material shall be noncombustible and not less than 2.54 mm (0.100 in.) in thickness, but they shall be permitted to be less than 2.54 mm (0.100 in.) in thickness if formed or reinforced to provide adequate mechanical strength.

## 404.10 Mounting of Snap Switches

**(A) Surface Type.** Snap switches used with open wiring on insulators shall be mounted on insulating material that separates the conductors at least 13 mm (½ in.) from the surface wired over.

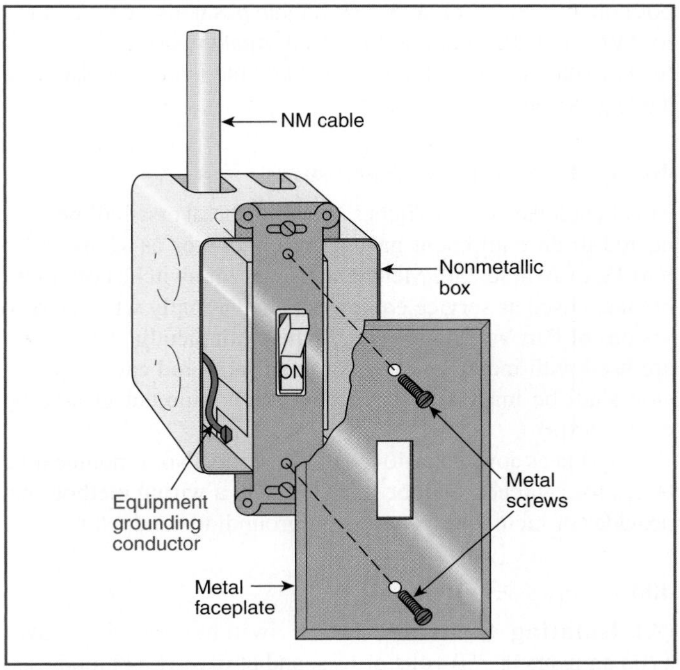


***EXHIBIT 404.1*** *A metal faceplate grounded through attachment to the grounded yoke of a snap switch.*

**(B) Box Mounted.** Flush-type snap switches mounted in boxes that are set back of the finished surface as permitted in 314.20 shall be installed so that the extension plaster ears are seated against the surface. Flush-type snap switches mounted in boxes that are flush with the finished surface or project from it shall be installed so that the mounting yoke or strap of the switch is seated against the box. Screws used for the purpose of attaching a snap switch to a box shall be of the type provided with a listed snap switch, or shall be machine screws having 32 threads per inch or part of listed assemblies or systems, in accordance with the manufacturer's instructions.

Regardless of where a box for a flush-type switch is installed (in a wall or in a ceiling), the switch yoke or strap must be seated against the box, or where the box is set back, the yoke or strap must be against the finished surface of the wall or ceiling. Requirements for mounting screws are provided to ensure proper mounting of the device and to preclude the use of drywall screws.

## 404.11 Circuit Breakers as Switches

A hand-operable circuit breaker equipped with a lever or handle, or a power-operated circuit breaker capable of being opened by hand in the event of a power failure, shall be permitted to serve as a switch if it has the required number of poles.

Informational Note: See the provisions contained in 240.81 and 240.83.

Circuit breakers that are capable of being hand operated must clearly indicate whether they are in the open (off) or closed (on) position. See 404.7 for details on handle positions. See 240.83(D) for SWD and HID marking for circuit breakers used as switches for 120-volt and 277-volt fluorescent and high-intensity discharge lighting circuits.

## 404.12 Grounding of Enclosures

Metal enclosures for switches or circuit breakers shall be connected to an equipment grounding conductor as specified in Part IV of Article 250. Metal enclosures for switches or circuit breakers used as service equipment shall comply with the provisions of Part V of Article 250. Where nonmetallic enclosures are used with metal raceways or metal-armored cables, provision shall be made for connecting the equipment grounding conductor(s).

Except as covered in 404.9(B), Exception No. 1, nonmetallic boxes for switches shall be installed with a wiring method that provides or includes an equipment grounding conductor.

## 404.13 Knife Switches

**(A) Isolating Switches.** Knife switches rated at over 1200 amperes at 250 volts or less, and at over 1000 amperes at 251 to 1000 volts, shall be used only as isolating switches and shall not be opened under load.

**(B) To Interrupt Currents.** To interrupt currents over 1200 amperes at 250 volts, nominal, or less, or over 600 amperes at 251 to 600 volts, nominal, a circuit breaker or a switch of special design listed for such purpose shall be used.

**(C) General-Use Switches.** Knife switches of ratings less than specified in 404.13(A) and (B) shall be considered general-use switches.

Informational Note: See the definition of *General-Use Switch* in Article 100.

**(D) Motor-Circuit Switches.** Motor-circuit switches shall be permitted to be of the knife-switch type.

Informational Note: See the definition of a *Motor-Circuit Switch* in Article 100.

## 404.14 Rating and Use of Snap Switches

Snap switches shall be used within their ratings and as indicated in 404.14(A) through (F).

Informational Note No. 1: For switches on signs and outline lighting, see 600.6.

Informational Note No. 2: For switches controlling motors, see 430.83, 430.109, and 430.110.

**(A) Alternating-Current General-Use Snap Switch.** A form of general-use snap switch suitable only for use on ac circuits for controlling the following:

(1) Resistive and inductive loads not exceeding the ampere rating of the switch at the voltage applied
(2) Tungsten-filament lamp loads not exceeding the ampere rating of the switch at 120 volts
(3) Motor loads not exceeding 80 percent of the ampere rating of the switch at its rated voltage

**(B) Alternating-Current or Direct-Current General-Use Snap Switch.** A form of general-use snap switch suitable for use on either ac or dc circuits for controlling the following:

(1) Resistive loads not exceeding the ampere rating of the switch at the voltage applied.
(2) Inductive loads not exceeding 50 percent of the ampere rating of the switch at the applied voltage. Switches rated in horsepower are suitable for controlling motor loads within their rating at the voltage applied.
(3) Tungsten-filament lamp loads not exceeding the ampere rating of the switch at the applied voltage if T-rated.

**(C) CO/ALR Snap Switches.** Snap switches rated 20 amperes or less directly connected to aluminum conductors shall be listed and marked CO/ALR.

**(D) Alternating-Current Specific-Use Snap Switches Rated for 347 Volts.** Snap switches rated 347 volts ac shall be listed and shall be used only for controlling the loads permitted by (D)(1) and (D)(2).

**(1) Noninductive Loads.** Noninductive loads other than tungsten-filament lamps not exceeding the ampere and voltage ratings of the switch.

**(2) Inductive Loads.** Inductive loads not exceeding the ampere and voltage ratings of the switch. Where particular load characteristics or limitations are specified as a condition of the listing, those restrictions shall be observed regardless of the ampere rating of the load.

The ampere rating of the switch shall not be less than 15 amperes at a voltage rating of 347 volts ac. Flush-type snap switches rated 347 volts ac shall not be readily interchangeable in box mounting with switches identified in 404.14(A) and (B).

Although not commonly used in the United States, 600Y/347-V systems are permitted by the *Code*. Sections 210.6 and 225.7(D) permit these systems to be used to supply outdoor lighting installations. An ac specific-use snap switch that is 347-volt rated is permitted for the purposes of controlling lighting circuits on these systems. Such switches, unless specifically restricted, are permitted to be used on circuits of a lower voltage, such as 277-volt and 120-volt circuits.

**(E) Dimmer Switches.** General-use dimmer switches shall be used only to control permanently installed incandescent luminaires unless listed for the control of other loads and installed accordingly.

General-use dimmers are not permitted to control receptacles or cord-and-plug-connected table and floor lamps. If a dimmer evaluated only for the control of incandescent luminaires is used, the potential for connecting incompatible equipment such as a cord-and-plug-connected motor-operated appliance or a portable fluorescent lamp is increased by using the dimmer to control a receptacle. Section 404.14(E) does not apply to commercial dimmers or theater dimmers that can be used for fluorescent lighting and portable lighting.

**(F) Cord- and Plug-Connected Loads.** Where a snap switch is used to control cord- and plug-connected equipment on a general-purpose branch circuit, each snap switch controlling receptacle outlets or cord connectors that are supplied by permanently connected cord pendants shall be rated at not less than the rating of the maximum permitted ampere rating or setting of the overcurrent device protecting the receptacles or cord connectors, as provided in 210.21(B).

Informational Note: See 210.50(A) and 400.7(A)(1) for equivalency to a receptacle outlet of a cord connector that is supplied by a permanently connected cord pendant.

*Exception: Where a snap switch is used to control not more than one receptacle on a branch circuit, the switch shall be permitted to be rated at not less than the rating of the receptacle.*

Where the load is cord-and-plug-connected through a receptacle or cord connector on a cord pendant (i.e., interchangeable), the load varies based on what is connected. A switch supplying cord-and-plug-connected loads is required to be rated at the rating or setting of the OCPD supplying the branch circuit. The exception recognizes circuits that may have multiple outlets where a switch controls one receptacle on the circuit and that such a switch be rated not less than the receptacle's rating.

## II. Construction Specifications

### 404.15 Marking

**(A) Ratings.** Switches shall be marked with the current, voltage, and, if horsepower rated, the maximum rating for which they are designed.

**(B) Off Indication.** Where in the off position, a switching device with a marked OFF position shall completely disconnect all ungrounded conductors to the load it controls.

### 404.16 Knife Switches Rated 600 to 1000 Volts

Auxiliary contacts of a renewable or quick-break type or the equivalent shall be provided on all knife switches rated 600 to 1000 volts and designed for use in breaking current over 200 amperes.

### 404.17 Fused Switches

A fused switch shall not have fuses in parallel except as permitted in 240.8.

### 404.18 Wire-Bending Space

The wire-bending space required by 404.3 shall meet Table 312.6(B) spacings to the enclosure wall opposite the line and load terminals.

# ARTICLE 406
# Receptacles, Cord Connectors, and Attachment Plugs (Caps)

### 406.1 Scope

This article covers the rating, type, and installation of receptacles, cord connectors, and attachment plugs (cord caps).

### 406.2 Definition

**Child Care Facility.** A building or structure, or portion thereof, for educational, supervisory, or personal care services for more than four children 7 years old or less.

The definition of *child care facility* takes into consideration various aspects of the current definitions for such facilities from building codes as well as from the GSA Guide.

## 406.3 Receptacle Rating and Type

**(A) Receptacles.** Receptacles shall be listed and marked with the manufacturer's name or identification and voltage and ampere ratings.

**(B) Rating.** Receptacles and cord connectors shall be rated not less than 15 amperes, 125 volts, or 15 amperes, 250 volts, and shall be of a type not suitable for use as lampholders.

Informational Note: See 210.21(B) for receptacle ratings where installed on branch circuits.

**(C) Receptacles for Aluminum Conductors.** Receptacles rated 20 amperes or less and designed for the direct connection of aluminum conductors shall be marked CO/ALR.

If the receptacle is not of the CO/ALR type, it can be connected with a copper pigtail to an aluminum branch-circuit conductor only if the wire connector is suitable for such a connection and is marked with the letters AL and CU. The commentary following 110.14(B) further explains the suitability of wire connectors used to join copper and aluminum conductors.

**(D) Isolated Ground Receptacles.** Receptacles incorporating an isolated grounding conductor connection intended for the reduction of electrical noise (electromagnetic interference) as permitted in 250.146(D) shall be identified by an orange triangle located on the face of the receptacle.

**(1) Isolated Equipment Grounding Conductor Required.** Receptacles so identified shall be used only with equipment grounding conductors that are isolated in accordance with 250.146(D).

**(2) Installation in Nonmetallic Boxes.** Isolated ground receptacles installed in nonmetallic boxes shall be covered with a nonmetallic faceplate.

*Exception: Where an isolated ground receptacle is installed in a nonmetallic box, a metal faceplate shall be permitted if the box contains a feature or accessory that permits the effective grounding of the faceplate.*

**(E) Controlled Receptacle Marking.** All nonlocking-type, 125-volt, 15- and 20-ampere receptacles that are controlled by an automatic control device, or that incorporate control features that remove power from the outlet for the purpose of energy management or building automation, shall be marked with the symbol shown in Figure 406.3(E) and located on the controlled receptacle outlet where visible after installation.

*Exception: The marking is not required for receptacles controlled by a wall switch that provide the required room lighting outlets as permitted by 210.70.*

Many energy efficiency codes require that a percentage of installed 125-volt, 15- and 20-ampere receptacles be automatically controlled. These receptacles are required to be marked to indicate to users which loads will be automatically de-energized by the controller. This allows the user to select a different receptacle if, for instance, the load must be supplied during overnight hours.

***FIGURE 406.3(E)*** *Controlled Receptacle Marking Symbol.*

## 406.4 General Installation Requirements

Receptacle outlets shall be located in branch circuits in accordance with Part III of Article 210. General installation requirements shall be in accordance with 406.4(A) through (F).

**(A) Grounding Type.** Except as provided in 406.4(D), receptacles installed on 15- and 20-ampere branch circuits shall be of the grounding type. Grounding-type receptacles shall be installed only on circuits of the voltage class and current for which they are rated, except as provided in Table 210.21(B)(2) and Table 210.21(B)(3).

•

**(B) To Be Grounded.** Receptacles and cord connectors that have equipment grounding conductor contacts shall have those contacts connected to an equipment grounding conductor.

*Exception No. 1: Receptacles mounted on portable and vehicle-mounted generators in accordance with 250.34.*

*Exception No. 2: Replacement receptacles as permitted by 406.4(D).*

**(C) Methods of Grounding.** The equipment grounding conductor contacts of receptacles and cord connectors shall be grounded by connection to the equipment grounding conductor of the circuit supplying the receptacle or cord connector.

Informational Note: For installation requirements for the reduction of electrical noise, see 250.146(D).

The branch-circuit wiring method shall include or provide an equipment grounding conductor to which the equipment grounding conductor contacts of the receptacle or cord connector are connected.

Informational Note No. 1: See 250.118 for acceptable grounding means.

Informational Note No. 2: For extensions of existing branch circuits, see 250.130.

**(D) Replacements.** Replacement of receptacles shall comply with 406.4(D)(1) through (D)(6), as applicable. Arc-fault circuit-interrupter type and ground-fault circuit-interrupter type receptacles shall be installed in a readily accessible location.

**(1) Grounding-Type Receptacles.** Where a grounding means exists in the receptacle enclosure or an equipment grounding conductor is installed in accordance with 250.130(C), grounding-type receptacles shall be used and shall be connected to the equipment grounding conductor in accordance with 406.4(C) or 250.130(C).

**(2) Non–Grounding-Type Receptacles.** Where attachment to an equipment grounding conductor does not exist in the receptacle enclosure, the installation shall comply with (D)(2)(a), (D)(2)(b), or (D)(2)(c).

(a) A non–grounding-type receptacle(s) shall be permitted to be replaced with another non–grounding-type receptacle(s).

(b) A non–grounding-type receptacle(s) shall be permitted to be replaced with a ground-fault circuit interrupter-type of receptacle(s). These receptacles shall be marked "No Equipment Ground." An equipment grounding conductor shall not be connected from the ground-fault circuit-interrupter-type receptacle to any outlet supplied from the ground-fault circuit-interrupter receptacle.

(c) A non–grounding-type receptacle(s) shall be permitted to be replaced with a grounding-type receptacle(s) where supplied through a ground-fault circuit interrupter. Grounding-type receptacles supplied through the ground-fault circuit interrupter shall be marked "GFCI Protected" and "No Equipment Ground." An equipment grounding conductor shall not be connected between the grounding-type receptacles.

**(3) Ground-Fault Circuit Interrupters.** Ground-fault circuit-interrupter protected receptacles shall be provided where replacements are made at receptacle outlets that are required to be so protected elsewhere in this *Code*.

*Exception: Where replacement of the receptacle type is impracticable, such as where the outlet box size will not permit the installation of the GFCI receptacle, the receptacle shall be permitted to be replaced with a new receptacle of the existing type, where GFCI protection is provided and the receptacle is marked "GFCI protected" and "no equipment ground," in accordance with 406.4(D)(2) (a), (b), or (c).*

**(4) Arc-Fault Circuit-Interrupter Protection.** Where a receptacle outlet is supplied by a branch circuit that requires arc-fault circuit-interrupter protection as specified elsewhere in this *Code*, a replacement receptacle at this outlet shall be one of the following:

(1) A listed outlet branch-circuit type arc-fault circuit-interrupter receptacle
(2) A receptacle protected by a listed outlet branch-circuit type arc-fault circuit-interrupter type receptacle
(3) A receptacle protected by a listed combination type arc-fault circuit-interrupter type circuit breaker

This requirement becomes effective January 1, 2014.

Older homes are statistically more vulnerable to electrical fires. Extra protection for older homes is provided by the gradual replacement, over time, of non-AFCI-protected receptacles with new AFCI-protected ones.

**(5) Tamper-Resistant Receptacles.** Listed tamper-resistant receptacles shall be provided where replacements are made at receptacle outlets that are required to be tamper-resistant elsewhere in this *Code*.

This requirement does not mandate receptacle replacement. It merely institutes a requirement for the receptacle if it is replaced. For example, an ordinary 15-ampere receptacle in a bedroom of a 10-year-old one-family dwelling would be required to be replaced with a tamper-resistant receptacle because tamper-resistant receptacles are required in a bedroom of a new home constructed under the current edition of the *Code*.

**(6) Weather-Resistant Receptacles.** Weather-resistant receptacles shall be provided where replacements are made at receptacle outlets that are required to be so protected elsewhere in this *Code*.

Without the requirement for weather-resistant receptacles to be installed at the time of replacement, ordinary receptacles may be installed and subjected to the same failures as the receptacles they replaced.

**(E) Cord- and Plug-Connected Equipment.** The installation of grounding-type receptacles shall not be used as a requirement that all cord-and plug-connected equipment be of the grounded type.

Informational Note: See 250.114 for types of cord-and plug-connected equipment to be grounded.

**(F) Noninterchangeable Types.** Receptacles connected to circuits that have different voltages, frequencies, or types of current (ac or dc) on the same premises shall be of such design that the attachment plugs used on these circuits are not interchangeable.

## 406.5 Receptacle Mounting

Receptacles shall be mounted in identified boxes or assemblies. The boxes or assemblies shall be securely fastened in place unless otherwise permitted elsewhere in this *Code*. Screws used for the purpose of attaching receptacles to a box shall be of the type provided with a listed receptacle, or shall be machine screws having 32 threads per inch or part of listed assemblies or systems, in accordance with the manufacturer's instructions.

Receptacles in pendant boxes are permitted, provided the box is supported from the flexible cord in accordance with 314.23(H)(1). A pendant box that is properly suspended is not required to be securely fastened in place.

**(A) Boxes That Are Set Back.** Receptacles mounted in boxes that are set back from the finished surface as permitted in 314.20 shall be installed such that the mounting yoke or strap of the receptacle is held rigidly at the finished surface.

**(B) Boxes That Are Flush.** Receptacles mounted in boxes that are flush with the finished surface or project therefrom shall be installed such that the mounting yoke or strap of the receptacle is held rigidly against the box or box cover.

An outlet box used to enclose a receptacle must be rigidly and securely supported according to 314.23(B) or (C). In addition, mounting outlet boxes with the proper setback, according to 314.20, requires the cooperation of other construction trades (drywall installers, plasterers, and carpenters) and the building designers.

The intent of 406.5(A) through (C) is to allow attachment plugs to be inserted or removed without moving the receptacle. Additionally, by restricting movement of the receptacle, effective grounding continuity can be maintained for contact devices or receptacle yokes where the box is installed flush with the wall surface or where it projects from it. The proper installation of receptacles helps ensure that attachment plugs can be fully inserted, thus providing a better contact.

**(C) Receptacles Mounted on Covers.** Receptacles mounted to and supported by a cover shall be held rigidly against the cover by more than one screw or shall be a device assembly or box cover listed and identified for securing by a single screw.

Receptacles mounted on raised covers, such as the receptacle illustrated in Exhibit 406.1, are not permitted to be secured by a single screw unless the device is listed and identified for that method.

**(D) Position of Receptacle Faces.** After installation, receptacle faces shall be flush with or project from faceplates of insulating material and shall project a minimum of 0.4 mm (0.015 in.) from metal faceplates.

Requiring receptacles to project from metal faceplates prevents faults between the blades of attachment plugs and metal faceplates.

***EXHIBIT 406.1*** *A receptacle mounted on a raised cover secured with three screws.*

Proper faceplate mounting ensures that attachment plugs can be fully inserted, thus providing a better contact. The *NEC* does not specify the position (blades up or blades down) of a common vertically mounted 15- or 20-ampere duplex receptacle. Although many drawings in this handbook show the slots for blades up, the receptacle may be installed with the slots for blades down. Receptacles can also be installed horizontally as well as vertically.

*Exception: Listed kits or assemblies encompassing receptacles and nonmetallic faceplates that cover the receptacle face, where the plate cannot be installed on any other receptacle, shall be permitted.*

This exception allows the use of listed kits, which include the receptacle and a nonmetallic faceplate, that have been evaluated to ensure that sufficient blade contact is achieved by the attachment plug when inserted in the receptacle. The kit's nonmetallic faceplate cannot fit standard style receptacles.

**(E) Receptacles in Countertops and Similar Work Surfaces.** Receptacles, unless listed as receptacle assemblies for countertop applications, shall not be installed in a face-up position in countertops or similar work surfaces. Where receptacle assemblies for countertop applications are required to provide ground-fault circuit-interrupter protection for personnel in accordance with 210.8, such assemblies shall be permitted to be listed as GFCI receptacle assemblies for countertop applications.

**(F) Receptacles in Seating Areas and Other Similar Surfaces.** In seating areas or similar surfaces, receptacles shall not be installed in a face-up position unless the receptacle is any of the following:

(1) Part of an assembly listed as a furniture power distribution unit, if cord-and plug-connected
(2) Part of an assembly listed either as household furnishings or as commercial furnishings
(3) Listed either as a receptacle assembly for countertop applications or as a GFCI receptacle assembly for countertop applications
(4) Installed in a listed floor box

**(G) Exposed Terminals.** Receptacles shall be enclosed so that live wiring terminals are not exposed to contact.

**(H) Voltage Between Adjacent Devices.** A receptacle shall not be grouped or ganged in enclosures with other receptacles, snap switches, or similar devices, unless they are arranged so that the voltage between adjacent devices does not exceed 300 volts, or unless they are installed in enclosures equipped with identified, securely installed barriers between adjacent devices.

## 406.6 Receptacle Faceplates (Cover Plates)

Receptacle faceplates shall be installed so as to completely cover the opening and seat against the mounting surface.

Receptacle faceplates mounted inside a box having a recess-mounted receptacle shall effectively close the opening and seat against the mounting surface.

The faceplates/cover plates for recessed receptacles must fit the inside dimensions of the box. A small tolerance is required to facilitate installation of the cover plate, which is typically 1/32 inch or less. The recessed design and the very small opening provide the necessary protection against access to live parts. That the receptacle is required to be "effectively" closed allows some tolerance in application of the requirement.

**(A) Thickness of Metal Faceplates.** Metal faceplates shall be of ferrous metal not less than 0.76 mm (0.030 in.) in thickness or of nonferrous metal not less than 1.02 mm (0.040 in.) in thickness.

**(B) Grounding.** Metal faceplates shall be grounded.

Generally, this requirement is easily met by grounding the metal box. However, isolated ground receptacles installed in nonmetallic boxes are problematic because grounding the receptacle in this case does not ground the faceplate. Section 406.3(D)(2) contains two solutions concerning the receptacle faceplate. First, the general solution is to use only nonmetallic faceplates. Second, the exception to 406.3(D)(2) allows a nonmetallic box manufacturer to add a feature or accessory to accomplish effective grounding of a metal faceplate.

**(C) Faceplates of Insulating Material.** Faceplates of insulating material shall be noncombustible and not less than 2.54 mm (0.10 in.) in thickness but shall be permitted to be less than 2.54 mm (0.10 in.) in thickness if formed or reinforced to provide adequate mechanical strength.

## 406.7 Attachment Plugs, Cord Connectors, and Flanged Surface Devices

All attachment plugs, cord connectors, and flanged surface devices (inlets and outlets) shall be listed and marked with the manufacturer's name or identification and voltage and ampere ratings.

An energized cord cap is often improperly used to supply power to a building from a portable generator when a power failure occurs. Section 406.7 prohibits the improper use of a cord cap, where the blades are exposed and energized, to supply power to a cord body or to plug into a receptacle to backfeed it. Prongs or blades that are exposed to contact by persons must not be energized. Exhibit 406.2 illustrates a flanged inlet device that must not be energized until the cord connector is installed.

**(A) Construction of Attachment Plugs and Cord Connectors.** Attachment plugs and cord connectors shall be constructed so that there are no exposed current-carrying parts except the prongs, blades, or pins. The cover for wire terminations shall be a part that is essential for the operation of an attachment plug or connector (dead-front construction).

**(B) Connection of Attachment Plugs.** Attachment plugs shall be installed so that their prongs, blades, or pins are not energized unless inserted into an energized receptacle or cord connectors.

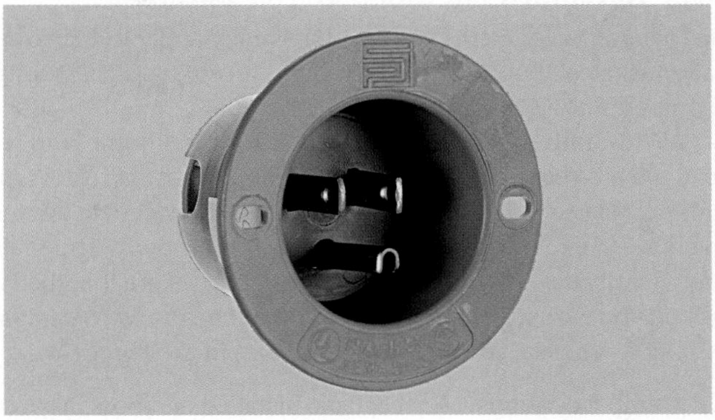

***EXHIBIT 406.2*** *Flanged inlet device. (Courtesy of Legrand/Pass & Seymour®)*

No receptacle shall be installed so as to require the insertion of an energized attachment plug as its source of supply.

A live attachment plug cap can be dangerous. Exposed, energized blades pose a serious shock hazard to anyone handling the attachment plug. Energized blades could also make contact with metal faceplates or screws exposed on the faceplate during insertion. Attachment plug caps should never be installed in such a way that the blades can be energized without being plugged into a device.

**(C) Attachment Plug Ejector Mechanisms.** Attachment plug ejector mechanisms shall not adversely affect engagement of the blades of the attachment plug with the contacts of the receptacle.

This device, designed for use by persons with mobility or visual impairment, reduces the likelihood of damage to the cord when the cord is pulled to remove the plug.

**(D) Flanged Surface Inlet.** A flanged surface inlet shall be installed such that the prongs, blades, or pins are not energized unless an energized cord connector is inserted into it.

## 406.8 Noninterchangeability

Receptacles, cord connectors, and attachment plugs shall be constructed such that receptacle or cord connectors do not accept an attachment plug with a different voltage or current rating from that for which the device is intended. However, a 20-ampere T-slot receptacle or cord connector shall be permitted to accept a 15-ampere attachment plug of the same voltage rating. Non–grounding-type receptacles and connectors shall not accept grounding-type attachment plugs.

For information on receptacle and attachment cap configurations, see NEMA WD 6, *Wiring Devices – Dimensional Requirements*, available for download at www.nema.org.

## 406.9 Receptacles in Damp or Wet Locations

The requirements of 406.9(A) and (B) as they apply to the covers that are typically used with lower rated receptacles (15 through 60 amperes) are summarized in Exhibit 406.3.

All 15- and 20-ampere receptacles for both damp and wet locations are required to be a listed weather-resistant (WR) type. Studies indicated that normal receptacles were inadequate because covers were either broken off or not closed properly. The major differences between WR and non-WR receptacles are that the WR has additional corrosion protection, UV resistance, and cold impact resistance. A typical WR receptacle is shown in Exhibit 406.4.

**(A) Damp Locations.** A receptacle installed outdoors in a location protected from the weather or in other damp locations shall have an enclosure for the receptacle that is weatherproof when the receptacle is covered (attachment plug cap not inserted and receptacle covers closed).

An installation suitable for wet locations shall also be considered suitable for damp locations.

A receptacle shall be considered to be in a location protected from the weather where located under roofed open porches, canopies, marquees, and the like, and will not be subjected to a beating rain or water runoff. All 15- and 20-ampere, 125- and 250-volt nonlocking receptacles shall be a listed weather-resistant type.

Informational Note: The types of receptacles covered by this requirement are identified as 5-15, 5-20, 6-15, and 6-20 in ANSI/NEMA WD 6-2002, National Electrical Manufacturers Association *Standard for Dimensions of Attachment Plugs and Receptacles.*

**(B) Wet Locations.**

**(1) Receptacles of 15 and 20 Amperes in a Wet Location.** Receptacles of 15 and 20 amperes installed in a wet location shall have an enclosure that is weatherproof whether or not the attachment plug cap is inserted. An outlet box hood installed for this purpose shall be listed and shall be identified as "extra duty." All 15- and 20-ampere, 125- and 250-volt nonlocking-type receptacles shall be listed weather-resistant type.

Informational Note No. 1: Requirements for extra-duty outlet box hoods are found in ANSI/UL 514D-2000, *Cover Plates for Flush-Mounted Wiring Devices.*

***EXHIBIT 406.3*** *Requirements for receptacle cover (enclosure) types.*

| Damp and Wet Receptacle Locations | Receptacle Cover (Enclosure) Type Requirements | |
|---|---|---|
| | Cover that *is not* weatherproof, with attachment plug cap inserted into receptacle  | Cover that *is* weatherproof, with attachment plug cap inserted into receptacle ("in-use" type) 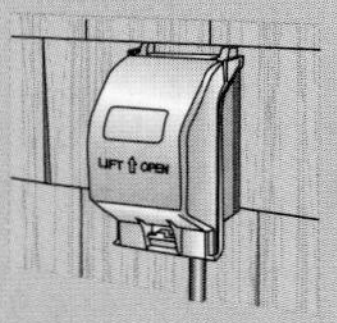 |
| 406.9(A): Outdoor damp locations | Minimum type required<br>*Note: "In-use" type covers permitted* | Permitted |
| 406.9(A): Indoor damp locations | Minimum type required<br>*Note: "In-use" type covers permitted* | Permitted |
| 406.9(B)(1)&(2): Outdoor wet locations | Required for receptacle types other than those rated 15 and 20 amperes, 125 and 250 volts, where the tool, appliance, or other utilization equipment plugged into the receptacle *is* attended while in use<br>*Note: "In-use" type covers permitted* | (a) Required for receptacles rated 15 and 20 amperes, 125 and 250 volts<br>(b) Required for receptacles other than those rated 15 and 20 amperes, 125 and 250 volts, where the tool, appliance, or other utilization equipment plugged into the receptacle *is not* attended while in use |
| 406.9(B)(2): Indoor wet locations | Required for receptacle types other than those rated 15 and 20 amperes, 125 and 250 volts, where the tool, appliance, or other utilization equipment plugged into the receptacle *is* attended while in use<br>*Note: "In-use" type covers permitted* | (a) Required for receptacles rated 15 and 20 amperes, 125 and 250 volts<br>(b) Required for receptacles other than those rated 15 and 20 amperes, 125 and 250 volts, where the tool, appliance, or other utilization equipment plugged into the receptacle *is not* attended while in use |

**EXHIBIT 406.4** *Two WR receptacles enclosed by a two-gang weatherproof cover suitable for use in wet locations. (Courtesy of Legrand/Pass & Seymour®)*

**EXHIBIT 406.5** *A single-gang weatherproof cover suitable for use in wet locations. (Courtesy of Thomas & Betts Corp.)*

Informational Note No. 2: The types of receptacles covered by this requirement are identified as 5-15, 5-20, 6-15, and 6-20 in ANSI/NEMA WD 6-2002, *Standard for Dimensions of Attachment Plugs and Receptacles.*

*Exception: 15- and 20-ampere, 125- through 250-volt receptacles installed in a wet location and subject to routine high-pressure spray washing shall be permitted to have an enclosure that is weatherproof when the attachment plug is removed.*

Where the cord-and-plug connection to receptacles is in a wet location, the enclosure is required to be weatherproof regardless of whether the plug is inserted. The requirement for this type of cover is not contingent on the anticipated use of the receptacle. This requirement applies to all 15- and 20-ampere, 125- and 250-volt receptacles that are installed in wet locations, including those receptacle outlets at dwelling units. Exhibit 406.5 is an example of the type of receptacle enclosure required by 406.9(B)(1).

This section requires outlet box hoods that are part of a weatherproof enclosure to have "extra duty" durability to retain a degree of protection for the receptacles.

**(2) Other Receptacles.** All other receptacles installed in a wet location shall comply with (B)(2)(a) or (B)(2)(b).

(a) A receptacle installed in a wet location, where the product intended to be plugged into it is not attended while in use, shall have an enclosure that is weatherproof with the attachment plug cap inserted or removed.

(b) A receptacle installed in a wet location where the product intended to be plugged into it will be attended while in use (e.g., portable tools) shall have an enclosure that is weatherproof when the attachment plug is removed.

Section 406.9(B)(2) does not apply to receptacles rated 15 and 20 amperes, 125 and 250 volts. Section 406.9(B)(2)(a) applies to receptacles of other ratings that supply cord-and-plug-connected equipment likely to be used outdoors or in a wet location for long periods of time. A portable pump motor is an example of such equipment. Receptacles for this application should remain weatherproof while they are in use.

Section 406.9(B)(2)(b) applies to receptacles of other ratings that supply cord-and-plug-connected portable tools or other portable equipment likely to be used outdoors for a specific purpose and then removed.

**(C) Bathtub and Shower Space.** Receptacles shall not be installed within or directly over a bathtub or shower stall.

The installation of receptacles inside bathtub and shower spaces or above their footprint is prohibited, even if the receptacle is installed in a weatherproof enclosure or is GFCI protected. The unprotected line-side of GFCI-protected receptacles installed in bathtub and shower spaces could possibly become wet and therefore create a shock hazard by energizing surrounding wet surfaces. Prohibiting such installation helps minimize the use of shavers, radios, hair dryers, and so forth, in these areas

**(D) Protection for Floor Receptacles.** Standpipes of floor receptacles shall allow floor-cleaning equipment to be operated without damage to receptacles.

**(E) Flush Mounting with Faceplate.** The enclosure for a receptacle installed in an outlet box flush-mounted in a finished surface shall be made weatherproof by means of a weatherproof faceplate assembly that provides a watertight connection between the plate and the finished surface.

## 406.10 Grounding-Type Receptacles, Adapters, Cord Connectors, and Attachment Plugs

**(A) Grounding Poles.** Grounding-type receptacles, cord connectors, and attachment plugs shall be provided with one fixed grounding pole in addition to the circuit poles. The grounding contacting pole of grounding-type plug-in ground-fault circuit interrupters shall be permitted to be of the movable, self-restoring type on circuits operating at not over 150 volts between any two conductors or any conductor and ground.

**(B) Grounding-Pole Identification.** Grounding-type receptacles, adapters, cord connections, and attachment plugs shall have a means for connection of an equipment grounding conductor to the grounding pole.

A terminal for connection to the grounding pole shall be designated by one of the following:

(1) A green-colored hexagonal-headed or -shaped terminal screw or nut, not readily removable.
(2) A green-colored pressure wire connector body (a wire barrel).
(3) A similar green-colored connection device, in the case of adapters. The grounding terminal of a grounding adapter shall be a green-colored rigid ear, lug, or similar device. The equipment grounding connection shall be so designed that it cannot make contact with current-carrying parts of the receptacle, adapter, or attachment plug. The adapter shall be polarized.
(4) If the terminal for the equipment grounding conductor is not visible, the conductor entrance hole shall be marked with the word *green* or *ground*, the letters *G* or *GR*, a grounding symbol, or otherwise identified by a distinctive green color. If the terminal for the equipment grounding conductor is readily removable, the area adjacent to the terminal shall be similarly marked.

Informational Note: See Informational Note Figure 406.10(B)(4).

***INFORMATIONAL NOTE FIGURE 406.10(B)(4)** One Example of a Symbol Used to Identify the Termination Point for an Equipment Grounding Conductor.*

Section 406.10(B)(3) requires the grounding terminal of an adapter to be a green-colored ear, lug, or similar device, thereby prohibiting use of an adapter with an attached pigtail grounding wire, which had been used for many years.

**(C) Grounding Terminal Use.** A grounding terminal shall not be used for purposes other than grounding.

**(D) Grounding-Pole Requirements.** Grounding-type attachment plugs and mating cord connectors and receptacles shall be designed such that the equipment grounding connection is made before the current-carrying connections. Grounding-type devices shall be so designed that grounding poles of attachment plugs cannot be brought into contact with current-carrying parts of receptacles or cord connectors.

The grounding blade of the attachment plug cap of most grounding-type combinations is longer than the circuit conductor blades and is used to ensure a "make-first, break-last" grounding connection. In some non-ANSI, pin-and-sleeve-type configurations, the grounding contact of the receptacle is closer to the face of the receptacle than it is to other contacts, serving the same purpose.

**(E) Use.** Grounding-type attachment plugs shall be used only with a cord having an equipment grounding conductor.

Informational Note: See 250.126 for identification of grounding conductor terminals.

## 406.11 Connecting Receptacle Grounding Terminal to Box

The connection of the receptacle grounding terminal shall comply with 250.146.

## 406.12 Tamper-Resistant Receptacles

Tamper-resistant receptacles shall be installed as specified in 406.12(A) through (C).

Statistics from 2005 indicated that approximately 89 percent of electrical burn and shock incidents occurred among children 6 years of age or less. The requirements for tamper-resistant receptacles ensure that children will be protected in all types of environments — in closely supervised areas, such as pediatric care locations and child care facilities, and in less structured, residential environments. Tamper-resistant construction provides the most effective and permanent means of preventing children from inserting foreign objects into receptacles. These receptacles are recognized in the U.S. General Services Administration (GSA) Child Care Center Design Guide (PBS100-July 1993) as a critical design feature for child care areas.

**(A) Dwelling Units.** In all areas specified in 210.52, all nonlocking-type 125-volt, 15- and 20-ampere receptacles shall be listed tamper-resistant receptacles.

**(B) Guest Rooms and Guest Suites of Hotels and Motels.** All nonlocking-type 125-volt, 15- and 20-ampere receptacles located

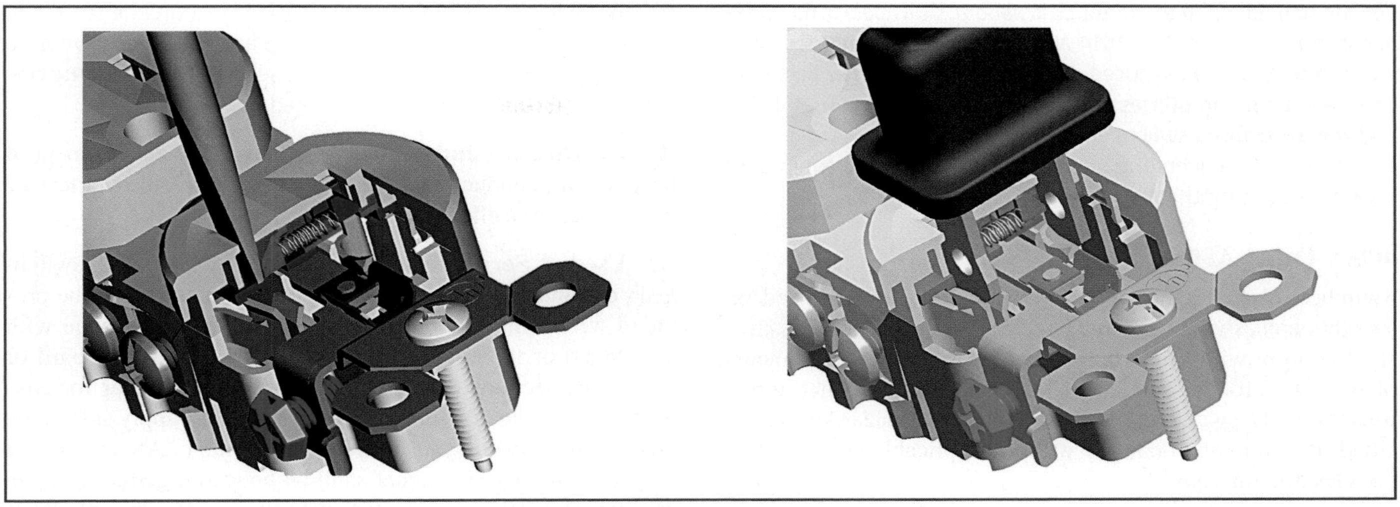

***EXHIBIT 406.6*** *Tamper-resistant receptacle. Insertion of an object in any one side does not open the shutter (left), but a two-bladed plug or grounding plug compresses the spring and simultaneously opens both shutters (right). (Courtesy of Legrand/Pass & Seymour®)*

in guest rooms and guest suites of hotels and motels shall be listed tamper-resistant receptacles.

**(C) Child Care Facilities.** In all child care facilities, all nonlocking-type 125-volt, 15- and 20-ampere receptacles shall be listed tamper-resistant receptacles.

*Exception to (A), (B), and (C): Receptacles in the following locations shall not be required to be tamper resistant:*

1. *Receptacles located more than 1.7 m (5½ ft) above the floor.*
2. *Receptacles that are part of a luminaire or appliance.*
3. *A single receptacle or a duplex receptacle for two appliances located within dedicated space for each appliance that, in normal use, is not easily moved from one place to another and that is cord-and plug-connected in accordance with 400.7(A)(6), (A)(7), or (A)(8).*
4. *Nongrounding receptacles used for replacements as permitted in 406.4(D)(2)(a).*

All areas specified in 210.52, as well as the pediatric areas in Article 517, need tamper-resistant receptacles. Lodging facilities require tamper-resistant receptacles to provide the same level of protection for children as they would have at home. Likewise, the receptacles in a child care facility are required to be tamper-resistant. Exhibit 406.6 shows a typical tamper-resistant receptacle.

Locking-type receptacles are not required to be tamper resistant. Only those receptacles installed at a height below 1.7 meters (5½ feet) can meet the requirements in 210.52 for wall spacing. Receptacles installed above 1.7 meters (5½ feet) are not accessible and well out of reach of small children. Allowing the exception for a single receptacle or duplex receptacle located within dedicated space eliminates the need for tamper-resistant receptacles to be installed behind dishwashers, refrigerators, washing machines, and so forth. Nongrounding-type receptacles were exempted because there are no known 125-volt, 15- or 20-ampere nongrounding-type receptacles listed as tamper-resistant receptacles.

### 406.15 Dimmer-Controlled Receptacles

A receptacle supplying lighting loads shall not be connected to a dimmer unless the plug/receptacle combination is a nonstandard configuration type that is specifically listed and identified for each such unique combination.

A cord-connected lamp may be removed and replaced with other equipment not suitable for use with a dimmer. The installation of a dimmer to control a standard receptacle is prohibited. To prevent misuse, a non-standard receptacle is required for dimmer controlled lighting.

# ARTICLE 408 Switchboards, Switchgear, and Panelboards

## I. General

### 408.1 Scope

This article covers switchboards, switchgear, and panelboards. It does not apply to equipment operating at over 1000 volts, except as specifically referenced elsewhere in the *Code*.

See the definitions of the terms *switchboard, switchgear,* and *panelboard* in Article 100. The 2014 *NEC* modifies and renames what was formerly "metal-enclosed power switchgear" to "switchgear." The new definition utilizes this generic term for both metal-clad and metal-enclosed switchgear.

Article 490 covers the general requirements for equipment operating over 1000 volts, nominal.

## 408.2 Other Articles

Switches, circuit breakers, and overcurrent devices used on switchboards, switchgear, and panelboards and their enclosures shall comply with this article and also with the requirements of Articles 240, 250, 312, 404, and other articles that apply. Switchboards, switchgear, and panelboards in hazardous (classified) locations shall comply with the applicable provisions of Articles 500 through 517.

## 408.3 Support and Arrangement of Busbars and Conductors

**(A) Conductors and Busbars on a Switchboard, Switchgear, or Panelboard.** Conductors and busbars on a switchboard, switchgear, or panelboard shall comply with 408.3(A)(1), (A)(2), and (A)(3) as applicable.

**(1) Location.** Conductors and busbars shall be located so as to be free from physical damage and shall be held firmly in place.

**(2) Service Switchboards and Switchgear.** Barriers shall be placed in all service switchboards and switchgear such that no uninsulated, ungrounded service busbar or service terminal is exposed to inadvertent contact by persons or maintenance equipment while servicing load terminations.

Where disconnecting or de-energizing the service conductors supplying a service switchboard or switchgear is not feasible, qualified electricians may be required to work on these switchboards or switchgear with the load terminals de-energized but with the service bus energized. Barriers are required in service switchboards and switchgear to provide physical separation (adequate distance or an obstacle) between load terminals and the service busbars and terminals, thus providing some measure of safety against inadvertent contact with line-energized parts during maintenance and installation of new feeders or branch circuits. In most multisection switchboards and switchgear, barriers are not required, because the line-side conductors and busbars are not in the same sections that contain the load terminals.

**(3) Same Vertical Section.** Other than the required interconnections and control wiring, only those conductors that are intended for termination in a vertical section of a switchboard or switchgear shall be located in that section.

*Exception: Conductors shall be permitted to travel horizontally through vertical sections of switchboards and switchgear where such conductors are isolated from busbars by a barrier.*

Horizontal travel of conductors through more than one section of a multisection switchboard or switchgear is necessary where a raceway or cable entry is made into a section other than the one at which the conductors are terminated.

**(B) Overheating and Inductive Effects.** The arrangement of busbars and conductors shall be such as to avoid overheating due to inductive effects.

**(C) Used as Service Equipment.** Each switchboard, switchgear, or panelboard, if used as service equipment, shall be provided with a main bonding jumper sized in accordance with 250.28(D) or the equivalent placed within the panelboard or one of the sections of the switchboard or switchgear for connecting the grounded service conductor on its supply side to the switchboard, switchgear, or panelboard frame. All sections of a switchboard or switchgear shall be bonded together using an equipment bonding conductor sized in accordance with Table 250.122 or Table 250.66 as appropriate.

*Exception: Switchboards, switchgear, and panelboards used as service equipment on high-impedance grounded neutral systems in accordance with 250.36 shall not be required to be provided with a main bonding jumper.*

**(D) Terminals.** In switchboards, switchgear, and panelboards, load terminals for field wiring, including grounded circuit conductor load terminals and connections to the equipment grounding conductor bus for load equipment grounding conductors, shall be so located that it is not necessary to reach across or beyond an uninsulated ungrounded line bus in order to make connections.

**(E) Bus Arrangement.**

The high leg is common on a 240/120-volt, 3-phase, 4-wire center-tap grounded delta system and is typically designated as "B phase." Section 110.15 requires the high-leg marking to be the color orange or another effective means of identification.

The exception to 408.3(E)(1) permits the phase leg having the higher voltage to ground to be located at the right-hand position (C phase), making it unnecessary to transpose the panelboard, switchgear, or switchboard busbar arrangement ahead of and beyond a metering compartment. The exception recognizes the fact that metering compartments have been standardized with the high leg at the right position (C phase) rather than in the center on B phase.

See also 110.15 and 230.56 for further information on identifying conductors with the higher voltage to ground. Other busbar arrangements for making additions to existing installations are permitted by 408.3(E).

**(1) AC Phase Arrangement.** Alternating-current phase arrangement on 3-phase buses shall be A, B, C from front to back, top to bottom, or left to right, as viewed from the front of the switchboard, switchgear, or panelboard. The B phase shall be that phase having the higher voltage to ground on 3-phase, 4-wire, delta-connected systems. Other busbar arrangements shall be permitted for additions to existing installations and shall be marked.

*Exception: Equipment within the same single section or multisection switchboard, switchgear, or panelboard as the meter on 3-phase, 4-wire, delta-connected systems shall be permitted to have the same phase configuration as the metering equipment.*

Informational Note: See 110.15 for requirements on marking the busbar or phase conductor having the higher voltage to ground where supplied from a 4-wire, delta-connected system.

**(2) DC Bus Arrangement.** Direct-current ungrounded buses shall be permitted to be in any order. Arrangement of dc buses shall be field marked as to polarity, grounding system, and nominal voltage.

**(F) Switchboard, Switchgear, or Panelboard Identification.** A caution sign(s) or a label(s) provided in accordance with 408.3(F)(1) through (F)(5) shall comply with 110.21(B).

**(1) High-Leg Identification.** A switchboard, switchgear, or panelboard containing a 4-wire, delta-connected system where the midpoint of one phase winding is grounded shall be legibly and permanently field marked as follows:

"Caution _____ Phase Has _____ Volts to Ground"

The requirement for legible marking of a switchboard, switchgear, or panelboard that contains a 3-phase, 4-wire center-tap grounded delta system results from injury and property damage caused by people not recognizing there is a high leg in the switchboard, switchgear, or panelboard. This requirement eliminates some of the hazards of accidentally connecting outlets to the high leg and causing injury to people and damage to equipment.

**(2) Ungrounded AC Systems.** A switchboard, switchgear, or panelboard containing an ungrounded ac electrical system as permitted in 250.21 shall be legibly and permanently field marked as follows:

"Caution Ungrounded System Operating — _____
Volts Between Conductors"

The intent of this requirement is to delineate grounded from ungrounded electrical systems. When a ground fault occurs on a 3-phase ungrounded system, the voltage to ground on the ungrounded system may equal the line-to-line voltage. The operational advantage of using an ungrounded system is continuity of operation, which in some processes might create a safer condition than would be achieved by automatic or unplanned opening of the supply circuit. Section 250.21(B) requires ungrounded systems of not less than 120 volts and not more than 1000 volts to be provided with ground detection. Ground detection will warn of the ground fault to permit an orderly shutdown of a process.

**(3) High-Impedance Grounded Neutral AC System.** A switchboard, switchgear, or panelboard containing a high-impedance grounded neutral ac system in accordance with 250.36 shall be legibly and permanently field marked as follows:

CAUTION: HIGH-IMPEDANCE GROUNDED
NEUTRAL AC SYSTEM OPERATING — _____
VOLTS BETWEEN CONDUCTORS AND MAY
OPERATE — _____ VOLTS TO GROUND FOR
INDEFINITE PERIODS UNDER FAULT CONDITIONS

**(4) Ungrounded DC Systems.** A switchboard, switchgear, or panelboard containing an ungrounded dc electrical system in accordance with 250.169 shall be legibly and permanently field marked as follows:

CAUTION: UNGROUNDED DC SYSTEM OPERATING —
_____ VOLTS BETWEEN CONDUCTORS

**(5) Resistively Grounded DC Systems.** A switchboard, switchgear, or panelboard containing a resistive connection between current-carrying conductors and the grounding system to stabilize voltage to ground shall be legibly and permanently field marked as follows:

CAUTION: DC SYSTEM OPERATING — _____
VOLTS BETWEEN CONDUCTORS AND MAY
OPERATE — _____ VOLTS TO GROUND FOR
INDEFINITE PERIODS UNDER FAULT CONDITIONS

**(G) Minimum Wire-Bending Space.** The minimum wire-bending space at terminals and minimum gutter space provided in switchboards, switchgear, and panelboards shall be as required in 312.6.

## 408.4 Field Identification Required

**(A) Circuit Directory or Circuit Identification.** Every circuit and circuit modification shall be legibly identified as to its clear, evident, and specific purpose or use. The identification shall include an approved degree of detail that allows each circuit to be distinguished from all others. Spare positions that contain unused overcurrent devices or switches shall be described accordingly. The identification shall be included in a circuit directory that is located on the face or inside of the panel door in the case of a panelboard and at each switch or circuit breaker in a switchboard or switchgear. No circuit shall be described in a manner that depends on transient conditions of occupancy.

The circuit directory is an important feature for the safe operation of an electrical system under normal and emergency conditions. The purpose of an accurate and legible circuit directory in these types of equipment is to provide clear identification of circuit breakers and switches that may need to be operated by service personnel or others responding who need to operate a switch or circuit breaker in an emergency. This requirement is specific to switchboards, switchgear, and panelboards; however, the identification requirements of 110.22 apply to all disconnecting means.

Circuits used for the same purpose must be identified by their location. For example, small-appliance branch circuits can supply outlets in the kitchen, dining room, and kitchen countertops. Identifying these circuits as small-appliance branch circuits is not acceptable; instead, they should be identified as "kitchen wall receptacles," "dining room floor receptacle," or "kitchen countertop receptacles left of sink." Circuit directories containing multiple entries with only "lights" or "outlets" do not provide the sufficient detail required by this section.

Spare devices are required to be marked to indicate that they are spares. Markings are required to indicate permanent features and not temporary conditions of occupancy. For example, for a circuit breaker supplying an office, a label with the employee's name is no longer useful when the employee no longer occupies that office.

**(B) Source of Supply.** All switchboards, switchgear, and panelboards supplied by a feeder(s) in other than one- or two-family dwellings shall be marked to indicate each device or equipment where the power originates.

Tracing a feeder circuit back to its originating switchboard, switchgear, panelboard, or other source can be a time-consuming and inaccurate process. Accurate identification of circuits promotes more efficient lockout/tagout processes, which provide a safer work environment for employees. Identification of the feeder circuit when the new feeder is being added is also more economical than the time-consuming process of tracing a circuit.

### 408.5 Clearance for Conductor Entering Bus Enclosures

Where conduits or other raceways enter a switchboard, switchgear, floor-standing panelboard, or similar enclosure at the bottom, approved space shall be provided to permit installation of conductors in the enclosure. The wiring space shall not be less than shown in Table 408.5 where the conduit or raceways enter or leave the enclosure below the busbars, their supports, or other obstructions. The conduit or raceways, including their end fittings, shall not rise more than 75 mm (3 in.) above the bottom of the enclosure.

***TABLE 408.5*** *Clearance for Conductors Entering Bus Enclosures*

| Conductor | Minimum Spacing Between Bottom of Enclosure and Busbars, Their Supports, or Other Obstructions: mm | in. |
|---|---|---|
| Insulated busbars, their supports, or other obstructions | 200 | 8 |
| Noninsulated busbars | 250 | 10 |

### 408.7 Unused Openings

Unused openings for circuit breakers and switches shall be closed using identified closures, or other approved means that provide protection substantially equivalent to the wall of the enclosure.

The requirement of 110.12(A) for closing unused openings (other than those provided for equipment mounting or drainage) applies to all electrical enclosures, including panelboard cabinets, switchgear, and switchboard enclosures. An unused opening may exist as a result of a renovation or an alteration of existing equipment. These two requirements are necessary to restore the electrical equipment enclosure integrity to a condition that minimizes the possibility of an escaping arc, spark, or molten metal igniting surrounding combustible material and also minimizes the potential for accidental contact with live parts.

## II. Switchboards and Switchgear

### 408.16 Switchboards and Switchgear in Damp or Wet Locations

Switchboards and switchgear in damp or wet locations shall be installed in accordance with 312.2.

### 408.17 Location Relative to Easily Ignitible Material

Switchboards and switchgear shall be placed so as to reduce to a minimum the probability of communicating fire to adjacent combustible materials. Where installed over a combustible floor, suitable protection thereto shall be provided.

Where flooring is combustible, one means of complying with this requirement is to form and attach a piece of sheet steel or other suitable noncombustible material to the floor under the electrical equipment.

### 408.18 Clearances

**(A) From Ceiling.** For other than a totally enclosed switchboard or switchgear, a space not less than 900 mm (3 ft) shall be provided between the top of the switchboard or switchgear and any combustible ceiling, unless a noncombustible shield is provided between the switchboard or switchgear and the ceiling.

**(B) Around Switchboards and Switchgear.** Clearances around switchboards and switchgear shall comply with the provisions of 110.26.

Sufficient access and working space permit safe operation and maintenance of switchboards and switchgear. Table 110.26(A)(1) indicates minimum working clearances from 0 to 600 volts, and Table 110.34(A) is used for voltages over 600 volts.

## 408.19 Conductor Insulation

An insulated conductor used within a switchboard or switchgear shall be listed, shall be flame retardant, and shall be rated not less than the voltage applied to it and not less than the voltage applied to other conductors or busbars with which it may come into contact.

## 408.20 Location of Switchboards and Switchgear

Switchboards and switchgear that have any exposed live parts shall be located in permanently dry locations and then only where under competent supervision and accessible only to qualified persons. Switchboards and switchgear shall be located such that the probability of damage from equipment or processes is reduced to a minimum.

## 408.22 Grounding of Instruments, Relays, Meters, and Instrument Transformers on Switchboards and Switchgear

Instruments, relays, meters, and instrument transformers located on switchboards and switchgear shall be grounded as specified in 250.170 through 250.178.

# III. Panelboards

## 408.30 General

All panelboards shall have a rating not less than the minimum feeder capacity required for the load calculated in accordance with Part III, IV, or V of Article 220, as applicable.

•

Many panelboards are suitable for use as service equipment and are so marked by the manufacturer. Listed panelboards are used with copper conductors, unless they are marked to indicate which terminals are suitable for use with aluminum conductors. Such marking must be independent of any marking on terminal connectors and must appear on a wiring diagram or other readily visible location. If all terminals are suitable for use with aluminum conductors as well as with copper conductors, the panelboard is marked "Use Copper or Aluminum Wire." A panelboard using terminals or main or branch-circuit units individually marked AL-CU is marked "Use Copper or Aluminum Wire" or "Use Copper Wire Only." The latter marking indicates that wiring space or other factors make the panelboard unsuitable for aluminum conductors. [See 110.14(C).]

Unless a panelboard is marked to indicate otherwise, the termination provisions are based on the use of 60°C ampacities for wire sizes 14 AWG through 1 AWG and 75°C ampacities for wire sizes 1/0 AWG and larger.

The terms *lighting* and *appliance branch-circuit panelboards* and *power panelboards* are no longer used. In addition, the requirement for a maximum of 42 overcurrent devices applies only with Exception No. 2 of 408.36. All panelboards need a single overcurrent device that protects the panelboard bus unless either of the exceptions of 408.36 applies.

## 408.36 Overcurrent Protection

In addition to the requirement of 408.30, a panelboard shall be protected by an overcurrent protective device having a rating not greater than that of the panelboard. This overcurrent protective device shall be located within or at any point on the supply side of the panelboard.

The main OCPD may be an integral part of a panelboard or may be located remote from the panelboard. Exhibit 408.1 shows a panelboard with a main circuit breaker and provisions for inserting 60 circuit breakers. Exhibit 408.2 illustrates overcurrent protection for the panelboard feeders having a rating not greater than the rating of the panelboard.

If a panelboard is required to have overcurrent protection, such protection can be provided by an OCPD in the panelboard or by an OCPD protecting the conductors that supply the panelboard. In either case, the OCPD rating is not permitted to exceed the panelboard rating. For example, a feeder protected by a 450-ampere OCPD supplies a panelboard with a 600-ampere rating. Because the panelboard is large enough to supply the

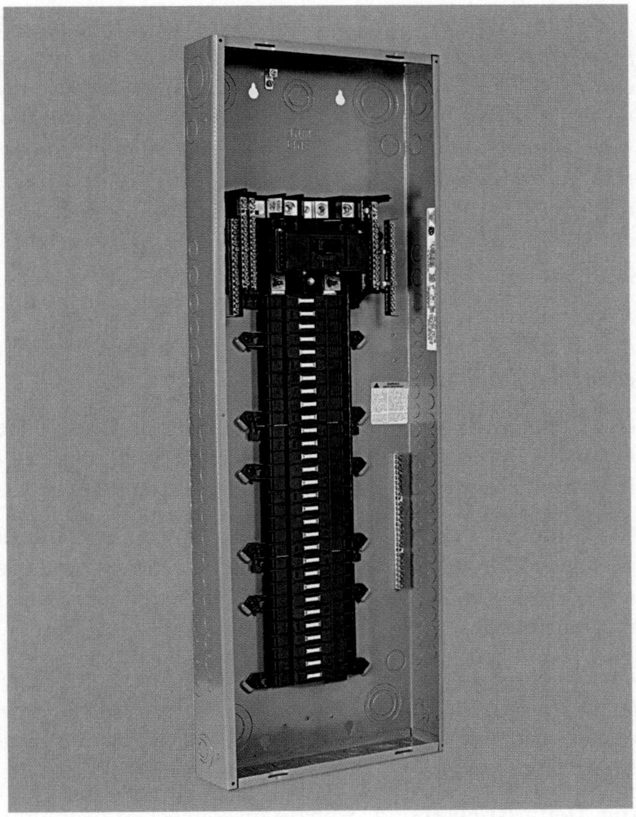

***EXHIBIT 408.1*** *A panelboard with main circuit breaker disconnect suitable for use as service equipment. (Courtesy of Schneider Electric)*

***EXHIBIT 408.2*** *An arrangement of three individual panelboards with main overcurrent protection per 408.36 remote from the panelboards.*

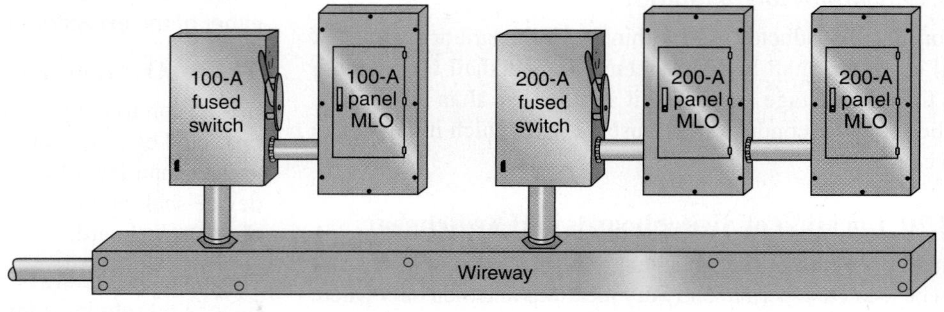


calculated load and the OCPD protecting the feeder does not exceed the panelboard rating, an individual OCPD in the panelboard is not required.

*Exception No. 1: Individual protection shall not be required for a panelboard used as service equipment with multiple disconnecting means in accordance with 230.71. In panelboards protected by three or more main circuit breakers or sets of fuses, the circuit breakers or sets of fuses shall not supply a second bus structure within the same panelboard assembly.*

*Exception No. 2: Individual protection shall not be required for a panelboard protected on its supply side by two main circuit breakers or two sets of fuses having a combined rating not greater than that of the panelboard. A panelboard constructed or wired under this exception shall not contain more than 42 overcurrent devices. For the purposes of determining the maximum of 42 overcurrent devices, a 2-pole or a 3-pole circuit breaker shall be considered as two or three overcurrent devices, respectively.*

*Exception No. 3: For existing panelboards, individual protection shall not be required for a panelboard used as service equipment for an individual residential occupancy.*

**(A) Snap Switches Rated at 30 Amperes or Less.** Panelboards equipped with snap switches rated at 30 amperes or less shall have overcurrent protection of 200 amperes or less.

**(B) Supplied Through a Transformer.** Where a panelboard is supplied through a transformer, the overcurrent protection required by 408.36 shall be located on the secondary side of the transformer.

*Exception: A panelboard supplied by the secondary side of a transformer shall be considered as protected by the overcurrent protection provided on the primary side of the transformer where that protection is in accordance with 240.21(C)(1).*

**(C) Delta Breakers.** A 3-phase disconnect or overcurrent device shall not be connected to the bus of any panelboard that has less than 3-phase buses. Delta breakers shall not be installed in panelboards.

**(D) Back-Fed Devices.** Plug-in-type overcurrent protection devices or plug-in type main lug assemblies that are backfed and used to terminate field-installed ungrounded supply conductors shall be secured in place by an additional fastener that requires other than a pull to release the device from the mounting means on the panel.

## 408.37 Panelboards in Damp or Wet Locations

Panelboards in damp or wet locations shall be installed to comply with 312.2.

## 408.38 Enclosure

Panelboards shall be mounted in cabinets, cutout boxes, or identified enclosures and shall be dead-front.

*Exception: Panelboards other than of the dead-front, externally operable type shall be permitted where accessible only to qualified persons.*

## 408.39 Relative Arrangement of Switches and Fuses

In panelboards, fuses of any type shall be installed on the load side of any switches.

*Exception: Fuses installed as part of service equipment in accordance with the provisions of 230.94 shall be permitted on the line side of the service switch.*

Sections 230.82 and 230.94 permit the service switch to be located on either the supply side or the load side of fuses such as cable limiters and other current-limiting devices. Where fuses of panelboards are accessible to other than qualified persons, such as occupants of a multifamily dwelling, 240.40 requires that disconnecting means be located on the supply side of all fuses in circuits of over 150 volts to ground and in cartridge-type fuses in circuits of any voltage. Thus, when the disconnect switch is opened, the fuses are de-energized, and danger from shock is reduced.

## 408.40 Grounding of Panelboards

Panelboard cabinets and panelboard frames, if of metal, shall be in physical contact with each other and shall be connected to an equipment grounding conductor. Where the panelboard is used with nonmetallic raceway or cable or where separate equipment grounding conductors are provided, a terminal bar for the equipment grounding conductors shall be secured inside the cabinet. The terminal bar shall be bonded to the cabinet and panelboard frame, if of metal; otherwise it shall be connected to the equipment grounding conductor that is run with the conductors feeding the panelboard.

A separate equipment grounding conductor (EGC) terminal bar must be installed and bonded to the panelboard for the termination of feeder and branch-circuit EGCs. Where installed within service equipment, this terminal is bonded to the neutral terminal bar. Any other connection between the equipment grounding terminal bar and the neutral bar (other than that allowed in 250.32) is not permitted. If this downstream connection occurs, current in the neutral or grounded conductor would take parallel paths through the EGCs (the raceway, the building structure, or earth, for example) back to the service equipment. Normal load currents on the EGCs could create a shock hazard. Exposed metal parts of equipment could have a potential difference of several volts created by the load current on the grounding conductors. Another safety hazard created by this effect, where subpanels are used, is arcing or loose connections at connectors and raceway fittings, creating a potential fire hazard.

*Exception: Where an isolated equipment grounding conductor is provided as permitted by 250.146(D), the insulated equipment grounding conductor that is run with the circuit conductors shall be permitted to pass through the panelboard without being connected to the panelboard's equipment grounding terminal bar.*

Equipment grounding conductors shall not be connected to a terminal bar provided for grounded conductors or neutral conductors unless the bar is identified for the purpose and is located where interconnection between equipment grounding conductors and grounded circuit conductors is permitted or required by Article 250.

Sensitive electronic equipment used in industrial and commercial power systems may fail to perform properly if electrical noise is present in the EGC.

The exception permits an isolated equipment grounding terminal, if it is necessary for the reduction of electrical noise on the grounding circuit. This equipment grounding terminal must be grounded by an insulated EGC that is run with the circuit conductors. The isolated EGC is also permitted to pass through one or more panelboards (without connection to the panelboard grounding terminal), but it is important that the EGC terminate directly at the applicable separately derived system or service grounding terminal. If the isolated EGC is run in a separate building, however, 250.146(D) requires the isolated EGC to terminate at a panelboard within the same building.

A connection only to a separate grounding electrode that places the earth in the fault return path prevents an insufficient level of ground-fault current to open the OCPD when a ground fault occurs. See the commentary following the informational note to 250.146(D) and 250.54.

### 408.41 Grounded Conductor Terminations

Each grounded conductor shall terminate within the panelboard in an individual terminal that is not also used for another conductor.

In accordance with 110.14(A), conductor terminations are only suitable for a single conductor unless the terminal is marked or otherwise identified as suitable for more than one conductor. The use of a single termination point within a panelboard to connect more than one grounded conductor or to connect a grounded conductor and an EGC can be problematic when it is necessary to isolate a particular grounded conductor for testing purposes. For example, if the grounded conductors of two branch circuits were terminated at a single connection point, and it were necessary to isolate one branch circuit for the purposes of troubleshooting, the fact that the circuit not being tested remained energized could create an unsafe working condition for service personnel disconnecting the grounded conductor of the circuit being tested. In some cases, panelboard instructions are provided that permit the use of a single-conductor termination for more than one EGC. See 408.40 for the requirements on panelboard terminations for EGCs and 408.41 for terminations of grounded conductors.

*Exception: Grounded conductors of circuits with parallel conductors shall be permitted to terminate in a single terminal if the terminal is identified for connection of more than one conductor.*

## IV. Construction Specifications

### 408.50 Panels

The panels of switchboards and switchgear shall be made of moisture-resistant, noncombustible material.

### 408.51 Busbars

Insulated or bare busbars shall be rigidly mounted.

### 408.52 Protection of Instrument Circuits

Instruments, pilot lights, voltage (potential) transformers, and other switchboard or switchgear devices with potential coils shall be supplied by a circuit that is protected by standard overcurrent devices rated 15 amperes or less.

*Exception No. 1: Overcurrent devices rated more than 15 amperes shall be permitted where the interruption of the circuit could create a hazard. Short-circuit protection shall be provided.*

*Exception No. 2: For ratings of 2 amperes or less, special types of enclosed fuses shall be permitted.*

### 408.53 Component Parts

Switches, fuses, and fuseholders used on panelboards shall comply with the applicable requirements of Articles 240 and 404.

### 408.54 Maximum Number of Overcurrent Devices

A panelboard shall be provided with physical means to prevent the installation of more overcurrent devices than that number for which the panelboard was designed, rated, and listed.

For the purposes of this section, a 2-pole circuit breaker or fusible switch shall be considered two overcurrent devices; a 3-pole circuit breaker or fusible switch shall be considered three overcurrent devices.

### 408.55 Wire-Bending Space Within an Enclosure Containing a Panelboard

**(A) Top and Bottom Wire-Bending Space.** The enclosure for a panelboard shall have the top and bottom wire-bending space sized in accordance with Table 312.6(B) for the largest conductor entering or leaving the enclosure.

*Exception No. 1: Either the top or bottom wire-bending space shall be permitted to be sized in accordance with Table 312.6(A) for a panelboard rated 225 amperes or less and designed to contain not over 42 overcurrent devices. For the purposes of this exception, a 2-pole or a 3-pole circuit breaker shall be considered as two or three overcurrent devices, respectively.*

*Exception No. 2: Either the top or bottom wire-bending space for any panelboard shall be permitted to be sized in accordance with Table 312.6(A) where at least one side wire-bending space is sized in accordance with Table 312.6(B) for the largest conductor to be terminated in any side wire-bending space.*

*Exception No. 3: The top and bottom wire-bending space shall be permitted to be sized in accordance with Table 312.6(A) spacings if the panelboard is designed and constructed for wiring using only a single 90-degree bend for each conductor, including the grounded circuit conductor, and the wiring diagram shows and specifies the method of wiring that shall be used.*

*Exception No. 4: Either the top or the bottom wire-bending space, but not both, shall be permitted to be sized in accordance with Table 312.6(A) where there are no conductors terminated in that space.*

Using Exhibit 408.3 as a reference (see 312.6), the general rule calls for wire-bending spaces $T_1$ and $T_4$ to be in accordance with Table 312.6(B) for size $M$ conductors (assuming these are the largest conductors entering the enclosure). Side wire-bending space $T_2$ must be in accordance with Table 312.6(A) for the wire size to be used with the largest-rated unit facing that side space, and $T_3$ must be similarly sized for the largest-rated unit facing the enclosure's right side.

Exception No. 1 to 408.55 permits either $T_1$ or $T_4$ (not both) to be reduced to the space required by Table 312.6(A) for size $M$ conductors for a panelboard rated 225 amperes or less and designed to contain not over 42 overcurrent devices.

Exception No. 2 to 408.55 permits either $T_1$ or $T_4$ (not both) to be reduced to the space required by Table 312.6(A) for size $M$ conductors for any panelboard. Exception No. 2 is valid where either $T_2$ or $T_3$ (or both) is sized in accordance with Table 312.6(B) for the largest conductor to be terminated in either the left or the right side space. Under the construction rules of 408.55, a panelboard enclosure might not be of adequate size for all manner of wiring; therefore, 312.6 must be considered when wiring is planned.

Exception No. 3 to 408.55 permits both the top and the bottom wire-bending space to be reduced as noted. A single 90-degree bend, meaning one and only one 90-degree bend, must be present for the ungrounded conductors. A grounded conductor is permitted to be wired straight in if spacing is provided per Table 312.6(B) for the grounded conductor.

Exception No. 4 to 408.55 permits a reduction to the Table 312.6(A) spacing for the top or bottom space where no terminals face that space. In this case, the space is a gutter space, and measurement is on a line perpendicular to the wall of the enclosure and to the closest barrier post or side of a switch, fuse, or circuit breaker unit that is, or may be, installed. Exhibit 408.3 illustrates that exception.

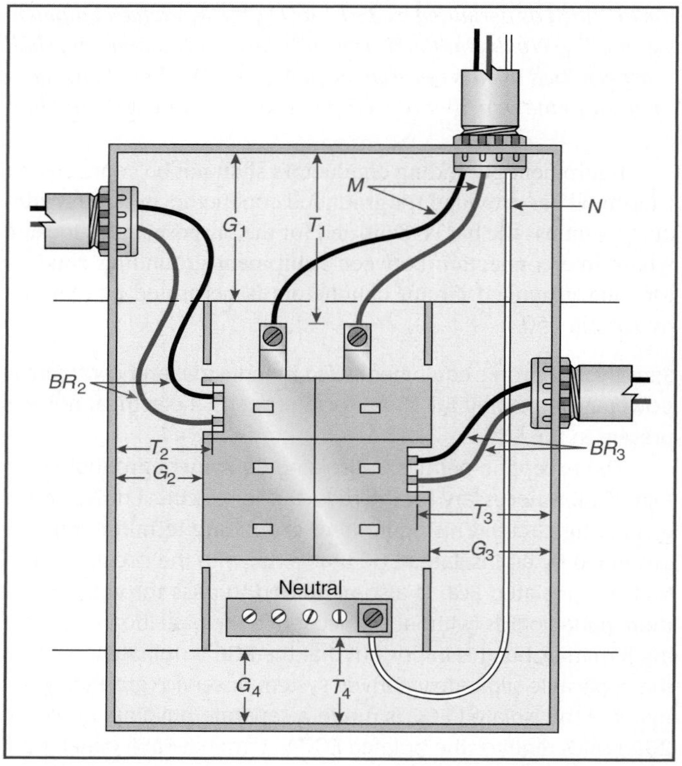


***EXHIBIT 408.3*** *Panelboard wire-bending space per 312.6 and 408.55.*

**(B) Side Wire-Bending Space.** Side wire-bending space shall be in accordance with Table 312.6(A) for the largest conductor to be terminated in that space.

**(C) Back Wire-Bending Space.** Where a raceway or cable entry is in the wall of the enclosure opposite a removable cover, the distance from that wall to the cover shall be permitted to comply with the distance required for one wire per terminal in Table 312.6(A). The distance between the center of the rear entry and the nearest termination for the entering conductors shall not be less than the distance given in Table 312.6(B).

### 408.56 Minimum Spacings

The distance between bare metal parts, busbars, and so forth shall not be less than specified in Table 408.56.

Where close proximity does not cause excessive heating, parts of the same polarity at switches, enclosed fuses, and so forth shall be permitted to be placed as close together as convenience in handling will allow.

*Exception: The distance shall be permitted to be less than that specified in Table 408.56 at circuit breakers and switches and in listed components installed in switchboards, switchgear, and panelboards.*

### 408.58 Panelboard Marking

Panelboards shall be durably marked by the manufacturer with the voltage and the current rating and the number of ac phases or dc buses for which they are designed and with the manufacturer's name or trademark in such a manner so as to be visible after installation, without disturbing the interior parts or wiring.

***TABLE 408.56*** *Minimum Spacings Between Bare Metal Parts*

| AC or DC Voltage | Opposite Polarity Where Mounted on the Same Surface | | Opposite Polarity Where Held Free in Air | | Live Parts to Ground* | |
|---|---|---|---|---|---|---|
| | mm | in. | mm | in. | mm | in. |
| Not over 125 volts, nominal | 19.1 | ¾ | 12.7 | ½ | 12.7 | ½ |
| Not over 250 volts, nominal | 31.8 | 1¼ | 19.1 | ¾ | 12.7 | ½ |
| Not over 1000 volts, nominal | 50.8 | 2 | 25.4 | 1 | 25.4 | 1 |

*For spacing between live parts and doors of cabinets, see 312.11(A)(1), (2), and (3).

# ARTICLE 409 Industrial Control Panels

## I. General

### 409.1 Scope

This article covers industrial control panels intended for general use and operating at 1000 volts or less.

Informational Note: ANSI/UL 508, *Standard for Industrial Control Panels*, is a safety standard for industrial control panels.

Field- and factory-assembled control panels are used for the control and operation of a multitude of processes – from a sewerage pump station to an industrial process line. Similar in function to motor control centers in some regards, control panels also contain control, overcurrent protection, and power distribution equipment for operation of industrial heating processes, robotics, spray painting and powder coating lines, and countless other processes.

•

### 409.3 Other Articles

In addition to the requirements of Article 409, industrial control panels that contain branch circuits for specific loads or components, or are for control of specific types of equipment addressed in other articles of this *Code*, shall be constructed and installed in accordance with the applicable requirements from the specific articles in Table 409.3.

## II. Installation

### 409.20 Conductor — Minimum Size and Ampacity

The size of the industrial control panel supply conductor shall have an ampacity not less than 125 percent of the full-load current rating of all heating loads plus 125 percent of the full-load current rating of the highest rated motor plus the sum of the full-load current ratings of all other connected motors and apparatus based on their duty cycle that may be in operation at the same time.

### 409.21 Overcurrent Protection

**(A) General.** Industrial control panels shall be provided with overcurrent protection in accordance with Parts I, II, and IX of Article 240.

**(B) Location.** This protection shall be provided for each incoming supply circuit by either of the following:

(1) An overcurrent protective device located ahead of the industrial control panel.

***TABLE 409.3*** *Other Articles*

| Equipment/Occupancy | Article | Section |
|---|---|---|
| Branch circuits | 210 | |
| Luminaires | 410 | |
| Motors, motor circuits, and controllers | 430 | |
| Air-conditioning and refrigerating equipment | 440 | |
| Capacitors | | 460.8, 460.9 |
| Hazardous (classified) locations | 500, 501, 502, 503, 504, 505 | |
| Commercial garages; aircraft hangars; motor fuel dispensing facilities; bulk storage plants; spray application, dipping, and coating processes; and inhalation anesthetizing locations | 511, 513, 514, 515, 516, and 517 Part IV | |
| Cranes and hoists | 610 | |
| Electrically driven or controlled irrigation machines | 675 | |
| Elevators, dumbwaiters, escalators, moving walks, wheelchair lifts, and stairway chair lifts | 620 | |
| Industrial machinery | 670 | |
| Resistors and reactors | 470 | |
| Transformers | 450 | |
| Class 1, Class 2, and Class 3 remote-control, signaling, and power-limited circuits | 725 | |

(2) A single main overcurrent protective device located within the industrial control panel. Where overcurrent protection is provided as part of the industrial control panel, the supply conductors shall be considered as either feeders or taps as covered by 240.21.

Each incoming power supply for an industrial control panel must be protected. See the commentary following 409.110(3).

**(C) Rating.** The rating or setting of the overcurrent protective device for the circuit supplying the industrial control panel shall not be greater than the sum of the largest rating or setting of the branch-circuit short-circuit and ground-fault protective device provided with the industrial control panel, plus 125 percent of the full-load current rating of all resistance heating loads, plus the sum of the full-load currents of all other motors and apparatus that could be in operation at the same time.

*Exception: Where one or more instantaneous trip circuit breakers or motor short-circuit protectors are used for motor branch-circuit short-circuit and ground-fault protection as permitted by 430.52(C), the procedure specified above for determining the maximum rating of the protective device for the circuit supplying the industrial control panel shall apply with the following provision: For the purpose of the calculation, each instantaneous trip circuit breaker or motor short-circuit protector shall be assumed to have a rating not exceeding the maximum percentage of motor full-load current permitted by Table 430.52 for the type of control panel supply circuit protective device employed.*

Where no branch-circuit short-circuit and ground-fault protective device is provided with the industrial control panel for motor or combination of motor and non-motor loads, the rating or setting of the overcurrent protective device shall be based on 430.52 and 430.53, as applicable.

### 409.22 Short-Circuit Current Rating

An industrial control panel shall not be installed where the available fault current exceeds its short-circuit current rating as marked in accordance with 409.110(4).

### 409.30 Disconnecting Means

Disconnecting means that supply motor loads shall comply with Part IX of Article 430.

### 409.60 Grounding

Multisection industrial control panels shall be bonded together with an equipment grounding conductor or an equivalent equipment grounding bus sized in accordance with Table 250.122. Equipment grounding conductors shall be connected to this equipment grounding bus or to an equipment grounding termination point provided in a single-section industrial control panel.

## III. Construction Specifications

Part III provides the AHJ with a set of requirements that can be used as a benchmark for approval of a field-constructed control panel.

### 409.100 Enclosures

Table 110.28 shall be used as the basis for selecting industrial control panel enclosures for use in specific locations other than hazardous (classified) locations. The enclosures are not intended to protect against conditions such as condensation, icing, corrosion, or contamination that may occur within the enclosure or enter via the conduit or unsealed openings.

### 409.102 Busbars and Conductors

Industrial control panels utilizing busbars shall comply with 409.102(A) and (B).

**(A) Support and Arrangement.** Busbars shall be protected from physical damage and be held firmly in place.

**(B) Phase Arrangement.** The phase arrangement on 3-phase horizontal common power and vertical buses shall be A, B, C from front to back, top to bottom, or left to right, as viewed from the front of the industrial control panel. The B phase shall be that phase having the higher voltage to ground on 3-phase, 4-wire, delta-connected systems. Other busbar arrangements shall be permitted for additions to existing installations, and the phases shall be permanently marked.

## 409.104 Wiring Space

**(A) General.** Industrial control panel enclosures shall not be used as junction boxes, auxiliary gutters, or raceways for conductors feeding through or tapping off to other switches or overcurrent devices or other equipment, unless the conductors fill less than 40 percent of the cross-sectional area of the wiring space. In addition, the conductors, splices, and taps shall not fill the wiring space at any cross section to more than 75 percent of the cross-sectional area of that space.

**(B) Wire Bending Space.** Wire bending space within industrial control panels for field wiring terminals shall be in accordance with the requirements in 430.10(B).

## 409.106 Spacings

Spacings in feeder circuits between uninsulated live parts of adjacent components, between uninsulated live parts of components and grounded or accessible non–current-carrying metal parts, between uninsulated live parts of components and the enclosure, and at field wiring terminals shall be as shown in Table 430.97(D).

*Exception: Spacings shall be permitted to be less than those specified in Table 430.97(D) at circuit breakers and switches and in listed components installed in industrial control panels.*

## 409.108 Service Equipment

Where used as service equipment, each industrial control panel shall be of the type that is suitable for use as service equipment.

Where a grounded conductor is provided, the industrial control panel shall be provided with a main bonding jumper, sized in accordance with 250.28(D), for connecting the grounded conductor, on its supply side, to the industrial control panel equipment ground bus or equipment ground terminal.

## 409.110 Marking

An industrial control panel shall be marked with the following information that is plainly visible after installation:

(1) Manufacturer's name, trademark, or other descriptive marking by which the organization responsible for the product can be identified.
(2) Supply voltage, number of phases, frequency, and full-load current for each incoming supply circuit.
(3) Industrial control panels supplied by more than one power source such that more than one disconnecting means is required to disconnect all power within the control panel shall be marked to indicate that more than one disconnecting means is required to de-energize the equipment.

The person servicing the industrial control panel may not realize that more than one power source supplies the panel. This requirement is similar to the requirement in Section 55.4 of UL 508A, which is the standard for listed industrial control panels.

(4) Short-circuit current rating of the industrial control panel based on one of the following:
    a. Short-circuit current rating of a listed and labeled assembly
    b. Short-circuit current rating established utilizing an approved method

Informational Note: ANSI/UL 508, *Standard for Industrial Control Panels*, Supplement SB, is an example of an approved method.

*Exception to (4): Short-circuit current rating markings are not required for industrial control panels containing only control circuit components.*

A group of components assembled in a common enclosure for the purposes of operation, control, and overcurrent protection should be able to limit and contain the effects of an internal fault (such as a short circuit or ground fault) so that the internal fault does not pose an external threat.

Without a short-circuit current rating and compliance therewith, a failure could occur that extends beyond the control panel enclosure. In many control panel installations, the available fault energy at the line terminals of components within the control panel is significant. In addition to possible high levels of short-circuit current available at the line terminals, there is also an interaction of the protective and control components under fault conditions that can only be assessed as part of the control panels evaluation under strict conformity assessment guidelines. The determination of the panel's short-circuit current rating should not be confused with the calculation of available short-circuit current at the panel terminals. The former determines what current the panel can withstand, and the latter is to ensure that the available short-circuit current is less than the panel's short-circuit current rating.

(5) If the industrial control panel is intended as service equipment, it shall be marked to identify it as being suitable for use as service equipment.
(6) Electrical wiring diagram or the identification number of a separate electrical wiring diagram or a designation referenced in a separate wiring diagram.
(7) An enclosure type number shall be marked on the industrial control panel enclosure.

# ARTICLE 410 Luminaires, Lampholders, and Lamps

## I. General

### 410.1 Scope

This article covers luminaires, portable luminaires, lampholders, pendants, incandescent filament lamps, arc lamps, electric-discharge lamps, decorative lighting products, lighting accessories for temporary seasonal and holiday use, portable flexible lighting products, and the wiring and equipment forming part of such products and lighting installations.

### 410.2 Definition

**Closet Storage Space.** The volume bounded by the sides and back closet walls and planes extending from the closet floor vertically to a height of 1.8 m (6 ft) or to the highest clothes-hanging rod and parallel to the walls at a horizontal distance of 600 mm (24 in.) from the sides and back of the closet walls, respectively, and continuing vertically to the closet ceiling parallel to the walls at a horizontal distance of 300 mm (12 in.) or the width of the shelf, whichever is greater; for a closet that permits access to both sides of a hanging rod, this space includes the volume below the highest rod extending 300 mm (12 in.) on either side of the rod on a plane horizontal to the floor extending the entire length of the rod. See Figure 410.2.

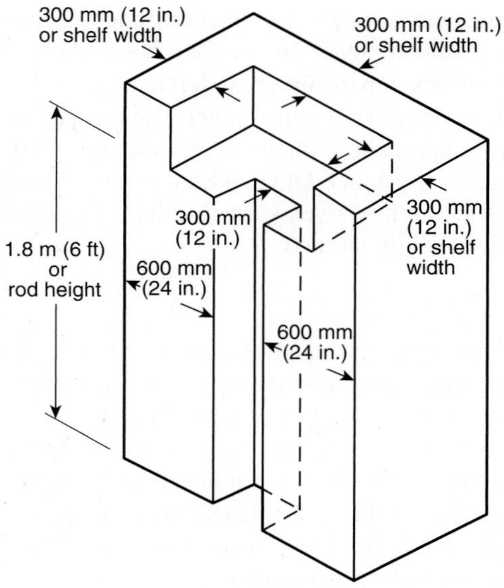


*FIGURE 410.2* *Closet Storage Space.*

The 24-inch dimension in the definition of the term *closet storage space* is intended to cover the clothes-hanging space, even if no clothes-hanging rod is installed. If such a rod is installed, the space extends from the floor to the top of the highest rod. If no clothes-hanging rod is installed, the space extends from the floor to a height of 6 feet.

In addition to the space in which clothing is hung from the closet pole or rod, this definition also establishes a 12-inch-wide shelf space to cover those installations where shelving is not in place at the time of fixture installation. If shelving is installed and the shelves are wider than 12 inches, the greater width must be applied in establishing this space.

The storage space for closets that permit access to both sides of the clothes-hanging rod is based on a horizontal plane extending 12 inches from both sides of the rod, from the rod down to the floor. This equates to the 24-inch space required for the closet rod where there is only one direction of access.

•

### 410.5 Live Parts

Luminaires, portable luminaires, lampholders, and lamps shall have no live parts normally exposed to contact. Exposed accessible terminals in lampholders and switches shall not be installed in metal luminaire canopies or in open bases of portable table or floor luminaires.

*Exception: Cleat-type lampholders located at least 2.5 m (8 ft) above the floor shall be permitted to have exposed terminals.*

### 410.6 Listing Required

All luminaires, lampholders, and retrofit kits shall be listed.

### 410.8 Inspection

Luminaires shall be installed such that the connections between the luminaire conductors and the circuit conductors can be inspected without requiring the disconnection of any part of the wiring unless the luminaires are connected by attachment plugs and receptacles.

## II. Luminaire Locations

### 410.10 Luminaires in Specific Locations

An Underwriters Laboratories Inc. pamphlet entitled *Luminaires Marking and Application Guide,* 2012 provides information on markings to help the AHJ quickly determine whether common types of listed fluorescent, high-intensity discharge, and incandescent fixtures are suitable for the application and are installed correctly.

**(A) Wet and Damp Locations.** Luminaires installed in wet or damp locations shall be installed such that water cannot enter or accumulate in wiring compartments, lampholders, or other electrical parts. All luminaires installed in wet locations shall be marked, "Suitable for Wet Locations." All luminaires installed in damp locations shall be marked "Suitable for Wet Locations" or "Suitable for Damp Locations."

Correct design, construction, and installation of these luminaires will prevent the entrance of rain, snow, ice, and dust. Outdoor parks and parking lots, outdoor recreational areas (tennis, golf, baseball, etc.), car wash areas, and building exteriors are examples of wet locations.

Luminaires in locations protected from the weather and not subject to water saturation but still exposed to moisture must be marked "Suitable for Damp Locations" or "Suitable for Wet Locations." The following are examples of damp locations:

1. The underside of store or gasoline station canopies or theater marquees
2. Some cold-storage warehouses
3. Some agricultural buildings
4. Some basements
5. Roofed open porches and carports

See the definitions of the terms *location, damp*; *location, dry*; and *location, wet* in Article 100.

**(B) Corrosive Locations.** Luminaires installed in corrosive locations shall be of a type suitable for such locations.

**(C) In Ducts or Hoods.** Luminaires shall be permitted to be installed in commercial cooking hoods where all of the following conditions are met:

(1) The luminaire shall be identified for use within commercial cooking hoods and installed such that the temperature limits of the materials used are not exceeded.
(2) The luminaire shall be constructed so that all exhaust vapors, grease, oil, or cooking vapors are excluded from the lamp and wiring compartment. Diffusers shall be resistant to thermal shock.
(3) Parts of the luminaire exposed within the hood shall be corrosion resistant or protected against corrosion, and the surface shall be smooth so as not to collect deposits and to facilitate cleaning.
(4) Wiring methods and materials supplying the luminaire(s) shall not be exposed within the cooking hood.

Informational Note: See 110.11 for conductors and equipment exposed to deteriorating agents.

NFPA 96, *Standard for Ventilation Control and Fire Protection of Commercial Cooking Operations,* provides the minimum fire safety requirements (preventive and operative) related to the design, installation, operation, inspection, and maintenance of all public and private cooking operations, except in single-family residential dwellings. NFPA 96 covers residential cooking equipment where used for purposes other than residential family use — such as employee kitchens or break areas and church and meeting hall kitchens — regardless of frequency of use.

Grease may cause the deterioration of conductor insulation, resulting in short circuits or ground faults in wiring, hence the requirement prohibiting wiring methods and materials (raceways, cables, lampholders) within ducts or hoods. Conventional enclosed and gasketed-type luminaires located in the path of travel of exhaust products are not permitted because a fire could result from the high temperatures on grease-coated glass bowls or globes enclosing the lamps. Recessed or surface gasketed-type luminaires intended for location within hoods must be identified as suitable for the specific purpose and should be installed with the required clearances maintained. Note that wiring systems, including rigid metal conduit, are not permitted to be run exposed within the cooking hood.

For further information, refer to UL 710, *Standard for Safety for Exhaust Hoods for Commercial Cooking Equipment*.

**(D) Bathtub and Shower Areas.** No parts of cord-connected luminaires, chain-, cable-, or cord-suspended luminaires, lighting track, pendants, or ceiling-suspended (paddle) fans shall be located within a zone measured 900 mm (3 ft) horizontally and 2.5 m (8 ft) vertically from the top of the bathtub rim or shower stall threshold. This zone is all encompassing and includes the space directly over the tub or shower stall. Luminaires located within the actual outside dimension of the bathtub or shower to a height of 2.5 m (8 ft) vertically from the top of the bathtub rim or shower threshold shall be marked for damp locations, or marked for wet locations where subject to shower spray.

Where luminaires are subject to shower spray, they must be listed for a wet location. Luminaires installed in the tub or shower zone and not subject to shower spray are required to be listed for use in a damp location. GFCI protection is required only where specified in the installation instructions for the luminaire.

The intent is to keep cord-connected, chain-hanging, or pendant luminaires and suspended fans out of the reach of an individual standing on a bathtub rim. The list of prohibited items recognizes that the same risk of electric shock is present for each one.

Exhibit 410.1 illustrates the restricted zone in which the specified luminaires, lighting track, and paddle fans are prohibited. This requirement applies to hydromassage bathtubs, as defined in 680.2, as well as other bathtub types and shower areas. See 680.43 for installation requirements for spas and hot tubs (as defined in 680.2) installed indoors.

**(E) Luminaires in Indoor Sports, Mixed-Use, and All-Purpose Facilities.** Luminaires subject to physical damage, using a mercury vapor or metal halide lamp, installed in playing and spectator seating areas of indoor sports, mixed-use, or all-purpose facilities shall be of the type that protects the lamp with a glass or plastic lens. Such luminaires shall be permitted to have an additional guard.

Accidental breakage of mercury or metal halide lamp outer jackets in open luminaires has occurred in sports facilities and other

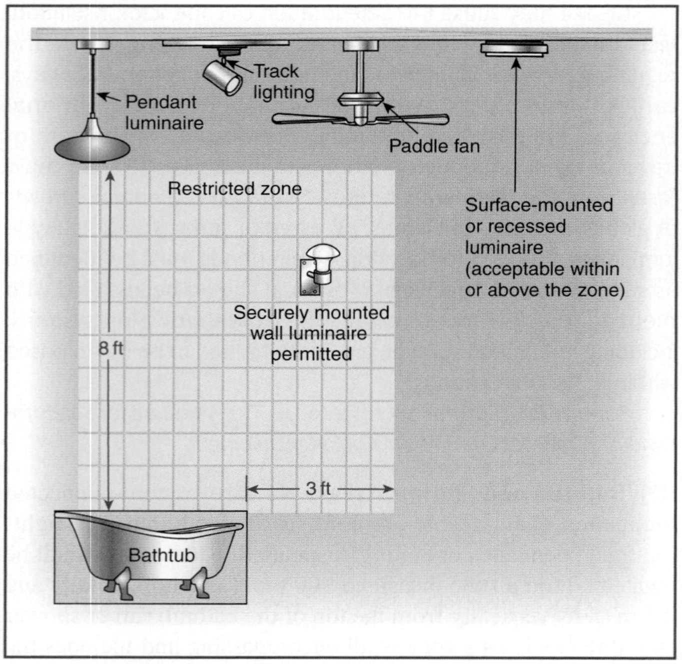


*EXHIBIT 410.1 Luminaires, lighting track, and suspended (paddle) fan located near a bathtub.*

similar locations. If the lamp is damaged, glass shards can fall on players or spectators. If the envelope is damaged, the arc tube may continue to operate even though the outer jacket may be cracked or missing.

**(F) Luminaires Installed in or Under Roof Decking.** Luminaires installed in exposed or concealed locations under metal-corrugated sheet roof decking shall be installed and supported so there is not less than 38 mm (1½ in.) measured from the lowest surface of the roof decking to the top of the luminaire.

This requirement correlates with 300.4(E), which prohibits cables, raceways, and boxes to be installed under metal-corrugated sheet roof decking. See the commentary for 300.4(E). Exhibit 410.2 illustrates an installation where the minimum clearance is not provided between the roof deck and the luminaire.

## 410.11 Luminaires Near Combustible Material

Luminaires shall be constructed, installed, or equipped with shades or guards so that combustible material is not subjected to temperatures in excess of 90°C (194°F).

The requirements of 410.11, 410.12, 410.14, and 410.16 regulate the placement of luminaires near combustible materials so that the luminaires do not become a heat source that could ignite the combustible material.

## 410.12 Luminaires over Combustible Material

Lampholders installed over highly combustible material shall be of the unswitched type. Unless an individual switch is provided for each luminaire, lampholders shall be located at least 2.5 m (8 ft) above the floor or shall be located or guarded so that the lamps cannot be readily removed or damaged.

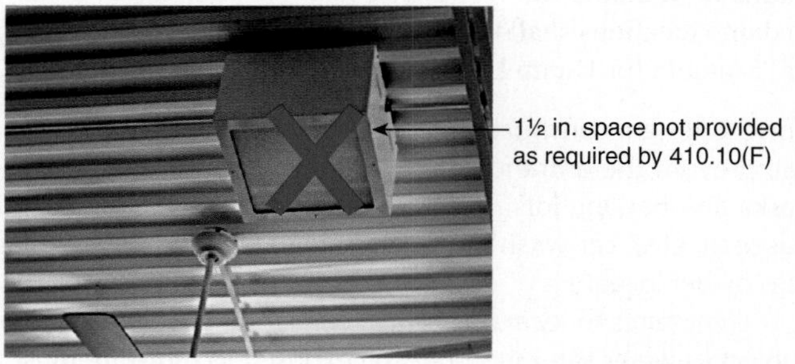


Similar to 300.4(E), 410.10(F) requires luminaires installed under roof decking to have a minimum 1½ in. between the top of the luminare and the decking.

*EXHIBIT 410.2 A minimum 1½ inch clearance is necessary to prevent damage from nail penetration.*

Pendants and fixed lighting equipment may be installed above highly combustible material. If a lamp cannot be located out of reach, the requirement can be met by equipping the lamp with a suitable guard. Section 410.12 does not apply to portable lamps.

## 410.14 Luminaires in Show Windows

Chain-supported luminaires used in a show window shall be permitted to be externally wired. No other externally wired luminaires shall be used.

## 410.16 Luminaires in Clothes Closets

**(A) Luminaire Types Permitted.** Only luminaires of the following types shall be permitted in a closet:

(1) Surface-mounted or recessed incandescent or LED luminaires with completely enclosed light sources
(2) Surface-mounted or recessed fluorescent luminaires
(3) Surface-mounted fluorescent or LED luminaires identified as suitable for installation within the closet storage space

**(B) Luminaire Types Not Permitted.** Incandescent luminaires with open or partially enclosed lamps and pendant luminaires or lampholders shall not be permitted.

**(C) Location.** The minimum clearance between luminaires installed in clothes closets and the nearest point of a closet storage space shall be as follows:

(1) 300 mm (12 in.) for surface-mounted incandescent or LED luminaires with a completely enclosed light source installed on the wall above the door or on the ceiling.
(2) 150 mm (6 in.) for surface-mounted fluorescent luminaires installed on the wall above the door or on the ceiling.

(3) 150 mm (6 in.) for recessed incandescent or LED luminaires with a completely enclosed light source installed in the wall or the ceiling.

A hot filament falling from a broken incandescent lamp can ignite combustible material below the luminaire in which the lamp is installed.

(4) 150 mm (6 in.) for recessed fluorescent luminaires installed in the wall or the ceiling.
(5) Surface-mounted fluorescent or LED luminaires shall be permitted to be installed within the closet storage space where identified for this use.

These requirements are intended to prevent hot lamps or parts of broken lamps from coming in contact with items such as boxes, cartons, and blankets stored on shelves and with clothing hung in closets. The clearance measurement for each requirement in 410.16(C) is to the luminaire, not to the lamp itself.

A luminaire in a clothes closet is not mandatory. If one is installed, however, the conditions for installation are as required by 410.16(C).

### 410.18 Space for Cove Lighting

Coves shall have adequate space and shall be located so that lamps and equipment can be properly installed and maintained.

Adequate space is necessary to allow easy access for relamping luminaires or replacing lampholders, ballasts, and so forth. Adequate space also improves ventilation.

## III. Provisions at Luminaire Outlet Boxes, Canopies, and Pans

### 410.20 Space for Conductors

Canopies and outlet boxes taken together shall provide sufficient space so that luminaire conductors and their connecting devices are capable of being installed in accordance with 314.16.

UL 1598, *Standard for Safety for Luminaires*, allows junction boxes and splice compartments that are integral to luminaires to have reduced free volume in respect to the requirements of 314.16. The reason is that 314.16 applies to general purpose boxes and conduit bodies where installed conductors and devices are variable, while the conductors and devices that will be contained within a luminaire junction box are known.

### 410.21 Temperature Limit of Conductors in Outlet Boxes

Luminaires shall be of such construction or installed so that the conductors in outlet boxes shall not be subjected to temperatures greater than that for which the conductors are rated.

Branch-circuit wiring, other than 2-wire or multiwire branch circuits supplying power to luminaires connected together, shall not be passed through an outlet box that is an integral part of a luminaire unless the luminaire is identified for through-wiring.

Informational Note: See 410.64(C) for wiring supplying power to luminaires connected together.

Branch-circuit conductors run to a lighting outlet box are not permitted to be subjected to temperatures higher than those for which they are rated. Examples of these are conductors that are rated 75°C and that supply a ceiling outlet box for the connection of a surface-mounted luminaire or are attached to the outlet box of a recessed luminaire. The design and installation of the luminaire should be such that the heat of the lamps does not subject the conductors to a temperature greater than 75°C. Recessed luminaires with integral boxes, listed by Underwriters Laboratories Inc., will specify the minimum wire gauge and insulation temperature and maximum number of conductors to account for the heat-contributing factor of the supply conductors.

Exhibit 410.3 illustrates recessed luminaires for one set of supply conductors, and Exhibit 410.4 illustrates luminaires listed for a feed-through installation.

### 410.22 Outlet Boxes to Be Covered

In a completed installation, each outlet box shall be provided with a cover unless covered by means of a luminaire canopy, lampholder, receptacle, or similar device.

### 410.23 Covering of Combustible Material at Outlet Boxes

Any combustible wall or ceiling finish exposed between the edge of a luminaire canopy or pan and an outlet box having a surface area of 1160 mm$^2$ (180 in.$^2$) or more shall be covered with noncombustible material.

Luminaires must be designed and installed not only to prevent overheating of conductors but also to prevent overheating of

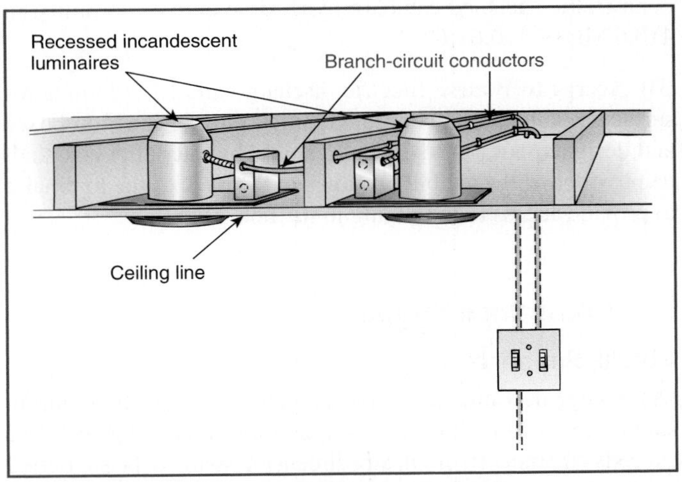


*EXHIBIT 410.3 Recessed luminaires designed for branch-circuit conductors terminating at each luminaire (no feed-through).*

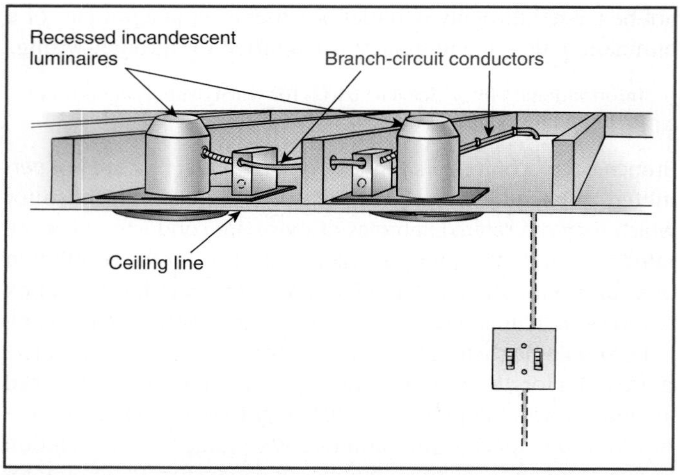


***EXHIBIT 410.4*** *Recessed luminaires designed for feed-through branch-circuit conductors.*

adjacent combustible wall or ceiling finishes. Canopy-style surface or ceiling-mounted luminaires listed to UL 1598 are not required to have a back-plate or back-cover where the total area of the surface being covered by the canopy is less than 180 in.$^2$, because they are evaluated to ensure that temperatures on wall or ceiling surfaces on which the luminaire is mounted do not exceed 90°C. See 314.20 for the requirements covering combustible finishes. Where luminaires are not directly mounted on outlet boxes, suitable outlet box covers are required.

### 410.24 Connection of Electric-Discharge and LED Luminaires

**(A) Independent of the Outlet Box.** Electric-discharge and LED luminaires supported independently of the outlet box shall be connected to the branch circuit through metal raceway, nonmetallic raceway, Type MC cable, Type AC cable, Type MI cable, nonmetallic sheathed cable, or by flexible cord as permitted in 410.62(B) or 410.62(C).

**(B) Access to Boxes.** Electric-discharge and LED luminaires surface mounted over concealed outlet, pull, or junction boxes and designed not to be supported solely by the outlet box shall be provided with suitable openings in the back of the luminaire to provide access to the wiring in the box.

## IV. Luminaire Supports

### 410.30 Supports

**(A) General.** Luminaires and lampholders shall be securely supported. A luminaire that weighs more than 3 kg (6 lb) or exceeds 400 mm (16 in.) in any dimension shall not be supported by the screw shell of a lampholder.

**(B) Metal or Nonmetallic Poles Supporting Luminaires.** Metal or nonmetallic poles shall be permitted to be used to support luminaires and as a raceway to enclose supply conductors, provided the following conditions are met:

(1) A pole shall have a handhole not less than 50 mm × 100 mm (2 in. × 4 in.) with a cover suitable for use in wet locations to provide access to the supply terminations within the pole or pole base.

*Exception No. 1: No handhole shall be required in a pole 2.5 m (8 ft) or less in height abovegrade where the supply wiring method continues without splice or pull point, and where the interior of the pole and any splices are accessible by removing the luminaire.*

This exception typically applies to both landscape (bollard-type) lighting and pole lights at residential dwellings.

*Exception No. 2: No handhole shall be required in a pole 6.0 m (20 ft) or less in height abovegrade that is provided with a hinged base.*

This exception recognizes metal light poles that do not have a handhole but instead use a hinged-base pole to permit access to splices made in the pole base. The height of the pole is limited to 20 feet. The pole and the base must be bonded in accordance with 250.96 (systems operating at 250 volts or less) or 250.97 (circuits operating at over 250 volts), depending on the system voltage.

Exhibit 410.5 illustrates a metal light pole with a hinged baseplate that meets the requirements of Exception No. 2. A handhole is not necessary because the pole can be tilted to allow access to terminations in the base.

(2) Where raceway risers or cable is not installed within the pole, a threaded fitting or nipple shall be brazed, welded, or attached to the pole opposite the handhole for the supply connection.

(3) A metal pole shall be provided with an equipment grounding terminal as follows:

a. A pole with a handhole shall have the equipment grounding terminal accessible from the handhole.

b. A pole with a hinged base shall have the equipment grounding terminal accessible within the base.

*Exception to (3): No grounding terminal shall be required in a pole 2.5 m (8 ft) or less in height abovegrade where the supply wiring method continues without splice or pull, and where the interior of the pole and any splices are accessible by removing the luminaire.*

(4) A metal pole with a hinged base shall have the hinged base and pole bonded together.

(5) Metal raceways or other equipment grounding conductors shall be bonded to the metal pole with an equipment

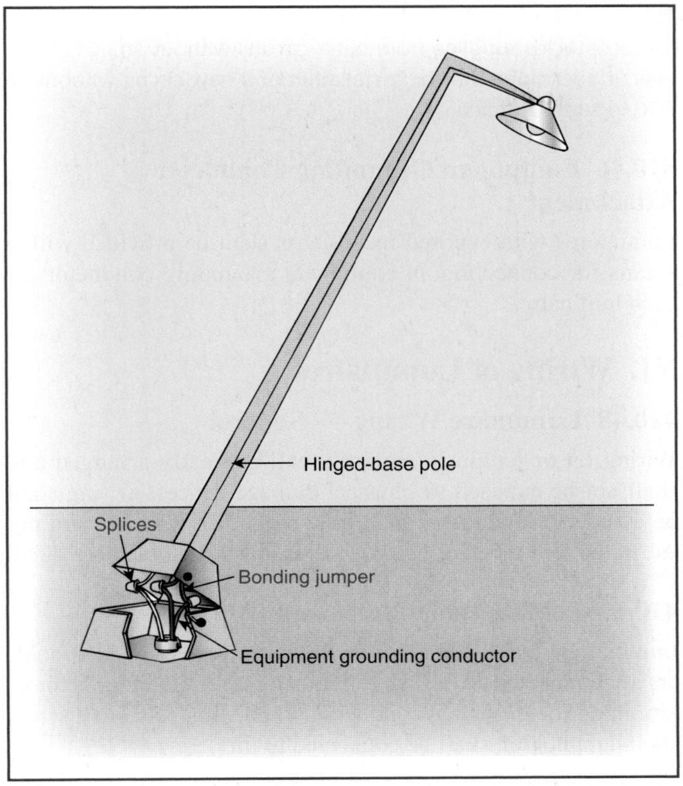


***EXHIBIT 410.5*** *A hinged-base metal pole supporting a luminaire.*

grounding conductor recognized by 250.118 and sized in accordance with 250.122.

(6) Conductors in vertical poles used as raceway shall be supported as provided in 300.19.

Metal poles are permitted to be used as raceways, and individual single conductors can be installed to supply the luminaire supported by the pole. Where individual single conductors are installed in a luminaire pole, separation is required between the power wiring and any communications, signaling, and power-limited circuits that are also installed within the pole. See 725.139, 800.133(A)(1)(d), or 820.133(A)(1)(c), as appropriate.

Where a light pole supports a luminaire and a security camera, the security camera signaling and power-limited wiring may be installed within the pole cavity if kept separated. Because the pole contains open circuit conductors (e.g., THW or XHHW conductors) that supply the luminaire, the separation requirement can be fulfilled by enclosing the camera conductors within a flexible raceway and installing that raceway within the pole. The use of a cable or cord assembly for the lighting circuit conductors is also an option. Section 410.30(B)(6) requires that conductors installed within poles be supported the same as in a vertical raceway in accordance with 300.19.

Section 410.30(B) is not intended to necessarily require the placement of a raceway for communications cables on the exterior of a lighting pole.

### 410.36 Means of Support

**(A) Outlet Boxes.** Outlet boxes or fittings installed as required by 314.23 and complying with the provisions of 314.27(A)(1) and 314.27(A)(2) shall be permitted to support luminaires.

Regardless of whether a luminaire is attached to an outlet box or is supported independently of the outlet box, care should be taken to securely fasten the outlet box or support the independent rod or hanger to ensure that the luminaire is securely mounted in place. The luminaire may be securely mounted to the box; however, if the box is not secured, it becomes the weak link in the luminaire support.

**(B) Suspended Ceilings.** Framing members of suspended ceiling systems used to support luminaires shall be securely fastened to each other and shall be securely attached to the building structure at appropriate intervals. Luminaires shall be securely fastened to the ceiling framing member by mechanical means such as bolts, screws, or rivets. Listed clips identified for use with the type of ceiling framing member(s) and luminaire(s) shall also be permitted.

Clips that are used to support a luminaire to the framing members of a suspended ceiling must be of a type listed for the application. However, the use of listed clips for luminaire support does not complete the requirements of this section. The ceiling framing members must be securely attached to each other and to the building structure. These requirements apply to all luminaires supported by a suspended ceiling assembly, including lay-in and surface-mounted types.

For the support of wiring that is located in the cavity of floor–ceiling assemblies, see 300.11(A).

**(C) Luminaire Studs.** Luminaire studs that are not a part of outlet boxes, hickeys, tripods, and crowfeet shall be made of steel, malleable iron, or other material suitable for the application.

**(D) Insulating Joints.** Insulating joints that are not designed to be mounted with screws or bolts shall have an exterior metal casing, insulated from both screw connections.

**(E) Raceway Fittings.** Raceway fittings used to support a luminaire(s) shall be capable of supporting the weight of the complete fixture assembly and lamp(s).

**(F) Busways.** Luminaires shall be permitted to be connected to busways in accordance with 368.17(C).

**(G) Trees.** Outdoor luminaires and associated equipment shall be permitted to be supported by trees.

Informational Note No. 1: See 225.26 for restrictions for support of overhead conductors.

The support of overhead conductor spans on trees is prohibited by 225.26.

Informational Note No. 2: See 300.5(D) for protection of conductors.

Section 300.5(D) requires buried conductors and cables to be protected from physical damage by the use of raceways from a specified point below grade to a point at least 8 feet above finish grade.

# V. Grounding

## 410.40 General

Luminaires and lighting equipment shall be grounded as required in Article 250 and Part V of this article.

## 410.42 Luminaire(s) with Exposed Conductive Parts

Exposed metal parts shall be connected to an equipment grounding conductor or insulated from the equipment grounding conductor and other conducting surfaces or be inaccessible to unqualified personnel. Lamp tie wires, mounting screws, clips, and decorative bands on glass spaced at least 38 mm (1½ in.) from lamp terminals shall not be required to be grounded.

## 410.44 Methods of Grounding

Luminaires and equipment shall be mechanically connected to an equipment grounding conductor as specified in 250.118 and sized in accordance with 250.122.

*Exception No. 1: Luminaires made of insulating material that is directly wired or attached to outlets supplied by a wiring method that does not provide a ready means for grounding attachment to an equipment grounding conductor shall be made of insulating material and shall have no exposed conductive parts.*

*Exception No. 2: Replacement luminaires shall be permitted to connect an equipment grounding conductor from the outlet in compliance with 250.130(C). The luminaire shall then comply with 410.42.*

Exception No. 2 provides a method by which a luminaire with exposed conductive parts can be installed at an outlet where the wiring method is not an EGC per 250.118, or where the wiring does not provide an EGC. In older installations where luminaires are replaced, the requirement to ground exposed metal parts of the luminaire is not negated simply because no means of grounding is provided by the existing wiring system. The means allowed by the exception is the same as is permitted for receptacles installed at outlets where no grounding means exists. A single grounding conductor can be run independently of the circuit conductors, from the outlet to a point on the wiring system where an effective grounding connection can be made. The acceptable termination points for this separate grounding conductor are specified by 250.130(C).

*Exception No. 3: Where no equipment grounding conductor exists at the outlet, replacement luminaires that are GFCI protected shall not be required to be connected to an equipment grounding conductor.*

This exception provides added protection similar to that provided for receptacles supplied from older circuits without an EGC. However, it does not allow the installation of a new circuit without an EGC to supply luminaires.

## 410.46 Equipment Grounding Conductor Attachment

Luminaires with exposed metal parts shall be provided with a means for connecting an equipment grounding conductor for such luminaires.

# VI. Wiring of Luminaires

## 410.48 Luminaire Wiring — General

Wiring on or within luminaires shall be neatly arranged and shall not be exposed to physical damage. Excess wiring shall be avoided. Conductors shall be arranged so that they are not subjected to temperatures above those for which they are rated.

## 410.50 Polarization of Luminaires

Luminaires shall be wired so that the screw shells of lampholders are connected to the same luminaire or circuit conductor or terminal. The grounded conductor, where connected to a screw shell lampholder, shall be connected to the screw shell.

## 410.52 Conductor Insulation

Luminaires shall be wired with conductors having insulation suitable for the environmental conditions, current, voltage, and temperature to which the conductors will be subjected.

Informational Note: For ampacity of fixture wire, maximum operating temperature, voltage limitations, minimum wire size, and other information, see Article 402.

## 410.54 Pendant Conductors for Incandescent Filament Lamps

**(A) Support.** Pendant lampholders with permanently attached leads, where used for other than festoon wiring, shall be hung from separate stranded rubber-covered conductors that are soldered directly to the circuit conductors but supported independently thereof.

**(B) Size.** Unless part of listed decorative lighting assemblies, pendant conductors shall not be smaller than 14 AWG for mogul-base or medium-base screw shell lampholders or smaller than 18 AWG for intermediate or candelabra-base lampholders.

**(C) Twisted or Cabled.** Pendant conductors longer than 900 mm (3 ft) shall be twisted together where not cabled in a listed assembly.

## 410.56 Protection of Conductors and Insulation

**(A) Properly Secured.** Conductors shall be secured in a manner that does not tend to cut or abrade the insulation.

**(B) Protection Through Metal.** Conductor insulation shall be protected from abrasion where it passes through metal.

**(C) Luminaire Stems.** Splices and taps shall not be located within luminaire arms or stems.

**(D) Splices and Taps.** No unnecessary splices or taps shall be made within or on a luminaire.

Informational Note: For approved means of making connections, see 110.14.

**(E) Stranding.** Stranded conductors shall be used for wiring on luminaire chains and on other movable or flexible parts.

**(F) Tension.** Conductors shall be arranged so that the weight of the luminaire or movable parts does not put tension on the conductors.

## 410.59 Cord-Connected Showcases

Individual showcases, other than fixed, shall be permitted to be connected by flexible cord to permanently installed receptacles, and groups of not more than six such showcases shall be permitted to be coupled together by flexible cord and separable locking-type connectors with one of the group connected by flexible cord to a permanently installed receptacle.

The installation shall comply with 410.59(A) through (E).

**(A) Cord Requirements.** Flexible cord shall be of the hard-service type, having conductors not smaller than the branch-circuit conductors, having ampacity at least equal to the branch-circuit overcurrent device, and having an equipment grounding conductor.

Informational Note: See Table 250.122 for size of equipment grounding conductor.

**(B) Receptacles, Connectors, and Attachment Plugs.** Receptacles, connectors, and attachment plugs shall be of a listed grounding type rated 15 or 20 amperes.

**(C) Support.** Flexible cords shall be secured to the undersides of showcases such that all of the following conditions are ensured:

(1) The wiring is not exposed to physical damage.
(2) The separation between cases is not in excess of 50 mm (2 in.), or more than 300 mm (12 in.) between the first case and the supply receptacle.
(3) The free lead at the end of a group of showcases has a female fitting not extending beyond the case.

**(D) No Other Equipment.** Equipment other than showcases shall not be electrically connected to showcases.

**(E) Secondary Circuit(s).** Where showcases are cord-connected, the secondary circuit(s) of each electric-discharge lighting ballast shall be limited to one showcase.

## 410.62 Cord-Connected Lampholders and Luminaires

**(A) Lampholders.** Where a metal lampholder is attached to a flexible cord, the inlet shall be equipped with an insulating bushing that, if threaded, is not smaller than metric designator 12 (trade size ⅜) pipe size. The cord hole shall be of a size appropriate for the cord, and all burrs and fins shall be removed in order to provide a smooth bearing surface for the cord.

Bushing having holes 7 mm (9⁄32 in.) in diameter shall be permitted for use with plain pendant cord and holes 11 mm (13⁄32 in.) in diameter with reinforced cord.

Metal lampholders (brass- and aluminum-shell type) used with flexible-cord pendants are required to be equipped with smooth and permanently secured insulating bushings. Nonmetallic-type lampholders do not require a bushing, because the material and design afford equivalent protection.

**(B) Adjustable Luminaires.** Luminaires that require adjusting or aiming after installation shall not be required to be equipped with an attachment plug or cord connector, provided the exposed cord is of the hard-usage or extra-hard-usage type and is not longer than that required for maximum adjustment. The cord shall not be subject to strain or physical damage.

**(C) Electric-Discharge and LED Luminaires.**

**(1) Cord-Connected Installation.** A luminaire or a listed assembly shall be permitted to be cord connected if the following conditions apply:

(1) The luminaire is located directly below the outlet or busway.
(2) The flexible cord meets all the following:
   a. Is visible for its entire length outside the luminaire
   b. Is not subject to strain or physical damage
   c. Is terminated in a grounding-type attachment plug cap or busway plug, or is a part of a listed assembly incorporating a manufactured wiring system connector in accordance with 604.6(C), or has a luminaire assembly with a strain relief and canopy having a maximum 152 mm (6 in.) long section of raceway for attachment to an outlet box above a suspended ceiling

Section 410.62(C)(1) applies to listed cord-and-plug-connected LED and electric-discharge luminaires, such as the luminaire illustrated in Exhibit 410.6. The supply cord is not permitted to penetrate a suspended ceiling, because the cord is required to be visible along its entire length.

Section 410.62(C)(1)(2)c permits a listed manufactured wiring system connector that is part of a fabricated assembly to supply the luminaires in place of a grounding-type attachment plug. The last phrase of this paragraph permits certain listed assemblies with a 6-inch maximum section of raceway to be installed above a suspended ceiling.

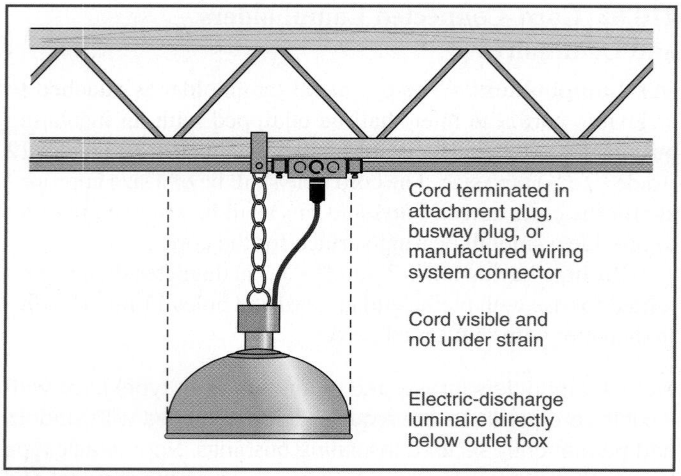


***EXHIBIT 410.6*** *A listed cord-and-plug-connected electric-discharge luminaire.*

Supply cords cannot be used as a supporting means, and the luminaires must be suspended directly below the outlet boxes supplying each luminaire. The luminaires are not permitted to be supplied by cord if they are installed in lift-out-type suspended ceilings. If the luminaire is suspended below the lift-out-type ceiling, the cord is not permitted to penetrate the ceiling, unless it is part of a listed luminaire assembly as described in 410.62(C)(1)(2)c.

Section 400.8 further explains the uses not permitted for cords. According to 368.56(B), luminaires are permitted to be connected to busways by cords plugged directly into the busway.

**(2) Provided with Mogul-Base, Screw Shell Lampholders.** Electric-discharge luminaires provided with mogul-base, screw shell lampholders shall be permitted to be connected to branch circuits of 50 amperes or less by cords complying with 240.5. Receptacles and attachment plugs shall be permitted to be of a lower ampere rating than the branch circuit but not less than 125 percent of the luminaire full-load current.

**(3) Equipped with Flanged Surface Inlet.** Electric-discharge luminaires equipped with a flanged surface inlet shall be permitted to be supplied by cord pendants equipped with cord connectors. Inlets and connectors shall be permitted to be of a lower ampere rating than the branch circuit but not less than 125 percent of the luminaire load current.

## 410.64 Luminaires as Raceways

Luminaires shall not be used as a raceway for circuit conductors unless they comply with 410.64(A), (B), or (C).

This section does not permit luminaires to be used as raceways for circuit conductors unless specifically listed and marked for this use. According to the *UL Luminaires Marking and Application Guide,* 2012, luminaires listed for use as raceways are marked "Suitable for Use as a Raceway" and also with the maximum number, size and type of conductor permitted in raceway. Without these markings, a row of luminaires connected end to end cannot be used as a raceway for circuit conductors other than the 2-wire or multiwire circuit supplying the luminaires. Luminaires identified for use as a raceway have been evaluated for the heat contribution caused by additional current-carrying conductors.

**(A) Listed.** Luminaires listed and marked for use as a raceway shall be permitted to be used as a raceway.

**(B) Through-Wiring.** Luminaires identified for through-wiring, as permitted by 410.21, shall be permitted to be used as a raceway.

**(C) Luminaires Connected Together.** Luminaires designed for end-to-end connection to form a continuous assembly, or luminaires connected together by recognized wiring methods, shall be permitted to contain the conductors of a 2-wire branch circuit, or one multiwire branch circuit, supplying the connected luminaires and shall not be required to be listed as a raceway. One additional 2-wire branch circuit separately supplying one or more of the connected luminaires shall also be permitted.

Informational Note: See Article 100 for the definition of *Multiwire Branch Circuit.*

Section 410.64(C) facilitates convenient switching and supply circuit arrangements for a physically continuous row of luminaires or a row that is made continuous via the wiring method. A single 2-wire or a single multiwire branch circuit supplying the luminaires is permitted to be run through the continuous row(s), and the luminaires are not required to be listed for use as a raceway. An additional 2-wire branch circuit is permitted to be run through these luminaires. This circuit may supply only luminaires in the connected row(s) and is commonly employed to switch night lighting as an energy conservation method.

## 410.68 Feeder and Branch-Circuit Conductors and Ballasts

Feeder and branch-circuit conductors within 75 mm (3 in.) of a ballast, LED driver, power supply, or transformer shall have an insulation temperature rating not lower than 90°C (194°F), unless supplying a luminaire marked as suitable for a different insulation temperature.

Temperature ratings, along with other insulated conductor specifications, are found in Table 310.104(A). Listed LED drivers (including the Class 2 output type) are limited to either 75°C or 90°C, depending on which standard was used to evaluate the device. In many ways, the installation rules established for discharge lighting ballasts over the years carried over to LED drivers. "LED driver" is a common industry term referring to the power supply for the LED.

# VII. Construction of Luminaires

## 410.70 Combustible Shades and Enclosures

Adequate airspace shall be provided between lamps and shades or other enclosures of combustible material.

### 410.74 Luminaire Rating

**(A) Marking.** All luminaires shall be marked with the maximum lamp wattage or electrical rating, manufacturer's name, trademark, or other suitable means of identification. A luminaire requiring supply wire rated higher than 60°C (140°F) shall be marked with the minimum supply wire temperature rating on the luminaire and shipping carton or equivalent.

**(B) Electrical Rating.** The electrical rating shall include the voltage and frequency and shall indicate the current rating of the unit, including the ballast, transformer, LED driver, power supply, or autotransformer.

### 410.82 Portable Luminaires

**(A) General.** Portable luminaires shall be wired with flexible cord recognized by 400.4 and an attachment plug of the polarized or grounding type. Where used with Edison-base lampholders, the grounded conductor shall be identified and attached to the screw shell and the identified blade of the attachment plug.

**(B) Portable Handlamps.** In addition to the provisions of 410.82(A), portable handlamps shall comply with the following:

(1) Metal shell, paper-lined lampholders shall not be used.
(2) Handlamps shall be equipped with a handle of molded composition or other insulating material.
(3) Handlamps shall be equipped with a substantial guard attached to the lampholder or handle.
(4) Metallic guards shall be grounded by means of an equipment grounding conductor run with circuit conductors within the power-supply cord.
(5) Portable handlamps shall not be required to be grounded where supplied through an isolating transformer with an ungrounded secondary of not over 50 volts.

### 410.84 Cord Bushings

A bushing or the equivalent shall be provided where flexible cord enters the base or stem of a portable luminaire. The bushing shall be of insulating material unless a jacketed type of cord is used.

## VIII. Installation of Lampholders

### 410.90 Screw Shell Type

Lampholders of the screw shell type shall be installed for use as lampholders only. Where supplied by a circuit having a grounded conductor, the grounded conductor shall be connected to the screw shell.

The common practice once was to install screw shell lampholders with screw shell adapters in baseboards and walls to connect cord-connected appliances and lighting equipment. This practice (now prohibited) permitted exposed live parts to be contacted by persons when the adapters were removed. See 406.3(B) for permitted uses of receptacles.

### 410.93 Double-Pole Switched Lampholders

Where supplied by the ungrounded conductors of a circuit, the switching device of lampholders of the switched type shall simultaneously disconnect both conductors of the circuit.

Single-pole switching may be used to interrupt the ungrounded conductor of a 2-wire circuit in which one conductor is grounded. The grounded conductor must be connected to the screw shell of the lampholder.

Where a 2-wire circuit is derived from the two ungrounded conductors of a multiwire circuit (3- or 4-wire system) or from the two ungrounded conductors of a 2-wire circuit (3-wire system) and is used with switched lampholders, the switching device is required to be double-pole and to simultaneously disconnect both ungrounded conductors of the circuit.

### 410.96 Lampholders in Wet or Damp Locations

Lampholders installed in wet locations shall be listed for use in wet locations. Lampholders installed in damp locations shall be listed for damp locations or shall be listed for wet locations.

### 410.97 Lampholders Near Combustible Material

Lampholders shall be constructed, installed, or equipped with shades or guards so that combustible material is not subjected to temperatures in excess of 90°C (194°F).

## IX. Lamps and Auxiliary Equipment

### 410.103 Bases, Incandescent Lamps

An incandescent lamp for general use on lighting branch circuits shall not be equipped with a medium base if rated over 300 watts, or with a mogul base if rated over 1500 watts. Special bases or other devices shall be used for over 1500 watts.

### 410.104 Electric-Discharge Lamp Auxiliary Equipment

**(A) Enclosures.** Auxiliary equipment for electric-discharge lamps shall be enclosed in noncombustible cases and treated as sources of heat.

The UL *Guide Information for Electrical Equipment – The White Book* contains two categories for ballasts under Electric Discharge Lamp Control Equipment (FKOT): fluorescent ballasts (FKVS) and HID (high-intensity discharge) ballasts (FLCR).

Fluorescent ballast enclosures are categorized by UL as indoor, outdoor, and weatherproof. Fluorescent ballasts may be an open-type that must be installed within an enclosure or may be enclosed. HID ballasts are categorized the same, except there is no open-type HID ballasts.

**Indoor ballasts** are suitable for use in an indoor, dry location only.

**Outdoor ballasts** are designated as Type 1 or Type 2. Type 2 ballasts are provided with their own enclosure. Both types are suitable for use in outdoor equipment, wet or damp location luminaires, or outdoor signs if the ballasts are within the overall electrical enclosure.

**Weatherproof ballasts** are suitable for use where exposed to the weather without an additional enclosure.

**(B) Switching.** Where supplied by the ungrounded conductors of a circuit, the switching device of auxiliary equipment shall simultaneously disconnect all conductors.

## X. Special Provisions for Flush and Recessed Luminaires

### 410.110 General

Luminaires installed in recessed cavities in walls or ceilings, including suspended ceilings, shall comply with 410.115 through 410.122.

### 410.115 Temperature

**(A) Combustible Material.** Luminaires shall be installed so that adjacent combustible material will not be subjected to temperatures in excess of 90°C (194°F).

**(B) Fire-Resistant Construction.** Where a luminaire is recessed in fire-resistant material in a building of fire-resistant construction, a temperature higher than 90°C (194°F) but not higher than 150°C (302°F) shall be considered acceptable if the luminaire is plainly marked for that service.

**(C) Recessed Incandescent Luminaires.** Incandescent luminaires shall have thermal protection and shall be identified as thermally protected.

Because many recessed incandescent luminaires are suitable for a wide variety of lamp sizes and types and finish trims, the temperature close to the lamp can vary widely. Therefore, many manufacturers have chosen to locate thermal protectors away from the source of heat – such as in the outlet box – and to design the protector so that it detects a change in temperature resulting from the addition of thermal insulation around the luminaire. This design prevents nuisance tripping of the protector (as a result of changing lamp wattage, for example) but still provides protection against overheating arising from thermal insulation around a recessed luminaire not designed for such use.

*Exception No. 1: Thermal protection shall not be required in a recessed luminaire identified for use and installed in poured concrete.*

*Exception No. 2: Thermal protection shall not be required in a recessed luminaire whose design, construction, and thermal performance characteristics are equivalent to a thermally protected luminaire and are identified as inherently protected.*

Recessed incandescent luminaires without thermal protection are permitted by Exception No. 2 if they are listed and identified as providing equivalent temperature protection by construction design.

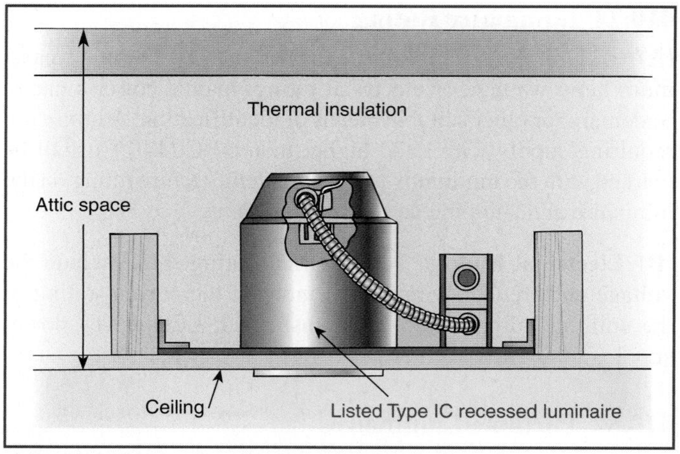


***EXHIBIT 410.7*** *A listed Type IC recessed luminaire suitable for use in insulated ceilings and installed in direct contact with thermal insulation.*

Exhibit 410.7 illustrates a listed Type IC recessed luminaire installed in direct contact with thermal insulation. Thermal protection is provided to deactivate the lamp should the luminaire be mislamped so that it overheats.

### 410.116 Clearance and Installation

**(A) Clearance.**

**(1) Non-Type IC.** A recessed luminaire that is not identified for contact with insulation shall have all recessed parts spaced not less than 13 mm (½ in.) from combustible materials. The points of support and the trim finishing off the openings in the ceiling, wall, or other finished surface shall be permitted to be in contact with combustible materials.

**(2) Type IC.** A recessed luminaire that is identified for contact with insulation, Type IC, shall be permitted to be in contact with combustible materials at recessed parts, points of support, and portions passing through or finishing off the opening in the building structure.

**(B) Installation.** Thermal insulation shall not be installed above a recessed luminaire or within 75 mm (3 in.) of the recessed luminaire's enclosure, wiring compartment, ballast, transformer, LED driver, or power supply unless the luminaire is identified as Type IC for insulation contact.

LED luminaires for installation in contact with thermal insulation must be identified as Type IC, which is similar to the requirements for other luminaires.

### 410.117 Wiring

**(A) General.** Conductors that have insulation suitable for the temperature encountered shall be used.

**(B) Circuit Conductors.** Branch-circuit conductors that have an insulation suitable for the temperature encountered shall be permitted to terminate in the luminaire.

**(C) Tap Conductors.** Tap conductors of a type suitable for the temperature encountered shall be permitted to run from the luminaire terminal connection to an outlet box placed at least 300 mm (1 ft) from the luminaire. Such tap conductors shall be in suitable raceway or Type AC or MC cable of at least 450 mm (18 in.) but not more than 1.8 m (6 ft) in length.

## XI. Construction of Flush and Recessed Luminaires

### 410.118 Temperature

Luminaires shall be constructed such that adjacent combustible material is not subject to temperatures in excess of 90°C (194°F).

### 410.120 Lamp Wattage Marking

Incandescent lamp luminaires shall be marked to indicate the maximum allowable wattage of lamps. The markings shall be permanently installed, in letters at least 6 mm (¼ in.) high, and shall be located where visible during relamping.

### 410.121 Solder Prohibited

No solder shall be used in the construction of a luminaire recessed housing.

### 410.122 Lampholders

Lampholders of the screw shell type shall be of porcelain or other suitable insulating materials.

## XII. Special Provisions for Electric-Discharge Lighting Systems of 1000 Volts or Less

### 410.130 General

**(A) Open-Circuit Voltage of 1000 Volts or Less.** Equipment for use with electric-discharge lighting systems and designed for an open-circuit voltage of 1000 volts or less shall be of a type identified for such service.

**(B) Considered as Energized.** The terminals of an electric-discharge lamp shall be considered as energized where any lamp terminal is connected to a circuit of over 300 volts.

**(C) Transformers of the Oil-Filled Type.** Transformers of the oil-filled type shall not be used.

**(D) Additional Requirements.** In addition to complying with the general requirements for luminaires, such equipment shall comply with Part XII of this article.

**(E) Thermal Protection — Fluorescent Luminaires.**

**(1) Integral Thermal Protection.** The ballast of a fluorescent luminaire installed indoors shall have integral thermal protection. Replacement ballasts shall also have thermal protection integral with the ballast.

**(2) Simple Reactance Ballasts.** A simple reactance ballast in a fluorescent luminaire with straight tubular lamps shall not be required to be thermally protected.

**(3) Exit Luminaires.** A ballast in a fluorescent exit luminaire shall not have thermal protection.

**(4) Egress Luminaires.** A ballast in a fluorescent luminaire that is used for egress lighting and energized only during a failure of the normal supply shall not have thermal protection.

Thermal protection that is integral with the ballast is required for fluorescent luminaires installed indoors. Thermally protected ballasts are also required as replacements for nonthermally protected ballasts in older fixtures. Thermally protected fluorescent lamp ballasts intended for use in accordance with 410.130(E) are marked "Class P." LED drivers are the LED lighting system's equivalent to ballasts. "LED driver" is a common industry term referring to the power supply for the LED.

Because different Class P ballasts have different heating characteristics, the heating characteristics should be considered when selecting replacements for nonthermally protected ballasts. This type of ballast protection is set to open the circuit at a predetermined temperature, to prevent abnormal ballast heat buildup caused by a fault in one or more of the ballast components or by some lampholder or wiring fault.

Exit sign fixtures are exempt from the thermal protection requirement, because overheating during high ambient conditions could cause the thermal protection to operate. This action could impair evacuation during a fire. Egress lighting is also exempt from the thermal protection requirement for the same reason that exit signs are exempt. However, this exemption applies to egress lighting that is energized only during the emergency condition.

Exhibit 410.8 illustrates a reactance-type ballast used in series with a preheat-type fluorescent lamp 30 watts or less. This type of ballast does not require thermal protection, and the luminaire may be equipped with automatic-type starters (such as used with medicine cabinet luminaires) or a manual momentary contact starter (such as used with desk lamps and some small under-cabinet luminaires).

**(F) High-Intensity Discharge Luminaires.**

**(1) Recessed.** Recessed high-intensity luminaires designed to be installed in wall or ceiling cavities shall have thermal protection and be identified as thermally protected.

**(2) Inherently Protected.** Thermal protection shall not be required in a recessed high-intensity luminaire whose design, construction, and thermal performance characteristics are equivalent to a thermally protected luminaire and are identified as inherently protected.

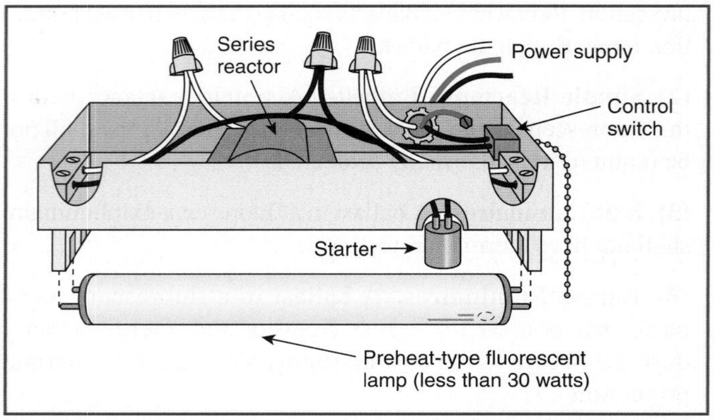


***EXHIBIT 410.8*** *The circuitry for a simple reactance-type ballast for fluorescent lighting.*

**(3) Installed in Poured Concrete.** Thermal protection shall not be required in a recessed high-intensity discharge luminaire identified for use and installed in poured concrete.

**(4) Recessed Remote Ballasts.** A recessed remote ballast for a high-intensity discharge luminaire shall have thermal protection that is integral with the ballast and shall be identified as thermally protected.

**(5) Metal Halide Lamp Containment.** Luminaires that use a metal halide lamp other than a thick-glass parabolic reflector lamp (PAR) shall be provided with a containment barrier that encloses the lamp, or shall be provided with a physical means that only allows the use of a lamp that is Type O.

Metal halide lamps have been identified as the likely cause of ignition in several major fires by insurers of large industrial facilities. HID luminaires can create an ignition source when physical damage occurs to the arc tube in open luminaires. Lamp types that are suitable for use in open luminaires (those that do not require lamp enclosures) are classified by the lamp manufacturers as Type O or Type S. Type-O lamps are provided with a shroud around the arc tube and are containment-tested in accordance with ANSI C78.387, *Electric Lamps – Metal-Halide Lamps – Methods of Measuring Characteristics*, and rated for use in open luminaires. The *Code* requires luminaires that use a metal halide lamp to have either a containment barrier that encloses the lamp or some means that allows only a Type-O lamp to be installed.

Informational Note: See ANSI Standard C78.389, *American National Standard for Electric Lamps — High Intensity Discharge, Methods of Measuring Characteristics.*

**(G) Disconnecting Means.**

**(1) General.** In indoor locations other than dwellings and associated accessory structures, fluorescent luminaires that utilize double-ended lamps and contain ballast(s) that can be serviced in place shall have a disconnecting means either internal or external to each luminaire. For existing installed luminaires without disconnecting means, at the time a ballast is replaced, a disconnecting means shall be installed. The line side terminals of the disconnecting means shall be guarded.

*Exception No. 1: A disconnecting means shall not be required for luminaires installed in hazardous (classified) location(s).*

*Exception No. 2: A disconnecting means shall not be required for emergency illumination required in 700.16.*

*Exception No. 3: For cord-and-plug-connected luminaires, an accessible separable connector or an accessible plug and receptacle shall be permitted to serve as the disconnecting means.*

•

*Exception No. 4: Where more than one luminaire is installed and supplied by other than a multiwire branch circuit, a disconnecting means shall not be required for every luminaire when the design of the installation includes disconnecting means, such that the illuminated space cannot be left in total darkness.*

**(2) Multiwire Branch Circuits.** When connected to multiwire branch circuits, the disconnecting means shall simultaneously break all the supply conductors to the ballast, including the grounded conductor.

**(3) Location.** The disconnecting means shall be located so as to be accessible to qualified persons before servicing or maintaining the ballast. Where the disconnecting means is external to the luminaire, it shall be a single device, and shall be attached to the luminaire or the luminaire shall be located within sight of the disconnecting means.

The disconnect can be either inside or outside the luminaire and must disconnect all supply conductors simultaneously, including the grounded conductor. Where the disconnecting means is external to the luminaire, it must be a single device either attached to the luminaire or within sight of the luminaire. Four exceptions to the required disconnect are provided. In addition to the ungrounded conductors, the grounded or neutral conductor of a multiwire branch circuit is required to be disconnected. A disconnecting means is required be installed when a ballast is replaced in a luminaire that does not have either an internal or an external disconnecting means.

## 410.134 Direct-Current Equipment

Luminaires installed on dc circuits shall be equipped with auxiliary equipment and resistors designed for dc operation. The luminaires shall be marked for dc operation.

## 410.135 Open-Circuit Voltage Exceeding 300 Volts

Equipment having an open-circuit voltage exceeding 300 volts shall not be installed in dwelling occupancies unless such equipment is designed so that there will be no exposed live parts when lamps are being inserted, are in place, or are being removed.

Luminaires intended for use in nondwelling occupancies are so marked, usually indicating that the luminaire has maintenance features beyond the capabilities of the ordinary homeowner or that the luminaire involves voltages in excess of those permitted by this *Code* for dwelling occupancies. For other references to voltage limitations within dwelling units, see 210.6(A) and 410.140(B).

### 410.136 Luminaire Mounting

**(A) Exposed Components.** Luminaires that have exposed ballasts, transformers, LED drivers, or power supplies shall be installed such that ballasts, transformers, LED drivers, or power supplies shall not be in contact with combustible material unless listed for such condition.

**(B) Combustible Low-Density Cellulose Fiberboard.** Where a surface-mounted luminaire containing a ballast, transformer, LED driver, or power supply is to be installed on combustible low-density cellulose fiberboard, it shall be marked for this condition or shall be spaced not less than 38 mm (1½ in.) from the surface of the fiberboard. Where such luminaires are partially or wholly recessed, the provisions of 410.110 through 410.122 shall apply.

Informational Note: Combustible low-density cellulose fiberboard includes sheets, panels, and tiles that have a density of 320 kg/m$^3$ (20 lb/ft$^3$) or less and that are formed of bonded plant fiber material but does not include solid or laminated wood or fiberboard that has a density in excess of 320 kg/m$^3$ (20 lb/ft$^3$) or is a material that has been integrally treated with fire-retarding chemicals to the degree that the flame spread index in any plane of the material will not exceed 25, determined in accordance with tests for surface burning characteristics of building materials. See ANSI/ASTM E84-2011b, *Test Method for Surface Burning Characteristics of Building Materials.*

Luminaires intended for mounting on combustible low-density cellulose fiberboard ceilings have been evaluated with thermal insulation above the ceiling in the vicinity of the luminaire and are marked "Suitable for Surface Mounting on Combustible Low-Density Cellulose Fiberboard."

Luminaires not so marked may be directly mounted against a ceiling surface constructed of a material other than combustible low-density fiberboard or may be spaced not less than 1½ inches from the surface of the low-density fiberboard.

### 410.137 Equipment Not Integral with Luminaire

**(A) Metal Cabinets.** Auxiliary equipment, including reactors, capacitors, resistors, and similar equipment, where not installed as part of a luminaire assembly, shall be enclosed in accessible, permanently installed metal cabinets.

**(B) Separate Mounting.** Separately mounted ballasts, transformers, LED drivers, or power supplies that are listed for direct connection to a wiring system shall not be required to be additionally enclosed.

**(C) Wired Luminaire Sections.** Wired luminaire sections are paired, with a ballast(s) supplying a lamp or lamps in both. For interconnection between paired units, it shall be permissible to use metric designator 12 (trade size ⅜) flexible metal conduit in lengths not exceeding 7.5 m (25 ft), in conformance with Article 348. Luminaire wire operating at line voltage, supplying only the ballast(s) of one of the paired luminaires shall be permitted in the same raceway as the lamp supply wires of the paired luminaires.

Wired luminaire sections are shipped in pairs and marked for use in pairs. Each individual unit includes lamps in odd-numbered quantities (one or three is most common), with the odd lamp in each luminaire supplied by a two-lamp ballast located in one luminaire of the pair. Two-lamp ballasts are more energy efficient than single-lamp or three-lamp ballasts.

### 410.138 Autotransformers

An autotransformer that is used to raise the voltage to more than 300 volts, as part of a ballast for supplying lighting units, shall be supplied only by a grounded system.

### 410.139 Switches

Snap switches shall comply with 404.14.

## XIII. Special Provisions for Electric-Discharge Lighting Systems of More Than 1000 Volts

### 410.140 General

**(A) Listing.** Electric-discharge lighting systems with an open-circuit voltage exceeding 1000 volts shall be listed and installed in conformance with that listing.

**(B) Dwelling Occupancies.** Equipment that has an open-circuit voltage exceeding 1000 volts shall not be installed in or on dwelling occupancies.

**(C) Live Parts.** The terminal of an electric-discharge lamp shall be considered as a live part.

**(D) Additional Requirements.** In addition to complying with the general requirements for luminaires, such equipment shall comply with Part XIII of this article.

Informational Note: For signs and outline lighting, see Article 600.

### 410.141 Control

**(A) Disconnection.** Luminaires or lamp installation shall be controlled either singly or in groups by an externally operable switch or circuit breaker that opens all ungrounded primary conductors.

**(B) Within Sight or Locked Type.** The switch or circuit breaker shall be located within sight from the luminaires or lamps, or it shall be permitted to be located elsewhere if it is lockable in accordance with 110.25.

### 410.142 Lamp Terminals and Lampholders

Parts that must be removed for lamp replacement shall be hinged or held captive. Lamps or lampholders shall be designed so that there are no exposed live parts when lamps are being inserted or removed.

### 410.143 Transformers

**(A) Type.** Transformers shall be enclosed, identified for the use, and listed.

**(B) Voltage.** The secondary circuit voltage shall not exceed 15,000 volts, nominal, under any load condition. The voltage to ground of any output terminals of the secondary circuit shall not exceed 7500 volts under any load conditions.

**(C) Rating.** Transformers shall have a secondary short-circuit current rating of not more than 150 mA if the open-circuit voltage is over 7500 volts, and not more than 300 mA if the open-circuit voltage rating is 7500 volts or less.

**(D) Secondary Connections.** Secondary circuit outputs shall not be connected in parallel or in series.

### 410.144 Transformer Locations

**(A) Accessible.** Transformers shall be accessible after installation.

**(B) Secondary Conductors.** Transformers shall be installed as near to the lamps as practicable to keep the secondary conductors as short as possible.

**(C) Adjacent to Combustible Materials.** Transformers shall be located so that adjacent combustible materials are not subjected to temperatures in excess of 90°C (194°F).

### 410.145 Exposure to Damage

Lamps shall not be located where normally exposed to physical damage.

### 410.146 Marking

Each luminaire or each secondary circuit of tubing having an open-circuit voltage of over 1000 volts shall have a clearly legible marking in letters not less than 6 mm (¼ in.) high reading "Caution ____ volts." The voltage indicated shall be the rated open-circuit voltage. The caution sign(s) or label(s) shall comply with 110.21(B).

## XIV. Lighting Track

### 410.151 Installation

**(A) Lighting Track.** Lighting track shall be permanently installed and permanently connected to a branch circuit. Only lighting track fittings shall be installed on lighting track. Lighting track fittings shall not be equipped with general-purpose receptacles.

A lighting track fitting differs from a fitting as defined in Article 100, in that it usually performs both an electrical and a mechanical function. Such assemblies are not intended to be used for locating convenience receptacles or as an alternative for required receptacle outlets such as those required in 210.62 for show windows. Lighting track can be removed and relocated and, therefore, is not a substitute for required receptacles.

**(B) Connected Load.** The connected load on lighting track shall not exceed the rating of the track. Lighting track shall be supplied by a branch circuit having a rating not more than that of the track. The load calculation in 220.43(B) shall not be required to limit the length of track on a single branch circuit, and it shall not be required to limit the number of luminaires on a single track.

Section 220.43(B) is intended to be used for load calculations of feeders and services. It does not limit the length of track or number of installed luminaires. See the example following 220.43(B) for load calculation method.

•

**(C) Locations Not Permitted.** Lighting track shall not be installed in the following locations:

(1) Where likely to be subjected to physical damage
(2) In wet or damp locations
(3) Where subject to corrosive vapors
(4) In storage battery rooms
(5) In hazardous (classified) locations
(6) Where concealed
(7) Where extended through walls or partitions
(8) Less than 1.5 m (5 ft) above the finished floor except where protected from physical damage or track operating at less than 30 volts rms open-circuit voltage
(9) Where prohibited by 410.10(D)

**(D) Support.** Fittings identified for use on lighting track shall be designed specifically for the track on which they are to be installed. They shall be securely fastened to the track, shall maintain polarization and connections to the equipment grounding conductor, and shall be designed to be suspended directly from the track.

### 410.153 Heavy-Duty Lighting Track

Heavy-duty lighting track is lighting track identified for use exceeding 20 amperes. Each fitting attached to a heavy-duty lighting track shall have individual overcurrent protection.

### 410.154 Fastening

Lighting track shall be securely mounted so that each fastening is suitable for supporting the maximum weight of luminaires that can be installed. Unless identified for supports at greater intervals, a single section 1.2 m (4 ft) or shorter in length shall have two supports, and, where installed in a continuous row, each individual section of not more than 1.2 m (4 ft) in length shall have one additional support.

### 410.155 Construction Requirements

**(A) Construction.** The housing for the lighting track system shall be of substantial construction to maintain rigidity. The conductors shall be installed within the track housing, permitting insertion of a luminaire, and designed to prevent tampering and accidental contact with live parts. Components of lighting track systems of different voltages shall not be interchangeable. The track conductors shall be a minimum 12 AWG or equal and shall be copper. The track system ends shall be insulated and capped.

**(B) Grounding.** Lighting track shall be grounded in accordance with Article 250, and the track sections shall be securely coupled to maintain continuity of the circuitry, polarization, and grounding throughout.

## XV. Decorative Lighting and Similar Accessories

### 410.160 Listing of Decorative Lighting

Decorative lighting and similar accessories used for holiday lighting and similar purposes, in accordance with 590.3(B), shall be listed.

# ARTICLE 411 Lighting Systems Operating at 30 Volts or Less and Lighting Equipment Connected to Class-2 Power Sources

### 411.1 Scope

This article covers lighting systems operating at 30 volts or less and their associated components. This article also covers lighting equipment connected to a Class 2 power source.

Article 411 covers low-voltage interior and exterior (landscape) lighting systems consisting of a maximum 30-volt isolating power supply or a Class 2 power supply rated in conformance with Chapter 9, Tables 11(A) or 11(B).

### 411.3 Low-Voltage Lighting Systems

**(A) General.** Lighting systems operating at 30 volts or less shall consist of an isolating power supply, low-voltage luminaires, and associated equipment that are all identified for the use. The output circuits of the power supply shall be rated for 25 amperes and 30 volts (42.4 volts peak) maximum under all load conditions.

**(B) Class 2.** Listed Class 2 lighting equipment shall be rated in conformance with Chapter 9, Table 11(A) or Table 11(B).

### 411.4 Listing Required

Lighting systems operating at 30 volts or less shall comply with 411.4(A) or 411.4(B). Class 2 power sources and lighting equipment connected to Class 2 power sources shall be listed.

**(A) Listed System.** Lighting systems operating at 30 volts or less shall be listed as a complete system. The luminaires, power supply, and luminaire fittings (including the exposed bare conductors) of an exposed bare conductor lighting system shall be listed for the use as part of the same identified lighting system.

**(B) Assembly of Listed Parts.** A lighting system assembled from the following listed parts shall be permitted:

(1) Low-voltage luminaires
(2) Low-voltage luminaire power supply

•

(3) Low-voltage luminaire fittings
(4) Cord (secondary circuit) for which the luminaires and power supply are listed for use
(5) Cable, conductors in conduit, or other fixed wiring method for the secondary circuit

The luminaires, power supply, and luminaire fittings (including the exposed bare conductors) of an exposed bare conductor lighting system shall be listed for use as part of the same identified lighting system.

A lighting system may be a complete listed system or an assembly of listed parts. Lighting systems operating at 30 volts or less have long been field assembled from individually listed low-voltage luminaires, listed luminaire power units, listed cord, and any involved listed luminaire fittings. Installers typically verify that individually listed lighting system parts (regularly from multiple manufacturers) are intended for the use and have the needed ratings to create and assemble a low-voltage lighting system.

### 411.5 Specific Location Requirements

**(A) Walls, Floors, and Ceilings.** Conductors concealed or extended through a wall, floor, or ceiling shall be in accordance with (1) or (2):

(1) Installed using any of the wiring methods specified in Chapter 3
(2) Installed using wiring supplied by a listed Class 2 power source and installed in accordance with 725.130

**(B) Pools, Spas, Fountains, and Similar Locations.** Lighting systems shall be installed not less than 3 m (10 ft) horizontally from the nearest edge of the water, unless permitted by Article 680.

The installation requirements of 411.5 recognize that shock and fire hazards still exist, even with low-voltage systems.

### 411.6 Secondary Circuits

**(A) Grounding.** Secondary circuits shall not be grounded.

**(B) Isolation.** The secondary circuit shall be insulated from the branch circuit by an isolating transformer.

**(C) Bare Conductors.** Exposed bare conductors and current-carrying parts shall be permitted for indoor installations only. Bare conductors shall not be installed less than 2.1 m (7 ft) above the finished floor, unless specifically listed for a lower installation height.

**(D) Insulated Conductors.** Exposed insulated secondary circuit conductors shall be of the type, and installed as, described in (1), (2), or (3):

(1) Class 2 cable supplied by a Class 2 power source and installed in accordance with Parts I and III of Article 725.
(2) Conductors, cord, or cable of the listed system and installed not less than 2.1 m (7 ft) above the finished floor unless the system is specifically listed for a lower installation height.
(3) Wiring methods described in Chapter 3.

### 411.7 Branch Circuit

Lighting systems covered by this article shall be supplied from a maximum 20-ampere branch circuit.

### 411.8 Hazardous (Classified) Locations

Where installed in hazardous (classified) locations, these systems shall conform with Articles 500 through 517 in addition to this article.

# ARTICLE 422 Appliances

## I. General

### 422.1 Scope

This article covers electrical appliances used in any occupancy.

Article 422 covers appliances that may be fastened in place or cord-and-plug-connected, such as air-conditioning units, dishwashers, heating appliances, water heaters, and infrared heating lamps. See 422.3 for the requirements of other articles as well as Article 100 for the definition of *appliance*.

### 422.2 Definition

**Vending Machine.** Any self-service device that dispenses products or merchandise without the necessity of replenishing the device between each vending operation and is designed to require insertion of coin, paper currency, token, card, key, or receipt of payment by other means.

### 422.3 Other Articles

The requirements of Article 430 shall apply to the installation of motor-operated appliances, and the requirements of Article 440 shall apply to the installation of appliances containing a hermetic refrigerant motor-compressor(s), except as specifically amended in this article.

### 422.4 Live Parts

Appliances shall have no live parts normally exposed to contact other than those parts functioning as open-resistance heating elements, such as the heating element of a toaster, which are necessarily exposed.

### 422.5 Ground-Fault Circuit-Interrupter (GFCI) Protection

The device providing GFCI protection required in this article shall be readily accessible.

GFCI-type receptacles require monthly testing. This requirement facilitates periodic testing as well as resetting a tripped device.

## II. Installation

### 422.10 Branch-Circuit Rating

This section specifies the ratings of branch circuits capable of carrying appliance current without overheating under the conditions specified.

Conductors that form integral parts of appliances are tested as part of the listing or labeling process.

**(A) Individual Circuits.** The rating of an individual branch circuit shall not be less than the marked rating of the appliance or the marked rating of an appliance having combined loads as provided in 422.62.

The rating of an individual branch circuit for motor-operated appliances not having a marked rating shall be in accordance with Part II of Article 430.

The branch-circuit rating for an appliance that is a continuous load, other than a motor-operated appliance, shall not be less than 125 percent of the marked rating, or not less than 100 percent of the marked rating if the branch-circuit device and its assembly are listed for continuous loading at 100 percent of its rating.

Branch circuits and branch-circuit conductors for household ranges and cooking appliances shall be permitted to be in accordance with Table 220.55 and shall be sized in accordance with 210.19(A)(3).

**(B) Circuits Supplying Two or More Loads.** For branch circuits supplying appliance and other loads, the rating shall be determined in accordance with 210.23.

## 422.11 Overcurrent Protection

Appliances shall be protected against overcurrent in accordance with 422.11(A) through (G) and 422.10.

**(A) Branch-Circuit Overcurrent Protection.** Branch circuits shall be protected in accordance with 240.4.

If a protective device rating is marked on an appliance, the branch-circuit overcurrent device rating shall not exceed the protective device rating marked on the appliance.

A labeled or listed appliance is provided with installation instructions from the manufacturer. The branch-circuit size is not permitted to be less than the minimum size stated in the installation instructions. See 110.3(B) and its related commentary regarding the installation and use of listed or labeled equipment.

**(B) Household-Type Appliances with Surface Heating Elements.** Household-type appliances with surface heating elements having a maximum demand of more than 60 amperes calculated in accordance with Table 220.55 shall have their power supply subdivided into two or more circuits, each of which shall be provided with overcurrent protection rated at not over 50 amperes.

**(C) Infrared Lamp Commercial and Industrial Heating Appliances.** Infrared lamp commercial and industrial heating appliances shall have overcurrent protection not exceeding 50 amperes.

**(D) Open-Coil or Exposed Sheathed-Coil Types of Surface Heating Elements in Commercial-Type Heating Appliances.** Open-coil or exposed sheathed-coil types of surface heating elements in commercial-type heating appliances shall be protected by overcurrent protective devices rated at not over 50 amperes.

**(E) Single Non–Motor-Operated Appliance.** If the branch circuit supplies a single non–motor-operated appliance, the rating of overcurrent protection shall comply with the following:

(1) Not exceed that marked on the appliance.
(2) Not exceed 20 amperes if the overcurrent protection rating is not marked and the appliance is rated 13.3 amperes or less; or
(3) Not exceed 150 percent of the appliance rated current if the overcurrent protection rating is not marked and the appliance is rated over 13.3 amperes. Where 150 percent of the appliance rating does not correspond to a standard overcurrent device ampere rating, the next higher standard rating shall be permitted.

**(F) Electric Heating Appliances Employing Resistance-Type Heating Elements Rated More Than 48 Amperes.**

**(1) Electric Heating Appliances.** Electric heating appliances employing resistance-type heating elements rated more than 48 amperes, other than household appliances with surface heating elements covered by 422.11(B), and commercial-type heating appliances covered by 422.11(D), shall have the heating elements subdivided. Each subdivided load shall not exceed 48 amperes and shall be protected at not more than 60 amperes.

These supplementary overcurrent protective devices shall be (1) factory-installed within or on the heater enclosure or provided as a separate assembly by the heater manufacturer; (2) accessible; and (3) suitable for branch-circuit protection.

The main conductors supplying these overcurrent protective devices shall be considered branch-circuit conductors.

**(2) Commercial Kitchen and Cooking Appliances.** Commercial kitchen and cooking appliances using sheathed-type heating elements not covered in 422.11(D) shall be permitted to be subdivided into circuits not exceeding 120 amperes and protected at not more than 150 amperes where one of the following is met:

(1) Elements are integral with and enclosed within a cooking surface.
(2) Elements are completely contained within an enclosure identified as suitable for this use.
(3) Elements are contained within an ASME-rated and stamped vessel.

**(3) Water Heaters and Steam Boilers.** Resistance-type immersion electric heating elements shall be permitted to be subdivided into circuits not exceeding 120 amperes and protected at not more than 150 amperes as follows:

(1) Where contained in ASME-rated and stamped vessels
(2) Where included in listed instantaneous water heaters
(3) Where installed in low-pressure water heater tanks or open-outlet water heater vessels

Informational Note: Low-pressure and open-outlet heaters are atmospheric pressure water heaters as defined in IEC 60335-2-21, *Household and similar electrical appliances — Safety — Particular requirements for storage water heaters.*

**(G) Motor-Operated Appliances.** Motors of motor-operated appliances shall be provided with overload protection in accordance with Part III of Article 430. Hermetic refrigerant motor-compressors in air-conditioning or refrigerating equipment shall be provided with overload protection in accordance with Part VI of Article 440. Where appliance overcurrent protective devices that are separate from the appliance are required, data for selection of these devices shall be marked on the appliance. The minimum marking shall be that specified in 430.7 and 440.4.

## 422.12 Central Heating Equipment

Central heating equipment other than fixed electric space-heating equipment shall be supplied by an individual branch circuit.

*Exception No. 1: Auxiliary equipment, such as a pump, valve, humidifier, or electrostatic air cleaner directly associated with the heating equipment, shall be permitted to be connected to the same branch circuit.*

*Exception No. 2: Permanently connected air-conditioning equipment shall be permitted to be connected to the same branch circuit.*

Exception No. 1 permits electric motors, ignition systems, controls, and so forth of fossil-fuel-fired central heating equipment to be connected to the same individual branch circuit as defined in Article 100 under the term *branch circuit, individual.*

Exception No. 2 allows a permanently connected air-conditioning unit to be supplied from a branch circuit that supplies central heating equipment other than fixed electric space-heating equipment, because central heating equipment and air-conditioning equipment are considered unlikely to operate at the same time.

## 422.13 Storage-Type Water Heaters

A fixed storage-type water heater that has a capacity of 450 L (120 gal) or less shall be considered a continuous load for the purposes of sizing branch circuits.

Informational Note: For branch-circuit rating, see 422.10.

Because certain water heaters are a continuous load, the branch-circuit overcurrent device and conductors are required to be sized based on 125 percent of the water heater nameplate rating unless the overcurrent device and the assembly it is installed in are listed to be used at 100 percent of its continuous current rating.

## 422.14 Infrared Lamp Industrial Heating Appliances

In industrial occupancies, infrared heating appliance lampholders shall be permitted to be operated in series on circuits of over 150 volts to ground, provided the voltage rating of the lampholders is not less than the circuit voltage.

Each section, panel, or strip carrying a number of infrared lampholders (including the internal wiring of such section, panel, or strip) shall be considered an appliance. The terminal connection block of each such assembly shall be considered an individual outlet.

## 422.15 Central Vacuum Outlet Assemblies

**(A)** Listed central vacuum outlet assemblies shall be permitted to be connected to a branch circuit in accordance with 210.23(A).

**(B)** The ampacity of the connecting conductors shall not be less than the ampacity of the branch circuit conductors to which they are connected.

**(C)** Accessible non–current-carrying metal parts of the central vacuum outlet assembly likely to become energized shall be connected to an equipment grounding conductor in accordance with 250.110. Incidental metal parts such as screws or rivets installed into or on insulating material shall not be considered likely to become energized.

Section 422.15 permits listed central vacuum outlet devices to be connected to the ordinary 15- or 20-ampere general-purpose branch circuits located in the same area in which the vacuum outlet is installed. Starting and stopping of the central vacuum system is achieved by a Class 2 control circuit that originates at the main unit of the central vacuum system. The circuit is switched at each outlet by the insertion or removal of the matching vacuum hose in the outlet.

"Incidental metal parts" is consistent with text in UL 1017, *Vacuum Cleaners, Blower Cleaners and Household Floor Finishing Machines*, which states: "Parts that are not considered likely to be energized are metal screws or rivets in polymeric enclosures or faceplates, external metal springs used on a self-closing polymeric cover, and the like. Electrified wall valves connected to an extra-low voltage circuit are excluded from this requirement."

## 422.16 Flexible Cords

**(A) General.** Flexible cord shall be permitted (1) for the connection of appliances to facilitate their frequent interchange or to prevent the transmission of noise or vibration or (2) to facilitate the removal or disconnection of appliances that are fastened in place, where the fastening means and mechanical connections are specifically designed to permit ready removal for maintenance or repair and the appliance is intended or identified for flexible cord connection.

**(B) Specific Appliances.**

**(1) Electrically Operated In-Sink Waste Disposers.** Electrically operated in-sink waste disposers shall be permitted to be cord-and plug-connected with a flexible cord identified as suitable in the installation instructions of the appliance manufacturer where all of the following conditions are met:

(1) The flexible cord shall be terminated with a grounding-type attachment plug.

*Exception: A listed in-sink waste disposer distinctly marked to identify it as protected by a system of double insulation, or its equivalent, shall not be required to be terminated with a grounding-type attachment plug.*

(2) The length of the cord shall not be less than 450 mm (18 in.) and not over 900 mm (36 in.).
(3) Receptacles shall be located to avoid physical damage to the flexible cord.
(4) The receptacle shall be accessible.

*EXHIBIT 422.1 A cord-and-plug-connected kitchen waste disposer. (Courtesy of the National Electrical Contractors Association)*

All cord-connected waste disposers are covered by this requirement whether they are in a sink at a kitchen, prep area at a deli counter, or a bar. The kitchen waste disposer illustrated in Exhibit 422.1 is an example of a cord-and-plug-connected appliance with mechanical connections designed to permit removal. The cord and receptacle are designed and installed in accordance with 422.16(B)(1). To facilitate control of the waste disposer from a wall location, this receptacle is permitted to be switched by a general-use snap switch as long as the load does not exceed the requirements of 404.14(A)(3).

**(2) Built-in Dishwashers and Trash Compactors.** Built-in dishwashers and trash compactors shall be permitted to be cord-and-plug-connected with a flexible cord identified as suitable for the purpose in the installation instructions of the appliance manufacturer where all of the following conditions are met:

(1) The flexible cord shall be terminated with a grounding-type attachment plug.

*Exception: A listed dishwasher or trash compactor distinctly marked to identify it as protected by a system of double insulation, or its equivalent, shall not be required to be terminated with a grounding-type attachment plug.*

(2) The length of the cord shall be 0.9 m to 1.2 m (3 ft to 4 ft) measured from the face of the attachment plug to the plane of the rear of the appliance.
(3) Receptacles shall be located to avoid physical damage to the flexible cord.
(4) The receptacle shall be located in the space occupied by the appliance or adjacent thereto.
(5) The receptacle shall be accessible.

**(3) Wall-Mounted Ovens and Counter-Mounted Cooking Units.** Wall-mounted ovens and counter-mounted cooking units complete with provisions for mounting and for making electrical connections shall be permitted to be permanently connected or, only for ease in servicing or for installation, cord-and-plug-connected.

A separable connector or a plug and receptacle combination in the supply line to an oven or cooking unit shall be approved for the temperature of the space in which it is located.

**(4) Range Hoods.** Range hoods shall be permitted to be cord-and-plug-connected with a flexible cord identified as suitable for use on range hoods in the installation instructions of the appliance manufacturer, where all of the following conditions are met:

(1) The flexible cord is terminated with a grounding-type attachment plug.

*Exception: A listed range hood distinctly marked to identify it as protected by a system of double insulation, or its equivalent, shall not be required to be terminated with a grounding-type attachment plug.*

(2) The length of the cord is not less than 450 mm (18 in.) and not over 900 mm (36 in.).
(3) Receptacles are located to avoid physical damage to the flexible cord.
(4) The receptacle is accessible.
(5) The receptacle is supplied by an individual branch circuit.

## 422.17 Protection of Combustible Material

Each electrically heated appliance that is intended by size, weight, and service to be located in a fixed position shall be placed so as to provide ample protection between the appliance and adjacent combustible material.

## 422.18 Support of Ceiling-Suspended (Paddle) Fans

Ceiling-suspended (paddle) fans shall be supported independently of an outlet box or by listed outlet box or outlet box systems identified for the use and installed in accordance with 314.27(C).

Paddle fans must be supported independently of the outlet box or be supported by a listed box identified for fan support. The fan must be supported from the building structure where the outlet box is not identified to support a paddle fan. An outlet box identified for support of a paddle fan must be marked with the amount of weight it is allowed to support if the weight exceeds 35 pounds. Exhibit 422.2 shows an example of a listed outlet box used to support a paddle fan.

## 422.19 Space for Conductors

Canopies of ceiling-suspended (paddle) fans and outlet boxes taken together shall provide sufficient space so that conductors and their connecting devices are capable of being installed in accordance with 314.16.

## 422.20 Outlet Boxes to Be Covered

In a completed installation, each outlet box shall be provided with a cover unless covered by means of a ceiling-suspended (paddle) fan canopy.

**EXHIBIT 422.2** *Supporting a ceiling-suspended (paddle) fan (35 lb or less) with a box identified for such use. (Courtesy of Hubbell RACO)*

### 422.21 Covering of Combustible Material at Outlet Boxes

Any combustible ceiling finish exposed between the edge of a ceiling-suspended (paddle) fan canopy or pan and an outlet box shall be covered with noncombustible material.

### 422.22 Other Installation Methods

Appliances employing methods of installation other than covered by this article shall be permitted to be used only by special permission.

### 422.23 Tire Inflation and Automotive Vacuum Machines

Tire inflation machines and automotive vacuum machines provided for public use shall be protected by a ground-fault circuit interrupter.

Tire inflation and automotive vacuum machines at service stations and car washes are often subject to exposure to the elements and damage by the public. In addition, they are used under all types of environmental conditions. Regardless of the operating voltage, GFCI protection is required for hard-wired and cord-and-plug connections.

## III. Disconnecting Means

### 422.30 General

A means shall be provided to simultaneously disconnect each appliance from all ungrounded conductors in accordance with the following sections of Part III. If an appliance is supplied by more than one branch-circuit or feeder, these disconnecting means shall be grouped and identified as the appliance disconnect.

### 422.31 Disconnection of Permanently Connected Appliances

**(A) Rated at Not over 300 Volt-Amperes or ⅛ Horsepower.** For permanently connected appliances rated at not over 300 volt-amperes or ⅛ hp, the branch-circuit overcurrent device shall be permitted to serve as the disconnecting means.

**(B) Appliances Rated over 300 Volt-Amperes.** For permanently connected appliances rated over 300 volt-amperes, the branch-circuit switch or circuit breaker shall be permitted to serve as the disconnecting means where the switch or circuit breaker is within sight from the appliance or is lockable in accordance with 110.25.

Informational Note: For appliances employing unit switches, see 422.34.

**(C) Motor-Operated Appliances Rated over ⅛ Horsepower.** The disconnecting means shall comply with 430.109 and 430.110. For permanently connected motor-operated appliances with motors rated over ⅛ hp, the disconnecting means shall meet 422.31(C)(1) or (2).

Since this section references an appliance's horsepower rating, a motor-driven appliance is implied. The disconnecting means requirement is more restrictive for appliances rated over ⅛ horsepower. The use of a lockable disconnecting means located out of sight of the appliance is not permitted.

(1) The branch-circuit switch or circuit breaker shall be permitted to serve as the disconnecting means where the switch or circuit breaker is within sight from the appliance.
(2) The disconnecting means shall be installed within sight of the appliance.

*Exception: If an appliance of more than ⅛ hp is provided with a unit switch that complies with 422.34(A), (B), (C), or (D), the switch or circuit breaker serving as the other disconnecting means shall be permitted to be out of sight from the appliance.*

### 422.33 Disconnection of Cord-and-Plug-Connected Appliances

**(A) Separable Connector or an Attachment Plug and Receptacle.** For cord-and-plug-connected appliances, an accessible separable connector or an accessible plug and receptacle shall be permitted to serve as the disconnecting means. Where the separable connector or plug and receptacle are not accessible, cord-and-plug-connected appliances shall be provided with disconnecting means in accordance with 422.31.

**(B) Connection at the Rear Base of a Range.** For cord-and-plug-connected household electric ranges, an attachment plug and receptacle connection at the rear base of a range, if it is accessible from the front by removal of a drawer, shall be considered as meeting the intent of 422.33(A).

**(C) Rating.** The rating of a receptacle or of a separable connector shall not be less than the rating of any appliance connected thereto.

*Exception: Demand factors authorized elsewhere in this Code shall be permitted to be applied to the rating of a receptacle or of a separable connector.*

### 422.34 Unit Switch(es) as Disconnecting Means

A unit switch(es) with a marked-off position that is a part of an appliance and disconnects all ungrounded conductors shall be permitted as the disconnecting means required by this article where other means for disconnection are provided in occupancies specified in 422.34(A) through (D).

**(A) Multifamily Dwellings.** In multifamily dwellings, the other disconnecting means shall be within the dwelling unit, or on the same floor as the dwelling unit in which the appliance is installed, and shall be permitted to control lamps and other appliances.

**(B) Two-Family Dwellings.** In two-family dwellings, the other disconnecting means shall be permitted either inside or outside of the dwelling unit in which the appliance is installed. In this case, an individual switch or circuit breaker for the dwelling unit shall be permitted and shall also be permitted to control lamps and other appliances.

**(C) One-Family Dwellings.** In one-family dwellings, the service disconnecting means shall be permitted to be the other disconnecting means.

**(D) Other Occupancies.** In other occupancies, the branch-circuit switch or circuit breaker, where readily accessible for servicing of the appliance, shall be permitted as the other disconnecting means.

### 422.35 Switch and Circuit Breaker to Be Indicating

Switches and circuit breakers used as disconnecting means shall be of the indicating type.

## IV. Construction

### 422.40 Polarity in Cord-and Plug-Connected Appliances

If the appliance is provided with a manually operated, line-connected, single-pole switch for appliance on–off operation, an Edison-base lampholder, or a 15- or 20-ampere receptacle, the attachment plug shall be of the polarized or grounding type.

A 2-wire, nonpolarized attachment plug shall be permitted to be used on a listed double-insulated shaver.

Informational Note: For polarity of Edison-base lampholders, see 410.82(A).

### 422.41 Cord-and Plug-Connected Appliances Subject to Immersion

Cord-and plug-connected portable, freestanding hydromassage units and hand-held hair dryers shall be constructed to provide protection for personnel against electrocution when immersed while in the "on" or "off" position.

Although receptacles in bathrooms of dwelling units have been required to be protected by GFCIs since the 1975 edition of the *Code*, many receptacles in existing bathrooms are not so protected. Cord-and-plug-connected appliances such as hand-held hair dryers and curling irons, which can accidentally fall into bathtubs and cause fatalities, are required to be provided with some form of protective device that is part of the appliance. Three types of protectors comply with this requirement:

1. Appliance-leakage circuit interrupters (ALCIs)
2. Immersion-detector circuit interrupters (IDCIs)
3. Ground-fault circuit interrupters (GFCIs)

ALCIs de-energize the supply to the appliance when leakage current exceeds a predetermined value. IDCIs de-energize the supply when a liquid causes a conductive path between a live part and a sensor, and GFCIs de-energize the supply when the current to ground exceeds a predetermined value.

### 422.42 Signals for Heated Appliances

In other than dwelling-type occupancies, each electrically heated appliance or group of appliances intended to be applied to combustible material shall be provided with a signal or an integral temperature-limiting device.

For electrically heated appliances in commercial or industrial locations, a common practice is to use a red light, connected to and within sight of the appliance, to indicate that the appliance is energized and operating. No signal is required for an electrically heated appliance provided with an integral high-temperature limiting device, such as a thermostat, that limits the temperature to which the appliance can heat.

### 422.43 Flexible Cords

**(A) Heater Cords.** All cord-and plug-connected smoothing irons and electrically heated appliances that are rated at more than 50 watts and produce temperatures in excess of 121°C (250°F) on surfaces with which the cord is likely to be in contact shall be provided with one of the types of approved heater cords listed in Table 400.4.

**(B) Other Heating Appliances.** All other cord-and plug-connected electrically heated appliances shall be connected with one of the approved types of cord listed in Table 400.4, selected in accordance with the usage specified in that table.

### 422.44 Cord-and Plug-Connected Immersion Heaters

Electric heaters of the cord-and plug-connected immersion type shall be constructed and installed so that current-carrying parts are effectively insulated from electrical contact with the substance in which they are immersed.

### 422.45 Stands for Cord-and Plug-Connected Appliances

Each smoothing iron and other cord-and plug-connected electrically heated appliance intended to be applied to combustible material shall be equipped with an approved stand, which shall be permitted to be a separate piece of equipment or a part of the appliance.

### 422.46 Flatirons

Electrically heated smoothing irons shall be equipped with an identified temperature-limiting means.

### 422.47 Water Heater Controls

All storage or instantaneous-type water heaters shall be equipped with a temperature-limiting means in addition to its control thermostat to disconnect all ungrounded conductors. Such means shall comply with both of the following:

(1) Installed to sense maximum water temperature.
(2) Be either a trip-free, manually reset type or a type having a replacement element. Such water heaters shall be marked to require the installation of a temperature and pressure relief valve.

*Exception No. 1: Storage water heaters that are identified as being suitable for use with a supply water temperature of 82°C (180°F) or above and a capacity of 60 kW or above.*

*Exception No. 2: Instantaneous-type water heaters that are identified as being suitable for such use, with a capacity of 4 L (1 gal) or less.*

Informational Note: See ANSI Z21.22-1999/CSA 4.4-M99, *Relief Valves for Hot Water Supply Systems*.

### 422.48 Infrared Lamp Industrial Heating Appliances

**(A) 300 Watts or Less.** Infrared heating lamps rated at 300 watts or less shall be permitted with lampholders of the medium-base, unswitched porcelain type or other types identified as suitable for use with infrared heating lamps rated 300 watts or less.

**(B) Over 300 Watts.** Screw shell lampholders shall not be used with infrared lamps rated over 300 watts, unless the lampholders are identified as being suitable for use with infrared heating lamps rated over 300 watts.

Infrared (heat) radiation lamps are tungsten-filament incandescent lamps similar in appearance to lighting lamps. However, they are designed to operate at a lower temperature, thus transferring more heat radiation and less light intensity. Infrared lamps are used for a variety of heating and drying purposes in industrial locations.

### 422.49 High-Pressure Spray Washers

Cord-and plug-connected high-pressure spray washing machines as specified in 422.49(1) or (2) shall be provided with factory-installed ground-fault circuit-interrupter protection for personnel that is an integral part of the attachment plug or that is located in the supply cord within 300 mm (12 in.) of the attachment plug.

(1) All single-phase equipment rated 250 volts or less
(2) All 3-phase equipment rated 208Y/120 volts and 60 amperes or less

### 422.50 Cord-and Plug-Connected Pipe Heating Assemblies

Cord-and plug-connected pipe heating assemblies intended to prevent freezing of piping shall be listed.

This listing requirement is a result of data that substantiated numerous fires initiated by heat tapes. Additional requirements for ground-fault protection of equipment are located in 427.22.

### 422.51 Vending Machines

**(A) Cord-and Plug-Connected.** Cord-and plug-connected vending machines manufactured or remanufactured on or after January 1, 2005, shall include a ground-fault circuit interrupter as an integral part of the attachment plug or be located within 300 mm (12 in.) of the attachment plug. Older vending machines manufactured or remanufactured prior to January 1, 2005, shall be connected to a GFCI-protected outlet.

**(B) Other Than Cord-and Plug-Connected.** Vending machines not utilizing a cord and plug connection shall be connected to a ground-fault circuit-interrupter protected circuit.

Informational Note: For further information, see ANSI/UL 541-2010, *Standard for Refrigerated Vending Machines*, or ANSI/UL 751-2010, *Standard for Vending Machines*.

Prior to 2005, the U.S. Consumer Product Safety Commission (CPSC) had investigated four separate electrocution incidents and three nonfatal shock incidents involving vending machines. Those investigations spurred this requirement that new and remanufactured vending machines be provided with a GFCI as an integral part of the power-supply cord within 12 inches of the attachment plug.

The GFCI requirement applies to all hard-wired and cord-and-plug-connected vending machines regardless of the voltage, current, or frequency rating of the appliance.

### 422.52 Electric Drinking Fountains

Electric drinking fountains shall be protected with ground-fault circuit-interrupter protection.

The GFCI protection must be either part of the fountain, included in the receptacle for the fountain, or provided on the branch circuit feeding the fountain. Bottled water coolers are not considered electric drinking fountains.

## V. Marking

### 422.60 Nameplate

**(A) Nameplate Marking.** Each electrical appliance shall be provided with a nameplate giving the identifying name and the rating in volts and amperes, or in volts and watts. If the appliance is to be used on a specific frequency or frequencies, it shall be so marked.

Where motor overload protection external to the appliance is required, the appliance shall be so marked.

Informational Note: See 422.11 for overcurrent protection requirements.

**(B) To Be Visible.** Marking shall be located so as to be visible or easily accessible after installation.

### 422.61 Marking of Heating Elements

All heating elements that are rated over one ampere, replaceable in the field, and a part of an appliance shall be legibly marked with the ratings in volts and amperes, or in volts and watts, or with the manufacturer's part number.

### 422.62 Appliances Consisting of Motors and Other Loads

**(A) Nameplate Horsepower Markings.** Where a motor-operated appliance nameplate includes a horsepower rating, that rating shall not be less than the horsepower rating on the motor nameplate. Where an appliance consists of multiple motors, or one or more motors and other loads, the nameplate value shall not be less than the equivalent horsepower of the combined loads, calculated in accordance with 430.110(C)(1).

**(B) Additional Nameplate Markings.** Appliances, other than those factory-equipped with cords and attachment plugs and with nameplates in compliance with 422.60, shall be marked in accordance with 422.62(B)(1) or (B)(2).

**(1) Marking.** In addition to the marking required in 422.60, the marking on an appliance consisting of a motor with other load(s) or motors with or without other load(s) shall specify the minimum supply circuit conductor ampacity and the maximum rating of the circuit overcurrent protective device. This requirement shall not apply to an appliance with a nameplate in compliance with 422.60 where both the minimum supply circuit conductor ampacity and maximum rating of the circuit overcurrent protective device are not more than 15 amperes.

**(2) Alternate Marking Method.** An alternative marking method shall be permitted to specify the rating of the largest motor in volts and amperes, and the additional load(s) in volts and amperes, or volts and watts in addition to the marking required in 422.60. The ampere rating of a motor ⅛ horsepower or less or a nonmotor load 1 ampere or less shall be permitted to be omitted unless such loads constitute the principal load.

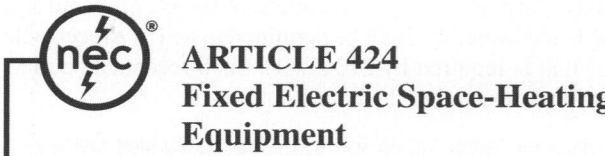

# ARTICLE 424 Fixed Electric Space-Heating Equipment

## I. General

### 424.1 Scope

This article covers fixed electric equipment used for space heating. For the purpose of this article, heating equipment shall include heating cable, unit heaters, boilers, central systems, or other approved fixed electric space-heating equipment. This article shall not apply to process heating and room air conditioning.

### 424.2 Other Articles

Fixed electric space-heating equipment incorporating a hermetic refrigerant motor-compressor shall also comply with Article 440.

### 424.3 Branch Circuits

**(A) Branch-Circuit Requirements.** Individual branch circuits shall be permitted to supply any volt-ampere or wattage rating of fixed electric space-heating equipment for which they are rated.

Branch circuits supplying two or more outlets for fixed electric space-heating equipment shall be rated 15, 20, 25, or 30 amperes. In other than a dwelling unit, fixed infrared heating equipment shall be permitted to be supplied from branch circuits rated not over 50 amperes.

**(B) Branch-Circuit Sizing.** Fixed electric space-heating equipment and motors shall be considered continuous load.

Branch-circuit conductors and overcurrent devices supplying fixed electric space-heating equipment must be sized at 125 percent of the total load of the heaters (and motors) to protect overcurrent devices – particularly those in panelboards – and conductors from overheating during periods of prolonged operation.

The requirement eliminates any question about whether the heating equipment is a continuous load, which is defined in Article 100. This requirement also impacts feeders and services that supply fixed electric space-heating equipment branch circuits.

### 424.6 Listed Equipment

Electric baseboard heaters, heating cables, duct heaters, and radiant heating systems shall be listed and labeled.

# II. Installation

## 424.9 General

All fixed electric space-heating equipment shall be installed in an approved manner.

Permanently installed electric baseboard heaters equipped with factory-installed receptacle outlets, or outlets provided as a separate listed assembly, shall be permitted in lieu of a receptacle outlet(s) that is required by 210.50(B). Such receptacle outlets shall not be connected to the heater circuits.

Informational Note: Listed baseboard heaters include instructions that may not permit their installation below receptacle outlets.

The second paragraph of 424.9 restates the permission granted in the second paragraph of 210.52 — that is, it allows factory-installed receptacle outlets in electric baseboard heaters to satisfy the spacing requirements for receptacle outlets in dwelling units according to 210.52(A).

Heating equipment and systems often have special installation instructions for spacings, types of supply wires, or special control equipment, which must be considered in determining the installation's suitability.

## 424.10 Special Permission

Fixed electric space-heating equipment and systems installed by methods other than covered by this article shall be permitted only by special permission.

## 424.11 Supply Conductors

Fixed electric space-heating equipment requiring supply conductors with over 60°C insulation shall be clearly and permanently marked. This marking shall be plainly visible after installation and shall be permitted to be adjacent to the field connection box.

Fixed electric space-heating equipment may require supply conductors with a temperature rating greater than 60°C, due to their proximity to the heating elements and the installation instructions provided with a listed product.

## 424.12 Locations

**(A) Exposed to Physical Damage.** Where subject to physical damage, fixed electric space-heating equipment shall be protected in an approved manner.

**(B) Damp or Wet Locations.** Heaters and related equipment installed in damp or wet locations shall be listed for such locations and shall be constructed and installed so that water or other liquids cannot enter or accumulate in or on wired sections, electrical components, or ductwork.

Informational Note No. 1: See 110.11 for equipment exposed to deteriorating agents.

Informational Note No. 2: See 680.27(C) for pool deck areas.

## 424.13 Spacing from Combustible Materials

Fixed electric space-heating equipment shall be installed to provide the required spacing between the equipment and adjacent combustible material, unless it is listed to be installed in direct contact with combustible material.

# III. Control and Protection of Fixed Electric Space-Heating Equipment

## 424.19 Disconnecting Means

Means shall be provided to simultaneously disconnect the heater, motor controller(s), and supplementary overcurrent protective device(s) of all fixed electric space-heating equipment from all ungrounded conductors. Where heating equipment is supplied by more than one source, feeder, or branch circuit, the disconnecting means shall be grouped and marked. The disconnecting means specified in 424.19(A) and (B) shall have an ampere rating not less than 125 percent of the total load of the motors and the heaters and shall be lockable in accordance with 110.25.

This section requires that the disconnecting means simultaneously open all the ungrounded conductors to prevent the practice of disconnecting one conductor at a time at terminal blocks or similar devices. The disconnect switch must have a rating of 125 percent of the heater's total load.

Section 424.19(A)(2) correlates with 422.31(C) by requiring disconnecting means for appliances that consist of heaters with motors of over ⅛ horsepower to comply with Part IX of Article 430.

A unit switch is permitted by 424.19(C) to serve as the disconnecting means, provided that it has a marked "off" position and disconnects all ungrounded conductors. In addition, other means must be provided in accordance with 424.19(C)(1) through (C)(4). Such other means are not required to be capable of being locked in the open position.

**(A) Heating Equipment with Supplementary Overcurrent Protection.** The disconnecting means for fixed electric space-heating equipment with supplementary overcurrent protection shall be within sight from the supplementary overcurrent protective device(s), on the supply side of these devices, if fuses, and, in addition, shall comply with either 424.19(A)(1) or (A)(2).

**(1) Heater Containing No Motor Rated over ⅛ Horsepower.** The disconnecting means specified in 424.19 or unit switches complying with 424.19(C) shall be permitted to serve as the required disconnecting means for both the motor controller(s) and heater under either of the following conditions:

(1) The disconnecting means provided is also within sight from the motor controller(s) and the heater.
(2) The disconnecting means is lockable in accordance with 110.25.

**(2) Heater Containing a Motor(s) Rated over ⅛ Horsepower.** The above disconnecting means shall be permitted to serve as the required disconnecting means for both the motor controller(s) and heater under either of the following conditions:

(1) Where the disconnecting means is in sight from the motor controller(s) and the heater and complies with Part IX of Article 430.
(2) Where a motor(s) of more than ⅛ hp and the heater are provided with a single unit switch that complies with 422.34(A), (B), (C), or (D), the disconnecting means shall be permitted to be out of sight from the motor controller.

**(B) Heating Equipment Without Supplementary Overcurrent Protection.**

**(1) Without Motor or with Motor Not over ⅛ Horsepower.** For fixed electric space-heating equipment without a motor rated over ⅛ hp, the branch-circuit switch or circuit breaker shall be permitted to serve as the disconnecting means where the switch or circuit breaker is within sight from the heater or is lockable in accordance with 110.25.

**(2) Over ⅛ Horsepower.** For motor-driven electric space-heating equipment with a motor rated over ⅛ hp, a disconnecting means shall be located within sight from the motor controller or shall be permitted to comply with the requirements in 424.19(A)(2).

**(C) Unit Switch(es) as Disconnecting Means.** A unit switch(es) with a marked "off" position that is part of a fixed heater and disconnects all ungrounded conductors shall be permitted as the disconnecting means required by this article where other means for disconnection are provided in the types of occupancies in 424.19(C)(1) through (C)(4).

**(1) Multifamily Dwellings.** In multifamily dwellings, the other disconnecting means shall be within the dwelling unit, or on the same floor as the dwelling unit in which the fixed heater is installed, and shall also be permitted to control lamps and appliances.

**(2) Two-Family Dwellings.** In two-family dwellings, the other disconnecting means shall be permitted either inside or outside of the dwelling unit in which the fixed heater is installed. In this case, an individual switch or circuit breaker for the dwelling unit shall be permitted and shall also be permitted to control lamps and appliances.

**(3) One-Family Dwellings.** In one-family dwellings, the service disconnecting means shall be permitted to be the other disconnecting means.

**(4) Other Occupancies.** In other occupancies, the branch-circuit switch or circuit breaker, where readily accessible for servicing of the fixed heater, shall be permitted as the other disconnecting means.

## 424.20 Thermostatically Controlled Switching Devices

**(A) Serving as Both Controllers and Disconnecting Means.** Thermostatically controlled switching devices and combination thermostats and manually controlled switches shall be permitted to serve as both controllers and disconnecting means, provided they meet all of the following conditions:

(1) Provided with a marked "off" position
(2) Directly open all ungrounded conductors when manually placed in the "off" position
(3) Designed so that the circuit cannot be energized automatically after the device has been manually placed in the "off" position
(4) Located as specified in 424.19

**(B) Thermostats That Do Not Directly Interrupt All Ungrounded Conductors.** Thermostats that do not directly interrupt all ungrounded conductors and thermostats that operate remote-control circuits shall not be required to meet the requirements of 424.20(A). These devices shall not be permitted as the disconnecting means.

## 424.21 Switch and Circuit Breaker to Be Indicating

Switches and circuit breakers used as disconnecting means shall be of the indicating type.

## 424.22 Overcurrent Protection

**(A) Branch-Circuit Devices.** Electric space-heating equipment, other than such motor-operated equipment as required by Articles 430 and 440 to have additional overcurrent protection, shall be permitted to be protected against overcurrent where supplied by one of the branch circuits in Article 210.

**(B) Resistance Elements.** Resistance-type heating elements in electric space-heating equipment shall be protected at not more than 60 amperes. Equipment rated more than 48 amperes and employing such elements shall have the heating elements subdivided, and each subdivided load shall not exceed 48 amperes. Where a subdivided load is less than 48 amperes, the rating of the supplementary overcurrent protective device shall comply with 424.3(B). A boiler employing resistance-type immersion heating elements contained in an ASME-rated and stamped vessel shall be permitted to comply with 424.72(A).

The reason for subdividing the overcurrent protection is to minimize the amount of damaging energy — in the form of both heat and magnetic energy — released into the heating elements during a short circuit. The damaging short-circuit energy released at the element is greatly reduced by limiting the size of the overcurrent device protecting the individual heating elements, thereby greatly reducing the risk of fire. In addition, a second benefit may be continuity of service if equipment is only partially affected.

Historically, the subdivision size of 60 amperes was selected to use the maximum fuseholder size of 60 amperes while maintaining up to a 48-ampere heating element (48 A × 125% = 60 A).

**(C) Overcurrent Protective Devices.** The supplementary overcurrent protective devices for the subdivided loads specified in 424.22(B) shall be (1) factory-installed within or on the heater enclosure or supplied for use with the heater as a separate assembly by the heater manufacturer; (2) accessible, but shall not be required to be readily accessible; and (3) suitable for branch-circuit protection.

Informational Note: See 240.10.

Where cartridge fuses are used to provide this overcurrent protection, a single disconnecting means shall be permitted to be used for the several subdivided loads.

Informational Note No. 1: For supplementary overcurrent protection, see 240.10.

Informational Note No. 2: For disconnecting means for cartridge fuses in circuits of any voltage, see 240.40.

**(D) Branch-Circuit Conductors.** The conductors supplying the supplementary overcurrent protective devices shall be considered branch-circuit conductors.

Where the heaters are rated 50 kW or more, the conductors supplying the supplementary overcurrent protective devices specified in 424.22(C) shall be permitted to be sized at not less than 100 percent of the nameplate rating of the heater, provided all of the following conditions are met:

(1) The heater is marked with a minimum conductor size.
(2) The conductors are not smaller than the marked minimum size.
(3) A temperature-actuated device controls the cyclic operation of the equipment.

**(E) Conductors for Subdivided Loads.** Field-wired conductors between the heater and the supplementary overcurrent protective devices shall be sized at not less than 125 percent of the load served. The supplementary overcurrent protective devices specified in 424.22(C) shall protect these conductors in accordance with 240.4.

Where the heaters are rated 50 kW or more, the ampacity of field-wired conductors between the heater and the supplementary overcurrent protective devices shall be permitted to be not less than 100 percent of the load of their respective subdivided circuits, provided all of the following conditions are met:

(1) The heater is marked with a minimum conductor size.
(2) The conductors are not smaller than the marked minimum size.
(3) A temperature-activated device controls the cyclic operation of the equipment.

# IV. Marking of Heating Equipment

## 424.28 Nameplate

**(A) Marking Required.** Each unit of fixed electric space-heating equipment shall be provided with a nameplate giving the identifying name and the normal rating in volts and watts or in volts and amperes.

Electric space-heating equipment intended for use on alternating current only, direct current only, or both shall be marked to so indicate. The marking of equipment consisting of motors over ⅛ hp and other loads shall specify the rating of the motor in volts, amperes, and frequency, and the heating load in volts and watts or in volts and amperes.

**(B) Location.** This nameplate shall be located so as to be visible or easily accessible after installation.

## 424.29 Marking of Heating Elements

All heating elements that are replaceable in the field and are part of an electric heater shall be legibly marked with the ratings in volts and watts or in volts and amperes.

# V. Electric Space-Heating Cables

## 424.34 Heating Cable Construction

Heating cables shall be furnished complete with factory-assembled nonheating leads at least 2.1 m (7 ft) in length.

## 424.35 Marking of Heating Cables

Each unit shall be marked with the identifying name or identification symbol, catalog number, and ratings in volts and watts or in volts and amperes.

Each unit length of heating cable shall have a permanent legible marking on each nonheating lead located within 75 mm (3 in.) of the terminal end. The lead wire shall have the following color identification to indicate the circuit voltage on which it is to be used:

(1) 120 volt, nominal — yellow
(2) 208 volt, nominal — blue
(3) 240 volt, nominal — red
(4) 277 volt, nominal — brown
(5) 480 volt, nominal — orange

## 424.36 Clearances of Wiring in Ceilings

Wiring located above heated ceilings shall be spaced not less than 50 mm (2 in.) above the heated ceiling and shall be considered as operating at an ambient temperature of 50°C (122°F). The ampacity of conductors shall be calculated on the basis of the correction factors shown in the 0–2000 volt ampacity tables of Article 310. If this wiring is located above thermal insulation having a minimum thickness of 50 mm (2 in.), the wiring shall not require correction for temperature.

## 424.38 Area Restrictions

**(A) Shall Not Extend Beyond the Room or Area.** Heating cables shall not extend beyond the room or area in which they originate.

**(B) Uses Prohibited.** Heating cables shall not be installed in the following:

(1) In closets
(2) Over walls
(3) Over partitions that extend to the ceiling, unless they are isolated single runs of embedded cable
(4) Over cabinets whose clearance from the ceiling is less than the minimum horizontal dimension of the cabinet to the nearest cabinet edge that is open to the room or area

**(C) In Closet Ceilings as Low-Temperature Heat Sources to Control Relative Humidity.** The provisions of 424.38(B) shall not prevent the use of cable in closet ceilings as low-temperature heat sources to control relative humidity, provided they are used only in those portions of the ceiling that are unobstructed to the floor by shelves or other permanent luminaires.

## 424.39 Clearance from Other Objects and Openings

Heating elements of cables shall be separated at least 200 mm (8 in.) from the edge of outlet boxes and junction boxes that are to be used for mounting surface luminaires. A clearance of not less than 50 mm (2 in.) shall be provided from recessed luminaires and their trims, ventilating openings, and other such openings in room surfaces. No heating cable shall be covered by any surface-mounted equipment.

## 424.40 Splices

Embedded cables shall be spliced only where necessary and only by approved means, and in no case shall the length of the heating cable be altered.

## 424.41 Installation of Heating Cables on Dry Board, in Plaster, and on Concrete Ceilings

**(A) In Walls.** Cables shall not be installed in walls unless it is necessary for an isolated single run of cable to be installed down a vertical surface to reach a dropped ceiling.

**(B) Adjacent Runs.** Adjacent runs of cable not exceeding 9 watts/m (2⅔ watts/ft) shall not be installed less than 38 mm (1½ in.) on centers.

**(C) Surfaces to Be Applied.** Heating cables shall be applied only to gypsum board, plaster lath, or other fire-resistant material. With metal lath or other electrically conductive surfaces, a coat of plaster shall be applied to completely separate the metal lath or conductive surface from the cable.

Informational Note: See also 424.41(F).

**(D) Splices.** All heating cables, the splice between the heating cable and nonheating leads, and 75-mm (3-in.) minimum of the nonheating lead at the splice shall be embedded in plaster or dry board in the same manner as the heating cable.

**(E) Ceiling Surface.** The entire ceiling surface shall have a finish of thermally noninsulating sand plaster that has a nominal thickness of 13 mm (½ in.), or other noninsulating material identified as suitable for this use and applied according to specified thickness and directions.

**(F) Secured.** Cables shall be secured by means of approved stapling, tape, plaster, nonmetallic spreaders, or other approved means either at intervals not exceeding 400 mm (16 in.) or at intervals not exceeding 1.8 m (6 ft) for cables identified for such use. Staples or metal fasteners that straddle the cable shall not be used with metal lath or other electrically conductive surfaces.

**(G) Dry Board Installations.** In dry board installations, the entire ceiling below the heating cable shall be covered with gypsum board not exceeding 13 mm (½ in.) thickness. The void between the upper layer of gypsum board, plaster lath, or other fire-resistant material and the surface layer of gypsum board shall be completely filled with thermally conductive, nonshrinking plaster or other approved material or equivalent thermal conductivity.

**(H) Free from Contact with Conductive Surfaces.** Cables shall be kept free from contact with metal or other electrically conductive surfaces.

**(I) Joists.** In dry board applications, cable shall be installed parallel to the joist, leaving a clear space centered under the joist of 65 mm (2½ in.) (width) between centers of adjacent runs of cable. A surface layer of gypsum board shall be mounted so that the nails or other fasteners do not pierce the heating cable.

**(J) Crossing Joists.** Cables shall cross joists only at the ends of the room unless the cable is required to cross joists elsewhere in order to satisfy the manufacturer's instructions that the installer avoid placing the cable too close to ceiling penetrations and luminaires.

## 424.42 Finished Ceilings

Finished ceilings shall not be covered with decorative panels or beams constructed of materials that have thermal insulating properties, such as wood, fiber, or plastic. Finished ceilings shall be permitted to be covered with paint, wallpaper, or other approved surface finishes.

## 424.43 Installation of Nonheating Leads of Cables

**(A) Free Nonheating Leads.** Free nonheating leads of cables shall be installed in accordance with approved wiring methods from the junction box to a location within the ceiling. Such installations shall be permitted to be single conductors in approved raceways, single or multiconductor Type UF, Type NMC, Type MI, or other approved conductors.

**(B) Leads in Junction Box.** Not less than 150 mm (6 in.) of free nonheating lead shall be within the junction box.

**(C) Excess Leads.** Excess leads of heating cables shall not be cut but shall be secured to the underside of the ceiling and embedded in plaster or other approved material, leaving only a length sufficient to reach the junction box with not less than 150 mm (6 in.) of free lead within the box.

### 424.44 Installation of Cables in Concrete or Poured Masonry Floors

**(A) Watts per Linear Meter (Foot).** Constant wattage heating cables shall not exceed 54 watts per linear meter (16½ watts per linear foot) of cable.

**(B) Spacing Between Adjacent Runs.** The spacing between adjacent runs of cable shall not be less than 25 mm (1 in.) on centers.

**(C) Secured in Place.** Cables shall be secured in place by nonmetallic frames or spreaders or other approved means while the concrete or other finish is applied.

Cables shall not be installed where they bridge expansion joints unless protected from expansion and contraction.

**(D) Spacing Between Heating Cable and Metal Embedded in the Floor.** Spacing shall be maintained between the heating cable and metal embedded in the floor, unless the cable is a grounded metal-clad cable.

**(E) Leads Protected.** Leads shall be protected where they leave the floor by rigid metal conduit, intermediate metal conduit, rigid nonmetallic conduit, electrical metallic tubing, or by other approved means.

**(F) Bushings or Approved Fittings.** Bushings or approved fittings shall be used where the leads emerge within the floor slab.

**(G) Ground-Fault Circuit-Interrupter Protection.** Ground-fault circuit-interrupter protection for personnel shall be provided for cables installed in electrically heated floors of bathrooms, kitchens, and in hydromassage bathtub locations.

GFCI protection is required where cables are installed in concrete or poured masonry floors, thereby reducing shock hazards to persons with bare feet in these areas. This requirement applies regardless of the type of floor covering over the concrete or poured masonry. Electric radiant heating cables are not permitted in spa, hot tub, or pool locations in accordance with 680.27(C)(3).

### 424.45 Inspection and Tests

Cable installations shall be made with due care to prevent damage to the cable assembly and shall be inspected and approved before cables are covered or concealed.

## VI. Duct Heaters

### 424.57 General

Part VI shall apply to any heater mounted in the airstream of a forced-air system where the air-moving unit is not provided as an integral part of the equipment.

### 424.58 Identification

Heaters installed in an air duct shall be identified as suitable for the installation.

### 424.59 Airflow

Means shall be provided to ensure uniform airflow over the face of the heater in accordance with the manufacturer's instructions.

Informational Note: Heaters installed within 1.2 m (4 ft) of the outlet of an air-moving device, heat pump, air conditioner, elbows, baffle plates, or other obstructions in ductwork may require turning vanes, pressure plates, or other devices on the inlet side of the duct heater to ensure an even distribution of air over the face of the heater.

### 424.60 Elevated Inlet Temperature

Duct heaters intended for use with elevated inlet air temperature shall be identified as suitable for use at the elevated temperatures.

### 424.61 Installation of Duct Heaters with Heat Pumps and Air Conditioners

Heat pumps and air conditioners having duct heaters closer than 1.2 m (4 ft) to the heat pump or air conditioner shall have both the duct heater and heat pump or air conditioner identified as suitable for such installation and so marked.

### 424.62 Condensation

Duct heaters used with air conditioners or other air-cooling equipment that could result in condensation of moisture shall be identified as suitable for use with air conditioners.

### 424.63 Fan Circuit Interlock

Means shall be provided to ensure that the fan circuit is energized when any heater circuit is energized. However, time- or temperature-controlled delay in energizing the fan motor shall be permitted.

### 424.64 Limit Controls

Each duct heater shall be provided with an approved, integral, automatic-reset temperature-limiting control or controllers to de-energize the circuit or circuits.

In addition, an integral independent supplementary control or controllers shall be provided in each duct heater that disconnects a sufficient number of conductors to interrupt current flow. This device shall be manually resettable or replaceable.

### 424.65 Location of Disconnecting Means

Duct heater controller equipment shall be either accessible with the disconnecting means installed at or within sight from the controller or as permitted by 424.19(A).

### 424.66 Installation

**(A) General.** Duct heaters shall be installed in accordance with the manufacturer's instructions in such a manner that operation does not create a hazard to persons or property. Furthermore, duct heaters shall be located with respect to building construction and other equipment so as to permit access to the heater. Sufficient clearance shall be maintained to permit replacement of controls and heating elements and for adjusting and cleaning of controls and other parts requiring such attention. See 110.26.

Working space about electrical enclosures for resistance heating element–type duct heaters that are mounted on duct systems and contain equipment that requires examination, adjustment, servicing, or maintenance while energized shall comply with 424.66(B).

**(B) Limited Access.** Where the enclosure is located in a space above a ceiling, all of the following shall apply:

(1) The enclosure shall be accessible through a lay-in type ceiling or an access panel(s).
(2) The width of the working space shall be the width of the enclosure or a minimum of 762 mm (30 in.), whichever is greater.
(3) All doors or hinged panels shall open to at least 90 degrees.
(4) The space in front of the enclosure shall comply with the depth requirements of Table 110.26(A)(1). A horizontal ceiling T-bar shall be permitted in this space.

Informational Note: For additional installation information, see NFPA 90A-2012, *Standard for the Installation of Air-Conditioning and Ventilating Systems*, and NFPA 90B-2012, *Standard for the Installation of Warm Air Heating and Air-Conditioning Systems*.

## VII. Resistance-Type Boilers

### 424.70 Scope

The provisions in Part VII of this article shall apply to boilers employing resistance-type heating elements. Electrode-type boilers shall not be considered as employing resistance-type heating elements. See Part VIII of this article.

### 424.71 Identification

Resistance-type boilers shall be identified as suitable for the installation.

### 424.72 Overcurrent Protection

**(A) Boiler Employing Resistance-Type Immersion Heating Elements in an ASME-Rated and Stamped Vessel.** A boiler employing resistance-type immersion heating elements contained in an ASME-rated and stamped vessel shall have the heating elements protected at not more than 150 amperes. Such a boiler rated more than 120 amperes shall have the heating elements subdivided into loads not exceeding 120 amperes.

Where a subdivided load is less than 120 amperes, the rating of the overcurrent protective device shall comply with 424.3(B).

**(B) Boiler Employing Resistance-Type Heating Elements Rated More Than 48 Amperes and Not Contained in an ASME-Rated and Stamped Vessel.** A boiler employing resistance-type heating elements not contained in an ASME-rated and stamped vessel shall have the heating elements protected at not more than 60 amperes. Such a boiler rated more than 48 amperes shall have the heating elements subdivided into loads not exceeding 48 amperes.

Where a subdivided load is less than 48 amperes, the rating of the overcurrent protective device shall comply with 424.3(B).

See the commentary following 424.22(B) for an explanation of the subdivision requirement.

**(C) Supplementary Overcurrent Protective Devices.** The supplementary overcurrent protective devices for the subdivided loads as required by 424.72(A) and (B) shall be as follows:

(1) Factory-installed within or on the boiler enclosure or provided as a separate assembly by the boiler manufacturer
(2) Accessible, but need not be readily accessible
(3) Suitable for branch-circuit protection

Where cartridge fuses are used to provide this overcurrent protection, a single disconnecting means shall be permitted for the several subdivided circuits. See 240.40.

**(D) Conductors Supplying Supplementary Overcurrent Protective Devices.** The conductors supplying these supplementary overcurrent protective devices shall be considered branch-circuit conductors.

Where the heaters are rated 50 kW or more, the conductors supplying the overcurrent protective device specified in 424.72(C) shall be permitted to be sized at not less than 100 percent of the nameplate rating of the heater, provided all of the following conditions are met:

(1) The heater is marked with a minimum conductor size.
(2) The conductors are not smaller than the marked minimum size.
(3) A temperature- or pressure-actuated device controls the cyclic operation of the equipment.

**(E) Conductors for Subdivided Loads.** Field-wired conductors between the heater and the supplementary overcurrent protective devices shall be sized at not less than 125 percent of the load served. The supplementary overcurrent protective devices specified in 424.72(C) shall protect these conductors in accordance with 240.4.

Where the heaters are rated 50 kW or more, the ampacity of field-wired conductors between the heater and the supplementary overcurrent protective devices shall be permitted to be not less than 100 percent of the load of their respective subdivided circuits, provided all of the following conditions are met:

(1) The heater is marked with a minimum conductor size.
(2) The conductors are not smaller than the marked minimum size.
(3) A temperature-activated device controls the cyclic operation of the equipment.

### 424.73 Overtemperature Limit Control

Each boiler designed so that in normal operation there is no change in state of the heat transfer medium shall be equipped with a temperature-sensitive limiting means. It shall be installed to limit maximum liquid temperature and shall directly or indirectly disconnect all ungrounded conductors to the heating elements. Such means shall be in addition to a temperature-regulating system and other devices protecting the tank against excessive pressure.

### 424.74 Overpressure Limit Control

Each boiler designed so that in normal operation there is a change in state of the heat transfer medium from liquid to vapor shall be equipped with a pressure-sensitive limiting means. It shall be installed to limit maximum pressure and shall directly or indirectly disconnect all ungrounded conductors to the heating elements. Such means shall be in addition to a pressure-regulating system and other devices protecting the tank against excessive pressure.

## VIII. Electrode-Type Boilers

### 424.80 Scope

The provisions in Part VIII of this article shall apply to boilers for operation at 600 volts, nominal, or less, in which heat is generated by the passage of current between electrodes through the liquid being heated.

Informational Note: For over 600 volts, see Part V of Article 490.

### 424.81 Identification

Electrode-type boilers shall be identified as suitable for the installation.

### 424.82 Branch-Circuit Requirements

The size of branch-circuit conductors and overcurrent protective devices shall be calculated on the basis of 125 percent of the total load (motors not included). A contactor, relay, or other device, approved for continuous operation at 100 percent of its rating, shall be permitted to supply its full-rated load. See 210.19(A), Exception. The provisions of this section shall not apply to conductors that form an integral part of an approved boiler.

Where an electrode boiler is rated 50 kW or more, the conductors supplying the boiler electrode(s) shall be permitted to be sized at not less than 100 percent of the nameplate rating of the electrode boiler, provided all the following conditions are met:

(1) The electrode boiler is marked with a minimum conductor size.
(2) The conductors are not smaller than the marked minimum size.
(3) A temperature- or pressure-actuated device controls the cyclic operation of the equipment.

### 424.83 Overtemperature Limit Control

Each boiler, designed so that in normal operation there is no change in state of the heat transfer medium, shall be equipped with a temperature-sensitive limiting means. It shall be installed to limit maximum liquid temperature and shall directly or indirectly interrupt all current flow through the electrodes. Such means shall be in addition to the temperature-regulating system and other devices protecting the tank against excessive pressure.

### 424.84 Overpressure Limit Control

Each boiler, designed so that in normal operation there is a change in state of the heat transfer medium from liquid to vapor, shall be equipped with a pressure-sensitive limiting means. It shall be installed to limit maximum pressure and shall directly or indirectly interrupt all current flow through the electrodes. Such means shall be in addition to a pressure-regulating system and other devices protecting the tank against excessive pressure.

### 424.85 Grounding

For those boilers designed such that fault currents do not pass through the pressure vessel, and the pressure vessel is electrically isolated from the electrodes, all exposed non–current-carrying metal parts, including the pressure vessel, supply, and return connecting piping, shall be grounded.

For all other designs, the pressure vessel containing the electrodes shall be isolated and electrically insulated from ground.

### 424.86 Markings

All electrode-type boilers shall be marked to show the following:

(1) The manufacturer's name.
(2) The normal rating in volts, amperes, and kilowatts.
(3) The electrical supply required specifying frequency, number of phases, and number of wires.
(4) The marking "Electrode-Type Boiler."
(5) A warning marking, "All Power Supplies Shall Be Disconnected Before Servicing, Including Servicing the Pressure Vessel." A field-applied warning marking or label shall comply with 110.21(B).

The nameplate shall be located so as to be visible after installation.

## IX. Electric Radiant Heating Panels and Heating Panel Sets

### 424.90 Scope

The provisions of Part IX of this article shall apply to radiant heating panels and heating panel sets.

### 424.91 Definitions

**Heating Panel.** A complete assembly provided with a junction box or a length of flexible conduit for connection to a branch circuit.

**Heating Panel Set.** A rigid or nonrigid assembly provided with nonheating leads or a terminal junction assembly identified as being suitable for connection to a wiring system.

### 424.92 Markings

**(A) Location.** Markings shall be permanent and in a location that is visible prior to application of panel finish.

**(B) Identified as Suitable.** Each unit shall be identified as suitable for the installation.

**(C) Required Markings.** Each unit shall be marked with the identifying name or identification symbol, catalog number, and rating in volts and watts or in volts and amperes.

**(D) Labels Provided by Manufacturer.** The manufacturers of heating panels or heating panel sets shall provide marking labels that indicate that the space-heating installation incorporates heating panels or heating panel sets and instructions that the labels shall be affixed to the panelboards to identify which branch circuits supply the circuits to those space-heating installations. If the heating panels and heating panel set installations are visible and distinguishable after installation, the labels shall not be required to be provided and affixed to the panelboards.

### 424.93 Installation

**(A) General.**

**(1) Manufacturer's Instructions.** Heating panels and heating panel sets shall be installed in accordance with the manufacturer's instructions.

**(2) Locations Not Permitted.** The heating portion shall not be installed as follows:

(1) In or behind surfaces where subject to physical damage
(2) Run through or above walls, partitions, cupboards, or similar portions of structures that extend to the ceiling
(3) Run in or through thermal insulation, but shall be permitted to be in contact with the surface of thermal insulation

**(3) Separation from Outlets for Luminaires.** Edges of panels and panel sets shall be separated by not less than 200 mm (8 in.) from the edges of any outlet boxes and junction boxes that are to be used for mounting surface luminaires. A clearance of not less than 50 mm (2 in.) shall be provided from recessed luminaires and their trims, ventilating openings, and other such openings in room surfaces, unless the heating panels and panel sets are listed and marked for lesser clearances, in which case they shall be permitted to be installed at the marked clearances. Sufficient area shall be provided to ensure that no heating panel or heating panel set is to be covered by any surface-mounted units.

**(4) Surfaces Covering Heating Panels.** After the heating panels or heating panel sets are installed and inspected, it shall be permitted to install a surface that has been identified by the manufacturer's instructions as being suitable for the installation. The surface shall be secured so that the nails or other fastenings do not pierce the heating panels or heating panel sets.

**(5) Surface Coverings.** Surfaces permitted by 424.93(A)(4) shall be permitted to be covered with paint, wallpaper, or other approved surfaces identified in the manufacturer's instructions as being suitable.

**(B) Heating Panel Sets.**

**(1) Mounting Location.** Heating panel sets shall be permitted to be secured to the lower face of joists or mounted in between joists, headers, or nailing strips.

**(2) Parallel to Joists or Nailing Strips.** Heating panel sets shall be installed parallel to joists or nailing strips.

**(3) Installation of Nails, Staples, or Other Fasteners.** Nailing or stapling of heating panel sets shall be done only through the unheated portions provided for this purpose. Heating panel sets shall not be cut through or nailed through any point closer than 6 mm (¼ in.) to the element. Nails, staples, or other fasteners shall not be used where they penetrate current-carrying parts.

**(4) Installed as Complete Unit.** Heating panel sets shall be installed as complete units unless identified as suitable for field cutting in an approved manner.

### 424.94 Clearances of Wiring in Ceilings

Wiring located above heated ceilings shall be spaced not less than 50 mm (2 in.) above the heated ceiling and shall be considered as operating at an ambient of 50°C (122°F). The ampacity shall be calculated on the basis of the correction factors given in the 0–2000 volt ampacity tables of Article 310. If this wiring is located above thermal insulations having a minimum thickness of 50 mm (2 in.), the wiring shall not require correction for temperature.

### 424.95 Location of Branch-Circuit and Feeder Wiring in Walls

**(A) Exterior Walls.** Wiring methods shall comply with Article 300 and 310.15(A)(3).

**(B) Interior Walls.** Any wiring behind heating panels or heating panel sets located in interior walls or partitions shall be considered as operating at an ambient temperature of 40°C (104°F), and the ampacity shall be calculated on the basis of the correction factors given in the 0–2000 volt ampacity tables of Article 310.

### 424.96 Connection to Branch-Circuit Conductors

**(A) General.** Heating panels or heating panel sets assembled together in the field to form a heating installation in one room or area shall be connected in accordance with the manufacturer's instructions.

**(B) Heating Panels.** Heating panels shall be connected to branch-circuit wiring by an approved wiring method.

**(C) Heating Panel Sets.**

**(1) Connection to Branch-Circuit Wiring.** Heating panel sets shall be connected to branch-circuit wiring by a method identified as being suitable for the purpose.

**(2) Panel Sets with Terminal Junction Assembly.** A heating panel set provided with terminal junction assembly shall be permitted to have the nonheating leads attached at the time of installation in accordance with the manufacturer's instructions.

### 424.97 Nonheating Leads

Excess nonheating leads of heating panels or heating panel sets shall be permitted to be cut to the required length. They shall meet the installation requirements of the wiring method employed in accordance with 424.96. Nonheating leads shall be an integral part of a heating panel and a heating panel set and shall not be subjected to the ampacity requirements of 424.3(B) for branch circuits.

### 424.98 Installation in Concrete or Poured Masonry

**(A) Maximum Heated Area.** Heating panels or heating panel sets shall not exceed 355 watts/m$^2$ (33 watts/ft$^2$) of heated area.

**(B) Secured in Place and Identified as Suitable.** Heating panels or heating panel sets shall be secured in place by means specified in the manufacturer's instructions and identified as suitable for the installation.

**(C) Expansion Joints.** Heating panels or heating panel sets shall not be installed where they bridge expansion joints unless provision is made for expansion and contraction.

**(D) Spacings.** Spacings shall be maintained between heating panels or heating panel sets and metal embedded in the floor. Grounded metal-clad heating panels shall be permitted to be in contact with metal embedded in the floor.

**(E) Protection of Leads.** Leads shall be protected where they leave the floor by rigid metal conduit, intermediate metal conduit, rigid nonmetallic conduit, or electrical metallic tubing, or by other approved means.

**(F) Bushings or Fittings Required.** Bushings or approved fittings shall be used where the leads emerge within the floor slabs.

### 424.99 Installation Under Floor Covering

A system that uses conductive-film heating elements is an example of a heating system that could be installed under a floor covering.

**(A) Identification.** Heating panels or heating panel sets for installation under floor covering shall be identified as suitable for installation under floor covering.

**(B) Maximum Heated Area.** Heating panels or panel sets installed under floor covering shall not exceed 160 watts/m$^2$ (15 watts/ft$^2$) of heated area.

**(C) Installation.** Listed heating panels or panel sets, if installed under floor covering, shall be installed on floor surfaces that are smooth and flat in accordance with the manufacturer's instructions and shall also comply with 424.99(C)(1) through (C)(5).

**(1) Expansion Joints.** Heating panels or heating panel sets shall not be installed where they bridge expansion joints unless protected from expansion and contraction.

**(2) Connection to Conductors.** Heating panels and heating panel sets shall be connected to branch-circuit and supply wiring by wiring methods recognized in Chapter 3.

**(3) Anchoring.** Heating panels and heating panel sets shall be firmly anchored to the floor using an adhesive or anchoring system identified for this use.

**(4) Coverings.** After heating panels or heating panel sets are installed and inspected, they shall be permitted to be covered by a floor covering that has been identified by the manufacturer as being suitable for the installation. The covering shall be secured to the heating panel or heating panel sets with release-type adhesives or by means identified for this use.

**(5) Fault Protection.** A device to open all ungrounded conductors supplying the heating panels or heating panel sets, provided by the manufacturer, shall function when a low- or high-resistance line-to-line, line-to-grounded conductor, or line-to-ground fault occurs, such as the result of a penetration of the element or element assembly.

Informational Note: An integral grounding shield may be required to provide this protection.

# ARTICLE 426 Fixed Outdoor Electric Deicing and Snow-Melting Equipment

## I. General

### 426.1 Scope

The requirements of this article shall apply to electrically energized heating systems and the installation of these systems.

**(A) Embedded.** Embedded in driveways, walks, steps, and other areas.

**(B) Exposed.** Exposed on drainage systems, bridge structures, roofs, and other structures.

Article 426 includes requirements for resistance heating elements, impedance heating systems, or skin-effect heating systems used for deicing and snow melting.

### 426.2 Definitions

**Heating System.** A complete system consisting of components such as heating elements, fastening devices, nonheating circuit wiring, leads, temperature controllers, safety signs, junction boxes, raceways, and fittings.

**Impedance Heating System.** A system in which heat is generated in a pipe or rod, or combination of pipes and rods, by causing current to flow through the pipe or rod by direct connection to an ac voltage source from an isolating transformer. The pipe or rod shall be permitted to be embedded in the surface to be heated, or constitute the exposed components to be heated.

**Resistance Heating Element.** A specific separate element to generate heat that is embedded in or fastened to the surface to be heated.

Informational Note: Tubular heaters, strip heaters, heating cable, heating tape, and heating panels are examples of resistance heaters.

**Skin-Effect Heating System.** A system in which heat is generated on the inner surface of a ferromagnetic envelope embedded in or fastened to the surface to be heated.

Informational Note: Typically, an electrically insulated conductor is routed through and connected to the envelope at the other end. The envelope and the electrically insulated conductor are connected to an ac voltage source from an isolating transformer.

### 426.3 Application of Other Articles

Cord-and-plug-connected fixed outdoor electric deicing and snow-melting equipment intended for specific use and identified as suitable for this use shall be installed according to Article 422.

### 426.4 Continuous Load

Fixed outdoor electric deicing and snow-melting equipment shall be considered as a continuous load.

Fixed outdoor electric deicing and snow-melting equipment can operate over 3 hours, so it is considered a continuous load when sizing branch circuits, feeders, service conductors, and OCPDs.

## II. Installation

### 426.10 General

Equipment for outdoor electric deicing and snow melting shall be identified as being suitable for the following:

(1) The chemical, thermal, and physical environment
(2) Installation in accordance with the manufacturer's drawings and instructions

### 426.11 Use

Electric heating equipment shall be installed in such a manner as to be afforded protection from physical damage.

Underwriters Laboratories Inc. requires that manufacturers of UL-listed mat or cable deicing and snow-melting equipment provide specific installation instructions for their products. These instructions supplement the requirements contained in Article 426. For example, if the equipment can be installed only in concrete that is double poured (poured in two parts), the installation instructions are to specifically require that installation technique. Where the instructions do not specify an installation process, the use of either a single- or double-pour installation method is acceptable. See 110.3(B) regarding the installation and use of listed or labeled equipment.

### 426.12 Thermal Protection

External surfaces of outdoor electric deicing and snow-melting equipment that operate at temperatures exceeding 60°C (140°F) shall be physically guarded, isolated, or thermally insulated to protect against contact by personnel in the area.

### 426.13 Identification

The presence of outdoor electric deicing and snow-melting equipment shall be evident by the posting of appropriate caution signs or markings where clearly visible.

### 426.14 Special Permission

Fixed outdoor deicing and snow-melting equipment employing methods of construction or installation other than covered by this article shall be permitted only by special permission.

See the definition of *special permission* in Article 100.

## III. Resistance Heating Elements

### 426.20 Embedded Deicing and Snow-Melting Equipment.

**(A) Watt Density.** Panels or units shall not exceed 1300 watts/m$^2$ (120 watts/ft$^2$) of heated area.

**(B) Spacing.** The spacing between adjacent cable runs is dependent upon the rating of the cable and shall be not less than 25 mm (1 in.) on centers.

**(C) Cover.** Units, panels, or cables shall be installed as follows:

(1) On a substantial asphalt or masonry base at least 50 mm (2 in.) thick and have at least 38 mm (1½ in.) of asphalt or masonry applied over the units, panels, or cables; or

(2) They shall be permitted to be installed over other approved bases and embedded within 90 mm (3½ in.) of masonry or asphalt but not less than 38 mm (1½ in.) from the top surface; or

(3) Equipment that has been listed for other forms of installation shall be installed only in the manner for which it has been identified.

**(D) Secured.** Cables, units, and panels shall be secured in place by frames or spreaders or other approved means while the masonry or asphalt finish is applied.

**(E) Expansion and Contraction.** Cables, units, and panels shall not be installed where they bridge expansion joints unless provision is made for expansion and contraction.

## 426.21 Exposed Deicing and Snow-Melting Equipment

**(A) Secured.** Heating element assemblies shall be secured to the surface being heated by approved means.

**(B) Overtemperature.** Where the heating element is not in direct contact with the surface being heated, the design of the heater assembly shall be such that its temperature limitations shall not be exceeded.

**(C) Expansion and Contraction.** Heating elements and assemblies shall not be installed where they bridge expansion joints unless provision is made for expansion and contraction.

**(D) Flexural Capability.** Where installed on flexible structures, the heating elements and assemblies shall have a flexural capability that is compatible with the structure.

## 426.22 Installation of Nonheating Leads for Embedded Equipment

**(A) Grounding Sheath or Braid.** Nonheating leads having a grounding sheath or braid shall be permitted to be embedded in the masonry or asphalt in the same manner as the heating cable without additional physical protection.

**(B) Raceways.** All but 25 mm to 150 mm (1 in. to 6 in.) of nonheating leads not having a grounding sheath shall be enclosed in a rigid metal conduit, electrical metallic tubing, intermediate metal conduit, or other raceways within asphalt or masonry. The distance from the factory splice to raceway shall not be less than 25 mm (1 in.) or more than 150 mm (6 in.).

**(C) Bushings.** Insulating bushings shall be used in the asphalt or masonry where leads enter conduit or tubing.

See 300.4(G) and the associated commentary for further information regarding insulating bushings.

**(D) Expansion and Contraction.** Leads shall be protected in expansion joints and where they emerge from masonry or asphalt by rigid conduit, electrical metallic tubing, intermediate metal conduit, other raceways, or other approved means.

**(E) Leads in Junction Boxes.** Not less than 150 mm (6 in.) of free nonheating lead shall be within the junction box.

## 426.23 Installation of Nonheating Leads for Exposed Equipment

**(A) Nonheating Leads.** Power supply nonheating leads (cold leads) for resistance elements shall be identified for the temperature encountered. Not less than 150 mm (6 in.) of nonheating leads shall be provided within the junction box. Preassembled factory-supplied and field-assembled nonheating leads on approved heaters shall be permitted to be shortened if the markings specified in 426.25 are retained.

**(B) Protection.** Nonheating power supply leads shall be enclosed in a rigid conduit, intermediate metal conduit, electrical metallic tubing, or other approved means.

## 426.24 Electrical Connection

**(A) Heating Element Connections.** Electrical connections, other than factory connections of heating elements to nonheating elements embedded in masonry or asphalt or on exposed surfaces, shall be made with insulated connectors identified for the use.

**(B) Circuit Connections.** Splices and terminations at the end of the nonheating leads, other than the heating element end, shall be installed in a box or fitting in accordance with 110.14 and 300.15.

## 426.25 Marking

Each factory-assembled heating unit shall be legibly marked within 75 mm (3 in.) of each end of the nonheating leads with the permanent identification symbol, catalog number, and ratings in volts and watts or in volts and amperes.

## 426.26 Corrosion Protection

Ferrous and nonferrous metal raceways, cable armor, cable sheaths, boxes, fittings, supports, and support hardware shall be permitted to be installed in concrete or in direct contact with the earth, or in areas subject to severe corrosive influences, where made of material suitable for the condition, or where provided with corrosion protection identified as suitable for the condition.

## 426.27 Grounding Braid or Sheath

Grounding means, such as copper braid, metal sheath, or other approved means, shall be provided as part of the heated section of the cable, panel, or unit.

## 426.28 Ground-Fault Protection of Equipment

Ground-fault protection of equipment shall be provided for fixed outdoor electric deicing and snow-melting equipment.

Rather than protecting the entire branch circuit, the ground-fault protection requirement is focused on protecting just the equipment itself. The manufacturer and the user have an option of providing both circuit and equipment protection or just the required equipment protection. This required protection may be accomplished by using circuit breakers equipped with equipment ground-fault protection or an integral device supplied as part of the deicing or snow-melting equipment that is sensitive to leakage currents of 6 mA to 50 mA. These protection devices, if applied properly, will substantially reduce the risk of a fire being started by low-level electrical arcing.

Note that the required equipment protection is not the same as a GFCI used for personal protection that trips at 5 mA (±1 mA).

## IV. Impedance Heating

### 426.30 Personnel Protection

Exposed elements of impedance heating systems shall be physically guarded, isolated, or thermally insulated with a weatherproof jacket to protect against contact by personnel in the area.

### 426.31 Isolation Transformer

An isolation transformer with a grounded shield between the primary and secondary windings shall be used to isolate the distribution system from the heating system.

### 426.32 Voltage Limitations

Unless protected by ground-fault circuit-interrupter protection for personnel, the secondary winding of the isolation transformer connected to the impedance heating elements shall not have an output voltage greater than 30 volts ac.

Where ground-fault circuit-interrupter protection for personnel is provided, the voltage shall be permitted to be greater than 30 but not more than 80 volts.

### 426.33 Induced Currents

All current-carrying components shall be installed in accordance with 300.20.

### 426.34 Grounding

An impedance heating system that is operating at a voltage greater than 30 but not more than 80 shall be grounded at a designated point(s).

## V. Skin-Effect Heating

### 426.40 Conductor Ampacity

The current through the electrically insulated conductor inside the ferromagnetic envelope shall be permitted to exceed the ampacity values shown in Article 310, provided it is identified as suitable for this use.

### 426.41 Pull Boxes

Where pull boxes are used, they shall be accessible without excavation by location in suitable vaults or abovegrade. Outdoor pull boxes shall be of watertight construction.

### 426.42 Single Conductor in Enclosure

The provisions of 300.20 shall not apply to the installation of a single conductor in a ferromagnetic envelope (metal enclosure).

### 426.43 Corrosion Protection

Ferromagnetic envelopes, ferrous or nonferrous metal raceways, boxes, fittings, supports, and support hardware shall be permitted to be installed in concrete or in direct contact with the earth, or in areas subjected to severe corrosive influences, where made of material suitable for the condition, or where provided with corrosion protection identified as suitable for the condition. Corrosion protection shall maintain the original wall thickness of the ferromagnetic envelope.

### 426.44 Grounding

The ferromagnetic envelope shall be connected to an equipment grounding conductor at both ends; and, in addition, it shall be permitted to be connected to an equipment grounding conductor at intermediate points as required by its design.

The provisions of 250.30 shall not apply to the installation of skin-effect heating systems.

Informational Note: For grounding methods, see Article 250.

## VI. Control and Protection

### 426.50 Disconnecting Means

**(A) Disconnection.** All fixed outdoor deicing and snow-melting equipment shall be provided with a means for simultaneous disconnection from all ungrounded conductors. Where readily accessible to the user of the equipment, the branch-circuit switch or circuit breaker shall be permitted to serve as the disconnecting means. The disconnecting means shall be of the indicating type and be capable of being locked in the open (off) position.

The disconnecting means must simultaneously open all ungrounded conductors to prevent the practice of disconnecting one conductor at a time at terminal blocks or similar devices.

**(B) Cord-and-Plug-Connected Equipment.** The factory-installed attachment plug of cord-and-plug-connected equipment rated 20 amperes or less and 150 volts or less to ground shall be permitted to be the disconnecting means.

### 426.51 Controllers

**(A) Temperature Controller with "Off" Position.** Temperature controlled switching devices that indicate an "off" position and that interrupt line current shall open all ungrounded conductors when the control device is in the "off" position. These devices shall not be permitted to serve as the disconnecting means unless they are lockable in accordance with 110.25.

**(B) Temperature Controller Without "Off" Position.** Temperature controlled switching devices that do not have an "off" position shall not be required to open all ungrounded conductors and shall not be permitted to serve as the disconnecting means.

**(C) Remote Temperature Controller.** Remote controlled temperature-actuated devices shall not be required to meet the requirements of 426.51(A). These devices shall not be permitted to serve as the disconnecting means.

**(D) Combined Switching Devices.** Switching devices consisting of combined temperature-actuated devices and manually controlled switches that serve both as the controller and the disconnecting means shall comply with all of the following conditions:

(1) Open all ungrounded conductors when manually placed in the "off" position
(2) Be so designed that the circuit cannot be energized automatically if the device has been manually placed in the "off" position
(3) Be lockable in accordance with 110.25

### 426.54 Cord-and-Plug-Connected Deicing and Snow-Melting Equipment

Cord-and-plug-connected deicing and snow-melting equipment shall be listed.

According to the UL *Guide Information for Electrical Equipment – The White Book*, category KOBQ, deicing and snow-melting equipment is provided with means for permanent wiring connection, except for equipment rated 20 amperes or less and 150 volts or less to ground, which may be of cord-and-plug-connected construction. See the definition of *listed* in Article 100.

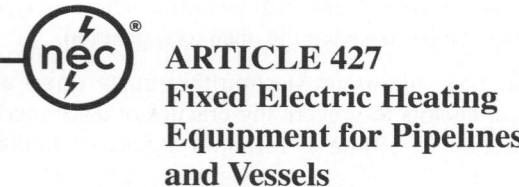

# ARTICLE 427 Fixed Electric Heating Equipment for Pipelines and Vessels

## I. General

### 427.1 Scope

The requirements of this article shall apply to electrically energized heating systems and the installation of these systems used with pipelines or vessels or both.

Informational Note: For further information, see ANSI/IEEE 515-2002, *Standard for the Testing, Design, Installation and Maintenance of Electrical Resistance Heat Tracing for Industrial Applications;* ANSI/IEEE 844-2000, *Recommended Practice for Electrical Impedance, Induction, and Skin Effect Heating of Pipelines and Vessels*; and ANSI/NECA 202-2006, *Standard for Installing and Maintaining Industrial Heat Tracing Systems.*

Article 427 includes requirements for impedance heating, induction heating, and skin-effect heating, in addition to resistance heating elements. Definitions of the various systems are provided in 427.2.

### 427.2 Definitions

**Impedance Heating System.** A system in which heat is generated in a pipeline or vessel wall by causing current to flow through the pipeline or vessel wall by direct connection to an ac voltage source from a dual-winding transformer.

**Induction Heating System.** A system in which heat is generated in a pipeline or vessel wall by inducing current and hysteresis effect in the pipeline or vessel wall from an external isolated ac field source.

•

**Pipeline.** A length of pipe including pumps, valves, flanges, control devices, strainers, and/or similar equipment for conveying fluids.

**Resistance Heating Element.** A specific separate element to generate heat that is applied to the pipeline or vessel externally or internally.

Informational Note: Tubular heaters, strip heaters, heating cable, heating tape, heating blankets, and immersion heaters are examples of resistance heaters.

**Skin-Effect Heating System.** A system in which heat is generated on the inner surface of a ferromagnetic envelope attached to a pipeline or vessel, or both.

Informational Note: Typically, an electrically insulated conductor is routed through and connected to the envelope at the other end. The envelope and the electrically insulated conductor are connected to an ac voltage source from a dual-winding transformer.

**Vessel.** A container such as a barrel, drum, or tank for holding fluids or other material.

### 427.3 Application of Other Articles

Cord-connected pipe heating assemblies intended for specific use and identified as suitable for this use shall be installed according to Article 422.

### 427.4 Continuous Load

Fixed electric heating equipment for pipelines and vessels shall be considered continuous load.

Fixed electric heating equipment is considered a continuous load for the purpose of sizing branch circuits, feeders, service conductors, and OCPDs.

## II. Installation

### 427.10 General

Equipment for pipeline and vessel electric heating shall be identified as being suitable for (1) the chemical, thermal, and physical environment and (2) installation in accordance with the manufacturer's drawings and instructions.

### 427.11 Use

Electric heating equipment shall be installed in such a manner as to be afforded protection from physical damage.

### 427.12 Thermal Protection

External surfaces of pipeline and vessel heating equipment that operate at temperatures exceeding 60°C (140°F) shall be physically guarded, isolated, or thermally insulated to protect against contact by personnel in the area.

### 427.13 Identification

The presence of electrically heated pipelines, vessels, or both, shall be evident by the posting of appropriate caution signs or markings at intervals not exceeding 6 m (20 ft) along the pipeline or vessel and on or adjacent to equipment in the piping system that requires periodic servicing.

## III. Resistance Heating Elements

### 427.14 Secured

Heating element assemblies shall be secured to the surface being heated by means other than the thermal insulation.

### 427.15 Not in Direct Contact

Where the heating element is not in direct contact with the pipeline or vessel being heated, means shall be provided to prevent overtemperature of the heating element unless the design of the heater assembly is such that its temperature limitations will not be exceeded.

### 427.16 Expansion and Contraction

Heating elements and assemblies shall not be installed where they bridge expansion joints unless provisions are made for expansion and contraction.

### 427.17 Flexural Capability

Where installed on flexible pipelines, the heating elements and assemblies shall have a flexural capability that is compatible with the pipeline.

### 427.18 Power Supply Leads

**(A) Nonheating Leads.** Power supply nonheating leads (cold leads) for resistance elements shall be suitable for the temperature encountered. Not less than 150 mm (6 in.) of nonheating leads shall be provided within the junction box. Preassembled factory-supplied and field-assembled nonheating leads on approved heaters shall be permitted to be shortened if the markings specified in 427.20 are retained.

**(B) Power Supply Leads Protection.** Nonheating power supply leads shall be protected where they emerge from electrically heated pipeline or vessel heating units by rigid metal conduit, intermediate metal conduit, electrical metallic tubing, or other raceways identified as suitable for the application.

**(C) Interconnecting Leads.** Interconnecting nonheating leads connecting portions of the heating system shall be permitted to be covered by thermal insulation in the same manner as the heaters.

### 427.19 Electrical Connections

**(A) Nonheating Interconnections.** Nonheating interconnections, where required under thermal insulation, shall be made with insulated connectors identified as suitable for this use.

**(B) Circuit Connections.** Splices and terminations outside the thermal insulation shall be installed in a box or fitting in accordance with 110.14 and 300.15.

### 427.20 Marking

Each factory-assembled heating unit shall be legibly marked within 75 mm (3 in.) of each end of the nonheating leads with the permanent identification symbol, catalog number, and ratings in volts and watts or in volts and amperes.

### 427.22 Ground-Fault Protection of Equipment

Ground-fault protection of equipment shall be provided for electric heat tracing and heating panels. This requirement shall not apply in industrial establishments where there is alarm indication of ground faults and the following conditions apply:

(1) Conditions of maintenance and supervision ensure that only qualified persons service the installed systems.
(2) Continued circuit operation is necessary for safe operation of equipment or processes.

Rather than protecting the entire branch circuit, the ground-fault protection requirement is focused on protecting just the equipment itself. Such protection affords the manufacturer and the user the option of providing both circuit and equipment protection or just the required equipment protection. Circuit breakers equipped with equipment ground-fault protection or an integral device supplied as part of the pipeline or vessel heating equipment that is sensitive to leakage currents from 6 mA to 50 mA will provide the required protection. These protective devices, if applied properly, substantially reduce the risk of fire being started by low-level electrical arcing.

The required equipment protection is not the same as that provided by a GFCI used for personal protection that trips at 5 mA (±1 mA).

### 427.23 Grounded Conductive Covering

Electric heating equipment shall be listed and have a grounded conductive covering in accordance with 427.23(A) or (B). The conductive covering shall provide an effective ground path for equipment protection.

The grounded conductive covering is intended to provide a ground-fault current path in order to trip circuit or ground-fault protective devices, thus reducing the potential for fire and electric shock. It also provides added mechanical protection of the heating cable or panel.

**(A) Heating Wires or Cables.** Heating wires or cables shall have a grounded conductive covering that surrounds the heating element and bus wires, if any, and their electrical insulation.

**(B) Heating Panels.** Heating panels shall have a grounded conductive covering over the heating element and its electrical insulation on the side opposite the side attached to the surface to be heated.

## IV. Impedance Heating

### 427.25 Personnel Protection

All accessible external surfaces of the pipeline, vessel, or both, being heated shall be physically guarded, isolated, or thermally insulated (with a weatherproof jacket for outside installations) to protect against contact by personnel in the area.

### 427.26 Isolation Transformer

A dual-winding transformer with a grounded shield between the primary and secondary windings shall be used to isolate the distribution system from the heating system.

### 427.27 Voltage Limitations

Unless protected by ground-fault circuit-interrupter protection for personnel, the secondary winding of the isolation transformer connected to the pipeline or vessel being heated shall not have an output voltage greater than 30 volts ac.

Where ground-fault circuit-interrupter protection for personnel is provided, the voltage shall be permitted to be greater than 30 but not more than 80 volts.

*Exception: In industrial establishments, the isolation transformer connected to the pipeline or vessel being heated shall be permitted to have an output voltage not greater than 132 volts ac to ground where all of the following conditions apply:*

*(1) Conditions of maintenance and supervision ensure that only qualified persons service the installed systems.*
*(2) Ground-fault protection of equipment is provided.*
*(3) The pipeline or vessel being heated is completely enclosed in a grounded metal enclosure.*
*(4) The transformer secondary connections to the pipeline or vessel being heated are completely enclosed in a grounded metal mesh or metal enclosure.*

The exception allows a voltage of not more than 132 volts for impedance heating where installed in industrial establishments, provided four prerequisites are followed. For other than industrial establishments, the maximum operating voltage allowed is 30 V ac; where GFCI protection for personnel is used, a maximum of 80 V ac is permitted.

### 427.28 Induced Currents

All current-carrying components shall be installed in accordance with 300.20.

### 427.29 Grounding

The pipeline, vessel, or both, that is being heated and operating at a voltage greater than 30 but not more than 80 shall be grounded at designated points.

### 427.30 Secondary Conductor Sizing

The ampacity of the conductors connected to the secondary of the transformer shall be at least 100 percent of the total load of the heater.

## V. Induction Heating

### 427.35 Scope

This part covers the installation of line frequency induction heating equipment and accessories for pipelines and vessels.

Informational Note: See Article 665 for other applications.

### 427.36 Personnel Protection

Induction coils that operate or may operate at a voltage greater than 30 volts ac shall be enclosed in a nonmetallic or split metallic enclosure, isolated, or made inaccessible by location to protect personnel in the area.

### 427.37 Induced Current

Induction coils shall be prevented from inducing circulating currents in surrounding metallic equipment, supports, or structures by shielding, isolation, or insulation of the current paths. Stray current paths shall be bonded to prevent arcing.

## VI. Skin-Effect Heating

### 427.45 Conductor Ampacity

The ampacity of the electrically insulated conductor inside the ferromagnetic envelope shall be permitted to exceed the values given in Article 310, provided it is identified as suitable for this use.

### 427.46 Pull Boxes

Pull boxes for pulling the electrically insulated conductor in the ferromagnetic envelope shall be permitted to be buried under the thermal insulation, provided their locations are indicated by permanent markings on the insulation jacket surface and on drawings. For outdoor installations, pull boxes shall be of watertight construction.

### 427.47 Single Conductor in Enclosure

The provisions of 300.20 shall not apply to the installation of a single conductor in a ferromagnetic envelope (metal enclosure).

### 427.48 Grounding

The ferromagnetic envelope shall be grounded at both ends, and, in addition, it shall be permitted to be grounded at intermediate points as required by its design. The ferromagnetic envelope shall be bonded at all joints to ensure electrical continuity.

The provisions of 250.30 shall not apply to the installation of skin-effect heating systems.

Informational Note: See Article 250 for grounding methods.

## VII. Control and Protection

### 427.55 Disconnecting Means

**(A) Switch or Circuit Breaker.** Means shall be provided to simultaneously disconnect all fixed electric pipeline or vessel heating equipment from all ungrounded conductors. The branch-circuit switch or circuit breaker, where readily accessible to the user of the equipment, shall be permitted to serve as the disconnecting means. The disconnecting means shall be of the indicating type and shall be capable of being locked in the open (off) position. The disconnecting means shall be installed in accordance with 110.25.

The simultaneous opening of all the ungrounded conductors by the disconnecting means is intended to prevent the practice of disconnecting one conductor at a time at terminal blocks or similar devices.

**(B) Cord-and-Plug-Connected Equipment.** The factory-installed attachment plug of cord-and-plug-connected equipment rated 20 amperes or less and 150 volts or less to ground shall be permitted to be the disconnecting means.

### 427.56 Controls

**(A) Temperature Control with "Off" Position.** Temperature-controlled switching devices that indicate an "off" position and that interrupt line current shall open all ungrounded conductors when the control device is in this "off" position. These devices shall not be permitted to serve as the disconnecting means unless capable of being locked in the open position.

**(B) Temperature Control Without "Off" Position.** Temperature controlled switching devices that do not have an "off" position shall not be required to open all ungrounded conductors and shall not be permitted to serve as the disconnecting means.

**(C) Remote Temperature Controller.** Remote controlled temperature-actuated devices shall not be required to meet the requirements of 427.56(A) and (B). These devices shall not be permitted to serve as the disconnecting means.

**(D) Combined Switching Devices.** Switching devices consisting of combined temperature-actuated devices and manually controlled switches that serve both as the controllers and the disconnecting means shall comply with all the following conditions:

(1) Open all ungrounded conductors when manually placed in the "off" position
(2) Be designed so that the circuit cannot be energized automatically if the device has been manually placed in the "off" position
(3) Be capable of being locked in the open position

### 427.57 Overcurrent Protection

Heating equipment shall be considered as protected against overcurrent where supplied by a branch circuit as specified in 210.3 and 210.23.

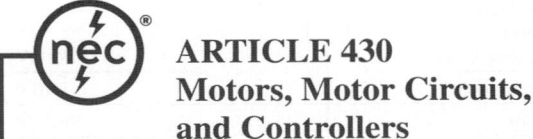

# ARTICLE 430 Motors, Motor Circuits, and Controllers

## I. General

Most electrical equipment is rated in volt-amperes (VA) or watt input, but motors traditionally have been rated in horsepower output. Circuits supplying motors are sized according to the input to the motor. The input includes the motor losses and the power factor of the motor. The losses are not the type of information found on the nameplate of a motor. Tables 430.249 and 430.250 contain accurate industry-wide input ampere ratings for motors.

Some motors are available with their output ratings expressed in watts and kilowatts. (One horsepower equals approximately 746 watts.) Circuits that supply motors not rated in horsepower still must be sized according to the input of the motor, rated in amperes. Sizing circuits based solely on kilowatt output results in seriously undersized conductors (because the current requirements of the losses and the power factor are neglected) and in the improper application of overcurrent devices. See 430.6 for ampacity and motor rating determination.

### 430.1 Scope

This article covers motors, motor branch-circuit and feeder conductors and their protection, motor overload protection, motor control circuits, motor controllers, and motor control centers.

Informational Note No. 1: Installation requirements for motor control centers are covered in 110.26(E). Air-conditioning and refrigerating equipment are covered in Article 440.

Informational Note No. 2: Figure 430.1 is for information only.

| | |
|---|---|
| General, 430.1 through 430.18 | Part I |
| Motor Circuit Conductors, 430.21 through 430.29 | Part II |
| Motor and Branch-Circuit Overload Protection, 430.31 through 430.44 | Part III |
| Motor Branch-Circuit Short-Circuit and Ground-Fault Protection, 430.51 through 430.58 | Part IV |
| Motor Feeder Short-Circuit and Ground-Fault Protection, 430.61 through 430.63 | Part V |
| Motor Control Circuits, 430.71 through 430.75 | Part VI |
| Motor Controllers, 430.81 through 430.90 | Part VII |
| Motor Control Centers, 430.92 through 430.98 | Part VIII |
| Disconnecting Means, 430.101 through 430.113 | Part IX |
| Adjustable-Speed Drive Systems, 430.120 through 430.131 | Part X |
| Over 1000 Volts, Nominal, 430.221 through 430.227 | Part XI |
| Protection of Live Parts—All Voltages, 430.231 through 430.233 | Part XII |
| Grounding—All Voltages, 430.241 through 430.245 | Part XIII |
| Tables, Tables 430.247 through 430.251(B) | Part XIV |

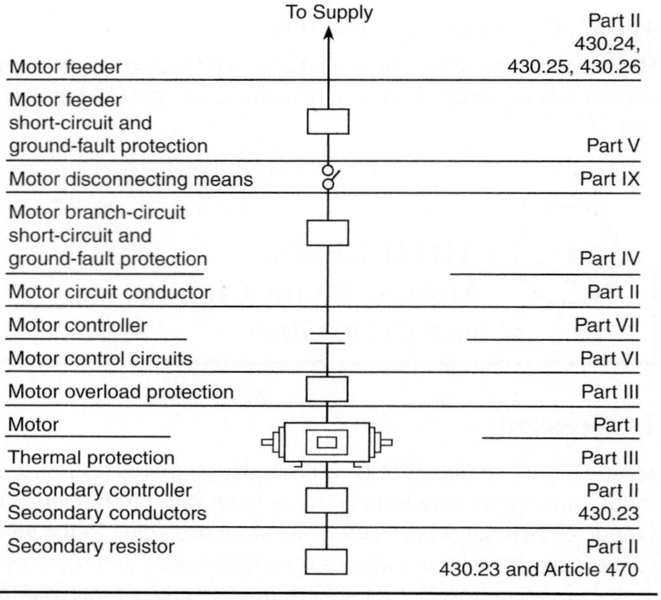

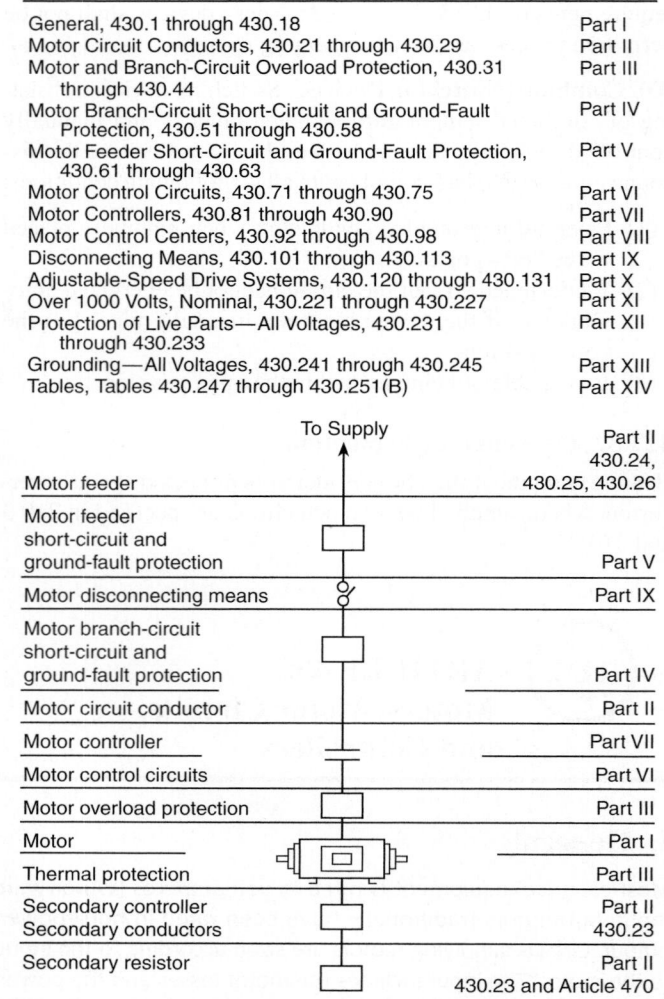

***FIGURE 430.1*** *Article 430 Contents.*

## 430.2 Definitions

•

**Controller.** For the purpose of this article, a controller is any switch or device that is normally used to start and stop a motor by making and breaking the motor circuit current.

•

**System Isolation Equipment.** A redundantly monitored, remotely operated contactor-isolating system, packaged to provide the disconnection/isolation function, capable of verifiable operation from multiple remote locations by means of lockout switches, each having the capability of being padlocked in the "off" (open) position.

**Valve Actuator Motor (VAM) Assemblies.** A manufactured assembly, used to operate a valve, consisting of an actuator motor and other components such as controllers, torque switches, limit switches, and overload protection.

Informational Note: VAMs typically have short-time duty and high-torque characteristics.

## 430.4 Part-Winding Motors

A part-winding start induction or synchronous motor is one that is arranged for starting by first energizing part of its primary (armature) winding and, subsequently, energizing the remainder of this winding in one or more steps. A standard part-winding start induction motor is arranged so that one-half of its primary winding can be energized initially, and, subsequently, the remaining half can be energized, both halves then carrying equal current. A hermetic refrigerant compressor motor shall not be considered a standard part-winding start induction motor.

Where separate overload devices are used with a standard part-winding start induction motor, each half of the motor winding shall be individually protected in accordance with 430.32 and 430.37 with a trip current one-half that specified.

Each motor-winding connection shall have branch-circuit short-circuit and ground-fault protection rated at not more than one-half that specified by 430.52.

*Exception: A short-circuit and ground-fault protective device shall be permitted for both windings if the device will allow the motor to start. Where time-delay (dual-element) fuses are used, they shall be permitted to have a rating not exceeding 150 percent of the motor full-load current.*

## 430.5 Other Articles

Motors and controllers shall also comply with the applicable provisions of Table 430.5.

## 430.6 Ampacity and Motor Rating Determination

The size of conductors supplying equipment covered by Article 430 shall be selected from the allowable ampacity tables in accordance with 310.15(B) or shall be calculated in accordance with 310.15(C). Where flexible cord is used, the size of the conductor shall be selected in accordance with 400.5. The required ampacity and motor ratings shall be determined as specified in 430.6(A), (B), (C), and (D).

**(A) General Motor Applications.** For general motor applications, current ratings shall be determined based on (A)(1) and (A)(2).

**(1) Table Values.** Other than for motors built for low speeds (less than 1200 RPM) or high torques, and for multispeed motors, the values given in Table 430.247, Table 430.248, Table 430.249, and Table 430.250 shall be used to determine the ampacity of conductors or ampere ratings of switches, branch-circuit short-circuit and ground-fault protection, instead of the actual current rating marked on the motor nameplate. Where a motor is marked in amperes, but not horsepower, the horsepower rating shall be assumed to be that corresponding to the value given in Table 430.247, Table 430.248, Table 430.249, and Table 430.250, interpolated if necessary. Motors built for low speeds (less than 1200 RPM) or high torques may have higher full-load currents, and multispeed motors will have

*TABLE 430.5* Other Articles

| Equipment/Occupancy | Article | Section |
|---|---|---|
| Air-conditioning and refrigerating equipment | 440 | |
| Capacitors | | 460.8, 460.9 |
| Commercial garages; aircraft hangars; motor fuel dispensing facilities; bulk storage plants; spray application, dipping, and coating processes; and inhalation anesthetizing locations | 511, 513, 514, 515, 516, and 517 Part IV | |
| Cranes and hoists | 610 | |
| Electrically driven or controlled irrigation machines | 675 | |
| Elevators, dumbwaiters, escalators, moving walks, wheelchair lifts, and stairway chair lifts | 620 | |
| Fire pumps | 695 | |
| Hazardous (classified) locations | 500–503, 505, and 506 | |
| Industrial machinery | 670 | |
| Motion picture projectors | | 540.11 and 540.20 |
| Motion picture and television studios and similar locations | 530 | |
| Resistors and reactors | 470 | |
| Theaters, audience areas of motion picture and television studios, and similar locations | | 520.48 |
| Transformers and transformer vaults | 450 | |

full-load current varying with speed, in which case the nameplate current ratings shall be used.

*Exception No. 1: Multispeed motors shall be in accordance with 430.22(A) and 430.52.*

*Exception No. 2: For equipment that employs a shaded-pole or permanent-split capacitor-type fan or blower motor that is marked with the motor type, the full load current for such motor marked on the nameplate of the equipment in which the fan or blower motor is employed shall be used instead of the horsepower rating to determine the ampacity or rating of the disconnecting means, the branch-circuit conductors, the controller, the branch-circuit short-circuit and ground-fault protection, and the separate overload protection. This marking on the equipment nameplate shall not be less than the current marked on the fan or blower motor nameplate.*

*Exception No. 3: For a listed motor-operated appliance that is marked with both motor horsepower and full-load current, the motor full-load current marked on the nameplate of the appliance shall be used instead of the horsepower rating on the appliance nameplate to determine the ampacity or rating of the disconnecting means, the branch-circuit conductors, the controller, the branch-circuit short-circuit and ground-fault protection, and any separate overload protection.*

Exception No. 3 is intended to resolve confusion that may result when motor-operated appliances are labeled with both horsepower and ampere ratings. The nameplate current rating in amperes is more accurate than the marked horsepower rating for motor-operated appliances.

**(2) Nameplate Values.** Separate motor overload protection shall be based on the motor nameplate current rating.

For general motor applications other than motors built for low speeds (less than 1200 rpm), multispeed motors, or high-torque motors, the ampacity of motor branch-circuit conductors, branch-circuit and ground-fault protection, and ampere rating of the motor disconnecting means are determined by the ampere values listed in Tables 430.247 through 430.250 rather than the ampere values marked on the motor nameplate. The ampere values are based on the horsepower rating and nominal voltage listed on the motor nameplate. Where the motor is marked with an ampere rating rather than a horsepower rating, the horsepower rating is assumed to be that in Tables 430.247 through 430.250, based on the motor ampere rating from the nameplate.

The ampere rating provided on the motor nameplate is used to size the overload protective devices that protect the motor, motor control apparatus, and motor branch-circuit conductors.

**(B) Torque Motors.** For torque motors, the rated current shall be locked-rotor current, and this nameplate current shall be used to determine the ampacity of the branch-circuit conductors covered in 430.22 and 430.24, the ampere rating of the motor overload protection, and the ampere rating of motor branch-circuit short-circuit and ground-fault protection in accordance with 430.52(B).

Informational Note: For motor controllers and disconnecting means, see 430.83(D) and 430.110.

**(C) Alternating-Current Adjustable Voltage Motors.** For motors used in alternating-current, adjustable voltage, variable torque drive systems, the ampacity of conductors, or ampere ratings of switches, branch-circuit short-circuit and ground-fault protection, and so forth, shall be based on the maximum operating current marked on the motor or control nameplate, or both. If the maximum operating current does not appear on the nameplate, the ampacity determination shall be based on 150 percent of the values given in Table 430.249 and Table 430.250.

**(D) Valve Actuator Motor Assemblies.** For valve actuator motor assemblies (VAMs), the rated current shall be the nameplate full-load current, and this current shall be used to determine the maximum rating or setting of the motor branch-circuit short-circuit and ground-fault protective device and the ampacity of the conductors.

A valve actuator motor assembly (VAM) typically has a full-load current (FLC) rating marked on the nameplate in addition to the locked-rotor current (LRC). Section 430.6(B) requires the LRC to be used to determine conductor sizes and overcurrent protection for torque motors. However, a VAM is a not a torque motor. The nameplate FLC is a more appropriate value to use for this purpose.

## 430.7 Marking on Motors and Multimotor Equipment

**(A) Usual Motor Applications.** A motor shall be marked with the following information:

(1) Manufacturer's name.
(2) Rated volts and full-load current. For a multispeed motor, full-load current for each speed, except shaded-pole and permanent-split capacitor motors where amperes are required only for maximum speed.
(3) Rated frequency and number of phases if an ac motor.
(4) Rated full-load speed.
(5) Rated temperature rise or the insulation system class and rated ambient temperature.
(6) Time rating. The time rating shall be 5, 15, 30, or 60 minutes, or continuous.
(7) Rated horsepower if ⅛ hp or more. For a multispeed motor ⅛ hp or more, rated horsepower for each speed, except shaded-pole and permanent-split capacitor motors ⅛ hp or more where rated horsepower is required only for maximum speed. Motors of arc welders are not required to be marked with the horsepower rating.
(8) Code letter or locked-rotor amperes if an alternating-current motor rated ½ hp or more. On polyphase wound-rotor motors, the code letter shall be omitted.

Informational Note: See 430.7(B).

(9) Design letter for design B, C, or D motors.

Informational Note: Motor design letter definitions are found in ANSI/NEMA MG 1-1993, *Motors and Generators, Part 1, Definitions,* and in IEEE 100-1996, *Standard Dictionary of Electrical and Electronic Terms*.

The design letters referred to in 430.7(A) indicate a motor's speed/torque characteristic curve and are not to be confused with code letters used in Table 430.7(B). For technical accuracy, code letters should be referred to as "locked-rotor indicating code letters," which are explained in 430.7(B). Design letters reflect characteristics inherent in motor design, such as locked-rotor current, slip at rated load, and locked-rotor and breakdown torque.

(10) Secondary volts and full-load current if a wound-rotor induction motor.
(11) Field current and voltage for dc excited synchronous motors.
(12) Winding — straight shunt, stabilized shunt, compound, or series, if a dc motor. Fractional horsepower dc motors 175 mm (7 in.) or less in diameter shall not be required to be marked.
(13) A motor provided with a thermal protector complying with 430.32(A)(2) or (B)(2) shall be marked "Thermally Protected." Thermally protected motors rated 100 watts or less and complying with 430.32(B)(2) shall be permitted to use the abbreviated marking "T.P."
(14) A motor complying with 430.32(B)(4) shall be marked "Impedance Protected." Impedance-protected motors rated 100 watts or less and complying with 430.32(B)(4) shall be permitted to use the abbreviated marking "Z.P."
(15) Motors equipped with electrically powered condensation prevention heaters shall be marked with the rated heater voltage, number of phases, and the rated power in watts.

Motors for outdoor installation or for use in locations where condensation might occur often come equipped with condensation prevention heaters. These heaters are energized when the motor is turned off. Section 430.7(A)(15) requires the manufacturer to mark a motor equipped with a condensation heater to alert the installer to provide the proper electrical supply to the heater. See 430.113 for heater disconnecting means.

**(B) Locked-Rotor Indicating Code Letters.** Code letters marked on motor nameplates to show motor input with locked rotor shall be in accordance with Table 430.7(B).

***TABLE 430.7(B)*** *Locked-Rotor Indicating Code Letters*

| Code Letter | Kilovolt-Amperes per Horsepower with Locked Rotor |
|---|---|
| A | 0–3.14 |
| B | 3.15–3.54 |
| C | 3.55–3.99 |
| D | 4.0–4.49 |
| E | 4.5–4.99 |
| F | 5.0–5.59 |
| G | 5.6–6.29 |
| H | 6.3–7.09 |
| J | 7.1–7.99 |
| K | 8.0–8.99 |
| L | 9.0–9.99 |
| M | 10.0–11.19 |
| N | 11.2–12.49 |
| P | 12.5–13.99 |
| R | 14.0–15.99 |
| S | 16.0–17.99 |
| T | 18.0–19.99 |
| U | 20.0–22.39 |
| V | 22.4 and up |

The code letter indicating motor input with locked rotor shall be in an individual block on the nameplate, properly designated.

**(1) Multispeed Motors.** Multispeed motors shall be marked with the code letter designating the locked-rotor kilovolt-ampere (kVA) per horsepower (hp) for the highest speed at which the motor can be started.

*Exception: Constant horsepower multispeed motors shall be marked with the code letter giving the highest locked-rotor kilovolt-ampere (kVA) per horsepower (hp).*

**(2) Single-Speed Motors.** Single-speed motors starting on wye connection and running on delta connections shall be marked with a code letter corresponding to the locked-rotor kilovolt-ampere (kVA) per horsepower (hp) for the wye connection.

**(3) Dual-Voltage Motors.** Dual-voltage motors that have a different locked-rotor kilovolt-ampere (kVA) per horsepower (hp) on the two voltages shall be marked with the code letter for the voltage giving the highest locked-rotor kilovolt-ampere (kVA) per horsepower (hp).

**(4) 50/60 Hz Motors.** Motors with 50- and 60-Hz ratings shall be marked with a code letter designating the locked-rotor kilovolt-ampere (kVA) per horsepower (hp) on 60 Hz.

**(5) Part-Winding Motors.** Part-winding start motors shall be marked with a code letter designating the locked-rotor kilovolt-ampere (kVA) per horsepower (hp) that is based on the locked-rotor current for the full winding of the motor.

The following example shows how to determine the locked-rotor current for a specific motor using Table 430.7(B).

**Calculation Example**

A 20-hp, 460-V, 3-phase motor has a nameplate kilovolt-ampere code letter G. Determine the maximum locked-rotor current for this motor.

*Solution*

STEP 1. Use Table 430.7(B) to find the maximum value in the range for code letter G, which is 6.29 kVA per horsepower.

STEP 2. Use the following formula to find the maximum locked-rotor current:

$$\begin{aligned}\text{Locked-rotor kVA} &= \text{motor hp} \\ &\quad \times \text{maximum code letter value} \\ &= 20 \times 6.29 = 125.8\end{aligned}$$

$$\text{Locked-rotor current} = \frac{\text{locked-rotor kVA}}{\sqrt{3} \times \text{kV}}$$

For 460 V,

$$\frac{125.8}{1.73 \times 0.46} = 158 \text{ amperes}$$

*Conclusion:* The maximum locked-rotor current for a 20-hp, 460-V motor with code letter G is 158 amperes when the system voltage is 460 volts.

**(C) Torque Motors.** Torque motors are rated for operation at standstill and shall be marked in accordance with 430.7(A), except that locked-rotor torque shall replace horsepower.

**(D) Multimotor and Combination-Load Equipment.**

**(1) Factory-Wired.** Multimotor and combination-load equipment shall be provided with a visible nameplate marked with the manufacturer's name, the rating in volts, frequency, number of phases, minimum supply circuit conductor ampacity, and the maximum ampere rating of the circuit short-circuit and ground-fault protective device. The conductor ampacity shall be calculated in accordance with 430.24 and counting all of the motors and other loads that will be operated at the same time. The short-circuit and ground-fault protective device rating shall not exceed the value calculated in accordance with 430.53. Multimotor equipment for use on two or more circuits shall be marked with the preceding information for each circuit.

Section 110.3(B) requires listed and labeled equipment to be used and installed in accordance with the manufacturer's instructions accompanying the equipment or marked on the nameplate. The nameplate marking for the maximum ampere rating of the branch-circuit short-circuit and ground-fault protective device may limit the type of protective device to a fuse by stipulating "fuse" without reference to a circuit breaker. That means that the circuit to the equipment must be protected by fuses, such as by a fused disconnect switch. A circuit breaker located in a panelboard, switchboard, or similar distribution equipment is permitted to supply the switch or other equipment in which the fuses are installed.

**(2) Not Factory-Wired.** Where the equipment is not factory-wired and the individual nameplates of motors and other loads are visible after assembly of the equipment, the individual nameplates shall be permitted to serve as the required marking.

## 430.8 Marking on Controllers

A controller shall be marked with the manufacturer's name or identification, the voltage, the current or horsepower rating, the short-circuit current rating, and other necessary data to properly indicate the applications for which it is suitable.

*Exception No. 1: The short-circuit current rating is not required for controllers applied in accordance with 430.81(A) or (B).*

*Exception No. 2: The short-circuit rating is not required to be marked on the controller when the short-circuit current rating of the controller is marked elsewhere on the assembly.*

*Exception No. 3: The short-circuit rating is not required to be marked on the controller when the assembly into which it is installed has a marked short-circuit current rating.*

*Exception No. 4: Short-circuit ratings are not required for controllers rated less than 2 hp at 300 V or less and listed exclusively for general-purpose branch circuits.*

A controller that includes motor overload protection suitable for group motor application shall be marked with the motor overload protection and the maximum branch-circuit short-circuit and ground-fault protection for such applications.

Combination controllers that employ adjustable instantaneous trip circuit breakers shall be clearly marked to indicate the ampere settings of the adjustable trip element.

Where a controller is built in as an integral part of a motor or of a motor-generator set, individual marking of the controller shall not be required if the necessary data are on the nameplate. For controllers that are an integral part of equipment approved as a unit, the above marking shall be permitted on the equipment nameplate.

Informational Note: See 110.10 for information on circuit impedance and other characteristics.

## 430.9 Terminals

**(A) Markings.** Terminals of motors and controllers shall be suitably marked or colored where necessary to indicate the proper connections.

**(B) Conductors.** Motor controllers and terminals of control circuit devices shall be connected with copper conductors unless identified for use with a different conductor.

**(C) Torque Requirements.** Control circuit devices with screw-type pressure terminals used with 14 AWG or smaller copper conductors shall be torqued to a minimum of 0.8 N·m (7 lb-in.) unless identified for a different torque value.

Proper torque is essential for safe and reliable connections. A screw-type pressure terminal that has not been torqued may loosen during motor operation, resulting in overheating. Safety is enhanced by providing a minimum torque value for screw-type pressure terminals. See the commentary following 110.14(B) for more information on electrical connections.

## 430.10 Wiring Space in Enclosures

**(A) General.** Enclosures for motor controllers and disconnecting means shall not be used as junction boxes, auxiliary gutters, or raceways for conductors feeding through or tapping off to the other apparatus unless designs are employed that provide adequate space for this purpose.

Informational Note: See 312.8 for switch and overcurrent-device enclosures.

**(B) Wire-Bending Space in Enclosures.** Minimum wire-bending space within the enclosures for motor controllers shall be in accordance with Table 430.10(B) where measured in a straight line from the end of the lug or wire connector (in the direction the wire leaves the terminal) to the wall or barrier. Where alternate wire termination means are substituted for that supplied by the manufacturer of the controller, they shall be of a type identified by the manufacturer for use with the controller and shall not reduce the minimum wire-bending space.

Exhibit 430.1 illustrates the application of the wire-bending space requirements of either 430.10(B) or 312.6(B) within an enclosure for a motor controller.

***TABLE 430.10(B)*** *Minimum Wire-Bending Space at the Terminals of Enclosed Motor Controllers*

| Size of Wire (AWG or kcmil) | Wires per Terminal* | | | |
|---|---|---|---|---|
| | 1 | | 2 | |
| | mm | in. | mm | in. |
| 14–10 | Not specified | | — | — |
| 8–6 | 38 | 1½ | — | — |
| 4–3 | 50 | 2 | — | — |
| 2 | 65 | 2½ | — | — |
| 1 | 75 | 3 | — | — |
| 1/0 | 125 | 5 | 125 | 5 |
| 2/0 | 150 | 6 | 150 | 6 |
| 3/0–4/0 | 175 | 7 | 175 | 7 |
| 250 | 200 | 8 | 200 | 8 |
| 300 | 250 | 10 | 250 | 10 |
| 350–500 | 300 | 12 | 300 | 12 |
| 600–700 | 350 | 14 | 400 | 16 |
| 750–900 | 450 | 18 | 475 | 19 |

*Where provision for three or more wires per terminal exists, the minimum wire-bending space shall be in accordance with the requirements of Article 312.

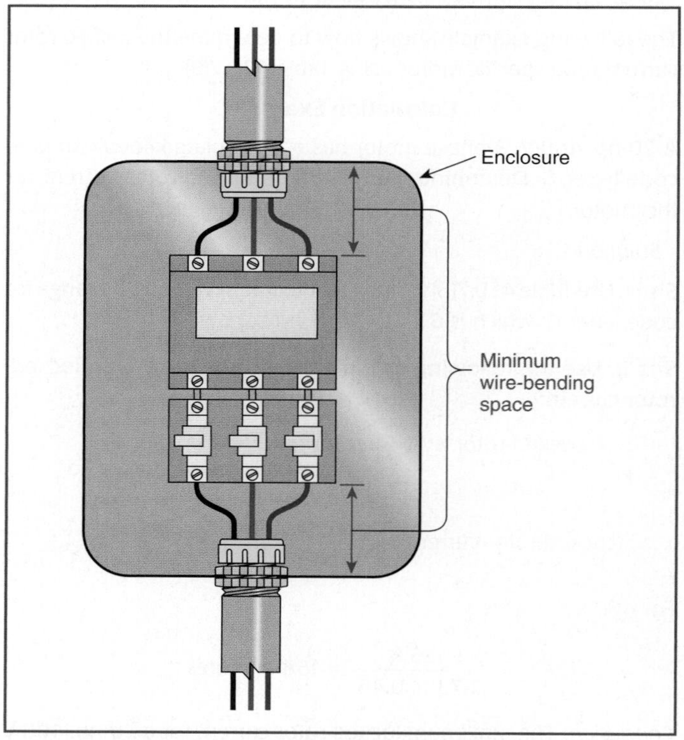


***EXHIBIT 430.1*** *Wire-bending space in enclosures for motor controllers.*

## 430.11 Protection Against Liquids

Suitable guards or enclosures shall be provided to protect exposed current-carrying parts of motors and the insulation of motor leads where installed directly under equipment, or in other locations where dripping or spraying oil, water, or other liquid is capable of occurring, unless the motor is designed for the existing conditions.

The presence of liquids may cause deterioration and insulation breakdown. Excess lubricants in the motor can collect dirt and clog the cooling passages of the motor, causing the motor to overheat.

## 430.12 Motor Terminal Housings

**(A) Material.** Where motors are provided with terminal housings, the housings shall be of metal and of substantial construction.

*Exception: In other than hazardous (classified) locations, substantial, nonmetallic, noncombustible housings shall be permitted, provided an internal grounding means between the motor frame and the equipment grounding connection is incorporated within the housing.*

Nonmetallic terminal housings are permitted on motors of any size, provided the housing material has been determined to be noncombustible with, for example, a flammability rating of at least 94-5V in accordance with UL 746C, *Polymeric Materials – Use in Electrical Equipment Evaluations.*

**(B) Dimensions and Space — Wire-to-Wire Connections.** Where these terminal housings enclose wire-to-wire connections, they shall have minimum dimensions and usable volumes in accordance with Table 430.12(B)

**(C) Dimensions and Space — Fixed Terminal Connections.** Where these terminal housings enclose rigidly mounted motor terminals, the terminal housing shall be of sufficient size to provide minimum terminal spacings and usable volumes in accordance with Table 430.12(C)(1) and Table 430.12(C)(2).

**(D) Large Wire or Factory Connections.** For motors with larger ratings, greater number of leads, or larger wire sizes, or where motors are installed as a part of factory-wired equipment, without additional connection being required at the motor terminal housing during equipment installation, the terminal housing shall be of ample size to make connections, but the foregoing provisions for the volumes of terminal housings shall not be considered applicable.

**(E) Equipment Grounding Connections.** A means for attachment of an equipment grounding conductor termination in accordance with 250.8 shall be provided at motor terminal housings for wire-to-wire connections or fixed terminal connections. The means for such connections shall be permitted to be located either inside or outside the motor terminal housing.

***TABLE 430.12(B)*** *Terminal Housings — Wire-to-Wire Connections*

**Motors 275 mm (11 in.) in Diameter or Less**

| | Cover Opening Minimum Dimension | | Usable Volume Minimum | |
|---|---|---|---|---|
| **Horsepower** | **mm** | **in.** | **cm$^3$** | **in.$^3$** |
| 1 and smaller[a] | 41 | $1\frac{5}{8}$ | 170 | 10.5 |
| $1\frac{1}{2}$, 2, and 3[b] | 45 | $1\frac{3}{4}$ | 275 | 16.8 |
| 5 and $7\frac{1}{2}$ | 50 | 2 | 365 | 22.4 |
| 10 and 15 | 65 | $2\frac{1}{2}$ | 595 | 36.4 |

**Motors Over 275 mm (11 in.) in Diameter — Alternating-Current Motors**

| Maximum Full Load Current for 3-Phase Motors with Maximum of 12 Leads (Amperes) | Terminal Box Cover Opening Minimum Dimension | | Usable Volume Minimum | | Typical Maximum Horsepower 3-Phase | |
|---|---|---|---|---|---|---|
| | **mm** | **in.** | **cm$^3$** | **in.$^3$** | **230 Volt** | **460 Volt** |
| 45 | 65 | 2.5 | 595 | 36.4 | 15 | 30 |
| 70 | 84 | 3.3 | 1,265 | 77 | 25 | 50 |
| 110 | 100 | 4.0 | 2,295 | 140 | 40 | 75 |
| 160 | 125 | 5.0 | 4,135 | 252 | 60 | 125 |
| 250 | 150 | 6.0 | 7,380 | 450 | 100 | 200 |
| 400 | 175 | 7.0 | 13,775 | 840 | 150 | 300 |
| 600 | 200 | 8.0 | 25,255 | 1540 | 250 | 500 |

**Direct-Current Motors**

| Maximum Full-Load Current for Motors with Maximum of 6 Leads (Amperes) | Terminal Box Minimum Dimensions | | Usable Volume Minimum | |
|---|---|---|---|---|
| | **mm** | **in.** | **cm$^3$** | **in.$^3$** |
| 68 | 65 | 2.5 | 425 | 26 |
| 105 | 84 | 3.3 | 900 | 55 |
| 165 | 100 | 4.0 | 1,640 | 100 |
| 240 | 125 | 5.0 | 2,950 | 180 |
| 375 | 150 | 6.0 | 5,410 | 330 |
| 600 | 175 | 7.0 | 9,840 | 600 |
| 900 | 200 | 8.0 | 18,040 | 1,100 |

Note: Auxiliary leads for such items as brakes, thermostats, space heaters, and exciting fields shall be permitted to be neglected if their current-carrying area does not exceed 25 percent of the current-carrying area of the machine power leads.

[a]For motors rated 1 hp and smaller, and with the terminal housing partially or wholly integral with the frame or end shield, the volume of the terminal housing shall not be less than 18.0 cm$^3$ (1.1 in.$^3$) per wire-to-wire connection. The minimum cover opening dimension is not specified.

[b]For motors rated $1\frac{1}{2}$, 2, and 3 hp, and with the terminal housing partially or wholly integral with the frame or end shield, the volume of the terminal housing shall not be less than 23.0 cm$^3$ (1.4 in.$^3$) per wire-to-wire connection. The minimum cover opening dimension is not specified.

*TABLE 430.12(C)(1) Terminal Spacings — Fixed Terminals*

| Nominal Volts | Minimum Spacing: Between Line Terminals (mm) | Minimum Spacing: Between Line Terminals (in.) | Minimum Spacing: Between Line Terminals and Other Uninsulated Metal Parts (mm) | Minimum Spacing: Between Line Terminals and Other Uninsulated Metal Parts (in.) |
|---|---|---|---|---|
| 240 or less | 6 | ¼ | 6 | ¼ |
| Over 250 – 1000 | 10 | ⅜ | 10 | ⅜ |

*TABLE 430.12(C)(2) Usable Volumes — Fixed Terminals*

| Power-Supply Conductor Size (AWG) | Minimum Usable Volume per Power-Supply Conductor: $cm^3$ | Minimum Usable Volume per Power-Supply Conductor: $in.^3$ |
|---|---|---|
| 14 | 16 | 1 |
| 12 and 10 | 20 | 1¼ |
| 8 and 6 | 37 | 2¼ |

*Exception: Where a motor is installed as a part of factory-wired equipment that is required to be grounded and without additional connection being required at the motor terminal housing during equipment installation, a separate means for motor grounding at the motor terminal housing shall not be required.*

### 430.13 Bushing

Where wires pass through an opening in an enclosure, conduit box, or barrier, a bushing shall be used to protect the conductors from the edges of openings having sharp edges. The bushing shall have smooth, well-rounded surfaces where it may be in contact with the conductors. If used where oils, greases, or other contaminants may be present, the bushing shall be made of material not deleteriously affected.

Informational Note: For conductors exposed to deteriorating agents, see 310.10(G).

### 430.14 Location of Motors

**(A) Ventilation and Maintenance.** Motors shall be located so that adequate ventilation is provided and so that maintenance, such as lubrication of bearings and replacing of brushes, can be readily accomplished.

*Exception: Ventilation shall not be required for submersible types of motors.*

**(B) Open Motors.** Open motors that have commutators or collector rings shall be located or protected so that sparks cannot reach adjacent combustible material.

*Exception: Installation of these motors on wooden floors or supports shall be permitted.*

### 430.16 Exposure to Dust Accumulations

In locations where dust or flying material collects on or in motors in such quantities as to seriously interfere with the ventilation or cooling of motors and thereby cause dangerous temperatures, suitable types of enclosed motors that do not overheat under the prevailing conditions shall be used.

Informational Note: Especially severe conditions may require the use of enclosed pipe-ventilated motors, or enclosure in separate dusttight rooms, properly ventilated from a source of clean air.

For motors exposed to combustible dust or readily ignitible flying material, see the requirements of 502.125(B) (Class II, Divisions 1 and 2) and 503.125 (Class III, Divisions 1 and 2). For classification of locations, see 500.5(C) (Class II locations) and 500.5(D) (Class III locations).

### 430.17 Highest Rated or Smallest Rated Motor

In determining compliance with 430.24, 430.53(B), and 430.53(C), the highest rated or smallest rated motor shall be based on the rated full-load current as selected from Table 430.247, Table 430.248, Table 430.249, and Table 430.250.

### 430.18 Nominal Voltage of Rectifier Systems

The nominal value of the ac voltage being rectified shall be used to determine the voltage of a rectifier derived system.

*Exception: The nominal dc voltage of the rectifier shall be used if it exceeds the peak value of the ac voltage being rectified.*

## II. Motor Circuit Conductors

### 430.21 General

Part II specifies ampacities of conductors that are capable of carrying the motor current without overheating under the conditions specified.

The provisions of Part II shall not apply to motor circuits rated over 1000 volts, nominal.

Informational Note: For over 1000 volts, nominal, see Part XI.

The provisions of Articles 250, 300, and 310 shall not apply to conductors that form an integral part of equipment, such as motors, motor controllers, motor control centers, or other factory-assembled control equipment.

•

Informational Note: See 110.14(C) and 430.9(B) for equipment device terminal requirements.

•

## 430.22 Single Motor

Conductors that supply a single motor used in a continuous duty application shall have an ampacity of not less than 125 percent of the motor full-load current rating, as determined by 430.6(A)(1), or not less than specified in 430.22(A) through (G).

The requirement that a conductor have an ampacity of at least 125 percent of the motor full-load current (FLC) rating does not constitute a conductor derating; rather, it is based on the need to provide for a sustained running current that is greater than the rated FLC and for protection of the conductors by the motor overload protective device set above the motor FLC rating.

The ampacity of the motor branch-circuit conductors is based on the FLC rating values provided in Tables 430.248 through 430.250. Motor nameplate FLC is not to be used to size branch-circuit conductors.

Exhibit 430.2 illustrates each motor on an individual branch circuit; the branch circuits are tapped from a feeder at a convenient location, such as a junction box or wireway, or from open wiring. The tap conductors are required to terminate in a branch-circuit protective device located not more than 25 feet from where the taps are connected to the feeder, in accordance with 430.28. Also see 430.28, Exception, which permits a 100-foot tap under some conditions in high-bay manufacturing facilities.

Exhibit 430.3 illustrates the following essential parts of a motor branch circuit:

1. Branch-circuit conductors
2. Disconnecting means
3. Branch-circuit short-circuit and ground-fault protective devices
4. Motor-controller
5. Motor overload protective devices

The branch-circuit short-circuit and ground-fault protective device may be a fuse or a circuit breaker and must be capable of carrying the starting current of the motor without opening the circuit. See Table 430.52.

In general, every motor must be provided with overload protective devices intended to protect the motor windings, motor-control apparatus, and motor branch-circuit conductors against excessive heating due to motor overloads and failure to start. Overload in equipment is defined as operation in excess of normal full-load rating, which, when it persists for a sufficient length of time, causes damage or dangerous overheating. Overload in a motor includes a stalled rotor but does not include fault currents due to short circuits or ground faults. See 430.44 for conditions where providing automatic opening of a motor circuit due to overload may be objectionable.

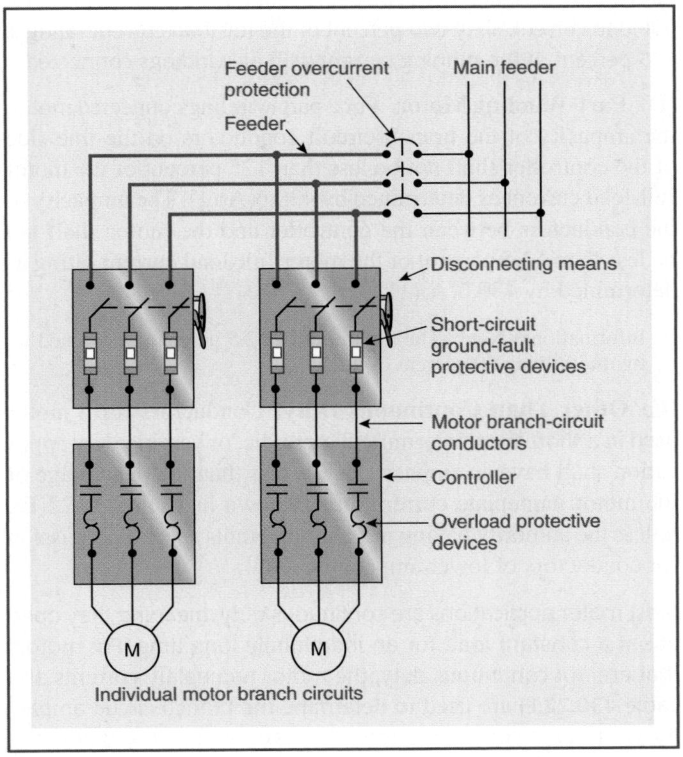


***EXHIBIT 430.2*** *A feeder supplying individual branch circuits to each motor.*

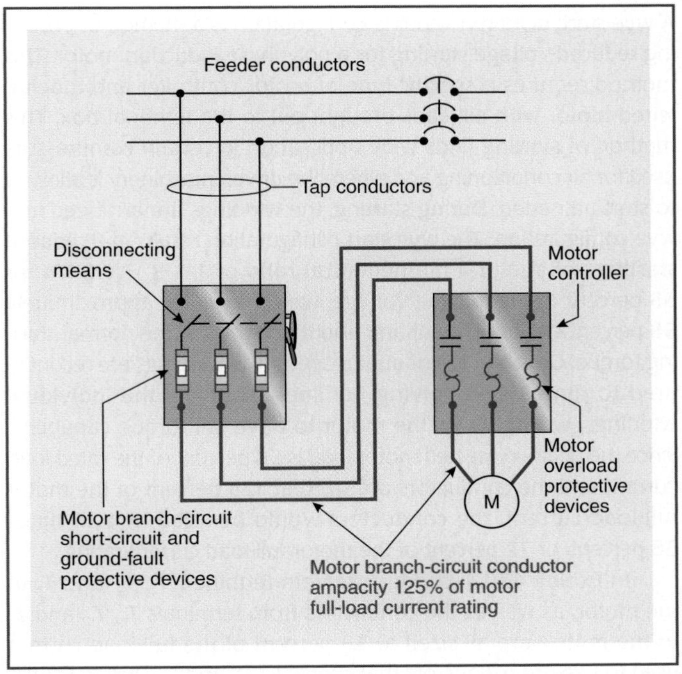


***EXHIBIT 430.3*** *A motor branch circuit showing the essential parts where both the controller and motor are assumed to be within sight of the disconnecting means.*

**(A) Direct-Current Motor-Rectifier Supplied.** For dc motors operating from a rectified power supply, the conductor ampacity on the input of the rectifier shall not be less than 125 percent of the rated input current to the rectifier. For dc motors operating from a rectified single-phase power supply, the conductors between the field wiring output terminals of the rectifier and the motor shall have an ampacity of not less than the following percentages of the motor full-load current rating:

(1) Where a rectifier bridge of the single-phase, half-wave type is used, 190 percent.
(2) Where a rectifier bridge of the single-phase, full-wave type is used, 150 percent.

**(B) Multispeed Motor.** For a multispeed motor, the selection of branch-circuit conductors on the line side of the controller shall be based on the highest of the full-load current ratings shown on the motor nameplate. The ampacity of the branch-circuit conductors between the controller and the motor shall not be less than 125 percent of the current rating of the winding(s) that the conductors energize.

**(C) Wye-Start, Delta-Run Motor.** For a wye-start, delta-run connected motor, the ampacity of the branch-circuit conductors on the line side of the controller shall not be less than 125 percent of the motor full-load current as determined by 430.6(A)(1). The ampacity of the conductors between the controller and the motor shall not be less than 72 percent of the motor full-load current rating as determined by 430.6(A)(1).

A wye-start, delta-run winding configuration is a method of providing reduced-voltage starting for a polyphase induction motor. This method requires a specific type of motor controller and a delta-wired motor with all leads brought out to the terminal box. This method of starting finds wide application in certain compressors used for air conditioning and where the driven machinery is allowed to start unloaded. During starting, the windings are arranged in a wye configuration. The wye-start configuration results in a reduced starting voltage of a mathematical ratio of $1/\sqrt{3} = 0.5774$, or 58 percent of the full line voltage, which results in approximately 58 percent starting current and about one-third of the normal starting torque. Once the motor attains speed, the windings are reconfigured to run as delta, giving full line voltage to the individual windings, which allows the motor to have full torque capability. Since the delta-connected motor load is 58 percent of the rated load current and the conductors are sized at 125 percent of the motor full-load current, the conductors would be sized at 1.25 times 58 percent, or 72 percent of the motor full-load current rating.

In Exhibit 430.4, conductors from terminals $T_1$, $T_2$, and $T_3$ to the motor, as well as the conductors from terminals $T_4$, $T_5$, and $T_6$ to the motor, are all sized at 58 percent of the full-load current used to size the conductors that supply $L_1$, $L_2$, and $L_3$. During START, contacts 1M and S are closed and contacts 2M are open. During RUN, contacts 1M and 2M are closed and contacts S are open.

Informational Note: The individual motor circuit conductors of a wye-start, delta-run connected motor carry 58 percent of the rated load current. The multiplier of 72 percent is obtained by multiplying 58 percent by 1.25.

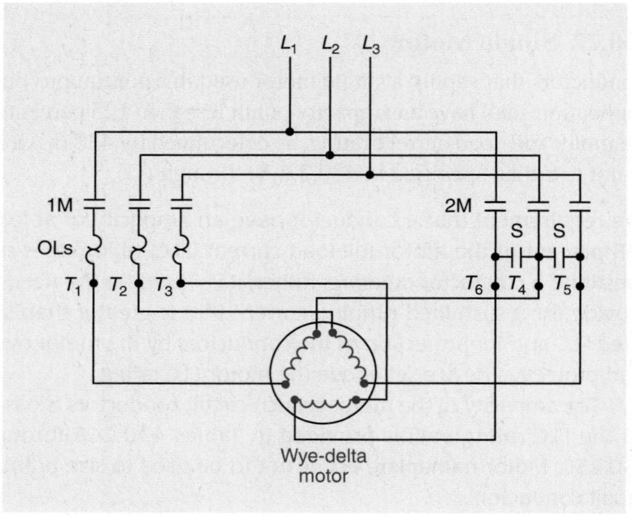


***EXHIBIT 430.4*** *An elementary wiring diagram of a typical wye-start, delta-run motor and controller. (Courtesy of Schneider Electric)*

A part-winding motor starter supplies power to partial sections of the primary winding of the motor. When the motor is running, load on the conductors will be approximately 50 percent of motor full-load current. Sixty-two percent of the full-load current rating is 125 percent of the running current with all windings connected.

**(D) Part-Winding Motor.** For a part-winding connected motor, the ampacity of the branch-circuit conductors on the line side of the controller shall not be less than 125 percent of the motor full-load current as determined by 430.6(A)(1). The ampacity of the conductors between the controller and the motor shall not be less than 62.5 percent of the motor full-load current rating as determined by 430.6(A)(1).

Informational Note: The multiplier of 62.5 percent is obtained by multiplying 50 percent by 1.25.

**(E) Other Than Continuous Duty.** Conductors for a motor used in a short-time, intermittent, periodic, or varying duty application shall have an ampacity of not less than the percentage of the motor nameplate current rating shown in Table 430.22(E), unless the authority having jurisdiction grants special permission for conductors of lower ampacity.

Most motor applications are continuous duty, meaning they operate at a constant load for an indefinitely long time. For motors that are not continuous duty, the motor nameplate currents and Table 430.22(E) are used to determine the branch-circuit ampacity. A motor is considered to be for continuous duty unless the nature of the apparatus it drives is such that the motor cannot operate continuously with load under any condition of use. Conductors for a motor used for short-time, intermittent, periodic, or varying duty are required to have an ampacity in accordance with

*TABLE 430.22(E) Duty-Cycle Service*

| Classification of Service | Nameplate Current Rating Percentages: 5-Minute Rated Motor | 15-Minute Rated Motor | 30- & 60-Minute Rated Motor | Continuous Rated Motor |
|---|---|---|---|---|
| Short-time duty operating valves, raising or lowering rolls, etc. | 110 | 120 | 150 | — |
| Intermittent duty freight and passenger elevators, tool heads, pumps, drawbridges, turntables, etc. (for arc welders, see 630.11) | 85 | 85 | 90 | 140 |
| Periodic duty rolls, ore- and coal-handling machines, etc. | 85 | 90 | 95 | 140 |
| Varying duty | 110 | 120 | 150 | 200 |

Note: Any motor application shall be considered as continuous duty unless the nature of the apparatus it drives is such that the motor will not operate continuously with load under any condition of use.

Table 430.22(E). Branch-circuit conductors for a motor with a rated horsepower used for 5-minute short-time duty service are permitted to be sized smaller than for the same motor with a 60-minute rating, due to the cooling intervals between operating periods. For example, a 5-minute rated motor will run for 5 minutes and then be off for 55 minutes. The terms *continuous duty, intermittent duty, periodic duty, short-time duty,* and *varying duty* are defined in Article 100.

**(F) Separate Terminal Enclosure.** The conductors between a stationary motor rated 1 hp or less and the separate terminal enclosure permitted in 430.245(B) shall be permitted to be smaller than 14 AWG but not smaller than 18 AWG, provided they have an ampacity as specified in 430.22(A).

**(G) Conductors for Small Motors.** Conductors for small motors shall not be smaller than 14 AWG unless otherwise permitted in 430.22(G)(1) or (G)(2).

**(1) 18 AWG Copper.** Where installed in a cabinet or enclosure, 18 AWG individual copper conductors, copper conductors that are part of a jacketed multiconductor cable assembly, or copper conductors in a flexible cord shall be permitted, under either of the following sets of conditions:

(1) The circuit supplies a motor with a full-load current rating, as determined by 430.6(A)(1), of greater than 3.5 amperes, and less than or equal to 5 amperes, and all the following conditions are met:

a. The circuit is protected in accordance with 430.52.
b. The circuit is provided with maximum Class 10 or Class 10A overload protection in accordance with 430.32.
c. Overcurrent protection is provided in accordance with 240.4(D)(1)(2).

(2) The circuit supplies a motor with a full-load current rating, as determined by 430.6(A)(1), of 3.5 amperes or less, and all the following conditions are met:

a. The circuit is protected in accordance with 430.52.
b. The circuit is provided with maximum Class 20 overload protection in accordance with 430.32.
c. Overcurrent protection is provided in accordance with 240.4(D)(1)(2).

**(2) 16 AWG Copper.** Where installed in a cabinet or enclosure, 16 AWG individual copper conductors, copper conductors that are part of a jacketed multiconductor cable assembly, or copper conductors in a flexible cord shall be permitted under either of the following sets of conditions:

(1) The circuit supplies a motor with a full-load current rating, as determined by 430.6(A)(1), of greater than 5.5 amperes, and less than or equal to 8 amperes, and all the following conditions are met:

a. The circuit is protected in accordance with 430.52.
b. The circuit is provided with maximum Class 10 or Class 10A overload protection in accordance with 430.32.
c. Overcurrent protection is provided in accordance with 240.4(D)(2)(2).

(2) The circuit supplies a motor with a full-load current rating, as determined by 430.6(A)(1), of 5.5 amperes or less, and all the following conditions are met:

a. The circuit is protected in accordance with 430.52.
b. The circuit is provided with maximum Class 20 overload protection in accordance with 430.32.
c. Overcurrent protection is provided in accordance with 240.4(D)(2)(2).

This section correlates the detailed requirements for applying small motor circuit conductors in Article 430 with those found in NFPA 79, *Electrical Standard for Industrial Machinery*. This requirement limits the type of conductors that can be used or limits their location to a "protected" area such as a cabinet or enclosure. It then specifies the classes of overload relays and the sizes of OCPDs, based upon the motor full-load amperes.

## 430.23 Wound-Rotor Secondary

**(A) Continuous Duty.** For continuous duty, the conductors connecting the secondary of a wound-rotor ac motor to its controller

*TABLE 430.23(C) Secondary Conductor*

| Resistor Duty Classification | Ampacity of Conductor in Percent of Full-Load Secondary Current |
|---|---|
| Light starting duty | 35 |
| Heavy starting duty | 45 |
| Extra-heavy starting duty | 55 |
| Light intermittent duty | 65 |
| Medium intermittent duty | 75 |
| Heavy intermittent duty | 85 |
| Continuous duty | 110 |

shall have an ampacity not less than 125 percent of the full-load secondary current of the motor.

**(B) Other Than Continuous Duty.** For other than continuous duty, these conductors shall have an ampacity, in percent of full-load secondary current, not less than that specified in Table 430.22(E).

**(C) Resistor Separate from Controller.** Where the secondary resistor is separate from the controller, the ampacity of the conductors between controller and resistor shall not be less than that shown in Table 430.23(C).

## 430.24 Several Motors or a Motor(s) and Other Load(s)

Conductors supplying several motors, or a motor(s) and other load(s), shall have an ampacity not less than the sum of each of the following:

(1) 125 percent of the full-load current rating of the highest rated motor, as determined by 430.6(A)
(2) Sum of the full-load current ratings of all the other motors in the group, as determined by 430.6(A)
(3) 100 percent of the noncontinuous non-motor load
(4) 125 percent of the continuous non-motor load.

Informational Note: See Informative Annex D, Example No. D8.

As illustrated in Exhibit 430.5, the requirements of Article 210 and Article 430 apply where motors are connected to a 15- or 20-ampere branch circuit that also supplies lighting or other appliance loads. Motors rated less than 1 hp may be connected to these circuits, and they must be provided with overload protective devices unless the motors are not permanently installed, are started manually, and are within sight from the controller location. For additional information on the installation of motors (1 hp or less), see 430.32(B) and (C) and 430.53(A).

Where branch circuits or feeders serve motors and/or other electrical loads, the highest rating or setting of the branch circuit or feeder short-circuit and ground-fault protective devices for the minimum-size branch circuit or feeder conductor permitted by 430.24 is specified in 430.62.

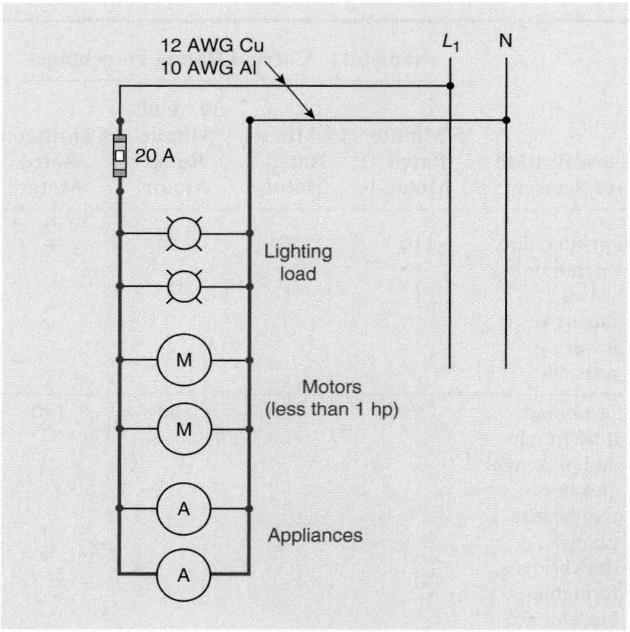


***EXHIBIT 430.5*** *A 20-ampere branch circuit supplying lighting, small motors, and appliances.*

Where the selection of a feeder protective device of higher rating or setting is based on the simultaneous starting of two or more motors, the size of the feeder conductors is required to be increased accordingly. Except where two or more motors may be started simultaneously, the heaviest load that a feeder will ever be required to carry occurs when the largest motor is started and all the other motors supplied by the same feeder are running and delivering their full-rated horsepower.

This requirement and those of 430.62 for the short-circuit and ground-fault protection of the branch circuit or feeder are based on the principle that the conductors should be sized to have an ampacity equal to 125 percent of the full-load current of the largest motor plus the full-load currents of all other motors and all other loads supplied by the feeder.

Where the conductors are branch-circuit conductors to multimotor equipment, 430.53 specifies the maximum rating of the branch-circuit short-circuit and ground-fault protective device, and 430.7(D)(1) requires the maximum ampere rating of the short-circuit and ground-fault protective device to be marked on multimotor equipment.

*Exception No. 1: Where one or more of the motors of the group are used for short-time, intermittent, periodic, or varying duty, the ampere rating of such motors to be used in the summation shall be determined in accordance with 430.22(E). For the highest rated motor, the greater of either the ampere rating from 430.22(E) or the largest continuous duty motor full-load current multiplied by 1.25 shall be used in the summation.*

*Exception No. 2: The ampacity of conductors supplying motor-operated fixed electric space-heating equipment shall comply with 424.3(B).*

*Exception No. 3: Where the circuitry is interlocked so as to prevent simultaneous operation of selected motors or other loads, the conductor ampacity shall be permitted to be based on the summation of the currents of the motors and other loads to be operated simultaneously that results in the highest total current.*

## 430.25 Multimotor and Combination-Load Equipment

The ampacity of the conductors supplying multimotor and combination-load equipment shall not be less than the minimum circuit ampacity marked on the equipment in accordance with 430.7(D). Where the equipment is not factory-wired and the individual nameplates are visible in accordance with 430.7(D)(2), the conductor ampacity shall be determined in accordance with 430.24.

When computing the load for the minimum allowable conductor size for a combination lighting (or lighting and appliance) load and motor load, the lighting load is determined in accordance with Article 220 (and other applicable articles and sections), the appliance load in accordance with Article 422, and the motor load in accordance with 430.22 (single motor) or 430.24 (two or more motors). The lighting load and the motor load are added together to determine the minimum conductor ampacity.

## 430.26 Feeder Demand Factor

Where reduced heating of the conductors results from motors operating on duty-cycle, intermittently, or from all motors not operating at one time, the authority having jurisdiction may grant permission for feeder conductors to have an ampacity less than specified in 430.24, provided the conductors have sufficient ampacity for the maximum load determined in accordance with the sizes and number of motors supplied and the character of their loads and duties.

The AHJ may grant permission to allow a demand factor of less than 100 percent if operational procedures, production demands, or the nature of the work is such that not all the motors are running at one time. Engineering study or evaluation of motor operation may provide information that allows a demand factor of less than 100 percent. Application of demand factors is subject to the approval of the AHJ.

Informational Note: Demand factors determined in the design of new facilities can often be validated against actual historical experience from similar installations. Refer to ANSI/IEEE Std. 141, *IEEE Recommended Practice for Electric Power Distribution for Industrial Plants,* and ANSI/IEEE Std. 241, *Recommended Practice for Electric Power Systems in Commercial Buildings,* for information on the calculation of loads and demand factor.

## 430.27 Capacitors with Motors

Where capacitors are installed in motor circuits, conductors shall comply with 460.8 and 460.9.

## 430.28 Feeder Taps

Feeder tap conductors shall have an ampacity not less than that required by Part II, shall terminate in a branch-circuit protective device, and, in addition, shall meet one of the following requirements:

(1) Be enclosed either by an enclosed controller or by a raceway, be not more than 3.0 m (10 ft) in length, and, for field installation, be protected by an overcurrent device on the line side of the tap conductor, the rating or setting of which shall not exceed 1000 percent of the tap conductor ampacity
(2) Have an ampacity of at least one-third that of the feeder conductors, be suitably protected from physical damage or enclosed in a raceway, and be not more than 7.5 m (25 ft) in length
(3) Have an ampacity not less than the feeder conductors

*Exception: Feeder taps over 7.5 m (25 ft) long. In high-bay manufacturing buildings [over 11 m (35 ft) high at walls], where conditions of maintenance and supervision ensure that only qualified persons service the systems, conductors tapped to a feeder shall be permitted to be not over 7.5 m (25 ft) long horizontally and not over 30.0 m (100 ft) in total length where all of the following conditions are met:*

*(1) The ampacity of the tap conductors is not less than one-third that of the feeder conductors.*
*(2) The tap conductors terminate with a single circuit breaker or a single set of fuses complying with (1) Part IV, where the load-side conductors are a branch circuit, or (2) Part V, where the load-side conductors are a feeder.*
*(3) The tap conductors are suitably protected from physical damage and are installed in raceways.*
*(4) The tap conductors are continuous from end-to-end and contain no splices.*
*(5) The tap conductors shall be 6 AWG copper or 4 AWG aluminum or larger.*
*(6) The tap conductors shall not penetrate walls, floors, or ceilings.*
*(7) The tap shall not be made less than 9.0 m (30 ft) from the floor.*

For a single motor load, the tap conductors are sized the same as the motor branch-circuit conductors — that is, according to 430.22, which requires that motor branch-circuit conductors be sized at least 125 percent of the full-load current (FLC) value for the motor given in Tables 430.248 through 430.250. The table value, rather than the nameplate value, is the FLC used for conductor sizing according to 430.6(A).

The tap conductors must terminate in a set of fuses or a circuit breaker, thus limiting the load on the tap conductors. The reduced-size tap conductors are protected from overload by the terminal overcurrent device but protected from short circuit (and ground fault) only from the feeder overcurrent device.

Where the tap conductor ampacity is less than the ampacity of the feeder, the tap conductor installation must meet the additional requirements associated with their tap conductor distance limits, that is, 10 feet, 25 feet, or, by exception, 100 feet.

The requirements for tap conductors that supply motor loads are somewhat similar to the basic tap requirements found in 240.21. For example, where tap conductors supply a motor load and do not exceed 10 feet, the tap conductors must be sized for the load, terminate in a set of fuses or a circuit breaker, be enclosed by a controller or a raceway, and be protected by a feeder overcurrent device not exceeding 10 times the tap conductor ampacity.

Where the tap conductors supply a motor load and do not exceed 25 feet, the tap conductors must be sized for the load, terminate in a set of fuses or a circuit breaker, be protected from physical damage or be enclosed in a raceway, and have an ampacity at least one-third that of the feeder conductor.

In a high-bay manufacturing building, feeder taps up to 100 feet long are conditionally permitted under the exception to 430.28.

For additional information concerning taps supplying motor circuits for group installations, see 430.53(D) and the associated commentary.

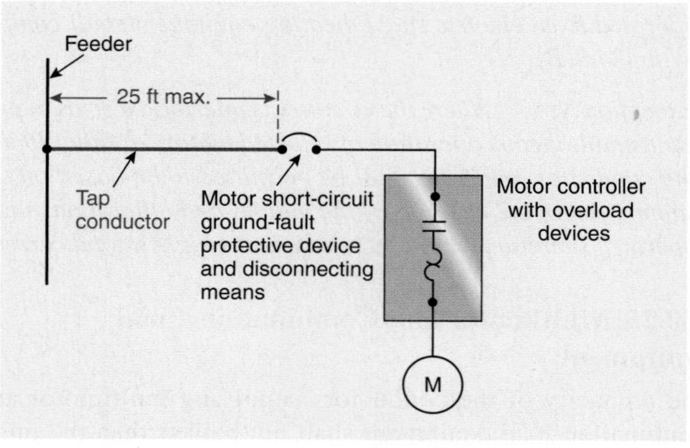


***EXHIBIT 430.6*** *Protective devices (branch-circuit short-circuit and ground-fault) for a motor branch circuit located not more than 25 feet from the point where the conductors are tapped to a feeder.*

### Calculation Example

A 15-hp, 230-V, 3-phase, NEMA Design B, squirrel-cage induction motor with a service factor of 1.15 and a nameplate FLC of 40 amperes is to be supplied by a tap from a 250 kcmil feeder. Assuming three Type THWN copper conductors in a raceway and no ambient correction factor, the feeder has an ampacity of 255 amperes [from Table 310.15(B)(16), 75°C column]. The tap conductors are not over 25 feet long (see Exhibit 430.6). Tap conductors with an ampacity of 85 amperes (4 AWG) are permitted ($\frac{1}{3} \times 255$ A $= 85$ A). Determine the required branch-circuit short-circuit and ground-fault protection and the overload protection required.

*Solution*

Step 1. Determine FLC based on 430.6(A) and Table 430.250. The FLC of the motor is 42 amperes.

Step 2. Determine the branch-circuit short-circuit and ground-fault protective device rating. According to 430.52(C)(1), the protective device cannot exceed the values given in Table 430.52. The maximum time-delay fuse value is $42 \times 1.75 = 73.5$ A. The maximum inverse time circuit breaker value is $42 \times 2.50 = 105$ A.

Section 430.52(C)(1), Exception No. 1, allows the next higher standard size – 80 and 110 amperes, respectively. A higher size, based on Exception No. 2, is allowed if the 80- or 110-ampere size is not adequate to start the motor.

Step 3. Determine overload protection rating. Based on 430.32, for a motor with a service factor of 1.15, the motor overload protective devices (heaters) are required to be set at a value not greater than 125 percent of the FLC marked on the motor nameplate and not at the FLC value from the table. A setting of up to 140 percent may be used according to the permissive rules in 430.32(C). With the motor overload protection set at 50 amperes, the 4 AWG THWN copper motor branch-circuit tap conductors are protected from overload.

## 430.29 Constant Voltage Direct-Current Motors — Power Resistors

Conductors connecting the motor controller to separately mounted power accelerating and dynamic braking resistors in the armature circuit shall have an ampacity not less than the value calculated from Table 430.29 using motor full-load current. If an armature shunt resistor is used, the power accelerating resistor conductor ampacity shall be calculated using the total of motor full-load current and armature shunt resistor current.

Armature shunt resistor conductors shall have an ampacity of not less than that calculated from Table 430.29 using rated shunt resistor current as full-load current.

***TABLE 430.29*** *Conductor Rating Factors for Power Resistors*

| Time in Seconds | | Ampacity of Conductor in Percent of Full-Load Current |
|---|---|---|
| **On** | **Off** | |
| 5 | 75 | 35 |
| 10 | 70 | 45 |
| 15 | 75 | 55 |
| 15 | 45 | 65 |
| 15 | 30 | 75 |
| 15 | 15 | 85 |
| Continuous Duty | | 110 |

## III. Motor and Branch-Circuit Overload Protection

### 430.31 General

Part III specifies overload devices intended to protect motors, motor-control apparatus, and motor branch-circuit conductors against excessive heating due to motor overloads and failure to start.

Informational Note No. 1: See Informative Annex D, Example No. D8.

Informational Note No. 2: See the definition of *Overload* in Article 100.

The purpose of motor and branch-circuit overload protection is to guard against abnormal operating conditions such as failure to start from a locked rotor, a single-phase condition, added friction on the driven load, or actual mechanical overloading of the driven load. The overload protection also guards against excessive heating in the motor caused by an overload condition or from a loss of phase condition.

Adequately applied overload protection should protect the motor from any overload condition prior to damage occurring in the motor. Overload protection is not designed or may not be capable of breaking short-circuit current or ground-fault current. Overload protection is not permitted to be installed where it could cause increased hazards as identified for fire pumps.

These provisions shall not require overload protection where a power loss would cause a hazard, such as in the case of fire pumps.

Informational Note: For protection of fire pump supply conductors, see 695.7.

The provisions of Part III shall not apply to motor circuits rated over 1000 volts, nominal.

Informational Note: For over 1000 volts, nominal, see Part XI.

•

### 430.32 Continuous-Duty Motors

**(A) More Than 1 Horsepower.** Each motor used in a continuous duty application and rated more than 1 hp shall be protected against overload by one of the means in 430.32(A)(1) through (A)(4).

The basic premise of 430.32(A) through (E) is that the operation of a motor in excess of its normal full-load rating for a prolonged period of time causes damage or dangerous overheating that may start a fire. Overload protection is intended to protect the motor and the system components from damaging overload currents.

A continuous-duty motor with a marked service factor of 1.15 or greater or with a marked temperature rise of 40°C or less can carry a 25-percent overload for an extended period without damage to the motor. Motors with a service factor of less than 1.15 or those with a marked temperature rise greater than 40°C may be incapable of withstanding a prolonged overload, where the motor overload protective device opens the circuit if the motor continues to draw 115 percent of its rated full-load current.

A "continuous-duty motor" is not the same as a "continuous load." The duty of a motor is determined by the application of the motor as defined in Article 100 under the term *duty*.

**(1) Separate Overload Device.** A separate overload device that is responsive to motor current. This device shall be selected to trip or shall be rated at no more than the following percent of the motor nameplate full-load current rating:

| | |
|---|---|
| Motors with a marked service factor 1.15 or greater | 125% |
| Motors with a marked temperature rise 40°C or less | 125% |
| All other motors | 115% |

Modification of this value shall be permitted as provided in 430.32(C). For a multispeed motor, each winding connection shall be considered separately.

Where a separate motor overload device is connected so that it does not carry the total current designated on the motor nameplate, such as for wye-delta starting, the proper percentage of nameplate current applying to the selection or setting of the overload device shall be clearly designated on the equipment, or the manufacturer's selection table shall take this into account.

Informational Note: Where power factor correction capacitors are installed on the load side of the motor overload device, see 460.9.

To protect a motor from an overload, the motor nameplate full-load current is used to select the overload protection rather than the full-load current values from Tables 430.248 through 430.250, which are used to select the feeder and branch-circuit wiring.

**(2) Thermal Protector.** A thermal protector integral with the motor, approved for use with the motor it protects on the basis that it will prevent dangerous overheating of the motor due to overload and failure to start. The ultimate trip current of a thermally protected motor shall not exceed the following percentage of motor full-load current given in Table 430.248, Table 430.249, and Table 430.250:

| | |
|---|---|
| Motor full-load current 9 amperes or less | 170% |
| Motor full-load current from 9.1 to, and including, 20 amperes | 156% |
| Motor full-load current greater than 20 amperes | 140% |

If the motor current-interrupting device is separate from the motor and its control circuit is operated by a protective device integral with the motor, it shall be arranged so that the opening of the control circuit will result in interruption of current to the motor.

**(3) Integral with Motor.** A protective device integral with a motor that will protect the motor against damage due to failure to start shall be permitted if the motor is part of an approved assembly that does not normally subject the motor to overloads.

**(4) Larger Than 1500 Horsepower.** For motors larger than 1500 hp, a protective device having embedded temperature detectors that cause current to the motor to be interrupted when the motor attains a temperature rise greater than marked on the nameplate in an ambient temperature of 40°C.

**(B) One Horsepower or Less, Automatically Started.** Any motor of 1 hp or less that is started automatically shall be protected against overload by one of the following means.

**(1) Separate Overload Device.** By a separate overload device following the requirements of 430.32(A)(1).

For a multispeed motor, each winding connection shall be considered separately. Modification of this value shall be permitted as provided in 430.32(C).

**(2) Thermal Protector.** A thermal protector integral with the motor, approved for use with the motor that it protects on the basis that it will prevent dangerous overheating of the motor due to overload and failure to start. Where the motor current-interrupting device is separate from the motor and its control circuit is operated by a protective device integral with the motor, it shall be arranged so that the opening of the control circuit results in interruption of current to the motor.

The thermal protector shown in Exhibit 430.7 is located inside the motor housing and is connected in series with the motor winding by a set of normally closed contacts attached to a bimetallic disk. The thermal protector heating coil causes the disk to heat rapidly. The heat-actuated disk snaps the contacts open to protect the motor windings from overheating due to failure to start, a sudden heavy overload, or a prolonged overload. After the circuit opens and the motor has cooled to a normal temperature, the contacts automatically close and restart the motor, which may not be desirable in some cases. For such applications, the protective device is designed so that it must be returned to the closed position by a manually controlled reset as required by 430.43.

For larger motors (usually over 1 hp), a similar device is used. This device, upon abnormal overload, acts as a control-circuit switch and operates the control circuit of a motor current-interrupting device, usually a motor contactor or starter, located separately from the motor. A thermal protector and circuit-interrupting device should be approved for use with the motor it protects and is required to open the circuit on an overcurrent as specified in 430.32(A)(2).

Section 501.125 permits open induction motors in Class I, Division 2 areas if they have no arc-producing parts. Because this thermal contact is an arc-producing part, a motor with a thermal protector, unless it is listed for a Class I, Division 2 area, is prohibited from use in a Class I, Division 2 area.

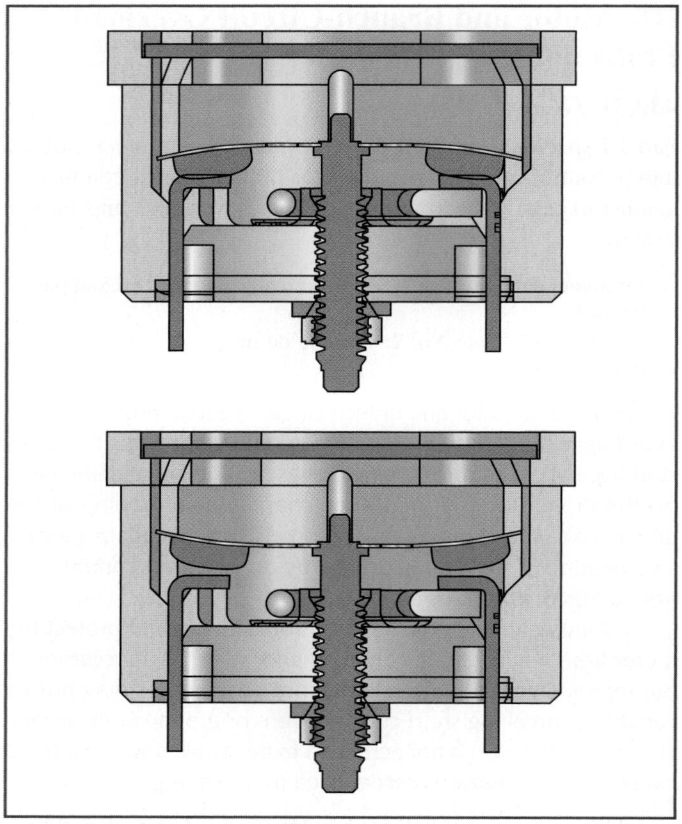

***EXHIBIT 430.7** A thermal protector for a motor, in which a heat-sensitive snap-action disk opens contacts and protects the motor in which it is mounted against dangerous overheating. (Courtesy of Sensata Technologies)*

**(3) Integral with Motor.** A protective device integral with a motor that protects the motor against damage due to failure to start shall be permitted (1) if the motor is part of an approved assembly that does not subject the motor to overloads, or (2) if the assembly is also equipped with other safety controls (such as the safety combustion controls on a domestic oil burner) that protect the motor against damage due to failure to start. Where the assembly has safety controls that protect the motor, it shall be so indicated on the nameplate of the assembly where it will be visible after installation.

**(4) Impedance-Protected.** If the impedance of the motor windings is sufficient to prevent overheating due to failure to start, the motor shall be permitted to be protected as specified in 430.32(D)(2)(a) for manually started motors if the motor is part of an approved assembly in which the motor will limit itself so that it will not be dangerously overheated.

Informational Note: Many ac motors of less than 1⁄20 hp, such as clock motors, series motors, and so forth, and also some larger motors such as torque motors, come within this classification. It does not include split-phase motors having automatic switches that disconnect the starting windings.

**(C) Selection of Overload Device.** Where the sensing element or setting or sizing of the overload device selected in accordance with 430.32(A)(1) and 430.32(B)(1) is not sufficient to start the motor or to carry the load, higher size sensing elements or incremental settings or sizing shall be permitted to be used, provided the trip current of the overload device does not exceed the following percentage of motor nameplate full-load current rating:

| | |
|---|---|
| Motors with marked service factor 1.15 or greater | 140% |
| Motors with a marked temperature rise 40°C or less | 140% |
| All other motors | 130% |

If not shunted during the starting period of the motor as provided in 430.35, the overload device shall have sufficient time delay to permit the motor to start and accelerate its load.

Informational Note: A Class 20 overload relay will provide a longer motor acceleration time than a Class 10 or Class 10A overload relay. A Class 30 overload relay will provide a longer motor acceleration time than a Class 20 overload relay. Use of a higher class overload relay may preclude the need for selection of a higher trip current.

**(D) One Horsepower or Less, Nonautomatically Started.**

**(1) Permanently Installed.** Overload protection shall be in accordance with 430.32(B).

**(2) Not Permanently Installed.**

(a) *Within Sight from Controller.* Overload protection shall be permitted to be furnished by the branch-circuit short-circuit and ground-fault protective device; such device, however, shall not be larger than that specified in Part IV of Article 430.

*Exception: Any such motor shall be permitted on a nominal 120-volt branch circuit protected at not over 20 amperes.*

(b) *Not Within Sight from Controller.* Overload protection shall be in accordance with 430.32(B).

Motors that are rated 1 hp or less and are not permanently installed and not automatically started, such as motors for bench grinders, drill presses, and portable electric tools, are not required to have overload protection and may be protected by the branch-circuit short-circuit fuse or circuit breaker. This type of equipment is usually attended by the operator, who can immediately shut off power to the motor if it overheats.

**(E) Wound-Rotor Secondaries.** The secondary circuits of wound-rotor ac motors, including conductors, controllers, resistors, and so forth, shall be permitted to be protected against overload by the motor-overload device.

## 430.33 Intermittent and Similar Duty

A motor used for a condition of service that is inherently short-time, intermittent, periodic, or varying duty, as illustrated by Table 430.22(E), shall be permitted to be protected against overload by the branch-circuit short-circuit and ground-fault protective device, provided the protective device rating or setting does not exceed that specified in Table 430.52.

Any motor application shall be considered to be for continuous duty unless the nature of the apparatus it drives is such that the motor cannot operate continuously with load under any condition of use.

Because duty-cycle service motors (short-time, intermittent, periodic, or varying) will not operate continuously, prolonged overloads are rare unless mechanical failure in the driven apparatus stalls the motor, in which case, the branch-circuit protective device would open the circuit. The omission of overload protective devices for such motors is based on the type of duty, not on the time rating of the motor.

## 430.35 Shunting During Starting Period

**(A) Nonautomatically Started.** For a nonautomatically started motor, the overload protection shall be permitted to be shunted or cut out of the circuit during the starting period of the motor if the device by which the overload protection is shunted or cut out cannot be left in the starting position and if fuses or inverse time circuit breakers rated or set at not over 400 percent of the full-load current of the motor are located in the circuit so as to be operative during the starting period of the motor.

**(B) Automatically Started.** The motor overload protection shall not be shunted or cut out during the starting period if the motor is automatically started.

*Exception: The motor overload protection shall be permitted to be shunted or cut out during the starting period on an automatically started motor where the following apply:*

*(a) The motor starting period exceeds the time delay of available motor overload protective devices, and*

*(b) Listed means are provided to perform the following:*

*(1) Sense motor rotation and automatically prevent the shunting or cutout in the event that the motor fails to start, and*

*(2) Limit the time of overload protection shunting or cutout to less than the locked rotor time rating of the protected motor, and*

*(3) Provide for shutdown and manual restart if motor running condition is not reached.*

## 430.36 Fuses — In Which Conductor

Where fuses are used for motor overload protection, a fuse shall be inserted in each ungrounded conductor and also in the grounded conductor if the supply system is 3-wire, 3-phase ac with one conductor grounded.

## 430.37 Devices Other Than Fuses — In Which Conductor

Where devices other than fuses are used for motor overload protection, Table 430.37 shall govern the minimum allowable number and location of overload units such as trip coils or relays.

All 3-phase motors, except those protected by other approved means, must be provided with three overload units, one in each phase. Examples of those motors protected by other means

*TABLE 430.37* *Overload Units*

| Kind of Motor | Supply System | Number and Location of Overload Units, Such as Trip Coils or Relays |
|---|---|---|
| 1-phase ac or dc | 2-wire, 1-phase ac or dc ungrounded | 1 in either conductor |
| 1-phase ac or dc | 2-wire, 1-phase ac or dc, one conductor grounded | 1 in ungrounded conductor |
| 1-phase ac or dc | 3-wire, 1-phase ac or dc, grounded neutral conductor | 1 in either ungrounded conductor |
| 1-phase ac | Any 3-phase | 1 in ungrounded conductor |
| 2-phase ac | 3-wire, 2-phase ac, ungrounded | 2, one in each phase |
| 2-phase ac | 3-wire, 2-phase ac, one conductor grounded | 2 in ungrounded conductors |
| 2-phase ac | 4-wire, 2-phase ac, grounded or ungrounded | 2, one for each phase in ungrounded conductors |
| 2-phase ac | Grounded neutral or 5-wire, 2-phase ac, ungrounded | 2, one for each phase in any ungrounded phase wire |
| 3-phase ac | Any 3-phase | 3, one in each phase* |

**Exception: An overload unit in each phase shall not be required where overload protection is provided by other approved means.*

include specially designed or integral-type detectors, with or without supplementary external protective devices. See 430.36 for instances in which fuses used as overloads are required even in the grounded conductor.

## 430.38 Number of Conductors Opened by Overload Device

Motor overload devices, other than fuses or thermal protectors, shall simultaneously open a sufficient number of ungrounded conductors to interrupt current flow to the motor.

## 430.39 Motor Controller as Overload Protection

A motor controller shall also be permitted to serve as an overload device if the number of overload units complies with Table 430.37 and if these units are operative in both the starting and running position in the case of a dc motor, and in the running position in the case of an ac motor.

A controller may be a switch, a circuit breaker, a contactor, or any other device that starts and stops a motor by making and breaking the motor circuit current. The controller must be capable of interrupting the stalled-rotor current of the motor and must have a horsepower rating that is not lower than the horsepower rating of the motor. Motor controllers are covered in Part VII of Article 430.

Dual-element fuses can be sized to provide motor overload protection (see 430.36). Automatically operated contactors or circuit breakers (with trip units) are governed by the requirements of 430.37 and 430.38 where these devices are used to provide overload protection.

## 430.40 Overload Relays

Overload relays and other devices for motor overload protection that are not capable of opening short circuits or ground faults shall be protected by fuses or circuit breakers with ratings or settings in accordance with 430.52 or by a motor short-circuit protector in accordance with 430.52.

*Exception: Where approved for group installation and marked to indicate the maximum size of fuse or inverse time circuit breaker by which they must be protected, the overload devices shall be protected in accordance with this marking.*

Some overload devices are marked with a maximum short-circuit and ground-fault protective device rating or setting. This rating sets the limit on the maximum rating or setting of a fuse or a circuit breaker that may be upstream from the overload device. The rating also notifies the user that coordination between the overload device and the short-circuit and ground-fault device is required, which is most often the case for group motor installation.

•

## 430.42 Motors on General-Purpose Branch Circuits

Overload protection for motors used on general-purpose branch circuits as permitted in Article 210 shall be provided as specified in 430.42(A), (B), (C), or (D).

**(A) Not over 1 Horsepower.** One or more motors without individual overload protection shall be permitted to be connected to a general-purpose branch circuit only where the installation complies with the limiting conditions specified in 430.32(B) and 430.32(D) and 430.53(A)(1) and (A)(2).

**(B) Over 1 Horsepower.** Motors of ratings larger than specified in 430.53(A) shall be permitted to be connected to general-purpose branch circuits only where each motor is protected by overload protection selected to protect the motor as specified in 430.32. Both the controller and the motor overload device shall be approved for group installation with the short-circuit and ground-fault protective device selected in accordance with 430.53.

**(C) Cord-and Plug-Connected.** Where a motor is connected to a branch circuit by means of an attachment plug and a receptacle or a cord connector, and individual overload protection is omitted as provided in 430.42(A), the rating of the attachment plug and receptacle or cord connector shall not exceed 15 amperes at 125 volts or 250 volts. Where individual overload protection is required as provided in 430.42(B) for a motor or motor-operated appliance that is attached to the branch circuit through an attachment plug and a receptacle or a cord connector, the overload device shall be an integral part of the motor or of the appliance. The rating of the attachment plug and receptacle or the cord connector shall determine the rating of the circuit to which the motor may be connected, as provided in 210.21(B).

**(D) Time Delay.** The branch-circuit short-circuit and ground-fault protective device protecting a circuit to which a motor or motor-operated appliance is connected shall have sufficient time delay to permit the motor to start and accelerate its load.

### 430.43 Automatic Restarting

A motor overload device that can restart a motor automatically after overload tripping shall not be installed unless approved for use with the motor it protects. A motor overload device that can restart a motor automatically after overload tripping shall not be installed if automatic restarting of the motor can result in injury to persons.

An integral motor overload protective device may be of the type that, after tripping and sufficiently cooling, automatically restarts the motor, or it may be of the type that, after tripping, can only be reset by using a manually operated reset button. See the commentary following 430.32(B)(2).

### 430.44 Orderly Shutdown

If immediate automatic shutdown of a motor by a motor overload protective device(s) would introduce additional or increased hazard(s) to a person(s) and continued motor operation is necessary for safe shutdown of equipment or process, a motor overload sensing device(s) complying with the provisions of Part III of this article shall be permitted to be connected to a supervised alarm instead of causing immediate interruption of the motor circuit, so that corrective action or an orderly shutdown can be initiated.

## IV. Motor Branch-Circuit Short-Circuit and Ground-Fault Protection

### 430.51 General

Part IV specifies devices intended to protect the motor branch-circuit conductors, the motor control apparatus, and the motors against overcurrent due to short circuits or ground faults. These rules add to or amend the provisions of Article 240. The devices specified in Part IV do not include the types of devices required by 210.8, 230.95, and 590.6.

Informational Note: See Informative Annex D, Example D8.

The provisions of Part IV shall not apply to motor circuits rated over 1000 volts, nominal.

Informational Note: For over 1000 volts, nominal, see Part XI.

•

### 430.52 Rating or Setting for Individual Motor Circuit

**(A) General.** The motor branch-circuit short-circuit and ground-fault protective device shall comply with 430.52(B) and either 430.52(C) or (D), as applicable.

For certain exceptions to the maximum rating or setting of these motor branch-circuit protective devices, as specified in Table 430.52, see 430.52, 430.53, and 430.54. Section 430.6 requires that if the current rating of a motor is used to determine the ampacity of conductors or ampere ratings of switches, branch-circuit overcurrent devices, and so forth, the values given in Tables 430.248 through 430.250 (including notes) must be used instead of the actual motor nameplate current rating. Separate motor overload protection must be based on the motor nameplate current rating.

Exhibit 430.3 illustrates a typical motor circuit in which the branch-circuit short-circuit and ground-fault protective fuse or circuit breaker rating must carry the starting current and may be sized 150 to 300 percent of the motor full-load current (depending on the type of motor).

The rules for short-circuit and ground-fault protection are specific for particular situations. A "short circuit" is a fault between two conductors or between phases. A "ground fault" is a fault between an ungrounded conductor and ground. During a short-circuit or phase-to-ground condition, the extremely high current causes the protective fuses or circuit breakers to open the circuit. Excess current flow caused by an overload condition passes through the overload protective device at the motor controller, thereby causing the device to open the control-circuit or motor-circuit conductors.

Branch-circuit conductors with an ampacity of 125 percent (not 150 to 300 percent) of the motor full-load current are reasonably protected by motor-protective devices set to operate at nearly the same current as the ampacity of the conductors. Branch-circuit short-circuit and ground-fault protective devices provide protection for both the motor and overload protective device; however, the overload protective device is not intended to open short circuits or ground faults.

The selected rating or setting of the branch-circuit short-circuit and ground-fault protective device should be as low as possible for maximum protection. However, if the rating or setting specified in Table 430.52 or permitted by 430.52(C)(1), Exception No. 1, is not sufficient for the starting current of the motor, a higher rating or setting is allowed per 430.52(C)(1), Exception No. 2. For example, a higher rating would be allowed for a motor under severe starting conditions in which the motor and its driven machinery require an extended period of time to reach the desired speed.

***TABLE 430.52*** *Maximum Rating or Setting of Motor Branch-Circuit Short-Circuit and Ground-Fault Protective Devices*

| | Percentage of Full-Load Current | | | |
|---|---|---|---|---|
| **Type of Motor** | **Nontime Delay Fuse[1]** | **Dual Element (Time-Delay) Fuse[1]** | **Instantaneous Trip Breaker** | **Inverse Time Breaker[2]** |
| Single-phase motors | 300 | 175 | 800 | 250 |
| AC polyphase motors other than wound-rotor | 300 | 175 | 800 | 250 |
| Squirrel cage — other than Design B energy-efficient | 300 | 175 | 800 | 250 |
| Design B energy-efficient | 300 | 175 | 1100 | 250 |
| Synchronous[3] | 300 | 175 | 800 | 250 |
| Wound-rotor | 150 | 150 | 800 | 150 |
| DC (constant voltage) | 150 | 150 | 250 | 150 |

Note: For certain exceptions to the values specified, see 430.54.

[1]The values in the Nontime Delay Fuse column apply to time-delay Class CC fuses.

[2]The values given in the last column also cover the ratings of nonadjustable inverse time types of circuit breakers that may be modified as in 430.52(C)(1), Exceptions No. 1 and No. 2.

[3]Synchronous motors of the low-torque, low-speed type (usually 450 rpm or lower), such as are used to drive reciprocating compressors, pumps, and so forth, that start unloaded, do not require a fuse rating or circuit-breaker setting in excess of 200 percent of full-load current.

**(B) All Motors.** The motor branch-circuit short-circuit and ground-fault protective device shall be capable of carrying the starting current of the motor.

**(C) Rating or Setting.**

**(1) In Accordance with Table 430.52.** A protective device that has a rating or setting not exceeding the value calculated according to the values given in Table 430.52 shall be used.

Although Class CC fuses are rated as time delay, they are permitted to be sized according to the requirements of non-time-delay-rated fuses because they are so fast acting. Examples of Class CC fuses are shown in Exhibit 430.8.

*Exception No. 1: Where the values for branch-circuit short-circuit and ground-fault protective devices determined by Table 430.52 do not correspond to the standard sizes or ratings of fuses, nonadjustable circuit breakers, thermal protective devices, or possible settings of adjustable circuit breakers, a higher size, rating, or possible setting that does not exceed the next higher standard ampere rating shall be permitted.*

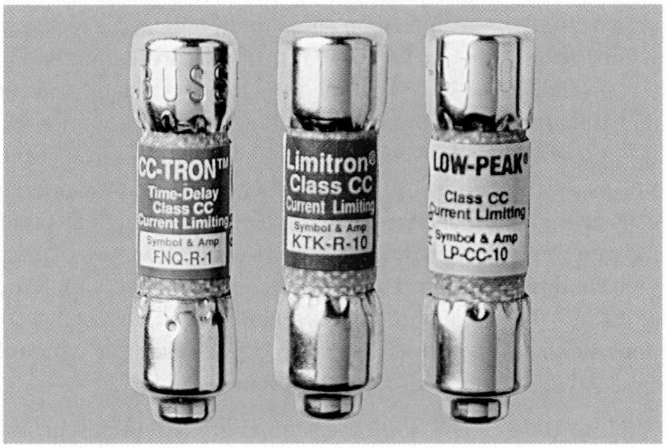


***EXHIBIT 430.8*** *Class CC fuses. (Courtesy of Cooper Bussmann, a division of Cooper Industries PLC)*

*Exception No. 2: Where the rating specified in Table 430.52, or the rating modified by Exception No. 1, is not sufficient for the starting current of the motor:*

*(a) The rating of a nontime-delay fuse not exceeding 600 amperes or a time-delay Class CC fuse shall be permitted to be increased but shall in no case exceed 400 percent of the full-load current.*

*(b) The rating of a time-delay (dual-element) fuse shall be permitted to be increased but shall in no case exceed 225 percent of the full-load current.*

*(c) The rating of an inverse time circuit breaker shall be permitted to be increased but shall in no case exceed 400 percent for full-load currents of 100 amperes or less or 300 percent for full-load currents greater than 100 amperes.*

*(d) The rating of a fuse of 601–6000 ampere classification shall be permitted to be increased but shall in no case exceed 300 percent of the full-load current.*

Informational Note: See Informative Annex D, Example D8, and Figure 430.1.

**(2) Overload Relay Table.** Where maximum branch-circuit short-circuit and ground-fault protective device ratings are shown in the manufacturer's overload relay table for use with a motor controller or are otherwise marked on the equipment, they shall not be exceeded even if higher values are allowed as shown above.

**(3) Instantaneous Trip Circuit Breaker.** An instantaneous trip circuit breaker shall be used only if adjustable and if part of a listed combination motor controller having coordinated motor overload and short-circuit and ground-fault protection in each conductor, and the setting is adjusted to no more than the value specified in Table 430.52.

Informational Note No. 1: Instantaneous trip circuit breakers are also known as motor-circuit protectors (MCPs).

Informational Note No. 2: For the purpose of this article, instantaneous trip circuit breakers may include a damping means to accommodate a transient motor inrush current without nuisance tripping of the circuit breaker.

*Exception No. 1: Where the setting specified in Table 430.52 is not sufficient for the starting current of the motor, the setting of an instantaneous trip circuit breaker shall be permitted to be increased but shall in no case exceed 1300 percent of the motor full-load current for other than Design B energy-efficient motors and no more than 1700 percent of full-load motor current for Design B energy-efficient motors. Trip settings above 800 percent for other than Design B energy-efficient motors and above 1100 percent for Design B energy-efficient motors shall be permitted where the need has been demonstrated by engineering evaluation. In such cases, it shall not be necessary to first apply an instantaneous-trip circuit breaker at 800 percent or 1100 percent.*

Informational Note: For additional information on the requirements for a motor to be classified "energy efficient," see NEMA Standards Publication No. MG1-1993, Revision, *Motors and Generators,* Part 12.59.

*Exception No. 2: Where the motor full-load current is 8 amperes or less, the setting of the instantaneous-trip circuit breaker with a continuous current rating of 15 amperes or less in a listed combination motor controller that provides coordinated motor branch-circuit overload and short-circuit and ground-fault protection shall be permitted to be increased to the value marked on the controller.*

**(4) Multispeed Motor.** For a multispeed motor, a single short-circuit and ground-fault protective device shall be permitted for two or more windings of the motor, provided the rating of the protective device does not exceed the above applicable percentage of the nameplate rating of the smallest winding protected.

*Exception: For a multispeed motor, a single short-circuit and ground-fault protective device shall be permitted to be used and sized according to the full-load current of the highest current winding, where all of the following conditions are met:*

*(a) Each winding is equipped with individual overload protection sized according to its full-load current.*

*(b) The branch-circuit conductors supplying each winding are sized according to the full-load current of the highest full-load current winding.*

*(c) The controller for each winding has a horsepower rating not less than that required for the winding having the highest horsepower rating.*

**(5) Power Electronic Devices.** Semiconductor fuses intended for the protection of electronic devices shall be permitted in lieu of devices listed in Table 430.52 for power electronic devices, associated electromechanical devices (such as bypass contactors and isolation contactors), and conductors in a solid-state motor controller system, provided that the marking for replacement fuses is provided adjacent to the fuses.

**(6) Self-Protected Combination Controller.** A listed self-protected combination controller shall be permitted in lieu of the devices specified in Table 430.52. Adjustable instantaneous-trip settings shall not exceed 1300 percent of full-load motor current for other than Design B energy-efficient motors and not more than 1700 percent of full-load motor current for Design B energy-efficient motors.

A self-protected combination controller combines the functions of short-circuit protection, disconnect, controller, and overload protection into a single unit. See Exhibit 430.9 for an example of a combination controller.

Informational Note: Proper application of self-protected combination controllers on 3-phase systems, other than solidly grounded wye, particularly on corner grounded delta systems, considers the self-protected combination controllers' individual pole-interrupting capability.

**(7) Motor Short-Circuit Protector.** A motor short-circuit protector shall be permitted in lieu of devices listed in Table 430.52 if the motor short-circuit protector is part of a listed combination motor controller having coordinated motor overload protection and short-circuit and ground-fault protection in each conductor and it will open the circuit at currents exceeding 1300 percent of motor full-load current for other

***EXHIBIT 430.9*** *A listed self-protected combination motor controller. (Courtesy of Eaton Corporation)*

than Design B energy-efficient motors and 1700 percent of motor full-load motor current for Design B energy-efficient motors.

Informational Note: A motor short-circuit protector, as used in this section, is a fused device and is not an instantaneous trip circuit breaker.

**(D) Torque Motors.** Torque motor branch circuits shall be protected at the motor nameplate current rating in accordance with 240.4(B).

### 430.53 Several Motors or Loads on One Branch Circuit

Two or more motors or one or more motors and other loads shall be permitted to be connected to the same branch circuit under conditions specified in 430.53(D) and in 430.53(A), (B), or (C). The branch-circuit protective device shall be fuses or inverse time circuit breakers.

**(A) Not Over 1 Horsepower.** Several motors, each not exceeding 1 hp in rating, shall be permitted on a nominal 120-volt branch circuit protected at not over 20 amperes or a branch circuit of 1000 volts, nominal, or less, protected at not over 15 amperes, if all of the following conditions are met:

(1) The full-load rating of each motor does not exceed 6 amperes.
(2) The rating of the branch-circuit short-circuit and ground-fault protective device marked on any of the controllers is not exceeded.
(3) Individual overload protection conforms to 430.32.

Two or more motors – or one or more motors and other loads – may be connected to the same 120-volt, 15- or 20-ampere, single-phase lighting circuit, as long as each motor is rated not more than 1 hp, the full-load rating of each motor does not exceed 6 amperes, and the rating of the branch-circuit protective device is not exceeded.

The requirements for overload protection provided in 430.32 must be applied in all cases, regardless of the number (one or more) of motors or the type of branch circuit.

**(B) If Smallest Rated Motor Protected.** If the branch-circuit short-circuit and ground-fault protective device is selected not to exceed that allowed by 430.52 for the smallest rated motor, two or more motors or one or more motors and other load(s), with each motor having individual overload protection, shall be permitted to be connected to a branch circuit where it can be determined that the branch-circuit short-circuit and ground-fault protective device will not open under the most severe normal conditions of service that might be encountered.

**(C) Other Group Installations.** Two or more motors of any rating or one or more motors and other load(s), with each motor having individual overload protection, shall be permitted to be connected to one branch circuit where the motor controller(s) and overload device(s) are (1) installed as a listed factory assembly and the motor branch-circuit short-circuit and ground-fault protective device either is provided as part of the assembly or is specified by a marking on the assembly, or (2) the motor branch-circuit short-circuit and ground-fault protective device, the motor controller(s), and overload device(s) are field-installed as separate assemblies listed for such use and provided with manufacturers' instructions for use with each other, and (3) all of the following conditions are complied with:

(1) Each motor overload device is either (a) listed for group installation with a specified maximum rating of fuse, inverse time circuit breaker, or both, or (b) selected such that the ampere rating of the motor-branch short-circuit and ground-fault protective device does not exceed that permitted by 430.52 for that individual motor overload device and corresponding motor load.
(2) Each motor controller is either (a) listed for group installation with a specified maximum rating of fuse, circuit breaker, or both, or (b) selected such that the ampere rating of the motor-branch short-circuit and ground-fault protective device does not exceed that permitted by 430.52 for that individual controller and corresponding motor load.
(3) Each circuit breaker is listed and is of the inverse time type.
(4) The branch circuit shall be protected by fuses or inverse time circuit breakers having a rating not exceeding that specified in 430.52 for the highest rated motor connected to the branch circuit plus an amount equal to the sum of the full-load current ratings of all other motors and the ratings of other loads connected to the circuit. Where this calculation results in a rating less than the ampacity of the branch-circuit conductors, it shall be permitted to increase the maximum rating of the fuses or circuit breaker to a value not exceeding that permitted by 240.4(B).
(5) The branch-circuit fuses or inverse time circuit breakers are not larger than allowed by 430.40 for the overload relay protecting the smallest rated motor of the group.
(6) Overcurrent protection for loads other than motor loads shall be in accordance with Parts I through VII of Article 240.

The ground-fault short-circuit protection for motors might be greater than is permitted for other loads in accordance with Article 240. Devices with the same ampere rating might have significantly different short-circuit current ratings. Section 110.10 addresses properly selecting components considering the characteristics of all the components, with the goal in mind that a fault will not cause unacceptable damage.

A motor controller or overload device need not be marked for group motor installation – since it is protected within its listing requirements – where it is applied within a group installation in which the branch-circuit protection for the group is within the same size limit as what would be permitted for a single motor installation of that device. Prior to the 2011 *NEC,* all motor controllers and overload devices were required to be listed for group installation.

Informational Note: See 110.10 for circuit impedance and other characteristics.

**(D) Single Motor Taps.** For group installations described above, the conductors of any tap supplying a single motor shall not be required to have an individual branch-circuit short-circuit and ground-fault protective device, provided they comply with one of the following:

(1) No conductor to the motor shall have an ampacity less than that of the branch-circuit conductors.
(2) No conductor to the motor shall have an ampacity less than one-third that of the branch-circuit conductors, with a minimum in accordance with 430.22. The conductors from the point of the tap to the motor overload device shall be not more than 7.5 m (25 ft) long and be protected from physical damage by being enclosed in an approved raceway or by use of other approved means.
(3) Conductors from the branch-circuit short-circuit and ground-fault protective device to a listed manual motor controller additionally marked "Suitable for Tap Conductor Protection in Group Installations," or to a branch-circuit protective device, shall be permitted to have an ampacity not less than one-tenth the rating or setting of the branch-circuit short-circuit and ground-fault protective device. The conductors from the controller to the motor shall have an ampacity in accordance with 430.22. The conductors from the point of the tap to the controller(s) shall (1) be suitably protected from physical damage and enclosed either by an enclosed controller or by a raceway and be not more than 3 m (10 ft) long or (2) have an ampacity not less than that of the branch-circuit conductors.

The conditions for applying this tap rule are similar to those in 430.28 covering motor supply conductors tapped to a feeder. The short-circuit ground-fault device on the line side of the tap conductors protects more than one set of conductors that supply individual motors, which eliminates the need for an individual short-circuit ground-fault device for each set of conductors that supply a motor. Additional branch-circuit protective devices (such as fuses, inverse time circuit breakers, and listed self-protected combination motor controllers) may be used in the same location in the circuit of a group installation as a manual motor controller additionally marked "Suitable for Tap Conductor Protection in Group Installations."

This approach requires that the tap conductors meet certain size, physical protection, length, and termination conditions. The tap conductors always have to meet the conductor size requirements of 430.22 and must have an ampacity not less than one-tenth the rating of the upstream short-circuit ground-fault protective device. Where the conductors are sized according to this provision, their length cannot exceed 10 feet and they have to be enclosed in a raceway or by the motor controller. If the conductor ampacity is not less than the rating of the upstream short-circuit ground-fault protective device, the length of the conductor is not limited.

The tap conductors are permitted to terminate in a listed manual motor controller that is marked "Suitable for Tap Conductor Protection in Group Installations." This controller provides an instantaneous trip mechanism, motor overload protection, and provisions for disconnecting the motor.

Exhibit 430.10 illustrates main branch-circuit conductors supplying a motor that is part of a group installation. The tap conductors have an ampacity equal to the ampacity of the main branch-circuit conductors. Therefore, branch-circuit short-circuit and ground-fault protective devices, fuses, or circuit breakers for the conductors in the tap are not required at the point of connection of the tap conductors to the main conductors, provided that the motor controller and motor overload protective device are listed for group installation with the size of the main branch-circuit short-circuit and ground-fault protective device used.

Exhibit 430.11 also illustrates main branch-circuit conductors supplying a motor that is part of a group installation. Here, the tap conductors have an ampacity at least one-third the ampacity of the main branch-circuit conductors, are not more than 25 feet in length, and are suitably protected from physical damage. The motor controller and motor overload protective device must be listed for group installation with the size of the main branch-circuit short-circuit and ground-fault protective device used.

In both examples, the main branch-circuit fuses or circuit breakers would operate in the event of a short circuit, and the

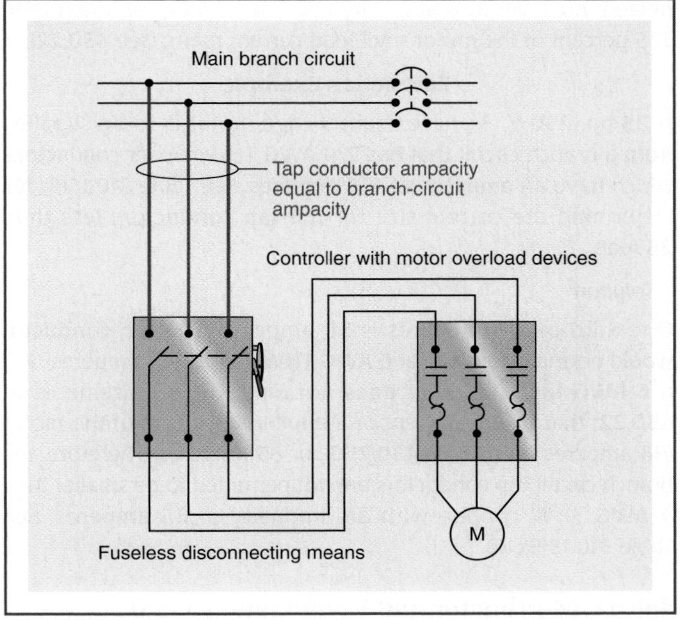


***EXHIBIT 430.10*** *An example of the permissible omission of motor branch-circuit protective devices for tap conductors per 430.53(D)(1).*

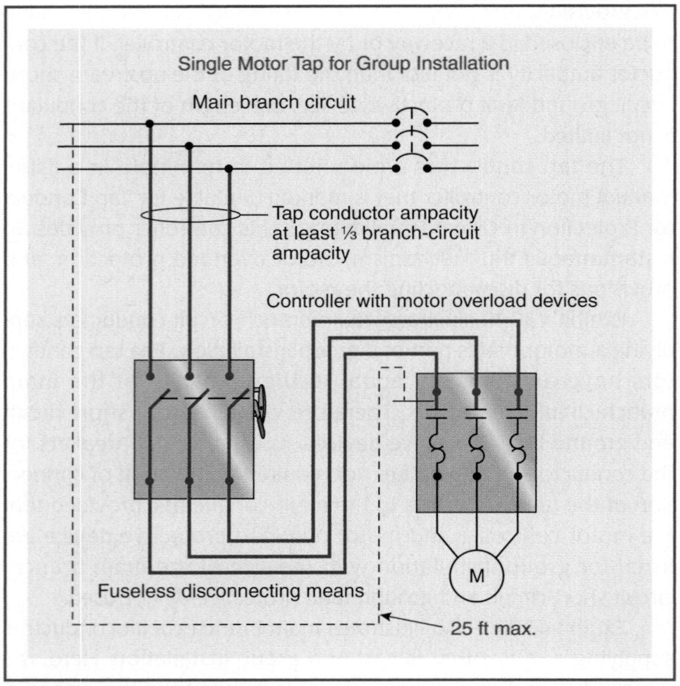


***EXHIBIT 430.11*** *An example of the permissible omission of motor branch-circuit protective devices for tap conductors per 430.53(D)(2).*

overload protective device would operate to protect the motor and tap conductors under overload conditions.

The tap conductors should never be of a smaller size and ampacity than the branch-circuit conductors required by 430.22. That is, a tap conductor (25 feet or less) may be one-third the ampacity of the main branch-circuit conductor to which it is connected; however, this ampacity must be equal to or larger than 125 percent of the motor's full-load current rating (see 430.22).

**Calculation Example**

A 25-hp, 230-V, 3-phase squirrel-cage motor is to be supplied from a branch circuit that has 2/0 AWG THW copper conductors, which have an ampacity of 175 amperes. See Table 310.15(B)(16). Determine the correct size for the tap conductors less than 25 feet.

*Solution*

One-third of 175 amperes is 58 amperes. The tap conductor would normally be sized at 6 AWG THW copper (65 amperes). But a 6 AWG tap conductor does not meet the requirements of 430.22: that is, 125 percent of the full-load current of the motor (68 amperes from Table 430.250), or 85 amperes. Therefore, the branch-circuit tap conductors are not permitted to be smaller than 4 AWG THW copper, with an ampacity of 85 amperes. See Table 310.15(B)(16).

## 430.54 Multimotor and Combination-Load Equipment

The rating of the branch-circuit short-circuit and ground-fault protective device for multimotor and combination-load equipment shall not exceed the rating marked on the equipment in accordance with 430.7(D).

## 430.55 Combined Overcurrent Protection

Motor branch-circuit short-circuit and ground-fault protection and motor overload protection shall be permitted to be combined in a single protective device where the rating or setting of the device provides the overload protection specified in 430.32.

Either a circuit breaker with inverse time characteristics or a dual-element (time-delay) fuse is permitted to serve both as motor overload protection and as the branch-circuit short-circuit and ground-fault protection, if the requirements of 430.32 are met. These devices are not permitted to be sized as overload protection according to the values of 430.32(C). Rather, fuses are permitted to be sized as overload protection only, according to the values found in 430.32(A)(1), 430.32(B)(1), and 430.32(D)(1).

One-time, time-delay dual-element and Type S dual-element fuses and adapters are available with up to a 30-ampere rating. Type S fuses are designed to prevent oversize fusing. See 240.50 through 240.54 for more information about these fuses and adapters.

Exhibits 430.12 and 430.13 are examples of time-delay, cartridge-type dual-element fuses that are able to withstand the normal motor starting current if sized at or near the motor full-load rating but that open when subjected to prolonged overload or blow quickly during a short circuit or ground fault. The dual-element characteristics are the thermal cutout element, which permits harmless high-inrush currents to flow for short periods (but which would open the circuit during a prolonged period), and the fuse link element, which has current-limiting ability for short-circuit currents (and which would blow quickly). Dual-element fuses may be used in larger sizes to provide only short-circuit and ground-fault protection.

## 430.56 Branch-Circuit Protective Devices — In Which Conductor

Branch-circuit protective devices shall comply with the provisions of 240.15.

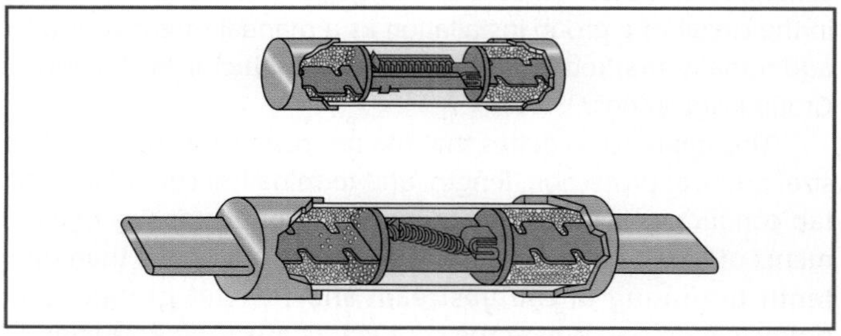

***EXHIBIT 430.12*** *Fusetron cartridge-type fuses. (Courtesy of Cooper Bussmann, a division of Cooper Industries PLC)*

*EXHIBIT 430.13 A fuse block rejection pin (red arrow), which ensures that circuit protection will be afforded by only fuses providing a specific level of overcurrent protection. (Courtesy of Cooper Bussmann, a division of Cooper Industries PLC)*

### 430.57 Size of Fuseholder

Where fuses are used for motor branch-circuit short-circuit and ground-fault protection, the fuseholders shall not be of a smaller size than required to accommodate the fuses specified by Table 430.52.

*Exception: Where fuses having time delay appropriate for the starting characteristics of the motor are used, it shall be permitted to use fuseholders sized to fit the fuses that are used.*

The use of dual-element (time-delay) fuses makes it possible to use smaller fuses, thereby providing better protection because of the smaller fuses' lower ratings. Dual-element fuses also save in installation cost by allowing smaller-size switches and panels, and they allow for easier arrangement of equipment where space is at a premium at motor control centers.

### 430.58 Rating of Circuit Breaker

A circuit breaker for motor branch-circuit short-circuit and ground-fault protection shall have a current rating in accordance with 430.52 and 430.110.

## V. Motor Feeder Short-Circuit and Ground-Fault Protection

### 430.61 General

Part V specifies protective devices intended to protect feeder conductors supplying motors against overcurrents due to short circuits or grounds.

Informational Note: See Informative Annex D, Example D8.

### 430.62 Rating or Setting — Motor Load

**(A) Specific Load.** A feeder supplying a specific fixed motor load(s) and consisting of conductor sizes based on 430.24 shall be provided with a protective device having a rating or setting not greater than the largest rating or setting of the branch-circuit short-circuit and ground-fault protective device for any motor supplied by the feeder [based on the maximum permitted value for the specific type of a protective device in accordance with 430.52, or 440.22(A) for hermetic refrigerant motor-compressors], plus the sum of the full-load currents of the other motors of the group.

Where the same rating or setting of the branch-circuit short-circuit and ground-fault protective device is used on two or more of the branch circuits supplied by the feeder, one of the protective devices shall be considered the largest for the above calculations.

The rating of a motor feeder short-circuit ground-fault protective device is determined by adding the rating of the largest branch-circuit short-circuit and ground-fault protective device for any motor supplied by the feeder to the sum of the full-load currents of all of the other motors supplied by that feeder. The largest branch-circuit short-circuit and ground-fault protective device is based on 430.52 and Table 430.52. The largest rating can be based on either of the exceptions to 430.52(C)(1). For the purposes of sizing the feeder protective device, it is assumed that the same type of protective device is being used for the feeder and the branch circuits. This assumption is necessary if the feeder protective device and the largest branch-circuit protective device are different types; for example, one is a fuse and the other is a circuit breaker.

Section 430.62(A) recognizes the lower setting for motor overload devices that is required for hermetic refrigerant motor-compressors.

*Exception No. 1: Where one or more instantaneous trip circuit breakers or motor short-circuit protectors are used for motor branch-circuit short-circuit and ground-fault protection as permitted in 430.52(C), the procedure provided above for determining the maximum rating of the feeder protective device shall apply with the following provision: For the purpose of the calculation, each instantaneous trip circuit breaker or motor short-circuit protector shall be assumed to have a rating not exceeding the maximum percentage of motor full-load current permitted by Table 430.52 for the type of feeder protective device employed.*

*Exception No. 2: Where the feeder overcurrent protective device also provides overcurrent protection for a motor control center, the provisions of 430.94 shall apply.*

Informational Note: See Informative Annex D, Example D8.

**(B) Other Installations.** Where feeder conductors have an ampacity greater than required by 430.24, the rating or setting of the feeder overcurrent protective device shall be permitted to be based on the ampacity of the feeder conductors.

Exception No. 2 to 430.62(A) correlates the requirement of 430.62(B) for determining feeder short-circuit ground-fault protection with the requirements of 430.94 covering overcurrent protection for motor control centers. Where the motor feeder short-circuit ground-fault protective device is also the OCPD for a motor control center, its rating cannot exceed that allowed for protecting the common power bus of the motor control center.

### 430.63 Rating or Setting — Motor Load and Other Load(s)

Where a feeder supplies a motor load and other load(s), the feeder protective device shall have a rating not less than that required for the sum of the other load(s) plus the following:

(1) For a single motor, the rating permitted by 430.52
(2) For a single hermetic refrigerant motor-compressor, the rating permitted by 440.22
(3) For two or more motors, the rating permitted by 430.62

*Exception: Where the feeder overcurrent device provides the overcurrent protection for a motor control center, the provisions of 430.94 shall apply.*

See the commentary following 430.62(B).

## VI. Motor Control Circuits

### 430.71 General

Part VI contains modifications of the general requirements and applies to the particular conditions of motor control circuits.

•

### 430.72 Overcurrent Protection

**(A) General.** A motor control circuit tapped from the load side of a motor branch-circuit short-circuit and ground-fault protective device(s) and functioning to control the motor(s) connected to that branch circuit shall be protected against overcurrent in accordance with 430.72. Such a tapped control circuit shall not be considered to be a branch circuit and shall be permitted to be protected by either a supplementary or branch-circuit overcurrent protective device(s). A motor control circuit other than such a tapped control circuit shall be protected against overcurrent in accordance with 725.43 or the notes to Table 11(A) and Table 11(B) in Chapter 9, as applicable.

**(B) Conductor Protection.** The overcurrent protection for conductors shall be provided as specified in 430.72(B) or (B)(2).

*Exception No. 1: Where the opening of the control circuit would create a hazard as, for example, the control circuit of a fire pump motor, and the like, conductors of control circuits shall require only short-circuit and ground-fault protection and shall be permitted to be protected by the motor branch-circuit short-circuit and ground-fault protective device(s).*

*Exception No. 2: Conductors supplied by the secondary side of a single-phase transformer having only a two-wire (single-voltage) secondary shall be permitted to be protected by overcurrent protection provided on the primary (supply) side of the transformer, provided this protection does not exceed the value determined by multiplying the appropriate maximum rating of the overcurrent device for the secondary conductor from Table 430.72(B) by the secondary-to-primary voltage ratio. Transformer secondary conductors (other than two-wire) shall not be considered to be protected by the primary overcurrent protection.*

**(1) Separate Overcurrent Protection.** Where the motor branch-circuit short-circuit and ground-fault protective device does not provide protection in accordance with 430.72(B)(2), separate overcurrent protection shall be provided. The overcurrent protection shall not exceed the values specified in Column A of Table 430.72(B).

**(2) Branch-Circuit Overcurrent Protective Device.** Conductors shall be permitted to be protected by the motor branch-circuit short-circuit and ground-fault protective device and shall require only short-circuit and ground-fault protection. Where the conductors do not extend beyond the motor control equipment enclosure, the rating of the protective device(s) shall not exceed the value specified in Column B of Table 430.72(B). Where the conductors extend beyond the motor control equipment enclosure, the rating of the protective device(s) shall not exceed the value specified in Column C of Table 430.72(B).

**(C) Control Circuit Transformer.** Where a motor control circuit transformer is provided, the transformer shall be protected in accordance with 430.72(C)(1), (C)(2), (C)(3), (C)(4), or (C)(5).

*Exception: Overcurrent protection shall be omitted where the opening of the control circuit would create a hazard as, for example, the control circuit of a fire pump motor and the like.*

**(1) Compliance with Article 725.** Where the transformer supplies a Class 1 power-limited circuit, Class 2, or Class 3 remote-control circuit complying with the requirements of Article 725, protection shall comply with Article 725.

**(2) Compliance with Article 450.** Protection shall be permitted to be provided in accordance with 450.3.

**(3) Less Than 50 Volt-Amperes.** Control circuit transformers rated less than 50 volt-amperes (VA) and that are an integral part of the motor controller and located within the motor controller enclosure shall be permitted to be protected by primary overcurrent devices, impedance limiting means, or other inherent protective means.

**(4) Primary Less Than 2 Amperes.** Where the control circuit transformer rated primary current is less than 2 amperes, an overcurrent device rated or set at not more than 500 percent of the rated primary current shall be permitted in the primary circuit.

**(5) Other Means.** Protection shall be permitted to be provided by other approved means.

Motor control circuits may receive their power either from the load side of the motor short-circuit and ground-fault protective device or from a separate source, such as a panelboard.

***TABLE 430.72(B)*** *Maximum Rating of Overcurrent Protective Device in Amperes*

| Control Circuit Conductor Size (AWG) | Protection Provided by Motor Branch-Circuit Protective Device(s) | | | | | |
|---|---|---|---|---|---|---|
| | Column A Separate Protection Provided | | Column B Conductors Within Enclosure | | Column C Conductors Extend Beyond Enclosure | |
| | Copper | Aluminum or Copper-Clad Aluminum | Copper | Aluminum or Copper-Clad Aluminum | Copper | Aluminum or Copper-Clad Aluminum |
| 18 | 7 | — | 25 | — | 7 | — |
| 16 | 10 | — | 40 | — | 10 | — |
| 14 | (Note 1) | — | 100 | — | 45 | — |
| 12 | (Note 1) | (Note 1) | 120 | 100 | 60 | 45 |
| 10 | (Note 1) | (Note 1) | 160 | 140 | 90 | 75 |
| Larger than 10 | (Note 1) | (Note 1) | (Note 2) | (Note 2) | (Note 3) | (Note 3) |

Notes:
1. Value specified in 310.15 as applicable.
2. 400 percent of value specified in Table 310.15(B)(17) for 60°C conductors.
3. 300 percent of value specified in Table 310.15(B)(16) for 60°C conductors.

Motor control circuits that receive their power from a separate source must be protected against overcurrent in accordance with 725.43 for Class 1 circuits. Conductor sizes 14 AWG and larger must be protected according to their ampacity listed in Tables 310.15(B)(16) through 310.15(B)(20). Conductor sizes 16 and 18 AWG must be protected at not more than 10 and 7 amperes, respectively, as specified in Table 430.72(B).

If a motor control circuit is tapped from the load side of the motor branch-circuit short-circuit and ground-fault protective device, the size of the tapped conductor and the rating of the overcurrent device are based on whether the conductor stays within the motor control enclosure or leaves it. The load on a motor control circuit is similar to a motor branch-circuit load in that there is a predetermined connected load. An initial high inrush of current also occurs until the armature of the relay is seated and the current decreases to a steady state. Therefore, the overcurrent protection is similar to the short-circuit and ground-fault protection provided for a motor and is allowed to be greater than the ampacity of the control circuit conductor.

## 430.73 Protection of Conductors from Physical Damage

Where damage to a motor control circuit would constitute a hazard, all conductors of such a remote motor control circuit that are outside the control device itself shall be installed in a raceway or be otherwise protected from physical damage.

If damage to the motor control circuit conductors would constitute a hazard, physical protection of the motor control circuit conductors is required. If damage to the control circuit conductors could result in an accidental ground fault or short circuit, causing the device to operate or rendering the device inoperative (either condition could constitute a hazard to persons or property), conductors must be installed in a raceway. Where boilers or furnaces are equipped with an automatic safety control device, damage to the conductors of the low-voltage control circuit (for example, a thermostat) does not constitute a hazard (see Article 725, Part III).

## 430.74 Electrical Arrangement of Control Circuits

Where one conductor of the motor control circuit is grounded, the motor control circuit shall be arranged so that a ground fault in the control circuit remote from the motor controller will (1) not start the motor and (2) not bypass manually operated shutdown devices or automatic safety shutdown devices.

The inadvertent grounding of control circuits is a significant safety issue. Section 430.74 requires that if one side of the motor control circuit is grounded, the circuit must be arranged so that a ground fault in the remote-control device will not start the motor. For example, in the control wiring illustrated in Exhibit 430.14, the control circuit is a 120-volt, single-phase circuit derived from a 208-volt, 3-phase wye system supplying the motor, and one side of the control circuit is the grounded neutral. If the start button of the motor control circuit is connected to the grounded neutral, a ground fault on the coil side of the start button can start the motor. As shown in Exhibit 430.15, the same condition exists if the ground fault is in the wiring rather than in the control device itself. This hazardous condition can be alleviated by locating the start button in the ungrounded side of the control circuit as shown in the correct arrangement in Exhibit 430.15.

Combinations of ground faults in motor and motor control circuits can also result in inadvertent motor starting. If the circuit is ungrounded, the first fault may go undetected. One solution is

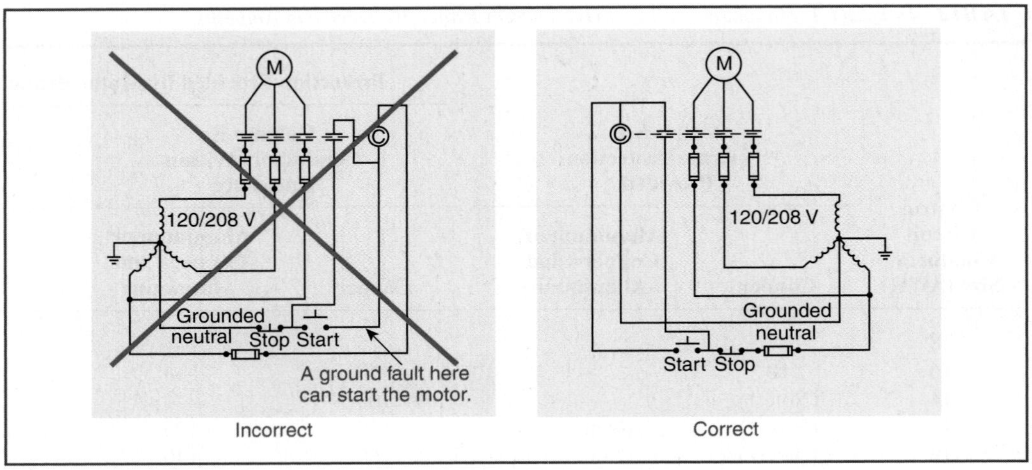


***EXHIBIT 430.14*** *An example of control wiring in violation of 430.74 (left), and in compliance with 430.74 (right). (For simplification, motor overload elements and disconnecting means are not shown.)*

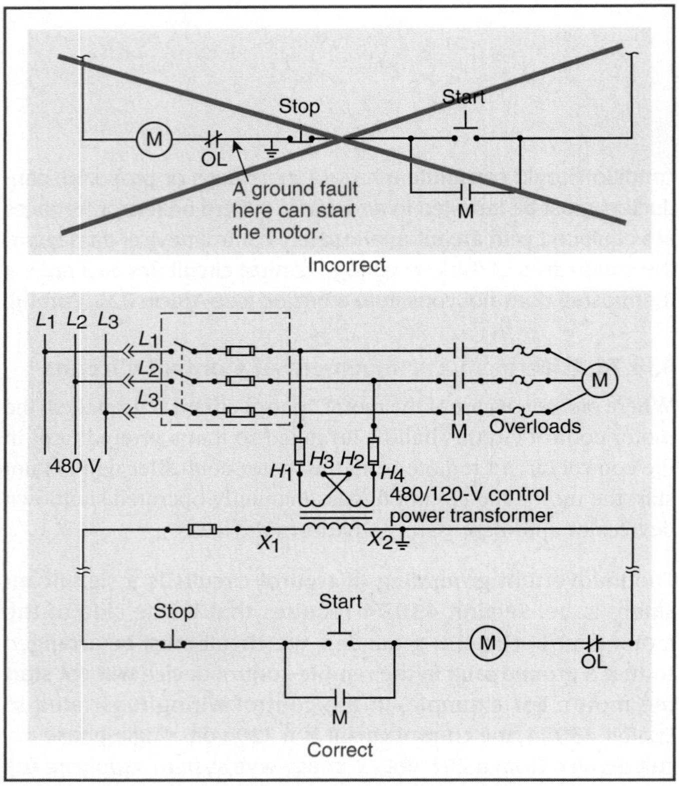


***EXHIBIT 430.15*** *An example of control wiring using a 480/120-V control power transformer. (The upper control circuit is not in compliance with 430.74.)*

to use double-pole control devices with one pole in each of the two control lines.

## 430.75 Disconnection

**(A) General.** Motor control circuits shall be arranged so that they will be disconnected from all sources of supply when the disconnecting means is in the open position. The disconnecting means shall be permitted to consist of two or more separate devices, one of which disconnects the motor and the controller from the source(s) of power supply for the motor, and the other(s), the motor control circuit(s) from its power supply. Where separate devices are used, they shall be located immediately adjacent to each other.

*Exception No. 1: Where more than 12 motor control circuit conductors are required to be disconnected, the disconnecting means shall be permitted to be located other than immediately adjacent to each other where all of the following conditions are complied with:*

*(a) Access to energized parts is limited to qualified persons in accordance with Part XII of this article.*

*(b) A warning sign is permanently located on the outside of each equipment enclosure door or cover permitting access to the live parts in the motor control circuit(s), warning that motor control circuit disconnecting means are remotely located and specifying the location and identification of each disconnect. Where energized parts are not in an equipment enclosure as permitted by 430.232 and 430.233, an additional warning sign(s) shall be located where visible to persons who may be working in the area of the energized parts.*

*Exception No. 2: The motor control circuit disconnecting means shall be permitted to be remote from the motor controller power supply disconnecting means where the opening of one or more motor control circuit disconnecting means is capable of resulting in potentially unsafe conditions for personnel or property and the conditions of items (a) and (b) of Exception No. 1 are complied with.*

**(B) Control Transformer in Controller Enclosure.** Where a transformer or other device is used to obtain a reduced voltage for the motor control circuit and is located in the controller enclosure, such transformer or other device shall be connected to the load side of the disconnecting means for the motor control circuit.

# VII. Motor Controllers

## 430.81 General

Part VII is intended to require suitable controllers for all motors.

**(A) Stationary Motor of ⅛ Horsepower or Less.** For a stationary motor rated at ⅛ hp or less that is normally left running and is constructed so that it cannot be damaged by overload or failure to start, such as clock motors and the like, the branch-circuit disconnecting means shall be permitted to serve as the controller.

**(B) Portable Motor of ⅓ Horsepower or Less.** For a portable motor rated at ⅓ hp or less, the controller shall be permitted to be an attachment plug and receptacle or cord connector.

## 430.82 Controller Design

**(A) Starting and Stopping.** Each controller shall be capable of starting and stopping the motor it controls and shall be capable of interrupting the locked-rotor current of the motor.

**(B) Autotransformer.** An autotransformer starter shall provide an "off" position, a running position, and at least one starting position. It shall be designed so that it cannot rest in the starting position or in any position that will render the overload device in the circuit inoperative.

**(C) Rheostats.** Rheostats shall be in compliance with the following:

(1) Motor-starting rheostats shall be designed so that the contact arm cannot be left on intermediate segments. The point or plate on which the arm rests when in the starting position shall have no electrical connection with the resistor.
(2) Motor-starting rheostats for dc motors operated from a constant voltage supply shall be equipped with automatic devices that will interrupt the supply before the speed of the motor has fallen to less than one-third its normal rate.

## 430.83 Ratings

The controller shall have a rating as specified in 430.83(A), unless otherwise permitted in 430.83(B) or (C), or as specified in (D), under the conditions specified.

**(A) General.**

**(1) Horsepower Ratings.** Controllers, other than inverse time circuit breakers and molded case switches, shall have horsepower ratings at the application voltage not lower than the horsepower rating of the motor.

**(2) Circuit Breaker.** A branch-circuit inverse time circuit breaker rated in amperes shall be permitted as a controller for all motors. Where this circuit breaker is also used for overload protection, it shall conform to the appropriate provisions of this article governing overload protection.

**(3) Molded Case Switch.** A molded case switch rated in amperes shall be permitted as a controller for all motors.

A molded case switch has the same frame appearance as a molded case circuit breaker and is designed to fit in circuit-breaker enclosures. However, the device is marked with only a short-circuit current withstand rating, which indicates that the switch does not provide overcurrent protection. Those fused molded case switches that do provide overcurrent protection are marked with a short-circuit current interrupting rating. Both fused and unfused molded case switches can be used in motor circuits.

Molded case switches are permitted as motor disconnecting means per 430.109. In general, molded case switches are rated only in amperes and, where used in a motor circuit, must be sized at 115 percent of the motor full-load current rating. Disconnecting means assemblies are available that employ molded case switches marked with horsepower ratings that can be used, instead of the ampere rating of the molded case switch.

**(B) Small Motors.** Devices as specified in 430.81(A) and (B) shall be permitted as a controller.

**(C) Stationary Motors of 2 Horsepower or Less.** For stationary motors rated at 2 hp or less and 300 volts or less, the controller shall be permitted to be either of the following:

(1) A general-use switch having an ampere rating not less than twice the full-load current rating of the motor
(2) On ac circuits, a general-use snap switch suitable only for use on ac (not general-use ac–dc snap switches) where the motor full-load current rating is not more than 80 percent of the ampere rating of the switch

**(D) Torque Motors.** For torque motors, the controller shall have a continuous-duty, full-load current rating not less than the nameplate current rating of the motor. For a motor controller rated in horsepower but not marked with the foregoing current rating, the equivalent current rating shall be determined from the horsepower rating by using Table 430.247, Table 430.248, Table 430.249, or Table 430.250.

**(E) Voltage Rating.** A controller with a straight voltage rating, for example, 240 volts or 480 volts, shall be permitted to be applied in a circuit in which the nominal voltage between any two conductors does not exceed the controller's voltage rating. A controller with a slash rating, for example, 120/240 volts or 480Y/277 volts, shall only be applied in a solidly grounded circuit in which the nominal voltage to ground from any conductor does not exceed the lower of the two values of the controller's voltage rating and the nominal voltage between any two conductors does not exceed the higher value of the controller's voltage rating.

Controllers identified with slash voltage ratings, such as 120/240 V and 480/277 V, may be used only on electrical systems in which the nominal voltage to ground does not exceed the lower voltage rating of the controller, and where the nominal voltage between any two phases of the electrical system is not greater than the higher value of the controller voltage rating.

## 430.84 Need Not Open All Conductors

The controller shall not be required to open all conductors to the motor.

*Exception: Where the controller serves also as a disconnecting means, it shall open all ungrounded conductors to the motor as provided in 430.111.*

A controller that does not serve as a disconnecting means must open only as many motor circuit conductors as are necessary to stop the motor – that is, one conductor for a dc or single-phase motor circuit, two conductors for a 3-phase motor circuit, and three conductors for a 2-phase motor circuit.

## 430.85 In Grounded Conductors

One pole of the controller shall be permitted to be placed in a permanently grounded conductor, provided the controller is designed so that the pole in the grounded conductor cannot be opened without simultaneously opening all conductors of the circuit.

Generally, one conductor of a 120-volt circuit is grounded, and a single-pole device must be connected in the ungrounded conductor to serve as a controller. A 2-pole controller is permitted for such a circuit, where both conductors (grounded and ungrounded) are opened simultaneously. The same requirement can be applied to other circuits, such as 240-volt, 3-wire circuits with one conductor grounded.

## 430.87 Number of Motors Served by Each Controller

Each motor shall be provided with an individual controller.

*Exception No. 1: For motors rated 1000 volts or less, a single controller rated at not less than the equivalent horsepower, as determined in accordance with 430.110(C)(1), of all the motors in the group shall be permitted to serve the group under any of the following conditions:*

*(a) Where a number of motors drive several parts of a single machine or piece of apparatus, such as metal and woodworking machines, cranes, hoists, and similar apparatus*
*(b) Where a group of motors is under the protection of one overcurrent device as permitted in 430.53(A)*
*(c) Where a group of motors is located in a single room within sight from the controller location*

*Exception No. 2: A branch-circuit disconnecting means serving as the controller as allowed in 430.81(A) shall be permitted to serve more than one motor.*

The conditions stated in Exception No. 1 are similar to those specified in the exception to 430.112, which permit the use of a single disconnecting means for a group of motors.

## 430.88 Adjustable-Speed Motors

Adjustable-speed motors that are controlled by means of field regulation shall be equipped and connected so that they cannot be started under a weakened field.

*Exception: Starting under a weakened field shall be permitted where the motor is designed for such starting.*

The torque and speed of a dc motor depend on the amount of current passing through the armature. This current is a function of shunt field strength and rpm of the armature. A reduction of the shunt field magnetic flux causes a reduction of the counterelectromotive force in the armature, resulting in an increase in armature current, thereby increasing torque and thus increasing speed.

Because of excessive armature starting currents, field-regulated, adjustable-speed motors are not permitted to be started under a weakened field condition unless some means is provided to limit the speed to within safe limits.

## 430.89 Speed Limitation

Machines of the following types shall be provided with speed-limiting devices or other speed-limiting means:

(1) Separately excited dc motors
(2) Series motors
(3) Motor-generators and converters that can be driven at excessive speed from the dc end, as by a reversal of current or decrease in load

*Exception: Separate speed-limiting devices or means shall not be required under either of the following conditions:*

*(1) Where the inherent characteristics of the machines, the system, or the load and the mechanical connection thereto are such as to safely limit the speed*
*(2) Where the machine is always under the manual control of a qualified operator*

Use of dc motors is common where speed control is essential, such as in the case of electric railways and elevators, where a smooth start, controlled acceleration, and a smooth stop are necessary. If the load is removed from a series motor when it is running, the speed of the motor will increase until it is dangerously high. To produce the necessary counterelectromotive force with a weakened field, the armature must turn correspondingly faster. Series motors are commonly used as gear-drive traction motors of electric locomotives and, thus, are continuously loaded.

Unless the exception applies, the motors, motor (compound-wound dc) generators, and (synchronous) converters must be provided with speed-limiting devices, such as a centrifugal device on the shaft of the machine or a remotely located overspeed device. This device may be set to operate a set of contacts at a predetermined speed and thereby trip a circuit breaker and de-energize the machine.

## 430.90 Combination Fuseholder and Switch as Controller

The rating of a combination fuseholder and switch used as a motor controller shall be such that the fuseholder will accommodate the size of the fuse specified in Part III of this article for motor overload protection.

*Exception: Where fuses having time delay appropriate for the starting characteristics of the motor are used, fuseholders of smaller size than specified in Part III of this article shall be permitted.*

Time-delay (dual-element) fuses can be used for both motor overload and branch-circuit short-circuit and ground-fault protection and can be sized in accordance with 430.32. See also 430.36, 430.55, and 430.57 for other requirements regarding fuses and fuseholders.

# VIII. Motor Control Centers

## 430.92 General

Part VIII covers motor control centers installed for the control of motors, lighting, and power circuits.

Motor control centers are made up of a number of motor starters, controls, and disconnect switches. Motor control centers are allowed to be used as service equipment if provided with a single main disconnecting means. A second service disconnecting means, however, is permitted in the motor control center if it is provided to serve other loads.

In addition to Part VIII, installation requirements, including access and working space clearances, for motor control centers are contained in 110.26. The requirements of 110.26(E) specify dedicated space for a motor control center and physical protection from mechanical systems that might leak or otherwise adversely affect a motor control center.

## 430.94 Overcurrent Protection

Motor control centers shall be provided with overcurrent protection in accordance with Parts I, II, and VIII of Article 240. The ampere rating or setting of the overcurrent protective device shall not exceed the rating of the common power bus. This protection shall be provided by (1) an overcurrent protective device located ahead of the motor control center or (2) a main overcurrent protective device located within the motor control center.

Overcurrent protection must not exceed the rating of the common power bus of a motor control center. Use of an OCPD with a rating less than the main bus is permitted, provided it is of sufficient size to carry the load determined in accordance with Part II of Article 430.

## 430.95 Service Equipment

Where used as service equipment, each motor control center shall be provided with a single main disconnecting means to disconnect all ungrounded service conductors.

*Exception: A second service disconnect shall be permitted to supply additional equipment.*

Where a grounded conductor is provided, the motor control center shall be provided with a main bonding jumper, sized in accordance with 250.28(D), within one of the sections for connecting the grounded conductor, on its supply side, to the motor control center equipment ground bus.

*Exception: High-impedance grounded neutral systems shall be permitted to be connected as provided in 250.36.*

## 430.96 Grounding

Multisection motor control centers shall be connected together with an equipment grounding conductor or an equivalent equipment grounding bus sized in accordance with Table 250.122. Equipment grounding conductors shall be connected to this equipment grounding bus or to a grounding termination point provided in a single-section motor control center.

## 430.97 Busbars and Conductors

**(A) Support and Arrangement.** Busbars shall be protected from physical damage and be held firmly in place. Other than for required interconnections and control wiring, only those conductors that are intended for termination in a vertical section shall be located in that section.

*Exception: Conductors shall be permitted to travel horizontally through vertical sections where such conductors are isolated from the busbars by a barrier.*

**(B) Phase Arrangement.** The phase arrangement on 3-phase horizontal common power and vertical buses shall be A, B, C from front to back, top to bottom, or left to right, as viewed from the front of the motor control center. The B phase shall be that phase having the higher voltage to ground on 3-phase, 4-wire, delta-connected systems. Other busbar arrangements shall be permitted for additions to existing installations and shall be marked.

*Exception: Rear-mounted units connected to a vertical bus that is common to front-mounted units shall be permitted to have a C, B, A phase arrangement where properly identified.*

**(C) Minimum Wire-Bending Space.** The minimum wire-bending space at the motor control center terminals and minimum gutter space shall be as required in Article 312(D).

**(D) Spacings.** Spacings between motor control center bus terminals and other bare metal parts shall not be less than specified in Table 430.97(D).

**(E) Barriers.** Barriers shall be placed in all service-entrance motor control centers to isolate service busbars and terminals from the remainder of the motor control center.

## 430.98 Marking

**(A) Motor Control Centers.** Motor control centers shall be marked according to 110.21, and the marking shall be plainly visible after installation. Marking shall also include common power bus current rating and motor control center short-circuit rating.

**(B) Motor Control Units.** Motor control units in a motor control center shall comply with 430.8.

*TABLE 430.97(D) Minimum Spacing Between Bare Metal Parts*

| | Opposite Polarity Where Mounted on the Same Surface | | Opposite Polarity Where Held Free in Air | | Live Parts to Ground | |
|---|---|---|---|---|---|---|
| **Nominal Voltage** | **mm** | **in.** | **mm** | **in.** | **mm** | **in.** |
| Not over 125 volts, nominal | 19.1 | ¾ | 12.7 | ½ | 12.7 | ½ |
| Not over 250 volts, nominal | 31.8 | 1¼ | 19.1 | ¾ | 12.7 | ½ |
| Not over 600 volts, nominal | 50.8 | 2 | 25.4 | 1 | 25.4 | 1 |

# IX. Disconnecting Means

## 430.101 General

Part IX is intended to require disconnecting means capable of disconnecting motors and controllers from the circuit.

•

## 430.102 Location

**(A) Controller.** An individual disconnecting means shall be provided for each controller and shall disconnect the controller. The disconnecting means shall be located in sight from the controller location.

*Exception No. 1: For motor circuits over 1000 volts, nominal, a controller disconnecting means lockable in accordance with 110.25 shall be permitted to be out of sight of the controller, provided that the controller is marked with a warning label giving the location of the disconnecting means.*

*Exception No. 2: A single disconnecting means shall be permitted for a group of coordinated controllers that drive several parts of a single machine or piece of apparatus. The disconnecting means shall be located in sight from the controllers, and both the disconnecting means and the controllers shall be located in sight from the machine or apparatus.*

*Exception No. 3: The disconnecting means shall not be required to be in sight from valve actuator motor (VAM) assemblies containing the controller where such a location introduces additional or increased hazards to persons or property and conditions (a) and (b) are met.*

*(a) The valve actuator motor assembly is marked with a warning label giving the location of the disconnecting means.*

*(b) The disconnecting means is lockable in accordance with 110.25.*

**(B) Motor.** A disconnecting means shall be provided for a motor in accordance with (B)(1) or (B)(2).

**(1) Separate Motor Disconnect.** A disconnecting means for the motor shall be located in sight from the motor location and the driven machinery location.

**(2) Controller Disconnect.** The controller disconnecting means required in accordance with 430.102(A) shall be permitted to serve as the disconnecting means for the motor if it is in sight from the motor location and the driven machinery location.

*Exception to (1) and (2): The disconnecting means for the motor shall not be required under either condition (a) or condition (b), which follow, provided that the controller disconnecting means required in 430.102(A) is lockable in accordance with 110.25.*

*(a) Where such a location of the disconnecting means for the motor is impracticable or introduces additional or increased hazards to persons or property*

Informational Note: Some examples of increased or additional hazards include, but are not limited to, motors rated in excess of 100 hp, multimotor equipment, submersible motors, motors associated with adjustable speed drives, and motors located in hazardous (classified) locations.

*(b) In industrial installations, with written safety procedures, where conditions of maintenance and supervision ensure that only qualified persons service the equipment*

•

Informational Note: For information on lockout/tagout procedures, see *NFPA 70E-2012, Standard for Electrical Safety in the Workplace.*

The main rules of 430.102(A) and (B) require that the disconnecting means be in sight of the controller, the motor location, and the driven-machinery location. For motors over 1000 volts, the controller disconnecting means is permitted to be out of sight of the controller, provided the controller has a warning label indicating the location and identification of the disconnecting means, which must be capable of being locked in the open position.

A single disconnecting means may be located adjacent to a group of coordinated controllers, as illustrated in Exhibit 430.16, where the controllers are mounted on a multimotor continuous process machine.

According to the exception, the disconnecting means is only permitted to be out of sight of the motor — as illustrated in Exhibit 430.17 — if the controller disconnecting means is individually capable of being locked in the open position and the criterion of either (a) or (b) is met. Disconnect switches or circuit breakers that are located only behind the locked door of a panelboard or within locked rooms do not comply with the requirements of 430.102.

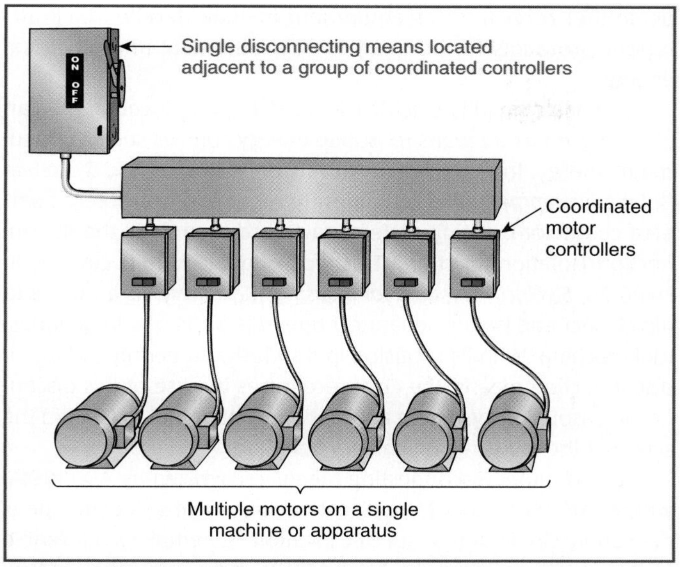


***EXHIBIT 430.16*** *A single disconnecting means located adjacent to a group of coordinated controllers mounted on a multimotor continuous process machine.*

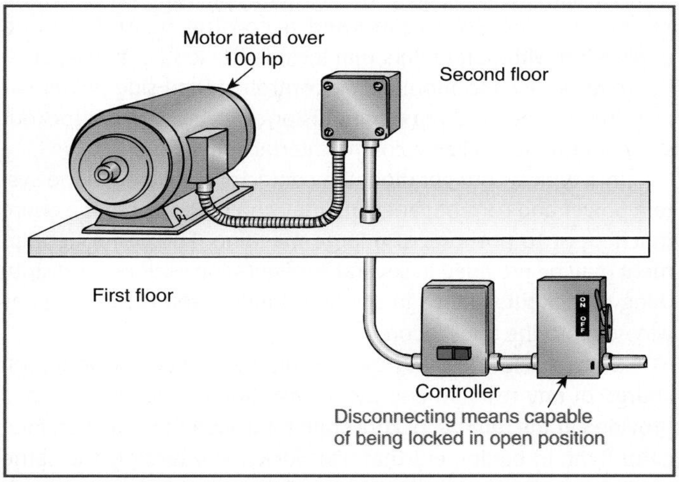


***EXHIBIT 430.17*** *A controller disconnecting means that is out of sight of the motor – only for cases that meet the requirements of (a) or (b) of 430.102(B), Exception.*

If locating the disconnecting means close to the motor location and driven machinery is impracticable due to the type of machinery, the type of facility, lack of space for locating large equipment such as disconnecting means rated over 600 volts, or any increased hazard to persons or property, the disconnecting means is permitted to be located remotely. Industrial facilities that comply with OSHA 29 CFR 1910.147, *The Control of Hazardous Energy (Lockout/Tagout),* are permitted to have the disconnecting means located remotely.

*NFPA 70E, Standard for Electrical Safety in the Workplace,* 120.2(A), requires in part that "All electrical conductors and circuit parts shall not be considered to be in an electrically safe work condition until all of the applicable requirements of Article 120 have been met." The work process specified in Article 120 includes removing the sources of energy, locking and tagging out the disconnecting means, and verifying the absence of voltage through the use of an approved voltage tester. Further, it states, "Lockout/tagout requirements shall apply to fixed, permanently installed equipment; to temporarily installed equipment; and to portable equipment." The principles and procedures in *NFPA 70E* establish strict work rules requiring locking off (out) and tagging out of disconnect switches.

## 430.103 Operation

The disconnecting means shall open all ungrounded supply conductors and shall be designed so that no pole can be operated independently. The disconnecting means shall be permitted in the same enclosure with the controller. The disconnecting means shall be designed so that it cannot be closed automatically.

Informational Note: See 430.113 for equipment receiving energy from more than one source.

A switch, circuit breaker, or other device serves as a disconnecting means for both the controller and the motor, thereby providing safety during maintenance and inspection shutdown periods. The disconnecting means also disconnects the controller; therefore, it cannot be a part of the controller.

However, separate disconnects and controllers may be mounted on the same panel or contained in the same enclosure, such as combination fused-switch, magnetic-starter units.

Depending on the size of the motor and other conditions, the type of disconnecting means required may be a motor circuit switch, a circuit breaker, a general-use switch, an isolating switch, an attachment plug and receptacle, or a branch-circuit short-circuit and ground-fault protective device, as specified in 430.109.

If a motor stalls or is under heavy overload and the motor controller fails to properly open the circuit, the disconnecting means, which must be rated to interrupt locked-rotor current, can be used to open the circuit. In accordance with 430.109(E), for motors larger than 100 hp ac or 40 hp dc, the disconnecting means is permitted to be a general-use or an isolating switch that is plainly marked "Do not operate under load."

## 430.104 To Be Indicating

The disconnecting means shall plainly indicate whether it is in the open (off) or closed (on) position.

## 430.105 Grounded Conductors

One pole of the disconnecting means shall be permitted to disconnect a permanently grounded conductor, provided the disconnecting means is designed so that the pole in the grounded conductor cannot be opened without simultaneously disconnecting all conductors of the circuit.

## 430.107 Readily Accessible

At least one of the disconnecting means shall be readily accessible.

## 430.108 Every Disconnecting Means

Every disconnecting means in the motor circuit between the point of attachment to the feeder or branch circuit and the point of connection to the motor shall comply with the requirements of 430.109 and 430.110.

## 430.109 Type

The disconnecting means shall be a type specified in 430.109(A), unless otherwise permitted in 430.109(B) through (G), under the conditions specified.

**(A) General.**

**(1) Motor Circuit Switch.** A listed motor-circuit switch rated in horsepower.

**(2) Molded Case Circuit Breaker.** A listed molded case circuit breaker.

**(3) Molded Case Switch.** A listed molded case switch.

**(4) Instantaneous Trip Circuit Breaker.** An instantaneous trip circuit breaker that is part of a listed combination motor controller.

**(5) Self-Protected Combination Controller.** Listed self-protected combination controller.

**(6) Manual Motor Controller.** Listed manual motor controllers additionally marked "Suitable as Motor Disconnect" shall be permitted as a disconnecting means where installed between the final motor branch-circuit short-circuit protective device and the motor. Listed manual motor controllers additionally marked "Suitable as Motor Disconnect" shall be permitted as disconnecting means on the line side of the fuses permitted in 430.52(C)(5). In this case, the fuses permitted in 430.52(C)(5) shall be considered supplementary fuses, and suitable branch-circuit short-circuit and ground-fault protective devices shall be installed on the line side of the manual motor controller additionally marked "Suitable as Motor Disconnect."

**(7) System Isolation Equipment.** System isolation equipment shall be listed for disconnection purposes. System isolation equipment shall be installed on the load side of the overcurrent protection and its disconnecting means. The disconnecting means shall be one of the types permitted by 430.109(A)(1) through (A)(3).

In large and often complex machines, repeated operation of disconnecting means for maintenance or servicing is inherent to the process, and the risk of injury to personnel is increased due to moving parts and multiple points of entry. This risk drives development of system isolation equipment (SIE). Safety procedures for personnel servicing this equipment include detailed lockout/tagout protocols for all sources of mechanical and electrical energy.

SIE helps simplify electrical lockout/tagout procedures; it can also be used as a means to isolate energy sources such as pneumatic energy. In accordance with its definition in 430.2, *system isolation equipment* is "a redundantly monitored, remotely operated contactor-isolating system, packaged to provide the disconnection/isolation function." This type of equipment is covered in NFPA 79, *Electrical Standard for Industrial Machinery,* as a means to disconnect and isolate separately operable parts of a large industrial machine. With its inclusion in 430.109 as a permitted type of disconnecting means, the *Code* recognizes the use of this disconnection/isolation system in applications that do not fall within the scope of the industrial machinery standard.

Unlike other disconnecting means recognized by 430.109(A) where the operation of the disconnecting means directly opens the supply circuit at that specific location, SIE employs a lockable control circuit switch(es) (lockout switch) and a verification indication at the disconnecting means location (lockout station). Also, operation of the lockout switch causes power components such as a monitored magnetic contactor to open and isolate the electrical equipment associated with the machine from its power supply circuit. The SIE is classified according to its intended application with parameters that include the load characteristics, the method used to monitor the controlled load-side power circuit, the number and maximum distance to the farthest lockout station, and the available control interface functions.

In a typical configuration, the contactor is located in the system power and control panel and may control power to the entire machine or to portions of a large machine. The control equipment may be provided in several configuration options for distributing to lockout stations in single or multiplexed radial schemes according to the application.

Once an electrically safe condition is achieved (including discharge of any residual energy), verification of such condition is provided at the remote lockout station through the use of an indicator light. In equipment that uses lockable guarding, the same verification signal could also be used as part of the guard access system.

In contrast to a simple start/stop station and control circuit operating a magnetic contactor, the control panel for this system provides a sophisticated level of monitoring upon actuation of the remote lockout switch. If any portion of the safety system cannot be verified for proper operation, the safe condition indicator light will not illuminate at the remote lockout station. As part of the standard operating procedure, the failure to receive the safe condition signal has to be considered an indication of an unsafe condition.

Among the critical safety elements that are provided by the control panel for the isolation system is the diversity and redundancy that is integrated into the safe condition verification logic. Another element is reducing the possibility of externally induced failure modes through the electrical isolation of the internal safety-related control circuits and the physical isolation of the equipment's internal components. The control panel modules are sealed as are the circuits between the SIE component enclosures, to discourage tampering that could compromise the safe operation of the equipment and endanger personnel. Where the system includes multiple lockout stations, the controlled equipment cannot be re-energized until all of the lockout switches are returned to the "on" position. Nominal configurations of the SIE include provisions to prevent power from unexpectedly reaching the machine upon the restoration of power from the utility source. To re-energize the machine, all lockout switches must be in the closed or "on" position while at least one lockout switch must have been in the "off" or open position or placed in the "off" or open position (and then moved to the "on" or closed position after the utility power had been restored).

The SIE is required to be listed for disconnection purposes and must be installed on the load side of the disconnecting means (could be a motor circuit switch, a molded case circuit breaker, or a molded case switch) and overcurrent protection for the supply circuit to the equipment.

**(B) Stationary Motors of ⅛ Horsepower or Less.** For stationary motors of ⅛ hp or less, the branch-circuit overcurrent device shall be permitted to serve as the disconnecting means.

**(C) Stationary Motors of 2 Horsepower or Less.** For stationary motors rated at 2 hp or less and 300 volts or less, the disconnecting means shall be permitted to be one of the devices specified in (1), (2), or (3):

(1) A general-use switch having an ampere rating not less than twice the full-load current rating of the motor
(2) On ac circuits, a general-use snap switch suitable only for use on ac (not general-use ac–dc snap switches) where the motor full-load current rating is not more than 80 percent of the ampere rating of the switch
(3) A listed manual motor controller having a horsepower rating not less than the rating of the motor and marked "Suitable as Motor Disconnect"

**(D) Autotransformer-Type Controlled Motors.** For motors of over 2 hp to and including 100 hp, the separate disconnecting means required for a motor with an autotransformer-type controller shall be permitted to be a general-use switch where all of the following provisions are met:

(1) The motor drives a generator that is provided with overload protection.
(2) The controller is capable of interrupting the locked-rotor current of the motors, is provided with a no voltage release, and is provided with running overload protection not exceeding 125 percent of the motor full-load current rating.
(3) Separate fuses or an inverse time circuit breaker rated or set at not more than 150 percent of the motor full-load current is provided in the motor branch circuit.

**(E) Isolating Switches.** For stationary motors rated at more than 40 hp dc or 100 hp ac, the disconnecting means shall be permitted to be a general-use or isolating switch where plainly marked "Do not operate under load."

**(F) Cord-and-Plug-Connected Motors.** For a cord-and-plug-connected motor, a horsepower-rated attachment plug and receptacle, flanged surface inlet and cord connector, or attachment plug and cord connector having ratings no less than the motor ratings shall be permitted to serve as the disconnecting means. Horsepower-rated attachment plugs, flanged surface inlets, receptacles, or cord connectors shall not be required for cord-and-plug-connected appliances in accordance with 422.33, room air conditioners in accordance with 440.63, or portable motors rated ⅓ hp or less.

A motor circuit switch is a horsepower-rated switch capable of interrupting the maximum overload current of a motor (see the definition of *switch, motor-circuit* in Article 100). A molded case switch (nonautomatic circuit interrupter) is a circuit-breaker-like device without the overcurrent element and automatic-trip mechanism. It is rated in amperes and is suitable for use as a motor circuit disconnect based on its ampere rating, as is a circuit breaker. The disconnecting means must be listed.

Exhibits 430.18, 430.19, 430.20, and 430.21 illustrate various methods of providing motor disconnecting means as permitted by 430.109(B), 430.109(C), 430.109(E), and 430.109(F), respectively.

Where horsepower-rated fused switches are required, marking within the enclosure usually permits a dual horsepower rating. The standard horsepower rating is based on the largest non-time-delay (non-dual-element) fuse rating that can be used in the switch and that will permit the motor to start. The maximum horsepower rating is based on the largest rated time-delay (dual-element) fuse that can be used in the switch and that will permit the motor to

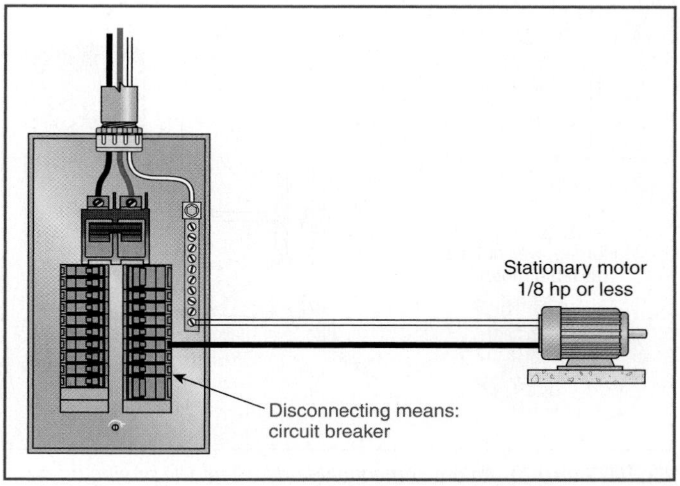


***EXHIBIT 430.18*** *A branch-circuit overcurrent device serving as the disconnecting means for a stationary motor of ⅛ hp or less according to 430.109(B).*

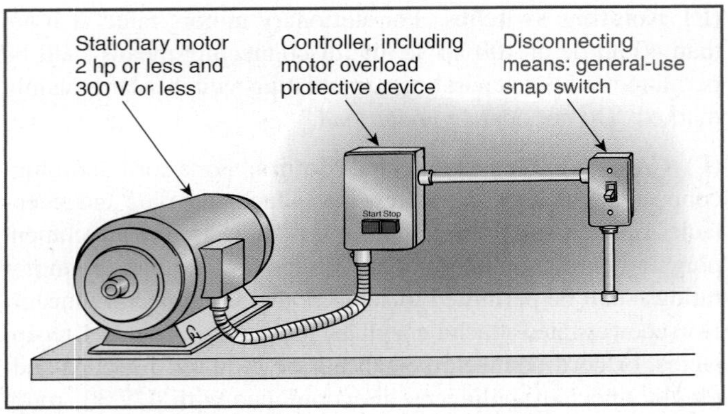


*EXHIBIT 430.19 A general-use snap switch serving as the disconnecting means for a stationary motor rated at 2 hp or less and at 300 volts or less according to 430.109(C).*

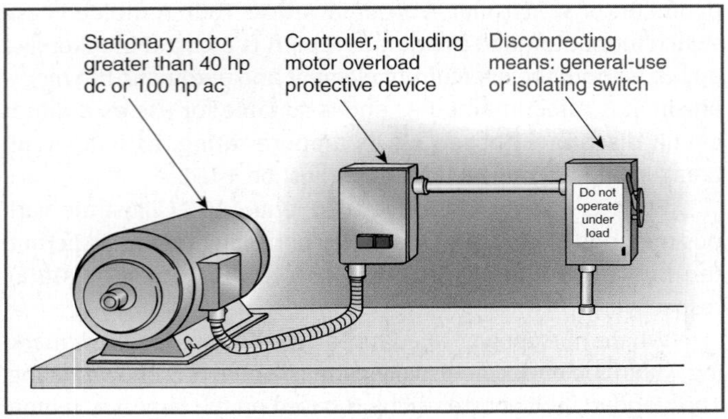


*EXHIBIT 430.20 A general-use or an isolating switch serving as the disconnecting means for a stationary motor rated at more than 40 hp dc or 100 hp ac according to 430.109(E).*

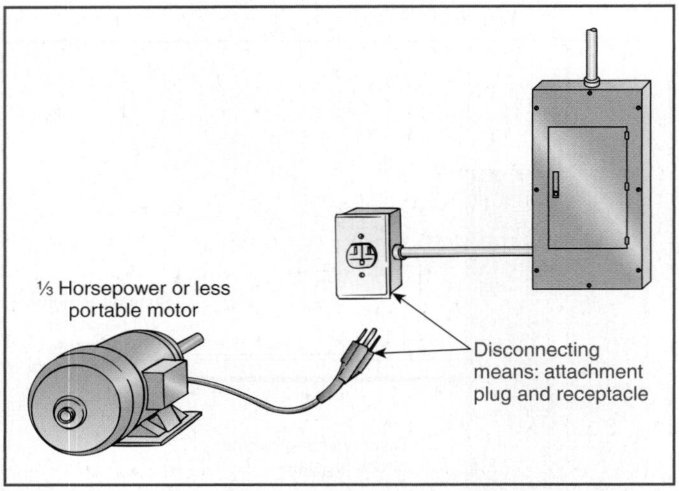


*EXHIBIT 430.21 An attachment plug and receptacle serving as the disconnecting means for a certain cord-and-plug-connected motor according to 430.109(F).*

start. Thus, where time-delay fuses are used, smaller-size switches and fuseholders can be used (see 430.57, Exception).

**(G) Torque Motors.** For torque motors, the disconnecting means shall be permitted to be a general-use switch.

## 430.110 Ampere Rating and Interrupting Capacity

**(A) General.** The disconnecting means for motor circuits rated 1000 volts, nominal, or less shall have an ampere rating not less than 115 percent of the full-load current rating of the motor.

*Exception: A listed unfused motor-circuit switch having a horsepower rating not less than the motor horsepower shall be permitted to have an ampere rating less than 115 percent of the full-load current rating of the motor.*

**(B) For Torque Motors.** Disconnecting means for a torque motor shall have an ampere rating of at least 115 percent of the motor nameplate current.

**(C) For Combination Loads.** Where two or more motors are used together or where one or more motors are used in combination with other loads, such as resistance heaters, and where the combined load may be simultaneous on a single disconnecting means, the ampere and horsepower ratings of the combined load shall be determined as follows.

**(1) Horsepower Rating.** The rating of the disconnecting means shall be determined from the sum of all currents, including resistance loads, at the full-load condition and also at the locked-rotor condition. The combined full-load current and the combined locked-rotor current so obtained shall be considered as a single motor for the purpose of this requirement as follows.

The full-load current equivalent to the horsepower rating of each motor shall be selected from Table 430.247, Table 430.248, Table 430.249, or Table 430.250. These full-load currents shall be added to the rating in amperes of other loads to obtain an equivalent full-load current for the combined load.

The locked-rotor current equivalent to the horsepower rating of each motor shall be selected from Table 430.251(A) or Table 430.251(B). The locked-rotor currents shall be added to the rating in amperes of other loads to obtain an equivalent locked-rotor current for the combined load. Where two or more motors or other loads cannot be started simultaneously, the largest sum of locked-rotor currents of a motor or group of motors that can be started simultaneously and the full-load currents of other concurrent loads shall be permitted to be used to determine the equivalent locked-rotor current for the simultaneous combined loads. In cases where different current ratings are obtained when applying these tables, the largest value obtained shall be used.

*Exception: Where part of the concurrent load is resistance load, and where the disconnecting means is a switch rated in horsepower and amperes, the switch used shall be permitted to have a horsepower rating that is not less than the combined load of the motor(s), if the ampere rating of the switch is not less than the locked-rotor current of the motor(s) plus the resistance load.*

**(2) Ampere Rating.** The ampere rating of the disconnecting means shall not be less than 115 percent of the sum of all currents at the full-load condition determined in accordance with 430.110(C)(1).

*Exception: A listed nonfused motor-circuit switch having a horsepower rating equal to or greater than the equivalent horsepower of the combined loads, determined in accordance with 430.110(C)(1), shall be permitted to have an ampere rating less than 115 percent of the sum of all currents at the full-load condition.*

**(3) Small Motors.** For small motors not covered by Table 430.247, Table 430.248, Table 430.249, or Table 430.250, the locked-rotor current shall be assumed to be six times the full-load current.

Listed circuit breakers and molded case switches are tested under overload conditions at six times their rating, to cover motor circuit applications, and are suitable for use as a motor disconnecting means.

**Calculation Example**

An installation consists of one 5-hp, one 3-hp, and two ½-hp motors, plus a 10-kW heater, all rated 240 V, 3-phase. All motors are Design B motors. Determine the size of the disconnect required for this combination load.

*Solution*

Use the appropriate tables to select the full-load and locked-rotor current equivalents from the tables in Article 430 as follows:

**Equivalent Full-Load and Locked-Rotor Current Rating**

| Motor or Other Load | Full-Load Current Amperes from Table 430.250 | Locked-Rotor Current Amperes from Table 430.251(B) |
|---|---|---|
| 5-hp motor | 15.2 | 92 |
| 3-hp motor | 9.6 | 64 |
| ½-hp motor | 2.2 | 20 |
| ½-hp motor | 2.2 | 20 |
| 10-kW heater | 24.1 | 24.1 |
| $\left[\frac{10 \times 1000}{(240 \times 1.732)} = 24.1\text{ A}\right]$ | | |
| | Total 53.3 | Total 220.1 |

STEP 1. The disconnecting means for a motor must have an ampere rating not less than 115 percent of the combined load. Therefore, the minimum ampere rating of the disconnecting means is 61.3 amperes (1.15 × 53.3 A = 61.3 A).

STEP 2. The disconnecting means must have a horsepower rating not less than the combined load. Use Table 430.251(B) to obtain an equivalent horsepower from the LRC rating. The closest value equal to or greater than 220.1 amperes under the 230-volt column is 232 amperes, which equates to 15 hp.

A 15-hp switch satisfies the minimum horsepower requirement but fails to satisfy the minimum current requirement of 61.3 amperes. Therefore, the next larger size disconnect switch must be used, which in horsepower rating is 20 hp. The LRC for a 20-hp, 3-phase, 230-volt motor is 290 amperes (over the needed 220.1 amperes), and the FLC is 54 amperes (over the needed 53.3 amperes).

## 430.111 Switch or Circuit Breaker as Both Controller and Disconnecting Means

A switch or circuit breaker shall be permitted to be used as both the controller and disconnecting means if it complies with 430.111(A) and is one of the types specified in 430.111(B).

**(A) General.** The switch or circuit breaker complies with the requirements for controllers specified in 430.83, opens all ungrounded conductors to the motor, and is protected by an overcurrent device in each ungrounded conductor (which shall be permitted to be the branch-circuit fuses). The overcurrent device protecting the controller shall be permitted to be part of the controller assembly or shall be permitted to be separate. An autotransformer-type controller shall be provided with a separate disconnecting means.

**(B) Type.** The device shall be one of the types specified in 430.111(B)(1), (B)(2), or (B)(3).

**(1) Air-Break Switch.** An air-break switch, operable directly by applying the hand to a lever or handle.

**(2) Inverse Time Circuit Breaker.** An inverse time circuit breaker operable directly by applying the hand to a lever or handle. The circuit breaker shall be permitted to be both power and manually operable.

**(3) Oil Switch.** An oil switch used on a circuit whose rating does not exceed 1000 volts or 100 amperes, or by special permission on a circuit exceeding this capacity where under expert supervision. The oil switch shall be permitted to be both power and manually operable.

If used as a controller, a switch or circuit breaker must meet all the requirements for controllers and be protected by branch-circuit short-circuit and ground-fault protective devices (fuses or a circuit breaker), which ensure that all ungrounded conductors will be opened.

If the controller consists of a manually operable air-break switch, an inverse time circuit breaker, or a 100-ampere maximum oil switch (higher rating by special permission), the controller is considered a satisfactory disconnecting means. The intent of 430.111 is to permit omission of an additional device to serve as a disconnecting means.

A separate disconnecting means must be provided if the controller is of the autotransformer or compensator type. (This switch may be combined in the same enclosure with a motor overload protective device.)

### 430.112 Motors Served by Single Disconnecting Means

Each motor shall be provided with an individual disconnecting means.

*Exception: A single disconnecting means shall be permitted to serve a group of motors under any one of the conditions of (a), (b), and (c). The single disconnecting means shall be rated in accordance with 430.110(C).*

*(a) Where a number of motors drive several parts of a single machine or piece of apparatus, such as metal- and woodworking machines, cranes, and hoists.*

*(b) Where a group of motors is under the protection of one set of branch-circuit protective devices as permitted by 430.53(A).*

*(c) Where a group of motors is in a single room within sight from the location of the disconnecting means.*

A single disconnecting means must have a rating equal to the sum of the horsepower or current of each motor in the group. If the sum is over 2 hp, a motor circuit switch (horsepower-rated) must be used; thus, for five 2-hp motors, the disconnecting means should be a motor-circuit switch rated at not less than 10 hp.

Part (a) of the exception indicates that a single disconnecting means may be used where a number of motors drive several parts of a single machine, such as cranes (see 610.31 through 610.33), metal or woodworking machines, steel rolling mill machinery, and so forth. The single disconnecting means for multimotor machinery provides a positive means of simultaneously de-energizing all motor branch circuits, including remote-control circuits, interlocking circuits, limit-switch circuits, and operator control stations.

Part (b) of the exception refers to 430.53(A), which permits a group of motors under the protection of the same branch-circuit device, provided the device is rated not more than 20 amperes at 125 volts, or 15 amperes at more than 125 volts, but not more than 1000 volts. The motors must be rated 1 hp or less, and the full-load current for each motor is not permitted to exceed 6 amperes. A single disconnecting means is both practical and economical for a group of such small motors.

Part (c) of the exception covers the common situation in which a group of motors is located in one room, such as a pump room, compressor room, mixer room, and so forth. It is possible to design the layout of a single disconnecting means with an unobstructed view (not more than 50 feet) from each motor.

These conditions for an individual disconnecting means are similar to those specified in 430.87, which permits the use of a single controller for a group of motors.

### 430.113 Energy from More Than One Source

Motor and motor-operated equipment receiving electric energy from more than one source shall be provided with disconnecting means from each source of electric energy immediately adjacent to the equipment served. Each source shall be permitted to have a separate disconnecting means. Where multiple disconnecting means are provided, a permanent warning sign shall be provided on or adjacent to each disconnecting means.

*Exception No. 1: Where a motor receives electric energy from more than one source, the disconnecting means for the main power supply to the motor shall not be required to be immediately adjacent to the motor, provided that the controller disconnecting means is lockable in accordance with 110.25.*

*Exception No. 2: A separate disconnecting means shall not be required for a Class 2 remote-control circuit conforming with Article 725, rated not more than 30 volts, and isolated and ungrounded.*

Some motors require multiple separate sources of power to operate properly. Power for a motor space heater is one example. Power for a speed switch is another. Synchronous motors commonly use dc power for excitation purposes. Section 430.113 could also apply to circuits that supply power to speed or vibration sensors mounted within or otherwise attached to the motor.

Where the individual sources have multiple disconnecting means, a permanent warning sign is required to warn the user that other power sources are present. Exception No. 2 removes the disconnect requirement only for Class 2 circuits.

## X. Adjustable-Speed Drive Systems

### 430.120 General

The installation provisions of Part I through Part IX are applicable unless modified or supplemented by Part X.

Adjustable-speed drives are used extensively in commercial, institutional, and industrial motor applications. Exhibit 430.22 shows 480 V adjustable-speed drives, also known as variable-frequency drives or VFDs.

Part X consolidates requirements that are unique to these drives. However, Parts I through IX must be followed unless modified or supplemented in Part X. The text provides rules regarding methods of overtemperature protection in motors. This is a critical area, because motors operating at reduced speed do not provide adequate air circulation over windings from a fan integral with the motor. An overload device that actuates on current in excess of full-load amperes will not operate, because the operating current at slower speeds is reduced. A thermal-sensing device integral with the motor will sense a temperature rise in the motor windings.

•

***EXHIBIT 430.22*** *Adjustable-speed drives for air-handling units provide significant power savings for fan appliances. (Courtesy of International Association of Electrical Inspectors)*

## 430.122 Conductors — Minimum Size and Ampacity

**(A) Branch/Feeder Circuit Conductors.** Circuit conductors supplying power conversion equipment included as part of an adjustable-speed drive system shall have an ampacity not less than 125 percent of the rated input current to the power conversion equipment.

Informational Note: Power conversion equipment can have multiple power ratings and corresponding input currents.

**(B) Bypass Device.** For an adjustable-speed drive system that utilizes a bypass device, the conductor ampacity shall not be less than required by 430.6. The ampacity of circuit conductors supplying power conversion equipment included as part of an adjustable-speed drive system that utilizes a bypass device shall be the larger of either of the following:

(1) 125 percent of the rated input current to the power conversion equipment
(2) 125 percent of the motor full-load current rating as determined by 430.6

## 430.124 Overload Protection

Overload protection of the motor shall be provided.

**(A) Included in Power Conversion Equipment.** Where the power conversion equipment is marked to indicate that motor overload protection is included, additional overload protection shall not be required.

**(B) Bypass Circuits.** For adjustable-speed drive systems that utilize a bypass device to allow motor operation at rated full-load speed, motor overload protection as described in Article 430, Part III, shall be provided in the bypass circuit.

**(C) Multiple Motor Applications.** For multiple motor application, individual motor overload protection shall be provided in accordance with Article 430, Part III.

## 430.126 Motor Overtemperature Protection

**(A) General.** Adjustable-speed drive systems shall protect against motor overtemperature conditions where the motor is not rated to operate at the nameplate rated current over the speed range required by the application. This protection shall be provided in addition to the conductor protection required in 430.32. Protection shall be provided by one of the following means.

(1) Motor thermal protector in accordance with 430.32
(2) Adjustable-speed drive system with load and speed-sensitive overload protection and thermal memory retention upon shutdown or power loss

*Exception to (2): Thermal memory retention upon shutdown or power loss is not required for continuous duty loads.*

(3) Overtemperature protection relay utilizing thermal sensors embedded in the motor and meeting the requirements of 430.32(A)(2) or (B)(2)
(4) Thermal sensor embedded in the motor whose communications are received and acted upon by an adjustable-speed drive system

Informational Note: The relationship between motor current and motor temperature changes when the motor is operated by an adjustable-speed drive. In certain applications, overheating of motors can occur when operated at reduced speed, even at current levels less than a motor's rated full-load current. The overheating can be the result of reduced motor cooling when its shaft-mounted fan is operating less than rated nameplate RPM. As part of the analysis to determine whether overheating will occur, it is necessary to consider the continuous torque capability curves for the motor given the application requirements. This will assist in determining whether the motor overload protection will be able, on its own, to provide protection against overheating. These overheating protection requirements are only intended to apply to applications where an adjustable-speed drive, as defined in Article 100, is used.

For motors that utilize external forced air or liquid cooling systems, overtemperature can occur if the cooling system is not operating. Although this issue is not unique to adjustable speed applications, externally cooled motors are most often encountered with such applications. In these instances, overtemperature protection using direct temperature sensing is recommended [i.e., 430.126(A)(1), (A)(3), or (A)(4)], or additional means should be provided to ensure that the cooling system is operating (flow or pressure sensing, interlocking of adjustable-speed drive system and cooling system, etc.).

**(B) Multiple Motor Applications.** For multiple motor applications, individual motor overtemperature protection shall be provided as required in 430.126(A).

**(C) Automatic Restarting and Orderly Shutdown.** The provisions of 430.43 and 430.44 shall apply to the motor overtemperature protection means.

### 430.128 Disconnecting Means

The disconnecting means shall be permitted to be in the incoming line to the conversion equipment and shall have a rating not less than 115 percent of the rated input current of the conversion unit.

### 430.130 Branch-Circuit Short-Circuit and Ground-Fault Protection for Single Motor Circuits Containing Power Conversion Equipment

**(A) Circuits Containing Power Conversion Equipment.** Circuits containing power conversion equipment shall be protected by a branch-circuit short-circuit and ground-fault protective device in accordance with the following:

(1) The rating and type of protection shall be determined by 430.52(C)(1), (C)(3), (C)(5), or (C)(6), using the full-load current rating of the motor load as determined by 430.6.

(2) Where maximum branch-circuit short-circuit and ground-fault protective ratings are stipulated for specific device types in the manufacturer's instructions for the power conversion equipment or are otherwise marked on the equipment, they shall not be exceeded even if higher values are permitted by 430.130(A)(1).

(3) A self-protected combination controller shall only be permitted where specifically identified in the manufacturer's instructions for the power conversion equipment or if otherwise marked on the equipment.

Informational Note: The type of protective device, its rating, and its setting are often marked on or provided with the power conversion equipment.

**(B) Bypass Circuit/Device.** Branch-circuit short-circuit and ground-fault protection shall also be provided for a bypass circuit/device(s). Where a single branch-circuit short-circuit and ground-fault protective device is provided for circuits containing both power conversion equipment and a bypass circuit, the branch-circuit protective device type and its rating or setting shall be in accordance with those determined for the power conversion equipment and for the bypass circuit/device(s) equipment.

### 430.131 Several Motors or Loads on One Branch Circuit Including Power Conversion Equipment

For installations meeting all the requirements of 430.53 that include one or more power converters, the branch-circuit short-circuit and ground-fault protective fuses or inverse time circuit breakers shall be of a type and rating or setting permitted for use with the power conversion equipment using the full-load current rating of the connected motor load in accordance with 430.53. For the purposes of 430.53 and 430.131, power conversion equipment shall be considered to be a motor controller.

## XI. Over 1000 Volts, Nominal

### 430.221 General

Part XI recognizes the additional hazard due to the use of higher voltages. It adds to or amends the other provisions of this article.

### 430.222 Marking on Controllers

In addition to the marking required by 430.8, a controller shall be marked with the control voltage.

### 430.223 Raceway Connection to Motors

Flexible metal conduit or liquidtight flexible metal conduit not exceeding 1.8 m (6 ft) in length shall be permitted to be employed for raceway connection to a motor terminal enclosure.

### 430.224 Size of Conductors

Conductors supplying motors shall have an ampacity not less than the current at which the motor overload protective device(s) is selected to trip.

### 430.225 Motor-Circuit Overcurrent Protection

**(A) General.** Each motor circuit shall include coordinated protection to automatically interrupt overload and fault currents in the motor, the motor-circuit conductors, and the motor control apparatus.

*Exception: Where a motor is critical to an operation and the motor should operate to failure if necessary to prevent a greater hazard to persons, the sensing device(s) shall be permitted to be connected to a supervised annunciator or alarm instead of interrupting the motor circuit.*

**(B) Overload Protection.**

**(1) Type of Overload Device.** Each motor shall be protected against dangerous heating due to motor overloads and failure to start by a thermal protector integral with the motor or external current-sensing devices, or both. Protective device settings for each motor circuit shall be determined under engineering supervision.

Selecting the proper overload and short-circuit protection for medium-voltage motor circuits is more complex than for low-voltage circuits. For medium-voltage motor circuits, it becomes very critical for the overload relay to coordinate with the short-circuit protection, because some short-circuit protective devices cannot safely open below certain multiples of their rating. In these overload cases, the overload relay must open before the short-circuit protective device is asked to open. At the same time, the overload relay may not safely open beyond certain multiples of its rating, requiring the short-circuit protective device to open. The curves of both the overload relay and the short-circuit protective device must be correlated to ensure that each opens only on levels of current for which it can safely open.

**(2) Wound-Rotor Alternating-Current Motors.** The secondary circuits of wound-rotor ac motors, including conductors, controllers, and resistors rated for the application, shall be considered as protected against overcurrent by the motor overload protection means.

**(3) Operation.** Operation of the overload interrupting device shall simultaneously disconnect all ungrounded conductors.

**(4) Automatic Reset.** Overload sensing devices shall not automatically reset after trip unless resetting of the overload sensing device does not cause automatic restarting of the motor or there is no hazard to persons created by automatic restarting of the motor and its connected machinery.

**(C) Fault-Current Protection.**

**(1) Type of Protection.** Fault-current protection shall be provided in each motor circuit as specified by either (1)(a) or (1)(b).

(a) A circuit breaker of suitable type and rating arranged so that it can be serviced without hazard. The circuit breaker shall simultaneously disconnect all ungrounded conductors. The circuit breaker shall be permitted to sense the fault current by means of integral or external sensing elements.

(b) Fuses of a suitable type and rating placed in each ungrounded conductor. Fuses shall be used with suitable disconnecting means, or they shall be of a type that can also serve as the disconnecting means. They shall be arranged so that they cannot be serviced while they are energized.

**(2) Reclosing.** Fault-current interrupting devices shall not automatically reclose the circuit.

*Exception: Automatic reclosing of a circuit shall be permitted where the circuit is exposed to transient faults and where such automatic reclosing does not create a hazard to persons.*

**(3) Combination Protection.** Overload protection and fault-current protection shall be permitted to be provided by the same device.

### 430.226 Rating of Motor Control Apparatus

The ultimate trip current of overcurrent (overload) relays or other motor-protective devices used shall not exceed 115 percent of the controller's continuous current rating. Where the motor branch-circuit disconnecting means is separate from the controller, the disconnecting means current rating shall not be less than the ultimate trip setting of the overcurrent relays in the circuit.

### 430.227 Disconnecting Means

The controller disconnecting means shall be lockable in accordance with 110.25.

## XII. Protection of Live Parts — All Voltages

### 430.231 General

Part XII specifies that live parts shall be protected in a manner judged adequate for the hazard involved.

### 430.232 Where Required

Exposed live parts of motors and controllers operating at 50 volts or more between terminals shall be guarded against accidental contact by enclosure or by location as follows:

(1) By installation in a room or enclosure that is accessible only to qualified persons
(2) By installation on a suitable balcony, gallery, or platform, elevated and arranged so as to exclude unqualified persons
(3) By elevation 2.5 m (8 ft) or more above the floor

*Exception: Live parts of motors operating at more than 50 volts between terminals shall not require additional guarding for stationary motors that have commutators, collectors, and brush rigging located inside of motor-end brackets and not conductively connected to supply circuits operating at more than 150 volts to ground.*

### 430.233 Guards for Attendants

Where live parts of motors or controllers operating at over 50 volts to ground are guarded against accidental contact only by location as specified in 430.232, and where adjustment or other attendance may be necessary during the operation of the apparatus, suitable insulating mats or platforms shall be provided so that the attendant cannot readily touch live parts unless standing on the mats or platforms.

Informational Note: For working space, see 110.26 and 110.34.

## XIII. Grounding — All Voltages

### 430.241 General

Part XIII specifies the grounding of exposed non–current-carrying metal parts, likely to become energized, of motor and controller frames to prevent a voltage aboveground in the event of accidental contact between energized parts and frames. Insulation, isolation, or guarding are suitable alternatives to grounding of motors under certain conditions.

### 430.242 Stationary Motors

The frames of stationary motors shall be grounded under any of the following conditions:

(1) Where supplied by metal-enclosed wiring
(2) Where in a wet location and not isolated or guarded

(3) If in a hazardous (classified) location
(4) If the motor operates with any terminal at over 150 volts to ground

Where the frame of the motor is not grounded, it shall be permanently and effectively insulated from the ground.

Stationary motors are usually supplied by wiring that is enclosed in metal raceways, flexible metal conduit, or cables with metal sheaths. When effectively attached to the motor junction box or frame, the metal raceway or cable armor serves as the EGC. See 250.118 for more information on types of EGCs.

### 430.243 Portable Motors

The frames of portable motors that operate over 150 volts to ground shall be guarded or grounded.

Informational Note No. 1: See 250.114(4) for grounding of portable appliances in other than residential occupancies.

Informational Note No. 2: See 250.119(C) for color of equipment grounding conductor.

*Exception No. 1: Listed motor-operated tools, listed motor-operated appliances, and listed motor-operated equipment shall not be required to be grounded where protected by a system of double insulation or its equivalent. Double-insulated equipment shall be distinctively marked.*

*Exception No. 2: Listed motor-operated tools, listed motor-operated appliances, and listed motor-operated equipment connected by a cord and attachment plug other than those required to be grounded in accordance with 250.114.*

### 430.244 Controllers

Controller enclosures shall be connected to the equipment grounding conductor regardless of voltage. Controller enclosures shall have means for attachment of an equipment grounding conductor termination in accordance with 250.8.

*Exception: Enclosures attached to ungrounded portable equipment shall not be required to be grounded.*

### 430.245 Method of Grounding

Connection to the equipment grounding conductor shall be done in the manner specified in Part VI of Article 250.

Most motors are subject to vibration. This may require that the wiring to motors that are fixed be installed with a short section (not more than 6 feet) of liquidtight flexible metal or nonmetallic conduit or of flexible metal conduit to the motor terminal housing to minimize the impact of the vibration. Such use of flexible conduit requires an EGC.

**(A) Grounding Through Terminal Housings.** Where the wiring to motors is metal-enclosed cable or in metal raceways, junction boxes to house motor terminals shall be provided, and the armor of the cable or the metal raceways shall be connected to them in the manner specified in 250.96(A) and 250.97.

•

**(B) Separation of Junction Box from Motor.** The junction box required by 430.245(A) shall be permitted to be separated from the motor by not more than 1.8 m (6 ft), provided the leads to the motor are stranded conductors within Type AC cable, interlocked metal tape Type MC cable where listed and identified in accordance with 250.118(10)(a), or armored cord or are stranded leads enclosed in liquidtight flexible metal conduit, flexible metal conduit, intermediate metal conduit, rigid metal conduit, or electrical metallic tubing not smaller than metric designator 12 (trade size ⅜), the armor or raceway being connected both to the motor and to the box.

Liquidtight flexible nonmetallic conduit and rigid nonmetallic conduit shall be permitted to enclose the leads to the motor, provided the leads are stranded and the required equipment grounding conductor is connected to both the motor and to the box.

Where stranded leads are used, protected as specified above, each strand within the conductor shall be not larger than 10 AWG and shall comply with other requirements of this *Code* for conductors to be used in raceways.

**(C) Grounding of Controller-Mounted Devices.** Instrument transformer secondaries and exposed non–current-carrying metal or other conductive parts or cases of instrument transformers, meters, instruments, and relays shall be grounded as specified in 250.170 through 250.178.

## XIV. Tables

Tables 430.248 through 430.250 reflect the typical and most used 4-pole and 2-pole induction motors in use.

***TABLE 430.247*** *Full-Load Current in Amperes, Direct-Current Motors*
The following values of full-load currents* are for motors running at base speed.

| | Armature Voltage Rating* | | | | | |
|---|---|---|---|---|---|---|
| **Horsepower** | **90 Volts** | **120 Volts** | **180 Volts** | **240 Volts** | **500 Volts** | **550 Volts** |
| ¼ | 4.0 | 3.1 | 2.0 | 1.6 | — | — |
| ⅓ | 5.2 | 4.1 | 2.6 | 2.0 | — | — |
| ½ | 6.8 | 5.4 | 3.4 | 2.7 | — | — |
| ¾ | 9.6 | 7.6 | 4.8 | 3.8 | — | — |
| 1 | 12.2 | 9.5 | 6.1 | 4.7 | — | — |
| 1½ | — | 13.2 | 8.3 | 6.6 | — | — |
| 2 | — | 17 | 10.8 | 8.5 | — | — |
| 3 | — | 25 | 16 | 12.2 | — | — |
| 5 | — | 40 | 27 | 20 | — | — |
| 7½ | — | 58 | — | 29 | 13.6 | 12.2 |
| 10 | — | 76 | — | 38 | 18 | 16 |
| 15 | — | — | — | 55 | 27 | 24 |
| 20 | — | — | — | 72 | 34 | 31 |
| 25 | — | — | — | 89 | 43 | 38 |
| 30 | — | — | — | 106 | 51 | 46 |
| 40 | — | — | — | 140 | 67 | 61 |
| 50 | — | — | — | 173 | 83 | 75 |
| 60 | — | — | — | 206 | 99 | 90 |
| 75 | — | — | — | 255 | 123 | 111 |
| 100 | — | — | — | 341 | 164 | 148 |
| 125 | — | — | — | 425 | 205 | 185 |
| 150 | — | — | — | 506 | 246 | 222 |
| 200 | — | — | — | 675 | 330 | 294 |

*These are average dc quantities.

***TABLE 430.248*** *Full-Load Currents in Amperes, Single-Phase Alternating-Current Motors*
The following values of full-load currents are for motors running at usual speeds and motors with normal torque characteristics. The voltages listed are rated motor voltages. The currents listed shall be permitted for system voltage ranges of 110 to 120 and 220 to 240 volts.

| Horsepower | 115 Volts | 200 Volts | 208 Volts | 230 Volts |
|---|---|---|---|---|
| 1/6 | 4.4 | 2.5 | 2.4 | 2.2 |
| 1/4 | 5.8 | 3.3 | 3.2 | 2.9 |
| 1/3 | 7.2 | 4.1 | 4.0 | 3.6 |
| 1/2 | 9.8 | 5.6 | 5.4 | 4.9 |
| 3/4 | 13.8 | 7.9 | 7.6 | 6.9 |
| 1 | 16 | 9.2 | 8.8 | 8.0 |
| 1½ | 20 | 11.5 | 11.0 | 10 |
| 2 | 24 | 13.8 | 13.2 | 12 |
| 3 | 34 | 19.6 | 18.7 | 17 |
| 5 | 56 | 32.2 | 30.8 | 28 |
| 7½ | 80 | 46.0 | 44.0 | 40 |
| 10 | 100 | 57.5 | 55.0 | 50 |

***TABLE 430.249*** *Full-Load Current, Two-Phase Alternating-Current Motors (4-Wire)*
The following values of full-load current are for motors running at speeds usual for belted motors and motors with normal torque characteristics. Current in the common conductor of a 2-phase, 3-wire system will be 1.41 times the value given. The voltages listed are rated motor voltages. The currents listed shall be permitted for system voltage ranges of 110 to 120, 220 to 240, 440 to 480, and 550 to 1000 volts.

| | Induction-Type Squirrel Cage and Wound Rotor (Amperes) | | | | |
|---|---|---|---|---|---|
| Horsepower | 115 Volts | 230 Volts | 460 Volts | 575 Volts | 2300 Volts |
| 1/2 | 4.0 | 2.0 | 1.0 | 0.8 | — |
| 3/4 | 4.8 | 2.4 | 1.2 | 1.0 | — |
| 1 | 6.4 | 3.2 | 1.6 | 1.3 | — |
| 1½ | 9.0 | 4.5 | 2.3 | 1.8 | — |
| 2 | 11.8 | 5.9 | 3.0 | 2.4 | — |
| 3 | — | 8.3 | 4.2 | 3.3 | — |
| 5 | — | 13.2 | 6.6 | 5.3 | — |
| 7½ | — | 19 | 9.0 | 8.0 | — |
| 10 | — | 24 | 12 | 10 | — |
| 15 | — | 36 | 18 | 14 | — |
| 20 | — | 47 | 23 | 19 | — |
| 25 | — | 59 | 29 | 24 | — |
| 30 | — | 69 | 35 | 28 | — |
| 40 | — | 90 | 45 | 36 | — |
| 50 | — | 113 | 56 | 45 | — |
| 60 | — | 133 | 67 | 53 | 14 |
| 75 | — | 166 | 83 | 66 | 18 |
| 100 | — | 218 | 109 | 87 | 23 |
| 125 | — | 270 | 135 | 108 | 28 |
| 150 | — | 312 | 156 | 125 | 32 |
| 200 | — | 416 | 208 | 167 | 43 |

*TABLE 430.250* *Full-Load Current, Three-Phase Alternating-Current Motors*

The following values of full-load currents are typical for motors running at speeds usual for belted motors and motors with normal torque characteristics.

The voltages listed are rated motor voltages. The currents listed shall be permitted for system voltage ranges of 110 to 120, 220 to 240, 440 to 480, and 550 to 1000 volts.

| | Induction-Type Squirrel Cage and Wound Rotor (Amperes) | | | | | | | Synchronous-Type Unity Power Factor* (Amperes) | | | |
|---|---|---|---|---|---|---|---|---|---|---|---|
| **Horsepower** | **115 Volts** | **200 Volts** | **208 Volts** | **230 Volts** | **460 Volts** | **575 Volts** | **2300 Volts** | **230 Volts** | **460 Volts** | **575 Volts** | **2300 Volts** |
| ½ | 4.4 | 2.5 | 2.4 | 2.2 | 1.1 | 0.9 | — | — | — | — | — |
| ¾ | 6.4 | 3.7 | 3.5 | 3.2 | 1.6 | 1.3 | — | — | — | — | — |
| 1 | 8.4 | 4.8 | 4.6 | 4.2 | 2.1 | 1.7 | — | — | — | — | — |
| 1½ | 12.0 | 6.9 | 6.6 | 6.0 | 3.0 | 2.4 | — | — | — | — | — |
| 2 | 13.6 | 7.8 | 7.5 | 6.8 | 3.4 | 2.7 | — | — | — | — | — |
| 3 | — | 11.0 | 10.6 | 9.6 | 4.8 | 3.9 | — | — | — | — | — |
| 5 | — | 17.5 | 16.7 | 15.2 | 7.6 | 6.1 | — | — | — | — | — |
| 7½ | — | 25.3 | 24.2 | 22 | 11 | 9 | — | — | — | — | — |
| 10 | — | 32.2 | 30.8 | 28 | 14 | 11 | — | — | — | — | — |
| 15 | — | 48.3 | 46.2 | 42 | 21 | 17 | — | — | — | — | — |
| 20 | — | 62.1 | 59.4 | 54 | 27 | 22 | — | — | — | — | — |
| 25 | — | 78.2 | 74.8 | 68 | 34 | 27 | — | 53 | 26 | 21 | — |
| 30 | — | 92 | 88 | 80 | 40 | 32 | — | 63 | 32 | 26 | — |
| 40 | — | 120 | 114 | 104 | 52 | 41 | — | 83 | 41 | 33 | — |
| 50 | — | 150 | 143 | 130 | 65 | 52 | — | 104 | 52 | 42 | — |
| 60 | — | 177 | 169 | 154 | 77 | 62 | 16 | 123 | 61 | 49 | 12 |
| 75 | — | 221 | 211 | 192 | 96 | 77 | 20 | 155 | 78 | 62 | 15 |
| 100 | — | 285 | 273 | 248 | 124 | 99 | 26 | 202 | 101 | 81 | 20 |
| 125 | — | 359 | 343 | 312 | 156 | 125 | 31 | 253 | 126 | 101 | 25 |
| 150 | — | 414 | 396 | 360 | 180 | 144 | 37 | 302 | 151 | 121 | 30 |
| 200 | | 552 | 528 | 480 | 240 | 192 | 49 | 400 | 201 | 161 | 40 |
| 250 | — | — | — | — | 302 | 242 | 60 | — | — | — | — |
| 300 | — | — | — | — | 361 | 289 | 72 | — | — | — | — |
| 350 | — | — | — | — | 414 | 336 | 83 | — | — | — | — |
| 400 | — | — | — | — | 477 | 382 | 95 | — | — | — | — |
| 450 | — | — | — | — | 515 | 412 | 103 | — | — | — | — |
| 500 | — | — | — | — | 590 | 472 | 118 | — | — | — | — |

*For 90 and 80 percent power factor, the figures shall be multiplied by 1.1 and 1.25, respectively.

***TABLE 430.251(A)*** *Conversion Table of Single-Phase Locked- Rotor Currents for Selection of Disconnecting Means and Controllers as Determined from Horsepower and Voltage Rating*
For use only with 430.110, 440.12, 440.41, and 455.8(C).

| Rated Horsepower | Maximum Locked-Rotor Current in Amperes, Single Phase | | |
|---|---|---|---|
| | 115 Volts | 208 Volts | 230 Volts |
| ½ | 58.8 | 32.5 | 29.4 |
| ¾ | 82.8 | 45.8 | 41.4 |
| 1 | 96 | 53 | 48 |
| 1½ | 120 | 66 | 60 |
| 2 | 144 | 80 | 72 |
| 3 | 204 | 113 | 102 |
| 5 | 336 | 186 | 168 |
| 7½ | 480 | 265 | 240 |
| 10 | 1000 | 332 | 300 |

***TABLE 430.251(B)*** *Conversion Table of Polyphase Design B, C, and D Maximum Locked-Rotor Currents for Selection of Disconnecting Means and Controllers as Determined from Horsepower and Voltage Rating and Design Letter*
For use only with 430.110, 440.12, 440.41 and 455.8(C).

| Rated Horsepower | Maximum Motor Locked-Rotor Current in Amperes, Two- and Three-Phase, Design B, C, and D* | | | | | |
|---|---|---|---|---|---|---|
| | 115 Volts | 200 Volts | 208 Volts | 230 Volts | 460 Volts | 575 Volts |
| | B, C, D | B, C, D | B, C, D | B, C, D | B, C, D | B, C, D |
| ½ | 40 | 23 | 22.1 | 20 | 10 | 8 |
| ¾ | 50 | 28.8 | 27.6 | 25 | 12.5 | 10 |
| 1 | 60 | 34.5 | 33 | 30 | 15 | 12 |
| 1½ | 80 | 46 | 44 | 40 | 20 | 16 |
| 2 | 100 | 57.5 | 55 | 50 | 25 | 20 |
| 3 | — | 73.6 | 71 | 64 | 32 | 25.6 |
| 5 | — | 105.8 | 102 | 92 | 46 | 36.8 |
| 7½ | — | 146 | 140 | 127 | 63.5 | 50.8 |
| 10 | — | 186.3 | 179 | 162 | 81 | 64.8 |
| 15 | — | 267 | 257 | 232 | 116 | 93 |
| 20 | — | 334 | 321 | 290 | 145 | 116 |
| 25 | — | 420 | 404 | 365 | 183 | 146 |
| 30 | — | 500 | 481 | 435 | 218 | 174 |
| 40 | — | 667 | 641 | 580 | 290 | 232 |
| 50 | — | 834 | 802 | 725 | 363 | 290 |
| 60 | — | 1001 | 962 | 870 | 435 | 348 |
| 75 | — | 1248 | 1200 | 1085 | 543 | 434 |
| 100 | — | 1668 | 1603 | 1450 | 725 | 580 |
| 125 | — | 2087 | 2007 | 1815 | 908 | 726 |
| 150 | — | 2496 | 2400 | 2170 | 1085 | 868 |
| 200 | — | 3335 | 3207 | 2900 | 1450 | 1160 |
| 250 | — | — | — | — | 1825 | 1460 |
| 300 | — | — | — | — | 2200 | 1760 |
| 350 | — | — | — | — | 2550 | 2040 |
| 400 | — | — | — | — | 2900 | 2320 |
| 450 | — | — | — | — | 3250 | 2600 |
| 500 | — | — | — | — | 3625 | 2900 |

*Design A motors are not limited to a maximum starting current or locked rotor current.

# ARTICLE 440
# Air-Conditioning and Refrigerating Equipment

## I. General

### 440.1 Scope

The provisions of this article apply to electric motor-driven air-conditioning and refrigerating equipment and to the branch circuits and controllers for such equipment. It provides for the special considerations necessary for circuits supplying hermetic refrigerant motor-compressors and for any air-conditioning or refrigerating equipment that is supplied from a branch circuit that supplies a hermetic refrigerant motor-compressor.

Article 440 provides special considerations necessary for circuits supplying hermetic refrigerant motor-compressors and is in addition to or amendatory of the provisions of Article 430 and other applicable articles. However, many requirements for disconnecting means, controllers, single or group installations, and sizing of conductors, for example, are the same as or similar to those applied in Article 430.

Article 440 does not apply unless a hermetic refrigerant motor-compressor is supplied. Article 440 must be applied in conjunction with Article 430.

### 440.2 Definitions

**Branch-Circuit Selection Current.** The value in amperes to be used instead of the rated-load current in determining the ratings of motor branch-circuit conductors, disconnecting means, controllers, and branch-circuit short-circuit and ground-fault protective devices wherever the running overload protective device permits a sustained current greater than the specified percentage of the rated-load current. The value of branch-circuit selection current will always be equal to or greater than the marked rated-load current.

•

**Leakage-Current Detector-Interrupter (LCDI).** A device provided in a power supply cord or cord set that senses leakage current flowing between or from the cord conductors and interrupts the circuit at a predetermined level of leakage current.

Opening of the circuit is accomplished through the use of electronic switching or by "air-break" contacts. The circuit remains open until the cause of the leakage current is eliminated or the protection device is manually reset. Leakage-current detection and interrupter protection is one of the protection methods for the power supply cord or cord set of a room air conditioner specified in 440.65.

**Rated-Load Current.** The rated-load current for a hermetic refrigerant motor-compressor is the current resulting when the motor-compressor is operated at the rated load, rated voltage, and rated frequency of the equipment it serves.

### 440.3 Other Articles

**(A) Article 430.** These provisions are in addition to, or amendatory of, the provisions of Article 430 and other articles in this *Code*, which apply except as modified in this article.

**(B) Articles 422, 424, or 430.** The rules of Articles 422, 424, or 430, as applicable, shall apply to air-conditioning and refrigerating equipment that does not incorporate a hermetic refrigerant motor-compressor. This equipment includes devices that employ refrigeration compressors driven by conventional motors, furnaces with air-conditioning evaporator coils installed, fan-coil units, remote forced air-cooled condensers, remote commercial refrigerators, and so forth.

**(C) Article 422.** Equipment such as room air conditioners, household refrigerators and freezers, drinking water coolers, and beverage dispensers shall be considered appliances, and the provisions of Article 422 shall also apply.

**(D) Other Applicable Articles.** Hermetic refrigerant motor-compressors, circuits, controllers, and equipment shall also comply with the applicable provisions of Table 440.3(D).

### 440.4 Marking on Hermetic Refrigerant Motor-Compressors and Equipment

**(A) Hermetic Refrigerant Motor-Compressor Nameplate.** A hermetic refrigerant motor-compressor shall be provided with a nameplate that shall indicate the manufacturer's name, trademark, or symbol; identifying designation; phase; voltage; and frequency. The rated-load current in amperes of the motor-compressor shall be marked by the equipment manufacturer on either or both the motor-compressor nameplate and the nameplate of the equipment in which the motor-compressor is used. The locked-rotor current of each single-phase motor-compressor having a rated-load current

***TABLE 440.3(D)*** *Other Articles*

| Equipment/Occupancy | Article | Section |
|---|---|---|
| Capacitors | | 460.9 |
| Commercial garages, aircraft hangars, motor fuel dispensing facilities, bulk storage plants, spray application, dipping, and coating processes, and inhalation anesthetizing locations | 511, 513, 514, 515, 516, and 517 Part IV | |
| Hazardous (classified) locations | 500–503, 505, and 506 | |
| Motion picture and television studios and similar locations | 530 | |
| Resistors and reactors | 470 | |

of more than 9 amperes at 115 volts, or more than 4.5 amperes at 230 volts, and each polyphase motor-compressor shall be marked on the motor-compressor nameplate. Where a thermal protector complying with 440.52(A)(2) and (B)(2) is used, the motor-compressor nameplate or the equipment nameplate shall be marked with the words "thermally protected." Where a protective system complying with 440.52(A)(4) and (B)(4) is used and is furnished with the equipment, the equipment nameplate shall be marked with the words, "thermally protected system." Where a protective system complying with 440.52(A)(4) and (B)(4) is specified, the equipment nameplate shall be appropriately marked.

**(B) Multimotor and Combination-Load Equipment.** Multimotor and combination-load equipment shall be provided with a visible nameplate marked with the maker's name, the rating in volts, frequency and number of phases, minimum supply circuit conductor ampacity, the maximum rating of the branch-circuit short-circuit and ground-fault protective device, and the short-circuit current rating of the motor controllers or industrial control panel. The ampacity shall be calculated by using Part IV and counting all the motors and other loads that will be operated at the same time. The branch-circuit short-circuit and ground-fault protective device rating shall not exceed the value calculated by using Part III. Multimotor or combination-load equipment for use on two or more circuits shall be marked with the above information for each circuit.

*Exception No. 1: Multimotor and combination-load equipment that is suitable under the provisions of this article for connection to a single 15- or 20-ampere, 120-volt, or a 15-ampere, 208- or 240-volt, single-phase branch circuit shall be permitted to be marked as a single load.*

*Exception No. 2: The minimum supply circuit conductor ampacity and the maximum rating of the branch-circuit short-circuit and ground-fault protective device shall not be required to be marked on a room air conditioner complying with 440.62(A).*

*Exception No. 3: Multimotor and combination-load equipment used in one- and two-family dwellings, cord-and-attachment-plug-connected equipment, or equipment supplied from a branch circuit protected at 60 A or less shall not be required to be marked with a short-circuit current rating.*

Motor controllers or control panels associated with multimotor and combination-load air-conditioning and refrigeration equipment are required by 440.4(B) to be marked with the short-circuit current rating. Where air-conditioning equipment is installed at large commercial, institutional, and industrial complexes, the controllers and control panels of air-conditioning and refrigeration equipment are often supplied from a point on the electrical distribution system where significant short-circuit current is available.

As is the case with any electrical installation where high levels of short-circuit current are available, the short-circuit current rating marked on the air-conditioning equipment controllers and control panels provides those responsible for designing and approving the electrical installation with the necessary information to ensure compliance with the requirements of 110.10. Multimotor and combination-load air-conditioning equipment used in one- and two-family dwellings, cord-and-plug-connected air-conditioning equipment, and air-conditioning equipment supplied by branch circuits rated 60 amperes or less are not required to be marked with their short-circuit current rating. A similar requirement for marking the short-circuit current rating on equipment is specified by 430.8 for motor controllers.

**(C) Branch-Circuit Selection Current.** A hermetic refrigerant motor-compressor, or equipment containing such a compressor, having a protection system that is approved for use with the motor-compressor that it protects and that permits continuous current in excess of the specified percentage of nameplate rated-load current given in 440.52(B)(2) or (B)(4) shall also be marked with a branch-circuit selection current that complies with 440.52(B)(2) or (B)(4). This marking shall be provided by the equipment manufacturer and shall be on the nameplate(s) where the rated-load current(s) appears.

## 440.5 Marking on Controllers

A controller shall be marked with the manufacturer's name, trademark, or symbol; identifying designation; voltage; phase; full-load and locked-rotor current (or horsepower) rating; and other data as may be needed to properly indicate the motor-compressor for which it is suitable.

## 440.6 Ampacity and Rating

The size of conductors for equipment covered by this article shall be selected from Table 310.15(B)(16) through Table 310.15(B)(19) or calculated in accordance with 310.15 as applicable. The required ampacity of conductors and rating of equipment shall be determined according to 440.6(A) and 440.6(B).

**(A) Hermetic Refrigerant Motor-Compressor.** For a hermetic refrigerant motor-compressor, the rated-load current marked on the nameplate of the equipment in which the motor-compressor is employed shall be used in determining the rating or ampacity of the disconnecting means, the branch-circuit conductors, the controller, the branch-circuit short-circuit and ground-fault protection, and the separate motor overload protection. Where no rated-load current is shown on the equipment nameplate, the rated-load current shown on the compressor nameplate shall be used.

*Exception No. 1: Where so marked, the branch-circuit selection current shall be used instead of the rated-load current to determine the rating or ampacity of the disconnecting means, the branch-circuit conductors, the controller, and the branch-circuit short-circuit and ground-fault protection.*

*Exception No. 2: For cord-and-plug-connected equipment, the nameplate marking shall be used in accordance with 440.22(B), Exception No. 2.*

•

**(B) Multimotor Equipment.** For multimotor equipment employing a shaded-pole or permanent split-capacitor-type fan or blower motor, the full-load current for such motor marked on the nameplate of the equipment in which the fan or blower motor is employed shall be used instead of the horsepower rating to determine the ampacity or rating of the disconnecting means, the branch-circuit conductors, the controller, the branch-circuit short-circuit and ground-fault protection, and the separate overload protection. This marking on the equipment nameplate shall not be less than the current marked on the fan or blower motor nameplate.

### 440.7 Highest Rated (Largest) Motor

In determining compliance with this article and with 430.24, 430.53(B) and 430.53(C), and 430.62(A), the highest rated (largest) motor shall be considered to be the motor that has the highest rated-load current. Where two or more motors have the same highest rated-load current, only one of them shall be considered as the highest rated (largest) motor. For other than hermetic refrigerant motor-compressors, and fan or blower motors as covered in 440.6(B), the full-load current used to determine the highest rated motor shall be the equivalent value corresponding to the motor horsepower rating selected from Table 430.248, Table 430.249, or Table 430.250.

*Exception: Where so marked, the branch-circuit selection current shall be used instead of the rated-load current in determining the highest rated (largest) motor-compressor.*

### 440.8 Single Machine

An air-conditioning or refrigerating system shall be considered to be a single machine under the provisions of 430.87, Exception No. 1, and 430.112, Exception. The motors shall be permitted to be located remotely from each other.

## II. Disconnecting Means

### 440.11 General

The provisions of Part II are intended to require disconnecting means capable of disconnecting air-conditioning and refrigerating equipment, including motor-compressors and controllers from the circuit conductors.

### 440.12 Rating and Interrupting Capacity

**(A) Hermetic Refrigerant Motor-Compressor.** A disconnecting means serving a hermetic refrigerant motor-compressor shall be selected on the basis of the nameplate rated-load current or branch-circuit selection current, whichever is greater, and locked-rotor current, respectively, of the motor-compressor as follows.

**(1) Ampere Rating.** The ampere rating shall be at least 115 percent of the nameplate rated-load current or branch-circuit selection current, whichever is greater.

*Exception: A listed unfused motor circuit switch, without fuseholders, having a horsepower rating not less than the equivalent horsepower determined in accordance with 440.12(A)(2) shall be permitted to have an ampere rating less than 115 percent of the specified current.*

**(2) Equivalent Horsepower.** To determine the equivalent horsepower in complying with the requirements of 430.109, the horsepower rating shall be selected from Table 430.248, Table 430.249, or Table 430.250 corresponding to the rated-load current or branch-circuit selection current, whichever is greater, and also the horsepower rating from Table 430.251(A) or Table 430.251(B) corresponding to the locked-rotor current. In case the nameplate rated-load current or branch-circuit selection current and locked-rotor current do not correspond to the currents shown in Table 430.248, Table 430.249, Table 430.250, Table 430.251(A), or Table 430.251(B), the horsepower rating corresponding to the next higher value shall be selected. In case different horsepower ratings are obtained when applying these tables, a horsepower rating at least equal to the larger of the values obtained shall be selected.

**(B) Combination Loads.** Where the combined load of two or more hermetic refrigerant motor-compressors or one or more hermetic refrigerant motor-compressor with other motors or loads may be simultaneous on a single disconnecting means, the rating for the disconnecting means shall be determined in accordance with 440.12(B)(1) and (B)(2).

**(1) Horsepower Rating.** The horsepower rating of the disconnecting means shall be determined from the sum of all currents, including resistance loads, at the rated-load condition and also at the locked-rotor condition. The combined rated-load current and the combined locked-rotor current so obtained shall be considered as a single motor for the purpose of this requirement as required by (1)(a) and (1)(b).

(a) The full-load current equivalent to the horsepower rating of each motor, other than a hermetic refrigerant motor-compressor, and fan or blower motors as covered in 440.6(B) shall be selected from Table 430.248, Table 430.249, or Table 430.250. These full-load currents shall be added to the motor-compressor rated-load current(s) or branch-circuit selection current(s), whichever is greater, and to the rating in amperes of other loads to obtain an equivalent full-load current for the combined load.

(b) The locked-rotor current equivalent to the horsepower rating of each motor, other than a hermetic refrigerant motor-compressor, shall be selected from Table 430.251(A) or Table 430.251(B), and, for fan and blower motors of the shaded-pole or permanent split-capacitor type marked with the locked-rotor current, the marked value shall be used. The locked-rotor currents shall be added to the motor-compressor locked-rotor current(s) and to the rating in amperes of other loads to obtain an equivalent locked-rotor current for the combined load. Where two or more motors or other loads such as resistance heaters, or both, cannot be started simultaneously, appropriate combinations of locked-rotor and rated-load current or branch-circuit selection current, whichever is greater, shall be an acceptable means of determining the equivalent locked-rotor current for the simultaneous combined load.

*Exception: Where part of the concurrent load is a resistance load and the disconnecting means is a switch rated in horsepower and amperes, the switch used shall be permitted to have a horsepower rating not less than the combined load to the motor-compressor(s) and other motor(s) at the locked-rotor condition, if the ampere rating of the switch is not less than this locked-rotor load plus the resistance load.*

**(2) Full-Load Current Equivalent.** The ampere rating of the disconnecting means shall be at least 115 percent of the sum of all currents at the rated-load condition determined in accordance with 440.12(B)(1).

*Exception: A listed unfused motor circuit switch, without fuseholders, having a horsepower rating not less than the equivalent horsepower determined by 440.12(B)(1) shall be permitted to have an ampere rating less than 115 percent of the sum of all currents.*

**(C) Small Motor-Compressors.** For small motor-compressors not having the locked-rotor current marked on the nameplate, or for small motors not covered by Table 430.247, Table 430.248, Table 430.249, or Table 430.250, the locked-rotor current shall be assumed to be six times the rated-load current.

**(D) Disconnecting Means.** Every disconnecting means in the refrigerant motor-compressor circuit between the point of attachment to the feeder and the point of connection to the refrigerant motor-compressor shall comply with the requirements of 440.12.

**(E) Disconnecting Means Rated in Excess of 100 Horsepower.** Where the rated-load or locked-rotor current as determined above would indicate a disconnecting means rated in excess of 100 hp, the provisions of 430.109(E) shall apply.

## 440.13 Cord-Connected Equipment

For cord-connected equipment such as room air conditioners, household refrigerators and freezers, drinking water coolers, and beverage dispensers, a separable connector or an attachment plug and receptacle shall be permitted to serve as the disconnecting means.

Informational Note: For room air conditioners, see 440.63.

## 440.14 Location

Disconnecting means shall be located within sight from and readily accessible from the air-conditioning or refrigerating equipment. The disconnecting means shall be permitted to be installed on or within the air-conditioning or refrigerating equipment.

The disconnecting means shall not be located on panels that are designed to allow access to the air-conditioning or refrigeration equipment or to obscure the equipment nameplate(s).

*Exception No. 1: Where the disconnecting means provided in accordance with 430.102(A) is lockable in accordance with 110.25 and the refrigerating or air-conditioning equipment is essential to an industrial process in a facility with written safety procedures, and where the conditions of maintenance and supervision ensure that only qualified persons service the equipment, a disconnecting means within sight from the equipment shall not be required.*

Exception No. 1 accommodates special conditions associated with process refrigeration equipment. Typically, this equipment is very large, so rated disconnects may not be available. Additionally, this equipment may be in hazardous locations, and locating disconnecting means within sight of the motor may introduce additional hazards. The provision for locking or attaching a lock to the disconnecting means must be a permanent part of the disconnect. An example of this type of locking hardware is shown in Exhibit 440.1.

*Exception No. 2: Where an attachment plug and receptacle serve as the disconnecting means in accordance with 440.13, their location shall be accessible but shall not be required to be readily accessible.*

Informational Note No. 1: See Parts VII and IX of Article 430 for additional requirements.

Informational Note No. 2: See 110.26.

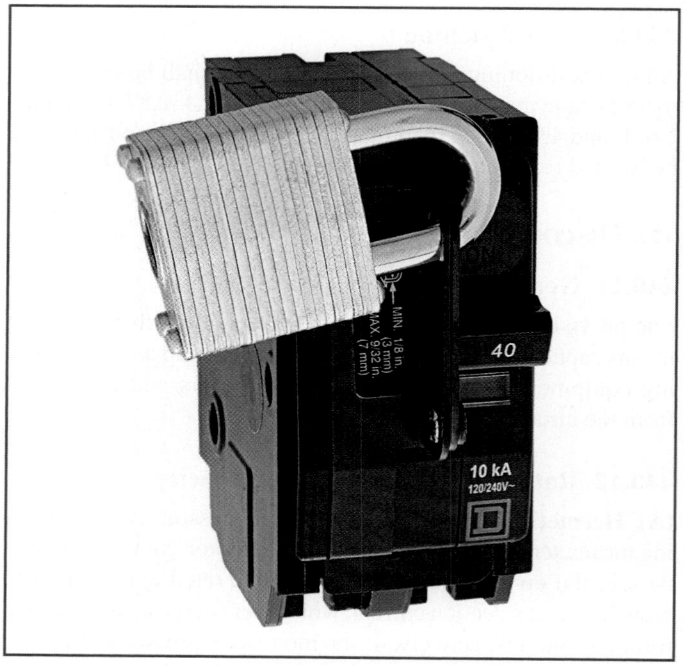


***EXHIBIT 440.1*** *An example of locking hardware (with lock installed) that is not readily removable or transferable. (Courtesy of Schneider Electric)*

The references to Parts VII and IX of Article 430 in Informational Note No. 1 are intended to call attention to the additional disconnect location requirements in 430.102, 430.107, and 430.113. Because 440.3(A) makes the requirements in Article 440 in addition to or amendatory of the provisions of Article 430, the requirement of 440.14 mandates that the equipment disconnecting means be within sight from and readily accessible from the equipment, even if a remote disconnect is capable of being locked in the "open" position under the provision of the exception to 430.102(B).

This special requirement for air-conditioning and refrigeration equipment covered by Article 440 is more stringent than the provisions in Article 430 and provides protection for service personnel working on equipment located in attics, on roofs, or outside in a remote location where it is difficult to gain access to a remote lockable disconnect. See Exception No. 1 to 440.14.

## III. Branch-Circuit Short-Circuit and Ground-Fault Protection

### 440.21 General

The provisions of Part III specify devices intended to protect the branch-circuit conductors, control apparatus, and motors in circuits supplying hermetic refrigerant motor-compressors against overcurrent due to short circuits and ground faults. They are in addition to or amendatory of the provisions of Article 240.

Where an air conditioner is listed by a qualified electrical testing laboratory with a nameplate that reads "maximum fuse size," the listing restricts the use of this unit to fuse protection only and does not cover its use with circuit breakers. If the air conditioner has been evaluated for both fuses and ordinary circuit breakers or both fuses and HACR-type circuit breakers, it is marked to indicate the acceptable type(s) of protective devices. UL-listed circuit breakers that have been found suitable for use with heating, air-conditioning, and refrigeration equipment comprising multimotor or combination loads are marked "Listed HACR Type." Section 430.53(C) permits the use of any listed inverse time circuit breaker for the short-circuit, ground-fault protection of group motor installations. Section 110.3(B) requires that the installation and use instructions provided by the manufacturer of listed equipment be followed, and in the case of air-conditioning equipment it is important to carefully read the nameplate so that the correct type of short-circuit, ground-fault protective device is selected.

Exhibit 440.2 illustrates three wiring configurations where the equipment is under fuse protection only as specified on the nameplate.

Current-limiting overcurrent devices, which may reduce the amount of fault current to which the equipment is subjected, can be installed in the branch circuit supplying the equipment. See 240.2 for the definition of *current-limiting overcurrent protective device* and the accompanying commentary explaining short-circuit damage, which is particularly important for larger commercial and industrial installations. See also 110.3(B) and 110.10 and associated commentary regarding the installation and use of listed or labeled equipment and the selection of OCPDs (such as fuses and circuit breakers).

### 440.22 Application and Selection

**(A) Rating or Setting for Individual Motor-Compressor.** The motor-compressor branch-circuit short-circuit and ground-fault protective device shall be capable of carrying the starting current of the motor. A protective device having a rating or setting not exceeding 175 percent of the motor-compressor rated-load current or branch-circuit selection current, whichever is greater, shall be permitted, provided that, where the protection specified is not sufficient for the starting current of the motor, the rating or setting shall be permitted to be increased but shall not exceed 225 percent of the motor rated-load current or branch-circuit selection current, whichever is greater.

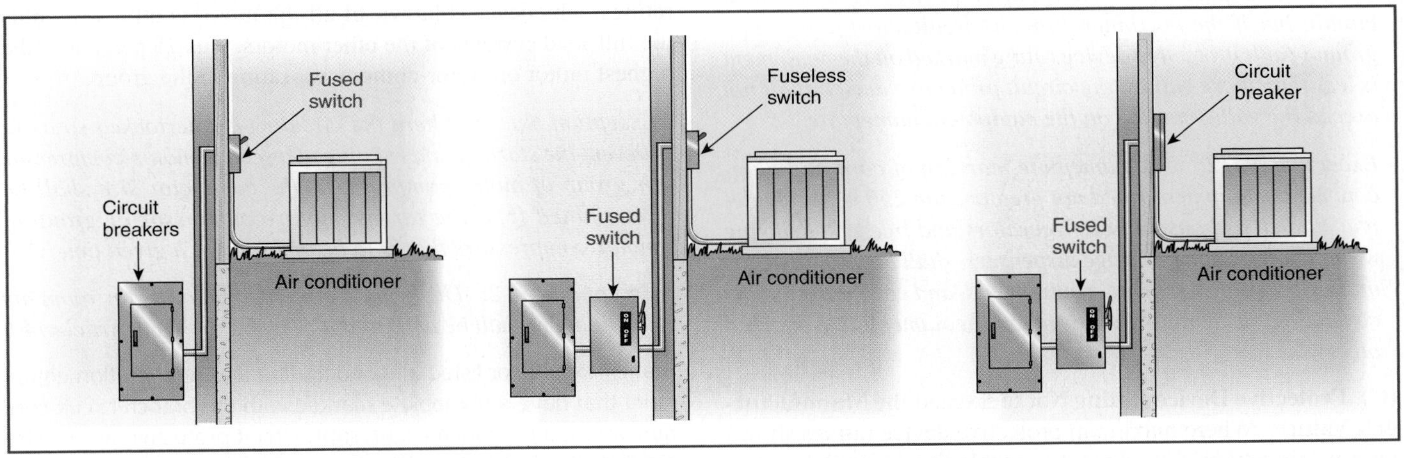


***EXHIBIT 440.2*** *Three correct alternate wiring configurations satisfying the restriction that the equipment be protected by fuses only. (Note that the fuse rating cannot exceed the maximum fuse size specified on the air-conditioner nameplate.)*

*Exception: The rating of the branch-circuit short-circuit and ground-fault protective device shall not be required to be less than 15 amperes.*

**(B) Rating or Setting for Equipment.** The equipment branch-circuit short-circuit and ground-fault protective device shall be capable of carrying the starting current of the equipment. Where the hermetic refrigerant motor-compressor is the only load on the circuit, the protection shall comply with 440.22(A). Where the equipment incorporates more than one hermetic refrigerant motor-compressor or a hermetic refrigerant motor-compressor and other motors or other loads, the equipment short-circuit and ground-fault protection shall comply with 430.53 and 440.22(B)(1) and (B)(2).

**(1) Motor-Compressor Largest Load.** Where a hermetic refrigerant motor-compressor is the largest load connected to the circuit, the rating or setting of the branch-circuit short-circuit and ground-fault protective device shall not exceed the value specified in 440.22(A) for the largest motor-compressor plus the sum of the rated-load current or branch-circuit selection current, whichever is greater, of the other motor-compressor(s) and the ratings of the other loads supplied.

**(2) Motor-Compressor Not Largest Load.** Where a hermetic refrigerant motor-compressor is not the largest load connected to the circuit, the rating or setting of the branch-circuit short-circuit and ground-fault protective device shall not exceed a value equal to the sum of the rated-load current or branch-circuit selection current, whichever is greater, rating(s) for the motor-compressor(s) plus the value specified in 430.53(C)(4) where other motor loads are supplied, or the value specified in 240.4 where only nonmotor loads are supplied in addition to the motor-compressor(s).

*Exception No. 1: Equipment that starts and operates on a 15- or 20-ampere 120-volt, or 15-ampere 208- or 240-volt single-phase branch circuit, shall be permitted to be protected by the 15- or 20-ampere overcurrent device protecting the branch circuit, but if the maximum branch-circuit short-circuit and ground-fault protective device rating marked on the equipment is less than these values, the circuit protective device shall not exceed the value marked on the equipment nameplate.*

*Exception No. 2: The nameplate marking of cord-and-plug-connected equipment rated not greater than 250 volts, single-phase, such as household refrigerators and freezers, drinking water coolers, and beverage dispensers, shall be used in determining the branch-circuit requirements, and each unit shall be considered as a single motor unless the nameplate is marked otherwise.*

**(C) Protective Device Rating Not to Exceed the Manufacturer's Values.** Where maximum protective device ratings shown on a manufacturer's overload relay table for use with a motor controller are less than the rating or setting selected in accordance with 440.22(A) and (B), the protective device rating shall not exceed the manufacturer's values marked on the equipment.

## IV. Branch-Circuit Conductors

### 440.31 General

The provisions of Part IV and Article 310 specify ampacities of conductors required to carry the motor current without overheating under the conditions specified, except as modified in 440.6(A), Exception No. 1.

The provisions of these articles shall not apply to integral conductors of motors, to motor controllers and the like, or to conductors that form an integral part of approved equipment.

•

### 440.32 Single Motor-Compressor

Branch-circuit conductors supplying a single motor-compressor shall have an ampacity not less than 125 percent of either the motor-compressor rated-load current or the branch-circuit selection current, whichever is greater.

For a wye-start, delta-run connected motor-compressor, the selection of branch-circuit conductors between the controller and the motor-compressor shall be permitted to be based on 72 percent of either the motor-compressor rated-load current or the branch-circuit selection current, whichever is greater.

For more information on wye-start, delta-run motors, see the commentary following 430.22(C).

Informational Note: The individual motor circuit conductors of wye-start, delta-run connected motor-compressors carry 58 percent of the rated load current. The multiplier of 72 percent is obtained by multiplying 58 percent by 1.25.

### 440.33 Motor-Compressor(s) With or Without Additional Motor Loads

Conductors supplying one or more motor-compressor(s) with or without an additional load(s) shall have an ampacity not less than the sum of the rated-load or branch-circuit selection current ratings, whichever is larger, of all the motor-compressors plus the full-load currents of the other motors, plus 25 percent of the highest motor or motor-compressor rating in the group.

*Exception No. 1: Where the circuitry is interlocked so as to prevent the starting and running of a second motor-compressor or group of motor-compressors, the conductor size shall be determined from the largest motor-compressor or group of motor-compressors that is to be operated at a given time.*

*Exception No. 2: The branch-circuit conductors for room air conditioners shall be in accordance with Part VII of Article 440.*

Branch circuits for listed air-conditioning and refrigeration equipment that have a nameplate marked with the branch-circuit conductor size and branch-circuit short-circuit protective device size are not required to have the branch-circuit conductors sized in accordance with 440.33. The standard includes the 25-percent increase for the largest motor or compressor in the group plus the other nonmotor or noncompressor load; therefore, the actual nameplate full-load amperes for the complete assembly can be used to size the branch-circuit conductors.

### 440.34 Combination Load

Conductors supplying a motor-compressor load in addition to other load(s) as calculated from Article 220 and other applicable articles shall have an ampacity sufficient for the other load(s) plus the required ampacity for the motor-compressor load determined in accordance with 440.33 or, for a single motor-compressor, in accordance with 440.32.

*Exception: Where the circuitry is interlocked so as to prevent simultaneous operation of the motor-compressor(s) and all other loads connected, the conductor size shall be determined from the largest size required for the motor-compressor(s) and other loads to be operated at a given time.*

### 440.35 Multimotor and Combination-Load Equipment

The ampacity of the conductors supplying multimotor and combination-load equipment shall not be less than the minimum circuit ampacity marked on the equipment in accordance with 440.4(B).

## V. Controllers for Motor-Compressors

### 440.41 Rating

**(A) Motor-Compressor Controller.** A motor-compressor controller shall have both a continuous-duty full-load current rating and a locked-rotor current rating not less than the nameplate rated-load current or branch-circuit selection current, whichever is greater, and locked-rotor current, respectively, of the compressor. In case the motor controller is rated in horsepower but is without one or both of the foregoing current ratings, equivalent currents shall be determined from the ratings as follows. Table 430.248, Table 430.249, and Table 430.250 shall be used to determine the equivalent full-load current rating. Table 430.251(A) and Table 430.251(B) shall be used to determine the equivalent locked-rotor current ratings.

**(B) Controller Serving More Than One Load.** A controller serving more than one motor-compressor or a motor-compressor and other loads shall have a continuous-duty full-load current rating and a locked-rotor current rating not less than the combined load as determined in accordance with 440.12(B).

## VI. Motor-Compressor and Branch-Circuit Overload Protection

### 440.51 General

The provisions of Part VI specify devices intended to protect the motor-compressor, the motor-control apparatus, and the branch-circuit conductors against excessive heating due to motor overload and failure to start.

Informational Note: See 240.4(G) for application of Parts III and VI of Article 440.

### 440.52 Application and Selection

**(A) Protection of Motor-Compressor.** Each motor-compressor shall be protected against overload and failure to start by one of the following means:

(1) A separate overload relay that is responsive to motor-compressor current. This device shall be selected to trip at not more than 140 percent of the motor-compressor rated-load current.
(2) A thermal protector integral with the motor-compressor, approved for use with the motor-compressor that it protects on the basis that it will prevent dangerous overheating of the motor-compressor due to overload and failure to start. If the current-interrupting device is separate from the motor-compressor and its control circuit is operated by a protective device integral with the motor-compressor, it shall be arranged so that the opening of the control circuit will result in interruption of current to the motor-compressor.
(3) A fuse or inverse time circuit breaker responsive to motor current, which shall also be permitted to serve as the branch-circuit short-circuit and ground-fault protective device. This device shall be rated at not more than 125 percent of the motor-compressor rated-load current. It shall have sufficient time delay to permit the motor-compressor to start and accelerate its load. The equipment or the motor-compressor shall be marked with this maximum branch-circuit fuse or inverse time circuit breaker rating.
(4) A protective system, furnished or specified and approved for use with the motor-compressor that it protects on the basis that it will prevent dangerous overheating of the motor-compressor due to overload and failure to start. If the current-interrupting device is separate from the motor-compressor and its control circuit is operated by a protective device that is not integral with the current-interrupting device, it shall be arranged so that the opening of the control circuit will result in interruption of current to the motor-compressor.

**(B) Protection of Motor-Compressor Control Apparatus and Branch-Circuit Conductors.** The motor-compressor controller(s), the disconnecting means, and the branch-circuit conductors shall be protected against overcurrent due to motor overload and failure to start by one of the following means, which shall be permitted to be the same device or system protecting the motor-compressor in accordance with 440.52(A):

*Exception: Overload protection of motor-compressors and equipment on 15- and 20-ampere, single-phase, branch circuits shall be permitted to be in accordance with 440.54 and 440.55.*

(1) An overload relay selected in accordance with 440.52(A)(1)
(2) A thermal protector applied in accordance with 440.52(A)(2), that will not permit a continuous current in excess of 156 percent of the marked rated-load current or branch-circuit selection current

(3) A fuse or inverse time circuit breaker selected in accordance with 440.52(A)(3)
(4) A protective system, in accordance with 440.52(A)(4), that will not permit a continuous current in excess of 156 percent of the marked rated-load current or branch-circuit selection current

### 440.53 Overload Relays

Overload relays and other devices for motor overload protection that are not capable of opening short circuits shall be protected by fuses or inverse time circuit breakers with ratings or settings in accordance with Part III unless identified for group installation or for part-winding motors and marked to indicate the maximum size of fuse or inverse time circuit breaker by which they shall be protected.

*Exception: The fuse or inverse time circuit breaker size marking shall be permitted on the nameplate of the equipment in which the overload relay or other overload device is used.*

### 440.54 Motor-Compressors and Equipment on 15- or 20-Ampere Branch Circuits — Not Cord- and Attachment-Plug-Connected

Overload protection for motor-compressors and equipment used on 15- or 20-ampere 120-volt, or 15-ampere 208- or 240-volt single-phase branch circuits as permitted in Article 210 shall be permitted as indicated in 440.54(A) and 440.54(B).

**(A) Overload Protection.** The motor-compressor shall be provided with overload protection selected as specified in 440.52(A). Both the controller and motor overload protective device shall be identified for installation with the short-circuit and ground-fault protective device for the branch circuit to which the equipment is connected.

**(B) Time Delay.** The short-circuit and ground-fault protective device protecting the branch circuit shall have sufficient time delay to permit the motor-compressor and other motors to start and accelerate their loads.

### 440.55 Cord- and Attachment-Plug-Connected Motor-Compressors and Equipment on 15- or 20-Ampere Branch Circuits

Overload protection for motor-compressors and equipment that are cord- and attachment-plug-connected and used on 15- or 20-ampere 120-volt, or 15-ampere 208- or 240-volt, single-phase branch circuits as permitted in Article 210 shall be permitted as indicated in 440.55(A), (B), and (C).

**(A) Overload Protection.** The motor-compressor shall be provided with overload protection as specified in 440.52(A). Both the controller and the motor overload protective device shall be identified for installation with the short-circuit and ground-fault protective device for the branch circuit to which the equipment is connected.

**(B) Attachment Plug and Receptacle or Cord Connector Rating.** The rating of the attachment plug and receptacle or cord connector shall not exceed 20 amperes at 125 volts or 15 amperes at 250 volts.

**(C) Time Delay.** The short-circuit and ground-fault protective device protecting the branch circuit shall have sufficient time delay to permit the motor-compressor and other motors to start and accelerate their loads.

## VII. Provisions for Room Air Conditioners

### 440.60 General

The provisions of Part VII shall apply to electrically energized room air conditioners that control temperature and humidity. For the purpose of Part VII, a room air conditioner (with or without provisions for heating) shall be considered as an ac appliance of the air-cooled window, console, or in-wall type that is installed in the conditioned room and that incorporates a hermetic refrigerant motor-compressor(s). The provisions of Part VII cover equipment rated not over 250 volts, single phase, and the equipment shall be permitted to be cord- and attachment-plug-connected.

A room air conditioner that is rated 3-phase or rated over 250 volts shall be directly connected to a wiring method recognized in Chapter 3, and provisions of Part VII shall not apply.

### 440.61 Grounding

The enclosures of room air conditioners shall be connected to the equipment grounding conductor in accordance with 250.110, 250.112, and 250.114.

### 440.62 Branch-Circuit Requirements

**(A) Room Air Conditioner as a Single Motor Unit.** A room air conditioner shall be considered as a single motor unit in determining its branch-circuit requirements where all the following conditions are met:

(1) It is cord- and attachment-plug-connected.
(2) Its rating is not more than 40 amperes and 250 volts, single phase.
(3) Total rated-load current is shown on the room air-conditioner nameplate rather than individual motor currents.
(4) The rating of the branch-circuit short-circuit and ground-fault protective device does not exceed the ampacity of the branch-circuit conductors or the rating of the receptacle, whichever is less.

**(B) Where No Other Loads Are Supplied.** The total marked rating of a cord- and attachment-plug-connected room air conditioner shall not exceed 80 percent of the rating of a branch circuit where no other loads are supplied.

**(C) Where Lighting Units or Other Appliances Are Also Supplied.** The total marked rating of a cord- and attachment-plug-connected room air conditioner shall not exceed 50 percent of the rating of a branch circuit where lighting outlets, other appliances, or general-use receptacles are also supplied. Where the circuitry is interlocked to prevent simultaneous operation of the room air conditioner and energization of other outlets on the same branch circuit, a cord- and attachment-plug-connected room air conditioner shall not exceed 80 percent of the branch-circuit rating.

### 440.63 Disconnecting Means

An attachment plug and receptacle or cord connector shall be permitted to serve as the disconnecting means for a single-phase room air conditioner rated 250 volts or less if (1) the manual controls on the room air conditioner are readily accessible and located within 1.8 m (6 ft) of the floor, or (2) an approved manually operable disconnecting means is installed in a readily accessible location within sight from the room air conditioner.

### 440.64 Supply Cords

Where a flexible cord is used to supply a room air conditioner, the length of such cord shall not exceed 3.0 m (10 ft) for a nominal, 120-volt rating or 1.8 m (6 ft) for a nominal, 208- or 240-volt rating.

### 440.65 Leakage-Current Detector-Interrupter (LCDI) and Arc-Fault Circuit Interrupter (AFCI)

Single-phase cord- and plug-connected room air conditioners shall be provided with factory-installed LCDI or AFCI protection. The LCDI or AFCI protection shall be an integral part of the attachment plug or be located in the power supply cord within 300 mm (12 in.) of the attachment plug.

## ARTICLE 445
## Generators

### 445.1 Scope

This article contains installation and other requirements for generators.

Generators are typically associated with fire pumps (see Article 695), emergency systems (see Article 700), legally required standby systems (see Article 701), optional standby systems (see Article 702), interconnected electric power production sources (see Article 705), and critical operations power systems (see Article 708). Article 445 covers the installation of generators; the aforementioned articles cover the use of generators in particular situations and applications. Exhibit 445.1 shows an example of a diesel engine–driven generator.

***EXHIBIT 445.1*** *A diesel engine–driven generator that may be used in an emergency system, legally required standby system, optional standby system, or critical operations standby system.*

### 445.10 Location

Generators shall be of a type suitable for the locations in which they are installed. They shall also meet the requirements for motors in 430.14.

### 445.11 Marking

Each generator shall be provided with a nameplate giving the manufacturer's name, the rated frequency, the number of phases if of ac, the rating in kilowatts or kilovolt-amperes, the normal volts and amperes corresponding to the rating, the rated revolutions per minute, and the rated ambient temperature or rated temperature rise.

Nameplates for all stationary generators and portable generators rated more than 15 kW shall also give the power factor, the subtransient and transient impedances, the insulation system class, and the time rating.

Marking shall be provided by the manufacturer to indicate whether or not the generator neutral is bonded to the generator frame. Where the bonding of a generator is modified in the field, additional marking shall be required to indicate whether the generator neutral is bonded to the generator frame.

### 445.12 Overcurrent Protection

**(A) Constant-Voltage Generators.** Constant-voltage generators, except ac generator exciters, shall be protected from overload by inherent design, circuit breakers, fuses, protective relays, or other identified overcurrent protective means suitable for the conditions of use.

**(B) Two-Wire Generators.** Two-wire, dc generators shall be permitted to have overcurrent protection in one conductor only if the overcurrent device is actuated by the entire current generated other than the current in the shunt field. The overcurrent device shall not open the shunt field.

**(C) 65 Volts or Less.** Generators operating at 65 volts or less and driven by individual motors shall be considered as protected by the overcurrent device protecting the motor if these devices will operate when the generators are delivering not more than 150 percent of their full-load rated current.

**(D) Balancer Sets.** Two-wire, dc generators used in conjunction with balancer sets to obtain neutral points for 3-wire systems shall be equipped with overcurrent devices that disconnect the 3-wire system in case of excessive unbalancing of voltages or currents.

**(E) Three-Wire, Direct-Current Generators.** Three-wire, dc generators, whether compound or shunt wound, shall be equipped with overcurrent devices, one in each armature lead, and connected so as to be actuated by the entire current from the armature. Such overcurrent devices shall consist either of a double-pole, double-coil circuit breaker or of a 4-pole circuit breaker connected in the main and equalizer leads and tripped by two overcurrent devices, one in each armature lead. Such protective devices shall be interlocked so that no one pole can be opened without simultaneously disconnecting both leads of the armature from the system.

*Exception to (A) through (E): Where deemed by the authority having jurisdiction that a generator is vital to the operation of an electrical system and the generator should operate to failure to prevent a greater hazard to persons, the overload sensing device(s) shall be permitted to be connected to an annunciator or alarm supervised by authorized personnel instead of interrupting the generator circuit.*

## 445.13 Ampacity of Conductors

The ampacity of the conductors from the generator terminals to the first distribution device(s) containing overcurrent protection shall not be less than 115 percent of the nameplate current rating of the generator. It shall be permitted to size the neutral conductors in accordance with 220.61. Conductors that must carry ground-fault currents shall not be smaller than required by 250.30(A). Neutral conductors of dc generators that must carry ground-fault currents shall not be smaller than the minimum required size of the largest conductor.

*Exception: Where the design and operation of the generator prevent overloading, the ampacity of the conductors shall not be less than 100 percent of the nameplate current rating of the generator.*

## 445.14 Protection of Live Parts

Live parts of generators operated at more than 50 volts to ground shall not be exposed to accidental contact where accessible to unqualified persons.

## 445.15 Guards for Attendants

Where necessary for the safety of attendants, the requirements of 430.233 shall apply.

## 445.16 Bushings

Where field-installed wiring passes through an opening in an enclosure, a conduit box, or a barrier, a bushing shall be used to protect the conductors from the edges of an opening having sharp edges. The bushing shall have smooth, well-rounded surfaces where it may be in contact with the conductors. If used where oils, grease, or other contaminants may be present, the bushing shall be made of a material not deleteriously affected.

## 445.17 Generator Terminal Housings

Generator terminal housings shall comply with 430.12. Where a horsepower rating is required to determine the required minimum size of the generator terminal housing, the full-load current of the generator shall be compared with comparable motors in Table 430.247 through Table 430.250. The higher horsepower rating of Table 430.247 and Table 430.250 shall be used whenever the generator selection is between two ratings.

*Exception: This section shall not apply to generators rated over 600 volts.*

## 445.18 Disconnecting Means Required for Generators

Generators shall be equipped with a disconnect(s), lockable in the open position by means of which the generator and all protective devices and control apparatus are able to be disconnected entirely from the circuits supplied by the generator except where the following conditions apply:

(1) Portable generators are cord- and plug-connected, or
(2) Both of the following conditions apply:
   a. The driving means for the generator can be readily shut down, is rendered incapable of restarting, and is lockable in the OFF position in accordance with 110.25.
   b. The generator is not arranged to operate in parallel with another generator or other source of voltage.

Informational Note: See UL 2200-2012, *Standard for Safety of Stationary Engine Generator Assemblies.*

•

## 445.20 Ground-Fault Circuit-Interrupter Protection for Receptacles on 15-kW or Smaller Portable Generators

All 125-volt, single-phase, 15- and 20-ampere receptacle outlets that are a part of a 15-kW or smaller portable generator either shall have ground-fault circuit-interrupter protection for personnel integral to the generator or receptacle or shall not be available for use when the 125/250-volt locking-type receptacle is in use. If the generator does not have a 125/250-volt locking-type receptacle, this requirement shall not apply.

# ARTICLE 450 Transformers and Transformer Vaults (Including Secondary Ties)

### 450.1 Scope

This article covers the installation of all transformers.

*Exception No. 1: Current transformers.*

See 110.23 for the requirement on energized current transformers that are not in use.

*Exception No. 2: Dry-type transformers that constitute a component part of other apparatus and comply with the requirements for such apparatus.*

*Exception No. 3: Transformers that are an integral part of an X-ray, high-frequency, or electrostatic-coating apparatus.*

*Exception No. 4: Transformers used with Class 2 and Class 3 circuits that comply with Article 725.*

*Exception No. 5: Transformers for sign and outline lighting that comply with Article 600.*

*Exception No. 6: Transformers for electric-discharge lighting that comply with Article 410.*

*Exception No. 7: Transformers used for power-limited fire alarm circuits that comply with Part III of Article 760.*

*Exception No. 8: Transformers used for research, development, or testing, where effective arrangements are provided to safeguard persons from contacting energized parts.*

This article covers the installation of transformers dedicated to supplying power to a fire pump installation as modified by Article 695.

This article also covers the installation of transformers in hazardous (classified) locations as modified by Articles 501 through 504.

## I. General Provisions

### 450.2 Definition

For the purpose of this article, the following definition shall apply.

**Transformer.** An individual transformer, single- or polyphase, identified by a single nameplate, unless otherwise indicated in this article.

### 450.3 Overcurrent Protection

Overcurrent protection of transformers shall comply with 450.3(A), (B), or (C). As used in this section, the word *transformer* shall mean a transformer or polyphase bank of two or more single-phase transformers operating as a unit.

Informational Note No. 1: See 240.4, 240.21, 240.100, and 240.101 for overcurrent protection of conductors.

The requirements for overcurrent protection of transformer secondaries apply only to the protection of transformers, not to the protection of conductors. The sections in Article 240 referenced in the informational note apply only to the protection of conductors, not to the protection of transformers. The overcurrent protection required by Article 450 may also satisfy the requirements in Article 240 for conductor protection, and vice versa, but it is also possible that they do not.

The overcurrent protection required for transformers may not provide satisfactory protection for the primary and secondary conductors. Where polyphase transformers are involved, primary and secondary conductors are usually not properly protected. The primary overcurrent device provides short-circuit protection for the primary conductors and a degree of overload protection for the transformer, and secondary overcurrent devices prevent the transformer and secondary conductors from being overloaded.

A transformer is considered the point of supply, and the conductors it supplies must be protected in accordance with their ampacity. Section 240.4(F) permits the secondary circuit conductors from a transformer to be protected by overcurrent devices in the primary circuit conductors only in two special cases: a transformer with a 2-wire primary and a 2-wire secondary; and a 3-phase, delta-delta-connected transformer having a 3-wire, single-voltage secondary. Either case requires transformer primary protection in accordance with 450.3. Where the primary feeder to the transformer incorporates OCPD rated (or set) at a level not to exceed those prescribed herein, it is not necessary to duplicate them at the transformer.

Requirements for the overcurrent protection of transformer conductors are found in 240.4(F) and 240.21(B). Also, Article 240, Part VIII and Part IX contain overcurrent protection requirements for feeders and feeder taps associated with transformers.

Informational Note No. 2: Nonlinear loads can increase heat in a transformer without operating its overcurrent protective device.

The increased heating effects of nonlinear load currents must be taken into account when determining the load on a transformer. Methods for handling these heating effects include derating equipment, oversizing equipment, increasing insulation ratings, installing thermal protection systems, and using K-factor transformers. The optimum method for dealing with transformer overheating varies, depending on several technical and economic factors, and is best determined during the design phase of the electrical system.

**(A) Transformers Over 1000 Volts, Nominal.** Overcurrent protection shall be provided in accordance with Table 450.3(A).

Unlike the information contained in informational notes, which are explanatory in nature and not enforceable, table notes are part of the requirements of the table.

***TABLE 450.3(A)*** *Maximum Rating or Setting of Overcurrent Protection for Transformers Over 1000 Volts (as a Percentage of Transformer-Rated Current)*

| | | Primary Protection over 1000 Volts | | Secondary Protection (See Note 2.) | | |
|---|---|---|---|---|---|---|
| | | | | Over 1000 Volts | | 1000 Volts or Less |
| **Location Limitations** | **Transformer Rated Impedance** | **Circuit Breaker (See Note 4.)** | **Fuse Rating** | **Circuit Breaker (See Note 4.)** | **Fuse Rating** | **Circuit Breaker or Fuse Rating** |
| Any location | Not more than 6% | 600% (See Note 1.) | 300% (See Note 1.) | 300% (See Note 1.) | 250% (See Note 1.) | 125% (See Note 1.) |
| | More than 6% and not more than 10% | 400% (See Note 1.) | 300% (See Note 1.) | 250% (See Note 1.) | 225% (See Note 1.) | 125% (See Note 1.) |
| Supervised locations only (See Note 3.) | Any | 300% (See Note 1.) | 250% (See Note 1.) | Not required | Not required | Not required |
| | Not more than 6% | 600% | 300% | 300% (See Note 5.) | 250% (See Note 5.) | 250% (See Note 5.) |
| | More than 6% and not more than 10% | 400% | 300% | 250% (See Note 5.) | 225% (See Note 5.) | 250% (See Note 5.) |

Notes:

1. Where the required fuse rating or circuit breaker setting does not correspond to a standard rating or setting, a higher rating or setting that does not exceed the following shall be permitted:

a. The next higher standard rating or setting for fuses and circuit breakers 1000 volts and below, or

b. The next higher commercially available rating or setting for fuses and circuit breakers above 1000 volts.

2. Where secondary overcurrent protection is required, the secondary overcurrent device shall be permitted to consist of not more than six circuit breakers or six sets of fuses grouped in one location. Where multiple overcurrent devices are utilized, the total of all the device ratings shall not exceed the allowed value of a single overcurrent device. If both circuit breakers and fuses are used as the overcurrent device, the total of the device ratings shall not exceed that allowed for fuses.

3. A supervised location is a location where conditions of maintenance and supervision ensure that only qualified persons monitor and service the transformer installation.

4. Electronically actuated fuses that may be set to open at a specific current shall be set in accordance with settings for circuit breakers.

5. A transformer equipped with a coordinated thermal overload protection by the manufacturer shall be permitted to have separate secondary protection omitted.

For Note 1 of Table 450.3(A), concerning standard ratings of circuit breakers and fuses, see 240.6. For Note 2, overcurrent protection of the secondary of a transformer is allowed to consist of not more than six sets of fuses or six circuit breakers. For Note 3, equipment maintenance is performed by personnel who have received safety training and are familiar with proper operation of the equipment and aware of the hazards associated with it. See Article 100 for the definition of a *qualified person*.

For Note 4, an *electronically actuated fuse* responds to a signal from an electronic control rather than heat from a current. See the definition in Article 100, Part II.

Exhibits 450.1 and 450.2 illustrate the conditions given in Note 2, which also appears in Table 450.3(B) for transformers rated 1000 volts and less.

The ratings or settings obtained from Table 450.3(A) are based on the type of protective device (fuse, electronic fuse, or circuit breaker), transformer-rated current and impedance, and primary and secondary voltages. The maximum ratings or settings of an OCPD for transformers rated over 1000 volts are separated into two broad categories: *any location* (or unsupervised) and *supervised locations only*.

The first category is not limited by location and is referred to as *any location*. The maximum ratings or settings for overcurrent devices permitted are applicable to all unsupervised locations. An *any location* transformer installation must be provided with overcurrent protection in both the primary and secondary circuit. See Exhibit 450.3 for an example of an installation using circuit breakers on the primary and the secondary for an over 1000-volt transformer with 6-percent impedance.

The second category for over 1000-volt transformers is *supervised locations only*. The maximum ratings or settings for overcurrent devices permitted are strictly limited to the supervised location conditions explained in Note 3. The installation shown in

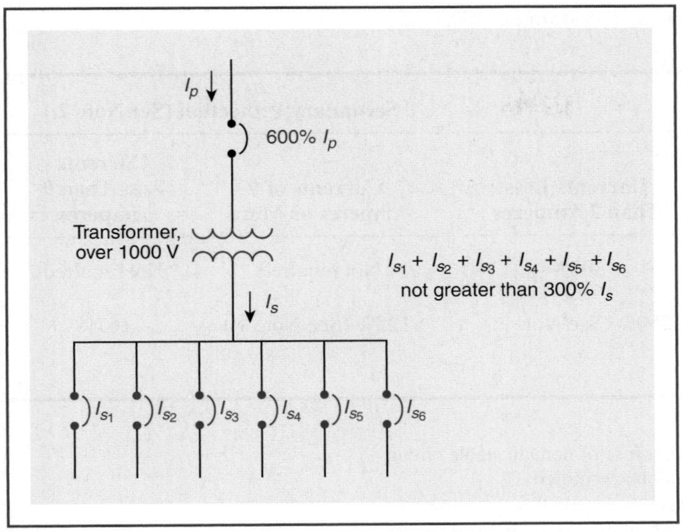


***EXHIBIT 450.1*** *A transformer rated over 1000 volts with a secondary rated over 1000 volts, with secondary protection consisting of six circuit breakers. The sum of the ratings of the circuit breakers is not permitted to exceed 300 percent of the rated secondary current.*

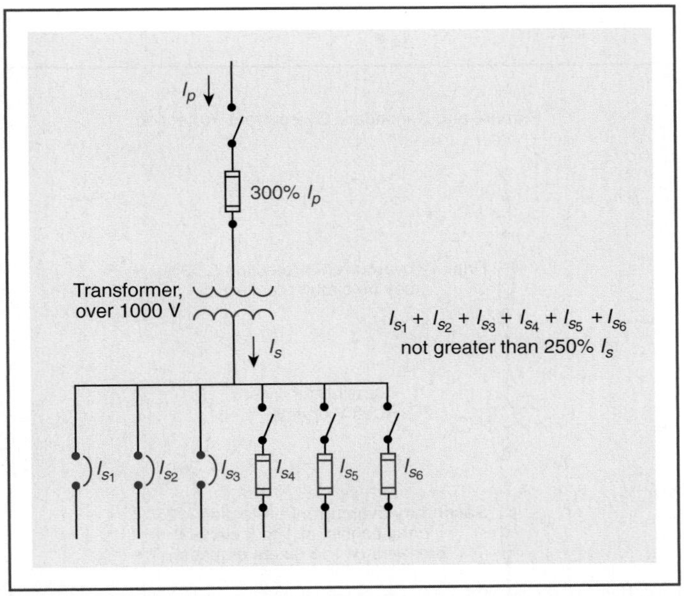


***EXHIBIT 450.2*** *A transformer rated over 1000 volts with a secondary rated over 1000 volts, with secondary protection consisting of fuses and circuit breakers. The sum of the ratings of all the overcurrent devices is not permitted to exceed the rating permitted for fuses.*

Exhibit 450.3 fulfills the requirements of both *any location* and *supervised locations only*.

See the commentary following 450.3 regarding the protection of transformer primary and secondary conductors.

**(B) Transformers 1000 Volts, Nominal, or Less.** Overcurrent protection shall be provided in accordance with Table 450.3(B).

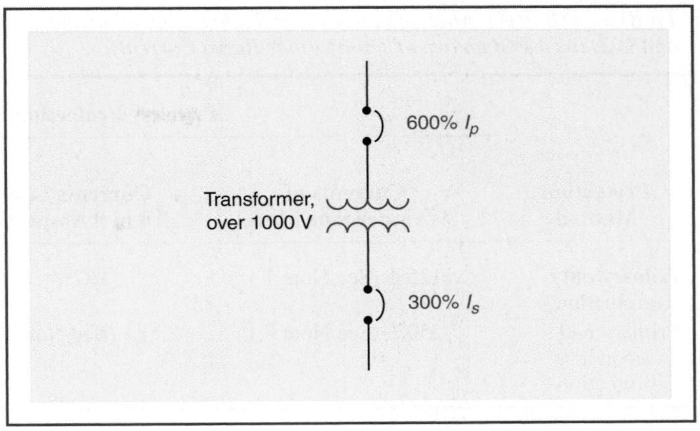


***EXHIBIT 450.3*** *A transformer with 6-percent impedance and rated over 1000 volts using circuit-breaker protection for both the primary and the secondary. Both the primary and the secondary voltages are over 1000 volts.*

The ratings or settings of the OCPD obtained from Table 450.3(B) are based on the transformer-rated current and whether secondary protection is provided. According to Table 450.3(B), the maximum ratings or settings of OCPDs for transformers rated 1000 volts and less are separated into two categories: *primary only protection* and *primary and secondary protection*.

Transformers must be protected by either of two methods. Method 1 requires primary protection only. Method 2 requires secondary side overcurrent protection at not more than 125 or 167 percent (depending on secondary current rating), provided the primary side overcurrent protection is not more than 250 percent of the primary side current rating.

An example of *primary only protection* is shown in Exhibit 450.4. An example of *primary and secondary protection* is shown in Exhibit 450.5. For overcurrent protection of motor control circuit transformers, see 430.72(C).

See the commentary following 450.3 regarding the protection of transformer primary and secondary conductors.

*Exception: Where the transformer is installed as a motor control circuit transformer in accordance with 430.72(C)(1) through (C)(5).*

**(C) Voltage (Potential) Transformers.** Voltage (potential) transformers installed indoors or enclosed shall be protected with primary fuses.

Informational Note: For protection of instrument circuits including voltage transformers, see 408.52.

## 450.4 Autotransformers 1000 Volts, Nominal, or Less

**(A) Overcurrent Protection.** Each autotransformer 1000 volts, nominal, or less shall be protected by an individual overcurrent device installed in series with each ungrounded input conductor. Such overcurrent device shall be rated or set at not more than 125 percent of the rated full-load input current of the autotransformer. Where this calculation does not correspond to a standard

***TABLE 450.3(B)*** *Maximum Rating or Setting of Overcurrent Protection for Transformers 1000 Volts and Less (as a Percentage of Transformer-Rated Current)*

| | Primary Protection | | | Secondary Protection (See Note 2.) | |
|---|---|---|---|---|---|
| **Protection Method** | **Currents of 9 Amperes or More** | **Currents Less Than 9 Amperes** | **Currents Less Than 2 Amperes** | **Currents of 9 Amperes or More** | **Currents Less Than 9 Amperes** |
| Primary only protection | 125% (See Note 1.) | 167% | 300% | Not required | Not required |
| Primary and secondary protection | 250% (See Note 3.) | 250% (See Note 3.) | 250% (See Note 3.) | 125% (See Note 1.) | 167% |

Notes:

1. Where 125 percent of this current does not correspond to a standard rating of a fuse or nonadjustable circuit breaker, a higher rating that does not exceed the next higher standard rating shall be permitted.
2. Where secondary overcurrent protection is required, the secondary overcurrent device shall be permitted to consist of not more than six circuit breakers or six sets of fuses grouped in one location. Where multiple overcurrent devices are utilized, the total of all the device ratings shall not exceed the allowed value of a single overcurrent device.
3. A transformer equipped with coordinated thermal overload protection by the manufacturer and arranged to interrupt the primary current shall be permitted to have primary overcurrent protection rated or set at a current value that is not more than six times the rated current of the transformer for transformers having not more than 6 percent impedance and not more than four times the rated current of the transformer for transformers having more than 6 percent but not more than 10 percent impedance.

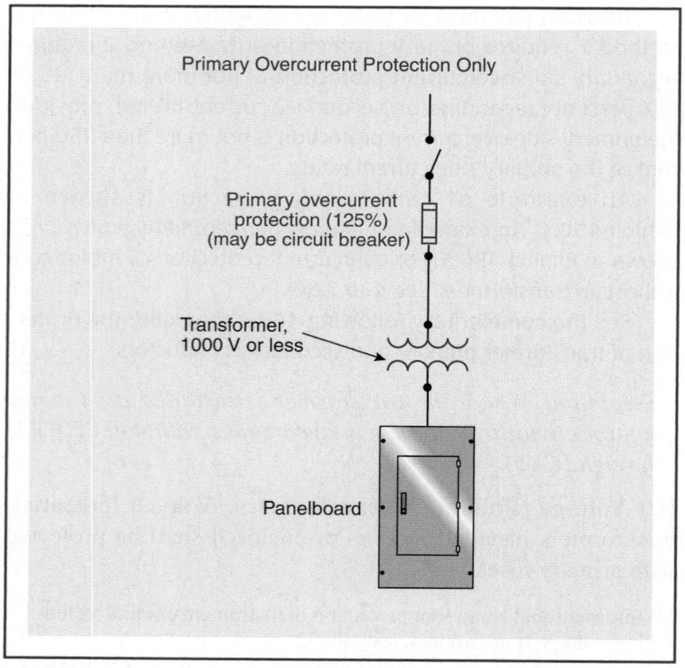


***EXHIBIT 450.4*** *A transformer (with currents of 9 amperes or more) rated 1000 volts or less with only primary overcurrent protection.*

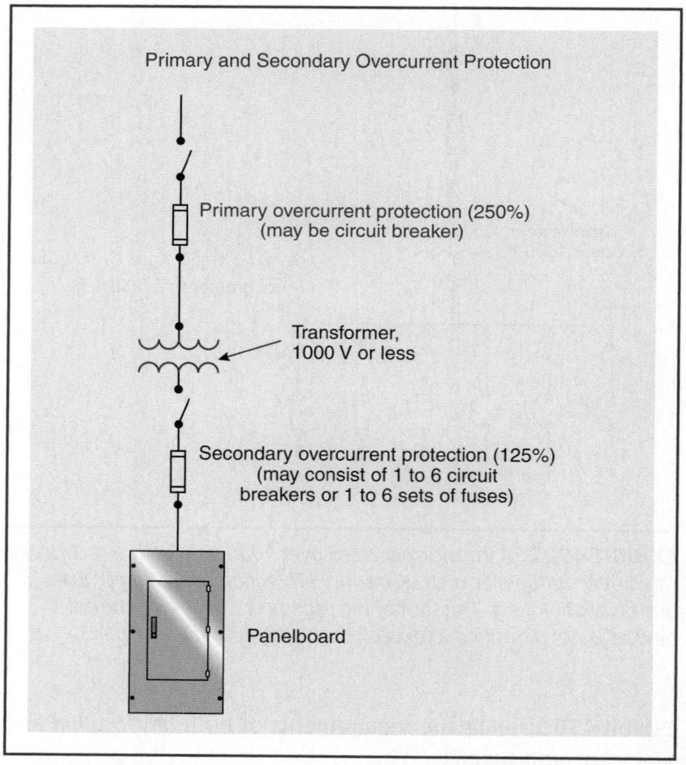


***EXHIBIT 450.5*** *A transformer (9 amperes or more) rated 1000 volts or less and protected by a combination of primary and secondary overcurrent protection.*

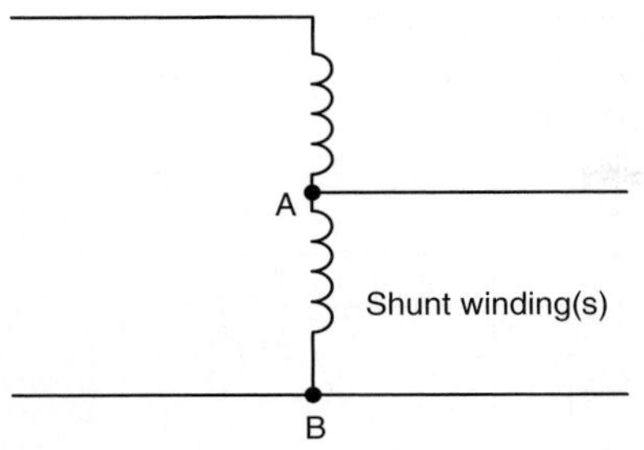


***FIGURE 450.4(A)*** *Autotransformer.*

rating of a fuse or nonadjustable circuit breaker and the rated input current is 9 amperes or more, the next higher standard rating described in 240.6 shall be permitted. An overcurrent device shall not be installed in series with the shunt winding (the winding common to both the input and the output circuits) of the autotransformer between Points A and B as shown in Figure 450.4(A).

Because of the voltage feedback problem that may occur, an overcurrent device is not permitted between points A and B in Figure 450.4.

Exhibit 450.6 provides an example of overcurrent protection for an autotransformer. It shows a 2-winding, single-phase transformer connected to boost a 208-volt supply to 240 volts. The autotransformer is provided with a 2-pole disconnect switch with both overcurrent devices (OC-1a and OC-1b) located on the supply side of the autotransformer. If an overcurrent device were located in series with the shunt winding and this overcurrent device opened, the full 208-volt supply voltage would be applied across the 32-volt secondary winding. Under those conditions, a higher-than-normal voltage would appear across the primary winding. If the load impedance were very low, this voltage could approach 208/32 × 208 = 1352 V.

*Exception: Where the rated input current of the autotransformer is less than 9 amperes, an overcurrent device rated or set at not more than 167 percent of the input current shall be permitted.*

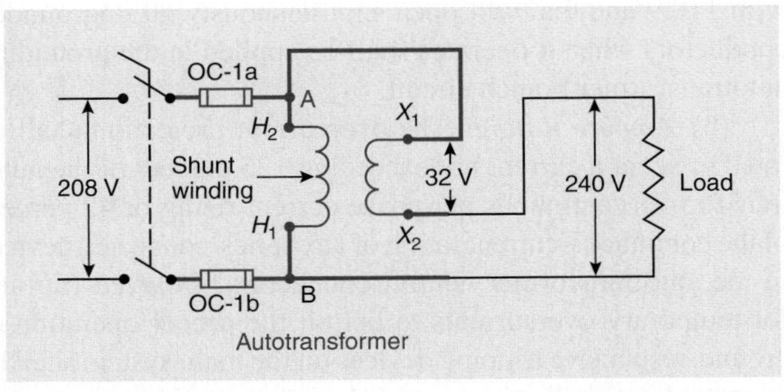


***EXHIBIT 450.6*** *A disconnect switch with overcurrent devices properly connected to protect an autotransformer and located to meet the requirements of 450.4(A), last sentence.*

**(B) Transformer Field-Connected as an Autotransformer.** A transformer field-connected as an autotransformer shall be identified for use at elevated voltage.

This requirement is necessary because of the dielectric voltage withstand test requirements applied to transformers. The test is conducted at 2500 volts for windings rated 250 volts or less and at 4000 volts for higher-rated windings. A transformer intended for buck or boost operation would require that the test for the low-voltage winding be based on the sum of the primary and secondary voltage ratings.

Informational Note: For information on permitted uses of autotransformers, see 210.9 and 215.11.

## 450.5 Grounding Autotransformers

Grounding autotransformers covered in this section are zigzag or T-connected transformers connected to 3-phase, 3-wire ungrounded systems for the purpose of creating a 3-phase, 4-wire distribution system or providing a neutral point for grounding purposes. Such transformers shall have a continuous per-phase current rating and a continuous neutral current rating. Zigzag-connected transformers shall not be installed on the load side of any system grounding connection, including those made in accordance with 250.24(B), 250.30(A)(1), or 250.32(B), Exception.

The installation of grounding autotransformers on the load side of a supply system grounding connection is prohibited. This restriction applies to services, to separately derived systems, and to feeders and branch circuits that supply separate buildings or structures. Where a zigzag transformer is used to create a neutral reference point on a circuit that is supplied from a grounded system, the current from a line-to-ground fault is shared through the supply system transformer and the zigzag transformer.

Where the rating of the circuit in which the line-to-ground fault occurs exceeds the rating of the circuit in which the zigzag transformer is used, the shared ground-fault current through the zigzag transformer has the potential to cause serious damage to the transformer. For instance, a zigzag transformer is installed on an existing 50-ampere, 3-phase, 3-wire branch circuit to create a neutral. The branch circuit is derived from a grounded wye service, from which large capacity, 800-ampere and 1000-ampere feeders are also supplied. A line-to-ground fault in one of these feeder circuits can result in serious damage to the zigzag transformer as a result of its sharing the fault current with the system supply transformer.

Informational Note: The phase current in a grounding autotransformer is one-third the neutral current.

*Exception: An auto transformer with a wye configuration on its line side and a zigzag configuration on its load side that does not permit neutral or ground-fault current to return over the line connection shall be permitted on the load side of a system grounding connection. This exception shall not apply to a connection made from a high-resistance grounded system applied in accordance with 250.36.*

**(A) Three-Phase, 4-Wire System.** A grounding autotransformer used to create a 3-phase, 4-wire distribution system from a 3-phase, 3-wire ungrounded system shall conform to 450.5(A)(1) through (A)(4).

**(1) Connections.** The transformer shall be directly connected to the ungrounded phase conductors and shall not be switched or provided with overcurrent protection that is independent of the main switch and common-trip overcurrent protection for the 3-phase, 4-wire system.

**(2) Overcurrent Protection.** An overcurrent sensing device shall be provided that will cause the main switch or common-trip overcurrent protection referred to in 450.5(A)(1) to open if the load on the autotransformer reaches or exceeds 125 percent of its continuous current per-phase or neutral rating. Delayed tripping for temporary overcurrents sensed at the autotransformer overcurrent device shall be permitted for the purpose of allowing proper operation of branch or feeder protective devices on the 4-wire system.

**(3) Transformer Fault Sensing.** A fault-sensing system that causes the opening of a main switch or common-trip overcurrent device for the 3-phase, 4-wire system shall be provided to guard against single-phasing or internal faults.

Informational Note: This can be accomplished by the use of two subtractive-connected donut-type current transformers installed to sense and signal when an unbalance occurs in the line current to the autotransformer of 50 percent or more of rated current.

**(4) Rating.** The autotransformer shall have a continuous neutral-current rating that is not less than the maximum possible neutral unbalanced load current of the 4-wire system.

Exhibit 450.7 illustrates the proper method of protecting a grounding autotransformer used to provide a neutral for a 3-phase system where necessary to supply a group of single-phase, line-to-neutral loads. Separate overcurrent protection is not provided for the autotransformer because there will be no control of the system line-to-neutral voltages if the autotransformer becomes disconnected. Consequently, simultaneous interruption of the power supply to all the line-to-neutral loads is necessary whenever the grounding autotransformer is switched off.

CT-1 is connected to an overload relay responsive to excess neutral current being supplied. See 450.5(A)(2). CT-2 and CT-3 are connected to differential-type fault-current sensing relays responsive to an unbalance of neutral current among the three phases of the grounding autotransformer (indicating an internal fault). All three relays are to be arranged to trip the circuit breaker located upstream of both the autotransformer and the line-to-neutral connected loads.

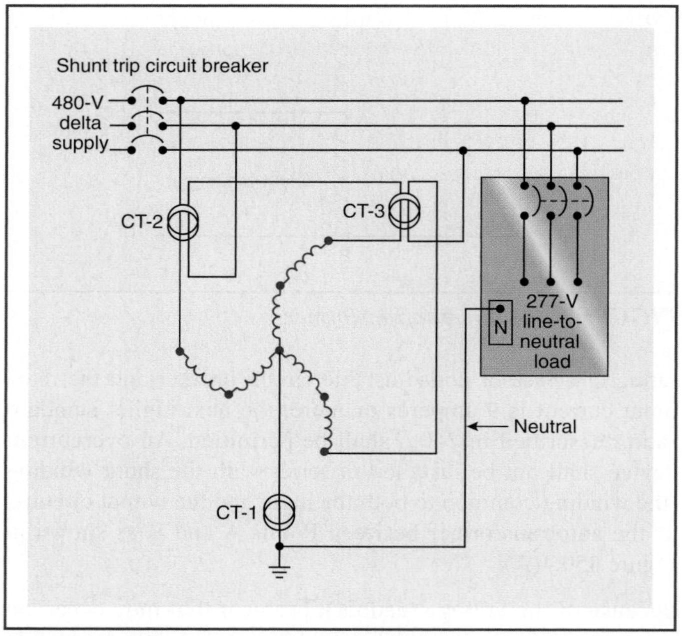


***EXHIBIT 450.7*** *A zigzag autotransformer used to establish a neutral connection for a 480Y/277-V, 3-phase ungrounded system to supply single-phase line-to-neutral loads.*

**(B) Ground Reference for Fault Protection Devices.** A grounding autotransformer used to make available a specified magnitude of ground-fault current for operation of a ground-responsive protective device on a 3-phase, 3-wire ungrounded system shall conform to 450.5(B)(1) and (B)(2).

**(1) Rating.** The autotransformer shall have a continuous neutral-current rating not less than the specified ground-fault current.

**(2) Overcurrent Protection.** Overcurrent protection shall comply with (a) and (b).

(a) *Operation and Interrupting Rating.* An overcurrent protective device having an interrupting rating in compliance with 110.9 and that will open simultaneously all ungrounded conductors when it operates shall be applied in the grounding autotransformer branch circuit.

(b) *Ampere Rating.* The overcurrent protection shall be rated or set at a current not exceeding 125 percent of the autotransformer continuous per-phase current rating or 42 percent of the continuous-current rating of any series-connected devices in the autotransformer neutral connection. Delayed tripping for temporary overcurrents to permit the proper operation of ground-responsive tripping devices on the main system shall be permitted but shall not exceed values that would be more than the short-time current rating of the grounding autotransformer or any series connected devices in the neutral connection thereto.

*Exception: For high-impedance grounded systems covered in 250.36, where the maximum ground-fault current is designed to be not more than 10 amperes, and where the grounding autotransformer and the grounding impedance are rated for continuous duty, an overcurrent device rated not more than 20 amperes that will simultaneously open all ungrounded conductors shall be permitted to be installed on the line side of the grounding autotransformer.*

In high-impedance grounded systems, the currents are low enough that finding overcurrent devices rated at 125 percent of a typical 5-ampere system was not practical.

Exhibit 450.8 illustrates the proper method of protecting a grounding autotransformer where it is used as a ground reference for fault protective devices. The overcurrent protective device is to have a rating (or setting) not in excess of 125 percent of the rated phase current of the autotransformer (42 percent of the neutral current rating) and not more than 42 percent of the continuous current rating of the neutral grounding resistor or other current-carrying device in the neutral connection, as specified in 450.5(B)(2).

**(C) Ground Reference for Damping Transitory Overvoltages.** A grounding autotransformer used to limit transitory overvoltages shall be of suitable rating and connected in accordance with 450.5(A)(1).

For installations involving a high-resistance grounding package, the functional performance of the installation parallels that described in 450.5(B), differing only in that the magnitude of available ground-fault current would likely be a lower value. It would be appropriate to employ the connections displayed in Exhibit 450.8 and to conform to the overcurrent protection requirements prescribed in 450.5(B)(2).

With any of the grounding autotransformer applications covered by 450.5(A), (B), or (C), the use of a ganged 3-pole switching interrupter for connecting and disconnecting the autotransformer accomplishes simultaneous connection (and disconnection) of the three line terminals. If, at any time, one or two of the line connections to the autotransformer were to open, which could occur if the protective devices were single pole, the grounding autotransformer would cease to function in the desired fashion and would act as a high-inductive-reactance connection between the electrical system and ground. The latter connection is prone to create high-value transitory overvoltages, line-to-ground.

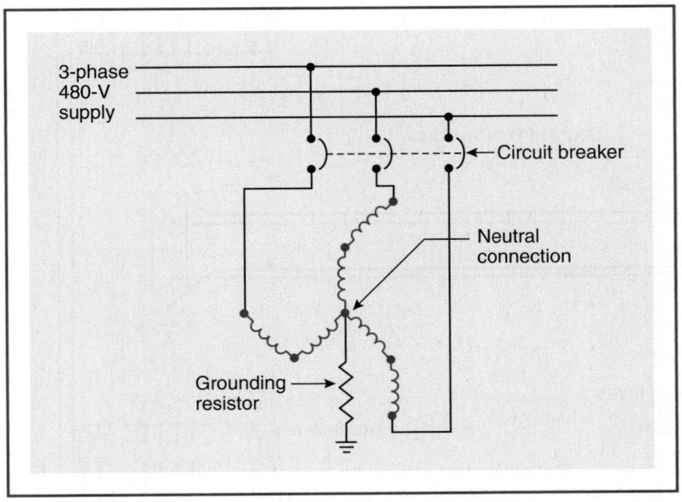


***EXHIBIT 450.8*** *A zigzag autotransformer used to establish a reference ground-fault current for fault-protective-device operation or for damping transitory overvoltage surges.*

## 450.6 Secondary Ties

As used in this article, a secondary tie is a circuit operating at 1000 volts, nominal, or less between phases that connects two power sources or power supply points, such as the secondaries of two transformers. The tie shall be permitted to consist of one or more conductors per phase or neutral. Conductors connecting the secondaries of transformers in accordance with 450.7 shall not be considered secondary ties.

As used in this section, the word *transformer* means a transformer or a bank of transformers operating as a unit.

**(A) Tie Circuits.** Tie circuits shall be provided with overcurrent protection at each end as required in Parts I, II, and VIII of Article 240.

Under the conditions described in 450.6(A)(1) and 450.6(A)(2), the overcurrent protection shall be permitted to be in accordance with 450.6(A)(3).

**(1) Loads at Transformer Supply Points Only.** Where all loads are connected at the transformer supply points at each end of the tie and overcurrent protection is not provided in accordance with Parts I, II, and VIII of Article 240, the rated ampacity of the tie shall not be less than 67 percent of the rated secondary current of the highest rated transformer supplying the secondary tie system.

**(2) Loads Connected Between Transformer Supply Points.** Where load is connected to the tie at any point between transformer supply points and overcurrent protection is not provided in accordance with Parts I, II, and VIII of Article 240, the rated ampacity of the tie shall not be less than 100 percent of the rated secondary current of the highest rated transformer supplying the secondary tie system.

*Exception: Tie circuits comprised of multiple conductors per phase shall be permitted to be sized and protected in accordance with 450.6(A)(4).*

**(3) Tie Circuit Protection.** Under the conditions described in 450.6(A)(1) and (A)(2), both supply ends of each ungrounded tie conductor shall be equipped with a protective device that opens at a predetermined temperature of the tie conductor under short-circuit conditions. This protection shall consist of one of the following: (1) a fusible link cable connector, terminal, or lug, commonly known as a limiter, each being of a size corresponding with that of the conductor and of construction and characteristics according to the operating voltage and the type of insulation on the tie conductors or

(2) automatic circuit breakers actuated by devices having comparable time–current characteristics.

**(4) Interconnection of Phase Conductors Between Transformer Supply Points.** Where the tie consists of more than one conductor per phase or neutral, the conductors of each phase or neutral shall comply with one of the following provisions.

(a) *Interconnected.* The conductors shall be interconnected in order to establish a load supply point, and the protective device specified in 450.6(A)(3) shall be provided in each ungrounded tie conductor at this point on both sides of the interconnection. The means of interconnection shall have an ampacity not less than the load to be served.

(b) *Not Interconnected.* The loads shall be connected to one or more individual conductors of a paralleled conductor tie without interconnecting the conductors of each phase or neutral and without the protection specified in 450.6(A)(3) at load connection points. Where this is done, the tie conductors of each phase or neutral shall have a combined capacity ampacity of not less than 133 percent of the rated secondary current of the highest rated transformer supplying the secondary tie system, the total load of such taps shall not exceed the rated secondary current of the highest rated transformer, and the loads shall be equally divided on each phase and on the individual conductors of each phase as far as practicable.

**(5) Tie Circuit Control.** Where the operating voltage exceeds 150 volts to ground, secondary ties provided with limiters shall have a switch at each end that, when open, de-energizes the associated tie conductors and limiters. The current rating of the switch shall not be less than the rated current ampacity of the conductors connected to the switch. It shall be capable of interrupting its rated current, and it shall be constructed so that it will not open under the magnetic forces resulting from short-circuit current.

**(B) Overcurrent Protection for Secondary Connections.** Where secondary ties are used, an overcurrent device rated or set at not more than 250 percent of the rated secondary current of the transformers shall be provided in the secondary connections of each transformer supplying the tie system. In addition, an automatic circuit breaker actuated by a reverse-current relay set to open the circuit at not more than the rated secondary current of the transformer shall be provided in the secondary connection of each transformer.

The requirements of 450.6 apply specifically to network systems for power distribution commonly employed where the load density is high and service reliability is important. This type of distribution system introduces a variety of problems not encountered in the more common radial-type distribution system and must be designed by experienced electrical engineers. Exhibit 450.9 illustrates a typical 3-phase network system for an industrial plant fed by two primary feeders, preferably from separate substations, energized at any standard voltage up to 34,500 volts. Each of the transformers is supplied by the two primary feeders via a double-throw switch at the transformer so that the transformer may be supplied by either feeder.

Each of the network transformers is rated in the range of 300 to 1000 kVA and is required to be protected as illustrated in Exhibit 450.10. The primary and secondary protection is in accordance with 450.3, but an additional protective device must be provided

***EXHIBIT 450.9*** *A typical 3-phase network system for an industrial plant fed by two primary feeders.*

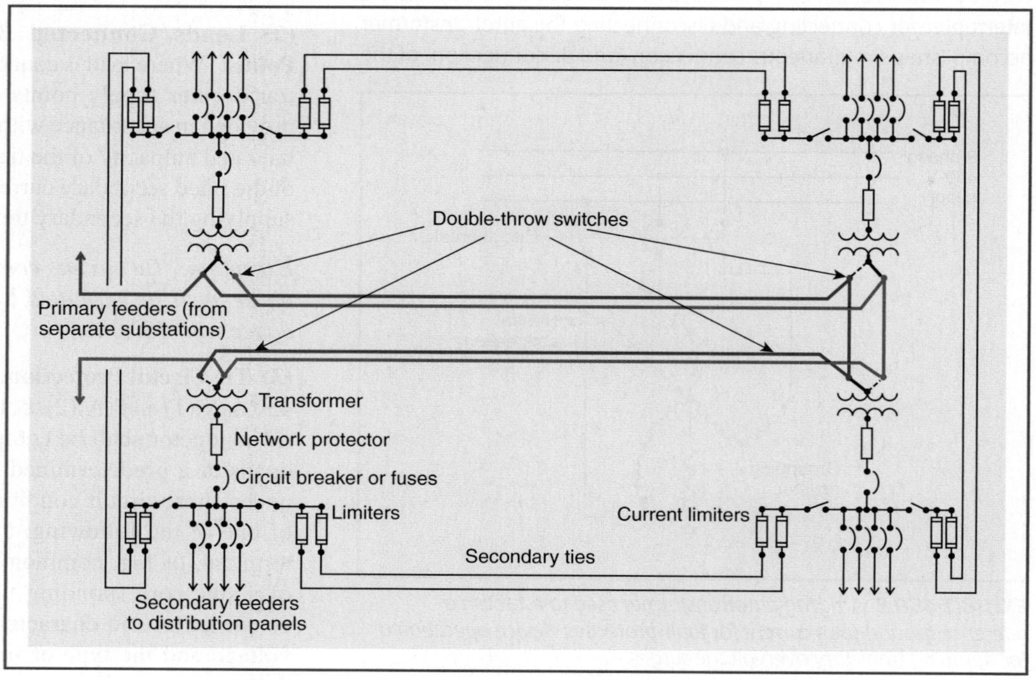

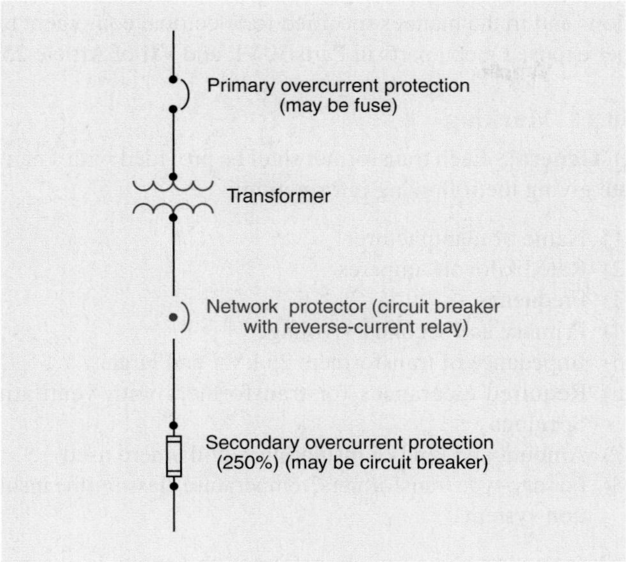


***EXHIBIT 450.10*** *Primary and secondary overcurrent protection for a transformer in a network system, showing a network protector (an automatic circuit breaker actuated by a reverse-current relay).*

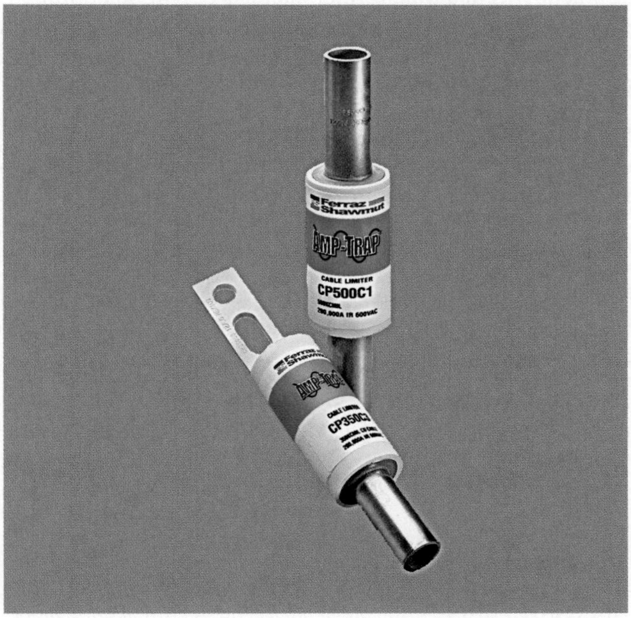


***EXHIBIT 450.11*** *A current limiter (a special type of high-interrupting-capacity fuse). (Courtesy of Ferraz Shawmut)*

on the secondary side. This protective device, known as a *network protector,* consists of a circuit breaker and a reverse-current relay. The network protector operates on reverse current to prevent power from being fed back into the transformer through the secondary ties if a fault were to occur in the transformer or a primary feeder. The reverse-current relay is set to trip the circuit breaker at a current value not more than the rated secondary current of the transformer. The relay is not designed to trip the circuit breaker in the event of an overload on the secondary of the transformer.

The secondary ties shown in Exhibit 450.9 must be protected at each end with an overcurrent device, in accordance with 450.6(A)(3). The overcurrent device most commonly provided is a special type of fuse known as a *current limiter,* shown in Exhibit 450.11. This high-interrupting-capacity device provides short-circuit protection only for the secondary ties by opening safely before temperatures damaging to the cable insulation are reached. See 240.2 for the definition of *current-limiting overcurrent protective device* and its accompanying commentary. The secondary ties form a closed loop equipped with switching devices so that any part of the loop may be isolated when repairs are needed or a current limiter must be replaced.

**(C) Grounding.** Where the secondary tie system is grounded, each transformer secondary supplying the tie system shall be grounded in accordance with the requirements of 250.30 for separately derived systems.

## 450.7 Parallel Operation

Transformers shall be permitted to be operated in parallel and switched as a unit, provided the overcurrent protection for each transformer meets the requirements of 450.3(A) for primary and secondary protective devices over 1000 volts, or 450.3(B) for primary and secondary protective devices 1000 volts or less.

Parallel operation of transformers that are not switched as a unit can present dangerous backfeed situations for workers performing electrical maintenance. Appropriate lockout/tagout procedures must be implemented during maintenance of electrical equipment operated or connected in parallel. See *NFPA 70E, Standard for Electrical Safety in the Workplace,* for safety-related work practices and appropriate lockout/tagout procedures.

## 450.8 Guarding

Transformers shall be guarded as specified in 450.8(A) through (D).

**(A) Mechanical Protection.** Appropriate provisions shall be made to minimize the possibility of damage to transformers from external causes where the transformers are exposed to physical damage.

One method of providing mechanical protection is to strategically place bollards around the transformer. This practice provides a degree of protection from vehicles.

**(B) Case or Enclosure.** Dry-type transformers shall be provided with a noncombustible moisture-resistant case or enclosure that provides protection against the accidental insertion of foreign objects.

**(C) Exposed Energized Parts.** Switches or other equipment operating at 1000 volts, nominal, or less and serving only equipment within a transformer enclosure shall be permitted to be installed in the transformer enclosure if accessible to qualified persons only. All energized parts shall be guarded in accordance with 110.27 and 110.34.

**(D) Voltage Warning.** The operating voltage of exposed live parts of transformer installations shall be indicated by signs or visible markings on the equipment or structures.

## 450.9 Ventilation

The ventilation shall dispose of the transformer full-load heat losses without creating a temperature rise that is in excess of the transformer rating.

Informational Note No. 1: See ANSI/IEEE C57.12.00-1993, *General Requirements for Liquid-Immersed Distribution, Power, and Regulating Transformers*, and ANSI/IEEE C57.12.01-1989, *General Requirements for Dry-Type Distribution and Power Transformers*.

Informational Note No. 2: Additional losses may occur in some transformers where nonsinusoidal currents are present, resulting in increased heat in the transformer above its rating. See ANSI/IEEE C57.110-1993, *Recommended Practice for Establishing Transformer Capability When Supplying Nonsinusoidal Load Currents*, where transformers are utilized with nonlinear loads.

Transformers with ventilating openings shall be installed so that the ventilating openings are not blocked by walls or other obstructions. The required clearances shall be clearly marked on the transformer.

Informational Note No. 2 warns of increased heating of transformers. See the commentary following 450.3, Informational Note No. 2, and the commentary following 310.15(B)(4) for additional information concerning nonlinear loads.

## 450.10 Grounding

**(A) Dry-Type Transformer Enclosures.** Where separate equipment grounding conductors and supply-side bonding jumpers are installed, a terminal bar for all grounding and bonding conductor connections shall be secured inside the transformer enclosure. The terminal bar shall be bonded to the enclosure in accordance with 250.12 and shall not be installed on or over any vented portion of the enclosure.

*Exception: Where a dry-type transformer is equipped with wire-type connections (leads), the grounding and bonding connections shall be permitted to be connected together using any of the methods in 250.8 and shall be bonded to the enclosure if of metal.*

An enclosure is typically not evaluated as a grounding and bonding device. The required busbar for EGCs and bonding jumpers prohibits the practice of using the transformer metal enclosure as a connection point for these conductors.

**(B) Other Metal Parts.** Where grounded, exposed non–current-carrying metal parts of transformer installations, including fences, guards, and so forth, shall be grounded and bonded under the conditions and in the manner specified for electrical equipment and other exposed metal parts in Parts V, VI, and VII of Article 250.

## 450.11 Marking

**(A) General.** Each transformer shall be provided with a nameplate giving the following information:

(1) Name of manufacturer
(2) Rated kilovolt-amperes
(3) Frequency
(4) Primary and secondary voltage
(5) Impedance of transformers 25 kVA and larger
(6) Required clearances for transformers with ventilating openings
(7) Amount and kind of insulating liquid where used
(8) For dry-type transformers, temperature class for the insulation system

The information given on a transformer nameplate is necessary to determine whether special precautions must be used pertaining to clearances for ventilation, overcurrent protection, or liquid confinement.

**(B) Source Marking.** A transformer shall be permitted to be supplied at the marked secondary voltage, provided that the installation is in accordance with the manufacturer's instructions.

Not all transformers are designed to be back-fed. Transformers are typically designed and evaluated with the supply on the primary side and the load on the secondary side. Back-feeding is only permitted when the manufacturer has indicated so in the instructions.

## 450.12 Terminal Wiring Space

The minimum wire-bending space at fixed, 1000-volt and below terminals of transformer line and load connections shall be as required in 312.6. Wiring space for pigtail connections shall conform to Table 314.16(B).

## 450.13 Accessibility

All transformers and transformer vaults shall be readily accessible to qualified personnel for inspection and maintenance or shall meet the requirements of 450.13(A) or 450.13(B).

Transformers are not accessible if wiring methods or other equipment obstruct the access of a worker or prevent removal of the covers for inspection or maintenance. Practical clearance considerations required for removal and replacement of the transformer are also important.

**(A) Open Installations.** Dry-type transformers 1000 volts, nominal, or less, located in the open on walls, columns, or structures, shall not be required to be readily accessible.

**(B) Hollow Space Installations.** Dry-type transformers 1000 volts, nominal, or less and not exceeding 50 kVA shall be permitted in hollow spaces of buildings not permanently closed in by structure, provided they meet the ventilation requirements of 450.9 and separation from combustible materials requirements of 450.21(A). Transformers so installed shall not be required to be readily accessible.

Transformers are permitted by 300.22(C)(3) to be installed in hollow spaces where the space is used for environmental air, provided the transformer is in a metal enclosure (ventilated or nonventilated) and the transformer is suitable for the ambient air temperature within the hollow space.

### 450.14 Disconnecting Means

Transformers, other than Class 2 or Class 3 transformers, shall have a disconnecting means located either in sight of the transformer or in a remote location. Where located in a remote location, the disconnecting means shall be lockable in accordance with 110.25, and its location shall be field marked on the transformer.

The requirement for a disconnecting means is especially important in installations utilizing the requirements of 240.21(B)(3) where several transformers in different locations may all be tapped from one feeder, and it may be impractical to de-energize the feeder to work on one of the transformers. The disconnect is required to be located within sight from the transformer but may be in a remote location if it is lockable. The location of any remote disconnect is required to be marked on the transformer.

## II. Specific Provisions Applicable to Different Types of Transformers

### 450.21 Dry-Type Transformers Installed Indoors

**(A) Not over 112½ kVA.** Dry-type transformers installed indoors and rated 112½ kVA or less shall have a separation of at least 300 mm (12 in.) from combustible material unless separated from the combustible material by a fire-resistant, heat-insulated barrier.

*Exception: This rule shall not apply to transformers rated for 1000 volts, nominal, or less that are completely enclosed, except for ventilating openings.*

**(B) Over 112½kVA.** Individual dry-type transformers of more than 112½ kVA rating shall be installed in a transformer room of fire-resistant construction. Unless specified otherwise in this article, the term *fire resistant* means a construction having a minimum fire rating of 1 hour.

*Exception No. 1: Transformers with Class 155 or higher insulation systems and separated from combustible material by a fire-resistant, heat-insulating barrier or by not less than 1.83 m (6 ft) horizontally and 3.7 m (12 ft) vertically.*

*Exception No. 2: Transformers with Class 155 or higher insulation systems and completely enclosed except for ventilating openings.*

Dry-type transformers with a Class 155 or higher insulation system rating are not required to be installed in transformer rooms or vaults if space separation or a fire-resistant heat-insulating barrier is provided. Although these units are designed for higher operating temperatures, the need for a transformer vault is mitigated by the fire-resistant characteristics of high-temperature insulations.

Further information on specific transformer class insulation systems may be found in UL 1561, *Dry-Type General Purpose and Power Transformers*.

Informational Note: See ANSI/ASTM E119-2012a, *Method for Fire Tests of Building Construction and Materials.*

**(C) Over 35,000 Volts.** Dry-type transformers rated over 35,000 volts shall be installed in a vault complying with Part III of this article.

Dry-type transformers depend on the surrounding air for adequate ventilation and must comply with 450.9. Where rated 112½ kVA or less, dry-type transformers are not required to be installed in a fire-resistant transformer room. For this reason, dry-type transformers or gas-filled or less-flammable liquid-insulated transformers (see 450.23), with a primary voltage of not more than 35,000 volts, are commonly used indoors.

Exhibit 450.12 shows a typical dry-type power transformer rated at 1000 kVA, 13,800 V to 480 V, 3-phase, 60 Hz. This transformer has

***EXHIBIT 450.12*** *A dry-type transformer with a core and coil design rated at 1000 kVA, 13,800 V to 480 V, 3-phase, 60 Hz. Note cooling fans beneath each winding. (Courtesy of Schneider Electric)*

a high-voltage and low-voltage flange for connection to switchgear and a high-voltage, 2-position (double-throw), 3-pole-load air-break switch that may be attached to the case and arranged as a selector switch for connection of the transformer primary to either of two feeder sources.

Dry-type transformers rated 112½ kVA or less require 12 inches of separation from combustible material or separation by fire-resistant barriers. Transformers rated less than 1000 volts and completely enclosed, except for ventilating openings, are exempt from this requirement unless the manufacturer's installation instructions specify clearance distances. Transformers rated over 112½ kVA must be located in fire-resistant transformer rooms or vaults unless the transformers have Class 155 or higher insulation ratings.

## 450.22 Dry-Type Transformers Installed Outdoors

Dry-type transformers installed outdoors shall have a weatherproof enclosure.

Transformers exceeding 112½ kVA shall not be located within 300 mm (12 in.) of combustible materials of buildings unless the transformer has Class 155 insulation systems or higher and is completely enclosed except for ventilating openings.

## 450.23 Less-Flammable Liquid-Insulated Transformers

Transformers insulated with listed less-flammable liquids that have a fire point of not less than 300°C shall be permitted to be installed in accordance with 450.23(A) or 450.23(B).

**(A) Indoor Installations.** Indoor installations shall be permitted in accordance with one of the following:

(1) In Type I or Type II buildings, in areas where all of the following requirements are met:

a. The transformer is rated 35,000 volts or less.
b. No combustible materials are stored.
c. A liquid confinement area is provided.
d. The installation complies with all restrictions provided for in the listing of the liquid.

(2) With an automatic fire extinguishing system and a liquid confinement area, provided the transformer is rated 35,000 volts or less

(3) In accordance with 450.26

**(B) Outdoor Installations.** Less-flammable liquid-filled transformers shall be permitted to be installed outdoors, attached to, adjacent to, or on the roof of buildings, where installed in accordance with (1) or (2):

(1) For Type I and Type II buildings, the installation shall comply with all restrictions provided for in the listing of the liquid.

Informational Note: Installations adjacent to combustible material, fire escapes, or door and window openings may require additional safeguards such as those listed in 450.27.

(2) In accordance with 450.27.

Informational Note No. 1: As used in this section, *Type I and Type II buildings* refers to Type I and Type II building construction as defined in NFPA 220-2012, *Standard on Types of Building Construction*. *Combustible materials* refers to those materials not classified as noncombustible or limited-combustible as defined in NFPA 220-2012.

Informational Note No. 2: See definition of *Listed* in Article 100.

Restrictions required by the listing of a less-flammable liquid are illustrated by the use of an FM Approvals LLC–approved liquid in a transformer tank. Pressure-relief devices must be provided. Spacing from adjacent buildings or transformers must also be provided. The spacing, as illustrated in Exhibit 450.13, is based not only on the fluid capacity of the transformer tank but also on the listing of the transformer and the building construction. In the event of a leak, the liquid confinement area prevents transformer dielectric fluid from spreading beyond the vicinity of the transformer. Further information on applications may be found in the Factory Mutual Loss Prevention Data Sheet 5-4.

The requirements in 450.23 refer to buildings of Types I and II construction. Table E.1 in Informative Annex E is a summary of the requirements for construction types. The arabic numbers at the top of the fire resistance rating columns reflect the fire resistance ratings of the following building elements: exterior bearing walls; columns, beams, girders, trusses and arches, supporting bearing walls, columns, or loads from more than one floor; and the floor construction.

For example, a building of Type I, 443 construction has 4-hour fire-resistance-rated exterior bearing walls; 4-hour fire-resistance-rated columns, beams, girders, trusses, or arches; and 3-hour fire-resistance-rated floor construction. Whether a building is of Type I, Type II, or other type is determined by the requirements of the building construction code adopted by a jurisdiction.

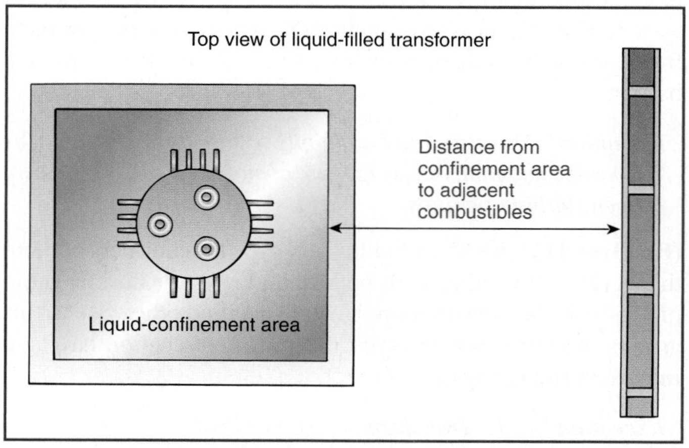


***EXHIBIT 450.13*** *A transformer tank containing a less-flammable fluid listed by FM Approvals, where the spacing from adjacent combustibles to the liquid confinement area is based on the capacity of the tank.*

## 450.24 Nonflammable Fluid-Insulated Transformers

Transformers insulated with a dielectric fluid identified as nonflammable shall be permitted to be installed indoors or outdoors. Such transformers installed indoors and rated over 35,000 volts shall be installed in a vault. Such transformers installed indoors shall be furnished with a liquid confinement area and a pressure-relief vent. The transformers shall be furnished with a means for absorbing any gases generated by arcing inside the tank, or the pressure-relief vent shall be connected to a chimney or flue that will carry such gases to an environmentally safe area.

Informational Note: Safety may be increased if fire hazard analyses are performed for such transformer installations.

For the purposes of this section, a nonflammable dielectric fluid is one that does not have a flash point or fire point and is not flammable in air.

A liquid confinement area and a pressure-relief vent are required for nonflammable fluid-insulated transformers. The liquid confinement area limits the extent of a spill if the tank leaks or ruptures. If a means for absorbing gases generated by arcing within the transformer is not provided, the pressure-relief vent must be connected to a chimney or flue that vents to an environmentally safe area.

The need for a gas absorption system or a chimney or flue that vents to an environmentally safe area is due to concerns about products generated during arcing. The high arc temperatures may cause the insulating medium to break down, resulting in the formation and emission of toxic or corrosive compounds.

## 450.25 Askarel-Insulated Transformers Installed Indoors

Askarel-insulated transformers installed indoors and rated over 25 kVA shall be furnished with a pressure-relief vent. Where installed in a poorly ventilated place, they shall be furnished with a means for absorbing any gases generated by arcing inside the case, or the pressure-relief vent shall be connected to a chimney or flue that carries such gases outside the building. Askarel-insulated transformers rated over 35,000 volts shall be installed in a vault.

Askarel-insulated transformers are no longer manufactured. The information in the *Code* is for reference and for the modification of existing askarel-insulated installations. Existing askarel-insulated transformers of 35,000 volts or less are not required to be installed in vaults because askarel is considered a noncombustible fluid.

## 450.26 Oil-Insulated Transformers Installed Indoors

Oil-insulated transformers installed indoors shall be installed in a vault constructed as specified in Part III of this article.

*Exception No. 1: Where the total capacity does not exceed 112½ kVA, the vault specified in Part III of this article shall be permitted to be constructed of reinforced concrete that is not less than 100 mm (4 in.) thick.*

*Exception No. 2: Where the nominal voltage does not exceed 1000, a vault shall not be required if suitable arrangements are made to prevent a transformer oil fire from igniting other materials and the total capacity in one location does not exceed 10 kVA in a section of the building classified as combustible or 75 kVA where the surrounding structure is classified as fire-resistant construction.*

*Exception No. 3: Electric furnace transformers that have a total rating not exceeding 75 kVA shall be permitted to be installed without a vault in a building or room of fire-resistant construction, provided suitable arrangements are made to prevent a transformer oil fire from spreading to other combustible material.*

*Exception No. 4: A transformer that has a total rating not exceeding 75 kVA and a supply voltage of 1000 volts or less that is an integral part of charged-particle-accelerating equipment shall be permitted to be installed without a vault in a building or room of noncombustible or fire-resistant construction, provided suitable arrangements are made to prevent a transformer oil fire from spreading to other combustible material.*

*Exception No. 5: Transformers shall be permitted to be installed in a detached building that does not comply with Part III of this article if neither the building nor its contents present a fire hazard to any other building or property, and if the building is used only in supplying electric service and the interior is accessible only to qualified persons.*

*Exception No. 6: Oil-insulated transformers shall be permitted to be used without a vault in portable and mobile surface mining equipment (such as electric excavators) if each of the following conditions is met:*

*(a) Provision is made for draining leaking fluid to the ground.*
*(b) Safe egress is provided for personnel.*
*(c) A minimum 6-mm (¼-in.) steel barrier is provided for personnel protection.*

## 450.27 Oil-Insulated Transformers Installed Outdoors

Combustible material, combustible buildings, and parts of buildings, fire escapes, and door and window openings shall be safeguarded from fires originating in oil-insulated transformers installed on roofs, attached to or adjacent to a building or combustible material.

In cases where the transformer installation presents a fire hazard, one or more of the following safeguards shall be applied according to the degree of hazard involved:

(1) Space separations
(2) Fire-resistant barriers
(3) Automatic fire suppression systems
(4) Enclosures that confine the oil of a ruptured transformer tank

Oil enclosures shall be permitted to consist of fire-resistant dikes, curbed areas or basins, or trenches filled with coarse, crushed stone. Oil enclosures shall be provided with trapped drains where the exposure and the quantity of oil involved are such that removal of oil is important.

Informational Note: For additional information on transformers installed on poles or structures or under ground, see ANSI C2-2007, *National Electrical Safety Code*.

### 450.28 Modification of Transformers

When modifications are made to a transformer in an existing installation that change the type of the transformer with respect to Part II of this article, such transformer shall be marked to show the type of insulating liquid installed, and the modified transformer installation shall comply with the applicable requirements for that type of transformer.

An existing askarel-insulated transformer may have the askarel replaced with either oil or a less flammable liquid. Where such a modification takes place, the completed installation must have the same degree of safety as a new installation. For example, replacement of askarel with oil in an indoor installation without a vault may not be acceptable (see 450.26 and its exceptions). The same is true if the replacement liquid is a less flammable liquid (see 450.23). Additional safety precautions may be necessary to compensate for the different fire characteristics of the new dielectric fluid.

## III. Transformer Vaults

### 450.41 Location

Vaults shall be located where they can be ventilated to the outside air without using flues or ducts wherever such an arrangement is practicable.

### 450.42 Walls, Roofs, and Floors

The walls and roofs of vaults shall be constructed of materials that have approved structural strength for the conditions with a minimum fire resistance of 3 hours. The floors of vaults in contact with the earth shall be of concrete that is not less than 100 mm (4 in.) thick, but, where the vault is constructed with a vacant space or other stories below it, the floor shall have approved structural strength for the load imposed thereon and a minimum fire resistance of 3 hours. For the purposes of this section, studs and wallboard construction shall not be permitted.

*Exception: Where transformers are protected with automatic sprinkler, water spray, carbon dioxide, or halon, construction of 1-hour rating shall be permitted.*

Informational Note No. 1: For additional information, see ANSI/ASTM E119-2012a, *Method for Fire Tests of Building Construction and Materials.*

Informational Note No. 2: A typical 3-hour construction is 150 mm (6 in.) thick reinforced concrete.

Vaults are intended primarily as passive fire protection. The need for vaults is dictated by the combustibility of the dielectric media and the size of the transformer. Transformers insulated with mineral oil have the greatest need for passive protection to prevent the spread of burning oil to other combustible materials.

Although construction of a 3-hour-rated wall may be possible using studs and wallboard, this construction method is not permitted for transformer vaults because of the concern for containing projectiles created in a transformer explosion. A reduction in fire resistance rating from 3 hours to 1 hour is permitted for vaults equipped with an automatic fire suppression system.

See the commentary following 450.23(B)(2), which relates to Type I and Type II building construction.

### 450.43 Doorways

Vault doorways shall be protected in accordance with 450.43(A), (B), and (C).

**(A) Type of Door.** Each doorway leading into a vault from the building interior shall be provided with a tight-fitting door that has a minimum fire rating of 3 hours. The authority having jurisdiction shall be permitted to require such a door for an exterior wall opening where conditions warrant.

*Exception: Where transformers are protected with automatic sprinkler, water spray, carbon dioxide, or halon, construction of 1-hour rating shall be permitted.*

Informational Note: For additional information, see NFPA 80-2013, *Standard for Fire Doors and Other Opening Protectives.*

**(B) Sills.** A door sill or curb that is of an approved height that will confine the oil from the largest transformer within the vault shall be provided, and in no case shall the height be less than 100 mm (4 in.).

**(C) Locks.** Doors shall be equipped with locks, and doors shall be kept locked, access being allowed only to qualified persons. Personnel doors shall swing out and be equipped with panic bars, pressure plates, or other devices that are normally latched but open under simple pressure.

Section 450.43 requires transformer vault doors to be locked but prohibits the use of conventional rotation-type door knobs. An injured worker attempting to escape from a transformer vault may not be able to operate a rotating-type door knob but would be able to escape through a door equipped with panic-type hardware.

### 450.45 Ventilation Openings

Where required by 450.9, openings for ventilation shall be provided in accordance with 450.45(A) through (F).

**(A) Location.** Ventilation openings shall be located as far as possible from doors, windows, fire escapes, and combustible material.

**(B) Arrangement.** A vault ventilated by natural circulation of air shall be permitted to have roughly half of the total area of openings required for ventilation in one or more openings near the floor and the remainder in one or more openings in the roof or in the sidewalls near the roof, or all of the area required for ventilation shall be permitted in one or more openings in or near the roof.

**(C) Size.** For a vault ventilated by natural circulation of air to an outdoor area, the combined net area of all ventilating openings, after deducting the area occupied by screens, gratings, or louvers, shall not be less than 1900 $mm^2$ (3 $in.^2$) per kVA of transformer capacity in service, and in no case shall the net area be less than 0.1 $m^2$ (1 $ft^2$) for any capacity under 50 kVA.

**(D) Covering.** Ventilation openings shall be covered with durable gratings, screens, or louvers, according to the treatment required in order to avoid unsafe conditions.

**(E) Dampers.** All ventilation openings to the indoors shall be provided with automatic closing fire dampers that operate in response to a vault fire. Such dampers shall possess a standard fire rating of not less than 1½ hours.

Informational Note: See ANSI/UL 555-2011, *Standard for Fire Dampers.*

**(F) Ducts.** Ventilating ducts shall be constructed of fire-resistant material.

### 450.46 Drainage

Where practicable, vaults containing more than 100 kVA transformer capacity shall be provided with a drain or other means that will carry off any accumulation of oil or water in the vault unless local conditions make this impracticable. The floor shall be pitched to the drain where provided.

### 450.47 Water Pipes and Accessories

Any pipe or duct system foreign to the electrical installation shall not enter or pass through a transformer vault. Piping or other facilities provided for vault fire protection, or for transformer cooling, shall not be considered foreign to the electrical installation.

Automatic sprinkler protection is permitted for transformer vaults. Piping or ductwork for cooling of the transformer is also permitted to be installed in a transformer vault. No other piping or ductwork is permitted to enter or pass through a transformer vault.

### 450.48 Storage in Vaults

Materials shall not be stored in transformer vaults.

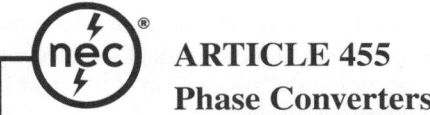

# ARTICLE 455
# Phase Converters

## I. General

### 455.1 Scope

This article covers the installation and use of phase converters.

A *phase converter* is an electrical device that converts single-phase electrical power to 3-phase for the operation of equipment that normally operates from a 3-phase electrical supply. Phase converters are of two types: *static,* with no moving parts, and *rotary,* with an internal rotor that must be rotating before a load is applied (see 455.2 for definitions of *rotary-phase converter* and *static-phase converter*).

Phase converters are most commonly used to supply 3-phase motor loads in locations where only single-phase power is available from the local utility. Electrical installations on farms and in other remote or rural areas are examples of such locations. Although their most common loads are motors, phase converters are increasingly used to supply such loads as cellular telephone and other communication transmitter sites.

### 455.2 Definitions

**Manufactured Phase.** The manufactured or derived phase originates at the phase converter and is not solidly connected to either of the single-phase input conductors.

**Phase Converter.** An electrical device that converts single-phase power to 3-phase electric power.

Informational Note: Phase converters have characteristics that modify the starting torque and locked-rotor current of motors served, and consideration is required in selecting a phase converter for a specific load.

**Rotary-Phase Converter.** A device that consists of a rotary transformer and capacitor panel(s) that permits the operation of 3-phase loads from a single-phase supply.

**Static-Phase Converter.** A device without rotating parts, sized for a given 3-phase load to permit operation from a single-phase supply.

### 455.3 Other Articles

Phase converters shall comply with this article and with the applicable provisions of other articles of this *Code.*

### 455.4 Marking

Each phase converter shall be provided with a permanent nameplate indicating the following:

(1) Manufacturer's name
(2) Rated input and output voltages
(3) Frequency
(4) Rated single-phase input full-load amperes
(5) Rated minimum and maximum single load in kilovolt-amperes (kVA) or horsepower
(6) Maximum total load in kilovolt-amperes (kVA) or horsepower
(7) For a rotary-phase converter, 3-phase amperes at full load

### 455.5 Equipment Grounding Connection

A means for attachment of an equipment grounding conductor termination in accordance with 250.8 shall be provided.

## 455.6 Conductors

**(A) Ampacity.** The ampacity of the single-phase supply conductors shall be determined by 455.6(A)(1) or (A)(2).

Informational Note: Single-phase conductors sized to prevent a voltage drop not exceeding 3 percent from the source of supply to the phase converter may help ensure proper starting and operation of motor loads.

**(1) Variable Loads.** Where the loads to be supplied are variable, the conductor ampacity shall not be less than 125 percent of the phase converter nameplate single-phase input full-load amperes.

**(2) Fixed Loads.** Where the phase converter supplies specific fixed loads, and the conductor ampacity is less than 125 percent of the phase converter nameplate single-phase input full-load amperes, the conductors shall have an ampacity not less than 250 percent of the sum of the full-load, 3-phase current rating of the motors and other loads served where the input and output voltages of the phase converter are identical. Where the input and output voltages of the phase converter are different, the current as determined by this section shall be multiplied by the ratio of output to input voltage.

**(B) Manufactured Phase Marking.** The manufactured phase conductors shall be identified in all accessible locations with a distinctive marking. The marking shall be consistent throughout the system and premises.

## 455.7 Overcurrent Protection

The single-phase supply conductors and phase converter shall be protected from overcurrent by 455.7(A) or (B). Where the required fuse or nonadjustable circuit breaker rating or settings of adjustable circuit breakers do not correspond to a standard rating or setting, a higher rating or setting that does not exceed the next higher standard rating shall be permitted.

**(A) Variable Loads.** Where the loads to be supplied are variable, overcurrent protection shall be set at not more than 125 percent of the phase converter nameplate single-phase input full-load amperes.

**(B) Fixed Loads.** Where the phase converter supplies specific fixed loads and the conductors are sized in accordance with 455.6(A)(2), the conductors shall be protected in accordance with their ampacity. The overcurrent protection determined from this section shall not exceed 125 percent of the phase converter nameplate single-phase input amperes.

## 455.8 Disconnecting Means

Means shall be provided to disconnect simultaneously all ungrounded single-phase supply conductors to the phase converter.

**(A) Location.** The disconnecting means shall be readily accessible and located in sight from the phase converter.

**(B) Type.** The disconnecting means shall be a switch rated in horsepower, a circuit breaker, or a molded-case switch. Where only nonmotor loads are served, an ampere-rated switch shall be permitted.

**(C) Rating.** The ampere rating of the disconnecting means shall not be less than 115 percent of the rated maximum single-phase input full-load amperes or, for specific fixed loads, shall be permitted to be selected from 455.8(C)(1) or (C)(2).

**(1) Current Rated Disconnect.** The disconnecting means shall be a circuit breaker or molded-case switch with an ampere rating not less than 250 percent of the sum of the following:

(1) Full-load, 3-phase current ratings of the motors
(2) Other loads served

**(2) Horsepower Rated Disconnect.** The disconnecting means shall be a switch with a horsepower rating. The equivalent locked rotor current of the horsepower rating of the switch shall not be less than 200 percent of the sum of the following:

(1) Nonmotor loads
(2) The 3-phase, locked-rotor current of the largest motor as determined from Table 430.251(B)
(3) The full-load current of all other 3-phase motors operating at the same time

**(D) Voltage Ratios.** The calculations in 455.8(C) shall apply directly where the input and output voltages of the phase converter are identical. Where the input and output voltages of the phase converter are different, the current shall be multiplied by the ratio of the output to input voltage.

## 455.9 Connection of Single-Phase Loads

Where single-phase loads are connected on the load side of a phase converter, they shall not be connected to the manufactured phase.

## 455.10 Terminal Housings

A terminal housing in accordance with the provisions of 430.12 shall be provided on a phase converter.

# II. Specific Provisions Applicable to Different Types of Phase Converters

## 455.20 Disconnecting Means

The single-phase disconnecting means for the input of a static phase converter shall be permitted to serve as the disconnecting means for the phase converter and a single load if the load is within sight of the disconnecting means.

## 455.21 Start-Up

Power to the utilization equipment shall not be supplied until the rotary-phase converter has been started.

### 455.22 Power Interruption

Utilization equipment supplied by a rotary-phase converter shall be controlled in such a manner that power to the equipment will be disconnected in the event of a power interruption.

Informational Note: Magnetic motor starters, magnetic contactors, and similar devices, with manual or time delay restarting for the load, provide restarting after power interruption.

### 455.23 Capacitors

Capacitors that are not an integral part of the rotary-phase conversion system but are installed for a motor load shall be connected to the line side of that motor overload protective device.

# ARTICLE 460 Capacitors

### 460.1 Scope

This article covers the installation of capacitors on electrical circuits.

Surge capacitors or capacitors included as a component part of other apparatus and conforming with the requirements of such apparatus are excluded from these requirements.

This article also covers the installation of capacitors in hazardous (classified) locations as modified by Articles 501 through 503.

### 460.2 Enclosing and Guarding

**(A) Containing More Than 11 L (3 gal) of Flammable Liquid.** Capacitors containing more than 11 L (3 gal) of flammable liquid shall be enclosed in vaults or outdoor fenced enclosures complying with Article 110, Part III. This limit shall apply to any single unit in an installation of capacitors.

**(B) Accidental Contact.** Where capacitors are accessible to unauthorized and unqualified persons, they shall be enclosed, located, or guarded so that persons cannot come into accidental contact or bring conducting materials into accidental contact with exposed energized parts, terminals, or buses associated with them. However, no additional guarding is required for enclosures accessible only to authorized and qualified persons.

## I. 1000 Volts, Nominal, and Under

### 460.6 Discharge of Stored Energy

Capacitors shall be provided with a means of discharging stored energy.

**(A) Time of Discharge.** The residual voltage of a capacitor shall be reduced to 50 volts, nominal, or less within 1 minute after the capacitor is disconnected from the source of supply.

**(B) Means of Discharge.** The discharge circuit shall be either permanently connected to the terminals of the capacitor or capacitor bank or provided with automatic means of connecting it to the terminals of the capacitor bank on removal of voltage from the line. Manual means of switching or connecting the discharge circuit shall not be used.

### 460.8 Conductors

**(A) Ampacity.** The ampacity of capacitor circuit conductors shall not be less than 135 percent of the rated current of the capacitor. The ampacity of conductors that connect a capacitor to the terminals of a motor or to motor circuit conductors shall not be less than one-third the ampacity of the motor circuit conductors and in no case less than 135 percent of the rated current of the capacitor.

Capacitors are rated in reactive kilovolt-amperes (kilovars or kVAr) or kilovolt-amperes capacitive (kVAc). Both ratings are synonymous. The kVAr rating shows how many reactive kilovolt-amperes the capacitor will supply to cancel out the reactive kilovolt-amperes caused by inductance. For example, a 20-kVAr capacitor will cancel out 20 kVAr of inductive reactive kilovolt-amperes.

The capacitor circuit conductors and disconnecting means must have an ampacity not less than 135 percent of the rated current of the capacitor. Capacitors are manufactured with a tolerance of zero percent to 15 percent, so a 100-kVAr capacitor may draw a current equivalent to that of a 115-kVAr capacitor. In addition, the current draw varies directly with the line voltage, and any variation in the line voltage from a pure sine wave form causes the capacitor to draw an increased current. Considering these factors, the increased current can amount to 135 percent of the rated current of the capacitor.

The current corresponding to the kVAr rating of a 3-phase capacitor, $i_c$, is computed from the following formula:

$$i_c = \frac{\text{kVAr} \times 1000}{\sqrt{3} \times \text{V}}$$

The ampacity of the conductors and the disconnecting device is then determined by multiplying $i_c$ by 1.35.

Where harmonic-producing loads are present, adding capacitors to the electrical system can place the system in a harmonic resonance condition. The harmonic loads can excite the electrical system at the harmonic resonance frequency and cause overcurrent and overvoltage conditions. If capacitors are to be placed on electrical systems with harmonic loads, an engineering study should be conducted that evaluates the size and placement of capacitors and the reactive impedance and load of the system. Capacitors may need a reactor placed in series with them to help detune the electrical system from a harmonic resonance condition.

**(B) Overcurrent Protection.** An overcurrent device shall be provided in each ungrounded conductor for each capacitor bank.

The rating or setting of the overcurrent device shall be as low as practicable.

*Exception: A separate overcurrent device shall not be required for a capacitor connected on the load side of a motor overload protective device.*

Unless the exception applies, the overcurrent device must be separate from the overcurrent device protecting any other equipment or conductor. See Exhibit 460.1, diagrams (a) and (b).

**(C) Disconnecting Means.** A disconnecting means shall be provided in each ungrounded conductor for each capacitor bank and shall meet the following requirements:

(1) The disconnecting means shall open all ungrounded conductors simultaneously.
(2) The disconnecting means shall be permitted to disconnect the capacitor from the line as a regular operating procedure.
(3) The rating of the disconnecting means shall not be less than 135 percent of the rated current of the capacitor.

*Exception: A separate disconnecting means shall not be required where a capacitor is connected on the load side of a motor controller.*

### 460.9 Rating or Setting of Motor Overload Device

Where a motor installation includes a capacitor connected on the load side of the motor overload device, the rating or setting of the motor overload device shall be based on the improved power factor of the motor circuit.

The effect of the capacitor shall be disregarded in determining the motor circuit conductor rating in accordance with 430.22.

Where a capacitor is connected on the load side of the motor overload relays [see Exhibit 460.1, diagram (a)], the line current will be reduced due to an improved power factor, which must be taken into account when selecting the rating of a motor overload device. A value lower than that specified in 430.32 should be used for proper protection of the motor.

The most effective power factor correction is obtained where the individual capacitors are connected closest to the inductive load. Capacitor manufacturers publish tables in which the required capacitor value is obtained by referring to the speed and the horsepower of the motor. These values improve the motor power factor to approximately 95 percent. To improve a plant power factor, capacitor manufacturers also publish tables to assist in calculating the total kVAr rating of capacitors required to improve the power factor to any desired value.

### 460.10 Grounding

Capacitor cases shall be connected to the equipment grounding conductor.

*Exception: Capacitor cases shall not be connected to the equipment grounding conductor where the capacitor units are supported on a structure designed to operate at other than ground potential.*

### 460.12 Marking

Each capacitor shall be provided with a nameplate giving the name of the manufacturer, rated voltage, frequency, kilovar or amperes, number of phases, and, if filled with a combustible liquid, the volume of liquid. Where filled with a nonflammable liquid, the nameplate shall so state. The nameplate shall also indicate whether a capacitor has a discharge device inside the case.

## II. Over 1000 Volts, Nominal

### 460.24 Switching

**(A) Load Current.** Group-operated switches shall be used for capacitor switching and shall be capable of the following:

(1) Carrying continuously not less than 135 percent of the rated current of the capacitor installation
(2) Interrupting the maximum continuous load current of each capacitor, capacitor bank, or capacitor installation that will be switched as a unit
(3) Withstanding the maximum inrush current, including contributions from adjacent capacitor installations
(4) Carrying currents due to faults on capacitor side of switch

**(B) Isolation.**

**(1) General.** A means shall be installed to isolate from all sources of voltage each capacitor, capacitor bank, or capacitor installation that will be removed from service as a unit. The isolating means shall provide a visible gap in the electrical circuit adequate for the operating voltage.

**(2) Isolating or Disconnecting Switches with No Interrupting Rating.** Isolating or disconnecting switches (with no interrupting rating) shall be interlocked with the load-interrupting device or shall be provided with prominently displayed caution signs in accordance with 490.22 to prevent switching load current.

**(C) Additional Requirements for Series Capacitors.** The proper switching sequence shall be ensured by use of one of the following:

(1) Mechanically sequenced isolating and bypass switches
(2) Interlocks
(3) Switching procedure prominently displayed at the switching location

### 460.25 Overcurrent Protection

**(A) Provided to Detect and Interrupt Fault Current.** A means shall be provided to detect and interrupt fault current likely to cause dangerous pressure within an individual capacitor.

**(B) Single Pole or Multipole Devices.** Single-pole or multipole devices shall be permitted for this purpose.

**(C) Protected Individually or in Groups.** Capacitors shall be permitted to be protected individually or in groups.

**(D) Protective Devices Rated or Adjusted.** Protective devices for capacitors or capacitor equipment shall be rated or adjusted to operate within the limits of the safe zone for individual capacitors. If the protective devices are rated or adjusted to operate within the limits for Zone 1 or Zone 2, the capacitors shall be enclosed or isolated.

In no event shall the rating or adjustment of the protective devices exceed the maximum limit of Zone 2.

Informational Note: For definitions of *Safe Zone, Zone 1,* and *Zone 2,* see ANSI/IEEE 18-1992, *Shunt Power Capacitors*.

## 460.26 Identification

Each capacitor shall be provided with a permanent nameplate giving the manufacturer's name, rated voltage, frequency, kilovar or amperes, number of phases, and the volume of liquid identified as flammable, if such is the case.

## 460.27 Grounding

Capacitor cases shall be connected to the equipment grounding conductor. If the capacitor neutral point is connected to a grounding electrode conductor, the connection shall be made in accordance with Part III of Article 250.

*Exception: Capacitor cases shall not be connected to the equipment grounding conductor where the capacitor units are supported on a structure designed to operate at other than ground potential.*

## 460.28 Means for Discharge

**(A) Means to Reduce the Residual Voltage.** A means shall be provided to reduce the residual voltage of a capacitor to 50 volts or less within 5 minutes after the capacitor is disconnected from the source of supply.

**(B) Connection to Terminals.** A discharge circuit shall be either permanently connected to the terminals of the capacitor or provided with automatic means of connecting it to the terminals of the capacitor bank after disconnection of the capacitor from the source of supply. The windings of motors, transformers, or other equipment directly connected to capacitors without a switch or overcurrent device interposed shall meet the requirements of 460.28(A).

Means are required to drain off the stored charge in a capacitor after the supply circuit has been opened. Otherwise, a person servicing the equipment could receive a severe shock, or damage could occur to the equipment.

Exhibit 460.1, diagram (a), shows a method in which capacitors are connected in a motor circuit so that they may be switched with the motor. In this arrangement, the stored charge drains off through the windings when the circuit is opened. Diagram (b) shows another arrangement in which the capacitor is connected to the line side of the motor starter contacts. An automatic discharge device and a separate disconnecting means are required. As shown in Exhibit 460.2, capacitors are often equipped with built-in resistors to drain off the stored charge; however, this type of capacitor is not needed where connected as shown in Exhibit 460.1, diagram (a).

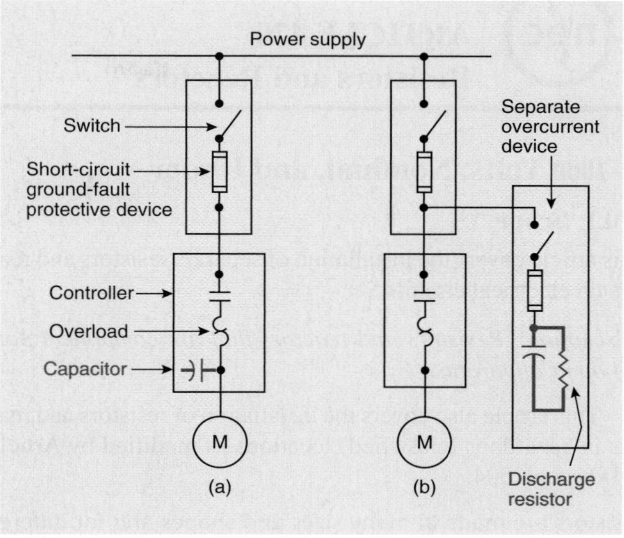


***EXHIBIT 460.1*** *Methods of connecting capacitors in induction motor circuit for power factor correction.*

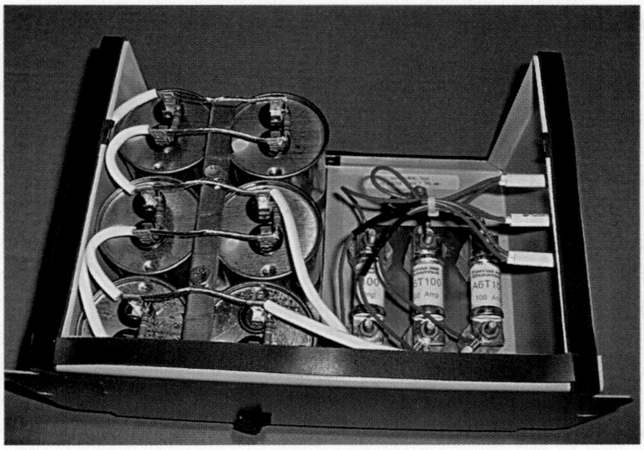

***EXHIBIT 460.2*** *Power factor correction capacitors with internal discharge resistors (blue) and overcurrent protection. (Courtesy of GE Energy)*

# ARTICLE 470
# Resistors and Reactors

## I. 1000 Volts, Nominal, and Under

### 470.1 Scope

This article covers the installation of separate resistors and reactors on electrical circuits.

*Exception: Resistors and reactors that are component parts of other apparatus.*

This article also covers the installation of resistors and reactors in hazardous (classified) locations as modified by Articles 501 through 504.

Resistors are made in many sizes and shapes and for different purposes. They may be wire, ribbon, form or edgewise wound, cast or punched steel grid, or box resistors. They may be mounted in the open or in ventilated metal boxes or cabinets, depending on their use and location. Because they give off heat, resistors must be guarded and located at safe distances from combustible materials. Where mounted on switchboards or installed in control panels, they are not required to have additional guards.

Current-limiting reactors are installed to limit the amount of current that can flow in a circuit when a short circuit occurs. Reactors can be divided into two classes: those with iron cores and those with no magnetic materials in the windings. Either type may be air cooled or oil immersed.

Mechanical stresses exist between adjacent air-core reactors due to their external fields, and the manufacturer's recommendations should be followed in spacing and bracing units and fastening supporting insulators.

Saturable reactors may be used for theater dimming [see 520.25(B) and commentary]. These reactors have, in addition to the ac winding, an auxiliary winding connected line-to-line or line-to-ground to neutralize charging current and prevent a voltage rise. Those reactors used on high-voltage systems may be oil immersed.

### 470.2 Location

Resistors and reactors shall not be placed where exposed to physical damage.

### 470.3 Space Separation

A thermal barrier shall be required if the space between the resistors and reactors and any combustible material is less than 305 mm (12 in.).

### 470.4 Conductor Insulation

Insulated conductors used for connections between resistance elements and controllers shall be suitable for an operating temperature of not less than 90°C (194°F).

*Exception: Other conductor insulations shall be permitted for motor starting service.*

## II. Over 1000 Volts, Nominal

### 470.18 General

**(A) Protected Against Physical Damage.** Resistors and reactors shall be protected against physical damage.

**(B) Isolated by Enclosure or Elevation.** Resistors and reactors shall be isolated by enclosure or elevation to protect personnel from accidental contact with energized parts.

**(C) Combustible Materials.** Resistors and reactors shall not be installed in close enough proximity to combustible materials to constitute a fire hazard and shall have a clearance of not less than 305 mm (12 in.) from combustible materials.

**(D) Clearances.** Clearances from resistors and reactors to grounded surfaces shall be adequate for the voltage involved.

•

**(E) Temperature Rise from Induced Circulating Currents.** Metallic enclosures of reactors and adjacent metal parts shall be installed so that the temperature rise from induced circulating currents is not hazardous to personnel or does not constitute a fire hazard.

### 470.19 Grounding

Resistor and reactor cases or enclosures shall be connected to the equipment grounding conductor.

*Exception: Resistor or reactor cases or enclosures supported on a structure designed to operate at other than ground potential shall not be connected to the equipment grounding conductor.*

### 470.20 Oil-Filled Reactors

Installation of oil-filled reactors, in addition to the above requirements, shall comply with applicable requirements of Article 450.

# ARTICLE 480
# Storage Batteries

### 480.1 Scope

The provisions of this article shall apply to all stationary installations of storage batteries.

Informational Note: The following standards are frequently referenced for the installation of stationary batteries:

(1) IEEE 484-2008, *Recommended Practice for Installation Design and Installation of Vented Lead-Acid Batteries for Stationary Applications*

(2) IEEE 485-1997, *Recommended Practice for Sizing Vented Lead-Acid Storage Batteries for Stationary Applications*

(3) IEEE 1145-2007, *Recommended Practice for Installation and Maintenance of Nickel-Cadmium Batteries for Photovoltaic (PV) Systems*
(4) IEEE 1187-2002, *Recommended Practice for Installation Design, and Installation of Valve-Regulated Lead-Acid Batteries for Stationary Applications*
(5) IEEE 1375-1996 (Rev. 2003), *IEEE Guide for the Protection of Stationary Battery Systems*
(6) IEEE 1578-2007, *Recommended Practice for Stationary Battery Spill Containment and Management*
(7) IEEE 1635/ASHRAE 21-2012, *Guide for the Ventilation and Thermal Management of Stationary Battery Installations*

There are many types of batteries. The most common types of storage cells are the lead-acid type, the alkali (nickel-cadmium) type, and the lithium-ion type. A lead-acid cell consists of a positive plate, usually lead peroxide (a semisolid compound) mounted on a framework or grid for support, and a negative plate, made of sponge lead mounted on a grid. Grids are generally made of a lead alloy, such as lead-calcium, lead-antimony, or lead-selenium. The electrolyte is sulfuric acid and distilled water. Lithium-ion batteries use a variety of different chemistries. These batteries are used in a variety of consumer electronics products. However, they are increasingly being used in large-scale applications because they have a very high energy density.

Lead-acid cells may be of the vented or sealed (valve-regulated) type. Under normal charging conditions, the vented type will liberate gases: hydrogen at the negative plate and oxygen at the positive plate. The valve-regulated type provides a means to recombine this gas, thus minimizing emissions from the cell.

In the alkali, or nickel-cadmium, battery, the principal active material in the positive plate is nickelous hydroxide; in the negative plate, it is cadmium hydroxide. The electrolyte is potassium hydroxide (an alkali).

In stationary installations, nickel-cadmium cells are generally of the vented type and liberate hydrogen and oxygen during normal charging. Hermetically sealed nickel-cadmium cells are sometimes used, but they require special charging equipment to prevent gas emissions.

Although some of the newer technology batteries do not ventilate hydrogen under normal operation, they may generate hydrogen during fault conditions.

## 480.2 Definitions

•

**Cell.** The basic electrochemical unit, characterized by an anode and a cathode, used to receive, store, and deliver electrical energy.

**Container.** A vessel that holds the plates, electrolyte, and other elements of a single unit in a battery.

Informational Note: A container may be single-cell or multi-cell and is sometimes referred to in the industry as a "jar."

**Electrolyte.** The medium that provides the ion transport mechanism between the positive and negative electrodes of a cell.

**Intercell Connector.** An electrically conductive bar or cable used to connect adjacent cells.

**Intertier Connector.** An electrical conductor used to connect two cells on different tiers of the same rack or different shelves of the same rack.

**Nominal Voltage (Battery or Cell).** The value assigned to a cell or battery of a given voltage class for the purpose of convenient designation. The operating voltage of the cell or battery may vary above or below this value.

Informational Note: The most common nominal cell voltages are 2 volts per cell for the lead-acid systems, 1.2 volts per cell for alkali systems, and 3.6 to 3.8 volts per cell for Li-ion systems. Nominal voltages might vary with different chemistries.

**Sealed Cell or Battery.** A cell or battery that has no provision for the routine addition of water or electrolyte or for external measurement of electrolyte specific gravity and might contain pressure relief venting.

**Storage Battery.** A battery comprised of one or more rechargeable cells of the lead-acid, nickel-cadmium, or other rechargeable electrochemical types.

**Terminal.** That part of a cell, container, or battery to which an external connection is made (commonly identified as post, pillar, pole, or terminal post).

## 480.3 Battery and Cell Terminations

**(A) Dissimilar Metals.** Where mating dissimilar metals, antioxidant material suitable for the battery connection shall be used.

Informational Note: The battery manufacturer's installation and instruction manual can be used for guidance for acceptable materials.

**(B) Intercell and Intertier Conductors and Connections.** The ampacity of field-assembled intercell and intertier connectors and conductors shall be of such cross-sectional area that the temperature rise under maximum load conditions and at maximum ambient temperature shall not exceed the safe operating temperature of the conductor insulation or of the material of the conductor supports.

Informational Note: Conductors sized to prevent a voltage drop exceeding 3 percent of maximum anticipated load, and where the maximum total voltage drop to the furthest point of connection does not exceed 5 percent, may not be appropriate for all battery applications. IEEE 1375-2003, *Guide for the Protection of Stationary Battery Systems*, provides guidance for overcurrent protection and associated cable sizing.

**(C) Battery Terminals.** Electrical connections to the battery, and the cable(s) between cells on separate levels or racks, shall not put mechanical strain on the battery terminals. Terminal plates shall be used where practicable.

## 480.4 Wiring and Equipment Supplied from Batteries

Wiring and equipment supplied from storage batteries shall be subject to the applicable provisions of this *Code* applying to wiring and equipment operating at the same voltage, unless otherwise permitted by 480.5.

The requirement to use single conductors only in conjunction with a Chapter 3 wiring method is not applicable to battery-powered conductors. For example, if it were necessary to extend the conductors from the battery to the prime mover starting solenoid at a generator location, these conductors would not be required to have overcurrent protection and could be run as open, single conductors.

## 480.5 Overcurrent Protection for Prime Movers

Overcurrent protection shall not be required for conductors from a battery with a nominal voltage of 50 volts or less if the battery provides power for starting, ignition, or control of prime movers. Section 300.3 shall not apply to these conductors.

## 480.6 DC Disconnect Methods

**(A) Disconnecting Means.** A disconnecting means shall be provided for all ungrounded conductors derived from a stationary battery system with a nominal voltage over 50 volts. A disconnecting means shall be readily accessible and located within sight of the battery system.

Informational Note: See 240.21(H) for information on the location of the overcurrent device for battery conductors.

A battery disconnecting means is needed to allow the battery to be disconnected for system maintenance. This requirement also correlates with 240.21(H), which addresses the location of overcurrent protection devices for battery conductors.

**(B) Remote Actuation.** Where controls to activate the disconnecting means of a battery are not located within sight of a stationary battery system, the disconnecting means shall be capable of being locked in the open position, in accordance with 110.25, and the location of the controls shall be field marked on the disconnecting means.

**(C) Busway.** Where a DC busway system is installed, the disconnecting means shall be permitted to be incorporated into the busway.

**(D) Notification.** The disconnecting means shall be legibly marked in the field. A label with the marking shall be placed in a conspicuous location near the battery if a disconnecting means is not provided. The marking shall be of sufficient durability to withstand the environment involved and shall include the following:

(1) Nominal battery voltage
(2) Maximum available short-circuit current derived from the stationary battery system
(3) Date the calculation was performed

Informational Note: Battery equipment suppliers can provide information about short-circuit current on any particular battery model.

## 480.7 Insulation of Batteries Not Over 250 Volts

This section shall apply to storage batteries having cells connected so as to operate at a nominal battery voltage of not over 250 volts.

**(A) Vented Lead-Acid Batteries.** Cells and multi-cell batteries with covers sealed to containers of nonconductive, heat-resistant material shall not require additional insulating support.

**(B) Vented Alkaline-Type Batteries.** Cells with covers sealed to containers of nonconductive, heat-resistant material shall require no additional insulation support. Cells in containers of conductive material shall be installed in trays of nonconductive material with not more than 20 cells (24 volts, nominal) in the series circuit in any one tray.

**(C) Rubber Containers.** Cells in rubber or composition containers shall require no additional insulating support where the total nominal voltage of all cells in series does not exceed 150 volts. Where the total voltage exceeds 150 volts, batteries shall be sectionalized into groups of 150 volts or less, and each group shall have the individual cells installed in trays or on racks.

**(D) Sealed Cells or Batteries.** Sealed cells and multicompartment sealed batteries constructed of nonconductive, heat-resistant material shall not require additional insulating support. Batteries constructed of a conducting container shall have insulating support if a voltage is present between the container and ground.

•

## 480.8 Racks and Trays

Racks and trays shall comply with 480.8(A) and (B).

**(A) Racks.** Racks, as required in this article, are rigid frames designed to support cells or trays. They shall be substantial and be made of one of the following:

(1) Metal, treated so as to be resistant to deteriorating action by the electrolyte and provided with nonconducting members directly supporting the cells or with continuous insulating material other than paint on conducting members
(2) Other construction such as fiberglass or other suitable nonconductive materials

**(B) Trays.** Trays are frames, such as crates or shallow boxes usually of wood or other nonconductive material, constructed or treated so as to be resistant to deteriorating action by the electrolyte.

**(C) Accessibility.** The terminals of all cells or multi-cell units shall be readily accessible for readings, inspection, and cleaning where required by the equipment design. One side of transparent battery containers shall be readily accessible for inspection of the internal components.

## 480.9 Battery Locations

Battery locations shall conform to 480.9(A), (B), and (C).

**(A) Ventilation.** Provisions appropriate to the battery technology shall be made for sufficient diffusion and ventilation of gases from the battery, if present, to prevent the accumulation of an explosive mixture.

Ventilation is necessary to prevent classification of a battery location as a hazardous (classified) location, in accordance with Article 500.

Mechanical ventilation is not mandated. Hydrogen disperses rapidly and requires little air movement to prevent accumulation. Unrestricted natural air movement in the vicinity of the battery, together with normal air changes for occupied spaces or heat removal, normally is sufficient. If the space is confined, mechanical ventilation may be required in the vicinity of the battery.

Hydrogen is lighter than air and tends to concentrate at ceiling level, so some form of ventilation should be provided at the upper portion of the structure. Ventilation can be a fan, roof ridge vent, or louvered area.

Although valve-regulated batteries are often referred to as "sealed," they actually emit very small quantities of hydrogen gas under normal operation and are capable of liberating large quantities of explosive gases if overcharged. These batteries therefore require the same amount of ventilation as their vented counterparts.

Informational Note No. 1: See NFPA 1, *Fire Code*, Chapter 52, for ventilation considerations for specific battery chemistries.

Informational Note No. 2: Some battery technologies do not require ventilation.

**(B) Live Parts.** Guarding of live parts shall comply with 110.27.

**(C) Spaces About Battery Systems.** Spaces about battery systems shall comply with 110.26. Working space shall be measured from the edge of the battery cabinet, racks, or trays.

For battery racks, there shall be a minimum clearance of 25 mm (1 in.) between a cell container and any wall or structure on the side not requiring access for maintenance. Battery stands shall be permitted to contact adjacent walls or structures, provided that the battery shelf has a free air space for not less than 90 percent of its length.

Informational Note: Additional space is often needed to accommodate battery hoisting equipment, tray removal, or spill containment.

Batteries should be located in clean, dry rooms and be arranged to provide sufficient work space for inspection and maintenance. Provision must also be made for adequate ventilation to prevent an accumulation of an explosive mixture of the gases from batteries that generate hydrogen.

The fumes given off by some storage batteries are very corrosive; therefore, wiring and its insulation must be of a type that withstands corrosion [see 310.10(G)]. Special precautions are necessary to ensure that all metalwork (such as metal raceways or metal racks) is designed or treated to be corrosion resistant. The battery racks shown in Exhibit 480.1 are coated with a nonmetallic outer covering as required by 480.8(A) that insulates and provides protection against the corrosive action of fumes from charging batteries and any electrolyte that may escape from the cells. Manufacturers sometimes suggest that aluminum or plastic conduit be used to withstand corrosive battery fumes, or, if steel conduit is used, that it be zinc coated and corrosion protected with a coating of an asphaltum-type paint (see 300.6).

***EXHIBIT 480.1*** *A well-arranged battery room with batteries installed on corrosion-resistant racks. (Courtesy of the International Association of Electrical Inspectors)*

Overcharging heats a battery and causes gassing and loss of water. A battery should not be allowed to reach temperatures over 110°F, because heat causes a shedding of active materials from the plates, which will eventually form a sediment buildup in the bottom of the case and short-circuit the plates and the cell. Because mixtures of oxygen and hydrogen are highly explosive, flame or sparks should never be allowed near a cell, especially if the filler cap is removed.

**(D) Top Terminal Batteries.** Where top terminal batteries are installed on tiered racks, working space in accordance with the battery manufacturer's instructions shall be provided between the highest point on a cell and the row or ceiling above that point.

Informational Note: Battery manufacturer's installation instructions typically define how much top working space is necessary for a particular battery model.

**(E) Egress.** A personnel door(s) intended for entrance to, and egress from, rooms designated as battery rooms shall open in the direction of egress and shall be equipped with listed panic hardware.

**(F) Piping in Battery Rooms.** Gas piping shall not be permitted in dedicated battery rooms.

**(G) Illumination.** Illumination shall be provided for working spaces containing battery systems. The lighting outlets shall not be controlled by automatic means only. Additional lighting outlets shall not be required where the work space is illuminated by an adjacent light source. The location of luminaires shall not:

(1) Expose personnel to energized battery components while performing maintenance on the luminaires in the battery space; or
(2) Create a hazard to the battery upon failure of the luminaire.

### 480.10 Vents

**(A) Vented Cells.** Each vented cell shall be equipped with a flame arrester that is designed to prevent destruction of the cell due to ignition of gases within the cell by an external spark or flame under normal operating conditions.

**(B) Sealed Cells.** Sealed battery or cells shall be equipped with a pressure-release vent to prevent excessive accumulation of gas pressure, or the battery or cell shall be designed to prevent scatter of cell parts in event of a cell explosion.

# ARTICLE 490 Equipment Over 1000 Volts, Nominal

## I. General

### 490.1 Scope

This article covers the general requirements for equipment operating at more than 1000 volts, nominal.

Informational Note No. 1: See *NFPA 70E-2012*, *Standard for Electrical Safety in the Workplace*, for electrical safety requirements for employee workplaces.

Informational Note No. 2: For further information on hazard signs and labels, see ANSI Z535.4-1998, *Product Signs and Safety Labels*.

For the 2014 edition, the threshold voltage for higher voltage equipment and requirements has been revised from 600 to 1000 V in conjunction with a coordinated effort throughout the *NEC* to recognize that standard configurations commonly used in alternative energy systems operate at over 600 V to increase efficiency and performance.

### 490.2 Definition

**High Voltage.** For the purposes of this article, more than 1000 volts, nominal.

### 490.3 Oil-Filled Equipment

Installation of electrical equipment, other than transformers covered in Article 450, containing more than 38 L (10 gal) of flammable oil per unit shall meet the requirements of Parts II and III of Article 450.

## II. Equipment — Specific Provisions

### 490.21 Circuit-Interrupting Devices

**(A) Circuit Breakers.**

**(1) Location.**

(a) Circuit breakers installed indoors shall be mounted either in metal-enclosed units or fire-resistant cell-mounted units, or they shall be permitted to be open-mounted in locations accessible to qualified persons only.

(b) Circuit breakers used to control oil-filled transformers in a vault shall either be located outside the transformer vault or be capable of operation from outside the vault.

(c) Oil circuit breakers shall be arranged or located so that adjacent readily combustible structures or materials are safeguarded in an approved manner.

**(2) Operating Characteristics.** Circuit breakers shall have the following equipment or operating characteristics:

(1) An accessible mechanical or other identified means for manual tripping, independent of control power
(2) Be release free (trip free)
(3) If capable of being opened or closed manually while energized, main contacts that operate independently of the speed of the manual operation
(4) A mechanical position indicator at the circuit breaker to show the open or closed position of the main contacts
(5) A means of indicating the open and closed position of the breaker at the point(s) from which they may be operated

**(3) Nameplate.** A circuit breaker shall have a permanent and legible nameplate showing manufacturer's name or trademark, manufacturer's type or identification number, continuous current rating, interrupting rating in megavolt-amperes (MVA) or amperes, and maximum voltage rating. Modification of a circuit breaker affecting its rating(s) shall be accompanied by an appropriate change of nameplate information.

**(4) Rating.** Circuit breakers shall have the following ratings:

(1) The continuous current rating of a circuit breaker shall not be less than the maximum continuous current through the circuit breaker.
(2) The interrupting rating of a circuit breaker shall not be less than the maximum fault current the circuit breaker will be required to interrupt, including contributions from all connected sources of energy.
(3) The closing rating of a circuit breaker shall not be less than the maximum asymmetrical fault current into which the circuit breaker can be closed.
(4) The momentary rating of a circuit breaker shall not be less than the maximum asymmetrical fault current at the point of installation.
(5) The rated maximum voltage of a circuit breaker shall not be less than the maximum circuit voltage.

**(B) Power Fuses and Fuseholders.**

**(1) Use.** Where fuses are used to protect conductors and equipment, a fuse shall be placed in each ungrounded conductor. Two power fuses shall be permitted to be used in parallel to protect the same load if both fuses have identical ratings and both fuses are installed in an identified common mounting with electrical connections that divide the current equally. Power fuses of the vented type shall not be used indoors, underground, or in metal enclosures unless identified for the use.

**(2) Interrupting Rating.** The interrupting rating of power fuses shall not be less than the maximum fault current the fuse is required to interrupt, including contributions from all connected sources of energy.

**(3) Voltage Rating.** The maximum voltage rating of power fuses shall not be less than the maximum circuit voltage. Fuses having a minimum recommended operating voltage shall not be applied below this voltage.

**(4) Identification of Fuse Mountings and Fuse Units.** Fuse mountings and fuse units shall have permanent and legible nameplates showing the manufacturer's type or designation, continuous current rating, interrupting current rating, and maximum voltage rating.

**(5) Fuses.** Fuses that expel flame in opening the circuit shall be designed or arranged so that they function properly without hazard to persons or property.

**(6) Fuseholders.** Fuseholders shall be designed or installed so that they are de-energized while a fuse is being replaced. A field-applied permanent and legible sign, in accordance with 110.21(B), shall be installed immediately adjacent to the fuseholders and shall be worded as follows:

DANGER — DISCONNECT CIRCUIT BEFORE REPLACING FUSES.

*Exception: Fuses and fuseholders designed to permit fuse replacement by qualified persons using identified equipment without de-energizing the fuseholder shall be permitted.*

**(7) High-Voltage Fuses.** Switchgear and substations that utilize high-voltage fuses shall be provided with a gang-operated disconnecting switch. Isolation of the fuses from the circuit shall be provided by either connecting a switch between the source and the fuses or providing roll-out switch and fuse-type construction. The switch shall be of the load-interrupter type, unless mechanically or electrically interlocked with a load-interrupting device arranged to reduce the load to the interrupting capability of the switch.

*Exception: More than one switch shall be permitted as the disconnecting means for one set of fuses where the switches are installed to provide connection to more than one set of supply conductors. The switches shall be mechanically or electrically interlocked to permit access to the fuses only when all switches are open. A conspicuous sign shall be placed at the fuses identifying the presence of more than one source.*

**(C) Distribution Cutouts and Fuse Links — Expulsion Type.**

**(1) Installation.** Cutouts shall be located so that they may be readily and safely operated and re-fused, and so that the exhaust of the fuses does not endanger persons. Distribution cutouts shall not be used indoors, underground, or in metal enclosures.

**(2) Operation.** Where fused cutouts are not suitable to interrupt the circuit manually while carrying full load, an approved means shall be installed to interrupt the entire load. Unless the fused cutouts are interlocked with the switch to prevent opening of the cutouts under load, a conspicuous sign shall be placed at such cutouts identifying that they shall not be operated under load.

**(3) Interrupting Rating.** The interrupting rating of distribution cutouts shall not be less than the maximum fault current the cutout is required to interrupt, including contributions from all connected sources of energy.

**(4) Voltage Rating.** The maximum voltage rating of cutouts shall not be less than the maximum circuit voltage.

**(5) Identification.** Distribution cutouts shall have on their body, door, or fuse tube a permanent and legible nameplate or identification showing the manufacturer's type or designation, continuous current rating, maximum voltage rating, and interrupting rating.

**(6) Fuse Links.** Fuse links shall have a permanent and legible identification showing continuous current rating and type.

**(7) Structure Mounted Outdoors.** The height of cutouts mounted outdoors on structures shall provide safe clearance between lowest energized parts (open or closed position) and standing surfaces, in accordance with 110.34(E).

**(D) Oil-Filled Cutouts.**

**(1) Continuous Current Rating.** The continuous current rating of oil-filled cutouts shall not be less than the maximum continuous current through the cutout.

**(2) Interrupting Rating.** The interrupting rating of oil-filled cutouts shall not be less than the maximum fault current the oil-filled cutout is required to interrupt, including contributions from all connected sources of energy.

**(3) Voltage Rating.** The maximum voltage rating of oil-filled cutouts shall not be less than the maximum circuit voltage.

**(4) Fault Closing Rating.** Oil-filled cutouts shall have a fault closing rating not less than the maximum asymmetrical fault current that can occur at the cutout location, unless suitable interlocks or operating procedures preclude the possibility of closing into a fault.

**(5) Identification.** Oil-filled cutouts shall have a permanent and legible nameplate showing the rated continuous current, rated maximum voltage, and rated interrupting current.

**(6) Fuse Links.** Fuse links shall have a permanent and legible identification showing the rated continuous current.

**(7) Location.** Cutouts shall be located so that they are readily and safely accessible for re-fusing, with the top of the cutout not over 1.5 m (5 ft) above the floor or platform.

**(8) Enclosure.** Suitable barriers or enclosures shall be provided to prevent contact with nonshielded cables or energized parts of oil-filled cutouts.

**(E) Load Interrupters.** Load-interrupter switches shall be permitted if suitable fuses or circuit breakers are used in conjunction with these devices to interrupt fault currents. Where these devices are used in combination, they shall be coordinated electrically so that they will safely withstand the effects of closing, carrying, or interrupting all possible currents up to the assigned maximum short-circuit rating.

Where more than one switch is installed with interconnected load terminals to provide for alternate connection to different supply conductors, each switch shall be provided with a conspicuous sign identifying this hazard.

**(1) Continuous Current Rating.** The continuous current rating of interrupter switches shall equal or exceed the maximum continuous current at the point of installation.

**(2) Voltage Rating.** The maximum voltage rating of interrupter switches shall equal or exceed the maximum circuit voltage.

**(3) Identification.** Interrupter switches shall have a permanent and legible nameplate including the following information: manufacturer's type or designation, continuous current rating, interrupting current rating, fault closing rating, maximum voltage rating.

**(4) Switching of Conductors.** The switching mechanism shall be arranged to be operated from a location where the operator is not exposed to energized parts and shall be arranged to open all ungrounded conductors of the circuit simultaneously with one operation. Switches shall be arranged to be locked in the open position. Metal-enclosed switches shall be operable from outside the enclosure.

**(5) Stored Energy for Opening.** The stored-energy operator shall be permitted to be left in the uncharged position after the switch has been closed if a single movement of the operating handle charges the operator and opens the switch.

**(6) Supply Terminals.** The supply terminals of fused interrupter switches shall be installed at the top of the switch enclosure, or, if the terminals are located elsewhere, the equipment

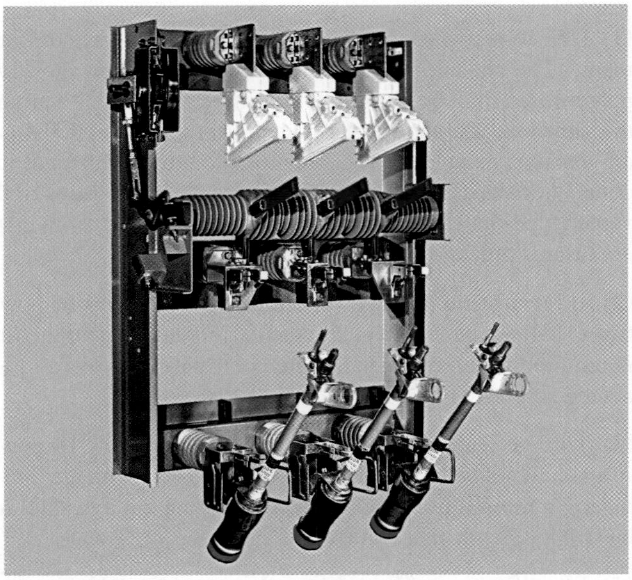

***EXHIBIT 490.1*** *Group-operated interrupter-switch and power fuse combination rated at 13.8 kV, 600 A continuous and interrupting, 40,000 A momentary, 40,000 A fault closing. (Courtesy of S&C Electric Co.)*

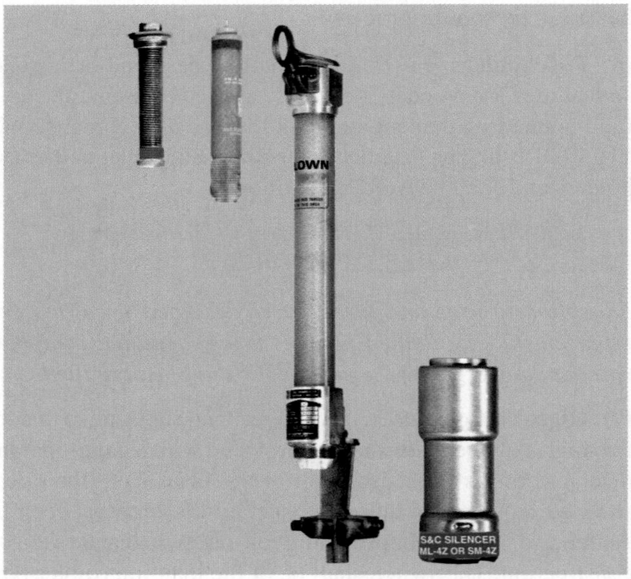


***EXHIBIT 490.2*** *Components of the indoor solid-material (SM) power fuseholder (boric-acid arc-extinguishing type) with a 14.4-kV, 400E-A maximum, 40,000-A rms asymmetrical interrupting rating. (Courtesy of S&C Electric Co.)*

*TABLE 490.24* *Minimum Clearance of Live Parts*

| Nominal Voltage Rating (kV) | Impulse Withstand, Basic Impulse Level B.I.L (kV) | | Minimum Clearance of Live Parts | | | | | | | |
|---|---|---|---|---|---|---|---|---|---|---|
| | | | Phase-to-Phase | | | | Phase-to-Ground | | | |
| | | | Indoors | | Outdoors | | Indoors | | Outdoors | |
| | Indoors | Outdoors | mm | in. | mm | in. | mm | in. | mm | in. |
| 2.4–4.16 | 60 | 95 | 115 | 4.5 | 180 | 7 | 80 | 3.0 | 155 | 6 |
| 7.2 | 75 | 95 | 140 | 5.5 | 180 | 7 | 105 | 4.0 | 155 | 6 |
| 13.8 | 95 | 110 | 195 | 7.5 | 305 | 12 | 130 | 5.0 | 180 | 7 |
| 14.4 | 110 | 110 | 230 | 9.0 | 305 | 12 | 170 | 6.5 | 180 | 7 |
| 23 | 125 | 150 | 270 | 10.5 | 385 | 15 | 190 | 7.5 | 255 | 10 |
| 34.5 | 150 | 150 | 320 | 12.5 | 385 | 15 | 245 | 9.5 | 255 | 10 |
| | 200 | 200 | 460 | 18.0 | 460 | 18 | 335 | 13.0 | 335 | 13 |
| 46 | — | 200 | — | — | 460 | 18 | — | — | 335 | 13 |
| | — | 250 | — | — | 535 | 21 | — | — | 435 | 17 |
| 69 | — | 250 | — | — | 535 | 21 | — | — | 435 | 17 |
| | — | 350 | — | — | 790 | 31 | — | — | 635 | 25 |
| 115 | — | 550 | — | — | 1350 | 53 | — | — | 1070 | 42 |
| 138 | — | 550 | — | — | 1350 | 53 | — | — | 1070 | 42 |
| | — | 650 | — | — | 1605 | 63 | — | — | 1270 | 50 |
| 161 | — | 650 | — | — | 1605 | 63 | — | — | 1270 | 50 |
| | — | 750 | — | — | 1830 | 72 | — | — | 1475 | 58 |
| 230 | — | 750 | — | — | 1830 | 72 | — | — | 1475 | 58 |
| | — | 900 | — | — | 2265 | 89 | — | — | 1805 | 71 |
| | — | 1050 | — | — | 2670 | 105 | — | — | 2110 | 83 |

Note: The values given are the minimum clearance for rigid parts and bare conductors under favorable service conditions. They shall be increased for conductor movement or under unfavorable service conditions or wherever space limitations permit. The selection of the associated impulse withstand voltage for a particular system voltage is determined by the characteristics of the surge protective equipment.

shall have barriers installed so as to prevent persons from accidentally contacting energized parts or dropping tools or fuses into energized parts.

See Exhibits 490.1 and 490.2 for examples of a fused interrupter switch and the fuseholder components. The components shown include the spring and cable assembly, refill unit, holder, and snuffler.

## 490.22 Isolating Means

Means shall be provided to completely isolate an item of equipment from all ungrounded conductors. The use of isolating switches shall not be required where there are other ways of de-energizing the equipment for inspection and repairs, such as draw-out-type switchgear units and removable truck panels.

Isolating switches not interlocked with an approved circuit-interrupting device shall be provided with a sign warning against opening them under load. The warning sign(s) or label(s) shall comply with 110.21(B).

An identified fuseholder and fuse shall be permitted as an isolating switch.

## 490.23 Voltage Regulators

Proper switching sequence for regulators shall be ensured by use of one of the following:

(1) Mechanically sequenced regulator bypass switch(es)
(2) Mechanical interlocks
(3) Switching procedure prominently displayed at the switching location

## 490.24 Minimum Space Separation.

In field-fabricated installations, the minimum air separation between bare live conductors and between such conductors and adjacent grounded surfaces shall not be less than the values given in Table 490.24. These values shall not apply to interior portions or exterior terminals of equipment designed, manufactured, and tested in accordance with accepted national standards.

### 490.25 Backfeed

Installations where the possibility of backfeed exists shall comply with (a) and (b), which follow.

(a) A permanent sign in accordance with 110.21(B) shall be installed on the disconnecting means enclosure or immediately adjacent to open disconnecting means with the following words or equivalent: DANGER — CONTACTS ON EITHER SIDE OF THIS DEVICE MAY BE ENERGIZED BY BACKFEED.

(b) A permanent and legible single-line diagram of the local switching arrangement, clearly identifying each point of connection to the high-voltage section, shall be provided within sight of each point of connection.

## III. Equipment — Switchgear and Industrial Control Assemblies

### 490.30 General

Part III covers assemblies of switchgear and industrial control equipment including, but not limited to, switches and interrupting devices and their control, metering, protection, and regulating equipment where they are an integral part of the assembly, with associated interconnections and supporting structures. Part III also includes switchgear assemblies that form a part of unit substations, power centers, or similar equipment.

Indicator instruments, such as voltmeters, ammeters, wattmeters, and protective relays, may be mounted on the panel doors as desired. This switchgear affords a high degree of safety because all live parts are metal-enclosed, and interlocks are provided for safe operation.

### 490.31 Arrangement of Devices in Assemblies

Arrangement of devices in assemblies shall be such that individual components can safely perform their intended function without adversely affecting the safe operation of other components in the assembly.

### 490.32 Guarding of High-Voltage Energized Parts Within a Compartment

Where access for other than visual inspection is required to a compartment that contains energized high-voltage parts, barriers shall be provided to prevent accidental contact by persons, tools, or other equipment with energized parts. Exposed live parts shall only be permitted in compartments accessible to qualified persons. Fuses and fuseholders designed to enable future replacement without de-energizing the fuseholder shall only be permitted for use by qualified persons.

***EXHIBIT 490.3*** *Open view of a tamperproof pad-mount transformer for loop-feed application with load-break elbow connectors. (Courtesy of Schneider Electric)*

An example of a high-voltage pad-mounted transformer and enclosure that may contain primary and secondary switches or circuit breakers is shown in Exhibit 490.3. The high-voltage compartment on the left of Exhibit 490.3 has bayonet fusing and a load-break transformer on/off switch, which are both hot-stick operated.

### 490.33 Guarding of Energized Parts Operating at 1000 Volts, Nominal, or Less Within Compartments

Energized bare parts mounted on doors shall be guarded where the door must be opened for maintenance of equipment or removal of draw-out equipment.

### 490.34 Clearance for Cable Conductors Entering Enclosure

The unobstructed space opposite terminals or opposite raceways or cables entering a switchgear or control assembly shall be approved for the type of conductor and method of termination.

### 490.35 Accessibility of Energized Parts

**(A) High-Voltage Equipment.** Doors that would provide unqualified persons access to high-voltage energized parts shall be locked. Permanent signs in accordance with 110.21(B) shall be installed on panels or doors that provide access to live parts over 1000 volts and shall read DANGER — HIGH VOLTAGE — KEEP OUT.

**(B) Control Equipment.** Where operating at 1000 volts, nominal, or less, control equipment, relays, motors, and the like shall not be installed in compartments with exposed high-voltage energized parts or high-voltage wiring, unless either of the following conditions is met:

(1) The access means is interlocked with the high-voltage switch or disconnecting means to prevent the access means from being opened or removed.
(2) The high-voltage switch or disconnecting means is in the isolating position.

**(C) High-Voltage Instruments or Control Transformers and Space Heaters.** High-voltage instrument or control transformers and space heaters shall be permitted to be installed in the high-voltage compartment without access restrictions beyond those that apply to the high-voltage compartment generally.

## 490.36 Grounding

Frames of switchgear and control assemblies shall be connected to an equipment grounding conductor or, where permitted, the grounded conductor.

## 490.37 Grounding of Devices

The metal cases or frames, or both, such as those of instruments, relays, meters, and instrument and control transformers, located in or on switchgear or control assemblies, shall be connected to an equipment grounding conductor or, where permitted, the grounded conductor.

## 490.38 Door Stops and Cover Plates

External hinged doors or covers shall be provided with stops to hold them in the open position. Cover plates intended to be removed for inspection of energized parts or wiring shall be equipped with lifting handles and shall not exceed 1.1 $m^2$ (12 $ft^2$) in area or 27 kg (60 lb) in weight, unless they are hinged and bolted or locked.

## 490.39 Gas Discharge from Interrupting Devices

Gas discharged during operating of interrupting devices shall be directed so as not to endanger personnel.

## 490.40 Visual Inspection Windows

Windows intended for visual inspection of disconnecting switches or other devices shall be of suitable transparent material.

## 490.41 Location of Industrial Control Equipment

Routinely operated industrial control equipment shall meet the requirements of (A) unless infrequently operated, as covered in 490.41(B).

**(A) Control and Instrument Transfer Switch Handles or Push Buttons.** Control and instrument transfer switch handles or push buttons shall be in a readily accessible location at an elevation of not over 2.0 m (6 ft 7 in.).

*Exception: Operating handles requiring more than 23 kg (50 lb) of force shall be located no higher than 1.7 m (66 in.) in either the open or closed position.*

**(B) Infrequently Operated Devices.** Where operating handles for such devices as draw-out fuses, fused potential or control transformers and their primary disconnects, and bus transfer and isolating switches are only operated infrequently, the handles shall be permitted to be located where they are safely operable and serviceable from a portable platform.

## 490.42 Interlocks — Interrupter Switches

Interrupter switches equipped with stored energy mechanisms shall have mechanical interlocks to prevent access to the switch compartment unless the stored energy mechanism is in the discharged or blocked position.

## 490.43 Stored Energy for Opening

The stored energy operator shall be permitted to be left in the uncharged position after the switch has been closed if a single movement of the operating handle charges the operator and opens the switch.

## 490.44 Fused Interrupter Switches

**(A) Supply Terminals.** The supply terminals of fused interrupter switches shall be installed at the top of the switch enclosure or, if the terminals are located elsewhere, the equipment shall have barriers installed so as to prevent persons from accidentally contacting energized parts or dropping tools or fuses into energized parts.

**(B) Backfeed.** Where fuses can be energized by backfeed, a sign shall be placed on the enclosure door identifying this hazard.

**(C) Switching Mechanism.** The switching mechanism shall be arranged to be operated from a location outside the enclosure where the operator is not exposed to energized parts and shall be arranged to open all ungrounded conductors of the circuit simultaneously with one operation. Switches shall be lockable in accordance with 110.25.

## 490.45 Circuit Breakers — Interlocks

**(A) Circuit Breakers.** Circuit breakers equipped with stored energy mechanisms shall be designed to prevent the release of the stored energy unless the mechanism has been fully charged.

**(B) Mechanical Interlocks.** Mechanical interlocks shall be provided in the housing to prevent the complete withdrawal of the circuit breaker from the housing when the stored energy mechanism is in the fully charged position, unless a suitable device is provided to block the closing function of the circuit breaker before complete withdrawal.

### 490.46 Circuit Breaker Locking

Circuit breakers shall be capable of being locked in the open position or, if they are installed in a drawout mechanism, that mechanism shall be capable of being locked in such a position that the mechanism cannot be moved into the connected position. In either case, the provision for locking shall be lockable in accordance with 110.25.

### 490.47 Switchgear Used as Service Equipment

Switchgear installed as high-voltage service equipment shall include a ground bus for the connection of service cable shields and to facilitate the attachment of safety grounds for personnel protection. This bus shall be extended into the compartment where the service conductors are terminated. Where the compartment door or panel provides access to parts that can only be de-energized and visibly isolated by the serving utility, the warning sign required by 490.35(A) shall include a notice that access is limited to the serving utility or is permitted only following an authorization of the serving utility.

Switchgear must include a ground bus for the service cable shields. It also provides a location for the connection of safety grounds for personnel protection during servicing of the equipment. The bus must extend to the compartment where the service conductor terminals are located.

### 490.48 Substation Design, Documentation, and Required Diagram

**(A) Design and Documentation.** Substations shall be designed by a qualified licensed professional engineer. Where components or the entirety of the substation are listed by a qualified electrical testing laboratory, documentation of internal design features subject to the listing investigation shall not be required. The design shall address but not be limited to the following topics and the documentation of this design shall be made available to the authority having jurisdiction.

(1) Clearances and exits
(2) Electrical enclosures
(3) Securing and support of electrical equipment
(4) Fire protection
(5) Safety ground connection provisions
(6) Guarding live parts
(7) Transformers and voltage regulation equipment
(8) Conductor insulation, electrical and mechanical protection, isolation, and terminations
(9) Application, arrangement, and disconnection of circuit breakers, switches, and fuses
(10) Provisions for oil filled equipment
(11) Switchgear
(12) Surge arrestors

**(B) Diagram.** A permanent single-line diagram of the switchgear shall be provided in a readily visible location within the same room or enclosed area with the switchgear and this diagram shall clearly identify interlocks, isolation means, and all possible sources of voltage to the installation under normal or emergency conditions, and the marking on the switchgear shall cross-reference the diagram.

*Exception: Where the equipment consists solely of a single cubicle or metal enclosed unit substation containing only one set of high-voltage switching devices, diagrams shall not be required.*

## IV. Mobile and Portable Equipment

### 490.51 General

**(A) Covered.** The provisions of this part shall apply to installations and use of high-voltage power distribution and utilization equipment that is portable, mobile, or both, such as substations and switch houses mounted on skids, trailers, or cars; mobile shovels; draglines; cranes; hoists; drills; dredges; compressors; pumps; conveyors; underground excavators; and the like.

**(B) Other Requirements.** The requirements of this part shall be additional to, or amendatory of, those prescribed in Articles 100 through 725 of this *Code*. Special attention shall be paid to Article 250.

**(C) Protection.** Approved enclosures or guarding, or both, shall be provided to protect portable and mobile equipment from physical damage.

**(D) Disconnecting Means.** Disconnecting means shall be installed for mobile and portable high-voltage equipment according to the requirements of Part VIII of Article 230 and shall disconnect all ungrounded conductors.

### 490.52 Overcurrent Protection

Motors driving single or multiple dc generators supplying a system operating on a cyclic load basis do not require overload protection, provided that the thermal rating of the ac drive motor cannot be exceeded under any operating condition. The branch-circuit protective device(s) shall provide short-circuit and locked-rotor protection and shall be permitted to be external to the equipment.

### 490.53 Enclosures

All energized switching and control parts shall be enclosed in grounded metal cabinets or enclosures. These cabinets or enclosures shall be marked DANGER — HIGH VOLTAGE — KEEP OUT and shall be locked so that only authorized and qualified persons can enter. The danger marking(s) or label(s) shall comply with 110.21(B). Circuit breakers and protective equipment shall have the operating means projecting through the metal cabinet or enclosure so these units can be reset without opening locked doors. With doors closed, safe access for normal operation of these units shall be provided.

### 490.54 Collector Rings

The collector ring assemblies on revolving-type machines (shovels, draglines, etc.) shall be guarded to prevent accidental contact with energized parts by personnel on or off the machine.

### 490.55 Power Cable Connections to Mobile Machines

A metallic enclosure shall be provided on the mobile machine for enclosing the terminals of the power cable. The enclosure shall include terminal connections to the machine frame for the equipment grounding conductor. Ungrounded conductors shall be attached to insulators or be terminated in approved high-voltage cable couplers (which include equipment grounding conductor connectors) of proper voltage and ampere rating. The method of cable termination used shall prevent any strain or pull on the cable from stressing the electrical connections. The enclosure shall have provision for locking so that only authorized and qualified persons may open it and shall be marked as follows:

DANGER — HIGH VOLTAGE — KEEP OUT.

The danger marking(s) or label(s) shall comply with 110.21(B).

### 490.56 High-Voltage Portable Cable for Main Power Supply

Flexible high-voltage cable supplying power to portable or mobile equipment shall comply with Article 250 and Article 400, Part III.

## V. Electrode-Type Boilers

### 490.70 General

The provisions of Part V shall apply to boilers operating over 1000 volts, nominal, in which heat is generated by the passage of current between electrodes through the liquid being heated.

### 490.71 Electrical Supply System

Electrode-type boilers shall be supplied only from a 3-phase, 4-wire solidly grounded wye system, or from isolating transformers arranged to provide such a system. Control circuit voltages shall not exceed 150 volts, shall be supplied from a grounded system, and shall have the controls in the ungrounded conductor.

### 490.72 Branch-Circuit Requirements.

**(A) Rating.** Each boiler shall be supplied from an individual branch circuit rated not less than 100 percent of the total load.

**(B) Common-Trip Fault-Interrupting Device.** The circuit shall be protected by a 3-phase, common-trip fault-interrupting device, which shall be permitted to automatically reclose the circuit upon removal of an overload condition but shall not reclose after a fault condition.

**(C) Phase-Fault Protection.** Phase-fault protection shall be provided in each phase, consisting of a separate phase-overcurrent relay connected to a separate current transformer in the phase.

**(D) Ground Current Detection.** Means shall be provided for detection of the sum of the neutral conductor and equipment grounding conductor currents and shall trip the circuit-interrupting device if the sum of those currents exceeds the greater of 5 amperes or 7½ percent of the boiler full-load current for 10 seconds or exceeds an instantaneous value of 25 percent of the boiler full-load current.

**(E) Grounded Neutral Conductor.** The grounded neutral conductor shall be as follows:

(1) Connected to the pressure vessel containing the electrodes
(2) Insulated for not less than 1000 volts
(3) Have not less than the ampacity of the largest ungrounded branch-circuit conductor
(4) Installed with the ungrounded conductors in the same raceway, cable, or cable tray, or, where installed as open conductors, in close proximity to the ungrounded conductors
(5) Not used for any other circuit

### 490.73 Pressure and Temperature Limit Control

Each boiler shall be equipped with a means to limit the maximum temperature, pressure, or both, by directly or indirectly interrupting all current flow through the electrodes. Such means shall be in addition to the temperature, pressure, or both, regulating systems and pressure relief or safety valves.

### 490.74 Bonding

All exposed non–current-carrying metal parts of the boiler and associated exposed metal structures or equipment shall be bonded to the pressure vessel or to the neutral conductor to which the vessel is connected in accordance with 250.102, except the ampacity of the bonding jumper shall not be less than the ampacity of the neutral conductor.

# 5 Special Occupancies

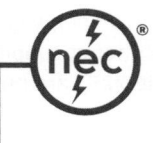

## ARTICLE 500 Hazardous (Classified) Locations, Classes I, II, and III, Divisions 1 and 2

Informational Note: Text that is followed by a reference in brackets has been extracted from NFPA 497-2008, *Recommended Practice for the Classification of Flammable Liquids, Gases, or Vapors and of Hazardous (Classified) Locations for Electrical Installations in Chemical Process Areas*, and NFPA 499-2008, *Recommended Practice for the Classification of Combustible Dusts and of Hazardous (Classified) Locations for Electrical Installation in Chemical Process Areas.* Only editorial changes were made to the extracted text to make it consistent with this *Code*.

### 500.1 Scope — Articles 500 Through 504

Articles 500 through 504 cover the requirements for electrical and electronic equipment and wiring for all voltages in Class I, Divisions 1 and 2; Class II, Divisions 1 and 2; and Class III, Divisions 1 and 2 locations where fire or explosion hazards may exist due to flammable gases, flammable liquid–produced vapors, combustible liquid–produced vapors, combustible dusts, or ignitible fibers/flyings.

Articles 500 through 516 cover the requirements for electrical installations in locations classified as hazardous due to the materials handled, processed, or stored in those locations. Hazardous (classified) locations, if properly treated, are not necessarily any more dangerous to work in than other areas or locations. The term *hazardous* is maintained in the *NEC* because it provides, by its very name, a degree of warning. Hazardous locations are sometimes referred to as classified locations. As used in NFPA codes and standards and the *NEC*, the terms are interchangeable.

Some of the most common materials encountered in hazardous locations are flammable and combustible liquids. A flammable liquid is one that has a flash point below 100°F, while a combustible liquid has a flash point at or above 100°F. A flammable or combustible liquid must be at its flash point before an explosion can occur. For example, No. 1-D diesel fuel oil and kerosene, with flash points higher than 100°F, are combustible liquids and do not emit flammable vapors unless heated above their flash points.

Article 500 is limited to locations classified as Class I, Class II, or Class III, which are further addressed in Articles 501, 502, and 503, respectively. Article 505 and Article 506 contain the requirements using the classification of Zones. Article 504 on intrinsically safe systems is used for both methods (Division and Zone) of area classification.

The *Code* does not classify areas where explosive materials, such as ammunition, dynamite, and blasting powder are present. Areas where such materials are present are not considered hazardous locations in accordance with Article 500. However, many organizations responsible for the safety of such areas require equipment and wiring methods suitable for hazardous locations as part of many safety precautions, even though the equipment and wiring have not been investigated for such locations. Further information on these locations can be found in NFPA 495, *Explosive Materials Code*.

For information on hazardous locations in general, including background on the classification of areas, equipment protection systems, ignition sources, static electricity and lightning, and requirements that apply outside the United States, see *Electrical Installations in Hazardous Locations*, by Peter J. Schram, Mark W. Earley, and Robert J. Benedetti. This book is available from NFPA.

Informational Note No. 1: The unique hazards associated with explosives, pyrotechnics, and blasting agents are not addressed in this article.

Informational Note No. 2: For the requirements for electrical and electronic equipment and wiring for all voltages in Zone 0, Zone 1, and Zone 2 hazardous (classified) locations where fire or explosion hazards may exist due to flammable gases or vapors or flammable liquids, refer to Article 505.

Informational Note No. 3: For the requirements for electrical and electronic equipment and wiring for all voltages in Zone 20, Zone 21, and Zone 22 hazardous (classified) locations where fire or explosion hazards may exist due to combustible dusts or ignitible fibers/flyings, refer to Article 506.

### 500.2 Definitions

For purposes of Articles 500 through 504 and Articles 510 through 516, the following definitions apply.

•

**Combustible Dust.** Dust particles that are 500 microns or smaller (material passing a U.S. No. 35 Standard Sieve as defined in ASTM E 11-09, *Standard Specification for Wire Cloth and Sieves for Testing Purposes*) and present a fire or explosion hazard when dispersed and ignited in air.

Informational Note: See ASTM E 1226–12a, *Standard Test Method for Explosibility of Dust Clouds*, or ISO 6184-1, *Explosion protection systems — Part 1: Determination of explosion indices of combustible dusts in air*, for procedures for determining the explosibility of dusts.

**Combustible Gas Detection System.** A protection technique utilizing stationary gas detectors in industrial establishments.

**Control Drawing.** A drawing or other document provided by the manufacturer of the intrinsically safe or associated apparatus, or of the nonincendive field wiring apparatus or associated nonincendive field wiring apparatus, that details the allowed interconnections between the intrinsically safe and associated apparatus or between the nonincendive field wiring apparatus or associated nonincendive field wiring apparatus.

**Dust-Ignitionproof.** Equipment enclosed in a manner that excludes dusts and does not permit arcs, sparks, or heat otherwise generated or liberated inside of the enclosure to cause ignition of exterior accumulations or atmospheric suspensions of a specified dust on or in the vicinity of the enclosure.

Informational Note: For further information on dustignitionproof enclosures, see Type 9 enclosure in ANSI/NEMA 250-2008, *Enclosures for Electrical Equipment,* and ANSI/UL 1203-2009, *Explosionproof and Dust-Ignitionproof Electrical Equipment for Hazardous (Classified) Locations*.

**Dusttight.** Enclosures constructed so that dust will not enter under specified test conditions.

Informational Note: See ANSI/ISA-12.12.01-2012, *Nonincendive Electrical Equipment for Use in Class I and II, Division 2, and Class III, Divisions 1 and 2 Hazardous (Classified) Locations*.

A dusttight enclosure has been determined to exclude dust under specified test conditions. Combustible dust is 500 microns or smaller as defined above. The AHJ determines the suitability of a dusttight enclosure or the acceptance of a specific standard, test, or listing organization.

•

**Hermetically Sealed.** Equipment sealed against the entrance of an external atmosphere where the seal is made by fusion, for example, soldering, brazing, welding, or the fusion of glass to metal.

Informational Note: For further information, see ANSI/ISA-12.12.01-2012, *Nonincendive Electrical Equipment for Use in Class I and II, Division 2, and Class III, Divisions 1 and 2 Hazardous (Classified) Locations*.

**Nonincendive Circuit.** A circuit, other than field wiring, in which any arc or thermal effect produced under intended operating conditions of the equipment, is not capable, under specified test conditions, of igniting the flammable gas–air, vapor–air, or dust–air mixture.

Informational Note: Conditions are described in ANSI/ISA-12.12.01-2012, *Nonincendive Electrical Equipment for Use in Class I and II, Division 2, and Class III, Divisions 1 and 2 Hazardous (Classified) Locations*.

A nonincendive circuit employs a protection technique that prevents electrical circuits from causing a fire or explosion in a hazardous location under normal conditions. This is in contrast to an intrinsic safety circuit, whose evaluation is conducted under abnormal conditions. Because of its definition, a nonincendive circuit is a low-energy circuit, but many low-voltage, low-energy circuits, including some communications circuits and thermocouple circuits (or Class 2 or 3 circuits as defined in Article 725), are not necessarily nonincendive.

**Nonincendive Component.** A component having contacts for making or breaking an incendive circuit and the contacting mechanism is constructed so that the component is incapable of igniting the specified flammable gas–air or vapor–air mixture. The housing of a nonincendive component is not intended to exclude the flammable atmosphere or contain an explosion.

Informational Note: For further information, see ANSI/ISA-12.12.01-2012, *Nonincendive Electrical Equipment for Use in Class I and II, Division 2, and Class III, Divisions 1 and 2 Hazardous (Classified) Locations*.

**Nonincendive Equipment.** Equipment having electrical/electronic circuitry that is incapable, under normal operating conditions, of causing ignition of a specified flammable gas–air, vapor–air, or dust–air mixture due to arcing or thermal means.

Informational Note: For further information, see ANSI/ISA-12.12.01-2012, *Nonincendive Electrical Equipment for Use in Class I and II, Division 2, and Class III, Divisions 1 and 2 Hazardous (Classified) Locations*.

**Nonincendive Field Wiring.** Wiring that enters or leaves an equipment enclosure and, under normal operating conditions of the equipment, is not capable, due to arcing or thermal effects, of igniting the flammable gas–air, vapor–air, or dust–air mixture. Normal operation includes opening, shorting, or grounding the field wiring.

Field wiring meeting this definition requires limitations of energy under normally expected conditions of operation, such as opening, shorting, or grounding. For example, stored energy in the form of mutual inductance or capacitance could be released during an opening, shorting, or grounding of nonincendive field wiring, thus defeating the purpose of this protection technique.

**Nonincendive Field Wiring Apparatus.** Apparatus intended to be connected to nonincendive field wiring.

Informational Note: For further information, see ANSI/ISA-12.12.01-2012, *Nonincendive Electrical Equipment for Use in Class I and II, Division 2, and Class III, Divisions 1 and 2 Hazardous (Classified) Locations.*

**Oil Immersion.** Electrical equipment immersed in a protective liquid in such a way that an explosive atmosphere that may be above the liquid or outside the enclosure cannot be ignited.

**Purged and Pressurized.** The process of (1) purging, supplying an enclosure with a protective gas at a sufficient flow and positive pressure to reduce the concentration of any flammable gas or vapor initially present to an acceptable level; and (2) pressurization, supplying an enclosure with a protective gas with or without continuous flow at sufficient pressure to prevent the entrance of a flammable gas or vapor, a combustible dust, or an ignitible fiber.

Informational Note: For further information, see ANSI/NFPA 496-2013, *Purged and Pressurized Enclosures for Electrical Equipment*.

**Unclassified Locations.** Locations determined to be neither Class I, Division 1; Class I, Division 2; Class I, Zone 0; Class I, Zone 1; Class I, Zone 2; Class II, Division 1; Class II, Division 2; Class III, Division 1; Class III, Division 2; Zone 20; Zone 21; Zone 22; or any combination thereof.

## 500.3 Other Articles

Except as modified in Articles 500 through 504, all other applicable rules contained in this *Code* shall apply to electrical equipment and wiring installed in hazardous (classified) locations.

The first four chapters of the *Code* cover general installation requirements for all electrical equipment and wiring (see 90.3). Materials and equipment installed in a hazardous location must also be suitable for environmental conditions such as rain, snow, ice, altitude, and heat; protected from any deteriorating effects on conductors and equipment; and have an interrupting rating sufficient for the nominal circuit voltage and available fault current, just to name a few.

Requirements in Articles 500 through 517 amend or modify the general installation rules to ensure the integrity of electrical installations and to minimize the possibility of electrical equipment being an ignition source in the environments covered in these articles. Examples of how the hazardous location articles modify the general requirements are found in the wiring method requirements of 501.10, 502.10, and 503.10. These sections limit the wiring methods that can be used in Class I, Class II, and Class III locations in order to provide the highest degree of protection against physical damage. The installation of any permitted wiring method must be in accordance with the Chapter 3 article that governs that particular wiring method and with any modifications in Articles 500 through 517.

## 500.4 General

**(A) Documentation.** All areas designated as hazardous (classified) locations shall be properly documented. This documentation shall be available to those authorized to design, install, inspect, maintain, or operate electrical equipment at the location.

One type of documentation consists of area classification drawings. Once the hazardous area has been classified and the hazardous area documentation has been developed, the materials and installation methods of the *NEC* are used to construct the electrical system in the classified area. This approach provides the necessary information for installers, service personnel, and AHJs to ensure that electrical equipment installed in classified areas is of the proper type. See Exhibit 500.1 for an example of an area classification drawing.

**(B) Reference Standards.** Important information relating to topics covered in Chapter 5 may be found in other publications.

Informational Note No. 1: It is important that the authority having jurisdiction be familiar with recorded industrial experience

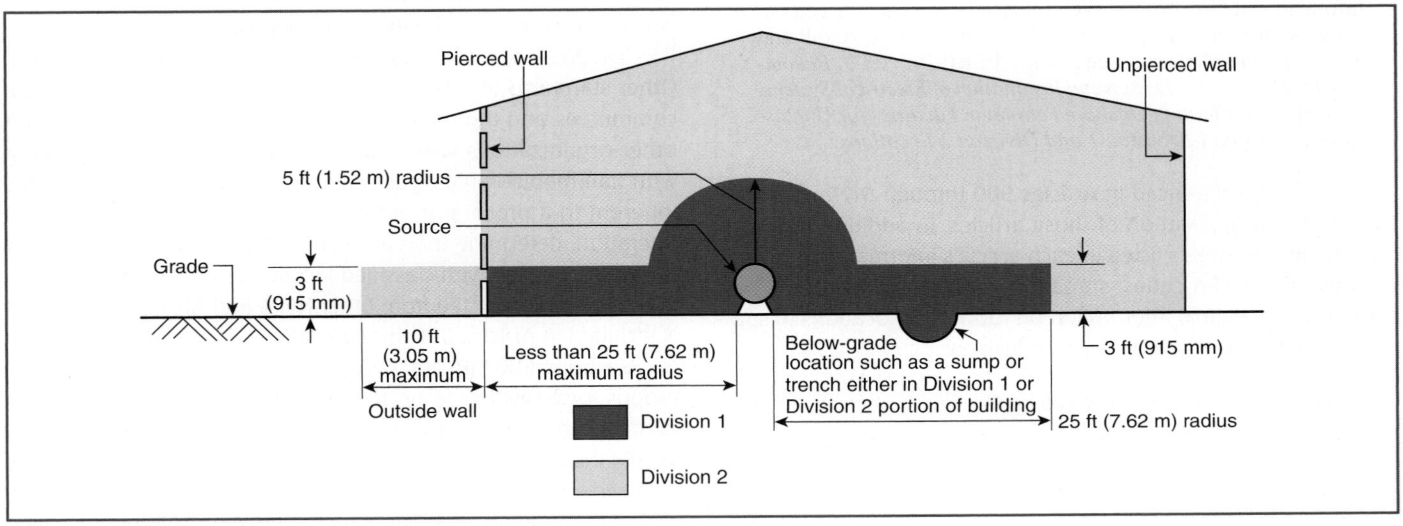


***EXHIBIT 500.1*** *An example of an area classification drawing.*

as well as with the standards of the National Fire Protection Association (NFPA), the American Petroleum Institute (API), and the International Society of Automation (ISA), that may be of use in the classification of various locations, the determination of adequate ventilation, and the protection against static electricity and lightning hazards.

Informational Note No. 2: For further information on the classification of locations, see NFPA 30-2012, *Flammable and Combustible Liquids Code*; NFPA 32-2011, *Standard for Drycleaning Plants*; NFPA 33-2011, *Standard for Spray Application Using Flammable or Combustible Materials*; NFPA 34-2011, *Standard for Dipping and Coating Processes Using Flammable or Combustible Liquids*; NFPA 35-2011, *Standard for the Manufacture of Organic Coatings*; NFPA 36-2013, *Standard for Solvent Extraction Plants*; NFPA 45-2011, *Standard on Fire Protection for Laboratories Using Chemicals*; NFPA 55-2013, *Compressed Gases and Cryogenic Fluids Code*; NFPA 58-2014, *Liquefied Petroleum Gas Code*; NFPA 59-2012, *Utility LP-Gas Plant Code*; NFPA 497-2012, *Recommended Practice for the Classification of Flammable Liquids, Gases, or Vapors and of Hazardous (Classified) Locations for Electrical Installations in Chemical Process Areas*; NFPA 499-2013, *Recommended Practice for the Classification of Combustible Dusts and of Hazardous (Classified) Locations for Electrical Installations in Chemical Process Areas*; NFPA 820-2012, *Standard for Fire Protection in Wastewater Treatment and Collection Facilities*; ANSI/API RP 500-2012, *Recommended Practice for Classification of Locations of Electrical Installations at Petroleum Facilities Classified as Class I, Division 1 and Division 2*; ISA-12.10-1988, *Area Classification in Hazardous (Classified) Dust Locations*.

Informational Note No. 3: For further information on protection against static electricity and lightning hazards in hazardous (classified) locations, see NFPA 77-2014, *Recommended Practice on Static Electricity*; NFPA 780-2014, *Standard for the Installation of Lightning Protection Systems*; and API RP 2003-1998, *Protection Against Ignitions Arising Out of Static Lightning and Stray Currents*.

Informational Note No. 4: For further information on ventilation, see NFPA 30-2012, *Flammable and Combustible Liquids Code*; and ANSI/API RP 500-2012, *Recommended Practice for Classification of Locations for Electrical Installations at Petroleum Facilities Classified as Class I, Division 1 and Division 2*.

Informational Note No. 5: For further information on electrical systems for hazardous (classified) locations on offshore oil- and gas-producing platforms, see ANSI/API RP 14F-1999, *Recommended Practice for Design and Installation of Electrical Systems for Fixed and Floating Offshore Petroleum Facilities for Unclassified and Class I, Division 1 and Division 2 Locations*.

The standards referenced in Articles 500 through 517 are essential for proper application of those articles. In addition to those documents and others listed in each article's Informational Notes, the following NFPA codes, standards, and recommended practices include valuable information on hazardous locations in specific applications, occupancies, or industries:

NFPA 32, *Standard for Drycleaning Plants*

NFPA 35, *Standard for the Manufacture of Organic Coatings*

NFPA 36, *Standard for Solvent Extraction Plants*

NFPA 51, *Standard for the Design and Installation of Oxygen–Fuel Gas Systems for Welding, Cutting, and Allied Processes*

NFPA 51A, *Standard for Acetylene Cylinder Charging Plants*

NFPA 52, *Vehicular Gaseous Fuel Systems Code*

NFPA 54, *Natural Fuel Gas Code*

NFPA 59A, *Standard for the Production, Storage, and Handling of Liquefied Natural Gas (LNG)*

NFPA 61, *Standard for the Prevention of Fires and Dust Explosions in Agricultural and Food Processing Facilities*

NFPA 85, *Boiler and Combustion Systems Hazards Code*

NFPA 99, *Health Care Facilities Code*

NFPA 407, *Standard for Aircraft Fuel Servicing*

NFPA 496, *Standard for Purged and Pressurized Enclosures for Electrical Equipment*

NFPA 654, *Standard for the Prevention of Fire and Dust Explosions from the Manufacturing, Processing, and Handling of Combustible Particulate Solids*

NFPA 655, *Standard for Prevention of Sulfur Fires and Explosions*

## 500.5 Classifications of Locations

**(A) Classifications of Locations.** Locations shall be classified depending on the properties of the flammable gas, flammable liquid–produced vapor, combustible liquid–produced vapors, combustible dusts, or fibers/flyings that may be present, and the likelihood that a flammable or combustible concentration or quantity is present. Each room, section, or area shall be considered individually in determining its classification. Where pyrophoric materials are the only materials used or handled, these locations are outside the scope of this article.

The definitions of classified locations include language such as "ignitible concentrations" or "quantities sufficient to produce explosive or ignitible mixtures." Determining the appropriate classification requires an understanding of the type and amount of material present, of the potential for that presence under normal and abnormal conditions, and of the entire process that the material will or might undergo in given circumstances.

The *NEC* does not classify specific Class I, II, and III locations. Other standards and recommended practices of NFPA technical committees and the American Petroleum Institute (API), among other organizations with experience and expertise in working with flammable liquids, gases, vapors, dusts, fibers, and flyings inherent to a process or present under abnormal conditions of operation, determine the parameters, distances, and degrees of hazard associated with classified locations. Some of this information has been extracted from other NFPA and API documents and is included in Articles 511 through 516.

Sections 500.5(B), (C), and (D) describe three classes of hazardous locations based on the type of material involved (gases and vapors — Class I; dusts — Class II; fibers/flyings — Class III). Within each class are varying degrees of hazard, so each class is subdivided into two divisions that are based on the likelihood the material will be present. The requirements for Division 1 of each class are more stringent than those for Division 2.

***EXHIBIT 500.2*** *A coal-handling operation classified as both a Class I and a Class II location. (Courtesy of Noel Williams)*

Classification requires consideration of the specific equipment and process and depends on the materials and physical equipment and location (for example, possible leaks at flanges and machinery seals). Common sense and judgment must prevail in classifying an area that is likely to become hazardous and in determining those portions of the premises to be classified Division 1 or Division 2.

If different types of material – such as a flammable gas and a combustible dust – exist in a process or location, the area must be classified as both a Class I and a Class II location, and protection must be provided for both hazards. Equipment that is identified for a Class I location may not be suitable for a Class II location and vice versa. Exhibit 500.2 shows a coal-handling operation classified as both a Class I (methane) and Class II (coal dust) location.

Because pyrophoric materials ignite spontaneously upon contact with air, locations containing these are not classified within the scope of the *NEC*. The use of electrical equipment suitable for hazardous locations will not prevent ignition of these materials. Instead, the process containment system should be designed to prevent contact between pyrophoric material and air.

Informational Note: Through the exercise of ingenuity in the layout of electrical installations for hazardous (classified) locations, it is frequently possible to locate much of the equipment in a reduced level of classification or in an unclassified location and, thus, to reduce the amount of special equipment required.

Rooms and areas containing ammonia refrigeration systems that are equipped with adequate mechanical ventilation may be classified as "unclassified" locations.

Informational Note: For further information regarding classification and ventilation of areas involving ammonia, see ANSI/ASHRAE 15-1994, *Safety Code for Mechanical Refrigeration*, and ANSI/CGA G2.1-1989, *Safety Requirements for the Storage and Handling of Anhydrous Ammonia*.

**(B) Class I Locations.** Class I locations are those in which flammable gases, flammable liquid–produced vapors, or combustible

***EXHIBIT 500.3*** *A CNG filling station, which is a Class I location.*

liquid–produced vapors are or may be present in the air in quantities sufficient to produce explosive or ignitible mixtures. Class I locations shall include those specified in 500.5(B)(1) and (B)(2).

For a given vapor, the vapor-in-air ratio must be within the flammable limits in order to be deemed a hazard. Many chemicals reach the lower limit within a few percent of vapor-to-air ratio, and some are lower than 1 percent. The flammable range may be very narrow or very wide (such as for acetylene, which has a flammable range of 2.5 percent to nearly 100 percent).

Class I locations, which involve the handling of a volatile flammable liquid or a flammable gas, are common in many cities and towns. Gas stations and propane filling stations are two examples of this type of location. The public interacts daily with these Class I locations without incident. Exhibit 500.3 shows another type of a Class I location, which is a station that dispenses compressed natural gas (CNG). Each of these types of dispensing facilities is covered under Article 514.

**(1) Class I, Division 1.** A Class I, Division 1 location is a location

(1) In which ignitible concentrations of flammable gases, flammable liquid–produced vapors, or combustible liquid–produced vapors can exist under normal operating conditions, or

Release of combustible materials during maintenance could require a Division 1 classification, which would be determined by the extent and frequency of the equipment maintenance.

(2) In which ignitible concentrations of such flammable gases, flammable liquid–produced vapors, or combustible liquids above their flash points may exist frequently because of repair or maintenance operations or because of leakage, or
(3) In which breakdown or faulty operation of equipment or processes might release ignitible concentrations of

flammable gases, flammable liquid–produced vapors, or combustible liquid–produced vapors and might also cause simultaneous failure of electrical equipment in such a way as to directly cause the electrical equipment to become a source of ignition.

Informational Note No. 1: This classification usually includes the following locations:

(1) Where volatile flammable liquids or liquefied flammable gases are transferred from one container to another
(2) Interiors of spray booths and areas in the vicinity of spraying and painting operations where volatile flammable solvents are used
(3) Locations containing open tanks or vats of volatile flammable liquids
(4) Drying rooms or compartments for the evaporation of flammable solvents
(5) Locations containing fat- and oil-extraction equipment using volatile flammable solvents
(6) Portions of cleaning and dyeing plants where flammable liquids are used
(7) Gas generator rooms and other portions of gas manufacturing plants where flammable gas may escape
(8) Inadequately ventilated pump rooms for flammable gas or for volatile flammable liquids
(9) The interiors of refrigerators and freezers in which volatile flammable materials are stored in open, lightly stoppered, or easily ruptured containers
(10) All other locations where ignitible concentrations of flammable vapors or gases are likely to occur in the course of normal operations

Informational Note No. 2: In some Division 1 locations, ignitible concentrations of flammable gases or vapors may be present continuously or for long periods of time. Examples include the following:

(1) The inside of inadequately vented enclosures containing instruments normally venting flammable gases or vapors to the interior of the enclosure
(2) The inside of vented tanks containing volatile flammable liquids
(3) The area between the inner and outer roof sections of a floating roof tank containing volatile flammable fluids
(4) Inadequately ventilated areas within spraying or coating operations using volatile flammable fluids
(5) The interior of an exhaust duct that is used to vent ignitible concentrations of gases or vapors

Experience has demonstrated the prudence of avoiding the installation of instrumentation or other electrical equipment in these particular areas altogether or where it cannot be avoided because it is essential to the process and other locations are not feasible [see 500.5(A), Informational Note] using electrical equipment or instrumentation approved for the specific application or consisting of intrinsically safe systems as described in Article 504.

**(2) Class I, Division 2.** A Class I, Division 2 location is a location

(1) In which volatile flammable gases, flammable liquid–produced vapors, or combustible liquid–produced vapors are handled, processed, or used, but in which the liquids, vapors, or gases will normally be confined within closed containers or closed systems from which they can escape only in case of accidental rupture or breakdown of such containers or systems or in case of abnormal operation of equipment, or
(2) In which ignitible concentrations of flammable gases, flammable liquid–produced vapors, or combustible liquid–produced vapors are normally prevented by positive mechanical ventilation and which might become hazardous through failure or abnormal operation of the ventilating equipment, or
(3) That is adjacent to a Class I, Division 1 location, and to which ignitible concentrations of flammable gases, flammable liquid–produced vapors, or combustible liquid–produced vapors above their flash points might occasionally be communicated unless such communication is prevented by adequate positive-pressure ventilation from a source of clean air and effective safeguards against ventilation failure are provided.

Informational Note No. 1: This classification usually includes locations where volatile flammable liquids or flammable gases or vapors are used but that, in the judgment of the authority having jurisdiction, would become hazardous only in case of an accident or of some unusual operating condition. The quantity of flammable material that might escape in case of accident, the adequacy of ventilating equipment, the total area involved, and the record of the industry or business with respect to explosions or fires are all factors that merit consideration in determining the classification and extent of each location.

Informational Note No. 2: Piping without valves, checks, meters, and similar devices would not ordinarily introduce a hazardous condition even though used for flammable liquids or gases. Depending on factors such as the quantity and size of the containers and ventilation, locations used for the storage of flammable liquids or liquefied or compressed gases in sealed containers may be considered either hazardous (classified) or unclassified locations. See NFPA 30-2012, *Flammable and Combustible Liquids Code*, and NFPA 58-2014, *Liquefied Petroleum Gas Code*.

**(C) Class II Locations.** Class II locations are those that are hazardous because of the presence of combustible dust. Class II locations shall include those specified in 500.5(C)(1) and (C)(2).

Housekeeping, settlement rates, and air velocity are all factors in determining the need for or extent of the Class II classified location. A layer of dust could ignite at a different temperature than that same dust dispersed into a cloud. Classification of dust layers is based on the thickness of the dust or the amount of dust expected to settle out, usually over a set period of time. One or more of the following four hazards may be present in a Class II location:

1. An explosive mixture of air and dust in suspension
2. Accumulation of dust that acts as a thermal blanket and interferes with the safe dissipation of heat from electrical equipment
3. Accumulation of electrically conductive dust lodged between terminals that have a difference of potential, thereby

causing tracking and glowing hot particles, short circuits, or ground faults that could ignite dust accumulated in the vicinity
4. Deposits of dust that could be ignited by arcs or sparks

**(1) Class II, Division 1.** A Class II, Division 1 location is a location

(1) In which combustible dust is in the air under normal operating conditions in quantities sufficient to produce explosive or ignitible mixtures, or
(2) Where mechanical failure or abnormal operation of machinery or equipment might cause such explosive or ignitible mixtures to be produced, and might also provide a source of ignition through simultaneous failure of electrical equipment, through operation of protection devices, or from other causes, or
(3) In which Group E combustible dusts may be present in quantities sufficient to be hazardous.

Informational Note: Dusts containing magnesium or aluminum are particularly hazardous, and the use of extreme precaution is necessary to avoid ignition and explosion.

Group E dusts are particularly hazardous. For example, current through the dust may cause a sufficient temperature rise or electrical arc to cause ignition. The classification of an area where a Group E dust is or may be present will be Division 1.

**(2) Class II, Division 2.** A Class II, Division 2 location is a location

(1) In which combustible dust due to abnormal operations may be present in the air in quantities sufficient to produce explosive or ignitible mixtures; or
(2) Where combustible dust accumulations are present but are normally insufficient to interfere with the normal operation of electrical equipment or other apparatus, but could as a result of infrequent malfunctioning of handling or processing equipment become suspended in the air; or
(3) In which combustible dust accumulations on, in, or in the vicinity of the electrical equipment could be sufficient to interfere with the safe dissipation of heat from electrical equipment, or could be ignitible by abnormal operation or failure of electrical equipment.

Informational Note No. 1: The quantity of combustible dust that may be present and the adequacy of dust removal systems are factors that merit consideration in determining the classification and may result in an unclassified area.

Informational Note No. 2: Where products such as seed are handled in a manner that produces low quantities of dust, the amount of dust deposited may not warrant classification.

**(D) Class III Locations.** Class III locations are those that are hazardous because of the presence of easily ignitible fibers or where materials producing combustible flyings are handled, manufactured, or used, but in which such fibers/flyings are not likely to be in suspension in the air in quantities sufficient to produce ignitible mixtures. Class III locations shall include those specified in 500.5(D)(1) and (D)(2).

**(1) Class III, Division 1.** A Class III, Division 1 location is a location in which easily ignitible fibers/flyings are handled, manufactured, or used.

Informational Note No. 1: Such locations usually include some parts of rayon, cotton, and other textile mills; combustible fibers/flyings manufacturing and processing plants; cotton gins and cotton-seed mills; flax-processing plants; clothing manufacturing plants; woodworking plants; and establishments and industries involving similar hazardous processes or conditions.

Informational Note No. 2: Easily ignitible fibers/flyings include rayon, cotton (including cotton linters and cotton waste), sisal or henequen, istle, jute, hemp, tow, cocoa fiber, oakum, baled waste kapok, Spanish moss, excelsior, and other materials of similar nature.

**(2) Class III, Division 2.** A Class III, Division 2 location is a location in which easily ignitible fibers/flyings are stored or handled other than in the process of manufacture.

## 500.6 Material Groups

For purposes of testing, approval, and area classification, various air mixtures (not oxygen-enriched) shall be grouped in accordance with 500.6(A) and (B).

Oxygen enrichment can drastically change the explosion characteristics of materials. It lowers minimum ignition energies, increases explosion pressures, and can reduce the maximum experimental safe gap, rendering both intrinsically safe and explosionproof equipment unsafe unless the equipment has been tested for the conditions involved.

Some metals burn freely in an oxygen-enriched atmosphere (in varying degrees depending on the concentration and pressure of the oxygen). Electrical contacts likewise can burn and initiate fires unless they are suitable for the particular pressure and oxygen. Equipment found to be safe in ordinary atmospheric conditions is not necessarily safe in oxygen concentrations or pressures higher than those of ordinary atmospheres. NFPA 53, *Recommended Practice on Materials, Equipment, and Systems Used in Oxygen-Enriched Atmospheres,* recommends that no electrical equipment be used in these atmospheres unless approved for use in the specific hazardous atmospheres at the maximum proposed pressure and oxygen concentration.

*Exception: Equipment identified for a specific gas, vapor, or dust.*

Informational Note: This grouping is based on the characteristics of the materials. Facilities are available for testing and identifying equipment for use in the various atmospheric groups.

NFPA 497, *Recommended Practice for the Classification of Flammable Liquids, Gases, or Vapors and of Hazardous (Classified) Locations for Electrical Installations in Chemical Process Areas,* contains a list of materials that have been categorized by group. For the complete table, refer to NFPA 497 or view the table at www.nfpa.org/nech.

For dust materials, NFPA 499, *Recommended Practice for the Classification of Combustible Dusts and of Hazardous (Classified) Locations for Electrical Installations in Chemical Process Areas,* contains a list of materials that have been categorized by group. For the complete table, refer to NFPA 497 or view the table at www.nfpa.org/nech.

If the group assignment of a material has not been determined, testing the material might be necessary to ensure that suitable electrical equipment can be selected for an installation.

**(A) Class I Group Classifications.** Class I groups shall be according to 500.6(A)(1) through (A)(4).

Informational Note No. 1: Informational Note Nos. 2 and 3 apply to 500.6(A).

Informational Note No. 2: The explosion characteristics of air mixtures of gases or vapors vary with the specific material involved. For Class I locations, Groups A, B, C, and D, the classification involves determinations of maximum explosion pressure and maximum safe clearance between parts of a clamped joint in an enclosure. It is necessary, therefore, that equipment be identified not only for class but also for the specific group of the gas or vapor that will be present.

Informational Note No. 3: Certain chemical atmospheres may have characteristics that require safeguards beyond those required for any of the Class I groups. Carbon disulfide is one of these chemicals because of its low autoignition temperature (90°C) and the small joint clearance permitted to arrest its flame.

**(1) Group A. Acetylene.** [**497:**3.3.5.1.1]

**(2) Group B.** Flammable gas, flammable liquid–produced vapor, or combustible liquid–produced vapor mixed with air that may burn or explode, having either a maximum experimental safe gap (MESG) value less than or equal to 0.45 mm or a minimum igniting current ratio (MIC ratio) less than or equal to 0.40. [**497:**3.3.5.1.2]

Informational Note: A typical Class I, Group B material is hydrogen.

*Exception No. 1: Group D equipment shall be permitted to be used for atmospheres containing butadiene, provided all conduit runs into explosionproof equipment are provided with explosionproof seals installed within 450 mm (18 in.) of the enclosure.*

*Exception No. 2: Group C equipment shall be permitted to be used for atmospheres containing allyl glycidyl ether, n-butyl glycidyl ether, ethylene oxide, propylene oxide, and acrolein, provided all conduit runs into explosionproof equipment are provided with explosionproof seals installed within 450 mm (18 in.) of the enclosure.*

The specific materials identified in Exception No. 1 and Exception No. 2 produce high pressures due to pressure piling in unsealed conduits. Pressure piling in conduits or pipes is a phenomenon in which the pressure caused by ignition of gas propels unburned gas ahead of the moving flame front. When the unburned gas is ignited, a more powerful explosion can occur. If all conduits are sealed, the volume of air and gas between the seal fittings and the arcing contacts is limited, thereby minimizing the pressure that can build within the enclosure and the raceway.

**(3) Group C.** Flammable gas, flammable liquid–produced vapor, or combustible liquid–produced vapor mixed with air that may burn or explode, having either a maximum experimental safe gap (MESG) value greater than 0.45 mm and less than or equal to 0.75 mm, or a minimum igniting current ratio (MIC ratio) greater than 0.40 and less than or equal to 0.80. [**497:**3.3.5.1.3]

Informational Note: A typical Class I, Group C material is ethylene.

**(4) Group D.** Flammable gas, flammable liquid–produced vapor, or combustible liquid–produced vapor mixed with air that may burn or explode, having either a maximum experimental safe gap (MESG) value greater than 0.75 mm or a minimum igniting current ratio (MIC ratio) greater than 0.80. [**497:**3.3.5.1.4]

Informational Note No. 1: A typical Class I, Group D material is propane.

Informational Note No. 2: For classification of areas involving ammonia atmospheres, see ANSI/ASHRAE 15-1994, *Safety Code for Mechanical Refrigeration*, and ANSI/CGA G2.1-1989, *Safety Requirements for the Storage and Handling of Anhydrous Ammonia*.

Flammable gases, flammable liquid–produced vapors, and combustible liquid–produced vapors are separated into four Class I groups – A, B, C, and D (or three Class I zone groups – IIC, IIB, and IIA; see 505.6), depending on their properties. By grouping explosive mixtures that have similar igniting current ratios and maximum safe clearances between parts of a joint in an enclosure, equipment can be designed for the entire group rather than an individual chemical. The *Code* requirements for Class I locations do not vary for different kinds of gas or vapor contained in a group, except as in Exceptions No. 1 and No. 2 to 500.6(A)(2). This method makes it easier to properly select equipment designed for use in the particular group involved.

Selected combustible materials have been evaluated for the purpose of designating the appropriate gas group – A, B, C, or D (or IIC, IIB, or IIA) – and this information is used to properly select electrical equipment for use in Class I locations. These materials, with their group classification and relevant physical properties, are listed in NFPA 497, *Recommended Practice for the Classification of Flammable Liquids, Gases, or Vapors and of Hazardous (Classified) Locations for Electrical Installations in Chemical Process Areas*. For the complete table, refer to NFPA 497 or view the table at www.nfpa.org/nech.

Many documents used to determine an area's classification do not require a hazardous classification unless the location contains a concentration over 25 percent of the material's lower flammable limit (LFL).

**(B) Class II Group Classifications.** Class II groups shall be in accordance with 500.6(B)(1) through (B)(3).

**(1) Group E.** Atmospheres containing combustible metal dusts, including aluminum, magnesium, and their commercial alloys, or other combustible dusts whose particle size, abrasiveness, and conductivity present similar hazards in the use of electrical equipment. [**499:**3.3.4.1]

Informational Note: Certain metal dusts may have characteristics that require safeguards beyond those required for atmospheres containing the dusts of aluminum, magnesium, and their commercial alloys. For example, zirconium, thorium, and uranium dusts have extremely low ignition temperatures [as low as 20°C (68°F)] and minimum ignition energies lower than any material classified in any of the Class I or Class II groups.

**(2) Group F.** Atmospheres containing combustible carbonaceous dusts that have more than 8 percent total entrapped volatiles (see ASTM D 3175-02, *Standard Test Method for Volatile Matter in the Analysis Sample for Coal and Coke*, for coal and coke dusts) or that have been sensitized by other materials so that they present an explosion hazard. Coal, carbon black, charcoal, and coke dusts are examples of carbonaceous dusts. [**499:**3.3.4.2]

Informational Note: Testing of specific dust samples, following established ASTM testing procedures, is a method used to identify the combustibility of a specific dust and the need to classify those locations containing that material as Group F.

**(3) Group G.** Atmospheres containing combustible dusts not included in Group E or Group F, including flour, grain, wood, plastic, and chemicals. [**499:**3.3.4.3]

NFPA 664, *Standard for the Prevention of Fires and Explosions in Wood Processing and Woodworking Facilities,* establishes minimum requirements for industrial, commercial, or institutional facilities that process wood or that manufacture wood products, creating wood chips, particles, or dust.

Informational Note No. 1: For additional information on group classification of Class II materials, see NFPA 499-2013, *Recommended Practice for the Classification of Combustible Dusts and of Hazardous (Classified) Locations for Electrical Installations in Chemical Process Areas*.

Informational Note No. 2: The explosion characteristics of air mixtures of dust vary with the materials involved. For Class II locations, Groups E, F, and G, the classification involves the tightness of the joints of assembly and shaft openings to prevent the entrance of dust in the dust-ignitionproof enclosure, the blanketing effect of layers of dust on the equipment that may cause overheating, and the ignition temperature of the dust. It is necessary, therefore, that equipment be identified not only for the class but also for the specific group of dust that will be present.

Informational Note No. 3: Certain dusts may require additional precautions due to chemical phenomena that can result in the generation of ignitible gases. See ANSI/IEEE C2-2012, *National Electrical Safety Code*, Section 127A, Coal Handling Areas.

Section 500.6(B) separates combustible dusts into three Class II groups — E, F, and G — depending on their properties.

Selected combustible dusts have been evaluated for the purpose of designating the appropriate dust group — E, F, or G — and this information is used to select electrical equipment for use in Class II locations. These selected materials, with their group classification and relevant physical properties, are listed in NFPA 499, *Recommended Practice for the Classification of Combustible Dusts and of Hazardous (Classified) Locations for Electrical Installations in Chemical Process Areas*. For the complete table, refer to NFPA 499 or view the table at www.nfpa.org/nech.

## 500.7 Protection Techniques

Section 500.7(A) through (L) shall be acceptable protection techniques for electrical and electronic equipment in hazardous (classified) locations.

**(A) Explosionproof Equipment.** This protection technique shall be permitted for equipment in Class I, Division 1 or 2 locations.

**(B) Dust Ignitionproof.** This protection technique shall be permitted for equipment in Class II, Division 1 or 2 locations.

**(C) Dusttight.** This protection technique shall be permitted for equipment in Class II, Division 2 or Class III, Division 1 or 2 locations.

**(D) Purged and Pressurized.** This protection technique shall be permitted for equipment in any hazardous (classified) location for which it is identified.

NFPA 496, *Standard for Purged and Pressurized Enclosures for Electrical Equipment*, provides requirements for these enclosures in Class I and Class II hazardous locations. In Class I locations, purged and pressurized enclosures are used to eliminate or reduce, within the enclosure, a Class I hazardous location classification. In Class II locations, pressurized enclosures prevent the entrance of dusts into an enclosure. Purged and pressurized enclosures make it possible for equipment that is not otherwise acceptable for Class I and Class II locations to be used in these locations.

Purging is the process of supplying an enclosure with a protective gas at a sufficient flow and positive pressure to reduce the initial concentration of any flammable gases, flammable liquid–produced vapors, or combustible liquid–produced vapors to an acceptable level.

Pressurization is the process of supplying an enclosure with a protective gas, with or without continuous flow, at sufficient pressure to prevent the entrance of a material.

A combustible dust inside an enclosure cannot be reduced to a safe level by purging with a flow of protective gas in the same manner as with gases or vapors. Supplying a flow of air into the enclosure could stir up the dust that has accumulated inside the enclosure and therein create a dust cloud that could explode if an ignition source occurs. The enclosure must be opened, and the dust must be removed. Positive pressure then prevents dust from re-entering the clean enclosure. The types of pressurizing are as follows:

1. Type X reduces the classification within a protected enclosure from Division 1 or Zone 1 to unclassified.

***EXHIBIT 500.4*** *A panel using Type X pressurization. (Courtesy of Pepperl and Fuchs, Inc.)*

2. Type Y reduces the classification within a protected enclosure from Division 1 to Division 2 or from Zone 1 to Zone 2.
3. Type Z reduces the classification within a protected enclosure from Division 2 or Zone 2 to unclassified.

Exhibit 500.4 shows a panel that uses the Type X purge and pressurization technique. The pressure regulator/filter on the side of the enclosure is for the purge gas supply connection. The integrated controller and vent are mounted on the side and bottom of the panel, respectively.

**(E) Intrinsic Safety.** This protection technique shall be permitted for equipment in Class I, Division 1 or 2; or Class II, Division 1 or 2; or Class III, Division 1 or 2 locations. The provisions of Articles 501 through 503 and Articles 510 through 516 shall not be considered applicable to such installations, except as required by Article 504, and installation of intrinsically safe apparatus and wiring shall be in accordance with the requirements of Article 504.

**(F) Nonincendive Circuit.** This protection technique shall be permitted for equipment in Class I, Division 2; Class II, Division 2; or Class III, Division 1 or 2 locations.

**(G) Nonincendive Equipment.** This protection technique shall be permitted for equipment in Class I, Division 2; Class II, Division 2; or Class III, Division 1 or 2 locations.

**(H) Nonincendive Component.** This protection technique shall be permitted for equipment in Class I, Division 2; Class II, Division 2; or Class III, Division 1 or 2 locations.

**(I) Oil Immersion.** This protection technique shall be permitted for current-interrupting contacts in Class I, Division 2 locations as described in 501.115(B)(1)(2).

**(J) Hermetically Sealed.** This protection technique shall be permitted for equipment in Class I, Division 2; Class II, Division 2; or Class III, Division 1 or 2 locations.

**(K) Combustible Gas Detection System.** A combustible gas detection system shall be permitted as a means of protection in industrial establishments with restricted public access and where the conditions of maintenance and supervision ensure that only qualified persons service the installation. Where such a system is installed, equipment specified in 500.7(K)(1), (K)(2), or (K)(3) shall be permitted.

The type of detection equipment, its listing, installation location(s), alarm and shutdown criteria, and calibration frequency shall be documented when combustible gas detectors are used as a protection technique.

Informational Note No. 1: For further information, see ANSI/ISA-60079-29-1, *Explosive Atmospheres - Part 29-1: Gas detectors - Performance requirements of detectors for flammable gases*, and ANSI/UL 2075, *Gas and Vapor Detectors and Sensors*.

Informational Note No. 2: For further information, see ANSI/API RP 500–Revised 2002, *Recommended Practice for Classification of Locations for Electrical Installations at Petroleum Facilities Classified as Class I, Division I or Division 2*.

Informational Note No. 3: For further information, see ANSI/ISA-60079-29-2009, *Explosive Atmospheres - Part 29-2: Gas detectors - Selection, installation, use and maintenance of detectors for flammable gases and oxygen*.

Informational Note No. 4: For further information, see ISA-TR12.13.03-2009, *Guide for Combustible Gas Detection as a Method of Protection*.

**(1) Inadequate Ventilation.** In a Class I, Division 1 location that is so classified due to inadequate ventilation, electrical equipment suitable for Class I, Division 2 locations shall be permitted. Combustible gas detection equipment shall be listed for Class I, Division 1, for the appropriate material group, and for the detection of the specific gas or vapor to be encountered.

**(2) Interior of a Building.** In a building located in, or with an opening into, a Class I, Division 2 location where the interior does not contain a source of flammable gas or vapor, electrical equipment for unclassified locations shall be permitted. Combustible gas detection equipment shall be listed for Class I, Division 1 or Class I, Division 2, for the appropriate material group, and for the detection of the specific gas or vapor to be encountered.

**(3) Interior of a Control Panel.** In the interior of a control panel containing instrumentation utilizing or measuring flammable liquids, gases, or vapors, electrical equipment suitable for Class I, Division 2 locations shall be permitted. Combustible gas detection equipment shall be listed for Class I, Division 1, for the appropriate material group, and for the detection of the specific gas or vapor to be encountered.

The gas detection system must be suitable for the original division classification of the area, even though the remainder of installed equipment is permitted to be suitable for the next lower division.

Section 17.11 of NFPA 30, *Flammable and Combustible Liquids Code*, and 3.3.1 of NFPA 497 provide information on what is considered adequate ventilation.

**(L) Other Protection Techniques.** Other protection techniques used in equipment identified for use in hazardous (classified) locations.

## 500.8 Equipment

Articles 500 through 504 require equipment construction and installation that ensure safe performance under conditions of proper use and maintenance.

Informational Note No. 1: It is important that inspection authorities and users exercise more than ordinary care with regard to installation and maintenance.

Informational Note No. 2: Since there is no consistent relationship between explosion properties and ignition temperature, the two are independent requirements.

Informational Note No. 3: Low ambient conditions require special consideration. Explosionproof or dust-ignitionproof equipment may not be suitable for use at temperatures lower than −25°C (−13°F) unless they are identified for low-temperature service. However, at low ambient temperatures, flammable concentrations of vapors may not exist in a location classified as Class I, Division 1 at normal ambient temperature.

At low ambient temperatures, such as those encountered in the Arctic, explosion pressures increase and the strengths of materials change, allowing the explosion pressure in explosionproof equipment to exceed the safe operating strength of the material. In addition, some sealing materials for fittings may become brittle. However, the extent of the hazardous location may also change at low ambient temperatures. The material may be used in a location where the temperature range is so low that no vapors are produced, based on the flash point of the material involved.

**(A) Suitability.** Suitability of identified equipment shall be determined by one of the following:

(1) Equipment listing or labeling

Many AHJs and local requirements call for listing of hazardous location equipment or for field evaluation to facilitate the approval process. Without a listing, every jurisdiction could require samples for destructive testing or copies of proprietary equipment drawings, or could conduct an evaluation that takes several months. Therefore, even if the *Code* does not require it, manufacturers often get equipment listed for hazardous locations to facilitate its installation and approval.

(2) Evidence of equipment evaluation from a qualified testing laboratory or inspection agency concerned with product evaluation

(3) Evidence acceptable to the authority having jurisdiction such as a manufacturer's self-evaluation or an owner's engineering judgment.

Informational Note: Additional documentation for equipment may include certificates demonstrating compliance with applicable equipment standards, indicating special conditions of use, and other pertinent information. Guidelines for certificates may be found in ANSI/ISA 12.00.02, *Certificate Standard for AEx Equipment for Hazardous (Classified) Locations*.

Several testing and product evaluation agencies list electrical equipment that is suitable for use in hazardous locations. The acceptance of the listing agency is a responsibility of the AHJ.

Testing laboratories outside the United States provide listing of equipment for use in hazardous locations, but they may not be testing and investigating the equipment for use in hazardous locations as defined in Article 500. These laboratories certify equipment for installation according to an IEC classification scheme similar to that in Articles 505 and 506. However, equipment certified to a product standard used in another country may require modification to be in compliance with the *National Electrical Code*. Some modifications can compromise safety.

**(B) Approval for Class and Properties.**

**(1)** Equipment shall be identified not only for the class of location but also for the explosive, combustible, or ignitible properties of the specific gas, vapor, dust, or fibers/flyings that will be present. In addition, Class I equipment shall not have any exposed surface that operates at a temperature in excess of the autoignition temperature of the specific gas or vapor. Class II equipment shall not have an external temperature higher than that specified in 500.8(D)(2). Class III equipment shall not exceed the maximum surface temperatures specified in 503.5.

Informational Note: Luminaires and other heat-producing apparatus, switches, circuit breakers, and plugs and receptacles are potential sources of ignition and are investigated for suitability in classified locations. Such types of equipment, as well as cable terminations for entry into explosionproof enclosures, are available as listed for Class I, Division 2 locations. Fixed wiring, however, may utilize wiring methods that are not evaluated with respect to classified locations. Wiring products such as cable, raceways, boxes, and fittings, therefore, are not marked as being suitable for Class I, Division 2 locations. Also see 500.8(C)(6)(a).

Where installed in a Class I or Class II location, equipment must be suitable for the specific group indicated on the classification document (see 500.4). An explosionproof enclosure suitable only for Group D, for example, is not acceptable for Group B. Enclosures are often rated for more than one Class or Group.

Some portable devices — cameras, multimeters, and flashlights — have the capacity to cause ignition of a hazardous location. Although this electrical equipment is outside the scope of the *Code*, all equipment used should be suitable for the specific hazardous location.

Powered forklifts are also capable of causing an ignition. Tables 4.2(a) and 4.2(b) from NFPA 505, *Fire Safety Standard for*

*Powered Industrial Trucks Including Type Designations, Areas of Use, Conversions, Maintenance, and Operations,* provide a summary of industrial truck types suitable for hazardous locations.

**(2)** Equipment that has been identified for a Division 1 location shall be permitted in a Division 2 location of the same class, group, and temperature class and shall comply with (a) or (b) as applicable.

(a) Intrinsically safe apparatus having a control drawing requiring the installation of associated apparatus for a Division 1 installation shall be permitted to be installed in a Division 2 location if the same associated apparatus is used for the Division 2 installation.

(b) Equipment that is required to be explosionproof shall incorporate seals in accordance with 501.15(A) or (D) when the wiring methods of 501.10(B) are employed.

**(3)** Where specifically permitted in Articles 501 through 503, general-purpose equipment or equipment in general-purpose enclosures shall be permitted to be installed in Division 2 locations if the equipment does not constitute a source of ignition under normal operating conditions.

**(4)** Equipment that depends on a single compression seal, diaphragm, or tube to prevent flammable or combustible fluids from entering the equipment shall be identified for a Class I, Division 2 location even if installed in an unclassified location. Equipment installed in a Class I, Division 1 location shall be identified for the Class I, Division 1 location.

Informational Note: Equipment used for flow measurement is an example of equipment having a single compression seal, diaphragm, or tube.

**(5)** Unless otherwise specified, normal operating conditions for motors shall be assumed to be rated full-load steady conditions.

Normal condition for a motor is considered to be at steady full-load. Locked-rotor or other abnormal motor conditions, such as single phasing, are not considered when evaluating motor-operating temperatures (internal and external) in Class I, Division 2 locations. However, such abnormal load conditions must be considered when evaluating the external temperatures of explosionproof motors for Class I, Division 1 locations and dust-ignitionproof motors for Class II, Division 1 locations. Awareness of the increase in temperature in some variable-speed motors is important when they are operated at a lower speed and are dependent on the fan for cooling.

**(6)** Where flammable gases, flammable liquid–produced vapors, combustible liquid–produced vapors, or combustible dusts are or may be present at the same time, the simultaneous presence of both shall be considered when determining the safe operating temperature of the electrical equipment.

Informational Note: The characteristics of various atmospheric mixtures of gases, vapors, and dusts depend on the specific material involved.

Examples of where flammable liquid and dust might be simultaneously present are at a coal-handling facility, where there is methane gas and coal dust; and in an automotive paint spray shop, where flammable paint and powdered metal flecks are sprayed. In the presence of this combination of simultaneous hazards, less energy may be needed and the accumulation of gas need not be in the flammable range for an ignition to occur.

**(C) Marking.** Equipment shall be marked to show the environment for which it has been evaluated. Unless otherwise specified or allowed in (C)(6), the marking shall include the information specified in (C)(1) through (C)(5).

**(1) Class.** The marking shall specify the class(es) for which the equipment is suitable.

**(2) Division.** The marking shall specify the division if the equipment is suitable for Division 2 only. Equipment suitable for Division 1 shall be permitted to omit the division marking.

Informational Note: Equipment not marked to indicate a division, or marked "Division 1" or "Div. 1," is suitable for both Division 1 and 2 locations; see 500.8(B)(2). Equipment marked "Division 2" or "Div. 2" is suitable for Division 2 locations only.

**(3) Material Classification Group.** The marking shall specify the applicable material classification group(s) in accordance with 500.6.

*Exception: Fixed luminaires marked for use only in Class I, Division 2 or Class II, Division 2 locations shall not be required to indicate the group.*

**(4) Equipment Temperature.** The marking shall specify the temperature class or operating temperature at a 40°C ambient temperature, or at the higher ambient temperature if the equipment is rated and marked for an ambient temperature of greater than 40°C. For equipment installed in a Class II, Division 1 location, the temperature class or operating temperature shall be based on operation of the equipment when blanketed with the maximum amount of dust that can accumulate on the equipment. The temperature class, if provided, shall be indicated using the temperature class (T codes) shown in Table 500.8(C). Equipment for Class I and Class II shall be marked with the maximum safe operating temperature, as determined by simultaneous exposure to the combinations of Class I and Class II conditions.

*Exception: Equipment of the non–heat-producing type, such as junction boxes, conduit, and fittings, and equipment of the heat-producing type having a maximum temperature not more than 1008C shall not be required to have a marked operating temperature or temperature class.*

Informational Note: More than one marked temperature class or operating temperature, for gases and vapors, dusts, and different ambient temperatures, may appear.

**(5) Ambient Temperature Range.** Electrical equipment designed for use in the ambient temperature range between −25°C to +40°C shall require no ambient temperature marking. For equipment rated for a temperature range other than

***TABLE 500.8(C)*** *Classification of Maximum Surface Temperature*

| Maximum Temperature | | Temperature Class (T Code) |
|---|---|---|
| °C | °F | |
| 450 | 842 | T1 |
| 300 | 572 | T2 |
| 280 | 536 | T2A |
| 260 | 500 | T2B |
| 230 | 446 | T2C |
| 215 | 419 | T2D |
| 200 | 392 | T3 |
| 180 | 356 | T3A |
| 165 | 329 | T3B |
| 160 | 320 | T3C |
| 135 | 275 | T4 |
| 120 | 248 | T4A |
| 100 | 212 | T5 |
| 85 | 185 | T6 |

−25°C to +40°C, the marking shall specify the special range of ambient temperatures in degrees Celsius. The marking shall include either the symbol "Ta" or "Tamb."

Informational Note: As an example, such a marking might be "−30°C ≤ Ta ≤ +40°C."

**(6) Special Allowances.**

(a) *General-Purpose Equipment.* Fixed general-purpose equipment in Class I locations, other than fixed luminaires, that is acceptable for use in Class I, Division 2 locations shall not be required to be marked with the class, division, group, temperature class, or ambient temperature range.

Part III of Article 501 specifies which equipment is permitted to be installed in a general-purpose enclosure for Class I, Division 2 locations.

(b) *Dusttight Equipment.* Fixed dusttight equipment, other than fixed luminaires, that is acceptable for use in Class II, Division 2 and Class III locations shall not be required to be marked with the class, division, group, temperature class, or ambient temperature range.

(c) *Associated Apparatus.* Associated intrinsically safe apparatus and associated nonincendive field wiring apparatus that are not protected by an alternative type of protection shall not be marked with the class, division, group, or temperature class. Associated intrinsically safe apparatus and associated nonincendive field wiring apparatus shall be marked with the class, division, and group of the apparatus to which it is to be connected.

(d) *Simple Apparatus.* "Simple apparatus" as defined in Article 504, shall not be required to be marked with class, division, group, temperature class, or ambient temperature range.

**(D) Temperature.**

**(1) Class I Temperature.** The temperature marking specified in 500.8(C) shall not exceed the autoignition temperature of the specific gas or vapor to be encountered.

Informational Note: For information regarding autoignition temperatures of gases and vapors, see NFPA 497-2013, *Recommended Practice for the Classification of Flammable Liquids, Gases, or Vapors, and of Hazardous (Classified) Locations for Electrical Installations in Chemical Process Areas.*

Electrical equipment intended for installation in a hazardous location is evaluated for maximum temperatures regardless of the type of protection afforded the equipment. Gases and vapors are qualified with an autoignition temperature and dusts are qualified with a layer or cloud ignition temperature, each of which is the temperature at which the material ignites. The equipment temperature is compared to these material temperatures to determine if a potential for a thermal ignition exists.

The autoignition temperature of a solid, liquid, or gaseous substance is the minimum temperature required to initiate or cause self-sustained combustion independent of the heating or heated element. The flash point is the minimum temperature at which a liquid gives off enough vapor to form an ignitible mixture with air. The ignition temperature and the flash point are unrelated properties, except that the flash point is always lower than the ignition temperature.

**(2) Class II Temperature.** The temperature marking specified in 500.8(C) shall be less than the ignition temperature of the specific dust to be encountered. For organic dusts that may dehydrate or carbonize, the temperature marking shall not exceed the lower of either the ignition temperature or 165°C (329°F).

Informational Note: See NFPA 499-2013, *Recommended Practice for the Classification of Combustible Dusts and of Hazardous (Classified) Locations for Electrical Installations in Chemical Process Areas*, for minimum ignition temperatures of specific dusts.

The ignition temperature for which equipment was approved prior to this requirement shall be assumed to be as shown in Table 500.8(D)(2).

**(E) Threading.** The supply connection entry thread form shall be NPT or metric. Conduit and fittings shall be made wrenchtight to prevent sparking when fault current flows through the conduit system, and to ensure the explosionproof integrity of the conduit system where applicable. Equipment provided with threaded entries for field wiring connections shall be installed in accordance with 500.8(E)(1) or (E)(2) and with (E)(3).

To ensure the integrity of the ground-fault current path of the conduit system, all conduit joints must be made up wrenchtight to prevent ignition-capable arcing between the conduit and the coupling, fitting, or enclosure under ground-fault conditions. The use of a bonding jumper in lieu of a wrenchtight connection is not permitted.

*TABLE 500.8(D)(2) Class II Temperatures*

| Class II Group | Equipment Not Subject to Overloading | | Equipment (Such as Motors or Power Transformers) That May Be Overloaded | | | |
|---|---|---|---|---|---|---|
| | | | Normal Operation | | Abnormal Operation | |
| | °C | °F | °C | °F | °C | °F |
| E | 200 | 392 | 200 | 392 | 200 | 392 |
| F | 200 | 392 | 150 | 302 | 200 | 392 |
| G | 165 | 329 | 120 | 248 | 165 | 329 |

**(1) Equipment Provided with Threaded Entries for NPT-Threaded Conduit or Fittings.** For equipment provided with threaded entries for NPT-threaded conduit or fittings, listed conduit, listed conduit fittings, or listed cable fittings shall be used. All NPT-threaded conduit and fittings shall be threaded with a National (American) Standard Pipe Taper (NPT) thread.

NPT-threaded entries into explosionproof equipment shall be made up with at least five threads fully engaged.

*Exception: For listed explosionproof equipment, joints with factory-threaded NPT entries shall be made up with at least four and one-half threads fully engaged.*

Informational Note No. 1: Thread specifications for male NPT threads are located in ANSI/ASME B1.20.1-1983, *Pipe Threads, General Purpose (Inch).*

Informational Note No. 2: Female NPT-threaded entries use a modified National Standard Pipe Taper (NPT) thread with thread form per ANSI/ASME B1.20.1-1983, *Pipe Threads, General Purpose (Inch).* See ANSI/UL 1203-2009, *Explosionproof and Dust-Ignition-Proof Electrical Equipment for Use in Hazardous (Classified) Locations.*

**(2) Equipment Provided with Threaded Entries for Metric-Threaded Conduit or Fittings.** For equipment with metric-threaded entries, listed conduit fittings or listed cable fittings shall be used. Such entries shall be identified as being metric, or listed adapters to permit connection to conduit or NPT-threaded fittings shall be provided with the equipment and shall be used for connection to conduit or NPT-threaded fittings.

Metric-threaded entries into explosionproof equipment shall have a class of fit of at least 6g/6H and shall be made up with at least five threads fully engaged for Group C and Group D, and at least eight threads fully engaged for Group A and Group B.

Informational Note: Threading specifications for metric-threaded entries are located in ISO 965-1-1998, *ISO general purpose metric screw threads — Tolerances — Part 1: Principles and basic data*, and ISO 965-3-1998, *ISO general purpose metric screw threads — Tolerances — Part 3: Deviations for constructional screw threads.*

**(3) Unused Openings.** All unused openings shall be closed with listed metal close-up plugs. The plug engagement shall comply with 500.8(E)(1) or (E)(2).

**(F) Optical Fiber Cables.** Where an optical fiber cable contains conductors that are capable of carrying current (composite optical fiber cable), the optical fiber cable shall be installed in accordance with the requirements of Article 500, 501, 502, or 503, as applicable.

The requirements of Articles 500, 501, 502, or 503 apply even if the conductor is grounded.

### 500.9 Specific Occupancies

Articles 510 through 517 cover garages, aircraft hangars, motor fuel dispensing facilities, bulk storage plants, spray application, dipping and coating processes, and health care facilities.

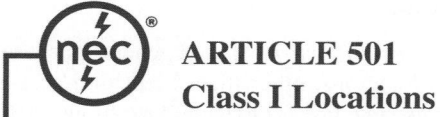

# ARTICLE 501
# Class I Locations

## I. General

### 501.1 Scope

Article 501 covers the requirements for electrical and electronic equipment and wiring for all voltages in Class I, Division 1 and 2 locations where fire or explosion hazards may exist due to flammable gases or vapors or flammable liquids.

Informational Note: For the requirements for electrical and electronic equipment and wiring for all voltages in Zone 0, Zone 1, or Zone 2 hazardous (classified) locations where fire or explosion hazards may exist due to flammable gases or vapors or flammable liquids, refer to Article 505.

Where ignitible concentrations (concentrations within flammable or explosive limits) are present, the atmosphere can be ignited by an arc, a spark, or high temperature. NFPA 497 includes information on the flammable limits of liquids and gases.

Hermetic sealing of all electrical equipment is impractical because equipment such as motors, conventional switches, and circuit breakers have movable parts that must be operated through the enclosing case. In addition, access to the inside of enclosures is often necessary for installation, servicing, or alterations. Therefore,

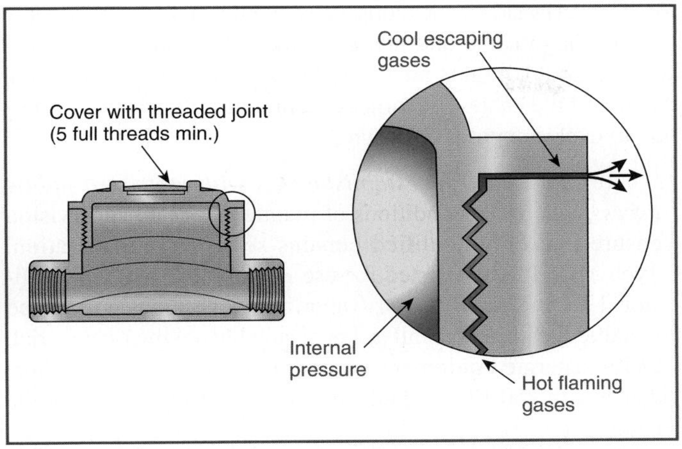


*EXHIBIT 501.1 Cooling of hot gases as they pass through the threads of a screw-type cover of an explosionproof junction box.*

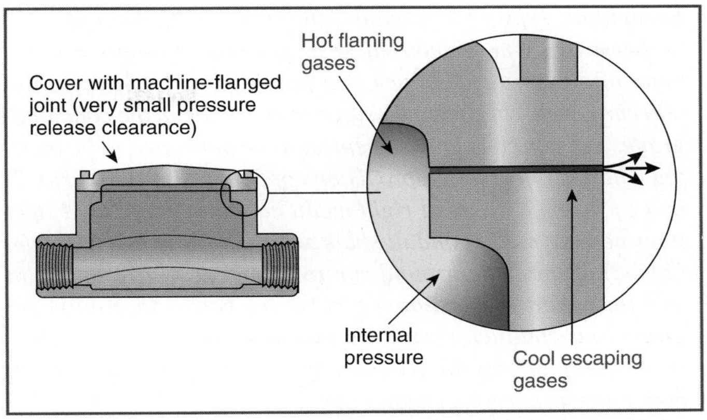


*EXHIBIT 501.2 Cooling of hot gases as they pass across a machine-flanged joint. The clearance between the machined surfaces is kept very small.*

it may be necessary to keep equipment in explosionproof enclosures or to use intrinsically safe equipment.

Electrical equipment that may cause ignition-capable arcs or sparks should be kept out of Class I locations, or the equipment must be identified for the appropriate hazardous location. It is practically impossible to make threaded joints gastight. The conduit system and apparatus enclosure "breathe" due to temperature changes, and any flammable gases or vapors in the room may slowly enter the conduit or enclosure, creating an explosive mixture. Should an arc occur, an explosion could occur.

When an explosion occurs within the enclosure or conduit system, the burning mixture or hot gases must be sufficiently confined within the system to prevent ignition of any explosive mixture outside of the system. An enclosure must be designed with sufficient strength to withstand the pressure generated by an internal explosion in order to prevent rupture and the release of burning or hot gases. During an explosion within an enclosure, gases escape through any paths or openings that exist, but the gases will be sufficiently cooled if they are carried out through an opening that is long in proportion to its width. Two examples of this are the screw-on type junction box cover, shown in Exhibit 501.1, and the tight tolerance, wide-machined flange between the body of the enclosure and its cover, illustrated in Exhibit 501.2. The function of the joint is the same whether it is flanged, threaded, rabbeted, or any other type designed for this purpose.

The clearance between flat surfaces may increase somewhat under explosion conditions, because the internal pressures created by the explosion tend to force the surfaces apart, as shown in Exhibit 501.3. The amount of increase in the joint clearance depends on the stiffness of the enclosure parts; the size, strength, and spacing of the bolts; and the explosion pressure. When there are no internal pressures, measuring the joint width and clearance does not indicate the actual clearances under the dynamic conditions of an explosion. Explosion tests are usually needed to demonstrate the acceptability of the design. Exhibit 501.3 illustrates the need to properly install all provided bolts, screws, fittings, and covers. If bolts are missing, it is essential that the manufacturer's specified bolt is used for replacement.

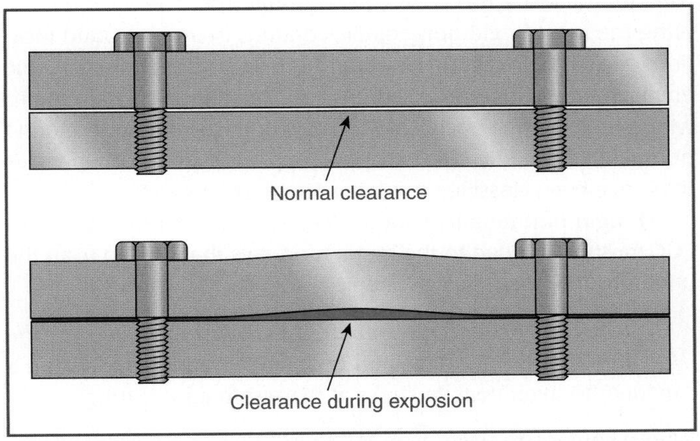


*EXHIBIT 501.3 Effect of internal explosion (bottom) on cover-to-body joint clearance in an explosionproof enclosure. (Courtesy of Underwriters Laboratories Inc.)*

### 501.5 Zone Equipment

Equipment listed and marked in accordance with 505.9(C)(2) for use in Zone 0, 1, or 2 locations shall be permitted in Class I, Division 2 locations for the same gas and with a suitable temperature class. Equipment listed and marked in accordance with 505.9(C)(2) for use in Zone 0 locations shall be permitted in Class I, Division 1 or Division 2 locations for the same gas and with a suitable temperature class.

## II. Wiring

### 501.10 Wiring Methods

Wiring methods shall comply with 501.10(A) or (B).

**(A) Class I, Division 1.**

**(1) General.** In Class I, Division 1 locations, the wiring methods in (a) through (e) shall be permitted.

(a) *Threaded rigid metal conduit or threaded steel intermediate metal conduit.*

*Exception: Type PVC conduit and Type RTRC conduit shall be permitted where encased in a concrete envelope a minimum of 50 mm (2 in.) thick and provided with not less than 600 mm (24 in.) of cover measured from the top of the conduit to grade. The concrete encasement shall be permitted to be omitted where subject to the provisions of 514.8, Exception No. 2, and 515.8(A). Threaded rigid metal conduit or threaded steel intermediate metal conduit shall be used for the last 600 mm (24 in.) of the underground run to emergence or to the point of connection to the aboveground raceway. An equipment grounding conductor shall be included to provide for electrical continuity of the raceway system and for grounding of non–current-carrying metal parts.*

This exception permits the use of Type PVC or RTRC conduit in some underground installations where encased in not less than 2 inches of concrete and threaded RMC or threaded steel IMC is used for the last 2 feet of the underground run. The sections covering the use of rigid nonmetallic conduits in underground locations per 514.8, Exception No. 2, and 515.8 do not require concrete encasement. Those requirements are for specific occupancies where there has been considerable experience with underground nonmetallic conduit. This exception applies to other occupancies that have been classified as Class I, Division 1 locations.

If rigid nonmetallic conduit (Type PVC or RTRC) is used, an EGC must be bonded to the metal raceways that extend from the underground nonmetallic conduit.

(b) *Type MI cable terminated with fittings listed for the location.* Type MI cable shall be installed and supported in a manner to avoid tensile stress at the termination fittings.

Where cables are used as a wiring method in hazardous locations, the termination fitting must be listed also [see 501.10(B)]. A termination fitting used in a Division 1 location with Type MI cable must be specifically listed for use in Class I, Division 1 hazardous locations.

Type MI cable fittings not investigated for use in hazardous locations may not be explosionproof and hence are not suitable for Class I, Division 1 locations. See Exhibit 501.4 for an example of this type of cable and fitting; in the installation pictured, the screw-on pot contains field-installed sealing compound to seal the end of the cable. The threaded gland has threads for connection to explosionproof enclosures.

(c) *In industrial establishments with restricted public access,* where the conditions of maintenance and supervision ensure that only qualified persons service the installation, Type MC-HL cable listed for use in Class I, Zone 1 or Division 1 locations, with a gas/vaportight continuous corrugated metallic sheath, an overall jacket of suitable polymeric material, and a separate equipment grounding conductor(s) in accordance with 250.122, and terminated with fittings listed for the application.

Type MC-HL cable shall be installed in accordance with the provisions of Article 330, Part II.

(d) *In industrial establishments with restricted public access,* where the conditions of maintenance and supervision ensure that only qualified persons service the installation, Type ITC-HL cable listed for use in Class I, Zone 1 or Division 1 locations, with a gas/vaportight continuous corrugated metallic sheath and an overall jacket of suitable polymeric material, and terminated with fittings listed for the application, and installed in accordance with the provisions of Article 727.

(e) *Optical fiber cable Types OFNP, OFCP, OFNR, OFCR, OFNG, OFCG, OFN, and OFC* shall be permitted to be installed in raceways in accordance with 501.10(A). These optical fiber cables shall be sealed in accordance with 501.15.

**(2) Flexible Connections.** Where necessary to employ flexible connections, as at motor terminals, the following shall be permitted:

(1) Flexible fittings listed for the location, or
(2) Flexible cord in accordance with the provisions of 501.140, terminated with cord connectors listed for the location, or
(3) In industrial establishments with restricted public access, where the conditions of maintenance and supervision

***EXHIBIT 501.4*** *Type MI cable and fitting listed for use in hazardous locations. (Courtesy of Tyco Thermal Controls)*

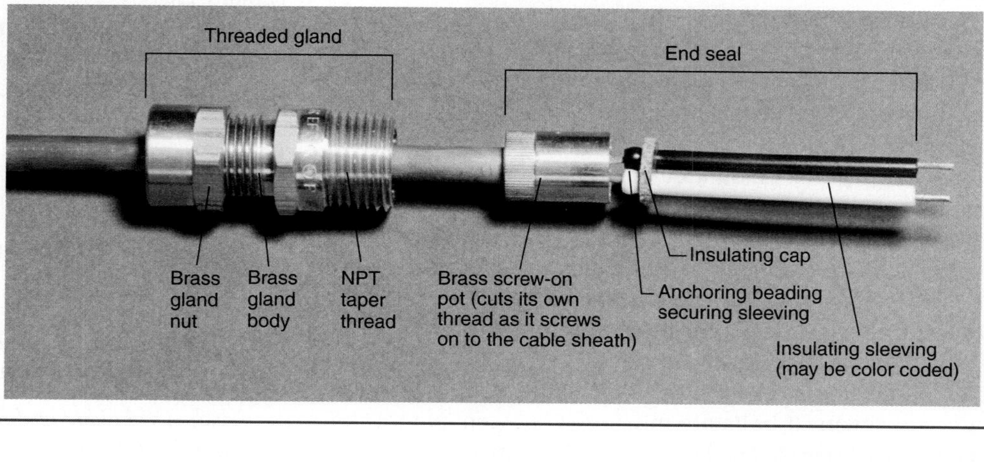

ensure that only qualified persons service the installation, for applications limited to 600 volts, nominal, or less, and where protected from damage by location or a suitable guard, listed Type TC-ER-HL cable with an overall jacket and a separate equipment grounding conductor(s) in accordance with 250.122 that is terminated with fittings listed for the location

"Flexible connections" refers only to the fittings. The intent of this section is not to permit a flexible wiring method. The fitting should be no longer than is needed; flexible connection fittings for use in Class I, Division 1 locations are available in lengths up to 3 feet. Explosionproof flexible fittings commonly used at motor connections can withstand continuous vibration for long periods and provide maximum protection to enclosed conductors.

Limited use of flexible cord is permitted in accordance with 501.140 for specific applications where flexibility of the wiring method is made necessary by the type of equipment being supplied.

The requirement to provide a seal within 18 inches of an explosionproof enclosure applies where flexible connections are used.

**(3) Boxes and Fittings.** All boxes and fittings shall be approved for Class I, Division 1.

Informational Note: For entry into enclosures required to be explosionproof, see the information on construction, testing, and marking of cables, explosionproof cable fittings, and explosionproof cord connectors in ANSI/UL 2225-2011, *Cables and Cable-Fittings for Use in Hazardous (Classified) Locations*.

Exhibit 501.5 shows an explosionproof junction box with two hubs and a threaded opening for the screw-type cover. Unused openings must be effectively closed by inserting threaded metal plugs that engage at least five full threads [4½ permitted in

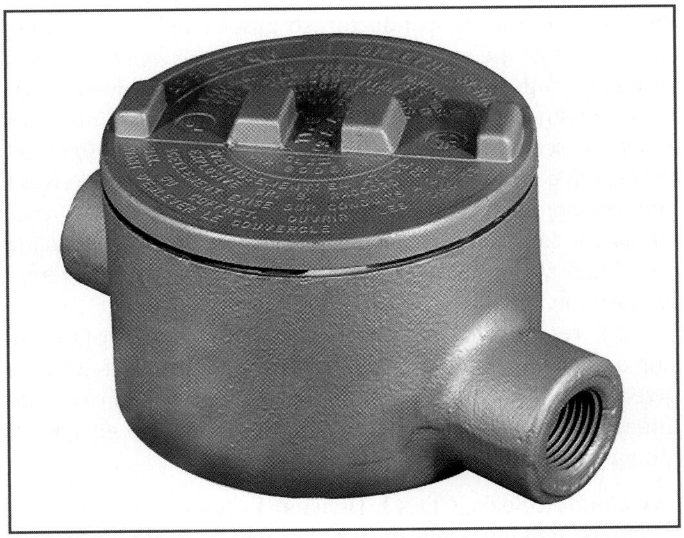

***EXHIBIT 501.5*** *An explosionproof junction box with a screw-type cover. (Courtesy of O-Z/Gedney, a division of EGS Electrical Group)*

accordance with the exception to 500.8(E)] and afford protection equivalent to that of the box wall.

**(B) Class I, Division 2.**

**(1) General.** In Class I, Division 2 locations, the following wiring methods shall be permitted:

(1) All wiring methods permitted in 501.10(A).
(2) Enclosed gasketed busways and enclosed gasketed wireways.
(3) Type PLTC and Type PLTC-ER cable in accordance with the provisions of Article 725, including installation in cable tray systems. The cable shall be terminated with listed fittings.
(4) Type ITC and Type ITC-ER cable as permitted in 727.4 and terminated with listed fittings.
(5) Type MC, MV, TC, or TC-ER cable, including installation in cable tray systems. The cable shall be terminated with listed fittings.
(6) In industrial establishments with restricted public access, where the conditions of maintenance and supervision ensure that only qualified persons service the installation and where metallic conduit does not provide sufficient corrosion resistance, listed reinforced thermosetting resin conduit (RTRC), factory elbows, and associated fittings, all marked with the suffix -XW, and Schedule 80 PVC conduit, factory elbows, and associated fittings shall be permitted.

    Where seals are required for boundary conditions as defined in 501.15(A)(4), the Division 1 wiring method shall extend into the Division 2 area to the seal, which shall be located on the Division 2 side of the Division 1–Division 2 boundary.
(7) Optical fiber cable Types OFNP, OFCP, OFNR, OFCR, OFNG, OFCG, OFN, and OFC shall be permitted to be installed in cable trays or any other raceway in accordance with 501.10(B). Optical fiber cables shall be sealed in accordance with 501.15.

**(2) Flexible Connections.** Where provision must be made for flexibility, one or more of the following shall be permitted:

(1) Listed flexible metal fittings.
(2) Flexible metal conduit with listed fittings.
(3) Interlocked armor Type MC cable with listed fittings.
(4) Liquidtight flexible metal conduit with listed fittings.
(5) Liquidtight flexible nonmetallic conduit with listed fittings.
(6) Flexible cord listed for extra-hard usage and terminated with listed fittings. A conductor for use as an equipment grounding conductor shall be included in the flexible cord.
(7) For elevator use, an identified elevator cable of Type EO, ETP, or ETT, shown under the "use" column in Table 400.4 for "hazardous (classified) locations" and terminated with listed fittings.

Informational Note: See 501.30(B) for grounding requirements where flexible conduit is used.

Type AC cable is not a permitted wiring method for this use because arcing can occur between convolutions during ground-fault conditions. See the commentary following 501.10(A)(2) concerning flexible connections.

**(3) Nonincendive Field Wiring.** Nonincendive field wiring shall be permitted using any of the wiring methods permitted for unclassified locations. Nonincendive field wiring systems shall be installed in accordance with the control drawing(s). Simple apparatus, not shown on the control drawing, shall be permitted in a nonincendive field wiring circuit, provided the simple apparatus does not interconnect the nonincendive field wiring circuit to any other circuit.

Informational Note: Simple apparatus is defined in 504.2.

Separate nonincendive field wiring circuits shall be installed in accordance with one of the following:

(1) In separate cables
(2) In multiconductor cables where the conductors of each circuit are within a grounded metal shield
(3) In multiconductor cables or in raceways, where the conductors of each circuit have insulation with a minimum thickness of 0.25 mm (0.01 in.)

Any wiring method suitable for ordinary locations may be used for nonincendive field wiring. Although many low-voltage, low-energy circuits, including some communications circuits and thermocouple circuits, are of the nonincendive type, a Class 2 or Class 3 circuit, as defined in Article 725, is not necessarily nonincendive. For additional information, see the defined terms *nonincendive circuit* and *nonincendive field wiring* in 500.2.

**(4) Boxes and Fittings.** Boxes and fittings shall not be required to be explosionproof except as required by 501.105(B)(1), 501.115(B)(1), and 501.150(B)(1).

In Class I, Division 2 locations, boxes, fittings, and joints are not required to be explosionproof if they contain no arcing devices; or for lighting outlets or enclosures containing non-arcing devices (such as solid-state relays, solenoids, and control transformers), if the maximum operating temperature of any exposed surface does not exceed 80 percent of the ignition temperature. However, when a general-purpose enclosure is installed using rigid, flexible, or intermediate metal conduit, a bonding jumper with proper fittings or bonding-type locknuts is required to be used between the enclosure and the raceway to ensure adequate bonding from the hazardous area to the point of grounding at the service equipment or separately derived system. See 501.30 for grounding and bonding requirements.

Where necessary, limited flexibility is provided through use of FMC, LFMC, and extra-hard-usage flexible cord, and the fittings are not required to be specifically approved for Class I locations. Section 501.10(B)(1) permits a variety of cable types, cable tray systems, enclosed gasketed wireways, and enclosed gasketed busways, each with associated fittings.

Informational Note: For entry into enclosures required to be explosionproof, see the information on construction, testing, and marking of cables, explosionproof cable fittings, and explosionproof cord connectors in ANSI/UL 2225-2011, *Cables and Cable-Fittings for Use in Hazardous (Classified) Locations*.

## 501.15 Sealing and Drainage

Seals in conduit and cable systems shall comply with 501.15(A) through (F). Sealing compound shall be used in Type MI cable termination fittings to exclude moisture and other fluids from the cable insulation.

Commentary Table 501.1 summarizes the sealing requirements of 501.15(A) through (F).

Informational Note No. 1: Seals are provided in conduit and cable systems to minimize the passage of gases and vapors and prevent the passage of flames from one portion of the electrical installation to another through the conduit. Such communication through Type MI cable is inherently prevented by construction of the cable. Unless specifically designed and tested for the purpose, conduit and cable seals are not intended to prevent the passage of liquids, gases, or vapors at a continuous pressure differential across the seal. Even at differences in pressure across the seal equivalent to a few inches of water, there may be a slow passage of gas or vapor through a seal and through conductors passing through the seal. Temperature extremes and highly corrosive liquids and vapors can affect the ability of seals to perform their intended function.

Informational Note No. 2: Gas or vapor leakage and propagation of flames may occur through the interstices between the strands of standard stranded conductors larger than 2 AWG. Special conductor constructions, such as compacted strands or sealing of the individual strands, are means of reducing leakage and preventing the propagation of flames.

Because the sealing compound used in conduit seal fittings is typically somewhat porous, gases, particularly those under slight pressure and those with small molecules such as hydrogen, may pass slowly through the compound. As well, the seal is around the insulation on the conductor, so gases can be transmitted slowly through the air spaces (the interstices) between strands of stranded conductors. Under normal conditions for smaller conductors with only normal atmospheric pressure differentials across the seal, the passage of gas through a seal is not sufficient to result in a hazard. For larger conductors, gas or vapor leakage and flame propagation may occur through the interstices. Special conductor constructions, such as compacted strands or sealing individual strands, may reduce leakage and prevent flame propagation.

Different sealing compounds have different rates of expansion and contraction that may affect their performance within a given fitting. Teflon™ tapes or joint compounds on conduit threads may weaken the seal fitting and interrupt the equipment grounding path.

**(A) Conduit Seals, Class I, Division 1.** In Class I, Division 1 locations, conduit seals shall be located in accordance with 501.15(A)(1) through (A)(4).

**COMMENTARY TABLE 501.1** Conduit and Cable Sealing Requirements

| Classification | Application | Location of Seal |
|---|---|---|
| **Conduit Seals**<br>Class I, Division 1 | Switch enclosure<br>Circuit breaker enclosure<br>Fuse enclosure<br>Relay enclosure<br>Resistor enclosure<br>Arcing or sparking apparatus<br>High-temperature apparatus | In conduit run within 18 in. of enclosure. |
| | Explosionproof enclosure containing arcing or sparking contacts that are hermetically sealed against gas or vapor entry<br>Explosionproof enclosure containing arcing or sparking contacts that are immersed in oil, in accordance with 501.115(B)(1)(2) | In conduit runs smaller than trade size 2, no seal is required. If conduit is trade size 2 or larger, in conduit run within 18 in. of enclosure. |
| | Enclosure containing terminals, splices, or taps; fitting containing terminals, splices, or taps | In conduit runs smaller than trade size 2, no seal is required. If conduit is trade size 2 or larger, in conduit run within 18 in. of enclosure. |
| | Two explosionproof enclosures with a conduit run between them of 36 in. or less | In conduit run within 18 in. of each enclosure. Permitted to use a single seal in each run as long as the seal is within 18 in. of each enclosure. |
| | Two explosionproof enclosures with a conduit run between them greater than 36 in. | In conduit run within 18 in. of each enclosure. |
| | Conduit run leaving Division 1 location | Within 10 ft of either side of boundary. No unions, couplings, boxes, or fittings (other than explosionproof reducers) permitted between the seal fitting and the point where the conduit leaves the Division 1 location. |
| | Metal conduit containing no unions, couplings, boxes, or fittings that passes completely through a Class I, Division 1 location, with no fittings less than 12 in. beyond each boundary | Not required to be sealed if the termination points of the unbroken conduit are in unclassified locations. |
| Class I, Division 2 | Enclosure required to be explosionproof | Seal as required for similar equipment in Division 1 location. |
| | Conduit run leaving Division 2 location | Within 10 ft of either side of boundary. No unions, couplings, boxes, or fittings (other than explosionproof reducers) permitted between the seal fitting and the point where the conduit leaves the Division 2 location. Not required to be explosionproof seal. |
| | Metal conduit containing no unions, couplings, boxes, or fittings that passes completely through a Division 2 location with no fittings less than 12 in. beyond each boundary | Not required to be sealed if the termination points of the unbroken conduit are in unclassified locations. |
| | Conduit systems terminating at an unclassified location where a wiring method transition is made to cable tray, cablebus, ventilated busway, Type MI cable, or open wiring | Not required to be sealed if passing from the Class I, Division 2 location into an outdoor unclassified location or an indoor location if the conduit system is all in one room. The conduits do not terminate at an enclosure containing an ignition source in normal operation. |
| | Conduit leaving a purged enclosure or room that is unclassified due to pressurization and entering a Division 2 location | Not required to be sealed at the boundary. |

*(continues)*

**COMMENTARY TABLE 501.1** *Continued*

| Classification | Application | Location of Seal |
|---|---|---|
| **Cable Seals** Class I, Division 1 | Enclosure with integral seal | Conduit seal fitting not required. |
| | Multiconductor Type MC-HL cables with a gastight/vaportight continuous corrugated metallic sheath and an overall jacket of suitable polymeric material | Seal at all terminations with a listed fitting after removing the jacket and any other covering, so that the sealing compound surrounds each individual insulated conductor. |
| | Cables in conduit with a gastight/vaportight continuous sheath capable of transmitting gases or vapors through the cable core | Seal in the Division 1 location after removing the jacket and any other coverings, so that the sealing compound surrounds each individual insulated conductor and the outer jacket. |
| | Multiconductor cables with a gastight/vaportight continuous sheath capable of transmitting gases or vapors through the cable core | Permitted to be considered a single conductor by sealing the cable in the conduit within 18 in. of the enclosure and the cable end within the enclosure by an approved means, to minimize the entrance of gases or vapors and prevent the propagation of flame into the cable core, or by other approved methods. |
| | For shielded cables and twisted pair cables | Removal of the shielding material or separation of the twisted pair is not required. Sealing the cable in the conduit and the cable end within the enclosure by an approved means, to minimize the entrance of gases or vapors and prevent the propagation of flame into the cable core, or by other approved methods. |
| | Each multiconductor cable in conduit if the cable is incapable of transmitting gases or vapors through the cable core | Considered a single conductor. These cables are sealed in accordance with 501.15(A). |
| Class I, Division 2 | Cables entering enclosures that are required to be explosionproof for Class I locations | Sealed at the point of entrance. These cables are sealed in accordance with the requirements of Division 1 locations. |
| | Multiconductor cables with a gastight/vaportight continuous sheath capable of transmitting gases or vapors through the cable core | Sealed in a listed fitting in the Division 2 location after removing the jacket and any other coverings, so that the sealing compound surrounds each individual insulated conductor. |
| | Multiconductor cables in conduit | Sealed in accordance with the requirements for Division 1 locations. |
| | Cables with a gastight/vaportight continuous sheath that will not transmit gases or vapors through the cable core in excess of the quantity permitted for seal fittings. The minimum length of such cable run is not less than that length that limits gas or vapor flow through the cable core to the rate permitted for seal fittings (0.007 $ft^3$ per hour of air at a pressure of 6 in. of water). | Not required to be sealed unless entering an enclosure that is required to be explosionproof. |
| | Cables with a gastight/vaportight continuous sheath capable of transmitting gases or vapors through the cable core | Not required to be sealed unless entering an enclosure that is required to be explosionproof or unless the cable is attached to process equipment or devices that may cause a pressure in excess of 6 in. of water to be exerted at a cable end, in which case a seal, barrier, or other means is provided to prevent migration of flammables into an unclassified area. |
| | Cables with an unbroken gastight/vaportight continuous sheath that pass through a Class I, Division 2 location | No seal required. |
| | Cables that do not have a gastight/vaportight continuous sheath | Sealed at the boundary of the Division 2 and unclassified location in such a manner as to minimize the passage of gases or vapors into an unclassified location. |

**(1) Entering Enclosures.** Each conduit entry into an explosionproof enclosure shall have a conduit seal where either of the following conditions applies:

(1) The enclosure contains apparatus, such as switches, circuit breakers, fuses, relays, or resistors, that may produce arcs, sparks, or temperatures that exceed 80 percent of the autoignition temperature, in degrees Celsius, of the gas or vapor involved in normal operation.

*Exception: Seals shall not be required for conduit entering an enclosure under any one of the following conditions:*

*a. The switch, circuit breaker, fuse, relay, or resistor is enclosed within a chamber hermetically sealed against the entrance of gases or vapors.*
*b. The switch, circuit breaker, fuse, relay, or resistor is immersed in oil in accordance with 501.115(B)(1)(2).*
*c. The switch, circuit breaker, fuse, relay, or resistor is enclosed within a factory-sealed explosionproof chamber located within the enclosure, identified for the location, and marked "factory sealed" or equivalent.*
*d. The switch, circuit breaker, fuse, relay, or resistor is part of a nonincendive circuit.*

(2) The entry is metric designator 53 (trade size 2) or larger, and the enclosure contains terminals, splices, or taps.

A seal fitting must be placed within 18 inches of the entrance of trade size 2 or larger conduit into any explosionproof enclosure, regardless of whether the enclosure contains arcing or sparking equipment or only splices, taps, or terminals.

•

Factory-sealed enclosures shall not be considered to serve as a seal for another adjacent explosionproof enclosure that is required to have a conduit seal.

Conduit seals shall be installed within 450 mm (18 in.) from the enclosure. Only explosionproof unions, couplings, reducers, elbows, capped elbows, and conduit bodies similar to L, T, and Cross types that are not larger than the trade size of the conduit shall be permitted between the sealing fitting and the explosionproof enclosure.

Conduit seals are to prevent an explosion from traveling through the conduit to another enclosure and to minimize the passage of gases or vapors from hazardous locations to nonhazardous locations. In Class 1, Division 1 locations, if the conduit enters an enclosure that contains arcing or high-temperature equipment, a seal fitting must be placed within 18 inches of the enclosure it isolates. Only conduit bodies ("L," "T," etc.), couplings, unions, and elbows are permitted between the seal and the enclosure. Exhibit 501.6 illustrates an explosionproof type of union.

**(2) Pressurized Enclosures.** Conduit seals shall be installed within 450 mm (18 in.) of the enclosure in each conduit entry into a pressurized enclosure where the conduit is not pressurized as part of the protection system.

***EXHIBIT 501.6*** *An explosionproof union. (Courtesy of Thomas & Betts Corp.)*

Informational Note No. 1: Installing the seal as close as possible to the enclosure will reduce problems with purging the dead airspace in the pressurized conduit.

Informational Note No. 2: For further information, see NFPA 496-2013, *Standard for Purged and Pressurized Enclosures for Electrical Equipment.*

**(3) Two or More Explosionproof Enclosures.** Where two or more explosionproof enclosures that require conduit seals are connected by nipples or runs of conduit not more than 900 mm (36 in.) long, a single conduit seal in each such nipple connection or run of conduit shall be considered sufficient if the seal is located not more than 450 mm (18 in.) from either enclosure.

If two enclosures are spaced within 36 inches as illustrated by Enclosures No. 1 and No. 2, in Exhibit 501.7, a single seal may be placed between two connecting nipples if the seal is located not more than 18 inches from either enclosure. Even if Enclosure No. 3 was not required to be sealed, the seal within 18 inches of Enclosure No. 1 in the vertical run of conduit to Enclosure No. 3 is required because the conduit run to the "T" fitting is a conduit run from Enclosure No. 1.

**(4) Class I, Division 1 Boundary.** A conduit seal shall be required in each conduit run leaving a Division 1 location. The sealing fitting shall be permitted to be installed on either side of the boundary within 3.05 m (10 ft) of the boundary, and it shall be designed and installed to minimize the amount of gas or vapor within the portion of the conduit installed in the Division 1 location that can be communicated beyond the seal. The conduit run between the conduit seal and the point at which the conduit leaves the Division 1 location shall contain no union, coupling, box, or other fitting except for a listed explosionproof reducer installed at the conduit seal.

A seal fitting is required at the boundary where the conduit leaves a Division 1 location or passes from a Division 2 location to an unclassified location per 501.15(B)(2). The seal is permitted on either side of the boundary, and no union, coupling, box, or similar fitting

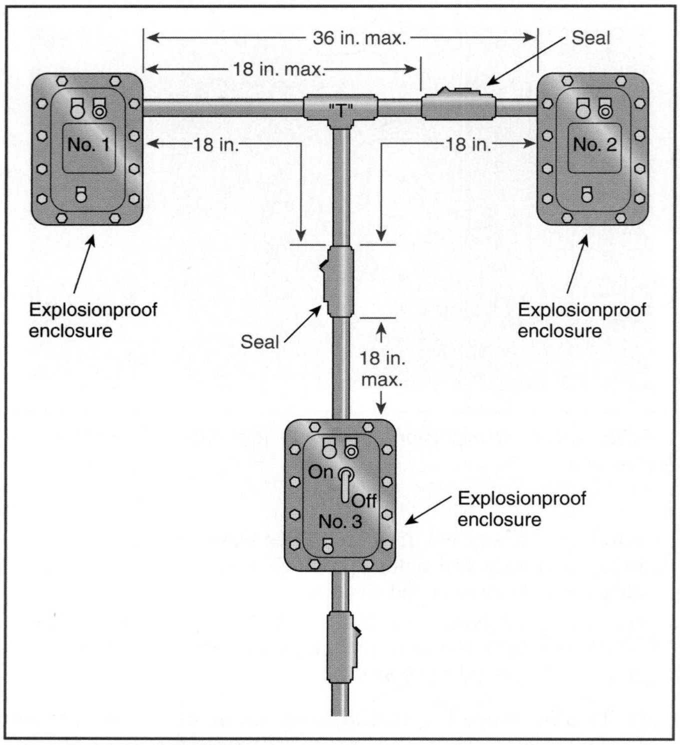


***EXHIBIT 501.7*** *Two seals required so that each run of conduit from Enclosure No. 1 is sealed.*

is permitted between the seal and the boundary. However, approved explosionproof reducers are permitted to be installed at the conduit seal.

The seal is best located on the nonhazardous side of the boundary where it serves two purposes: completion of the explosionproof wiring method and completion of the explosionproof enclosure system. For example, even though a conduit seal is not required for ½-inch conduit connected to an explosionproof box that contains only splices, the required seal at the boundary of the Division 1 location serves to complete the explosionproof system. The seal at the boundary also prevents the conduit system from serving as a pipe to transmit flammable mixtures from either a Division 1 or a Division 2 location to an unclassified location.

*Exception No. 1: Metal conduit that contains no unions, couplings, boxes, or fittings, that passes completely through a Division 1 location with no fittings installed within 300 mm (12 in.) of either side of the boundary, shall not require a conduit seal if the termination points of the unbroken conduit are located in unclassified locations.*

*Exception No. 2: For underground conduit installed in accordance with 300.5 where the boundary is below grade, the sealing fitting shall be permitted to be installed after the conduit emerges from below grade, but there shall be no union, coupling, box, or fitting, other than listed explosionproof reducers at the sealing fitting, in the conduit between the sealing fitting and the point at which the conduit emerges from below grade.*

Exhibit 501.8 illustrates a Class I, Division 1 location where the enclosures for the disconnecting means and motor controller for the motor (right portion of the drawing) are placed on the other

***EXHIBIT 501.8*** *A Class I, Division 1 location where threaded metal conduits, seal fittings, explosionproof fittings, and equipment for power and lighting are used.*

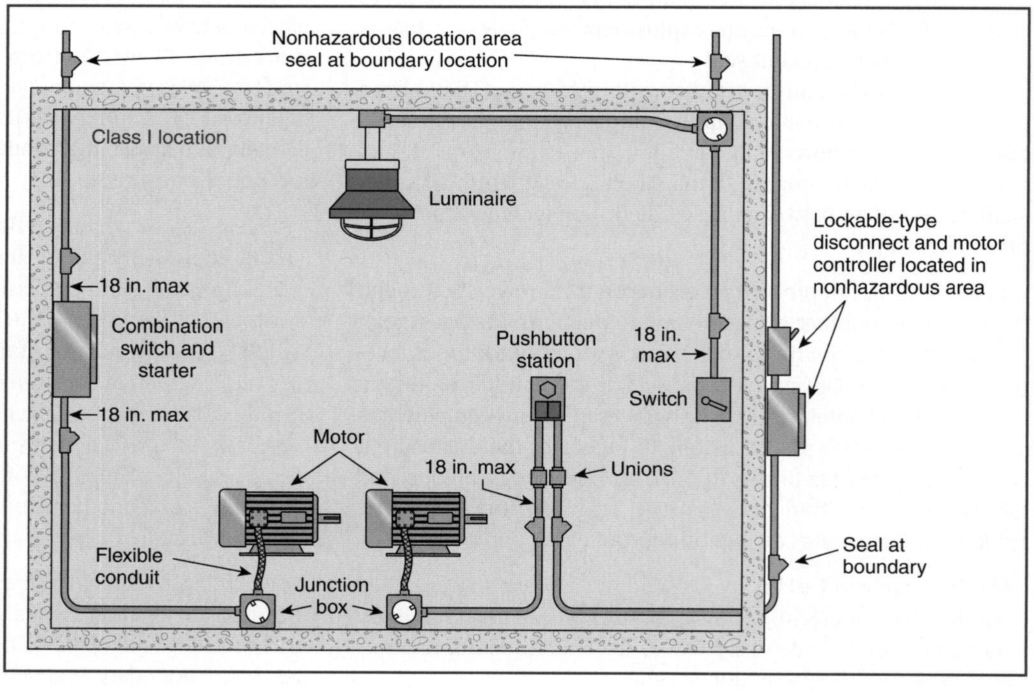

side of the wall in a nonhazardous location and are thus not required to be explosionproof.

Each of the three conduits shown is sealed on the nonhazardous side before entering the hazardous location in accordance with 501.15(A)(4). The pigtail leads of both motors are factory sealed at the motor-terminal housing, and, unless the size of the flexible fitting entering the motor-terminal housing is trade size 2 or larger, no other seals are needed at that point. Because the pushbutton control station and the combination switch and starter are considered arc-producing devices, conduits are sealed within 18 inches of the entrance to these enclosures. Seals are required even though the contacts may be immersed in oil if the conduit is trade size 2 or larger.

Additionally, Exhibit 501.8 shows a seal provided within 18 inches of the switch controlling the lighting. The explosionproof luminaire has an explosionproof chamber for the wiring that is separated or sealed from the lamp compartment. Hence, a separate seal is not required adjacent to this luminaire. See 501.130 for luminaire requirements.

**(B) Conduit Seals, Class I, Division 2.** In Class I, Division 2 locations, conduit seals shall be located in accordance with 501.15(B)(1) and (B)(2).

**(1) Entering Enclosures.** For connections to enclosures that are required to be explosionproof, a conduit seal shall be provided in accordance with 501.15(A)(1)(1) and (A)(3). All portions of the conduit run or nipple between the seal and enclosure shall comply with 501.10(A).

An enclosure will not be explosionproof if the necessary conduit seals are not provided. Seals complete the explosionproof enclosure.

**(2) Class I, Division 2 Boundary.** A conduit seal shall be required in each conduit run leaving a Class I, Division 2 location. The sealing fitting shall be permitted to be installed on either side of the boundary within 3.05 m (10 ft) of the boundary and it shall be designed and installed to minimize the amount of gas or vapor within the portion of the conduit installed in the Division 2 location that can be communicated beyond the seal. Rigid metal conduit or threaded steel intermediate metal conduit shall be used between the sealing fitting and the point at which the conduit leaves the Division 2 location, and a threaded connection shall be used at the sealing fitting. The conduit run between the conduit seal and the point at which the conduit leaves the Division 2 location shall contain no union, coupling, box, or other fitting except for a listed explosionproof reducer installed at the conduit seal. Such seals shall not be required to be explosionproof but shall be identified for the purpose of minimizing the passage of gases permitted under normal operating conditions and shall be accessible.

Informational Note: For further information, refer to ANSI/UL 514B-2012, *Conduit, Tubing, and Cable Fittings*.

*Exception No. 1: Metal conduit that contains no unions, couplings, boxes, or fittings, that passes completely through a Division 2 location with no fittings installed within 300 mm (12 in.) of either side of the boundary, shall not be required to be sealed if the termination points of the unbroken conduit are located in unclassified locations.*

*Exception No. 2: Conduit systems terminating in an unclassified location where the metal conduit transitions to cable tray, cablebus, ventilated busway, or Type MI cable, or to cable not installed in any cable tray or raceway system, shall not be required to be sealed where passing from the Division 2 location into the unclassified location under the following conditions:*

*(1) The unclassified location is outdoors located or the unclassified location is indoors and the conduit system is entirely in one room.*

*(2) The conduits shall not terminate at an enclosure containing an ignition source in normal operation.*

*Exception No. 3: Conduit systems passing from an enclosure or a room that is unclassified, as a result of pressurization, into a Division 2 location shall not require a seal at the boundary.*

Informational Note: For further information, refer to NFPA 496-2013, *Standard for Purged and Pressurized Enclosures for Electrical Equipment*.

*Exception No. 4: Segments of aboveground conduit systems shall not be required to be sealed where passing from a Division 2 location into an unclassified location if all of the following conditions are met:*

*(1) No part of the conduit system segment passes through a Division 1 location where the conduit segment contains unions, couplings, boxes, or fittings that are located within 300 mm (12 in.) of the Division 1 location.*

*(2) The conduit system segment is located entirely in outdoor locations.*

*(3) The conduit system segment is not directly connected to canned pumps, process or service connections for flow, pressure, or analysis measurement, and so forth, that depend on a single compression seal, diaphragm, or tube to prevent flammable or combustible fluids from entering the conduit system.*

*(4) The conduit system segment contains only threaded metal conduit, unions, couplings, conduit bodies, and fittings in the unclassified location.*

*(5) The conduit system segment is sealed at its entry to each enclosure or fitting located in the Division 2 location that contains terminals, splices, or taps.*

**(C) Class I, Divisions 1 and 2.** Seals installed in Class I, Division 1 and Division 2 locations shall comply with 501.15(C)(1) through (C)(6).

*Exception: Seals that are not required to be explosionproof by 501.15(B)(2) or 504.70 shall not be required to comply with 501.15(C).*

**(1) Fittings.** Enclosures that contain connections or equipment shall be provided with an integral sealing means, or sealing fittings listed for the location shall be used. Sealing fittings shall be listed for use with one or more specific compounds and shall be accessible.

Fittings should be sealed only with the sealing compound or compounds specified by the manufacturer's instructions furnished with the fitting.

**(2) Compound.** The compound shall provide a seal to minimize the passage of gas and/or vapors through the sealing fitting and shall not be affected by the surrounding atmosphere or liquids. The melting point of the compound shall not be less than 93°C (200°F).

**(3) Thickness of Compounds.** The thickness of the sealing compound installed in completed seals, other than listed cable sealing fittings, shall not be less than the metric designator (trade size) of the sealing fitting expressed in the units of measurement employed; however, in no case shall the thickness of the compound be less than 16 mm (⅝ in.).

Exhibit 501.9 is a cutaway of a seal fitting containing sealing compound and conductors. A dam must be provided to prevent the compound, while still in the liquid state, from running out of the fitting. All conductors must be separated to permit the compound to run between them. The compound must have a minimum thickness of not less than the trade size of the conduit and never less than ⅝ inch.

To eliminate the time-consuming task of field-poured seals, a factory-sealed device with the seal designed into the device is permissible. A wide selection of factory-sealed devices is available for a variety of installations in hazardous locations. For example, explosionproof motors are normally factory-sealed and therefore require no additional seal. Factory-sealed devices are usually marked as such.

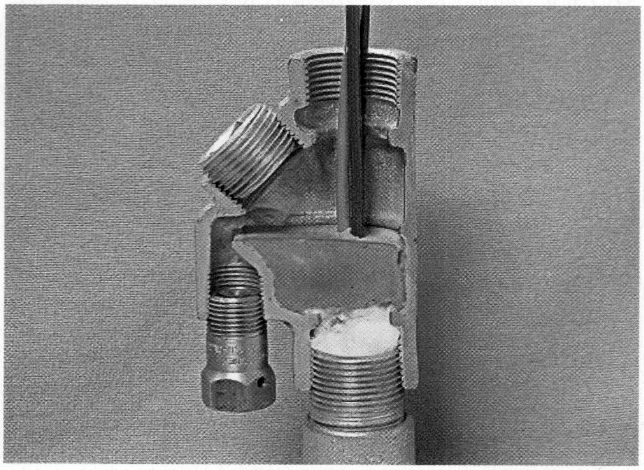

***EXHIBIT 501.9*** *A seal fitting to minimize the passage of gases from one portion of the electrical installation to another. (Courtesy of Appleton Electric Co., EGS Electrical Group)*

**(4) Splices and Taps.** Splices and taps shall not be made in fittings intended only for sealing with compound; nor shall other fittings in which splices or taps are made be filled with compound.

**(5) Assemblies.** An entire assembly shall be identified for the location where the equipment that may produce arcs, sparks, or high temperatures is located in a compartment that is separate from the compartment containing splices or taps, and an integral seal is provided where conductors pass from one compartment to the other. In Division 1 locations, seals shall be provided in conduit connecting to the compartment containing splices or taps where required by 501.15(A)(1)(2).

**(6) Conductor or Optical Fiber Fill.** The cross-sectional area of the conductors or optical fiber tubes (metallic or nonmetallic) permitted in a seal shall not exceed 25 percent of the cross-sectional area of a rigid metal conduit of the same trade size unless the seal is specifically identified for a higher percentage of fill.

The maximum permitted fill for most conduit seals is 25 percent, which is less than permitted for most conduit applications. If the conduit fill exceeds 25 percent of the cross-sectional area of the seal fitting, a larger trade size or expanded seal may be required. Reducers are allowed for connection of a larger trade size seal fitting to the conduit per 501.15(A)(1).

**(D) Cable Seals, Class I, Division 1.** In Division 1 locations, cable seals shall be located according to 501.15(D)(1) through (D)(3).

**(1) At Terminations.** Cables shall be sealed with sealing fittings that comply with 501.15(C) at all terminations. Type MC-HL cables with a gas/vaportight continuous corrugated metallic sheath and an overall jacket of suitable polymeric material shall be sealed with a listed fitting after the jacket and any other covering have been removed so that the sealing compound can surround each individual insulated conductor in such a manner as to minimize the passage of gases and vapors.

The sealing requirements for cables installed in Class I, Division 1 locations differ from the requirements for sealing conduits. As required by 501.15(A), conduits entering explosionproof enclosures must be sealed if the enclosure contains equipment that produces arcs, sparks, or high temperatures or if the conduit entering the enclosure is trade size 2 or larger. Section 501.15(D)(1) requires cables to be sealed at all terminations regardless of the type of equipment contained in the enclosure or the cable diameter.

Type MC-HL cable is specifically listed for use as a wiring method in Class I, Division 1 locations. Exhibit 501.10 shows an example of a cable seal fitting for Type MC-HL cable.

*Exception: Shielded cables and twisted pair cables shall not require the removal of the shielding material or separation*

**EXHIBIT 501.10** *An explosionproof seal fitting for Type MC-HL cable. (Courtesy of Cooper Crouse-Hinds)*

*of the twisted pairs, provided the termination is sealed by an approved means to minimize the entrance of gases or vapors and prevent propagation of flame into the cable core.*

**(2) Cables Capable of Transmitting Gases or Vapors.** Cables with a gas/vaportight continuous sheath capable of transmitting gases or vapors through the cable core, installed in conduit, shall be sealed in the Class 1, Division 1 location after the jacket and any other coverings have been removed so that the sealing compound can surround each individual insulated conductor or optical fiber tube and the outer jacket.

In addition to the conduit seal, the cable within the conduit must also be sealed to prevent gases from passing through the cable. A single conduit seal can serve both purposes, sealing the conduit and sealing the cable.

*Exception: Multiconductor cables with a gas/vaportight continuous sheath capable of transmitting gases or vapors through the cable core shall be permitted to be considered as a single conductor by sealing the cable in the conduit within 450 mm (18 in.) of the enclosure and the cable end within the enclosure by an approved means to minimize the entrance of gases or vapors and prevent the propagation of flame into the cable core, or by other approved methods. It shall not be required to remove the shielding material or separate the twisted pairs of shielded cables and twisted pair cables.*

**(3) Cables Incapable of Transmitting Gases or Vapors.** Each multiconductor cable installed in conduit shall be considered as a single conductor if the cable is incapable of transmitting gases or vapors through the cable core. These cables shall be sealed in accordance with 501.15(A).

**(E) Cable Seals, Class I, Division 2.** In Division 2 locations, cable seals shall be located in accordance with 501.15(E)(1) through (E)(4).

*Exception: Cables with an unbroken gas/vaportight continuous sheath shall be permitted to pass through a Division 2 location without seals.*

**(1) Terminations.** Cables entering enclosures that are required to be explosionproof shall be sealed at the point of entrance. The sealing fitting shall comply with 501.15(B)(1) or be explosionproof. Multiconductor or optical multifiber cables with a gas/vaportight continuous sheath capable of transmitting gases or vapors through the cable core that are installed in a Division 2 location shall be sealed with a listed fitting after the jacket and any other coverings have been removed so that the sealing compound can surround each individual insulated conductor or optical fiber tube in such a manner as to minimize the passage of gases and vapors. Multiconductor or optical multifiber cables installed in conduit shall be sealed as described in 501.15(D).

*Exception No. 1: Cables leaving an enclosure or room that is unclassified as a result of Type Z pressurization and entering into a Division 2 location shall not require a seal at the boundary.*

If cables are run from a Type Z pressurized (unclassified interior) room or enclosure into a Class I, Division 2 location, Exception No. 1 allows the cables to be installed without a seal fitting at the enclosure boundary. This exception correlates with a similar allowance for conduit systems found in 501.15(B)(2), Exception No. 3.

*Exception No. 2: Shielded cables and twisted pair cables shall not require the removal of the shielding material or separation of the twisted pairs, provided the termination is by an approved means to minimize the entrance of gases or vapors and prevent propagation of flame into the cable core.*

**(2) Cables That Do Not Transmit Gases or Vapors.** Cables that have a gas/vaportight continuous sheath and do not transmit gases or vapors through the cable core in excess of the quantity permitted for seal fittings shall not be required to be sealed except as required in 501.15(E)(1). The minimum length of such a cable run shall not be less than the length needed to limit gas or vapor flow through the cable core, excluding the interstices of the conductor strands, to the rate permitted for seal fittings [200 $cm^3$/hr (0.007 $ft^3$/hr) of air at a pressure of 1500 pascals (6 in. of water)].

The ability of a cable to transmit gases or vapors through the core (primarily between insulated conductors) depends not only on how tightly packed the conductors are within the outer sheaths and the location and composition of fillers but also on how the cable has been handled and the geometry of the cable run. If any concern that the cable run is capable of transmitting gases or vapors through the core exists, a seal fitting should be installed. See the commentary following 501.15, Informational Note No. 2.

To conduct a leak rate test to measure the gas flow through a cable core, the ends of each individual conductor in the cable are sealed to prevent migration of gases or vapors between the individual strands of wire. Sealing can be achieved by dipping the cable end in hot wax. The rate of flow through the filler between the insulated conductors can then be accurately measured, excluding any leakage through the conductor strands. If this sealing is done, the wax should be removed before the connections are made and the system is placed in service.

•

**(3) Cables Capable of Transmitting Gases or Vapors.** Cables with a gas/vaportight continuous sheath capable of transmitting gases or vapors through the cable core shall be sealed as required in 501.15(E)(1), unless the cable is attached to process equipment or devices that may cause a pressure in excess of 1500 pascals (6 in. of water) to be exerted at a cable end, in which case a seal, a barrier, or other means shall be provided to prevent migration of flammables into an unclassified location.

•

**(4) Cables Without Gas/Vaportight Sheath.** Cables that do not have a gas/vaportight continuous sheath shall be sealed at the boundary of the Division 2 and unclassified location in such a manner as to minimize the passage of gases or vapors into an unclassified location.

**(F) Drainage.**

Exhibit 501.11 shows a seal designed for use in a vertical run of conduit to provide drainage for any condensation of moisture trapped above the enclosure by the seal. Any accumulation of water runs down over the surface of the sealing compound, flowing through an explosionproof drain.

Exhibit 501.12 shows a combination drain and breather fitting. This fitting is specifically designed to permit the escape of accumulated water through its drain and allow the continuous circulation of air through the breather, preventing condensation of any moisture that may be present while still providing explosionproof protection. A good practice is to consider the installation of a drain, breather, or combination fitting to guard against

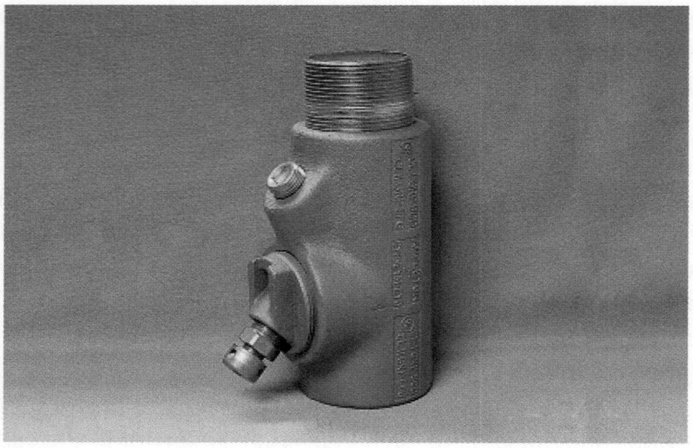

***EXHIBIT 501.11*** *A seal with an automatic drain plug. (Courtesy of Appleton Electric Co., EGS Electrical Group)*

***EXHIBIT 501.12*** *A combination breather-drainage fitting. (Courtesy of Appleton Electric Co., EGS Electrical Group)*

water accumulation, which can cause future insulation failures, even though prevalent conditions may not indicate a need.

**(1) Control Equipment.** Where there is a probability that liquid or other condensed vapor may be trapped within enclosures for control equipment or at any point in the raceway system, approved means shall be provided to prevent accumulation or to permit periodic draining of such liquid or condensed vapor.

**(2) Motors and Generators.** Where liquid or condensed vapor may accumulate within motors or generators, joints and conduit systems shall be arranged to minimize the entrance of liquid. If means to prevent accumulation or to permit periodic draining are necessary, such means shall be provided at the time of manufacture and shall be considered an integral part of the machine.

## 501.17 Process Sealing

This section shall apply to process-connected equipment, which includes, but is not limited to, canned pumps, submersible pumps, flow, pressure, temperature, or analysis measurement instruments. A process seal is a device to prevent the migration of process fluids from the designed containment into the external electrical system. Process-connected electrical equipment that incorporates a single process seal, such as a single compression seal, diaphragm, or tube to prevent flammable or combustible fluids from entering a conduit or cable system capable of transmitting fluids, shall be provided with an additional means to mitigate a single process seal failure, The additional means may include, but is not limited to, the following:

(1) A suitable barrier meeting the process temperature and pressure conditions that the barrier will be subjected to upon failure of the single process seal. There shall be a vent or drain between the single process seal and the suitable barrier. Indication of the single process seal failure shall be provided by visible leakage, an audible whistle, or other means of monitoring.

(2) A listed Type MI cable assembly, rated at not less than 125 percent of the process pressure and not less than 125 percent of the maximum process temperature (in degrees Celsius), installed between the cable or conduit and the single process seal.

(3) A drain or vent located between the single process seal and a conduit or cable seal. The drain or vent shall be sufficiently sized to prevent overpressuring the conduit or cable seal above 6 in. water column (1493 Pa). Indication of the single process seal failure shall be provided by visible leakage, an audible whistle, or other means of monitoring.

(4) An add-on secondary seal marked "secondary seal" and rated for the pressure and temperature conditions to which it will be subjected upon failure of the single process seal.

Process-connected electrical equipment that does not rely on a single process seal or is listed and marked "single seal" or "dual seal" shall not be required to be provided with an additional means of sealing.

Informational Note: For construction and testing requirements for process sealing for listed and marked single seal, dual seal, or secondary seal equipment, refer to ANSI/ISA-12.27.01-2011, *Requirements for Process Sealing Between Electrical Systems and Flammable or Combustible Process Fluids.*

In addition to the primary seal provided with canned pumps and other process equipment that operate above atmospheric pressure, 501.17 requires a second means to prevent fluid from entering the electrical conduit or cable system if the primary seal fails. The method employed is not limited to the four common techniques described.

The application of pressure or exposure to extreme temperatures must be prevented at the additional seal or barrier so that the process fluid will not enter the conduit system if the primary seal fails. This protection may be accomplished through the use of MI cable assembly rated 125 percent of the process temperature and pressure or through the use of a vent or drain. If the process fluid is a gas or can become a gas under ordinary atmospheric conditions (liquefied natural gas, for example), the drain should be a vent. In addition to this seal or barrier, a drain, vent, or other device that indicates failure of the primary seal must be provided.

This redundant protection system may be achieved by installing a vented junction (box) enclosure within the classified area where the conductors terminate on busbars. Terminating the conductors in this manner allows any fluid that escapes through the primary seal and that has traveled through the stranding of the conductors to vent at the terminations. The circuit continues through the vented enclosure at normal atmospheric pressure to another set of conductors, which also must be sealed with a seal fitting if they travel into a different classified area. A gas detector can be installed in the vicinity of the vented termination box to signal that the primary seal has failed and allow an orderly shutdown of the process system either automatically or manually.

## 501.20 Conductor Insulation, Class I, Divisions 1 and 2

Where condensed vapors or liquids may collect on, or come in contact with, the insulation on conductors, such insulation shall be of a type identified for use under such conditions; or the insulation shall be protected by a sheath of lead or by other approved means.

Type THWN conductors are commonly used in areas where they may be exposed to gasoline because of their ease of handling. Not all Type THWN conductors are suitable where they could be exposed to gasoline. THWN wire suitable for exposure to liquid gasoline and gasoline vapors at ordinary ambient temperature is marked "Gasoline and Oil Resistant I" (or "GR1") or "Gasoline and Oil Resistant II" (or "GR2"). See the UL White Book for further information on these cables.

## 501.25 Uninsulated Exposed Parts, Class I, Divisions 1 and 2

There shall be no uninsulated exposed parts, such as electrical conductors, buses, terminals, or components, that operate at more than 30 volts (15 volts in wet locations). These parts shall additionally be protected by a protection technique according to 500.7(E), (F), or (G) that is suitable for the location.

Exposed live parts are permitted in Class I, Division 1 and 2 locations provided the voltage does not exceed 30 volts in dry locations or 15 volts in wet locations, and the parts are protected by either intrinsically safe or nonincendive techniques (as applicable). These techniques limit the circuit's energy to a level incapable of causing ignition of the hazardous area.

## 501.30 Grounding and Bonding, Class I, Divisions 1 and 2

Regardless of the voltage of the electrical system, wiring and equipment in Class I, Division 1 and 2 locations shall be grounded as specified in Article 250 and in accordance with the requirements of 501.30(A) and (B).

**(A) Bonding.** The locknut-bushing and double-locknut types of contacts shall not be depended on for bonding purposes, but bonding jumpers with proper fittings or other approved means of bonding shall be used. Such means of bonding shall apply to all intervening raceways, fittings, boxes, enclosures, and so forth between Class I locations and the point of grounding for service equipment or point of grounding of a separately derived system.

*Exception: The specific bonding means shall be required only to the nearest point where the grounded circuit conductor and the grounding electrode are connected together on the line side of the building or structure disconnecting means as specified in 250.32(B), provided the branch-circuit overcurrent protection is located on the load side of the disconnecting means.*

The specific bonding methods mentioned in this section are intended to provide a mechanical/electrical connection that is low impedance and free from accidental arcing due to loose connections; they apply to raceways and raceway-to-enclosure connections both inside and outside the hazardous location. Section 250.100 specifies this enhanced level of bonding for all raceways and enclosures, and the requirement is not contingent on the circuit voltage. This includes metal raceways and enclosures containing signaling, communications, or other power-limited circuits.

To be effective, proper grounding and bonding apply to all interconnected raceways, fittings, enclosures, and so forth, between hazardous locations and the point of grounding for service equipment or the point of grounding of the building disconnecting means that supplies the branch-circuit overcurrent protection. If conduit is used in hazardous locations, it is preferable that threaded connections also be employed in the nonhazardous location.

Section 250.100 clarifies that the installation of a wire-type EGC in a metal raceway does not negate the special raceway and enclosure bonding requirements. Unless it is an isolated EGC as permitted by 250.96(B) or 250.146(D), the wire-type EGC and the metal raceway will be electrically in parallel and will share ground-fault current based on the impedance of the wire and of the metal raceway. For that reason, the electrical continuity of raceways and raceway-to-enclosure connections must always be ensured through compliance with 250.100 and 501.30(A), regardless of whether a supplementary EGC has been installed in the raceway.

The exception covers the grounding and bonding requirements that are specific to hazardous locations where the installation occurs at a multibuilding or multistructure setting. If the service equipment and the electrical equipment supplying the hazardous location are not located in the same building or structure, applying the bonding requirement of 501.30(A) from the hazardous location back to the service equipment is not necessary. It is necessary only to apply the bonding requirement from the hazardous location back to the grounding electrode on the line side of the building or structure disconnecting means. This connection must be ahead of the branch circuits that are on the load side of the disconnecting means for the building or structure.

•

**(B) Types of Equipment Grounding Conductors.** Flexible metal conduit and liquidtight flexible metal conduit shall include an equipment bonding jumper of the wire type in compliance with 250.102.

Although flexible metal conduit is permitted by 250.118 to function as an EGC, it is generally required to be installed with a wire-type bonding jumper if used in a hazardous location. The intent is to ensure permanent and effective mechanical and electrical connections in order to prevent the possibility of arcs or sparks caused by ineffective or poor grounding methods. Exhibit 501.13 illustrates an equipment bonding jumper routed outside of LFMC for a length not to exceed 6 feet in accordance with 250.102(E). The exception permits omission of the bonding jumper under certain conditions.

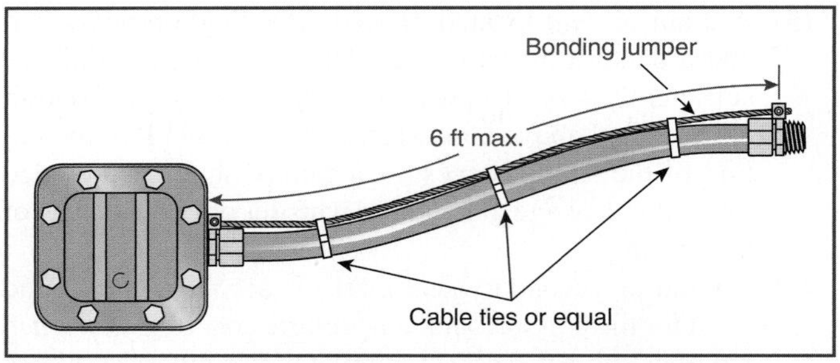


***EXHIBIT 501.13*** *The connection of an external bonding jumper used with LFMC.*

*Exception: In Class I, Division 2 locations, the bonding jumper shall be permitted to be deleted where all of the following conditions are met:*

*(1) Listed liquidtight flexible metal conduit 1.8 m (6 ft) or less in length, with fittings listed for grounding, is used.*

*(2) Overcurrent protection in the circuit is limited to 10 amperes or less.*

*(3) The load is not a power utilization load.*

### 501.35 Surge Protection

**(A) Class I, Division 1.** Surge arresters, surge-protective devices, and capacitors shall be installed in enclosures identified for Class I, Division 1 locations. Surge-protective capacitors shall be of a type designed for specific duty.

**(B) Class I, Division 2.** Surge arresters and surge-protective devices shall be nonarcing, such as metal-oxide varistor (MOV) sealed type, and surge-protective capacitors shall be of a type designed for specific duty. Enclosures shall be permitted to be of the general-purpose type. Surge protection of types other than described in this paragraph shall be installed in enclosures identified for Class I, Division 1 locations.

In Class I, Division 2 locations, only spark-producing types of surge arresters require installation in an enclosure identified for the location. Nonarcing, sealed, and solid state–type SPDs are permitted where installed in a general purpose–type enclosure. Refer to Article 285 for further information on SPDs and see Article 280 for surge arresters. Surge arresters can also be installed in oil-filled enclosures or have the arcing or sparking contacts enclosed in hermetically sealed chambers.

•

## III. Equipment

### 501.100 Transformers and Capacitors

**(A) Class I, Division 1.** In Class I, Division 1 locations, transformers and capacitors shall comply with 501.100(A)(1) and (A)(2).

**(1) Containing Liquid That Will Burn.** Transformers and capacitors containing a liquid that will burn shall be installed only in vaults that comply with 450.41 through 450.48 and with (1) through (4) as follows:

(1) There shall be no door or other communicating opening between the vault and the Division 1 location.
(2) Ample ventilation shall be provided for the continuous removal of flammable gases or vapors.
(3) Vent openings or ducts shall lead to a safe location outside of buildings.
(4) Vent ducts and openings shall be of sufficient area to relieve explosion pressures within the vault, and all portions of vent ducts within the buildings shall be of reinforced concrete construction.

**(2) Not Containing Liquid That Will Burn.** Transformers and capacitors that do not contain a liquid that will burn shall be installed in vaults complying with 501.100(A)(1) or be identified for Class I locations.

**(B) Class I, Division 2.** In Class I, Division 2 locations, transformers shall comply with 450.21 through 450.27, and capacitors shall comply with 460.2 through 460.28.

## 501.105 Meters, Instruments, and Relays

**(A) Class I, Division 1.** In Class I, Division 1 locations, meters, instruments, and relays, including kilowatt-hour meters, instrument transformers, resistors, rectifiers, and thermionic tubes, shall be provided with enclosures identified for Class I, Division 1 locations. Enclosures for Class I, Division 1 locations include explosionproof enclosures and purged and pressurized enclosures.

Informational Note: See NFPA 496-2013, *Standard for Purged and Pressurized Enclosures for Electrical Equipment.*

**(B) Class I, Division 2.** In Class I, Division 2 locations, meters, instruments, and relays shall comply with 501.105(B)(1) through (B)(6).

**(1) Contacts.** Switches, circuit breakers, and make-and-break contacts of pushbuttons, relays, alarm bells, and horns shall have enclosures identified for Class I, Division 1 locations in accordance with 501.105(A).

*Exception: General-purpose enclosures shall be permitted if current-interrupting contacts comply with one of the following:*

*(1) Are immersed in oil*
*(2) Are enclosed within a chamber that is hermetically sealed against the entrance of gases or vapors*
*(3) Are in nonincendive circuits*
*(4) Are listed for Division 2*

This exception identifies the conditions under which general-purpose enclosures are permitted for certain equipment in Class 1, Division 2 locations. One such condition is the use of hermetic seals, which include fusion seals such as the glass-to-metal seals in mercury-tube switches and some reed switches, welded seals, and soldered seals. Seals of the glass-to-metal-fusion type are usually the most reliable. Soft-soldered seals can be relatively porous, and their effectiveness is highly dependent on workmanship. Although gasketed seals can be very effective, depending on the gasket material used, gasket materials can be damaged and deteriorate rapidly if exposed to atmospheres that contain solvent vapors. Gasketed enclosures may be considered hermetically sealed under some conditions; however, in accordance with the 500.2 definition of *hermetically sealed*, such enclosures cannot be used to satisfy those requirements in which hermetic sealing is recognized as a protection technique.

This does not mean that the entire enclosure or circuit is required to qualify as being protected by one of these techniques, but that only the contacts need such protection.

**(2) Resistors and Similar Equipment.** Resistors, resistance devices, thermionic tubes, rectifiers, and similar equipment that are used in or in connection with meters, instruments, and relays shall comply with 501.105(A).

*Exception: General-purpose-type enclosures shall be permitted if such equipment is without make-and-break or sliding contacts [other than as provided in 501.105(B)(1)] and if the maximum operating temperature of any exposed surface will not exceed 80 percent of the autoignition temperature in degrees Celsius of the gas or vapor involved or has been tested and found incapable of igniting the gas or vapor. This exception shall not apply to thermionic tubes.*

**(3) Without Make-or-Break Contacts.** Transformer windings, impedance coils, solenoids, and other windings that do not incorporate sliding or make-or-break contacts shall be provided with enclosures. General-purpose-type enclosures shall be permitted.

**(4) General-Purpose Assemblies.** Where an assembly is made up of components for which general-purpose enclosures are acceptable as provided in 501.105(B)(1), (B)(2), and (B)(3), a single general-purpose enclosure shall be acceptable for the assembly. Where such an assembly includes any of the equipment described in 501.105(B)(2), the maximum obtainable surface temperature of any component of the assembly shall be clearly and permanently indicated on the outside of the enclosure. Alternatively, equipment shall be permitted to be marked to indicate the temperature class for which it is suitable, using the temperature class (T Code) of Table 500.8(C).

**(5) Fuses.** Where general-purpose enclosures are permitted in 501.105(B)(1) through (B)(4), fuses for overcurrent protection of instrument circuits not subject to overloading in normal use shall be permitted to be mounted in general-purpose enclosures if each such fuse is preceded by a switch complying with 501.105(B)(1).

**(6) Connections.** To facilitate replacements, process control instruments shall be permitted to be connected through flexible cord, attachment plug, and receptacle, provided all of the following conditions apply:

(1) A switch complying with 501.105(B)(1) is provided so that the attachment plug is not depended on to interrupt current.

*Exception: The switch is not required if the circuit is nonincendive field wiring.*

(2) The current does not exceed 3 amperes at 120 volts, nominal.
(3) The power-supply cord does not exceed 900 mm (3 ft), is of a type listed for extra-hard usage or for hard usage if protected by location, and is supplied through an attachment plug and receptacle of the locking and grounding type.
(4) Only necessary receptacles are provided.
(5) The receptacle carries a label warning against unplugging under load.

## 501.115 Switches, Circuit Breakers, Motor Controllers, and Fuses

**(A) Class I, Division 1.** In Class I, Division 1 locations, switches, circuit breakers, motor controllers, and fuses, including pushbuttons, relays, and similar devices, shall be provided with enclosures, and the enclosure in each case, together with the enclosed apparatus, shall be identified as a complete assembly for use in Class I locations.

Exhibit 501.14 shows an explosionproof panelboard that consists of an assembly of branch-circuit devices enclosed in a cast metal explosionproof housing. Explosionproof panelboards are provided with bolted access covers and threaded conduit-entry hubs designed to withstand the force of an internal explosion.

Exhibit 501.15 shows a cylindrical-type (spin-top) combination motor controller, motor control starter, and circuit breaker in an explosionproof enclosure. The top and bottom covers are threaded on for quick removal for installation and servicing. Exhibit 501.16 shows the same type of equipment in a rectangular enclosure with a hinged, bolted-on cover. These types of housings are designed to accommodate a wide range of manually or magnetically operated across-the-line types of motor starters in a variety of ratings.

Exhibit 501.17 illustrates a standard toggle switch in an explosionproof enclosure.

**(B) Class I, Division 2.** Switches, circuit breakers, motor controllers, and fuses in Class I, Division 2 locations shall comply with 501.115(B)(1) through (B)(4).

**(1) Type Required.** Circuit breakers, motor controllers, and switches intended to interrupt current in the normal performance of the function for which they are installed shall be provided with enclosures identified for Class I, Division 1 locations in accordance with 501.105(A), unless general-purpose enclosures are provided and any of the following apply:

***EXHIBIT 501.14*** *An explosionproof panelboard. (Courtesy of Appleton Electric Co., EGS Electrical Group)*

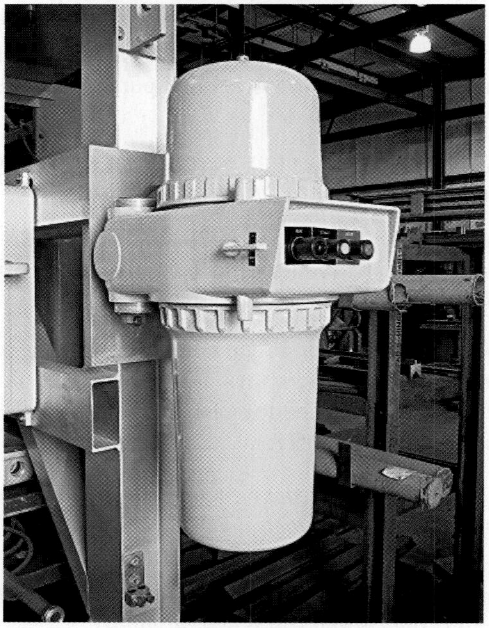

***EXHIBIT 501.15*** *An explosionproof enclosure for a motor control starter and circuit breaker. (Courtesy of Appleton Electric Co., EGS Electrical Group)*

(1) The interruption of current occurs within a chamber hermetically sealed against the entrance of gases and vapors.
(2) The current make-and-break contacts are oil-immersed and of the general-purpose type having a 50-mm (2-in.)

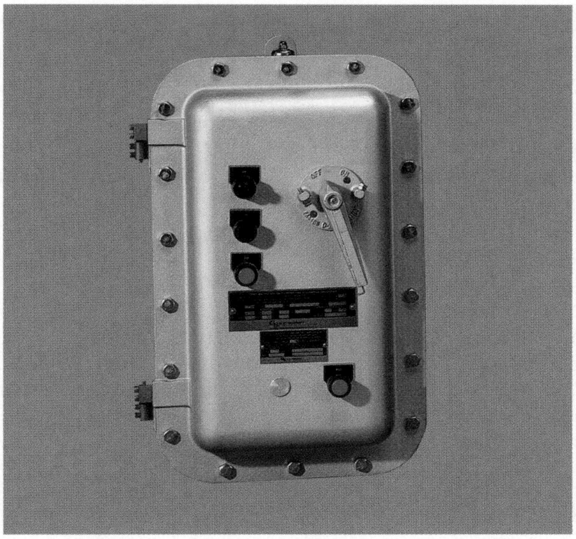

***EXHIBIT 501.16*** *A magnetic motor starter for use in a Class I, Group D location. Note the number of securing bolts and the width of the flange. (Courtesy of O-Z/Gedney, a division of EGS Electrical Group)*

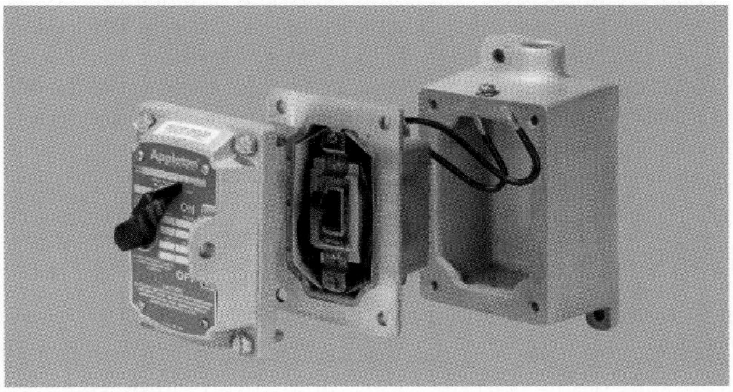

***EXHIBIT 501.17*** *A standard toggle switch in an explosionproof enclosure. (Courtesy of Appleton Electric Co., EGS Electrical Group)*

minimum immersion for power contacts and a 25-mm (1-in.) minimum immersion for control contacts.

(3) The interruption of current occurs within a factory-sealed explosionproof chamber.
(4) The device is a solid state, switching control without contacts, where the surface temperature does not exceed 80 percent of the autoignition temperature in degrees Celsius of the gas or vapor involved.

**(2) Isolating Switches.** Fused or unfused disconnect and isolating switches for transformers or capacitor banks that are not intended to interrupt current in the normal performance of the function for which they are installed shall be permitted to be installed in general-purpose enclosures.

**(3) Fuses.** For the protection of motors, appliances, and lamps, other than as provided in 501.115(B)(4), standard plug or cartridge fuses shall be permitted, provided they are placed within enclosures identified for the location; or fuses shall be permitted if they are within general-purpose enclosures, and if they are of a type in which the operating element is immersed in oil or other approved liquid, or the operating element is enclosed within a chamber hermetically sealed against the entrance of gases and vapors, or the fuse is a nonindicating, filled, current-limiting type.

**(4) Fuses Internal to Luminaires.** Listed cartridge fuses shall be permitted as supplementary protection within luminaires.

## 501.120 Control Transformers and Resistors

Transformers, impedance coils, and resistors used as, or in conjunction with, control equipment for motors, generators, and appliances shall comply with 501.120(A) and (B).

**(A) Class I, Division 1.** In Class I, Division 1 locations, transformers, impedance coils, and resistors, together with any switching mechanism associated with them, shall be provided with enclosures identified for Class I, Division 1 locations in accordance with 501.105(A).

**(B) Class I, Division 2.** In Class I, Division 2 locations, control transformers and resistors shall comply with 501.120(B)(1) through (B)(3).

**(1) Switching Mechanisms.** Switching mechanisms used in conjunction with transformers, impedance coils, and resistors shall comply with 501.115(B).

**(2) Coils and Windings.** Enclosures for windings of transformers, solenoids, or impedance coils shall be permitted to be of the general-purpose type.

**(3) Resistors.** Resistors shall be provided with enclosures; and the assembly shall be identified for Class I locations, unless resistance is nonvariable and maximum operating temperature, in degrees Celsius, will not exceed 80 percent of the autoignition temperature of the gas or vapor involved or the resistor has been tested and found incapable of igniting the gas or vapor.

## 501.125 Motors and Generators

**(A) Class I, Division 1.** In Class I, Division 1 locations, motors, generators, and other rotating electrical machinery shall be one of the following:

(1) Identified for Class I, Division 1 locations
(2) Of the totally enclosed type supplied with positive-pressure ventilation from a source of clean air with discharge to a safe area, so arranged to prevent energizing of the machine until ventilation has been established and the enclosure has been purged with at least 10 volumes of air, and also arranged to automatically de-energize the equipment when the air supply fails
(3) Of the totally enclosed inert gas-filled type supplied with a suitable reliable source of inert gas for pressurizing the

enclosure, with devices provided to ensure a positive pressure in the enclosure and arranged to automatically de-energize the equipment when the gas supply fails

(4) Of a type designed to be submerged in a liquid that is flammable only when vaporized and mixed with air, or in a gas or vapor at a pressure greater than atmospheric and that is flammable only when mixed with air; and the machine is so arranged to prevent energizing it until it has been purged with the liquid or gas to exclude air, and also arranged to automatically de-energize the equipment when the supply of liquid or gas or vapor fails or the pressure is reduced to atmospheric

Totally enclosed motors of the types specified in 501.125(A)(2) or (A)(3) shall have no external surface with an operating temperature in degrees Celsius in excess of 80 percent of the autoignition temperature of the gas or vapor involved. Appropriate devices shall be provided to detect and automatically de-energize the motor or provide an adequate alarm if there is any increase in temperature of the motor beyond designed limits. Auxiliary equipment shall be of a type identified for the location in which it is installed.

•

**(B) Class I, Division 2.** In Class I, Division 2 locations, motors, generators, and other rotating electrical machinery in which are employed sliding contacts, centrifugal or other types of switching mechanism (including motor overcurrent, overloading, and overtemperature devices), or integral resistance devices, either while starting or while running, shall be identified for Class I, Division 1 locations, unless such sliding contacts, switching mechanisms, and resistance devices are provided with enclosures identified for Class I, Division 2 locations in accordance with 501.105(B). The exposed surface of space heaters used to prevent condensation of moisture during shutdown periods shall not exceed 80 percent of the autoignition temperature in degrees Celsius of the gas or vapor involved when operated at rated voltage, and the maximum space heater surface temperature [based on a 40°C or higher marked ambient] shall be permanently marked on a visible nameplate mounted on the motor. Otherwise, space heaters shall be identified for Class I, Division 2 locations. In Class I, Division 2 locations, the installation of open or nonexplosionproof enclosed motors, such as squirrel-cage induction motors without brushes, switching mechanisms, or similar arc-producing devices that are not identified for use in a Class I, Division 2 location, shall be permitted.

Motors with arcing devices, such as commutators, are required to be provided with an enclosure identified for the location, such as an explosionproof enclosure. Other motor types without arcing devices, such as a squirrel-cage induction motor, are permitted without special enclosures. Additionally, indication of the maximum temperature present in a motor is critical to proper installation. Many motor heaters are de-energized automatically when the motor is running. However, the heater ratings are usually low when compared with the normal heat generated during motor operation. Unless otherwise indicated on the motor wiring diagram or in instructions provided with the motor, there is no need to de-energize the heater except to save energy. The heater temperature must be marked on the motor, or the heater must be identified for the location.

Informational Note No. 1: It is important to consider the temperature of internal and external surfaces that may be exposed to the flammable atmosphere.

Informational Note No. 2: It is important to consider the risk of ignition due to currents arcing across discontinuities and overheating of parts in multisection enclosures of large motors and generators. Such motors and generators may need equipotential bonding jumpers across joints in the enclosure and from enclosure to ground. Where the presence of ignitible gases or vapors is suspected, clean-air purging may be needed immediately prior to and during start-up periods.

Informational Note No. 3: For further information on the application of electric motors in Class I, Division 2 hazardous (classified) locations, see IEEE 1349-2011, *IEEE Guide for the Application of Electric Motors in Class I, Division 2 and Class I, Zone 2 Hazardous (Classified) Locations.*

Informational Note No. 4: Reciprocating engine-driven generators, compressors, and other equipment installed in Class I, Division 2 locations may present a risk of ignition of flammable materials associated with fuel, starting, compression, and so forth, due to inadvertent release or equipment malfunction by the engine ignition system and controls. For further information on the requirements for ignition systems for reciprocating engines installed in Class I, Division 2 hazardous (classified) locations, see ANSI/ISA-12.20.01-2009, *General Requirements for Electrical Ignition Systems for Internal Combustion Engines in Class I, Division 2 or Zone 2, Hazardous (Classified) Locations.*

High-inertia loads can cause increased rotor heating during starting, and sparking can occur between motor housing assemblies when starting. Motor types used where flammable gases or vapors with very low ignition temperatures may be present should be carefully selected. Modern motors with high-temperature insulation systems, such as Class H [180°C (356°F)], may operate close to or above the ignition temperature of the flammable mixture.

Exhibits 501.18 and 501.19 show a totally enclosed fan-cooled motor listed for use in explosive atmospheres. The main frame and end-bells are designed with sufficient strength to withstand an internal explosion. Flames or hot gases are cooled while escaping because of the wide metal-to-metal joints between the frame and the end-bells and the long, close-tolerance clearance provided for the free turn of the shaft. Air circulation outside the motor is maintained by a nonsparking (aluminum, bronze, or non-static-generating-type plastic) fan on the end opposite the shaft end of the motor. A sheet metal housing surrounds the fan to reduce the likelihood of an individual or object coming into contact with the moving blades and to direct the flow of air. An internal fan on the shaft, as shown in Exhibit 501.19, circulates air around the windings.

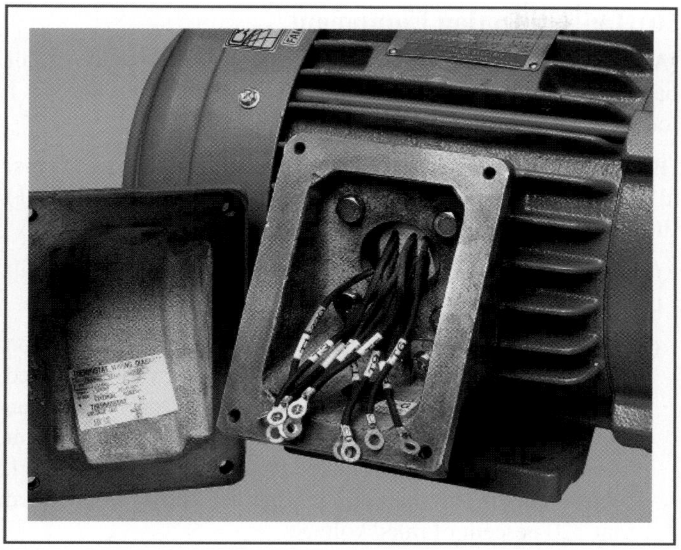

**EXHIBIT 501.18** *Terminal housing of a motor listed for use in specific hazardous locations. Note integral sealing of the motor. (Courtesy of General Electric Co.)*

**EXHIBIT 501.19** *View showing internal fan of motor in Exhibit 501.18. (Courtesy of General Electric Co.)*

## 501.130 Luminaires

Luminaires shall comply with 501.130(A) or (B).

**(A) Class I, Division 1.** In Class I, Division 1 locations, luminaires shall comply with 501.130(A)(1) through (A)(4).

**(1) Luminaires.** Each luminaire shall be identified as a complete assembly for the Class I, Division 1 location and shall be clearly marked to indicate the maximum wattage of lamps for which it is identified. Luminaires intended for portable use shall be specifically listed as a complete assembly for that use.

**(2) Physical Damage.** Each luminaire shall be protected against physical damage by a suitable guard or by location.

**(3) Pendant Luminaires.** Pendant luminaires shall be suspended by and supplied through threaded rigid metal conduit stems or threaded steel intermediate conduit stems, and threaded joints shall be provided with set-screws or other effective means to prevent loosening. For stems longer than 300 mm (12 in.), permanent and effective bracing against lateral displacement shall be provided at a level not more than 300 mm (12 in.) above the lower end of the stem, or flexibility in the form of a fitting or flexible connector identified for the Class I, Division 1 location shall be provided not more than 300 mm (12 in.) from the point of attachment to the supporting box or fitting.

**(4) Supports.** Boxes, box assemblies, or fittings used for the support of luminaires shall be identified for Class I locations.

**(B) Class I, Division 2.** In Class I, Division 2 locations, luminaires shall comply with 501.130(B)(1) through (B)(6).

**(1) Luminaires.** Where lamps are of a size or type that may, under normal operating conditions, reach surface temperatures exceeding 80 percent of the autoignition temperature in degrees Celsius of the gas or vapor involved, luminaires shall comply with 501.130(A)(1) or shall be of a type that has been tested in order to determine the marked operating temperature or temperature class (T code).

**(2) Physical Damage.** Luminaires shall be protected from physical damage by suitable guards or by location. Where there is danger that falling sparks or hot metal from lamps or luminaires might ignite localized concentrations of flammable vapors or gases, suitable enclosures or other effective protective means shall be provided.

**(3) Pendant Luminaires.** Pendant luminaires shall be suspended by threaded rigid metal conduit stems, threaded steel intermediate metal conduit stems, or other approved means. For rigid stems longer than 300 mm (12 in.), permanent and effective bracing against lateral displacement shall be provided at a level not more than 300 mm (12 in.) above the lower end of the stem, or flexibility in the form of an identified fitting or flexible connector shall be provided not more than 300 mm (12 in.) from the point of attachment to the supporting box or fitting.

**(4) Portable Lighting Equipment.** Portable lighting equipment shall comply with 501.130(A)(1).

*Exception: Where portable lighting equipment is mounted on movable stands and is connected by flexible cords, as covered in 501.140, it shall be permitted to comply with 501.130(B)(1), where mounted in any position, provided that it also complies with 501.130(B)(2).*

**(5) Switches.** Switches that are a part of an assembled fixture or of an individual lampholder shall comply with 501.115(B)(1).

**(6) Starting Equipment.** Starting and control equipment for electric-discharge lamps shall comply with 501.120(B).

Exhibit 501.20 shows a typical luminaire for Class I, Group C and D locations. The outlet boxes have an internally threaded opening designed to receive the cover. A pendant luminaire is attached to the cover by threaded RMC or threaded IMC. To prevent loosening from vibration or lamp changing, threaded joints must be provided with set-screws. The set-screws should not interrupt the explosionproof joint. RMC or IMC stems longer than 12 inches require effective bracing or a flexible fitting identified for the purpose, placed not more than 12 inches from the point of attachment to the supporting box, cover, or fitting.

A globe holder is threaded onto the body of the luminaire housing to support a heavy glass globe, guard, and reflector. In designing any hazardous location lighting system, operating temperatures must be considered. If the area is Class I, Division 1, luminaires that are identified for this location and are properly marked must be used. Generally, enclosed and gasketed luminaires – without guards, if breakage is unlikely – or luminaires identified for Class I, Division 2 locations are required in Division 2 locations.

Portable luminaires are required to be specifically listed as a complete assembly for use in Class 1, Division 1 or 2 locations. Exhibit 501.21 shows an explosionproof hand lamp. Lamp compartments must be sealed from the terminal compartment. Provisions must be made for the connection of 3-conductor (one must be a grounding conductor) flexible, extra-hard-usage cord. See 501.140(A)(1).

*Exception: A thermal protector potted into a thermally protected fluorescent lamp ballast if the luminaire is identified for the location.*

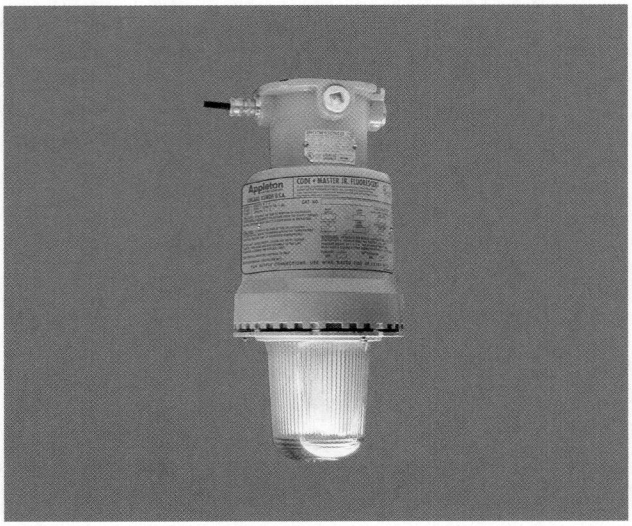

***EXHIBIT 501.20*** *A typical luminaire for use in Class I, Group C and D locations. (Courtesy of Appleton Electric Co., EGS Electrical Group)*

### 501.135 Utilization Equipment

**(A) Class I, Division 1.** In Class I, Division 1 locations, all utilization equipment shall be identified for Class I, Division 1 locations.

**(B) Class I, Division 2.** In Class I, Division 2 locations, all utilization equipment shall comply with 501.135(B)(1) through (B)(3).

**(1) Heaters.** Electrically heated utilization equipment shall conform with either item (1) or item (2):

(1) The heater shall not exceed 80 percent of the autoignition temperature in degrees Celsius of the gas or vapor involved on any surface that is exposed to the gas or vapor when continuously energized at the maximum rated ambient temperature. If a temperature controller is not provided, these conditions shall apply when the heater is operated at 120 percent of rated voltage.

*Exception No. 1: For motor-mounted anticondensation space heaters, see 501.125.*

*Exception No. 2: Where a current-limiting device is applied to the circuit serving the heater to limit the current in the heater to a value less than that required to raise the heater surface temperature to 80 percent of the autoignition temperature.*

(2) The heater shall be identified for Class I, Division 1 locations.

*Exception to (2): Electrical resistance heat tracing identified for Class I, Division 2 locations.*

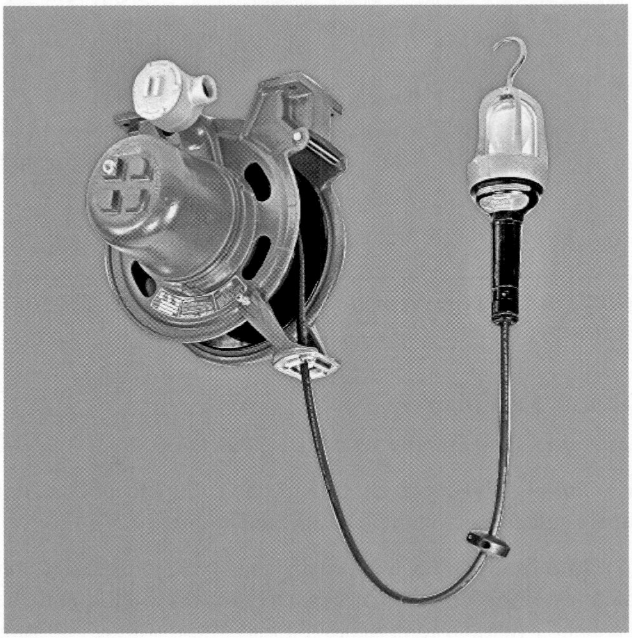

***EXHIBIT 501.21*** *An explosionproof hand lamp for use in Class I locations. (Courtesy of Appleton Electric Co., EGS Electrical Group)*

**(2) Motors.** Motors of motor-driven utilization equipment shall comply with 501.125(B).

**(3) Switches, Circuit Breakers, and Fuses.** Switches, circuit breakers, and fuses shall comply with 501.115(B).

## 501.140 Flexible Cords, Class I, Divisions 1 and 2

**(A) Permitted Uses.** Flexible cord shall be permitted:

(1) For connection between portable lighting equipment or other portable utilization equipment and the fixed portion of their supply circuit. The flexible cord shall be attached to the utilization equipment with a cord connector listed for the protection technique of the equipment wiring compartment. An attachment plug in accordance with 501.140(B)(4) shall be employed.

(2) For that portion of the circuit where the fixed wiring methods of 501.10(A) cannot provide the necessary degree of movement for fixed and mobile electrical utilization equipment, and the flexible cord is protected by location or by a suitable guard from damage and only in an industrial establishment where conditions of maintenance and engineering supervision ensure that only qualified persons install and service the installation.

(3) For electric submersible pumps with means for removal without entering the wet-pit. The extension of the flexible cord within a suitable raceway between the wet-pit and the power source shall be permitted.

(4) For electric mixers intended for travel into and out of open-type mixing tanks or vats.

Because electric mixers used in mixing tanks are frequently inserted and removed, the mixers are considered portable utilization equipment and are permitted to be wired with flexible cord per 501.140(A)(4).

(5) For temporary portable assemblies consisting of receptacles, switches, and other devices that are not considered portable utilization equipment but are individually listed for the location.

An example of a portable assembly that is not considered utilization equipment is a power cart (see Exhibit 501.22) that provides power during servicing or maintenance. The flexible cord is required to be continuous from the power source to the assembly and from the assembly to the utilization equipment.

**(B) Installation.** Where flexible cords are used, the cords shall comply with all of the following:

(1) Be of a type listed for extra-hard usage

(2) Contain, in addition to the conductors of the circuit, an equipment grounding conductor complying with 400.23

(3) Be supported by clamps or by other suitable means in such a manner that there is no tension on the terminal connections

(4) In Division 1 locations or in Division 2 locations where the boxes, fittings, or enclosures are required to be explosionproof, the cord shall be terminated with a cord connector or attachment plug listed for the location or a listed cord connector installed with a seal listed for the location. In Division 2 locations where explosionproof equipment is not required, the cord shall be terminated with a listed cord connector or listed attachment plug.

(5) Be of continuous length. Where 501.140(A)(5) is applied, cords shall be of continuous length from the power source to the temporary portable assembly and from the temporary portable assembly to the utilization equipment.

***EXHIBIT 501.22*** *A temporary portable assembly. (Courtesy of Killark, a division of Hubbell Incorporated)*

Informational Note: See 501.20 for flexible cords exposed to liquids having a deleterious effect on the conductor insulation.

## 501.145 Receptacles and Attachment Plugs, Class I, Divisions 1 and 2

**(A) Receptacles.** Receptacles shall be part of the premises wiring, except as permitted by 501.140(A).

**(B) Attachment Plugs.** Attachment plugs shall be of the type providing for connection to the equipment grounding conductor of a flexible cord and shall be identified for the location.

*Exception: Receptacles and attachment plugs as provided in 501.105(B)(6).*

Exhibit 501.23 shows an explosionproof receptacle and attachment plug with an interlocking switch. The design of this device is such that when the switch is in the "on" position, the plug cannot be removed. Also, the switch cannot be placed in the "on" position when the plug has been removed; the receptacle is factory sealed, with a provision for threaded-conduit entry to the switch compartment. The plug is to be used with Type S or equivalent extra-hard-service flexible cord having an EGC.

Exhibit 501.24 shows a 30-ampere, 4-pole receptacle and attachment plug assembly that is suitable for use without a switch. The design is such that the mating parts of the receptacle and plug are enclosed in a chamber that seals the arc and, by delayed-action construction, prevents complete removal of the plug until the arc or hot metal has cooled. The receptacle is

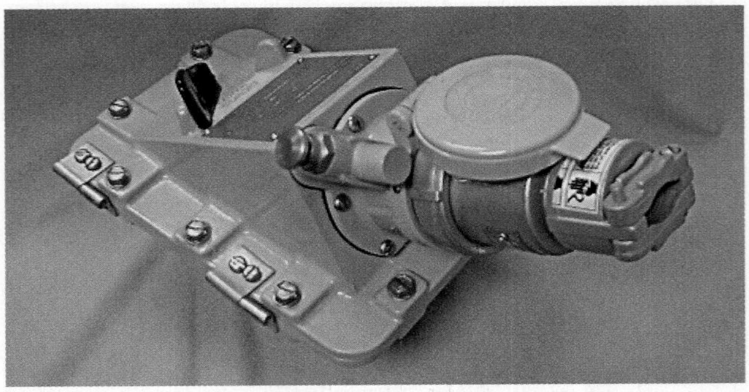

***EXHIBIT 501.23*** *A receptacle and attachment plug of the explosionproof type with an interlocking switch. The switch must be in the "off" position before the attachment plug can be inserted or removed. (Courtesy of Appleton Electric Co., EGS Electrical Group)*

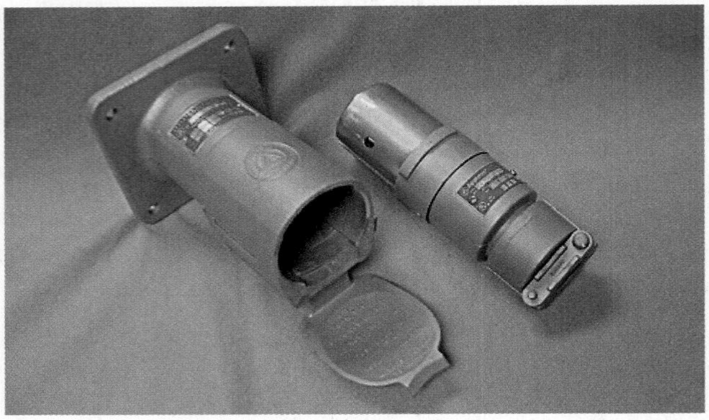

***EXHIBIT 501.24*** *A 4-pole (delayed action) explosionproof receptacle and attachment plug suitable for use without a switch. (Courtesy of Appleton Electric Co., EGS Electrical Group)*

factory sealed, and the attachment plug is designed for use with a 4-conductor cord (3-conductor, 3-phase circuit with one EGC) or a 3-conductor cord (two circuit conductors and one EGC).

## 501.150 Signaling, Alarm, Remote-Control, and Communications Systems

**(A) Class I, Division 1** In Class I, Division 1 locations, all apparatus and equipment of signaling, alarm, remote-control, and communications systems, regardless of voltage, shall be identified for Class I, Division 1 locations, and all wiring shall comply with 501.10(A), 501.15(A), and 501.15(C).

**(B) Class I, Division 2.** In Class I, Division 2 locations, signaling, alarm, remote-control, and communications systems shall comply with 501.150(B)(1) through (B)(4).

**(1) Contacts.** Switches, circuit breakers, and make-and-break contacts of pushbuttons, relays, alarm bells, and horns shall have enclosures identified for Class I, Division 1 locations in accordance with 501.105(A).

*Exception: General-purpose enclosures shall be permitted if current-interrupting contacts are one of the following:*

*(1) Immersed in oil*
*(2) Enclosed within a chamber hermetically sealed against the entrance of gases or vapors*
*(3) In nonincendive circuits*
*(4) Part of a listed nonincendive component*

Explosionproof devices or explosionproof enclosures may prove more practical than oil-immersed contacts because maintaining the condition and level of the oil can be a problem. Hermetically sealed enclosures, such as float-operated mercury-tube switches, are available for some applications.

See the commentary regarding arcing contacts and the use of general-purpose enclosures in Class I, Division 2 locations following the exception to 501.105(B)(1).

**(2) Resistors and Similar Equipment.** Resistors, resistance devices, thermionic tubes, rectifiers, and similar equipment shall comply with 501.105(B)(2).

**(3) Protectors.** Enclosures shall be provided for lightning protective devices and for fuses. Such enclosures shall be permitted to be of the general-purpose type.

**(4) Wiring and Sealing.** All wiring shall comply with 501.10(B), 501.15(B), and 501.15(C).

Some audible signaling devices may contain make-and-break contacts that are capable of producing a spark of sufficient energy to cause ignition of a hazardous atmospheric mixture. If used in Class I locations, this type of equipment must be contained in explosionproof or purged and pressurized enclosures, wiring methods must comply with 501.10, and seal fittings must be provided in accordance with 501.15. (See Exhibit 501.25.) Electronic signal devices without make-and-break contacts usually do not require explosionproof enclosures in Division 2 locations.

***EXHIBIT 501.25*** *An audible signaling device for use in hazardous locations. (Courtesy of Cooper Crouse-Hinds)*

# ARTICLE 502 Class II Locations

## I. General

### 502.1 Scope

Article 502 covers the requirements for electrical and electronic equipment and wiring for all voltages in Class II, Division 1 and 2 locations where fire or explosion hazards may exist due to combustible dust.

Class II, Division 1 and 2 locations are defined in 500.5(C) as "hazardous because of the presence of combustible dust." These locations are separated into three groups: Group E, Group F, and Group G [see 500.6(B)]. Two different types of dust environments typically warrant a Class II, Division 1 area classification. The first is where a cloud of combustible dust is likely to be present continuously or intermittently under normal operating conditions or as a result of repair or maintenance operations or leakage. The other environment is one in which a dust layer is likely to accumulate to a depth greater than ⅛ inch on major horizontal surfaces over a defined period of time, usually 24 hours. A Class II, Division 2 location is typically one where these conditions exist infrequently or under abnormal conditions.

The size of the dust particle is the primary factor in determining whether it should be classified as combustible. Combustible dust, as defined in 500.2, is any finely divided solid material 500 microns or smaller in diameter (material passing through a U.S. No. 35 standard sieve) that presents a fire or explosion hazard when dispersed and ignited in air.

NFPA 484, *Standard for Combustible Metals*, provides requirements for means to minimize the occurrence of, and resulting damage from, fire or explosion in areas where combustible metals or metal dusts are produced, processed, finished, handled, stored, and used.

### 502.5 Explosionproof Equipment

Explosionproof equipment and wiring shall not be required and shall not be acceptable in Class II locations unless also identified for such locations.

The electrical equipment required in Class II locations is different from that required for Class I locations. Class II equipment is designed to prevent the ignition of layers of dust, which may also cause an increase in equipment operating temperature, while Class I equipment does not address this concern. To protect against explosions in hazardous locations, all electrical equipment exposed to the hazardous atmosphere must be suitable for such locations. Equipment suitable for one class and group is not necessarily suitable for any other class and group.

Class I equipment is not necessarily suitable for a Class II location because the hazard contemplated in the equipment design is different. Grain dust, for example, ignites at a temperature lower than that of most flammable vapors. Motors listed for use in Class I locations may not have dust shields on the bearings to prevent entrance of dust into the bearing race, thereby causing overheating of the bearing and resulting in ignition of dust on the motor. Class I equipment is not designed for dust layering unless it is also designed and identified for Class II locations.

Dust-ignitionproof enclosures are not required to be explosionproof. Explosionproof enclosures are not necessarily dust-ignitionproof. However, they are allowed to be used in Class II locations if the equipment is dual rated and identified as suitable for the Class II division and group.

These dual-rated enclosures, where used in an environment that is only a Class II location, are not required to be sealed to complete the explosionproof assembly as required in a Class I environment. However, they must be provided with seals to prevent the entrance of dust into the enclosure where a raceway provides a means for dust to enter the system, such as in a conduit run between the enclosure and a general-purpose junction box.

### 502.6 Zone Equipment

Equipment listed and marked in accordance with 506.9(C)(2) for Zone 20 locations shall be permitted in Class II, Division 1 locations for the same dust atmosphere; and with a suitable temperature class.

Equipment listed and marked in accordance with 506.9(C)(2) for Zone 20, 21, or 22 locations shall be permitted in Class II, Division 2 locations for the same dust atmosphere and with a suitable temperature class.

## II. Wiring

### 502.10 Wiring Methods

Wiring methods shall comply with 502.10(A) or (B).

**(A) Class II, Division 1.**

**(1) General.** In Class II, Division 1 locations, the wiring methods in (1) through (4) shall be permitted:

(1) Threaded rigid metal conduit, or threaded steel intermediate metal conduit.

(2) Type MI cable with termination fittings listed for the location. Type MI cable shall be installed and supported in a manner to avoid tensile stress at the termination fittings.

(3) In industrial establishments with limited public access, where the conditions of maintenance and supervision ensure that only qualified persons service the installation, Type MC-HL cable, listed for use in Class II, Division 1 locations, with a gas/vaportight continuous corrugated metallic sheath, an overall jacket of suitable polymeric material, a separate equipment grounding conductor(s) in accordance with 250.122, and provided with termination fittings listed for the location, shall be permitted.

•

(4) Optical fiber cables Types OFNP, OFCP, OFNR, OFCR, OFNG, OFCG, OFN, and OFC shall be permitted to be installed in raceways in accordance with 502.10(A). Optical fiber cables shall be sealed in accordance with 502.15.

**(2) Flexible Connections.** Where necessary to employ flexible connections, one or more of the following shall also be permitted:

(1) Dusttight flexible connectors.
(2) Liquidtight flexible metal conduit with listed fittings.
(3) Liquidtight flexible nonmetallic conduit with listed fittings.
(4) Interlocked armor Type MC cable having an overall jacket of suitable polymeric material and provided with termination fittings listed for Class II, Division 1 locations.
(5) Flexible cord listed for extra-hard usage and terminated with listed dusttight cord connectors. Where flexible cords are used, they shall comply with 502.140.
(6) For elevator use, an identified elevator cable of Type EO, ETP, or ETT, shown under the "use" column in Table 400.4 for "hazardous (classified) locations" and terminated with listed dusttight fittings.

Informational Note: See 502.30(B) for grounding requirements where flexible conduit is used.

Where it is necessary to use flexible connections, liquidtight flexible conduit or extra-hard-usage flexible cord is permitted. Where liquidtight flexible conduit is used, a bonding jumper (internal or external) must be provided (see 502.30). An alternative method is to use a flexible fitting as described in the commentary following 501.10(A)(2).

**(3) Boxes and Fittings.** Boxes and fittings shall be provided with threaded bosses for connection to conduit or cable terminations and shall be dusttight. Boxes and fittings in which taps, joints, or terminal connections are made, or that are used in Group E locations, shall be identified for Class II locations.

Informational Note: For entry into enclosures required to be dust-ignitionproof, see the information on construction, testing, and marking of cables, dust-ignitionproof cable fittings, and dust-ignitionproof cord connectors in ANSI/UL 2225-2011, *Cables and Cable-Fittings for Use in Hazardous (Classified) Locations.*

**(B) Class II, Division 2.**

**(1) General.** In Class II, Division 2 locations, the following wiring methods shall be permitted:

(1) All wiring methods permitted in 502.10(A).
(2) Rigid metal conduit, intermediate metal conduit, electrical metallic tubing, dusttight wireways.
(3) Type MC or MI cable with listed termination fittings.
(4) Type PLTC and Type PLTC-ER cable in accordance with the provisions of Article 725, including installation in cable tray systems. The cable shall be terminated with listed fittings.
(5) Type ITC and Type ITC-ER cable as permitted in 727.4 and terminated with listed fittings.
(6) Type MC, MI, or TC cable installed in ladder, ventilated trough, or ventilated channel cable trays in a single layer, with a space not less than the larger cable diameter between the two adjacent cables, shall be the wiring method employed.

*Exception to (6): Type MC cable listed for use in Class II, Division 1 locations shall be permitted to be installed without the spacings required by (6).*

(7) In industrial establishments with restricted public access where the conditions of maintenance and supervision ensure that only qualified persons service the installation and where metallic conduit does not provide sufficient corrosion resistance, reinforced thermosetting resin conduit (RTRC) factory elbows, and associated fittings, all marked with suffix -XW, and Schedule 80 PVC conduit, factory elbows and associated fittings shall be permitted.
(8) Optical fiber cable Types OFNP, OFCP, OFNR, OFCR, OFNG, OFCG, OFN, and OFC shall be permitted to be installed in cable trays or any other raceway in accordance with 502.10(B). Optical fiber cables shall be sealed in accordance with 502.15.

**(2) Flexible Connections.** Where provision must be made for flexibility, 502.10(A)(2) shall apply.

**(3) Nonincendive Field Wiring.** Nonincendive field wiring shall be permitted using any of the wiring methods permitted for unclassified locations. Nonincendive field wiring systems shall be installed in accordance with the control drawing(s). Simple apparatus, not shown on the control drawing, shall be permitted in a nonincendive field wiring circuit, provided the simple apparatus does not interconnect the nonincendive field wiring circuit to any other circuit.

Informational Note: Simple apparatus is defined in 504.2.

Separate nonincendive field wiring circuits shall be installed in accordance with one of the following:

(1) In separate cables
(2) In multiconductor cables where the conductors of each circuit are within a grounded metal shield
(3) In multiconductor cables or in raceways where the conductors of each circuit have insulation with a minimum thickness of 0.25 mm (0.01 in.)

**(4) Boxes and Fittings.** All boxes and fittings shall be dusttight.

Boxes and fittings in a Class II, Division 2 location need only be dusttight. Whereas, in Division 1 locations, boxes containing taps, joints, or terminal connections, in addition to being dusttight, must be provided with threaded hubs and must be identified for use in Class II locations. Threaded hubs also provide adequate

*EXHIBIT 502.1 Junction box with threaded hubs, suitable for use in Class II, Group E hazardous atmospheres. (Courtesy of Appleton Electric Co., EGS Electrical Group)*

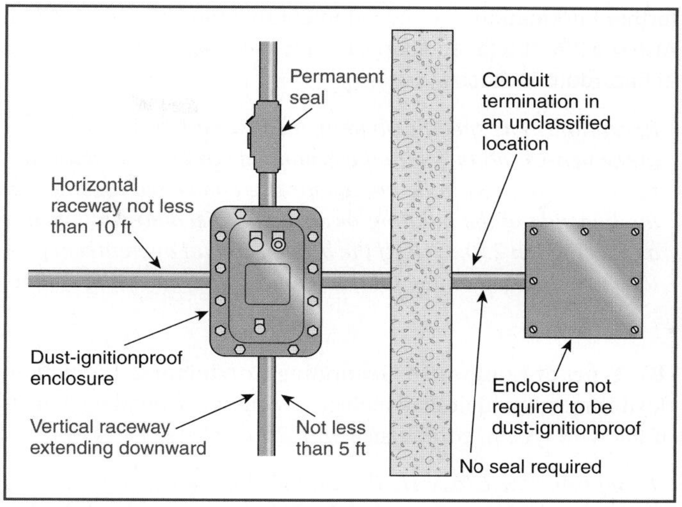


*EXHIBIT 502.2 Three methods for preventing dust from entering a dust-ignitionproof enclosure through the raceway.*

bonding in Division 2 locations. Exhibit 502.1 shows a dusttight cover, which is necessary for Class II locations.

## 502.15 Sealing, Class II, Divisions 1 and 2

Where a raceway provides communication between an enclosure that is required to be dust-ignitionproof and one that is not, suitable means shall be provided to prevent the entrance of dust into the dust-ignitionproof enclosure through the raceway. One of the following means shall be permitted:

(1) A permanent and effective seal
(2) A horizontal raceway not less than 3.05 m (10 ft) long
(3) A vertical raceway not less than 1.5 m (5 ft) long and extending downward from the dust-ignitionproof enclosure
(4) A raceway installed in a manner equivalent to (2) or (3) that extends only horizontally and downward from the dust-ignition proof enclosures

Where a raceway provides communication between an enclosure that is required to be dust-ignitionproof and an enclosure in an unclassified location, seals shall not be required.

Sealing fittings shall be accessible.

Seals shall not be required to be explosionproof.

Informational Note: Electrical sealing putty is a method of sealing.

Four suitable ways are provided to prevent dust from entering dust-ignitionproof enclosures through the raceway. Sealing methods (1) through (3) in Class II locations are shown in Exhibit 502.2.

The requirement to provide a seal applies if a raceway connects two enclosures in a hazardous location — from one enclosure that is required to be dust-ignitionproof to one that is not required to be dust-ignitionproof. Dust may enter the system through the other enclosure that is not dust-ignitionproof. If a raceway extends from a dust-ignitionproof enclosure to an enclosure in an unclassified location, a seal in that raceway is not required since dust will not enter through the conduit system.

Seal fittings designed for use in Class I locations are acceptable for Class II locations. However, because the Class I location pressure-piling considerations are not inherent in Class II locations, conduit seals are not required to be explosionproof. Conduit seals are expected only to prevent the migration of dust into dust-ignitionproof enclosures.

## 502.25 Uninsulated Exposed Parts, Class II, Divisions 1 and 2

There shall be no uninsulated exposed parts, such as electrical conductors, buses, terminals, or components, that operate at more than 30 volts (15 volts in wet locations). These parts shall additionally be protected by a protection technique according to 500.7(E), (F), or (G) that is suitable for the location.

## 502.30 Grounding and Bonding, Class II, Divisions 1 and 2

Regardless of the voltage of the electrical system, wiring and equipment in Class II, Division 1 and 2 locations shall be grounded as specified in Article 250 and in accordance with the requirements of 502.30(A) and (B).

**(A) Bonding.** The locknut-bushing and double-locknut types of contact shall not be depended on for bonding purposes, but bonding jumpers with proper fittings or other approved means of bonding shall be used. Such means of bonding shall apply to all intervening raceways, fittings, boxes, enclosures, and so forth, between Class II locations and the point of grounding for service equipment or point of grounding of a separately derived system.

The requirements for enhanced bonding in Class II locations are the same as those given in 501.30(A) for Class I locations. For further information, see the commentary following that section. Also see 250.100 for additional requirements applying to bonding in hazardous locations.

*Exception: The specific bonding means shall only be required to the nearest point where the grounded circuit conductor and the grounding electrode conductor are connected together on the line side of the building or structure disconnecting means as specified in 250.32(B) if the branch-circuit overcurrent protection is located on the load side of the disconnecting means.*

•

**(B) Types of Equipment Grounding Conductors.** Liquidtight flexible metal conduit shall include an equipment bonding jumper of the wire type in compliance with 250.102.

*Exception: In Class II, Division 2 locations, the bonding jumper shall be permitted to be deleted where all of the following conditions are met:*

*(1) Listed liquidtight flexible metal conduit 1.8 m (6 ft) or less in length, with fittings listed for grounding, is used.*
*(2) Overcurrent protection in the circuit is limited to 10 amperes or less.*
*(3) The load is not a power utilization load.*

### 502.35 Surge Protection — Class II, Divisions 1 and 2

Surge arresters and surge-protective devices installed in a Class II, Division 1 location shall be in suitable enclosures. Surge-protective capacitors shall be of a type designed for specific duty.

•

## III. Equipment

In the layout of electrical installations for hazardous locations, the preferred location of service equipment, switchboards, panelboards, and much of the electrical equipment is in less hazardous areas, usually in a separate room. The use of pressurized rooms, as described in NFPA 496, is a common method of protecting panelboards and switchboards in grain elevators and similar locations.

### 502.100 Transformers and Capacitors

**(A) Class II, Division 1.** In Class II, Division 1 locations, transformers and capacitors shall comply with 502.100(A)(1) through (A)(3).

**(1) Containing Liquid That Will Burn.** Transformers and capacitors containing a liquid that will burn shall be installed only in vaults complying with 450.41 through 450.48, and, in addition, (1), (2), and (3) shall apply.

(1) Doors or other openings communicating with the Division 1 location shall have self-closing fire doors on both sides of the wall, and the doors shall be carefully fitted and provided with suitable seals (such as weather stripping) to minimize the entrance of dust into the vault.
(2) Vent openings and ducts shall communicate only with the outside air.
(3) Suitable pressure-relief openings communicating with the outside air shall be provided.

**(2) Not Containing Liquid That Will Burn.** Transformers and capacitors that do not contain a liquid that will burn shall be installed in vaults complying with 450.41 through 450.48 or be identified as a complete assembly, including terminal connections.

**(3) Group E.** No transformer or capacitor shall be installed in a Class II, Division 1, Group E location.

**(B) Class II, Division 2.** In Class II, Division 2 locations, transformers and capacitors shall comply with 502.100(B)(1) through (B)(3).

**(1) Containing Liquid That Will Burn.** Transformers and capacitors containing a liquid that will burn shall be installed in vaults that comply with 450.41 through 450.48.

**(2) Containing Askarel.** Transformers containing askarel and rated in excess of 25 kVA shall be as follows:

(1) Provided with pressure-relief vents
(2) Provided with a means for absorbing any gases generated by arcing inside the case, or the pressure-relief vents shall be connected to a chimney or flue that will carry such gases outside the building
(3) Have an airspace of not less than 150 mm (6 in.) between the transformer cases and any adjacent combustible material

**(3) Dry-Type Transformers.** Dry-type transformers shall be installed in vaults or shall have their windings and terminal connections enclosed in tight metal housings without ventilating or other openings and shall operate at not over 600 volts, nominal.

See Section 460.2 for requirements on enclosing and guarding of capacitors.

### 502.115 Switches, Circuit Breakers, Motor Controllers, and Fuses

**(A) Class II, Division 1.** In Class II, Division 1 locations, switches, circuit breakers, motor controllers, fuses, push buttons, relays, and similar devices shall be provided with enclosures identified for the location.

**(B) Class II, Division 2.** In Class II, Division 2 locations, enclosures for fuses, switches, circuit breakers, and motor controllers, including push buttons, relays, and similar devices, shall be dust-tight or otherwise identified for the location.

Exhibits 502.3 and 502.4 show dust-ignitionproof equipment that is suitable for use in Class II, Division 1 locations. Dust-ignitionproof equipment enclosures can be used in Class II,

***EXHIBIT 502.3*** *A dust-ignitionproof pushbutton control station suitable for use in Class II, Group E, F, and G locations. (Courtesy of Appleton Electric Co., EGS Electrical Group)*

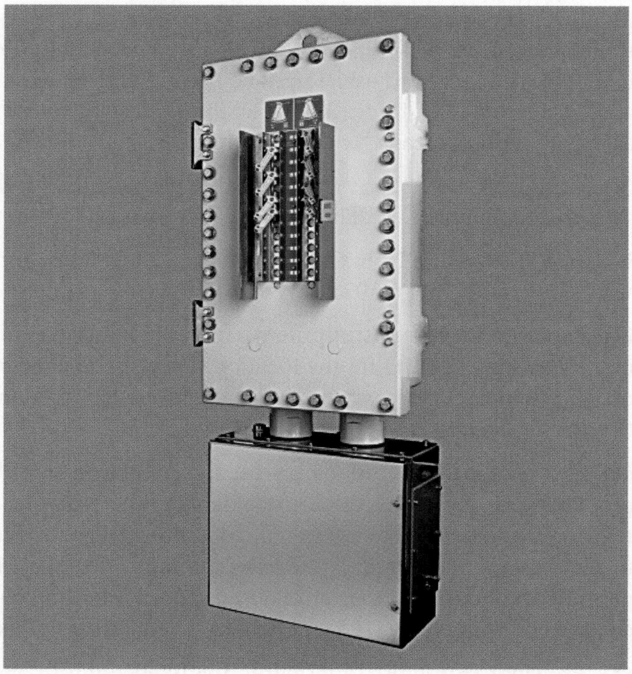

***EXHIBIT 502.4*** *A dust-ignitionproof panelboard for use in Class II, Division 2, Group F and G locations. (Courtesy of Cooper Crouse-Hinds)*

Division 2 locations, but because of the reduced level of hazard associated with Division 2, dusttight equipment enclosures are also permitted. In addition to being suitable for Class II and the specific division, the equipment must also be suitable for the dust group(s) (i.e., Groups E, F, and G) present in a specific hazardous location.

## 502.120 Control Transformers and Resistors

**(A) Class II, Division 1.** In Class II, Division 1 locations, control transformers, solenoids, impedance coils, resistors, and any overcurrent devices or switching mechanisms associated with them shall be provided with enclosures identified for the location.

Switching and heat-generating equipment must have enclosures identified specifically for use in Class II, Division 1 environments.

**(B) Class II, Division 2.** In Class II, Division 2 locations, transformers and resistors shall comply with 502.120(B)(1) through (B)(3).

**(1) Switching Mechanisms.** Switching mechanisms (including overcurrent devices) associated with control transformers, solenoids, impedance coils, and resistors shall be provided with enclosures that are dusttight or otherwise identified for the location.

**(2) Coils and Windings.** Where not located in the same enclosure with switching mechanisms, control transformers, solenoids, and impedance coils shall be provided with enclosures that are dusttight or otherwise identified for the location.

**(3) Resistors.** Resistors and resistance devices shall have dust-ignitionproof enclosures that are dusttight or otherwise identified for the location.

## 502.125 Motors and Generators

**(A) Class II, Division 1.** In Class II, Division 1 locations, motors, generators, and other rotating electrical machinery shall be in conformance with either of the following:

(1) Identified for the location
(2) Totally enclosed pipe-ventilated

Markings on the motor must identify its suitability for Class II, Division 1 locations, or a totally enclosed pipe-ventilated installation is required. In a pipe-ventilated system, the supply and exhaust of air from/to outside the Class II area is through a closed pipe system. This prevents accumulation of dust inside the totally enclosed equipment.

**(B) Class II, Division 2.** In Class II, Division 2 locations, motors, generators, and other rotating electrical equipment shall be totally enclosed nonventilated, totally enclosed pipe-ventilated, totally enclosed water-air-cooled, totally enclosed fan-cooled or dust-ignitionproof for which maximum full-load external temperature shall be in accordance with 500.8(D)(2) for normal operation when operating in free air (not dust blanketed) and shall have no external openings.

*Exception: If the authority having jurisdiction believes accumulations of nonconductive, nonabrasive dust will be moderate and if machines can be easily reached for routine cleaning and maintenance, the following shall be permitted to be installed:*

*(1) Standard open-type machines without sliding contacts, centrifugal or other types of switching mechanism (including motor overcurrent, overloading, and overtemperature devices), or integral resistance devices*
*(2) Standard open-type machines with such contacts, switching mechanisms, or resistance devices enclosed within dust-tight housings without ventilating or other openings*
*(3) Self-cleaning textile motors of the squirrel-cage type*

Totally enclosed motors are permitted in Class II, Division 2 locations if the external surface temperatures, without a dust blanket, do not exceed the temperatures indicated under the maximum full-load (normal operation) conditions in 500.8(D)(2). Totally enclosed fan-cooled (TEFC) motors should be examined carefully to verify that there are no external openings, even though the motor may be marked TEFC.

## 502.128 Ventilating Piping

Ventilating pipes for motors, generators, or other rotating electrical machinery, or for enclosures for electrical equipment, shall be of metal not less than 0.53 mm (0.021 in.) in thickness or of equally substantial noncombustible material and shall comply with all of the following:

(1) Lead directly to a source of clean air outside of buildings
(2) Be screened at the outer ends to prevent the entrance of small animals or birds
(3) Be protected against physical damage and against rusting or other corrosive influences

Ventilating pipes shall also comply with 502.128(A) and (B).

**(A) Class II, Division 1.** In Class II, Division 1 locations, ventilating pipes, including their connections to motors or to the dust-ignitionproof enclosures for other equipment, shall be dusttight throughout their length. For metal pipes, seams and joints shall comply with one of the following:

(1) Be riveted and soldered
(2) Be bolted and soldered
(3) Be welded
(4) Be rendered dusttight by some other equally effective means

**(B) Class II, Division 2.** In Class II, Division 2 locations, ventilating pipes and their connections shall be sufficiently tight to prevent the entrance of appreciable quantities of dust into the ventilated equipment or enclosure and to prevent the escape of sparks, flame, or burning material that might ignite dust accumulations or combustible material in the vicinity. For metal pipes, lock seams and riveted or welded joints shall be permitted; and tight-fitting slip joints shall be permitted where some flexibility is necessary, as at connections to motors.

## 502.130 Luminaires

**(A) Class II, Division 1.** In Class II, Division 1 locations, luminaires for fixed and portable lighting shall comply with 502.130(A)(1) through (A)(4).

**(1) Marking.** Each luminaire shall be identified for the location and shall be clearly marked to indicate the type and maximum wattage of the lamp for which it is designed.

**(2) Physical Damage.** Each luminaire shall be protected against physical damage by a suitable guard or by location.

**(3) Pendant Luminaires.** Pendant luminaires shall be suspended by threaded rigid metal conduit stems, by threaded steel intermediate metal conduit stems, by chains with approved fittings, or by other approved means. For rigid stems longer than 300 mm (12 in.), permanent and effective bracing against lateral displacement shall be provided at a level not more than 300 mm (12 in.) above the lower end of the stem, or flexibility in the form of a fitting or a flexible connector listed for the location shall be provided not more than 300 mm (12 in.) from the point of attachment to the supporting box or fitting. Threaded joints shall be provided with set screws or other effective means to prevent loosening. Where wiring between an outlet box or fitting and a pendant luminaire is not enclosed in conduit, flexible cord listed for hard usage shall be permitted to be used in accordance with 502.10(A)(2)(5). Flexible cord shall not serve as the supporting means for a luminaire.

**(4) Supports.** Boxes, box assemblies, or fittings used for the support of luminaires shall be identified for Class II locations.

**(B) Class II, Division 2.** In Class II, Division 2 locations, luminaires shall comply with 502.130(B)(1) through (B)(5).

**(1) Portable Lighting Equipment.** Portable lighting equipment shall be identified for the location. They shall be clearly marked to indicate the maximum wattage of lamps for which they are designed.

**(2) Fixed Lighting.** Luminaires for fixed lighting shall be provided with enclosures that are dusttight or otherwise identified for the location. Each luminaire shall be clearly marked to indicate the maximum wattage of the lamp that shall be permitted without exceeding an exposed surface temperature in accordance with 500.8(D)(2) under normal conditions of use.

**(3) Physical Damage.** Luminaires for fixed lighting shall be protected from physical damage by suitable guards or by location.

**(4) Pendant Luminaires.** Pendant luminaires shall be suspended by threaded rigid metal conduit stems, by threaded steel intermediate metal conduit stems, by chains with approved fittings, or by other approved means. For rigid stems longer than 300 mm (12 in.), permanent and effective bracing against lateral displacement shall be provided at a level not more than 300 mm (12 in.) above the lower end of the stem, or flexibility in the form of an identified fitting or a flexible connector shall be provided not more than 300 mm (12 in.) from the point of attachment to the supporting box or fitting. Where wiring between an outlet box or fitting and a pendant luminaire is not enclosed in conduit, flexible cord listed for hard usage shall be permitted if terminated with a listed cord connector that maintains the protection technique. Flexible cord shall not serve as the supporting means for a luminaire.

**(5) Electric-Discharge Lamps.** Starting and control equipment for electric-discharge lamps shall comply with the requirements of 502.120(B).

All lighting equipment in Class II, Division 1 locations and all portable lighting equipment in a Class II, Division 2 location must also be identified for such use. If not identified for use in a Class II location, all fixed lighting equipment in a Class II, Division 2 location must be dusttight.

Section 502.130 requires luminaires in Class II locations to be marked to indicate maximum lamp wattage. In addition, lamps must be guarded or protected by their location to prevent damage that could allow the escape of sparks or burning material.

Flexible cord of the hard-usage type is permitted with approved sealed connections for the wiring of chain-suspended or hook-and-eye–suspended luminaires as long as they are not used as a means of support.

Exhibit 502.5 shows a listed luminaire suitable for use in Class II, Group E, F, and G locations. The portable hand lamp shown in Exhibit 501.21 is listed as a complete assembly for use in Class I locations and also in any Class II, Group F or G location.

## 502.135 Utilization Equipment

**(A) Class II, Division 1.** In Class II, Division 1 locations, all utilization equipment shall be identified for the location.

***EXHIBIT 502.5*** *A typical luminaire for use in Class II, Division 1 locations. (Courtesy of Cooper Crouse-Hinds)*

**(B) Class II, Division 2.** In Class II, Division 2 locations, all utilization equipment shall comply with 502.135(B)(1) through (B)(4).

**(1) Heaters.** Electrically heated utilization equipment shall be identified for the location.

*Exception: Metal-enclosed radiant heating panel equipment shall be permitted to be dusttight and marked in accordance with 500.8(C).*

**(2) Motors.** Motors of motor-driven utilization equipment shall comply with 502.125(B).

**(3) Switches, Circuit Breakers, and Fuses.** Enclosures for switches, circuit breakers, and fuses shall comply with 502.115(B).

**(4) Transformers, Solenoids, Impedance Coils, and Resistors.** Transformers, solenoids, impedance coils, and resistors shall comply with 502.120(B).

## 502.140 Flexible Cords — Class II, Divisions 1 and 2

**(A) Permitted Uses.** Flexible cords used in Class II locations shall comply with all of the following:

(1) For connection between portable lighting equipment or other portable utilization equipment and the fixed portion of its supply circuit. The flexible cord shall be attached to the utilization equipment with a cord connector listed for the protection technique of the equipment wiring compartment. An attachment plug in accordance with 502.145 shall be employed.
(2) Where flexible cord is permitted by 502.10(A)(2) for fixed and mobile electrical utilization equipment; where the flexible cord is protected by location or by a suitable guard from damage; and only in an industrial establishment where conditions of maintenance and engineering supervision ensure that only qualified persons install and service the installation.
(3) For electric submersible pumps with means for removal without entering the wet-pit. The extension of the flexible cord within a suitable raceway between the wet-pit and the power source shall be permitted.
(4) For electric mixers intended for travel into and out of open-type mixing tanks or vats.
(5) For temporary portable assemblies consisting of receptacles, switches, and other devices that are not considered portable utilization equipment but are individually listed for the location.

**(B) Installation.** Where flexible cords are used, the cords shall comply with all of the following:

(1) Be of a type listed for extra-hard usage.

*Exception: Flexible cord listed for hard usage as permitted by 502.130(A)(3) and (B)(4).*

(2) Contain, in addition to the conductors of the circuit, an equipment grounding conductor complying with 400.23.
(3) Be supported by clamps or by other suitable means in such a manner that there will be no tension on the terminal connections.
(4) In Division 1 locations, the cord shall be terminated with a cord connector listed for the location or a listed cord connector installed with a seal listed for the location. In Division 2 locations, the cord shall be terminated with a listed dusttight cord connector.
(5) Be of continuous length. Where 502.140(A)(5) is applied, cords shall be of continuous length from the power source to the temporary portable assembly and from the temporary portable assembly to the utilization equipment.

### 502.145 Receptacles and Attachment Plugs

Receptacles and attachment plugs shall be identified for the location.

**(A) Class II, Division 1**

**(1) Receptacles.** In Class II, Division 1 locations, receptacles shall be part of the premises wiring.

**(2) Attachment Plugs.** Attachment plugs shall be of the type that provides for connection to the equipment grounding conductor of the flexible cord.

**(B) Class II, Division 2.**

**(1) Receptacles.** In Class II, Division 2 locations, receptacles shall be part of the premises wiring.

**(2) Attachment Plugs.** Attachment plugs shall be of the type that provides for connection to the equipment grounding conductor of the flexible cord.

### 502.150 Signaling, Alarm, Remote-Control, and Communications Systems; and Meters, Instruments, and Relays

Informational Note: See Article 800 for rules governing the installation of communications circuits.

**(A) Class II, Division 1.** In Class II, Division 1 locations, signaling, alarm, remote-control, and communications systems; and meters, instruments, and relays shall comply with 502.150(A)(1) through (A)(3).

**(1) Contacts.** Switches, circuit breakers, relays, contactors, fuses and current-breaking contacts for bells, horns, howlers, sirens, and other devices in which sparks or arcs may be produced shall be provided with enclosures identified for the location.

*Exception: Where current-breaking contacts are immersed in oil or where the interruption of current occurs within a chamber sealed against the entrance of dust, enclosures shall be permitted to be of the general-purpose type.*

**(2) Resistors and Similar Equipment.** Resistors, transformers, choke coils, rectifiers, thermionic tubes, and other heat-generating equipment shall be provided with enclosures identified for the location.

*Exception: Where resistors or similar equipment are immersed in oil or enclosed in a chamber sealed against the entrance of dust, enclosures shall be permitted to be of the general-purpose type.*

**(3) Rotating Machinery.** Motors, generators, and other rotating electrical machinery shall comply with 502.125(A).

**(B) Class II, Division 2.** In Class II, Division 2 locations, signaling, alarm, remote-control, and communications systems; and meters, instruments, and relays shall comply with 502.150(B)(1) through (B)(4).

**(1) Contacts.** Contacts shall comply with 502.150(A)(1) or shall be installed in enclosures that are dusttight or otherwise identified for the location.

*Exception: In nonincendive circuits, enclosures shall be permitted to be of the general-purpose type.*

**(2) Transformers and Similar Equipment.** The windings and terminal connections of transformers, choke coils, and similar equipment shall comply with 502.120(B)(2).

**(3) Resistors and Similar Equipment.** Resistors, resistance devices, thermionic tubes, rectifiers, and similar equipment shall comply with 502.120(B)(3).

**(4) Rotating Machinery.** Motors, generators, and other rotating electrical machinery shall comply with 502.125(B).

# ARTICLE 503
# Class III Locations

## I. General

### 503.1 Scope

Article 503 covers the requirements for electrical and electronic equipment and wiring for all voltages in Class III, Division 1 and 2 locations where fire or explosion hazards may exist due to ignitible fibers/flyings.

Class III locations usually include textile mills that process cotton, rayon, and other fabrics, where easily ignitible fibers/flyings are present in the manufacturing process. Sawmills and other woodworking plants, where sawdust, wood shavings, and combustible fibers/flyings are present, may also become hazardous. However, if wood flour (dust) is present, the location is a Class II, Group G location and not a Class III location.

Fibers/flyings are hazardous not only because they are easily ignited, but also because flames spread through them quickly.

Such fires travel with a rapidity approaching an explosion and are commonly called flash fires.

Class III, Division 1 applies to locations where material is handled, manufactured, or used. Division 2 applies to locations where material is stored or handled but where no manufacturing processes are performed. Unlike Class I locations (Groups A, B, C, and D) and Class II locations (Groups E, F, and G), Class III locations do not have material group designations.

## 503.5 General

Equipment installed in Class III locations shall be able to function at full rating without developing surface temperatures high enough to cause excessive dehydration or gradual carbonization of accumulated fibers/flyings. Organic material that is carbonized or excessively dry is highly susceptible to spontaneous ignition. The maximum surface temperatures under operating conditions shall not exceed 165°C (329°F) for equipment that is not subject to overloading, and 120°C (248°F) for equipment (such as motors or power transformers) that may be overloaded. In a Class III, Division 1 location, the operating temperature shall be the temperature of the equipment when blanketed with the maximum amount of dust (simulating fibers/flyings) that can accumulate on the equipment.

Informational Note: For electric trucks, see NFPA 505-2013, *Fire Safety Standard for Powered Industrial Trucks Including Type Designations, Areas of Use, Conversions, Maintenance, and Operation.*

## 503.6 Zone Equipment

Equipment listed and marked in accordance with 506.9(C)(2) for Zone 20 locations and with a temperature class of not greater than T120°C (for equipment that may be overloaded) or not greater than T165°C (for equipment not subject to overloading) shall be permitted in Class III, Division 1 locations.

Equipment listed and marked in accordance with 506.9(C)(2) for Zone 20, 21, or 22 locations and with a temperature class of not greater than T120°C (for equipment that may be overloaded) or not greater than T165°C (for equipment not subject to overloading) shall be permitted in Class III, Division 2 locations.

# II. Wiring

## 503.10 Wiring Methods

Wiring methods shall comply with 503.10(A) or (B).

### (A) Class III, Division 1

**(1) General.** In Class III, Division 1 locations, the wiring method shall be in accordance with (1) through (4):

(1) Rigid metal conduit, Type PVC conduit, Type RTRC conduit, intermediate metal conduit, electrical metallic tubing, dusttight wireways, or Type MC or MI cable with listed termination fittings.
(2) Type PLTC and Type PLTC-ER cable in accordance with the provisions of Article 725 including installation in cable tray systems. The cable shall be terminated with listed fittings.
(3) Type ITC and Type ITC-ER cable as permitted in 727.4 and terminated with listed fittings.
(4) Type MC, MI, TC, or TC-ER cable installed in ladder, ventilated trough, or ventilated channel cable trays in a single layer, with a space not less than the larger cable diameter between the two adjacent cables, shall be the wiring method employed. The cable shall be terminated with listed fittings.

*Exception to (4): Type MC cable listed for use in Class II, Division 1 locations shall be permitted to be installed without the spacings required by 503.10(A)(1)(4).*

**(2) Boxes and Fittings.** All boxes and fittings shall be dusttight.

**(3) Flexible Connections.** Where necessary to employ flexible connections, one or more of the following shall be permitted:

(1) Dusttight flexible connectors
(2) Liquidtight flexible metal conduit with listed fittings,
(3) Liquidtight flexible nonmetallic conduit with listed fittings,
(4) Interlocked armor Type MC cable having an overall jacket of suitable polymeric material and installed with listed dusttight termination fittings
(5) Flexible cord in compliance with 503.140

Informational Note: See 503.30(B) for grounding requirements where flexible conduit is used.

(6) For elevator use, an identified elevator cable of Type EO, ETP, or ETT, shown under the "use" column in Table 400.4 for "hazardous (classified) locations" and terminated with listed dusttight fittings.

**(4) Nonincendive Field Wiring.** Nonincendive field wiring shall be permitted using any of the wiring methods permitted for unclassified locations. Nonincendive field wiring systems shall be installed in accordance with the control drawing(s). Simple apparatus, not shown on the control drawing, shall be permitted in a nonincendive field wiring circuit, provided the simple apparatus does not interconnect the nonincendive field wiring circuit to any other circuit.

Informational Note: Simple apparatus is defined in 504.2.

Separate nonincendive field wiring circuits shall be installed in accordance with one of the following:

(1) In separate cables
(2) In multiconductor cables where the conductors of each circuit are within a grounded metal shield
(3) In multiconductor cables where the conductors of each circuit have insulation with a minimum thickness of 0.25 mm (0.01 in.)

Except as permitted in the exception to 503.10(B), the wiring methods used for Class III, Division 2 locations are identical to those required for Division 1 locations.

**(B) Class III, Division 2.** In Class III, Division 2 locations, the wiring method shall comply with 503.10(A).

*Exception: In sections, compartments, or areas used solely for storage and containing no machinery, open wiring on insulators shall be permitted where installed in accordance with Article 398, but only on condition that protection as required by 398.15(C) be provided where conductors are not run in roof spaces and are well out of reach of sources of physical damage.*

## 503.25 Uninsulated Exposed Parts, Class III, Divisions 1 and 2

There shall be no uninsulated exposed parts, such as electrical conductors, buses, terminals, or components, that operate at more than 30 volts (15 volts in wet locations). These parts shall additionally be protected by a protection technique according to 500.7(E), (F), or (G) that is suitable for the location.

*Exception: As provided in 503.155.*

Exposed live parts are permitted in Class III, Division 1 and 2 locations provided the voltage does not exceed 30 volts in dry locations or 15 volts in wet locations. Protection techniques permitted for these parts are intrinsically safe or nonincendive. These techniques limit the circuit's energy to a level incapable of causing ignition of the hazardous area.

## 503.30 Grounding and Bonding — Class III, Divisions 1 and 2

Regardless of the voltage of the electrical system, wiring and equipment in Class III, Division 1 and 2 locations shall be grounded as specified in Article 250 and with the following additional requirements in 503.30(A) and (B).

**(A) Bonding.** The locknut-bushing and double-locknut types of contacts shall not be depended on for bonding purposes, but bonding jumpers with proper fittings or other approved means of bonding shall be used. Such means of bonding shall apply to all intervening raceways, fittings, boxes, enclosures, and so forth, between Class III locations and the point of grounding for service equipment or point of grounding of a separately derived system.

The requirements for enhanced bonding in Class III locations are the same as those for Class I locations given in 501.30(A) and for Class II locations in 502.30(A). For further information, see the commentary following 501.30(A). Also see 250.100 for additional bonding requirements applying in hazardous locations.

*Exception: The specific bonding means shall only be required to the nearest point where the grounded circuit conductor and the grounding electrode conductor are connected together on the line side of the building or structure disconnecting means as specified in 250.32(B) if the branch-circuit overcurrent protection is located on the load side of the disconnecting means.*

•

**(B) Types of Equipment Bonding Conductors.** Liquidtight flexible metal conduit shall include an equipment bonding jumper of the wire type in compliance with 250.102.

*Exception: In Class III, Division 1 and 2 locations, the bonding jumper shall be permitted to be deleted where all of the following conditions are met:*

*(1) Listed liquidtight flexible metal conduit 1.8 m (6 ft) or less in length, with fittings listed for grounding, is used.*
*(2) Overcurrent protection in the circuit is limited to 10 amperes or less.*
*(3) The load is not a power utilization load.*

# III. Equipment

## 503.100 Transformers and Capacitors — Class III, Divisions 1 and 2

Transformers and capacitors shall comply with 502.100(B).

## 503.115 Switches, Circuit Breakers, Motor Controllers, and Fuses — Class III, Divisions 1 and 2

Switches, circuit breakers, motor controllers, and fuses, including pushbuttons, relays, and similar devices, shall be provided with dusttight enclosures.

## 503.120 Control Transformers and Resistors — Class III, Divisions 1 and 2

Transformers, impedance coils, and resistors used as, or in conjunction with, control equipment for motors, generators, and appliances shall be provided with dusttight enclosures complying with the temperature limitations in 503.5.

## 503.125 Motors and Generators — Class III, Divisions 1 and 2

In Class III, Divisions 1 and 2 locations, motors, generators, and other rotating machinery shall be totally enclosed nonventilated, totally enclosed pipe ventilated, or totally enclosed fan cooled.

*Exception: In locations where, in the judgment of the authority having jurisdiction, only moderate accumulations of lint or flyings are likely to collect on, in, or in the vicinity of a rotating electrical machine and where such machine is readily accessible for routine cleaning and maintenance, one of the following shall be permitted:*

*(1) Self-cleaning textile motors of the squirrel-cage type*
*(2) Standard open-type machines without sliding contacts, centrifugal or other types of switching mechanisms, including motor overload devices*

*(3) Standard open-type machines having such contacts, switching mechanisms, or resistance devices enclosed within tight housings without ventilating or other openings*

### 503.128 Ventilating Piping — Class III, Divisions 1 and 2

Ventilating pipes for motors, generators, or other rotating electrical machinery, or for enclosures for electric equipment, shall be of metal not less than 0.53 mm (0.021 in.) in thickness, or of equally substantial noncombustible material, and shall comply with the following:

(1) Lead directly to a source of clean air outside of buildings
(2) Be screened at the outer ends to prevent the entrance of small animals or birds
(3) Be protected against physical damage and against rusting or other corrosive influences

Ventilating pipes shall be sufficiently tight, including their connections, to prevent the entrance of appreciable quantities of fibers/flyings into the ventilated equipment or enclosure and to prevent the escape of sparks, flame, or burning material that might ignite accumulations of fibers/flyings or combustible material in the vicinity. For metal pipes, lock seams and riveted or welded joints shall be permitted; and tight-fitting slip joints shall be permitted where some flexibility is necessary, as at connections to motors.

### 503.130 Luminaires — Class III, Divisions 1 and 2

**(A) Fixed Lighting.** Luminaires for fixed lighting shall provide enclosures for lamps and lampholders that are designed to minimize entrance of fibers/flyings and to prevent the escape of sparks, burning material, or hot metal. Each luminaire shall be clearly marked to show the maximum wattage of the lamps that shall be permitted without exceeding an exposed surface temperature of 165°C (329°F) under normal conditions of use.

**(B) Physical Damage.** A luminaire that may be exposed to physical damage shall be protected by a suitable guard.

**(C) Pendant Luminaires.** Pendant luminaires shall be suspended by stems of threaded rigid metal conduit, threaded intermediate metal conduit, threaded metal tubing of equivalent thickness, or by chains with approved fittings. For stems longer than 300 mm (12 in.), permanent and effective bracing against lateral displacement shall be provided at a level not more than 300 mm (12 in.) above the lower end of the stem, or flexibility in the form of an identified fitting or a flexible connector shall be provided not more than 300 mm (12 in.) from the point of attachment to the supporting box or fitting.

**(D) Portable Lighting Equipment.** Portable lighting equipment shall be equipped with handles and protected with substantial guards. Lampholders shall be of the unswitched type with no provision for receiving attachment plugs. There shall be no exposed current-carrying metal parts, and all exposed non–current-carrying metal parts shall be grounded. In all other respects, portable lighting equipment shall comply with 503.130(A).

### 503.135 Utilization Equipment — Class III, Divisions 1 and 2

**(A) Heaters.** Electrically heated utilization equipment shall be identified for Class III locations.

**(B) Motors.** Motors of motor-driven utilization equipment shall comply with 503.125.

**(C) Switches, Circuit Breakers, Motor Controllers, and Fuses.** Switches, circuit breakers, motor controllers, and fuses shall comply with 503.115.

### 503.140 Flexible Cords — Class III, Divisions 1 and 2

Flexible cords shall comply with the following:

(1) Be of a type listed for extra-hard usage
(2) Contain, in addition to the conductors of the circuit, an equipment grounding conductor complying with 400.23
(3) Be supported by clamps or other suitable means in such a manner that there will be no tension on the terminal connections
(4) Be terminated with a listed dusttight cord connector.

### 503.145 Receptacles and Attachment Plugs — Class III, Divisions 1 and 2

Receptacles and attachment plugs shall be of the grounding type, shall be designed so as to minimize the accumulation or the entry of fibers/flyings, and shall prevent the escape of sparks or molten particles.

*Exception: In locations where, in the judgment of the authority having jurisdiction, only moderate accumulations of lint or flyings are likely to collect in the vicinity of a receptacle, and where such receptacle is readily accessible for routine cleaning, general-purpose grounding-type receptacles mounted so as to minimize the entry of fibers/flyings shall be permitted.*

### 503.150 Signaling, Alarm, Remote-Control, and Local Loudspeaker Intercommunications Systems — Class III, Divisions 1 and 2

Signaling, alarm, remote-control, and local loudspeaker intercommunications systems shall comply with the requirements of Article 503 regarding wiring methods, switches, transformers, resistors, motors, luminaires, and related components.

### 503.155 Electric Cranes, Hoists, and Similar Equipment — Class III, Divisions 1 and 2

Where installed for operation over combustible fibers or accumulations of flyings, traveling cranes and hoists for material handling, traveling cleaners for textile machinery, and similar equipment shall comply with 503.155(A) through (D).

**(A) Power Supply.** The power supply to contact conductors shall be electrically isolated from all other systems, ungrounded, and shall be equipped with an acceptable ground detector that gives an alarm and automatically de-energizes the contact conductors in case of a fault to ground or gives a visual and audible alarm as long as power is supplied to the contact conductors and the ground fault remains.

**(B) Contact Conductors.** Contact conductors shall be located or guarded so as to be inaccessible to other than authorized persons and shall be protected against accidental contact with foreign objects.

**(C) Current Collectors.** Current collectors shall be arranged or guarded so as to confine normal sparking and prevent escape of sparks or hot particles. To reduce sparking, two or more separate surfaces of contact shall be provided for each contact conductor. Reliable means shall be provided to keep contact conductors and current collectors free of accumulations of lint or flyings.

**(D) Control Equipment.** Control equipment shall comply with 503.115 and 503.120.

In Class III locations, cranes that are installed over accumulations of fibers/flyings and equipped with rolling or sliding collectors that make contact with bare conductors introduce two hazards.

The first hazard results from arcing between a conductor and a collector rail igniting combustible fibers or lint that has accumulated on or near the bare conductor. This hazard may be prevented by maintaining the proper alignment of the bare conductor, by using a collector designed so that proper contact is always maintained, and by using guards or shields to confine hot metal particles that result from arcing.

The second hazard occurs if enough moisture is present and fibers/flyings accumulating on the insulating supports of the bare conductors form a conductive path between the conductors or from one conductor to ground, permitting enough current to flow to ignite the fibers. If the system is ungrounded, a current flow to ground is unlikely to start a fire.

A suitable recording ground detector sounds an alarm and automatically de-energizes contact conductors when the insulation resistance is lowered by an accumulation of fibers on the insulators or in case of a fault to ground. A ground-fault indicator that maintains an alarm until the system is de-energized or the ground fault is cleared is permitted.

## 503.160 Storage Battery Charging Equipment — Class III, Divisions 1 and 2

Storage battery charging equipment shall be located in separate rooms built or lined with substantial noncombustible materials. The rooms shall be constructed to prevent the entrance of ignitible amounts of flyings or lint and shall be well ventilated.

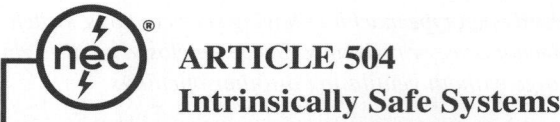

# ARTICLE 504 Intrinsically Safe Systems

## 504.1 Scope

This article covers the installation of intrinsically safe (I.S.) apparatus, wiring, and systems for Class I, II, and III locations.

Informational Note: For further information, see ANSI/ISA-RP 12.06.01-2003, *Recommended Practice for Wiring Methods for Hazardous (Classified) Locations Instrumentation — Part 1: Intrinsic Safety.*

The standard used in the United States for construction and performance requirements for intrinsically safe (IS) systems is ANSI/UL 913, *Standard for Intrinsically Safe Apparatus and Associated Apparatus for Use in Class I, II, and III, Division 1, Hazardous (Classified) Locations.* ANSI/UL 913 is similar to standards for IS equipment in other countries; all these standards are based on the IEC (International Electrotechnical Commission) standard. ANSI/UL 60079-11, *Electrical Apparatus for Explosive Gas Atmospheres – Part 11: Intrinsic Safety "I,"* is based on the IEC 60079-11 standard, which contains the U.S. deviations that allow it to be compatible for installations in the United States. The *NEC* offers the choice of designating hazardous locations as two divisions (1 and 2) or three zones (0, 1, and 2). ANSI/UL 60079-11 requirements, however, are based on the zone requirements, including Zone 0, which is the most stringent. Equipment certified by a testing laboratory for Zone 1 would not necessarily meet ANSI/UL 913 requirements for Division 1.

## 504.2 Definitions

**Associated Apparatus.** Apparatus in which the circuits are not necessarily intrinsically safe themselves but that affects the energy in the intrinsically safe circuits and is relied on to maintain intrinsic safety. Such apparatus is one of the following:

(1) Electrical apparatus that has an alternative type of protection for use in the appropriate hazardous (classified) location
(2) Electrical apparatus not so protected that shall not be used within a hazardous (classified) location

Informational Note No. 1: Associated apparatus has identified intrinsically safe connections for intrinsically safe apparatus and also may have connections for nonintrinsically safe apparatus.

Informational Note No. 2: An example of associated apparatus is an intrinsic safety barrier, which is a network designed to limit the energy (voltage and current) available to the protected circuit in the hazardous (classified) location, under specified fault conditions.

•

**Different Intrinsically Safe Circuits.** Intrinsically safe circuits in which the possible interconnections have not been evaluated and identified as intrinsically safe.

**Intrinsically Safe Apparatus.** Apparatus in which all the circuits are intrinsically safe.

**Intrinsically Safe Circuit.** A circuit in which any spark or thermal effect is incapable of causing ignition of a mixture of flammable or combustible material in air under prescribed test conditions.

Informational Note: Test conditions are described in ANSI/UL 913-2006, *Standard for Safety, Intrinsically Safe Apparatus and Associated Apparatus for Use in Class I, II, and III, Division 1, Hazardous (Classified) Locations.*

Due to its physical and electrical characteristics, an IS circuit does not develop sufficient electrical energy (millijoules) in an arc or spark to cause ignition, or sufficient thermal energy resulting from an overload condition to cause the temperature of the installed circuit to exceed the ignition temperature of a specified gas or vapor under normal or abnormal operating conditions.

An abnormal condition may occur due to accidental damage, failure of electrical components, excessive voltage, or improper adjustment or maintenance of the equipment.

**Intrinsically Safe System.** An assembly of interconnected intrinsically safe apparatus, associated apparatus, and interconnecting cables, in that those parts of the system that may be used in hazardous (classified) locations are intrinsically safe circuits.

Although low-energy devices, such as thermocouples, crystal strain transducers, or pressure transducers generate millivolts and currents in the microampere range, they are not necessarily intrinsically safe. Low-energy devices are normally connected to amplifiers and power supplies that are in turn connected to circuits that are 120 volts or higher. Should a fault occur within the amplifier or power supply, or a voltage surge occur in the electrical supply system, high-energy arcing, sparking, or overheating of the low-energy portion of the circuit could occur. Therefore, these devices must be installed as part of an IS system.

Informational Note No. 2 to the definition of associated apparatus identifies an intrinsic safety barrier as a type of associated apparatus that limits energy to a protected circuit. For an example of an IS barrier, see Exhibit 504.1. Note the marking indicating that the installation is to be in accordance with a specific control drawing.

Informational Note: An intrinsically safe system may include more than one intrinsically safe circuit.

**Simple Apparatus.** An electrical component or combination of components of simple construction with well-defined electrical parameters that does not generate more than 1.5 volts, 100 mA, and 25 mW, or a passive component that does not dissipate more than 1.3 watts and is compatible with the intrinsic safety of the circuit in which it is used.

Informational Note: The following apparatus are examples of simple apparatus:

(1) Passive components; for example, switches, junction boxes, resistance temperature devices, and simple semiconductor devices such as LEDs

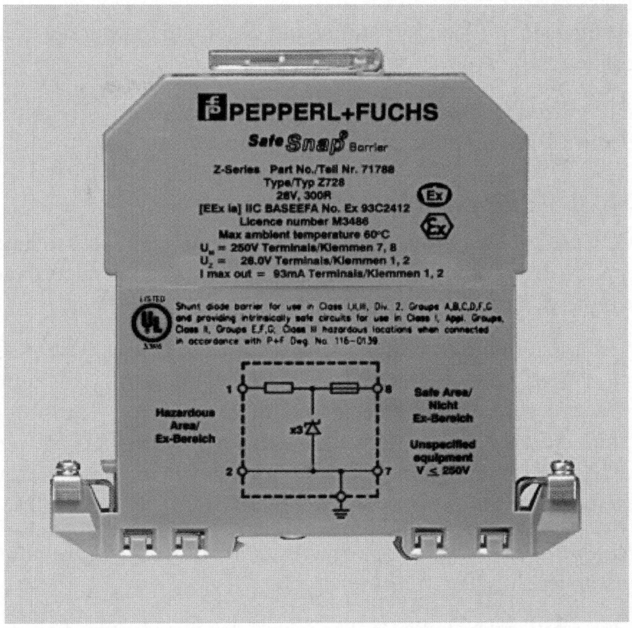


***EXHIBIT 504.1*** *A typical IS barrier that limits the energy available to the hazardous location. (Courtesy of Pepperl and Fuchs, Inc.)*

(2) Sources of stored energy consisting of single components in simple circuits with well-defined parameters; for example, capacitors or inductors, whose values are considered when determining the overall safety of the system

(3) Sources of generated energy; for example, thermocouples and photocells, that do not generate more than 1.5 volts, 100 mA, and 25 mW

## 504.3 Application of Other Articles

Except as modified by this article, all applicable articles of this *Code* shall apply.

Because IS wiring must be low energy, the wiring itself is most likely to be a Class 2 or a power-limited fire-protective signaling circuit. See Article 725 or 760, as appropriate, for the requirements for such wiring. The installation may also fall under the scope of Article 800. The associated apparatus, on the other hand, may be supplied by ordinary power circuits, in which case other *Code* rules may apply.

Often IS apparatus is supplied by associated apparatus located in a hazardous location. The associated apparatus is not normally suitable for a hazardous location. Therefore another protection technique, such as installing the associated apparatus in an explosionproof enclosure, is commonly used. In this case, requirements for an explosionproof installation would apply to the associated apparatus. The additional requirements that would apply would be governed by the protection technique that is used. IS systems are not exempt from the grounding and bonding requirements of 501.30, 502.30, 503.30, and 505.25.

## 504.4 Equipment

All intrinsically safe apparatus and associated apparatus shall be listed.

*Exception: Simple apparatus, as described on the control drawing, shall not be required to be listed.*

## 504.10 Equipment Installation

**(A) Control Drawing.** Intrinsically safe apparatus, associated apparatus, and other equipment shall be installed in accordance with the control drawing(s).

An example of the control drawing required to be followed to correctly install an IS system is shown in Exhibit 504.2. This represents the drawing provided by the associated equipment manufacturer. A similar drawing is provided by the IS equipment manufacturer. Compliance with the requirements of both drawings is required to properly install an IS system.

*Exception: A simple apparatus that does not interconnect intrinsically safe circuits.*

Informational Note No. 1: The control drawing identification is marked on the apparatus.

Informational Note No. 2: Associated apparatus with a marked Um of less than 250 V may require additional overvoltage protection at the inputs to limit any possible fault voltages to less than the Um marked on the product.

An IS system is required to be installed according to the control drawing, which may put limitations on cables and on the separation of circuits in the system. The control drawing also illustrates what is permitted to be connected in the system. Compliance with all the provisions in the control drawing is essential if intrinsic safety is to be maintained. The investigation of the equipment by third-party testing laboratories is based on installation in accordance with the control drawing. See Exhibit 504.2 for an example of a control drawing.

**(B) Location.** Intrinsically safe apparatus shall be permitted to be installed in any hazardous (classified) location for which it has been identified.

Associated apparatus shall be permitted to be installed in any hazardous (classified) location for which it has been identified.

Simple apparatus shall be permitted to be installed in any hazardous (classified) location in which the maximum surface temperature of the simple apparatus does not exceed the ignition temperature of the flammable gases or vapors, flammable liquids, combustible dusts, or ignitible fibers/flyings present.

**(C) Enclosures.** General-purpose enclosures shall be permitted for intrinsically safe apparatus and associated apparatus unless otherwise specified in the manufacturer's documentation.

**(D) Simple Apparatus.** For simple apparatus, the maximum surface temperature can be determined from the values of the output power from the associated apparatus or apparatus to which it is connected to obtain the temperature class. The temperature class can be determined by:

(1) Reference to Table 504.10(D)
(2) Calculation using the following equation:

$$T = P_O R_{th} + T_{amb}$$

***EXHIBIT 504.2*** *A sample zener barrier control drawing.*

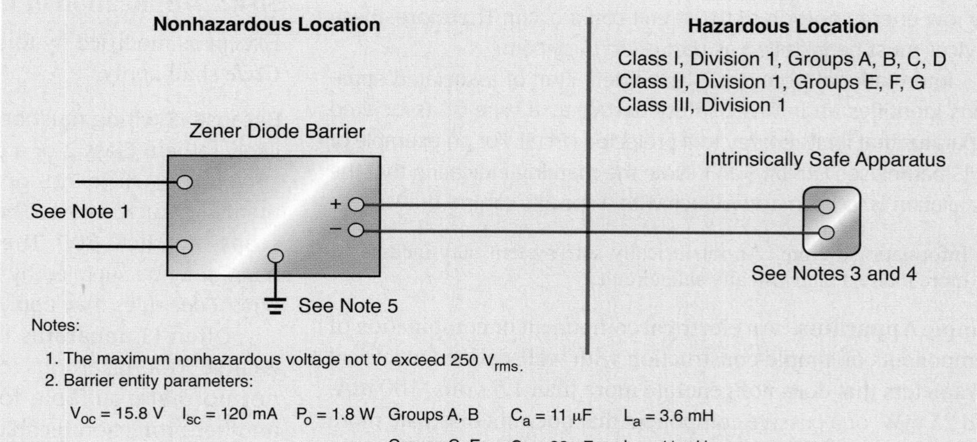


Notes:

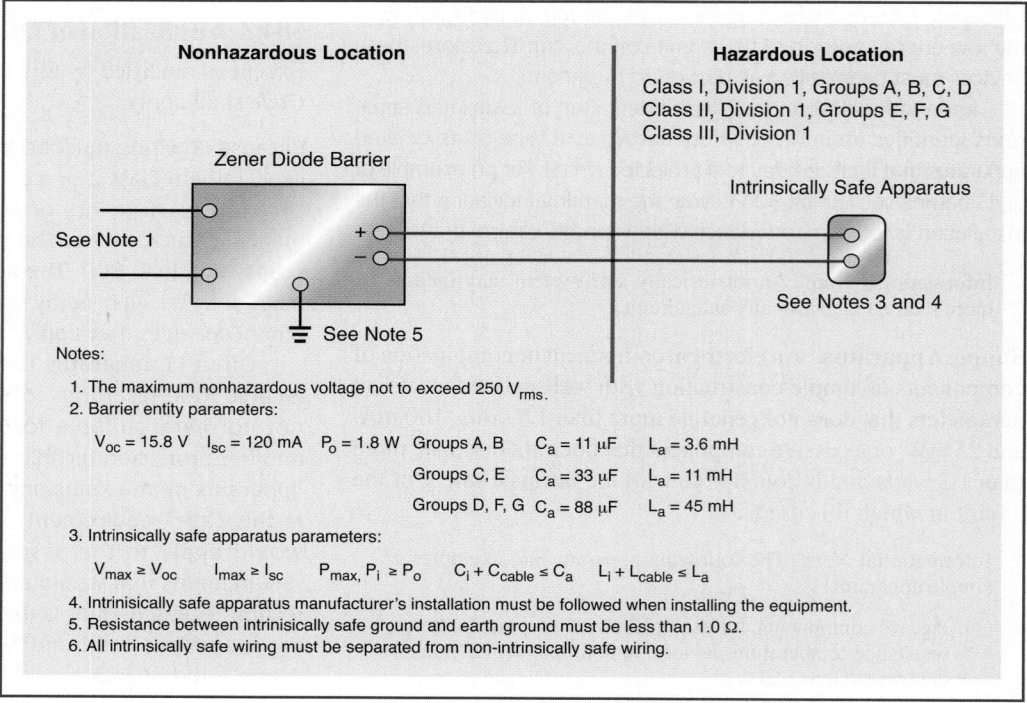

1. The maximum nonhazardous voltage not to exceed 250 $V_{rms}$.
2. Barrier entity parameters:

| | | | | | |
|---|---|---|---|---|---|
| $V_{oc}$ = 15.8 V | $I_{sc}$ = 120 mA | $P_o$ = 1.8 W | Groups A, B | $C_a$ = 11 µF | $L_a$ = 3.6 mH |
| | | | Groups C, E | $C_a$ = 33 µF | $L_a$ = 11 mH |
| | | | Groups D, F, G | $C_a$ = 88 µF | $L_a$ = 45 mH |

3. Intrinsically safe apparatus parameters:

$V_{max} \geq V_{oc}$ $I_{max} \geq I_{sc}$ $P_{max}, P_i \geq P_o$ $C_i + C_{cable} \leq C_a$ $L_i + L_{cable} \leq L_a$

4. Intrinisically safe apparatus manufacturer's installation must be followed when installing the equipment.
5. Resistance between intrinisically safe ground and earth ground must be less than 1.0 Ω.
6. All intrinsically safe wiring must be separated from non-intrinsically safe wiring.

***TABLE 504.10(D)*** *Assessment for T4 Classification According to Component Size and Temperature*

| Total Surface Area Excluding Lead Wires | Requirement for T4 Classification |
|---|---|
| <20 mm² | Surface temperature ≤275°C |
| ≥20 mm² ≤10 cm² | Surface temperature ≤200°C |
| ≥20 mm² | Power not exceeding 1.3 W* |

*Based on 40°C ambient temperature. Reduce to 1.2 W with an ambient of 60°C or 1.0 W with 80°C ambient temperature.

where:

$T$ = surface temperature

$P_O$ = output power marked on the associated apparatus or intrinsically safe apparatus

$R_{th}$ = thermal resistance of the simple apparatus

$T_{amb}$ = ambient temperature (normally 40°C) and reference Table 500.8(C)

In addition, components with a surface area smaller than 10 cm² (excluding lead wires) may be classified as T5 if their surface temperature does not exceed 150°C.

Simple apparatus stores little or no energy without requiring the apparatus to be listed or to be specifically mentioned on the control drawing. The informational note accompanying the definition of the term in 504.2 provides examples of simple apparatus.

•

## 504.20 Wiring Methods

Any of the wiring methods suitable for unclassified locations, including those covered by Chapter 7 and Chapter 8, shall be permitted for installing intrinsically safe apparatus. Sealing shall be as provided in 504.70, and separation shall be as provided in 504.30.

An IS system evaluation also includes wiring faults and cable parameters (e.g., short circuits and cable capacitance); therefore, any of the wiring methods for unclassified locations may be applied to IS systems. See the commentary following 504.3 for more information on the types of circuits typically used as part of an IS system.

## 504.30 Separation of Intrinsically Safe Conductors

### (A) From Nonintrinsically Safe Circuit Conductors.

**(1) In Raceways, Cable Trays, and Cables.** Conductors of intrinsically safe circuits shall not be placed in any raceway, cable tray, or cable with conductors of any nonintrinsically safe circuit.

*Exception No. 1: Where conductors of intrinsically safe circuits are separated from conductors of nonintrinsically safe circuits by a distance of at least 50 mm (2 in.) and secured, or by a grounded metal partition or an approved insulating partition.*

Informational Note: No. 20 gauge sheet metal partitions 0.91 mm (0.0359 in.) or thicker are generally considered acceptable.

*Exception No. 2: Where either (1) all of the intrinsically safe circuit conductors or (2) all of the nonintrinsically safe circuit conductors are in grounded metal-sheathed or metal-clad cables where the sheathing or cladding is capable of carrying fault current to ground.*

Informational Note: Cables meeting the requirements of Articles 330 and 332 are typical of those considered acceptable.

Type MI cable with a copper sheath and Type MC cable of smooth or corrugated metallic sheath construction meet the conditions prescribed in Exception No. 2. The metallic sheath of interlocked-tape Type MC cable is generally not investigated as an EGC and therefore may not be capable of carrying a fault current to ground as required by Exception No. 2.

*Exception No. 3: Intrinsically safe circuits in a Division 2 or Zone 2 location shall be permitted to be installed in a raceway, cable tray, or cable along with nonincendive field wiring circuits when installed in accordance with 504.30(B).*

*Exception No. 4: Intrinsically safe circuits passing through a Division 2 or Zone 2 location to supply apparatus that is located in a Division 1, Zone 0 or Zone 1 location shall be permitted to be installed in a raceway, cable tray, or cable along with nonincendive field wiring circuits when installed in accordance with 504.30(B).*

Informational Note: Nonincendive field wiring circuits are described in 501.10(B)(3), 502.10(B)(3), and 503.10(A)(4).

Exceptions No. 3 and No. 4 permit IS circuits in a Division 2 or Zone 2 location to be installed with nonincendive field wiring circuits when the circuits are installed as required for two separate IS circuits.

**(2) Within Enclosures.** Conductors of intrinsically safe circuits shall be secured so that any conductor that might come loose from a terminal is unlikely to come into contact with another terminal. The conductors shall be separated from conductors of nonintrinsically safe circuits by one of the methods in (1) through (4).

(1) Separation by at least 50 mm (2 in.) from conductors of any nonintrinsically safe circuits
(2) Separation from conductors of nonintrinsically safe circuits by use of a grounded metal partition 0.91 mm (0.0359 in.) or thicker
(3) Separation from conductors of nonintrinsically safe circuits by use of an approved insulating partition that extends to within 1.5 mm (0.0625 in.) of the enclosure walls
(4) Where either (1) all of the intrinsically safe circuit conductors or (2) all of the nonintrinsically safe circuit conductors are in grounded metal-sheathed or metal-clad cables where the sheathing or cladding is capable of carrying fault current to ground

Informational Note No. 1: Cables meeting the requirements of Articles 330 and 332 are typical of those considered acceptable.

Informational Note No. 2: The use of separate wiring compartments for the intrinsically safe and nonintrinsically safe terminals is a typical method of complying with this requirement.

Informational Note No. 3: Physical barriers such as grounded metal partitions or approved insulating partitions or approved restricted access wiring ducts separated from other such ducts by at least 19 mm (¾ in.) can be used to help ensure the required separation of the wiring.

The intent of 504.30(A) is to prevent unsafe energy from being introduced into the IS system by a wiring fault. Other low-voltage, low-energy circuits, such as Class 2 and communications circuits, are not IS circuits and must not be installed in the same raceways or cables as IS circuits in either a hazardous or a nonhazardous location. It is essential that non-IS circuits and IS circuits be physically and electrically separated.

**(3) Other (Not in Raceway or Cable Tray Systems).** Conductors and cables of intrinsically safe circuits run in other than raceway or cable tray systems shall be separated by at least 50 mm (2 in.) and secured from conductors and cables of any nonintrinsically safe circuits.

Even where not installed in an enclosure, raceway, or cable tray, IS circuit conductors are required to be securely separated from other conductors.

*Exception: Where either (1) all of the intrinsically safe circuit conductors are in Type MI or MC cables or (2) all of the nonintrinsically safe circuit conductors are in raceways or Type MI or MC cables where the sheathing or cladding is capable of carrying fault current to ground.*

**(B) From Different Intrinsically Safe Circuit Conductors.** The clearance between two terminals for connection of field wiring of different intrinsically safe circuits shall be at least 6 mm (0.25 in.), unless this clearance is permitted to be reduced by the control drawing. Different intrinsically safe circuits shall be separated from each other by one of the following means:

(1) The conductors of each circuit are within a grounded metal shield.

(2) The conductors of each circuit have insulation with a minimum thickness of 0.25 mm (0.01 in.).

*Exception: Unless otherwise identified.*

The minimum required clearance provides a safeguard against an inadvertent connection between adjacent terminals (with different IS circuits) that could occur during maintenance or connection of a new circuit to an existing terminal block for IS circuits.

**(C) From Grounded Metal.** The clearance between the uninsulated parts of field wiring conductors connected to terminals and grounded metal or other conducting parts shall be at least 3 mm (0.125 in.).

## 504.50 Grounding

**(A) Intrinsically Safe Apparatus, Enclosures, and Raceways.** Intrinsically safe apparatus, enclosures, and raceways, if of metal, shall be connected to the equipment grounding conductor.

Informational Note: In addition to an equipment grounding conductor connection, a connection to a grounding electrode may be needed for some associated apparatus; for example, zener diode barriers, if specified in the control drawing. See ANSI/ISA-RP 12.06.01-2003, *Recommended Practice for Wiring Methods for Hazardous (Classified) Locations Instrumentation — Part 1: Intrinsic Safety.*

**(B) Associated Apparatus and Cable Shields.** Associated apparatus and cable shields shall be grounded in accordance with the required control drawing. See 504.10(A).

Informational Note: Supplementary connection(s) to the grounding electrode may be needed for some associated apparatus; for example, zener diode barriers, if specified in the control drawing. See ANSI/ISA RP 12.06.01-2003, *Recommended Practice for Wiring Methods for Hazardous (Classified) Locations Instrumentation — Part 1: Intrinsic Safety.*

Exhibit 504.3 illustrates a common type of zener diode barrier, which is often called a shunt diode barrier. Maintaining a low-impedance path to ground for zener diode barrier systems is important because such systems shunt fault currents to ground.

Through the use of fuses, resistors, and zener diodes, the shunt diode barrier limits the fault energy delivered to the hazardous location to a level below the ignition energy of a given classified material or group. The basic concept of intrinsic safety is that safety be maintained in the event of two simultaneous faults. For example, although diodes $Z_1$ and $Z_2$ themselves can be considered subject to open-circuit fault, they are redundant components, and either one alone provides the necessary protection in the event of a first fault, that is, high voltage across input terminals 1 and 2. Terminals 2 and 4 and the ends of the two diodes are connected to a ground bus that, in turn, is connected to a ground system to which all grounds in the IS system are connected. A very low impedance (1 ohm or less is usually recommended) is necessary so that the voltage level on the ground bus will not be raised to an unsafe level under high-current fault conditions.

The basic operation of a shunt diode barrier is that under normal operating conditions, the input voltage is below that of the diodes. With 250 volts ac across terminals 1 and 2, representing a

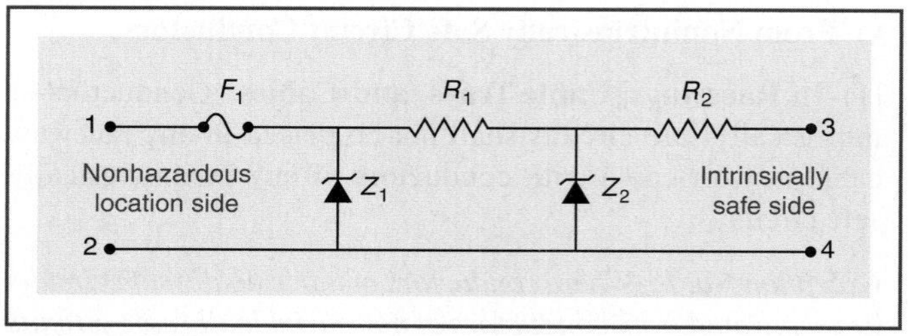


***EXHIBIT 504.3*** *Zener, or shunt, diode barrier.*

fault in equipment on the nonhazardous location side, diode $Z_1$ conducts at its rated voltage, thus limiting the output voltage of the circuit. The voltage at which this diode conducts is designed to be higher than the rated input voltage (under no-fault conditions) at terminals 1 and 2. Fuse $F_1$ is selected so that it opens before the power rating of $Z_1$ is exceeded. Diode $Z_2$ usually conducts within 1 or 2 volts of $Z_1$ and serves as a backup in the event $Z_1$ fails in an open-circuit condition for any reason. At the output voltage of $Z_1$, resistor $R_2$ limits the current in the IS circuit. Therefore, with up to 250 V ac applied to the input of the circuit (terminals 1 and 2) as a result of a fault, the output at terminals 3 and 4 cannot exceed the voltage and current permitted by the diodes and resistor. The rating of the components and therefore the barrier output level is selected to be below the ignition energy of the material or group for which the barrier is designed. By adjusting the values of the components, the barrier can be designed for a variety of uses, including for different groups or for different types of instrument systems.

Shunt diode barriers normally have maximum allowable inductance and capacitance ratings on the intrinsically safe side even though both the output voltage and current are limited. Too much inductance in the circuit could result in the release of an ignition-capable spark when the circuit is opened. In a like manner, too much capacitance could result in the release of an ignition-capable spark if a short-circuit occurs between wires or between a wire and ground.

Wiring always has inductance and capacitance associated with it, depending on the spacing between conductors, size of conductors, and length of conductors. Therefore, limiting the length of conductors connected to an intrinsic safety barrier is necessary, just as limiting the inductance and capacitance of connected equipment is necessary. The length limitation is usually on the order of thousands of feet. The control drawing provides information on installation limitations.

**(C) Connection to Grounding Electrodes.** Where connection to a grounding electrode is required, the grounding electrode shall be as specified in 250.52(A)(1), (A)(2), (A)(3), and (A)(4) and shall comply with 250.30(A)(4). Sections 250.52(A)(5), (A)(7), and (A)(8) shall not be used if any of the electrodes specified in 250.52(A)(1), (A)(2), (A)(3), or (A)(4) are present.

Where a grounding electrode is necessary for associated apparatus, the electrodes specified in 250.52(A)(1) through (A)(4) (metal underground water pipes, the metal frame of a building, concrete-encased electrodes, and ground rings) are required to be used if present. These electrodes usually provide lower resistance grounds than ground rods and plate electrodes, which are covered in 250.52(A)(5) and (A)(7).

## 504.60 Bonding

**(A) Hazardous Locations.** In hazardous (classified) locations, intrinsically safe apparatus shall be bonded in the hazardous (classified) location in accordance with 250.100.

**(B) Unclassified.** In unclassified locations, where metal raceways are used for intrinsically safe system wiring in hazardous (classified) locations, associated apparatus shall be bonded in accordance with 501.30(A), 502.30(A), 503.30(A), 505.25, or 506.25 as applicable.

## 504.70 Sealing

Conduits and cables that are required to be sealed by 501.15, 502.15, 505.16, and 506.16 shall be sealed to minimize the passage of gases, vapors, or dusts. Such seals shall not be required to be explosionproof or flameproof but shall be identified for the purpose of minimizing passage of gases, vapors, or dusts under normal operating conditions and shall be accessible.

The use of an IS system does not remove the need to seal interconnecting cables. Any cable capable of transmitting material to another location must be sealed. These seals are not required to be explosionproof or flameproof, but they must be identified to minimize the passage of gases or dust and must be accessible.

*Exception: Seals shall not be required for enclosures that contain only intrinsically safe apparatus, except as required by 501.17.*

## 504.80 Identification

Labels required by this section shall be suitable for the environment where they are installed with consideration given to exposure to chemicals and sunlight.

**(A) Terminals.** Intrinsically safe circuits shall be identified at terminal and junction locations in a manner that is intended to prevent unintentional interference with the circuits during testing and servicing.

**(B) Wiring.** Raceways, cable trays, and other wiring methods for intrinsically safe system wiring shall be identified with permanently affixed labels with the wording "Intrinsic Safety Wiring" or equivalent. The labels shall be located so as to be visible after installation and placed so that they may be readily traced through the entire length of the installation. Intrinsic safety circuit labels shall appear in every section of the wiring system that is separated by enclosures, walls, partitions, or floors. Spacing between labels shall not be more than 7.5 m (25 ft).

*Exception: Circuits run underground shall be permitted to be identified where they become accessible after emergence from the ground.*

Informational Note No. 1: Wiring methods permitted in unclassified locations may be used for intrinsically safe systems in hazardous (classified) locations. Without labels to identify the application of the wiring, enforcement authorities cannot determine that an installation is in compliance with this *Code*.

Informational Note No. 2: In unclassified locations, identification is necessary to ensure that nonintrinsically safe wire will not be inadvertently added to existing raceways at a later date.

**(C) Color Coding.** Color coding shall be permitted to identify intrinsically safe conductors where they are colored light blue and where no other conductors colored light blue are used. Likewise, color coding shall be permitted to identify raceways, cable trays, and junction boxes where they are colored light blue and contain only intrinsically safe wiring.

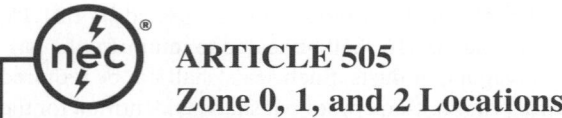

# ARTICLE 505
# Zone 0, 1, and 2 Locations

Informational Note: Text that is followed by a reference in brackets has been extracted from NFPA 497-2012, *Recommended Practice for the Classification of Flammable Liquids, Gases, or Vapors and of Hazardous (Classified) Locations for Electrical Installations in Chemical Process Areas*. Only editorial changes were made to the extracted text to make it consistent with this *Code*.

## 505.1 Scope

This article covers the requirements for the zone classification system as an alternative to the division classification system covered in Article 500 for electrical and electronic equipment and wiring for all voltages in Class I, Zone 0, Zone 1, and Zone 2 hazardous (classified) locations where fire or explosion hazards may exist due to flammable gases, vapors, or liquids.

Informational Note: For the requirements for electrical and electronic equipment and wiring for all voltages in Class I, Division 1 or Division 2; Class II, Division 1 or Division 2; and Class III, Division 1 or Division 2 hazardous (classified) locations where fire or explosion hazards may exist due to flammable gases or vapors, flammable liquids, or combustible dusts or fibers, refer to Articles 500 through 504.

The requirements in this article parallel those in Articles 500 and 501. The zone classification concept, based on the standards developed by the International Electrotechnical Commission (IEC), offers an alternative method of classifying Class I hazardous locations. The IEC classification scheme includes underground mines, whereas in the United States, underground mines are under the jurisdiction of the Mine Safety and Health Administration (MSHA) and are outside the scope of the *NEC*.

## 505.2 Definitions

**Combustible Gas Detection System.** A protection technique utilizing stationary gas detectors in industrial establishments.

**Electrical and Electronic Equipment.** Materials, fittings, devices, appliances, and the like that are part of, or in connection with, an electrical installation.

Informational Note: Portable or transportable equipment having self-contained power supplies, such as battery-operated equipment, could potentially become an ignition source in hazardous (classified) locations. See ANSI/ISA-12.12.03-2011, *Standard for Portable Electronic Products Suitable for Use in Class I and II, Division 2, Class I Zone 2 and Class III, Division 1 and 2 Hazardous (Classified) Locations.*

**Encapsulation "m."** Type of protection where electrical parts that could ignite an explosive atmosphere by either sparking or heating are enclosed in a compound in such a way that this explosive atmosphere cannot be ignited.

Informational Note No. 1: See ANSI/ISA-60079-18 (12.23.01)-2009, *Explosive atmospheres — Part 18: Equipment protection by encapsulation "m"*; and ANSI/UL 60079-18-2009, *Explosive atmospheres — Part 18: Equipment protection by encapsulation "m."*

Informational Note No. 2: Encapsulation is designated type of protection "ma" for use in Zone 0 locations. Encapsulation is designated type of protection "m" or "mb" for use in Zone 1 locations. Encapsulation is designated type of protection "mc" for use in Zone 2 locations.

**Flameproof "d."** Type of protection where the enclosure will withstand an internal explosion of a flammable mixture that has penetrated into the interior, without suffering damage and without causing ignition, through any joints or structural openings in the enclosure, of an external explosive gas atmosphere consisting of one or more of the gases or vapors for which it is designed.

Informational Note: See ANSI/ISA-60079-1 (12.22.01)-2009, *Explosive Atmospheres, Part 1: Equipment protection by flameproof enclosures "d"*; and ANSI/UL 60079-1-2009, *Electrical Apparatus for Explosive Gas Atmospheres — Part 1: Flameproof Enclosures "d."*

**Increased Safety "e."** Type of protection applied to electrical equipment that does not produce arcs or sparks in normal service and under specified abnormal conditions, in which additional measures are applied so as to give increased security against the possibility of excessive temperatures and of the occurrence of arcs and sparks.

Informational Note: See ANSI/ISA-60079-7 (12.16.01)-2008, *Explosive Atmospheres, Part 7: Equipment protection by increased safety "e"*; and ANSI/UL 60079-7–2008, *Electrical Apparatus for Explosive Gas Atmospheres — Part 7: Increased Safety "e."*

**Intrinsic Safety "i."** Type of protection where any spark or thermal effect is incapable of causing ignition of a mixture of flammable or combustible material in air under prescribed test conditions.

Informational Note No. 1: See ANSI/UL 913-2006, *Intrinsically Safe Apparatus and Associated Apparatus for Use in Class I, II, and III, Hazardous Locations*; ANSI/ISA-60079-11 (12.02.01)-2011, *Explosive Atmospheres: Part 11: Equipment protection by intrinsic safety "i"*; and ANSI/UL 60079-11-2011, *Explosive Atmospheres, Part 11: Equipment protection by intrinsic safety "i."*

Informational Note No. 2: Intrinsic safety is designated type of protection "ia" for use in Zone 0 locations. Intrinsic safety is designated type of protection "ib" for use in Zone 1 locations. Intrinsic safety is designated type of protection "ic" for use in Zone 2 locations.

Informational Note No. 3: Intrinsically safe associated apparatus, designated by [ia], [ib], or [ic], is connected to intrinsically safe apparatus ("ia," "ib," or "ic," respectively) but is located outside the hazardous (classified) location unless also protected by another type of protection (such as flameproof).

**Oil Immersion "o."** Type of protection where electrical equipment is immersed in a protective liquid in such a way that an explosive atmosphere that may be above the liquid or outside the enclosure cannot be ignited.

Informational Note: See ANSI/ISA-60079-6 (12.00.05)-2009, *Explosive Atmospheres, Part 6: Equipment protection by oil immersion "o"*; and ANSI/UL 60079-6-2009, *Electrical Apparatus for Explosive Gas Atmospheres — Part 6: Oil-Immersion "o."*

**Powder Filling "q."** Type of protection where electrical parts capable of igniting an explosive atmosphere are fixed in position and completely surrounded by filling material (glass or quartz powder) to prevent the ignition of an external explosive atmosphere.

Informational Note: See ANSI/ISA-60079-5 (12.00.04)-2009, *Explosive Atmospheres, Part 5: Equipment protection by powder filling "q"*; and ANSI/UL 60079-5, *Electrical Apparatus for Explosive Gas Atmospheres — Part 5: Powder Filling "q."*

**Pressurization "p."** Type of protection for electrical equipment that uses the technique of guarding against the ingress of the external atmosphere, which may be explosive, into an enclosure by maintaining a protective gas therein at a pressure above that of the external atmosphere.

Informational Note: See ANSI/ISA-60079-2 (12.04.01)-2010, *Explosive Atmospheres, Part 2: Equipment protection by pressurized enclosures "p"*; and IEC 60079-13-2010, *Electrical apparatus for explosive gas atmospheres — Part 13: Construction and use of rooms or buildings protected by pressurization.*

**Type of Protection "n."** Type of protection where electrical equipment, in normal operation, is not capable of igniting a surrounding explosive gas atmosphere and a fault capable of causing ignition is not likely to occur.

Informational Note: See ANSI/UL 60079-15-2009, *Electrical Apparatus for Explosive Gas Atmospheres — Part 15: Type of Protection "n"*; and ANSI/ISA-60079-15 (12.12.02)-2009, *Explosive Atmospheres — Part 15: Equipment protection by type of protection "n."*

**Unclassified Locations.** Locations determined to be neither Class I, Division 1; Class I, Division 2; Class I, Zone 0; Class I, Zone 1; Class I, Zone 2; Class II, Division 1; Class II, Division 2; Class III, Division 1; Class III, Division 2; Zone 20; Zone 21; Zone 22; or any combination thereof.

## 505.3 Other Articles

All other applicable rules contained in this *Code* shall apply to electrical equipment and wiring installed in hazardous (classified) locations.

*Exception: As modified by Article 504 and this article.*

## 505.4 General

**(A) Documentation for Industrial Occupancies.** All areas in industrial occupancies designated as hazardous (classified) locations shall be properly documented. This documentation shall be available to those authorized to design, install, inspect, maintain, or operate electrical equipment at the location.

Informational Note: For examples of area classification drawings, see ANSI/API RP 505-1997, *Recommended Practice for Classification of Locations for Electrical Installations at Petroleum Facilities Classified as Class I, Zone 0, Zone 1, or Zone 2*; ANSI/ISA-TR(12.24.01)-1998 (IEC 60079-10 Mod), *Recommended Practice for Classification of Locations for Electrical Installations Classified as Class I, Zone 0, Zone 1, or Zone 2*; IEC 60079-10-1995, *Electrical Apparatus for Explosive Gas Atmospheres, Classification of Hazardous Areas*; and *Model Code of Safe Practice in the Petroleum Industry, Part 15: Area Classification Code for Petroleum Installations*, IP 15, The Institute of Petroleum, London.

**(B) Reference Standards.** Important information relating to topics covered in Chapter 5 may be found in other publications.

Informational Note No. 1: It is important that the authority having jurisdiction be familiar with recorded industrial experience as well as with standards of the National Fire Protection Association (NFPA), the American Petroleum Institute (API), the International Society of Automation (ISA), and the International Electrotechnical Commission (IEC) that may be of use in the classification of various locations, the determination of adequate ventilation, and the protection against static electricity and lightning hazards.

Informational Note No. 2: For further information on the classification of locations, see NFPA 497-2012, *Recommended Practice for the Classification of Flammable Liquids, Gases, or Vapors and of Hazardous (Classified) Locations for Electrical Installations in Chemical Process Areas*; ANSI/API RP 505-1997, *Recommended Practice for Classification of Locations for Electrical Installations at Petroleum Facilities Classified as Class I, Zone 0, Zone 1, or Zone 2*; ANSI/ISA-TR(12.24.01)-1998 (IEC 60079-10-Mod), *Recommended Practice for Classification of Locations for Electrical Installations Classified as Class I, Zone 0, Zone 1, or Zone 2*; IEC 60079-10-1995, *Electrical Apparatus for Explosive Gas Atmospheres, Classification of Hazardous Areas*; and *Model Code of Safe Practice in the Petroleum Industry, Part 15: Area Classification Code for Petroleum Installations*, IP 15, The Institute of Petroleum, London.

Informational Note No. 3: For further information on protection against static electricity and lightning hazards in hazardous (classified) locations, see NFPA 77-2014, *Recommended Practice on Static Electricity*; NFPA 780-2014, *Standard for the Installation of Lightning Protection Systems*; and API RP 2003-1998, *Protection Against Ignitions Arising Out of Static Lightning and Stray Currents.*

Informational Note No. 4: For further information on ventilation, see NFPA 30-2012, *Flammable and Combustible Liquids Code*, and ANSI/API RP 505-1997, *Recommended Practice for Classification of Locations for Electrical Installations at Petroleum Facilities Classified as Class I, Zone 0, Zone 1, or Zone 2.*

Informational Note No. 5: For further information on electrical systems for hazardous (classified) locations on offshore oil and gas producing platforms, see ANSI/API RP 14FZ-2000, *Recommended Practice for Design and Installation of Electrical*

*Systems for Fixed and Floating Offshore Petroleum Facilities for Unclassified and Class I, Zone 0, Zone 1, and Zone 2 Locations.*

Informational Note No. 6: For further information on the installation of electrical equipment in hazardous (classified) locations in general, see IEC 60079-14-1996, *Electrical apparatus for explosive gas atmospheres — Part 14: Electrical installations in explosive gas atmospheres (other than mines)*, and IEC 60079-16-1990, *Electrical apparatus for explosive gas atmospheres — Part 16: Artificial ventilation for the protection of analyzer(s) houses.*

Informational Note No. 7: For further information on application of electrical equipment in hazardous (classified) locations in general, see ANSI/ISA-60079-0 (12.00.01)-2009, *Explosive Atmospheres — Part 0: Equipment — General Requirements*; ANSI/ISA-12.01.01-1999, *Definitions and Information Pertaining to Electrical Apparatus in Hazardous (Classified) Locations*; and ANSI/UL 60079-0, *Electrical Apparatus for Explosive Gas Atmospheres — Part 0: General Requirements.*

## 505.5 Classifications of Locations

**(A) Classification of Locations.** Locations shall be classified depending on the properties of the flammable gases, flammable liquid–produced vapors, combustible liquid–produced vapors, combustible dusts, or fibers/flyings that may be present and the likelihood that a flammable or combustible concentration or quantity is present. Each room, section, or area shall be considered individually in determining its classification. Where pyrophoric materials are the only materials used or handled, these locations are outside the scope of this article.

Informational Note No. 1: See 505.7 for restrictions on area classification.

Informational Note No. 2: Through the exercise of ingenuity in the layout of electrical installations for hazardous (classified) locations, it is frequently possible to locate much of the equipment in reduced level of classification or in an unclassified location and, thus, to reduce the amount of special equipment required.

Rooms and areas containing ammonia refrigeration systems that are equipped with adequate mechanical ventilation may be classified as "unclassified" locations.

Informational Note: For further information regarding classification and ventilation of areas involving ammonia, see ANSI/ASHRAE 15-1994, *Safety Code for Mechanical Refrigeration*; and ANSI/CGA G2.1-1989 (14-39), *Safety Requirements for the Storage and Handling of Anhydrous Ammonia.*

**(B) Class I, Zone 0, 1, and 2 Locations.** Class I, Zone 0, 1, and 2 locations are those in which flammable gases or vapors are or may be present in the air in quantities sufficient to produce explosive or ignitible mixtures. Class I, Zone 0, 1, and 2 locations shall include those specified in 505(B)(1), (B)(2), and (B)(3).

**(1) Class I, Zone 0.** A Class I, Zone 0 location is a location in which

(1) Ignitible concentrations of flammable gases or vapors are present continuously, or
(2) Ignitible concentrations of flammable gases or vapors are present for long periods of time.

Informational Note No. 1: As a guide in determining when flammable gases or vapors are present continuously or for long periods of time, refer to ANSI/API RP 505-1997, *Recommended Practice for Classification of Locations for Electrical Installations of Petroleum Facilities Classified as Class I, Zone 0, Zone 1 or Zone 2*; ANSI/ISA-TR12.24.01-1998 (IEC 60079-10 Mod), *Recommended Practice for Classification of Locations for Electrical Installations Classified as Class I, Zone 0, Zone 1, or Zone 2*; IEC 60079-10-1995, *Electrical apparatus for explosive gas atmospheres, classifications of hazardous areas*; and *Area Classification Code for Petroleum Installations, Model Code, Part 15*, Institute of Petroleum.

Informational Note No. 2: This classification includes locations inside vented tanks or vessels that contain volatile flammable liquids; inside inadequately vented spraying or coating enclosures, where volatile flammable solvents are used; between the inner and outer roof sections of a floating roof tank containing volatile flammable liquids; inside open vessels, tanks and pits containing volatile flammable liquids; the interior of an exhaust duct that is used to vent ignitible concentrations of gases or vapors; and inside inadequately ventilated enclosures that contain normally venting instruments utilizing or analyzing flammable fluids and venting to the inside of the enclosures.

**(2) Class I, Zone 1.** A Class I, Zone 1 location is a location

(1) In which ignitible concentrations of flammable gases or vapors are likely to exist under normal operating conditions; or
(2) In which ignitible concentrations of flammable gases or vapors may exist frequently because of repair or maintenance operations or because of leakage; or
(3) In which equipment is operated or processes are carried on, of such a nature that equipment breakdown or faulty operations could result in the release of ignitible concentrations of flammable gases or vapors and also cause simultaneous failure of electrical equipment in a mode to cause the electrical equipment to become a source of ignition; or
(4) That is adjacent to a Class I, Zone 0 location from which ignitible concentrations of vapors could be communicated, unless communication is prevented by adequate positive pressure ventilation from a source of clean air and effective safeguards against ventilation failure are provided.

Informational Note No. 1: Normal operation is considered the situation when plant equipment is operating within its design parameters. Minor releases of flammable material may be part of normal operations. Minor releases include the releases from mechanical packings on pumps. Failures that involve repair or shutdown (such as the breakdown of pump seals and flange gaskets, and spillage caused by accidents) are not considered normal operation.

Informational Note No. 2: This classification usually includes locations where volatile flammable liquids or liquefied flammable gases are transferred from one container to another. In areas in the vicinity of spraying and painting operations where flammable solvents are used; adequately ventilated drying rooms or compartments for evaporation of flammable solvents; adequately ventilated locations containing fat and oil extraction equipment using volatile flammable solvents; portions of cleaning and dyeing plants where volatile flammable liquids are used; adequately ventilated gas generator rooms and other portions of gas manufacturing plants where flammable gas may escape; inadequately ventilated pump rooms for flammable gas or for volatile flammable liquids; the interiors of refrigerators and freezers in which volatile flammable materials are stored in the open, lightly stoppered, or in easily ruptured containers; and other locations where ignitible concentrations of flammable vapors or gases are likely to occur in the course of normal operation but not classified Zone 0.

**(3) Class I, Zone 2.** A Class I, Zone 2 location is a location

(1) In which ignitible concentrations of flammable gases or vapors are not likely to occur in normal operation and, if they do occur, will exist only for a short period; or
(2) In which volatile flammable liquids, flammable gases, or flammable vapors are handled, processed, or used but in which the liquids, gases, or vapors normally are confined within closed containers of closed systems from which they can escape, only as a result of accidental rupture or breakdown of the containers or system, or as a result of the abnormal operation of the equipment with which the liquids or gases are handled, processed, or used; or
(3) In which ignitible concentrations of flammable gases or vapors normally are prevented by positive mechanical ventilation but which may become hazardous as a result of failure or abnormal operation of the ventilation equipment; or
(4) That is adjacent to a Class I, Zone 1 location, from which ignitible concentrations of flammable gases or vapors could be communicated, unless such communication is prevented by adequate positive-pressure ventilation from a source of clean air and effective safeguards against ventilation failure are provided.

Informational Note: The Zone 2 classification usually includes locations where volatile flammable liquids or flammable gases or vapors are used but which would become hazardous only in case of an accident or of some unusual operating condition.

## 505.6 Material Groups

For purposes of testing, approval, and area classification, various air mixtures (not oxygen enriched) shall be grouped as required in 505.6(A), (B), and (C).

Informational Note No. 1: Group I is intended for use in describing atmospheres that contain firedamp (a mixture of gases, composed mostly of methane, found underground, usually in mines). This *Code* does not apply to installations underground in mines. See 90.2(B).

•

Informational Note No. 2: The gas and vapor subdivision as described above is based on the maximum experimental safe gap (MESG), minimum igniting current (MIC), or both. Test equipment for determining the MESG is described in IEC 60079-1A-1975, Amendment No. 1 (1993), *Construction and verification tests of flameproof enclosures of electrical apparatus*; and *UL Technical Report No. 58* (1993). The test equipment for determining MIC is described in IEC 60079-11-1999, *Electrical apparatus for explosive gas atmospheres — Part 11: Intrinsic safety "i."* The classification of gases or vapors according to their maximum experimental safe gaps and minimum igniting currents is described in IEC 60079-12-1978, *Classification of mixtures of gases or vapours with air according to their maximum experimental safe gaps and minimum igniting currents.*

•

Informational Note No. 3: Group II is currently subdivided into Group IIA, Group IIB, and Group IIC. Prior marking requirements permitted some types of protection to be marked without a subdivision, showing only Group II.

Informational Note No. 4: It is necessary that the meanings of the different equipment markings and Group II classifications be carefully observed to avoid confusion with Class I, Divisions 1 and 2, Groups A, B, C, and D.

Class I, Zone 0, 1, and 2, groups shall be as follows:

**(A) Group IIC.** Atmospheres containing acetylene, hydrogen, or flammable gas, flammable liquid–produced vapor, or combustible liquid–produced vapor mixed with air that may burn or explode, having either a maximum experimental safe gap (MESG) value less than or equal to 0.50 mm or minimum igniting current ratio (MIC ratio) less than or equal to 0.45. [**497:**3.3.5.2.1]

Informational Note: Group IIC is equivalent to a combination of Class I, Group A, and Class I, Group B, as described in 500.6(A)(1) and (A)(2).

**(B) Group IIB.** Atmospheres containing acetaldehyde, ethylene, or flammable gas, flammable liquid–produced vapor, or combustible liquid–produced vapor mixed with air that may burn or explode, having either maximum experimental safe gap (MESG) values greater than 0.50 mm and less than or equal to 0.90 mm or minimum igniting current ratio (MIC ratio) greater than 0.45 and less than or equal to 0.80. [**497:**3.3.5.2.2]

Informational Note: Group IIB is equivalent to Class I, Group C, as described in 500.6(A)(3).

**(C) Group IIA.** Atmospheres containing acetone, ammonia, ethyl alcohol, gasoline, methane, propane, or flammable gas, flammable liquid–produced vapor, or combustible liquid–produced vapor mixed with air that may burn or explode, having either a maximum experimental safe gap (MESG) value greater than 0.90 mm or minimum igniting current ratio (MIC ratio) greater than 0.80. [**497:**3.3.5.2.3]

Informational Note: Group IIA is equivalent to Class I, Group D as described in 500.6(A)(4).

The zone classification system for gases is different from the division system used in Articles 500 and 501. Commentary Table 505.1 contrasts the classification of the two systems. Note that Group I is used for the classification of gases normally encountered in mining applications. The group consists primarily of methane, and is known internationally as firedamp.

**COMMENTARY TABLE 505.1** Comparison of Zone and Division Classification Systems

| Zone Classification | Division Classification |
|---|---|
| Group IIC | Groups A and B |
| Group IIB | Group C |
| Group IIA | Group D |
| Group I | Group D |

## 505.7 Special Precaution

Article 505 requires equipment construction and installation that ensures safe performance under conditions of proper use and maintenance.

Informational Note No. 1: It is important that inspection authorities and users exercise more than ordinary care with regard to the installation and maintenance of electrical equipment in hazardous (classified) locations.

Informational Note No. 2: Low ambient conditions require special consideration. Electrical equipment depending on the protection techniques described by 505.8(A) may not be suitable for use at temperatures lower than −20°C (−4°F) unless they are identified for use at lower temperatures. However, at low ambient temperatures, flammable concentrations of vapors may not exist in a location classified Class I, Zones 0, 1, or 2 at normal ambient temperature.

**(A) Implementation of Zone Classification System.** Classification of areas, engineering and design, selection of equipment and wiring methods, installation, and inspection shall be performed by qualified persons.

**(B) Dual Classification.** In instances of areas within the same facility classified separately, Class I, Zone 2 locations shall be permitted to abut, but not overlap, Class I, Division 2 locations. Class I, Zone 0 or Zone 1 locations shall not abut Class I, Division 1 or Division 2 locations.

An installation is permitted to be designed using either the classification scheme of Article 500 or the classification scheme of Article 505. Both schemes cannot be used for classifying the same area. In areas within the same facility, Class I, Zone 2 locations are allowed to be adjacent to and share the same border, but they are not allowed to overlap Class I, Division 2 locations. However, Class I, Zone 0 or Zone 1 locations are not allowed to be adjacent to and share the same border with Class I, Division 1 or Division 2 locations.

**(C) Reclassification Permitted.** A Class I, Division 1 or Division 2 location shall be permitted to be reclassified as a Class I, Zone 0, Zone 1, or Zone 2 location, provided all of the space that is classified because of a single flammable gas or vapor source is reclassified under the requirements of this article.

**(D) Solid Obstacles.** Flameproof equipment with flanged joints shall not be installed such that the flange openings are closer than the distances shown in Table 505.7(D) to any solid obstacle that is not a part of the equipment (such as steelworks, walls, weather guards, mounting brackets, pipes, or other electrical equipment) unless the equipment is listed for a smaller distance of separation.

***TABLE 505.7(D)*** *Minimum Distance of Obstructions from Flameproof "d" Flange Openings*

| Gas Group | Minimum Distance | |
|---|---|---|
| | mm | in. |
| IIC | 40 | 1 37/64 |
| IIB | 30 | 13/16 |
| IIA | 10 | 25/64 |

**(E) Simultaneous Presence of Flammable Gases and Combustible Dusts or Fibers/Flyings.** Where flammable gases, combustible dusts, or fibers/flyings are or may be present at the same time, the simultaneous presence shall be considered during the selection and installation of the electrical equipment and the wiring methods, including the determination of the safe operating temperature of the electrical equipment.

**(F) Available Short-Circuit Current for Type of Protection "e".** The available short-circuit current for electrical equipment using type of protection "e" for the field wiring connections in Zone 1 locations shall be limited to 10,000 rms symmetrical amperes to reduce the likelihood of ignition of a flammable atmosphere by an arc during a short-circuit event.

Informational Note: Limitation of the available short-circuit current to this level may require the application of current-limiting fuses or current-limiting circuit breakers.

The limit on available short-circuit current is due to the rating of terminals and terminal blocks of equipment evaluated under ANSI/UL 508A, *Standard for Industrial Control Panels,* and using the Type "e" protection technique.

## 505.8 Protection Techniques

Acceptable protection techniques for electrical and electronic equipment in hazardous (classified) locations shall be as described in 505.8(A) through (I).

Informational Note: For additional information, see ANSI/ISA-60079-0 (12.00.01)-2009, *Explosive Atmospheres — Part 0: Equipment — General Requirements*; ANSI/ISA-12.01.01-1999, *Definitions and Information Pertaining to Electrical Apparatus in Hazardous (Classified) Locations*; and ANSI/UL 60079–0, *Electrical Apparatus for Explosive Gas Atmospheres — Part 0: General Requirements.*

Where the area is classified in accordance with the zone method, 505.8(A) through (K) identifies the methods for protecting electrical and electronic equipment. Many of the protection methods defined in 505.2 are different than those in Article 501, and only those specified are suitable for zone installations.

***EXHIBIT 505.1*** *Typical control stations with the combination of flameproof and increased safety types of protection suitable for use in Class I, Zone 1 areas. (Courtesy of Cooper Crouse-Hinds)*

See 505.9(C) for marking requirements. Of the many protection techniques, intrinsic safety, flameproof, and increased safety are the most common for Zone 1 locations.

**(A) Flameproof "d".** This protection technique shall be permitted for equipment in Class I, Zone 1 or Zone 2 locations.

Equipment identified as flameproof is similar to explosionproof equipment. Flameproof protection is commonly combined with increased safety protection. For example, motor control and other switching contacts are commonly protected by flameproof enclosures, with the field wiring terminals protected in a separate but attached enclosure by increased safety. The conductors between the enclosures are protected by flameproof feed-through insulators. The equipment shown in Exhibit 505.1 employs this combination of protection techniques.

**(B) Purged and Pressurized.** This protection technique shall be permitted for equipment in those Class I, Zone 1 or Zone 2 locations for which it is identified.

**(C) Intrinsic Safety.** This protection technique shall be permitted for apparatus and associated apparatus in Class I, Zone 0, Zone 1, or Zone 2 locations for which it is listed.

The identifying letter for intrinsic safety is "i" followed by either "a," "b," or "c" indicating whether the equipment is suitable for Zone 0 (ia), Zone 1 (ib), or Zone 2 (ic). The associated apparatus is identified by the same letters in brackets, that is, [ia], [ib], or [ic].

The "ic" classification extends the intrinsic safety concept to Zone 2 locations. The concept is similar to the nonincendive technique in that faults to the circuit or system are not applied.

**(D) Type of Protection "n".** This protection technique shall be permitted for equipment in Class I, Zone 2 locations. Type of protection "n" is further subdivided into nA, nC, and nR.

Informational Note: See Table 505.9(C)(2)(4) for the descriptions of subdivisions for type of protection "n".

**(E) Oil Immersion "o".** This protection technique shall be permitted for equipment in Class I, Zone 1 or Zone 2 locations.

**(F) Increased Safety "e".** This protection technique shall be permitted for equipment in Class I, Zone 1 or Zone 2 locations.

The increased safety protection technique is commonly used for terminal boxes, fluorescent luminaires, motors, and generators (see 505.22).

**(G) Encapsulation "m".** This protection technique shall be permitted for equipment in Class I, Zone 0, Zone 1, or Zone 2 locations for which it is identified.

Informational Note: See Table 505.9(C)(2)(4) for the descriptions of subdivisions for encapsulation.

**(H) Powder Filling "q".** This protection technique shall be permitted for equipment in Class I, Zone 1 or Zone 2 locations.

**(I) Combustible Gas Detection System.** A combustible gas detection system shall be permitted as a means of protection in industrial establishments with restricted public access and where the conditions of maintenance and supervision ensure that only qualified persons service the installation. Where such a system is installed, equipment specified in 505.8(I)(1), (I)(2), or (I)(3) shall be permitted. The type of detection equipment, its listing, installation location(s), alarm and shutdown criteria, and calibration frequency shall be documented when combustible gas detectors are used as a protection technique.

Informational Note No. 1: For further information, see ANSI/API RP 505-1997, *Recommended Practice for Classification of Locations for Electrical Installations at Petroleum Facilities Classified as Class I, Zone 0, Zone 1, and Zone 2*.

Informational Note No. 2: For further information, see ANSI/ISA-60079-29-2, *Explosive Atmospheres — Part 29-2: Gas detectors — Selection, installation, use and maintenance of detectors for flammable gases and oxygen*.

Informational Note No. 3: For further information, see ANSI/ISA-TR12.13.03-2009, *Guide for Combustible Gas Detection as a Method of Protection*.

**(1) Inadequate Ventilation.** In a Class I, Zone 1 location that is so classified due to inadequate ventilation, electrical equipment suitable for Class I, Zone 2 locations shall be permitted. Combustible gas detection equipment shall be listed for Class I, Zone 1, for the appropriate material group, and for the detection of the specific gas or vapor to be encountered.

**(2) Interior of a Building.** In a building located in, or with an opening into, a Class I, Zone 2 location where the interior does not contain a source of flammable gas or vapor, electrical equipment for unclassified locations shall be permitted. Combustible gas detection equipment shall be listed for Class I, Zone 1 or Class I, Zone 2, for the appropriate material group, and for the detection of the specific gas or vapor to be encountered.

**(3) Interior of a Control Panel.** In the interior of a control panel containing instrumentation utilizing or measuring flammable liquids, gases, or vapors, electrical equipment suitable for Class I, Zone 2 locations shall be permitted. Combustible gas detection equipment shall be listed for Class I, Zone 1, for the appropriate material group, and for the detection of the specific gas or vapor to be encountered.

The gas detection system must be suitable for the original zone classification of the area even though the remainder of installed equipment is permitted to be suitable for one zone lower.

## 505.9 Equipment

**(A) Suitability.** Suitability of identified equipment shall be determined by one of the following:

(1) Equipment listing or labeling
(2) Evidence of equipment evaluation from a qualified testing laboratory or inspection agency concerned with product evaluation
(3) Evidence acceptable to the authority having jurisdiction such as a manufacturer's self-evaluation or an owner's engineering judgment

Informational Note: Additional documentation for equipment may include certificates demonstrating compliance with applicable equipment standards, indicating special conditions of use, and other pertinent information.

**(B) Listing.**

(1) Equipment that is listed for a Zone 0 location shall be permitted in a Zone 1 or Zone 2 location of the same gas or vapor, provided that it is installed in accordance with the requirements for the marked type of protection. Equipment that is listed for a Zone 1 location shall be permitted in a Zone 2 location of the same gas or vapor, provided that it is installed in accordance with the requirements for the marked type of protection.
(2) Equipment shall be permitted to be listed for a specific gas or vapor, specific mixtures of gases or vapors, or any specific combination of gases or vapors.

Informational Note: One common example is equipment marked for "IIB. + H2."

**(C) Marking.** Equipment shall be marked in accordance with 505.9(C)(1) or (C)(2).

**(1) Division Equipment.** Equipment identified for Class I, Division 1 or Class I, Division 2 shall, in addition to being marked in accordance with 500.8(C), be permitted to be marked with all of the following:

(1) Class I, Zone 1 or Class I, Zone 2 (as applicable)
(2) Applicable gas classification group(s) in accordance with Table 505.9(C)(1)(2)
(3) Temperature classification in accordance with 505.9(D)(1)

**(2) Zone Equipment.** Equipment meeting one or more of the protection techniques described in 505.8 shall be marked with all of the following in the order shown:

(1) Class
(2) Zone

***TABLE 505.9(C)(1)(2)*** *Material Groups*

| Material Group | Comment |
|---|---|
| IIC | See 505.6(A) |
| IIB | See 505.6(B) |
| IIA | See 505.6(C) |

(3) Symbol "AEx"
(4) Protection technique(s) in accordance with Table 505.9(C)(2)(4)
(5) Applicable material group in accordance with Table 505.9(C)(1)(2) or a specific gas or vapor
(6) Temperature classification in accordance with 505.9(D)(1)

The symbol AEx identifies the equipment as meeting American national standards. In European Union countries, the symbol is EEx. In the IEC standards, on which American and European standards are based, the symbol is Ex. Only equipment marked AEx has been evaluated for use in electrical systems and hazardous locations covered under the *NEC*.

*Exception No. 1: Associated apparatus NOT suitable for installation in a hazardous (classified) location shall be required to be marked only with (3), (4), and (5), but BOTH the symbol AEx (3) and the symbol for the type of protection (4) shall be enclosed within the same square brackets, for example, [AEx ia] IIC.*

*Exception No. 2: Simple apparatus as defined in 504.2 shall not be required to have a marked operating temperature or temperature class.*

*Exception No. 3: Fittings for the termination of cables shall not be required to have a marked operating temperature or temperature class.*

•

Informational Note No. 1: An example of the required marking for intrinsically safe apparatus for installation in Class I, Zone 0 is "Class I, Zone 0, AEx ia IIC T6." An explanation of the marking that is required is shown in Informational Note Figure 505.9(C)(2), No.1.

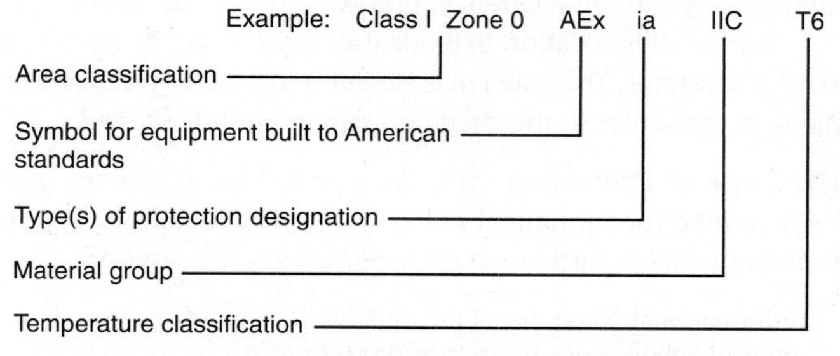


***INFORMATIONAL NOTE FIGURE 505.9(C)(2), NO. 1,*** *Zone Equipment Marking.*

***TABLE 505.9(C)(2)(4)*** *Types of Protection Designation*

| Designation | Technique | Zone* |
|---|---|---|
| d | Flameproof enclosure | 1 |
| db | Flameproof enclosure | 1 |
| e | Increased safety | 1 |
| eb | Increased safety | 1 |
| ia | Intrinsic safety | 0 |
| ib | Intrinsic safety | 1 |
| ic | Intrinsic safety | 2 |
| [ia] | Associated apparatus | Unclassified** |
| [ib] | Associated apparatus | Unclassified** |
| [ic] | Associated apparatus | Unclassified** |
| ma | Encapsulation | 0 |
| m | Encapsulation | 1 |
| mb | Encapsulation | 1 |
| mc | Encapsulation | 2 |
| nA | Nonsparking equipment | 2 |
| nAc | Nonsparking equipment | 2 |
| nC | Sparking equipment in which the contacts are suitably protected other than by restricted breathing enclosure | 2 |
| nCc | Sparking equipment in which the contacts are suitably protected other than by restricted breathing enclosure | 2 |
| nR | Restricted breathing enclosure | 2 |
| nRc | Restricted breathing enclosure | 2 |
| o | Oil immersion | 1 |
| ob | Oil immersion | 1 |
| px | Pressurization | 1 |
| pxb | Pressurization | 1 |
| py | Pressurization | 1 |
| pyb | Pressurization | 1 |
| pz | Pressurization | 2 |
| pzc | Pressurization | 2 |
| q | Powder filled | 1 |
| qb | Powder filled | 1 |

*Does not address use where a combination of techniques is used.

**Associated apparatus is permitted to be installed in a hazardous (classified) location if suitably protected using another type of protection.

Informational Note No. 2: An example of the required marking for intrinsically safe associated apparatus mounted in a flameproof enclosure for installation in Class I, Zone 1 is "Class I, Zone 1 AEx d[ia] IIC T4."

Informational Note No. 3: An example of the required marking for intrinsically safe associated apparatus NOT for installation in a hazardous (classified) location is "[AEx ia] IIC."

Informational Note No. 4: The EPL (or equipment protection level) may appear in the product marking. EPLs are designated as G for gas, D for dust, or M for mining and are then followed by a letter (a, b, or c) to give the user a better understanding as to whether the equipment provides either (a) a "very high," (b) a "high," or (c) an "enhanced" level of protection against ignition of an explosive atmosphere. For example, an AEx d IIC T4 motor (which is suitable by protection concept for application in Zone 1) may additionally be marked with an EPL of "Gb" to indicate that it was provided with a high level of protection, such as AEx d IIC T4 Gb.

Informational Note No. 5: Equipment installed outside a Zone 0 location, electrically connected to equipment located inside a Zone 0 location, may be marked Class I, Zone 0/1. The "/" indicates that equipment contains a separation element and can be installed at the boundary between a Zone 0 and a Zone 1 location. See ANSI/ISA-60079-26, *Electrical Apparatus for Use in Class I, Zone 0 Hazardous (Classified) Locations*.

**(D) Class I Temperature.** The temperature marking specified in 505.9(D)(1) shall not exceed the autoignition temperature of the specific gas or vapor to be encountered.

Informational Note: For information regarding autoignition temperatures of gases and vapors, see NFPA 497-2012, *Recommended Practice for the Classification of Flammable Liquids, Gases, or Vapors and of Hazardous (Classified) Locations for Electrical Installations in Chemical Process Areas*; and IEC 60079-20-1996, *Electrical Apparatus for Explosive Gas Atmospheres, Data for Flammable Gases and Vapours, Relating to the Use of Electrical Apparatus*.

**(1) Temperature Classifications.** Equipment shall be marked to show the operating temperature or temperature class referenced to a 40°C ambient, or at the higher ambient temperature if the equipment is rated and marked for an ambient temperature of greater than 40°C. The temperature class, if provided, shall be indicated using the temperature class (T code) shown in Table 505.9(D)(1).

Electrical equipment designed for use in the ambient temperature range between −20°C and +40°C shall require no ambient temperature marking.

Electrical equipment that is designed for use in a range of ambient temperatures other than −20°C to +40°C is considered to be special; and the ambient temperature range shall then be marked on the equipment, including either the symbol "Ta" or

***TABLE 505.9(D)(1)*** *Classification of Maximum Surface Temperature for Group II Electrical Equipment*

| Temperature Class (T Code) | Maximum Surface Temperature (°C) |
|---|---|
| T1 | ≤450 |
| T2 | ≤300 |
| T3 | ≤200 |
| T4 | ≤135 |
| T5 | ≤100 |
| T6 | ≤85 |

"Tamb" together with the special range of ambient temperatures, in degrees Celsius.

Informational Note: As an example, such a marking might be "–30°C to +40°C."

*Exception No. 1: Equipment of the non–heat-producing type, such as conduit fittings, and equipment of the heat-producing type having a maximum temperature of not more than 100°C (212°F) shall not be required to have a marked operating temperature or temperature class.*

*Exception No. 2: Equipment identified for Class 1, Division 1 or Division 2 locations as permitted by 505.20(A), (B), and (C) shall be permitted to be marked in accordance with 505.8(C) and Table 500.8(C).*

**(E) Threading.** The supply connection entry thread form shall be NPT or metric. Conduit and fittings shall be made wrenchtight to prevent sparking when fault current flows through the conduit system, and to ensure the explosionproof or flameproof integrity of the conduit system where applicable. Equipment provided with threaded entries for field wiring connections shall be installed in accordance with 505.9(E)(1) or (E)(2) and with (E)(3).

**(1) Equipment Provided with Threaded Entries for NPT Threaded Conduit or Fittings.** For equipment provided with threaded entries for NPT threaded conduit or fittings, listed conduit, listed conduit fittings, or listed cable fittings shall be used.

All NPT threaded conduit and fittings referred to herein shall be threaded with a National (American) Standard Pipe Taper (NPT) thread.

NPT threaded entries into explosionproof or flameproof equipment shall be made up with at least five threads fully engaged.

*Exception: For listed explosionproof or flameproof equipment, factory threaded NPT entries shall be made up with at least 4½ threads fully engaged.*

Informational Note No. 1: Thread specifications for male NPT threads are located in ANSI/ASME B1.20.1-1983, *Pipe Threads, General Purpose (Inch).*

Informational Note No. 2: Female NPT threaded entries use a modified National Standard Pipe Taper (NPT) thread with thread form per ANSI/ASME B1.20.1-1983, *Pipe Threads, General Purpose (Inch).* See ANSI UL/ISA 60079-1, *Electrical Apparatus for Explosive Gas Atmospheres — Part 1: Flameproof Enclosures "d."*

**(2) Equipment Provided with Threaded Entries for Metric Threaded Conduit or Fittings.** For equipment with metric threaded entries, listed conduit fittings or listed cable fittings shall be used. Such entries shall be identified as being metric, or listed adapters to permit connection to conduit or NPT threaded fittings shall be provided with the equipment and shall be used for connection to conduit or NPT threaded fittings.

Metric threaded entries into explosionproof or flameproof equipment shall have a class of fit of at least 6g/6H and be made up with at least five threads fully engaged for Groups C, D, IIB, or IIA and not less than eight threads fully engaged for Groups A, B, IIC, or IIB + $H_2$.

Informational Note: Threading specifications for metric threaded entries are located in ISO 965/1-1980, *ISO general pupose metric screw threads — Tolerances — Part 1: Principles and basic data*; and ISO 965-3-1998, *ISO general purpose metric screw threads — Tolerances — Part 3: Deviations for constructional screw threads*; and ISO 965/3-1980, *Metric Screw Threads*.

Listed fittings must be used with metric threaded entries to assure the integrity of the conduit system. In addition, either the entries must be identified as being metric or adapters must be provided for making connection to NPT fittings. Exhibit 505.2 is an example of an adapter that provides a means of connecting conduit or fitting with NPT threads to a Type "e" (increased safety) enclosure that has metric threads.

**(3) Unused Openings.** All unused openings shall be closed with close-up plugs listed for the location and shall maintain the type of protection. The plug engagement shall comply with 505.9(E)(1) or 505.9(E)(2).

**(F) Optical Fiber Cables.** Where an optical fiber cable contains conductors that are capable of carrying current (composite optical fiber cable), the optical fiber cable shall be installed in accordance with the requirements of Articles 505.15 and 505.16.

## 505.15 Wiring Methods

Wiring methods shall maintain the integrity of protection techniques and shall comply with 505.15(A) through (C).

**(A) Class I, Zone 0.** In Class I, Zone 0 locations, only intrinsically safe wiring methods in accordance with Article 504 shall be permitted.

This requirement is one of the most significant differences between the zone and division area classification requirements. The degree of hazard within a Zone 0 area is considered so severe that all wiring in this area must be intrinsically safe (technique "ia"). In

***EXHIBIT 505.2*** *A typical hub providing an NPT-threaded entry for conduit or cable into an increased safety enclosure. (Courtesy of Cooper Crouse-Hinds)*

general, only instrumentation and signaling circuits installed in accordance with Article 504 can be used in a Zone 0 area.

•

**(B) Class I, Zone 1.**

**(1) General.** In Class I, Zone 1 locations, the wiring methods in (B)(1)(a) through (B)(1)(i) shall be permitted.

(a) All wiring methods permitted by 505.15(A).

(b) In industrial establishments with restricted public access, where the conditions of maintenance and supervision ensure that only qualified persons service the installation, and where the cable is not subject to physical damage, Type MC-HL cable listed for use in Class I, Zone 1 or Division 1 locations, with a gas/vaportight continuous corrugated metallic sheath, an overall jacket of suitable polymeric material, and a separate equipment grounding conductor(s) in accordance with 250.122, and terminated with fittings listed for the application. Type MC-HL cable shall be installed in accordance with the provisions of Article 330, Part II.

(c) In industrial establishments with restricted public access, where the conditions of maintenance and supervision ensure that only qualified persons service the installation, and where the cable is not subject to physical damage, Type ITC-HL cable listed for use in Class I, Zone 1 or Division 1 locations, with a gas/vaportight continuous corrugated metallic sheath and an overall jacket of suitable polymeric material, and terminated with fittings listed for the application. Type ITC-HL cable shall be installed in accordance with the provisions of Article 727.

Informational Note: See 727.4 and 727.5 for restrictions on use of Type ITC cable.

(d) Type MI cable terminated with fittings listed for Class I, Zone 1 or Division 1 locations. Type MI cable shall be installed and supported in a manner to avoid tensile stress at the termination fittings.

(e) Threaded rigid metal conduit, or threaded steel intermediate metal conduit.

(f) Type PVC conduit and Type RTRC conduit shall be permitted where encased in a concrete envelope a minimum of 50 mm (2 in.) thick and provided with not less than 600 mm (24 in.) of cover measured from the top of the conduit to grade. Threaded rigid metal conduit or threaded steel intermediate metal conduit shall be used for the last 600 mm (24 in.) of the underground run to emergence or to the point of connection to the aboveground raceway. An equipment grounding conductor shall be included to provide for electrical continuity of the raceway system and for grounding of non–current-carrying metal parts.

(g) Intrinsic safety type of protection "ib" shall be permitted using the wiring methods specified in Article 504.

Informational Note: For entry into enclosures required to be flameproof, explosionproof, or of increased safety, see the information on construction, testing, and marking of cables; flameproof and increased safety cable fittings; and flameproof and increased safety cord connectors in ANSI/UL 2225-2011, *Cables and Cable-Fittings for Use in Hazardous (Classified) Locations*.

(h) Optical fiber cable Types OFNP, OFCP, OFNR, OFCR, OFNG, OFCG, OFN, and OFC shall be permitted to be installed in raceways in accordance with 505.15(B). Optical fiber cable shall be sealed in accordance with 505.16.

Informational Note: For entry into enclosures required to be flameproof, explosionproof, or of increased safety, see the information on construction, testing, and marking of cables; flameproof and increased safety cable fittings; and flameproof and increased safety cord connectors in ANSI/UL 2225-2011, *Cables and Cable-Fittings for Use in Hazardous (Classified) Locations*.

(i) In industrial establishments with restricted public access, where the conditions of maintenance and supervision ensure that only qualified persons service the installation, for applications limited to 600 volts nominal or less, for cable diameters 25 mm (1 in.) or less, and where the cable is not subject to physical damage, Type TC-ER-HL cable listed for use in Class I, Zone 1 locations, with an overall jacket and a separate equipment grounding conductor(s) in accordance with 250.122, and terminated with fittings listed for the location, Type TC-ER-HL cable shall be installed in accordance with the provisions of Article 336, including the restrictions of 336.10(7).

**(2) Flexible Connections.** Where necessary to employ flexible connections, flexible fittings listed for Class I, Zone 1 or Division 1 locations, or flexible cord in accordance with the provisions of 505.17(A) terminated with a listed cord connector that maintains the type of protection of the terminal compartment, shall be permitted.

**(C) Class I, Zone 2.**

**(1) General.** In Class I, Zone 2 locations, the following wiring methods shall be permitted.

(a) All wiring methods permitted by 505.15(B).

(b) Types MC, MV, TC, or TC-ER cable, including installation in cable tray systems. The cable shall be terminated with listed fittings. Single conductor Type MV cables shall be shielded or metallic-armored.

(c) Type ITC and Type ITC-ER cable as permitted in 727.4 and terminated with listed fittings.

(d) Type PLTC and Type PLTC-ER cable in accordance with the provisions of Article 725, including installation in cable tray systems. The cable shall be terminated with listed fittings.

(e) Enclosed gasketed busways, enclosed gasketed wireways.

(f) In industrial establishments with restricted public access, where the conditions of maintenance and supervision ensure that only qualified persons service the installation, and where metallic conduit does not provide sufficient corrosion resistance, listed reinforced thermosetting resin conduit (RTRC), factory elbows, and associated fittings, all marked with the suffix -XW, and Schedule 80 PVC conduit, factory elbows, and associated fittings shall be permitted. Where seals are required for boundary conditions as defined in 505.16(C)(1)(b), the Zone 1 wiring method shall extend into the Zone 2 area to the seal, which shall be located on the Zone 2 side of the Zone 1/Zone 2 boundary.

(g) Intrinsic safety type of protection "ic" shall be permitted using any of the wiring methods permitted for unclassified locations. Intrinsic safety type of protection "ic" systems shall be installed in accordance with the control drawing(s). Simple apparatus, not shown on the control drawing, shall be permitted in an intrinsic safety type of protection "ic" circuit, provided that the simple apparatus does not interconnect the intrinsic safety type of protection "ic" systems to any other circuit.

Informational Note: Simple apparatus is defined in 504.2.

(h) Optical fiber cable of Types OFNP, OFCP, OFNR, OFCR, OFNG, OFCG, OFN, and OFC shall be permitted to be installed in cable trays or any other raceway in accordance with 505.15(C). Optical fiber cable shall be sealed in accordance with 505.16.

Separate intrinsic safety type of protection "ic" systems shall be installed in accordance with one of the following:

(1) In separate cables
(2) In multiconductor cables where the conductors of each circuit are within a grounded metal shield
(3) In multiconductor cables where the conductors of each circuit have insulation with a minimum thickness of 0.25 mm (0.01 in.)

**(2) Flexible Connections.** Where provision must be made for flexibility, flexible metal fittings, flexible metal conduit with listed fittings, liquidtight flexible metal conduit with listed fittings, liquidtight flexible nonmetallic conduit with listed fittings, or flexible cord in accordance with the provisions of 505.17 terminated with a listed cord connector that maintains the type of protection of the terminal compartment shall be permitted.

Informational Note: See 505.25(B) for grounding requirements where flexible conduit is used.

*Exception: For elevator use, an identified elevator cable of Type EO, ETP, or ETT, shown under the "use" column in Table 400.4 for "hazardous (classified) locations," that is terminated with listed connectors that maintain the type of protection of the terminal compartment, shall be permitted.*

## 505.16 Sealing and Drainage

Seals in conduit and cable systems shall comply with 505.16(A) through (E). Sealing compound shall be used in Type MI cable termination fittings to exclude moisture and other fluids from the cable insulation.

See the associated commentary to 501.15 for further information regarding conduit and cable sealing for hazardous locations.

Informational Note No. 1: Seals are provided in conduit and cable systems to minimize the passage of gases and vapors and prevent the passage of flames from one portion of the electrical installation to another through the conduit. Such communication through Type MI cable is inherently prevented by construction of the cable. Unless specifically designed and tested for the purpose, conduit and cable seals are not intended to prevent the passage of liquids, gases, or vapors at a continuous pressure differential across the seal. Even at differences in pressure across the seal equivalent to a few inches of water, there may be a slow passage of gas or vapor through a seal and through conductors passing through the seal. See 505.16(C)(2)(b). Temperature extremes and highly corrosive liquids and vapors can affect the ability of seals to perform their intended function. See 505.16(D)(2).

Informational Note No. 2: Gas or vapor leakage and propagation of flames may occur through the interstices between the strands of standard stranded conductors larger than 2 AWG. Special conductor constructions, for example, compacted strands or sealing of the individual strands, are means of reducing leakage and preventing the propagation of flames.

**(A) Zone 0.** In Class I, Zone 0 locations, seals shall be located according to 505.16(A)(1), (A)(2), and (A)(3).

**(1) Conduit Seals.** Seals shall be provided within 3.05 m (10 ft) of where a conduit leaves a Zone 0 location. There shall be no unions, couplings, boxes, or fittings, except listed reducers at the seal, in the conduit run between the seal and the point at which the conduit leaves the location.

*Exception: A rigid unbroken conduit that passes completely through the Zone 0 location with no fittings less than 300 mm (12 in.) beyond each boundary shall not be required to be sealed if the termination points of the unbroken conduit are in unclassified locations.*

**(2) Cable Seals.** Seals shall be provided on cables at the first point of termination after entry into the Zone 0 location.

**(3) Not Required to Be Explosionproof or Flameproof.** Seals shall not be required to be explosionproof or flameproof.

**(B) Zone 1.** In Class I, Zone 1 locations, seals shall be located in accordance with 505.16(B)(1) through (B)(8).

**(1) Type of Protection "d" or "e" Enclosures.** Conduit seals shall be provided within 50 mm (2 in.) for each conduit entering enclosures having type of protection "d" or "e."

*Exception No. 1: Where the enclosure having type of protection "d" is marked to indicate that a seal is not required.*

*Exception No. 2: For type of protection "e," conduit and fittings employing only NPT to NPT raceway joints or fittings listed for type of protection "e" shall be permitted between the enclosure and the seal, and the seal shall not be required to be within 50 mm (2 in.) of the entry.*

Informational Note: Examples of fittings employing other than NPT threads include conduit couplings, capped elbows, unions, and breather drains.

*Exception No. 3: For conduit installed between type of protection "e" enclosures employing only NPT to NPT raceway joints or conduit fittings listed for type of protection "e," a seal shall not be required.*

**(2) Explosionproof Equipment.** Conduit seals shall be provided for each conduit entering explosionproof equipment according to 505.16(B)(2)(a), (B)(2)(b), and (B)(2)(c).

(a) In each conduit entry into an explosionproof enclosure where either (1) the enclosure contains apparatus, such as switches, circuit breakers, fuses, relays, or resistors, that may produce arcs, sparks, or high temperatures that are considered to be an ignition source in normal operation, or (2) the entry is metric designator 53 (trade size 2) or larger and the enclosure contains terminals, splices, or taps. For the purposes of this section, high temperatures shall be considered to be any temperatures exceeding 80 percent of the autoignition temperature in degrees Celsius of the gas or vapor involved.

*Exception: Conduit entering an enclosure where such switches, circuit breakers, fuses, relays, or resistors comply with one of the following:*

*(1) Are enclosed within a chamber hermetically sealed against the entrance of gases or vapors.*
*(2) Are immersed in oil.*
*(3) Are enclosed within a factory-sealed explosionproof chamber located within the enclosure, identified for the location, and marked "factory sealed" or equivalent, unless the entry is metric designator 53 (trade size 2) or larger. Factory-sealed enclosures shall not be considered to serve as a seal for another adjacent explosionproof enclosure that is required to have a conduit seal.*

(b) Conduit seals shall be installed within 450 mm (18 in.) from the enclosure. Only explosionproof unions, couplings, reducers, elbows, capped elbows, and conduit bodies similar to L, T, and cross types that are not larger than the trade size of the conduit shall be permitted between the sealing fitting and the explosionproof enclosure.

(c) Where two or more explosionproof enclosures for which conduit seals are required under 505.16(B)(2) are connected by nipples or by runs of conduit not more than 900 mm (36 in.) long, a single conduit seal in each such nipple connection or run of conduit shall be considered sufficient if located not more than 450 mm (18 in.) from either enclosure.

**(3) Pressurized Enclosures.** Conduit seals shall be provided in each conduit entry into a pressurized enclosure where the conduit is not pressurized as part of the protection system. Conduit seals shall be installed within 450 mm (18 in.) from the pressurized enclosure.

Informational Note No. 1: Installing the seal as close as possible to the enclosure reduces problems with purging the dead airspace in the pressurized conduit.

Informational Note No. 2: For further information, see NFPA 496-2013, *Standard for Purged and Pressurized Enclosures for Electrical Equipment.*

**(4) Class I, Zone 1 Boundary.** Conduit seals shall be provided in each conduit run leaving a Class I, Zone 1 location. The sealing fitting shall be permitted on either side of the boundary of such location within 3.05 m (10 ft) of the boundary and shall be designed and installed so as to minimize the amount of gas or vapor within the Zone 1 portion of the conduit from being communicated to the conduit beyond the seal. Except for listed explosionproof reducers at the conduit seal, there shall be no union, coupling, box, or fitting between the conduit seal and the point at which the conduit leaves the Zone 1 location.

*Exception: Metal conduit containing no unions, couplings, boxes, or fittings and passing completely through a Class I, Zone 1 location with no fittings less than 300 mm (12 in.) beyond each boundary shall not require a conduit seal if the termination points of the unbroken conduit are in unclassified locations.*

**(5) Cables Capable of Transmitting Gases or Vapors.** Conduits containing cables with a gas/vaportight continuous sheath capable of transmitting gases or vapors through the cable core shall be sealed in the Zone 1 location after removing the jacket and any other coverings so that the sealing compound surrounds each individual insulated conductor or optical fiber tube and the outer jacket.

*Exception: Multiconductor cables with a gas/vaportight continuous sheath capable of transmitting gases or vapors through the cable core shall be permitted to be considered as a single conductor by sealing the cable in the conduit within 450 mm (18 in.) of the enclosure and the cable end within the enclosure by an approved means to minimize the entrance of gases or vapors and prevent the propagation of flame into the cable core, or by other approved methods. For shielded cables and twisted pair cables, it shall not be required to remove the shielding material or separate the twisted pair.*

**(6) Cables Incapable of Transmitting Gases or Vapors.** Each multiconductor or optical multifiber cable in conduit shall be considered as a single conductor or single optical fiber tube if the cable is incapable of transmitting gases or vapors through the cable core. These cables shall be sealed in accordance with 505.16(D).

**(7) Cables Entering Enclosures.** Cable seals shall be provided for each cable entering flameproof or explosionproof enclosures. The seal shall comply with 505.16(D).

**(8) Class I, Zone 1 Boundary.** Cables shall be sealed at the point at which they leave the Zone 1 location.

*Exception: Where cable is sealed at the termination point.*

**(C) Zone 2.** In Class I, Zone 2 locations, seals shall be located in accordance with 505.16(C)(1) and (C)(2).

**(1) Conduit Seals.** Conduit seals shall be located in accordance with (C)(1)(a) and (C)(1)(b).

(a) For connections to enclosures that are required to be flameproof or explosionproof, a conduit seal shall be provided in accordance with 505.16(B)(1) and (B)(2). All portions of the conduit run or nipple between the seal and enclosure shall comply with 505.16(B).

(b) In each conduit run passing from a Class I, Zone 2 location into an unclassified location. The sealing fitting shall be permitted on either side of the boundary of such location within 3.05 m (10 ft) of the boundary and shall be designed and installed so as to minimize the amount of gas or vapor within the Zone 2 portion of the conduit from being communicated to the conduit beyond the seal. Rigid metal conduit or threaded steel intermediate metal conduit shall be used between the sealing fitting and the point at which the conduit leaves the Zone 2 location, and a threaded connection shall be used at the sealing fitting. Except for listed explosionproof reducers at the conduit seal, there shall be no union, coupling, box, or fitting between the conduit seal and the point at which the conduit leaves the Zone 2 location. Conduits shall be sealed to minimize the amount of gas or vapor within the Class I, Zone 2 portion of the conduit from being communicated to the conduit beyond the seal. Such seals shall not be required to be flameproof or explosionproof but shall be identified for the purpose of minimizing passage of gases under normal operating conditions and shall be accessible.

*Exception No. 1: Metal conduit containing no unions, couplings, boxes, or fittings and passing completely through a Class I, Zone 2 location with no fittings less than 300 mm (12 in.) beyond each boundary shall not be required to be sealed if the termination points of the unbroken conduit are in unclassified locations.*

*Exception No. 2: Conduit systems terminating at an unclassified location where a wiring method transition is made to cable tray, cablebus, ventilated busway, Type MI cable, or cable that is not installed in a raceway or cable tray system shall not be required to be sealed where passing from the Class I, Zone 2 location into the unclassified location. The unclassified location shall be outdoors or, if the conduit system is all in one room, it shall be permitted to be indoors. The conduits shall not terminate at an enclosure containing an ignition source in normal operation.*

*Exception No. 3: Conduit systems passing from an enclosure or room that is unclassified as a result of pressurization into a Class I, Zone 2 location shall not require a seal at the boundary.*

Informational Note: For further information, refer to NFPA 496-2013, *Standard for Purged and Pressurized Enclosures for Electrical Equipment.*

*Exception No. 4: Segments of aboveground conduit systems shall not be required to be sealed where passing from a Class I, Zone 2 location into an unclassified location if all the following conditions are met:*

*(1) No part of the conduit system segment passes through a Zone 0 or Zone 1 location where the conduit contains unions, couplings, boxes, or fittings within 300 mm (12 in.) of the Zone 0 or Zone 1 location.*
*(2) The conduit system segment is located entirely in outdoor locations.*
*(3) The conduit system segment is not directly connected to canned pumps, process or service connections for flow, pressure, or analysis measurement, and so forth, that depend on a single compression seal, diaphragm, or tube to prevent flammable or combustible fluids from entering the conduit system.*
*(4) The conduit system segment contains only threaded metal conduit, unions, couplings, conduit bodies, and fittings in the unclassified location.*
*(5) The conduit system segment is sealed at its entry to each enclosure or fitting housing terminals, splices, or taps in Zone 2 locations.*

**(2) Cable Seals.** Cable seals shall be located in accordance with (C)(2)(a), (C)(2)(b), and (C)(2)(c).

(a) *Explosionproof and Flameproof Enclosures.* Cables entering enclosures required to be flameproof or explosionproof shall be sealed at the point of entrance. The seal shall comply with 505.16(D). Multiconductor or optical multifiber cables with a gas/vaportight continuous sheath capable of transmitting gases or vapors through the cable core shall be sealed in the Zone 2 location after removing the jacket and any other coverings so that the sealing compound surrounds each individual insulated conductor or optical fiber tube in such a manner as to minimize the passage of gases and vapors. Multiconductor or optical multifiber cables in conduit shall be sealed as described in 505.16(B)(4).

*Exception No. 1: Cables passing from an enclosure or room that is unclassified as a result of Type Z pressurization into a Zone 2 location shall not require a seal at the boundary.*

*Exception No. 2: Shielded cables and twisted pair cables shall not require the removal of the shielding material or separation of the twisted pairs, provided the termination is by an approved means to minimize the entrance of gases or vapors and prevent propagation of flame into the cable core.*

(b) *Cables That Will Not Transmit Gases or Vapors.* Cables with a gas/vaportight continuous sheath and that will not transmit gases or vapors through the cable core in excess of the quantity permitted for seal fittings shall not be required to be sealed except as required in 505.16(C)(2)(a). The minimum length of such cable run shall not be less than the length that limits gas or vapor flow through the cable core to the rate permitted for seal fittings [200 $cm^3$/hr (0.007 $ft^3$/hr) of air at a pressure of 1500 pascals (6 in. of water)].

Informational Note No. 1: For further information on construction, testing, and marking of cables, cable fittings, and cord connectors, see ANSI/UL 2225-2011, *Cables and Cable-Fittings for Use in Hazardous (Classified) Locations*.

Informational Note No. 2: The cable core does not include the interstices of the conductor strands.

(c) *Cables Capable of Transmitting Gases or Vapors.* Cables with a gas/vaportight continuous sheath capable of transmitting gases or vapors through the cable core shall not be required to be sealed except as required in 505.16(C)(2)(a), unless the cable is attached to process equipment or devices that may cause a pressure in excess of 1500 pascals (6 in. of water) to be exerted at a cable end, in which case a seal, barrier, or other means shall be provided to prevent migration of flammables into an unclassified area.

*Exception: Cables with an unbroken gas/vaportight continuous sheath shall be permitted to pass through a Class I, Zone 2 location without seals.*

(d) *Cables Without Gas/Vaportight Continuous Sheath.* Cables that do not have gas/vaportight continuous sheath shall be sealed at the boundary of the Zone 2 and unclassified location in such a manner as to minimize the passage of gases or vapors into an unclassified location.

Informational Note: The cable sheath may be either metal or a nonmetallic material.

**(D) Class I, Zones 0, 1, and 2.** Where required, seals in Class I, Zones 0, 1, and 2 locations shall comply with 505.16(D)(1) through (D)(5).

**(1) Fittings.** Enclosures for connections or equipment shall be provided with an integral means for sealing, or sealing fittings listed for the location shall be used. Sealing fittings shall be listed for use with one or more specific compounds and shall be accessible.

**(2) Compound.** The compound shall provide a seal against passage of gas or vapors through the seal fitting, shall not be affected by the surrounding atmosphere or liquids, and shall not have a melting point less than 93°C (200°F).

**(3) Thickness of Compounds.** In a completed seal, the minimum thickness of the sealing compound shall not be less than the trade size of the sealing fitting and, in no case, less than 16 mm (⅝ in.).

*Exception: Listed cable sealing fittings shall not be required to have a minimum thickness equal to the trade size of the fitting.*

**(4) Splices and Taps.** Splices and taps shall not be made in fittings intended only for sealing with compound, nor shall other fittings in which splices or taps are made be filled with compound.

**(5) Conductor or Optical Fiber Fill.** The cross-sectional area of the conductors or optical fiber tubes (metallic or nonmetallic) permitted in a seal shall not exceed 25 percent of the cross-sectional area of a rigid metal conduit of the same trade size unless it is specifically listed for a higher percentage of fill.

**(E) Drainage.**

**(1) Control Equipment.** Where there is a probability that liquid or other condensed vapor may be trapped within enclosures for control equipment or at any point in the raceway system, approved means shall be provided to prevent accumulation or to permit periodic draining of such liquid or condensed vapor.

**(2) Motors and Generators.** Where liquid or condensed vapor may accumulate within motors or generators, joints and conduit systems shall be arranged to minimize entrance of liquid. If means to prevent accumulation or to permit periodic draining are necessary, such means shall be provided at the time of manufacture and shall be considered an integral part of the machine.

## 505.17 Flexible Cords and Connections

**(A) Flexible Cords, Class I, Zones 1 and 2.** A flexible cord shall be permitted for connection between portable lighting equipment or other portable utilization equipment and the fixed portion of their supply circuit. Flexible cord shall also be permitted for that portion of the circuit where the fixed wiring methods of 505.15(B) and (C) cannot provide the necessary degree of movement for fixed and mobile electrical utilization equipment in an industrial establishment where conditions of maintenance and engineering supervision ensure that only qualified persons install and service the installation, and where the flexible cord is protected by location or by a suitable guard from damage. The length of the flexible cord shall be continuous. Where flexible cords are used, the cords shall comply with the following:

(1) Be of a type listed for extra-hard usage
(2) Contain, in addition to the conductors of the circuit, an equipment grounding conductor complying with 400.23
(3) Be connected to terminals or to supply conductors in an approved manner
(4) Be supported by clamps or by other suitable means in such a manner that there will be no tension on the terminal connections
(5) Be terminated with a listed cord connector that maintains the type of protection where the flexible cord enters boxes, fittings, or enclosures that are required to be explosionproof or flameproof
(6) Cord entering an increased safety "e" enclosure shall be terminated with a listed increased safety "e" cord connector.

Informational Note: See 400.7 for permitted uses of flexible cords.

Electric submersible pumps with means for removal without entering the wet-pit shall be considered portable utilization equipment. The extension of the flexible cord within a suitable raceway between the wet-pit and the power source shall be permitted.

Electric mixers intended for travel into and out of open-type mixing tanks or vats shall be considered portable utilization equipment.

Informational Note: See 505.18 for flexible cords exposed to liquids having a deleterious effect on the conductor insulation.

**(B) Instrumentation Connections for Zone 2.** To facilitate replacements, process control instruments shall be permitted to be connected through flexible cords, attachment plugs, and receptacles, provided that all of the following conditions apply:

(1) A switch listed for Zone 2 is provided so that the attachment plug is not depended on to interrupt current, unless the circuit is type "ia," "ib," or "ic" protection, in which case the switch is not required.
(2) The current does not exceed 3 amperes at 120 volts, nominal.
(3) The power-supply cord does not exceed 900 mm (3 ft), is of a type listed for extra-hard usage or for hard usage if protected by location, and is supplied through an attachment plug and receptacle of the locking and grounding type.
(4) Only necessary receptacles are provided.
(5) The receptacle carries a label warning against unplugging under load.

## 505.18 Conductors and Conductor Insulation

**(A) Conductors.** For type of protection "e," field wiring conductors shall be copper. Every conductor (including spares) that enters Type "e" equipment shall be terminated at a Type "e" terminal.

**(B) Conductor Insulation.** Where condensed vapors or liquids may collect on, or come in contact with, the insulation on conductors, such insulation shall be of a type identified for use under such conditions, or the insulation shall be protected by a sheath of lead or by other approved means.

## 505.19 Uninsulated Exposed Parts

There shall be no uninsulated exposed parts, such as electrical conductors, buses, terminals, or components that operate at more than 30 volts (15 volts in wet locations). These parts shall additionally be protected by type of protection "ia," "ib," or "nA" that is suitable for the location.

## 505.20 Equipment Requirements

**(A) Zone 0.** In Class I, Zone 0 locations, only equipment specifically listed and marked as suitable for the location shall be permitted.

*Exception: Intrinsically safe apparatus listed for use in Class I, Division 1 locations for the same gas, or as permitted by 505.9(B)(2), and with a suitable temperature class shall be permitted.*

**(B) Zone 1.** In Class I, Zone 1 locations, only equipment specifically listed and marked as suitable for the location shall be permitted.

*Exception No. 1: Equipment identified for use in Class I, Division 1 or listed for use in Zone 0 locations for the same gas, or as permitted by 505.9(B)(2), and with a suitable temperature class shall be permitted.*

*Exception No. 2: Equipment identified for Class I, Zone 1 or Zone 2 type of protection "p" shall be permitted.*

**(C) Zone 2.** In Class I, Zone 2 locations, only equipment specifically listed and marked as suitable for the location shall be permitted.

*Exception No. 1: Equipment listed for use in Zone 0 or Zone 1 locations for the same gas, or as permitted by 505.9(B)(2), and with a suitable temperature class, shall be permitted.*

*Exception No. 2: Equipment identified for Class I, Zone 1 or Zone 2 type of protection "p" shall be permitted.*

*Exception No. 3: Equipment identified for use in Class I, Division 1 or Division 2 locations for the same gas, or as permitted by 505.9(B)(2), and with a suitable temperature class shall be permitted.*

*Exception No. 4: In Class I, Zone 2 locations, the installation of open or nonexplosionproof or nonflameproof enclosed motors, such as squirrel-cage induction motors without brushes, switching mechanisms, or similar arc-producing devices that are not identified for use in a Class I, Zone 2 location shall be permitted.*

Informational Note No. 1: It is important to consider the temperature of internal and external surfaces that may be exposed to the flammable atmosphere.

Informational Note No. 2: It is important to consider the risk of ignition due to currents arcing across discontinuities and overheating of parts in multisection enclosures of large motors and generators. Such motors and generators may need equipotential bonding jumpers across joints in the enclosure and from enclosure to ground. Where the presence of ignitible gases or vapors is suspected, clean air purging may be needed immediately prior to and during start-up periods.

Informational Note No. 3: For further information on the application of electric motors in Class I, Zone 2 hazardous (classified) locations, see IEEE 1349-2011, *IEEE Guide for the Application of Electric Motors in Class I, Division 2 and Class I, Zone 2 Hazardous (Classified) Locations*.

Informational Notes 1 and 2 identify unique issues that apply to motors in Class I, Zone 2 locations. High-inertia loads can cause increased rotor heating during starting. In addition, sparking can occur between motor housing assemblies when starting. See Exhibits 501.18 and 501.19 and the accompanying commentary for more information on electric motors installed in hazardous locations.

**(D) Materials.** Equipment marked Group IIC shall be permitted for applications requiring Group IIA or Group IIB equipment.

Similarly, equipment marked Group IIB shall be permitted for applications requiring Group IIA equipment.

Equipment marked for a specific gas or vapor shall be permitted for applications where the specific gas or vapor may be encountered.

Informational Note: One common example combines these markings with equipment marked IIB +H2. This equipment is suitable for applications requiring Group IIA equipment, Group IIB equipment, or equipment for hydrogen atmospheres.

**(E) Manufacturer's Instructions.** Electrical equipment installed in hazardous (classified) locations shall be installed in accordance with the instructions (if any) provided by the manufacturer.

•

## 505.22 Increased Safety "e" Motors and Generators

In Class I, Zone 1 locations, Increased Safety "e" motors and generators of all voltage ratings shall be listed for Zone 1 locations, and shall comply with all of the following:

(1) Motors shall be marked with the current ratio, *IA/IN*, and time, *tE*.
(2) Motors shall have controllers marked with the model or identification number, output rating (horsepower or kilowatt), full-load amperes, starting current ratio (*IA/IN*), and time (*tE*) of the motors that they are intended to protect; the controller marking shall also include the specific overload protection type (and setting, if applicable) that is listed with the motor or generator.
(3) Connections shall be made with the specific terminals listed with the motor or generator.
(4) Terminal housings shall be permitted to be of substantial, nonmetallic, nonburning material, provided an internal grounding means between the motor frame and the equipment grounding connection is incorporated within the housing.
(5) The provisions of Part III of Article 430 shall apply regardless of the voltage rating of the motor.
(6) The motors shall be protected against overload by a separate overload device that is responsive to motor current. This device shall be selected to trip or shall be rated in accordance with the listing of the motor and its overload protection.
(7) Sections 430.32(C) and 430.44 shall not apply to such motors.
(8) The motor overload protection shall not be shunted or cut out during the starting period.

Informational Note: Reciprocating engine-driven generators, compressors, and other equipment installed in Class I, Zone 2 locations may present a risk of ignition of flammable materials associated with fuel, starting, compression, and so forth, due to inadvertent release or equipment malfunction by the engine ignition system and controls. For further information on the requirements for ignition systems for reciprocating engines installed in Class I, Zone 2 hazardous (classified) locations, see ANSI/ISA-12.20.01-2009, *General Requirements for Electrical Ignition Systems for Internal Combustion Engines in Class I, Division 2 or Zone 2, Hazardous (Classified) Locations.*

## 505.25 Grounding and Bonding

Regardless of the voltage of the electrical system, grounding and bonding shall comply with Article 250 and the requirements in 505.25(A) and (B).

**(A) Bonding.** The locknut-bushing and double-locknut types of contacts shall not be depended on for bonding purposes, but bonding jumpers with proper fittings or other approved means of bonding shall be used. Such means of bonding shall apply to all intervening raceways, fittings, boxes, enclosures, and so forth, between Class I locations and the point of grounding for service equipment or point of grounding of a separately derived system.

*Exception: The specific bonding means shall be required only to the nearest point where the grounded circuit conductor and the grounding electrode are connected together on the line side of the building or structure disconnecting means as specified in 250.32(B), provided the branch-circuit overcurrent protection is located on the load side of the disconnecting means.*

•

**(B) Types of Equipment Grounding Conductors.** Flexible metal conduit and liquidtight flexible metal conduit shall include an equipment bonding jumper of the wire type in compliance with 250.102.

*Exception: In Class I, Zone 2 locations, the bonding jumper shall be permitted to be deleted where all of the following conditions are met:*

*(a) Listed liquidtight flexible metal conduit 1.8 m (6 ft) or less in length, with fittings listed for grounding, is used.*
*(b) Overcurrent protection in the circuit is limited to 10 amperes or less.*
*(c) The load is not a power utilization load.*

The grounding and bonding requirements for the zone classification system are identical to those for the division classification system. For information on grounding and bonding requirements in Class I locations, see the commentary following 501.30(A) and (B).

## 505.26 Process Sealing

This section shall apply to process-connected equipment, which includes, but is not limited to, canned pumps, submersible pumps, flow, pressure, temperature, or analysis measurement instruments. A process seal is a device to prevent the migration of process fluids from the designed containment into the external electrical system. Process connected electrical equipment that incorporates a single process seal, such as a single compression seal, diaphragm, or tube to prevent flammable or combustible fluids from entering a conduit or cable system capable of transmitting fluids, shall be provided with an additional means to mitigate a single process seal failure. The additional means may include, but is not limited to the following:

(1) A suitable barrier meeting the process temperature and pressure conditions that the barrier is subjected to upon failure of the single process seal. There shall be a vent or drain between the single process seal and the suitable barrier. Indication of the single process seal failure shall be provided by visible leakage, an audible whistle, or other means of monitoring.
(2) A listed Type MI cable assembly, rated at not less than 125 percent of the process pressure and not less than 125 percent of the maximum process temperature (in degrees Celsius), installed between the cable or conduit and the single process seal.
(3) A drain or vent located between the single process seal and a conduit or cable seal. The drain or vent shall be sufficiently sized to prevent overpressuring the conduit or cable seal above 6 in. water column (1493 Pa). Indication of the single process seal failure shall be provided by visible leakage, an audible whistle, or other means of monitoring.
(4) An add-on secondary seal marked "secondary seal" and rated for the pressure and temperature conditions to which it will be subjected upon failure of the single process seal.

Process-connected electrical equipment that does not rely on a single process seal or is listed and marked "single seal" or "dual seal" shall not be required to be provided with an additional means of sealing.

Informational Note: For construction and testing requirements for process sealing for listed and marked single seal, dual seal, or secondary seal equipment, refer to ANSI/ISA-12.27.01-2011, *Requirements for Process Sealing Between Electrical Systems and Flammable or Combustible Process Fluids.*

The requirements for process sealing clarify the sealing, venting, and primary seal failure indication methods for the additional process seal. See the commentary following 501.17 for further information on process sealing.

# ARTICLE 506
# Zone 20, 21, and 22 Locations for Combustible Dusts or Ignitible Fibers/Flyings

Informational Note: Text that is followed by a reference in brackets has been extracted from NFPA 499-2013, *Recommended Practice for the Classification of Combustible Dusts and of Hazardous (Classified) Locations for Electrical Installation in Chemical Process Areas.* Only editorial changes were made to the extracted text to make it consistent with this *Code.*

## 506.1 Scope

This article covers the requirements for the zone classification system as an alternative to the division classification system covered in Article 500, Article 502, and Article 503 for electrical and electronic equipment and wiring for all voltages in Zone 20, Zone 21, and Zone 22 hazardous (classified) locations where fire and explosion hazards may exist due to combustible dusts or ignitible fibers/flyings.

Just as in the division hazardous location classification scheme, the zone classification scheme also addresses combustible dusts and ignitible fibers/flyings. Zones 20, 21, and 22 are analogous to both Class II and III, Division 1 and 2 hazardous locations in Articles 502 and 503.

Hazardous locations containing combustible dusts or ignitible fibers/flyings are not subdivided in classes of material under the zone method. Similar to Class II, Division 1 and 2 locations, combustible dusts for zone classifications also include group designations. Group IIIC for combustible metal dusts is equivalent to Group E, and Group IIIB for all other combustible dusts is equivalent to Groups F and G. However, unlike Class III locations, the zone classification includes Group IIIA for fibers/flyings and is equivalent to Class III.

Informational Note No. 1: For the requirements for electrical and electronic equipment and wiring for all voltages in Class I, Division 1 or Division 2; Class II, Division 1 or Division 2; Class III, Division 1 or Division 2; and Class I, Zone 0 or Zone 1 or Zone 2 hazardous (classified) locations where fire or explosion hazards may exist due to flammable gases or vapors, flammable liquids, or combustible dusts or fibers, refer to Articles 500 through 505.

Informational Note No. 2: Zone 20, Zone 21, and Zone 22 area classifications are based on the modified IEC area classification system as defined in ANSI/ISA-61241-10 (12.10.05)-2004, *Electrical Apparatus for Use in Zone 20, Zone 21, and Zone 22 Hazardous (Classified) Locations — Classification of Zone 20, Zone 21, and Zone 22 Hazardous (Classified) Locations.*

Informational Note No. 3: The unique hazards associated with explosives, pyrotechnics, and blasting agents are not addressed in this article.

## 506.2 Definitions

For purposes of this article, the following definitions apply.

**Associated Nonincendive Field Wiring Apparatus.** Apparatus in which the circuits are not necessarily nonincendive themselves but that affect the energy in nonincendive field wiring circuits and are relied upon to maintain nonincendive energy levels. Such apparatus are one of the following:

(1) Electrical apparatus that has an alternative type of protection for use in the appropriate hazardous (classified) location
(2) Electrical apparatus not so protected that shall not be used in a hazardous (classified) location

Informational Note: Associated nonincendive field wiring apparatus has designated associated nonincendive field wiring apparatus connections for nonincendive field wiring apparatus and may also have connections for other electrical apparatus.

**Combustible Dust.** Dust particles that are 500 microns or smaller (material passing a U.S. No. 35 Standard Sieve as defined in ASTM E 11-09, Standard Specification for Wire Cloth and Sieves for Testing Purposes) and present a fire or explosion hazard when dispersed and ignited in air.

Informational Note: See ASTM E 1226–12a, *Standard Test Method for Explosibility of Dust Clouds*, or ISO 6184-1, *Explosion protection systems — Part 1: Determination of explosion indices of combustible dusts in air*, for procedures for determining the explosibility of dusts.

**Dust-Ignitionproof.** Equipment enclosed in a manner that excludes dusts and does not permit arcs, sparks, or heat otherwise generated or liberated inside of the enclosure to cause ignition of exterior accumulations or atmospheric suspensions of a specified dust on or in the vicinity of the enclosure.

Informational Note: For further information on dust-ignitionproof enclosures, see Type 9 enclosure in ANSI/NEMA 250-2008, *Enclosures for Electrical Equipment*, and ANSI/UL 1203-2009, *Explosionproof and Dust-Ignitionproof Electrical Equipment for Hazardous (Classified) Locations.*

**Dusttight.** Enclosures constructed so that dust will not enter under specified test conditions.

Informational Note: See ANSI/ISA-12.12.01-2012, *Nonincendive Electrical Equipment for Use in Class I and II, Division 2, and Class III, Divisions 1 and 2 Hazardous (Classified) Locations.*

**Nonincendive Circuit.** A circuit, other than field wiring, in which any arc or thermal effect produced under intended operating conditions of the equipment is not capable, under specified test conditions, of igniting the flammable gas–air, vapor–air, or dust–air mixture.

Informational Note: Conditions are described in ANSI/ISA-12.12.01-2012, *Nonincendive Electrical Equipment for Use in Class I and II, Division 2, and Class III, Divisions 1 and 2 Hazardous (Classified) Locations.*

**Nonincendive Equipment.** Equipment having electrical/electronic circuitry that is incapable, under normal operating conditions, of causing ignition of a specified flammable gas–air, vapor–air, or dust–air mixture due to arcing or thermal means.

Informational Note: For further information, see ANSI/ISA-12.12.01-2012, *Nonincendive Electrical Equipment for Use in Class I and II, Division 2, and Class III, Divisions 1 and 2 Hazardous (Classified) Locations.*

**Nonincendive Field Wiring.** Wiring that enters or leaves an equipment enclosure and, under normal operating conditions of the equipment, is not capable, due to arcing or thermal effects, of igniting the flammable gas–air, vapor–air, or dust–air mixture. Normal operation includes opening, shorting, or grounding the field wiring.

**Nonincendive Field Wiring Apparatus.** Apparatus intended to be connected to nonincendive field wiring.

Informational Note: For further information, see ANSI/ISA-12.12.01-2012, *Nonincendive Electrical Equipment for Use in Class I and II, Division 2, and Class III, Divisions 1 and 2 Hazardous (Classified) Locations.*

**Pressurized.** The process of supplying an enclosure with a protective gas with or without continuous flow, at sufficient pressure to prevent the entrance of combustible dust or ignitible fibers/flyings.

Informational Note: Informational Note: For further information, see ANSI/NFPA 496-2013, *Standard for Purged and Pressurized Enclosures for Electrical Equipment.*

**Protection by Encapsulation "m."** Type of protection where electrical parts that could cause ignition of a mixture of combustible dust or fibers/flyings in air are protected by enclosing them in a compound in such a way that the explosive atmosphere cannot be ignited.

Informational Note No. 1: For additional information, see ANSI/ISA-60079-18 (12.23.01)-2009, *Explosive atmospheres — Part 18: Equipment protection by encapsulation "m"*; ANSI/UL 60079-18-2009, *Explosive atmospheres — Part 18: Equipment protection by encapsulation "m"*; and ANSI/ISA-61241-18 (12.10.07)-2011, *Electrical Apparatus for Use in Zone 20, Zone 21 and Zone 22 Hazardous (Classified) Locations — Protection by Encapsulation "m."*

Informational Note No. 2: Encapsulation is designated level of protection "maD" or "ma" for use in Zone 20 locations. Encapsulation is designated level of protection "mbD" or "mb" for use in Zone 21 locations. Encapsulation is designated type of protection "mc" for use in Zone 22 locations.

**Protection by Enclosure "t."** Type of protection for explosive dust atmospheres where electrical apparatus is provided with an enclosure providing dust ingress protection and a means to limit surface temperatures.

Informational Note No. 1: For additional information, see ANSI/ISA-60079-31 (12.10.03)-2009, *Explosive Atmospheres — Part 31: Equipment Dust Ignition Protection by Enclosure "t"*; and ANSI/ISA-61241-1 (12.10.03)-2011, *Electrical Apparatus for Use in Zone 21 and Zone 22 Hazardous (Classified) Locations — Protection by Enclosure "t."*

Informational Note No. 2: Protection by enclosure is designated level of protection "ta" for use in Zone 20 locations. Protection by enclosure is designated level of protection "tb" or "tD" for use in Zone 21 locations. Protection by enclosure is designated level of protection "tc" or "tD" for use in Zone 22 locations.

**Protection by Intrinsic Safety "iD."** Type of protection where any spark or thermal effect is incapable of causing ignition of a mixture of combustible dust, fibers, or flyings in air under prescribed test conditions.

Informational Note No. 1: For additional information, see ANSI/ISA-60079-11 (12.01.01)-2011, *Electrical Apparatus for Explosive Gas Atmospheres — Part 11: intrinsic safety "i"*; ANSI/UL 60079-11-2011, *Electrical Apparatus for Explosive Gas Atmospheres — Part 11: Intrinsic safety "i"*; and ANSI/ISA-61241-11 (12.10.04)-2011, *Electrical Apparatus for Use in Zone 20, Zone 21 and Zone 22 Hazardous (Classified) Locations — Protection by Intrinsic Safety "i."*

Informational Note No. 2: Intrinsic safety is designated level of protection "iaD" or "ia" for use in Zone 20 locations. Intrinsic safety is designated level of protection "ibD" or "ib" for use in Zone 21 locations. Intrinsic safety is designated type of protection "ic" for use in Zone 22 locations.

**Protection by Pressurization "pD."** Type of protection that guards against the ingress of a mixture of combustible dust or fibers/flyings in air into an enclosure containing electrical equipment by providing and maintaining a protective gas atmosphere inside the enclosure at a pressure above that of the external atmosphere.

Informational Note: For additional information, see ANSI/ISA-61241-2 (12.10.06)-2006, *Electrical Apparatus for Use in Zone 21 and Zone 22 Hazardous (Classified) Locations — Protection by Pressurization "pD."*

**Zone 20 Hazardous (Classified) Location.** An area where combustible dust or ignitible fibers/flyings are present continuously or for long periods of time in quantities sufficient to be hazardous, as classified by 506.5(B)(1).

**Zone 21 Hazardous (Classified) Location.** An area where combustible dust or ignitible fibers/flyings are likely to exist occasionally under normal operation in quantities sufficient to be hazardous, as classified by 506.5(B)(2).

**Zone 22 Hazardous (Classified) Location.** An area where combustible dust or ignitible fibers/flyings are not likely to occur under normal operation in quantities sufficient to be hazardous, as classified by 506.5(B)(3).

## 506.3 Other Articles

All other applicable rules contained in this Code shall apply to electrical equipment and wiring installed in hazardous (classified) locations.

*Exception: As modified by Article 504 and this article.*

## 506.4 General

**(A) Documentation for Industrial Occupancies.** Areas designated as hazardous (classified) locations shall be properly documented. This documentation shall be available to those authorized to design, install, inspect, maintain, or operate electrical equipment.

**(B) Reference Standards.** Important information relating to topics covered in Chapter 5 are found in other publications.

Informational Note: It is important that the authority having jurisdiction be familiar with the recorded industrial experience as well as with standards of the National Fire Protection Association (NFPA), the International Society of Automation (ISA), and the International Electrotechnical Commission (IEC) that may be of use in the classification of various locations, the determination of adequate ventilation, and the protection against static electricity and lightning hazards.

## 506.5 Classification of Locations

**(A) Classifications of Locations.** Locations shall be classified on the basis of the properties of the combustible dust or ignitible fibers/flyings that may be present, and the likelihood that a combustible or combustible concentration or quantity is present. Each room, section, or area shall be considered individually in determining its classification. Where pyrophoric materials are the only materials used or handled, these locations are outside of the scope of this article.

**(B) Zone 20, Zone 21, and Zone 22 Locations.** Zone 20, Zone 21, and Zone 22 locations are those in which combustible dust or ignitible fibers/flyings are or may be present in the air or in layers, in quantities sufficient to produce explosive or ignitible mixtures. Zone 20, Zone 21, and Zone 22 locations shall include those specified in 506.5(B)(1), (B)(2), and (B)(3).

Informational Note: Through the exercise of ingenuity in the layout of electrical installations for hazardous (classified) locations, it is frequently possible to locate much of the equipment in a reduced level of classification and, thus, to reduce the amount of special equipment required.

**(1) Zone 20.** A Zone 20 location is a location in which

(a) Ignitible concentrations of combustible dust or ignitible fibers/flyings are present continuously.

(b) Ignitible concentrations of combustible dust or ignitible fibers/flyings are present for long periods of time.

Informational Note No. 1: As a guide to classification of Zone 20 locations, refer to ANSI/ISA-61241-10 (12.10.05)-2004, *Electrical Apparatus for Use in Zone 20, Zone 21, and Zone 22 Hazardous (Classified) Locations — Classification of Zone 20, Zone 21, and Zone 22 Hazardous (Classified) Locations.*

Informational Note No. 2: Zone 20 classification includes locations inside dust containment systems; hoppers, silos, etc., cyclones and filters, dust transport systems, except some parts of belt and chain conveyors, etc.; blenders, mills, dryers, bagging equipment, etc.

**(2) Zone 21.** A Zone 21 location is a location

(a) In which ignitible concentrations of combustible dust or ignitible fibers/flyings are likely to exist occasionally under normal operating conditions; or

(b) In which ignitible concentrations of combustible dust or ignitible fibers/flyings may exist frequently because of repair or maintenance operations or because of leakage; or

(c) In which equipment is operated or processes are carried on, of such a nature that equipment breakdown or faulty operations could result in the release of ignitible concentrations of combustible dust or ignitible fibers/flyings and also cause simultaneous failure of electrical equipment in a mode to cause the electrical equipment to become a source of ignition; or

(d) That is adjacent to a Zone 20 location from which ignitible concentrations of dust or ignitible fibers/flyings could be communicated, unless communication is prevented by adequate positive pressure ventilation from a source of clean air and effective safeguards against ventilation failure are provided.

Informational Note No. 1: As a guide to classification of Zone 21 locations, refer to ANSI/ISA-61241-10 (12.10.05)-2004, *Electrical Apparatus for Use in Zone 20, Zone 21, and Zone 22*

*Hazardous (Classified) Locations — Classification of Zone 20, Zone 21, and Zone 22 Hazardous (Classified) Locations.*

Informational Note No. 2: This classification usually includes locations outside dust containment and in the immediate vicinity of access doors subject to frequent removal or opening for operation purposes when internal combustible mixtures are present; locations outside dust containment in the proximity of filling and emptying points, feed belts, sampling points, truck dump stations, belt dump over points, etc. where no measures are employed to prevent the formation of combustible mixtures; locations outside dust containment where dust accumulates and where due to process operations the dust layer is likely to be disturbed and form combustible mixtures; locations inside dust containment where explosive dust clouds are likely to occur (but neither continuously, nor for long periods, nor frequently) as, for example, silos (if filled and/or emptied only occasionally) and the dirty side of filters if large self-cleaning intervals are occurring.

**(3) Zone 22.** A Zone 22 location is a location

(a) In which ignitible concentrations of combustible dust or ignitible fibers/flyings are not likely to occur in normal operation and, if they do occur, will only persist for a short period; or

(b) In which combustible dust or fibers/flyings are handled, processed, or used but in which the dust or fibers/flyings are normally confined within closed containers of closed systems from which they can escape only as a result of the abnormal operation of the equipment with which the dust or fibers/flyings are handled, processed, or used; or

(c) That is adjacent to a Zone 21 location, from which ignitible concentrations of dust or fibers/flyings could be communicated, unless such communication is prevented by adequate positive pressure ventilation from a source of clean air and effective safeguards against ventilation failure are provided.

Informational Note No. 1: As a guide to classification of Zone 22 locations, refer to ANSI/ISA-61241-10 (12.10.05)-2004, *Electrical Apparatus for Use in Zone 20, Zone 21, and Zone 22 Hazardous (Classified) Locations — Classification of Zone 20, Zone 21, and Zone 22 Hazardous (Classified) Locations.*

Informational Note No. 2: Zone 22 locations usually include outlets from bag filter vents, because in the event of a malfunction there can be emission of combustible mixtures; locations near equipment that has to be opened at infrequent intervals or equipment that from experience can easily form leaks where, due to pressure above atmospheric, dust will blow out; pneumatic equipment, flexible connections that can become damaged, etc.; storage locations for bags containing dusty product, since failure of bags can occur during handling, causing dust leakage; and locations where controllable dust layers are formed that are likely to be raised into explosive dust–air mixtures. Only if the layer is removed by cleaning before hazardous dust–air mixtures can be formed is the area designated unclassified.

Informational Note No. 3: Locations that normally are classified as Zone 21 can fall into Zone 22 when measures are employed to prevent the formation of explosive dust–air mixtures. Such measures include exhaust ventilation. The measures should be used in the vicinity of (bag) filling and emptying points, feed belts, sampling points, truck dump stations, belt dump over points, etc.

## 506.6 Material Groups

For the purposes of testing, approval, and area classification, various air mixtures (not oxygen enriched) shall be grouped as required in 506.6(A), (B), and (C).

**(A) Group IIIC.** Combustible metal dust.

Informational Note: Group IIIC is equivalent to Class II, Group E as described in 500.6(B)(1).

**(B) Group IIIB.** Combustible dust other than combustible metal dust.

Informational Note: Group IIIB is equivalent to Class II, Groups F and G as described in 500.6(B)(2) and 500.6(B)(3), respectively.

**(C) Group IIIA.** Solid particles, including fibers, greater than 500 μm in nominal size, which may be suspended in air and could settle out of the atmosphere under their own weight.

Informational Note No. 1: Group IIIA is equivalent to Class III.

Informational Note No. 2: Examples of flyings include rayon, cotton (including cotton linters and cotton waste), sisal, jute, hemp, cocoa fiber, oakum, and baled waste kapok.

The zone classification system for dusts also includes combustible fibers and flyings. The grouping of material by hazard also differs from the system used in Articles 500, 502, and 503. Commentary Table 506.1 contrasts the classification in the two systems.

## 506.7 Special Precaution

Article 506 requires equipment construction and installation that ensures safe performance under conditions of proper use and maintenance.

Informational Note: It is important that inspection authorities and users exercise more than ordinary care with regard to the installation and maintenance of electrical equipment in hazardous (classified) locations.

**COMMENTARY TABLE 506.1** Comparison of Zone and Division Classification Systems

| Hazard | Zone Classification | Division Classification |
|---|---|---|
| Combustible metal dusts | Group IIIC | Class II, Group E |
| Coal, coke, and other carbonaceous dusts | Group IIIB | Class II, Group F |
| Combustible dusts that are nonmetallic, and not carbonaceous. This includes dusts such as flour, grain, wood, plastic, and chemicals | Group IIIB | Class II, Group G |
| Combustible fibers and flyings | Group IIIA | Class III |

**(A) Implementation of Zone Classification System.** Classification of areas, engineering and design, selection of equipment and wiring methods, installation, and inspection shall be performed by qualified persons.

**(B) Dual Classification.** In instances of areas within the same facility classified separately, Zone 22 locations shall be permitted to abut, but not overlap, Class II or Class III, Division 2 locations. Zone 20 or Zone 21 locations shall not abut Class II or Class III, Division 1 or Division 2 locations.

**(C) Reclassification Permitted.** A Class II or Class III, Division 1 or Division 2 location shall be permitted to be reclassified as a Zone 20, Zone 21, or Zone 22 location, provided that all of the space that is classified because of a single combustible dust or ignitible fiber/flying source is reclassified under the requirements of this article.

**(D) Simultaneous Presence of Flammable Gases and Combustible Dusts or Fibers/Flyings.** Where flammable gases, combustible dusts, or fibers/flyings are or may be present at the same time, the simultaneous presence shall be considered during the selection and installation of the electrical equipment and the wiring methods, including the determination of the safe operating temperature of the electrical equipment.

## 506.8 Protection Techniques

Acceptable protection techniques for electrical and electronic equipment in hazardous (classified) locations shall be as described in 506.8(A) through (J).

**(A) Dust Ignitionproof.** This protection technique shall be permitted for equipment in Zone 20, Zone 21, and Zone 22 locations for which it is identified.

**(B) Pressurized.** This protection technique shall be permitted for equipment in Zone 21 and Zone 22 locations for which it is identified.

**(C) Intrinsic Safety.** This protection technique shall be permitted for equipment in Zone 20, Zone 21, and Zone 22 locations for which it is identified.

**(D) Dusttight.** This protection technique shall be permitted for equipment in Zone 22 locations for which it is identified.

**(E) Protection by Encapsulation "m".** This protection technique shall be permitted for equipment in Zone 20, Zone 21, and Zone 22 locations for which it is identified.

Informational Note: See Table 506.9(C)(2)(3) for the descriptions of subdivisions for encapsulation.

•

**(F) Nonincendive Equipment.** This protection technique shall be permitted for equipment in Zone 22 locations for which it is identified.

**(G) Protection by Enclosure "t".** This protection technique shall be permitted for equipment in Zone 20, Zone 21, and Zone 22 locations for which it is identified.

Informational Note: See Table 506.9(C)(2)(3) for the descriptions of subdivisions for protection by enclosure "t."

**(H) Protection by Pressurization "pD".** This protection technique shall be permitted for equipment in Zone 21 and Zone 22 locations for which it is identified.

**(I) Protection by Intrinsic Safety "iD".** This protection technique shall be permitted for equipment in Zone 20, Zone 21, and Zone 22 locations for which it is listed.

## 506.9 Equipment Requirements

**(A) Suitability.** Suitability of identified equipment shall be determined by one of the following:

(1) Equipment listing or labeling
(2) Evidence of equipment evaluation from a qualified testing laboratory or inspection agency concerned with product evaluation
(3) Evidence acceptable to the authority having jurisdiction such as a manufacturer's self-evaluation or an owner's engineering judgment

Informational Note: Additional documentation for equipment may include certificates demonstrating compliance with applicable equipment standards, indicating special conditions of use, and other pertinent information.

**(B) Listing.**

(1) Equipment that is listed for Zone 20 shall be permitted in a Zone 21 or Zone 22 location of the same dust or ignitible fiber/flying. Equipment that is listed for Zone 21 may be used in a Zone 22 location of the same dust fiber/flying.
(2) Equipment shall be permitted to be listed for a specific dust or ignitible fiber/flying or any specific combination of dusts fibers/flyings.

**(C) Marking.**

**(1) Division Equipment.** Equipment identified for Class II, Division 1 or Class II, Division 2 shall, in addition to being marked in accordance with 500.8(C), be permitted to be marked with all of the following:

(1) Zone 20, 21, or 22 (as applicable)
(2) Material group in accordance with 506.6
(3) Maximum surface temperature in accordance with 506.9(D), marked as a temperature value in degrees C, preceded by "T" and followed by the symbol "°C"

**(2) Zone Equipment.** Equipment meeting one or more of the protection techniques described in 506.8 shall be marked with the following in the order shown:

(1) Zone
(2) Symbol "AEx"

(3) Protection technique(s) in accordance with Table 506.9(C)(2)(3)
(4) Material group in accordance with 506.6
(5) Maximum surface temperature in accordance with 506.9(D), marked as a temperature value in degrees C, preceded by "T" and followed by the symbol "°C"
(6) Ambient temperature marking in accordance with 506.9(D)

Informational Note: The EPL (or equipment protection level) may appear in the product marking. EPLs are designated as G for gas, D for dust, or M for mining, and are then followed by a letter (a, b, or c) to give the user a better understanding as to whether the equipment provides either (a) a "very high," (b) "high," or (c) an "enhanced" level of protection against ignition of an explosive atmosphere. For example, an AEx pb IIIB T165°C motor (which is suitable by protection concept for application in Zone 21) may additionally be marked with an EPL of "Db", AEx p IIIB T165°C Db.

*Exception: Associated apparatus NOT suitable for installation in a hazardous (classified) location shall be required to be marked only with 506.9(C)(2)(2), (3), and (5), but BOTH the symbol AEx in 506.9(C)(2)(2) and the symbol for the type of protection in 506.9(C)(2)(3) shall be enclosed within the same square brackets; for example, [AEx iaD] or [AEx ia] IIIC.*

***TABLE 506.9(C)(2)(3)*** *Types of Protection Designation*

| Designation | Technique | Zone* |
|---|---|---|
| iaD | Protection by intrinsic safety | 20 |
| ia | Protection by intrinsic safety | 20 |
| ibD | Protection by intrinsic safety | 21 |
| ib | Protection by intrinsic safety | 21 |
| ic | Protection by intrinsic safety | 22 |
| [iaD] | Associated apparatus | Unclassified** |
| [ia] | Associated apparatus | Unclassified** |
| [ibD] | Associated apparatus | Unclassified** |
| [ib] | Associated apparatus | Unclassified** |
| [ic] | Associated apparatus | Unclassified** |
| maD | Protection by encapsulation | 20 |
| ma | Protection by encapsulation | 20 |
| mbD | Protection by encapsulation | 21 |
| mb | Protection by encapsulation | 21 |
| mc | Protection by encapsulation | 22 |
| pD | Protection by pressurization | 21 |
| p | Protection by pressurization | 21 |
| pb | Protection by pressurization | 21 |
| tD | Protection by enclosures | 21 |
| ta | Protection by enclosures | 20 |
| tb | Protection by enclosures | 21 |
| tc | Protection by enclosures | 22 |

*Does not address use where a combination of techniques is used.

**Associated apparatus is permitted to be installed in a hazardous (classified) location if suitably protected using another type of protection.

The symbol AEx identifies the equipment as meeting American National Standards. In European Union countries, the symbol is EEx. In the IEC standards, on which American and European standards are based, the symbol is Ex. Only equipment marked AEx has been evaluated for use in electrical systems and hazardous locations covered under the *NEC*.

Informational Note: The "D" suffix on the type of protection designation was employed prior to the introduction of Group IIIA, IIIB, and IIIC; which is now used to distinguish between the type of protection employed for Group II (Gases) or Group III (Dusts).

**(D) Temperature Classifications.** Equipment shall be marked to show the maximum surface temperature referenced to a 40°C ambient, or at the higher marked ambient temperature if the equipment is rated and marked for an ambient temperature of greater than 40°C. For equipment installed in a Zone 20 or Zone 21 location, the operating temperature shall be based on operation of the equipment when blanketed with the maximum amount of dust (or with dust-simulating fibers/flyings) that can accumulate on the equipment. Electrical equipment designed for use in the ambient temperature range between −20°C and +40°C shall require no additional ambient temperature marking. Electrical equipment that is designed for use in a range of ambient temperatures other than −20°C and +40°C is considered to be special; and the ambient temperature range shall then be marked on the equipment, including either the symbol "Ta" or "Tamb" together with the special range of ambient temperatures.

Informational Note: As an example, such a marking might be "−30°C ≤ Ta ≤ +40°C."

*Exception No. 1: Equipment of the non–heat-producing type, such as conduit fittings, shall not be required to have a marked operating temperature.*

*Exception No. 2: Equipment identified for Class II, Division 1 or Class II, Division 2 locations as permitted by 506.20(B) and (C) shall be permitted to be marked in accordance with 500.8(C) and Table 500.8(C).*

**(E) Threading.** The supply connection entry thread form shall be NPT or metric. Conduit and fittings shall be made wrenchtight to prevent sparking when the fault current flows through the conduit system and to ensure the integrity of the conduit system. Equipment provided with threaded entries for field wiring connections shall be installed in accordance with 506.9(E)(1) or (E)(2) and with (E)(3).

**(1) Equipment Provided with Threaded Entries for NPT-Threaded Conduit or Fittings.** For equipment provided with threaded entries for NPT-threaded conduit or fittings, listed conduit fittings or listed cable fittings shall be used. All NPT-threaded conduit and fittings referred to herein shall be threaded with a National (American) Standard Pipe Taper (NPT) thread.

Informational Note: Thread specifications for NPT threads are located in ANSI/ASME B1.20.1-1983, *Pipe Threads, General Purpose (Inch)*.

**(2) Equipment Provided with Threaded Entries for Metric-Threaded Conduit or Fittings.** For equipment with metric-threaded entries, listed conduit fittings or listed cable fittings shall be used. Such entries shall be identified as being metric, or listed adapters to permit connection to conduit or NPT-threaded fittings shall be provided with the equipment and shall be used for connection to conduit or NPT-threaded fittings. Metric-threaded entries shall be made up with at least five threads fully engaged.

**(3) Unused Openings.** All unused openings shall be closed with listed metal close-up plugs. The plug engagement shall comply with 506.9(E)(1) or (E)(2).

**(F) Optical Fiber Cables.** Where an optical fiber cable contains conductors that are capable of carrying current (composite optical fiber cable), the optical fiber cable shall be installed in accordance with the requirements of Articles 506.15 and 506.16.

The requirements for fiber optic cables with conductors capable of carrying current are required to follow the general wiring and sealing methods for Zone 20, 21, or 22 locations.

## 506.15 Wiring Methods

Wiring methods shall maintain the integrity of the protection techniques and shall comply with 506.15(A), (B), or (C).

**(A) Zone 20.** In Zone 20 locations, the following wiring methods shall be permitted.

(1) Threaded rigid metal conduit or threaded steel intermediate metal conduit.
(2) Type MI cable terminated with fittings listed for the location. Type MI cable shall be installed and supported in a manner to avoid tensile stress at the termination fittings.

*Exception No. 1: MI cable and fittings listed for Class II, Division 1 locations shall be permitted to be used.*

*Exception No. 2: Equipment identified as intrinsically safe "iaD" or "ia" shall be permitted to be connected using the wiring methods identified in 504.20.*

(3) In industrial establishments with limited public access, where the conditions of maintenance and supervision ensure that only qualified persons service the installation, Type MC-HL cable listed for use in Zone 20 locations, with a continuous corrugated metallic sheath, an overall jacket of suitable polymeric material, and a separate equipment grounding conductor(s) in accordance with 250.122, and terminated with fittings listed for the application, shall be permitted. Type MC-HL cable shall be installed in accordance with the provisions of Article 330, Part II.

*Exception: Type MC-HL cable and fittings listed for Class II, Division 1 locations shall be permitted to be used.*

(4) In industrial establishments with restricted public access, where the conditions of maintenance and supervision ensure that only qualified persons service the installation, and where the cable is not subject to physical damage, Type ITC-HL cable listed for use in Zone 1 or Class I, Division 1 locations, with a gas/vaportight continuous corrugated metallic sheath and an overall jacket of suitable polymeric material, and terminated with fittings listed for the application. Type ITC-HL cable shall be installed in accordance with the provisions of Article 727.
(5) Fittings and boxes shall be identified for use in Zone 20 locations.

*Exception: Boxes and fittings listed for Class II, Division 1 locations shall be permitted to be used.*

(6) Where necessary to employ flexible connections, liquidtight flexible metal conduit with listed fittings, liquidtight flexible nonmetallic conduit with listed fittings, or flexible cord listed for extra-hard usage and provided with listed fittings shall be used. Where flexible cords are used, they shall also comply with 506.17 and shall be terminated with a listed cord connector that maintains the type of protection of the terminal compartment. Where flexible connections are subject to oil or other corrosive conditions, the insulation of the conductors shall be of a type listed for the condition or shall be protected by means of a suitable sheath.

*Exception No. 1: Flexible conduit and flexible conduit and cord fittings listed for Class II, Division 1 locations shall be permitted to be used.*

*Exception No. 2: For elevator use, an identified elevator cable of Type EO, ETP, or ETT, shown under the "use" column in Table 400.4 for "hazardous (classified) locations," and terminated with listed connectors that maintain the type of protection of the terminal compartment shall be permitted.*

Informational Note No. 1: See 506.25 for grounding requirements where flexible conduit is used.

Informational Note No. 2: For further information on construction, testing, and marking of cables, cable fittings, and cord connectors, see ANSI/UL 2225-2011, *Cables and Cable-Fittings for Use in Hazardous (Classified) Locations*.

(7) Optical fiber cable Types OFNP, OFCP, OFNR, OFCR, OFNG, OFCG, OFN, and OFC shall be permitted to be installed in raceways in accordance with 506.15(A). Optical fiber cables shall be sealed in accordance with 506.16.

**(B) Zone 21.** In Zone 21 locations, the wiring methods in (B)(1) and (B)(2) shall be permitted.

(1) All wiring methods permitted in 506.15(A).
(2) Fittings and boxes that are dusttight, provided with threaded bosses for connection to conduit, in which taps, joints, or terminal connections are not made, and are not used in locations where metal dust is present, may be used.

Informational Note: For further information on construction, testing, and marking of cables, cable fittings, and cord connectors, see ANSI/UL 2225-2011, *Cables and Cable-Fittings for Use in Hazardous (Classified) Locations.*

*Exception: Equipment identified as intrinsically safe "ibD" or "ib" shall be permitted to be connected using the wiring methods identified in 504.20.*

**(C) Zone 22.** In Zone 22 locations, the following wiring methods shall be permitted.

(1) All wiring methods permitted in 506.15(B).
(2) Rigid metal conduit, intermediate metal conduit, electrical metallic tubing, dusttight wireways.
(3) Type MC or MI cable with listed termination fittings.
(4) Type PLTC and Type PLTC-ER cable in accordance with the provisions of Article 725, including installation in cable tray systems. The cable shall be terminated with listed fittings.
(5) Type ITC and Type ITC-ER cable as permitted in 727.4 and terminated with listed fittings.
(6) Type MC, MI, MV, TC, or TC-ER cable installed in ladder, ventilated trough, or ventilated channel cable trays in a single layer, with a space not less than the larger cable diameter between two adjacent cables, shall be the wiring method employed. Single-conductor Type MV cables shall be shielded or metallic armored. The cable shall be terminated with listed fittings.
(7) Intrinsic safety type of protection "ic" shall be permitted using any of the wiring methods permitted for unclassified locations. Intrinsic safety type of protection "ic" systems shall be installed in accordance with the control drawing(s). Simple apparatus, not shown on the control drawing, shall be permitted in a circuit of intrinsic safety type of protection "ic", provided that the simple apparatus does not interconnect the intrinsic safety type of protection "ic" circuit to any other circuit.

Informational Note: The term *Simple Apparatus* is defined in 504.2.

Separation of circuits of intrinsic safety type of protection "ic" shall be in accordance with one of the following:

a. Be in separate cables
b. Be in multiconductor cables where the conductors of each circuit are within a grounded metal shield
c. Be in multiconductor cables where the conductors have insulation with a minimum thickness of 0.25 mm (0.01 in.)

(8) Boxes and fittings shall be dusttight.
(9) Optical fiber cable Types OFNP, OFCP, OFNR, OFCR, OFNG, OFCG, OFN, and OFC shall be permitted to be installed in cable trays or any raceway in accordance with 506.15(C). Optical fiber cables shall be sealed in accordance with 506.16.

## 506.16 Sealing

Where necessary to protect the ingress of combustible dust or ignitible fibers/flyings, or to maintain the type of protection, seals shall be provided. The seal shall be identified as capable of preventing the ingress of combustible dust or ignitible fibers/flyings and maintaining the type of protection but need not be explosionproof or flameproof.

## 506.17 Flexible Cords

Flexible cords used in Zone 20, Zone 21, and Zone 22 locations shall comply with all of the following:

(1) Be of a type listed for extra-hard usage
(2) Contain, in addition to the conductors of the circuit, an equipment grounding conductor complying with 400.23
(3) Be connected to terminals or to supply conductors in an approved manner
(4) Be supported by clamps or by other suitable means in such a manner to minimize tension on the terminal connections
(5) Be terminated with a listed cord connector that maintains the protection technique of the terminal compartment

Informational Note: For further information on construction, testing, and marking of cables, cable fittings, and cord connectors, see ANSI/UL 2225-2011, *Cables and Cable-Fittings for Use in Hazardous (Classified) Locations.*

## 506.20 Equipment Installation

**(A) Zone 20.** In Zone 20 locations, only equipment listed and marked as suitable for the location shall be permitted.

*Exception: Equipment listed for use in Class II, Division 1 locations with a suitable temperature class shall be permitted.*

**(B) Zone 21.** In Zone 21 locations, only equipment listed and marked as suitable for the location shall be permitted.

*Exception No. 1: Apparatus listed for use in Class II, Division 1 locations with a suitable temperature class shall be permitted.*

*Exception No. 2: Pressurized equipment identified for Class II, Division 1 shall be permitted.*

**(C) Zone 22.** In Zone 22 locations, only equipment listed and marked as suitable for the location shall be permitted.

*Exception No. 1: Apparatus listed for use in Class II, Division 1 locations with a suitable temperature class shall be permitted.*

*Exception No. 2: Pressurized equipment identified for Class II, Division 1 or Division 2 shall be permitted.*

**(D) Material Group.** Equipment marked Group IIIC shall be permitted for applications requiring IIIA or IIIB equipment. Similarly, equipment marked Group IIIB shall be permitted for applications requiring IIIA equipment.

**(E) Manufacturer's Instructions.** Electrical equipment installed in hazardous (classified) locations shall be installed in accordance with the instructions (if any) provided by the manufacturer.

**(F) Temperature.** The temperature marking specified in 506.9(C)(2)(5) shall comply with (E)(1) or (E)(2):

(1) For combustible dusts, less than the lower of either the layer or cloud ignition temperature of the specific combustible dust. For organic dusts that may dehydrate or carbonize, the temperature marking shall not exceed the lower of either the ignition temperature or 165°C (329°F).
(2) For ignitible fibers/flyings, less than 165°C (329°F) for equipment that is not subject to overloading, or 120°C (248°F) for equipment (such as motors or power transformers) that may be overloaded.

Informational Note: See NFPA 499-2013, *Recommended Practice for the Classification of Combustible Dusts and of Hazardous (Classified) Locations for Electrical Installations in Chemical Processing Areas*, for minimum ignition temperatures of specific dusts.

•

## 506.25 Grounding and Bonding

Regardless of the voltage of the electrical system, grounding and bonding shall comply with Article 250 and the requirements in 506.25(A) and (B).

**(A) Bonding.** The locknut-bushing and double-locknut types of contacts shall not be depended on for bonding purposes, but bonding jumpers with proper fittings or other approved means of bonding shall be used. Such means of bonding shall apply to all intervening raceways, fittings, boxes, enclosures, and so forth, between Zone 20, Zone 21, and Zone 22 locations and the point of grounding for service equipment or point of grounding of a separately derived system.

*Exception: The specific bonding means shall be required only to the nearest point where the grounded circuit conductor and the grounding electrode conductor are connected together on the line side of the building or structure disconnecting means as specified in 250.32(B) if the branch side overcurrent protection is located on the load side of the disconnecting means.*

•

**(B) Types of Equipment Grounding Conductors.** Liquidtight flexible metal conduit shall include an equipment bonding jumper of the wire type in compliance with 250.102.

*Exception: In Zone 22 locations, the bonding jumper shall be permitted to be deleted where all of the following conditions are met:*

*(1) Listed liquidtight flexible metal conduit 1.8 m (6 ft) or less in length, with fittings listed for grounding, is used.*
*(2) Overcurrent protection in the circuit is limited to 10 amperes or less.*
*(3) The load is not a power utilization load.*

# ARTICLE 510 Hazardous (Classified) Locations — Specific

## 510.1 Scope

Articles 511 through 517 cover occupancies or parts of occupancies that are or may be hazardous because of atmospheric concentrations of flammable liquids, gases, or vapors, or because of deposits or accumulations of materials that may be readily ignitible.

## 510.2 General

The general rules of this *Code* and the provisions of Articles 500 through 504 shall apply to electrical wiring and equipment in occupancies within the scope of Articles 511 through 517, except as such rules are modified in Articles 511 through 517. Where unusual conditions exist in a specific occupancy, the authority having jurisdiction shall judge with respect to the application of specific rules.

# ARTICLE 511 Commercial Garages, Repair and Storage

Informational Note: Text that is followed by a reference in brackets has been extracted from NFPA 30A-2012, *Code for Motor Fuel Dispensing Facilities and Repair Garages*. Only editorial changes were made to the extracted text to make it consistent with this *Code*.

## 511.1 Scope

These occupancies shall include locations used for service and repair operations in connection with self-propelled vehicles (including, but not limited to, passenger automobiles, buses, trucks, and tractors) in which volatile flammable liquids or flammable gases are used for fuel or power.

Article 100 defines *garage* as "a building or portion of a building in which one or more self-propelled vehicles can be kept for use, sale, storage, rental, repair, exhibition, or demonstration purposes." Article 511 applies to commercial garages in which the primary operation is the service and repair of self-propelled vehicles that use flammable gases or liquids for fuel. These commercial garages include automotive service centers; repair garages for commercial vehicles, such as trucks and tractors; and service garages for fleet vehicles, such as buses, cars, and trucks.

The requirements of Article 511 are intended to mitigate the potential for an ignition-capable arc or spark from electrical wiring or equipment used in or above hazardous locations. Also covered are requirements for personnel protection in occupancies that are frequently wet or damp in which service personnel are subject to contact with large grounded surfaces, such as concrete slabs in direct contact with the earth. Service operations in which minor repairs, such as oil changes, occur are covered under the requirements of this article. See 511.3 and its associated commentary.

Parking, storage, and similar occupancies are not required to be classified, provided that any repair that occurs is minor and does not involve the use of electrical equipment. In accordance with NFPA 88A, *Standard for Parking Structures*, a mechanical ventilating system that is capable of continuously providing a ventilation rate of 1 ft$^3$ per minute for each square foot of floor area is required for all enclosed, basement, and underground parking garages.

Operations that involve open flames or electric arcs, including fusion gas welding and electric welding, as well as the requirements for heat-producing appliances, are found in NFPA 30A. Repair work that involves an open flame or electric arcs must be restricted to areas specifically provided for such purposes.

Section 555.22 requires that the repair facilities for boats and other marine craft comply with the requirements of Article 511.

## 511.2 Definitions

**Major Repair Garage.** A building or portions of a building where major repairs, such as engine overhauls, painting, body and fender work, and repairs that require draining of the motor vehicle fuel tank are performed on motor vehicles, including associated floor space used for offices, parking, or showrooms. [**30A:**3.3.12.1]

**Minor Repair Garage.** A building or portions of a building used for lubrication, inspection, and minor automotive maintenance work, such as engine tune-ups, replacement of parts, fluid changes (e.g., oil, antifreeze, transmission fluid, brake fluid, air-conditioning refrigerants), brake system repairs, tire rotation, and similar routine maintenance work, including associated floor space used for offices, parking, or showrooms. [**30A:**3.3.12.2]

## 511.3 Area Classification, General

Where Class I liquids or gaseous fuels are stored, handled, or transferred, electrical wiring and electrical utilization equipment shall be designed in accordance with the requirements for Class I, Division 1 or 2 hazardous (classified) locations as classified in accordance with 500.5 and 500.6, and this article. A Class I location shall not extend beyond an unpierced wall, roof, or other solid partition that has no openings. [**30A:**8.3.5, 8.3.2]

The classification of areas in a garage is dependent upon the level of repair (major or minor) being conducted and the handling of flammable liquids or gaseous fuels other than for dispensing purposes. As stated in 511.3(B), the classification of dispensing areas of a repair garage is addressed in Article 514.

The term *transferred* is used in determining the requirements for classifying locations where a significant quantity of flammable or gaseous liquids are exposed to the atmosphere by a motor vehicle repair operation (e.g., major engine overhauls or repairs that require draining of the motor vehicle fuel tank). Minor repair garages, by definition under NFPA 30A, would not be permitted to conduct such types of repair operations involving the transfer of flammable or gaseous liquids.

The term *Class I liquids* refers to *flammable liquids* as defined in NFPA 30, *Flammable and Combustible Liquids Code*, and are liquids with a flash point below 100°F (38°C). Gasoline is a common Class I liquid, whereas diesel fuel, with its flash point being above 100°F, is classified as a Class II combustible liquid. The use of Class I, Class II, and Class III in NFPA 30 for the classification of liquids has no direct correlation to the use of Class I, Class II, and Class III in the *NEC* to designate hazardous locations.

The need to establish hazardous locations can be mitigated through the use of mechanical ventilation that meets the specified air exchange parameters. The *Code* indicates different classification areas dependent on whether ventilation is or is not provided. Where it is established that there will be hazardous locations within a commercial garage, all applicable requirements of Article 501 for installing wiring and equipment in Class I, Division 1 and 2 locations must be followed.

Many steps are required to properly classify a hazardous location. Although the *NEC* provides general area classifications, it does not classify specific locations. The *NEC* classifications have been extracted from other NFPA documents. The classifications from those documents are based on the premise that all applicable requirements of the document have been met. Deviations in on-site conditions, such as process conditions, area ventilation, and room construction, from those assumed by the document may alter the general classification. Those responsible for the specific area classification must consider the basis for the general classifications to determine the applicability to their specific location.

NFPA 30A contains the specific construction and installation requirements used to develop the area classifications in Article 511.

**(A) Parking Garages.** Parking garages used for parking or storage shall be permitted to be unclassified.

Informational Note: For further information, see NFPA 88A-2011, *Standard for Parking Structures*, and NFPA 30A-2012, *Code for Motor Fuel Dispensing Facilities and Repair Garages.*

**(B) Repair Garages, with Dispensing.** Major and minor repair garages that dispense motor fuels into the fuel tanks of vehicles, including flammable liquids having a flash point below 38°C (100°F) such as gasoline, or gaseous fuels such as natural gas, hydrogen, or LPG, shall have the dispensing functions and components classified in accordance with Table 514.3(B)(1) in addition to any classification required by this section. Where Class I liquids, other than fuels, are dispensed, the area within

900 mm (3 ft) of any fill or dispensing point, extending in all directions, shall be a Class I, Division 2 location.

**(C) Major Repair Garages.** Where flammable liquids having a flash point below 38°C (100°F) such as gasoline, or gaseous fuels such as natural gas, hydrogen, or LPG, will not be dispensed, but repair activities that involve the transfer of such fluids or gases are performed, the classification rules in (1), (2), and (3) shall apply.

**(1) Floor Areas.**

(a) *Ventilation Provided.* The floor area shall be unclassified where there is mechanical ventilation providing a minimum of four air changes per hour or 0.3 $m^3/min/m^2$ (1 $cfm/ft^2$) of exchanged air for each square meter (foot) of floor area. Ventilation shall provide for air exchange across the entire floor area, and exhaust air shall be taken at a point within 0.3 m (12 in.) of the floor.

(b) *Ventilation Not Provided.* The entire floor area up to a level of 450 mm (18 in.) above the floor shall be classified as Class I, Division 2 if the ventilation does not comply with 511.3(C)(1)(a).

**(2) Ceiling Areas.** Where lighter-than-air gaseous fueled vehicles, such as vehicles fueled by natural gas or hydrogen, are repaired or stored, the area within 450 mm (18 in.) of the ceiling shall be considered for classification in accordance with (a) and (b).

(a) *Ventilation Provided.* The ceiling area shall be unclassified where ventilation is provided, from a point not more than 450 mm (18 in.) from the highest point in the ceiling, to exhaust the ceiling area at a rate of not less than 0.3 $m^3/min/m^2$ (1 $cfm/ft^2$) of ceiling area at all times that the building is occupied or when vehicles using lighter-than-air gaseous fuels are parked below this area.

(b) *Ventilation Not Provided.* Ceiling areas that are not ventilated in accordance with 511.3(C)(2)(a) shall be classified as Class I, Division 2.

**(3) Pit Areas in Lubrication or Service Room.** Any pit, belowgrade work area, or subfloor work area shall be classified as provided in (a) or (b).

The Class I, Division 2 location above grade within a commercial garage in which Class I liquids or gaseous fuels are transferred extends 18 inches above floor level, unless mechanical ventilation provides at least four air changes per hour. The same 18 inches applies down from ceiling areas when lighter-than-air-fuel vehicles are maintained, unless ventilation similar to the floor area of the ceiling area is provided. Areas suitably cut off and areas adjacent to unclassified, ventilated garages are not classified as hazardous.

The Class I, Division 1 location below grade extends from the floor of the pit or depression to floor level, unless the pit or depression is classified as Class I, Division 2, because ventilation providing at least six air changes per hour exhausts air at the floor level of the pit or depression. See Exhibit 511.1 for an illustration of classified and unclassified locations in a major repair garage.

(a) *Ventilation Provided.* The pit area shall be a Class I, Division 2 location where there is mechanical ventilation providing a minimum of six air changes per hour.

(b) *Ventilation Not Provided.* Where ventilation is not provided in accordance with 511.3(C)(3)(a), any pit or depression below floor level shall be a Class I, Division 1 location that extends up to the floor level.

**(D) Minor Repair Garages.** Where flammable liquids having a flash point below 38°C (100°F) such as gasoline, or gaseous fuels such as natural gas or hydrogen, will not be dispensed or transferred, the classification rules in (D)(1), (D)(2), and (D)(3) shall apply to the lubrication and service rooms.

Most lubritoriums primarily offer oil and filter change and lubrication-type services, but do not transfer fuel. If the lower-level work area of a lubritorium is provided with exhaust ventilation at the rate specified, the lower level is not classified as a hazardous location.

**(1) Floor Areas.** Floor areas in minor repair garages without pits, belowgrade work areas, or subfloor work areas shall be unclassified. Where floor areas include pits, belowgrade work areas, or subfloor work areas in lubrication or service rooms, the classification rules in (a) or (b) shall apply.

(a) *Ventilation Provided.* The entire floor area shall be unclassified where there is mechanical ventilation providing a minimum of four air changes per hour or 0.3 $m^3/min/m^2$ (1 $cfm/ft^2$) of exchanged air for each square meter (foot) of floor area. Ventilation shall provide for air exchange across the entire floor area, and exhaust air shall be taken at a point within 0.3 m (12 in.) of the floor.

(b) *Ventilation Not Provided.* The floor area up to a level of 450 mm (18 in.) above any unventilated pit, belowgrade work area, or subfloor work area and extending a distance of 900 mm (3 ft) horizontally from the edge of any such pit, belowgrade work area, or subfloor work area, shall be classified as Class I, Division 2.

**(2) Ceiling Areas.** Where lighter-than-air gaseous fuels (such as natural gas or hydrogen) will not be transferred, such locations shall be unclassified.

**(3) Pit Areas in Lubrication or Service Room.** Any pit, belowgrade work area, or subfloor work area shall be classified as provided in (a) or (b).

(a) *Ventilation Provided.* Where ventilation is provided to exhaust the pit area at a rate of not less than 0.3 $m^3/min/m^2$ (1 $cfm/ft^2$) of floor area at all times that the building is occupied, or when vehicles are parked in or over this area and where exhaust air is taken from a point within 300 mm (12 in.) of the

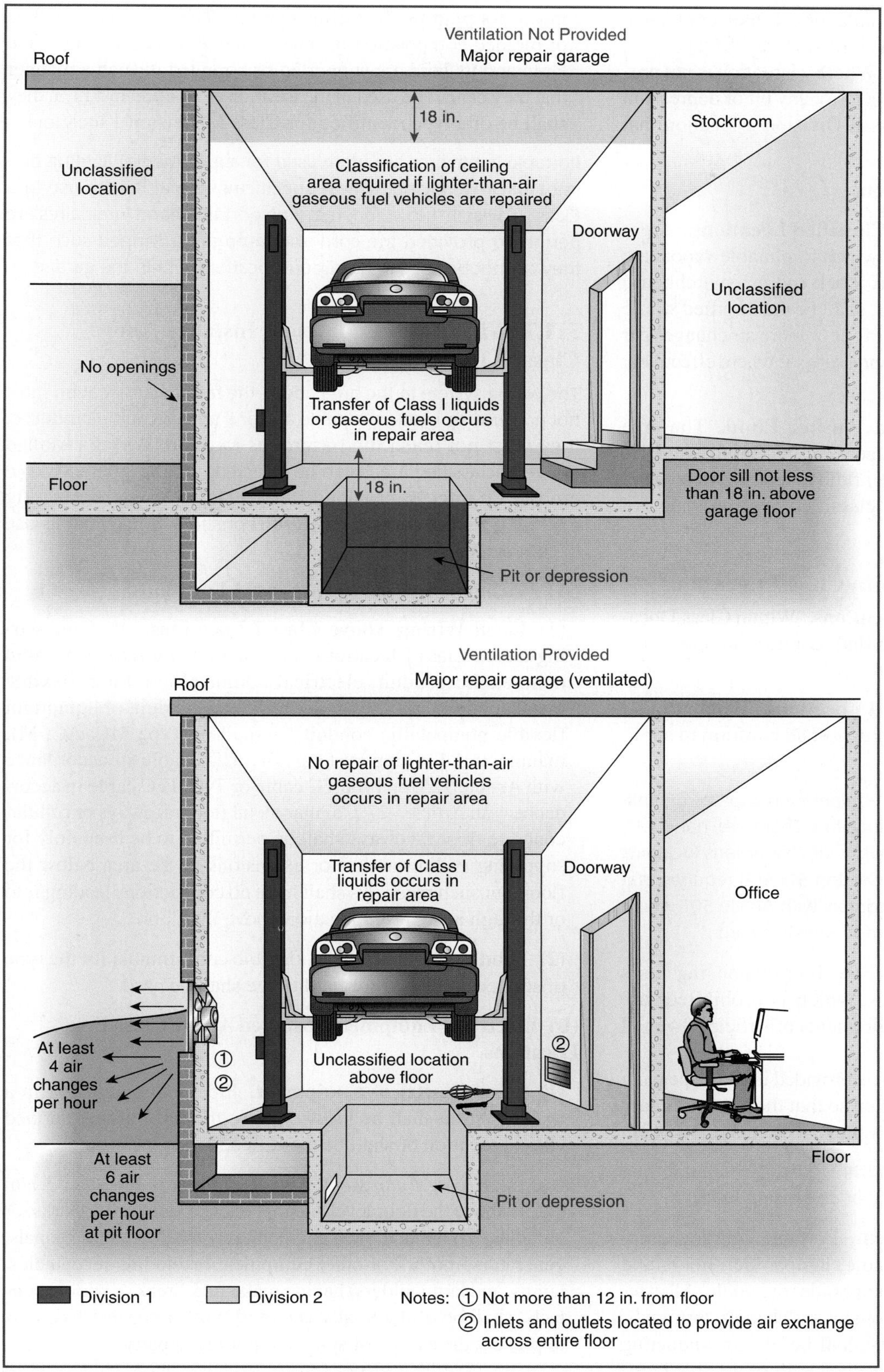
Ventilation Not Provided
Major repair garage
Roof
18 in.
Stockroom
Unclassified location
Classification of ceiling area required if lighter-than-air gaseous fuel vehicles are repaired
Doorway
Unclassified location
No openings
Transfer of Class I liquids or gaseous fuels occurs in repair area
Floor
18 in.
Door sill not less than 18 in. above garage floor
Pit or depression
Ventilation Provided
Major repair garage (ventilated)
Roof
No repair of lighter-than-air gaseous fuel vehicles occurs in repair area
Transfer of Class I liquids occurs in repair area
Doorway
Office
At least 4 air changes per hour
①
②
Unclassified location above floor
②
Floor
At least 6 air changes per hour at pit floor
Pit or depression
Division 1
Division 2
Notes: ① Not more than 12 in. from floor
② Inlets and outlets located to provide air exchange across entire floor

***EXHIBIT 511.1*** *Classification of locations in commercial garages.*

floor of the pit, belowgrade work area, or subfloor work area, the pit shall be unclassified. [**30A:**7.4.5.4. Table 8.3.1]

(b) *Ventilation Not Provided.* Where ventilation is not provided in accordance with 511.3(D)(3)(a), any pit or depression below floor level shall be a Class I, Division 2 location that extends up to the floor level.

**(E) Modifications to Classification.**

**(1) Specific Areas Adjacent to Classified Locations.** Areas adjacent to classified locations in which flammable vapors are not likely to be released, such as stock rooms, switchboard rooms, and other similar locations, shall be unclassified where mechanically ventilated at a rate of four or more air changes per hour, or designed with positive air pressure, or where effectively cut off by walls or partitions.

**(2) Alcohol-Based Windshield Washer Fluid.** The area used for storage, handling, or dispensing into motor vehicles of alcohol-based windshield washer fluid in repair garages shall be unclassified unless otherwise classified by a provision of 511.3. [**30A:**8.3.5, Exception]

## 511.4 Wiring and Equipment in Class I Locations

**(A) Wiring Located in Class I Locations.** Within Class I locations as classified in 511.3, wiring shall conform to applicable provisions of Article 501.

**(B) Equipment Located in Class I Locations.** Within Class I locations as defined in 511.3, equipment shall conform to applicable provisions of Article 501.

Most battery-operated and portable equipment is capable of igniting the atmosphere in a hazardous location. Therefore, handheld, portable, and mobile equipment brought into hazardous locations should be suitable for the location. Section 511.4(A) requires that flexible cords supplying equipment comply with Article 501 where 501.140 details specific requirements for a flexible cord.

**(1) Fuel-Dispensing Units.** Where fuel-dispensing units (other than liquid petroleum gas, which is prohibited) are located within buildings, the requirements of Article 514 shall govern.

Where mechanical ventilation is provided in the dispensing area, the control shall be interlocked so that the dispenser cannot operate without ventilation, as prescribed in 500.5(B)(2).

Figure 514.3 in the *Code* and Exhibit 514.1 provide more information on classified areas in the vicinity of dispensing units.

**(2) Portable Lighting Equipment.** Portable lighting equipment shall be equipped with handle, lampholder, hook, and substantial guard attached to the lampholder or handle. All exterior surfaces that might come in contact with battery terminals, wiring terminals, or other objects shall be of nonconducting material or shall be effectively protected with insulation. Lampholders shall be of an unswitched type and shall not provide means for plug-in of attachment plugs. The outer shell shall be of molded composition or other suitable material. Unless the lamp and its cord are supported or arranged in such a manner that they cannot be used in the locations classified in 511.3, they shall be of a type identified for Class I, Division 1 locations.

Portable luminaires are often used for supplemental lighting during vehicle servicing. Unless specifically identified for use in a Class I, Division 1 location, reel-type portable hand luminaires are permitted provided the cord and lamp are arranged such that they cannot be used in hazardous locations within the garage.

## 511.7 Wiring and Equipment Installed Above Class I Locations

The wiring system in the area above the hazardous location must not produce an ignition-capable arc, or if sparks can be produced, they must not reach the hazardous location. Wiring installed above unclassified areas can be selected from the methods covered in Chapter 3, provided the article covering that wiring method does not contain any restrictions that would limit its use in commercial garages.

**(A) Wiring in Spaces Above Class I Locations.**

**(1) Fixed Wiring Above Class I Locations.** All fixed wiring above Class I locations shall be in metal raceways, rigid nonmetallic conduit, electrical nonmetallic tubing, flexible metal conduit, liquidtight flexible metal conduit, or liquidtight flexible nonmetallic conduit, or shall be Type MC, AC, MI, manufactured wiring systems, or PLTC cable in accordance with Article 725, or Type TC cable or Type ITC cable in accordance with Article 727. Cellular metal floor raceways or cellular concrete floor raceways shall be permitted to be used only for supplying ceiling outlets or extensions to the area below the floor, but such raceways shall have no connections leading into or through any Class I location above the floor.

**(2) Pendant.** For pendants, flexible cord suitable for the type of service and listed for hard usage shall be used.

**(B) Electrical Equipment Installed Above Class I Locations.**

**(1) Fixed Electrical Equipment.** Electrical equipment in a fixed position shall be located above the level of any defined Class I location or shall be identified for the location.

(a) *Arcing Equipment.* Equipment that is less than 3.7 m (12 ft) above the floor level and that may produce arcs, sparks, or particles of hot metal, such as cutouts, switches, charging panels, generators, motors, or other equipment (excluding receptacles, lamps, and lampholders) having make-and-break or sliding contacts, shall be of the totally enclosed type or constructed so as to prevent the escape of sparks or hot metal particles.

(b) *Fixed Lighting.* Lamps and lampholders for fixed lighting that is located over lanes through which vehicles are commonly driven or that may otherwise be exposed to physical damage shall be located not less than 3.7 m (12 ft) above floor level, unless of the totally enclosed type or constructed so as to prevent escape of sparks or hot metal particles.

## 511.9 Sealing

Seals complying with the requirements of 501.15 and 501.15(B)(2) shall be provided and shall apply to horizontal as well as vertical boundaries of the defined Class I locations.

Note that the general rules of 501.15(A)(4) and (B)(2) on providing seals at hazardous location boundaries apply where raceway installations in a commercial garage pass from classified to unclassified locations and where conduit fittings, outlet boxes, or both are installed less than 12 inches from either side of the boundary. In accordance with 501.15(A)(4), Exception No. 2, where a raceway runs from a Class I, Division 1 location into an underground unclassified location and then emerges from below ground into an unclassified location, the boundary seal is permitted to be located more than 10 feet from the actual boundary, provided the seal is located at the point the conduit emerges from below grade into the unclassified location. The seal and, if necessary, an associated explosionproof union are required to be the first fitting(s) at the point the conduit emerges from below ground into the unclassified location.

Exhibit 511.2 depicts two receptacle outlet enclosures that are located at least 12 inches above an area that has been classified as Class I. The rigid metal conduit passes unbroken from the outlet boxes through the Class I location into the unclassified underground location beneath the floor. The conduit coupling is located 12 inches or more from the penetration into the hazardous location. No seals are required for this installation in accordance with 501.15(A)(4), Exception No. 1.

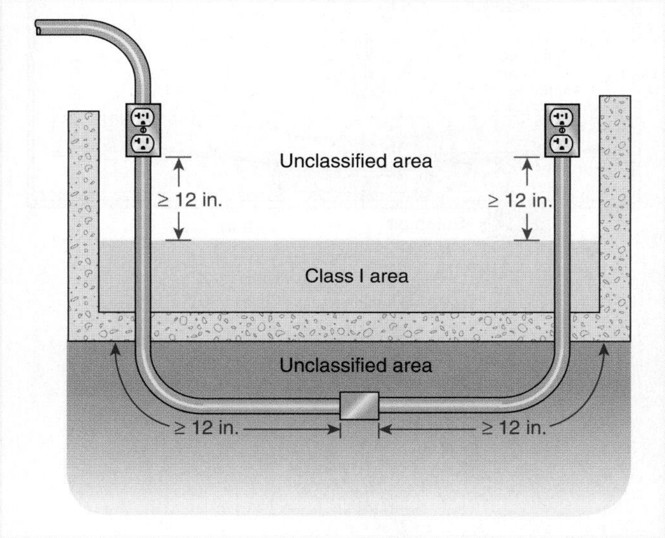


***EXHIBIT 511.2*** *Seals not required for conduits that pass unbroken through the Class I location.*

## 511.10 Special Equipment

**(A) Battery Charging Equipment.** Battery chargers and their control equipment, and batteries being charged, shall not be located within locations classified in 511.3.

**(B) Electric Vehicle Charging Equipment.**

**(1) General.** All electrical equipment and wiring shall be installed in accordance with Article 625, except as noted in 511.10(B)(2) and (B)(3). Flexible cords shall be of a type identified for extra-hard usage.

**(2) Connector Location.** No connector shall be located within a Class I location as defined in 511.3.

**(3) Plug Connections to Vehicles.** Where the cord is suspended from overhead, it shall be arranged so that the lowest point of sag is at least 150 mm (6 in.) above the floor. Where an automatic arrangement is provided to pull both cord and plug beyond the range of physical damage, no additional connector shall be required in the cable or at the outlet.

## 511.12 Ground-Fault Circuit-Interrupter Protection for Personnel

All 125-volt, single-phase, 15- and 20-ampere receptacles installed in areas where electrical diagnostic equipment, electrical hand tools, or portable lighting equipment are to be used shall have ground-fault circuit-interrupter protection for personnel.

## 511.16 Grounding and Bonding Requirements

**(A) General Grounding Requirements.** All metal raceways, the metal armor or metallic sheath on cables, and all non–current-carrying metal parts of fixed or portable electrical equipment, regardless of voltage, shall be grounded.

**(B) Supplying Circuits with Grounded and Grounding Conductors in Class I Locations.** Grounding in Class I locations shall comply with 501.30.

**(1) Circuits Supplying Portable Equipment or Pendants.** Where a circuit supplies portables or pendants and includes a grounded conductor as provided in Article 200, receptacles, attachment plugs, connectors, and similar devices shall be of the grounding type, and the grounded conductor of the flexible cord shall be connected to the screw shell of any lampholder or to the grounded terminal of any utilization equipment supplied.

**(2) Approved Means.** Approved means shall be provided for maintaining continuity of the equipment grounding conductor between the fixed wiring system and the non–current-carrying metal portions of pendant luminaires, portable luminaires, and portable utilization equipment.

# ARTICLE 513
# Aircraft Hangars

## 513.1 Scope

This article shall apply to buildings or structures in any part of which aircraft containing Class I (flammable) liquids or Class II (combustible) liquids whose temperatures are above their flash points are housed or stored and in which aircraft might undergo service, repairs, or alterations. It shall not apply to locations used exclusively for aircraft that have never contained fuel or unfueled aircraft.

Article 513 does not apply to areas in which the only fuel contained in the aircraft is a Class II combustible liquid, unless the fuel will be used or stored above its flash point. A Class II liquid has a closed-cup flash point at or above 100°F. Some aviation fuel, such as Jet-A, is a Class II combustible liquid. See the definition of *combustible liquid* in NFPA 30. An aircraft manufacturing plant in which the aircraft under construction have never contained fuel is an example of a facility not covered by the requirements of Article 513.

Many steps are required to properly classify a hazardous location. Although the *NEC* provides general area classifications, it does not classify specific locations. The *NEC* classifications have been extracted from other NFPA documents. The classifications from those documents are based on the premise that all applicable requirements of the document have been met. Deviations in on-site conditions, such as process conditions, area ventilation, and room construction, from those assumed by the document may alter the general classification. Those responsible for the specific area classification must consider the basis for the general classifications to determine the applicability to their specific location.

NFPA 409 contains the specific construction and installation requirements used to develop the area classifications in Article 513.

Informational Note No. 1: For definitions of aircraft hangar and unfueled aircraft, see NFPA 409-2011, *Standard on Aircraft Hangars*.

Informational Note No. 2: For further information on fuel classification see NFPA 30-2012, *Flammable and Combustible Liquids Code*.

## 513.2 Definitions

For the purpose of this article, the following definitions shall apply.

**Aircraft Painting Hangar.** An aircraft hangar constructed for the express purpose of spray/coating/dipping applications and provided with dedicated ventilation supply and exhaust.

**Mobile Equipment.** Equipment with electrical components suitable to be moved only with mechanical aids or is provided with wheels for movement by person(s) or powered devices.

**Portable Equipment (as applied to Article 513).** Equipment with electrical components suitable to be moved by a single person without mechanical aids.

## 513.3 Classification of Locations

**(A) Below Floor Level.** Any pit or depression below the level of the hangar floor shall be classified as a Class I, Division 1 or Zone 1 location that shall extend up to said floor level.

**(B) Areas Not Cut Off or Ventilated.** The entire area of the hangar, including any adjacent and communicating areas not suitably cut off from the hangar, shall be classified as a Class I, Division 2 or Zone 2 location up to a level 450 mm (18 in.) above the floor.

**(C) Vicinity of Aircraft.**

**(1) Aircraft Maintenance and Storage Hangars.** The area within 1.5 m (5 ft) horizontally from aircraft power plants or aircraft fuel tanks shall be classified as a Class I, Division 2 or Zone 2 location that shall extend upward from the floor to a level 1.5 m (5 ft) above the upper surface of wings and of engine enclosures.

To properly classify hangar areas, information on the aircraft parking patterns, the types of aircraft, and the operations to be performed in the hangar must be obtained. Much of the hangar area classification is dependent on the dimensional outline of the aircraft; therefore, consideration of future aircraft and parking patterns is appropriate to avoid costs associated with changes in the area classification. Exhibit 513.1 illustrates the area classifications

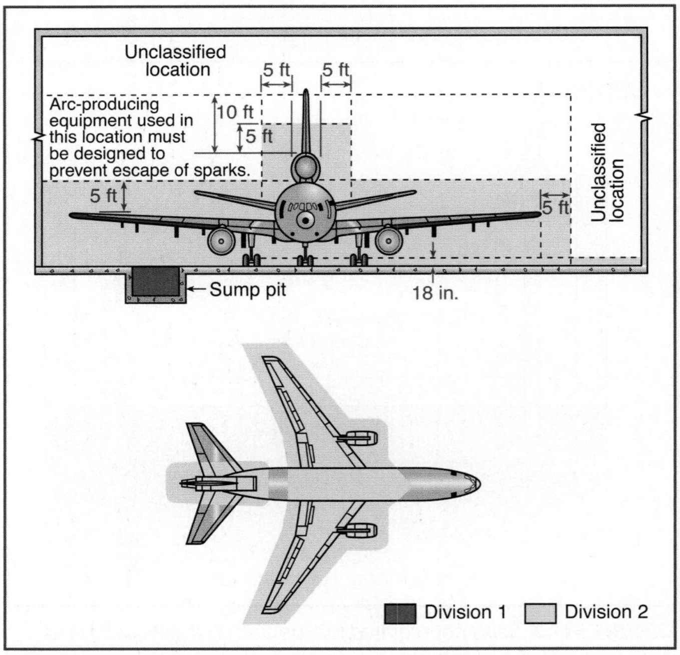


***EXHIBIT 513.1*** *Area classification in aircraft hangars.*

in aircraft hangars. In addition to illustrating the requirements of 513.3(A), (B), and (C)(1), the provision in 513.7(C) to afford protection from arcing equipment is also depicted.

**(2) Aircraft Painting Hangars.** The area within 3 m (10 ft) horizontally from aircraft surfaces from the floor to 3 m (10 ft) above the aircraft shall be classified as Class I, Division 1 or Class I, Zone 1. The area horizontally from aircraft surfaces between 3.0 m (10 ft) and 9.0 m (30 ft) from the floor to 9.0 m (30 ft) above the aircraft surface shall be classified as Class I, Division 2 or Class I, Zone 2.

Informational Note: See NFPA 33-2011, *Standard for Spray Application Using Flammable or Combustible Materials*, for information on ventilation and grounding for static protection in spray painting areas.

**(D) Areas Suitably Cut Off and Ventilated.** Adjacent areas in which flammable liquids or vapors are not likely to be released, such as stock rooms, electrical control rooms, and other similar locations, shall be unclassified where adequately ventilated and where effectively cut off from the hangar itself by walls or partitions.

## 513.4 Wiring and Equipment in Class I Locations

**(A) General.** All wiring and equipment that is or may be installed or operated within any of the Class I locations defined in 513.3 shall comply with the applicable provisions of Article 501 or Article 505 for the division or zone in which they are used.

Attachment plugs and receptacles in Class I locations shall be identified for Class I locations or shall be designed such that they cannot be energized while the connections are being made or broken.

**(B) Stanchions, Rostrums, and Docks.** Electrical wiring, outlets, and equipment (including lamps) on or attached to stanchions, rostrums, or docks that are located or likely to be located in a Class I location, as defined in 513.3(C), shall comply with the applicable provisions of Article 501 or Article 505 for the division or zone in which they are used.

## 513.7 Wiring and Equipment Not Installed in Class I Locations

**(A) Fixed Wiring.** All fixed wiring in a hangar but not installed in a Class I location as classified in 513.3 shall be installed in metal raceways or shall be Type MI, TC, or MC cable.

*Exception: Wiring in unclassified locations, as described in 513.3(D), shall be permitted to be any suitable type wiring method recognized in Chapter 3.*

**(B) Pendants.** For pendants, flexible cord suitable for the type of service and identified for hard usage or extra-hard usage shall be used. Each such cord shall include a separate equipment grounding conductor.

**(C) Arcing Equipment.** In locations above those described in 513.3, equipment that is less than 3.0 m (10 ft) above wings and engine enclosures of aircraft and that may produce arcs, sparks, or particles of hot metal, such as lamps and lampholders for fixed lighting, cutouts, switches, receptacles, charging panels, generators, motors, or other equipment having make-and-break or sliding contacts, shall be of the totally enclosed type or constructed so as to prevent the escape of sparks or hot metal particles.

*Exception: Equipment in areas described in 513.3(D) shall be permitted to be of the general-purpose type.*

**(D) Lampholders.** Lampholders of metal-shell, fiber-lined types shall not be used for fixed incandescent lighting.

**(E) Stanchions, Rostrums, or Docks.** Where stanchions, rostrums, or docks are not located or likely to be located in a Class I location, as defined in 513.3(C), wiring and equipment shall comply with 513.7, except that such wiring and equipment not more than 457 mm (18 in.) above the floor in any position shall comply with 513.4(B). Receptacles and attachment plugs shall be of a locking type that will not readily disconnect.

**(F) Mobile Stanchions.** Mobile stanchions with electrical equipment complying with 513.7(E) shall carry at least one permanently affixed warning sign with the following words or equivalent:

WARNING
KEEP 5 FT CLEAR OF AIRCRAFT
ENGINES AND FUEL TANK AREAS

or

WARNING
KEEP 1.5 METERS CLEAR OF AIRCRAFT
ENGINES AND FUEL TANK AREAS

## 513.8 Underground Wiring

**(A) Wiring and Equipment Embedded, Under Slab, or Underground.** All wiring installed in or under the hangar floor shall comply with the requirements for Class I, Division 1 locations. Where such wiring is located in vaults, pits, or ducts, adequate drainage shall be provided.

**(B) Uninterrupted Raceways, Embedded, Under Slab, or Underground.** Uninterrupted raceways that are embedded in a hangar floor or buried beneath the hangar floor shall be considered to be within the Class I location above the floor, regardless of the point at which the raceway descends below or rises above the floor.

Wiring and equipment embedded in or buried below the hangar floor is considered to be in a Class I, Division 1 location, whereas uninterrupted raceways that are embedded in or buried below the hangar floor are considered to be in the same hazardous location that exists above the floor. Raceways that rise out of the floor in unclassified locations must be provided with boundary seals in accordance with 501.15.

## 513.9 Sealing

Seals shall be provided in accordance with 501.15 or 505.16, as applicable. Sealing requirements specified shall apply to horizontal as well as to vertical boundaries of the defined Class I locations.

## 513.10 Special Equipment

**(A) Aircraft Electrical Systems.**

**(1) De-energizing Aircraft Electrical Systems.** Aircraft electrical systems shall be de-energized when the aircraft is stored in a hangar and, whenever possible, while the aircraft is undergoing maintenance.

**(2) Aircraft Batteries.** Aircraft batteries shall not be charged where installed in an aircraft located inside or partially inside a hangar.

**(B) Aircraft Battery Charging and Equipment.** Battery chargers and their control equipment shall not be located or operated within any of the Class I locations defined in 513.3 and shall preferably be located in a separate building or in an area such as defined in 513.3(D). Mobile chargers shall carry at least one permanently affixed warning sign with the following words or equivalent:

WARNING
KEEP 5 FT CLEAR OF AIRCRAFT ENGINES
AND FUEL TANK AREAS

or

WARNING
KEEP 1.5 METERS CLEAR OF AIRCRAFT
ENGINES AND FUEL TANK AREAS

Tables, racks, trays, and wiring shall not be located within a Class I location and, in addition, shall comply with Article 480.

**(C) External Power Sources for Energizing Aircraft.**

**(1) Not Less Than 450 mm (18 in.) Above Floor.** Aircraft energizers shall be designed and mounted such that all electrical equipment and fixed wiring will be at least 450 mm (18 in.) above floor level and shall not be operated in a Class I location as defined in 513.3(C).

**(2) Marking for Mobile Units.** Mobile energizers shall carry at least one permanently affixed warning sign with the following words or equivalent:

WARNING
KEEP 5 FT CLEAR OF AIRCRAFT
ENGINES AND FUEL TANK AREAS

or

WARNING
KEEP 1.5 METERS CLEAR OF AIRCRAFT
ENGINES AND FUEL TANK AREAS

**(3) Cords.** Flexible cords for aircraft energizers and ground support equipment shall be identified for the type of service and extra-hard usage and shall include an equipment grounding conductor.

**(D) Mobile Servicing Equipment with Electrical Components.**

**(1) General.** Mobile servicing equipment (such as vacuum cleaners, air compressors, air movers) having electrical wiring and equipment not suitable for Class I, Division 2 or Zone 2 locations shall be so designed and mounted that all such fixed wiring and equipment will be at least 450 mm (18 in.) above the floor. Such mobile equipment shall not be operated within the Class I location defined in 513.3(C) and shall carry at least one permanently affixed warning sign with the following words or equivalent:

WARNING
KEEP 5 FT CLEAR OF AIRCRAFT ENGINES
AND FUEL TANK AREAS

or

WARNING
KEEP 1.5 METERS CLEAR OF AIRCRAFT ENGINES
AND FUEL TANK AREAS

**(2) Cords and Connectors.** Flexible cords for mobile equipment shall be suitable for the type of service and identified for extra-hard usage and shall include an equipment grounding conductor. Attachment plugs and receptacles shall be identified for the location in which they are installed and shall provide for connection of the equipment grounding conductor.

**(3) Restricted Use.** Equipment that is not identified as suitable for Class I, Division 2 locations shall not be operated in locations where maintenance operations likely to release flammable liquids or vapors are in progress.

**(E) Portable Equipment.**

**(1) Portable Lighting Equipment.** Portable lighting equipment that is used within a hangar shall be identified for the location in which they are used. For portable luminaires, flexible cord suitable for the type of service and identified for extra-hard usage shall be used. Each such cord shall include a separate equipment grounding conductor.

**(2) Portable Utilization Equipment.** Portable utilization equipment that is or may be used within a hangar shall be of a type suitable for use in Class I, Division 2 or Zone 2 locations. For portable utilization equipment, flexible cord suitable for the type of service and approved for extra-hard usage shall be used. Each such cord shall include a separate equipment grounding conductor.

## 513.12 Ground-Fault Circuit-Interrupter Protection for Personnel

All 125-volt, 50/60-Hz, single-phase, 15- and 20-ampere receptacles installed in areas where electrical diagnostic equipment, electrical hand tools, or portable lighting equipment are to be used shall have ground-fault circuit-interrupter protection for personnel.

GFCI protection is not required on circuits supplied at 400 Hz and higher.

## 513.16 Grounding and Bonding Requirements

**(A) General Grounding Requirements.** All metal raceways, the metal armor or metallic sheath on cables, and all non–current-carrying metal parts of fixed or portable electrical equipment, regardless of voltage, shall be grounded. Grounding in Class I locations shall comply with 501.30 for Class I, Division 1 and 2 locations and 505.25 for Class I, Zone 0, 1, and 2 locations.

**(B) Supplying Circuits with Grounded and Grounding Conductors in Class I Locations.**

**(1) Circuits Supplying Portable Equipment or Pendants.** Where a circuit supplies portables or pendants and includes a grounded conductor as provided in Article 200, receptacles, attachment plugs, connectors, and similar devices shall be of the grounding type, and the grounded conductor of the flexible cord shall be connected to the screw shell of any lampholder or to the grounded terminal of any utilization equipment supplied.

**(2) Approved Means.** Approved means shall be provided for maintaining continuity of the grounding conductor between the fixed wiring system and the non–current-carrying metal portions of pendant luminaires, portable luminaires, and portable utilization equipment.

# ARTICLE 514 Motor Fuel Dispensing Facilities

Informational Note: Text that is followed by a reference in brackets has been extracted from NFPA 30A-2012, *Code for Motor Fuel Dispensing Facilities and Repair Garages*. Only editorial changes were made to the extracted text to make it consistent with this *Code*.

## 514.1 Scope

This article shall apply to motor fuel dispensing facilities, marine/motor fuel dispensing facilities, motor fuel dispensing facilities located inside buildings, and fleet vehicle motor fuel dispensing facilities.

Article 514 encompasses all locations where volatile flammable liquids or gases are dispensed into the fuel tanks of self-propelled vehicles or other approved fuel tanks. It also includes dispensing at marine facilities such as marinas and boatyards. The phrase *approved containers* in the 514.2 definition covers portable gasoline containers and also applies to dispensing locations for liquefied petroleum gas (LPG), including those locations that do not serve self-propelled vehicles. The popularity of outdoor cooking appliances using LPG has resulted in a marked increase in portable-container dispensing sites. Electrical area classification for LPG sites is specified in Table 514.3(B)(2).

Informational Note: For further information regarding safeguards for motor fuel dispensing facilities, see NFPA 30A-2012, *Code for Motor Fuel Dispensing Facilities and Repair Garages*.

## 514.2 Definition

**Motor Fuel Dispensing Facility.** That portion of a property where motor fuels are stored and dispensed from fixed equipment into the fuel tanks of motor vehicles or marine craft or into approved containers, including all equipment used in connection therewith. [**30A:**3.3.11]

Informational Note: Refer to Articles 510 and 511 with respect to electrical wiring and equipment for other areas used as lubritoriums, service rooms, repair rooms, offices, salesrooms, compressor rooms, and similar locations.

## 514.3 Classification of Locations (See Figure 514.3)

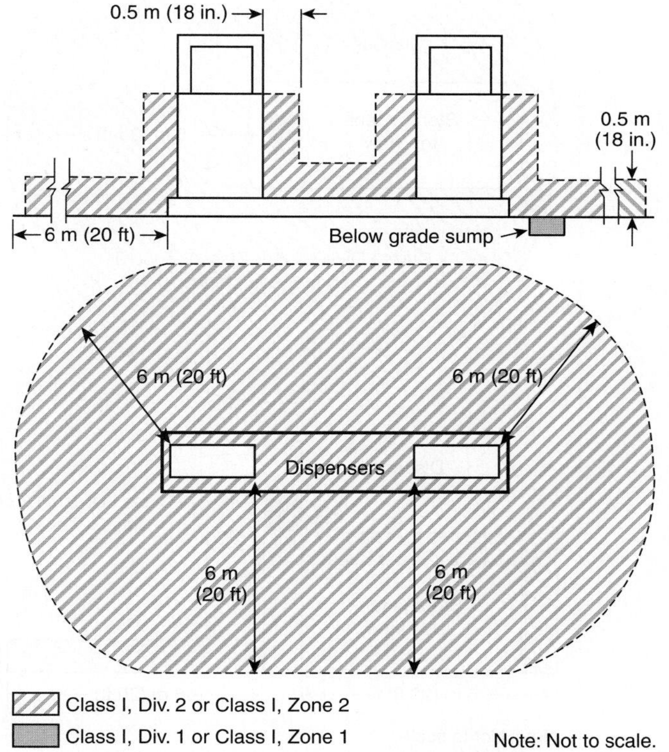


***FIGURE 514.3*** *Classified Areas Adjacent to Dispensers. [**30A:** Figure 8.3.2(a)]*

**(A) Unclassified Locations.** Where the authority having jurisdiction can satisfactorily determine that flammable liquids having a flash point below 38°C (100°F), such as gasoline, will not be handled, such location shall not be required to be classified.

**(B) Classified Locations. [See Figure 514.3(B).]**

Many steps are required to properly classify a hazardous location. Although the *NEC* provides general area classifications, it does not classify specific locations. The *NEC* classifications have been extracted from other NFPA documents. The classifications from those documents are based on the premise that all applicable requirements of the document have been met. Deviations in on-site conditions, such as process conditions, area ventilation, and room construction, from those assumed by the document may alter the general classification. Those responsible for the specific area classification must consider the basis for the general classifications to determine the applicability to their specific location.

NFPA 30A contains the specific construction and installation requirements used to develop the area classifications in Article 514.

**(1) Class I Locations.** Table 514.3(B)(1) shall be applied where Class I liquids are stored, handled, or dispensed and shall be used to delineate and classify motor fuel dispensing facilities and commercial garages as defined in Article 511. Table 515.3 shall be used for the purpose of delineating and classifying aboveground tanks. A Class I location shall not extend beyond an unpierced wall, roof, or other solid partition. [**30A:**8.1, 8.3]

**(2) Compressed Natural Gas, Liquefied Natural Gas, and Liquefied Petroleum Gas Areas.** Table 514.3(B)(2) shall be used to delineate and classify areas where compressed natural gas (CNG), liquefied natural gas (LNG), or liquefied petroleum gas (LPG) is stored, handled, or dispensed. Where CNG or LNG dispensers are installed beneath a canopy or enclosure, either the canopy or the enclosure shall be designed to prevent accumulation or entrapment of ignitible vapors, or all electrical equipment installed beneath the canopy or enclosure shall be suitable for Class I, Division 2 hazardous (classified) locations. Dispensing devices for liquefied petroleum gas shall be located

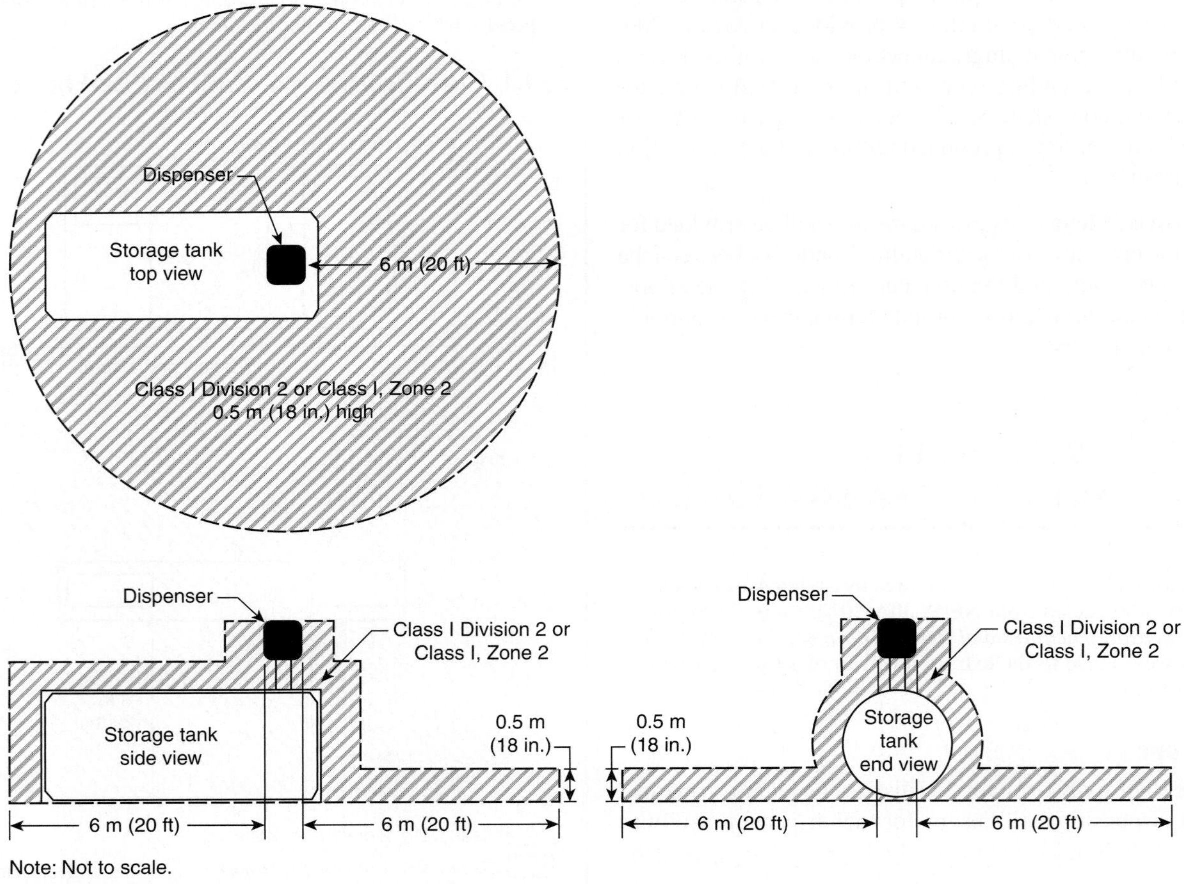


***FIGURE 514.3(B)*** *Classified Areas Adjacent to Dispenser Mounted on Aboveground Storage Tank. [**30A:** Figure 8.3.2(b)]*

***TABLE 514.3(B)(1)*** *Class I Locations — Motor Fuel Dispensing Facilities*

| Location | Division (Group D) | Zone (Group IIA) | Extent of Classified Location[1] |
|---|---|---|---|
| **Dispensing Device (except Overhead Type)**[2, 3] | | | |
| Under dispenser containment | 1 | 1 | Entire space within and under dispenser pit or containment |
| Dispenser | 2 | 2 | Within 450 mm (18 in.) of dispenser enclosure or that portion of dispenser enclosure containing liquid handling components, extending horizontally in all directions and down to grade level |
| Outdoor | 2 | 2 | Up to 450 mm (18 in.) above grade level, extending 6 m (20 ft) horizontally in all directions from dispenser enclosure |
| Indoor | | | |
| - with mechanical ventilation | 2 | 2 | Up to 450 mm (18 in.) above floor level, extending 6 m (20 ft) horizontally in all directions from dispenser enclosure |
| - with gravity ventilation | 2 | 2 | Up to 450 mm (18 in.) above floor level, extending 7.5 m (25 ft) horizontally in all directions from dispenser enclosure |
| **Dispensing Device — Overhead Type**[4] | 1 | 1 | Space within dispenser enclosure and all electrical equipment integral with dispensing hose or nozzle |
| | 2 | 2 | Within 450 mm (18 in.) of dispenser enclosure, extending horizontally in all directions and down to grade level |
| | 2 | 2 | Up to 450 mm (18 in.) above grade level, extending 6 m (20 ft) horizontally in all directions from a point vertically below edge of dispenser enclosure |
| **Remote Pump —** | | | |
| Outdoor | 1 | 1 | Entire space within any pit or box below grade level, any part of which is within 3 m (10 ft) horizontally from any edge of pump |
| | 2 | 2 | Within 900 mm (3 ft) of any edge of pump, extending horizontally in all directions |
| | 2 | 2 | Up to 450 mm (18 in.) above grade level, extending 3 m (10 ft) horizontally in all directions from any edge of pump |
| Indoor | 1 | 1 | Entire space within any pit |
| | 2 | 2 | Within 1.5 m (5 ft) of any edge of pump, extending in all directions |
| | 2 | 2 | Up to 900 mm (3 ft) above floor level, extending 7.5 m (25 ft) horizontally in all directions from any edge of pump |
| **Sales, Storage, Rest Rooms** | unclassified | unclassified | Except as noted below |
| including structures (such as the attendant's kiosk) on or adjacent to dispensers | 1 | 1 | Entire volume, if there is any opening to room within the extent of a Division 1 or Zone 1 location |
| | 2 | 2 | Entire volume, if there is any opening to room within the extent of a Division 2 or Zone 2 location |
| **Tank, Aboveground** | | | |
| Inside tank | 1 | 0 | Entire inside volume |
| Shell, ends, roof, dike area | 1 | 1 | Entire space within dike, where dike height exceeds distance from tank shell to inside of dike wall for more than 50 percent of tank circumference |
| | 2 | 2 | Entire space within dike, where dike height does not exceed distance from tank shell to inside of dike wall for more than 50 percent of tank circumference |
| Vent | 2 | 2 | Within 3 m (10 ft) of shell, ends, or roof of tank |
| | 1 | 1 | Within 1.5 m (5 ft) of open end of vent, extending in all directions |
| | 2 | 2 | Between 1.5 m and 3 m (5 ft and 10 ft) from open end of vent, extending in all directions |

*(continues)*

*TABLE 514.3(B)(1) Continued*

| Location | Division (Group D) | Zone (Group IIA) | Extent of Classified Location[1] |
|---|---|---|---|
| **Tank, Underground** | | | |
| Inside tank | 1 | 0 | Entire inside volume |
| Fill Opening | 1 | 1 | Entire space within any pit or box below grade level, any part of which is within a Division 1 or Division 2 classified location or within a Zone 1 or Zone 2 classified location |
| | 2 | 2 | Up to 450 mm (18 in.) above grade level, extending 1.5 m (5 ft) horizontally in all directions from any tight-fill connection and extending 3 m (10 ft) horizontally in all directions from any loose-fill connection |
| Vent | 1 | 1 | Within 1.5 m (5 ft) of open end of vent, extending in all directions |
| | 2 | 2 | Between 1.5 m and 3 m (5 ft and 10 ft) from open end of vent, extending in all directions |
| **Vapor Processing System** | | | |
| Pits | 1 | 1 | Entire space within any pit or box below grade level, any part of which: (1) is within a Division 1 or Division 2 classified location; (2) is within a Zone 1 or Zone 2 classified location; (3) houses any equipment used to transfer or process vapors |
| Equipment in protective enclosures | 2 | 2 | Entire space within enclosure |
| Equipment *not* within protective enclosure | 2 | 2 | Within 450 mm (18 in.) of equipment containing flammable vapors or liquid, extending horizontally in all directions and down to grade level |
| | 2 | 2 | Up to 450 mm (18 in.) above grade level within 3 m (10 ft) horizontally of the vapor processing equipment |
| - Equipment enclosure | 1 | 1 | Entire space within enclosure, if flammable vapor or liquid is present under normal operating conditions |
| | 2 | 2 | Entire space within enclosure, if flammable vapor or liquid is not present under normal operating conditions |
| - Vacuum assist blower | 2 | 2 | Within 450 mm (18 in.) of blower, extending horizontally in all directions and down to grade level |
| | 2 | 2 | Up to 450 mm (18 in.) above grade level, extending 3 m (10 ft) horizontally in all directions |
| **Vault** | 1 | 1 | Entire interior space, if Class I liquids are stored within |

[1]For marine application, *grade level* means the surface of a pier, extending down to water level.
[2]Refer to Figure 514.3 and Figure 514.3(B) for an illustration of classified location around dispensing devices.
[3]Area classification inside the dispenser enclosure is covered in UL 87, *Standard for Power-Operated Dispensing Devices for Petroleum Products.*
[4]Ceiling-mounted hose reel. [**30A:** Table 8.3.1]

not less than 1.5 m (5 ft) from any dispensing device for Class I liquids. [**30A:**12.1, 12.4, 12.5]

Informational Note No. 1: For information on area classification where liquefied petroleum gases are dispensed, see NFPA 58-2014, *Liquefied Petroleum Gas Code.*

Informational Note No. 2: For information on classified areas pertaining to LP-Gas systems other than residential or commercial, see NFPA 58-2014, *Liquefied Petroleum Gas Code,* and NFPA 59-2012, *Utility LP-Gas Plant Code.*

Informational Note No. 3: See 514.3(C) for motor fuel dispensing stations in marinas and boatyards.

Tables 514.3(B)(1) and (B)(2) are extracted from NFPA 30A. See the commentary following 511.3(D) for information regarding the classification of facilities that offer primarily lubrication-type services. Also see Exhibit 514.1 for an illustration of the Class I location around overhead motor fuel dispensing units.

Aboveground tanks with dispensing equipment are frequently used at fleet motor fuel dispensing facilities. Hazardous area classification for aboveground tank installations is also performed in accordance with Table 514.3(B)(1).

### (C) Motor Fuel Dispensing Stations in Boatyards and Marinas.

**(1) General.** Electrical wiring and equipment located at or serving motor fuel dispensing locations shall be installed on the side of the wharf, pier, or dock opposite from the liquid piping system.

*TABLE 514.3(B)(2) Electrical Equipment Classified Areas for Dispensing Devices*

| Dispensing Device | Extent of Classified Area: Class I, Division 1 | Extent of Classified Area: Class I, Division 2 |
|---|---|---|
| Compressed natural gas | Entire space within the dispenser enclosure | 1.5 m (5 ft) in all directions from dispenser enclosure |
| Liquefied natural gas | Entire space within the dispenser enclosure and 1.5 m (5 ft) in all directions from the dispenser enclosure | From 1.5 m to 3.0 m (5 ft to 10 ft) in all directions from the dispenser enclosure |
| Liquefied petroleum gas | Entire space within the dispenser enclosure; 450 mm (18 in.) from the exterior surface of the dispenser enclosure to an elevation of 1.2 m (4 ft) above the base of the dispenser; the entire pit or open space beneath the dispenser and within 6.0 m (20 ft) horizontally from any edge of the dispenser when the pit or trench is not mechanically ventilated. | Up to 450 mm (18 in.) aboveground and within 6.0 m (20 ft) horizontally from any edge of the dispenser enclosure, including pits or trenches within this area when provided with adequate mechanical ventilation |

[**30A:** Table 12.6.2]

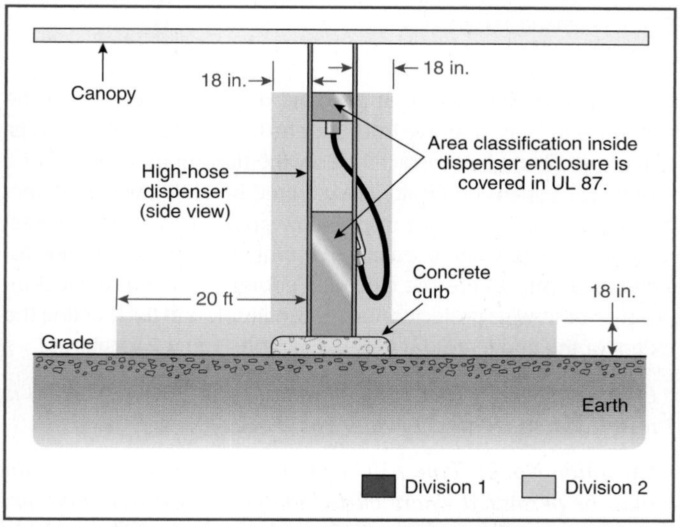


***EXHIBIT 514.1*** *Extent of Class I location around overhead motor fuel dispensing units in accordance with Table 514.3(B)(1).*

Informational Note: For additional information, see NFPA 303-2011, *Fire Protection Standard for Marinas and Boatyards*, and NFPA 30A-2012, *Motor Fuel Dispensing Facilities and Repair Garages*.

**(2) Classification of Class I, Division 1 and 2 Areas.** The following criteria shall be used for the purposes of applying Table 514.3(B)(1) and Table 514.3(B)(2) to motor fuel dispensing equipment on floating or fixed piers, wharfs, or docks.

**(D) Closed Construction.** Where the construction of floating docks, piers, or wharfs is closed so that there is no space between the bottom of the dock, pier, or wharf and the water, as in the case of concrete-enclosed expanded foam or similar construction, and the construction includes integral service boxes with supply chases, the following shall apply:

(1) The space above the surface of the floating dock, pier, or wharf shall be a Class I, Division 2 location with distances as specified in Table 514.3(B)(1)(1) for dispenser and outdoor locations.

(2) Spaces below the surface of the floating dock, pier, or wharf that have areas or enclosures, such as tubs, voids, pits, vaults, boxes, depressions, fuel piping chases, or similar spaces, where flammable liquid or vapor can accumulate shall be a Class I, Division 1 location.

*Exception No. 1: Dock, pier, or wharf sections that do not support fuel dispensers and abut, but are located 6.0 m (20 ft) or more from, dock sections that support a fuel dispenser(s) shall be permitted to be Class I, Division 2 locations where documented air space is provided between dock sections to allow flammable liquids or vapors to dissipate without traveling to such dock sections. The documentation shall comply with 500.4(A).*

*Exception No. 2: Dock, pier, or wharf sections that do not support fuel dispensers and do not directly abut sections that support fuel dispensers shall be permitted to be unclassified where documented air space is provided and where flammable liquids or vapors cannot travel to such dock sections. The documentation shall comply with 500.4(A).*

**(E) Open Construction.** Where the construction of piers, wharfs, or docks is open, as in the case of decks built on stringers supported by pilings, floats, pontoons, or similar construction, the following shall apply:

(1) The area 450 mm (18 in.) above the surface of the dock, pier, or wharf and extending 6.0 m (20 ft) horizontally in all directions from the outside edge of the dispenser and down to the water level shall be a Class 1, Division 2 location.

(2) Enclosures such as tubs, voids, pits, vaults, boxes, depressions, piping chases, or similar spaces where flammable liquids or vapors can accumulate within 6.0 m (20 ft) of the dispenser shall be a Class I, Division 1 location.

EXHIBIT 514.2 *A typical explosionproof junction box used in a dispenser application. (Courtesy of Cooper Crouse-Hinds)*

## 514.4 Wiring and Equipment Installed in Class I Locations

All electrical equipment and wiring installed in Class I locations as classified in 514.3 shall comply with the applicable provisions of Article 501.

*Exception: As permitted in 514.8.*

Informational Note: For special requirements for conductor insulation, see 501.20.

The applicable wiring methods and equipment in Article 501 must be used within the Class I areas at a motor fuel dispensing facility. An explosionproof junction box of the type frequently used in gasoline dispensing units is shown in Exhibit 514.2. The branch-circuit conductors for dispenser power, lighting, or both connect to the internal wiring of the dispenser in the explosionproof junction box.

## 514.7 Wiring and Equipment Above Class I Locations

Wiring and equipment above the Class I locations as classified in 514.3 shall comply with 511.7.

## 514.8 Underground Wiring

Underground wiring shall be installed in threaded rigid metal conduit or threaded steel intermediate metal conduit. Any portion of electrical wiring that is below the surface of a Class I, Division 1, or a Class I, Division 2, location [as classified in Table 514.3(B)(1) and Table 514.3(B)(2)] shall be sealed within 3.05 m (10 ft) of the point of emergence above grade. Except for listed explosionproof reducers at the conduit seal, there shall be no union, coupling, box, or fitting between the conduit seal and the point of emergence above grade. Refer to Table 300.5.

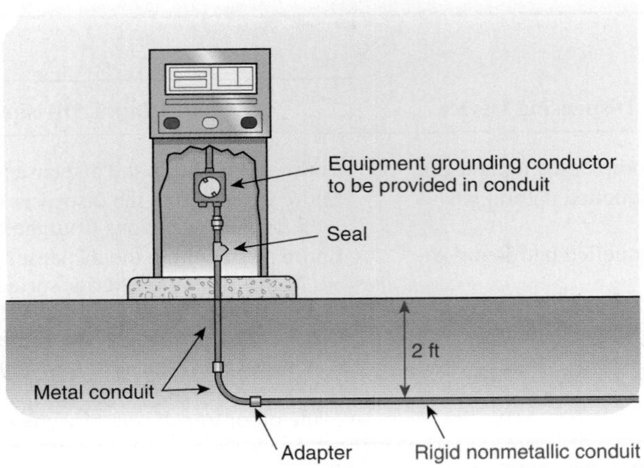


EXHIBIT 514.3 *Use of rigid nonmetallic conduit in accordance with 514.8, Exception No. 2.*

Fuel spilled in the vicinity of gasoline dispensers seeps into the ground and may migrate into underground electrical conduits. Therefore, all conduits installed below the hazardous locations of a motor fuel dispensing facility are required be sealed within 10 feet of the point of emergence from below grade. This boundary seal minimizes the passage of gasoline or other fuel vapors into unclassified locations where the electrical equipment is not explosionproof or otherwise protected. Tables 514.3(B)(1) and (B)(2) define the extent of the aboveground Class I, Divisions 1 and 2 locations.

*Exception No. 1: Type MI cable shall be permitted where it is installed in accordance with Article 332.*

*Exception No. 2: Type PVC conduit and Type RTRC conduit shall be permitted where buried under not less than 600 mm (2 ft) of cover. Where Type PVC conduit or Type RTRC conduit is used, threaded rigid metal conduit or threaded steel intermediate metal conduit shall be used for the last 600 mm (2 ft) of the underground run to emergence or to the point of connection to the aboveground raceway, and an equipment grounding conductor shall be included to provide electrical continuity of the raceway system and for grounding of non–current-carrying metal parts.*

Where Type PVC or RTRC conduit is used for underground wiring, threaded rigid metal conduit or threaded steel intermediate metal conduit must be used for the last 2 feet of the underground run to the point of emergence or to the point of connection to the aboveground raceway. These rigid nonmetallic conduits, including any nonmetallic conduit elbows and fittings, must be located not less than 2 feet below grade, as shown in Exhibit 514.3.

If rigid nonmetallic conduit (Type PVC or RTRC) is used, an EGC must be included and must be bonded to the explosionproof raceway system inside the dispenser. Installation is accomplished by terminating the EGC on the ground screw (or other means) provided in the dispenser junction box.

## 514.9 Sealing

**(A) At Dispenser.** A listed seal shall be provided in each conduit run entering or leaving a dispenser or any cavities or enclosures in direct communication therewith. The sealing fitting shall be the first fitting after the conduit emerges from the earth or concrete.

**(B) At Boundary.** Additional seals shall be provided in accordance with 501.15. Sections 501.15(A)(4) and (B)(2) shall apply to horizontal as well as to vertical boundaries of the defined Class I locations.

Seal fittings are required in all conduits entering or leaving a dispenser and leaving a Class I location. Even though a conduit runs from dispenser to dispenser and does not leave the hazardous location, a seal is necessary where the conduit leaves, and again where it enters, a dispenser. Conduits passing under the boundaries of the hazardous locations (20-foot radius from dispenser, 10-foot radius from a loose-fill tank connection, and 5-foot radius from a tight-fill tank connection) are considered to be in a Class I location, and the seal is to be the first fitting at the point of emergence.

Panelboards are generally located in a room classified as a nonhazardous location; however, any conduit coming from the dispenser or passing under the hazardous location boundaries would require a seal at the panelboard location to minimize the likelihood of gas migration into the remote location. If the panelboard is located in the lube or repair room, all conduits emerging into the 18-inch hazardous location would require seals. See Exhibits 514.4 and 514.5.

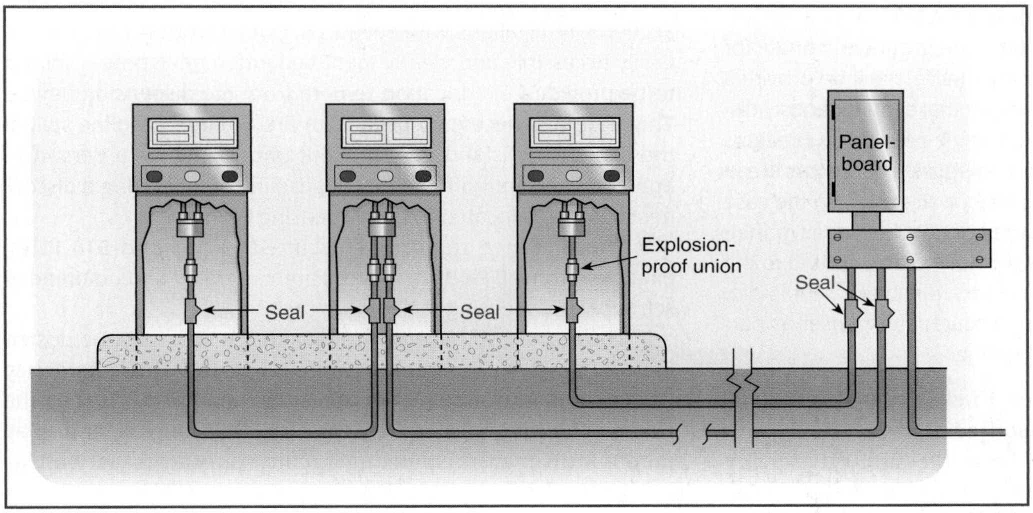


***EXHIBIT 514.4*** *A gasoline dispenser installation indicating locations for seal fittings.*

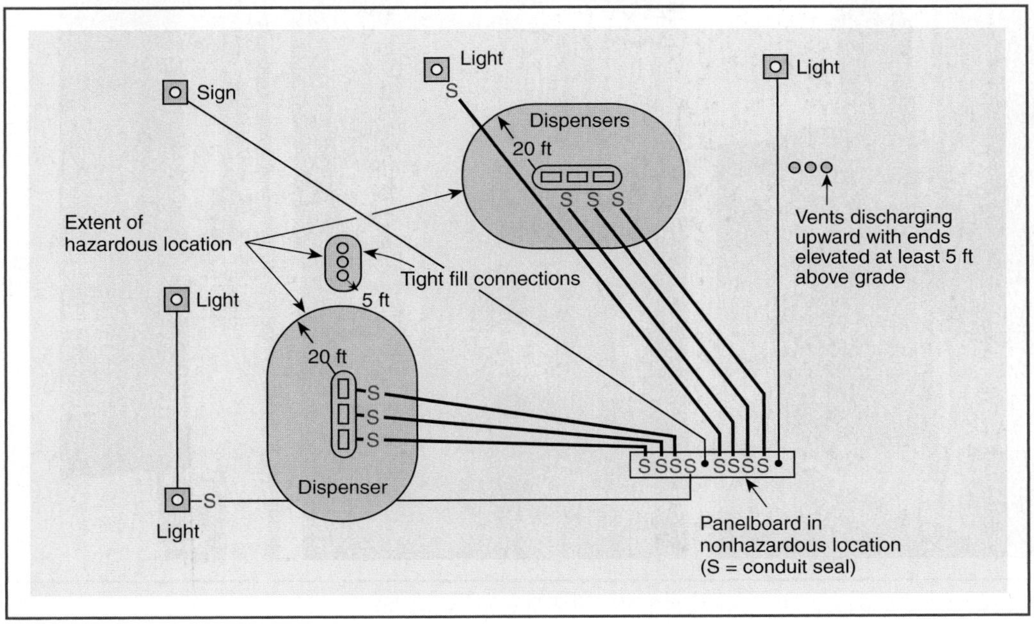


***EXHIBIT 514.5*** *Required seals at points marked "S." Seals are not required at the sign and two of the lights because conduit runs do not pass through a hazardous location.*

## 514.11 Circuit Disconnects

**(A) General.** Each circuit leading to or through dispensing equipment, including all associated power, communications, data, and video circuits, and equipment for remote pumping systems, shall be provided with a clearly identified and readily accessible switch or other approved means, located remote from the dispensing devices, to disconnect simultaneously from the source of supply, all conductors of the circuits, including the grounded conductor, if any.

Single-pole breakers utilizing handle ties shall not be permitted.

The disconnecting means must be clearly marked and readily accessible. The disconnecting means also must be remote from the dispensing device, so that if an emergency occurs that requires rapid shutdown of the dispensing equipment, the person operating the disconnecting means is not exposed to the hazard.

All conductors of a circuit, including the grounded conductor that may be present within a dispensing device, must be provided with a switch or special-type circuit breaker that simultaneously disconnects all conductors. Handle ties on single-pole circuit breakers are not permitted. The intent is that no energized conductors are in the dispenser vicinity during maintenance or alteration. In the case of accidental reversal of the polarities of conductors at panelboards, the grounded conductor must be able to be switched to the open or "off" position. Grounded conductors may be present in old-style pump motors, or they may pass through a dispenser as part of a circuit for the dispensing island lighting.

**(B) Attended Self-Service Motor Fuel Dispensing Facilities.** Emergency controls as specified in 514.11(A) shall be installed at a location acceptable to the authority having jurisdiction, but controls shall not be more than 30 m (100 ft) from dispensers. [**30A:**6.7.1]

**(C) Unattended Self-Service Motor Fuel Dispensing Facilities.** Emergency controls as specified in 514.11(A) shall be installed at a location acceptable to the authority having jurisdiction, but the control shall be more than 6 m (20 ft) but less than 30 m (100 ft) from the dispensers. Additional emergency controls shall be installed on each group of dispensers or the outdoor equipment used to control the dispensers. Emergency controls shall shut off all power to all dispensing equipment at the station. Controls shall be manually reset only in a manner approved by the authority having jurisdiction. [**30A:**6.7.2]

Informational Note: For additional information, see 6.7.1 and 6.7.2 of NFPA 30A-2012, *Code for Motor Fuel Dispensing Facilities and Repair Garages*.

Section 514.11 aligns with Section 6.7 of NFPA 30A requiring an easily accessible and clearly identified emergency power shutoff to be provided at a location remote from the dispensing device. This shutoff is necessary because a fire or large gasoline spill at the dispensing island may make it impossible for a person to approach and shut off the flow of gasoline by operating a disconnecting means located at the dispensing island.

The distance requirements in 514.11(B) and 514.11(C), extracted from NFPA 30A, address both attended and unattended self-service dispensing facilities.

The term *clearly identified* means that a sign must be posted indicating where the shutoff switch is located. This emergency power shutoff must be readily accessible and not blocked by the storage of such things as tires, cases of lubricating oil, or display merchandise. All dispensing facility operators as well as

***EXHIBIT 514.6*** *A self-service dispensing island (left) and associated emergency disconnecting means (right).*

responding fire fighters should know the location of the emergency power shutoff.

Exhibit 514.6 shows a typical dispensing island at a self-service gas station (left) and the required clearly identified emergency disconnect (right). During the evening hours, this location is an unattended self-service dispensing facility. In accordance with 514.11(C), the disconnect must be located where acceptable to the AHJ and is required to be located between 20 feet and 100 feet from the dispensers.

### 514.13 Provisions for Maintenance and Service of Dispensing Equipment

Each dispensing device shall be provided with a means to remove all external voltage sources, including power, communications, data, and video circuits and including feedback, during periods of maintenance and service of the dispensing equipment. The location of this means shall be permitted to be other than inside or adjacent to the dispensing device. The means shall be capable of being locked in the open position in accordance with 110.25.

As more sophisticated control circuitry is integrated into dispensing equipment, simply shutting off the main power source to the dispenser or remote pump does not necessarily ensure that the equipment has been isolated from all sources of voltage. Ensuring that the equipment is completely isolated from all voltage sources by a means to remove all external voltage sources from each dispensing device, including sources that may backfeed into the dispenser, enhances the level of safety for personnel servicing dispensing equipment. Exhibit 514.7 shows the interior of an electronic dispensing unit. Power is required to be removed from printers, displays, keypads, and card readers.

***EXHIBIT 514.7*** *An example of the complexity of the interior of electronic dispensing units.*

### 514.16 Grounding and Bonding

All metal raceways, the metal armor or metallic sheath on cables, and all non–current-carrying metal parts of fixed and portable electrical equipment, regardless of voltage, shall be grounded and bonded. Grounding and bonding in Class I locations shall comply with 501.30.

## ARTICLE 515
## Bulk Storage Plants

Informational Note: Text that is followed by a reference in brackets has been extracted from NFPA 30-2012, *Flammable and Combustible Liquids Code*. Only editorial changes were made to the extracted text to make it consistent with this *Code*.

### 515.1 Scope

This article covers a property or portion of a property where flammable liquids are received by tank vessel, pipelines, tank car, or tank vehicle and are stored or blended in bulk for the purpose of distributing such liquids by tank vessel, pipeline, tank car, tank vehicle, portable tank, or container.

Article 515 covers facilities that store (in bulk) and distribute flammable liquids as opposed to dispensing liquids into fuel tanks of vehicles. Flammable liquid dispensing locations, including those within the bulk storage facility, are covered under Article 514. Bulk storage tanks may be located inside buildings or outside either aboveground or underground. This article addresses the hazardous locations in the vicinity of the storage tank, and the tank vehicle, pier, or wharf from which the liquids are loaded and offloaded. This article also covers the classification of the areas around drum storage containers.

•

### 515.3 Class I Locations

Table 515.3 shall be applied where Class I liquids are stored, handled, or dispensed and shall be used to delineate and classify bulk storage plants. The class location shall not extend beyond a floor, wall, roof, or other solid partition that has no communicating openings. [**30:**8.1, 8.2.2]

Many steps are required to properly classify a hazardous location. Although the *NEC* provides general area classifications, it does not classify specific locations. The *NEC* classifications have been extracted from other NFPA documents. The classifications from those documents are based on the premise that all applicable requirements of the document have been met. Deviations in on-site conditions, such as process conditions, area ventilation, and room construction, from those assumed by the document may alter the general classification. Those responsible for the specific area classification must consider the basis for the general classifications to determine the applicability to their specific location.

NFPA 30 contains the specific construction and installation requirements used to develop the area classifications in Article 515. Table 515.3 is extracted from Table 7.3.3 of NFPA 30. The area classifications listed in the table are based on the premise that all applicable requirements of NFPA 30 have been met as conveyed by Informational Note No. 1 to 515.3.

Exhibits 515.1 through 515.5 illustrate the hazardous locations associated with several types of flammable liquid containers and operations. Exhibits 515.1 and 515.2 depict the

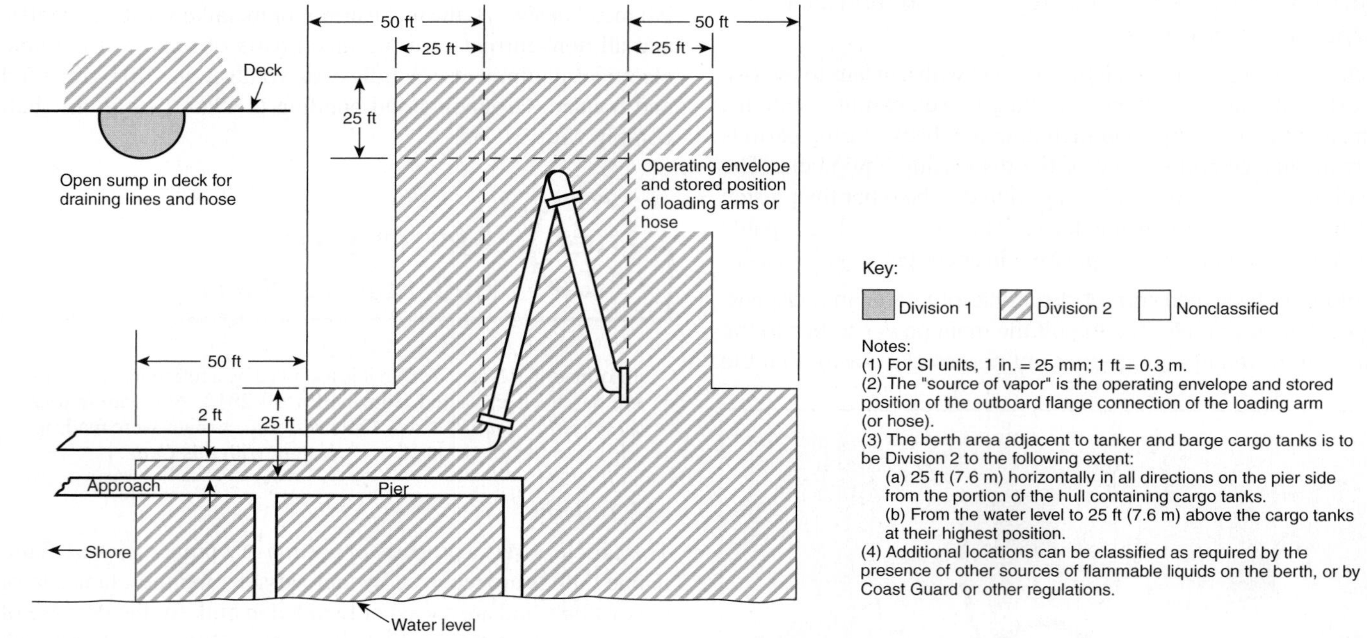


***FIGURE 515.3*** *Marine Terminal Handling Flammable Liquids. [**30:**Figure 29.3.22]*

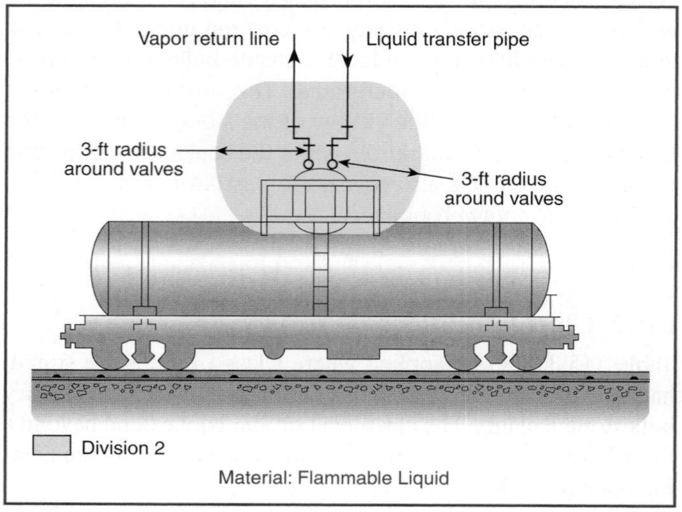


***EXHIBIT 515.1*** *Tank car/tank truck loading and unloading via closed system with transfer through dome only.*

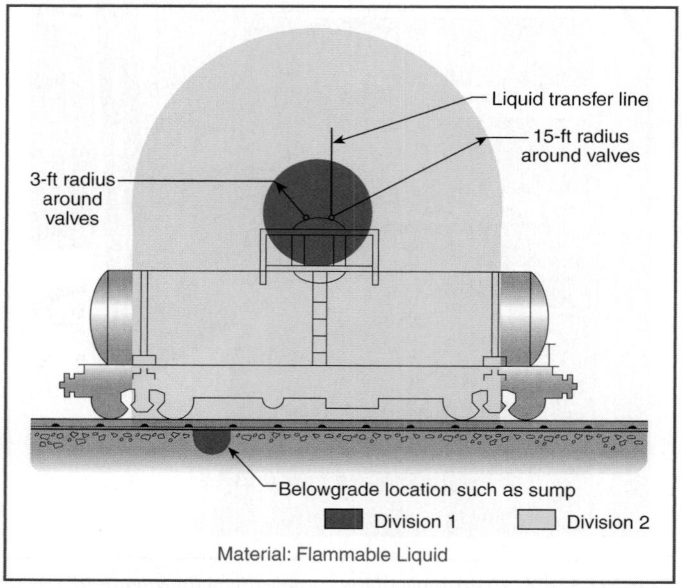


***EXHIBIT 515.2*** *Open system with top or bottom product transfer.*

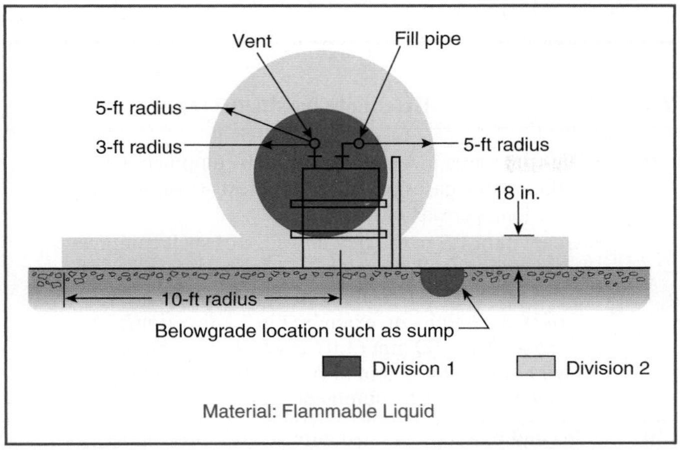


***EXHIBIT 515.3*** *Drum filling station, outdoors or indoors, with adequate ventilation.*

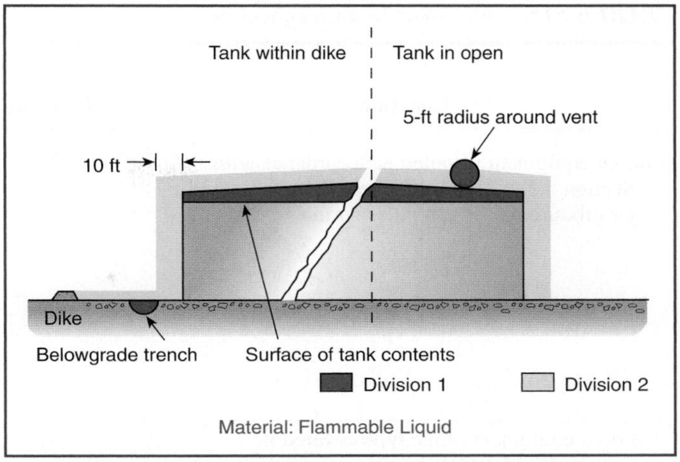


***EXHIBIT 515.5*** *Fixed roof storage tank, outdoors at grade.*

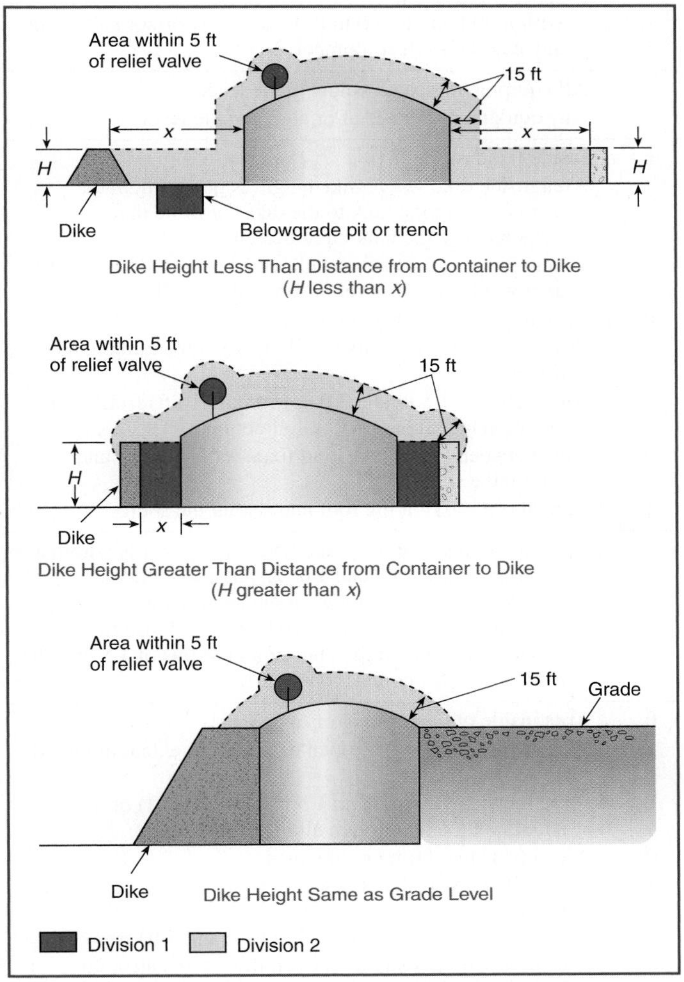


***EXHIBIT 515.4*** *Storage tanks for cryogenic liquids. [Source: Adapted from NFPA 59A-2009, Figures 10.7.2(a) through 10.7.2(c)]*

classification difference between using a closed and open transfer system on a tank car.

Informational Note No. 1: The area classifications listed in Table 515.3 are based on the premise that the installation meets the applicable requirements of NFPA 30-2012, *Flammable and Combustible Liquids Code*, Chapter 5, in all respects. Should this not be the case, the authority having jurisdiction has the authority to classify the extent of the classified space.

Informational Note No. 2: See 555.21 for gasoline dispensing stations in marinas and boatyards.

## 515.4 Wiring and Equipment Located in Class I Locations

All electrical wiring and equipment within the Class I locations defined in 515.3 shall comply with the applicable provisions of Article 501 or Article 505 for the division or zone in which they are used.

*Exception: As permitted in 515.8.*

## 515.7 Wiring and Equipment Above Class I Locations

**(A) Fixed Wiring.** All fixed wiring above Class I locations shall be in metal raceways, Schedule 80 PVC conduit, Type RTRC marked with the suffix -XW, or Type MI, Type TC, or Type MC cable, or Type PLTC and Type PLTC-ER cable in accordance with the provisions of Article 725, including installation in cable tray systems or Type ITC and Type ITC-ER cable as permitted in 727.4. The cable shall be terminated with listed fittings.

**(B) Fixed Equipment.** Fixed equipment that may produce arcs, sparks, or particles of hot metal, such as lamps and lampholders for fixed lighting, cutouts, switches, receptacles, motors, or other equipment having make-and-break or sliding contacts, shall be of the totally enclosed type or be constructed so as to prevent the escape of sparks or hot metal particles.

***TABLE 515.3*** *Electrical Area Classifications*

| Location | NEC Class I Division | Zone | Extent of Classified Area |
|---|---|---|---|
| Indoor equipment installed in accordance with Section 5.3 of NFPA 30 where flammable vapor–air mixtures can exist under normal operation | 1 | 0 | The entire area associated with such equipment where flammable gases or vapors are present continuously or for long periods of time |
| | 1 | 1 | Area within 1.5 m (5 ft) of any edge of such equipment, extending in all directions |
| | 2 | 2 | Area between 1.5 m and 2.5 m (5 ft and 8 ft) of any edge of such equipment, extending in all directions; also, space up to 900 mm (3 ft) above floor or grade level within 1.5 m to 7.5 m (5 ft to 25 ft) horizontally from any edge of such equipment[1] |
| Outdoor equipment of the type covered in Section 5.3 of NFPA 30 where flammable vapor–air mixtures may exist under normal operation | 1 | 0 | The entire area associated with such equipment where flammable gases or vapors are present continuously or for long periods of time |
| | 1 | 1 | Area within 900 mm (3 ft) of any edge of such equipment, extending in all directions |
| | 2 | 2 | Area between 900 mm (3 ft) and 2.5 m (8 ft) of any edge of such equipment, extending in all directions; also, space up to 900 mm (3 ft) above floor or grade level within 900 mm to 3.0 m (3 ft to 10 ft) horizontally from any edge of such equipment |
| Tank storage installations inside buildings | 1 | 1 | All equipment located below grade level |
| | 2 | 2 | Any equipment located at or above grade level |
| Tank – aboveground | 1 | 0 | Inside fixed roof tank |
| | 1 | 1 | Area inside dike where dike height is greater than the distance from the tank to the dike for more than 50 percent of the tank circumference |
| Shell, ends, or roof and dike area | 2 | 2 | Within 3.0 m (10 ft) from shell, ends, or roof of tank; also, area inside dike to level of top of dike wall |
| Vent | 1 | 0 | Area inside of vent piping or opening |
| | 1 | 1 | Within 1.5 m (5 ft) of open end of vent, extending in all directions |
| | 2 | 2 | Area between 1.5 m and 3.0 m (5 ft and 10 ft) from open end of vent, extending in all directions |
| Floating roof with fixed outer roof | 1 | 0 | Area between the floating and fixed roof sections and within the shell |
| Floating roof with no fixed outer roof | 1 | 1 | Area above the floating roof and within the shell |
| Underground tank fill opening | 1 | 1 | Any pit, or space below grade level, if any part is within a Division 1 or 2, or Zone 1 or 2, classified location |
| | 2 | 2 | Up to 450 mm (18 in.) above grade level within a horizontal radius of 3.0 m (10 ft) from a loose fill connection, and within a horizontal radius of 1.5 m (5 ft) from a tight fill connection |
| Vent – discharging upward | 1 | 0 | Area inside of vent piping or opening |
| | 1 | 1 | Within 900 mm (3 ft) of open end of vent, extending in all directions |
| | 2 | 2 | Area between 900 mm and 1.5 m (3 ft and 5 ft) of open end of vent, extending in all directions |
| Drum and container filling – outdoors or indoors | 1 | 0 | Area inside the drum or container |
| | 1 | 1 | Within 900 mm (3 ft) of vent and fill openings, extending in all directions |
| | 2 | 2 | Area between 900 mm and 1.5 m (3 ft and 5 ft) from vent or fill opening, extending in all directions; also, up to 450 mm (18 in.) above floor or grade level within a horizontal radius of 3.0 m (10 ft) from vent or fill opening |

*TABLE 515.3 Continued*

| Location | NEC Class I Division | Zone | Extent of Classified Area |
|---|---|---|---|
| Pumps, bleeders, withdrawal fittings | | | |
| Indoors | 2 | 2 | Within 1.5 m (5 ft) of any edge of such devices, extending in all directions; also, up to 900 mm (3 ft) above floor or grade level within 7.5 m (25 ft) horizontally from any edge of such devices |
| Outdoors | 2 | 2 | Within 900 mm (3 ft) of any edge of such devices, extending in all directions. Also, up to 450 mm (18 in.) above grade level within 3.0 m (10 ft) horizontally from any edge of such devices |
| Pits and sumps | | | |
| Without mechanical ventilation | 1 | 1 | Entire area within a pit or sump if any part is within a Division 1 or 2, or Zone 1 or 2, classified location |
| With adequate mechanical ventilation | 2 | 2 | Entire area within a pit or sump if any part is within a Division 1 or 2, or Zone 1 or 2, classified location |
| Containing valves, fittings, or piping, and not within a Division 1 or 2, or Zone 1 or 2, classified location | 2 | 2 | Entire pit or sump |
| Drainage ditches, separators, impounding basins | | | |
| Outdoors | 2 | 2 | Area up to 450 mm (18 in.) above ditch, separator, or basin; also, area up to 450 mm (18 in.) above grade within 4.5 m (15 ft) horizontally from any edge |
| Indoors | | | Same classified area as pits |
| Tank vehicle and tank car[2] loading through open dome | 1 | 0 | Area inside of the tank |
| | 1 | 1 | Within 900 mm (3 ft) of edge of dome, extending in all directions |
| | 2 | 2 | Area between 900 mm and 4.5 m (3 ft and 15 ft) from edge of dome, extending in all directions |
| Loading through bottom connections with atmospheric venting | 1 | 0 | Area inside of the tank |
| | 1 | 1 | Within 900 mm (3 ft) of point of venting to atmosphere, extending in all directions |
| | 2 | 2 | Area between 900 mm and 4.5 m (3 ft and 15 ft) from point of venting to atmosphere, extending in all directions; also, up to 450 mm (18 in.) above grade within a horizontal radius of 3.0 m (10 ft) from point of loading connection |
| Office and rest rooms | Unclassified | | If there is any opening to these rooms within the extent of an indoor classified location, the room shall be classified the same as if the wall, curb, or partition did not exist. |
| Loading through closed dome with atmospheric venting | 1 | 1 | Within 900 mm (3 ft) of open end of vent, extending in all directions |
| | 2 | 2 | Area between 900 mm and 4.5 m (3 ft and 15 ft) from open end of vent, extending in all directions; also, within 900 mm (3 ft) of edge of dome, extending in all directions |
| Loading through closed dome with vapor control | 2 | 2 | Within 900 mm (3 ft) of point of connection of both fill and vapor lines extending in all directions |
| Bottom loading with vapor control or any bottom unloading | 2 | 2 | Within 900 mm (3 ft) of point of connections, extending in all directions; also up to 450 mm (18 in.) above grade within a horizontal radius of 3.0 m (10 ft) from point of connections |

*(continues)*

*TABLE 515.3 Continued*

| Location | NEC Class I Division | Zone | Extent of Classified Area |
|---|---|---|---|
| Storage and repair garage for tank vehicles | 1 | 1 | All pits or spaces below floor level |
| | 2 | 2 | Area up to 450 mm (18 in.) above floor or grade level for entire storage or repair garage |
| Garages for other than tank vehicles | Unclassified | | If there is any opening to these rooms within the extent of an outdoor classified location, the entire room shall be classified the same as the area classification at the point of the opening. |
| Outdoor drum storage | Unclassified | | |
| Inside rooms or storage lockers used for the storage of Class I liquids | 2 | 2 | Entire room |
| Indoor warehousing where there is no flammable liquid transfer | Unclassified | | If there is any opening to these rooms within the extent of an indoor classified location, the room shall be classified the same as if the wall, curb, or partition did not exist. |
| Piers and wharves | | | See Figure 515.3. |

[1]The release of Class I liquids may generate vapors to the extent that the entire building, and possibly an area surrounding it, should be considered a Class I, Division 2 or Zone 2 location.

[2]When classifying extent of area, consideration shall be given to fact that tank cars or tank vehicles may be spotted at varying points. Therefore, the extremities of the loading or unloading positions shall be used. [**30:**Table 8.2.2]

**(C) Portable Luminaires or Other Utilization Equipment.** Portable luminaires or other utilization equipment and their flexible cords shall comply with the provisions of Article 501 or Article 505 for the class of location above which they are connected or used.

## 515.8 Underground Wiring

**(A) Wiring Method.** Underground wiring shall be installed in threaded rigid metal conduit or threaded steel intermediate metal conduit or, where buried under not less than 600 mm (2 ft) of cover, shall be permitted in Type PVC conduit, Type RTRC conduit, or a listed cable. Where Type PVC conduit or Type RTRC conduit is used, threaded rigid metal conduit or threaded steel intermediate metal conduit shall be used for not less than the last 600 mm (2 ft) of the conduit run to the conduit point of emergence from the underground location or to the point of connection to an aboveground raceway. Where cable is used, it shall be enclosed in threaded rigid metal conduit or threaded steel intermediate metal conduit from the point of lowest buried cable level to the point of connection to the aboveground raceway.

See the commentary following 514.8 regarding underground wiring.

**(B) Insulation.** Conductor insulation shall comply with 501.20.

**(C) Nonmetallic Wiring.** Where Type PVC conduit, Type RTRC conduit, or cable with a nonmetallic sheath is used, an equipment grounding conductor shall be included to provide for electrical continuity of the raceway system and for grounding of non–current-carrying metal parts.

## 515.9 Sealing

Sealing requirements shall apply to horizontal as well as to vertical boundaries of the defined Class I locations. Buried raceways and cables under defined Class I locations shall be considered to be within a Class I, Division 1 or Zone 1 location.

## 515.10 Special Equipment — Gasoline Dispensers

Where gasoline or other volatile flammable liquids or liquefied flammable gases are dispensed at bulk stations, the applicable provisions of Article 514 shall apply.

## 515.16 Grounding and Bonding

All metal raceways, the metal armor or metallic sheath on cables, and all non–current-carrying metal parts of fixed or portable electrical equipment, regardless of voltage, shall be grounded and bonded as provided in Article 250.

Grounding and bonding in Class I locations shall comply with 501.30 for Class I, Division 1 and 2 locations and 505.25 for Class I, Zone 0, 1, and 2 locations.

Informational Note: For information on grounding for static protection, see 4.5.3.4 and 4.5.3.5 of NFPA 30-2012, *Flammable and Combustible Liquids Code*.

# ARTICLE 516
# Spray Application, Dipping, Coating, and Printing Processes Using Flammable or Combustible Materials

Informational Note: Text that is followed by a reference in brackets has been extracted from NFPA 33-2011, *Standard for Spray Application Using Flammable and Combustible Materials*, or NFPA 34-2011, *Standard for Dipping, Coating, and Printing Processes Using Flammable or Combustible Liquids*. Only editorial changes were made to the extracted text to make it consistent with this *Code*.

## 516.1 Scope

This article covers the regular or frequent application of flammable liquids, combustible liquids, and combustible powders by spray operations and the application of flammable liquids, or combustible liquids at temperatures above their flashpoint, by dipping, coating, printing, or other means.

NFPA 33 covers the spray application of flammable or combustible materials by means of compressed air atomization, airless or hydraulic atomization, electrostatic application methods, or any other means in continuous or intermittent processes. NFPA 33 also covers the application of combustible powders applied by powder spray guns, electrostatic powder spray guns, and the fluidized bed application method or electrostatic fluidized bed application method. NFPA 33 contains requirements for the maintenance of safe conditions as well as personal safety.

NFPA 34 covers the fire and explosion hazards of dipping and coating processes that use flammable or combustible liquids, or use water-borne, water-based, and water-reducible materials that contain flammable or combustible liquids or that produce combustible deposits or residues.

Informational Note: For further information regarding safeguards for these processes, such as fire protection, posting of warning signs, and maintenance, see NFPA 33-2011, *Standard for Spray Application Using Flammable and Combustible Materials*, and NFPA 34-2011, *Standard for Dipping, Coating, and Printing Processes Using Flammable or Combustible Liquids*. For additional information regarding ventilation, see NFPA 91-2010, *Standard for Exhaust Systems for Air Conveying of Vapors, Gases, Mists, and Noncombustible Particulate Solids*.

Maintenance and correct operation of equipment where flammable and combustible materials are present are critical to protect life and property from fire or explosion. Industry experience has shown that the largest fire losses and frequency of fires occur where the proper codes and standards have not been used or applied.

Flammable or combustible liquids or powders and their vapors or mists, and highly combustible residues or powders, are the principal hazards of these spray application locations. Careful consideration of the location and installation of extinguishing equipment can reduce the possibility of fire spreading. NFPA 33 and NFPA 34 address these additional, nonelectrical hazards.

## 516.2 Definitions

For the purpose of this article, the following definitions shall apply.

**Flash-Off Area.** An open or enclosed area after a spray application process where vapors are released due to exposure to ambient air or a heated atmosphere. [**33:**3.3.1.1]

**Limited Finishing Workstation.** An apparatus that is capable of confining the vapors, mists, residues, dusts, or deposits that are generated by a spray application process and that meets the requirements of Section 14.3 of NFPA 33, *Standard for Spray Application Using Flammable or Combustible Materials*, but does not meet the requirements of a spray booth or spray room, as herein defined. [**33:**3.3.15.1]

**Resin Application Area.** Any area in which polyester resins or gelcoats are spray applied. [**33:**3.3.1.2]

**Spray Area.** Any fully enclosed, partly enclosed, or unenclosed area in which ignitible quantities of flammable or combustible vapors, mists, residues, dusts, or deposits are present due to the operation of spray processes, including (1) any area in the direct path of a spray application process; (2) the interior of a spray booth or spray room or limited finishing workstation, as herein defined; (3) the interior of any exhaust plenum, eliminator section, or scrubber section; (4) the interior of any exhaust duct or exhaust stack leading from a spray application process; (5) the interior of any air recirculation filter house or enclosure, including secondary recirculation particulate filters; (6) any solvent concentrator (pollution abatement) unit or solvent recovery (distillation) unit. The following are not considered to be a part of the spray area: (1) fresh air make-up units; (2) air supply ducts and air supply plenums; (3) recirculation air supply ducts downstream of secondary filters; (4) exhaust ducts from solvent concentrator (pollution abatement) units. [**33:**3.3.2.3]

Informational Note: Unenclosed spray areas are locations outside of buildings or are localized operations within a larger room or space. Such are normally provided with some local vapor extraction/ventilation system. In automated operations, the area limits are the maximum area in the direct path of spray operations. In manual operations, the area limits are the maximum area of spray when aimed at 90 degrees to the application surface.

**Spray Booth.** A power-ventilated enclosure for a spray application operation or process that confines and limits the escape of the material being sprayed, including vapors, mists, dusts, and residues that are produced by the spraying operation and conducts or directs these materials to an exhaust system. [**33:**3.3.14]

Informational Note: A spray booth is an enclosure or insert within a larger room used for spray/coating/dipping applications. A spray booth may be fully enclosed or have open front or face and may include a separate conveyor entrance and exit. The spray booth is provided with a dedicated ventilation exhaust but may draw supply air from the larger room or have a dedicated air supply.

**Spray Room.** A power-ventilated fully enclosed room used exclusively for open spraying of flammable or combustible materials. A spray room is a purposefully enclosed room built for spray/coating/dipping applications provided with dedicated ventilation supply and exhaust. Normally the room is configured to house the item to be painted, providing reasonable access around the item/process. Depending on the size of the item being painted, such rooms may actually be the entire building or the major portion thereof. [**33:**3.3.15]

**Unenclosed Spray Area.** Any spray area that is not confined by a limited finishing workstation, spray booth, or spray room, as herein defined. [**33:**3.3.2.3.2]

Subsection 516.3(A)(1)2 was added by a tentative interim amendments (TIA).

## 516.3 Classification of Locations

Classification is based on quantities of flammable vapors, combustible mists, residues, dusts, or deposits that are present or might be present in quantities sufficient to produce ignitible or explosive mixtures with air.

Many steps are required to properly classify a hazardous location. Although the *NEC* provides general area classifications, it does not classify specific locations. The *NEC* classifications have been extracted from other NFPA documents. The classifications from those documents are based on the premise that all applicable requirements of the document have been met. Deviations in on-site conditions, such as process conditions, area ventilation, and room construction, from those assumed by the document may alter the general classification. Those responsible for the specific area classification must consider the basis for the general classifications to determine the applicability to their specific location.

NFPA 33 and NFPA 34 contain the specific construction and installation requirements used to develop the area classifications in Article 516.

The determination of the extent of hazardous areas involved in spray application requires an understanding of the multiple hazards of flammable vapors, mists, powders, and highly combustible deposits applied at each location.

**(A) Zone Classification of Locations.**

(1) For the purposes of this article, the zone system of electrical area classification shall be applied as follows:
  a. The inside of open or closed containers or vessels shall be considered a Class I, Zone 0 location.
  b. A Class I, Division 1 location shall be permitted to be alternatively classified as a Class I, Zone 1 location.
  c. A Class I, Division 2 location shall be permitted to be alternatively classified as a Class I, Zone 2 location.
  d. A Class II, Division 1 location shall be permitted to be alternatively classified as a Zone 21 location.
  e. A Class II, Division 2 location shall be permitted to be alternatively classified as a Zone 22 location. [**33:**6.2.2]

(2) For the purposes of electrical area classification, the division system and the zone system shall not be intermixed for any given source of release. [**33:**6.2.3]

(3) In instances of areas within the same facility classified separately, Class I, Zone 2 locations shall be permitted to abut, but not overlap, Class I, Division 2 locations. Class I, Zone 0 or Zone 1 locations shall not abut Class I, Division 1 or Division 2 locations. [**33:**6.2.4]

(4) Open flames, spark-producing equipment or processes, and equipment whose exposed surfaces exceed the autoignition temperature of the material being sprayed shall not be located in a spray area or in any surrounding area that is classified as Division 2, Zone 2, or Zone 22.

*Exception: This requirement shall not apply to drying, curing, or fusing apparatus. [**33:**6.2.5]*

(5) Any utilization equipment or apparatus that is capable of producing sparks or particles of hot metal and that is located above or adjacent to either the spray area or the surrounding Division 2, Zone 2, or Zone 22 areas shall be of the totally enclosed type or shall be constructed to prevent the escape of sparks or particles of hot metal. [**33:**6.2.6]

**(B) Class I, Division 1 or Class I, Zone 0 Locations.** The following spaces shall be considered Class I, Division 1, or Class I, Zone 0, as applicable:

(1) The interior of any open or closed container or vessel of a flammable liquid
(2) The interior of any dip tank or coating tank
(3) The interior of any ink fountain, ink reservoir, or ink tank

Informational Note: For additional guidance, see Chapter 6 of NFPA 33-2011, *Standard for Spray Application Using Flammable or Combustible Materials*, and Chapter 6 of NFPA 34-2011, *Standard for Dipping, Coating, and Printing Processes Using Flammable or Combustible Liquids*.

**(C) Class I, Division 1; Class I, Zone 1; Class II, Division 1; or Zone 21 Locations.** The following spaces shall be considered Class I, Division 1, or Class I, Zone 1, Class II, Division 1, or Zone 21 locations, as applicable:

Where spraying operations are confined to adequately ventilated spray booths or rooms, deposits of combustible residues or dangerous concentrations of flammable vapors, mists, or dusts outside the spray booth are not likely under normal operating conditions. Adjacent to the Division 1 spraying area, there is usually a Division 2 area. This Division 2 location should contain no equipment that produces ignition-capable sparks under normal operation. Within this Division 2 location, electric lamps must be enclosed to prevent hot particles from falling on freshly painted stock or other readily ignitible material and, if subject to physical damage, the lamps must be properly guarded. See 516.7(B).

(1) The interior of spray booths and rooms except as specifically provided in 516.3(D)(7).
(2) The interior of exhaust ducts.
(3) Any area in the direct path of spray operations.
(4) For open dipping and coating operations, all spaces within a 1.5-m (5-ft) radial distance from the vapor sources extending from these surfaces to the floor. The vapor source shall be the liquid exposed in the process and the drainboard, and any dipped or coated object from which it is possible to measure vapor concentrations exceeding 25 percent of the lower flammable limit at a distance of 300 mm (1 ft), in any direction, from the object as in Figure 516.3(D)(1).
(5) Sumps, pits, or belowgrade channels within 7.5 m (25 ft) horizontally of a vapor source. If the sump, pit, or channel extends beyond 7.5 m (25 ft) from the vapor source, it shall be provided with a vapor stop or it shall be classified as Class I, Division 1 for its entire length.
(6) All space in all directions outside of but within 900 mm (3 ft) of open containers, supply containers, spray gun cleaners, and solvent distillation units containing flammable liquids.
(7) For limited finishing workstations, the area inside the curtains or partitions. See Figure 516.3(D)(5).

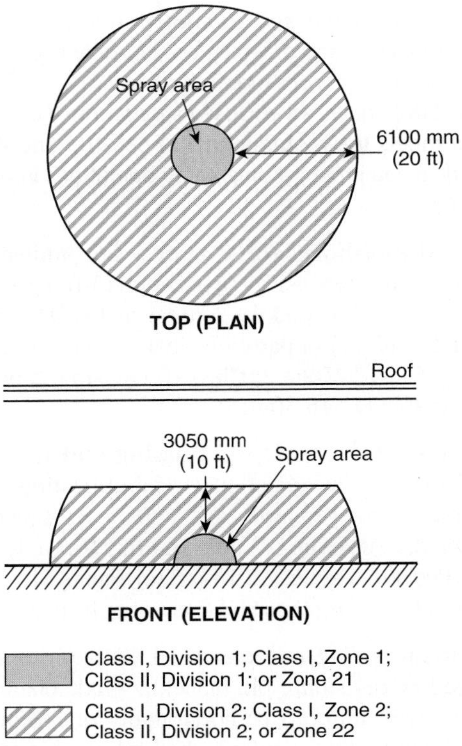


***FIGURE 516.3(D)(1)** Electrical Area Classification for Unenclosed Spray Areas. [**33:**Figure 6.5.1]*

**(D) Class I, Division 2; Class I, Zone 2; Class II, Division 2; or Zone 22 Locations.** The following spaces shall be considered Class I, Division 2; Class I, Zone 2; Class II, Division 2; or Zone 22, as applicable.

**(1) Unenclosed Spray Processes.** For unenclosed spraying, all space outside of but within 6 m (20 ft) horizontally and 3 m (10 ft) vertically of the Class I, Division 1 or Class I, Zone 1 location as defined in 516.3(A) and not separated from it by partitions. See Figure 516.3(D)(1). [**33:**6.5.1]

**(2) Closed-Top, Open-Face, and Open-Front Spray Booths and Spray Rooms.** If spray application operations are conducted within a closed-top, open-face, or open-front booth or room, as shown in Figure 516.3(D)(2), any electrical wiring or utilization equipment located outside of the booth or room but within 915 mm (3 ft) of any opening shall be suitable for Class I, Division 2; Class I, Zone 2; Class II, Division 2; or Zone 22 locations, whichever is applicable. The Class I, Division 2; Class I, Zone 2; Class II, Division 2; or Zone 22 locations shown in Figure 516.3(D)(2) shall extend from the edges of the open face or open front of the booth or room in accordance with the following:

Informational Note: For both interlocked and non-interlocked exhaust ventilation systems, the Division 2, Zone 2 or Zone 22

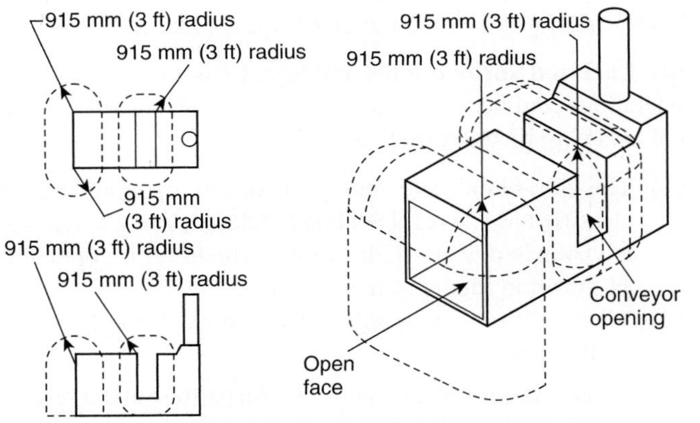


***FIGURE 516.3(D)(2)** Class I, Division 2; Class I, Zone 2; Class II, Division 2; or Zone 22 Locations Adjacent to a Closed Top, Open Face, or Open Front Spray Booth or Room.*

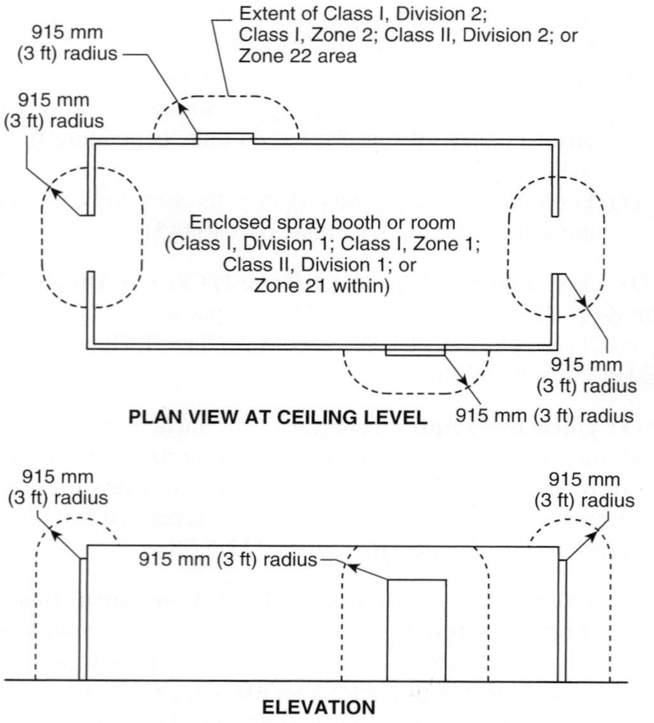


*FIGURE 516.3(D)(4)(1) Class I, Division 2; Class I, Zone 2; Class II, Division 2; or Zone 22 Locations Adjacent to an Enclosed Spray Booth or Spray Room. [33:Figure 6.5.4]*

location extends 915 mm (3 ft) horizontally and 915 mm (3 ft) vertically from the open face or open front of the booth or room, as shown in Figure 516.3(D)(2).

**(3) Open-Top Spray Booths.** For spraying operations conducted within an open top spray booth, the space 915 mm (3 ft) vertically above the booth and within 915 mm (3 ft) of other booth openings shall be considered Class I, Division 2; Class I, Zone 2; Class II, Division 2; or Zone 22. [**33:**6.5.3]

**(4) Enclosed Spray Booths and Spray Rooms.** For spraying operations confined to an enclosed spray booth or room, electrical area classification shall be as follows: [**33:**6.5.4]

(1) The area within 915 mm (3 ft) of any opening shall be classified as Class I, Division 2; Class I, Zone 2; Class II, Division 2; or Zone 22 locations, whichever is applicable, as shown in Figure 516.3(D)(4)(1).
(2) Where exhaust air is recirculated, both of the following shall apply:
  a. The interior of any recirculation path from the secondary particulate filters up to and including the air supply plenum shall be classified as Class I, Division 2; Class I, Zone 2; Class II, Division 2; or Zone 22 locations, whichever is applicable.
  b. The interior of fresh air supply ducts shall be unclassified.

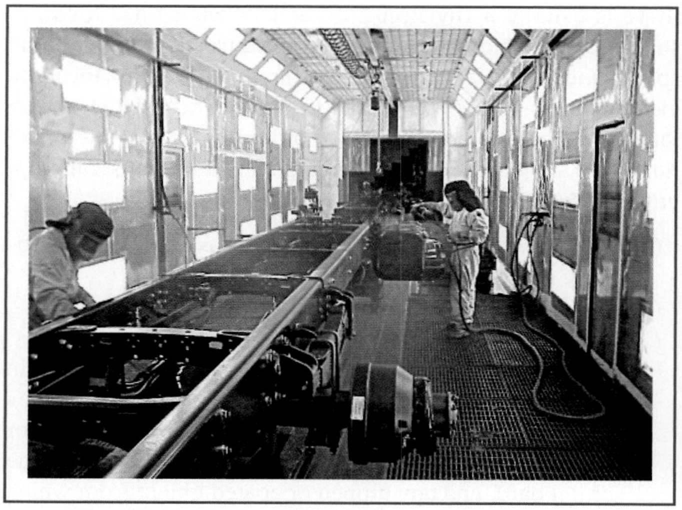

*EXHIBIT 516.1 An enclosed paint spray room.*

(3) Where exhaust air is not recirculated, the interior of fresh air supply ducts and fresh air supply plenums shall be unclassified.

Spray booths with adequate mechanical ventilation must discharge the vapors or powder to a safe location to reduce the possibility of an explosion and to control the accumulation of overspray residues, many of which are not only highly combustible but also subject to spontaneous ignition.

Exhibit 516.1 shows an enclosed spray room. Note the illumination that meets the requirements of 516.4(C) and the lack of other electrical equipment during spray operations in accordance with 516.4(D).

**(5) Limited Finishing Workstations.** For limited finishing workstations, the area inside the 915-mm (3-ft) space horizontally and vertically beyond the volume enclosed by the outside surface of the curtains or partitions shall be classified as Class I, Division 2; Class I, Zone 2; Class II, Division 2; or Zone 22, as shown in Figure 516.3(D)(5).

**(6) Areas Adjacent to Open Dipping and Coating Processes.** Electrical wiring and electrical utilization equipment located adjacent to open processes shall meet the requirements of 516(D)(6)(1) through (4) and Figure 516.3(D)(6)(a), Figure 516.3(D)(6)(b), Figure 516.3(D)(6)(c), Figure 516.3(D)(6)(d), or Figure 516.3(D)(6)(e), whichever is applicable. [**34:**6.4]

(1) Electrical wiring and electrical utilization equipment located in any sump, pit, or below grade channel that is within 7620 mm (25 ft) horizontally of a vapor source, as defined by this standard, shall be suitable for Class I, Division 1 or Class I, Zone 1 locations. If the sump, pit, or channel extends beyond 7620 mm (25 ft) of the vapor source, it shall be provided with a vapor stop, or it shall

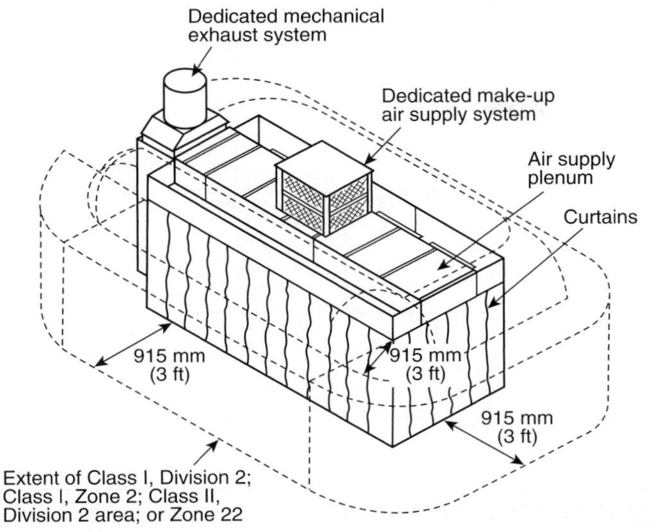


*FIGURE 516.3(D)(5) Class I, Division 2; Class I, Zone 2; Class II, Division 2; or Zone 22 Locations Adjacent to a Limited Finishing Workstation. [33:Figure 14.3.5.1]*

be classified as Class I, Division 1 or Class I, Zone 1 for its entire length. [**34:**6.4.1]

(2) Electrical wiring and electrical utilization equipment located within 1525 mm (5 ft) of a vapor source shall be suitable for Class I, Division 1 or Class I, Zone 1 locations. The space inside a dip tank, ink fountain, ink reservoir, or ink tank shall be classified as Class I, Division 1 or Class I, Zone 0, whichever is applicable. [**34:**6.4.2]

(3) Electrical wiring and electrical utilization equipment located within 915 mm (3 ft) of the Class I, Division 1 or Class I, Zone 1 location described in 516.3(D)(6)(2) shall be suitable for Class I, Division 2 or Class I, Zone 2 locations, whichever is applicable. [**34:**6.4.3]

(4) The space 915 mm (3 ft) above the floor and extending 6100 mm (20 ft) horizontally in all directions from the Class I, Division 1 or Class I, Zone 1 location described in 6.4.3 shall be classified as Class I, Division 2 or Class I, Zone 2, and electrical wiring and electrical utilization equipment located within this space shall be suitable for Class I, Division 2 or Class I, Zone 2 locations, whichever is applicable. [**34:**6.4.4]

*Exception: This space shall be permitted to be unclassified for purposes of electrical installations if the surface area of the vapor source does not exceed 0.5 m² (5 ft²), the contents of the dip tank, ink fountain, ink reservoir, or ink tank do not exceed 19 L (5 gal), and the vapor concentration during operating and shutdown periods does not exceed 25 percent of the lower flammable limit.*

**(7) Enclosed Coating and Dipping Operations.** Areas adjacent to enclosed dipping and coating processes shall be classified in accordance with 516.3(D)(7) and Figure 516.3(D)(7). The space adjacent to an enclosed dipping or coating process or apparatus shall be considered unclassified. [**34:**6.5.3]

*Exception: The space within 915 mm (3 ft) in all directions from any opening in the enclosures shall be classified as Class I, Division 2 or Class I, Zone 2, as applicable. [34:6.5.2]*

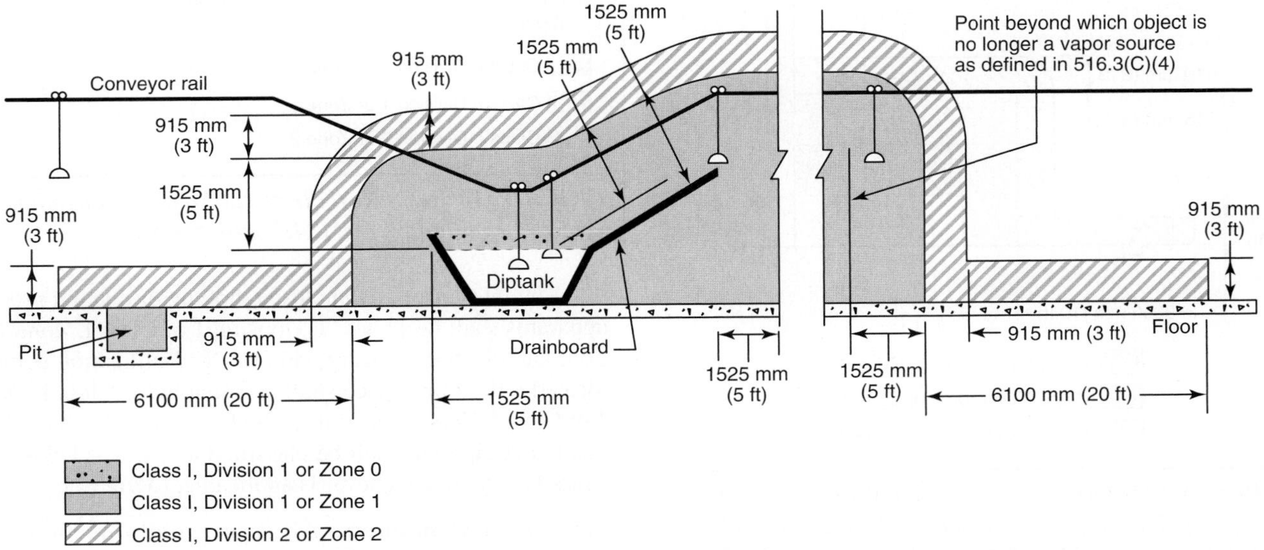


*FIGURE 516.3(D)(6)(a) Electrical Area Classification for Open Dipping and Coating Processes Without Vapor Containment or Ventilation. [34:Figure 6.4(a)]*

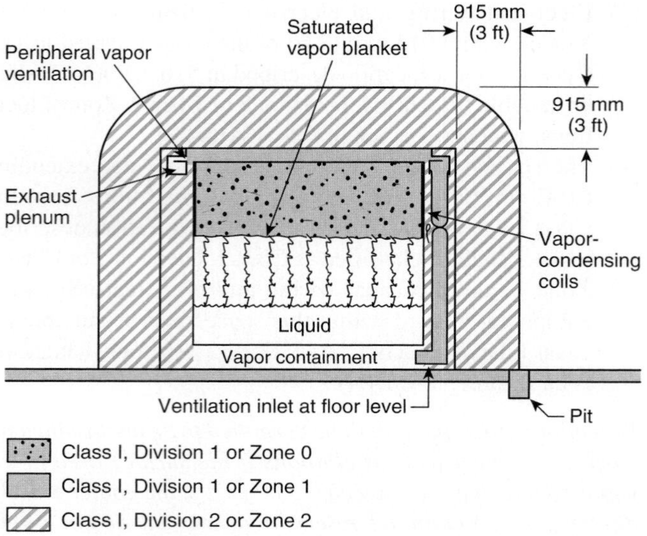


Definitions

Freeboard: The distance from the maximum solvent or coating material level to the top of the tank

Freeboard ratio: The freeboard height divided by the smaller of the interior length or interior width of the tank

***FIGURE 516.3(D)(6)(b)** Electrical Area Classification for Open Dipping and Coating Processes with Peripheral Vapor Containment and Ventilation — Vapors Confined to Process Equipment. [**34:**Figure 6.4(b)]*

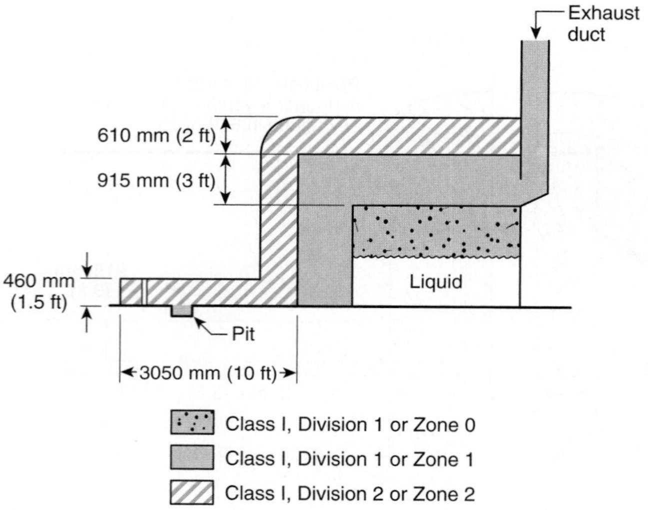


***FIGURE 516.3(D)(6)(c)** Electrical Area Classification for Open Dipping and Coating Processes with Partial Peripheral Vapor Containment and Ventilation — Vapors NOT Confined to Process Equipment. [**34:**Figure 6.4(c)]*

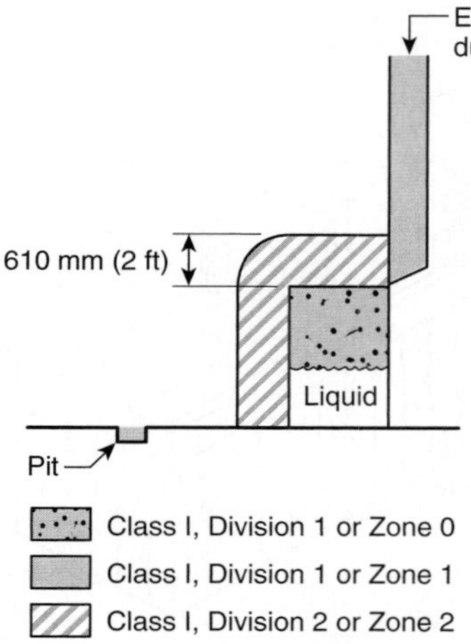


***FIGURE 516.3(D)(6)(d)** Electrical Area Classification for Open Dipping and Coating Processes with Partial Peripheral Vapor Containment and Ventilation — Vapors Confined to Process Equipment. [**34:**Figure 6.4(d)]*

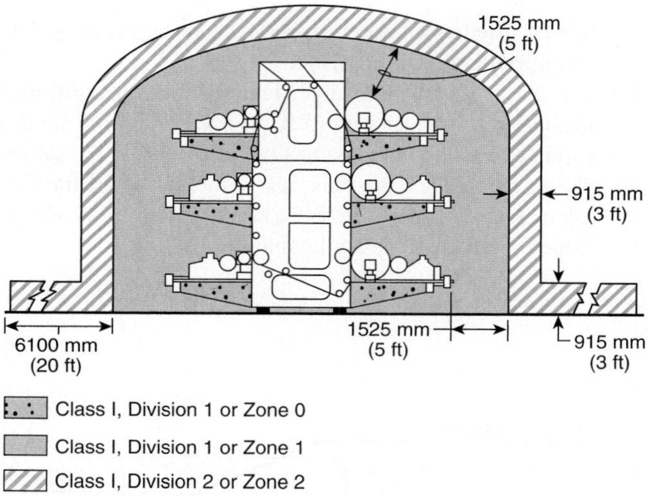


***FIGURE 516.3(D)(6)(e)** Electrical Area Classification for a Typical Printing Process. [**34:**Figure 6.4(e)]*

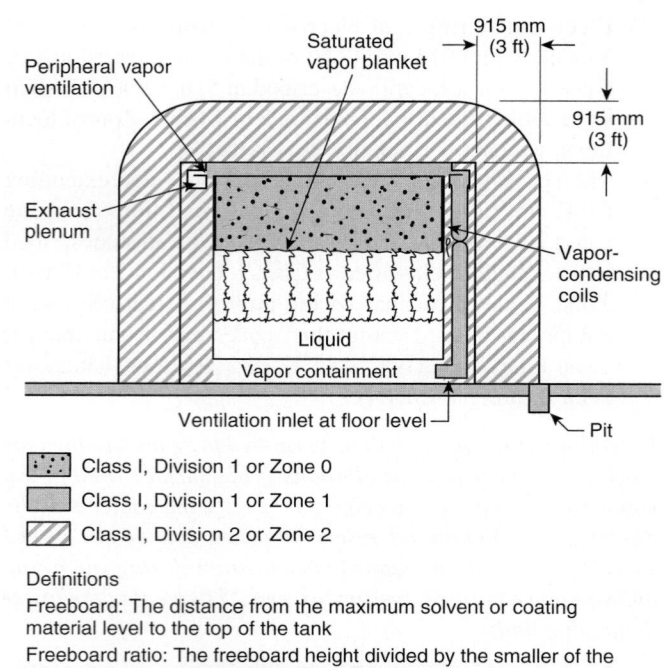

The interior of any enclosed dipping or coating process or apparatus shall be a Class I, Division 1 or Class I, Zone 1 location, and electrical wiring and electrical utilization equipment located within this space shall be suitable for Class I, Division 1 or Class I, Zone 1 locations, whichever is applicable. The area inside the dip tank shall be classified as Class I, Division 1 or Class I, Zone 0, whichever is applicable. [**34:**6.5.1]

**(8) Open Containers.** All space in all directions within 600 mm (2 ft) of the Division 1 or Zone 1 area surrounding open containers, supply containers, spray gun cleaners, and

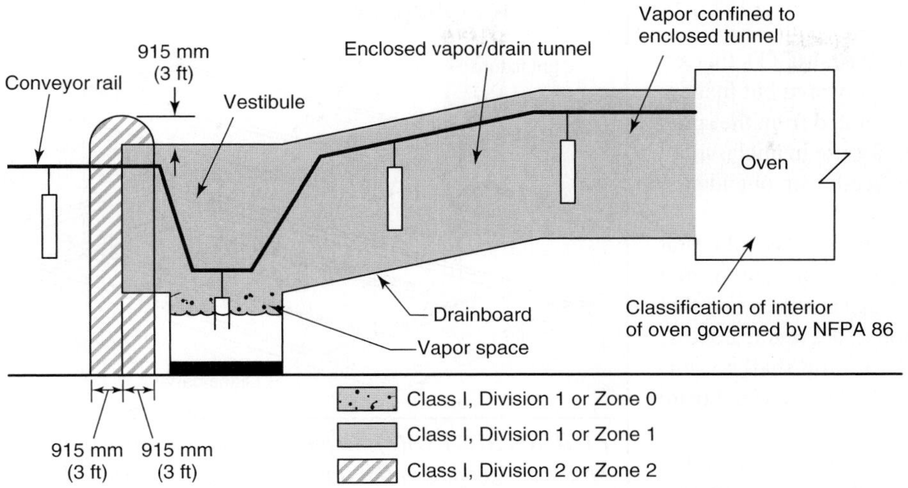


***FIGURE 516.3(D)(7)*** *Electrical Area Classification Around an Enclosed Dipping or Coating Process. [**34:**Figure 6.5]*

solvent distillation units containing flammable liquids, as well as the area extending 1.5 m (5 ft) beyond the Division 1 or Zone 1 area up to a height of 460 mm (18 in.) above the floor or grade level. [**33:**6.5.5]

The parameters of the hazardous location defined around open containers containing flammable liquids are extracted from 6.5.5.1(2) and (3) of NFPA 33.

•

**(E) Adjacent Locations.** Adjacent locations that are cut off from the defined Class I or Class II locations by tight partitions without communicating openings, and within which flammable vapors or combustible powders are not likely to be released, shall be unclassified.

The adjacent unpartitioned area, which may be safe under normal operating conditions, could become dangerous due to accident or careless operation. Equipment known to produce sparks or flames under normal operating conditions should not be installed in these adjacent unpartitioned areas.

**(F) Unclassified Locations.** Locations using drying, curing, or fusion apparatus and provided with positive mechanical ventilation adequate to prevent accumulation of flammable concentrations of vapors, and provided with effective interlocks to de-energize all electrical equipment (other than equipment identified for Class I locations) in case the ventilating equipment is inoperative, shall be permitted to be unclassified where the authority having jurisdiction so judges.

Inadequate ventilation can permit explosive vapor–air mixtures to exist. No open flames or spark-producing equipment should be located in an area without adequate ventilation. Some residues may be ignited at very low temperatures. Therefore, consideration must be given to the operating temperatures of equipment subject to residue deposits. Some deposits may be ignited at temperatures produced by steam pipes or incandescent light bulbs or globes, even luminaires of the explosionproof type.

Informational Note: For further information regarding safeguards, see NFPA 86-2011, *Standard for Ovens and Furnaces*.

## 516.4 Wiring and Equipment in Class I Locations

**(A) Wiring and Equipment — Vapors.** All electrical wiring and equipment within the Class I location (containing vapor only — not residues) defined in 516.3 shall comply with the applicable provisions of Article 501 or Article 505, as applicable.

**(B) Wiring and Equipment — Vapors and Residues.** Unless specifically listed for locations containing deposits of dangerous quantities of flammable or combustible vapors, mists, residues, dusts, or deposits (as applicable), there shall be no electrical equipment in any spray area as herein defined whereon deposits of combustible residue may readily accumulate. All electrical wiring shall be comply with 516.4(A).

Electrical equipment is generally not permitted inside any spray booth, in the exhaust duct from a spray booth, in the entrained air of an exhaust system from a spraying operation, or in the direct path of spray, unless such equipment is specifically listed for readily ignitible deposits and flammable vapor.

**(C) Illumination.**

(1) Luminaires, like that shown in Figure 516.4(C)(1), that are attached to the walls or ceiling of a spray area but that are outside any classified area and are separated from the spray area by glass panels shall be suitable for use in unclassified locations. Such fixtures shall be serviced from outside the spray area. [**33:**6.6.1]

(2) Luminaires, like that shown in Figure 516.4(C)(1), that are attached to the walls or ceiling of a spray area; that are separated from the spray area by glass panels and that are located within a Class I, Division 2; a Class I, Zone 2; a Class II, Division 2; or a Zone 22 location shall be suitable for such location. Such fixtures shall be serviced from outside the spray area. [**33:**6.6.2]

(3) Luminaires, like that shown in Figure 516.4(C)(3), that are an integral part of the walls or ceiling of a spray area shall be permitted to be separated from the spray area by glass panels that are an integral part of the fixture. Such fixtures shall be listed for use in Class I, Division 2; Class I, Zone 2; Class II, Division 2; or Zone 22 locations, whichever is applicable, and also shall be listed for accumulations of deposits of combustible residues. Such fixtures shall be permitted to be serviced from inside the spray area. [**33:**6.6.3]

(4) Glass panels used to separate luminaires from the spray area or that are an integral part of the luminaire shall meet the following requirements.

a. Panels for light fixtures or for observation shall be of heat-treated glass, laminated glass, wired glass, or hammered-wired glass and shall be sealed to confine vapors, mists, residues, dusts, and deposits to the spray area. [**33:**5.5.1]

*Exception: Listed spray booth assemblies that have vision panels constructed of other materials shall be permitted.*

b. Panels for light fixtures shall be separated from the fixture to prevent the surface temperature of the panel from exceeding 938C (2008F). [**33:**5.5.2]

c. The panel frame and method of attachment shall be designed to not fail under fire exposure before the vision panel fails. [**33:**5.5.3]

Sufficient permanent illumination should be provided at the time the equipment is installed to avoid the use of temporary or emergency luminaires connected to ordinary extension cords in any spray area. See 516.4(D). A practical method of lighting is the use of wired or tempered glass panels in the top or sides of the spray booth, with electrical luminaires located outside the booth, avoiding the direct path of the spray.

Even where a panel of glass or translucent material separates a lighting unit from the readily ignitible location, the unit

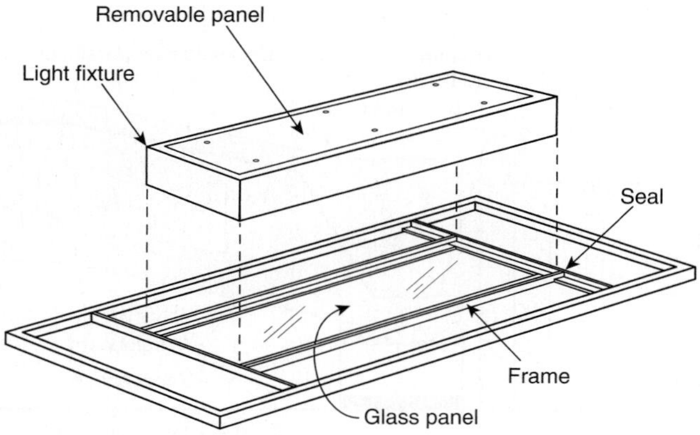


***FIGURE 516.4(C)(1)*** *Example of a Luminaire that is Mounted Outside of the Spray Area and is Serviced from Outside the Spray Area. [33:Figure 6.6.1]*

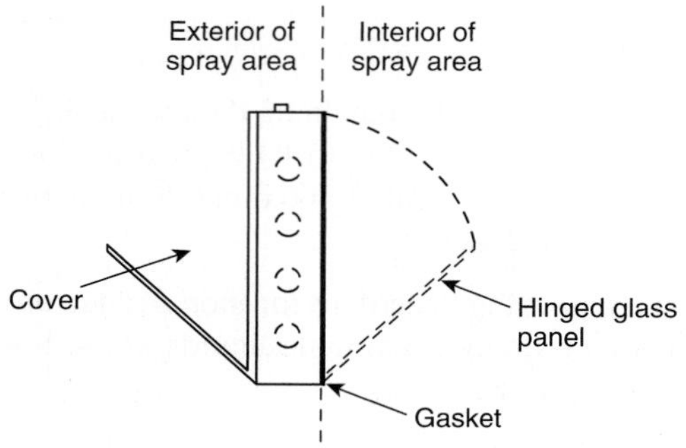


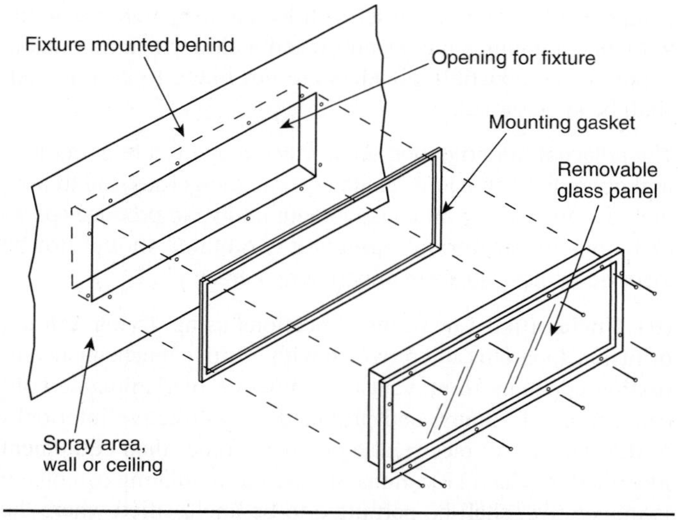


***FIGURE 516.4(C)(3)*** *Example of a Luminaire that is an Integral Part of the Spray Area and is Serviced from Inside the Spray Area. [33:Figure 6.6.3]*

may be in a hazardous area and must be suitable for use in the specific location. Lighting units that are integral to the paint spray booth must be listed both for the location and for deposits of readily combustible paint residues on the side of the luminaire that forms part of the interior surface of the spray booth.

**(D) Portable Equipment.** Portable electric luminaires or other utilization equipment shall not be used in a spray area during spray operations.

*Exception No. 1: Where portable electric luminaires are required for operations in spaces not readily illuminated by fixed lighting within the spraying area, they shall be of the type identified for Class I, Division 1 or Class 1, Zone 1 locations where readily ignitible residues may be present. [**33:**6.9 Exception]*

*Exception No. 2: Where portable electric drying apparatus is used in spray booths and the following requirements are met:*

*(a) The apparatus and its electrical connections are not located within the spray enclosure during spray operations.*
*(b) Electrical equipment within 450 mm (18 in.) of the floor is identified for Class I, Division 2 or Class I, Zone 2 locations.*
*(c) All metallic parts of the drying apparatus are electrically bonded and grounded.*
*(d) Interlocks are provided to prevent the operation of spray equipment while drying apparatus is within the spray enclosure, to allow for a 3-minute purge of the enclosure before energizing the drying apparatus and to shut off drying apparatus on failure of ventilation system.*

**(E) Electrostatic Equipment.** Electrostatic spraying or detearing equipment shall be installed and used only as provided in 516.10.

Informational Note: For further information, see NFPA 33-2011, *Standard for Spray Application Using Flammable or Combustible Materials.*

**(F) Static Electric Discharges.**

(1) All persons and all electrically conductive objects, including any metal parts of the process equipment or apparatus, containers of material, exhaust ducts, and piping systems that convey flammable or combustible liquids, shall be electrically grounded. [**34:**6.8.1]
(2) Provision shall be made to dissipate static electric charges from all nonconductive substrates in printing processes.

## 516.7 Wiring and Equipment Not Within Classified Locations

**(A) Wiring.** All fixed wiring above the Class I and II locations shall be in metal raceways, Type PVC conduit, Type RTRC conduit, or electrical nonmetallic tubing; where cables are used, they shall be Type MI, Type TC, or Type MC cable. Cellular metal floor raceways shall only be permitted to supply ceiling outlets or as extensions to the area below the floor of a Class I or II location. Where cellular metal raceways are used, they shall not have connections leading into or passing through the Class I or II location unless suitable seals are provided.

**(B) Equipment.** Equipment that may produce arcs, sparks, or particles of hot metal, such as lamps and lampholders for fixed lighting, cutouts, switches, receptacles, motors, or other equipment having make-and-break or sliding contacts, where installed above a Classified location or above a location where freshly finished goods are handled, shall be of the totally enclosed type or be constructed so as to prevent the escape of sparks or hot metal particles.

Subsection 5.16.10(A) was revised by a tentative interim amendment (TIA).

Even though areas adjacent to spray booths (particularly where coating-material stocks are located) are assumed to have ventilation sufficient to prevent the presence of flammable vapors or deposits, luminaires should be totally enclosed to prevent hot particles from falling in any area that may have freshly painted stock, accidentally spilled flammable or combustible materials, readily ignitible refuse, or flammable or combustible liquid containers that have been left open accidentally. Where luminaires are in areas subject to atmospheres of flammable vapor, lamps should be replaced while electricity is off; otherwise, sparking might occur.

## 516.10 Special Equipment

**(A) Fixed Electrostatic Equipment.** This section shall apply to any equipment using electrostatically charged elements for the atomization, charging, and/or precipitation of hazardous materials for coatings on articles or for other similar purposes in which the charging or atomizing device is attached to a mechanical support or manipulator. This shall include robotic devices. This section shall not apply to devices that are held or manipulated by hand. Where robot or programming procedures involve manual manipulation of the robot arm while spraying with the high voltage on, the provisions of 516.10(B) shall apply. The installation of electrostatic spraying equipment shall comply with 516.10(A)(1) through (A)(10). Spray equipment shall be listed except as otherwise permitted. All automatic electrostatic equipment systems shall comply with 516.4(A)(1) through (A)(9). [**33:**11.5]

Informational Note: For more information on listing and approval of electrostatic spray equipment, see NFPA 33-2011, *Standard for Spray Application Using Flammable or Combustible Materials*, Section 11.5. NFPA 33 permits certain electrostatic spray equipment to be approved for use when additional mitigation equipment is employed.

**(1) Power and Control Equipment.** Transformers, high-voltage supplies, control apparatus, and all other electrical portions of the equipment shall be installed outside of the Class I location as defined in 516.3 or be of a type identified for the location.

*Exception: High-voltage grids, electrodes, electrostatic atomizing heads, and their connections shall be permitted within the Class I location.*

**(2) Electrostatic Equipment.** Electrodes and electrostatic atomizing heads shall be adequately supported in permanent locations and shall be effectively insulated from ground. Electrodes and electrostatic atomizing heads that are permanently attached to their bases, supports, reciprocators, or robots shall be deemed to comply with this section.

**(3) High-Voltage Leads.** High-voltage leads shall be properly insulated and protected from mechanical damage or exposure to destructive chemicals. Any exposed element at high voltage shall be effectively and permanently supported on suitable insulators and shall be effectively guarded against accidental contact or grounding.

**(4) Support of Goods.** Goods being coated using this process shall be supported on conveyors or hangers. The conveyors or hangers shall be arranged (1) to ensure that the parts being coated are electrically connected to ground with a resistance of 1 megohm or less and (2) to prevent parts from swinging.

**(5) Automatic Controls.** Electrostatic apparatus shall be equipped with automatic means that will rapidly de-energize the high-voltage elements under any of the following conditions:

(1) Stoppage of ventilating fans or failure of ventilating equipment from any cause
(2) Stoppage of the conveyor carrying goods through the high-voltage field unless stoppage is required by the spray process
(3) Occurrence of excessive current leakage at any point in the high-voltage system
(4) De-energizing the primary voltage input to the power supply

**(6) Grounding.** All electrically conductive objects in the spray area, except those objects required by the process to be at high voltage, shall be adequately grounded. This requirement shall apply to paint containers, wash cans, guards, hose connectors, brackets, and any other electrically conductive objects or devices in the area.

Informational Note: For more information on grounding and bonding for static electricity purposes, see NFPA 33-2011, *Standard for Spray Application Using Flammable or Combustible Materials*; NFPA 34-2011, *Standard for Dipping, Coating, and Printing Processes Using Flammable or Combustible Liquids*; and NFPA 77-2014, *Recommended Practice on Static Electricity*.

All electrically conductive objects, including metal parts of spray booths, exhaust ducts, piping systems conveying flammable or combustible liquids or paint, solvent tanks, and canisters should be properly grounded to prevent sparks from the accumulation of static electricity. See Section 6.7 of NFPA 33 and Section 9.3 of NFPA 77, *Recommended Practice on Static Electricity*.

The same informational note appears in 516.10(B)(4) and (C)(4) to increase awareness of three standards dealing with spray applications, dipping and coating processes, and static electricity for grounding and bonding.

**(7) Isolation.** Safeguards such as adequate booths, fencing, railings, interlocks, or other means shall be placed about the equipment or incorporated therein so that they, either by their location, character, or both, ensure that a safe separation of the process is maintained.

**(8) Signs.** Signs shall be conspicuously posted to convey the following:

(1) Designate the process zone as dangerous with regard to fire and accident
(2) Identify the grounding requirements for all electrically conductive objects in the spray area
(3) Restrict access to qualified personnel only

**(9) Insulators.** All insulators shall be kept clean and dry.

**(10) Other Than Nonincendive Equipment.** Spray equipment that cannot be classified as nonincendive shall comply with (A)(10)(a) and (A)(10)(b).

(a) Conveyors, hangers, and application equipment shall be arranged so that a minimum separation of at least twice the sparking distance is maintained between the workpiece or material being sprayed and electrodes, electrostatic atomizing heads, or charged conductors. Warnings defining this safe distance shall be posted. [**33:**11.4.1]

(b) The equipment shall provide an automatic means of rapidly de-energizing the high-voltage elements in the event the distance between the goods being painted and the electrodes or electrostatic atomizing heads falls below that specified in (a). [**33:**11.3.8]

**(B) Electrostatic Hand-Spraying Equipment.** This section shall apply to any equipment using electrostatically charged elements for the atomization, charging, or precipitation of flammable and combustible materials for coatings on articles, or for other similar purposes in which the charging or atomizing device is hand-held and manipulated during the spraying operation. Electrostatic hand-spraying equipment and devices used in connection with paint-spraying operations shall be of listed types and shall comply with 516.10(B)(1) through (B)(5).

**(1) General.** The high-voltage circuits shall be designed so as not to produce a spark of sufficient intensity to ignite the most readily ignitible of those vapor–air mixtures likely to be encountered, or result in appreciable shock hazard upon coming in contact with a grounded object under all normal operating conditions. The electrostatically charged exposed elements of the handgun shall be capable of being energized only by an actuator that also controls the coating material supply.

**(2) Power Equipment.** Transformers, power packs, control apparatus, and all other electrical portions of the equipment shall be located outside of the Class I location or be identified for the location.

*Exception: The handgun itself and its connections to the power supply shall be permitted within the Class I location.*

**(3) Handle.** The handle of the spraying gun shall be electrically connected to ground by a conductive material and be constructed so that the operator in normal operating position is in direct electrical contact with the grounded handle with a resistance of not more than 1 megohm to prevent buildup of a static charge on the operator's body. Signs indicating the necessity for grounding other persons entering the spray area shall be conspicuously posted.

**(4) Electrostatic Equipment.** All electrically conductive objects in the spraying area, except those objects required by the process to be at high voltage, shall be electrically connected to ground with a resistance of not more than 1 megohm. This requirement shall apply to paint containers, wash cans, and any other electrical conductive objects or devices in the area. The equipment shall carry a prominent, permanently installed warning regarding the necessity for this grounding feature.

Informational Note: For more information on grounding and bonding for static electricity purposes, see NFPA 33-2011, *Standard for Spray Application Using Flammable or Combustible Materials*; NFPA 34-2011, *Standard for Dipping, Coating, and Printing Processes Using Flammable or Combustible Liquids*; and NFPA 77-2014, *Recommended Practice on Static Electricity*.

**(5) Support of Objects.** Objects being painted shall be maintained in electrical contact with the conveyor or other grounded support. Hooks shall be regularly cleaned to ensure adequate grounding of 1 megohm or less. Areas of contact shall be sharp points or knife edges where possible. Points of support of the object shall be concealed from random spray where feasible; and, where the objects being sprayed are supported from a conveyor, the point of attachment to the conveyor shall be located so as to not collect spray material during normal operation. [**33:**Chapter 12]

**(C) Powder Coating.** This section shall apply to processes in which combustible dry powders are applied. The hazards associated with combustible dusts are present in such a process to a degree, depending on the chemical composition of the material, particle size, shape, and distribution.

**(1) Electrical Equipment and Sources of Ignition.** Electrical equipment and other sources of ignition shall comply with the requirements of Article 502. Portable electric luminaires and other utilization equipment shall not be used within a Class II location during operation of the finishing processes. Where such luminaires or utilization equipment are used during cleaning or repairing operations, they shall be of a type identified for Class II, Division 1 locations, and all exposed metal parts shall be connected to an equipment grounding conductor.

*Exception: Where portable electric luminaires are required for operations in spaces not readily illuminated by fixed lighting within the spraying area, they shall be of the type listed for Class II, Division 1 locations where readily ignitible residues may be present.*

**(2) Fixed Electrostatic Spraying Equipment.** The provisions of 516.10(A) and 516.10(C)(1) shall apply to fixed electrostatic spraying equipment.

**(3) Electrostatic Hand-Spraying Equipment.** The provisions of 516.10(B) and 516.10(C)(1) shall apply to electrostatic hand-spraying equipment.

**(4) Electrostatic Fluidized Beds.** Electrostatic fluidized beds and associated equipment shall be of identified types. The high-voltage circuits shall be designed such that any discharge produced when the charging electrodes of the bed are approached or contacted by a grounded object shall not be of sufficient intensity to ignite any powder–air mixture likely to be encountered or to result in an appreciable shock hazard.

(a) Transformers, power packs, control apparatus, and all other electrical portions of the equipment shall be located outside the powder-coating area or shall otherwise comply with the requirements of 516.10(C)(1).

*Exception: The charging electrodes and their connections to the power supply shall be permitted within the powder-coating area.*

(b) All electrically conductive objects within the powder-coating area shall be adequately grounded. The powder-coating equipment shall carry a prominent, permanently installed warning regarding the necessity for grounding these objects.

Informational Note: For more information on grounding and bonding for static electricity purposes, see NFPA 33-2011, *Standard for Spray Application Using Flammable or Combustible Materials*; NFPA 34-2011, *Standard for Dipping, Coating, and Printing Processes Using Flammable or Combustible Liquids*; and NFPA 77-2014, *Recommended Practice on Static Electricity*.

(c) Objects being coated shall be maintained in electrical contact (less than 1 megohm) with the conveyor or other support in order to ensure proper grounding. Hangers shall be regularly cleaned to ensure effective electrical contact. Areas of electrical contact shall be sharp points or knife edges where possible.

(d) The electrical equipment and compressed air supplies shall be interlocked with a ventilation system so that the equipment cannot be operated unless the ventilating fans are in operation. [**33:**Chapter 15]

## 516.16 Grounding

All metal raceways, the metal armors or metallic sheath on cables, and all non–current-carrying metal parts of fixed or portable electrical equipment, regardless of voltage, shall be grounded and bonded. Grounding and bonding shall comply with 501.30, 502.30, or 505.25, as applicable.

# ARTICLE 517
# Health Care Facilities

Informational Note: Text that is followed by a reference in brackets has been extracted from NFPA 99-2012, *Health Care Facilities Code*, and NFPA *101*-2012, *Life Safety Code*. Only editorial changes were made to the extracted text to make it consistent with this *Code*.

## I. General

### 517.1 Scope

The provisions of this article shall apply to electrical construction and installation criteria in health care facilities that provide services to human beings.

The requirements in Parts II and III not only apply to single-function buildings but are also intended to be individually applied to their respective forms of occupancy within a multifunction building (e.g., a doctor's examining room located within a limited care facility would be required to meet the provisions of 517.10).

Informational Note: For information concerning performance, maintenance, and testing criteria, refer to the appropriate health care facilities documents.

The requirements of Article 517 apply to all types of health care facilities. In some instances, parts of health care facilities are not directly used for the treatment of patients. For example, in a suite of doctors' offices within an office building, a doctor's business office is treated as an ordinary occupancy and the electrical installation is required to comply with the applicable requirements of Chapters 1 through 4. The wiring and electrical equipment in the examining rooms and any other patient care areas within the office suite are required to be installed per the applicable rules of Article 517.

The scope of Article 517 also includes health care facilities that are mobile or that supply outpatient service, but it does not include veterinary offices or animal hospitals.

Other standards referenced in Article 517 include NFPA 99, NFPA *101*®, and NFPA 20, *Standard for the Installation of Stationary Pumps for Fire Protection*.

### 517.2 Definitions

**Alternate Power Source.** One or more generator sets, or battery systems where permitted, intended to provide power during the interruption of the normal electrical service; or the public utility electrical service intended to provide power during interruption of service normally provided by the generating facilities on the premises. [**99:**3.3.5]

Section 517.30(B)(4) permits alternate power sources to serve essential electrical systems of contiguous or multiple building facilities such as a health care campus with a centrally located alternate power plant.

**Ambulatory Health Care Occupancy.** A building or portion thereof used to provide services or treatment simultaneously to four or more patients that provides, on an outpatient basis, one or more of the following:

(1) Treatment for patients that renders the patients incapable of taking action for self-preservation under emergency conditions without assistance of others.
(2) Anesthesia that renders the patients incapable of taking action for self-preservation under emergency conditions without the assistance of others.
(3) Emergency or urgent care for patients who, due to the nature of their injury or illness, are incapable of taking action for self-preservation under emergency conditions without the assistance of others. [***101:***3.3.188.1]

Ambulatory health care occupancies, including outpatient surgery centers, freestanding emergency medical centers, and hemodialysis units, are subject to the requirements of Part II of Article 517 and 517.45. This definition, which correlates with the definition of the same term in NFPA 99, recognizes that some emergency or urgent care may be performed at ambulatory health care occupancies.

**Anesthetizing Location.** Any area of a facility that has been designated to be used for the administration of any flammable or nonflammable inhalation anesthetic agent in the course of examination or treatment, including the use of such agents for relative analgesia.

In an emergency, it may be necessary to administer an anesthetic almost anywhere in a health care facility. Only those areas specifically set aside for administering anesthetics are required to meet the requirements of Part IV of Article 517. The provisions of Part IV do not apply to administering analgesic or local anesthetics, such as might be used in minor medical or dental procedures.

The definition of *anesthetizing location* applies to areas where inhalation anesthetics are used for relative analgesia. The term *relative analgesia* (see definition in 517.2) is sometimes referred to as "conscious sedation" and is a state of sedation in which the perception of pain is partially blocked but the patient does not lose consciousness. This form of anesthesia is commonly used by oral surgeons. For guidance on flammable anesthetizing locations, see Annex E of NFPA 99.

**Battery-Powered Lighting Units.** Individual unit equipment for backup illumination consisting of the following:

(1) Rechargeable battery
(2) Battery-charging means
(3) Provisions for one or more lamps mounted on the equipment, or with terminals for remote lamps, or both
(4) Relaying device arranged to energize the lamps automatically upon failure of the supply to the unit equipment

**Critical Branch.** A system of feeders and branch circuits supplying power for task illumination, fixed equipment, select receptacles, and select power circuits serving areas and functions related to patient care and that is automatically connected to alternate power sources by one or more transfer switches during interruption of normal power source. [**99:**3.3.30]

**Electrical Life-Support Equipment.** Electrically powered equipment whose continuous operation is necessary to maintain a patient's life. [**99:**3.3.37]

•

**Equipment Branch.** A system of feeders and branch circuits arranged for delayed, automatic, or manual connection to the alternate power source and that serves primarily 3-phase power equipment. [**99:**3.3.46]

**Essential Electrical System.** A system comprised of alternate sources of power and all connected distribution systems and ancillary equipment, designed to ensure continuity of electrical power to designated areas and functions of a health care facility during disruption of normal power sources, and also to minimize disruption within the internal wiring system. [**99:**3.3.48]

Emergency systems in occupancies other than health care are installed primarily for life safety and building evacuation. The essential systems in hospitals are for life safety systems as well as for task illumination, fixed equipment, select receptacles, and select power circuits serving areas and functions related to patient care.

**Exposed Conductive Surfaces.** Those surfaces that are capable of carrying electric current and that are unprotected, unenclosed, or unguarded, permitting personal contact. Paint, anodizing, and similar coatings are not considered suitable insulation, unless they are listed for such use.

**Fault Hazard Current.** See *Hazard Current*.

**Flammable Anesthetics.** Gases or vapors, such as fluroxene, cyclopropane, divinyl ether, ethyl chloride, ethyl ether, and ethylene, which may form flammable or explosive mixtures with air, oxygen, or reducing gases such as nitrous oxide.

**Flammable Anesthetizing Location.** Any area of the facility that has been designated to be used for the administration of any flammable inhalation anesthetic agents in the normal course of examination or treatment.

**Hazard Current.** For a given set of connections in an isolated power system, the total current that would flow through a low impedance if it were connected between either isolated conductor and ground.

*Fault Hazard Current.* The hazard current of a given isolated system with all devices connected except the line isolation monitor.

*Monitor Hazard Current.* The hazard current of the line isolation monitor alone.

*Total Hazard Current.* The hazard current of a given isolated system with all devices, including the line isolation monitor, connected.

**Health Care Facilities.** Buildings or portions of buildings in which medical, dental, psychiatric, nursing, obstetrical, or surgical care are provided. Health care facilities include, but are not limited to, hospitals, nursing homes, limited care facilities, clinics, medical and dental offices, and ambulatory care centers, whether permanent or movable.

NFPA *101* defines a health care occupancy as one used to provide medical or other treatment or care simultaneously to four or more patients on an inpatient basis, where such patients are mostly incapable of self-preservation due to age, physical or mental disability, or security measures not under the occupants' control. NFPA 99 defines health care facilities as buildings, portions of buildings, or mobile enclosures in which medical, dental, psychiatric, nursing, obstetrical, or surgical care is provided.

The term *health care facility* should not be confused with the term *health care occupancy*. All health care occupancies, including ambulatory health care occupancies, are considered health care facilities; however, not all health care facilities are considered health care occupancies. A medical office building can be a medical facility, but under NFPA *101* it would typically be a business occupancy. A hospital is a health care facility, but it can be several different occupancies (health care, assembly, business, ambulatory health care, and so forth).

The *NEC* does not designate the type of facility or level of care provided. The governing body of the facility determines the type of services being provided at a facility. For example, depending on the jurisdiction, a chiropractic office may or may not be a facility covered under Article 517.

**Hospital.** A building or portion thereof used on a 24-hour basis for the medical, psychiatric, obstetrical, or surgical care of four or more inpatients. [***101:***3.3.142]

**Isolated Power System.** A system comprising an isolating transformer or its equivalent, a line isolation monitor, and its ungrounded circuit conductors.

**Isolation Transformer.** A transformer of the multiple-winding type, with the primary and secondary windings physically separated, which inductively couples its secondary winding(s) to circuit conductors connected to its primary winding(s).

**Life Safety Branch.** A system of feeders and branch circuits supplying power for lighting, receptacles, and equipment essential for life safety that is automatically connected to alternate power sources by one or more transfer switches during interruption of the normal power source. [**99:**3.3.94]

**Limited Care Facility.** A building or portion thereof used on a 24-hour basis for the housing of four or more persons who are incapable of self-preservation because of age; physical limitation due to accident or illness; or limitations such as mental retardation/developmental disability, mental illness, or chemical dependency. [**99:**3.3.97]

**Line Isolation Monitor.** A test instrument designed to continually check the balanced and unbalanced impedance from each line of an isolated circuit to ground and equipped with a built-in test circuit to exercise the alarm without adding to the leakage current hazard.

**Monitor Hazard Current.** See *Hazard Current*.

**Nurses' Stations.** Areas intended to provide a center of nursing activity for a group of nurses serving bed patients, where the patient calls are received, nurses are dispatched, nurses' notes written, inpatient charts prepared, and medications prepared for distribution to patients. Where such activities are carried on in more than one location within a nursing unit, all such separate areas are considered a part of the nurses' station.

**Nursing Home.** A building or portion of a building used on a 24-hour basis for the housing and nursing care of four or more persons who, because of mental or physical incapacity, might be unable to provide for their own needs and safety without the assistance of another person. [**99:**3.3.127]

**Patient Bed Location.** The location of a patient sleeping bed, or the bed or procedure table of a critical care area. [**99:**3.3.136]

**Patient Care Space.** Space within a health care facility wherein patients are intended to be examined or treated.

*Basic Care Space.* Space in which failure of equipment or a system is not likely to cause injury to the patients or caregivers but may cause patient discomfort.

*General Care Space.* Space in which failure of equipment or a system is likely to cause minor injury to patients or caregivers.

*Critical Care Space.* Space in which failure of equipment or a system is likely to cause major injury or death to patients or caregivers.

*Support Space.* Space in which failure of equipment or a system is not likely to have a physical impact on patients or caregivers.

The word "space" conveys that the patient care area is often not defined by the four walls of a room. Formerly called the patient care area, patient care spaces are defined by four varying degrees of possible injury to a patient or caregiver due to an electrical system failure.

An operating room, where patients are subject to invasive procedures, is a critical care space. See Exhibit 517.1.

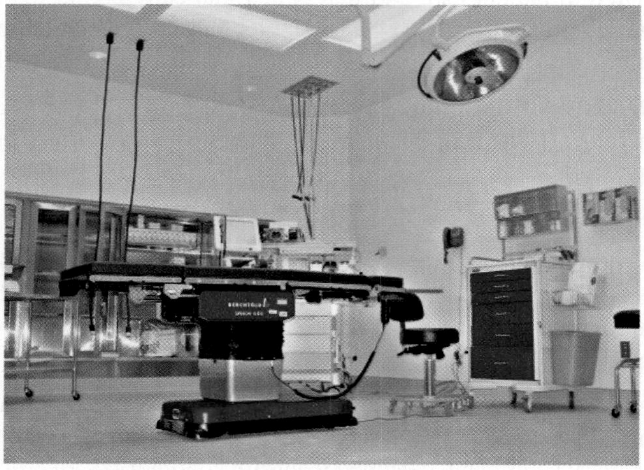

***EXHIBIT 517.1*** *An operating room where patients are connected to line-operated electromedical devices while undergoing invasive procedures. (Courtesy of the International Association of Electrical Inspectors)*

Informational Note No. 1: The governing body of the facility designates patient care space in accordance with the type of patient care anticipated and with the definitions of the area classification. Business offices, corridors, lounges, day rooms, dining rooms, or similar areas typically are not classified as patient care space.

Informational Note No. 2: Basic care space is typically a location where basic medical or dental care, treatment, or examinations are performed. Examples include, but are not limited to, examination or treatment rooms in clinics, medical and dental offices, nursing homes, and limited care facilities.

Informational Note No. 3: General care space includes areas such as patient bedrooms, examining rooms, treatment rooms, clinics, and similar areas where the patient may come into contact with electromedical devices or ordinary appliances such as a nurse call system, electric beds, examining lamps, telephones, and entertainment devices.

Informational Note No. 4: Critical care space includes special care units, intensive care units, coronary care units, angiography laboratories, cardiac catheterization laboratories, delivery rooms, operating rooms, and similar areas in which are patients are intended to be subjected to invasive procedures and are connected to line-operated, electromedical devices.

Informational Note No. 5: Spaces where a procedure is performed that subjects patients or staff to wet conditions are considered as wet procedure areas. Wet conditions include standing fluids on the floor or drenching of the work area. Routine housekeeping procedures and incidental spillage of liquids do not define wet procedure areas. It is the responsibility of the governing body of the health care facility to designate the wet procedure areas.

**Patient Care Vicinity.** A space, within a location intended for the examination and treatment of patients, extending 1.8 m (6 ft) beyond the normal location of the patient bed, chair, table, treadmill, or other device that supports the patient during examination and treatment and extending vertically to 2.3 m (7 ft 6 in.) above the floor. [**99:**3.3.139]

The patient care vicinity is defined not only by a patient bed but also by other equipment that supports a patient during examination or treatment. The vicinity is determined by equipment in its normal location, that is, the location called for in the architect's plans, rather than the temporary location of equipment subject to movement by housekeeping staff or for the convenience of the medical staff.

**Patient Equipment Grounding Point.** A jack or terminal that serves as the collection point for redundant grounding of electrical appliances serving a patient care vicinity or for grounding other items in order to eliminate electromagnetic interference problems. [**99:**3.3.140]

**Psychiatric Hospital.** A building used exclusively for the psychiatric care, on a 24-hour basis, of four or more inpatients.

**Reference Grounding Point.** The ground bus of the panelboard or isolated power system panel supplying the patient care area.

**Relative Analgesia.** A state of sedation and partial block of pain perception produced in a patient by the inhalation of concentrations of nitrous oxide insufficient to produce loss of consciousness (conscious sedation).

**Selected Receptacles.** A minimum number of electrical receptacles to accommodate appliances ordinarily required for local tasks or likely to be used in patient care emergencies.

**Task Illumination.** Provision for the minimum lighting required to carry out necessary tasks in the described areas, including safe access to supplies and equipment, and access to exits.

**Total Hazard Current.** See *Hazard Current.*

•

**Wet Procedure Location.** The area in a patient care space where a procedure is performed that is normally subject to wet conditions while patients are present, including standing fluids on the floor or drenching of the work area, where either such condition is intimate to the patient or staff.

*Wet procedure locations* may also include such areas as hydrotherapy areas, dialysis laboratories, and certain wet laboratories at the discretion of the governing body of the facility. The definition excludes areas such as lavatories or bathrooms within a health care facility. For infection control purposes, many patient and treatment areas have a sink for hand washing, which also is not a wet procedure location.

Informational Note: Routine housekeeping procedures and incidental spillage of liquids do not define a wet procedure location.

**X-Ray Installations, Long-Time Rating.** A rating based on an operating interval of 5 minutes or longer.

**X-Ray Installations, Mobile.** X-ray equipment mounted on a permanent base with wheels, casters, or a combination of both to facilitate moving the equipment while completely assembled.

**X-Ray Installations, Momentary Rating.** A rating based on an operating interval that does not exceed 5 seconds.

**X-Ray Installations, Portable.** X-ray equipment designed to be hand carried.

**X-Ray Installations, Transportable.** X-ray equipment to be conveyed by a vehicle or that is readily disassembled for transport by a vehicle.

## II. Wiring and Protection

**Formal Interpretation 99-1**

*Reference:* Article 517, Part II

*Question:* Does Part II of Article 517 of the NEC apply to patient sleeping rooms of nursing homes or limited care facilities where patient care activities do not involve the use of electrical or electronic life support systems; or invasive procedures where patients are electrically connected to line connected electromedical devices?

*Answer:* No

*Issue Edition:* 1999

*Reference:* Article 517

*Issue Date:* August 1, 2000

*Effective Date:* August 21, 2000

### 517.10 Applicability

**(A) Applicability.** Part II shall apply to patient care space of all health care facilities.

The designation of the types of patient care spaces is the responsibility of the governing body of the health care facility. Both the design and the inspection of a patient care space are based on the patient care anticipated in the area.

**(B) Not Covered.** Part II shall not apply to the following:

(1) Business offices, corridors, waiting rooms, and the like in clinics, medical and dental offices, and outpatient facilities
(2) Areas of nursing homes and limited care facilities wired in accordance with Chapters 1 through 4 of this *Code* where these areas are used exclusively as patient sleeping rooms

Informational Note: See NFPA *101*-2012, *Life Safety Code®*.

Areas in nursing homes that are designated as patient sleeping rooms are not considered to be patient care spaces, even though residents may require assistance to attend to their personal needs and safety. However, areas of nursing homes, including patient bedrooms, in which residents are intended to be examined or treated can be considered examining rooms and are classified in the broader category as general care space. See the defined term *patient care space* in 517.2. See the Formal Interpretation on Article 517, Part II applicability.

### 517.11 General Installation — Construction Criteria

The purpose of this article is to specify the installation criteria and wiring methods that minimize electrical hazards by the maintenance of adequately low potential differences only between exposed conductive surfaces that are likely to become energized and could be contacted by a patient.

Informational Note: In a health care facility, it is difficult to prevent the occurrence of a conductive or capacitive path from the patient's body to some grounded object, because that path may be established accidentally or through instrumentation directly connected to the patient. Other electrically conductive surfaces that may make an additional contact with the patient, or instruments that may be connected to the patient, then become possible sources of electric currents that can traverse the patient's body. The hazard is increased as more apparatus is associated with the patient, and, therefore, more intensive precautions are needed. Control of electric shock hazard requires the limitation of electric current that might flow in an electrical circuit involving the patient's body by raising the resistance of the conductive circuit that includes the patient, or by insulating exposed surfaces that might become energized, in addition to reducing the potential difference that can appear between exposed conductive surfaces in the patient care vicinity, or by combinations of these methods. A special problem is presented by the patient with an externalized direct conductive path to the heart muscle. The patient may be electrocuted at current levels so low that additional protection in the design of appliances, insulation of the catheter, and control of medical practice is required.

Sensitivity to electric shock may be increased for patients whose body resistance is compromised either accidentally or by a necessary medical procedure. For example, incontinence or the insertion of a catheter may render a patient much more vulnerable to the effects of an electric current. Therefore, those responsible for the design, installation, and maintenance of the electrical system in patient care spaces must be well acquainted with at least the rudiments of the hazard as explained in this note.

## 517.12 Wiring Methods

Except as modified in this article, wiring methods shall comply with the applicable provisions of Chapters 1 through 4 of this *Code*.

## 517.13 Grounding of Receptacles and Fixed Electrical Equipment in Patient Care Areas

Wiring in patient care areas shall comply with 517.13(A) and (B).

**(A) Wiring Methods.** All branch circuits serving patient care areas shall be provided with an effective ground-fault current path by installation in a metal raceway system, or a cable having a metallic armor or sheath assembly. The metal raceway system, or metallic cable armor, or sheath assembly shall itself qualify as an equipment grounding conductor in accordance with 250.118.

These wiring method requirements apply to the branch circuits in areas used for patient care and are not limited to patient rooms. Additional areas, such as examining rooms, therapy areas, recreational areas, solaria, and certain patient corridors, are also included. The branch circuit wiring method used in these areas is one component of a two-part redundant grounding scheme unique to patient care spaces. The metal raceway or metal cable armor or sheath must qualify as an EGC in accordance with 250.118, independent of the second component of this grounding scheme required by 517.13(B).

Metal-sheathed cable assemblies are not permitted as a general wiring method for life safety and critical branch circuits, because 517.30(C)(3) requires such wiring to be protected by installation in metal raceways. However, listed flexible raceway and metal-sheathed cable assemblies are allowed for limited application for these circuits where used in listed prefabricated headwalls, in office furnishings, or where they are fished into existing walls or ceilings and are not subject to physical damage.

**(B) Insulated Equipment Grounding Conductor.**

**(1) General.** The following shall be directly connected to an insulated copper equipment grounding conductor that is installed with the branch circuit conductors in the wiring methods as provided in 517.13(A).

(1) The grounding terminals of all receptacles.
(2) Metal boxes and enclosures containing receptacles.
(3) All non–current-carrying conductive surfaces of fixed electrical equipment likely to become energized that are subject to personal contact, operating at over 100 volts.

These requirements cover the second component of the redundant grounding scheme. An insulated copper EGC, either solid or stranded, sized in accordance with 250.122, must be installed with the branch-circuit conductors in a wiring method that meets the provisions of 517.13(A). This does not require an additional insulated EGC to be run with the feeder.

*Exception: An insulated equipment bonding jumper that directly connects to the equipment grounding conductor is permitted to connect the box and receptacle(s) to the equipment grounding conductor.*

*Exception No. 1 to (3): Metal faceplates shall be permitted to be connected to the equipment grounding conductor by means of a metal mounting screw(s) securing the faceplate to a grounded outlet box or grounded wiring device.*

The installation of a separate EGC or bonding jumper run to the metal faceplate is not necessary. See 404.9(B), which requires switches and their metal faceplates to be effectively grounded.

*Exception No. 2 to (3): Luminaires more than 2.3 m (7½ ft) above the floor and switches located outside of the patient care vicinity shall be permitted to be connected to an equipment grounding return path complying with 517.13(A).*

Luminaires mounted 7½ feet above the floor and switches located outside the patient vicinity are excluded from having a separate insulated EGC; it is unlikely that a patient would contact these items or that an attendant would contact these items and a patient at the same time. The patient vicinity space consists of a volume 6 feet horizontally in all directions from the bed and vertically to 7½ feet above the floor.

**(2) Sizing.** Equipment grounding conductors and equipment bonding jumpers shall be sized in accordance with 250.122.

## 517.14 Panelboard Bonding

The equipment grounding terminal buses of the normal and essential branch-circuit panelboards serving the same individual patient care vicinity shall be connected together with an insulated continuous copper conductor not smaller than 10 AWG. Where two or more panelboards serving the same individual patient care vicinity are served from separate transfer switches on the essential electrical system, the equipment grounding terminal buses of those panelboards shall be connected together with an insulated continuous copper conductor not smaller than 10 AWG. This conductor shall be permitted to be broken in order to terminate on the equipment grounding terminal bus in each panelboard.

The requirement for bonding panelboards applies to multiple panelboards that may be supplied from the same system or from different systems serving the patient care vicinity.

## 517.16 Use of Isolated Ground Receptacles

An isolated ground receptacle shall not be installed within a patient care vicinity. [**99:**6.3.2.2.7.1(B)]

This section prohibits the use of isolated grounding receptacles, which are intended to provide for the reduction of noise as noted in 406.3(D). Because the equipment grounding terminal is isolated, it does not provide the functional benefit of the multiple equipment grounding paths specified in 517.13. These receptacles are identified by an orange triangle located on the face of the receptacle.

## 517.17 Ground-Fault Protection

Wherever ground-fault protection of equipment (GFPE) is applied to the service providing power to a health care facility, an additional level of ground-fault protection is required downstream. Section 517.17(B) requires ground-fault protection for every feeder.

This requirement is unlike the requirements of 215.10 and 230.95 where mandatory GFPE is based on the rating of the disconnecting means (1000 amperes or more). The second level of GFPE is based on the need to provide selectivity between the feeder protective devices and the service or building supply protective devices. With proper ground-fault coordination per 517.17(C), this additional level of protection is intended to limit a ground fault to a single feeder and thereby prevent a total power outage of the entire health facility. Coordination includes consideration of the trip setting, the time setting, and the time required for operation (opening time) of each level of the ground-fault protection system.

**(A) Applicability.** The requirements of 517.17 shall apply to hospitals, and other buildings (including multiple-occupancy buildings) with critical care space or utilizing electrical life-support equipment, and buildings that provide the required essential utilities or services for the operation of critical care space or electrical life-support equipment.

The requirement for providing two levels of ground-fault protection includes health care facilities that are located in multiple-occupancy buildings. These multiple-occupancy buildings may be multiple medical office or clinic-type occupancies or may be multiple occupancies of mixed use in which one or more of the occupancies are health care facilities. The selectivity required by 517.17(C) is accomplished through the installation of GFPE for all feeder disconnecting means in the first level of distribution downstream of the GFPE-protected service equipment or building disconnecting means specified in 215.10 or 230.95. Therefore, any feeder supplied from this level of distribution will be required to have GFPE regardless of the occupancy type or use group.

Note that the requirement for the second level of GFPE does not apply to all health care facilities located in multiple-occupancy buildings. This requirement applies only where the health care facility governing authority has classified certain portions of the facility as critical care space or where the types of procedures for which the facility is approved or licensed require the use of life support equipment. If there are no designated critical care spaces or life support equipment is not used, second-level GFPE is not required. Based on these criteria, most general medical and dental practices located in multiple-occupancy buildings are not impacted by this requirement. In addition, if the service is not provided with GFPE (either because it is not required by the *Code* or, if optional, has not been incorporated as part of the design), then, of course, the second level of GFPE becomes moot.

In the case of existing multiple-occupancy buildings that have GFPE for the service equipment, a tenant build-out or a renovation for a new health care occupancy may result in the need to also provide second-level GFPE for the feeders supplying all other occupancy types in order to provide the selectivity of GFPE operation required by this section. Careful analysis of the impact of this requirement on the existing service equipment may warrant an alternative approach such as installation of another service if permitted under the provisions of 230.2(A) through (D).

In accordance with 517.45, health care facilities, such as clinics or ambulatory care facilities, that use life support equipment or have areas designated as critical care are required to be provided with an essential electrical system that includes an alternate power source. It is not intended that ground-fault protection be installed between the on-site generator and the transfer switch or on the load side of the essential electrical system transfer switch. Informational Note No. 3 to 230.95(C) calls attention to problems that may arise when ground-fault-protected systems are transferred to another supply system.

**(B) Feeders.** Where ground-fault protection is provided for operation of the service disconnecting means or feeder disconnecting means as specified by 230.95 or 215.10, an additional step of ground-fault protection shall be provided in all next level feeder disconnecting means downstream toward the load. Such protection shall consist of overcurrent devices and current transformers or other equivalent protective equipment that shall cause the feeder disconnecting means to open.

The additional levels of ground-fault protection shall not be installed on the load side of an essential electrical system transfer switch.

**(C) Selectivity.** Ground-fault protection for operation of the service and feeder disconnecting means shall be fully selective such that the feeder device, but not the service device, shall open on ground faults on the load side of the feeder device. Separation of ground-fault protection time-current characteristics shall conform to manufacturer's recommendations and shall consider all required tolerances and disconnect operating time to achieve 100 percent selectivity.

See 230.95(C) and its commentary for information on ground-fault selectivity and performance testing.

Informational Note: See 230.95, informational note, for transfer of alternate source where ground-fault protection is applied.

**(D) Testing.** When equipment ground-fault protection is first installed, each level shall be performance tested to ensure compliance with 517.17(C).

## 517.18 General Care Areas

**(A) Patient Bed Location.** Each patient bed location shall be supplied by at least two branch circuits, one from the critical branch and one from the normal system. All branch circuits from the normal system shall originate in the same panelboard. The electrical receptacles or the cover plate for the electrical receptacles supplied from the critical branch shall have a distinctive color or marking so as to be readily identifiable and shall also indicate the panelboard and branch-circuit number supplying them.

Branch circuits serving patient bed locations shall not be part of a multiwire branch circuit.

*Exception No. 1: Branch circuits serving only special purpose outlets or receptacles, such as portable X-ray outlets, shall not be required to be served from the same distribution panel or panels.*

*Exception No. 2: The requirements of 517.18(A) shall not apply to patient bed locations in clinics, medical and dental offices, and outpatient facilities; psychiatric, substance abuse, and rehabilitation hospitals; sleeping rooms of nursing home; and limited care facilities meeting the requirements of 517.18(B)(2).*

*Exception No. 3: A general care patient bed location served from two separate transfer switches on the critical branch shall not be required to have circuits from the normal system.*

Patient bed locations in general care spaces are prohibited from deriving all their branch circuits from the essential electrical system, and at least one branch circuit for each location must originate in a normal system panelboard. This supply circuit arrangement correlates with the limitations imposed by 517.33 regarding the types of loads permitted to be supplied by the critical branch of the essential electrical system.

Exception No. 3 allows both of the required branch circuits for a general care patient bed location to be supplied by the critical branch, provided they are supplied by two separate transfer switches. Two critical branch circuits have a higher reliability than one normal and one critical branch circuit.

Multiwire branch circuits are required by 210.4(B) to be simultaneously disconnected from all ungrounded conductors. This could result in unintended and potentially dangerous interruption of power to lighting or receptacle loads at a patient bed location. Because of this concern, multiwire branch circuits are not permitted to be used to meet the requirements of 517.18(A).

**(B) Patient Bed Location Receptacles.** Each patient bed location shall be provided with a minimum of eight receptacles. They shall be permitted to be of the single, duplex, or quadruplex type or any combination of the three. All receptacles shall be listed "hospital grade" and shall be so identified. The grounding terminal of each receptacle shall be connected to an insulated copper equipment grounding conductor sized in accordance with Table 250.122.

*Exception No. 1: The requirements of 517.18(B) shall not apply to psychiatric, substance abuse, and rehabilitation hospitals meeting the requirements of 517.10(B)(2).*

*Exception No. 2: Psychiatric security rooms shall not be required to have receptacle outlets installed in the room.*

Informational Note: It is not intended that there be a total, immediate replacement of existing non–hospital grade receptacles. It is intended, however, that non–hospital grade receptacles be replaced with hospital grade receptacles upon modification of use, renovation, or as existing receptacles need replacement.

**(C) Designated General Care Pediatric Locations.** Receptacles that are located within the patient rooms, bathrooms, playrooms, and activity rooms of pediatric units, other than nurseries, shall be listed tamper-resistant or shall employ a listed tamper-resistant cover. [**99:**6.3.2.2.6.2(F)]

Unlike the requirement in 406.12, this requirement covers all receptacles installed in specified rooms of pediatric locations. Safeguarding can be achieved through the use of either listed tamper-resistant receptacles or listed tamper-resistant covers. The use of locking covers over ordinary receptacles does not meet this requirement. Only 125-volt, 15- and 20-ampere tamper-resistant receptacles are available; therefore, other receptacle types are likely to require a tamper-resistant cover. Exhibit 517.2 shows a listed hospital-grade tamper-resistant receptacle, identified by a green dot on its face, that can be used to comply with 517.18(C). Exhibit 517.3 shows a listed tamper-resistant, hospital-grade, GFCI receptacle installed adjacent to a bathroom sink in a pediatric ward.

## 517.19 Critical Care Areas

**(A) Patient Bed Location Branch Circuits.** Each patient bed location shall be supplied by at least two branch circuits, one or more from the critical branch and one or more circuits from

**EXHIBIT 517.2** *A tamper-resistant, hospital-grade receptacle. (Courtesy of Legrand/Pass & Seymour®)*

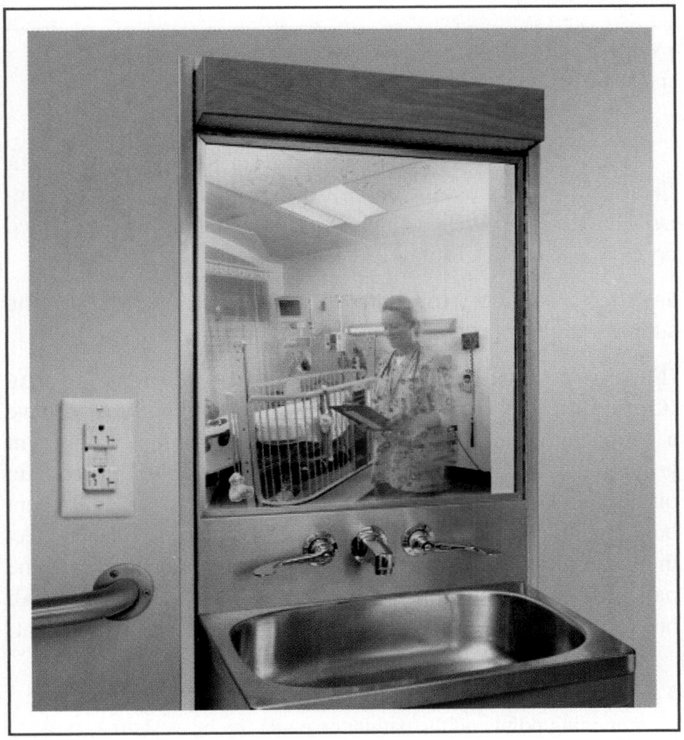

**EXHIBIT 517.3** *Hospital-grade, tamper-resistant GFCI in a hospital environment. (Courtesy of Legrand/Pass & Seymour®)*

**EXHIBIT 517.4** *A receptacle from the normal system (left) and the essential electrical system (right). Note the label that indicates the panelboard and the circuit number. (Courtesy of the International Association of Electrical Inspectors)*

the normal system. At least one branch circuit from the critical branch shall supply an outlet(s) only at that bed location. All branch circuits from the normal system shall be from a single panelboard. Critical branch receptacles shall be identified and shall also indicate the panelboard and circuit number supplying them.

The branch circuit serving patient bed locations shall not be part of a multiwire branch circuit.

*Exception No. 1: Branch circuits serving only special-purpose receptacles or equipment in critical care spaces shall be permitted to be served by other panelboards.*

*Exception No. 2: Critical care space served from two separate critical branch transfer switches shall not be required to have circuits from the normal system.*

Each patient bed location must be served by receptacles supplied from the normal system as well as the essential electrical system. The critical branch receptacles are required to be labeled with the panelboard and the circuit supplying them. They are also required to be identified as being supplied from the essential electrical system, which is often accomplished through the use of a color code for receptacles established for that facility (the color red is used in many health care facilities to identify such receptacles), as illustrated in Exhibit 517.4.

Multiwire branch circuits are required by 210.4(B) to be simultaneously disconnected from all ungrounded conductors. This could result in unintended and potentially dangerous interruption of power to lighting or receptacle loads at a patient bed location. Because of this concern, these circuits are not permitted to be used to meet the requirements of 517.19(A). Exception No. 2 covers the special case in which two separate transfer switches supply a single patient care area. Critical branch circuits supplied from two separate transfer switches provide an equivalent level of redundancy to that specified by the main requirement.

**(B) Patient Bed Location Receptacles.**

**(1) Minimum Number and Supply.** Each patient bed location shall be provided with a minimum of 14 receptacles, at least one of which shall be connected to either of the following:

(1) The normal system branch circuit required in 517.19(A)
(2) A critical branch circuit supplied by a different transfer switch than the other receptacles at the same patient bed location

**(2) Receptacle Requirements.** The receptacles required in 517.19(B)(1) shall be permitted to be single, duplex, or quadruplex type or any combination thereof. All receptacles shall be listed "hospital grade" and shall be so identified. The grounding terminal of each receptacle shall be connected to the reference grounding point by means of an insulated copper equipment grounding conductor.

Each patient bed location must be provided with at least fourteen receptacles that may be of the single, duplex, or quadruplex type, provided they are listed hospital-grade type and are identified as such, typically with a green dot (shown in Exhibit 517.2). A duplex receptacle is two receptacles on one yoke. A quadruplex receptacle is considered four receptacles.

Each patient bed location in critical care spaces must be supplied by at least two branch circuits, one from the normal panel and one from the essential electrical panel, as shown in Exhibit 517.5. The normal circuits must be supplied from the same panel (L-1). The critical branch circuits are permitted to be supplied from different panels (EES-1 and EES-2). However, the critical branch circuit to patient bed location A cannot supply receptacles for patient bed location B.

Patient bed location receptacles can also be supplied by two different critical branch circuits, instead of one critical branch and one normal, provided the critical branch circuits are supplied from two different transfer switches. The requirements for the number and type of branch circuits in critical care spaces are intended to ensure that critical care patients will not be without electrical power regardless of whether the equipment, the branch circuits, or the normal system itself is at fault.

**(C) Operating Room Receptacles.**

**(1) Minimum Number and Supply.** Each operating room shall be provided with a minimum of 36 receptacles, at least 12 of which shall be connected to either of the following:

(1) The normal system branch circuit required in 517.19(A)
(2) A critical branch circuit supplied by a different transfer switch than the other receptacles at the same location

**(2) Receptacle Requirements.** The receptacles required in C(517.19)(1) shall be permitted to be of the single or duplex types or a combination of both.

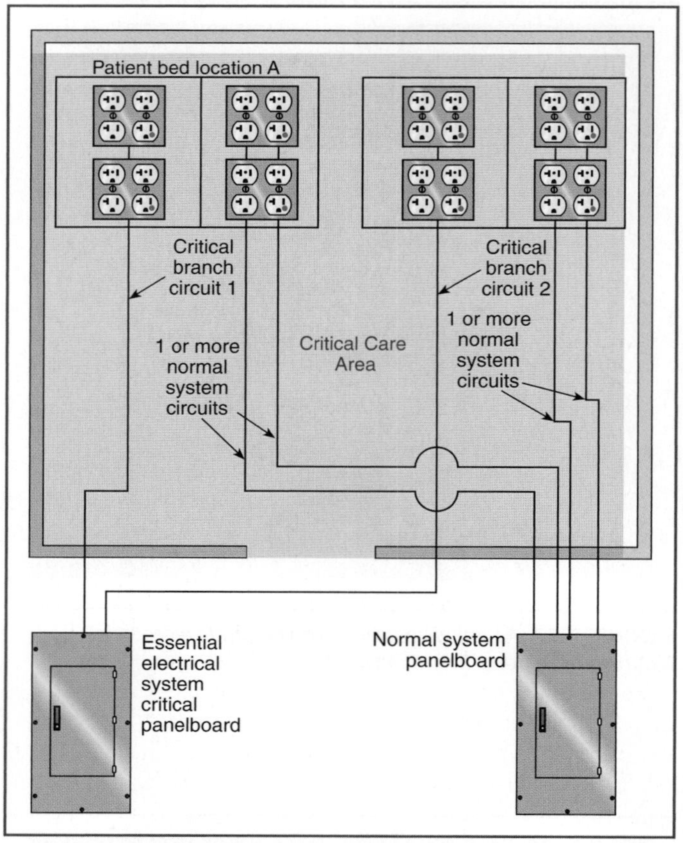


***EXHIBIT 517.5*** *Examples of normal and critical branch circuits supplying patient bed locations in a critical care space.*

All receptacles shall be listed hospital grade and so identified. The grounding terminal of each receptacle shall be connected to the reference grounding point by means of an insulated copper equipment grounding conductor.

See the commentary following 517.19(B) for information on the supply and receptacle requirements.

**(D) Patient Care Vicinity Grounding and Bonding (Optional).** A patient care vicinity shall be permitted to have a patient equipment grounding point. The patient equipment grounding point, where supplied, shall be permitted to contain one or more listed grounding and bonding jacks. An equipment bonding jumper not smaller than 10 AWG shall be used to connect the grounding terminal of all grounding-type receptacles to the patient equipment grounding point. The bonding conductor shall be permitted to be arranged centrically or looped as convenient.

Informational Note: Where there is no patient equipment grounding point, it is important that the distance between the reference grounding point and the patient care vicinity be as short as possible to minimize any potential differences.

A patient vicinity is permitted to have a patient equipment grounding point with multiple grounding or bonding jacks. See Exhibit 517.6.

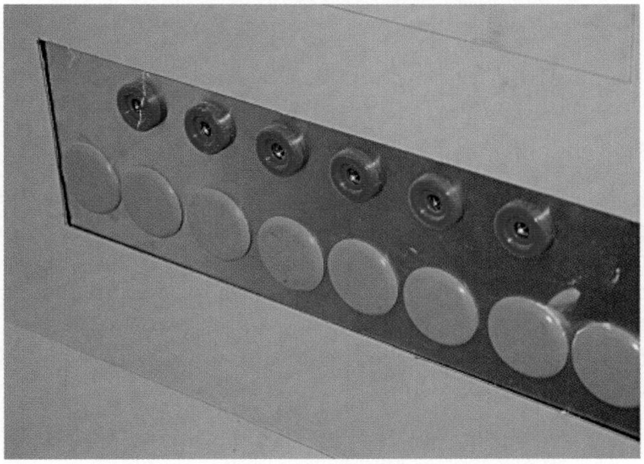

*EXHIBIT 517.6 A patient equipment grounding point. (Courtesy of the International Association of Electrical Inspectors)*

**(E) Equipment Grounding and Bonding.** Where a grounded electrical distribution system is used and metal feeder raceway or Type MC or MI cable that qualifies as an equipment grounding conductor in accordance with 250.118 is installed, grounding of enclosures and equipment, such as panelboards, switchboards, and switchgear, shall be ensured by one of the following bonding means at each termination or junction point of the metal raceway or Type MC or MI cable:

(1) A grounding bushing and a continuous copper bonding jumper, sized in accordance with 250.122, with the bonding jumper connected to the junction enclosure or the ground bus of the panel
(2) Connection of feeder raceways or Type MC or MI cable to threaded hubs or bosses on terminating enclosures
(3) Other approved devices such as bonding-type locknuts or bushings

**(F) Additional Protective Techniques in Critical Care Spaces (Optional).** Isolated power systems shall be permitted to be used for critical care spaces, and, if used, the isolated power system equipment shall be listed as isolated power equipment. The isolated power system shall be designed and installed in accordance with 517.160.

*Exception: The audible and visual indicators of the line isolation monitor shall be permitted to be located at the nursing station for the area being served.*

**(G) Isolated Power System Equipment Grounding.** Where an isolated ungrounded power source is used and limits the first-fault current to a low magnitude, the equipment grounding conductor associated with the secondary circuit shall be permitted to be run outside of the enclosure of the power conductors in the same circuit.

Informational Note: Although it is permitted to run the grounding conductor outside of the conduit, it is safer to run it with the power conductors to provide better protection in case of a second ground fault.

Installing the EGC inside the raceway with the conductors delivering the fault current reduces the impedance of the grounding path.

**(H) Special-Purpose Receptacle Grounding.** The equipment grounding conductor for special-purpose receptacles, such as the operation of mobile X-ray equipment, shall be extended to the reference grounding points of branch circuits for all locations likely to be served from such receptacles. Where such a circuit is served from an isolated ungrounded system, the grounding conductor shall not be required to be run with the power conductors; however, the equipment grounding terminal of the special-purpose receptacle shall be connected to the reference grounding point.

## 517.20 Wet Procedure Locations

**(A) Receptacles and Fixed Equipment.** Wet procedure location patient care areas shall be provided with special protection against electric shock by one of the following means:

(1) Power distribution system that inherently limits the possible ground-fault current due to a first fault to a low value, without interrupting the power supply
(2) Power distribution system in which the power supply is interrupted if the ground-fault current does, in fact, exceed a value of 6 mA

See the definition of *wet procedure location* in 517.2.

*Exception: Branch circuits supplying only listed, fixed, therapeutic and diagnostic equipment shall be permitted to be supplied from a grounded service, single- or 3-phase system, provided that*

*(a) Wiring for grounded and isolated circuits does not occupy the same raceway, and*
*(b) All conductive surfaces of the equipment are connected to an insulated copper equipment grounding conductor.*

**(B) Isolated Power Systems.** Where an isolated power system is utilized, the isolated power equipment shall be listed as isolated power equipment, and the isolated power system shall be designed and installed in accordance with 517.160.

Informational Note: For requirements for installation of therapeutic pools and tubs, see Part VI of Article 680.

## 517.21 Ground-Fault Circuit-Interrupter Protection for Personnel

Ground-fault circuit-interrupter protection for personnel shall not be required for receptacles installed in those critical care areas where the toilet and basin are installed within the patient room.

This requirement does not exempt receptacles installed in bathrooms for critical care spaces from the GFCI requirements in 210.8(B). It also does not exempt receptacles in other bathrooms for patients, staff, or the public from the requirements of 210.8(B). Patients in critical care areas are often bedridden. Therefore, a basin and toilet may be provided within the patient room or as part of the bed assembly. Although the presence of a basin and toilet meets the definition of a bathroom, the receptacles are exempt from the GFCI requirement because of the specialized use of a critical care space.

# III. Essential Electrical System

## 517.25 Scope

The essential electrical system for these facilities shall comprise a system capable of supplying a limited amount of lighting and power service, which is considered essential for life safety and orderly cessation of procedures during the time normal electrical service is interrupted for any reason. This includes clinics, medical and dental offices, outpatient facilities, nursing homes, limited care facilities, hospitals, and other health care facilities serving patients.

Informational Note: For information on the need for an essential electrical system, see NFPA 99-2012, *Health Care Facilities Code*.

## 517.26 Application of Other Articles

The life safety branch of the essential electrical system shall meet the requirements of Article 700, except as amended by Article 517.

In larger health care facilities, the requirements of 700.5(D) and 517.30(B)(2) correlate from the perspective that one or more transfer switches supply only essential loads. However, 517.30(B)(2) also differs from 700.5 in smaller facilities with a maximum demand on the essential electrical system of 150 kVA or less. In those cases, a single transfer switch is permitted to supply the entire essential electrical system. This example is where a requirement in Article 517 differs from a requirement in Article 700, and Article 517 takes precedence.

Similar to the physical separation requirements for emergency systems specified in 700.10(B), the physical separation requirements for circuits supplied by essential electrical systems are covered in 517.30(C)(1) for hospitals and 517.41(D) for nursing homes and limited care facilities.

Informational Note No. 1: For additional information, see NFPA 110-2013, *Standard for Emergency and Standby Power Systems*.

Informational Note No. 2: For additional information, see 517.30 and NFPA 99-2012, *Health Care Facilities Code*.

## 517.30 Essential Electrical Systems for Hospitals

**(A) Applicability.** The requirements of Part III, 517.30 through 517.35, shall apply to hospitals where an essential electrical system is required.

Informational Note No. 1: For performance, maintenance, and testing requirements of essential electrical systems in hospitals, see NFPA 99-2012, *Health Care Facilities Code*. For installation of centrifugal fire pumps, see NFPA 20-2013, *Standard for the Installation of Stationary Fire Pumps for Fire Protection*.

Informational Note No. 2: For additional information, see NFPA 99-2012, *Health Care Facilities Code*.

**(B) General.**

**(1) Separate Branches.** Essential electrical systems for hospitals shall be comprised of three separate branches capable of supplying a limited amount of lighting and power service that is considered essential for life safety and effective hospital operation during the time the normal electrical service is interrupted for any reason. The three branches are life safety, critical, and equipment.

•

**(2) Transfer Switches.** The number of transfer switches to be used shall be based on reliability, design, and load considerations. Each branch of the essential electrical system shall have one or more transfer switches. One transfer switch and downstream distribution system shall be permitted to serve one or more branches in a facility with a maximum demand on the essential electrical system of 150 kVA.

Informational Note No. 1: See NFPA 99-2012, *Health Care Facilities Code*, 6.4.3.2, Transfer Switches; 6.4.2.1.5, Automatic Transfer Switch Features; 6.4.2.1.5.15, Nonautomatic Transfer Switch Features; and 6.4.2.1.7, Nonautomatic Transfer Device Features.

Informational Note No. 2: See Informational Note Figure 517.30, No. 1.

Informational Note No. 3: See Informational Note Figure 517.30, No. 2.

A small load can be served by a single transfer switch that can handle all loads associated with the essential electrical system. This is based on the assumption that the alternate source of

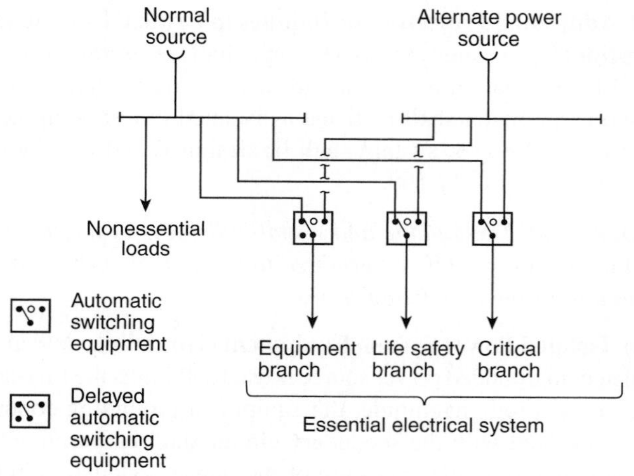


***INFORMATIONAL NOTE FIGURE 517.30, NO. 1***
*Hospital — Minimum Requirement (greater than 150 kVA) for Transfer Switch Arrangement.*

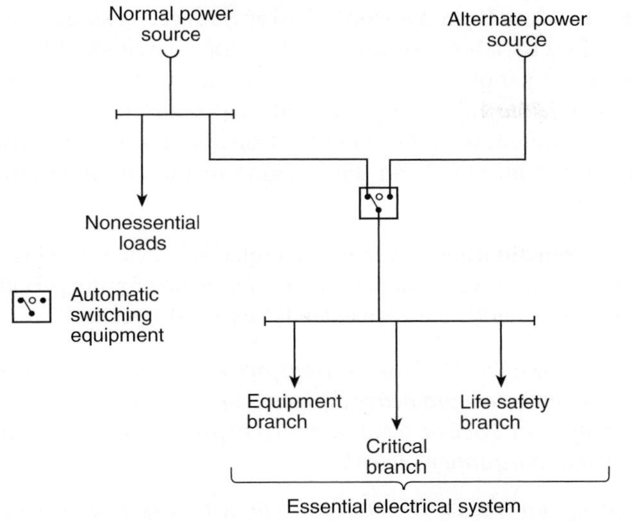


***INFORMATIONAL NOTE FIGURE 517.30, NO. 2***
*Hospital — Minimum Requirement (150 kVA or less) for Transfer Switch Arrangement.*

power is sufficiently large to handle the simultaneous transfer of all systems in the event of a normal power loss. For further explanation of loads permitted on an essential electrical system, see NFPA 99.

**(3) Optional Loads.** Loads served by the generating equipment not specifically named in Article 517 shall be served by their own transfer switches such that the following conditions apply:

(1) These loads shall not be transferred if the transfer will overload the generating equipment.
(2) These loads shall be automatically shed upon generating equipment overloading.

**(4) Contiguous Facilities.** Hospital power sources and alternate power sources shall be permitted to serve the essential electrical systems of contiguous or same site facilities.

**(C) Wiring Requirements.**

**(1) Separation from Other Circuits.** The life safety branch and critical branch of the essential electrical system shall be kept entirely independent of all other wiring and equipment and shall not enter the same raceways, boxes, or cabinets with each other or other wiring.

Where general care locations are served from two separate transfer switches on the essential electrical system in accordance with 517.18(A), Exception No. 3, the general care circuits from the two separate systems shall be kept independent of each other.

Where critical care locations are served from two separate transfer switches on the essential electrical system in accordance with 517.19(A), Exception No. 2, the critical care circuits from the two separate systems shall be kept independent of each other.

Wiring of the life safety branch and the critical branch shall be permitted to occupy the same raceways, boxes, or cabinets of other circuits not part of the branch where such wiring complies with one of the following:

(1) Is in transfer equipment enclosures
(2) Is in exit or emergency luminaires supplied from two sources
(3) Is in a common junction box attached to exit or emergency luminaires supplied from two sources
(4) Is for two or more circuits supplied from the same branch and same transfer switch

The wiring of the equipment branch shall be permitted to occupy the same raceways, boxes, or cabinets of other circuits that are not part of the essential electrical system.

Although the life safety branch and critical branch of the essential electrical system are not to enter the same raceways, boxes, or cabinets with each other or other wiring, circuits on the life safety or critical branch are allowed with other circuits of the same branch.

Where general care or critical care locations are supplied from two transfer switches on an essential electrical system, the separate feeder and branch circuits are to be kept independent of each other. The issue of a failure in one circuit affecting the other is the same whether there is a normal and an essential circuit or two essential circuits.

**(2) Isolated Power Systems.** Where isolated power systems are installed in any of the areas in 517.33(A)(1) and (A)(2), each system shall be supplied by an individual circuit serving no other load.

**(3) Mechanical Protection of the Essential Electrical System.** The wiring of the life safety and critical branches shall be mechanically protected. Where installed as branch circuits in patient care spaces, the installation shall comply with the requirements of 517.13(A) and (B). The following wiring methods shall be permitted:

To increase reliability of power delivery to life safety and patient care equipment, the wiring of the life safety and critical branch requires additional protection against mechanical damage that is not normally mandated for other occupancies. The wiring methods for branch circuits in patient care spaces are limited by 517.13.

(1) Nonflexible metal raceways, Type MI cable, Type RTRC marked with the suffix –XW, or Schedule 80 PVC conduit. Nonmetallic raceways shall not be used for branch circuits that supply patient care areas.
(2) Where encased in not less than 50 mm (2 in.) of concrete, Schedule 40 PVC conduit, flexible nonmetallic or jacketed metallic raceways, or jacketed metallic cable assemblies listed for installation in concrete. Nonmetallic raceways shall not be used for branch circuits that supply patient care areas.

(3) Listed flexible metal raceways and listed metal sheathed cable assemblies in any of the following:
   a. Where used in listed prefabricated medical headwalls
   b. In listed office furnishings
   c. Where fished into existing walls or ceilings, not otherwise accessible and not subject to physical damage
   d. Where necessary for flexible connection to equipment
(4) Flexible power cords of appliances or other utilization equipment connected to the emergency system.
(5) Cables for Class 2 or Class 3 systems permitted by Part VI of this Article, with or without raceways.

Informational Note: See 517.13 for additional grounding requirements in patient care areas.

Some conditions do not require the installation of conductors in nonflexible metal raceways. Section 517.30(C)(3)(3) permits fishing flexible metal raceways and metal sheathed cables in existing installations. This facilitates installations in renovated areas where the existing walls or ceilings remain intact. The secondary conductors of limited energy systems, such as nurse call, telephone, and alarm circuits, are exempt from being run in raceways, provided they comply with their applicable articles. Although this requirement allows substantial latitude in the wiring method, the restrictions of 300.22 (ducts, plenums, and other air-handling spaces) apply, unless cables specifically listed for use in these environments are used.

**(D) Capacity of Systems.** The essential electrical system shall have the capacity and rating to meet the maximum actual demand likely to be produced by the connected load.

Feeders shall be sized in accordance with 215.2 and Part III of Article 220. The generator set(s) shall have the capacity and rating to meet the demand produced by the load at any given time.

Demand calculations for sizing of the generator set(s) shall be based on any of the following:

(1) Prudent demand factors and historical data
(2) Connected load
(3) Feeder calculation procedures described in Article 220
(4) Any combination of the above

The sizing requirements in 700.4 and 701.4 shall not apply to hospital generator set(s).

The sizing of generators is permitted to be based on actual demand likely to be produced by the connected load of the system at any one time. This method of calculation facilitates practical sizing of generators in health care facilities and helps eliminate prime mover operational problems associated with lightly loaded generators.

**(E) Receptacle Identification.** The cover plates for the electrical receptacles or the electrical receptacles themselves supplied from the essential electrical system shall have a distinctive color or marking so as to be readily identifiable. [**99:**6.4.2.2.6.2(C)]

**(F) Feeders from Alternate Power Source.** A single feeder supplied by a local or remote alternate source shall be permitted to supply the essential electrical system to the point at which the life safety, critical, and equipment branches are separated. Installation of the transfer equipment shall be permitted at other than the location of the alternate power source.

**(G) Coordination.** Overcurrent protective devices serving the essential electrical system shall be coordinated for the period of time that a fault's duration extends beyond 0.1 second.

*Exception No. 1: Between transformer primary and secondary overcurrent protective devices, where only one overcurrent protective device or set of overcurrent protective devices exists on the transformer secondary.*

*Exception No. 2: Between overcurrent protective devices of the same size (ampere rating) in series.*

Informational Note: The terms *coordination* and *coordinated* as used in this section do not cover the full range of overcurrent conditions.

## 517.31 Branches Requiring Automatic Connection

Those functions of patient care depending on lighting or appliances that are connected to the essential electrical system shall be divided into the life safety branch and the critical branch, as described in 517.32 and 517.33.

The life safety and critical branches shall be installed and connected to the alternate power source so that all functions supplied by these branches specified here shall be automatically restored to operation within 10 seconds after interruption of the normal source. [**99:**6.4.3.1]

## 517.32 Life Safety Branch

No functions other than those listed in 517.32(A) through (H) shall be connected to the life safety branch. The life safety branch of the essential electrical system shall supply power for the following lighting, receptacles, and equipment.

**(A) Illumination of Means of Egress.** Illumination of means of egress, such as lighting required for corridors, passageways, stairways, and landings at exit doors, and all necessary ways of approach to exits. Switching arrangements to transfer patient corridor lighting in hospitals from general illumination circuits to night illumination circuits shall be permitted, provided only one of two circuits can be selected and both circuits cannot be extinguished at the same time.

Informational Note: See NFPA *101*-2012, *Life Safety Code*, Sections 7.8 and 7.9.

**(B) Exit Signs.** Exit signs and exit directional signs.

Informational Note: See NFPA *101*-2012, *Life Safety Code*, Section 7.10.

**(C) Alarm and Alerting Systems.** Alarm and alerting systems including the following:

(1) Fire alarms

Informational Note: See NFPA *101*-2012, *Life Safety Code*, Section 9.6 and 18.3.4.

(2) Alarms required for systems used for the piping of nonflammable medical gases

Informational Note: See NFPA 99-2012, *Health Care Facilities Code*, 6.4.2.2.3.3.

(3) Mechanical, control, and other accessories required for effective life safety systems operation shall be permitted to be connected to the life safety branch.

HVAC controls are permitted to be on the life safety branch because the operation of an HVAC system and associated dampers can impact smoke control and life safety.

**(D) Communications Systems.** Hospital communications systems, where used for issuing instructions during emergency conditions.

**(E) Generator Set and Transfer Switch Locations.** Task illumination battery charger for battery-powered lighting unit(s) and selected receptacles at the generator set and essential transfer switch locations. [**99:**6.4.2.2.3.2(4)]

**(F) Generator Set Accessories.** Generator set accessories as required for generator performance. Loads dedicated to a specific generator, including the fuel transfer pump(s), ventilation fans, electrically operated louvers, controls, cooling system, and other generator accessories essential for generator operation, shall be connected to the life safety branch or to the output terminals of the generator with overcurrent protective devices.

An example of an accessory is task illumination. The connection of a battery charger for task lighting for transfer switch locations, in addition to generator locations, allows for servicing when the building is being supplied from the alternate source of the essential electrical system.

**(G) Elevators.** Elevator cab lighting, control, communications, and signal systems.

**(H) Automatic Doors.** Automatically operated doors used for building egress. [**99:**4.4.2.2.2.2(7)]

## 517.33 Critical Branch

**(A) Task Illumination and Selected Receptacles.** The critical branch of the essential electrical system shall supply power for task illumination, fixed equipment, selected receptacles, and special power circuits serving the following areas and functions related to patient care:

(1) Critical care areas that utilize anesthetizing gases — task illumination, selected receptacles, and fixed equipment
(2) The isolated power systems in special environments
(3) Patient care areas — task illumination and selected receptacles in the following:
   a. Infant nurseries
   b. Medication preparation areas
   c. Pharmacy dispensing areas
   d. Selected acute nursing areas
   e. Psychiatric bed areas (omit receptacles)
   f. Ward treatment rooms
   g. Nurses' stations (unless adequately lighted by corridor luminaires)
(4) Additional specialized patient care task illumination and receptacles, where needed
(5) Nurse call systems
(6) Blood, bone, and tissue banks
(7) Telephone and data equipment rooms and closets
(8) Task illumination, selected receptacles, and selected power circuits for the following:
   a. General care beds (at least one duplex receptacle in each patient bedroom)
   b. Angiographic labs
   c. Cardiac catheterization labs
   d. Coronary care units
   e. Hemodialysis rooms or areas
   f. Emergency room treatment areas (selected)
   g. Human physiology labs
   h. Intensive care units
   i. Postoperative recovery rooms (selected)
(9) Additional task illumination, receptacles, and selected power circuits needed for effective hospital operation. Single-phase fractional horsepower motors shall be permitted to be connected to the critical branch. [**99:**6.4.2.2.4.2(9)]

The critical branch is intended to serve a limited number of receptacles and locations in order to reduce the load and minimize the chances of a fault condition. Receptacles in general patient care area corridors are permitted on the critical branch, but they must be identified in some manner (color-coded or labeled) as part of the essential electrical system, in accordance with 517.30(E).

**(B) Subdivision of the Critical Branch.** It shall be permitted to subdivide the critical branch into two or more branches.

Informational Note: It is important to analyze the consequences of supplying an area with only critical care branch power when failure occurs between the area and the transfer switch. Some proportion of normal and critical power or critical power from separate transfer switches may be appropriate.

## 517.34 Equipment Branch Connection to Alternate Power Source

The equipment branch shall be installed and connected to the alternate power source such that the equipment described in 517.34(A) is automatically restored to operation at appropriate time-lag intervals following the energizing of the essential electrical system. Its arrangement shall also provide for the subsequent connection of equipment described in 517.34(B). [**99:**6.4.2.2.5.2]

*Exception: For essential electrical systems under 150 kVA, deletion of the time-lag intervals feature for delayed automatic connection to the equipment system shall be permitted.*

**(A) Equipment for Delayed Automatic Connection.** The following equipment shall be permitted to be arranged for delayed automatic connection to the alternate power source:

(1) Central suction systems serving medical and surgical functions, including controls. Such suction systems shall be permitted on the critical branch.
(2) Sump pumps and other equipment required to operate for the safety of major apparatus, including associated control systems and alarms.
(3) Compressed air systems serving medical and surgical functions, including controls. Such air systems shall be permitted on the critical branch.
(4) Smoke control and stair pressurization systems, or both.
(5) Kitchen hood supply or exhaust systems, or both, if required to operate during a fire in or under the hood.
(6) Supply, return, and exhaust ventilating systems for airborne infectious/isolation rooms, protective environment rooms, exhaust fans for laboratory fume hoods, nuclear medicine areas where radioactive material is used, ethylene oxide evacuation, and anesthesia evacuation. Where delayed automatic connection is not appropriate, such ventilation systems shall be permitted to be placed on the critical branch. [**99:**6.4.2.2.5.3(A)(6) and (B)]
(7) Supply, return, and exhaust ventilating systems for operating and delivery rooms.
(8) Supply, return, exhaust ventilating systems and/or air-conditioning systems serving telephone equipment rooms and closets and data equipment rooms and closets.

*Exception: Sequential delayed automatic connection to the alternate power source to prevent overloading the generator shall be permitted where engineering studies indicate it is necessary.*

**(B) Equipment for Delayed Automatic or Manual Connection.** The following equipment shall be permitted to be arranged for either delayed automatic or manual connection to the alternate power source:

(1) Heating equipment to provide heating for operating, delivery, labor, recovery, intensive care, coronary care, nurseries, infection/isolation rooms, emergency treatment spaces, and general patient rooms and pressure maintenance (jockey or make-up) pump(s) for water-based fire protection systems.

*Exception: Heating of general patient rooms and infection/isolation rooms during disruption of the normal source shall not be required under any of the following conditions:*

*(1) The outside design temperature is higher than −6.7°C (20°F).*
*(2) The outside design temperature is lower than −6.7°C (20°F), and where a selected room(s) is provided for the needs of all confined patients, only such room(s) need be heated.*
*(3) The facility is served by a dual source of normal power.*

Informational Note No. 1: The design temperature is based on the 97½ percent design value as shown in Chapter 24 of the ASHRAE *Handbook of Fundamentals* (1997).

A common practice in some areas is to install individual room heating/air conditioning rather than to have a central heating/air-conditioning plant. If these individual units are electrically powered, applying their high-demand load to the generator may not be practical. Providing limited heating during emergency conditions is based on consideration of outside design temperatures. Delayed connection of the loads listed in this section is optional, rather than mandatory. Many health care facilities have generators with sufficient capacity to pick up these loads during startup.

Informational Note No. 2: For a description of a dual source of normal power, see 517.35(C), Informational Note.

(2) An elevator(s) selected to provide service to patient, surgical, obstetrical, and ground floors during interruption of normal power. In instances where interruption of normal power would result in other elevators stopping between floors, throw-over facilities shall be provided to allow the temporary operation of any elevator for the release of patients or other persons who may be confined between floors.
(3) Hyperbaric facilities.
(4) Hypobaric facilities.
(5) Automatically operated doors
(6) Minimal electrically heated autoclaving equipment shall be permitted to be arranged for either automatic or manual connection to the alternate source.
(7) Controls for equipment listed in 517.34.
(8) Other selected equipment shall be permitted to be served by the equipment system. [**99:**6.4.2.2.5.4(9)]

**(C) AC Equipment for Nondelayed Automatic Connection.** Generator accessories, including but not limited to, the transfer fuel pump, electrically operated louvers, and other generator accessories essential for generator operation, shall be arranged for automatic connection to the alternate power source. [**99:**6.5.2.2.3.2]

### 517.35 Sources of Power

**(A) Two Independent Sources of Power.** Essential electrical systems shall have a minimum of two independent sources of power: a normal source generally supplying the entire electrical system and one or more alternate sources for use when the normal source is interrupted. [**99:**6.4.1.1.4]

**(B) Alternate Source of Power.** The alternate source of power shall be one of the following:

(1) Generator(s) driven by some form of prime mover(s) and located on the premises
(2) Another generating unit(s) where the normal source consists of a generating unit(s) located on the premises
(3) An external utility service when the normal source consists of a generating unit(s) located on the premises
(4) A battery system located on the premises [**99:**6.4.1.2]

**(C) Location of Essential Electrical System Components.** Careful consideration shall be given to the location of the spaces housing the components of the essential electrical system to minimize interruptions caused by natural forces common to the area (e.g., storms, floods, earthquakes, or hazards created by adjoining structures or activities). Consideration shall also be given to the possible interruption of normal electrical services resulting from similar causes as well as possible disruption of normal electrical service due to internal wiring and equipment failures. Consideration shall be given to the physical separation of the main feeders of the alternate source from the main feeders of the normal electrical source to prevent possible simultaneous interruption.

Informational Note: Facilities in which the normal source of power is supplied by two or more separate central station-fed services experience greater than normal electrical service reliability than those with only a single feed. Such a dual source of normal power consists of two or more electrical services fed from separate generator sets or a utility distribution network that has multiple power input sources and is arranged to provide mechanical and electrical separation so that a fault between the facility and the generating sources is not likely to cause an interruption of more than one of the facility service feeders.

### 517.40 Essential Electrical Systems for Nursing Homes and Limited Care Facilities

NFPA 99 recognizes two classes of nursing homes or limited care facilities. For the smaller, less complex facility, only a minimum alternate lighting and alarm service needs to be furnished. At nursing homes or limited care facilities where patients are sustained by electrical life-support equipment or inpatient hospital care is provided, the requirements of 517.41 through 517.44 apply. Because the level of care is comparable to that provided in a hospital, an essential electrical system is required for this type of nursing home.

**(A) Applicability.** The requirements of Part III, 517.40(C) through 517.44, shall apply to nursing homes and limited care facilities.

*Exception: The requirements of Part III, 517.40(C) through 517.44, shall not apply to freestanding buildings used as nursing homes and limited care facilities, provided that the following apply:*

*(a) Admitting and discharge policies are maintained that preclude the provision of care for any patient or resident who may need to be sustained by electrical life-support equipment.*
*(b) No surgical treatment requiring general anesthesia is offered.*
*(c) An automatic battery-operated system(s) or equipment is provided that shall be effective for at least 1½ hours and is otherwise in accordance with 700.12 and that shall be capable of supplying lighting for exit lights, exit corridors, stairways, nursing stations, medical preparation areas, boiler rooms, and communications areas. This system shall also supply power to operate all alarm systems.*

Informational Note: See NFPA *101*-2012, *Life Safety Code*.

**(B) Inpatient Hospital Care Facilities.** For those nursing homes and limited care facilities that admit patients who need to be sustained by electrical life support equipment, the essential electrical system from the source to the portion of the facility where such patients are treated shall comply with the requirements of Part III, 517.30 through 517.35.

Regardless of the name given to a facility, the type of electrical system required corresponds with the level of patient care provided. If inpatient care requires the use of life support equipment, a hospital-type essential electrical system must be installed. The type of care that can be provided at a nursing home or limited care facility is generally controlled through the administrative agency that licenses and regulates the facility.

**(C) Facilities Contiguous or Located on the Same Site with Hospitals.** Nursing homes and limited care facilities that are contiguous or located on the same site with a hospital shall be permitted to have their essential electrical systems supplied by that of the hospital.

A single alternate power supply is permitted to serve a single building or campus with multiple types of health care occupancies. This is the same allowance specified in 517.30(B)(4) for a campus having only hospital occupancies. The use of multiple alternate sources is permitted and may be desirable to ensure reliability.

Informational Note: For performance, maintenance, and testing requirements of essential electrical systems in nursing homes and limited care facilities, see NFPA 99-2012, *Health Care Facilities Code*.

### 517.41 Essential Electrical Systems

**(A) General.** Essential electrical systems for nursing homes and limited care facilities shall be comprised of two separate branches capable of supplying a limited amount of lighting and power service, which is considered essential for the protection of life safety and effective operation of the institution during the time normal electrical service is interrupted for any reason. These two separate branches shall be the life safety branch and the critical branch. [**99:**A.6.5.2.1.1]

**(B) Transfer Switches.** The number of transfer switches to be used shall be based on reliability, design, and load considerations. Each branch of the essential electrical system shall be served by one or more transfer switches. One transfer switch shall be permitted to serve one or more branches or systems in a facility with a maximum demand on the essential electrical system of 150 kVA. [**99:**6.5.2.2.1]

Informational Note No. 1: See NFPA 99-2012, *Health Care Facilities Code*, 6.5.3.2, Transfer Switch Operation Type II; 6.4.2.1.5, Automatic Transfer Switch Features; and 6.4.2.1.7, Nonautomatic Transfer Device Features.

Informational Note No. 2: See Informational Note Figure 517.41, No. 1.

Informational Note No. 3: See Informational Note Figure 517.41, No. 2.

**(C) Capacity of System.** The essential electrical system shall have adequate capacity to meet the demand for the operation of all functions and equipment to be served by each branch at one time.

**(D) Separation from Other Circuits.** The life safety branch shall be kept entirely independent of all other wiring and equipment and shall not enter the same raceways, boxes, or cabinets with other wiring except as follows:

(1) In transfer switches
(2) In exit or emergency luminaires supplied from two sources
(3) In a common junction box attached to exit or emergency luminaires supplied from two sources

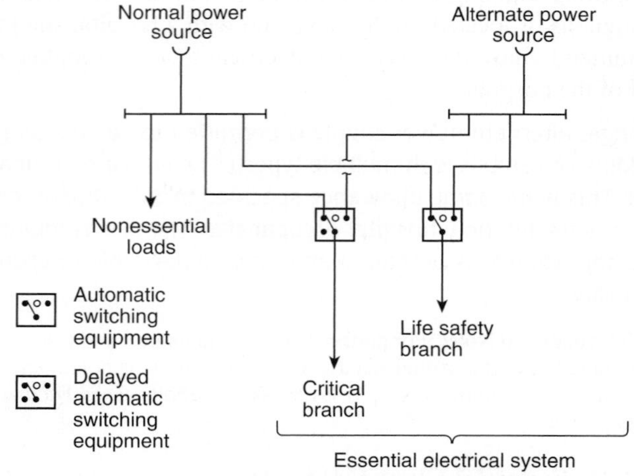


*INFORMATIONAL NOTE FIGURE 517.41, NO. 1 Nursing Home and Limited Health Care Facilities — Minimum Requirement (greater than 150 kVA) for Transfer Switch Arrangement.*

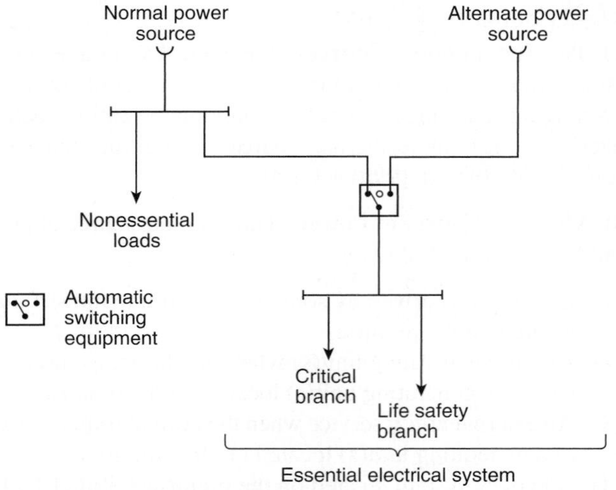


*INFORMATIONAL NOTE FIGURE 517.41, NO. 2 Nursing Home and Limited Health Care Facilities — Minimum Requirement (150 kVA or less) for Transfer Switch Arrangement.*

The wiring of the critical branch shall be permitted to occupy the same raceways, boxes, or cabinets of other circuits that are not part of the life safety branch.

**(E) Receptacle Identification.** The cover plates for the electrical receptacles or the electrical receptacles themselves supplied from the essential electrical system shall have a distinctive color or marking so as to be readily identifiable. [**99:**6.5.2.2.4.2]

Nonlocking-type, 125-volt, 15- and 20-ampere receptacles shall have an illuminated face or an indicator light to indicate that there is power to the receptacle.

## 517.42 Automatic Connection to Life Safety Branch

The life safety branch shall be installed and connected to the alternate source of power so that all functions specified herein shall be automatically restored to operation within 10 seconds after the interruption of the normal source. No functions other than those listed in 517.42(A) through (G) shall be connected to the life safety branch. The life safety branch shall supply power for the following lighting, receptacles, and equipment.

•

**(A) Illumination of Means of Egress.** Illumination of means of egress as is necessary for corridors, passageways, stairways, landings, and exit doors and all ways of approach to exits. Switching arrangement to transfer patient corridor lighting from general illumination circuits shall be permitted, providing only one of two circuits can be selected and both circuits cannot be extinguished at the same time.

Informational Note: See NFPA *101*-2012, *Life Safety Code*, Sections 7.8 and 7.9.

**(B) Exit Signs.** Exit signs and exit directional signs.

Informational Note: See NFPA *101-2012*, *Life Safety Code*, Section 7.10.

**(C) Alarm and Alerting Systems.** Alarm and alerting systems, including the following:

(1) Fire alarms

Informational Note: See NFPA *101-2012*, *Life Safety Code*, Sections 9.6 and 18.3.4.

(2) Alarms required for systems used for the piping of nonflammable medical gases

Informational Note: See NFPA 99-2012, *Health Care Facilities Code*, 6.5.2.2.2.1(3).

**(D) Communications Systems.** Communications systems, where used for issuing instructions during emergency conditions.

**(E) Dining and Recreation Areas.** Sufficient lighting in dining and recreation areas to provide illumination to exit ways.

**(F) Generator Set Location.** Task illumination and selected receptacles in the generator set location.

**(G) Elevators.** Elevator cab lighting, control, communications, and signal systems. [**99:**6.4.2.2.3.2(5)]

## 517.43 Connection to Critical Branch

The critical branch shall be installed and connected to the alternate power source so that the equipment listed in 517.43(A) shall be automatically restored to operation at appropriate time-lag intervals following the restoration of the life safety branch to operation. Its arrangement shall also provide for the additional connection of equipment listed in 517.43(B) by either delayed automatic or manual operation. [**99:**6.5.2.2.3.1(A) and (B)]

*Exception: For essential electrical systems under 150 kVA, deletion of the time-lag intervals feature for delayed automatic connection to the equipment branch shall be permitted.*

**(A) Delayed Automatic Connection.** The following equipment shall be permitted to be connected to the critical branch and shall be arranged for delayed automatic connection to the alternate power source:

(1) Patient care areas — task illumination and selected receptacles in the following:
 a. Medication preparation areas
 b. Pharmacy dispensing areas
 c. Nurses' stations (unless adequately lighted by corridor luminaires)
(2) Sump pumps and other equipment required to operate for the safety of major apparatus and associated control systems and alarms
(3) Smoke control and stair pressurization systems
(4) Kitchen hood supply and/or exhaust systems, if required to operate during a fire in or under the hood
(5) Supply, return, and exhaust ventilating systems for airborne infectious isolation rooms [**99:**6.5.2.2.3.3]

**(B) Delayed Automatic or Manual Connection.** The following equipment shall be permitted to be connected to the critical branch and shall be arranged for either delayed automatic or manual connection to the alternate power source:

(1) Heating equipment to provide heating for patient rooms.

*Exception: Heating of general patient rooms during disruption of the normal source shall not be required under any of the following conditions:*

*(1) The outside design temperature is higher than −6.7°C (20°F).*
*(2) The outside design temperature is lower than −6.7°C (20°F) and where a selected room(s) is provided for the needs of all confined patients, only such room(s) need be heated.*
*(3) The facility is served by a dual source of normal power as described in 517.44(C), Informational Note.*

Informational Note: The outside design temperature is based on the 97½ percent design values as shown in Chapter 24 of the ASHRAE Handbook of Fundamentals (1997).

(2) Elevator service — in instances where disruption of power would result in elevators stopping between floors, throwover facilities shall be provided to allow the temporary operation of any elevator for the release of passengers. For elevator cab lighting, control, and signal system requirements, see 517.42(G).
(3) Additional illumination, receptacles, and equipment shall be permitted to be connected only to the critical branch.

[**99:**6.5.2.2.3.4(A), (B), and (C)]

## 517.44 Sources of Power

**(A) Two Independent Sources of Power.** Essential electrical systems shall have a minimum of two independent sources of power: a normal source generally supplying the entire electrical system and one or more alternate sources for use when the normal source is interrupted. [**99:**6.5.1]

**(B) Alternate Source of Power.** The alternate source of power shall be a generator(s) driven by some form of prime mover(s) and located on the premises.

*Exception No. 1: Where the normal source consists of generating units on the premises, the alternate source shall be either another generator set or an external utility service.*

*Exception No. 2: Nursing homes or limited care facilities meeting the requirement of 517.40(A) and other health care facilities meeting the requirement of 517.45 shall be permitted to use a battery system or self-contained battery integral with the equipment.*

**(C) Location of Essential Electrical System Components.** Careful consideration shall be given to the location of the spaces housing the components of the essential electrical system to minimize interruptions caused by natural forces common to the area (e.g., storms, floods, earthquakes, or hazards created by adjoining structures or activities). Consideration shall also be given to the possible interruption of normal electrical services resulting from similar causes as well as possible disruption of normal electrical service due to internal wiring and equipment failures.

Informational Note: Facilities in which the normal source of power is supplied by two or more separate central station-fed services experience greater than normal electrical service reliability than those with only a single feed. Such a dual source of normal power consists of two or more electrical services fed from separate generator sets or a utility distribution network that has multiple power input sources and is arranged to provide mechanical and electrical separation so that a fault between the facility and the generating sources will not likely cause an interruption of more than one of the facility service feeders.

### 517.45 Essential Electrical Systems for Other Health Care Facilities

**(A) Essential Electrical Distribution.** The essential electrical distribution system shall be a battery or generator system.

Informational Note: See NFPA 99-2012, *Health Care Facilities Code*.

**(B) Electrical Life Support Equipment.** Where electrical life support equipment is required, the essential electrical distribution system shall be as described in 517.30 through 517.35.

**(C) Critical Care Areas.** Where critical care areas are present, the essential electrical distribution system shall be as described in 517.30 through 517.35.

**(D) Power Systems.** Battery systems shall be installed in accordance with the requirements of Article 700, and generator systems shall be as described in 517.30 through 517.35.

Other health care facilities include medical and dental offices and ambulatory health care facilities. Depending on the type and level of patient care, these facilities may require an alternate source of power similar to that provided at a hospital because the level of patient care is essentially the same, even though the facility is not called or licensed as a hospital.

## IV. Inhalation Anesthetizing Locations

Informational Note: For further information regarding safeguards for anesthetizing locations, see NFPA 99-2012, *Health Care Facilities Code*.

### 517.60 Anesthetizing Location Classification

Informational Note: If either of the anesthetizing locations in 517.60(A) or 517.60(B) is designated a wet procedure location, refer to 517.20.

**(A) Hazardous (Classified) Location.**

**(1) Use Location.** In a location where flammable anesthetics are employed, the entire area shall be considered to be a Class I, Division 1 location that extends upward to a level 1.52 m (5 ft) above the floor. The remaining volume up to the structural ceiling is considered to be above a hazardous (classified) location. [**99:**Annex E, E.1, and E.2]

**(2) Storage Location.** Any room or location in which flammable anesthetics or volatile flammable disinfecting agents are stored shall be considered to be a Class I, Division 1 location from floor to ceiling.

Some countries outside the United States still use flammable anesthetics and rely on these safety measures; and although there are no known medical schools in the United States still teaching the use of flammable anesthetics or health care facilities in the United States using flammable anesthetics, use of these precautions would be necessary should flammable anesthetics be re-instituted.

Section 517.60 designates anesthetizing locations either as hazardous locations, where flammable or nonflammable anesthetics may be interchangeably employed [517.60(A)], or as other-than-hazardous locations, where only nonflammable anesthetics are used [517.60(B)]. In the case of the flammable anesthetizing location, the volume of the room extending upward from a level 5 feet above the floor to the surface of the structural ceiling, including the space between a drop ceiling and the structural ceiling, is considered to be above a hazardous location.

**(B) Other-Than-Hazardous (Classified) Location.** Any inhalation anesthetizing location designated for the exclusive use of nonflammable anesthetizing agents shall be considered to be an other-than-hazardous (classified) location.

### 517.61 Wiring and Equipment

**(A) Within Hazardous (Classified) Anesthetizing Locations.**

**(1) Isolation.** Except as permitted in 517.160, each power circuit within, or partially within, a flammable anesthetizing location as referred to in 517.60 shall be isolated from any distribution system by the use of an isolated power system.

**(2) Design and Installation.** Where an isolated power system is utilized, the isolated power equipment shall be listed as isolated power equipment, and the isolated power system shall be designed and installed in accordance with 517.160.

**(3) Equipment Operating at More Than 10 Volts.** In hazardous (classified) locations referred to in 517.60, all fixed wiring and equipment and all portable equipment, including lamps and other utilization equipment, operating at more than 10 volts between conductors shall comply with the requirements of 501.1 through 501.25, and 501.100 through 501.150, and 501.30(A) and 501.30(B) for Class I, Division 1 locations. All such equipment shall be specifically approved for the hazardous atmospheres involved.

**(4) Extent of Location.** Where a box, fitting, or enclosure is partially, but not entirely, within a hazardous (classified) location(s), the hazardous (classified) location(s) shall be considered to be extended to include the entire box, fitting, or enclosure.

**(5) Receptacles and Attachment Plugs.** Receptacles and attachment plugs in a hazardous (classified) location(s) shall be listed for use in Class I, Group C hazardous (classified) locations and shall have provision for the connection of a grounding conductor.

See the commentary following 517.19(B)(2) regarding receptacles listed for hospital use.

**(6) Flexible Cord Type.** Flexible cords used in hazardous (classified) locations for connection to portable utilization equipment, including lamps operating at more than 8 volts between conductors, shall be of a type approved for extra-hard usage in accordance with Table 400.4 and shall include an additional conductor for grounding.

**(7) Flexible Cord Storage.** A storage device for the flexible cord shall be provided and shall not subject the cord to bending at a radius of less than 75 mm (3 in.).

**(B) Above Hazardous (Classified) Anesthetizing Locations.**

**(1) Wiring Methods.** Wiring above a hazardous (classified) location referred to in 517.60 shall be installed in rigid metal conduit, electrical metallic tubing, intermediate metal conduit, Type MI cable, or Type MC cable that employs a continuous, gas/vaportight metal sheath.

**(2) Equipment Enclosure.** Installed equipment that may produce arcs, sparks, or particles of hot metal, such as lamps and lampholders for fixed lighting, cutouts, switches, generators, motors, or other equipment having make-and-break or sliding contacts, shall be of the totally enclosed type or be constructed so as to prevent escape of sparks or hot metal particles.

*Exception: Wall-mounted receptacles installed above the hazardous (classified) location in flammable anesthetizing locations shall not be required to be totally enclosed or have openings guarded or screened to prevent dispersion of particles.*

**(3) Luminaires.** Surgical and other luminaires shall conform to 501.130(B).

*Exception No. 1: The surface temperature limitations set forth in 501.130(B)(1) shall not apply.*

*Exception No. 2: Integral or pendant switches that are located above and cannot be lowered into the hazardous (classified) location(s) shall not be required to be explosionproof.*

**(4) Seals.** Listed seals shall be provided in conformance with 501.15, and 501.15(A)(4) shall apply to horizontal as well as to vertical boundaries of the defined hazardous (classified) locations.

**(5) Receptacles and Attachment Plugs.** Receptacles and attachment plugs located above hazardous (classified) anesthetizing locations shall be listed for hospital use for services of prescribed voltage, frequency, rating, and number of conductors with provision for the connection of the grounding conductor. This requirement shall apply to attachment plugs and receptacles of the 2-pole, 3-wire grounding type for single-phase, 120-volt, nominal, ac service.

**(6) 250-Volt Receptacles and Attachment Plugs Rated 50 and 60 Amperes.** Receptacles and attachment plugs rated 250 volts, for connection of 50-ampere and 60-ampere ac medical equipment for use above hazardous (classified) locations, shall be arranged so that the 60-ampere receptacle will accept either the 50-ampere or the 60-ampere plug. Fifty-ampere receptacles shall be designed so as not to accept the 60-ampere attachment plug. The attachment plugs shall be of the 2-pole, 3-wire design with a third contact connecting to the insulated (green or green with yellow stripe) equipment grounding conductor of the electrical system.

**(C) Other-Than-Hazardous (Classified) Anesthetizing Locations.**

**(1) Wiring Methods.** Wiring serving other-than-hazardous (classified) locations, as defined in 517.60, shall be installed in a metal raceway system or cable assembly. The metal raceway system or cable armor or sheath assembly shall qualify as an equipment grounding conductor in accordance with 250.118. Type MC and Type MI cable shall have an outer metal armor, sheath, or sheath assembly that is identified as an acceptable equipment grounding conductor.

*Exception: Pendant receptacle installations that employ listed Type SJO, or equivalent hard usage or extra-hard usage, flexible cords suspended not less than 1.8 m (6 ft) from the floor shall not be required to be installed in a metal raceway or cable assembly.*

**(2) Receptacles and Attachment Plugs.** Receptacles and attachment plugs installed and used in other-than-hazardous (classified) locations shall be listed "hospital grade" for services of prescribed voltage, frequency, rating, and number of conductors with provision for connection of the grounding conductor. This requirement shall apply to 2-pole, 3-wire grounding type for single-phase, 120-, 208-, or 240-volt, nominal, ac service.

See the commentary following 517.19(B)(2) regarding receptacles listed for hospital use.

**(3) 250-Volt Receptacles and Attachment Plugs Rated 50 Amperes and 60 Amperes.** Receptacles and attachment plugs rated 250 volts, for connection of 50-ampere and 60-ampere ac medical equipment for use in other-than-hazardous (classified) locations, shall be arranged so that the 60-ampere receptacle will accept either the 50-ampere or the 60-ampere plug. Fifty-ampere receptacles shall be designed so as not to accept the 60-ampere attachment plug. The attachment plugs shall be of the 2-pole, 3-wire design with a third contact connecting to the insulated (green or green with yellow stripe) equipment grounding conductor of the electrical system.

## 517.62 Grounding

In any anesthetizing area, all metal raceways and metal-sheathed cables and all normally non–current-carrying conductive portions of fixed electrical equipment shall be connected to an equipment grounding conductor. Grounding and bonding in Class I locations shall comply with 501.30.

*Exception: Equipment operating at not more than 10 volts between conductors shall not be required to be connected to an equipment grounding conductor.*

The grounding requirements for anesthetizing locations apply to metal raceways, metal-sheathed cables, and electrical equipment. Carts, tables, and other nonelectrical items are not required to be grounded. In flammable anesthetizing locations, however, portable carts and tables with conductive tires and wheels and conductive flooring are usually employed to avoid the buildup of static electrical charges. See NFPA 99, E.6.6.8, for electrostatic safeguards.

## 517.63 Grounded Power Systems in Anesthetizing Locations

**(A) Battery-Powered Lighting Units.** One or more battery-powered lighting units shall be provided and shall be permitted to be wired to the critical lighting circuit in the area and connected ahead of any local switches.

Failure of the emergency circuit feeder that supplies the operating room will ordinarily plunge the room into darkness. Unless some form of uninterruptible power supply is installed, a delay in the restoration of illumination may occur until the alternate source of the essential electrical system comes on-line. Even though this delay is limited to 10 seconds (per 517.31), loss of illumination at a critical point in a surgical procedure could result in danger to the patient or operating room personnel. To safeguard against being thrust into complete darkness upon interruption of normal power, at least one battery-operated emergency lighting unit is required to be installed. This type of unit provides immediate illumination upon loss of power, helping to mitigate the impact of sudden interruption of the normal illumination. It is permitted to connect the lighting unit to the critical branch circuit.

**(B) Branch-Circuit Wiring.** Branch circuits supplying only listed, fixed, therapeutic and diagnostic equipment, permanently installed above the hazardous (classified) location and in other-than-hazardous (classified) locations, shall be permitted to be supplied from a normal grounded service, single- or three-phase system, provided the following apply:

(1) Wiring for grounded and isolated circuits does not occupy the same raceway or cable.
(2) All conductive surfaces of the equipment are connected to an equipment grounding conductor.
(3) Equipment (except enclosed X-ray tubes and the leads to the tubes) is located at least 2.5 m (8 ft) above the floor or outside the anesthetizing location.
(4) Switches for the grounded branch circuit are located outside the hazardous (classified) location.

*Exception: Sections 517.63(B)(3) and (B)(4) shall not apply in other-than-hazardous (classified) locations.*

**(C) Fixed Lighting Branch Circuits.** Branch circuits supplying only fixed lighting shall be permitted to be supplied by a normal grounded service, provided the following apply:

(1) Such luminaires are located at least 2.5 m (8 ft) above the floor.
(2) All conductive surfaces of luminaires are connected to an equipment grounding conductor.
(3) Wiring for circuits supplying power to luminaires does not occupy the same raceway or cable for circuits supplying isolated power.
(4) Switches are wall-mounted and located above hazardous (classified) locations.

*Exception: Sections 517.63(C)(1) and (C)(4) shall not apply in other-than-hazardous (classified) locations.*

**(D) Remote-Control Stations.** Wall-mounted remote-control stations for remote-control switches operating at 24 volts or less shall be permitted to be installed in any anesthetizing location.

**(E) Location of Isolated Power Systems.** Where an isolated power system is utilized, the isolated power equipment shall be listed as isolated power equipment. Isolated power system equipment and its supply circuit shall be permitted to be located in an anesthetizing location, provided it is installed above a hazardous (classified) location or in an other-than-hazardous (classified) location.

**(F) Circuits in Anesthetizing Locations.** Except as permitted above, each power circuit within, or partially within, a flammable anesthetizing location as referred to in 517.60 shall be isolated from any distribution system supplying other-than-anesthetizing locations.

## 517.64 Low-Voltage Equipment and Instruments

**(A) Equipment Requirements.** Low-voltage equipment that is frequently in contact with the bodies of persons or has exposed current-carrying elements shall comply with one of the following:

(1) Operate on an electrical potential of 10 volts or less
(2) Be approved as intrinsically safe or double-insulated equipment
(3) Be moisture resistant

**(B) Power Supplies.** Power shall be supplied to low-voltage equipment from one of the following:

(1) An individual portable isolating transformer (autotransformers shall not be used) connected to an isolated power circuit receptacle by means of an appropriate cord and attachment plug
(2) A common low-voltage isolating transformer installed in an other-than-hazardous (classified) location
(3) Individual dry-cell batteries
(4) Common batteries made up of storage cells located in an other-than-hazardous (classified) location

**(C) Isolated Circuits.** Isolating-type transformers for supplying low-voltage circuits shall have both of the following:

(1) Approved means for insulating the secondary circuit from the primary circuit
(2) The core and case connected to an equipment grounding conductor

**(D) Controls.** Resistance or impedance devices shall be permitted to control low-voltage equipment but shall not be used to limit the maximum available voltage to the equipment.

**(E) Battery-Powered Appliances.** Battery-powered appliances shall not be capable of being charged while in operation unless their charging circuitry incorporates an integral isolating-type transformer.

**(F) Receptacles or Attachment Plugs.** Any receptacle or attachment plug used on low-voltage circuits shall be of a type that does not permit interchangeable connection with circuits of higher voltage.

Informational Note: Any interruption of the circuit, even circuits as low as 10 volts, either by any switch or loose or defective connections anywhere in the circuit, may produce a spark that is sufficient to ignite flammable anesthetic agents.

## V. X-Ray Installations

### 517.70 Applicability

Nothing in this part shall be construed as specifying safeguards against the useful beam or stray X-ray radiation.

Informational Note No. 1: Radiation safety and performance requirements of several classes of X-ray equipment are regulated under Public Law 90-602 and are enforced by the Department of Health and Human Services.

Informational Note No. 2: In addition, information on radiation protection by the National Council on Radiation Protection and Measurements is published as *Reports of the National Council on Radiation Protection and Measurement*. These reports are obtainable from NCRP Publications, P.O. Box 30175, Washington, DC 20014.

### 517.71 Connection to Supply Circuit

**(A) Fixed and Stationary Equipment.** Fixed and stationary X-ray equipment shall be connected to the power supply by means of a wiring method complying with applicable requirements of Chapters 1 through 4 of this *Code*, as modified by this article.

*Exception: Equipment properly supplied by a branch circuit rated at not over 30 amperes shall be permitted to be supplied through a suitable attachment plug and hard-service cable or cord.*

**(B) Portable, Mobile, and Transportable Equipment.** Individual branch circuits shall not be required for portable, mobile, and transportable medical X-ray equipment requiring a capacity of not over 60 amperes.

**(C) Over 1000-Volt Supply.** Circuits and equipment operated on a supply circuit of over 1000 volts shall comply with Article 490.

### 517.72 Disconnecting Means

**(A)** Capacity. A disconnecting means of adequate capacity for at least 50 percent of the input required for the momentary rating or 100 percent of the input required for the long-time rating of the X-ray equipment, whichever is greater, shall be provided in the supply circuit.

**(B) Location.** The disconnecting means shall be operable from a location readily accessible from the X-ray control.

**(C) Portable Equipment.** For equipment connected to a 120-volt branch circuit of 30 amperes or less, a grounding-type attachment plug and receptacle of proper rating shall be permitted to serve as a disconnecting means.

### 517.73 Rating of Supply Conductors and Overcurrent Protection

**(A) Diagnostic Equipment.**

**(1) Branch Circuits.** The ampacity of supply branch-circuit conductors and the current rating of overcurrent protective devices shall not be less than 50 percent of the momentary rating or 100 percent of the long-time rating, whichever is greater.

**(2) Feeders.** The ampacity of supply feeders and the current rating of overcurrent protective devices supplying two or more branch circuits supplying X-ray units shall not be less than 50 percent of the momentary demand rating of the largest unit plus 25 percent of the momentary demand rating of the next largest unit plus 10 percent of the momentary demand rating of each additional unit. Where simultaneous biplane examinations are undertaken with the X-ray units, the supply conductors and overcurrent protective devices shall be 100 percent of the momentary demand rating of each X-ray unit.

Informational Note: The minimum conductor size for branch and feeder circuits is also governed by voltage regulation requirements. For a specific installation, the manufacturer usually specifies minimum distribution transformer and conductor sizes, rating of disconnecting means, and overcurrent protection.

**(B) Therapeutic Equipment.** The ampacity of conductors and rating of overcurrent protective devices shall not be less than 100 percent of the current rating of medical X-ray therapy equipment.

Informational Note: The ampacity of the branch-circuit conductors and the ratings of disconnecting means and overcurrent protection for X-ray equipment are usually designated by the manufacturer for the specific installation.

### 517.74 Control Circuit Conductors

**(A) Number of Conductors in Raceway.** The number of control circuit conductors installed in a raceway shall be determined in accordance with 300.17.

**(B) Minimum Size of Conductors.** Size 18 AWG or 16 AWG fixture wires as specified in 725.49 and flexible cords shall be permitted for the control and operating circuits of X-ray and auxiliary equipment where protected by not larger than 20-ampere overcurrent devices.

### 517.75 Equipment Installations

All equipment for new X-ray installations and all used or reconditioned X-ray equipment moved to and reinstalled at a new location shall be of an approved type.

### 517.76 Transformers and Capacitors

Transformers and capacitors that are part of X-ray equipment shall not be required to comply with Articles 450 and 460.

Capacitors shall be mounted within enclosures of insulating material or grounded metal.

### 517.77 Installation of High-Tension X-Ray Cables

Cables with grounded shields connecting X-ray tubes and image intensifiers shall be permitted to be installed in cable trays or cable troughs along with X-ray equipment control and power supply conductors without the need for barriers to separate the wiring.

### 517.78 Guarding and Grounding

**(A) High-Voltage Parts.** All high-voltage parts, including X-ray tubes, shall be mounted within grounded enclosures. Air, oil, gas, or other suitable insulating media shall be used to insulate the high-voltage from the grounded enclosure. The connection from the high-voltage equipment to X-ray tubes and other high-voltage components shall be made with high-voltage shielded cables.

**(B) Low-Voltage Cables.** Low-voltage cables connecting to oil-filled units that are not completely sealed, such as transformers, condensers, oil coolers, and high-voltage switches, shall have insulation of the oil-resistant type.

**(C) Non–Current-Carrying Metal Parts.** Non–current-carrying metal parts of X-ray and associated equipment (controls, tables, X-ray tube supports, transformer tanks, shielded cables, X-ray tube heads, etc.) shall be connected to an equipment grounding conductor in the manner specified in Part VII of Article 250, as modified by 517.13(A) and (B).

## VI. Communications, Signaling Systems, Data Systems, Fire Alarm Systems, and Systems Less Than 120 Volts, Nominal

### 517.80 Patient Care Areas

Equivalent insulation and isolation to that required for the electrical distribution systems in patient care areas shall be provided for communications, signaling systems, data system circuits, fire alarm systems, and systems less than 120 volts, nominal.

Class 2 and Class 3 signaling and communications systems and power-limited fire alarm systems shall not be required to comply with the grounding requirements of 517.13, to comply with the mechanical protection requirements of 517.30(C)(3)(5), or to be enclosed in raceways, unless otherwise specified by Chapter 7 or 8.

Secondary circuits of transformer-powered communications or signaling systems shall not be required to be enclosed in raceways unless otherwise specified by Chapter 7 or 8. [**99:**6.4.2.2.6.6]

One of the major objectives of the requirements in Article 517 is to minimize patients' exposure to any level of current that could compromise their well-being. In general, circuits covered under Part VI are required to provide insulation and isolation equivalent to that required for the electrical power distribution system. The equivalent insulation and isolation required by 517.80 is for protection of patients from any shock hazard that could result from inadvertent contact with energized circuit conductors or parts associated with the limited energy systems. See the informational note to 517.11. Nurse call, intercom, speaker, cable television, and fire alarm systems are examples of the types of circuits covered by this requirement.

Some classes of limited energy circuits, including Class 2 and Class 3 remote control and signaling circuits as well as power-limited fire alarm circuits that are installed in patient care spaces, are not required to comply with the same grounding or mechanical protection requirements as power and lighting circuits. This does not remove applicable installation requirements, such as the use of a raceway, if specified in Chapter 7 or Chapter 8.

### 517.81 Other-Than-Patient-Care Areas

In other-than-patient-care areas, installations shall be in accordance with the applicable provisions of other parts of this *Code*.

### 517.82 Signal Transmission Between Appliances

**(A) General.** Permanently installed signal cabling from an appliance in a patient location to remote appliances shall employ a signal transmission system that prevents hazardous grounding interconnection of the appliances.

Informational Note: See 517.13(A) for additional grounding requirements in patient care areas.

**(B) Common Signal Grounding Wire.** Common signal grounding wires (i.e., the chassis ground for single-ended transmission) shall be permitted to be used between appliances all located within the patient care vicinity, provided the appliances are served from the same reference grounding point.

## VII. Isolated Power Systems

### 517.160 Isolated Power Systems

**(A) Installations.**

**(1) Isolated Power Circuits.** Each isolated power circuit shall be controlled by a switch or circuit breaker that has a disconnecting pole in each isolated circuit conductor to simultaneously disconnect all power. Such isolation shall be accomplished by means of one or more isolation transformers, by means of generator sets, or by means of electrically isolated batteries. Conductors of isolated power circuits shall not be installed in cables, raceways, or other enclosures containing conductors of another system.

**(2) Circuit Characteristics.** Circuits supplying primaries of isolating transformers shall operate at not more than 600 volts between conductors and shall be provided with proper overcurrent protection. The secondary voltage of such transformers shall not exceed 600 volts between conductors of each circuit. All circuits supplied from such secondaries shall be ungrounded and shall have an approved overcurrent device of proper ratings in each conductor. Circuits supplied directly from batteries or from motor generator sets shall be ungrounded and shall be protected against overcurrent in the same manner as transformer-fed secondary circuits. If an electrostatic shield is present, it shall be connected to the reference grounding point. [**99:**6.3.2.6.1]

**(3) Equipment Location.** The isolating transformers, motor generator sets, batteries and battery chargers, and associated primary or secondary overcurrent devices shall not be installed in hazardous (classified) locations. The isolated secondary circuit wiring extending into a hazardous anesthetizing location shall be installed in accordance with 501.10.

**(4) Isolation Transformers.** An isolation transformer shall not serve more than one operating room except as covered in (A)(4)(a) and (A)(4)(b).

For purposes of this section, anesthetic induction rooms are considered part of the operating room or rooms served by the induction rooms.

(a) *Induction Rooms.* Where an induction room serves more than one operating room, the isolated circuits of the induction room shall be permitted to be supplied from the isolation transformer of any one of the operating rooms served by that induction room.

(b) *Higher Voltages.* Isolation transformers shall be permitted to serve single receptacles in several patient areas where the following apply:

(1) The receptacles are reserved for supplying power to equipment requiring 150 volts or higher, such as portable X-ray units.
(2) The receptacles and mating plugs are not interchangeable with the receptacles on the local isolated power system. [**99:**13.4.1.2.6.6]

**(5) Conductor Identification.** The isolated circuit conductors shall be identified as follows:

(1) Isolated Conductor No. 1 — Orange with at least one distinctive colored stripe other than white, green, or gray along the entire length of the conductor
(2) Isolated Conductor No. 2 — Brown with at least one distinctive colored stripe other than white, green, or gray along the entire length of the conductor

For 3-phase systems, the third conductor shall be identified as yellow with at least one distinctive colored stripe other than white, green, or gray along the entire length of the conductor. Where isolated circuit conductors supply 125-volt, single-phase, 15- and 20-ampere receptacles, the striped orange conductor(s) shall be connected to the terminal(s) on the receptacles that are identified in accordance with 200.10(B) for connection to the grounded circuit conductor.

**(6) Wire-Pulling Compounds.** Wire-pulling compounds that increase the dielectric constant shall not be used on the secondary conductors of the isolated power supply.

Informational Note No. 1: It is desirable to limit the size of the isolation transformer to 10 kVA or less and to use conductor insulation with low leakage to meet impedance requirements.

Informational Note No. 2: Minimizing the length of branch-circuit conductors and using conductor insulations with a dielectric constant less than 3.5 and insulation resistance constant greater than 6100 megohm-meters (20,000 megohm-feet) at 16°C (60°F) reduces leakage from line to ground, reducing the hazard current.

See Exhibit 517.7 for an example of a hospital isolated power system panel.

**(B) Line Isolation Monitor.**

**(1) Characteristics.** In addition to the usual control and overcurrent protective devices, each isolated power system shall be provided with a continually operating line isolation monitor that indicates total hazard current. The monitor shall be designed such that a green signal lamp, conspicuously visible

***EXHIBIT 517.7*** *An example of a hospital isolated power system panel with built-in isolation transformer, line isolation monitor, load center, and grounded busbar. (Courtesy of Schneider Electric)*

to persons in each area served by the isolated power system, remains lighted when the system is adequately isolated from ground. An adjacent red signal lamp and an audible warning signal (remote if desired) shall be energized when the total hazard current (consisting of possible resistive and capacitive leakage currents) from either isolated conductor to ground reaches a threshold value of 5 mA under nominal line voltage conditions. The line monitor shall not alarm for a fault hazard of less than 3.7 mA or for a total hazard current of less than 5 mA.

*Exception: A system shall be permitted to be designed to operate at a lower threshold value of total hazard current. A line isolation monitor for such a system shall be permitted to be approved, with the provision that the fault hazard current shall be permitted to be reduced but not to less than 35 percent of the corresponding threshold value of the total hazard current, and the monitor hazard current is to be correspondingly reduced to not more than 50 percent of the alarm threshold value of the total hazard current.*

**(2) Impedance.** The line isolation monitor shall be designed to have sufficient internal impedance such that, when properly connected to the isolated system, the maximum internal current that can flow through the line isolation monitor, when any point of the isolated system is grounded, shall be 1 mA.

*Exception: The line isolation monitor shall be permitted to be of the low-impedance type such that the current through the line isolation monitor, when any point of the isolated system is grounded, will not exceed twice the alarm threshold value for a period not exceeding 5 milliseconds.*

Informational Note: Reduction of the monitor hazard current, provided this reduction results in an increased "not alarm" threshold value for the fault hazard current, will increase circuit capacity.

**(3) Ammeter.** An ammeter calibrated in the total hazard current of the system (contribution of the fault hazard current plus monitor hazard current) shall be mounted in a plainly visible place on the line isolation monitor with the "alarm on" zone at approximately the center of the scale.

*Exception: The line isolation monitor shall be permitted to be a composite unit, with a sensing section cabled to a separate display panel section on which the alarm or test functions are located.*

Informational Note: It is desirable to locate the ammeter so that it is conspicuously visible to persons in the anesthetizing location.

# ARTICLE 518
# Assembly Occupancies

## 518.1 Scope

Except for the assembly occupancies explicitly covered by 520.1, this article covers all buildings or portions of buildings or structures designed or intended for the gathering together of 100 or more persons for such purposes as deliberation, worship, entertainment, eating, drinking, amusement, awaiting transportation, or similar purposes.

Article 518 applies to assembly occupancies designed or intended for 100 or more persons with the population capacity determined by methods utilized in NFPA *101, Life Safety Code*. Article 518 would apply, for example, to a church chapel or an auditorium for occupancy of 100 or more persons but not to a supermarket. Even though a supermarket may contain 100 or more persons, it is not specifically designed or intended for the assembly of persons. Article 518 does not apply to office buildings or schools, even though such buildings, as a rule, are designed for occupancy by 100 or more persons. The article does, however, apply to assembly halls, restaurants, and so forth, within an office or school building if these parts of the building are designed or intended for the assembly of 100 or more persons.

The following information for determining new assembly occupancy capacity is extracted from NFPA *101, Life Safety Code*:

**12.1.7 Occupant Load.**

**12.1.7.1 General.** The occupant load, in number of persons for whom means of egress and other provisions are required, shall be determined on the basis of the occupant load factors of Table 7.3.1.2 [Commentary Table 518.1] that are characteristic of the use of the space or shall be determined as the maximum probable population of the space under consideration, whichever is greater.

⋮

**7.3.1.1.2** For other than existing means of egress, where more than one means of egress is required, the means of egress shall be of such width and capacity that the loss of any one means of egress leaves available not less than 50 percent of the required capacity.

**7.3.1.2 Occupant Load Factor.** The occupant load in any building or portion thereof shall be not less than the number of persons determined by dividing the floor area assigned to that use by the occupant load factor for that use as specified in Table 7.3.1.2 [Commentary Table 518.1]. Where both gross and net area figures are given for the same occupancy, calculations shall be made by applying the gross area figure to the gross area of the portion of the building devoted to the use for which the gross area figure is specified and by applying the net area figure to the net area of the portion of the building devoted to the use for which the net area figure is specified.

## 518.2 General Classification

**(A) Examples.** Assembly occupancies shall include, but not be limited to, the following:

| | |
|---|---|
| Armories | Bowling lanes |
| Assembly halls | Club rooms |
| Auditoriums | Conference rooms |

**COMMENTARY TABLE 518.1** Occupant Load Factor

| Use | $m^2$ (per person)[1] | $ft^2$ (per person)[1] |
|---|---|---|
| **Assembly Use** | | |
| Concentrated use, without fixed seating | 0.65 net | 7 net |
| Less concentrated use, without fixed seating | 1.4 net | 15 net |
| Bench-type seating | 1 person/455 linear mm | 1 person/18 linear in. |
| Fixed seating | Number of fixed seats | Number of fixed seats |
| Waiting spaces | See 12.1.7.2 and 13.1.7.2 [of NFPA *101*] | See 12.1.7.2 and 13.1.7.2 [of NFPA *101*] |
| Kitchens | 9.3 | 100 |
| Library stack areas | 9.3 | 100 |
| Library reading rooms | 4.6 net | 50 net |
| Swimming pools | 4.6 (water surface) | 50 (water surface) |
| Swimming pool decks | 2.8 | 30 |
| Exercise rooms with equipment | 4.6 | 50 |
| Exercise rooms without equipment | 1.4 | 15 |
| Stages | 1.4 net | 15 net |
| Lighting and access catwalks, galleries, gridirons | 9.3 net | 100 net |
| Casinos and similar gaming areas | 1 | 11 |
| Skating rinks | 4.6 | 50 |
| **Educational Use** | | |
| Classrooms | 1.9 net | 20 net |
| Shops, laboratories, vocational rooms | 4.6 net | 50 net |
| **Day-Care Use** | 3.3 net | 35 net |
| **Health Care Use** | | |
| Inpatient treatment departments | 22.3 | 240 |
| Sleeping departments | 11.1 | 120 |
| Ambulatory health care | 9.3 | 100 |
| **Detention and Correctional Use** | 11.1 | 120 |
| **Residential Use** | | |
| Hotels and dormitories | 18.6 | 200 |
| Apartment buildings | 18.6 | 200 |
| Board and care, large | 18.6 | 200 |
| **Industrial Use** | | |
| General and high hazard industrial | 9.3 | 100 |
| Special purpose industrial | NA | NA |
| **Business Use (other than below)** | 9.3 | 100 |
| Air traffic control tower observation levels | 3.7 | 40 |
| **Storage Use** | | |
| In storage occupancies | NA | NA |
| In mercantile occupancies | 27.9 | 300 |
| In other than storage and mercantile occupancies | 46.5 | 500 |
| **Mercantile Use** | | |
| Sales area on street floor[2,3] | 2.8 | 30 |
| Sales area on two or more street floors[3] | 3.7 | 40 |
| Sales area on floor below street floor[3] | 2.8 | 30 |
| Sales area on floors above street floor[3] | 5.6 | 60 |
| Floors or portions of floors used only for offices | See business use | See business use |
| Floors or portions of floors used only for storage, receiving, and shipping, and not open to general public | 27.9 | 300 |
| Mall buildings[4] | Per factors applicable to use of space[5] | Per factors applicable to use of space[5] |

Note: NA = not applicable. The occupant load is the maximum probable number of occupants present at any time.

[1]All factors are expressed in gross area unless marked "net."

Notes 2 through 5 contain specific load or egress considerations, such as when no direct egress to a street is available, for the occupancies referenced. Refer to NFPA *101* for further information.

Courtrooms
Dance halls
Dining and drinking facilities
Exhibition halls
Gymnasiums
Mortuary chapels
Multipurpose rooms
Museums
Places of awaiting transportation
Places of religious worship
Pool rooms
Restaurants
Skating rinks

**(B) Multiple Occupancies.** Where an assembly occupancy forms a portion of a building containing other occupancies, Article 518 applies only to that portion of the building considered an assembly occupancy. Occupancy of any room or space for assembly purposes by less than 100 persons in a building of other occupancy, and incidental to such other occupancy, shall be classified as part of the other occupancy and subject to the provisions applicable thereto.

**(C) Theatrical Areas.** Where any such building structure, or portion thereof, contains a projection booth or stage platform or area for the presentation of theatrical or musical productions, either fixed or portable, the wiring for that area, including associated audience seating areas, and all equipment that is used in the referenced area, and portable equipment and wiring for use in the production that will not be connected to permanently installed wiring, shall comply with Article 520.

The requirements in Article 520 apply to theatrical areas within assembly occupancies. See the commentary following 520.1.

Informational Note: For methods of determining population capacity, see local building code or, in its absence, NFPA *101*-2012, *Life Safety Code*.

## 518.3 Other Articles

**(A) Hazardous** (Classified) Areas. Electrical installations in hazardous (classified) areas located in assembly occupancies shall comply with Article 500.

**(B) Temporary Wiring.** In exhibition halls used for display booths, as in trade shows, the temporary wiring shall be permitted to be installed in accordance with Article 590. Flexible cables and cords approved for hard or extra-hard usage shall be permitted to be laid on floors where protected from contact by the general public. The ground-fault circuit-interrupter requirements of 590.6 shall not apply. All other ground-fault circuit-interrupter requirements of this *Code* shall apply.

Where ground-fault circuit interrupter protection for personnel is supplied by plug-and-cord-connection to the branch circuit or to the feeder, the ground fault circuit interrupter protection shall be listed as portable ground fault circuit interrupter protection or provide a level of protection equivalent to a portable ground fault circuit interrupter, whether assembled in the field or at the factory.

GFCI requirements are present in 590.6, but those requirements do not apply to temporary installations under Article 518. Temporary installations in exhibition halls must meet all other applicable GFCI protection requirements. Although trade show booths are connected to temporary wiring, GFCI protection is required for installations of water features at garden shows and for receptacles near sinks for food vendors for example.

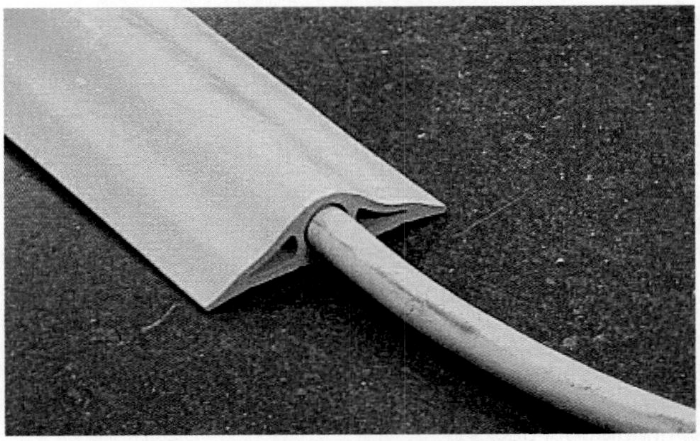

***EXHIBIT 518.1*** *A treadle used to protect temporary cords. (Courtesy of Woodhead Industries, a division of Molex)*

Permanent GFCI protection differs from portable GFCI protection. Product standards for GFCI equipment require portable devices to de-energize the contacts when the grounded conductor is open, a grounded and grounded conductor is transposed (miswired), and/or an ungrounded conductor is open. For this reason, GFCI protection for temporary wiring must be listed as portable or provide equivalent protection to a portable GFCI device.

A treadle, such as the one shown in Exhibit 518.1, is an example of a protection technique used to protect temporary cords from abuse in areas where the cords must be laid across pedestrian ways, such as in exhibition halls.

*Exception: Where conditions of supervision and maintenance ensure that only qualified persons will service the installation, flexible cords or cables identified in Table 400.4 for hard usage or extra-hard usage shall be permitted in cable trays used only for temporary wiring. All cords or cables shall be installed in a single layer. A permanent sign shall be attached to the cable tray at intervals not to exceed 7.5 m (25 ft). The sign shall read*

*CABLE TRAY FOR TEMPORARY WIRING ONLY*

**(C) Emergency Systems.** Control of emergency systems shall comply with Article 700.

## 518.4 Wiring Methods

**(A) General.** The fixed wiring methods shall be metal raceways, flexible metal raceways, nonmetallic raceways encased in not less than 50 mm (2 in.) of concrete, Type MI, MC, or AC cable. The wiring method shall itself qualify as an equipment grounding conductor according to 250.118 or shall contain an insulated equipment grounding conductor sized in accordance with Table 250.122.

*Exception: Fixed wiring methods shall be as provided in*

*(a) Audio signal processing, amplification, and reproduction equipment — Article 640*

(b) *Communications circuits — Article 800*

(c) *Class 2 and Class 3 remote-control and signaling circuits — Article 725*

(d) *Fire alarm circuits — Article 760*

**(B) Nonrated Construction.** In addition to the wiring methods of 518.4(A), nonmetallic-sheathed cable, Type AC cable, electrical nonmetallic tubing, and rigid nonmetallic conduit shall be permitted to be installed in those buildings or portions thereof that are not required to be of fire-rated construction by the applicable building code.

Informational Note: Fire-rated construction is the fire-resistive classification used in building codes.

**(C) Spaces with Finish Rating.** Electrical nonmetallic tubing and rigid nonmetallic conduit shall be permitted to be installed in club rooms, conference and meeting rooms in hotels or motels, courtrooms, dining facilities, restaurants, mortuary chapels, museums, libraries, and places of religious worship where the following apply:

(1) The electrical nonmetallic tubing or rigid nonmetallic conduit is installed concealed within walls, floors, and ceilings where the walls, floors, and ceilings provide a thermal barrier of material that has at least a 15-minute finish rating as identified in listings of fire-rated assemblies.

(2) The electrical nonmetallic tubing or rigid nonmetallic conduit is installed above suspended ceilings where the suspended ceilings provide a thermal barrier of material that has at least a 15-minute finish rating as identified in listings of fire-rated assemblies.

Electrical nonmetallic tubing and rigid nonmetallic conduit are not recognized for use in other space used for environmental air in accordance with 300.22(C).

Informational Note: A finish rating is established for assemblies containing combustible (wood) supports. The finish rating is defined as the time at which the wood stud or wood joist reaches an average temperature rise of 121°C (250°F) or an individual temperature rise of 163°C (325°F) as measured on the plane of the wood nearest the fire. A finish rating is not intended to represent a rating for a membrane ceiling.

The wiring methods identified in 518.4(A) and its exception apply to any wall, floor, or ceiling within an assembly occupancy, as classified in 518.2. The requirements of 518.4(B) apply to those portions of the building and those assembly occupancies not required to be fire rated. The use of electrical nonmetallic tubing and rigid nonmetallic conduit as permitted in 518.4(C) applies only to the specific occupancies described, provided these wiring methods are installed concealed behind a surface that has a 15-minute finish rating.

Exhibit 518.2 illustrates a single-story facility in which the washrooms and office area are not assembly occupancies, as defined in 518.2, and therefore require no special wiring methods. Ordinary wiring methods may be used on the inside surface of

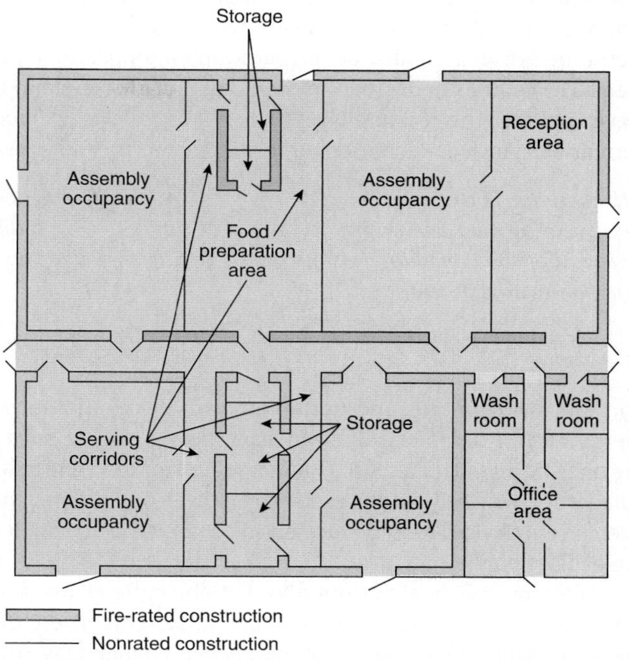


***EXHIBIT 518.2*** *Floor plan of a single-story facility. The walls (represented by wide yellow lines) are required by the local building code to be of fire-rated construction; the thin black lines represent walls not required by the local building code to be of fire-rated construction.*

the storage area walls and on or in the partitions between storage areas, because those areas are not assembly occupancies. Inside any hollow spaces of the fire-rated storage area walls, however, the main requirements of 518.4 apply, because the serving corridors are part of the assembly occupancies as a result of this particular building design. If the hollow spaces of fire-rated walls or ceiling also provide a 15-minute finish rating and are not "other space for environmental air" as described in 300.22(C), electrical nonmetallic tubing as well as rigid nonmetallic conduit is permitted for specifically described occupancies. (See 362.10.) Also, wiring in ceilings or floors that are required to be of fire-rated construction in the assembly occupancy must also comply with 518.4, except as noted in 518.4(A), Exception.

Assembly occupancies frequently require emergency wiring, particularly for emergency illumination and exit lighting. The requirements of 700.10(D) contain special fire protection requirements for emergency circuits in assembly occupancies with an occupant capacity of 1000 or more.

## 518.5 Supply

Portable switchboards and portable power distribution equipment shall be supplied only from listed power outlets of sufficient voltage and ampere rating. Such power outlets shall be protected by overcurrent devices. Such overcurrent devices and power outlets shall not be accessible to the general public. Provisions for connection of an equipment grounding conductor shall be provided.

The neutral conductor of feeders supplying solid-state phase control, 3-phase, 4-wire dimmer systems shall be considered a current-carrying conductor for purposes of ampacity adjustment. The neutral conductor of feeders supplying solid-state sine wave, 3-phase, 4-wire dimming systems shall not be considered a current-carrying conductor for purposes of ampacity adjustment.

*Exception: The neutral conductor of feeders supplying systems that use or may use both phase-control and sine-wave dimmers shall be considered as current-carrying for purposes of ampacity adjustment.*

Informational Note: For definitions of solid-state dimmer types, see 520.2.

Portable switchboards and portable power distribution equipment must be supplied only from listed power outlets, such as the one shown in Exhibit 518.3, that are rated for the voltage and current for which they are used. The power outlets and their overcurrent devices must be located so as not to be accessible to the general public.

Some professional performance lighting systems use solid-state, sine wave, 3-phase, 4-wire dimming systems. These dimmers vary with the amplitude of the applied voltage waveform, without any of the nonlinear switching found in phase-control solid-state dimmers. Because solid-state sine wave dimmers are linear loads, they do not require the neutral to be considered as a current-carrying conductor.

***EXHIBIT 518.3*** *A listed power outlet for connection of portable switchboards in an assembly occupancy. (Courtesy of Union Connector Co., Inc.)*

# ARTICLE 520
# Theaters, Audience Areas of Motion Picture and Television Studios, Performance Areas, and Similar Locations

## I. General

### 520.1 Scope

This article covers all buildings or that part of a building or structure, indoor or outdoor, designed or used for presentation, dramatic, musical, motion picture projection, or similar purposes and to specific audience seating areas within motion picture or television studios.

The special requirements of Article 520 apply only to that part of a building used as a theater or for a similar purpose and do not necessarily apply to the entire building. In a school building, for example, the requirements of Article 520 apply to an auditorium used for dramatic or other performances. The special requirements of this article apply to the stage, auditorium, dressing rooms, and main corridors leading to the auditorium, but not to other parts of the building that are not involved in the use of the auditorium for performances or entertainment. The theater space may be a traditional theater, where the audience sits in the auditorium (house) facing the proscenium arch and views the performance on the stage on the other side of the arch. It may house other spaces, such as a simple stage platform, either indoors or outdoors, with seats on three or four sides facing the platform. Audience areas of motion picture and television studios, as defined and covered in Article 530, are also covered by the requirements of Article 520.

### 520.2 Definitions

**Border Light.** A permanently installed overhead strip light.

**Breakout Assembly.** An adapter used to connect a multipole connector containing two or more branch circuits to multiple individual branch-circuit connectors.

**Bundled.** Cables or conductors that are tied, wrapped, taped, or otherwise periodically bound together.

**Connector Strip.** A metal wireway containing pendant or flush receptacles.

**Drop Box.** A box containing pendant- or flush-mounted receptacles attached to a multiconductor cable via strain relief or a multipole connector.

**Footlight.** A border light installed on or in the stage.

**Grouped.** Cables or conductors positioned adjacent to one another but not in continuous contact with each other.

**Performance Area.** The stage and audience seating area associated with a temporary stage structure, whether indoors or outdoors, constructed of scaffolding, truss, platforms, or similar devices, that is used for the presentation of theatrical or musical productions or for public presentations.

**Portable Equipment.** Equipment fed with portable cords or cables intended to be moved from one place to another.

**Portable Power Distribution Unit.** A power distribution box containing receptacles and overcurrent devices.

**Proscenium.** The wall and arch that separates the stage from the auditorium (house).

**Solid-State Phase-Control Dimmer.** A solid-state dimmer where the wave shape of the steady-state current does not follow the wave shape of the applied voltage, such that the wave shape is nonlinear.

Solid-state phase-control dimmers are nonlinear devices. Where they are used, nonlinear loading of the neutral conductor occurs, which may necessitate increasing the size of the neutral conductor.

**Solid-State Sine Wave Dimmer.** A solid-state dimmer where the wave shape of the steady-state current follows the wave shape of the applied voltage such that the wave shape is linear.

Some professional performance lighting systems use a solid-state, 3-phase, 4-wire dimming system whose wave shape varies with the amplitude of the applied voltage wave-form without any of the nonlinear switching found in phase-control solid-state dimmers. Because solid-state sine wave dimmers are linear loads, they do not require the neutral to be considered as a current-carrying conductor. See Exhibit 520.1.

**Stage Equipment.** Equipment at any location on the premises integral to the stage production including, but not limited to, equipment for lighting, audio, special effects, rigging, motion control, projection, or video.

**Stage Lighting Hoist.** A motorized lifting device that contains a mounting position for one or more luminaires, with wiring devices for connection of luminaires to branch circuits, and integral flexible cables to allow the luminaires to travel over the lifting range of the hoist while energized.

**Stage Switchboard.** A switchboard, panelboard, or rack containing dimmers or relays with associated overcurrent protective devices, or overcurrent protective devices alone, used primarily to feed stage equipment.

**Stand Lamp (Work Light).** A portable stand that contains a general-purpose luminaire or lampholder with guard for the purpose of providing general illumination on the stage or in the auditorium.

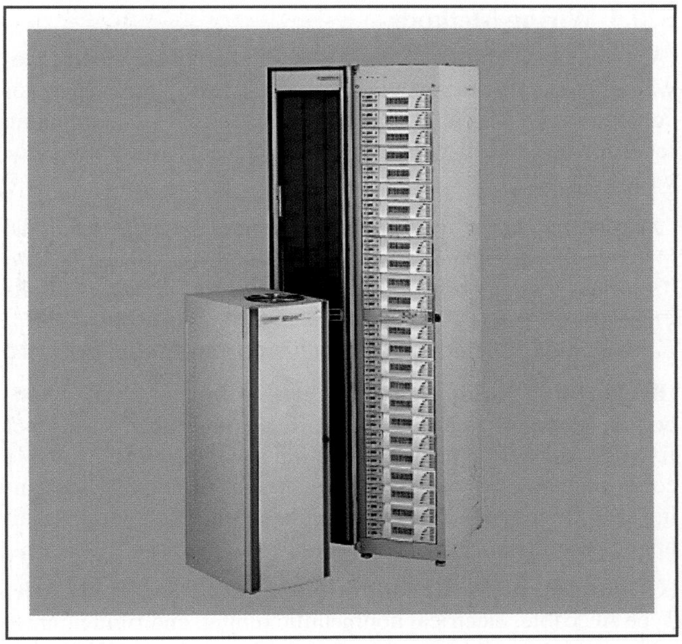

***EXHIBIT 520.1*** *Sine wave dimmer racks of 24 and 48 dimmers (2.4 kW). (Courtesy of Electronic Theatre Controls, Inc.)*

**Strip Light.** A luminaire with multiple lamps arranged in a row.

**Two-Fer.** An adapter cable containing one male plug and two female cord connectors used to connect two loads to one branch circuit.

A two-fer, as shown in Exhibit 520.2, consists of two cord connectors on separate cords connected to a single supply cord.

## 520.3 Motion Picture Projectors

Motion picture equipment and its installation and use shall comply with Article 540.

## 520.4 Audio Signal Processing, Amplification, and Reproduction Equipment

Audio signal processing, amplification, and reproduction equipment and its installation shall comply with Article 640.

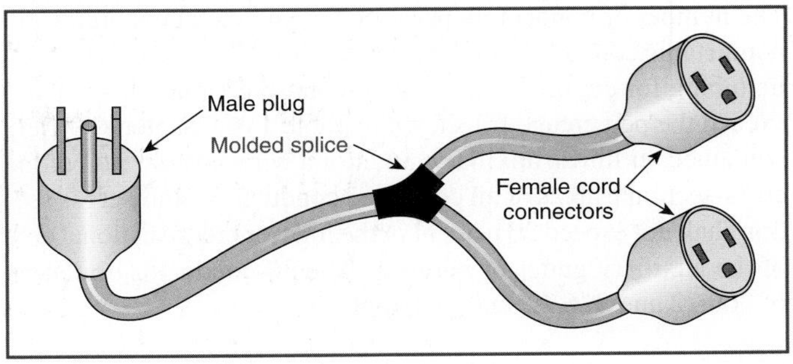


***EXHIBIT 520.2*** *A two-fer.*

## 520.5 Wiring Methods

**(A) General.** The fixed wiring method shall be metal raceways, nonmetallic raceways encased in at least 50 mm (2 in.) of concrete, Type MI cable, MC cable, or AC cable containing an insulated equipment grounding conductor sized in accordance with Table 250.122.

*Exception: Fixed wiring methods shall be as provided in Article 640 for audio signal processing, amplification, and reproduction equipment, in Article 800 for communications circuits, in Article 725 for Class 2 and Class 3 remote-control and signaling circuits, and in Article 760 for fire alarm circuits.*

**(B) Portable Equipment.** The wiring for portable switchboards, stage set lighting, stage effects, and other wiring not fixed as to location shall be permitted with approved flexible cords and cables as provided elsewhere in Article 520. Fastening such cables and cords by uninsulated staples or nailing shall not be permitted.

**(C) Nonrated Construction.** Nonmetallic-sheathed cable, Type AC cable, electrical nonmetallic tubing, and rigid nonmetallic conduit shall be permitted to be installed in those buildings or portions thereof that are not required to be of fire-rated construction by the applicable building code.

Because theaters and similar buildings are usually required to be of fire-rated construction as determined by applicable building codes, the fixed wiring methods in 520.5 are limited. See the commentary for 518.4 on wiring methods where fire-rated construction is required.

The exception to 520.5(A) permits the installation of communications circuits, Class 2 and Class 3 remote-control and signaling circuits, sound-reproduction wiring, and fire alarm circuits using wiring methods from the respective articles covering these systems. Where portability, flexibility, and adjustments are necessary, suitable cords and cables are permitted. Section 520.5(C) permits Type AC cable as one of the wiring methods in buildings or portions of buildings that are not required to be of fire-rated construction. In this application, Type AC cable is not required to contain an insulated EGC.

## 520.6 Number of Conductors in Raceway

The number of conductors permitted in any metal conduit, rigid nonmetallic conduit as permitted in this article, or electrical metallic tubing for circuits or for remote-control conductors shall not exceed the percentage fill shown in Table 1 of Chapter 9. Where contained within an auxiliary gutter or a wireway, the sum of the cross-sectional areas of all contained conductors at any cross section shall not exceed 20 percent of the interior cross-sectional area of the auxiliary gutter or wireway. The 30-conductor limitation of 366.22 and 376.22 shall not apply.

## 520.7 Enclosing and Guarding Live Parts

Live parts shall be enclosed or guarded to prevent accidental contact by persons and objects. All switches shall be of the externally operable type. Dimmers, including rheostats, shall be placed in cases or cabinets that enclose all live parts.

## 520.8 Emergency Systems

Control of emergency systems shall comply with Article 700.

## 520.9 Branch Circuits

A branch circuit of any size supplying one or more receptacles shall be permitted to supply stage set lighting. The voltage rating of the receptacles shall be not less than the circuit voltage. Receptacle ampere ratings and branch-circuit conductor ampacity shall be not less than the branch-circuit overcurrent device ampere rating. Table 210.21(B)(2) shall not apply.

The occupancies referenced in Article 520 are excluded from all the general requirements relating to connector rating and branch-circuit loading found elsewhere in the *Code*, such as in Table 210.21(B)(2). These requirements modify several other sections, such as 210.23(C) and (D), which would disallow 40-ampere and larger branch circuits from serving 5000-watt and larger portable stage lighting equipment found in theaters. They require only that connectors be rated sufficiently for the parameters involved, thus permitting connectors with voltage and current ratings higher than the branch-circuit rating to be used.

The stage set lighting and associated equipment, such as stage effects, both fixed and portable, must be as flexible as possible. Connectors are often used for different purposes and are therefore marked on a show-by-show basis to designate the voltage, current, and type of current actually employed. Stage set lighting is usually planned in advance, and the loads on each receptacle are known. Loads are not casually connected as they might be at a typical general-use wall receptacle. Care is taken to ensure that circuits are not overloaded, thereby avoiding nuisance tripping during a performance.

## 520.10 Portable Equipment Used Outdoors

Portable stage and studio lighting equipment and portable power distribution equipment not identified for outdoor use shall be permitted for temporary use outdoors, provided the equipment is supervised by qualified personnel while energized and barriered from the general public.

Portable indoor stage or studio equipment that is not marked as suitable for wet or damp locations is permitted to be used temporarily in outdoor locations. If rain occurs, this equipment is typically de-energized, and a protective cover is installed before it is re-energized. At the end of the day, this equipment is either de-energized and protected or dismantled and stored.

# II. Fixed Stage Switchboards

## 520.21 General

Fixed stage switchboards shall comply with 520.21(1) through (4):

(1) Fixed stage switchboards shall be listed.

(2) Fixed stage switchboards shall be readily accessible but shall not be required to be located on or adjacent to the stage. Multiple fixed stage switchboards shall be permitted at different locations.
(3) A fixed stage switchboard shall contain overcurrent protective devices for all branch circuits supplied by that switchboard.
(4) A fixed stage switchboard shall be permitted to supply both stage and non-stage equipment.

•

## 520.25 Dimmers

Dimmers shall comply with 520.25(A) through (D).

A high-density digital dimmer rack typically contains one dimmer (usually of 20-, 50-, or 100-ampere capacity) for each branch circuit connected to it. The rack is usually supplied by a 3-phase, 4-wire-plus-ground feeder, which is distributed via buses to all dimmers in the rack. Typical dimmer racks contain between 12 and 96 dimmers and may have total power capacities of up to 288 kW. In large theatrical systems, many racks may be bused together. A central control electronics module drives multiple dimmers in the rack. A digital data link may connect the dimmer rack to a remotely located computer control console.

Exhibit 520.3 shows a high-density digital SCR dimmer switchboard. Dimmers for individual circuits are contained in dual plug-in dimmer modules. These modules also contain circuit breakers for overcurrent protection and filter chokes to eliminate acoustic noise from the lamp filaments. The digital control electronics are contained in a plug-in module with front-panel controls for configuration and testing.

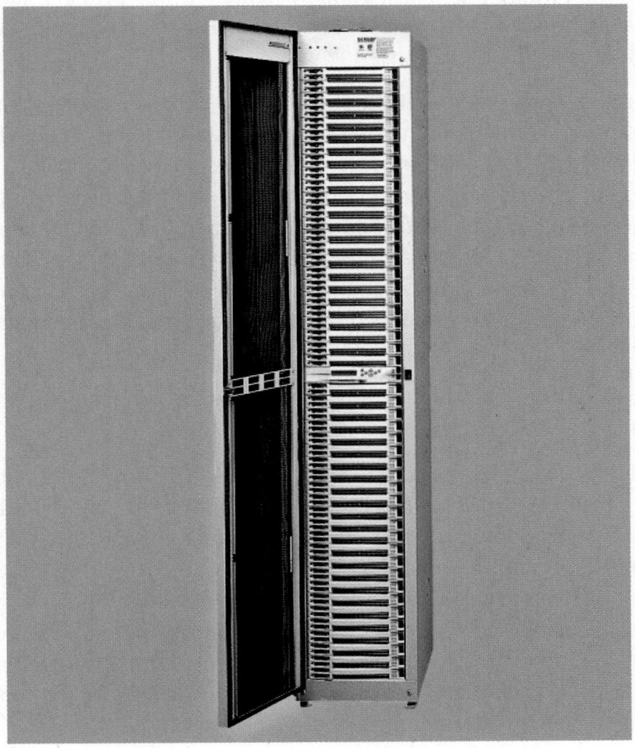

***EXHIBIT 520.3*** *A typical high-density digital SCR dimmer switchboard. (Courtesy of Electronic Theatre Controls, Inc.)*

**(A) Disconnection and Overcurrent Protection.** Where dimmers are installed in ungrounded conductors, each dimmer shall have overcurrent protection not greater than 125 percent of the dimmer rating and shall be disconnected from all ungrounded conductors when the master or individual switch or circuit breaker supplying such dimmer is in the open position.

**(B) Resistance- or Reactor-Type Dimmers.** Resistance- or series reactor-type dimmers shall be permitted to be placed in either the grounded or the ungrounded conductor of the circuit. Where designed to open either the supply circuit to the dimmer or the circuit controlled by it, the dimmer shall then comply with 404.2(B). Resistance- or reactor-type dimmers placed in the grounded neutral conductor of the circuit shall not open the circuit.

**(C) Autotransformer-Type Dimmers.** The circuit supplying an autotransformer-type dimmer shall not exceed 150 volts between conductors. The grounded conductor shall be common to the input and output circuits.

Informational Note: See 210.9 for circuits derived from autotransformers.

Any desired voltage may be applied to the lamps, from full-line voltage to voltage so low that the lamps provide no illumination, by means of a movable contact tap. This type of dimmer produces very little heat and operates at high efficiency. Its dimming effect, within its maximum rating, is independent of the wattage of the load. See the commentary that follows 470.1 regarding saturable reactors that are sometimes used for stage dimmers.

**(D) Solid-State-Type Dimmers.** The circuit supplying a solid-state dimmer shall not exceed 150 volts between conductors unless the dimmer is listed specifically for higher voltage operation. Where a grounded conductor supplies a dimmer, it shall be common to the input and output circuits. Dimmer chassis shall be connected to the equipment grounding conductor.

Solid-state stage dimmers are often used since stage switchboards are usually remote-controlled. The switchboard or dimmer rack is normally located offstage in a dimmer room, where proper climate control can be furnished and noise from the rack cooling fans does not interfere with the performance onstage. Branch circuits are usually connected to the dimmer rack on a dimmer-per-circuit basis. A digital control cable connects the dimmer rack to a remote computer lighting control console, such as the one shown in Exhibit 520.4, which can be located on stage or in the auditorium in view of the stage.

## 520.26 Type of Switchboard

A stage switchboard shall be either one or a combination of the types specified in 520.26(A), (B), (C), and (D).

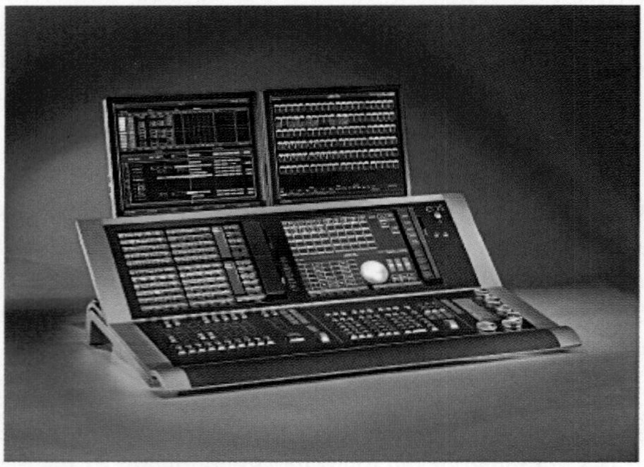

*EXHIBIT 520.4 An electronic computer lighting control console for remotely controlling solid-state-type dimmers. (Courtesy of Electronic Theatre Controls, Inc.)*

**(A) Manual.** Dimmers and switches are operated by handles mechanically linked to the control devices.

**(B) Remotely Controlled.** Devices are operated electrically from a pilot-type control console or panel. Pilot control panels either shall be part of the switchboard or shall be permitted to be at another location.

**(C) Intermediate.** A stage switchboard with circuit interconnections is a secondary switchboard (patch panel) or panelboard remote to the primary stage switchboard. It shall contain overcurrent protection. Where the required branch-circuit overcurrent protection is provided in the dimmer panel, it shall be permitted to be omitted from the intermediate switchboard.

The intermediate-stage switchboard located between the dimmer switchboard and the branch circuits is usually called a patch panel. Its purpose is to break down larger dimmer circuits to smaller branch circuits, to select the branch circuits to be controlled by a dimmer, or both.

**(D) Constant Power.** A stage switchboard containing only overcurrent protective devices and no control elements.

## 520.27 Stage Switchboard Feeders

**(A) Type of Feeder.** Feeders supplying stage switchboards shall be one of the types in 520.27(A)(1) through (A)(3).

**(1) Single Feeder.** A single feeder disconnected by a single disconnect device.

**(2) Multiple Feeders to Intermediate Stage Switchboard (Patch Panel).** Multiple feeders of unlimited quantity shall be permitted, provided that all multiple feeders are part of a single system. Where combined, neutral conductors in a given raceway shall be of sufficient ampacity to carry the maximum unbalanced current supplied by multiple feeder conductors in the same raceway, but they need not be greater than the ampacity of the neutral conductor supplying the primary stage switchboard. Parallel neutral conductors shall comply with 310.10(H).

The feeders are often many dimmer-controlled circuits at 100 amperes or less, single phase, so they can be distributed to different combinations of the same size or smaller branch circuits. This type of installation usually requires a common neutral, and because of the quantity of circuits, many installations require several parallel neutrals running in several raceways. Generally, these parallel neutrals are sized as follows:

1. Size the common neutral to the feeder of the primary switchboard
2. Split this neutral into multiple parallel conductors, one per raceway
3. Equally divide, per phase, and size each ungrounded conductor of the many single-phase circuits among the raceways

In no case are the ungrounded conductors permitted to be installed in one raceway and the common neutral installed in another.

**(3) Separate Feeders to Single Primary Stage Switchboard (Dimmer Bank).** Installations with separate feeders to a single primary stage switchboard shall have a disconnecting means for each feeder. The primary stage switchboard shall have a permanent and obvious label stating the number and location of disconnecting means. If the disconnecting means are located in more than one distribution switchboard, the primary stage switchboard shall be provided with barriers to correspond with these multiple locations.

Large primary stage switchboards usually consist of several sections, often called dimmer racks, that form a dimmer bank. The dimmer racks may be fed separately or may be bused together to accept one or more feeder circuits. In older theaters where an intermediate stage switchboard is connected to a primary stage switchboard, a single large feeder usually supplies the primary stage switchboard, because the intermediate stage switchboard patches only the ungrounded conductors and requires a common neutral.

**(B) Neutral Conductor.** For the purpose of ampacity adjustment, the following shall apply:

(1) The neutral conductor of feeders supplying solid-state, phase-control 3-phase, 4-wire dimming systems shall be considered a current-carrying conductor.
(2) The neutral conductor of feeders supplying solid-state, sine wave 3-phase, 4-wire dimming systems shall not be considered a current-carrying conductor.
(3) The neutral conductor of feeders supplying systems that use or may use both phase-control and sine wave dimmers shall be considered as current-carrying.

It is not necessary to consider the neutral as a current-carrying conductor in every instance with the use of solid-state dimmers. If the sine-wave–type dimmer is the only type in use, the neutral of the circuits supplying it need not be considered as a current-carrying conductor. However, if phase-control dimmers are used, or if combinations of phase-control and sine-wave–type dimmers are connected to the same feeder or branch circuit, the neutral conductor must be considered as a current-carrying conductor. The neutral of feeders supplying solid-state, 3-phase, 4-wire dimming systems carries third-harmonic currents that are present even under balanced load conditions.

**(C) Supply Capacity.** For the purposes of calculating supply capacity to switchboards, it shall be permissible to consider the maximum load that the switchboard is intended to control in a given installation, provided that the following apply:

(1) All feeders supplying the switchboard shall be protected by an overcurrent device with a rating not greater than the ampacity of the feeder.
(2) The opening of the overcurrent device shall not affect the proper operation of the egress or emergency lighting systems.

Informational Note: For calculation of stage switchboard feeder loads, see 220.40.

The feeder for a single, primary stage switchboard is permitted to be sized for the maximum load the switchboard controls for a specific location, rather than for the rating of the switchboard. The feeder(s) must be protected by an overcurrent device that has a rating not greater than the feeder ampacity.

## III. Fixed Stage Equipment Other Than Switchboards

### 520.40 Stage Lighting Hoists

Where a stage lighting hoist is listed as a complete assembly and contains an integral cable-handling system and cable to connect a moving wiring device to a fixed junction box for connection to permanent wiring, the extra-hard usage requirement of 520.44(C)(1) shall not apply.

### 520.41 Circuit Loads

**(A) Circuits Rated 20 Amperes or Less.** Footlights, border lights, and proscenium sidelights shall be arranged so that no branch circuit supplying such equipment carries a load exceeding 20 amperes.

**(B) Circuits Rated Greater Than 20 Amperes.** Where only heavy-duty lampholders are used, such circuits shall be permitted to comply with Article 210 for circuits supplying heavy-duty lampholders.

In accordance with 210.23(B) and (C), 30-, 40-, or 50-ampere branch circuits are permitted if heavy-duty lampholders, such as medium- or mogul-base Edison screw shell types, are used for fixed lighting.

### 520.42 Conductor Insulation

Foot, border, proscenium, or portable strip lights and connector strips shall be wired with conductors that have insulation suitable for the temperature at which the conductors are operated, but not less than 125°C (257°F). The ampacity of the 125°C (257°F) conductors shall be that of 60°C (140°F) conductors. All drops from connector strips shall be 90°C (194°F) wire sized to the ampacity of 60°C (140°F) cords and cables with no more than 150 mm (6 in.) of conductor extending into the connector strip. Section 310.15(B)(3)(a) shall not apply.

Informational Note: See Table 310.104(A) for conductor types.

The 125°C minimum temperature rating is due to the heat from the lamps raising the ambient temperature where the wiring is located. Drops from connector strips are usually flexible cord. The derating factor for more than three current-carrying conductors may not be necessary if the conductors are not all energized at one time, or are not often energized at full intensity (dimmed), and are not energized continuously.

### 520.43 Footlights

**(A) Metal Trough Construction.** Where metal trough construction is employed for footlights, the trough containing the circuit conductors shall be made of sheet metal not lighter than 0.81 mm (0.032 in.) and treated to prevent oxidation. Lampholder terminals shall be kept at least 13 mm (½ in.) from the metal of the trough. The circuit conductors shall be soldered to the lampholder terminals.

**(B) Other-Than-Metal Trough Construction.** Where the metal trough construction specified in 520.43(A) is not used, footlights shall consist of individual outlets with lampholders wired with rigid metal conduit, intermediate metal conduit, or flexible metal conduit, Type MC cable, or mineral-insulated, metal-sheathed cable. The circuit conductors shall be soldered to the lampholder terminals.

**(C) Disappearing Footlights.** Disappearing footlights shall be arranged so that the current supply is automatically disconnected when the footlights are replaced in the storage recesses designed for them.

The footlights described in 520.43(A) and (B) were historically built in the field. Modern footlights are compartmentalized, factory-wired assemblies for field installation, as shown in Exhibit 520.5, and may be permanently exposed or be of the disappearing type. Disappearing footlights must automatically disconnect the current supply when the footlights are in the closed position, thereby preventing heat entrapment that could cause a fire. Disconnection is accomplished by mercury switches in the terminal compartment.

### 520.44 Borders, Proscenium Sidelights, Drop Boxes, and Connector Strips

**(A) General.** Borders and proscenium sidelights shall be as follows:

(1) Constructed as specified in 520.43
(2) Suitably stayed and supported

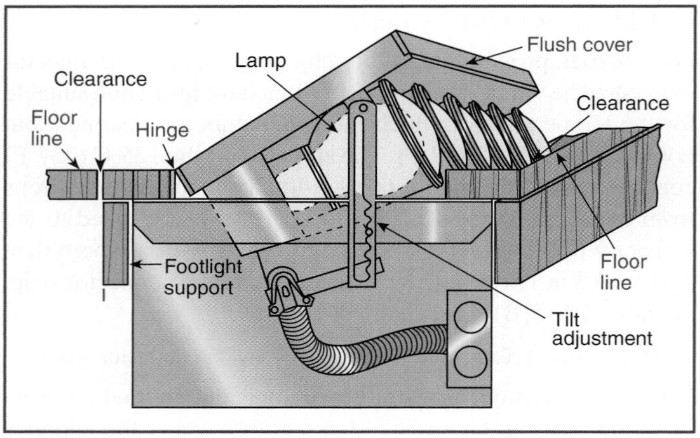


*EXHIBIT 520.5 Disappearing footlights.*

*EXHIBIT 520.6 A suspended border light assembly over a stage.*

(3) Designed so that the flanges of the reflectors or other adequate guards protect the lamps from mechanical damage and from accidental contact with scenery or other combustible material

Both of these types of stage lighting instruments must be suitably supported and protected from mechanical damage. Exhibit 520.6 shows a suspended border light installed over a stage. Commonly, lampholders are wired alternately on three or four circuits. A splice box is provided on top of the housing for enclosing connections between the cable supplying the border light and the border light's internal wiring, which consists of wiring from the splice box to the lamp sockets in a trough extending the length of the border.

**(B) Connector Strips and Drop Boxes.** Connector strips and drop boxes shall be as follows:

(1) Suitably stayed and supported
(2) Listed as stage and studio wiring devices

**(C) Cords and Cables for Border Lights, Drop Boxes, and Connector Strips.**

**(1) General.** Cords and cables for supply to border lights, drop boxes, and connector strips shall be listed for extra-hard usage. The cords and cables shall be suitably supported. Such cords and cables shall be employed only where flexible conductors are necessary. Ampacity of the conductors shall be as provided in 400.5.

Border lights are typically supported by steel cables to facilitate height adjustment for cleaning and lamp replacement, and the circuit conductors supplying the border lights are carried to each border light in flexible cable. Each of these flexible cables usually contains many circuits; however, its overall length is limited by its ability to travel up and down without getting tangled. See Exhibit 520.7.

**(2) Cords and Cables Not in Contact with Heat-Producing Equipment.** Listed multiconductor extra-hard-usage-type cords and cables not in direct contact with equipment containing heat-producing elements shall be permitted to have their ampacity determined by Table 520.44. Maximum load current in any conductor with an ampacity determined by Table 520.44 shall not exceed the values in Table 520.44.

Extra-hard-usage cords and cables not in direct contact with heat-producing equipment are permitted to have their ampacity determined by Table 520.44 instead of 400.5(A).

Table 520.44 is based on a minimum 50-percent diversity factor. It also makes allowance for the fact that not all circuits are on at the same time, not all circuits are at full intensity (dimmed), and not all circuits are on for a long period of time. If the load diversity does not follow this pattern, such as border lights that are all left on at full intensity to light the stage for rehearsal, lecture, or classroom purposes, Table 520.44 must not be used.

Flexible cords and cables are only permitted to be used where necessary, such as for border lights requiring height adjustment, whereas a Chapter 3 wiring method is required for a fixed connection. These fixed conductors are required to follow the ampacity calculations in Article 310, but the adjustment factors in 310.15(B)(3) do not take into account load diversity. Informational Note No. 1 to 310.15(B)(3) refers to Table B.310.15(B)(2)(11) for adjustment factors with at least a 50-percent load diversity. This annex table for Chapter 3 wiring methods correlates with the flexible cord adjustment factors in Table 520.44 for a load diversity of 50 percent.

**(3) Identification of Conductors in Multiconductor Extra-hard Usage Cords and Cables.** Grounded (neutral) conductors shall be white without stripe or shall be identified by a distinctive white marking at their terminations. Grounding conductors shall be green with or without yellow stripe or shall be identified by a distinctive green marking at their terminations.

## 520.45 Receptacles

Receptacles for electrical equipment on stages shall be rated in amperes. Conductors supplying receptacles shall be in accordance with Articles 310 and 400.

***TABLE 520.44*** *Ampacity of Listed Extra-Hard-Usage Cords and Cables with Temperature Ratings of 75°C (167°F) and 90°C (194°F)* [Based on Ambient Temperature of 30°C (86°F)]*

| | Temperature Rating of Cords and Cables | | |
|---|---|---|---|
| **Size (AWG)** | **75°C (167°F)** | **90°C (194°F)** | **Maximum Rating of Overcurrent Device** |
| 14 | 24 | 28 | 15 |
| 12 | 32 | 35 | 20 |
| 10 | 41 | 47 | 25 |
| 8 | 57 | 65 | 35 |
| 6 | 77 | 87 | 45 |
| 4 | 101 | 114 | 60 |
| 2 | 133 | 152 | 80 |

*Ampacity shown is the ampacity for multiconductor cords and cables where only three copper conductors are current-carrying as described in 400.5. If the number of current-carrying conductors in a cord or cable exceeds three and the load diversity factor is a minimum of 50 percent, the ampacity of each conductor shall be reduced as shown in the following table:

| Number of Conductors | Percent of Ampacity |
|---|---|
| 4–6 | 80 |
| 7–24 | 70 |
| 25–42 | 60 |
| 43 and above | 50 |

Note: Ultimate insulation temperature. In no case shall conductors be associated together in such a way with respect to the kind of circuit, the wiring method used, or the number of conductors such that the temperature limit of the conductors is exceeded.

A neutral conductor that carries only the unbalanced current from other conductors of the same circuit need not be considered as a current-carrying conductor.

In a 3-wire circuit consisting of two phase conductors and the neutral conductor of a 4-wire, 3-phase, wye-connected system, the neutral conductor carries approximately the same current as the line-to-neutral currents of the other conductors and shall be considered to be a current-carrying conductor.

On a 4-wire, 3-phase wye circuit where the major portion of the load consists of nonlinear loads, there are harmonic currents in the neutral conductor. Therefore, the neutral conductor shall be considered to be a current-carrying conductor.

## 520.46 Connector Strips, Drop Boxes, Floor Pockets, and Other Outlet Enclosures

Receptacles for the connection of portable stage-lighting equipment shall be pendant or mounted in suitable pockets or enclosures and shall comply with 520.45. Supply cables for connector strips and drop boxes shall be as specified in 520.44(C).

Exhibit 520.7 shows a hanging connector strip with its associated hardware and flexible cable allowing for height adjustment. Exhibits 520.8 and 520.9 illustrate two other types of connections for portable stage lighting equipment.

## 520.47 Backstage Lamps (Bare Bulbs)

Lamps (bare bulbs) installed in backstage and ancillary areas where they can come in contact with scenery shall be located and guarded so as to be free from physical damage and shall provide an air space of not less than 50 mm (2 in.) between such lamps and any combustible material.

*Exception: Decorative lamps installed in scenery shall not be considered to be backstage lamps for the purpose of this section.*

## 520.48 Curtain Machines

Curtain machines shall be listed.

## 520.49 Smoke Ventilator Control

Where stage smoke ventilators are released by an electrical device, the circuit operating the device shall be normally closed and shall be controlled by at least two externally operable switches, one switch being placed at a readily accessible location on stage and the other where designated by the authority having jurisdiction. The device shall be designed for the full voltage of the circuit to which it is connected, no resistance being inserted. The device shall be located in the loft above the scenery and shall be enclosed in a suitable metal box having a tight, self-closing door.

In addition to the two externally operable switches at different locations, the design of a normally closed circuit ensures that smoke ventilators operate when the circuit opens for any reason, such as a circuit breaker tripping or a fuse blowing.

# IV. Portable Switchboards on Stage

## 520.50 Road Show Connection Panel (A Type of Patch Panel)

A panel designed to allow for road show connection of portable stage switchboards to fixed lighting outlets by means of permanently installed supplementary circuits. The panel, supplementary circuits, and outlets shall comply with 520.50(A) through (D).

Also known as a road show interconnect or intercept panel, this panel is designed to connect the load side of a portable switchboard to the fixed building branch circuits and associated outlets. It may also provide for the fixed branch circuits to be connected to a fixed switchboard when the portable switchboard is not installed.

**(A) Load Circuits.** Circuits shall originate from grounding-type polarized inlets of current and voltage rating that match the fixed-load receptacle.

The required grounding-type polarized inlets may be flush or pendant. The fixed-load receptacle is where the portable switchboard connects to the house circuits to control the theater lights.

**(B) Circuit Transfer.** Circuits that are transferred between fixed and portable switchboards shall have all circuit conductors transferred simultaneously.

*EXHIBIT 520.7 A suspended connector strip with border lights attached.*

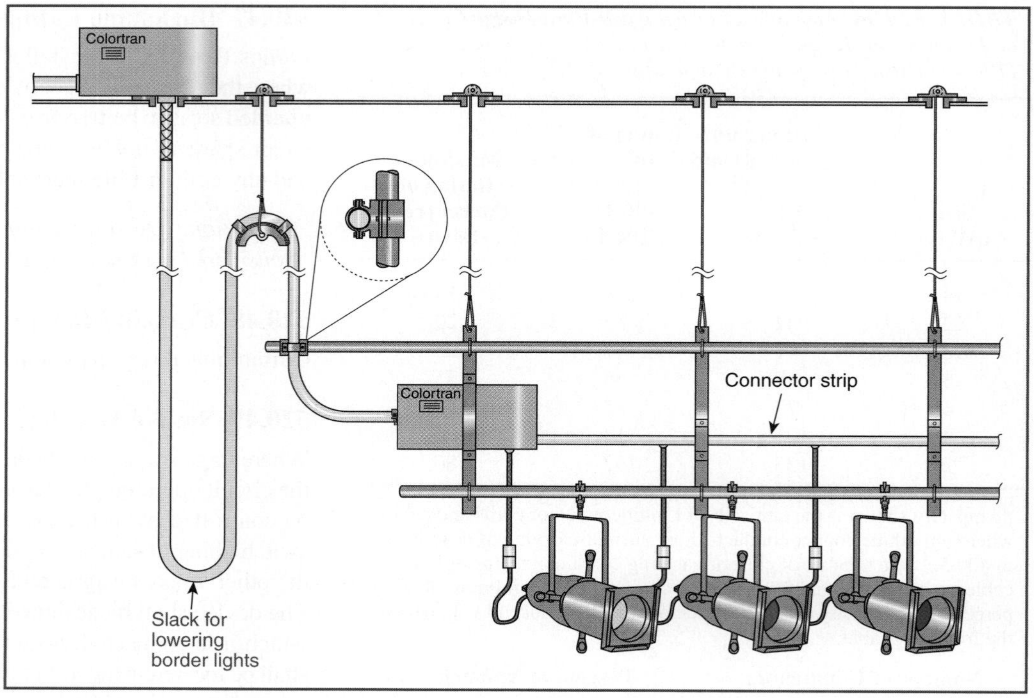

*EXHIBIT 520.8 A 4-gang, 4-receptacle pin-plug outlet box designed for flush mounting. (Courtesy of Electronic Theatre Controls, Inc.)*

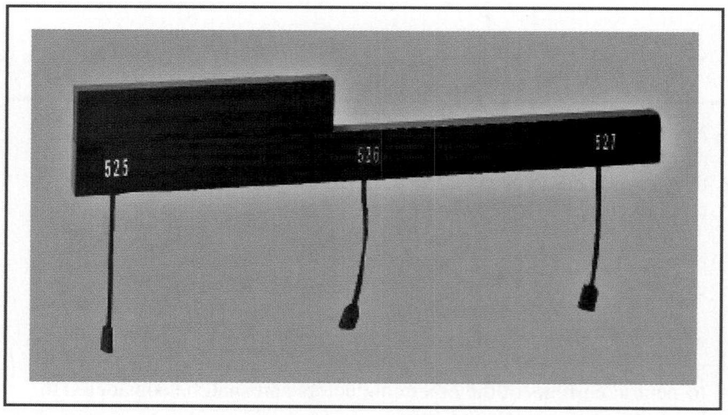


*EXHIBIT 520.9 A typical three-circuit connector strip designed for wall or pipe mounting. (Courtesy of Electronic Theatre Controls, Inc.)*

The required simultaneous transfer of all conductors of the circuit includes any grounded conductors.

**(C) Overcurrent Protection.** The supply devices of these supplementary circuits shall be protected by branch-circuit overcurrent protective devices. Each supplementary circuit, within the road show connection panel and theater, shall be protected by branch-circuit overcurrent protective devices installed within the road show connection panel.

The branch-circuit overcurrent protection normally should be in the switchboard, but because some older units do not have this protection, backup overcurrent protection is provided in the road show connection panel.

**(D) Enclosure.** Panel construction shall be in accordance with Article 408.

## 520.51 Supply

Portable switchboards shall be supplied only from power outlets of sufficient voltage and ampere rating. Such power outlets shall include only externally operable, enclosed fused switches or circuit breakers mounted on stage or at the permanent switchboard in locations readily accessible from the stage floor. Provisions for connection of an equipment grounding conductor shall be provided. For the purposes of conductor derating, the requirements of 520.27(B) shall apply.

Power outlets, known in the entertainment industry as company switches or bull switches, are the point in the wiring system where portable feeder cables connect to the fixed building wiring. They may be as simple as an overcurrent-protected multipole receptacle designed to accept the supply cable described in 520.53(P), Exception, or they may be multiple sets of parallel single-conductor feeder cables. These single-conductor feeder cables, as described in 520.53(H), may be terminated via single-pole separable connectors, as described in 520.53(K), or directly to busbars, fused disconnect switches, or circuit breakers with wire connectors (lugs).

## 520.52 Overcurrent Protection for Branch Circuits

Portable switchboards shall contain overcurrent protection for branch circuits. The requirements of 210.23 shall not apply.

## 520.53 Construction and Feeders

Portable switchboards and feeders for use on stages shall comply with 520.53(A) through (P).

Exhibit 520.10 illustrates a portable switchboard known as a rolling rack.

***EXHIBIT 520.10*** *A large, portable SCR dimmer switchboard (rolling rack). (Courtesy of Electronic Theatre Controls, Inc.)*

**(A) Enclosure.** Portable switchboards shall be placed within an enclosure of substantial construction, which shall be permitted to be arranged so that the enclosure is open during operation. Enclosures of wood shall be completely lined with sheet metal of not less than 0.51 mm (0.020 in.) and shall be well galvanized, enameled, or otherwise properly coated to prevent corrosion or be of a corrosion-resistant material.

**(B) Energized Parts.** There shall not be exposed energized parts within the enclosure.

**(C) Switches and Circuit Breakers.** All switches and circuit breakers shall be of the externally operable, enclosed type.

**(D) Circuit Protection.** Overcurrent devices shall be provided in each ungrounded conductor of every circuit supplied through the switchboard. Enclosures shall be provided for all overcurrent devices in addition to the switchboard enclosure.

**(E) Dimmers.** The terminals of dimmers shall be provided with enclosures, and dimmer faceplates shall be arranged such that accidental contact cannot be readily made with the faceplate contacts.

**(F) Interior Conductors.**

**(1) Type.** All conductors other than busbars within the switchboard enclosure shall be stranded. Conductors shall be approved for an operating temperature at least equal to the approved operating temperature of the dimming devices used in the switchboard and in no case less than the following:

(1) Resistance-type dimmers — 200°C (392°F); or
(2) Reactor-type, autotransformer, and solid-state dimmers — 125°C (257°F)

**(2) Protection.** Each conductor shall have an ampacity not less than the rating of the circuit breaker, switch, or fuse that it supplies. Circuit interrupting and bus bracing shall be in accordance with 110.9 and 110.10. The short-circuit current rating shall be marked on the switchboard.

Conductors shall be enclosed in metal wireways or shall be securely fastened in position and shall be bushed where they pass through metal.

**(G) Pilot Light.** A pilot light shall be provided within the enclosure and shall be connected to the circuit supplying the board so that the opening of the master switch does not cut off the supply to the lamp. This lamp shall be on an individual branch circuit having overcurrent protection rated or set at not over 15 amperes.

This requirement applies only to switchboards with a main disconnect provided on the switchboard. The pilot light serves as a warning at the switchboard to indicate the presence of power before the main disconnect is activated.

**(H) Supply Conductors.**

**(1) General.** The supply to a portable switchboard shall be by means of listed extra-hard usage cords or cables. The supply cords or cables shall terminate within the switchboard enclosure in an externally operable fused master switch or circuit breaker or in an identified connector assembly. The supply cords or cable (and connector assembly) shall have current ratings not less than the total load connected to the switchboard and shall be protected by overcurrent devices.

As with the supply end described in 520.51, the termination connection required in 520.53(H)(1) could be as simple as a permanently terminated multiconductor supply cord or multipole connector assembly (inlet) or as complex as a set of parallel single-conductor feeder cables. These cables may be field-connected to an assembly of single-pole connectors (inlet) or directly connected, with wire connectors, to busbars or a fused switch or breaker.

Road shows with fixed lighting plans are permitted to size the feeder to the actual connected load rather than sizing it based on the overcurrent protection rating.

**(2) Single-Conductor Cables.** Single-conductor portable supply cable sets shall be not smaller than 2 AWG conductors. The equipment grounding conductor shall not be smaller than 6 AWG conductor. Single-conductor grounded neutral cables for a supply shall be sized in accordance with 520.53(O)(2). Where single conductors are paralleled for increased ampacity, the paralleled conductors shall be of the same length and size. Single-conductor supply cables shall be grouped together but not bundled. The equipment grounding conductor shall be permitted to be of a different type, provided it meets the other requirements of this section, and it shall be permitted to be reduced in size as permitted by 250.122. Grounded (neutral) and equipment grounding conductors shall be identified in accordance with 200.6, 250.119, and 310.110. Grounded conductors shall be permitted to be identified by marking at least the first 150 mm (6 in.) from both ends of each length of conductor with white or gray. Equipment grounding conductors shall be permitted to be identified by marking at least the first 150 mm (6 in.) from both ends of each length of conductor with green or green with yellow stripes. Where more than one nominal voltage exists within the same premises, each ungrounded conductor shall be identified by system.

**(3) Supply Conductors Not Over 3.0 m (10 ft) Long.** Where supply conductors do not exceed 3.0 m (10 ft) in length between supply and switchboard or supply and a subsequent overcurrent device, the supply conductors shall be permitted to be reduced in size where all of the following conditions are met:

(1) The ampacity of the supply conductors shall be at least one-quarter of the current rating of the supply overcurrent protective device.
(2) The supply conductors shall terminate in a single overcurrent protective device that will limit the load to the ampacity of the supply conductors. This single overcurrent device shall be permitted to supply additional overcurrent devices on its load side.
(3) The supply conductors shall not penetrate walls, floors, or ceilings or be run through doors or traffic areas. The supply conductors shall be adequately protected from physical damage.
(4) The supply conductors shall be suitably terminated in an approved manner.
(5) Conductors shall be continuous without splices or connectors.
(6) Conductors shall not be bundled.
(7) Conductors shall be supported above the floor in an approved manner.

**(4) Supply Conductors Not Over 6.0 m (20 ft) Long.** Where supply conductors do not exceed 6.0 m (20 ft) in length between supply and switchboard or supply and a subsequent overcurrent protection device, the supply conductors shall be permitted to be reduced in size where all of the following conditions are met:

(1) The ampacity of the supply conductors shall be at least one-half of the current rating of the supply overcurrent protective device.
(2) The supply conductors shall terminate in a single overcurrent protective device that limits the load to the ampacity of the supply conductors. This single overcurrent device shall be permitted to supply additional overcurrent devices on its load side.
(3) The supply conductors shall not penetrate walls, floors, or ceilings or be run through doors or traffic areas. The supply conductors shall be adequately protected from physical damage.
(4) The supply conductors shall be suitably terminated in an approved manner.
(5) The supply conductors shall be supported in an approved manner at least 2.1 m (7 ft) above the floor except at terminations.
(6) The supply conductors shall not be bundled.
(7) Tap conductors shall be in unbroken lengths.

Similar to the requirements for taps in 240.21, 520.53(H)(3) and (H)(4) permit supply conductors for portable switchboards to be sized according to their overcurrent protection, not by the total connected load. Loads of 144 kVA and greater are not uncommon, even on portable switchboard equipment. Installations in the field include lighting for theatrical-type productions with large numbers of stage lighting fixtures. However, only a fraction of the many fixtures installed are used at any one time.

These tap rules are designed to allow one or more switchboards with smaller feeders to be connected to larger supplies (company switches). If these "rules" are not complied with, proper fixed or portable overcurrent protection devices must be provided for each of the smaller switchboards.

Column D of Table 400.5(A)(2) is able to be employed if the conductors are not bundled. If the conductors were bundled, column F and all applicable derating factors would apply. Most devices used in the theater to terminate single-conductor cables are rated for use at 90°C ampacity. However, if single-conductor cables are terminated directly to a circuit breaker or fused switch, a 75°C ampacity or lower most likely would apply.

**(5) Supply Conductors Not Reduced in Size.** Supply conductors not reduced in size under provisions of 520.53(H)(3) or (H)(4) shall be permitted to pass through holes in walls specifically designed for the purpose. If penetration is through the fire-resistant–rated wall, it shall be in accordance with 300.21.

**(I) Cable Arrangement.** Cables shall be protected by bushings where they pass through enclosures and shall be arranged so that tension on the cable is not transmitted to the connections. Where power conductors pass through metal, the requirements of 300.20 shall apply.

Tension on the connections can be removed by using conventional strain relief devices or, often, by lashing the cable to the enclosure with rope.

**(J) Number of Supply Interconnections.** Where connectors are used in a supply conductor, there shall be a maximum number of three interconnections (mated connector pairs) where the total length from supply to switchboard does not exceed 30 m (100 ft). In cases where the total length from supply to switchboard exceeds 30 m (100 ft), one additional interconnection shall be permitted for each additional 30 m (100 ft) of supply conductor.

The intent is to prevent the addition of excessive numbers of interconnections that could jeopardize the mechanical and electrical integrity of the supply conductors.

**(K) Single-Pole Separable Connectors.** Where single-pole portable cable connectors are used, they shall be listed and of the locking type. Sections 400.10, 406.7, and 406.8 shall not apply to listed single-pole separable connectors and single-conductor cable assemblies utilizing listed single-pole separable connectors. Where paralleled sets of current-carrying, single-pole separable connectors are provided as input devices, they shall be prominently labeled with a warning indicating the presence of internal parallel connections. The use of single-pole separable connectors shall comply with at least one of the following conditions:

(1) Connection and disconnection of connectors are possible only where the supply connectors are interlocked to the source and it is not possible to connect or disconnect connectors when the supply is energized.
(2) Line connectors are of the listed sequential-interlocking type so that load connectors shall be connected in the following sequence:
  a. Equipment grounding conductor connection
  b. Grounded circuit conductor connection, if provided
  c. Ungrounded conductor connection, and that disconnection shall be in the reverse order
(3) A caution notice shall be provided adjacent to the line connectors indicating that plug connection shall be in the following order:
  a. Equipment grounding conductor connectors
  b. Grounded circuit conductor connectors, if provided
  c. Ungrounded conductor connectors, and that disconnection shall be in the reverse order

A listed, special type of connection device suitable for connecting single-conductor feeder cables must be of the locking type to reduce the likelihood of its separating while under load. The connectors must be used in sets because they are only single-pole types. The connector sets must be arranged to reduce the likelihood that connections are made in the incorrect order, in accordance with one of the following methods:

1. The main disconnect cannot be energized until all conductors are connected.
2. The connectors are precluded from being connected in any order other than the proper one (first make/last break of the grounding conductor and connect next-to-first and disconnect next-to-last for the grounded conductor).
3. The individual connectors, free of any special electromechanical intervention, are marked with instructions to the user regarding proper connection.

Single-pole separable connectors are quick-connect feeder splicing and terminating devices, not attachment plugs or receptacles. They are designed to be sized, terminated, and inspected by a qualified person before being energized and are to be guarded from accidental disconnection before being de-energized.

The warning sign(s) or label(s) shall comply with 110.21(B).

**(L) Protection of Supply Conductors and Connectors.** All supply conductors and connectors shall be protected against physical damage by an approved means. This protection shall not be required to be raceways.

Rubber mats and commercially available rubber bridges are often used to protect supply conductors and connectors.

**(M) Flanged Surface Inlets.** Flanged surface inlets (recessed plugs) that are used to accept the power shall be rated in amperes.

**(N) Terminals.** Terminals to which stage cables are connected shall be located so as to permit convenient access to the terminals.

This requirement facilitates the field connection and disconnection of the large feeder cables as a show travels from place to place.

**(O) Neutral Conductor.**

**(1) Neutral Terminal.** In portable switchboard equipment designed for use with 3-phase, 4-wire with ground supply, the current rating of the supply neutral terminal, and the ampacity of its associated busbar or equivalent wiring, or both, shall have an ampacity equal to at least twice the ampacity of the largest ungrounded supply terminal.

*Exception: Where portable switchboard equipment is specifically constructed and identified to be internally converted in the field, in an approved manner, from use with a balanced 3-phase, 4-wire with ground supply to a balanced single-phase, 3-wire with ground supply, the supply neutral terminal and its associated busbar, equivalent wiring, or both, shall have an ampacity equal to at least that of the largest ungrounded single-phase supply terminal.*

**(2) Supply Neutral Conductor.** The power supply conductors for portable switchboards utilizing solid-state phase-control dimmers shall be sized considering the neutral conductor as a current-carrying conductor for ampacity adjustment purposes. The power supply conductors for portable switchboards utilizing only solid-state sine wave dimmers shall be sized considering the neutral conductor as a non–current-carrying conductor for ampacity adjustment purposes. Where single-conductor feeder cables, not installed in raceways, are used on multiphase circuits feeding portable switchboards containing solid-state phase-control dimmers, the neutral conductor shall have an ampacity of at least 130 percent of the ungrounded circuit conductors feeding the portable switchboard. Where such feeders are supplying only solid-state sine wave dimmers, the neutral conductor shall have an ampacity of at least 100 percent of the ungrounded circuit conductors feeding the portable switchboard.

Section 520.53(O) involves overlapping concepts regarding the neutral conductor in power supplies for portable switchboards. If a 3-phase, 4-wire switchboard of any kind is brought into a space that has only single-phase, 3-wire service, the switchboard most likely will be connected with two phases to one leg and one phase to the other. This connection could double the current flowing through the neutral, so the neutral terminal and busbar must be rated for double size to allow for that possibility. The exception to 520.53(O)(1) provides for a smaller neutral sized for the single-phase feed where a switchboard contains switching devices that can divide the B-phase load equally between the A-phase and C-phase buses for single-phase operation.

Additionally from 520.53(O)(2), 3-phase, 4-wire switchboards that contain solid-state phase-control dimming devices must, when connected to a 3-phase, 4-wire supply, be connected to that supply with a multiconductor cable sized by counting the neutral as a current-carrying conductor or with a set of single-conductor cables where the neutral is sized 130 percent greater than the phases.

The double-neutral requirement covers the terminal and associated busbar or wiring. This requirement begins at the main input terminals or busing, main input inlet connector, or attached main input cord-and-plug set and includes all wiring on the load side of that point.

Power supply feeders easily detached at the terminals or inlet connector need not adhere to the 200-percent neutral rule, because they can easily be sized on a show-by-show basis for the type of supply encountered. These cables must, however, adhere to the requirements of the neutral as a current-carrying conductor or to the requirements of the 130-percent single-conductor-cable neutral.

Solid-state sine-wave dimmers are linear devices that do not add nonlinear loads to the neutral conductor. Where feeders supply solid-state sine-wave dimmers, the neutral conductor is sized by considering it as a non–current-carrying conductor. However, it must have an ampacity of at least 100 percent of the ampacity of the phase conductors.

**(P) Qualified Personnel.** The routing of portable supply conductors, the making and breaking of supply connectors and other supply connections, and the energization and de-energization of supply services shall be performed by qualified personnel, and portable switchboards shall be so marked, indicating this requirement in a permanent and conspicuous manner.

*Exception: A portable switchboard shall be permitted to be connected to a permanently installed supply receptacle by other than qualified personnel, provided that the supply receptacle is protected for its current rating by an overcurrent device of not greater than 150 amperes, and where the receptacle, interconnection, and switchboard comply with all of the following:*

*(a) Employ listed multipole connectors suitable for the purpose for every supply interconnection*

*(b) Prevent access to all supply connections by the general public*

*(c) Employ listed extra-hard usage multiconductor cords or cables with an ampacity not less than the load and not less than the ampere rating of the connectors.*

This divides the acceptable practices for professional and professional-grade educational venues (qualified) from those in amateur or amateur-grade educational venues (other than qualified). The requirements allow for such things as single-conductor feeder systems, feeders sized for the current-connected load, tap rules, and so forth, and require the services of a qualified person. The exception provides for a conventional feeder system suitable for use by an untrained person.

## V. Portable Stage Equipment Other Than Switchboards

### 520.61 Arc Lamps

Arc lamps, including enclosed arc lamps and associated ballasts, shall be listed. Interconnecting cord sets and interconnecting cords and cables shall be extra-hard usage type and listed.

### 520.62 Portable Power Distribution Units

Portable power distribution units shall comply with 520.62(A) through (E).

**(A) Enclosure.** The construction shall be such that no current-carrying part will be exposed.

**(B) Receptacles and Overcurrent Protection.** Receptacles shall comply with 520.45 and shall have branch-circuit overcurrent protection in the box. Fuses and circuit breakers shall be protected against physical damage. Flexible cords or cables supplying pendant receptacles or cord connectors shall be listed for extra-hard usage.

**(C) Busbars and Terminals.** Busbars shall have an ampacity equal to the sum of the ampere ratings of all the circuits connected to the busbar. Lugs shall be provided for the connection of the master cable.

**(D) Flanged Surface Inlets.** Flanged surface inlets (recessed plugs) that are used to accept the power shall be rated in amperes.

**(E) Cable Arrangement.** Cables shall be adequately protected where they pass through enclosures and be arranged so that tension on the cable is not transmitted to the terminations.

## 520.63 Bracket Fixture Wiring

**(A) Bracket Wiring.** Brackets for use on scenery shall be wired internally, and the fixture stem shall be carried through to the back of the scenery where a bushing shall be placed on the end of the stem. Externally wired brackets or other fixtures shall be permitted where wired with cords designed for hard usage that extend through scenery and without joint or splice in canopy of fixture back and terminate in an approved-type stage connector located, where practical, within 450 mm (18 in.) of the fixture.

**(B) Mounting.** Fixtures shall be securely fastened in place.

## 520.64 Portable Strips

Portable strips shall be constructed in accordance with the requirements for border lights and proscenium sidelights in 520.44(A). The supply cable shall be protected by bushings where it passes through metal and shall be arranged so that tension on the cable will not be transmitted to the connections.

Informational Note No. 1: See 520.42 for wiring of portable strips.

Informational Note No. 2: See 520.68(A)(3) for insulation types required on single conductors.

## 520.65 Festoons

Joints in festoon wiring shall be staggered. Where such lampholders have terminals of a type that puncture the insulation and make contact with the conductors, they shall be attached only to conductors of the stranded type. Lamps enclosed in lanterns or similar devices of combustible material shall be equipped with guards.

Staggering joints in festoon wiring ensures that connections are not opposite one another. Non-staggered joints could cause sparking due to improper insulation or unraveling of insulation, which, in turn, could ignite lanterns or other combustible material enclosing lamps. Where lampholders have terminals that puncture the conductor insulation to make contact with the conductors, stranded conductors must be used. (See the definition of *festoon lighting* in Article 100.)

## 520.66 Special Effects

Electrical devices used for simulating lightning, waterfalls, and the like shall be constructed and located so that flames, sparks, or hot particles cannot come in contact with combustible material.

## 520.67 Multipole Branch-Circuit Cable Connectors

Multipole branch-circuit cable connectors, male and female, for flexible conductors shall be constructed so that tension on the cord or cable is not transmitted to the connections. The female half shall be attached to the load end of the power supply cord or cable. The connector shall be rated in amperes and designed so that differently rated devices cannot be connected together; however, a 20-ampere T-slot receptacle shall be permitted to accept a 15-ampere attachment plug of the same voltage rating. Alternating-current multipole connectors shall be polarized and comply with 406.7 and 406.10.

Informational Note: See 400.10 for pull at terminals.

## 520.68 Conductors for Portables

**(A) Conductor Type.**

**(1) General.** Flexible conductors, including cable extensions, used to supply portable stage equipment shall be listed extra-hard usage cords or cables.

**(2) Stand Lamps.** Listed, hard usage cord shall be permitted to supply stand lamps where the cord is not subject to physical damage and is protected by an overcurrent device rated at not over 20 amperes.

**(3) Luminaire Supply Cords.** Listed hard usage supply cords shall be permitted to supply luminaires when all of the following conditions are met:

(1) The supply cord is not longer than 1.0 m (3.3 ft).
(2) The supply cord is attached at one end to the luminaire or a luminaire-specific listed connector that mates with a panel-mounted inlet on the body of the luminaire.
(3) The supply cord is protected by an overcurrent protective device of not more than 20 amperes.
(4) The luminaire is listed.
(5) The supply cord is not subject to physical damage.

**(4) High-Temperature Applications.** A special assembly of conductors in sleeving not longer than 1.0 m (3.3 ft) shall be permitted to be employed in lieu of flexible cord if the individual wires are stranded and rated not less than 125°C (257°F) and the outer sleeve is glass fiber with a wall thickness of at least 0.635 mm (0.025 in.).

Portable stage equipment requiring flexible supply conductors with a higher temperature rating where one end is permanently attached to the equipment shall be permitted to employ alternate, suitable conductors as determined by a qualified testing laboratory and recognized test standards.

Stage equipment, such as stage lighting fixtures, often operate at elevated temperatures. High-temperature (150°C to 250°C), extra-hard-usage cords are not generally available. The alternate use of conductors in a glass fiber sleeve is limited to 3.3 feet in length to reduce the likelihood that they would be placed on the floor or other area where they might be damaged by traffic or moving scenery.

**(5) Breakouts.** Listed, hard usage (junior hard service) cords shall be permitted in breakout assemblies where all of the following conditions are met:

(1) The cords are utilized to connect between a single multipole connector containing two or more branch circuits and multiple 2-pole, 3-wire connectors.
(2) The longest cord in the breakout assembly does not exceed 6.0 m (20 ft).
(3) The breakout assembly is protected from physical damage by attachment over its entire length to a pipe, truss, tower, scaffold, or other substantial support structure.
(4) All branch circuits feeding the breakout assembly are protected by overcurrent devices rated at not over 20 amperes.

These requirements apply to multiconductor cable assemblies with multipole connectors that contain more than one branch circuit. The breakout assembly is a multipole connector with several pendant receptacles connected to it, separating the multiple branch circuits into individual branch circuits. The use of a similar arrangement of pendant plugs to form a breaking assembly on the other end of the multiconductor cable is also possible.

**(B) Conductor Ampacity.** The ampacity of conductors shall be as given in 400.5, except multiconductor, listed, extra-hard usage portable cords that are not in direct contact with equipment containing heat-producing elements shall be permitted to have their ampacity determined by Table 520.44. Maximum load current in any conductor with an ampacity determined by Table 520.44 shall not exceed the values in Table 520.44.

Listed portable, multiconductor cable is permitted to be sized in accordance with Table 520.44, similar to the method used for border light cable. A cable, used in lieu of a connector strip, directly above heat-producing equipment should be spaced sufficiently above that equipment to avoid the elevated temperatures or should be sized in accordance with 400.5.

*Exception: Where alternate conductors are allowed in 520.68(A)(3), their ampacity shall be as given in the appropriate table in this Code for the types of conductors employed.*

### 520.69 Adapters

Adapters, two-fers, and other single- and multiple-circuit outlet devices shall comply with 520.69(A), (B), and (C).

**(A) No Reduction in Current Rating.** Each receptacle and its corresponding cable shall have the same current and voltage rating as the plug supplying it. It shall not be utilized in a stage circuit with a greater current rating.

**(B) Connectors.** All connectors shall be wired in accordance with 520.67.

Plugs and receptacles must be of the same rating even though available adapters allow connector bodies to be connected to a plug of a larger rating. For example, a 12 AWG conductor with an ampacity of 20 amperes could be connected to a 100-ampere circuit via an adapter. An overload could result in a fire because the circuit breaker or fuse would not provide adequate protection.

**(C) Conductor Type.** Conductors for adapters and two-fers shall be listed extra-hard usage or listed hard usage (junior hard service) cord. Hard usage (junior hard service) cord shall be restricted in overall length to 1.0 m (3.3 ft).

## VI. Dressing Rooms

### 520.71 Pendant Lampholders

Pendant lampholders shall not be installed in dressing rooms.

### 520.72 Lamp Guards

All exposed incandescent lamps in dressing rooms, where less than 2.5 m (8 ft) from the floor, shall be equipped with open-end guards riveted to the outlet box cover or otherwise sealed or locked in place.

Lamps in dressing rooms are required to be provided with suitable open-end guards that permit relamping and that are not easily removed. Guards make it difficult to circumvent their purpose of preventing contact between the lamps and flammable materials, such as costumes and wigs, typically found in dressing rooms.

### 520.73 Switches Required

All lights and any receptacles adjacent to the mirror(s) and above the dressing table counter(s) installed in dressing rooms shall be controlled by wall switches installed in the dressing room(s). Each switch controlling receptacles adjacent to the mirror(s) and above the dressing table counter(s) shall be provided with a pilot light located outside the dressing room, adjacent to the door to indicate when the receptacles are energized. Other outlets installed in the dressing room shall not be required to be switched.

Only receptacles located adjacent to the mirror and on the countertop are subject to the disconnect and pilot light requirements. The purpose of the switching requirement is to ensure that all coffee pots, curling irons, hair dryers, and other similar countertop appliances can be readily disconnected at the end of a performance.

## VII. Grounding

### 520.81 Grounding

All metal raceways and metal-sheathed cables shall be connected to an equipment grounding conductor. The metal frames and enclosures of all equipment, including border lights and portable luminaires, shall be connected to an equipment grounding conductor.

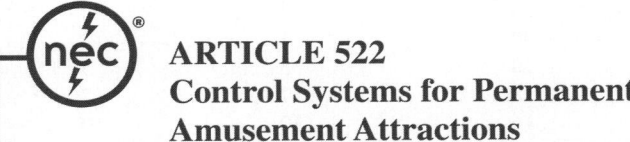

# ARTICLE 522
# Control Systems for Permanent Amusement Attractions

## I. General

### 522.1 Scope

This article covers the installation of control circuit power sources and control circuit conductors for electrical equipment, including associated control wiring in or on all structures, that are an integral part of a permanent amusement attraction.

Article 522 provides requirements for permanent amusement attractions and theme parks. Article 525 applies to temporary attractions, such as carnivals, circuses, and fairs, where most of the attractions consist of portable modules that are moved from place to place. In contrast, theme parks are permanent facilities that have entertainment features fixed in place so that they are not readily portable. In the United States, more than 450 parks operate a wide variety of permanent entertainment features, such as the one shown in Exhibit 522.1.

Article 522 addresses the unique applications and installations utilized in the theme park and amusement industry and covers the wiring requirements for the control circuit power source and control circuit conductors, allowing for wiring methods not available under the provisions of Article 725. The control voltage used is a maximum of 150 volts ac to ground or 300 volts dc to ground.

### 522.2 Definitions

•

**Entertainment Device.** A mechanical or electromechanical device that provides an entertainment experience.

Informational Note: These devices may include animated props, show action equipment, animated figures, and special effects, coordinated with audio and lighting to provide an entertainment experience.

**Permanent Amusement Attraction.** Ride devices, entertainment devices, or combination thereof, that are installed so that portability or relocation is impracticable.

***EXHIBIT 522.1*** *An amusement park facility covered by the requirements of Article 522.*

**Ride Device.** A device or combination of devices that carry, convey, or direct a person(s) over or through a fixed or restricted course within a defined area for the primary purpose of amusement or entertainment.

### 522.5 Voltage Limitations

Control voltage shall be a maximum of 150 volts, nominal, ac to ground or 300 volts dc to ground.

### 522.7 Maintenance

The conditions of maintenance and supervision shall ensure that only qualified persons service the permanent amusement attraction.

## II. Control Circuits

### 522.10 Power Sources for Control Circuits

**(A) Power-Limited Control Circuits.** Power-limited control circuits shall be supplied from a source that has a rated output of not more than 30 volts and 1000 volt-amperes.

**(1) Control Transformers.** Transformers used to supply power-limited control circuits shall comply with the applicable sections within Parts I and II of Article 450.

**(2) Other Power-Limited Control Power Sources.** Power-limited control power sources, other than transformers, shall be protected by overcurrent devices rated at not more than

167 percent of the volt-ampere rating of the source divided by the rated voltage. The fusible overcurrent devices shall not be interchangeable with fusible overcurrent devices of higher ratings. The overcurrent device shall be permitted to be an integral part of the power source.

To comply with the 1000 volt-ampere limitation of 522.10(A), the maximum output of power sources, other than transformers, shall be limited to 2500 volt-amperes, and the product of the maximum current and maximum voltage shall not exceed 10,000 volt-amperes. These ratings shall be determined with any overcurrent-protective device bypassed.

**(B) Non–Power-Limited Control Circuits.** Non–power-limited control circuits shall not exceed 300 volts. The power output of the source shall not be required to be limited.

**(1) Control Transformers.** Transformers used to supply non–power-limited control circuits shall comply with the applicable sections within Parts I and II of Article 450.

**(2) Other Non–Power-Limited Control Power Sources.** Non–power-limited control power sources, other than transformers, shall be protected by overcurrent devices rated at not more than 125 percent of the volt-ampere rating of the source divided by the rated voltage. The fusible overcurrent devices shall not be interchangeable with fusible overcurrent devices of higher ratings. The overcurrent device shall be permitted to be an integral part of the power source.

## III. Control Circuit Wiring Methods

### 522.20 Conductors, Busbars, and Slip Rings

Insulated control circuit conductors shall be copper and shall be permitted to be stranded or solid. Listed multiconductor cable assemblies shall be permitted.

*Exception No. 1: Busbars and slip rings shall be permitted to be materials other than copper.*

*Exception No. 2: Conductors used as specific-purpose devices, such as thermocouples and resistive thermal devices, shall be permitted to be materials other than copper.*

### 522.21 Conductor Sizing

**(A) Conductors Within a Listed Component or Assembly.** Conductors of size 30 AWG or larger shall be permitted within a listed component or as part of the wiring of a listed assembly.

**(B) Conductors Within an Enclosure or Operator Station.** Conductors of size 30 AWG or larger shall be permitted in a listed and jacketed multiconductor cable within an enclosure or operator station. Conductors in a non-jacketed multiconductor cable, such as ribbon cable, shall not be smaller than 26 AWG. Single conductors shall not be smaller than 24 AWG.

*Exception: Single conductors 30 AWG or larger shall be permitted for jumpers and special wiring applications.*

***TABLE 522.22*** *Conductor Ampacity Based on Copper Conductors with 60°C and 75°C Insulation in an Ambient Temperature of 30°C*

| Conductor Size (AWG) | Ampacity | |
|---|---|---|
| | 60°C | 75°C |
| 30 | – | 0.5 |
| 28 | – | 0.8 |
| 26 | – | 1 |
| 24 | 2 | 2 |
| 22 | 3 | 3 |
| 20 | 5 | 5 |
| 18 | 7 | 7 |
| 16 | 10 | 10 |

Notes:
1. For ambient temperatures other than 30°C, use Table 310.15(B)(16) temperature correction factors.
2. Ampacity adjustment for conductors with 90°C or greater insulation shall be based on ampacities in the 75°C column.

**(C) Conductors Outside of an Enclosure or Operator Station.** The size of conductors in a listed and jacketed, multiconductor cable shall not be smaller than 26 AWG. Single conductors shall not be smaller than 18 AWG and shall be installed only where part of a recognized wiring method of Chapter 3.

### 522.22 Conductor Ampacity

Conductors sized 16 AWG and smaller shall not exceed the continuous current values provided in Table 522.22.

### 522.23 Overcurrent Protection for Conductors

Conductors 30 AWG through 16 AWG shall have overcurrent protection in accordance with the appropriate conductor ampacity in Table 522.22. Conductors larger than 16 AWG shall have overcurrent protection in accordance with the appropriate conductor ampacity in Table 310.15(B)(16).

### 522.24 Conductors of Different Circuits in the Same Cable, Cable Tray, Enclosure, or Raceway

Control circuits shall be permitted to be installed with other circuits as specified in 522.24(A) and (B).

**(A) Two or More Control Circuits.** Control circuits shall be permitted to occupy the same cable, cable tray, enclosure, or raceway without regard to whether the individual circuits are alternating current or direct current, provided all conductors are insulated for the maximum voltage of any conductor in the cable, cable tray, enclosure, or raceway.

**(B) Control Circuits with Power Circuits.** Control circuits shall be permitted to be installed with power conductors as specified in 522.24(B)(1) through (B)(3).

**(1) In a Cable, Enclosure, or Raceway.** Control circuits and power circuits shall be permitted to occupy the same cable, enclosure, or raceway only where the equipment powered is functionally associated.

**(2) In Factory- or Field-Assembled Control Centers.** Control circuits and power circuits shall be permitted to be installed in factory- or field-assembled control centers.

**(3) In a Manhole.** Control circuits and power circuits shall be permitted to be installed as underground conductors in a manhole in accordance with one of the following:

(1) The power or control circuit conductors are in a metal-enclosed cable or Type UF cable.
(2) The conductors are permanently separated from the power conductors by a continuous firmly fixed nonconductor, such as flexible tubing, in addition to the insulation on the wire.
(3) The conductors are permanently and effectively separated from the power conductors and securely fastened to racks, insulators, or other approved supports.
(4) In cable trays, where the control circuit conductors and power conductors not functionally associated with them are separated by a solid fixed barrier of a material compatible with the cable tray, or where the power or control circuit conductors are in a metal-enclosed cable.

### 522.25 Ungrounded Control Circuits

Separately derived ac and 2-wire dc circuits and systems 50 volts or greater shall be permitted to be ungrounded, provided that all the following conditions are met:

(1) Continuity of control power is required for orderly shutdown.
(2) Ground detectors are installed on the control system.

### 522.28 Control Circuits in Wet Locations

Where wet contact is likely to occur, ungrounded 2-wire direct-current control circuits shall be limited to 30 volts maximum for continuous dc or 12.4 volts peak for direct current that is interrupted at a rate of 10 to 200 Hz.

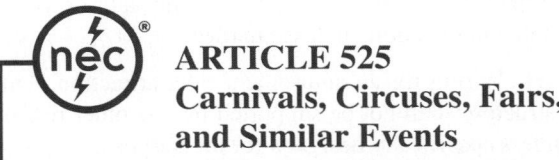

# ARTICLE 525
# Carnivals, Circuses, Fairs, and Similar Events

## I. General Requirements

### 525.1 Scope

This article covers the installation of portable wiring and equipment for carnivals, circuses, fairs, and similar functions, including wiring in or on all structures.

Article 525 addresses the installation of portable wiring and equipment for temporary attractions, such as carnivals, circuses, and fairs. Article 525 is intended to apply to all wiring in or on portable structures, whereas Articles 518, 520, and 522 apply to permanent structures.

### 525.2 Definitions

**Operator.** The individual responsible for starting, stopping, and controlling an amusement ride or supervising a concession.

**Portable Structures.** Units designed to be moved including, but not limited to, amusement rides, attractions, concessions, tents, trailers, trucks, and similar units.

### 525.3 Other Articles

**(A) Portable Wiring and Equipment.** Wherever the requirements of other articles of this *Code* and Article 525 differ, the requirements of Article 525 shall apply to the portable wiring and equipment.

**(B) Permanent Structures.** Articles 518 and 520 shall apply to wiring in permanent structures.

**(C) Audio Signal Processing, Amplification, and Reproduction Equipment.** Article 640 shall apply to the wiring and installation of audio signal processing, amplification, and reproduction equipment.

**(D) Attractions Utilizing Pools, Fountains, and Similar Installations with Contained Volumes of Water.** This equipment shall be installed to comply with the applicable requirements of Article 680.

### 525.5 Overhead Conductor Clearances

**(A) Vertical Clearances.** Conductors shall have a vertical clearance to ground in accordance with 225.18. These clearances shall apply only to wiring installed outside of tents and concessions.

**(B) Clearance to Portable Structures.**

**(1) 600 Volts (or Less).** Portable structures shall be maintained not less than 4.5 m (15 ft) in any direction from overhead conductors operating at 600 volts or less, except for the conductors supplying the portable structure. Portable structures included in 525.3(D) shall comply with Table 680.8(A).

**(2) Over 600 Volts.** Portable structures shall not be located under or within a space that is located 4.5 m (15 ft) horizontally and extending vertically to grade of conductors operating in excess of 600 volts.

Portable structures, which include rides, attractions, and vendor booths, are not permitted in the area defined by a square that extends 15 feet horizontally from the overhead conductors and down to grade level. Exhibit 525.1 depicts the restricted area.

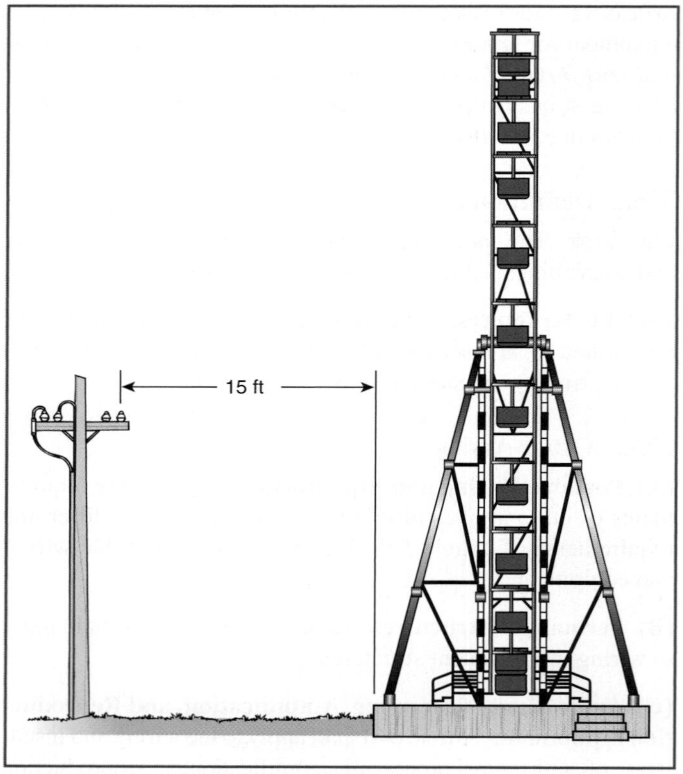


*EXHIBIT 525.1 The area restricted around overhead conductors.*

### 525.6 Protection of Electrical Equipment

Electrical equipment and wiring methods in or on portable structures shall be provided with mechanical protection where such equipment or wiring methods are subject to physical damage.

## II. Power Sources

Power sources include services and multiple sources of supply, such as generators and transformers, that are separately derived systems. In addition to the requirements in 525.10(A) and (B), the requirements for services in Article 230 are applicable.

Service equipment must be installed in accordance with Article 230 and must be lockable where accessible to unqualified persons. Fairs, carnivals, and similar events generate significant pedestrian traffic throughout the sites, including those areas where electrical equipment is located. This requirement helps safeguard the general public from accidentally coming in contact with energized service equipment.

### 525.10 Services

Services shall comply with 525.10(A) and (B).

**(A) Guarding.** Service equipment shall not be installed in a location that is accessible to unqualified persons, unless the equipment is lockable.

**(B) Mounting and Location.** Service equipment shall be securely fastened to a solid backing and be installed so as to be protected from the weather, unless of weatherproof construction.

### 525.11 Multiple Sources of Supply

Where multiple services or separately derived systems, or both, supply portable structures, the equipment grounding conductors of all the sources of supply that serve such structures separated by less than 3.7 m (12 ft) shall be bonded together at the portable structures. The bonding conductor shall be copper and sized in accordance with Table 250.122 based on the largest overcurrent device supplying the portable structures, but not smaller than 6 AWG.

To maintain an equal potential between exposed, non-current-carrying metal parts of portable structures that have a physical separation less than 12 feet, they must be bonded to each other using a copper conductor sized per Table 250.122, but not smaller than 6 AWG.

## III. Wiring Methods

### 525.20 Wiring Methods

**(A) Type.** Where flexible cords or cables are used, they shall be listed for extra-hard usage. Where flexible cords or cables are used and are not subject to physical damage, they shall be permitted to be listed for hard usage. Where used outdoors, flexible cords and cables shall also be listed for wet locations and shall be sunlight resistant. Extra-hard usage flexible cords or cables shall be permitted for use as permanent wiring on portable amusement rides and attractions where not subject to physical damage.

**(B) Single-Conductor.** Single-conductor cable shall be permitted only in sizes 2 AWG or larger.

**(C) Open Conductors.** Open conductors are prohibited except as part of a listed assembly or festoon lighting installed in accordance with Article 225.

**(D) Splices.** Flexible cords or cables shall be continuous without splice or tap between boxes or fittings.

**(E) Cord Connectors.** Cord connectors shall not be laid on the ground unless listed for wet locations. Connectors and cable connections shall not be placed in audience traffic paths or within areas accessible to the public unless guarded.

**(F) Support.** Wiring for an amusement ride, attraction, tent, or similar structure shall not be supported by any other ride or structure unless specifically designed for the purpose.

**(G) Protection.** Flexible cords or cables accessible to the public shall be arranged to minimize the tripping hazard and shall be permitted to be covered with nonconductive matting, provided that the matting does not constitute a greater tripping hazard than the uncovered cables. It shall be permitted to bury cables. The requirements of 300.5 shall not apply.

**(H) Boxes and Fittings.** A box or fitting shall be installed at each connection point, outlet, switchpoint, or junction point.

### 525.21 Rides, Tents, and Concessions

**(A) Disconnecting Means.** A means to disconnect each portable structure from all ungrounded conductors shall be provided. The disconnecting means shall be located within sight of and within 1.8 m (6 ft) of the operator's station. The disconnecting means shall be readily accessible to the operator, including when the ride is in operation. Where accessible to unqualified persons, the disconnecting means shall be lockable. A shunt trip device that opens the fused disconnect or circuit breaker when a switch located in the ride operator's console is closed shall be a permissible method of opening the circuit.

**(B) Portable Wiring Inside Tents and Concessions.** Electrical wiring for lighting, where installed inside of tents and concessions, shall be securely installed and, where subject to physical damage, shall be provided with mechanical protection. All lamps for general illumination shall be protected from accidental breakage by a suitable luminaire or lampholder with a guard.

### 525.22 Portable Distribution or Termination Boxes

Portable distribution or termination boxes shall comply with 525.22(A) through (D).

**(A) Construction.** Boxes shall be designed so that no live parts are exposed except when necessary for examination, adjustment, servicing, or maintenance by qualified persons. Where installed outdoors, the box shall be of weatherproof construction and mounted so that the bottom of the enclosure is not less than 150 mm (6 in.) above the ground.

Requiring equipment to be mounted so that the bottom of the enclosure is at least 6 inches above the ground prevents excessive moisture from entering the equipment and allows for proper radius of bend on conductors entering and exiting the equipment from below.

**(B) Busbars and Terminals.** Busbars shall have an ampere rating not less than the overcurrent device supplying the feeder supplying the box. Where conductors terminate directly on busbars, busbar connectors shall be provided.

**(C) Receptacles and Overcurrent Protection.** Receptacles shall have overcurrent protection installed within the box. The overcurrent protection shall not exceed the ampere rating of the receptacle, except as permitted in Article 430 for motor loads.

**(D) Single-Pole Connectors.** Where single-pole connectors are used, they shall comply with 530.22.

### 525.23 Ground-Fault Circuit-Interrupter (GFCI) Protection

**(A) Where GFCI Protection Is Required.** GFCI protection for personnel shall be provided for the following:

(1) All 125-volt, single-phase, 15- and 20-ampere non-locking-type receptacles used for disassembly and reassembly or readily accessible to the general public
(2) Equipment that is readily accessible to the general public and supplied from a 125-volt, single-phase, 15- or 20-ampere branch circuit

The ground-fault circuit-interrupter shall be permitted to be an integral part of the attachment plug or located in the power-supply cord within 300 mm (12 in.) of the attachment plug. Listed cord sets incorporating ground-fault circuit-interrupter for personnel shall be permitted.

**(B) Where GFCI Protection Is Not Required.** Receptacles that are not accessible from grade level and that only facilitate quick disconnecting and reconnecting of electrical equipment shall not be required to be provided with GFCI protection. These receptacles shall be of the locking type.

**(C) Where GFCI Protection Is Not Permitted.** Egress lighting shall not be protected by a GFCI.

Section 525.23 provides three categories: where GFCIs are required, where GFCIs are not required, and where GFCIs are not permitted to be installed. The application where GFCI protection is not required is very specific. The receptacles must be locking, quick disconnect/reconnect and must not be accessible from grade. GFCI protection is not allowed on circuits that supply means-of-egress illumination.

## IV. Grounding and Bonding

### 525.30 Equipment Bonding

The following equipment connected to the same source shall be bonded:

(1) Metal raceways and metal-sheathed cable
(2) Metal enclosures of electrical equipment
(3) Metal frames and metal parts of portable structures, trailers, trucks, or other equipment that contain or support electrical equipment

The equipment grounding conductor of the circuit supplying the equipment in items (1), (2) or (3) that is likely to energize the metal frame or part shall be permitted to serve as the bonding means.

### 525.31 Equipment Grounding

All equipment to be grounded shall be connected to an equipment grounding conductor of a type recognized by 250.118 and installed in accordance with Parts VI and VII of Article 250. The equipment grounding conductor shall be connected to the system grounded conductor at the service disconnecting means or, in the case of a separately derived system such as a generator, at the generator or first disconnecting means supplied by the generator.

The grounded circuit conductor shall not be connected to the equipment grounding conductor on the load side of the service disconnecting means or on the load side of a separately derived system disconnecting means.

### 525.32 Equipment Grounding Conductor Continuity Assurance

The continuity of the equipment grounding conductor system used to reduce electrical shock hazards as required by 250.114, 250.138, 406.4(C), and 590.4(D) shall be verified each time that portable electrical equipment is connected.

The transient nature of the events covered under Article 525 and, in some cases, the entire associated electrical distribution system increases the possibility that continuity of the EGC system could be interrupted. Verification of the grounding system continuity each time equipment is reconnected helps ensure the safety of workers and members of the general public who may come in contact with electrical equipment.

# ARTICLE 530 Motion Picture and Television Studios and Similar Locations

## I. General

### 530.1 Scope

The requirements of this article shall apply to television studios and motion picture studios using either film or electronic cameras, except as provided in 520.1, and exchanges, factories, laboratories, stages, or a portion of the building in which film or tape more than 22 mm (⅞ in.) in width is exposed, developed, printed, cut, edited, rewound, repaired, or stored.

Informational Note: For methods of protecting against cellulose nitrate film hazards, see NFPA 40-2011, *Standard for the Storage and Handling of Cellulose Nitrate Film.*

The requirements for motion picture studios and for television studios are virtually the same and are intended to apply only to those locations presenting special hazards, for example, where film is handled or for temporary structures constructed of wood or other combustible material. In other areas of the facilities, the conditions are similar to those for theater stages, and the provisions of Article 520 apply for such areas as stages and dressing rooms.

NFPA 140, *Standard on Motion Picture and Television Production Studio Soundstages, Approved Production Facilities, and Production Locations,* addresses additional aspects of these facilities.

### 530.2 Definitions

**Alternating-Current Power Distribution Box (Alternating-Current Plugging Box, Scatter Box).** An ac distribution center or box that contains one or more grounding-type polarized receptacles that may contain overcurrent protective devices.

**Bull Switch.** An externally operated wall-mounted safety switch that may or may not contain overcurrent protection and is designed for the connection of portable cables and cords.

**Location (Shooting Location).** A place outside a motion picture studio where a production or part of it is filmed or recorded.

**Location Board (Deuce Board).** Portable equipment containing a lighting contactor or contactors and overcurrent protection designed for remote control of stage lighting.

**Motion Picture Studio (Lot).** A building or group of buildings and other structures designed, constructed, or permanently altered for use by the entertainment industry for the purpose of motion picture or television production.

**Plugging Box.** A dc device consisting of one or more 2-pole, 2-wire, nonpolarized, nongrounding-type receptacles intended to be used on dc circuits only.

**Portable Equipment.** Equipment intended to be moved from one place to another.

**Single-Pole Separable Connector.** A device that is installed at the ends of portable, flexible, single-conductor cable that is used to establish connection or disconnection between two cables or one cable and a single-pole, panel-mounted separable connector.

**Spider (Cable Splicing Block).** A device that contains busbars that are insulated from each other for the purpose of splicing or distributing power to portable cables and cords that are terminated with single-pole busbar connectors.

**Stage Effect (Special Effect).** An electrical or electromechanical piece of equipment used to simulate a distinctive visual or audible effect such as wind machines, lightning simulators, sunset projectors, and the like.

**Stage Property.** An article or object used as a visual element in a motion picture or television production, except painted backgrounds (scenery) and costumes.

**Stage Set.** A specific area set up with temporary scenery and properties designed and arranged for a particular scene in a motion picture or television production.

**Stand Lamp (Work Light).** A portable stand that contains a general-purpose luminaire or lampholder with guard for the purpose of providing general illumination in the studio or stage.

**Television Studio or Motion Picture Stage (Sound Stage).** A building or portion of a building usually insulated from the outside noise and natural light for use by the entertainment industry for the purpose of motion picture, television, or commercial production.

### 530.6 Portable Equipment

Portable stage and studio lighting equipment and portable power distribution equipment shall be permitted for temporary use outdoors if the equipment is supervised by qualified personnel while energized and barriered from the general public.

Portable indoor stage or studio equipment is permitted to be temporarily used outdoors. If it rains, the equipment is typically de-energized and covered. At the end of the day, the equipment is either de-energized and protected, or dismantled and stored.

## II. Stage or Set

### 530.11 Permanent Wiring

The permanent wiring shall be Type MC cable, Type AC cable containing an insulated equipment grounding conductor sized in accordance with Table 250.122, Type MI cable, or in approved raceways.

*Exception: Communications circuits; audio signal processing, amplification, and reproduction circuits; Class 1, Class 2, and Class 3 remote-control or signaling circuits and power-limited fire alarm circuits shall be permitted to be wired in accordance with Articles 640, 725, 760, and 800.*

### 530.12 Portable Wiring

**(A) Stage Set Wiring.** The wiring for stage set lighting and other supply wiring not fixed as to location shall be done with listed hard usage flexible cords and cables. Where subject to physical damage, such wiring shall be listed extra-hard usage flexible cords and cables. Splices or taps in cables shall be permitted if the total connected load does not exceed the maximum ampacity of the cable.

**(B) Stage Effects and Electrical Equipment Used as Stage Properties.** The wiring for stage effects and electrical equipment used as stage properties shall be permitted to be wired with single- or multiconductor listed flexible cords or cables if the conductors are protected from physical damage and secured to the scenery by approved cable ties or by insulated staples. Splices or taps shall be permitted where such are made with listed devices and the circuit is protected at not more than 20 amperes.

**(C) Other Electrical Equipment.** Cords and cables other than extra-hard usage, where supplied as a part of a listed assembly, shall be permitted.

### 530.13 Stage Lighting and Effects Control

Switches used for studio stage set lighting and effects (on the stages and lots and on location) shall be of the externally operable type. Where contactors are used as the disconnecting means for fuses, an individual externally operable switch, suitably rated, for the control of each contactor shall be located at a distance of not more than 1.8 m (6 ft) from the contactor, in addition to remote-control switches. A single externally operable switch shall be permitted to simultaneously disconnect all the contactors on any one location board, where located at a distance of not more than 1.8 m (6 ft) from the location board.

### 530.14 Plugging Boxes

Each receptacle of dc plugging boxes shall be rated at not less than 30 amperes.

### 530.15 Enclosing and Guarding Live Parts

**(A) Live Parts.** Live parts shall be enclosed or guarded to prevent accidental contact by persons and objects.

**(B) Switches.** All switches shall be of the externally operable type.

**(C) Rheostats.** Rheostats shall be placed in approved cases or cabinets that enclose all live parts, having only the operating handles exposed.

**(D) Current-Carrying Parts.** Current-carrying parts of bull switches, location boards, spiders, and plugging boxes shall be enclosed, guarded, or located so that persons cannot accidentally come into contact with them or bring conductive material into contact with them.

### 530.16 Portable Luminaires

Portable luminaires and work lights shall be equipped with flexible cords, composition or metal-sheathed porcelain sockets, and substantial guards.

*Exception: Portable luminaires used as properties in a motion picture set or television stage set, on a studio stage or lot, or on location shall not be considered to be portable luminaires for the purpose of this section.*

### 530.17 Portable Arc Lamps

**(A) Portable Carbon Arc Lamps.** Portable carbon arc lamps shall be substantially constructed. The arc shall be provided with an enclosure designed to retain sparks and carbons and to prevent persons or materials from coming into contact with the arc or bare live parts. The enclosures shall be ventilated. All switches shall be of the externally operable type.

**(B) Portable Noncarbon Arc Electric-Discharge Lamps.** Portable noncarbon arc lamps, including enclosed arc lamps, and associated ballasts shall be listed. Interconnecting cord sets and interconnecting cords and cables shall be extra-hard usage type and listed.

### 530.18 Overcurrent Protection — General

Automatic overcurrent protective devices (circuit breakers or fuses) for motion picture studio stage set lighting and the stage cables for such stage set lighting shall be as given in 530.18(A) through (G). The maximum ampacity allowed on a given conductor, cable, or cord size shall be as given in the applicable tables of Articles 310 and 400.

**(A) Stage Cables.** Stage cables for stage set lighting shall be protected by means of overcurrent devices set at not more than 400 percent of the ampacity given in the applicable tables of Articles 310 and 400.

**(B) Feeders.** In buildings used primarily for motion picture production, the feeders from the substations to the stages shall be protected by means of overcurrent devices (generally located in the substation) having a suitable ampere rating. The overcurrent devices shall be permitted to be multipole or single-pole gang operated. No pole shall be required in the neutral conductor. The overcurrent device setting for each feeder shall not exceed 400 percent of the ampacity of the feeder, as given in the applicable tables of Article 310.

**(C) Cable Protection.** Cables shall be protected by bushings where they pass through enclosures and shall be arranged so that tension on the cable is not transmitted to the connections. Where power conductors pass through metal, the requirements of 300.20 shall apply.

Portable feeder cables shall be permitted to temporarily penetrate fire-rated walls, floors, or ceilings provided that all of the following apply:

(1) The opening is of noncombustible material.
(2) When in use, the penetration is sealed with a temporary seal of a listed firestop material.
(3) When not in use, the opening shall be capped with a material of equivalent fire rating.

**(D) Location Boards.** Overcurrent protection (fuses or circuit breakers) shall be provided at the location boards. Fuses in the location boards shall have an ampere rating of not over 400 percent of the ampacity of the cables between the location boards and the plugging boxes.

**(E) Plugging Boxes.** Cables and cords supplied through plugging boxes shall be of copper. Cables and cords smaller than 8 AWG shall be attached to the plugging box by means of a plug containing two cartridge fuses or a 2-pole circuit breaker. The rating of the fuses or the setting of the circuit breaker shall not be over 400 percent of the rated ampacity of the cables or cords as given in the applicable tables of Articles 310 and 400. Plugging boxes shall not be permitted on ac systems.

**(F) Alternating-Current Power Distribution Boxes.** Alternating-current power distribution boxes used on sound stages and shooting locations shall contain connection receptacles of a polarized, grounding type.

**(G) Lighting.** Work lights, stand lamps, and luminaires rated 1000 watts or less and connected to dc plugging boxes shall be by means of plugs containing two cartridge fuses not larger than 20 amperes, or they shall be permitted to be connected to special outlets on circuits protected by fuses or circuit breakers rated at not over 20 amperes. Plug fuses shall not be used unless they are on the load side of the fuse or circuit breakers on the location boards.

## 530.19 Sizing of Feeder Conductors for Television Studio Sets

**(A) General.** It shall be permissible to apply the demand factors listed in Table 530.19(A) to that portion of the maximum possible connected load for studio or stage set lighting for all permanently installed feeders between substations and stages and to all permanently installed feeders between the main stage switchboard and stage distribution centers or location boards.

**(B) Portable Feeders.** A demand factor of 50 percent of maximum possible connected load shall be permitted for all portable feeders.

## 530.20 Grounding

Type MC cable, Type MI cable, Type AC cable containing an insulated equipment grounding conductor, metal raceways, and all non–current-carrying metal parts of appliances, devices, and equipment shall be connected to an equipment grounding conductor. This shall not apply to pendant and portable lamps, to portable stage lighting and stage sound equipment, or to other portable and special stage equipment operating at not over 150 volts dc to ground.

## 530.21 Plugs and Receptacles

**(A) Rating.** Plugs and receptacles, including cord connectors and flanged surface devices, shall be rated in amperes. The voltage rating of the plugs and receptacles shall be not less than the nominal circuit voltage. Plug and receptacle ampere ratings for ac circuits shall not be less than the feeder or branch-circuit overcurrent device ampere rating. Table 210.21(B)(2) shall not apply.

**(B) Interchangeability.** Plugs and receptacles used in portable professional motion picture and television equipment shall be permitted to be interchangeable for ac or dc use on the same premises, provided they are listed for ac/dc use and marked in a suitable manner to identify the system to which they are connected.

## 530.22 Single-Pole Separable Connectors

**(A) General.** Where ac single-pole portable cable connectors are used, they shall be listed and of the locking type. Sections 400.10, 406.7, and 406.8 shall not apply to listed single-pole

***TABLE 530.19(A)*** *Demand Factors for Stage Set Lighting*

| Portion of Stage Set Lighting Load to Which Demand Factor Applied (volt-amperes) | Feeder Demand Factor (%) |
|---|---|
| First 50,000 or less at | 100 |
| From 50,001 to 100,000 at | 75 |
| From 100,001 to 200,000 at | 60 |
| Remaining over 200,000 at | 50 |

separable connections and single-conductor cable assemblies utilizing listed single-pole separable connectors. Where paralleled sets of current-carrying single-pole separable connectors are provided as input devices, they shall be prominently labeled with a warning indicating the presence of internal parallel connections. The use of single-pole separable connectors shall comply with at least one of the following conditions:

(1) Connection and disconnection of connectors are only possible where the supply connectors are interlocked to the source and it is not possible to connect or disconnect connectors when the supply is energized.
(2) Line connectors are of the listed sequential-interlocking type so that load connectors shall be connected in the following sequence:
   a. Equipment grounding conductor connection
   b. Grounded circuit conductor connection, if provided
   c. Ungrounded conductor connection, and that disconnection shall be in the reverse order
(3) A caution notice shall be provided adjacent to the line connectors, indicating that plug connection shall be in the following order:
   a. Equipment grounding conductor connectors
   b. Grounded circuit-conductor connectors, if provided
   c. Ungrounded conductor connectors, and that disconnection shall be in the reverse order

The warning sign(s) or label(s) shall comply with 110.21(B).

**(B) Interchangeability.** Single-pole separable connectors used in portable professional motion picture and television equipment shall be permitted to be interchangeable for ac or dc use or for different current ratings on the same premises, provided they are listed for ac/dc use and marked in a suitable manner to identify the system to which they are connected.

### 530.23 Branch Circuits

A branch circuit of any size supplying one or more receptacles shall be permitted to supply stage set lighting loads.

## III. Dressing Rooms

### 530.31 Dressing Rooms

Fixed wiring in dressing rooms shall be installed in accordance with the wiring methods covered in Chapter 3. Wiring for portable dressing rooms shall be approved.

## IV. Viewing, Cutting, and Patching Tables

### 530.41 Lamps at Tables

Only composition or metal-sheathed, porcelain, keyless lampholders equipped with suitable means to guard lamps from physical damage and from film and film scrap shall be used at patching, viewing, and cutting tables.

## V. Cellulose Nitrate Film Storage Vaults

### 530.51 Lamps in Cellulose Nitrate Film Storage Vaults

Lamps in cellulose nitrate film storage vaults shall be installed in rigid luminaires of the glass-enclosed and gasketed type. Lamps shall be controlled by a switch having a pole in each ungrounded conductor. This switch shall be located outside of the vault and provided with a pilot light to indicate whether the switch is on or off. This switch shall disconnect from all sources of supply all ungrounded conductors terminating in any outlet in the vault.

NFPA 40, *Standard for the Storage and Handling of Cellulose Nitrate Film,* requires that luminaires in vaults be suitable for Class I, Division 2, Group D with a temperature rating of T6.

### 530.52 Electrical Equipment in Cellulose Nitrate Film Storage Vaults

Except as permitted in 530.51, no receptacles, outlets, heaters, portable lights, or other portable electrical equipment shall be located in cellulose nitrate film storage vaults. Electric motors shall be permitted, provided they are listed for the application and comply with Article 500, Class I, Division 2.

## VI. Substations

### 530.61 Substations

Wiring and equipment of over 1000 volts, nominal, shall comply with Article 490.

### 530.62 Portable Substations

Wiring and equipment in portable substations shall conform to the sections applying to installations in permanently fixed substations, but, due to the limited space available, the working spaces shall be permitted to be reduced, provided that the equipment shall be arranged so that the operator can work safely and so that other persons in the vicinity cannot accidentally come into contact with current-carrying parts or bring conducting objects into contact with them while they are energized.

### 530.63 Overcurrent Protection of Direct-Current Generators

Three-wire generators shall have overcurrent protection in accordance with 445.12(E).

### 530.64 Direct-Current Switchboards

**(A) General.** Switchboards of not over 250 volts dc between conductors, where located in substations or switchboard rooms accessible to qualified persons only, shall not be required to be dead-front.

**(B) Circuit Breaker Frames.** Frames of dc circuit breakers installed on switchboards shall not be required to be connected to an equipment grounding conductor.

# ARTICLE 540 Motion Picture Projection Rooms

## I. General

### 540.1 Scope

The provisions of this article apply to motion picture projection rooms, motion picture projectors, and associated equipment of the professional and nonprofessional types using incandescent, carbon arc, xenon, or other light source equipment that develops hazardous gases, dust, or radiation.

Informational Note: For further information, see NFPA 40-2011, *Standard for the Storage and Handling of Cellulose Nitrate Film.*

Motion picture projection rooms are not hazardous locations as defined in Article 500. Some older types of film, such as cellulose nitrate film, are highly flammable and rarely used today. The more commonly used cellulose acetate film is not volatile at ordinary temperatures and does not emit flammable gases. Therefore, wiring methods are not required to be suitable for hazardous locations.

### 540.2 Definitions

Nonprofessional Projector. Nonprofessional projectors are those types other than as described in 540.2.

**Professional Projector.** A type of projector using 35- or 70-mm film that has a minimum width of 35 mm (1⅜ in.) and has on each edge 212 perforations per meter (5.4 perforations per inch), or a type using carbon arc, xenon, or other light source equipment that develops hazardous gases, dust, or radiation.

## II. Equipment and Projectors of the Professional Type

### 540.10 Motion Picture Projection Room Required

Every professional-type projector shall be located within a projection room. Every projection room shall be of permanent construction, approved for the type of building in which the projection room is located. All projection ports, spotlight ports, viewing ports, and similar openings shall be provided with glass or other approved material so as to completely close the opening. Such rooms shall not be considered as hazardous (classified) locations as defined in Article 500.

Informational Note: For further information on protecting openings in projection rooms handling cellulose nitrate motion picture film, see NFPA *101*-2012, *Life Safety Code.*

The audience area of a motion picture theater is covered by Article 520. Article 540 addresses projection rooms for theaters. Every professional projector is required to be located within a permanent projection room that is closed from the theater by glass or other material over all projection, spotlight, viewing, or other ports. Modern projectors contain rectifiers as an integral part of their equipment, thereby eliminating generators and other associated equipment described in 540.11. In accordance with 540.31, nonprofessional projectors are not required to be installed within a projection room.

### 540.11 Location of Associated Electrical Equipment

**(A) Motor Generator Sets, Transformers, Rectifiers, Rheostats, and Similar Equipment.** Motor-generator sets, transformers, rectifiers, rheostats, and similar equipment for the supply or control of current to projection or spotlight equipment shall, where nitrate film is used, be located in a separate room. Where placed in the projection room, they shall be located or guarded so that arcs or sparks cannot come in contact with film, and the commutator end or ends of motor generator sets shall comply with one of the conditions in 540.11(A)(1) through (A)(6).

**(1) Types.** Be of the totally enclosed, enclosed fan-cooled, or enclosed pipe-ventilated type.

**(2) Separate Rooms or Housings.** Be enclosed in separate rooms or housings built of noncombustible material constructed so as to exclude flyings or lint with approved ventilation from a source of clean air.

**(3) Solid Metal Covers.** Have the brush or sliding-contact end of motor-generator enclosed by solid metal covers.

**(4) Tight Metal Housings.** Have brushes or sliding contacts enclosed in substantial, tight metal housings.

**(5) Upper and Lower Half Enclosures.** Have the upper half of the brush or sliding-contact end of the motor-generator enclosed by a wire screen or perforated metal and the lower half enclosed by solid metal covers.

**(6) Wire Screens or Perforated Metal.** Have wire screens or perforated metal placed at the commutator of brush ends. No dimension of any opening in the wire screen or perforated metal shall exceed 1.27 mm (0.05 in.), regardless of the shape of the opening and of the material used.

**(B) Switches, Overcurrent Devices, or Other Equipment.** Switches, overcurrent devices, or other equipment not normally required or used for projectors, sound reproduction, flood or other special effect lamps, or other equipment shall not be installed in projection rooms.

*Exception No. 1: In projection rooms approved for use only with cellulose acetate (safety) film, the installation of appurtenant electrical equipment used in conjunction with the operation of the projection equipment and the control of lights, curtains, and audio equipment, and so forth, shall be permitted. In such projection rooms, a sign reading "Safety Film Only Permitted in This Room" shall be posted on the outside of each projection room door and within the projection room itself in a conspicuous location.*

*Exception No. 2: Remote-control switches for the control of auditorium lights or switches for the control of motors operating curtains and masking of the motion picture screen shall be permitted to be installed in projection rooms.*

**(C) Emergency Systems.** Control of emergency systems shall comply with Article 700.

### 540.12 Work Space

Each motion picture projector, floodlight, spotlight, or similar equipment shall have clear working space not less than 750 mm (30 in.) wide on each side and at the rear thereof.

*Exception: One such space shall be permitted between adjacent pieces of equipment.*

### 540.13 Conductor Size

Conductors supplying outlets for arc and xenon projectors of the professional type shall not be smaller than 8 AWG and shall have an ampacity not less than the projector current rating. Conductors for incandescent-type projectors shall conform to normal wiring standards as provided in 210.24.

### 540.14 Conductors on Lamps and Hot Equipment

Insulated conductors having a rated operating temperature of not less than 200°C (392°F) shall be used on all lamps or other equipment where the ambient temperature at the conductors as installed will exceed 50°C (122°F).

### 540.15 Flexible Cords

Cords approved for hard usage, as provided in Table 400.4, shall be used on portable equipment.

### 540.20 Listing Requirements

Projectors and enclosures for arc, xenon, and incandescent lamps and rectifiers, transformers, rheostats, and similar equipment shall be listed.

### 540.21 Marking

Projectors and other equipment shall be marked with the manufacturer's name or trademark and with the voltage and current for which they are designed in accordance with 110.21.

## III. Nonprofessional Projectors

### 540.31 Motion Picture Projection Room Not Required

Projectors of the nonprofessional or miniature type, where employing cellulose acetate (safety) film, shall be permitted to be operated without a projection room.

### 540.32 Listing Requirements

Projection equipment shall be listed.

## IV. Audio Signal Processing, Amplification, and Reproduction Equipment

### 540.50 Audio Signal Processing, Amplification, and Reproduction Equipment

Audio signal processing, amplification, and reproduction equipment shall be installed as provided in Article 640.

# ARTICLE 545 Manufactured Buildings

### 545.1 Scope

This article covers requirements for a manufactured building and building components as herein defined.

The term *manufactured building* is defined in 545.2. The distinction between manufactured buildings covered in Article 545 and manufactured homes covered and defined in Article 550 is important. The most distinguishing feature between the two types of structures is how they are placed on the building site. Manufactured homes are built on a chassis and installed on site with or without a permanent foundation. Manufactured buildings are generally constructed within a factory or assembly plant and then transported to the building site. They are not built on a chassis and are designed to be installed on a permanent foundation.

In addition, the organizations responsible for construction standards for these units differ. In the case of manufactured homes, the U.S. Department of Housing and Urban Development, 24 CFR 3280, *Manufactured Home Construction and Safety Standards*, contains construction requirements for manufactured homes. Manufactured homes bear a nameplate documenting that the unit was constructed in accordance with the federal standard in force at the time of manufacture. In accordance with federal law, this identifying mark is universally recognized throughout the United States.

Manufactured building construction standards generally are promulgated through state or local units of government. Manufactured building construction can be affected by differences in building construction regulations among the jurisdictions where the buildings will be delivered. The building will typically have an information sheet (often inside the cabinet below the kitchen sink or on a closet wall) indicating the applicable building, electrical, plumbing, and mechanical codes to which the building was constructed.

### 545.2 Definitions

**Building Component.** Any subsystem, subassembly, or other system designed for use in or integral with or as part of a structure, which can include structural, electrical, mechanical, plumbing, and fire protection systems, and other systems affecting health and safety.

**Building System.** Plans, specifications, and documentation for a system of manufactured building or for a type or a system of building components, which can include structural, electrical, mechanical, plumbing, and fire protection systems, and other systems affecting health and safety, and including such variations thereof as are specifically permitted by regulation, and which variations are submitted as part of the building system or amendment thereto.

**Closed Construction.** Any building, building component, assembly, or system manufactured in such a manner that all concealed parts of processes of manufacture cannot be inspected after installation at the building site without disassembly, damage, or destruction.

**Manufactured Building.** Any building that is of closed construction and is made or assembled in manufacturing facilities on or off the building site for installation, or for assembly and installation on the building site, other than manufactured homes, mobile homes, park trailers, or recreational vehicles.

## 545.4 Wiring Methods

**(A) Methods Permitted.** All raceway and cable wiring methods included in this *Code* and other wiring systems specifically intended and listed for use in manufactured buildings shall be permitted with listed fittings and with fittings listed and identified for manufactured buildings.

**(B) Securing Cables.** In closed construction, cables shall be permitted to be secured only at cabinets, boxes, or fittings where 10 AWG or smaller conductors are used and protection against physical damage is provided.

## 545.5 Supply Conductors

Provisions shall be made to route the service-entrance conductors, underground service conductors, service-lateral, feeder, or branch-circuit supply to the service or building disconnecting means conductors.

## 545.6 Installation of Service-Entrance Conductors

Service-entrance conductors shall be installed after erection at the building site.

*Exception: Where point of attachment is known prior to manufacture.*

## 545.7 Service Equipment

Service equipment shall be installed in accordance with 230.70.

## 545.8 Protection of Conductors and Equipment

Protection shall be provided for exposed conductors and equipment during processes of manufacturing, packaging, in transit, and erection at the building site.

## 545.9 Boxes

**(A) Other Dimensions.** Boxes of dimensions other than those required in Table 314.16(A) shall be permitted to be installed where tested, identified, and listed to applicable standards.

**(B) Not Over 1650 cm$^3$ (100 in.$^3$).** Any box not over 1650 cm$^3$ (100 in.$^3$) in size, intended for mounting in closed construction, shall be affixed with anchors or clamps so as to provide a rigid and secure installation.

## 545.10 Receptacle or Switch with Integral Enclosure

A receptacle or switch with integral enclosure and mounting means, where tested, identified, and listed to applicable standards, shall be permitted to be installed.

See the commentary following 300.15(E) for additional discussion about wiring devices with integral enclosures.

## 545.11 Bonding and Grounding

Prewired panels and building components shall provide for the bonding, or bonding and grounding, of all exposed metals likely to become energized, in accordance with Article 250, Parts V, VI, and VII.

## 545.12 Grounding Electrode Conductor

Provisions shall be made to route a grounding electrode conductor from the service, feeder, or branch-circuit supply to the point of attachment to the grounding electrode.

## 545.13 Component Interconnections

Fittings and connectors that are intended to be concealed at the time of on-site assembly, where tested, identified, and listed to applicable standards, shall be permitted for on-site interconnection of modules or other building components. Such fittings and connectors shall be equal to the wiring method employed in insulation, temperature rise, and fault-current withstand and shall be capable of enduring the vibration and minor relative motions occurring in the components of manufactured buildings.

The structural components or modules are usually constructed in manufacturing facilities and then transported over the road to a building site for complete assembly of a structure, such as a dwelling unit, motel, or office building. Each module may be prewired at the factory and supplied with fittings and connectors. At the on-site location, these connectors are used to interconnect two or more modules.

# ARTICLE 547
# Agricultural Buildings

## 547.1 Scope

The provisions of this article shall apply to the following agricultural buildings or that part of a building or adjacent areas of similar or like nature as specified in 547.1(A) or (B).

**(A) Excessive Dust and Dust with Water.** Agricultural buildings where excessive dust and dust with water may accumulate, including all areas of poultry, livestock, and fish confinement systems, where litter dust or feed dust, including mineral feed particles, may accumulate.

**(B) Corrosive Atmosphere.** Agricultural buildings where a corrosive atmosphere exists. Such buildings include areas where the following conditions exist:

(1) Poultry and animal excrement may cause corrosive vapors.
(2) Corrosive particles may combine with water.
(3) The area is damp and wet by reason of periodic washing for cleaning and sanitizing with water and cleansing agents.
(4) Similar conditions exist.

Article 547 applies not only to agricultural buildings but also to adjacent areas similar in nature. The requirements address the severe environmental conditions that regularly exist on agricultural premises. Damp and wet conditions, dust from feed and litter, and corrosive agents from livestock excrement are all present in these settings as part of normal operating conditions.

## 547.2 Definitions

**Distribution Point.** An electrical supply point from which service drops, service conductors, feeders, or branch circuits to buildings or structures utilized under single management are supplied.

Informational Note No. 1: Distribution points are also known as the center yard pole, meterpole, or the common distribution point.

Informational Note No. 2: The service point as defined in Article 100 is typically at the distribution point.

**Equipotential Plane.** An area where wire mesh or other conductive elements are embedded in or placed under concrete, bonded to all metal structures and fixed nonelectrical equipment that may become energized, and connected to the electrical grounding system to minimize voltage potentials within the plane and between the planes, the grounded equipment, and the earth.

**Site-Isolating Device.** A disconnecting means installed at the distribution point for the purposes of isolation, system maintenance, emergency disconnection, or connection of optional standby systems.

The site-isolating device provides a means to disconnect and isolate the agricultural premises wiring system from the serving utility under emergency conditions for maintenance of the load-side wiring system or to allow for the connection of an alternate power source when there is a power outage. In accordance with 547.9(A)(2), the site-isolating device must be pole-mounted, and, as its name implies, it is an isolating switch and is not considered to be the service disconnecting means for the agricultural premises.

## 547.3 Other Articles

For buildings and structures not having conditions as specified in 547.1, the electrical installations shall be made in accordance with the applicable articles in this *Code*.

## 547.4 Surface Temperatures

Electrical equipment or devices installed in accordance with the provisions of this article shall be installed in a manner such that they will function at full rating without developing surface temperatures in excess of the specified normal safe operating range of the equipment or device.

## 547.5 Wiring Methods

**(A) Wiring Systems.** Types UF, NMC, copper SE cables, jacketed Type MC cable, rigid nonmetallic conduit, liquidtight flexible nonmetallic conduit, or other cables or raceways suitable for the location, with approved termination fittings, shall be the wiring methods employed. The wiring methods of Article 502, Part II, shall be permitted for areas described in 547.1(A).

Informational Note: See 300.7, 352.44, and 355.44 for installation of raceway systems exposed to widely different temperatures.

**(B) Mounting.** All cables shall be secured within 200 mm (8 in.) of each cabinet, box, or fitting. Nonmetallic boxes, fittings, conduit, and cables shall be permitted to be mounted directly to any building surface covered by this article without maintaining the 6 mm (¼ in.) airspace in accordance with 300.6(D).

The 8-inch support distance is less than that required for cables in other types of occupancies. The ¼-inch airspace is unnecessary provided nonmetallic wiring methods are used. Also, locating the wiring method directly on the interior surface of the building allows a sealant to be placed along the wiring method to facilitate cleaning. Decreasing the support spacing along with eliminating the ¼-inch airspace reduces the potential for physical damage to cable-type wiring methods in agricultural buildings. See also 300.6(D), Exception.

**(C) Equipment Enclosures, Boxes, Conduit Bodies, and Fittings.**

**(1) Excessive Dust.** Equipment enclosures, boxes, conduit bodies, and fittings installed in areas of buildings where excessive dust may be present shall be designed to minimize the entrance of dust and shall have no openings (such as holes for attachment screws) through which dust could enter the enclosure.

**(2) Damp or Wet Locations.** In damp or wet locations, equipment enclosures, boxes, conduit bodies, and fittings shall be placed or equipped so as to prevent moisture from entering or accumulating within the enclosure, box, conduit body, or fitting. In wet locations, including normally dry or damp locations where surfaces are periodically washed or sprayed with water, boxes, conduit bodies, and fittings shall be listed for use in wet locations and equipment enclosures shall be weatherproof.

**(3) Corrosive Atmosphere.** Where wet dust, excessive moisture, corrosive gases or vapors, or other corrosive conditions may be present, equipment enclosures, boxes, conduit bodies, and fittings shall have corrosion resistance properties suitable for the conditions.

Informational Note No. 1: See Table 110.28 for appropriate enclosure type designations.

Informational Note No. 2: Aluminum and magnetic ferrous materials may corrode in agricultural environments.

**(D) Flexible Connections.** Where necessary to employ flexible connections, dusttight flexible connectors, liquidtight flexible metal conduit, liquidtight flexible nonmetallic conduit, or flexible cord listed and identified for hard usage shall be used.

**(E) Physical Protection.** All electrical wiring and equipment subject to physical damage shall be protected.

**(F) Separate Equipment Grounding Conductor.** Where an equipment grounding conductor is installed underground within a location falling under the scope of Article 547, it shall be insulated or covered.

This requirement improves the longevity of EGCs installed above ground and underground in the highly corrosive locations that are typical of many farm buildings.

**(G) Receptacles.** All 125-volt, single-phase, 15- and 20-ampere general-purpose receptacles installed in the locations listed in (1) through (4) shall have ground-fault circuit-interrupter protection:

(1) Areas having an equipotential plane
(2) Outdoors
(3) Damp or wet locations
(4) Dirt confinement areas for livestock

## 547.6 Switches, Receptacles, Circuit Breakers, Controllers, and Fuses

Switches, including pushbuttons, relays, and similar devices, receptacles, circuit breakers, controllers, and fuses, shall be provided with enclosures as specified in 547.5(C).

## 547.7 Motors

Motors and other rotating electrical machinery shall be totally enclosed or designed so as to minimize the entrance of dust, moisture, or corrosive particles.

## 547.8 Luminaires

Luminaires shall comply with 547.8(A) through (C).

**(A) Minimize the Entrance of Dust.** Luminaires shall be installed to minimize the entrance of dust, foreign matter, moisture, and corrosive material.

**(B) Exposed to Physical Damage.** Luminaires exposed to physical damage shall be protected by a suitable guard.

**(C) Exposed to Water.** Luminaires exposed to water from condensation, building cleansing water, or solution shall be listed as suitable for use in wet locations.

## 547.9 Electrical Supply to Building(s) or Structure(s) from a Distribution Point

A distribution point shall be permitted to supply any building or structure located on the same premises. The overhead electrical supply shall comply with 547.9(A) and (B), or with 547.9(C). The underground electrical supply shall comply with 547.9(C).

**(A) Site-Isolating Device.** Site-isolating devices shall comply with 547.9(A)(1) through (A)(10).

**(1) Where Required.** A site-isolating device shall be installed at the distribution point where two or more buildings or structures are supplied from the distribution point.

**(2) Location.** The site-isolating device shall be pole-mounted and be not less than the height above grade required by 230.24 for the conductors it supplies.

**(3) Operation.** The site-isolating device shall simultaneously disconnect all ungrounded service conductors from the premises wiring.

**(4) Bonding Provisions.** The site-isolating device enclosure shall be connected to the grounded circuit conductor and the grounding electrode system.

**(5) Grounding.** At the site-isolating device, the system grounded conductor shall be connected to a grounding electrode system via a grounding electrode conductor.

**(6) Rating.** The site-isolating device shall be rated for the calculated load as determined by Part V of Article 220.

**(7) Overcurrent Protection.** The site-isolating device shall not be required to provide overcurrent protection.

**(8) Accessibility.** The site-isolating device shall be capable of being remotely operated by an operating handle installed at a readily accessible location. The operating handle of the site-isolating device, when in its highest position, shall not be more than 2.0 m (6 ft 7 in.) above grade or a working platform.

**(9) Series Devices.** An additional site-isolating device for the premises wiring system shall not be required where a site-isolating device meeting all applicable requirements of this section is provided by the serving utility as part of their service requirements.

**(10) Marking.** A site-isolating device shall be permanently marked to identify it as a site-isolating device. This marking shall be located on the operating handle or immediately adjacent thereto.

**(B) Service Disconnecting Means and Overcurrent Protection at the Building(s) or Structure(s).** Where the service disconnecting means and overcurrent protection are located at the building(s) or structure(s), the requirements of 547.9(B)(1) through (B)(3) shall apply.

**(1) Conductor Sizing.** The supply conductors shall be sized in accordance with Part V of Article 220.

**(2) Conductor Installation.** The supply conductors shall be installed in accordance with the requirements of Part II of Article 225.

**(3) Grounding and Bonding.** For each building or structure, grounding and bonding of the supply conductors shall be in accordance with the requirements of 250.32, and the following conditions shall be met:

(1) The equipment grounding conductor is not smaller than the largest supply conductor if of the same material, or is adjusted in size in accordance with the equivalent size columns of Table 250.122 if of different materials.
(2) The equipment grounding conductor is connected to the grounded circuit conductor and the site-isolating device enclosure at the distribution point.

**(C) Service Disconnecting Means and Overcurrent Protection at the Distribution Point.** Where the service disconnecting means and overcurrent protection for each set of feeders or branch circuits are located at the distribution point, the feeders or branch circuits to buildings or structures shall comply with the provisions of 250.32 and Article 225, Parts I and II.

Informational Note: Methods to reduce neutral-to-earth voltages in livestock facilities include supplying buildings or structures with 4-wire single-phase services, sizing 3-wire single-phase service and feeder conductors to limit voltage drop to 2 percent, and connecting loads line-to-line.

**(D) Identification.** Where a site is supplied by more than one distribution point, a permanent plaque or directory shall be installed at each of these distribution points denoting the location of each of the other distribution points and the buildings or structures served by each.

The requirements in 547.9 cover the installation of conductors that originate from an electrical distribution point and supply agricultural buildings. The term *distribution point* is defined in 547.2. A distribution point, sometimes referred to as the center yard pole, is often used as a means of centrally locating the origin of the electrical distribution system.

Many agricultural sites consist of multiple buildings that are directly related to or support the operation. A distribution point often supplies multiple buildings via an overhead distribution system. A means to disconnect all ungrounded conductors run to the buildings and structures is required. This disconnecting means is referred to as the site-isolating device. This device provides a means to disconnect all power to the buildings from a single location in the event of an emergency, for the purposes of maintenance, or for connection to a standby power source. It is not considered the service disconnecting means.

The site-isolating device is required to be pole-mounted to the height required for the conductors it supplies, thereby rendering the device inaccessible. The remote operating handle for the device must be readily accessible to personnel, be located not more than 6 feet 7 inches above finished grade, and be permanently marked to identify it as the site-isolating device. If the supply system includes a grounded conductor, it must be connected to a grounding electrode system at the site-isolating device.

The site-isolating device is not required to provide overcurrent protection, nor is overcurrent protection for the load-side conductors required to be located immediately adjacent to this device. Based on the requirements in 547.9(B) and (C), the location of the service disconnecting means and overcurrent protection is on the load side of the site-isolating device, either at the distribution point or at the building or structure supplied. Where the site is supplied by more than one distribution point, a plaque or directory is required at each site-isolating device that provides information about the location of each of the distribution points and the buildings served by each distribution point.

In Exhibit 547.1, the pole-mounted site-isolating device is located at the distribution point. A set of overhead service conductors is run to each of the three structures, and a service disconnecting means and overcurrent protection are installed at each building. The supply conductors are considered service conductors because there is no overcurrent protection at the site-isolating device. A grounding electrode system is required at the distribution point, and a grounding electrode conductor connection to the supply system grounded conductor must be made at the site-isolating device. A grounding electrode system is also required at each of the buildings.

Any portion of an EGC installed underground must be insulated or covered copper in accordance with 547.5(F). This is to protect the EGC from the corrosive influences inherent to agricultural premises and to reduce leakage current in those areas where livestock are kept, because prevention of stray voltage at agricultural premises is important.

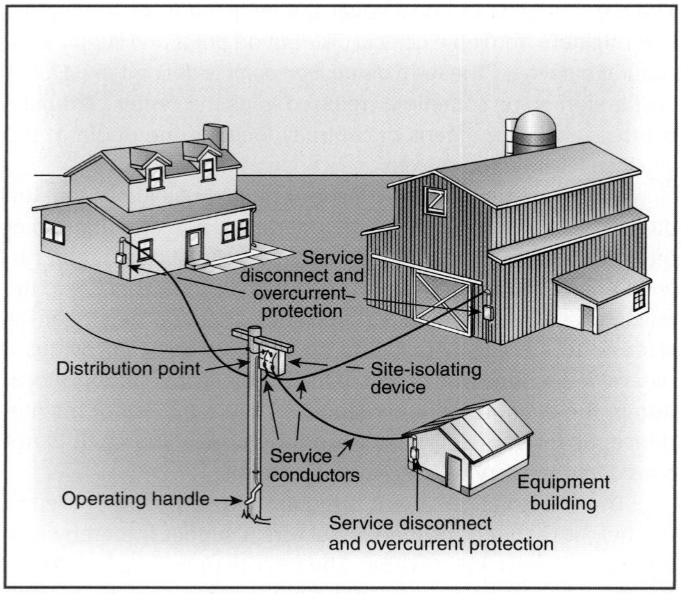


***EXHIBIT 547.1*** *Site-isolating device located at the distribution point with service conductors run to each building. A service disconnecting means and overcurrent protection is installed at each building.*

### 547.10 Equipotential Planes and Bonding of Equipotential Planes

The installation and bonding of equipotential planes shall comply with 547.10(A) and (B). For the purposes of this section, the term *livestock* shall not include poultry.

**(A) Where Required.** Equipotential planes shall be installed where required in (A)(1) and (A)(2).

**(1) Indoors.** Equipotential planes shall be installed in confinement areas with concrete floors where metallic equipment is located that may become energized and is accessible to livestock.

**(2) Outdoors.** Equipotential planes shall be installed in concrete slabs where metallic equipment is located that may become energized and is accessible to livestock.

The equipotential plane shall encompass the area where the livestock stands while accessing metallic equipment that may become energized.

**(B) Bonding.** Equipotential planes shall be connected to the electrical grounding system. The bonding conductor shall be solid copper, insulated, covered or bare, and not smaller than 8 AWG. The means of bonding to wire mesh or conductive elements shall be by pressure connectors or clamps of brass, copper, copper alloy, or an equally substantial approved means. Slatted floors that are supported by structures that are a part of an equipotential plane shall not require bonding.

Grounding and bonding requirements unique to agricultural settings are necessary due to the sensitivity of livestock to slight differences in potential between surfaces with which they are in direct contact. The wet or damp concrete common to animal confinement areas enhances this sensitivity.

Informational Note No. 1: Methods to establish equipotential planes are described in American Society of Agricultural and Biological Engineers (ASABE) EP473.2-2001, *Equipotential Planes in Animal Containment Areas*.

Informational Note No. 2: Methods for safe installation of livestock waterers are described in American Society of Agricultural and Biological Engineers (ASABE) EP342.3-2010, *Safety for Electrically Heated Livestock Waterers*.

Electrically heated livestock watering troughs could pose an electric shock hazard for livestock and personnel. The referenced document provides information on the proper installation of this equipment.

Informational Note No. 3: Low grounding electrode system resistances may reduce potential differences in livestock facilities.

# ARTICLE 550 Mobile Homes, Manufactured Homes, and Mobile Home Parks

## I. General

### 550.1 Scope

The provisions of this article cover the electrical conductors and equipment installed within or on mobile and manufactured homes, the conductors that connect mobile and manufactured homes to a supply of electricity, and the installation of electrical wiring, luminaires, equipment, and appurtenances related to electrical installations within a mobile home park up to the mobile home service-entrance conductors or, if none, the mobile home service equipment.

Informational Note: For additional information on manufactured housing see NFPA 501-2013, *Standard on Manufactured Housing*, and Part 3280, *Manufactured Home Construction and Safety Standards*, of the Federal Department of Housing and Urban Development.

The *Federal Mobile Home Construction and Safety Standard*, issued by the Federal Housing and Urban Development Administration (HUD), incorporates many of the requirements of Article 550 of the *NEC*. The federal standard contains the requirements for electrical systems, conductors, and equipment installed within or on mobile homes and the conductors that connect mobile homes to a supply of electricity. Mobile homes are defined as manufactured homes in the HUD regulations. For the purposes of this *Code*, and unless otherwise indicated, the term *mobile home* includes manufactured homes.

The regulations pertaining to electrical systems are located in 24 CFR 3280.801–3280.816. They require that new manufactured homes comply with the federal standard. In some cases, HUD has delegated the enforcement of this standard to state and private inspection agencies and qualified testing laboratories. The service equipment and feeders installed at the mobile or manufactured home site are covered by the requirements in Part III of this article.

For information on the distinction between manufactured homes and manufactured buildings, see the commentary following 545.1.

## 550.2 Definitions

**Appliance, Fixed.** An appliance that is fastened or otherwise secured at a specific location.

**Appliance, Portable.** An appliance that is actually moved or can easily be moved from one place to another in normal use.

Informational Note: For the purpose of this article, the following major appliances, other than built-in, are considered portable if cord connected: refrigerators, range equipment, clothes washers, dishwashers without booster heaters, or other similar appliances.

•

**Feeder Assembly.** The overhead or under-chassis feeder conductors, including the grounding conductor, together with the necessary fittings and equipment or a power-supply cord listed for mobile home use, identified for the delivery of energy from the source of electrical supply to the panelboard within the mobile home.

**Laundry Area.** An area containing or designed to contain a laundry tray, clothes washer, or a clothes dryer.

**Manufactured Home.** A structure, transportable in one or more sections, that, in the traveling mode, is 2.4 m (8 body-ft) or more in width or 12.2 m (40 body-ft) or more in length, or, when erected on site, is 29.7 $m^2$ (320 $ft^2$) or more and that is built on a permanent chassis and designed to be used as a dwelling, with or without a permanent foundation, when connected therein. The term *manufactured home* includes any structure that meets all the provisions of this paragraph except the size requirements and with respect to which the manufacturer voluntarily files a certification required by the regulatory agency, and except that such term does not include any self-propelled recreational vehicle. Calculations used to determine the number of square meters (square feet) in a structure are based on the structure's exterior dimensions, measured at the largest horizontal projections when erected on site. These dimensions include all expandable rooms, cabinets, and other projections containing interior space but do not include bay windows.

For the purpose of this *Code* and unless otherwise indicated, the term *mobile home* includes manufactured homes.

Informational Note No. 1: See the applicable building code for definition of the term *permanent foundation*.

Informational Note No. 2: See Part 3280, *Manufactured Home Construction and Safety Standards, of the Federal Department of Housing and Urban Development*, for additional information on the definition.

**Mobile Home.** A factory-assembled structure or structures transportable in one or more sections that are built on a permanent chassis and designed to be used as a dwelling without a permanent foundation where connected to the required utilities and that include the plumbing, heating, air-conditioning, and electrical systems contained therein.

For the purpose of this *Code* and unless otherwise indicated, the term *mobile home* includes manufactured homes.

**Mobile Home Accessory Building or Structure.** Any awning, cabana, ramada, storage cabinet, carport, fence, windbreak, or porch established for the use of the occupant of the mobile home on a mobile home lot.

*Mobile home* is the original term covering a structure that is built on a chassis, designed to be transportable and intended for installation on a site with or without a permanent foundation. Manufactured homes (not to be confused with manufactured buildings, covered in Article 545) are also covered by Article 550 and, for the purposes of this article, are considered mobile homes. The requirements in Article 550 treat mobile and manufactured homes the same unless specifically stated otherwise. An example of a distinction between the two is found in 550.32(A) and (B), which cover the location of service equipment for each structure.

The requirements contained in Article 550, Part III, cover the installation of service equipment and feeders at mobile and manufactured home sites. Homes constructed in accordance with the requirements of Article 550, Parts I and II (or under the HUD 24 CFR 3280 regulations), are intended to be installed at their sites in accordance with the requirements of Part III.

**Mobile Home Lot.** A designated portion of a mobile home park designed for the accommodation of one mobile home and its accessory buildings or structures for the exclusive use of its occupants.

**Mobile Home Park.** A contiguous parcel of land that is used for the accommodation of occupied mobile homes.

**Mobile Home Service Equipment.** The equipment containing the disconnecting means, overcurrent protective devices, and receptacles or other means for connecting a mobile home feeder assembly.

**Park Electrical Wiring Systems.** All of the electrical wiring, luminaires, equipment, and appurtenances related to electrical installations within a mobile home park, including the mobile home service equipment.

## 550.4 General Requirements

**(A) Mobile Home Not Intended as a Dwelling Unit.** A mobile home not intended as a dwelling unit — for example, those equipped for sleeping purposes only, contractor's on-site offices, construction job dormitories, mobile studio dressing rooms, banks, clinics, mobile stores, or intended for the display or demonstration of merchandise or machinery — shall not be required to meet the provisions of this article pertaining to the number or capacity of circuits required. It shall, however, meet all other applicable requirements of this article if provided with an electrical installation intended to be energized from a 120-volt or 120/240-volt ac power supply system. Where different voltage is required by either design or available power supply system, adjustment shall be made in accordance with other articles and sections for the voltage used.

**(B) In Other Than Mobile Home Parks.** Mobile homes installed in other than mobile home parks shall comply with the provisions of this article.

**(C) Connection to Wiring System.** The provisions of this article shall apply to mobile homes intended for connection to a wiring system rated 120/240 volts, nominal, 3-wire ac, with a grounded neutral conductor.

**(D) Listed or Labeled.** All electrical materials, devices, appliances, fittings, and other equipment shall be listed or labeled by a qualified testing agency and shall be connected in an approved manner when installed.

## II. Mobile and Manufactured Homes

Manufactured and mobile homes are required to meet federal standards. See the commentary following 550.1. Manufacturers attach a label certifying that the construction is in accordance with those standards. This certification facilitates the transport of these homes nationwide by allowing AHJ approval without having to dismantle the home to verify the construction and electrical installation. Although this federal process supersedes local codes for the home construction, local codes are applicable for the on-site work. Refer to Exhibit 550.1 for an example of the HUD label.

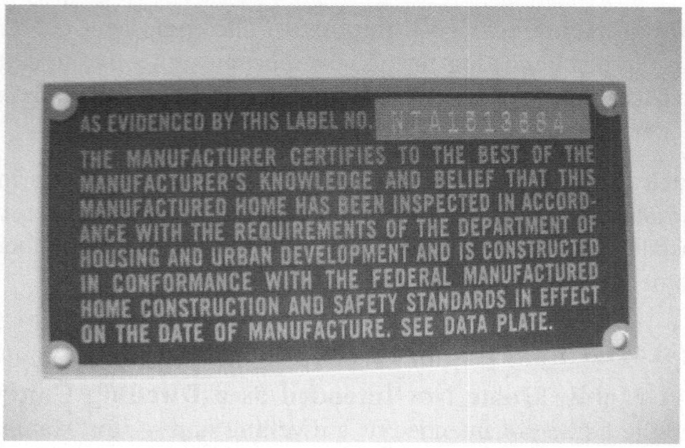


***EXHIBIT 550.1*** *A HUD label required for mobile homes.*

### 550.10 Power Supply

**(A) Feeder.** The power supply to the mobile home shall be a feeder assembly consisting of not more than one listed 50-ampere mobile home power-supply cord or a permanently installed feeder.

*Exception No. 1: A mobile home that is factory equipped with gas or oil-fired central heating equipment and cooking appliances shall be permitted to be provided with a listed mobile home power-supply cord rated 40 amperes.*

*Exception No. 2: A feeder assembly shall not be required for manufactured homes constructed in accordance with 550.32(B).*

Exception No. 2 modifies the requirement only for manufactured homes. The installation of service equipment is permitted in or on a manufactured home per 550.32(B).

**(B) Power-Supply Cord.** If the mobile home has a power-supply cord, it shall be permanently attached to the panelboard, or to a junction box permanently connected to the panelboard, with the free end terminating in an attachment plug cap.

Cords with adapters and pigtail ends, extension cords, and similar items shall not be attached to, or shipped with, a mobile home.

A suitable clamp or the equivalent shall be provided at the panelboard knockout to afford strain relief for the cord to prevent strain from being transmitted to the terminals when the power-supply cord is handled in its intended manner.

The cord shall be a listed type with four conductors, one of which shall be identified by a continuous green color or a continuous green color with one or more yellow stripes for use as the grounding conductor.

**(C) Attachment Plug Cap.** The attachment plug cap shall be a 3-pole, 4-wire, grounding type, rated 50 amperes, 125/250 volts with a configuration as shown in Figure 550.10(C) and intended for use with the 50-ampere, 125/250-volt receptacle configuration shown in Figure 550.10(C). It shall be listed, by itself or as part of a power-supply cord assembly, for the purpose and shall be molded to or installed on the flexible cord so that it is secured tightly to the cord at the point where the cord enters the attachment plug cap. If a right-angle cap is used, the configuration shall be oriented so that the grounding member is farthest from the cord.

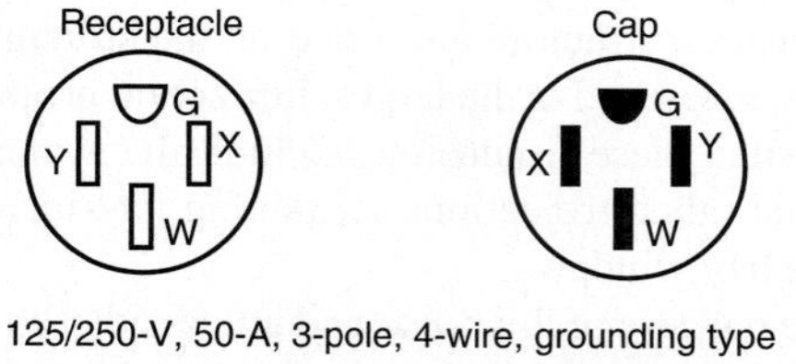


***FIGURE 550.10(C)*** *50-Ampere, 125/250-Volt Receptacle and Attachment Plug Cap Configurations, 3-Pole, 4-Wire, Grounding-Types, Used for Mobile Home Supply Cords and Mobile Home Parks.*

Informational Note: Complete details of the 50-ampere plug and receptacle configuration can be found in ANSI/NEMA WD 6-2002 (R2008), *Standard for Dimensions of Attachment Plugs and Receptacles*, Figure 14-50.

**(D) Overall Length of a Power-Supply Cord.** The overall length of a power-supply cord, measured from the end of the cord, including bared leads, to the face of the attachment plug cap shall not be less than 6.4 m (21 ft) and shall not exceed 11 m (36½ ft). The length of the cord from the face of the attachment plug cap to the point where the cord enters the mobile home shall not be less than 6.0 m (20 ft).

**(E) Marking.** The power-supply cord shall bear the following marking:

FOR USE WITH MOBILE HOMES — 40 AMPERES

or

FOR USE WITH MOBILE HOMES — 50 AMPERES

**(F) Point of Entrance.** The point of entrance of the feeder assembly to the mobile home shall be in the exterior wall, floor, or roof.

**(G) Protected.** Where the cord passes through walls or floors, it shall be protected by means of conduits and bushings or equivalent. The cord shall be permitted to be installed within the mobile home walls, provided a continuous raceway having a maximum size of 32 mm (1¼ in.) is installed from the branch-circuit panelboard to the underside of the mobile home floor.

**(H) Protection Against Corrosion and Mechanical Damage.** Permanent provisions shall be made for the protection of the attachment plug cap of the power-supply cord and any connector cord assembly or receptacle against corrosion and mechanical damage if such devices are in an exterior location while the mobile home is in transit.

**(I) Mast Weatherhead or Raceway.** Where the calculated load exceeds 50 amperes or where a permanent feeder is used, the supply shall be by means of either of the following:

(1) One mast weatherhead installation, installed in accordance with Article 230, containing four continuous, insulated, color-coded feeder conductors, one of which shall be an equipment grounding conductor
(2) A metal raceway or rigid nonmetallic conduit from the disconnecting means in the mobile home to the underside of the mobile home, with provisions for the attachment to a suitable junction box or fitting to the raceway on the underside of the mobile home [with or without conductors as in 550.10(I)(1)]. The manufacturer shall provide written installation instructions stating the proper feeder conductor sizes for the raceway and the size of the junction box to be used.

Cord-and-plug connection is permitted for loads that do not exceed 50 amperes, but many larger mobile and manufactured homes often contain much electrical equipment and often exceed this rating. A permanently connected feeder is required if the calculated load exceeds 50 amperes.

A raceway is required from the distribution panelboard in the mobile home to the underside of the mobile home. Typically, the feeder conductors in this raceway are installed when the mobile home is located at its site. The raceway provides a means to install the feeder conductors to the mobile home panelboard without damaging the interior finish. The feeder assembly must comprise four continuous, insulated, color-coded conductors, as indicated in 550.10(I)(1) and 550.33(A).

## 550.11 Disconnecting Means and Branch-Circuit Protective Equipment

The branch-circuit equipment shall be permitted to be combined with the disconnecting means as a single assembly. Such a combination shall be permitted to be designated as a panelboard. If a fused panelboard is used, the maximum fuse size for the mains shall be plainly marked with lettering at least 6 mm (¼ in.) high and visible when fuses are changed.

Where plug fuses and fuseholders are used, they shall be tamper-resistant Type S, enclosed in dead-front fuse panelboards. Electrical panelboards containing circuit breakers shall also be dead-front type.

Informational Note: See 110.22 concerning identification of each disconnecting means and each service, feeder, or branch circuit at the point where it originated and the type marking needed.

**(A) Disconnecting Means.** A single disconnecting means shall be provided in each mobile home consisting of a circuit breaker, or a switch and fuses and its accessories installed in a readily accessible location near the point of entrance of the supply cord or conductors into the mobile home. The main circuit breakers or fuses shall be plainly marked "Main." This equipment shall contain a solderless type of grounding connector or bar for the purposes of grounding, with sufficient terminals for all grounding conductors. The terminations of the grounded circuit conductors shall be insulated in accordance with 550.16(A). The disconnecting equipment shall have a rating not less than the calculated load. The distribution equipment, either circuit breaker or fused type, shall be located a minimum of 600 mm (24 in.) from the bottom of such equipment to the floor level of the mobile home.

Informational Note: See 550.20(B) for information on disconnecting means for branch circuits designed to energize heating or air-conditioning equipment, or both, located outside the mobile home, other than room air conditioners.

A panelboard shall be rated not less than 50 amperes and employ a 2-pole circuit breaker rated 40 amperes for a 40-ampere supply cord, or 50 amperes for a 50-ampere supply cord. A panelboard employing a disconnect switch and fuses shall be rated 60 amperes and shall employ a single 2-pole, 60-ampere fuseholder with 40- or 50-ampere main fuses for 40- or 50-ampere supply cords, respectively. The outside of the panelboard shall be plainly marked with the fuse size.

The panelboard shall be located in an accessible location but shall not be located in a bathroom or a clothes closet. A clear working space at least 750 mm (30 in.) wide and 750 mm (30 in.) in front of the panelboard shall be provided. This space shall extend from the floor to the top of the panelboard.

**(B) Branch-Circuit Protective Equipment.** Branch-circuit distribution equipment shall be installed in each mobile home and shall include overcurrent protection for each branch circuit consisting of either circuit breakers or fuses.

The branch-circuit overcurrent devices shall be rated as follows:

(1) Not more than the circuit conductors; and
(2) Not more than 150 percent of the rating of a single appliance rated 13.3 amperes or more that is supplied by an individual branch circuit; but
(3) Not more than the overcurrent protection size and of the type marked on the air conditioner or other motor-operated appliance.

**(C) Two-Pole Circuit Breakers.** Where circuit breakers are provided for branch-circuit protection, 240-volt circuits shall be protected by a 2-pole common or companion trip, or by circuit breakers with identified handle ties.

**(D) Electrical Nameplates.** A metal nameplate on the outside adjacent to the feeder assembly entrance shall read as follows:

THIS CONNECTION FOR 120/240-VOLT,
3-POLE, 4-WIRE, 60-HERTZ,
_____ AMPERE SUPPLY

The correct ampere rating shall be marked in the blank space.

*Exception: For manufactured homes, the manufacturer shall provide in its written installation instructions or in the data plate the minimum ampere rating of the feeder assembly or, where provided, the service-entrance conductors intended for connection to the manufactured home. The rating provided shall not be less than the minimum load calculated in accordance with 550.18.*

## 550.12 Branch Circuits

The number of branch circuits required shall be determined in accordance with 550.12(A) through (E).

**(A) Lighting.** The number of branch circuits shall be based on 33 volt-amperes/m$^2$ (3 VA/ft$^2$) times outside dimensions of the mobile home (coupler excluded) divided by 120 volts to determine the number of 15- or 20-ampere lighting area circuits, for example,

$$\frac{3 \times \text{length} \times \text{width}}{120 \times 15 \text{ (or 20)}} = \text{No. of 15- (or 20-) ampere circuits}$$

**(B) Small Appliances.** In kitchens, pantries, dining rooms, and breakfast rooms, two or more 20-ampere small-appliance circuits, in addition to the number of circuits required elsewhere in this section, shall be provided for all receptacle outlets required by 550.13(D) in these rooms. Such circuits shall have no other outlets.

*Exception No. 1: Receptacle outlets installed solely for the electrical supply and support of an electric clock in any the rooms specified in 550.12(B) shall be permitted.*

*Exception No. 2: Receptacle outlets installed to provide power for supplemental equipment and lighting on gas-fired ranges, ovens, or counter-mounted cooking units shall be permitted.*

*Exception No. 3: A single receptacle for refrigeration equipment shall be permitted to be supplied from an individual branch circuit rated 15 amperes or greater.*

Countertop receptacle outlets installed in the kitchen shall be supplied by not less than two small-appliance circuit branch circuits, either or both of which shall be permitted to supply receptacle outlets in the kitchen and other locations specified in 550.12(B).

**(C) Laundry Area.** Where a laundry area is provided, a 20-ampere branch circuit shall be provided to supply the laundry receptacle outlet(s). This circuit shall have no other outlets.

**(D) General Appliances.** (Including furnace, water heater, range, and central or room air conditioner, etc.). There shall be one or more circuits of adequate rating in accordance with the following:

Informational Note: For central air conditioning, see Article 440.

(1) The ampere rating of fixed appliances shall be not over 50 percent of the circuit rating if lighting outlets (receptacles, other than kitchen, dining area, and laundry, considered as lighting outlets) are on the same circuit.
(2) For fixed appliances on a circuit without lighting outlets, the sum of rated amperes shall not exceed the branch-circuit rating. Motor loads or continuous loads shall not exceed 80 percent of the branch-circuit rating.
(3) The rating of a single cord-and-plug-connected appliance on a circuit having no other outlets shall not exceed 80 percent of the circuit rating.
(4) The rating of a range branch circuit shall be based on the range demand as specified for ranges in 550.18(B)(5).

**(E) Bathrooms.** Bathroom receptacle outlets shall be supplied by at least one 20-ampere branch circuit. Such circuits shall have no outlets other than as provided for in 550.13(E)(2).

## 550.13 Receptacle Outlets

**(A) Grounding-Type Receptacle Outlets.** All receptacle outlets shall comply with the following:

(1) Be of grounding type

(2) Be installed according to 406.4
(3) Except where supplying specific appliances, be 15- or 20-ampere, 125-volt, either single or multiple type, and accept parallel-blade attachment plugs

**(B) Ground-Fault Circuit Interrupters (GFCI).** All 125-volt, single-phase, 15- and 20-ampere receptacle outlets installed outdoors, in compartments accessible from outside the unit, or in bathrooms, including receptacles in luminaires, shall have GFCI protection. GFCI protection shall be provided for receptacle outlets serving countertops in kitchens and receptacle outlets located within 1.8 m (6 ft) of a wet bar sink. The exceptions in 210.8(A) shall be permitted.

Feeders supplying branch circuits shall be permitted to be protected by a ground-fault circuit-interrupter in lieu of the provision for such interrupters specified herein.

The locations where GFCI protection is required for receptacles installed in a mobile or manufactured home parallel those specified in 210.8(A) for site-built dwelling units. The exception to 210.8(A)(5) is limited to a basement receptacle installed to supply a fire or burglar control panel. Because mobile and manufactured homes typically do not have basements, only the 210.8(A)(3) exception covering an outdoor receptacle for deicing or snow-melting equipment has any relevance within the scope of Article 550.

**(C) Cord-Connected Fixed Appliance.** A grounding-type receptacle outlet shall be provided for each cord-connected fixed appliance installed.

**(D) Receptacle Outlets Required.** Except in the bath, closet, and hallway areas, receptacle outlets shall be installed at wall spaces 600 mm (2 ft) wide or more so that no point along the floor line is more than 1.8 m (6 ft) measured horizontally from an outlet in that space. In addition, a receptacle outlet shall be installed in the following locations:

(1) Over or adjacent to countertops in the kitchen [at least one on each side of the sink if countertops are on each side and are 300 mm (12 in.) or over in width].
(2) Adjacent to the refrigerator and freestanding gas-range space. A multiple-type receptacle shall be permitted to serve as the outlet for a countertop and a refrigerator.
(3) At countertop spaces for built-in vanities.
(4) At countertop spaces under wall-mounted cabinets.
(5) In the wall at the nearest point to where a bar-type counter attaches to the wall.
(6) In the wall at the nearest point to where a fixed room divider attaches to the wall.
(7) In laundry areas within 1.8 m (6 ft) of the intended location of the laundry appliance(s).
(8) At least one receptacle outlet located outdoors and accessible at grade level and not more than 2.0 m (6½ ft) above grade. A receptacle outlet located in a compartment accessible from the outside of the unit shall be considered an outdoor receptacle.
(9) At least one receptacle outlet shall be installed in bathrooms within 900 mm (36 in.) of the outside edge of each basin. The receptacle outlet shall be located above or adjacent to the basin location. This receptacle shall be in addition to any receptacle that is a part of a luminaire or appliance. The receptacle shall not be enclosed within a bathroom cabinet or vanity.

**(E) Pipe Heating Cable(s) Outlet.** For the connection of pipe heating cable(s), a receptacle outlet shall be located on the underside of the unit as follows:

(1) Within 600 mm (2 ft) of the cold water inlet.
(2) Connected to an interior branch circuit, other than a small-appliance branch circuit. It shall be permitted to use a bathroom receptacle circuit for this purpose.
(3) On a circuit where all of the outlets are on the load side of the ground-fault circuit-interrupter.
(4) This outlet shall not be considered as the receptacle required by 550.13(D)(8).

A receptacle outlet (sometimes referred to as a heat tape outlet) is required on the underside of mobile homes to supply cord-and-plug-connected pipe-heating cables. The receptacle must be GFCI protected and connected to a branch circuit that serves the interior of the mobile home. Requiring all outlets on this branch circuit to be on the load (downstream) side of the GFCI is to allow supervision of the power supply and GFCI for this outlet from inside the mobile home. If the OCPD or GFCI device opens, the occupants of the home are more likely to notice it than if the heating outlet were supplied by a dedicated circuit.

**(F) Receptacle Outlets Not Permitted.** Receptacle outlets shall not be permitted in the following locations:

(1) Receptacle outlets shall not be installed within or directly over a bathtub or shower space.
(2) A receptacle shall not be installed in a face-up position in any countertop.
(3) Receptacle outlets shall not be installed above electric baseboard heaters, unless provided for in the listing or manufacturer's instructions.

**(G) Receptacle Outlets Not Required.** Receptacle outlets shall not be required in the following locations:

(1) In the wall space occupied by built-in kitchen or wardrobe cabinets
(2) In the wall space behind doors that can be opened fully against a wall surface
(3) In room dividers of the lattice type that are less than 2.5 m (8 ft) long, not solid, and within 150 mm (6 in.) of the floor
(4) In the wall space afforded by bar-type counters

## 550.14 Luminaires and Appliances

**(A) Fasten Appliances in Transit.** Means shall be provided to securely fasten appliances when the mobile home is in transit. (See 550.16 for provisions on grounding.)

**(B) Accessibility.** Every appliance shall be accessible for inspection, service, repair, or replacement without removal of permanent construction.

**(C) Pendants.** Listed pendant-type luminaires or pendant cords shall be permitted.

**(D) Bathtub and Shower Luminaires.** Where a luminaire is installed over a bathtub or in a shower stall, it shall be of the enclosed and gasketed type listed for wet locations.

## 550.15 Wiring Methods and Materials

Except as specifically limited in this section, the wiring methods and materials included in this *Code* shall be used in mobile homes. Aluminum conductors, aluminum alloy conductors, and aluminum core conductors such as copper-clad aluminum shall not be acceptable for use as branch-circuit wiring.

**(A) Nonmetallic Boxes.** Nonmetallic boxes shall be permitted only with nonmetallic cable or nonmetallic raceways.

**(B) Nonmetallic Cable Protection.** Nonmetallic cable located 380 mm (15 in.) or less above the floor, if exposed, shall be protected from physical damage by covering boards, guard strips, or raceways. Cable likely to be damaged by stowage shall be so protected in all cases.

**(C) Metal-Covered and Nonmetallic Cable Protection.** Metal-covered and nonmetallic cables shall be permitted to pass through the centers of the wide side of 2 by 4 studs. However, they shall be protected where they pass through 2 by 2 studs or at other studs or frames where the cable or armor would be less than 32 mm (1¼ in.) from the inside or outside surface of the studs where the wall covering materials are in contact with the studs. Steel plates on each side of the cable, or a tube, with not less than 1.35 mm (0.053 in.) wall thickness shall be required to protect the cable. These plates or tubes shall be securely held in place.

**(D) Metal Faceplates.** Where metal faceplates are used, they shall be grounded.

**(E) Installation Requirements.** Where a range, clothes dryer, or other appliance is connected by metal-covered cable or flexible metal conduit, a length of not less than 900 mm (3 ft) of unsupported cable or conduit shall be provided to service the appliance. The cable or flexible metal conduit shall be secured to the wall. Type NM or Type SE cable shall not be used to connect a range or dryer. This shall not prohibit the use of Type NM or Type SE cable between the branch-circuit overcurrent protective device and a junction box or range or dryer receptacle.

**(F) Raceways.** Where rigid metal conduit or intermediate metal conduit is terminated at an enclosure with a locknut and bushing connection, two locknuts shall be provided, one inside and one outside of the enclosure. Rigid nonmetallic conduit, electrical nonmetallic tubing, or surface raceway shall be permitted. All cut ends of conduit and tubing shall be reamed or otherwise finished to remove rough edges.

**(G) Switches.** Switches shall be rated as follows:

(1) For lighting circuits, switches shall be rated not less than 10 amperes, 120 to 125 volts, and in no case less than the connected load.
(2) Switches for motor or other loads shall comply with the provisions of 404.14.

**(H) Under-Chassis Wiring (Exposed to Weather).** Where outdoor or under-chassis line-voltage (120 volts, nominal, or higher) wiring is exposed to moisture or physical damage, it shall be protected by a conduit or raceway approved for use in wet locations or where subject to physical damage. The conductors shall be listed for use in wet locations.

The under-chassis wiring method is not restricted to rigid or intermediate conduit as long as the raceway is suitable for installation in wet locations and, if necessary, for locations where it is subject to physical damage. This allows for the standard practice in manufactured home construction of installing PVC conduit or RTRC under the chassis.

•

**(I) Boxes, Fittings, and Cabinets.** Boxes, fittings, and cabinets shall be securely fastened in place and shall be supported from a structural member of the home, either directly or by using a substantial brace.

*Exception: Snap-in-type boxes. Boxes provided with special wall or ceiling brackets and wiring devices with integral enclosures that securely fasten to walls or ceilings and are identified for the use shall be permitted without support from a structural member or brace. The testing and approval shall include the wall and ceiling construction systems for which the boxes and devices are intended to be used.*

**(J) Appliance Terminal Connections.** Appliances having branch-circuit terminal connections that operate at temperatures higher than 60°C (140°F) shall have circuit conductors as described in the following:

(1) Branch-circuit conductors having an insulation suitable for the temperature encountered shall be permitted to be run directly to the appliance.
(2) Conductors having an insulation suitable for the temperature encountered shall be run from the appliance terminal connection to a readily accessible outlet box placed at least 300 mm (1 ft) from the appliance. These conductors shall be in a suitable raceway or Type AC or MC cable of at least 450 mm (18 in.) but not more than 1.8 m (6 ft) in length.

**(K) Component Interconnections.** Fittings and connectors that are intended to be concealed at the time of assembly shall be listed and identified for the interconnection of building components. Such fittings and connectors shall be equal to the wiring method employed in insulation, temperature rise, and fault-current withstanding and shall be capable of enduring the vibration and shock occurring in mobile home transportation.

Informational Note: See 550.19 for interconnection of multiple section units.

## 550.16 Grounding

Grounding of both electrical and nonelectrical metal parts in a mobile home shall be through connection to a grounding bus in the mobile home panelboard and shall be connected through the green-colored insulated conductor in the supply cord or the feeder wiring to the grounding bus in the service-entrance equipment located adjacent to the mobile home location. Neither the frame of the mobile home nor the frame of any appliance shall be connected to the grounded circuit conductor in the mobile home. Where the panelboard is the service equipment as permitted by 550.32(B), the neutral conductors and the equipment grounding bus shall be connected.

**(A) Grounded Conductor.**

**(1) Insulated.** The grounded circuit conductor shall be insulated from the grounding conductors and from equipment enclosures and other grounded parts. The grounded circuit conductor terminals in the panelboard and in ranges, clothes dryers, counter-mounted cooking units, and wall-mounted ovens shall be insulated from the equipment enclosure. Bonding screws, straps, or buses in the panelboard or in appliances shall be removed and discarded. Where the panelboard is the service equipment as permitted by 550.32(B), the neutral conductors and the equipment grounding bus shall be connected.

The feeder assembly must consist of a listed cord or four color-coded insulated conductors, one of which is the grounded conductor (white) and one of which is used for grounding purposes (green). Thus, the grounded and grounding conductors are kept independent of each other and are connected only at the service equipment (at the point of connection of the grounding electrode conductor). Grounding of metal parts, including the frame of the mobile home or the frame of any appliance, is accomplished by connection to the equipment grounding bus and never to the grounded conductor (neutral bus). This prevents incidental contact between the grounded conductor and non–current-carrying metal parts of electrical equipment. Without the separation of the grounded and grounding conductors, this contact could result in the metal structure or metal sheathing of the mobile home becoming a parallel path for neutral current.

Many ranges and clothes dryers have a factory-installed bonding jumper. Removing this jumper does not compromise or void the listing of the product, because isolation of the metal appliance frame from the grounded circuit conductor is required by the *Code*.

**(2) Connections of Ranges and Clothes Dryers.** Connections of ranges and clothes dryers with 120/240-volt, 3-wire ratings shall be made with 4-conductor cord and 3-pole, 4-wire, grounding-type plugs or by Type AC cable, Type MC cable, or conductors enclosed in flexible metal conduit.

**(B) Equipment Grounding Means.**

**(1) Supply Cord or Permanent Feeder.** The green-colored insulated grounding wire in the supply cord or permanent feeder wiring shall be connected to the grounding bus in the panelboard or disconnecting means.

**(2) Electrical System.** In the electrical system, all exposed metal parts, enclosures, frames, luminaire canopies, and so forth, shall be effectively bonded to the grounding terminal or enclosure of the panelboard.

**(3) Cord-Connected Appliances.** Cord-connected appliances, such as washing machines, clothes dryers, and refrigerators, and the electrical system of gas ranges and so forth, shall be grounded by means of a cord with an equipment grounding conductor and grounding-type attachment plug.

**(C) Bonding of Non–Current-Carrying Metal Parts.**

**(1) Exposed Non–Current-Carrying Metal Parts.** All exposed non–current-carrying metal parts that are likely to become energized shall be effectively bonded to the grounding terminal or enclosure of the panelboard. A bonding conductor shall be connected between the panelboard and an accessible terminal on the chassis.

**(2) Grounding Terminals.** Grounding terminals shall be of the solderless type and listed as pressure-terminal connectors recognized for the wire size used. The bonding conductor shall be solid or stranded, insulated or bare, and shall be 8 AWG copper minimum, or equivalent. The bonding conductor shall be routed so as not to be exposed to physical damage.

**(3) Metallic Piping and Ducts.** Metallic gas, water, and waste pipes and metallic air-circulating ducts shall be considered bonded if they are connected to the terminal on the chassis [see 550.16(C)(1)] by clamps, solderless connectors, or by suitable grounding-type straps.

**(4) Metallic Roof and Exterior Coverings.** Any metallic roof and exterior covering shall be considered bonded if the following conditions are met:

(1) The metal panels overlap one another and are securely attached to the wood or metal frame parts by metallic fasteners.
(2) The lower panel of the metallic exterior covering is secured by metallic fasteners at a cross member of the chassis by two metal straps per mobile home unit or section at opposite ends.

The bonding strap material shall be a minimum of 100 mm (4 in.) in width of material equivalent to the skin or a material of equal or better electrical conductivity. The straps shall be fastened with paint-penetrating fittings such as screws and starwashers or equivalent.

### 550.17 Testing

**(A) Dielectric Strength Test.** The wiring of each mobile home shall be subjected to a 1-minute, 900-volt, dielectric strength test (with all switches closed) between live parts (including neutral conductor) and the mobile home ground. Alternatively, the test shall be permitted to be performed at 1080 volts for 1 second. This test shall be performed after branch circuits are complete and after luminaires or appliances are installed.

*Exception: Listed luminaires or appliances shall not be required to withstand the dielectric strength test.*

**(B) Continuity and Operational Tests and Polarity Checks.** Each mobile home shall be subjected to all of the following:

(1) An electrical continuity test to ensure that all exposed electrically conductive parts are properly bonded
(2) An electrical operational test to demonstrate that all equipment, except water heaters and electric furnaces, is connected and in working order
(3) Electrical polarity checks of permanently wired equipment and receptacle outlets to determine that connections have been properly made

### 550.18 Calculations

The following method shall be employed in calculating the supply-cord and distribution-panelboard load for each feeder assembly for each mobile home in lieu of the procedure shown in Article 220 and shall be based on a 3-wire, 120/240-volt supply with 120-volt loads balanced between the two ungrounded conductors of the 3-wire system.

**(A) Lighting, Small-Appliance, and Laundry Load.**

**(1) Lighting Volt-Amperes.** Length times width of mobile home floor (outside dimensions) times 33 volt-amperes/m$^2$ (3 VA/ft$^2$), for example, length $\times$ width $\times$ 3 = lighting volt-amperes.

**(2) Small-Appliance Volt-Amperes.** Number of circuits times 1500 volt-amperes for each 20-ampere appliance receptacle circuit (see definition of *Appliance, Portable*, with a fine print note in 550.2), for example, number of circuits $\times$ 1500 = small-appliance volt-amperes.

**(3) Laundry Area Circuit Volt-Amperes.** 1500 volt-amperes.

**(4) Total Volt-Amperes.** Lighting volt-amperes plus small-appliance volt-amperes plus laundry area volt-amperes equals total volt-amperes.

**(5) Net Volt-Amperes.** First 3000 total volt-amperes at 100 percent plus remainder at 35 percent equals volt-amperes to be divided by 240 volts to obtain current (amperes) per leg.

**(B) Total Load for Determining Power Supply.** Total load for determining power supply is the sum of the following:

(1) Lighting and small-appliance load as calculated in 550.18(A)(5).
(2) Nameplate amperes for motors and heater loads (exhaust fans, air conditioners, electric, gas, or oil heating). Omit smaller of the heating and cooling loads, except include blower motor if used as air-conditioner evaporator motor. Where an air conditioner is not installed and a 40-ampere power-supply cord is provided, allow 15 amperes per leg for air conditioning.
(3) Twenty-five percent of current of largest motor in (2).
(4) Total of nameplate amperes for waste disposer, dishwasher, water heater, clothes dryer, wall-mounted oven, cooking units. Where the number of these appliances exceeds three, use 75 percent of total.
(5) Derive amperes for freestanding range (as distinguished from separate ovens and cooking units) by dividing the following values by 240 volts:

| Nameplate Rating (watts) | Use (volt-amperes) |
|---|---|
| 0–10,000 | 80 percent of rating |
| Over 10,000–12,500 | 8,000 |
| Over 12,500–13,500 | 8,400 |
| Over 13,500–14,500 | 8,800 |
| Over 14,500–15,500 | 9,200 |
| Over 15,500–16,500 | 9,600 |
| Over 16,500–17,500 | 10,000 |

(6) If outlets or circuits are provided for other than factory-installed appliances, include the anticipated load.

Informational Note: Refer to Informative Annex D, Example D11, for an illustration of the application of this calculation.

**(C) Optional Method of Calculation for Lighting and Appliance Load.** The optional method for calculating lighting and appliance load shown in 220.82 shall be permitted.

### 550.19 Interconnection of Multiple-Section Mobile or Manufactured Home Units

**(A) Wiring Methods.** Approved and listed fixed-type wiring methods shall be used to join portions of a circuit that must be electrically joined and are located in adjacent sections after the home is installed on its support foundation. The circuit's junction shall be accessible for disassembly when the home is prepared for relocation.

Informational Note: See 550.15(K) for component interconnections.

**(B) Disconnecting Means.** Expandable or multiunit manufactured homes, not having permanently installed feeders, that are to be moved from one location to another shall be permitted to have disconnecting means with branch-circuit protective equipment in each unit when so located that after assembly or joining together of units, the requirements of 550.10 will be met.

### 550.20 Outdoor Outlets, Luminaires, Air-Cooling Equipment, and So Forth

**(A) Listed for Outdoor Use.** Outdoor luminaires and equipment shall be listed for wet locations or outdoor use. Outdoor receptacles shall comply with 406.9. Where located on the underside of the home or located under roof extensions or similarly protected locations, outdoor luminaires and equipment shall be listed for use in damp locations.

See the commentary following 406.9 for information on receptacles in damp or wet outdoor locations.

**(B) Outside Heating Equipment, Air-Conditioning Equipment, or Both.** A mobile home provided with a branch circuit designed to energize outside heating equipment, air-conditioning equipment, or both, located outside the mobile home, other than room air conditioners, shall have such branch-circuit conductors terminate in a listed outlet box, or disconnecting means, located on the outside of the mobile home. A label shall be permanently affixed adjacent to the outlet box and shall contain the following information:

THIS CONNECTION IS FOR HEATING
AND/OR AIR-CONDITIONING EQUIPMENT.
THE BRANCH CIRCUIT IS RATED AT NOT
MORE THAN _____ AMPERES, AT ______ VOLTS,
60 HERTZ, _____ CONDUCTOR AMPACITY.
A DISCONNECTING MEANS SHALL BE LOCATED
WITHIN SIGHT OF THE EQUIPMENT.

The correct voltage and ampere rating shall be given. The tag shall be not less than 0.51 mm (0.020 in.) thick etched brass, stainless steel, anodized or alclad aluminum, or equivalent. The tag shall not be less than 75 mm by 45 mm (3 in. by 1¾ in.) minimum size.

### 550.25 Arc-Fault Circuit-Interrupter Protection

**(A) Definition.** Arc-fault circuit interrupters are defined in Article 100.

**(B) Mobile Homes and Manufactured Homes.** All 120-volt branch circuits that supply 15- and 20-ampere outlets installed in family rooms, dining rooms, living rooms, parlors, libraries, dens, bedrooms, sunrooms, recreation rooms, closets, hallways, or similar rooms or areas of mobile homes and manufactured homes shall comply with 210.12.

AFCI protection is required in accordance with 210.12. The branch circuits covered by this requirement include those that fall within the voltage and current ratings specified and those that supply lighting outlets, receptacle outlets, smoke alarm outlets, and other power outlets. This requirement does not supersede the current HUD 24 CFR 3280 requirements for factory-installed wiring in manufactured homes.

## III. Services and Feeders

### 550.30 Distribution System

The mobile home park secondary electrical distribution system to mobile home lots shall be single-phase, 120/240 volts, nominal. For the purpose of Part III, where the park service exceeds 240 volts, nominal, transformers and secondary panelboards shall be treated as services.

The distribution systems at mobile home parks must supply 120/240 volts to the mobile home lot. Because appliances and other equipment, nominally rated 120/240 volts, are usually installed during the manufacturing process of mobile homes, the home is intended to connect to a 120/240-volt, 3-wire ac, grounded neutral system. A 120/208-volt supply derived from a 4-wire, 120/208-volt wye system is unsuitable.

### 550.31 Allowable Demand Factors

Park electrical wiring systems shall be calculated (at 120/240 volts) on the larger of the following:

(1) 16,000 volt-amperes for each mobile home lot
(2) The load calculated in accordance with 550.18 for the largest typical mobile home that each lot will accept

It shall be permissible to calculate the feeder or service load in accordance with Table 550.31. No demand factor shall be allowed for any other load, except as provided in this *Code*.

Mobile home park electrical wiring systems must be calculated on the basis of the larger of (1) not less than 16,000 VA (at 120/240 volts)

***TABLE 550.31*** *Demand Factors for Services and Feeders*

| Number of Mobile Homes | Demand Factor (%) |
|---|---|
| 1 | 100 |
| 2 | 55 |
| 3 | 44 |
| 4 | 39 |
| 5 | 33 |
| 6 | 29 |
| 7–9 | 28 |
| 10–12 | 27 |
| 13–15 | 26 |
| 16–21 | 25 |
| 22–40 | 24 |
| 41–60 | 23 |
| 61 and over | 22 |

for each mobile home lot or (2) the calculated load of the largest typical mobile home the lot accommodates. However, the ampacity of the feeder-circuit conductors to each mobile home lot cannot be less than 100 amperes (at 120/240 volts), per 550.33(B).

## 550.32 Service Equipment

**(A) Mobile Home Service Equipment.** The mobile home service equipment shall be located adjacent to the mobile home and not mounted in or on the mobile home. The service equipment shall be located in sight from and not more than 9.0 m (30 ft) from the exterior wall of the mobile home it serves. The service equipment shall be permitted to be located elsewhere on the premises, provided that a disconnecting means suitable for use as service equipment is located within sight from and not more than 9.0 m (30 ft) from the exterior wall of the mobile home it serves and is rated not less than that required for service equipment in accordance with 550.32(C). Grounding at the disconnecting means shall be in accordance with 250.32.

Mobile home service equipment must be located in sight of the mobile home, but the equipment can be up to 30 feet from any point on the exterior wall of the mobile home. Service equipment may be located more than 30 feet from the mobile home if an additional disconnecting means is located within 30 feet of the mobile home, and grounding and bonding of this additional disconnecting means are performed in accordance with the requirements of 250.32. In a mobile home park, this arrangement facilitates locating service equipment at one or more centralized locations. Feeders are then installed from this service equipment to the mobile home site disconnecting means located within 30 feet of the mobile home.

**(B) Manufactured Home Service Equipment.** The manufactured home service equipment shall be permitted to be installed in or on a manufactured home, provided that all of the following conditions are met:

(1) The manufacturer shall include in its written installation instructions information indicating that the home shall be secured in place by an anchoring system or installed on and secured to a permanent foundation.
(2) The installation of the service shall comply with Part I through Part VII of Article 230.
(3) Means shall be provided for the connection of a grounding electrode conductor to the service equipment and routing it outside the structure.
(4) Bonding and grounding of the service shall be in accordance with Part I through Part V of Article 250.
(5) The manufacturer shall include in its written installation instructions one method of grounding the service equipment at the installation site. The instructions shall clearly state that other methods of grounding are found in Article 250.
(6) The minimum size grounding electrode conductor shall be specified in the instructions.
(7) A red warning label shall be mounted on or adjacent to the service equipment. The label shall state the following:

WARNING
DO NOT PROVIDE ELECTRICAL POWER
UNTIL THE GROUNDING ELECTRODE(S)
IS INSTALLED AND CONNECTED
(SEE INSTALLATION INSTRUCTIONS).

Where the service equipment is not installed in or on the unit, the installation shall comply with the other provisions of this section.

This section specifies the conditions required for installing the service equipment in or on a manufactured home. The concern over the unit being moved off site without the ability to disconnect the electrical supply is addressed in condition (1). A manufactured home with a service in or on the unit must be anchored in place or secured to a permanent foundation.

The other specified conditions cover the need to provide proper grounding and bonding conductors, systems, and connections and the need to install the service equipment in accordance with the applicable requirements in Article 230. These requirements apply only to manufactured homes as defined in 550.2.

**(C) Rating.** Mobile home service equipment shall be rated at not less than 100 amperes at 120/240 volts, and provisions shall be made for connecting a mobile home feeder assembly by a permanent wiring method. Power outlets used as mobile home service equipment shall also be permitted to contain receptacles rated up to 50 amperes with appropriate overcurrent protection. Fifty-ampere receptacles shall conform to the configuration shown in Figure 550.10(C).

Informational Note: Complete details of the 50-ampere plug and receptacle configuration can be found in ANSI/NEMA WD 6-2002 (Rev. 2008), *Standard for Wiring Devices — Dimensional Requirements*, Figure 14-50.

**(D) Additional Outside Electrical Equipment.** Means for connecting a mobile home accessory building or structure or additional electrical equipment located outside a mobile home by a fixed wiring method shall be provided in either the mobile home service equipment or the local external disconnecting means permitted in 550.32(A).

**(E) Additional Receptacles.** Additional receptacles shall be permitted for connection of electrical equipment located outside the mobile home, and all such 125-volt, single-phase, 15- and 20-ampere receptacles shall be protected by a listed ground-fault circuit interrupter.

**(F) Mounting Height.** Outdoor mobile home disconnecting means shall be installed so the bottom of the enclosure containing the disconnecting means is not less than 600 mm (2 ft) above finished grade or working platform. The disconnecting means shall be installed so that the center of the grip of the operating handle, when in the highest position, is not more than 2.0 m (6 ft 7 in.) above the finished grade or working platform.

**(G) Marking.** Where a 125/250-volt receptacle is used in mobile home service equipment, the service equipment shall be marked as follows:

TURN DISCONNECTING SWITCH OR CIRCUIT BREAKER OFF BEFORE INSERTING OR REMOVING PLUG. PLUG MUST BE FULLY INSERTED OR REMOVED.

The marking shall be located on the service equipment adjacent to the receptacle outlet.

### 550.33 Feeder

**(A) Feeder Conductors.** Feeder conductors shall comply with the following:

(1) Feeder conductors shall consist of either a listed cord, factory installed in accordance with 550.10(B), or a permanently installed feeder consisting of four insulated, color-coded conductors that shall be identified by the factory or field marking of the conductors in compliance with 310.110. Equipment grounding conductors shall not be identified by stripping the insulation.
(2) Feeder conductors shall be installed in compliance with 250.32(B).

*Exception: For an existing feeder that is installed between the service equipment and a disconnecting means as covered in 550.32(A), it shall be permitted to omit the equipment grounding conductor where the grounded circuit conductor is grounded at the disconnecting means in accordance with 250.32(B) Exception.*

**(B) Feeder Capacity.** Mobile home and manufactured home lot feeder circuit conductors shall have a capacity not less than the loads supplied, shall be rated at not less than 100 amperes, and shall be permitted to be sized in accordance with 310.15(B)(7).

# ARTICLE 551 Recreational Vehicles and Recreational Vehicle Parks

## I. General

### 551.1 Scope

The provisions of this article cover the electrical conductors and equipment other than low-voltage and automotive vehicle circuits or extensions thereof, installed within or on recreational vehicles, the conductors that connect recreational vehicles to a supply of electricity, and the installation of equipment and devices related to electrical installations within a recreational vehicle park.

Laws in many states require a factory inspection of recreational vehicles by either a governmental or a private inspection agency. NFPA 1192, *Standard on Recreational Vehicles*, is widely accepted by the recreational vehicle (RV) industry and AHJs who are responsible for ensuring that RVs are built to a recognized safety standard. Section 4.4 of NFPA 1192 requires compliance with Parts I, III, IV, V, and VI of Article 551 and with ANSI/RVIA 12V, *Low Voltage Systems in Conversion and Recreational Vehicles*, for the RV electrical systems rated 24 volts, nominal, or less.

Informational Note: For information on low-voltage systems, refer to NFPA 1192-2011, *Standard on Recreational Vehicles*, and ANSI/RVIA 12V-2011, *Standard for Low Voltage Systems in Conversion and Recreational Vehicles*.

### 551.2 Definitions

(See Article 100 for additional definitions.)

**Air-Conditioning or Comfort-Cooling Equipment.** All of that equipment intended or installed for the purpose of processing the treatment of air so as to control simultaneously or individually its temperature, humidity, cleanliness, and distribution to meet the requirements of the conditioned space.

**Appliance, Fixed.** An appliance that is fastened or otherwise secured at a specific location.

•

**Camping Trailer.** A vehicular portable unit mounted on wheels and constructed with collapsible partial side walls that fold for towing by another vehicle and unfold at the campsite to provide temporary living quarters for recreational, camping, or travel use. *(See Recreational Vehicle.)*

**Converter.** A device that changes electrical energy from one form to another, as from alternating current to direct current.

**Dead Front (as applied to switches, circuit breakers, switchboards, and panelboards).** Designed, constructed, and installed so that no current-carrying parts are normally exposed on the front.

**Disconnecting Means.** The necessary equipment usually consisting of a circuit breaker or switch and fuses, and their accessories, located near the point of entrance of supply conductors in a recreational vehicle and intended to constitute the means of cutoff for the supply to that recreational vehicle.

•

**Frame.** Chassis rail and any welded addition thereto of metal thickness of 1.35 mm (0.053 in.) or greater.

**Low Voltage.** An electromotive force rated 24 volts, nominal, or less.

**Motor Home.** A vehicular unit designed to provide temporary living quarters for recreational, camping, or travel use built on or permanently attached to a self-propelled motor vehicle chassis or on a chassis cab or van that is an integral part of the completed vehicle. *(See Recreational Vehicle.)*

**Power-Supply Assembly.** The conductors, including ungrounded, grounded, and equipment grounding conductors, the connectors, attachment plug caps, and all other fittings, grommets, or devices installed for the purpose of delivering energy from the source of electrical supply to the distribution panel within the recreational vehicle.

**Recreational Vehicle.** A vehicular-type unit primarily designed as temporary living quarters for recreational, camping, or travel use, which either has its own motive power or is mounted on or drawn by another vehicle.

Informational Note: The basic entities are travel trailer, camping trailer, truck camper, and motor home as referenced in NFPA 1192-2011, *Standard on Recreational Vehicles*. See 3.3.46, *Recreational Vehicle*, and A.3.3.46 of NFPA 1192.

**Recreational Vehicle Park.** A plot of land upon which two or more recreational vehicle sites are located, established, or maintained for occupancy by recreational vehicles of the general public as temporary living quarters for recreation or vacation purposes.

**Recreational Vehicle Site.** A plot of ground within a recreational vehicle park set aside for the accommodation of a recreational vehicle on a temporary basis or used as a camping unit site.

**Recreational Vehicle Site Feeder Circuit Conductors.** The conductors from the park service equipment to the recreational vehicle site supply equipment.

**Recreational Vehicle Site Supply Equipment.** The necessary equipment, usually a power outlet, consisting of a circuit breaker or switch and fuse and their accessories, located near the point of entrance of supply conductors to a recreational vehicle site and intended to constitute the disconnecting means for the supply to that site.

**Recreational Vehicle Stand.** That area of a recreational vehicle site intended for the placement of a recreational vehicle.

**Travel Trailer.** A vehicular unit, mounted on wheels, designed to provide temporary living quarters for recreational, camping, or travel use, of such size or weight as not to require special highway movement permits when towed by a motorized vehicle, and of gross trailer area less than 30 $m^2$ (320 $ft^2$). *(See Recreational Vehicle.)*

**Truck Camper.** A portable unit constructed to provide temporary living quarters for recreational, travel, or camping use, consisting of a roof, floor, and sides, designed to be loaded onto and unloaded from the bed of a pickup truck. *(See Recreational Vehicle.)*

### 551.4 General Requirements

**(A) Not Covered.** A recreational vehicle not used for the purposes as defined in 551.2 shall not be required to meet the provisions of Part IV pertaining to the number or capacity of circuits required. It shall, however, meet all other applicable requirements of this article if the recreational vehicle is provided with an electrical installation intended to be energized from a 120-volt, 208Y/120-volt, or 120/240-volt, nominal, ac power-supply system.

**(B) Systems.** This article covers combination electrical systems, generator installations, and 120-volt, 208Y/120-volt, or 120/240-volt, nominal, systems.

Informational Note: For information on low-voltage systems, refer to NFPA 1192-2011, *Standard on Recreational Vehicles*, and ANSI/RVIA 12V-2011, *Standard for Low Voltage Systems in Conversion and Recreational Vehicles*.

**(C) Labels.** Labels required by Article 551 shall be made of etched, metal-stamped, or embossed brass; stainless steel; plastic laminates not less than 0.13 mm (0.005 in.) thick; or anodized or alclad aluminum not less than 0.5 mm (0.020 in.) thick or the equivalent.

Informational Note: For guidance on other label criteria used in the recreational vehicle industry, refer to ANSI Z535.4-2011, *Product Safety Signs and Labels*.

## II. Combination Electrical Systems

### 551.20 Combination Electrical Systems

**(A) General.** Vehicle wiring suitable for connection to a battery or dc supply source shall be permitted to be connected to a 120-volt source, provided the entire wiring system and equipment are rated and installed in full conformity with Parts I, II, III, IV, and V requirements of this article covering 120-volt electrical systems. Circuits fed from ac transformers shall not supply dc appliances.

**(B) Voltage Converters (120-Volt Alternating Current to Low-Voltage Direct Current).** The 120-volt ac side of the voltage converter shall be wired in full conformity with the requirements of Parts I, II, and IV of this article for 120-volt electrical systems.

*Exception: Converters supplied as an integral part of a listed appliance shall not be subject to 551.20(B).*

All converters and transformers shall be listed for use in recreational vehicles and designed or equipped to provide over-temperature protection. To determine the converter rating, the following percentages shall be applied to the total connected load, including average battery-charging rate, of all 12-volt equipment:

The first 20 amperes of load at 100 percent plus
The second 20 amperes of load at 50 percent plus
All load above 40 amperes at 25 percent

*Exception: A low-voltage appliance that is controlled by a momentary switch (normally open) that has no means for holding in the closed position or refrigerators with a 120-volt function shall not be considered as a connected load when determining the required converter rating. Momentarily energized appliances shall be limited to those used to prepare the vehicle for occupancy or travel.*

**(C) Bonding Voltage Converter Enclosures.** The non–current-carrying metal enclosure of the voltage converter shall be connected to the frame of the vehicle with a minimum 8 AWG copper conductor. The voltage converter shall be provided with a separate chassis bonding conductor that shall not be used as a current-carrying conductor.

This requirement reduces the possibility of damage to the power-supply cord by large dc fault currents that may find their way back to the vehicle frame or battery through the ac grounding conductor of the converter. Metal enclosures of listed converters are provided with an external pressure terminal connector for this purpose.

**(D) Dual-Voltage Fixtures, Including Luminaires or Appliances.** Fixtures, including luminaires, or appliances having both 120-volt and low-voltage connections shall be listed for dual voltage.

In the dual-voltage fixtures, barriers are used to separate the 120-volt and the 12-volt wiring connections.

**(E) Autotransformers.** Autotransformers shall not be used.

**(F) Receptacles and Plug Caps.** Where a recreational vehicle is equipped with an ac system, a low-voltage system, or both, receptacles and plug caps of the low-voltage system shall differ in configuration from those of the ac system. Where a vehicle equipped with a battery or other low-voltage system has an external connection for low-voltage power, the connector shall have a configuration that will not accept ac power.

## III. Other Power Sources

### 551.30 Generator Installations

**(A) Mounting.** Generators shall be mounted in such a manner as to be effectively bonded to the recreational vehicle chassis.

**(B) Generator Protection.** Equipment shall be installed to ensure that the current-carrying conductors from the engine generator and from an outside source are not connected to a vehicle circuit at the same time. Automatic transfer switches in such applications shall be listed for use in one of the following:

(1) Emergency systems
(2) Optional standby systems

Receptacles used as disconnecting means shall be accessible (as applied to wiring methods) and capable of interrupting their rated current without hazard to the operator.

**(C) Installation of Storage Batteries and Generators.** Storage batteries and internal-combustion-driven generator units (subject to the provisions of this *Code*) shall be secured in place to avoid displacement from vibration and road shock.

**(D) Ventilation of Generator Compartments.** Compartments accommodating internal-combustion-driven generator units shall be provided with ventilation in accordance with instructions provided by the manufacturer of the generator unit.

Informational Note: For generator compartment construction requirements, see NFPA 1192-2011, *Standard on Recreational Vehicles*.

**(E) Supply Conductors.** The supply conductors from the engine generator to the first termination on the vehicle shall be of the stranded type and be installed in listed flexible conduit or listed liquidtight flexible conduit. The point of first termination shall be in one of the following:

(1) Panelboard
(2) Junction box with a blank cover
(3) Junction box with a receptacle
(4) Enclosed transfer switch
(5) Receptacle assembly listed in conjunction with the generator

The panelboard, enclosed transfer switch, or junction box with a receptacle shall be installed within 450 mm (18 in.) of the point of entry of the supply conductors into the vehicle. A junction box with a blank cover shall be mounted on the compartment wall inside or outside the compartment; to any part of the generator-supporting structure (but not to the generator); to the vehicle floor on the outside of the vehicle; or within 450 mm (18 in.) of the point of entry of the supply conductors into the vehicle. A receptacle assembly listed in conjunction with the generator shall be mounted in accordance with its listing.

### 551.31 Multiple Supply Source

**(A) Multiple Supply Sources.** Where a multiple supply system consisting of an alternate power source and a power-supply cord is installed, the feeder from the alternate power source shall be protected by an overcurrent protective device. Installation shall be in accordance with 551.30(A), 551.30(B), and 551.40.

**(B) Multiple Supply Sources Capacity.** The multiple supply sources shall not be required to be of the same capacity.

**(C) Alternate Power Sources Exceeding 30 Amperes.** If an alternate power source exceeds 30 amperes, 120 volts, nominal, it shall be permissible to wire it as a 120-volt, nominal, system, a 208Y/120-volt, nominal, system, or a 120/240-volt, nominal, system, provided an overcurrent protective device of the proper rating is installed in the feeder.

**(D) Power-Supply Assembly Not Less Than 30 Amperes.** The external power-supply assembly shall be permitted to be less than the calculated load but not less than 30 amperes and shall have overcurrent protection not greater than the capacity of the external power-supply assembly.

### 551.32 Other Sources

Other sources of ac power, such as inverters, motor generators, or engine generators, shall be listed for use in recreational vehicles and shall be installed in accordance with the terms of the listing. Other sources of ac power shall be wired in full conformity with the requirements in Parts I, II, III, IV, and V of this article covering 120-volt electrical systems.

### 551.33 Alternate Source Restrictions

Transfer equipment, if not integral with the listed power source, shall be installed to ensure that the current-carrying conductors from other sources of ac power and from an outside source are not connected to the vehicle circuit at the same time. Automatic transfer switches in such applications shall be listed for use in one of the following:

(1) Emergency systems
(2) Optional standby systems

Automatic transfer switches with relays may, under a fault condition such as a relay failure, simultaneously connect multiple sources. Therefore, the transfer switch must be listed for use in emergency or standby systems.

## IV. Nominal 120-Volt or 120/240-Volt Systems

### 551.40 120-Volt or 120/240-Volt, Nominal, Systems

**(A) General Requirements.** The electrical equipment and material of recreational vehicles indicated for connection to a wiring system rated 120 volts, nominal, 2-wire with equipment grounding conductor, or a wiring system rated 120/240 volts, nominal, 3-wire with equipment grounding conductor, shall be listed and installed in accordance with the requirements of Parts I, II, III, IV, and V of this article. Electrical equipment connected line-to-line shall have a voltage rating of 208–230 volts.

Electrical equipment is required to be rated 208–230 volts when it is to be connected line-to-line. This rating allows for compatibility with RV parks that have 208Y/120-volt, 3-phase, 4-wire electrical distribution systems.

**(B) Materials and Equipment.** Electrical materials, devices, appliances, fittings, and other equipment installed in, intended for use in, or attached to the recreational vehicle shall be listed. All products shall be used only in the manner in which they have been tested and found suitable for the intended use.

**(C) Ground-Fault Circuit-Interrupter Protection.** The internal wiring of a recreational vehicle having only one 15- or 20-ampere branch circuit as permitted in 551.42(A) and (B) shall have ground-fault circuit-interrupter protection for personnel. The ground-fault circuit interrupter shall be installed at the point where the power supply assembly terminates within the recreational vehicle. Where a separable cord set is not employed, the ground-fault circuit interrupter shall be permitted to be an integral part of the attachment plug of the power supply assembly. The ground-fault circuit interrupter shall provide protection also under the conditions of an open grounded circuit conductor, interchanged circuit conductors, or both.

### 551.41 Receptacle Outlets Required

**(A) Spacing.** Receptacle outlets shall be installed at wall spaces 600 mm (2 ft) wide or more so that no point along the floor line is more than 1.8 m (6 ft), measured horizontally, from an outlet in that space.

*Exception No. 1: Bath and hallway areas.*

*Exception No. 2: Wall spaces occupied by kitchen cabinets, wardrobe cabinets, built-in furniture, behind doors that may open fully against a wall surface, or similar facilities.*

**(B) Location.** Receptacle outlets shall be installed as follows:

(1) Adjacent to countertops in the kitchen [at least one on each side of the sink if countertops are on each side and are 300 mm (12 in.) or over in width and depth].
(2) Adjacent to the refrigerator and gas range space, except where a gas-fired refrigerator or cooking appliance, requiring no external electrical connection, is factory installed.
(3) Adjacent to countertop spaces of 300 mm (12 in.) or more in width and depth that cannot be reached from a receptacle required in 551.41(B)(1) by a cord of 1.8 m (6 ft) without crossing a traffic area, cooking appliance, or sink.
(4) Rooftop decks that are accessible from inside the RV shall have at least one receptacle installed within the perimeter of the rooftop deck. The receptacle shall not be located more than 1.2 m (4 ft) above the balcony, deck, or porch surface. The receptacle shall comply with the requirements of 406.9(B) for wet locations.

**(C) Ground-Fault Circuit-Interrupter Protection.** Where provided, each 125-volt, single-phase, 15- or 20-ampere receptacle outlet shall have ground-fault circuit-interrupter protection for personnel in the following locations:

(1) Adjacent to a bathroom lavatory

The walls of an RV often do not provide the necessary depth for the installation of a GFCI receptacle. This requirement does not prohibit a bathroom receptacle from being mounted in the side of a lavatory cabinet.

(2) Where the receptacles are installed to serve the countertop surfaces and are within 1.8 m (6 ft) of any lavatory or sink

*Exception No. 1: Receptacles installed for appliances in dedicated spaces, such as for dishwashers, disposals, refrigerators, freezers, and laundry equipment.*

*Exception No. 2: Single receptacles for interior connections of expandable room sections.*

*Exception No. 3: De-energized receptacles that are within 1.8 m (6 ft) of any sink or lavatory due to the retraction of the expandable room section.*

(3) In the area occupied by a toilet, shower, tub, or any combination thereof
(4) On the exterior of the vehicle

*Exception: Receptacles that are located inside of an access panel that is installed on the exterior of the vehicle to supply power for an installed appliance shall not be required to have ground-fault circuit-interrupter protection.*

The receptacle outlet shall be permitted in a listed luminaire. A receptacle outlet shall not be installed in a tub or combination tub–shower compartment.

**(D) Face-Up Position.** A receptacle shall not be installed in a face-up position in any countertop or similar horizontal surface.

## 551.42 Branch Circuits Required

Each recreational vehicle containing an ac electrical system shall contain one of the circuit arrangements in 551.42(A) through (D).

**(A) One 15-Ampere Circuit.** One 15-ampere circuit to supply lights, receptacle outlets, and fixed appliances. Such recreational vehicles shall be equipped with one 15-ampere switch and fuse or one 15-ampere circuit breaker.

**(B) One 20-Ampere Circuit.** One 20-ampere circuit to supply lights, receptacle outlets, and fixed appliances. Such recreational vehicles shall be equipped with one 20-ampere switch and fuse or one 20-ampere circuit breaker.

**(C) Two to Five 15- or 20-Ampere Circuits.** A maximum of five 15- or 20-ampere circuits to supply lights, receptacle outlets, and fixed appliances shall be permitted. Such recreational vehicles shall be permitted to be equipped with panelboards rated 120 V maximum or 120/240 V maximum and listed for 30-ampere application supplied by the appropriate power-supply assemblies. Not more than two 120-volt thermostatically controlled appliances (e.g., air conditioner and water heater) shall be installed in such systems unless appliance isolation switching, energy management systems, or similar methods are used.

*Exception No. 1: Additional 15- or 20-ampere circuits shall be permitted where a listed energy management system rated at 30-ampere maximum is employed within the system.*

*Exception No. 2: Six 15- or 20-ampere circuits shall be permitted without employing an energy management system, provided that the added sixth circuit serves only the power converter; and the combined load of all six circuits does not exceed the allowable load that was designed for use by the original five circuits.*

Informational Note: See 210.23(A) for permissible loads. See 551.45(C) for main disconnect and overcurrent protection requirements.

**(D) More Than Five Circuits Without a Listed Energy Management System.** A 50-ampere, 120/208–240-volt power-supply assembly and a minimum 50-ampere-rated panelboard shall be used where six or more circuits are employed. The load distribution shall ensure a reasonable current balance between phases.

In RVs with six or more circuits, the power-supply assembly must have a minimum rating of 50 amperes. Reasonable balancing of the electrical load between phases is required. In addition, the panelboard is required to be rated 120/208-240 volt to allow connection to a 120/240-volt single-phase service or a 120/208-volt 3-phase distribution. Based on experience and field data, supply calculations, which in some cases have resulted in oversized power supplies, are considered unnecessary.

## 551.43 Branch-Circuit Protection

**(A) Rating.** The branch-circuit overcurrent devices shall be rated as follows:

(1) Not more than the circuit conductors, and
(2) Not more than 150 percent of the rating of a single appliance rated 13.3 amperes or more and supplied by an individual branch circuit, but
(3) Not more than the overcurrent protection size marked on an air conditioner or other motor-operated appliances

**(B) Protection for Smaller Conductors.** A 20-ampere fuse or circuit breaker shall be permitted for protection for fixtures, including luminaires, leads, cords, or small appliances, and 14 AWG tap conductors, not over 1.8 m (6 ft) long for recessed luminaires.

**(C) Fifteen-Ampere Receptacle Considered Protected by 20 Amperes.** If more than one receptacle or load is on a branch circuit, a 15-ampere receptacle shall be permitted to be protected by a 20-ampere fuse or circuit breaker.

## 551.44 Power-Supply Assembly

Each recreational vehicle shall have only one of the main power-supply assemblies covered in 551.44(A) through (D).

**(A) Fifteen-Ampere Main Power-Supply Assembly.** Recreational vehicles wired in accordance with 551.42(A) shall use a listed 15-ampere or larger main power-supply assembly.

**(B) Twenty-Ampere Main Power-Supply Assembly.** Recreational vehicles wired in accordance with 551.42(B) shall use a listed 20-ampere or larger main power-supply assembly.

**(C) Thirty-Ampere Main Power-Supply Assembly.** Recreational vehicles wired in accordance with 551.42(C) shall use a listed 30-ampere or larger main power-supply assembly.

**(D) Fifty-Ampere Power-Supply Assembly.** Recreational vehicles wired in accordance with 551.42(D) shall use a listed 50-ampere, 120/208–240-volt main power-supply assembly.

## 551.45 Panelboard

**(A) Listed and Appropriately Rated.** A listed and appropriately rated panelboard or other equipment specifically listed for this purpose shall be used. The grounded conductor termination bar shall be insulated from the enclosure as provided in 551.54(C). An equipment grounding terminal bar shall be attached inside the enclosure of the panelboard.

**(B) Location.** The panelboard shall be installed in a readily accessible location with the RV in the setup mode. Working clearance for the panelboard with the RV in the setup mode shall be not less than 600 mm (24 in.) wide and 750 mm (30 in.) deep.

Some RVs have expandable room sections (or slide-out rooms) that are in a stowed position when the vehicle travels from place to place. The specified working clearance in front of equipment in an expanded section must be maintained in the setup mode.

*Exception No. 1: Where the panelboard cover is exposed to the inside aisle space, one of the working clearance dimensions shall be permitted to be reduced to a minimum of 550 mm (22 in.). A panelboard is considered exposed where the panelboard cover is within 50 mm (2 in.) of the aisle's finished surface or not more than 25 mm (1 in.) from the backside of doors that enclose the space.*

*Exception No. 2: Compartment doors used for access to a generator shall be permitted to be equipped with a locking system.*

**(C) Dead-Front Type.** The panelboard shall be of the dead-front type and shall consist of one or more circuit breakers or Type S fuseholders. A main disconnecting means shall be provided where fuses are used or where more than two circuit breakers are employed. A main overcurrent protective device not exceeding the power-supply assembly rating shall be provided where more than two branch circuits are employed.

## 551.46 Means for Connecting to Power Supply

**(A) Assembly.** The power-supply assembly or assemblies shall be factory supplied or factory installed and be of one of the types specified herein.

**(1) Separable.** Where a separable power-supply assembly consisting of a cord with a female connector and molded attachment plug cap is provided, the vehicle shall be equipped with a permanently mounted, flanged surface inlet (male, recessed-type motor-base attachment plug) wired directly to the panelboard by an approved wiring method. The attachment plug cap shall be of a listed type.

**(2) Permanently Connected.** Each power-supply assembly shall be connected directly to the terminals of the panelboard or conductors within a junction box and provided with means to prevent strain from being transmitted to the terminals. The ampacity of the conductors between each junction box and the terminals of each panelboard shall be at least equal to the ampacity of the power-supply cord. The supply end of the assembly shall be equipped with an attachment plug of the type described in 551.46(C). Where the cord passes through the walls or floors, it shall be protected by means of conduit and bushings or equivalent. The cord assembly shall have permanent provisions for protection against corrosion and mechanical damage while the vehicle is in transit and while the cord assembly is being stored or removed for use.

**(B) Cord.** The cord exposed usable length shall be measured from the point of entrance to the recreational vehicle or the face of the flanged surface inlet (motor-base attachment plug) to the face of the attachment plug at the supply end.

The cord exposed usable length, measured to the point of entry on the vehicle exterior, shall be a minimum of 7.5 m (25 ft) where the point of entrance is at the side of the vehicle or shall be a minimum 9.0 m (30 ft) where the point of entrance is at the rear of the vehicle.

Where the cord entrance into the vehicle is more than 900 mm (3 ft) above the ground, the minimum cord lengths above shall be increased by the vertical distance of the cord entrance heights above 900 mm (3 ft).

Informational Note: See 551.46(E) for location of point of entrance of a power-supply assembly on the recreational vehicle exterior.

**(C) Attachment Plugs.**

**(1) Units with One 15-Ampere Branch Circuit.** Recreational vehicles having only one 15-ampere branch circuit as permitted by 551.42(A) shall have an attachment plug that shall be 2-pole, 3-wire grounding type, rated 15 amperes, 125 volts, conforming to the configuration shown in Figure 551.46(C)(1).

Informational Note: Complete details of this configuration can be found in ANSI/NEMA WD 6-2002, *Standard for Dimensions of Attachment Plugs and Receptacle*, Figure 5.15.

**(2) Units with One 20-Ampere Branch Circuit.** Recreational vehicles having only one 20-ampere branch circuit as permitted in 551.42(B) shall have an attachment plug that shall be 2-pole, 3-wire grounding type, rated 20 amperes, 125 volts, conforming to the configuration shown in Figure 551.46(C)(1).

Informational Note: Complete details of this configuration can be found in ANSI/NEMA WD 6-2002, National Electrical Manufacturers Association's *Standard for Dimensions of Attachment Plugs and Receptacles*, Figure 5.20.

**(3) Units with Two to Five 15- or 20-Ampere Branch Circuits.** Recreational vehicles wired in accordance with 551.42(C) shall have an attachment plug that shall be 2-pole, 3-wire grounding type, rated 30 amperes, 125 volts, conforming to the configuration shown in Figure 551.46(C)(1) intended for use with units rated at 30 amperes, 125 volts.

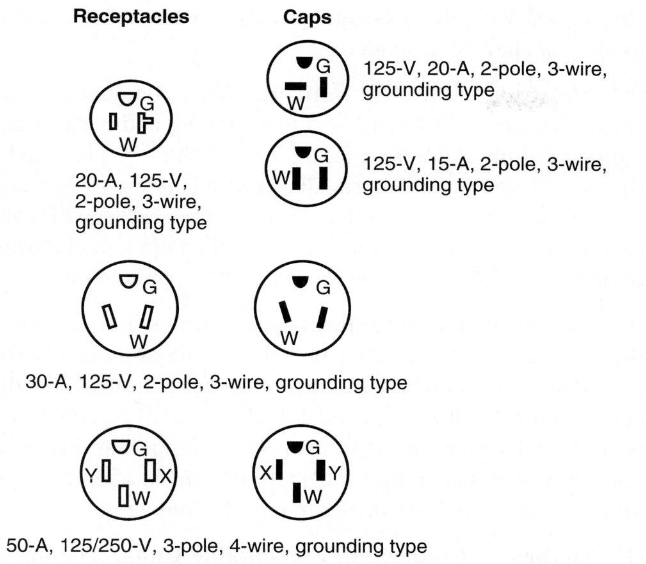


***FIGURE 551.46(C)(1)*** *Configurations for Grounding-Type Receptacles and Attachment Plug Caps Used for Recreational Vehicle Supply Cords and Recreational Vehicle Lots.*

Informational Note: Complete details of this configuration can be found in ANSI/NEMA WD 6-2002, National Electrical Manufacturers Association's *Standard for Dimensions of Attachment Plugs and Receptacles*, Figure 5.15.

The 30-ampere plug and receptacle configuration in Figure 551.46(C)(1) is not a standard 5-30P plug and 5-30R receptacle. The configuration is unique to RVs.

**(4) Units with 50-Ampere Power-Supply Assembly.** Recreational vehicles having a power-supply assembly rated 50 amperes as permitted by 551.42(D) shall have a 3-pole, 4-wire grounding-type attachment plug rated 50 amperes, 125/250 volts, conforming to the configuration shown in Figure 551.46(C)(1).

Informational Note: Complete details of this configuration can be found in ANSI/NEMA WD 6-2002, *Standard for Dimensions of Attachment Plugs and Receptacles*, Figure 14.50.

**(D) Labeling at Electrical Entrance.** Each recreational vehicle shall have a safety label with the signal word WARNING in minimum 6-mm (¼-in.) high letters and body text in minimum 3-mm (⅛-in.) high letters on a contrasting background. The safety label shall be affixed to the exterior skin, at or near the point of entrance of the power-supply cord(s), and shall read, using one of the following warnings, as appropriate:

WARNING
THIS CONNECTION IS FOR 110–125-VOLT AC,
60 HZ, ____ AMPERE SUPPLY.

or

THIS CONNECTION IS FOR 208Y/120-VOLT or
120/240-VOLT AC, 3-POLE, 4-WIRE,
60 HZ, ______ AMPERE SUPPLY.
DO NOT EXCEED CIRCUIT RATING.
EXCEEDING THE CIRCUIT RATING MAY CAUSE A
FIRE AND RESULT IN DEATH OR SERIOUS INJURY.

The correct ampere rating shall be marked in the blank space.

**(E) Location.** The point of entrance of a power-supply assembly shall be located within 4.5 m (15 ft) of the rear, on the left (road) side or at the rear, left of the longitudinal center of the vehicle, within 450 mm (18 in.) of the outside wall.

*Exception No. 1: A recreational vehicle equipped with only a listed flexible drain system or a side-vent drain system shall be permitted to have the electrical point of entrance located on either side, provided the drain(s) for the plumbing system is (are) located on the same side.*

*Exception No. 2: A recreational vehicle shall be permitted to have the electrical point of entrance located more than 4.5 m (15 ft) from the rear. Where this occurs, the distance beyond the 4.5-m (15-ft) dimension shall be added to the cord's minimum length as specified in 551.46(B).*

*Exception No. 3: Recreational vehicles designed for transporting livestock shall be permitted to have the electrical point of entrance located on either side or the front.*

## 551.47 Wiring Methods

**(A) Wiring Systems.** Cables and raceways installed in accordance with Articles 320, 322, 330 through 340, 342 through 362, 386, and 388 shall be permitted in accordance with their applicable article, except as otherwise specified in this article. An equipment grounding means shall be provided in accordance with 250.118.

See the commentary following 348.60 for information regarding the use of flexible metal conduit as an EGC.

**(B) Conduit and Tubing.** Where rigid metal conduit or intermediate metal conduit is terminated at an enclosure with a locknut and bushing connection, two locknuts shall be provided, one inside and one outside of the enclosure. All cut ends of conduit and tubing shall be reamed or otherwise finished to remove rough edges.

See the commentary following 344.28, 358.28(A), and 300.4(G), which covers the protection of conductor insulation against abrasion at conduit and tubing terminations.

**(C) Nonmetallic Boxes.** Nonmetallic boxes shall be acceptable only with nonmetallic-sheathed cable or nonmetallic raceways.

**(D) Boxes.** In walls and ceilings constructed of wood or other combustible material, boxes and fittings shall be flush with the finished surface or project therefrom.

**(E) Mounting.** Wall and ceiling boxes shall be mounted in accordance with Article 314.

*Exception No. 1: Snap-in-type boxes or boxes provided with special wall or ceiling brackets that securely fasten boxes in walls or ceilings shall be permitted.*

*Exception No. 2: A wooden plate providing a 38-mm (1½- in.) minimum width backing around the box and of a thickness of 13 mm (½ in.) or greater (actual) attached directly to the wall panel shall be considered as approved means for mounting outlet boxes.*

This exception permits the mounting of outlet boxes by screws to a wooden plate that is secured directly to the back of the wall panel. The wooden plate must extend at least 1½ inches around the box. This requirement recognizes the special construction of RV walls, which often makes it difficult or impossible to attach an outlet box to a structural member, as required by 314.23(B).

**(F) Raceway and Cable Continuity.** Raceways and cable sheaths shall be continuous between boxes and other enclosures.

**(G) Protected.** Metal-clad, Type AC, or nonmetallic-sheathed cables and electrical nonmetallic tubing shall be permitted to pass through the centers of the wide side of 2 by 4 wood studs. However, they shall be protected where they pass through 2 by 2 wood studs or at other wood studs or frames where the cable or tubing would be less than 32 mm (1¼ in.) from the inside or outside surface. Steel plates on each side of the cable or tubing or a steel tube, with not less than 1.35 mm (0.053 in.) wall thickness, shall be installed to protect the cable or tubing. These plates or tubes shall be securely held in place. Where nonmetallic-sheathed cables pass through punched, cut, or drilled slots or holes in metal members, the cable shall be protected by bushings or grommets securely fastened in the opening prior to installation of the cable.

**(H) Bends.** No bend shall have a radius of less than five times the cable diameter.

**(I) Cable Supports.** Where connected with cable connectors or clamps, cables shall be secured and supported within 300 mm (12 in.) of outlet boxes, panelboards, and splice boxes on appliances. Supports and securing shall be provided at intervals not exceeding 1.4 m (4½ ft) at other places.

**(J) Nonmetallic Box Without Cable Clamps.** Nonmetallic-sheathed cables shall be secured and supported within 200 mm (8 in.) of a nonmetallic outlet box without cable clamps. Where wiring devices with integral enclosures are employed with a loop of extra cable to permit future replacement of the device, the cable loop shall be considered as an integral portion of the device.

**(K) Physical Damage.** Where subject to physical damage, exposed nonmetallic cable shall be protected by covering boards, guard strips, raceways, or other means.

**(L) Receptacle Faceplates.** Metal faceplates shall comply with Section 406.5(A). Nonmetallic faceplates shall comply with Section 406.5(C).

**(M) Metal Faceplates Grounded.** Where metal faceplates are used, they shall be grounded.

**(N) Moisture or Physical Damage.** Where outdoor or under-chassis wiring is 120 volts, nominal, or over and is exposed to moisture or physical damage, the wiring shall be protected by rigid metal conduit, by intermediate metal conduit, or by electrical metallic tubing, rigid nonmetallic conduit, or Type MI cable, that is closely routed against frames and equipment enclosures or other raceway or cable identified for the application.

**(O) Component Interconnections.** Fittings and connectors that are intended to be concealed at the time of assembly shall be listed and identified for the interconnection of building components. Such fittings and connectors shall be equal to the wiring method employed in insulation, temperature rise, and fault-current withstanding and shall be capable of enduring the vibration and shock occurring in recreational vehicles.

**(P) Method of Connecting Expandable Units.** The method of connecting expandable units to the main body of the vehicle shall comply with 551.47(P)(1) or (P)(2):

**(1) Cord-and-Plug-Connected.** Cord-and-plug connections shall comply with (a) through (d).

(a) That portion of a branch circuit that is installed in an expandable unit shall be permitted to be connected to the portion of the branch circuit in the main body of the vehicle by means of an attachment plug and cord listed for hard usage. The cord and its connections shall comply with all provisions of Article 400 and shall be considered as a permitted use under 400.7. Where the attachment plug and cord are located within the vehicle's interior, use of plastic thermoset or elastomer parallel cord Type SPT-3, SP-3, or SPE shall be permitted.

(b) Where the receptacle provided for connection of the cord to the main circuit is located on the outside of the vehicle, it shall be protected with a ground-fault circuit interrupter for personnel and be listed for wet locations. A cord located on the outside of a vehicle shall be identified for outdoor use.

(c) Unless removable or stored within the vehicle interior, the cord assembly shall have permanent provisions for protection against corrosion and mechanical damage while the vehicle is in transit.

(d) The attachment plug and cord shall be installed so as not to permit exposed live attachment plug pins.

**(2) Direct Wired.** That portion of a branch circuit that is installed in an expandable unit shall be permitted to be connected to the portion of the branch circuit in the main body of the vehicle by means of flexible cord installed in accordance with 551.47(P)(2)(a) through (P)(2)(e) or other approved wiring method.

This section covers the interconnection between the main body of a vehicle and an expandable unit. Two methods of interconnection are permitted: one by means of cord-and-plug connections with the cord listed for hard usage; the other by means of a direct-wired connection, using flexible cord with the outer jacket intact, installed in nonflexible conduit or tubing. Where subject to physical damage, RMC, IMC, Schedule 80 PVC, RTRC, or other approved means must be used to protect the cord.

(a) The flexible cord shall be listed for hard usage and for use in wet locations.

(b) The flexible cord shall be permitted to be exposed on the underside of the vehicle.

(c) The flexible cord shall be permitted to pass through the interior of a wall or floor assembly or both a maximum concealed length of 600 mm (24 in.) before terminating at an outlet or junction box.

(d) Where concealed, the flexible cord shall be installed in nonflexible conduit or tubing that is continuous from the outlet or junction box inside the recreational vehicle to a weatherproof outlet box, junction box, or strain relief fitting listed for use in wet locations that is located on the underside of the recreational vehicle. The outer jacket of the flexible cord shall be continuous into the outlet or junction box.

(e) Where the flexible cord passes through the floor to an exposed area inside of the recreational vehicle, it shall be protected by means of conduit and bushings or equivalent.

Where subject to physical damage, the flexible cord shall be protected with RMC, IMC, Schedule 80 PVC, reinforced thermosetting resin conduit (RTRC) listed for exposure to physical damage, or other approved means and shall extend at least 150 mm (6 in.) above the floor. A means shall be provided to secure the flexible cord where it enters the recreational vehicle.

**(Q) Prewiring for Air-Conditioning Installation.** Prewiring installed for the purpose of facilitating future air-conditioning installation shall comply with the applicable portions of this article and the following:

(1) An overcurrent protective device with a rating compatible with the circuit conductors shall be installed in the panelboard and wiring connections completed.
(2) The load end of the circuit shall terminate in a junction box with a blank cover or other listed enclosure. Where a junction box with a blank cover is used, the free ends of the conductors shall be adequately capped or taped.
(3) A safety label with the signal word WARNING in minimum 6-mm (¼-in.) high letters and body text in minimum 3-mm (⅛-in.) high letters on a contrasting background shall be affixed on or adjacent to the junction box and shall read as follows:

WARNING
AIR-CONDITIONING CIRCUIT.
THIS CONNECTION IS FOR AIR CONDITIONERS
RATED 110–125-VOLT AC, 60 HZ,
___ AMPERES MAXIMUM.
DO NOT EXCEED CIRCUIT RATING.
EXCEEDING THE CIRCUIT RATING MAY
CAUSE A FIRE AND RESULT IN DEATH
OR SERIOUS INJURY.

An ampere rating, not to exceed 80 percent of the circuit rating, shall be legibly marked in the blank space.

(4) The circuit shall serve no other purpose.

**(R) Prewiring for Generator Installation.** Prewiring installed for the purpose of facilitating future generator installation shall comply with the other applicable portions of this article and the following:

(1) Circuit conductors shall be appropriately sized in relation to the anticipated load as stated on the label required in (R)(4).
(2) Where junction boxes are utilized at either of the circuit originating or terminus points, free ends of the conductors shall be adequately capped or taped.
(3) Where devices such as receptacle outlet, transfer switch, and so forth, are installed, the installation shall be complete, including circuit conductor connections.
(4) A safety label with the signal word WARNING in minimum 6-mm (¼-in.) high letters and body text in minimum 3-mm (⅛-in.) high letters on a contrasting background shall be affixed on the cover of each junction box containing incomplete circuitry and shall read, using one of the following warnings, as appropriate:

WARNING
GENERATOR
ONLY INSTALL A GENERATOR LISTED
SPECIFICALLY FOR RV USE
HAVING OVERCURRENT PROTECTION
RATED 110–125-VOLT AC,
60 HZ, ______ AMPERES MAXIMUM.

or

GENERATOR
ONLY INSTALL A GENERATOR LISTED
SPECIFICALLY FOR RV USE
HAVING OVERCURRENT PROTECTION
RATED 120–240-VOLT AC,
60 HZ, _______ AMPERES MAXIMUM.

The correct ampere rating shall be legibly marked in the blank space.

**(S) Prewiring for Other Circuits.** Prewiring installed for the purpose of installing other appliances or devices shall comply with the applicable portions of this article and the following:

(1) An overcurrent protection device with a rating compatible with the circuit conductors shall be installed in the panelboard with wiring connections completed.
(2) The load end of the circuit shall terminate in a junction box with a blank cover or a device listed for the purpose. Where a junction box with blank cover is used, the free ends of the conductors shall be adequately capped or taped.

(3) A safety label with the signal word WARNING in minimum 6-mm (¼-in.) high letters and body text in minimum 3-mm (⅛-in.) high letters on a contrasting background shall be affixed on or adjacent to the junction box or device listed for the purpose and shall read as follows:

WARNING
THIS CONNECTION IS FOR _____ RATED
_____ VOLT AC, 60 HZ, _____ AMPERES
MAXIMUM. DO NOT EXCEED CIRCUIT
RATING.

EXCEEDING THE CIRCUIT RATING MAY
CAUSE A FIRE AND RESULT IN DEATH OR
SERIOUS INJURY.

An ampere rating not to exceed 80 percent of the circuit rating shall be legibly marked in the blank space.

### 551.48 Conductors and Boxes

The maximum number of conductors permitted in boxes shall be in accordance with 314.16.

### 551.49 Grounded Conductors

The identification of grounded conductors shall be in accordance with 200.6.

### 551.50 Connection of Terminals and Splices

Conductor splices and connections at terminals shall be in accordance with 110.14.

### 551.51 Switches

**(A) Rating.** Switches shall be rated in accordance with 551.51(A)(1) and (A)(2).

**(1) Lighting Circuits.** For lighting circuits, switches shall be rated not less than 10 amperes, 120–125 volts and in no case less than the connected load.

**(2) Motors or Other Loads.** Switches for motor or other loads shall comply with the provisions of 404.14.

**(B) Location.** Switches shall not be installed within wet locations in tub or shower spaces unless installed as part of a listed tub or shower assembly.

### 551.52 Receptacles

All receptacle outlets shall be of the grounding type and installed in accordance with 406.4 and 210.21.

### 551.53 Luminaires and Other Equipment

**(A) General.** Any combustible wall or ceiling finish exposed between the edge of a canopy or pan of a luminaire or ceiling-suspended (paddle) fan and the outlet box shall be covered with noncombustible material.

**(B) Shower Luminaires.** If a luminaire is provided over a bathtub or in a shower stall, it shall be of the enclosed and gasketed type and listed for the type of installation, and it shall be ground-fault circuit-interrupter protected.

Due to the low ceilings in RVs, luminaires installed above a tub or shower enclosure may be easily reached by most persons standing in the enclosure. Accordingly, only luminaires that are listed for wet locations and have GFCI protection are permitted to be installed above a tub or shower enclosure.

**(C) Outdoor Outlets, Luminaires, Air-Cooling Equipment, and So On.** Outdoor luminaires and other equipment shall be listed for outdoor use.

### 551.54 Grounding

(See also 551.56 on bonding of non–current-carrying metal parts.)

**(A) Power-Supply Grounding.** The grounding conductor in the supply cord or feeder shall be connected to the grounding bus or other approved grounding means in the panelboard.

**(B) Panelboard.** The panelboard shall have a grounding bus with terminals for all grounding conductors or other approved grounding means.

**(C) Insulated Grounded Conductor (Neutral Conductor).** The grounded circuit conductor (neutral conductor) shall be insulated from the equipment grounding conductors and from equipment enclosures and other grounded parts. The grounded circuit conductor (neutral conductor) terminals in the panelboard and in ranges, clothes dryers, counter-mounted cooking units, and wall-mounted ovens shall be insulated from the equipment enclosure. Bonding screws, straps, or buses in the panelboard or in appliances shall be removed and discarded. Connection of electric ranges and electric clothes dryers utilizing a grounded conductor, if cord-connected, shall be made with 4-conductor cord and 3-pole, 4-wire grounding-type plug caps and receptacles.

### 551.55 Interior Equipment Grounding

**(A) Exposed Metal Parts.** In the electrical system, all exposed metal parts, enclosures, frames, luminaire canopies, and so forth, shall be effectively bonded to the grounding terminals or enclosure of the panelboard.

**(B) Equipment Grounding and Bonding Conductors.** Bare wires, insulated wire with an outer finish that is green or green with one or more yellow stripes, shall be used for equipment grounding or bonding conductors only.

**(C) Grounding of Electrical Equipment.** Grounding of electrical equipment shall be accomplished by one or more of the following methods:

(1) Connection of metal raceway, the sheath of Type MC and Type MI cable where the sheath is identified for grounding, or the armor of Type AC cable to metal enclosures.

(2) A connection between the one or more equipment grounding conductors and a metal enclosure by means of a grounding screw, which shall be used for no other purpose, or a listed grounding device.
(3) The equipment grounding conductor in nonmetallic-sheathed cable shall be permitted to be secured under a screw threaded into the luminaire canopy other than a mounting screw or cover screw, or attached to a listed grounding means (plate) in a nonmetallic outlet box for luminaire mounting. [Grounding means shall also be permitted for luminaire attachment screws.]

**(D) Grounding Connection in Nonmetallic Box.** A connection between the one or more equipment grounding conductors brought into a nonmetallic outlet box shall be so arranged that a connection of the equipment grounding conductor can be made to any fitting or device in that box that requires grounding.

**(E) Grounding Continuity.** Where more than one equipment grounding or bonding conductor of a branch circuit enters a box, all such conductors shall be in good electrical contact with each other, and the arrangement shall be such that the disconnection or removal of a receptacle, luminaire, or other device fed from the box will not interfere with or interrupt the grounding continuity.

**(F) Cord-Connected Appliances.** Cord-connected appliances, such as washing machines, clothes dryers, refrigerators, and the electrical system of gas ranges, and so forth, shall be grounded by means of an approved cord with equipment grounding conductor and grounding-type attachment plug.

### 551.56 Bonding of Non–Current-Carrying Metal Parts

**(A) Required Bonding.** All exposed non–current-carrying metal parts that are likely to become energized shall be effectively bonded to the grounding terminal or enclosure of the panelboard.

**(B) Bonding Chassis.** A bonding conductor shall be connected between any panelboard and an accessible terminal on the chassis. Aluminum or copper-clad aluminum conductors shall not be used for bonding if such conductors or their terminals are exposed to corrosive elements.

*Exception: Any recreational vehicle that employs a unitized metal chassis-frame construction to which the panelboard is securely fastened with a bolt(s) and nut(s) or by welding or riveting shall be considered to be bonded.*

**(C) Bonding Conductor Requirements.** Grounding terminals shall be of the solderless type and listed as pressure terminal connectors recognized for the wire size used. The bonding conductor shall be solid or stranded, insulated or bare, and shall be 8 AWG copper minimum, or equal.

**(D) Metallic Roof and Exterior Bonding.** The metal roof and exterior covering shall be considered bonded where both of the following conditions apply:

(1) The metal panels overlap one another and are securely attached to the wood or metal frame parts by metal fasteners.
(2) The lower panel of the metal exterior covering is secured by metal fasteners at each cross member of the chassis, or the lower panel is connected to the chassis by a metal strap.

**(E) Gas, Water, and Waste Pipe Bonding.** The gas, water, and waste pipes shall be considered grounded if they are bonded to the chassis.

**(F) Furnace and Metal Air Duct Bonding.** Furnace and metal circulating air ducts shall be bonded.

### 551.57 Appliance Accessibility and Fastening

Every appliance shall be accessible for inspection, service, repair, and replacement without removal of permanent construction. Means shall be provided to securely fasten appliances in place when the recreational vehicle is in transit.

## V. Factory Tests

### 551.60 Factory Tests (Electrical)

Each recreational vehicle designed with a 120-volt or a 120/240-volt electrical system shall withstand the applied potential without electrical breakdown of a 1-minute, 900-volt ac or 1280-volt dc dielectric strength test, or a 1-second, 1080-volt ac or 1530-volt dc dielectric strength test, with all switches closed, between ungrounded and grounded conductors and the recreational vehicle ground. During the test, all switches and other controls shall be in the "on" position. Fixtures, including luminaires and permanently installed appliances, shall not be required to withstand this test. The test shall be performed after branch circuits are complete prior to energizing the system and again after all outer coverings and cabinetry have been secured. The dielectric test shall be performed in accordance with the test equipment manufacturer's written instructions.

Each recreational vehicle shall be subjected to all of the following:

(1) A continuity test to ensure that all metal parts are properly bonded
(2) Operational tests to demonstrate that all equipment is properly connected and in working order
(3) Polarity checks to determine that connections have been properly made
(4) GFCI test to demonstrate that the ground fault protection device(s) installed on the recreational vehicle are operating properly.

## VI. Recreational Vehicle Parks

### 551.71 Type Receptacles Provided

Every recreational vehicle site with electrical supply shall be equipped with at least one 20-ampere, 125-volt receptacle. A minimum of 20 percent of all recreational vehicle sites, with electrical supply, shall each be equipped with a 50-ampere, 125/250-volt receptacle conforming to the configuration as identified in Figure 551.46(C)(1). Every recreational vehicle site equipped with a 50-ampere receptacle shall also be equipped with a 30-ampere, 125-volt receptacle conforming to Figure 551.46(C)(1). These electrical supplies shall be permitted to include additional receptacles that have configurations in accordance with 551.81. A minimum of 70 percent of all recreational vehicle sites with electrical supply shall each be equipped with a 30-ampere, 125-volt receptacle conforming to Figure 551.46(C)(1). This supply shall be permitted to include additional receptacle configurations conforming to 551.81. The remainder of all recreational vehicle sites with electrical supply shall be equipped with one or more of the receptacle configurations conforming to 551.81. Dedicated tent sites with a 15- or 20-ampere electrical supply shall be permitted to be excluded when determining the percentage of recreational vehicle sites with 30- or 50-ampere receptacles.

Additional receptacles shall be permitted for the connection of electrical equipment outside the recreational vehicle within the recreational vehicle park.

At least one 20-ampere, 125-volt receptacle must be installed at each RV campsite. Many RVs require a 30-ampere connection, and 70 percent of sites must also provide a 30-ampere receptacle.

Some RVs have a 50-ampere, 120/240-volt supply installed, and 20 percent of RV sites must be provided with a 50-ampere receptacle to accommodate the larger electrical system. This receptacle is in addition to the 20- and 30-ampere receptacles required for the site. This requirement increases the load capacity for RV park services and feeders. Receptacle configurations are shown in Figure 551.46(C), and receptacle ratings are described in 551.81.

Adapter plugs or "cheater" cords are often used to connect an RV with a 20-ampere supply cord to a 30-ampere receptacle outlet or a 30-ampere supply cord to a 50-ampere receptacle. This practice does not provide adequate overload protection for the cord or the connected load. The requirement ensures the availability of the properly rated receptacle at new campgrounds.

All 125-volt, single-phase, 15- and 20-ampere receptacles shall have listed ground-fault circuit-interrupter protection for personnel.

Informational Note: The percentage of 50 ampere sites required by 551.71 may be inadequate for seasonal recreational vehicle sites serving a higher percentage of recreational vehicles with 50 ampere electrical systems. In that type of recreational vehicle park, the percentage of 50 ampere sites could approach 100 percent.

## 551.72 Distribution System

Receptacles rated at 50 amperes shall be supplied from a branch circuit of the voltage class and rating of the receptacle. Other recreational vehicle sites with 125-volt, 20- and 30-ampere receptacles shall be permitted to be derived from any grounded distribution system that supplies 120-volt single-phase power. The neutral conductors shall not be reduced in size below the size of the ungrounded conductors for the site distribution. The neutral conductors shall be permitted to be reduced in size below the minimum required size of the ungrounded conductors for 240-volt, line-to-line, permanently connected loads only.

The distribution system of an RV park must supply the sites by a 120-volt, 2-wire circuit; a 120/240-volt, 3-wire circuit; or a 120/208-volt, 3-wire circuit derived from a 208Y/120-volt, 3-phase, 4-wire system. See the commentary following 551.40(A) for information regarding the voltage rating of line-to-line connected appliances in RVs.

## 551.73 Calculated Load

**(A) Basis of Calculations.** Electrical services and feeders shall be calculated on the basis of not less than 9600 volt-amperes per site equipped with 50-ampere, 208Y/120 or 120/240-volt supply facilities; 3600 volt-amperes per site equipped with both 20-ampere and 30-ampere supply facilities; 2400 volt-amperes per site equipped with only 20-ampere supply facilities; and 600 volt-amperes per site equipped with only 20-ampere supply facilities that are dedicated to tent sites. The demand factors set forth in Table 551.73(A) shall be the minimum allowable demand factors that shall be permitted in calculating load for service and feeders. Where the electrical supply for a recreational vehicle site has more than one receptacle, the calculated load shall be calculated only for the highest rated receptacle.

The calculated load for dedicated tent sites supplied with electricity can be smaller since these sites are not intended to accommodate recreational vehicles.

Where the electrical supply is in a location that serves two recreational vehicles, the equipment for both sites shall comply with 551.77, and the calculated load shall only be calculated for the two receptacles with the highest rating.

**(B) Transformers and Secondary Panelboards.** For the purpose of this *Code*, where the park service exceeds 240 volts, transformers and secondary panelboards shall be treated as services.

**(C) Demand Factors.** The demand factor for a given number of sites shall apply to all sites indicated. For example, 20 sites calculated at 45 percent of 3600 volt-amperes results in a permissible demand of 1620 volt-amperes per site or a total of 32,400 volt-amperes for 20 sites.

Informational Note: These demand factors may be inadequate in areas of extreme hot or cold temperature with loaded circuits for heating or air conditioning.

**(D) Feeder-Circuit Capacity.** Recreational vehicle site feeder-circuit conductors shall have an ampacity not less than the loads supplied and shall be rated not less than 30 amperes. The neutral conductors shall have an ampacity not less than the ungrounded conductors.

*TABLE 551.73(A)* *Demand Factors for Site Feeders and Service-Entrance Conductors for Park Sites*

| Number of Recreational Vehicle Sites | Demand Factor (%) |
|---|---|
| 1 | 100 |
| 2 | 90 |
| 3 | 80 |
| 4 | 75 |
| 5 | 65 |
| 6 | 60 |
| 7–9 | 55 |
| 10–12 | 50 |
| 13–15 | 48 |
| 16–18 | 47 |
| 19–21 | 45 |
| 22–24 | 43 |
| 25–35 | 42 |
| 36 plus | 41 |

Informational Note: Due to the long circuit lengths typical in most recreational vehicle parks, feeder conductor sizes found in the ampacity tables of Article 310 may be inadequate to maintain the voltage regulation suggested in the fine print note to 210.19. Total circuit voltage drop is a sum of the voltage drops of each serial circuit segment, where the load for each segment is calculated using the load that segment sees and the demand factors of 551.73(A).

Loads for other amenities such as, but not limited to, service buildings, recreational buildings, and swimming pools shall be calculated separately and then be added to the value calculated for the recreational vehicle sites where they are all supplied by a common service.

## 551.74 Overcurrent Protection

Overcurrent protection shall be provided in accordance with Article 240.

## 551.75 Grounding

All electrical equipment and installations in recreational vehicle parks shall be grounded as required by Article 250.

Informational Note: See 250.32(A), Exception, for single branch circuits.

## 551.76 Grounding — Recreational Vehicle Site Supply Equipment

**(A) Exposed Non–Current-Carrying Metal Parts.** Exposed non–current-carrying metal parts of fixed equipment, metal boxes, cabinets, and fittings that are not electrically connected to grounded equipment shall be grounded by an equipment grounding conductor run with the circuit conductors from the service equipment or from the transformer of a secondary distribution system. Equipment grounding conductors shall be sized in accordance with 250.122 and shall be permitted to be spliced by listed means.

The arrangement of equipment grounding connections shall be such that the disconnection or removal of a receptacle or other device will not interfere with, or interrupt, the grounding continuity.

**(B) Secondary Distribution System.** Each secondary distribution system shall be grounded at the transformer.

**(C) Grounded Conductor Not to Be Used as an Equipment Ground.** The grounded conductor shall not be used as an equipment grounding conductor for recreational vehicles or equipment within the recreational vehicle park.

**(D) No Connection on the Load Side.** No connection to a grounding electrode shall be made to the grounded conductor on the load side of the service disconnecting means except as covered in 250.30(A) for separately derived systems, and 250.32(B) Exception for separate buildings.

## 551.77 Recreational Vehicle Site Supply Equipment

**(A) Location.** Where provided on back-in sites, the recreational vehicle site electrical supply equipment shall be located on the left (road) side of the parked vehicle, on a line that is 1.5 m to 2.1 m (5 ft to 7 ft) from the left edge (driver's side of the parked RV) of the stand and shall be located at any point on this line from the rear of the stand to 4.5 m (15 ft) forward of the rear of the stand.

For pull-through sites, the electrical supply equipment shall be permitted to be located at any point along the line that is 1.5 m to 2.1 m (5 ft to 7 ft) from the left edge (driver's side of the parked RV) from 4.9 m (16 ft) forward of the rear of the stand to the center point between the two roads that gives access to and egress from the pull-through sites.

The left edge (driver's side of the parked RV) of the stand shall be marked.

These requirements accommodate vehicles towing boats or other trailers. The location of the site supply equipment permitted for pull-through sites reduces the use of extension cords.

**(B) Disconnecting Means.** A disconnecting switch or circuit breaker shall be provided in the site supply equipment for disconnecting the power supply to the recreational vehicle.

**(C) Access.** All site supply equipment shall be accessible by an unobstructed entrance or passageway not less than 600 mm (2 ft) wide and 2.0 m (6 ft 6 in.) high.

**(D) Mounting Height.** Site supply equipment shall be located not less than 600 mm (2 ft) or more than 2.0 m (6 ft 6 in.) above the ground.

**(E) Working Space.** Sufficient space shall be provided and maintained about all electrical equipment to permit ready and safe operation, in accordance with 110.26.

**(F) Marking.** Where the site supply equipment contains a 125/250-volt receptacle, the equipment shall be marked as follows: "Turn disconnecting switch or circuit breaker off before inserting or removing plug. Plug must be fully inserted or removed." The marking shall be located on the equipment adjacent to the receptacle outlet.

The marking is to reduce the possibility of a partially engaged attachment plug, which could result in intermittent neutral (grounded conductor) contact. Loss of the neutral could momentarily apply the line-to-line voltage (240 volts) across 125-volt equipment, causing malfunction or damage.

### 551.78 Protection of Outdoor Equipment

**(A) Wet Locations.** All switches, circuit breakers, receptacles, control equipment, and metering devices located in wet locations shall be weatherproof.

**(B) Meters.** If secondary meters are installed, meter sockets without meters installed shall be blanked off with an approved blanking plate.

### 551.79 Clearance for Overhead Conductors

Open conductors of not over 1000 volts, nominal, shall have a vertical clearance of not less than 5.5 m (18 ft) and a horizontal clearance of not less than 900 mm (3 ft) in all areas subject to recreational vehicle movement. In all other areas, clearances shall conform to 225.18 and 225.19.

Informational Note: For clearances of conductors over 600 volts, nominal, see 225.60 and 225.61.

### 551.80 Underground Service, Feeder, Branch-Circuit, and Recreational Vehicle Site Feeder-Circuit Conductors

**(A) General.** All direct-burial conductors, including the equipment grounding conductor if of aluminum, shall be insulated and identified for the use. All conductors shall be continuous from equipment to equipment. All splices and taps shall be made in approved junction boxes or by use of listed material.

**(B) Protection Against Physical Damage.** Direct-buried conductors and cables entering or leaving a trench shall be protected by rigid metal conduit, intermediate metal conduit, electrical metallic tubing with supplementary corrosion protection, rigid polyvinyl chloride conduit (PVC), nonmetallic underground conduit with conductors (NUCC), high density polyethylene conduit (HDPE), reinforced thermosetting resin conduit (RTRC), liquidtight flexible nonmetallic conduit, liquidtight flexible metal conduit, or other approved raceways or enclosures. Where subject to physical damage, the conductors or cables shall be protected by rigid metal conduit, intermediate metal conduit, Schedule 80 PVC conduit, or RTRC listed for exposure to physical damage. All such protection shall extend at least 450 mm (18 in.) into the trench from finished grade.

Informational Note: See 300.5 and Article 340 for conductors or Type UF cable used underground or in direct burial in earth.

### 551.81 Receptacles

A receptacle to supply electric power to a recreational vehicle shall be one of the configurations shown in Figure 551.46(C)(1) in the following ratings:

(1) 50-ampere — 125/250-volt, 50-ampere, 3-pole, 4-wire grounding type for 120/240-volt systems
(2) 30-ampere — 125-volt, 30-ampere, 2-pole, 3-wire grounding type for 120-volt systems
(3) 20-ampere — 125-volt, 20-ampere, 2-pole, 3-wire grounding type for 120-volt systems

Informational Note: Complete details of these configurations can be found in ANSI/NEMA WD 6-2002, National Electrical Manufacturers Association's *Standard for Dimensions of Attachment Plugs and Receptacles*, Figures 14-50, TT, and 5-20.

# ARTICLE 552 Park Trailers

## I. General

### 552.1 Scope

The provisions of this article cover the electrical conductors and equipment installed within or on park trailers not covered fully under Articles 550 and 551.

This article covers park trailers that have a single chassis and wheels, that do not exceed 400 $ft^2$ (set up), and that are not used as permanent residences. Additionally, Article 552 does not apply to units that meet the definition of *park trailer* (see 552.2) but are used for commercial purposes (see 552.4).

Park trailers equipped with electrical loads similar to those used in mobile homes are not uncommon. It is also not uncommon for a park trailer to be located in the same park trailer community for several years without relocation.

Park trailers are somewhat similar to mobile homes and recreational vehicles, and many requirements in Article 552 parallel those contained in Articles 550 and 551. Article 552, therefore, is similar in structure to Articles 550 and 551.

### 552.2 Definition

(See Articles 100, 550, and 551 for additional definitions.)

**Park Trailer.** A unit that is built on a single chassis mounted on wheels and has a gross trailer area not exceeding 37 $m^2$ (400 $ft^2$) in the set-up mode.

### 552.4 General Requirements

A park trailer as specified in 552.2 is intended for seasonal use. It is not intended as a permanent dwelling unit or for commercial uses such as banks, clinics, offices, or similar.

## II. Low-Voltage Systems

In some park trailers, 12-volt systems are used for interior lighting and other small loads. The 12-volt system is often supplied from an on-board battery or through a transfer switch from a 120/12-volt transformer in conjunction with a full-wave rectifier.

### 552.10 Low-Voltage Systems

**(A) Low-Voltage Circuits.** Low-voltage circuits furnished and installed by the park trailer manufacturer, other than those related to braking, are subject to this *Code*. Circuits supplying lights subject to federal or state regulations shall comply with applicable government regulations and this *Code*.

The requirements of Part II apply to the low-voltage wiring within the park trailer that would be used in place of 120-volt ac supplies. These requirements do not apply to the trailer braking circuits.

**(B) Low-Voltage Wiring.**

**(1) Material.** Copper conductors shall be used for low-voltage circuits.

*Exception: A metal chassis or frame shall be permitted as the return path to the source of supply.*

The intent is to not permit the sidewalls or the roof of a park trailer to serve as the ground return path. See the definition of *frame* in 551.2.

**(2) Conductor Types.** Conductors shall conform to the requirements for Type GXL, HDT, SGT, SGR, or Type SXL or shall have insulation in accordance with Table 310.104(A) or the equivalent. Conductor sizes 6 AWG through 18 AWG or SAE shall be listed. Single-wire, low-voltage conductors shall be of the stranded type.

Informational Note: See SAE J1128-2011, *Low Tension Primary Cable,* for Types GXL, HDT, and SXL, and SAE J1127-2010, *Battery Cable,* for Types SGT and SGR.

**(3) Marking.** All insulated low-voltage conductors shall be surface marked at intervals not greater than 1.2 m (4 ft) as follows:

(1) Listed conductors shall be marked as required by the listing agency.
(2) SAE conductors shall be marked with the name or logo of the manufacturer, specification designation, and wire gauge.
(3) Other conductors shall be marked with the name or logo of the manufacturer, temperature rating, wire gauge, conductor material, and insulation thickness.

**(C) Low-Voltage Wiring Methods.**

**(1) Physical Protection.** Conductors shall be protected against physical damage and shall be secured. Where insulated conductors are clamped to the structure, the conductor insulation shall be supplemented by an additional wrap or layer of equivalent material, except that jacketed cables shall not be required to be so protected. Wiring shall be routed away from sharp edges, moving parts, or heat sources.

**(2) Splices.** Conductors shall be spliced or joined with splicing devices that provide a secure connection or by brazing, welding, or soldering with a fusible metal or alloy. Soldered splices shall first be spliced or joined to be mechanically and electrically secure without solder, and then soldered. All splices, joints, and free ends of conductors shall be covered with an insulation equivalent to that on the conductors.

**(3) Separation.** Battery and other low-voltage circuits shall be physically separated by at least a 13-mm (½-in.) gap or other approved means from circuits of a different power source. Acceptable methods shall be by clamping, routing, or equivalent means that ensure permanent total separation. Where circuits of different power sources cross, the external jacket of the nonmetallic-sheathed cables shall be deemed adequate separation.

**(4) Ground Connections.** Ground connections to the chassis or frame shall be made in an accessible location and shall be mechanically secure. Ground connections shall be by means of copper conductors and copper or copper-alloy terminals of the solderless type identified for the size of wire used. The surface on which ground terminals make contact shall be cleaned and be free from oxide or paint or shall be electrically connected through the use of a cadmium, tin, or zinc-plated internal/external-toothed lockwasher or locking terminals. Ground terminal attaching screws, rivets or bolts, nuts, and lockwashers shall be cadmium, tin, or zinc-plated except rivets shall be permitted to be unanodized aluminum where attaching to aluminum structures.

The chassis-grounding terminal of the battery shall be connected to the unit chassis with a minimum 8 AWG copper conductor. In the event the unbonded lead from the battery exceeds 8 AWG, the bonding conductor size shall be not less than that of the unbonded lead.

In a combination ac/dc appliance, this ground connection arrangement minimizes the possibility of low-voltage circuit-fault currents passing through the ac panelboard bonding conductor and the EGC, and subsequently passing through the negative dc conductor feeding the appliance; that conductor may also be bonded to the external metal cover of the appliance. The ac EGC of the appliance may not have sufficient ampacity to safely conduct the dc fault current, which would necessitate installation of the battery bonding conductor. Some recreational vehicles already have one side of the battery circuit bonded to the frame by an 8 AWG or larger copper conductor.

**(D) Battery Installations.** Storage batteries subject to the provisions of this *Code* shall be securely attached to the unit and installed in an area vaportight to the interior and ventilated directly to the exterior of the unit. Where batteries are installed in a compartment, the compartment shall be ventilated

*TABLE 552.10(E)(1) Low-Voltage Overcurrent Protection*

| Wire Size (AWG) | Ampacity | Wire Type |
|---|---|---|
| 18 | 6 | Stranded only |
| 16 | 8 | Stranded only |
| 14 | 15 | Stranded or solid |
| 12 | 20 | Stranded or solid |
| 10 | 30 | Stranded or solid |

with openings having a minimum area of 1100 mm$^2$ (1.7 in.$^2$) at both the top and at the bottom. Where compartment doors are equipped for ventilation, the openings shall be within 50 mm (2 in.) of the top and bottom. Batteries shall not be installed in a compartment containing spark- or flame-producing equipment.

**(E) Overcurrent Protection.**

**(1) Rating.** Low-voltage circuit wiring shall be protected by overcurrent protective devices rated not in excess of the ampacity of copper conductors, in accordance with Table 552.10(E)(1).

**(2) Type.** Circuit breakers or fuses shall be of an approved type, including automotive types. Fuseholders shall be clearly marked with maximum fuse size and shall be protected against shorting and physical damage by a cover or equivalent means.

Informational Note: For further information, see ANSI/SAE J554-1987, *Standard for Electric Fuses (Cartridge Type)*; SAE J1284-1988, *Standard for Blade Type Electric Fuses*; and UL 275-2005, *Standard for Automotive Glass Tube Fuses*.

Protection of fuseholders by a cover or equivalent means reduces the possibility of the low-voltage system shorting to ground.

**(3) Appliances.** Appliances such as pumps, compressors, heater blowers, and similar motor-driven appliances shall be installed in accordance with the manufacturer's instructions.

Motors that are controlled by automatic switching or by latching-type manual switches shall be protected in accordance with 430.32(B).

**(4) Location.** The overcurrent protective device shall be installed in an accessible location on the unit within 450 mm (18 in.) of the point where the power supply connects to the unit circuits. If located outside the park trailer, the device shall be protected against weather and physical damage.

*Exception: External low-voltage supply shall be permitted to have the overcurrent protective device within 450 mm (18 in.) after entering the unit or after leaving a metal raceway.*

**(F) Switches.** Switches shall have a dc rating not less than the connected load.

**(G) Luminaires.** All low-voltage interior luminaires rated more than 4 watts, employing lamps rated more than 1.2 watts, shall be listed.

Twelve-volt systems for running and signal lights, similar to those used in conventional automobiles, are covered in 552.10 and 552.20.

# III. Combination Electrical Systems

## 552.20 Combination Electrical Systems

**(A) General.** Unit wiring suitable for connection to a battery or other low-voltage supply source shall be permitted to be connected to a 120-volt source, provided that the entire wiring system and equipment are rated and installed in full conformity with Parts I, III, IV, and V requirements of this article covering 120-volt electrical systems. Circuits fed from ac transformers shall not supply dc appliances.

**(B) Voltage Converters (120-Volt Alternating Current to Low-Voltage Direct Current).** The 120-volt ac side of the voltage converter shall be wired in full conformity with the requirements of Parts I and IV of this article for 120-volt electrical systems.

*Exception: Converters supplied as an integral part of a listed appliance shall not be subject to 552.20(B).*

All converters and transformers shall be listed for use in recreation units and designed or equipped to provide over-temperature protection. To determine the converter rating, the following percentages shall be applied to the total connected load, including average battery-charging rate, of all 12-volt equipment:

The first 20 amperes of load at 100 percent plus

The second 20 amperes of load at 50 percent plus

All load above 40 amperes at 25 percent

*Exception: A low-voltage appliance that is controlled by a momentary switch (normally open) that has no means for holding in the closed position shall not be considered as a connected load when determining the required converter rating. Momentarily energized appliances shall be limited to those used to prepare the unit for occupancy or travel.*

**(C) Bonding Voltage Converter Enclosures.** The non–current-carrying metal enclosure of the voltage converter shall be connected to the frame of the unit with an 8 AWG copper conductor minimum. The grounding conductor for the battery and the metal enclosure shall be permitted to be the same conductor.

**(D) Dual-Voltage Fixtures Including Luminaires or Appliances.** Fixtures, including luminaires, or appliances having both 120-volt and low-voltage connections shall be listed for dual voltage.

In the dual-voltage fixtures, barriers are used to separate the 120-volt and the 12-volt wiring connections.

**(E) Autotransformers.** Autotransformers shall not be used.

**(F) Receptacles and Plug Caps.** Where a park trailer is equipped with a 120-volt or 120/240-volt ac system, a low-voltage system, or both, receptacles and plug caps of the low-voltage system shall differ in configuration from those of the 120-volt or 120/240-volt system. Where a unit equipped with a battery or dc system has an external connection for low-voltage power, the connector shall have a configuration that will not accept 120-volt power.

## IV. Nominal 120-Volt or 120/240-Volt Systems

### 552.40 120-Volt or 120/240-Volt, Nominal, Systems

**(A) General Requirements.** The electrical equipment and material of park trailers indicated for connection to a wiring system rated 120 volts, nominal, 2-wire with an equipment grounding conductor, or a wiring system rated 120/240 volts, nominal, 3-wire with an equipment grounding conductor, shall be listed and installed in accordance with the requirements of Parts I, III, IV, and V of this article.

**(B) Materials and Equipment.** Electrical materials, devices, appliances, fittings, and other equipment installed, intended for use in, or attached to the park trailer shall be listed. All products shall be used only in the manner in which they have been tested and found suitable for the intended use.

### 552.41 Receptacle Outlets Required

**(A) Spacing.** Receptacle outlets shall be installed at wall spaces 600 mm (2 ft) wide or more so that no point along the floor line is more than 1.8 m (6 ft), measured horizontally, from an outlet in that space.

*Exception No. 1: Bath and hallway areas.*

*Exception No. 2: Wall spaces occupied by kitchen cabinets, wardrobe cabinets, built-in furniture; behind doors that may open fully against a wall surface; or similar facilities.*

**(B) Location.** Receptacle outlets shall be installed as follows:

(1) Adjacent to countertops in the kitchen [at least one on each side of the sink if countertops are on each side and are 300 mm (12 in.) or over in width]
(2) Adjacent to the refrigerator and gas range space, except where a gas-fired refrigerator or cooking appliance, requiring no external electrical connection, is factory-installed
(3) Adjacent to countertop spaces of 300 mm (12 in.) or more in width that cannot be reached from a receptacle required in 552.41(B)(1) by a cord of 1.8 m (6 ft) without crossing a traffic area, cooking appliance, or sink

**(C) Ground-Fault Circuit-Interrupter Protection.** Each 125-volt, single-phase, 15- or 20-ampere receptacle shall have ground-fault circuit-interrupter protection for personnel in the following locations:

(1) Where the receptacles are installed to serve kitchen countertop surfaces
(2) Within 1.8 m (6 ft) of any lavatory or sink

*Exception: Receptacles installed for appliances in dedicated spaces, such as for dishwashers, disposals, refrigerators, freezers, and laundry equipment.*

(3) In the area occupied by a toilet, shower, tub, or any combination thereof
(4) On the exterior of the unit

*Exception: Receptacles that are located inside of an access panel that is installed on the exterior of the unit to supply power for an installed appliance shall not be required to have ground-fault circuit-interrupter protection.*

The receptacle outlet shall be permitted in a listed luminaire. A receptacle outlet shall not be installed in a tub or combination tub–shower compartment.

The walls of a park trailer often do not provide the necessary depth for the installation of a GFCI receptacle. This requirement does not prohibit a bathroom receptacle from being mounted in the side of a lavatory cabinet. Receptacles of any type are not permitted in a tub or tub–shower compartment.

**(D) Pipe Heating Cable Outlet.** Where a pipe heating cable outlet is installed, the outlet shall be as follows:

(1) Located within 600 mm (2 ft) of the cold water inlet
(2) Connected to an interior branch circuit, other than a small-appliance branch circuit
(3) On a circuit where all of the outlets are on the load side of the ground-fault circuit-interrupter protection for personnel
(4) Mounted on the underside of the park trailer and shall not be considered to be the outdoor receptacle outlet required in 552.41(E)

**(E) Outdoor Receptacle Outlets.** At least one receptacle outlet shall be installed outdoors. A receptacle outlet located in a compartment accessible from the outside of the park trailer shall be considered an outdoor receptacle. Outdoor receptacle outlets shall be protected as required in 552.41(C)(4).

**(F) Receptacle Outlets Not Permitted.**

**(1) Shower or Bathtub Space.** Receptacle outlets shall not be installed in or within reach [750 mm (30 in.)] of a shower or bathtub space.

**(2) Face-Up Position.** A receptacle shall not be installed in a face-up position in any countertop.

### 552.43 Power Supply

**(A) Feeder.** The power supply to the park trailer shall be a feeder assembly consisting of not more than one listed 30-ampere or 50-ampere park trailer power-supply cord, with an integrally molded or securely attached cap, or a permanently installed feeder.

**(B) Power-Supply Cord.** If the park trailer has a power-supply cord, it shall be permanently attached to the panelboard, or to a junction box permanently connected to the panelboard, with the free end terminating in a molded-on attachment plug cap.

Cords with adapters and pigtail ends, extension cords, and similar items shall not be attached to, or shipped with, a park trailer.

A suitable clamp or the equivalent shall be provided at the panelboard knockout to afford strain relief for the cord to prevent strain from being transmitted to the terminals when the power-supply cord is handled in its intended manner.

The cord shall be a listed type with 3-wire, 120-volt or 4-wire, 120/240-volt conductors, one of which shall be identified by a continuous green color or a continuous green color with one or more yellow stripes for use as the grounding conductor.

**(C) Mast Weatherhead or Raceway.** Where the calculated load exceeds 50 amperes or where a permanent feeder is used, the supply shall be by means of one of the following:

(1) One mast weatherhead installation, installed in accordance with Article 230, containing four continuous, insulated, color-coded feeder conductors, one of which shall be an equipment grounding conductor

(2) A metal raceway, rigid nonmetallic conduit, or liquidtight flexible nonmetallic conduit from the disconnecting means in the park trailer to the underside of the park trailer, with provisions for the attachment to a suitable junction box or fitting to the raceway on the underside of the park trailer [with or without conductors as in 550.10(I)(1)]

## 552.44 Cord

**(A) Permanently Connected.** Each power-supply assembly shall be factory supplied or factory installed and connected directly to the terminals of the panelboard or conductors within a junction box and provided with means to prevent strain from being transmitted to the terminals. The ampacity of the conductors between each junction box and the terminals of each panelboard shall be at least equal to the ampacity of the power-supply cord. The supply end of the assembly shall be equipped with an attachment plug of the type described in 552.44(C). Where the cord passes through the walls or floors, it shall be protected by means of conduit and bushings or equivalent. The cord assembly shall have permanent provisions for protection against corrosion and mechanical damage while the unit is in transit.

**(B) Cord Length.** The cord-exposed usable length shall be measured from the point of entrance to the park trailer or the face of the flanged surface inlet (motor-base attachment plug) to the face of the attachment plug at the supply end.

The cord-exposed usable length, measured to the point of entry on the unit exterior, shall be a minimum of 7.0 m (23 ft) where the point of entrance is at the side of the unit, or shall be a minimum 8.5 m (28 ft) where the point of entrance is at the rear of the unit. The maximum length shall not exceed 11 m (36½ ft).

Where the cord entrance into the unit is more than 900 mm (3 ft) above the ground, the minimum cord lengths above shall be increased by the vertical distance of the cord entrance heights above 900 mm (3 ft).

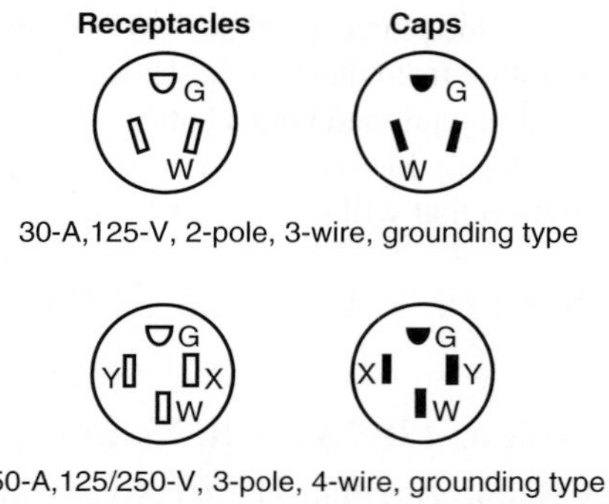


*FIGURE 552.44(C)(1) Attachment Cap and Receptacle Configurations.*

**(C) Attachment Plugs.**

**(1) Units with Two to Five 15- or 20-Ampere Branch Circuits.** Park trailers wired in accordance with 552.46(A) shall have an attachment plug that shall be 2-pole, 3-wire grounding type, rated 30 amperes, 125 volts, conforming to the configuration shown in Figure 552.44(C)(1) intended for use with units rated at 30 amperes, 125 volts.

Informational Note: Complete details of this configuration can be found in ANSI/NEMA WD 6-2002 (Rev. 2008), *Standard for Dimensions of Attachment Plugs and Receptacles*, Figure TT.

**(2) Units with 50-Ampere Power Supply Assembly.** Park trailers having a power-supply assembly rated 50 amperes as permitted by 552.43(B) shall have a 3-pole, 4-wire grounding-type attachment plug rated 50 amperes, 125/250 volts, conforming to the configuration shown in Figure 552.44(C)(1).

Informational Note: Complete details of this configuration can be found in ANSI/NEMA WD 6-2002 (Rev. 2008), *Standard for Dimensions of Attachment Plugs and Receptacles*, Figure 14-50.

The 30-ampere plug and receptacle configuration in Figure 552.44(C)(1) is unique to RVs. They are not a standard 5-30P plug or 5-30R receptacle.

**(D) Labeling at Electrical Entrance.** Each park trailer shall have permanently affixed to the exterior skin, at or near the point of entrance of the power-supply assembly, a label 75 mm × 45 mm (3 in. × 1¾ in.) minimum size, made of etched, metal-stamped, or embossed brass, stainless steel, or anodized or alclad aluminum not less than 0.51 mm (0.020 in.) thick, or other suitable material [e.g., 0.13 mm (0.005 in.) thick plastic laminate], that reads, as appropriate, either

THIS CONNECTION IS FOR 110–125-VOLT AC,
60 HZ, 30 AMPERE SUPPLY

or

THIS CONNECTION IS FOR 208Y/120-VOLT
OR 120/240-VOLT AC, 3-POLE, 4-WIRE,
60 HZ, ______ AMPERE SUPPLY.

The correct ampere rating shall be marked in the blank space.

**(E) Location.** The point of entrance of a power-supply assembly shall be located within 4.5 m (15 ft) of the rear, on the left (road) side or at the rear, left of the longitudinal center of the unit, within 450 mm (18 in.) of the outside wall.

*Exception: A park trailer shall be permitted to have the electrical point of entrance located more than 4.5 m (15 ft) from the rear. Where this occurs, the distance beyond the 4.5-m (15-ft) dimension shall be added to the cord's minimum length as specified in 551.46(B).*

## 552.45 Panelboard

**(A) Listed and Appropriately Rated.** A listed and appropriately rated panelboard shall be used. The grounded conductor termination bar shall be insulated from the enclosure as provided in 552.55(C). An equipment grounding terminal bar shall be attached inside the metal enclosure of the panelboard.

**(B) Location.** The panelboard shall be installed in a readily accessible location. Working clearance for the panelboard shall be not less than 600 mm (24 in.) wide and 750 mm (30 in.) deep.

*Exception: Where the panelboard cover is exposed to the inside aisle space, one of the working clearance dimensions shall be permitted to be reduced to a minimum of 550 mm (22 in.). A panelboard shall be considered exposed where the panelboard cover is within 50 mm (2 in.) of the aisle's finished surface.*

**(C) Dead-Front Type.** The panelboard shall be of the dead-front type. A main disconnecting means shall be provided where fuses are used or where more than two circuit breakers are employed. A main overcurrent protective device not exceeding the power-supply assembly rating shall be provided where more than two branch circuits are employed.

## 552.46 Branch Circuits

Branch circuits shall be determined in accordance with 552.46(A) and (B).

**(A) Two to Five 15- or 20-Ampere Circuits.** Two to five 15- or 20-ampere circuits to supply lights, receptacle outlets, and fixed appliances shall be permitted. Such park trailers shall be equipped with a panelboard rated at 120 V maximum with a 30-ampere-rated main power supply assembly. Not more than two 120-volt thermostatically controlled appliances (i.e., air conditioner and water heater) shall be installed in such systems unless appliance isolation switching, energy management systems, or similar methods are used.

*Exception: Additional 15- or 20-ampere circuits shall be permitted where a listed energy management system rated at 30 amperes maximum is employed within the system.*

**(B) More Than Five Circuits.** Where more than five circuits are needed, they shall be determined in accordance with 552.46(B)(1), (B)(2), and (B)(3).

**(1) Lighting.** Based on 33 volt-amperes/m$^2$ (3 VA/ft$^2$) multiplied by the outside dimensions of the park trailer (coupler excluded) divided by 120 volts to determine the number of 15- or 20-ampere lighting area circuits, for example,

$$\frac{3 \times \text{length} \times \text{width}}{120 \times 15 \text{ (or 20)}} = \text{No. of 15- (or 20-) ampere circuits}$$

The lighting circuits shall be permitted to serve listed cord-connected kitchen waste disposers and to provide power for supplemental equipment and lighting on gas-fired ranges, ovens, or counter-mounted cooking units.

**(2) Small Appliances.** Small-appliance branch circuits shall be installed in accordance with 210.11(C)(1).

**(3) General Appliances.** (including furnace, water heater, space heater, range, and central or room air conditioner, etc.) An individual branch circuit shall be permitted to supply any load for which it is rated. There shall be one or more circuits of adequate rating in accordance with (a) through (d).

Informational Note No. 1: For the laundry branch circuit, see 210.11(C)(2).

Informational Note No. 2: For central air conditioning, see Article 440.

(a) The total rating of fixed appliances shall not exceed 50 percent of the circuit rating if lighting outlets, general-use receptacles, or both are also supplied.

(b) For fixed appliances with a motor(s) larger than ⅛ horsepower, the total calculated load shall be based on 125 percent of the largest motor plus the sum of the other loads. Where a branch circuit supplies continuous load(s) or any combination of continuous and noncontinuous loads, the branch-circuit conductor size shall be in accordance with 210.19(A).

(c) The rating of a single cord-and-plug-connected appliance supplied by other than an individual branch circuit shall not exceed 80 percent of the circuit rating.

(d) The rating of a range branch circuit shall be based on the range demand as specified for ranges in 552.47(B)(5).

## 552.47 Calculations

The following method shall be employed in computing the supply-cord and distribution-panelboard load for each feeder assembly for each park trailer in lieu of the procedure shown in Article 220 and shall be based on a 3-wire, 208Y/120-volt or 120/240-volt supply with 120-volt loads balanced between the two phases of the 3-wire system.

**(A) Lighting and Small-Appliance Load.** Lighting Volt-Amperes: Length times width of park trailer floor (outside dimensions) times 33 volt-amperes/m$^2$ (3 VA/ft$^2$). For example,

Length × width × 3 = lighting volt-amperes

Small-Appliance Volt-Amperes: Number of circuits times 1500 volt-amperes for each 20-ampere appliance receptacle circuit (see definition of *Appliance, Portable* with fine print note) including 1500 volt-amperes for laundry circuit. For example,

No. of circuits × 1500 = small-appliance volt-amperes

Total: Lighting volt-amperes plus small-appliance volt-amperes = total volt-amperes

First 3000 total volt-amperes at 100 percent plus remainder at 35 percent = volt-amperes to be divided by 240 volts to obtain current (amperes) per leg.

**(B) Total Load for Determining Power Supply.** Total load for determining power supply is the sum of the following:

(1) Lighting and small-appliance load as calculated in 552.47(A).
(2) Nameplate amperes for motors and heater loads (exhaust fans, air conditioners, electric, gas, or oil heating). Omit smaller of the heating and cooling loads, except include blower motor if used as air-conditioner evaporator motor. Where an air conditioner is not installed and a 50-ampere power-supply cord is provided, allow 15 amperes per phase for air conditioning.
(3) Twenty-five percent of current of largest motor in (B)(2).
(4) Total of nameplate amperes for disposal, dishwasher, water heater, clothes dryer, wall-mounted oven, cooking units. Where the number of these appliances exceeds three, use 75 percent of total.
(5) Derive amperes for freestanding range (as distinguished from separate ovens and cooking units) by dividing the following values by 240 volts:

| Nameplate Rating (watts) | Use (volt-amperes) |
|---|---|
| 0–10,000 | 80 percent of rating |
| Over 10,000–12,500 | 8,000 |
| Over 12,500–13,500 | 8,400 |
| Over 13,500–14,500 | 8,800 |
| Over 14,500–15,500 | 9,200 |
| Over 15,500–16,500 | 9,600 |
| Over 16,500–17,500 | 10,000 |

(6) If outlets or circuits are provided for other than factory-installed appliances, include the anticipated load.

Informational Note: Refer to Informative Annex D, Example D12, for an illustration of the application of this calculation.

**(C) Optional Method of Calculation for Lighting and Appliance Load.** For park trailers, the optional method for calculating lighting and appliance load shown in 220.82 shall be permitted.

## 552.48 Wiring Methods

**(A) Wiring Systems.** Cables and raceways installed in accordance with Articles 320, 322, 330 through 340, 342 through 362, 386, and 388 shall be permitted in accordance with their applicable article, except as otherwise specified in this article. An equipment grounding means shall be provided in accordance with 250.118.

See the commentary following 348.60 for information regarding the use of flexible metal conduit as an EGC.

**(B) Conduit and Tubing.** Where rigid metal conduit or intermediate metal conduit is terminated at an enclosure with a locknut and bushing connection, two locknuts shall be provided, one inside and one outside of the enclosure. All cut ends of conduit and tubing shall be reamed or otherwise finished to remove rough edges.

For more information on the protection of conductor insulation against abrasion at conduit and tubing terminations, see the commentary following 344.28, 358.28(A), and 300.4(G).

**(C) Nonmetallic Boxes.** Nonmetallic boxes shall be acceptable only with nonmetallic-sheathed cable or nonmetallic raceways.

**(D) Boxes.** In walls and ceilings constructed of wood or other combustible material, boxes and fittings shall be flush with the finished surface or project therefrom.

**(E) Mounting.** Wall and ceiling boxes shall be mounted in accordance with Article 314.

*Exception No. 1: Snap-in-type boxes or boxes provided with special wall or ceiling brackets that securely fasten boxes in walls or ceilings shall be permitted.*

*Exception No. 2: A wooden plate providing a 38-mm (1½-in.) minimum width backing around the box and of a thickness of 13 mm (½ in.) or greater (actual) attached directly to the wall panel shall be considered as approved means for mounting outlet boxes.*

Exception No. 2 permits the mounting of outlet boxes by screws to a wooden plate that is secured directly to the back of a wall panel. The wooden plate must extend at least 1½ inches around the box. This requirement recognizes the special construction of recreational vehicle walls, which often makes it difficult or impossible to attach an outlet box to a structural member, as required by 314.23(B).

**(F) Cable Sheath.** The sheath of nonmetallic-sheathed cable, and the armor of metal-clad cable and Type AC cable, shall be continuous between outlet boxes and other enclosures.

**(G) Protected.** Metal-clad, Type AC, or nonmetallic-sheathed cables and electrical nonmetallic tubing shall be permitted to pass through the centers of the wide side of 2 by 4 wood studs. However, they shall be protected where they pass through 2 by 2 wood studs or at other wood studs or frames where the cable or tubing would be less than 32 mm (1¼ in.) from the inside or outside surface. Steel plates on each side of the cable or tubing, or a steel tube, with not less than 1.35 mm (0.053 in.) wall thickness, shall be installed to protect the cable or tubing. These plates or tubes shall be securely held in place. Where nonmetallic-sheathed cables pass through punched, cut, or drilled slots or holes in metal members, the cable shall be protected by bushings or grommets securely fastened in the opening prior to installation of the cable.

**(H) Cable Supports.** Where connected with cable connectors or clamps, cables shall be supported within 300 mm (12 in.) of outlet boxes, panelboards, and splice boxes on appliances. Supports shall be provided at intervals not exceeding 1.4 m (4½ ft) at other places.

**(I) Nonmetallic Box Without Cable Clamps.** Nonmetallic-sheathed cables shall be supported within 200 mm (8 in.) of a nonmetallic outlet box without cable clamps.

*Exception: Where wiring devices with integral enclosures are employed with a loop of extra cable to permit future replacement of the device, the cable loop shall be considered as an integral portion of the device.*

**(J) Physical Damage.** Where subject to physical damage, exposed nonmetallic cable shall be protected by covering boards, guard strips, raceways, or other means.

**(K) Receptacle Faceplates.** Metal faceplates shall comply with 406.5(A). Nonmetallic faceplates shall comply with 406.5(C).

**(L) Metal Faceplates Grounded.** Where metal faceplates are used, they shall be grounded.

**(M) Moisture or Physical Damage.** Where outdoor or under-chassis wiring is 120 volts, nominal, or over and is exposed to moisture or physical damage, the wiring shall be protected by rigid metal conduit, by intermediate metal conduit, by electrical metallic tubing, by rigid nonmetallic conduit, or by Type MI cable that is closely routed against frames and equipment enclosures or other raceway or cable identified for the application.

**(N) Component Interconnections.** Fittings and connectors that are intended to be concealed at the time of assembly shall be listed and identified for the interconnection of building components. Such fittings and connectors shall be equal to the wiring method employed in insulation, temperature rise, and fault-current withstanding, and shall be capable of enduring the vibration and shock occurring in park trailers.

**(O) Method of Connecting Expandable Units.** The method of connecting expandable units to the main body of the vehicle shall comply with the following as applicable:

(1) That portion of a branch circuit that is installed in an expandable unit shall be permitted to be connected to the branch circuit in the main body of the vehicle by means of a flexible cord or attachment plug and cord listed for hard usage. The cord and its connections shall conform to all provisions of Article 400 and shall be considered as a permitted use under 400.7.
(2) If the receptacle provided for connection of the cord to the main circuit is located on the outside of the unit, it shall be protected with a ground-fault circuit interrupter for personnel and be listed for wet locations. A cord located on the outside of a unit shall be identified for outdoor use.
(3) Unless removable or stored within the unit interior, the cord assembly shall have permanent provisions for protection against corrosion and mechanical damage while the unit is in transit.
(4) If an attachment plug and cord is used, it shall be installed so as not to permit exposed live attachment plug pins.

**(P) Prewiring for Air-Conditioning Installation.** Prewiring installed for the purpose of facilitating future air-conditioning installation shall comply with the applicable portions of this article and the following:

(1) An overcurrent protective device with a rating compatible with the circuit conductors shall be installed in the panelboard and wiring connections completed.
(2) The load end of the circuit shall terminate in a junction box with a blank cover or other listed enclosure. Where a junction box with a blank cover is used, the free ends of the conductors shall be adequately capped or taped.
(3) A label conforming to 552.44(D) shall be placed on or adjacent to the junction box and shall read as follows:

AIR-CONDITIONING CIRCUIT.
THIS CONNECTION IS FOR AIR CONDITIONERS
RATED 110–125-VOLT AC, 60 HZ,
____ AMPERES MAXIMUM.
DO NOT EXCEED CIRCUIT RATING.

An ampere rating, not to exceed 80 percent of the circuit rating, shall be legibly marked in the blank space.

(4) The circuit shall serve no other purpose.

## 552.49 Maximum Number of Conductors in Boxes

The maximum number of conductors permitted in boxes shall be in accordance with 314.16.

## 552.50 Grounded Conductors

The identification of grounded conductors shall be in accordance with 200.6.

## 552.51 Connection of Terminals and Splices

Conductor splices and connections at terminals shall be in accordance with 110.14.

## 552.52 Switches

Switches shall be rated as required by 552.52(A) and (B).

**(A) Lighting Circuits.** For lighting circuits, switches shall be rated not less than 10 amperes, 120/125 volts, and in no case less than the connected load.

**(B) Motors or Other Loads.** For motors or other loads, switches shall have ampere or horsepower ratings, or both, adequate for loads controlled. (An ac general-use snap switch shall be permitted to control a motor 2 hp or less with full-load current not over 80 percent of the switch ampere rating.)

## 552.53 Receptacles

All receptacle outlets shall be of the grounding type and installed in accordance with 210.21 and 406.4.

## 552.54 Luminaires

**(A) General.** Any combustible wall or ceiling finish exposed between the edge of a luminaire canopy or pan and the outlet box shall be covered with noncombustible material or a material identified for the purpose.

**(B) Shower Luminaires.** If a luminaire is provided over a bathtub or in a shower stall, it shall be of the enclosed and gasketed type and listed for the type of installation, and it shall be ground-fault circuit-interrupter protected.

The switch for shower luminaires and exhaust fans, located over a tub or in a shower stall, shall be located outside the tub or shower space.

Due to the low ceilings in park trailers, luminaires installed above a tub or shower enclosure may be easily reached by most persons standing in the enclosure. Accordingly, only luminaires that are listed for wet locations and have GFCI protection are permitted to be installed above a tub or shower enclosure.

**(C) Outdoor Outlets, Luminaires, Air-Cooling Equipment, and So On.** Outdoor luminaires and other equipment shall be listed for outdoor use or wet locations.

## 552.55 Grounding

(See also 552.57 on bonding of non–current-carrying metal parts.)

**(A) Power-Supply Grounding.** The grounding conductor in the supply cord or feeder shall be connected to the grounding bus or other approved grounding means in the panelboard.

**(B) Panelboard.** The panelboard shall have a grounding bus with sufficient terminals for all grounding conductors or other approved grounding means.

**(C) Insulated Grounded Conductor.** The grounded circuit conductor shall be insulated from the equipment grounding conductors and from equipment enclosures and other grounded parts. The grounded circuit conductor terminals in the panelboard and in ranges, clothes dryers, counter-mounted cooking units, and wall-mounted ovens shall be insulated from the equipment enclosure. Bonding screws, straps, or buses in the panelboard or in appliances shall be removed and discarded. Connection of electric ranges and electric clothes dryers utilizing a grounded conductor, if cord-connected, shall be made with 4-conductor cord and 3-pole, 4-wire, grounding-type plug caps and receptacles.

## 552.56 Interior Equipment Grounding

**(A) Exposed Metal Parts.** In the electrical system, all exposed metal parts, enclosures, frames, luminaire canopies, and so forth, shall be effectively bonded to the grounding terminals or enclosure of the panelboard.

**(B) Equipment Grounding Conductors.** Bare conductors or conductors with insulation or individual covering that is green or green with one or more yellow stripes shall be used for equipment grounding conductors only.

**(C) Grounding of Electrical Equipment.** Where grounding of electrical equipment is specified, it shall be permitted as follows:

(1) Connection of metal raceway (conduit or electrical metallic tubing), the sheath of Type MC and Type MI cable where the sheath is identified for grounding, or the armor of Type AC cable to metal enclosures.
(2) A connection between the one or more equipment grounding conductors and a metal box by means of a grounding screw, which shall be used for no other purpose, or a listed grounding device.
(3) The equipment grounding conductor in nonmetallic-sheathed cable shall be permitted to be secured under a screw threaded into the luminaire canopy other than a mounting screw or cover screw or attached to a listed grounding means (plate) in a nonmetallic outlet box for luminaire mounting (grounding means shall also be permitted for luminaire attachment screws).

**(D) Grounding Connection in Nonmetallic Box.** A connection between the one or more grounding conductors brought into a nonmetallic outlet box shall be arranged so that a connection can be made to any fitting or device in that box that requires grounding.

**(E) Grounding Continuity.** Where more than one equipment grounding conductor of a branch circuit enters a box, all such conductors shall be in good electrical contact with each other, and the arrangement shall be such that the disconnection or removal of a receptacle, fixture, including a luminaire, or other device fed from the box will not interfere with or interrupt the grounding continuity.

**(F) Cord-Connected Appliances.** Cord-connected appliances, such as washing machines, clothes dryers, refrigerators, and the electrical system of gas ranges, and so on, shall be grounded by means of an approved cord with equipment grounding conductor and grounding-type attachment plug.

## 552.57 Bonding of Non–Current-Carrying Metal Parts

**(A) Required Bonding.** All exposed non–current-carrying metal parts that are likely to become energized shall be effectively bonded to the grounding terminal or enclosure of the panelboard.

**(B) Bonding Chassis.** A bonding conductor shall be connected between any panelboard and an accessible terminal on the chassis. Aluminum or copper-clad aluminum conductors shall not be used for bonding if such conductors or their terminals are exposed to corrosive elements.

*Exception: Any park trailer that employs a unitized metal chassis-frame construction to which the panelboard is securely fastened with a bolt(s) and nut(s) or by welding or riveting shall be considered to be bonded.*

**(C) Bonding Conductor Requirements.** Grounding terminals shall be of the solderless type and listed as pressure terminal connectors recognized for the wire size used. The bonding conductor shall be solid or stranded, insulated or bare, and shall be 8 AWG copper minimum or equivalent.

**(D) Metallic Roof and Exterior Bonding.** The metal roof and exterior covering shall be considered bonded where both of the following conditions apply:

(1) The metal panels overlap one another and are securely attached to the wood or metal frame parts by metal fasteners.
(2) The lower panel of the metal exterior covering is secured by metal fasteners at each cross member of the chassis, or the lower panel is connected to the chassis by a metal strap.

**(E) Gas, Water, and Waste Pipe Bonding.** The gas, water, and waste pipes shall be considered grounded if they are bonded to the chassis.

**(F) Furnace and Metal Air Duct Bonding.** Furnace and metal circulating air ducts shall be bonded.

## 552.58 Appliance Accessibility and Fastening

Every appliance shall be accessible for inspection, service, repair, and replacement without removal of permanent construction. Means shall be provided to securely fasten appliances in place when the park trailer is in transit.

## 552.59 Outdoor Outlets, Fixtures, Including Luminaires, Air-Cooling Equipment, and So On

**(A) Listed for Outdoor Use.** Outdoor fixtures, including luminaires, and equipment shall be listed for outdoor use. Outdoor receptacle outlets shall be in accordance with 406.9(A) and (B). Switches and circuit breakers installed outdoors shall comply with 404.4.

**(B) Outside Heating Equipment, Air-Conditioning Equipment, or Both.** A park trailer provided with a branch circuit designed to energize outside heating equipment or air-conditioning equipment, or both, located outside the park trailer, other than room air conditioners, shall have such branch-circuit conductors terminate in a listed outlet box or disconnecting means located on the outside of the park trailer. A label shall be permanently affixed within 150 mm (6 in.) from the listed box or disconnecting means and shall contain the following information:

THIS CONNECTION IS FOR HEATING
AND/OR AIR-CONDITIONING EQUIPMENT.
THE BRANCH CIRCUIT IS RATED AT NOT MORE
THAN ______ AMPERES, AT ______ VOLTS, 60 HZ,
_______ CONDUCTOR AMPACITY.
A DISCONNECTING MEANS SHALL BE
LOCATED WITHIN SIGHT OF THE EQUIPMENT.

The correct voltage and ampere rating shall be given. The tag shall not be less than 0.51 mm (0.020 in.) thick etched brass, stainless steel, anodized or alclad aluminum, or equivalent. The tag shall not be less than 75 mm × 45 mm (3 in. × 1¾ in.) minimum size.

# V. Factory Tests

## 552.60 Factory Tests (Electrical)

Each park trailer shall be subjected to the tests required by 552.60(A) and (B).

**(A) Circuits of 120 Volts or 120/240 Volts.** Each park trailer designed with a 120-volt or a 120/240-volt electrical system shall withstand the applied potential without electrical breakdown of a 1-minute, 900-volt dielectric strength test, or a 1-second, 1080-volt dielectric strength test, with all switches closed, between ungrounded and grounded conductors and the park trailer ground. During the test, all switches and other controls shall be in the on position. Fixtures, including luminaires, and permanently installed appliances shall not be required to withstand this test.

Each park trailer shall be subjected to the following:

(1) A continuity test to ensure that all metal parts are properly bonded
(2) Operational tests to demonstrate that all equipment is properly connected and in working order
(3) Polarity checks to determine that connections have been properly made
(4) Receptacles requiring GFCI protection shall be tested for correct function by the use of a GFCI testing device

**(B) Low-Voltage Circuits.** An operational test of low-voltage circuits shall be conducted to demonstrate that all equipment is connected and in electrical working order. This test shall be performed in the final stages of production after all outer coverings and cabinetry have been secured.

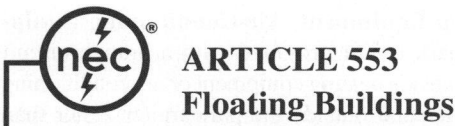

# ARTICLE 553 Floating Buildings

## I. General

### 553.1 Scope

This article covers wiring, services, feeders, and grounding for floating buildings.

Although the *Code* does not cover electrical installations on ships or watercraft, it does cover installations for floating buildings. Floating buildings may be restaurants, aquariums, dwelling units, or many other occupancies that are permanently moored in one location. All other applicable articles of the *Code* apply to these floating buildings.

### 553.2 Definition

**Floating Building.** A building unit, as defined in Article 100, that floats on water, is moored in a permanent location, and has a premises wiring system served through connection by permanent wiring to an electrical supply system not located on the premises.

## II. Services and Feeders

### 553.4 Location of Service Equipment

The service equipment for a floating building shall be located adjacent to, but not in or on, the building or any floating structure. The main overcurrent protective device that feeds the floating structure shall have ground fault protection not exceeding 100 mA. Ground fault protection of each individual branch or feeder circuit shall be permitted as a suitable alternative.

This requirement ensures that supply conductors to a floating building can be disconnected in an emergency, such as during a storm, when the floating building has to be moved quickly. Service equipment is not permitted to be installed on the floating building and any other floating structure such as a wharf or pier.

Overcurrent protection for supply conductors is provided by the service equipment, and since these conductors may develop leakage, ground fault protection is required at this main device or, alternatively, for each feeder or branch circuit. Factors such as corrosion or lack of maintenance may cause ground faults to the metal surfaces of floating buildings or shore-powered vessels. Persons in contact with these metal surfaces, in proximity to the water surrounding the metal surface, or attempting to exit the water via a metal swim platform or ladder may be subjected to an electrical shock. While branch-circuit GFCI devices, which may trip as low as 4 mA, are permitted to be used, this is not practical for all floating buildings. Therefore, the ground-fault current level of the device is not permitted to exceed 100 mA. Devices operating at current levels higher than those specified for a Class A GFCI in UL 943, *Standard for Ground-Fault Circuit Interrupters,* do not provide GFCI protection of personnel. See Article 100 for the definition of *ground-fault circuit interrupter.*

### 553.5 Service Conductors

One set of service conductors shall be permitted to serve more than one set of service equipment.

### 553.6 Feeder Conductors

Each floating building shall be supplied by a single set of feeder conductors from its service equipment.

*Exception: Where the floating building has multiple occupancy, each occupant shall be permitted to be supplied by a single set of feeder conductors extended from the occupant's service equipment to the occupant's panelboard.*

### 553.7 Installation of Services and Feeders

**(A) Flexibility.** Flexibility of the wiring system shall be maintained between floating buildings and the supply conductors. All wiring shall be installed so that motion of the water surface and changes in the water level will not result in unsafe conditions.

**(B) Wiring Methods.** Liquidtight flexible metal conduit or liquidtight flexible nonmetallic conduit with approved fittings shall be permitted for feeders and where flexible connections are required for services. Extra-hard usage portable power cable listed for both wet locations and sunlight resistance shall be permitted for a feeder to a floating building where flexibility is required. Other raceways suitable for the location shall be permitted to be installed where flexibility is not required.

Type LFMC with approved fittings is permitted where flexible connections are required. Where Type W cables from Table 400.4 are used, they must be listed for use in wet locations. Not all Type W cables are listed for wet location applications.

•

## III. Grounding

### 553.8 General Requirements

Grounding at floating buildings shall comply with 553.8(A) through (D).

**(A) Grounding of Electrical and Nonelectrical Parts.** Grounding of both electrical and nonelectrical parts in a floating building shall be through connection to a grounding bus in the building panelboard.

**(B) Installation and Connection of Equipment Grounding Conductor.** The equipment grounding conductor shall be installed with the feeder conductors and connected to a grounding terminal in the service equipment.

**(C) Identification of Equipment Grounding Conductor.** The equipment grounding conductor shall be an insulated copper conductor with a continuous outer finish that is either green or green with one or more yellow stripes. For conductors larger than 6 AWG, or where multiconductor cables are used, re-identification of conductors as allowed in 250.119(A)(2)(b) and (A)(2)(c) or 250.119(B)(b) and (B)(c) shall be permitted.

**(D) Grounding Electrode Conductor Connection.** The grounding terminal in the service equipment shall be grounded by connection through an insulated grounding electrode conductor to a grounding electrode on shore.

This section provides four requirements for grounding a floating building. Grounding of electrical and nonelectrical parts must be connected to a grounding bus at the building panelboard. An EGC must also be included in the feeder supplying the building and connected to the grounding terminal at the service equipment. For conductor sizes 6 AWG and smaller, the EGC must be provided with green insulation or green insulation with a yellow tracer. Finally, an insulated EGC must be installed between the service equipment and the grounding electrode on shore.

### 553.9 Insulated Neutral

The grounded circuit conductor (neutral) shall be an insulated conductor identified in compliance with 200.6. The neutral conductor shall be connected to the equipment grounding terminal in the service equipment, and, except for that connection, it shall be insulated from the equipment grounding conductors, equipment enclosures, and all other grounded parts. The neutral conductor terminals in the panelboard and in ranges, clothes dryers, counter-mounted cooking units, and the like shall be insulated from the enclosures.

### 553.10 Equipment Grounding

**(A) Electrical Systems.** All enclosures and exposed metal parts of electrical systems shall be connected to the grounding bus.

**(B) Cord-Connected Appliances.** Where required to be grounded, cord-connected appliances shall be grounded by means of an equipment grounding conductor in the cord and a grounding-type attachment plug.

### 553.11 Bonding of Non–Current-Carryingand Metal Parts

All metal parts in contact with the water, all metal piping, and all non–current-carrying metal parts that are likely to become energized shall be connected to the grounding bus in the panelboard.

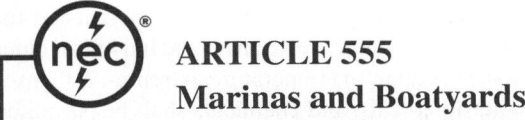

## ARTICLE 555 Marinas and Boatyards

### 555.1 Scope

This article covers the installation of wiring and equipment in the areas comprising fixed or floating piers, wharves, docks, and other areas in marinas, boatyards, boat basins, boathouses, yacht clubs, boat condominiums, docking facilities associated with residential condominiums, any multiple docking facility, or similar occupancies, and facilities that are used, or intended for use, for the purpose of repair, berthing, launching, storage, or fueling of small craft and the moorage of floating buildings.

Private, noncommercial docking facilities constructed or occupied for the use of the owner or residents of the associated single-family dwelling are not covered by this article.

Informational Note: See NFPA 303-2011, *Fire Protection Standard for Marinas and Boatyards*, for additional information.

The requirements of Article 555 apply to public and private docking, storage, repair, and fueling facilities for small craft. The term *small craft* is not defined in the *Code*. However, based on the scope of NFPA 303, *Fire Protection Standard for Marinas and Boatyards*, small craft includes recreational and commercial boats, yachts, and other craft that do not exceed 300 gross tons. For facilities that serve larger craft and ships, see NFPA 307, *Standard for the Construction and Fire Protection of Marine Terminals, Piers, and Wharves*. See Article 553 for requirements for floating buildings, including floating dwelling units.

The requirements of this article apply to stand-alone boathouses, except those constructed and used in association with a single-family dwelling. Electrical installations on docks and piers located at a single-family dwelling are not subject to the requirements of Article 555. However, all requirements in Chapters 1 through 4 for these outdoor, wet locations are applicable.

### 555.2 Definitions

**Electrical Datum Plane.** The electrical datum plane is defined as follows:

(1) In land areas subject to tidal fluctuation, the electrical datum plane is a horizontal plane 606 mm (2 ft) above the highest tide level for the area occurring under normal circumstances, that is, highest high tide.
(2) In land areas not subject to tidal fluctuation, the electrical datum plane is a horizontal plane 606 mm (2 ft) above the highest water level for the area occurring under normal circumstances.
(3) The electrical datum plane for floating piers and landing stages that are (a) installed to permit rise and fall response to water level, without lateral movement, and (b) that are so equipped that they can rise to the datum plane established for (1) or (2), is a horizontal plane 762 mm (30 in.) above the water level at the floating pier or landing stage and a minimum of 305 mm (12 in.) above the level of the deck.

Throughout Article 555, the physical location of electrical equipment is referenced to the electrical datum plane, which is used as a horizontal benchmark on land and on floating piers. The definition of *electrical datum plane* encompasses areas subject to tidal movement and areas in which the water level is affected only by conditions such as climate (rain or snowfall) or by human intervention (the opening or closing of dams and floodgates). In either case, the term covers the normal highest water level, such as astronomical high tides. The term does not cover extremes due to natural or manmade disasters.

**Marine Power Outlet.** An enclosed assembly that can include equipment such as receptacles, circuit breakers, fused switches, fuses, a watt-hour meter(s), panelboards, and monitoring means approved for marine use.

## 555.3 Ground-Fault Protection

The main overcurrent protective device that feeds the marina shall have ground fault protection not exceeding 100 mA. Ground-fault protection of each individual branch or feeder circuit shall be permitted as a suitable alternative.

See the commentary following 553.4 regarding ground-fault protection of the main overcurrent device.

## 555.4 Distribution System

Yard and pier distribution systems shall not exceed 1000 volts phase to phase.

## 555.5 Transformers

Transformers and enclosures shall be specifically approved for the intended location. The bottom of enclosures for transformers shall not be located below the electrical datum plane.

## 555.7 Location of Service Equipment

The service equipment for floating docks or marinas shall be located adjacent to, but not on or in, the floating structure.

The requirement covering service equipment location is similar to that in 553.4 for service equipment supplying floating buildings.

## 555.9 Electrical Connections

Electrical connections shall be located at least 305 mm (12 in.) above the deck of a floating pier. Conductor splices, within approved junction boxes, utilizing sealed wire connector systems listed and identified for submersion shall be permitted where located above the waterline but below the electrical datum plane for floating piers.

All electrical connections shall be located at least 305 mm (12 in.) above the deck of a fixed pier but not below the electrical datum plane.

The use of sealed, waterproof wire-to-wire splices in approved junction boxes that are below the datum plane but above the waterline are permitted where the connector systems are listed and identified for submersion.

## 555.10 Electrical Equipment Enclosures

**(A) Securing and Supporting.** Electrical equipment enclosures installed on piers above deck level shall be securely and substantially supported by structural members, independent of any conduit connected to them. If enclosures are not attached to mounting surfaces by means of external ears or lugs, the internal screw heads shall be sealed to prevent seepage of water through mounting holes.

**(B) Location.** Electrical equipment enclosures on piers shall be located so as not to interfere with mooring lines.

## 555.11 Circuit Breakers, Switches, Panelboards, and Marine Power Outlets

Circuit breakers and switches installed in gasketed enclosures shall be arranged to permit required manual operation without exposing the interior of the enclosure. All such enclosures shall be arranged with a weep hole to discharge condensation.

## 555.12 Load Calculations for Service and Feeder Conductors

General lighting and other loads shall be calculated in accordance with Part III of Article 220, and, in addition, the demand factors set forth in Table 555.12 shall be permitted for each service and/or feeder circuit supplying receptacles that provide shore power for boats. These calculations shall be permitted to be modified as indicated in notes (1) and (2) to Table 555.12. Where demand factors of Table 555.12 are applied, the demand factor specified in 220.61(B) shall not be permitted.

Informational Note: These demand factors may be inadequate in areas of extreme hot or cold temperatures with loaded circuits for heating, air-conditioning, or refrigerating equipment.

The demand factors from Table 555.12 are permitted but not mandatory. For receptacles that supply shore power for boats, the demand factors of Table 555.12 are the only ones permitted by the *Code*. The demand factor from 220.61(B) is not permitted to be added, because then conductors could be undersized.

## 555.13 Wiring Methods and Installation

**(A) Wiring Methods.**

**(1) General.** Wiring methods of Chapter 3 shall be permitted where identified for use in wet locations.

**(2) Portable Power Cables.** Extra-hard usage portable power cables rated not less than 167°F (75°C), 600 volts; listed for both wet locations and sunlight resistance; and having an outer jacket rated to be resistant to temperature extremes, oil, gasoline, ozone, abrasion, acids, and chemicals shall be permitted as follows:

(1) As permanent wiring on the underside of piers (floating or fixed)

*TABLE 555.12 Demand Factors*

| Number of Shore Power Receptacles | Sum of the Rating of the Receptacles (%) |
|---|---|
| 1–4 | 100 |
| 5–8 | 90 |
| 9–14 | 80 |
| 15–30 | 70 |
| 31–40 | 60 |
| 41–50 | 50 |
| 51–70 | 40 |
| ≥71 | 30 |

Notes:
1. Where shore power accommodations provide two receptacles specifically for an individual boat slip and these receptacles have different voltages (for example, one 30 ampere, 125 volt and one 50 ampere, 125/250 volt), only the receptacle with the larger kilowatt demand shall be required to be calculated.
2. If the facility being installed includes individual kilowatt-hour submeters for each slip and is being calculated using the criteria listed in Table 555.12, the total demand amperes may be multiplied by 0.9 to achieve the final demand amperes.

(2) Where flexibility is necessary as on piers composed of floating sections

The cable construction requirements are necessary due to the cables' exposure to extremes in weather conditions and to operational hazards such as oil and gasoline spills. Table 400.4 identifies Types G, PPE, and W as portable power cables suitable for extra-hard usage. The use of these cables on floating piers and docks provides the necessary degree of flexibility to compensate for tidal and wave action.

**(3) Temporary Wiring.** Temporary wiring, except as permitted by Article 590, shall not be used to supply power to boats.

**(B) Installation.**

**(1) Overhead Wiring.** Overhead wiring shall be installed to avoid possible contact with masts and other parts of boats being moved in the yard.

Conductors and cables shall be routed to avoid wiring closer than 6.0 m (20 ft) from the outer edge or any portion of the yard that can be used for moving vessels or stepping or unstepping masts.

**(2) Outside Branch Circuits and Feeders.** Outside branch circuits and feeders shall comply with Article 225 except that clearances for overhead wiring in portions of the yard other than those described in 555.13(B)(1) shall not be less than 5.49 m (18 ft) abovegrade.

**(3) Wiring Over and Under Navigable Water.** Wiring over and under navigable water shall be subject to approval by the authority having jurisdiction.

Approval for wiring over and under navigable water by the AHJ may include federal and local agencies, such as the Army Corps of Engineers, the Coast Guard, or local harbormasters, who have specific authority over the waterways.

Informational Note: See NFPA 303-2011, *Fire Protection Standard for Marinas and Boatyards*, for warning sign requirements.

**(4) Portable Power Cables.**

(a) Where portable power cables are permitted by 555.13(A)(2), the installation shall comply with the following:

(1) Cables shall be properly supported.
(2) Cables shall be located on the underside of the pier.
(3) Cables shall be securely fastened by nonmetallic clips to structural members other than the deck planking.
(4) Cables shall not be installed where subject to physical damage.
(5) Where cables pass through structural members, they shall be protected against chafing by a permanently installed oversized sleeve of nonmetallic material.

(b) Where portable power cables are used as permitted in 555.13(A)(2)(2), there shall be an approved junction box of corrosion-resistant construction with permanently installed terminal blocks on each pier section to which the feeder and feeder extensions are to be connected. A listed marine power outlet employing terminal blocks/bars shall be permitted in lieu of a junction box. Metal junction boxes and their covers, and metal screws and parts that are exposed externally to the boxes, shall be of corrosion-resistant materials or protected by material resistant to corrosion.

**(5) Protection.** Rigid metal conduit, reinforced thermosetting resin conduit (RTRC) listed for aboveground use, or rigid polyvinyl chloride (PVC) conduit suitable for the location, shall be installed to protect wiring above decks of piers and landing stages and below the enclosure that it serves. The conduit shall be connected to the enclosure by full standard threads or fittings listed for use in damp or wet locations, as applicable.

## 555.15 Grounding

Wiring and equipment within the scope of this article shall be grounded as specified in Article 250 and as required by 555.15(A) through (E).

**(A) Equipment to Be Grounded.** The following items shall be connected to an equipment grounding conductor run with the circuit conductors in the same raceway, cable, or trench:

(1) Metal boxes, metal cabinets, and all other metal enclosures
(2) Metal frames of utilization equipment
(3) Grounding terminals of grounding-type receptacles

**(B) Type of Equipment Grounding Conductor.** The equipment grounding conductor shall be an insulated conductor with a continuous outer finish that is either green or green with one or more yellow stripes. The equipment grounding conductor of Type MI cable shall be permitted to be identified at terminations. For conductors larger than 6 AWG, or where multiconductor cables are used, re-identification of conductors as allowed in 250.119(A)(2)(b) and (A)(2)(c) or 250.119(B)(2) and (B)(3) shall be permitted.

The use of an insulated aluminum or copper EGC ensures a high-integrity path for ground-fault current. Because of the corrosive conditions in marinas and boatyards, metal raceways are not permitted to serve as the sole EGC.

**(C) Size of Equipment Grounding Conductor.** The insulated equipment grounding conductor shall be sized in accordance with 250.122 but not smaller than 12 AWG.

**(D) Branch-Circuit Equipment Grounding Conductor.** The insulated equipment grounding conductor for branch circuits shall terminate at a grounding terminal in a remote panelboard or the grounding terminal in the main service equipment.

**(E) Feeder Equipment Grounding Conductors.** Where a feeder supplies a remote panelboard, an insulated equipment grounding conductor shall extend from a grounding terminal in the service equipment to a grounding terminal in the remote panelboard.

## 555.17 Disconnecting Means for Shore Power Connection(s)

Disconnecting means shall be provided to isolate each boat from its supply connection(s).

**(A) Type.** The disconnecting means shall consist of a circuit breaker, switch, or both, and shall be properly identified as to which receptacle it controls.

**(B) Location.** The disconnecting means shall be readily accessible, located not more than 762 mm (30 in.) from the receptacle it controls, and shall be located in the supply circuit ahead of the receptacle. Circuit breakers or switches located in marine power outlets complying with this section shall be permitted as the disconnecting means.

## 555.19 Receptacles

Receptacles shall be mounted not less than 305 mm (12 in.) above the deck surface of the pier and not below the electrical datum plane on a fixed pier.

The location of enclosures for receptacles on fixed and floating piers is based on the electrical datum plane as defined in 555.2. For floating piers, the datum plane is 12 inches above the deck of the pier. The purpose of this requirement is to prevent submersion of receptacle enclosures.

The requirements for enclosures in 555.19(A)(1) address their exposure to the severe weather (wind-driven rain) and environmental conditions (splashing from breaking waves or wakes) frequently encountered at marine locations.

**(A) Shore Power Receptacles.**

**(1) Enclosures.** Receptacles intended to supply shore power to boats shall be housed in marine power outlets listed as marina power outlets or listed for set locations, or shall be installed in listed enclosures protected from the weather or in listed weatherproof enclosures. The integrity of the assembly shall not be affected when the receptacles are in use with any type of booted or nonbooted attachment plug/cap inserted.

**(2) Strain Relief.** Means shall be provided where necessary to reduce the strain on the plug and receptacle caused by the weight and catenary angle of the shore power cord.

**(3) Branch Circuits.** Each single receptacle that supplies shore power to boats shall be supplied from a marine power outlet or panelboard by an individual branch circuit of the voltage class and rating corresponding to the rating of the receptacle.

Informational Note: Supplying receptacles at voltages other than the voltages marked on the receptacle may cause overheating or malfunctioning of connected equipment, for example, supplying single-phase, 120/240-volt, 3-wire loads from a 208Y/120-volt, 3-wire source.

The requirement that each single receptacle that supplies shore power to boats be supplied from an individual branch circuit can be met through the use of multiwire branch circuits derived from single-phase, 3-wire systems or from 3-phase, 4-wire systems. Although the ungrounded conductors of a multiwire branch circuit share the same grounded (neutral) conductor, this configuration can be considered multiple branch circuits in accordance with 210.4(A). See the commentary following 300.13(B) regarding device removal for multiwire branch circuits.

**(4) Ratings.** Shore power for boats shall be provided by single receptacles rated not less than 30 amperes.

Informational Note: For locking- and grounding-type receptacles for auxiliary power to boats, see NFPA 303-2011, *Fire Protection Standard for Marinas and Boatyards*.

(a) Receptacles rated 30 amperes and 50 amperes shall be of the locking and grounding type.

Informational Note: For various configurations and ratings of locking- and grounding-type receptacles and caps, see ANSI/NEMA WD 6-2002 (Rev. 2008), *Standard for Dimensions of Attachment Plugs and Receptacles*.

(b) Receptacles rated 60 amperes and 100 amperes shall be of the pin and sleeve type.

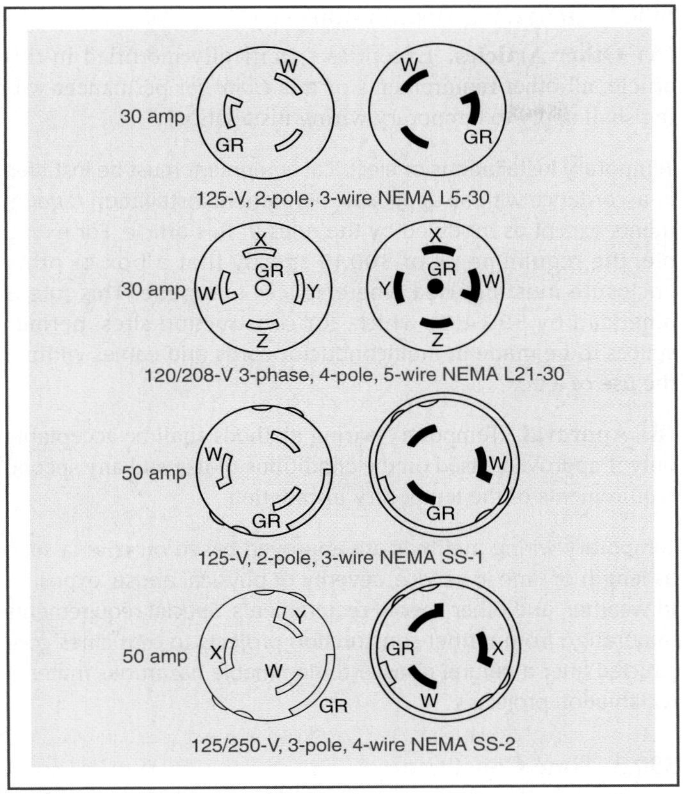


***EXHIBIT 555.1*** *Typical configurations from 30 A to 50 A for single locking- and grounding-type receptacles and attachment plug caps used to provide shore power for boats in marinas and boatyards.*

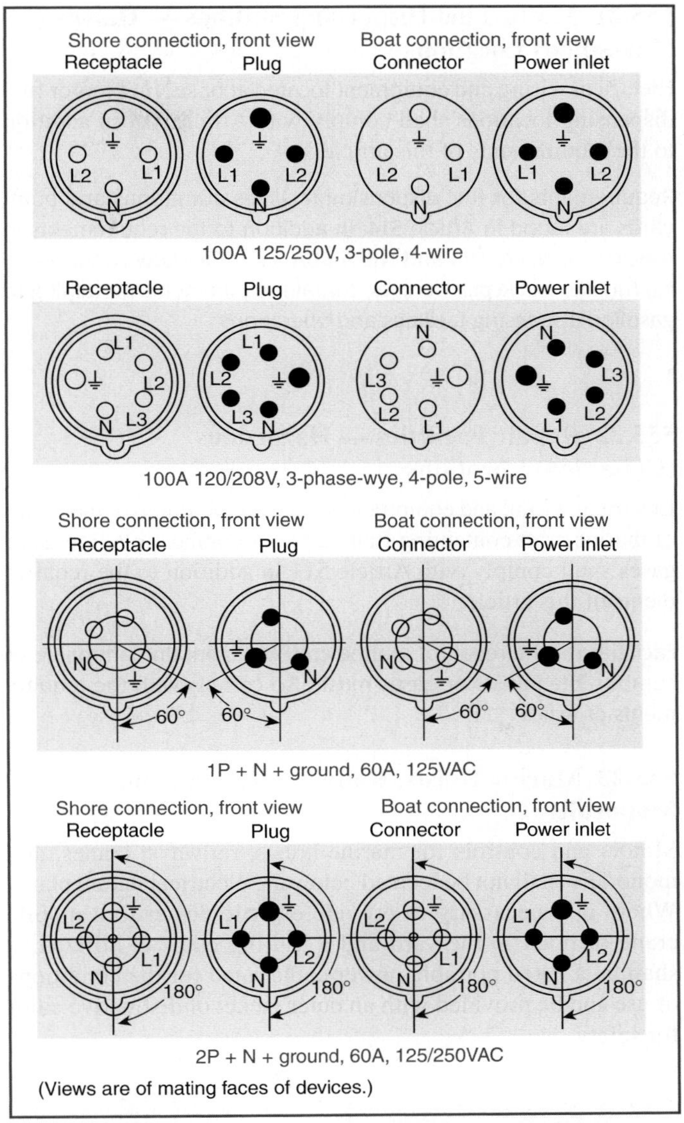


***EXHIBIT 555.2*** *Typical configurations of 60 A or 100 A for safety pin-and-sleeve–type receptacles, plugs, connectors, and power inlets used to provide shore power for boats in marinas and boatyards.*

Informational Note: For various configurations and ratings of pin and sleeve receptacles, see ANSI/UL 1686, *UL Standard for Safety Pin and Sleeve Configurations.*

Single locking- and grounding-type receptacles and attachment caps are required for providing shore power to boats. This facilitates proper connections and prevents unintentional disconnection of on-board equipment, such as bilge pumps, refrigerators, and so forth. Exhibit 555.1 illustrates a chart of grounding-type locking plug and receptacle configurations. Exhibit 555.2 shows pin-and-sleeve–type receptacle configurations.

**(B) Other Than Shore Power.**

**(1) Ground-Fault Circuit-Interrupter (GFCI) Protection for Personnel.** Fifteen- and 20-ampere, single-phase, 125-volt receptacles installed outdoors, in boathouses, in buildings or structures used for storage, maintenance, or repair where portable electrical hand tools, electrical diagnostic equipment, or portable lighting equipment are to be used shall be provided with GFCI protection for personnel. Receptacles in other locations shall be protected in accordance with 210.8(B).

Fifteen- and 20-ampere, single-phase, 125-volt receptacles, other than those supplying shore power to boats and used for maintenance or other purposes at piers, wharves, and so forth, may be of the general-purpose, nonlocking type and are required to be protected by GFCIs. See Exhibit 210.8 for an example of a GFCI receptacle.

**(2) Marking.** Receptacles other than those supplying shore power to boats shall be permitted to be housed in marine power outlets with the receptacles that provide shore power to boats, provided they are marked to clearly indicate that they are not to be used to supply power to boats.

### 555.21 Motor Fuel Dispensing Stations — Hazardous (Classified) Locations

Electrical wiring and equipment located at or serving motor fuel dispensing locations shall comply with Article 514 in addition to the requirements of this article.

Requirements for fuel dispensing facilities at marinas and boatyards are found in Article 514. In addition to the requirements in Article 514, NFPA 303 and NFPA 30A, *Code for Motor Fuel Dispensing Facilities and Repair Garages*, contain requirements pertaining to gasoline dispensing facilities and operations.

•

### 555.22 Repair Facilities — Hazardous (Classified) Locations

Electrical wiring and equipment located at facilities for the repair of marine craft containing flammable or combustible liquids or gases shall comply with Article 511 in addition to the requirements of this article.

Facilities for the repair of marine craft that contain flammable or combustible liquids or gases must also comply with the requirements of Article 511.

### 555.23 Marine Hoists, Railways, Cranes, and Monorails

Motors and controls for marine hoists, railways, cranes, and monorails shall not be located below the electrical datum plane. Where it is necessary to provide electric power to a mobile crane or hoist in the yard and a trailing cable is utilized, it shall be a listed portable power cable rated for the conditions of use and be provided with an outer jacket of distinctive color for safety.

## ARTICLE 590 Temporary Installations

### 590.1 Scope

The provisions of this article apply to temporary electric power and lighting installations.

Temporary installations are temporary as approved by the AHJ. Article 590 applies to any temporary installation whether it is at a transient or permanent location. The installation could be at a construction site, a box store parking lot, or the local craft fair in a field.

### 590.2 All Wiring Installations

**(A) Other Articles.** Except as specifically modified in this article, all other requirements of this *Code* for permanent wiring shall apply to temporary wiring installations.

Temporary installations of electrical equipment must be installed in accordance with all applicable permanent installation requirements except as modified by the rules in this article. For example, the requirements of 300.15 specify that a box or other enclosure must be used where splices are made. This rule is amended by 590.4(G), which, for construction sites, permits splices to be made in multiconductor cords and cables without the use of a box.

**(B) Approval.** Temporary wiring methods shall be acceptable only if approved based on the conditions of use and any special requirements of the temporary installation.

Temporary wiring methods are approved based on criteria such as length of time in service, severity of physical abuse, exposure to weather, and other special requirements. Special requirements may range from tunnel construction projects to tent cities constructed after a natural disaster to flammable hazardous material reclamation projects.

### 590.3 Time Constraints

**(A) During the Period of Construction.** Temporary electric power and lighting installations shali be permitted during the period of construction, remodeling, maintenance, repair, or demolition of buildings, structures, equipment, or similar activities.

**(B) 90 Days.** Temporary electric power and lighting installations shall be permitted for a period not to exceed 90 days for holiday decorative lighting and similar purposes.

The 90-day time limit applies only to temporary electrical installations associated with holiday displays. Other installations are not bound by this time limit.

**(C) Emergencies and Tests.** Temporary electric power and lighting installations shall be permitted during emergencies and for tests, experiments, and developmental work.

**(D) Removal.** Temporary wiring shall be removed immediately upon completion of construction or purpose for which the wiring was installed.

Because temporary wiring installations may not meet all of the requirements for a permanent installation due to the modifications permitted by Article 590, all temporary wiring not only must be disconnected but also removed from the building, structure, or other location of installation.

## 590.4 General

**(A) Services.** Services shall be installed in conformance with Parts I through VIII of Article 230, as applicable.

**(B) Feeders.** Overcurrent protection shall be provided in accordance with 240.4, 240.5, 240.100, and 240.101. Conductors shall be permitted within cable assemblies or within multiconductor cords or cables of a type identified in Table 400.4 for hard usage or extra-hard usage. For the purpose of this section, Type NM and Type NMC cables shall be permitted to be used in any dwelling, building, or structure without any height limitation or limitation by building construction type and without concealment within walls, floors, or ceilings.

*Exception: Single insulated conductors shall be permitted where installed for the purpose(s) specified in 590.3(C), where accessible only to qualified persons.*

**(C) Branch Circuits.** All branch circuits shall originate in an approved power outlet, switchgear, switchboard or panelboard, motor control center, or fused switch enclosure. Conductors shall be permitted within cable assemblies or within multiconductor cord or cable of a type identified in Table 400.4 for hard usage or extra-hard usage. Conductors shall be protected from overcurrent as provided in 240.4, 240.5, and 240.100. For the purposes of this section, Type NM and Type NMC cables shall be permitted to be used in any dwelling, building, or structure without any height limitation or limitation by building construction type and without concealment within walls, floors, or ceilings.

Types NM and NMC cable may be used in any building or structure regardless of building height and construction type.

Temporary feeders and branch circuits are permitted to be cable assemblies, multiconductor cords, or single-conductor cords. Cords must be identified for hard or extra-hard usage according to Table 400.4. Individual conductors, as described in Table 310.104(A), are not permitted as open conductors but may be part of a cable assembly or used in a raceway system. Open or individual conductor feeders are permitted only during emergencies or tests by the exception to 590.4(B).

The basic requirement is that temporary wiring be located and installed so that it will not be physically damaged. Note that hard-usage or extra-hard-usage extension cords are permitted to be laid on the floor.

*Exception: Branch circuits installed for the purposes specified in 590.3(B) or 590.3(C) shall be permitted to be run as single insulated conductors. Where the wiring is installed in accordance with 590.3(B), the voltage to ground shall not exceed 150 volts, the wiring shall not be subject to physical damage, and the conductors shall be supported on insulators at intervals of not more than 3.0 m (10 ft); or, for festoon lighting, the conductors shall be so arranged that excessive strain is not transmitted to the lampholders.*

**(D) Receptacles.**

**(1) All Receptacles.** All receptacles shall be of the grounding type. Unless installed in a continuous metal raceway that qualifies as an equipment grounding conductor in accordance with 250.118 or a continuous metal-covered cable that qualifies as an equipment grounding conductor in accordance with 250.118, all branch circuits shall include a separate equipment grounding conductor, and all receptacles shall be electrically connected to the equipment grounding conductor(s). Receptacles on construction sites shall not be installed on any branch circuit that supplies temporary lighting.

Conductors for lighting and receptacle loads are required to be separate so that the activation of an overcurrent device or GFCI does not de-energize the lighting circuit. Metal cables or raceways must be continuous and qualify as an EGC. If the metal raceway or metal cable is not continuous or does not qualify as an EGC, a separate EGC must be installed.

**(2) Receptacles in Wet Locations.** All 15- and 20-ampere, 125- and 250-volt receptacles installed in a wet location shall comply with 406.9(B)(1).

**(E) Disconnecting Means.** Suitable disconnecting switches or plug connectors shall be installed to permit the disconnection of all ungrounded conductors of each temporary circuit. Multiwire branch circuits shall be provided with a means to disconnect simultaneously all ungrounded conductors at the power outlet or panelboard where the branch circuit originated. Identified handle ties shall be permitted.

**(F) Lamp Protection.** All lamps for general illumination shall be protected from accidental contact or breakage by a suitable luminaire or lampholder with a guard.

Brass shell, paper-lined sockets, or other metal-cased sockets shall not be used unless the shell is grounded.

**(G) Splices.** On construction sites, a box shall not be required for splices or junction connections where the circuit conductors are multiconductor cord or cable assemblies, provided that the equipment grounding continuity is maintained with or without the box. See 110.14(B) and 400.9. A box, conduit body, or terminal fitting having a separately bushed hole for each conductor shall be used wherever a change is made to a conduit or tubing system or a metal-sheathed cable system.

**(H) Protection from Accidental Damage.** Flexible cords and cables shall be protected from accidental damage. Sharp corners and projections shall be avoided. Where passing through doorways or other pinch points, protection shall be provided to avoid damage.

Flexible cords and cables, because of the nature of their temporary use, are permitted to pass through doorways, unlike the requirement in 400.8(3).

**(I) Termination(s) at Devices.** Flexible cords and cables entering enclosures containing devices requiring termination shall be secured to the box with fittings listed for connecting flexible cords and cables to boxes designed for the purpose.

**(J) Support.** Cable assemblies and flexible cords and cables shall be supported in place at intervals that ensure that they will be protected from physical damage. Support shall be in the form of staples, cable ties, straps, or similar type fittings installed so as not to cause damage. Cable assemblies and flexible cords and cables installed as branch circuits or feeders shall not be installed on the floor or on the ground. Extension cords shall not be required to comply with 590.4(J). Vegetation shall not be used for support of overhead spans of branch circuits or feeders.

*Exception: For holiday lighting in accordance with 590.3(B), where the conductors or cables are arranged with strain relief devices, tension take-up devices, or other approved means to avoid damage from the movement of the live vegetation, trees shall be permitted to be used for support of overhead spans of branch-circuit conductors or cables.*

Temporary wiring methods do not have to be supported in accordance with the permanent installation requirements (from Chapter 3) for the particular wiring method. Adequate support is needed only to minimize the possibility of damage to the wiring method during its temporary period of use. The use of vegetation as a support structure for overhead spans of branch-circuit and feeder conductors is not permitted.

The exception allows holiday lighting to be installed and supported by trees for a period of not more than 90 days, provided the wiring is arranged with proper strain relief devices, tension take-up devices, or other means to prevent damage to the conductors from the tree swaying. Note that all temporary wiring must be removed at the end of the temporary period or project.

## 590.5 Listing of Decorative Lighting

Decorative lighting used for holiday lighting and similar purposes, in accordance with 590.3(B), shall be listed.

## 590.6 Ground-Fault Protection for Personnel

Ground-fault protection for personnel for all temporary wiring installations shall be provided to comply with 590.6(A) and (B). This section shall apply only to temporary wiring installations used to supply temporary power to equipment used by personnel during construction, remodeling, maintenance, repair, or demolition of buildings, structures, equipment, or similar activities. This section shall apply to power derived from an electric utility company or from an on-site-generated power source.

**(A) Receptacle Outlets.** Temporary receptacle installations used to supply temporary power to equipment used by personnel during construction, remodeling, maintenance, repair, or demolition of buildings, structures, equipment, or similar activities shall comply with the requirements of 590.6(A)(1) through (A)(3), as applicable.

*Exception: In industrial establishments only, where conditions of maintenance and supervision ensure that only qualified personnel are involved, an assured equipment grounding conductor program as specified in 590.6(B)(2) shall be permitted for only those receptacle outlets used to supply equipment that would create a greater hazard if power were interrupted or having a design that is not compatible with GFCI protection.*

**(1) Receptacle Outlets Not Part of Permanent Wiring.** All 125-volt, single-phase, 15-, 20-, and 30-ampere receptacle outlets that are not a part of the permanent wiring of the building or structure and that are in use by personnel shall have ground-fault circuit-interrupter protection for personnel. Listed cord sets or devices incorporating listed ground-fault circuit-interrupter protection for personnel identified for portable use shall be permitted.

**(2) Receptacle Outlets Existing or Installed as Permanent Wiring.** Ground-fault circuit-interrupter protection for personnel shall be provided for all 125-volt, single-phase, 15-, 20-, and 30-ampere receptacle outlets installed or existing as part of the permanent wiring of the building or structure and used for temporary electric power. Listed cord sets or devices incorporating listed ground-fault circuit-interrupter protection for personnel identified for portable use shall be permitted.

**(3) Receptacles on 15-kW or less Portable Generators.** All 125-volt and 125/250-volt, single-phase, 15-, 20-, and 30-ampere receptacle outlets that are a part of a 15-kW or smaller portable generator shall have listed ground-fault circuit-interrupter protection for personnel. All 15- and 20-ampere, 125- and 250-volt receptacles, including those that are part of a portable generator, used in a damp or wet location shall comply with 406.9(A) and (B). Listed cord sets or devices incorporating listed ground-fault circuit-interrupter protection for personnel identified for portable use shall be permitted for use with 15-kW or less portable generators manufactured or remanufactured prior to January 1, 2011.

The requirements for GFCI protection are based on the type of wiring being used to provide temporary power. For example, receptacles at a construction site may provide power via temporary wiring or via the permanent wiring of the structure. The latter may occur when the premises wiring is available prior to project completion. This requirement applies even where the final occupancy would not require GFCI protection for the receptacle being utilized.

Section 590.6(A)(3) specifically addresses GFCI protection for small generators that are common at construction sites. Generators manufactured prior to January 1, 2011, were not required to provide this protection. Therefore, listed cord sets or other devices are permitted to provide GFCI protection. Exhibits 590.1 and 590.2 show some examples of ways to implement the GFCI requirements for temporary installations.

**(B) Use of Other Outlets.** For temporary wiring installations, receptacles, other than those covered by 590.6(A)(1) through (A)(3) used to supply temporary power to equipment used by personnel during construction, remodeling, maintenance, repair, or demolition of buildings, structures, or equipment, or similar activities, shall have protection in accordance with (B)(1) or the

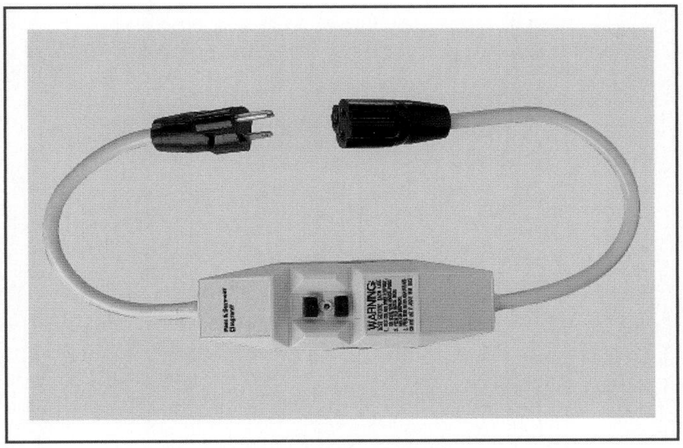

***EXHIBIT 590.1*** *A raintight, portable GFCI with open neutral protection that is designed for use on the line end of a flexible cord. (Courtesy of Legrand/Pass & Seymour®)*

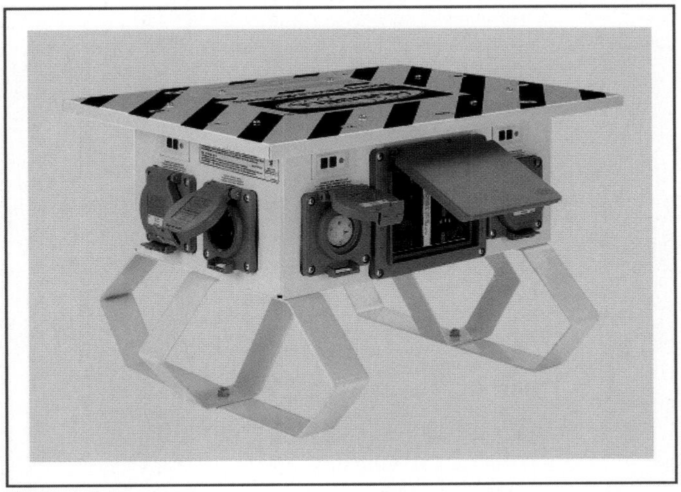

***EXHIBIT 590.2*** *A temporary power outlet unit commonly used on construction sites with a variety of configurations, including GFCI protection. (Courtesy of Hubbell Wiring Device - Kellems)*

assured equipment grounding conductor program in accordance with (B)(2).

**(1) GFCI Protection.** Ground-fault circuit-interrupter protection for personnel.

**(2) Assured Equipment Grounding Conductor Program.** A written assured equipment grounding conductor program continuously enforced at the site by one or more designated persons to ensure that equipment grounding conductors for all cord sets, receptacles that are not a part of the permanent wiring of the building or structure, and equipment connected by cord and plug are installed and maintained in accordance with the applicable requirements of 250.114, 250.138, 406.4(C), and 590.4(D).

(a) The following tests shall be performed on all cord sets, receptacles that are not part of the permanent wiring of the building or structure, and cord-and-plug-connected equipment required to be connected to an equipment grounding conductor:

(1) All equipment grounding conductors shall be tested for continuity and shall be electrically continuous.
(2) Each receptacle and attachment plug shall be tested for correct attachment of the equipment grounding conductor. The equipment grounding conductor shall be connected to its proper terminal.
(3) All required tests shall be performed as follows:
   a. Before first use on site
   b. When there is evidence of damage
   c. Before equipment is returned to service following any repairs
   d. At intervals not exceeding 3 months

(b) The tests required in item (2)(a) shall be recorded and made available to the authority having jurisdiction.

The environmental conditions encountered during activities such as construction or demolition subject personnel to an elevated exposure to electrical shock hazards, including the possibility of electrocution. Requiring GFCI protection of all temporarily installed, 125-volt, single-phase, 15-, 20-, and 30-ampere receptacles is intended to protect personnel using these receptacles from shock hazards that may be encountered during construction and maintenance activities. Other receptacle configurations supplying temporary power must also be GFCI protected or be installed and maintained in accordance with the assured EGC program specified in 590.6(B)(2).

The exception to 590.6(A) is limited in scope and application. The exception applies only to industrial occupancies in which qualified persons will be using 125-volt, single-phase, 15-, 20-, and 30-ampere receptacles. Additionally, either the nature of the equipment being supplied by these receptacles has to be of such importance that the hazard of power interruption outweighs the benefits of GFCI protection, or the equipment has been demonstrated to be incompatible with the proper operation of GFCI protective devices. In instances where those conditions are present, the use of the assured EGC program is permitted. An electrically operated air supply for personnel working in toxic environments is an example of where the loss of power is the greater hazard. Some electrically operated testing equipment has proved to be incompatible with GFCI protection.

The OSHA test requirements are very similar to the *NEC* requirements for an assured grounding program.

According to OSHA 29 CFR 1926.404(b)(1)(iii):

> The employer shall establish and implement an assured equipment grounding conductor program on construction sites covering all cord sets, receptacles which are not a part of the building or structure, and equipment connected by cord and plug which are available for use or used by employees. This program shall comply with the following minimum requirements:
>
> (A) A written description of the program, including the specific procedures adopted by the employer, shall be available at the jobsite for inspection and copying by the Assistant Secretary and any affected employee.
>
> (B) The employer shall designate one or more competent persons to implement the program.

## 590.7 Guarding

For wiring over 600 volts, nominal, suitable fencing, barriers, or other effective means shall be provided to limit access only to authorized and qualified personnel.

# 6 Special Equipment

## ARTICLE 600 Electric Signs and Outline Lighting

## I. General

### 600.1 Scope

This article covers the installation of conductors, equipment, and field wiring for electric signs and outline lighting, regardless of voltage. All installations and equipment using neon tubing, such as signs, decorative elements, skeleton tubing, or art forms, are covered by this article.

Informational Note: Sign and outline lighting illumination systems include, but are not limited to, cold cathode neon tubing, high-intensity discharge lamps (HID), fluorescent or incandescent lamps, light-emitting diodes (LEDs), and electroluminescent and inductance lighting.

Covered under these requirements are signs of the fixed, stationary, and portable self-contained type. Electric signs and outline lighting frequently include sources of illumination identical to those of luminaires; however, the structure and electrical operation of many electric signs are far more complex than simply a set of fluorescent lamps within an enclosure. The terms *electric sign* and *outline lighting* as defined in Article 100 distinguish the function and use of equipment covered by these requirements from the equipment covered by the requirements of Article 410.

Neon tubing is used extensively in the sign industry, and its uses go far beyond the typical electric sign or outline lighting applications. Neon tubing is used in decorative and artistic applications to enhance the indoor and outdoor appearance of buildings and structures. Neon art forms are mounted on enclosures, sign bodies, and other support structures, or they are field-installed skeleton tubing. Depending on how these neon art forms are constructed and installed, they are subject to the requirements of either Part I or Parts I and II.

### 600.2 Definitions

•

**LED Sign Illumination System.** A complete lighting system for use in signs and outline lighting consisting of light-emitting diode (LED) light sources, power supplies, wire, and connectors to complete the installation.

**Neon Tubing.** Electric-discharge luminous tubing, including cold cathode luminous tubing, that is manufactured into shapes to illuminate signs, form letters, parts of letters, skeleton tubing, outline lighting, other decorative elements, or art forms and filled with various inert gases.

•

**Section Sign.** A sign or outline lighting system, shipped as subassemblies, that requires field-installed wiring between the subassemblies to complete the overall sign. The subassemblies are either physically joined to form a single sign unit or are installed as separate remote parts of an overall sign.

A large electric sign may be constructed in multiple factory-wired subassemblies that can be assembled at the sign-installation location. The definition of *section sign* clarifies that the multiple parts are referred to as subassemblies and the only field wiring involved is the connections between subassemblies and connection of the subassemblies to the power source.

The power source may be a branch circuit or the secondary wiring from a sign power supply. In accordance with 600.3, section signs are required to be listed. In accordance with UL 48, *Standard for Electric Signs*, each subassembly is provided with installation instructions containing detailed information on the mechanical and electrical connections that are performed when the subassemblies are installed to form the completed section sign (see Exhibit 600.1).

**Sign Body.** A portion of a sign that may provide protection from the weather but is not an electrical enclosure.

**Skeleton Tubing.** Neon tubing that is itself the sign or outline lighting and is not attached to an enclosure or sign body.

**EXHIBIT 600.1** *Example of a section sign where the subassemblies are field assembled to form a single sign. (Courtesy of Kieffer & Co. Inc.)*

## 600.3 Listing

Fixed, mobile, or portable electric signs, section signs, outline lighting, and retrofit kits, regardless of voltage, shall be listed, provided with installation instructions, and installed in conformance with that listing, unless otherwise approved by special permission.

**(A) Field-Installed Skeleton Tubing.** Field-installed skeleton tubing shall not be required to be listed where installed in conformance with this *Code*.

**(B) Outline Lighting.** Outline lighting shall not be required to be listed as a system when it consists of listed luminaires wired in accordance with Chapter 3.

Electric signs and outline lighting are occasionally designed for a very specific purpose and may be a one-time construction for which the manufacturer might not obtain listing. Listing or approval by special permission helps ensure that electrical equipment does not pose a shock or fire hazard.

## 600.4 Markings

**(A) Signs and Outline Lighting Systems.** Signs and outline lighting systems shall be marked with the manufacturer's name, trademark, or other means of identification; and input voltage and current rating.

**(B) Signs with Lampholders for Incandescent Lamps.** Signs and outline lighting systems with lampholders for incandescent lamps shall be marked to indicate the maximum allowable lamp wattage per lampholder. The markings shall be permanently installed, in letters at least 6 mm (¼ in.) high, and shall be located where visible during relamping.

**(C) Visibility.** The markings required in 600.4(A) and listing labels shall not be required to be visible after installation but shall be permanently applied in a location visible during servicing.

The required markings are only required to be visible during servicing of the sign. The markings may be placed within the interior of a sign body or sign equipment enclosure.

**(D) Durability.** Marking labels shall be permanent, durable and, when in wet locations, shall be weatherproof.

**(E) Installation Instructions.** All signs, outline lighting, skeleton tubing systems, and retrofit kits shall be marked to indicate that field wiring and installation instructions are required.

*Exception: Portable, cord-connected signs are not required to be marked.*

## 600.5 Branch Circuits

**(A) Required Branch Circuit.** Each commercial building and each commercial occupancy accessible to pedestrians shall be provided with at least one outlet in an accessible location at each entrance to each tenant space for sign or outline lighting system use. The outlet(s) shall be supplied by a branch circuit rated at least 20 amperes that supplies no other load. Service hallways or corridors shall not be considered accessible to pedestrians.

A 20-ampere outlet on a circuit dedicated to the purpose of supplying an electric sign is required to be installed at the entrance to single-occupant commercial buildings and at the entrance to each occupancy of multiple-occupant commercial buildings (e.g., shopping malls). This requirement is not contingent on whether an electric sign will be installed at the time an occupant moves in, since it is not uncommon to install an electric sign after the space is occupied or when a new occupant moves into an existing space.

**(B) Rating.** Branch circuits that supply signs shall be rated in accordance with 600.5(B)(1) or (B)(2) and shall be considered to be continuous loads for the purposes of calculations.

**(1) Neon Signs.** Branch circuits that supply neon tubing installations shall not be rated in excess of 30 amperes.

**(2) All Other Signs.** Branch circuits that supply all other signs and outline lighting systems shall be rated not to exceed 20 amperes.

Large signs often have load requirements that exceed the ratings permitted by 600.5(B). These signs are typically supplied by a feeder that in turn supplies branch circuits rated within the parameters of this requirement. In some cases, particularly for signs installed along highways, a utility service dedicated to the sign is provided. The rating of the feeder or service is not limited by this requirement. Because sign loads are continuous, the conductors and OCPDs for circuits supplying sign loads have to be sized in accordance with the rules for continuous loads contained in Articles 210, 215, and 230.

**(C) Wiring Methods.** Wiring methods used to supply signs shall comply with 600.5(C)(1), (C)(2), and (C)(3).

**(1) Supply.** The wiring method used to supply signs and outline lighting systems shall terminate within a sign, an outline lighting system enclosure, a suitable box, or a conduit body.

**(2) Enclosures as Pull Boxes.** Signs and transformer enclosures shall be permitted to be used as pull or junction boxes for conductors supplying other adjacent signs, outline lighting systems, or floodlights that are part of a sign and shall be permitted to contain both branch and secondary circuit conductors.

**(3) Metal or Nonmetallic Poles.** Metal or nonmetallic poles used to support signs shall be permitted to enclose supply conductors, provided the poles and conductors are installed in accordance with 410.30(B).

## 600.6 Disconnects

Each sign and outline lighting system, feeder circuit or branch circuit supplying a sign, outline lighting system, or skeleton tubing shall be controlled by an externally operable switch or circuit breaker that opens all ungrounded conductors and controls no other load. The switch or circuit breaker shall open all ungrounded conductors simultaneously on multi-wire branch circuits in accordance with 210.4(B). Signs and outline lighting systems located within fountains shall have the disconnect located in accordance with 680.12.

*Exception No. 1: A disconnecting means shall not be required for an exit directional sign located within a building.*

*Exception No. 2: A disconnecting means shall not be required for cord-connected signs with an attachment plug.*

**(A) Location.**

**(1) At Point of Entry to a Sign Enclosure.** The disconnect shall be located at the point the feeder circuit or branch circuit(s) supplying a sign or outline lighting system enters a sign enclosure or a pole in accordance with 600.5(C)(3) and shall disconnect all wiring where it enters the enclosure of the sign or pole.

*Exception: A disconnect shall not be required for branch or feeder circuits passing through the sign where enclosed in a Chapter 3 listed raceway.*

**(2) Within Sight of the Sign.** The disconnecting means shall be within sight of the sign or outline lighting system that it controls. Where the disconnecting means is out of the line of sight from any section that is able to be energized, the disconnecting means shall be lockable in accordance with 110.25.

**(3) Within Sight of the Controller.** The following shall apply for signs or outline lighting systems operated by electronic or electromechanical controllers located external to the sign or outline lighting system:

(1) The disconnecting means shall be located within sight of the controller or in the same enclosure with the controller.
(2) The disconnecting means shall disconnect the sign or outline lighting system and the controller from all ungrounded supply conductors.
(3) The disconnecting means shall be designed such that no pole can be operated independently and shall be lockable in accordance with 110.25.

For signs or outline lighting systems operated by mechanical or electromechanical controllers located external to the sign, the disconnecting means is required to be located within sight of or in the same enclosure as the controller and must be capable of being locked in the open position. This requirement enhances safe working conditions for persons servicing the controller or the sign.

**(B) Control Switch Rating.** Switches, flashers, and similar devices controlling transformers and electronic power supplies shall be rated for controlling inductive loads or have a current rating not less than twice the current rating of the transformer.

A switching device that controls the primary circuit of a transformer supplying a luminous gas tube is subject to a highly inductive load that causes severe arcing of its contacts. Therefore, the switch or flasher is required to be rated for the inductive load, or it must have a current rating that is at least twice the current rating of the transformer it controls.

•

## 600.7 Grounding and Bonding

**(A) Grounding.**

**(1) Equipment Grounding.** Metal equipment of signs, outline lighting, and skeleton tubing systems shall be grounded by connection to the equipment grounding conductor of the supply branch circuit(s) or feeder using the types of equipment grounding conductors specified in 250.118.

*Exception: Portable cord-connected signs shall not be required to be connected to the equipment grounding conductor where protected by a system of double insulation or its equivalent. Double insulated equipment shall be distinctively marked.*

**(2) Size of Equipment Grounding Conductor.** The equipment grounding conductor size shall be in accordance with 250.122 based on the rating of the overcurrent device protecting the branch circuit or feeder conductors supplying the sign or equipment.

**(3) Connections.** Equipment grounding conductor connections shall be made in accordance with 250.130 and in a method specified in 250.8.

**(4) Auxiliary Grounding Electrode.** Auxiliary grounding electrode(s) shall be permitted for electric signs and outline lighting systems covered by this article and shall meet the requirements of 250.54.

**(5) Metal Building Parts.** Metal parts of a building shall not be permitted as a secondary return conductor or an equipment grounding conductor.

**(B) Bonding.**

**(1) Bonding of Metal Parts.** Metal parts and equipment of signs and outline lighting systems shall be bonded together and to the associated transformer or power-supply equipment grounding conductor of the branch circuit or feeder supplying the sign or outline lighting system and shall meet the requirements of 250.90.

*Exception: Remote metal parts of a section sign or outline lighting system only supplied by a remote Class 2 power supply shall not be required to be bonded to an equipment grounding conductor.*

**(2) Bonding Connections.** Bonding connections shall be made in accordance with 250.8.

**(3) Metal Building Parts.** Metal parts of a building shall not be permitted to be used as a means for bonding metal parts and equipment of signs or outline lighting systems together or to the transformer or power-supply equipment grounding conductor of the supply circuit.

**(4) Flexible Metal Conduit Length.** Listed flexible metal conduit or listed liquidtight flexible metal conduit that encloses the secondary circuit conductor from a transformer or power supply for use with neon tubing shall be permitted as a bonding means if the total accumulative length of the conduit in the secondary circuit does not exceed 30 m (100 ft).

**(5) Small Metal Parts.** Small metal parts not exceeding 50 mm (2 in.) in any dimension, not likely to be energized, and spaced at least 19 mm (¾ in.) from neon tubing, shall not require bonding.

**(6) Nonmetallic Conduit.** Where listed nonmetallic conduit is used to enclose the secondary circuit conductor from a transformer or power supply and a bonding conductor is required, the bonding conductor shall be installed separate and remote from the nonmetallic conduit and be spaced at least 38 mm (1½ in.) from the conduit when the circuit is operated at 100 Hz or less or 45 mm (1¾ in.) when the circuit is operated at over 100 Hz.

**(7) Bonding Conductors.** Bonding conductors shall comply with (1) and (2).

(1) Bonding conductors shall be copper and not smaller than 14 AWG.
(2) Bonding conductors installed externally of a sign or raceway shall be protected from physical damage.

**(8) Signs in Fountains.** Signs or outline lighting installed inside a fountain shall have all metal parts bonded to the equipment grounding conductor of the branch circuit for the fountain recirculating system. The bonding connection shall be as near as practicable to the fountain and shall be permitted to be made to metal piping systems that are bonded in accordance with 680.53.

Informational Note: Refer to 600.32(J) for restrictions on length of high-voltage secondary conductors.

All metal parts larger than 2 inches are required to be bonded. A common practice in the sign industry is to use flexible metal raceways to enclose the conductors supplied from the secondary circuit of a transformer or electronic power supply. In addition to providing protection from physical damage, the FMC or LFMC is permitted as the bonding means for non–current-carrying metal parts where the total length of the conduit in the secondary circuit does not exceed 100 feet.

Secondary circuit raceways normally contain only one conductor, which is connected to one side of the neon tube. Where PVC conduit or LFNC is used and any sign parts are required to be bonded, the bonding conductor(s) must be run outside of and be separated from the nonmetallic conduit. Installing bonding conductors inside the nonmetallic conduit with secondary power-supply conductors could increase the chance of failure of the conductor or nonmetallic tubing.

## 600.8 Enclosures

Live parts, other than lamps, and neon tubing shall be enclosed. Transformers and power supplies provided with an integral enclosure, including a primary and secondary circuit splice enclosure, shall not require an additional enclosure.

**(A) Strength.** Enclosures shall have ample structural strength and rigidity.

**(B) Material.** Sign and outline lighting system enclosures shall be constructed of metal or shall be listed.

**(C) Minimum Thickness of Enclosure Metal.** Sheet copper or aluminum shall be at least 0.51 mm (0.020 in.) thick. Sheet steel shall be at least 0.41 mm (0.016 in.) thick.

**(D) Protection of Metal.** Metal parts of equipment shall be protected from corrosion.

## 600.9 Location

**(A) Vehicles.** Sign or outline lighting system equipment shall be at least 4.3 m (14 ft) above areas accessible to vehicles unless protected from physical damage.

**(B) Pedestrians.** Neon tubing, other than listed, dry-location, portable signs, readily accessible to pedestrians shall be protected from physical damage.

Informational Note: See 600.41(D) for additional requirements.

**(C) Adjacent to Combustible Materials.** Signs and outline lighting systems shall be installed so that adjacent combustible materials are not subjected to temperatures in excess of 90°C (194°F).

The spacing between wood or other combustible materials and an incandescent or HID lamp or lampholder shall not be less than 50 mm (2 in.).

**(D) Wet Location.** Signs and outline lighting system equipment for wet location use, other than listed watertight type, shall be weatherproof and have drain holes, as necessary, in accordance with the following:

(1) Drain holes shall not be larger than 13 mm (½ in.) or smaller than 6 mm (¼ in.).
(2) Every low point or isolated section of the equipment shall have at least one drain hole.
(3) Drain holes shall be positioned such that there will be no external obstructions.

## 600.10 Portable or Mobile Signs

These requirements address the safety concerns associated with signs that are frequently moved and that may be used in damp or wet environments.

**(A) Support.** Portable or mobile signs shall be adequately supported and readily movable without the use of tools.

**(B) Attachment Plug.** An attachment plug shall be provided for each portable or mobile sign.

**(C) Wet or Damp Location.** Portable or mobile signs in wet or damp locations shall comply with 600.10(C)(1) and (C)(2).

**(1) Cords.** All cords shall be junior hard-service or hard-service types as designated in Table 400.4 and have an equipment grounding conductor.

**(2) Ground-Fault Circuit Interrupter.** The manufacturer of portable or mobile signs shall provide listed ground-fault circuit-interrupter protection for personnel. The ground-fault circuit interrupter shall be an integral part of the attachment plug or shall be located in the power-supply cord within 300 mm (12 in.) of the attachment plug.

The GFCIs required for portable electric signs must have integral *open-neutral* protection in accordance with UL 48, *Standard for Electric Signs.* An interruption of the neutral conductor on the supply side of the GFCI disables the protection circuitry. Open-neutral protection ensures that if damage to the supply cord causes a break in the grounded conductor, both conductors on the load-side circuit to the portable sign will be opened and no voltage will be present at the sign. These protective devices are required to be original equipment installed by the manufacturer as an integrated GFCI device in the attachment plug or as part of the cord as shown in Exhibit 600.2.

**(D) Dry Location.** Portable or mobile signs in dry locations shall meet the following:

(1) Cords shall be SP-2, SPE-2, SPT-2, or heavier, as designated in Table 400.4.
(2) The cord shall not exceed 4.5 m (15 ft) in length.

## 600.12 Field-Installed Secondary Wiring

Field-installed secondary circuit wiring for electric signs, retrofit kits, outline lighting systems, and skeleton tubing systems shall be in accordance with their installation instructions and 600.12(A), (B), or (C).

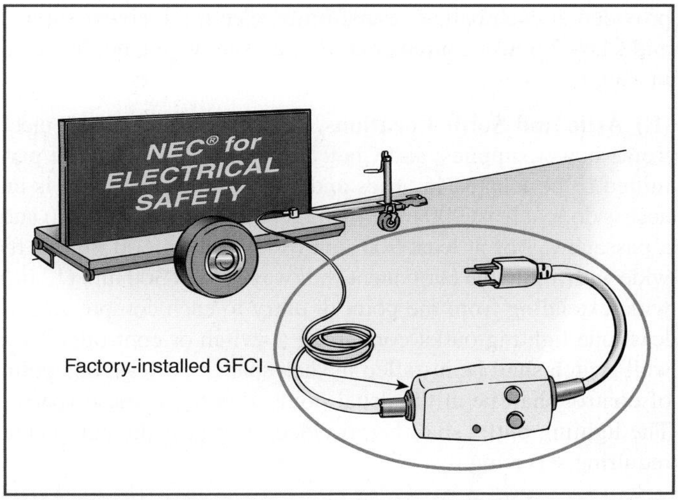


***EXHIBIT 600.2*** *A factory-installed GFCI device located in the power-supply cord within 12 inches of the attachment plug.*

**(A) 1000 Volts or Less.** Neon and secondary circuit wiring of 1000 volts or less shall comply with 600.31.

**(B) Over 1000 Volts.** Neon secondary circuit wiring of over 1000 volts shall comply with 600.32.

**(C) Class 2.** Where the installation complies with 600.33 and the power source provides a Class 2 output that complies with 600.24, either of the following wiring methods shall be permitted as determined by the installation instructions and conditions.

(1) Wiring methods identified in Chapter 3
(2) Class 2 cables complying with Part III of Article 725

## 600.21 Ballasts, Transformers, Electronic Power Supplies, and Class 2 Power Sources

Ballasts, transformers, electronic power supplies, and Class 2 power sources shall be of the self-contained type or be enclosed by placement in a listed sign body or listed separate enclosure.

**(A) Accessibility.** Ballasts, transformers, electronic power supplies, and Class 2 power sources shall be located where accessible and shall be securely fastened in place.

**(B) Location.** Ballasts, transformers, electronic power supplies, and Class 2 power sources shall be installed as near to the lamps or neon tubing as practicable to keep the secondary conductors as short as possible.

**(C) Wet Location.** Ballasts, transformers, electronic power supplies, and Class 2 power sources used in wet locations shall be of the weatherproof type or be of the outdoor type and protected from the weather by placement in a sign body or separate enclosure.

**(D) Working Space.** A working space at least 900 mm (3 ft) high × 900 mm (3 ft) wide × 900 mm (3 ft) deep shall be provided at each ballast, transformer, electronic power supply, and Class 2 power source or at its enclosure where not installed in a sign.

**(E) Attic and Soffit Locations.** Ballasts, transformers, electronic power supplies, and Class 2 power sources shall be permitted to be located in attics and soffits, provided there is an access door at least 900 mm × 562.5 mm (36 in. × 22½ in.) and a passageway of at least 900 mm (3 ft) high × 600 mm (2 ft) wide with a suitable permanent walkway at least 300 mm (12 in.) wide extending from the point of entry to each component. At least one lighting outlet containing a switch or controlled by a wall switch shall be installed in such spaces. At least one point of control shall be at the usual point of entry to these spaces. The lighting outlet shall be provided at or near the equipment requiring servicing.

**(F) Suspended Ceilings.** Ballasts, transformers, electronic power supplies, and Class 2 power sources shall be permitted to be located above suspended ceilings, provided that their enclosures are securely fastened in place and not dependent on the suspended-ceiling grid for support. Ballasts, transformers, and electronic power supplies installed in suspended ceilings shall not be connected to the branch circuit by flexible cord.

### 600.22 Ballasts

**(A) Type.** Ballasts shall be identified for the use and shall be listed.

**(B) Thermal Protection.** Ballasts shall be thermally protected.

### 600.23 Transformers and Electronic Power Supplies

**(A) Type.** Transformers and electronic power supplies shall be identified for the use and shall be listed.

**(B) Secondary-Circuit Ground-Fault Protection.** Transformers and electronic power supplies other than the following shall have secondary-circuit ground-fault protection:

(1) Transformers with isolated ungrounded secondaries and with a maximum open circuit voltage of 7500 volts or less
(2) Transformers with integral porcelain or glass secondary housing for the neon tubing and requiring no field wiring of the secondary circuit

**(C) Voltage.** Secondary-circuit voltage shall not exceed 15,000 volts, nominal, under any load condition. The voltage to ground of any output terminals of the secondary circuit shall not exceed 7500 volts, under any load condition.

**(D) Rating.** Transformers and electronic power supplies shall have a secondary-circuit current rating of not more than 300 mA.

**(E) Secondary Connections.** Secondary circuit outputs shall not be connected in parallel or in series.

**(F) Marking.** Transformers and electronic power supplies that are equipped with secondary-circuit ground-fault protection shall be so marked.

### 600.24 Class 2 Power Sources

Signs and outline lighting systems supplied by Class 2 transformers, power supplies, and power sources shall comply with the requirements of Class 2 circuits and 600.24(A), (B), (C), and (D).

**(A) Listing.** Class 2 power supplies and power sources shall be listed for use with electric signs and outline lighting systems or shall be a component in a listed electric sign.

**(B) Grounding.** Metal parts of signs and outline lighting systems shall be grounded and bonded in accordance with 600.7.

**(C) Wiring Methods on the Supply Side of the Class 2 Power Supply.** Conductors and equipment on the supply side of the power source shall be installed in accordance with the appropriate requirements of Chapter 3.

**(D) Secondary Wiring.** Secondary wiring from Class 2 power sources shall comply with 600.12(C) and 600.33.

## II. Field-Installed Skeleton Tubing, Outline Lighting, and Secondary Wiring

### 600.30 Applicability

Part II of this article shall apply to all of the following:

(1) Field-installed skeleton tubing
(2) Field-installed secondary circuits
(3) Outline lighting

These requirements are in addition to the requirements of Part I.

### 600.31 Neon Secondary-Circuit Wiring, 1000 Volts or Less, Nominal

**(A) Wiring Method.** Conductors shall be installed using any wiring method included in Chapter 3 suitable for the conditions.

**(B) Insulation and Size.** Conductors shall be listed, insulated, and not smaller than 18 AWG.

**(C) Number of Conductors in Raceway.** The number of conductors in a raceway shall be in accordance with Table 1 of Chapter 9.

**(D) Installation.** Conductors shall be installed so they are not subject to physical damage.

**(E) Protection of Leads.** Bushings shall be used to protect wires passing through an opening in metal.

### 600.32 Neon Secondary-Circuit Wiring, over 1000 Volts, Nominal

**(A) Wiring Methods.**

**(1) Installation.** Conductors shall be installed in rigid metal conduit, intermediate metal conduit, liquidtight flexible nonmetallic conduit, flexible metal conduit, liquidtight flexible metal conduit, electrical metallic tubing, metal enclosures; on insulators in metal raceways; or in other equipment listed for use with neon secondary circuits over 1000 volts.

**(2) Number of Conductors.** Conduit or tubing shall contain only one conductor.

**(3) Size.** Conduit or tubing shall be a minimum of metric designator 16 (trade size ½).

**(4) Spacing from Grounded Parts.** Other than at the location of connection to a metal enclosure or sign body, nonmetallic conduit or flexible nonmetallic conduit shall be spaced no less than 38 mm (1½ in.) from grounded or bonded parts when the conduit contains a conductor operating at 100 Hz or less, and shall be spaced no less than 45 mm (1¾ in.) from grounded or bonded parts when the conduit contains a conductor operating at more than 100 Hz.

Where installed in nonmetallic conduit, gas tubing sign (GTO) cable located in close proximity to a grounded surface may result in damaging stress to the cable insulation due to capacitive coupling and the resulting production of ozone.

**(5) Metal Building Parts.** Metal parts of a building shall not be permitted as a secondary return conductor or an equipment grounding conductor.

**(B) Insulation and Size.** Conductors shall be insulated, listed as gas tube sign and ignition cable type GTO, rated for 5, 10, or 15 kV, not smaller than 18 AWG, and have a minimum temperature rating of 105°C (221°F).

Informative Annex A identifies the product standard for GTO cable as UL 814, *Gas-Tube-Sign Cable*.

**(C) Installation.** Conductors shall be so installed that they are not subject to physical damage.

**(D) Bends in Conductors.** Sharp bends in insulated conductors shall be avoided.

**(E) Spacing.** Secondary conductors shall be separated from each other and from all objects other than insulators or neon tubing by a spacing of not less than 38 mm (1½ in.). GTO cable installed in metal conduit or tubing requires no spacing between the cable insulation and the conduit or tubing.

**(F) Insulators and Bushings.** Insulators and bushings for conductors shall be listed for use with neon secondary circuits over 1000 volts.

**(G) Conductors in Raceways.** The insulation on all conductors shall extend not less than 65 mm (2½ in.) beyond the metal conduit or tubing.

**(H) Between Neon Tubing and Midpoint Return.** Conductors shall be permitted to run between the ends of neon tubing or to the secondary circuit midpoint return of listed transformers or listed electronic power supplies and provided with terminals or leads at the midpoint.

**(I) Dwelling Occupancies.** Equipment having an open circuit voltage exceeding 1000 volts shall not be installed in or on dwelling occupancies.

**(J) Length of Secondary Circuit Conductors.**

**(1) Secondary Conductor to the First Electrode.** The length of secondary circuit conductors from a high-voltage terminal or lead of a transformer or electronic power supply to the first neon tube electrode shall not exceed the following:

(1) 6 m (20 ft) where installed in metal conduit or tubing
(2) 15 m (50 ft) where installed in nonmetallic conduit

**(2) Other Secondary Circuit Conductors.** All other sections of secondary circuit conductor in a neon tube circuit shall be as short as practicable.

**(K) Splices.** Splices in high-voltage secondary circuit conductors shall be made in listed enclosures rated over 1000 volts. Splice enclosures shall be accessible after installation and listed for the location where they are installed.

## 600.33 LED Sign Illumination Systems, Secondary Wiring

The wiring methods and materials shall be installed in accordance with the sign manufacturer's installation instructions using any applicable wiring methods from Chapter 3 and the requirements for Class 2 circuits contained in Part III of Article 725, as applicable.

**(A) Insulation and Sizing of Class 2 Conductors.** Listed Class 2 cable that complies with Table shall be installed on the load side of the Class 2 power source. The conductors shall have an ampacity not less than the load to be supplied and shall not be sized smaller than 22 AWG.

**(1) Wet Locations.** Class 2 cable used in a wet location shall be identified for use in wet locations or have a moisture-impervious metal sheath.

**(2) Other Locations.** In other locations, any applicable cable permitted in Table 725.154 shall be permitted to be used.

**(B) Installation.** Secondary wiring shall be installed in accordance with (B)(1) and (B)(2).

(1) Support wiring shall be installed in a neat and workmanlike manner. Cables and conductors installed exposed on the surface of ceilings and sidewalls shall be supported by the building structure in such a manner that the cable is not be damaged by normal building use. Such cables shall be supported by straps, staples, hangers, cable ties, or similar fittings designed and installed so as not to damage the cable. The installation shall also comply with 300.4(D).
(2) Connections in cable and conductors shall be made with listed insulating devices and be accessible after installation. Where made in a wall, connections shall be enclosed in a listed box.

**(C) Protection Against Physical Damage.** Where subject to physical damage, the conductors shall be protected and installed in accordance with 300.4.

**(D) Grounding and Bonding.** Grounding and bonding shall be in accordance with 600.7.

Section 600.7(B)(1), Exception, does not require remote metal parts supplied by a Class 2 power supply to be bonded.

## 600.41 Neon Tubing

A tube that is too long or too small in diameter increases the impedance of the load and thus stresses the transformer insulation. Gas tube sign transformers are designed to operate at or near short-circuit current. Generally, the primary voltage of the transformers is 120 volts, and proper installation and maintenance of transformers and high-voltage secondary conductors minimize the possibility of injury or fire. Secondary conductors should be properly terminated to the tube electrodes and the connections protected from contact by unauthorized persons or contact with any flammable or combustible material. Broken tubes should be replaced or de-energized.

**(A) Design.** The length and design of the tubing shall not cause a continuous overcurrent beyond the design loading of the transformer or electronic power supply.

**(B) Support.** Tubing shall be supported by listed tube supports. The neon tubing shall be supported within 150 mm (6 in.) from the electrode connection.

**(C) Spacing.** A spacing of not less than 6 mm (¼ in.) shall be maintained between the tubing and the nearest surface, other than its support.

**(D) Protection.** Field-installed skeleton tubing shall not be subject to physical damage. Where the tubing is readily accessible to other than qualified persons, field-installed skeleton tubing shall be provided with suitable guards or protected by other approved means.

## 600.42 Electrode Connections

**(A) Points of Transition.** Where the high-voltage secondary circuit conductors emerge from the wiring methods specified in 600.32(A), they shall be enclosed in a listed assembly.

**(B) Accessibility.** Terminals of the electrode shall not be accessible to unqualified persons.

**(C) Electrode Connections.** Connections shall be made by use of a connection device, twisting of the wires together, or use of an electrode receptacle. Connections shall be electrically and mechanically secure and shall be in an enclosure listed for the purpose.

**(D) Support.** Neon secondary conductor(s) shall be supported not more than 150 mm (6 in.) from the electrode connection to the tubing.

**(E) Receptacles.** Electrode receptacles shall be listed.

**(F) Bushings.** Where electrodes penetrate an enclosure, bushings listed for the purpose shall be used unless receptacles are provided.

**(G) Wet Locations.** A listed cap shall be used to close the opening between neon tubing and a receptacle where the receptacle penetrates a building. Where a bushing or neon tubing penetrates a building, the opening between neon tubing and the bushing shall be sealed.

**(H) Electrode Enclosures.** Electrode enclosures shall be listed.

**(1) Dry Locations.** Electrode enclosures that are listed for use in dry, damp, or wet locations shall be permitted to be installed and used in such locations.

**(2) Damp and Wet Locations.** Electrode enclosures installed in damp and wet locations shall be specifically listed and identified for use in such locations.

Informational Note: See 110.3(B) covering installation and use of electrical equipment.

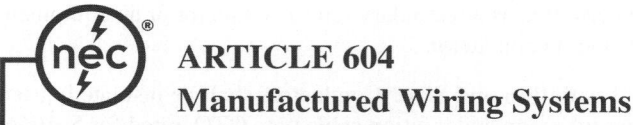

# ARTICLE 604
# Manufactured Wiring Systems

## 604.1 Scope

The provisions of this article apply to field-installed wiring using off-site manufactured subassemblies for branch circuits, remote-control circuits, signaling circuits, and communications circuits in accessible areas.

## 604.2 Definition

**Manufactured Wiring System.** A system containing component parts that are assembled in the process of manufacture and cannot be inspected at the building site without damage or destruction to the assembly and used for the connection of luminaires, utilization equipment, continuous plug-in type busways, and other devices.

## 604.4 Uses Permitted

Manufactured wiring systems shall be permitted in accessible and dry locations and in ducts, plenums, and other air-handling spaces where listed for this application and installed in accordance with 300.22.

Manufactured wiring systems are typically constructed of Type AC or Type MC cable and are provided with factory connectors and receptacles. The connection devices used with these systems facilitate ease of initial installation and future relocation of equipment. These systems are used extensively for the installation of branch-circuit and tap conductors supplying luminaires in accessible locations, including open and suspended-ceiling construction. Manufactured wiring systems employing flexible conduits, flexible cords, busways, and surface-mounted raceways are also permitted by this article.

*Exception No. 1: In concealed spaces, one end of tapped cable shall be permitted to extend into hollow walls for direct termination at switch and outlet points.*

*Exception No. 2: Manufactured wiring system assemblies installed outdoors shall be listed for use in outdoor locations.*

## 604.5 Uses Not Permitted

Manufactured wiring system types shall not be permitted where limited by the applicable article in Chapter 3 for the wiring method used in its construction.

## 604.6 Construction

**(A) Cable or Conduit Types.**

**(1) Cables.** Cable shall be one of the following:

(1) Listed Type AC cable containing nominal 600-volt, 8 to 12 AWG insulated copper conductors with a bare or insulated copper equipment grounding conductor equivalent in size to the ungrounded conductor.
(2) Listed Type MC cable containing nominal 600-volt, 8 to 12 AWG insulated copper conductors with a bare or insulated copper equipment grounding conductor equivalent in size to the ungrounded conductor.
(3) Listed Type MC cable containing nominal 600-volt, 8 to 12 AWG insulated copper conductors with a grounding conductor and armor assembly listed and identified for grounding in accordance with 250.118(10). The combined metallic sheath and grounding conductor shall have a current-carrying capacity equivalent to that of the ungrounded copper conductor.

Other cables as listed in 725.154, 800.113, 820.113, and 830.179 shall be permitted in manufactured wiring systems for wiring of equipment within the scope of their respective articles.

**(2) Conduits.** Conduit shall be listed flexible metal conduit or listed liquidtight flexible conduit containing nominal 600-volt, 8 to 12 AWG insulated copper conductors with a bare or insulated copper equipment grounding conductor equivalent in size to the ungrounded conductor.

*Exception No. 1 to (1) and (2): A luminaire tap, no longer than 1.8 m (6 ft) and intended for connection to a single luminaire, shall be permitted to contain conductors smaller than 12 AWG but not smaller than 18 AWG.*

*Exception No. 2 to (1) and (2): Listed manufactured wiring assemblies containing conductors smaller than 12 AWG shall be permitted for remote-control, signaling, or communication circuits.*

*Exception No. 3 to (2): Listed manufactured wiring systems containing unlisted flexible metal conduit of noncircular cross section or trade sizes smaller than permitted by 348.20(A), or both, shall be permitted where the wiring systems are supplied with fittings and conductors at the time of manufacture.*

**(3) Flexible Cord.** Flexible cord suitable for hard usage, with minimum 12 AWG conductors, shall be permitted as part of a listed factory-made assembly not exceeding 1.8 m (6 ft) in length when making a transition between components of a manufactured wiring system and utilization equipment not permanently secured to the building structure. The cord shall be visible for the entire length, shall not be subject to physical damage, and shall be provided with identified strain relief.

Flexible cord facilitates a transition between manufactured wiring systems and utilization equipment found in display cases, merchandise racks, temporary workstations, and the like. This transition is limited, however, to hard-usage cord not over 6 feet in length, to minimize damage, as illustrated in Exhibit 604.1. Examples of polarized receptacles and connectors are shown in Exhibit 604.2.

*Exception: Listed electric-discharge luminaires that comply with 410.62(C) shall be permitted with conductors smaller than 12 AWG.*

This exception and the requirements in 410.62(C)(1) permit the use of flexible cord equipped with a manufactured wiring system connector as a means to supply listed electric-discharge luminaires such as fluorescent or high-intensity discharge types. In this application, the cord-equipped luminaires are supplied from branch-circuit conductors installed using a manufactured wiring system. This method of supplying luminaires is permitted only where the cord is visible for its entire length, from its attachment to the luminaire to its interface with the branch-circuit conductors of the manufactured wiring system. Where used for connection of listed electric-discharge luminaires, listed manufactured wiring system cord assemblies not longer than 6 feet and containing conductors smaller than 12 AWG copper are permitted.

**(4) Busways.** Busways shall be listed continuous plug-in type containing factory-mounted, bare or insulated conductors, which shall be copper or aluminum bars, rods, or tubes. The busway shall be provided with an equipment ground. The busway shall be rated nominal 600 volts, 20, 30, or 40 amperes. Busways shall be installed in accordance with 368.12, 368.17(D), and 368.30.

**(5) Raceway.** Prewired, modular, surface-mounted raceways shall be listed for the use, rated nominal 600 volts, 20 amperes, and installed in accordance with 386.12, 386.30, 386.60, and 386.100.

Metal and nonmetallic surface-mounted raceways prewired as a manufactured wiring system are required to be listed specifically for the application. ANSI/UL 183, *Manufactured Wiring Systems,* covers the construction of these systems. Article 380 contains the

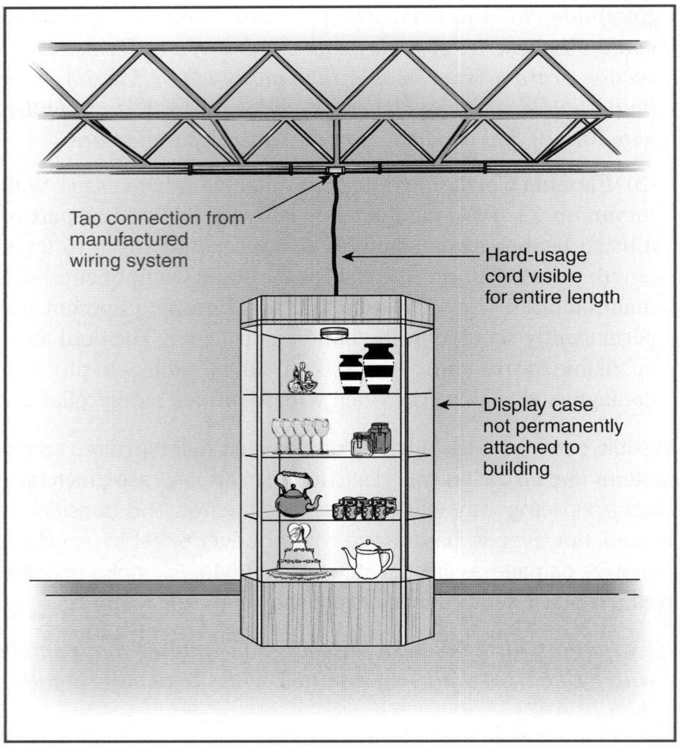


**EXHIBIT 604.1** *Transition wiring between a manufactured wiring system and utilization equipment.*

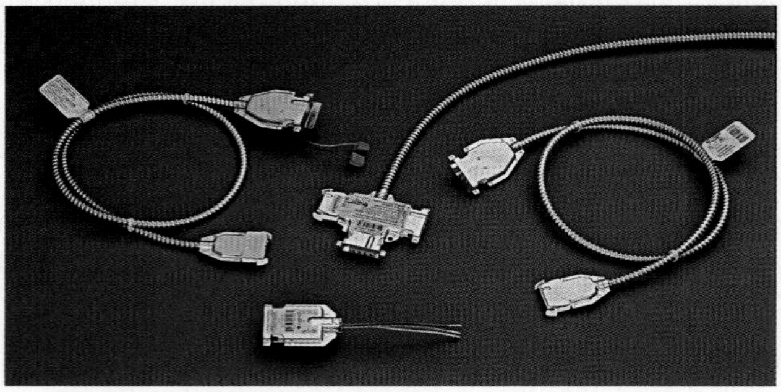

**EXHIBIT 604.2** *Polarized receptacles and connectors of a manufactured wiring system. (Courtesy of RELOC® Wiring Solutions, an Acuity Brands Company)*

requirements for the installation of surface, flush, or freestanding raceways that contain conductors and receptacles. Although multioutlet assemblies are permitted to be assembled in the field, listed multioutlet assemblies are covered by ANSI/UL 5, *Surface Metal Raceways and Fittings,* and ANSI UL 5A, *Nonmetallic Surface Raceways and Fittings.*

**(B) Marking.** Each section shall be marked to identify the type of cable, flexible cord, or conduit.

**(C) Receptacles and Connectors.** Receptacles and connectors shall be of the locking type, uniquely polarized and identified for the purpose, and shall be part of a listed assembly for the appropriate system. All connector openings shall be designed to prevent inadvertent contact with live parts or capped to effectively close the connector openings.

**(D) Other Component Parts.** Other component parts shall be listed for the appropriate system.

### 604.7 Installation

Manufactured wiring systems shall be secured and supported in accordance with the applicable cable or conduit article for the cable or conduit type employed.

The securing and supporting requirements for manufactured wiring systems are taken from the specific article in Chapter 3 that covers the wiring method employed in the system construction.

## ARTICLE 605
## Office Furnishings

### 605.1 Scope

This article covers electrical equipment, lighting accessories, and wiring systems used to connect, contained within, or installed on office furnishings.

This article covers electrical equipment and conductors installed in office furnishings. Office furnishings may be partitions that are freestanding or fixed, but are not as permanent as a conventional stud-and-wallboard type of partition. They may also be storage units, desks, and workstations that are interconnected much in the way traditional office partitions are connected. The electrical equipment and devices shown in Exhibit 605.1 are components of a modular office furnishing system.

### 605.2 Definition

**Office Furnishing.** Cubicle panels, partitions, study carrels, workstations, desks, shelving systems, and storage units that may be mechanically and electrically interconnected to form an office furnishing system.

### 605.3 General

Wiring systems shall be identified as suitable for providing power for lighting accessories and utilization equipment used within office furnishings. A wired partition shall not extend from floor to ceiling.

*Exception: Where permitted by the authority having jurisdiction, these relocatable wired partitions shall be permitted to extend to, but shall not penetrate, the ceiling.*

**EXHIBIT 605.1** *Example of a fixed-type office partition where the branch-circuit wiring is run through the partition.*

**(A) Use.** These assemblies shall be installed and used only as provided for by this article.

**(B) Hazardous (Classified) Locations.** Where used in hazardous (classified) locations, these assemblies shall comply with Articles 500 through 517 in addition to this article.

## 605.4 Wireways

All conductors and connections shall be contained within wiring channels of metal or other material identified as suitable for the conditions of use. Wiring channels shall be free of projections or other conditions that might damage conductor insulation.

A wiring channel that is separate from the channel containing the branch circuits for light and power may be provided within the system components for the routing of communications, signaling, and optical fiber cables.

## 605.5 Office Furnishing Interconnections

The electrical connection between office furnishings shall be a flexible assembly identified for use with office furnishings or shall be permitted to be installed using flexible cord, provided that all the following conditions are met:

(1) The cord is extra-hard usage type with 12 AWG or larger conductors, with an insulated equipment grounding conductor.
(2) The office furnishings are mechanically contiguous.
(3) The cord is not longer than necessary for maximum positioning of the office furnishing but is in no case to exceed 600 mm (2 ft).
(4) The cord is terminated at an attachment plug-and-cord connector with strain relief.

## 605.6 Lighting Accessories

Lighting equipment shall be listed and identified for use with office furnishings and shall comply with 605.6(A), (B), and (C).

**(A) Support.** A means for secure attachment or support shall be provided.

**(B) Connection.** Where cord and plug connection is provided, it shall comply with all of the following:

(1) The cord length shall be suitable for the intended application but shall not exceed 2.7 m (9 ft) in length.
(2) The cord shall not be smaller than 18 AWG.
(3) The cord shall contain an equipment grounding conductor, except as specified in 605.6(B)(4).
(4) Cords on the load side of a listed Class 2 power source shall not be required to contain an equipment grounding conductor.
(5) The cord shall be of the hard usage type, except as specified in 605.6(B)(6).
(6) A cord provided on a listed Class 2 power source shall be of the type provided with the listed luminaire assembly or of the type specified in 725.130 and 725.127.
(7) Connection by other means shall be identified as suitable for the conditions of use.

**(C) Receptacle Outlet.** Receptacles shall not be permitted in lighting accessories.

## 605.7 Fixed-Type Office Furnishings

Office furnishings that are fixed (secured to building surfaces) shall be permanently connected to the building electrical system by one of the wiring methods of Chapter 3.

## 605.8 Freestanding-Type Office Furnishings

Office furnishings of the freestanding type (not fixed) shall be permitted to be connected to the building electrical system by one of the wiring methods of Chapter 3.

Office furnishings that are attached to the building are required by 605.7 to be connected to the premises wiring with a permanent wiring method. Freestanding furnishings are permitted by 605.8 to connect with a permanent wiring method or by 605.9 to connect with a flexible cord. Although multiwire branch circuits are not permitted in cord-connected furnishings by 605.9(D), they may be supplied by multiple branch circuits.

### 605.9 Freestanding-Type Office Furnishings, Cord- and Plug-Connected

Individual office furnishings of the freestanding type, or groups of individual office furnishings that are electrically connected, are mechanically contiguous, and do not exceed 9.0 m (30 ft) when assembled, shall be permitted to be connected to the building electrical system by a single flexible cord and plug, provided that all of the conditions of 605.9(A) through (D) are met.

**(A) Flexible Power-Supply Cord.** The flexible power supply cord shall be extra-hard usage type with 12 AWG or larger conductors with an insulated equipment grounding conductor and shall not exceed 600 mm (2 ft) in length.

**(B) Receptacle Supplying Power.** The receptacle(s) supplying power shall be on a separate circuit serving only the office furnishing and no other loads and shall be located not more than 300 mm (12 in.) from the office furnishing that is connected to it.

**(C) Receptacle Outlets, Maximum.** An individual office furnishing or groups of interconnected individual office furnishings shall not contain more than 13 15-ampere, 125-volt receptacle outlets.

**(D) Multiwire Circuits, Not Permitted.** An individual office furnishing or groups of interconnected office furnishings shall not contain multiwire circuits.

Informational Note: See 210.4 for circuits supplying office furnishings in 605.7 and 605.8.

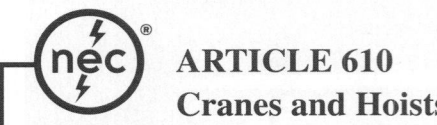

# ARTICLE 610 Cranes and Hoists

## I. General

### 610.1 Scope

This article covers the installation of electrical equipment and wiring used in connection with cranes, monorail hoists, hoists, and all runways.

Informational Note: For further information, see ANSI B30, *Safety Code for Cranes, Derricks, Hoists, Jacks, and Slings*.

Electric cranes, such as the ones shown in Exhibit 610.1, present unique challenges to ensuring that electrical safety is maintained. Constant movement of the crane requires flexibility of power and control wiring or the use of contact conductors installed along the crane runway or bridge. The duty cycle of crane motors is addressed in Table 610.14(A), which covers conductor ampacities for short-time–rated crane and hoist motors.

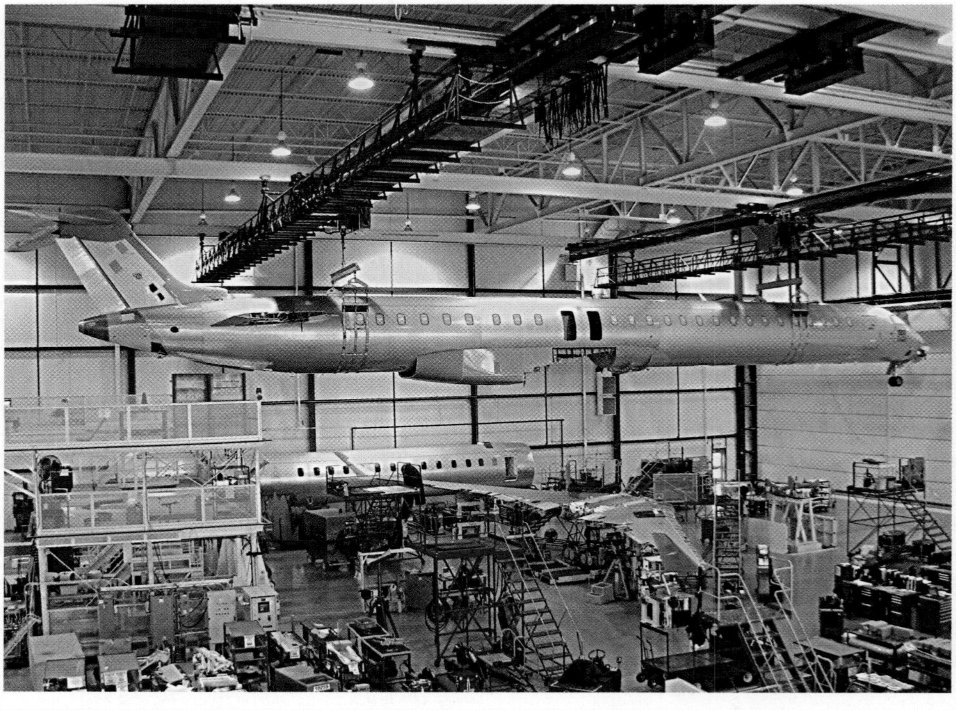

***EXHIBIT 610.1*** *New CRJ1000 NextGen aircraft fuselage being hoisted at Bombardier Aerospace's Mirabel Manufacturing Plant. (Courtesy of Bombardier Aerospace)*

### 610.2 Definition

**Festoon Cable.** Single- and multiple-conductor cable intended for use and installation in accordance with Article 610 where flexibility is required.

Informational Note: Festoon cable consists of one or more insulated conductors cabled together with an overall jacket. It is rated 60°C (140°F), 75°C (167°F), 90°C (194°F), or 105°C (221°F) and 600 V.

### 610.3 Special Requirements for Particular Locations

**(A) Hazardous (Classified) Locations.** All equipment that operates in a hazardous (classified) location shall conform to Article 500.

**(1) Class I Locations.** Equipment used in locations that are hazardous because of the presence of flammable gases or vapors shall conform to Article 501.

**(2) Class II Locations.** Equipment used in locations that are hazardous because of combustible dust shall conform to Article 502.

**(3) Class III Locations.** Equipment used in locations that are hazardous because of the presence of easily ignitible fibers or flyings shall conform to Article 503.

For more details on cranes and hoists in Class III locations, see the commentary following 503.155(D).

**(B) Combustible Materials.** Where a crane, hoist, or monorail hoist operates over readily combustible material, the resistors shall be located as permitted in the following:

(1) A well ventilated cabinet composed of noncombustible material constructed so that it does not emit flames or molten metal
(2) A cage or cab constructed of noncombustible material that encloses the sides of the cage or cab from the floor to a point at least 150 mm (6 in.) above the top of the resistors

**(C) Electrolytic Cell Lines.** See 668.32.

Special precautions are necessary on electrolytic cell lines to prevent the introduction of exposed grounded parts. Conductive surfaces of cranes in the cell line work zone are to be insulated from ground as described in 668.32.

## II. Wiring

### 610.11 Wiring Method

Conductors shall be enclosed in raceways or be Type AC cable with insulated grounding conductor, Type MC cable, or Type MI cable unless otherwise permitted or required in 610.11(A) through (E).

Type AC cable with an insulated wire-type EGC, terminated on the grounding terminals of crane- and hoist-associated equipment, is required to ensure the continuity of the grounding and bonding connection to equipment that is frequently subject to vibration.

**(A) Contact Conductor.** Contact conductors shall not be required to be enclosed in raceways.

**(B) Exposed Conductors.** Short lengths of exposed conductors at resistors, collectors, and other equipment shall not be required to be enclosed in raceways.

Short runs of open conductors facilitate connection to resistors, collectors, and similar equipment. Each conductor is required by 610.12 to be provided with separately bushed holes in boxes as well as in cable and raceway fittings used where the transition to open wiring is made.

**(C) Flexible Connections to Motors and Similar Equipment.** Where flexible connections are necessary, flexible stranded conductors shall be used. Conductors shall be in flexible metal conduit, liquidtight flexible metal conduit, liquidtight flexible nonmetallic conduit, multiconductor cable, or an approved nonmetallic flexible raceway.

**(D) Pushbutton Station Multiconductor Cable.** Where multiconductor cable is used with a suspended pushbutton station, the station shall be supported in some satisfactory manner that protects the electrical conductors against strain.

Exhibit 610.2 shows an example of suitable strain relief for a cord that supports a control pushbutton station for an overhead crane.

***EXHIBIT 610.2*** *A suitable strain relief grip for a cord-suspended pushbutton station.*

**(E) Flexibility to Moving Parts.** Where flexibility is required for power or control to moving parts, listed festoon cable or a cord suitable for the purpose shall be permitted, provided the following apply:

(1) Suitable strain relief and protection from physical damage is provided.
(2) In Class I, Division 2 locations, the cord is approved for extra-hard usage.

## 610.12 Raceway or Cable Terminal Fittings

Conductors leaving raceways or cables shall comply with either 610.12(A) or (B).

**(A) Separately Bushed Hole.** A box or terminal fitting that has a separately bushed hole for each conductor shall be used wherever a change is made from a raceway or cable to exposed wiring. A fitting used for this purpose shall not contain taps or splices and shall not be used at luminaire outlets.

**(B) Bushing in Lieu of a Box.** A bushing shall be permitted to be used in lieu of a box at the end of a rigid metal conduit, intermediate metal conduit, or electrical metallic tubing where the raceway terminates at unenclosed controls or similar equipment, including contact conductors, collectors, resistors, brakes, power-circuit limit switches, and dc split-frame motors.

## 610.13 Types of Conductors

Conductors shall comply with Table 310.104(A) unless otherwise permitted in 610.13(A) through (D).

**(A) Exposed to External Heat or Connected to Resistors.** A conductor(s) exposed to external heat or connected to resistors shall have a flame-resistant outer covering or be covered with flame-resistant tape individually or as a group.

**(B) Contact Conductors.** Contact conductors along runways, crane bridges, and monorails shall be permitted to be bare and shall be copper, aluminum, steel, or other alloys or combinations thereof in the form of hard-drawn wire, tees, angles, tee rails, or other stiff shapes.

**(C) Flexibility.** Where flexibility is required, listed flexible cord or cable, or listed festoon cable, shall be permitted to be used and, where necessary, cable reels or take-up devices shall be used.

**(D) Class 1, Class 2, and Class 3 Circuits.** Conductors for Class 1, Class 2, and Class 3 remote-control, signaling, and power-limited circuits, installed in accordance with Article 725, shall be permitted.

This cross-reference to Article 725 is necessary because Article 725 is not permitted to modify Article 610, as explained in 90.3. This section allows the methods of wiring control and signaling circuits permitted in Article 725 to be used for crane- and hoist-control circuits.

## 610.14 Rating and Size of Conductors

**(A) Ampacity.** The allowable ampacities of conductors shall be as shown in Table 610.14(A).

Informational Note: For the ampacities of conductors between controllers and resistors, see 430.23.

**(B) Secondary Resistor Conductors.** Where the secondary resistor is separate from the controller, the minimum size of the conductors between controller and resistor shall be calculated by multiplying the motor secondary current by the appropriate factor from Table 610.14(B) and selecting a wire from Table 610.14(A).

**(C) Minimum Size.** Conductors external to motors and controls shall be not smaller than 16 AWG unless otherwise permitted in (1) or (2):

(1) 18 AWG wire in multiconductor cord shall be permitted for control circuits not exceeding 7 amperes.
(2) Wires not smaller than 20 AWG shall be permitted for electronic circuits.

**(D) Contact Conductors.** Contact wires shall have an ampacity not less than that required by Table 610.14(A) for 75°C (167°F) wire, and in no case shall they be smaller than as shown in Table 610.14(D).

**(E) Calculation of Motor Load.**

**(1) Single Motor.** For one motor, 100 percent of motor nameplate full-load ampere rating shall be used.

**(2) Multiple Motors on Single Crane or Hoist.** For multiple motors on a single crane or hoist, the minimum ampacity of the power supply conductors shall be the nameplate full-load ampere rating of the largest motor or group of motors for any single crane motion, plus 50 percent of the nameplate full-load ampere rating of the next largest motor or group of motors, using that column of Table 610.14(A) that applies to the longest time-rated motor.

**(3) Multiple Cranes or Hoists on a Common Conductor System.** For multiple cranes, hoists, or both, supplied by a common conductor system, calculate the motor minimum ampacity for each crane as defined in 610.14(E), add them together, and multiply the sum by the appropriate demand factor from Table 610.14(E)

**(F) Other Loads.** Additional loads, such as heating, lighting, and air conditioning, shall be provided for by application of the appropriate sections of this *Code*.

**(G) Nameplate.** Each crane, monorail, or hoist shall be provided with a visible nameplate marked with the manufacturer's name, rating in volts, frequency, number of phases, and circuit amperes as calculated in 610.14(E) and (F).

***TABLE 610.14(A)*** *Ampacities of Insulated Copper Conductors Used with Short-Time Rated Crane and Hoist Motors. Based on Ambient Temperature of 30°C (86°F).*

| Maximum Operating Temperature | Up to Four Simultaneously Energized Conductors in Raceway or Cable[1] | | | | Up to Three ac[2] or Four dc[1] Simultaneously Energized Conductors in Raceway or Cable | | Maximum Operating Temperature |
|---|---|---|---|---|---|---|---|
| | 75°C (167°F) | | 90°C (194°F) | | 125°C (257°F) | | |
| | Types MTW, RHW, THW, THWN, XHHW, USE, ZW | | Types TA, TBS, SA, SIS, PFA, FEP, FEPB, RHH, THHN, XHHW, Z, ZW | | Types FEP, FEPB, PFA, PFAH, SA, TFE, Z, ZW | | |
| **Size (AWG or kcmil)** | **60 Min** | **30 Min** | **60 Min** | **30 Min** | **60 Min** | **30 Min** | **Size (AWG or kcmil)** |
| 16 | 10 | 12 | — | — | — | — | 16 |
| 14 | 25 | 26 | 31 | 32 | 38 | 40 | 14 |
| 12 | 30 | 33 | 36 | 40 | 45 | 50 | 12 |
| 10 | 40 | 43 | 49 | 52 | 60 | 65 | 10 |
| 8 | 55 | 60 | 63 | 69 | 73 | 80 | 8 |
| 6 | 76 | 86 | 83 | 94 | 101 | 119 | 6 |
| 5 | 85 | 95 | 95 | 106 | 115 | 134 | 5 |
| 4 | 100 | 117 | 111 | 130 | 133 | 157 | 4 |
| 3 | 120 | 141 | 131 | 153 | 153 | 183 | 3 |
| 2 | 137 | 160 | 148 | 173 | 178 | 214 | 2 |
| 1 | 143 | 175 | 158 | 192 | 210 | 253 | 1 |
| 1/0 | 190 | 233 | 211 | 259 | 253 | 304 | 1/0 |
| 2/0 | 222 | 267 | 245 | 294 | 303 | 369 | 2/0 |
| 3/0 | 280 | 341 | 305 | 372 | 370 | 452 | 3/0 |
| 4/0 | 300 | 369 | 319 | 399 | 451 | 555 | 4/0 |
| 250 | 364 | 420 | 400 | 461 | 510 | 635 | 250 |
| 300 | 455 | 582 | 497 | 636 | 587 | 737 | 300 |
| 350 | 486 | 646 | 542 | 716 | 663 | 837 | 350 |
| 400 | 538 | 688 | 593 | 760 | 742 | 941 | 400 |
| 450 | 600 | 765 | 660 | 836 | 818 | 1042 | 450 |
| 500 | 660 | 847 | 726 | 914 | 896 | 1143 | 500 |

**AMPACITY CORRECTION FACTORS**

| Ambient Temperature (°C) | For ambient temperatures other than 30°C (86°F), multiply the ampacities shown above by the appropriate factor shown below. | | | | | | Ambient Temperature (°F) |
|---|---|---|---|---|---|---|---|
| 21–25 | 1.05 | 1.05 | 1.04 | 1.04 | 1.02 | 1.02 | 70–77 |
| 26–30 | 1.00 | 1.00 | 1.00 | 1.00 | 1.00 | 1.00 | 79–86 |
| 31–35 | 0.94 | 0.94 | 0.96 | 0.96 | 0.97 | 0.97 | 88–95 |
| 36–40 | 0.88 | 0.88 | 0.91 | 0.91 | 0.95 | 0.95 | 97–104 |
| 41–45 | 0.82 | 0.82 | 0.87 | 0.87 | 0.92 | 0.92 | 106–113 |
| 46–50 | 0.75 | 0.75 | 0.82 | 0.82 | 0.89 | 0.89 | 115–122 |
| 51–55 | 0.67 | 0.67 | 0.76 | 0.76 | 0.86 | 0.86 | 124–131 |
| 56–60 | 0.58 | 0.58 | 0.71 | 0.71 | 0.83 | 0.83 | 133–140 |
| 61–70 | 0.33 | 0.33 | 0.58 | 0.58 | 0.76 | 0.76 | 142–158 |
| 71–80 | — | — | 0.41 | 0.41 | 0.69 | 0.69 | 160–176 |
| 81–90 | — | — | — | — | 0.61 | 0.61 | 177–194 |
| 91–100 | — | — | — | — | 0.51 | 0.51 | 195–212 |
| 101–120 | — | — | — | — | 0.40 | 0.40 | 213–248 |

Note: Other insulations shown in Table 310.104(A) and approved for the temperature and location shall be permitted to be substituted for those shown in Table 610.14(A). The allowable ampacities of conductors used with 15-minute motors shall be the 30-minute ratings increased by 12 percent.

[1]For 5 to 8 simultaneously energized power conductors in raceway or cable, the ampacity of each power conductor shall be reduced to a value of 80 percent of that shown in this table.

[2]For 4 to 6 simultaneously energized 125°C (257°F) ac power conductors in raceway or cable, the ampacity of each power conductor shall be reduced to a value of 80 percent of that shown in this table.

**TABLE 610.14(B)** *Secondary Conductor Rating Factors*

| Time in Seconds | | Ampacity of Wire in Percent of Full-Load Secondary Current |
|---|---|---|
| On | Off | |
| 5 | 75 | 35 |
| 10 | 70 | 45 |
| 15 | 75 | 55 |
| 15 | 45 | 65 |
| 15 | 30 | 75 |
| 15 | 15 | 85 |
| Continuous Duty | | 110 |

**TABLE 610.14(D)** *Minimum Contact Conductor Size Based on Distance Between Supports*

| Minimum Size of Wire (AWG) | Maximum Distance Between End Strain Insulators or Clamp-Type Intermediate Supports |
|---|---|
| 6 | 9.0 m (30 ft) or less |
| 4 | 18 m (60 ft) or less |
| 2 | Over 18 m (60 ft) |

**TABLE 610.14(E)** *Demand Factors*

| Number of Cranes or Hoists | Demand Factor |
|---|---|
| 2 | 0.95 |
| 3 | 0.91 |
| 4 | 0.87 |
| 5 | 0.84 |
| 6 | 0.81 |
| 7 | 0.78 |

### 610.15 Common Return

Where a crane or hoist is operated by more than one motor, a common-return conductor of proper ampacity shall be permitted.

## III. Contact Conductors

### 610.21 Installation of Contact Conductors

Contact conductors shall comply with 610.21(A) through (H).

**(A) Locating or Guarding Contact Conductors.** Runway contact conductors shall be guarded, and bridge contact conductors shall be located or guarded in such a manner that persons cannot inadvertently touch energized current-carrying parts.

**(B) Contact Wires.** Wires that are used as contact conductors shall be secured at the ends by means of approved strain insulators and shall be mounted on approved insulators so that the extreme limit of displacement of the wire does not bring the latter within less than 38 mm (1½ in.) from the surface wired over.

**(C) Supports Along Runways.** Main contact conductors carried along runways shall be supported on insulating supports placed at intervals not exceeding 6.0 m (20 ft) unless otherwise permitted in 610.21(F).

Such conductors shall be separated at not less than 150 mm (6 in.), other than for monorail hoists where a spacing of not less than 75 mm (3 in.) shall be permitted. Where necessary, intervals between insulating supports shall be permitted to be increased up to 12 m (40 ft), the separation between conductors being increased proportionately.

**(D) Supports on Bridges.** Bridge wire contact conductors shall be kept at least 65 mm (2½ in.) apart, and, where the span exceeds 25 m (80 ft), insulating saddles shall be placed at intervals not exceeding 15 m (50 ft).

**(E) Supports for Rigid Conductors.** Conductors along runways and crane bridges, that are of the rigid type specified in 610.13(B) and not contained within an approved enclosed assembly, shall be carried on insulating supports spaced at intervals of not more than 80 times the vertical dimension of the conductor, but in no case greater than 4.5 m (15 ft), and spaced apart sufficiently to give a clear electrical separation of conductors or adjacent collectors of not less than 25 mm (1 in.).

**(F) Track as Circuit Conductor.** Monorail, tram rail, or crane runway tracks shall be permitted as a conductor of current for one phase of a 3-phase, ac system furnishing power to the carrier, crane, or trolley, provided all of the following conditions are met:

(1) The conductors supplying the other two phases of the power supply are insulated.
(2) The power for all phases is obtained from an insulating transformer.
(3) The voltage does not exceed 300 volts.
(4) The rail serving as a conductor shall be bonded to the equipment grounding conductor at the transformer and also shall be permitted to be grounded by the fittings used for the suspension or attachment of the rail to a building or structure.

Crane runway tracks are permitted as a current-carrying conductor where part of a 3-phase system is furnishing power to the crane. The track is also permitted to be grounded through the metal supporting means attached to the building's metal frame.

**(G) Electrical Continuity of Contact Conductors.** All sections of contact conductors shall be mechanically joined to provide a continuous electrical connection.

**(H) Not to Supply Other Equipment.** Contact conductors shall not be used as feeders for any equipment other than the crane(s) or hoist(s) that they are primarily designed to serve.

### 610.22 Collectors

Collectors shall be designed so as to reduce to a minimum sparking between them and the contact conductor; and, where operated in rooms used for the storage of easily ignitible combustible fibers and materials, they shall comply with 503.155.

## IV. Disconnecting Means

### 610.31 Runway Conductor Disconnecting Means

A disconnecting means that has a continuous ampere rating not less than that calculated in 610.14(E) and (F) shall be provided between the runway contact conductors and the power supply. The disconnecting means shall comply with 430.109. This disconnecting means shall be as follows:

(1) Readily accessible and operable from the ground or floor level
(2) Lockable open in accordance with 110.25
(3) Open all ungrounded conductors simultaneously
(4) Placed within view of the runway contact conductors

*Exception: The runway conductor disconnecting means for electrolytic cell lines shall be permitted to be placed out of view of the runway contact conductors where either of the following conditions are met:*

*(1) Where a location in view of the contact conductors is impracticable or introduces additional or increased hazards to persons or property*
*(2) In industrial installations, with written safety procedures, where conditions of maintenance and supervision ensure that only qualified persons service the equipment*

### 610.32 Disconnecting Means for Cranes and Monorail Hoists

A disconnecting means in compliance with 430.109 shall be provided in the leads from the runway contact conductors or other power supply on all cranes and monorail hoists. The disconnecting means shall be lockable open in accordance with 110.25.

Where a monorail hoist or hand-propelled crane bridge installation meets all of the following, the disconnecting means shall be permitted to be omitted:

(1) The unit is controlled from the ground or floor level.
(2) The unit is within view of the power supply disconnecting means.
(3) No fixed work platform has been provided for servicing the unit.

Where the disconnecting means is not readily accessible from the crane or monorail hoist operating station, means shall be provided at the operating station to open the power circuit to all motors of the crane or monorail hoist.

Many crane installations are not arranged so that the unit is within view of the power-supply disconnecting means. When one crane is being serviced, another unit on the same system could remain energized, and could be run into the person performing maintenance on the crane. Therefore, a disconnecting means (lock-open type) must be provided in the contact conductors to disconnect all power to the system.

### 610.33 Rating of Disconnecting Means

The continuous ampere rating of the switch or circuit breaker required by 610.32 shall not be less than 50 percent of the combined short-time ampere rating of the motors or less than 75 percent of the sum of the short-time ampere rating of the motors required for any single motion.

## V. Overcurrent Protection

### 610.41 Feeders, Runway Conductors

**(A) Single Feeder.** The runway supply conductors and main contact conductors of a crane or monorail shall be protected by an overcurrent device(s) that shall not be greater than the largest rating or setting of any branch-circuit protective device plus the sum of the nameplate ratings of all the other loads with application of the demand factors from Table 610.14(E).

**(B) More Than One Feeder Circuit.** Where more than one feeder circuit is installed to supply runway conductors, each feeder circuit shall be sized and protected in compliance with 610.41(A).

Multiple feeders are sometimes used to supply long runway conductors to minimize voltage drops on the runway conductors.

### 610.42 Branch-Circuit Short-Circuit and Ground-Fault Protection

Branch circuits shall be protected in accordance with 610.42(A). Branch-circuit taps, where made, shall comply with 610.42(B).

**(A) Fuse or Circuit Breaker Rating.** Crane, hoist, and monorail hoist motor branch circuits shall be protected by fuses or inverse-time circuit breakers that have a rating in accordance with Table 430.52. Where two or more motors operate a single motion, the sum of their nameplate current ratings shall be considered as that of a single motor.

**(B) Taps.**

**(1) Multiple Motors.** Where two or more motors are connected to the same branch circuit, each tap conductor to an individual motor shall have an ampacity not less than one-third that of the branch circuit. Each motor shall be protected from overload according to 610.43.

**(2) Control Circuits.** Where taps to control circuits originate on the load side of a branch-circuit protective device, each tap and piece of equipment shall be protected in accordance with 430.72.

**(3) Brake Coils.** Taps without separate overcurrent protection shall be permitted to brake coils.

### 610.43 Overload Protection

**(A) Motor and Branch-Circuit Overload Protection.** Each motor, motor controller, and branch-circuit conductor shall be protected from overload by one of the following means:

(1) A single motor shall be considered as protected where the branch-circuit overcurrent device meets the rating requirements of 610.42.
(2) Overload relay elements in each ungrounded circuit conductor, with all relay elements protected from short circuit by the branch-circuit protection.
(3) Thermal sensing devices, sensitive to motor temperature or to temperature and current, that are thermally in contact with the motor winding(s). A hoist or trolley shall be considered to be protected if the sensing device is connected in the hoist's upper limit switch circuit so as to prevent further hoisting during an overload condition of either motor.

**(B) Manually Controlled Motor.** If the motor is manually controlled, with spring return controls, the overload protective device shall not be required to protect the motor against stalled rotor conditions.

**(C) Multimotor.** Where two or more motors drive a single trolley, truck, or bridge and are controlled as a unit and protected by a single set of overload devices with a rating equal to the sum of their rated full-load currents, a hoist or trolley shall be considered to be protected if the sensing device is connected in the hoist's upper limit switch circuit so as to prevent further hoisting during an overtemperature condition of either motor.

**(D) Hoists and Monorail Hoists.** Hoists and monorail hoists and their trolleys that are not used as part of an overhead traveling crane shall not require individual motor overload protection, provided the largest motor does not exceed 7½ hp and all motors are under manual control of the operator.

## VI. Control

### 610.51 Separate Controllers

Each motor shall be provided with an individual controller unless otherwise permitted in 610.51(A) or (B).

**(A) Motions with More Than One Motor.** Where two or more motors drive a single hoist, carriage, truck, or bridge, they shall be permitted to be controlled by a single controller.

**(B) Multiple Motion Controller.** One controller shall be permitted to be switched between motors, under the following conditions:

(1) The controller has a horsepower rating that is not lower than the horsepower rating of the largest motor.
(2) Only one motor is operated at one time.

### 610.53 Overcurrent Protection

Conductors of control circuits shall be protected against overcurrent. Control circuits shall be considered as protected by overcurrent devices that are rated or set at not more than 300 percent of the ampacity of the control conductors, unless otherwise permitted in 610.53(A) or (B).

**(A) Taps to Control Transformers.** Taps to control transformers shall be considered as protected where the secondary circuit is protected by a device rated or set at not more than 200 percent of the rated secondary current of the transformer and not more than 200 percent of the ampacity of the control circuit conductors.

**(B) Continuity of Power.** Where the opening of the control circuit would create a hazard, as for example, the control circuit of a hot metal crane, the control circuit conductors shall be considered as being properly protected by the branch-circuit overcurrent devices.

### 610.55 Limit Switch

A limit switch or other device shall be provided to prevent the load block from passing the safe upper limit of travel of all hoisting mechanisms.

### 610.57 Clearance

The dimension of the working space in the direction of access to live parts that are likely to require examination, adjustment, servicing, or maintenance while energized shall be a minimum of 750 mm (2½ ft). Where controls are enclosed in cabinets, the door(s) shall either open at least 90 degrees or be removable.

## VII. Grounding

### 610.61 Grounding

All exposed non–current-carrying metal parts of cranes, monorail hoists, hoists, and accessories, including pendant controls, shall be bonded either by mechanical connections or bonding jumpers, where applicable, so that the entire crane or hoist is a ground-fault current path as required or permitted by Article 250, Parts V and VII.

Moving parts, other than removable accessories, or attachments that have metal-to-metal bearing surfaces, shall be considered to be electrically bonded to each other through bearing surfaces for grounding purposes. The trolley frame and bridge frame shall not be considered as electrically grounded through the bridge and trolley wheels and its respective tracks. A separate bonding conductor shall be provided.

These requirements are not intended to allow the trolley frame or bridge frame to serve as the EGC for electrical equipment (such as motors, motor controllers, luminaires, and transformers) on a crane. The EGCs that are run with the circuit conductors are required to be one of the types described in 250.118. Metal-to-metal bearing surfaces of moving parts are considered to be a suitable grounding and bonding connection. However, the bridge and trolley wheel contact with their tracks is not permitted to be used as a reliable grounding and bonding connection. Because dirt or other foreign surfaces could impede the effectiveness of the wheel-to-track contact as a reliable grounding and bonding connection, the bridge and trolley frames of an electric crane are required to be bonded through the use of a separate conductor.

# ARTICLE 620 Elevators, Dumbwaiters, Escalators, Moving Walks, Platform Lifts, and Stairway Chairlifts

## I. General

### 620.1 Scope

This article covers the installation of electrical equipment and wiring used in connection with elevators, dumbwaiters, escalators, moving walks, platform lifts, and stairway chairlifts.

Informational Note No. 1: For further information, see ASME A17.1-2010/CSA B44-10, *Safety Code for Elevators and Escalators*.

Informational Note No. 2: For further information, see CSA B44.1-11/ASME-A17.5-2011, *Elevator and Escalator Electrical Equipment Certification Standard*.

Informational Note No. 3: The term *wheelchair lift* has been changed to *platform lift*. For further information, see ASME A18.1-2008, *Safety Standard for Platform Lifts and Stairway Lifts*.

### 620.2 Definitions

Informational Note No. 1: The motor controller, motion controller, and operation controller are located in a single enclosure or a combination of enclosures.

Informational Note No. 2: Figure 620.2, No. 2 is for information only.

**Control Room (for Elevator, Dumbwaiter).** An enclosed control space outside the hoistway, intended for full bodily entry, that contains the elevator motor controller. The room could also contain electrical and/or mechanical equipment used directly in connection with the elevator or dumbwaiter but not the electric driving machine or the hydraulic machine.

**Control Space (for Elevator, Dumbwaiter).** A space inside or outside the hoistway, intended to be accessed with or without full bodily entry, that contains the elevator motor controller. This space could also contain electrical and/or mechanical equipment used directly in connection with the elevator or dumbwaiter but not the electrical driving machine or the hydraulic machine.

**Control System.** The overall system governing the starting, stopping, direction of motion, acceleration, speed, and retardation of the moving member.

**Controller, Motion.** The electrical device(s) for that part of the control system that governs the acceleration, speed, retardation, and stopping of the moving member.

**Controller, Motor.** The operative units of the control system comprised of the starter device(s) and power conversion equipment used to drive an electric motor, or the pumping unit used to power hydraulic control equipment.

**Controller, Operation.** The electrical device(s) for that part of the control system that initiates the starting, stopping, and direction of motion in response to a signal from an operating device.

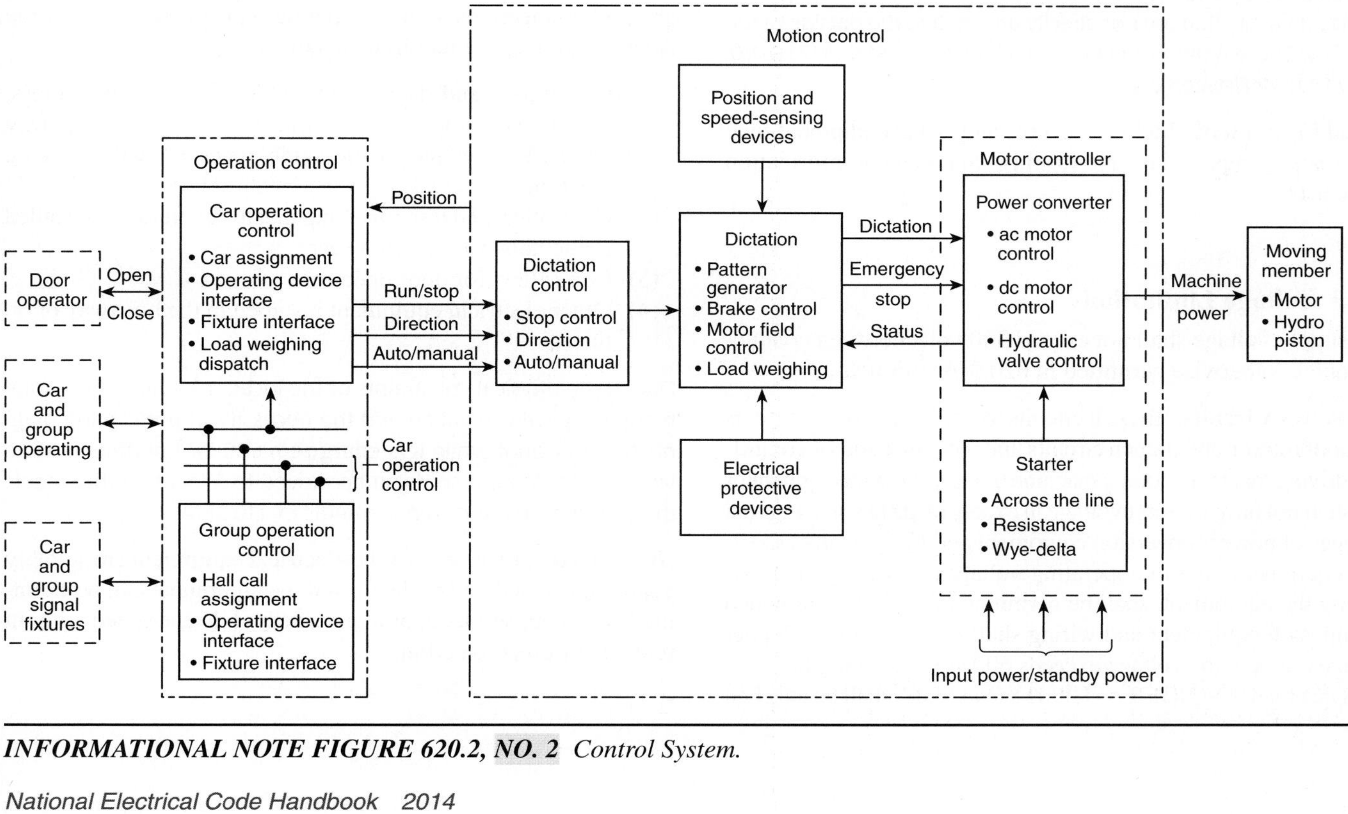


***INFORMATIONAL NOTE FIGURE 620.2, NO. 2*** *Control System.*

**Machine Room (for Elevator, Dumbwaiter).** An enclosed machinery space outside the hoistway, intended for full bodily entry, that contains the electrical driving machine or the hydraulic machine. The room could also contain electrical and/or mechanical equipment used directly in connection with the elevator or dumbwaiter.

**Machinery Space (for Elevator, Dumbwaiter).** A space inside or outside the hoistway, intended to be accessed with or without full bodily entry, that contains elevator or dumbwaiter mechanical equipment, and could also contain electrical equipment used directly in connection with the elevator or dumbwaiter. This space could also contain the electrical driving machine or the hydraulic machine.

**Operating Device.** The car switch, pushbuttons, key or toggle switch(s), or other devices used to activate the operation controller.

**Remote Machine Room and Control Room (for Elevator, Dumbwaiter).** A machine room or control room that is not attached to the outside perimeter or surface of the walls, ceiling, or floor of the hoistway.

**Remote Machinery Space and Control Space (for Elevator, Dumbwaiter).** A machinery space or control space that is not within the hoistway, machine room, or control room and that is not attached to the outside perimeter or surface of the walls, ceiling, or floor of the hoistway.

Definitions of *remote machine room and control room (for elevator, dumbwaiter)* and *remote machinery space and control space (for elevator, dumbwaiter)* describe elevator and dumbwaiter equipment rooms and areas that are not directly attached to the outside of the hoistway. These terms correlate with their use in ASME A17.1-2010, *Safety Code for Elevators and Escalators*.

**Signal Equipment.** Includes audible and visual equipment such as chimes, gongs, lights, and displays that convey information to the user.

•

## 620.3 Voltage Limitations

The supply voltage shall not exceed 300 volts between conductors unless otherwise permitted in 620.3(A) through (C).

**(A) Power Circuits.** Branch circuits to door operator controllers and door motors and branch circuits and feeders to motor controllers, driving machine motors, machine brakes, and motor-generator sets shall not have a circuit voltage in excess of 1000 volts. Internal voltages of power conversion equipment and functionally associated equipment, and the operating voltages of wiring interconnecting the equipment, shall be permitted to be higher, provided that all such equipment and wiring shall be listed for the higher voltages. Where the voltage exceeds 600 volts, warning labels or signs that read "DANGER — HIGH VOLTAGE" shall be attached to the equipment and shall be plainly visible. The danger sign(s) or label(s) shall comply with 110.21(B).

**(B) Lighting Circuits.** Lighting circuits shall comply with the requirements of Article 410.

**(C) Heating and Air-Conditioning Circuits.** Branch circuits for heating and air-conditioning equipment located on the elevator car shall not have a circuit voltage in excess of 1000 volts.

## 620.4 Live Parts Enclosed

All live parts of electrical apparatus in the hoistways, at the landings, in or on the cars of elevators and dumbwaiters, in the wellways or the landings of escalators or moving walks, or in the runways and machinery spaces of platform lifts and stairway chairlifts shall be enclosed to protect against accidental contact.

Informational Note: See 110.27 for guarding of live parts (1000 volts, nominal, or less).

## 620.5 Working Clearances

Working space shall be provided about controllers, disconnecting means, and other electrical equipment in accordance with 110.26(A).

Where conditions of maintenance and supervision ensure that only qualified persons examine, adjust, service, and maintain the equipment, the clearance requirements of 110.26(A) shall not be required where any of the conditions in 620.5(A) through (D) are met.

**(A) Flexible Connections to Equipment.** Electrical equipment in (A)(1) through (A)(4) is provided with flexible leads to all external connections so that it can be repositioned to meet the clear working space requirements of 110.26.

(1) Controllers and disconnecting means for dumbwaiters, escalators, moving walks, platform lifts, and stairway chairlifts installed in the same space with the driving machine
(2) Controllers and disconnecting means for elevators installed in the hoistway or on the car
(3) Controllers for door operators
(4) Other electrical equipment installed in the hoistway or on the car

Due to the physical constraints of the locations where this equipment is typically installed and the necessity of performing diagnostic work on it while it is energized, 620.5(A) permits flexible leads on equipment so it can be moved to a location that meets the working clearance requirements of 110.26(A).

**(B) Guards.** Live parts of the electrical equipment are suitably guarded, isolated, or insulated, and the equipment can be examined, adjusted, serviced, or maintained while energized without removal of this protection.

•

**(C) Examination, Adjusting, and Servicing.** Electrical equipment is not required to be examined, adjusted, serviced, or maintained while energized.

**(D) Low Voltage.** Uninsulated parts are at a voltage not greater than 30 volts rms, 42 volts peak, or 60 volts dc.

# II. Conductors

## 620.11 Insulation of Conductors

The insulation of conductors shall comply with 620.11(A) through (D).

Informational Note: One method of determining that conductors are flame retardant is by testing the conductors to the VW-1 (Vertical-Wire) Flame Test in ANSI/UL 1581-2011, *Reference Standard for Electrical Wires, Cables, and Flexible Cords.*

**(A) Hoistway Door Interlock Wiring.** The conductors to the hoistway door interlocks from the hoistway riser shall be flame retardant and suitable for a temperature of not less than 200°C (392°F). Conductors shall be Type SF or equivalent.

**(B) Traveling Cables.** Traveling cables used as flexible connections between the elevator or dumbwaiter car or counterweight and the raceway shall be of the types of elevator cable listed in Table 400.4 or other approved types.

**(C) Other Wiring.** All conductors in raceways shall have flame-retardant insulation.

Conductors shall be Type MTW, TF, TFF, TFN, TFFN, THHN, THW, THWN, TW, XHHW, hoistway cable, or any other conductor with insulation designated as flame retardant. Shielded conductors shall be permitted if such conductors are insulated for the maximum nominal circuit voltage applied to any conductor within the cable or raceway system.

**(D) Insulation.** All conductors shall have an insulation voltage rating equal to at least the maximum nominal circuit voltage applied to any conductor within the enclosure, cable, or raceway. Insulations and outer coverings that are marked for limited smoke and are so listed shall be permitted.

## 620.12 Minimum Size of Conductors

The minimum size of conductors, other than conductors that form an integral part of control equipment, shall be in accordance with 620.12(A) and (B).

**(A) Traveling Cables.**

**(1) Lighting Circuits.** For lighting circuits, 14 AWG copper, 20 AWG copper or larger conductors shall be permitted in parallel, provided the ampacity is equivalent to at least that of 14 AWG copper.

**(2) Other Circuits.** For other circuits, 20 AWG copper.

**(B) Other Wiring.** 24 AWG copper. Smaller size listed conductors shall be permitted.

Section 310.10(H) provides the conditions under which conductors can be installed in parallel for power and lighting circuits. One of those conditions stipulates that the minimum size for parallel conductors is 1/0 AWG. In high-rise structures, the length of the elevator traveling cables makes it hard to maintain an acceptable level of voltage drop for equipment on or within the car. To require compliance with the 310.10(H) rules for parallel conductors would result in exceptionally large traveling cables.

Section 620.12(A)(1) amends these general requirements for parallel conductors and permits 20 AWG and larger conductors to be installed in parallel for lighting circuits, provided that the combined ampacity of the paralleled conductors is not less than that of a 14 AWG copper conductor (for example, 15 amperes for 60°C). This provision is unique to Article 620 and is an example of the structure of the *Code* as set forth in 90.3.

In response to the extensive use of electronics with lower currents, conductors smaller than 24 AWG are permitted by 620.12(B), provided that they are listed and have the necessary strength and durability for the conditions to which they will be exposed. One application is the shielded cables interconnecting various microprocessors in an elevator distributed control system.

## 620.13 Feeder and Branch-Circuit Conductors

Conductors shall have an ampacity in accordance with 620.13(A) through (D). With generator field control, the conductor ampacity shall be based on the nameplate current rating of the driving motor of the motor-generator set that supplies power to the elevator motor.

Informational Note No. 1: The heating of conductors depends on root-mean-square current values, which, with generator field control, are reflected by the nameplate current rating of the motor-generator driving motor rather than by the rating of the elevator motor, which represents actual but short-time and intermittent full-load current values.

Informational Note No. 2: See Informational Note, Figure 620.13, No. 2.

**(A) Conductors Supplying Single Motor.** Conductors supplying a single motor shall have an ampacity not less than the percentage of motor nameplate current determined from 430.22(A) and (E).

Informational Note: Some elevator motor currents, or those motor currents of similar function, exceed the motor nameplate value. Heating of the motor and conductors is dependent on the root-mean square (rms) current value and the length of operation time. Because this motor application is inherently intermittent duty, conductors are sized for duty cycle service as shown in Table 430.22(E).

**(B) Conductors Supplying a Single Motor Controller.** Conductors supplying a single motor controller shall have an ampacity not less than the motor controller nameplate current rating, plus all other connected loads. Motor controller nameplate current ratings shall be permitted to be derived based on the rms

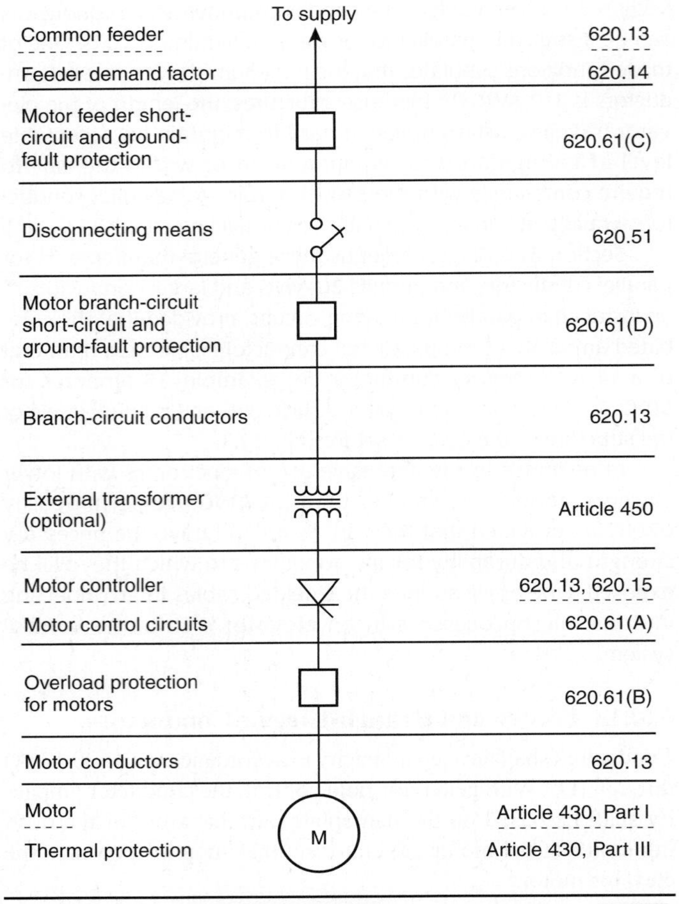


***FIGURE 620.13*** *Informational Note Single-Line Diagram, No. 2.*

value of the motor current using an intermittent duty cycle and other control system loads, if present.

**(C) Conductors Supplying a Single Power Transformer.** Conductors supplying a single power transformer shall have an ampacity not less than the nameplate current rating of the power transformer plus all other connected loads.

Informational Note No. 1: The nameplate current rating of a power transformer supplying a motor controller reflects the nameplate current rating of the motor controller at line voltage (transformer primary).

Informational Note No. 2: See Informative Annex D, Example No. D10.

**(D) Conductors Supplying More Than One Motor, Motor Controller, or Power Transformer.** Conductors supplying more than one motor, motor controller, or power transformer shall have an ampacity not less than the sum of the nameplate current ratings of the equipment plus all other connected loads. The ampere ratings of motors to be used in the summation shall be determined from Table 430.22(E), 430.24 and 430.24, Exception No. 1.

Informational Note: See Informative Annex D, Example Nos. D9 and D10.

***TABLE 620.14*** *Feeder Demand Factors for Elevators*

| Number of Elevators on a Single Feeder | Demand Factor* |
|---|---|
| 1 | 1.00 |
| 2 | 0.95 |
| 3 | 0.90 |
| 4 | 0.85 |
| 5 | 0.82 |
| 6 | 0.79 |
| 7 | 0.77 |
| 8 | 0.75 |
| 9 | 0.73 |
| 10 or more | 0.72 |

*Demand factors are based on 50 percent duty cycle (i.e., half time on and half time off).

### 620.14 Feeder Demand Factor

Feeder conductors of less ampacity than required by 620.13 shall be permitted, subject to the requirements of Table 620.14.

•

### 620.15 Motor Controller Rating

The motor controller rating shall comply with 430.83. The rating shall be permitted to be less than the nominal rating of the elevator motor, when the controller inherently limits the available power to the motor and is marked as power limited.

Informational Note: For controller markings, see 430.8.

The inherent power-limiting ability of certain adjustable-speed drive controllers is the basis for permitting the controller to have a lower current or horsepower rating than that of the motor. For a controller to be used in this manner, the manufacturer's marking must indicate that it is power limiting.

## III. Wiring

### 620.21 Wiring Methods

Conductors and optical fibers located in hoistways, in escalator and moving walk wellways, in platform lifts, stairway chairlift runways, machinery spaces, control spaces, in or on cars, in machine rooms and control rooms, not including the traveling cables connecting the car or counterweight and hoistway wiring, shall be installed in rigid metal conduit, intermediate metal conduit, electrical metallic tubing, rigid nonmetallic conduit, or wireways, or shall be Type MC, MI, or AC cable unless otherwise permitted in 620.21(A) through (C).

*Exception: Cords and cables of listed cord- and plug-connected equipment shall not be required to be installed in a raceway.*

**(A) Elevators.**

**(1) Hoistways.**

(a) Cables used in Class 2 power-limited circuits shall be permitted to be installed between risers and signal equipment and operating devices, provided the cables are supported and protected from physical damage and are of a jacketed and flame-retardant type.

(b) Flexible cords and cables that are components of listed equipment and used in circuits operating at 30 volts rms or less or 42 volts dc or less shall be permitted in lengths not to exceed 1.8 m (6 ft), provided the cords and cables are supported and protected from physical damage and are of a jacketed and flame-retardant type.

(c) The following wiring methods shall be permitted in the hoistway in lengths not to exceed 1.8 m (6 ft):

(1) Flexible metal conduit
(2) Liquidtight flexible metal conduit
(3) Liquidtight flexible nonmetallic conduit
(4) Flexible cords and cables, or conductors grouped together and taped or corded, shall be permitted to be installed without a raceway. They shall be located to be protected from physical damage and shall be of a flame-retardant type and shall be part of the following:
  a. Listed equipment
  b. A driving machine, or
  c. A driving machine brake

*Exception to A(620.21)(1)(c)(1), (2), and (3): The conduit length shall not be required to be limited between risers and limit switches, interlocks, operating buttons, and similar devices.*

(d) A sump pump or oil recovery pump located in the pit shall be permitted to be cord connected. The cord shall be a hard usage oil-resistant type, of a length not to exceed 1.8 m (6 ft), and shall be located to be protected from physical damage.

**(2) Cars.**

(a) Flexible metal conduit, liquidtight flexible metal conduit, or liquidtight flexible nonmetallic conduit of metric designator 12 (trade size ⅜), or larger, not exceeding 1.8 m (6 ft) in length, shall be permitted on cars where so located as to be free from oil and if securely fastened in place.

*Exception: Liquidtight flexible nonmetallic conduit of metric designator 12 (trade size ⅜), or larger, as defined by 356.2(2), shall be permitted in lengths in excess of 1.8 m (6 ft).*

(b) Hard-service cords and junior hard-service cords that conform to the requirements of Article 400 (Table 400.4) shall be permitted as flexible connections between the fixed wiring on the car and devices on the car doors or gates. Hard-service cords only shall be permitted as flexible connections for the top-of-car operating device or the car-top work light. Devices or luminaires shall be grounded by means of an equipment grounding conductor run with the circuit conductors. Cables with smaller conductors and other types and thicknesses of insulation and jackets shall be permitted as flexible connections between the fixed wiring on the car and devices on the car doors or gates, if listed for this use.

(c) Flexible cords and cables that are components of listed equipment and used in circuits operating at 30 volts rms or less or 42 volts dc or less shall be permitted in lengths not to exceed 1.8 m (6 ft), provided the cords and cables are supported and protected from physical damage and are of a jacketed and flame-retardant type.

(d) The following wiring methods shall be permitted on the car assembly in lengths not to exceed 1.8 m (6 ft):

(1) Flexible metal conduit
(2) Liquidtight flexible metal conduit
(3) Liquidtight flexible nonmetallic conduit
(4) Flexible cords and cables, or conductors grouped together and taped or corded, shall be permitted to be installed without a raceway. They shall be located to be protected from physical damage and shall be of a flame-retardant type and shall be part of the following:
  a. Listed equipment
  b. A driving machine, or
  c. A driving machine brake

The requirements of 620.21(A)(2)(d) describe the permitted wiring methods where a driving machine or driving machine brake is located on the car. In addition to flexible metal and nonmetallic conduits, the use of single conductors that are taped or corded together is permitted. This use includes rack-and-pinion or screw column drives located on cars. See also 620.71(B).

**(3) Within Machine Rooms, Control Rooms, and Machinery Spaces and Control Spaces.**

(a) Flexible metal conduit, liquidtight flexible metal conduit, or liquidtight flexible nonmetallic conduit of metric designator 12 (trade size ⅜), or larger, not exceeding 1.8 m (6 ft) in length, shall be permitted between control panels and machine motors, machine brakes, motor-generator sets, disconnecting means, and pumping unit motors and valves.

*Exception: Liquidtight flexible nonmetallic conduit metric designator 12 (trade size ⅜) or larger, as defined in 356.2(2), shall be permitted to be installed in lengths in excess of 1.8 m (6 ft).*

(b) Where motor-generators, machine motors, or pumping unit motors and valves are located adjacent to or underneath control equipment and are provided with extra-length terminal leads not exceeding 1.8 m (6 ft) in length, such leads shall be permitted to be extended to connect directly to controller terminal studs without regard to the carrying-capacity requirements of Articles 430 and 445. Auxiliary gutters shall be permitted in machine and control rooms between controllers, starters, and similar apparatus.

(c) Flexible cords and cables that are components of listed equipment and used in circuits operating at 30 volts rms or less or 42 volts dc or less shall be permitted in lengths not to exceed 1.8 m (6 ft), provided the cords and cables are supported and protected from physical damage and are of a jacketed and flame-retardant type.

(d) On existing or listed equipment, conductors shall also be permitted to be grouped together and taped or corded without being installed in a raceway. Such cable groups shall be supported at intervals not over 900 mm (3 ft) and located so as to be protected from physical damage.

(e) Flexible cords and cables in lengths not to exceed 1.8 m (6 ft) that are of a flame-retardant type and located to be protected from physical damage shall be permitted in these rooms and spaces without being installed in a raceway. They shall be part of the following:

(1) Listed equipment
(2) A driving machine, or
(3) A driving machine brake

**(4) Counterweight.** The following wiring methods shall be permitted on the counterweight assembly in lengths not to exceed 1.8 m (6 ft):

(1) Flexible metal conduit
(2) Liquidtight flexible metal conduit
(3) Liquidtight flexible nonmetallic conduit
(4) Flexible cords and cables, or conductors grouped together and taped or corded, shall be permitted to be installed without a raceway. They shall be located to be protected from physical damage, shall be of a flame-retardant type, and shall be part of the following:
    a. Listed equipment
    b. A driving machine, or
    c. A driving machine brake

**(B) Escalators.**

**(1) Wiring Methods.** Flexible metal conduit, liquidtight flexible metal conduit, or liquidtight flexible nonmetallic conduit shall be permitted in escalator and moving walk wellways. Flexible metal conduit or liquidtight flexible conduit of metric designator 12 (trade size ⅜) shall be permitted in lengths not in excess of 1.8 m (6 ft).

*Exception: Metric designator 12 (trade size ⅜), nominal, or larger liquidtight flexible nonmetallic conduit, as defined in 356.2(2), shall be permitted to be installed in lengths in excess of 1.8 m (6 ft).*

**(2) Class 2 Circuit Cables.** Cables used in Class 2 power-limited circuits shall be permitted to be installed within escalators and moving walkways, provided the cables are supported and protected from physical damage and are of a jacketed and flame-retardant type.

**(3) Flexible Cords.** Hard-service cords that conform to the requirements of Article 400 (Table 400.4) shall be permitted as flexible connections on escalators and moving walk control panels and disconnecting means where the entire control panel and disconnecting means are arranged for removal from machine spaces as permitted in 620.5.

**(C) Platform Lifts and Stairway Chairlift Raceways.**

**(1) Wiring Methods.** Flexible metal conduit or liquidtight flexible metal conduit shall be permitted in platform lifts and stairway chairlift runways and machinery spaces. Flexible metal conduit or liquidtight flexible conduit of metric designator 12 (trade size ⅜) shall be permitted in lengths not in excess of 1.8 m (6 ft).

*Exception: Metric designator 12 (trade size ⅜) or larger liquidtight flexible nonmetallic conduit, as defined in 356.2(2), shall be permitted to be installed in lengths in excess of 1.8 m (6 ft).*

**(2) Class 2 Circuit Cables.** Cables used in Class 2 power-limited circuits shall be permitted to be installed within platform lifts and stairway chairlift runways and machinery spaces, provided the cables are supported and protected from physical damage and are of a jacketed and flame-retardant type.

**(3) Flexible Cords and Cables.** Flexible cords and cables that are components of listed equipment and used in circuits operating at 30 volts rms or less or 42 volts dc or less shall be permitted in lengths not to exceed 1.8 m (6 ft), provided the cords and cables are supported and protected from physical damage and are of a jacketed and flame-retardant type.

## 620.22 Branch Circuits for Car Lighting, Receptacle(s), Ventilation, Heating, and Air-Conditioning

**(A) Car Light Source.** A separate branch circuit shall supply the car lights, receptacle(s), auxiliary lighting power source, and ventilation on each elevator car. The overcurrent device protecting the branch circuit shall be located in the elevator machine room or control room/machinery space or control space.

Required lighting shall not be connected to the load side of a ground-fault circuit interrupter.

A service receptacle installed on an elevator car top is required to be a GFCI-type device, and because the car lights and receptacle are supplied from the same branch circuit, this requirement ensures that GFCI operation will not also interrupt power to the car lighting. The same requirement is found in 620.23(A) and 620.24(A) for machine room lighting and hoistway pit lighting.

**(B) Air-Conditioning and Heating Source.** A separate branch circuit shall supply the air-conditioning and heating units on each elevator car. The overcurrent device protecting the branch circuit shall be located in the elevator machine room or control room/machinery space or control space.

### 620.23 Branch Circuits for Machine Room or Control Room/Machinery Space or Control Space Lighting and Receptacle(s)

**(A) Separate Branch Circuit.** A separate branch circuit shall supply the machine room or control room/machinery space or control space lighting and receptacle(s).

Required lighting shall not be connected to the load side of a ground-fault circuit interrupter.

Luminaires are not permitted to be connected to the load side of a GFCI device. This placement prevents power interruption to the machine room lighting if the GFCI operates.

**(B) Lighting Switch.** The machine room or control room/machinery space or control space lighting switch shall be located at the point of entry.

**(C) Duplex Receptacle.** At least one 125-volt, single-phase, 15- or 20-ampere duplex receptacle shall be provided in each machine room or control room and machinery space or control space.

Informational Note: See ASME A17.1-2010/CSA B44-10, *Safety Code for Elevators and Escalators*, for illumination levels.

The receptacles required by 620.23 and 620.24 are also required to be provided with GFCI protection for personnel (see 620.85). Luminaires are not permitted to be connected to the load side of GFCI devices. See also the commentary following 620.23(A). ASME A17.1 requires a minimum of 5 foot-candles (54 lux) at the pit floor and requires luminaires to be externally guarded to prevent accidental breakage. Luminaires in pits should be mounted so that the car or counterweight does not strike them when on fully compressed buffers.

### 620.24 Branch Circuit for Hoistway Pit Lighting and Receptacle(s)

**(A) Separate Branch Circuit.** A separate branch circuit shall supply the hoistway pit lighting and receptacle(s).

Required lighting shall not be connected to the load side of a ground-fault circuit interrupter.

**(B) Lighting Switch.** The lighting switch shall be so located as to be readily accessible from the pit access door.

**(C) Duplex Receptacle.** At least one 125-volt, single-phase, 15- or 20-ampere duplex receptacle shall be provided in the hoistway pit.

Informational Note: See ASME A17.1-2010/CSA B44-10, *Safety Code for Elevators and Escalators*, for illumination levels.

### 620.25 Branch Circuits for Other Utilization Equipment

**(A) Additional Branch Circuits.** Additional branch circuit(s) shall supply utilization equipment not identified in 620.22, 620.23, and 620.24. Other utilization equipment shall be restricted to that equipment identified in 620.1.

**(B) Overcurrent Devices.** The overcurrent devices protecting the branch circuit(s) shall be located in the elevator machinery room or control room/machinery space or control space.

## IV. Installation of Conductors

### 620.32 Metal Wireways and Nonmetallic Wireways

The sum of the cross-sectional area of the individual conductors in a wireway shall not be more than 50 percent of the interior cross-sectional area of the wireway.

Vertical runs of wireways shall be securely supported at intervals not exceeding 4.5 m (15 ft) and shall have not more than one joint between supports. Adjoining wireway sections shall be securely fastened together to provide a rigid joint.

### 620.33 Number of Conductors in Raceways

The sum of the cross-sectional area of the individual conductors in raceways shall not exceed 40 percent of the interior cross-sectional area of the raceway, except as permitted in 620.32 for wireways.

### 620.34 Supports

Supports for cables or raceways in a hoistway or in an escalator or moving walk wellway or platform lift and stairway chairlift runway shall be securely fastened to the guide rail; escalator or moving walk truss; or to the hoistway, wellway, or runway construction.

### 620.35 Auxiliary Gutters

Auxiliary gutters shall not be subject to the restrictions of 366.12(2) covering length or of 366.22 covering number of conductors.

### 620.36 Different Systems in One Raceway or Traveling Cable

Optical fiber cables and conductors for operating devices, operation and motion control, power, signaling, fire alarm, lighting, heating, and air-conditioning circuits of 1000 volts or less shall be permitted to be run in the same traveling cable or raceway system if all conductors are insulated for the maximum voltage applied to any conductor within the cables or raceway system and if all live parts of the equipment are insulated from ground for this maximum voltage. Such a traveling cable or raceway shall also be permitted to include shielded conductors and/or one or more coaxial cables if such conductors are insulated for the maximum voltage applied to any conductor within the cable or raceway system. Conductors shall be permitted to be covered with suitable shielding for telephone, audio, video, or higher frequency communications circuits.

The use of greater numbers of individual cables and of much longer cables in tall buildings increases the likelihood of the multiple cable loops becoming twisted. To prevent the practice of tying other cables to the traveling cable, one elevator cable or raceway is permitted to enclose optical fiber cables and all the conductors for power, control, lighting, video, fire alarm, and communications circuits if all conductors are insulated for the maximum voltage applied to any conductor within the cable or raceway and if all live parts are insulated from ground for the maximum voltage present.

Power-limited and non–power-limited fire alarm conductors are permitted in the same raceway or traveling cable as power and other types of signaling conductors. All power, signaling, and fire alarm conductors are to be insulated for the maximum voltage applied to any conductor in the raceway or cable.

### 620.37 Wiring in Hoistways, Machine Rooms, Control Rooms, Machinery Spaces, and Control Spaces

**(A) Uses Permitted.** Only such electrical wiring, raceways, and cables used directly in connection with the elevator or dumbwaiter, including wiring for signals, for communication with the car, for lighting, heating, air conditioning, and ventilating the elevator car, for fire detecting systems, for pit sump pumps, and for heating, lighting, and ventilating the hoistway, shall be permitted inside the hoistway, machine rooms, control rooms, machinery spaces, and control spaces.

**(B) Lightning Protection.** Bonding of elevator rails (car and/or counterweight) to a lightning protection system grounding down conductor(s) shall be permitted. The lightning protection system grounding down conductor(s) shall not be located within the hoistway. Elevator rails or other hoistway equipment shall not be used as the grounding down conductor for lightning protection systems.

Informational Note: See 250.106 for bonding requirements. For further information, see NFPA 780-2014, *Standard for the Installation of Lightning Protection Systems*.

Where a lightning protection system is provided with the system grounding "down" conductor(s) located outside the hoistway within a critical horizontal distance of the elevator rails, bonding of the rails to the lightning protection system grounding down conductor(s) is required by NFPA 780, *Standard for the Installation of Lightning Protection Systems*. The requirements of 620.37(B) provide the necessary correlation for this bonding of the elevator rails to occur. Bonding prevents a dangerous side flash between the lightning protection system grounding down conductor(s) and the elevator rails. A lightning strike on the building air terminal will be conducted through the lightning protection system grounding down conductor(s), and, if the elevator rails are not at the same potential as the lightning protection system grounding down conductor(s), a side flash may occur. Generally, down conductors are installed vertically near the structure's perimeter.

**(C) Main Feeders.** Main feeders for supplying power to elevators and dumbwaiters shall be installed outside the hoistway unless as follows:

(1) By special permission, feeders for elevators shall be permitted within an existing hoistway if no conductors are spliced within the hoistway.
(2) Feeders shall be permitted inside the hoistway for elevators with driving machine motors located in the hoistway or on the car or counterweight.

### 620.38 Electrical Equipment in Garages and Similar Occupancies

Electrical equipment and wiring used for elevators, dumbwaiters, escalators, moving walks, and platform lifts and stairway chairlifts in garages shall comply with the requirements of Article 511.

Informational Note: Garages used for parking or storage and where no repair work is done in accordance with 511.3(A) are not classified.

## V. Traveling Cables

### 620.41 Suspension of Traveling Cables

Traveling cables shall be suspended at the car and hoistways' ends, or counterweight end where applicable, so as to reduce the strain on the individual copper conductors to a minimum.

Traveling cables shall be supported by one of the following means:

(1) By their steel supporting member(s)
(2) By looping the cables around supports for unsupported lengths less than 30 m (100 ft)
(3) By suspending from the supports by a means that automatically tightens around the cable when tension is increased for unsupported lengths up to 60 m (200 ft)

Unsupported length for the hoistway suspension means shall be that length of cable measured from the point of suspension in the hoistway to the bottom of the loop, with the elevator car located at the bottom landing. Unsupported length for the car suspension means shall be that length of cable measured from the point of suspension on the car to the bottom of the loop, with the elevator car located at the top landing.

### 620.42 Hazardous (Classified) Locations

In hazardous (classified) locations, traveling cables shall be of a type approved for hazardous (classified) locations and shall comply with 501.140, 502.140, or 503.140, as applicable.

### 620.43 Location of and Protection for Cables

Traveling cable supports shall be located so as to reduce to a minimum the possibility of damage due to the cables coming in contact with the hoistway construction or equipment in the hoistway. Where necessary, suitable guards shall be provided to protect the cables against damage.

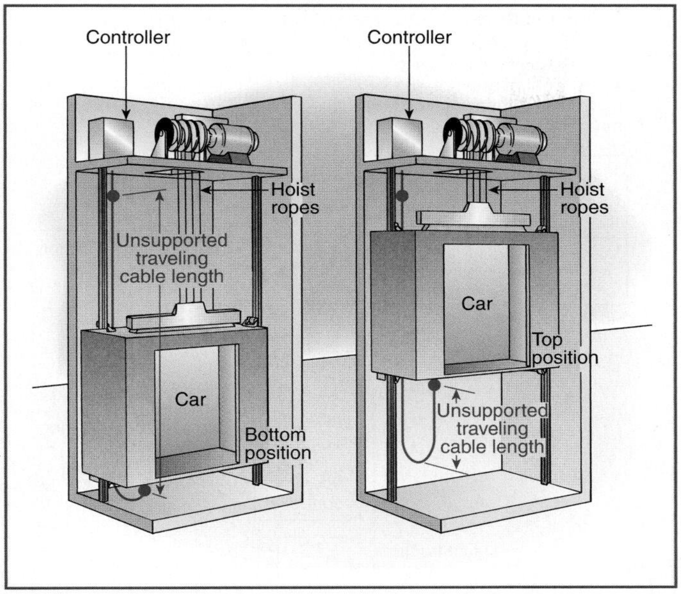


***EXHIBIT 620.1*** *Unsupported lengths of traveling cable.*

## 620.44 Installation of Traveling Cables

Traveling cables that are suitably supported and protected from physical damage shall be permitted to be run without the use of a raceway in either or both of the following:

(a) When used inside the hoistway, on the elevator car, hoistway wall, counterweight, or controllers and machinery that are located inside the hoistway, provided the cables are in the original sheath.

(b) From inside the hoistway, to elevator controller enclosures and to elevator car and machine room, control room, machinery space, and control space connections that are located outside the hoistway for a distance not exceeding 1.8 m (6 ft) in length as measured from the first point of support on the elevator car or hoistway wall, or counterweight where applicable, provided the conductors are grouped together and taped or corded, or in the original sheath. These traveling cables shall be permitted to be continued to this equipment.

Traveling cables between fixed suspension points are not required to be installed in a raceway. If the fixed suspension point is on top of the car, the cables on the side of the car might be exposed. If the suspension point is under the car, the cables might be run up the side of the car to the car's top junction box. Suitable guards may be necessary to protect these cables from damage. In order to connect to equipment, the traveling cable is permitted to continue out of the hoistway, up to a length of 6 feet from the first support, without the use of a raceway. Refer to 620.44 and Exhibit 620.1.

# VI. Disconnecting Means and Control

## 620.51 Disconnecting Means

A single means for disconnecting all ungrounded main power supply conductors for each unit shall be provided and be designed so that no pole can be operated independently. Where multiple driving machines are connected to a single elevator, escalator, moving walk, or pumping unit, there shall be one disconnecting means to disconnect the motor(s) and control valve operating magnets.

The disconnecting means for the main power supply conductors shall not disconnect the branch circuit required in 620.22, 620.23, and 620.24.

The branch circuits that supply elevator car lighting, receptacles, ventilation, air conditioning, and heating are required to be independent of the control portion of the elevator. In addition, the branch circuits supplying hoistway pit lighting and receptacles and machine room or control room lights and receptacles are not permitted to be disconnected by the main elevator power disconnect. This requirement provides for passenger safety and comfort and for the safety of elevator maintenance personnel during an inadvertent or emergency shutdown of the main power circuit to the elevator. This disconnecting means must always be located outside the hoistway.

**(A) Type.** The disconnecting means shall be an enclosed externally operable fused motor circuit switch or circuit breaker that is lockable open in accordance with 110.25.

The disconnecting means shall be a listed device.

Informational Note: For additional information, see ASME A17.1-2010/CSA B44-10, *Safety Code for Elevators and Escalators.*

*Exception No. 1: Where an individual branch circuit supplies a platform lift, the disconnecting means required by 620.51(C)(4) shall be permitted to comply with 430.109(C). This disconnecting means shall be listed and shall be lockable open in accordance with 110.25.*

*Exception No. 2: Where an individual branch circuit supplies a stairway chairlift, the stairway chairlift shall be permitted to be cord-and-plug-connected, provided it complies with 422.16(A) and the cord does not exceed 1.8 m (6 ft) in length.*

**(B) Operation.** No provision shall be made to open or close this disconnecting means from any other part of the premises. If sprinklers are installed in hoistways, machine rooms, control rooms, machinery spaces, or control spaces, the disconnecting means shall be permitted to automatically open the power supply to the affected elevator(s) prior to the application of water. No provision shall be made to automatically close this disconnecting means. Power shall only be restored by manual means.

Informational Note: To reduce hazards associated with water on live elevator electrical equipment.

Where sprinklers are installed in hoistways, machine rooms, or machinery spaces, a means must be provided to automatically disconnect the main line power supply to the affected elevator(s) upon or prior to the application of water in accordance with ASME A17.1-2010, Section 2.8.3.3.2. Water on elevator electrical equipment can result in hazards such as uncontrolled car movement (wet machine brakes), movement of elevator with open doors (water on safety circuits bypassing car and/or hoistway door interlocks), and shock hazards.

Automatic disconnection of the main line power supply is not required by ASME A17.1 where hoistways and machine rooms are not sprinklered. NFPA 13, *Standard for the Installation of Sprinkler Systems*, provides requirements for sprinkler installation in machine rooms, hoistways, and pits.

Elevator shutdown is generally accomplished through the use of heat detectors located near sprinkler heads. The heat detectors are designed to actuate and generate an alarm signal prior to water discharge from the sprinkler heads. An output control relay powered by the fire alarm system then provides a monitored output to the main line disconnecting means control circuit, which activates the shunt trip. This practice ensures that all components have secondary power and are monitored for integrity as required by *NFPA 72, National Fire Alarm and Signaling Code*. Stand-alone heat detectors connected directly to the elevator disconnecting means control circuit are not monitored for integrity, have no secondary power supply, and are not permitted by *NFPA 72*.

Elevator shutdown can occur even if the car is not at a landing. In order to avoid trapping occupants in elevator car(s) when the main power to the elevator driving machine is interrupted due to sprinkler activation, Section 2.27.3.2 of ASME A17.1 requires installation of a fire alarm initiating device(s) in sprinklered hoistways for the purposes of initiating elevator car(s) recall before the main line power is disconnected. This operation is referred to as "Phase I Emergency Recall" in the ASME code. See Section 21.4 of *NFPA 72* for additional requirements relating to the fire alarm system and elevator shutdown.

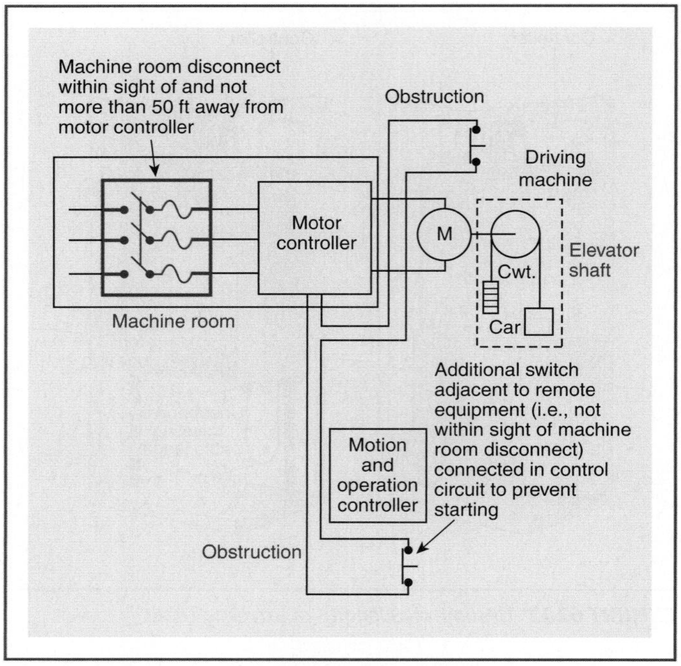


***EXHIBIT 620.2*** *Disconnecting means for driving machines or motion and operation controllers not within sight of the main line disconnecting means. (Courtesy of ASME)*

**(C) Location.** The disconnecting means shall be located where it is readily accessible to qualified persons.

**(1) On Elevators Without Generator Field Control.** On elevators without generator field control, the disconnecting means shall be located within sight of the motor field controller. Where the motor controller is located in the elevator hoistway, the disconnecting means required by 620.51(A) shall be located in a machinery space, machine room, control space, or control room outside the hoistway; and an additional fused or non-fused, enclosed, externally operable motor-circuit switch that is lockable open in accordance with 110.25 to disconnect all ungrounded main power-supply conductors shall be located within sight of the motor controller. The additional switch shall be a listed device and shall comply with 620.91(C).

A common installation is a machine room, containing the driving machine, motor controller, motion controller, and operation controller, located outside of the hoistway. A disconnecting means must be within sight of the motor controller. Any driving machine or motion and operation controller not within sight of this disconnecting means must be provided with a manual switch in its control circuit to prevent it from starting. Exhibit 620.2 illustrates the requirement on disconnecting means for driving machines or motion and operation controllers not within sight of the main line disconnecting means.

•

Driving machines or motion and operation controllers not within sight of the disconnecting means shall be provided with a manually operated switch installed in the control circuit to prevent starting. The manually operated switch(es) shall be installed adjacent to this equipment.

Where the driving machine of an electric elevator or the hydraulic machine of a hydraulic elevator is located in a remote machine room or remote machinery space, a single means for disconnecting all ungrounded main power-supply conductors shall be provided and be lockable open in accordance with 110.25.

**(2) On Elevators with Generator Field Control.** On elevators with generator field control, the disconnecting means shall be located within sight of the motor controller for the driving motor of the motor-generator set. Driving machines, motor-generator sets, or motion and operation controllers not within sight of the disconnecting means shall be provided with a manually operated switch installed in the control circuit to prevent starting. The manually operated switch(es) shall be installed adjacent to this equipment.

Where the driving machine or the motor-generator set is located in a remote machine room or remote machinery space,

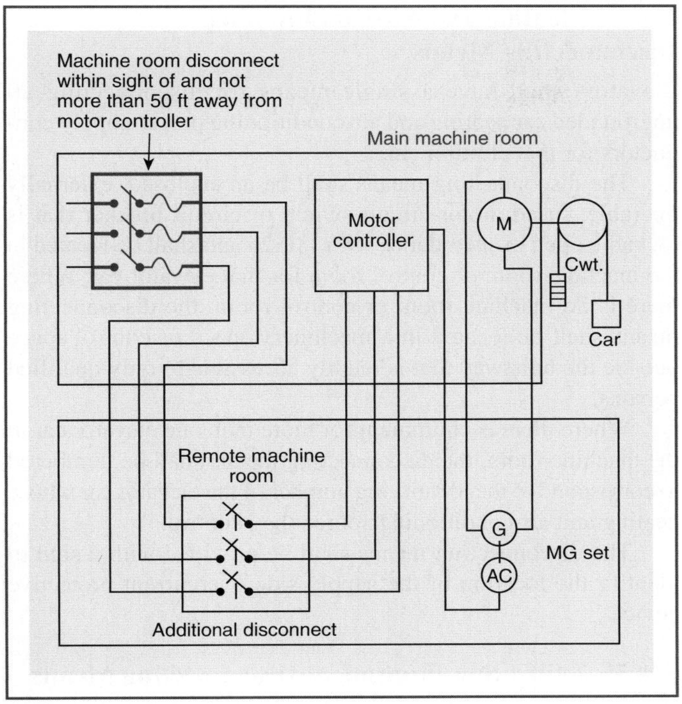


**EXHIBIT 620.3** *Disconnecting means for a motor-generator set in a remote location. (Courtesy of ASME)*

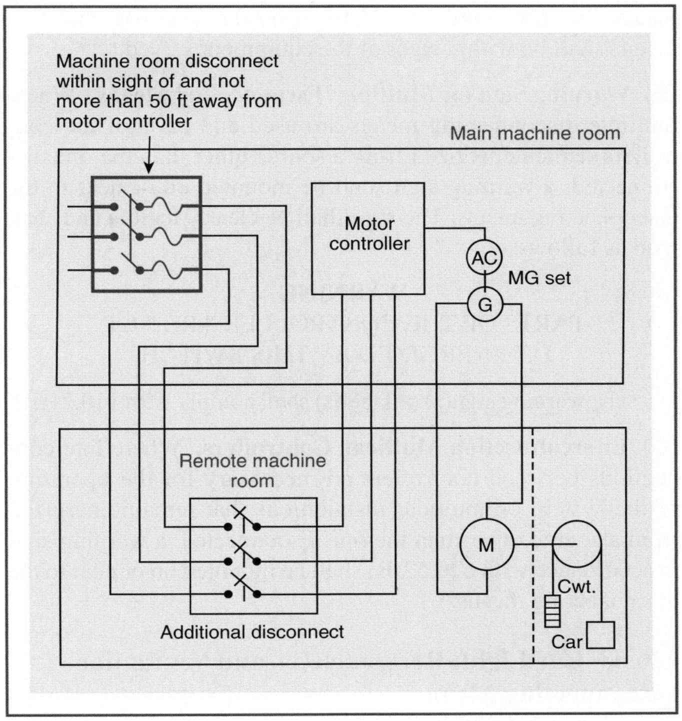


**EXHIBIT 620.4** *Disconnecting means for driving machines in a remote location. (Courtesy of ASME)*

a single means for disconnecting all ungrounded main power-supply conductors shall be provided and be lockable open in accordance with 110.25.

Where the driving machine is located in a remote machine room away from the control room, a disconnecting means must be within sight of the motor controller. Additionally, a means for disconnecting all ungrounded main power-supply conductors must be provided in the remote machine room. See Exhibits 620.3 and 620.4 for examples of disconnecting means for a motor-generator set and for driving machines in remote locations.

**(3) On Escalators and Moving Walks.** On escalators and moving walks, the disconnecting means shall be installed in the space where the controller is located.

The local emergency stop control at the escalator location shown in Exhibit 620.5, which is required by Section 6.1.6.3.1 of ASME A.17.1-2010 for passenger safety, cannot be used as the disconnecting means required by this section.

**(4) On Platform Lifts and Stairway Chairlifts.** On platform lifts and stairway chairlifts, the disconnecting means shall be located within sight of the motor controller.

**(D) Identification and Signs.** Where there is more than one driving machine in a machine room, the disconnecting means shall be numbered to correspond to the identifying number of the driving machine that they control.

**EXHIBIT 620.5** *The emergency stop button installed at the escalator location is not considered to be the required disconnecting means.*

The disconnecting means shall be provided with a sign to identify the location of the supply side overcurrent protective device.

## 620.52 Power from More Than One Source

**(A) Single-Car and Multicar Installations.** On single-car and multicar installations, equipment receiving electrical power from more than one source shall be provided with a disconnecting means for each source of electrical power. The disconnecting means shall be within sight of the equipment served.

**(B) Warning Sign for Multiple Disconnecting Means.** Where multiple disconnecting means are used and parts of the controllers remain energized from a source other than the one disconnected, a warning sign shall be mounted on or next to the disconnecting means. The sign shall be clearly legible and shall read as follows:

**WARNING**
PARTS OF THE CONTROLLER ARE NOT
DE-ENERGIZED BY THIS SWITCH.

The warning sign(s) or label(s) shall comply with 110.21(B).

**(C) Interconnection Multicar Controllers.** Where interconnections between controllers are necessary for the operation of the system on multicar installations that remain energized from a source other than the one disconnected, a warning sign in accordance with 620.52(B) shall be mounted on or next to the disconnecting means.

## 620.53 Car Light, Receptacle(s), and Ventilation Disconnecting Means

Elevators shall have a single means for disconnecting all ungrounded car light, receptacle(s), and ventilation power-supply conductors for that elevator car.

The disconnecting means shall be an enclosed, externally operable, fused motor-circuit switch or circuit breaker that is lockable open in accordance with 110.25 and shall be located in the machine room or control room for that elevator car. Where there is no machine room or control room, the disconnecting means shall be located in a machinery space or control space outside the hoistway that is readily accessible to only qualified persons.

Disconnecting means shall be numbered to correspond to the identifying number of the elevator car whose light source they control.

The disconnecting means shall be provided with a sign to identify the location of the supply side overcurrent protective device.

*Exception: Where a separate branch circuit supplies car lighting, a receptacle(s), and a ventilation motor not exceeding 2 hp, the disconnecting means required by 620.53 shall be permitted to comply with 430.109(C). This disconnecting means shall be listed and shall be lockable open in accordance with 110.25.*

Section 430.109(C) permits the use of general-use snap switches and listed manual motor controllers as disconnecting means for motors rated 2 hp or less and 300 volts or less. General-use snap switches used as a disconnecting means for motors not exceeding these parameters are not required to have a horsepower rating. All of the methods permitted by this section are required to be capable of being locked in the open position.

## 620.54 Heating and Air-Conditioning Disconnecting Means

Elevators shall have a single means for disconnecting all ungrounded car heating and air-conditioning power-supply conductors for that elevator car.

The disconnecting means shall be an enclosed, externally operable, fused motor-circuit switch or circuit breaker that is lockable open in accordance with 110.25 and shall be located in the machine room or control room for that elevator car. Where there is no machine room or control room, the disconnecting means shall be located in a machinery space or control space outside the hoistway that is readily accessible to only qualified persons.

Where there is equipment for more than one elevator car in the machine room, the disconnecting means shall be numbered to correspond to the identifying number of the elevator car whose heating and air-conditioning source they control.

The disconnecting means shall be provided with a sign to identify the location of the supply side overcurrent protective device.

## 620.55 Utilization Equipment Disconnecting Means

Each branch circuit for other utilization equipment shall have a single means for disconnecting all ungrounded conductors. The disconnecting means shall be lockable open in accordance with 110.25.

Where there is more than one branch circuit for other utilization equipment, the disconnecting means shall be numbered to correspond to the identifying number of the equipment served. The disconnecting means shall be provided with a sign to identify the location of the supply side overcurrent protective device.

# VII. Overcurrent Protection

## 620.61 Overcurrent Protection

Overcurrent protection shall be provided in accordance with 620.61(A) through (D)

**(A) Operating Devices and Control and Signaling Circuits.** Operating devices and control and signaling circuits shall be protected against overcurrent in accordance with the requirements of 725.43 and 725.45.

Class 2 power-limited circuits shall be protected against overcurrent in accordance with the requirements of Chapter 9, Notes to Tables 11(A) and 11(B).

**(B) Overload Protection for Motors** Motor and branch-circuit overload protection shall conform to Article 430, Part III, and (B)(1) through (B)(4).

**(1) Duty Rating on Elevator, Dumbwaiter, and Motor-Generator Sets Driving Motors.** Duty on elevator and dumbwaiter driving machine motors and driving motors of motor-generators used with generator field control shall be rated as intermittent. Such motors shall be permitted to be protected against overload in accordance with 430.33.

**(2) Duty Rating on Escalator Motors.** Duty on escalator and moving walk driving machine motors shall be rated as continuous. Such motors shall be protected against overload in accordance with 430.32.

**(3) Overload Protection.** Escalator and moving walk driving machine motors and driving motors of motor-generator sets shall be protected against running overload as provided in Table 430.37.

**(4) Duty Rating and Overload Protection on Platform Lift and Stairway Chairlift Motors.** Duty on platform lift and stairway chairlift driving machine motors shall be rated as intermittent. Such motors shall be permitted to be protected against overload in accordance with 430.33.

Informational Note: For further information, see 430.44 for orderly shutdown.

**(C) Motor Feeder Short-Circuit and Ground-Fault Protection.** Motor feeder short-circuit and ground-fault protection shall be as required in Article 430, Part V.

**(D) Motor Branch-Circuit Short-Circuit and Ground-Fault Protection.** Motor branch-circuit short-circuit and ground-fault protection shall be as required in Article 430, Part IV.

### 620.62 Selective Coordination

Where more than one driving machine disconnecting means is supplied by a single feeder, the overcurrent protective devices in each disconnecting means shall be selectively coordinated with any other supply side overcurrent protective devices.

Selective coordination shall be selected by a licensed professional engineer or other qualified person engaged primarily in the design, installation, or maintenance of electrical systems. The selection shall be documented and made available to those authorized to design, install, inspect, maintain, and operate the system.

Coordination of the OCPDs is important to ensure continuity of power where more than one elevator is supplied by a single feeder. For example, if a building contains three elevators and a fault occurs in the circuit conductors to one of the elevators, only the overcurrent device ahead of that faulted circuit should open. Coordination leaves the remaining two elevators in operation. This arrangement is especially important because elevators are commonly used to carry fire fighters and equipment closer to the fire during fire-fighting operations. Where the overcurrent devices in the elevator room do not have proper coordination with the upstream feeder overcurrent device, the potential for interruption of power to all three elevators is increased.

For selective coordination of OCPDs, the manufacturer's time–current curves and data on let-through and withstand capacity and unlatching times must be used for sizing or setting overcurrent devices. The one-line diagram on the right side of Exhibit 620.6 illustrates the overcurrent protection arrangement required by 620.62, while the left side shows the potential for unnecessary power interruption to all of the elevators where the overcurrent protection is not selectively coordinated. Feeder overcurrent devices in the main distribution panel should not open, so two elevators can remain in use. See 620.51(A) for the requirement regarding power-supply disconnecting means.

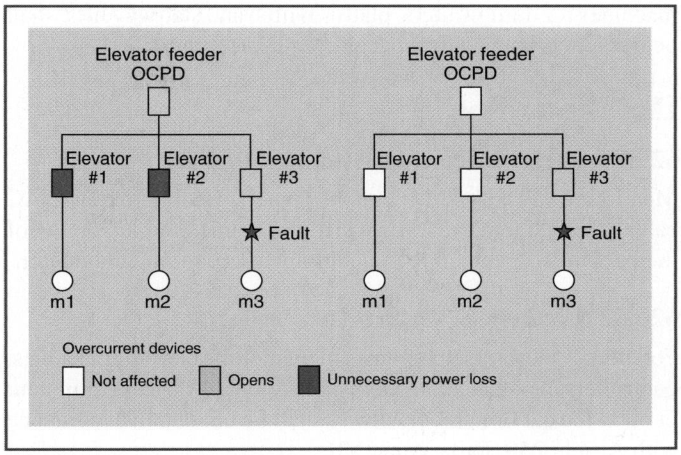


***EXHIBIT 620.6*** *Examples of a system of OCPDs that are not selectively coordinated (left) and a system where selectively coordinated overcurrent protection limits the power outage to only the elevator circuit in which the fault has occurred (right).*

## VIII. Machine Rooms, Control Rooms, Machinery Spaces, and Control Spaces

### 620.71 Guarding Equipment

Elevator, dumbwaiter, escalator, and moving walk driving machines; motor-generator sets; motor controllers; and disconnecting means shall be installed in a room or space set aside for that purpose unless otherwise permitted in 620.71(A) or (B). The room or space shall be secured against unauthorized access.

**(A) Motor Controllers.** Motor controllers shall be permitted outside the spaces herein specified, provided they are in enclosures with doors or removable panels that are capable of being locked in the closed position and the disconnecting means is located adjacent to or is an integral part of the motor controller. Motor controller enclosures for escalator or moving walks shall be permitted in the balustrade on the side located away from the moving steps or moving treadway. If the disconnecting means is an integral part of the motor controller, it shall be operable without opening the enclosure.

**(B) Driving Machines.** Elevators with driving machines located on the car, on the counterweight, or in the hoistway, and driving machines for dumbwaiters, platform lifts, and stairway lifts, shall be permitted outside the spaces herein specified.

## IX. Grounding

### 620.81 Metal Raceways Attached to Cars

Metal raceways, Type MC cable, Type MI cable, or Type AC cable attached to elevator cars shall be bonded to metal parts of the car that are bonded to the equipment grounding conductor.

### 620.82 Electric Elevators

For electric elevators, the frames of all motors, elevator machines, controllers, and the metal enclosures for all electrical equipment in or on the car or in the hoistway shall be bonded in accordance with Article 250, Parts V and VII.

### 620.83 Nonelectric Elevators

For elevators other than electric having any electrical conductors attached to the car, the metal frame of the car, where normally accessible to persons, shall be bonded in accordance with Article 250, Parts V and VII.

### 620.84 Escalators, Moving Walks, Platform Lifts, and Stairway Chairlifts

Escalators, moving walks, platform lifts, and stairway chairlifts shall comply with Article 250.

### 620.85 Ground-Fault Circuit-Interrupter Protection for Personnel

Each 125-volt, single-phase, 15- and 20-ampere receptacle installed in pits, in hoistways, on elevator car tops, and in escalator and moving walk wellways shall be of the ground-fault circuit-interrupter type.

All 125-volt, single-phase, 15- and 20-ampere receptacles installed in machine rooms and machinery spaces shall have ground-fault circuit-interrupter protection for personnel.

A single receptacle supplying a permanently installed sump pump shall not require ground-fault circuit-interrupter protection.

These GFCI requirements are intended to reduce the shock hazard to maintenance personnel who service elevator equipment using portable hand tools and temporary lighting.

The first paragraph requires a GFCI-type receptacle for each 15- and 20-ampere receptacle installed in pits, on elevator car tops, and in escalator and moving-walk wellways. This requirement is based on the premise that the reset pushbutton for a tripped GFCI receptacle should be within easy reach of an elevator mechanic working in confined spaces.

The second paragraph requires that all 15- and 20-ampere receptacles installed in machine rooms and machinery spaces have GFCI protection for personnel. This protection can be afforded by either a GFCI-type circuit breaker or a GFCI-type receptacle because machine spaces usually do not cause access hazards for service personnel.

## X. Emergency and Standby Power Systems

### 620.91 Emergency and Standby Power Systems

An elevator(s) shall be permitted to be powered by an emergency or standby power system.

Informational Note: See ASME A17.1-2010/CSA B44-10, *Safety Code for Elevators and Escalators*, 2.27.2, for additional information.

**(A) Regenerative Power.** For elevator systems that regenerate power back into the power source that is unable to absorb the regenerative power under overhauling elevator load conditions, a means shall be provided to absorb this power.

**(B) Other Building Loads.** Other building loads, such as power and lighting, shall be permitted as the energy absorption means required in 620.91(A), provided that such loads are automatically connected to the emergency or standby power system operating the elevators and are large enough to absorb the elevator regenerative power.

**(C) Disconnecting Means.** The disconnecting means required by 620.51 shall disconnect the elevator from both the emergency or standby power system and the normal power system.

Where an additional power source is connected to the load side of the disconnecting means, which allows automatic movement of the car to permit evacuation of passengers, the disconnecting means required in 620.51 shall be provided with an auxiliary contact that is positively opened mechanically, and the opening shall not be solely dependent on springs. This contact shall cause the additional power source to be disconnected from its load when the disconnecting means is in the open position.

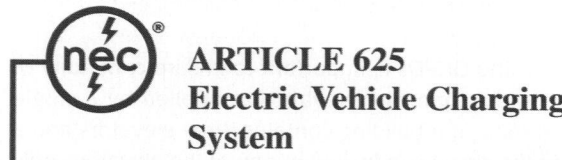

# ARTICLE 625 Electric Vehicle Charging System

## I. General

A variety of street- and highway-worthy electric and combination electric/fossil fuel or hybrid vehicles are available to consumers (see Exhibit 625.1). Plug-in hybrid and all-electric vehicle (EV) technology has evolved to the point that they are a viable option to the consumer. EV charging occurs in all occupancies, including residential, commercial, retail, and public sites. The National Institute of Standards and Technology projects that many homes and businesses will soon have EV charging infrastructure included as part of the premises wiring system.

***EXHIBIT 625.1*** *A plug-in electric vehicle.*

Article 625 sets forth installation safety requirements for charging equipment and for connecting the charging (supply) equipment to the EV through either a conductive or inductive connection. In addition, ventilation requirements aimed at preventing an ignitible air/hydrogen mixture are included in the requirements. The fundamental purpose of the *Code* to minimize fire and shock hazards is conveyed through the Article 625 requirements covering the charger–vehicle interface and the environment that is created by the charging of some types of batteries.

## 625.1 Scope

The provisions of this article cover the electrical conductors and equipment external to an electric vehicle that connect an electric vehicle to a supply of electricity by conductive or inductive means, and the installation of equipment and devices related to electric vehicle charging.

Informational Note No. 1: For industrial trucks, see NFPA 505-2011, *Fire Safety Standard for Powered Industrial Trucks Including Type Designations, Areas of Use, Conversions, Maintenance, and Operation.*

Informational Note No. 2: UL 2594-2013, *Standard for Electric Vehicle Supply Equipment*, is a safety standard for electric vehicle supply equipment. UL 2202-2009, *Standard for Electric Vehicle Charging System Equipment*, is a safety standard for electric vehicle charging equipment.

Article 625 covers all electrical wiring and equipment installed between the service point and the skin of an automotive-type EV. Automotive-type EVs are emphasized because they are much different from other commonly used electric vehicles, such as industrial forklifts, hoists, lifts, transports, golf carts, and airport personnel trams. The charging requirements and other exterior electrical connections for these off-road vehicles are usually serviced and maintained by trained mechanics or technicians. The *NEC* has adequate provisions to allow the AHJ to make interpretations that provide the safety levels needed for these installations.

## 625.2 Definitions

Several of the definitions in 625.2 correlate with industry standards such as the following documents from these organizations:

Society of Automotive Engineers

SAE J1772, *SAE Electric Vehicle Conductive Charge Coupler*

SAE J1773, *SAE Electric Vehicle Inductively Coupled Charging*

Underwriters Laboratories

UL 2231-1, *Standard for Personnel Protection Systems for Electric Vehicle Supply Circuits: General Requirements*

UL 2231-2, *Standard for Personnel Protection Systems for Electric Vehicle Supply Circuits: Particular Requirements for Protection Devices for Use in Charging Systems*

**Cable Management System (Electric Vehicle Supply Equipment).** An apparatus designed to control and organize unused lengths of output cable to the electric vehicle.

**Electric Vehicle.** An automotive-type vehicle for on-road use, such as passenger automobiles, buses, trucks, vans, neighborhood electric vehicles, electric motorcycles, and the like, primarily powered by an electric motor that draws current from a rechargeable storage battery, fuel cell, photovoltaic array, or other source of electric current. Plug-in hybrid electric vehicles (PHEV) are considered electric vehicles. For the purpose of this article, off-road, self-propelled electric vehicles, such as industrial trucks, hoists, lifts, transports, golf carts, airline ground support equipment, tractors, boats, and the like, are not included.

The primary difference between EVs as defined in Article 625 and EVs covered by other sections in the *NEC* is in their road and highway worthiness. Automotive EVs are comparable in performance and function to conventional automobiles and light trucks. Automotive EVs must be capable of complying with the Federal Motor Vehicle Safety Standards and other Department of Transportation, National Highway Traffic Safety Administration, and U.S. Environmental Protection Agency requirements.

The definition of *electric vehicle* includes neighborhood electric vehicles (NEVs), which are low-speed, limited-use EVs similar to golf carts but provided with automotive-grade headlights, seat belts, windshields, brakes, and other safety equipment that makes them street legal. NEVs are popular as low-cost, energy-efficient, zero-polluting alternatives to traditional automobiles. Under National Highway Traffic Safety Administration guidelines, the intended use for these vehicles is in inner-city areas and planned and retirement communities where the street speed limit is 35 mph or less. Electric vehicles such as lift trucks and golf carts are not covered by Article 625.

EVs can also be used as a power source for an optional standby system as covered in Article 702 or as an electric power production source as covered in Article 705. The EV supply equipment can either transfer power from the premises to the EV or from the EV to the premises wiring system through an interactive system covered in 625.48.

**Electric Vehicle Connector.** A device that, when electrically coupled (conductive or inductive) to an electric vehicle inlet, establishes an electrical connection to the electric vehicle for the purpose of power transfer and information exchange. This device is part of the electric vehicle coupler.

Informational Note: For further information, see 625.48 for interactive systems.

**Electric Vehicle Coupler.** A mating electric vehicle inlet and electric vehicle connector set.

**Electric Vehicle Inlet.** The device on the electric vehicle into which the electric vehicle connector is electrically coupled (conductive or inductive) for power transfer and information exchange. This device is part of the electric vehicle coupler. For the purposes of this *Code*, the electric vehicle inlet is considered to be part of the electric vehicle and not part of the electric vehicle supply equipment.

Informational Note: For further information, see 625.48 for interactive systems.

**Electric Vehicle Storage Battery.** A battery, comprised of one or more rechargeable electrochemical cells, that has no provision for the release of excessive gas pressure during normal charging and operation, or for the addition of water or electrolyte for external measurements of electrolyte-specific gravity.

**Electric Vehicle Supply Equipment.** The conductors, including the ungrounded, grounded, and equipment grounding conductors, and the electric vehicle connectors, attachment plugs, and all other fittings, devices, power outlets, or apparatus installed specifically for the purpose of transferring energy between the premises wiring and the electric vehicle.

EV supply equipment comprises the components between the skin of the EV and the premises wiring, including any flexible cable, disconnecting means, enclosures, power outlet, and EV connector. The defined term includes all off-vehicle charging equipment and does not include charging equipment installed on the vehicle. See Exhibit 625.2 for an EV charging station at a business and a vehicle connector plugged into a vehicle.

Informational Note No. 1: For further information, see 625.48 for interactive systems.

Informational Note No. 2: Within this article, the terms *electric vehicle supply equipment* and *electric vehicle charging system equipment* are considered to be equivalent.

**Output Cable to the Electric Vehicle.** An assembly consisting of a length of flexible EV cable and an electric vehicle connector (supplying power to the electric vehicle).

**Personnel Protection System.** A system of personnel protection devices and constructional features that when used together provide protection against electric shock of personnel.

**Plug-In Hybrid Electric Vehicle (PHEV).** A type of electric vehicle intended for on-road use with the ability to store and use off-vehicle electrical energy in the rechargeable energy storage system, and having a second source of motive power.

A *plug-in hybrid electric vehicle* can be charged through either its own rechargeable energy storage system or through connection to EV supply equipment located at a home, business, or other location.

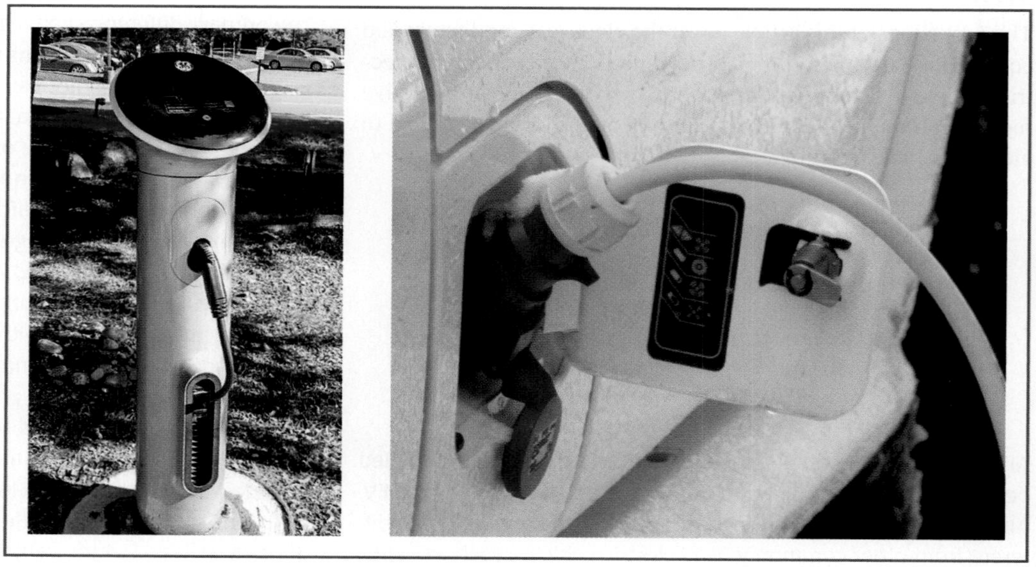

***EXHIBIT 625.2*** *EV supply equipment (left) and a connection to a vehicle (right).*

**Power-Supply Cord.** An assembly consisting of an attachment plug and length of flexible cord that connects the electric vehicle supply equipment (EVSE) to a receptacle.

**Rechargeable Energy Storage System.** Any power source that has the capability to be charged and discharged.

Informational Note: Batteries, capacitors, and electromechanical flywheels are examples of rechargeable energy storage systems.

### 625.4 Voltages

Unless other voltages are specified, the nominal ac system voltages of 120, 120/240, 208Y/120, 240, 480Y/277, 480, 600Y/347, and 600 volts and dc system voltages of up to 600 volts shall be used to supply equipment covered by this article.

### 625.5 Listed

All electrical materials, devices, fittings, and associated equipment shall be listed.

## II. Equipment Construction

### 625.10 Electric Vehicle Coupler

The electric vehicle coupler shall comply with 625.10(A) through (F).

The EV connector is the device that inserts into the EV inlet (charge port) of the vehicle. The EV inlet is not a premises wiring receptacle or an attachment cap. An EV coupler is the mating set of the EV connector and EV inlet. The coupler is required to be noninterchangeable to prevent equipment damage or personal injury, and it is not permitted to be a standard NEMA-configuration wiring device.

**(A) Polarization.** The electric vehicle coupler shall be polarized.

*Exception: A coupler that is part of a listed electric vehicle supply equipment.*

**(B) Noninterchangeability.** The electric vehicle coupler shall have a configuration that is noninterchangeable with wiring devices in other electrical systems. Nongrounding-type electric vehicle couplers shall not be interchangeable with grounding-type electric vehicle couplers.

**(C) Construction and Installation.** The electric vehicle coupler shall be constructed and installed so as to guard against inadvertent contact by persons with parts made live from the electric vehicle supply equipment or the electric vehicle battery.

The requirements for coupler construction provide a safe interface component for persons connecting the vehicle to or disconnecting the vehicle from the charging system. This connection/disconnection is generally performed by individuals who do not have any knowledge of the operation of the specific equipment or the associated hazards.

**(D) Unintentional Disconnection.** The electric vehicle coupler shall be provided with a positive means to prevent unintentional disconnection.

**(E) Grounding Pole.** The electric vehicle coupler shall be provided with a grounding pole, unless provided as part of a listed isolated electric vehicle supply equipment system.

**(F) Grounding Pole Requirements.** If a grounding pole is provided, the electric vehicle coupler shall be so designed that the grounding pole connection is the first to make and the last to break contact.

•

### 625.15 Markings

The electric vehicle supply equipment shall comply with 625.15(A) through (C).

**(A) General.** All electric vehicle supply equipment shall be marked by the manufacturer as follows:

FOR USE WITH ELECTRIC VEHICLES

**(B) Ventilation Not Required.** Where marking is required by 625.52(A), the electric vehicle supply equipment shall be clearly marked by the manufacturer as follows:

VENTILATION NOT REQUIRED

The marking shall be located so as to be clearly visible after installation.

**(C) Ventilation Required.** Where marking is required by 625.52(B), the electric vehicle supply equipment shall be clearly marked by the manufacturer, "Ventilation Required." The marking shall be located so as to be clearly visible after installation.

### 625.16 Means of Coupling

The means of coupling to the electric vehicle shall be either conductive or inductive. Attachment plugs, electric vehicle connectors, and electric vehicle inlets shall be listed or labeled for the purpose.

### 625.17 Cords and Cables

**(A) Power-Supply Cord.** The cable for cord-connected equipment shall comply with all of the following:

(1) Be any of the types specified in 625.17(B) or hard service cord, junior hard service cord, or portable power cable types in accordance with Table 400.4. Hard service cord, junior hard service cord, or portable power cable types shall be listed, as applicable, for exposure to oil and damp and wet locations.
(2) Have an ampacity as specified in Table 400.5(A)(1) or, for 8 AWG and larger, in the 60°C columns of Table 400.5(A)(2).

(3) Have an overall length as specified in 625.17(A)(3)a or b as follows:

a. When the interrupting device of the personnel protection system specified in 625.22 is located within the enclosure of the supply equipment or charging system, the power-supply cord shall be not more than 300 mm (12 in.) long,

b. When the interrupting device of the personnel protection system specified in 625.22 is located at the attachment plug, or within the first 300 mm (12 in.) of the power-supply cord, the overall cord length shall be a minimum of 1.8 m (6 ft) and shall be not greater than 4.6 m (15 ft).

**(B) Output Cable to the Electric Vehicle.** The output cable to the electric vehicle shall be Type EV, EVJ, EVE, EVJE, EVT, or EVJT flexible cable as specified in Table 400.4.

Informational Note: Listed electric vehicle supply equipment may incorporate output cables having ampacities greater than 60°C based on the permissible temperature limits for the components and the cable.

**(C) Overall Cord and Cable Length.** The overall usable length shall not exceed 7.5 m (25 ft) unless equipped with a cable management system that is part of the listed electric vehicle supply equipment.

**(1) Not Fastened in Place.** Where the electric vehicle supply equipment or charging system is not fastened in place, the cord-exposed useable length shall be measured from the face of the attachment plug to the face of the electric vehicle connector.

**(2) Fastened in Place.** Where the electric vehicle supply equipment or charging system is fastened in place, the useable length of the output cable shall be measured from the cable exit of the electric vehicle supply equipment or charging system to the face of the electric vehicle connector.

The maximum 25-foot cable length takes into account both the power supply cable and output cable to the EV. For example, 625.17(A)(3)(b) allows a 15-foot long power cord, which limits the length of the output cable to 10 feet. Cable lengths in excess of 25 feet are permitted where provided with a listed cable management system. This provides commercial parking areas with flexibility in the number of charging spaces they can provide.

### 625.18 Interlock

Electric vehicle supply equipment shall be provided with an interlock that de-energizes the electric vehicle connector whenever the electrical connector is uncoupled from the electric vehicle. An interlock shall not be required for portable cord-and-plug-connected electric vehicle supply equipment intended for connection to receptacle outlets rated at 125 volts, single phase, 15 and 20 amperes. An interlock shall not be required for dc supplies less than 50 volts dc.

To reduce shock hazard, a pilot or communications interlock establishes power through the EV supply equipment. Loss of the pilot or communications circuit locks out power, isolating possible hazardous situations in the EV supply equipment. See 625.52(B)(4) for mechanical ventilation interlock requirements.

### 625.19 Automatic De-Energization of Cable

The electric vehicle supply equipment or the cable-connector combination of the equipment shall be provided with an automatic means to de-energize the cable conductors and electric vehicle connector upon exposure to strain that could result in either cable rupture or separation of the cable from the electric connector and exposure of live parts. Automatic means to de-energize the cable conductors and electric vehicle connector shall not be required for portable cord-and-plug-connected electric vehicle supply equipment intended for connection to receptacle outlets rated at 125 volts, single phase, 15 and 20 amperes. An interlock shall not be required for dc supplies less than 50 volts dc.

•

### 625.22 Personnel Protection System

The electric vehicle supply equipment shall have a listed system of protection against electric shock of personnel. Where cord- and plug-connected electric vehicle supply equipment is used, the interrupting device of a listed personnel protection system shall be provided and shall be an integral part of the attachment plug or shall be located in the power-supply cord not more than 300 mm (12 in.) from the attachment plug.

The listed personnel protection system may consist of one or more components that provide protection against electric shock for different portions of the EV supply equipment circuitry, which may be operating at frequencies other than 50/60 Hz, at direct-current potentials, and/or voltages above 150 volts to ground. Standard GFCI devices do not provide the range of protection needed for the various types of charging systems available. For systems operating above 150 volts to ground, the protective system may include monitoring systems to ensure that proper grounding is provided and maintained during charging.

## III. Installation

### 625.40 Overcurrent Protection

Overcurrent protection for feeders and branch circuits supplying electric vehicle supply equipment shall be sized for continuous duty and shall have a rating of not less than 125 percent of the maximum load of the electric vehicle supply equipment. Where noncontinuous loads are supplied from the same feeder or branch circuit, the overcurrent device shall have a rating of not less than the sum of the noncontinuous loads plus 125 percent of the continuous loads.

## 625.41 Rating

Electric vehicle supply equipment shall have sufficient rating to supply the load served. Electric vehicle charging loads shall be considered to be continuous loads for the purposes of this article. Where an automatic load management system is used, the maximum electric vehicle supply equipment load on a service and feeder shall be the maximum load permitted by the automatic load management system.

Three methods for EV charging — referred to as Level 1, Level 2, and Level 3 EV charging — cover the range of power levels anticipated for charging EVs. EVs are treated as continuous loads.

*Level 1* allows broad access to charge an EV by plugging into a common, grounded 120-volt electrical receptacle (NEMA 5-15R or 5-20R). The maximum load on this receptacle is 12 A/1.4 kVA (15 ampere receptacle) or 16 A/1.9 kVA (20 ampere receptacle).

*Level 2* is the primary method of EV charging at both private and public facilities. It requires special equipment and connection to an electric power supply dedicated to EV charging. The voltage of this connection is either 240 volts or 208 volts. The maximum load is 32 amperes (7.7 kVA at 240 V or 6.7 kVA at 208 V), with a minimum circuit and overcurrent rating of 40 amperes. See 625.40 for sizing OCPDs.

*Level 3* is the EV equivalent of a commercial gasoline dispensing station. This high-speed, high-power method charges an EV in about the same time it takes to refuel a conventional vehicle. Because of individual supply requirements and available source voltages, exact voltage and load specifications for Level 3 charging have not been defined in the same way that Level 1 and Level 2 have. These power requirements are specified by the equipment manufacturer, but at present the maximum current is specified as 400 amperes with 240 kW of continuous power supplied.

Connection of EV supply equipment to automatic load management system can preclude the need for a service or feeder upgrade to an existing electrical installation.

## 625.42 Disconnecting Means

For electric vehicle supply equipment rated more than 60 amperes or more than 150 volts to ground, the disconnecting means shall be provided and installed in a readily accessible location. The disconnecting means shall be lockable open in accordance with 110.25.

## 625.44 Electric Vehicle Supply Equipment Connection

Electric vehicle supply equipment shall be permitted to be cord- and plug-connected to the premises wiring system in accordance with one of the following:

**(A) Connections to 125-Volt, Single-Phase, 15- and 20-Ampere Receptacle Outlets.** Electric vehicle supply equipment intended for connection to nonlocking, 2-pole, 3-wire grounding-type receptacle outlets rated at 125 V, single phase, 15 and 20 amperes or from a supply of less than 50 volts dc.

**(B) Connections to Other Receptacle Outlets.** Electric vehicle supply equipment that is rated 250 V maximum and complying with all of the following:

(1) It is intended for connection to nonlocking, 2-pole, 3-wire and 3-pole, 4-wire, grounding-type receptacle outlets rated not more than 50 amperes.
(2) EVSE is fastened in place to facilitate any of the following:
  a. Ready removal for interchange
  b. Facilitation of maintenance and repair
  c. Repositioning of portable, movable, or EVSE fastened in place
(3) Power-supply cord length for electric vehicle supply equipment fastened in place is limited to 1.8 m (6 ft).
(4) Receptacles are located to avoid physical damage to the flexible cord.

All other electric vehicle supply equipment shall be permanently wired and fastened in place to the supporting surface, a wall, a pole, or other structure. The electric vehicle supply equipment shall have no exposed live parts.

Some manufacturers produce 125-volt, single-phase, 15- or 20-ampere portable charging units for convenience charging. These charging units may be stored in the vehicle. Fastened EV supply equipment rated up to 250 volts and 50 amperes may be cord-and-plug-connected under the specified conditions. However, all other equipment must be mounted and permanently wired.

## 625.46 Loss of Primary Source

Means shall be provided such that, upon loss of voltage from the utility or other electrical system(s), energy cannot be back fed through the electric vehicle and the supply equipment to the premises wiring system unless permitted by 625.48.

## 625.48 Interactive Systems

Electric vehicle supply equipment and other parts of a system, either on board or off board the vehicle, that are intended to be interconnected to a vehicle and also serve as an optional standby system or an electric power production source or provide for bi-directional power feed shall be listed and marked as suitable for that purpose. When used as an optional standby system, the requirements of Article 702 shall apply, and when used as an electric power production source, the requirements of Article 705 shall apply.

The on-board power production system of some EVs is capable of operating as a stand-alone or interactive power supply for premises wiring systems. Such systems are required to be listed for this type of use and are required to comply with the requirements of Article 702 or Article 705, depending on how the system connects to premises wiring system and/or the primary source of electricity.

•

### 625.50 Location

The electric vehicle supply equipment shall be located for direct electrical coupling of the EV connector (conductive or inductive) to the electric vehicle. Unless specifically listed and marked for the location, the coupling means of the electric vehicle supply equipment shall be stored or located at a height of not less than 450 mm (18 in.) above the floor level for indoor locations and 600 mm (24 in.) above the grade level for outdoor locations.

### 625.52 Ventilation

The ventilation requirement for charging an electric vehicle in an indoor enclosed space shall be determined by 625.52(A) or (B).

Where the EV charging operation is conducted in outdoor or open locations, the resulting off-gassing of hydrogen does not pose the same risk of creating an ignitible environment as in indoor locations. The lighter-than-air hydrogen readily diffuses into the atmosphere. In addition to driveways and parking lots, structures with adequate natural ventilation – such as carports and open parking structures – do not require mechanical ventilation. NFPA 88A, *Standard for Parking Structures*, provides a quantifiable definition of the term *open parking structure*.

**(A) Ventilation Not Required.** Where electric vehicle storage batteries are used or where the electric vehicle supply equipment is listed for charging electric vehicles indoors without ventilation and marked in accordance with 625.15(B), mechanical ventilation shall not be required.

Most batteries used in EVs do not emit hydrogen gas in quantities that could cause an explosion. Preventive measures such as mechanical or passive ventilation are not required, because the EV batteries and charging systems are designed to prevent or limit the emission of hydrogen during charging. The Society of Automotive Engineers recommended practice SAE J-1718, *Measurement of Hydrogen Gas Emission from Battery-Powered Passenger Cars and Light Trucks During Battery Charging*, can be used to assess suitability for indoor charging. This standard includes provisions for tests during normal charging operations and potential equipment failure modes.

**(B) Ventilation Required.** Where the electric vehicle supply equipment is listed for charging electric vehicles that require ventilation for indoor charging, and is marked in accordance with 625.15(C), mechanical ventilation, such as a fan, shall be provided. The ventilation shall include both supply and exhaust equipment and shall be permanently installed and located to intake from, and vent directly to, the outdoors. Positive-pressure ventilation systems shall be permitted only in vehicle charging buildings or areas that have been specifically designed and approved for that application. Mechanical ventilation requirements shall be determined by one of the methods specified in 625.52(B)(1) through (B)(4).

**(1) Table Values.** For supply voltages and currents specified in Table 625.52(B)(1) or Table 625.52(B)(2), the minimum ventilation requirements shall be as specified in Table 625.52(B)(1) or Table 625.52(B)(2) for each of the total number of electric vehicles that can be charged at one time.

**(2) Other Values.** For supply voltages and currents other than specified in Table 625.52(B)(1) or Table 625.52(B)(2), the minimum ventilation requirements shall be calculated by means of the following general formulas, as applicable:

(1) Single-phase ac or dc:
Ventilation $_{\text{single-phase ac or dc}}$ in cubic meters per minute ($m^3$/min) =

$$\frac{(\text{volts})(\text{amperes})}{1718}$$

Ventilation $_{\text{single-phase ac or dc}}$ in cubic feet per minute (cfm) =

$$\frac{(\text{volts})(\text{amperes})}{48.7}$$

(2) Three-phase ac:
Ventilation $_{\text{3-phase}}$ in cubic meters per minute ($m^3$/min) =

$$\frac{1.732(\text{volts})(\text{amperes})}{1718}$$

Ventilation $_{\text{3-phase}}$ in cubic feet per minute (cfm) =

$$\frac{1.732(\text{volts})(\text{amperes})}{48.7}$$

**(3) Engineered Systems.** For an electric vehicle supply equipment ventilation system designed by a person qualified to perform such calculations as an integral part of a building's total ventilation system, the minimum ventilation requirements shall be permitted to be determined in accordance with calculations specified in the engineering study.

**(4) Supply Circuits.** The supply circuit to the mechanical ventilation equipment shall be electrically interlocked with the electric vehicle supply equipment and shall remain energized during the entire electric vehicle charging cycle. Electric vehicle supply equipment shall be marked in accordance with 625.15. Electric vehicle supply equipment receptacles rated at 125 volts, single phase, 15 and 20 amperes shall be marked in accordance with 625.15 and shall be switched, and the mechanical ventilation system shall be electrically interlocked through the switch supply power to the receptacle. Electric vehicle supply equipment supplied from less than 50 volts dc shall be marked in accordance with 625.15(C) and shall be switched, and the mechanical ventilation system shall be electrically interlocked through the switch supply power to the electric vehicle supply equipment.

•

***TABLE 625.52(B)(1)*** *Minimum Ventilation Required in Cubic Meters per Minute ($m^3$/min) for Each of the Total Number of Electric Vehicles That Can Be Charged at One Time*

| Branch-Circuit Ampere Rating | Branch-Circuit Voltage | | | | | | | |
|---|---|---|---|---|---|---|---|---|
| | | Single Phase | | | 3 Phase | | | |
| | DC ≥ 50 V | 120 V | 208 V | 240 V or 120/240 V | 208 V or 208Y/120 V | 240 V | 480 V or 480Y/277 V | 600 V or 600Y/347 V |
| 15 | 0.5 | 1.1 | 1.8 | 2.1 | — | — | — | — |
| 20 | 0.6 | 1.4 | 2.4 | 2.8 | 4.2 | 4.8 | 9.7 | 12 |
| 30 | 0.9 | 2.1 | 3.6 | 4.2 | 6.3 | 7.2 | 15 | 18 |
| 40 | 1.2 | 2.8 | 4.8 | 5.6 | 8.4 | 9.7 | 19 | 24 |
| 50 | 1.5 | 3.5 | 6.1 | 7.0 | 10 | 12 | 24 | 30 |
| 60 | 1.8 | 4.2 | 7.3 | 8.4 | 13 | 15 | 29 | 36 |
| 100 | 2.9 | 7.0 | 12 | 14 | 21 | 24 | 48 | 60 |
| 150 | — | — | — | — | 31 | 36 | 73 | 91 |
| 200 | — | — | — | — | 42 | 48 | 97 | 120 |
| 250 | — | — | — | — | 52 | 60 | 120 | 150 |
| 300 | — | — | — | — | 63 | 73 | 145 | 180 |
| 350 | — | — | — | — | 73 | 85 | 170 | 210 |
| 400 | — | — | — | — | 84 | 97 | 195 | 240 |

***TABLE 625.52(B)(2)*** *Minimum Ventilation Required in Cubic Feet per Minute (cfm) for Each of the Total Number of Electric Vehicles That Can Be Charged at One Time*

| Branch-Circuit Ampere Rating | Branch-Circuit Voltage | | | | | | | |
|---|---|---|---|---|---|---|---|---|
| | | Single Phase | | | 3 Phase | | | |
| | DC ≥ 50 V | 120 V | 208 V | 240 V or 120/240 V | 208 V or 208Y/120 V | 240 V | 480 V or 480Y/277 V | 600 V or 600Y/347 V |
| 15 | 15.4 | 37 | 64 | 74 | — | — | — | — |
| 20 | 20.4 | 49 | 85 | 99 | 148 | 171 | 342 | 427 |
| 30 | 30.8 | 74 | 128 | 148 | 222 | 256 | 512 | 641 |
| 40 | 41.3 | 99 | 171 | 197 | 296 | 342 | 683 | 854 |
| 50 | 51.3 | 123 | 214 | 246 | 370 | 427 | 854 | 1066 |
| 60 | 61.7 | 148 | 256 | 296 | 444 | 512 | 1025 | 1281 |
| 100 | 102.5 | 246 | 427 | 493 | 740 | 854 | 1708 | 2135 |
| 150 | — | — | — | — | 1110 | 1281 | 2562 | 3203 |
| 200 | — | — | — | — | 1480 | 1708 | 3416 | 4270 |
| 250 | — | — | — | — | 1850 | 2135 | 4270 | 5338 |
| 300 | — | — | — | — | 2221 | 2562 | 5125 | 6406 |
| 350 | — | — | — | — | 2591 | 2989 | 5979 | 7473 |
| 400 | — | — | — | — | 2961 | 3416 | 6832 | 8541 |

The sufficient diffusion and dilution of hydrogen gas from gas-emitting batteries prevents a hazardous condition. During the charging process, certain batteries used in some EVs emit hydrogen gas, which is colorless, odorless, tasteless, nontoxic, and flammable. At atmospheric pressure, the flammable range for hydrogen is 4 to 75 percent by volume in air.

NFPA 69, *Standard on Explosion Prevention Systems*, establishes requirements to ensure safety with flammable mixtures. Section 7.3, Design and Operating Requirements, of NFPA 69-2008 specifies that combustible gas concentrations be restricted to 25 percent of the lower flammable limit to provide a safety margin for personnel. Safety is accomplished by keeping the concentration of hydrogen below 25 percent of the lower flammability limit. That is 1 percent (25 percent × 4 percent = 1 percent) hydrogen by volume in air, or below 10,000 ppm hydrogen.

A ventilation system for a typical residential-type garage includes both supply and mechanical exhaust equipment and is permanently installed. The system brings outdoor air into the space, circulates the air through the space, and exhausts the air directly to the outdoors. Typically, the equipment includes a passive vent for intake on one side of the enclosed space and an exhaust fan vented to the outside on the other side.

In enclosed commercial garages and other structures, additional ventilation is not required if the exhaust, as required by the building code for carbon monoxide or other purposes, is greater than the quantity listed in the table. Other engineered EV ventilation systems are allowed as part of the building ventilation system.

The ventilation system and the charging system must be interlocked to prevent charging if the ventilation is not operating. This charging arrangement can be used with EVs equipped with a self-contained charging system in which activation of the charging system does not depend on a signal from the EV. A manually operated switch controls the receptacle used to supply the vehicle charging system, and it is also interlocked with the power supply to the ventilation fan. This arrangement ensures that the ventilation fan is operating whenever the vehicle charging receptacle is energized.

# ARTICLE 626 Electrified Truck Parking Spaces

## I. General

### 626.1 Scope

The provisions of this article cover the electrical conductors and equipment external to the truck or transport refrigerated unit that connect trucks or transport refrigerated units to a supply of electricity, and the installation of equipment and devices related to electrical installations within an electrified truck parking space.

Stringent federal and state mandates to reduce diesel engine emissions have led to using electric power for operation of transport truck heating and refrigeration equipment while the truck is parked. Because much of the transport industry is interstate commerce, this article provides for standardization of truck parking space equipment so that driver interface with electrical connection devices can be safely accomplished from coast to coast.

### 626.2 Definitions

**Cable Management System (Electrified Truck Parking Spaces).** An apparatus designed to control and organize unused lengths of cable or cord at electrified truck parking spaces.

**Cord Connector.** A device that, by inserting it into a truck flanged surface inlet, establishes an electrical connection to the truck for the purpose of providing power for the on-board electric loads and may provide a means for information exchange. This device is part of the truck coupler.

**Disconnecting Means, Parking Space.** The necessary equipment usually consisting of a circuit breaker or switch and fuses, and their accessories, located near the point of entrance of supply conductors in an electrified truck parking space and intended to constitute the means of cutoff for the supply to that truck.

**Electrified Truck Parking Space.** A truck parking space that has been provided with an electrical system that allows truck operators to connect their vehicles while stopped and to use off-board power sources in order to operate on-board systems such as air conditioning, heating, and appliances, without any engine idling.

Informational Note: An electrified truck parking space also includes dedicated parking areas for heavy-duty trucks at travel plazas, warehouses, shipper and consignee yards, depot facilities, and border crossings. It does not include areas such as the shoulders of highway ramps and access roads, camping and recreational vehicle sites, residential and commercial parking areas used for automotive parking or other areas where ac power is provided solely for the purpose of connecting automotive and other light electrical loads, such as engine block heaters, and at private residences.

**Electrified Truck Parking Space Wiring Systems.** All of the electrical wiring, equipment, and appurtenances related to electrical installations within an electrified truck parking space, including the electrified parking space supply equipment.

**Overhead Gantry.** A structure consisting of horizontal framework, supported by vertical columns spanning above electrified truck parking spaces, that supports equipment, appliances, raceway, and other necessary components for the purpose of supplying electrical, HVAC, internet, communications, and other services to the spaces.

**Separable Power Supply Cable Assembly.** A flexible cord or cable, including ungrounded, grounded, and equipment grounding conductors, provided with a cord connector, an attachment plug, and all other fittings, grommets, or devices installed for the purpose of delivering energy from the source of electrical supply to the truck or TRU flanged surface inlet.

**Transport Refrigerated Unit (TRU).** A trailer or container, with integrated cooling or heating, or both, used for the purpose of maintaining the desired environment of temperature-sensitive goods or products.

**Truck.** A motor vehicle designed for the transportation of goods, services, and equipment.

**Truck Coupler.** A truck flanged surface inlet and mating cord connector.

**Truck Flanged Surface Inlet.** The device(s) on the truck into which the connector(s) is inserted to provide electric energy and other services. This device is part of the truck coupler. For the purposes of this article, the truck flanged surface inlet is considered to be part of the truck and not part of the electrified truck parking space supply equipment.

### 626.3 Other Articles

Wherever the requirements of other articles of this *Code* and Article 626 differ, the requirements of Article 626 shall apply. Unless electrified truck parking space wiring systems are supported or arranged in such a manner that they cannot be used in or above locations classified in 511.3 or 514.3, or both, they shall comply with 626.3(A) and (B) in addition to the requirements of this article.

**(A) Vehicle Repair and Storage Facilities.** Electrified truck parking space electrical wiring systems located at facilities for the repair or storage of self-propelled vehicles that use volatile flammable liquids or flammable gases for fuel or power shall comply with Article 511.

**(B) Motor Fuel Dispensing Stations.** Electrified truck parking space electrical wiring systems located at or serving motor fuel dispensing stations shall comply with Article 514.

Informational Note: For additional information, see NFPA 88A-2011, *Standard for Parking Structures*, and NFPA 30A-2012, *Code for Fuel Dispensing Facilities and Repair Garages*.

### 626.4 General Requirements

**(A) Not Covered.** The provisions of this article shall not apply to that portion of other equipment in residential, commercial, or industrial facilities that requires electric power used to load and unload cargo, operate conveyors, and for other equipment used on the site or truck.

**(B) Distribution System Voltages.** Unless other voltages are specified, the nominal ac system voltages of 120, 120/240, 208Y/120, 240, or 480Y/277 shall be used to supply equipment covered by this article.

**(C) Connection to Wiring System.** The provisions of this article shall apply to the electrified truck parking space supply equipment intended for connection to a wiring system as defined in 626.4(B).

## II. Electrified Truck Parking Space Electrical Wiring Systems

### 626.10 Branch Circuits

Electrified truck parking space single-phase branch circuits shall be derived from a 208Y/120-volt, 3-phase, 4-wire system or a 120/240-volt, single-phase, 3-wire system.

*Exception: A 120-volt distribution system shall be permitted to supply existing electrified truck parking spaces.*

***TABLE 626.11(B)*** *Demand Factors for Services and Feeders*

| Climatic Temperature Zone (USDA Hardiness Zone) See Note | Demand Factor (%) |
|---|---|
| 1 | 70% |
| 2a | 67% |
| 2b | 62% |
| 3a | 59% |
| 3b | 57% |
| 4a | 55% |
| 4b | 51% |
| 5a | 47% |
| 5b | 43% |
| 6a | 39% |
| 6b | 34% |
| 7a | 29% |
| 7b | 24% |
| 8a | 21% |
| 8b | 20% |
| 9a | 20% |
| 9b | 20% |
| 10a | 21% |
| 10b | 23% |
| 11 | 24% |

Note: The climatic temperature zones shown in Table 626.11(B) correlate with those found on the "USDA Plant Hardiness Zone Map," and the climatic temperature zone selected for use with the table shall be determined through the use of this map based on the installation location.

### 626.11 Feeder and Service Load Calculations

**(A) Parking Space Load.** The calculated load of a feeder or service shall be not less than the sum of the loads on the branch circuits. Electrical service and feeders shall be calculated on the basis of not less than 11 kVA per electrified truck parking space.

**(B) Demand Factors.** Electrified truck parking space electrical wiring system demand factors shall be based upon the climatic temperature zone in which the equipment is installed. The demand factors set forth in Table 626.11(B) shall be the minimum allowable demand factors that shall be permitted for calculating load for service and feeders. No demand factor shall be allowed for any other load, except as provided in this article.

Informational Note: The U.S. Department of Agriculture (USDA) has developed a commonly used "Plant Hardiness Zone" map that is publicly available. The map provides guidance for determining the Climatic Temperature Zone. Data indicate that the HVAC has the highest power requirement in cold climates, with the heating demand representing the greatest load, which in turn is dependent on outside temperature. In very warm climates, where no heating load is necessary, the cooling load increases as the outdoor temperature rises.

**(C) Two or More Electrified Truck Parking Spaces.** Where the electrified truck parking space wiring system is in a location that serves two or more electrified truck parking spaces, the equipment for each space shall comply with 626.11(A), and the calculated load shall be calculated on the basis of each parking space.

**(D) Conductor Rating.** Truck space branch-circuit supplied loads shall be considered to be continuous.

## III. Electrified Truck Parking Space Supply Equipment

### 626.22 Wiring Methods and Materials

**(A) Electrified Truck Parking Space Supply Equipment Type.** The electrified truck parking space supply equipment shall be provided in one of the following forms:

(1) Pedestal
(2) Overhead gantry
(3) Raised concrete pad

**(B) Mounting Height.** Post, pedestal, and raised concrete pad types of electrified truck parking space supply equipment shall be not less than 600 mm (2 ft) aboveground or above the point identified as the prevailing highest water level mark or an equivalent benchmark based on seasonal or storm-driven flooding from the authority having jurisdiction.

**(C) Access to Working Space.** All electrified truck parking space supply equipment shall be accessible by an unobstructed entrance or passageway not less than 600 mm (2 ft) wide and not less than 2.0 m (6 ft 6 in.) high.

**(D) Disconnecting Means.** A disconnecting switch or circuit breaker shall be provided to disconnect one or more electrified truck parking space supply equipment sites from a remote location. The disconnecting means shall be provided and installed in a readily accessible location and shall be lockable open in accordance with 110.25.

### 626.23 Overhead Gantry or Cable Management System

**(A) Cable Management.** Electrified truck parking space equipment provided from either overhead gantry or cable management systems shall utilize a permanently attached power supply cable in electrified truck parking space supply equipment. Other cable types and assemblies listed as being suitable for the purpose, including optional hybrid communications, signal, and composite optical fiber cables, shall be permitted.

**(B) Strain Relief.** Means to prevent strain from being transmitted to the wiring terminals shall be provided. Permanently attached power supply cable(s) shall be provided with a means to de-energize the cable conductors and power service delivery device upon exposure to strain that could result in either cable damage or separation from the power service delivery device and exposure of live parts.

### 626.24 Electrified Truck Parking Space Supply Equipment Connection Means

**(A) General.** Each truck shall be supplied from electrified truck parking space supply equipment through suitable extra-hard service cables or cords. Each connection to the equipment shall be by a single separable power supply cable assembly.

**(B) Receptacle.** All receptacles shall be listed and of the grounding type. Every truck parking space with electrical supply shall be equipped with (B)(1) and (B)(2).

(1) A maximum of three receptacles, each 2-pole, 3-wire grounding type and rated 20 amperes, 125 volts, and two of the three connected to two separate branch circuits.

Informational Note: For the nonlocking-type and grounding-type 20-ampere receptacle configuration, see ANSI/NEMA WD6-2002, *Standard for Dimensions of Attachment Plugs and Receptacles*, Figure 5-20.

(2) One single receptacle, 3-pole, 4-wire grounding type, single phase rated either 30 amperes 208Y/120 volts or 125/250 volts. The 125/250-volt receptacle shall be permitted to be used on a 208Y/120-volt, single-phase circuit.

Informational Note: For various configurations of 30-ampere pin and sleeve receptacles, see ANSI/UL1686, *Standard for Pin and Sleeve Configurations*, Figure C2.9 or Part C3.

*Exception: Where electrified truck parking space supply equipment provides the heating, air-conditioning, and comfort-cooling function without requiring a direct electrical connection at the truck, only two receptacles identified in 626.24(B)(1) shall be required.*

**(C) Disconnecting Means, Parking Space.** The electrified truck parking space supply equipment shall be provided with a switch or circuit breaker for disconnecting the power supply to the electrified truck parking space. A disconnecting means shall be provided and installed in a readily accessible location and shall be lockable open in accordance with 110.25.

**(D) Ground-Fault Circuit-Interrupter Protection for Personnel.** The electrified truck parking space equipment shall be designed and constructed such that all receptacle outlets in 626.24 are provided with ground-fault circuit-interrupter protection for personnel.

### 626.25 Separable Power-Supply Cable Assembly

A separable power-supply cable assembly, consisting of a power-supply cord, a cord connector, and an attachment plug intended for connection with a truck flanged surface inlet, shall be of a listed type. The power-supply cable assembly or assemblies shall be identified and be one of the types and ratings specified in 626.25(A) and (B). Cords with adapters and pigtail ends, extension cords, and similar items shall not be used.

**(A) Rating(s).**

**(1) Twenty-Ampere Power-Supply Cable Assembly.** Equipment with a 20-ampere, 125-volt receptacle, in accordance with 626.24(B)(1), shall use a listed 20-ampere power-supply cable assembly.

*Exception: It shall be permitted to use a listed separable power-supply cable assembly, either hard service or extra-hard service and rated 15 amperes, 125 volts, for connection to an engine block heater for legacy vehicles.*

**(2) Thirty-Ampere Power-Supply Cable Assembly.** Equipment with a 30-ampere, 208Y/120-volt or 125/250-volt receptacle, in accordance with 626.24(B)(2), shall use a listed 30-ampere main power-supply cable assembly.

**(B) Power-Supply Cord.**

**(1) Conductors.** The cord shall be a listed type with three or four conductors, for single-phase connection, one conductor of which shall be identified in accordance with 400.23.

*Exception: It shall be permitted to use a separate listed three-conductor separable power-supply cable assembly, one conductor of which shall be identified in accordance with 400.23 and rated 15 amperes, 125 volts for connection to an engine block heater for existing vehicles.*

**(2) Cord.** Extra-hard usage flexible cords and cables rated not less than 90°C (194°F), 600 volts; listed for both wet locations and sunlight resistance; and having an outer jacket rated to be resistant to temperature extremes, oil, gasoline, ozone, abrasion, acids, and chemicals shall be permitted where flexibility is necessary between the electrified truck parking space supply equipment, the panel board and flanged surface inlet(s) on the truck.

*Exception: Cords for the separable power supply cable assembly for 15- and 20-ampere connections shall be permitted to be a hard service type.*

**(3) Cord Overall Length.** The exposed cord length shall be measured from the face of the attachment plug to the point of entrance to the truck or the face of the flanged surface inlet or to the point where the cord enters the truck. The overall length of the cable shall not exceed 7.5 m (25 ft) unless equipped with a cable management system that is listed as suitable for the purpose.

**(4) Attachment Plug.** The attachment plug(s) shall be listed, by itself or as part of a cord set, for the purpose and shall be molded to or installed on the flexible cord so that it is secured tightly to the cord at the point where the cord enters the attachment plug. If a right-angle cap is used, the configuration shall be oriented so that the grounding member is farthest from the cord. Where a flexible cord is provided, the attachment plug shall comply with 250.138(A).

(a) *Connection to a 20-Ampere Receptacle.* A separable power-supply cable assembly for connection to a truck flanged surface inlet, rated at 20 amperes, shall have a nonlocking-type attachment plug that shall be 2-pole, 3-wire grounding type rated 20 amperes, 125 volts and intended for use with the 20-ampere, 125-volt receptacle.

*Exception: A separable power-supply cable assembly, rated 15 amperes, provided for the connection of an engine block heater, only, shall have an attachment plug that shall be 2-pole, 3-wire grounding type rated 15 amperes, 125 volts.*

Informational Note: For nonlocking- and grounding-type 15- or 20-ampere plug and receptacle configurations, see ANSI/NEMA WD6-2002, *Standard for Dimensions of Attachment Plugs and Receptacles*, Figure 5-15 or Figure 5-20.

(b) *Connection to a 30-Ampere Receptacle.* A separable power-supply cable assembly for connection to a truck flanged surface inlet, rated at 30 amperes, shall have an attachment plug that shall be 3-pole, 4-wire grounding type rated 30-amperes, 208Y/120 volts or 125/250 volts, and intended for use with the receptacle in accordance with 626.24(B)(2). The 125/250-volt attachment plug shall be permitted to be used on a 208Y/120-volt, single-phase circuit.

Informational Note: For various configurations of 30-ampere pin and sleeve plugs, see ANSI/UL 1686-2012, *Standard for Pin and Sleeve Configurations*, Figure C2.10 or Part C3.

**(5) Cord Connector.** The cord connector for a separable power-supply cable assembly, as specified in 626.25(A)(1), shall be a 2-pole, 3-wire grounding type rated 20 amperes, 125 volts. The cord connector for a separable power-supply cable assembly, as specified in 626.25(A)(2), shall be a 3-pole, 4-wire grounding type rated 30 amperes, 208Y/120 volts or 125/250 volts. The 125/250-volt cord connector shall be permitted to be used on a 208Y/120-volt, single-phase circuit.

*Exception: The cord connector for a separable power supply cable assembly, rated 15 amperes, provided for the connection of an engine block heater for existing vehicles, shall have an attachment plug that shall be 2-pole, 3-wire grounding type rated 15 amperes, 125 volts.*

Informational Note: For various configurations of 30-ampere cord connectors, see ANSI/UL 1686-2012, *Standard for Pin and Sleeve Configurations*, Figure C2.9 or Part C3.

## 626.26 Loss of Primary Power

Means shall be provided such that, upon loss of voltage from the utility or other electric supply system(s), energy cannot be back-fed through the truck and the truck supply equipment to the electrified truck parking space wiring system unless permitted by 626.27.

### 626.27 Interactive Systems

Electrified truck parking space supply equipment and other parts of a system, either on-board or off-board the vehicle, that are identified for and intended to be interconnected to a vehicle and also serve as an optional standby system or an electric power production source or provide for bi-directional power feed shall be listed as suitable for that purpose. When used as an optional standby system, the requirements of Article 702 shall apply, and when used as an electric power production source, the requirements of Article 705 shall apply.

## IV. Transport Refrigerated Units (TRUs)

### 626.30 Transport Refrigerated Units

Electrified truck parking spaces intended to supply transport refrigerated units (TRUs) shall include an individual branch circuit and receptacle for operation of the refrigeration/heating units. The receptacle associated with the TRUs shall be provided in addition to the receptacles required in 626.24(B).

**(A) Branch Circuits.** TRU spaces shall be supplied from 208-volt, 3-phase or 480-volt, 3-phase branch circuits and with an equipment grounding conductor.

**(B) Electrified Truck Parking Space Supply Equipment.** The electrified truck parking space supply equipment, or portion thereof, providing electric power for the operation of TRUs shall be independent of the loads in Part III of Article 626.

### 626.31 Disconnecting Means and Receptacles

**(A) Disconnecting Means.** Disconnecting means shall be provided to isolate each refrigerated unit from its supply connection. A disconnecting means shall be provided and installed in a readily accessible location and shall be lockable open in accordance with 110.25.

**(B) Location.** The disconnecting means shall be readily accessible, located not more than 750 mm (30 in.) from the receptacle it controls, and located in the supply circuit ahead of the receptacle. Circuit breakers or switches located in power outlets complying with this section shall be permitted as the disconnecting means.

**(C) Receptacles.** All receptacles shall be listed and of the grounding type. Every electrified truck parking space intended to provide an electrical supply for transport refrigerated units shall be equipped with one or both of the following:

(1) A 30-ampere, 480-volt, 3-phase, 3-pole, 4-wire receptacle
(2) A 60-ampere, 208-volt, 3-phase, 3-pole, 4-wire receptacle

Informational Note: Complete details of the 30-ampere pin and sleeve receptacle configuration for refrigerated containers (transport refrigerated units) can be found in ANSI/UL 1686-2012, *Standard for Pin and Sleeve Configurations*, Figure C2.11. For various configurations of 60-ampere pin and sleeve receptacles, see ANSI/UL1686.

### 626.32 Separable Power Supply Cable Assembly

A separable power supply cable assembly, consisting of a cord with an attachment plug and cord connector, shall be one of the types and ratings specified in 626.32(A), (B), and (C). Cords with adapters and pigtail ends, extension cords, and similar items shall not be used.

**(A) Rating(s).** The power supply cable assembly shall be listed and be rated in accordance with (1) or (2).

(1) 30 ampere, 480-volt, 3-phase
(2) 60 ampere, 208-volt, 3-phase

**(B) Cord Assemblies.** The cord shall be a listed type with four conductors, for 3-phase connection, one of which shall be identified in accordance with 400.23 for use as the equipment grounding conductor. Extra-hard usage cables rated not less than 90°C (194°F), 600 volts, listed for both wet locations and sunlight resistance, and having an outer jacket rated to be resistant to temperature extremes, oil, gasoline, ozone, abrasion, acids, and chemicals, shall be permitted where flexibility is necessary between the electrified truck parking space supply equipment and the inlet(s) on the TRU.

**(C) Attachment Plug(s) and Cord Connector(s).** Where a flexible cord is provided with an attachment plug and cord connector, they shall comply with 250.138(A). The attachment plug(s) and cord connector(s) shall be listed, by itself or as part of the power-supply cable assembly, for the purpose and shall be molded to or installed on the flexible cord so that it is secured tightly to the cord at the point where the cord enters the attachment plug or cord connector. If a right-angle cap is used, the configuration shall be oriented so that the grounding member is farthest from the cord. An attachment plug and cord connector for the connection of a truck or trailer shall be rated in accordance with (1) or (2) as follows:

(1) 30-ampere, 480-volt, 3-phase, 3-pole, 4-wire and intended for use with a 30-ampere 480-volt, 3-phase, 3-pole, 4-wire receptacles and inlets, respectively, or
(2) 60-ampere, 208-volt, 3-phase, 3-pole, 4-wire and intended for use with a 60-ampere, 208-volt, 3-phase, 3-pole, 4-wire receptacles and inlets, respectively.

Informational Note: Complete details of the 30-ampere pin and sleeve attachment plug and cord connector configurations for refrigerated containers (transport refrigerated units) can be found in ANSI/UL 1686-2012, *Standard for Pin and Sleeve Configurations*, Figures C2.12 and C2.11. For various configurations of 60-ampere pin and sleeve attachment plugs and cord connectors, see ANSI/UL1686.

# ARTICLE 630
# Electric Welders

## I. General

### 630.1 Scope

This article covers apparatus for electric arc welding, resistance welding, plasma cutting, and other similar welding and cutting process equipment that is connected to an electrical supply system.

The two general types of electric welding are resistance welding and arc welding. Resistance welding, or "spot" welding, is the process of electrically fusing two or more metal sheets or parts. The metal parts are placed between two electrodes or welding points, and a high current at a low voltage is passed through the electrodes. The resistance of the metal parts to the flow of current heats them to a molten state, and a weld is made.

Arc welding is the butting of two metal parts, then striking an arc at the joint with a metal electrode (a flux-coated wire rod). The electrode itself is melted and supplies the extra metal necessary for joining the metal parts.

The scope of Article 630 also covers electrically supplied equipment associated with plasma cutting operations. This electrically powered equipment controls the flammable gas or gases used for cutting.

## II. Arc Welders

### 630.11 Ampacity of Supply Conductors

The ampacity of conductors for arc welders shall be in accordance with 630.11(A) and (B).

**(A) Individual Welders.** The ampacity of the supply conductors shall be not less than the $I_{1eff}$ value on the rating plate. Alternatively, if the $I_{1eff}$ is not given, the ampacity of the supply conductors shall not be less than the current value determined by multiplying the rated primary current in amperes given on the welder rating plate by the factor shown in Table 630.11(A) based on the duty cycle of the welder.

**(B) Group of Welders.** Minimum conductor ampacity shall be based on the individual currents determined in 630.11(A) as the sum of 100 percent of the two largest welders, plus 85 percent of the third largest welder, plus 70 percent of the fourth largest welder, plus 60 percent of all remaining welders.

*Exception: Percentage values lower than those given in 630.11(B) shall be permitted in cases where the work is such that a high-operating duty cycle for individual welders is impossible.*

***TABLE 630.11(A)*** *Duty Cycle Multiplication Factors for Arc Welders*

| | Multiplier for Arc Welders | |
|---|---|---|
| **Duty Cycle** | **Nonmotor Generator** | **Motor Generator** |
| 100 | 1.00 | 1.00 |
| 90 | 0.95 | 0.96 |
| 80 | 0.89 | 0.91 |
| 70 | 0.84 | 0.86 |
| 60 | 0.78 | 0.81 |
| 50 | 0.71 | 0.75 |
| 40 | 0.63 | 0.69 |
| 30 | 0.55 | 0.62 |
| 20 or less | 0.45 | 0.55 |

Informational Note: Duty cycle considers welder loading based on the use to be made of each welder and the number of welders supplied by the conductors that will be in use at the same time. The load value used for each welder considers both the magnitude and the duration of the load while the welder is in use.

Even under high-production conditions, the loads on transformer arc welders are considered intermittent. Therefore, the minimum ampacity of feeder conductors supplying several transformers (three or more) is permitted to be determined by applying the percentage values specified in 630.11(B). See also 630.31(B). The ampacity of the conductors is based on the $I_{1eff}$ rating on the welder rating plate. If the $I_{1eff}$ rating is not available, a calculation is done by selecting the appropriate factor from Table 630.11(A) based on the type of welder and the duty cycle of the welder. The selected factor is then multiplied by the primary current rating from the welder rating plate to determine the minimum ampacity of the supply conductors.

### 630.12 Overcurrent Protection

Overcurrent protection for arc welders shall be as provided in 630.12(A) and (B). Where the values as determined by this section do not correspond to the standard ampere ratings provided in 240.6 or where the rating or setting specified results in unnecessary opening of the overcurrent device, the next higher standard rating or setting shall be permitted.

**(A) For Welders.** Each welder shall have overcurrent protection rated or set at not more than 200 percent of $I_{1max}$. Alternatively, if the $I_{1max}$ is not given, the overcurrent protection shall be rated or set at not more than 200 percent of the rated primary current of the welder.

An overcurrent device shall not be required for a welder that has supply conductors protected by an overcurrent device rated or set at not more than 200 percent of $I_{1max}$ or at the rated primary current of the welder.

If the supply conductors for a welder are protected by an overcurrent device rated or set at not more than 200 percent of $I_{1max}$ or at the rated primary current of the welder, a separate overcurrent device shall not be required.

**(B) For Conductors.** Conductors that supply one or more welders shall be protected by an overcurrent device rated or set at not more than 200 percent of the conductor ampacity.

Informational Note: $I_{1max}$ is the maximum value of the rated supply current at maximum rated output. $I_{1eff}$ is the maximum value of the effective supply current, calculated from the rated supply current ($I_1$), the corresponding duty cycle (duty factor) ($X$), and the supply current at no-load ($I_0$) by the following equation:

$$I_{1eff} = \sqrt{I_1^2 X + I_0^2(1 - X)}$$

Some arc welding machines have a welding range involving an excess secondary-current output capacity beyond that indicated by the secondary rating marked on the machines. This excess capacity (generally not more than 150 percent of the marked output capacity) is usually supplied by means of secondary taps in addition to the tap(s) intended for normal output current; the higher currents thus available are intended to provide for heavier welding work, including the use of larger-sized electrodes. This excess capacity is somewhat analogous to the inherent overload capacity of motors and transformers. However, the use of this excess current capacity and the overloading of welding machines, except for relatively short periods of time, could be hazardous and should be undertaken with caution.

**Calculation Example**

A motor-generator-type electric arc welder has a nameplate primary current rating of 95 amperes and a duty cycle of 80 percent. Determine the minimum ampacity of the branch-circuit conductors and the maximum rating or setting for the branch-circuit OCPD.

*Solution*

Step 1. Determine the minimum ampacity for the supply circuit conductors [630.11(A)]:

$$95 \text{ A} \times 0.91 \text{ (duty cycle factor)} = 86.45 \text{ A}$$

Copper THWN conductor selected from the 75°C column of Table 310.15(B)(16): 3 AWG (100-A allowable ampacity).

Step 2. Determine the maximum rating of setting for the OCPD for the welder and the branch-circuit conductors [630.12(A) and (B)]:

$$95 \text{ A} \times 200\% = 190 \text{ A}$$

Next standard size OCPD: 200 A.

*Conclusion:* The minimum conductor ampacity is 86.45 amperes, and the maximum rating or setting for the branch-circuit OCPD is 200 amperes. This rating or setting is the maximum permitted for a circuit supplying a single welder. However, the *Code* does not prohibit the use of a smaller-size OCPD, because that is a performance consideration related to the intended use of the welder.

## 630.13 Disconnecting Means

A disconnecting means shall be provided in the supply circuit for each arc welder that is not equipped with a disconnect mounted as an integral part of the welder. The disconnecting means identity shall be marked in accordance with 110.22(A).

The disconnecting means shall be a switch or circuit breaker, and its rating shall be not less than that necessary to accommodate overcurrent protection as specified under 630.12.

## 630.14 Marking

A rating plate shall be provided for arc welders giving the following information:

(1) Name of manufacturer
(2) Frequency
(3) Number of phases
(4) Primary voltage
(5) $I_{1max}$ and $I_{1eff}$, or rated primary current
(6) Maximum open-circuit voltage
(7) Rated secondary current
(8) Basis of rating, such as the duty cycle

## 630.15 Grounding of Welder Secondary Circuit

The secondary circuit conductors of an arc welder, consisting of the electrode conductor and the work conductor, shall not be considered as premises wiring for the purpose of applying Article 250.

Informational Note: Connecting welder secondary circuits to grounded objects can create parallel paths and can cause objectionable current over equipment grounding conductors.

In theory and in accordance with the *NEC* definition, the secondary circuit of an arc welder could be viewed as a *separately derived system*. However, the function of a welder is to create a high-current circuit between the electrode and the work surface. In the normal operation of an ac power distribution system, such an event would be considered a fault, and the operation of an overcurrent device to open the circuit and clear the fault is a fundamental concept of Articles 240 and 250. In the case of an arc welder, the opening of an overcurrent device is not intended unless the welding operation significantly exceeds the operating parameters of the welder. Grounding of a welder secondary terminal has the potential to cause excessive and potentially degrading parallel currents on power system EGCs.

This requirement clarifies that for the purposes of Article 250 — specifically, the requirements covering grounding of separately derived systems — the secondary circuit of a welder is not treated as premises wiring and is not required to be grounded as such. This removes any potential conflict where grounding in the welder secondary circuit occurs at the work object.

# III. Resistance Welders

## 630.31 Ampacity of Supply Conductors

The ampacity of the supply conductors for resistance welders necessary to limit the voltage drop to a value permissible for the satisfactory performance of the welder is usually greater than that required to prevent overheating as covered in 630.31(A) and (B).

**(A) Individual Welders.** The rated ampacity for conductors for individual welders shall comply with the following:

(1) The ampacity of the supply conductors for a welder that may be operated at different times at different values of primary current or duty cycle shall not be less than 70 percent of the rated primary current for seam and automatically fed welders, and 50 percent of the rated primary current for manually operated nonautomatic welders.
(2) The ampacity of the supply conductors for a welder wired for a specific operation for which the actual primary current and duty cycle are known and remain unchanged shall not be less than the product of the actual primary current and the multiplier specified in Table 630.31(A)(2) for the duty cycle at which the welder will be operated.

**(B) Groups of Welders.** The ampacity of conductors that supply two or more welders shall not be less than the sum of the value obtained in accordance with 630.31(A) for the largest welder supplied and 60 percent of the values obtained for all the other welders supplied.

Informational Note: **Explanation of Terms**

(1) The *rated primary current* is the rated kilovolt-amperes (kVA) multiplied by 1000 and divided by the rated primary voltage, using values given on the nameplate.
(2) The *actual primary current* is the current drawn from the supply circuit during each welder operation at the particular heat tap and control setting used.
(3) The *duty cycle* is the percentage of the time during which the welder is loaded. For instance, a spot welder supplied by a 60-Hz system (216,000 cycles per hour) and making 400 15-cycle welds per hour would have a duty cycle of 2.8 percent (400 multiplied by 15, divided by 216,000, multiplied by 100). A seam welder operating 2 cycles "on" and 2 cycles "off" would have a duty cycle of 50 percent.

***TABLE 630.31(A)(2)*** *Duty Cycle Multiplication Factors for Resistance Welders*

| Duty Cycle (%) | Multiplier |
|---|---|
| 50 | 0.71 |
| 40 | 0.63 |
| 30 | 0.55 |
| 25 | 0.50 |
| 20 | 0.45 |
| 15 | 0.39 |
| 10 | 0.32 |
| 7.5 | 0.27 |
| 5 or less | 0.22 |

The ampacity of supply conductors for a welder that is not wired for a specific function (i.e., one operated at varying intervals for different applications, such as dissimilar metals or thicknesses) is permitted by 630.31(A)(1) to be 70 percent of the rated primary current for automatically fed welders and 50 percent of the rated primary current for manually operated welders. The rated primary current can be determined using the following equation with the values given on the welder nameplate:

$$\text{Rated primary voltage} = \frac{\text{welder kVA} \times 1000}{\text{rated primary voltage}}$$

Where the actual primary current and the duty cycle are known, such as for a welder wired for a specific operation, the ampacity of the supply conductors is not permitted to be less than the product of the actual primary current (current drawn during weld operation) and the multiplier, as provided in Table 630.31(A)(2), for the duty cycle at which the welder will be operated.

### Calculation Example

A seam welder is set to draw current for 3 cycles and to be off for 4 cycles during every 7-cycle period. The welder's duty cycle is calculated as follows:

$$\frac{3}{7} \times 100\% = 42.9\% \text{ (duty cycle)}$$

The duty cycle is set for a specific operation by adjusting the controller for the welder. An instrument capable of measuring current impulses for 3 cycles (1⁄20 second) is required to measure the actual primary current as required by 630.31(A)(2) in order to size the conductors. For the sizing of supply conductors, voltage drop should be limited to a value permissible for the satisfactory performance of the welder.

## 630.32 Overcurrent Protection

Overcurrent protection for resistance welders shall be as provided in 630.32(A) and (B). Where the values as determined by this section do not correspond with the standard ampere ratings provided in 240.6 or where the rating or setting specified results in unnecessary opening of the overcurrent device, a higher rating or setting that does not exceed the next higher standard ampere rating shall be permitted.

**(A) For Welders.** Each welder shall have an overcurrent device rated or set at not more than 300 percent of the rated primary current of the welder. If the supply conductors for a welder are protected by an overcurrent device rated or set at not more than

200 percent of the rated primary current of the welder, a separate overcurrent device shall not be required.

**(B) For Conductors.** Conductors that supply one or more welders shall be protected by an overcurrent device rated or set at not more than 300 percent of the conductor ampacity.

### 630.33 Disconnecting Means

A switch or circuit breaker shall be provided by which each resistance welder and its control equipment can be disconnected from the supply circuit. The ampere rating of this disconnecting means shall not be less than the supply conductor ampacity determined in accordance with 630.31. The supply circuit switch shall be permitted as the welder disconnecting means where the circuit supplies only one welder.

### 630.34 Marking

A nameplate shall be provided for each resistance welder, giving the following information:

(1) Name of manufacturer
(2) Frequency
(3) Primary voltage
(4) Rated kilovolt-amperes (kVA) at 50 percent duty cycle
(5) Maximum and minimum open-circuit secondary voltage
(6) Short-circuit secondary current at maximum secondary voltage
(7) Specified throat and gap setting

## IV. Welding Cable

### 630.41 Conductors

Insulation of conductors intended for use in the secondary circuit of electric welders shall be flame retardant.

Listed welding cable is intended to be used for the secondary circuits of electric welders and cannot be used as "building wire" for circuits operating at 600 volts or less unless the cable is also one of the types covered in Table 310.104(A). The fine stranding allows for the flexibility necessary in manual and automatic welding operations. Terminals used with this type of cable must be suitable for use with the fine stranding used in this type of cable construction. See 110.14 for more information regarding terminations used with conductors having other than Class B or C stranding.

### 630.42 Installation

Cables shall be permitted to be installed in a dedicated cable tray as provided in 630.42(A), (B), and (C).

**(A) Cable Support.** The cable tray shall provide support at not greater than 150-mm (6-in.) intervals.

**(B) Spread of Fire and Products of Combustion.** The installation shall comply with 300.21.

**(C) Signs.** A permanent sign shall be attached to the cable tray at intervals not greater than 6.0 m (20 ft). The sign shall read as follows:

CABLE TRAY FOR WELDING CABLES ONLY

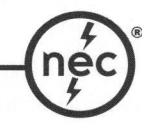

# ARTICLE 640 Audio Signal Processing, Amplification, and Reproduction Equipment

## I. General

### 640.1 Scope

**(A) Covered.** This article covers equipment and wiring for audio signal generation, recording, processing, amplification, and reproduction; distribution of sound; public address; speech input systems; temporary audio system installations; and electronic organs or other electronic musical instruments. This also includes audio systems subject to Article 517, Part VI, and Articles 518, 520, 525, and 530.

Informational Note: Examples of permanently installed distributed audio system locations include, but are not limited to, restaurant, hotel, business office, commercial and retail sales environments, churches, and schools. Both portable and permanently installed equipment locations include, but are not limited to, residences, auditoriums, theaters, stadiums, and movie and television studios. Temporary installations include, but are not limited to, auditoriums, theaters, stadiums (which use both temporary and permanently installed systems), and outdoor events such as fairs, festivals, circuses, public events, and concerts.

Equipment covered by Article 640 includes amplifiers; public address (PA) systems and centralized sound systems used in schools, factories, businesses, stadiums, and similar locations; intercommunications devices and systems; and devices used for recording or reproducing voice or music. The scope is limited to equipment whose main function is the processing, distribution, amplification, and reproduction of audio frequency bandwidth signals. This limitation does not preclude equipment that uses radio frequency or other forms of transmission between equipment components, such as wireless microphone systems.

Electronic organs are synthesizers, and synthesizers also generate audio signals. For the sake of clarity, electronic organs are uniquely cited in the scope, and electronic musical instruments are included to cover all other forms of electronic tone generation. Electronic musical instruments create an electronic signal as their sole or primary output and require amplification and reproduction equipment to be audible.

**(B) Not Covered.** This article does not cover the installation and wiring of fire and burglary alarm signaling devices.

## 640.2 Definitions

For purposes of this article, the following definitions apply.

**Abandoned Audio Distribution Cable.** Installed audio distribution cable that is not terminated at equipment and not identified for future use with a tag.

**Audio Amplifier or Pre-Amplifier.** Electronic equipment that increases the current or voltage potential, or both, of an audio signal intended for use by another piece of audio equipment. *Amplifier* is the term used to denote an audio amplifier within this article.

**Audio Autotransformer.** A transformer with a single winding and multiple taps intended for use with an amplifier loudspeaker signal output.

**Audio Signal Processing Equipment.** Electrically operated equipment that produces, processes, or both, electronic signals that, when appropriately amplified and reproduced by a loudspeaker, produce an acoustic signal within the range of normal human hearing (typically 20–20 kHz). Within this article, the terms *equipment* and *audio equipment* are assumed to be equivalent to audio signal processing equipment.

Informational Note: This equipment includes, but is not limited to, loudspeakers; headphones; pre-amplifiers; microphones and their power supplies; mixers; MIDI (musical instrument digital interface) equipment or other digital control systems; equalizers, compressors, and other audio signal processing equipment; and audio media recording and playback equipment, including turntables, tape decks and disk players (audio and multimedia), synthesizers, tone generators, and electronic organs. Electronic organs and synthesizers may have integral or separate amplification and loudspeakers. With the exception of amplifier outputs, virtually all such equipment is used to process signals (utilizing analog or digital techniques) that have nonhazardous levels of voltage or current potential.

The definition of *audio signal processing equipment* clarifies the limits of signal processing (frequency bandwidth), which falls under Article 640. The Informational Note enumerates the breadth of equipment considered to fall within the defined scope of Article 640 and provides a sufficiently broad list of current technology equipment to assist in determining the applicability of Article 640.

"MIDI (musical instrument digital interface) equipment or other digital control systems" is mentioned specifically because, while MIDI or similar digital control signals may issue from an electronic musical instrument, such signals also can be obtained from a computer that is appropriately configured to perform similar controlling functions.

**Audio System.** Within this article, the totality of all equipment and interconnecting wiring used to fabricate a fully functional audio signal processing, amplification, and reproduction system.

**Audio Transformer.** A transformer with two or more electrically isolated windings and multiple taps intended for use with an amplifier loudspeaker signal output.

The definition of *audio transformer* states that such transformers are intended only for use with audio signals, not light and power.

**Equipment Rack.** A framework for the support, enclosure, or both, of equipment; may be portable or stationary.

ANSI/EIA 310-D-1992, *Cabinets, Racks, Panels and Associated Equipment,* defines *commercial equipment racks.* Within Article 640, both the terms *equipment rack* and *rack* are used to refer to equipment enclosures that are conceptually similar in intended use to those defined by the ANSI/EIA standard.

Informational Note: See ANSI/EIA/310-D-1992, *Cabinets, Racks, Panels and Associated Equipment.*

**Loudspeaker.** Equipment that converts an ac electric signal into an acoustic signal. The term *speaker* is commonly used to mean *loudspeaker.*

**Maximum Output Power.** The maximum power delivered by an amplifier into its rated load as determined under specified test conditions.

Informational Note: The maximum output power can exceed the manufacturer's rated output power for the same amplifier.

**Mixer.** Equipment used to combine and level match a multiplicity of electronic signals, such as from microphones, electronic instruments, and recorded audio.

Typical peak signal operating voltages for such equipment vary from a few millivolts for microphones to 2 to 4 volts for disc players. A mixer's purpose is to balance these inputs to provide (typically) a 1-volt peak output signal to an amplifier.

•

**Portable Equipment.** Equipment fed with portable cords or cables intended to be moved from one place to another.

•

**Rated Output Power.** The amplifier manufacturer's stated or marked output power capability into its rated load.

•

**Technical Power System.** An electrical distribution system with grounding in accordance with 250.146(D), where the equipment grounding conductor is isolated from the premises grounded conductor except at a single grounded termination point within a branch-circuit panelboard, at the originating (main breaker) branch-circuit panelboard, or at the premises grounding electrode.

The terms *technical power* and *technical ground* are commonly used by audio/video technicians and electricians to designate a wiring system that is in compliance with 250.146(D). Including the definition of *technical power system* in Article 640 broadens the scope of this term to include the commonly employed distribution systems fabricated in compliance with 250.146(D).

**Temporary Equipment.** Portable wiring and equipment intended for use with events of a transient or temporary nature where all equipment is presumed to be removed at the conclusion of the event.

Temporary equipment may be used in permanent or temporary facilities, or in areas with no services other than a source of electrical power. Locations include indoor and outdoor areas such as athletic facilities, halls, auditoriums, concert shells, athletic fields, beaches, and other places designated for public assembly.

## 640.3 Locations and Other Articles

Circuits and equipment shall comply with 640.3(A) through (M), as applicable.

**(A) Spread of Fire or Products of Combustion.** Section 300.21 shall apply.

**(B) Ducts, Plenums, and Other Air-Handling Spaces.** See 300.22 for circuits and equipment installed in ducts or plenums or other space used for environmental air.

Informational Note: NFPA 90A-2012, *Standard for the Installation of Air Conditioning and Ventilation Systems*, 4.3.10.2.6.5, permits loudspeakers, loudspeaker assemblies, and their accessories listed in accordance with UL 2043-2008, *Fire Test for Heat and Visible Smoke Release for Discrete Products and Their Accessories Installed in Air-Handling Spaces*, to be installed in other spaces used for environmental air (ceiling cavity plenums).

**(C) Cable Trays.** Cable trays shall be used in accordance with Article 392.

Informational Note: See 725.154(C) for the use of Class 2, Class 3, and Type PLTC cable in cable trays.

**(D) Hazardous (Classified) Locations.** Equipment used in hazardous (classified) locations shall comply with the applicable requirements of Chapter 5.

**(E) Assembly Occupancies.** Equipment used in assembly occupancies shall comply with Article 518.

The examples of assembly occupancies described in 518.2 are some of the most common locations for the installation of both distributed audio systems (e.g., background music) and centralized systems (permanently installed sound reinforcement systems for meeting rooms, auditoriums, gymnasiums, and so forth).

**(F) Theaters, Audience Areas of Motion Picture and Television Studios, and Similar Locations.** Equipment used in theaters, audience areas of motion picture and television studios, and similar locations shall comply with Article 520.

**(G) Carnivals, Circuses, Fairs, and Similar Events.** Equipment used in carnivals, circuses, fairs, and similar events shall comply with Article 525.

**(H) Motion Picture and Television Studios.** Equipment used in motion picture and television studios shall comply with Article 530.

**(I) Swimming Pools, Fountains, and Similar Locations.** Audio equipment used in or near swimming pools, fountains, and similar locations shall comply with Article 680.

The underwater installation of audio equipment in swimming pools is covered in 680.27(A). The acceptable placement, wiring, and use of equipment used near (rather than immersed in) bodies of water, both natural and artificial, are covered in 640.10.

**(J) Combination Systems.** Where the authority having jurisdiction permits audio systems for paging or music, or both, to be combined with fire alarm systems, the wiring shall comply with Article 760.

Informational Note: For installation requirements for such combination systems, refer to *NFPA 72*-2013, *National Fire Alarm and Signaling Code*, and NFPA *101*-2012, *Life Safety Code*.

In addition to alarm tones, fire alarm systems frequently use loudspeakers for verbal announcements. All such systems must comply with Article 760 and *NFPA 72, National Fire Alarm and Signaling Code*. Audio systems that use a paging or background music system are permitted to be used as part of a fire alarm warning system, but they must comply with Article 760. The installation of fire alarm systems is governed by *NFPA 72*. Refer to NFPA *101*, *Life Safety Code*, for multipurpose systems.

**(K) Antennas.** Equipment used in audio systems that contain an audio or video tuner and an antenna input shall comply with Article 810. Wiring other than antenna wiring that connects such equipment to other audio equipment shall comply with this article.

The term *receiver* is commonly used in the consumer market to mean an amplifier combined with a radio tuner (typically AM/FM) and other signal processing and/or switching functions. Except for the tuner function and the antenna input, the signal processing functions are the same as those provided by equipment in Article 640. Article 810, Part II, covers the antenna installation for such equipment, and 810.3 references Article 640 as appropriate for wiring requirements (other than for the antenna).

**(L) Generators.** Generators shall be installed in accordance with 445.10 through 445.12, 445.14 through 445.16, and 445.18. Grounding of portable and vehicle-mounted generators shall be in accordance with 250.34.

**(M) Organ Pipes.** Additions of pipe organ pipes to an electronic organ shall be in accordance with 650.4 through 650.8.

## 640.4 Protection of Electrical Equipment

Amplifiers, loudspeakers, and other equipment shall be so located or protected as to guard against environmental exposure or physical damage, such as might result in fire, shock, or personal hazard.

## 640.5 Access to Electrical Equipment Behind Panels Designed to Allow Access

Access to equipment shall not be denied by an accumulation of wires and cables that prevents removal of panels, including suspended ceiling panels.

## 640.6 Mechanical Execution of Work

**(A) Neat and Workmanlike Manner.** Audio signal processing, amplification, and reproduction equipment, cables, and circuits shall be installed in a neat workmanlike manner.

**(B) Installation of Audio Distribution Cables.** Cables installed exposed on the surface of ceilings and sidewalls shall be supported in such a manner that the audio distribution cables will not be damaged by normal building use. Such cables shall be secured by straps, staples, cable ties, hangers, or similar fittings designed and installed so as not to damage the cable. The installation shall conform to 300.4 and 300.11(A).

Where installed in the hollow space above a suspended, dropped, or similar ceiling, cables of audio systems covered in Article 640 are required to be supported in accordance with 300.11. Without specific instructions permitting the use of the ceiling system support wires as a means to support wiring methods, an independent support system for the cables must be installed. Additional ceiling wires installed to support the audio system wiring are required to be secured in place. Attachment of the additional support wires to the ceiling system and to the building structure above the ceiling provides the secure support required by 300.11(A)(1) and (A)(2). The use of securing hardware such as straps, staples, cable ties, hangers, or other approved means is required, and this hardware must be installed so as not to damage the audio cable. The *Code* does not specify the distance between securing points.

**(C) Abandoned Audio Distribution Cables.** The accessible portion of abandoned audio distribution cables shall be removed.

**(D) Installed Audio Distribution Cable Identified for Future Use.**

**(1)** Cables identified for future use shall be marked with a tag of sufficient durability to withstand the environment involved.

**(2)** Cable tags shall have the following information:

(1) Date cable was identified for future use
(2) Date of intended use
(3) Information related to the intended future use of cable

## 640.7 Grounding

**(A) General.** Wireways and auxiliary gutters shall be connected to an equipment grounding conductor(s), to an equipment bonding jumper, or to the grounded conductor where permitted or required by 250.92(B)(1) or 250.142. Where the wireway or auxiliary gutter does not contain power-supply wires, the equipment grounding conductor shall not be required to be larger than 14 AWG copper or its equivalent. Where the wireway or auxiliary gutter contains power-supply wires, the equipment grounding conductor shall not be smaller than specified in 250.122.

**(B) Separately Derived Systems with 60 Volts to Ground.** Grounding of separately derived systems with 60 volts to ground shall be in accordance with 647.6.

**(C) Isolated Ground Receptacles.** Isolated grounding-type receptacles shall be permitted as described in 250.146(D), and for the implementation of other technical power systems in compliance with Article 250. For separately derived systems with 60 volts to ground, the branch-circuit equipment grounding conductor shall be terminated as required in 647.6(B).

Informational Note: See 406.3(D) for grounding-type receptacles and required identification.

The reference to 647.6 provides guidance for grounding separately derived systems operating at 60 volts to ground. Such separately derived systems are used for the reduction of electromagnetic noise in audio systems and in video systems.

Section 640.7(C) addresses the proper use of isolated ground receptacles when used with technical power systems of the type that are separately derived systems with 60 volts to ground.

## 640.8 Grouping of Conductors

Insulated conductors of different systems grouped or bundled so as to be in close physical contact with each other in the same raceway or other enclosure, or in portable cords or cables, shall comply with 300.3(C)(1).

## 640.9 Wiring Methods

**(A) Wiring to and Between Audio Equipment.**

**(1) Power Wiring.** Wiring and equipment from source of power to and between devices connected to the premises wiring systems shall comply with the requirements of Chapters 1 through 4, except as modified by this article.

**(2) Separately Derived Power Systems.** Separately derived systems shall comply with the applicable articles of this *Code*, except as modified by this article. Separately derived systems with 60 volts to ground shall be permitted for use in audio system installations as specified in Article 647.

**(3) Other Wiring.** All wiring not connected to the premises wiring system or to a wiring system separately derived from the premises wiring system shall comply with Article 725.

**(B) Auxiliary Power Supply Wiring.** Equipment that has a separate input for an auxiliary power supply shall be wired in compliance with Article 725. Battery installation shall be in accordance with Article 480. This section shall not apply to the use of uninterruptible power supply (UPS) equipment, or other sources of supply, that are intended to act as a direct replacement for the primary circuit power source and are connected to the primary circuit input.

Informational Note: Refer to *NFPA 72-2013*, *National Fire Alarm and Signaling Code*, where equipment is used for a fire alarm system.

Audio equipment with a separate input for an auxiliary power supply is typically used for emergency paging or fire alarm systems. These auxiliary power-supply inputs typically range from 12 to 48 V dc. Article 480 covers installation and overcurrent protection of battery circuits of this type. The term *auxiliary* is used to indicate that the equipment is also capable of being powered by the premises wiring system through an independent input connector, cord, or cable.

A replacement source for the premises wiring system such as a UPS or a standby generator is not covered by the requirements of this section unless it is directly connected to the auxiliary power-supply input and supplies the audio equipment with a dc voltage.

**(C) Output Wiring and Listing of Amplifiers.** Amplifiers with output circuits carrying audio program signals shall be permitted to employ Class 1, Class 2, or Class 3 wiring where the amplifier is listed and marked for use with the specific class of wiring method. Such listing shall ensure the energy output is equivalent to the shock and fire risk of the same class as stated in Article 725. Overcurrent protection shall be provided and shall be permitted to be inherent in the amplifier.

Audio amplifier output circuits wired using Class 1 wiring methods shall be considered equivalent to Class 1 circuits and shall be installed in accordance with 725.46, where applicable.

Audio amplifier output circuits wired using Class 2 or Class 3 wiring methods shall be considered equivalent to Class 2 or Class 3 circuits, respectively. They shall use conductors insulated at not less than the requirements of 725.179 and shall be installed in accordance with 725.133 and 725.154.

Informational Note No. 1: ANSI/UL 1711-2006, *Amplifiers for Fire Protective Signaling Systems*, contains requirements for the listing of amplifiers used for fire alarm systems in compliance with *NFPA 72-2013*, *National Fire Alarm and Signaling Code*.

Informational Note No. 2: Examples of requirements for listing amplifiers used in residential, commercial, and professional use are found in ANSI/UL 813-1996, *Commercial Audio Equipment*; ANSI/UL 1419-2011, *Professional Video and Audio Equipment*; ANSI/UL 1492-2010, *Audio-Video Products and Accessories*; ANSI/UL 6500-2006, *Audio/Video and Musical Instrument Apparatus for Household, Commercial, and Similar Use*; and UL 62368-1-2012, *Audio/Video, Information and Communication Technology Equipment — Part 1: Safety Requirements*.

**(D) Use of Audio Transformers and Autotransformers.** Audio transformers and autotransformers shall be used only for audio signals in a manner so as not to exceed the manufacturer's stated input or output voltage, impedance, or power limitations. The input or output wires of an audio transformer or autotransformer shall be allowed to connect directly to the amplifier or loudspeaker terminals. No electrical terminal or lead shall be required to be grounded or bonded.

Audio transformers and autotransformers are commonly used between the amplifier output and the loudspeaker input for the following reasons:

1. At the output of the amplifier to change the amplifier's operating voltage to match the design impedance of the loudspeaker
2. At the loudspeaker, where the inherently low voice coil impedance is raised, to match the output voltage of the amplifier (or autotransformer)
3. Between the amplifier output and loudspeaker input as an attenuating device (volume control)

Audio autotransformers are similar in concept to autotransformers used for light and power. Audio transformers are commonly used to provide electrical isolation of the speakers from the signal source. Either type of audio transformer (two windings or an autotransformer) is referred to as an "impedance matching transformer."

The last sentence of 640.9(D) specifically addresses the fact that electrical terminals are not to be treated in the same manner as transformers used for light and power might be (e.g., grounding the common terminal of an autotransformer). Some amplifier outputs are deliberately isolated from equipment ground, in which case such a connection could damage the amplifier and violate the manufacturer's recommended use. The frame of the transformer may or may not require bonding, depending on the manufacturer's installation instructions.

## 640.10 Audio Systems Near Bodies of Water

Audio systems near bodies of water, either natural or artificial, shall be subject to the restrictions specified in 640.10(A) and (B).

*Exception: This section does not include audio systems intended for use on boats, yachts, or other forms of land or water transportation used near bodies of water, whether or not supplied by branch-circuit power.*

Informational Note: See 680.27(A) for installation of underwater audio equipment.

**(A) Equipment Supplied by Branch-Circuit Power.** Audio system equipment supplied by branch-circuit power shall not be placed horizontally within 1.5 m (5 ft) of the inside wall of a pool, spa, hot tub, or fountain, or within 1.5 m (5 ft) of the prevailing or tidal high water mark. The equipment shall be provided with branch-circuit power protected by a ground-fault circuit interrupter where required by other articles.

The particular application of underwater loudspeakers is unique in construction and wiring to pools and is addressed in Article 680. Other locations where audio equipment might be used "near bodies of water" are not addressed in Article 680. Locations where audio equipment is used near bodies of water are covered by 640.10.

The term *prevailing or tidal high water mark* recognizes that the edges of natural bodies of water can advance or recede. Such water level changes must be anticipated.

Unless required by other sections of the *Code,* the requirement for GFCI protection does not apply where the equipment (specifically an amplifier or a receiver) is not installed near a body of water.

**(B) Equipment Not Supplied by Branch-Circuit Power.** Audio system equipment powered by a listed Class 2 power supply or by the output of an amplifier listed as permitting the use of Class 2 wiring shall be restricted in placement only by the manufacturer's recommendations.

Informational Note: See 640.10(A) for placement of the power supply or amplifier if supplied by branch-circuit power.

## II. Permanent Audio System Installations

Permanent audio systems are characterized by fixed locations for the wiring, signal processing equipment, and reproduction equipment. Wiring is attached to the building structure and is frequently concealed. Speakers in commercial, hospital, school, and restaurant areas are commonly recessed into ceiling or wall surfaces, or they are mounted to structure surfaces using brackets.

### 640.21 Use of Flexible Cords and Cables

**(A) Between Equipment and Branch-Circuit Power.** Power supply cords for audio equipment shall be suitable for the use and shall be permitted to be used where the interchange, maintenance, or repair of such equipment is facilitated through the use of a power-supply cord.

**(B) Between Loudspeakers and Amplifiers or Between Loudspeakers.** Cables used to connect loudspeakers to each other or to an amplifier shall comply with Article 725. Other listed cable types and assemblies, including optional hybrid communications, signal, and composite optical fiber cables, shall be permitted.

Some loudspeakers are identified as being for outdoor use. The requirements of 110.11, which specify electrical equipment and conductors be identified for use in the operating environment, apply to audio equipment and its conductors. Exhibit 640.1 shows loudspeakers identified for outdoor use. The conductors supplying these outdoor speakers must also be identified for the environment.

***EXHIBIT 640.1*** *Loudspeakers for outdoor use above ground or partially in ground. (Courtesy of Bose Corp.)*

**(C) Between Equipment.** Cables used for the distribution of audio signals between equipment shall comply with Article 725. Other listed cable types and assemblies, including optional hybrid communications, signal, and composite optical fiber cables, shall be permitted. Other cable types and assemblies specified by the equipment manufacturer as acceptable for the use shall be permitted in accordance with 110.3(B).

**(D) Between Equipment and Power Supplies Other Than Branch-Circuit Power.** The following power supplies, other than branch-circuit power supplies, shall be installed and wired between equipment in accordance with the requirements of this *Code* for the voltage and power delivered:

(1) Storage batteries
(2) Transformers
(3) Transformer rectifiers
(4) Other ac or dc power supplies

Informational Note: For some equipment, these sources such as in items (1) and (2) serve as the only source of power. These could, in turn, be supplied with intermittent or continuous branch-circuit power.

**(E) Between Equipment Racks and Premises Wiring System.** Flexible cords and cables shall be permitted for the electrical connection of permanently installed equipment racks to the premises wiring system to facilitate access to equipment or for the purpose of isolating the technical power system of the rack from the premises ground. Connection shall be made either by using approved plugs and receptacles or by direct connection within an approved enclosure. Flexible cords and cables shall not be subjected to physical manipulation or abuse while the rack is in use.

### 640.22 Wiring of Equipment Racks and Enclosures

Metal equipment racks and enclosures shall be grounded. Bonding shall not be required if the rack is connected to a technical power ground.

Equipment racks shall be wired in a neat and workmanlike manner. Wires, cables, structural components, or other equipment shall not be placed in such a manner as to prevent reasonable access to equipment power switches and resettable or replaceable circuit overcurrent protection devices.

Supply cords or cables, if used, shall terminate within the equipment rack enclosure in an identified connector assembly. The supply cords or cable (and connector assembly if used) shall have sufficient ampacity to carry the total load connected to the equipment rack and shall be protected by overcurrent devices.

### 640.23 Conduit or Tubing

**(A) Number of Conductors.** The number of conductors permitted in a single conduit or tubing shall not exceed the percentage fill specified in Table 1, Chapter 9.

**(B) Nonmetallic Conduit or Tubing and Insulating Bushings.** The use of nonmetallic conduit or tubing and insulating bushings shall be permitted where a technical power system is employed and shall comply with applicable articles.

### 640.24 Wireways, Gutters, and Auxiliary Gutters

The use of metallic and nonmetallic wireways, gutters, and auxiliary gutters shall be permitted for use with audio signal conductors and shall comply with applicable articles with respect to permitted locations, construction, and fill.

### 640.25 Loudspeaker Installation in Fire Resistance-Rated Partitions, Walls, and Ceilings

Loudspeakers installed in a fire resistance-rated partition, wall, or ceiling shall be listed for that purpose or installed in an enclosure or recess that maintains the fire resistance rating.

Informational Note: Fire-rated construction is the fire-resistive classification used in building codes. One method of determining fire rating is testing in accordance with NFPA 256-2003, *Standard Methods of Fire Tests of Roof Coverings*.

The enclosure must maintain the requisite fire resistance rating of the wall or ceiling in which a flush-mounted loudspeaker is installed. Listed enclosures are available for this purpose. Site-built enclosures may be installed with the approval of the AHJ and have been used as a method to maintain the fire resistance rating of the wall or ceiling.

## III. Portable and Temporary Audio System Installations

While the equipment used for portable and temporary audio systems may not differ fundamentally from that used in permanent installations, the enclosures that serve as portable equipment racks provide both transit protection and mechanical protection while the equipment is in use. Such enclosures may accommodate one or multiple pieces of equipment and may be of metal, wood, plastic, or reinforced plastic construction. The nonmetallic construction enclosures frequently do not comply with ANSI/EIA 310-D-1992, *Cabinets, Racks, Panels and Associated Equipment*, except for the mounting attachment points for the equipment.

### 640.41 Multipole Branch-Circuit Cable Connectors

Multipole branch-circuit cable connectors, male and female, for power-supply cords and cables shall be so constructed that tension on the cord or cable is not transmitted to the connections. The female half shall be attached to the load end of the power supply cord or cable. The connector shall be rated in amperes and designed so that differently rated devices cannot be connected together. Alternating-current multipole connectors shall be polarized and comply with 406.7(A) and (B) and 406.10. Alternating-current or direct-current multipole connectors utilized for connection between loudspeakers and amplifiers, or between loudspeakers, shall not be compatible with nonlocking 15- or 20-ampere rated connectors intended for branch-circuit power or with connectors rated 250 volts or greater and of either the locking or nonlocking type. Signal cabling not intended for such loudspeaker and amplifier interconnection shall not be permitted to be compatible with multipole branch-circuit cable connectors of any accepted configuration.

Informational Note: See 400.10 for pull at terminals.

### 640.42 Use of Flexible Cords and Cables

**(A) Between Equipment and Branch-Circuit Power.** Power supply cords for audio equipment shall be listed and shall be permitted to be used where the interchange, maintenance, or repair of such equipment is facilitated through the use of a power-supply cord.

**(B) Between Loudspeakers and Amplifiers, or Between Loudspeakers.** Flexible cords and cables used to connect loudspeakers to each other or to an amplifier shall comply with Article 400 and Article 725, respectively. Cords and cables listed for portable use, either hard or extra-hard usage as defined by Article 400, shall also be permitted. Other listed cable types and assemblies, including optional hybrid communications, signal, and optical fiber cables, shall be permitted.

**(C) Between Equipment and/or Between Equipment Racks.** Flexible cords and cables used for the distribution of audio signals between equipment shall comply with Article 400 and Article 725, respectively. Cords and cables listed for portable use, either hard or extra-hard service as defined by Article 400, shall also be permitted. Other listed cable types and assemblies, including optional hybrid communications, signal, and optical fiber cables, shall be permitted.

**(D) Between Equipment, Equipment Racks, and Power Supplies Other Than Branch-Circuit Power.** Wiring between the following power supplies, other than branch-circuit power supplies, shall be installed, connected, or wired in accordance with the requirements of this *Code* for the voltage and power required:

(1) Storage batteries
(2) Transformers
(3) Transformer rectifiers
(4) Other ac or dc power supplies

**(E) Between Equipment Racks and Branch-Circuit Power.** The supply to a portable equipment rack shall be by means of listed extra-hard usage cords or cables, as defined in Table 400.4. For outdoor portable or temporary use, the cords or cables shall be further listed as being suitable for wet locations and sunlight resistant. Sections 520.5, 520.10, and 525.3 shall apply as appropriate when the following conditions exist:

(1) Where equipment racks include audio and lighting and/or power equipment
(2) When using or constructing cable extensions, adapters, and breakout assemblies

## 640.43 Wiring of Equipment Racks

Equipment racks fabricated of metal l be grounded. Nonmetallic racks with covers (if provided) removed shall not allow access to Class 1, Class 3, or primary circuit power without the removal of covers over terminals or the use of tools.

Equipment racks shall be wired in a neat and workmanlike manner. Wires, cables, structural components, or other equipment shall not be placed in such a manner as to prevent reasonable access to equipment power switches and resettable or replaceable circuit overcurrent protection devices.

Wiring that exits the equipment rack for connection to other equipment or to a power supply shall be relieved of strain or otherwise suitably terminated such that a pull on the flexible cord or cable will not increase the risk of damage to the cable or connected equipment such as to cause an unreasonable risk of fire or electric shock.

## 640.44 Environmental Protection of Equipment

Portable equipment not listed for outdoor use shall be permitted only where appropriate protection of such equipment from adverse weather conditions is provided to prevent risk of fire or electric shock. Where the system is intended to remain operable during adverse weather, arrangements shall be made for maintaining operation and ventilation of heat-dissipating equipment.

Although most portable audio equipment used in temporary audio systems is not listed for outdoor use, such equipment may be used with an appropriate enclosure or other means to protect the equipment from anticipated adverse weather conditions.

## 640.45 Protection of Wiring

Where accessible to the public, flexible cords and cables laid or run on the ground or on the floor shall be covered with approved nonconductive mats. Cables and mats shall be arranged so as not to present a tripping hazard. The cover requirements of 300.5 shall not apply to wiring protected by burial.

## 640.46 Equipment Access

Equipment likely to present a risk of fire, electric shock, or physical injury to the public shall be protected by barriers or supervised by qualified personnel so as to prevent public access.

# ARTICLE 645 Information Technology Equipment

Informational Note: Text that is followed by a reference in brackets has been extracted from NFPA 75-2013, *Standard for the Protection of Information Technology Equipment*. Only editorial changes were made to the extracted text to make it consistent with this *Code*.

## 645.1 Scope

This article covers equipment, power-supply wiring, equipment interconnecting wiring, and grounding of information technology equipment and systems in an information technology equipment room.

Informational Note: For further information, see NFPA 75-2013, *Standard for the Protection of Information Technology Equipment*, which covers the requirements for the protection of information technology equipment and information technology equipment areas.

The term *information technology equipment* (ITE) is also used by UL 60950-1, *Safety of Information Technology, Part 1: General Requirements*, as well as by international standards, as a more inclusive term for the equipment addressed by Article 645.

Article 645 applies only to equipment and systems, including the associated wiring, located within the ITE room. An ITE room is an enclosed area that contains computer-based business and industrial equipment. It is designed to comply with the special construction and fire protection provisions of NFPA 75 as well as 645.4.

Small terminals, such as remote telephone terminal units, remote data terminals, personal computers, and cash registers in stores and supermarkets, are not covered by Article 645.

## 645.2 Definitions

**Abandoned Supply Circuits and Interconnecting Cables.** Installed supply circuits and interconnecting cables that are not terminated at equipment and not identified for future use with a tag.

**Critical Operations Data System.** An information technology equipment system that requires continuous operation for reasons of public safety, emergency management, national security, or business continuity.

Similar to the application of Article 708 covering critical operations power systems, the designation of which information technology systems are critical in function is the responsibility of the AHJ, who in many cases may be an emergency management director or similar person rather than the electrical AHJ. Once the system is designated as being "critical," the AHJ responsible for approving the installation ensures compliance with the applicable requirements of Article 645.

**Information Technology Equipment (ITE).** Equipment and systems rated 600 volts or less, normally found in offices or other business establishments and similar environments classified as ordinary locations, that are used for creation and manipulation of data, voice, video, and similar signals that are not communications equipment as defined in Part I of Article 100 and do not process communications circuits as defined in 800.2.

This definition delineates equipment that is subject to the requirements of this article from communications equipment covered in Article 800. ITE equipment and communications equipment are commonly installed in the same ITE room.

Informational Note: For information on listing requirements for both information technology equipment and communications equipment, see UL 60950-1-2007, *Information Technology Equipment — Safety — Part 1: General Requirements*.

**Information Technology Equipment Room.** A room within the information technology equipment area that contains the information technology equipment. [75:3.3.9]

**Remote Disconnect Control.** An electric device and circuit that controls a disconnecting means through a relay or equivalent device.

**Zone.** A physically identifiable area (such as barriers or separation by distance) within an information technology equipment room, with dedicated power and cooling systems for the information technology equipment or systems.

## 645.3 Other Articles

Circuits and equipment shall comply with 645.3(A) through (G), as applicable.

**(A) Spread of Fire or Products of Combustion.** Sections 300.21, 770.26, 800.26, and 820.26 shall apply to penetrations of the fire-resistant room boundary.

**(B) Plenums.** Sections 300.22(C)(1), 725.135(B), 760.53(B)(2), 760.135(B), 770.113(C), 800.113(C), and 820.113(C) and Table 725.154, Table 760.154, Table 770.154(a), Table 800.154(a), and Table 820.154(a) shall apply to wiring and cabling in a plenum (other space used for environmental air) above an information technology equipment room.

**(C) Grounding.** The non–current-carrying conductive members of optical fiber cables in an information technology equipment room shall be grounded in accordance with 770.114.

**(D) Electrical Classification of Data Circuits.** Section 725.121(A)(4) shall apply to the electrical classification of listed information technology equipment signaling circuits. Sections 725.139(D)(1) and 800.133(A)(1)(b) shall apply to the electrical classification of Class 2 and Class 3 circuits in the same cable with communications circuits.

**(E) Fire Alarm Equipment.** Parts I, II, and III of Article 760 shall apply to fire alarm systems equipment installed in an information technology equipment room.

**(F) Communications Equipment.** Parts I, II, III, IV, and V of Article 800 shall apply to communications equipment installed in an information technology equipment room. Article 645 shall apply to the powering of communications equipment in an information technology equipment room.

Informational Note: See Part I of Article 100, Definitions, for a definition of communications equipment.

**(G) Community Antenna Television and Radio Distribution Systems Equipment.** Parts I, II, III, IV, and V of Article 820 shall apply to community antenna television and radio distribution systems equipment installed in an information technology equipment room. Article 645 shall apply to the powering of community antenna television and radio distribution systems equipment installed in an information technology equipment room.

## 645.4 Special Requirements for Information Technology Equipment Room

This article shall be permitted to provide alternate wiring methods to the provisions of Chapter 3 and Article 708 for power wiring, Parts I and III of Article 725 for signaling wiring, and Parts I and V of Article 770 for optical fiber cabling where all of the following conditions are met:

(1) Disconnecting means complying with 645.10 are provided.
(2) A heating/ventilating/air-conditioning (HVAC) system is provided in one of the methods identified in 645.4(2) a or b.
  a. A separate HVAC system that is dedicated for information technology equipment use and is separated from other areas of occupancy; or
  b. An HVAC system that serves other occupancies and meets all of the following:
    i. Also serves the information technology equipment room
    ii. Provides fire/smoke dampers at the point of penetration of the room boundary
    iii. Activates the damper operation upon initiation by smoke detector alarms, by operation of the disconnecting means required by 645.10, or by both

Informational Note: For further information, see NFPA 75-2013, *Standard for the Protection of Information Technology Equipment*, Chapter 10, 10.1, 10.1.1, 10.1.2, and 10.1.3.

(3) All information technology and communications equipment installed in the room is listed.
(4) The room is occupied by, and accessible to, only those personnel needed for the maintenance and functional operation of the installed information technology equipment.
(5) The room is separated from other occupancies by fire-resistant-rated walls, floors, and ceilings with protected openings.

Informational Note: For further information on room construction requirements, see NFPA 75-2013, *Standard for the Protection of Information Technology Equipment*, Chapter 5.

(6) Only electrical equipment and wiring associated with the operation of the information technology room is installed in the room.

Informational Note: HVAC systems, communications systems, and monitoring systems such as telephone, fire alarm systems, security systems, water detection systems, and other related protective equipment are examples of equipment associated with the operation of the information technology room.

Use of Article 645 is based on the assumption that construction of the ITE room complies with NFPA 75. For those ITE rooms,

Article 645 contains electrical installation requirements – for example, requirements for wiring methods in the space beneath the raised floor used for environmental air of an ITE room – that are less stringent than those in Chapters 1 through 4 for the same type of space.

Application of these requirements is contingent on the ITE room construction and equipment meeting all six conditions specified in 645.4. These less restrictive provisions cannot be taken advantage of if any one of the six conditions is not met. Application of Article 645 is typically determined by the designer or facility manager who wants to use its requirements instead of the relevant requirements in Chapters 1 through 4 for the electrical installation in the ITE room.

## 645.5 Supply Circuits and Interconnecting Cables

**(A) Branch-Circuit Conductors.** The branch-circuit conductors supplying one or more units of information technology equipment shall have an ampacity not less than 125 percent of the total connected load.

**(B) Power-Supply Cords.** Information technology equipment shall be permitted to be connected to a branch circuit by a power-supply cord.

(1) Power-supply cords shall not exceed 4.5 m (15 ft).
(2) Power cords shall be listed and a type permitted for use on listed information technology equipment or shall be constructed of listed flexible cord and listed attachment plugs and cord connectors of a type permitted for information technology equipment.

Informational Note: One method of determining if cords are of a type permitted for the purpose is found in UL 60950-1-2007, *Safety of Information Technology Equipment — Safety — Part 1: General Requirements*; or UL 62368-1-2012, *Audio/Video, Information and Communication Technology Equipment — Part 1: Safety Requirements*.

**(C) Interconnecting Cables.** Separate information technology equipment units shall be permitted to be interconnected by means of listed cables and cable assemblies. The 4.5 m (15 ft) limitation in 645.5(B)(1) shall not apply to interconnecting cables.

**(D) Physical Protection.** Where exposed to physical damage, supply circuits and interconnecting cables shall be protected.

**(E) Under Raised Floors.** Power cables, communications cables, connecting cables, interconnecting cables, cord-and-plug connections, and receptacles associated with the information technology equipment shall be permitted under a raised floor, provided the following conditions are met:

(1) The raised floor is of approved construction, and the area under the floor is accessible.
(2) The branch-circuit supply conductors to receptacles or field-wired equipment are in rigid metal conduit, rigid nonmetallic conduit, intermediate metal conduit, electrical metallic tubing, electrical nonmetallic tubing, metal wireway, nonmetallic wireway, surface metal raceway with metal cover, surface nonmetallic raceway, flexible metal conduit, liquidtight flexible metal conduit, or liquidtight flexible nonmetallic conduit, Type MI cable, Type MC cable, or Type AC cable and associated metallic and nonmetallic boxes or enclosures. These supply conductors shall be installed in accordance with the requirements of 300.11.

Branch-circuit conductors installed under the raised floor of an ITE room using any of the wiring methods listed are required to conform to the specific article for the wiring method used. In addition, 300.11 requires raceways, cables, and boxes to be securely fastened in place, even though they are installed below a raised floor.

(3) Supply cords of listed information technology equipment are in accordance with 645.5(B).
(4) Ventilation in the underfloor area is used for the information technology equipment room only, except as provided in 645.4(2).

The underfloor area of an ITE room is required to be provided with a smoke detection device(s). Upon detection of smoke, the circulation of air in the underfloor area must be interrupted. In addition, the smoke detectors may provide other fire protection functions as part of a complete building fire alarm system. A ventilation system can serve the underfloor areas of an ITE room as well as other areas of a building if the ventilation system is equipped with smoke and fire dampers at the ITE room boundaries. Upon detection of smoke or activation of the ITE room disconnecting means, these fire protection features isolate the underfloor area from other areas served by the ventilation system.

•

(5) Openings in raised floors for cords and cables protect cords and cables against abrasion and minimize the entrance of debris beneath the floor.

Supply cords of ITE equipment are permitted to be run through holes in a raised floor to connect to receptacles located below the raised floor. Openings in a raised floor through which cords and cables are run must be made so the cords and cables are not subject to abrasion. Allowing cords through openings in a raised floor is a 90.3 amendment to the general prohibition of this practice found in 400.8.

(6) Cables, other than those covered in 645.5(E)(2) and (E)(3), are one of the following:
   a. Listed Type DP cable having adequate fire-resistant characteristics suitable for use under raised floors of an information technology equipment room
   b. Interconnecting cables enclosed in a raceway
   c. Cable type designations shown in Table 645.5(E)(6)
   d. Equipment grounding conductors

Informational Note: One method of defining *fire resistance* is by establishing that the cables do not spread fire to the top of the tray in the "UL Flame Exposure, Vertical Tray Flame Test"

*TABLE 645.5(E)(6) Cable Types Permitted Under Raised Floors*

| Article | Plenum | Riser | General Purpose |
|---|---|---|---|
| 336 | | | TC |
| 725 | CL2P & CL3P | CL2R & CL3R | CL2, CL3 & PLTC |
| 727 | | | ITC |
| 760 | NPLFP & FPLP | NPLFR & FPLR | NPLF & FPL |
| 770 | OFNP & OFCP | OFNR & OFCR | OFN & OFC |
| 800 | CMP | CMR | CM & CMG |
| 820 | CATVP | CATVR | CATV |

in UL 1685-2011, *Standard for Safety for Vertical-Tray Fire-Propagation and Smoke-Release Test for Electrical and Optical-Fiber Cables*. The smoke measurements in the test method are not applicable.

Another method of defining *fire resistance* is for the damage (char length) not to exceed 1.5 m (4 ft 11 in.) when performing the CSA "Vertical Flame Test — Cables in Cable Trays," as described in CSA C22.2 No. 0.3-M-2001, *Test Methods for Electrical Wires and Cables*.

Other than branch-circuit conductors and power supply cords, interconnecting cables used under raised floors are required to be enclosed in a raceway, be listed as Type DP (data processing) cables, or be the appropriate cable type from Table 645.5(E)(6).

**(F) Securing in Place.** Power cables; communications cables; connecting cables; interconnecting cables; and associated boxes, connectors, plugs, and receptacles that are listed as part of, or for, information technology equipment shall not be required to be secured in place.

**(G) Abandoned Supply Circuits and Interconnecting Cables.** The accessible portion of abandoned supply circuits and interconnecting cables shall be removed unless contained in a raceway.

**(H) Installed Supply Circuits and Interconnecting Cables Identified for Future Use.**

(1) Supply circuits and interconnecting cables identified for future use shall be marked with a tag of sufficient durability to withstand the environment involved.

(2) Supply circuit tags and interconnecting cable tags shall have the following information:

a. Date identified for future use

b. Date of intended use

c. Information relating to the intended future use

## 645.6 Cables Not in Information Technology Equipment Room

Cables extending beyond the information technology equipment room shall be subject to the applicable requirements of this *Code*.

•

## 645.10 Disconnecting Means

An approved means shall be provided to disconnect power to all electronic equipment in the information technology equipment room or in designated zones within the room. There shall also be a similar approved means to disconnect the power to all dedicated HVAC systems serving the room or designated zones and shall cause all required fire/smoke dampers to close. The disconnecting means shall comply with either 645.10(A) or (B).

*Exception: Installations qualifying under the provisions of Article 685.*

**(A) Remote Disconnect Controls.**

**(1)** Remote disconnect controls shall be located at approved locations readily accessible in case of fire to authorized personnel and emergency responders.

**(2)** The remote disconnect means for the control of electronic equipment power and HVAC systems shall be grouped and identified. A single means to control both systems shall be permitted.

**(3)** Where multiple zones are created, each zone shall have an approved means to confine fire or products of combustion to within the zone.

**(4)** Additional means to prevent unintentional operation of remote disconnect controls shall be permitted.

Typically, the circuit supplying the ITE and the circuit supplying the HVAC system will be controlled through separate disconnecting means. However, operation of these disconnecting means can be accomplished through the use of a single remote control, such as one pushbutton. The disconnecting means is required to disconnect the conductors of each circuit from their supply source and close all required fire/smoke dampers. (See the definition of *disconnecting means* in Article 100.) The disconnecting means is permitted to be remote-controlled switching devices, such as relays, with pushbutton stations at the principal exit doors.

The requirements of 645.10(A) and those of 645.3(A) for sealing penetrations are intended to minimize the passage of smoke or fire to other parts of the building.

Informational Note: For further information, see NFPA 75-2013, *Standard for the Protection of Information Technology Equipment*.

**(B) Critical Operations Data Systems.** Remote disconnecting controls shall not be required for critical operations data systems when all of the following conditions are met:

(1) An approved procedure has been established and maintained for removing power and air movement within the room or zone.

(2) Qualified personnel are continuously available to meet emergency responders and to advise them of disconnecting methods.
(3) A smoke-sensing fire detection system is in place.

Informational Note: For further information, see *NFPA 72-2013, National Fire Alarm and Signaling Code.*

(4) An approved fire suppression system suitable for the application is in place.
(5) Cables installed under a raised floor, other than branch-circuit wiring, and power cords are installed in compliance with 645.5(E)(2) or (E)(3), or in compliance with 300.22(C), 725.135(B), and Table 725.154; 770.113(C) and Table 770.154(a); 800.113(C) and Table 800.154(a); or 820.113(C) and Table 820.154(a).

See the commentary following the definition of *critical operations data system* in 645.2. Only those data systems designated as critical in function based on that definition are permitted to implement the provision for not installing the remote disconnect control covered in 645.10(A).

## 645.11 Uninterruptible Power Supplies (UPSs)

Except for installations and constructions covered in 645.11(1) or (2), UPS systems installed within the information technology equipment room, and their supply and output circuits, shall comply with 645.10. The disconnecting means shall also disconnect the battery from its load.

(1) Installations qualifying under the provisions of Article 685
(2) Power sources limited to 750 volt-amperes or less derived either from UPS equipment or from battery circuits integral to electronic equipment

## 645.14 System Grounding

Separately derived power systems shall be installed in accordance with the provisions of Parts I and II of Article 250. Power systems derived within listed information technology equipment that supply information technology systems through receptacles or cable assemblies supplied as part of this equipment shall not be considered separately derived for the purpose of applying 250.30.

## 645.15 Equipment Grounding and Bonding

All exposed non–current-carrying metal parts of an information technology system shall be bonded to the equipment grounding conductor in accordance with Parts I, V, VI, VII, and VIII of Article 250 or shall be double insulated. Where signal reference structures are installed, they shall be bonded to the equipment grounding conductor provided for the information technology equipment. Any auxiliary grounding electrode(s) installed for information technology equipment shall be installed in accordance with 250.54.

Properly bonded high-frequency signal reference structures provide additional safety measures.

Informational Note No. 1: The bonding requirements in the product standards governing this listed equipment ensure that it complies with Article 250.

Informational Note No. 2: Where isolated grounding-type receptacles are used, see 250.146(D) and 406.3(D).

## 645.16 Marking

Each unit of an information technology system supplied by a branch circuit shall be provided with a manufacturer's nameplate, which shall also include the input power requirements for voltage, frequency, and maximum rated load in amperes.

## 645.17 Power Distribution Units

Power distribution units that are used for information technology equipment shall be permitted to have multiple panelboards within a single cabinet if the power distribution unit is utilization equipment listed for information technology application.

Power distribution units (PDUs) are specialized electrical distribution equipment used to supply multiple bays of rack-mounted modules installed in an ITE room. Due to the large number of OCPDs used in this type of application, PDUs are built with multiple panelboards installed in a single cabinet.

## 645.25 Engineering Supervision

As an alternative to the feeder and service load calculations required by Parts III and IV of Article 220, feeder and service load calculations for new or existing loads shall be permitted to be used if provided by qualified persons under engineering supervision.

An engineered alternative to the load calculations in Parts III and IV of Article 220 recognizes that the loads associated with computer hardware vary according to the operating system and software being used. Therefore, identical pieces of equipment installed in a facility will each have a load based on how that equipment is being used and applied.

The engineered alternative provides for a customized load calculation based on how a facility or a specific industry applies its computer hardware.

## 645.27 Selective Coordination

Critical operations data system(s) overcurrent protective devices shall be selectively coordinated with all supply-side overcurrent protective devices.

# ARTICLE 646 Modular Data Centers

## I. General

### 646.1 Scope

This article covers modular data centers.

Informational Note No. 1: Modular data centers include the installed information technology equipment (ITE) and support equipment, electrical supply and distribution, wiring and protection, working space, grounding, HVAC, and the like, that are located in an equipment enclosure.

Informational Note No. 2: For further information, see NFPA 75-2013, *Standard for the Protection of Information Technology Equipment*, which covers the requirements for the protection of information technology equipment and systems in an information technology equipment room.

A modular data center (MDC) is similar to an information technology equipment (ITE) room covered under Article 645. The distinction between the two is that the MDC is a preassembled and packaged enclosure delivered with the ITE equipment installed, while the ITE room is built and equipped on site. The MDC may be supplied through the premises wiring system or through a separate MDC enclosure. Data, fire alarm, communications, control, audio, and visual circuits from the MDC are typically brought outside of the MDC and into the facility. An MDC is large enough for personnel to enter; therefore, access, working space, and emergency lighting are specifically addressed.

### 646.2 Definitions

The definitions in 645.2 shall apply. For the purposes of this article, the following additional definition applies.

**Modular Data Center (MDC).** Prefabricated units, rated 600 volts or less, consisting of an outer enclosure housing multiple racks or cabinets of information technology equipment (ITE) (e.g., servers) and various support equipment, such as electrical service and distribution equipment, HVAC systems, and the like.

Informational Note No. 1: A typical construction may use a standard ISO shipping container or other structure as the outer enclosure, racks or cabinets of ITE, service-entrance equipment and power distribution components, power storage such as a UPS, and an air or liquid cooling system. Modular data centers are intended for fixed installation, either indoors or outdoors, based on their construction and resistance to environmental conditions. MDCs can be configured as an all-in-one system housed in a single equipment enclosure or as a system with the support equipment housed in separate equipment enclosures.

Informational Note No. 2: For information on listing requirements for both information technology equipment and communications equipment, see UL 60950-1-2011, *Information Technology Equipment — Safety — Part 1: General Requirements*, and UL 62368-1-2012, *Audio/Video, Information and Communication Technology Equipment — Part 1: Safety Requirements.*

Informational Note No. 3: *Modular data centers* as defined in this article are sometimes referred to as containerized data centers.

Informational Note No. 4: Equipment enclosures housing only support equipment (e.g., HVAC or power distribution equipment) that are not part of a specific modular data center are not considered a modular data center as defined in this article.

### 646.3 Other Articles

Circuits and equipment shall comply with 646.3(A) through (N) as applicable. Wherever the requirements of other articles of this *Code* and Article 646 differ, the requirements of Article 646 shall apply.

**(A) Spread of Fire or Products of Combustion.** Sections 300.21, 770.26, 800.26, and 820.26 shall apply to penetrations of a fire-resistant room boundary, if provided.

**(B) Plenums.** Sections 300.22(C)(1), 725.154(A), 760.53(B)(2), 760.154(A), 770.113(C), 800.113(C), and Table 725.154, Table 760.154, Table 770.154(a), Table 800.154(a), and Table 820.154(a) shall apply to wiring and cabling in other spaces used for environmental air (plenums).

Informational Note: Environmentally controlled working spaces, aisles, and equipment areas in an MDC are not considered a plenum.

**(C) Grounding. Grounding and bonding of an MDC shall comply with Article 250.** The non–current-carrying conductive members of optical fiber cables in an MDC shall be grounded in accordance with 770.114. Grounding and bonding of communications protectors, cable shields, and non–current-carrying metallic members of cable shall comply with Part IV of Article 800.

**(D) Electrical Classification of Data Circuits.** Section 725.121(A)(4) shall apply to the electrical classification of listed information technology equipment signaling circuits. Sections 725.139(D)(1) and 800.133(A)(1)(b) shall apply to the electrical classification of Class 2 and Class 3 circuits in the same cable with communications circuits.

**(E) Fire Alarm Equipment.** The provisions of Parts I, II, and III of Article 760 shall apply to fire alarm system equipment installed in an MDC, where provided.

**(F) Communications Equipment.** Parts I, II, III, IV, and V of Article 800 shall apply to communications equipment installed in an MDC.

Informational Note: See Part I of Article 100 for a definition of *communications equipment*.

**(G) Community Antenna Television and Radio Distribution Systems Equipment.** Parts I, II, III, IV, and V of Article 820 shall apply to community antenna television and radio distribution systems equipment installed in an MDC.

**(H) Storage Batteries.** Installation of storage batteries shall comply with Article 480.

*Exception: Batteries that are part of listed and labeled equipment and installed in accordance with the listing requirements.*

**(I) Surge-Protective Devices (SPDs).** Where provided, surge-protective devices shall be listed and labeled and installed in accordance with Article 285.

**(J) Lighting.** Lighting shall be installed in accordance with Article 410.

**(K) Power Distribution Wiring and Wiring Protection.** Power distribution wiring and wiring protection within an MDC shall comply with Article 210 for branch circuits.

**(L) Wiring Methods and Materials.**

(1) Unless modified elsewhere in this article, wiring methods and materials for power distribution shall comply with Chapter 3. Wiring shall be suitable for its use and installation and shall be listed and labeled.

*Exception: This requirement shall not apply to wiring that is part of listed and labeled equipment.*

(2) The following wiring methods shall not be permitted:
   a. Integrated gas spacer cable: Type IGS (Article 326)
   b. Concealed knob-and-tube wiring (Article 394)
   c. Messenger-supported wiring (Article 396)
   d. Open wiring on insulators (Article 398)
   e. Outdoor overhead conductors over 600 volts (Article 399)

(3) Wiring in areas under a raised floor that are constructed and used for ventilation as described in 645.5(E) shall be permitted to use the wiring methods described in 645.5(E).

(4) Installation of wiring for remote-control, signaling, and power-limited circuits shall comply with Part III of Article 725.

(5) Installation of optical fiber cables shall comply with Part V of Article 770.

(6) Installation of wiring for fire alarm systems shall comply with Parts II and III of Article 760.

(7) Installation of communications wires and cables, raceways, and cable routing assemblies shall comply with Part V of Article 800.

(8) Alternate wiring methods as permitted by Article 645 shall be permitted for MDCs, provided that all of the conditions of 645.4 are met.

**(M) Service Equipment.** For an MDC that is designed such that it may be powered from a separate electrical service, the service equipment for control and protection of services and their installation shall comply with Article 230. The service equipment and their arrangement and installation shall permit the installation of the service-entrance conductors in accordance with Article 230. Service equipment shall be listed and labeled and marked as being suitable for use as service equipment.

**(N) Disconnecting Means.** An approved means shall be provided to disconnect power to all electronic equipment in the MDC in accordance with 645.10. There shall also be a similar approved means to disconnect the power to all dedicated HVAC systems serving the MDC that shall cause all required fire/smoke dampers to close.

## 646.4 Applicable Requirements

All MDCs shall:

(1) Be listed and labeled and comply with 646.3(N) and 646.5 through 646.9, or

Informational Note: One way to determine applicable listing requirements is to refer to UL Subject 2755, *Outline of Investigation for Modular Data Centers.*

(2) Comply with the provisions of this article.

An MDC may be listed, in which case an evaluation of the equipment, installed wiring, lighting, and work space is conducted as part of the listing. Any field-installed wiring — including supply circuits and data circuits — is required to comply with the appropriate *NEC* article. Non-listed MDCs may also be installed under the requirements of this article.

## 646.5 Nameplate Data

A permanent nameplate shall be attached to each equipment enclosure of an MDC and shall be plainly visible after installation. The nameplate shall include the information in 646.5(1) through (6), as applicable:

(1) Supply voltage, number of phases, frequency, and full-load current. The full-load current shown on the nameplate shall not be less than the sum of the full-load currents required for all motors and other equipment that may be in operation at the same time under normal conditions of use. Where unusual type loads, duty cycles, and so forth, require oversized conductors or permit reduced-size conductors, the required capacity shall be included in the marked full-load current. Where more than one incoming supply circuit is to be provided, the nameplate shall state the preceding information for each circuit.

Informational Note No. 1: See 430.22(E) and 430.26 for duty cycle requirements.

Informational Note No. 2: For listed equipment, the full-load current shown on the nameplate may be the maximum, measured, 15-minute, average full-load current.

(2) For MDCs powered by a separate service, the short-circuit current rating of the service equipment provided as part of the MDC.

Informational Note: This rating may be part of the service equipment marking.

(3) For MDCs powered by a separate service, if the required service as determined by Parts III and IV of Article 220 is less than the rating of the service panel used, the required service shall be included on the nameplate.

Informational Note: Branch circuits supplying ITE loads are assumed to be loaded not less than 80 percent of the branch-circuit rating with a 100 percent duty cycle. As an alternative to the feeder and service load calculations required by Parts III and IV of Article 220, feeder and service load calculations for new, future, or existing loads may be permitted to be used if performed by qualified persons under engineering supervision.

(4) Electrical diagram number(s) or the number of the index to the electrical drawings.
(5) For MDC equipment enclosures that are not powered by a separate service, feeder, or branch circuit, a reference to the powering equipment.
(6) Manufacturer's name or trademark.

## 646.6 Supply Conductors and Overcurrent Protection

Whether the MDC is listed or not, a permanent nameplate is required on each enclosure to indicate the required supply and the full-load current. The full-load current is not necessarily determined with all equipment operating simultaneously as indicated in 646.5(1). The MDC is considered a continuous load; therefore, supply conductors must be sized for 125 percent of the marked full-load current.

**(A) Size.** The size of the supply conductor shall be such as to have an ampacity not less than 125 percent of the full-load current rating.

Informational Note No. 1: See the 0–2000-volt ampacity tables of Article 310 for ampacity of conductors rated 600 V and below.

Informational Note No. 2: See 430.22(E) and 430.26 for duty cycle requirements.

**(B) Overcurrent Protection.** Where overcurrent protection for supply conductors is furnished as part of the MDC, overcurrent protection for each supply circuit shall comply with 646.6(B)(1) through (B)(2):

**(1) Service Equipment — Overcurrent Protection.** Service conductors shall be provided with overcurrent protection in accordance with 230.90 through 230.95.

**(2) Taps and Feeders.** Where overcurrent protection for supply conductors is furnished as part of the MDC as permitted by 240.21, the overcurrent protection shall comply with the following:

(1) The overcurrent protection shall consist of a single circuit breaker or set of fuses.
(2) The MDC shall be marked "OVERCURRENT PROTECTION PROVIDED AT MDC SUPPLY TERMINALS."
(3) The supply conductors shall be considered either as feeders or as taps and be provided with overcurrent protection complying with 240.21.

## 646.7 Short-Circuit Current Rating

**(A) Service Equipment.** The service equipment of an MDC that connects directly to a service shall have a short-circuit current rating not less than the available fault current of the service.

**(B) MDCs Connected to Branch Circuits and Feeders.** Modular data centers that connect to a branch circuit or a feeder circuit shall have a short-circuit current rating not less than the available fault current of the branch circuit or feeder. The short-circuit current rating of the MDC shall be based on the short-circuit current rating of a listed and labeled MDC or the short-circuit current rating established utilizing an approved method.

Informational Note No. 1: UL 508A-2001, *Standard for Industrial Control Panels, Supplement SB*, is an example of an approved method.

Informational Note No. 2: This requirement does not apply to listed and labeled equipment connected to branch circuits located inside of the MDC equipment enclosure.

**(C) MDCs Powered from Separate MDC System Enclosures.** Modular data center equipment enclosures, powered from a separate MDC system enclosure that is part of the specific MDC system, shall have a short-circuit current rating coordinated with the powering module in accordance with 110.10.

Informational Note: UL 508A-2001, *Standard for Industrial Control Panels, Supplement SB*, is an example of an approved method for determining short-circuit current ratings.

## 646.8 Field-Wiring Compartments

A field-wiring compartment in which service or feeder connections are to be made shall be readily accessible and comply with 646.8(1) through (3) as follows:

(1) Permit the connection of the supply wires after the MDC is installed
(2) Permit the connection to be introduced and readily connected
(3) Be located so that the connections may be readily inspected after the MDC is installed

## 646.9 Flexible Power Cords and Cables for Connecting Equipment Enclosures of an MDC System

**(A) Uses Permitted.** Flexible power cords and cables shall be permitted to be used for connections between equipment enclosures of an MDC system where not subject to physical damage.

Informational Note: One example of flexible power cord usage for connections between equipment enclosures of an MDC system is between an MDC enclosure containing only servers and one containing power distribution equipment.

**(B) Uses Not Permitted.** Flexible power cords and cables shall not be used for connection to external sources of power.

Informational Note: Examples of external sources of power are electrical services, feeders, and premises branch circuits.

**(C) Listing.** Where flexible power cords or cables are used, they shall be listed as suitable for extra-hard usage. Where used outdoors, flexible power cords and cables shall also be listed as suitable for wet locations and shall be sunlight resistant.

**(D) Single-Conductor Cable.** Single-conductor power cable shall be permitted to be used only in sizes 2 AWG or larger.

## II. Equipment

### 646.10 Electrical Supply and Distribution

Equipment used for electrical supply and distribution in an MDC, including fittings, devices, luminaires, apparatus, machinery, and the like, shall comply with Parts I and II of Article 110.

### 646.11 Distribution Transformers

**(A) Utility-Owned Transformers.** Utility-owned distribution transformers shall not be permitted in an MDC.

**(B) Non-Utility-Owned Premises Transformers.** Non-utility-owned premises distribution transformers installed in the vicinity of an MDC shall be of the dry type or the type filled with a noncombustible dielectric medium. Such transformers shall be installed in accordance with the requirements of Article 450. Non-utility-owned premises distribution transformers shall not be permitted in an MDC.

**(C) Power Transformers.** Power transformers that supply power only to the MDC shall be permitted to be installed in the MDC equipment enclosure. Only dry-type transformers shall be permitted to be installed in the MDC equipment enclosure. Such transformers shall be installed in accordance with the requirements of Article 450.

### 646.12 Receptacles

At least one 125-volt ac, 15- or 20-ampere-rated duplex convenience outlet shall be provided in each work area of the MDC to facilitate the powering of test and measurement equipment that may be required during routine maintenance and servicing without having to route flexible power cords through or across doorways or around line-ups of equipment, or the like.

### 646.13 Other Electrical Equipment

Electrical equipment that is an integral part of the MDC, including lighting, control, power, HVAC (heating, ventilation, and air-conditioning), emergency lighting, alarm circuits, and the like, shall comply with the requirements for its use and installation and shall be listed and labeled.

### 646.14 Installation and Use

Listed and labeled equipment shall be installed and used in accordance with any instructions or limitations included in the listing.

## III. Lighting

### 646.15 General Illumination

Illumination shall be provided for all workspaces and areas that are used for exit access and exit discharge. The illumination shall be arranged so that the failure of any single lighting unit does not result in a complete loss of illumination.

Informational Note: See NFPA *101*®-2012, *Life Safety Code*, Section 7.8, for information on illumination of means of egress.

### 646.16 Emergency Lighting

Areas that are used for exit access and exit discharge shall be provided with emergency lighting. Emergency lighting systems shall be listed and labeled equipment installed in accordance with the manufacturer's instructions.

Informational Note: See NFPA *101*®-2012, *Life Safety Code*, Section 7.9, for information on emergency lighting.

### 646.17 Emergency Lighting Circuits

No appliances or lamps, other than those specified as required for emergency use, shall be supplied by emergency lighting circuits. Branch circuits supplying emergency lighting shall be installed to provide service from storage batteries, generator sets, UPS, separate service, fuel cells, or unit equipment. No other equipment shall be connected to these circuits unless the emergency lighting system includes a backup system where only the lighting is supplied by battery circuits under power failure conditions. All boxes and enclosures (including transfer switches, generators, and power panels) for emergency circuits shall be marked to identify them as components of an emergency circuit or system.

## IV. Workspace

### 646.18 General

Space about electrical equipment shall comply with 110.26.

### 646.19 Entrance to and Egress from Working Space

For equipment over 1.8 m (6 ft) wide or deep, there shall be one entrance to and egress from the required working space not less than 610 mm (24 in.) wide and 2.0 m (6½ ft) high at each end of the working space. The door(s) shall open in the direction of egress and be equipped with panic bars, pressure plates, or other devices that are normally latched but open under simple pressure. A single entrance to and egress from the required working space shall be permitted where either of the conditions in 646.19(1) or (2) is met.

**(1) Unobstructed Egress.** Where the location permits a continuous and unobstructed way of egress travel, a single entrance to the working space shall be permitted.

**(2) Extra Working Space.** Where the depth of the working space is twice that required by 110.26(1), a single entrance shall be permitted. It shall be located such that the distance from the equipment to the nearest edge of the entrance is not less than the minimum clear distance specified in Table 110.26(1) for equipment operating at that voltage and in that condition.

### 646.20 Working Space for ITE

**(A) Low-Voltage Circuits.** The working space about ITE where any live parts that may be exposed during routine servicing operate at not greater than 30 volts rms, 42 volts peak, or 60 volts dc shall not be required to comply with the workspace requirements of 646.19.

**(B) Other Circuits.** Any areas of ITE that require servicing of parts that are greater than 30 volts rms, 42 volts peak, or 60 volts dc shall comply with the workspace requirements of 646.19.

Informational Note No. 1: For example, field-wiring compartments for ac mains connections, power distribution units, and the like.

Informational Note No. 2: It is assumed that ITE operates at voltages not exceeding 600 volts.

### 646.21 Work Areas and Working Space Around Batteries

Working space around a battery system shall comply with 110.26. Working clearance shall be measured from the edge of the battery rack.

### 646.22 Workspace for Routine Service and Maintenance

Workspace shall be provided to facilitate routine servicing and maintenance (those tasks involving operations that can be accomplished by employees and where extensive disassembly of equipment is not required). Routine servicing and maintenance shall be able to be performed without exposing the worker to a risk of electric shock or personal injury.

Informational Note: An example of such routine maintenance is cleaning or replacing an air filter.

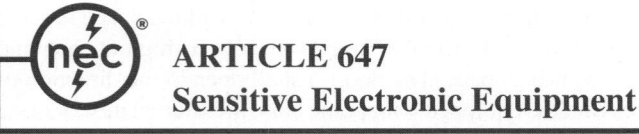

## ARTICLE 647 Sensitive Electronic Equipment

### 647.1 Scope

This article covers the installation and wiring of separately derived systems operating at 120 volts line-to-line and 60 volts to ground for sensitive electronic equipment.

This type of supply system is employed as a means to reduce objectionable noise and its adverse effect on the performance of electronic audio and video equipment. Article 647 permits the use of this type of supply system for all commercial and industrial applications where sensitive audio/video or similar electronic equipment is used. Such systems can only be used in areas that are under the close supervision of qualified individuals.

### 647.3 General

Use of a separately derived 120-volt single-phase 3-wire system with 60 volts on each of two ungrounded conductors to a grounded neutral conductor shall be permitted for the purpose of reducing objectionable noise in sensitive electronic equipment locations, provided the following conditions apply:

(1) The system is installed only in commercial or industrial occupancies.
(2) The system's use is restricted to areas under close supervision by qualified personnel.
(3) All of the requirements in 647.4 through 647.8 are met.

### 647.4 Wiring Methods

**(A) Panelboards and Overcurrent Protection.** Use of standard single-phase panelboards and distribution equipment with a higher voltage rating shall be permitted. The system shall be clearly marked on the face of the panel or on the inside of the panel doors. Common trip two-pole circuit breakers or a combination two-pole fused disconnecting means that are identified for use at the system voltage shall be provided for both ungrounded conductors in all feeders and branch circuits. Branch circuits and feeders shall be provided with a means to simultaneously disconnect all ungrounded conductors.

Circuit breakers and fuses are acceptable means of providing overcurrent protection for technical power circuits. Additionally, all technical power feeder and branch circuits are required to be provided with a disconnecting means that simultaneously opens all ungrounded conductors of the circuit.

**(B) Junction Boxes.** All junction box covers shall be clearly marked to indicate the distribution panel and the system voltage.

**(C) Conductor Identification.** All feeders and branch-circuit conductors installed under this section shall be identified as to system at all splices and terminations by color, marking, tagging, or equally effective means. The means of identification shall be posted at each branch-circuit panelboard and at the disconnecting means for the building.

**(D) Voltage Drop.** The voltage drop on any branch circuit shall not exceed 1.5 percent. The combined voltage drop of feeder and branch-circuit conductors shall not exceed 2.5 percent.

Unlike electrical distribution systems that supply lighting and appliance branch circuits, the supply systems covered by Article 647 are subject to mandatory voltage-drop requirements. These voltage-drop requirements are needed to ensure the operation of overcurrent devices in order to protect conductors and equipment supplied by these systems. Because the use of standard overcurrent devices and distribution equipment with higher voltage ratings is permitted, the impedance in circuits supplied by these systems under fault conditions is a primary concern.

**(1) Fixed Equipment.** The voltage drop on branch circuits supplying equipment connected using wiring methods in Chapter 3 shall not exceed 1.5 percent. The combined voltage drop of feeder and branch-circuit conductors shall not exceed 2.5 percent.

**(2) Cord-Connected Equipment.** The voltage drop on branch circuits supplying receptacles shall not exceed 1 percent. For the purposes of making this calculation, the load connected to the receptacle outlet shall be considered to be 50 percent of the branch-circuit rating. The combined voltage drop of feeder and branch-circuit conductors shall not exceed 2.0 percent.

Informational Note: The purpose of this provision is to limit voltage drop to 1.5 percent where portable cords may be used as a means of connecting equipment.

## 647.5 Three-Phase Systems

Where 3-phase power is supplied, a separately derived 6-phase "wye" system with 60 volts to ground installed under this article shall be configured as three separately derived 120-volt single-phase systems having a combined total of no more than six disconnects.

## 647.6 Grounding

**(A) General.** The transformer secondary center tap of the 60/120-volt, 3-wire system shall be grounded as provided in 250.30.

A technical power system has two ungrounded wires with 120 volts between them and a grounded reference wire at 60 volts with respect to the ungrounded wires.

**(B) Grounding Conductors Required.** Permanently wired utilization equipment and receptacles shall be grounded by means of an equipment grounding conductor run with the circuit conductors to an equipment grounding bus prominently marked "Technical Equipment Ground" in the originating branch-circuit panelboard. The grounding bus shall be connected to the grounded conductor on the line side of the separately derived system's disconnecting means. The grounding conductor shall not be smaller than that specified in Table 250.122 and run with the feeder conductors. The technical equipment grounding bus need not be bonded to the panelboard enclosure. Other grounding methods authorized elsewhere in this *Code* shall be permitted where the impedance of the grounding return path does not exceed the impedance of equipment grounding conductors sized and installed in accordance with this article.

Informational Note No. 1: See 250.122 for equipment grounding conductor sizing requirements where circuit conductors are adjusted in size to compensate for voltage drop.

Informational Note No. 2: These requirements limit the impedance of the ground fault path where only 60 volts apply to a fault condition instead of the usual 120 volts.

## 647.7 Receptacles

**(A) General.** Where receptacles are used as a means of connecting equipment, the following conditions shall be met:

(1) All 15- and 20-ampere receptacles shall be GFCI protected.
(2) All receptacle outlet strips, adapters, receptacle covers, and faceplates shall be marked with the following words or equivalent:

WARNING — TECHNICAL POWER
Do not connect to lighting equipment.
For electronic equipment use only.
60/120 V. 1ϕac
GFCI protected
The warning sign(s) or label(s) shall comply with 110.21(B).

(3) A 125-volt, single-phase, 15- or 20-ampere-rated receptacle having one of its current-carrying poles connected to a grounded circuit conductor shall be located within 1.8 m (6 ft) of all permanently installed 15- or 20-ampere-rated 60/120-volt technical power-system receptacles.
(4) All 125-volt receptacles used for 60/120-volt technical power shall have a unique configuration and be identified for use with this class of system.

*Exception: Receptacles and attachment plugs rated 125-volt, single-phase, 15- or 20-amperes, and that are identified for use with grounded circuit conductors, shall be permitted in machine rooms, control rooms, equipment rooms, equipment racks, and other similar locations that are restricted to use by qualified personnel.*

**(B) Isolated Ground Receptacles.** Isolated ground receptacles shall be permitted as described in 250.146(D); however, the branch-circuit equipment grounding conductor shall be terminated as required in 647.6(B).

## 647.8 Lighting Equipment

Lighting equipment installed under this article for the purpose of reducing electrical noise originating from lighting equipment shall meet the conditions of 647.8(A) through (C).

**(A) Disconnecting Means.** All luminaires connected to separately derived systems operating at 60 volts to ground, and associated control equipment if provided, shall have a disconnecting means that simultaneously opens all ungrounded conductors. The disconnecting means shall be located within sight of the luminaire or be lockable open in accordance with 110.25.

**(B) Luminaires.** All luminaires shall be permanently installed and listed for connection to a separately derived system at 120 volts line-to-line and 60 volts to ground.

**(C) Screw Shell.** Luminaires installed under this section shall not have an exposed lamp screw shell.

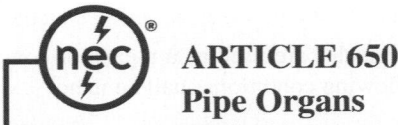

# ARTICLE 650
# Pipe Organs

## 650.1 Scope

This article covers those electrical circuits and parts of electrically operated pipe organs that are employed for the control of the sounding apparatus and keyboards.

## 650.3 Other Articles

**(A) Electronic Organ Equipment.** Installations of digital/analog–sampled sound production technology and associated audio signal processing, amplification, reproduction equipment, and wiring installed as part of a pipe organ shall be in accordance with Article 640.

Some pipe organ installations incorporate electronics for digital/analog sampled sound technology. The requirements in Article 640 are necessary for electronic sound production, amplification, signal processing, and other sound reproduction circuits and equipment installed as part of a pipe organ.

**(B) Optical Fiber Cable.** Installations of optical fiber cables shall be in accordance with Parts I and V of Article 770.

## 650.4 Source of Energy

The source of power shall be a transformer-type rectifier, the dc potential of which shall not exceed 30 volts dc.

## 650.5 Grounding

The rectifier shall be bonded to the equipment grounding conductor according to the provisions in Article 250, Parts V, VI, VII, and VIII.

## 650.6 Conductors

Conductors shall comply with 650.6(A) through (D).

**(A) Size.** Conductors shall be not less than 28 AWG for electronic signal circuits and not less than 26 AWG for electromagnetic valve supply and the like. A main common-return conductor in the electromagnetic supply shall not be less than 14 AWG.

**(B) Insulation.** Conductors shall have thermoplastic or thermosetting insulation.

**(C) Conductors to Be Cabled.** Except for the common-return conductor and conductors inside the organ proper, the organ sections and the organ console conductors shall be cabled. The common-return conductors shall be permitted under an additional covering enclosing both cable and return conductor, or they shall be permitted as a separate conductor and shall be permitted to be in contact with the cable.

**(D) Cable Covering.** Each cable shall be provided with an outer covering, either overall or on each of any subassemblies of grouped conductors. Tape shall be permitted in place of a covering. Where not installed in metal raceway, the covering shall be resistant to flame spread, or the cable or each cable subassembly shall be covered with a closely wound listed fireproof tape.

Informational Note: One method of determining that cable is resistant to flame spread is by testing the cable to the VW-1 (vertical-wire) flame test in ANSI/UL 1581-2011, *Reference Standard for Electrical Wires, Cables and Flexible Cords.*

## 650.7 Installation of Conductors

Cables shall be securely fastened in place and shall be permitted to be attached directly to the organ structure without insulating supports. Cables shall not be placed in contact with other conductors. Abandoned cables that are not terminated at equipment shall be identified with a tag.

## 650.8 Overcurrent Protection

Circuits shall be so arranged that 26 AWG and 28 AWG conductors shall be protected by an overcurrent device rated at not more than 6 amperes. Other conductor sizes shall be protected in accordance with their ampacity. A common return conductor shall not require overcurrent protection.

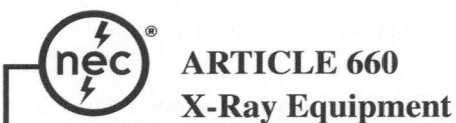

# ARTICLE 660
# X-Ray Equipment

## I. General

## 660.1 Scope

This article covers all X-ray equipment operating at any frequency or voltage for industrial or other nonmedical or nondental use.

Informational Note: See Article 517, Part V, for X-ray installations in health care facilities.

Nothing in this article shall be construed as specifying safeguards against the useful beam or stray X-ray radiation.

Informational Note No. 1: Radiation safety and performance requirements of several classes of X-ray equipment are regulated under Public Law 90-602 and are enforced by the Department of Health and Human Services.

Informational Note No. 2: In addition, information on radiation protection by the National Council on Radiation Protection and Measurements is published as *Reports of the National Council on Radiation Protection and Measurement.* These reports can be obtained from NCRP Publications, 7910 Woodmont Ave., Suite 1016, Bethesda, MD 20814.

Article 660 covers X-ray equipment in industrial facilities or similar locations, where it is commonly used for inspecting a process or product. This permits nondestructive testing without dismantling or applying stress to detect cracks, flaws, or structural defects. Welded joints are frequently inspected with X-ray equipment to detect hidden defects that can cause failure under stress.

The most common industrial application of X-rays is radiography, in which shadow pictures of the object are produced on photographic film. The type and thickness of the material involved govern the voltage to be employed, which can range from a few thousand volts (kV) to millions of volts (MV). Metal objects that are as much as 20 inches thick can be X-rayed.

Fluoroscopy is another X-ray technique used for industrial and commercial applications. Fluoroscopy is similar to radiography, but it operates at a much lower voltage (less than 250 kV). Instead of producing a film, it projects a shadow picture on a screen, similar to those used for security checks of luggage at airport terminals. Fluoroscopy is capable of detecting minute flaws or defects.

## 660.2 Definitions

**Long-Time Rating.** A rating based on an operating interval of 5 minutes or longer.

**Mobile.** X-ray equipment mounted on a permanent base with wheels and/or casters for moving while completely assembled.

**Momentary Rating.** A rating based on an operating interval that does not exceed 5 seconds.

**Portable.** X-ray equipment designed to be hand-carried.

**Transportable.** X-ray equipment that is to be installed in a vehicle or that may be readily disassembled for transport in a vehicle.

## 660.3 Hazardous (Classified) Locations

Unless identified for the location, X-ray and related equipment shall not be installed or operated in hazardous (classified) locations.

Informational Note: See Article 517, Part IV.

## 660.4 Connection to Supply Circuit

**(A) Fixed and Stationary Equipment.** Fixed and stationary X-ray equipment shall be connected to the power supply by means of a wiring method meeting the general requirements of this *Code*. Equipment properly supplied by a branch circuit rated at not over 30 amperes shall be permitted to be supplied through a suitable attachment plug cap and hard-service cable or cord.

**(B) Portable, Mobile, and Transportable Equipment.** Individual branch circuits shall not be required for portable, mobile, and transportable X-ray equipment requiring a capacity of not over 60 amperes. Portable and mobile types of X-ray equipment of any capacity shall be supplied through a suitable hard-service cable or cord. Transportable X-ray equipment of any capacity shall be permitted to be connected to its power supply by suitable connections and hard-service cable or cord.

**(C) Over 1000 Volts, Nominal.** Circuits and equipment operated at more than 1000 volts, nominal, shall comply with Article 490.

## 660.5 Disconnecting Means

A disconnecting means of adequate capacity for at least 50 percent of the input required for the momentary rating, or 100 percent of the input required for the long-time rating, of the X-ray equipment, whichever is greater, shall be provided in the supply circuit. The disconnecting means shall be operable from a location readily accessible from the X-ray control. For equipment connected to a 120-volt, nominal, branch circuit of 30 amperes or less, a grounding-type attachment plug cap and receptacle of proper rating shall be permitted to serve as a disconnecting means.

## 660.6 Rating of Supply Conductors and Overcurrent Protection

**(A) Branch-Circuit Conductors.** The ampacity of supply branch-circuit conductors and the overcurrent protective devices shall not be less than 50 percent of the momentary rating or 100 percent of the long-time rating, whichever is greater.

**(B) Feeder Conductors.** The rated ampacity of conductors and overcurrent devices of a feeder for two or more branch circuits supplying X-ray units shall not be less than 100 percent of the momentary demand rating [as determined by 660.6(A)] of the two largest X-ray apparatus plus 20 percent of the momentary ratings of other X-ray apparatus.

Informational Note: The minimum conductor size for branch and feeder circuits is also governed by voltage regulation requirements. For a specific installation, the manufacturer usually specifies minimum distribution transformer and conductor sizes, rating of disconnect means, and overcurrent protection.

## 660.7 Wiring Terminals

X-ray equipment not provided with a permanently attached cord or cord set shall be provided with suitable wiring terminals or leads for the connection of power-supply conductors of the size required by the rating of the branch circuit for the equipment.

## 660.9 Minimum Size of Conductors

Size 18 AWG or 16 AWG fixture wires, as specified in 725.49, and flexible cords shall be permitted for the control and operating circuits of X-ray and auxiliary equipment where protected by not larger than 20-ampere overcurrent devices.

## 660.10 Equipment Installations

All equipment for new X-ray installations and all used or reconditioned X-ray equipment moved to and reinstalled at a new location shall be of an approved type.

## II. Control

### 660.20 Fixed and Stationary Equipment

**(A) Separate Control Device.** A separate control device, in addition to the disconnecting means, shall be incorporated in the X-ray control supply or in the primary circuit to the high-voltage transformer. This device shall be a part of the X-ray equipment but shall be permitted in a separate enclosure immediately adjacent to the X-ray control unit.

A control device provides means for initiating and terminating X-ray exposures and automatically times their duration.

**(B) Protective Device.** A protective device, which shall be permitted to be incorporated into the separate control device, shall be provided to control the load resulting from failures in the high-voltage circuit.

### 660.21 Portable and Mobile Equipment

Portable and mobile equipment shall comply with 660.20, but the manually controlled device shall be located in or on the equipment.

### 660.23 Industrial and Commercial Laboratory Equipment

**(A) Radiographic and Fluoroscopic Types.** All radiographic- and fluoroscopic-type equipment shall be effectively enclosed or shall have interlocks that de-energize the equipment automatically to prevent ready access to live current-carrying parts.

**(B) Diffraction and Irradiation Types.** Diffraction- and irradiation-type equipment or installations not effectively enclosed or not provided with interlocks to prevent access to uninsulated live parts during operation shall be provided with a positive means to indicate when they are energized. The indicator shall be a pilot light, readable meter deflection, or equivalent means.

### 660.24 Independent Control

Where more than one piece of equipment is operated from the same high-voltage circuit, each piece or each group of equipment as a unit shall be provided with a high-voltage switch or equivalent disconnecting means. This disconnecting means shall be constructed, enclosed, or located so as to avoid contact by persons with its live parts.

## III. Transformers and Capacitors

### 660.35 General

Transformers and capacitors that are part of an X-ray equipment shall not be required to comply with Articles 450 and 460.

High-ratio step-up transformers that are an integral part of X-ray equipment are not required to comply with Article 450 and are generally used to provide the high voltage necessary for X-ray tubes. Because the fire hazard is lower due to the low primary voltage, X-ray transformers are not required to be installed in fire-resistant vaults.

### 660.36 Capacitors

Capacitors shall be mounted within enclosures of insulating material or grounded metal.

## IV. Guarding and Grounding

### 660.47 General

**(A) High-Voltage Parts.** All high-voltage parts, including X-ray tubes, shall be mounted within grounded enclosures. Air, oil, gas, or other suitable insulating media shall be used to insulate the high voltage from the grounded enclosure. The connection from the high-voltage equipment to X-ray tubes and other high-voltage components shall be made with high-voltage shielded cables.

**(B) Low-Voltage Cables.** Low-voltage cables connecting to oil-filled units that are not completely sealed, such as transformers, condensers, oil coolers, and high-voltage switches, shall have insulation of the oil-resistant type.

### 660.48 Grounding

Non–current-carrying metal parts of X-ray and associated equipment (controls, tables, X-ray tube supports, transformer tanks, shielded cables, X-ray tube heads, and so forth) shall be grounded in the manner specified in Article 250. Portable and mobile equipment shall be provided with an approved grounding-type attachment plug cap.

*Exception: Battery-operated equipment.*

Grounded enclosures are required to be provided for all high-voltage X-ray equipment, including X-ray tubes. Section 660.47(A) requires that high-voltage shielded cables be used to connect high-voltage equipment to X-ray tubes, and this section requires the shield to be grounded.

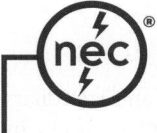

# ARTICLE 665 Induction and Dielectric Heating Equipment

## I. General

### 665.1 Scope

This article covers the construction and installation of dielectric heating, induction heating, induction melting, and induction welding equipment and accessories for industrial and scientific applications. Medical or dental applications, appliances, or line frequency pipeline and vessel heating are not covered in this article.

Informational Note: See Article 427, Part V, for line frequency induction heating of pipelines and vessels.

To prevent spurious radiation caused by induction and dielectric heating equipment, the Federal Communications Commission (FCC) has established rules that govern the use of this type of industrial heating equipment operating above 10 kHz (FCC, 47 CFR 18).

For further information on electric heating systems using an induction heater or a dielectric heater in ovens and furnaces, see NFPA 86, *Standard for Ovens and Furnaces*.

## 665.2 Definitions

**Applicator.** The device used to transfer energy between the output circuit and the object or mass to be heated.

**Converting Device.** That part of the heating equipment that converts input mechanical or electrical energy to the voltage, current, and frequency used for the heating applicator. A converting device consists of equipment using line frequency, all static multipliers, oscillator-type units using vacuum tubes, inverters using solid-state devices, or motor-generator equipment.

**Dielectric Heating.** Heating of a nominally insulating material due to its own dielectric losses when the material is placed in a varying electric field.

**Heating Equipment.** As used in this article, any equipment that is used for heating purposes and whose heat is generated by induction or dielectric methods.

•

**Induction Heating, Melting, and Welding.** The heating, melting, or welding of a nominally conductive material due to its own $I^2R$ losses when the material is placed in a varying electromagnetic field.

Induction and dielectric heating methods are used for ovens, furnaces, and industrial equipment where pieces of material are heated by a rapidly alternating magnetic or electric field.

### Theory of Operation – Solid-State Converter Power Circuit

A solid-state power converter consists of three sections: the rectifier section, the inverter section, and the output section, which includes the load coil and is usually located outside the power supply. Exhibit 665.1 is an example of an enclosed power supply for an induction heating process.

The rectifier section converts 3-phase, line frequency voltage to direct current. The output of the rectifier section supplies energy for the inverter section. The inverter section converts the energy to a variable frequency for the output circuit. The variable output frequency controls the power delivered to the load.

The output section consists of a capacitor in parallel (current fed) or in series (voltage fed) with a coil. Capacitance and inductance operate at a resonant frequency, and as the output frequency approaches this resonant frequency, the power to the load approaches its maximum. The output power is very low at minimum frequency.

***EXHIBIT 665.1*** *Enclosed power supply for an induction heating process. (Courtesy of Ajax Tocco Magnethermic, Park Ohio Industries)*

### Induction Heating

Induction heating occurs when an electrically conductive material (load) is placed in a varying magnetic field generated by a coil (inductor) around or adjacent to the workpiece to be heated. The varying magnetic field induces current in the load. Heat is generated by the resulting $I^2R$ losses in the load. Induction heating can be further subdivided into heating, melting, and welding.

Induction heating raises the temperature of the load to a temperature below its melting point, usually for the purposes of hardening, tempering, annealing, forging, extruding, or rolling. Frequencies used for heating range from about 50 Hz to 500 kHz. Power levels range from 5 kW to 42 MW.

Induction melting raises the temperature of the load to a temperature above its melting point, so the molten material can be alloyed, homogenized, and/or poured. Frequencies used for melting range from about 50 Hz to 10 kHz. Power levels range from 5 kW to 16.5 MW.

Induction welding is primarily used in the manufacture of welded pipe and tubing. In this process, a high-frequency current is passed through an induction coil in the proximity of the conducting metal surfaces to be joined. Selected portions are heated nearly instantaneously to the forging temperature, then are

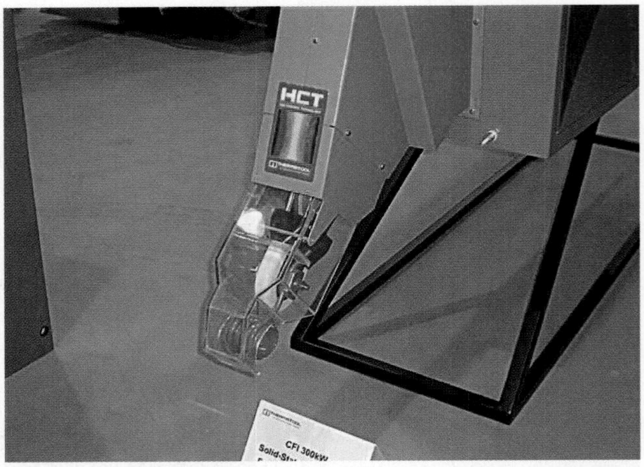


***EXHIBIT 665.2*** *A solid-state induction welding machine. (Courtesy of Thermatool Corp.)*

joined under pressure to produce a forge weld. Frequencies used for welding range from about 100 kHz to 800 kHz. Power levels range from 20 kW to 1 MW. Exhibit 665.2 shows an induction welding machine used in the manufacture of pipe and tubing.

**Dielectric Heating**

Dielectric heating equipment is similar to induction heating equipment, except that it is used to heat nonmetallic materials as opposed to metals. Typical applications include the drying of textiles after dyeing, drying of water-based coatings on paper, preheating of wood fibers for the medium-density fiberboard (MDF) industry, welding of plastic materials, and food processing.

At radio frequencies, the material to be heated forms a dielectric when placed between metal capacitor plates connected across the output of the generator. A high-frequency alternating electric field is created between the electrode plates. The molecules vibrate in the dielectric field, causing dissipation of energy through the material and frictional heating of the dielectric material. At higher (microwave) frequencies, a similar process occurs, but the generator is coupled to a resonant cavity into which the dielectric material is placed.

The frequency of operation of dielectric heating equipment is considerably higher than for induction heating. These machines operate at the assigned radio frequencies of 13.56 MHz, 27.12 MHz, and 40.68 MHz or at microwave frequencies of 915 MHz and 2450 MHz.

The majority of installed machines use vacuum tube generators, and powers range from 0.5 kW to 1 MW. Solid-state generators also have been installed, although power has been limited to 5 kW or less.

## 665.4 Hazardous (Classified) Locations

Heating equipment shall not be installed in hazardous (classified) locations as defined in Article 500 unless the equipment and wiring are designed and approved for the hazardous (classified) locations.

## 665.5 Output Circuit

The output circuit shall include all output components external to the converting device, including contactors, switches, busbars, and other conductors. The current flow from the output circuit to ground under operating and ground-fault conditions shall be limited to a value that does not cause 50 volts or more to ground to appear on any accessible part of the heating equipment and its load. The output circuit shall be permitted to be isolated from ground.

If the load (object being heated) accidentally comes in contact with the output coil, a voltage to ground will appear on the load, depending on the various impedances to ground of the coil and the load. If the voltage on the load is limited to less than 50 volts, guarding per 110.27(A) is not required. If the coil is isolated from ground and the load is grounded through an impedance that is low (less than 1 percent) relative to the coil impedance to ground, the voltage of the load to ground will be low no matter where the load contacts the coil.

In induction melting furnaces, an additional reason for isolating the coil from ground is to limit the fault current when a coil does go to ground. Limiting the fault current prevents severe damage to the water-cooled coil, resulting in a water leak and the potential for a water–molten metal explosion. If water is trapped under molten metal, the rapid transfer of heat to the water causes the water to turn almost instantly into steam. The resulting 1600-to-1 expansion of the steam results in the ejection of molten metal from the furnace.

## 665.7 Remote Control

**(A) Multiple Control Points.** Where multiple control points are used for applicator energization, a means shall be provided and interlocked so that the applicator can be energized from only one control point at a time. A means for de-energizing the applicator shall be provided at each control point.

**(B) Foot Switches.** Switches operated by foot pressure shall be provided with a shield over the contact button to avoid accidental closing of a foot switch.

## 665.10 Ampacity of Supply Conductors

The ampacity of supply conductors shall be determined by 665.10(A) or (B).

**(A) Nameplate Rating.** The ampacity of conductors supplying one or more pieces of equipment shall be not less than the sum of the nameplate ratings for the largest group of machines capable of simultaneous operation, plus 100 percent of the standby currents of the remaining machines. Where standby currents are not given on the nameplate, the nameplate rating shall be used as the standby current.

**(B) Motor-Generator Equipment.** The ampacity of supply conductors for motor-generator equipment shall be determined in accordance with Article 430, Part II.

### 665.11 Overcurrent Protection

Overcurrent protection for the heating equipment shall be provided as specified in Article 240. This overcurrent protection shall be permitted to be provided separately or as a part of the equipment.

### 665.12 Disconnecting Means

A readily accessible disconnecting means shall be provided to disconnect each heating equipment from its supply circuit. The disconnecting means shall be located within sight from the controller or be lockable open in accordance with 110.25.

The rating of this disconnecting means shall not be less than the nameplate rating of the heating equipment. Motor-generator equipment shall comply with Article 430, Part IX. The supply circuit disconnecting means shall be permitted to serve as the heating equipment disconnecting means where only one heating equipment is supplied.

## II. Guarding, Grounding, and Labeling

### 665.19 Component Interconnection

The interconnection components required for a complete heating equipment installation shall be guarded.

### 665.20 Enclosures

The converting device (excluding the component interconnections) shall be completely contained within an enclosure(s) of noncombustible material.

### 665.21 Control Panels

All control panels shall be of dead-front construction.

### 665.22 Access to Internal Equipment

Access doors or detachable access panels shall be employed for internal access to heating equipment. Access doors to internal compartments containing equipment employing voltages from 150 volts to 1000 volts ac or dc shall be capable of being locked closed or shall be interlocked to prevent the supply circuit from being energized while the door(s) is open. The provision for locking or adding a lock to the access doors shall be installed on or at the access door and shall remain in place with or without the lock installed.

Access doors to internal compartments containing equipment employing voltages exceeding 1000 volts ac or dc shall be provided with a disconnecting means equipped with mechanical lockouts to prevent access while the heating equipment is energized, or the access doors shall be capable of being locked closed and interlocked to prevent the supply circuit from being energized while the door(s) is open. Detachable panels not normally used for access to such parts shall be fastened in a manner that makes them inconvenient to remove.

### 665.23 Warning Labels or Signs

Warning labels or signs that read "DANGER — HIGH VOLTAGE — KEEP OUT" shall be attached to the equipment and shall be plainly visible where persons might come in contact with energized parts when doors are open or closed or when panels are removed from compartments containing over 150 volts ac or dc. The warning sign(s) or label(s) shall comply with 110.21(B).

### 665.24 Capacitors

The time and means of discharge shall be in accordance with 460.6 for capacitors rated 600 volts, nominal, and under. The time and means of discharge shall be in accordance with 460.28 for capacitors rated over 600 volts, nominal. Capacitor internal pressure switches connected to a circuit-interrupter device shall be permitted for capacitor overcurrent protection.

Enhanced protection against rupture of capacitor cases is needed when capacitors are operated at the higher frequencies used for induction and dielectric heating. A high-resistance fault condition can cause case pressure to build up inside the capacitor over a very short time. Capacitor internal pressure switches are the preferred method to detect this type of failure.

Consider a 5000-kVAR, 2500-V, 300-Hz capacitor. Nominal current is 2000 amperes. A "high-resistance" fault of 10 ohms results in 250 amperes of resistive current, or a total capacitor current of 2016 amperes rms. This small increase in rms current will not result in the opening of an overcurrent device even though 625 kW of thermal energy is being generated inside the capacitor, which is designed to dissipate about 1.5 kW of losses.

### 665.25 Dielectric Heating Applicator Shielding

Protective cages or adequate shielding shall be used to guard dielectric heating applicators. Interlock switches shall be used on all hinged access doors, sliding panels, or other easy means of access to the applicator. All interlock switches shall be connected in such a manner as to remove all power from the applicator when any one of the access doors or panels is open.

### 665.26 Grounding and Bonding

Bonding to the equipment grounding conductor or inter-unit bonding, or both, shall be used wherever required for circuit operation, and for limiting to a safe value radio frequency voltages between all exposed non–current-carrying parts of the equipment and earth ground, between all equipment parts and surrounding objects, and between such objects and earth ground. Such connection to the equipment grounding conductor and bonding shall be installed in accordance with Article 250, Parts II and V.

Informational Note: Under certain conditions, contact between the object being heated and the applicator results in an unsafe condition, such as eruption of heated materials. Grounding of the object being heated and ground detection can be used to prevent this unsafe condition.

Because of stray currents between units of equipment or between equipment and the ground, bonding presents special problems at radio frequencies. Special bonding requirements are particularly needed at dielectric heating frequencies (100 to 200 MHz) because of the differences in radio frequency potential that can exist between the equipment and surrounding metal units or other units of the installation. Bonding has been accomplished by placing all units of the equipment on a flooring or base consisting of a copper or aluminum sheet, then thoroughly bonding by soldering, welding, or bolting. Such special bonding holds the radio frequency resistance and reactance between units to a minimum, and any stray circulating currents flowing through the bonding will not cause a dangerous voltage drop.

The operator can be protected from high radio frequency potentials by shielding at dielectric heating frequencies. Interference with radio communications systems at such high frequencies can be eliminated by totally enclosing all components in a shielding of copper or aluminum.

### 665.27 Marking

Each heating equipment shall be provided with a nameplate giving the manufacturer's name and model identification and the following input data: line volts, frequency, number of phases, maximum current, full-load kilovolt-amperes (kVA), and full-load power factor. Additional data shall be permitted.

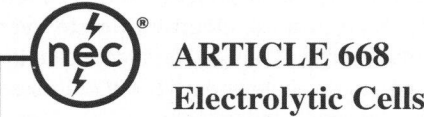

## ARTICLE 668 Electrolytic Cells

### 668.1 Scope

The provisions of this article apply to the installation of the electrical components and accessory equipment of electrolytic cells, electrolytic cell lines, and process power supply for the production of aluminum, cadmium, chlorine, copper, fluorine, hydrogen peroxide, magnesium, sodium, sodium chlorate, and zinc.

Not covered by this article are cells used as a source of electric energy and for electroplating processes and cells used for the production of hydrogen.

Informational Note No. 1: In general, any cell line or group of cell lines operated as a unit for the production of a particular metal, gas, or chemical compound may differ from any other cell lines producing the same product because of variations in the particular raw materials used, output capacity, use of proprietary methods or process practices, or other modifying factors to the extent that detailed *Code* requirements become overly restrictive and do not accomplish the stated purpose of this *Code*.

Informational Note No. 2: For further information, see IEEE 463-1993, *Standard for Electrical Safety Practices in Electrolytic Cell Line Working Zones*.

Within a cell line working zone, both an electrolytic cell line and its dc process power-supply circuit are treated as an individual machine supplied from a single source, even though they might cover acres of space, have a load current in excess of 400,000 amperes dc, or have a circuit voltage in excess of 1000 volts dc. The cell line process current passes through each cell in a series connection, and the load current cannot be subdivided the way it can in the heating circuit of a resistance-type electric furnace. Because a cell line is supplied by its individual dc rectifier system, the rectifier or the entire cell line circuit is de-energized by removing its primary power source.

In some electrolytic cell systems, the terminal voltage of the process supply can be appreciable. The voltage to ground of exposed live parts from one end of a cell line to the other is variable between the limits of the terminal voltage. Hence, operating and maintenance personnel and their tools are required to be insulated from ground.

### 668.2 Definitions

**Cell Line.** An assembly of electrically interconnected electrolytic cells supplied by a source of direct-current power.

**Cell Line Attachments and Auxiliary Equipment.** As applied to this article, a term that includes, but is not limited to, auxiliary tanks; process piping; ductwork; structural supports; exposed cell line conductors; conduits and other raceways; pumps, positioning equipment, and cell cutout or bypass electrical devices. Auxiliary equipment includes tools, welding machines, crucibles, and other portable equipment used for operation and maintenance within the electrolytic cell line working zone.

In the cell line working zone, auxiliary equipment includes the exposed conductive surfaces of ungrounded cranes and crane-mounted cell-servicing equipment.

**Electrically Connected.** A connection capable of carrying current as distinguished from connection through electromagnetic induction.

**Electrolytic Cell.** A tank or vat in which electrochemical reactions are caused by applying electric energy for the purpose of refining or producing usable materials.

**Electrolytic Cell Line Working Zone.** The space envelope wherein operation or maintenance is normally performed on or in the vicinity of exposed energized surfaces of electrolytic cell lines or their attachments.

### 668.3 Other Articles

**(A) Lighting, Ventilating, Material Handling.** Chapters 1 through 4 shall apply to services, feeders, branch circuits, and apparatus for supplying lighting, ventilating, material handling, and the like that are outside the electrolytic cell line working zone.

**(B) Systems Not Electrically Connected.** Those elements of a cell line power-supply system that are not electrically connected to the cell supply system, such as the primary winding of a two-winding transformer, the motor of a motor-generator set, feeders, branch circuits, disconnecting means, motor controllers, and overload protective equipment, shall be required to comply with all applicable provisions of this *Code*.

**(C) Electrolytic Cell Lines.** Electrolytic cell lines shall comply with the provisions of Chapters 1 through 4 except as amended in 668.3(C)(1) through (C)(4).

**(1) Conductors.** The electrolytic cell line conductors shall not be required to comply with the provisions of Articles 110, 210, 215, 220, and 225. See 668.11.

**(2) Overcurrent Protection.** Overcurrent protection of electrolytic cell dc process power circuits shall not be required to comply with the requirements of Article 240.

**(3) Grounding.** Equipment located or used within the electrolytic cell line working zone or associated with the cell line direct-current power circuits shall not be required to comply with the provisions of Article 250.

**(4) Working Zone.** The electrolytic cells, cell line attachments, and the wiring of auxiliary equipment and devices within the cell line working zone shall not be required to comply with the provisions of Articles 110, 210, 215, 220, and 225. See 668.30.

Informational Note: See 668.15 for equipment, apparatus, and structural component grounding.

## 668.10 Cell Line Working Zone

**(A) Area Covered.** The space envelope of the cell line working zone shall encompass spaces that meet any of the following conditions:

(1) Is within 2.5 m (96 in.) above energized surfaces of electrolytic cell lines or their energized attachments
(2) Is below energized surfaces of electrolytic cell lines or their energized attachments, provided the headroom in the space beneath is less than 2.5 m (96 in.)
(3) Is within 1.0 m (42 in.) horizontally from energized surfaces of electrolytic cell lines or their energized attachments or from the space envelope described in 668.10(A)(1) or (A)(2)

**(B) Area Not Covered.** The cell line working zone shall not be required to extend through or beyond walls, floors, roofs, partitions, barriers, or the like.

## 668.11 Direct-Current Cell Line Process Power Supply

**(A) Not Grounded.** The direct-current cell line process power-supply conductors shall not be required to be grounded.

**(B) Metal Enclosures Grounded.** All metal enclosures of power-supply apparatus for the direct-current cell line process operating at a power-supply potential between terminals of over 50 volts shall be grounded by either of the following means:

(1) Through protective relaying equipment
(2) By a minimum 2/0 AWG copper grounding conductor or a conductor of equal or greater conductance

**(C) Grounding Requirements.** The grounding connections required by 668.11(B) shall be installed in accordance with 250.8, 250.10, 250.12, 250.68, and 250.70.

## 668.12 Cell Line Conductors

**(A) Insulation and Material.** Cell line conductors shall be either bare, covered, or insulated and of copper, aluminum, copper-clad aluminum, steel, or other suitable material.

**(B) Size.** Cell line conductors shall be of such cross-sectional area that the temperature rise under maximum load conditions and at maximum ambient shall not exceed the safe operating temperature of the conductor insulation or the material of the conductor supports.

**(C) Connections.** Cell line conductors shall be joined by bolted, welded, clamped, or compression connectors.

## 668.13 Disconnecting Means

**(A) More Than One Process Power Supply.** Where more than one direct-current cell line process power supply serves the same cell line, a disconnecting means shall be provided on the cell line circuit side of each power supply to disconnect it from the cell line circuit.

**(B) Removable Links or Conductors.** Removable links or removable conductors shall be permitted to be used as the disconnecting means.

## 668.14 Shunting Means

**(A) Partial or Total Shunting.** Partial or total shunting of cell line circuit current around one or more cells shall be permitted.

**(B) Shunting One or More Cells.** The conductors, switches, or combination of conductors and switches used for shunting one or more cells shall comply with the applicable requirements of 668.12.

## 668.15 Grounding

For equipment, apparatus, and structural components that are required to be grounded by provisions of Article 668, the provisions of Article 250 shall apply, except a water pipe electrode shall not be required to be used. Any electrode or combination of electrodes described in 250.52 shall be permitted.

## 668.20 Portable Electrical Equipment

**(A) Portable Electrical Equipment Not to Be Grounded.** The frames and enclosures of portable electrical equipment used within the cell line working zone shall not be grounded.

*Exception No. 1: Where the cell line voltage does not exceed 200 volts dc, these frames and enclosures shall be permitted to be grounded.*

*Exception No. 2: These frames and enclosures shall be permitted to be grounded where guarded.*

**(B) Isolating Transformers.** Electrically powered, hand-held, cord-connected portable equipment with ungrounded frames or enclosures used within the cell line working zone shall be connected to receptacle circuits that have only ungrounded conductors such as a branch circuit supplied by an isolating transformer with an ungrounded secondary.

**(C) Marking.** Ungrounded portable electrical equipment shall be distinctively marked and shall employ plugs and receptacles of a configuration that prevents connection of this equipment to grounding receptacles and that prevents inadvertent interchange of ungrounded and grounded portable electrical equipments.

## 668.21 Power-Supply Circuits and Receptacles for Portable Electrical Equipment

**(A) Isolated Circuits.** Circuits supplying power to ungrounded receptacles for hand-held, cord-connected equipment shall be electrically isolated from any distribution system supplying areas other than the cell line working zone and shall be ungrounded. Power for these circuits shall be supplied through isolating transformers. Primaries of such transformers shall operate at not more than 1000 volts between conductors and shall be provided with proper overcurrent protection. The secondary voltage of such transformers shall not exceed 300 volts between conductors, and all circuits supplied from such secondaries shall be ungrounded and shall have an approved overcurrent device of proper rating in each conductor.

**(B) Noninterchangeability.** Receptacles and their mating plugs for ungrounded equipment shall not have provision for a grounding conductor and shall be of a configuration that prevents their use for equipment required to be grounded.

**(C) Marking.** Receptacles on circuits supplied by an isolating transformer with an ungrounded secondary shall be a distinctive configuration, shall be distinctively marked, and shall not be used in any other location in the plant.

## 668.30 Fixed and Portable Electrical Equipment

**(A) Electrical Equipment Not Required to Be Grounded.** Alternating-current systems supplying fixed and portable electrical equipment within the cell line working zone shall not be required to be grounded.

**(B) Exposed Conductive Surfaces Not Required to Be Grounded.** Exposed conductive surfaces, such as electrical equipment housings, cabinets, boxes, motors, raceways, and the like, that are within the cell line working zone shall not be required to be grounded.

**(C) Wiring Methods.** Auxiliary electrical equipment such as motors, transducers, sensors, control devices, and alarms, mounted on an electrolytic cell or other energized surface, shall be connected to premises wiring systems by any of the following means:

(1) Multiconductor hard usage cord.
(2) Wire or cable in suitable raceways or metal or nonmetallic cable trays. If metal conduit, cable tray, armored cable, or similar metallic systems are used, they shall be installed with insulating breaks such that they do not cause a potentially hazardous electrical condition.

**(D) Circuit Protection.** Circuit protection shall not be required for control and instrumentation that are totally within the cell line working zone.

**(E) Bonding.** Bonding of fixed electrical equipment to the energized conductive surfaces of the cell line, its attachments, or auxiliaries shall be permitted. Where fixed electrical equipment is mounted on an energized conductive surface, it shall be bonded to that surface.

## 668.31 Auxiliary Nonelectrical Connections

Auxiliary nonelectrical connections, such as air hoses, water hoses, and the like, to an electrolytic cell, its attachments, or auxiliary equipment shall not have continuous conductive reinforcing wire, armor, braids, and the like. Hoses shall be of a nonconductive material.

## 668.32 Cranes and Hoists

**(A) Conductive Surfaces to Be Insulated from Ground.** The conductive surfaces of cranes and hoists that enter the cell line working zone shall not be required to be grounded. The portion of an overhead crane or hoist that contacts an energized electrolytic cell or energized attachments shall be insulated from ground.

**(B) Hazardous Electrical Conditions.** Remote crane or hoist controls that could introduce hazardous electrical conditions into the cell line working zone shall employ one or more of the following systems:

(1) Isolated and ungrounded control circuit in accordance with 668.21(A)
(2) Nonconductive rope operator
(3) Pendant pushbutton with nonconductive supporting means and having nonconductive surfaces or ungrounded exposed conductive surfaces
(4) Radio

## 668.40 Enclosures

General-purpose electrical equipment enclosures shall be permitted where a natural draft ventilation system prevents the accumulation of gases.

# ARTICLE 669 Electroplating

## 669.1 Scope

The provisions of this article apply to the installation of the electrical components and accessory equipment that supply the power and controls for electroplating, anodizing, electropolishing, and electrostripping. For purposes of this article, the term *electroplating* shall be used to identify any or all of these processes.

Because of the extremely high currents and low voltages normally involved, conventional wiring methods cannot be used in electroplating, anodizing, electropolishing, and electrostripping processes. Section 669.6 permits the use of bare conductors even in systems exceeding 50 volts dc. Some systems in the aluminum anodizing process have potentials up to 240 volts. Warning signs are required to be posted to indicate the presence of bare conductors.

## 669.3 General

Equipment for use in electroplating processes shall be identified for such service.

## 669.5 Branch-Circuit Conductors

Branch-circuit conductors supplying one or more units of equipment shall have an ampacity of not less than 125 percent of the total connected load. The ampacities for busbars shall be in accordance with 366.23.

## 669.6 Wiring Methods

Conductors connecting the electrolyte tank equipment to the conversion equipment shall be in accordance with 669.6(A) and (B).

**(A) Systems Not Exceeding 50 Volts Direct Current.** Insulated conductors shall be permitted to be run without insulated support, provided they are protected from physical damage. Bare copper or aluminum conductors shall be permitted where supported on insulators.

**(B) Systems Exceeding 50 Volts Direct Current.** Insulated conductors shall be permitted to be run on insulated supports, provided they are protected from physical damage. Bare copper or aluminum conductors shall be permitted where supported on insulators and guarded against accidental contact up to the point of termination in accordance with 110.27.

## 669.7 Warning Signs

Warning signs shall be posted to indicate the presence of bare conductors. The warning sign(s) or label(s) shall comply with 110.21(B).

## 669.8 Disconnecting Means

**(A) More Than One Power Supply.** Where more than one power supply serves the same dc system, a disconnecting means shall be provided on the dc side of each power supply.

**(B) Removable Links or Conductors.** Removable links or removable conductors shall be permitted to be used as the disconnecting means.

## 669.9 Overcurrent Protection

Direct-current conductors shall be protected from overcurrent by one or more of the following:

(1) Fuses or circuit breakers
(2) A current-sensing device that operates a disconnecting means
(3) Other approved means

# ARTICLE 670 Industrial Machinery

## 670.1 Scope

This article covers the definition of, the nameplate data for, and the size and overcurrent protection of supply conductors to industrial machinery.

> Informational Note No. 1: For further information, see NFPA 79-2012, *Electrical Standard for Industrial Machinery*.

The equipment and wiring of industrial machinery, for which different component parts may be purchased and assembled at the location of use, must be installed in accordance with the applicable articles in the *NEC*. Machinery assembled by the manufacturer, in accordance with NFPA 79, *Electrical Standard for Industrial Machinery*, then disassembled for shipping and reassembled at its place of use, comes only under Article 670 and any *NEC* sections referenced herein. In this case, the machinery is treated as a package unit.

The information to be included on the nameplate allows for proper conductor sizing, overcurrent protection of the feeder or branch circuit supplying the industrial machine, and integration of the machine into the facility electrical system.

> Informational Note No. 2: For information on the workspace requirements for equipment containing supply conductor terminals, see 110.26. For information on the workspace requirements for machine power and control equipment, see NFPA 79-2012, *Electrical Standard for Industrial Machinery*.

Working clearances around control equipment enclosures and compartments containing equipment operating at 600 volts or less, that are an integral part of an industrial machine, are contained in Section 11.5 in NFPA 79. The requirements of NFPA 79 closely parallel those found in 110.26(A), but some provisions in NFPA 79 allow smaller clearances under very specific conditions of operation and equipment construction.

## 670.2 Definition

**Industrial Machinery (Machine).** A power-driven machine (or a group of machines working together in a coordinated manner), not portable by hand while working, that is used to process material by cutting; forming; pressure; electrical, thermal, or optical techniques; lamination; or a combination of these processes. It can include associated equipment used to transfer material or tooling, including fixtures, to assemble/disassemble, to inspect or test, or to package. [The associated electrical equipment, including the logic controller(s) and associated software or logic together with the machine actuators and sensors, are considered as part of the industrial machine.]

This definition permits the inclusion of other types of industrial machines without the need to continuously modify the scope of Article 670 and NFPA 79. Also, the scope and definition are in harmony with IEC 60204-1, *Safety of Machinery – Electrical Equipment of Machines – Part 1: General Requirements*. Exhibit 670.1 shows an example of an industrial machine.

## 670.3 Machine Nameplate Data

**(A) Permanent Nameplate.** A permanent nameplate shall be attached to the control equipment enclosure or machine and shall be plainly visible after installation. The nameplate shall include the following information:

(1) Supply voltage, number of phases, frequency, and full-load current
(2) Maximum ampere rating of the short-circuit and ground-fault protective device
(3) Ampere rating of largest motor, from the motor nameplate, or load
(4) Short-circuit current rating of the machine industrial control panel based on one of the following:

***EXHIBIT 670.1*** *Thread-spooling machine used in the textile industry. (Courtesy of the International Association of Electrical Inspectors)*

a. Short-circuit current rating of a listed and labeled machine control enclosure or assembly
b. Short-circuit current rating established utilizing an approved method

Informational Note: UL 508A-2001, Supplement SB, is an example of an approved method.

(5) Electrical diagram number(s) or the number of the index to the electrical drawings

The full-load current shown on the nameplate shall not be less than the sum of the full-load currents required for all motors and other equipment that may be in operation at the same time under normal conditions of use. Where unusual type loads, duty cycles, and so forth require oversized conductors or permit reduced-size conductors, the required capacity shall be included in the marked "full-load current." Where more than one incoming supply circuit is to be provided, the nameplate shall state the preceding information for each circuit.

Informational Note: See 430.22(E) and 430.26 for duty cycle requirements.

An industrial machine's nameplate must provide the short-circuit current rating of the machine's industrial control panel. That rating is established either as part of the listing of the control enclosure or assembly or, for assemblies that are not listed, by an approved method of determining the short-circuit current rating.

In the absence of product listing, Supplement SB to UL 508A, *Standard for Industrial Control Panels*, is referred to as one example of a method for determining the short-circuit current rating of a control panel or assembly that could be used as a basis for equipment approval.

The second paragraph of 670.3(A) recognizes that the operating characteristics of an industrial machine may permit the use of a feeder demand factor, which is covered in 430.26. An example of this is an industrial machine containing motors sized for high torque but, in normal operation, run at close to no-load current values. In this case, it may be appropriate to reduce the full-load current marking on the machine nameplate.

**(B) Overcurrent Protection.** Where overcurrent protection is provided in accordance with 670.4(C), the machine shall be marked "overcurrent protection provided at machine supply terminals."

## 670.4 Supply Conductors and Overcurrent Protection

**(A) Size.** The size of the supply conductor shall be such as to have an ampacity not less than 125 percent of the full-load current rating of all resistance heating loads plus 125 percent of the full-load current rating of the highest rated motor plus the sum of the full-load current ratings of all other connected motors and apparatus, based on their duty cycle, that may be in operation at the same time.

Informational Note No. 1: See Table 310.15(B)(16) through Table 310.15(B)(20) for ampacity of conductors rated 1000 V and below.

Informational Note No. 2: See 430.22(E) and 430.26 for duty cycle requirements.

The duty cycle of motors and apparatus must be considered when determining the minimum ampacity of a supply circuit conductor for an industrial machine. Depending on the operating characteristics of the motor, the duty cycle of the apparatus might not always result in reduction of the supply conductor ampacity. Where motors are used in other than a continuous-duty mode of operation, Table 430.22(E) provides percentages by which the full-load current of a given motor is increased or decreased for the purpose of sizing motor circuit conductors. A motor that is loaded continuously under any conditions of use is an example of a continuous-duty application.

**(B) Disconnecting Means.** A machine shall be considered as an individual unit and therefore shall be provided with disconnecting means. The disconnecting means shall be permitted to be supplied by branch circuits protected by either fuses or circuit breakers. The disconnecting means shall not be required to incorporate overcurrent protection.

In regard to the machine disconnecting means, NFPA 79 states, in part:

> The center of the grip of the operating handle of the disconnecting means, when in its highest position, shall be not more than 2.0 m (6 ft 7 in.) above the servicing level. A permanent operating platform, readily accessible by means of a permanent stair or ladder, shall be considered the servicing level for the purpose of this requirement.
>
> [The disconnecting means shall] be provided with a permanent means permitting it to be locked in the off (open) position only (e.g., by padlocks) independent of the door position. When so locked, remote as well as local closing shall be prevented.

**(C) Overcurrent Protection.** Where furnished as part of the machine, overcurrent protection for each supply circuit shall consist of a single circuit breaker or set of fuses, the machine shall bear the marking required in 670.3, and the supply conductors shall be considered either as feeders or as taps as covered by 240.21.

The rating or setting of the overcurrent protective device for the circuit supplying the machine shall not be greater than the sum of the largest rating or setting of the branch-circuit short-circuit and ground-fault protective device provided with the machine, plus 125 percent of the full-load current rating of all resistance heating loads, plus the sum of the full-load currents of all other motors and apparatus that could be in operation at the same time.

*Exception: Where one or more instantaneous trip circuit breakers or motor short-circuit protectors are used for motor branch-circuit short-circuit and ground-fault protection as permitted by 430.52(C), the procedure specified in 670.4(C) for determining the maximum rating of the protective device for the circuit supplying the machine shall apply with the following provision: For the purpose of the calculation, each instantaneous trip circuit breaker or motor short-circuit protector shall be assumed to have a rating not exceeding the maximum percentage of motor full-load current permitted by Table 430.52 for the type of machine supply circuit protective device employed.*

Where no branch-circuit short-circuit and ground-fault protective device is provided with the machine, the rating or setting of the overcurrent protective device shall be based on 430.52 and 430.53, as applicable.

The nameplate provides the necessary information to size the branch-circuit or feeder conductors, the machine disconnecting means, and overcurrent protection. The computation of motor and nonmotor loads is reflected on the nameplate as full-load amperes, and no further calculation is necessary. Sizing of circuit conductors and overcurrent protection beyond the machine disconnecting means is under the scope of NFPA 79.

### 670.5 Short-Circuit Current Rating

Industrial machinery shall not be installed where the available fault current exceeds its short-circuit current rating as marked in accordance with 670.3(A)(4).

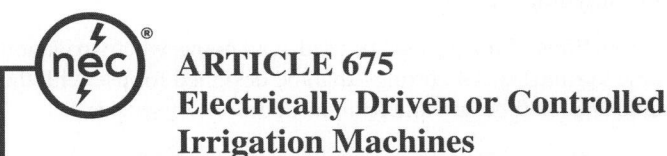

# ARTICLE 675 Electrically Driven or Controlled Irrigation Machines

## I. General

### 675.1 Scope

The provisions of this article apply to electrically driven or controlled irrigation machines, and to the branch circuits and controllers for such equipment.

Electric pump motors used to supply water to irrigation machines are governed by the general requirements of the *NEC* and not by Article 675.

### 675.2 Definitions

**Center Pivot Irrigation Machine.** A multimotored irrigation machine that revolves around a central pivot and employs alignment switches or similar devices to control individual motors.

**Collector Rings.** An assembly of slip rings for transferring electric energy from a stationary to a rotating member.

**Irrigation Machine.** An electrically driven or controlled machine, with one or more motors, not hand-portable, and used primarily to transport and distribute water for agricultural purposes.

## 675.4 Irrigation Cable

**(A) Construction.** The cable used to interconnect enclosures on the structure of an irrigation machine shall be an assembly of stranded, insulated conductors with nonhygroscopic and nonwicking filler in a core of moisture- and flame-resistant nonmetallic material overlaid with a metallic covering and jacketed with a moisture-, corrosion-, and sunlight-resistant nonmetallic material.

The conductor insulation shall be of a type listed in Table 310.104(A) for an operating temperature of 75°C (167°F) and for use in wet locations. The core insulating material thickness shall not be less than 0.76 mm (30 mils), and the metallic overlay thickness shall be not less than 0.20 mm (8 mils). The jacketing material thickness shall be not less than 1.27 mm (50 mils).

A composite of power, control, and grounding conductors in the cable shall be permitted.

**(B) Alternate Wiring Methods.** Installation of other listed cables complying with the construction requirements of 675.4(A) shall be permitted.

**(C) Supports.** Irrigation cable shall be secured by straps, hangers, or similar fittings identified for the purpose and so installed as not to damage the cable. Cable shall be supported at intervals not exceeding 1.2 m (4 ft).

**(D) Fittings.** Fittings shall be used at all points where irrigation cable terminates. The fittings shall be designed for use with the cable and shall be suitable for the conditions of service.

## 675.5 More Than Three Conductors in a Raceway or Cable

The signal and control conductors of a raceway or cable shall not be counted for the purpose of ampacity adjustment as required in 310.15(B)(3)(a).

## 675.6 Marking on Main Control Panel

The main control panel shall be provided with a nameplate that shall give the following information:

(1) The manufacturer's name, the rated voltage, the phase, and the frequency
(2) The current rating of the machine
(3) The rating of the main disconnecting means and size of overcurrent protection required

## 675.7 Equivalent Current Ratings

Where intermittent duty is not involved, the provisions of Article 430 shall be used for determining ratings for controllers, disconnecting means, conductors, and the like. Where irrigation machines have inherent intermittent duty, the determinations of equivalent current ratings in 675.7(A) and (B) shall be used.

**(A) Continuous-Current Rating.** The equivalent continuous-current rating for the selection of branch-circuit conductors and overcurrent protection shall be equal to 125 percent of the motor nameplate full-load current rating of the largest motor, plus a quantity equal to the sum of each of the motor nameplate full-load current ratings of all remaining motors on the circuit, multiplied by the maximum percent duty cycle at which they can continuously operate.

**(B) Locked-Rotor Current.** The equivalent locked-rotor current rating shall be equal to the numerical sum of the locked-rotor current of the two largest motors plus 100 percent of the sum of the motor nameplate full-load current ratings of all the remaining motors on the circuit.

## 675.8 Disconnecting Means

**(A) Main Controller.** A controller that is used to start and stop the complete machine shall meet all of the following requirements:

(1) An equivalent continuous current rating not less than specified in 675.7(A) or 675.22(A)
(2) A horsepower rating not less than the value from Table 430.251(A) and Table 430.251(B), based on the equivalent locked-rotor current specified in 675.7(B) or 675.22(B)

*Exception: A listed molded case switch shall not require a horsepower rating.*

A molded case switch used as a motor controller is not required to have a horsepower rating, but it is required to have a continuous current (ampere) rating not less than that specified by 675.7(A) or 675.22(A).

**(B) Main Disconnecting Means.** The main disconnecting means for the machine shall provide overcurrent protection, shall be at the point of connection of electric power to the machine, or shall be in sight from the machine, and it shall be readily accessible and lockable in accordance with 110.25. This disconnecting means shall have a horsepower and current rating not less than required for the main controller.

The main disconnecting means is permitted to be up to 50 feet from the machine but must be readily accessible and capable of being locked in the open position. This eliminates one set of OCPDs and one disconnecting means where the circuit originates at the motor control panel for the irrigation pump and the panel is located within 50 feet of the center pivot machine. It also alleviates some potential problems with machines designed to be towed to a second site.

*Exception No. 1: Circuit breakers without marked horsepower ratings shall be permitted in accordance with 430.109.*

*Exception No. 2: A listed molded case switch without marked horsepower ratings shall be permitted.*

**(C) Disconnecting Means for Individual Motors and Controllers.** A disconnecting means shall be provided to simultaneously disconnect all ungrounded conductors for each motor and controller and shall be located as required by Article 430, Part IX. The disconnecting means shall not be required to be readily accessible.

See the commentary following 430.103 regarding motor disconnects.

### 675.9 Branch-Circuit Conductors

The branch-circuit conductors shall have an ampacity not less than specified in 675.7(A) or 675.22(A).

### 675.10 Several Motors on One Branch Circuit

**(A) Protection Required.** Several motors, each not exceeding 2 hp rating, shall be permitted to be used on an irrigation machine circuit protected at not more than 30 amperes at 1000 volts, nominal, or less, provided all of the following conditions are met:

(1) The full-load rating of any motor in the circuit shall not exceed 6 amperes.
(2) Each motor in the circuit shall have individual overload protection in accordance with 430.32.
(3) Taps to individual motors shall not be smaller than 14 AWG copper and not more than 7.5 m (25 ft) in length.

The requirements for this special equipment application are a modified version of those in 430.53.

**(B) Individual Protection Not Required.** Individual branch-circuit short-circuit protection for motors and motor controllers shall not be required where the requirements of 675.10(A) are met.

### 675.11 Collector Rings

**(A) Transmitting Current for Power Purposes.** Collector rings shall have a current rating not less than 125 percent of the full-load current of the largest device served plus the full-load current of all other devices served, or as determined from 675.7(A) or 675.22(A).

**(B) Control and Signal Purposes.** Collector rings for control and signal purposes shall have a current rating not less than 125 percent of the full-load current of the largest device served plus the full-load current of all other devices served.

**(C) Grounding.** The collector ring used for grounding shall have a current rating not less than that sized in accordance with 675.11(A).

**(D) Protection.** Collector rings shall be protected from the expected environment and from accidental contact by means of a suitable enclosure.

### 675.12 Grounding

The following equipment shall be grounded:

(1) All electrical equipment on the irrigation machine
(2) All electrical equipment associated with the irrigation machine
(3) Metal junction boxes and enclosures
(4) Control panels or control equipment that supplies or controls electrical equipment to the irrigation machine

*Exception: Grounding shall not be required on machines where all of the following provisions are met:*

*(a) The machine is electrically controlled but not electrically driven.*
*(b) The control voltage is 30 volts or less.*
*(c) The control or signal circuits are current limited as specified in Chapter 9, Tables 11(A) and 11(B).*

### 675.13 Methods of Grounding

Machines that require grounding shall have a non–current-carrying equipment grounding conductor provided as an integral part of each cord, cable, or raceway. This grounding conductor shall be sized not less than the largest supply conductor in each cord, cable, or raceway. Feeder circuits supplying power to irrigation machines shall have an equipment grounding conductor sized according to Table 250.122.

### 675.14 Bonding

Where electrical grounding is required on an irrigation machine, the metallic structure of the machine, metallic conduit, or metallic sheath of cable shall be connected to the grounding conductor. Metal-to-metal contact with a part that is connected to the grounding conductor and the non–current-carrying parts of the machine shall be considered as an acceptable bonding path.

### 675.15 Lightning Protection

If an irrigation machine has a stationary point, a grounding electrode system in accordance with Article 250, Part III, shall be connected to the machine at the stationary point for lightning protection.

Where the electrical power supply to irrigation machine equipment is a service, the requirements of Article 250 for grounding the system and equipment are applicable. Due to the physical location of irrigation equipment, the most likely grounding electrode is a driven ground rod or ground plate. Where lightning protection is installed, NFPA 780, *Standard for the Installation of Lightning Protection Systems,* requires an electrode for that system. In accordance with 250.60, a common electrode is not permitted to serve the dual function of grounding the electric service and grounding the lightning protection system. The separate electrode systems are required to be bonded together.

### 675.16 Energy from More Than One Source

Equipment within an enclosure receiving electric energy from more than one source shall not be required to have a disconnecting means for the additional source, provided that its voltage is 30 volts or less and it meets the requirements of Part III of Article 725.

### 675.17 Connectors

External plugs and connectors on the equipment shall be of the weatherproof type.

Unless provided solely for the connection of circuits meeting the requirements of Part III of Article 725, external plugs and connectors shall be constructed as specified in 250.124(A).

## II. Center Pivot Irrigation Machines

### 675.21 General

The provisions of Part II are intended to cover additional special requirements that are peculiar to center pivot irrigation machines. See 675.2 for the definition of *Center Pivot Irrigation Machine*.

### 675.22 Equivalent Current Ratings

To establish ratings of controllers, disconnecting means, conductors, and the like, for the inherent intermittent duty of center pivot irrigation machines, the determinations in 675.22(A) and (B) shall be used.

**(A) Continuous-Current Rating.** The equivalent continuous-current rating for the selection of branch-circuit conductors and branch-circuit devices shall be equal to 125 percent of the motor nameplate full-load current rating of the largest motor plus 60 percent of the sum of the motor nameplate full-load current ratings of all remaining motors on the circuit.

**(B) Locked-Rotor Current.** The equivalent locked-rotor current rating shall be equal to the numerical sum of two times the locked-rotor current of the largest motor plus 80 percent of the sum of the motor nameplate full-load current ratings of all the remaining motors on the circuit.

# ARTICLE 680 Swimming Pools, Fountains, and Similar Installations

## I. General

### 680.1 Scope

The provisions of this article apply to the construction and installation of electrical wiring for, and equipment in or adjacent to, all swimming, wading, therapeutic, and decorative pools; fountains; hot tubs; spas; and hydromassage bathtubs, whether permanently installed or storable, and to metallic auxiliary equipment, such as pumps, filters, and similar equipment. The term *body of water* used throughout Part I applies to all bodies of water covered in this scope unless otherwise amended.

The installations covered by this article can be indoors or outdoors, permanent or storable, and may or may not be directly supplied by electrical circuits of any nature. This article also applies to pools used in religious services where participants are immersed in water. Requirements for natural and artificially made bodies of water not covered by Article 680 are contained in Article 682.

Studies conducted by Underwriters Laboratories, various manufacturers, and others indicate that a person in a swimming pool can receive a severe electric shock by reaching out and touching the energized casing of a faulty appliance — such as a radio or a hair dryer — because the immersed person's body, which has a lower resistance to electric current, establishes a conductive path through the water to earth. Also, a person not in contact with a faulty appliance or any grounded object can receive an electric shock and be rendered immobile by a potential gradient in the water itself. The level of electrical current necessary to cause immobilization may not cause death from electrical shock, but it could lead to accidental drowning. Shock hazards in and around a swimming pool can result from faulty electrical equipment directly associated with the pool or from faulty electrical equipment not associated with but in close proximity to the pool.

Accordingly, the requirements of Article 680 covering effective bonding and grounding, installation of receptacles and luminaires, use of GFCIs, modified wiring methods, and so forth, apply not only to the installation of the pool but also to installations and equipment adjacent to or associated with the pool.

Enhanced electric shock protection in this wet environment, where people are immersed in bodies of water that also contain electrical equipment, is provided through one or more of the following means:

- GFCI protection and low-voltage equipment
- Double-insulated equipment
- Insulation and isolation
- Equipotential bonding
- Physical separation and restricted locations
- Robust physical protection requirements for circuit conductors

### 680.2 Definitions

**Cord-and-Plug-Connected Lighting Assembly.** A lighting assembly consisting of a luminaire intended for installation in the wall of a spa, hot tub, or storable pool, and a cord-and-plug-connected transformer.

**Dry-Niche Luminaire.** A luminaire intended for installation in the floor or wall of a pool, spa, or fountain in a niche that is sealed against the entry of water.

**Fixed (as applied to equipment).** Equipment that is fastened or otherwise secured at a specific location.

**Forming Shell.** A structure designed to support a wet-niche luminaire assembly and intended for mounting in a pool or fountain structure.

**Fountain.** Fountains, ornamental pools, display pools, and reflection pools. The definition does not include drinking fountains.

**Hydromassage Bathtub.** A permanently installed bathtub equipped with a recirculating piping system, pump, and associated equipment. It is designed so it can accept, circulate, and discharge water upon each use.

See the commentary following 680.71, 680.73, and 680.74 for more information on installation requirements for hydromassage bath tubs.

**Low Voltage Contact Limit.** A voltage not exceeding the following values:

(1) 15 volts (RMS) for sinusoidal ac
(2) 21.2 volts peak for nonsinusoidal ac
(3) 30 volts for continuous dc
(4) 12.4 volts peak for dc that is interrupted at a rate of 10 to 200 Hz

The low-voltage contact limits are based on the wet contact limits specified in Tables 11(A) and 11(B) in Chapter 9. Before 2008, the use of isolated winding–type transformers that provided a sinusoidal ac voltage not exceeding 15 volts was used as the "low-voltage" operational threshold for underwater luminaires. Although the 15-volt limit in Article 680 was based on a transformer-type power supply, it applied to all "low-voltage" systems. Any luminaire operating at more than 15 volts was required to be protected by a GFCI.

In addition to sinusoidal alternating current, this definition also identifies the maximum acceptable safe levels for other voltage systems. Depending on the system, this maximum level is either higher or lower than that considered safe for sinusoidal ac voltage. Underwater lighting has been developed that integrates power supplies other than the traditional isolated winding–type transformer, and this definition and associated requirements in Article 680 ensure that those power supplies can be safely integrated into the swimming pool environment.

**Maximum Water Level.** The highest level that water can reach before it spills out.

The "maximum water level" is used as the benchmark for determining minimum distance between equipment such as luminaires or ceiling-suspended (paddle) fans and the water in a pool, spa, or hot tub. The maximum water level is also used in determining the clearance between overhead conductors and the pool water. The normal water level is used as the benchmark for determining location of underwater luminaires. Exhibit 680.1 illustrates the difference between the normal water level (typically controlled by the pool filtration system) and the maximum water level.

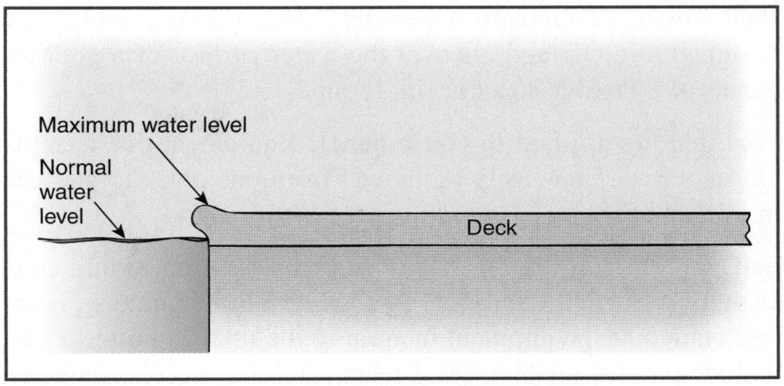


***EXHIBIT 680.1*** *A pool showing maximum water level.*

**No-Niche Luminaire.** A luminaire intended for installation above or below the water without a niche.

**Packaged Spa or Hot Tub Equipment Assembly.** A factory-fabricated unit consisting of water-circulating, heating, and control equipment mounted on a common base, intended to operate a spa or hot tub. Equipment can include pumps, air blowers, heaters, lights, controls, sanitizer generators, and so forth.

**Packaged Therapeutic Tub or Hydrotherapeutic Tank Equipment Assembly.** A factory-fabricated unit consisting of water-circulating, heating, and control equipment mounted on a common base, intended to operate a therapeutic tub or hydrotherapeutic tank. Equipment can include pumps, air blowers, heaters, lights, controls, sanitizer generators, and so forth.

**Permanently Installed Decorative Fountains and Reflection Pools.** Those that are constructed in the ground, on the ground, or in a building in such a manner that the fountain cannot be readily disassembled for storage, whether or not served by electrical circuits of any nature. These units are primarily constructed for their aesthetic value and are not intended for swimming or wading.

**Permanently Installed Swimming, Wading, Immersion, and Therapeutic Pools.** Those that are constructed in the ground or partially in the ground, and all others capable of holding water in a depth greater than 1.0 m (42 in.), and all pools installed inside of a building, regardless of water depth, whether or not served by electrical circuits of any nature.

The word *immersion* extends this definition to pools used in religious services or for some other function in which people become immersed as an inherent use or purpose of the pool. Also see the commentary following Part VI of Article 680.

**Pool.** Manufactured or field-constructed equipment designed to contain water on a permanent or semipermanent basis and used for swimming, wading, immersion, or therapeutic purposes.

**Pool Cover, Electrically Operated.** Motor-driven equipment designed to cover and uncover the water surface of a pool by means of a flexible sheet or rigid frame.

**Portable (as applied to equipment).** Equipment that is actually moved or can easily be moved from one place to another in normal use.

**Self-Contained Spa or Hot Tub.** Factory-fabricated unit consisting of a spa or hot tub vessel with all water-circulating, heating, and control equipment integral to the unit. Equipment can include pumps, air blowers, heaters, lights, controls, sanitizer generators, and so forth.

**Self-Contained Therapeutic Tubs or Hydrotherapeutic Tanks.** A factory-fabricated unit consisting of a therapeutic tub or hydrotherapeutic tank with all water-circulating, heating, and control equipment integral to the unit. Equipment may include pumps, air blowers, heaters, light controls, sanitizer generators, and so forth.

**Spa or Hot Tub.** A hydromassage pool, or tub for recreational or therapeutic use, not located in health care facilities, designed for immersion of users, and usually having a filter, heater, and motor-driven blower. It may be installed indoors or outdoors, on the ground or supporting structure, or in the ground or supporting structure. Generally, a spa or hot tub is not designed or intended to have its contents drained or discharged after each use.

**Stationary (as applied to equipment).** Equipment that is not moved from one place to another in normal use.

**Storable Swimming, Wading, or Immersion Pools; or Storable/ Portable Spas and Hot Tubs.** Those that are constructed on or above the ground and are capable of holding water to a maximum depth of 1.0 m (42 in.), or a pool, spa, or hot tub with nonmetallic, molded polymeric walls or inflatable fabric walls regardless of dimension.

Storable pools are intended to be temporary structures, without the need for special wiring or modification to the pool site. They are usually sold as a complete package, consisting of the pool walls, vinyl liner, plumbing kit, and pump/filter device. A storable pool is often disassembled and stored during the winter months.

The main difference between a storable and permanent pool is wall height. Generally, pools intended to be disassembled at season's end have wall heights of 42 inches or less, while those not intended for disassembly have wall heights of 48 inches or more. The surface area of the pool is not a factor. Inflatable pools are treated as storable pools regardless of their wall height.

Storable pools are supplied as two distinct types. One type is intended to be disassembled at the end of each swimming season. The second type, by the nature of its construction, can be disassembled, but manufacturers recommend leaving it assembled. The pools in the latter category frequently require special modification to and preparation of the pool site, making them impractical to disassemble.

**Through-Wall Lighting Assembly.** A lighting assembly intended for installation above grade, on or through the wall of a pool, consisting of two interconnected groups of components separated by the pool wall.

**Wet-Niche Luminaire.** A luminaire intended for installation in a forming shell mounted in a pool or fountain structure where the luminaire will be completely surrounded by water.

## 680.3 Other Articles

Except as modified by this article, wiring and equipment in or adjacent to pools and fountains shall comply with other applicable provisions of this *Code*, including those provisions identified in Table 680.3.

## 680.4 Approval of Equipment

All electrical equipment installed in the water, walls, or decks of pools, fountains, and similar installations shall comply with the provisions of this article.

## 680.5 Ground-Fault Circuit Interrupters

Ground-fault circuit interrupters (GFCIs) shall be self-contained units, circuit-breaker or receptacle types, or other listed types.

See the definition of *ground-fault circuit interrupter (GFCI)* in Article 100. A Class A GFCI trips where the current to ground is in the range of 4 through 6 mA. However, circuits supplying pool equipment, including underwater luminaires, that were installed before local adoption of the 1965 *NEC* may have sufficient leakage current to cause a Class A GFCI to trip. Prior to the 1965 *Code*, a Class B GFCI, which trips if the current to ground exceeds 20 mA, was suitable for use with underwater swimming pool lighting.

## 680.6 Grounding

Electrical equipment shall be grounded in accordance with Parts V, VI, and VII of Article 250 and connected by wiring methods of Chapter 3, except as modified by this article. The following equipment shall be grounded:

(1) Through-wall lighting assemblies and underwater luminaires, other than those low-voltage lighting products listed for the application without a grounding conductor

***TABLE 680.3*** *Other Articles*

| Topic | Section or Article |
|---|---|
| Site lighting systems operating at 30 volts or less | 411.4(B) |
| Audio equipment | Article 640, Parts I and II |
| Adjacent to pools and fountains | 640.10 |
| Underwater speakers* | |

*Underwater loudspeakers shall be installed in accordance with 680.27(A).

(2) All electrical equipment located within 1.5 m (5 ft) of the inside wall of the specified body of water
(3) All electrical equipment associated with the recirculating system of the specified body of water
(4) Junction boxes
(5) Transformer and power supply enclosures
(6) Ground-fault circuit interrupters
(7) Panelboards that are not part of the service equipment and that supply any electrical equipment associated with the specified body of water

An outdoor receptacle installed to meet the requirements of 680.22(A)(3) is permitted to be wired with Type UF cable containing an insulated or bare conductor for equipment grounding purposes. Although Type UF cable can be used for the receptacle and for some pool-related equipment, circuit conductors for underwater luminaires are required to be run in raceways. Circuit conductors, other than flexible cord, for pool-associated motors are required to be installed in raceways except in the interior of one-family dwelling units as allowed by 680.21(A)(4).

Equipment grounding requirements are contained in 680.6, 680.21(A)(1), 680.23(F)(2), and 680.25(B). These requirements specify that EGCs be connected to non–current-carrying metal parts of the specified equipment. These EGCs are required to be run in the raceway with the circuit conductors, and they must be terminated at the grounding terminal bus of the service panelboard, the source of the separately derived system, or the subpanel. This EGC provides a path of low impedance that limits the voltage to ground and facilitates operation of the circuit OCPD(s). The EGC is required to be an insulated copper conductor not smaller than 12 AWG.

The requirements of 680.26 are in addition to the EGC requirements. Bonding establishes an equipotential plane to limit the voltage between all non–current-carrying parts of electrical and nonelectrical equipment in the pool area. Bonding conductors may be insulated, covered, or bare; are required to be 8 AWG solid copper or larger; and may be direct buried. All the bonded parts form a common grid that establishes an equipotential bonding system, and the bonding conductors do not have to be run to the equipment grounding terminals of panelboards or service equipment.

## 680.7 Cord-and-Plug-Connected Equipment

Fixed or stationary equipment, other than underwater luminaires, for a permanently installed pool shall be permitted to be connected with a flexible cord and plug to facilitate the removal or disconnection for maintenance or repair.

**(A) Length.** For other than storable pools, the flexible cord shall not exceed 900 mm (3 ft) in length.

**(B) Equipment Grounding.** The flexible cord shall have a copper equipment grounding conductor sized in accordance with 250.122 but not smaller than 12 AWG. The cord shall terminate in a grounding-type attachment plug.

**(C) Construction.** The equipment grounding conductors shall be connected to a fixed metal part of the assembly. The removable part shall be mounted on or bonded to the fixed metal part.

In some climates, disconnecting and removing a permanent pool's filter pump during cold-weather months is preferable. A 3-foot cord is permitted to facilitate the removal of fixed or stationary equipment for maintenance and storage. The 3-foot cord limitation does not apply to cord-and-plug-connected filter pumps used with storable-type pools (covered in Part III), since these pumps are neither fixed nor stationary. Listed filter pumps for use with storable pools are considered portable and are permitted to be equipped with cords longer than 3 feet.

## 680.8 Overhead Conductor Clearances

Overhead conductors shall meet the clearae requirements in this section. Where a minimum clearance from the water level is given, the measurement shall be taken from the maximum water level of the specified body of water.

**(A) Power.** With respect to service-drop conductors, overhead service conductors, and open overhead wiring, swimming pool and similar installations shall comply with the minimum clearances given in Table 680.8(A) and illustrated in Figure 680.8(A).

Informational Note: Open overhead wiring as used in this article typically refers to conductor(s) not in an enclosed raceway.

**(B) Communications Systems.** Communications, radio, and television coaxial cables within the scope of Articles 800 through 820 shall be permitted at a height of not less than 3.0 m (10 ft) above swimming and wading pools, diving structures, and observation stands, towers, or platforms.

**(C) Network-Powered Broadband Communications Systems.** The minimum clearances for overhead network-powered broadband communications systems conductors from pools or fountains shall comply with the provisions in Table 680.8(A) for conductors operating at 0 to 750 volts to ground.

These clearances consider factors such as the use of skimmers with aluminum handles and provide sufficient separation between the conductors and the pool. In some instances, locating

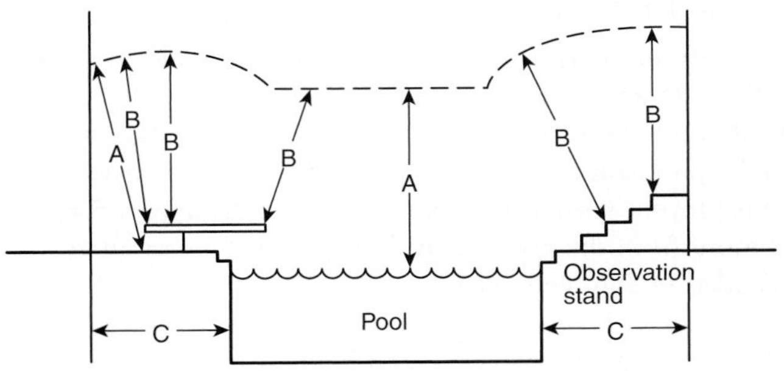


*FIGURE 680.8(A) Clearances from Pool Structures.*

*TABLE 680.8(A)* *Overhead Conductor Clearances*

| | | Insulated Cables, 0–750 Volts to Ground, Supported on and Cabled Together with a Solidly Grounded Bare Messenger or Solidly Grounded Neutral Conductor | | All Other Conductors Voltage to Ground | | | |
|---|---|---|---|---|---|---|---|
| | | | | 0 through 15 kV | | Over 15 through 50 kV | |
| | Clearance Parameters | m | ft | m | ft | m | ft |
| A. | Clearance in any direction to the water level, edge of water surface, base of diving platform, or permanently anchored raft | 6.9 | 22.5 | 7.5 | 25 | 8.0 | 27 |
| B. | Clearance in any direction to the observation stand, tower, or diving platform | 4.4 | 14.5 | 5.2 | 17 | 5.5 | 18 |
| C. | Horizontal limit of clearance measured from inside wall of the pool | This limit shall extend to the outer edge of the structures listed in A and B of this table but not less than 3 m (10 ft). | | | | | |

a swimming pool below electric fixed conductors is unavoidable; for example, on a building lot with limited area or an existing lot where the electric supply lines are already in place.

The clearances for conductors from pools and pool structures harmonize the *NEC* with ANSI C2, *National Electrical Safety Code (NESC)*. The maximum water level (see 680.2) of the body of water (pool, spa, hot tub, or other) is used to determine compliance with 680.8.

## 680.9 Electric Pool Water Heaters

All electric pool water heaters shall have the heating elements subdivided into loads not exceeding 48 amperes and protected at not over 60 amperes. The ampacity of the branch-circuit conductors and the rating or setting of overcurrent protective devices shall not be less than 125 percent of the total nameplate-rated load.

## 680.10 Underground Wiring Location

Underground wiring shall not be permitted under the pool or within the area extending 1.5 m (5 ft) horizontally from the inside wall of the pool unless this wiring is necessary to supply pool equipment permitted by this article. Where space limitations prevent wiring from being routed a distance 1.5 m (5 ft) or more from the pool, such wiring shall be permitted where installed in complete raceway systems of rigid metal conduit, intermediate metal conduit, or a nonmetallic raceway system. All metal conduit shall be corrosion resistant and suitable for the location. The minimum cover depth shall be as given in Table 680.10.

*TABLE 680.10* *Minimum Cover Depths*

| | Minimum Cover | |
|---|---|---|
| Wiring Method | mm | in. |
| Rigid metal conduit | 150 | 6 |
| Intermediate metal conduit | 150 | 6 |
| Nonmetallic raceways listed for direct burial under minimum of 102 mm (4 in.) thick concrete exterior slab and extending not less than 162 mm (6 in.) beyond the underground installation | 150 | 6 |
| Nonmetallic raceways listed for direct burial without concrete encasement | 450 | 18 |
| Other approved raceways* | 450 | 18 |

*Raceways approved for burial only where concrete encased shall require a concrete envelope not less than 50 mm (2 in.) thick.

Wiring is allowed within 5 feet of the inside walls of the swimming pool under two conditions. The first condition permits wiring to pool-associated equipment such as an underwater luminaire. The second condition permits wiring not associated with the pool within this area where spatial constraints such as property lines preclude the 5-foot minimum separation. Under the second condition, underground wiring located within the 5-foot zone is required to be installed in RMC, IMC, or a nonmetallic raceway and must be buried to a depth not less than that required by Table 680.10 for these permitted wiring methods. The raceway must be installed as a complete system between points of termination and cannot simply be a sleeve through the 5-foot zone. Beyond the 5-foot zone, the minimum cover requirements of Table 300.5 apply to the underground wiring methods used for circuits rated 600 volts and less.

The focus of 680.10 is to mitigate shock hazards that may occur as a result of a faulty or damaged underground installation that is in close proximity to the swimming pool. Due to water splashing out of the pool and water dripping off those who have been in the pool, the area within 5 feet of the inside walls is generally the wettest location; electrical leakage from underground installations presents a greater shock hazard in this continuously wet environment.

### 680.11 Equipment Rooms and Pits

Electrical equipment shall not be installed in rooms or pits that do not have drainage that prevents water accumulation during normal operation or filter maintenance.

### 680.12 Maintenance Disconnecting Means

One or more means to simultaneously disconnect all ungrounded conductors shall be provided for all utilization equipment other than lighting. Each means shall be readily accessible and within sight from its equipment and shall be located at least 1.5 m (5 ft) horizontally from the inside walls of a pool, spa, fountain, or hot tub unless separated from the open water by a permanently installed barrier that provides a 1.5 m (5 ft) reach path or greater. This horizontal distance is to be measured from the water's edge along the shortest path required to reach the disconnect.

A readily accessible disconnecting means is required to be located within sight of pool, spa, fountain, and hot tub equipment in order to provide service personnel with the ability to safely disconnect power while servicing equipment such as motors, heaters, and control panels. Underwater luminaires are not subject to this requirement. The proximity of the disconnecting means to the pool must be not less than 5 feet unless the disconnecting means is separated from the water by a permanent barrier.

## II. Permanently Installed Pools

### 680.20 General

Electrical installations at permanently installed pools shall comply with the provisions of Part I and Part II of this article.

### 680.21 Motors

**(A) Wiring Methods.** The wiring to a pool motor shall comply with (A)(1) unless modified for specific circumstances by (A)(2), (A)(3), (A)(4), or (A)(5).

**(1) General.** The branch circuits for pool-associated motors shall be installed in rigid metal conduit, intermediate metal conduit, rigid polyvinyl chloride conduit, reinforced thermosetting resin conduit, or Type MC cable listed for the location. Other wiring methods and materials shall be permitted in specific locations or applications as covered in this section. Any wiring method employed shall contain an insulated copper equipment grounding conductor sized in accordance with 250.122 but not smaller than 12 AWG.

Type MC cables listed for installation in direct sunlight or direct burial are marked to indicate suitability for such applications.

Other than cable assemblies installed on the interior of a one-family dwelling per 680.21(A)(4), wiring methods used for the supply circuit to a swimming pool pump motor must include an insulated copper EGC not less than 12 AWG.

**(2) On or Within Buildings.** Where installed on or within buildings, electrical metallic tubing shall be permitted.

**(3) Flexible Connections.** Where necessary to employ flexible connections at or adjacent to the motor, liquidtight flexible metal or liquidtight flexible nonmetallic conduit with approved fittings shall be permitted.

**(4) One-Family Dwellings.** In the interior of dwelling units, or in the interior of accessory buildings associated with a dwelling unit, any of the wiring methods recognized in Chapter 3 of this *Code* that comply with the provisions of this section shall be permitted. Where run in a cable assembly, the equipment grounding conductor shall be permitted to be uninsulated, but it shall be enclosed within the outer sheath of the cable assembly.

**(5) Cord-and-Plug Connections.** Pool-associated motors shall be permitted to employ cord-and-plug connections. The flexible cord shall not exceed 900 mm (3 ft) in length. The flexible cord shall include a copper equipment grounding conductor sized in accordance with 250.122 but not smaller than 12 AWG. The cord shall terminate in a grounding-type attachment plug.

**(B) Double Insulated Pool Pumps.** A listed cord-and-plug-connected pool pump incorporating an approved system of double insulation that provides a means for grounding only the internal and nonaccessible, non–current-carrying metal parts of the pump shall be connected to any wiring method recognized in Chapter 3 that is suitable for the location. Where the bonding grid is connected to the equipment grounding conductor of the motor circuit in accordance with the second sentence of 680.26(B)(6)(a), the branch-circuit wiring shall comply with 680.21(A).

Cord-and-plug-connected double-insulated swimming pool filter pumps have been used with permanently installed aboveground pools and some storable pools, regardless of the pool's size, for many years without any known field-related problems. The internal metal parts of a swimming pool filter pump incorporating a system of double insulation are grounded; however, they are not required to be incorporated into the bonding system required by 680.26(B), because the act of bonding compromises the double-insulation system.

**(C) GFCI Protection.** Outlets supplying pool pump motors connected to single-phase, 120-volt through 240-volt branch circuits, whether by receptacle or by direct connection, shall be provided with ground-fault circuit-interrupter protection for personnel.

Using the *Code* definition of the term *outlet* is important to correctly apply this requirement. The term *outlet* includes a point on the wiring system where a receptacle is installed to supply a cord-and-plug-connected pool pump motor and also includes a point on the wiring system from where the branch circuit is run directly to the pool pump motor, which is often referred to as a "hard-wired" installation.

## 680.22 Lighting, Receptacles, and Equipment

**(A) Receptacles.**

**(1) Required Receptacle, Location.** Where a permanently installed pool is installed, no fewer than one 125-volt, 15- or 20-ampere receptacle on a general-purpose branch circuit shall be located not less than 1.83 m (6 ft) from, and not more than 6.0 m (20 ft) from, the inside wall of the pool. This receptacle shall be located not more than 2.0 m (6 ft 6 in.) above the floor, platform, or grade level serving the pool.

**(2) Circulation and Sanitation System, Location.** Receptacles that provide power for water-pump motors or for other loads directly related to the circulation and sanitation system shall be located at least 3.0 m (10 ft) from the inside walls of the pool, or not less than 1.83 m (6 ft) from the inside walls of the pool if they meet all of the following conditions:

(1) Consist of single receptacles

•

(2) Are of the grounding type
(3) Have GFCI protection

**(3) Other Receptacles, Location.** Other receptacles shall be not less than 1.83 m (6 ft) from the inside walls of a pool.

**(4) GFCI Protection.** All 15- and 20-ampere, single-phase, 125-volt receptacles located within 6.0 m (20 ft) of the inside walls of a pool shall be protected by a ground-fault circuit interrupter.

**(5) Measurements.** In determining the dimensions in this section addressing receptacle spacings, the distance to be measured shall be the shortest path the supply cord of an appliance connected to the receptacle would follow without piercing a floor, wall, ceiling, doorway with hinged or sliding door, window opening, or other effective permanent barrier.

The requirements of 680.22(A) apply to receptacles located near a permanently installed pool or fountain. They do not apply to direct-connected equipment. Direct-connected pool pump motors are covered in 680.21(C).

As required by 680.22(A)(1), each permanently installed pool is required to have at least one receptacle that is located at least 6 feet from the pool and not more than 20 feet from the pool. This allows ordinary appliances to be safely plugged in and used near the pool, but avoids the need for extension cords in the pool vicinity. The 6-foot minimum dimension reduces the likelihood that an appliance with a 6-foot cord could be accidentally knocked into the pool.

Section 680.22(A)(4) applies to pools located outdoors or indoors, permanently installed or storable, and for residential or commercial use. Because people within 20 feet of a pool are normally subjected to dampness and moisture, the GFCI requirement within the 20-foot space is warranted.

Examples of receptacles meeting the requirements of 680.22(A) are shown in Exhibits 680.2 and 680.3. Receptacles within a structure are permitted to be less than 6 feet from the pool, because the structure is considered to provide a permanent barrier separating the receptacle from the pool. Where this installation is at a dwelling unit, at least one receptacle must be provided between 6 feet and 20 feet from the inside walls of the pool. This location precludes having to run the cord of an appliance used on the pool deck through a doorway.

**(B) Luminaires, Lighting Outlets, and Ceiling-Suspended (Paddle) Fans.**

**(1) New Outdoor Installation Clearances.** In outdoor pool areas, luminaires, lighting outlets, and ceiling-suspended (paddle) fans installed above the pool or the area extending 1.5 m (5 ft) horizontally from the inside walls of the pool shall be installed at a height not less than 3.7 m (12 ft) above the maximum water level of the pool.

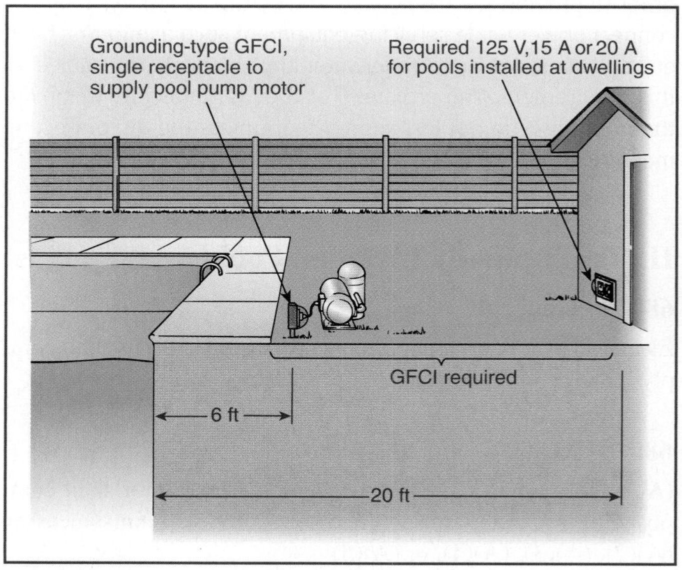


***EXHIBIT 680.2*** *An example of a receptacle installed according to 680.22(A).*

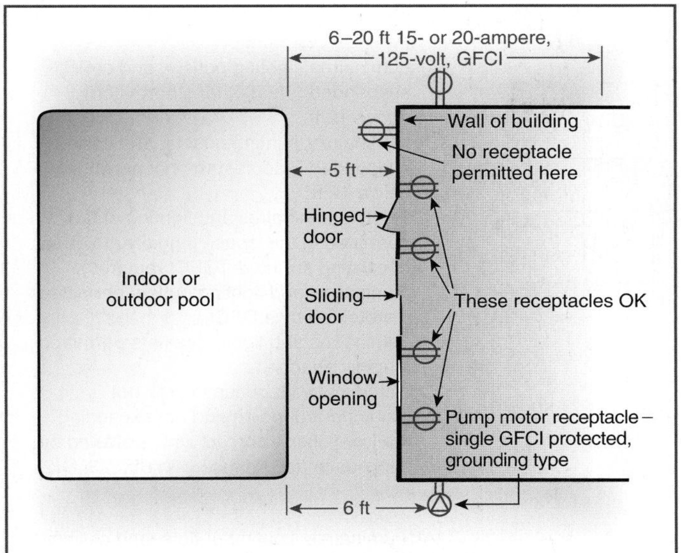


***EXHIBIT 680.3*** *Acceptable receptacle locations within 20 feet of a permanently installed swimming pool.*

**(2) Indoor Clearances.** For installations in indoor pool areas, the clearances shall be the same as for outdoor areas unless modified as provided in this paragraph. If the branch circuit supplying the equipment is protected by a ground-fault circuit interrupter, the following equipment shall be permitted at a height not less than 2.3 m (7 ft 6 in.) above the maximum pool water level:

(1) Totally enclosed luminaires
(2) Ceiling-suspended (paddle) fans identified for use beneath ceiling structures such as provided on porches or patios

**(3) Existing Installations.** Existing luminaires and lighting outlets located less than 1.5 m (5 ft) measured horizontally from the inside walls of a pool shall be not less than 1.5 m (5 ft) above the surface of the maximum water level, shall be rigidly attached to the existing structure, and shall be protected by a ground-fault circuit interrupter.

**(4) GFCI Protection in Adjacent Areas.** Luminaires, lighting outlets, and ceiling-suspended (paddle) fans installed in the area extending between 1.5 m (5 ft) and 3.0 m (10 ft) horizontally from the inside walls of a pool shall be protected by a ground-fault circuit interrupter unless installed not less than 1.5 m (5 ft) above the maximum water level and rigidly attached to the structure adjacent to or enclosing the pool.

**(5) Cord-and-Plug-Connected Luminaires.** Cord-and-plug-connected luminaires shall comply with the requirements of 680.7 where installed within 4.9 m (16 ft) of any point on the water surface, measured radially.

Exhibit 680.4 provides diagrams that clarify the limitations of 680.22(B) for the areas surrounding outdoor and indoor pools.

**(6) Low-Voltage Luminaires.** Listed low-voltage luminaires not requiring grounding, not exceeding the low-voltage contact limit, and supplied by listed transformers or power supplies that comply with 680.23(A)(2) shall be permitted to be located less than 1.5 m (5 ft) from the inside walls of the pool.

**(C) Switching Devices.** Switching devices shall be located at least 1.5 m (5 ft) horizontally from the inside walls of a pool unless separated from the pool by a solid fence, wall, or other permanent barrier. Alternatively, a switch that is listed as being acceptable for use within 1.5 m (5 ft) shall be permitted.

**(D) Other Outlets.** Other outlets shall be not less than 3.0 m (10 ft) from the inside walls of the pool. Measurements shall be determined in accordance with 680.22(A)(5).

Informational Note: Other outlets may include, but are not limited to, remote-control, signaling, fire alarm, and communications circuits.

## 680.23 Underwater Luminaires

This section covers all luminaires installed below the maximum water level of the pool.

**(A) General.**

**(1) Luminaire Design, Normal Operation.** The design of an underwater luminaire supplied from a branch circuit either directly or by way of a transformer or power supply meeting the requirements of this section shall be such that, where the luminaire is properly installed without a ground-fault circuit interrupter, there is no shock hazard with any likely combination of fault conditions during normal use (not relamping).

**(2) Transformers and Power Supplies.** Transformers and power supplies used for the supply of underwater luminaires, together with the transformer or power supply enclosure, shall be listed for swimming pool and spa use. The transformer or power supply shall incorporate either a transformer of the isolated winding type, with an ungrounded secondary that has a grounded metal barrier between the primary and secondary windings, or one that incorporates an approved system of double insulation between the primary and secondary windings.

Transformers and power supplies are required to be specifically listed for swimming pool use. Unless marked otherwise, swimming pool and spa transformers are not suitable for connection to a conduit that extends directly to an underwater pool light forming shell. Swimming pool and spa transformers are not permitted to be used outdoors unless marked "For Outdoor Use" or in an equivalent manner that signifies that they have been found acceptable for outdoor use. See 110.3(B) for requirements on the installation and use of listed electrical equipment.

**(3) GFCI Protection, Relamping.** A ground-fault circuit interrupter shall be installed in the branch circuit supplying luminaires operating at more than the low voltage contact

***EXHIBIT 680.4*** *Limitations of 680.22(B) for the areas surrounding outdoor and indoor pools.*

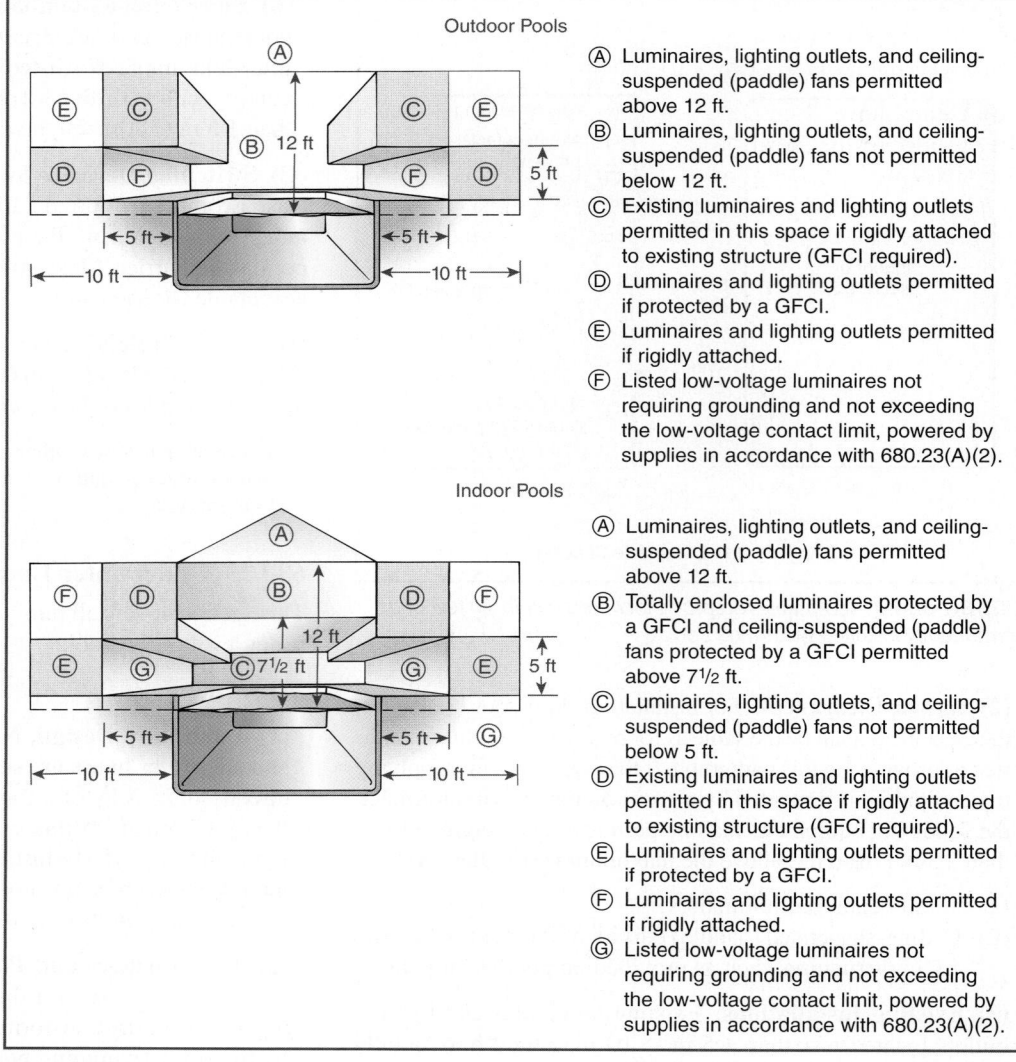

Outdoor Pools

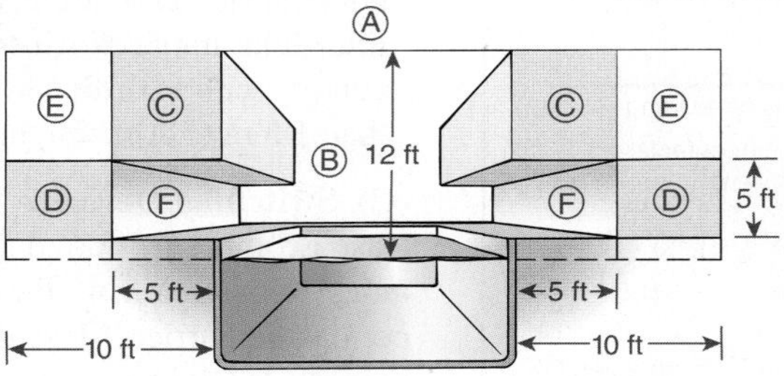


Ⓐ Luminaires, lighting outlets, and ceiling-suspended (paddle) fans permitted above 12 ft.
Ⓑ Luminaires, lighting outlets, and ceiling-suspended (paddle) fans not permitted below 12 ft.
Ⓒ Existing luminaires and lighting outlets permitted in this space if rigidly attached to existing structure (GFCI required).
Ⓓ Luminaires and lighting outlets permitted if protected by a GFCI.
Ⓔ Luminaires and lighting outlets permitted if rigidly attached.
Ⓕ Listed low-voltage luminaires not requiring grounding and not exceeding the low-voltage contact limit, powered by supplies in accordance with 680.23(A)(2).

Indoor Pools

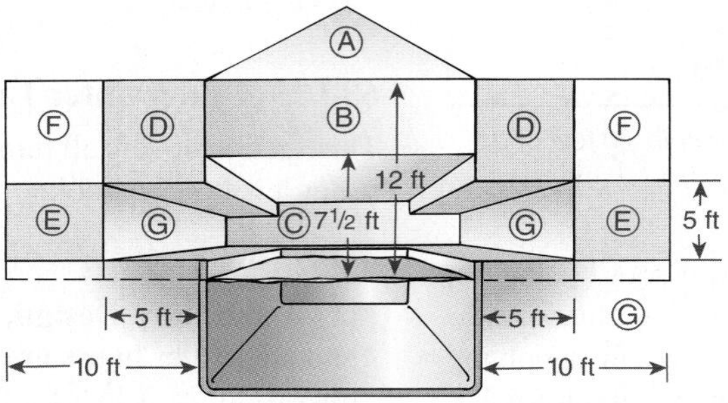


Ⓐ Luminaires, lighting outlets, and ceiling-suspended (paddle) fans permitted above 12 ft.
Ⓑ Totally enclosed luminaires protected by a GFCI and ceiling-suspended (paddle) fans protected by a GFCI permitted above 7½ ft.
Ⓒ Luminaires, lighting outlets, and ceiling-suspended (paddle) fans not permitted below 5 ft.
Ⓓ Existing luminaires and lighting outlets permitted in this space if rigidly attached to existing structure (GFCI required).
Ⓔ Luminaires and lighting outlets permitted if protected by a GFCI.
Ⓕ Luminaires and lighting outlets permitted if rigidly attached.
Ⓖ Listed low-voltage luminaires not requiring grounding and not exceeding the low-voltage contact limit, powered by supplies in accordance with 680.23(A)(2).

limit such that there is no shock hazard during relamping. The installation of the ground-fault circuit interrupter shall be such that there is no shock hazard with any likely fault-condition combination that involves a person in a conductive path from any ungrounded part of the branch circuit or the luminaire to ground.

Dry-niche, no-niche, or wet-niche underwater luminaires operating at more than 15 volts require GFCI protection. See the commentary following 680.5.

**(4) Voltage Limitation.** No luminaires shall be installed for operation on supply circuits over 150 volts between conductors.

**(5) Location, Wall-Mounted Luminaires.** Luminaires mounted in walls shall be installed with the top of the luminaire lens not less than 450 mm (18 in.) below the normal water level of the pool, unless the luminaire is listed and identified for use at lesser depths. No luminaire shall be installed less than 100 mm (4 in.) below the normal water level of the pool.

The 18-inch minimum submergence requirement for luminaires is to reduce the likelihood that a person in the water and hanging onto the side of the pool directly in front of a luminaire will have his or her chest in line with the luminaire. This section covers luminaires that have been investigated and found acceptable for use where a person's chest may be directly in front of a luminaire. The highest level of leakage current in a pool coming from a wet-niche luminaire with a broken lens and bulb is found directly in front of the luminaire.

**(6) Bottom-Mounted Luminaires.** A luminaire facing upward shall comply with either (1) or (2):

(1) Have the lens guarded to prevent contact by any person
(2) Be listed for use without a guard

**(7) Dependence on Submersion.** Luminaires that depend on submersion for safe operation shall be inherently protected against the hazards of overheating when not submerged.

Protection against overheating is required to be built into a luminaire or to be a part of it. A remotely located low-water cutoff switch does not provide the intended protection.

**(8) Compliance.** Compliance with these requirements shall be obtained by the use of a listed underwater luminaire and by installation of a listed ground-fault circuit interrupter in the branch circuit or a listed transformer or power supply for luminaires operating at not more than the low voltage contact limit.

**(B) Wet-Niche Luminaires.**

**(1) Forming Shells.** Forming shells shall be installed for the mounting of all wet-niche underwater luminaires and shall be equipped with provisions for conduit entries. Metal parts of the luminaire and forming shell in contact with the pool water shall be of brass or other approved corrosion-resistant metal. All forming shells used with nonmetallic conduit systems, other than those that are part of a listed low-voltage lighting system not requiring grounding, shall include provisions for terminating an 8 AWG copper conductor.

**(2) Wiring Extending Directly to the Forming Shell.** Conduit shall be installed from the forming shell to a junction box or other enclosure conforming to the requirements in 680.24. Conduit shall be rigid metal, intermediate metal, liquidtight flexible nonmetallic, or rigid nonmetallic.

(a) *Metal Conduit.* Metal conduit shall be approved and shall be of brass or other approved corrosion-resistant metal.

(b) *Nonmetallic Conduit.* Where a nonmetallic conduit is used, an 8 AWG insulated solid or stranded copper bonding jumper shall be installed in this conduit unless a listed low-voltage lighting system not requiring grounding is used. The bonding jumper shall be terminated in the forming shell, junction box or transformer enclosure, or ground-fault circuit-interrupter enclosure. The termination of the 8 AWG bonding jumper in the forming shell shall be covered with, or encapsulated in, a listed potting compound to protect the connection from the possible deteriorating effect of pool water.

An 8 AWG insulated copper bonding jumper is required to be installed in the conduit to provide electrical continuity between the forming shell and the junction box or other enclosure. This bonding conductor is in addition to the EGC required by 680.23(F)(2).

The function of this conductor is twofold. It permanently bonds all non–current-carrying metal surfaces of the forming shell to any non–current-carrying parts of the deck box and to the EGC of the circuit that supplies the wet-niche luminaire. Additionally, this conductor serves as the path for ground-fault current in the event of a ground fault when the wet-niche luminaire is removed from the forming shell, which is typically done during relamping. Damage to the wet-niche luminaire supply cord could result in this ground-fault scenario.

Low-voltage lighting systems that are listed for installation without an EGC or bonding conductor are exempt from this requirement.

**(3) Equipment Grounding Provisions for Cords.** Other than listed low-voltages lighting systems not requiring grounding wet-niche luminaires that are supplied by a flexible cord or cable shall have all exposed non–current-carrying metal parts grounded by an insulated copper equipment grounding conductor that is an integral part of the cord or cable. This grounding conductor shall be connected to a grounding terminal in the supply junction box, transformer enclosure, or other enclosure. The grounding conductor shall not be smaller than the supply conductors and not smaller than 16 AWG.

**(4) Luminaire Grounding Terminations.** The end of the flexible-cord jacket and the flexible-cord conductor terminations within a luminaire shall be covered with, or encapsulated in, a suitable potting compound to prevent the entry of water into the luminaire through the cord or its conductors. If present, the grounding connection within a luminaire shall be similarly treated to protect such connection from the deteriorating effect of pool water in the event of water entry into the luminaire.

**(5) Luminaire Bonding.** The luminaire shall be bonded to, and secured to, the forming shell by a positive locking device that ensures a low-resistance contact and requires a tool to remove the luminaire from the forming shell. Bonding shall not be required for luminaires that are listed for the application and have no non–current-carrying metal parts.

**(6) Servicing.** All wet-niche luminaires shall be removable from the water for inspection, relamping, or other maintenance. The forming shell location and length of cord in the forming shell shall permit personnel to place the removed luminaire on the deck or other dry location for such maintenance. The luminaire maintenance location shall be accessible without entering or going in the pool water.

Custom swimming pool installations where the pool is incorporated as an architectural feature of a building or structure can present access problems for those who have to change the lamps of an underwater luminaire. In some cases, the length of the flexible cord connected to a wet-niche luminaire does not permit the luminaire to be removed from the pool for relamping or servicing. To address the concern over a person having to be in the pool in order to change lamps, this requirement specifies that the underwater luminaire installation has to be made such that changing of the lamp occurs on the pool deck or other dry location outside of the pool, and that the location can be accessed without having to enter the pool water.

**(C) Dry-Niche Luminaires.**

**(1) Construction.** A dry-niche luminaire shall have provision for drainage of water. Other than listed low voltage luminaires not requiring grounding, a dry-niche luminaire shall have means for accommodating one equipment grounding conductor for each conduit entry.

**(2) Junction Box.** A junction box shall not be required but, if used, shall not be required to be elevated or located as specified in 680.24(A)(2) if the luminaire is specifically identified for the purpose.

**(D) No-Niche Luminaires.** A no-niche luminaire shall meet the construction requirements of 680.23(B)(3) and be installed in accordance with the requirements of 680.23(B). Where connection to a forming shell is specified, the connection shall be to the mounting bracket.

**(E) Through-Wall Lighting Assembly.** A through-wall lighting assembly shall be equipped with a threaded entry or hub, or a nonmetallic hub, for the purpose of accommodating the termination of the supply conduit. A through-wall lighting assembly shall meet the construction requirements of 680.23(B)(3) and be installed in accordance with the requirements of 680.23. Where connection to a forming shell is specified, the connection shall be to the conduit termination point.

**(F) Branch-Circuit Wiring.**

Section 680.23(B)(2) covers bonding from the junction box to a luminaire forming shell. Section 680.23(F)(1) covers the wiring method from the supply to the junction box, and 680.23(F)(2) covers the grounding of the entire branch circuit to the luminaire.

In addition to the bonding jumper and EGC of the cord or cable contained in the nonmetallic conduit between the forming shell and the deck box, the wiring method from the deck box to the power source is also required to contain a separate EGC regardless of the type of conduit installed. This EGC must be insulated, copper, and not smaller than 12 AWG. The grounding terminals within the deck (junction) box are used to terminate and bond together all of the conductors. See 680.24 for commentary on listed pool junction boxes.

Exhibit 680.5 illustrates an installation of a forming shell for a wet-niche luminaire and a flush junction (deck) box installed according to 680.24(A)(2). (Surface deck boxes are shown in Exhibit 680.5.)

**(1) Wiring Methods.** Branch-circuit wiring on the supply side of enclosures and junction boxes connected to conduits run to wet-niche and no-niche luminaires, and the field wiring compartments of dry-niche luminaires, shall be installed using rigid metal conduit, intermediate metal conduit, liquidtight flexible nonmetallic conduit, rigid polyvinyl chloride conduit, or reinforced thermosetting resin conduit. Where installed on buildings, electrical metallic tubing shall be permitted, and where installed within buildings, electrical nonmetallic tubing, Type MC cable, electrical metallic tubing, or Type AC cable shall be permitted. In all cases, an insulated equipment grounding conductor sized in accordance with Table 250.122 but not less than 12 AWG shall be required.

*Exception: Where connecting to transformers for pool lights, liquidtight flexible metal conduit shall be permitted. The length shall not exceed 1.8 m (6 ft) for any one length or exceed 3.0 m (10 ft) in total length used.*

**(2) Equipment Grounding.** Other than listed low-voltage luminaires not requiring grounding, all through-wall lighting assemblies, wet-niche, dry-niche, or no-niche luminaires shall be connected to an insulated copper equipment grounding conductor installed with the circuit conductors. The equipment grounding conductor shall be installed without joint or splice except as permitted in (F)(2)(a) and (F)(2)(b). The equipment grounding conductor shall be sized in accordance with Table 250.122 but shall not be smaller than 12 AWG.

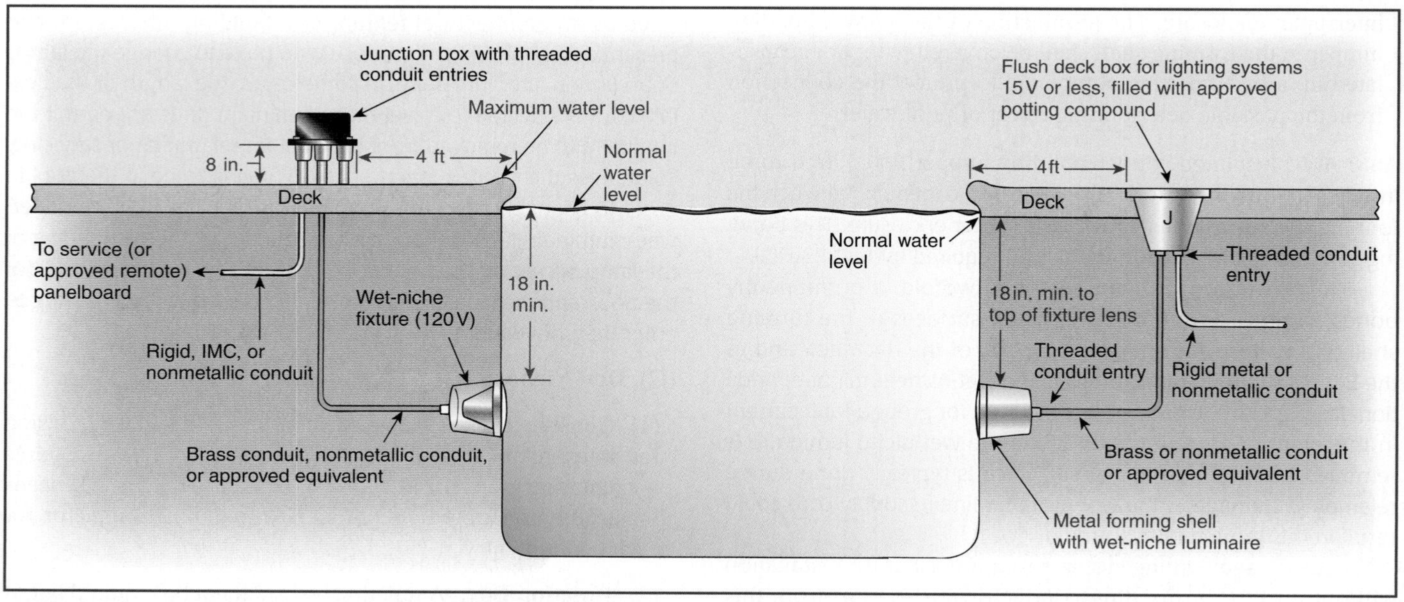


***EXHIBIT 680.5*** *A flush junction (deck) box and a forming shell for a wet-niche luminaire.*

*Exception: An equipment grounding conductor between the wiring chamber of the secondary winding of a transformer and a junction box shall be sized in accordance with the overcurrent device in this circuit.*

(a) If more than one underwater luminaire is supplied by the same branch circuit, the equipment grounding conductor, installed between the junction boxes, transformer enclosures, or other enclosures in the supply circuit to wet-niche luminaires, or between the field-wiring compartments of dry-niche luminaires, shall be permitted to be terminated on grounding terminals.

(b) If the underwater luminaire is supplied from a transformer, ground-fault circuit interrupter, clock-operated switch, or a manual snap switch that is located between the panelboard and a junction box connected to the conduit that extends directly to the underwater luminaire, the equipment grounding conductor shall be permitted to terminate on grounding terminals on the transformer, ground-fault circuit interrupter, clock-operated switch enclosure, or an outlet box used to enclose a snap switch.

**(3) Conductors.** Conductors on the load side of a ground-fault circuit interrupter or of a transformer, used to comply with the provisions of 680.23(A)(8), shall not occupy raceways, boxes, or enclosures containing other conductors unless one of the following conditions applies:

(1) The other conductors are protected by ground-fault circuit interrupters.
(2) The other conductors are grounding conductors.
(3) The other conductors are supply conductors to a feed-through-type ground-fault circuit interrupter.
(4) Ground-fault circuit interrupters shall be permitted in a panelboard that contains circuits protected by other than ground-fault circuit interrupters.

## 680.24 Junction Boxes and Electrical Enclosures for Transformers or Ground-Fault Circuit Interrupters

The requirements in 680.24(A) through (F) cover the construction and installation of boxes and enclosures associated with underwater luminaires. Boxes and enclosures used for the supply wiring to wet-niche and no-niche underwater luminaires must be listed for that purpose. The requirements of 680.24(D) ensure the availability of integral grounding terminals necessary for the grounding and bonding of underwater luminaires. A box that is not specifically listed for use with swimming pools does not provide the correct number of integral grounding and bonding terminals. The number of grounding terminals in a box or enclosure is required to be one more than the number of conduit entries for which the box is designed.

**(A) Junction Boxes.** A junction box connected to a conduit that extends directly to a forming shell or mounting bracket of a no-niche luminaire shall meet the requirements of this section.

**(1) Construction.** The junction box shall be listed as a swimming pool junction box and shall comply with the following conditions:

(1) Be equipped with threaded entries or hubs or a nonmetallic hub
(2) Be comprised of copper, brass, suitable plastic, or other approved corrosion-resistant material
(3) Be provided with electrical continuity between every connected metal conduit and the grounding terminals by means of copper, brass, or other approved corrosion-resistant metal that is integral with the box

**(2) Installation.** Where the luminaire operates over the low voltage contact limit, the junction box location shall comply with (A)(2)(a) and (A)(2)(b). Where the luminaire operates at the low voltage contact limit or less, the junction box location shall be permitted to comply with (A)(2)(c).

(a) *Vertical Spacing.* The junction box shall be located not less than 100 mm (4 in.), measured from the inside of the bottom of the box, above the ground level, or pool deck, or not less than 200 mm (8 in.) above the maximum pool water level, whichever provides the greater elevation.

(b) *Horizontal Spacing.* The junction box shall be located not less than 1.2 m (4 ft) from the inside wall of the pool, unless separated from the pool by a solid fence, wall, or other permanent barrier.

(c) *Flush Deck Box.* If used on a lighting system operating at the low voltage contact limit or less, a flush deck box shall be permitted if both of the following conditions are met:

(1) An approved potting compound is used to fill the box to prevent the entrance of moisture.
(2) The flush deck box is located not less than 1.2 m (4 ft) from the inside wall of the pool.

**(B) Other Enclosures.** An enclosure for a transformer, ground-fault circuit interrupter, or a similar device connected to a conduit that extends directly to a forming shell or mounting bracket of a no-niche luminaire shall meet the requirements of this section.

**(1) Construction.** The enclosure shall be listed and labeled for the purpose and meet the following requirements:

(1) Equipped with threaded entries or hubs or a nonmetallic hub
(2) Comprised of copper, brass, suitable plastic, or other approved corrosion-resistant material
(3) Provided with an approved seal, such as duct seal at the conduit connection, that prevents circulation of air between the conduit and the enclosures
(4) Provided with electrical continuity between every connected metal conduit and the grounding terminals by means of copper, brass, or other approved corrosion-resistant metal that is integral with the box

**(2) Installation.**

(a) *Vertical Spacing.* The enclosure shall be located not less than 100 mm (4 in.), measured from the inside of the bottom of the box, above the ground level, or pool deck, or not less than 200 mm (8 in.) above the maximum pool water level, whichever provides the greater elevation.

(b) *Horizontal Spacing.* The enclosure shall be located not less than 1.2 m (4 ft) from the inside wall of the pool, unless separated from the pool by a solid fence, wall, or other permanent barrier.

**(C) Protection.** Junction boxes and enclosures mounted above the grade of the finished walkway around the pool shall not be located in the walkway unless afforded additional protection, such as by location under diving boards, adjacent to fixed structures, and the like.

**(D) Grounding Terminals.** Junction boxes, transformer and power-supply enclosures, and ground-fault circuit-interrupter enclosures connected to a conduit that extends directly to a forming shell or mounting bracket of a no-niche luminaire shall be provided with a number of grounding terminals that shall be no fewer than one more than the number of conduit entries.

**(E) Strain Relief.** The termination of a flexible cord of an underwater luminaire within a junction box, transformer or power-supply enclosure, ground-fault circuit interrupter, or other enclosure shall be provided with a strain relief.

**(F) Grounding.** The equipment grounding conductor terminals of a junction box, transformer enclosure, or other enclosure in the supply circuit to a wet-niche or no-niche luminaire and the field-wiring chamber of a dry-niche luminaire shall be connected to the equipment grounding terminal of the panelboard. This terminal shall be directly connected to the panelboard enclosure.

## 680.25 Feeders

These provisions shall apply to any feeder on the supply side of panelboards supplying branch circuits for pool equipment covered in Part II of this article and on the load side of the service equipment or the source of a separately derived system.

**(A) Wiring Methods.**

**(1) Feeders.** Feeders shall be installed in rigid metal conduit or intermediate metal conduit. The following wiring methods shall be permitted if not subject to physical damage:

(1) Liquidtight flexible nonmetallic conduit
(2) Rigid polyvinyl chloride conduit
(3) Reinforced thermosetting resin conduit
(4) Electrical metallic tubing where installed on or within a building
(5) Electrical nonmetallic tubing where installed within a building
(6) Type MC cable where installed within a building and if not subject to corrosive environment

**(2) Aluminum Conduit.** Aluminum conduit shall not be permitted in the pool area where subject to corrosion.

**(B) Grounding.** An equipment grounding conductor shall be installed with the feeder conductors between the grounding terminal of the pool equipment panelboard and the grounding terminal of the applicable service equipment or source of a separately derived system. For other than (1) existing feeders, or (2) feeders to separate buildings that do not utilize an insulated equipment grounding conductor in accordance with 680.25(B)(2), this equipment grounding conductor shall be insulated.

**(1) Size.** This conductor shall be sized in accordance with 250.122 but not smaller than 12 AWG. On separately derived systems, this conductor shall be sized in accordance with 250.30(A)(3) but not smaller than 8 AWG.

**(2) Separate Buildings.** A feeder to a separate building or structure shall be permitted to supply swimming pool equipment branch circuits, or feeders supplying swimming pool equipment branch circuits, if the grounding arrangements in the separate building meet the requirements in 250.32(B).

The insulated EGC to the panelboard serving swimming pool equipment can be aluminum or copper, whereas the branch-circuit EGC is required to be copper. Where a remote panelboard supplying a pool is supplied by a separately derived system, the rules covering the EGC apply only to the feeder between the separately derived system and the panelboard, not all the way back to the service. The feeder is also required to be installed in a raceway.

The general rule in 680.25(B) requires an EGC to be installed with the feeder to a panelboard serving swimming pool equipment. Section 680.25(B)(2) allows feeder circuit conductors run to a remote panelboard in a separate building where the installation complies with the exception to 250.32(B). This allows for existing installations that were permitted to use the grounded conductor as the ground connection between the two buildings.

See Exhibit 680.6 for an illustration of applying the wiring method and grounding requirements to a feeder-supplied panelboard that provides branch circuits for swimming pool–related equipment. Also note the branch-circuit wiring methods shown in this illustration.

## 680.26 Equipotential Bonding

**(A) Performance.** The equipotential bonding required by this section shall be installed to reduce voltage gradients in the pool area.

Article 100 defines bonding as "connected to establish electrical continuity and conductivity." As described in 680.26(A), the function of equipotential bonding differs from the primary function of bonding to meet the requirements of Article 250 in that providing a path for ground-fault current is not the function of the equipotential bonding grid and associated bonding conductors.

Creating an electrically safe environment in and around permanently installed swimming pools requires the installation of a bonding system with the sole function of establishing equal

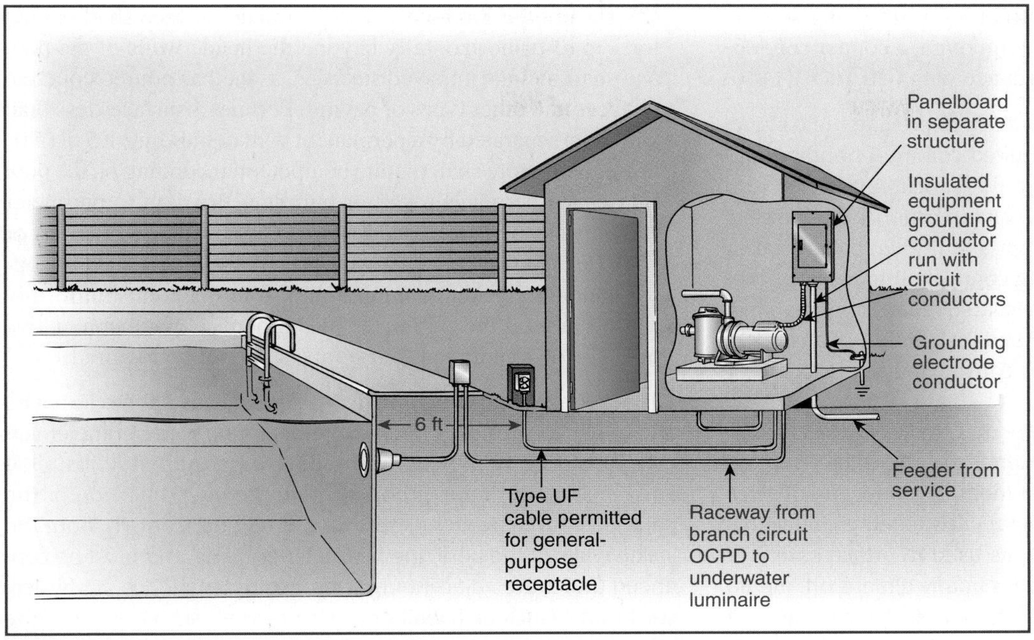


***EXHIBIT 680.6*** *Wiring method and grounding requirements for a feeder panelboard supplying swimming pool equipment.*

electrical potential (voltage) in the vicinity of the swimming pool. A person who is immersed in a pool or who is lying on or walking on a concrete deck or other conductive perimeter surface is vulnerable to any differences in electrical potential that may be present in the pool area.

The only function of the 8 AWG conductor required by 680.26(B) is equipotential bonding to eliminate the voltage gradient in the pool area. The bonding conductor is not required to extend or connect to any parts or equipment other than those covered in 680.26(B)(1) through (7) and to a pool water bonding element covered in 680.26(C).

The reason for electrically connecting all of the metal parts described in 680.26(B)(1) through (7) is to ensure that they all are at the same electrical potential. This bonding reduces possible injurious or disabling shock hazards created by stray currents in the ground or piping connected to the swimming pool. Stray currents can also exist in nonmetallic piping because of the low resistivity of chlorinated water. See Exhibit 680.7.

**(B) Bonded Parts.** The parts specified in 680.26(B)(1) through (B)(7) shall be bonded together using solid copper conductors, insulated covered, or bare, not smaller than 8 AWG or with rigid metal conduit of brass or other identified corrosion-resistant metal. Connections to bonded parts shall be made in accordance with 250.8. An 8 AWG or larger solid copper bonding conductor provided to reduce voltage gradients in the pool area shall not be required to be extended or attached to remote panelboards, service equipment, or electrodes.

**(1) Conductive Pool Shells.** Bonding to conductive pool shells shall be provided as specified in 680.26(B)(1)(a) or

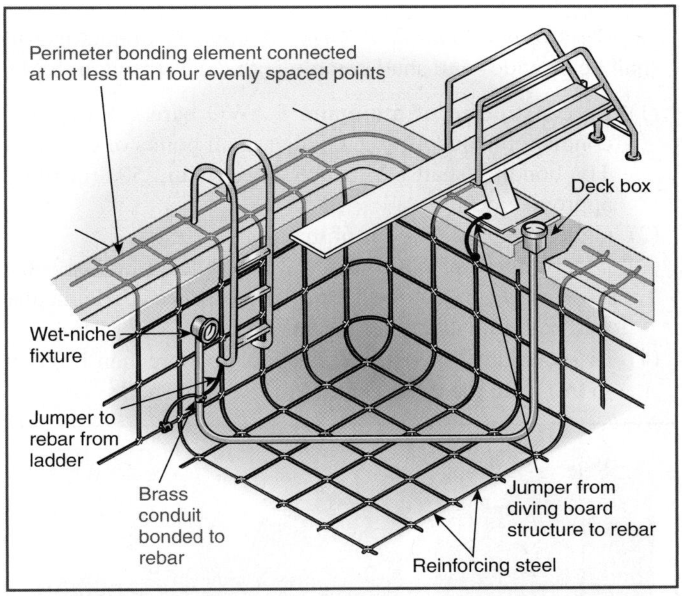


***EXHIBIT 680.7*** *Bonding of conductive metal equipment and parts associated with a swimming pool.*

(B)(1)(b). Poured concrete, pneumatically applied or sprayed concrete, and concrete block with painted or plastered coatings shall all be considered conductive materials due to water permeability and porosity. Vinyl liners and fiberglass composite shells shall be considered to be nonconductive materials.

(a) *Structural Reinforcing Steel.* Unencapsulated structural reinforcing steel shall be bonded together by steel tie wires or the equivalent. Where structural reinforcing steel is encapsulated in a nonconductive compound, a copper conductor grid shall be installed in accordance with 680.26(B)(1)(b).

Encapsulated reinforcing steel is not likely to provide the conductivity necessary to establish the required common bonding grid around the contour of a conductive pool shell. A bonding grid around the contour of the pool shell will not be formed if the steel is effectively encapsulated by a compound during installation or manufacturing. Therefore, a bonding connection to the encapsulated reinforcing steel, such as epoxy-coated rebar, is not required. However, a copper bonding grid around the contour of a conductive pool shell must be provided and constructed as prescribed in 680.26(B)(1)(b).

In Exhibit 680.8, structural reinforcing steel serves as a common point to which all metal appurtenances associated with the pool are connected. This connection method is one way of satisfying the requirement to bond all metal parts together. Individual pieces of hardware such as the hooks used to attach safety or lane ropes — that are less than 4 inches in any dimension and do not penetrate into the pool structure more than 1 inch — are not required to be bonded per 680.26(B)(5). The flush deck box meets the requirements of 680.24(A).

(b) *Copper Conductor Grid.* A copper conductor grid shall be provided and shall comply with (b)(1) through (b)(4).

(1) Be constructed of minimum 8 AWG bare solid copper conductors bonded to each other at all points of crossing. The bonding shall be in accordance with 250.8 or other approved means.
(2) Conform to the contour of the pool
(3) Be arranged in a 300-mm (12-in.) by 300-mm (12-in.) network of conductors in a uniformly spaced perpendicular grid pattern with a tolerance of 100 mm (4 in.)
(4) Be secured within or under the pool no more than 150 mm (6 in.) from the outer contour of the pool shell

***EXHIBIT 680.8*** *A poured-concrete pool with structural reinforcing steel that serves as the pool shell bonding grid.*

**(2) Perimeter Surfaces.** The perimeter surface shall extend for 1 m (3 ft) horizontally beyond the inside walls of the pool and shall include unpaved surfaces, as well as poured concrete surfaces and other types of paving. Perimeter surfaces less than 1 m (3 ft) separated by a permanent wall or building 1.5 m (5 ft) in height or more shall require equipotential bonding on the pool side of the permanent wall or building. Bonding to perimeter surfaces shall be provided as specified in 680.26(B)(2)(a) or (2)(b) and shall be attached to the pool reinforcing steel or copper conductor grid at a minimum of four (4) points uniformly spaced around the perimeter of the pool. For nonconductive pool shells, bonding at four points shall not be required.

The requirement for bonding perimeter surfaces applies to paved and unpaved surfaces. An example of an unpaved perimeter surface would be the lawn surrounding a permanently installed aboveground swimming pool. Where the paved portion of the perimeter surface extends less than 3 feet horizontally from the inside walls of the pool, the perimeter bonding grid must be continued under the adjacent unpaved perimeter surface. If physical constraints (such as a wall or other physical barrier) prevent the perimeter from extending 3 feet beyond the inside walls of the pool, the bonding grid is required only to extend under the available perimeter area.

The perimeter bonding grid can be comprised of structural reinforcing metal (rebar or welded wire mesh) that is conductive to the perimeter surface and installed in or under the perimeter surface. Where structural reinforcing steel is not available, a single, bare, solid 8 AWG or larger copper conductor can be installed around the pool's perimeter in an area measuring between 18 inches and 24 inches from the inside pool walls. This 8 AWG bonding conductor can be installed in the paving material (i.e., in the concrete), or it can be buried in the material (*subgrade*) below the paving material. Where buried, the bonding conductor is to be not less than 4 inches and not more than 6 inches below the surface level of the subgrade material.

The perimeter surface bonding medium has to be connected, at four evenly spaced points around the pool perimeter, to either the structural steel of a conductive pool shell or to the copper bonding grid provided for the conductive pool shell that has encapsulated rebar or no rebar at all. Connection between the perimeter bonding medium and nonconductive pool shells is not required.

(a) *Structural Reinforcing Steel.* Structural reinforcing steel shall be bonded in accordance with 680.26(B)(1)(a).

(b) *Alternate Means.* Where structural reinforcing steel is not available or is encapsulated in a nonconductive compound, a copper conductor(s) shall be utilized where the following requirements are met:

(1) At least one minimum 8 AWG bare solid copper conductor shall be provided.
(2) The conductors shall follow the contour of the perimeter surface.
(3) Only listed splices shall be permitted.

(4) The required conductor shall be 450 mm to 600 mm (18 in. to 24 in.) from the inside walls of the pool.
(5) The required conductor shall be secured within or under the perimeter surface 100 mm to 150 mm (4 in. to 6 in.) below the subgrade.

**(3) Metallic Components.** All metallic parts of the pool structure, including reinforcing metal not addressed in 680.26(B)(1)(a), shall be bonded. Where reinforcing steel is encapsulated with a nonconductive compound, the reinforcing steel shall not be required to be bonded.

**(4) Underwater Lighting.** All metal forming shells and mounting brackets of no-niche luminaires shall be bonded.

*Exception: Listed low-voltage lighting systems with nonmetallic forming shells shall not require bonding.*

**(5) Metal Fittings.** All metal fittings within or attached to the pool structure shall be bonded. Isolated parts that are not over 100 mm (4 in.) in any dimension and do not penetrate into the pool structure more than 25 mm (1 in.) shall not require bonding.

**(6) Electrical Equipment.** Metal parts of electrical equipment associated with the pool water circulating system, including pump motors and metal parts of equipment associated with pool covers, including electric motors, shall be bonded.

*Exception: Metal parts of listed equipment incorporating an approved system of double insulation shall not be bonded.*

(a) *Double-Insulated Water Pump Motors.* Where a double-insulated water pump motor is installed under the provisions of this rule, a solid 8 AWG copper conductor of sufficient length to make a bonding connection to a replacement motor shall be extended from the bonding grid to an accessible point in the vicinity of the pool pump motor. Where there is no connection between the swimming pool bonding grid and the equipment grounding system for the premises, this bonding conductor shall be connected to the equipment grounding conductor of the motor circuit.

(b) *Pool Water Heaters.* For pool water heaters rated at more than 50 amperes and having specific instructions regarding bonding and grounding, only those parts designated to be bonded shall be bonded and only those parts designated to be grounded shall be grounded.

**(7) Fixed Metal Parts.** All fixed metal parts shall be bonded including, but not limited to, metal-sheathed cables and raceways, metal piping, metal awnings, metal fences, and metal door and window frames.

*Exception No. 1: Those separated from the pool by a permanent barrier that prevents contact by a person shall not be required to be bonded.*

*Exception No. 2: Those greater than 1.5 m (5 ft) horizontally from the inside walls of the pool shall not be required to be bonded.*

*Exception No. 3: Those greater than 3.7 m (12 ft) measured vertically above the maximum water level of the pool, or as measured vertically above any observation stands, towers, or platforms, or any diving structures, shall not be required to be bonded.*

The metal parts required to be bonded include all metal parts of electrical equipment associated with the water-circulating system of the pool, all metal parts of the pool structure, and all fixed metal parts within 5 feet of the inside walls of the pool and not separated by a permanent barrier. The bonding of these parts can be accomplished by one or more of the following methods using a solid 8 AWG or larger, insulated, covered, or bare copper conductor:

- Connecting the parts directly to each other in series or parallel configurations
- Connecting the parts to the unencapsulated structural metal forming the shell of a conductive pool or connecting the parts to a copper conductor grid system used around the contour of a conductive pool shell
- Connecting the parts together using the pool shell constructed of bolted or welded steel as a common connection point
- Connecting the parts to the perimeter bonding grid consisting of either structural reinforcing steel (rebar or welded wire mesh) or a solid 8 AWG bare copper conductor encircling the pool's perimeter

Brass or other corrosion-resistant rigid metal conduit (RMC) can also be used as a bonding conductor for connecting metal parts together. See Exhibit 680.7 for an example of using brass RMC as the method of bonding two electrical enclosures and as a point to connect bonding jumpers run to the pool reinforcing steel and stainless steel ladder.

As specified in 250.7, exothermic welding, pressure connectors and clamps specifically listed for the purpose, and other listed means are permitted as the method of connecting bonding conductors to swimming pool equipment. Connections in pool areas must be suitable for wet conditions and high levels of chlorine. High concentrations of chlorine in swimming pool water make the wet locations in the vicinity of swimming pool areas (including many pool pump rooms) a corrosive environment. The integrity of the bonding connections should be periodically inspected, particularly those bonding connections between the 8 AWG copper conductor and an aluminum (or other dissimilar metal) ladder. See Exhibit 680.9 for an illustration of two acceptable methods of making swimming pool bonding connections.

**(C) Pool Water.** Where none of the bonded parts is in direct connection with the pool water, the pool water shall be in direct contact with an approved corrosion-resistant conductive surface that exposes not less than 5800 mm$^2$ (9 in.$^2$) of surface area to the pool water at all times. The conductive surface shall be located where it is not exposed to physical damage or dislodgement

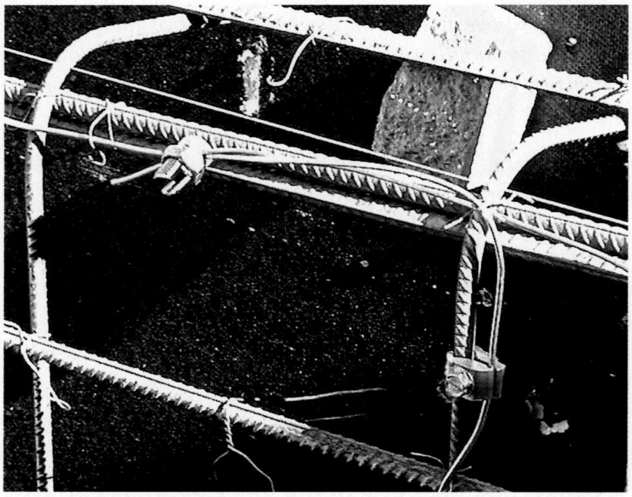

***EXHIBIT 680.9*** *Use of listed connectors for making bonding connections at a swimming pool.*

during usual pool activities, and it shall be bonded in accordance with 680.26(B).

A conductive element that is part of the pool bonding system must be in direct contact with the pool water. Where bonded items such as underwater luminaires, rails, or ladders are in direct contact with the pool water and provide the required surface area, it is not necessary to provide another conductive element. A conductive pool shell also satisfies this requirement. However, where the pool does not include any of these items, it is necessary to install one. Devices have been specifically listed as a means to provide this contact with the pool water. The requirement applies to all permanently installed swimming pools covered by the requirements in Part II of Article 680.

## 680.27 Specialized Pool Equipment

**(A) Underwater Audio Equipment.** All underwater audio equipment shall be identified.

**(1) Speakers.** Each speaker shall be mounted in an approved metal forming shell, the front of which is enclosed by a captive metal screen, or equivalent, that is bonded to, and secured to, the forming shell by a positive locking device that ensures a low-resistance contact and requires a tool to open for installation or servicing of the speaker. The forming shell shall be installed in a recess in the wall or floor of the pool.

**(2) Wiring Methods.** Rigid metal conduit of brass or other identified corrosion-resistant metal, liquidtight flexible nonmetallic conduit (LFNC-B), rigid polyvinyl chloride conduit, or reinforced thermosetting resin conduit shall extend from the forming shell to a listed junction box or other enclosure as provided in 680.24. Where rigid polyvinyl chloride conduit, reinforced thermosetting resin conduit, or liquidtight flexible nonmetallic conduit is used, an 8 AWG insulated solid or stranded copper bonding jumper shall be installed in this conduit. The bonding jumper shall be terminated in the forming shell and the junction box. The termination of the 8 AWG bonding jumper in the forming shell shall be covered with, or encapsulated in, a listed potting compound to protect such connection from the possible deteriorating effect of pool water.

**(3) Forming Shell and Metal Screen.** The forming shell and metal screen shall be of brass or other approved corrosion-resistant metal. All forming shells shall include provisions for terminating an 8 AWG copper conductor.

**(B) Electrically Operated Pool Covers.**

**(1) Motors and Controllers.** The electric motors, controllers, and wiring shall be located not less than 1.5 m (5 ft) from the inside wall of the pool unless separated from the pool by a wall, cover, or other permanent barrier. Electric motors installed below grade level shall be of the totally enclosed type. The device that controls the operation of the motor for an electrically operated pool cover shall be located such that the operator has full view of the pool.

Informational Note No. 1: For cabinets installed in damp and wet locations, see 312.2.

Informational Note No. 2: For switches or circuit breakers installed in wet locations, see 404.4.

Informational Note No. 3: For protection against liquids, see 430.11.

**(2) Protection.** The electric motor and controller shall be connected to a branch circuit protected by a ground-fault circuit interrupter.

**(C) Deck Area Heating.** The provisions of this section shall apply to all pool deck areas, including a covered pool, where electrically operated comfort heating units are installed within 6.0 m (20 ft) of the inside wall of the pool.

**(1) Unit Heaters.** Unit heaters shall be rigidly mounted to the structure and shall be of the totally enclosed or guarded type. Unit heaters shall not be mounted over the pool or within the area extending 1.5 m (5 ft) horizontally from the inside walls of a pool.

**(2) Permanently Wired Radiant Heaters.** Radiant electric heaters shall be suitably guarded and securely fastened to their mounting device(s). Heaters shall not be installed over a pool or within the area extending 1.5 m (5 ft) horizontally from the inside walls of the pool and shall be mounted at least 3.7 m (12 ft) vertically above the pool deck unless otherwise approved.

**(3) Radiant Heating Cables Not Permitted.** Radiant heating cables embedded in or below the deck shall not be permitted.

Only unit heaters and permanently connected radiant heaters are permitted in the area that extends 5 feet to 20 feet horizontally from the inside walls of a pool.

## III. Storable Pools, Storable Spas, and Storable Hot Tubs

### 680.30 General

Electrical installations at storable pools, storable spas, or storable hot tubs shall comply with the provisions of Part I and Part III of this article.

Pools, spas, and hot tubs of any dimension with inflatable walls are considered storable. Other storable units are those that can be readily disassembled and are limited to a maximum water depth of 42 inches. See the defined terms *storable swimming, wading, or immersion pools,* or *storable/portable spas and hot tubs* in 680.2. A storable pool, its associated equipment, and perimeter do not require equipotential bonding conductors. However, the filter pump must be double insulated, and an EGC that is an integral part of the flexible cord is required. The 3-foot length limitation for flexible cords in 680.7 does not apply to the power cord of equipment listed for use with a storable swimming pool. All electrical equipment used with a storable pool is required to have GFCI protection for personnel. Exhibit 680.10 illustrates an inflatable swimming pool that is considered to be a storable swimming pool regardless of the water depth.

Underwriters Laboratories developed testing and labeling criteria for listing the pump/filter units designed especially for storable pools. This equipment has the following characteristics:

1. It must have an approved system of double insulation or the equivalent.
2. It is permitted to have a flexible cord equipped with a parallel-blade, grounding-type attachment plug for electrical connection.
3. It must have a grounding conductor included in the flexible cord.
4. The flexible cord is not limited to 3 feet, as required by 680.7(A), but is specified by UL to be not less than 25 feet long. This length was chosen to discourage the use of extension cords.

***EXHIBIT 680.10** Example of an inflatable swimming pool subject to the requirements in Parts I and III of Article 680.*

The UL labeling requirement for these listed units includes the wording "Do Not Use with Permanently Installed Pools." Despite this marking, consumers and swimming pool installers often attempt to use these pump/filter units on any pool, regardless of the pool's dimensions or its "storability."

### 680.31 Pumps

A cord-connected pool filter pump shall incorporate an approved system of double insulation or its equivalent and shall be provided with means for grounding only the internal and nonaccessible non–current-carrying metal parts of the appliance.

The means for grounding shall be an equipment grounding conductor run with the power-supply conductors in the flexible cord that is properly terminated in a grounding-type attachment plug having a fixed grounding contact member.

Cord-connected pool filter pumps shall be provided with a ground-fault circuit interrupter that is an integral part of the attachment plug or located in the power supply cord within 300 mm (12 in.) of the attachment plug.

### 680.32 Ground-Fault Circuit Interrupters Required

All electrical equipment, including power-supply cords, used with storable pools shall be protected by ground-fault circuit interrupters.

All 125-volt, 15- and 20-ampere receptacles located within 6.0 m (20 ft) of the inside walls of a storable pool, storable spa, or storable hot tub shall be protected by a ground-fault circuit interrupter. In determining these dimensions, the distance to be measured shall be the shortest path the supply cord of an appliance connected to the receptacle would follow without piercing a floor, wall, ceiling, doorway with hinged or sliding door, window opening, or other effective permanent barrier.

Informational Note: For flexible cord usage, see 400.4.

### 680.33 Luminaires

An underwater luminaire, if installed, shall be installed in or on the wall of the storable pool, storable spa, or storable hot tub. It shall comply with either 680.33(A) or (B).

Listed luminaire assemblies are permitted to be cord-and-plug-connected to facilitate disconnection and removal when the storable pool is disassembled.

**(A) Within the Low Voltage Contact Limit.** A luminaire shall be part of a cord-and plug connected lighting assembly. This assembly shall be listed as an assembly for the purpose and have the following construction features:

(1) No exposed metal parts
(2) A luminaire lamp that is suitable for use at the supplied voltage
(3) An impact-resistant polymeric lens, luminaire body, and transformer enclosure
(4) A transformer or power supply meeting the requirements of 680.23(A)(2) with a primary rating not over 150 V

**(B) Over the Low Voltage Contact Limit But Not over 150 Volts.** A lighting assembly without a transformer or power supply and with the luminaire lamp(s) operating at not over 150 volts shall be permitted to be cord-and-plug-connected where the assembly is listed as an assembly for the purpose. The installation shall comply with 680.23(A)(5), and the assembly shall have the following construction features:

(1) No exposed metal parts
(2) An impact-resistant polymeric lens and luminaire body
(3) A ground-fault circuit interrupter with open neutral conductor protection as an integral part of the assembly
(4) The luminaire lamp permanently connected to the ground-fault circuit interrupter with open-neutral protection
(5) Compliance with the requirements of 680.23(A)

### 680.34 Receptacle Locations

Receptacles shall not be located less than 1.83 m (6 ft) from the inside walls of a storable pool, storable spa, or storable hot tub. In determining these dimensions, the distance to be measured shall be the shortest path the supply cord of an appliance connected to the receptacle would follow without piercing a floor, wall, ceiling, doorway with hinged or sliding door, window opening, or other effective permanent barrier.

## IV. Spas and Hot Tubs

### 680.40 General

Electrical installations at spas and hot tubs shall comply with the provisions of Part I and Part IV of this article.

### 680.41 Emergency Switch for Spas and Hot Tubs

A clearly labeled emergency shutoff or control switch for the purpose of stopping the motor(s) that provide power to the recirculation system and jet system shall be installed at a point readily accessible to the users and not less than 1.5 m (5 ft) away, adjacent to, and within sight of the spa or hot tub. This requirement shall not apply to single-family dwellings.

A local disconnecting device that is capable of being used in an emergency is required for spas and hot tubs. This addresses entrapment hazards associated with spas and hot tubs. For more information regarding this issue, visit the U.S. Consumer Product Safety Commission online at www.poolsafely.gov.

The emergency shutoff switch must be installed within sight of and at least 5 feet from the spa or hot tub and must be clearly labeled "Emergency Shutoff." See Exhibit 680.11 for an illustration of the switch location. The shutoff switch can be either a line-operated device or a remote-control circuit that causes the pump circuit to open. This requirement does not apply to one-family dwellings.

### 680.42 Outdoor Installations

A spa or hot tub installed outdoors shall comply with the provisions of Parts I and II of this article, except as permitted in

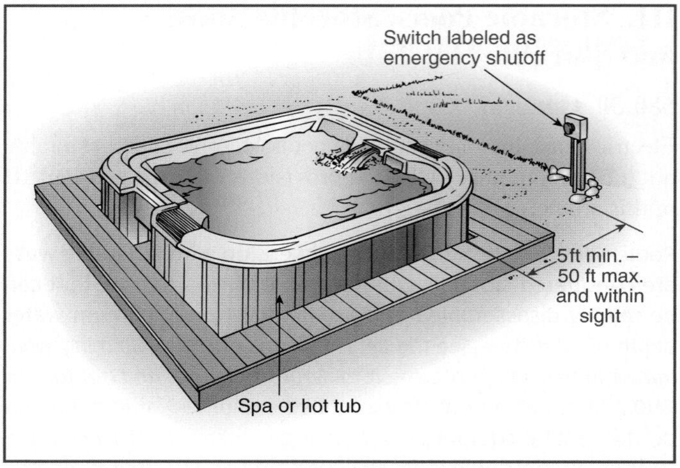


***EXHIBIT 680.11*** *Location of the required emergency shutoff device.*

680.42(A) and (B), that would otherwise apply to pools installed outdoors.

**(A) Flexible Connections.** Listed packaged spa or hot tub equipment assemblies or self-contained spas or hot tubs utilizing a factory-installed or assembled control panel or panelboard shall be permitted to use flexible connections as covered in 680.42(A)(1) and (A)(2).

**(1) Flexible Conduit.** Liquidtight flexible metal conduit or liquidtight flexible nonmetallic conduit shall be permitted.

**(2) Cord-and-Plug Connections.** Cord-and-plug connections with a cord not longer than 4.6 m (15 ft) shall be permitted where protected by a ground-fault circuit interrupter.

**(B) Bonding.** Bonding by metal-to-metal mounting on a common frame or base shall be permitted. The metal bands or hoops used to secure wooden staves shall not be required to be bonded as required in 680.26.

Equipotential bonding of perimeter surfaces in accordance with 680.26(B)(2) shall not be required to be provided for spas and hot tubs where all of the following conditions apply:

(1) The spa or hot tub shall be listed as a self-contained spa for aboveground use.
(2) The spa or hot tub shall not be identified as suitable only for indoor use.
(3) The installation shall be in accordance with the manufacturer's instructions and shall be located on or above grade.
(4) The top rim of the spa or hot tub shall be at least 710 mm (28 in.) above all perimeter surfaces that are within 760 mm (30 in.), measured horizontally from the spa or hot tub. The height of nonconductive external steps for entry to or exit from the self-contained spa shall not be used to reduce or increase this rim height measurement.

Informational Note: For information regarding listing requirements for self-contained spas and hot tubs, see ANSI/UL 1563-2010,

*Standard for Electric Spas, Equipment Assemblies, and Associated Equipment.*

The equipotential bonding requirements of 680.26 are amended by 680.42(B) only for three conditions:

1. Bonding of metal bands or hoops used to secure wooden staves is not required.
2. Bonding by metal-to-metal mounting on a common base is permitted.
3. Perimeter surface bonding of listed self-contained spas and tubs meeting the stated conditions is not required.

All the other bonding requirements in 680.26 apply to outdoor spas and hot tubs, including the packaged and self-contained types.

**(C) Interior Wiring to Outdoor Installations.** In the interior of a dwelling unit or in the interior of another building or structure associated with a dwelling unit, any of the wiring methods recognized or permitted in Chapter 3 of this *Code* that contains a copper equipment grounding conductor that is insulated or enclosed within the outer sheath of the wiring method and not smaller than 12 AWG shall be permitted to be used for the connection to motor, heating, and control loads that are part of a self-contained spa or hot tub or a packaged spa or hot tub equipment assembly. Wiring to an underwater luminaire shall comply with 680.23 or 680.33.

## 680.43 Indoor Installations

A spa or hot tub installed indoors shall comply with the provisions of Parts I and II of this article except as modified by this section and shall be connected by the wiring methods of Chapter 3.

*Exception No. 1: Listed spa and hot tub packaged units rated 20 amperes or less shall be permitted to be cord-and-plug-connected to facilitate the removal or disconnection of the unit for maintenance and repair.*

*Exception No. 2: The equipotential bonding requirements for perimeter surfaces in 680.26(B)(2) shall not apply to a listed self-contained spa or hot tub installed above a finished floor.*

*Exception No. 3: For a dwelling unit(s) only, where a listed spa or hot tub is installed indoors, the wiring method requirements of 680.42(C) shall also apply.*

The raceway requirement in 680.25 is modified by Exception No. 3, which permits any Chapter 3 wiring method containing a copper EGC for listed spa or tub installations in a dwelling unit.

**(A) Receptacles.** At least one 125-volt, 15- or 20-ampere receptacle on a general-purpose branch circuit shall be located not less than 1.83 m (6 ft) from, and not exceeding 3.0 m (10 ft) from, the inside wall of the spa or hot tub.

**(1) Location.** Receptacles shall be located at least 1.83 m (6 ft) measured horizontally from the inside walls of the spa or hot tub.

**(2) Protection, General.** Receptacles rated 125 volts and 30 amperes or less and located within 3.0 m (10 ft) of the inside walls of a spa or hot tub shall be protected by a ground-fault circuit interrupter.

**(3) Protection, Spa or Hot Tub Supply Receptacle.** Receptacles that provide power for a spa or hot tub shall be ground-fault circuit-interrupter protected.

**(4) Measurements.** In determining the dimensions in this section addressing receptacle spacings, the distance to be measured shall be the shortest path the supply cord of an appliance connected to the receptacle would follow without piercing a floor, wall, ceiling, doorway with hinged or sliding door, window opening, or other effective permanent barrier.

**(B) Installation of Luminaires, Lighting Outlets, and Ceiling-Suspended (Paddle) Fans.**

**(1) Elevation.** Luminaires, except as covered in 680.43(B)(2), lighting outlets, and ceiling-suspended (paddle) fans located over the spa or hot tub or within 1.5 m (5 ft) from the inside walls of the spa or hot tub shall comply with the clearances specified in (B)(1)(a), (B)(1)(b), and (B)(1)(c) above the maximum water level.

(a) *Without GFCI.* Where no GFCI protection is provided, the mounting height shall be not less than 3.7 m (12 ft).

(b) *With GFCI.* Where GFCI protection is provided, the mounting height shall be permitted to be not less than 2.3 m (7 ft 6 in.).

(c) *Below 2.3 m (7 ft 6 in.).* Luminaires meeting the requirements of item (1) or (2) and protected by a ground-fault circuit interrupter shall be permitted to be installed less than 2.3 m (7 ft 6 in.) over a spa or hot tub:

(1) Recessed luminaires with a glass or plastic lens, nonmetallic or electrically isolated metal trim, and suitable for use in damp locations
(2) Surface-mounted luminaires with a glass or plastic globe, a nonmetallic body, or a metallic body isolated from contact, and suitable for use in damp locations

**(2) Underwater Applications.** Underwater luminaires shall comply with the provisions of 680.23 or 680.33.

**(C) Switches.** Switches shall be located at least 1.5 m (5 ft), measured horizontally, from the inside walls of the spa or hot tub.

Receptacles, wall switches, and electrical devices and controls not associated with a spa or hot tub are required to be located at least 5 feet from the inside wall of the spa or hot tub. Receptacles within 10 feet are required to be protected by a GFCI. Receptacles supplying power to a spa or hot tub are also required to be protected by a GFCI unless the unit is a listed package unit with integral GFCI protection.

Luminaires, lighting outlets, and ceiling-suspended (paddle) fans located less than 12 feet above a spa or hot tub and within

5 feet horizontally from the inside walls of the spa or hot tub are required to be protected by a GFCI.

**(D) Bonding.** The following parts shall be bonded together:

(1) All metal fittings within or attached to the spa or hot tub structure
(2) Metal parts of electrical equipment associated with the spa or hot tub water circulating system, including pump motors, unless part of a listed self-contained spa or hot tub
(3) Metal raceway and metal piping that are within 1.5 m (5 ft) of the inside walls of the spa or hot tub and that are not separated from the spa or hot tub by a permanent barrier
(4) All metal surfaces that are within 1.5 m (5 ft) of the inside walls of the spa or hot tub and that are not separated from the spa or hot tub area by a permanent barrier

*Exception: Small conductive surfaces not likely to become energized, such as air and water jets and drain fittings, where not connected to metallic piping, towel bars, mirror frames, and similar nonelectrical equipment, shall not be required to be bonded.*

(5) Electrical devices and controls that are not associated with the spas or hot tubs and that are located less than 1.5 m (5 ft) from such units; otherwise, they shall be bonded to the spa or hot tub system

Bonding requirements for spas and hot tubs are similar to those in Parts I and II of Article 680, except that metal-to-metal mounting on a common frame or base is an acceptable bonding method.

Small conductive surfaces such as air and water jets, drain fittings, and towel bars are not required to be bonded. See 680.43(D)(4), Exception.

**(E) Methods of Bonding.** All metal parts associated with the spa or hot tub shall be bonded by any of the following methods:

(1) The interconnection of threaded metal piping and fittings
(2) Metal-to-metal mounting on a common frame or base
(3) The provisions of a solid copper bonding jumper, insulated, covered, or bare, not smaller than 8 AWG

**(F) Grounding.** The following equipment shall be grounded:

(1) All electrical equipment located within 1.5 m (5 ft) of the inside wall of the spa or hot tub
(2) All electrical equipment associated with the circulating system of the spa or hot tub

**(G) Underwater Audio Equipment.** Underwater audio equipment shall comply with the provisions of Part II of this article.

### 680.44 Protection

Except as otherwise provided in this section, the outlet(s) that supplies a self-contained spa or hot tub, a packaged spa or hot tub equipment assembly, or a field-assembled spa or hot tub shall be protected by a ground-fault circuit interrupter.

**(A) Listed Units.** If so marked, a listed self-contained unit or listed packaged equipment assembly that includes integral ground-fault circuit-interrupter protection for all electrical parts within the unit or assembly (pumps, air blowers, heaters, lights, controls, sanitizer generators, wiring, and so forth) shall be permitted without additional GFCI protection.

**(B) Other Units.** A field-assembled spa or hot tub rated 3 phase or rated over 250 volts or with a heater load of more than 50 amperes shall not require the supply to be protected by a ground-fault circuit interrupter.

Informational Note: See 680.2 for definitions of *self-contained spa or hot tub* and for *packaged spa or hot tub equipment assembly.*

Spas and hot tubs using voltages over 250 volts or 3-phase power are not required to have GFCI protection, because GFCI devices are not available in all voltage, amperage, and phasing arrangements.

## V. Fountains

Part V applies to permanently installed decorative fountains and reflecting pools in the ground, partially in the ground, or in a building. These units are primarily for aesthetic value and are not intended for swimming or wading.

Part V does not cover installations in natural lakes, rivers, or ponds. Such installations are covered by the requirements of Article 682.

### 680.50 General

The provisions of Part I and Part V of this article shall apply to all permanently installed fountains as defined in 680.2. Fountains that have water common to a pool shall additionally comply with the requirements in Part II of this article. Part V does not cover self-contained, portable fountains. Portable fountains shall comply with Parts II and III of Article 422.

### 680.51 Luminaires, Submersible Pumps, and Other Submersible Equipment

**(A) Ground-Fault Circuit Interrupter.** Luminaires, submersible pumps, and other submersible equipment, unless listed for operation at low voltage contact limit or less and supplied by a transformer or power supply that complies with 680.23(A)(2), shall be protected by a ground-fault circuit interrupter.

**(B) Operating Voltage.** No luminaires shall be installed for operation on supply circuits over 150 volts between conductors. Submersible pumps and other submersible equipment shall operate at 300 volts or less between conductors.

**(C) Luminaire Lenses.** Luminaires shall be installed with the top of the luminaire lens below the normal water level of the fountain unless listed for above-water locations. A luminaire facing upward shall comply with either (1) or (2):

(1) Have the lens guarded to prevent contact by any person
(2) Be listed for use without a guard

**(D) Overheating Protection.** Electrical equipment that depends on submersion for safe operation shall be protected against overheating by a low-water cutoff or other approved means when not submerged.

**(E) Wiring.** Equipment shall be equipped with provisions for threaded conduit entries or be provided with a suitable flexible cord. The maximum length of each exposed cord in the fountain shall be limited to 3.0 m (10 ft). Cords extending beyond the fountain perimeter shall be enclosed in approved wiring enclosures. Metal parts of equipment in contact with water shall be of brass or other approved corrosion-resistant metal.

**(F) Servicing.** All equipment shall be removable from the water for relamping or normal maintenance. Luminaires shall not be permanently embedded into the fountain structure such that the water level must be reduced or the fountain drained for relamping, maintenance, or inspection.

**(G) Stability.** Equipment shall be inherently stable or be securely fastened in place.

## 680.52 Junction Boxes and Other Enclosures

**(A) General.** Junction boxes and other enclosures used for other than underwater installation shall comply with 680.24.

**(B) Underwater Junction Boxes and Other Underwater Enclosures.** Junction boxes and other underwater enclosures shall meet the requirements of 680.52(B)(1) and (B)(2).

**(1) Construction.**

(a) Underwater enclosures shall be equipped with provisions for threaded conduit entries or compression glands or seals for cord entry.

(b) Underwater enclosures shall be submersible and made of copper, brass, or other approved corrosion-resistant material.

**(2) Installation.** Underwater enclosure installations shall comply with (a) and (b).

(a) Underwater enclosures shall be filled with an approved potting compound to prevent the entry of moisture.

(b) Underwater enclosures shall be firmly attached to the supports or directly to the fountain surface and bonded as required. Where the junction box is supported only by conduits in accordance with 314.23(E) and (F), the conduits shall be of copper, brass, stainless steel, or other approved corrosion-resistant metal. Where the box is fed by nonmetallic conduit, it shall have additional supports and fasteners of copper, brass, or other approved corrosion-resistant material.

## 680.53 Bonding

All metal piping systems associated with the fountain shall be bonded to the equipment grounding conductor of the branch circuit supplying the fountain.

Informational Note: See 250.122 for sizing of these conductors.

## 680.54 Grounding

The following equipment shall be grounded:

(1) Other than listed low-voltage luminaires not requiring grounding, all electrical equipment located within the fountain or within 1.5 m (5 ft) of the inside wall of the fountain
(2) All electrical equipment associated with the recirculating system of the fountain
(3) Panelboards that are not part of the service equipment and that supply any electrical equipment associated with the fountain

## 680.55 Methods of Grounding

**(A) Applied Provisions.** The provisions of 680.21(A), 680.23(B)(3), 680.23(F)(1) and (F)(2), 680.24(F), and 680.25 shall apply.

**(B) Supplied by a Flexible Cord.** Electrical equipment that is supplied by a flexible cord shall have all exposed non–current-carrying metal parts grounded by an insulated copper equipment grounding conductor that is an integral part of this cord. The equipment grounding conductor shall be connected to an equipment grounding terminal in the supply junction box, transformer enclosure, power supply enclosure, or other enclosure.

## 680.56 Cord-and-Plug-Connected Equipment

**(A) Ground-Fault Circuit Interrupter.** All electrical equipment, including power-supply cords, shall be protected by ground-fault circuit interrupters.

**(B) Cord Type.** Flexible cord immersed in or exposed to water shall be of a type for extra-hard usage, as designated in Table 400.4, and shall be a listed type with a "W" suffix.

**(C) Sealing.** The end of the flexible cord jacket and the flexible cord conductor termination within equipment shall be covered with, or encapsulated in, a suitable potting compound to prevent the entry of water into the equipment through the cord or its conductors. In addition, the ground connection within equipment shall be similarly treated to protect such connections from the deteriorating effect of water that may enter into the equipment.

**(D) Terminations.** Connections with flexible cord shall be permanent, except that grounding-type attachment plugs and receptacles shall be permitted to facilitate removal or disconnection for maintenance, repair, or storage of fixed or stationary equipment not located in any water-containing part of a fountain.

### 680.57 Signs

**(A) General.** This section covers electric signs installed within a fountain or within 3.0 m (10 ft) of the fountain edge.

**(B) Ground-Fault Circuit-Interrupter Protection for Personnel.** Branch circuits or feeders supplying the sign shall have ground-fault circuit-interrupter protection for personnel.

**(C) Location.**

**(1) Fixed or Stationary.** A fixed or stationary electric sign installed within a fountain shall be not less than 1.5 m (5 ft) inside the fountain measured from the outside edges of the fountain.

**(2) Portable.** A portable electric sign shall not be placed within a pool or fountain or within 1.5 m (5 ft) measured horizontally from the inside walls of the fountain.

**(D) Disconnect.** A sign shall have a local disconnecting means in accordance with 600.6 and 680.12.

**(E) Bonding and Grounding.** A sign shall be grounded and bonded in accordance with 600.7.

Electric signs in fountains are required to have GFCI protection. This protection may be provided in the feeder or branch circuit. To prevent contact by persons around the fountain, the sign must be at least 5 feet from the edge of the fountain (see Exhibit 680.12). Disconnecting and bonding requirements in Article 600 apply, and grounding must be provided in accordance with Article 250.

### 680.58 GFCI Protection for Adjacent Receptacle Outlets

All 15- or 20-ampere, single-phase 125-volt through 250-volt receptacles located within 6.0 m (20 ft) of a fountain edge shall be provided with GFCI protection.

## VI. Pools and Tubs for Therapeutic Use

Part VI recognizes therapeutic equipment in other locations, such as athletic training rooms and health care facilities. Portable therapeutic appliances, which are covered by Article 422, are required to provide protection from shock or electrocution while in the "on" or "off" position. The device used is an immersion-detection circuit interrupter (IDCI).

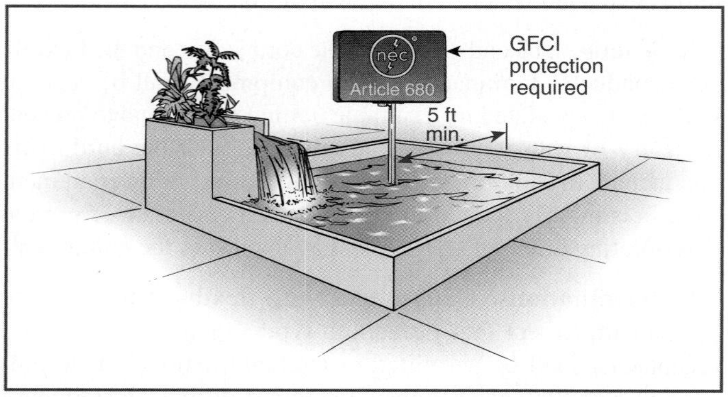


***EXHIBIT 680.12*** *Electric sign located in a fountain as described in 680.57.*

Permanently installed therapeutic pools that cannot be readily disassembled are required to comply with Parts I and II of Article 680. The limitations regarding luminaires over and around a swimming pool do not apply to therapeutic pools and tubs. Luminaires in the tub area are required to be totally enclosed, and therapeutic tubs that cannot easily be moved are subject to the same basic requirements.

Bonding and grounding requirements are similar to those in Parts I and II of Article 680, except that metal-to-metal mounting on a common frame or base is permitted. Where equipment is connected by a flexible cord, the EGC is required to be connected to a fixed metal part of the assembly.

### 680.60 General

The provisions of Part I and Part VI of this article shall apply to pools and tubs for therapeutic use in health care facilities, gymnasiums, athletic training rooms, and similar areas. Portable therapeutic appliances shall comply with Parts II and III of Article 422.

Informational Note: See 517.2 for definition of health care facilities.

### 680.61 Permanently Installed Therapeutic Pools

Therapeutic pools that are constructed in the ground, on the ground, or in a building in such a manner that the pool cannot be readily disassembled shall comply with Parts I and II of this article.

*Exception: The limitations of 680.22(B)(1) through (C)(4) shall not apply where all luminaires are of the totally enclosed type.*

### 680.62 Therapeutic Tubs (Hydrotherapeutic Tanks)

Therapeutic tubs, used for the submersion and treatment of patients, that are not easily moved from one place to another in normal use or that are fastened or otherwise secured at a specific location, including associated piping systems, shall conform to Part VI.

**(A) Protection.** Except as otherwise provided in this section, the outlet(s) that supplies a self-contained therapeutic tub or hydrotherapeutic tank, a packaged therapeutic tub or hydrotherapeutic tank, or a field-assembled therapeutic tub or hydrotherapeutic tank shall be protected by a ground-fault circuit interrupter.

**(1) Listed Units.** If so marked, a listed self-contained unit or listed packaged equipment assembly that includes integral ground-fault circuit-interrupter protection for all electrical parts within the unit or assembly (pumps, air blowers, heaters, lights, controls, sanitizer generators, wiring, and so forth) shall be permitted without additional GFCI protection.

**(2) Other Units.** A therapeutic tub or hydrotherapeutic tank rated 3 phase or rated over 250 volts or with a heater load of more than 50 amperes shall not require the supply to be protected by a ground-fault circuit interrupter.

The requirements in 680.62(A)(2) for large therapeutic tanks and therapeutic tubs are similar to the requirements for large hot tubs and spas. Large field-assembled therapeutic tubs are not required to have GFCI protection in their electrical supply when the heater load is over 50 amperes.

**(B) Bonding.** The following parts shall be bonded together:

(1) All metal fittings within or attached to the tub structure
(2) Metal parts of electrical equipment associated with the tub water circulating system, including pump motors
(3) Metal-sheathed cables and raceways and metal piping that are within 1.5 m (5 ft) of the inside walls of the tub and not separated from the tub by a permanent barrier
(4) All metal surfaces that are within 1.5 m (5 ft) of the inside walls of the tub and not separated from the tub area by a permanent barrier
(5) Electrical devices and controls that are not associated with the therapeutic tubs and located within 1.5 m (5 ft) from such units.

*Exception: Small conductive surfaces not likely to become energized, such as air and water jets and drain fittings not connected to metallic piping, and towel bars, mirror frames, and similar nonelectrical equipment not connected to metal framing, shall not be required to be bonded.*

**(C) Methods of Bonding.** All metal parts required to be bonded by this section shall be bonded by any of the following methods:

(1) The interconnection of threaded metal piping and fittings
(2) Metal-to-metal mounting on a common frame or base
(3) Connections by suitable metal clamps
(4) By the provisions of a solid copper bonding jumper, insulated, covered, or bare, not smaller than 8 AWG

**(D) Grounding.**

**(1) Fixed or Stationary Equipment.** The equipment specified in (a) and (b) shall be connected to the equipment grounding conductor.

(a) *Location.* All electrical equipment located within 1.5 m (5 ft) of the inside wall of the tub shall be connected to the equipment grounding conductor.

(b) *Circulation System.* All electrical equipment associated with the circulating system of the tub shall be connected to the equipment grounding conductor.

**(2) Portable Equipment.** Portable therapeutic appliances shall meet the grounding requirements in 250.114.

**(E) Receptacles.** All receptacles within 1.83 m (6 ft) of a therapeutic tub shall be protected by a ground-fault circuit interrupter.

**(F) Luminaires.** All luminaires used in therapeutic tub areas shall be of the totally enclosed type.

## VII. Hydromassage Bathtubs

### 680.70 General

Hydromassage bathtubs as defined in 680.2 shall comply with Part VII of this article. They shall not be required to comply with other parts of this article.

### 680.71 Protection

Hydromassage bathtubs and their associated electrical components shall be on an individual branch circuit(s) and protected by a readily accessible ground-fault circuit interrupter. All 125-volt, single-phase receptacles not exceeding 30 amperes and located within 1.83 m (6 ft) measured horizontally of the inside walls of a hydromassage tub shall be protected by a ground-fault circuit interrupter.

Hydromassage bathtubs are treated the same as ordinary bathtubs in regard to the installation of luminaires, switches, and other electrical equipment. See 410.10(D) for special requirements relating to cord-connected luminaires, hanging luminaires, and pendants near bathtubs. Also see 210.8(A)(1) and (B)(1) for requirements for GFCI protection of bathroom receptacles. The GFCI device protecting the hydromassage bathtub is required to be readily accessible. Where the GFCI device is installed in the space under a hydromassage bathtub, the opening to that space must provide ready access.

### 680.72 Other Electrical Equipment

Luminaires, switches, receptacles, and other electrical equipment located in the same room, and not directly associated with a hydromassage bathtub, shall be installed in accordance with the requirements of Chapters 1 through 4 in this *Code* covering the installation of that equipment in bathrooms.

### 680.73 Accessibility

Hydromassage bathtub electrical equipment shall be accessible without damaging the building structure or building finish. Where the hydromassage bathtub is cord- and plug-connected with the supply receptacle accessible only through a service access opening, the receptacle shall be installed so that its face is within direct view and not more than 300 mm (1 ft) of the opening.

Where a GFCI-type receptacle is installed under a hydromassage tub, the receptacle is required by 680.71 to be readily accessible. Access may be either an integral part of the tub or an access panel that is provided in the finish that encloses the tub. Where utilization equipment associated with a hydromassage bathtub is cord-and-plug-connected, the receptacle must be installed within 1 foot of the access opening and be positioned so that the face of the receptacle is visible. This requirement facilitates connection and disconnection of the cord-and-plug-connected equipment for

***EXHIBIT 680.13*** *Hydromassage bathtub circulating pump motor and supply receptacle. (Courtesy of the International Association of Electrical Inspectors)*

safety. Exhibit 680.13 shows a receptacle for a cord-and-plug-connected hydromassage bathtub positioned to meet these requirements.

### 680.74 Bonding

Both metal piping systems and grounded metal parts in contact with the circulating water shall be bonded together using a solid copper bonding jumper, insulated, covered, or bare, not smaller than 8 AWG. The bonding jumper shall be connected to the terminal on the circulating pump motor that is intended for this purpose. The bonding jumper shall not be required to be connected to a double insulated circulating pump motor. The 8 AWG or larger solid copper bonding jumper shall be required for equipotential bonding in the area of the hydromassage bathtub and shall not be required to be extended or attached to any remote panelboard, service equipment, or any electrode. The 8 AWG or larger solid copper bonding jumper shall be long enough to terminate on a replacement non-double-insulated pump motor and shall be terminated to the equipment grounding conductor of the branch circuit of the motor when a double-insulated circulating pump motor is used.

Equipotential bonding is necessary in the immediate vicinity of the hydromassage unit, and the EGC of the branch circuit supplying the hydromassage tub provides the path for ground-fault current. The 8 AWG copper bonding conductor is required to be a solid conductor. Solid conductors are required for Article 680 equipotential bonding applications in order to provide an added level of resistance to physical damage.

Double-insulated circulating pump motors are not permitted to be bonded. However, where a double-insulated water circulating pump motor is used and other metal objects associated with the water circulating system are required to be bonded, a solid, 8 AWG or larger copper bonding conductor is to be run to the pump location, with suitable length to connect it to a replacement pump that is not double insulated. In addition, during the time the double-insulated pump is in service, the bonding conductor is required to be connected to the EGC of the branch circuit supplying the hydromassage bathtub pump motor.

# ARTICLE 682 Natural and Artificially Made Bodies of Water

## I. General

### 682.1 Scope

This article applies to the installation of electrical wiring for, and equipment in and adjacent to, natural or artificially made bodies of water not covered by other articles in this *Code*, such as but not limited to aeration ponds, fish farm ponds, storm retention basins, treatment ponds, irrigation (channels) facilities.

Artificially made bodies of water do not include the pools, fountains, and spas that are specifically covered under the scope of Article 680. Electrical equipment such as pumps, luminaires, and their associated supply wiring are frequently installed in lakes, ponds, aeration and treatment basins, and similar bodies of water, and the requirements in Article 682 are designed to minimize the shock hazards inherent in those wet and damp locations.

### 682.2 Definitions

**Artificially Made Bodies of Water.** Bodies of water that have been constructed or modified to fit some decorative or commercial purpose such as, but not limited to, aeration ponds, fish farm ponds, storm retention basins, treatment ponds, and irrigation (channel) facilities. Water depths may vary seasonally or be controlled.

The term *artificially made bodies of water* includes all bodies of water that are not naturally created and that are not covered by the requirements of Article 680. The uses of artificially made bodies of water include decorative, agricultural, municipal infrastructure, and industrial. The decorative pond shown in Exhibit 682.1 is an example of an artificially made body of water because it was constructed and filled with water and did not occur naturally.

**Electrical Datum Plane.** The electrical datum plane as used in this article is defined as follows:

(1) In land areas subject to tidal fluctuation, the electrical datum plane is a horizontal plane 600 mm (2 ft) above the highest tide level for the area occurring under normal circumstances, that is, highest high tide.

*EXHIBIT 682.1 The electrical equipment associated with pumps used to circulate water in this artificial pond is subject to the requirements of Article 682.*

(2) In land areas not subject to tidal fluctuation, the electrical datum plane is a horizontal plane 600 mm (2 ft) above the highest water level for the area occurring under normal circumstances.

(3) In land areas subject to flooding, the electrical datum plane based on (1) or (2) above is a horizontal plane 600 mm (2 ft) above the point identified as the prevailing high water mark or an equivalent benchmark based on seasonal or storm-driven flooding from the authority having jurisdiction.

(4) The electrical datum plane for floating structures and landing stages that are (1) installed to permit rise and fall response to water level, without lateral movement, and (2) that are so equipped that they can rise to the datum plane established for (1) or (2) above, is a horizontal plane 750 mm (30 in.) above the water level at the floating structure or landing stage and a minimum of 300 mm (12 in.) above the level of the deck.

See the commentary on the electrical datum plane in 555.2.

**Equipotential Plane.** An area where wire mesh or other conductive elements are on, embedded in, or placed under the walk surface within 75 mm (3 in.), bonded to all metal structures and fixed nonelectrical equipment that may become energized, and connected to the electrical grounding system to prevent a difference in voltage from developing within the plane.

**Natural Bodies of Water.** Bodies of water such as lakes, streams, ponds, rivers, and other naturally occurring bodies of water, which may vary in depth throughout the year.

**Shoreline.** The farthest extent of standing water under the applicable conditions that determine the electrical datum plane for the specified body of water.

### 682.3 Other Articles

If the water is subject to boat traffic, the wiring shall comply with 555.13(B).

## II. Installation

### 682.10 Electrical Equipment and Transformers

Electrical equipment and transformers, including their enclosures, shall be specifically approved for the intended location. No portion of an enclosure for electrical equipment not identified for operation while submerged shall be located below the electrical datum plane.

See 640.10 for requirements covering the installation of audio system equipment near a body of water.

### 682.11 Location of Service Equipment

On land, the service equipment for floating structures and submersible electrical equipment shall be located no closer than 1.5 m (5 ft) horizontally from the shoreline and live parts shall be elevated a minimum of 300 mm (12 in.) above the electrical datum plane. Service equipment shall disconnect when the water level reaches the height of the established electrical datum plane.

### 682.12 Electrical Connections

All electrical connections not intended for operation while submerged shall be located at least 300 mm (12 in.) above the deck of a floating or fixed structure, but not below the electrical datum plane.

### 682.13 Wiring Methods and Installation

Liquidtight flexible metal conduit or liquidtight flexible nonmetallic conduit with approved fittings shall be permitted for feeders and where flexible connections are required for services. Extra-hard usage portable power cable listed for both wet locations and sunlight resistance shall be permitted for a feeder or a branch circuit where flexibility is required. Other wiring methods suitable for the location shall be permitted to be installed where flexibility is not required. Temporary wiring in accordance with 590.4 shall be permitted.

### 682.14 Submersible or Floating Equipment Power Connection(s)

Submersible or floating equipment shall be cord- and plug-connected, using extra-hard usage cord, as designated in Table 400.4, and listed with a "W" suffix. The plug and receptacle combination shall be arranged to be suitable for the location while in use. Disconnecting means shall be provided to isolate each submersible or floating electrical equipment from its supply connection(s) without requiring the plug to be removed from the receptacle.

*Exception: Equipment listed for direct connection and equipment anchored in place and incapable of routine movement caused by water currents or wind shall be permitted to be connected using wiring methods covered in 682.13.*

**(A) Type and Marking.** The disconnecting means shall consist of a circuit breaker, a switch, or both, or a molded case switch, and shall be specifically marked to designate which receptacle or other outlet it controls.

**(B) Location.** The disconnecting means shall be readily accessible on land, located not more than 750 mm (30 in.) from the receptacle it controls, and shall be located in the supply circuit ahead of the receptacle. The disconnecting means shall be located within sight of but not closer than 1.5 m (5 ft) from the shoreline and shall be elevated not less than 300 mm (12 in.) above the datum plane.

The general requirement on connecting submersible and floating equipment is that a cord-and-plug connection be provided; only under the conditions specified in the exception is equipment permitted to be direct-connected (hard-wired). However, the separable connection is not permitted to serve as the required disconnecting means. One of the types of disconnecting means described in 682.14(A) and located as specified in 682.14(B) must be provided for each piece of submersible or floating equipment so that power can be turned off without having to separate the cord cap from the receptacle. In the case of tidal waters, the amount of fluctuation between low and high tides has a direct impact on the datum plane and shoreline benchmarks and, consequently, on the length of the flexible cord from the receptacle to the floating or submersible equipment. See 682.2 for the definitions of *datum plane* and *shoreline.*

### 682.15 Ground-Fault Circuit-Interrupter (GFCI) Protection

Fifteen- and 20-ampere single-phase, 125-volt through 250-volt receptacles installed outdoors and in or on floating buildings or structures within the electrical datum plane area that are used for storage, maintenance, or repair where portable electric hand tools, electrical diagnostic equipment, or portable lighting equipment are to be used shall be provided with GFCI protection. The GFCI protection device shall be located not less than 300 mm (12 in.) above the established electrical datum plane.

## III. Grounding and Bonding

### 682.30 Grounding

Wiring and equipment within the scope of this article shall be grounded as specified in Part III of 553, 555.15, and with the requirements in Part III of this article.

### 682.31 Equipment Grounding Conductors

**(A) Type.** Equipment grounding conductors shall be insulated copper conductors sized in accordance with 250.122 but not smaller than 12 AWG.

**(B) Feeders.** Where a feeder supplies a remote panelboard or other distribution equipment, an insulated equipment grounding conductor shall extend from a grounding terminal in the service to a grounding terminal and busbar in the remote panelboard or other distribution equipment.

**(C) Branch Circuits.** The insulated equipment grounding conductor for branch circuits shall terminate at a grounding terminal in a remote panelboard or other distribution equipment or the grounding terminal in the main service equipment.

**(D) Cord-and-Plug-Connected Appliances.** Where grounded, cord-and-plug-connected appliances shall be grounded by means of an equipment grounding conductor in the cord and a grounding-type attachment plug.

### 682.32 Bonding of Non–Current-Carrying Metal Parts

All metal parts in contact with the water, all metal piping, tanks, and all non–current-carrying metal parts that are likely to become energized shall be bonded to the grounding terminal in the distribution equipment.

### 682.33 Equipotential Planes and Bonding of Equipotential Planes

An equipotential plane shall be installed where required in this section to mitigate step and touch voltages at electrical equipment.

**(A) Areas Requiring Equipotential Planes.** Equipotential planes shall be installed adjacent to all outdoor service equipment or disconnecting means that control equipment in or on water, that have a metallic enclosure and controls accessible to personnel, and that are likely to become energized. The equipotential plane shall encompass the area around the equipment and shall extend from the area directly below the equipment out not less than 900 mm (36 in.) in all directions from which a person would be able to stand and come in contact with the equipment.

**(B) Areas Not Requiring Equipotential Planes.** Equipotential planes shall not be required for the controlled equipment supplied by the service equipment or disconnecting means. All circuits rated not more than 60 amperes at 120 through 250 volts, single phase, shall have GFCI protection.

**(C) Bonding.** Equipotential planes shall be bonded to the electrical grounding system. The bonding conductor shall be solid copper, insulated, covered or bare, and not smaller than 8 AWG. Connections shall be made by exothermic welding or by listed pressure connectors or clamps that are labeled as being suitable for the purpose and are of stainless steel, brass, copper, or copper alloy.

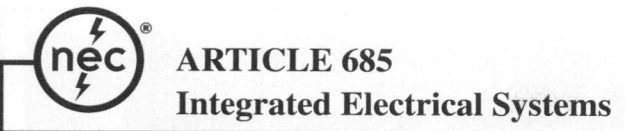

# ARTICLE 685 Integrated Electrical Systems

## I. General

### 685.1 Scope

This article covers integrated electrical systems, other than unit equipment, in which orderly shutdown is necessary to ensure safe operation. An *integrated electrical system* as used in this article is a unitized segment of an industrial wiring system where all of the following conditions are met:

(1) An orderly shutdown is required to minimize personnel hazard and equipment damage.

(2) The conditions of maintenance and supervision ensure that qualified persons service the system. The name(s) of the qualified person(s) shall be kept in a permanent record at the office of the establishment in charge of the completed installation.

A person designated as a qualified person shall possess the skills and knowledge related to the construction and operation of the electrical equipment and installation and shall have received documented safety training on the hazards involved. Documentation of their qualifications shall be on file with the office of the establishment in charge of the completed installation.

(3) Effective safeguards acceptable to the authority having jurisdiction are established and maintained.

The integrated electrical systems commonly used in large and complex industrial processes are designed, installed, and operated under stringent on-site engineering supervision. The control equipment, including overcurrent devices, is located so that it is accessible to qualified personnel, but that location might not meet – and is not required to meet – the conditions described in the Article 100 definition of *readily accessible.* Locating overcurrent devices and their associated disconnecting means so that they are not readily accessible to unqualified personnel is one of the preventive measures used to help maintain continuity of operation.

For some industrial processes, the sudden loss of electric power to vital equipment is an unacceptable level of risk, and an orderly shutdown procedure is necessary to prevent severe equipment damage, injury to personnel, or – in some extreme cases – catastrophic failure. Orderly shutdown is commonly employed in nuclear power–generating facilities, paper mills, and other areas with hazardous processes.

### 685.3 Application of Other Articles

The articles/sections in Table 685.3 apply to particular cases of installation of conductors and equipment, where there are orderly shutdown requirements that are in addition to those of this article or are modifications of them.

***TABLE 685.3*** *Application of Other Articles*

| Conductor/Equipment | Section |
|---|---|
| More than one building or other structure | 225, Part II |
| Ground-fault protection of equipment | 230.95, Exception |
| Protection of conductors | 240.4 |
| Electrical system coordination | 240.12 |
| Ground-fault protection of equipment | 240.13(1) |
| Grounding ac systems of 50 volts to less than 1000 volts | 250.21 |
| Equipment protection | 427.22 |
| Orderly shutdown | 430.44 |
| Disconnection | 430.74, Exception Nos. 1 and 2 |
| Disconnecting means in sight from controller | 430.102(A), Exception No. 2 |
| Energy from more than one source | 430.113, Exception Nos. 1 and 2 |
| Disconnecting means | 645.10, Exception |
| Uninterruptible power supplies (UPS) | 645.11(1) |
| Point of connection | 705.12(A) |

## II. Orderly Shutdown

### 685.10 Location of Overcurrent Devices in or on Premises

Location of overcurrent devices that are critical to integrated electrical systems shall be permitted to be accessible, with mounting heights permitted to ensure security from operation by unqualified personnel.

### 685.12 Direct-Current System Grounding

Two-wire dc circuits shall be permitted to be ungrounded.

### 685.14 Ungrounded Control Circuits

Where operational continuity is required, control circuits of 150 volts or less from separately derived systems shall be permitted to be ungrounded.

# ARTICLE 690 Solar Photovoltaic (PV) Systems

## I. General

### 690.1 Scope

The provisions of this article apply to solar PV electrical energy systems, including the array circuit(s), inverter(s), and

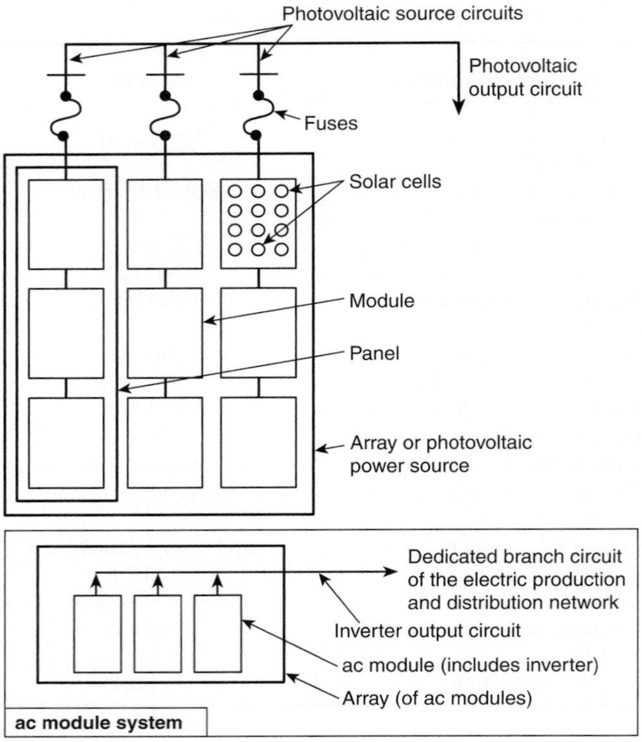


Notes:
1. These diagrams are intended to be a means of identification for photovoltaic system components, circuits, and connections.
2. Disconnecting means required by Article 690, Part III, are not shown.
3. System grounding and equipment grounding are not shown. See Article 690, Part V.

***FIGURE 690.1(a)*** *Identification of Solar Photovoltaic System Components.*

controller(s) for such systems. [See Figure 690.1(a) and Figure 690.1(b).] Solar PV systems covered by this article may be interactive with other electrical power production sources or stand-alone, with or without electrical energy storage such as batteries. These systems may have ac or dc output for utilization.

The use of photovoltaic (PV) systems as utility-interactive or stand-alone power-supply systems has steadily increased as the technology of PV equipment has evolved and its availability has improved. The requirements of Article 690 cover the use of stand-alone and utility-interactive PV systems. Utility-interactive photovoltaic systems are also subject to the requirements for interconnected electric power production sources contained in Article 705.

Exhibit 690.1 shows a typical installation of a PV array in a field.

## 690.2 Definitions

**Alternating-Current (ac) Module (Alternating-Current Photovoltaic Module).** A complete, environmentally protected unit consisting of solar cells, optics, inverter, and other components, exclusive of tracker, designed to generate ac power when exposed to sunlight.

***EXHIBIT 690.1*** *A PV array. (Courtesy of Solar Design Associates, LLC)*

An ac PV module consists of a single integrated mechanical unit. Because there is no accessible, field-installed dc wiring in this single unit, the dc PV source-circuit requirements in this *Code* are not applicable to the dc wiring in an ac PV module.

**Array.** A mechanically integrated assembly of modules or panels with a support structure and foundation, tracker, and other components, as required, to form a direct-current power-producing unit.

An array composed of multiple panels installed on a support structure is illustrated in Exhibit 690.2.

**Bipolar Photovoltaic Array.** A PV array that has two outputs, each having opposite polarity to a common reference point or center tap.

**Blocking Diode.** A diode used to block reverse flow of current into a PV source circuit.

Blocking diodes are not required by this *Code*, although the instructions or labels supplied with the PV module may require them. Blocking diodes are not overcurrent devices and may not be substituted for any overcurrent device required by the *NEC*.

**Building Integrated Photovoltaics.** Photovoltaic cells, devices, modules, or modular materials that are integrated into the outer surface or structure of a building and serve as the outer protective surface of that building.

•

**DC-to-DC Converter.** A device installed in the PV source circuit or PV output circuit that can provide an output dc voltage and current at a higher or lower value than the input dc voltage and current.

**Direct-Current (dc) Combiner.** A device used in the PV source and PV output circuits to combine two or more dc circuit inputs and provide one dc circuit output.

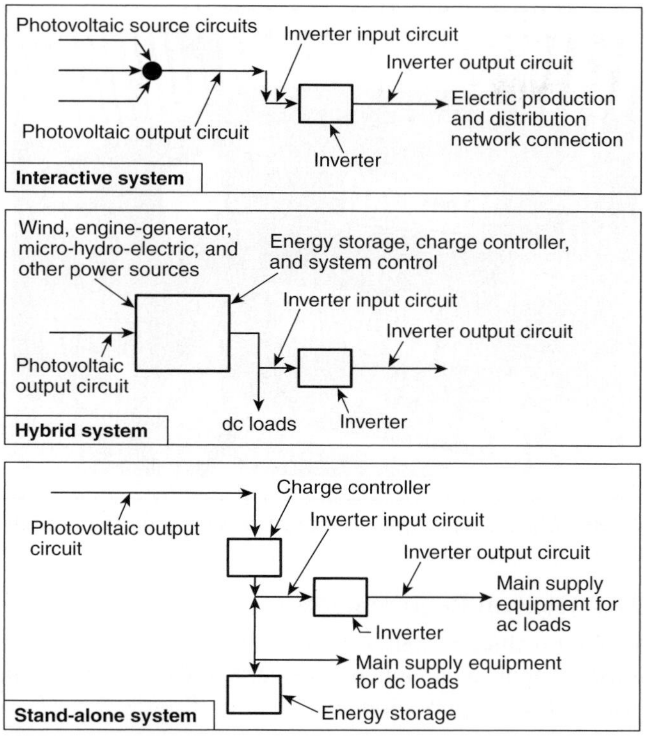


Notes:
1. These diagrams are intended to be a means of identification for photovoltaic system components, circuits, and connections.
2. Disconnecting means and overcurrent protection required by Article 690, Part III, are not shown.
3. System grounding and equipment grounding are not shown. See Article 690, Part V.
4. Custom designs occur in each configuration, and some components are optional.

***FIGURE 690.1(b)*** *Identification of Solar Photovoltaic System Components in Common System Configurations.*

***EXHIBIT 690.2*** *A PV array support structure that allows for continued use of the walkway. (Courtesy of Solar Design Associates, LLC)*

**Diversion Charge Controller.** Equipment that regulates the charging process of a battery by diverting power from energy storage to direct-current or alternating-current loads or to an interconnected utility service.

**Electrical Production and Distribution Network.** A power production, distribution, and utilization system, such as a utility system and connected loads, that is external to and not controlled by the PV power system.

•

**Interactive System.** A solar PV system that operates in parallel with and may deliver power to an electrical production and distribution network. For the purpose of this definition, an energy storage subsystem of a solar PV system, such as a battery, is not another electrical production source.

**Inverter.** Equipment that is used to change voltage level or waveform, or both, of electrical energy. Commonly, an inverter [also known as a power conditioning unit (PCU) or power conversion system (PCS)] is a device that changes dc input to an ac output. Inverters may also function as battery chargers that use alternating current from another source and convert it into direct current for charging batteries.

Exhibit 690.3 shows a utility-interactive inverter intended for use in parallel with an electric utility.

**Inverter Input Circuit.** Conductors between the inverter and the battery in stand-alone systems or the conductors between the inverter and the PV output circuits for electrical production and distribution network.

***EXHIBIT 690.3*** *A utility-interactive inverter. (Courtesy of SMA Technologies AG)*

**Inverter Output Circuit.** Conductors between the inverter and an ac panelboard for stand-alone systems or the conductors between the inverter and the service equipment or another electric power production source, such as a utility, for electrical production and distribution network.

**Module.** A complete, environmentally protected unit consisting of solar cells, optics, and other components, exclusive of tracker, designed to generate dc power when exposed to sunlight.

**Monopole Subarray.** A PV subarray that has two conductors in the output circuit, one positive (+) and one negative (−). Two monopole PV subarrays are used to form a bipolar PV array.

**Multimode Inverter.** Equipment having the capabilities of both the utility-interactive inverter and the stand-alone inverter.

**Panel.** A collection of modules mechanically fastened together, wired, and designed to provide a field-installable unit.

**Photovoltaic Output Circuit.** Circuit conductors between the PV source circuit(s) and the inverter or dc utilization equipment.

**Photovoltaic Power Source.** An array or aggregate of arrays that generates dc power at system voltage and current.

**Photovoltaic Source Circuit.** Circuits between modules and from modules to the common connection point(s) of the dc system.

**Photovoltaic System Voltage.** The direct current (dc) voltage of any PV source or PV output circuit. For multiwire installations, the PV system voltage is the highest voltage between any two dc conductors.

**Solar Cell.** The basic PV device that generates electricity when exposed to light.

•

**Stand-Alone System.** A solar PV system that supplies power independently of an electrical production and distribution network.

Exhibit 690.4 shows the controller and energy storage components of a stand-alone system.

**Subarray.** An electrical subset of a PV array.

### 690.3 Other Articles

Wherever the requirements of other articles of this *Code* and Article 690 differ, the requirements of Article 690 shall apply and, if the system is operated in parallel with a primary source(s) of electricity, the requirements in 705.14, 705.16, 705.32, and 705.143 shall apply.

*Exception: Solar PV systems, equipment, or wiring installed in a hazardous (classified) location shall also comply with the applicable portions of Articles 500 through 516.*

***EXHIBIT 690.4*** *Stand-alone system components. (Courtesy of Solar Design Associates, LLC)*

### 690.4 General Requirements

**(A) Photovoltaic Systems.** Photovoltaic systems shall be permitted to supply a building or other structure in addition to any other electrical supply system(s).

•

**(B) Equipment.** Inverters, motor generators, PV modules, PV panels, ac PV modules, dc combiners, dc-to-dc converters, and charge controllers intended for use in PV power systems shall be listed for the PV application.

The simplified circuit schematics in Exhibits 690.5 through 690.8 demonstrate the use of various components in a PV system. Specific requirements for overcurrent protection, disconnecting means, and grounding are covered in other sections of Article 690 and should not be assumed based on these drawings. Instructions for or labels on the PV module might require additional overcurrent devices that may not be shown.

Equipment listed for marine, mobile, telecommunications, or other applications may not be suitable for installation in permanent PV power systems.

•

**(C) Qualified Personnel.** The installation of equipment and all associated wiring and interconnections shall be performed only by qualified persons.

Informational Note: See Article 100 for the definition of *qualified person.*

**(D) Multiple Inverters.** A PV system shall be permitted to have multiple inverters installed in or on a single building or structure. Where the inverters are remotely located from each other, a directory in accordance with 705.10 shall be installed at each dc PV system disconnecting means, at each ac disconnecting means, and at the main service disconnecting means showing the location of all ac and dc PV system disconnecting means in the building.

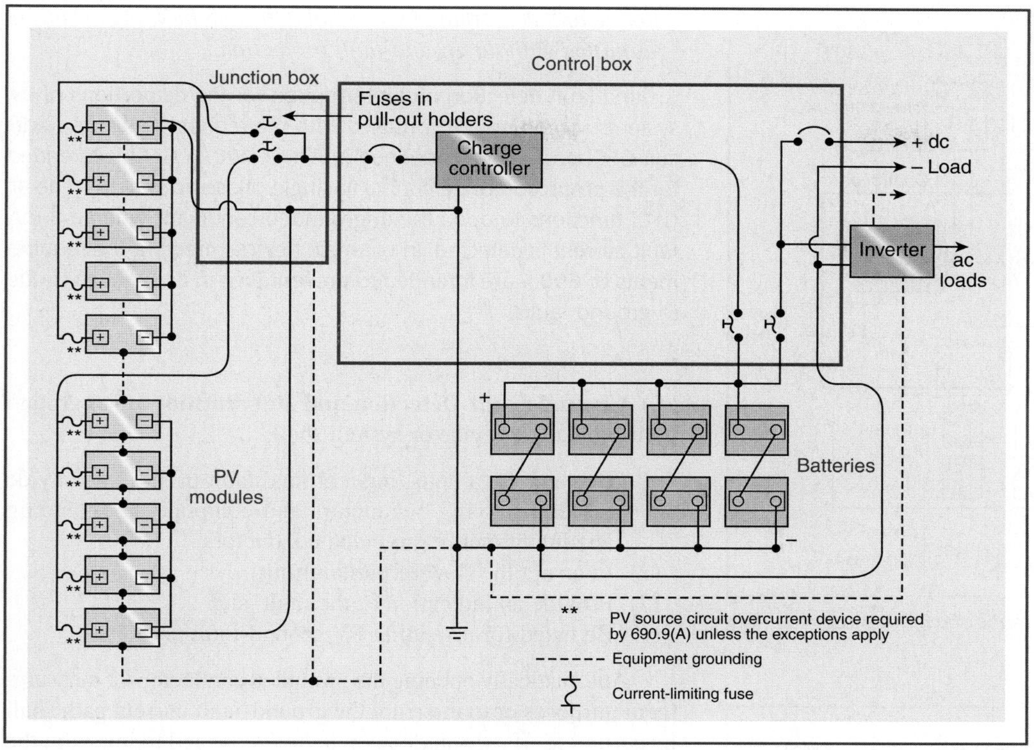


*EXHIBIT 690.5 Simplified circuit schematic of a small residential stand-alone system.*

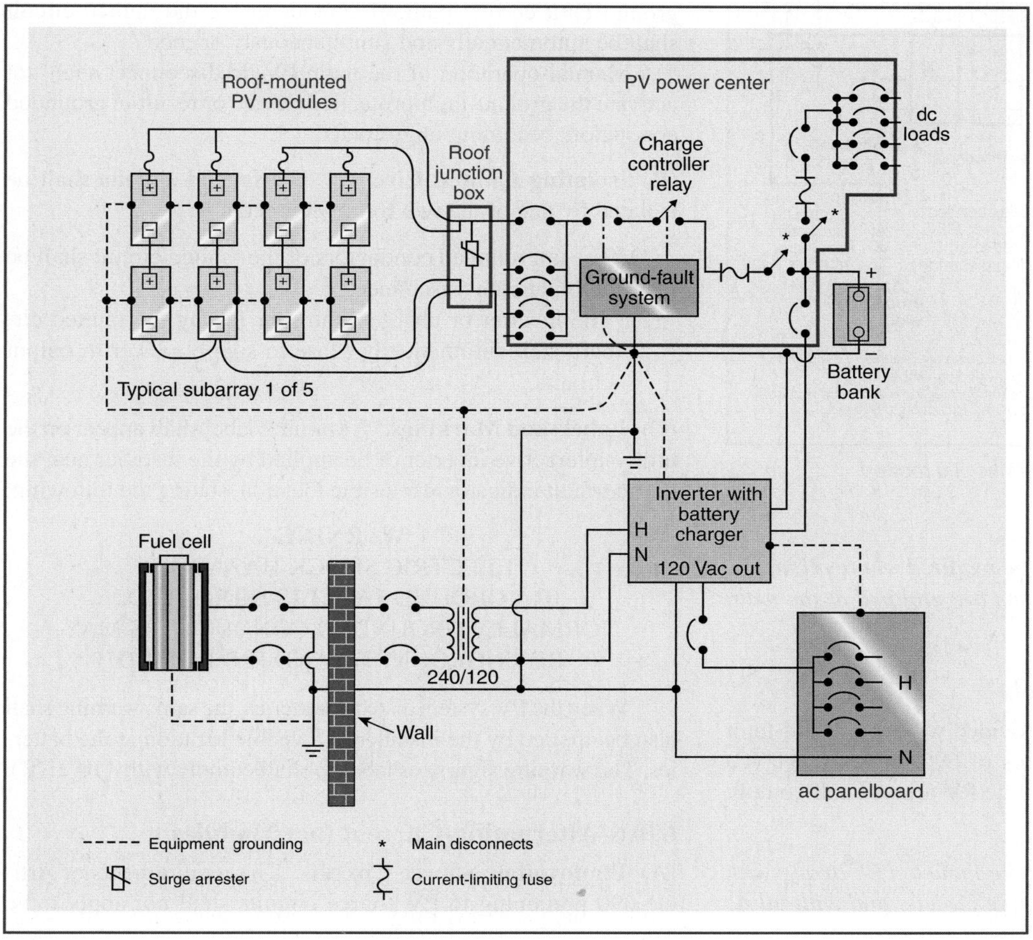


*EXHIBIT 690.6 Simplified circuit schematic of a medium-sized residential hybrid system.*

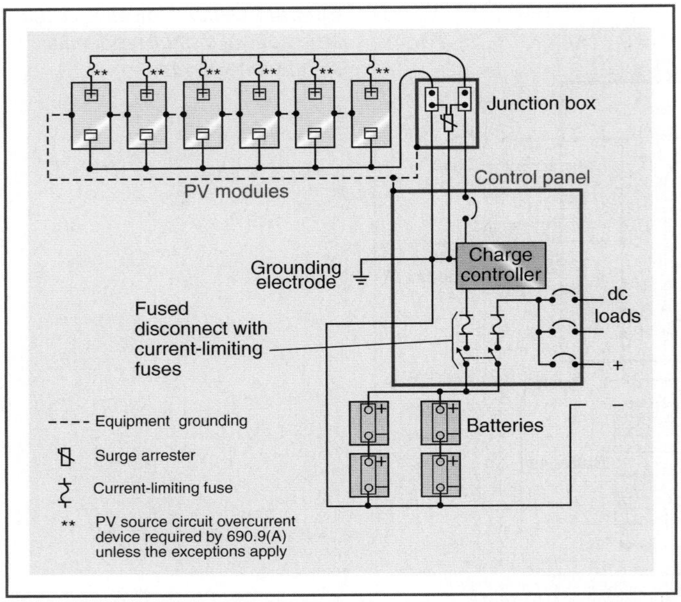


***EXHIBIT 690.7*** *Simplified circuit schematic of a remote-cabin dc-only system.*

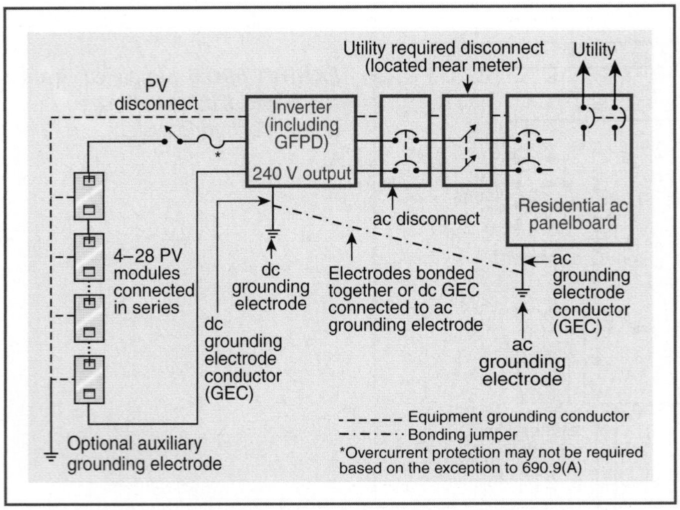


***EXHIBIT 690.8*** *Simplified circuit schematic of a rooftop grid-connected system.*

*Exception: A directory shall not be required where all inverters and PV dc disconnecting means are grouped at the main service disconnecting means.*

## 690.5 Ground-Fault Protection

Grounded dc PV arrays shall be provided with dc ground-fault protection meeting the requirements of 690.5(A) through (C) to reduce fire hazards. Ungrounded dc PV arrays shall comply with 690.35.

*Exception: Ground-mounted or pole-mounted PV arrays with not more than two paralleled source circuits and with all dc source and dc output circuits isolated from buildings shall be permitted without ground-fault protection.*

Ground-fault detection and interruption for the dc portions of PV systems should not be confused with the requirements for ac circuit GFCI protection as defined in Article 100. A GFCI is intended for the protection of personnel in single-phase ac systems. The ac GFCI functions to open the ungrounded conductor when a 5-mA fault current is detected. In contrast, devices meeting the requirements of 690.5 are intended to prevent fires in dc PV circuits due to ground faults.

•

**(A) Ground-Fault Detection and Interruption.** The ground fault protection device or system shall:

(1) Be capable of detecting a ground fault in the PV array dc current-carrying conductors and components, including any intentionally grounded conductors,
(2) Interrupt the flow of fault current
(3) Provide an indication of the fault, and
(4) Be listed for providing PV ground-fault protection

Automatically opening the grounded conductor for measurement purposes or to interrupt the ground-fault current path shall be permitted. If a grounded conductor is opened to interrupt the ground-fault current path, all conductors of the faulted circuit shall be automatically and simultaneously opened.

Manual operation of the main PV dc disconnect shall not activate the ground-fault protection device or result in grounded conductors becoming ungrounded.

**(B) Isolating Faulted Circuits.** The faulted circuits shall be isolated by one of the two following methods:

(1) The ungrounded conductors of the faulted circuit shall be automatically disconnected.
(2) The inverter or charge controller fed by the faulted circuit shall automatically cease to supply power to output circuits.

**(C) Labels and Markings.** A warning label shall appear on the utility-interactive inverter or be applied by the installer near the ground-fault indicator at a visible location, stating the following:

**WARNING**
ELECTRIC SHOCK HAZARD
IF A GROUND FAULT IS INDICATED,
NORMALLY GROUNDED CONDUCTORS MAY
BE UNGROUNDED AND ENERGIZED

When the PV system also has batteries, the same warning shall also be applied by the installer in a visible location at the batteries. The warning sign(s) or label(s) shall comply with 110.21(B).

## 690.6 Alternating-Current (ac) Modules

**(A) Photovoltaic Source Circuits.** The requirements of Article 690 pertaining to PV source circuits shall not apply to ac modules. The PV source circuit, conductors, and inverters shall be considered as internal wiring of an ac module.

**(B) Inverter Output Circuit.** The output of an ac module shall be considered an inverter output circuit.

**(C) Disconnecting Means.** A single disconnecting means, in accordance with 690.15 and 690.17, shall be permitted for the combined ac output of one or more ac modules. Additionally, each ac module in a multiple ac module system shall be provided with a connector, bolted, or terminal-type disconnecting means.

Utility-interactive ac PV modules are designed to produce power only when they are connected to an external power source at the correct voltage and frequency. A single disconnecting means removes the external source and turns off all ac PV modules connected to that disconnecting device.

•

**(D) Overcurrent Protection.** The output circuits of ac modules shall be permitted to have overcurrent protection and conductor sizing in accordance with 240.5(B)(2).

## II. Circuit Requirements

### 690.7 Maximum Voltage

**(A) Maximum Photovoltaic System Voltage.** In a dc PV source circuit or output circuit, the maximum PV system voltage for that circuit shall be calculated as the sum of the rated open-circuit voltage of the series-connected PV modules corrected for the lowest expected ambient temperature. For crystalline and multicrystalline silicon modules, the rated open-circuit voltage shall be multiplied by the correction factor provided in Table 690.7. This voltage shall be used to determine the voltage rating of cables, disconnects, overcurrent devices, and other equipment. Where the lowest expected ambient temperature is below −40°C (−40°F), or where other than crystalline or multicrystalline silicon PV modules are used, the system voltage adjustment shall be made in accordance with the manufacturer's instructions.

When open-circuit voltage temperature coefficients are supplied in the instructions for listed PV modules, they shall be used to calculate the maximum PV system voltage as required by 110.3(B) instead of using Table 690.7.

Informational Note: One source for statistically valid, lowest-expected, ambient temperature design data for various locations is the Extreme Annual Mean Minimum Design Dry Bulb Temperature found in the *ASHRAE Handbook — Fundamentals*. These temperature data can be used to calculate maximum voltage using the manufacturer's temperature coefficients relative to the rating temperature of 25°C.

A PV source is not a constant-voltage source, and the difference between the rated operating voltage determined under controlled laboratory conditions and the open-circuit voltage under field-installed conditions can be significant. Consequently, the higher-rated open-circuit voltage must be used to select circuit components with proper voltage ratings.

The voltage (both open circuit and operating) of a PV power source increases as the temperature decreases. The installer should note the temperature conditions for which the PV device was rated. If the anticipated lowest temperature at the installation site is lower than the rating condition (25°C), Table 690.7 must be used to adjust the maximum open-circuit voltage of crystalline systems before conductors, overcurrent devices, and switchgear are selected. For other than crystalline systems, see the manufacturer's instructions.

Where a listed PV module includes open-circuit voltage temperature coefficients in the installation instructions, these temperature coefficients provide a more accurate maximum system voltage than those from Table 690.7 and are required to be used instead of applying the table.

Bipolar PV systems (with positive and negative voltages) must use the sum of the absolute values of the open-circuit voltages to determine the rated open-circuit system voltage. See the definition of *photovoltaic system voltage* in 690.2. For example, a system with open-circuit voltages of +480 volts and −480 volts with respect to ground would have a system open-circuit voltage of 960 volts. This voltage should be multiplied by a temperature-dependent factor from Table 690.7, yielding a system design voltage of up to 1200 volts. The system design voltage should be used in the selection of cables and other equipment. Certain methods of connecting bipolar PV arrays meeting the requirements of 690.7(E) may have different requirements for calculating the maximum system voltage.

**(B) Direct-Current Utilization Circuits.** The voltage of dc utilization circuits shall conform to 210.6.

**(C) Photovoltaic Source and Output Circuits.** In one- and two-family dwellings, PV source circuits and PV output circuits

***TABLE 690.7*** *Voltage Correction Factors for Crystalline and Multicrystalline Silicon Modules*

| Correction Factors for Ambient Temperatures Below 25°C (77°F). (Multiply the rated open circuit voltage by the appropriate correction factor shown below.) | | |
|---|---|---|
| **Ambient Temperature (°C)** | **Factor** | **Ambient Temperature (°F)** |
| 24 to 20 | 1.02 | 76 to 68 |
| 19 to 15 | 1.04 | 67 to 59 |
| 14 to 10 | 1.06 | 58 to 50 |
| 9 to 5 | 1.08 | 49 to 41 |
| 4 to | 1.10 | 40 to 32 |
| −1 to −5 | 1.12 | 31 to 23 |
| −6 to −10 | 1.14 | 22 to 14 |
| −11 to −15 | 1.16 | 13 to 5 |
| −16 to −20 | 1.18 | 4 to −4 |
| −21 to −25 | 1.20 | −5 to −13 |
| −26 to −30 | 1.21 | −14 to −22 |
| −31 to −35 | 1.23 | −23 to −31 |
| −36 to −40 | 1.25 | −32 to −40 |

that do not include lampholders, fixtures, or receptacles shall be permitted to have a maximum PV system voltage up to 600 volts. Other installations with a maximum PV system voltage over 1000 volts shall comply with Article 690, Part IX.

**(D) Circuits over 150 Volts to Ground.** In one- and two-family dwellings, live parts in PV source circuits and PV output circuits over 150 volts to ground shall not be accessible to other than qualified persons while energized.

Informational Note: See 110.27 for guarding of live parts, and 210.6 for voltage to ground and between conductors.

Conduit, closed cabinets, or enclosures that require tools to open them are means of limiting access to qualified persons only.

**(E) Bipolar Source and Output Circuits.** For 2-wire circuits connected to bipolar systems, the maximum system voltage shall be the highest voltage between the conductors of the 2-wire circuit if all of the following conditions apply:

(1) One conductor of each circuit of a bipolar subarray is solidly grounded.

*Exception: The operation of ground-fault or arc-fault devices (abnormal operation) shall be permitted to interrupt this connection to ground when the entire bipolar array becomes two distinct arrays isolated from each other and the utilization equipment.*

(2) Each circuit is connected to a separate subarray.
(3) The equipment is clearly marked with a label as follows:

**WARNING**
BIPOLAR PHOTOVOLTAIC ARRAY.
DISCONNECTION OF NEUTRAL
OR GROUNDED CONDUCTORS
MAY RESULT IN OVERVOLTAGE
ON ARRAY OR INVERTER.

The warning sign(s) or label(s) shall comply with 110.21(B).

## 690.8 Circuit Sizing and Current

**(A) Calculation of Maximum Circuit Current.** The maximum current for the specific circuit shall be calculated in accordance with 690.8(A)(1) through (A)(5).

Informational Note: Where the requirements of 690.8(A)(1) and (B)(1) are both applied, the resulting multiplication factor is 156 percent.

**(1) Photovoltaic Source Circuit Currents.** The maximum current shall be the sum of parallel module rated short-circuit currents multiplied by 125 percent.

The use of the array short-circuit current allows for proper sizing of conductors to handle the current generated during extended periods of operation under a short-circuit current operating point.

The 125-percent factor is required because PV modules, PV source circuits, and PV output circuits can deliver output currents higher than the rated short-circuit currents for more than 3 hours near solar noon. This requirement is duplicated in the instructions provided with each listed module, but this factor only needs to be applied once. A second 125-percent factor is required by 690.8(B).

PV modules in hot climates operate at temperatures of 60°C to 80°C due to solar heating. Conductors with insulation types rated at least 90°C should be used, and these conductors should have the ampacity corrected in accordance with Table 310.15(B)(16) or Table 310.15(B)(17).

**(2) Photovoltaic Output Circuit Currents.** The maximum current shall be the sum of parallel source circuit maximum currents as calculated in 690.8(A)(1).

**(3) Inverter Output Circuit Current.** The maximum current shall be the inverter continuous output current rating.

Both stand-alone and utility-interactive inverters are power-limited devices. Output circuits connected to these devices are sized on the continuous-rated outputs of these devices and are not based on load calculations or reduced-size PV arrays or battery banks. Exhibit 690.9 shows an inverter label displaying the maximum output circuit current along with other necessary ratings.

**(4) Stand-Alone Inverter Input Circuit Current.** The maximum current shall be the stand-alone continuous inverter input current rating when the inverter is producing rated power at the lowest input voltage.

Stand-alone inverters are nearly constant-output voltage devices. As the input battery voltage decreases, the input battery current increases to maintain a constant ac output power. The input current for such inverters is calculated by taking the rated full-power output of the inverter in watts and dividing it by the lowest operating battery voltage and then by the rated efficiency of the inverter under those operating conditions.

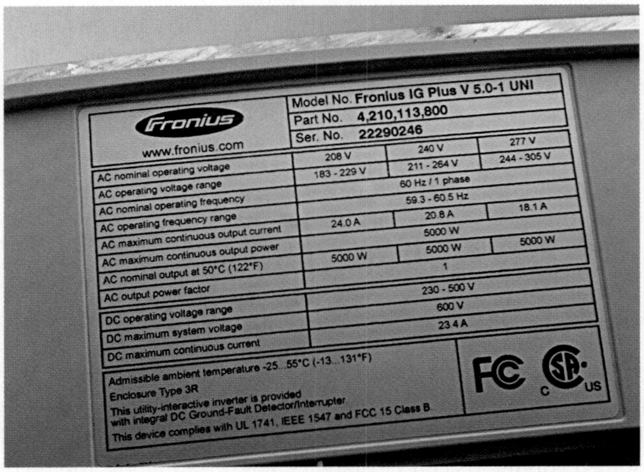


***EXHIBIT 690.9*** *The label of a utility-interactive inverter.*

For example, the input current for a 4000-W, 24-volt inverter that is 85 percent efficient at 22 volts can be calculated as follows:

$$\text{Ampere input} = \frac{\text{watt output}}{\text{voltage} \times \text{efficiency}}$$

$$= \frac{4000\ \text{W}}{22\ \text{V} \times 0.85} = 214\ \text{A}$$

Ripple currents might be present in the dc-input circuits of single-phase, stand-alone inverters. These ripple currents might cause nuisance operation of overcurrent devices at continuous high inverter outputs. In such cases, the measured maximum true rms value of the total (ac + dc) input current, which will be greater than the average current calculated here, should be used to determine conductor sizes and overcurrent device ratings.

**(5) DC-to-DC Converter Output Current.** The maximum current shall be the dc-to-dc converter continuous output current rating.

**(B) Conductor Ampacity.** PV system currents shall be considered to be continuous. Circuit conductors shall be sized to carry not less than the larger of 690.8(B)(1) or (2).

**(1)** One hundred and twenty-five percent of the maximum currents calculated in 690.8(A) before the application of adjustment and correction factors.

*Exception: Circuits containing an assembly, together with its overcurrent device(s), that is listed for continuous operation at 100 percent of its rating shall be permitted to be used at 100 percent of its rating.*

•

**(2)** The maximum currents calculated in 690.8(A) after the application of adjustment and correction factors.

•

**(C) Systems with Multiple Direct-Current Voltages.** For a PV power source that has multiple output circuit voltages and employs a common-return conductor, the ampacity of the common-return conductor shall not be less than the sum of the ampere ratings of the overcurrent devices of the individual output circuits.

**(D) Sizing of Module Interconnection Conductors.** Where a single overcurrent device is used to protect a set of two or more parallel-connected module circuits, the ampacity of each of the module interconnection conductors shall not be less than the sum of the rating of the single overcurrent device plus 125 percent of the short-circuit current from the other parallel-connected modules.

Normally, labels or module instructions require reverse overcurrent protection for each module or string of modules. In some cases, modules with low-rated short-circuit currents and high values of the required series protective fuse may allow the use of one overcurrent device to provide reverse-current protection for two modules or strings of modules and overcurrent protection for the conductors. The PV module manufacturer should be contacted for specific information.

## 690.9 Overcurrent Protection

**(A) Circuits and Equipment.** PV source circuit, PV output circuit, inverter output circuit, and storage battery circuit conductors and equipment shall be protected in accordance with the requirements of Article 240. Protection devices for PV source circuits and PV output circuits shall be in accordance with the requirements of 690.9(B) through (E). Circuits, either ac or dc, connected to current-limited supplies (e.g., PV modules, ac output of utility-interactive inverters), and also connected to sources having significantly higher current availability (e.g., parallel strings of modules, utility power), shall be protected at the source from overcurrent.

*Exception: An overcurrent device shall not be required for PV modules or PV source circuit conductors sized in accordance with 690.8(B) where one of the following applies:*

*(a) There are no external sources such as parallel-connected source circuits, batteries, or backfeed from inverters.*

*(b) The short-circuit currents from all sources do not exceed the ampacity of the conductors and the maximum overcurrent protective device size rating specified on the PV module nameplate.*

The conductors of the specified circuits and equipment are required to be protected at their source of supply in accordance with Article 240. The additional requirements of 690.9(B) through (E) apply to PV source and output circuit conductors. Because these circuits are subject to electrical and environmental stresses, the overcurrent devices are required to be specifically listed for use in PV systems. The overcurrent devices may be either supplemental or branch-circuit devices.

In the circuits illustrated in Exhibits 690.5 through 690.8, the PV source-circuit and output-circuit overcurrent devices are required to be rated at 125 percent of the source current calculated in 690.8(A). Possible backfeed currents from the other PV source circuits, other supply sources through the inverter, and storage-battery circuits, if any, have to be considered.

Each module or string of modules is a source circuit; therefore, reverse-current protection is required on nearly all PV modules (as indicated by the fuse requirement labeled on the back of each module).

Blocking diodes (possibly required by the module manufacturer for specific applications) can lose their blocking ability because of overtemperature conditions or internal breakdown. Therefore, overcurrent protection has to be considered with a condition of shorted blocking diodes if they are used in the circuit.

In typical one- and two-family dwelling unit utility-interactive PV systems, overcurrent devices may only be required in the dc PV source or output circuits where more than two strings of PV modules are connected in parallel.

The need for overcurrent protection at the inverter or battery/charge controller end of the PV output circuit may depend on the maximum backfeed fault current available from other sources.

•

**(B) Overcurrent Device Ratings.** Overcurrent device ratings shall be not less than 125 percent of the maximum currents calculated in 690.8(A).

*Exception: Circuits containing an assembly, together with its overcurrent device(s), that is listed for continuous operation at 100 percent of its rating shall be permitted to be used at 100 percent of its rating.*

Overcurrent devices may be installed in enclosures that are exposed to direct sunlight and so may be subject to operating temperatures higher than 40°C. Derating may be necessary.

**(C) Direct-Current Rating.** Overcurrent devices, either fuses or circuit breakers, used in any dc portion of a PV power system shall be listed and shall have the appropriate voltage, current, and interrupt ratings.

Direct-current fault currents are considerably harder to interrupt than ac faults. Overcurrent devices marked or listed only for ac use should not be used in dc circuits. Automotive- and marine-type fuses, although used in dc systems, might not be suitable for use in permanently wired residential or commercial electrical power systems meeting the requirements of the *Code.*

**(D) Photovoltaic Source and Output Circuits.** Listed PV overcurrent devices shall be required to provide overcurrent protection in PV source and output circuits. The overcurrent devices shall be accessible but shall not be required to be readily accessible.

**(E) Series Overcurrent Protection.** In grounded PV source circuits, a single overcurrent protection device, where required, shall be permitted to protect the PV modules and the interconnecting conductors. In ungrounded PV source circuits complying with 690.35, an overcurrent protection device, where required, shall be installed in each ungrounded circuit conductor and shall be permitted to protect the PV modules and the interconnecting cables.

The single overcurrent device may provide both the reverse-current protection required for the series-connected PV modules and the overcurrent protection required for the interconnecting conductors.

**(F) Power Transformers.** Overcurrent protection for a transformer with a source(s) on each side shall be provided in accordance with 450.3 by considering first one side of the transformer, then the other side of the transformer, as the primary.

*Exception: A power transformer with a current rating on the side connected toward the utility-interactive inverter output, not less than the rated continuous output current of the inverter, shall be permitted without overcurrent protection from the inverter.*

## 690.10 Stand-Alone Systems

The premises wiring system shall be adequate to meet the requirements of this *Code* for a similar installation connected to a service. The wiring on the supply side of the building or structure disconnecting means shall comply with the requirements of this *Code*, except as modified by 690.10(A) through (E).

**(A) Inverter Output.** The ac output from a stand-alone inverter(s) shall be permitted to supply ac power to the building or structure disconnecting means at current levels less than the calculated load connected to that disconnect. The inverter output rating or the rating of an alternate energy source shall be equal to or greater than the load posed by the largest single utilization equipment connected to the system. Calculated general lighting loads shall not be considered as a single load.

Even though a stand-alone installation may have service-entrance equipment rated at 100 or 200 amperes at 120/240 volts, the PV source is not required to provide either the full current rating or the dual voltages of the service equipment. A PV installation is usually designed so that the actual ac demands on the system are sized to the output rating of the PV system. The inverter output is required to have sufficient capacity to power the largest single piece of utilization equipment to be supplied by the PV system, but the inverter output does not have to be rated for potential multiple loads to be simultaneously connected to it.

**(B) Sizing and Protection.** The circuit conductors between the inverter output and the building or structure disconnecting means shall be sized based on the output rating of the inverter. These conductors shall be protected from overcurrents in accordance with Article 240. The overcurrent protection shall be located at the output of the inverter.

**(C) Single 120-Volt Supply.** The inverter output of a stand-alone solar PV system shall be permitted to supply 120 volts to single-phase, 3-wire, 120/240-volt service equipment or distribution panels where there are no 240-volt outlets and where there are no multiwire branch circuits. In all installations, the rating of the overcurrent device connected to the output of the inverter shall be less than the rating of the neutral bus in the service equipment. This equipment shall be marked with the following words or equivalent:

**WARNING**

SINGLE 120-VOLT SUPPLY. DO NOT CONNECT MULTIWIRE BRANCH CIRCUITS!

The warning sign(s) or label(s) shall comply with 110.21(B).

Multiwire branch circuits are common in one- and two-family dwelling units. When connected to a normal 120/240-volt ac service, the currents in the neutral conductors of these multiwire branch circuits (typically 14-3 AWG) subtract, or are at most no larger than the rating of the branch-circuit overcurrent device. If the electrical system consists of a single 120-volt PV power system inverter supplying the two buses in the panelboard, the currents in the grounded conductor for each multiwire branch circuit add, rather than subtract. Because the two buses are in phase, there is no neutral conductor. The currents in these conductors may be as high as twice the rating of the branch-circuit overcurrent device, and overloading is possible.

**(D) Energy Storage or Backup Power System Requirements.** Energy storage or backup power supplies are not required.

**(E) Back-Fed Circuit Breakers.** Plug-in type back-fed circuit breakers connected to a stand-alone or multimode inverter output in stand-alone systems shall be secured in accordance with 408.36(D). Circuit breakers marked "line" and "load" shall not be back-fed.

### 690.11 Arc-Fault Circuit Protection (Direct Current)

Photovoltaic systems with dc source circuits, dc output circuits, or both, operating at a PV system maximum system voltage of 80 volts or greater, shall be protected by a listed (dc) arc-fault circuit interrupter, PV type, or other system components listed to provide equivalent protection. The PV arc-fault protection means shall comply with the following requirements:

(1) The system shall detect and interrupt arcing faults resulting from a failure in the intended continuity of a conductor, connection, module, or other system component in the dc PV source and dc PV output circuits.

•

(2) The system shall require that the disabled or disconnected equipment be manually restarted.
(3) The system shall have an annunciator that provides a visual indication that the circuit interrupter has operated. This indication shall not reset automatically.

The arc-fault protective device used to meet this requirement must be listed for dc use and listed for use in PV systems. Listed components that provide protection equivalent to arc-fault protection are also permitted by this requirement.

### 690.12 Rapid Shutdown of PV Systems on Buildings

PV system circuits installed on or in buildings shall include a rapid shutdown function that controls specific conductors in accordance with 690.12(1) through (5) as follows.

(1) Requirements for controlled conductors shall apply only to PV system conductors of more than 1.5 m (5 ft) in length inside a building, or more than 3 m (10 ft) from a PV array.
(2) Controlled conductors shall be limited to not more than 30 volts and 240 volt-amperes within 10 seconds of rapid shutdown initiation.
(3) Voltage and power shall be measured between any two conductors and between any conductor and ground.
(4) The rapid shutdown initiation methods shall be labeled in accordance with 690.56(B).
(5) Equipment that performs the rapid shutdown shall be listed and identified.

First responders must contend with elements of a PV system that remain energized after the service disconnect is opened. This rapid shutdown requirement provides a zone outside of which the potential for shock has been mitigated. Conductors more than 5 feet inside a building or more than 10 feet from an array will be limited to a maximum of 30 V and 240 VA within 10 seconds of activation of shutdown. Ten seconds allows time for any dc capacitor banks to discharge. Methods and designs for achieving proper rapid shutdown are not addressed by the *NEC* but instead are addressed in the product standards for this type of equipment.

## III. Disconnecting Means

### 690.13 Building or Other Structure Supplied by a Photovoltaic System

Means shall be provided to disconnect all ungrounded dc conductors of a PV system from all other conductors in a building or other structure.

•

**(A) Location.** The PV disconnecting means shall be installed at a readily accessible location either on the outside of a building or structure or inside nearest the point of entrance of the system conductors.

*Exception: Installations that comply with 690.31(F) shall be permitted to have the disconnecting means located remote from the point of entry of the system conductors.*

The PV system disconnecting means shall not be installed in bathrooms.

These requirements generally prohibit long runs of PV source and output circuits inside a building before reaching the required PV disconnect. A short conductor run through a wall at the point of first penetration to reach a disconnect mounted inside the building is allowed. Section 690.31(G) permits these circuits to be run inside a building where installed in metal conduit from the point of entrance to the system disconnecting means.

**(B) Marking.** Each PV system disconnecting means shall be permanently marked to identify it as a PV system disconnect.

**(C) Suitable for Use.** Each PV system disconnecting means shall not be required to be suitable as service equipment.

**(D) Maximum Number of Disconnects.** The PV system disconnecting means shall consist of not more than six switches or six circuit breakers mounted in a single enclosure or in a group of separate enclosures.

**(E) Grouping.** The PV system disconnecting means shall be grouped with other disconnecting means for the system in accordance with 690.13(D). A PV disconnecting means shall not be required at the PV module or array location.

Where a building has multiple sources of power, such as the utility, the PV array, backup generator, and wind system, no more than six disconnects for each source of power to the building are permitted, and the disconnects for each source are required to be grouped together.

•

## 690.15 Disconnection of Photovoltaic Equipment

Means shall be provided to disconnect equipment, such as inverters, batteries, and charge controllers, from all ungrounded conductors of all sources. If the equipment is energized from more than one source, the disconnecting means shall be grouped and identified.

A single disconnecting means in accordance with 690.17 shall be permitted for the combined ac output of one or more inverters or ac modules in an interactive system.

**(A) Utility-Interactive Inverters Mounted in Not Readily Accessible Locations.** Utility-interactive inverters shall be permitted to be mounted on roofs or other exterior areas that are not readily accessible and shall comply with 690.15(A)(1) through (4):

(1) A dc PV disconnecting means shall be mounted within sight of or in each inverter.
(2) An ac disconnecting means shall be mounted within sight of or in each inverter.
(3) The ac output conductors from the inverter and an additional ac disconnecting means for the inverter shall comply with 690.13(A).
(4) A plaque shall be installed in accordance with 705.10.

Exhibit 690.10 shows utility interactive inverters with associated disconnects installed on a rooftop.

**(B) Equipment.** Equipment such as PV source circuit isolating switches, overcurrent devices, dc-to-dc converters, and blocking diodes shall be permitted on the PV side of the PV disconnecting means.

In general, equipment that needs servicing must be disconnected from sources of supply. In a PV system, however, some equipment is permitted to be located on the PV power source side of the disconnecting means. See Exhibit 690.11. Servicing the exempted equipment might require disabling all or portions of the array as explained in the commentary following 690.18.

A disconnecting means located in each PV source circuit or located physically at each PV module location is not required. Unlike load circuits (such as rooftop air conditioners), PV

*EXHIBIT 690.10 Rooftop utility interactive inverters. (Courtesy of Solar Design Associates, LLC)*

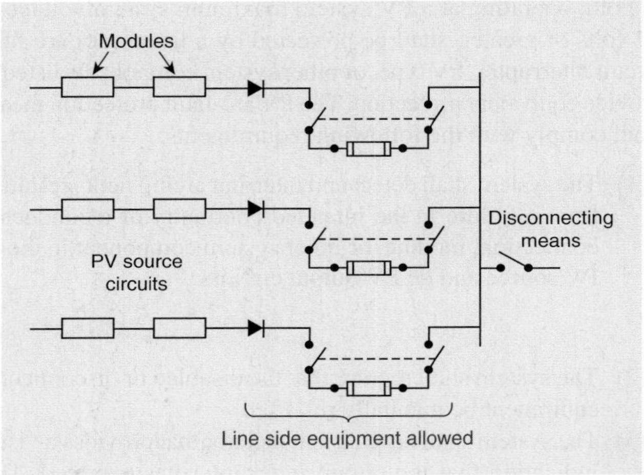


*EXHIBIT 690.11 Equipment permitted on the PV power source side of the disconnecting means.*

source-circuit conductors may be energized at any time from the PV modules. A centrally located disconnect meeting the requirements of 690.13(A) near the inverter or batteries serves to disconnect the PV source circuits from the other portions of the electric power system.

**(C) Direct-Current Combiner Disconnects.** The dc output of dc combiners mounted on roofs of dwellings or other buildings shall have a load break disconnecting means located in the combiner or within 1.8 m (6 ft) of the combiner. The disconnecting means shall be permitted to be remotely controlled but shall be manually operable locally when control power is not available.

**(D) Maximum Number of Disconnects.** The PV system disconnecting means shall consist of not more than six switches or six circuit breakers mounted in a single enclosure or in a group of separate enclosures.

### 690.16 Fuses

**(A) Disconnecting Means.** Disconnecting means shall be provided to disconnect a fuse from all sources of supply if the fuse is energized from both directions. Such a fuse in a PV source circuit shall be capable of being disconnected independently of fuses in other PV source circuits.

Switches, pullouts, or similar devices that have suitable ratings may serve as means to disconnect fuses from all sources of supply.

**(B) Fuse Servicing.** Disconnecting means shall be installed on PV output circuits where overcurrent devices (fuses) must be serviced that cannot be isolated from energized circuits. The disconnecting means shall be within sight of, and accessible to, the location of the fuse or integral with fuse holder and shall comply with 690.17. Where the disconnecting means are located more than 1.8 m (6 ft) from the overcurrent device, a directory showing the location of each disconnect shall be installed at the overcurrent device location.

Non-load-break-rated disconnecting means shall be marked "Do not open under load."

### 690.17 Disconnect Type

**(A) Manually Operable.** The disconnecting means for ungrounded PV conductors shall consist of a manually operable switch(es) or circuit breaker(s). The disconnecting means shall be permitted to be power operable with provisions for manual operation in the event of a power-supply failure. The disconnecting means shall be one of the following listed devices:

(1) A PV industrial control switch marked for use in PV systems
(2) A PV molded-case circuit breaker marked for use in PV systems
(3) A PV molded-case switch marked for use in PV systems
(4) A PV enclosed switch marked for use in PV systems
(5) A PV open-type switch marked for use in PV systems
(6) A dc-rated molded-case circuit breaker suitable for backfeed operation
(7) A dc-rated molded-case switch suitable for backfeed operation
(8) A dc-rated enclosed switch
(9) A dc-rated open-type switch
(10) A dc-rated rated low-voltage power circuit breaker

Informational Note: Devices marked with "line" and "load" are not suitable for backfeed or reverse current.

**(B) Simultaneous Opening of Poles.** The PV disconnecting means shall simultaneously disconnect all ungrounded supply conductors.

**(C) Externally Operable and Indicating.** The PV disconnecting means shall be externally operable without exposing the operator to contact with live parts and shall indicate whether in the open or closed position.

**(D) Disconnection of Grounded Conductor.** A switch, circuit breaker, or other device shall not be installed in a grounded conductor if operation of that switch, circuit breaker, or other device leaves the marked, grounded conductor in an ungrounded and energized state.

*Exception No. 1: A switch or circuit breaker that is part of a ground-fault detection system required by 690.5, or that is part of an arc-fault detection/interruption system required by 690.11, shall be permitted to open the grounded conductor when that switch or circuit breaker is automatically opened as a normal function of the device in responding to ground faults.*

*Exception No. 2: A disconnecting switch shall be permitted in a grounded conductor if all of the following conditions are met:*

*(1) The switch is used only for PV array maintenance.*
*(2) The switch is accessible only by qualified persons.*
*(3) The switch is rated for the maximum dc voltage and current that could be present during any operation, including ground-fault conditions.*

**(E) Interrupting Rating.** The building or structure disconnecting means shall have an interrupting rating sufficient for the maximum circuit voltage and current that is available at the line terminals of the equipment. Where all terminals of the disconnecting means may be energized in the open position, a warning sign shall be mounted on or adjacent to the disconnecting means. The sign shall be clearly legible and have the following words or equivalent:

**WARNING**
ELECTRIC SHOCK HAZARD
DO NOT TOUCH TERMINALS.
TERMINALS ON BOTH THE LINE
AND LOAD SIDES MAY BE ENERGIZED
IN THE OPEN POSITION.

The warning sign(s) or label(s) shall comply with 110.21(B).

*Exception: A connector shall be permitted to be used as an ac or a dc disconnecting means, provided that it complies with the requirements of 690.33 and is listed and identified for use with specific equipment.*

### 690.18 Installation and Service of an Array

Open circuiting, short circuiting, or opaque covering shall be used to disable an array or portions of an array for installation and service.

Informational Note: Photovoltaic modules are energized while exposed to light. Installation, replacement, or servicing of array components while a module(s) is energized may expose persons to electric shock.

To prevent contact with energized parts by personnel during installation, servicing, or other procedures, a number of methods can be used to disable an array or portions of an array. One method, used infrequently because of the expense in time and materials, is to cover all of the array or portions of it with an opaque material. Care must be taken that all of the area to be covered is shielded from light. Another method subdivides the array into smaller segments, which can be accomplished by switches or connectors. Also see 690.33. Short-circuiting all or portions of an array by means of switches or plug-in connectors, in conjunction with the bypass diodes internal to each module, can also disable the array. (Bypass diodes are incorporated in PV modules for performance purposes.)

# IV. Wiring Methods

## 690.31 Methods Permitted

**(A) Wiring Systems.** All raceway and cable wiring methods included in this *Code*, other wiring systems and fittings specifically listed for use on PV arrays, and wiring as part of a listed system shall be permitted. Where wiring devices with integral enclosures are used, sufficient length of cable shall be provided to facilitate replacement.

Where PV source and output circuits operating at maximum system voltages greater than 30 volts are installed in readily accessible locations, circuit conductors shall be guarded or installed in a raceway.

Most PV modules do not have means for attaching raceways. These circuits may have to be made "not readily accessible" by use of physical barriers such as wire screening.

Informational Note: Photovoltaic modules operate at elevated temperatures when exposed to high ambient temperatures and to bright sunlight. These temperatures routinely exceed 70°C (158°F) in many locations. Module interconnection conductors are available with insulation rated for wet locations and a temperature rating of 90°C (194°F) or greater.

**(B) Identification and Grouping.** PV source circuits and PV output circuits shall not be contained in the same raceway, cable tray, cable, outlet box, junction box, or similar fitting as conductors, feeders, branch circuits of other non-PV systems, or inverter output circuits, unless the conductors of the different systems are separated by a partition. PV system conductors shall be identified and grouped as required by 690.31(B)(1) through (4). The means of identification shall be permitted by separate color coding, marking tape, tagging, or other approved means.

Alternating-current branch-circuit conductors that supply an exterior luminaire installed near a roof-mounted PV array are examples of conductors that must not share the same raceway or cable with PV source or output circuit conductors.

Conductors directly related to a specific PV system, such as those in dc and ac output power circuits, are permitted in the same raceway as PV source and output conductors, provided they meet the requirements of 690.31(B)(1) through (B)(4) and 300.3(C).

**(1) PV Source Circuits.** PV source circuits shall be identified at all points of termination, connection, and splices.

**(2) PV Output and Inverter Circuits.** The conductors of PV output circuits and inverter input and output circuits shall be identified at all points of termination, connection, and splices.

**(3) Conductors of Multiple Systems.** Where the conductors of more than one PV system occupy the same junction box, raceway, or equipment, the conductors of each system shall be identified at all termination, connection, and splice points.

*Exception: Where the identification of the conductors is evident by spacing or arrangement, further identification shall not be required.*

**(4) Grouping.** Where the conductors of more than one PV system occupy the same junction box or raceway with a removable cover(s), the ac and dc conductors of each system shall be grouped separately by cable ties or similar means at least once and shall then be grouped at intervals not to exceed 1.8 m (6 ft).

*Exception: The requirement for grouping shall not apply if the circuit enters from a cable or raceway unique to the circuit that makes the grouping obvious.*

**(C) Single-Conductor Cable.**

(1) **General.** Single-conductor cable Type USE-2, and single-conductor cable listed and labeled as photovoltaic (PV) wire shall be permitted in exposed outdoor locations in PV source circuits for PV module interconnections within the PV array.

*Exception: Raceways shall be used when required by 690.31(A).*

Most PV modules are designed for a direct series connection by using factory-installed leads and connectors. To accommodate such a connection, use of a single-conductor Type USE-2 cable and single-conductor cable listed and labeled for PV applications is permitted in PV source circuits. Long runs of separated conductors (with loop inductance and distributed capacitance) and the resulting long time constants in dc circuits may result in improper operation of overcurrent devices. Running both positive and negative conductors of each circuit and the EGC as close together as possible minimizes the circuit time constant. It also decreases induced currents from nearby lightning strikes. Because PV modules may operate at high temperatures and are installed in outdoor, exposed locations, the use of high-temperature conductors rated for wet locations, such as USE-2, THWN-2, or RHW-2, is often necessary. See 310.15(B)(3)(c) for requirements on the ampacities of conductors in circular raceways installed on rooftops exposed to sunlight. Single-conductor cables listed and labeled for use in PV applications will be identified as "PV Wire," "PV Cable," "Photovoltaic Wire," or "Photovoltaic Cable."

(2) **Cable Tray.** PV source circuits and PV output circuits using single-conductor cable listed and labeled as photovoltaic (PV) wire of all sizes, with or without a cable tray

marking/rating, shall be permitted in cable trays installed in outdoor locations, provided that the cables are supported at intervals not to exceed 300 mm (12 in.) and secured at intervals not to exceed 1.4 m (4.5 ft).

Informational Note: Photovoltaic wire and PV cable have a non-standard outer diameter. See Table 1 of Chapter 9 for conduit fill calculations.

**(D) Multiconductor Cable.** Multiconductor cable Type TC-ER or Type USE-2 shall be permitted in outdoor locations in PV inverter output circuits where used with utility-interactive inverters mounted in locations that are not readily accessible. The cable shall be secured at intervals not exceeding 1.8 m (6 ft). Equipment grounding for the utilization equipment shall be provided by an equipment grounding conductor within the cable.

**(E) Flexible Cords and Cables.** Flexible cords and cables, where used to connect the moving parts of tracking PV modules, shall comply with Article 400 and shall be of a type identified as a hard service cord or portable power cable; they shall be suitable for extra-hard usage, listed for outdoor use, water resistant, and sunlight resistant. Allowable ampacities shall be in accordance with 400.5. For ambient temperatures exceeding 30°C (86°F), the ampacities shall be derated by the appropriate factors given in Table 690.31(E).

**(F) Small-Conductor Cables.** Single-conductor cables listed for outdoor use that are sunlight resistant and moisture resistant in sizes 16 AWG and 18 AWG shall be permitted for module interconnections where such cables meet the ampacity requirements of 400.5. Section 310.15 shall be used to determine the cable ampacity adjustment and correction factors.

Because the smaller cables are not normally marked with standard *Code*-recognized markings (such as USE-2), the PV module manufacturer or installer should verify that these cables are listed and labeled for PV use, which would indicate that they have the necessary sunlight and moisture resistance and are suitable for exposed, outdoor use.

In accordance with 200.6(A), grounded conductors that are smaller than 6 AWG and used in PV source circuits are permitted to be marked at the time of installation with a white marking at all terminations.

**(G) Direct-Current Photovoltaic Source and Direct-Current Output Circuits on or Inside a Building.** Where dc PV source or dc PV output circuits from building-integrated systems or other PV systems are run inside a building or structure, they shall be contained in metal raceways, Type MC metal-clad cable that complies with 250.118(10), or metal enclosures from the point of penetration of the surface of the building or structure to the first readily accessible disconnecting means. The disconnecting means shall comply with 690.13(B) and (C) and 690.15(A) and (B). The wiring methods shall comply with the additional installation requirements in 690.31(G)(1) through (4).

The use of metal raceways, Type MC cable, or metal enclosures inside a building provides additional physical protection for these circuits. Metal raceways also provide additional fire resistance should faults develop in the cable, and they provide an additional ground-fault detection path for the ground-fault protection device required by 690.5.

**(1) Embedded in Building Surfaces.** Where circuits are embedded in built-up, laminate, or membrane roofing materials in roof areas not covered by PV modules and associated equipment, the location of circuits shall be clearly marked using a marking protocol that is approved as being suitable for continuous exposure to sunlight and weather.

The distance between the array and the disconnecting means is not limited by the *Code*. The PV circuit conductors between the PV power source and the PV disconnecting means are energized whenever the source is producing power. Because of this potential exposure to energized conductors, a marking is required to warn roofers, other tradespersons, or first responders of the location of energized PV conductors where PV source or output circuits are "imbedded" or concealed by the roofing material. This

***TABLE 690.31(E)*** *Correction Factors*

| Ambient Temperature (°C) | Temperature Rating of Conductor | | | | Ambient Temperature (°F) |
|---|---|---|---|---|---|
| | 60°C (140°F) | 75°C (167°F) | 90°C (194°F) | 105°C (221°F) | |
| 30 | 1.00 | 1.00 | 1.00 | 1.00 | 86 |
| 31–35 | 0.91 | 0.94 | 0.96 | 0.97 | 87–95 |
| 36–40 | 0.82 | 0.88 | 0.91 | 0.93 | 96–104 |
| 41–45 | 0.71 | 0.82 | 0.87 | 0.89 | 105–113 |
| 46–50 | 0.58 | 0.75 | 0.82 | 0.86 | 114–122 |
| 51–55 | 0.41 | 0.67 | 0.76 | 0.82 | 123–131 |
| 56–60 | — | 0.58 | 0.71 | 0.77 | 132–140 |
| 61–70 | — | 0.33 | 0.58 | 0.68 | 141–158 |
| 71–80 | — | — | 0.41 | 0.58 | 159–176 |

requirement does not apply to conductors installed in areas of the roof covered by PV modules or other associated equipment.

•

**(2) Flexible Wiring Methods.** Where flexible metal conduit (FMC) smaller than metric designator 21 (trade size ¾) or Type MC cable smaller than 25 mm (1 in.) in diameter containing PV power circuit conductors is installed across ceilings or floor joists, the raceway or cable shall be protected by substantial guard strips that are at least as high as the raceway or cable. Where run exposed, other than within 1.8 m (6 ft) of their connection to equipment, these wiring methods shall closely follow the building surface or be protected from physical damage by an approved means.

**(3) Marking and Labeling Required.** The following wiring methods and enclosures that contain PV power source conductors shall be marked with the wording WARNING: PHOTOVOLTAIC POWER SOURCE by means of permanently affixed labels or other approved permanent marking:

(1) Exposed raceways, cable trays, and other wiring methods
(2) Covers or enclosures of pull boxes and junction boxes
(3) Conduit bodies in which any of the available conduit openings are unused

**(4) Marking and Labeling Methods and Locations.** The labels or markings shall be visible after installation. The labels shall be reflective, and all letters shall be capitalized and shall be a minimum height of 9.5 mm (⅜ in.) in white on a red background. PV power circuit labels shall appear on every section of the wiring system that is separated by enclosures, walls, partitions, ceilings, or floors. Spacing between labels or markings, or between a label and a marking, shall not be more than 3 m (10 ft). Labels required by this section shall be suitable for the environment where they are installed.

The objective of the requirements contained in 690.31(G) is to protect persons from inadvertently damaging PV source and output circuit conductors. Where the location of the PV circuit conductors is not obvious, fire fighters, other first responders, and maintenance personnel could be exposed to shock hazards. Ventilating roofs containing PV source or output circuits by cutting the membrane with saws could expose personnel to shock hazards and the building to further damage resulting from the ignition of combustible members due to arcing from damaged conductors.

**(H) Flexible, Fine-Stranded Cables.** Flexible, fine-stranded cables shall be terminated only with terminals, lugs, devices, or connectors in accordance with 110.14.

The terminals, connectors, and crimp-on lugs found on most electrical equipment used in PV systems are *not* suitable for use with fine-stranded conductors. The requirements for these conductors are covered in 110.14.

**(I) Bipolar Photovoltaic Systems.** Where the sum, without consideration of polarity, of the PV system voltages of the two monopole subarrays exceeds the rating of the conductors and connected equipment, monopole subarrays in a bipolar PV system shall be physically separated, and the electrical output circuits from each monopole subarray shall be installed in separate raceways until connected to the inverter. The disconnecting means and overcurrent protective devices for each monopole subarray output shall be in separate enclosures. All conductors from each separate monopole subarray shall be routed in the same raceway. Bipolar PV systems shall be clearly marked with a permanent, legible warning notice indicating that the disconnection of the grounded conductor(s) may result in overvoltage on the equipment.

*Exception: Listed switchgear rated for the maximum voltage between circuits and containing a physical barrier separating the disconnecting means for each monopole subarray shall be permitted to be used instead of disconnecting means in separate enclosures.*

**(J) Module Connection Arrangement.** The connection to a module or panel shall be arranged so that removal of a module or panel from a PV source circuit does not interrupt a grounded conductor connection to other PV source circuits.

If interrupted, grounded conductors operate at the system potential with respect to ground, and a shock hazard could result.

## 690.32 Component Interconnections

Fittings and connectors that are intended to be concealed at the time of on-site assembly, where listed for such use, shall be permitted for on-site interconnection of modules or other array components. Such fittings and connectors shall be equal to the wiring method employed in insulation, temperature rise, and fault-current withstand, and shall be capable of resisting the effects of the environment in which they are used.

## 690.33 Connectors

The connectors permitted by Article 690 shall comply with 690.33(A) through (E).

**(A) Configuration.** The connectors shall be polarized and shall have a configuration that is noninterchangeable with receptacles in other electrical systems on the premises.

**(B) Guarding.** The connectors shall be constructed and installed so as to guard against inadvertent contact with live parts by persons.

**(C) Type.** The connectors shall be of the latching or locking type. Connectors that are readily accessible and that are used in circuits operating at over 30 volts, nominal, maximum system voltage for dc circuits, or 30 volts for ac circuits, shall require a tool for opening.

**(D) Grounding Member.** The grounding member shall be the first to make and the last to break contact with the mating connector.

**(E) Interruption of Circuit.** Connectors shall be either (1) or (2):

(1) Be rated for interrupting current without hazard to the operator.
(2) Be a type that requires the use of a tool to open and marked "Do Not Disconnect Under Load" or "Not for Current Interrupting."

The two options for connectors in this requirement provide for safe disconnection of circuit connectors either by allowing them to be opened under load or by requiring a warning indicating that disconnection prior to opening the connector is necessary. Connectors that can be opened or disconnected using only the hands are not acceptable.

### 690.34 Access to Boxes

Junction, pull, and outlet boxes located behind modules or panels shall be so installed that the wiring contained in them can be rendered accessible directly or by displacement of a module(s) or panel(s) secured by removable fasteners and connected by a flexible wiring system.

### 690.35 Ungrounded Photovoltaic Power Systems

Photovoltaic power systems shall be permitted to operate with ungrounded PV source and output circuits where the system complies with 690.35(A) through (G).

**(A) Disconnects.** All PV source and output circuit conductors shall have disconnects complying with 690, Part III.

**(B) Overcurrent Protection.** All PV source and output circuit conductors shall have overcurrent protection complying with 690.9.

**(C) Ground-Fault Protection.** All PV source and output circuits shall be provided with a ground-fault protection device or system that complies with 690.35(1) through (4):

(1) Detects ground fault(s) in the PV array dc current-carrying conductors and components
(2) Indicates that a ground fault has occurred
(3) Automatically disconnects all conductors or causes the inverter or charge controller connected to the faulted circuit to automatically cease supplying power to output circuits
(4) Is listed for providing PV ground-fault protection

Aging of the conductors, dust and dirt infiltration, and moisture and water intrusion create leakage paths from the conductors to ground. These high-resistance leakage paths can result in leakage current values less than those detected by the required ground-fault detection device, but they can cause any ungrounded conductor to become a potential shock hazard with respect to ground.

**(D) Conductors.** The PV source conductors shall consist of the following:

(1) Metallic or nonmetallic jacketed multiconductor cables
(2) Conductors installed in raceways
(3) Conductors listed and identified as PV wire installed as exposed, single conductors, or
(4) Conductors that are direct-buried and identified for direct-burial use

All cables and conductors installed outdoors and exposed to direct sunlight and wet conditions must be suitable for these conditions. Conductors inside raceways installed in wet locations are required to be identified or listed as suitable for wet locations. See 310.10(C) for the requirements on conductors installed in wet locations. Open, single conductors are permitted where listed and identified as "Photovoltaic Wire," "Photovoltaic Cable," "PV Wire," or "PV Cable." These conductors are evaluated for use where exposed to direct sunlight and wet conditions.

**(E) Battery Systems.** The PV power system direct-current circuits shall be permitted to be used with ungrounded battery systems complying with 690.71(G).

**(F) Marking.** The PV power source shall be labeled with the following warning at each junction box, combiner box, disconnect, and device where energized, ungrounded circuits may be exposed during service:

WARNING
ELECTRIC SHOCK HAZARD. THE DC CONDUCTORS OF THIS PHOTOVOLTAIC SYSTEM ARE UNGROUNDED AND MAY BE ENERGIZED.

The warning sign(s) or label(s) shall comply with 110.21(B).

**(G) Equipment.** The inverters or charge controllers used in systems with ungrounded PV source and output circuits shall be listed for the purpose.

Many types of PV equipment are designed to operate only on grounded systems.

## V. Grounding

### 690.41 System Grounding

Photovoltaic systems shall comply with one of the following:

(1) Ungrounded systems shall comply with 690.35.
(2) Grounded two-wire systems shall have one conductor grounded or be impedance grounded, and the system shall comply with 690.5.
(3) Grounded bipolar systems shall have the reference (center tap) conductor grounded or be impedance grounded, and the system shall comply with 690.5.
(4) Other methods that accomplish equivalent system protection in accordance with 250.4(A) with equipment listed and identified for the use shall be permitted to be used.

•

## 690.42 Point of System Grounding Connection

The dc circuit grounding connection shall be made at any single point on the PV output circuit.

Informational Note: Locating the grounding connection point as close as practicable to the PV source better protects the system from voltage surges due to lightning.

*Exception: Systems with a 690.5 ground-fault protection device shall be permitted to have the required grounded conductor-to-ground bond made by the ground-fault protection device. This bond, where internal to the ground-fault equipment, shall not be duplicated with an external connection.*

## 690.43 Equipment Grounding

Equipment grounding conductors and devices shall comply with 690.43(A) through (F).

**(A) Equipment Grounding Required.** Exposed non–current-carrying metal parts of PV module frames, electrical equipment, and conductor enclosures shall be grounded in accordance with 250.134 or 250.136(A), regardless of voltage.

**(B) Equipment Grounding Conductor Required.** An equipment grounding conductor between a PV array and other equipment shall be required in accordance with 250.110.

**(C) Structure as Equipment Grounding Conductor.** Devices listed and identified for grounding the metallic frames of PV modules or other equipment shall be permitted to bond the exposed metal surfaces or other equipment to mounting structures. Metallic mounting structures, other than building steel, used for grounding purposes shall be identified as equipment-grounding conductors or shall have identified bonding jumpers or devices connected between the separate metallic sections and shall be bonded to the grounding system.

**(D) Photovoltaic Mounting Systems and Devices.** Devices and systems used for mounting PV modules that are also used to provide grounding of the module frames shall be identified for the purpose of grounding PV modules.

**(E) Adjacent Modules.** Devices identified and listed for bonding the metallic frames of PV modules shall be permitted to bond the exposed metallic frames of PV modules to the metallic frames of adjacent PV modules.

**(F) All Conductors Together.** Equipment grounding conductors for the PV array and structure (where installed) shall be contained within the same raceway or cable or otherwise run with the PV array circuit conductors when those circuit conductors leave the vicinity of the PV array.

To maintain the shortest electrical time constant in each dc circuit, the EGC should be routed as closely as possible to the circuit conductors. This routing facilitates the operation of overcurrent devices.

## 690.45 Size of Equipment Grounding Conductors

Equipment grounding conductors for PV source and PV output circuits shall be sized in accordance with 250.122. Where no overcurrent protective device is used in the circuit, an assumed overcurrent device rated at the PV maximum circuit current shall be used when applying Table 250.122. Increases in equipment grounding conductor size to address voltage drop considerations shall not be required. An equipment grounding conductor shall not be smaller than 14 AWG.

•

## 690.46 Array Equipment Grounding Conductors

For PV modules, equipment grounding conductors smaller than 6 AWG shall comply with 250.120(C).

Where installed in raceways, equipment grounding conductors and grounding electrode conductors not larger than 6 AWG shall be permitted to be solid.

## 690.47 Grounding Electrode System

**(A) Alternating-Current Systems.** If installing an ac system, a grounding electrode system shall be provided in accordance with 250.50 through 250.60. The grounding electrode conductor shall be installed in accordance with 250.64.

**(B) Direct-Current Systems.** If installing a dc system, a grounding electrode system shall be provided in accordance with 250.166 for grounded systems or 250.169 for ungrounded systems. The grounding electrode conductor shall be installed in accordance with 250.64.

A common dc grounding-electrode conductor shall be permitted to serve multiple inverters. The size of the common grounding electrode and the tap conductors shall be in accordance with 250.166. The tap conductors shall be connected to the common grounding-electrode conductor by exothermic welding or with connectors listed as grounding and bonding equipment in such a manner that the common grounding electrode conductor remains without a splice or joint.

An ac equipment grounding system shall be permitted to be used for equipment grounding of inverters and other equipment and for the ground-fault detection reference for ungrounded PV systems.

**(C) Systems with Alternating-Current and Direct-Current Grounding Requirements.** Photovoltaic systems having dc circuits and ac circuits with no direct connection between the dc grounded conductor and ac grounded conductor shall have a dc grounding system. The dc grounding system shall be bonded to the ac grounding system by one of the methods in (1), (2), or (3).

This section shall not apply to ac PV modules.

When using the methods of (C)(2) or (C)(3), the existing ac grounding electrode system shall meet the applicable requirements of Article 250, Part III.

Informational Note No. 1: ANSI/UL 1741, *Standard for Inverters, Converters, and Controllers for Use in Independent Power*

*Systems,* requires that any inverter or charge controller that has a bonding jumper between the grounded dc conductor and the grounding system connection point have that point marked as a grounding electrode conductor (GEC) connection point. In PV inverters, the terminals for the dc equipment grounding conductors and the terminals for ac equipment grounding conductors are generally connected to, or electrically in common with, a grounding busbar that has a marked dc GEC terminal.

Informational Note No. 2: For utility-interactive systems, the existing premises grounding system serves as the ac grounding system.

**(1) Separate Direct-Current Grounding Electrode System Bonded to the Alternating-Current Grounding Electrode System.** A separate dc grounding electrode or system shall be installed, and it shall be bonded directly to the ac grounding electrode system. The size of any bonding jumper(s) between the ac and dc systems shall be based on the larger size of the existing ac grounding electrode conductor or the size of the dc grounding electrode conductor specified by 250.166. The dc grounding electrode system conductor(s) or the bonding jumpers to the ac grounding electrode system shall not be used as a substitute for any required ac equipment grounding conductors.

**(2) Common Direct-Current and Alternating-Current Grounding Electrode.** A dc grounding electrode conductor of the size specified by 250.166 shall be run from the marked dc grounding electrode connection point to the ac grounding electrode. Where an ac grounding electrode is not accessible, the dc grounding electrode conductor shall be connected to the ac grounding electrode conductor in accordance with 250.64(C)(1) or 250.64(C)(C)(2) or by using a connector listed for grounding and bonding. This dc grounding electrode conductor shall not be used as a substitute for any required ac equipment grounding conductors.

**(3) Combined Direct-Current Grounding Electrode Conductor and Alternating-Current Equipment Grounding Conductor.** An unspliced, or irreversibly spliced, combined grounding conductor shall be run from the marked dc grounding electrode conductor connection point along with the ac circuit conductors to the grounding busbar in the associated ac equipment. This combined grounding conductor shall be the larger of the sizes specified by 250.122 or 250.166 and shall be installed in accordance with 250.64(E). For ungrounded systems, this conductor shall be sized in accordance with 250.122 and shall not be required to be larger than the largest ungrounded phase conductor.

Inverters used in PV power systems usually contain a transformer that isolates the dc grounded circuit conductor from the ac grounded circuit conductor. Isolation necessitates that both a dc and an ac grounding system be installed. The two grounding systems are to be bonded together or have a common grounding electrode so that all ac and dc grounded circuit conductors and EGCs have the same near-zero potential to earth.

**(D) Additional Auxiliary Electrodes for Array Grounding.** A grounding electrode shall be installed in accordance with 250.52 and 250.54 at the location of all ground- and pole-mounted PV arrays and as close as practicable to the location of roof-mounted PV arrays. The electrodes shall be connected directly to the array frame(s) or structure. The dc grounding electrode conductor shall be sized according to 250.166. Additional electrodes are not permitted to be used as a substitute for equipment bonding or equipment grounding conductor requirements. The structure of a ground- or pole-mounted PV array shall be permitted to be considered a grounding electrode if it meets the requirements of 250.52. Roof-mounted PV arrays shall be permitted to use the metal frame of a building or structure if the requirements of 250.52(A)(2) are met.

*Exception No. 1: An array grounding electrode(s) shall not be required where the load served by the array is integral with the array.*

*Exception No. 2: An additional array grounding electrode(s) shall not be required if located within 1.8 m (6 ft) of the premises wiring electrode.*

## 690.48 Continuity of Equipment Grounding Systems

Where the removal of equipment disconnects the bonding connection between the grounding electrode conductor and exposed conducting surfaces in the PV source or output circuit equipment, a bonding jumper shall be installed while the equipment is removed.

PV source and output circuits are energized anytime the PV modules are exposed to light. In many PV systems, the main bonding jumper and the equipment grounding busbar are located in the inverter or in a dc power center that may require removal for service. The continuity of the EGCs should be maintained even when the equipment is removed.

## 690.49 Continuity of Photovoltaic Source and Output Circuit Grounded Conductors

Where the removal of the utility-interactive inverter or other equipment disconnects the bonding connection between the grounding electrode conductor and the PV source and/or PV output circuit grounded conductor, a bonding jumper shall be installed to maintain the system grounding while the inverter or other equipment is removed.

PV source and output circuits are energized anytime the PV modules are exposed to light. The marked, grounded circuit conductors should always remain grounded because they may be energized daily and cannot be easily disconnected from the source. In many PV systems, the main bonding jumper is located in the inverter or in a dc power center that may require removal for service. The continuity to ground of the grounded circuit conductors should be maintained even when the equipment is removed.

## 690.50 Equipment Bonding Jumpers

Equipment bonding jumpers, if used, shall comply with 250.120(C).

# VI. Marking

## 690.51 Modules

Modules shall be marked with identification of terminals or leads as to polarity, maximum overcurrent device rating for module protection, and with the following ratings:

(1) Open-circuit voltage
(2) Operating voltage
(3) Maximum permissible system voltage
(4) Operating current
(5) Short-circuit current
(6) Maximum power

## 690.52 Alternating-Current Photovoltaic Modules

Alternating-current modules shall be marked with identification of terminals or leads and with identification of the following ratings:

(1) Nominal operating ac voltage
(2) Nominal operating ac frequency
(3) Maximum ac power
(4) Maximum ac current
(5) Maximum overcurrent device rating for ac module protection

## 690.53 Direct-Current Photovoltaic Power Source

A permanent label for the direct-current PV power source indicating the information specified in (1) through (5) shall be provided by the installer at the PV disconnecting means:

(1) Rated maximum power-point current.
(2) Rated maximum power-point voltage.
(3) Maximum system voltage.

Informational Note to (3): See 690.7(A) for maximum PV system voltage.

(4) Maximum circuit current. Where the PV power source has multiple outputs, 690.53(1) and (4) shall be specified for each output.

Informational Note to (4): See 690.8(A) for calculation of maximum circuit current.

(5) Maximum rated output current of the charge controller (if installed).

Informational Note: Reflecting systems used for irradiance enhancement may result in increased levels of output current and power.

The rated values for the PV power source can be calculated by adding voltage ratings of series-connected modules and adding current ratings of parallel-connected modules or PV source circuits.

Some charge controllers have higher-rated output currents than the input currents from the PV array. They reduce the input voltage from the PV array while increasing the output to the battery.

## 690.54 Interactive System Point of Interconnection

All interactive system(s) points of interconnection with other sources shall be marked at an accessible location at the disconnecting means as a power source and with the rated ac output current and the nominal operating ac voltage.

## 690.55 Photovoltaic Power Systems Employing Energy Storage

Photovoltaic power systems employing energy storage shall also be marked with the maximum operating voltage, including any equalization voltage and the polarity of the grounded circuit conductor.

## 690.56 Identification of Power Sources

**(A) Facilities with Stand-Alone Systems.** Any structure or building with a PV power system that is not connected to a utility service source and is a stand-alone system shall have a permanent plaque or directory installed on the exterior of the building or structure at a readily visible location acceptable to the authority having jurisdiction. The plaque or directory shall indicate the location of system disconnecting means and that the structure contains a stand-alone electrical power system. The marking shall be in accordance with 690.31(G).

**(B) Facilities with Utility Services and PV Systems.** Buildings or structures with both utility service and a PV system shall have a permanent plaque or directory providing the location of the service disconnecting means and the PV system disconnecting means if not located at the same location. The warning sign(s) or label(s) shall comply with 110.21(B).

**(C) Facilities with Rapid Shutdown.** Buildings or structures with both utility service and a PV system, complying with 690.12, shall have a permanent plaque or directory including the following wording:

PHOTOVOLTAIC SYSTEM EQUIPPED
WITH RAPID SHUTDOWN

The plaque or directory shall be reflective, with all letters capitalized and having a minimum height of 9.5 mm (3⁄8 in.), in white on red background.

# VII. Connection to Other Sources

## 690.57 Load Disconnect

A load disconnect that has multiple sources of power shall disconnect all sources when in the off position.

## 690.60 Identified Interactive Equipment

Only inverters and ac modules listed and identified as interactive shall be permitted in interactive systems.

### 690.61 Loss of Interactive System Power

An inverter or an ac module in an interactive solar PV system shall automatically de-energize its output to the connected electrical production and distribution network upon loss of voltage in that system and shall remain in that state until the electrical production and distribution network voltage has been restored.

A normally interactive solar PV system shall be permitted to operate as a stand-alone system to supply loads that have been disconnected from electrical production and distribution network sources.

This requirement prevents energizing of otherwise de-energized system conductors or output conductors of other off-site sources (such as an electrical utility) and is intended to prevent electric shock. The ability to automatically de-energize output upon loss of voltage is normally a feature of the utility-interactive inverter.

### 690.63 Unbalanced Interconnections

Unbalanced connections shall be in accordance with 705.100.

### 690.64 Point of Connection

Point of connection shall be in accordance with 705.12.

## VIII. Storage Batteries

### 690.71 Installation

**(A) General.** Storage batteries in a solar photovoltaic system shall be installed in accordance with the provisions of Article 480. The interconnected battery cells shall be considered grounded where the photovoltaic power source is installed in accordance with 690.41.

**(B) Dwellings.**

**(1) Operating Voltage.** Storage batteries for dwellings shall have the cells connected so as to operate at a voltage of 50 volts, nominal, or less.

*Exception: Where live parts are not accessible during routine battery maintenance, a battery system voltage in accordance with 690.7 shall be permitted.*

**(2) Guarding of Live Parts.** Live parts of battery systems for dwellings shall be guarded to prevent accidental contact by persons or objects, regardless of voltage or battery type.

Informational Note: Batteries in solar photovoltaic systems are subject to extensive charge–discharge cycles and typically require frequent maintenance, such as checking electrolyte and cleaning connections.

**(C) Current Limiting.** A listed, current-limiting, overcurrent device shall be installed in each circuit adjacent to the batteries where the available short-circuit current from a battery or battery bank exceeds the interrupting or withstand ratings of other equipment in that circuit. The installation of current-limiting fuses shall comply with 690.16.

**(D) Battery Nonconductive Cases and Conductive Racks.** Flooded, vented, lead-acid batteries with more than twenty-four 2-volt cells connected in series (48 volts, nominal) shall not use conductive cases or shall not be installed in conductive cases. Conductive racks used to support the nonconductive cases shall be permitted where no rack material is located within 150 mm (6 in.) of the tops of the nonconductive cases.

This requirement shall not apply to any type of valve-regulated lead-acid (VRLA) battery or any other types of sealed batteries that may require steel cases for proper operation.

**(E) Disconnection of Series Battery Circuits.** Battery circuits subject to field servicing, where more than twenty-four 2-volt cells are connected in series (48 volts, nominal), shall have provisions to disconnect the series-connected strings into segments of 24 cells or less for maintenance by qualified persons. Non–load-break bolted or plug-in disconnects shall be permitted.

**(F) Battery Maintenance Disconnecting Means.** Battery installations, where there are more than twenty-four 2-volt cells connected in series (48 volts, nominal), shall have a disconnecting means, accessible only to qualified persons, that disconnects the grounded circuit conductor(s) in the battery electrical system for maintenance. This disconnecting means shall not disconnect the grounded circuit conductor(s) for the remainder of the photovoltaic electrical system. A non–load-break-rated switch shall be permitted to be used as the disconnecting means.

**(G) Battery Systems of More Than 48 Volts.** On photovoltaic systems where the battery system consists of more than twenty-four 2-volt cells connected in series (more than 48 volts, nominal), the battery system shall be permitted to operate with ungrounded conductors, provided the following conditions are met:

(1) The photovoltaic array source and output circuits shall comply with 690.41.
(2) The dc and ac load circuits shall be solidly grounded.
(3) All main ungrounded battery input/output circuit conductors shall be provided with switched disconnects and overcurrent protection.
(4) A ground-fault detector and indicator shall be installed to monitor for ground faults in the battery bank.

**(H) Disconnects and Overcurrent Protection.** Where energy storage device input and output terminals are more than 1.5 m (5 ft) from connected equipment, or where the circuits from these terminals pass through a wall or partition, the installation shall comply with the following:

(1) A disconnecting means and overcurrent protection shall be provided at the energy storage device end of the circuit. Fused disconnecting means or circuit breakers shall be permitted to be used.
(2) Where fused disconnecting means are used, the line terminals of the disconnecting means shall be connected toward the energy storage device terminals.

(3) Overcurrent devices or disconnecting means shall not be installed in energy storage device enclosures where explosive atmospheres can exist.
(4) A second disconnecting means located at the connected equipment shall be installed where the disconnecting means required by 690.71(H)(1) is not within sight of the connected equipment.
(5) Where the energy storage device disconnecting means is not within sight of the PV system ac and dc disconnecting means, placards or directories shall be installed at the locations of all disconnecting means indicating the location of all disconnecting means.

Circuits between a PV system and an energy storage device are bidirectional; a supply source is present on both ends of the circuit. Many energy storage devices are capable of significant short-circuit currents. Therefore, overcurrent protection is needed for circuits connected to these devices.

A disconnecting means is required at the battery end to disconnect the batteries from the circuit during equipment failure or maintenance. Any penetration of a wall or partition necessitates the installation of an additional disconnecting means at the equipment end of the circuit.

## 690.72 Charge Control

**(A) General.** Equipment shall be provided to control the charging process of the battery. Charge control shall not be required where the design of the photovoltaic source circuit is matched to the voltage rating and charge current requirements of the interconnected battery cells and the maximum charging current multiplied by 1 hour is less than 3 percent of the rated battery capacity expressed in ampere-hours or as recommended by the battery manufacturer.

All adjusting means for control of the charging process shall be accessible only to qualified persons.

Informational Note: Certain battery types such as valve-regulated lead acid or nickel cadmium can experience thermal failure when overcharged.

**(B) Diversion Charge Controller.**

**(1) Sole Means of Regulating Charging.** A photovoltaic power system employing a diversion charge controller as the sole means of regulating the charging of a battery shall be equipped with a second independent means to prevent overcharging of the battery.

**(2) Circuits with Direct-Current Diversion Charge Controller and Diversion Load.** Circuits containing a dc diversion charge controller and a dc diversion load shall comply with the following:

(1) The current rating of the diversion load shall be less than or equal to the current rating of the diversion load charge controller. The voltage rating of the diversion load shall be greater than the maximum battery voltage. The power rating of the diversion load shall be at least 150 percent of the power rating of the photovoltaic array.
(2) The conductor ampacity and the rating of the overcurrent device for this circuit shall be at least 150 percent of the maximum current rating of the diversion charge controller.

Diversion loads are typically rated by the current that they will draw at some rated voltage. If the rated current of the diversion load exceeds the current rating of the diversion load controller, the controller may not function properly.

**(3) PV Systems Using Utility-Interactive Inverters.** Photovoltaic power systems using utility-interactive inverters to control battery state-of-charge by diverting excess power into the utility system shall comply with (1) and (2):

(1) These systems shall not be required to comply with 690.72(B)(2). The charge regulation circuits used shall comply with the requirements of 400.5.
(2) These systems shall have a second, independent means of controlling the battery charging process for use when the utility is not present or when the primary charge controller fails or is disabled.

**(C) Buck/Boost Direct-Current Converters.** When buck/boost charge controllers and other dc power converters that increase or decrease the output current or output voltage with respect to the input current or input voltage are installed, the requirements shall comply with 690.72(C)(1) and (C)(2).

(1) The ampacity of the conductors in output circuits shall be based on the maximum rated continuous output current of the charge controller or converter for the selected output voltage range.
(2) The voltage rating of the output circuits shall be based on the maximum voltage output of the charge controller or converter for the selected output voltage range.

## 690.74 Battery Interconnections

**(A) Flexible Cables.** Flexible cables, as identified in Article 400, in sizes 2/0 AWG and larger shall be permitted within the battery enclosure from battery terminals to a nearby junction box where they shall be connected to an approved wiring method. Flexible battery cables shall also be permitted between batteries and cells within the battery enclosure. Such cables shall be listed for hard-service use and identified as moisture resistant.

Flexible, fine-stranded cables shall be terminated only with terminals, lugs, devices, or connectors in accordance with 110.14.

# IX. Systems over 1000 Volts

## 690.80 General

Solar PV systems with a maximum system voltage over 1000 volts dc shall comply with Article 490 and other requirements applicable to installations rated over 1000 volts.

### 690.81 Listing

Products listed for PV systems shall be permitted to be used and installed in accordance with their listing. PV wire that is listed for direct burial at voltages above 600 volts, but not exceeding 2000 volts, shall be installed in accordance with Table 300.50, column 1.

### 690.85 Definitions

For the purposes of Part VIII of this article, the voltages used to determine cable and equipment ratings are as follows.

**Battery Circuits.** In battery circuits, the highest voltage experienced under charging or equalizing conditions.

**Photovoltaic Circuits.** In dc PV source circuits and PV output circuits, the maximum system voltage.

## X. Electric Vehicle Charging

### 690.90 General

Photovoltaic systems used directly to charge electric vehicles shall comply with Article 625 in addition to the requirements of this article.

### 690.91 Charging Equipment

Electric vehicle couplers shall comply with 625.10. Personnel protection systems in accordance with 625.22 and automatic de-energization of cables in accordance with 625.19 are not required for PV systems with maximum system voltages of less than 80 volts dc.

# ARTICLE 692
# Fuel Cell Systems

## I. General

### 692.1 Scope

This article identifies the requirements for the installation of fuel cell power systems, which may be stand-alone or interactive with other electric power production sources and may be with or without electric energy storage such as batteries. These systems may have ac or dc output for utilization.

The rising demand for electric power has led to the development of power sources that are viable alternatives to or can be interconnected with electric utility distribution systems. Article 692 covers the installation of on-premises electrical supply systems where the power is derived from an electrochemical system that consumes fuel to generate an electric current.

The principle of operation is that direct current is generated through a chemical reaction in which fuel such as natural gas or LP-Gas is consumed. As opposed to internal combustion prime movers, the consumption of the fuel gas is via an electrochemical process rather than a combustion process. A power inverter converts the dc to ac. The installation requirements of Article 692 allow power derived from fuel cells to be safely delivered into residential and light commercial occupancies as the sole source of electric power or as an integrated source with a utility or other power source.

### 692.2 Definitions

**Fuel Cell.** An electrochemical system that consumes fuel to produce an electric current. In such cells, the main chemical reaction used for producing electric power is not combustion. However, there may be sources of combustion used within the overall cell system, such as reformers/fuel processors.

**Fuel Cell System.** The complete aggregate of equipment used to convert chemical fuel into usable electricity and typically consisting of a reformer, stack, power inverter, and auxiliary equipment.

**Interactive System.** A fuel cell system that operates in parallel with and may deliver power to an electrical production and distribution network. For the purpose of this definition, an energy storage subsystem of a fuel cell system, such as a battery, is not another electrical production source.

**Maximum System Voltage.** The highest fuel cell inverter output voltage between any ungrounded conductors present at accessible output terminals.

**Output Circuit.** The conductors used to connect the fuel cell system to its electrical point of delivery.

> Informational Note: In the case of sites that have series- or parallel-connected multiple units, the term *output circuit* also refers to the conductors used to electrically interconnect the fuel cell system(s).

**Point of Common Coupling.** The point at which the power production and distribution network and the customer interface occurs in an interactive system. Typically, this is the load side of the power network meter.

**Stand-Alone System.** A fuel cell system that supplies power independently of an electrical production and distribution network.

### 692.3 Other Articles

Wherever the requirements of other articles of this *Code* and Article 692 differ, the requirements of Article 692 shall apply, and, if the system is operated in parallel with a primary source(s) of electricity, the requirements in 705.14, 705.16, 705.32, and 705.143 shall apply.

### 692.4 Installation

**(A) Fuel Cell System.** A fuel cell system shall be permitted to supply a building or other structure in addition to any service(s) of another electricity supply system(s).

**(B) Identification.** A permanent plaque or directory, denoting all electric power sources on or in the premises, shall be installed at each service equipment location.

**(C) System Installation.** Fuel cell systems including all associated wiring and interconnections shall be installed by only qualified persons.

Informational Note: See Article 100 for the definition of *qualified person*.

### 692.6 Listing Requirement

The fuel cell system shall be evaluated and listed for its intended application prior to installation.

## II. Circuit Requirements

### 692.8 Circuit Sizing and Current

**(A) Nameplate Rated Circuit Current.** The nameplate(s) rated circuit current shall be the rated current indicated on the fuel cell nameplate(s).

**(B) Conductor Ampacity and Overcurrent Device Ratings.** The ampacity of the feeder circuit conductors from the fuel cell system(s) to the premises wiring system shall not be less than the greater of (1) nameplate(s) rated circuit current or (2) the rating of the fuel cell system(s) overcurrent protective device(s).

**(C) Ampacity of Grounded or Neutral Conductor.** If an interactive single-phase, 2-wire fuel cell output(s) is connected to the grounded or neutral conductor and a single ungrounded conductor of a 3-wire system or of a 3-phase, 4-wire, wye-connected system, the maximum unbalanced neutral load current plus the fuel cell system(s) output rating shall not exceed the ampacity of the grounded or neutral conductor.

### 692.9 Overcurrent Protection

**(A) Circuits and Equipment.** If the fuel cell system is provided with overcurrent protection sufficient to protect the circuit conductors that supply the load, additional circuit overcurrent devices shall not be required. Equipment and conductors connected to more than one electrical source shall be protected.

**(B) Accessibility.** Overcurrent devices shall be readily accessible.

### 692.10 Stand-Alone Systems

The premises wiring system shall meet the requirements of this Code except as modified by 692.10(a), (B), and (C).

**(A) Fuel Cell System Output.** The fuel cell system output from a stand-alone system shall be permitted to supply ac power to the building or structure disconnecting means at current levels below the rating of that disconnecting means.

**(B) Sizing and Protection.** The circuit conductors between the fuel cell system(s) output and the building or structure disconnecting means shall be sized based on the output rating of the fuel cell system(s). These conductors shall be protected from overcurrents in accordance with 240.4. The overcurrent protection shall be located at the output of the fuel cell system(s).

**(C) Single 120-Volt Nominal Supply.** The inverter output of a stand-alone fuel cell system shall be permitted to supply 120 volts, nominal, to single-phase, 3-wire 120/240-volt service equipment or distribution panels where there are no 240-volt loads and where there are no multiwire branch circuits. In all installations, the rating of the overcurrent device connected to the output of the fuel cell system(s) shall be less than the rating of the service equipment. This equipment shall be marked as follows:

**WARNING**
SINGLE 120-VOLT SUPPLY.
DO NOT CONNECT MULTIWIRE
BRANCH CIRCUITS!

The warning sign(s) or label(s) shall comply with 110.21(B).

## III. Disconnecting Means

### 692.13 All Conductors

Means shall be provided to disconnect all current-carrying conductors of a fuel cell system power source from all other conductors in a building or other structure.

### 692.17 Switch or Circuit Breaker

The disconnecting means for ungrounded conductors shall consist of readily accessible, manually operable switch(es) or circuit breaker(s).

Where all terminals of the disconnecting means may be energized in the open position, a warning sign shall be mounted on or adjacent to the disconnecting means. The sign shall be clearly legible and shall have the following words or equivalent:

DANGER
ELECTRIC SHOCK HAZARD.
DO NOT TOUCH TERMINALS.
TERMINALS ON BOTH THE LINE AND
LOAD SIDES MAY BE ENERGIZED
IN THE OPEN POSITION.

The danger sign(s) or label(s) shall comply with 110.21(B).

## IV. Wiring Methods

### 692.31 Wiring Systems

All raceway and cable wiring methods included in Chapter 3 of this *Code* and other wiring systems and fittings specifically intended and identified for use with fuel cell systems shall be permitted. Where wiring devices with integral enclosures are used, sufficient length of cable shall be provided to facilitate replacement.

## V. Grounding

### 692.41 System Grounding

**(A) AC Systems.** Grounding of ac systems shall be in accordance with 250.20, and with 250.30 for stand-alone systems.

**(B) DC Systems.** Grounding of dc systems shall be in accordance with 250.160.

**(C) Systems with Alternating-Current and Direct-Current Grounding Requirements.** When fuel cell power systems have both alternating-current (ac) and direct-current (dc) grounding requirements, the dc grounding system shall be bonded to the ac grounding system. The bonding conductor shall be sized according to 692.45. A single common grounding electrode and grounding bar may be used for both systems, in which case the common grounding electrode conductor shall be sized to meet the requirements of both 250.66 (ac) and 250.166 (dc).

### 692.44 Equipment Grounding Conductor

A separate equipment grounding conductor shall be installed.

### 692.45 Size of Equipment Grounding Conductor

The equipment grounding conductor shall be sized in accordance with 250.122.

### 692.47 Grounding Electrode System

Any auxiliary grounding electrode(s) required by the manufacturer shall be connected to the equipment grounding conductor specified in 250.118.

## VI. Marking

### 692.53 Fuel Cell Power Sources

A marking specifying the fuel cell system, output voltage, output power rating, and continuous output current rating shall be provided at the disconnecting means for the fuel cell power source at an accessible location on the site.

### 692.54 Fuel Shut-Off

The location of the manual fuel shut-off valve shall be marked at the location of the primary disconnecting means of the building or circuits supplied.

### 692.56 Stored Energy

A fuel cell system that stores electric energy shall require the following warning sign, or equivalent, at the location of the service disconnecting means of the premises:

WARNING
FUEL CELL POWER SYSTEM CONTAINS
ELECTRICAL ENERGY STORAGE DEVICES.

The warning sign(s) or label(s) shall comply with 110.21(B).

## VII. Connection to Other Circuits

### 692.59 Transfer Switch

A transfer switch shall be required in non–grid-interactive systems that use utility grid backup. The transfer switch shall maintain isolation between the electrical production and distribution network and the fuel cell system. The transfer switch shall be permitted to be located externally or internally to the fuel cell system unit. Where the utility service conductors of the structure are connected to the transfer switch, the switch shall comply with Article 230, Part V.

### 692.60 Identified Interactive Equipment

Only fuel cell systems listed and marked as interactive shall be permitted in interactive systems.

### 692.61 Output Characteristics

Output characteristics shall be in accordance with 705.14.

### 692.62 Loss of Interactive System Power

The fuel cell system shall be provided with a means of detecting when the electrical production and distribution network has become de-energized and shall not feed the electrical production and distribution network side of the point of common coupling during this condition. The fuel cell system shall remain in that state until the electrical production and distribution network voltage has been restored.

A normally interactive fuel cell system shall be permitted to operate as a stand-alone system to supply loads that have been disconnected from electrical production and distribution network sources.

### 692.64 Unbalanced Interconnections

Unbalanced interconnections shall be in accordance with 705.100.

### 692.65 Utility-Interactive Point of Connection

Point of connection shall be in accordance with 705.12.

## VIII. Outputs over 1000 Volts

### 692.80 General

Fuel cell systems with a maximum output voltage over 1000 volts ac shall comply with the requirements of other articles applicable to such installations.

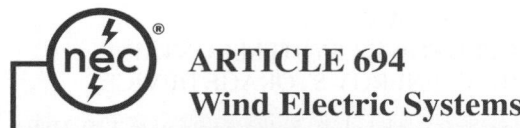

# ARTICLE 694
# Wind Electric Systems

## I. General

### 694.1 Scope

The provisions of this article apply to wind (turbine) electric systems that consist of one or more wind electric generators. These systems can include generators, alternators, inverters, and controllers.

Informational Note: Wind electric systems can be interactive with other electrical power production sources or might be stand-alone systems. Wind electric systems can have ac or dc output, with or without electrical energy storage, such as batteries. See Informational Note Figure 694.1(a) and Informational Note Figure 694.1(b).

Conversion of wind power to electricity is not new. It was used in rural farming areas prior to the Rural Electrification Act of 1936. Small-wattage wind-powered generating systems, generally under 200 W, were used to charge storage batteries that in turn powered appliances on rural farms. More recently, wind-driven turbines have been implemented into the utility power generation portfolio to generate power using clean, renewable sources.

Like photovoltaic and fuel cell systems, wind-driven turbines as a stand-alone or interconnected power production source are available for use as part of the premises wiring system, and these systems have seen a significant increase in use. According to the U.S. Department of Energy, although the United States reached 10 GW of wind power capacity in 25 years, it only took 4 years to add an additional 40 GW (2008–2012). These systems are anticipated to grow in popularity and use.

Wind turbine farms are more common in the utility sector, while a wind electric system consisting of a single wind turbine, such as the one shown in Exhibit 694.1, is the norm in most installations covered under the scope of this *Code*. These systems may be stand-alone or interconnected with a utility or other on-site source.

Article 694 contains the requirements for stand-alone and interconnected wind systems, and many of the requirements are similar to those contained in Articles 690 and 692. The requirements apply to all wind turbines within the scope of the *NEC*, regardless of the kW rating. This article also introduces specific terminology and requirements that address some of the safety concerns unique to this type of power generation system. For example, the elevated towers used to support the wind turbines present lightning protection concerns in areas susceptible to thunderstorm activity.

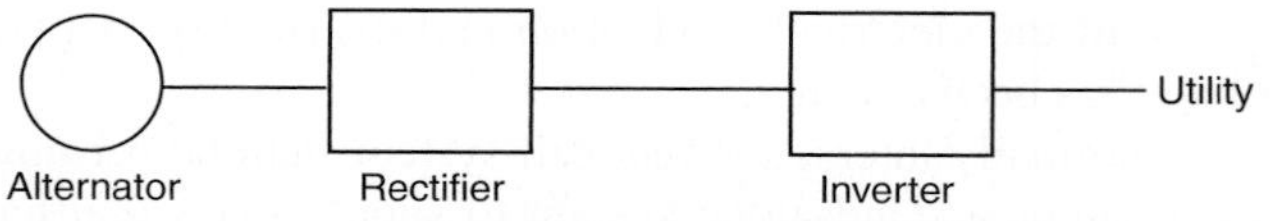


*INFORMATIONAL NOTE FIGURE 694.1(a) Identification of Wind Electric System Components — Interactive System.*

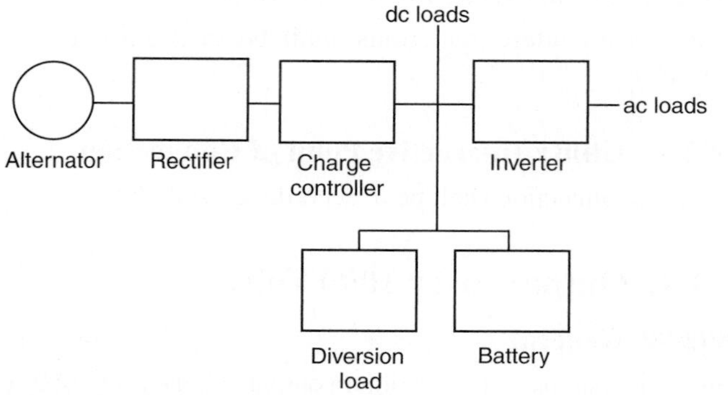


*INFORMATIONAL NOTE FIGURE 694.1(b) Identification of Wind Electric System Components — Stand-Alone System.*

### 694.2 Definitions

•

**Diversion Charge Controller.** Equipment that regulates the charging process of a battery or other energy storage device by diverting power from energy storage to dc or ac loads, or to an interconnected utility service.

**Diversion Load.** A load connected to a diversion charge controller or diversion load controller, also known as a dump load.

**Diversion Load Controller.** Equipment that regulates the output of a wind generator by diverting power from the generator to dc or ac loads or to an interconnected utility service.

**Guy.** A cable that mechanically supports a wind turbine tower.

**Inverter Output Circuit.** The conductors between an inverter and an ac panelboard for stand-alone systems, or the conductors

***EXHIBIT 694.1*** *A wind electric system consisting of a single wind turbine.*

between an inverter and service equipment or another electric power production source, such as a utility, for an electrical production and distribution network.

**Maximum Output Power.** The maximum 1 minute average power output a wind turbine produces in normal steady-state operation (instantaneous power output can be higher).

**Maximum Voltage.** The maximum voltage the wind turbine produces in operation including open circuit conditions.

**Nacelle.** An enclosure housing the alternator and other parts of a wind turbine.

**Rated Power.** The output power of a wind turbine at its rated wind speed.

Informational Note: The method for measuring wind turbine power output is specified in IEC 61400-12-1, *Power Performance Measurements of Electricity Producing Wind Turbines*.

**Tower.** A pole or other structure that supports a wind turbine.

**Wind Turbine.** A mechanical device that converts wind energy to electrical energy.

**Wind Turbine Output Circuit.** The circuit conductors between the internal components of a wind turbine (which might include an alternator, integrated rectifier, controller, and/or inverter) and other equipment.

Informational Note: See also definitions for interconnected systems in Article 705.

•

### 694.3 Other Articles

Where the system is operated in parallel with primary sources of electricity, the requirements of Article 705 shall apply.

*Exception: Wind electric systems, equipment, or wiring installed in a hazardous (classified) location shall also comply with the applicable portions of Articles 500 through 516.*

### 694.7 Installation

Systems covered by this article shall be installed only by qualified persons.

Informational Note: See Article 100 for the definition of *Qualified Person*.

Wind-powered systems present some unique hazards, including the danger of working in elevated, confined spaces. Therefore, personnel working on these systems have to be trained to recognize and avoid all hazards associated with installation and servicing this type of power generation system.

**(A) Wind Electric Systems.** A wind electric system(s) shall be permitted to supply a building or other structure in addition to other sources of supply.

**(B) Equipment.** Wind electric systems shall be listed and labeled for the application.

Three documents published by Underwriters Laboratories — Subject 6140, *Outline of Investigation for Wind Turbine Generating Systems;* Subject 6141, *Outline of Investigation for Wind Turbine Converters and Interconnection Systems;* and UL 6142, *Small Wind Turbine Systems* — provide the basis for certifying (product listing or classification) the overall wind turbine generator and its associated components, or individual components such as inverters and interconnection hardware, that are associated with a wind-generating system. All equipment of the wind electrical system requires listing, but field evaluations may be necessary for specific installations.

**(C) Diversion Load Controllers.** A small wind electric system employing a diversion load controller as the primary means of regulating the speed of a wind turbine rotor shall be equipped with an additional, independent, reliable means to prevent overspeed operation. An interconnected utility service shall not be considered to be a reliable diversion load.

**(D) Surge Protective Devices (SPD).** A surge protective device shall be installed between a small wind electric system and any loads served by the premises electrical system. The surge protective device shall be permitted to be a Type 3 SPD on a dedicated branch circuit serving a small wind electric system or a Type 2 SPD located anywhere on the load side of the service disconnect. Surge protective devices shall be installed in accordance with Part II of Article 285.

For most installations, the decision to install surge protective devices (SPDs) is made by the designer, installer, or building owner. Because the towers associated with wind electric systems will generally be the tallest structure in the vicinity, the use of surge protection devices covered in Article 285 is mandatory in order to help protect the premises wiring systems against the effects of lightning. Grounding of towers and guy wires is covered in 694.40(B)(3) and (4).

**(E) Receptacles.** A receptacle shall be permitted to be supplied by a wind electric system branch or feeder circuit for maintenance or data acquisition use. Receptacles shall be protected with an overcurrent device with a rating not to exceed the current rating of the receptacle. All 125-volt, single-phase, 15- and 20-ampere receptacles installed for maintenance of the wind turbine shall have ground-fault circuit-interrupter protection for personnel.

Any outdoor or indoor receptacle installed for the maintenance of a wind turbine is required to be GFCI protected. For example, a receptacle installed in a shed for system maintenance requires GFCI protection.

**(F) Metal or Nonmetallic Poles or Towers Supporting Wind Turbines Used as a Raceway.** A metallic or non-metallic pole or tower shall be permitted to be used as a raceway if evaluated as part of the listing for the wind turbine or otherwise shall be listed or evaluated for the purpose.

## II. Circuit Requirements

### 694.10 Maximum Voltage

**(A) Wind Turbine Output Circuits.** For wind turbines connected to one- and two-family dwellings, turbine output circuits shall be permitted to have a maximum voltage up to 600 volts. Other installations with a maximum voltage over 1000 volts shall comply with Part IX of Article 694.

**(B) Direct-Current Utilization Circuits.** The voltage of dc utilization circuits shall comply with 210.6.

**(C) Circuits over 150 Volts to Ground.** In one- and two-family dwellings, live parts in circuits over 150 volts to ground shall not be accessible to other than qualified persons while energized.

Informational Note: See 110.27 for guarding of live parts and 210.6 for branch circuit voltage limitations.

### 694.12 Circuit Sizing and Current

**(A) Calculation of Maximum Circuit Current.** The maximum current for a circuit shall be calculated in accordance with 694.12(A)(1) through (A)(3).

**(1) Turbine Output Circuit Currents.** The maximum current shall be based on the circuit current of the wind turbine operating at maximum output power.

**(2) Inverter Output Circuit Current.** The maximum output current shall be the inverter continuous output current rating.

**(3) Stand-Alone Inverter Input Circuit Current.** The maximum input current shall be the stand-alone continuous inverter input current rating of the inverter producing rated power at the lowest input voltage.

**(B) Ampacity and Overcurrent Device Ratings.**

**(1) Continuous Current.** Small wind turbine electric system currents shall be considered to be continuous.

**(2) Sizing of Conductors and Overcurrent Devices.** Circuit conductors and overcurrent devices shall be sized to carry not less than 125 percent of the maximum current as calculated in 694.12(A). The rating or setting of overcurrent devices shall be permitted in accordance with 240.4(B) and (C).

*Exception: Circuits containing an assembly, together with its overcurrent devices, listed for continuous operation at 100 percent of its rating shall be permitted to be used at 100 percent of its rating.*

### 694.15 Overcurrent Protection

**(A) Circuits and Equipment.** Turbine output circuits, inverter output circuits, and storage battery circuit conductors and equipment shall be protected in accordance with the requirements of Article 240. Circuits connected to more than one electrical source shall have overcurrent devices located so as to provide overcurrent protection from all sources.

*Exception: An overcurrent device shall not be required for circuit conductors sized in accordance with 694.12(B) where the maximum current from all sources does not exceed the ampacity of the conductors.*

Informational Note: Possible backfeed of current from any source of supply, including a supply through an inverter to the wind turbine output circuit, is a consideration in determining whether overcurrent protection from all sources is provided. Some wind electric systems rely on the turbine output circuit to regulate turbine speed. Inverters may also operate in reverse for turbine startup or speed control.

**(B) Power Transformers.** Overcurrent protection for a transformer with sources on each side shall be provided in accordance with 450.3 by considering first one side of the transformer, then the other side of the transformer, as the primary.

*Exception: A power transformer with a current rating on the side connected to the inverter output, which is not less than the rated continuous output current rating of the inverter, shall not be required to have overcurrent protection at the inverter.*

**(C) Direct-Current Rating.** Overcurrent devices, either fuses or circuit breakers, used in any dc portion of a small wind electric system shall be listed for use in dc circuits and shall have appropriate voltage, current, and interrupting ratings.

### 694.18 Stand-Alone Systems

The premises wiring system shall be adequate to meet the requirements of this *Code* for a similar installation connected to a service. The wiring on the supply side of the building or structure disconnecting means shall comply with this *Code*, except as modified by 694.18(A) through (D).

**(A) Inverter Output.** The ac output from stand-alone inverters shall be permitted to supply ac power to the building or structure disconnecting means at current levels less than the calculated load connected to that disconnect. The inverter output rating or the rating of a wind energy source shall be not less than the load of the largest single utilization equipment connected to the system. Calculated general lighting loads shall not be considered as a single load.

The total calculated ac load is not required to be used in sizing the inverter of a stand-alone wind electric system. The inverter of a stand-alone system is only required to be sized to supply the load of the largest single piece of utilization equipment. Larger inverters that will supply more of the ac load are, of course, permitted. This requirement parallels the rules for sizing the inverters of a stand-alone PV system specified in 690.10(A). Inrush currents should be taken into consideration when sizing the inverter.

**(B) Sizing and Protection.** The circuit conductors between the inverter output and the building or structure disconnecting means shall be sized based on the output rating of the inverter. These conductors shall be protected in accordance with Article 240. The overcurrent protection shall be located at the output of the inverter.

**(C) Single 120-Volt Supply.** The inverter output of a standalone small wind electric system shall be permitted to supply 120 volts to single-phase, 3-wire, 120/240-volt service equipment or distribution panels where there are no 240-volt outlets and where there are no multiwire branch circuits. In all installations, the rating of the overcurrent device connected to the output of the inverter shall be less than the rating of the neutral bus in the service equipment. This equipment shall be marked with the following words or equivalent:

WARNING.
SINGLE 120-VOLT SUPPLY.
DO NOT CONNECT.
MULTIWIRE BRANCH CIRCUITS!

The warning sign(s) or label(s) shall comply with 110.21(B).

Multiwire branch circuits are common in one- and two-family dwelling units. When connected to a normal 120/240-volt ac service, the currents in the neutral conductors of these multiwire branch circuits (typically 14-3 AWG) subtract or are at most no larger than the rating of the branch-circuit overcurrent device. If the electrical system consists of a single 120-volt wind electrical system inverter supplying the two buses in the panelboard, the currents in the grounded conductor for each multiwire branch circuit add rather than subtract. Because the two buses are in phase, there is no neutral conductor. The currents in these conductors may be as high as twice the rating of the branch-circuit overcurrent device, and overloading is possible.

**(D) Energy Storage or Backup Power System Requirements.** Energy storage or backup power supplies shall not be required.

## III. Disconnecting Means

### 694.20 All Conductors

Means shall be provided to disconnect all current-carrying conductors of a small wind electric power source from all other conductors in a building or other structure. A switch, circuit breaker, or other device, either ac or dc, shall not be installed in a grounded conductor if operation of that switch, circuit breaker, or other device leaves the marked, grounded conductor in an ungrounded and energized state.

*Exception: A wind turbine that uses the turbine output circuit for regulating turbine speed shall not require a turbine output circuit disconnecting means.*

### 694.22 Additional Provisions

Disconnecting means shall comply with 694.22(A) through (D).

**(A) Disconnecting Means.** The disconnecting means shall not be required to be suitable for use as service equipment. The disconnecting means for ungrounded conductors shall consist of manually operable switches or circuit breakers complying with all of the following requirements:

(1) They shall be located where readily accessible.
(2) They shall be externally operable without exposing the operator to contact with live parts.
(3) They shall plainly indicate whether in the open or closed position.
(4) They shall have an interrupting rating sufficient for the nominal circuit voltage and the current that is available at the line terminals of the equipment.

Where all terminals of the disconnecting means are capable of being energized in the open position, a warning sign shall be mounted on or adjacent to the disconnecting means. The sign shall be clearly legible and shall have the following words or equivalent:

WARNING.
ELECTRIC SHOCK HAZARD.
DO NOT TOUCH TERMINALS.
TERMINALS ON BOTH THE LINE
AND LOAD SIDES MAY BE
ENERGIZED IN THE OPEN POSITION.

The warning sign(s) or label(s) shall comply with 110.21(B).

**(B) Equipment.** Equipment such as rectifiers, controllers, output circuit isolating and shorting switches, and over-current devices shall be permitted on the wind turbine side of the disconnecting means.

**(C) Requirements for Disconnecting Means.**

**(1) Location.** The small wind electric system disconnecting means shall be installed at a readily accessible location either on or adjacent to the turbine tower, on the outside of a building or structure or inside, at the point of entrance of the wind system conductors.

*Exception: Installations that comply with 694.30(C) shall be permitted to have the disconnecting means located remotely from the point of entry of the wind system conductors.*

A wind turbine disconnecting means shall not be required to be located at the nacelle or tower.

The disconnecting means shall not be installed in bathrooms.

The general requirement for locating the wind electric system disconnecting means uses the same concept as found in 230.70(A) for services. The exception permits the system disconnecting means to be located at any readily accessible location within a building or structure except a bathroom, provided the wiring method between the conductor point of entry and the disconnecting means is a metal raceway or the conductors are protected by a metal enclosure. The supply conductors from a turbine to a building or structure are permitted to be installed without a disconnecting means until they reach the building or structure.

**(2) Marking.** Each turbine system disconnecting means shall be permanently marked to identify it as a small wind electric system disconnect. A plaque shall be installed in accordance with 705.10.

**(3) Suitable for Use.** Turbine system disconnecting means shall be suitable for the prevailing conditions.

**(4) Maximum Number of Disconnects.** The turbine disconnecting means shall consist of not more than six switches or six circuit breakers mounted in a single enclosure, in a group of separate enclosures, or in or on a switchgear.

**(D) Equipment That Is Not Readily Accessible.** Rectifiers, controllers, and inverters shall be permitted to be mounted in nacelles or other exterior areas that are not readily accessible.

### 694.23 Turbine Shutdown

**(A) Manual Shutdown.** Wind turbines shall be required to have a readily accessible manual shutdown button or switch. Operation of the button or switch shall result in a parked turbine state that shall either stop the turbine rotor or allow limited rotor speed combined with a means to de-energize the turbine output circuit.

*Exception: Turbines with a swept area of less than 50 $m^2$ (538 $ft^2$) shall not be required to have a manual shutdown button or switch.*

**(B) Shutdown Procedure.** The shutdown procedure for a wind turbine shall be defined and permanently posted at the location of a shutdown means and at the location of the turbine controller or disconnect, if the location is different.

Although the exception to 694.23(A) permits a smaller turbine without a manual shutdown switch, a shutdown procedure is required for all turbines.

### 694.24 Disconnection of Wind Electric System Equipment

Means shall be provided to disconnect equipment, such as inverters, batteries, and charge controllers, from all ungrounded conductors of all sources. If the equipment is energized from more than one source, the disconnecting means shall be grouped and identified.

A single disconnecting means in accordance with 694.22 shall be permitted for the combined ac output of one or more inverters in an interactive system.

A shorting switch or plug shall be permitted to be used as an alternative to a disconnect in systems that regulate turbine speed using the turbine output circuit.

*Exception: Equipment housed in a turbine nacelle shall not be required to have a disconnecting means.*

### 694.26 Fuses

Means shall be provided to disconnect a fuse from all sources of supply where the fuse is energized from both directions and is accessible to other than qualified persons. Switches, pullouts, or similar devices that are rated for the application shall be permitted to serve as a means to disconnect fuses from all sources of supply.

### 694.28 Installation and Service of a Wind Turbine

Open circuiting, short circuiting, or mechanical brakes shall be used to disable a turbine for installation and service.

Informational Note: Some wind turbines rely on the connection from the alternator to a remote controller for speed regulation. Opening turbine output circuit conductors may cause mechanical damage to a turbine and create excessive voltages that could damage equipment or expose persons to electric shock.

## IV. Wiring Methods

### 694.30 Permitted Methods

**(A) Wiring Systems.** All raceway and cable wiring methods included in this *Code*, and other wiring systems and fittings specifically intended for use on wind turbines, shall be permitted. In readily accessible locations, turbine output circuits that operate at voltages greater than 30 volts shall be installed in raceways.

**(B) Flexible Cords and Cables.** Flexible cords and cables, where used to connect the moving parts of turbines or where used for ready removal for maintenance and repair, shall comply with Article 400 and shall be of a type identified as hard service cord or portable power cable, shall be suitable for extra-hard usage, shall be listed for outdoor use, and shall be water resistant. Cables exposed to sunlight shall be sunlight resistant. Flexible, fine-stranded cables shall be terminated only with terminals, lugs, devices, or connectors in accordance with 110.14(A).

To provide a greater degree of flexibility, the conductors associated with flexible cords and cables are typically more finely stranded than conductors with Class B or C stranding. Terminals used with classes of stranding other than Class B or C are required to be identified for the class(es) of stranding for which they are suitable. See 110.14 and its associated commentary.

**(C) Direct-Current Turbine Output Circuits Inside a Building.** Direct-current turbine output circuits installed inside a building or structure shall be enclosed in metal raceways or installed in metal enclosures, or run in Type MC metal-clad cable that complies with 250.118(10), from the point of penetration of the surface of the building or structure to the first readily accessible disconnecting means.

## V. Grounding

### 694.40 Equipment Grounding

**(A) General.** Exposed non–current-carrying metal parts of towers, turbine nacelles, other equipment, and conductor enclosures shall be grounded in accordance with Parts IV, V, and VI of Article 250. Attached metal parts, such as turbine blades and tails that are not likely to become energized, shall not be required to be grounded or bonded.

**(B) Tower Grounding and Bonding.**

**(1) Grounding Electrodes and Grounding Electrode Conductors.** A wind turbine tower shall be connected to a grounding electrode system. Where installed in close proximity to galvanized foundation or tower anchor components, galvanized grounding electrodes shall be used.

Informational Note: Copper and copper-clad grounding electrodes, where used in highly conductive soils, can cause electrolytic corrosion of galvanized foundation and tower anchor components.

**(2) Bonding Conductor.** Equipment grounding conductors or supply-side bonding jumpers, as applicable, shall be required between turbines, towers, and the premises grounding system in accordance with Parts V and VI of Article 250.

**(3) Tower Connections.** Equipment grounding conductors and grounding electrode conductors, where used, shall be connected to metallic towers using listed means. All mechanical elements used to terminate these conductors shall be accessible.

**(4) Guy Wires.** Guy wires used to support turbine towers shall not be required to be connected to an equipment grounding or bonding conductor or to comply with the requirements of 250.110.

Informational Note: Guy wires supporting grounded towers are unlikely to become energized. Grounding of metallic guy wires may be required by lightning codes. For information on lightning protection systems, see NFPA 780-2014, *Standard for the Installation of Lightning Protection Systems*.

## VI. Marking

### 694.50 Interactive System Point of Interconnection

All interactive system points of interconnection with other sources shall be marked at an accessible location at the disconnecting means and with the rated ac output current and the nominal operating ac voltage.

### 694.52 Power Systems Employing Energy Storage

Wind electric systems employing energy storage shall be marked with the maximum operating voltage, any equalization voltage, and the polarity of the grounded circuit conductor.

### 694.54 Identification of Power Sources

**(A) Facilities with Stand-Alone Systems.** Any structure or building with a stand-alone system and not connected to a utility service source shall have a permanent plaque or directory installed on the exterior of the building or structure at a readily visible location. The plaque or directory shall indicate the location of system disconnecting means and shall indicate that the structure contains a stand-alone electrical power system.

**(B) Facilities with Utility Services and Wind Electric Systems.** Buildings or structures with both utility service and wind electric systems shall have a permanent plaque or directory providing the location of the service disconnecting means and the wind electric system disconnecting means.

### 694.56 Instructions for Disabling Turbine

A plaque shall be installed at or adjacent to the turbine location providing basic instructions for disabling the turbine.

## VII. Connection to Other Sources

### 694.60 Identified Interactive Equipment

Only inverters listed and identified as interactive shall be permitted in interactive systems.

### 694.62 Installation

Wind electric systems, where connected to utility electric sources, shall comply with the requirements of Article 705.

### 694.66 Operating Voltage Range

Wind electric systems connected to dedicated branch or feeder circuits shall be permitted to exceed normal voltage operating ranges on these circuits, provided that the voltage at any distribution equipment supplying other loads remains within normal ranges.

Informational Note: Wind turbines might use the electric grid to dump energy from short-term wind gusts. Normal operating voltages are defined in ANSI C84.1-2006, *Voltage Ratings for Electric Power Systems and Equipment (60 Hz)*.

### 694.68 Point of Connection

Points of connection to interconnected electric power sources shall comply with 705.12.

•

## VIII. Systems over 1000 Volts

### 694.80 General

Wind electric systems with a maximum system voltage exceeding 1000 volts ac or dc shall comply with Article 490 and other requirements applicable to installations rated over 1000 V.

### 694.85 Cable and Equipment Ratings

For the purposes of Part IX of this article, the voltages used to determine cable and equipment ratings shall be as specified in 694.85(A) and (B).

**(A) Battery Circuits.** In battery circuits, the voltage used shall be the highest voltage experienced under charging or equalizing conditions.

**(B) Other Circuits.** In other circuits, the voltage used shall be the maximum voltage experienced in normal operation.

# ARTICLE 695 Fire Pumps

## 695.1 Scope

Informational Note: Text that is followed by a reference in brackets has been extracted from NFPA 20-2013, *Standard for the Installation of Stationary Pumps for Fire Protection*. Only editorial changes were made to the extracted text to make it consistent with this *Code*.

**(A) Covered.** This article covers the installation of the following:

(1) Electric power sources and interconnecting circuits
(2) Switching and control equipment dedicated to fire pump drivers

**(B) Not Covered.** This article does not cover the following:

(1) The performance, maintenance, and acceptance testing of the fire pump system, and the internal wiring of the components of the system
(2) The installation of pressure maintenance (jockey or makeup) pumps

Informational Note: For the installation of pressure maintenance (jockey or makeup) pumps supplied by the fire pump circuit or another source, see Article 430.

(3) Transfer equipment upstream of the fire pump transfer switch(es)

Informational Note: See NFPA 20-2013, *Standard for the Installation of Stationary Pumps for Fire Protection*, for further information.

The requirements covering reliable power supplies for electric fire pump motors correlate with those in NFPA 20, *Standard for the Installation of Stationary Pumps for Fire Protection*. However, the *NEC* and NFPA 20 have a distinct division of responsibility for fire pump requirements. Performance issues, including the determination of power supply reliability, are under the jurisdiction of the NFPA Technical Committee on Fire Pumps, while electrical installation requirements are within the purview of the National Electrical Code Committee. The scope of Article 695 speaks to this division of responsibility.

An electric motor–driven fire pump such as the one shown in Exhibit 695.1 is covered by the requirements of Article 695. This article does not apply to pumps used to supply sprinkler systems in one- and two-family dwellings. NFPA 13D, *Standard for the Installation of Sprinkler Systems in One- and Two-Family Dwellings and Manufactured Homes*, does not require the use of a fire pump; thus, neither NFPA 20 nor Article 695 is applicable. Although the installation requirements for pressure maintenance (jockey)

***EXHIBIT 695.1*** *An electric motor specifically listed for fire pump service. (Courtesy of Liberty Mutual Insurance)*

pumps are not covered by Article 695, these pumps are permitted to be supplied by a fire pump service or feeder.

Generally the requirements of Article 695 are independent of those in Article 700 unless otherwise mandated by the AHJ. The only exception to this is the specific reference in 695.4(B)(3)(b).

## 695.2 Definitions

**Fault-Tolerant External Control Circuits.** Those control circuits either entering or leaving the fire pump controller enclosure, which if broken, disconnected, or shorted will not prevent the controller from starting the fire pump from all other internal or external means and may cause the controller to start the pump under these conditions.

**On-Site Power Production Facility.** The normal supply of electric power for the site that is expected to be constantly producing power.

**On-Site Standby Generator.** A facility producing electric power on site as the alternate supply of electric power. It differs from an on-site power production facility, in that it is not constantly producing power.

## 695.3 Power Source(s) for Electric Motor-Driven Fire Pumps

Electric motor-driven fire pumps shall have a reliable source of power.

**(A) Individual Sources.** Where reliable, and where capable of carrying indefinitely the sum of the locked-rotor current of the fire pump motor(s) and the pressure maintenance pump motor(s) and the full-load current of the associated fire pump accessory equipment when connected to this power supply, the power source for an electric motor driven fire pump shall be one or more of the following.

The power source for an electric motor–driven fire pump must be reliable and have adequate capacity to carry the locked-rotor currents of the fire pump motor and accessory equipment. These two main requirements ensure that the fire pump operates in the event of a fire without being accidentally disconnected, and that the fire pump continues to operate until the fire is extinguished, the fire pump is purposely shut down, or the pump itself is destroyed.

The determination of whether the serving electric utility is a reliable source of power is an issue for the AHJ. The following excerpt of A.9.3.2 in Annex A of NFPA 20 elaborates on several key characteristics of a reliable power supply:

> **A.9.3.2** A reliable power source possesses the following characteristics:
>
> (1) The source power plant has not experienced any shutdowns longer than 4 continuous hours in the year prior to plan submittal. NFPA 25, *Standard for the Inspection, Testing, and Maintenance of Water-Based Fire Protection Systems,* requires special undertakings (i.e., fire watches) when a water-based fire protection system is taken out of service for longer than 4 hours. If the normal source power plant has been intentionally shut down for longer than 4 hours in the past, it is reasonable to require a backup source of power.
>
> (2) Power outages have not routinely been experienced in the area of the protected facility caused by failures in generation or transmission. The standard is not intended to require that the normal source of power be infallible to deem the power reliable. NFPA 20 does not intend to require a back-up source of power for every installation using an electric motor–driven fire pump. Note that should the normal source of power fail in a rare event, the impairment procedures of NFPA 25 could be followed to mitigate the risk. If a fire does occur during the power loss, the fire protection system could be supplied through the fire department connection.
>
> (3) The normal source of power is not supplied by overhead conductors outside the protected facility. Fire departments responding to an incident at the protected facility will not operate aerial apparatus near live overhead power lines, without exception. A backup source of power is required in case this scenario occurs and the normal source of power must be shut off. Additionally, many utility providers will remove power to the protected facility by physically cutting the overhead conductors. If the normal source of power is provided by overhead conductors, which will not be identified, the utility provider could mistakenly cut the overhead conductor supplying the fire pump.
>
> (4) Only the disconnect switches and overcurrent protection devices permitted by 9.2.3 are installed in the normal source of power. Power disconnection and activated overcurrent protection should only occur in the fire pump controller. The provisions of 9.2.2 for the disconnect switch and overcurrent protection essentially require disconnection and overcurrent protection to occur in the fire pump controller. If unanticipated disconnect switches or overcurrent protection devices are installed in the normal source of power that do not meet the requirements of 9.2.2, the normal source of power must be considered not reliable and a back-up source of power is necessary.

Performance requirements for the alternate source of electric power can be found in NFPA 110, *Standard for Emergency and Standby Power Systems.*

For an on-site power production facility to be considered a reliable power source for an electric motor–driven fire pump(s), fire protection measures must be in place to protect the source and maintain a reliable power supply. In many cases, on-site power production sources are electric generating stations dedicated to a particular facility or to a particular facility's campus-style distribution system. Information on fire protection systems for on-site generating stations can be found in NFPA 850, *Recommended Practice for Fire Protection for Electric Generating Plants and High Voltage Direct Current Converter Stations.*

**(1) Electric Utility Service Connection.** A fire pump shall be permitted to be supplied by a separate service, or from a connection located ahead of and not within the same cabinet, enclosure, vertical switchgear section, or vertical switchboard section as the service disconnecting means. The connection shall be located and arranged so as to minimize the possibility of damage by fire from within the premises and from exposing hazards. A tap ahead of the service disconnecting means shall comply with 230.82(5). The service equipment shall comply with the labeling requirements in 230.2 and the location requirements in 230.72(B). [**20:**9.2.2(1)]

Configuration No. 1 of Exhibit 695.2 shows a single service where a dedicated set of service-entrance conductors to supply the fire pump is tapped to the incoming service conductors. The tap cannot be made in the section of the equipment that contains the service disconnecting means. This tap is permitted under the

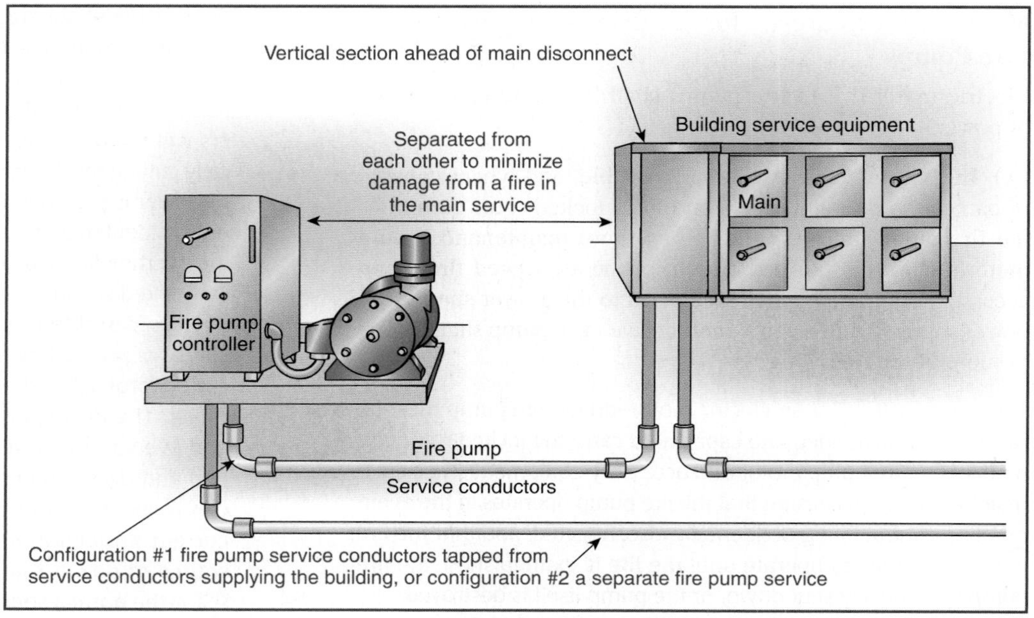


***EXHIBIT 695.2*** *Two permitted configurations for connecting to an electric utility–supplied service.*

conditions specified in 230.40, Exception No. 5, and in 230.82(5). Configuration No. 2 shows a dedicated service supplying the fire pump as permitted by 230.2(A)(1).

**(2) On-Site Power Production Facility.** A fire pump shall be permitted to be supplied by an on-site power production facility. The source facility shall be located and protected to minimize the possibility of damage by fire. [**20:**9.2.2(3)]

*On-site power production facilities* are defined in 695.2 and are permitted to be used as the sole power source for an electrically driven fire pump motor. An on-site power production facility differs in normal application from an on-site standby generator, in that it is the normal source of electrical supply for a structure and is not a utility-owned generating facility. For some small installations, the normal use of the generator is the feature that determines whether the equipment is defined as an on-site power production facility or an on-site standby generator. Exhibit 695.3 illustrates generating equipment that is the normal source of power for the premises wiring system and meets the definition of on-site power production facility. On-site power production is not restricted to a generator.

**(3) Dedicated Feeder.** A dedicated feeder shall be permitted where it is derived from a service connection as described in 695.3(A)(1). [**20:**9.2.2(3)]

**(B) Multiple Sources.** If reliable power cannot be obtained from a source described in 695.3(A), power shall be supplied by one of the following: [**20:**9.3.2]

**(1) Individual Sources.** An approved combination of two or more of the sources from 695.3(A).

**(2) Individual Source and On-site Standby Generator.** An approved combination of one or more of the sources in 695.3(A)

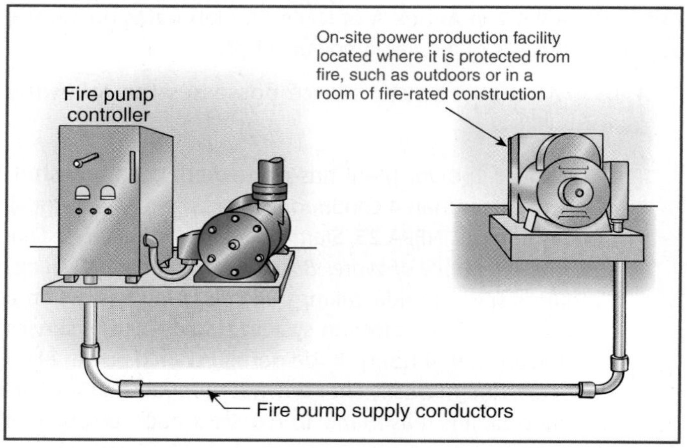


***EXHIBIT 695.3*** *On-site power production facility as a power source for a fire pump installation.*

and an on-site standby generator complying with 695.3(D). [**20:**9.3.4]

*Exception to (B)(1) and (B)(2): An alternate source of power shall not be required where a back-up engine-driven or back-up steam turbine-driven fire pump is installed. [**20:**9.3.3]*

If none of the power supply sources specified in 695.3(A)(1) through (3) can individually provide reliable power with adequate capacity, 695.3(B) permits an approved combination (two or more) of these sources or a combination of one or more of these sources with an on-site standby generator.

In lieu of installing an on-site standby generator, an engine- or steam turbine–driven fire pump may be provided as backup for an electric fire pump. In this instance, the electric fire pump is

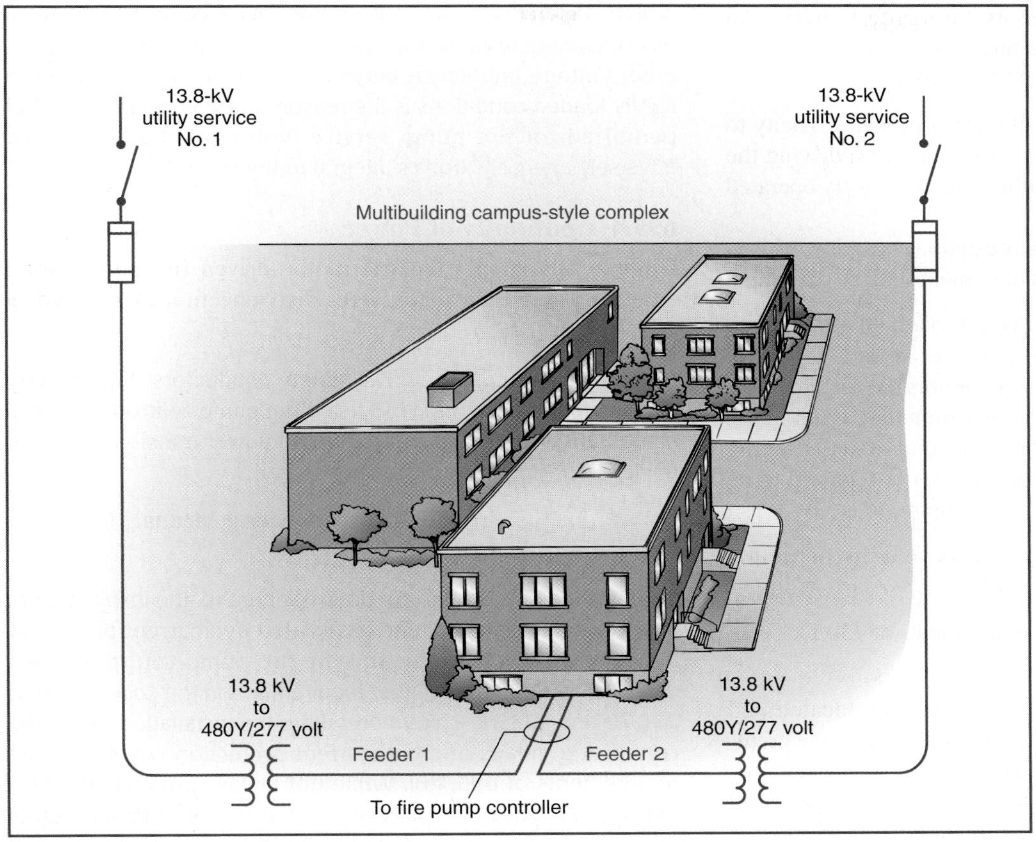


***EXHIBIT 695.4*** *Multiple feeder sources for campus-style application.*

permitted to be supplied by only a single power source. This allowance provides some design options for augmenting an electric fire pump that is supplied by an unreliable source.

**(C) Multibuilding Campus-Style Complexes.** If the sources in 695.3(A) are not practicable and the installation is part of a multibuilding campus-style complex, feeder sources shall be permitted if approved by the authority having jurisdiction and installed in accordance with either (C)(1) and (C)(3) or (C)(2) and (C)(3).

**(1) Feeder Sources.** Two or more feeders shall be permitted as more than one power source if such feeders are connected to, or derived from, separate utility services. The connection(s), overcurrent protective device(s), and disconnecting means for such feeders shall meet the requirements of 695.4(B).

**(2) Feeder and Alternate Source.** A feeder shall be permitted as a normal source of power if an alternate source of power independent from the feeder is provided. The connection(s), overcurrent protective device(s), and disconnecting means for such feeders shall meet the requirements of 695.4(B).

**(3) Selective Coordination.** The overcurrent protective device(s) in each disconnecting means shall be selectively coordinated with any other supply-side overcurrent protective device(s).

A fire pump supplied by a radial loop type of distribution system (commonly used for medium- and high-voltage distribution) where the two feeders originate from a single substation has to be augmented by an on-site standby generator. In the system shown in Exhibit 695.4, the two feeders originate from different utility substations, a distribution arrangement that allows the two feeders, without an on-site standby generator, to be multiple sources for the electric fire pump as permitted by 695.3(C)(1).

Section 695.3(C) allows for fire pumps to be supplied by feeder circuits that are part of a medium- or high-voltage premises wiring system. This distribution arrangement is common in industrial and institutional campus settings. The conductors supplied by the higher voltage level distribution systems are not service conductors, because the service point and service-disconnecting means is generally located at a campus distribution switchyard or distribution building. Also, all of the distribution conductors on the load side of the service equipment, even though they resemble electric utility–type distribution, are considered to be feeders or – in some cases where the circuit supplies a single piece of utilization equipment – branch circuits.

Where more than one overcurrent device is in series with the fire pump supply, 695.3(C)(3) requires each device to be selectively coordinated with supply-side OCPDs. See the commentary for and definition of *selective coordination* in Article 100.

**(D) On-Site Standby Generator as Alternate Source.** An on-site standby generator(s) used as an alternate source of power shall comply with (D)(1) through (D)(3). [**20:**9.6.2.1]

**(1) Capacity.** The generator shall have sufficient capacity to allow normal starting and running of the motor(s) driving the fire pump(s) while supplying all other simultaneously operated load(s). [**20:**9.6.1.1]

Automatic shedding of one or more optional standby loads in order to comply with this capacity requirement shall be permitted.

Only the sources specified in 695.3(A)(1) through (3) are required to be capable of indefinitely carrying the locked-rotor current of the fire pump motor. On-site standby generators are required only to be capable of carrying the starting and running current of the fire pump motor. The generator disconnecting means and the OCPD(s) for the electric-driven fire pump are not required to be sized for locked-rotor current of the fire pump motor(s).

**(2) Connection.** A tap ahead of the generator disconnecting means shall not be required. [**20:**9.6.1.2]

**(3) Adjacent Disconnects.** The requirements of 430.113 shall not apply.

**(E) Arrangement.** All power supplies shall be located and arranged to protect against damage by fire from within the premises and exposing hazards. [**20:**9.1.4]

Multiple power sources shall be arranged so that a fire at one source does not cause an interruption at the other source.

Determining compliance of the installation requires review of individual building or structure characteristics. The type of construction, type of content, proximity of building to other hazard exposures, and the location of the primary and alternate power sources for the fire pump should be considered.

**(F) Transfer of Power.** Transfer of power to the fire pump controller between the individual source and one alternate source shall take place within the pump room. [**20:**9.6.4]

**(1) Power Source Selection.** Selection of power source shall be performed by a transfer switch listed for fire pump service. [**20:**10.8.1.3.1]

**(2) Overcurrent Device Selection.** An instantaneous trip circuit breaker shall be permitted in lieu of the overcurrent devices specified in 695.4(B)(2)(a)(1), provided that it is part of a transfer switch assembly listed for fire pump service that complies with 695.4(B)(2)(a)(2).

A listed fire pump transfer switch with a factory-installed instantaneous circuit breaker provides ground-fault and short-circuit protection. Overload protection is provided by the circuit breaker in the fire pump controller. Selective coordination of the breakers is accomplished as part the equipment evaluation.

**(G) Phase Converters.** Phase converters shall not be permitted to be used for fire pump service. [**20:**9.1.7]

A phase converter used in a fire pump circuit would be in continuous operation, because the controller has to be constantly powered. Voltage imbalance between phases under unloaded or lightly loaded conditions is the reason phase converters are not permitted for fire pump service. Voltage imbalance could adversely affect electronics integral to the controller.

## 695.4 Continuity of Power

Circuits that supply electric motor–driven fire pumps shall be supervised from inadvertent disconnection as covered in 695.4(A) or (B).

**(A) Direct Connection.** The supply conductors shall directly connect the power source to a listed fire pump controller, a listed combination fire pump controller and power transfer switch, or a listed fire pump power transfer switch.

**(B) Connection Through Disconnecting Means and Overcurrent Device.**

Section 695.4(B) permits, but does not require, the installation of a disconnecting means and associated overcurrent protection between a power source and the fire pump control devices described in 695.4(B)(1). Other requirements in the *Code* – such as 230.70 and 225.31 – may necessitate the installation of the disconnecting means and overcurrent protection covered in the requirements of 695.4(B). While not always possible, the best method to provide continuity of power is the direct connection of the source to the fire pump control equipment in accordance with 695.4(A).

**(1) Number of Disconnecting Means.**

(a) *General.* A single disconnecting means and associated overcurrent protective device(s) shall be permitted to be installed between the fire pump power source(s) and one of the following: [**20:**9.1.2]

(1) A listed fire pump controller
(2) A listed fire pump power transfer switch
(3) A listed combination fire pump controller and power transfer switch

(b) *Feeder Sources.* For systems installed under the provisions of 695.3(C) only, additional disconnecting means and the associated overcurrent protective device(s) shall be permitted as required to comply with other provisions of this *Code*.

(c) *On-Site Standby Generator.* Where an on-site standby generator is used to supply a fire pump, an additional disconnecting means and an associated overcurrent protective device(s) shall be permitted.

An on-site standby generator equipped with an integral disconnecting means and overcurrent protection is allowed in addition to a disconnecting means and overcurrent protection installed elsewhere in the alternate supply circuit to the fire pump. The second disconnecting means and overcurrent device could be located in distribution equipment and is required to comply with the requirements of 695.4(B)(2)(b) and 695.4(B)(3)(b) through (e).

**(2) Overcurrent Device Selection.** Overcurrent devices shall comply with 695.4(B)(2)(a) or (b).

(a) *Individual Sources.* Overcurrent protection for individual sources shall comply with 695.4(B)(2)(a)(1) or (2).

(1) Overcurrent protective device(s) shall be rated to carry indefinitely the sum of the locked-rotor current of the largest fire pump motor and the pressure maintenance pump motor(s) and the full-load current of all of the other pump motors and associated fire pump accessory equipment when connected to this power supply. Where the locked-rotor current value does not correspond to a standard overcurrent device size, the next standard overcurrent device size shall be used in accordance with 240.6. The requirement to carry the locked-rotor currents indefinitely shall not apply to conductors or devices other than overcurrent devices in the fire pump motor circuit(s). [**20:**9.2.3.4]

A key factor in the reliable power source equation is sizing the overcurrent protection in a supervised fire pump disconnecting means so it is able to carry locked-rotor current (LRC) indefinitely. Opening of the circuit by an overcurrent device installed in a fire pump circuit cannot be tolerated, except under short circuits or ground faults. The circuit has to perform as if a direct connection exists to the power source. Sizing for LRC applies only to OCPDs and does not extend to conductors or other devices in the fire pump motor circuit. Similar requirements are contained in 695.5(B) and 695.5(C)(2). Alternately, a listed fire pump assembly complying with 695.4(B)(2) is permitted.

It is unlikely that all fire pumps on a circuit will be under simultaneous locked-rotor conditions. Therefore, only the LRC of largest fire pump and maintenance pump is required. Full-load current is used for all other pumps.

(2) Overcurrent protection shall be provided by an assembly listed for fire pump service and complying with the following:

a. The overcurrent protective device shall not open within 2 minutes at 600 percent of the full-load current of the fire pump motor(s).
b. The overcurrent protective device shall not open with a re-start transient of 24 times the full-load current of the fire pump motor(s).
c. The overcurrent protective device shall not open within 10 minutes at 300 percent of the full-load current of the fire pump motor(s).
d. The trip point for circuit breakers shall not be field adjustable. [**20:**9.2.3.4.1]

(b) *On-Site Standby Generators.* Overcurrent protective devices between an on-site standby generator and a fire pump controller shall be selected and sized to allow for instantaneous pickup of the full pump room load, but shall not be larger than the value selected to comply with 430.62 to provide short-circuit protection only. [**20:**9.6.1.1]

This requirement correlates with 695.3(D)(1) covering the required capacity of an on-site standby generator. OCPDs supplied by an on-site standby generator are not required to be sized to carry the locked-rotor current of the fire pump(s) indefinitely. The on-site standby generator is not limited to supplying only the fire pump. Where other pump room loads such as lights or fans are supplied, the generator must have sufficient capacity to instantaneously carry the entire load supplied. The OCPDs are not required to provide overload protection and are required to be sized per 430.62, which covers devices supplying multiple motors or motors and other loads.

**(3) Disconnecting Means.** All disconnecting devices that are unique to the fire pump loads shall comply with items (a) through (e).

(a) *Features and Location — Normal Power Source.* The disconnecting means for the normal power source shall comply with all of the following: [**20:**9.2.3.1]

(1) Be identified as suitable for use as service equipment.
(2) Be lockable in the closed position. The provision for locking or adding a lock to the disconnecting means shall be installed on or at the switch or circuit breaker used as the disconnecting means and shall remain in place with or without the lock installed.
(3) Not be located within the same enclosure, panelboard, switchboard, switchgear, or motor control center, with or without common bus, that supplies loads other than the fire pump.
(4) Be located sufficiently remote from other building or other fire pump source disconnecting means such that inadvertent operation at the same time would be unlikely.

The notion of keeping the fire pump disconnecting means "sufficiently remote" from other disconnecting means is to prevent interruption of fire pump supply through inadvertent operation of the disconnecting means. It is not practical to specify a distance as a rule; the AHJ must make this determination.

A disconnecting means supplied by one of the individual sources specified in 695.3(A) cannot be installed in distribution equipment that supplies other than fire pump loads. "Sufficiently remote" also cannot be interpreted as permitting the fire pump disconnecting means to be located in a separate switchboard section of equipment that supplies other than fire pump loads.

(b) *Features and Location — On-Site Standby Generator.* The disconnecting means for an on-site standby generator(s) used as the alternate power source shall be installed in accordance with 700.10(B)(5) for emergency circuits and shall be lockable in the closed position. The provision for locking or adding a lock to the disconnecting means shall be installed on or at the switch or circuit breaker used as the disconnecting means and shall remain in place with or without the lock installed.

A disconnecting means supplied by an on-site standby generator is permitted to be installed in equipment that supplies other loads. However, the requirements of 700.10(B)(5) must be followed. The effect of this requirement is that a fire pump feeder cannot be supplied from equipment in which the fire pump conductors are installed in the same enclosure or vertical switchboard section with conductors supplying loads that are designated or classed as legally required standby (Article 701) or optional standby (Article 702) loads. To help minimize inadvertent opening of the fire pump circuit, the disconnecting means is required to be capable of being locked in the closed (on) position.

(c) *Disconnect Marking.* The disconnecting means shall be marked "Fire Pump Disconnecting Means." The letters shall be at least 25 mm (1 in.) in height, and they shall be visible without opening enclosure doors or covers. [**20:**9.2.3.1(5)]

(d) *Controller Marking.* A placard shall be placed adjacent to the fire pump controller, stating the location of this disconnecting means and the location of the key (if the disconnecting means is locked). [**20:**9.2.3.2]

(e) *Supervision.* The disconnecting means shall be supervised in the closed position by one of the following methods:

Supervision of the disconnecting means is required to assure continued operation of the fire pump.

(1) Central station, proprietary, or remote station signal device
(2) Local signaling service that causes the sounding of an audible signal at a constantly attended point
(3) Locking the disconnecting means in the closed position
(4) Sealing of disconnecting means and approved weekly recorded inspections when the disconnecting means are located within fenced enclosures or in buildings under the control of the owner [**20:**9.2.3.3]

Ideally, power supply conductors are run directly to the listed fire pump control and/or transfer equipment without the need for an additional service disconnecting means and overcurrent protection. However, this arrangement is not always possible; therefore, the single disconnecting means in 695.4(B) is permitted, provided it is monitored to be in the closed position.

Supervision of the disconnecting means by a local (protected premises) fire alarm system, central station, proprietary supervising station, or remote supervising station requires a connection to the premises fire alarm system. A fire alarm system initiating device circuit is programmed to generate a supervisory signal at the fire alarm control unit on loss of voltage to the fire pump controller. A supervisory signal indicates that the suppression system is "off-normal." For more information on this interface with the fire alarm system, see *NFPA 72, National Fire Alarm and Signaling Code.*

**Calculation Example**

A fusible service disconnect switch supplies power to a 100-hp, 460-volt, 3-phase fire pump and to a 1½-hp, 460-volt, 3-phase jockey pump. Determine the sizes of the disconnecting means and OCPD for the system. Also determine the minimum ampacity of the feeder conductors.

*Solution*

STEP 1. Determine the minimum ratings of the disconnecting means and the OCPD.

According to the motor nameplates, the locked-rotor current (LRC) is 725 amperes for the 100-hp motor and 20 amperes for the 1½-hp motor. If the locked-rotor amperes are not on the nameplates, the LRCs found in Table 430.251(B) must be used. Calculate the size by summing the LRC of both motors and then going to the next larger standard-size OCPD, as follows:

100-hp, 3-phase LRC = 725 A
1½-hp, 3-phase LRC = 20 A
Total LRC = 745 A

The next larger standard-size disconnect switch and overcurrent device is 800 amperes. An adjustable-trip circuit breaker of 750 amperes is also permitted, because it, too, will carry the LRC indefinitely.

STEP 2. Determine the minimum ampacity for the fire pump feeder conductor.

Even though the disconnect switch and overcurrent device are sized according to LRCs, the feeder conductors to the fire pump and associated equipment are required to have an ampacity not less than 125 percent of the full-load current (FLC) rating of the fire pump motor(s) and pressure maintenance pump motor(s), plus 100 percent of associated accessory equipment. Calculate the size of the feeder to the fire pump controller using 430.6(A)(1) and Table 430.250 for the FLC of the motors:

100-hp, 3-phase FLC

$$124 \text{ A} \times 1.25 = 155.0 \text{ A}$$

1½-hp, 3-phase FLC

$$3 \text{ A} \times 1.25 = 3.75 \text{ A}$$

Total FLC = 158.75 A or 159 A

Thus, the minimum ampacity for the feeder conductors is 159 amperes. Using the 75°C column, per 110.14(C)(1)(b), from Table 310.15(B)(16), a 2/0 copper conductor is the minimum size required.

## 695.5 Transformers

Where the service or system voltage is different from the utilization voltage of the fire pump motor, transformer(s) protected by disconnecting means and overcurrent protective devices shall be permitted to be installed between the system supply and the fire pump controller in accordance with 695.5(A) and (B), or with (C). Only transformers covered in 695.5(C) shall be permitted to supply loads not directly associated with the fire pump system.

**(A) Size.** Where a transformer supplies an electric motor driven fire pump, it shall be rated at a minimum of 125 percent of the sum of the fire pump motor(s) and pressure maintenance pump(s) motor loads, and 100 percent of the associated fire pump accessory equipment supplied by the transformer.

**(B) Overcurrent Protection.** The primary overcurrent protective device(s) shall be selected or set to carry indefinitely the sum of the locked-rotor current of the fire pump motor(s) and the pressure maintenance pump motor(s) and the full-load current of the associated fire pump accessory equipment when connected to this power supply. Secondary overcurrent protection shall not be permitted. The requirement to carry the locked-rotor currents indefinitely shall not apply to conductors or devices other than overcurrent devices in the fire pump motor circuit(s).

The sizing of dedicated transformers and overcurrent protection can be broken down into three basic requirements. Generally stated, they are as follows:

1. The transformer must be sized to at least 125 percent of the sum of the loads.
2. The transformer primary overcurrent device must be at least a specified minimum size.
3. The transformer secondary must not contain any overcurrent devices whatsoever.

See Exhibit 695.5 for a simple one-line diagram on applying the dedicated fire pump transformer overcurrent protection requirements. The jockey pump could alternately be supplied by a separate panelboard.

### Calculation Example

A 4160/480-V, 3-phase, dedicated transformer supplies power to a 100-hp, 460-volt, 3-phase, code letter G fire pump and to a 1½-hp, 460-volt, 3-phase, code letter H jockey pump. Determine the sizes of the dedicated transformer and its primary overcurrent protection.

*Solution*

STEP 1. Determine the minimum standard-size transformer. First, to determine the minimum current value for use in the 3-phase power calculation, add the full-load currents (FLCs) of the fire pump motor(s) and jockey pump motor(s). The FLCs of the two motors, using the FLC values from Table 430.250, are as follows:

$$\begin{aligned} \text{100-hp, 3-phase FLC} &= 124\text{ A} \\ \text{1½-hp, 3-phase FLC} &= \underline{\phantom{12}3\text{ A}} \\ \text{Total FLC} &= 127\text{ A} \end{aligned}$$

Now, increase the sum of the fire pump motor and the jockey pump motor to 125 percent:

$$127\text{ A} \times 1.25 = 158.75\text{ A}$$

Then, size the transformer as follows:

$$\begin{aligned} \text{Transformer kVA} &= \frac{\text{volts} \times \text{amperes} \times \sqrt{3}}{1000} \\ &= \frac{480 \times 158.75 \times \sqrt{3}}{1000} \\ &= 131.98\text{ kVA} \end{aligned}$$

The minimum-size transformer permitted is 131.98 kVA. The next larger standard-size transformer available is 150 kVA, but any larger size is permitted.

STEP 2. Calculate the minimum-size primary OCPD permitted for this transformer. According to 695.5(B), the minimum primary OCPD must allow the transformer secondary to supply the locked-rotor current (LRC) to the fire pump and, in this case, the jockey pump. The LRC of each motor must be individually calculated if it is not available on the motor nameplate. In this example, however, only the kVA code letters are assumed to be available. According to 430.7(B) and using the maximum values for the individual code letters per Table 430.7(B), calculate the maximum LRCs, as follows.

For the 100-hp motor, code letter G:

$$\text{LRC} = \text{motor hp} \times \text{max. code letter value} \times \frac{1000}{\text{motor voltage} \times \text{3-phase factor}}$$

$$= 100\text{ hp} \times \frac{6.29\text{ kVa}}{\text{hp}} \times \frac{1000}{460 \times \sqrt{3}} = 789.49\text{ A}$$

For the 1½-hp motor, code letter H (using the same formula):

$$\text{LRC} = 1\tfrac{1}{2}\text{ hp} \times \frac{7.09\text{ kVa}}{\text{hp}} \times \frac{1000}{460 \times \sqrt{3}} = 13.35\text{ A}$$

For the total LRC:

$$\begin{aligned} \text{100-hp LRC} &= 789.49\text{ A} \\ \text{1½-hp LRC} &= \underline{\phantom{7}13.35\text{ A}} \\ \text{Total LRC} &= 802.84\text{ A or }803\text{ A} \end{aligned}$$

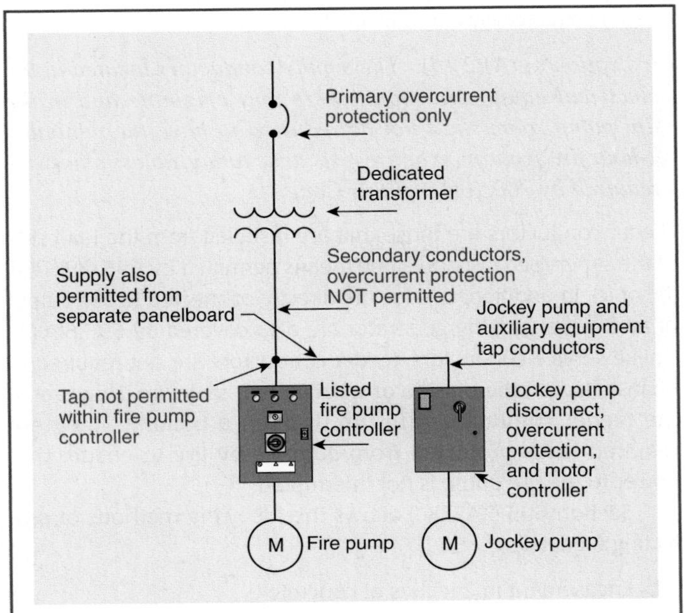


***EXHIBIT 695.5*** *Overcurrent protection for a transformer supplying a fire pump and associated equipment. The device must be capable of carrying LRCs of fire pump motor and jockey pump motor indefinitely.*

Now, calculate the equivalent LRC on the primary side of the transformer, based on the calculated LRC of the secondary of the transformer, as follows:

$$LRC_{primary} = \frac{\text{secondary voltage}}{\text{primary voltage}} \times LRC_{secondary}$$
$$= \frac{480\text{ V}}{4160\text{ V}} \times 803\text{ A}$$
$$= 92.65\text{ A or } 93\text{ A}$$

This 93 amperes represents the secondary LRC reflected to the primary side of the transformer. Because this value is the absolute smallest OCPD permitted, the next larger standard size, according to 240.6, is 100 amperes.

*Conclusion:*

1. The smallest standard-size transformer that is permitted is 150 kVA.
2. The smallest standard-size OCPD permitted on the primary of the transformer is 100 amperes.
3. A secondary OCPD is not permitted.

**(C) Feeder Source.** Where a feeder source is provided in accordance with 695.3(C), transformers supplying the fire pump system shall be permitted to supply other loads. All other loads shall be calculated in accordance with Article 220, including demand factors as applicable.

**(1) Size.** Transformers shall be rated at a minimum of 125 percent of the sum of the fire pump motor(s) and pressure maintenance pump(s) motor loads, and 100 percent of the remaining load supplied by the transformer.

**(2) Overcurrent Protection.** The transformer size, the feeder size, and the overcurrent protective device(s) shall be coordinated such that overcurrent protection is provided for the transformer in accordance with 450.3 and for the feeder in accordance with 215.3, and such that the overcurrent protective device(s) is selected or set to carry indefinitely the sum of the locked-rotor current of the fire pump motor(s), the pressure maintenance pump motor(s), the full-load current of the associated fire pump accessory equipment, and 100 percent of the remaining loads supplied by the transformer. The requirement to carry the locked-rotor currents indefinitely shall not apply to conductors or devices other than overcurrent devices in the fire pump motor circuit(s).

## 695.6 Power Wiring

Power circuits and wiring methods shall comply with the requirements in 695.6(A) through (J), and as permitted in 230.90(A), Exception No. 4; 230.94, Exception No. 4; 240.13; 230.208; 240.4(A); and 430.31.

**(A) Supply Conductors.**

**(1) Services and On-Site Power Production Facilities.** Service conductors and conductors supplied by on-site power production facilities shall be physically routed outside a building(s) and shall be installed as service-entrance conductors in accordance with 230.6, 230.9, and Parts III and IV of Article 230. Where supply conductors cannot be physically routed outside of buildings, the conductors shall be permitted to be routed through the building(s) where installed in accordance with 230.6(1) or (2).

**(2) Feeders.** Fire pump supply conductors on the load side of the final disconnecting means and overcurrent device(s) permitted by 695.4(B), or conductors that connect directly to an on-site standby generator, shall comply with all of the following:

(a) *Independent Routing.* The conductors shall be kept entirely independent of all other wiring.

(b) *Associated Fire Pump Loads.* The conductors shall supply only loads that are directly associated with the fire pump system.

(c) *Protection from Potential Damage.* The conductors shall be protected from potential damage by fire, structural failure, or operational accident.

(d) *Inside of a Building.* Where routed through a building, the conductors shall be installed using one of the following methods:

(1) Be encased in a minimum 50 mm (2 in.) of concrete
(2) Be protected by a fire-rated assembly listed to achieve a minimum fire rating of 2 hours and dedicated to the fire pump circuit(s)
(3) Be a listed electrical circuit protective system with a minimum 2-hour fire rating

Informational Note: UL guide information for electrical circuit protective systems (FHIT) contains information on proper installation requirements to maintain the fire rating.

*Exception to (A)(2)(d): The supply conductors located in the electrical equipment room where they originate and in the fire pump room shall not be required to have the minimum 2-hour fire separation or fire resistance rating, unless otherwise required by 700.10(D) of this Code.*

Feeder conductors are those that are installed from the load side of the supervised disconnecting means permitted by 695.4(B)(1)(a), (b), or (c). In addition, conductors directly connected to the output of an on-site standby generator are also covered by 695.6(A)(2). Unlike service conductors, feeder conductors are not required to be installed on the outside of a building or structure. However, if the feeder conductors are run through a building, they are required to be protected from damage by fire to ensure that power to the fire pump is not interrupted.

Section 695.6(A)(2)(d) allows the following methods of protecting feeders.

1. Encasement in 2 inches of concrete
2. Enclosure of the wiring method within 2-hour fire-resistant building construction
3. Use of a wiring method listed as an electrical circuit protective system with a minimum 2-hour fire rating

These wiring methods are recognized for protecting feeder conductors against fire damage.

The difference between a 2-hour fire rating of an electrical circuit, such as a conduit with wires, and a 2-hour fire resistance rating of a structural member, such as a wall, is that at the end of a 2-hour fire test on an electrical conduit with wires, the circuit must function electrically (no short circuits, grounds, or opens are permitted) and its insulation must be intact. A wall subjected to a 2-hour fire resistance test must only prevent a fire from passing through or past the wall, without regard to damage to the wall. All fire ratings and fire resistance ratings are based on the assumption that the structural supports for the assembly are not impaired by the effects of the fire.

The UL *Fire Resistance Directory,* Volume 2, describes three categories of products that can be used in the fire protection of electrical circuits for fire pumps: electrical circuit integrity systems (FHIT), electrical circuit protective materials (FHIY), and fire-resistive cables (FHJR). (The four-letter codes in parentheses are the UL product category guide designations.) For information on electrical circuit protective systems, see UL 1724, *Fire Tests for Electrical Circuit Protective Systems.*

**(B) Conductor Size.**

**(1) Fire Pump Motors and Other Equipment.** Conductors supplying a fire pump motor(s), pressure maintenance pumps, and associated fire pump accessory equipment shall have a rating not less than 125 percent of the sum of the fire pump motor(s) and pressure maintenance motor(s) full-load current(s), and 100 percent of the associated fire pump accessory equipment.

**(2) Fire Pump Motors Only.** Conductors supplying only a fire pump motor shall have a minimum ampacity in accordance with 430.22 and shall comply with the voltage drop requirements in 695.7.

Listed fire pump controller and pump combinations are available in a wye-start, delta-run configuration as well as variable speed drive configurations. In the wye-delta configuration, six circuit conductors are run from the controller to the motor; when the motor is in the run mode, the conductors that supply each winding are connected in parallel. See commentary following 430.22(C) for wye-start, delta-run operation. The minimum conductor ampacity for the controller and for each of the six leads between the controller and the motor is calculated as shown in the following example.

### Calculation Example

Determine the minimum size for the line and load side conductors of a controller with a fire pump with a 50-hp, 3-phase, 460-volt motor. The pump motor and controller are configured for a wye-start, delta-run operation.

- Table 430.250 specifies full-load current (FLC) for 50-hp motor as 65 amperes.
- Section 430.22(C) requires a controller line side minimum conductor ampacity based on 125 percent of motor FLC.
- Section 430.22(C) requires a controller load side minimum conductor ampacity based on 72 percent of motor FLC.

*Solution*

STEP 1. Determine minimum conductor ampacity.

(a) Load side: 65 A × 0.72 = 47 A

(b) Line side: 65 A × 1.25 = 81 A

STEP 2. Determine Type THWN copper conductor minimum size using Table 310.15(B)(16) and assuming 75°C terminations in the controller.

(a) Load side: 50 amperes requires 8 AWG conductors.

- The combined ampacity of the two 8 AWG circuit conductors connected in parallel to each winding in the run mode is 100 amperes.

(b) Line side: 81 A requires 4 AWG conductors.

- The minimum size for the conductors may have to be increased to comply with the mandatory voltage-drop performance requirements in 695.7.

**(C) Overload Protection.** Power circuits shall not have automatic protection against overloads. Except for protection of transformer primaries provided in 695.5(C)(2), branch-circuit and feeder conductors shall be protected against short circuit only. Where a tap is made to supply a fire pump, the wiring shall be treated as service conductors in accordance with 230.6. The applicable distance and size restrictions in 240.21 shall not apply.

*Exception No. 1: Conductors between storage batteries and the engine shall not require overcurrent protection or disconnecting means.*

*Exception No. 2: For an on-site standby generator(s) rated to produce continuous current in excess of 225 percent of the full-load amperes of the fire pump motor, the conductors between the on-site generator(s) and the combination fire pump transfer switch controller or separately mounted transfer switch shall be installed in accordance with 695.6(A)(2). The protection provided shall be in accordance with the short-circuit current rating of the combination fire pump transfer switch controller or separately mounted transfer switch.*

**(D) Pump Wiring.** All wiring from the controllers to the pump motors shall be in rigid metal conduit, intermediate metal conduit, electrical metallic tubing, liquidtight flexible metal conduit, or liquidtight flexible nonmetallic conduit Type LFNC-B, listed Type MC cable with an impervious covering, or Type MI cable. Electrical connections at motor terminal boxes shall be made with a listed means of connection. Twist-on, insulation-piercing–type, and soldered wire connectors shall not be permitted to be used for this purpose.

**(E) Loads Supplied by Controllers and Transfer Switches.** A fire pump controller and fire pump power transfer switch, if provided, shall not serve any load other than the fire pump for which it is intended.

**(F) Mechanical Protection.** All wiring from engine controllers and batteries shall be protected against physical damage and shall be installed in accordance with the controller and engine manufacturer's instructions.

**(G) Ground-Fault Protection of Equipment.** Ground-fault protection of equipment shall not be permitted for fire pumps.

Although ground-fault protection of equipment is a major safety concern elsewhere in the *Code*, the continued operation of the fire pumps until the fire is extinguished is essential. Ground-fault protection of equipment is not permitted to be used to protect components of a fire pump installation. The function of ground-fault protection of equipment protection should not be confused with the function of GFCI protection for personnel. See 240.13(3). Ground-fault detection that provides an alarm only is not prohibited by this requirement.

**(H) Listed Electrical Circuit Protective System to Controller Wiring.** Electrical circuit protective system installation shall comply with any restrictions provided in the listing of the electrical circuit protective system used and the following also shall apply:

(1) A junction box shall be installed ahead of the fire pump controller a minimum of 300 mm (12 in.) beyond the fire-rated wall or floor bounding the fire zone.

The required junction box allows for a transition between solid conductors that are used in some electrical circuit protective systems (Type MI cable, for example) and stranded conductors that are required at the supply terminals of the controller by the controller manufacturer and its listing, without having to make the splice in the controller enclosure. In addition, where an electrical circuit protective system employs single conductor cables, such as Type MI cable, the necessity to modify enclosures to prevent inductive heating can result in a compromise of the controller enclosure's resistance to water infiltration.

(2) Where required by the manufacturer of a listed electrical circuit protective system or by the listing, or as required elsewhere in this *Code*, the raceway between a junction box and the fire pump controller shall be sealed at the junction box end as required and in accordance with the instructions of the manufacturer. [**20:**9.8.2]

Sealing of the raceway between the junction box and the enclosure may be required by manufacturers of some electrical circuit protective systems. Sealing prevents any conductive material or gases that emanate in the electrical circuit protective system from entering the controller enclosure and compromising the controller operation.

(3) Standard wiring between the junction box and the controller shall be permitted. [**20:**9.8.3]

**(I) Junction Boxes.** Where fire pump wiring to or from a fire pump controller is routed through a junction box, the following requirements shall be met:

(1) The junction box shall be securely mounted. [**20:**9.7(1)]
(2) Mounting and installing of a junction box shall not violate the enclosure type rating of the fire pump controller(s). [**20:**9.7(2)]
(3) Mounting and installing of a junction box shall not violate the integrity of the fire pump controller(s) and shall not affect the short-circuit rating of the controller(s). [**20:**9.7(3)]
(4) As a minimum, a Type 2, drip-proof enclosure (junction box) shall be used where installed in the fire pump room. The enclosure shall be listed to match the fire pump controller enclosure type rating. [**20:**9.7(4)]

These requirements maintain the controller enclosure's environmental rating. Use of conduit hubs having the same environmental rating as the controller enclosure minimizes the entry of water or other liquids into the enclosure.

(5) Terminals, junction blocks, wire connectors, and splices, where used, shall be listed. [**20:**9.7(5)]
(6) A fire pump controller or fire pump power transfer switch, where provided, shall not be used as a junction box to supply other equipment, including a pressure maintenance (jockey) pump(s).

**(J) Raceway Terminations.** Where raceways are terminated at a fire pump controller, the following requirements shall be met: [**20:**9.9]

(1) Listed conduit hubs shall be used. [**20:**9.9.1]
(2) The type rating of the conduit hub(s) shall be at least equal to that of the fire pump controller. [**20:**9.9.2]
(3) The installation instructions of the manufacturer of the fire pump controller shall be followed. [**20:**9.9.3]
(4) Alterations to the fire pump controller, other than conduit entry as allowed elsewhere in this *Code*, shall be approved by the authority having jurisdiction. [**20:**9.9.4]

## 695.7 Voltage Drop

**(A) Starting.** The voltage at the fire pump controller line terminals shall not drop more than 15 percent below normal (controller-rated voltage) under motor starting conditions.

*Exception: This limitation shall not apply for emergency run mechanical starting. [**20:**9.4.2]*

**(B) Running.** The voltage at the load terminals of the fire pump controller shall not drop more than 5 percent below the voltage rating of the motor connected to those terminals when the motor is operating at 115 percent of the full-load current rating of the motor.

## 695.10 Listed Equipment

Diesel engine fire pump controllers, electric fire pump controllers, electric motors, fire pump power transfer switches, foam pump controllers, and limited service controllers shall be listed for fire pump service. [**20:**9.5.1.1, 10.1.2.1, 12.1.3.1]

Prior to being shipped to the installation site, listed fire pump controllers are matched with the listed electric motor(s) they will control, to ensure compatibility of the individually listed components. The fire pump controller and transfer switch shown in Exhibit 695.6 is an example of listed equipment.

## 695.12 Equipment Location

**(A) Controllers and Transfer Switches.** Electric motor-driven fire pump controllers and power transfer switches shall be located as close as practicable to, and within sight of, the motors that they control.

**(B) Engine-Drive Controllers.** Engine-drive fire pump controllers shall be located as close as is practical to, and within sight of, the engines that they control.

**(C) Storage Batteries.** Storage batteries for fire pump engine drives shall be supported above the floor, secured against displacement, and located where they are not subject to physical damage, flooding with water, excessive temperature, or excessive vibration.

**(D) Energized Equipment.** All energized equipment parts shall be located at least 300 mm (12 in.) above the floor level.

**(E) Protection Against Pump Water.** Fire pump controller and power transfer switches shall be located or protected so that they are not damaged by water escaping from pumps or pump connections.

***EXHIBIT 695.6*** *Listed fire pump controller and power transfer switch.*

**(F) Mounting.** All fire pump control equipment shall be mounted in a substantial manner on noncombustible supporting structures.

NFPA 20 specifies a suitable space for fire pump equipment. This space must be free from hazards that could impair the operation of the fire pump. Neither the *Code* nor NFPA 20 mandates a dedicated room for the fire pump.

Even though 695.12(A) requires fire pump controllers and transfer switches to be "as close as practicable" to their associated fire pump motor, the minimum working space required by 110.26 must be maintained.

Fire pump controllers are housed in enclosures suitable to protect the contents against limited amounts of falling water and dirt. In addition, all energized parts in the enclosure must be mounted at least 12 inches above the floor. Typically, the floor space for this area is equipped with a floor drain.

Section 695.12(F) does not permit fire pump control equipment to be mounted on combustible backboards (such as plywood).

## 695.14 Control Wiring

**(A) Control Circuit Failures.** External control circuits that extend outside the fire pump room shall be arranged so that failure of any external circuit (open or short circuit) shall not prevent the operation of a pump(s) from all other internal or external means. Breakage, disconnecting, shorting of the wires, or loss of power to these circuits could cause continuous running of the fire pump but shall not prevent the controller(s) from starting the fire pump(s) due to causes other than these external control circuits. All control conductors within the fire pump room that are not fault tolerant shall be protected against physical damage. [**20:**10.5.2.6, 12.5.2.5]

**(B) Sensor Functioning.** No undervoltage, phase-loss, frequency-sensitive, or other sensor(s) shall be installed that automatically or manually prohibits actuation of the motor contactor. [**20:**10.4.5.6]

*Exception: A phase loss sensor(s) shall be permitted only as a part of a listed fire pump controller.*

**(C) Remote Device(s).** No remote device(s) shall be installed that will prevent automatic operation of the transfer switch. [**20:**10.8.1.3]

**(D) Engine-Drive Control Wiring.** All wiring between the controller and the diesel engine shall be stranded and sized to continuously carry the charging or control currents as required by the controller manufacturer. Such wiring shall be protected against physical damage. Controller manufacturer's specifications for distance and wire size shall be followed. [**20:**12.3.5.1]

**(E) Electric Fire Pump Control Wiring Methods.** All electric motor–driven fire pump control wiring shall be in rigid metal conduit, intermediate metal conduit, liquidtight flexible metal conduit, liquidtight flexible nonmetallic conduit Type B

(LFNC-B), listed Type MC cable with an impervious covering, or Type MI cable.

The wiring methods described in 695.14(E) apply only to the control wiring for electric motor–driven fire pumps. These methods do not apply to the control wiring for engine-driven fire pumps.

**(F) Generator Control Wiring Methods.** Control conductors installed between the fire pump power transfer switch and the standby generator supplying the fire pump during normal power loss shall be kept entirely independent of all other wiring. They shall be protected to resist potential damage by fire or structural failure. They shall be permitted to be routed through a building(s) using one of the following methods:

(1) Be encased in a minimum 50 mm (2 in.) of concrete.
(2) Be protected by a fire-rated assembly listed to achieve a minimum fire rating of 2 hours and dedicated to the fire pump circuits.
(3) Be a listed electrical circuit protective system with a minimum 2-hour fire rating. The installation shall comply with any restrictions provided in the listing of the electrical circuit protective system used.

Informational Note: UL guide information for electrical circuit protective systems (FHIT) contains information on proper installation requirements to maintain the fire rating.

Having the power wiring protected against fire damage is only one reliability consideration. In order for the generator to provide power, it has to receive the necessary signal to start. It is also critical to protect the control circuit wiring between the fire pump transfer switch/controller and the on-site standby generator. Otherwise, the fire pump is subject to failure.

# 7 Special Conditions

## ARTICLE 700
## Emergency Systems

### I. General

#### 700.1 Scope

The provisions of this article apply to the electrical safety of the installation, operation, and maintenance of emergency systems consisting of circuits and equipment intended to supply, distribute, and control electricity for illumination, power, or both, to required facilities when the normal electrical supply or system is interrupted.

Informational Note No. 1: For further information regarding wiring and installation of emergency systems in health care facilities, see Article 517.

Informational Note No. 2: For further information regarding performance and maintenance of emergency systems in health care facilities, see NFPA 99-2012, *Health Care Facilities Code*.

Informational Note No. 3: For specification of locations where emergency lighting is considered essential to life safety, see NFPA *101*-2012, *Life Safety Code*.

Informational Note No. 4: For further information regarding performance of emergency and standby power systems, see NFPA 110-2013, *Standard for Emergency and Standby Power Systems*.

Emergency systems are designed and installed to maintain a specific degree of illumination or to provide power for essential equipment, such as emergency lighting for means of egress, if the normal power supply fails.

Article 700 applies to the installation of emergency systems that are essential for safety to human life and are legally required by municipal, state, federal, or other codes or by a governmental agency having jurisdiction. Article 700 does not dictate whether emergency systems are required or where emergency or exit lights should be located. These determinations may rely on NFPA *101*®, *Life Safety Code*®.

Article 708 provides requirements for power facilities that must be kept continuously operational throughout the duration of an emergency. Critical operations power systems (COPS) are generally installed in vital infrastructure facilities – those that, if destroyed or incapacitated, would disrupt national security, the economy, public health, or safety – and in areas where enhanced electrical infrastructure for continuity of operation has been deemed necessary by governmental authority.

#### 700.2 Definitions

**Emergency Systems.** Those systems legally required and classed as emergency by municipal, state, federal, or other codes, or by any governmental agency having jurisdiction. These systems are intended to automatically supply illumination, power, or both, to designated areas and equipment in the event of failure of the normal supply or in the event of accident to elements of a system intended to supply, distribute, and control power and illumination essential for safety to human life.

Informational Note: Emergency systems are generally installed in places of assembly where artificial illumination is required for safe exiting and for panic control in buildings subject to occupancy by large numbers of persons, such as hotels, theaters, sports arenas, health care facilities, and similar institutions. Emergency systems may also provide power for such functions as ventilation where essential to maintain life, fire detection and alarm systems, elevators, fire pumps, public safety communications systems, industrial processes where current interruption would produce serious life safety or health hazards, and similar functions.

**Relay, Automatic Load Control.** A device used to set normally dimmed or normally-off switched emergency lighting equipment to full power illumination levels in the event of a loss of the normal supply by bypassing the dimming/switching controls, and to return the emergency lighting equipment to normal status when the device senses the normal supply has been restored.

Informational Note: See ANSI/UL 924, *Emergency Lighting and Power Equipment*, for the requirements covering automatic load control relays.

One use of automatic load control relays is in a lighting branch circuit supplied by the emergency system where the load is controlled by an energy management system. The automatic load control relay functions to restore the required level of emergency lighting where the lighting has either been dimmed or completely turned off by an energy management system. When the emergency loads are transferred from the normal source to the alternate source, the relay overrides the energy management mode and provides full power to the load. Upon restoration of the normal source, the relay returns the load to the normal operating mode that is controlled by the energy management system.

## 700.3 Tests and Maintenance

**(A) Conduct or Witness Test.** The authority having jurisdiction shall conduct or witness a test of the complete system upon installation and periodically afterward.

**(B) Tested Periodically.** Systems shall be tested periodically on a schedule acceptable to the authority having jurisdiction to ensure the systems are maintained in proper operating condition.

**(C) Battery Systems Maintenance.** Where battery systems or unit equipments are involved, including batteries used for starting, control, or ignition in auxiliary engines, the authority having jurisdiction shall require periodic maintenance.

**(D) Written Record.** A written record shall be kept of such tests and maintenance.

**(E) Testing Under Load.** Means for testing all emergency lighting and power systems during maximum anticipated load conditions shall be provided.

Informational Note: For information on testing and maintenance of emergency power supply systems (EPSSs), see NFPA 110-2013, *Standard for Emergency and Standby Power Systems.*

Emergency system testing can be divided into two general categories – acceptance testing and operational testing. Section 700.3 requires both types of testing as well as written records of each and of maintenance performance.

Acceptance testing is performed after the emergency system has been installed but before the system is used. Acceptance testing ensures that the emergency system meets or exceeds the original installation specification.

Operational testing, which is performed during the life of the system, ensures that the emergency system remains functional and that maintenance is performed adequately. One method of operational testing is running the generating system to power the load of the facility. Generally, actual emergency system loads are smaller than the design capacity of the emergency generator system. Actual peak loads of the emergency system should be kept as part of the written record.

Further information on tests and maintenance may be found in NFPA 70B, *Recommended Practice for Electrical Equipment Maintenance*; NFPA 99, *Standard for Health Care Facilities*; NFPA *101*®, *Life Safety Code*®; NFPA 110, *Standard for Emergency and Standby Power Systems*; and NFPA 111, *Standard on Stored Electrical Energy Emergency and Standby Power Systems.*

## 700.4 Capacity

**(A) Capacity and Rating.** An emergency system shall have adequate capacity and rating for all loads to be operated simultaneously. The emergency system equipment shall be suitable for the maximum available fault current at its terminals.

The emergency system must be designed with adequate capacity and rating to safely carry, at one time, the entire load connected to the emergency system. It must be capable of restarting emergency loads that have been interrupted, such as motors that may have stopped, and it must be suitable for the available fault current.

**(B) Selective Load Pickup, Load Shedding, and Peak Load Shaving.** The alternate power source shall be permitted to supply emergency, legally required standby, and optional standby system loads where the source has adequate capacity or where automatic selective load pickup and load shedding is provided as needed to ensure adequate power to (1) the emergency circuits, (2) the legally required standby circuits, and (3) the optional standby circuits, in that order of priority. The alternate power source shall be permitted to be used for peak load shaving, provided these conditions are met.

Peak load shaving operation shall be permitted for satisfying the test requirement of 700.3(B), provided all other conditions of 700.3 are met.

A portable or temporary alternate source shall be available whenever the emergency generator is out of service for major maintenance or repair.

Where a generator is used for peak load shaving, supplying backup power, and other uses, priority loads must be properly and reliably served. Selective load pickup and load shedding are not required where the generator has the capacity to supply all loads served.

If a generator is used for peak load shaving or in a cogeneration system, the increase in wear and tear will likely result in an increase in downtime for maintenance. Also, using the emergency generator on a regular basis for nonemergency loads provides assurance that the emergency generator will supply emergency power when it is needed. The requirement for a portable or temporary alternate source is intended to provide emergency power when the generator set is out of service for major maintenance. A major maintenance or repair procedure is one that keeps the generator set out of service for more than a few hours.

## 700.5 Transfer Equipment

Double-throw automatic transfer switches (ATS) are typically used for emergency and standby power generation systems rated 600 volts or less. These transfer switches do not normally incorporate overcurrent protection. ATS are available in ratings up to 38 kV. For reliability, those used for emergency and legally required standby systems must be electrically operated and mechanically held. System grounding is determined by the type of transfer switch employed. See 250.30 and associated commentary regarding separately derived systems.

It is desirable to locate transfer switches close to the load and to keep the operation of the transfer switches independent of overcurrent protection. It may be advantageous to use multiple transfer switches of lower current rating located near the load rather than one large transfer switch at the point of incoming service.

**Time-Delay Devices on Automatic Transfer Switches**

The normal power source is usually a service, and the emergency power source is an automatically started engine generator set that starts when the normal source fails. Time-delay controls are essential to the operation of the ATS.

To avoid unnecessary starting and transfer to the alternate supply, a time delay can override momentary interruptions and temporary reductions in normal source voltage but still allow starting and transfer if the reduction or outage is sustained. However, the time delay should be set fast enough to effectively operate the transfer switch and provide backup power for long-term outages.

This delay is generally set at 1 second but may be set higher if reclosers or circuit breakers on the utility power lines take longer to operate or if momentary power dips exceed 1 second. If longer delay settings are used, care must be taken to ensure that sufficient time remains to meet 10-second power restoration requirements. The AHJ may determine that an outage is not a longer-term power failure until the utility automatic protective devices fail to restore power to the facility. For example, the 10-second power restoration requirements would become effective after the 2-second recloser cycle.

Once the load is transferred to the alternate source, another timer delays retransfer to the normal source until that source has time to stabilize. Another important function of this timer required by 700.12(B)(1) is to allow an engine generator to operate under load for at least 15 minutes to ensure continued good performance of the set and its starting system. This delay should be automatically nullified if the alternate source fails and the normal source is available.

Engine generator manufacturers often recommend a cool-down period for their sets that allows them to run unloaded after the load is retransferred to the normal source. A third time delay, usually 5 minutes, is provided for this purpose. Running an unloaded engine longer is usually not recommended, because it can cause deterioration in engine performance.

If more than one ATS is connected to the same engine generator, it is sometimes recommended that transfer of the loads be sequenced to the alternate source. Using a sequencing scheme can reduce starting kW capacity requirements of the generator. A fourth timer, adjustable from 0 to 5 minutes, will delay transfer to the emergency supply source for this and other similar requirements.

**(A) General.** Transfer equipment, including automatic transfer switches, shall be automatic, identified for emergency use, and approved by the authority having jurisdiction. Transfer equipment shall be designed and installed to prevent the inadvertent interconnection of normal and emergency sources of supply in any operation of the transfer equipment. Transfer equipment and electric power production systems installed to permit operation in parallel with the normal source shall meet the requirements of Article 705.

Traditional ATS are not designed to permit parallel operation of generation equipment and the normal source. Therefore, traditional ATS need not comply with Article 705. However, certain ATS configurations are intentionally designed to briefly (for a few cycles) parallel the generation equipment with the normal source upon load transfer from generator to normal source. This load transfer method may result in minimal disturbance or effect on the load. Transfer switches that employ this type of paralleling must comply with Article 705.

If continuous parallel operation of generation equipment and the source is desired, paralleling switchgear or paralleling equipment with appropriate protection is required. (See Article 705.)

**(B) Bypass Isolation Switches.** Means shall be permitted to bypass and isolate the transfer equipment. Where bypass isolation switches are used, inadvertent parallel operation shall be avoided.

**(C) Automatic Transfer Switches.** Automatic transfer switches shall be electrically operated and mechanically held. Automatic transfer switches, rated 1000 VAC and below, shall be listed for emergency system use.

Relay contacts are required to be mechanically held so that if a coil fails, the generator will not drop offline. Transfer switches rated 1000 volts ac and below are required to be listed for emergency systems use.

When emergency systems are tested, both the normal and the emergency system are energized. If the two sources are not synchronized, as much as twice the rated voltage may exist across the transfer switch contacts. Some listed transfer switches are designed and tested to be suitable for switching between out-of-phase power sources. Other protection methods may be employed, such as a mechanical interlock that prevents inadvertent interconnection, or an electronic method that prevents both systems from being interconnected. A typical emergency system transfer switch is shown in Exhibit 700.1.

**(D) Use.** Transfer equipment shall supply only emergency loads.

The alternate power source can supply emergency loads as well as other loads. However, the emergency system transfer switch is limited to supplying emergency loads. Legally required standby loads or optional standby loads (covered by Articles 701 and 702) require separate transfer switches.

## 700.6 Signals

Audible and visual signal devices shall be provided, where practicable, for the purpose described in 700.6(A) through (D).

**(A) Derangement.** To indicate derangement of the emergency source.

**(B) Carrying Load.** To indicate that the battery is carrying load.

*EXHIBIT 700.1* *An emergency system transfer switch. (Courtesy of the International Association of Electrical Inspectors)*

**(C) Not Functioning.** To indicate that the battery charger is not functioning.

The major causes of emergency equipment failure are inadequate testing and maintenance. Installing signal devices that annunciate trouble where attendants or other personnel familiar with the operation of the emergency equipment can see or hear them allows action to be taken to maintain system function.

Battery-operated unit equipment generally has a test switch that simulates failure of the normal system, and an indicating light that glows brightly while charging and dims when ready. Transparent cases for lead-acid batteries allow easy viewing of electrolyte levels.

**(D) Ground Fault.** To indicate a ground fault in solidly grounded wye emergency systems of more than 150 volts to ground and circuit-protective devices rated 1000 amperes or more. The sensor for the ground-fault signal devices shall be located at, or ahead of, the main system disconnecting means for the emergency source, and the maximum setting of the signal devices shall be for a ground-fault current of 1200 amperes. Instructions on the course of action to be taken in event of indicated ground fault shall be located at or near the sensor location.

Informational Note: For signals for generator sets, see NFPA 110-2013, *Standard for Emergency and Standby Power Systems*.

Automatic ground-fault protection is not required on emergency systems (see 700.27), because it could interrupt the system when it is needed. However, ground faults must be detected and indicated so that the ground fault can be cleared as soon as practical.

### 700.7 Signs

**(A) Emergency Sources.** A sign shall be placed at the service-entrance equipment, indicating type and location of on-site emergency power sources.

*Exception: A sign shall not be required for individual unit equipment as specified in 700.12(F).*

**(B) Grounding.** Where removal of a grounding or bonding connection in normal power source equipment interrupts the grounding electrode conductor connection to the alternate power source(s) grounded conductor, a warning sign shall be installed at the normal power source equipment stating:

WARNING
SHOCK HAZARD EXISTS IF GROUNDING
ELECTRODE CONDUCTOR OR BONDING JUMPER
CONNECTION IN THIS EQUIPMENT IS REMOVED
WHILE ALTERNATE SOURCE(S) IS ENERGIZED.

The warning sign(s) or label(s) shall comply with 110.21(B).

Emergency and standby systems that have a solid (unswitched) neutral in the transfer equipment (non-separately derived system) rely on the grounding and bonding connections in the normal source supply equipment to ensure that the ground-fault current path is completed from a ground fault to the alternate source. If a main or system bonding jumper is removed (for example, to perform testing on GFPE systems), an electrician or other service personnel could inadvertently become part of the current path if a ground fault occurs while the alternate source is supplying power to loads. This poses a significant shock hazard to personnel who may not be aware of the grounding and bonding configuration for the alternate source.

### 700.8 Surge Protection

A listed SPD shall be installed in or on all emergency systems switchboards and panelboards.

## II. Circuit Wiring

### 700.10 Wiring, Emergency System

**(A) Identification.** All boxes and enclosures (including transfer switches, generators, and power panels) for emergency circuits shall be permanently marked so they will be readily identified as a component of an emergency circuit or system.

The required marking can be by color code, the words "emergency system," or any other method that identifies the box or enclosure as a component of the emergency system.

**(B) Wiring.** Wiring of two or more emergency circuits supplied from the same source shall be permitted in the same raceway, cable, box, or cabinet. Wiring from an emergency source or emergency source distribution overcurrent protection to emergency loads shall be kept entirely independent of all other wiring and equipment, unless otherwise permitted in 700.10(B)(1) through (5):

(1) Wiring from the normal power source located in transfer equipment enclosures
(2) Wiring supplied from two sources in exit or emergency luminaires
(3) Wiring from two sources in a listed load control relay supplying exit or emergency luminaires, or in a common junction box, attached to exit or emergency luminaires
(4) Wiring within a common junction box attached to unit equipment, containing only the branch circuit supplying the unit equipment and the emergency circuit supplied by the unit equipment

Where an alternate power source supplies a switchboard from a single feeder or feeders in parallel, that switchboard may further distribute and provide power for the emergency, legally required, and optional standby systems, provided separate vertical switchboard sections are used.

Separate vertical switchboard sections provide the physical separation requirements of both system and wiring from a common power source. This physical separation cannot occur within a panelboard enclosure because of its open design. The supply tap box on generators equipped with disconnects with or without overcurrent protection is not generally designed or manufactured for the installation of multiple transfer switches to serve separate circuits for emergency systems, fire pump loads, legally required standby systems, and optional standby systems. In addition, large systems may employ multiple generators. Any combination of these systems may be supplied from a single feeder or multiple feeders, or from separate vertical sections of a switchboard that are either supplied by a common bus or supplied individually.

Where generators operate in parallel, frequency and voltage must be synchronized and can supply a common bus array. From this bus, emergency, legally required, and optional standby transfer switches can be supplied.

Except as noted in this section, wiring for the emergency circuits must be completely independent of all other wiring and equipment. This practice ensures that a fault in any other system wiring will not affect the performance of the emergency wiring or equipment.

Sections 700.10(B)(2) and (B)(3) permit the use of two-lamp exit or two-lamp emergency fixtures, where one lamp is connected to the normal supply and one lamp is connected to the alternate supply. Both lamps may be illuminated as part of the regular lighting operation.

Wiring on the load side of a transfer switch serves as both the emergency circuit wiring and the normal circuit wiring. Two sets of wiring are not required to supply emergency loads from the load side of the transfer switch to the emergency load distribution panel or from these emergency distribution panels to the emergency loads. See Exhibits 700.2, 700.3, and 700.4 for illustrations of feeder configurations.

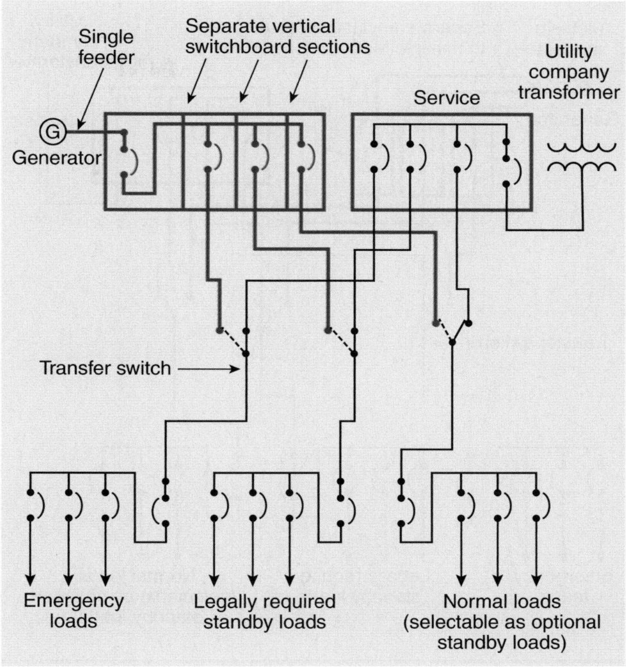


***EXHIBIT 700.2*** *Illustration of a single feeder that supplies separate vertical sections of the switchboard.*

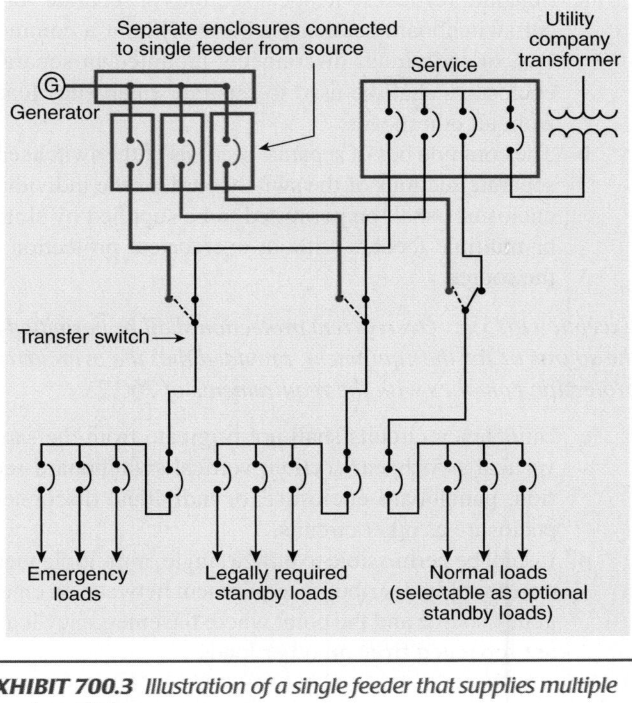


***EXHIBIT 700.3*** *Illustration of a single feeder that supplies multiple transfer switches.*

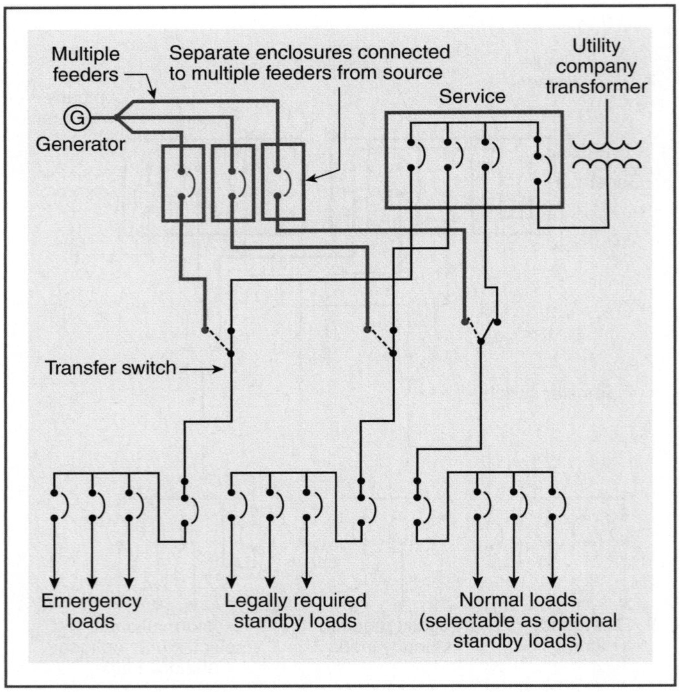


***EXHIBIT 700.4*** *Illustration of a generator that supplies multiple feeders at its terminals.*

(5) Wiring from an emergency source to supply emergency and other loads in accordance with 700.10(B)(5)a, b, c, and d as follows:

a. Separate vertical switchgear sections or separate vertical switchboard sections, with or without a common bus, or individual disconnects mounted in separate enclosures shall be used to separate emergency loads from all other loads.

b. The common bus of separate sections of the switchgear, separate sections of the switchboard, or the individual enclosures shall be permitted to be supplied by single or multiple feeders without overcurrent protection at the source.

*Exception to (5)b: Overcurrent protection shall be permitted at the source or for the equipment, provided that the overcurrent protection complies with the requirements of 700.28.*

c. Emergency circuits shall not originate from the same vertical switchgear section, vertical switchboard section, panelboard enclosure, or individual disconnect enclosure as other circuits.

d. It shall be permissible to utilize single or multiple feeders to supply distribution equipment between an emergency source and the point where the emergency loads are separated from all other loads.

A single common feeder is permitted to be installed between the alternate source and the point in the distribution system at which the physical separation of the emergency, legally required standby, and optional standby system conductors occurs, such as at a switchboard or other distribution equipment. Many large campus facilities with multiple buildings, such as medical centers, colleges or universities, prisons, and shopping malls, rely on central generation of emergency power. See the commentary following 700.10(B).

**(C) Wiring Design and Location.** Emergency wiring circuits shall be designed and located so as to minimize the hazards that might cause failure due to flooding, fire, icing, vandalism, and other adverse conditions.

**(D) Fire Protection.** Emergency systems shall meet the additional requirements in (D)(1) through (D)(3) in assembly occupancies for not less than 1000 persons or in buildings above 23 m (75 ft) in height.

•

**(1) Feeder-Circuit Wiring.** Feeder-circuit wiring shall meet one of the following conditions:

(1) Be installed in spaces or areas that are fully protected by an approved automatic fire suppression system

Where emergency system feeders are installed above a suspended ceiling, for the system to be fully protected by a fire suppression system, sprinklers must be provided above the suspended ceilings even though sprinklers might be installed below the ceiling.

(2) Be a listed electrical circuit protective system with a minimum 2-hour fire rating

Informational Note: UL guide information for electrical circuit protective systems (FHIT) contains information on proper installation requirements to maintain the fire rating.

•

(3) Be protected by a listed thermal barrier system for electrical system components with a minimum 2-hour fire rating

Where emergency system wiring is installed in a listed fire-rated assembly, no other wiring is permitted within the assembly. If a fire-rated assembly is needed for normal circuits, it must be a separate fire-rated assembly from the one used for the emergency system.

(4) Be protected by a listed fire-rated assembly that has a minimum fire rating of 2 hours and contains only emergency wiring circuits

(5) Be encased in a minimum of 50 mm (2 in.) of concrete

If feeders are not located in building spaces that are fully protected by a fire suppression system, other fire protection techniques to comply with 700.10(D)(1) include the following.

**Listed electrical circuit protective systems** are described in the UL *White Book*. The four-letter code (shown in parentheses) is the UL product category guide designation. Examples of these systems include electrical circuit protective systems (FHIT), electrical circuit protective materials (FHIY), and fire-resistive cables (FHJR). Circuit integrity cable is covered under category FHJR.

**Listed thermal barrier systems (XCLF)** are described in the UL *White Book*. An example of the thermal barrier protection technique is batts and blankets (XCLR) wrapped over the wiring method to achieve a predetermined fire rating.

**Fire-rated assemblies** are described in the UL *Fire Resistance Directory*, Volumes 1 and 2. Volume 1 includes hourly ratings for beams, floors, roofs, columns, and walls and partitions. Volume 2A and 2B include hourly ratings for joint, through-penetration firestops, and electrical circuit protective systems. All fire ratings and fire resistance ratings are based on the assumption that the structural supports for the assembly are not impaired by the fire.

**Encasement in concrete** has been successful for many years in protecting premises from faults in service conductors per 230.6. Encasement in 2 inches of concrete is possible after orginal construction.

There is a difference between a 2-hour fire rating of an electrical circuit, such as a conduit with wires, and a 2-hour fire resistance rating of a structural member, such as a wall. At the end of a 2-hour fire test on an electrical conduit with wires, its insulation must be intact and the circuit must function electrically; no short circuits, grounds, or opens are permitted. A wall subjected to a 2-hour fire resistance test must only prevent a fire from passing through or past the wall, regardless of damage to the wall.

**(2) Feeder-Circuit Equipment.** Equipment for feeder circuits (including transfer switches, transformers, and panelboards) shall be located either in spaces fully protected by approved automatic fire suppression systems (including sprinklers, carbon dioxide systems) or in spaces with a 2-hour fire resistance rating.

Fire protection requirements for both emergency system feeder circuits and equipment ensure the integrity as well as the performance of the emergency electrical system. If feeders and equipment are located in building spaces that are fully protected by an approved fire suppression system, no further fire protection techniques are generally required.

Sprinkler systems are the most common fire suppression systems. Building spaces that are fully protected by automatic sprinkler systems meet the requirements of 700.10(D). Requirements for fire suppression systems are included in the following standards:

NFPA 12, *Standard on Carbon Dioxide Extinguishing Systems*

NFPA 12A, *Standard on Halon 1301 Fire Extinguishing Systems*

NFPA 13, *Standard for the Installation of Sprinkler Systems*

NFPA 15, *Standard for Water Spray Fixed Systems for Fire Protection*

NFPA 17, *Standard for Dry Chemical Extinguishing Systems*

NFPA 2001, *Standard on Clean Agent Fire Extinguishing Systems*

If feeder-circuit equipment is not located in a space that is fully protected by a fire suppression system, the space must have a 2-hour fire resistance rating. See the commentary following 700.10(D)(1) regarding fire-rated assemblies.

**(3) Generator Control Wiring.** Control conductors installed between the transfer equipment and the emergency generator shall be kept entirely independent of all other wiring and shall meet the conditions of 700.10(D)(1).

## III. Sources of Power

### General Requirements for Emergency Lighting Systems

At least two sources of power must be provided – one normal supply and one or more of the emergency systems described in 700.12. The sources (see Exhibits 700.5 and 700.6.) may be one of the following:

1. Two services – one normal supply and one emergency supply (preferably from separate utility stations)
2. One normal service and a storage battery (or unit equipment) system
3. One normal service and a generator set

A means must be provided to transfer the emergency loads to the alternate supply when the normal source of supply is interrupted. If a separate service is used, both may operate normally, but equipment for emergency lighting and power must be arranged to be energized from either service.

If the alternate or emergency source of supply is a storage battery or generator set, the single emergency system is usually operated on the normal service, and the battery (or batteries) or generator operates only if the normal service fails. However, a generator may be used for peak load shaving and other standby systems in accordance with 700.4.

Two or more separate and complete systems may provide power for emergency lighting, but means must be provided for energizing one system if the other one fails.

Disconnecting means and overcurrent protection (see Exhibits 700.5 and 700.6) must be provided for emergency systems as required by Articles 225 and 230.

## 700.12 General Requirements

Current supply shall be such that, in the event of failure of the normal supply to, or within, the building or group of buildings concerned, emergency lighting, emergency power, or both shall be available within the time required for the application but not to exceed 10 seconds. The supply system for emergency purposes, in addition to the normal services to the building and meeting

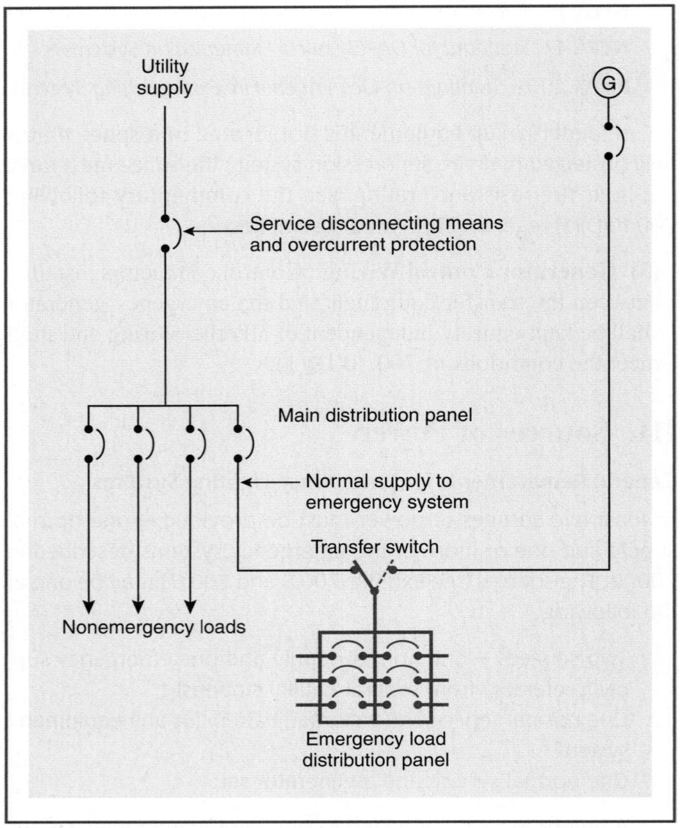


***EXHIBIT 700.5*** *Emergency load arranged to be supplied from a generator, as permitted by 700.12(B).*

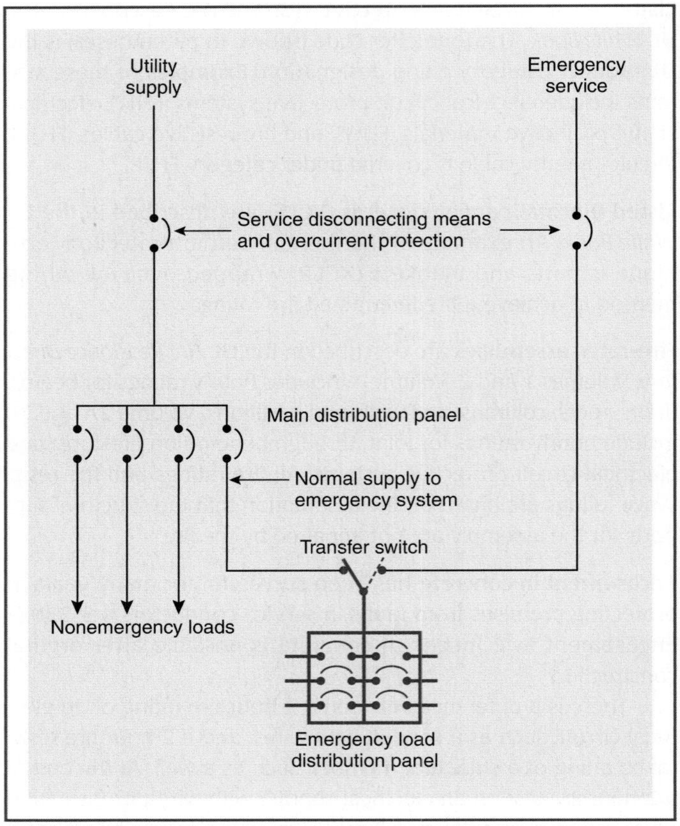


***EXHIBIT 700.6*** *Emergency load arranged to be supplied from two widely separated services, as permitted by 700.12(D).*

the general requirements of this section, shall be one or more of the types of systems described in 700.12(A) through (E). Unit equipment in accordance with 700.12(F) shall satisfy the applicable requirements of this article.

In selecting an emergency source of power, consideration shall be given to the occupancy and the type of service to be rendered, whether of minimum duration, as for evacuation of a theater, or longer duration, as for supplying emergency power and lighting due to an indefinite period of current failure from trouble either inside or outside the building.

Equipment shall be designed and located so as to minimize the hazards that might cause complete failure due to flooding, fires, icing, and vandalism.

The design and selection of a location must consider hazards that could impair reliability. See Exhibit 700.7.

Equipment for sources of power as described in 700.12(A) through (E) where located within assembly occupancies for greater than 1000 persons or in buildings above 23 m (75 ft) in height with any of the following occupancy classes — assembly, educational, residential, detention and correctional, business, and mercantile — shall be installed either in spaces fully protected

***EXHIBIT 700.7*** *Alternate source of power in a heated and secure enclosure that is located in an area not subject to flooding. (Courtesy of the International Association of Electrical Inspectors)*

by approved automatic fire suppression systems (sprinklers, carbon dioxide systems, and so forth) or in spaces with a 1-hour fire rating.

Informational Note No. 1: For the definition of *Occupancy Classification*, see Section 6.1 of NFPA *101-2012*, *Life Safety Code*.

Informational Note No. 2: For further information, see ANSI/IEEE 493-2007, *Recommended Practice for the Design of Reliable Industrial and Commercial Power Systems*.

**(A) Storage Battery.** Storage batteries used as a source of power for emergency systems shall be of suitable rating and capacity to supply and maintain the total load for a minimum period of 1½ hours, without the voltage applied to the load falling below 87½ percent of normal.

Batteries, whether of the acid or alkali type, shall be designed and constructed to meet the requirements of emergency service and shall be compatible with the charger for that particular installation.

For a sealed battery, the container shall not be required to be transparent. However, for the lead acid battery that requires water additions, transparent or translucent containers shall be furnished. Automotive-type batteries shall not be used.

An automatic battery charging means shall be provided.

**(B) Generator Set.**

**(1) Prime Mover-Driven.** For a generator set driven by a prime mover acceptable to the authority having jurisdiction and sized in accordance with 700.4, means shall be provided for automatically starting the prime mover on failure of the normal service and for automatic transfer and operation of all required electrical circuits. A time-delay feature permitting a 15-minute setting shall be provided to avoid retransfer in case of short-time reestablishment of the normal source.

**(2) Internal Combustion Engines as Prime Movers.** Where internal combustion engines are used as the prime mover, an on-site fuel supply shall be provided with an on-premises fuel supply sufficient for not less than 2 hours' full-demand operation of the system. Where power is needed for the operation of the fuel transfer pumps to deliver fuel to a generator set day tank, this pump shall be connected to the emergency power system.

Engine-driven generators that use an electric fuel transfer pump might not start or continue operating if the fuel pump is not operating. These pumps, which transfer fuel into a day tank, must be supplied by the emergency system.

**(3) Dual Supplies.** Prime movers shall not be solely dependent on a public utility gas system for their fuel supply or municipal water supply for their cooling systems. Means shall be provided for automatically transferring from one fuel supply to another where dual fuel supplies are used.

*Exception: Where acceptable to the authority having jurisdiction, the use of other than on-site fuels shall be permitted where there is a low probability of a simultaneous failure of both the off-site fuel delivery system and power from the outside electrical utility company.*

**(4) Battery Power and Dampers.** Where a storage battery is used for control or signal power or as the means of starting the prime mover, it shall be suitable for the purpose and shall be equipped with an automatic charging means independent of the generator set. Where the battery charger is required for the operation of the generator set, it shall be connected to the emergency system. Where power is required for the operation of dampers used to ventilate the generator set, the dampers shall be connected to the emergency system.

**(5) Auxiliary Power Supply.** Generator sets that require more than 10 seconds to develop power shall be permitted if an auxiliary power supply energizes the emergency system until the generator can pick up the load.

**(6) Outdoor Generator Sets.** Where an outdoor housed generator set is equipped with a readily accessible disconnecting means in accordance with 445.18, and the disconnecting means is located within sight of the building or structure supplied, an additional disconnecting means shall not be required where ungrounded conductors serve or pass through the building or structure. Where the generator supply conductors terminate at a disconnecting means in or on a building or structure, the disconnecting means shall meet the requirements of 225.36.

The disconnecting means on the generator can be used as the disconnecting means required in 225.31, provided the disconnecting means is readily accessible and is within sight of the building. (See the definitions of *readily accessible* and *in sight from* in Article 100.) Where an additional disconnecting means is necessary, it must be suitable for use as service equipment in accordance with 225.36.

*Exception: For installations under single management, where conditions of maintenance and supervision ensure that only qualified persons will monitor and service the installation and where documented safe switching procedures are established and maintained for disconnection, the generator set disconnecting means shall not be required to be located within sight of the building or structure served.*

The circuit between the generator and the building or structure is a feeder. Therefore, the requirements for outdoor feeders contained in Article 225 must be followed, including those covering disconnecting means for outdoor branch circuits and feeders. Section 700.12(B)(6) modifies the requirement in 225.32 for the location of disconnecting means. The feeder disconnecting means is permitted to be located at the generator location provided the disconnecting means is within sight and readily accessible from the building being supplied.

**(C) Uninterruptible Power Supplies.** Uninterruptible power supplies used to provide power for emergency systems shall comply with the applicable provisions of 700.12(A) and (B).

**(D) Separate Service.** Where approved by the authority having jurisdiction as suitable for use as an emergency source of power, an additional service shall be permitted. This service shall be in accordance with the applicable provisions of Article 230 and the following additional requirements:

(1) Separate overhead service conductors, service drops, underground service conductors, or service laterals shall be installed.
(2) The service conductors for the separate service shall be installed sufficiently remote electrically and physically from any other service conductors to minimize the possibility of simultaneous interruption of supply.

**(E) Fuel Cell System.** Fuel cell systems used as a source of power for emergency systems shall be of suitable rating and capacity to supply and maintain the total load for not less than 2 hours of full-demand operation.

Installation of a fuel cell system shall meet the requirements of Parts II through VIII of Article 692.

Where a single fuel cell system serves as the normal supply for the building or group of buildings concerned, it shall not serve as the sole source of power for the emergency standby system.

Emergency systems can be designed using one or more of the following systems:

1. *One storage battery or a group of storage batteries* provided with an automatic battery-charging means. (See Article 480.)
2. *A generator set driven by a prime mover,* acceptable to the AHJ, and with adequate capacity to carry the maximum load connected. Prime movers may be internal-combustion engines, steam or gas turbines, or other approved types of mechanical drivers. A storage battery used to start the prime mover must be provided with an automatic battery-charging means. An on-site fuel supply that is sufficient to operate internal-combustion engines at full load for 2 hours must also be available.

   Off-site fuel supplies such as natural gas or piped steam may be used where experience has demonstrated their reliability. Off-site fuel supplies may also be used where they provide greater reliability than gasoline or diesel engines or in isolated areas where maintenance or refueling could be a problem.

   Some types of drivers, particularly large ones, may take longer than 10 seconds to accelerate and develop generator voltage. Gas and steam turbines and large internal-combustion engines may have prolonged starting times. Depending on the specific loads, short-time supply could be provided by an uninterruptible power supply; a generator shared with other loads; or a generator with limited emergency supply, such as an expander, a steam turbine, or a waste heat system.
3. *Uninterruptible power supplies (UPS),* which generally include a rectifier, a storage battery, and an inverter to ac. Uninterruptible power supplies may be very complex systems with redundant components and high-speed solid-state switching. A common practice is to include an automatic bypass for UPS malfunction to permit maintenance.
4. *The use of a separate service,* which requires a judgment by the AHJ. Such judgment should be based on the nature of the emergency loads and the expected reliability of the other available sources.

**(F) Unit Equipment.**

**(1) Components of Unit Equipment.** Individual unit equipment for emergency illumination shall consist of the following:

(1) A rechargeable battery
(2) A battery charging means
(3) Provisions for one or more lamps mounted on the equipment, or shall be permitted to have terminals for remote lamps, or both
(4) A relaying device arranged to energize the lamps automatically upon failure of the supply to the unit equipment

Unit equipment must be permanently fixed in place, usually by mounting screws that are accessible only from within the unit. One or more lamps may be mounted on or remote from the unit. The unit should be located where it can be readily checked or tested for proper performance. See Exhibit 700.8.

Unit equipment is intended to provide illumination for the area where it is installed. For instance, if a unit is located in a corridor, it must be connected to the branch circuit supplying the

***EXHIBIT 700.8*** *Self-contained, fully automatic unit equipment for operating emergency lighting located on the unit or for remotely located exit signs or lighting heads. (Courtesy of the International Association of Electrical Inspectors)*

normal corridor lights (on the line side of any switching arrangements). If normal power fails, the unit automatically energizes the unit lamps, restoring illumination to the corridor. A separate circuit is not permitted for unit equipment [except as noted in the exception to 700.12(F)(2)(3)] because failure of the normal corridor circuit would not affect the unit equipment, and the corridor would remain dark. The branch circuit feeding the unit must be identified at the panelboard.

**(2) Installation of Unit Equipment.** Unit equipment shall be installed in accordance with 700.12(F)(2)(1) through (6).

(1) The batteries shall be of suitable rating and capacity to supply and maintain at not less than 87½ percent of the nominal battery voltage for the total lamp load associated with the unit for a period of at least 1½ hours, or the unit equipment shall supply and maintain not less than 60 percent of the initial emergency illumination for a period of at least 1½ hours. Storage batteries, whether of the acid or alkali type, shall be designed and constructed to meet the requirements of emergency service.
(2) Unit equipment shall be permanently fixed in place (i.e., not portable) and shall have all wiring to each unit installed in accordance with the requirements of any of the wiring methods in Chapter 3. Flexible cord-and-plug connection shall be permitted, provided that the cord does not exceed 900 mm (3 ft) in length.
(3) The branch circuit feeding the unit equipment shall be the same branch circuit as that serving the normal lighting in the area and connected ahead of any local switches.

*Exception: In a separate and uninterrupted area supplied by a minimum of three normal lighting circuits that are not part of a multiwire branch circuit, a separate branch circuit for unit equipment shall be permitted if it originates from the same panelboard as that of the normal lighting circuits and is provided with a lock-on feature.*

(4) The branch circuit that feeds unit equipment shall be clearly identified at the distribution panel.
(5) Emergency luminaires that obtain power from a unit equipment and are not part of the unit equipment shall be wired to the unit equipment as required by 700.10 and by one of the wiring methods of Chapter 3.
(6) Remote heads providing lighting for the exterior of an exit door shall be permitted to be supplied by the unit equipment serving the area immediately inside the exit door.

Prior to the 2011 *Code*, outside unit equipment was required to be supplied from the outside lighting circuit. Unit equipment serving the area immediately inside the exit door is allowed to supply remote emergency luminaires installed outside the exit door. The normal lighting branch circuit for the area inside the exit door can be used to supply unit equipment that in turn supplies emergency luminaires installed inside and outside the exit door. If the power to the normal lighting branch circuit for this area is interrupted, the indoor and outdoor emergency luminaires will activate, even if the normal branch circuit for exterior lighting remains energized.

## IV. Emergency System Circuits for Lighting and Power

### 700.15 Loads on Emergency Branch Circuits

No appliances and no lamps, other than those specified as required for emergency use, shall be supplied by emergency lighting circuits.

### 700.16 Emergency Illumination

Emergency illumination shall include all required means of egress lighting, illuminated exit signs, and all other lights specified as necessary to provide required illumination.

Emergency lighting systems shall be designed and installed so that the failure of any individual lighting element, such as the burning out of a lamp, cannot leave in total darkness any space that requires emergency illumination.

Where high-intensity discharge lighting such as high- and low-pressure sodium, mercury vapor, and metal halide is used as the sole source of normal illumination, the emergency lighting system shall be required to operate until normal illumination has been restored.

Where an emergency system is installed, emergency illumination shall be provided in the area of the disconnecting means required by 225.31 and 230.70, as applicable, where the disconnecting means are installed indoors.

*Exception: Alternative means that ensure that the emergency lighting illumination level is maintained shall be permitted.*

High-intensity discharge (HID) fixtures take some time to start once they are energized. Therefore, if HID fixtures are the sole source of normal illumination in an area, the *Code* requires that the emergency lighting system operate not only until the normal system is returned to service but also until the HID fixtures provide illumination. This does not apply if another type of fixture, such as an incandescent one, also normally illuminates the area.

For unit equipment, a second lamp ensures that the area is not left in total darkness. This section does not require redundant batteries or control circuitry.

### 700.17 Branch Circuits for Emergency Lighting

Branch circuits that supply emergency lighting shall be installed to provide service from a source complying with 700.12 when the normal supply for lighting is interrupted. Such installations shall provide either of the following:

(1) An emergency lighting supply, independent of the normal lighting supply, with provisions for automatically

transferring the emergency lights upon the event of failure of the normal lighting branch circuit

(2) Two or more branch circuits supplied from separate and complete systems with independent power sources. One of the two power sources and systems shall be part of the emergency system, and the other shall be permitted to be part of the normal power source and system. Each system shall provide sufficient power for emergency lighting purposes.

Unless both systems are used for regular lighting purposes and are both kept lighted, means shall be provided for automatically energizing either system upon failure of the other. Either or both systems shall be permitted to be a part of the general lighting of the protected occupancy if circuits supplying lights for emergency illumination are installed in accordance with other sections of this article.

The terms *normal* and *branch circuit* indicate that the emergency lighting supply must be independent of the normal lighting supply and that it must automatically operate when there is a failure of the branch circuit(s) supplying the normal lighting.

Section 700.17(2) requires emergency lighting to be supplied by a minimum of two branch circuits from separate systems with different power sources. Where a failure of the normal lighting branch circuit activates the emergency lighting supply, an area supplied by only one lighting branch circuit will be in total darkness if that branch circuit fails. For example, if a single branch circuit, supplied by an emergency circuit panelboard, supplies the lighting in a stairwell (means of egress), a failure of that branch circuit plunges that stairwell into total darkness. If two branch circuits from separate systems are run to the stairwell, it is unlikely that both circuits to the stairway would fail simultaneously; therefore the risk to occupants created by total darkness is minimized.

### 700.18 Circuits for Emergency Power

For branch circuits that supply equipment classed as emergency, there shall be an emergency supply source to which the load will be transferred automatically upon the failure of the normal supply.

### 700.19 Multiwire Branch Circuits

The branch circuit serving emergency lighting and power circuits shall not be part of a multiwire branch circuit.

## V. Control — Emergency Lighting Circuits

### 700.20 Switch Requirements

The switch or switches installed in emergency lighting circuits shall be arranged so that only authorized persons have control of emergency lighting.

*Exception No. 1: Where two or more single-throw switches are connected in parallel to control a single circuit, at least one of these switches shall be accessible only to authorized persons.*

*Exception No. 2: Additional switches that act only to put emergency lights into operation but not disconnect them shall be permissible.*

Switches connected in series or 3- and 4-way switches shall not be used.

### 700.21 Switch Location

All manual switches for controlling emergency circuits shall be in locations convenient to authorized persons responsible for their actuation. In facilities covered by Articles 518 and 520, a switch for controlling emergency lighting systems shall be located in the lobby or at a place conveniently accessible thereto.

In no case shall a control switch for emergency lighting be placed in a motion-picture projection booth or on a stage or platform.

*Exception: Where multiple switches are provided, one such switch shall be permitted in such locations where arranged so that it can only energize the circuit but cannot de-energize the circuit.*

### 700.22 Exterior Lights

Those lights on the exterior of a building that are not required for illumination when there is sufficient daylight shall be permitted to be controlled by an automatic light-actuated device.

### 700.23 Dimmer and Relay Systems

A dimmer or relay system containing more than one dimmer or relay and listed for use in emergency systems shall be permitted to be used as a control device for energizing emergency lighting circuits. Upon failure of normal power, the dimmer or relay system shall be permitted to selectively energize only those branch circuits required to provide minimum emergency illumination. All branch circuits supplied by the dimmer or relay system cabinet shall comply with the wiring methods of Article 700.

Dimmer systems that are listed for emergency system use include a method to sense failure of normal power and selectively energize branch circuits fed from the dimmer cabinet, regardless of the setting of control switches or panels normally used to control the dimmer system. Dimmer systems are usually supplied by a feeder that is transferred from the normal system to the emergency system by a transfer switch. See Exhibit 700.9.

### 700.24 Directly Controlled Luminaires

Where emergency illumination is provided by one or more directly controlled luminaires that respond to an external control input to bypass normal control upon loss of normal power, such luminaires and external bypass controls shall be individually listed for use in emergency systems.

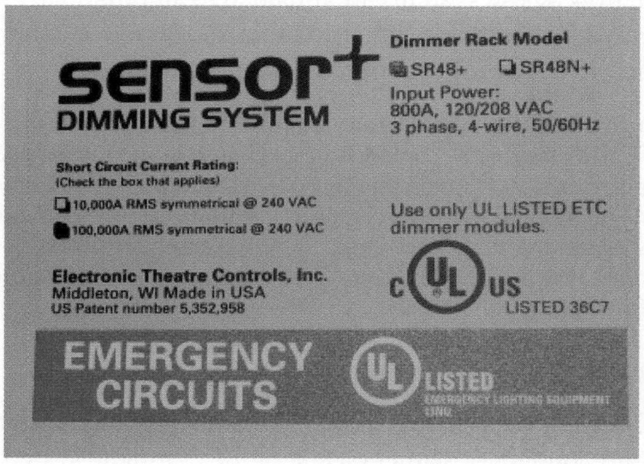


***EXHIBIT 700.9*** *An example of a label for a dimmer system that is listed for emergency system use. (Courtesy of Electronic Theatre Controls, Inc.)*

## 700.25 Automatic Load Control Relay

If an emergency lighting load is automatically energized upon loss of the normal supply, a listed automatic load control relay shall be permitted to energize the load. The load control relay shall not be used as transfer equipment.

Automatic load control relays were traditionally part of emergency unit equipment, but stand-alone devices are now listed under ANSI/UL 924, *Standard for Emergency Lighting and Power Equipment*. Proper application of these devices depends upon their function in an emergency circuit.

Load control relays listed to UL 924 are not to be used to transfer a load between two nonsynchronous power sources; only transfer equipment listed to UL 1008 is suitable for this application. These power sources might be out of phase with one another. Load control relays do not have mechanisms required by UL 1008 to prevent inadvertent connection of the normal and emergency sources, and they do not undergo the fault-current evaluation that is required of UL 1008 for transfer switches.

In addition, the UL *White Book* differentiates automatic transfer switches (product category WPWR) from automatic load control relays (product category FTBR).

# VI. Overcurrent Protection

## 700.26 Accessibility

The branch-circuit overcurrent devices in emergency circuits shall be accessible to authorized persons only.

## 700.27 Ground-Fault Protection of Equipment

The alternate source for emergency systems shall not be required to have ground-fault protection of equipment with automatic disconnecting means. Ground-fault indication of the emergency source shall be provided in accordance with 700.6(D) if ground-fault protection of equipment with automatic disconnecting means is not provided.

## 700.28 Selective Coordination

Emergency system(s) overcurrent devices shall be selectively coordinated with all supply-side overcurrent protective devices.

Selective coordination shall be selected by a licensed professional engineer or other qualified persons engaged primarily in the design, installation, or maintenance of electrical systems. The selection shall be documented and made available to those authorized to design, install, inspect, maintain, and operate the system.

*Exception: Selective coordination shall not be required between two overcurrent devices located in series if no loads are connected in parallel with the downstream device.*

The term *coordination (selective)*, as defined in Article 100, indicates that a selectively coordinated system is one where the operation of the overcurrent protective scheme localizes an overcurrent condition to the circuit conductors or equipment in which an overload or fault (short circuit or ground fault) has occurred. Because the purpose of an emergency system is to provide power to essential life safety systems in a building or facility, a selectively coordinated overcurrent protection scheme that localizes and minimizes the extent of an interruption of power due to the opening of a protective device is a critical safety element.

Continuity of operation of illumination for occupant evacuation or maintaining continuity of operation of essential safety equipment such as smoke evacuation systems is necessary for occupant safety. This prohibits an overload, short circuit, or ground fault in a 20-ampere branch circuit from causing the feeder protective device supplying the branch-circuit panelboard to open. Coordination must be carried through each level of distribution that supplies power to the emergency system.

Design and verification of electrical system coordination can be achieved only through a coordination study. A coordination study entails detailed analysis of electrical supply system fault-current characteristics. The design must integrate overcurrent protective devices that interact by localizing the overcurrent problem and isolating that part of the emergency system. Modifications to the electrical system after the initial design and installation can affect the original implementation of the coordinated system. For additional discussion on selective coordination, see the commentary for 620.62 and Exhibit 620.8.

The exception to 700.28 recognizes devices that are in series where the upstream device has no other loads. A typical example would be a series-rated system where the device immediately upstream is designed to open before the downstream device under short-circuit conditions.

# ARTICLE 701 Legally Required Standby Systems

## I. General

### 701.1 Scope

The provisions of this article apply to the electrical safety of the installation, operation, and maintenance of legally required standby systems consisting of circuits and equipment intended to supply, distribute, and control electricity to required facilities for illumination or power, or both, when the normal electrical supply or system is interrupted.

The systems covered by this article consist only of those that are permanently installed in their entirety, including the power source.

Informational Note No. 1: For additional information, see NFPA 99-2012, *Health Care Facilities Code*.

Informational Note No. 2: For further information regarding performance of emergency and standby power systems, see NFPA 110-2013, *Standard for Emergency and Standby Power Systems*.

Informational Note No. 3: For further information, see ANSI/IEEE 446-1995, *Recommended Practice for Emergency and Standby Power Systems for Industrial and Commercial Applications*.

Legally required standby systems are intended to provide electric power to aid in fire fighting, rescue operations, control of health hazards, and similar operations. In comparison, emergency systems (see Article 700) are those systems essential for safety to life. Optional standby systems (see Article 702) are those in which failure can cause physical discomfort, interruption of an industrial process, damage to process equipment, or disruption of business, for example.

The requirements for legally required standby systems are much the same as for emergency systems, except for a few differences. When normal power is lost, legally required systems must be able to supply standby power in 60 seconds or less, instead of the 10 seconds or less required of emergency systems. Wiring for legally required standby systems may occupy the same raceways, cables, boxes, and cabinets as other general wiring, whereas wiring for emergency systems must be kept entirely independent of other wiring. Legally required standby systems take second priority to emergency systems if they are involved in sharing an alternate supply and/or load shedding or peak shaving schemes.

### 701.2 Definition

**Legally Required Standby Systems.** Those systems required and so classed as legally required standby by municipal, state, federal, or other codes or by any governmental agency having jurisdiction. These systems are intended to automatically supply power to selected loads (other than those classed as emergency systems) in the event of failure of the normal source.

Informational Note: Legally required standby systems are typically installed to serve loads, such as heating and refrigeration systems, communications systems, ventilation and smoke removal systems, sewage disposal, lighting systems, and industrial processes, that, when stopped during any interruption of the normal electrical supply, could create hazards or hamper rescue or fire-fighting operations.

### 701.3 Tests and Maintenance

**(A) Conduct or Witness Test.** The authority having jurisdiction shall conduct or witness a test of the complete system upon installation.

**(B) Tested Periodically.** Systems shall be tested periodically on a schedule and in a manner acceptable to the authority having jurisdiction to ensure the systems are maintained in proper operating condition.

**(C) Battery Systems Maintenance.** Where batteries are used for control, starting, or ignition of prime movers, the authority having jurisdiction shall require periodic maintenance.

**(D) Written Record.** A written record shall be kept on such tests and maintenance.

**(E) Testing Under Load.** Means for testing legally required standby systems under load shall be provided.

Informational Note: For information on testing and maintenance of emergency power supply systems (EPSSs), see NFPA 110-2013, *Standard for Emergency and Standby Power Systems*.

### 701.4 Capacity and Rating

A legally required standby system shall have adequate capacity and rating for the supply of all equipment intended to be operated at one time. Legally required standby system equipment shall be suitable for the maximum available fault current at its terminals.

The legally required standby alternate power source shall be permitted to supply both legally required standby and optional standby system loads under either of the following conditions:

(1) Where the alternate source has adequate capacity to handle all connected loads
(2) Where automatic selective load pickup and load shedding is provided that will ensure adequate power to the legally required standby circuits

Selective load pickup and load shedding are not required if the generator has sufficient capacity to supply all connected loads.

### 701.5 Transfer Equipment

**(A) General.** Transfer equipment, including automatic transfer switches, shall be automatic and identified for standby use and approved by the authority having jurisdiction. Transfer equipment shall be designed and installed to prevent the inadvertent interconnection of normal and alternate sources of supply in any operation of the transfer equipment. Transfer equipment and electric power production systems installed to permit operation in parallel with the normal source shall meet the requirements of Article 705.

Parallel operation of the generation equipment with the normal source is permitted as long as the requirements of Article 705 are met. Traditional automatic transfer switches (ATS) are not designed to permit parallel operation of generation equipment and the normal source. Therefore, traditional ATS need not comply with Article 705. However, certain ATS configurations are intentionally designed to briefly (for a few cycles) parallel the generation equipment with the normal source upon load transfer. This load transfer can occur with minimal disturbance or effect on the load. Transfer switches that employ this type of paralleling must comply with Article 705.

**(B) Bypass Isolation Switches.** Means to bypass and isolate the transfer switch equipment shall be permitted. Where bypass isolation switches are used, inadvertent parallel operation shall be avoided.

**(C) Automatic Transfer Switches.** Automatic transfer switches shall be electrically operated and mechanically held. Automatic transfer switches, rated 1000 VAC and below, shall be listed for emergency use.

The intent is to ensure that relay contacts are mechanically held in the event of coil failure. This requirement also correlates with NFPA 110, *Standard for Emergency and Standby Power Systems*.

When standby systems are tested, both the normal and the standby system are energized. If the two sources are not synchronized, as much as twice the rated voltage may exist across the transfer switch contacts. Some listed transfer switches are designed and tested to be suitable for switching between out-of-phase power sources. Other protection methods may be employed, such as a mechanical interlock that prevents inadvertent interconnection or an electronic method that prevents both systems from being interconnected.

## 701.6 Signals

Audible and visual signal devices shall be provided, where practicable, for the purposes described in 701.6(A), (B), (C), and (D).

**(A) Derangement.** To indicate derangement of the standby source.

**(B) Carrying Load.** To indicate that the standby source is carrying load.

**(C) Not Functioning.** To indicate that the battery charger is not functioning.

Informational Note: For signals for generator sets, see NFPA 110-2013, *Standard for Emergency and Standby Power Systems*.

**(D) Ground Fault.** To indicate a ground fault in solidly grounded wye, legally required standby systems of more than 150 volts to ground and circuit-protective devices rated 1000 amperes or more. The sensor for the ground-fault signal devices shall be located at, or ahead of, the main system disconnecting means for the legally required standby source, and the maximum setting of the signal devices shall be for a ground-fault current of 1200 amperes. Instructions on the course of action to be taken in event of indicated ground fault shall be located at or near the sensor location.

Informational Note: For signals for generator sets, see NFPA 110-2013, *Standard for Emergency and Standby Power Systems*.

Ground-fault indication is required for legally required standby systems. Although 701.26 specifies that automatic ground-fault protection of equipment is not required to be provided on the alternate source, ground faults can occur on such systems, and they can result in equipment burndown. Because of the importance of legally required systems, automatic disconnect in the event of a ground fault is inappropriate. Detection of such a fault, however, is required so that the condition can be corrected.

## 701.7 Signs

**(A) Mandated Standby.** A sign shall be placed at the service entrance indicating type and location of on-site legally required standby power sources.

*Exception: A sign shall not be required for individual unit equipment as specified in 701.12(G).*

**(B) Grounding.** Where removal of a grounding or bonding connection in normal power source equipment interrupts the grounding electrode conductor connection to the alternate power source(s) grounded conductor, a warning sign shall be installed at the normal power source equipment stating:

WARNING
SHOCK HAZARD EXISTS IF GROUNDING
ELECTRODE CONDUCTOR OR BONDING JUMPER
CONNECTION IN THIS EQUIPMENT IS REMOVED
WHILE ALTERNATE SOURCE(S) IS ENERGIZED.

The warning sign(s) or label(s) shall comply with 110.21(B).

See the commentary following 700.7(B). Removal of grounding and bonding connections presents the same hazard in the normal supply equipment for legally required standby systems as in the normal supply equipment for an emergency system.

# II. Circuit Wiring

## 701.10 Wiring Legally Required Standby Systems

The legally required standby system wiring shall be permitted to occupy the same raceways, cables, boxes, and cabinets with other general wiring.

## III. Sources of Power

### 701.12 General Requirements

Current supply shall be such that, in the event of failure of the normal supply to, or within, the building or group of buildings concerned, legally required standby power will be available within the time required for the application but not to exceed 60 seconds. The supply system for legally required standby purposes, in addition to the normal services to the building, shall be permitted to comprise one or more of the types of systems described in 701.12(A) through (F). Unit equipment in accordance with 701.12(G) shall satisfy the applicable requirements of this article.

In selecting a legally required standby source of power, consideration shall be given to the type of service to be rendered, whether of short-time duration or long duration.

Consideration shall be given to the location or design, or both, of all equipment to minimize the hazards that might cause complete failure due to floods, fires, icing, and vandalism.

Informational Note: For further information, see ANSI/IEEE 493-2007, *Recommended Practice for the Design of Reliable Industrial and Commercial Power Systems*.

**(A) Storage Battery.** A storage battery shall be of suitable rating and capacity to supply and maintain at not less than 87½ percent of system voltage the total load of the circuits supplying legally required standby power for a period of at least 1½ hours.

Batteries, whether of the acid or alkali type, shall be designed and constructed to meet the service requirements of emergency service and shall be compatible with the charger for that particular installation.

For a sealed battery, the container shall not be required to be transparent. However, for the lead acid battery that requires water additions, transparent or translucent containers shall be furnished. Automotive-type batteries shall not be used.

An automatic battery charging means shall be provided.

**(B) Generator Set.**

**(1) Prime Mover-Driven.** For a generator set driven by a prime mover acceptable to the authority having jurisdiction and sized in accordance with 701.4, means shall be provided for automatically starting the prime mover upon failure of the normal service and for automatic transfer and operation of all required electrical circuits. A time-delay feature permitting a 15-minute setting shall be provided to avoid retransfer in case of short-time re-establishment of the normal source.

**(2) Internal Combustion Engines as Prime Mover.** Where internal combustion engines are used as the prime mover, an on-site fuel supply shall be provided with an on-premises fuel supply sufficient for not less than 2 hours of full-demand operation of the system. Where power is needed for the operation of the fuel transfer pumps to deliver fuel to a generator set day tank, the pumps shall be connected to the legally required standby power system.

When power is needed for the operation of the fuel transfer pumps, they must be connected to the legally required standby system for the continued delivery of fuel.

**(3) Dual Supplies.** Prime movers shall not be solely dependent on a public utility gas system for their fuel supply or on a municipal water supply for their cooling systems. Means shall be provided for automatically transferring one fuel supply to another where dual fuel supplies are used.

*Exception: Where acceptable to the authority having jurisdiction, the use of other than on-site fuels shall be permitted where there is a low probability of a simultaneous failure of both the off-site fuel delivery system and power from the outside electrical utility company.*

**(4) Battery Power.** Where a storage battery is used for control or signal power or as the means of starting the prime mover, it shall be suitable for the purpose and shall be equipped with an automatic charging means independent of the generator set.

**(5) Outdoor Generator Sets.** Where an outdoor housed generator set is equipped with a readily accessible disconnecting means in accordance with 445.18, and the disconnecting means is located within sight of the building or structure supplied, an additional disconnecting means shall not be required where ungrounded conductors serve or pass through the building or structure. Where the generator supply conductors terminate at a disconnecting means in or on a building or structure, the disconnecting means shall meet the requirements of 225.36.

The disconnecting means on an outdoor generator set can be used as the disconnecting means required in 225.31, provided the disconnecting means is readily accessible and is within sight of the building. When an additional disconnecting means is necessary, it must be suitable for use as service equipment in accordance with 225.36. See the definitions of the terms *readily accessible* and *in sight from* in Article 100.

**(C) Uninterruptible Power Supplies.** Uninterruptible power supplies used to provide power for legally required standby systems shall comply with the applicable provisions of 701.12(A) and (B).

**(D) Separate Service.** Where approved, a separate service shall be permitted as a legally required source of standby power. This service shall be in accordance with the applicable provisions of Article 230, with a separate service drop or lateral or a separate set of overhead or underground service conductors sufficiently remote electrically and physically from any other service to minimize the possibility of simultaneous interruption of supply from an occurrence in another service.

**(E) Connection Ahead of Service Disconnecting Means.** Where acceptable to the authority having jurisdiction, connections located ahead of and not within the same cabinet, enclosure, vertical switchgear section, or vertical switchboard section as the service disconnecting means shall be permitted.

The legally required standby service shall be sufficiently separated from the normal main service disconnecting means to minimize simultaneous interruption of supply through an occurrence within the building or groups of buildings served.

Informational Note: See 230.82 for equipment permitted on the supply side of a service disconnecting means.

Where a legally required standby system is supplied by conductors tapped to the normal service conductors (connection required to be on line side of the normal service disconnecting means), 230.82 requires that the tapped conductors be installed in accordance with all of the requirements for service-entrance conductors and that the conductors terminate in equipment suitable for use as service equipment. These requirements help ensure that the legally required standby system disconnecting means can safely interrupt the fault current available from the utility and that the tapped conductors, which do not have short-circuit and ground-fault protection, are not run through the interior of a building.

**(F) Fuel Cell System.** Fuel cell systems used as a source of power for legally required standby systems shall be of suitable rating and capacity to supply and maintain the total load for not less than 2 hours of full-demand operation.

Installation of a fuel cell system shall meet the requirements of Parts II through VIII of Article 692.

Where a single fuel cell system serves as the normal supply for the building or group of buildings concerned, it shall not serve as the sole source of power for the legally required standby system.

**(G) Unit Equipment.** Individual unit equipment for legally required standby illumination shall consist of the following:

(1) A rechargeable battery
(2) A battery charging means
(3) Provisions for one or more lamps mounted on the equipment and shall be permitted to have terminals for remote lamps
(4) A relaying device arranged to energize the lamps automatically upon failure of the supply to the unit equipment

The batteries shall be of suitable rating and capacity to supply and maintain at not less than 87½ percent of the nominal battery voltage for the total lamp load associated with the unit for a period of at least 1½ hours, or the unit equipment shall supply and maintain not less than 60 percent of the initial legally required standby illumination for a period of at least 1½ hours. Storage batteries, whether of the acid or alkali type, shall be designed and constructed to meet the requirements of emergency service.

Unit equipment shall be permanently fixed in place (i.e., not portable) and shall have all wiring to each unit installed in accordance with the requirements of any of the wiring methods in Chapter 3. Flexible cord-and-plug connection shall be permitted, provided that the cord does not exceed 900 mm (3 ft) in length. The branch circuit feeding the unit equipment shall be the same branch circuit as that serving the normal lighting in the area and connected ahead of any local switches. Legally required standby luminaires that obtain power from a unit equipment and are not part of the unit equipment shall be wired to the unit equipment by one of the wiring methods of Chapter 3.

*Exception: In a separate and uninterrupted area supplied by a minimum of three normal lighting circuits, a separate branch circuit for unit equipment shall be permitted if it originates from the same panelboard as that of the normal lighting circuits and is provided with a lock-on feature.*

## IV. Overcurrent Protection

### 701.25 Accessibility

The branch-circuit overcurrent devices in legally required standby circuits shall be accessible to authorized persons only.

### 701.26 Ground-Fault Protection of Equipment

The alternate source for legally required standby systems shall not be required to have ground-fault protection of equipment with automatic disconnecting means. Ground-fault indication of the legally required standby source shall be provided in accordance with 701.6(D) if ground-fault protection of equipment with automatic disconnecting means is not provided.

### 701.27 Selective Coordination

Legally required standby system(s) overcurrent devices shall be selectively coordinated with all supply-side overcurrent protective devices.

Selective coordination shall be selected by a licensed professional engineer or other qualified persons engaged primarily in the design, installation, or maintenance of electrical systems. The selection shall be documented and made available to those authorized to design, install, inspect, maintain, and operate the system.

*Exception: Selective coordination shall not be required between two overcurrent devices located in series if no loads are connected in parallel with the downstream device.*

See the commentary regarding selective coordination for emergency systems following 700.28.

# ARTICLE 702
# Optional Standby Systems

## I. General

### 702.1 Scope

The provisions of this article apply to the installation and operation of optional standby systems.

***EXHIBIT 702.1*** *A trailer (vehicle)-mounted portable generator. (Courtesy of the International Association of Electrical Inspectors)*

The systems covered by this article consist of those that are permanently installed in their entirety, including prime movers, and those that are arranged for a connection to a premises wiring system from a portable alternate power supply.

Article 702 applies not only to permanently installed generators and prime movers but also to portable alternate power supplies that can be connected to an optional standby system. For example, upon failure of an optional standby generator at a frozen food processing plant, a vehicle-mounted generator can be brought in and connected to the plant's optional standby system, which has provisions for such a connection. Events such as carnivals and fairs require the use of portable generators. See Exhibit 702.1.

Optional standby systems are those that upon failing can cause physical discomfort, interruption of an industrial process, damage to process equipment, or disruption of business, whereas emergency systems (see Article 700) are those systems essential for safety to life and legally required standby systems (see Article 701) are intended to provide electric power to aid fire fighting, rescue operations, control of health hazards, and similar operations.

The installation of optional standby systems at one-family dwelling units has increased in recent years. Spikes in the increased use of these systems can be directly related to natural disasters, resulting in prolonged outages of utility power.

## 702.2 Definition

**Optional Standby Systems.** Those systems intended to supply power to public or private facilities or property where life safety does not depend on the performance of the system. These systems are intended to supply on-site generated power to selected loads either automatically or manually.

Informational Note: Optional standby systems are typically installed to provide an alternate source of electric power for such facilities as industrial and commercial buildings, farms, and residences and to serve loads such as heating and refrigeration systems, data processing and communications systems, and industrial processes that, when stopped during any power outage, could cause discomfort, serious interruption of the process, damage to the product or process, or the like.

## 702.4 Capacity and Rating

**(A) Available Short-Circuit Current.** Optional standby system equipment shall be suitable for the maximum available short-circuit current at its terminals.

**(B) System Capacity.** The calculations of load on the standby source shall be made in accordance with Article 220 or by another approved method.

**(1) Manual Transfer Equipment.** Where manual transfer equipment is used, an optional standby system shall have adequate capacity and rating for the supply of all equipment intended to be operated at one time. The user of the optional standby system shall be permitted to select the load connected to the system.

**(2) Automatic Transfer Equipment.** Where automatic transfer equipment is used, an optional standby system shall comply with (2)(a) or (2)(b).

(a) *Full Load.* The standby source shall be capable of supplying the full load that is transferred by the automatic transfer equipment.

(b) *Load Management.* Where a system is employed that will automatically manage the connected load, the standby source shall have a capacity sufficient to supply the maximum load that will be connected by the load management system.

The standby source must have the capacity to supply all of the loads connected to it, unless an automatic load management system is used to ensure that the transferred load does not overload the source. This requirement applies only to systems where the switching between power sources occurs automatically.

## 702.5 Transfer Equipment

Transfer equipment shall be suitable for the intended use and designed and installed so as to prevent the inadvertent interconnection of normal and alternate sources of supply in any operation of the transfer equipment. Transfer equipment and electric power production systems installed to permit operation in parallel with the normal source shall meet the requirements of Article 705.

Transfer equipment, located on the load side of branch circuit protection, shall be permitted to contain supplemental overcurrent protection having an interrupting rating sufficient for the available fault current that the generator can deliver. The supplementary overcurrent protection devices shall be part of a listed transfer equipment.

Transfer equipment shall be required for all standby systems subject to the provisions of this article and for which an electric utility supply is either the normal or standby source.

Parallel operation of the generation equipment with the normal source is permitted as long as the requirements of Article 705 are met. Traditional automatic transfer switches (ATS) are not designed to permit parallel operation of generation equipment and the normal source. Therefore, traditional ATS need not comply with Article 705. However, certain ATS configurations are intentionally designed to briefly (for a few cycles) parallel the generation equipment with the normal source upon load transfer from generator to normal source. This load transfer can occur with minimal disturbance or effect on the load. Transfer switches that employ this type of paralleling must comply with Article 705.

*Exception: Temporary connection of a portable generator without transfer equipment shall be permitted where conditions of maintenance and supervision ensure that only qualified persons service the installation and where the normal supply is physically isolated by a lockable disconnecting means or by disconnection of the normal supply conductors.*

The exception provides requirements for the connection of loads to a generator without the use of a transfer switch. Such applications often occur when necessary for equipment maintenance or when an extended power outage occurs. In such instances, a portable generator can be brought to a facility and connected to the existing distribution system. Supervision by qualified personnel is critical to ensuring that a dangerous backfeed condition is not created by connecting the generator to the system without the benefit of transfer equipment.

### 702.6 Signals

Audible and visual signal devices shall be provided, where practicable, for the following purposes:

**(1) Derangement.** To indicate derangement of the optional standby source.

**(2) Carrying Load.** To indicate that the optional standby source is carrying load.

*Exception: Signals shall not be required for portable standby power sources.*

### 702.7 Signs

**(A) Standby.** A sign shall be placed at the service-entrance equipment that indicates the type and location of on-site optional standby power sources. A sign shall not be required for individual unit equipment for standby illumination.

**(B) Grounding.** Where removal of a grounding or bonding connection in normal power source equipment interrupts the grounding electrode conductor connection to the alternate power source(s) grounded conductor, a warning sign shall be installed at the normal power source equipment stating:

WARNING
SHOCK HAZARD EXISTS IF GROUNDING ELECTRODE CONDUCTOR OR BONDING JUMPER CONNECTION IN THIS EQUIPMENT IS REMOVED WHILE ALTERNATE SOURCE(S) IS ENERGIZED.

The warning sign(s) or label(s) shall comply with 110.21(B).

See the commentary following 700.7(B). Removal of grounding and bonding connections presents the same hazard in the normal supply equipment for optional standby systems as in the normal supply equipment for an emergency system.

**(C) Power Inlet.** Where a power inlet is used for a temporary connection to a portable generator, a warning sign shall be placed near the inlet to indicate the type of derived system that the system is capable of based on the wiring of the transfer equipment. The sign shall display one of the following warnings:

WARNING:
FOR CONNECTION OF A SEPARATELY DERIVED (BONDED NEUTRAL) SYSTEM ONLY

or

WARNING:
FOR CONNECTION OF A NONSEPARATELY DERIVED (FLOATING NEUTRAL) SYSTEM ONLY

## II. Wiring

### 702.10 Wiring Optional Standby Systems

The optional standby system wiring shall be permitted to occupy the same raceways, cables, boxes, and cabinets with other general wiring.

### 702.11 Portable Generator Grounding

**(A) Separately Derived System.** Where a portable optional standby source is used as a separately derived system, it shall be grounded to a grounding electrode in accordance with 250.30.

**(B) Nonseparately Derived System.** Where a portable optional standby source is used as a nonseparately derived system, the equipment grounding conductor shall be bonded to the system grounding electrode.

### 702.12 Outdoor Generator Sets

**(A) Permanently Installed Generators and Portable Generators Greater Than 15 kW.** Where an outdoor housed generator set is equipped with a readily accessible disconnecting means in accordance with 445.18, and the disconnecting means is located within sight of the building or structure supplied, an additional disconnecting means shall not be required where ungrounded conductors serve or pass through the building or structure. Where the generator supply conductors terminate at a disconnecting means in or on a building or structure, the disconnecting means shall meet the requirements of 225.36.

The disconnecting means on an outdoor generator set can be used as the disconnecting means required in 225.31, provided the disconnecting means is readily accessible and is within sight of the building. When an additional disconnecting means is necessary, it must be suitable for use as service equipment in accordance with 225.36.

**(B) Portable Generators 15 kW or Less.** Where a portable generator, rated 15 kW or less, is installed using a flanged inlet or other cord- and plug-type connection, a disconnecting means shall not be required where ungrounded conductors serve or pass through a building or structure.

# ARTICLE 705 Interconnected Electric Power Production Sources

## I. General

### 705.1 Scope

This article covers installation of one or more electric power production sources operating in parallel with a primary source(s) of electricity.

Informational Note: Examples of the types of primary sources include a utility supply or an on-site electric power source(s).

Article 705 sets forth basic safety requirements for interconnecting generators and other types of power production sources that operate in parallel as distributed generation. Power sources include any systems that produce electric power, including electric utility sources and on-premises sources ranging from rotating generators (see Article 445) to solar photovoltaic systems (see Article 690) to fuel cells (see Article 692) to wind power systems (see Article 694).

Article 705 addresses the basic safety requirements specifically related to parallel operation for the generators and other power sources, the power system that interconnects the power sources, and the equipment connected to these systems.

### 705.2 Definitions

•

**Multimode Inverter.** Equipment having the capabilities of both the utility-interactive inverter and the stand-alone inverter.

•

**Power Production Equipment.** The generating source, and all distribution equipment associated with it, that generates electricity from a source other than a utility supplied service.

Informational Note: Examples of power production equipment include such items as generators, solar photovoltaic systems, and fuel cell systems.

The power production equipment covered by this definition does not include the utility-supplied service.

**Utility-Interactive Inverter Output Circuit.** The conductors between the utility interactive inverter and the service equipment or another electric power production source, such as a utility, for electrical production and distribution network.

### 705.3 Other Articles

Interconnected electric power production sources shall comply with this article and also with the applicable requirements of the articles in Table 705.3.

### 705.4 Equipment Approval

All equipment shall be approved for the intended use. Utility-interactive inverters for interconnection systems shall be listed and identified for interconnection service.

### 705.6 System Installation

Installation of one or more electrical power production sources operating in parallel with a primary source(s) of electricity shall be installed only by qualified persons.

Interconnected power production sources introduce hazards unique to systems operating in parallel. A qualified person must have the skills and knowledge to recognize the hazards associated with these systems. Special training for persons working on interconnected systems is key to ensuring that personnel can work safely on these systems.

Informational Note: See Article 100 for the definition of *Qualified Person.*

***TABLE 705.3*** *Other Articles*

| Equipment/System | Article |
|---|---|
| Generators | 445 |
| Solar photovoltaic systems | 690 |
| Fuel cell systems | 692 |
| Wind electric systems | 694 |
| Emergency systems | 700 |
| Legally required standby systems | 701 |
| Optional standby systems | 702 |

### 705.10 Directory

A permanent plaque or directory, denoting all electric power sources on or in the premises, shall be installed at each service equipment location and at locations of all electric power production sources capable of being interconnected.

*Exception: Installations with large numbers of power production sources shall be permitted to be designated by groups.*

### 705.12 Point of Connection

The output of an interconnected electric power source shall be connected as specified in 705.12(A), (B), (C), or (D).

**(A) Supply Side.** An electric power production source shall be permitted to be connected to the supply side of the service disconnecting means as permitted in 230.82(6). The sum of the ratings of all overcurrent devices connected to power production sources shall not exceed the rating of the service.

**(B) Integrated Electrical Systems.** The outputs shall be permitted to be interconnected at a point or points elsewhere on the premises where the system qualifies as an integrated electrical system and incorporates protective equipment in accordance with all applicable sections of Article 685.

**(C) Greater Than 100 kW.** The outputs shall be permitted to be interconnected at a point or points elsewhere on the premises where all of the following conditions are met:

(1) The aggregate of non-utility sources of electricity has a capacity in excess of 100 kW, or the service is above 1000 volts.
(2) The conditions of maintenance and supervision ensure that qualified persons service and operate the system.
(3) Safeguards, documented procedures, and protective equipment are established and maintained.

Electric power production sources are permitted to be connected on the supply side of the utility-supplied service disconnecting means, or the source can be connected on the load side. This practice accommodates the safe work practices of many utilities, which provide a readily accessible disconnect for distributed generation. (See Exhibit 705.1.)

Sections 705.12(B) and (C) recognize that generators and other power sources can be safely connected elsewhere on the premises system. These locations include where the premises has an integrated electrical system as set forth in Article 685, where the total generator capacity on premises is greater than 100 kW, and where the service is greater than 1000 volts.

**(D) Utility-Interactive Inverters.** The output of a utility-interactive inverter shall be permitted to be connected to the load side of the service disconnecting means of the other source(s) at any distribution equipment on the premises. Where distribution equipment, including switchgear, switchboards, or panelboards, is fed simultaneously by a primary source(s) of electricity and one or more utility-interactive inverters, and where this distribution equipment is capable of supplying multiple branch circuits or feeders, or both, the interconnecting provisions for the utility-interactive inverter(s) shall comply with 705.12(D)(1) through (D)(6).

**(1) Dedicated Overcurrent and Disconnect.** The source interconnection of one or more inverters installed in one system shall be made at a dedicated circuit breaker or fusible disconnecting means.

**(2) Bus or Conductor Ampere Rating.** One hundred twenty-five percent of the inverter output circuit current shall be used in ampacity calculations for the following:

(1) *Feeders.* Where the inverter output connection is made to a feeder at a location other than the opposite end of the feeder from the primary source overcurrent device, that portion of the feeder on the load side of the inverter output connection shall be protected by one of the following:
   (a) The feeder ampacity shall be not less than the sum of the primary source overcurrent device and 125 percent of the inverter output circuit current.
   (b) An overcurrent device on the load side of the inverter connection shall be rated not greater than the ampacity of the feeder.
(2) *Taps.* In systems where inverter output connections are made at feeders, any taps shall be sized based on the sum of 125 percent of the inverter(s) output circuit current and the rating of the overcurrent device protecting the feeder conductors as calculated in 240.21(B).
(3) *Busbars.* One of the methods that follows shall be used to determine the ratings of busbars in panelboards.

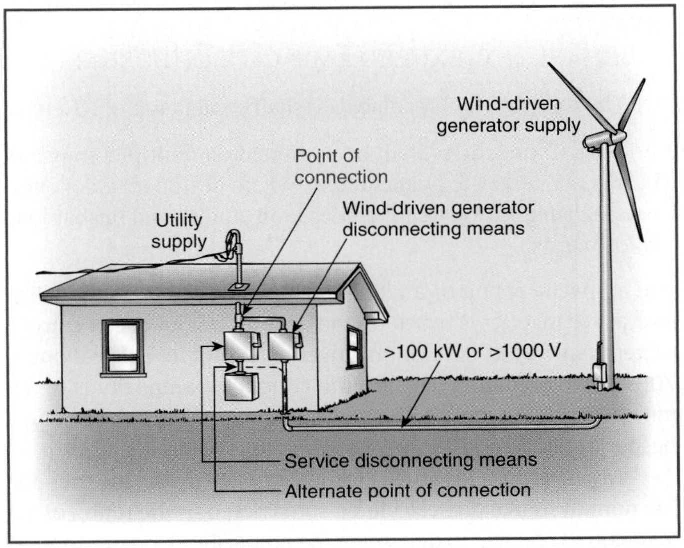


***EXHIBIT 705.1*** *The points of interconnection permitted by 705.12(A), (B), and (C).*

(a) The sum of 125 percent of the inverter(s) output circuit current and the rating of the overcurrent device protecting the busbar shall not exceed the ampacity of the busbar.

Informational Note: This general rule assumes no limitation in the number of the loads or sources applied to busbars or their locations.

(b) Where two sources, one a utility and the other an inverter, are located at opposite ends of a busbar that contains loads, the sum of 125 percent of the inverter(s) output circuit current and the rating of the overcurrent device protecting the busbar shall not exceed 120 percent of the ampacity of the busbar. The busbar shall be sized for the loads connected in accordance with Article 220. A permanent warning label shall be applied to the distribution equipment adjacent to the back-fed breaker from the inverter that displays the following or equivalent wording:

WARNING:
INVERTER OUTPUT CONNECTION;
DO NOT RELOCATE THIS OVERCURRENT DEVICE.

The warning sign(s) or label (s) shall comply with 110.21(B).

(c) The sum of the ampere ratings of all overcurrent devices on panelboards, both load and supply devices, excluding the rating of the overcurrent device protecting the busbar, shall not exceed the ampacity of the busbar. The rating of the overcurrent device protecting the busbar shall not exceed the rating of the busbar. Permanent warning labels shall be applied to distribution equipment that displays the following or equivalent wording:

WARNING:
THIS EQUIPMENT FED BY MULTIPLE SOURCES.
TOTAL RATING OF ALL OVERCURRENT DEVICES,
EXCLUDING MAIN SUPPLY OVERCURRENT DEVICE,
SHALL NOT EXCEED AMPACITY OF BUSBAR.

The warning sign(s) or label (s) shall comply with 110.21(B).

(d) Connections shall be permitted on multiple-ampacity busbars or center-fed panelboards where designed under engineering supervision that includes fault studies and busbar load calculations.

The ampacity rating of a feeder or bus connected to the utility-interactive inverter is based on the inverter output circuit current, rather than on the OCPD in the inverter. Except for calculations in 705.12(D)(2)(1)(b) and (D)(3)(c), the conductor ampacity is determined by adding the ampacity of the primary OCPD protecting a busbar or feeder and 125 percent of inverter output current.

Where a tap is made to a feeder supplied by the inverter and the normal source, the calculated sum is used as the rating of the "overcurrent device" to determine the ampacity of the tap conductors in 240.21(B).

Unlike in service equipment where the number or rating of overcurrent devices is not limited, 705.12(D)(3)(c) places a limit on panelboard overcurrent devices. The sum of the ratings of all overcurrent devices (excluding the main overcurrent device) supplying and/or being supplied by the panelboard is limited to the busbar rating. In addition, the main overcurrent device must also be limited to the ampacity of busbar.

•

**(3) Marking.** Equipment containing overcurrent devices in circuits supplying power to a busbar or conductor supplied from multiple sources shall be marked to indicate the presence of all sources.

**(4) Suitable for Backfeed.** Circuit breakers, if backfed, shall be suitable for such operation.

Informational Note: Fused disconnects, unless otherwise marked, are suitable for backfeeding.

**(5) Fastening.** Listed plug-in-type circuit breakers backfed from utility-interactive inverters that are listed and identified as interactive shall be permitted to omit the additional fastener normally required by 408.36(D) for such applications.

•

**(6) Wire Harness and Exposed Cable Arc-Fault Protection.** A utility-interactive inverter(s) that has a wire harness or cable output circuit rated 240 V, 30 amperes, or less, that is not installed within an enclosed raceway, shall be provided with listed ac AFCI protection.

## 705.14 Output Characteristics

The output of a generator or other electric power production source operating in parallel with an electrical supply system shall be compatible with the voltage, wave shape, and frequency of the system to which it is connected.

Informational Note: The term *compatible* does not necessarily mean matching the primary source wave shape.

Control of the generator or power production source should include real power, reactive power, and harmonic content of the output. The interconnected equipment must be compatible with the electric supply system in voltage, wave shape, and frequency. The output characteristics of a rotating generator are significantly different from those of a solid-state power source. Compatibility with other sources and with different types of loads is limited in different ways. Control of the driver speed causes real power (kW) to flow from an induction generator, whereas control of the prime mover torque causes real power (kW) to flow from a synchronous generator. Control of voltage causes reactive power (kVAr) to flow to or from a synchronous generator, while induction generators have no means to control reactive power (kVAr) flow and continuously draw reactive power.

The parallel operation of generators is a complex balance of several variables that are design parameters and therefore beyond the scope of the *Code*.

Some inverters, uninterruptible power supplies (UPS), or solid-state variable-speed drives may produce harmonic currents.

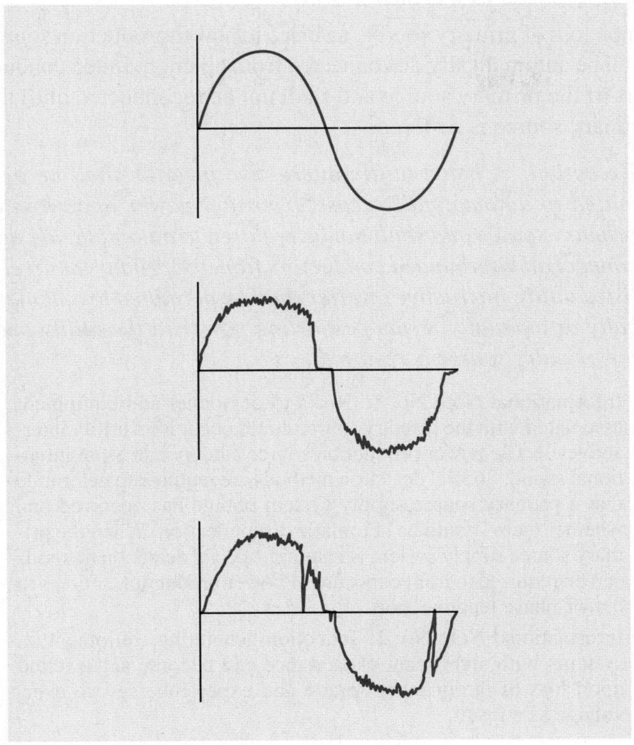

***EXHIBIT 705.2*** *Typical output wave shapes: (top) with rotating generator and system wave shape normally encountered with motor, lighting, and heating loads; (middle) with inverter source; and (bottom) with variable speed drive, rectifier, and uninterruptible power to supply loads.*

(See Exhibit 705.2.) The multiples of the basic supply frequency (usually 60 Hz) can cause additional heating, which may require derating of generators, transformers, cables, and motors. Special generator voltage control systems are required to avoid erratic operation or destruction of control devices. Circuit breakers may require derating if the higher harmonics become significant.

In Exhibit 705.2 (middle), motors and transformers will be driven by harmonic-rich voltage and may require derating. In Exhibit 705.2 (bottom), the source generator may require derating, and special voltage control may be needed.

Significant magnitudes of harmonics may be inadvertently matched to system resonance and result in the opening of capacitor fuses, overheating of circuits, and erratic operation of controls. The usual solution is to detune the system by rearrangement, installation of reactors, or both.

## 705.16 Interrupting and Short-Circuit Current Rating

Consideration shall be given to the contribution of fault currents from all interconnected power sources for the interrupting and short-circuit current ratings of equipment on interactive systems.

## 705.20 Disconnecting Means, Sources

Means shall be provided to disconnect all ungrounded conductors of an electric power production source(s) from all other conductors.

## 705.21 Disconnecting Means, Equipment

Means shall be provided to disconnect power production equipment, such as utility interactive inverters or transformers associated with a power production source, from all ungrounded conductors of all sources of supply. Equipment intended to be operated and maintained as an integral part of a power production source exceeding 1000 volts shall not be required to have a disconnecting means.

## 705.22 Disconnect Device

The disconnecting means for ungrounded conductors shall consist of a manually or power operable switch(es) or circuit breaker(s) with the following features:

(1) Located where readily accessible
(2) Externally operable without exposing the operator to contact with live parts and, if power operable, of a type that could be opened by hand in the event of a power-supply failure
(3) Plainly indicating whether in the open (off) or closed (on) position
(4) Having ratings not less than the load to be carried and the fault current to be interrupted. For disconnect equipment energized from both sides, a marking shall be provided to indicate that all contacts of the disconnect equipment might be energized.

Informational Note to (4): In parallel generation systems, some equipment, including knife blade switches and fuses, is likely to be energized from both directions. See 240.40.

(5) Simultaneous disconnect of all ungrounded conductors of the circuit
(6) Capable of being locked in the open (off) position

Each generating source must have a disconnecting means. This disconnecting means must be suitable for use as service equipment. Still another disconnecting means may be applied to separate the grounding systems.

The basic requirement in 705.22 recognizes the success of applying switches as well as circuit breakers as the disconnecting means for ungrounded conductors. Most safe work practices on the premises use these disconnect devices.

The disconnect at the service entrance is required for disconnecting the premises wiring system from the utility. The utility safe work practices may also use this disconnect device.

## 705.30 Overcurrent Protection

Conductors shall be protected in accordance with Article 240. Equipment and conductors connected to more than one electrical source shall have a sufficient number of overcurrent devices located so as to provide protection from all sources.

**(A) Solar Photovoltaic Systems.** Solar photovoltaic systems shall be protected in accordance with Article 690.

**(B) Transformers.** Overcurrent protection for a transformer with a source(s) on each side shall be provided in accordance with 450.3 by considering first one side of the transformer, then the other side of the transformer, as the primary.

**(C) Fuel Cell Systems.** Fuel cell systems shall be protected in accordance with Article 692.

**(D) Utility-Interactive Inverters.** Utility-interactive inverters shall be protected in accordance with 705.65.

**(E) Generators.** Generators shall be protected in accordance with 705.130.

## 705.31 Location of Overcurrent Protection

Overcurrent protection for electric power production source conductors, connected to the supply side of the service disconnecting means in accordance with 705.12(A), shall be located within 3 m (10 ft) of the point where the electric power production source conductors are connected to the service.

Informational Note: This overcurrent protection protects against short-circuit current supplied from the primary source(s) of electricity.

*Exception: Where the overcurrent protection for the power production source is located more than 3 m (10 ft) from the point of connection for the electric power production source to the service, cable limiters or current-limited circuit breakers for each ungrounded conductor shall be installed at the point where the electric power production conductors are connected to the service.*

## 705.32 Ground-Fault Protection

Where ground-fault protection is used, the output of an interactive system shall be connected to the supply side of the ground-fault protection.

*Exception: Connection shall be permitted to be made to the load side of ground-fault protection, provided that there is ground-fault protection for equipment from all ground-fault current sources.*

Large optional standby systems (see Article 702) are sophisticated and may involve many special protective relays. Protective relays include under- and overvoltage, under- and overfrequency, voltage-restrained overcurrent, anti-motoring, loss of excitation, overtemperature, and shutdown for derangement of the mechanical driver.

Small generator installations cannot justify the cost of protective relays. Therefore, small generator protection comprises more common devices with fewer features. Application guides for relay protection are available in manufacturers' technical literature.

## 705.40 Loss of Primary Source

Upon loss of primary source, an electric power production source shall be automatically disconnected from all ungrounded conductors of the primary source and shall not be reconnected until the primary source is restored.

*Exception: A listed utility-interactive inverter shall be permitted to automatically cease exporting power upon loss of primary source and shall not be required to automatically disconnect all ungrounded conductors from the primary source. A listed utility-interactive inverter shall be permitted to automatically or manually resume exporting power to the utility once the primary source is restored.*

Informational Note No. 1: Risks to personnel and equipment associated with the primary source could occur if an utility interactive electric power production source can operate as an intentional island. Special detection methods are required to determine that a primary source supply system outage has occurred and whether there should be automatic disconnection. When the primary source supply system is restored, special detection methods can be required to limit exposure of power production sources to out-of-phase reconnection.

Informational Note No. 2: Induction-generating equipment on systems with significant capacitance can become self-excited upon loss of the primary source and experience severe overvoltage as a result.

A utility-interactive inverter shall be permitted to operate as a stand-alone system to supply loads that have been disconnected from electrical production and distribution network sources.

When two interconnected power systems separate, they can drift out of synchronism. When the utility and the interconnected power system separate, there is a risk of damage to the system if restoration of the utility occurs out of phase. If the timing of the reconnection is random, violent electromechanical stresses can destroy mechanical components such as gears, couplings, and shafts and can displace coils. Therefore, the premises wiring system or generator must be disconnected from the primary source.

Many technical guides are available from which to select appropriate protective systems and equipment for large systems. A limited choice of low-cost devices is available for application to small systems. Induction generators are commonly used. Induction generators have characteristics quite different from those of synchronous machines. They are more rugged because of the construction of the rotor, and they are less expensive because of their basic design, availability, and type of starting and control equipment. Theoretically, an induction machine can continue to run on an isolated system if a large capacitor bank provides excitation. In reality, an induction machine will probably lose stability and be shut down quickly by one of the protective devices.

## 705.42 Loss of 3-Phase Primary Source

A 3-phase electric power production source shall be automatically disconnected from all ungrounded conductors of the interconnected systems when one of the phases of that source opens. This requirement shall not be applicable to an electric power production source providing power for an emergency or legally required standby system.

*Exception: A listed utility-interactive inverter shall be permitted to automatically cease exporting power when one of the phases of the source opens and shall not be required to automatically disconnect all ungrounded conductors from the primary source. A listed utility-interactive inverter shall be permitted to automatically or manually resume exporting power to the utility once all phases of the source are restored.*

### 705.50 Grounding

Interconnected electric power production sources shall be grounded in accordance with Article 250.

*Exception: For direct-current systems connected through an inverter directly to a grounded service, other methods that accomplish equivalent system protection and that utilize equipment listed and identified for the use shall be permitted.*

## II. Utility-Interactive Inverters

### 705.60 Circuit Sizing and Current

**(A) Calculation of Maximum Circuit Current.** The maximum current for the specific circuit shall be calculated in accordance with 705.60 (A)(1) and (A)(2).

**(1) Inverter Input Circuit Currents.** The maximum current shall be the maximum rated input current of the inverter.

**(2) Inverter Output Circuit Current.** The maximum current shall be the inverter continuous output current rating.

**(B) Ampacity and Overcurrent Device Ratings.** Inverter system currents shall be considered to be continuous. The circuit conductors and overcurrent devices shall be sized to carry not less than 125 percent of the maximum currents as calculated in 705.60(A). The rating or setting of overcurrent devices shall be permitted in accordance with 240.4(B) and (C).

*Exception: Circuits containing an assembly together with its overcurrent device(s) that is listed for continuous operation at 100 percent of its rating shall be permitted to be utilized at 100 percent of its rating.*

### 705.65 Overcurrent Protection

**(A) Circuits and Equipment.** Inverter input circuits, inverter output circuits, and storage battery circuit conductors and equipment shall be protected in accordance with the requirements of Article 240. Circuits connected to more than one electrical source shall have overcurrent devices located so as to provide overcurrent protection from all sources.

*Exception: An overcurrent device shall not be required for circuit conductors sized in accordance with 705.60(B) and located where one of the following applies:*

*(1) There are no external sources such as parallel-connected source circuits, batteries, or backfeed from inverters.*
*(2) The short-circuit currents from all sources do not exceed the ampacity of the conductors.*

Informational Note: Possible backfeed of current from any source of supply, including a supply through an inverter into the inverter output circuit and inverter source circuits, is a consideration in determining whether adequate overcurrent protection from all sources is provided for conductors and modules.

**(B) Power Transformers.** Overcurrent protection for a transformer with a source(s) on each side shall be provided in accordance with 450.3 by considering first one side of the transformer, then the other side of the transformer, as the primary.

*Exception: A power transformer with a current rating on the side connected toward the utility-interactive inverter output that is not less than the rated continuous output current of the inverter shall be permitted without overcurrent protection from that source.*

### 705.70 Utility-Interactive Inverters Mounted in Not-Readily-Accessible Locations

Utility-interactive inverters shall be permitted to be mounted on roofs or other exterior areas that are not readily accessible. These installations shall comply with (1) through (4):

(1) A direct-current disconnecting means shall be mounted within sight of or in the inverter.
(2) An alternating-current disconnecting means shall be mounted within sight of or in the inverter.
(3) An additional alternating-current disconnecting means for the inverter shall comply with 705.22.
(4) A plaque shall be installed in accordance with 705.10.

### 705.80 Utility-Interactive Power Systems Employing Energy Storage

Utility-interactive power systems employing energy storage shall also be marked with the maximum operating voltage, including any equalization voltage, and the polarity of the grounded circuit conductor.

### 705.82 Hybrid Systems

Hybrid systems shall be permitted to be interconnected with utility-interactive inverters.

### 705.95 Ampacity of Neutral Conductor

The ampacity of the neutral conductors shall comply with either (A) or (B).

**(A) Neutral Conductor for Single Phase, 2-Wire Inverter Output.** If a single-phase, 2-wire inverter output is connected to the neutral and one ungrounded conductor (only) of a 3-wire system or of a 3-phase, 4-wire, wye-connected system, the maximum load connected between the neutral and any one ungrounded conductor plus the inverter output rating shall not exceed the ampacity of the neutral conductor.

**(B) Neutral Conductor for Instrumentation, Voltage Detection or Phase Detection.** A conductor used solely for instrumentation, voltage detection, or phase detection and connected to a single-phase or 3-phase utility-interactive inverter, shall be permitted to be sized at less than the ampacity of the other current-carrying conductors and shall be sized equal to or larger than the equipment grounding conductor.

The operation of the inverter in the presence of load currents tends to decrease currents in the neutral. If a neutral conductor is associated with a circuit supplying both power and instrumentation loads, 705.95(A) applies to sizing of the neutral conductor.

### 705.100 Unbalanced Interconnections

**(A) Single Phase.** Single-phase inverters for hybrid systems and ac modules in interactive hybrid systems shall be connected to three-phase power systems in order to limit unbalanced voltages to not more than 3 percent.

Informational Note: For utility-interactive single-phase inverters, unbalanced voltages can be minimized by the same methods that are used for single-phase loads on a three-phase power system. See ANSI/C84.1-2011, *Electric Power Systems and Equipment — Voltage Ratings (60 Hertz).*

**(B) Three Phase.** Three-phase inverters and 3-phase ac modules in interactive systems shall have all phases automatically de-energized upon loss of, or unbalanced, voltage in one or more phases unless the interconnected system is designed so that significant unbalanced voltages will not result.

## III. Generators

### 705.130 Overcurrent Protection

Conductors shall be protected in accordance with Article 240. Equipment and conductors connected to more than one electrical source shall have overcurrent devices located so as to provide protection from all sources. Generators shall be protected in accordance with 445.12.

### 705.143 Synchronous Generators

Synchronous generators in a parallel system shall be provided with the necessary equipment to establish and maintain a synchronous condition.

# ARTICLE 708 Critical Operations Power Systems (COPS)

Informational Note: Text that is followed by a reference in brackets has been extracted from *NFPA 1600-2013*, *Standard on Disaster/Emergency Management and Business Continuity Programs*. Only editorial changes were made to the extracted text to make it consistent with this *Code*.

## I. General

### 708.1 Scope

The provisions of this article apply to the installation, operation, monitoring, control, and maintenance of the portions of the premises wiring system intended to supply, distribute, and control electricity to designated critical operations areas (DCOA) in the event of disruption to elements of the normal system.

Critical operations power systems are those systems so classed by municipal, state, federal, or other codes by any governmental agency having jurisdiction or by facility engineering documentation establishing the necessity for such a system. These systems include but are not limited to power systems, HVAC, fire alarm, security, communications, and signaling for designated critical operations areas.

Informational Note No. 1: Critical operations power systems are generally installed in vital infrastructure facilities that, if destroyed or incapacitated, would disrupt national security, the economy, public health or safety; and where enhanced electrical infrastructure for continuity of operation has been deemed necessary by governmental authority.

Informational Note No. 2: For further information on disaster and emergency management, see *NFPA 1600-2013*, *Standard on Disaster/Emergency Management and Business Continuity Programs.*

Informational Note No. 3: For further information regarding performance of emergency and standby power systems, see NFPA 110-2013, *Standard for Emergency and Standby Power Systems.*

Informational Note No. 4: For further information regarding performance and maintenance of emergency systems in health care facilities, see NFPA 99-2012, *Standard for Health Care Facilities.*

Informational Note No. 5: For specification of locations where emergency lighting is considered essential to life safety, see NFPA *101*-2012, *Life Safety Code*, or the applicable building code.

Informational Note No. 6: For further information regarding physical security, see NFPA 730-2011, *Guide for Premises Security.*

Informational Note No. 7: Threats to facilities that may require transfer of operation to the critical systems include both naturally occurring hazards and human-caused events. See also A.5.3.2 of *NFPA 1600-2013.*

Informational Note No. 8: See Informative Annex F, Availability and Reliability for Critical Operations Power Systems; and Development and Implementation of Functional Performance Tests (FPTs) for Critical Operations Power Systems.

Informational Note No. 9: See Informative Annex G, Supervisory Control and Data Acquisition (SCADA).

Article 708 addresses homeland security issues for facilities that are "mission critical." These requirements go beyond those of Article 700, in that these electrical systems must continue to operate during the full duration of an emergency and beyond. See Exhibit 708.1. Examples of facilities that would use a critical operations power system (COPS) include police stations, fire stations, and hospitals. It may not include every one of these facilities within an area. Only facilities that are designated as critical because power must operate continuously with a robust power supply would be included.

## 708.2 Definitions

**Commissioning.** The acceptance testing, integrated system testing, operational tune-up, and start-up testing is the process by which baseline test results verify the proper operation and sequence of operation of electrical equipment, in addition to developing baseline criteria by which future trend analysis can identify equipment deterioration.

***EXHIBIT 708.1*** *New Hampshire Department of Safety Emergency Management Center, a typical COPS facility: (top) facility and (bottom) facility signage.*

**Critical Operations Power Systems (COPS).** Power systems for facilities or parts of facilities that require continuous operation for the reasons of public safety, emergency management, national security, or business continuity.

**Designated Critical Operations Areas (DCOA).** Areas within a facility or site designated as requiring critical operations power.

**Supervisory Control and Data Acquisition (SCADA).** An electronic system that provides monitoring and controls for the operation of the critical operations power system. This can include the fire alarm system, security system, control of the HVAC, the start/stop/monitoring of the power supplies and electrical distribution system, annunciation and communications equipment to emergency personnel, facility occupants, and remote operators.

## 708.4 Risk Assessment

Risk assessment for critical operations power systems shall be documented and shall be conducted in accordance with 708.4(A) through (C).

Informational Note: Chapter 5 of *NFPA 1600-2013*, *Standard on Disaster/Emergency Management and Business Continuity Programs*, provides additional guidance concerning risk assessment and hazard analysis.

**(A) Conducting Risk Assessment.** In critical operations power systems, risk assessment shall be performed to identify hazards, the likelihood of their occurrence, and the vulnerability of the electrical system to those hazards.

**(B) Identification of Hazards.** Hazards to be considered at a minimum shall include, but shall not be limited to, the following:

(1) Naturally occurring hazards (geological, meteorological, and biological)
(2) Human-caused events (accidental and intentional) [**1600:**5.3.2]

**(C) Developing Mitigation Strategy.** Based on the results of the risk assessment, a strategy shall be developed and implemented to mitigate the hazards that have not been sufficiently mitigated by the prescriptive requirements of this *Code*.

## 708.5 Physical Security

Physical security shall be provided for critical operations power systems in accordance with 708.5(A) and (B).

**(A) Risk Assessment.** Based on the results of the risk assessment, a strategy for providing physical security for critical operations power systems shall be developed, documented, and implemented.

**(B) Restricted Access.** Electrical circuits and equipment for critical operations power systems shall be accessible to qualified personnel only.

### 708.6 Testing and Maintenance

**(A) Conduct or Witness Test.** The authority having jurisdiction shall conduct or witness a test of the complete system upon installation and periodically afterward.

**(B) Tested Periodically.** Systems shall be tested periodically on a schedule acceptable to the authority having jurisdiction to ensure the systems are maintained in proper operating condition.

**(C) Maintenance.** The authority having jurisdiction shall require a documented preventive maintenance program for critical operations power systems.

Informational Note: For information concerning maintenance, see NFPA 70B-2013, *Recommended Practice for Electrical Equipment Maintenance*.

**(D) Written Record.** A written record shall be kept of such tests and maintenance.

**(E) Testing Under Load.** Means for testing all critical power systems during maximum anticipated load conditions shall be provided.

Informational Note: For information concerning testing and maintenance of emergency power supply systems (EPSSs) that are also applicable to COPS, see NFPA 110-2013, *Standard for Emergency and Standby Power Systems*.

### 708.8 Commissioning

**(A) Commissioning Plan.** A commissioning plan shall be developed and documented.

Informational Note: For further information on developing a commissioning program see NFPA 70B-2013, *Recommended Practice for Electrical Equipment Maintenance*.

**(B) Component and System Tests.** The installation of the equipment shall undergo component and system tests to ensure that, when energized, the system will function properly.

**(C) Baseline Test Results.** A set of baseline test results shall be documented for comparison with future periodic maintenance testing to identify equipment deterioration.

**(D) Functional Performance Tests.** A functional performance test program shall be established, documented, and executed upon complete installation of the critical system in order to establish a baseline reference for future performance requirements.

Informational Note: See Informative Annex F for more information on developing and implementing a functional performance test program.

## II. Circuit Wiring and Equipment

### 708.10 Feeder and Branch Circuit Wiring

**(A) Identification.**

**(1) Boxes and Enclosures.** In a building or at a structure where a critical operations power system and any other type of power system are present, all boxes and enclosures (including transfer switches, generators, and power panels) for critical operations power system circuits shall be permanently marked so they will be readily identified as a component of the critical operations power system.

**(2) Receptacle Identification.** In a building in which COPS are present with other types of power systems described in other sections in this article, the cover plates for the receptacles or the receptacles themselves supplied from the COPS shall have a distinctive color or marking so as to be readily identifiable.

*Exception: If the COPS supplies power to a DCOA that is a stand-alone building, receptacle cover plates or the receptacles themselves shall not be required to have distinctive marking.*

In a building where a COPS and another type of power system are installed, those receptacles supplied by the COPS or their cover plates must have a distinctive cover or marking. The exception provides that where a COPS supplies power to a designated critical operations area (DCOA) that is a stand-alone building, the distinctive cover or marking is not required.

**(B) Wiring.** Wiring of two or more COPS circuits supplied from the same source shall be permitted in the same raceway, cable, box, or cabinet. Wiring from a COPS source or COPS source distribution overcurrent protection to critical loads shall be kept entirely independent of all other wiring and equipment.

*Exception: Where the COPS feeder is installed in transfer equipment enclosures.*

**(C) COPS Feeder Wiring Requirements.** COPS feeders shall comply with 708.10(C)(1) through (C)(3).

**(1) Protection Against Physical Damage.** The wiring of the COPS system shall be protected against physical damage. Wiring methods shall be permitted to be installed in accordance with the following:

(1) Rigid metal conduit, intermediate metal conduit, or Type MI cable.
(2) Where encased in not less than 50 mm (2 in.) of concrete, any of the following wiring methods shall be permitted:
    a. Schedule 40 or Schedule 80 rigid polyvinyl chloride conduit (Type PVC)
    b. Reinforced thermosetting resin conduit (Type RTRC)
    c. Electrical metallic tubing (Type EMT)
    d. Flexible nonmetallic or jacketed metallic raceways
    e. Jacketed metallic cable assemblies listed for installation in concrete
(3) Where provisions must be made for flexibility at equipment connection, one or more of the following shall also be permitted:
    a. Flexible metal fittings
    b. Flexible metal conduit with listed fittings
    c. Liquidtight flexible metal conduit with listed fittings

**(2) Fire Protection for Feeders.** Feeders shall meet one of the following conditions:

(1) Be a listed electrical circuit protective system with a minimum 2-hour fire rating

Informational Note: UL guide information for electrical circuit protection systems (FHIT) contains information on proper installation requirements to maintain the fire rating.

(2) Be protected by a listed fire-rated assembly that has a minimum fire rating of 2 hours
(3) Be encased in a minimum 50 mm (2 in.) of concrete

Unlike the emergency system feeders covered in Article 700, COPS feeders are required to employ another fire protection technique even where located in building spaces that are fully protected by a fire suppression system.

The feeder-circuit wiring requires a minimum 2-hour fire rating provided by a listed electrical circuit protective system or a listed fire-rated assembly unless encased in 2 inches of concrete.

It is important to understand the difference between a 2-hour fire rating of an electrical circuit, such as a conduit with wires, and a 2-hour fire resistance rating of a structural member, such as a wall. At the end of a 2-hour fire test on an electrical conduit with wires, its insulation must be intact and the circuit must function electrically; no short circuits, grounds, or opens are permitted. A wall subjected to a 2-hour fire resistance test must only prevent a fire from passing through or past the wall, without regard to damage to the wall. All fire ratings and fire resistance ratings are based on the assumption that the structural supports for the assembly are not impaired by the effects of the fire.

Listed electrical circuit protective systems are described in the UL *White Book*. The four-letter code (shown in parentheses) is the UL product category guide designation. Examples of these systems include electrical circuit protective systems (FHIT), electrical circuit protective materials (FHIY), and fire-resistive cables (FHJR). Circuit integrity cable is covered under category FHJR.

**(3) Floodplain Protection.** Where COPS feeders are installed below the level of the 100-year floodplain, the insulated circuit conductors shall be listed for use in a wet location and be installed in a wiring method that is permitted for use in wet locations.

**(D) COPS Branch Circuit Wiring.**

(a) *Outside the DCOA.* COPS branch circuits installed outside the DCOA shall comply with the physical and fire protection requirements of 708.10(C)(1) through (C)(3).

(b) *Within the DCOA.* Any of the wiring methods recognized in Chapter 3 of this *Code* shall be permitted within the DCOA.

## 708.11 Branch Circuit and Feeder Distribution Equipment

**(A) Branch Circuit Distribution Equipment.** COPS branch circuit distribution equipment shall be located within the same DCOA as the branch circuits it supplies.

**(B) Feeder Distribution Equipment.** Equipment for COPS feeder circuits (including transfer equipment, transformers, and panelboards) shall comply with (1) and (2).

(1) Be located in spaces with a 2-hour fire resistance rating
(2) Be located above the 100-year floodplain

## 708.12 Feeders and Branch Circuits Supplied by COPS

Feeders and branch circuits supplied by the COPS shall supply only equipment specified as required for critical operations use.

## 708.14 Wiring of HVAC, Fire Alarm, Security, Emergency Communications, and Signaling Systems

All conductors or cables shall be installed using any of the metal wiring methods permitted by 708.10(C)(1) and, in addition, shall comply with 708.14(1) through (8), as applicable.

(1) All cables for fire alarm, security, signaling systems, and emergency communications shall be shielded twisted pair cables or installed to comply with the performance requirements of the system.
(2) Shields of cables for fire alarm, security, signaling systems, and emergency communications shall be arranged in accordance with the manufacturer's published installation instructions.
(3) Optical fiber cables shall be used for connections between two or more buildings on the property and under single management.
(4) A listed primary protector shall be provided on all communications circuits. Listed secondary protectors shall be provided at the terminals of the communications circuits.
(5) Conductors for all control circuits rated above 50 volts shall be rated not less than 600 volts.
(6) Communications, fire alarm, and signaling circuits shall use relays with contact ratings that exceed circuit voltage and current ratings in the controlled circuit.
(7) All cables for fire alarm, security, and signaling systems shall be riser-rated and shall be a listed 2-hour electrical circuit protective system. Emergency communication cables shall be Type CMR-CI or shall be riser-rated and shall be a listed 2-hour electrical circuit protective system.
(8) Control, monitoring, and power wiring to HVAC systems shall be a listed 2-hour electrical circuit protective system.

# III. Power Sources and Connection

## 708.20 Sources of Power

**(A) General Requirements.** Current supply shall be such that, in the event of failure of the normal supply to the DCOA, critical operations power shall be available within the time required for the application. The supply system for critical operations power, in addition to the normal services to the building and meeting the general requirements of this section, shall be one or more of the types of systems described in 708.20(E) through (H).

Informational Note: Assignment of degree of reliability of the recognized critical operations power system depends on the careful evaluation in accordance with the risk assessment.

**(B) Fire Protection.** Where located within a building, equipment for sources of power as described in 708.20(E) through (H) shall be installed either in spaces fully protected by approved automatic fire suppression systems (sprinklers, carbon dioxide systems, and so forth) or in spaces with a 2-hour fire rating.

**(C) Grounding.** All sources of power shall be grounded as a separately derived source in accordance with 250.30.

*Exception: Where the equipment containing the main bonding jumper or system bonding jumper for the normal source and the feeder wiring to the transfer equipment are installed in accordance with 708.10(C) and 708.11(B).*

**(D) Surge Protection Devices.** Surge protection devices shall be provided at all facility distribution voltage levels.

**(E) Storage Battery.** An automatic battery charging means shall be provided. Batteries shall be compatible with the charger for that particular installation. For a sealed battery, the container shall not be required to be transparent. However, for the lead acid battery that requires water additions, transparent or translucent containers shall be furnished. Automotive-type batteries shall not be used.

**(F) Generator Set.**

**(1) Prime Mover-Driven.** Generator sets driven by a prime mover shall be provided with means for automatically starting the prime mover on failure of the normal service. A time-delay feature permitting a minimum 15-minute setting shall be provided to avoid retransfer in case of short-time reestablishment of the normal source.

**(2) Power for fuel transfer pumps.** Where power is needed for the operation of the fuel transfer pumps to deliver fuel to a generator set day tank, this pump shall be connected to the COPS.

**(3) Dual Supplies.** Prime movers shall not be solely dependent on a public utility gas system for their fuel supply or municipal water supply for their cooling systems. Means shall be provided for automatically transferring from one fuel supply to another where dual fuel supplies are used.

**(4) Battery Power and Dampers.** Where a storage battery is used for control or signal power or as the means of starting the prime mover, it shall be suitable for the purpose and shall be equipped with an automatic charging means independent of the generator set. Where the battery charger is required for the operation of the generator set, it shall be connected to the COPS. Where power is required for the operation of dampers used to ventilate the generator set, the dampers shall be connected to the COPS.

**(5) Outdoor Generator Sets.**

(a) *Permanently Installed Generators and Portable Generators Greater Than 15 kW.* Where an outdoor housed generator set is equipped with a readily accessible disconnecting means in accordance with 445.18, and the disconnecting means is located within sight of the building or structure supplied, an additional disconnecting means shall not be required where ungrounded conductors serve or pass through the building or structure. Where the generator supply conductors terminate at a disconnecting means in or on a building or structure, the disconnecting means shall meet the requirements of 225.36.

(b) *Portable Generators 15 kW or Less.* Where a portable generator, rated 15 kW or less, is installed using a flanged inlet or other cord-and plug-type connection, a disconnecting means shall not be required where ungrounded conductors serve or pass through a building or structure.

**(6) Means for Connecting Portable or Vehicle-Mounted Generator.** Where the COPS is supplied by a single generator, a means to connect a portable or vehicle-mounted generator shall be provided.

**(7) On-Site Fuel Supply.** Where internal combustion engines are used as the prime mover, an on-site fuel supply shall be provided. The on-site fuel supply shall be secured and protected in accordance with the risk assessment.

**(G) Uninterruptible Power Supplies.** Uninterruptible power supplies used as the sole source of power for COPS shall comply with the applicable provisions of 708.20(E) and (F).

**(H) Fuel Cell System.** Installation of a fuel cell system shall meet the requirements of Parts II through VIII of Article 692.

## 708.21 Ventilation

Adequate ventilation shall be provided for the alternate power source for continued operation under maximum anticipated ambient temperatures.

Informational Note: NFPA 110-2013, *Standard for Emergency and Standby Power Systems*, and NFPA 111-2013, *Standard for Stored Energy Emergency and Standby Power Systems*, include additional information on ventilation air for combustion and cooling.

Air-cooled radiators, air intake for combustion engines, and discharge of generated heat are some factors affecting adequate ventilation. In addition, internal combustion engines require proper exhaust ventilation to remove carbon monoxide.

## 708.22 Capacity of Power Sources

**(A) Capacity and Rating.** A COPS shall have capacity and rating for all loads to be operated simultaneously for continuous operation with variable load for an unlimited number of hours, except for required maintenance of the power source. A portable, temporary, or redundant alternate power source shall be available for use whenever the COPS power source is out of service for maintenance or repair.

**(B) Selective Load Pickup, Load Shedding, and Peak Load Shaving.** The alternate power source shall be permitted to supply COPS emergency, legally required standby, and optional loads where the source has adequate capacity or where automatic selective load pickup and load shedding is provided as needed to ensure adequate power to (1) the COPS and emergency circuits, (2) the legally required standby circuits, and (3) the optional standby circuits, in that order of priority. The alternate power source shall be permitted to be used for peak load shaving, provided these conditions are met.

Peak load-shaving operation shall be permitted for satisfying the test requirement of 708.6(B), provided all other conditions of 708.6 are met.

**(C) Duration of COPS Operation.** The alternate power source shall be capable of operating the COPS for a minimum of 72 hours at full load of DCOA with a steady-state voltage within ±10 percent of nominal utilization voltage.

### 708.24 Transfer Equipment

**(A) General.** Transfer equipment, including automatic transfer switches, shall be automatic and identified for emergency use. Transfer equipment shall be designed and installed to prevent the inadvertent interconnection of normal and critical operations sources of supply in any operation of the transfer equipment. Transfer equipment and electric power production systems installed to permit operation in parallel with the normal source shall meet the requirements of Article 705.

**(B) Bypass Isolation Switches.** Means shall be permitted to bypass and isolate the transfer equipment. Where bypass isolation switches are used, inadvertent parallel operation shall be avoided.

**(C) Automatic Transfer Switches.** Where used with sources that are not inherently synchronized, automatic transfer switches shall comply with (C)(1) and (C)(2).

(1) Automatic transfer switches shall be listed for emergency use.
(2) Automatic transfer switches shall be electrically operated and mechanically held.

**(D) Use.** Transfer equipment shall supply only COPS loads.

### 708.30 Branch Circuits Supplied by COPS

Branch circuits supplied by the COPS shall only supply equipment specified as required for critical operations use.

## IV. Overcurrent Protection

### 708.50 Accessibility

The feeder- and branch-circuit overcurrent devices shall be accessible to authorized persons only.

### 708.52 Ground-Fault Protection of Equipment

**(A) Applicability.** The requirements of 708.52 shall apply to critical operations (including multiple occupancy buildings) with critical operation areas.

**(B) Feeders.** Where ground-fault protection is provided for operation of the service disconnecting means or feeder disconnecting means as specified by 230.95 or 215.10, an additional step of ground-fault protection shall be provided in all next level feeder disconnecting means downstream toward the load. Such protection shall consist of overcurrent devices and current transformers or other equivalent protective equipment that causes the feeder disconnecting means to open.

The additional levels of ground-fault protection shall not be installed on electrical systems that are not solidly grounded wye systems with greater than 150 volts to ground but not exceeding 1000 volts phase-to-phase.

**(C) Testing.** When equipment ground-fault protection is first installed, each level shall be tested to ensure that ground-fault protection is operational.

Informational Note: Testing is intended to verify the ground-fault function is operational. The performance test is not intended to verify selectivity in 708.52(D), as this is often coordinated similarly to circuit breakers by reviewing time and current curves and properly setting the equipment. (Selectivity of fuses and circuit breakers is not performance tested for overload and short circuit.)

**(D) Selectivity.** Ground-fault protection for operation of the service and feeder disconnecting means shall be fully selective such that the feeder device, but not the service device, shall open on ground faults on the load side of the feeder device. Separation of ground-fault protection time-current characteristics shall conform to the manufacturer's recommendations and shall consider all required tolerances and disconnect operating time to achieve 100 percent selectivity.

Informational Note: See 230.95, Informational Note No. 4, for transfer of alternate source where ground-fault protection is applied.

### 708.54 Selective Coordination

Critical operations power system(s) overcurrent devices shall be selectively coordinated with all supply-side overcurrent protective devices.

Selective coordination shall be selected by a licensed professional engineer or other qualified persons engaged primarily in the design, installation, or maintenance of electrical systems.

The selection shall be documented and made available to those authorized to design, install, inspect, maintain, and operate the system.

*Exception: Selective coordination shall not be required between two overcurrent devices located in series if no loads are connected in parallel with the downstream device.*

## V. System Performance and Analysis

### 708.64 Emergency Operations Plan

A facility with a COPS shall have documented an emergency operations plan. The plan shall consider emergency operations and response, recovery, and continuity of operations.

Informational Note: *NFPA 1600-2013, Standard on Disaster/Emergency Management and Business Continuity Programs*, Section 5.7, provides guidance for the development and implementation of emergency plans.

# ARTICLE 720 Circuits and Equipment Operating at Less Than 50 Volts

### 720.1 Scope

This article covers installations operating at less than 50 volts, direct current or alternating current.

### 720.2 Other Articles

Direct current or alternating-current installations operating at less than 50 volts, as covered in 411.1 through 411.8; Part VI of Article 517; Part II of Article 551; Parts II and III and 552.60(B) of Article 552; 650.1 through 650.8; 669.1 through 669.9; Parts I and VIII of Article 690; Parts I and III of Article 725; or Parts I and III of Article 760 shall not be required to comply with this article.

Lighting systems operating at 30 volts or less are covered by Article 411, not Article 720.

### 720.3 Hazardous (Classified) Locations

Installations within the scope of this article and installed in hazardous (classified) locations shall also comply with the appropriate provisions for hazardous (classified) locations in other applicable articles of this *Code*.

Low voltage alone does not render a circuit incapable of igniting flammable atmospheres. Ordinary flashlights using two 1½-volt D-cell batteries, for example, can become a source of ignition in some hazardous (classified) locations.

### 720.4 Conductors

Conductors shall not be smaller than 12 AWG copper or equivalent. Conductors for appliance branch circuits supplying more than one appliance or appliance receptacle shall not be smaller than 10 AWG copper or equivalent.

### 720.5 Lampholders

Standard lampholders that have a rating of not less than 660 watts shall be used.

### 720.6 Receptacle Rating

Receptacles shall have a rating of not less than 15 amperes.

### 720.7 Receptacles Required

Receptacles of not less than 20-ampere rating shall be provided in kitchens, laundries, and other locations where portable appliances are likely to be used.

### 720.9 Batteries

Installations of storage batteries shall comply with 480.1 through 480.5 and 480.8 through 480.10.

### 720.11 Mechanical Execution of Work

Circuits operating at less than 50 volts shall be installed in a neat and workmanlike manner. Cables shall be supported by the building structure in such a manner that the cable will not be damaged by normal building use.

Cables are required to be installed in a manner that is consistent with standard industry practice.

# ARTICLE 725 Class 1, Class 2, and Class 3 Remote-Control, Signaling, and Power-Limited Circuits

## I. General

### 725.1 Scope

This article covers remote-control, signaling, and power-limited circuits that are not an integral part of a device or appliance.

Informational Note: The circuits described herein are characterized by usage and electrical power limitations that differentiate them from electric light and power circuits; therefore, alternative requirements to those of Chapters 1 through 4 are given with regard to minimum wire sizes, ampacity adjustment and correction factors, overcurrent protection, insulation requirements, and wiring methods and materials.

Article 725 includes systems such as security system circuits (see Exhibit 725.1), access control circuits, sound circuits, nurse call circuits, intercom circuits, some computer network systems, some control circuits for lighting dimmer systems, and some low-voltage control circuits that originate from listed appliances or from listed computer equipment.

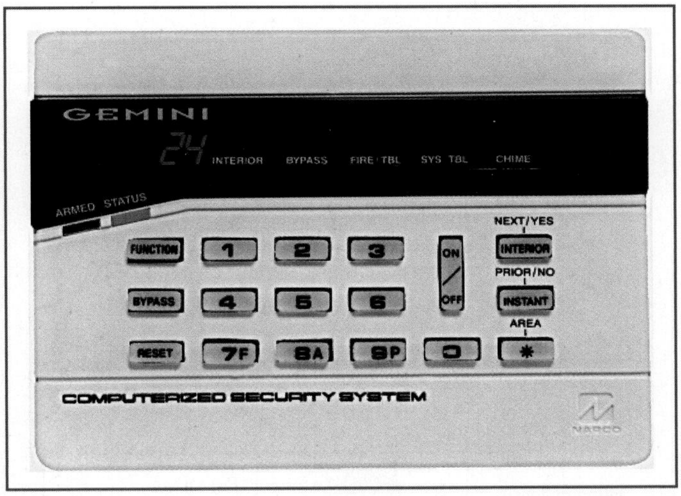


***EXHIBIT 725.1*** *Typical security system keypad. (Courtesy of NAPCO Security Technologies, Inc.)*

The installation requirements for the wiring of information technology equipment (electronic data processing and computer equipment) located within the confines of a room that is constructed according to the requirements of NFPA 75, *Standard for the Protection of Information Technology Equipment*, are not covered by Article 725. The wiring within these specially constructed rooms is covered in Article 645.

In addition, if listed computer equipment is interconnected and all the interconnected equipment is in close proximity, the wiring is considered an integral part of the equipment and, therefore, not subject to the requirements of Article 725. If the wiring leaves the group of equipment to connect to other devices in the same room or elsewhere in the building, the wiring is considered "wiring within buildings" and is subject to the requirements of Article 725.

The wiring methods required by Chapters 1 through 4 of the *Code* apply to remote-control, signaling, and power-limited circuits, except as amended by Article 725 for specified conditions.

A remote-control, signaling, or power-limited circuit is the portion of the wiring system between the load side of the overcurrent device or the power-limited supply and all connected equipment. The circuit is categorized as Class 1, Class 2, or Class 3.

Class 1 circuits are not permitted to exceed 600 volts. In many cases, Class 1 circuits are extensions of power systems and are subject to the requirements of the power systems, except under the following conditions:

1. Conductors size 16 AWG and 18 AWG may be used. (See 725.43.)
2. Where damage to the circuit would introduce a hazard, the circuit must be mechanically protected. [See 725.31(B).]
3. The adjustment factors of 310.15(B)(3) apply only if such conductors carry a continuous load. (See 725.51.)

Class 1 remote-control circuits are commonly used to operate motor controllers in conjunction with moving equipment or mechanical processes, elevators, conveyors, and other such equipment. Class 1 remote-control circuits may also be used as shunt trip circuits for circuit breakers. Class 1 signaling circuits often operate at 120 volts but are not limited to this value.

Conductors and equipment on the supply side of overcurrent protection, transformers, or current-limiting devices of Class 2 and Class 3 circuits must be installed according to the applicable requirements of Chapter 3. Load-side conductors and equipment must comply with Article 725. Class 2 and Class 3 conductors are required to be separated from and not occupy the same raceways, cable trays, cables, or enclosures as electric light, power, and Class 1 conductors, except as noted in 725.136.

Dry-cell batteries are considered Class 2 power supplies, provided the voltage is 30 volts or less and the capacity is equal to or less than that available from series-connected No. 6 carbon-zinc cells. See 725.121(A)(5).

Circuits originating from thermocouples are categorized as Class 2 circuits. Neither dry-cell batteries nor thermocouples are required to be listed.

## 725.2 Definitions

**Abandoned Class 2, Class 3, and PLTC Cable.** Installed Class 2, Class 3, and PLTC cable that is not terminated at equipment and not identified for future use with a tag.

**Circuit Integrity (CI) Cable.** Cable(s) used for remote-control, signaling, or power-limited systems that supply critical circuits to ensure survivability for continued circuit operation for a specified time under fire conditions.

**Class 1 Circuit.** The portion of the wiring system between the load side of the overcurrent device or power-limited supply and the connected equipment.

Informational Note: See 725.41 for voltage and power limitations of Class 1 circuits.

**Class 2 Circuit.** The portion of the wiring system between the load side of a Class 2 power source and the connected equipment. Due to its power limitations, a Class 2 circuit considers safety from a fire initiation standpoint and provides acceptable protection from electric shock.

**Class 3 Circuit.** The portion of the wiring system between the load side of a Class 3 power source and the connected equipment. Due to its power limitations, a Class 3 circuit considers safety from a fire initiation standpoint. Since higher levels of voltage and current than for Class 2 are permitted, additional safeguards are specified to provide protection from an electric shock hazard that could be encountered.

**Power-Limited Tray Cable (PLTC).** A factory assembly of two or more insulated conductors rated at 300 V, with or without associated bare or insulated equipment grounding conductors, under a nonmetallic jacket.

## 725.3 Other Articles

Circuits and equipment shall comply with the articles or sections listed in 725.3(A) through (L). Only those sections of Article 300 referenced in this article shall apply to Class 1, Class 2, and Class 3 circuits.

**(A) Number and Size of Conductors in Raceway.** Section 300.17.

**(B) Spread of Fire or Products of Combustion.** Installation of Class 1, Class 2, and Class 3 circuits shall comply with 300.21.

**(C) Ducts, Plenums, and Other Air-Handling Spaces.** Class 1, Class 2, and Class 3 circuits installed in ducts, plenums, or other space used for environmental air shall comply with 300.22.

*Exception: As permitted in Table 725.154.*

See the commentary following 300.22(B), 300.22(C), and informational note to 725.179(A) for information on wiring in ducts, plenums, and other air-handling spaces.

**(D) Hazardous (Classified) Locations.** Articles 500 through 516 and Article 517, Part IV, where installed in hazardous (classified) locations.

**(E) Cable Trays.** Article 392, where installed in cable tray.

**(F) Motor Control Circuits.** Article 430, Part VI, where tapped from the load side of the motor branch-circuit protective device(s) as specified in 430.72(A).

**(G) Instrumentation Tray Cable.** See Article 727.

**(H) Raceways Exposed to Different Temperatures.** Installations shall comply with 300.7(A).

Condensation often forms in conduit exposed to nonconditioned and conditioned spaces. Section 725.3(H) brings the requirements of 300.7(A) into Article 725.

**(I) Vertical Support for Fire-Rated Cables and Conductors.** Vertical installations of circuit integrity (CI) cables and conductors installed in a raceway or conductors and cables of electrical circuit protective systems shall be installed in accordance with 300.19.

The strength of cables and conductors decreases with heat, and they may break if not properly supported. This could adversely impact the operation of signaling systems that are important to public safety.

**(J) Bushing.** A bushing shall be installed where cables emerge from raceway used for mechanical support or protection in accordance with 300.15(C).

**(K) Installation of Conductors with Other Systems.** Installations shall comply with 300.8.

**(L) Corrosive, Damp, or Wet Locations.** Class 2 and Class 3 cables installed in corrosive, damp, or wet locations shall comply with the applicable requirements in 110.11, 300.5(B), 300.6, 300.9, and 310.10(G).

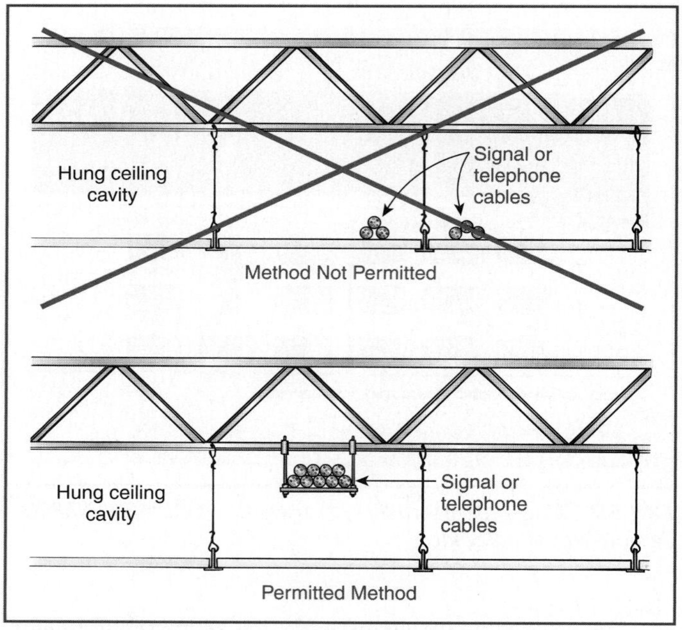


***EXHIBIT 725.2*** *Incorrect cable installation (upper diagram) and correct method (lower diagram).*

## 725.21 Access to Electrical Equipment Behind Panels Designed to Allow Access

Access to electrical equipment shall not be denied by an accumulation of wires and cables that prevents removal of panels, including suspended ceiling panels.

An excess accumulation of wires and cables can limit access to electrical equipment by preventing the removal of access panels. To safely service, rearrange, or install electrical equipment, the worker must have an accessible work space. Incorrect installation of conductors and cables can prevent access to equipment or cables. See Exhibit 725.2. See 300.11(A), which permits the use of support wires and approved fittings that are independent of the suspended ceiling support wires.

## 725.24 Mechanical Execution of Work

Class 1, Class 2, and Class 3 circuits shall be installed in a neat and workmanlike manner. Cables and conductors installed exposed on the surface of ceilings and sidewalls shall be supported by the building structure in such a manner that the cable will not be damaged by normal building use. Such cables shall be supported by straps, staples, hangers, cable ties, or similar fittings designed and installed so as not to damage the cable. The installation shall also comply with 300.4(D).

Cable must be attached to or supported by the building structure by cable ties, straps, clamps, hangers, and so forth. The installation method must not damage the cable. In addition, the location of the cable should be carefully evaluated to ensure that activities and processes within the building do not cause damage to the cable. (See 725.143 and Exhibit 725.2.)

Section 300.4(D) requires protection of cables that are installed on framing members. Such cables are required to be installed in a manner that protects them from nail or screw penetration. This section permits attachment to baseboards and non-load-bearing walls, which are not structural components.

### 725.25 Abandoned Cables

The accessible portion of abandoned Class 2, Class 3, and PLTC cables shall be removed. Where cables are identified for future use with a tag, the tag shall be of sufficient durability to withstand the environment involved.

### 725.30 Class 1, Class 2, and Class 3 Circuit Identification

Class 1, Class 2, and Class 3 circuits shall be identified at terminal and junction locations in a manner that prevents unintentional interference with other circuits during testing and servicing.

### 725.31 Safety-Control Equipment

**(A) Remote-Control Circuits.** Remote-control circuits for safety-control equipment shall be classified as Class 1 if the failure of the equipment to operate introduces a direct fire or life hazard. Room thermostats, water temperature regulating devices, and similar controls used in conjunction with electrically controlled household heating and air conditioning shall not be considered safety-control equipment.

The remote-control circuits to safety-control devices are required to be classified as Class 1 if failure of the safety-control circuit could cause a direct fire or life hazard. One example of the direct link between a failure and the initiation of a fire hazard is a boiler explosion caused by failure of the low-water cutoff circuit. See Exhibit 725.3.

Generally, signaling systems such as a nurse call system do not fit this category. These systems do not have a direct link to the initiation of fire or the initiation of a life hazard but, rather, serve as the reporting or warning link of a hazard initiated by some other (indirect) cause.

**(B) Physical Protection.** Where damage to remote-control circuits of safety-control equipment would introduce a hazard, as covered in 725.31(A), all conductors of such remote-control circuits shall be installed in rigid metal conduit, intermediate metal conduit, rigid nonmetallic conduit, electrical metallic tubing, Type MI cable, Type MC cable, or be otherwise suitably protected from physical damage.

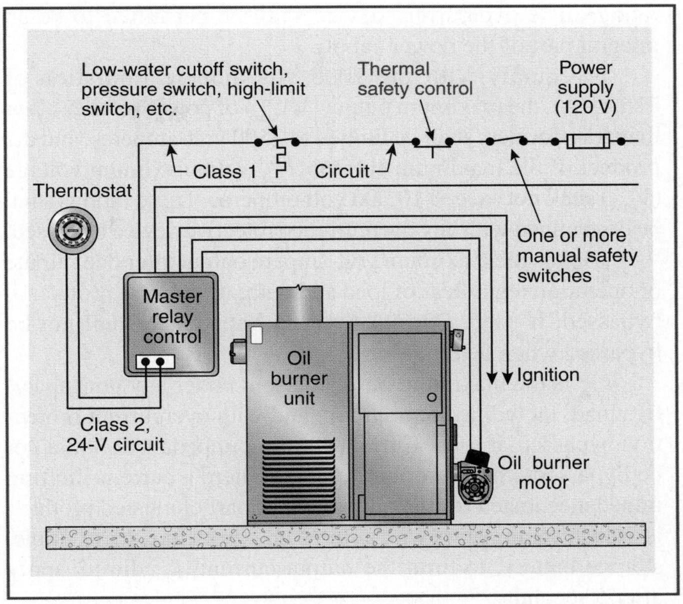


***EXHIBIT 725.3*** *Typical installation of an automatic oil burner unit for a boiler employing a safety shutdown circuit required to be Class 1.*

### 725.35 Class 1, Class 2, and Class 3 Circuit Requirements

A remote-control, signaling, or power-limited circuit shall comply with the following parts of this article:

(1) Class 1 Circuits: Parts I and II
(2) Class 2 and Class 3 Circuits: Parts I and III

## II. Class 1 Circuits

### 725.41 Class 1 Circuit Classifications and Power Source Requirements

Class 1 circuits shall be classified as either Class 1 power-limited circuits where they comply with the power limitations of 725.41(A) or as Class 1 remote-control and signaling circuits where they are used for remote-control or signaling purposes and comply with the power limitations of 725.41(B).

**(A) Class 1 Power-Limited Circuits.** These circuits shall be supplied from a source that has a rated output of not more than 30 volts and 1000 volt-amperes.

**(1) Class 1 Transformers.** Transformers used to supply power-limited Class 1 circuits shall comply with the applicable sections within Parts I and II of Article 450.

**(2) Other Class 1 Power Sources.** Power sources other than transformers shall be protected by overcurrent devices rated at not more than 167 percent of the volt-ampere rating of the source divided by the rated voltage. The overcurrent devices shall not be interchangeable with overcurrent devices of higher ratings. The overcurrent device shall be permitted to be an integral part of the power supply.

To comply with the 1000 volt-ampere limitation of 725.41(A), the maximum output ($VA_{max}$) of power sources other than transformers shall be limited to 2500 volt-amperes, and the product of the maximum current ($I_{max}$) and maximum voltage ($V_{max}$) shall not exceed 10,000 volt-amperes. These ratings shall be determined with any overcurrent-protective device bypassed.

$VA_{max}$ is the maximum volt-ampere output after one minute of operation regardless of load and with overcurrent protection bypassed, if used. Current-limiting impedance shall not be bypassed when determining $VA_{max}$.

$I_{max}$ is the maximum output current under any noncapacitive load, including short circuit, and with overcurrent protection bypassed, if used. Current-limiting impedance should not be bypassed when determining $I_{max}$. Where a current-limiting impedance, listed for the purpose or as part of a listed product, is used in combination with a stored energy source, for example, storage battery, to limit the output current, $I_{max}$ limits apply after 5 seconds.

$V_{max}$ is the maximum output voltage regardless of load with rated input applied.

**(B) Class 1 Remote-Control and Signaling Circuits.** These circuits shall not exceed 600 volts. The power output of the source shall not be required to be limited.

## 725.43 Class 1 Circuit Overcurrent Protection

Overcurrent protection for conductors 14 AWG and larger shall be provided in accordance with the conductor ampacity, without applying the ampacity adjustment and correction factors of 310.15 to the ampacity calculation. Overcurrent protection shall not exceed 7 amperes for 18 AWG conductors and 10 amperes for 16 AWG.

*Exception: Where other articles of this Code permit or require other overcurrent protection.*

Informational Note: For example, see 430.72 for motors, 610.53 for cranes and hoists, and 517.74(B) and 660.9 for X-ray equipment.

## 725.45 Class 1 Circuit Overcurrent Device Location

Overcurrent devices shall be located as specified in 725.45(A), (B), (C), (D), or (E).

**(A) Point of Supply.** Overcurrent devices shall be located at the point where the conductor to be protected receives its supply.

**(B) Feeder Taps.** Class 1 circuit conductors shall be permitted to be tapped, without overcurrent protection at the tap, where the overcurrent device protecting the circuit conductor is sized to protect the tap conductor.

**(C) Branch-Circuit Taps.** Class 1 circuit conductors 14 AWG and larger that are tapped from the load side of the overcurrent protective device(s) of a controlled light and power circuit shall require only short-circuit and ground-fault protection and shall be permitted to be protected by the branch-circuit overcurrent protective device(s) where the rating of the protective device(s) is not more than 300 percent of the ampacity of the Class 1 circuit conductor.

**(D) Primary Side of Transformer.** Class 1 circuit conductors supplied by the secondary of a single-phase transformer having only a 2-wire (single-voltage) secondary shall be permitted to be protected by overcurrent protection provided on the primary side of the transformer, provided this protection is in accordance with 450.3 and does not exceed the value determined by multiplying the secondary conductor ampacity by the secondary-to-primary transformer voltage ratio. Transformer secondary conductors other than 2-wire shall not be considered to be protected by the primary overcurrent protection.

**(E) Input Side of Electronic Power Source.** Class 1 circuit conductors supplied by the output of a single-phase, listed electronic power source, other than a transformer, having only a 2-wire (single-voltage) output for connection to Class 1 circuits shall be permitted to be protected by overcurrent protection provided on the input side of the electronic power source, provided this protection does not exceed the value determined by multiplying the Class 1 circuit conductor ampacity by the output-to-input voltage ratio. Electronic power source outputs, other than 2-wire (single voltage), shall not be considered to be protected by the primary overcurrent protection.

## 725.46 Class 1 Circuit Wiring Methods

Class 1 circuits shall be installed in accordance with Part I of Article 300 and with the wiring methods from the appropriate articles in Chapter 3.

*Exception No. 1: The provisions of 725.48 through 725.51 shall be permitted to apply in installations of Class 1 circuits.*

*Exception No. 2: Methods permitted or required by other articles of this Code shall apply to installations of Class 1 circuits.*

## 725.48 Conductors of Different Circuits in the Same Cable, Cable Tray, Enclosure, or Raceway

Class 1 circuits shall be permitted to be installed with other circuits as specified in 725.48(A) and (B).

**(A) Two or More Class 1 Circuits.** Class 1 circuits shall be permitted to occupy the same cable, cable tray, enclosure, or raceway without regard to whether the individual circuits are alternating current or direct current, provided all conductors are insulated for the maximum voltage of any conductor in the cable, cable tray, enclosure, or raceway.

**(B) Class 1 Circuits with Power-Supply Circuits.** Class 1 circuits shall be permitted to be installed with power-supply conductors as specified in 725.48(B)(1) through (B)(4).

**(1) In a Cable, Enclosure, or Raceway.** Class 1 circuits and power-supply circuits shall be permitted to occupy the same cable, enclosure, or raceway only where the equipment powered is functionally associated.

**(2) In Factory- or Field-Assembled Control Centers.** Class 1 circuits and power-supply circuits shall be permitted to be installed in factory- or field-assembled control centers.

**(3) In a Manhole.** Class 1 circuits and power-supply circuits shall be permitted to be installed as underground conductors in a manhole in accordance with one of the following:

(1) The power-supply or Class 1 circuit conductors are in a metal-enclosed cable or Type UF cable.
(2) The conductors are permanently separated from the power-supply conductors by a continuous firmly fixed nonconductor, such as flexible tubing, in addition to the insulation on the wire.
(3) The conductors are permanently and effectively separated from the power supply conductors and securely fastened to racks, insulators, or other approved supports.

Class 1 power-limited circuit conductors are permitted to be installed in manholes with wiring of non–power-limited systems where permanent separation requirements comply with the following, as applicable:

1. Class 2 and Class 3 power-limited circuits in 725.136
2. Communications circuits in 800.133(A)
3. Radio/television antennas and lead-in conductors in 810.18
4. CATV conductors in 820.133(A)

**(4) In Cable Trays.** Installations in cable trays shall comply with 725.48(B)(4)(1) or (B)(4)(2).

(1) Class 1 circuit conductors and power-supply conductors not functionally associated with the Class 1 circuit conductors shall be separated by a solid fixed barrier of a material compatible with the cable tray.
(2) Class 1 circuit conductors and power-supply conductors not functionally associated with the Class 1 circuit conductors shall be permitted to be installed in a cable tray without barriers where all of the conductors are installed with separate multiconductor Type AC, Type MC, Type MI, or Type TC cables and all the conductors in the cables are insulated at 600 volts or greater.

### 725.49 Class 1 Circuit Conductors

**(A) Sizes and Use.** Conductors of sizes 18 AWG and 16 AWG shall be permitted to be used, provided they supply loads that do not exceed the ampacities given in 402.5 and are installed in a raceway, an approved enclosure, or a listed cable. Conductors larger than 16 AWG shall not supply loads greater than the ampacities given in 310.15. Flexible cords shall comply with Article 400.

**(B) Insulation.** Insulation on conductors shall be rated for the system voltage and not less than 600 volts. Conductors larger than 16 AWG shall comply with Article 310. Conductors in sizes 18 AWG and 16 AWG shall be Type FFH-2, KF-2, KFF-2, PAF, PAFF, PF, PFF, PGF, PGFF, PTF, PTFF, RFH-2, RFHH-2, RFHH-3, SF-2, SFF-2, TF, TFF, TFFN, TFN, ZF, or ZFF. Conductors with other types and thicknesses of insulation shall be permitted if listed for Class 1 circuit use.

Class 1 circuit conductors are required to be rated at 600 volts. This effectively requires Class 1 circuits to be wired using the wiring methods found in Chapter 3 or the use of conductors specifically listed for Class 1 circuit use.

### 725.51 Number of Conductors in Cable Trays and Raceway, and Ampacity Adjustment

**(A) Class 1 Circuit Conductors.** Where only Class 1 circuit conductors are in a raceway, the number of conductors shall be determined in accordance with 300.17. The ampacity adjustment factors given in 310.15(B)(3)(a) shall apply only if such conductors carry continuous loads in excess of 10 percent of the ampacity of each conductor.

**(B) Power-Supply Conductors and Class 1 Circuit Conductors.** Where power-supply conductors and Class 1 circuit conductors are permitted in a raceway in accordance with 725.48, the number of conductors shall be determined in accordance with 300.17. The ampacity adjustment factors given in 310.15(B)(3)(a) shall apply as follows:

(1) To all conductors where the Class 1 circuit conductors carry continuous loads in excess of 10 percent of the ampacity of each conductor and where the total number of conductors is more than three
(2) To the power-supply conductors only, where the Class 1 circuit conductors do not carry continuous loads in excess of 10 percent of the ampacity of each conductor and where the number of power-supply conductors is more than three

**(C) Class 1 Circuit Conductors in Cable Trays.** Where Class 1 circuit conductors are installed in cable trays, they shall comply with the provisions of 392.22 and 392.80(A).

### 725.52 Circuits Extending Beyond One Building

Class 1 circuits that extend aerially beyond one building shall also meet the requirements of Article 225.

## III. Class 2 and Class 3 Circuits

### 725.121 Power Sources for Class 2 and Class 3 Circuits

**(A) Power Source.** The power source for a Class 2 or a Class 3 circuit shall be as specified in 725.121(A)(1), (A)(2), (A)(3), (A)(4), or (A)(5):

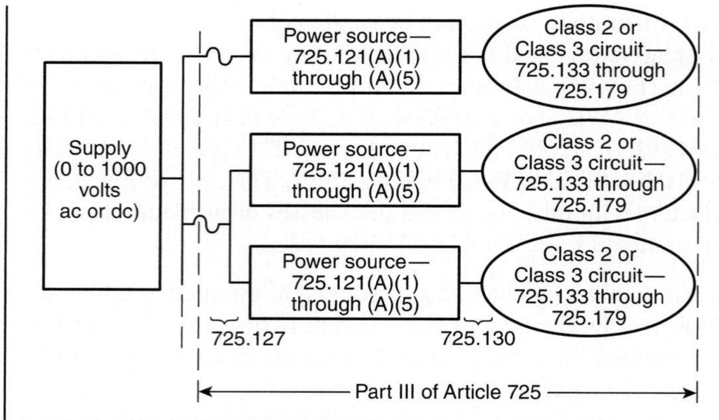


***INFORMATIONAL NOTE FIGURE 725.121, NO. 1***
*Class 2 and Class 3 Circuits.*

Informational Note No. 1: Informational Note Figure 725.121, No. 1 illustrates the relationships between Class 2 or Class 3 power sources, their supply, and the Class 2 or Class 3 circuits.

Informational Note No. 2: Table 11(A) and Table 11(B) in Chapter 9 provide the requirements for listed Class 2 and Class 3 power sources.

(1) A listed Class 2 or Class 3 transformer
(2) A listed Class 2 or Class 3 power supply
(3) Other listed equipment marked to identify the Class 2 or Class 3 power source

*Exception No. 1 to (3): Thermocouples shall not require listing as a Class 2 power source.*

*Exception No. 2 to (3): Limited power circuits of listed equipment where these circuits have energy levels rated at or below the limits established in Chapter 9, Table 11(A) and Table 11(B).*

Informational Note: Examples of other listed equipment are as follows:

(1) A circuit card listed for use as a Class 2 or Class 3 power source where used as part of a listed assembly
(2) A current-limiting impedance, listed for the purpose, or part of a listed product, used in conjunction with a non–power-limited transformer or a stored energy source, for example, storage battery, to limit the output current
(3) A thermocouple
(4) Limited voltage/current or limited impedance secondary communications circuits of listed industrial control equipment

(4) Listed information technology (computer) equipment limited-power circuits.

Informational Note: One way to determine applicable requirements for listing of information technology (computer) equipment is to refer to UL 60950-1-2011, *Standard for Safety of Information Technology Equipment*. Typically such circuits are used to interconnect information technology equipment for the purpose of exchanging information (data).

(5) A dry cell battery shall be considered an inherently limited Class 2 power source, provided the voltage is 30 volts or less and the capacity is equal to or less than that available from series connected No. 6 carbon zinc cells.

**(B) Interconnection of Power Sources.** Class 2 or Class 3 power sources shall not have the output connections paralleled or otherwise interconnected unless listed for such interconnection.

## 725.124 Circuit Marking

The equipment supplying the circuits shall be durably marked where plainly visible to indicate each circuit that is a Class 2 or Class 3 circuit.

This section requires the power source to be marked, but not the individual raceways, cables, and fittings containing the circuits.

## 725.127 Wiring Methods on Supply Side of the Class 2 or Class 3 Power Source

Conductors and equipment on the supply side of the power source shall be installed in accordance with the appropriate requirements of Chapters 1 through 4. Transformers or other devices supplied from electric light or power circuits shall be protected by an overcurrent device rated not over 20 amperes.

*Exception: The input leads of a transformer or other power source supplying Class 2 and Class 3 circuits shall be permitted to be smaller than 14 AWG, but not smaller than 18 AWG if they are not over 12 in. (305 mm) long and if they have insulation that complies with 725.49(B).*

Listed Class 2 and Class 3 transformers must be protected by an overcurrent device not exceeding 20 amperes, unless the transformers are fed from circuits other than power or lighting.

## 725.130 Wiring Methods and Materials on Load Side of the Class 2 or Class 3 Power Source

Class 2 and Class 3 circuits on the load side of the power source shall be permitted to be installed using wiring methods and materials in accordance with either 725.130(A) or (B).

**(A) Class 1 Wiring Methods and Materials.** Installation shall be in accordance with 725.46.

*Exception No. 1: The ampacity adjustment factors given in 310.15(B)(3)(a) shall not apply.*

*Exception No. 2: Class 2 and Class 3 circuits shall be permitted to be reclassified and installed as Class 1 circuits if the Class 2 and Class 3 markings required in 725.124 are eliminated and the entire circuit is installed using the wiring methods and materials in accordance with Part II, Class 1 circuits.*

Informational Note: Class 2 and Class 3 circuits reclassified and installed as Class 1 circuits are no longer Class 2 or Class 3 circuits, regardless of the continued connection to a Class 2 or Class 3 power source.

Where it is necessary to locate Class 2 or Class 3 circuits inside the same cable or raceway as a Class 1 circuit, Exception No. 2

permits a Class 2 or Class 3 circuit to be reclassified and installed as Class 1, provided the Class 2 or Class 3 marking is removed, that overcurrent protection complies with 725.43, and that the reclassified circuit maintains separation from other Class 2 and Class 3 circuits in accordance with 725.136.

**(B) Class 2 and Class 3 Wiring Methods.** Conductors on the load side of the power source shall be insulated at not less than the requirements of 725.179 and shall be installed in accordance with 725.133 and 725.154.

*Exception No. 1: As provided for in 620.21 for elevators and similar equipment.*

*Exception No. 2: Other wiring methods and materials installed in accordance with the requirements of 725.3 shall be permitted to extend or replace the conductors and cables described in 725.179 and permitted by 725.130(B).*

*Exception No. 3: Bare Class 2 conductors shall be permitted as part of a listed intrusion protection system where installed in accordance with the listing instructions for the system.*

### 725.133 Installation of Conductors and Equipment in Cables, Compartments, Cable Trays, Enclosures, Manholes, Outlet Boxes, Device Boxes, Raceways, and Cable Routing Assemblies for Class 2 and Class 3 Circuits

Conductors and equipment for Class 2 and Class 3 circuits shall be installed in accordance with 725.135 through 725.143.

### 725.135 Installation of Class 2, Class 3, and PLTC Cables

Installation of Class 2, Class 3, and PLTC cables shall comply with 725.135(A) through (M).

**(A) Listing.** Class 2, Class 3, and PLTC cables installed in buildings shall be listed.

**(B) Fabricated Ducts Used for Environmental Air.** The following wires and cables shall be permitted in ducts used for environmental air as described in 300.22(B) if they are directly associated with the air distribution system:

(1) Types CL2P and CL3P cables in lengths as short as practicable to perform the required function
(2) Types CL2P, CL3P, CL2R, CL3R, CL2, CL3, CL2X, CL3X, and PLTC cables installed in raceways that are installed in compliance with 300.22(B)

Informational Note: For information on fire protection of wiring installed in fabricated ducts, see 4.3.4.1 and 4.3.11.3.3 of NFPA 90A-2012, *Standard for the Installation of Air-Conditioning and Ventilating Systems*.

**(C) Other Spaces Used for Environmental Air (Plenums).** The following cables shall be permitted in other spaces used for environmental air as described in 300.22(C):

(1) Types CL2P and CL3P cables
(2) Types CL2P and CL3P cables installed in plenum communications raceways
(3) Types CL2P and CL3P cables and plenum communications raceways supported by open metallic cable trays or cable tray systems
(4) Types CL2P, CL3P, CL2R, CL3R, CL2, CL3, CL2X, CL3X, and PLTC cables installed in raceways that are installed in compliance with 300.22(C)
(5) Types CL2P, CL3P, CL2R, CL3R, CL2, CL3, CL2X, CL3X, and PLTC cables supported by solid bottom metal cable trays with solid metal covers in other spaces used for environmental air (plenums) as described in 300.22(C)
(6) Types CL2P, CL3P, CL2R, CL3R, CL2, CL3, CL2X, CL3X, and PLTC cables installed in plenum communications raceways, riser communications raceways, and general-purpose communications raceways supported by solid bottom metal cable trays with solid metal covers in other spaces used for environmental air (plenums) as described in 300.22(C)

**(D) Risers — Cables in Vertical Runs.** The following cables shall be permitted in vertical runs penetrating one or more floors and in vertical runs in a shaft:

(1) Types CL2P, CL3P, CL2R, and CL3R cables
(2) Types CL2P, CL3P, CL2R, and CL3R cables installed in the following:
   a. Plenum communications raceways
   b. Plenum cable routing assemblies
   c. Riser communications raceways
   d. Riser cable routing assemblies

Informational Note: See 300.21 for firestop requirements for floor penetrations.

**(E) Risers — Cables in Metal Raceways.** The following cables shall be permitted in metal raceways in a riser having firestops at each floor:

(1) Types CL2P, CL3P, CL2R, CL3R, CL2, CL3, CL2X, CL3X, and PLTC cables
(2) Types CL2P, CL3P, CL2R, CL3R, CL2, CL3, CL2X, CL3X, and PLTC cables installed in the following:
   a. Plenum communications raceways
   b. Riser communications raceways
   c. General-purpose communications raceways

Informational Note: See 300.21 for firestop requirements for floor penetrations.

**(F) Risers — Cables in Fireproof Shafts.** The following shall be permitted to be installed in fireproof riser shafts having firestops at each floor:

(1) Types CL2P, CL3P, CL2R, CL3R, CL2, CL3, CL2X, CL3X, and PLTC cables

(2) Types CL2P, CL3P, CL2R, CL3R, CL2, CL3, and PLTC cables installed in the following:
   a. Plenum communications raceways
   b. Plenum cable routing assemblies
   c. Riser communications raceways
   d. Riser cable routing assemblies
   e. General-purpose communications raceways
   f. General-purpose cable routing assemblies

Informational Note: See 300.21 for firestop requirements for floor penetrations.

**(G) Risers — One- and Two-Family Dwellings.** The following cables shall be permitted in one- and two-family dwellings:

(1) Types CL2P, CL3P, CL2R, CL3R, CL2, CL3, and PLTC cables
(2) Types CL2X and CL3X cables less than 6 mm (0.25 in.) in diameter
(3) Types CL2P, CL3P, CL2R, CL3R, CL2, CL3, and PLTC cables installed in the following:
   a. Plenum communications raceways
   b. Plenum cable routing assemblies
   c. Riser communications raceways
   d. Riser cable routing assemblies
   e. General-purpose communications raceways
   f. General-purpose cable routing assemblies

**(H) Cable Trays.** Cables installed in cable trays outdoors shall be Type PLTC. The following cables shall be permitted to be supported by cable trays in buildings:

(1) Types CM CL2P, CL3P, CL2R, CL3R, CL2, CL3, and PLTC cables
(2) Types CL2P, CL3P, CL2R, CL3R, CL2, CL3, and PLTC cables installed in the following:
   a. Plenum communications raceways
   b. Riser communications raceways
   c. General-purpose communications raceways

**(I) Cross-Connect Arrays.** The following cables shall be permitted to be installed in cross-connect arrays:

(1) Types CL2P, CL3P, CL2R, CL3R, CL2, CL3, and PLTC cables
(2) Types CL2P, CL3P, CL2R, CL3R, CL2, CL3, and PLTC cables installed in the following:
   a. Plenum communications raceways
   b. Plenum cable routing assemblies
   c. Riser communications raceways
   d. Riser cable routing assemblies
   e. General-purpose communications raceways
   f. General-purpose cable routing assemblies

**(J) Industrial Establishments.** In industrial establishments where the conditions of maintenance and supervision ensure that only qualified persons service the installation, Type PLTC cable shall be permitted in accordance with either (1) or (2) as follows:

(1) Where the cable is not subject to physical damage, Type PLTC cable that complies with the crush and impact requirements of Type MC cable and is identified as PLTC-ER for such use shall be permitted to be exposed between the cable tray and the utilization equipment or device. The cable shall be continuously supported and protected against physical damage using mechanical protection such as dedicated struts, angles, or channels. The cable shall be supported and secured at intervals not exceeding 1.8 m (6 ft).

Type PLTC cable that complies with the crush and impact requirements of Type MC cable is identified as PLTC-ER and is permitted to be run between a cable tray and utilization equipment or a device.

(2) Type PLTC cable, with a metallic sheath or armor in accordance with 725.179(E), shall be permitted to be installed exposed. The cable shall be continuously supported and protected against physical damage using mechanical protection such as dedicated struts, angles, or channels. The cable shall be secured at intervals not exceeding 1.8 m (6 ft).

**(K) Other Building Locations.** The following wires and cables shall be permitted to be installed in building locations other than the locations covered in 725.135(B) through (I):

(1) Types CL2P, CL3P, CL2R, CL3R, CL2, CL3, and PLTC cables
(2) A maximum of 3 m (10 ft) of exposed Type CL2X wires and cables in nonconcealed spaces
(3) A maximum of 3 m (10 ft) of exposed Type CL3X wires and cables in nonconcealed spaces
(4) Types CL2P, CL3P, CL2R, CL3R, CL2, CL3, and PLTC cables installed in the following:
   a. Plenum communications raceways
   b. Plenum cable routing assemblies
   c. Riser communications raceways
   d. Riser cable routing assemblies
   e. General-purpose communications raceways
   f. General-purpose cable routing assemblies
(5) Types CL2P, CL3P, CL2R, CL3R, CL2, CL3, CL2X, CL3X, and PLTC cables installed in raceways recognized in Chapter 3
(6) Type CMUC undercarpet communications wires and cables installed under carpet

**(L) Multifamily Dwellings.** The following wires and cables shall be permitted to be installed in multifamily dwellings in locations other than the locations covered in 725.135(B) through (I):

(1) Types CL2P, CL3P, CL2R, CL3R, CL2, CL3, and PLTC wires and cables
(2) Type CL2X wires and cables less than 6 mm (0.25 in.) in diameter in nonconcealed spaces

(3) Type CL3X wires and cables less than 6 mm (0.25 in.) in diameter in nonconcealed spaces
(4) Types CL2P, CL3P, CL2R, CL3R, CL2, CL3, and PLTC wires and cables installed in the following:
   a. Plenum communications raceways
   b. Plenum cable routing assemblies
   c. Riser communications raceways
   d. Riser cable routing assemblies
   e. General-purpose communications raceways
   f. General-purpose cable routing assemblies
(5) Types CL2P, CL3P, CL2R, CL3R, CL2, CL3, CL2X, CL3X, and PLTC wires and cables installed in raceways recognized in Chapter 3
(6) Type CMUC undercarpet communications wires and cables installed under carpet

**(M) One- and Two-Family Dwellings.** The following wires and cables shall be permitted to be installed in one- and two-family dwellings in locations other than the locations covered in 725.135(B) through (I):

(1) Types CL2P, CL3P, CL2R, CL3R, CL2, CL3, and PLTC wires and cables
(2) Type CL2X wires and cables less than 6 mm (0.25 in.) in diameter
(3) Type CL3X wires and cables less than 6 mm (0.25 in.) in diameter
(4) Communications wires and Types CL2P, CL3P, CL2R, CL3R, CL2, CL3, and PLTC cables installed in the following:
   a. Plenum communications raceways
   b. Plenum cable routing assemblies
   c. Riser communications raceways
   d. Riser cable routing assemblies
   e. General-purpose communications raceways
   f. General-purpose cable routing assemblies
(5) Types CL2P, CL3P, CL2R, CL3R, CL2, CL3, CL2X, CL3X, and PLTC wires and cables installed in raceways recognized in Chapter 3
(6) Type CMUC undercarpet communications wires and cables installed under carpet

## 725.136 Separation from Electric Light, Power, Class 1, Non–Power-Limited Fire Alarm Circuit Conductors, and Medium-Power Network-Powered Broadband Communications Cables

**(A) General.** Cables and conductors of Class 2 and Class 3 circuits shall not be placed in any cable, cable tray, compartment, enclosure, manhole, outlet box, device box, raceway, or similar fitting with conductors of electric light, power, Class 1, non–power-limited fire alarm circuits, and medium-power network-powered broadband communications circuits unless permitted by 725.136(B) through (I).

The lower voltage ratings of listed Class 2 and Class 3 cables do not allow them to be installed with electric light, power, Class 1, non–power-limited fire alarm circuits, and medium-power network-powered broadband communications cables. Failure of the cable insulation due to a fault could lead to hazardous voltages being imposed on the Class 2 or Class 3 circuit conductors.

**(B) Separated by Barriers.** Class 2 and Class 3 circuits shall be permitted to be installed together with the conductors of electric light, power, Class 1, non–power-limited fire alarm and medium power network-powered broadband communications circuits where they are separated by a barrier.

**(C) Raceways Within Enclosures.** In enclosures, Class 2 and Class 3 circuits shall be permitted to be installed in a raceway to separate them from Class 1, non–power-limited fire alarm and medium-power network-powered broadband communications circuits.

**(D) Associated Systems Within Enclosures.** Class 2 and Class 3 circuit conductors in compartments, enclosures, device boxes, outlet boxes, or similar fittings shall be permitted to be installed with electric light, power, Class 1, non–power-limited fire alarm, and medium-power network-powered broadband communications circuits where they are introduced solely to connect the equipment connected to Class 2 and Class 3 circuits, and where (1) or (2) applies:

(1) The electric light, power, Class 1, non–power-limited fire alarm, and medium-power network-powered broadband communications circuit conductors are routed to maintain a minimum of 6 mm (0.25 in.) separation from the conductors and cables of Class 2 and Class 3 circuits.
(2) The circuit conductors operate at 150 volts or less to ground and also comply with one of the following:
   a. The Class 2 and Class 3 circuits are installed using Type CL3, CL3R, or CL3P or permitted substitute cables, provided these Class 3 cable conductors extending beyond the jacket are separated by a minimum of 6 mm (0.25 in.) or by a nonconductive sleeve or nonconductive barrier from all other conductors.
   b. The Class 2 and Class 3 circuit conductors are installed as a Class 1 circuit in accordance with 725.41.

An example of associated systems is where the Class 2 circuit source is the secondary of a control transformer in the same motor-starter enclosure. In such an installation, the Class 2 conductor insulation is not required to have the same voltage rating as the insulation on the power conductors in the same enclosure.

**(E) Enclosures with Single Opening.** Class 2 and Class 3 circuit conductors entering compartments, enclosures, device boxes, outlet boxes, or similar fittings shall be permitted to be installed with Class 1, non–power-limited fire alarm and medium-power network-powered broadband communications circuits where they are introduced solely to connect the equipment connected to Class 2 and Class 3 circuits. Where Class 2 and Class 3 circuit conductors must enter an enclosure that is provided with a single opening, they shall be permitted to enter through a single fitting (such as a tee), provided the conductors are separated from the conductors of the other circuits by a continuous and firmly fixed nonconductor, such as flexible tubing.

**(F) Manholes.** Underground Class 2 and Class 3 circuit conductors in a manhole shall be permitted to be installed with Class 1, non–power-limited fire alarm and medium-power network-powered broadband communications circuits where one of the following conditions is met:

(1) The electric light, power, Class 1, non–power-limited fire alarm and medium-power network-powered broadband communications circuit conductors are in a metal-enclosed cable or Type UF cable.
(2) The Class 2 and Class 3 circuit conductors are permanently and effectively separated from the conductors of other circuits by a continuous and firmly fixed nonconductor, such as flexible tubing, in addition to the insulation or covering on the wire.
(3) The Class 2 and Class 3 circuit conductors are permanently and effectively separated from conductors of the other circuits and securely fastened to racks, insulators, or other approved supports.

See the commentary following 725.48(B)(3).

**(G) Cable Trays.** Class 2 and Class 3 circuit conductors shall be permitted to be installed in cable trays, where the conductors of the electric light, Class 1, and non–power-limited fire alarm circuits are separated by a solid fixed barrier of a material compatible with the cable tray or where the Class 2 or Class 3 circuits are installed in Type MC cable.

**(H) In Hoistways.** In hoistways, Class 2 or Class 3 circuit conductors shall be installed in rigid metal conduit, rigid nonmetallic conduit, intermediate metal conduit, liquidtight flexible nonmetallic conduit, or electrical metallic tubing. For elevators or similar equipment, these conductors shall be permitted to be installed as provided in 620.21.

**(I) Other Applications.** For other applications, conductors of Class 2 and Class 3 circuits shall be separated by at least 50 mm (2 in.) from conductors of any electric light, power, Class 1 non–power-limited fire alarm or medium power network-powered broadband communications circuits unless one of the following conditions is met:

(1) Either (a) all of the electric light, power, Class 1, non–power-limited fire alarm and medium-power network-powered broadband communications circuit conductors or (b) all of the Class 2 and Class 3 circuit conductors are in a raceway or in metal-sheathed, metal-clad, non–metallic-sheathed, or Type UF cables.
(2) All of the electric light, power, Class 1 non–power-limited fire alarm, and medium-power network-powered broadband communications circuit conductors are permanently separated from all of the Class 2 and Class 3 circuit conductors by a continuous and firmly fixed nonconductor, such as porcelain tubes or flexible tubing, in addition to the insulation on the conductors.

## 725.139 Installation of Conductors of Different Circuits in the Same Cable, Enclosure, Cable Tray, Raceway, or Cable Routing Assembly

**(A) Two or More Class 2 Circuits.** Conductors of two or more Class 2 circuits shall be permitted within the same cable, enclosure, raceway, or cable routing assembly.

**(B) Two or More Class 3 Circuits.** Conductors of two or more Class 3 circuits shall be permitted within the same cable, enclosure, raceway, or cable routing assembly.

**(C) Class 2 Circuits with Class 3 Circuits.** Conductors of one or more Class 2 circuits shall be permitted within the same cable, enclosure, raceway, or cable routing assembly with conductors of Class 3 circuits, provided that the insulation of the Class 2 circuit conductors in the cable, enclosure, raceway, or cable routing assembly is at least that required for Class 3 circuits.

**(D) Class 2 and Class 3 Circuits with Communications Circuits.**

**(1) Classified as Communications Circuits.** Class 2 and Class 3 circuit conductors shall be permitted in the same cable with communications circuits, in which case the Class 2 and Class 3 circuits shall be classified as communications circuits and shall be installed in accordance with the requirements of Article 800. The cables shall be listed as communications cables.

**(2) Composite Cables.** Cables constructed of individually listed Class 2, Class 3, and communications cables under a common jacket shall be permitted to be classified as communications cables. The fire resistance rating of the composite cable shall be determined by the performance of the composite cable.

**(E) Class 2 or Class 3 Cables with Other Circuit Cables.** Jacketed cables of Class 2 or Class 3 circuits shall be permitted in the same enclosure, cable tray, raceway, or cable routing assembly with jacketed cables of any of the following:

(1) Power-limited fire alarm systems in compliance with Parts I and III of Article 760
(2) Nonconductive and conductive optical fiber cables in compliance with Parts I and IV of Article 770
(3) Communications circuits in compliance with Parts I and IV of Article 800
(4) Community antenna television and radio distribution systems in compliance with Parts I and IV of Article 820
(5) Low-power, network-powered broadband communications in compliance with Parts I and IV of Article 830

**(F) Class 2 or Class 3 Conductors or Cables and Audio System Circuits.** Audio system circuits described in 640.9(C), and installed using Class 2 or Class 3 wiring methods in compliance with 725.133 and 725.154, shall not be permitted to be installed in the same cable, raceway, or cable routing assembly with Class 2 or Class 3 conductors or cables.

### 725.141 Installation of Circuit Conductors Extending Beyond One Building

Where Class 2 or Class 3 circuit conductors extend beyond one building and are run so as to be subject to accidental contact with electric light or power conductors operating over 300 volts to ground, or are exposed to lightning on interbuilding circuits on the same premises, the requirements of the following shall also apply:

(1) Sections 800.44, 800.50, 800.53, 800.93, 800.100, 800.170(A), and 800.170(B) for other than coaxial conductors

(2) Sections 820.44, 820.93, and 820.100 for coaxial conductors

### 725.143 Support of Conductors

Class 2 or Class 3 circuit conductors shall not be strapped, taped, or attached by any means to the exterior of any conduit or other raceway as a means of support. These conductors shall be permitted to be installed as permitted by 300.11(B)(2).

See the commentary following 725.24 for more information on the support of conductors.

### 725.154 Applications of Listed Class 2, Class 3, and PLTC Cables

Class 2, Class 3, and PLTC cables shall comply with any of the requirements described in 725.154(A) through (C) and as indicated in Table 725.154.

•

**(A) Class 2 and Class 3 Cable Substitutions.** The substitutions for Class 2 and Class 3 cables illustrated in Figure 725.154(A) shall be permitted. Where substitute cables are installed, the wiring requirements of Article 725, Parts I and III, shall apply.

Informational Note: For information on Types CMP, CMR, CM, and CMX, see 800.179.

**(B) Class 2, Class 3, PLTC Circuit Integrity (CI) Cable or Electrical Circuit Protective System.** Circuit integrity (CI) cable or a listed electrical circuit protective system shall be permitted for use in remote control, signaling, or power-limited systems that supply critical circuits to ensure survivability for continued circuit operation for a specified time under fire conditions.

**(C) Thermocouple Circuits.** Conductors in Type PLTC cables used for Class 2 thermocouple circuits shall be permitted to be any of the materials used for thermocouple extension wire.

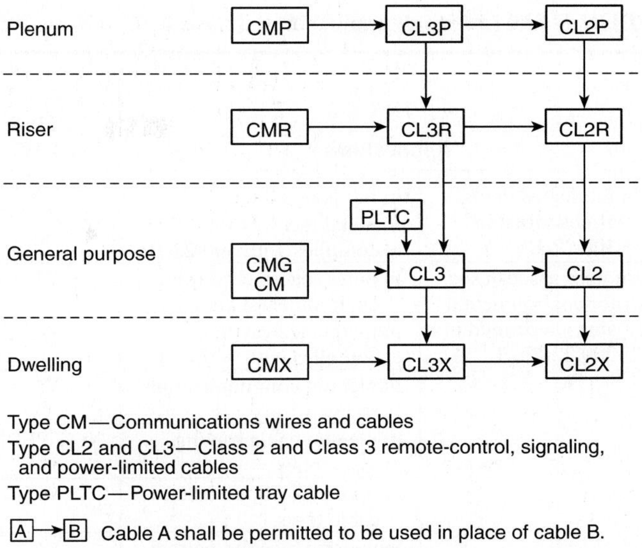


***FIGURE 725.154(A)** Cable Substitution Hierarchy.*

## IV. Listing Requirements

### 725.179 Listing and Marking of Class 2, Class 3, and Type PLTC Cables; Communications Raceways; and Cable Routing Assemblies

Class 2, Class 3, and Type PLTC cables, nonmetallic signaling raceways and cable routing assemblies installed as wiring methods within buildings shall be listed as being resistant to the spread of fire and other criteria in accordance with 725.179(A) through (J) and shall be marked in accordance with 725.179 (K).

**(A) Types CL2P and CL3P.** Types CL2P and CL3P plenum cable shall be listed as being suitable for use in ducts, plenums, and other space for environmental air and shall also be listed as having adequate fire-resistant and low-smoke producing characteristics.

Informational Note: One method of defining a cable that is low-smoke producing cable and fire-resistant cable is that the cable exhibits a maximum peak optical density of 0.50 or less, an average optical density of 0.15 or less, and a maximum flame spread distance of 1.52 m (5 ft) or less when tested in accordance with NFPA 262-2011, *Standard Method of Test for Flame Travel and Smoke of Wires and Cables for Use in Air-Handling Spaces.*

NFPA 262, *Standard Method of Test for Flame Travel and Smoke of Wires and Cables for Use in Air-Handling Spaces,* is a test method for electrical wires and cables that are to be installed without raceways in plenums and other spaces used for environmental air. NFPA 262 was originally developed as an adaptation of the Steiner Tunnel test (ASTM E 84/UL 723, *Standard Method of Test of Surface Burning Characteristics of Building Materials*).

NFPA 262 does not list pass/fail criteria. The criteria for acceptance of a given application are given in the appropriate

***TABLE 725.154*** *Applications of Listed Class 2, Class 3, and PLTC Cables in Buildings*

| Applications | | Wire and Cable Type | | | | | |
|---|---|---|---|---|---|---|---|
| | | **CL2P & CL3P** | **CL2R & CL3R** | **CL2 & CL3** | **CL2X & CL3X** | **CMUC** | **PLTC** |
| In fabricated ducts as described in 300.22(B) | In fabricated ducts | Y* | N | N | N | N | N |
| | In metal raceway that complies with 300.22(B) | Y* | Y* | Y* | Y* | N | Y* |
| In other spaces used for environmental air as described in 300.22(C) | In other spaces used for environmental air | Y* | N | N | N | N | N |
| | In metal raceway that complies with 300.22(C) | Y* | Y* | Y* | Y* | N | Y* |
| | In plenum communications raceways | Y* | N | N | N | N | N |
| | In plenum cable routing assemblies | NOT PERMITTED | | | | | |
| | Supported by open metal cable trays | Y* | N | N | N | N | N |
| | Supported by solid bottom metal cable trays with solid metal covers | Y* | Y* | Y* | Y* | N | N |
| In risers | In vertical runs | Y* | Y* | N | N | N | N |
| | In metal raceways | Y* | Y* | Y* | Y* | N | Y* |
| | In fireproof shafts | Y* | Y* | Y* | Y* | N | Y* |
| | In plenum communications raceways | Y* | Y* | N | N | N | N |
| | In plenum cable routing assemblies | Y* | Y* | N | N | N | N |
| | In riser communications raceways | Y* | Y* | N | N | N | N |
| | In riser cable routing assemblies | Y* | Y* | N | N | N | N |
| | In one- and two-family dwellings | Y* | Y* | Y* | Y* | N | Y* |
| Within buildings in other than air-handling spaces and risers | General | Y* | Y* | Y* | Y* | N | Y* |
| | In one- and two-family dwellings | Y* | Y* | Y* | Y* | Y* | Y* |
| | In multifamily dwellings | Y* | Y* | Y* | Y* | Y* | Y* |
| | In nonconcealed spaces | Y* | Y* | Y* | Y* | Y* | Y* |
| | Supported by cable trays | Y* | Y* | Y* | N | N | Y* |
| | Under carpet | N | N | N | N | Y* | N |
| | In cross-connect arrays | Y* | Y* | Y* | N | N | Y* |
| | In any raceway recognized in Chapter 3 | Y* | Y* | Y* | Y* | N | Y* |
| | In plenum communications raceways | Y* | Y* | Y* | N | N | Y* |
| | In plenum cable routing assemblies | Y* | Y* | Y* | N | N | Y* |
| | In riser communications raceways | Y* | Y* | Y* | N | N | Y* |
| | In riser cable routing assemblies | Y* | Y* | Y* | N | N | Y* |
| | In general-purpose communications raceways | Y* | Y* | Y* | N | N | Y* |
| | In general-purpose cable routing assemblies | Y* | Y* | Y* | N | N | Y* |

Note: An "N" in the table indicates that the cable type shall not be permitted to be installed in the application. A "Y*" indicates that the cable shall be permitted to be installed in the application, subject to the limitations described in 725.130 through 725.143.

sections of the *Code*. See the informational notes that follow 725.179(A), 760.179(D), 770.179(A), 800.182, and 820.179(A).

A Class 2 or Class 3 cable that has passed the requirements of this test may be used in ducts, plenums, or other air-handling spaces. In addition, such cable may be used anywhere in a building where Class 2 or Class 3 cable is permitted. [See Table 725.154(G).]

**(B) Types CL2R and CL3R.** Types CL2R and CL3R riser cables shall be marked as Type CL2R or CL3R, respectively, and be listed as suitable for use in a vertical run in a shaft or from floor to floor and shall also be listed as having fire-resistant characteristics capable of preventing the carrying of fire from floor to floor.

Informational Note: One method of defining fire-resistant characteristics capable of preventing the carrying of fire from floor to floor is that the cables pass the requirements of ANSI/UL 1666-2012, *Test for Flame Propagation Height of Electrical and Optical-Fiber Cable Installed Vertically in Shafts.*

In the fire test covered in UL 1666, *Test for Flame Propagation Height of Electrical and Optical-Fiber Cables Installed Vertically in Shafts*, cables are arranged in a simulated vertical shaft and subjected to an ignition source. The shaft is a 19-foot-high concrete shaft divided into two compartments by a 1-foot by 2-foot opening at the 12-foot level. To pass, cables must not propagate flame to the top of the 12-foot-high compartment during the 30-minute test.

**(C) Types CL2 and CL3.** Types CL2 and CL3 cables shall be marked as Type CL2 or CL3, respectively, and be listed as suitable for general-purpose use, with the exception of risers, ducts, plenums, and other space used for environmental air, and shall also be listed as being resistant to the spread of fire.

Informational Note: One method of defining *resistant to the spread of fire* is that the cables do not spread fire to the top of the tray in the "UL Flame Exposure, Vertical Tray Flame Test" in ANSI/UL 1685-2010, *Standard for Safety for Vertical-Tray Fire-Propagation and Smoke-Release Test for Electrical and Optical-Fiber Cables*. The smoke measurements in the test method are not applicable.

Another method of defining *resistant to the spread of fire* is for the damage (char length) not to exceed 1.5 m (4 ft 11 in.) when performing the CSA "Vertical Flame Test — Cables in Cable Trays," as described in CSA C22.2 No. 0.3-M-2001, *Test Methods for Electrical Wires and Cables*.

The UL vertical tray flame test determines whether cables installed in a ladder-type cable tray will propagate fire from a given exposure. The samples are considered to have passed the test if flame has not propagated to the top of the cable tray by the conclusion of the 20-minute test. A cable that passes this testing may be listed as a Type CL2 or Type CL3 cable. [See also the commentary following the informational note to 725.179(D).]

**(D) Types CL2X and CL3X.** Types CL2X and CL3X limited-use cables shall be marked as Type CL2X or CL3X, respectively, and be listed as being suitable for use in dwellings and for use in raceway and shall also be listed as being resistant to flame spread.

Informational Note: One method of determining that cable is resistant to flame spread is by testing the cable to the VW-1 (vertical wire) flame test in ANSI/UL 1581-2011, *Reference Standard for Electrical Wires, Cables and Flexible Cords.*

UL 1581, *Reference Standard for Electrical Wires, Cables and Flexible Cords*, contains basic requirements for conductors, insulation, jackets, and other coverings and the methods of sample preparation, specimen selection and conditioning, and measurements and calculations required in UL 44, *Thermoset-Insulated Wires and Cables*; UL 83, *Thermoplastic-Insulated Wires and Cables*; and UL 62, *Flexible Cord and Fixture Wire*. The flame test methods of these standards include the vertical wire flame test (VW-1) and the vertical tray flame test. [Also see the commentary following the informational note to 725.179(C).]

A sample that passes the test may be listed as Type CL2X or Type CL3X, suitable for use as a limited-use, power-limited cable. This cable may be used in one-family, two-family, and multifamily dwellings.

**(E) Type PLTC.** Type PLTC nonmetallic-sheathed, power-limited tray cable shall be listed as being suitable for cable trays and shall consist of a factory assembly of two or more insulated conductors under a nonmetallic jacket. The insulated conductors shall be 22 AWG through 12 AWG. The conductor material shall be copper (solid or stranded). Insulation on conductors shall be rated for 300 volts. The cable core shall be either (1) two or more parallel conductors, (2) one or more group assemblies of twisted or parallel conductors, or (3) a combination thereof. A metallic shield or a metallized foil shield with drain wire(s) shall be permitted to be applied either over the cable core, over groups of conductors, or both. The cable shall be listed as being resistant to the spread of fire. The outer jacket shall be a sunlight- and moisture-resistant nonmetallic material. Type PLTC cable used in a wet location shall be listed for use in wet locations or have a moisture-impervious metal sheath.

*Exception No. 1: Where a smooth metallic sheath, continuous corrugated metallic sheath, or interlocking tape armor is applied over the nonmetallic jacket, an overall nonmetallic jacket shall not be required. On metallic-sheathed cable without an overall nonmetallic jacket, the information required in 310.120 shall be located on the nonmetallic jacket under the sheath.*

*Exception No. 2: Conductors in PLTC cables used for Class 2 thermocouple circuits shall be permitted to be any of the materials used for thermocouple extension wire.*

Informational Note: One method of defining *resistant to the spread of fire* is that the cables do not spread fire to the top of the tray in the "UL Flame Exposure, Vertical Tray Flame Test" in ANSI/UL 1685-2010, *Standard for Safety for Vertical-Tray Fire-Propagation and Smoke-Release Test for Electrical and Optical-Fiber Cables*. The smoke measurements in the test method are not applicable.

Another method of defining *resistant to the spread of fire* is for the damage (char length) not to exceed 1.5 m (4 ft 11 in.)

when performing the CSA "Vertical Flame Test — Cables in Cable Trays," as described in CSA C22.2 No. 0.3-M-2001, *Test Methods for Electrical Wires and Cables*.

**(F) Circuit Integrity (CI) Cable or Electrical Circuit Protective System.** Cables that are used for survivability of critical circuits under fire conditions shall meet either 725.179(F)(1) or (F)(2) as follows:

**(1) Circuit Integrity (CI) Cables.** Circuit Integrity (CI) cables, specified in 725.154(A), (B), and used for survivability of critical circuits, shall have the additional classification using the suffix "CI." Circuit integrity (CI) cables shall only be permitted to be installed in a raceway where specifically listed and marked as part of an electrical circuit protective system as covered in 725.179(F)(2).

**(2) Electrical Circuit Protective System.** Cables specified in 725.154(A), (B), and (F)(1) that are part of an electrical circuit protective system shall be identified with the protective system number and hourly rating printed on the outer jacket of the cable and installed in accordance with the listing of the protective system.

Informational Note No. 1: One method of defining circuit integrity (CI) cable or an electrical circuit protective system is by establishing a minimum 2-hour fire-resistive rating when tested in accordance with UL 2196-2012, *Standard for Tests of Fire Resistive Cables*.

Informational Note No. 2: UL guide information for electrical circuit protective systems (FHIT) contains information on proper installation requirements to maintain the fire rating.

Section 725.179(F) permits the use of circuit integrity (CI) cable for applications where continuity of the operations of critical circuits is needed during a fire. Such circuits could be essential to firefighting operations or could be circuits whose interruption could cause a more dangerous condition to occur. A smoke removal system is an example of where it may be necessary to use CI cables for control circuits to ensure that the dampers operate during a fire.

**(G) Class 2 and Class 3 Cable Voltage Ratings.** Class 2 cables shall have a voltage rating of not less than 150 volts. Class 3 cables shall have a voltage rating of not less than 300 volts.

**(H) Class 3 Single Conductors.** Class 3 single conductors used as other wiring within buildings shall not be smaller than 18 AWG and shall be Type CL3. Conductor types described in 725.49(B) that are also listed as Type CL3 shall be permitted.

Informational Note: One method of defining *resistant to the spread of fire* is that the cables do not spread fire to the top of the tray in the "UL Flame Exposure, Vertical Tray Flame Test" in ANSI/UL 1685-2010, *Standard for Safety for Vertical-Tray Fire-Propagation and Smoke-Release Test for Electrical and Optical-Fiber Cables*. The smoke measurements in the test method are not applicable.

Another method of defining *resistant to the spread of fire* is for the damage (char length) not to exceed 1.5 m (4 ft 11 in.) when performing the CSA "Vertical Flame Test — Cables in Cable Trays," as described in CSA C22.2 No. 0.3-M-2001, *Test Methods for Electrical Wires and Cables*.

•

**(I) Riser Cable Routing Assemblies.** Riser cable routing assemblies shall be listed as having fire-resistant characteristics capable of preventing the carrying of fire from floor to floor.

Informational Note: One method of defining fire-resistant characteristics capable of preventing the carrying of fire from floor to floor is that the cable routing assemblies pass the requirements of the test for flame propagation (riser) in Subject 2024A, *UL Outline of Investigation for Cable Routing Assemblies*.

Raceways listed for risers must pass fire tests similar to those required of cables listed for risers.

**(J) General-Use Cable Routing Assemblies.** General-use cable routing assemblies shall be listed as being resistant to the spread of fire.

Informational Note: One method of defining resistance to the spread of fire is that the cable routing assemblies pass the requirements of the vertical tray flame test (general use) in Subject 2024A, *UL Outline of Investigation for Cable Routing Assemblies*.

Raceways listed for general-purpose applications must pass fire tests similar to those required of cables listed for general-purpose applications.

**(K) Marking.** Cables shall be marked in accordance with 310.120(A)(2), (A)(3), (A)(4), and (A)(5) and Table 725.179(K). Voltage ratings shall not be marked on the cables.

Informational Note: Voltage markings on cables may be misinterpreted to suggest that the cables may be suitable for Class 1 electric light and power applications.

***TABLE 725.179(K)*** *Cable Marking*

| Cable Marking | Type |
|---|---|
| CL3P | Class 3 plenum cable |
| CL2P | Class 2 plenum cable |
| CL3R | Class 3 riser cable |
| CL2R | Class 2 riser cable |
| PLTC | Power-limited tray cable |
| CL3 | Class 3 cable |
| CL2 | Class 2 cable |
| CL3X | Class 3 cable, limited use |
| CL2X | Class 2 cable, limited use |

Informational Note: Class 2 and Class 3 cable types are listed in descending order of fire resistance rating, and Class 3 cables are listed above Class 2 cables because Class 3 cables can substitute for Class 2 cables.

*Exception: Voltage markings shall be permitted where the cable has multiple listings and a voltage marking is required for one or more of the listings.*

# ARTICLE 727 Instrumentation Tray Cable: Type ITC

## 727.1 Scope

This article covers the use, installation, and construction specifications of instrumentation tray cable for application to instrumentation and control circuits operating at 150 volts or less and 5 amperes or less.

Article 727 permits an alternate wiring method for circuits that do not exceed 5 amperes and 150 volts. Instrument tray cable is particularly suited for instrumentation circuits in industrial establishments where qualified persons perform service and maintenance.

## 727.2 Definition

**Type ITC Instrumentation Tray Cable.** A factory assembly of two or more insulated conductors, with or without a grounding conductor(s), enclosed in a nonmetallic sheath.

## 727.3 Other Articles

In addition to the provisions of this article, installation of Type ITC cable shall comply with other applicable articles of this *Code*.

## 727.4 Uses Permitted

Type ITC cable shall be permitted to be used as follows in industrial establishments where the conditions of maintenance and supervision ensure that only qualified persons service the installation:

(1) In cable trays.
(2) In raceways.
(3) In hazardous locations as permitted in 501.10, 502.10, 503.10, 504.20, 504.30, 504.80, and 505.15.
(4) Enclosed in a smooth metallic sheath, continuous corrugated metallic sheath, or interlocking tape armor applied over the nonmetallic sheath in accordance with 727.6. The cable shall be supported and secured at intervals not exceeding 1.8 m (6 ft).
(5) Cable, without a metallic sheath or armor, that complies with the crush and impact requirements of Type MC cable and is identified for such use with the marking *ITC-ER* shall be permitted to be installed exposed. The cable shall be continuously supported and protected against physical damage using mechanical protection such as dedicated struts, angles, or channels. The cable shall be secured at intervals not exceeding 1.8 m (6 ft).
(6) As aerial cable on a messenger.
(7) Direct buried where identified for the use.
(8) Under raised floors in rooms containing industrial process control equipment and rack rooms where arranged to prevent damage to the cable.
(9) Under raised floors in information technology equipment rooms in accordance with 645.5(E)(5)(b).

## 727.5 Uses Not Permitted

Type ITC cable shall not be installed on circuits operating at more than 150 volts or more than 5 amperes.

Installation of Type ITC cable with other cables shall be subject to the stated provisions of the specific articles for the other cables. Where the governing articles do not contain stated provisions for installation with Type ITC cable, the installation of Type ITC cable with the other cables shall not be permitted.

Type ITC cable shall not be installed with power, lighting, Class 1 circuits that are not power limited, or non–power-limited circuits.

*Exception No. 1: Where terminated within equipment or junction boxes and separations are maintained by insulating barriers or other means.*

*Exception No. 2: Where a metallic sheath or armor is applied over the nonmetallic sheath of the Type ITC cable.*

## 727.6 Construction

The insulated conductors of Type ITC cable shall be in sizes 22 AWG through 12 AWG. The conductor material shall be copper or thermocouple alloy. Insulation on the conductors shall be rated for 300 volts. Shielding shall be permitted.

The cable shall be listed as being resistant to the spread of fire. The outer jacket shall be sunlight and moisture resistant.

Where a smooth metallic sheath, continuous corrugated metallic sheath, or interlocking tape armor is applied over the nonmetallic sheath, an overall nonmetallic jacket shall not be required.

Informational Note: One method of defining *resistant to the spread of fire* is that the cables do not spread fire to the top of the tray in the "UL Flame Exposure, Vertical Tray Flame Test" in ANSI/UL 1685-2010, *Standard for Safety for Vertical-Tray Fire-Propagation and Smoke-Release Test for Electrical and Optical-Fiber Cables*. The smoke measurements in the test method are not applicable.

Another method of defining *resistant to the spread of fire* is for the damage (char length) not to exceed 1.5 m (4 ft 11 in.) when performing the CSA "Vertical Flame Test — Cables in Cable Trays," as described in CSA C22.2 No. 0.3-M-2001, *Test Methods for Electrical Wires and Cables*.

## 727.7 Marking

The cable shall be marked in accordance with 310.120(A)(2), (A)(3), (A)(4), and (A)(5). Voltage ratings shall not be marked on the cable.

### 727.8 Allowable Ampacity

The allowable ampacity of the conductors shall be 5 amperes, except for 22 AWG conductors, which shall have an allowable ampacity of 3 amperes.

### 727.9 Overcurrent Protection

Overcurrent protection shall not exceed 5 amperes for 20 AWG and larger conductors, and 3 amperes for 22 AWG conductors.

### 727.10 Bends

Bends in Type ITC cables shall be made so as not to damage the cable.

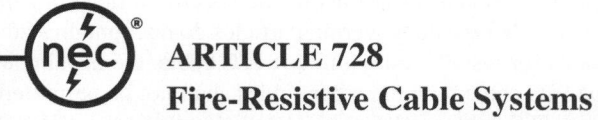

## ARTICLE 728 Fire-Resistive Cable Systems

### 728.1 Scope

This article covers the installation of fire-resistive cables, fire-resistive conductors, and other system components used for survivability of critical circuits to ensure continued operation during a specified time under fire conditions as required in this *Code*.

### 728.2 Definition

**Fire-Resistive Cable System.** A cable and components used to ensure survivability of critical circuits for a specified time under fire conditions.

### 728.3 Other Articles

Wherever the requirements of other articles of this *Code* and Article 728 differ, the requirements of Article 728 shall apply.

### 728.4 General

Fire-resistive cables, fire-resistive conductors, and components shall be tested and listed as a complete system, shall be designated for use in a specific fire-rated system, and shall not be interchangeable between systems. Fire-resistive cables, conductors, and components shall be approved.

Informational Note No. 1: One method of defining the fire rating is by testing the system in accordance with UL 2196-2012, *Standard for Tests of Fire Resistive Cables*.

Informational Note No. 2: Fire-resistive cable systems are considered part of an electrical circuit protective system.

### 728.5 Installations

Fire-resistive cable systems installed outside the fire-rated rooms that they serve, such as the electrical room or the fire pump room, shall comply with the requirements of 728.5(A) through (H) and all other installation instructions provided in the listing.

**(A) Mounting.** The fire-resistive cable system shall be secured to the building structure in accordance with the listing and the manufacturer's installation instructions.

**(B) Supports.** The fire-resistive system shall be supported in accordance with the listing and the manufacturer's installation instructions.

Informational Note: The supports are critical for survivability of the system. Each system has its specific support requirements.

**(C) Raceways and Couplings.** Where the fire-resistive system is listed to be installed in a raceway, the raceways enclosing the system, any couplings, and connectors shall be listed as part of the fire-rated system.

**(D) Cable Trays.** Cable trays used as part of a fire-resistive system shall be listed as part of the fire-resistive system.

**(E) Boxes.** Boxes or enclosures used as part of a fire-resistive system shall be listed as part of the fire-resistive system and shall be secured to the building structure independently of the raceways or cables listed in the system.

**(F) Pulling Lubricants.** Fire-resistive cable systems installed in a raceway shall only use pulling lubricants listed as part of the fire-resistive cable system.

**(G) Vertical Supports.** Cables and conductors installed in vertical raceways shall be supported in accordance with the listing of the fire-resistive cable system.

**(H) Splices.** Only splices that are part of the listing for the fire-resistive cable system shall be used. Splices shall have manufacturer's installation instructions.

### 728.60 Grounding

Fire-resistive systems installed in a raceway requiring an equipment grounding conductor shall use the same fire-rated cable described in the system, unless alternative equipment grounding conductors are listed with the system. Any alternative equipment grounding conductor shall be marked with the system number. The system shall specify a permissible equipment grounding conductor. If not specified, the equipment grounding conductor shall be the same as the fire-rated cable described in the system.

### 728.120 Marking

In addition to the marking required in 310.120, system cables and conductors shall be surface marked with the suffix "FRR" (fire-resistive rating), along with the circuit integrity duration in hours, and with the system identifier.

# ARTICLE 750
# Energy Management Systems

## 750.1 Scope

This article applies to the installation and operation of energy management systems.

Informational Note: Performance provisions in other codes establish prescriptive requirements that may further restrict the requirements contained in this article.

This new article addresses the installation and operation of energy management systems. Energy management systems have become integral elements of the electrical infrastructure through the control of utilization equipment, energy storage, and power protection. Energy management has two basic aspects, monitoring the system and controlling some part of the system. These two elements must be separated in order to allow the system to monitor and possibly restrict those areas of control that would adversely impact the electrical system. These requirements ensure that an energy management system does not overload a branch circuit, feeder, or service or override a load-shedding system for an alternate power source for fire pumps and other emergency systems.

## 750.2 Definitions

For the purpose of this article, the following definitions shall apply.

**Control.** The predetermined process of connecting, disconnecting, increasing, or reducing electric power.

**Energy Management System.** A system consisting of any of the following: a monitor(s), communications equipment, a controller(s), a timer(s), or other device(s) that monitors and /or controls an electrical load or a power production or storage source.

**Monitor.** An electrical or electronic means to observe, record, or detect the operation or condition of the electric power system or apparatus.

## 750.20 Alternate Power Sources

An energy management system shall not override any control necessary to ensure continuity of an alternate power source for the following:

(1) Fire pumps
(2) Health care facilities
(3) Emergency systems
(4) Legally required standby systems
(5) Critical operations power systems

## 750.30 Load Management

Energy management systems shall be permitted to monitor and control electrical loads unless restricted in accordance with 750.30(A) through (C).

**(A) Load Shedding Controls.** An energy management system shall not override the load shedding controls put in place to ensure the minimum electrical capacity for the following:

(1) Fire pumps
(2) Emergency systems
(3) Legally required standby systems
(4) Critical operations power systems

**(B) Disconnection of Power.** An energy management system shall not be permitted to cause disconnection of power to the following:

(1) Elevators, escalators, moving walks, or stairway lift chairs
(2) Positive mechanical ventilation for hazardous (classified) locations
(3) Ventilation used to exhaust hazardous gas or reclassify an area
(4) Circuits supplying emergency lighting
(5) Essential electrical systems in health care facilities

Systems necessary for life safety, fire protection, and critical operations must continue to operate even with the loss of primary power. Load shedding is often employed for the alternate power source of a backup system to give priority to needed equipment. Disconnecting some equipment – such as a ventilation system that prevents an explosive concentration from being reached – may introduce safety hazards and must be avoided. Therefore, an energy management system must not disconnect or override the control of these systems.

**(C) Capacity of Branch Circuit, Feeder, or Service.** An energy management system shall not cause a branch circuit, feeder, or service to be overloaded at any time.

## 750.50 Field Markings

Where an energy management system is employed to control electrical power through the use of a remote means, a directory identifying the controlled device(s) and circuit(s) shall be posted on the enclosure of the controller, disconnect, or branch-circuit overcurrent device.

Informational Note: The use of the term *remote* is intended to convey that a controller can be operated via another means or location through communications without a direct operator interface with the controlled device.

# ARTICLE 760
# Fire Alarm Systems

## I. General

### 760.1 Scope

This article covers the installation of wiring and equipment of fire alarm systems including all circuits controlled and powered by the fire alarm system.

Informational Note No. 1: Fire alarm systems include fire detection and alarm notification, guard's tour, sprinkler waterflow, and sprinkler supervisory systems. Circuits controlled and powered by the fire alarm system include circuits for the control of building systems safety functions, elevator capture, elevator shutdown, door release, smoke doors and damper control, fire doors and damper control and fan shutdown, but only where these circuits are powered by and controlled by the fire alarm system. For further information on the installation and monitoring for integrity requirements for fire alarm systems, refer to the *NFPA 72*-2013, *National Fire Alarm and Signaling Code*.

Informational Note No. 2: Class 1, 2, and 3 circuits are defined in Article 725.

Article 760 covers only circuits that are powered and controlled by the fire alarm system, including fire safety features such as smoke door control, damper control, fan shutdown, and elevator recall. Circuits powered and controlled by other building systems such as heating, ventilating, and air conditioning (HVAC); security; lighting controls; and time recording are covered by Article 725.

*NFPA 72, National Fire Alarm and Signaling Code,* requires that all wiring, cable, and equipment be in accordance with *NFPA 70, National Electrical Code,* and specifically with Article 760. Article 760 covers the wiring between the devices and equipment required by *NFPA 72. NFPA 72* provides the requirements for the listing, selection, installation, performance, use, testing, and maintenance of fire alarm system components. For whether a specific occupancy is required to have a fire alarm system, see NFPA *101, Life Safety Code,* or other local codes.

Examples of fire alarm equipment and devices are shown in Exhibits 760.1 and 760.2. The installation of these system components is covered by *NFPA 72,* but the circuit wiring associated with these components must be installed in accordance with the requirements of Article 760. Single- and multiple-station smoke alarms, such as those commonly installed in dwelling units, are supplied through 120-volt branch circuits rather than through a fire alarm signaling circuit that is powered and controlled by a fire alarm control panel. Branch circuits supplying power to single- and multiple-station smoke alarms are not subject to the requirements of Article 760.

***EXHIBIT 760.1*** *Typical fire alarm control unit. (Courtesy of the International Association of Electrical Inspectors)*

***EXHIBIT 760.2*** *Typical spot-type smoke detector.*

### 760.2 Definitions

**Abandoned Fire Alarm Cable.** Installed fire alarm cable that is not terminated at equipment other than a connector and not identified for future use with a tag.

**Fire Alarm Circuit.** The portion of the wiring system between the load side of the overcurrent device or the power-limited supply and the connected equipment of all circuits powered and controlled by the fire alarm system. Fire alarm circuits are classified as either non–power-limited or power-limited.

**Fire Alarm Circuit Integrity (CI) Cable.** Cable used in fire alarm systems to ensure continued operation of critical circuits during a specified time under fire conditions.

**Non–Power-Limited Fire Alarm Circuit (NPLFA).** A fire alarm circuit powered by a source that complies with 760.41 and 760.43.

**Power-Limited Fire Alarm Circuit (PLFA).** A fire alarm circuit powered by a source that complies with 760.121.

## 760.3 Other Articles

Circuits and equipment shall comply with 760.3(A) through (K). Only those sections of Article 300 referenced in this article shall apply to fire alarm systems.

**(A) Spread of Fire or Products of Combustion.** See 300.21.

**(B) Ducts, Plenums, and Other Air-Handling Spaces.** Section 300.22, where installed in ducts or plenums or other spaces used for environmental air.

*Exception: As permitted in 760.53(B)(1) and (B)(2) and Table 760.154.*

See the commentary following 300.22(B) and (C) for more information on wiring installed in ducts, plenums, or other air-handling spaces.

**(C) Hazardous (Classified) Locations.** Articles 500 through 516 and Article 517, Part IV, where installed in hazardous (classified) locations.

**(D) Corrosive, Damp, or Wet Locations.** Sections 110.11, 300.5(B), 300.6, 300.9, and 310.10(G), where installed in corrosive, damp, or wet locations.

Cables and equipment used in wet or damp locations, high ambient temperature areas, or corrosive locations must be identified as suitable for the particular use. Underground installations are considered wet locations.

**(E) Building Control Circuits.** Article 725, where building control circuits (e.g., elevator capture, fan shutdown) are associated with the fire alarm system.

**(F) Optical Fiber Cables.** Where optical fiber cables are utilized for fire alarm circuits, the cables shall be installed in accordance with Article 770.

**(G) Installation of Conductors with Other Systems.** Installations shall comply with 300.8.

**(H) Raceways or Sleeves Exposed to Different Temperatures.** Installations shall comply with 300.7(A).

Condensation often forms in conduit exposed to nonconditioned and conditioned spaces. Section 300.7(A) requires blocking the circulation of warm air into the colder section of the raceway.

**(I) Vertical Support for Fire Rated Cables and Conductors.** Vertical installations of circuit integrity (CI) cables and conductors installed in a raceway or conductors and cables of electrical circuit protective systems shall be installed in accordance with 300.19.

Support requirements for fire-rated cable are critical and contained in 300.19(B). The strength of cables and conductors decreases with heat, and they may break if not properly supported. Support of vertical runs of fire-rated cables helps ensure continued performance if the cable is exposed directly to a fire or to high temperatures caused by a fire.

**(J) Number and Size of Cables and Conductors in Raceway.** Installations shall comply with 300.17.

Compliance with the raceway fill requirement protects conductors against abrasion or other physical damage resulting from too many conductors being pulled through a raceway. Because of the critical life safety function of the fire alarm system conductors, protecting the conductors against damage increases overall system reliability. In addition, system modifications are facilitated by providing the ability to install and withdraw conductors without causing damage to the conductors that remain in the raceway. This requirement applies to non–power-limited and power-limited fire alarm circuit conductors.

**(K) Bushing.** A bushing shall be installed where cables emerge from raceway used for mechanical support or protection in accordance with 300.15(C).

Conduits and other raceways are often used for mechanical support or protection of cables. A bushing is needed to protect cables from damage.

## 760.21 Access to Electrical Equipment Behind Panels Designed to Allow Access

Access to electrical equipment shall not be denied by an accumulation of conductors and cables that prevents removal of panels, including suspended ceiling panels.

An excess accumulation of wires and cables can limit access to equipment by preventing the removal of access panels. See Exhibit 760.3.

## 760.24 Mechanical Execution of Work

**(A) General.** Fire alarm circuits shall be installed in a neat workmanlike manner. Cables and conductors installed exposed on the surface of ceilings and sidewalls shall be supported by the building structure in such a manner that the cable will not be damaged by normal building use. Such cables shall be supported by straps, staples, cable ties, hangers, or similar fittings designed and installed so as not to damage the cable. The installation shall also comply with 300.4(D).

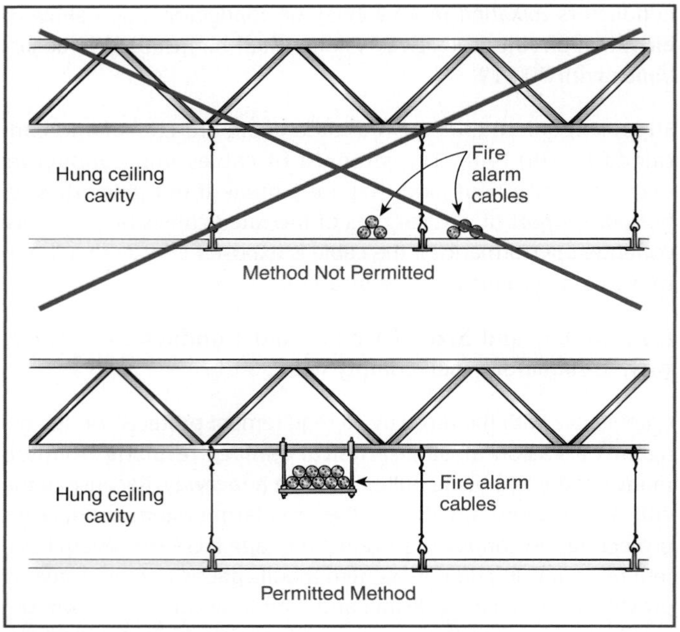


***EXHIBIT 760.3*** *Incorrect cable installation (upper diagram) and correct method (lower diagram).*

Cable must be attached to or supported by the structure by cable ties, straps, clamps, hangers, and so forth. The installation method must not damage the cable. In addition, the location of the cable should be carefully evaluated to ensure that activities and processes within the building do not cause damage to the cable.

The reference to 300.4(D) calls attention to the hazards to which cables are exposed where they are installed on framing members. Such cables are required to be installed in a manner that protects them from nail or screw penetration. This section permits attachment to baseboards and non-load-bearing walls, which are not structural components.

**(B) Circuit Integrity (CI) Cable.** Circuit integrity (CI) cables shall be supported at a distance not exceeding 610 mm (24 in.). Where located within 2.1 m (7 ft) of the floor, as covered in 760.53(A)(1) and 760.130(1), as applicable, the cable shall be fastened in an approved manner at intervals of not more than 450 mm (18 in.). Cable supports and fasteners shall be steel.

### 760.25 Abandoned Cables

The accessible portion of abandoned fire alarm cables shall be removed. Where cables are identified for future use with a tag, the tag shall be of sufficient durability to withstand the environment involved.

Accessible abandoned fire alarm cable must be removed. Abandoned cable increases fire loading unnecessarily, and, where installed in plenums, it can affect airflow. See the definition of *abandoned fire alarm cable* in 760.2.

### 760.30 Fire Alarm Circuit Identification

Fire alarm circuits shall be identified at terminal and junction locations in a manner that helps to prevent unintentional signals on fire alarm system circuit(s) during testing and servicing of other systems.

### 760.32 Fire Alarm Circuits Extending Beyond One Building

Non–power-limited fire alarm circuits and power-limited fire alarm circuits that extend beyond one building and run outdoors shall meet the installation requirements of Parts II, III, and IV of Article 800 and shall meet the installation requirements of Part I of Article 300.

Informational Note: An example of a protective device suitable to provide protection is a device tested to the requirements of ANSI/UL 497B, *Protectors for Data Communications*.

### 760.35 Fire Alarm Circuit Requirements

Fire alarm circuits shall comply with 760.35(A) and (B).

**(A) Non–Power-Limited Fire Alarm (NPLFA) Circuits.** See Parts I and II.

**(B) Power-Limited Fire Alarm (PLFA) Circuits.** See Parts I and III.

Exact power source limitations for power-limited fire alarm circuits used by testing laboratories are found in Chapter 9, Tables 12(A) and 12(B). Table 12(A) covers ac source limitations, and Table 12(B) covers dc source limitations.

## II. Non–Power-Limited Fire Alarm (NPLFA) Circuits

### 760.41 NPLFA Circuit Power Source Requirements

**(A) Power Source.** The power source of non–power-limited fire alarm circuits shall comply with Chapters 1 through 4, and the output voltage shall be not more than 600 volts, nominal. The fire alarm circuit disconnect shall be permitted to be secured in the "on" position.

The fire alarm circuit disconnect is permitted to be secured in the "on" position. This provides correlation with *NFPA 72* that requires the circuit disconnecting means to be accessible only to authorized personnel. By limiting access, the chance that the power to the fire alarm system is turned off decreases.

**(B) Branch Circuit.** The branch circuit supplying the fire alarm equipment(s) shall supply no other loads. The location of the branch-circuit overcurrent protective device shall be permanently identified at the fire alarm control unit. The circuit disconnecting means shall have red identification, shall be accessible only to qualified personnel, and shall be identified as "FIRE ALARM CIRCUIT." The red identification shall not damage

***EXHIBIT 760.4*** *Duplex receptacle supplied by circuit dedicated to fire alarm equipment. (Courtesy of Merton Bunker)*

the overcurrent protective devices or obscure the manufacturer's markings. This branch circuit shall not be supplied through ground-fault circuit interrupters or arc-fault circuit-interrupters.

Informational Note: See 210.8(A)(5), Exception, for receptacles in dwelling-unit unfinished basements that supply power for fire alarm systems.

*NFPA 72* requires that the power to the fire alarm system be supplied from a branch circuit dedicated to the fire alarm system. The dedicated circuit may be used to power other equipment that is part of the system, but this power circuit cannot be used to power other equipment, such as phone switches, computer stations, and other equipment that is not directly associated with the fire alarm system functions. Further, *NFPA 72* requires the location of the branch circuit disconnecting means to be permanently identified at the control unit. The circuit disconnecting means is required to be identified as "FIRE ALARM CIRCUIT" and have a red marking.

Exhibit 760.4 illustrates a plug-in power supply for a fire alarm control unit. The other half of the duplex receptacle shown is permitted only to supply other fire alarm–related equipment.

In order to minimize interruption of the normal ac power, non–power-limited fire alarm equipment is not permitted to be supplied by a branch circuit protected with a GFCI or with an AFCI device. The branch circuit is required to have overcurrent protection in accordance with the general provisions of Articles 210 and 240 and to use the wiring methods of 210.12(A), Exception, if the fire alarm equipment outlet is installed in an area of a dwelling unit that is subject to the AFCI protection requirements.

The AFCI protection requirements apply to outlets supplying single- or multiple-station smoke alarms where the smoke alarm outlet is located in an area covered by the requirements of 210.12(A). These smoke alarms are supplied by a branch circuit covered by the requirements of Article 210, not by the fire alarm control panel. In new construction, single- and multiple-station smoke alarms are required by *NFPA 72* to have a backup battery that will supply power in the event that the branch circuit power is interrupted due to the operation of an AFCI device.

## 760.43 NPLFA Circuit Overcurrent Protection

Overcurrent protection for conductors 14 AWG and larger shall be provided in accordance with the conductor ampacity without applying the ampacity adjustment and correction factors of 310.15 to the ampacity calculation. Overcurrent protection shall not exceed 7 amperes for 18 AWG conductors and 10 amperes for 16 AWG conductors.

*Exception: Where other articles of this Code permit or require other overcurrent protection.*

## 760.45 NPLFA Circuit Overcurrent Device Location

Overcurrent devices shall be located at the point where the conductor to be protected receives its supply.

*Exception No. 1: Where the overcurrent device protecting the larger conductor also protects the smaller conductor.*

*Exception No. 2: Transformer secondary conductors. Non–power-limited fire alarm circuit conductors supplied by the secondary of a single-phase transformer that has only a 2-wire (single-voltage) secondary shall be permitted to be protected by overcurrent protection provided by the primary (supply) side of the transformer, provided the protection is in accordance with 450.3 and does not exceed the value determined by multiplying the secondary conductor ampacity by the secondary-to-primary transformer voltage ratio. Transformer secondary conductors other than 2-wire shall not be considered to be protected by the primary overcurrent protection.*

*Exception No. 3: Electronic power source output conductors. Non–power-limited circuit conductors supplied by the output of a single-phase, listed electronic power source, other than a transformer, having only a 2-wire (single-voltage) output for connection to non–power-limited circuits shall be permitted to be protected by overcurrent protection provided on the input side of the electronic power source, provided this protection does not exceed the value determined by multiplying the non–power-limited circuit conductor ampacity by the output-to-input voltage ratio. Electronic power source outputs, other than 2-wire (single voltage), connected to non–power-limited circuits shall not be considered to be protected by overcurrent protection on the input of the electronic power source.*

Informational Note: A single-phase, listed electronic power supply whose output supplies a 2-wire (single-voltage) circuit is an example of a non–power-limited power source that meets the requirements of 760.41.

Non–power-limited electronic power supplies that do not supply energy directly through the use of transformers are covered by Exception No. 3, which permits overcurrent protection for the non–power-limited circuit conductors to be installed on the input side of the electronic power source rather than on the output side for 2-wire circuits only.

## 760.46 NPLFA Circuit Wiring

Installation of non–power-limited fire alarm circuits shall be in accordance with 110.3(B), 300.7, 300.11, 300.15, 300.17, 300.19(B), and other appropriate articles of Chapter 3.

*Exception No. 1: As provided in 760.48 through 760.53.*

*Exception No. 2: Where other articles of this Code require other methods.*

The appropriate wiring methods in Chapter 3 are required to be used for non–power-limited circuits. However, Exception No. 1 permits special non–power-limited cable types to be used in place of Chapter 3 wiring methods.

Section 300.11(A) requires devices and equipment to be securely mounted. Section 300.15 is referenced to require non–power-limited circuit terminations to be made in a box or conduit body. However, 300.15(E) permits devices with integral terminal enclosures and mounting brackets to be used without a box. Devices must be mounted on a box or conduit body where the instructions or listing require the use of a box. Fire alarm system components such as manual fire alarm boxes are frequently tested. Therefore, secure mounting of the back box is necessary to ensure that the manual fire alarm device will remain in place. (See Exhibit 760.5.)

## 760.48 Conductors of Different Circuits in Same Cable, Enclosure, or Raceway

**(A) Class 1 with NPLFA Circuits.** Class 1 and non–power-limited fire alarm circuits shall be permitted to occupy the same cable, enclosure, or raceway without regard to whether the individual circuits are alternating current or direct current, provided all conductors are insulated for the maximum voltage of any conductor in the enclosure or raceway.

**(B) Fire Alarm with Power-Supply Circuits.** Power-supply and fire alarm circuit conductors shall be permitted in the same cable, enclosure, or raceway only where connected to the same equipment.

## 760.49 NPLFA Circuit Conductors

**(A) Sizes and Use.** Only copper conductors shall be permitted to be used for fire alarm systems. Size 18 AWG and 16 AWG conductors shall be permitted to be used, provided they supply loads that do not exceed the ampacities given in Table 402.5 and are installed in a raceway, an approved enclosure, or a listed cable. Conductors larger than 16 AWG shall not supply loads greater than the ampacities given in 310.15, as applicable.

***EXHIBIT 760.5*** *Typical manual fire alarm box. (Courtesy of the Protectowire Fire Systems)*

The minimum size of conductors permitted to be used on non–power-limited fire protective signaling circuits is 18 AWG. The load must not exceed the conductor ampacities specified in Table 402.5.

*NFPA 72* requires fire alarm device and appliance voltages to be between 85 and 110 percent of nominal rated voltage. Calculations should be made to ensure that all devices or appliances will be operating within these limits at full circuit load. Where future circuit extensions are anticipated, larger conductors should be considered. Some manufacturers specify maximum circuit loop resistances. The equipment specifications should be consulted to ensure that maximum allowable loop resistances are not exceeded.

**(B) Insulation.** Insulation on conductors shall be rated for the system voltage and not less than 600 volts. Conductors larger than 16 AWG shall comply with Article 310. Conductors 18 AWG and 16 AWG shall be Type KF-2, KFF-2, PAFF, PTFF, PF, PFF, PGF, PGFF, RFH-2, RFHH-2, RFHH-3, SF-2, SFF-2, TF, TFF, TFN, TFFN, ZF, or ZFF. Conductors with other types and thickness of insulation shall be permitted if listed for non–power-limited fire alarm circuit use.

Informational Note: For application provisions, see Table 402.3.

**(C) Conductor Materials.** Conductors shall be solid or stranded copper.

*Exception to (B) and (C): Wire Types PAF and PTF shall be permitted only for high-temperature applications between 90°C (194°F) and 250°C (482°F).*

## 760.51 Number of Conductors in Cable Trays and Raceways, and Ampacity Adjustment Factors

**(A) NPLFA Circuits and Class 1 Circuits.** Where only non–power-limited fire alarm circuit and Class 1 circuit conductors are in a raceway, the number of conductors shall be determined in accordance with 300.17. The ampacity adjustment factors given in 310.15(B)(3)(a) shall apply if such conductors carry continuous load in excess of 10 percent of the ampacity of each conductor.

**(B) Power-Supply Conductors and NPLFA Circuit Conductors.** Where power-supply conductors and non–power-limited fire alarm circuit conductors are permitted in a raceway in accordance with 760.48, the number of conductors shall be determined in accordance with 300.17. The ampacity adjustment factors given in 310.15(B)(3)(a) shall apply as follows:

(1) To all conductors where the fire alarm circuit conductors carry continuous loads in excess of 10 percent of the ampacity of each conductor and where the total number of conductors is more than three
(2) To the power-supply conductors only, where the fire alarm circuit conductors do not carry continuous loads in excess of 10 percent of the ampacity of each conductor and where the number of power-supply conductors is more than three

**(C) Cable Trays.** Where fire alarm circuit conductors are installed in cable trays, they shall comply with 392.22 and 392.80(A).

## 760.53 Multiconductor NPLFA Cables

Multiconductor non–power-limited fire alarm cables that meet the requirements of 760.176 shall be permitted to be used on fire alarm circuits operating at 150 volts or less and shall be installed in accordance with 760.53(A) and (B).

**(A) NPLFA Wiring Method.** Multiconductor non–power-limited fire alarm circuit cables shall be installed in accordance with 760.53(A)(1), (A)(2), and (A)(3).

**(1) In Raceways, Exposed on Ceilings or Sidewalls, or Fished in Concealed Spaces.** Cable splices or terminations shall be made in listed fittings, boxes, enclosures, fire alarm devices, or utilization equipment. Where installed exposed, cables shall be adequately supported and installed in such a way that maximum protection against physical damage is afforded by building construction such as baseboards, door frames, ledges, and so forth. Where located within 2.1 m (7 ft) of the floor, cables shall be securely fastened in an approved manner at intervals of not more than 450 mm (18 in.).

**(2) Passing Through a Floor or Wall.** Cables shall be installed in metal raceway or rigid nonmetallic conduit where passing through a floor or wall to a height of 2.1 m (7 ft) above the floor, unless adequate protection can be afforded by building construction such as detailed in 760.53(A)(1), or unless an equivalent solid guard is provided.

**(3) In Hoistways.** Cables shall be installed in rigid metal conduit, rigid nonmetallic conduit, intermediate metal conduit, liquidtight flexible nonmetallic conduit, or electrical metallic tubing where installed in hoistways.

*Exception: As provided for in 620.21 for elevators and similar equipment.*

**(B) Applications of Listed NPLFA Cables.** The use of non–power-limited fire alarm circuit cables shall comply with 760.53(B)(1) through (B)(4).

**(1) Ducts.** Multiconductor non–power-limited fire alarm circuit cables, Types NPLFP, NPLFR, and NPLF, shall not be installed exposed in ducts.

Informational Note: See 300.22(B).

Wiring methods for non–power-limited circuits in ducts and plenums must be in accordance with 300.22(B). Cables marked NPLFP may not be installed in ducts or plenums, but they are permitted in other spaces for environmental air. See 760.53(B)(2). The higher possible voltages and currents of non–power-limited fire alarm circuits preclude the use of the listed cables inside plenums.

**(2) Other Spaces Used for Environmental Air.** Cables installed in other spaces used for environmental air shall be Type NPLFP.

*Exception No. 1: Types NPLFR and NPLF cables installed in compliance with 300.22(C).*

*Exception No. 2: Other wiring methods in accordance with 300.22(C) and conductors in compliance with 760.49(C).*

Other spaces used for environmental air are covered by 300.22(C) and the related informational note. Spaces over suspended ceilings used as an environmental air-handling return are considered by the *Code* as "other spaces used for environmental air." Non–power-limited cables in other spaces used for environmental air must be marked NPLFP. [See 760.176(C).]

*Exception No. 3: Type NPLFP-CI cable shall be permitted to be installed to provide a 2-hour circuit integrity rated cable.*

**(3) Riser.** Cables installed in vertical runs and penetrating one or more floors, or cables installed in vertical runs in a shaft, shall be Type NPLFR. Floor penetrations requiring Type NPLFR shall contain only cables suitable for riser or plenum use.

*Exception No. 1: Type NPLF or other cables that are specified in Chapter 3 and are in compliance with 760.49(C) and encased in metal raceway.*

*Exception No. 2: Type NPLF cables located in a fireproof shaft having firestops at each floor.*

Informational Note: See 300.21 for firestop requirements for floor penetrations.

*Exception No. 3: Type NPLF-CI cable shall be permitted to be installed to provide a 2-hour circuit integrity rated cable.*

**(4) Other Wiring Within Buildings.** Cables installed in building locations other than the locations covered in 760.53(B)(1), (B)(2), and (B)(3) shall be Type NPLF.

*Exception No. 1: Chapter 3 wiring methods with conductors in compliance with 760.49(C).*

*Exception No. 2: Type NPLFP or Type NPLFR cables shall be permitted.*

*Exception No. 3: Type NPLFR-CI cable shall be permitted to be installed to provide a 2-hour circuit integrity rated cable.*

## III. Power-Limited Fire Alarm (PLFA) Circuits

### 760.121 Power Sources for PLFA Circuits

**(A) Power Source.** The power source for a power-limited fire alarm circuit shall be as specified in 760.121(A)(1), (A)(2), or (A)(3).

Informational Note No. 1: Tables 12(A) and 12(B) in Chapter 9 provide the listing requirements for power-limited fire alarm circuit sources.

Informational Note No. 2: See 210.8(A)(5), Exception, for receptacles in dwelling-unit unfinished basements that supply power for fire alarm systems.

(1) A listed PLFA or Class 3 transformer.
(2) A listed PLFA or Class 3 power supply.
(3) Listed equipment marked to identify the PLFA power source.

Informational Note: Examples of listed equipment are a fire alarm control panel with integral power source; a circuit card listed for use as a PLFA source, where used as part of a listed assembly; a current-limiting impedance, listed for the purpose or part of a listed product, used in conjunction with a non–power-limited transformer or a stored energy source, for example, storage battery, to limit the output current.

**(B) Branch Circuit.** The branch circuit supplying the fire alarm equipment(s) shall supply no other loads. The location of the branch-circuit overcurrent protective device shall be permanently identified at the fire alarm control unit. The circuit disconnecting means shall have red identification, shall be accessible only to qualified personnel, and shall be identified as "FIRE ALARM CIRCUIT." The red identification shall not damage the overcurrent protective devices or obscure the manufacturer's markings. This branch circuit shall not be supplied through ground-fault circuit interrupters or arc-fault circuit interrupters.

For more information on branch-circuit requirements for fire alarm systems, see the commentary following 760.41(B). The requirements covering branch circuits for non–power-limited circuits and for power-limited circuits are the same.

### 760.124 Circuit Marking

The equipment supplying PLFA circuits shall be durably marked where plainly visible to indicate each circuit that is a power-limited fire alarm circuit.

Informational Note: See 760.130(A), Exception No. 3, where a power-limited circuit is to be reclassified as a non–power-limited circuit.

### 760.127 Wiring Methods on Supply Side of the PLFA Power Source

Conductors and equipment on the supply side of the power source shall be installed in accordance with the appropriate requirements of Part II and Chapters 1 through 4. Transformers or other devices supplied from power-supply conductors shall be protected by an overcurrent device rated not over 20 amperes.

*Exception: The input leads of a transformer or other power source supplying power-limited fire alarm circuits shall be permitted to be smaller than 14 AWG, but not smaller than 18 AWG, if they are not over 300 mm (12 in.) long and if they have insulation that complies with 760.49(B).*

### 760.130 Wiring Methods and Materials on Load Side of the PLFA Power Source

Fire alarm circuits on the load side of the power source shall be permitted to be installed using wiring methods and materials in accordance with 760.130(A), (B), or a combination of (A) and (B).

Individual power-limited circuits are permitted to be installed using Chapter 3 wiring methods, non–power-limited fire alarm circuit wiring methods, power-limited circuit wiring methods, or a combination. If it is desirable to run power-limited circuits in the same cable or raceway with non–power-limited circuits, the power-limited circuits may be reclassified as permitted by 760.130(A), Exception No. 3. Also refer to the informational note that follows the exception regarding circuit classification.

**(A) NPLFA Wiring Methods and Materials.** Installation shall be in accordance with 760.46, and conductors shall be solid or stranded copper.

*Exception No. 1: The ampacity adjustment factors given in 310.15(B)(3)(a) shall not apply.*

*Exception No. 2: Conductors and multiconductor cables described in and installed in accordance with 760.49 and 760.53 shall be permitted.*

*Exception No. 3: Power-limited circuits shall be permitted to be reclassified and installed as non–power-limited circuits if the power-limited fire alarm circuit markings required by 760.124 are eliminated and the entire circuit is installed using the wiring methods and materials in accordance with Part II, Non–Power-Limited Fire Alarm Circuits.*

Informational Note: Power-limited circuits reclassified and installed as non–power-limited circuits are no longer power-limited circuits, regardless of the continued connection to a power-limited source.

Exception No. 3 allows power-limited circuits to be reclassified and installed in accordance with the requirements for non–power-limited circuits. Where installed as non–power-limited circuits, the power-limited marking must be removed from equipment, overcurrent protection must be provided in accordance with 760.43, and reclassified circuits must maintain separation from power-limited circuits in accordance with 760.48 and 760.133.

**(B) PLFA Wiring Methods and Materials.** Power-limited fire alarm conductors and cables described in 760.179 shall be installed as detailed in 760.130(B)(1), (B)(2), or (B)(3) of this section and 300.7. Devices shall be installed in accordance with 110.3(B), 300.11(A), and 300.15.

Mechanical protection is required at splices and termination points. Because failure of a circuit often occurs at splices or termination points, this requirement offers more protection and strain relief for these cable connections.

**(1) In Raceways, Exposed on Ceilings or Sidewalls, or Fished in Concealed Spaces.** Cable splices or terminations shall be made in listed fittings, boxes, enclosures, fire alarm devices, or utilization equipment. Where installed exposed, cables shall be adequately supported and installed in such a way that maximum protection against physical damage is afforded by building construction such as baseboards, door frames, ledges, and so forth. Where located within 2.1 m (7 ft) of the floor, cables shall be securely fastened in an approved manner at intervals of not more than 450 mm (18 in.).

**(2) Passing Through a Floor or Wall.** Cables shall be installed in metal raceways or rigid nonmetallic conduit where passing through a floor or wall to a height of 2.1 m (7 ft) above the floor, unless adequate protection can be afforded by building construction such as detailed in 760.130(B)(1), or unless an equivalent solid guard is provided.

**(3) In Hoistways.** Cables shall be installed in rigid metal conduit, rigid nonmetallic conduit, intermediate metal conduit, or electrical metallic tubing where installed in hoistways.

*Exception: As provided for in 620.21 for elevators and similar equipment.*

### 760.133 Installation of Conductors and Equipment in Cables, Compartments, Cable Trays, Enclosures, Manholes, Outlet Boxes, Device Boxes, and Raceways for Power-Limited Circuits

Conductors and equipment for power-limited fire alarm circuits shall be installed in accordance with 760.135 through 760.143.

### 760.135 Installation of PLFA Cables in Buildings

Installation of power-limited fire alarm cables in buildings shall comply with 760.135(A) through (J).

**(A) Listing.** PLFA cables installed in buildings shall be listed.

**(B) Fabricated Ducts Used for Environmental Air.** The following cables shall be permitted in ducts, as described in 300.22(B), if they are directly associated with the air distribution system:

(1) Types FPLP and FPLP-CI cables in lengths as short as practicable to perform the required function
(2) Types FPLP, FPLP-CI, FPLR, FPLR-CI, FPL, and FPL-CI cables installed in raceways that are installed in compliance with 300.22(B)

Informational Note: For information on fire protection of wiring installed in fabricated ducts, see 4.3.4.1 and 4.3.11.3.3 of NFPA 90A-2012, *Standard for the Installation of Air-Conditioning and Ventilating Systems.*

**(C) Other Spaces Used For Environmental Air (Plenums).** The following cables shall be permitted in other spaces used for environmental air as described in 300.22(C):

(1) Type FPLP cables
(2) Type FPLP cables installed in plenum communications raceways
(3) Types FPLP and FPLP-CI cables supported by open metallic cable trays or cable tray systems
(4) Types FPLP, FPLR, and FPL cables installed in raceways that are installed in compliance with 300.22(C)
(5) Types FPLP, FPLR, and FPL cables supported by solid bottom metal cable trays with solid metal covers in other spaces used for environmental air (plenums) as described in 300.22(C)
(6) Types FPLP, FPLR, and FPL cables installed in plenum communications raceways, riser communications raceways, or general-purpose communications raceways supported by solid bottom metal cable trays with solid metal covers in other spaces used for environmental air (plenums) as described in 300.22(C).

**(D) Risers — Cables in Vertical Runs.** The following cables shall be permitted in vertical runs penetrating one or more floors and in vertical runs in a shaft:

(1) Types FPLP and FPLR cables
(2) Types FPLP and FPLR cables installed in the following:
  a. Plenum communications raceways
  b. Plenum cable routing assemblies
  c. Riser communications raceways
  d. Riser cable routing assemblies

Informational Note: See 300.21 for firestop requirements for floor penetrations.

**(E) Risers — Cables in Metal Raceways.** The following cables shall be permitted in metal raceways in a riser having firestops at each floor:

(1) Types FPLP, FPLR, and FPL cables
(2) Types FPLP, FPLR, and FPL cables installed in the following:
  a. Plenum communications raceways
  b. Riser communications raceways
  c. General-purpose communications raceways

Informational Note: See 300.21 for firestop requirements for floor penetrations.

**(F) Risers — Cables in Fireproof Shafts.** The following cables shall be permitted to be installed in fireproof riser shafts having firestops at each floor:

(1) Types FPLP, FPLR, and FPL cables
(2) Types FPLP, FPLR, and FPL cables installed in the following:
  a. Plenum communications raceways
  b. Plenum cable routing assemblies
  c. Riser communications raceways
  d. Riser cable routing assemblies
  e. General-purpose communications raceways
  f. General-purpose cable routing assemblies

Informational Note: See 300.21 for firestop requirements for floor penetrations.

**(G) Risers — One- and Two-Family Dwellings.** The following cables shall be permitted in one- and two-family dwellings:

(1) Types FPLP, FPLR, and FPL cables
(2) Types FPLP, FPLR, and FPL cables installed in the following:
  a. Plenum communications raceways
  b. Plenum cable routing assemblies
  c. Riser communications raceways
  d. Riser cable routing assemblies
  e. General-purpose communications raceways
  f. General-purpose cable routing assemblies

**(H) Other Building Locations.** The following cables shall be permitted to be installed in building locations other than the locations covered in 770.113(B) through (H):

(1) Types FPLP, FPLR, and FPL cables
(2) Types FPLP, FPLR, and FPL cables installed in the following:
  a. Plenum communications raceways
  b. Plenum cable routing assemblies
  c. Riser communications raceways
  d. Riser cable routing assemblies
  e. General-purpose communications raceways
  f. General-purpose cable routing assemblies
(3) Types FPLP, FPLR, and FPL cables installed in a raceway of a type recognized in Chapter 3

**(I) Nonconcealed Spaces.** Cables specified in Chapter 3 and meeting the requirements of 760.179(A) and (B) shall be permitted to be installed in nonconcealed spaces where the exposed length of cable does not exceed 3 m (10 ft).

**(J) Portable Fire Alarm System.** A portable fire alarm system provided to protect a stage or set when not in use shall be permitted to use wiring methods in accordance with 530.12.

## 760.136 Separation from Electric Light, Power, Class 1, NPLFA, and Medium-Power Network-Powered Broadband Communications Circuit Conductors

**(A) General.** Power-limited fire alarm circuit cables and conductors shall not be placed in any cable, cable tray, compartment, enclosure, manhole, outlet box, device box, raceway, or similar fitting with conductors of electric light, power, Class 1, non–power-limited fire alarm circuits, and medium-power network-powered broadband communications circuits unless permitted by 760.136(B) through (G).

Jackets of listed power-limited fire alarm cables do not have sufficient construction specifications to permit them to be installed with electric light, power, Class 1, non–power-limited fire alarm circuits, and medium-power network-powered broadband communications cables. Failure of the cable insulation due to a fault could lead to hazardous voltages being imposed on the power-limited fire alarm circuit conductors.

**(B) Separated by Barriers.** Power-limited fire alarm circuit cables shall be permitted to be installed together with Class 1, non–power-limited fire alarm, and medium-power network-powered broadband communications circuits where they are separated by a barrier.

**(C) Raceways Within Enclosures.** In enclosures, power-limited fire alarm circuits shall be permitted to be installed in a raceway within the enclosure to separate them from Class 1, non–power-limited fire alarm, and medium-power network-powered broadband communications circuits.

**(D) Associated Systems Within Enclosures.** Power-limited fire alarm conductors in compartments, enclosures, device boxes, outlet boxes, or similar fittings shall be permitted to be installed with electric light, power, Class 1, non–power-limited fire alarm, and medium power network-powered broadband communications circuits where they are introduced solely to connect the equipment connected to power-limited fire alarm circuits, and comply with either of the following conditions:

(1) The electric light, power, Class 1, non–power-limited fire alarm, and medium-power network-powered broadband communications circuit conductors are routed to maintain a minimum of 6 mm (0.25 in.) separation from the conductors and cables of power-limited fire alarm circuits.
(2) The circuit conductors operate at 150 volts or less to ground and also comply with one of the following:
   a. The fire alarm power-limited circuits are installed using Type FPL, FPLR, FPLP, or permitted substitute cables, provided these power-limited cable conductors extending beyond the jacket are separated by a minimum of 6 mm (0.25 in.) or by a nonconductive sleeve or nonconductive barrier from all other conductors.
   b. The power-limited fire alarm circuit conductors are installed as non–power-limited circuits in accordance with 760.46.

**(E) Enclosures with Single Opening.** Power-limited fire alarm circuit conductors entering compartments, enclosures, device boxes, outlet boxes, or similar fittings shall be permitted to be installed with electric light, power, Class 1, non–power-limited fire alarm, and medium-power network-powered broadband communications circuits where they are introduced solely to connect the equipment connected to power-limited fire alarm circuits or to other circuits controlled by the fire alarm system to which the other conductors in the enclosure are connected. Where power-limited fire alarm circuit conductors must enter an enclosure that is provided with a single opening, they shall be permitted to enter through a single fitting (such as a tee), provided the conductors are separated from the conductors of the other circuits by a continuous and firmly fixed nonconductor, such as flexible tubing.

**(F) In Hoistways.** In hoistways, power-limited fire alarm circuit conductors shall be installed in rigid metal conduit, rigid nonmetallic conduit, intermediate metal conduit, liquidtight flexible nonmetallic conduit, or electrical metallic tubing. For elevators or similar equipment, these conductors shall be permitted to be installed as provided in 620.21.

**(G) Other Applications.** For other applications, power-limited fire alarm circuit conductors shall be separated by at least 50 mm (2 in.) from conductors of any electric light, power, Class 1, non–power-limited fire alarm, or medium-power network-powered broadband communications circuits unless one of the following conditions is met:

(1) Either (a) all of the electric light, power, Class 1, non–power-limited fire alarm, and medium-power network-powered broadband communications circuit conductors or (b) all of the power-limited fire alarm circuit conductors are in a raceway or in metal-sheathed, metal-clad, nonmetallic-sheathed, or Type UF cables.
(2) All of the electric light, power, Class 1, non–power-limited fire alarm, and medium-power network-powered broadband communications circuit conductors are permanently separated from all of the power-limited fire alarm circuit conductors by a continuous and firmly fixed nonconductor, such as porcelain tubes or flexible tubing, in addition to the insulation on the conductors.

### 760.139 Installation of Conductors of Different PLFA Circuits, Class 2, Class 3, and Communications Circuits in the Same Cable, Enclosure, Cable Tray, Raceway, or Cable Routing Assembly

**(A) Two or More PLFA Circuits.** Cable and conductors of two or more power-limited fire alarm circuits, communications circuits, or Class 3 circuits shall be permitted within the same cable, enclosure, cable tray, raceway, or cable routing assembly.

**(B) Class 2 Circuits with PLFA Circuits.** Conductors of one or more Class 2 circuits shall be permitted within the same cable, enclosure, cable tray, raceway, or cable routing assembly with conductors of power-limited fire alarm circuits, provided that the insulation of the Class 2 circuit conductors in the cable, enclosure, raceway, or cable routing assembly is at least that required by the power-limited fire alarm circuits.

**(C) Low-Power Network-Powered Broadband Communications Cables and PLFA Cables.** Low-power network-powered broadband communications circuits shall be permitted in the same enclosure, cable tray, raceway, or cable routing assembly with PLFA cables.

**(D) Audio System Circuits and PLFA Circuits.** Audio system circuits described in 640.9(C) and installed using Class 2 or Class 3 wiring methods in compliance with 725.133 and 725.154 shall not be permitted to be installed in the same cable, cable tray, raceway, or cable routing assembly with power-limited conductors or cables.

Audio circuits that are installed as Class 2 or Class 3 circuits are prohibited from being installed in the same cable or raceway with power-limited fire alarm wiring. A fault between audio amplifier circuits and power-limited fire alarm circuits has the potential to impair the fire alarm system.

### 760.142 Conductor Size

Conductors of 26 AWG shall be permitted only where spliced with a connector listed as suitable for 26 AWG to 24 AWG or larger conductors that are terminated on equipment or where the 26 AWG conductors are terminated on equipment listed as suitable for 26 AWG conductors. Single conductors shall not be smaller than 18 AWG.

Due to a signaling method called "multiplexing" used with digitally addressable fire alarm systems, power-limited fire alarm cable is permitted to contain circuit conductors as small as 26 AWG, provided they are used as specified and as permitted by the listing or installation instructions of the fire alarm equipment.

### 760.143 Support of Conductors

Power-limited fire alarm circuit conductors shall not be strapped, taped, or attached by any means to the exterior of any conduit or other raceway as a means of support.

See the commentary following 760.24(A) for more information on the support of conductors.

### 760.145 Current-Carrying Continuous Line-Type Fire Detectors

**(A) Application.** Listed continuous line-type fire detectors, including insulated copper tubing of pneumatically operated detectors, employed for both detection and carrying signaling currents shall be permitted to be used in power-limited circuits.

**(B) Installation.** Continuous line-type fire detectors shall be installed in accordance with 760.124 through 760.130 and 760.133.

### 760.154 Applications of Listed PLFA Cables

PLFA cables shall comply with the requirements described in Table 760.154 or where cable substitutions are made as shown in 760.154(A). Where substitute cables are installed, the wiring requirements of Article 760, Parts I and III, shall apply. Types FPLP-CI, FPLR-CI, and FPL-CI cables shall be permitted to be installed to provide 2-hour circuit integrity rated cables.

Type CI cable is permitted for applications where survivability of fire alarm circuits is needed during a fire. Such circuits could be essential to communicating evacuation or relocation instructions to building occupants under fire or other emergency conditions.

•

**(A) Fire Alarm Cable Substitutions.** The substitutions for fire alarm cables listed in Table 760.154(A) and illustrated in Figure 760.154(A) shall be permitted. Where substitute cables are installed, the wiring requirements of Article 760, Parts I and III, shall apply.

Informational Note: For information on communications cables (CMP, CMR, CMG, CM), see 800.179.

## IV. Listing Requirements

### 760.176 Listing and Marking of NPLFA Cables

Non–power-limited fire alarm cables installed as wiring within buildings shall be listed in accordance with 760.176(A) and (B) and as being resistant to the spread of fire in accordance with 760.176(C) through (F), and shall be marked in accordance with 760.176(G). Cable used in a wet location shall be listed for use in wet locations or have a moisture-impervious metal sheath.

**(A) NPLFA Conductor Materials.** Conductors shall be 18 AWG or larger solid or stranded copper.

**(B) Insulated Conductors.** Insulation on conductors shall be rated for the system voltage and not less than 600 V. Insulated conductors 14 AWG and larger shall be one of the types listed in Table 310.104(A) or one that is identified for this use. Insulated conductors 18 AWG and 16 AWG shall be in accordance with 760.49.

**(C) Type NPLFP.** Type NPLFP non–power-limited fire alarm cable for use in other space used for environmental air shall be listed as being suitable for use in other space used for environmental air as described in 300.22(C) and shall also be listed as having adequate fire-resistant and low smoke–producing characteristics.

Informational Note: One method of defining a cable that is low-smoke producing cable and fire-resistant cable is that the cable exhibits a maximum peak optical density of 0.50 or less, an average optical density of 0.15 or less, and a maximum flame spread distance of 1.52 m (5 ft) or less when tested in accordance with NFPA 262-2011, *Standard Method of Test for Flame Travel and Smoke of Wires and Cables for Use in Air-Handling Spaces*.

See the commentary following the informational note to 725.179(A) for more information on this test and the commentary following 760.53(B)(2), Exception No. 2, which discusses other spaces used for environmental air.

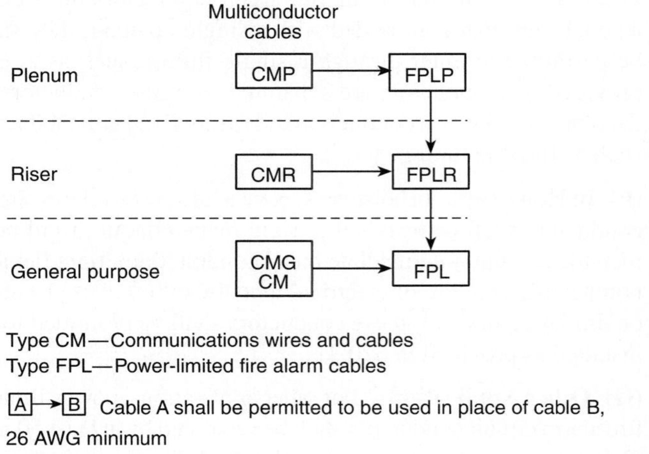


***FIGURE 760.154(A)*** *Cable Substitution Hierarchy.*

*TABLE 760.154 Applications of Listed PLFA Cables in Buildings*

| Applications | | Cable Type: FPLP & FPLP-CI | Cable Type: FPLR & FPLR-CI | Cable Type: FPL & FPL-CI |
|---|---|---|---|---|
| In fabricated ducts as described in 300.22(B) | In fabricated ducts | Y* | N | N |
| | In metal raceway that complies with 300.22(B) | Y* | Y* | Y* |
| In other spaces used for environmental air as described in 300.22(C) | In other spaces used for environmental air | Y* | N | N |
| | In metal raceway that complies with 300.22(C) | Y* | Y* | Y* |
| | In plenum communications raceways | Y* | N | N |
| | In plenum cable routing assemblies | NOT PERMITTED | | |
| | Supported by open metal cable trays | Y* | N | N |
| | Supported by solid bottom metal cable trays with solid metal covers | Y* | Y* | Y* |
| In risers | In vertical runs | Y* | Y* | N |
| | In metal raceways | Y* | Y* | Y* |
| | In fireproof shafts | Y* | Y* | Y* |
| | In plenum communications raceways | Y* | Y* | N |
| | In plenum cable routing assemblies | Y* | Y* | N |
| | In riser communications raceways | Y* | Y* | N |
| | In riser cable routing assemblies | Y* | Y* | N |
| | In one- and two-family dwellings | Y* | Y* | Y* |
| Within buildings in other than air-handling spaces and risers | General | Y* | Y* | Y* |
| | Supported by cable trays | Y* | Y* | Y* |
| | In any raceway recognized in Chapter 3 | Y* | Y* | Y* |
| | In plenum communications raceway | Y* | Y* | Y* |
| | In plenum cable routing assemblies | Y* | Y* | Y* |
| | In riser communications raceways | Y* | Y* | Y* |
| | In riser cable routing assemblies | Y* | Y* | Y* |
| | In general-purpose communications raceways | Y* | Y* | Y* |
| | In general-purpose cable routing assemblies | Y* | Y* | Y* |

Note: An "N" in the table indicates that the cable type shall not be permitted to be installed in the application. A "Y*" indicates that the cable shall be permitted to be installed in the application subject to the limitations described in 760.130 through 760.145.

*TABLE 760.154(A) Cable Substitutions*

| Cable Type | Permitted Substitutions |
|---|---|
| FPLP | CMP |
| FPLR | CMP, FPLP, CMR |
| FPL | CMP, FPLP, CMR, FPLR, CMG, CM |

**(D) Type NPLFR.** Type NPLFR non–power-limited fire alarm riser cable shall be listed as being suitable for use in a vertical run in a shaft or from floor to floor and shall also be listed as having fire-resistant characteristics capable of preventing the carrying of fire from floor to floor.

Informational Note: One method of defining fire-resistant characteristics capable of preventing the carrying of fire from floor to floor is that the cables pass ANSI/UL 1666-2012, *Test for Flame Propagation Height of Electrical and Optical-Fiber Cables Installed Vertically in Shafts*.

See the commentary following the informational note to 725.179(B) for more information on this test.

**(E) Type NPLF.** Type NPLF non–power-limited fire alarm cable shall be listed as being suitable for general-purpose fire alarm use, with the exception of risers, ducts, plenums, and other space used for environmental air, and shall also be listed as being resistant to the spread of fire.

Informational Note: One method of defining *resistant to the spread of fire* is that the cables do not spread fire to the top of the tray in the "UL Flame Exposure, Vertical Tray Flame Test" in ANSI/UL 1685-2010, *Standard for Safety for Vertical-Tray Fire-Propagation and Smoke-Release Test for Electrical and Optical-Fiber Cables*. The smoke measurements in the test method are not applicable.

Another method of defining *resistant to the spread of fire* is for the damage (char length) not to exceed 1.5 m (4 ft 11 in.) when performing the CSA "Vertical Flame Test — Cables in Cable Trays," as described in CSA C22.2 No. 0.3-M-2001, *Test Methods for Electrical Wires and Cables*.

For further information on the fire test method for cables used as other wiring within buildings, see the commentary following the informational note to 725.179(C).

**(F) Fire Alarm Circuit Integrity (CI) Cable or Electrical Circuit Protective System.** Cables that are used for survivability of critical circuits under fire conditions shall meet either 760.176(F)(1) or (F)(2) as follows:

**(1) Circuit Integrity (CI) Cables.** Circuit integrity (CI) cables, specified in 760.176(C), (D), and (E), and used for survivability of critical circuits, shall have an additional classification using the suffix "CI." Circuit integrity (CI) cables shall only be permitted to be installed in a raceway where specifically listed and marked as part of an electrical circuit protective system as covered in 760.176(F)(2).

*TABLE 760.176(G) NPLFA Cable Markings*

| Cable Marking | Type | Reference |
|---|---|---|
| NPLFP | Non–power-limited fire alarm circuit cable for use in "other space used for environmental air" | 760.176(C) and (G) |
| NPLFR | Non–power-limited fire alarm circuit riser cable | 760.176(D) and (G) |
| NPLF | Non–power-limited fire alarm circuit cable | 760.176(E) and (G) |

Note: Cables identified in 760.176(C), (D), and (E) and meeting the requirements for circuit integrity shall have the additional classification using the suffix "CI" (for example, NPLFP-CI, NPLFR-CI, and NPLF-CI).

Type CI cable is intended to meet the performance requirements for survivability required by *NFPA 72*. This type of cable is designed to retain vital electrical performance during and immediately after fire exposure. CI cable is considered a 2-hour-rated cable assembly and is an alternative to fire-rated mineral-insulated cable (Type MI).

**(2) Electrical Circuit Protective System.** Cables specified in 760.176(C), (D), (E), and (F)(1), that are part of an electrical circuit protective system, shall be identified with the protective system number and hourly rating printed on the outer jacket of the cable and installed in accordance with the listing of the protective system.

Informational Note No. 1: Fire alarm circuit integrity (CI) cable and electrical circuit protective systems may be used for fire alarm circuits to comply with the survivability requirements of *NFPA 72-2013*, *National Fire Alarm and Signaling Code*, 12.4.3 and 12.4.4, that the circuit maintain its electrical function during fire conditions for a defined period of time.

Informational Note No. 2: One method of defining circuit integrity (CI) cable or an electrical circuit protective system is by establishing a minimum 2-hour fire-resistive rating for the cable when tested in accordance with UL 2196-2012, *Standard for Tests of Fire Resistive Cables*.

Informational Note No. 3: UL guide information for electrical circuit protective systems (FHIT) contains information on proper installation requirements for maintaining the fire rating.

**(G) NPLFA Cable Markings.** Multiconductor non–power-limited fire alarm cables shall be marked in accordance with Table 760.176(G). Non–power-limited fire alarm circuit cables shall be permitted to be marked with a maximum usage voltage rating of 150 volts. Cables that are listed for circuit integrity shall be identified with the suffix "CI" as defined in 760.176(F).

Informational Note: Cable types are listed in descending order of fire resistance rating.

### 760.179 Listing and Marking of PLFA Cables and Insulated Continuous Line-Type Fire Detectors

PLFA cables installed as wiring within buildings shall be listed as being resistant to the spread of fire and other criteria in accordance with 760.179(A) through (H) and shall be marked in accordance with 760.179(I). Insulated continuous line-type fire detectors shall be listed in accordance with 760.179(J). Cable used in a wet location shall be listed for use in wet locations or have a moisture-impervious metal sheath.

**(A) Conductor Materials.** Conductors shall be solid or stranded copper.

**(B) Conductor Size.** The size of conductors in a multiconductor cable shall not be smaller than 26 AWG. Single conductors shall not be smaller than 18 AWG.

**(C) Ratings.** The cable shall have a voltage rating of not less than 300 volts.

**(D) Type FPLP.** Type FPLP power-limited fire alarm plenum cable shall be listed as being suitable for use in ducts, plenums, and other space used for environmental air and shall also be listed as having adequate fire-resistant and low smoke–producing characteristics.

Informational Note: One method of defining a cable that is low-smoke producing cable and fire-resistant cable is that the cable exhibits a maximum peak optical density of 0.50 or less, an average optical density of 0.15 or less, and a maximum flame spread distance of 1.52 m (5 ft) or less when tested in accordance with NFPA 262-2011, *Standard Method of Test for Flame Travel and Smoke of Wires and Cables for Use in Air-Handling Spaces.*

For further information on the fire test method for plenum cables, see the commentary following the informational note to 725.179(A).

**(E) Type FPLR.** Type FPLR power-limited fire alarm riser cable shall be listed as being suitable for use in a vertical run in a shaft or from floor to floor and shall also be listed as having fire-resistant characteristics capable of preventing the carrying of fire from floor to floor.

Informational Note: One method of defining fire-resistant characteristics capable of preventing the carrying of fire from floor to floor is that the cables pass the requirements of ANSI/UL 1666-2012, *Standard Test for Flame Propagation Height of Electrical and Optical-Fiber Cable Installed Vertically in Shafts.*

For further information on the fire test method for riser cables, see the commentary following the informational note to 725.179(B).

**(F) Type FPL.** Type FPL power-limited fire alarm cable shall be listed as being suitable for general-purpose fire alarm use, with the exception of risers, ducts, plenums, and other spaces used for environmental air, and shall also be listed as being resistant to the spread of fire.

Informational Note: One method of defining *resistant to the spread of fire* is that the cables do not spread fire to the top of the tray in the "UL Flame Exposure, Vertical Tray Flame Test" in ANSI/UL 1685-2012, *Standard for Safety for Vertical-Tray Fire-Propagation and Smoke-Release Test for Electrical and Optical-Fiber Cables.* The smoke measurements in the test method are not applicable.

Another method of defining *resistant to the spread of fire* is for the damage (char length) not to exceed 1.5 m (4 ft 11 in.) when performing the CSA "Vertical Flame Test — Cables in Cable Trays," as described in CSA C22.2 No. 0.3-M-2001, *Test Methods for Electrical Wires and Cables.*

For further information on the fire test method for cables, see the commentary following the informational note to 725.179(C).

**(G) Fire Alarm Circuit Integrity (CI) Cable or Electrical Circuit Protective System.** Cables that are used for survivability of critical circuits under fire conditions shall meet either 760.179(G)(1) or (G)(2) as follows:

**(1) Circuit Integrity (CI) Cables.** Circuit integrity (CI) cables specified in 760.179(D), (E), (F), and (H), and used for survivability of critical circuits, shall have an additional classification using the suffix "CI." Circuit integrity (CI) cables shall only be permitted to be installed in a raceway where specifically listed and marked as part of an electrical circuit protective system as covered in 760.179(G)(2).

**(2) Electrical Circuit Protective System.** Cables specified in 760.179(D), (E), (F), (H), and (G)(1), that are part of an electrical circuit protective system, shall be identified with the protective system number and hourly rating printed on the outer jacket of the cable and installed in accordance with the listing of the protective system.

Informational Note No. 1: Fire alarm circuit integrity (CI) cable and electrical circuit protective systems may be used for fire alarm circuits to comply with the survivability requirements of *NFPA 72-2013, National Fire Alarm and Signaling Code*, 12.4.3 and 12.4.4, that the circuit maintain its electrical function during fire conditions for a defined period of time.

Informational Note No. 2: One method of defining circuit integrity (CI) cable or an electrical circuit protective system is by establishing a minimum 2-hour fire-resistive rating for the cable when tested in accordance with UL 2196-2012, *Standard for Tests of Fire Resistive Cables.*

In coordination with certain requirements in *NFPA 72*, 760.179(G) requires listed CI cable or an electrical circuit protective system for critical circuits of the fire alarm system, which must operate under severe conditions such as attack by fire. Informational Note No. 2 references cables tested in accordance with UL 2196, *Standard for Tests for Fire Resistive Cables*, to provide the 2-hour fire-rated CI cable recognized by *NFPA 72*. The construction of typical CI cable is illustrated in Exhibit 760.6.

Informational Note No. 3: UL guide information for electrical circuit protective systems (FHIT) contains information on proper installation requirements for maintaining the fire rating.

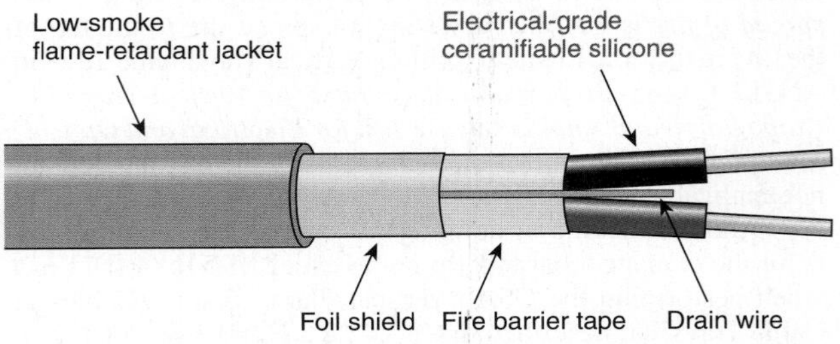


*EXHIBIT 760.6 Type CI cable. (Courtesy of RSCC Wire & Cable, LLC)*

**(H) Coaxial Cables.** Coaxial cables shall be permitted to use 30 percent conductivity copper-covered steel center conductor wire and shall be listed as Type FPLP, FPLR, or FPL cable.

**(I) Cable Marking.** The cable shall be marked in accordance with Table 760.179(I). The voltage rating shall not be marked on the cable. Cables that are listed for circuit integrity shall be identified with the suffix CI as defined in 760.179(G).

Informational Note: Voltage ratings on cables may be misinterpreted to suggest that the cables may be suitable for Class 1, electric light, and power applications.

*Exception: Voltage markings shall be permitted where the cable has multiple listings and voltage marking is required for one or more of the listings.*

Informational Note: Cable types are listed in descending order of fire resistance rating.

**(J) Insulated Continuous Line-Type Fire Detectors.** Insulated continuous line-type fire detectors shall be rated in accordance with 760.179(C), listed as being resistant to the spread of fire in accordance with 760.179(D) through (F), marked in accordance with 760.179(I), and the jacket compound shall have a high degree of abrasion resistance.

Section 760.179(A) is not intended to require conductors made of 100-percent copper. Some listed line-type fire detectors may not be made exclusively of copper.

***TABLE 760.179(I)*** *Cable Markings*

| Cable Marking | Type |
|---|---|
| FPLP | Power-limited fire alarm plenum cable |
| FPLR | Power-limited fire alarm riser cable |
| FPL | Power-limited fire alarm cable |

Note: Cables identified in 760.179(D), (E), and (F) as meeting the requirements for circuit integrity shall have the additional classification using the suffix "CI" (for example, FPLP-CI, FPLR-CI, and FPL-CI).

# ARTICLE 770 Optical Fiber Cables and Raceways

Informational Note: The general term *grounding conductor* as previously used in this article is replaced by either the term *bonding conductor* or the term *grounding electrode conductor* (GEC), where applicable, to more accurately reflect the application and function of the conductor.

See Informational Note Figure 800(a) and Informational Note Figure 800(b) for illustrative application of a bonding conductor or grounding electrode conductor.

## I. General

### 770.1 Scope

The provisions of this article apply to the installation of optical fiber cables, raceways, and cable routing assemblies. This article does not cover the construction of optical fiber cables and raceways.

Article 770 covers the installation of optical fiber cables, raceways, and cable routing assemblies. It provides applications and listing requirements for these raceways. These raceways and routing assemblies are listed to ANSI/UL 2024, *Standard for Safety of Signaling, Optical Fiber and Communications Raceways and Cable Routing Assemblies.* Routing assemblies are U-shaped wiring troughs that may or may not have covers. The significant difference between optical fiber or communications cable routing assemblies and optical fiber raceways is that the routing assemblies are larger and open and, therefore, may present a greater fire load.

Article 770 permits the use of optical fiber technology in conjunction with electrical conductors for communications, signaling, and control circuits in lieu of metallic conductors. The most common optical fiber cable used in buildings is nonconductive.

Because they are not affected by electrical noise, optical fiber cables may be desirable in some circumstances to transmit data or other communications where electrical noise is a problem. Optical fiber cables may be nonconductive, or they may be composite, containing electrical conductors. See Exhibits 770.1 and 770.2.

### 770.2 Definitions

See Part I of Article 100. For purposes of this article, the following additional definitions apply.

**Abandoned Optical Fiber Cable.** Installed optical fiber cable that is not terminated at equipment other than a connector and not identified for future use with a tag.

Informational Note: See Part I of Article 100 for a definition of *Equipment*.

•

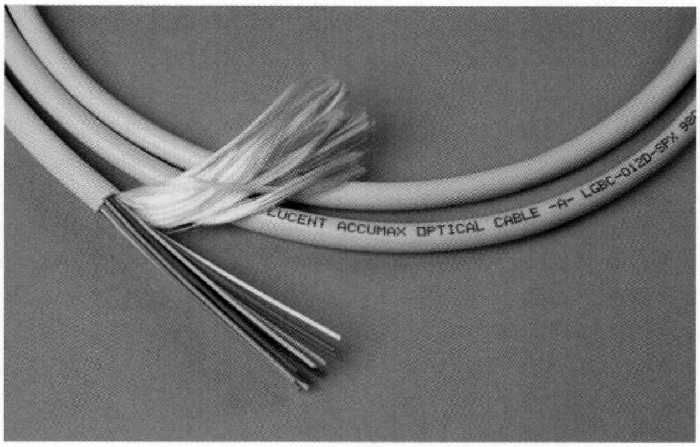

***EXHIBIT 770.1*** *An example of a nonconductive optical fiber cable.*

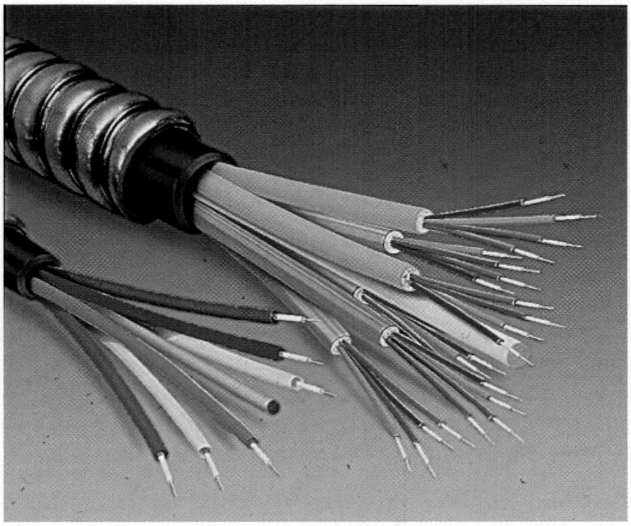

***EXHIBIT 770.2*** *An example of a composite optical fiber cable that also meets the requirements of Article 330 and is referred to as Type MC cable. (Courtesy of AFC Cable Systems, Inc.)*

**Cable Sheath.** A covering over the optical fiber assembly that includes one or more jackets and may include one or more metallic members or strength members.

**Composite Optical Fiber Cable.** A cable containing optical fibers and current-carrying electrical conductors.

**Conductive Optical Fiber Cable.** A factory assembly of one or more optical fibers having an overall covering and containing non–current-carrying conductive member(s) such as metallic strength member(s), metallic vapor barrier(s), metallic armor or metallic sheath.

**Electrical Circuit Protective System.** A system consisting of components and materials intended for installation as protection for specific electrical wiring systems with respect to the disruption of electrical circuit integrity upon exterior fire exposure.

**Exposed (to Accidental Contact).** A conductive optical fiber cable in such a position that, in case of failure of supports or insulation, contact between the cable's non–current-carrying conductive members and an electrical circuit may result.

Informational Note: See Part I of Article 100 for two other definitions of *Exposed.*

**Innerduct.** A nonmetallic raceway placed within a larger raceway.

**Nonconductive Optical Fiber Cable.** A factory assembly of one or more optical fibers having an overall covering and containing no electrically conductive materials.

**Optical Fiber Cable.** A factory assembly or field assembly of one or more optical fibers having an overall covering.

Informational Note: A field-assembled optical fiber cable is an assembly of one or more optical fibers within a jacket. The jacket, without optical fibers, is installed in a manner similar to conduit or raceway. Once the jacket is installed, the optical fibers are inserted into the jacket, completing the cable assembly.

•

**Point of Entrance.** The point within a building at which the optical fiber cable emerges from an external wall, from a concrete floor slab, from rigid metal conduit (RMC), or from intermediate metal conduit (IMC).

## 770.3 Other Articles

Installations of optical fiber cables and raceways shall comply with 770.3(A) and (B). Only those sections of Chapter 2 and Article 300 referenced in this article shall apply to optical fiber cables and raceways.

**(A) Hazardous (Classified) Locations.** Listed optical fiber cables shall be permitted to be installed in hazardous (classified) locations. The cables shall be sealed in accordance with the requirements of 501.15, 502.15, 505.16, or 506.16, as applicable.

**(B) Cables in Ducts for Dust, Loose Stock, or Vapor Removal.** The requirements of 300.22(A) for wiring systems shall apply to conductive optical fiber cables.

**(C) Composite Cables.** Composite optical fiber cables shall be classified as electrical cables in accordance with the type of electrical conductors. They shall be constructed, listed, and marked in accordance with the appropriate article for each type of electrical cable.

## 770.12 Innerduct for Optical Fiber Cables

Listed plenum communications raceway, listed riser communications raceway, and listed general-purpose communications raceway selected in accordance with the provisions of Table 800.154(b) shall be permitted to be installed as innerduct in any type of listed raceway permitted in Chapter 3.

### 770.21 Access to Electrical Equipment Behind Panels Designed to Allow Access

Access to electrical equipment shall not be denied by an accumulation of optical fiber cables that prevents removal of panels, including suspended ceiling panels.

An excess accumulation of wires and cables can limit access to equipment by preventing the removal of access panels. (See Exhibit 800.2.)

### 770.24 Mechanical Execution of Work

Optical fiber cables shall be installed in a neat and workmanlike manner. Cables installed exposed on the surface of ceilings and sidewalls shall be supported by the building structure in such a manner that the cable will not be damaged by normal building use. Such cables shall be secured by hardware including straps, staples, cable ties, hangers, or similar fittings designed and installed so as not to damage the cable. The installation shall also conform with 300.4(D) through (G) and 300.11. Nonmetallic cable ties and other nonmetallic cable accessories used to secure and support cables in other spaces used for environmental air (plenums) shall be listed as having low smoke and heat release properties.

Informational Note No. 1: Accepted industry practices are described in ANSI/NECA/BICSI 568-2006, *Standard for Installing Commercial Building Telecommunications Cabling*; ANSI/NECA/FOA 301-2009, *Standard for Installing and Testing Fiber Optic Cables*; and other ANSI-approved installation standards.

Optical fiber cables must be installed in a neat and workmanlike manner. Cable must be attached to or supported by the structure by cable ties, straps, clamps, hangers, and so forth. The installation method must not damage the cable.

This requirement does not contain specific supporting and securing intervals. It does reference 300.11 for general requirements on securing equipment and cables and 300.4(D) through (G) for protection of cables.

Informational Note No. 2: See 4.3.11.2.6.5 and 4.3.11.5.5.6 of NFPA 90A-2012, *Standard for the Installation of Air-Conditioning and Ventilating Systems*, for discrete combustible components installed in accordance with 300.22(C).

Informational Note No. 2 addresses requirements in NFPA 90A, *Standard for the Installation of Air-Conditioning and Ventilating Systems*, that have an influence on installations covered in this *Code*. The intent is not to require that products in this section be listed for more than their smoke and heat properties, which would result in additional requirements not included in NFPA 90A.

The NFPA 90A requirements are focused on smoke and heat generated from a fire in an air-handling plenum. Where discrete combustible components are installed in the ceiling cavity plenum, NFPA 90A requires speakers, fixtures, and other electrical equipment with combustible enclosures, including their assemblies and accessories, cable ties, and other discrete products, to be listed for their smoke density and heat release in accordance with ANSI/UL 2043, *Standard for Safety Fire Test for Heat and Visible Smoke Release for Discrete Products and Their Accessories Installed in Air-Handling Spaces.* Similar requirements in NFPA 90A apply for discrete combustible products installed in a raised floor plenum.

None of these requirements pertain to noncombustible products. Many metallic products, including metallic cable ties, used to support power, data, and communications raceways and cables are not required to be listed.

### 770.25 Abandoned Cables

The accessible portion of abandoned optical fiber cables shall be removed. Where cables are identified for future use with a tag, the tag shall be of sufficient durability to withstand the environment involved.

See Article 100 for the definition of *accessible* as applied to wire methods. Abandoned cable increases fire loading unnecessarily, and, where installed in plenums, it can affect airflow. See the definition of *abandoned optical fiber cable* in 770.2.

### 770.26 Spread of Fire or Products of Combustion

Installations of optical fiber cables and communications raceways in hollow spaces, vertical shafts, and ventilation or air-handling ducts shall be made so that the possible spread of fire or products of combustion will not be substantially increased. Openings around penetrations of optical fiber cables and communications raceways through fire-resistant–rated walls, partitions, floors, or ceilings shall be firestopped using approved methods to maintain the fire resistance rating.

Informational Note: Directories of electrical construction materials published by qualified testing laboratories contain many listing installation restrictions necessary to maintain the fire-resistive rating of assemblies where penetrations or openings are made. Building codes also contain restrictions on membrane penetrations on opposite sides of a fire resistance–rated wall assembly. An example is the 600-mm (24-in.) minimum horizontal separation that usually applies between boxes installed on opposite sides of the wall. Assistance in complying with 770.26 can be found in building codes, fire resistance directories, and product listings.

## II. Cables Outside and Entering Buildings

### 770.47 Underground Optical Fiber Cables Entering Buildings

Underground optical fiber cables entering buildings shall comply with 770.47(A) and (B).

**(A) Underground Systems with Electric Light, Power, Class 1, or Non–Power-Limited Fire Alarm Circuit**

**Conductors.** Underground conductive optical fiber cables entering buildings with electric light, power, Class 1, or non–power-limited fire alarm circuit conductors in a raceway, handhole enclosure, or manhole shall be located in a section separated from such conductors by means of brick, concrete, or tile partitions or by means of a suitable barrier.

**(B) Direct-Buried Cables and Raceways.** Direct-buried conductive optical fiber cables shall be separated by at least 300 mm (12 in.) from conductors of any electric light, power, or non–power-limited fire alarm circuit conductors or Class 1 circuit.

*Exception No. 1: Direct-buried conductive optical fiber cables shall not be required to be separated by at least 300 mm (12 in.) from electric service conductors where electric service conductors are installed in raceways or have metal cable armor.*

*Exception No. 2: Direct-buried conductive optical fiber cables shall not be required to be separated by at least 300 mm (12 in.) from electric light or power branch-circuit or feeder conductors, non–power-limited fire alarm circuit conductors, or Class 1 circuit conductors where electric light or power branch-circuit or feeder conductors, non–power-limited fire alarm circuit conductors, or Class 1 circuit conductors are installed in a raceway or in metal-sheathed, metal-clad, or Type UF or Type USE cables.*

### 770.48 Unlisted Cables and Raceways Entering Buildings

**(A) Conductive and Nonconductive Cables.** Unlisted conductive and nonconductive outside plant optical fiber cables shall be permitted to be installed in building spaces, other than risers, ducts used for environmental air, plenums used for environmental air, and other spaces used for environmental air, where the length of the cable within the building, measured from its point of entrance, does not exceed 15 m (50 ft) and the cable enters the building from the outside and is terminated in an enclosure.

Informational Note No. 1: Splice cases or terminal boxes, both metallic and plastic types, typically are used as enclosures for splicing or terminating optical fiber cables.

Informational Note No. 2: See 770.2 for the definition of *Point of Entrance*.

**(B) Nonconductive Cables in Raceway.** Unlisted nonconductive outside plant optical fiber cables shall be permitted to enter the building from the outside and shall be permitted to be installed in any of the following raceways:

(1) Intermediate metal conduit (IMC)
(2) Rigid metal conduit (RMC)
(3) Rigid polyvinyl chloride conduit (PVC)
(4) Electrical metallic tubing (EMT)

### 770.49 Metallic Entrance Conduit Grounding

Rigid metal conduit (RMC) or intermediate metal conduit (IMC) containing optical fiber entrance cable shall be connected by a bonding conductor or grounding electrode conductor to a grounding electrode in accordance with 770.100(B).

## III. Protection

### 770.93 Grounding or Interruption of Non–Current-Carrying Metallic Members of Optical Fiber Cables

Optical fiber cables entering the building or terminating on the outside of the building shall comply with 770.93(A) or (B).

**(A) Entering Buildings.** In installations where an optical fiber cable is exposed to contact with electric light or power conductors and the cable enters the building, the non–current-carrying metallic members shall be either grounded as specified in 770.100, or interrupted by an insulating joint or equivalent device. The grounding or interruption shall be as close as practicable to the point of entrance.

Informational Note: See 770.100(B) for a definition of *Point of Entrance*.

**(B) Terminating On the Outside of Buildings.** In installations where an optical fiber cable is exposed to contact with electric light or power conductors and the cable is terminated on the outside of the building, the non–current-carrying metallic members shall be either grounded as specified in 770.100, or interrupted by an insulating joint or equivalent device. The grounding or interruption shall be as close as practicable to the point of termination of the cable.

## IV. Grounding Methods

### 770.100 Entrance Cable Bonding and Grounding

Where required, the non–current-carrying metallic members of optical fiber cables entering buildings shall be bonded or grounded as specified in 770.100(A) through (D).

**(A) Bonding Conductor or Grounding Electrode Conductor**

**(1) Insulation.** The bonding conductor or grounding electrode conductor shall be listed and shall be permitted to be insulated, covered, or bare.

**(2) Material.** The bonding conductor or grounding electrode conductor shall be copper or other corrosion-resistant conductive material, stranded or solid.

**(3) Size.** The bonding conductor or grounding electrode conductor shall not be smaller than 14 AWG. It shall have a current-carrying capacity not less than that of the grounded metallic member(s). The bonding conductor or grounding electrode conductor shall not be required to exceed 6 AWG.

**(4) Length.** The bonding conductor or grounding electrode conductor shall be as short as practicable. In one- and two-family dwellings, the bonding conductor or grounding electrode conductor shall be as short as practicable not to exceed 6.0 m (20 ft) in length.

Informational Note: Similar bonding conductor or grounding electrode conductor length limitations applied at apartment buildings and commercial buildings help to reduce voltages that may develop between the building's power and communications systems during lightning events.

*Exception: In one- and two-family dwellings where it is not practicable to achieve an overall maximum bonding conductor or grounding electrode conductor length of 6.0 m (20 ft), a separate ground rod meeting the minimum dimensional criteria of 770.100(B)(3)(2) shall be driven, the grounding electrode conductor shall be connected to the separate ground rod in accordance with 770.100(C), and the separate ground rod shall be bonded to the power grounding electrode system in accordance with 770.100(D).*

**(5) Run in Straight Line.** The bonding conductor or grounding electrode conductor shall be run in as straight a line as practicable.

**(6) Physical Protection.** Bonding conductors and grounding electrode conductors shall be protected where exposed to physical damage. Where the bonding conductor or grounding electrode conductor is installed in a metal raceway, both ends of the raceway shall be bonded to the contained conductor or to the same terminal or electrode to which the bonding conductor or grounding electrode conductor is connected.

**(B) Electrode.** The bonding conductor and grounding electrode conductor shall be connected in accordance with 770.100(B)(1), (B)(2), or (B)(3).

**(1) In Buildings or Structures with an Intersystem Bonding Termination.** If the building or structure served has an intersystem bonding termination as required by 250.94, the bonding conductor shall be connected to the intersystem bonding termination.

Informational Note: See Part I of Article 100 for the definition of *Intersystem Bonding Termination*.

**(2) In Buildings or Structures with Grounding Means.** If the building or structure served has no intersystem bonding termination, the bonding conductor or grounding electrode conductor shall be connected to the nearest accessible location on the following:

(1) The building or structure grounding electrode system as covered in 250.50
(2) The grounded interior metal water piping system, within 1.5 m (5 ft) from its point of entrance to the building, as covered in 250.52
(3) The power service accessible means external to enclosures as covered in 250.94
(4) The nonflexible metallic power service raceway
(5) The service equipment enclosure
(6) The grounding electrode conductor or the grounding electrode conductor metal enclosure of the power service, or
(7) The grounding electrode conductor or the grounding electrode of a building or structure disconnecting means that is grounded to an electrode as covered in 250.32

**(3) In Buildings or Structures Without Intersystem Bonding Termination or Grounding Means.** If the building or structure served has no intersystem bonding termination or grounding means, as described in 770.100(B)(2), the grounding electrode conductor shall be connected to either of the following:

(1) To any one of the individual grounding electrodes described in 250.52(A)(1), (A)(2), (A)(3), or (A)(4).
(2) If the building or structure served has no grounding means, as described in 770.100(B)(2) or (B)(3)(1), to any one of the individual grounding electrodes described in 250.52(A)(7) and (A)(8) or to a ground rod or pipe not less than 1.5 m (5 ft) in length and 12.7 mm (½ in.) in diameter, driven, where practicable, into permanently damp earth and separated from lightning conductors as covered in 800.53 and at least 1.8 m (6 ft) from electrodes of other systems. Steam or hot water pipes or air terminal conductors (lightning-rod conductors) shall not be employed as electrodes for non–current-carrying metallic members.

**(C) Electrode Connection.** Connections to grounding electrodes shall comply with 250.70.

**(D) Bonding of Electrodes.** A bonding jumper not smaller than 6 AWG copper or equivalent shall be connected between the grounding electrode and power grounding electrode system at the building or structure served where separate electrodes are used.

*Exception: At mobile homes as covered in 770.106.*

Informational Note No. 1: See 250.60 for use of air terminals (lightning rods).

Informational Note No. 2: Bonding together of all separate electrodes limits potential differences between them and between their associated wiring systems.

## 770.106 Grounding and Bonding of Entrance Cables at Mobile Homes

**(A) Grounding.** Grounding shall comply with 770.106(A)(1) and (A)(2).

(1) Where there is no mobile home service equipment located within 9.0 m (30 ft) of the exterior wall of the mobile home it serves, the non–current-carrying metallic members of optical fiber cables entering the mobile home shall be grounded in accordance with 770.100(B)(3).
(2) Where there is no mobile home disconnecting means grounded in accordance with 250.32 and located within 9.0 m (30 ft) of the exterior wall of the mobile home it serves, the non–current-carrying metallic members of optical fiber cables entering the mobile home shall be grounded in accordance with 770.100(B)(3).

**(B) Bonding.** The grounding electrode shall be bonded to the metal frame or available grounding terminal of the mobile home with a copper conductor not smaller than 12 AWG under either of the following conditions:

(1) Where there is no mobile home service equipment or disconnecting means as in 770.106(A)
(2) Where the mobile home is supplied by cord and plug

## V. Installation Methods Within Buildings

### 770.110 Raceways and Cable Routing Assemblies for Optical Fiber Cables

**(A) Types of Raceways.** Optical fiber cables shall be permitted to be installed in any raceway that complies with either 770.110(A)(1) or (A)(2) and in cable routing assemblies installed in compliance with 770.110(C).

**(1) Raceways Recognized in Chapter 3.** Optical fiber cables shall be permitted to be installed in any raceway included in Chapter 3. The raceways shall be installed in accordance with the requirements of Chapter 3.

**(2) Communications Raceways.** Optical fiber cables shall be permitted to be installed in listed plenum communications raceways, listed riser communications raceways, and listed general-purpose communications raceways selected in accordance with the provisions of 770.113, 800.110, and 800.113, and installed in accordance with 362.24 through 362.56, where the requirements applicable to electrical nonmetallic tubing (ENT) apply.

See the commentary following 800.182(C) for information on listed communications raceways.

**(B) Raceway Fill for Optical Fiber Cables.** Raceway fill for optical fiber cables shall comply with either 770.110(B)(1) or (B)(2).

**(1) Without Electric Light or Power Conductors.** Where optical fiber cables are installed in raceway without electric light or power conductors, the raceway fill requirements of Chapters 3 and 9 shall not apply.

**(2) Nonconductive Optical Fiber Cables with Electric Light or Power Conductors.** Where nonconductive optical fiber cables are installed with electric light or power conductors in a raceway, the raceway fill requirements of Chapters 3 and 9 shall apply.

Conduit fill requirements apply where optical fiber cables are installed in a raceway with electrical conductors. Chapter 3 raceway articles refer to Chapter 9 for the cylindrical raceway fill tables.

**(C) Cable Routing Assemblies.** Optical fiber cables shall be permitted to be installed in plenum cable routing assemblies, riser cable routing assemblies, and general-purpose cable routing assemblies selected in accordance with the provisions of 800.113 and Table 800.154(c) and installed in accordance with 770.110(C)(1) and (C)(2).

See the commentary following 800.182(C) for information on listed cable routing assemblies.

**(1) Horizontal Support.** Cable routing assemblies shall be supported where run horizontally at intervals not to exceed 900 mm (3 ft), and at each end or joint, unless listed for other support intervals. In no case shall the distance between supports exceed 3 m (10 ft).

**(2) Vertical Support.** Vertical runs of cable routing assemblies shall be supported at intervals not exceeding 1.2 m (4 ft), unless listed for other support intervals, and shall not have more than one joint between supports.

### 770.113 Installation of Optical Fiber Cables

Installation of optical fiber cables shall comply with 770.113(A) through (J). Installation of raceways shall comply with 770.110.

**(A) Listing.** Optical fiber cables installed in buildings shall be listed.

*Exception: Optical fiber cables that comply with 770.48 shall not be required to be listed.*

**(B) Fabricated Ducts Used for Environmental Air.** The following cables shall be permitted in ducts, as described in 300.22(B) if they are directly associated with the air distribution system:

(1) Up to 1.22 m (4 ft) of Types OFNP and OFCP cables
(2) Types OFNP, OFCP, OFNR, OFCR, OFNG, OFCG, OFN, and OFC cables installed in raceways that are installed in compliance with 300.22(B)

Informational Note: For information on fire protection of wiring installed in fabricated ducts, see 4.3.4.1 and 4.3.11.3.3 of NFPA 90A-2012, *Standard for the Installation of Air-Conditioning and Ventilating Systems.*

**(C) Other Spaces Used For Environmental Air (Plenums).** The following cables shall be permitted in other spaces used for environmental air as described in 300.22(C):

(1) Types OFNP and OFCP cables

•

(2) Types OFNP and OFCP cables installed in plenum communications raceways
(3) Types OFNP and OFCP cables supported by open metallic cable trays or cable tray systems
(4) Types OFNP, OFCP, OFNR, OFCR, OFNG, OFCG, OFN, and OFC cables installed in raceways that are installed in compliance with 300.22(C)
(5) Types OFNP, OFCP, OFNR, OFCR, OFNG, OFCG, OFN, and OFC cables supported by solid bottom metal cable

trays with solid metal covers in other spaces used for environmental air (plenums), as described in 300.22(C)

(6) Types OFNP, OFCP, OFNR, OFCR, OFNG, OFCG, OFN, and OFC cables installed in plenum communications raceways, riser communications raceways, or general-purpose communications raceways supported by solid bottom metal cable trays with solid metal covers in other spaces used for environmental air (plenums), as described in 300.22(C)

Informational Note No. 1: For information on fire protection of wiring installed in other spaces used for environmental air, see 4.3.11.2, 4.3.11.4, and 4.3.11.5 of NFPA 90A-2012, *Standard for the Installation of Air-Conditioning and Ventilating Systems.*

Informational Note No. 2: See 800.110 and 800.113 for installation requirements for cable routing assemblies and communications raceways.

**(D) Risers — Cables, Raceways, and Cable Routing Assemblies in Vertical Runs.** The following cables shall be permitted in vertical runs penetrating one or more floors and in vertical runs in a shaft:

(1) Types OFNP, OFCP, OFNR, and OFCR cables

•

(2) Types OFNP, OFCP, OFNR, and OFCR cables installed in:

a. Plenum communications raceways
b. Plenum cable routing assemblies

•

c. Riser communications raceways
d. Riser cable routing assemblies

Informational Note: See 770.26 for firestop requirements for floor penetrations.

**(E) Risers — Cables in Metal Raceways.** The following cables shall be permitted in metal raceways in a riser having firestops at each floor:

(1) Types OFNP, OFCP, OFNR, OFCR, OFNG, OFCG, OFN, and OFC cables

•

(2) Types OFNP, OFCP, OFNR, OFCR, OFNG, OFCG, OFN, and OFC cables installed in:

•

a. Plenum communications raceways

•

b. Riser communications raceways

•

c. General-purpose communications raceways

Informational Note: See 770.26 for firestop requirements for floor penetrations.

**(F) Risers — Cables in Fireproof Shafts.** The following cables shall be permitted to be installed in fireproof riser shafts having firestops at each floor:

(1) Types OFNP, OFCP, OFNR, OFCR, OFNG, OFCG, OFN, and OFC cables

•

(2) Types OFNP, OFCP, OFNR, OFCR, OFNG, OFCG, OFN, and OFC cables installed in:

•

a. Plenum communications raceways
b. Plenum cable routing assemblies

•

c. Riser communications raceways

•

d. Riser cable routing assemblies

•

e. General-purpose communications raceways

•

f. General-purpose cable routing assemblies

Informational Note: See 770.26 for firestop requirements for floor penetrations.

**(G) Risers — One- and Two-Family Dwellings.** The following cables shall be permitted in one- and two-family dwellings:

(1) Types OFNP, OFCP, OFNR, OFCR, OFNG, OFCG, OFN, and OFC cables

•

(2) Types OFNP, OFCP, OFNR, OFCR, OFNG, OFCG, OFN, and OFC cables installed in:

•

a. Plenum communications raceways

•

b. Plenum cable routing assemblies

•

c. Riser communications raceways

•

d. Riser cable routing assemblies

•

e. General-purpose communications raceways

•

f. General-purpose cable routing assemblies

**(H) Cable Trays.** The following cables shall be permitted to be supported by cable trays:

(1) Types OFNP, OFCP, OFNR, OFCR, OFNG, OFCG, OFN, and OFC cables

•

(2) Types OFNP, OFCP, OFNR, OFCR, OFNG, OFCG, OFN, and OFC cables installed in:

•

a. Plenum communications raceways

•

b. Riser communications raceways

•

c. General-purpose communications raceways

**(I) Distributing Frames and Cross-Connect Arrays.** The following cables shall be permitted to be installed in distributing frames and cross-connect arrays:

(1) Types OFNP, OFCP, OFNR, OFCR, OFNG, OFCG, OFN, and OFC cables

•

(2) Types OFNP, OFCP, OFNR, OFCR, OFNG, OFCG, OFN, and OFC cables installed in:

•

a. Plenum communications raceways
b. Plenum cable routing assemblies

•

c. Riser communications raceways
d. Riser cable routing assemblies

•

e. General-purpose communications raceways

•

f. General-purpose cable routing assemblies

**(J) Other Building Locations.** The following cables shall be permitted to be installed in building locations other than the locations covered in 770.113(B) through (I):

(1) Types OFNP, OFCP, OFNR, OFCR, OFNG, OFCG, OFN, and OFC cables

•

(2) Types OFNP, OFCP, OFNR, OFCR, OFNG, OFCG, OFN, and OFC cables installed in:

•

a. Plenum communications raceways
b. Plenum cable routing assemblies

•

c. Riser communications raceways

•

d. Riser cable routing assemblies

•

e. General-purpose communications raceways

•

f. General-purpose cable routing assemblies

(3) Types OFNP, OFCP, OFNR, OFCR, OFNG, OFCG, OFN, and OFC cables installed in a raceway of a type recognized in Chapter 3

## 770.114 Grounding

Non–current-carrying conductive members of optical fiber cables shall be bonded to a grounded equipment rack or enclosure, or grounded in accordance with the grounding methods specified by 770.100(B)(2).

## 770.133 Installation of Optical Fibers and Electrical Conductors

**(A) With Conductors for Electric Light, Power, Class 1, Non–Power-Limited Fire Alarm, or Medium Power Network-Powered Broadband Communications Circuits.** When optical fibers are within the same composite cable for electric light, power, Class 1, non–power-limited fire alarm, or medium-power network-powered broadband communications circuits operating at 1000 volts or less, they shall be permitted to be installed only where the functions of the optical fibers and the electrical conductors are associated.

Nonconductive optical fiber cables shall be permitted to occupy the same cable tray or raceway with conductors for electric light, power, Class 1, non–power-limited fire alarm, Type ITC, or medium-power network-powered broadband communications circuits operating at 1000 volts or less. Conductive optical fiber cables shall not be permitted to occupy the same cable tray or raceway with conductors for electric light, power, Class 1, non–power-limited fire alarm, Type ITC, or medium-power network-powered broadband communications circuits.

Optical fibers in composite optical fiber cables containing only current-carrying conductors for electric light, power, or Class 1 circuits rated 1000 volts or less shall be permitted to occupy the same cabinet, cable tray, outlet box, panel, raceway, or other termination enclosure with conductors for electric light, power, or Class 1 circuits operating at 1000 volts or less.

Nonconductive optical fiber cables shall not be permitted to occupy the same cabinet, outlet box, panel, or similar enclosure housing the electrical terminations of an electric light, power, Class 1, non–power-limited fire alarm, or medium-power network-powered broadband communications circuit.

*Exception No. 1: Occupancy of the same cabinet, outlet box, panel, or similar enclosure shall be permitted where nonconductive optical fiber cable is functionally associated with the electric light, power, Class 1, non–power-limited fire alarm, or medium-power network-powered broadband communications circuit.*

*Exception No. 2: Occupancy of the same cabinet, outlet box, panel, or similar enclosure shall be permitted where nonconductive optical fiber cables are installed in factory- or field-assembled control centers.*

*Exception No. 3: In industrial establishments only, where conditions of maintenance and supervision ensure that only qualified persons service the installation, nonconductive optical fiber cables shall be permitted with circuits exceeding 1000 volts.*

*Exception No. 4: In industrial establishments only, where conditions of maintenance and supervision ensure that only qualified persons service the installation, optical fibers in composite optical fiber cables containing current-carrying conductors operating over 1000 volts shall be permitted to be installed.*

*Exception No. 5: Where all of the conductors of electric light, power, Class 1, nonpower-limited fire alarm, and medium-power network-powered broadband communications circuits are separated from all of the optical fiber cables by a permanent barrier or listed divider.*

**(B) With Communications Cables.** Optical fibers shall be permitted in the same cable, and conductive and nonconductive optical fiber cables shall be permitted in the same raceway, cable tray, box, enclosure, or cable routing assembly, with conductors of any of the following:

(1) Communications circuits in compliance with Parts I and V of Article 800
(2) Community antenna television and radio distribution systems in compliance with Parts I and V of Article 820
(3) Low-power network-powered broadband communications circuits in compliance with Parts I and V of Article 830

**(C) With Other Circuits.** Optical fibers shall be permitted in the same cable, and conductive and nonconductive optical fiber cables shall be permitted in the same raceway, cable tray, box, enclosure, or cable routing assembly, with conductors of any of the following:

(1) Class 2 and Class 3 remote-control, signaling, and power-limited circuits in compliance with Article 645 or Parts I and III of Article 725.
(2) Power-limited fire alarm systems in compliance with Parts I and III of Article 760.

**(D) Support of Cables.** Raceways shall be used for their intended purpose. Optical fiber cables shall not be strapped, taped, or attached by any means to the exterior of any conduit or raceway as a means of support.

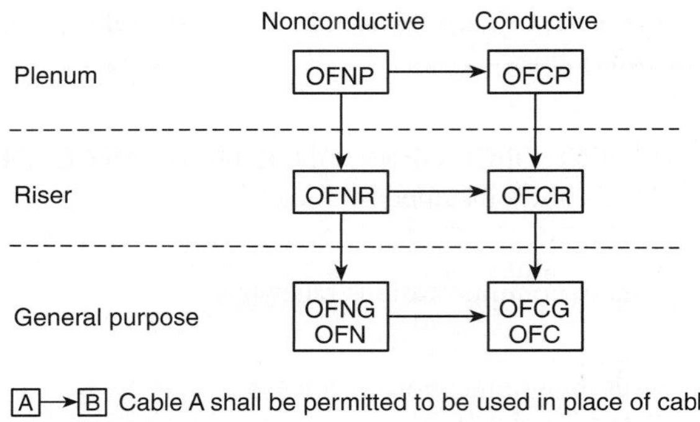


***FIGURE 770.154*** *Cable Substitution Hierarchy.*

*Exception: Overhead (aerial) spans of optical fiber cables shall be permitted to be attached to the exterior of a raceway-type mast intended for the attachment and support of such cables.*

### 770.154 Applications of Listed Optical Fiber Cables

Permitted and nonpermitted applications of listed optical fiber cables shall be as indicated in Table 770.154(a). The permitted applications shall be subject to the installation requirements of 770.110 and 770.113. The substitutions for optical fiber cables in Table 770.154(b) and illustrated in Figure 770.154 shall be permitted.

The application of optical fiber cables, communications raceway, and cable routing assemblies are summarized in Table 770.154(a). The installation location dictates the type of optical fiber cable permitted within the raceway or assembly and is subject to the installation requirements of 770.110 and 770.113.

## VI. Listing Requirements

### 770.179 Optical Fiber Cables

Optical fiber cables shall be listed in accordance with 770.179(A) through (F) and shall be marked in accordance with Table 770.179. In addition, the overall covering of a field-assembled optical fiber cable shall have a surface marking indicating the specific optical fiber conductors with which it is listed, and the optical fiber conductors shall have a permanent marking, such as a marker tape, indicating the overall covering with which they are listed. The overall covering of a field-assembled optical fiber cable shall meet the listing requirements for optical fiber raceways. Optical fiber cables shall have a temperature rating of not less than 60°C (140°F).

Optical fiber cables must have a temperature rating of not less than 140°F to correlate with requirements for communications wires and cables that are addressed in 800.179.

***TABLE 770.154(a)*** *Applications of Listed Optical Fiber Cables in Buildings*

| **Applications** | | **Cable Type** | | |
|---|---|---|---|---|
| | | **OFNP, OFCP** | **OFNR, OFCR** | **OFNG, OFCG, OFN, OFC** |
| In specifically fabricated ducts as described in 300.22(B) | In fabricated ducts | Y* | N | N |
| | In metal raceway that complies with 300.22(B) | Y* | Y* | Y* |
| In other spaces used for environmental air as described in 300.22(C) | In other spaces used for environmental air | Y* | N | N |
| | In metal raceway that complies with 300.22(C) | Y* | Y* | Y* |
| | In plenum communications raceways | Y* | N | N |
| | In plenum cable routing assemblies | NOT PERMITTED | | |
| | Supported by open metal cable trays | Y* | N | N |
| | Supported by solid bottom metal cable trays with solid metal covers | Y* | Y* | Y* |
| In risers | In vertical runs | Y* | Y* | N |
| | In metal raceways | Y* | Y* | Y* |
| | In fireproof shafts | Y* | Y* | Y* |
| | In plenum communications raceways | Y* | Y* | N |
| | In plenum cable routing assemblies | Y* | Y* | N |
| | In riser communications raceways | Y* | Y* | N |
| | In riser cable routing assemblies | Y* | Y* | N |
| | In one- and two-family dwellings | Y* | Y* | Y* |
| Within buildings in other than air-handling spaces and risers | General | Y* | Y* | Y* |
| | Supported by cable trays | Y* | Y* | Y* |
| | In distributing frames and cross-connect arrays | Y* | Y* | Y* |
| | In any raceway recognized in Chapter 3 | Y* | Y* | Y* |
| | In plenum communications raceways | Y* | Y* | Y* |
| | In plenum cable routing assemblies | Y* | Y* | Y* |
| | In riser communications raceways | Y* | Y* | Y* |
| | In riser cable routing assemblies | Y* | Y* | Y* |
| | In general-purpose communications raceways | Y* | Y* | Y* |
| | In general-purpose cable routing assemblies | Y* | Y* | Y* |

Note: An "N" in the table indicates that the cable type shall not be permitted to be installed in the application. A "Y*" indicates that the cable shall be permitted to be installed in the application subject to the limitations described in 770.110 and 770.113.

Informational Note No. 1: Part V of Article 770 covers installation methods within buildings. This table covers the applications of listed optical fiber cables and raceways and cable routing assemblies in buildings. The definition of *Point of Entrance* is in 770.2. Optical fiber entrance cables that have not emerged from the rigid metal conduit (RMC) or intermediate metal conduit (IMC) are not considered to be in the building.

Informational Note No. 2: For information on the restrictions to the installation of optical fiber cables in fabricated ducts, see 770.113(B).

Informational Note No. 3: Cable routing assemblies are not addressed in NFPA 90A-2012, *Standard for the Installation of Air Conditioning and Ventilation Systems.*

***TABLE 770.154(b)*** *Cable Substitutions*

| Cable Type | Permitted Substitutions |
|---|---|
| OFNP | None |
| OFCP | OFNP |
| OFNR | OFNP |
| OFCR | OFNP, OFCP, OFNR |
| OFNG, OFN | OFNP, OFNR |
| OFCG, OFC | OFNP, OFCP, OFNR, OFCR, OFNG, OFN |

***TABLE 770.179*** *Cable Markings*

| Cable Marking | Type |
|---|---|
| OFNP | Nonconductive optical fiber plenum cable |
| OFCP | Conductive optical fiber plenum cable |
| OFNR | Nonconductive optical fiber riser cable |
| OFCR | Conductive optical fiber riser cable |
| OFNG | Nonconductive optical fiber general-purpose cable |
| OFCG | Conductive optical fiber general-purpose cable |
| OFN | Nonconductive optical fiber general-purpose cable |
| OFC | Conductive optical fiber general-purpose cable |

**(A) Types OFNP and OFCP.** Types OFNP and OFCP nonconductive and conductive optical fiber plenum cables shall be listed as being suitable for use in ducts, plenums, and other space used for environmental air and shall also be listed as having adequate fire resistant and low smoke producing characteristics.

Informational Note: One method of defining a cable that is low-smoke producing cable and fire-resistant cable is that the cable exhibits a maximum peak optical density of 0.50 or less, an average optical density of 0.15 or less, and a maximum flame spread distance of 1.52 m (5 ft) or less when tested in accordance with NFPA 262-2011, *Standard Method of Test for Flame Travel and Smoke of Wires and Cables for Use in Air-Handling Spaces.*

For further information on the fire test method for plenum cables, see the commentary following the informational note to 725.179(A).

**(B) Types OFNR and OFCR.** Types OFNR and OFCR nonconductive and conductive optical fiber riser cables shall be listed as being suitable for use in a vertical run in a shaft or from floor to floor and shall also be listed as having the fire-resistant characteristics capable of preventing the carrying of fire from floor to floor.

Informational Note: One method of defining fire-resistant characteristics capable of preventing the carrying of fire from floor to floor is that the cables pass the requirements of ANSI/UL 1666-2011, *Standard Test for Flame Propagation Height of Electrical and Optical-Fiber Cable Installed Vertically in Shafts.*

For further information on the fire test method for riser cables, see the commentary following the informational note to 725.179(B).

**(C) Types OFNG and OFCG.** Types OFNG and OFCG nonconductive and conductive general-purpose optical fiber cables shall be listed as being suitable for general-purpose use, with the exception of risers and plenums, and shall also be listed as being resistant to the spread of fire.

Informational Note: One method of defining *resistant to the spread of fire* is for the damage (char length) not to exceed 1.5 m (4 ft 11 in.) when performing the CSA "Vertical Flame Test — Cables in Cable Trays," as described in CSA C22.2 No. 0.3-M-2001, *Test Methods for Electrical Wires and Cables.*

For further information on the fire test method for cables used as other wiring within buildings, see the commentary following the informational note to 725.179(C).

**(D) Types OFN and OFC.** Types OFN and OFC nonconductive and conductive optical fiber cables shall be listed as being suitable for general-purpose use, with the exception of risers, plenums, and other spaces used for environmental air, and shall also be listed as being resistant to the spread of fire.

Informational Note: One method of defining *resistant to the spread of fire* is that the cables do not spread fire to the top of the tray in the "UL Flame Exposure, Vertical Tray Flame Test" in ANSI/UL 1685-2010, *Standard for Safety for Vertical-Tray Fire-Propagation and Smoke-Release Test for Electrical and Optical-Fiber Cables.* The smoke measurements in the test method are not applicable.

Another method of defining *resistant to the spread of fire* is for the damage (char length) not to exceed 1.5 m (4 ft 11 in.) when performing the CSA "Vertical Flame Test — Cables in Cable Trays," as described in CSA C22.2 No. 0.3-M-2001, *Test Methods for Electrical Wires and Cables.*

Informational Note: Cable types are listed in descending order of fire resistance rating. Within each fire resistance rating, nonconductive cable is listed first because it may substitute for the conductive cable.

**(E) Circuit Integrity (CI) Cable or Electrical Circuit Protective System.** Cables that are used for survivability of critical circuits under fire conditions shall be listed and meet either 770.179(E)(1) or (E)(2).

Informational Note: The listing organization provides information for circuit integrity (CI) cable and electrical circuit protective systems, including installation requirements necessary to maintain the fire rating.

This correlates with *NFPA 72, National Fire Alarm and Signaling Code,* Section 12.4, Pathway Survivability.

**(1) Circuit Integrity (CI) Cables.** Circuit integrity (CI) cables specified in 770.179(A) through (D), and used for survivability of critical circuits, shall have an additional classification using the suffix "CI." In order to maintain its listed fire rating, circuit integrity (CI) cable shall only be installed in free air.

Informational Note: One method of defining circuit integrity (CI) cable is by establishing a minimum 2-hour fire resistance rating for the cable when tested in accordance with ANSI/UL 2196-2006, Standard for Tests of Fire-Resistive Cable.

**(2) Fire-Resistive Cables.** Cables specified in 770.179(A) through (D) and 770.179(E)(1), that are part of an electrical circuit protective system, shall be fire-resistive cable and identified with the protective system number on the product or on the smallest unit container in which the product is packaged and installed in accordance with the listing of the protective system.

Informational Note No. 1: One method of defining an electrical circuit protective system is by establishing a minimum 2-hour fire resistance rating for the system when tested in accordance with UL Subject 1724, Outline of Investigation for Fire Tests for Electrical Circuit Protective Systems.

Informational Note No. 2: The listing organization provides information for electrical circuit protective systems (FHIT), including installation requirements for maintaining the fire rating.

**(F) Field-Assembled Optical Fiber Cables.** Field-assembled optical fiber cable shall comply with 770.179(F)(1) or (2).

**(1) Marking and Listing of Combination of Jacket and Optical Fibers.** The specific combination of jacket and optical fibers intended to be installed as a field-assembled optical fiber cable shall be listed in accordance with 770.179(A), (B), or (D) and shall be marked in accordance with Table 770.179.

a. The jacket of a field-assembled optical fiber cable shall have a surface marking indicating the specific optical fibers with which it is listed for use.

b. The optical fibers shall have a permanent marking, such as a marker tape, indicating the jacket with which they are listed for use.

**(2) Listing of Jacket Without Fibers.** The jacket without fibers shall meet the listing requirements for communications raceways in 800.182(A), (B), or (C) in accordance with the cable marking.

## 770.180 Grounding Devices

Where bonding or grounding is required, devices used to connect a shield, a sheath, or non–current-carrying metallic members of a cable to a bonding conductor or grounding electrode conductor shall be listed or be part of listed equipment.

•

# 8 Communications Systems

## ARTICLE 800 Communications Circuits

Informational Note: The general term *grounding conductor* as previously used in this article is replaced by either the term *bonding conductor* or the term *grounding electrode conductor* (GEC), where applicable, to more accurately reflect the application and function of the conductor. See Informational Note Figure 800(a) and Informational Note Figure 800(b).

## I. General

### 800.1 Scope

This article covers communications circuits and equipment.

Informational Note No. 1: See 90.2(B)(4) for installations of communications circuits and equipment that are not covered.

Informational Note No. 2: For further information for remote-control, signaling, and power-limited circuits, see Article 725.

Informational Note No. 3: For further information for fire alarm systems, see Article 760.

Section 90.3, covering the structure of the *NEC*, specifies that Chapter 8 (which comprises Articles 800, 810, 820, 830, and 840) covers communications systems and is not subject to the requirements of Chapters 1 through 7, other than where a requirement from these chapters is specifically cited by a Chapter 8 requirement. As an example, 800.24 references 300.4(D) and 300.11, 800.44(A)(3) references 225.14(D), and 800.90(C) references 500.5 and 505.5.

Although information technology equipment systems are often used for or with communications systems, Article 800 does not cover wiring of this equipment. Article 645 provides requirements for wiring contained solely within an information technology equipment room. (See 645.4 for a description of information technology equipment room.) Article 725 provides requirements for wiring that extends beyond a computer room, and also covers wiring of local area networks within buildings. Article 760 covers wiring requirements for fire alarm systems.

In some cases, telephone system wiring is also used for data transmission, which is covered by Article 800. Telephone company

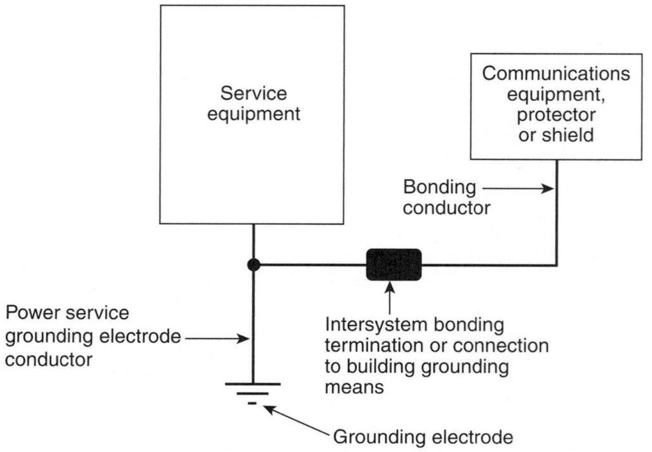


***INFORMATIONAL NOTE FIGURE 800(a)*** *Example of the Use of the Term Bonding Conductor Used in a Communications Installation.*

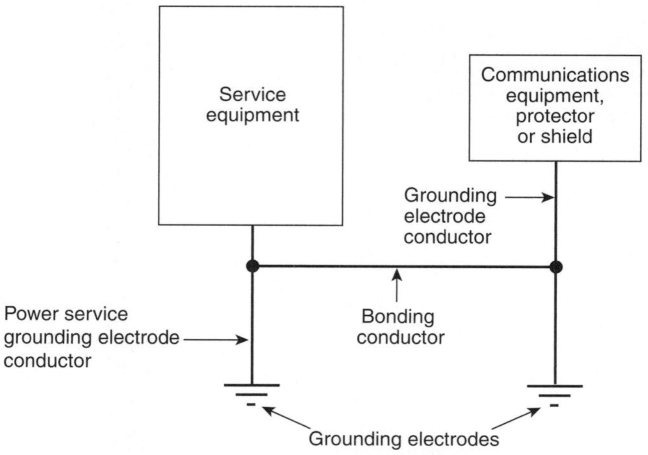


***INFORMATIONAL NOTE FIGURE 800(b)*** *Example of the Use of the Term Grounding Electrode Conductor Used in a Communications Installation.*

*EXHIBIT 800.1 A private automatic branch exchange, one part of telecommunications equipment.*

central offices are exempt from the requirements of Article 800 by 90.2(B)(4). The arrangement of the requirements in Article 800 is similar to those of Articles 725, 760, 770, 820, 830, and 840. *Communications equipment* (see Article 100 for definition), such as the private automatic branch exchange shown in Exhibit 800.1, and all of the premises wiring for a *communications circuit* (see 800.2 for definition), are subject to the requirements of Article 800.

## 800.2 Definitions

See Part I of Article 100. For the purposes of this article, the following additional definitions apply.

**Abandoned Communications Cable.** Installed communications cable that is not terminated at both ends at a connector or other equipment and not identified for future use with a tag.

Informational Note: See Part I of Article 100 for a definition of *Equipment*.

**Block** A square or portion of a city, town, or village enclosed by streets and including the alleys so enclosed, but not any street.

**Cable.** A factory assembly of two or more conductors having an overall covering.

**Cable Sheath.** A covering over the conductor assembly that may include one or more metallic members, strength members, or jackets.

**Communications Circuit.** The circuit that extends voice, audio, video, data, interactive services, telegraph (except radio), outside wiring for fire alarm and burglar alarm from the communications utility to the customer's communications equipment up to and including terminal equipment such as a telephone, fax machine, or answering machine.

**Communications Circuit Integrity (CI) Cable.** Cable used in communications systems to ensure continued operation of critical circuits during a specified time under fire conditions.

•

**Electrical Circuit Protective System.** A system consisting of components and materials intended for installation as protection for specific electrical wiring systems with respect to the disruption of electrical circuit integrity upon exterior fire exposure.

**Exposed (to Accidental Contact).** A circuit that is in such a position that, in case of failure of supports or insulation, contact with another circuit may result.

Informational Note: See Part I of Article 100 for two other definitions of *Exposed*.

**Innerduct.** A nonmetallic raceway placed within a larger raceway.

**Point of Entrance.** The point within a building at which the communications wire or cable emerges from an external wall, from a concrete floor slab, from rigid metal conduit (RMC), or from intermediate metal conduit (IMC).

**Premises.** The land and buildings of a user located on the user side of the utility-user network point of demarcation.

**Wire.** A factory assembly of one or more insulated conductors without an overall covering.

## 800.3 Other Articles

**(A) Hazardous (Classified) Locations.** Communications circuits and equipment installed in a location that is classified in accordance with 500.5 and 505.5 shall comply with the applicable requirements of Chapter 5.

**(B) Wiring in Ducts for Dust, Loose Stock, or Vapor Removal.** The requirements of 300.22(A) shall apply.

**(C) Equipment in Other Space Used for Environmental Air.** The requirements of 300.22(C) (3) shall apply.

**(D) Installation and Use.** The requirements of 110.3(B) shall apply.

**(E) Network-Powered Broadband Communications Systems.** Article 830 shall apply to network-powered broadband communications systems.

**(F) Premises-Powered Broadband Communications Systems.** Article 840 shall apply to premises-powered broadband communications systems.

•

**(G) Optical Fiber Cable.** Where optical fiber cable is used, either in whole or in part, to provide a communications circuit within a building, Article 770 shall apply to the installation of the optical fiber portion of the communications circuit.

•

## 800.12 Innerduct

Listed plenum communications raceway, listed riser communications raceway, and listed general-purpose communications raceway selected in accordance with the provisions of Table 800.154(b) shall be permitted to be installed as innerduct in any type of listed raceway permitted in Chapter 3.

## 800.18 Installation of Equipment

Equipment electrically connected to a communications network shall be listed in accordance with 800.170.

*Exception: This listing requirement shall not apply to test equipment that is intended for temporary connection to a telecommunications network by qualified persons during the course of installation, maintenance, or repair of telecommunications equipment or systems.*

UL 1863, *Communications Circuit Accessories*, and UL 60950, *Safety of Information Technology Equipment, Part 1: General Requirements*, are two safety standards that contain requirements for determining whether equipment connected to a telecommunications network is suitable for the intended purpose. Listed equipment that is connected to the telecommunications network and evaluated according to other U.S. safety standards is also subject to telecommunications requirements appropriate for the equipment. Examples include information technology equipment, audio-video equipment, and signaling equipment connected to a central station. The appropriate requirements contained within the applicable safety standard are extracted from UL 1863, UL 60950, or both.

## 800.21 Access to Electrical Equipment Behind Panels Designed to Allow Access

Access to electrical equipment shall not be denied by an accumulation of communications wires and cables that prevents removal of panels, including suspended ceiling panels.

An excess accumulation of wires and cables can limit access to equipment by preventing the removal of access panels, as shown in Exhibit 800.2.

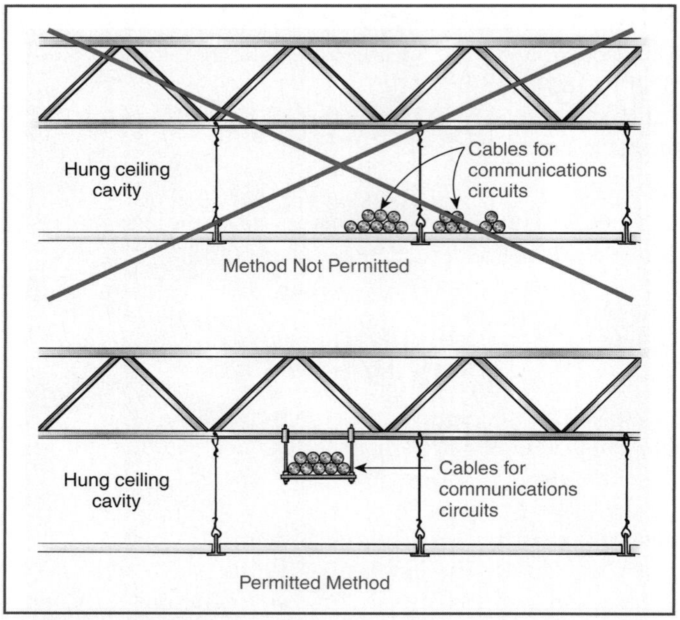


***EXHIBIT 800.2*** *Incorrect installation of cables (upper diagram) and correct method (lower diagram).*

## 800.24 Mechanical Execution of Work

Communications circuits and equipment shall be installed in a neat and workmanlike manner. Cables installed exposed on the surface of ceilings and sidewalls shall be supported by the building structure in such a manner that the cable will not be damaged by normal building use. Such cables shall be secured by hardware, including straps, staples, cable ties, hangers, or similar fittings designed and installed so as not to damage the cable. The installation shall also conform to 300.4(D) and 300.11. Nonmetallic cable ties and other nonmetallic cable accessories used to secure and support cables in other spaces used for environmental air (plenums) shall be listed as having low smoke and heat release properties.

Cable must be attached to or supported by the structure by cable ties, straps, clamps, hangers, and so forth. This requirement does not contain specific supporting and securing intervals. It does reference 300.11 as a general rule on securing equipment and cables, as well as 300.4(D) for protection of cables installed parallel to framing methods. See 800.106(C) for support of cable routing assemblies.

Informational Note No. 1: Accepted industry practices are described in ANSI/NECA/BICSI 568-2006, *Standard for Installing Commercial Building Telecommunications Cabling*; ANSI/TIA/EIA-568-B.1-2004 — Part 1, General Requirements Commercial Building Telecommunications Cabling Standard; ANSI/TIA-569-B-2004, *Commercial Building Standard for Telecommunications Pathways and Spaces*; ANSI/TIA-570-B, *Residential Telecommunications Infrastructure*, and other ANSI-approved installation standards.

Informational Note No. 2: See 4.3.11.2.6.5 and 4.3.11.5.5.6 of NFPA 90A-2012, *Standard for the Installation of Air-Conditioning and Ventilating Systems*, for discrete combustible components installed in accordance with 300.22(C).

## 800.25 Abandoned Cables

The accessible portion of abandoned communications cables shall be removed. Where cables are identified for future use with a tag, the tag shall be of sufficient durability to withstand the environment involved.

See Article 100 for the definition of *accessible* as applied to wiring methods. Abandoned cable increases fire loading unnecessarily, and, where installed in plenums, can affect airflow. See the definition of *abandoned communications cable* in 800.2.

## 800.26 Spread of Fire or Products of Combustion

Installations of communications cables, communications raceways, cable routing assemblies in hollow spaces, vertical shafts, and ventilation or air-handling ducts shall be made so that the possible spread of fire or products of combustion will not be substantially increased. Openings around penetrations of communications cables, communications raceways, and cable routing assemblies through fire-resistant-rated walls, partitions, floors, or ceilings shall be firestopped using approved methods to maintain the fire resistance rating.

Informational Note: Directories of electrical construction materials published by qualified testing laboratories contain many listing installation restrictions necessary to maintain the fire-resistive rating of assemblies where penetrations or openings are made. Building codes also contain restrictions on membrane penetrations on opposite sides of a fire resistance–rated wall assembly. An example is the 600 mm (24 in.) minimum horizontal separation that usually applies between boxes installed on opposite sides of the wall. Assistance in complying with 800.26 can be found in building codes, fire resistance directories, and product listings.

# II. Wires and Cables Outside and Entering Buildings

## 800.44 Overhead (Aerial) Communications Wires and Cables

Overhead (aerial) communications wires and cables entering buildings shall comply with 800.44(A) and (B).

**(A) On Poles and In-Span.** Where communications wires and cables and electric light or power conductors are supported by the same pole or are run parallel to each other in-span, the conditions described in 800.44(A)(1) through (A)(4) shall be met.

**(1) Relative Location.** Where practicable, the communications wires and cables shall be located below the electric light or power conductors.

**(2) Attachment to Cross-Arms.** Communications wires and cables shall not be attached to a cross-arm that carries electric light or power conductors.

**(3) Climbing Space.** The climbing space through communications wires and cables shall comply with the requirements of 225.14(D).

**(4) Clearance.** Supply service drops and sets of overhead service conductors of 0 to 750 volts running above and parallel to communications service drops shall have a minimum separation of 300 mm (12 in.) at any point in the span, including the point of and at their attachment to the building, provided that the ungrounded conductors are insulated and that a clearance of not less than 1.0 m (40 in.) is maintained between the two services at the pole.

**(B) Above Roofs.** Communications wires and cables shall have a vertical clearance of not less than 2.5 m (8 ft) from all points of roofs above which they pass.

*Exception No. 1: Auxiliary buildings, such as garages and the like.*

*Exception No. 2: A reduction in clearance above only the overhanging portion of the roof to not less than 450 mm (18 in.) shall be permitted if (a) not more than 1.2 m (4 ft) of communications service-drop conductors pass above the roof overhang and (b) they are terminated at a through- or above-the-roof raceway or approved support.*

*Exception No. 3: Where the roof has a slope of not less than 100 mm in 300 mm (4 in. in 12 in.), a reduction in clearance to not less than 900 mm (3 ft) shall be permitted.*

Informational Note: For additional information regarding overhead (aerial) wires and cables, see ANSI C2-2007, *National Electric Safety Code*, Part 2, Safety Rules for Overhead Lines.

## 800.47 Underground Communications Wires and Cables Entering Buildings

Underground communications wires and cables entering buildings shall comply with 800.47(A) and (B). The requirements of 310.10(C) shall not apply to communications wires and cables.

**(A) With Electric Light or Power Conductors.** Underground communications wires and cables in a raceway, handhole enclosure, or manhole containing electric light, power, Class 1, or non–power-limited fire alarm circuit conductors shall be in a section separated from such conductors by means of brick, concrete, or tile partitions or by means of a suitable barrier.

**(B) Underground Block Distribution.** Where the entire street circuit is run underground and the circuit within the block is placed so as to be free from the likelihood of accidental contact with electric light or power circuits of over 300 volts to ground, the insulation requirements of 800.50(A) and (C) shall not apply, insulating supports shall not be required for the conductors, and bushings shall not be required where the conductors enter the building.

## 800.48 Unlisted Cables Entering Buildings

Unlisted outside plant communications cables shall be permitted to be installed in building spaces other than risers, ducts used for environmental air, plenums used for environmental air, and other spaces used for environmental air, where the length of the cable within the building, measured from its point of entrance, does not exceed 15 m (50 ft) and the cable enters the building from the outside and is terminated in an enclosure or on a listed primary protector.

Informational Note No. 1: Splice cases or terminal boxes, both metallic and plastic types, are typically used as enclosures for splicing or terminating telephone cables.

Informational Note No. 2: This section limits the length of unlisted outside plant cable to 15 m (50 ft), while 800.90(B) requires that the primary protector be located as close as practicable to the point at which the cable enters the building. Therefore, in installations requiring a primary protector, the outside plant cable may not be permitted to extend 15 m (50 ft) into the building if it is practicable to place the primary protector closer than 15 m (50 ft) to the point of entrance.

Informational Note No. 3: See 800.2 for the definition of *point of entrance*.

### 800.49 Metallic Entrance Conduit Grounding

Rigid metal conduit (RMC) or intermediate metal conduit (IMC) containing communications entrance wire or cable shall be connected by a bonding conductor or grounding electrode conductor to a grounding electrode in accordance with 800.100(B).

### 800.50 Circuits Requiring Primary Protectors

Circuits that require primary protectors as provided in 800.90 shall comply with 800.50(A), (B), and (C).

**(A) Insulation, Wires, and Cables.** Communications wires and cables without a metallic shield, running from the last outdoor support to the primary protector, shall be listed in accordance with 800.173.

**(B) On Buildings.** Communications wires and cables in accordance with 800.50(A) shall be separated at least 100 mm (4 in.) from electric light or power conductors not in a raceway or cable or be permanently separated from conductors of the other systems by a continuous and firmly fixed nonconductor in addition to the insulation on the wires, such as porcelain tubes or flexible tubing. Communications wires and cables in accordance with 800.50(A) exposed to accidental contact with electric light and power conductors operating at over 300 volts to ground and attached to buildings shall be separated from woodwork by being supported on glass, porcelain, or other insulating material.

*Exception: Separation from woodwork shall not be required where fuses are omitted as provided for in 800.90(A)(1), or where conductors are used to extend circuits to a building from a cable having a grounded metal sheath.*

**(C) Entering Buildings.** Where a primary protector is installed inside the building, the communications wires and cables shall enter the building either through a noncombustible, nonabsorbent insulating bushing or through a metal raceway. The insulating bushing shall not be required where the entering communications wires and cables (1) are in metal-sheathed cable, (2) pass through masonry, (3) meet the requirements of 800.50(A) and fuses are omitted as provided in 800.90(A)(1), or (4) meet the requirements of 800.50(A) and are used to extend circuits to a building from a cable having a grounded metallic sheath. Raceways or bushings shall slope upward from the outside or, where this cannot be done, drip loops shall be formed in the communications wires and cables immediately before they enter the building.

Raceways shall be equipped with an approved service head. More than one communications wire and cable shall be permitted to enter through a single raceway or bushing. Conduits or other metal raceways located ahead of the primary protector shall be grounded.

### 800.53 Lightning Conductors

Where practicable, a separation of at least 1.8 m (6 ft) shall be maintained between communications wires and cables on buildings and lightning conductors.

## III. Protection

### 800.90 Protective Devices

**(A) Application.** A listed primary protector shall be provided on each circuit run partly or entirely in aerial wire or aerial cable not confined within a block. Also, a listed primary protector shall be provided on each circuit, aerial or underground, located within the block containing the building served so as to be exposed to accidental contact with electric light or power conductors operating at over 300 volts to ground. In addition, where there exists a lightning exposure, each interbuilding circuit on a premises shall be protected by a listed primary protector at each end of the interbuilding circuit. Installation of primary protectors shall also comply with 110.3(B).

Informational Note No. 1: On a circuit not exposed to accidental contact with power conductors, providing a listed primary protector in accordance with this article helps protect against other hazards, such as lightning and above-normal voltages induced by fault currents on power circuits in proximity to the communications circuit.

Informational Note No. 2: Interbuilding circuits are considered to have a lightning exposure unless one or more of the following conditions exist:

(1) Circuits in large metropolitan areas where buildings are close together and sufficiently high to intercept lightning.
(2) Interbuilding cable runs of 42 m (140 ft) or less, directly buried or in underground conduit, where a continuous metallic cable shield or a continuous metallic conduit containing the cable is connected to each building grounding electrode system.
(3) Areas having an average of five or fewer thunderstorm days per year and earth resistivity of less than 100 ohm-meters. Such areas are found along the Pacific coast.

Telephone utility companies ordinarily provide primary protectors if telephone lines are exposed to lightning. Installers of private networks that include interbuilding cable should also install primary protectors where cables are exposed to lightning. A primary protector is required at each end of an interbuilding communications circuit where lightning exposure exists.

**(1) Fuseless Primary Protectors.** Fuseless-type primary protectors shall be permitted under any of the conditions given in (A)(1)(a) through (A)(1)(e).

(a) Where conductors enter a building through a cable with grounded metallic sheath member(s) and where the conductors in the cable safely fuse on all currents greater than the current-carrying capacity of the primary protector and of the primary protector bonding conductor or grounding electrode conductor

(b) Where insulated conductors in accordance with 800.50(A) are used to extend circuits to a building from a cable with an effectively grounded metallic sheath member(s) and where the conductors in the cable or cable stub, or the connections between the insulated conductors and the plant exposed to accidental contact with electric light or power conductors operating at greater than 300 volts to ground, safely fuse on all currents greater than the current-carrying capacity of the primary protector, or the associated insulated conductors and of the primary protector bonding conductor or grounding electrode conductor

(c) Where insulated conductors in accordance with 800.50(A) or (B) are used to extend circuits to a building from other than a cable with metallic sheath member(s), where (1) the primary protector is listed as being suitable for this purpose for application with circuits extending from other than a cable with metallic sheath members, and (2) the connections of the insulated conductors to the plant exposed to accidental contact with electric light or power conductors operating at greater than 300 volts to ground or the conductors of the plant exposed to accidental contact with electric light or power conductors operating at greater than 300 volts to ground safely fuse on all currents greater than the current-carrying capacity of the primary protector, or associated insulated conductors and of the primary protector bonding conductor or grounding electrode conductor

(d) Where insulated conductors in accordance with 800.50(A) are used to extend circuits aerially to a building from a buried or underground circuit that is unexposed to accidental contact with electric light or power conductors operating at greater than 300 volts to ground

(e) Where insulated conductors in accordance with 800.50(A) are used to extend circuits to a building from cable with an effectively grounded metallic sheath member(s), and where (1) the combination of the primary protector and insulated conductors is listed as being suitable for this purpose for application with circuits extending from a cable with an effectively grounded metallic sheath member(s), and (2) the insulated conductors safely fuse on all currents greater than the current-carrying capacity of the primary protector and of the primary protector bonding conductor or grounding electrode conductor

Informational Note: Section 9 of ANSI C2-2007, *National Electrical Safety Code*, provides an example of methods of protective grounding that can achieve effective grounding of communications cable sheaths for cables from which communications circuits are extended.

**(2) Fused Primary Protectors.** Where the requirements listed under 800.90(A)(1)(a) through (A)(1)(e) are not met, fused-type primary protectors shall be used. Fused-type primary protectors shall consist of an arrester connected between each line conductor and ground, a fuse in series with each line conductor, and an appropriate mounting arrangement. Primary protector terminals shall be marked to indicate line, instrument, and ground, as applicable.

**(B) Location.** The primary protector shall be located in, on, or immediately adjacent to the structure or building served and as close as practicable to the point of entrance.

Informational Note: See 800.2 for the definition of *Point of Entrance*.

For purposes of this section, primary protectors located at mobile home service equipment within 9.0 m (30 ft) of the exterior wall of the mobile home it serves, or at a mobile home disconnecting means connected to an electrode by a grounding electrode conductor in accordance with 250.32 and located within 9.0 m (30 ft) of the exterior wall of the mobile home it serves, shall be considered to meet the requirements of this section.

Informational Note: Selecting a primary protector location to achieve the shortest practicable primary protector bonding conductor or grounding electrode conductor helps limit potential differences between communications circuits and other metallic systems.

Exhibit 800.3 shows an example of a primary protector unit typically installed in commercial buildings.

**(C) Hazardous (Classified) Locations.** The primary protector shall not be located in any hazardous (classified) locations, as defined in 500.5 and 505.5, or in the vicinity of easily ignitible material.

*Exception: As permitted in 501.150, 502.150, and 503.150.*

**(D) Secondary Protectors.** Where a secondary protector is installed in series with the indoor communications wire and cable between the primary protector and the equipment, it shall be listed for the purpose in accordance with 800.170(B).

Informational Note: Secondary protectors on circuits exposed to accidental contact with electric light or power conductors operating at greater than 300 volts to ground are not intended for use without primary protectors.

## 800.93 Grounding or Interruption of Metallic Sheath Members of Communications Cables

Communications cables entering the building or terminating on the outside of the building shall comply with 800.93(A) or (B).

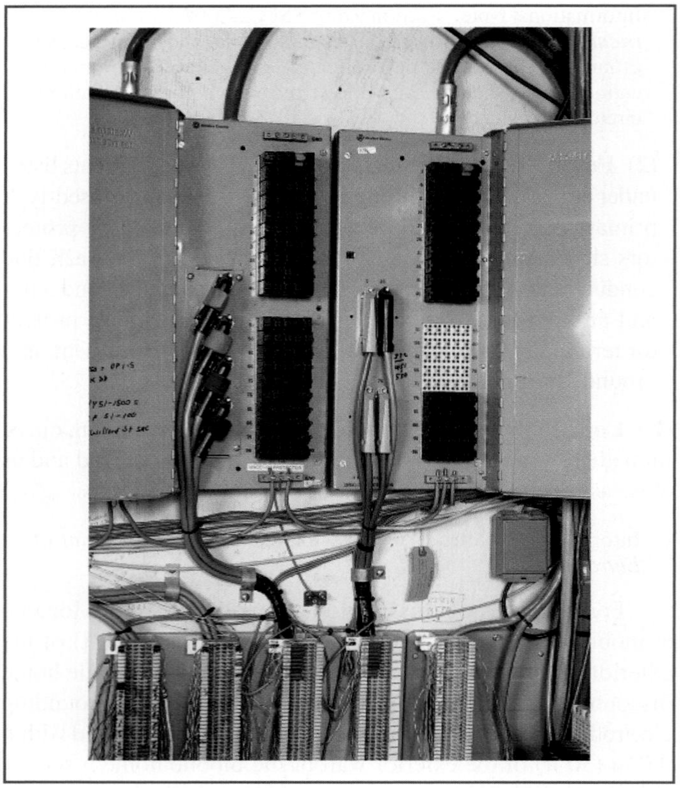

***EXHIBIT 800.3*** *A primary protector unit installed in a commercial building that is the interface to the outside plant cable.*

**(A) Entering Buildings.** In installations where the communications cable enters a building, the metallic sheath members of the cable shall be either grounded as specified in 800.100 or interrupted by an insulating joint or equivalent device. The grounding or interruption shall be as close as practicable to the point of entrance.

**(B) Terminating on the Outside of Buildings.** In installations where the communications cable is terminated on the outside of the building, the metallic sheath members of the cable shall be either grounded as specified in 800.100 or interrupted by an insulating joint or equivalent device. The grounding or interruption shall be as close as practicable to the point of termination of the cable.

Informational Note: See 800.2 for a definition of *Point of Entrance*.

## IV. Grounding Methods

### 800.100 Cable and Primary Protector Bonding and Grounding

The primary protector and the metallic member(s) of the cable sheath shall be bonded or grounded as specified in 800.100(A) through (D).

**(A) Bonding Conductor or Grounding Electrode Conductor.**

**(1) Insulation.** The bonding conductor or grounding electrode conductor shall be listed and shall be permitted to be insulated, covered, or bare.

**(2) Material.** The bonding conductor or grounding electrode conductor shall be copper or other corrosion-resistant conductive material, stranded or solid.

**(3) Size.** The bonding conductor or grounding electrode conductor shall not be smaller than 14 AWG. It shall have a current-carrying capacity not less than the grounded metallic sheath member(s) and protected conductor(s) of the communications cable. The bonding conductor or grounding electrode conductor shall not be required to exceed 6 AWG.

**(4) Length.** The primary protector bonding conductor or grounding electrode conductor shall be as short as practicable. In one- and two-family dwellings, the primary protector bonding conductor or grounding electrode conductor shall be as short as practicable, not to exceed 6.0 m (20 ft) in length.

Informational Note: Similar bonding conductor or grounding electrode conductor length limitations applied at apartment buildings and commercial buildings help to reduce voltages that may be developed between the building's power and communications systems during lightning events.

*Exception: In one- and two-family dwellings where it is not practicable to achieve an overall maximum primary protector bonding conductor or grounding electrode conductor length of 6.0 m (20 ft), a separate communications ground rod meeting the minimum dimensional criteria of 800.100(B)(3)(2) shall be driven, the primary protector shall be connected to the communications ground rod in accordance with 800.100(C), and the communications ground rod shall be connected to the power grounding electrode system in accordance with 800.100(D).*

The restricted conductor length reduces the impedance of the bonding conductor, resulting in a lower potential difference between the communications system conductors and equipment and the electrical conductors and equipment in the building. The low impedance bonding connection reduces the fire hazard and shock hazard to persons in the event that electric utility power lines come in contact with communications conductors. Section 800.100(D) requires bonding of communications and power grounding electrodes at the same building or structure.

The informational note to 800.100(A)(4) provides guidance for the treatment of the cable and primary protector grounding conductor length at apartment and commercial buildings that is consistent with the 20-foot rule for one- and two-family dwellings. However, a specific length is not specified in the *Code* because such a limitation may not be practical in some installations.

**(5) Run in Straight Line.** The bonding conductor or grounding electrode conductor shall be run in as straight a line as practicable.

**(6) Physical Protection.** Bonding conductors and grounding electrode conductors shall be protected where exposed to physical damage. Where the bonding conductor or grounding electrode conductor is installed in a metal raceway, both ends of the raceway shall be bonded to the contained conductor or to the same terminal or electrode to which the bonding conductor or grounding electrode conductor is connected.

**(B) Electrode.** The bonding conductor or grounding electrode conductor shall be connected in accordance with 800.100(B)(1), (B)(2), or (B)(3).

**(1) In Buildings or Structures with an Intersystem Bonding Termination.** If the building or structure served has an intersystem bonding termination as required by 250.94, the bonding conductor shall be connected to the intersystem bonding termination.

Informational Note: See Part I of Article 100 for the definition of *Intersystem Bonding Termination.*

**(2) In Buildings or Structures with Grounding Means.** If the building or structure served has no intersystem bonding termination, the bonding conductor or grounding electrode conductor shall be connected to the nearest accessible location on one of the following:

(1) The building or structure grounding electrode system as covered in 250.50
(2) The grounded interior metal water piping system, within 1.5 m (5 ft) from its point of entrance to the building, as covered in 250.52

For more information on the use of a metal water piping system as a grounding electrode, see the commentary following 250.52(A)(1).

(3) The power service accessible means external to enclosures as covered in 250.94
(4) The nonflexible metallic power service raceway
(5) The service equipment enclosure
(6) The grounding electrode conductor or the grounding electrode conductor metal enclosure of the power service, or
(7) The grounding electrode conductor or the grounding electrode of a building or structure disconnecting means that is grounded to an electrode as covered in 250.32

A bonding device intended to provide a termination point for the bonding conductor (intersystem bonding) shall not interfere with the opening of an equipment enclosure. A bonding device shall be mounted on nonremovable parts. A bonding device shall not be mounted on a door or cover even if the door or cover is nonremovable.

For purposes of this section, the mobile home service equipment or the mobile home disconnecting means, as described in 800.90(B), shall be considered accessible.

**(3) In Buildings or Structures Without an Intersystem Bonding Termination or Grounding Means.** If the building or structure served has no intersystem bonding termination or grounding means, as described in 800.100(B)(2), the grounding electrode conductor shall be connected to either of the following:

(1) To any one of the individual grounding electrodes described in 250.52(A)(1), (A)(2), (A)(3), or (A)(4).
(2) If the building or structure served has no intersystem bonding termination or has no grounding means, as described in 800.100(B)(2) or (B)(3)(1), to any one of the individual grounding electrodes described in 250.52(A)(7) and (A)(8) or to a ground rod or pipe not less than 1.5 m (5 ft) in length and 12.7 mm (½ in.) in diameter, driven, where practicable, into permanently damp earth and separated from lightning conductors as covered in 800.53 and at least 1.8 m (6 ft) from electrodes of other systems. Steam or hot water pipes or air terminal conductors (lightning-rod conductors) shall not be employed as electrodes for protectors and grounded metallic members.

**(C) Electrode Connection.** Connections to grounding electrodes shall comply with 250.70.

**(D) Bonding of Electrodes.** A bonding jumper not smaller than 6 AWG copper or equivalent shall be connected between the communications grounding electrode and power grounding electrode system at the building or structure served where separate electrodes are used.

*Exception: At mobile homes as covered in 800.106.*

Informational Note No. 1: See 250.60 for use of air terminals (lightning rods).

Informational Note No. 2: Bonding together of all separate electrodes limits potential differences between them and between their associated wiring systems.

## 800.106 Primary Protector Grounding and Bonding at Mobile Homes

**(A) Grounding.** Grounding shall comply with 800.106(A)(1) and (A)(2).

(1) Where there is no mobile home service equipment located within 9.0 m (30 ft) of the exterior wall of the mobile home it serves, the primary protector grounding terminal shall be connected to a grounding electrode conductor or grounding electrode in accordance with 800.100(B)(3).
(2) Where there is no mobile home disconnecting means grounded in accordance with 250.32 and located within 9.0 m (30 ft) of the exterior wall of the mobile home it serves, the primary protector grounding terminal shall be connected to a grounding electrode in accordance with 800.100(B)(3).

**(B) Bonding.** The primary protector grounding terminal or grounding electrode shall be connected to the metal frame or available grounding terminal of the mobile home with a copper conductor not smaller than 12 AWG under either of the following conditions:

(1) Where there is no mobile home service equipment or disconnecting means as in 800.106(A)
(2) Where the mobile home is supplied by cord and plug

## V. Installation Methods Within Buildings

Data circuits between computers are classified as Class 2 circuits. In a typical office environment consisting of a group of computers connected to a local area network, data wiring is as prevalent as telephone wiring. One common way to minimize the amount of cabling is to run the telephone and data circuits in the same cable. Section 725.139(D) requires that a listed communications cable be used for this purpose.

### 800.110 Raceways and Cable Routing Assemblies for Communications Wires and Cables

**(A) Types of Raceways.** Communications wires and cables shall be permitted to be installed in any raceway that complies with either (A)(1) or (A)(2) and in cable routing assemblies installed in compliance with 800.110(C).

**(1) Raceways Recognized in Chapter 3.** Communications wires and cables shall be permitted to be installed in any raceway included in Chapter 3. The raceways shall be installed in accordance with the requirements of Chapter 3.

**(2) Communications Raceways.** Communications wires and cables shall be permitted to be installed in listed plenum communications raceways, listed riser communications raceways, and listed general-purpose communications raceways selected in accordance with the provisions of 800.113, and installed in accordance with 362.24 through 362.56, where the requirements applicable to electrical nonmetallic tubing (ENT) apply.

**(B) Raceway Fill for Communications Wires and Cables.** The raceway fill requirements of Chapters 3 and 9 shall not apply to communications wires and cables.

**(C) Cable Routing Assemblies.** Communications wires and cables shall be permitted to be installed in plenum cable routing assemblies, riser cable routing assemblies, and general-purpose cable routing assemblies selected in accordance with the provisions of 800.113 and installed in accordance with 800.110(C)(1) and (2).

**(1) Horizontal Support.** Cable routing assemblies shall be supported where run horizontally at intervals not to exceed 900 mm (3 ft), and at each end or joint, unless listed for other support intervals. In no case shall the distance between supports exceed 3 m (10 ft).

**(2) Vertical Support.** Vertical runs of cable routing assemblies shall be supported at intervals not exceeding 1.2 m (4 ft), unless listed for other support intervals, and shall not have more than one joint between supports.

### 800.113 Installation of Communications Wires, Cables and Raceways, and Cable Routing Assemblies

Installation of communications wires, cables and raceways, and cable routing assemblies shall comply with 800.113(A) through (L). Installation of raceways and cable routing assemblies shall also comply with 800.110.

**(A) Listing.** Communications wires, communications cables, communications raceways, and cable routing assemblies installed in buildings shall be listed.

*Exception: Communications cables that comply with 800.48 shall not be required to be listed.*

**(B) Fabricated Ducts Used for Environmental Air.** The following wires and cables shall be permitted in ducts used for environmental air as described in 300.22(B) if they are directly associated with the air distribution system:

(1) Up to 1.22 m (4 ft) of Type CMP cable
(2) Types CMP, CMR, CMG, CM, and CMX cables and communications wires installed in raceways that are installed in compliance with 300.22(B)

Informational Note: For information on fire protection of wiring installed in fabricated ducts see 4.3.4.1 and 4.3.11.3.3 of NFPA 90A-2012, *Standard for the Installation of Air-Conditioning and Ventilating Systems.*

**(C) Other Spaces Used for Environmental Air (Plenums).** The following wires, cables, and raceways shall be permitted in other spaces used for environmental air as described in 300.22(C):

(1) Type CMP cables
(2) Plenum communications raceways
(3) Type CMP cables installed in plenum communications raceways
(4) Type CMP cables and plenum communications raceways supported by open metallic cable trays or cable tray systems
(5) Types CMP, CMR, CMG, CM, and CMX cables and communications wires installed in raceways that are installed in compliance with 300.22(C)
(6) Types CMP, CMR, CMG, CM, and CMX cables, plenum communications raceways, riser communications raceways, and general-purpose communications raceways supported by solid bottom metal cable trays with solid metal covers in other spaces used for environmental air (plenums) as described in 300.22(C)
(7) Types CMP, CMR, CMG, CM, and CMX cables installed in plenum communications raceways, riser communications raceways, and general-purpose communications raceways supported by solid bottom metal cable trays with solid metal covers in other spaces used for environmental air (plenums) as described in 300.22(C)

Informational Note: For information on fire protection of wiring installed in other spaces used for environmental air, see 4.3.11.2, 4.3.11.4, and 4.3.11.5 of NFPA 90A-2012, *Standard for the Installation of Air-Conditioning and Ventilating Systems.*

**(D) Risers — Cables and Raceways in Vertical Runs.** The following cables, raceways, and cable routing assemblies shall be permitted in vertical runs penetrating one or more floors and in vertical runs in a shaft:

(1) Types CMP and CMR cables
(2) Plenum and riser communications raceways
(3) Plenum and riser cable routing assemblies
(4) Types CMP and CMR cables installed in:

a. Plenum communications raceways
b. Riser communications raceways
c. Plenum cable routing assemblies
d. Riser cable routing assemblies

Informational Note: See 800.26 for firestop requirements for floor penetrations.

**(E) Risers — Cables and Raceways in Metal Raceways.** The following cables and raceways shall be permitted in metal raceways in a riser having firestops at each floor:

(1) Types CMP, CMR, CMG, CM, and CMX cables
(2) Plenum, riser, and general-purpose communications raceways
(3) Types CMP, CMR, CMG, CM, and CMX cables installed in:

a. Plenum communications raceways
b. Riser communications raceways
c. General-purpose communications raceways

Informational Note: See 800.26 for firestop requirements for floor penetrations.

**(F) Risers — Cables, Raceways, and Cable Routing Assemblies in Fireproof Shafts.** The following cables, raceways, and cable routing assemblies shall be permitted to be installed in fireproof riser shafts having firestops at each floor:

(1) Types CMP, CMR, CMG, CM, and CMX cables
(2) Plenum, riser, and general-purpose communications raceways
(3) Plenum, riser, and general-purpose cable routing assemblies
(4) Types CMP, CMR, CMG, and CM cables installed in:

a. Plenum communications raceways
b. Riser communications raceways
c. General-purpose communications raceways
d. Plenum cable routing assemblies
e. Riser cable routing assemblies
f. General-purpose cable routing assemblies

Informational Note: See 800.26 for firestop requirements for floor penetrations.

**(G) Risers — One- and Two-Family Dwellings.** The following cables, raceways, and cable routing assemblies shall be permitted in one- and two-family dwellings:

(1) Types CMP, CMR, CMG, and CM cables
(2) Type CMX cables less than 6 mm (0.25 in.) in diameter
(3) Plenum, riser, and general-purpose communications raceways
(4) Plenum, riser, and general-purpose cable routing assemblies
(5) Types CMP, CMR, CMG, and CM cables installed in:

a. Plenum communications raceways
b. Riser communications raceways
c. General-purpose communications raceways
d. Plenum cable routing assemblies
e. Riser cable routing assemblies
f. General-purpose cable routing assemblies

**(H) Cable Trays.** The following wires, cables, and raceways shall be permitted to be supported by cable trays:

(1) Types CMP, CMR, CMG, and CM cables
(2) Plenum, riser, and general-purpose communications raceways
(3) Communications wires and Types CMP, CMR, CMG, and CM cables installed in:

a. Plenum communications raceways
b. Riser communications raceways
c. General-purpose communications raceways

**(I) Distributing Frames and Cross-Connect Arrays.** The following wires, cables, raceways, and cable routing assemblies shall be permitted to be installed in distributing frames and cross-connect arrays:

(1) Types CMP, CMR, CMG, and CM cables and communications wires
(2) Plenum, riser, and general-purpose communications raceways
(3) Plenum, riser, and general-purpose cable routing assemblies
(4) Communications wires and Types CMP, CMR, CMG, and CM cables installed in:

a. Plenum communications raceways
b. Riser communications raceways
c. General-purpose communications raceways
d. Plenum cable routing assemblies
e. Riser cable routing assemblies
f. General-purpose cable routing assemblies

**(J) Other Building Locations.** The following wires, cables, raceways, and cable routing assemblies shall be permitted to be installed in building locations other than the locations covered in 800.113(B) through (I):

(1) Types CMP, CMR, CMG, and CM cables
(2) A maximum of 3 m (10 ft) of exposed Type CMX in nonconcealed spaces
(3) Plenum, riser, and general-purpose communications raceways
(4) Plenum, riser, and general-purpose cable routing assemblies

(5) Communications wires and Types CMP, CMR, CMG, and CM cables installed in:

a. Plenum communications raceways
b. Riser communications raceways
c. General-purpose communications raceways

(6) Types CMP, CMR, CMG, and CM cables installed in:

a. Plenum cable routing assemblies
b. Riser cable routing assemblies
c. General-purpose cable routing assemblies

(7) Communications wires and Types CMP, CMR, CMG, CM, and CMX cables installed in raceways recognized in Chapter 3
(8) Type CMUC under-carpet communications wires and cables installed under carpet

**(K) Multifamily Dwellings.** The following cables, raceways, and cable routing assemblies shall be permitted to be installed in multifamily dwellings in locations other than the locations covered in 800.113(B) through (G):

(1) Types CMP, CMR, CMG, and CM cables
(2) Type CMX cables less than 6 mm (0.25 in.) in diameter in nonconcealed spaces
(3) Plenum, riser, and general-purpose communications raceways
(4) Plenum, riser, and general-purpose cable routing assemblies
(5) Communications wires and Types CMP, CMR, CMG, and CM cables installed in:

a. Plenum communications raceways
b. Riser communications raceways
c. General-purpose communications raceways

(6) Types CMP, CMR, CMG, and CM cables installed in:

a. Plenum cable routing assemblies
b. Riser cable routing assemblies
c. General-purpose cable routing assemblies

(7) Communications wires and Types CMP, CMR, CMG, CM, and CMX cables installed in raceways recognized in Chapter 3
(8) Type CMUC under-carpet communications wires and cables installed under carpet

**(L) One- and Two-Family Dwellings.** The following cables, raceways, and cable routing assemblies shall be permitted to be installed in one- and two-family dwellings in locations other than the locations covered in 800.113(B) through (F):

(1) Types CMP, CMR, CMG, and CM cables
(2) Type CMX cables less than 6 mm (0.25 in.) in diameter
(3) Plenum, riser, and general-purpose communications raceways
(4) Plenum, riser, and general-purpose cable routing assemblies
(5) Communications wires and Types CMP, CMR, CMG, and CM cables installed in:

a. Plenum communications raceways
b. Riser communications raceways
c. General-purpose communications raceways

(6) Types CMP, CMR, CMG, and CM cables installed in:

a. Plenum cable routing assemblies
b. Riser cable routing assemblies
c. General-purpose cable routing assemblies

(7) Communications wires and Types CMP, CMR, CMG, CM, and CMX cables installed in raceways recognized in Chapter 3
(8) Type CMUC under-carpet communications wires and cables installed under carpet
(9) Hybrid power and communications cable listed in accordance with 800.179(I)

## 800.133 Installation of Communications Wires, Cables, and Equipment

Communications wires and cables from the protector to the equipment or, where no protector is required, communications wires and cables attached to the outside or inside of the building shall comply with 800.133(A) and (B).

**(A) Separation from Other Conductors.**

**(1) In Raceways, Cable Trays, Boxes, Cables, Enclosures, and Cable Routing Assemblies.**

(a) *Optical Fiber and Communications Cables.* Communications cables shall be permitted in the same raceway, cable tray, box, enclosure, or cable routing assembly with cables of any of the following:

(1) Nonconductive and conductive optical fiber cables in compliance with Parts I and V of Article 770
(2) Community antenna television and radio distribution systems in compliance with Parts I and V of Article 820
(3) Low-power network-powered broadband communications circuits in compliance with Parts I and V of Article 830

(b) *Other Circuits.* Communications cables shall be permitted in the same raceway, cable tray, box, enclosure, or cable routing assembly with cables of any of the following:

(1) Class 2 and Class 3 remote-control, signaling, and power-limited circuits in compliance with Article 645 or Parts I and III of Article 725
(2) Power-limited fire alarm systems in compliance with Parts I and III of Article 760

(c) *Class 2 and Class 3 Circuits.* Class 1 circuits shall not be run in the same cable with communications circuits. Class 2 and Class 3 circuit conductors shall be permitted in the same cable with communications circuits, in which case the Class 2 and Class 3 circuits shall be classified as communications circuits and shall meet the requirements of this article. The cables shall be listed as communications cables.

*Exception: Cables constructed of individually listed Class 2, Class 3, and communications cables under a common jacket shall not be required to be classified as communications cable. The fire-resistance rating of the composite cable shall be determined by the performance of the composite cable.*

(d) *Electric Light, Power, Class 1, Non–Power-Limited Fire Alarm, and Medium-Power Network-Powered Broadband Communications Circuits in Raceways, Compartments, and Boxes.* Communications conductors shall not be placed in any raceway, compartment, outlet box, junction box, or similar fitting with conductors of electric light, power, Class 1, non–power-limited fire alarm, or medium-power network-powered broadband communications circuits.

*Exception No. 1: Where all of the conductors of electric light, power, Class 1, non–power-limited fire alarm, and medium-power network-powered broadband communications circuits are separated from all of the conductors of communications circuits by a permanent barrier or listed divider.*

*Exception No. 2: Power conductors in outlet boxes, junction boxes, or similar fittings or compartments where such conductors are introduced solely for power supply to communications equipment. The power circuit conductors shall be routed within the enclosure to maintain a minimum of 6 mm (0.25 in.) separation from the communications circuit conductors.*

*Exception No. 3: As permitted by 620.36.*

**(2) Other Applications.** Communications wires and cables shall be separated at least 50 mm (2 in.) from conductors of any electric light, power, Class 1, non–power-limited fire alarm, or medium-power network-powered broadband communications circuits.

*Exception No. 1: Where either (1) all of the conductors of the electric light, power, Class 1, non–power-limited fire alarm, and medium-power network-powered broadband communications circuits are in a raceway or in metal-sheathed, metal-clad, nonmetallic-sheathed, Type AC, or Type UF cables, or (2) all of the conductors of communications circuits are encased in raceway.*

*Exception No. 2: Where the communications wires and cables are permanently separated from the conductors of electric light, power, Class 1, non–power-limited fire alarm, and medium-power network-powered broadband communications circuits by a continuous and firmly fixed nonconductor, such as porcelain tubes or flexible tubing, in addition to the insulation on the wire.*

**(B) Support of Communications Wires and Cables.** Raceways shall be used for their intended purpose. Communications wires and cables shall not be strapped, taped, or attached by any means to the exterior of any raceway as a means of support.

*Exception: Overhead (aerial) spans of communications wires and cables shall be permitted to be attached to the exterior of a raceway-type mast intended for the attachment and support of such wires and cables.*

In some instances, the only way to achieve the proper clearance above roadways, driveways, or structures is by use of a mast. The exception permits overhead spans of communications cable to be attached to the exterior of a raceway-type mast only if the mast is installed to support communications cable. The attachment of communications cable to a service mast or a feeder and/or branch-circuit mast is prohibited by 230.28 and 225.17, respectively.

## 800.154 Applications of Listed Communications Wires, Cables and Raceways, and Listed Cable Routing Assemblies

Permitted and nonpermitted applications of listed communications wires, cables, and raceways, and listed cable routing assemblies, shall be in accordance with one of the following:

(1) Listed communications wires and cables as indicated in Table 800.154(a)
(2) Listed communications raceways as indicated in Table 800.154(b)
(3) Listed cable routing assemblies as indicated in Table 800.154(c)

The permitted applications shall be subject to the installation requirements of 800.110 and 800.113. The substitutions for communications cables listed in Table 800.154(d) and illustrated in Figure 800.154 shall be permitted.

The length of unlisted outside plant cable permitted in a building depends on the location of the primary protector in accordance with 800.48 and 800.90(B).

Exhibit 800.4 illustrates applications of listed communications cables.

## 800.156 Dwelling Unit Communications Outlet

For new construction, a minimum of one communications outlet shall be installed within the dwelling in a readily accessible area and cabled to the service provider demarcation point.

The location of the communications outlet within a dwelling is not specified, but cable must be installed between the outlet location and the point at which the communication services provider installs its equipment or the point at which it connects to owner-supplied equipment. Although an increasing number of dwelling unit owners or occupants may use cellular or PCS telephones exclusively, having at least one wired communications outlet in every dwelling facilitates connection of dial-up devices used in home fire detection and security systems. Exhibit 800.5 illustrates a communications outlet provided during the "rough-in" phase of a dwelling's electrical installation.

*EXHIBIT 800.4* *Applications of listed communications cables within various locations of a building.*

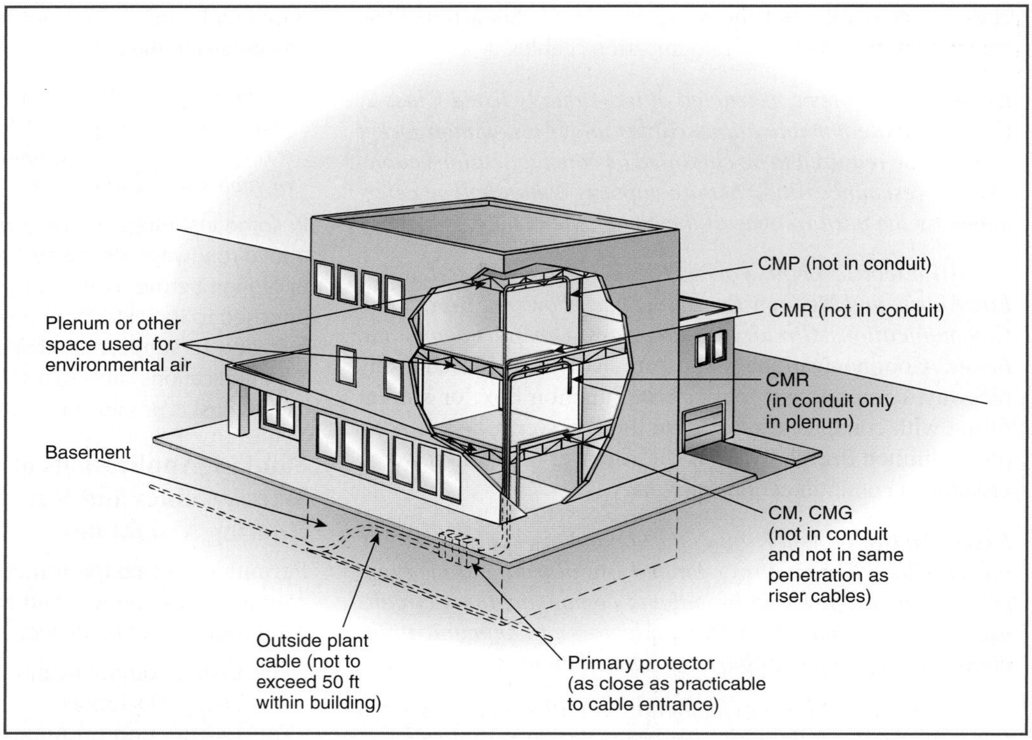


***TABLE 800.154(a)*** *Applications of Listed Communications Wires and Cables in Buildings*

| Applications | | Wire and Cable Type: CMP | CMR | CMG CM | CMX | CMUC | Hybrid Power and Communications Cables | Communications Wires |
|---|---|---|---|---|---|---|---|---|
| In specifically fabricated ducts as described in 300.22(B) | In fabricated ducts | Y* | N | N | N | N | N | N |
| | In metal raceway that complies with 300.22(B) | Y* | Y* | Y* | Y* | N | N | Y* |
| In other spaces used for environmental air as described in 300.22(C) | In other spaces used for environmental air | Y* | N | N | N | N | N | N |
| | In metal raceway that complies with 300.22(C) | Y* | Y* | Y* | Y* | N | N | Y* |
| | In plenum communications raceways | Y* | N | N | N | N | N | Y* |
| | In plenum cable routing assemblies | NOT PERMITTED | | | | | | |
| | Supported by open metal cable trays | Y* | N | N | N | N | N | N |
| | Supported by solid bottom metal cable trays with solid metal covers | Y* | Y* | Y* | Y* | N | N | N |

*TABLE 800.154(a) Continued*

| Applications | | Wire and Cable Type | | | | | | |
|---|---|---|---|---|---|---|---|---|
| | | CMP | CMR | CMG CM | CMX | CMUC | Hybrid Power and Communications Cables | Communications Wires |
| In risers | In vertical runs | Y* | Y* | N | N | N | N | N |
| | In metal raceways | Y* | Y* | Y* | Y* | N | N | N |
| | In fireproof shafts | Y* | Y* | Y* | Y* | N | N | N |
| | In plenum communications raceways | Y* | Y* | N | N | N | N | N |
| | In plenum cable routing assemblies | Y* | Y* | N | N | N | N | N |
| | In riser communications raceways | Y* | Y* | N | N | N | N | N |
| | In riser cable routing assemblies | Y* | Y* | N | N | N | N | N |
| | In one- and two-family dwellings | Y* | Y* | Y* | Y* | N | Y* | N |
| Within buildings in other than air-handling spaces and risers | General | Y* | Y* | Y* | Y* | N | N | N |
| | In one- and two-family dwellings | Y* | Y* | Y* | Y* | Y* | Y* | N |
| | In multifamily dwellings | Y* | Y* | Y* | Y* | Y* | N | N |
| | In nonconcealed spaces | Y* | Y* | Y* | Y* | Y* | N | N |
| | Supported by cable trays | Y* | Y* | Y* | N | N | N | N |
| | Under carpet | N | N | N | N | Y* | N | N |
| | In distributing frames and cross-connect arrays | Y* | Y* | Y* | N | N | N | Y* |
| | In any raceway recognized in Chapter 3 | Y* | Y* | Y* | Y* | N | N | Y* |
| | In plenum communications raceways | Y* | Y* | Y* | N | N | N | Y* |
| | In plenum cable routing assemblies | Y* | Y* | Y* | N | N | N | Y* |
| | In riser communications raceways | Y* | Y* | Y* | N | N | N | Y* |
| | In riser cable routing assemblies | Y* | Y* | Y* | N | N | N | Y* |
| | In general-purpose communications raceways | Y* | Y* | Y* | N | N | N | Y* |
| | In general-purpose cable routing assemblies | Y* | Y* | Y* | N | N | N | Y* |

Note: An "N" in the table indicates that the cable type is not permitted to be installed in the application. A "Y*" indicates that the cable is permitted to be installed in the application subject to the limitations described in 800.113.

Informational Note No. 1: Part V of Article 800 covers installation methods within buildings. This table covers the applications of listed communications wires, cables, and raceways in buildings. The definition of *Point of Entrance* is in 800.2. Communications entrance cables that have not emerged from the rigid metal conduit (RMC) or intermediate metal conduit (IMC) are not considered to be in the building.

Informational Note No. 2: For information on the restrictions to the installation of communications cables in fabricated ducts, see 800.113(B).

Informational Note No. 3: Cable routing assemblies are not addressed in NFPA 90A-2012, *Standard for the Installation of Air-Conditioning and Ventilating Systems*.

*TABLE 800.154(b) Applications of Listed Communications Raceways in Buildings*

| Applications | | Plenum Communications Raceways | Riser Communications Raceways | General-Purpose Communications Raceways |
|---|---|---|---|---|
| | | **Raceway Type** | | |
| In specifically fabricated ducts as described in 300.22(B) | In fabricated ducts | N | N | N |
| | In metal raceway that complies with 300.22(B) | Y* | Y* | Y* |
| In other spaces used for environmental air as described in 300.22(C) | In other spaces used for environmental air | Y* | N | N |
| | In metal raceway that complies with 300.22(C) | Y* | Y* | Y* |
| | In plenum cable routing assemblies | NOT PERMITTED | | |
| | Supported by open metal cable trays | Y* | N | N |
| | Supported by solid bottom metal cable trays with solid metal covers | Y* | Y* | Y* |
| In risers | In vertical runs | Y* | Y* | N |
| | In metal raceways | Y* | Y* | Y* |
| | In fireproof shafts | Y* | Y* | Y* |
| | In plenum cable routing assemblies | N | N | N |
| | In riser cable routing assemblies | N | N | N |
| | In one- and two-family dwellings | Y* | Y* | Y* |
| Within buildings in other than air-handling spaces and risers | General | Y* | Y* | Y* |
| | In one- and two-family dwellings | Y* | Y* | Y* |
| | In multifamily dwellings | Y* | Y* | Y* |
| | In nonconcealed spaces | Y* | Y* | Y* |
| | Supported by cable trays | Y* | Y* | Y* |
| | Under carpet | N | N | N |
| | In distributing frames and cross-connect arrays | Y* | Y* | Y* |
| | In any raceway recognized in Chapter 3 | Y* | Y* | Y* |
| | In plenum cable routing assemblies | N | N | N |
| | In riser cable routing assemblies | N | N | N |
| | In general-purpose cable routing assemblies | N | N | N |

Note: An "N" in the table indicates that the cable type shall not be permitted to be installed in the application. A "Y*" indicates that the cable shall be permitted to be installed in the application subject to the limitations described in 800.110 and 800.113.

Informational Note: Cable routing assemblies are not addressed in NFPA 90A-2012, *Standard for the Installation of Air-Conditioning and Ventilating Systems.*

***TABLE 800.154(c)*** *Applications of Listed Cable Routing Assemblies in Buildings*

| Applications | | Cable Routing Assembly Type: Plenum Cable Routing Assembly | Cable Routing Assembly Type: Riser Cable Routing Assembly | Cable Routing Assembly Type: General-Purpose Cable Routing Assembly |
|---|---|---|---|---|
| In specifically fabricated ducts as described in 300.22(B) | In fabricated ducts | N | N | N |
| | In metal raceway that complies with 300.22(B) | N | N | N |
| In other spaces used for environmental air as described in 300.22(C) | In other spaces used for environmental air | N | N | N |
| | In metal raceway that complies with 300.22(C) | N | N | N |
| | In plenum communications raceways | N | N | N |
| | Supported by open metal cable trays | N | N | N |
| | Supported by solid bottom metal cable trays with solid metal covers | N | N | N |
| In risers | In vertical runs | Y* | Y* | N |
| | In metal raceways | N | N | N |
| | In fireproof shafts | Y* | Y* | Y* |
| | In plenum communications raceways | N | N | N |
| | In riser communications raceways | N | N | N |
| | In one- and two-family dwellings | Y* | Y* | Y* |
| Within buildings in other than air-handling spaces and risers | General | Y* | Y* | Y* |
| | In one- and two-family dwellings | Y* | Y* | Y* |
| | In multifamily dwellings | Y* | Y* | Y* |
| | In nonconcealed spaces | Y* | Y* | Y* |
| | Supported by cable trays | N | N | N |
| | Under carpet | N | N | N |
| | In distributing frames and cross-connect arrays | Y* | Y* | Y* |
| | In any raceway recognized in Chapter 3 | N | N | N |
| | In plenum communications raceways | N | N | N |
| | In riser communications raceways | N | N | N |
| | In general-purpose communications raceways | N | N | N |

Note: An "N" in the table indicates that the cable type shall not be permitted to be installed in the application. A "Y*" indicates that the cable shall be permitted to be installed in the application subject to the limitations described in 800.113.

***TABLE 800.154(d)*** *Cable Substitutions*

| Cable Type | Permitted Substitutions |
|---|---|
| CMR | CMP |
| CMG, CM | CMP, CMR |
| CMX | CMP, CMR, CMG, CM |

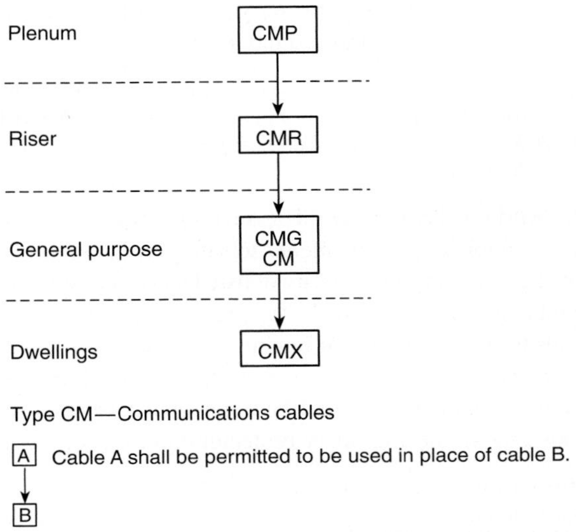


***FIGURE 800.154*** *Cable Substitution Hierarchy.*

***EXHIBIT 800.5*** *Communications outlet provision installed at "rough-in." (Courtesy of the International Association of Electrical Inspectors)*

# VI. Listing Requirements

## 800.170 Equipment

Communications equipment shall be listed as being suitable for electrical connection to a communications network.

Informational Note: One way to determine applicable requirements is to refer to UL 60950-1-2007, *Standard for Safety of Information Technology Equipment*; UL 1459-1999, *Standard for Safety Telephone Equipment*; or UL 1863-2004, *Standard for Safety Communications Circuit Accessories*. For information on listing requirements for cable routing assemblies and communications raceways, see UL 2024-2011, *Standard for Signaling, Optical Fiber and Communications Cable Raceways and Cable Routing Assemblies*.

**(A) Primary Protectors.** The primary protector shall consist of an arrester connected between each line conductor and ground in an appropriate mounting. Primary protector terminals shall be marked to indicate line and ground as applicable.

Informational Note: One way to determine applicable requirements for a listed primary protector is to refer to ANSI/UL 497-2009, *Standard for Protectors for Paired Conductor Communications Circuits*.

**(B) Secondary Protectors.** The secondary protector shall be listed as suitable to provide means to safely limit currents to less than the current-carrying capacity of listed indoor communications wire and cable, listed telephone set line cords, and listed communications terminal equipment having ports for external wire line communications circuits. Any overvoltage protection, arresters, or grounding connection shall be connected on the equipment terminals side of the secondary protector current-limiting means.

Informational Note: One way to determine applicable requirements for a listed secondary protector is to refer to ANSI/UL 497A-2008, *Standard for Secondary Protectors for Communications Circuits*.

**(C) Plenum Grade Cable Ties.** Cable ties intended for use in other space used for environmental air (plenums) shall be listed as having low smoke and heat release properties.

Informational Note: See NFPA 90A-2012, *Standard for the Installation of Air-Conditioning and Ventilating Systems*, and ANSI/UL 2043, *Standard for Safety Fire Test for Heat and Visible Smoke Release for Discrete Products and Their Accessories Installed in Air-Handling Spaces*, for information on listing discrete products as having low smoke and heat release properties.

## 800.173 Drop Wire and Cable

Communications wires and cables without a metallic shield, running from the last outdoor support to the primary protector, shall be listed as being suitable for the purpose and shall have current-carrying capacity as specified in 800.90(A)(1)(b) or (A)(1)(c).

## 800.179 Communications Wires and Cables

Communications wires and cables shall be listed in accordance with 800.179(A) through (I) and marked in accordance with Table 800.179. Conductors in communications cables, other than in a coaxial cable, shall be copper.

Communications wires and cables shall have a voltage rating of not less than 300 volts. The insulation for the individual conductors, other than the outer conductor of a coaxial cable, shall be rated for 300 volts minimum. The cable voltage rating shall not be marked on the cable or on the undercarpet communications wire. Communications wires and cables shall have a temperature rating of not less than 60°C.

Conductor insulation rating of at least 300 volts is required for the following reasons:

1. To coordinate with protector installation requirements (i.e., protectors are not required within a block unless the cable is exposed to over 300 volts)
2. To recognize the fact that primary protectors are designed to allow voltages below 300 to pass
3. To accommodate the voltages ordinarily found on a telephone line (48 volts dc plus ringing voltage up to 130 volts rms)
4. To permit communications cable to substitute for 300-volt power-limited fire-protective signaling cable

*Exception: Voltage markings shall be permitted where the cable has multiple listings and voltage marking is required for one or more of the listings.*

Informational Note No. 1: Voltage markings on cables may be misinterpreted to suggest that the cables may be suitable for Class 1, electric light, and power applications.

Informational Note No. 2: See 800.170 for listing requirement for equipment.

**(A) Type CMP.** Type CMP communications plenum cables shall be listed as being suitable for use in ducts, plenums, and other spaces used for environmental air and shall also be listed

*TABLE 800.179 Cable Markings*

| Cable Marking | Type |
|---|---|
| CMP | Communications plenum cable |
| CMR | Communications riser cable |
| CMG | Communications general-purpose cable |
| CM | Communications general-purpose cable |
| CMX | Communications cable, limited use |
| CMUC | Undercarpet communications wire and cable |

Informational Note: Cable types are listed in descending order of fire resistance rating.

as having adequate fire-resistant and low smoke-producing characteristics.

Informational Note: One method of defining a cable that is low-smoke producing cable and fire-resistant cable is that the cable exhibits a maximum peak optical density of 0.50 or less, an average optical density of 0.15 or less, and a maximum flame spread distance of 1.52 m (5 ft) or less when tested in accordance with NFPA 262-2011, *Standard Method of Test for Flame Travel and Smoke of Wires and Cables for Use in Air-Handling Spaces.*

See the commentary following the informational note to 725.179(A).

**(B) Type CMR.** Type CMR communications riser cables shall be listed as being suitable for use in a vertical run in a shaft or from floor to floor and shall also be listed as having fire-resistant characteristics capable of preventing the carrying of fire from floor to floor.

Informational Note: One method of defining fire-resistant characteristics capable of preventing the carrying of fire from floor to floor is that the cables pass the requirements of ANSI/UL 1666-2011, *Standard Test for Flame Propagation Height of Electrical and Optical-Fiber Cable Installed Vertically in Shafts.*

See the commentary following the informational note to 725.179(B).

**(C) Type CMG.** Type CMG general-purpose communications cables shall be listed as being suitable for general-purpose communications use, with the exception of risers and plenums, and shall also be listed as being resistant to the spread of fire.

Informational Note: One method of defining *resistant to the spread of fire* is for the damage (char length) not to exceed 1.5 m (4 ft 11 in.) when performing the CSA "Vertical Flame Test — Cables in Cable Trays," as described in CSA C22.2 No. 0.3-M-2001, *Test Methods for Electrical Wires and Cables.*

See the commentary following the informational note to 725.179(C).

**(D) Type CM.** Type CM communications cables shall be listed as being suitable for general-purpose communications use, with the exception of risers and plenums, and shall also be listed as being resistant to the spread of fire.

Informational Note: One method of defining *resistant to the spread of fire* is that the cables do not spread fire to the top of the tray in the "UL Flame Exposure, Vertical Flame Tray Test" in ANSI/UL 1685-2011, *Standard for Safety for Vertical-Tray Fire-Propagation and Smoke-Release Test for Electrical and Optical-Fiber Cables.* The smoke measurements in the test method are not applicable.

Another method of defining *resistant to the spread of fire* is for the damage (char length) not to exceed 1.5 m (4 ft 11 in.) when performing the CSA "Vertical Flame Test— Cables in Cable Trays," as described in CSA C22.2 No. 0.3-M-2001, *Test Methods for Electrical Wires and Cables.*

See the commentary following the informational note to 725.179(D).

**(E) Type CMX.** Type CMX limited-use communications cables shall be listed as being suitable for use in dwellings and for use in raceway and shall also be listed as being resistant to flame spread.

Informational Note: One method of determining that cable is resistant to flame spread is by testing the cable to the VW-1 (vertical-wire) flame test in ANSI/UL 1581-2011, *Reference Standard for Electrical Wires, Cables and Flexible Cords.*

**(F) Type CMUC Undercarpet Wires and Cables.** Type CMUC undercarpet communications wires and cables shall be listed as being suitable for undercarpet use and shall also be listed as being resistant to flame spread.

Informational Note: One method of determining that cable is resistant to flame spread is by testing the cable to the VW-1 (vertical-wire) flame test in ANSI/UL 1581-2011, *Reference Standard for Electrical Wires, Cables and Flexible Cords.*

**(G) Circuit Integrity (CI) Cable or Electrical Circuit Protective System.** Cables that are used for survivability of critical circuits under fire conditions shall be listed and meet either 800.179(G)(1) or (2) as follows:

Informational Note: The listing organization provides information for circuit integrity (CI) cable and electrical circuit protective systems, including installation requirements required to maintain the fire rating.

**(1) Circuit Integrity (CI) Cables.** Circuit integrity (CI) cables specified in 800.179(A) through (E), and used for survivability of critical circuits, shall have an additional classification using the suffix "CI." In order to maintain its listed fire rating, circuit integrity (CI) cable shall only be installed in free air.

Informational Note: One method of defining circuit integrity (CI) cable is by establishing a minimum 2-hour fire resistance rating for the cable when tested in accordance with ANSI/UL 2196-2006, *Standard for Tests of Fire-Resistive Cable.*

**(2) Fire-Resistive Cables.** Cables specified in 800.179(A) through (E) and 800.179(G)(1), that are part of an electrical circuit protective system, shall be fire-resistive cable identified with the protective system number on the product, or on the smallest unit container in which the product is packaged, and shall be installed in accordance with the listing of the protective system.

Informational Note No. 1: One method of defining an electrical circuit protective system is by establishing a minimum 2-hour fire resistance rating for the system when tested in accordance with UL Subject 1724, *Outline of Investigation for Fire Tests for Electrical Circuit Protective Systems.*

Informational Note No. 2: The listing organization provides information for electrical circuit protective systems (FHIT), including installation requirements for maintaining the fire rating.

**(H) Communications Wires.** Communications wires, such as distributing frame wire and jumper wire, shall be listed as being resistant to the spread of fire.

Informational Note: One method of defining *resistant to the spread of fire* is that the cables do not spread fire to the top of the tray in the "UL Flame Exposure, Vertical Flame Tray Test" in ANSI/XUL 1685-2010, *Standard for Safety for Vertical-Tray Fire-Propagation and Smoke-Release Test for Electrical and Optical-Fiber Cables.* The smoke measurements in the test method are not applicable.

Another method of defining *resistant to the spread of fire* is for the damage (char length) not to exceed 1.5 m (4 ft 11 in.) when performing the CSA "Vertical Flame Test— Cables in Cable Trays," as described in CSA C22.2 No. 0.3-M-2001, *Test Methods for Electrical Wires and Cables.*

**(I) Hybrid Power and Communications Cables.** Listed hybrid power and communications cables shall be permitted where the power cable is a listed Type NM or NM-B, conforming to the provisions of Part III of Article 334, and the communications cable is a listed Type CM, the jackets on the listed NM or NM-B, and listed CM cables are rated for 600 volts minimum, and the hybrid cable is listed as being resistant to the spread of fire.

Informational Note: One method of defining *resistant to the spread of fire* is that the cables do not spread fire to the top of the tray in the "UL Flame Exposure, Vertical Flame Tray Test" in ANSI/UL 1685-2010, *Standard for Safety for Vertical-Tray Fire-Propagation and Smoke-Release Test for Electrical and Optical-Fiber Cables.* The smoke measurements in the test method are not applicable.

Another method of defining *resistant to the spread of fire* is for the damage (char length) not to exceed 1.5 m (4 ft 11 in.) when performing the CSA "Vertical Flame Test — Cables in Cable Trays," as described in CSA C22.2 No. 0.3-M-2001, *Test Methods for Electrical Wires and Cables.*

## 800.180 Grounding Devices

Where bonding or grounding is required, devices used to connect a shield, a sheath, or non–current-carrying metallic members of a cable to a bonding conductor or grounding electrode conductor shall be listed or be part of listed equipment.

## 800.182 Communications Raceways and Cable Routing Assemblies

Communications raceways and cable routing assemblies shall be listed in accordance with 800.182(A) through (C).

Informational Note: For information on listing requirements for both communications raceways and cable routing assemblies, see ANSI/UL 2024-4-2011, *Signaling, Optical Fiber and Communications Raceways and Cable Routing Assemblies.*

**(A) Plenum Communications Raceways and Plenum Cable Routing Assemblies.** Plenum communications raceways and plenum cable routing assemblies shall be listed as having adequate fire-resistant and low-smoke producing characteristics.

•

**(B) Riser Communications Raceways and Riser Cable Routing Assemblies.** Riser communications raceways and riser cable routing assemblies shall be listed as having adequate fire-resistant characteristics capable of preventing the carrying of fire from floor to floor.

•

**(C) General-Purpose Communications Raceways and General-Purpose Cable Routing Assemblies.** General-purpose communications raceways and general-purpose cable routing assemblies shall be listed as being resistant to the spread of fire.

The application of communications raceway and cable routing assemblies are summarized in Tables 800.154(b) and (c). The installation location will dictate the type of cable permitted within the raceway or assembly as summarized in Table 800.154(a).

A raceway or routing assembly marked "plenum" is suitable for use in ducts, plenums, or other spaces used for environmental air in accordance with 800.154. These are identified by a marking on its surface or on a marker tape indicating "plenum." A "plenum" raceway or routing assembly is also suitable for installation in risers, for general-purpose use, and for dwellings.

A raceway or routing assembly marked "riser" is suitable for installation in risers in accordance with 800.154. These are identified by a marking on its surface or on a marker tape indicating "riser." A "riser" raceway or routing assembly is also suitable for general-purpose use and for dwellings.

A raceway or routing assembly marked "general purpose" is suitable for installation in general-purpose areas in accordance with 800.154 and for dwellings.

Pliable raceway is raceway that can be bent by hand without the use of tools. The smallest radius of the curve of the inner edge of any bend to which the raceway can be bent without cracking either on the outer surface or internally is not less than 2½ times the outside diameter of the raceway.

•

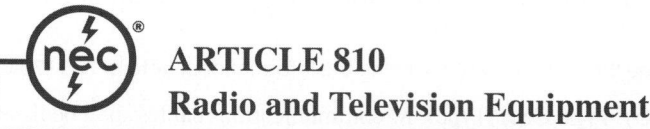

# ARTICLE 810
# Radio and Television Equipment

Informational Note: The general term *grounding conductor* as previously used in this article is replaced by either the term *bonding conductor* or the term *grounding electrode conductor* (GEC), where applicable, to more accurately reflect the application and function of the conductor.

# I. General

## 810.1 Scope

This article covers antenna systems for radio and television receiving equipment, amateur and citizen band radio transmitting and receiving equipment, and certain features of transmitter safety. This article covers antennas such as wire-strung type, multi-element, vertical rod, flat, or parabolic and also covers the wiring and cabling that connect them to equipment. This article does not cover equipment and antennas used for coupling carrier current to power line conductors.

Article 810 covers wiring requirements for television and radio receiving equipment, specifically including digital satellite receiving equipment for television signals and wiring for amateur radio equipment and citizens band (CB) radio equipment. Chapters 1 through 4 cover wiring for the power supply. Article 640 contains requirements for sound distribution systems. The interior wiring using coaxial cable is covered by Article 820.

## 810.2 Definitions

For definitions applicable to this article, see Part I of Article 100.

## 810.3 Other Articles

Wiring from the source of power to and between devices connected to the interior wiring system shall comply with Chapters 1 through 4 other than as modified by Parts I and II of Article 640. Wiring for audio signal processing, amplification, and reproduction equipment shall comply with Article 640. Coaxial cables that connect antennas to equipment shall comply with Article 820.

## 810.4 Community Television Antenna

The antenna shall comply with this article. The distribution system shall comply with Article 820.

## 810.5 Radio Noise Suppressors

Radio interference eliminators, interference capacitors, or noise suppressors connected to power-supply leads shall be of a listed type. They shall not be exposed to physical damage.

## 810.6 Antenna Lead-In Protectors

Where an antenna lead-in surge protector is installed, it shall be listed as being suitable for limiting surges on the cable that connects the antenna to the receiver/transmitter electronics and shall be connected between the conductors and the grounded shield or other ground connection. The antenna lead-in protector shall be grounded using a bonding conductor or grounding electrode conductor installed in accordance with 810.21(F).

Informational Note: For requirements covering protectors for antenna lead-in conductors, refer to UL Subject 497E, *Outline of Investigation for Protectors for Antenna Lead-In Conductors*.

## 810.7 Grounding Devices

Where bonding or grounding is required, devices used to connect a shield, a sheath, non–current-carrying metallic members of a cable, or metal parts of equipment or antennas to a bonding conductor or grounding electrode conductor shall be listed or be part of listed equipment.

# II. Receiving Equipment — Antenna Systems

## 810.11 Material

Antennas and lead-in conductors shall be of hard-drawn copper, bronze, aluminum alloy, copper-clad steel, or other high-strength, corrosion-resistant material.

*Exception: Soft-drawn or medium-drawn copper shall be permitted for lead-in conductors where the maximum span between points of support is less than 11 m (35 ft).*

## 810.12 Supports

Outdoor antennas and lead-in conductors shall be securely supported. The antennas or lead-in conductors shall not be attached to the electric service mast. They shall not be attached to poles or similar structures carrying open electric light or power wires or trolley wires of over 250 volts between conductors. Insulators supporting the antenna conductors shall have sufficient mechanical strength to safely support the conductors. Lead-in conductors shall be securely attached to the antennas.

## 810.13 Avoidance of Contacts with Conductors of Other Systems

Outdoor antennas and lead-in conductors from an antenna to a building shall not cross over open conductors of electric light or power circuits and shall be kept well away from all such circuits so as to avoid the possibility of accidental contact. Where proximity to open electric light or power service conductors of less than 250 volts between conductors cannot be avoided, the installation shall be such as to provide a clearance of at least 600 mm (2 ft).

Where practicable, antenna conductors shall be installed so as not to cross under open electric light or power conductors.

One of the leading causes of electric shock and electrocution, according to statistical reports, is the accidental contact of antennas, ladders, and other equipment with overhead light or power conductors. Extreme caution should be exercised during installation. Antennas should be visually inspected periodically to ensure that they continue to be safe from exposure to overhead power conductors.

## 810.14 Splices

Splices and joints in antenna spans shall be made mechanically secure with approved splicing devices or by such other means as will not appreciably weaken the conductors.

## 810.15 Grounding

Masts and metal structures supporting antennas shall be grounded in accordance with 810.21.

## 810.16 Size of Wire-Strung Antenna — Receiving Station

**(A) Size of Antenna Conductors.** Outdoor antenna conductors for receiving stations shall be of a size not less than given in Table 810.16(A).

**(B) Self-Supporting Antennas.** Outdoor antennas, such as vertical rods and flat, parabolic, or dipole structures, shall be of corrosion-resistant materials and of strength suitable to withstand ice and wind loading conditions and shall be located well away from overhead conductors of electric light and power circuits of over 150 volts to ground, so as to avoid the possibility of the antenna or structure falling into or making accidental contact with such circuits.

## 810.17 Size of Lead-in — Receiving Station

Lead-in conductors from outside antennas for receiving stations shall, for various maximum open span lengths, be of such size as to have a tensile strength at least as great as that of the conductors for antennas as specified in 810.16. Where the lead-in consists of two or more conductors that are twisted together, are enclosed in the same covering, or are concentric, the conductor size shall, for various maximum open span lengths, be such that the tensile strength of the combination is at least as great as that of the conductors for antennas as specified in 810.16.

## 810.18 Clearances — Receiving Stations

**(A) Outside of Buildings.** Lead-in conductors attached to buildings shall be installed so that they cannot swing closer than 600 mm (2 ft) to the conductors of circuits of 250 volts or less between conductors, or 3.0 m (10 ft) to the conductors of circuits of over 250 volts between conductors, except that in the case of circuits not over 150 volts between conductors, where all conductors involved are supported so as to ensure permanent separation, the clearance shall be permitted to be reduced but shall not be less than 100 mm (4 in.). The clearance between lead-in conductors and any conductor forming a part of a lightning protection system shall not be less than 1.8 m (6 ft). Underground conductors shall be separated at least 300 mm (12 in.) from conductors of any light or power circuits or Class 1 circuits.

*Exception: Where the electric light or power conductors, Class 1 conductors, or lead-in conductors are installed in raceways or metal cable armor.*

Informational Note No. 1: See 250.60 for use of air terminals. For further information, see NFPA 780-2014, *Standard for the Installation of Lightning Protection Systems*, which contains detailed information on grounding, bonding, and spacing from lightning protection systems.

Informational Note No. 2: Metal raceways, enclosures, frames, and other non–current-carrying metal parts of electrical equipment installed on a building equipped with a lightning protection system may require bonding or spacing from the lightning protection conductors in accordance with NFPA 780-2011, *Standard for the Installation of Lightning Protection Systems*. Separation from lightning protection conductors is typically 1.8 m (6 ft) through air or 900 mm (3 ft) through dense materials such as concrete, brick, or wood.

**(B) Antennas and Lead-ins — Indoors.** Indoor antennas and indoor lead-ins shall not be run nearer than 50 mm (2 in.) to conductors of other wiring systems in the premises.

*Exception No. 1: Where such other conductors are in metal raceways or cable armor.*

*Exception No. 2: Where permanently separated from such other conductors by a continuous and firmly fixed nonconductor, such as porcelain tubes or flexible tubing.*

**(C) In Boxes or Other Enclosures.** Indoor antennas and indoor lead-ins shall be permitted to occupy the same box or enclosure with conductors of other wiring systems where separated from such other conductors by an effective permanently installed barrier.

***TABLE 810.16(A)*** *Size of Receiving Station Outdoor Antenna Conductors*

| | Minimum Size of Conductors (AWG) Where Maximum Open Span Length Is | | |
|---|---|---|---|
| **Material** | **Less Than 11 m (35 ft)** | **11 m to 45 m (35 ft to 150 ft)** | **Over 45 m (150 ft)** |
| Aluminum alloy, hard-drawn copper | 19 | 14 | 12 |
| Copper-clad steel, bronze, or other high-strength material | 20 | 17 | 14 |

## 810.19 Electrical Supply Circuits Used in Lieu of Antenna — Receiving Stations

Where an electrical supply circuit is used in lieu of an antenna, the device by which the radio receiving set is connected to the supply circuit shall be listed.

The connecting device is usually a small, fixed capacitor connecting the antenna terminal of the receiver and one wire of the supply circuit. As is the case with most receivers, the capacitor should be designed for operation at not less than 300 volts. This voltage rating ensures a high degree of safety and minimizes the possibility of a breakdown in the capacitor, thereby avoiding a short circuit to ground through the antenna coil of the set.

## 810.20 Antenna Discharge Units — Receiving Stations

**(A) Where Required.** Each conductor of a lead-in from an outdoor antenna shall be provided with a listed antenna discharge unit.

*Exception: Where the lead-in conductors are enclosed in a continuous metallic shield that either is grounded with a conductor in accordance with 810.21 or is protected by an antenna discharge unit.*

An antenna discharge unit (lightning arrester) is not required if the lead-in conductors are enclosed in a continuous metal shield, such as rigid or intermediate metal conduit, electrical metallic tubing, or any metal raceway or metal-shielded cable that is effectively grounded. A lightning discharge will take the path of lower impedance and jump from the lead-in conductors to the metal raceway or shield rather than take the path through the antenna coil of the receiver.

**(B) Location.** Antenna discharge units shall be located outside the building or inside the building between the point of entrance of the lead-in and the radio set or transformers and as near as practicable to the entrance of the conductors to the building. The antenna discharge unit shall not be located near combustible material or in a hazardous (classified) location as defined in Article 500.

**(C) Grounding.** The antenna discharge unit shall be grounded in accordance with 810.21.

## 810.21 Bonding Conductors and Grounding Electrode Conductors — Receiving Stations

Bonding conductors and grounding electrode conductors shall comply with 810.21(A) through (K).

**(A) Material.** The bonding conductor or grounding electrode conductor shall be of copper, aluminum, copper-clad steel, bronze, or similar corrosion-resistant material. Aluminum or copper-clad aluminum bonding conductors or grounding electrode conductors shall not be used where in direct contact with masonry or the earth or where subject to corrosive conditions. Where used outside, aluminum or copper-clad aluminum conductors shall not be installed within 450 mm (18 in.) of the earth.

**(B) Insulation.** Insulation on bonding conductors or grounding electrode conductors shall not be required.

**(C) Supports.** The bonding conductor or grounding electrode conductor shall be securely fastened in place and shall be permitted to be directly attached to the surface wired over without the use of insulating supports.

*Exception: Where proper support cannot be provided, the size of the bonding conductors or grounding electrode conductors shall be increased proportionately.*

**(D) Physical Protection.** Bonding conductors and grounding electrode conductors shall be protected where exposed to physical damage. Where the bonding conductor or grounding electrode conductor is installed in a metal raceway, both ends of the raceway shall be bonded to the contained conductor or to the same terminal or electrode to which the bonding conductor or grounding electrode conductor is connected.

Where metal raceways are used to enclose the grounding electrode conductor, a connection between the grounding electrode conductor and the metal conduit must be provided at both ends of the conduit to provide an adequate low-impedance current path to ground.

**(E) Run in Straight Line.** The bonding conductor or grounding electrode conductor for an antenna mast or antenna discharge unit shall be run in as straight a line as practicable.

**(F) Electrode.** The bonding conductor or grounding electrode conductor shall be connected as required in (F)(1) through (F)(3).

**(1) In Buildings or Structures with an Intersystem Bonding Termination.** If the building or structure served has an intersystem bonding termination as required by 250.94, the bonding conductor shall be connected to the intersystem bonding termination.

Informational Note: See Article 100 for the definition of *Intersystem Bonding Termination.*

**(2) In Buildings or Structures with Grounding Means.** If the building or structure served has no intersystem bonding termination, the bonding conductor or grounding electrode conductor shall be connected to the nearest accessible location on the following:

(1) The building or structure grounding electrode system as covered in 250.50
(2) The grounded interior metal water piping systems, within 1.52 m (5 ft) from its point of entrance to the building, as covered in 250.52

For more information on the use of a metal water piping system, see the commentary following 250.52(A)(1).

(3) The power service accessible means external to the building, as covered in 250.94
(4) The nonflexible metallic power service raceway
(5) The service equipment enclosure, or
(6) The grounding electrode conductor or the grounding electrode conductor metal enclosures of the power service

A bonding device intended to provide a termination point for the bonding conductor (intersystem bonding) shall not interfere with the opening of an equipment enclosure. A bonding device shall be mounted on non-removable parts. A bonding device shall not be mounted on a door or cover even if the door or cover is non-removable.

**(3) In Buildings or Structures Without an Intersystem Bonding Termination or Grounding Means.** If the building or structure served has no intersystem bonding termination or grounding means as described in 810.21(F)(2), the grounding electrode conductor shall be connected to a grounding electrode as described in 250.52.

**(G) Inside or Outside Building.** The bonding conductor or grounding electrode conductor shall be permitted to be run either inside or outside the building.

**(H) Size.** The bonding conductor or grounding electrode conductor shall not be smaller than 10 AWG copper, 8 AWG aluminum, or 17 AWG copper-clad steel or bronze.

**(I) Common Ground.** A single bonding conductor or grounding electrode conductor shall be permitted for both protective and operating purposes.

**(J) Bonding of Electrodes.** A bonding jumper not smaller than 6 AWG copper or equivalent shall be connected between the radio and television equipment grounding electrode and the power grounding electrode system at the building or structure served where separate electrodes are used.

Antenna masts must be grounded to the same grounding electrode used for the building's electrical system to ensure that all exposed, non–current-carrying metal parts are at the same potential. In many cases, masts are connected incorrectly to conveniently located vent pipes, metal gutters, or downspouts. Such a connection could create potential differences between lead-in conductors and various metal parts located in or on buildings, resulting in possible shock and fire hazards. An underground gas piping system is not permitted to be used as a grounding electrode.

The use of separate radio/television grounding electrodes is not required. However, where they are provided, 800.21(J) requires the radio/television system grounding electrode to be connected, via a bonding jumper, to the grounding electrode of the electrical distribution system of the building or structure.

**(K) Electrode Connection.** Connections to grounding electrodes shall comply with 250.70.

This requirement is similar to the requirements for grounding electrode conductors in Articles 800, 820, and 830.

## III. Amateur and Citizen Band Transmitting and Receiving Stations — Antenna Systems

### 810.51 Other Sections

In addition to complying with Part III, antenna systems for amateur and citizen band transmitting and receiving stations shall also comply with 810.11 through 810.15.

Amateur radio and citizens band (CB) radio are two different noncommercial radio services. From the standpoint of the *NEC*, the installation requirements are similar.

***TABLE 810.52*** *Size of Outdoor Antenna Conductors*

| Material | Minimum Size of Conductors (AWG) Where Maximum Open Span Length Is: Less Than 45 m (150 ft) | Over 45 m (150 ft) |
|---|---|---|
| Hard-drawn copper | 14 | 10 |
| Copper-clad steel, bronze, or other high-strength material | 14 | 12 |

### 810.52 Size of Antenna

Antenna conductors for transmitting and receiving stations shall be of a size not less than given in Table 810.52.

### 810.53 Size of Lead-in Conductors

Lead-in conductors for transmitting stations shall, for various maximum span lengths, be of a size at least as great as that of conductors for antennas as specified in 810.52.

### 810.54 Clearance on Building

Antenna conductors for transmitting stations, attached to buildings, shall be firmly mounted at least 75 mm (3 in.) clear of the surface of the building on nonabsorbent insulating supports, such as treated pins or brackets equipped with insulators having not less than 75-mm (3-in.) creepage and airgap distances. Lead-in conductors attached to buildings shall also comply with these requirements.

*Exception: Where the lead-in conductors are enclosed in a continuous metallic shield that is grounded with a conductor in accordance with 810.58, they shall not be required to comply with these requirements. Where grounded, the metallic shield shall also be permitted to be used as a conductor.*

Creepage distance is measured from the conductor across the face of the supporting insulator to the building surface. Air gap distance is measured from the conductor (at its closest point) across the air space (not necessarily in a straight line) to the surface of the building.

### 810.55 Entrance to Building

Except where protected with a continuous metallic shield that is grounded with a conductor in accordance with 810.58, lead-in conductors for transmitting stations shall enter buildings by one of the following methods:

(1) Through a rigid, noncombustible, nonabsorbent insulating tube or bushing
(2) Through an opening provided for the purpose in which the entrance conductors are firmly secured so as to provide a clearance of at least 50 mm (2 in.)
(3) Through a drilled window pane

### 810.56 Protection Against Accidental Contact

Lead-in conductors to radio transmitters shall be located or installed so as to make accidental contact with them difficult.

### 810.57 Antenna Discharge Units — Transmitting Stations

Each conductor of a lead-in for outdoor antennas shall be provided with an antenna discharge unit or other suitable means that drain static charges from the antenna system.

*Exception No. 1: Where the lead-in is protected by a continuous metallic shield that is grounded with a conductor in accordance with 810.58, an antenna discharge unit or other suitable means shall not be required.*

*Exception No. 2: Where the antenna is grounded with a conductor in accordance with 810.58, an antenna discharge unit or other suitable means shall not be required.*

If an antenna discharge unit is not installed at a transmitting station, protection against lightning may be provided by a switch that connects the lead-in conductors to ground during the times the station is not in operation.

### 810.58 Bonding Conductors and Grounding Electrode Conductors — Amateur and Citizen Band Transmitting and Receiving Stations

Bonding conductors and grounding electrode conductors shall comply with 810.58(A) through (C).

**(A) Other Sections.** All bonding conductors and grounding electrode conductors for amateur and citizen band transmitting and receiving stations shall comply with 810.21(A) through (K).

**(B) Size of Protective Bonding Conductor or Grounding Electrode Conductor.** The protective bonding conductor or grounding electrode conductor for transmitting stations shall be as large as the lead-in but not smaller than 10 AWG copper, bronze, or copper-clad steel.

**(C) Size of Operating Bonding Conductor or Grounding Electrode Conductor.** The operating bonding conductor or grounding electrode conductor for transmitting stations shall not be less than 14 AWG copper or its equivalent.

## IV. Interior Installation — Transmitting Stations

### 810.70 Clearance from Other Conductors

All conductors inside the building shall be separated at least 100 mm (4 in.) from the conductors of any electric light, power, or signaling circuit.

*Exception No. 1: As provided in Article 640.*

*Exception No. 2: Where separated from other conductors by raceway or some firmly fixed nonconductor, such as porcelain tubes or flexible tubing.*

### 810.71 General

Transmitters shall comply with 810.71(A) through (C).

**(A) Enclosing.** The transmitter shall be enclosed in a metal frame or grille or separated from the operating space by a barrier or other equivalent means, all metallic parts of which are effectively connected to a bonding conductor or grounding electrode conductor.

**(B) Grounding of Controls.** All external metal handles and controls accessible to the operating personnel shall be effectively connected to an equipment grounding conductor if the transmitter is powered by the premises wiring system or grounded with a conductor in accordance with 810.21.

**(C) Interlocks on Doors.** All access doors shall be provided with interlocks that disconnect all voltages of over 350 volts between conductors when any access door is opened.

# ARTICLE 820
# Community Antenna Television and Radio Distribution Systems

Informational Note: The general term *grounding conductor* as previously used in this article is replaced by either the term *bonding conductor* or the term *grounding electrode conductor* (GEC), where applicable, to more accurately reflect the application and function of the conductor.

See Informational Note Figure 800(a) and Informational Note Figure 800(b) for an illustrative application of a bonding conductor or grounding electrode conductor.

## I. General

### 820.1 Scope

This article covers coaxial cable distribution of radio frequency signals typically employed in community antenna television (CATV) systems.

Informational Note: See 90.2(B)(4) for installations of CATV and radio distribution systems that are not covered.

Article 820 covers the installation of coaxial cable for closed-circuit television, cable television, and security television cameras. This article also covers coaxial cable for radio and television receiving equipment. Article 830 covers network-powered broadband system installations.

### 820.2 Definitions

See Part I of Article 100. For the purposes of this article, the following additional definitions apply.

**Abandoned Coaxial Cable.** Installed coaxial cable that is not terminated at equipment other than a coaxial connector and not identified for future use with a tag.

Informational Note: See Part I of Article 100 for a definition of *Equipment*.

**Coaxial Cable.** A cylindrical assembly composed of a conductor centered inside a metallic tube or shield, separated by a dielectric material, and usually covered by an insulating jacket.

**Exposed (to Accidental Contact).** A circuit in such a position that, in case of failure of supports and or insulation, contact with another circuit may result.

Informational Note: See Part I of Article 100 for two other definitions of *Exposed*.

**Point of Entrance.** The point within a building at which the coaxial cable emerges from an external wall, from a concrete floor slab, from rigid metal conduit (RMC), or from intermediate metal conduit (IMC).

**Premises.** The land and buildings of a user located on the user side of utility-user network point of demarcation.

## 820.3 Other Articles

Circuits and equipment shall comply with 820.3(A) through (J).

**(A) Hazardous (Classified) Locations.** CATV equipment installed in a location that is classified in accordance with 500.5 and 505.5 shall comply with the applicable requirements of Chapter 5.

**(B) Wiring in Ducts for Dust, Loose Stock, or Vapor Removal.** The requirements of 300.22(A) shall apply.

**(C) Equipment in Other Space Used for Environmental Air.** The requirements of 300.22(C)(3) shall apply.

**(D) Installation and Use.** The requirements of 110.3(B) shall apply.

**(E) Installations of Conductive and Nonconductive Optical Fiber Cables.** The requirements of Article 770 shall apply.

**(F) Communications Circuits.** The requirements of Article 800 shall apply.

**(G) Network-Powered Broadband Communications Systems.** The requirements of Article 830 shall apply.

**(H) Premises-Powered Broadband Communications Systems.** The requirements of Article 840 shall apply.

**(I) Alternate Wiring Methods.** The wiring methods of Article 830 shall be permitted to substitute for the wiring methods of Article 820.

Informational Note: Use of Article 830 wiring methods will facilitate the upgrading of Article 820 installations to network-powered broadband applications.

**(J) Cable Routing Assemblies.** The definition in Article 100, the applications in Table 800.154(c) , and the installation requirements in 800.110 shall apply to Article 820 and 800.113.

## 820.15 Power Limitations

Coaxial cable shall be permitted to deliver power to equipment that is directly associated with the radio frequency distribution system if the voltage is not over 60 volts and if the current is supplied by a transformer or other device that has power-limiting characteristics.

Power shall be blocked from premises devices on the network that are not intended to be powered via the coaxial cable.

## 820.21 Access to Electrical Equipment Behind Panels Designed to Allow Access

Access to electrical equipment shall not be denied by an accumulation of coaxial cables that prevents removal of panels, including suspended ceiling panels.

An excess accumulation of wires and cables can limit access to equipment by preventing the removal of access panels. (See Exhibit 820.1.)

## 820.24 Mechanical Execution of Work

Community television and radio distribution systems shall be installed in a neat and workmanlike manner. Coaxial cables installed exposed on the surface of ceiling and sidewalls shall be supported by the building structure in such a manner that the cables will not be damaged by normal building use. Such cables shall be secured by hardware including straps, staples, cable ties, hangers, or similar fittings designed and installed so as not to damage the cable. The installation shall also conform to 300.4(D) and 300.11. Nonmetallic cable ties and other nonmetallic cable

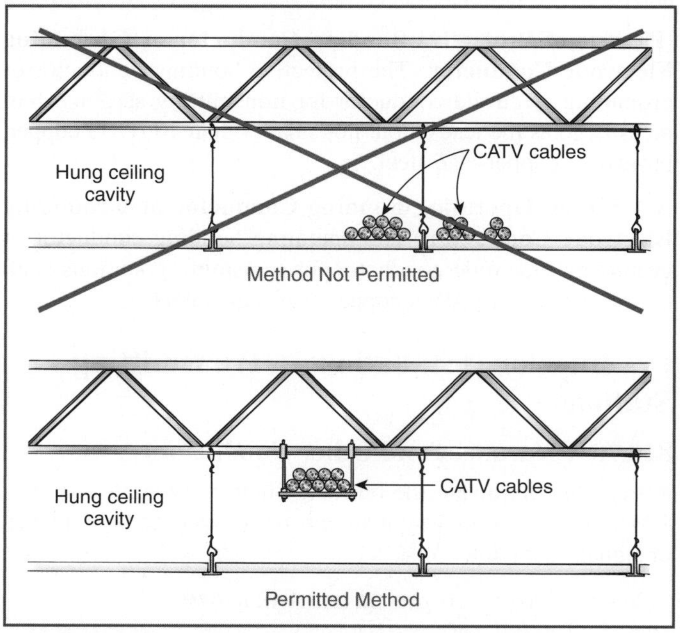


***EXHIBIT 820.1*** *Incorrect installation of cables (upper diagram), and correct method (lower diagram).*

accessories used to secure and support cables in other spaces used for environmental air (plenums) shall be listed as having low smoke and heat release properties.

Cable must be attached to or supported by the structure by cable ties, straps, clamps, hangers, and so forth. The installation method must not damage the cable. Although this requirement does not contain specific supporting and securing intervals, it does reference 300.11 as a general rule on securing equipment and cables and 300.4(D) for protection of cables installed parallel to framing methods.

Informational Note No. 1: Accepted industry practices are described in ANSI/NECA/BICSI 568–2006, *Standard for Installing Commercial Building Telecommunications Cabling*; ANSI/TIA/EIA-568-B.1 2004 — *Part 1, General Requirements Commercial Building Telecommunications Cabling Standard*; ANSI/TIA-569-B 2004, *Commercial Building Standard for Telecommunications Pathways and Spaces*; ANSI/TIA-570-B, *Residential Telecommunications Infrastructure*; and other ANSI-approved installation standards.

Informational Note No. 2: See 4.3.11.2.6.5 and 4.3.11.5.5.6 of NFPA 90A-2012, *Standard for the Installation of Air-Conditioning and Ventilating Systems*, for discrete combustible components installed in accordance with 300.22(C).

### 820.25 Abandoned Cables

The accessible portion of abandoned coaxial cables shall be removed. Where cables are identified for future use with a tag, the tag shall be of sufficient durability to withstand the environment involved.

See Article 100 for the definition of *accessible* (as applied to wiring methods). Abandoned cable increases fire loading unnecessarily and, where installed in plenums, can affect airflow.

### 820.26 Spread of Fire or Products of Combustion

Installations of coaxial cables and communications raceways in hollow spaces, vertical shafts, and ventilation or air-handling ducts shall be made so that the possible spread of fire or products of combustion will not be substantially increased. Openings around penetrations of coaxial cables and communications raceways through fire-resistant-rated walls, partitions, floors, or ceilings shall be firestopped using approved methods to maintain the fire resistance rating.

Informational Note: Directories of electrical construction materials published by qualified testing laboratories contain many listing installation restrictions necessary to maintain the fire-resistive rating of assemblies where penetrations or openings are made. Building codes also contain restrictions on membrane penetrations on opposite sides of a fire resistance–rated wall assembly. An example is the 600-mm (24-in.) minimum horizontal separation that usually applies between boxes installed on opposite sides of the wall. Assistance in complying with 820.26 can be found in building codes, fire resistance directories, and product listings.

## II. Coaxial Cables Outside and Entering Buildings

### 820.44 Overhead (Aerial) Coaxial Cables

Overhead (aerial) coaxial cables, prior to the point of grounding, as specified in 820.93, shall comply with 820.44(A) through (E).

**(A) On Poles and In-Span.** Where coaxial cables and electric light or power conductors are supported by the same pole or are run parallel to each other in-span, the conditions described in 820.44(A)(1) through (A)(4) shall be met.

**(1) Relative Location.** Where practicable, the coaxial cables shall be located below the electric light or power conductors.

**(2) Attachment to Cross-Arms.** Coaxial cables shall not be attached to cross-arm that carries electric light or power conductors.

**(3) Climbing Space.** The climbing space through coaxial cables shall comply with the requirements of 225.14(D).

**(4) Clearance.** Lead-in or overhead (aerial)-drop coaxial cables from a pole or other support, including the point of initial attachment to a building or structure, shall be kept away from electric light, power, Class 1, or non–power-limited fire alarm circuit conductors so as to avoid the possibility of accidental contact.

*Exception: Where proximity to electric light, power, Class 1, or non–power-limited fire alarm circuit conductors cannot be avoided, the installation shall provide clearances of not less than 300 mm (12 in.) from electric light, power, Class 1, or non–power-limited fire alarm circuit conductors. The clearance requirement shall apply at all points along the drop, and it shall increase to 1.0 m (40 in.) at the pole.*

**(B) Above Roofs.** Coaxial cables shall have a vertical clearance of not less than 2.5 m (8 ft) from all points of roofs above which they pass.

*Exception No. 1: Auxiliary buildings such as garages and the like.*

*Exception No. 2: A reduction in clearance above only the overhanging portion of the roof to not less than 450 mm (18 in.) shall be permitted if (1) not more than 1.2 m (4 ft) of communications service drop conductors pass above the roof overhang, and (2) they are terminated at a raceway mast or other approved support.*

*Exception No. 3: Where the roof has a slope of not less than 100 mm in 300 mm (4 in. in 12 in.), a reduction in clearance to not less than 900 mm (3 ft) shall be permitted.*

**(C) On Masts.** Overhead (aerial) coaxial cables shall be permitted to be attached to an above-the-roof raceway mast that does not enclose or support conductors of electric light or power circuits.

**(D) Between Buildings.** Coaxial cables extending between buildings or structures, and also the supports or attachment fixtures, shall be identified and shall have sufficient strength to withstand the loads to which they might be subjected.

Wind and ice loads should be considered because they can damage cables and attachment points.

*Exception: Where a coaxial cable does not have sufficient strength to be self-supporting, it shall be attached to a supporting messenger cable that, together with the attachment fixtures or supports, shall be acceptable for the purpose and shall have sufficient strength to withstand the loads to which they may be subjected.*

**(E) On Buildings.** Where attached to buildings, coaxial cables shall be securely fastened in such a manner that they will be separated from other conductors in accordance with 820.44(E)(1), (E)(2), and (E)(3).

**(1) Electric Light or Power.** The coaxial cable shall have a separation of at least 100 mm (4 in.) from electric light, power, Class 1, or non–power-limited fire alarm circuit conductors not in raceway or cable, or shall be permanently separated from conductors of the other system by a continuous and firmly fixed nonconductor in addition to the insulation on the wires.

**(2) Other Communications Systems.** Coaxial cable shall be installed so that there will be no unnecessary interference in the maintenance of the separate systems. In no case shall the conductors, cables, messenger strand, or equipment of one system cause abrasion to the conductors, cable, messenger strand, or equipment of any other system.

**(3) Lightning Conductors.** Where practicable, a separation of at least 1.8 m (6 ft) shall be maintained between any coaxial cable and lightning conductors.

Informational Note: For additional information regarding overhead (aerial) wires and cables, see ANSI C2-2007, *National Electric Safety Code*, Part 2, Safety Rules for Overhead Lines.

## 820.47 Underground Coaxial Cables Entering Buildings

Underground coaxial cables entering buildings shall comply with 820.47(A) and (B).

**(A) Underground Systems with Electric Light, Power, Class 1, or Non–Power-Limited Fire Alarm Circuit Conductors.** Underground coaxial cables in a duct, pedestal, handhole enclosure, or manhole that contains electric light, power, or Class 1 or non–power-limited fire alarm circuit conductors shall be in a section permanently separated from such conductors by means of a suitable barrier.

**(B) Direct-Buried Cables and Raceways.** Direct-buried coaxial cable shall be separated at least 300 mm (12 in.) from conductors of any light or power or Class 1 circuit.

*Exception No. 1: Where electric service conductors or coaxial cables are installed in raceways or have metal cable armor.*

*Exception No. 2: Where electric light or power branch-circuit or feeder conductors or Class 1 circuit conductors are installed in a raceway or in metal-sheathed, metal-clad, or Type UF or Type USE cables; or the coaxial cables have metal cable armor or are installed in a raceway.*

## 820.48 Unlisted Cables Entering Buildings

Unlisted outside plant coaxial cables shall be permitted to be installed in building spaces other than risers, ducts used for environmental air, plenums used for environmental air, and other spaces used for environmental air, where the length of the cable within the building, measured from its point of entrance, does not exceed 15 m (50 ft) and the cable enters the building from the outside and is terminated at a grounding block.

## 820.49 Metallic Entrance Conduit Grounding

Rigid metal conduit (RMC) or intermediate metal conduit (IMC) containing entrance coaxial cable shall be connected by a bonding conductor or grounding electrode conductor to a grounding electrode in accordance with 820.100(B).

# III. Protection

## 820.93 Grounding of the Outer Conductive Shield of Coaxial Cables

Coaxial cables entering buildings or attached to buildings shall comply with 820.93(A) or (B). Where the outer conductive shield of a coaxial cable is grounded, no other protective devices shall be required. For purposes of this section, grounding located at mobile home service equipment located within 9.0 m (30 ft) of the exterior wall of the mobile home it serves, or at a mobile home disconnecting means grounded in accordance with 250.32 and located within 9.0 m (30 ft) of the exterior wall of the mobile home it serves, shall be considered to meet the requirements of this section.

Informational Note: Selecting a grounding block location to achieve the shortest practicable bonding conductor or grounding electrode conductor helps limit potential differences between CATV and other metallic systems.

Proper bonding of the CATV system coaxial cable sheath to the electrical power grounding electrode is needed to prevent potential fire and shock hazards.

**(A) Entering Buildings.** In installations where the coaxial cable enters the building, the outer conductive shield shall be grounded in accordance with 820.100. The grounding shall be as close as practicable to the point of entrance.

Informational Note: See 820.2 for a definition of *Point of Entrance.*

**(B) Terminating Outside of the Building.** In installations where the coaxial cable is terminated outside of the building, the outer conductive shield shall be grounded in accordance with 820.100. The grounding shall be as close as practicable to the point of attachment or termination.

**(C) Location.** Where installed, a listed primary protector shall be applied on each community antenna and radio distribution (CATV) cable external to the premises. The listed primary protector shall be located as close as practicable to the entrance point of the cable on either side or integral to the ground block.

**(D) Hazardous (Classified) Locations.** Where a primary protector or equipment providing the primary protection function is used, it shall not be located in any hazardous (classified) location as defined in 500.5 and 505.5 or in the vicinity of easily ignitible material.

*Exception: As permitted in 501.150, 502.150, and 503.150.*

# IV. Grounding Methods

## 820.100 Cable Bonding and Grounding

The shield of the coaxial cable shall be bonded or grounded as specified in 820.100(A) through (D).

*Exception: For communications systems using coaxial cable confined within the premises and isolated from outside cable plant, the shield shall be permitted to be grounded by a connection to an equipment grounding conductor as described in 250.118. Connecting to an equipment grounding conductor through a grounded receptacle using a dedicated bonding jumper and a permanently connected listed device shall be permitted. Use of a cord and plug for the connection to an equipment grounding conductor shall not be permitted.*

**(A) Bonding Conductor or Grounding Electrode Conductor.**

**(1) Insulation.** The bonding conductor or grounding electrode conductor shall be listed and shall be permitted to be insulated, covered, or bare.

**(2) Material.** The bonding conductor or grounding electrode conductor shall be copper or other corrosion-resistant conductive material, stranded or solid.

**(3) Size.** The bonding conductor or grounding electrode conductor shall not be smaller than 14 AWG. It shall have a current-carrying capacity not less than the outer sheath of the coaxial cable. The bonding conductor or grounding electrode conductor shall not be required to exceed 6 AWG.

**(4) Length.** The bonding conductor or grounding electrode conductor shall be as short as practicable. In one- and two-family dwellings, the bonding conductor or grounding electrode conductor shall be as short as practicable, not to exceed 6.0 m (20 ft) in length.

Informational Note: Similar bonding conductor or grounding electrode conductor length limitations applied at apartment buildings and commercial buildings help to reduce voltages that may be developed between the building's power and communications systems during lightning events.

*Exception: In one- and two-family dwellings where it is not practicable to achieve an overall maximum bonding conductor or grounding electrode conductor length of 6.0 m (20 ft), a separate grounding electrode as specified in 250.52(A)(5), (A)(6), or (A)(7) shall be used, the grounding electrode conductor shall be connected to the separate grounding electrode in accordance with 250.70, and the separate grounding electrode shall be connected to the power grounding electrode system in accordance with 820.100(D).*

The 20-foot limitation on length results in a lower impedance, which in turn limits the potential difference between CATV systems and other systems during a lightning strike. Large potential differences between grounding conductors can result in damage if a lightning strike were to occur.

The informational note provides guidance for the treatment of the cable and primary protector grounding conductor length at apartment and commercial buildings that is consistent with the 20-foot rule for one- and two-family dwellings. However, a specific length is not specified in the *Code*, because such a limitation may not be practical in some installations.

**(5) Run in Straight Line.** The bonding conductor or grounding electrode conductor shall be run in as straight a line as practicable.

**(6) Physical Protection.** Bonding conductors and grounding electrode conductors shall be protected where exposed to physical damage. Where the bonding conductor or grounding electrode conductor is installed in a metal raceway, both ends of the raceway shall be bonded to the contained conductor or to the same terminal or electrode to which the bonding conductor or grounding electrode conductor is connected.

**(B) Electrode.** The bonding conductor or grounding electrode conductor shall be connected in accordance with 820.100(B)(2), or (B)(3).

**(1) In Buildings or Structures with an Intersystem Bonding Termination.** If the building or structure served has an intersystem bonding termination as required by 250.94, the bonding conductor shall be connected to the intersystem bonding termination.

Informational Note: See Part I of Article 100 for the definition of *Intersystem Bonding Termination.*

**(2) In Buildings or Structures with Grounding Means.** If the building or structure served has no intersystem bonding termination, the bonding conductor or grounding electrode conductor shall be connected to the nearest accessible location on one of the following:

(1) The building or structure grounding electrode system as covered in 250.50
(2) The grounded interior metal water piping system, within 1.5 m (5 ft) from its point of entrance to the building, as covered in 250.52

For more information on the use of a metal water piping system as a grounding electrode, see the commentary following 250.52(A)(1).

(3) The power service accessible means external to enclosures as covered in 250.94
(4) The nonflexible metallic power service raceway
(5) The service equipment enclosure
(6) The grounding electrode conductor or the grounding electrode conductor metal enclosure of the power service, or
(7) The grounding electrode conductor or the grounding electrode of a building or structure disconnecting means that is connected to an electrode as covered in 250.32

A bonding device intended to provide a termination point for the bonding conductor (intersystem bonding) shall not interfere with the opening of an equipment enclosure. A bonding device shall be mounted on non-removable parts. A bonding device shall not be mounted on a door or cover even if the door or cover is nonremovable.

For purposes of this section, the mobile home service equipment or the mobile home disconnecting means, as described in 820.93, shall be considered accessible.

**(3) In Buildings or Structures Without an Intersystem Bonding Termination or Grounding Means.** If the building or structure served has no intersystem bonding termination or grounding means, as described in 820.100(B)(2), the grounding electrode conductor shall be connected to either of the following:

(1) To any one of the individual grounding electrodes described in 250.52(A)(1), (A)(2), (A)(3), or (A)(4).
(2) If the building or structure served has no intersystem bonding termination or grounding means, as described in 820.100(B)(2) or (B)(3)(1), to any one of the individual grounding electrodes described in 250.52(A)(5), (A)(7), and (A)(8). Steam or hot water pipes or air terminal conductors (lightning-rod conductors) shall not be employed as grounding electrodes for bonding conductors or grounding electrode conductors.

**(C) Electrode Connection.** Connections to grounding electrodes shall comply with 250.70.

**(D) Bonding of Electrodes.** A bonding jumper not smaller than 6 AWG copper or equivalent shall be connected between the community antenna television system's grounding electrode and the power grounding electrode system at the building or structure served where separate electrodes are used.

*Exception: At mobile homes as covered in 820.106.*

Informational Note No. 1: See 250.60 for use of air terminals (lightning rods).

Informational Note No. 2: Bonding together of all separate electrodes limits potential differences between them and between their associated wiring systems.

Bonding of CATV and power grounding electrodes is required at the same building or structure. A common error made in grounding CATV systems is connecting the coaxial cable sheath to a rod-type grounding electrode driven by the CATV installer at a convenient location near the point of cable entry to the building, instead of bonding it to the electrical service grounding electrode system.

Section 250.94 requires that a bonding means with not less than three termination points that is accessible and external to the service equipment be provided for making the bonding and grounding connection for other systems. For existing installations where an intersystem bonding termination is not available, alternate bonding means are described in 820.100(B)(2). A separate grounding electrode is permitted by 800.100(B)(3) only if the building or structure has neither an intersystem bonding termination nor a grounding means, which is rare. The earth cannot be used as the bonding conductor, because it does not have the required low-impedance path. (See 250.54.)

Both CATV systems and power systems are subject to current surges as a result of, for example, induced voltages from lightning in the vicinity of the outside distribution systems. Surges also result from switching operations on power systems. If the grounded conductors and parts of the two systems are not bonded by a low-impedance path, such line surges can raise the potential difference between the two systems to many thousands of volts. This can result in arcing between the two systems – for example, wherever the coaxial cable jacket contacts a grounded part, such as a metal water pipe or metal structural member – inside the building.

If a person is the interface between two systems that are not bonded in accordance with the *Code*, a high-voltage surge could result in electric shock. More common, however, is burnout of a television tuner, a part that is almost always an interface between the two systems. The tuner is connected to the power system ground through the grounded neutral of the power supply, even if the television set itself is not provided with an EGC.

Also see the commentary following 250.92(B) and 820.100(E).

**(E) Shield Protection Devices.** Grounding of a coaxial drop cable shield by means of a protective device that does not interrupt the grounding system within the premises shall be permitted.

The electric utility supply, the CATV system, and the premises wiring are all grounded. When a ground fault occurs at the premises, the current tries to return to its source and follows the multiple paths through the different premises metal water piping systems or earth. Such ground faults can cause current on the CATV shield, whose primary function is to prevent RF leakage out of the cable. The fault current can cause the cable shield to burn open and also damage the cable insulation. A device that can safely conduct current at 60 Hz and block current at the higher frequencies can be connected between the cable shield and ground, thereby maintaining grounding integrity. An ordinary fuse, for example, would not be suitable.

### 820.103 Equipment Grounding

Unpowered equipment and enclosures or equipment powered by the coaxial cable shall be considered grounded where connected to the metallic cable shield.

### 820.106 Grounding and Bonding at Mobile Homes

**(A) Grounding.** Grounding shall comply with 820.106(A)(1) and (A)(2).

(1) Where there is no mobile home service equipment located within 9.0 m (30 ft) of the exterior wall of the mobile home it serves, the coaxial cable shield ground, or surge arrester grounding terminal, shall be connected to a grounding electrode conductor or grounding electrode in accordance with 820.100(B)(3).
(2) Where there is no mobile home disconnecting means grounded in accordance with 250.32 and located within 9.0 m (30 ft) of the exterior wall of the mobile home it serves, the coaxial cable shield ground, or surge arrester grounding terminal, shall be connected to a grounding electrode in accordance with 820.100(B)(3).

**(B) Bonding.** The coaxial cable shield grounding terminal, surge arrester grounding terminal, or grounding electrode shall be connected to the metal frame or available grounding terminal of the mobile home with a copper conductor not smaller than 12 AWG under any of the following conditions:

(1) Where there is no mobile home service equipment or disconnecting means as in 820.106(A)
(2) Where the mobile home is supplied by cord and plug

## V. Installation Methods Within Buildings

### 820.110 Raceways and Cable Routing Assemblies for Coaxial Cables

**(A) Types of Raceways.** Coaxial cables shall be permitted to be installed in any raceway that complies with either (A)(1) or (A)(2) and in cable routing assemblies installed in compliance with 820.110(C).

**(1) Raceways Recognized in Chapter 3.** Coaxial cables shall be permitted to be installed in any raceway included in Chapter 3. The raceways shall be installed in accordance with the requirements of Chapter 3.

**(2) Communications Raceways.** Coaxial cables shall be permitted to be installed in listed plenum communications raceways, listed riser communications raceways, and listed general-purpose communications raceways, selected in accordance with the provisions of 800.110, 800.113, and 820.113 and installed in accordance with 362.24 through 362.56, where the requirements applicable to electrical nonmetallic tubing (ENT) apply.

See the commentary following 800.182(C) for information on listed communications raceways.

**(B) Raceway Fill for Coaxial Cables.** The raceway fill requirements of Chapters 3 and 9 shall not apply to coaxial cables.

**(C) Cable Routing Assemblies.** Coaxial cables shall be permitted to be installed in plenum cable routing assemblies, riser cable routing assemblies, and general-purpose cable routing assemblies selected in accordance with the provisions of 800.113 and installed in accordance with 820.110(C)(1) and (2).

See the commentary following 800.182(C) for information on listed cable routing assemblies.

**(1) Horizontal Support.** Cable routing assemblies shall be supported where run horizontally at intervals not to exceed 900 mm (3 ft), and at each end or joint, unless listed for other support intervals. In no case shall the distance between supports exceed 3 m (10 ft).

**(2) Vertical Support.** Vertical runs of cable routing assemblies shall be supported at intervals not exceeding 1.2 m (4 ft), unless listed for other support intervals, and shall not have more than one joint between supports.

### 820.113 Installation of Coaxial Cables

Installation of coaxial cables shall comply with 820.113(A) through (K). Installation of raceways shall comply with 820.110.

**(A) Listing.** Coaxial cables installed in buildings shall be listed.

*Exception: Coaxial cables that comply with 820.48 shall not be required to be listed.*

**(B) Fabricated Ducts Used for Environmental Air.** The following cables shall be permitted in ducts as described in 300.22(B) if they are directly associated with the air distribution system:

(1) Up to 1.22 m (4 ft) of Type CATVP cable
(2) Types CATVP, CATVR, CATV, and CATVX cables installed in raceways that are installed in compliance with 300.22(B)

Informational Note: For information on fire protection of wiring installed in fabricated ducts see 4.3.4.1 and 4.3.11.3.3 of NFPA 90A-2012, *Standard for the Installation of Air-Conditioning and Ventilating Systems.*

**(C) Other Spaces Used For Environmental Air (Plenums).** The following cables shall be permitted in other spaces used for environmental air as described in 300.22(C):

(1) Type CATVP cable
(2) Type CATVP cable installed in plenum communications raceways
(3) Type CATVP cable supported by open metallic cable trays or cable tray systems
(4) Types CATVP, CATVR, CATV, and CATVX cables installed in raceways that are installed in compliance with 300.22(C)

(5) Types CATVP, CATVR, CATV, and CATVX cables supported by solid bottom metal cable trays with solid metal covers in other spaces used for environmental air (plenums) as described in 300.22(C)
(6) Types CATVP, CATVR, CATV, and CATVX cables installed in plenum communications raceways, riser communications raceways, or general-purpose communications raceways supported by solid bottom metal cable trays with solid metal covers in other spaces used for environmental air (plenums) as described in 300.22(C)

Informational Note: For information on fire protection of wiring installed in other spaces used for environmental air, see 4.3.11.2, 4.3.11.4, and 4.3.11.5 of NFPA 90A-2012, *Standard for the Installation of Air-Conditioning and Ventilating Systems*.

**(D) Risers — Cables in Vertical Runs.** The following cables shall be permitted in vertical runs penetrating one or more floors and in vertical runs in a shaft:

(1) Types CATVP and CATVR cables
(2) Types CATVP and CATVR cables installed in:
 a. Plenum communications raceways
 b. Plenum cable routing assemblies
 c. Riser communications raceways
 d. Riser cable routing assemblies

Informational Note: See 820.26 for firestop requirements for floor penetrations.

**(E) Risers — Cables in Metal Raceways.** The following cables shall be permitted in metal raceways in a riser having firestops at each floor:

(1) Types CATVP, CATVR, CATV, and CATVX cables
(2) Types CATVP, CATVR, CATV, and CATVX cables installed in:
 a. Plenum communications raceways
 b. Riser communications raceways
 c. General-purpose communications raceways

Informational Note: See 820.26 for firestop requirements for floor penetrations.

**(F) Risers — Cables in Fireproof Shafts.** The following cables shall be permitted to be installed in fireproof riser shafts with firestops at each floor:

(1) Types CATVP, CATVR, CATV, and CATVX cables
(2) Types CATVP, CATVR, and CATV cables installed in:
 a. Plenum communications raceways
 b. Plenum cable routing assemblies
 c. Riser communications raceways
 d. Riser cable routing assemblies
 e. General-purpose communications raceways
 f. General-purpose cable routing assemblies

Informational Note: See 820.26 for firestop requirements for floor penetrations.

**(G) Risers — One- and Two-Family Dwellings.** The following cables shall be permitted in one- and two-family dwellings:

(1) Types CATVP, CATVR, and CATV cables
(2) Type CATVX cable less than 10 mm (0.375 in.) in diameter
(3) Types CATVP, CATVR, and CATV cables installed in:
 a. Plenum communications raceways
 b. Plenum cable routing assemblies
 c. Riser communications raceways
 d. Riser cable routing assemblies
 e. General-purpose communications raceways
 f. General-purpose cable routing assemblies

Informational Note: See 820.26 for firestop requirements for floor penetrations.

**(H) Cable Trays.** The following cables shall be permitted to be supported by cable trays:

(1) Types CATVP, CATVR, and CATV cables
(2) Types CATVP, CATVR, and CATV cables installed in:
 a. Plenum communications raceways
 b. Riser communications raceways
 c. General-purpose communications raceways

**(I) Distributing Frames and Cross-Connect Arrays.** The following cables shall be permitted to be installed in distributing frames and cross-connect arrays:

(1) Types CATVP, CATVR, and CATV cables
(2) Types CATVP, CATVR, and CATV cables installed in:
 a. Plenum communications raceways
 b. Plenum cable routing assemblies
 c. Riser communications raceways
 d. Riser cable routing assemblies
 e. General-purpose communications raceways
 f. General-purpose cable routing assemblies

**(J) Other Building Locations.** The following cables shall be permitted to be installed in building locations other than the locations covered in 820.113(B) through (I):

(1) Types CATVP, CATVR, and CATV cables
(2) A maximum of 3 m (10 ft) of exposed Type CATVX cables in nonconcealed spaces
(3) Types CATVP, CATVR, and CATV cables installed in:
 a. Plenum communications raceways
 b. Plenum cable routing assemblies
 c. Riser communications raceways
 d. Riser cable routing assemblies
 e. General-purpose communications raceways
 f. General-purpose cable routing assemblies
(4) Types CATVP, CATVR, CATV, and Type CATVX cables installed in a raceway of a type recognized in Chapter 3

**(K) One- and Two-Family and Multifamily Dwellings.** The following cables and cable routing assemblies shall be permitted to be installed in one- and two-family and multifamily dwellings in locations other than those locations covered in 820.113(B) through (I):

(1) Types CATVP, CATVR, and CATV cables
(2) Type CATVX cable less than 10 mm (0.375 in.) in diameter
(3) Types CATVP, CATVR, and CATV cables installed in:
  a. Plenum communications raceways
  b. Plenum cable routing assemblies
  c. Riser communications raceways
  d. Riser cable routing assemblies
  e. General-purpose communications raceways
  f. General-purpose cable routing assemblies
(4) Types CATVP, CATVR, CATV, and Type CATVX cables installed in a raceway of a type recognized in Chapter 3

## 820.133 Installation of Coaxial Cables and Equipment

Beyond the point of grounding, as defined in 820.93, the coaxial cable installation shall comply with 820.133(A) and (B).

Jackets of coaxial cable do not have sufficient construction specifications to permit them to be installed with electric light, power, Class 1, non–power-limited fire alarm circuits, and medium- and high-power network-powered broadband communications cable. Failure of the cable insulation due to a fault could lead to hazardous voltages being imposed on the Class 2 or Class 3 circuit conductors.

**(A) Separation from Other Conductors.**

**(1) In Raceways, Cable Trays, Boxes, Enclosures, and Cable Routing Assemblies.**

(a) *Optical Fiber and Communications Cables.* Coaxial cables shall be permitted in the same raceway, cable tray, box, enclosure, or cable routing assembly with jacketed cables of any of the following:

(1) Nonconductive and conductive optical fiber cables in compliance with Parts I and V of Article 770
(2) Communications circuits in compliance with Parts I and V of Article 800
(3) Low-power network-powered broadband communications circuits in compliance with Parts I and V of Article 830

(b) *Other Circuits.* Coaxial cables shall be permitted in the same raceway, cable tray, box, enclosure, or cable routing assembly with jacketed cables of any of the following:

(1) Class 2 and Class 3 remote-control, signaling, and power-limited circuits in compliance with Article 645 or Parts I and III of Article 725
(2) Power-limited fire alarm systems in compliance with Parts I and III of Article 760

(c) *Electric Light, Power, Class 1, Non–Power-Limited Fire Alarm, and Medium-Power Network-Powered Broadband Communications Circuits.* Coaxial cable shall not be placed in any raceway, compartment, outlet box, junction box, or other enclosures with conductors of electric light, power, Class 1, non–power-limited fire alarm, or medium-power network-powered broadband communications circuits.

*Exception No. 1: Where all of the conductors of electric light, power, Class 1, non–power-limited fire alarm, and medium-power network-powered broadband communications circuits are separated from all of the coaxial cables by a permanent barrier or listed divider.*

This exception recognizes the use of a listed field-installed divider to separate the communications circuits from the power circuits.

*Exception No. 2: Power circuit conductors in outlet boxes, junction boxes, or similar fittings or compartments where such conductors are introduced solely for power supply to the coaxial cable system distribution equipment. The power circuit conductors shall be routed within the enclosure to maintain a minimum 6-mm (0.25-in.) separation from coaxial cables.*

**(2) Other Applications.** Coaxial cable shall be separated at least 50 mm (2 in.) from conductors of any electric light, power, Class 1, non–power-limited fire alarm, or medium-power network-powered broadband communications circuits.

*Exception No. 1: Where either (1) all of the conductors of electric light, power, Class 1, non–power-limited fire alarm, and medium-power network-powered broadband communications circuits are in a raceway, or in metal-sheathed, metal-clad, nonmetallic-sheathed, Type AC or Type UF cables, or (2) all of the coaxial cables are encased in raceway.*

*Exception No. 2: Where the coaxial cables are permanently separated from the conductors of electric light, power, Class 1, non–power-limited fire alarm, and medium-power network-powered broadband communications circuits by a continuous and firmly fixed nonconductor, such as porcelain tubes or flexible tubing, in addition to the insulation on the wire.*

**(B) Support of Coaxial Cables.** Raceways shall be used for their intended purpose. Coaxial cables shall not be strapped, taped, or attached by any means to the exterior of any conduit or raceway as a means of support.

*Exception: Overhead (aerial) spans of coaxial cables shall be permitted to be attached to the exterior of a raceway-type mast intended for the attachment and support of such cables.*

## 820.154 Applications of Listed CATV Cables

Permitted and nonpermitted applications of listed coaxial cables shall be as indicated in Table 820.154(a). The permitted applications shall be subject to the installation requirements of 820.110 and 820.113. The substitutions for coaxial cables in Table 820.154(b) and illustrated in Figure 820.154 shall be permitted.

***TABLE 820.154(a)*** *Applications of Listed Coaxial Cables in Buildings*

| Applications | | CATVP | CATVR | CATV | CATVX |
|---|---|---|---|---|---|
| | | **Cable Type** | | | |
| In specifically fabricated ducts as described in 300.22(B) | In fabricated ducts as described in 300.22(B) | Y* | N | N | N |
| | In metal raceway that complies with 300.22(B) | Y* | Y* | Y* | Y* |
| In other spaces used for environmental air as described in 300.22(C) | In other spaces used for environmental air (plenums) as described in 300.22(C) | Y* | N | N | N |
| | In metal raceway that complies with 300.22(C) | Y* | Y* | Y* | Y* |
| | In plenum communications raceways | NOT PERMITTED | | | |
| | In plenum cable routing assemblies | Y* | N | N | N |
| | Supported by open metal cable trays | Y* | N | N | N |
| | Supported by solid bottom metal cable trays with solid metal covers | Y* | Y* | Y* | Y* |
| In risers | In vertical runs | Y* | Y* | N | N |
| | In metal raceways | Y* | Y* | Y* | Y* |
| | In fireproof shafts | Y* | Y* | Y* | Y* |
| | In plenum communications raceways | Y* | Y* | N | N |
| | In plenum cable routing assemblies | Y* | Y* | N | N |
| | In riser communications raceways | Y* | Y* | N | N |
| | In riser cable routing assemblies | Y* | Y* | N | N |
| | In one- and two- family dwellings | Y* | Y* | Y* | Y* |
| Within buildings in other than air-handling spaces and risers | General | Y* | Y* | Y* | Y* |
| | In one- and two-family dwellings | Y* | Y* | Y* | Y* |
| | In multifamily dwellings | Y* | Y* | Y* | Y* |
| | In nonconcealed spaces | Y* | Y* | Y* | Y* |
| | Supported by cable trays | Y* | Y* | Y* | N |
| | In distributing frames and cross-connect arrays | Y* | Y* | Y* | N |
| | In any raceway recognized in Chapter 3 | Y* | Y* | Y* | Y* |
| | In plenum communications raceways | Y* | Y* | Y* | N |
| | In plenum cable routing assemblies | Y* | Y* | Y* | N |
| | In riser communications raceways | Y* | Y* | Y* | N |
| | In riser cable routing assemblies | Y* | Y* | Y* | N |
| | In general-purpose communications raceways | Y* | Y* | Y* | N |
| | In general-purpose cable routing assemblies | Y* | Y* | Y* | N |

Note: An "N" in the table indicates that the cable type is not permitted to be installed in the application. A "Y*" indicates that the cable is permitted to be installed in the application, subject to the limitations described in 820.113.

Informational Note No. 1: Part V of Article 820 covers installation methods within buildings. This table covers the applications of listed coaxial cables in buildings. The definition of *Point of Entrance* is in 820.2. Coaxial entrance cables that have not emerged from the rigid metal conduit (RMC) or intermediate metal conduit (IMC) are not considered to be in the building.

Informational Note No. 2: For information on the restrictions to the installation of communications cables in fabricated ducts, see 820.113(B).

Informational Note No. 3: Cable routing assemblies are not addressed in NFPA 90A-2012, *Standard for the Installation of Air-Conditioning and Ventilating Systems*.

***TABLE 820.154(b)*** *Coaxial Cable Uses and Permitted Substitutions*

| Cable Type | Permitted Substitutions |
|---|---|
| CATVP | CMP, BLP |
| CATVR | CATVP, CMP, CMR, BMR, BLP, BLR |
| CATV | CATVP, CMP, CATVR, CMR, CMG, CM, BMR, BM, BLP, BLR, BL |
| CATVX | CATVP, CMP, CATVR, CMR, CATV, CMG, CM, BMR, BM, BLP, BLR, BL, BLX |

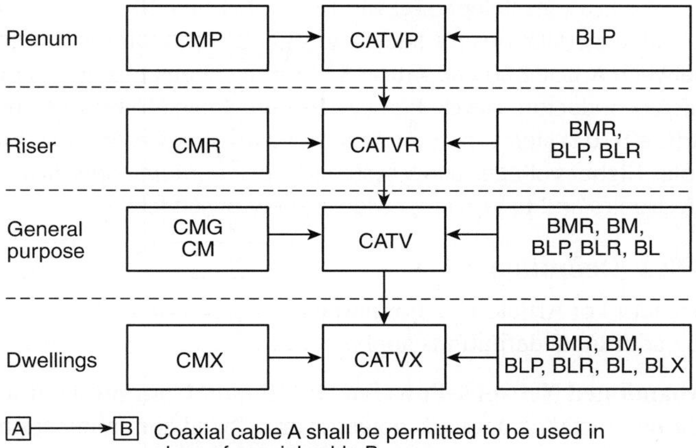


Coaxial cable A shall be permitted to be used in place of coaxial cable B.

Type BL—Network-powered broadband communications low-power cables

Type BM—Network-powered broadband communications medium-power cables

Type CATV—Community antenna television cables

Type CM—Communications cables

***FIGURE 820.154*** *Cable Substitution Hierarchy.*

The application of coaxial cables, communications raceways, and cable routing assemblies are summarized in Table 820.154(a). The installation location will dictate the type of coaxial cable permitted within the raceway or assembly and is subject to the installation requirements of 820.110 and 820.113.

Informational Note: The substitute cables in Table 820.154(b) and Figure 820.154 are only coaxial-type cables.

# VI. Listing Requirements

## 820.179 Coaxial Cables

Cables shall be listed in accordance with 820.179(A) through (D) and marked in accordance with Table 820.179. The cable voltage rating shall not be marked on the cable.

Informational Note: Voltage markings on cables could be misinterpreted to suggest that the cables may be suitable for Class 1, electric light, and power applications.

See the commentary following the informational note to 725.179(A).

*Exception: Voltage markings shall be permitted where the cable has multiple listings and voltage marking is required for one or more of the listings.*

**(A) Type CATVP.** Type CATVP community antenna television plenum coaxial cables shall be listed as being suitable for use in ducts, plenums, and other spaces used for environmental air and shall also be listed as having adequate fire-resistant and low smoke-producing characteristics.

Informational Note: One method of defining a cable that is low smoke-producing cable and fire-resistant cable is that the cable exhibits a maximum peak optical density of 0.50 or less, an average optical density of 0.15 or less, and a maximum flame spread distance of 1.52 m (5 ft) or less when tested in accordance with NFPA 262-2011, *Standard Method of Test for Flame Travel and Smoke of Wires and Cables for Use in Air-Handling Spaces.*

**(B) Type CATVR.** Type CATVR community antenna television riser coaxial cables shall be listed as being suitable for use in a vertical run in a shaft or from floor to floor and shall also be listed as having fire-resistant characteristics capable of preventing the carrying of fire from floor to floor.

Informational Note: One method of defining fire-resistant characteristics capable of preventing the carrying of fire from floor to floor is that the cables pass the requirements of ANSI/UL 1666-2011, *Standard Test for Flame Propagation Height of Electrical and Optical-Fiber Cable Installed Vertically in Shafts.*

See the commentary following the informational note to 725.179(B).

**(C) Type CATV.** Type CATV community antenna television coaxial cables shall be listed as being suitable for general-purpose CATV use, with the exception of risers and plenums, and shall also be listed as being resistant to the spread of fire.

Informational Note: One method of defining *resistant to the spread of fire* is that the cables do not spread fire to the top of the tray in the "UL Flame Exposure, Vertical Tray Flame Test" in ANSI/UL 1685-2010, *Standard for Safety for Vertical-Tray Fire-Propagation and Smoke-Release Test for Electrical and Optical-Fiber Cables.* The smoke measurements in the test method are not applicable.

Another method of defining *resistant to the spread of fire* is for the damage (char length) not to exceed 1.5 m (4 ft 11 in.) when performing the CSA "Vertical Flame Test — Cables in Cable Trays," as described in CSA C22.2 No. 0.3-M-2001, *Test Methods for Electrical Wires and Cables.*

See the commentary following the informational note to 725.179(C).

**(D) Type CATVX.** Type CATVX limited-use community antenna television coaxial cables shall be listed as being suitable for use in dwellings and for use in raceway and shall also be listed as being resistant to flame spread.

Informational Note: One method of determining that cable is resistant to flame spread is by testing the cable to the VW-1

*TABLE 820.179* *Coaxial Cable Markings*

| Cable Marking | Type |
|---|---|
| CATVP | CATV plenum cable |
| CATVR | CATV riser cable |
| CATV | CATV cable |
| CATVX | CATV cable, limited use |

Informational Note: Cable types are listed in descending order of fire resistance rating.

(vertical-wire) flame test in ANSI/UL 1581-2011, *Reference Standard for Electrical Wires, Cables and Flexible Cords.*

See the commentary following the informational note to 725.179(D).

### 820.180 Grounding Devices

Where bonding or grounding is required, devices used to connect a shield, a sheath, or non–current-carrying metallic members of a cable to a bonding conductor, or grounding electrode conductor, shall be listed or be part of listed equipment.

# ARTICLE 830 Network-Powered Broadband Communications Systems

Informational Note: The general term *grounding conductor* as previously used in this article is replaced by either the term *bonding conductor* or the term *grounding electrode conductor* (GEC), where applicable, to more accurately reflect the application and function of the conductor.

See Informational Note Figure 800(a) and Informational Note Figure 800(b) for an illustrative application of a bonding conductor or grounding electrode conductor.

## I. General

### 830.1 Scope

This article covers network-powered broadband communications systems that provide any combination of voice, audio, video, data, and interactive services through a network interface unit.

Informational Note No. 1: A typical basic system configuration includes a cable supplying power and broadband signal to a network interface unit that converts the broadband signal to the component signals. Typical cables are coaxial cable with both broadband signal and power on the center conductor, composite metallic cable with a coaxial member for the broadband signal and a twisted pair for power, and composite optical fiber cable with a pair of conductors for power. Larger systems may also include network components such as amplifiers that require network power.

Informational Note No. 2: See 90.2(B)(4) for installations of broadband communications systems that are not covered.

Network-powered broadband communications circuits provide a wide array of subscriber services, including voice, data (such as Internet access), interactive services, and television signals.

Article 830 contains requirements for wiring both the inside and the outside of buildings. Other articles cover the wiring derived from the network interface unit (NIU) into the premises. For example, Article 725 covers wiring of Class 2 and Class 3 circuits, Article 760 covers wiring of fire alarm systems, Article 770 covers the installation of optical fiber cable, Article 800 covers communications (telephone) wiring, and Article 820 covers coaxial cable installations for television signals. The major difference between Article 820 and Article 830 is the voltage present on the circuit conductors. Article 820 systems are limited to 60 volts, but Article 830 systems are permitted to have ratings as high as 150 volts. Higher voltages allow systems to power more sophisticated electronics and to provide a wider variety of services.

### 830.2 Definitions

See Part I of Article 100. For purposes of this article, the following additional definitions apply.

**Abandoned Network-Powered Broadband Communications Cable.** Installed network-powered broadband communications cable that is not terminated at equipment other than a connector and not identified for future use with a tag.

Informational Note: See Part I of Article 100 for a definition of *Equipment*.

**Block.** A square or portion of a city, town, or village enclosed by streets, including the alleys so enclosed but not any street.

**Exposed (to Accidental Contact).** A circuit in such a position that, in case of failure of supports or insulation, contact with another circuit may result.

Informational Note: See Part I of Article 100 for two other definitions of *Exposed*.

**Fault Protection Device.** An electronic device that is intended for the protection of personnel and functions under fault conditions, such as network-powered broadband communications cable short or open circuit, to limit the current or voltage, or both, for a low-power network-powered broadband communications circuit and provide acceptable protection from electric shock.

**Network Interface Unit (NIU).** A device that converts a broadband signal into component voice, audio, video, data, and interactive services signals and provides isolation between the network power and the premises signal circuits. These devices often contain primary and secondary protectors.

Exhibit 830.1 illustrates an NIU with derived circuits.

**Network-Powered Broadband Communications Circuit.** The circuit extending from the communications utility's serving terminal or tap up to and including the NIU.

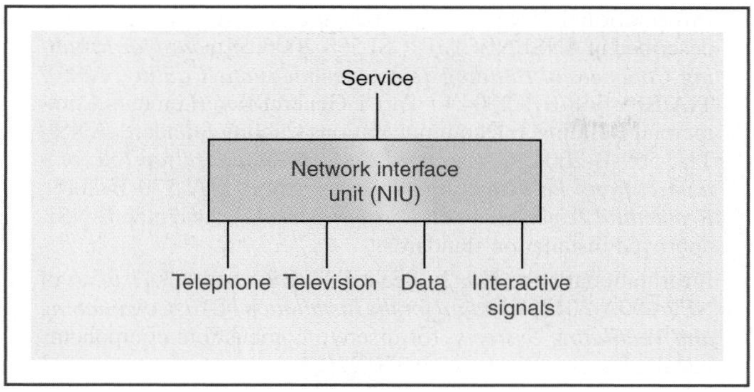


***EXHIBIT 830.1*** *An NIU diagram showing derived circuits.*

Informational Note: A typical single-family network-powered communications circuit consists of a communications drop or communications service cable and an NIU and includes the communications utility's serving terminal or tap where it is not under the exclusive control of the communications utility.

**Point of Entrance.** The point within a building at which the network-powered broadband communications cable emerges from an external wall, from a concrete floor slab, from rigid metal conduit (RMC), or from intermediate metal conduit (IMC).

## 830.3 Other Articles

Circuits and equipment shall comply with 830.3(A) through (G).

**(A) Hazardous (Classified) Locations.** Network-powered broadband communications circuits and equipment installed in a location that is classified in accordance with 500.5 and 505.5 shall comply with the applicable requirements of Chapter 5.

**(B) Wiring in Ducts for Dust, Loose Stock, or Vapor Removal.** The requirements of 300.22(A) shall apply.

**(C) Equipment in Other Space Used for Environmental Air.** The requirements of 300.22(C)(3) shall apply.

**(D) Installation and Use.** The requirements of 110.3(B) shall apply.

**(E) Output Circuits.** As appropriate for the services provided, the output circuits derived from the optical network terminal shall comply with the requirements of the following:

(1) Installations of communications circuits — Part V of Article 800
(2) Installations of community antenna television and radio distribution circuits — Part V of Article 820

*Exception: 830.90(B)(3) shall apply where protection is provided in the output of the NIU.*

(3) Installations of optical fiber cables — Part V of Article 770
(4) Installations of Class 2 and Class 3 circuits — Part III of Article 725
(5) Installations of power-limited fire alarm circuits — Part III of Article 760

**(F) Protection Against Physical Damage.** The requirements of 300.4 shall apply.

**(G) Cable Routing Assemblies.** The definition in Article 100, the applications in Table 800.154(c), and the installation requirements in 800.110 and 800.113 shall apply to Article 830.

## 830.15 Power Limitations

Network-powered broadband communications systems shall be classified as having low- or medium-power sources as specified in 830.15(1) or (2).

(1) Sources shall be classified as defined in Table 830.15.
(2) Direct-current power sources exceeding 150 volts to ground, but no more than 200 volts to ground, with the current to ground limited to 10 mA dc, that meet the current and power limitation for medium-power sources in Table 830.15 shall be classified as medium-power sources.

Informational Note: One way to determine compliance with 830.15(2) is listed information technology equipment intended to supply power via a communications network that complies with the requirements for RFT-V circuits as defined in UL 60950-21-2007, *Standard for Safety for Information Technology Equipment — Safety — Part 21: Remote Power Feeding.*

Only network-powered broadband systems that operate within the voltage, current, and power parameters specified in Table 830.15 or a dc system operating at not more than 200 volts and 10 mA to ground are permitted. These dc systems must meet the current and power limitations for medium-power systems specified in Table 830.15.

***TABLE 830.15*** *Limitations for Network-Powered Broadband Communications Systems*

| Network Power Source | Low | Medium |
|---|---|---|
| Circuit voltage, $V_{max}$ (volts)[1] | 0–100 | 0–150 |
| Power limitation, $VA_{max}$(volt-amperes)[1] | 250 | 250 |
| Current limitation, $I_{max}$ (amperes)[1] | $1000/V_{max}$ | $1000/V_{max}$ |
| Maximum power rating (volt-amperes) | 100 | 100 |
| Maximum voltage rating (volts) | 100 | 150 |
| Maximum overcurrent protection (amperes)[2] | $100/V_{max}$ | NA |

[1]$V_{max}$, $I_{max}$, and $VA_{max}$ are determined with the current-limiting impedance in the circuit (not bypassed) as follows:

$V_{max}$ — Maximum system voltage regardless of load with rated input applied.

$I_{max}$ — Maximum system current under any noncapacitive load, including short circuit, and with overcurrent protection bypassed if used. $I_{max}$ limits apply after 1 minute of operation.

$VA_{max}$ — Maximum volt-ampere output after 1 minute of operation regardless of load and overcurrent protection bypassed if used.

[2]Overcurrent protection is not required where the current-limiting device provides equivalent current limitation and the current-limiting device does not reset until power or the load is removed.

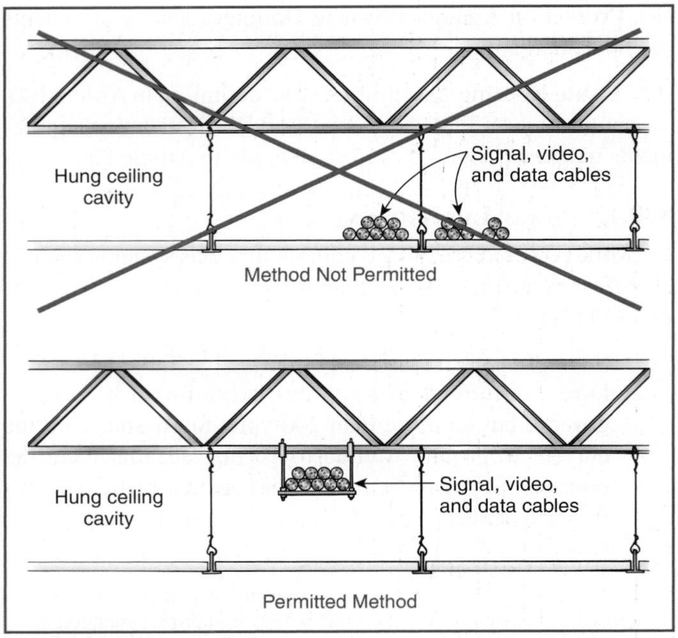


***EXHIBIT 830.2*** *Incorrect installation of conductors and cables (upper diagram), which can prevent access to equipment or cables. Correct method is shown in the lower diagram.*

## 830.21 Access to Electrical Equipment Behind Panels Designed to Allow Access

Access to electrical equipment shall not be denied by an accumulation of network-powered broadband communications cables that prevents removal of panels, including suspended ceiling panels.

An excess accumulation of wires and cables can limit access to equipment by preventing the removal of access panels. (See Exhibit 830.2.)

## 830.24 Mechanical Execution of Work

Network-powered broadband communications circuits and equipment shall be installed in a neat and workmanlike manner. Cables installed exposed on the surface of ceilings and sidewalls shall be supported by the building structure in such a manner that the cable will not be damaged by normal building use. Such cables shall be secured by hardware including straps, staples, cable ties, hangers, or similar fittings designed and installed so as not to damage the cable. The installation shall also conform to 300.11. Nonmetallic cable ties and other nonmetallic cable accessories used to secure and support cables in other spaces used for environmental air (plenums) shall be listed as having low smoke and heat release properties.

Cable must be attached to or supported by the structure by cable ties, straps, clamps, hangers, and so forth. The installation method must not damage the cable. Although this requirement does not contain specific supporting and securing intervals, it does reference 300.11 as a general rule on securing equipment and cables.

Informational Note No. 1: Accepted industry practices are described in ANSI/NECA/BICSI 568-2006, *Standard for Installing Commercial Building Telecommunications Cabling*; ANSI/TIA/EIA-568-B.1-2004 — Part 1 General Requirements Commercial Building Telecommunications Cabling Standard; ANSI/TIA-569-B-2004, *Commercial Building Standard for Telecommunications Pathways and Spaces*; ANSI/TIA-570-B-2009, *Residential Telecommunications Infrastructure*; and other ANSI-approved installation standards.

Informational Note No. 2: See 4.3.11.2.6.5 and 4.3.11.5.5.6 of NFPA 90A-2012, *Standard for the Installation of Air-Conditioning and Ventilating Systems*, for discrete combustible components installed in accordance with 300.22(C).

## 830.25 Abandoned Cables

The accessible portion of abandoned network-powered broadband cables shall be removed. Where cables are identified for future use with a tag, the tag shall be of sufficient durability to withstand the environment involved.

See Article 100 for the definition of *accessible* (as applied to wiring methods). Abandoned cable increases fire loading unnecessarily and, where installed in plenums, can affect airflow. See the definition of *abandoned network-powered broadband communications cable* in 830.2.

## 830.26 Spread of Fire or Products of Combustion

Installations of network-powered broadband cables in hollow spaces, vertical shafts, and ventilation or air-handling ducts shall be made so that the possible spread of fire or products of combustion will not be substantially increased. Openings around penetrations of network-powered broadband cables through fire-resistant-rated walls, partitions, floors, or ceilings shall be firestopped using approved methods to maintain the fire resistance rating.

Informational Note: Directories of electrical construction materials published by qualified testing laboratories contain many listing installation restrictions necessary to maintain the fire-resistive rating of assemblies where penetrations or openings are made. Building codes also contain restrictions on membrane penetrations on opposite sides of a fire resistance–rated wall assembly. An example is the 600-mm (24-in.) minimum horizontal separation that usually applies between boxes installed on opposite sides of the wall. Assistance in complying with 830.26 can be found in building codes, fire resistance directories, and product listings.

# II. Cables Outside and Entering Buildings

## 830.40 Entrance Cables

Network-powered broadband communications cables located outside and entering buildings shall comply with 830.40(A) and (B).

**(A) Medium-Power Circuits.** Medium-power network-powered broadband communications circuits located outside and entering buildings shall be installed using Type BMU, Type BM, or Type BMR network-powered broadband communications medium-power cables.

**(B) Low-Power Circuits.** Low-power network-powered broadband communications circuits located outside and entering buildings shall be installed using Type BLU or Type BLX low-power network-powered broadband communications cables. Cables shown in Table 830.154(b) shall be permitted to substitute.

*Exception: Outdoor community antenna television and radio distribution system coaxial cables installed prior to January 1, 2000, and installed in accordance with Article 820, shall be permitted for low-power-type, network-powered broadband communications circuits.*

## 830.44 Overhead (Aerial) Cables

Overhead (aerial) network-powered broadband communications cables shall comply with 830.44(A) through (G).

Informational Note: For additional information regarding overhead (aerial) wires and cables, see ANSI C2-2007, *National Electric Safety Code*, Part 2, Safety Rules for Overhead Lines.

Network-powered broadband communications systems may contain sufficient energy to pose an electric shock hazard. For that reason, they are subject to requirements similar to those for other higher-powered circuits.

Section 830.44 requires that conductor spans be of sufficient size and strength to maintain clearances and avoid possible contact with light or power conductors. Splices and joints must be made with approved connectors or other means that provide sufficient mechanical strength so that conductors are not weakened, which could cause them to break and come into contact with higher-voltage conductors.

**(A) On Poles and In-Span.** Where network-powered broadband communications cables and electric light or power conductors are supported by the same pole or are run parallel to each other in-span, the conditions described in 830.44(A)(1) through (A)(4) shall be met.

**(1) Relative Location.** Where practicable, the network-powered broadband communications cables shall be located below the electric light or power conductors.

**(2) Attachment to Cross-Arms.** Network-powered broadband communications cables shall not be attached to a cross-arm that carries electric light or power conductors.

**(3) Climbing Space.** The climbing space through network-powered broadband communications wires and cables shall comply with the requirements of 225.14(D).

**(4) Clearance.** Lead-in or overhead (aerial)-drop network-powered broadband communications cables from a pole or other support, including the point of initial attachment to a building or structure, shall be kept away from electric light, power, Class 1, or non–power-limited fire alarm circuit conductors so as to avoid the possibility of accidental contact.

*Exception: Where proximity to electric light, power, Class 1, or non–power-limited fire alarm circuit conductors cannot be avoided, the installation shall provide clearances of not less than 300 mm (12 in.) from electric light, power, Class 1, or non–power-limited fire alarm circuit conductors. The clearance requirement shall apply to all points along the drop, and it shall increase to 1.02 m (40 in.) at the pole.*

**(B) Above Roofs.** Network-powered broadband communications cables shall have a vertical clearance of not less than 2.5 m (8 ft) from all points of roofs above which they pass.

*Exception No. 1: Auxiliary buildings such as garages and the like.*

*Exception No. 2: A reduction in clearance above only the overhanging portion of the roof to not less than 450 mm (18 in.) shall be permitted if (1) not more than 1.2 m (4 ft) of the broadband communications drop cables pass above the roof overhang, and (2) they are terminated at a through-the-roof raceway or support.*

*Exception No. 3: Where the roof has a slope of not less than 100 mm in 300 mm (4 in. in 12 in.), a reduction in clearance to not less than 900 mm (3 ft) shall be permitted.*

**(C) Clearance from Ground.** Overhead (aerial) spans of network-powered broadband communications cables shall conform to not less than the following:

(1) 2.9 m (9.5 ft) — above finished grade, sidewalks, or from any platform or projection from which they might be reached and accessible to pedestrians only
(2) 3.5 m (11.5 ft) — over residential property and driveways, and those commercial areas not subject to truck traffic
(3) 4.7 m (15.5 ft) — over public streets, alleys, roads, parking areas subject to truck traffic, driveways on other than residential property, and other land traversed by vehicles such as cultivated, grazing, forest, and orchard

Informational Note: These clearances have been specifically chosen to correlate with ANSI C2-2007, *National Electrical Safety Code*, Table 232-1, which provides for clearances of wires, conductors, and cables above ground and roadways, rather than using the clearances referenced in 225.18. Because Article 800 and Article 820 have had no required clearances, the communications industry has used the clearances from the NESC for their installed cable plant.

**(D) Over Pools.** Clearance of network-powered broadband communications cable in any direction from the water level, edge of pool, base of diving platform, or anchored raft shall comply with those clearances in 680.8.

**(E) Final Spans.** Final spans of network-powered broadband communications cables without an outer jacket shall be permitted to be attached to the building, but they shall be kept not less than 900 mm (3 ft) from windows that are designed to be opened, doors, porches, balconies, ladders, stairs, fire escapes, or similar locations.

*Exception: Conductors run above the top level of a window shall be permitted to be less than the 900-mm (3-ft) requirement above.*

Overhead (aerial) network-powered broadband communications cables shall not be installed beneath openings through which materials might be moved, such as openings in farm and commercial buildings, and shall not be installed where they obstruct entrance to these building openings.

**(F) Between Buildings.** Network-powered broadband communications cables extending between buildings or structures, and also the supports or attachment fixtures, shall be identified as suitable for outdoor aerial applications and shall have sufficient strength to withstand the loads to which they may be subjected.

Wind and ice loads must be considered because they can damage cables and attachment points.

*Exception: Where a network-powered broadband communications cable does not have sufficient strength to be self-supporting, it shall be attached to a supporting messenger cable that, together with the attachment fixtures or supports, shall be acceptable for the purpose and shall have sufficient strength to withstand the loads to which they may be subjected.*

**(G) On Buildings.** Where attached to buildings, network-powered broadband communications cables shall be securely fastened in such a manner that they are separated from other conductors in accordance with 830.44(I)(1) through (I)(4).

**(1) Electric Light or Power.** The network-powered broadband communications cable shall have a separation of at least 100 mm (4 in.) from electric light, power, Class 1, or non–power-limited fire alarm circuit conductors not in raceway or cable, or be permanently separated from conductors of the other system by a continuous and firmly fixed nonconductor in addition to the insulation on the wires.

**(2) Other Communications Systems.** Network-powered broadband communications cables shall be installed so that there will be no unnecessary interference in the maintenance of the separate systems. In no case shall the conductors, cables, messenger strand, or equipment of one system cause abrasion to the conductors, cables, messenger strand, or equipment of any other system.

**(3) Lightning Conductors.** Where practicable, a separation of at least 1.8 m (6 ft) shall be maintained between any network-powered broadband communications cable and lightning conductors.

**(4) Protection from Damage.** Network-powered broadband communications cables attached to buildings or structures and located within 2.5 m (8 ft) of finished grade shall be protected by enclosures, raceways, or other approved means.

*Exception: A low-power network-powered broadband communications circuit that is equipped with a listed fault protection device, appropriate to the network-powered broadband communications cable used, and located on the network side of the network-powered broadband communications cable being protected.*

## 830.47 Underground Network-Powered Broadband Communications Cables Entering Buildings

Underground network-powered broadband communications cables entering buildings shall comply with 830.47(A) through (D).

**(A) Underground Systems with Electric Light and Power Conductors.** Underground network-powered broadband communications cables in a duct, pedestal, handhole enclosure, or manhole that contains electric light, power conductors, non–power-limited fire alarm circuit conductors, or Class 1 circuits shall be in a section permanently separated from such conductors by means of a suitable barrier.

**(B) Direct-Buried Cables and Raceways.** Direct-buried network-powered broadband communications cables shall be separated by at least 300 mm (12 in.) from conductors of any light, power, non–power-limited fire alarm circuit conductors or Class 1 circuit.

*Exception No. 1: Where electric service conductors or network-powered broadband communications cables are installed in raceways or have metal cable armor.*

*Exception No. 2: Where electric light or power branch-circuit or feeder conductors, non–power-limited fire alarm circuit conductors, or Class 1 circuit conductors are installed in a raceway or in metal-sheathed, metal-clad, or Type UF or Type USE cables; or the network-powered broadband communications cables have metal cable armor or are installed in a raceway.*

**(C) Mechanical Protection.** Direct-buried cable, conduit, or other raceways shall be installed to meet the minimum cover requirements of Table 830.47(C). In addition, direct-buried cables emerging from the ground shall be protected by enclosures, raceways, or other approved means extending from the minimum cover distance required by Table 830.47(C) below grade to a point at least 2.5 m (8 ft) above finished grade. In no case shall the protection be required to exceed 450 mm (18 in.) below finished grade. Types BMU and BLU direct-buried cables emerging from the ground shall be installed in rigid metal conduit (RMC), intermediate metal conduit (IMC), rigid nonmetallic conduit, or other approved means extending from the minimum cover distance required by Table 830.47(C) below grade to the point of entrance.

*Exception: A low-power network-powered broadband communications circuit that is equipped with a listed fault protection device, appropriate to the network-powered broadband communications cable used, and located on the network side of the network-powered broadband communication cable being protected.*

**(D) Pools.** Cables located under the pool or within the area extending 1.5 m (5 ft) horizontally from the inside wall of the pool shall meet those clearances and requirements specified in 680.10.

***TABLE 830.47(C)*** *Network-Powered Broadband Communications Systems Minimum Cover Requirements (Cover is the shortest distance measured between a point on the top surface of any direct-buried cable, conduit, or other raceway and the top surface of finished grade, concrete, or similar cover.)*

| Location of Wiring Method or Circuit | Direct Burial Cables | | Rigid Metal Conduit (RMC) or Intermediate Metal Conduit (IMC) | | Nonmetallic Raceways Listed for Direct Burial; Without Concrete Encasement or Other Approved Raceways | |
|---|---|---|---|---|---|---|
| | mm | in. | mm | in. | mm | in. |
| All locations not specified below | 450 | 18 | 150 | 6 | 300 | 12 |
| In trench below 50-mm (2-in.) thick concrete or equivalent | 300 | 12 | 150 | 6 | 150 | 6 |
| Under a building (in raceway only) | 0 | 0 | 0 | 0 | 0 | 0 |
| Under minimum of 100-mm (4-in.) thick concrete exterior slab with no vehicular traffic and the slab extending not less than 150 mm (6 in.) beyond the underground installation | 300 | 12 | 100 | 4 | 100 | 4 |
| One- and two-family dwelling driveways and outdoor parking areas and used only for dwelling-related purposes | 300 | 12 | 300 | 12 | 300 | 12 |

Notes:

1. Raceways approved for burial only where concrete encased shall require a concrete envelope not less than 50 mm (2 in.) thick.
2. Lesser depths shall be permitted where cables rise for terminations or splices or where access is otherwise required.
3. Where solid rock is encountered, all wiring shall be installed in metal or nonmetallic raceway permitted for direct burial. The raceways shall be covered by a minimum of 50 mm (2 in.) of concrete extending down to rock.
4. Low-power network-powered broadband communications circuits using directly buried community antenna television and radio distribution system coaxial cables that were installed outside and entering buildings prior to January 1, 2000, in accordance with Article 820 shall be permitted where buried to a minimum depth of 300 mm (12 in.).

### 830.49 Metallic Entrance Conduit Grounding

Rigid metal conduit (RMC) or intermediate metal conduit (IMC) containing network-powered broadband communications entrance cable shall be connected by a bonding conductor or grounding electrode conductor to a grounding electrode in accordance with 830.100(B).

## III. Protection

### 830.90 Primary Electrical Protection

**(A) Application.** Primary electrical protection shall be provided on all network-powered broadband communications conductors that are neither grounded nor interrupted and are run partly or entirely in aerial cable not confined within a block. Also, primary electrical protection shall be provided on all aerial or underground network-powered broadband communications conductors that are neither grounded nor interrupted and are located within the block containing the building served so as to be exposed to lightning or accidental contact with electric light or power conductors operating at over 300 volts to ground.

*Exception: Where electrical protection is provided on the derived circuit(s) (output side of the NIU) in accordance with 830.90(B)(3).*

Informational Note No. 1: On network-powered broadband communications conductors not exposed to lightning or accidental contact with power conductors, providing primary electrical protection in accordance with this article helps protect against other hazards, such as ground potential rise caused by power fault currents, and above-normal voltages induced by fault currents on power circuits in proximity to the network-powered broadband communications conductors.

Informational Note No. 2: Network-powered broadband communications circuits are considered to have a lightning exposure unless one or more of the following conditions exist:

(1) Circuits in large metropolitan areas where buildings are close together and sufficiently high to intercept lightning.
(2) Areas having an average of five or fewer thunderstorm days each year and earth resistivity of less than 100 ohm-meters. Such areas are found along the Pacific coast.

Utility companies may provide primary protectors if conductors are exposed to lightning. Cable is not considered to be exposed to lightning unless one or both of the conditions in Informational

Note No. 2 exist. A primary protector is required at each end of a communications circuit where lightning exposure exists, unless protection is provided on the output side of the NIU.

**(1) Fuseless Primary Protectors.** Fuseless-type primary protectors shall be permitted where power fault currents on all protected conductors in the cable are safely limited to a value no greater than the current-carrying capacity of the primary protector and of the primary protector bonding conductor or grounding electrode conductor.

**(2) Fused Primary Protectors.** Where the requirements listed in 830.90(A)(1) are not met, fused-type primary protectors shall be used. Fused-type primary protectors shall consist of an arrester connected between each conductor to be protected and ground, a fuse in series with each conductor to be protected, and an appropriate mounting arrangement. Fused primary protector terminals shall be marked to indicate line, instrument, and ground, as applicable.

**(B) Location.** The location of the primary protector, where required, shall comply with (B)(1), (B)(2), or (B)(3):

(1) A listed primary protector shall be applied on each network-powered broadband communications cable external to and on the network side of the network interface unit.
(2) The primary protector function shall be an integral part of and contained in the network interface unit. The network interface unit shall be listed as being suitable for application with network-powered broadband communications systems and shall have an external marking indicating that it contains primary electrical protection.
(3) The primary protector(s) shall be provided on the derived circuit(s) (output side of the NIU), and the combination of the NIU and the protector(s) shall be listed as being suitable for application with network-powered broadband communications systems.

A primary protector, whether provided integrally or external to the network interface unit, shall be located as close as practicable to the point of entrance.

For purposes of this section, a network interface unit and any externally provided primary protectors located at mobile home service equipment located in sight from and not more than 9.0 m (30 ft) from the exterior wall of the mobile home it serves, or at a mobile home disconnecting means grounded in accordance with 250.32 and located in sight from and not more than 9.0 m (30 ft) from the exterior wall of the mobile home it serves, shall be considered to meet the requirements of this section.

Informational Note: Selecting a network interface unit and primary protector location to achieve the shortest practicable primary protector bonding conductor or grounding electrode conductor helps limit potential differences between communications circuits and other metallic systems.

**(C) Hazardous (Classified) Locations.** The primary protector or equipment providing the primary protection function shall not be located in any hazardous (classified) location as defined in 500.5 and 505.5 or in the vicinity of easily ignitible material.

*Exception: As permitted in 501.150, 502.150, and 503.150.*

### 830.93 Grounding or Interruption of Metallic Members of Network-Powered Broadband Communications Cables

Network-powered communications cables entering buildings or attaching to buildings shall comply with 830.93(A) or (B).

For purposes of this section, grounding located at mobile home service equipment located within 9.0 m (30 ft) of the exterior wall of the mobile home it serves, or at a mobile home disconnecting means grounded in accordance with 250.32 and located within 9.0 m (30 ft) of the exterior wall of the mobile home it serves, shall be considered to meet the requirements of this section.

Informational Note: Selecting a grounding location to achieve the shortest practicable bonding conductor or grounding electrode conductor helps limit potential differences between the network-powered broadband communications circuits and other metallic systems.

**(A) Entering Buildings.** In installations where the network-powered communications cable enters the building, the shield shall be grounded in accordance with 830.100 and metallic members of the cable not used for communications or powering shall be grounded in accordance with 830.100, or interrupted by an insulating joint or equivalent device. The grounding or interruption shall be as close as practicable to the point of entrance.

Informational Note: See 830.2 for the definition of *Point of Entrance*.

Proper bonding of the network-powered broadband communications system cable sheath to the electrical power grounding electrode is needed to prevent potential fire and shock hazards.

**(B) Terminating Outside of the Building.** In installations where the network-powered communications cable is terminated outside of the building, the shield shall be grounded in accordance with 830.100, and metallic members of the cable not used for communications or powering shall be grounded in accordance with 830.100 or interrupted by an insulating joint or equivalent device. The grounding or interruption shall be as close as practicable to the point of attachment of the NIU.

## IV. Grounding Methods

### 830.100 Cable, Network Interface Unit, and Primary Protector Bonding and Grounding

Network interface units containing protectors, NIUs with metallic enclosures, primary protectors, and the metallic members of the network-powered broadband communications cable that are intended to be bonded or grounded shall be connected as specified in 830.100(A) through (D).

**(A) Bonding Conductor or Grounding Electrode Conductor.**

**(1) Insulation.** The bonding conductor or grounding electrode conductor shall be listed and shall be permitted to be insulated, covered, or bare.

**(2) Material.** The bonding conductor or grounding electrode conductor shall be copper or other corrosion-resistant conductive material, stranded or solid.

**(3) Size.** The bonding conductor or grounding electrode conductor shall not be smaller than 14 AWG and shall have a current-carrying capacity not less than that of the grounded metallic member(s) and protected conductor(s) of the network-powered broadband communications cable. The bonding conductor or grounding electrode conductor shall not be required to exceed 6 AWG.

**(4) Length.** The bonding conductor or grounding electrode conductor shall be as short as practicable. In one- and two-family dwellings, the bonding conductor or grounding electrode conductor shall be as short as practicable, not to exceed 6.0 m (20 ft) in length.

Informational Note: Similar bonding conductor or grounding electrode conductor length limitations applied at apartment buildings and commercial buildings help to reduce voltages that may be developed between the building's power and communications systems during lightning events.

*Exception: In one- and two-family dwellings where it is not practicable to achieve an overall maximum bonding conductor or grounding electrode conductor length of 6.0 m (20 ft), a separate communications ground rod meeting the minimum dimensional criteria of 830.100(B)(3)(2) shall be driven, and the grounding electrode conductor shall be connected to the communications ground rod in accordance with 830.100(C). The communications ground rod shall be bonded to the power grounding electrode system in accordance with 830.100(D).*

The limitation on length will result in a lower impedance, which will limit the potential difference between network-powered broadband communications systems and other systems during a lightning strike. Large potential differences between grounding conductors can result in increased damage if a lightning strike were to occur.

The informational note provides guidance for the treatment of the cable and primary protector grounding conductor length at apartment and commercial buildings that is consistent with the 20-foot rule for one- and two-family dwellings. However, a specific length is not specified in the *Code*, because such a limitation may not be practical in some installations.

**(5) Run in Straight Line.** The bonding conductor or grounding electrode conductor shall be run in as straight a line as practicable.

**(6) Physical Protection.** Bonding conductors and grounding electrode conductors shall be protected where exposed to physical damage. Where the bonding conductor or grounding electrode conductor is installed in a metal raceway, both ends of the raceway shall be bonded to the contained conductor or to the same terminal or electrode to which the bonding conductor or grounding electrode conductor is connected.

**(B) Electrode.** The bonding conductor or grounding electrode conductor shall be connected in accordance with 830.100(B)(1), (B)(2), or (B)(3).

**(1) In Buildings or Structures with an Intersystem Bonding Termination.** If the building or structure served has an intersystem bonding termination as required by 250.94, the bonding conductor shall be connected to the intersystem bonding termination.

Informational Note: See Part I of Article 100 for the definition of *Intersystem Bonding Termination.*

**(2) In Buildings or Structures with Grounding Means.** If the building or structure served has no intersystem bonding termination, the bonding conductor or grounding electrode conductor shall be connected to the nearest accessible location on one of the following:

(1) The building or structure grounding electrode system as covered in 250.50
(2) The grounded interior metal water piping system, within 1.5 m (5 ft) from its point of entrance to the building, as covered in 250.52

For more information on the use of a metal water piping system as a grounding electrode, see the commentary following 250.52(A)(1).

(3) The power service accessible means external to enclosures as covered in 250.94
(4) The nonflexible metallic power service raceway
(5) The service equipment enclosure
(6) The grounding electrode conductor or the grounding electrode conductor metal enclosure of the power service, or
(7) The grounding electrode conductor or the grounding electrode of a building or structure disconnecting means that is connected to an electrode as covered in 250.32

A bonding device intended to provide a termination point for the bonding conductor (intersystem bonding) shall not interfere with the opening of an equipment enclosure. A bonding device shall be mounted on nonremovable parts. A bonding device shall not be mounted on a door or cover even if the door or cover is nonremovable.

For purposes of this section, the mobile home service equipment or the mobile home disconnecting means, as described in 830.93, shall be considered accessible.

**(3) In Buildings or Structures Without an Intersystem Bonding Termination or Grounding Means.** If the building or structure served has no intersystem bonding termination or grounding means, as described in 830.100(B)(2), the grounding electrode conductor shall be connected to either of the following:

(1) To any one of the individual grounding electrodes described in 250.52(A)(1), (A)(2), (A)(3), or (A)(4).
(2) If the building or structure served has no intersystem bonding termination or has no grounding means, as described in 830.100(B)(2) or (B)(3)(1), to any one of the individual grounding electrodes described in 250.52(A)(7) and (A)(8), or to a ground rod or pipe not less than 1.5 m (5 ft) in length and 12.7 mm (½ in.) in diameter, driven, where practicable, into permanently damp earth and separated from lightning conductors as covered in 800.53 and at least 1.8 m (6 ft) from electrodes of other systems. Steam or hot water pipes or lightning-rod conductors shall not be employed as grounding electrodes for protectors, NIUs with integral protection, grounded metallic members, NIUs with metallic enclosures, and other equipment.

**(C) Electrode Connection.** Connections to grounding electrodes shall comply with 250.70.

**(D) Bonding of Electrodes.** A bonding jumper not smaller than 6 AWG copper or equivalent shall be connected between the network-powered broadband communications system grounding electrode and the power grounding electrode system at the building or structure served where separate electrodes are used.

*Exception: At mobile homes as covered in 830.106.*

Informational Note No. 1: See 250.60 for use of air terminals (lightning rods).

Informational Note No. 2: Bonding together of all separate electrodes limits potential differences between them and between their associated wiring systems.

A common error made in grounding network-powered broadband communications systems is connecting the cable sheath to a separate rod-type grounding electrode driven by the communications utility installer at a convenient location near the point of cable entry to the building, instead of bonding it to the electrical service grounding electrode system.

Section 250.94 requires that a bonding means with not less than three termination points that is accessible and external to the service equipment be provided for making the bonding and grounding connection for other systems. For existing installations where an intersystem bonding termination is not available, alternate bonding means are described in 820.100(B)(2). A separate grounding electrode is permitted by 800.100(B)(3) only if the building or structure has neither an intersystem bonding termination nor a grounding means, which is rare. The earth cannot be used as the bonding conductor, because it does not have the required low-impedance path. (See 250.54.)

Both network-powered broadband communications systems and power systems are subject to current surges as a result of, for example, induced voltages from lightning in the vicinity of the outside distribution systems. Surges also result from switching operations on power systems. If the grounded conductors and parts of the two systems are not bonded by a low-impedance path, line surges can raise the potential difference between the two systems to many thousands of volts. This can result in arcing between the two systems wherever the network-powered broadband communications system cable jacket contacts a grounded part, such as a metal water pipe or metal structural member, inside the building.

If a person is the interface between two systems not bonded in accordance with the *Code*, the high-voltage surge could result in electric shock. More common, however, is burnout of a television tuner, a part that is almost always an interface between the two systems. The tuner is connected to the power system ground through the grounded neutral of the power supply, even if the television set itself is not provided with an EGC.

Also see the commentary following 250.92(B) and 820.100(E).

## 830.106 Grounding and Bonding at Mobile Homes

**(A) Grounding.** Grounding shall comply with 830.106(A)(1) or (A)(2).

(1) Where there is no mobile home service equipment located within 9.0 m (30 ft) of the exterior wall of the mobile home it serves, the network-powered broadband communications cable shield, network-powered broadband communications cable metallic members not used for communications or powering, network interface unit, and primary protector grounding terminal shall be connected to a grounding electrode conductor or grounding electrode in accordance with 830.100(B)(3).
(2) Where there is no mobile home disconnecting means grounded in accordance with 250.32 and located within 9.0 m (30 ft) of the exterior wall of the mobile home it serves, the network-powered broadband communications cable shield, network-powered broadband communications cable metallic members not used for communications or powering, network interface unit, and primary protector grounding terminal shall be connected to a grounding electrode in accordance with 830.100(B)(3).

**(B) Bonding.** The network-powered broadband communications cable grounding terminal, network interface unit grounding terminal, if present, and primary protector grounding terminal shall be bonded together with a copper bonding conductor not smaller than 12 AWG. The network-powered broadband communications cable grounding terminal, network interface unit grounding terminal, primary protector grounding terminal, or the grounding electrode shall be bonded to the metal frame or available grounding terminal of the mobile home with a copper bonding conductor not smaller than 12 AWG under any of the following conditions:

(1) Where there is no mobile home service equipment or disconnecting means as in 830.106(A)
(2) Where the mobile home is supplied by cord and plug

## V. Installation Methods Within Buildings

### 830.110 Raceways and Cable Routing Assemblies for Network-Powered Broadband Communications Cables

**(A) Types of Raceways.** Low-power network-powered broadband communications cables shall be permitted to be installed in any raceway that complies with either 830.110(A)(1) or (A)(2) and in cable routing assemblies installed in compliance with 830.110(C). Medium-power network-powered broadband communications cables shall be permitted to be installed in any raceway that complies with 830.110(A)(1).

**(1) Raceways Recognized in Chapter 3.** Low- and medium-power network-powered broadband communications cables shall be permitted to be installed in any raceway included in Chapter 3. The raceways shall be installed in accordance with the requirements of Chapter 3.

**(2) Communications Raceways.** Low-power network-powered broadband communications cables shall be permitted to be installed in listed plenum communications raceways, listed riser communications raceways, and listed general-purpose communications raceways, selected in accordance with the provisions of 800.113 and 830.113 and installed in accordance with 362.24 through 362.56, where the requirements applicable to electrical nonmetallic tubing apply.

**(B) Raceway Fill for Network-Powered Broadband Communications Cables.** Raceway fill for network-powered broadband communications cables shall comply with either (B)(1) or (B)(2).

**(1) Low-Power Network-Powered Broadband Communications Cables.** The raceway fill requirements of Chapters 3 and 9 shall not apply to low-power network-powered broadband communications cables.

**(2) Medium-Power Network-Powered Broadband Communications Cables.** Where medium-power network-powered broadband communications cables are installed in a raceway, the raceway fill requirements of Chapters 3 and 9 shall apply.

**(C) Cable Routing Assemblies.** Network-powered broadband communications cables shall be permitted to be installed in plenum cable routing assemblies, riser cable routing assemblies, and general-purpose cable routing assemblies selected in accordance with the provisions of 800.113 and installed in accordance with 830.110(C)(1) and (2).

**(1) Horizontal Support.** Cable routing assemblies shall be supported where run horizontally at intervals not to exceed 900 mm (3 ft), and at each end or joint, unless listed for other support intervals. In no case shall the distance between supports exceed 3 m (10 ft).

**(2) Vertical Support.** Vertical runs of cable routing assemblies shall be supported at intervals not exceeding 1.2 m (4 ft), unless listed for other support intervals, and shall not have more than one joint between supports.

### 830.113 Installation of Network-Powered Broadband Communications Cables

Installation of network-powered broadband communications cables shall comply with 830.113(A) through (H).

**(A) Listing.** Network-powered broadband communications cables installed in buildings shall be listed.

**(B) Fabricated Ducts Used for Environmental Air.** The following cables shall be permitted in ducts as described in 300.22(B) if they are directly associated with the air distribution system:

(1) Up to 1.22 m (4 ft) of Type BLP cable
(2) Types BLP, BMR, BLR, BM, BL, and BLX cables installed in raceways that are installed in compliance with 300.22(B)

Informational Note: For information on fire protection of wiring installed in fabricated ducts, see 4.3.4.1 and 4.3.11.3.3 in NFPA 90A-2012, *Standard for the Installation of Air-Conditioning and Ventilating Systems.*

**(C) Other Spaces Used For Environmental Air (Plenums).** The following cables shall be permitted in other spaces used for environmental air as described in 300.22(C):

(1) Type BLP cable
(2) Type BLP cable installed in plenum communications raceways
(3) Type BLP cable supported by open metallic cable trays or cable tray systems
(4) Types BLP, BMR, BLR, BM, BL, and BLX cables installed in raceways that are installed in compliance with 300.22(C)
(5) Types BLP, BMR, BLR, BM, BL, and BLX cables supported by solid bottom metal cable trays with solid metal covers in other spaces used for environmental air (plenums) as described in 300.22(C)
(6) Types BLP, BMR, BLR, BM, BL, and BLX cables installed in plenum communications raceways, riser communications raceways, or general-purpose communications raceways supported by solid bottom metal cable trays with

solid metal covers in other spaces used for environmental air (plenums) as described in 300.22(C)

Informational Note: For information on fire protection of wiring installed in other spaces used for environmental air, see 4.3.11.2, 4.3.11.4, and 4.3.11.5 of NFPA 90A-2012, *Standard for the Installation of Air-Conditioning and Ventilating Systems.*

**(D) Risers — Cables in Vertical Runs.** The following cables shall be permitted in vertical runs penetrating one or more floors and in vertical runs in a shaft:

(1) Types BLP, BMR, and BLR cables
(2) Types BLP and BLR cables installed in:

a. Plenum communications raceways
b. Plenum cable routing assemblies
c. Riser communications raceways
d. Riser cable routing assemblies

Informational Note: See 830.26 for firestop requirements for floor penetrations.

**(E) Risers — Cables in Metal Raceways.** The following cables shall be permitted in a metal raceway in a riser with firestops at each floor:

(1) Types BLP, BMR, BLR, BM, BL, and BLX cables
(2) Types BLP, BLR, and BL cables installed in:

a. Plenum communications raceways
b. Riser communications raceways
c. General-purpose communications raceways

Informational Note: See 830.26 for firestop requirements for floor penetrations.

**(F) Risers — Cables in Fireproof Shafts.** The following cables shall be permitted to be installed in fireproof riser shafts with firestops at each floor:

(1) Types BLP, BMR, BLR, BM, BL, and BLX cables
(2) Types BLP, BLR, and BL cables installed in:

a. Plenum communications raceways
b. Plenum cable routing assemblies
c. Riser communications raceways
d. Riser cable routing assemblies
e. General-purpose communications raceways
f. General-purpose cable routing assemblies

Informational Note: See 830.26 for firestop requirements for floor penetrations.

**(G) Risers — One- and Two-Family Dwellings.** The following cables shall be permitted in one- and two-family dwellings:

(1) Types BLP, BMR, BLR, BM, BL, and BLX cables less than 10 mm (0.375 in.) in diameter
(2) Types BLP, BLR, and BL cables installed in:

a. Plenum communications raceways
b. Plenum cable routing assemblies
c. Riser communications raceways
d. Riser cable routing assemblies
e. General-purpose communications raceways
f. General-purpose cable routing assemblies

Informational Note: See 830.26 for firestop requirements for floor penetrations.

**(H) Other Building Locations.** The following cables and raceways shall be permitted to be installed in building locations other than those covered in 830.113(B) through (G):

(1) Types BLP, BMR, BLR, BM, and BL cables
(2) Types BLP, BMR, BLR, BM, BL, and BLX cables installed in a raceway
(3) Types BLP, BLR, and BL cables installed in:

a. Plenum communications raceways
b. Plenum cable routing assemblies
c. Riser communications raceways
d. Riser cable routing assemblies
e. General-purpose communications raceways
f. General-purpose cable routing assemblies

(4) Type BLX cables less than 10 mm (0.375 in.) in diameter in one- and two-family dwellings
(5) Types BMU and BLU cables entering the building from outside and run in rigid metal conduit (RMC) or intermediate metal conduit (IMC) where the conduit is connected by a bonding conductor or grounding electrode conductor in accordance with 830.100(B)

Informational Note: This provision limits the length of Type BLX cable to 15 m (50 ft), while 830.90(B) requires that the primary protector, or NIU with integral protection, be located as close as practicable to the point at which the cable enters the building. Therefore, in installations requiring a primary protector, or NIU with integral protection, Type BLX cable may not be permitted to extend 15 m (50 ft) into the building if it is practicable to place the primary protector closer than 15 m (50 ft) to the entrance point.

(6) A maximum length of 15 m (50 ft), within the building, of Type BLX cable entering the building from outside and terminating at an NIU or a primary protection location

## 830.133 Installation of Network-Powered Broadband Communications Cables and Equipment

Cable and equipment installations within buildings shall comply with 830.133(A) through (C), as applicable.

**(A) Separation of Conductors.**

**(1) In Raceways, Cable Trays, Boxes, Enclosures, and Cable Routing Assemblies.**

(a) *Low- and Medium-Power Network-Powered Broadband Communications Circuit Cables.* Low- and medium-power network-powered broadband communications cables shall be permitted in the same raceway, cable tray, box, enclosure, or cable routing assembly.

(b) *Low-Power Network-Powered Broadband Communications Circuit Cables with Optical Fiber Cables and Other Communications Cables.* Low-power network-powered broadband communications cables shall be permitted in the same raceway, cable tray, box, enclosure, or cable routing assembly with jacketed cables of any of the following circuits:

(1) Communications circuits in compliance with Parts I and V of Article 800
(2) Nonconductive and conductive optical fiber cables in compliance with Parts I and V of Article 770
(3) Community antenna television and radio distribution systems in compliance with Parts I and V of Article 820

(c) *Low-Power Network-Powered Broadband Communications Circuit Cables with Other Circuits.* Low-power network-powered broadband communications cables shall be permitted in the same raceway, cable tray, box, enclosure, or cable routing assembly with jacketed cables of any of the following circuits:

(1) Class 2 and Class 3 remote-control, signaling, and power-limited circuits in compliance with Parts I and III of Article 725
(2) Power-limited fire alarm systems in compliance with Parts I and III of Article 760

(d) *Medium-Power Network-Powered Broadband Communications Circuit Cables with Optical Fiber Cables and Other Communications Cables.* Medium-power network-powered broadband communications cables shall not be permitted in the same raceway, cable tray, box, enclosure, or cable routing assembly with conductors of any of the following circuits:

(1) Communications circuits in compliance with Parts I and V of Article 800
(2) Conductive optical fiber cables in compliance with Parts I and V of Article 770
(3) Community antenna television and radio distribution systems in compliance with Parts I and V of Article 820

(e) *Medium-Power Network-Powered Broadband Communications Circuit Cables with Other Circuits.* Medium-power network-powered broadband communications cables shall not be permitted in the same raceway, cable tray, box, enclosure, or cable routing assembly with conductors of any of the following circuits:

(1) Class 2 and Class 3 remote-control, signaling, and power-limited circuits in compliance with Parts I and III of Article 725
(2) Power-limited fire alarm systems in compliance with Parts I and III of Article 760

(f) *Electric Light, Power, Class 1, Non–Powered Broadband Communications Circuit Cables.* Network-powered broadband communications cable shall not be placed in any raceway, cable tray, compartment, outlet box, junction box, or similar fittings with conductors of electric light, power, Class 1, or non–power-limited fire alarm circuit cables.

*Exception No. 1: Where all of the conductors of electric light, power, Class 1, non–power-limited fire alarm circuits are separated from all of the network-powered broadband communications cables by a permanent barrier or listed divider.*

This exception recognizes the use of a listed field-installed divider to provide physical separation between the communication circuits and the power circuits.

*Exception No. 2: Power circuit conductors in outlet boxes, junction boxes, or similar fittings or compartments where such conductors are introduced solely for power supply to the network-powered broadband communications system distribution equipment. The power circuit conductors shall be routed within the enclosure to maintain a minimum 6-mm (0.25-in.) separation from network-powered broadband communications cables.*

**(2) Other Applications.** Network-powered broadband communications cable shall be separated at least 50 mm (2 in.) from conductors of any electric light, power, Class 1, and non–power-limited fire alarm circuits.

*Exception No. 1: Where either (1) all of the conductors of electric light, power, Class 1, and non–power-limited fire alarm circuits are in a raceway, or in metal-sheathed, metal-clad, nonmetallic-sheathed, Type AC, or Type UF cables, or (2) all of the network-powered broadband communications cables are encased in raceway.*

*Exception No. 2: Where the network-powered broadband communications cables are permanently separated from the conductors of electric light, power, Class 1, and non-power-limited fire alarm circuits by a continuous and firmly fixed nonconductor, such as porcelain tubes or flexible tubing, in addition to the insulation on the wire.*

**(B) Support of Network-Powered Broadband Communications Cables.** Raceways shall be used for their intended purpose. Network-powered broadband communications cables shall not be strapped, taped, or attached by any means to the exterior of any conduit or raceway as a means of support.

## 830.154 Applications of Network-Powered Broadband Communications System Cables

Permitted and nonpermitted applications of listed network-powered broadband communications system cables shall be as indicated in Table 830.154(a). The permitted applications shall be subject to the installation requirements of 830.40, 830.110, and 830.113. The substitutions for network-powered broadband system cables listed in Table 830.154(b) shall be permitted.

The applications for the cable, communications raceways, and cable routing assemblies are summarized in Table 830.154(a). The installation location dictates the type of cable permitted within

***TABLE 830.154(a)*** *Applications of Network-Powered Broadband Cables in Buildings*

| Applications | | Cable Types | | | | | | |
|---|---|---|---|---|---|---|---|---|
| | | BLP | BLR | BL | BMR | BM | BLX | BMU, BLU |
| In specifically fabricated ducts as described in 300.22(B) | In fabricated ducts as described in 300.22(B) | Y* | N | N | N | N | N | N |
| | In metal raceway that complies with 300.22(B) | Y* | Y* | Y* | Y* | Y* | Y* | N |
| In other spaces used for environmental air as described in 300.22(C) | In other spaces used for environmental air as described in 300.22(C) | Y* | N | N | N | N | N | N |
| | In metal raceway that complies with 300.22(C) | Y* | Y* | Y* | Y* | Y* | Y* | N |
| | In plenum communications raceways | Y* | N | N | N | N | N | N |
| | In plenum cable routing assemblies | NOT PERMITTED | | | | | | |
| | Supported by open metal cable trays | Y* | N | N | N | N | N | N |
| | Supported by solid bottom metal cable trays with solid metal covers | Y* | Y* | Y* | Y* | Y* | Y* | N |
| In risers | In vertical runs | Y* | Y* | N | Y* | N | N | N |
| | In metal raceways | Y* | Y* | Y* | Y* | Y* | Y* | N |
| | In fireproof shafts | Y* | Y* | Y* | Y* | Y* | Y* | N |
| | In plenum communications raceways | Y* | Y* | N | N | N | N | N |
| | In plenum cable routing assemblies | Y* | Y* | N | N | N | N | N |
| | In riser communications raceways | Y* | Y* | N | N | N | N | N |
| | In riser cable routing assemblies | Y* | Y* | N | N | N | N | N |
| | In one- and two-family dwellings | Y* | Y* | Y* | Y* | Y* | Y* | N |
| Within buildings in other than air-handling spaces and risers | General | Y* | Y* | Y* | Y* | Y* | Y* | N |
| | In one- and two-family dwellings | Y* | Y* | Y* | Y* | Y* | Y* | N |
| | Supported by cable trays | Y* | Y* | Y* | Y* | Y* | N | N |
| | In rigid metal conduit (RMC) and intermediate metal conduit (IMC) | Y* | Y* | Y* | Y* | Y* | Y* | Y* |
| | In any raceway recognized in Chapter 3 | Y* | Y* | Y* | Y* | Y* | Y* | N |
| | In plenum communications raceways | Y* | Y* | Y* | N | N | N | N |
| | In plenum cable routing assemblies | Y* | Y* | Y* | N | N | N | N |
| | In riser communications raceways | Y* | Y* | Y* | N | N | N | N |
| | In riser cable routing assemblies | Y* | Y* | Y* | N | N | N | N |
| | In general-purpose communications raceways | Y* | Y* | Y* | N | N | N | N |
| | In general-purpose cable routing assemblies | Y* | Y* | Y* | N | N | N | N |

Note: An "N" in the table indicates that the cable type shall not be permitted to be installed in the application. A "Y*" indicates that the cable shall be permitted to be installed in the application subject to the limitations described in 830.113.

Informational Note No. 1: Part V of Article 830 covers installation methods within buildings. This table covers the applications of listed network-powered broadband communications cables in buildings. The definition of *Point of Entrance* is in 830.2. Network-powered broadband communications cables entrance cables that have not emerged from the rigid metal conduit (RMC) or intermediate metal conduit (IMC) are not considered to be in the building.

Informational Note No. 2: For information on the restrictions to the installation of network-powered broadband communications cables in fabricated ducts, see 830.113(B).

Informational Note No. 3: Cable routing assemblies are not addressed in NFPA 90A-2012, *Standard for the Installation of Air-Conditioning and Ventilating Systems*.

*TABLE 830.154(b)* *Cable Substitutions*

| Cable Type | Permitted Cable Substitutions |
|---|---|
| BM | BMR |
| BLP | CMP, CL3P |
| BLR | CMP, CL3P, CMR, CL3R, BLP, BMR |
| BL | CMP, CMR, CM, CMG, CL3P, CL3R, CL3, BMR, BM, BLP, BLR |
| BLX | CMP, CMR, CM, CMG, CMX, CL3P, CL3R, CL3, CL3X, BMR, BM, BLP, BRP, BL |

the raceway or assembly and is subject to the installation requirements of 830.40, 830.110, and 830.113.

### 830.160 Bends

Bends in network broadband cable shall be made so as not to damage the cable.

## VI. Listing Requirements

### 830.179 Network-Powered Broadband Communications Equipment and Cables

Network-powered broadband communications equipment and cables shall be listed and marked in accordance with 830.179(A) or (B).

See the informational note following 830.1 regarding three types of cable: coaxial, coaxial with a twisted pair of conductors, and fiber optical with a twisted pair of conductors, typically used in network-powered systems.

*Exception No. 1: This listing requirement shall not apply to community antenna television and radio distribution system coaxial cables that were installed prior to January 1, 2000, in accordance with Article 820 and are used for low-power network-powered broadband communications circuits.*

*Exception No. 2: Substitute cables for network-powered broadband communications cables shall be permitted as shown in Table 830.154(b).*

**(A) Network-Powered Broadband Communications Medium-Power Cables.** Network-powered broadband communications medium-power cables shall be factory-assembled cables consisting of a jacketed coaxial cable, a jacketed combination of coaxial cable and multiple individual conductors, or a jacketed combination of an optical fiber cable and multiple individual conductors. The insulation for the individual conductors shall be rated for 300 volts minimum. Cables intended for outdoor use shall be listed as suitable for the application. Cables shall be marked in accordance with 310.120.

A rating of 300 volts is necessary for the following reasons:

1. To coordinate with protector installation requirements (i.e., protectors are not required within a block unless the cable is exposed to over 300 volts)
2. To recognize the fact that primary protectors are designed to allow voltages below 300 to pass
3. To accommodate the voltages ordinarily found on a network-powered broadband communications circuit (voltage up to 150 volts rms)

**(1) Type BMR.** Type BMR cables shall be listed as being suitable for use in a vertical run in a shaft or from floor to floor and shall also be listed as having fire-resistant characteristics capable of preventing the carrying of fire from floor to floor.

Informational Note: One method of defining fire-resistant characteristics capable of preventing the carrying of fire from floor to floor is that the cables pass the requirements of ANSI/UL 1666-2011, *Standard Test for Flame Propagation Height of Electrical and Optical-Fiber Cable Installed Vertically in Shafts.*

See the commentary following the informational note to 725.179(B).

**(2) Type BM.** Type BM cables shall be listed as being suitable for general-purpose use, with the exception of risers and plenums, and shall also be listed as being resistant to the spread of fire.

Informational Note: One method of defining resistant to the spread of fire is that the cables do not spread fire to the top of the tray in the UL Flame Exposure, Vertical Tray Flame Test in ANSI/UL 1685-2010, *Standard for Safety for Vertical-Tray Fire-Propagation and Smoke-Release Test for Electrical and Optical-Fiber Cables.* The smoke measurements in the test method are not applicable.

Another method of defining resistant to the spread of fire is for the damage (char length) not to exceed 1.5 m (4 ft 11 in.) when performing the CSA Vertical Flame Test — Cables in Cable Trays, as described in CSA C22.2 No. 0.3-M-2001, *Test Methods for Electrical Wires and Cables.*

See the commentary following the informational note to 725.179(C).

**(3) Type BMU.** Type BMU cables shall be jacketed and listed as being suitable for outdoor underground use.

**(B) Network-Powered Broadband Communication Low-Power Cables.** Network-powered broadband communications low-power cables shall be factory-assembled cables consisting of a jacketed coaxial cable, a jacketed combination of coaxial cable and multiple individual conductors, or a jacketed combination of an optical fiber cable and multiple individual conductors. The insulation for the individual conductors shall be rated for 300 volts minimum. Cables intended for outdoor use shall be listed as suitable for the application. Cables shall be marked in accordance with 310.120.

**(1) Type BLP.** Type BLP cables shall be listed as being suitable for use in ducts, plenums, and other spaces used for environmental air and shall also be listed as having adequate fire-resistant and low-smoke producing characteristics.

Informational Note: One method of defining a cable that is low-smoke producing cable and fire-resistant cable is that the cable exhibits a maximum peak optical density of 0.50 or less, an average optical density of 0.15 or less, and a maximum flame spread distance of 1.52 m (5 ft) or less when tested in accordance with NFPA 262-2011, *Standard Method of Test for Flame Travel and Smoke of Wires and Cables for Use in Air-Handling Spaces*.

See the commentary following the informational note to 725.179(A).

**(2) Type BLR.** Type BLR cables shall be listed as being suitable for use in a vertical run in a shaft, or from floor to floor, and shall also be listed as having fire-resistant characteristics capable of preventing the carrying of fire from floor to floor.

Informational Note: One method of defining fire-resistant characteristics capable of preventing the carrying of fire from floor to floor is that the cables pass the requirements of ANSI/UL 1666-2011, *Standard Test for Flame Propagation Height of Electrical and Optical-Fiber Cable Installed Vertically in Shafts*.

See the commentary following the informational note to 725.179(B).

**(3) Type BL.** Type BL cables shall be listed as being suitable for general-purpose use, with the exception of risers and plenums, and shall also be listed as being resistant to the spread of fire.

Informational Note: One method of defining resistant to the spread of fire is that the cables do not spread fire to the top of the tray in the UL Flame Exposure, Vertical Tray Flame Test in ANSI/UL 1685-2010, *Standard for Safety for Vertical-Tray Fire-Propagation and Smoke-Release Test for Electrical and Optical-Fiber Cables*. The smoke measurements in the test method are not applicable.

Another method of defining resistant to the spread of fire is for the damage (char length) not to exceed 1.5 m (4 ft 11 in.) when performing the CSA Vertical Flame Test — Cables in Cable Trays, as described in CSA C22.2 No. 0.3-2001, *Test Methods for Electrical Wires and Cables*.

See the commentary following the informational note to 725.179(C).

**(4) Type BLX.** Type BLX limited-use cables shall be listed as being suitable for use outside, for use in dwellings, and for use in raceways and shall also be listed as being resistant to flame spread.

Informational Note: One method of determining that cable is resistant to flame spread is by testing the cable to VW-1 (vertical-wire) flame test in ANSI/UL 1581-2011, *Reference Standard for Electrical Wires, Cables and Flexible Cords*.

**(5) Type BLU.** Type BLU cables shall be jacketed and listed as being suitable for outdoor underground use.

### 830.180 Grounding Devices

Where bonding or grounding is required, devices used to connect a shield, a sheath, or non–current-carrying metallic members of a cable to a bonding conductor, or grounding electrode conductor, shall be listed or be part of listed equipment.

# ARTICLE 840 Premises-Powered Broadband Communications Systems

## I. General

### 840.1 Scope

This article covers premises-powered optical fiber-based broadband communications systems that provide any combination of voice, video, data, and interactive services through an optical network terminal (ONT).

Informational Note No. 1: A typical basic system configuration consists of an optical fiber cable to the premises (FTTP) supplying a broadband signal to an ONT that converts the broadband optical signal into component electrical signals, such as traditional telephone, video, high-speed internet, and interactive services. Powering of the ONT is typically accomplished through an ONT power supply unit (OPSU) and battery backup unit (BBU) that derive their power input from the available ac at the premises. The optical fiber cable is unpowered and may be nonconductive or conductive.

Informational Note No. 2: See 90.2(B)(4) for installations of premises-powered broadband communications systems that are not covered in this article.

Although similar to Article 830, which addresses network-powered broadband communications systems, Article 840 covers premises-powered optical fiber–based broadband communications systems.

Premises-powered optical fiber–based broadband communications systems provide a wide array of subscriber services, including voice, video, data (such as Internet access), and interactive services through an optical network terminal (ONT).

Article 840 contains requirements for wiring both the inside and the outside of buildings. Other articles cover the wiring derived from the ONT into the premises. They include the following:

- Article 725 covers wiring of Class 2 and Class 3 circuits.
- Article 760 covers the wiring of fire alarm systems.
- Article 770 covers the installation of optical fiber cable.
- Article 800 covers communications (telephone) wiring.
- Article 820 covers coaxial cable installations for television signals.

### 840.2 Definitions

The definitions in Part I of Article 100 and 770.2, 800.2, and 820.2 shall apply. For purposes of this article, the following additional definitions apply.

**Fiber-to-the-Premises (FTTP).** Conductive or nonconductive optical fiber cable that is brought to the premises is terminated at an optical network terminal (ONT) and establishes a connection to a communications network.

**Optical Network Terminal (ONT).** A device that converts an optical signal into component signals, including voice, audio, video, data, wireless, and interactive service electrical, and is considered to be network interface equipment.

**Premises Communications Circuit.** The circuit that extends voice, audio, video, data, interactive services, telegraph (except radio), and outside wiring for fire alarm and burglar alarm from the service provider's ONT to the customer's communications equipment up to and including terminal equipment, such as a telephone, a fax machine, or an answering machine.

**Premises Community Antenna Television (CATV) Circuit.** The circuit that extends community antenna television (CATV) systems for audio, video, data, and interactive services from the service provider's ONT to the appropriate customer equipment.

### 840.3 Other Articles

**(A) Hazardous (Classified) Locations.** Premises-powered broadband communications circuits and equipment installed in a location that is classified in accordance with 500.5 and 505.5 shall comply with the applicable requirements of Chapter 5.

**(B) Cables in Ducts for Dust, Loose Stock, or Vapor Removal.** The requirements of 300.22(A) for wiring systems shall apply to conductive optical fiber cables.

**(C) Equipment in Other Space Used for Environmental Air.** The requirements of 300.22(C)(3) shall apply.

**(D) Installation and Use.** The requirements of 110.3(B) shall apply.

**(E) Output Circuits.** As appropriate for the services provided, the output circuits derived from the optical network terminal shall comply with the requirements of the following:

(1) Installations of communications circuits — Part V of Article 800
(2) Installations of community antenna television and radio distribution circuits — Part V of Article 820
(3) Installations of optical fiber cables — Part V of Article 770
(4) Installations of Class 2 and Class 3 circuits — Part III of Article 725
(5) Installations of power-limited fire alarm circuits — Part III of Article 760

### 840.21 Access to Electrical Equipment Behind Panels Designed to Allow Access

Access to electrical equipment shall not be denied by an accumulation of premises-powered broadband cables that prevents removal of panels, including suspended ceiling panels.

An excess accumulation of wires and cables can limit access to equipment by preventing the removal of access panels.

### 840.24 Mechanical Execution of Work

The requirements of 770.24, 800.24, and 820.24 shall apply.

Cable must be attached to or supported by the structure by cable ties, straps, clamps, hangers, and so forth. The installation method must not damage the cable. Although the referenced sections do not contain specific supporting and securing intervals, they reference 300.11 as a general rule on securing equipment and cables and 300.4(D) through (G) for protection of cables.

### 840.25 Abandoned Cables

The requirements of 770.25, 800.25, and 820.25 shall apply.

### 840.26 Spread of Fire or Products of Combustion

The requirements of 770.26, 800.26, and 820.26 shall apply.

## II. Cables Outside and Entering Buildings

### 840.44 Overhead Optical Fiber Cables

Overhead optical fiber cables containing a non–current-carrying metallic member entering buildings shall comply with 840.44(A) and (B).

Composite premises-powered optical fiber–based broadband communications systems may contain sufficient energy to pose an electric shock hazard. For that reason, those types of systems are subject to requirements similar to those for other high-powered circuits.

Section 840.44 requires that conductor spans be of sufficient size and strength to maintain clearances and avoid possible contact with light or power conductors. Splices and joints must be made with approved connectors or other means that provide sufficient mechanical strength so that conductors are not weakened, which could cause them to break and come into contact with higher-voltage conductors.

**(A) On Poles and In-Span.** Where outside plant optical fiber cables and electric light or power conductors are supported by the same pole or are run parallel to each other in-span, the conditions described in 840.44(A)(1) through (A)(4) shall be met.

**(1) Relative Location.** Where practicable, the outside plant optical fiber cables shall be located below the electric light or power conductors.

**(2) Attachment to Cross-Arms.** Attachment of outside plant optical fiber cables to a cross-arm that carries electric light or power conductors shall not be permitted.

**(3) Climbing Space.** The climbing space through outside plant optical fiber cables shall comply with the requirements of 225.14(D).

**(4) Clearance.** Supply service drops and sets of overhead service conductors of 0 to 750 volts running above and parallel to broadband communications service drops shall have a minimum separation of 300 mm (12 in.) at any point in the span, including the point of and at their attachment to the building. Clearance of not less than 1.0 m (40 in.) shall be maintained between the two services at the pole.

**(B) Above Roofs.** Outside plant optical fiber cables shall have a vertical clearance of not less than 2.5 m (8 ft) from all points of roofs above which they pass.

*Exception No. 1: The requirement of 840.44(B) shall not apply to auxiliary buildings, such as garages and the like.*

*Exception No. 2: A reduction in clearance above only the overhanging portion of the roof, to not less than 450 mm (18 in.), shall be permitted if (a) not more than 1.2 m (4 ft) of premises-powered broadband communications service-drop cable passes above the roof overhang, and (b) the cable is terminated at a through- or above-the-roof raceway or approved support.*

*Exception No. 3: Where the roof has a slope of not less than 100 mm in 300 mm (4 in. in 12 in.), a reduction in clearance to not less than 900 mm (3 ft) shall be permitted.*

Informational Note: For additional information regarding overhead wires and cables, see ANSI C2-2007, *National Electric Safety Code, Part 2, Safety Rules for Overhead Lines.*

### 840.47 Underground Optical Fiber Cables Entering Buildings

Underground optical fiber cables entering buildings shall comply with 840.47(A) through (C).

**(A) Class 1 or Non–Power-Limited Fire Alarm Circuits.** Underground optical fiber cables with a non–current-carrying metallic member entering buildings with electric light, power, Class 1, or non–power-limited fire alarm circuit conductors in a raceway, handhole enclosure, or manhole shall be located in a section separated from such conductors by means of brick, concrete, or tile partitions or by means of a suitable barrier.

**(B) Direct-Buried Cables and Raceways.** Direct-buried optical fiber cables with a non–current-carrying metallic member shall be separated by at least 300 mm (12 in.) from conductors of any electric light, power, or non–power-limited fire alarm circuit conductors or Class 1 circuit.

*Exception No. 1: Where electric service conductors are installed in raceways or have metal cable armor.*

*Exception No. 2: Where electric light or power branch-circuit or feeder conductors, non–power-limited fire alarm circuit conductors, or Class 1 circuit conductors are installed in a raceway or in metal-sheathed, metal-clad, or Type UF or Type USE cables.*

**(C) Mechanical Protection.** Direct-buried cable, conduit, or other raceway shall be installed to have a minimum cover of 150 mm (6 in.).

### 840.48 Unlisted Cables and Raceways Entering Buildings

The requirements of 770.48 shall apply.

### 840.49 Metallic Entrance Conduit Grounding

The requirements of 770.49 shall apply.

## III. Protection

### 840.90 Protective Devices

The requirements of 800.90 shall apply.

### 840.93 Grounding or Interruption

Non–current-carrying metallic members of optical fiber cables, communications cables, or coaxial cables entering buildings or attaching to buildings shall comply with 840.93(A), (B), or (C), respectively.

**(A) Non–Current-Carrying Metallic Members of Optical Fiber Cables.** Non–current-carrying metallic members of optical fiber cables entering a building or terminating on the outside of a building shall comply with 770.93(A) or (B).

**(B) Communications Cables.** The grounding or interruption of the metallic sheath of communications cable shall comply with 800.93.

**(C) Coaxial Cables.** Where the ONT is installed inside or outside of the building, with coaxial cables terminating at the ONT, and is either entering, exiting, or attached to the outside of the building, 820.93 shall apply.

## IV. Grounding Methods

### 840.100 ONT and Optical Fiber Cable Grounding

Grounding required for protection of the ONT and optical fiber cable shall comply with 770.100, 800.100, or 820.100, as applicable.

### 840.101 Premises Circuits Not Leaving the Building

Where the ONT is served by a nonconductive optical fiber cable, or where any non–current-carrying metallic member is interrupted by an insulating joint or equivalent device, and circuits that terminate at the ONT and are completely contained within the building (i.e., they do not exit the building), 840.101(A), (B), and (C) shall apply, as applicable.

**(A) Coaxial Cable Shield Grounding.** The shield of coaxial cable shall be grounded by one of the following:

(1) Any of the methods described in 820.100 or 820.106
(2) A fixed connection to an equipment grounding conductor as described in 250.118
(3) Connection to the ONT grounding terminal provided that the terminal is connected to ground by one of the methods described in 820.100 or 820.106, or to an equipment grounding conductor through a listed grounding device that will retain the ground connection if the ONT is unplugged

The coaxial shield is permitted to be grounded through the ONT as long as the ONT grounding connection is permanent, or the connection is to an EGC through a listed grounding device that will retain the grounding connection if the ONT is unplugged.

**(B) Communications Circuit Grounding.** Communications circuits shall not be required to be grounded.

**(C) ONT Grounding.** The ONT shall not be required to be grounded unless required by its listing. If the coaxial cable shield is separately grounded as described in 840.101(A)(1) or 840.101(A)(2), the use of a cord and plug for the connection to the ONT grounding connection shall be permitted.

Informational Note: Where required to be grounded, a listed device that extends the equipment grounding conductor from the receptacle to the ONT equipment grounding terminal is permitted. Sizing of the extended equipment grounding conductor is covered in Table 250.122.

•

### 840.106 Grounding and Bonding at Mobile Homes

**(A) Grounding.** Grounding shall comply with (1) and (2).

**(1)** Where there is no mobile home service equipment located within 9.0 m (30 ft) of the exterior wall of the mobile home it serves, the non–current-carrying metallic members of optical fiber cables shall be connected to a grounding electrode in accordance with 770.106(A)(1). The ONT, if required to be grounded, shall be connected to a grounding electrode in accordance with 800.106(A)(1). Premises CATV circuits shall be grounded in accordance with 820.106(A)(1), unless the ONT is listed to provide the grounding path for the shield of the coaxial cable. The grounding electrode shall be bonded in accordance with 770.106(B).

**(2)** Where there is no mobile home disconnecting means grounded in accordance with 250.32 and located within 9.0 m (30 ft) of the exterior wall of the mobile home it serves, the non–current-carrying metallic members of optical fiber cables shall be connected to a grounding electrode in accordance with 770.106(A)(2). The ONT, if required to be grounded, shall be connected to a grounding electrode in accordance with 800.106(A)(2). Premises CATV circuits shall be grounded in accordance with 820.106(A)(2), unless the ONT is listed to provide the grounding path for the shield of the coaxial cable. The grounding electrode shall be bonded in accordance with 770.106(B).

**(B) Bonding.** The ONT grounding terminal or grounding electrode shall be connected to the metal frame or available grounding terminal of the mobile home with a copper conductor not smaller than 12 AWG under any of the following conditions:

**(1)** Where there is no mobile home service equipment or disconnecting means as specified in 840.106(A).

**(2)** Where the mobile home is supplied by cord and plug.

## V. Installation Methods Within Buildings

### 840.110 Raceways for Premises-Powered Broadband Communications Optical Fiber Cables

The requirements of 770.110 shall apply.

### 840.113 Installation Past the ONT

Installation of premises communications circuits and premises coaxial circuits shall comply with 840.113(A) and (B).

**(A) Premises Communications Circuits.** Premises communications wires and cables installed in a building from the ONT shall be listed in accordance with 800.179, and the installation shall comply with 800.113 and 800.133.

**(B) Premises Community Antenna Television (CATV) Circuits.** Premises CATV coaxial cables installed in a building from the ONT shall be listed in accordance with 820.179, and the installation shall comply with 820.113 and 820.133.

### 840.133 Installation of Optical Fibers and Electrical Conductors Associated with Premises-Powered Broadband Communications Systems

The requirements of 770.133 shall apply.

### 840.154 Applications of Listed Optical Fiber Cables and Raceways

The requirements of 770.154 shall apply.

## VI. Listing Requirements

### 840.170 Equipment and Cables

Premises-powered broadband communications systems equipment and cables shall comply with 840.170(A) through (D).

**(A) Optical Network Terminal.** The ONT and applicable grounding means shall be listed for application with premises-powered broadband communications systems.

Informational Note No. 1: One way to determine applicable requirements is to refer to UL 60950-1-2007, *Standard for Safety of Information Technology Equipment*; UL 498A-2008, *Current Taps and Adapters*; or UL 467-2007, *Grounding and Bonding Equipment.*

Informational Note No. 2: There are no requirements on the ONT and its grounding methodologies except for those covered by the listing of the product.

**(B) Optical Fiber Cables.** Optical fiber cables shall be listed in accordance with 770.179(A) through (D) and shall be marked in accordance with Table 770.179.

**(C) Premises Communications Circuits.** Premises communications wires and cables connecting to the ONT shall be listed in accordance with 800.179. Communications raceways associated with the premises-powered broadband communications system shall be listed in accordance with 800.182.

**(D) Premises Community Antenna Television (CATV) Circuits.** Premises community antenna television (CATV) coaxial cables connecting to the ONT shall be listed in accordance with 820.179. Applicable grounding means shall be listed for application with premises-powered broadband communications systems.

The ONT grounding means must be listed for use with premises-powered broadband communications systems.

## 840.180 Grounding Devices

Where bonding or grounding is required, devices used to connect a shield, a sheath, or non–current-carrying metallic members of a cable to a bonding conductor, or grounding electrode conductor, shall be listed or be part of listed equipment.

# 9 Tables

The tables in Chapter 9 are part of the mandatory requirements of the *Code*. Tables 1 through 10 deal with conductors and raceways. The last four tables provide parameters for power limitations for Class 2 and 3 power-limited circuits and for power-limited fire alarm circuits. The tables are as follows:

Table 1 Percent of Cross Section of Conduit and Tubing for Conductors and Cables

Table 2 Radius of Conduit and Tubing Bends

Table 4 Dimensions and Percent Area of Conduit and Tubing (Areas of Conduit or Tubing for the Combinations of Wires Permitted in Table 1, Chapter 9)

Table 5 Dimensions of Insulated Conductors and Fixture Wires

Table 5A Compact Copper and Aluminum Building Wire Nominal Dimensions and Areas

Table 8 Conductor Properties

Table 9 Alternating-Current Resistance and Reactance for 600-Volt Cables, 3-Phase, 60 Hz, 75°C (167°F) — Three Single Conductors in Conduit

Table 10 Conductor Stranding

Table 11(A) Class 2 and Class 3 Alternating-Current Power Source Limitations

Table 11(B) Class 2 and Class 3 Direct-Current Power Source Limitations

Table 12(A) PLFA Alternating-Current Power Source Limitations

Table 12(B) PLFA Direct-Current Power Source Limitations

As the adoption and use of the *NEC* has increased in areas of the world where the metric system is the standard, providing a means to allow for the use of electrical products with metric measurements or designations has become necessary for assimilation of *NEC* requirements in the metric world. For every provision that specifies the size of a conduit or tubing with a cylindrical cross-section size, two size designations, referred to as metric designator and trade size, are given per Table 300.1(C). This designation affects all metal, nonmetallic, rigid, and flexible conduit and tubing types that have a cylindrical cross section. For example, ½ -inch conduit and tubing are referred to as metric designator 16 or trade size ½; ¾ inch is metric designator 21 or trade size ¾; 4 inches is metric designator 103 or trade size 4, and so forth.

For further information on metrication and the revised way of giving conduit and tubing sizes in the *NEC*, see 90.9 and its associated commentary and the commentary following Table 300.1(C).

Because conduits and tubing from different manufacturers have different internal diameters for the same trade size, Table 4 provides the diameter and the actual area of different conduit and tubing types at fill percentages of 100, 60, 53 (one wire), 31 (two wires), and 40 (more than two wires). The 60-percent fill is provided in Table 4 to correlate with Note 4 (found in the Notes to Tables section of this chapter) to the conduit and tubing fill tables, which permits conduit or tubing nipples 24 inches or less in length to have a conductor fill of up to 60 percent. Separate sections in Table 4 cover metal, nonmetallic, rigid, and flexible conduit and tubing types. Examples of how to use the conduit and tubing conductor fill tables are included in the commentary both here and in Informative Annex C.

Informative Annex C contains conductor fill tables for each of 12 types of conduit and tubing. The Informative Annex C tables — which are based on the dimensions given in Tables 1 and 4 of Chapter 9 for conduit and tubing fill and on the dimensions for conductors in Table 5 of Chapter 9 — provide conductor fill information based on the specific conduit or tubing and on the conductor insulation type, size, and stranding characteristics. Examples of how to use these tables are included in the commentary both here and in Informative Annex C.

***TABLE 1*** *Percent of Cross Section of Conduit and Tubing for Conductors and Cables*

| Number of Conductors and/or Cables | Cross-Sectional Area (%) |
|---|---|
| 1 | 53 |
| 2 | 31 |
| Over 2 | 40 |

Informational Note No. 1: Table 1 is based on common conditions of proper cabling and alignment of conductors where the length of the pull and the number of bends are within reasonable limits. It should be recognized that, for certain conditions, a larger size conduit or a lesser conduit fill should be considered.

Table 1 establishes the maximum fill permitted for the circular conduit and tubing types. It is the basis for Table 4 and for the information on conduit and tubing fill provided in the Informative Annex C tables. Informational Note No. 1 advises that factors such as the length of the run or the number and total radius of bends can increase the difficulty of pulling conductors into the raceway and in extreme cases could result in damage to conductor insulation. To mitigate such adverse effects and to facilitate the ease of installing the conductors in the conduit or tubing, it is recommended that where a difficult installation is anticipated, the maximum number of conductors permitted not be installed, or the size of the conduit or tubing be increased by at least one trade size larger than the minimum required by the *Code*. Experienced personnel can attest to the wisdom of this advice.

Informational Note No. 2: When pulling three conductors or cables into a raceway, if the ratio of the raceway (inside diameter) to the conductor or cable (outside diameter) is between 2.8 and 3.2, jamming can occur. While jamming can occur when pulling four or more conductors or cables into a raceway, the probability is very low.

Informational Note No. 2 warns of another potential pitfall associated with pulling conductors into conduit or tubing. Conductor jamming may occur during the installation (pulling) of conductors into a conduit even if fill allowances of 40 percent are observed. During the installation of three conductors or cables into the raceway, one conductor could slip between the other two conductors. This is more likely to take place at bends, where the raceway may be slightly oval.

As an example, Table C.1 in Informative Annex C permits three 8 AWG conductors in trade size ½ electrical metallic tubing (EMT). An 8 AWG conductor has an outside diameter (OD) of 0.216 inch (from Table 5), and a ½ inch EMT has an internal diameter (ID) of 0.622 inch (from Table 4).

The EMT in a straight run has an internal diameter of 0.622 inch, but because it may not be round at a bend, one conductor may slip between the other two and cause a jam as the conductors exit the bend. In a straight run, assuming no variation in the EMT's internal diameter or in a conductor's outside diameter, one conductor usually cannot slip between the other two, because the total of the outside diameters of the conductors (3 × 0.216 inch = 0.648 inch) is greater than the EMT's internal diameter of 0.622 inch At a bend, however, the major internal diameter of the raceway may increase due to bending, particularly in tubing, to a diameter slightly larger than 0.648 inch, permitting the middle conductor to be pulled between the outer two conductors. As the conductors exit the bend and the raceway returns to its normal shape with an internal diameter of 0.622 inch, the conductors may jam. This can also occur in straight runs where the ratio of the raceway's internal diameter to the conductor's outside diameter approaches 3. The jam ratio is calculated as follows:

$$\text{Jam ratio} = \frac{\text{ID of raceway}}{\text{OD of conductor}} = \frac{0.622}{0.216} = 2.88$$

To avoid difficult conductor installations and potential conductor insulation damage due to jamming within the conduit or tubing, a jam ratio between 2.8 and 3.2 should be avoided.

**Notes to Tables**

(1) See Informative Annex C for the maximum number of conductors and fixture wires, all of the same size (total cross-sectional area including insulation) permitted in trade sizes of the applicable conduit or tubing.

(2) Table 1 applies only to complete conduit or tubing systems and is not intended to apply to sections of conduit or tubing used to protect exposed wiring from physical damage.

The maximum fill requirements do not apply to short sections of conduit or tubing used for the physical protection of conductors and cables. Cables are commonly protected from physical damage by conduit or tubing sleeves sized to enable the cable to be passed through with relative ease without injuring or abrading the protective jacket of the cable. The requirement of 300.5(D)(1) regarding physical protection of direct-buried cables and conductors as they emerge from below grade is an example of conduit or tubing being used as a protective sleeve and not as a continuous raceway system per 300.12. However, a fitting is required on the end(s) of the conduit or tubing to protect the conductors or cables from abrasion. [See 300.15(C).]

(3) Equipment grounding or bonding conductors, where installed, shall be included when calculating conduit or tubing fill. The actual dimensions of the equipment grounding or bonding conductor (insulated or bare) shall be used in the calculation.

All insulated, covered, and bare conductors occupy space within a raceway. Therefore, all installed conductors must be included in the raceway fill calculation, including non–current-carrying conductors such as equipment grounding conductors, bonding conductors, and bonding jumpers. The only exception to this rule is the addition of an equipment grounding conductor permitted in trade size ⅜ flexible metal conduit (see the note to Table 348.22). The dimensions of bare conductors are given in Table 8.

(4) Where conduit or tubing nipples having a maximum length not to exceed 600 mm (24 in.) are installed between boxes, cabinets, and similar enclosures, the nipples shall be permitted to be filled to 60 percent of their total cross-sectional area, and 310.15(B)(3)(a) adjustment factors need not apply to this condition.

(5) For conductors not included in Chapter 9, such as multiconductor cables and optical fiber cables, the actual dimensions shall be used.

For conductors not included in Chapter 9, such as high-voltage types, the cross-sectional area can be calculated in the following manner, using the actual dimensions of each conductor:

$$\text{cross-sectional area} = d^2 \text{ cmil}$$

Where:

$d$ = outside diameter of a conductor (including insulation) [1 in. = 1000 mil (1 mil = 0.001 in.)]

cmil = circular mil, a unit measure of area equal to $\pi/4$ (3.1416/4 = 0.7854) square mil. In other words, 1 cmil = 0.7854 square mil.

### Calculation Example

Three 15-kV single conductors are to be installed in rigid metal conduit (RMC). The outside diameter of each conductor measures 1⅝ inch or 1.625 inch. What size RMC will accommodate the three conductors?

*Solution*

Step 1. Find the cross-sectional area within the conduit to be displaced by the three conductors:

$$(1.625 \text{ in.})^2 \times 0.7854 \times 3 = 6.2218 \text{ in.}^2 \text{ or } 6.222 \text{ in.}^2$$

Step 2. Determine the correct conduit size to accommodate the three conductors. Table 1 allows 40-percent conduit fill for three or more conductors, and Table 4 indicates that 40 percent of trade size 5 RMC is 8.085 in.$^2$ Thus, trade size 5 RMC will accommodate three 15-kV single conductors.

(6) For combinations of conductors of different sizes, use actual dimensions or Table 5 and Table 5A for dimensions of conductors and Table 4 for the applicable conduit or tubing dimensions.

The following two examples demonstrate how to calculate the minimum trade size conduit or tubing required for conductors of different sizes.

### Calculation Example 1

A 200-ampere feeder is routed in various wiring methods [EMT; PVC (Schedule 40); and RMC] from the main switchboard in one building to a distribution panelboard in another building. The circuit consists of four 4/0 AWG XHHW copper conductors and one 6 AWG XHHW copper conductor. Select the proper trade size for the various types of conduit and tubing to be used for the feeder.

*Solution*

All the raceways for this example require conduit fill to be calculated according to Table 1 in Chapter 9, which permits conduit fill to a maximum of 40 percent where more than two conductors are installed. (See 344.22 for RMC, 352.22 for PVC, and 358.22 for EMT.) Table 1, Note 6 refers to Table 5 for the area required for each insulated conductor. Note 6 also refers to Table 4 for selection of the appropriate trade size conduit or tubing. Table 4 contains the allowable cross-sectional area for conduit and tubing based on conductor-occupied space (40 percent maximum in this example).

Step 1. Calculate the total area occupied by the conductors, using the approximate areas listed in Table 5:

Four 4/0 AWG:

$$4 \times 0.3197 \text{ in.}^2 = 1.2788 \text{ in.}^2$$

One 6 AWG XHHW:

$$1 \times 0.0590 \text{ in.}^2 = 0.0590 \text{ in.}^2$$

Total area = 1.3378 in.$^2$ or 1.338 in.$^2$

Step 2. Determine the proper trade size EMT, RMC, and PVC (Schedule 40) from Table 4. The portion of this feeder installed in EMT requires a minimum trade size 2, which has 1.342 in.$^2$ of available space for 40-percent fill. RMC also requires a minimum trade size 2, because trade size 2 RMC has 1.363 in.$^2$ of available space for the conductors. PVC (Schedule 40), however, requires a minimum trade size 2½. Trade size 2 PVC has 1.316 in.$^2$ allowable space, which is less than the 1.338 in.$^2$ required for this combination of conductors. Therefore, it is necessary to increase the PVC size to 2½ trade size, the next standard size increment.

### Calculation Example 2

Determine the minimum size RMC allowed for the 10 mixed conductor sizes and types described as follows.

| Quantity | Wire Size and Type | Cross-Sectional Area of Each Wire (from Table 5) | Cross-Sectional Area |
|---|---|---|---|
| 4 | 12 AWG THWN | 0.0133 | 0.0532 |
| 3 | 8 AWG TW | 0.0437 | 0.1311 |
| 3 | 6 AWG THW | 0.0726 | 0.2178 |
| | | Total | 0.4021 |

*Solution*

The "Over 2 Wires" column in Table 4 indicates that 40 percent of a trade size 1¼ RMC is 0.610 in.$^2$ Therefore, trade size 1¼ is the minimum size RMC allowed for this combination of 10 conductors.

(7) When calculating the maximum number of conductors or cables permitted in a conduit or tubing, all of the same size (total cross-sectional area including insulation), the next higher whole number shall be used to determine the maximum number of conductors permitted when the calculation results in a decimal greater than or equal to 0.8. When calculating the size for conduit or tubing permitted for a single conductor, one conductor shall be permitted when the calculation results in a decimal greater than or equal to 0.8.

**Calculation Example**

Determine how many 10 AWG THHN conductors are permitted in a trade size 1¼ RMC.

*Solution*

Table 1 permits 40-percent fill for over two conductors. From Table 4, 40-percent fill for trade size 1¼ RMC is 0.610 inch, and from Table 5, the cross-sectional area of a 10 AWG THHN conductor is 0.0211 in.$^2$. The number of conductors permitted is calculated as follows:

$$\frac{0.610 \text{ in.}^2}{0.0211 \text{ in.}^2 \text{ per conductor}} = 28.910 \text{ conductors}$$

Based on the maximum allowable fill, the number of 10 AWG THHN conductors in trade size 1¼ RMC cannot exceed 28. However, in accordance with Note 7, an increase to the next whole number of 29 conductors is permitted in this case, because 0.910 is greater than 0.8, which is the benchmark for determining whether an increase to the next whole number for the maximum number of conductors is permitted. Although increasing the total to 29 conductors results in the raceway fill exceeding 40 percent, the amount by which it is exceeded is a fraction of 1 percent and will not adversely affect the installation of the conductors. This number of conductors does have to be addressed from a mutual heating effect in accordance with 310.15(B)(3)(a). Bear in mind that this is the maximum number of conductors permitted, and in accordance with Informational Note No. 1 to Table 1 in Chapter 9, an installation with fewer than the maximum number of conductors allowed may be prudent.

Verification of this solution can be found in Informative Annex C, Table C.8, which lists twenty-nine 10 AWG THWN conductors as the maximum number permitted in trade size 1¼ RMC.

Several examples of the application of Note 7 are found in the Informative Annex C tables. Where the calculation results in a decimal value less than 0.8, the maximum number of conductors permitted is based on the next lower whole number.

(8) Where bare conductors are permitted by other sections of this *Code*, the dimensions for bare conductors in Table 8 shall be permitted.

(9) A multiconductor cable, optical fiber cable, or flexible cord of two or more conductors shall be treated as a single conductor for calculating percentage conduit fill area. For cables that have elliptical cross sections, the cross-sectional area calculation shall be based on using the major diameter of the ellipse as a circle diameter.

(10) The values for approximate conductor diameter and area shown in Table 5 are based on worst-case scenario and indicate round concentric-lay-stranded conductors. Solid and round concentric-lay-stranded conductor values are grouped together for the purpose of Table 5. Round compact-stranded conductor values are shown in Table 5A. If the actual values of the conductor diameter and area are known, they shall be permitted to be used.

Prior to the 2005 *Code*, Table 2 was included in Article 344 as Table 344.24. The requirements on minimum bending radius (depending on the type of bending equipment employed) apply to all the rigid, flexible, metallic, and nonmetallic conduit and tubing types. Because of its widespread application for all the circular conduit and tubing types (refer to Section __.24 of applicable article), this table is more appropriately located in Chapter 9 and referred to in the respective conduit and tubing articles.

***TABLE 2*** *Radius of Conduit and Tubing Bends*

| Conduit or Tubing Size | | One Shot and Full Shoe Benders | | Other Bends | |
|---|---|---|---|---|---|
| **Metric Designator** | **Trade Size** | **mm** | **in.** | **mm** | **in.** |
| 16 | ½ | 101.6 | 4 | 101.6 | 4 |
| 21 | ¾ | 114.3 | 4½ | 127 | 5 |
| 27 | 1 | 146.05 | 5¾ | 152.4 | 6 |
| 35 | 1¼ | 184.15 | 7¼ | 203.2 | 8 |
| 41 | 1½ | 209.55 | 8¼ | 254 | 10 |
| 53 | 2 | 241.3 | 9½ | 304.8 | 12 |
| 63 | 2½ | 266.7 | 10½ | 381 | 15 |
| 78 | 3 | 330.2 | 13 | 457.2 | 18 |
| 91 | 3½ | 381 | 15 | 533.4 | 21 |
| 103 | 4 | 406.4 | 16 | 609.6 | 24 |
| 129 | 5 | 609.6 | 24 | 762 | 30 |
| 155 | 6 | 762 | 30 | 914.4 | 36 |

***TABLE 4*** *Dimensions and Percent Area of Conduit and Tubing (Areas of Conduit or Tubing for the Combinations of Wires Permitted in Table 1, Chapter 9)*

**Article 358 — Electrical Metallic Tubing (EMT)**

| | | Over 2 Wires 40% | | 60% | | 1 Wire 53% | | 2 Wires 31% | | Nominal Internal Diameter | | Total Area 100% | |
|---|---|---|---|---|---|---|---|---|---|---|---|---|---|
| **Metric Designator** | **Trade Size** | **mm²** | **in.²** | **mm²** | **in.²** | **mm²** | **in.²** | **mm²** | **in.²** | **mm** | **in.** | **mm²** | **in.²** |
| 16 | ½ | 78 | 0.122 | 118 | 0.182 | 104 | 0.161 | 61 | 0.094 | 15.8 | 0.622 | 196 | 0.304 |
| 21 | ¾ | 137 | 0.213 | 206 | 0.320 | 182 | 0.283 | 106 | 0.165 | 20.9 | 0.824 | 343 | 0.533 |
| 27 | 1 | 222 | 0.346 | 333 | 0.519 | 295 | 0.458 | 172 | 0.268 | 26.6 | 1.049 | 556 | 0.864 |
| 35 | 1¼ | 387 | 0.598 | 581 | 0.897 | 513 | 0.793 | 300 | 0.464 | 35.1 | 1.380 | 968 | 1.496 |
| 41 | 1½ | 526 | 0.814 | 788 | 1.221 | 696 | 1.079 | 407 | 0.631 | 40.9 | 1.610 | 1314 | 2.036 |
| 53 | 2 | 866 | 1.342 | 1299 | 2.013 | 1147 | 1.778 | 671 | 1.040 | 52.5 | 2.067 | 2165 | 3.356 |
| 63 | 2½ | 1513 | 2.343 | 2270 | 3.515 | 2005 | 3.105 | 1173 | 1.816 | 69.4 | 2.731 | 3783 | 5.858 |
| 78 | 3 | 2280 | 3.538 | 3421 | 5.307 | 3022 | 4.688 | 1767 | 2.742 | 85.2 | 3.356 | 5701 | 8.846 |
| 91 | 3½ | 2980 | 4.618 | 4471 | 6.927 | 3949 | 6.119 | 2310 | 3.579 | 97.4 | 3.834 | 7451 | 11.545 |
| 103 | 4 | 3808 | 5.901 | 5712 | 8.852 | 5046 | 7.819 | 2951 | 4.573 | 110.1 | 4.334 | 9521 | 14.753 |

**Article 362 — Electrical Nonmetallic Tubing (ENT)**

| | | Over 2 Wires 40% | | 60% | | 1 Wire 53% | | 2 Wires 31% | | Nominal Internal Diameter | | Total Area 100% | |
|---|---|---|---|---|---|---|---|---|---|---|---|---|---|
| **Metric Designator** | **Trade Size** | **mm²** | **in.²** | **mm²** | **in.²** | **mm²** | **in.²** | **mm²** | **in.²** | **mm** | **in.** | **mm²** | **in.²** |
| 16 | ½ | 73 | 0.114 | 110 | 0.171 | 97 | 0.151 | 57 | 0.088 | 15.3 | 0.602 | 184 | 0.285 |
| 21 | ¾ | 131 | 0.203 | 197 | 0.305 | 174 | 0.269 | 102 | 0.157 | 20.4 | 0.804 | 328 | 0.508 |
| 27 | 1 | 215 | 0.333 | 322 | 0.499 | 284 | 0.441 | 166 | 0.258 | 26.1 | 1.029 | 537 | 0.832 |
| 35 | 1¼ | 375 | 0.581 | 562 | 0.872 | 497 | 0.770 | 291 | 0.450 | 34.5 | 1.36 | 937 | 1.453 |
| 41 | 1½ | 512 | 0.794 | 769 | 1.191 | 679 | 1.052 | 397 | 0.616 | 40.4 | 1.59 | 1281 | 1.986 |
| 53 | 2 | 849 | 1.316 | 1274 | 1.975 | 1125 | 1.744 | 658 | 1.020 | 52 | 2.047 | 2123 | 3.291 |
| 63 | 2½ | — | — | — | — | — | — | — | — | — | — | — | — |
| 78 | 3 | — | — | — | — | — | — | — | — | — | — | — | — |
| 91 | 3½ | — | — | — | — | — | — | — | — | — | — | — | — |

*(continues)*

***TABLE 4*** *Continued*

**Article 348 — Flexible Metal Conduit (FMC)**

| Metric Designator | Trade Size | Over 2 Wires 40% mm² | Over 2 Wires 40% in.² | 60% mm² | 60% in.² | 1 Wire 53% mm² | 1 Wire 53% in.² | 2 Wires 31% mm² | 2 Wires 31% in.² | Nominal Internal Diameter mm | Nominal Internal Diameter in. | Total Area 100% mm² | Total Area 100% in.² |
|---|---|---|---|---|---|---|---|---|---|---|---|---|---|
| 12 | ⅜ | 30 | 0.046 | 44 | 0.069 | 39 | 0.061 | 23 | 0.036 | 9.7 | 0.384 | 74 | 0.116 |
| 16 | ½ | 81 | 0.127 | 122 | 0.190 | 108 | 0.168 | 63 | 0.098 | 16.1 | 0.635 | 204 | 0.317 |
| 21 | ¾ | 137 | 0.213 | 206 | 0.320 | 182 | 0.283 | 106 | 0.165 | 20.9 | 0.824 | 343 | 0.533 |
| 27 | 1 | 211 | 0.327 | 316 | 0.490 | 279 | 0.433 | 163 | 0.253 | 25.9 | 1.020 | 527 | 0.817 |
| 35 | 1¼ | 330 | 0.511 | 495 | 0.766 | 437 | 0.677 | 256 | 0.396 | 32.4 | 1.275 | 824 | 1.277 |
| 41 | 1½ | 480 | 0.743 | 720 | 1.115 | 636 | 0.985 | 372 | 0.576 | 39.1 | 1.538 | 1201 | 1.858 |
| 53 | 2 | 843 | 1.307 | 1264 | 1.961 | 1117 | 1.732 | 653 | 1.013 | 51.8 | 2.040 | 2107 | 3.269 |
| 63 | 2½ | 1267 | 1.963 | 1900 | 2.945 | 1678 | 2.602 | 982 | 1.522 | 63.5 | 2.500 | 3167 | 4.909 |
| 78 | 3 | 1824 | 2.827 | 2736 | 4.241 | 2417 | 3.746 | 1414 | 2.191 | 76.2 | 3.000 | 4560 | 7.069 |
| 91 | 3½ | 2483 | 3.848 | 3724 | 5.773 | 3290 | 5.099 | 1924 | 2.983 | 88.9 | 3.500 | 6207 | 9.621 |
| 103 | 4 | 3243 | 5.027 | 4864 | 7.540 | 4297 | 6.660 | 2513 | 3.896 | 101.6 | 4.000 | 8107 | 12.566 |

**Article 342 — Intermediate Metal Conduit (IMC)**

| Metric Designator | Trade Size | Over 2 Wires 40% mm² | Over 2 Wires 40% in.² | 60% mm² | 60% in.² | 1 Wire 53% mm² | 1 Wire 53% in.² | 2 Wires 31% mm² | 2 Wires 31% in.² | Nominal Internal Diameter mm | Nominal Internal Diameter in. | Total Area 100% mm² | Total Area 100% in.² |
|---|---|---|---|---|---|---|---|---|---|---|---|---|---|
| 12 | ⅜ | — | — | — | — | — | — | — | — | — | — | — | — |
| 16 | ½ | 89 | 0.137 | 133 | 0.205 | 117 | 0.181 | 69 | 0.106 | 16.8 | 0.660 | 222 | 0.342 |
| 21 | ¾ | 151 | 0.235 | 226 | 0.352 | 200 | 0.311 | 117 | 0.182 | 21.9 | 0.864 | 377 | 0.586 |
| 27 | 1 | 248 | 0.384 | 372 | 0.575 | 329 | 0.508 | 192 | 0.297 | 28.1 | 1.105 | 620 | 0.959 |
| 35 | 1¼ | 425 | 0.659 | 638 | 0.988 | 564 | 0.873 | 330 | 0.510 | 36.8 | 1.448 | 1064 | 1.647 |
| 41 | 1½ | 573 | 0.890 | 859 | 1.335 | 759 | 1.179 | 444 | 0.690 | 42.7 | 1.683 | 1432 | 2.225 |
| 53 | 2 | 937 | 1.452 | 1405 | 2.178 | 1241 | 1.924 | 726 | 1.125 | 54.6 | 2.150 | 2341 | 3.630 |
| 63 | 2½ | 1323 | 2.054 | 1985 | 3.081 | 1753 | 2.722 | 1026 | 1.592 | 64.9 | 2.557 | 3308 | 5.135 |
| 78 | 3 | 2046 | 3.169 | 3069 | 4.753 | 2711 | 4.199 | 1586 | 2.456 | 80.7 | 3.176 | 5115 | 7.922 |
| 91 | 3½ | 2729 | 4.234 | 4093 | 6.351 | 3616 | 5.610 | 2115 | 3.281 | 93.2 | 3.671 | 6822 | 10.584 |
| 103 | 4 | 3490 | 5.452 | 5235 | 8.179 | 4624 | 7.224 | 2705 | 4.226 | 105.4 | 4.166 | 8725 | 13.631 |

**Article 356 — Liquidtight Flexible Nonmetallic Conduit (LFNC-B*)**

| Metric Designator | Trade Size | Over 2 Wires 40% mm² | Over 2 Wires 40% in.² | 60% mm² | 60% in.² | 1 Wire 53% mm² | 1 Wire 53% in.² | 2 Wires 31% mm² | 2 Wires 31% in.² | Nominal Internal Diameter mm | Nominal Internal Diameter in. | Total Area 100% mm² | Total Area 100% in.² |
|---|---|---|---|---|---|---|---|---|---|---|---|---|---|
| 12 | ⅜ | 49 | 0.077 | 74 | 0.115 | 65 | 0.102 | 38 | 0.059 | 12.5 | 0.494 | 123 | 0.192 |
| 16 | ½ | 81 | 0.125 | 122 | 0.188 | 108 | 0.166 | 63 | 0.097 | 16.1 | 0.632 | 204 | 0.314 |
| 21 | ¾ | 140 | 0.216 | 210 | 0.325 | 185 | 0.287 | 108 | 0.168 | 21.1 | 0.830 | 350 | 0.541 |
| 27 | 1 | 226 | 0.349 | 338 | 0.524 | 299 | 0.462 | 175 | 0.270 | 26.8 | 1.054 | 564 | 0.873 |
| 35 | 1¼ | 394 | 0.611 | 591 | 0.917 | 522 | 0.810 | 305 | 0.474 | 35.4 | 1.395 | 984 | 1.528 |
| 41 | 1½ | 510 | 0.792 | 765 | 1.188 | 676 | 1.050 | 395 | 0.614 | 40.3 | 1.588 | 1276 | 1.981 |
| 53 | 2 | 836 | 1.298 | 1255 | 1.948 | 1108 | 1.720 | 648 | 1.006 | 51.6 | 2.033 | 2091 | 3.246 |

*Corresponds to 356.2(2).

***TABLE 4*** *Continued*

**Article 356 — Liquidtight Flexible Nonmetallic Conduit (LFNC-A*)**

| Metric Designator | Trade Size | Over 2 Wires 40% | | 60% | | 1 Wire 53% | | 2 Wires 31% | | Nominal Internal Diameter | | Total Area 100% | |
|---|---|---|---|---|---|---|---|---|---|---|---|---|---|
| | | mm² | in.² | mm² | in.² | mm² | in.² | mm² | in.² | mm | in. | mm² | in.² |
| 12 | ⅜ | 50 | 0.077 | 75 | 0.115 | 66 | 0.102 | 39 | 0.060 | 12.6 | 0.495 | 125 | 0.192 |
| 16 | ½ | 80 | 0.125 | 121 | 0.187 | 107 | 0.165 | 62 | 0.097 | 16.0 | 0.630 | 201 | 0.312 |
| 21 | ¾ | 139 | 0.214 | 208 | 0.321 | 184 | 0.283 | 107 | 0.166 | 21.0 | 0.825 | 346 | 0.535 |
| 27 | 1 | 221 | 0.342 | 331 | 0.513 | 292 | 0.453 | 171 | 0.265 | 26.5 | 1.043 | 552 | 0.854 |
| 35 | 1¼ | 387 | 0.601 | 581 | 0.901 | 513 | 0.796 | 300 | 0.466 | 35.1 | 1.383 | 968 | 1.502 |
| 41 | 1½ | 520 | 0.807 | 781 | 1.211 | 690 | 1.070 | 403 | 0.626 | 40.7 | 1.603 | 1301 | 2.018 |
| 53 | 2 | 863 | 1.337 | 1294 | 2.006 | 1143 | 1.772 | 669 | 1.036 | 52.4 | 2.063 | 2157 | 3.343 |

*Corresponds to 356.2(1).

**Article 350 — Liquidtight Flexible Metal Conduit (LFMC)**

| Metric Designator | Trade Size | Over 2 Wires 40% | | 60% | | 1 Wire 53% | | 2 Wires 31% | | Nominal Internal Diameter | | Total Area 100% | |
|---|---|---|---|---|---|---|---|---|---|---|---|---|---|
| | | mm² | in.² | mm² | in.² | mm² | in.² | mm² | in.² | mm | in. | mm² | in.² |
| 12 | ⅜ | 49 | 0.077 | 74 | 0.115 | 65 | 0.102 | 38 | 0.059 | 12.5 | 0.494 | 123 | 0.192 |
| 16 | ½ | 81 | 0.125 | 122 | 0.188 | 108 | 0.166 | 63 | 0.097 | 16.1 | 0.632 | 204 | 0.314 |
| 21 | ¾ | 140 | 0.216 | 210 | 0.325 | 185 | 0.287 | 108 | 0.168 | 21.1 | 0.830 | 350 | 0.541 |
| 27 | 1 | 226 | 0.349 | 338 | 0.524 | 299 | 0.462 | 175 | 0.270 | 26.8 | 1.054 | 564 | 0.873 |
| 35 | 1¼ | 394 | 0.611 | 591 | 0.917 | 522 | 0.810 | 305 | 0.474 | 35.4 | 1.395 | 984 | 1.528 |
| 41 | 1½ | 510 | 0.792 | 765 | 1.188 | 676 | 1.050 | 395 | 0.614 | 40.3 | 1.588 | 1276 | 1.981 |
| 53 | 2 | 836 | 1.298 | 1255 | 1.948 | 1108 | 1.720 | 648 | 1.006 | 51.6 | 2.033 | 2091 | 3.246 |
| 63 | 2½ | 1259 | 1.953 | 1888 | 2.929 | 1668 | 2.587 | 976 | 1.513 | 63.3 | 2.493 | 3147 | 4.881 |
| 78 | 3 | 1931 | 2.990 | 2896 | 4.485 | 2559 | 3.962 | 1497 | 2.317 | 78.4 | 3.085 | 4827 | 7.475 |
| 91 | 3½ | 2511 | 3.893 | 3766 | 5.839 | 3327 | 5.158 | 1946 | 3.017 | 89.4 | 3.520 | 6277 | 9.731 |
| 103 | 4 | 3275 | 5.077 | 4912 | 7.615 | 4339 | 6.727 | 2538 | 3.935 | 102.1 | 4.020 | 8187 | 12.692 |
| 129 | 5 | — | — | — | — | — | — | — | — | — | — | — | — |
| 155 | 6 | — | — | — | — | — | — | — | — | — | — | — | — |

**Article 344 — Rigid Metal Conduit (RMC)**

| Metric Designator | Trade Size | Over 2 Wires 40% | | 60% | | 1 Wire 53% | | 2 Wires 31% | | Nominal Internal Diameter | | Total Area 100% | |
|---|---|---|---|---|---|---|---|---|---|---|---|---|---|
| | | mm² | in.² | mm² | in.² | mm² | in.² | mm² | in.² | mm | in. | mm² | in.² |
| 12 | ⅜ | — | — | — | — | — | — | — | — | — | — | — | — |
| 16 | ½ | 81 | 0.125 | 122 | 0.188 | 108 | 0.166 | 63 | 0.097 | 16.1 | 0.632 | 204 | 0.314 |
| 21 | ¾ | 141 | 0.220 | 212 | 0.329 | 187 | 0.291 | 109 | 0.170 | 21.2 | 0.836 | 353 | 0.549 |
| 27 | 1 | 229 | 0.355 | 344 | 0.532 | 303 | 0.470 | 177 | 0.275 | 27.0 | 1.063 | 573 | 0.887 |
| 35 | 1¼ | 394 | 0.610 | 591 | 0.916 | 522 | 0.809 | 305 | 0.473 | 35.4 | 1.394 | 984 | 1.526 |
| 41 | 1½ | 533 | 0.829 | 800 | 1.243 | 707 | 1.098 | 413 | 0.642 | 41.2 | 1.624 | 1333 | 2.071 |
| 53 | 2 | 879 | 1.363 | 1319 | 2.045 | 1165 | 1.806 | 681 | 1.056 | 52.9 | 2.083 | 2198 | 3.408 |
| 63 | 2½ | 1255 | 1.946 | 1882 | 2.919 | 1663 | 2.579 | 972 | 1.508 | 63.2 | 2.489 | 3137 | 4.866 |
| 78 | 3 | 1936 | 3.000 | 2904 | 4.499 | 2565 | 3.974 | 1500 | 2.325 | 78.5 | 3.090 | 4840 | 7.499 |
| 91 | 3½ | 2584 | 4.004 | 3877 | 6.006 | 3424 | 5.305 | 2003 | 3.103 | 90.7 | 3.570 | 6461 | 10.010 |
| 103 | 4 | 3326 | 5.153 | 4990 | 7.729 | 4408 | 6.828 | 2578 | 3.994 | 102.9 | 4.050 | 8316 | 12.882 |
| 129 | 5 | 5220 | 8.085 | 7830 | 12.127 | 6916 | 10.713 | 4045 | 6.266 | 128.9 | 5.073 | 13050 | 20.212 |
| 155 | 6 | 7528 | 11.663 | 11292 | 17.495 | 9975 | 15.454 | 5834 | 9.039 | 154.8 | 6.093 | 18821 | 29.158 |

*(continues)*

***TABLE 4*** *Continued*

**Article 352 — Rigid PVC Conduit (PVC), Schedule 80**

| Metric Designator | Trade Size | Over 2 Wires 40% | | 60% | | 1 Wire 53% | | 2 Wires 31% | | Nominal Internal Diameter | | Total Area 100% | |
|---|---|---|---|---|---|---|---|---|---|---|---|---|---|
| | | mm² | in.² | mm² | in.² | mm² | in.² | mm² | in.² | mm | in. | mm² | in.² |
| 12 | 3/8 | — | — | — | — | — | — | — | — | — | — | — | — |
| 16 | 1/2 | 56 | 0.087 | 85 | 0.130 | 75 | 0.115 | 44 | 0.067 | 13.4 | 0.526 | 141 | 0.217 |
| 21 | 3/4 | 105 | 0.164 | 158 | 0.246 | 139 | 0.217 | 82 | 0.127 | 18.3 | 0.722 | 263 | 0.409 |
| 27 | 1 | 178 | 0.275 | 267 | 0.413 | 236 | 0.365 | 138 | 0.213 | 23.8 | 0.936 | 445 | 0.688 |
| 35 | 1¼ | 320 | 0.495 | 480 | 0.742 | 424 | 0.656 | 248 | 0.383 | 31.9 | 1.255 | 799 | 1.237 |
| 41 | 1½ | 442 | 0.684 | 663 | 1.027 | 585 | 0.907 | 342 | 0.530 | 37.5 | 1.476 | 1104 | 1.711 |
| 53 | 2 | 742 | 1.150 | 1113 | 1.725 | 983 | 1.523 | 575 | 0.891 | 48.6 | 1.913 | 1855 | 2.874 |
| 63 | 2½ | 1064 | 1.647 | 1596 | 2.471 | 1410 | 2.183 | 825 | 1.277 | 58.2 | 2.290 | 2660 | 4.119 |
| 78 | 3 | 1660 | 2.577 | 2491 | 3.865 | 2200 | 3.414 | 1287 | 1.997 | 72.7 | 2.864 | 4151 | 6.442 |
| 91 | 3½ | 2243 | 3.475 | 3365 | 5.213 | 2972 | 4.605 | 1738 | 2.693 | 84.5 | 3.326 | 5608 | 8.688 |
| 103 | 4 | 2907 | 4.503 | 4361 | 6.755 | 3852 | 5.967 | 2253 | 3.490 | 96.2 | 3.786 | 7268 | 11.258 |
| 129 | 5 | 4607 | 7.142 | 6911 | 10.713 | 6105 | 9.463 | 3571 | 5.535 | 121.1 | 4.768 | 11518 | 17.855 |
| 155 | 6 | 6605 | 10.239 | 9908 | 15.359 | 8752 | 13.567 | 5119 | 7.935 | 145.0 | 5.709 | 16513 | 25.598 |

**Articles 352 and 353 — Rigid PVC Conduit (PVC), Schedule 40, and HDPE Conduit (HDPE)**

| Metric Designator | Trade Size | Over 2 Wires 40% | | 60% | | 1 Wire 53% | | 2 Wires 31% | | Nominal Internal Diameter | | Total Area 100% | |
|---|---|---|---|---|---|---|---|---|---|---|---|---|---|
| | | mm² | in.² | mm² | in.² | mm² | in.² | mm² | in.² | mm | in. | mm² | in.² |
| 12 | 3/8 | — | — | — | — | — | — | — | — | — | — | — | — |
| 16 | 1/2 | 74 | 0.114 | 110 | 0.171 | 97 | 0.151 | 57 | 0.088 | 15.3 | 0.602 | 184 | 0.285 |
| 21 | 3/4 | 131 | 0.203 | 196 | 0.305 | 173 | 0.269 | 101 | 0.157 | 20.4 | 0.804 | 327 | 0.508 |
| 27 | 1 | 214 | 0.333 | 321 | 0.499 | 284 | 0.441 | 166 | 0.258 | 26.1 | 1.029 | 535 | 0.832 |
| 35 | 1¼ | 374 | 0.581 | 561 | 0.872 | 495 | 0.770 | 290 | 0.450 | 34.5 | 1.360 | 935 | 1.453 |
| 41 | 1½ | 513 | 0.794 | 769 | 1.191 | 679 | 1.052 | 397 | 0.616 | 40.4 | 1.590 | 1282 | 1.986 |
| 53 | 2 | 849 | 1.316 | 1274 | 1.975 | 1126 | 1.744 | 658 | 1.020 | 52.0 | 2.047 | 2124 | 3.291 |
| 63 | 2½ | 1212 | 1.878 | 1817 | 2.817 | 1605 | 2.488 | 939 | 1.455 | 62.1 | 2.445 | 3029 | 4.695 |
| 78 | 3 | 1877 | 2.907 | 2816 | 4.361 | 2487 | 3.852 | 1455 | 2.253 | 77.3 | 3.042 | 4693 | 7.268 |
| 91 | 3½ | 2511 | 3.895 | 3766 | 5.842 | 3327 | 5.161 | 1946 | 3.018 | 89.4 | 3.521 | 6277 | 9.737 |
| 103 | 4 | 3237 | 5.022 | 4855 | 7.532 | 4288 | 6.654 | 2508 | 3.892 | 101.5 | 3.998 | 8091 | 12.554 |
| 129 | 5 | 5099 | 7.904 | 7649 | 11.856 | 6756 | 10.473 | 3952 | 6.126 | 127.4 | 5.016 | 12748 | 19.761 |
| 155 | 6 | 7373 | 11.427 | 11060 | 17.140 | 9770 | 15.141 | 5714 | 8.856 | 153.2 | 6.031 | 18433 | 28.567 |

***TABLE 4** Continued*

**Article 352 — Type A, Rigid PVC Conduit (PVC)**

| Metric Designator | Trade Size | Over 2 Wires 40% | | 60% | | 1 Wire 53% | | 2 Wires 31% | | Nominal Internal Diameter | | Total Area 100% | |
|---|---|---|---|---|---|---|---|---|---|---|---|---|---|
| | | mm² | in.² | mm² | in.² | mm² | in.² | mm² | in.² | mm | in. | mm² | in.² |
| 16 | ½ | 100 | 0.154 | 149 | 0.231 | 132 | 0.204 | 77 | 0.119 | 17.8 | 0.700 | 249 | 0.385 |
| 21 | ¾ | 168 | 0.260 | 251 | 0.390 | 222 | 0.345 | 130 | 0.202 | 23.1 | 0.910 | 419 | 0.650 |
| 27 | 1 | 279 | 0.434 | 418 | 0.651 | 370 | 0.575 | 216 | 0.336 | 29.8 | 1.175 | 697 | 1.084 |
| 35 | 1¼ | 456 | 0.707 | 684 | 1.060 | 604 | 0.937 | 353 | 0.548 | 38.1 | 1.500 | 1140 | 1.767 |
| 41 | 1½ | 600 | 0.929 | 900 | 1.394 | 795 | 1.231 | 465 | 0.720 | 43.7 | 1.720 | 1500 | 2.324 |
| 53 | 2 | 940 | 1.459 | 1410 | 2.188 | 1245 | 1.933 | 728 | 1.131 | 54.7 | 2.155 | 2350 | 3.647 |
| 63 | 2½ | 1406 | 2.181 | 2109 | 3.272 | 1863 | 2.890 | 1090 | 1.690 | 66.9 | 2.635 | 3515 | 5.453 |
| 78 | 3 | 2112 | 3.278 | 3169 | 4.916 | 2799 | 4.343 | 1637 | 2.540 | 82.0 | 3.230 | 5281 | 8.194 |
| 91 | 3½ | 2758 | 4.278 | 4137 | 6.416 | 3655 | 5.668 | 2138 | 3.315 | 93.7 | 3.690 | 6896 | 10.694 |
| 103 | 4 | 3543 | 5.489 | 5315 | 8.234 | 4695 | 7.273 | 2746 | 4.254 | 106.2 | 4.180 | 8858 | 13.723 |
| 129 | 5 | — | — | — | — | — | — | — | — | — | — | — | — |
| 155 | 6 | — | — | — | — | — | — | — | — | — | — | — | — |

**Article 352 — Type EB, Rigid PVC Conduit (PVC)**

| Metric Designator | Trade Size | Over 2 Wires 40% | | 60% | | 1 Wire 53% | | 2 Wires 31% | | Nominal Internal Diameter | | Total Area 100% | |
|---|---|---|---|---|---|---|---|---|---|---|---|---|---|
| | | mm² | in.² | mm² | in.² | mm² | in.² | mm² | in.² | mm | in. | mm² | in.² |
| 16 | ½ | — | — | — | — | — | — | — | — | — | — | — | — |
| 21 | ¾ | — | — | — | — | — | — | — | — | — | — | — | — |
| 27 | 1 | — | — | — | — | — | — | — | — | — | — | — | — |
| 35 | 1¼ | — | — | — | — | — | — | — | — | — | — | — | — |
| 41 | 1½ | — | — | — | — | — | — | — | — | — | — | — | — |
| 53 | 2 | 999 | 1.550 | 1499 | 2.325 | 1324 | 2.053 | 774 | 1.201 | 56.4 | 2.221 | 2498 | 3.874 |
| 63 | 2½ | — | — | — | — | — | — | — | — | — | — | — | — |
| 78 | 3 | 2248 | 3.484 | 3373 | 5.226 | 2979 | 4.616 | 1743 | 2.700 | 84.6 | 3.330 | 5621 | 8.709 |
| 91 | 3½ | 2932 | 4.546 | 4397 | 6.819 | 3884 | 6.023 | 2272 | 3.523 | 96.6 | 3.804 | 7329 | 11.365 |
| 103 | 4 | 3726 | 5.779 | 5589 | 8.669 | 4937 | 7.657 | 2887 | 4.479 | 108.9 | 4.289 | 9314 | 14.448 |
| 129 | 5 | 5726 | 8.878 | 8588 | 13.317 | 7586 | 11.763 | 4437 | 6.881 | 135.0 | 5.316 | 14314 | 22.195 |
| 155 | 6 | 8133 | 12.612 | 12200 | 18.918 | 10776 | 16.711 | 6303 | 9.774 | 160.9 | 6.336 | 20333 | 31.530 |

***TABLE 5*** *Dimensions of Insulated Conductors and Fixture Wires*

| Type | Size (AWG or kcmil) | Approximate Area mm² | Approximate Area in.² | Approximate Diameter mm | Approximate Diameter in. |
|---|---|---|---|---|---|
| **Type: FFH-2, RFH-1, RFH-2, RFHH-2, RHH*, RHW*, RHW-2*, RHH, RHW, RHW-2, SF-1, SF-2, SFF-1, SFF-2, TF, TFF, THHW, THW, THW-2, TW, XF, XFF** | | | | | |
| RFH-2, FFH-2, RFHH-2 | 18 | 9.355 | 0.0145 | 3.454 | 0.136 |
| | 16 | 11.10 | 0.0172 | 3.759 | 0.148 |
| RHH, RHW, RHW-2 | 14 | 18.90 | 0.0293 | 4.902 | 0.193 |
| | 12 | 22.77 | 0.0353 | 5.385 | 0.212 |
| | 10 | 28.19 | 0.0437 | 5.994 | 0.236 |
| | 8 | 53.87 | 0.0835 | 8.280 | 0.326 |
| | 6 | 67.16 | 0.1041 | 9.246 | 0.364 |
| | 4 | 86.00 | 0.1333 | 10.46 | 0.412 |
| | 3 | 98.13 | 0.1521 | 11.18 | 0.440 |
| | 2 | 112.9 | 0.1750 | 11.99 | 0.472 |
| | 1 | 171.6 | 0.2660 | 14.78 | 0.582 |
| | 1/0 | 196.1 | 0.3039 | 15.80 | 0.622 |
| | 2/0 | 226.1 | 0.3505 | 16.97 | 0.668 |
| | 3/0 | 262.7 | 0.4072 | 18.29 | 0.720 |
| | 4/0 | 306.7 | 0.4754 | 19.76 | 0.778 |
| | 250 | 405.9 | 0.6291 | 22.73 | 0.895 |
| | 300 | 457.3 | 0.7088 | 24.13 | 0.950 |
| | 350 | 507.7 | 0.7870 | 25.43 | 1.001 |
| | 400 | 556.5 | 0.8626 | 26.62 | 1.048 |
| | 500 | 650.5 | 1.0082 | 28.78 | 1.133 |
| | 600 | 782.9 | 1.2135 | 31.57 | 1.243 |
| | 700 | 874.9 | 1.3561 | 33.38 | 1.314 |
| | 750 | 920.8 | 1.4272 | 34.24 | 1.348 |
| | 800 | 965.0 | 1.4957 | 35.05 | 1.380 |
| | 900 | 1057 | 1.6377 | 36.68 | 1.444 |
| | 1000 | 1143 | 1.7719 | 38.15 | 1.502 |
| | 1250 | 1515 | 2.3479 | 43.92 | 1.729 |
| | 1500 | 1738 | 2.6938 | 47.04 | 1.852 |
| | 1750 | 1959 | 3.0357 | 49.94 | 1.966 |
| | 2000 | 2175 | 3.3719 | 52.63 | 2.072 |
| SF-2, SFF-2 | 18 | 7.419 | 0.0115 | 3.073 | 0.121 |
| | 16 | 8.968 | 0.0139 | 3.378 | 0.133 |
| | 14 | 11.10 | 0.0172 | 3.759 | 0.148 |
| SF-1, SFF-1 | 18 | 4.194 | 0.0065 | 2.311 | 0.091 |
| RFH-1, TF, TFF, XF, XFF | 18 | 5.161 | 0.0088 | 2.692 | 0.106 |
| TF, TFF, XF, XFF | 16 | 7.032 | 0.0109 | 2.997 | 0.118 |
| TW, XF, XFF, THHW, THW, THW-2 | 14 | 8.968 | 0.0139 | 3.378 | 0.133 |
| TW, THHW, THW, THW-2 | 12 | 11.68 | 0.0181 | 3.861 | 0.152 |
| | 10 | 15.68 | 0.0243 | 4.470 | 0.176 |
| | 8 | 28.19 | 0.0437 | 5.994 | 0.236 |
| RHH*, RHW*, RHW-2* | 14 | 13.48 | 0.0209 | 4.140 | 0.163 |

***TABLE 5*** *Continued*

| Type | Size (AWG or kcmil) | Approximate Area mm² | Approximate Area in.² | Approximate Diameter mm | Approximate Diameter in. |
|---|---|---|---|---|---|
| RHH*, RHW*, RHW-2*, XF, XFF | 12 | 16.77 | 0.0260 | 4.623 | 0.182 |
| **Type: RHH*, RHW*, RHW-2*, THHN, THHW, THW, THW-2, TFN, TFFN, THWN, THWN-2, XF, XFF** | | | | | |
| RHH,* RHW,* RHW-2,* XF, XFF | 10 | 21.48 | 0.0333 | 5.232 | 0.206 |
| RHH*, RHW*, RHW-2* | 8 | 35.87 | 0.0556 | 6.756 | 0.266 |
| TW, THW, THHW, THW-2, RHH*, RHW*, RHW-2* | 6 | 46.84 | 0.0726 | 7.722 | 0.304 |
| | 4 | 62.77 | 0.0973 | 8.941 | 0.352 |
| | 3 | 73.16 | 0.1134 | 9.652 | 0.380 |
| | 2 | 86.00 | 0.1333 | 10.46 | 0.412 |
| | 1 | 122.6 | 0.1901 | 12.50 | 0.492 |
| | 1/0 | 143.4 | 0.2223 | 13.51 | 0.532 |
| | 2/0 | 169.3 | 0.2624 | 14.68 | 0.578 |
| | 3/0 | 201.1 | 0.3117 | 16.00 | 0.630 |
| | 4/0 | 239.9 | 0.3718 | 17.48 | 0.688 |
| | 250 | 296.5 | 0.4596 | 19.43 | 0.765 |
| | 300 | 340.7 | 0.5281 | 20.83 | 0.820 |
| | 350 | 384.4 | 0.5958 | 22.12 | 0.871 |
| | 400 | 427.0 | 0.6619 | 23.32 | 0.918 |
| | 500 | 509.7 | 0.7901 | 25.48 | 1.003 |
| | 600 | 627.7 | 0.9729 | 28.27 | 1.113 |
| | 700 | 710.3 | 1.1010 | 30.07 | 1.184 |
| | 750 | 751.7 | 1.1652 | 30.94 | 1.218 |
| | 800 | 791.7 | 1.2272 | 31.75 | 1.250 |
| | 900 | 874.9 | 1.3561 | 33.38 | 1.314 |
| | 1000 | 953.8 | 1.4784 | 34.85 | 1.372 |
| | 1250 | 1200 | 1.8602 | 39.09 | 1.539 |
| | 1500 | 1400 | 2.1695 | 42.21 | 1.662 |
| | 1750 | 1598 | 2.4773 | 45.11 | 1.776 |
| | 2000 | 1795 | 2.7818 | 47.80 | 1.882 |
| TFN, TFFN | 18 | 3.548 | 0.0055 | 2.134 | 0.084 |
| | 16 | 4.645 | 0.0072 | 2.438 | 0.096 |
| THHN, THWN, THWN-2 | 14 | 6.258 | 0.0097 | 2.819 | 0.111 |
| | 12 | 8.581 | 0.0133 | 3.302 | 0.130 |
| | 10 | 13.61 | 0.0211 | 4.166 | 0.164 |
| | 8 | 23.61 | 0.0366 | 5.486 | 0.216 |
| | 6 | 32.71 | 0.0507 | 6.452 | 0.254 |
| | 4 | 53.16 | 0.0824 | 8.230 | 0.324 |
| | 3 | 62.77 | 0.0973 | 8.941 | 0.352 |
| | 2 | 74.71 | 0.1158 | 9.754 | 0.384 |
| | 1 | 100.8 | 0.1562 | 11.33 | 0.446 |
| | 1/0 | 119.7 | 0.1855 | 12.34 | 0.486 |
| | 2/0 | 143.4 | 0.2223 | 13.51 | 0.532 |
| | 3/0 | 172.8 | 0.2679 | 14.83 | 0.584 |
| | 4/0 | 208.8 | 0.3237 | 16.31 | 0.642 |

*(continues)*

***TABLE 5*** *Continued*

| Type | Size (AWG or kcmil) | Approximate Area | | Approximate Diameter | |
|---|---|---|---|---|---|
| | | mm² | in.² | mm | in. |
| | 250 | 256.1 | 0.3970 | 18.06 | 0.711 |
| | 300 | 297.3 | 0.4608 | 19.46 | 0.766 |
| **Type: FEP, FEPB, PAF, PAFF, PF, PFA, PFAH, PFF, PGF, PGFF, PTF, PTFF, TFE, THHN, THWN, THWN-2, Z, ZF, ZFF, ZHF** | | | | | |
| THHN, THWN, THWN-2 | 350 | 338.2 | 0.5242 | 20.75 | 0.817 |
| | 400 | 378.3 | 0.5863 | 21.95 | 0.864 |
| | 500 | 456.3 | 0.7073 | 24.10 | 0.949 |
| | 600 | 559.7 | 0.8676 | 26.70 | 1.051 |
| | 700 | 637.9 | 0.9887 | 28.50 | 1.122 |
| | 750 | 677.2 | 1.0496 | 29.36 | 1.156 |
| | 800 | 715.2 | 1.1085 | 30.18 | 1.188 |
| | 900 | 794.3 | 1.2311 | 31.80 | 1.252 |
| | 1000 | 869.5 | 1.3478 | 33.27 | 1.310 |
| PF, PGFF, PGF, PFF, PTF, PAF, PTFF, PAFF | 18 | 3.742 | 0.0058 | 2.184 | 0.086 |
| | 16 | 4.839 | 0.0075 | 2.489 | 0.098 |
| PF, PGFF, PGF, PFF, PTF, PAF, PTFF, PAFF, TFE, FEP, PFA, FEPB, PFAH | 14 | 6.452 | 0.0100 | 2.870 | 0.113 |
| TFE, FEP, PFA, FEPB, PFAH | 12 | 8.839 | 0.0137 | 3.353 | 0.132 |
| | 10 | 12.32 | 0.0191 | 3.962 | 0.156 |
| | 8 | 21.48 | 0.0333 | 5.232 | 0.206 |
| | 6 | 30.19 | 0.0468 | 6.198 | 0.244 |
| | 4 | 43.23 | 0.0670 | 7.417 | 0.292 |
| | 3 | 51.87 | 0.0804 | 8.128 | 0.320 |
| | 2 | 62.77 | 0.0973 | 8.941 | 0.352 |
| TFE, PFAH, PFA | 1 | 90.26 | 0.1399 | 10.72 | 0.422 |
| TFE, PFA, PFAH, Z | 1/0 | 108.1 | 0.1676 | 11.73 | 0.462 |
| | 2/0 | 130.8 | 0.2027 | 12.90 | 0.508 |
| | 3/0 | 158.9 | 0.2463 | 14.22 | 0.560 |
| | 4/0 | 193.5 | 0.3000 | 15.70 | 0.618 |
| ZF, ZFF, ZHF | 18 | 2.903 | 0.0045 | 1.930 | 0.076 |
| | 16 | 3.935 | 0.0061 | 2.235 | 0.088 |
| Z, ZF, ZFF, ZHF | 14 | 5.355 | 0.0083 | 2.616 | 0.103 |
| Z | 12 | 7.548 | 0.0117 | 3.099 | 0.122 |
| | 10 | 12.32 | 0.0191 | 3.962 | 0.156 |
| | 8 | 19.48 | 0.0302 | 4.978 | 0.196 |
| | 6 | 27.74 | 0.0430 | 5.944 | 0.234 |
| | 4 | 40.32 | 0.0625 | 7.163 | 0.282 |
| | 3 | 55.16 | 0.0855 | 8.382 | 0.330 |
| | 2 | 66.39 | 0.1029 | 9.195 | 0.362 |
| | 1 | 81.87 | 0.1269 | 10.21 | 0.402 |

***TABLE 5*** *Continued*

| Type | Size (AWG or kcmil) | Approximate Area | | Approximate Diameter | |
|---|---|---|---|---|---|
| | | mm² | in.² | mm | in. |
| **Type: KF-1, KF-2, KFF-1, KFF-2, XHH, XHHW, XHHW-2, ZW** | | | | | |
| XHHW, ZW, XHHW-2, XHH | 14 | 8.968 | 0.0139 | 3.378 | 0.133 |
| | 12 | 11.68 | 0.0181 | 3.861 | 0.152 |
| | 10 | 15.68 | 0.0243 | 4.470 | 0.176 |
| | 8 | 28.19 | 0.0437 | 5.994 | 0.236 |
| | 6 | 38.06 | 0.0590 | 6.960 | 0.274 |
| | 4 | 52.52 | 0.0814 | 8.179 | 0.322 |
| | 3 | 62.06 | 0.0962 | 8.890 | 0.350 |
| | 2 | 73.94 | 0.1146 | 9.703 | 0.382 |
| XHHW, XHHW-2, XHH | 1 | 98.97 | 0.1534 | 11.23 | 0.442 |
| | 1/0 | 117.7 | 0.1825 | 12.24 | 0.482 |
| | 2/0 | 141.3 | 0.2190 | 13.41 | 0.528 |
| | 3/0 | 170.5 | 0.2642 | 14.73 | 0.58 |
| | 4/0 | 206.3 | 0.3197 | 16.21 | 0.638 |
| | 250 | 251.9 | 0.3904 | 17.91 | 0.705 |
| | 300 | 292.6 | 0.4536 | 19.30 | 0.76 |
| | 350 | 333.3 | 0.5166 | 20.60 | 0.811 |
| | 400 | 373.0 | 0.5782 | 21.79 | 0.858 |
| | 500 | 450.6 | 0.6984 | 23.95 | 0.943 |
| | 600 | 561.9 | 0.8709 | 26.75 | 1.053 |
| | 700 | 640.2 | 0.9923 | 28.55 | 1.124 |
| | 750 | 679.5 | 1.0532 | 29.41 | 1.158 |
| | 800 | 717.5 | 1.1122 | 30.23 | 1.190 |
| | 900 | 796.8 | 1.2351 | 31.85 | 1.254 |
| | 1000 | 872.2 | 1.3519 | 33.32 | 1.312 |
| | 1250 | 1108 | 1.7180 | 37.57 | 1.479 |
| | 1500 | 1300 | 2.0156 | 40.69 | 1.602 |
| | 1750 | 1492 | 2.3127 | 43.59 | 1.716 |
| | 2000 | 1682 | 2.6073 | 46.28 | 1.822 |
| KF-2, KFF-2 | 18 | 2.000 | 0.003 | 1.575 | 0.062 |
| | 16 | 2.839 | 0.0043 | 1.88 | 0.074 |
| | 14 | 4.129 | 0.0064 | 2.286 | 0.090 |
| | 12 | 6.000 | 0.0092 | 2.743 | 0.108 |
| | 10 | 8.968 | 0.0139 | 3.378 | 0.133 |
| KF-1, KFF-1 | 18 | 1.677 | 0.0026 | 1.448 | 0.057 |
| | 16 | 2.387 | 0.0037 | 1.753 | 0.069 |
| | 14 | 3.548 | 0.0055 | 2.134 | 0.084 |
| | 12 | 5.355 | 0.0083 | 2.616 | 0.103 |
| | 10 | 8.194 | 0.0127 | 3.226 | 0.127 |

*Types RHH, RHW, and RHW-2 without outer covering.

***TABLE 5A*** *Compact Copper and Aluminum Building Wire Nominal Dimensions* and Areas*

| Size (AWG or kcmil) | Bare Conductor | | Types RHH**, RHW**, or USE | | | | Types THW and THHW | | | | Type THHN | | | | Type XHHW | | | | Size (AWG or kcmil) |
|---|---|---|---|---|---|---|---|---|---|---|---|---|---|---|---|---|---|---|---|
| | Diameter | | Approximate Diameter | | Approximate Area | | Approximate Diameter | | Approximate Area | | Approximate Diameter | | Approximate Area | | Approximate Diameter | | Approximate Area | | |
| | mm | in. | mm | in. | $mm^2$ | $in.^2$ | mm | in. | $mm^2$ | $in.^2$ | mm | in. | $mm^2$ | $in.^2$ | mm | in. | $mm^2$ | $in.^2$ | |
| 8 | 3.404 | 0.134 | 6.604 | 0.260 | 34.25 | 0.0531 | 6.477 | 0.255 | 32.90 | 0.0510 | — | — | — | — | 5.690 | 0.224 | 25.42 | 0.0394 | 8 |
| 6 | 4.293 | 0.169 | 7.493 | 0.295 | 44.10 | 0.0683 | 7.366 | 0.290 | 42.58 | 0.0660 | 6.096 | 0.240 | 29.16 | 0.0452 | 6.604 | 0.260 | 34.19 | 0.0530 | 6 |
| 4 | 5.410 | 0.213 | 8.509 | 0.335 | 56.84 | 0.0881 | 8.509 | 0.335 | 56.84 | 0.0881 | 7.747 | 0.305 | 47.10 | 0.0730 | 7.747 | 0.305 | 47.10 | 0.0730 | 4 |
| 2 | 6.807 | 0.268 | 9.906 | 0.390 | 77.03 | 0.1194 | 9.906 | 0.390 | 77.03 | 0.1194 | 9.144 | 0.360 | 65.61 | 0.1017 | 9.144 | 0.360 | 65.61 | 0.1017 | 2 |
| 1 | 7.595 | 0.299 | 11.81 | 0.465 | 109.5 | 0.1698 | 11.81 | 0.465 | 109.5 | 0.1698 | 10.54 | 0.415 | 87.23 | 0.1352 | 10.54 | 0.415 | 87.23 | 0.1352 | 1 |
| 1/0 | 8.534 | 0.336 | 12.70 | 0.500 | 126.6 | 0.1963 | 12.70 | 0.500 | 126.6 | 0.1963 | 11.43 | 0.450 | 102.6 | 0.1590 | 11.43 | 0.450 | 102.60 | 0.1590 | 1/0 |
| 2/0 | 9.550 | 0.376 | 13.72 | 0.540 | 147.8 | 0.2290 | 13.84 | 0.545 | 150.5 | 0.2332 | 12.57 | 0.495 | 124.1 | 0.1924 | 12.45 | 0.490 | 121.6 | 0.1885 | 2/0 |
| 3/0 | 10.74 | 0.423 | 14.99 | 0.590 | 176.3 | 0.2733 | 14.99 | 0.590 | 176.3 | 0.2733 | 13.72 | 0.540 | 147.7 | 0.2290 | 13.72 | 0.540 | 147.7 | 0.2290 | 3/0 |
| 4/0 | 12.07 | 0.475 | 16.26 | 0.640 | 207.6 | 0.3217 | 16.38 | 0.645 | 210.8 | 0.3267 | 15.11 | 0.595 | 179.4 | 0.2780 | 14.99 | 0.590 | 176.3 | 0.2733 | 4/0 |
| 250 | 13.21 | 0.520 | 18.16 | 0.715 | 259.0 | 0.4015 | 18.42 | 0.725 | 266.3 | 0.4128 | 17.02 | 0.670 | 227.4 | 0.3525 | 16.76 | 0.660 | 220.7 | 0.3421 | 250 |
| 300 | 14.48 | 0.570 | 19.43 | 0.765 | 296.5 | 0.4596 | 19.69 | 0.775 | 304.3 | 0.4717 | 18.29 | 0.720 | 262.6 | 0.4071 | 18.16 | 0.715 | 259.0 | 0.4015 | 300 |
| 350 | 15.65 | 0.616 | 20.57 | 0.810 | 332.3 | 0.5153 | 20.83 | 0.820 | 340.7 | 0.5281 | 19.56 | 0.770 | 300.4 | 0.4656 | 19.30 | 0.760 | 292.6 | 0.4536 | 350 |
| 400 | 16.74 | 0.659 | 21.72 | 0.855 | 370.5 | 0.5741 | 21.97 | 0.865 | 379.1 | 0.5876 | 20.70 | 0.815 | 336.5 | 0.5216 | 20.32 | 0.800 | 324.3 | 0.5026 | 400 |
| 500 | 18.69 | 0.736 | 23.62 | 0.930 | 438.2 | 0.6793 | 23.88 | 0.940 | 447.7 | 0.6939 | 22.48 | 0.885 | 396.8 | 0.6151 | 22.35 | 0.880 | 392.4 | 0.6082 | 500 |
| 600 | 20.65 | 0.813 | 26.29 | 1.035 | 542.8 | 0.8413 | 26.67 | 1.050 | 558.6 | 0.8659 | 25.02 | 0.985 | 491.6 | 0.7620 | 24.89 | 0.980 | 486.6 | 0.7542 | 600 |
| 700 | 22.28 | 0.877 | 27.94 | 1.100 | 613.1 | 0.9503 | 28.19 | 1.110 | 624.3 | 0.9676 | 26.67 | 1.050 | 558.6 | 0.8659 | 26.67 | 1.050 | 558.6 | 0.8659 | 700 |
| 750 | 23.06 | 0.908 | 28.83 | 1.135 | 652.8 | 1.0118 | 29.21 | 1.150 | 670.1 | 1.0386 | 27.31 | 1.075 | 585.5 | 0.9076 | 27.69 | 1.090 | 602.0 | 0.9331 | 750 |
| 900 | 25.37 | 0.999 | 31.50 | 1.240 | 779.3 | 1.2076 | 31.09 | 1.224 | 759.1 | 1.1766 | 30.33 | 1.194 | 722.5 | 1.1196 | 29.69 | 1.169 | 692.3 | 1.0733 | 900 |
| 1000 | 26.92 | 1.060 | 32.64 | 1.285 | 836.6 | 1.2968 | 32.64 | 1.285 | 836.6 | 1.2968 | 31.88 | 1.255 | 798.1 | 1.2370 | 31.24 | 1.230 | 766.6 | 1.1882 | 1000 |

*Dimensions are from industry sources.

**Types RHH and RHW without outer coverings.

Most aluminum building wire in Types THW, THHW, THWN/THHN, and XHHW conductors is compact stranded. Table 5A provides appropriate dimensions for these types of wire.

***TABLE 8*** *Conductor Properties*

| | | | Conductors | | | | | | | Direct-Current Resistance at 75°C (167°F) | | | | | |
|---|---|---|---|---|---|---|---|---|---|---|---|---|---|---|---|
| | | | Stranding | | | Overall | | | | Copper | | | | | |
| Size (AWG or kcmil) | Area | | | Diameter | | Diameter | | Area | | Uncoated | | Coated | | Aluminum | |
| | mm² | Circular mils | Quantity | mm | in. | mm | in. | mm² | in.² | ohm/km | ohm/kFT | ohm/km | ohm/kFT | ohm/km | ohm/kFT |
| 18 | 0.823 | 1620 | 1 | — | — | 1.02 | 0.040 | 0.823 | 0.001 | 25.5 | 7.77 | 26.5 | 8.08 | 42.0 | 12.8 |
| 18 | 0.823 | 1620 | 7 | 0.39 | 0.015 | 1.16 | 0.046 | 1.06 | 0.002 | 26.1 | 7.95 | 27.7 | 8.45 | 42.8 | 13.1 |
| 16 | 1.31 | 2580 | 1 | — | — | 1.29 | 0.051 | 1.31 | 0.002 | 16.0 | 4.89 | 16.7 | 5.08 | 26.4 | 8.05 |
| 16 | 1.31 | 2580 | 7 | 0.49 | 0.019 | 1.46 | 0.058 | 1.68 | 0.003 | 16.4 | 4.99 | 17.3 | 5.29 | 26.9 | 8.21 |
| 14 | 2.08 | 4110 | 1 | — | — | 1.63 | 0.064 | 2.08 | 0.003 | 10.1 | 3.07 | 10.4 | 3.19 | 16.6 | 5.06 |
| 14 | 2.08 | 4110 | 7 | 0.62 | 0.024 | 1.85 | 0.073 | 2.68 | 0.004 | 10.3 | 3.14 | 10.7 | 3.26 | 16.9 | 5.17 |
| 12 | 3.31 | 6530 | 1 | — | — | 2.05 | 0.081 | 3.31 | 0.005 | 6.34 | 1.93 | 6.57 | 2.01 | 10.45 | 3.18 |
| 12 | 3.31 | 6530 | 7 | 0.78 | 0.030 | 2.32 | 0.092 | 4.25 | 0.006 | 6.50 | 1.98 | 6.73 | 2.05 | 10.69 | 3.25 |
| 10 | 5.261 | 10380 | 1 | — | — | 2.588 | 0.102 | 5.26 | 0.008 | 3.984 | 1.21 | 4.148 | 1.26 | 6.561 | 2.00 |
| 10 | 5.261 | 10380 | 7 | 0.98 | 0.038 | 2.95 | 0.116 | 6.76 | 0.011 | 4.070 | 1.24 | 4.226 | 1.29 | 6.679 | 2.04 |
| 8 | 8.367 | 16510 | 1 | — | — | 3.264 | 0.128 | 8.37 | 0.013 | 2.506 | 0.764 | 2.579 | 0.786 | 4.125 | 1.26 |
| 8 | 8.367 | 16510 | 7 | 1.23 | 0.049 | 3.71 | 0.146 | 10.76 | 0.017 | 2.551 | 0.778 | 2.653 | 0.809 | 4.204 | 1.28 |
| 6 | 13.30 | 26240 | 7 | 1.56 | 0.061 | 4.67 | 0.184 | 17.09 | 0.027 | 1.608 | 0.491 | 1.671 | 0.510 | 2.652 | 0.808 |
| 4 | 21.15 | 41740 | 7 | 1.96 | 0.077 | 5.89 | 0.232 | 27.19 | 0.042 | 1.010 | 0.308 | 1.053 | 0.321 | 1.666 | 0.508 |
| 3 | 26.67 | 52620 | 7 | 2.20 | 0.087 | 6.60 | 0.260 | 34.28 | 0.053 | 0.802 | 0.245 | 0.833 | 0.254 | 1.320 | 0.403 |
| 2 | 33.62 | 66360 | 7 | 2.47 | 0.097 | 7.42 | 0.292 | 43.23 | 0.067 | 0.634 | 0.194 | 0.661 | 0.201 | 1.045 | 0.319 |
| 1 | 42.41 | 83690 | 19 | 1.69 | 0.066 | 8.43 | 0.332 | 55.80 | 0.087 | 0.505 | 0.154 | 0.524 | 0.160 | 0.829 | 0.253 |
| 1/0 | 53.49 | 105600 | 19 | 1.89 | 0.074 | 9.45 | 0.372 | 70.41 | 0.109 | 0.399 | 0.122 | 0.415 | 0.127 | 0.660 | 0.201 |
| 2/0 | 67.43 | 133100 | 19 | 2.13 | 0.084 | 10.62 | 0.418 | 88.74 | 0.137 | 0.3170 | 0.0967 | 0.329 | 0.101 | 0.523 | 0.159 |
| 3/0 | 85.01 | 167800 | 19 | 2.39 | 0.094 | 11.94 | 0.470 | 111.9 | 0.173 | 0.2512 | 0.0766 | 0.2610 | 0.0797 | 0.413 | 0.126 |
| 4/0 | 107.2 | 211600 | 19 | 2.68 | 0.106 | 13.41 | 0.528 | 141.1 | 0.219 | 0.1996 | 0.0608 | 0.2050 | 0.0626 | 0.328 | 0.100 |
| 250 | 127 | — | 37 | 2.09 | 0.082 | 14.61 | 0.575 | 168 | 0.260 | 0.1687 | 0.0515 | 0.1753 | 0.0535 | 0.2778 | 0.0847 |
| 300 | 152 | — | 37 | 2.29 | 0.090 | 16.00 | 0.630 | 201 | 0.312 | 0.1409 | 0.0429 | 0.1463 | 0.0446 | 0.2318 | 0.0707 |
| 350 | 177 | — | 37 | 2.47 | 0.097 | 17.30 | 0.681 | 235 | 0.364 | 0.1205 | 0.0367 | 0.1252 | 0.0382 | 0.1984 | 0.0605 |
| 400 | 203 | — | 37 | 2.64 | 0.104 | 18.49 | 0.728 | 268 | 0.416 | 0.1053 | 0.0321 | 0.1084 | 0.0331 | 0.1737 | 0.0529 |
| 500 | 253 | — | 37 | 2.95 | 0.116 | 20.65 | 0.813 | 336 | 0.519 | 0.0845 | 0.0258 | 0.0869 | 0.0265 | 0.1391 | 0.0424 |
| 600 | 304 | — | 61 | 2.52 | 0.099 | 22.68 | 0.893 | 404 | 0.626 | 0.0704 | 0.0214 | 0.0732 | 0.0223 | 0.1159 | 0.0353 |
| 700 | 355 | — | 61 | 2.72 | 0.107 | 24.49 | 0.964 | 471 | 0.730 | 0.0603 | 0.0184 | 0.0622 | 0.0189 | 0.0994 | 0.0303 |
| 750 | 380 | — | 61 | 2.82 | 0.111 | 25.35 | 0.998 | 505 | 0.782 | 0.0563 | 0.0171 | 0.0579 | 0.0176 | 0.0927 | 0.0282 |
| 800 | 405 | — | 61 | 2.91 | 0.114 | 26.16 | 1.030 | 538 | 0.834 | 0.0528 | 0.0161 | 0.0544 | 0.0166 | 0.0868 | 0.0265 |
| 900 | 456 | — | 61 | 3.09 | 0.122 | 27.79 | 1.094 | 606 | 0.940 | 0.0470 | 0.0143 | 0.0481 | 0.0147 | 0.0770 | 0.0235 |
| 1000 | 507 | — | 61 | 3.25 | 0.128 | 29.26 | 1.152 | 673 | 1.042 | 0.0423 | 0.0129 | 0.0434 | 0.0132 | 0.0695 | 0.0212 |
| 1250 | 633 | — | 91 | 2.98 | 0.117 | 32.74 | 1.289 | 842 | 1.305 | 0.0338 | 0.0103 | 0.0347 | 0.0106 | 0.0554 | 0.0169 |
| 1500 | 760 | — | 91 | 3.26 | 0.128 | 35.86 | 1.412 | 1011 | 1.566 | 0.02814 | 0.00858 | 0.02814 | 0.00883 | 0.0464 | 0.0141 |
| 1750 | 887 | — | 127 | 2.98 | 0.117 | 38.76 | 1.526 | 1180 | 1.829 | 0.02410 | 0.00735 | 0.02410 | 0.00756 | 0.0397 | 0.0121 |
| 2000 | 1013 | — | 127 | 3.19 | 0.126 | 41.45 | 1.632 | 1349 | 2.092 | 0.02109 | 0.00643 | 0.02109 | 0.00662 | 0.0348 | 0.0106 |

Notes:

1. These resistance values are valid **only** for the parameters as given. Using conductors having coated strands, different stranding type, and, especially, other temperatures changes the resistance.1.
2. Equation for temperature change: $R_2 = R_1\ [1 + \alpha\ (T_2 - 75)]$ where $\alpha_{cu} = 0.00323$, $\alpha_{AL} = 0.00330$ at 75°C.
3. Conductors with compact and compressed stranding have about 9 percent and 3 percent, respectively, smaller bare conductor diameters than those shown. See Table 5A for actual compact cable dimensions.
4. The IACS conductivities used: bare copper = 100%, aluminum = 61%.
5. Class B stranding is listed as well as solid for some sizes. Its overall diameter and area are those of its circumscribing circle.

Informational Note: The construction information is in accordance with NEMA WC/70-2009 or ANSI/UL 1581-2011. The resistance is calculated in accordance with National Bureau of Standards Handbook 100, dated 1966, and Handbook 109, dated 1972.

In addition to traditional wire sizes expressed as American Wire Gage (AWG), circular mil (cmil) area, or thousands of circular mil (kcmil) area, wire is available with its cross-sectional area expressed in square millimeters ($mm^2$).

The *Code* requires that insulated conductors be marked with their sizes and that the sizes be expressed in either AWG or circular mil area. [See 110.6 and 310.120(A)(4).] The *Code* allows no exceptions to either of these two requirements. Because Article 310 does not specifically prohibit optional marking on insulated conductors, the *Code* permits square millimeter ($mm^2$) markings on conductors, but only if they are in addition to the required traditional markings of AWG or circular mil area.

According to IEEE/ASTM SI 10-2002, *Standard for Use of the International System of Units (SI): The Modern Metric System*, conversion from circular mils to square meters is done by multiplying circular mils by $5.067075 \times 10^{-10}$. However, because square millimeters, rather than square meters, is the standard marking for wire size and because the reciprocal is more appropriate for this conversion, a simpler conversion factor to convert from square millimeters to circular mils (approximately) follows:

$$k = 1973.53 \frac{\text{circular mils}}{\text{mm}^2}$$

The following example provides a comparison of the square millimeter wire gauge to traditional wire sizes.

### Calculation Example

What traditional wire size does the size 125 $mm^2$ represent (approximately)?

*Solution*

$$\begin{aligned}\text{Circular mil area} &= \text{wire size(mm}^2) \times \text{conversion factor}\\ &= 125\ \text{mm}^2 \times 1973.53 \frac{\text{circular mils}}{\text{mm}^2}\\ &= 246{,}691 \text{ circular mils or } 246.691 \text{ kcmil}\end{aligned}$$

Therefore, the 125 $mm^2$ wire is larger than 4/0 AWG (211.6 kcmil) but smaller than a 250 kcmil conductor.

*Conclusion:* If a 125 $mm^2$ wire is determined to be the minimum or recommended size conductor, it is important to understand that size 250 kcmil would be the only Table 8 conductor with equivalent cross-sectional area because 4/0 AWG is simply not enough metal. It is important, however, to note that the 125 $mm^2$ conductor ampacity could not be used for a 250 kcmil conductor, because the metric conductor size is smaller. The ampacity of a 4/0 AWG can be used, or the ampacity can be calculated under engineering supervision.

***TABLE 9*** *Alternating-Current Resistance and Reactance for 600-Volt Cables, 3-Phase, 60 Hz, 75°C (167°F) — Three Single Conductors in Conduit*

| | Ohms to Neutral per Kilometer / Ohms to Neutral per 1000 Feet | | | | | | | | | | | | | | |
|---|---|---|---|---|---|---|---|---|---|---|---|---|---|---|---|
| | $X_L$ (Reactance) for All Wires | | Alternating-Current Resistance for Uncoated Copper Wires | | | Alternating-Current Resistance for Aluminum Wires | | | Effective *Z* at 0.85 *PF* for Uncoated Copper Wires | | | Effective *Z* at 0.85 *PF* for Aluminum Wires | | | |
| Size (AWG or kcmil) | PVC, Aluminum Conduits | Steel Conduit | PVC Conduit | Aluminum Conduit | Steel Conduit | PVC Conduit | Aluminum Conduit | Steel Conduit | PVC Conduit | Aluminum Conduit | Steel Conduit | PVC Conduit | Aluminum Conduit | Steel Conduit | Size (AWG or kcmil) |
| 14 | 0.190<br>0.058 | 0.240<br>0.073 | 10.2<br>3.1 | 10.2<br>3.1 | 10.2<br>3.1 | —<br>— | —<br>— | —<br>— | 8.9<br>2.7 | 8.9<br>2.7 | 8.9<br>2.7 | —<br>— | —<br>— | —<br>— | 14 |
| 12 | 0.177<br>0.054 | 0.223<br>0.068 | 6.6<br>2.0 | 6.6<br>2.0 | 6.6<br>2.0 | 10.5<br>3.2 | 10.5<br>3.2 | 10.5<br>3.2 | 5.6<br>1.7 | 5.6<br>1.7 | 5.6<br>1.7 | 9.2<br>2.8 | 9.2<br>2.8 | 9.2<br>2.8 | 12 |
| 10 | 0.164<br>0.050 | 0.207<br>0.063 | 3.9<br>1.2 | 3.9<br>1.2 | 3.9<br>1.2 | 6.6<br>2.0 | 6.6<br>2.0 | 6.6<br>2.0 | 3.6<br>1.1 | 3.6<br>1.1 | 3.6<br>1.1 | 5.9<br>1.8 | 5.9<br>1.8 | 5.9<br>1.8 | 10 |
| 8 | 0.171<br>0.052 | 0.213<br>0.065 | 2.56<br>0.78 | 2.56<br>0.78 | 2.56<br>0.78 | 4.3<br>1.3 | 4.3<br>1.3 | 4.3<br>1.3 | 2.26<br>0.69 | 2.26<br>0.69 | 2.30<br>0.70 | 3.6<br>1.1 | 3.6<br>1.1 | 3.6<br>1.1 | 8 |
| 6 | 0.167<br>0.051 | 0.210<br>0.064 | 1.61<br>0.49 | 1.61<br>0.49 | 1.61<br>0.49 | 2.66<br>0.81 | 2.66<br>0.81 | 2.66<br>0.81 | 1.44<br>0.44 | 1.48<br>0.45 | 1.48<br>0.45 | 2.33<br>0.71 | 2.36<br>0.72 | 2.36<br>0.72 | 6 |
| 4 | 0.157<br>0.048 | 0.197<br>0.060 | 1.02<br>0.31 | 1.02<br>0.31 | 1.02<br>0.31 | 1.67<br>0.51 | 1.67<br>0.51 | 1.67<br>0.51 | 0.95<br>0.29 | 0.95<br>0.29 | 0.98<br>0.30 | 1.51<br>0.46 | 1.51<br>0.46 | 1.51<br>0.46 | 4 |
| 3 | 0.154<br>0.047 | 0.194<br>0.059 | 0.82<br>0.25 | 0.82<br>0.25 | 0.82<br>0.25 | 1.31<br>0.40 | 1.35<br>0.41 | 1.31<br>0.40 | 0.75<br>0.23 | 0.79<br>0.24 | 0.79<br>0.24 | 1.21<br>0.37 | 1.21<br>0.37 | 1.21<br>0.37 | 3 |
| 2 | 0.148<br>0.045 | 0.187<br>0.057 | 0.62<br>0.19 | 0.66<br>0.20 | 0.66<br>0.20 | 1.05<br>0.32 | 1.05<br>0.32 | 1.05<br>0.32 | 0.62<br>0.19 | 0.62<br>0.19 | 0.66<br>0.20 | 0.98<br>0.30 | 0.98<br>0.30 | 0.98<br>0.30 | 2 |
| 1 | 0.151<br>0.046 | 0.187<br>0.057 | 0.49<br>0.15 | 0.52<br>0.16 | 0.52<br>0.16 | 0.82<br>0.25 | 0.85<br>0.26 | 0.82<br>0.25 | 0.52<br>0.16 | 0.52<br>0.16 | 0.52<br>0.16 | 0.79<br>0.24 | 0.79<br>0.24 | 0.82<br>0.25 | 1 |

***TABLE 9*** *Continued*

| | Ohms to Neutral per Kilometer<br>Ohms to Neutral per 1000 Feet | | | | | | | | | | | | | | |
|---|---|---|---|---|---|---|---|---|---|---|---|---|---|---|---|
| | $X_L$ (Reactance) for All Wires | | Alternating-Current Resistance for Uncoated Copper Wires | | | Alternating-Current Resistance for Aluminum Wires | | | Effective *Z* at 0.85 *PF* for Uncoated Copper Wires | | | Effective *Z* at 0.85 *PF* for Aluminum Wires | | | |
| **Size (AWG or kcmil)** | **PVC, Aluminum Conduits** | **Steel Conduit** | **PVC Conduit** | **Aluminum Conduit** | **Steel Conduit** | **PVC Conduit** | **Aluminum Conduit** | **Steel Conduit** | **PVC Conduit** | **Aluminum Conduit** | **Steel Conduit** | **PVC Conduit** | **Aluminum Conduit** | **Steel Conduit** | **Size (AWG or kcmil)** |
| 1/0 | 0.144<br>0.044 | 0.180<br>0.055 | 0.39<br>0.12 | 0.43<br>0.13 | 0.39<br>0.12 | 0.66<br>0.20 | 0.69<br>0.21 | 0.66<br>0.20 | 0.43<br>0.13 | 0.43<br>0.13 | 0.43<br>0.13 | 0.62<br>0.19 | 0.66<br>0.20 | 0.66<br>0.20 | 1/0 |
| 2/0 | 0.141<br>0.043 | 0.177<br>0.054 | 0.33<br>0.10 | 0.33<br>0.10 | 0.33<br>0.10 | 0.52<br>0.16 | 0.52<br>0.16 | 0.52<br>0.16 | 0.36<br>0.11 | 0.36<br>0.11 | 0.36<br>0.11 | 0.52<br>0.16 | 0.52<br>0.16 | 0.52<br>0.16 | 2/0 |
| 3/0 | 0.138<br>0.042 | 0.171<br>0.052 | 0.253<br>0.077 | 0.269<br>0.082 | 0.259<br>0.079 | 0.43<br>0.13 | 0.43<br>0.13 | 0.43<br>0.13 | 0.289<br>0.088 | 0.302<br>0.092 | 0.308<br>0.094 | 0.43<br>0.13 | 0.43<br>0.13 | 0.46<br>0.14 | 3/0 |
| 4/0 | 0.135<br>0.041 | 0.167<br>0.051 | 0.203<br>0.062 | 0.220<br>0.067 | 0.207<br>0.063 | 0.33<br>0.10 | 0.36<br>0.11 | 0.33<br>0.10 | 0.243<br>0.074 | 0.256<br>0.078 | 0.262<br>0.080 | 0.36<br>0.11 | 0.36<br>0.11 | 0.36<br>0.11 | 4/0 |
| 250 | 0.135<br>0.041 | 0.171<br>0.052 | 0.171<br>0.052 | 0.187<br>0.057 | 0.177<br>0.054 | 0.279<br>0.085 | 0.295<br>0.090 | 0.282<br>0.086 | 0.217<br>0.066 | 0.230<br>0.070 | 0.240<br>0.073 | 0.308<br>0.094 | 0.322<br>0.098 | 0.33<br>0.10 | 250 |
| 300 | 0.135<br>0.041 | 0.167<br>0.051 | 0.144<br>0.044 | 0.161<br>0.049 | 0.148<br>0.045 | 0.233<br>0.071 | 0.249<br>0.076 | 0.236<br>0.072 | 0.194<br>0.059 | 0.207<br>0.063 | 0.213<br>0.065 | 0.269<br>0.082 | 0.282<br>0.086 | 0.289<br>0.088 | 300 |
| 350 | 0.131<br>0.040 | 0.164<br>0.050 | 0.125<br>0.038 | 0.141<br>0.043 | 0.128<br>0.039 | 0.200<br>0.061 | 0.217<br>0.066 | 0.207<br>0.063 | 0.174<br>0.053 | 0.190<br>0.058 | 0.197<br>0.060 | 0.240<br>0.073 | 0.253<br>0.077 | 0.262<br>0.080 | 350 |
| 400 | 0.131<br>0.040 | 0.161<br>0.049 | 0.108<br>0.033 | 0.125<br>0.038 | 0.115<br>0.035 | 0.177<br>0.054 | 0.194<br>0.059 | 0.180<br>0.055 | 0.161<br>0.049 | 0.174<br>0.053 | 0.184<br>0.056 | 0.217<br>0.066 | 0.233<br>0.071 | 0.240<br>0.073 | 400 |
| 500 | 0.128<br>0.039 | 0.157<br>0.048 | 0.089<br>0.027 | 0.105<br>0.032 | 0.095<br>0.029 | 0.141<br>0.043 | 0.157<br>0.048 | 0.148<br>0.045 | 0.141<br>0.043 | 0.157<br>0.048 | 0.164<br>0.050 | 0.187<br>0.057 | 0.200<br>0.061 | 0.210<br>0.064 | 500 |
| 600 | 0.128<br>0.039 | 0.157<br>0.048 | 0.075<br>0.023 | 0.092<br>0.028 | 0.082<br>0.025 | 0.118<br>0.036 | 0.135<br>0.041 | 0.125<br>0.038 | 0.131<br>0.040 | 0.144<br>0.044 | 0.154<br>0.047 | 0.167<br>0.051 | 0.180<br>0.055 | 0.190<br>0.058 | 600 |
| 750 | 0.125<br>0.038 | 0.157<br>0.048 | 0.062<br>0.019 | 0.079<br>0.024 | 0.069<br>0.021 | 0.095<br>0.029 | 0.112<br>0.034 | 0.102<br>0.031 | 0.118<br>0.036 | 0.131<br>0.040 | 0.141<br>0.043 | 0.148<br>0.045 | 0.161<br>0.049 | 0.171<br>0.052 | 750 |
| 1000 | 0.121<br>0.037 | 0.151<br>0.046 | 0.049<br>0.015 | 0.062<br>0.019 | 0.059<br>0.018 | 0.075<br>0.023 | 0.089<br>0.027 | 0.082<br>0.025 | 0.105<br>0.032 | 0.118<br>0.036 | 0.131<br>0.040 | 0.128<br>0.039 | 0.138<br>0.042 | 0.151<br>0.046 | 1000 |

Notes:

1. These values are based on the following constants: UL-Type RHH wires with Class B stranding, in cradled configuration. Wire conductivities are 100 percent IACS copper and 61 percent IACS aluminum, and aluminum conduit is 45 percent IACS. Capacitive reactance is ignored, since it is negligible at these voltages. These resistance values are valid only at 75°C (167°F) and for the parameters as given, but are representative for 600-volt wire types operating at 60 Hz.

2. *Effective Z* is defined as $R \cos(\theta) + X \sin(\theta)$, where θ is the power factor angle of the circuit. Multiplying current by effective impedance gives a good approximation for line-to-neutral voltage drop. Effective impedance values shown in this table are valid only at 0.85 power factor. For another circuit power factor (*PF*), effective impedance (*Ze*) can be calculated from *R* and $X_L$ values given in this table as follows: $Ze = R \times PF + X_L \sin[\arccos(PF)]$.

Voltage-drop calculations using the dc-resistance formula are not always accurate for ac circuits, especially for those with a less-than-unity power factor or for those that use conductors larger than 2 AWG. Table 9 allows *Code* users to perform simple ac voltage-drop calculations. Table 9 was compiled using the Neher–McGrath ac-resistance calculation method, and the values presented are both reliable and conservative. This table contains completed calculations of effective impedance ($Z$) for the average ac circuit with an 85-percent power factor (see Calculation Example 1). If calculations with a different power factor are necessary, Table 9 also contains the appropriate values of inductive reactance and ac resistance (see Example 2). The basic assumptions and the limitations of Table 9 are as follows:

1. Capacitive reactance is ignored.
2. Three conductors are in a raceway.
3. The calculated voltage-drop values are approximate.
4. For circuits with other parameters, the Neher–McGrath ac-resistance calculation method is used.

### Calculation Example 1

A feeder has a 100-ampere continuous load. The system source is 240 volts, 3 phase, and the supplying circuit breaker is 125 amperes. The feeder is in a trade size 1¼ aluminum conduit with three 1 AWG THHN copper conductors operating at their maximum temperature rating of 75°C. The circuit length is 150 feet, and the power factor is 85 percent. Using Table 9, determine the approximate voltage drop of this circuit.

*Solution*

STEP 1. Find the approximate line-to-neutral voltage drop. Using the Table 9 column "Effective $Z$ at 0.85 *PF* for Uncoated Copper Wires," select aluminum conduit and size 1 AWG copper wire. Use the given value of 0.16 ohm per 1000 ft in the following formula:

$$\begin{aligned}\text{Voltage drop}_{\text{(line-to-neutral)}} &= \text{table value} \times \frac{\text{circuit length}}{1000\text{ ft}} \times \text{circuit load}\\ &= 0.16\text{ ohm} \times \frac{150\text{ ft}}{1000\text{ ft}} \times 100\text{ A}\\ &= 2.40\text{ V}\end{aligned}$$

STEP 2. Find the line-to-line voltage drop:

$$\begin{aligned}\text{Voltage drop}_{\text{(line-to-line)}} &= \text{voltage drop}_{\text{(line-to-neutral)}} \times \sqrt{3}\\ &= 2.40\text{ V} \times 1.732\\ &= 4.157\text{ V}\end{aligned}$$

STEP 3. Find the voltage present at the load end of the circuit:

$$240\text{ V} - 4.157\text{ V} = 235.84\text{ V}$$

### Calculation Example 2

A 270-ampere continuous load is present on a feeder. The circuit consists of a single 4-inch PVC conduit with three 600 kcmil XHHW/USE aluminum conductors fed from a 480-volt, 3-phase, 3-wire source. The conductors are operating at their maximum rated temperature of 75°C. If the power factor is 0.7 and the circuit length is 250 feet, is the voltage drop excessive?

*Solution*

STEP 1. Using the Table 9 column "$X_L$ (Reactance) for All Wires," select PVC conduit and the row for size 600 kcmil. A value of 0.039 ohm per 1000 ft is given as this $X_L$. Next, using the column "Alternating-Current Resistance for Aluminum Wires," select PVC conduit and the row for size 600 kcmil. A value of 0.036 ohm per 1000 feet is given as this $R$.

STEP 2. Find the angle representing a power factor of 0.7. Using a calculator with trigonometric functions or a trigonometric function table, find the arccosine ($\cos^{-1}$) $\theta$ of 0.7, which is 45.57 degrees. For this example, call this angle $\theta$.

STEP 3. Find the impedance ($Z$) corrected to 0.7 power factor ($Z_c$):

$$\begin{aligned}Z_c &= (R \times \cos\theta) + (X_L \times \sin\theta)\\ &= (0.036 \times 0.7) + (0.039 \times 0.7141)\\ &= 0.0252 + 0.0279\\ &= 0.0531\text{ ohm to neutral}\end{aligned}$$

STEP 4. As in Calculation Example 1, find the approximate line-to-neutral voltage drop:

$$\begin{aligned}\text{Voltage drop}_{\text{(line-to-neutral)}} &= Z_C \times \frac{\text{circuit length}}{1000\text{ ft}} \times \text{circuit load}\\ &= 0.0531 \times \frac{250\text{ ft}}{1000\text{ ft}} \times 270\text{ A}\\ &= 3.584\text{ V}\end{aligned}$$

STEP 5. Find the approximate line-to-line voltage drop:

$$\begin{aligned}\text{Voltage drop}_{\text{(line-to-line)}} &= \text{voltage drop}_{\text{(line-to-neutral)}} \times \sqrt{3}\\ &= 3.584\text{ V} \times 1.732\\ &= 6.208\text{ V}\end{aligned}$$

STEP 6. Find the approximate voltage drop expressed as a percentage of the circuit voltage:

$$\begin{aligned}\text{Percentage voltage drop}_{\text{(line-to-line)}} &= \frac{6.208\text{ V}}{480\text{ V}} \times 100\\ &= 1.29\%\text{ VD}\end{aligned}$$

STEP 7. Find the voltage present at the load end of the circuit:

$$480\text{ V} - 6.208\text{ V} = 473.8\text{ V}$$

*Conclusion:* According to 210.19(A)(1), Informational Note No. 4, this voltage drop does not appear to be excessive.

*TABLE 10 Conductor Stranding*

| Conductor Size | | Number of Strands | | |
|---|---|---|---|---|
| | | Copper | | Aluminum |
| AWG or kcmil | mm² | Class B[a] | Class C | Class B[a] |
| 24–30 | 0.20–0.05 | [b] | — | — |
| 22 | 0.32 | 7 | — | — |
| 20 | 0.52 | 10 | — | — |
| 18 | 0.82 | 16 | — | — |
| 16 | 1.3 | 26 | — | — |
| 14–2 | 2.1–33.6 | 7 | 19 | 7[c] |
| 1–4/0 | 42.4–107 | 19 | 37 | 19 |
| 250–500 | 127–253 | 37 | 61 | 37 |
| 600–1000 | 304–508 | 61 | 91 | 61 |
| 1250–1500 | 635–759 | 91 | 127 | 91 |
| 1750–2000 | 886–1016 | 127 | 271 | 127 |

[a]Conductors with a lesser number of strands shall be permitted based on an evaluation for connectability and bending.

[b]Number of strands vary.

[c]Aluminum 14 AWG (2.1 mm²) is not available.



## Table 11(A) and Table 11(B)

For listing purposes, Table 11(A) and Table 11(B) provide the required power source limitations for Class 2 and Class 3 power sources. Table 11(A) applies for alternating-current sources, and Table 11(B) applies for direct-current sources.

The power for Class 2 and Class 3 circuits shall be either (1) inherently limited, requiring no overcurrent protection, or (2) not inherently limited, requiring a combination of power source and overcurrent protection. Power sources designed for interconnection shall be listed for the purpose.

As part of the listing, the Class 2 or Class 3 power source shall be durably marked where plainly visible to indicate the class of supply and its electrical rating. A Class 2 power source not suitable for wet location use shall be so marked.

*Exception: Limited power circuits used by listed information technology equipment.*

Overcurrent devices, where required, shall be located at the point where the conductor to be protected receives its supply and shall not be interchangeable with devices of higher ratings. The overcurrent device shall be permitted as an integral part of the power source.

***TABLE 11(A)*** *Class 2 and Class 3 Alternating-Current Power Source Limitations*

| Power Source | Inherently Limited Power Source (Overcurrent Protection Not Required) | | | | Not Inherently Limited Power Source (Overcurrent Protection Required) | | | |
|---|---|---|---|---|---|---|---|---|
| | Class 2 | | | Class 3 | Class 2 | | Class 3 | |
| Source voltage $V_{max}$ (volts) (see Note 1) | 0 through 20* | Over 20 and through 30* | Over 30 and through 150 | Over 30 and through 100 | 0 through 20* | Over 20 and through 30* | Over 30 and through 100 | Over 100 and through 150 |
| Power limitations $VA_{max}$ (volt-amperes) (see Note 1) | — | — | — | — | 250 (see Note 3) | 250 | 250 | N.A. |
| Current limitations $I_{max}$ (amperes) (see Note 1) | 8.0 | 8.0 | 0.005 | $150/V_{max}$ | $1000/V_{max}$ | $1000/V_{max}$ | $1000/V_{max}$ | 1.0 |
| Maximum overcurrent protection (amperes) | — | — | — | — | 5.0 | $100/V_{max}$ | $100/ V_{max}$ | 1.0 |
| Power source maximum nameplate rating: VA (volt-amperes) | $5.0 \times V_{max}$ | 100 | $0.005 \times V_{max}$ | 100 | $5.0 \times V_{max}$ | 100 | 100 | 100 |
| Power source maximum nameplate rating: Current (amperes) | 5.0 | $100/V_{max}$ | 0.005 | $100/V_{max}$ | 5.0 | $100/V_{max}$ | $100/V_{max}$ | $100/V_{max}$ |

Note: Notes for this table can be found following Table 11(B).

*Voltage ranges shown are for sinusoidal ac in indoor locations or where wet contact is not likely to occur. For nonsinusoidal or wet contact conditions, see Note 2.

***TABLE 11(B)*** *Class 2 and Class 3 Direct-Current Power Source Limitations*

| Power Source | Inherently Limited Power Source (Overcurrent Protection Not Required) | | | | | Not Inherently Limited Power Source (Overcurrent Protection Required) | | | |
|---|---|---|---|---|---|---|---|---|---|
| | Class 2 | | | | Class 3 | Class 2 | | Class 3 | |
| Source voltage $V_{max}$ (volts) (see Note 1) | 0 through 20* | Over 20 and through 30* | Over 30 and through 60* | Over 60 and through 150 | Over 60 and through 100 | 0 through 20* | Over 20 and through 60* | Over 60 and through 100 | Over 100 and through 150 |
| Power limitations $VA_{max}$ (volt-amperes) (see Note 1) | — | — | — | — | — | 250 (see Note 3) | 250 | 250 | N.A. |
| Current limitations $I_{max}$ (amperes) (see Note 1) | 8.0 | 8.0 | $150/V_{max}$ | 0.005 | $150/V_{max}$ | $1000/V_{max}$ | $1000/V_{max}$ | $1000/V_{max}$ | 1.0 |
| Maximum overcurrent protection (amperes) | — | — | — | — | — | 5.0 | $100/V_{max}$ | $100/V_{max}$ | 1.0 |
| Power source maximum nameplate rating: VA (volt-amperes) | $5.0 \times V_{max}$ | 100 | 100 | $0.005 \times V_{max}$ | 100 | $5.0 \times V_{max}$ | 100 | 100 | 100 |
| Power source maximum nameplate rating: Current (amperes) | 5.0 | $100/V_{max}$ | $100/V_{max}$ | 0.005 | $100/V_{max}$ | 5.0 | $100/V_{max}$ | $100/V_{max}$ | $100/V_{max}$ |

*Voltage ranges shown are for continuous dc in indoor locations or where wet contact is not likely to occur. For interrupted dc or wet contact conditions, see Note 4.

**Notes for Table 11(A) and Table 11(B)**

1. $V_{max}$, $I_{max}$, and $VA_{max}$ are determined with the current-limiting impedance in the circuit (not bypassed) as follows:

$V_{max}$: Maximum output voltage regardless of load with rated input applied.

$I_{max}$: Maximum output current under any noncapacitive load, including short circuit, and with overcurrent protection bypassed if used. Where a transformer limits the output current, $I_{max}$ limits apply after 1 minute of operation. Where a current-limiting impedance, listed for the purpose, or as part of a listed product, is used in combination with a nonpower-limited transformer or a stored energy source, e.g., storage battery, to limit the output current, $I_{max}$ limits apply after 5 seconds.

$VA_{max}$: Maximum volt-ampere output after 1 minute of operation regardless of load and overcurrent protection bypassed if used.

2. For nonsinusoidal ac, $V_{max}$ shall not be greater than 42.4 volts peak. Where wet contact (immersion not included) is likely to occur, Class 3 wiring methods shall be used or $V_{max}$ shall not be greater than 15 volts for sinusoidal ac and 21.2 volts peak for nonsinusoidal ac.

3. If the power source is a transformer, $VA_{max}$ is 350 or less when $V_{max}$ is 15 or less.

4. For dc interrupted at a rate of 10 to 200 Hz, $V_{max}$ shall not be greater than 24.8 volts peak. Where wet contact (immersion not included) is likely to occur, Class 3 wiring methods shall be used, or $V_{max}$ shall not be greater than 30 volts for continuous dc; 12.4 volts peak for dc that is interrupted at a rate of 10 to 200 Hz.

Because Class 2 and Class 3 power supplies have a listing requirement in 725.121, Tables 11(A) and 11(B) are not directly referenced by the typical installer. The information has been retained in Chapter 9 to provide direction for organizations properly equipped and qualified to evaluate and list these products.

### Table 12(A) and Table 12(B)

For listing purposes, Table 12(A) and Table 12(B) provide the required power source limitations for power-limited fire alarm sources. Table 12(A) applies for alternating-current sources, and Table 12(B) applies for direct-current sources. The power for power-limited fire alarm circuits shall be either (1) inherently limited, requiring no overcurrent protection, or (2) not inherently limited, requiring the power to be limited by a combination of power source and overcurrent protection.

As part of the listing, the PLFA power source shall be durably marked where plainly visible to indicate that it is a power-limited fire alarm power source. The overcurrent device, where required, shall be located at the point where the conductor to be protected receives its supply and shall not be interchangeable with devices of higher ratings. The overcurrent device shall be permitted as an integral part of the power source.

Because power-limited fire alarm (PLFA) power supplies have a listing requirement in 760.121, this information is not directly referenced by the typical installer. Tables 12(A) and 12(B) provide direction for organizations properly equipped and qualified to evaluate and list these products.

***TABLE 12(A)*** *PLFA Alternating-Current Power Source Limitations*

| Power Source | | Inherently Limited Power Source (Overcurrent Protection Not Required) | | | Not Inherently Limited Power Source (Overcurrent Protection Required) | | |
|---|---|---|---|---|---|---|---|
| Circuit voltage $V_{max}$ (volts) (see Note 1) | | 0 through 20 | Over 20 and through 30 | Over 30 and through 100 | 0 through 20 | Over 20 and through 100 | Over 100 and through 150 |
| Power limitations $VA_{max}$ (volt-amperes) (see Note 1) | | — | — | — | 250 (see Note 2) | 250 | N.A. |
| Current limitations $I_{max}$ (amperes) (see Note 1) | | 8.0 | 8.0 | $150/V_{max}$ | $1000/V_{max}$ | $1000/V_{max}$ | 1.0 |
| Maximum overcurrent protection (amperes) | | — | — | — | 5.0 | $100/V_{max}$ | 1.0 |
| Power source maximum nameplate ratings | VA (volt-amperes) | $5.0 \times V_{max}$ | 100 | 100 | $5.0 \times V_{max}$ | 100 | 100 |
| | Current (amperes) | 5.0 | $100/V_{max}$ | $100/V_{max}$ | 5.0 | $100/V_{max}$ | $100/V_{max}$ |

Note: Notes for this table can be found following Table 12(B).

***TABLE 12(B)*** *PLFA Direct-Current Power Source Limitations*

| **Power Source** | | **Inherently Limited Power Source (Overcurrent Protection Not Required)** | | | **Not Inherently Limited Power Source (Overcurrent Protection Required)** | | |
|---|---|---|---|---|---|---|---|
| Circuit voltage $V_{max}$ (volts) (see Note 1) | | 0 through 20 | Over 20 and through 30 | Over 30 and through 100 | 0 through 20 | Over 20 and through 100 | Over 100 and through 150 |
| Power limitations $VA_{max}$ (volt-amperes) (see Note 1) | | — | — | — | 250 (see Note 2) | 250 | N.A. |
| Current limitations $I_{max}$ (amperes) (see Note 1) | | 8.0 | 8.0 | $150/V_{max}$ | $1000/V_{max}$ | $1000/V_{max}$ | 1.0 |
| Maximum overcurrent protection (amperes) | | — | — | — | 5.0 | $100/V_{max}$ | 1.0 |
| Power source maximum nameplate ratings | VA (volt-amperes) | $5.0 \times V_{max}$ | 100 | 100 | $5.0 \times V_{max}$ | 100 | 100 |
| | Current (amperes) | 5.0 | $100/V_{max}$ | $100/V_{max}$ | 5.0 | $100/V_{max}$ | $100/V_{max}$ |

**Notes for Table 12(A) and Table 12(B)**

1. $V_{max}$, $I_{max}$, and $VA_{max}$ are determined as follows:

$V_{max}$: Maximum output voltage regardless of load with rated input applied.

$I_{max}$: Maximum output current under any noncapacitive load, including short circuit, and with overcurrent protection bypassed if used. Where a transformer limits the output current, $I_{max}$ limits apply after 1 minute of operation. Where a current-limiting impedance, listed for the purpose, is used in combination with a nonpower-limited transformer or a stored energy source, e.g., storage battery, to limit the output current, $I_{max}$ limits apply after 5 seconds.

$VA_{max}$: Maximum volt-ampere output after 1 minute of operation regardless of load and overcurrent protection bypassed if used. Current limiting impedance shall not be bypassed when determining $I_{max}$ and $VA_{max}$.

2. If the power source is a transformer, $VA_{max}$ is 350 or less when $V_{max}$ is 15 or less.

# A Informative Annex
## Product Safety Standards

*Informative Annex A is not a part of the requirements of this NFPA document but is included for informational purposes only.*

This informative annex provides a list of product safety standards used for product listing where that listing is required by this *Code*. It is recognized that this list is current at the time of publication but that new standards or modifications to existing standards can occur at any time while this edition of the *Code* is in effect.

A key element of a safe and *Code*-compliant electrical installation is adherence to product installation requirements imposed by product-testing organizations as part of their evaluation of an electrical product. Section 110.3(B) requires compliance with the installation and use instructions included with listed and labeled products. Numerous requirements in the *NEC* specify the use of listed products. For those *Code* requirements where product listing is mandatory, Informative Annex A is a compilation of applicable product safety standards. The product safety standards included in Informative Annex A are only those for which the *Code* has a mandatory listing requirement. Many other safety standards are associated with products that do not have a mandatory listing or labeling requirement in the *NEC*. For more information on product standards, consult the product directories available from testing organizations.

Product safety standards, installation codes such as the *NEC*, and qualified electrical inspection are separate but not mutually exclusive components of the North American electrical safety system. The effectiveness of this system strongly depends on a close working relationship among the organizations responsible for the development of product standards and installation codes and the electrical inspection community. In addition, proper electrical installations by *qualified persons* (see the definition of this term in Article 100) trained on the installation codes and application of listed products is a key component of this interdependent system that places public safety as its primary objective. All of these components must be in place for the electrical safety system to be effective.

The function of Informative Annex A is to provide users of the *NEC* with the name, number, and developing organization for all product standards related to mandatory *Code* requirements for the use of listed products. For more information on listing and labeling, see the commentary following 110.3(B).

This informative annex does not form a mandatory part of the requirements of this *Code* but is intended only to provide *Code* users with informational guidance about the product characteristics about which *Code* requirements have been based.

| Product Standard Name | Product Standard Number |
|---|---|
| Aboveground Reinforced Thermosetting Resin Conduit (RTRC) and Fittings | UL 2515 |
| Antenna-Discharge Units | UL 452 |
| Arc-Fault Circuit-Interrupters | UL 1699 |
| Armored Cable | UL 4 |
| Attachment Plugs and Receptacles | UL 498 |
| Audio, Video and Similar Electronic Apparatus — Safety Requirements | UL 60065 |
| Audio/Video, Information and Communication Technology Equipment — Part 1: Safety Requirements | UL 62368-1 |
| Batteries for Use in Electric Vehicles | UL 2580 |
| Batteries for Use in Light Electric Rail (LER) Applications and Stationary Applications | Subject 1973 |
| Belowground Reinforced Thermosetting Resin Conduit (RTRC) and Fittings | UL 2420 |
| Busways | UL 857 |
| Cables — Thermoplastic-Insulated Underground Feeder and Branch-Circuit Cables | UL 493 |
| Cables — Thermoplastic-Insulated Wires and Cables | UL 83 |
| Cables — Thermoset-Insulated Wires and Cables | UL 44 |
| Cable and Cable Fittings for Use in Hazardous (Classified) Locations | UL 2225 |
| Cables for Non–Power-Limited Fire-Alarm Circuits | UL 1425 |
| Cables for Power-Limited Fire-Alarm Circuits | UL 1424 |
| Capacitors | UL 810 |
| Cellular Metal Floor Raceways and Fittings | UL 209 |
| Circuit Breakers for Use in Communication Equipment | UL 489A |
| Circuit Integrity (CI) Cable — Fire Tests for Electrical Circuit Protective Systems | Subject 1724 |
| Circuit Integrity (CI) Cable — Tests for Fire Resistive Cables | UL 2196 |
| Class 2 Power Units | UL 1310 |
| Communication Circuit Accessories | UL 1863 |
| Communications Cables | UL 444 |
| Community-Antenna Television Cables | UL 1655 |
| Concentrator Photovoltaic Modules and Assemblies | Subject 8703 |
| Conduit, Tubing, and Cable Fittings | UL 514B |
| • | |
| Connectors for Use in Photovoltaic Systems | Subject 6703 |
| Cord Sets and Power-Supply Cords | UL 817 |
| Cover Plates for Flush-Mounted Wiring Devices | UL 514D |
| Data-Processing Cable | UL 1690 |
| Distributed Wiring Harnesses | Subject 9703 |
| Electric Generators | UL 1004-4 |
| Electric Heating Appliances | UL 499 |
| Electric Sign Components | UL 879 |
| Electric Signs | UL 48 |
| Electric Spas, Equipment Assemblies, and Associated Equipment | UL 1563 |
| Electric Vehicle (EV) Charging System Equipment | UL 2202 |
| Electric Vehicle Supply Equipment | UL 2594 |
| Electric Water Heaters for Pools and Tubs | UL 1261 |
| • | |
| Electrical Apparatus for Explosive Gas Atmospheres — Part 15: Type of Protection "n" | ANSI/ISA-60079-15/ANSI/UL 60079-15 |
| Electrical Apparatus for Use in Class I, Zone 1 Hazardous (Classified) Locations Type of Protection — Encapsulation "m" | ANSI/ISA-60079-18/ANSI/UL 60079-18 |
| Electrical Apparatus for Use in Zone 20, Zone 21, and Zone 22 Hazardous (Classified) Locations — Protection by Encapsulation "mD" | ANSI/ISA-61241-18 |
| Electrical Apparatus for Use in Zone 21 and Zone 22 Hazardous (Classified) Locations — Protection by Enclosure "tD" | ANSI/ISA-61241-1 |
| Electrical Apparatus for Use in Zone 20, Zone 21, and Zone 22 Hazardous (Classified) Locations — General Requirements | ANSI/ISA-61241-0 |
| Electrical Apparatus for Use in Zone 20, Zone 21, and Zone 22 Hazardous (Classified) Locations — Protection by Intrinsic Safety "iD" | ANSI/ISA-61241-11 |
| Electrical Apparatus for Use in Zone 21 and Zone 22 Hazardous (Classified) Locations — Protection by Pressurization "pD" | ANSI/ISA-61241-2 |

| Product Standard Name | Product Standard Number |
|---|---|
| • | |
| Electrical Intermediate Metal Conduit — Steel | UL 1242 |
| Electrical Metallic Tubing — Aluminum | UL 797A |
| Electrical Metallic Tubing — Steel | UL 797 |
| Electrical Nonmetallic Tubing | UL 1653 |
| Electrical Resistance Heat Tracing for Industrial Applications | IEEE 515 |
| Electrical Rigid Metal Conduit — Steel | UL 6 |
| Electric-Battery-Powered Industrial Trucks | UL 583 |
| Electrochemical Capacitors | UL 810A |
| • | |
| Emergency Lighting and Power Equipment | UL 924 |
| Enclosed and Dead-Front Switches | UL 98 |
| Enclosed and Dead-Front Switches for Use in Photovoltaic Systems | Subject 98B |
| Enclosures for Electrical Equipment | UL 50 |
| Enclosures for Electrical Equipment, Environmental Considerations | UL 50E |
| Energy Management Equipment | UL 916 |
| Explosionproof and Dust-Ignition-Proof Electrical Equipment for Use in Hazardous (Classified) Locations | UL 1203 |
| Explosive Gas Atmospheres — Part 0: Equipment- General requirements | ANSI/ISA-60079-0/ANSI/UL 60079-0 |
| Explosive Gas Atmospheres — Part 7: Increased safety "e" | ANSI/ISA-60079-7/ANSI/UL 60079-7 |
| Explosive Gas Atmospheres — Part 1: Type of protection – Flameproof "d" | ANSI/ISA-60079-1/ANSI/UL 60079-1 |
| Explosive Gas Atmospheres — Part 5: Type of protection – Powder filling "q" | ANSI/ISA-60079-5/ANSI/UL 60079-5 |
| Explosive Gas Atmospheres — Part 6: Type of protection – Oil immersion "o" | ANSI/ISA-60079-6/ANSI/UL 60079-6 |
| Fire Pump Controllers | UL 218 |
| Fire Pump Motors | UL 1004-5 |
| Fire Resistive Cables, Test for | UL 2196 |
| Fixture Wire | UL 66 |
| Flame Propagation Height of Electrical and Optical-Fiber Cables Installed Vertically in Shafts, Test for | UL 1666 |
| Flat-Plate Photovoltaic Modules and Panels | UL 1703 |
| Flexible Cords and Cables | UL 62 |
| Flexible Lighting Products | UL 2388 |
| Flexible Metal Conduit | UL 1 |
| Fluorescent-Lamp Ballasts | UL 935 |
| Gas and Vapor Detectors and Sensors | UL 2075 |
| Gas-Burning Heating Appliances for Manufactured Homes and Recreational Vehicles | UL 307B |
| Gas-Fired Cooking Appliances for Recreational Vehicles | UL 1075 |
| Gas-Tube-Sign Cable | UL 814 |
| General-Use Snap Switches | UL 20 |
| Ground-Fault Circuit-Interrupters | UL 943 |
| Ground-Fault Sensing and Relaying Equipment | UL 1053 |
| Grounding and Bonding Equipment | UL 467 |
| Hardware for the Support of Conduit, Tubing and Cable | UL 2239 |
| Heating and Cooling Equipment | UL 1995 |
| High-Intensity-Discharge Lamp Ballasts | UL 1029 |
| • | |
| Household and Similar Electrical Appliances, Part 2: Particular Requirements for Heating and Cooling | UL 60335-2-40 |
| Household and Similar Electrical Appliances, Part 2: Particular Requirements for Refrigerating Appliances, Ice-Cream Appliances, and Ice-makers | UL 60335-2-24 |
| Household Refrigerators and Freezers | UL 250 |
| Impedance Protected Motors | UL 1004-2 |
| Industrial Battery Chargers | UL 1564 |
| Industrial Control Equipment | UL 508 |
| Industrial Control Panels | UL 508A |
| Information Technology Equipment Safety — Part 1: General Requirements | UL 60950-1 |

| Product Standard Name | Product Standard Number |
|---|---|
| Information Technology Equipment Safety — Part 21: Remote Power Feeding | UL 60950-21 |
| Information Technology Equipment Safety — Part 22: Equipment to be Installed Outdoors | UL 60950-22 |
| Information Technology Equipment Safety — Part 23: Large Data Storage Equipment | UL 60950-23 |
| Instrumentation Tray Cable | UL 2250 |
| Insulated Multi-Pole Splicing Wire Connectors | UL 2459 |
| Inverters, Converters, Controllers and Interconnection System Equipment for Use with Distributed Energy Resources | UL 1741 |
| Isolated Power Systems Equipment | UL 1047 |
| Junction Boxes for Swimming Pool Luminaires | UL 1241 |
| Light Emitting Diode (LED) Equipment for Use in Lighting Products | UL 8750 |
| Liquid Fuel-Burning Heating Appliances for Manufactured Homes and Recreational Vehicles | UL 307A |
| Liquid-Tight Flexible Nonmetallic Conduit | UL 1660 |
| Liquid-Tight Flexible Steel Conduit | UL 360 |
| Lithium Batteries | UL 1642 |
| Low-Voltage Fuses — Fuses for Photovoltaic Systems | Subject 2579 |
| Low-Voltage Fuses — Part 1: General Requirements | UL 248-1 |
| Low-Voltage Fuses — Part 2: Class C Fuses | UL 248-2 |
| Low-Voltage Fuses — Part 3: Class CA and CB Fuses | UL 248-3 |
| Low-Voltage Fuses — Part 4: Class CC Fuses | UL 248-4 |
| Low-Voltage Fuses — Part 5: Class G Fuses | UL 248-5 |
| Low-Voltage Fuses — Part 6: Class H Renewable Fuses | UL 248-6 |
| Low-Voltage Fuses — Part 7: Class H Renewable Fuses | UL 248-7 |
| Low-Voltage Fuses — Part 8: Class J Fuses | UL 248-8 |
| Low-Voltage Fuses — Part 9: Class K Fuses | UL 248-9 |
| Low-Voltage Fuses — Part 10: Class L Fuses | UL 249-10 |
| Low-Voltage Fuses — Part 11: Plug Fuses | UL 248-11 |
| Low-Voltage Fuses — Part 12: Class R Fuses | UL 248-12 |
| Low-Voltage Fuses — Part 13: Semiconductor Fuses | UL 248–13 |
| Low-Voltage Fuses — Part 14: Supplemental Fuses | UL 248–14 |
| Low-Voltage Fuses — Part 15: Class T Fuses | UL 248-15 |
| Low-Voltage Fuses — Part 16: Test Limiters | UL 248-16 |
| Low-Voltage Landscape Lighting Systems | UL 1838 |
| Low-Voltage Lighting Fixtures for Use in Recreational Vehicles | UL 234 |
| Low-Voltage Lighting Systems | UL 2108 |
| Low-Voltage Switchgear and Controlgear — Part 4-1A: Contactors and Motor-Starters — Electromechanical Contactors and Motor-Starters | UL 60947-4-1A |
| Low-Voltage Switchgear and Controlgear — Part 5-2: Control Circuit Devices and Switching Elements —— Proximity Switches | UL 60947-5-2 |
| Low Voltage Transformers — Part 1: General Requirements | UL 5085-1 |
| Low Voltage Transformers — Part 3: Class 2 and Class 3 Transformers | UL 5085-3 |
| Luminaire Reflector Kits for Installation on Previously Installed Fluorescent Luminaires, Supplemental Requirements | UL 1598B |
| Luminaires | UL 1598 |
| Machine-Tool Wires and Cables | UL 1063 |
| Manufactured Wiring Systems | UL 183 |
| Medical Electrical Equipment — Part 1: General Requirements for Safety | UL 60601–1 |
| Medium-Voltage AC Contactors, Controllers, and Control Centers | UL 347 |
| Medium-Voltage Power Cables | UL 1072 |
| Metal-Clad Cables | UL 1569 |
| • | |
| Metallic Outlet Boxes | UL 514A |
| Mobile Home Pipe Heating Cable | Subject 1462 |
| Molded-Case Circuit Breakers, Molded-Case Switches, and Circuit-Breaker Enclosures | UL 489 |
| Molded-Case Circuit Breakers, Molded-Case Switches, and Circuit-Breaker Enclosures for Use with Photovoltaic (PV) Systems | Subject 489B |
| Motor Control Centers | UL 845 |

| Product Standard Name | Product Standard Number |
|---|---|
| Motor-Operated Appliances | UL 73 |
| Multi-Pole Connectors for Use in Photovoltaic Systems | Subject 6703A |
| Neon Transformers and Power Supplies | UL 2161 |
| Nonincendive Electrical Equipment for Use in Class I and II, Division 2 and Class III, Divisions 1 and 2 Hazardous (Classified) Locations | ANSI/ISA-12.12.01 |
| Nonmetallic Outlet Boxes, Flush-Device Boxes, and Covers | UL 514C |
| Nonmetallic Surface Raceways and Fittings | UL 5A |
| Nonmetallic Underground Conduit with Conductors | UL 1990 |
| Office Furnishings | UL 1286 |
| Optical Fiber Cable | UL 1651 |
| Optical Fiber and Communication Cable Raceway | UL 2024 |
| Panelboards | UL 67 |
| Performance Requirements of Detectors for Flammable Gases | ANSI/ISA-60079-29-1 |
| Personnel Protection Systems for Electric Vehicle (EV) Supply Circuits; Part 1: General Requirements | UL 2231–1 |
| Personnel Protection Systems for Electric Vehicle (EV) Supply Circuits; Part 2: Particular Requirements for Protection Devices for Use in Charging Systems | UL 2231–2 |
| Photovoltaic DC Arc-Fault Circuit Protection | Subject 1699B |
| Photovoltaic Junction Boxes | Subject 3730 |
| Photovoltaic Wire | Subject 4703 |
| Plugs, Receptacles and Couplers for Electrical Vehicles | UL 2251 |
| Portable Electric Luminaires | UL 153 |
| Portable Power-Distribution Equipment | UL 1640 |
| Potting Compounds for Swimming Pool, Fountain, and Spa Equipment | UL 676A |
| Power Conversion Equipment | UL 508C |
| Power Outlets | UL 231 |
| Power Units Other Than Class 2 | UL 1012 |
| Power-Limited Circuit Cables | UL 13 |
| Professional Video and Audio Equipment | UL 1419 |
| Programmable Controllers – Part 2: Equipment Requirements and Tests | UL 61131-2 |
| Protectors for Coaxial Communications Circuits | UL 497C |
| Protectors for Data Communication and Fire Alarm Circuits | UL 497B |
| Protectors for Paired-Conductor Communications Circuits | UL 497 |
| • | |
| Reference Standard for Electrical Wires, Cables, and Flexible Cords | UL 1581 |
| • | |
| Requirements for Process Sealing Between Electrical Systems and Potentially Flammable or Combustible Process Fluids | ANSI/ISA-12.27.01 |
| Residential Pipe Heating Cable | Subject 2049 |
| Roof and Gutter De-Icing Cable Units | Subject 1588 |
| Room Air Conditioners | UL 484 |
| Rotating Electrical Machines — General Requirements | UL 1004-1 |
| • | |
| Schedule 40, 80, Type EB and A Rigid PVC Conduit and Fittings | UL 651 |
| Schedule 40 and 80 High Density Polyethylene (HDPE) Conduit | UL 651A |
| Sealed Wire Connector Systems | UL 486D |
| Seasonal and Holiday Decorative Products | UL 588 |
| Secondary Protectors for Communications Circuits | UL 497A |
| Self-Ballasted Lamps and Lamp Adapters | UL 1993 |
| Service-Entrance Cables | UL 854 |
| Smoke Detectors for Fire Alarm Signaling Systems | UL 268 |
| Solar Trackers | Subject 3703 |
| Specialty Transformers | UL 506 |
| Splicing Wire Connectors | UL 486C |
| Stage and Studio Luminaires and Connector Strips | UL 1573 |
| Standby Batteries | UL 1989 |

| Product Standard Name | Product Standard Number |
|---|---|
| Stationary Engine Generator Assemblies | UL 2200 |
| Strut-Type Channel Raceways and Fittings | UL 5B |
| Supplemental Requirements for Extra-Heavy Wall Reinforced Thermosetting Resin Conduit (RTRC) and Fittings | UL 2515A |
| Surface Metal Raceways and Fittings | UL 5 |
| Surface Raceways and Fittings for Use with Data, Signal and Control Circuits | UL 5C |
| Surge Arresters — Gapped Silicon-Carbide Surge Arresters for AC Power Circuits | IEEE C62.1 |
| Surge Arresters — Metal-Oxide Surge Arresters for AC Power Circuits | IEEE C62.11 |
| Surge Protective Devices | UL 1449 |
| Swimming Pool Pumps, Filters, and Chlorinators | UL 1081 |
| Switchboards | UL 891 |
| • | |
| Thermally Protected Motors | UL 1004-3 |
| Transfer Switch Equipment | UL 1008 |
| • | |
| Underfloor Raceways and Fittings | UL 884 |
| Underwater Luminaires and Submersible Junction Boxes | UL 676 |
| Uninterruptible Power Systems | UL 1778 |
| Vacuum Cleaners, Blower Cleaners, and Household Floor Finishing Machines | UL 1017 |
| Waste Disposers | UL 430 |
| Wind Turbine Generating Systems | Subject 6140 |
| Wind Turbine Generating Systems — Small | Subject 6142 |
| Wire Connectors | UL 486A, UL 486B |
| Wireways, Auxiliary Gutters, and Associated Fittings | UL 870 |

# B Informative Annex
## Application Information for Ampacity Calculation

*This informative annex is not a part of the requirements of this NFPA document but is included for informational purposes only.*

### B.310.15(B)(1) Equation Application Information

This informative annex provides application information for ampacities calculated under engineering supervision.

Informative Annex B is based on calculations using the Neher–McGrath conductor ampacity formula. The use of this method to calculate conductor ampacity is permitted under engineering supervision per 310.15(A)(1) and 310.15(C). The formula is found in 310.15(C).

Although conductor ampacities calculated using this formula may exceed those found in a table of allowable ampacities, such as Table 310.15(B)(16), the limitations for connecting to equipment terminals specified in 110.14(C) must be followed. For equipment 600 volts and under, the ampacities used with equipment terminals are based on Table 310.15(B)(16). See the commentary following 110.14(C) and 310.15(C).

### B.310.15(B)(2) Typical Applications Covered by Tables

Typical ampacities for conductors rated 0 through 2000 volts are shown in Table B.310.15(B)(2)(1) through Table B.310.15(B)(2)(10). Table B.310.15(B)(2)(11) provides the adjustment factors for more than three current-carrying conductors in a raceway or cable with load diversity. Underground electrical duct bank configurations, as detailed in Figure B.310.15(B)(2)(3), Figure B.310.15(B)(2)(4), and Figure B.310.15(B)(2)(5), are utilized for conductors rated 0 through 5000 volts. In Figure B.310.15(B)(2)(2) through Figure B.310.15(B)(2)(5), where adjacent duct banks are used, a separation of 1.5 m (5 ft) between the centerlines of the closest ducts in each bank or 1.2 m (4 ft) between the extremities of the concrete envelopes is sufficient to prevent derating of the conductors due to mutual heating. These ampacities were calculated as detailed in the basic ampacity paper, AIEE Paper 57-660, *The Calculation of the Temperature Rise and Load Capability of Cable Systems,* by J. H. Neher and M. H. McGrath. For additional information concerning the application of these ampacities, see IEEE/ICEA Standard S-135/P-46-426, *Power Cable Ampacities*, and IEEE Standard 835-1994, *Standard Power Cable Ampacity Tables*.

Typical values of thermal resistivity (Rho) are as follows:

Average soil (90 percent of USA) = 90

Concrete = 55

Damp soil (coastal areas, high water table) = 60

Paper insulation = 550

Polyethylene (PE) = 450

Polyvinyl chloride (PVC) = 650

Rubber and rubber-like = 500

Very dry soil (rocky or sandy) = 120

*Thermal resistivity*, as used in this informative annex, refers to the heat transfer capability through a substance by conduction. It is the reciprocal of thermal conductivity and is normally expressed in the units°C-cm/watt. For additional information on determining soil thermal resistivity (Rho), see ANSI/IEEE Standard 442-1996, *Guide for Soil Thermal Resistivity Measurements*.

If other factors remain the same, a soil resistivity higher than 90 reduces ampacities to values below those listed in Tables B.310.15(B)(2)(5) through B.310.15(B)(2)(10). Conversely, a load factor less than 100 percent increases ampacities if other factors remain the same. See B.310.15(B)(7) for allowable adjustments if the load factor is less than 100 percent. Reduced load factors are used in Informational Note Figures B.310.15(B)(2)(3) through B.310.15(B)(2)(5).

### B.310.15(B)(3) Criteria Modifications

Where values of load factor and Rho are known for a particular electrical duct bank installation and they are different from those shown in a specific table or figure, the ampacities shown in the table or figure can be modified by the application of factors derived from the use of Figure B.310.15(B)(2)(1).

Where two different ampacities apply to adjacent portions of a circuit, the higher ampacity can be used beyond the point of transition, a distance equal to 3 m (10 ft) or 10 percent of the circuit length calculated at the higher ampacity, whichever is less.

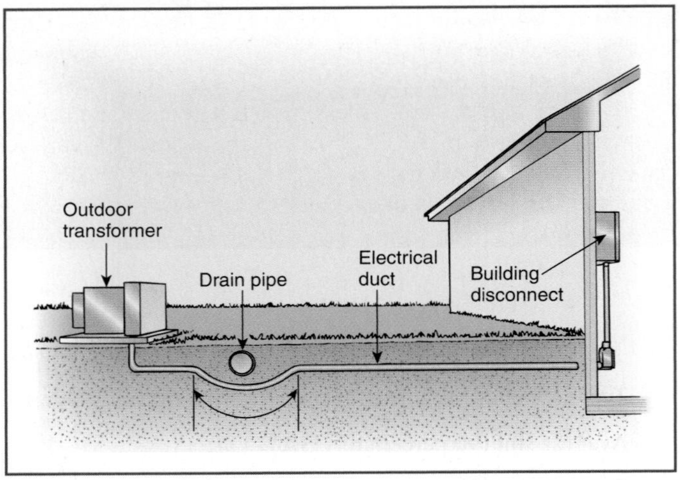


***EXHIBIT B.1*** *The presence of a drain pipe results in a change of burial depth for that portion of the electrical installation under the drain pipe.*

Where the burial depth of direct burial or electrical duct bank circuits are modified from the values shown in a figure or table, ampacities can be modified as shown in (a) and (b) as follows.

The information provided in B.310.15(B)(3) on how to address different conductor ampacities that apply to adjacent portions of a circuit is also specified in the exception to 310.15(A)(2). See the commentary following that exception.

Exhibit B.1 illustrates an installation where a change in the burial depth results in an analysis of whether a modification to the conductor ampacity is necessary. If that portion deeper than 30 inches does not exceed 25 percent of the total run length, no decrease in ampacity is required, even if part of the run is more than 30 inches deep, which is the maximum depth assumed in Figure B.310.15(B)(2)(2), Note 1, to maintain the accuracy of the tables.

(a) Where burial depths are increased in part(s) of an electrical duct run to avoid underground obstructions, no decrease in ampacity of the conductors is needed, provided the total length of parts of the duct run increased in depth to avoid obstructions is less than 25 percent of the total run length.

(b) Where burial depths are deeper than shown in a specific underground ampacity table or figure, an ampacity derating factor of 6 percent per increased 300 mm (foot) of depth for all values of Rho can be utilized. No rating change is needed where the burial depth is decreased.

For example, in accordance with Table B.310.15(B)(2)(7) and Figure B.310.15(B)(2)(2), the ampacity of six parallel runs of 500 kcmil, Type XHHW copper conductors [as shown in Detail 3 in Figure B.310.15(B)(2)(2)], where Rho = 90, is 6 × 273 A = 1638 A at a depth measuring no more than 30 inches to the top duct in the bank. If the burial depth is 6 feet, the ampacity is calculated as follows: 1638 A − (3.5 × 0.06 × 1638 A) = 1294 A.

## B.310.15(B)(4) Electrical Ducts

The term *electrical duct(s)* is defined in 310.60.

## B.310.15(B)(5) Table B.310.15(B)(2)(6) and Table B.310.15(B)(2)(7)

(a) To obtain the ampacity of cables installed in two electrical ducts in one horizontal row with 190-mm (7.5-in.) center-to-center spacing between electrical ducts, similar to Figure B.310.15(B)(2)(2), Detail 1, multiply the ampacity shown for one duct in Table B.310.15(B)(2)(6) and Table B.310.15(B)(2)(7) by 0.88.

(b) To obtain the ampacity of cables installed in four electrical ducts in one horizontal row with 190-mm (7.5-in.) center-to-center spacing between electrical ducts, similar to Figure B.310.15(B)(2)(2), Detail 2, multiply the ampacity shown for three electrical ducts in Table B.310.15(B)(2)(6) and Table B.310.15(B)(2)(7) by 0.94.

The underground ampacity tables [Tables B.310.15(B)(2)(5) through B.310.15(B)(2)(10)] are based on the 7.5-inch center-to-center spacing illustrated in Figure B.310.15(B)(2)(2). Although moving directly buried cables or electrical ducts farther apart will increase ampacities, the effect is surprisingly small. One calculation indicates that two side-by-side electrical ducts buried 30 inches below grade would have to be spaced about 5 feet apart before they could be considered single electrical ducts as shown in Detail 1, Figure B.310.15(B)(2)(2). Each of these factors – decreasing the burial depth, decreasing the thermal resistivity of the earth or other surrounding medium, and decreasing the load factor – has a much greater effect in increasing ampacity than does increasing the horizontal spacing.

## B.310.15(B)(6) Electrical Ducts Used in Figure B.310.15(B)(2)(2)

If spacing between electrical ducts, as shown in Figure B.310.15(B)(2)(2), is less than as specified where electrical ducts enter equipment enclosures from underground, the ampacity of conductors contained within such electrical ducts need not be reduced.

## B.310.15(B)(7) Examples Showing Use of Figure B.310.15(B)(2)(1) for Electrical Duct Bank Ampacity Modifications

Figure B.310.15(B)(2)(1) is used for interpolation or extrapolation for values of Rho and load factor for cables installed in electrical ducts. The upper family of curves shows the variation in ampacity and Rho at unity load factor in terms of $I_1$, the ampacity for Rho = 60, and 50 percent load factor. Each curve is designated for a particular ratio $I_2/I_1$, where $I_2$ is the ampacity at Rho = 120 and 100 percent load factor.

The lower family of curves shows the relationship between Rho and load factor that will give substantially the same ampacity as the indicated value of Rho at 100 percent load factor.

As an example, to find the ampacity of a 500-kcmil copper cable circuit for six electrical ducts as shown in Table B.310.15(B)(2)(5): At the Rho = 60, LF = 50, $I_1$ = 583; for Rho = 120 and LF = 100, $I_2$ = 400. The ratio $I_2/I_1$ = 0.686. Locate Rho = 90 at the bottom of the chart and follow the 90 Rho line to the intersection with 100 percent load factor where the equivalent Rho = 90. Then follow the 90 Rho line to $I_2/I_1$ ratio of 0.686 where $F = 0.74$. The desired ampacity = 0.74 × 583 = 431, which agrees with the table for Rho = 90, LF = 100.

To determine the ampacity for the same circuit where Rho = 80 and LF = 75, using Figure B.310.15(B)(2)(1), the equivalent Rho = 43, F = 0.855, and the desired ampacity = 0.855 × 583 = 498 amperes. Values for using Figure B.310.15(B)(2)(1) are found in the electrical duct bank ampacity tables of this informative annex.

Where the load factor is less than 100 percent and can be verified by measurement or calculation, the ampacity of electrical duct bank installations can be modified as shown. Different values of Rho can be accommodated in the same manner.

***TABLE B.310.15(B)(2)(1)*** *Ampacities of Two or Three Insulated Conductors, Rated 0 Through 2000 Volts, Within an Overall Covering (Multiconductor Cable), in Raceway in Free Air Based on Ambient Air Temperature of 30°C (86°F)**

| | Temperature Rating of Conductor. [See Table 310.104(A).] | | | | | | |
|---|---|---|---|---|---|---|---|
| | **60°C (140°F)** | **75°C (167°F)** | **90°C (194°F)** | **60°C (140°F)** | **75°C (167°F)** | **90°C (194°F)** | |
| | **Types TW, UF** | **Types RHW, THHW, THW, THWN, XHHW, ZW** | **Types THHN, THHW, THW-2, THWN-2, RHH, RWH-2, USE-2, XHHW, XHHW-2, ZW-2** | **Type TW** | **Types RHW, THHW, THW, THWN, XHHW** | **Types THHN, THHW, THW-2, THWN-2, RHH, RWH-2, USE-2, XHHW, XHHW-2, ZW-2** | |
| **Size (AWG or kcmil)** | **COPPER** | | | **ALUMINUM OR COPPER-CLAD ALUMINUM** | | | **Size (AWG or kcmil)** |
| 14 | 16** | 18** | 21** | — | — | — | 14 |
| 12 | 20** | 24** | 27** | 16** | 18** | 21** | 12 |
| 10 | 27** | 33** | 36** | 21** | 25** | 28** | 10 |
| 8 | 36 | 43 | 48 | 28 | 33 | 37 | 8 |
| 6 | 48 | 58 | 65 | 38 | 45 | 51 | 6 |
| 4 | 66 | 79 | 89 | 51 | 61 | 69 | 4 |
| 3 | 76 | 90 | 102 | 59 | 70 | 79 | 3 |
| 2 | 88 | 105 | 119 | 69 | 83 | 93 | 2 |
| 1 | 102 | 121 | 137 | 80 | 95 | 106 | 1 |
| 1/0 | 121 | 145 | 163 | 94 | 113 | 127 | 1/0 |
| 2/0 | 138 | 166 | 186 | 108 | 129 | 146 | 2/0 |
| 3/0 | 158 | 189 | 214 | 124 | 147 | 167 | 3/0 |
| 4/0 | 187 | 223 | 253 | 147 | 176 | 197 | 4/0 |
| 250 | 205 | 245 | 276 | 160 | 192 | 217 | 250 |
| 300 | 234 | 281 | 317 | 185 | 221 | 250 | 300 |
| 350 | 255 | 305 | 345 | 202 | 242 | 273 | 350 |
| 400 | 274 | 328 | 371 | 218 | 261 | 295 | 400 |
| 500 | 315 | 378 | 427 | 254 | 303 | 342 | 500 |
| 600 | 343 | 413 | 468 | 279 | 335 | 378 | 600 |
| 700 | 376 | 452 | 514 | 310 | 371 | 420 | 700 |
| 750 | 387 | 466 | 529 | 321 | 384 | 435 | 750 |
| 800 | 397 | 479 | 543 | 331 | 397 | 450 | 800 |
| 900 | 415 | 500 | 570 | 350 | 421 | 477 | 900 |
| 1000 | 448 | 542 | 617 | 382 | 460 | 521 | 1000 |

*Refer to 310.15(B)(2) for the ampacity correction factors where the ambient temperature is other than 30°C (86°F).

**Unless otherwise specifically permitted elsewhere in this *Code*, the overcurrent protection for these conductor types shall not exceed 15 amperes for 14 AWG, 20 amperes for 12 AWG, and 30 amperes for 10 AWG copper; or 15 amperes for 12 AWG and 25 AWG amperes for 10 AWG aluminum and copper-clad aluminum.

***TABLE B.310.15(B)(2)(3)*** *Ampacities of Multiconductor Cables with Not More Than Three Insulated Conductors, Rated 0 Through 2000 Volts, in Free Air Based on Ambient Air Temperature of 40°C (104°F) (for Types TC, MC, MI, UF, and USE Cables)**

| Size (AWG or kcmil) | Temperature Rating of Conductor. [See Table 310.104(A).] | | | | | | | | Size (AWG or kcmil) |
|---|---|---|---|---|---|---|---|---|---|
| | 60°C (140°F) | 75°C (167°F) | 85°C (185°F) | 90°C (194°F) | 60°C (140°F) | 75°C (167°F) | 85°C (185°F) | 90°C (194°F) | |
| | COPPER | | | | ALUMINUM OR COPPER-CLAD ALUMINUM | | | | |
| 18 | — | — | — | 11 | — | — | — | — | 18 |
| 16 | — | — | — | 16 | — | — | — | — | 16 |
| 14 | 18** | 21** | 24** | 25** | — | — | — | — | 14 |
| 12 | 21** | 28** | 30** | 32** | 18** | 21** | 24** | 25** | 12 |
| 10 | 28** | 36** | 41** | 43** | 21** | 28** | 30** | 32** | 10 |
| 8 | 39 | 50 | 56 | 59 | 30 | 39 | 44 | 46 | 8 |
| 6 | 52 | 68 | 75 | 79 | 41 | 53 | 59 | 61 | 6 |
| 4 | 69 | 89 | 100 | 104 | 54 | 70 | 78 | 81 | 4 |
| 3 | 81 | 104 | 116 | 121 | 63 | 81 | 91 | 95 | 3 |
| 2 | 92 | 118 | 132 | 138 | 72 | 92 | 103 | 108 | 2 |
| 1 | 107 | 138 | 154 | 161 | 84 | 108 | 120 | 126 | 1 |
| 1/0 | 124 | 160 | 178 | 186 | 97 | 125 | 139 | 145 | 1/0 |
| 2/0 | 143 | 184 | 206 | 215 | 111 | 144 | 160 | 168 | 2/0 |
| 3/0 | 165 | 213 | 238 | 249 | 129 | 166 | 185 | 194 | 3/0 |
| 4/0 | 190 | 245 | 274 | 287 | 149 | 192 | 214 | 224 | 4/0 |
| 250 | 212 | 274 | 305 | 320 | 166 | 214 | 239 | 250 | 250 |
| 300 | 237 | 306 | 341 | 357 | 186 | 240 | 268 | 280 | 300 |
| 350 | 261 | 337 | 377 | 394 | 205 | 265 | 296 | 309 | 350 |
| 400 | 281 | 363 | 406 | 425 | 222 | 287 | 317 | 334 | 400 |
| 500 | 321 | 416 | 465 | 487 | 255 | 330 | 368 | 385 | 500 |
| 600 | 354 | 459 | 513 | 538 | 284 | 368 | 410 | 429 | 600 |
| 700 | 387 | 502 | 562 | 589 | 306 | 405 | 462 | 473 | 700 |
| 750 | 404 | 523 | 586 | 615 | 328 | 424 | 473 | 495 | 750 |
| 800 | 415 | 539 | 604 | 633 | 339 | 439 | 490 | 513 | 800 |
| 900 | 438 | 570 | 639 | 670 | 362 | 469 | 514 | 548 | 900 |
| 1000 | 461 | 601 | 674 | 707 | 385 | 499 | 558 | 584 | 1000 |

*Refer to 310.15(B)(2) for the ampacity correction factors where the ambient temperature is other than 40°C (104°F).

**Unless otherwise specifically permitted elsewhere in this *Code*, the overcurrent protection for these conductor types shall not exceed 15 amperes for 14 AWG, 20 amperes for 12 AWG, and 30 amperes for 10 AWG copper; or 15 amperes for 12 AWG and 25 amperes for 10 AWG aluminum and copper-clad aluminum.

***TABLE B.310.15(B)(2)(5)*** *Ampacities of Single Insulated Conductors, Rated 0 through 2000 Volts, in Nonmagnetic Underground Electrical Ducts (One Conductor per Electrical Duct), Based on Ambient Earth Temperature of 20°C (68°F), Electrical Duct Arrangement in Accordance with Figure B.310.15(B)(2)(2), Conductor Temperature 75°C (167°F)*

| Size (kcmil) | 3 Electrical Ducts (Fig. B.310.15(B)(2)(2), Detail 2) | | | 6 Electrical Ducts (Fig. B.310.15(B)(2)(2), Detail 3) | | | 9 Electrical Ducts (Fig. B.310.15(B)(2)(2), Detail 4) | | | 3 Electrical Ducts (Fig. B.310.15(B)(2)(2), Detail 2) | | | 6 Electrical Ducts (Fig. B.310.15(B)(2)(2), Detail 3) | | | 9 Electrical Ducts (Fig. B.310.15(B)(2)(2), Detail 4) | | | Size (kcmil) |
|---|---|---|---|---|---|---|---|---|---|---|---|---|---|---|---|---|---|---|---|
| | Types RHW, THHW, THW, THWN, XHHW, USE | | | Types RHW, THHW, THW, THWN, XHHW, USE | | | Types RHW, THHW, THW, THWN, XHHW, USE | | | Types RHW, THHW, THW, THWN, XHHW, USE | | | Types RHW, THHW, THW, THWN, XHHW, USE | | | Types RHW, THHW, THW, THWN, XHHW, USE | | | |
| | COPPER | | | | | | | | | ALUMINUM OR COPPER-CLAD ALUMINUM | | | | | | | | | |
| | RHO 60 LF 50 | RHO 90 LF 100 | RHO 120 LF 100 | RHO 60 LF 50 | RHO 90 LF 100 | RHO 120 LF 100 | RHO 60 LF 50 | RHO 90 LF 100 | RHO 120 LF 100 | RHO 60 LF 50 | RHO 90 LF 100 | RHO 120 LF 100 | RHO 60 LF 50 | RHO 90 LF 100 | RHO 120 LF 100 | RHO 60 LF 50 | RHO 90 LF 100 | RHO 120 LF 100 | |
| 250 | 410 | 344 | 327 | 386 | 295 | 275 | 369 | 270 | 252 | 320 | 269 | 256 | 302 | 230 | 214 | 288 | 211 | 197 | 250 |
| 350 | 503 | 418 | 396 | 472 | 355 | 330 | 446 | 322 | 299 | 393 | 327 | 310 | 369 | 277 | 258 | 350 | 252 | 235 | 350 |
| 500 | 624 | 511 | 484 | 583 | 431 | 400 | 545 | 387 | 360 | 489 | 401 | 379 | 457 | 337 | 313 | 430 | 305 | 284 | 500 |
| 750 | 794 | 640 | 603 | 736 | 534 | 494 | 674 | 469 | 434 | 626 | 505 | 475 | 581 | 421 | 389 | 538 | 375 | 347 | 750 |
| 1000 | 936 | 745 | 700 | 864 | 617 | 570 | 776 | 533 | 493 | 744 | 593 | 557 | 687 | 491 | 453 | 629 | 432 | 399 | 1000 |
| 1250 | 1055 | 832 | 781 | 970 | 686 | 632 | 854 | 581 | 536 | 848 | 668 | 627 | 779 | 551 | 508 | 703 | 478 | 441 | 1250 |
| 1500 | 1160 | 907 | 849 | 1063 | 744 | 685 | 918 | 619 | 571 | 941 | 736 | 689 | 863 | 604 | 556 | 767 | 517 | 477 | 1500 |
| 1750 | 1250 | 970 | 907 | 1142 | 793 | 729 | 975 | 651 | 599 | 1026 | 796 | 745 | 937 | 651 | 598 | 823 | 550 | 507 | 1750 |
| 2000 | 1332 | 1027 | 959 | 1213 | 836 | 768 | 1030 | 683 | 628 | 1103 | 850 | 794 | 1005 | 693 | 636 | 877 | 581 | 535 | 2000 |
| Ambient Temp. (°C) | Correction Factors | | | | | | | | | | | | | | | | | | Ambient Temp. (°F) |
| 6–10 | 1.09 | | | 1.09 | | | 1.09 | | | 1.09 | | | 1.09 | | | 1.09 | | | 43–50 |
| 11–15 | 1.04 | | | 1.04 | | | 1.04 | | | 1.04 | | | 1.04 | | | 1.04 | | | 52–59 |
| 16–20 | 1.00 | | | 1.00 | | | 1.00 | | | 1.00 | | | 1.00 | | | 1.00 | | | 61–68 |
| 21–25 | 0.95 | | | 0.95 | | | 0.95 | | | 0.95 | | | 0.95 | | | 0.95 | | | 70–77 |
| 26–30 | 0.90 | | | 0.90 | | | 0.90 | | | 0.90 | | | 0.90 | | | 0.90 | | | 79–86 |

***TABLE B.310.15(B)(2)(6)*** *Ampacities of Three Insulated Conductors, Rated 0 through 2000 Volts, Within an Overall Covering (Three-Conductor Cable) in Underground Electrical Ducts (One Cable per Electrical Duct) Based on Ambient Earth Temperature of 20°C (68°F), Electrical Duct Arrangement in Accordance with Figure B.310.15(B)(2)(2), Conductor Temperature 75°C (167°F)*

| Size (AWG or kcmil) | 1 Electrical Duct (Fig. B.310.15(B)(2)(2), Detail 1) | | | 3 Electrical Ducts (Fig. B.310.15(B)(2)(2), Detail 2) | | | 6 Electrical Ducts (Fig. B.310.15(B)(2)(2), Detail 3) | | | 1 Electrical Duct (Fig. B.310.15(B)(2)(2), Detail 1) | | | 3 Electrical Ducts (Fig. B.310.15(B)(2)(2), Detail 2) | | | 6 Electrical Ducts (Fig. B.310.15(B)(2)(2), Detail 3) | | | Size (AWG or kcmil) |
|---|---|---|---|---|---|---|---|---|---|---|---|---|---|---|---|---|---|---|---|
| | Types RHW, THHW, THW, THWN, XHHW, USE | | | Types RHW, THHW, THW, THWN, XHHW, USE | | | Types RHW, THHW, THW, THWN, XHHW, USE | | | Types RHW, THHW, THW, THWN, XHHW, USE | | | Types RHW, THHW, THW, THWN, XHHW, USE | | | Types RHW, THHW, THW, THWN, XHHW, USE | | | |
| | COPPER | | | | | | | | | ALUMINUM OR COPPER-CLAD ALUMINUM | | | | | | | | | |
| | RHO 60 LF 50 | RHO 90 LF 100 | RHO 120 LF 100 | RHO 60 LF 50 | RHO 90 LF 100 | RHO 120 LF 100 | RHO 60 LF 50 | RHO 90 LF 100 | RHO 120 LF 100 | RHO 60 LF 50 | RHO 90 LF 100 | RHO 120 LF 100 | RHO 60 LF 50 | RHO 90 LF 100 | RHO 120 LF 100 | RHO 60 LF 50 | RHO 90 LF 100 | RHO 120 LF 100 | |
| 8 | 58 | 54 | 53 | 56 | 48 | 46 | 53 | 42 | 39 | 45 | 42 | 41 | 43 | 37 | 36 | 41 | 32 | 30 | 8 |
| 6 | 77 | 71 | 69 | 74 | 63 | 60 | 70 | 54 | 51 | 60 | 55 | 54 | 57 | 49 | 47 | 54 | 42 | 39 | 6 |
| 4 | 101 | 93 | 91 | 96 | 81 | 77 | 91 | 69 | 65 | 78 | 72 | 71 | 75 | 63 | 60 | 71 | 54 | 51 | 4 |
| 2 | 132 | 121 | 118 | 126 | 105 | 100 | 119 | 89 | 83 | 103 | 94 | 92 | 98 | 82 | 78 | 92 | 70 | 65 | 2 |
| 1 | 154 | 140 | 136 | 146 | 121 | 114 | 137 | 102 | 95 | 120 | 109 | 106 | 114 | 94 | 89 | 107 | 79 | 74 | 1 |
| 1/0 | 177 | 160 | 156 | 168 | 137 | 130 | 157 | 116 | 107 | 138 | 125 | 122 | 131 | 107 | 101 | 122 | 90 | 84 | 1/0 |
| 2/0 | 203 | 183 | 178 | 192 | 156 | 147 | 179 | 131 | 121 | 158 | 143 | 139 | 150 | 122 | 115 | 140 | 102 | 95 | 2/0 |
| 3/0 | 233 | 210 | 204 | 221 | 178 | 158 | 205 | 148 | 137 | 182 | 164 | 159 | 172 | 139 | 131 | 160 | 116 | 107 | 3/0 |
| 4/0 | 268 | 240 | 232 | 253 | 202 | 190 | 234 | 168 | 155 | 209 | 187 | 182 | 198 | 158 | 149 | 183 | 131 | 121 | 4/0 |
| 250 | 297 | 265 | 256 | 280 | 222 | 209 | 258 | 184 | 169 | 233 | 207 | 201 | 219 | 174 | 163 | 202 | 144 | 132 | 250 |
| 350 | 363 | 321 | 310 | 340 | 267 | 250 | 312 | 219 | 202 | 285 | 252 | 244 | 267 | 209 | 196 | 245 | 172 | 158 | 350 |
| 500 | 444 | 389 | 375 | 414 | 320 | 299 | 377 | 261 | 240 | 352 | 308 | 297 | 328 | 254 | 237 | 299 | 207 | 190 | 500 |
| 750 | 552 | 478 | 459 | 511 | 388 | 362 | 462 | 314 | 288 | 446 | 386 | 372 | 413 | 314 | 293 | 374 | 254 | 233 | 750 |
| 1000 | 628 | 539 | 518 | 579 | 435 | 405 | 522 | 351 | 321 | 521 | 447 | 430 | 480 | 361 | 336 | 433 | 291 | 266 | 1000 |
| **Ambient Temp. (°C)** | **Correction Factors** | | | | | | | | | | | | | | | | | | **Ambient Temp. (°F)** |
| 6–10 | 1.09 | | | 1.09 | | | 1.09 | | | 1.09 | | | 1.09 | | | 1.09 | | | 43–50 |
| 11–15 | 1.04 | | | 1.04 | | | 1.04 | | | 1.04 | | | 1.04 | | | 1.04 | | | 52–59 |
| 16–20 | 1.00 | | | 1.00 | | | 1.00 | | | 1.00 | | | 1.00 | | | 1.00 | | | 61–68 |
| 21–25 | 0.95 | | | 0.95 | | | 0.95 | | | 0.95 | | | 0.95 | | | 0.95 | | | 70–77 |
| 26–30 | 0.90 | | | 0.90 | | | 0.90 | | | 0.90 | | | 0.90 | | | 0.90 | | | 79–86 |

***TABLE B.310.15(B)(2)(7)*** *Ampacities of Three Single Insulated Conductors, Rated 0 Through 2000 Volts, in Underground Electrical Ducts (Three Conductors per Electrical Duct) Based on Ambient Earth Temperature of 20°C (68°F), Electrical Duct Arrangement in Accordance with Figure B.310.15(B)(2)(2), Conductor Temperature 75°C (167°F)*

| Size (AWG or kcmil) | 1 Electrical Duct (Fig. B.310.15(B)(2)(2), Detail 1) | | | 3 Electrical Ducts (Fig. B.310.15(B)(2)(2), Detail 2) | | | 6 Electrical Ducts (Fig. B.310.15(B)(2)(2), Detail 3) | | | 1 Electrical Duct (Fig. B.310.15(B)(2)(2), Detail 1) | | | 3 Electrical Ducts (Fig. B.310.15(B)(2)(2), Detail 2) | | | 6 Electrical Ducts (Fig. B.310.15(B)(2)(2), Detail 3) | | | Size (AWG or kcmil) |
|---|---|---|---|---|---|---|---|---|---|---|---|---|---|---|---|---|---|---|---|
| | Types RHW, THHW, THW, THWN, XHHW, USE | | | Types RHW, THHW, THW, THWN, XHHW, USE | | | Types RHW, THHW, THW, THWN, XHHW, USE | | | Types RHW, THHW, THW, THWN, XHHW, USE | | | Types RHW, THHW, THW, THWN, XHHW, USE | | | Types RHW, THHW, THW, THWN, XHHW, USE | | | |
| | COPPER | | | | | | | | | ALUMINUM OR COPPER-CLAD ALUMINUM | | | | | | | | | |
| | RHO 60 LF 50 | RHO 90 LF 100 | RHO 120 LF 100 | RHO 60 LF 50 | RHO 90 LF 100 | RHO 120 LF 100 | RHO 60 LF 50 | RHO 90 LF 100 | RHO 120 LF 100 | RHO 60 LF 50 | RHO 90 LF 100 | RHO 120 LF 100 | RHO 60 LF 50 | RHO 90 LF 100 | RHO 120 LF 100 | RHO 60 LF 50 | RHO 90 LF 100 | RHO 120 LF 100 | |
| 8 | 63 | 58 | 57 | 61 | 51 | 49 | 57 | 44 | 41 | 49 | 45 | 44 | 47 | 40 | 38 | 45 | 34 | 32 | 8 |
| 6 | 84 | 77 | 75 | 80 | 67 | 63 | 75 | 56 | 53 | 66 | 60 | 58 | 63 | 52 | 49 | 59 | 44 | 41 | 6 |
| 4 | 111 | 100 | 98 | 105 | 86 | 81 | 98 | 73 | 67 | 86 | 78 | 76 | 79 | 67 | 63 | 77 | 57 | 52 | 4 |
| 3 | 129 | 116 | 113 | 122 | 99 | 94 | 113 | 83 | 77 | 101 | 91 | 89 | 83 | 77 | 73 | 84 | 65 | 60 | 3 |
| 2 | 147 | 132 | 128 | 139 | 112 | 106 | 129 | 93 | 86 | 115 | 103 | 100 | 108 | 87 | 82 | 101 | 73 | 67 | 2 |
| 1 | 171 | 153 | 148 | 161 | 128 | 121 | 149 | 106 | 98 | 133 | 119 | 115 | 126 | 100 | 94 | 116 | 83 | 77 | 1 |
| 1/0 | 197 | 175 | 169 | 185 | 146 | 137 | 170 | 121 | 111 | 153 | 136 | 132 | 144 | 114 | 107 | 133 | 94 | 87 | 1/0 |
| 2/0 | 226 | 200 | 193 | 212 | 166 | 156 | 194 | 136 | 126 | 176 | 156 | 151 | 165 | 130 | 121 | 151 | 106 | 98 | 2/0 |
| 3/0 | 260 | 228 | 220 | 243 | 189 | 177 | 222 | 154 | 142 | 203 | 178 | 172 | 189 | 147 | 138 | 173 | 121 | 111 | 3/0 |
| 4/0 | 301 | 263 | 253 | 280 | 215 | 201 | 255 | 175 | 161 | 235 | 205 | 198 | 219 | 168 | 157 | 199 | 137 | 126 | 4/0 |
| 250 | 334 | 290 | 279 | 310 | 236 | 220 | 281 | 192 | 176 | 261 | 227 | 218 | 242 | 185 | 172 | 220 | 150 | 137 | 250 |
| 300 | 373 | 321 | 308 | 344 | 260 | 242 | 310 | 210 | 192 | 293 | 252 | 242 | 272 | 204 | 190 | 245 | 165 | 151 | 300 |
| 350 | 409 | 351 | 337 | 377 | 283 | 264 | 340 | 228 | 209 | 321 | 276 | 265 | 296 | 222 | 207 | 266 | 179 | 164 | 350 |
| 400 | 442 | 376 | 361 | 394 | 302 | 280 | 368 | 243 | 223 | 349 | 297 | 284 | 321 | 238 | 220 | 288 | 191 | 174 | 400 |
| 500 | 503 | 427 | 409 | 460 | 341 | 316 | 412 | 273 | 249 | 397 | 338 | 323 | 364 | 270 | 250 | 326 | 216 | 197 | 500 |
| 600 | 552 | 468 | 447 | 511 | 371 | 343 | 457 | 296 | 270 | 446 | 373 | 356 | 408 | 296 | 274 | 365 | 236 | 215 | 600 |
| 700 | 602 | 509 | 486 | 553 | 402 | 371 | 492 | 319 | 291 | 488 | 408 | 389 | 443 | 321 | 297 | 394 | 255 | 232 | 700 |
| 750 | 632 | 529 | 505 | 574 | 417 | 385 | 509 | 330 | 301 | 508 | 425 | 405 | 461 | 334 | 309 | 409 | 265 | 241 | 750 |
| 800 | 654 | 544 | 520 | 597 | 428 | 395 | 527 | 338 | 308 | 530 | 439 | 418 | 481 | 344 | 318 | 427 | 273 | 247 | 800 |
| 900 | 692 | 575 | 549 | 628 | 450 | 415 | 554 | 355 | 323 | 563 | 466 | 444 | 510 | 365 | 337 | 450 | 288 | 261 | 900 |
| 1000 | 730 | 605 | 576 | 659 | 472 | 435 | 581 | 372 | 338 | 597 | 494 | 471 | 538 | 385 | 355 | 475 | 304 | 276 | 1000 |
| Ambient Temp. (°C) | Correction Factors | | | | | | | | | | | | | | | | | | Ambient Temp. (°F) |
| 6–10 | 1.09 | | | 1.09 | | | 1.09 | | | 1.09 | | | 1.09 | | | 1.09 | | | 43–50 |
| 11–15 | 1.04 | | | 1.04 | | | 1.04 | | | 1.04 | | | 1.04 | | | 1.04 | | | 52–59 |
| 16–20 | 1.00 | | | 1.00 | | | 1.00 | | | 1.00 | | | 1.00 | | | 1.00 | | | 61–68 |
| 21–25 | 0.95 | | | 0.95 | | | 0.95 | | | 0.95 | | | 0.95 | | | 0.95 | | | 70–77 |
| 26–30 | 0.90 | | | 0.90 | | | 0.90 | | | 0.90 | | | 0.90 | | | 0.90 | | | 79–86 |

***TABLE B.310.15(B)(2)(8)*** *Ampacities of Two or Three Insulated Conductors, Rated 0 Through 2000 Volts, Cabled Within an Overall (Two- or Three-Conductor) Covering, Directly Buried in Earth, Based on Ambient Earth Temperature of 20°C (68°F), Electrical Duct Arrangement in Accordance with Figure B.310.15(B)(2)(2), 100 Percent Load Factor, Thermal Resistance (Rho) of 90*

| | 1 Cable (Fig. B.310.15(B)(2)(2), Detail 5) | | 2 Cables (Fig. B.310.15(B)(2)(2), Detail 6) | | 1 Cable (Fig. B.310.15(B)(2)(2), Detail 5) | | 2 Cables (Fig. B.310.15(B)(2)(2), Detail 6) | | |
|---|---|---|---|---|---|---|---|---|---|
| | 60°C (140°F) | 75°C (167°F) | 60°C (140°F) | 75°C (167°F) | 60°C (140°F) | 75°C (167°F) | 60°C (140°F) | 75°C (167°F) | |
| | TYPES | | | | TYPES | | | | |
| | UF | RHW, THHW, THW, THWN, XHHW, USE | UF | RHW, THHW, THW, THWN, XHHW, USE | UF | RHW, THHW, THW, THWN, XHHW, USE | UF | RHW, THHW, THW, THWN, XHHW, USE | |
| Size (AWG or kcmil) | COPPER | | | | ALUMINUM OR COPPER-CLAD ALUMINUM | | | | Size (AWG or kcmil) |
| 8 | 64 | 75 | 60 | 70 | 51 | 59 | 47 | 55 | 8 |
| 6 | 85 | 100 | 81 | 95 | 68 | 75 | 60 | 70 | 6 |
| 4 | 107 | 125 | 100 | 117 | 83 | 97 | 78 | 91 | 4 |
| 2 | 137 | 161 | 128 | 150 | 107 | 126 | 110 | 117 | 2 |
| 1 | 155 | 182 | 145 | 170 | 121 | 142 | 113 | 132 | 1 |
| 1/0 | 177 | 208 | 165 | 193 | 138 | 162 | 129 | 151 | 1/0 |
| 2/0 | 201 | 236 | 188 | 220 | 157 | 184 | 146 | 171 | 2/0 |
| 3/0 | 229 | 269 | 213 | 250 | 179 | 210 | 166 | 195 | 3/0 |
| 4/0 | 259 | 304 | 241 | 282 | 203 | 238 | 188 | 220 | 4/0 |
| 250 | — | 333 | — | 308 | — | 261 | — | 241 | 250 |
| 350 | — | 401 | — | 370 | — | 315 | — | 290 | 350 |
| 500 | — | 481 | — | 442 | — | 381 | — | 350 | 500 |
| 750 | — | 585 | — | 535 | — | 473 | — | 433 | 750 |
| 1000 | — | 657 | — | 600 | — | 545 | — | 497 | 1000 |
| Ambient Temp. (°C) | Correction Factors | | | | | | | | Ambient Temp. (°F) |
| 6–10 | 1.12 | 1.09 | 1.12 | 1.09 | 1.12 | 1.09 | 1.12 | 1.09 | 43–50 |
| 11–15 | 1.06 | 1.04 | 1.06 | 1.04 | 1.06 | 1.04 | 1.06 | 1.04 | 52–59 |
| 16–20 | 1.00 | 1.00 | 1.00 | 1.00 | 1.00 | 1.00 | 1.00 | 1.00 | 61–68 |
| 21–25 | 0.94 | 0.95 | 0.94 | 0.95 | 0.94 | 0.95 | 0.94 | 0.95 | 70–77 |
| 26–30 | 0.87 | 0.90 | 0.87 | 0.90 | 0.87 | 0.90 | 0.87 | 0.90 | 79–86 |

Note: For ampacities of Type UF cable in underground electrical ducts, multiply the ampacities shown in the table by 0.74.

***TABLE B.310.15(B)(2)(9)*** *Ampacities of Three Triplexed Single Insulated Conductors, Rated 0 Through 2000 Volts, Directly Buried in Earth Based on Ambient Earth Temperature of 20°C (68°F), Electrical Duct Arrangement in Accordance with Figure B.310.15(B)(2)(2), 100 Percent Load Factor, Thermal Resistance (Rho) of 90*

| | See Fig. B.310.15(B)(2)(2), Detail 7 | | See Fig. B.310.15(B)(2)(2), Detail 8 | | See Fig. B.310.15(B)(2)(2), Detail 7 | | See Fig. B.310.15(B)(2)(2), Detail 8 | | |
|---|---|---|---|---|---|---|---|---|---|
| | 60°C (140°F) | 75°C (167°F) | 60°C (140°F) | 75°C (167°F) | 60°C (140°F) | 75°C (167°F) | 60°C (140°F) | 75°C (167°F) | |
| | TYPES | | | | TYPES | | | | |
| | UF | USE | UF | USE | UF | USE | UF | USE | |
| Size (AWG or kcmil) | COPPER | | | | ALUMINUM OR COPPER-CLAD ALUMINUM | | | | Size (AWG or kcmil) |
| 8 | 72 | 84 | 66 | 77 | 55 | 65 | 51 | 60 | 8 |
| 6 | 91 | 107 | 84 | 99 | 72 | 84 | 66 | 77 | 6 |
| 4 | 119 | 139 | 109 | 128 | 92 | 108 | 85 | 100 | 4 |
| 2 | 153 | 179 | 140 | 164 | 119 | 139 | 109 | 128 | 2 |
| 1 | 173 | 203 | 159 | 186 | 135 | 158 | 124 | 145 | 1 |
| 1/0 | 197 | 231 | 181 | 212 | 154 | 180 | 141 | 165 | 1/0 |
| 2/0 | 223 | 262 | 205 | 240 | 175 | 205 | 159 | 187 | 2/0 |
| 3/0 | 254 | 298 | 232 | 272 | 199 | 233 | 181 | 212 | 3/0 |
| 4/0 | 289 | 339 | 263 | 308 | 226 | 265 | 206 | 241 | 4/0 |
| 250 | — | 370 | — | 336 | — | 289 | — | 263 | 250 |
| 350 | — | 445 | — | 403 | — | 349 | — | 316 | 350 |
| 500 | — | 536 | — | 483 | — | 424 | — | 382 | 500 |
| 750 | — | 654 | — | 587 | — | 525 | — | 471 | 750 |
| 1000 | — | 744 | — | 665 | — | 608 | — | 544 | 1000 |
| **Ambient Temp. (°C)** | **Correction Factors** | | | | | | | | **Ambient Temp. (°F)** |
| 6–10 | 1.12 | 1.09 | 1.12 | 1.09 | 1.12 | 1.09 | 1.12 | 1.09 | 43–50 |
| 11–15 | 1.06 | 1.04 | 1.06 | 1.04 | 1.06 | 1.04 | 1.06 | 1.04 | 52–59 |
| 16–20 | 1.00 | 1.00 | 1.00 | 1.00 | 1.00 | 1.00 | 1.00 | 1.00 | 61–68 |
| 21–25 | 0.94 | 0.95 | 0.94 | 0.95 | 0.94 | 0.95 | 0.94 | 0.95 | 70–77 |
| 26–30 | 0.87 | 0.90 | 0.87 | 0.90 | 0.87 | 0.90 | 0.87 | 0.90 | 79–86 |

***TABLE B.310.15(B)(2)(10)*** *Ampacities of Three Single Insulated Conductors, Rated 0 Through 2000 Volts, Directly Buried in Earth Based on Ambient Earth Temperature of 20°C (68°F), Electrical Duct Arrangement in Accordance with Figure B.310.15(B)(2)(2), 100 Percent Load Factor, Thermal Resistance (Rho) of 90*

| Size (AWG or kcmil) | See Fig. B.310.15(B)(2)(2), Detail 9 | | See Fig. B.310.15(B)(2)(2), Detail 10 | | See Fig. B.310.15(B)(2)(2), Detail 9 | | See Fig. B.310.15(B)(2)(2), Detail 10 | | Size (AWG or kcmil) |
|---|---|---|---|---|---|---|---|---|---|
| | 60°C (140°F) | 75°C (167°F) | 60°C (140°F) | 75°C (167°F) | 60°C (140°F) | 75°C (167°F) | 60°C (140°F) | 75°C (167°F) | |
| | TYPES | | | | TYPES | | | | |
| | UF | USE | UF | USE | UF | USE | UF | USE | |
| | COPPER | | | | ALUMINUM OR COPPER-CLAD ALUMINUM | | | | |
| 8 | 84 | 98 | 78 | 92 | 66 | 77 | 61 | 72 | 8 |
| 6 | 107 | 126 | 101 | 118 | 84 | 98 | 78 | 92 | 6 |
| 4 | 139 | 163 | 130 | 152 | 108 | 127 | 101 | 118 | 4 |
| 2 | 178 | 209 | 165 | 194 | 139 | 163 | 129 | 151 | 2 |
| 1 | 201 | 236 | 187 | 219 | 157 | 184 | 146 | 171 | 1 |
| 1/0 | 230 | 270 | 212 | 249 | 179 | 210 | 165 | 194 | 1/0 |
| 2/0 | 261 | 306 | 241 | 283 | 204 | 239 | 188 | 220 | 2/0 |
| 3/0 | 297 | 348 | 274 | 321 | 232 | 272 | 213 | 250 | 3/0 |
| 4/0 | 336 | 394 | 309 | 362 | 262 | 307 | 241 | 283 | 4/0 |
| 250 | — | 429 | — | 394 | — | 335 | — | 308 | 250 |
| 350 | — | 516 | — | 474 | — | 403 | — | 370 | 350 |
| 500 | — | 626 | — | 572 | — | 490 | — | 448 | 500 |
| 750 | — | 767 | — | 700 | — | 605 | — | 552 | 750 |
| 1000 | — | 887 | — | 808 | — | 706 | — | 642 | 1000 |
| 1250 | — | 979 | — | 891 | — | 787 | — | 716 | 1250 |
| 1500 | — | 1063 | — | 965 | — | 862 | — | 783 | 1500 |
| 1750 | — | 1133 | — | 1027 | — | 930 | — | 843 | 1750 |
| 2000 | — | 1195 | — | 1082 | — | 990 | — | 897 | 2000 |
| Ambient Temp.(°C) | Correction Factors | | | | | | | | Ambient Temp.(°F) |
| 6–10 | 1.12 | 1.09 | 1.12 | 1.09 | 1.12 | 1.09 | 1.12 | 1.09 | 43–50 |
| 11–15 | 1.06 | 1.04 | 1.06 | 1.04 | 1.06 | 1.04 | 1.06 | 1.04 | 52–59 |
| 16–20 | 1.00 | 1.00 | 1.00 | 1.00 | 1.00 | 1.00 | 1.00 | 1.00 | 61–68 |
| 21–25 | 0.94 | 0.95 | 0.94 | 0.95 | 0.94 | 0.95 | 0.94 | 0.95 | 70–77 |
| 26–30 | 0.87 | 0.90 | 0.87 | 0.90 | 0.87 | 0.90 | 0.87 | 0.90 | 79–86 |

***TABLE B.310.15(B)(2)(11)*** *Adjustment Factors for More Than Three Current-Carrying Conductors in a Raceway or Cable with Load Diversity*

| Number of Conductors* | Percent of Values in Tables as Adjusted for Ambient Temperature if Necessary |
|---|---|
| 4 – 6 | 80 |
| 7 – 9 | 70 |
| 10 – 24 | 70** |
| 25 – 42 | 60** |
| 43– 85 | 50** |

*Number of conductors is the total number of conductors in the raceway or cable adjusted in accordance with 310.15(B)(4) and (5).

**These factors include the effects of a load diversity of 50 percent.

Informational Note: The ampacity limit for 10 through 85 current-carrying conductors is based on the following equation. For more than 85 conductors, special calculations are required that are beyond the scope of this table.

$$A_2 = \left[\sqrt{\frac{0.5N}{E}} \times (A_1)\right] \text{ or } A_1\text{, whichever is less}$$

where:

$A_1$ = ampacity from Table 310.15(B)(16), Table 310.15(B)(18), Table B.310.15(B)(2)(1), Table B.310.15(B)(2)(6), or Table B.310.15(B)(2)(7) multiplied by the appropriate adjustment factor from Table B.310.15(B)(2)(11).

$N$ = total number of conductors used to select adjustment factor from Table B.310.15(B)(2)(11)

$E$ = number of conductors carrying current simultaneously in the raceway or cable

$A_2$ = ampacity limit for the current-carrying conductors in the raceway or cable

**Example 1**

Calculate the ampacity limit for twelve 14 AWG THWN current-carrying conductors (75°C) in a raceway that contains 24 conductors that may, at different times, be current-carrying.

$$A_2 = \sqrt{\frac{(0.5)(24)}{12}} \times 20(0.7)$$

$$= 14 \text{ amperes (i.e., 50 percent diversity)}$$

**Example 2**

Calculate the ampacity limit for eighteen 14 AWG THWN current-carrying conductors (75°C) in a raceway that contains 24 conductors that may, at different times, be current-carrying.

$$A_2 = \sqrt{\frac{(0.5)(24)}{18}} \times 20(0.7) = 11.5 \text{ amperes}$$

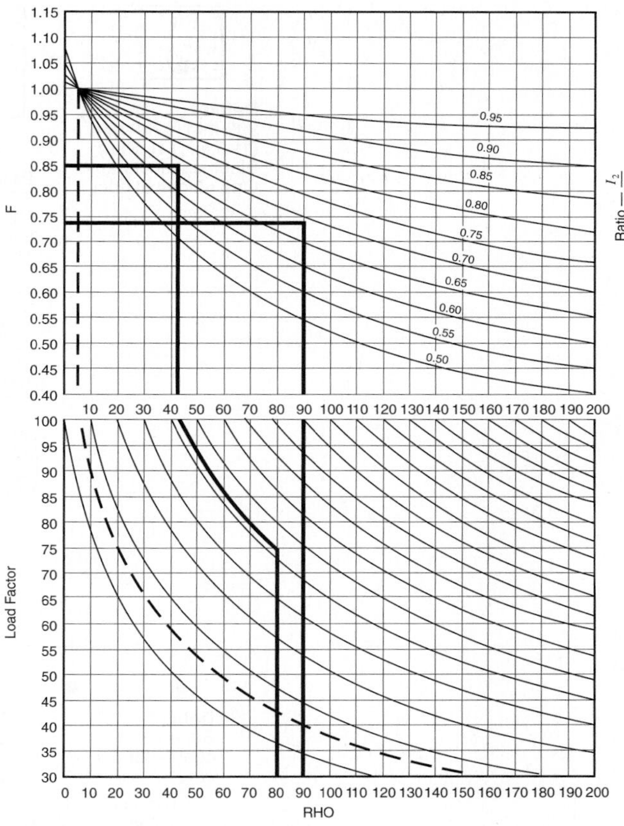


***FIGURE B.310.15(B)(2)(1)*** *Interpolation Chart for Cables in a Duct Bank $I_1$ = ampacity for Rho = 60, 50 LF; $I_2$ = ampacity for Rho = 120, 100 LF (load factor); desired ampacity = F × $I_1$.*

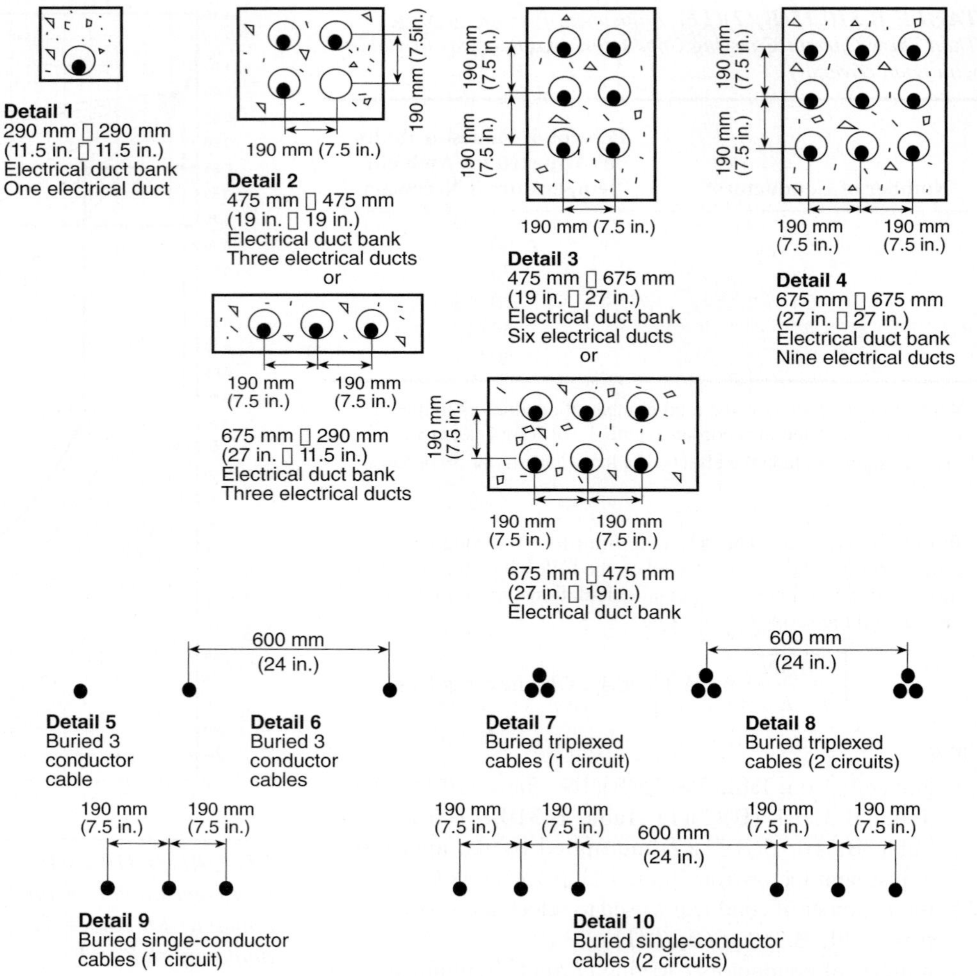


Note 1: Minimum burial depths to top electrical ducts or cables shall be in accordance with 300.5. Maximum depth to the top of electrical duct banks shall be 750 mm (30 in.) and maximum depth to the top of direct-buried cables shall be 900 mm (36 in.)

Note 2: For two and four electrical duct installations with electrical ducts installed in a single row, see B.310.15(B)(5).

**Legend**

- Backfill (earth or concrete)
- Electrical duct
- Cable or cables

***FIGURE B.310.15(B)(2)(2)*** *Cable Installation Dimensions for Use with Table B.310.15(B)(2)(5) Through Table B.310.15(B)(2)(10)*

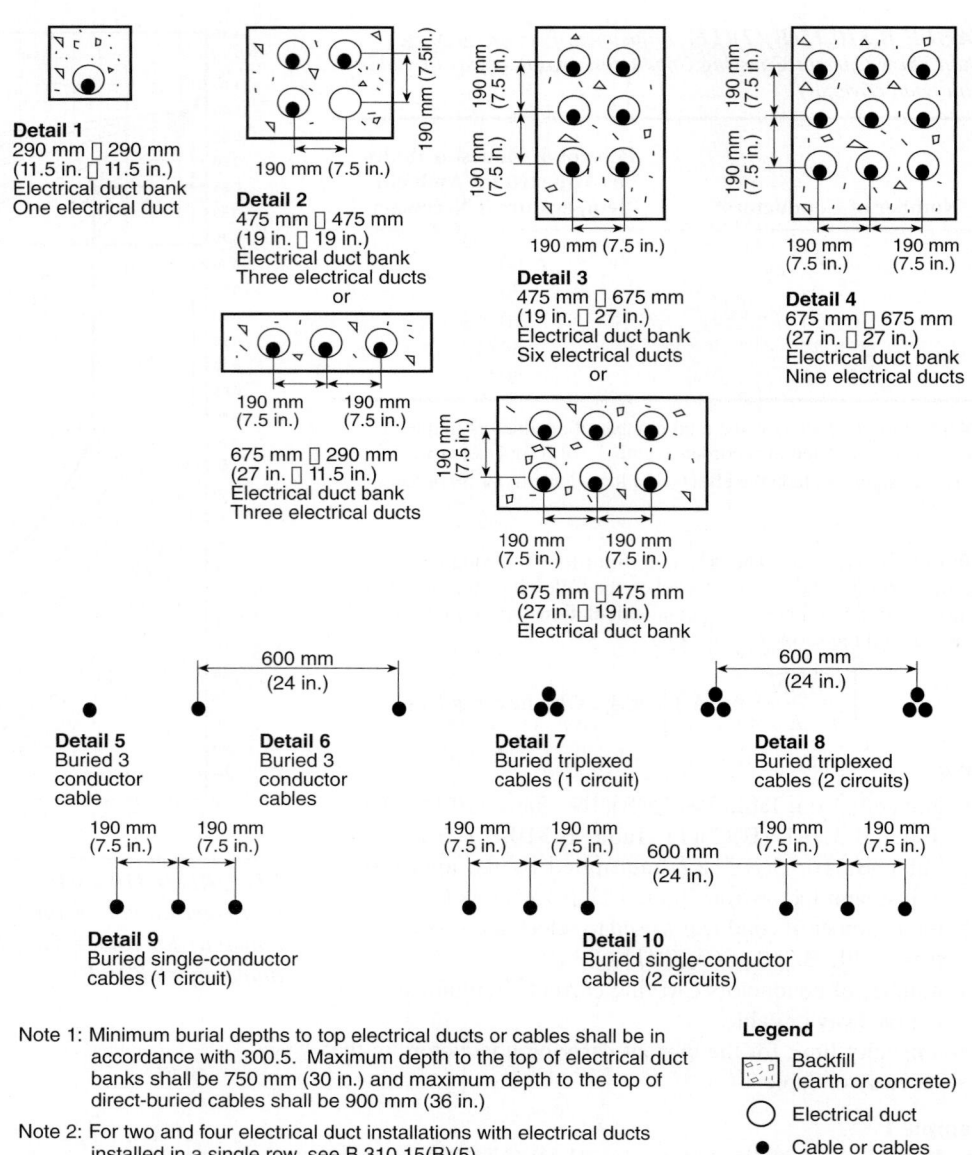

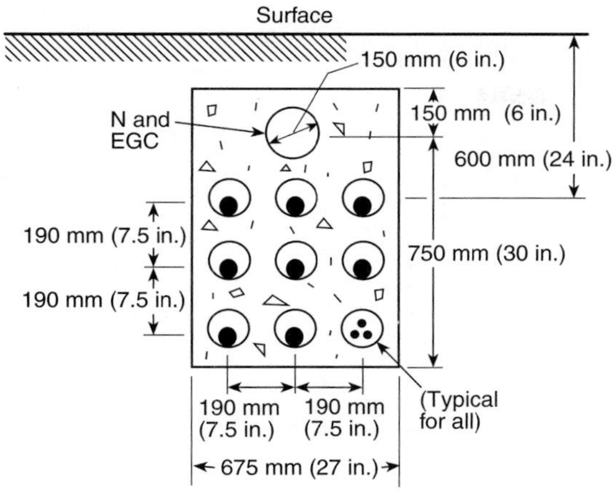


Design Criteria
Neutral and Equipment
Grounding conductor (EGC)
Duct = 150 mm (6 in.)
Phase Ducts = 75 to 125 mm (3 to 5 in.)
Conductor Material = Copper
Number of Cables per Duct = 3

Number of Cables per Phase = 9
Rho Concrete = Rho Earth – 5
Rho PVC Duct = 650
Rho Cable Insulation = 500
Rho Cable Jacket = 650

Notes:
1. Neutral configuration per 300.5(I), Exception No. 2, for isolated phase installations in nonmagnetic ducts.
2. Phasing is A, B, C in rows or columns. Where magnetic electrical ducts are used, conductors are installed A, B, C per electrical duct with the neutral and all equipment grounding conductors in the same electrical duct. In this case, the 6-in. trade size neutral duct is eliminated.
3. Maximum harmonic loading on the neutral conductor cannot exceed 50 percent of the phase current for the ampacities shown in the table below.
4. Metallic shields of Type MV-90 cable shall be grounded at one point only where using A, B, C phasing in rows or columns.

| Size kcmil | TYPES RHW, THHW, THW, THWN, XHHW, USE, OR MV-90* — Total per Phase Ampere Rating | | | Size kcmil |
|---|---|---|---|---|
| | RHO EARTH 60 LF 50 | RHO EARTH 90 LF 100 | RHO EARTH 120 LF 100 | |
| 250 | 2340 (260A/Cable) | 1530 (170A/Cable) | 1395 (155A/Cable) | 250 |
| 350 | 2790 (310A/Cable) | 1800 (200A/Cable) | 1665 (185A/Cable) | 350 |
| 500 | 3375 (375A/Cable) | 2160 (240A/Cable) | 1980 (220A/Cable) | 500 |

| Ambient Temp. (∞C) | For ambient temperatures other than 20∞C (68∞F), multiply the ampacities shown above by the appropriate factor shown below. | | | | | Ambient Temp. (∞F) |
|---|---|---|---|---|---|---|
| 6–10 | 1.09 | 1.09 | 1.09 | 1.09 | 1.09 | 43–50 |
| 11–15 | 1.04 | 1.04 | 1.04 | 1.04 | 1.04 | 52–59 |
| 16–20 | 1.00 | 1.00 | 1.00 | 1.00 | 1.00 | 61–68 |
| 21–25 | 0.95 | 0.95 | 0.95 | 0.95 | 0.95 | 70–77 |
| 26–30 | 0.90 | 0.90 | 0.90 | 0.90 | 0.90 | 79–86 |

*Limited to 75∞C conductor temperature.

***INFORMATIONAL NOTE FIGURE B.310.15(B)(2)(3)***
*Ampacities of Single Insulated Conductors Rated 0 through 5000 Volts in Underground Electrical Ducts (Three Conductors per Electrical Duct), Nine Single-Conductor Cables per Phase Based on Ambient Earth Temperature of 20°C (68°F), Conductor Temperature 75°C (167°F).*

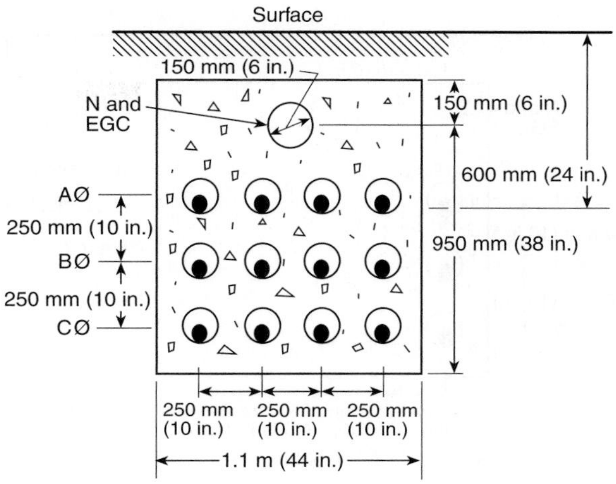


Design Criteria
Neutral and Equipment
Grounding conductor (EGC)
Duct = 150 mm (6 in.)
Phase Ducts = 75 mm (3 in.)
Conductor Material = Copper
Number of Cables per Duct = 1

Number of Cables per Phase = 4
Rho Concrete = Rho Earth – 5
Rho PVC Duct = 650
Rho Cable Insulation = 500
Rho Cable Jacket = 650

Notes:
1. Neutral configuration per 300.5(I), Exception No 2.
2. Maximum harmonic loading on the neutral conductor cannot exceed 50 percent of the phase current for the ampacities shown in the table below.
3. Metallic shields of Type MV-90 cable shall be grounded at one point only.

| Size kcmil | TYPES RHW,THHW,THW,THWN, XHHW,USE,OR MV-90* — Total per Phase Ampere Rating | | | Size kcmil |
|---|---|---|---|---|
| | RHO EARTH 60 LF 50 | RHO EARTH 90 LF 100 | RHO EARTH 120 LF 100 | |
| 750 | 2820 (705A/Cable) | 1860 (465A/Cable) | 1680 (420A/Cable) | 750 |
| 1000 | 3300 (825A/Cable) | 2140 (535A/Cable) | 1920 (480A/Cable) | 1000 |
| 1250 | 3700 (925A/Cable) | 2380 (595A/Cable) | 2120 (530A/Cable) | 1250 |
| 1500 | 4060 (1015A/Cable) | 2580 (645A/Cable) | 2300 (575A/Cable) | 1500 |
| 1750 | 4360 (1090A/Cable) | 2740 (685A/Cable) | 2460 (615A/Cable) | 1750 |

| Ambient Temp. (∞C) | For ambient temperatures other than 20∞C (68∞F), multiply the ampacities shown above by the appropriate factor shown below. | | | | | Ambient Temp. (∞F) |
|---|---|---|---|---|---|---|
| 6–10 | 1.09 | 1.09 | 1.09 | 1.09 | 1.09 | 43–50 |
| 11–15 | 1.04 | 1.04 | 1.04 | 1.04 | 1.04 | 52–59 |
| 16–20 | 1.00 | 1.00 | 1.00 | 1.00 | 1.00 | 61–68 |
| 21–25 | 0.95 | 0.95 | 0.95 | 0.95 | 0.95 | 70–77 |
| 26–30 | 0.90 | 0.90 | 0.90 | 0.90 | 0.90 | 79–86 |

*Limited to 75∞C conductor temperature.

***INFORMATIONAL NOTE FIGURE B.310.15(B)(2)(4)***
*Ampacities of Single Insulated Conductors Rated 0 through 5000 Volts in Nonmagnetic Underground Electrical Ducts (One Conductor per Electrical Duct), Four Single-Conductor Cables per Phase Based on Ambient Earth Temperature of 20°C (68°F), Conductor Temperature 75°C (167°F).*

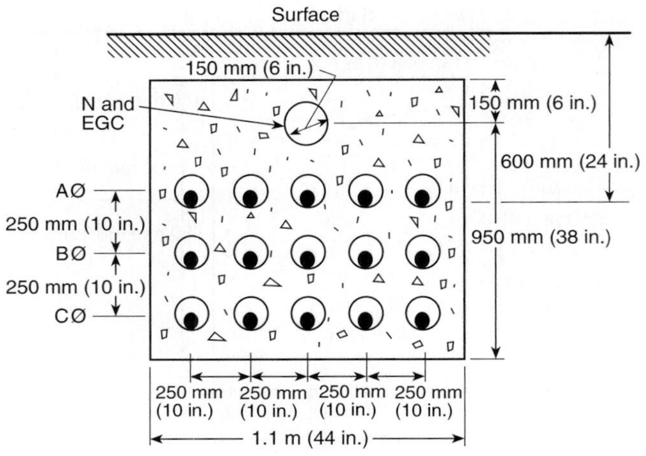


Design Criteria
Neutral and Equipment
Grounding conductor (EGC)
Duct = 150 mm (6 in.)
Phase Ducts = 75 mm (3 in.)
Conductor Material = Copper
Number of Cables per Duct = 1
Number of Cables per Phase = 5
Rho Concrete = Rho Earth – 5
Rho PVC Duct = 650
Rho Cable Insulation = 500
Rho Cable Jacket = 650

Notes:
1. Neutral configuration per 300.5(I), Exception No. 2.
2. Maximum harmonic loading on the neutral conductor cannot exceed 50 percent of the phase current for the ampacities shown in the table below.
3. Metallic shields of Type MV-90 cable shall be grounded at one point only.

| Size kcmil | TYPES RHW, THHW, THW, THWN, XHHW, USE, OR MV-90* — Total per Phase Ampere Rating | | | Size kcmil |
|---|---|---|---|---|
| | RHO EARTH 60 LF 50 | RHO EARTH 90 LF 100 | RHO EARTH 120 LF 100 | |
| 2000 | 5575 (1115A/Cable) | 3375 (675A/Cable) | 3000 (600A/Cable) | 2000 |

| Ambient Temp. (∞C) | For ambient temperatures other than 20∞C (68∞F), multiply the ampacities shown above by the appropriate factor shown below. | | | | | Ambient Temp. (∞F) |
|---|---|---|---|---|---|---|
| 6–10 | 1.09 | 1.09 | 1.09 | 1.09 | 1.09 | 43–50 |
| 11–15 | 1.04 | 1.04 | 1.04 | 1.04 | 1.04 | 52–59 |
| 16–20 | 1.00 | 1.00 | 1.00 | 1.00 | 1.00 | 61–68 |
| 21–25 | 0.95 | 0.95 | 0.95 | 0.95 | 0.95 | 70–77 |
| 26–30 | 0.90 | 0.90 | 0.90 | 0.90 | 0.90 | 79–86 |

*Limited to 75∞C conductor temperature.

***INFORMATIONAL NOTE FIGURE B.310.15(B)(2)(5)***
*Ampacities of Single Insulated Conductors Rated 0 through 5000 Volts in Nonmagnetic Underground Electrical Ducts (One Conductor per Electrical Duct), Five Single-Conductor Cables per Phase Based on Ambient Earth Temperature of 20°C (68°F), Conductor Temperature 75°C (167°F).*

# Informative Annex C
## Conduit and Tubing Fill Tables for Conductors and Fixture Wires of the Same Size

*This informative annex is not a part of the requirements of this NFPA document but is included for informational purposes only.*



*Where this table is used in conjunction with Tables C.1 through C.12, the conductors installed must be of the compact type.

***TABLE C.1*** *Maximum Number of Conductors or Fixture Wires in Electrical Metallic Tubing (EMT)*
(Based on Chapter 9: Table 1, Table 4, and Table 5)

| Type | Conductor Size (AWG/kcmil) | Trade Size (Metric Designator) ⅜ (12) | ½ (16) | ¾ (21) | 1 (27) | 1¼ (35) | 1½ (41) | 2 (53) | 2½ (63) | 3 (78) | 3½ (91) | 4 (103) | 5 (129) | 6 (155) |
|---|---|---|---|---|---|---|---|---|---|---|---|---|---|---|
| **CONDUCTORS** | | | | | | | | | | | | | | |
| RHH, RHW, RHW-2 | 14 | — | 4 | 7 | 11 | 20 | 27 | 46 | 80 | 120 | 157 | 201 | — | — |
| | 12 | — | 3 | 6 | 9 | 17 | 23 | 38 | 66 | 100 | 131 | 167 | — | — |
| | 10 | — | 2 | 5 | 8 | 13 | 18 | 30 | 53 | 81 | 105 | 135 | — | — |
| | 8 | — | 1 | 2 | 4 | 7 | 9 | 16 | 28 | 42 | 55 | 70 | — | — |
| | 6 | — | 1 | 1 | 3 | 5 | 8 | 13 | 22 | 34 | 44 | 56 | — | — |
| | 4 | — | 1 | 1 | 2 | 4 | 6 | 10 | 17 | 26 | 34 | 44 | — | — |
| | 3 | — | 1 | 1 | 1 | 4 | 5 | 9 | 15 | 23 | 30 | 38 | — | — |
| | 2 | — | 1 | 1 | 1 | 3 | 4 | 7 | 13 | 20 | 26 | 33 | — | — |
| | 1 | — | 0 | 1 | 1 | 1 | 3 | 5 | 9 | 13 | 17 | 22 | — | — |

*(continues)*

***TABLE C.1*** *Continued*

| Type | Conductor Size (AWG/ kcmil) | Trade Size (Metric Designator) | | | | | | | | | | | | |
|---|---|---|---|---|---|---|---|---|---|---|---|---|---|---|
| | | ⅜ (12) | ½ (16) | ¾ (21) | 1 (27) | 1¼ (35) | 1½ (41) | 2 (53) | 2½ (63) | 3 (78) | 3½ (91) | 4 (103) | 5 (129) | 6 (155) |
| | 1/0 | — | 0 | 1 | 1 | 1 | 2 | 4 | 7 | 11 | 15 | 19 | — | — |
| | 2/0 | — | 0 | 1 | 1 | 1 | 2 | 4 | 6 | 10 | 13 | 17 | — | — |
| | 3/0 | — | 0 | 0 | 1 | 1 | 1 | 3 | 5 | 8 | 11 | 14 | — | — |
| | 4/0 | — | 0 | 0 | 1 | 1 | 1 | 3 | 5 | 7 | 9 | 12 | — | — |
| | 250 | — | 0 | 0 | 0 | 1 | 1 | 1 | 3 | 5 | 7 | 9 | — | — |
| | 300 | — | 0 | 0 | 0 | 1 | 1 | 1 | 3 | 5 | 6 | 8 | — | — |
| | 350 | — | 0 | 0 | 0 | 1 | 1 | 1 | 3 | 4 | 6 | 7 | — | — |
| | 400 | — | 0 | 0 | 0 | 1 | 1 | 1 | 2 | 4 | 5 | 7 | — | — |
| | 500 | — | 0 | 0 | 0 | 0 | 1 | 1 | 2 | 3 | 4 | 6 | — | — |
| | 600 | — | 0 | 0 | 0 | 0 | 1 | 1 | 1 | 3 | 4 | 5 | — | — |
| | 700 | — | 0 | 0 | 0 | 0 | 0 | 1 | 1 | 2 | 3 | 4 | — | — |
| | 750 | — | 0 | 0 | 0 | 0 | 0 | 1 | 1 | 2 | 3 | 4 | — | — |
| | 800 | — | 0 | 0 | 0 | 0 | 0 | 1 | 1 | 2 | 3 | 4 | — | — |
| | 900 | — | 0 | 0 | 0 | 0 | 0 | 1 | 1 | 1 | 3 | 3 | — | — |
| | 1000 | — | 0 | 0 | 0 | 0 | 0 | 1 | 1 | 1 | 2 | 3 | — | — |
| | 1250 | — | 0 | 0 | 0 | 0 | 0 | 0 | 1 | 1 | 1 | 2 | — | — |
| | 1500 | — | 0 | 0 | 0 | 0 | 0 | 0 | 1 | 1 | 1 | 1 | — | — |
| | 1750 | — | 0 | 0 | 0 | 0 | 0 | 0 | 1 | 1 | 1 | 1 | — | — |
| | 2000 | — | 0 | 0 | 0 | 0 | 0 | 0 | 1 | 1 | 1 | 1 | — | — |
| • TW, THHW, THW, THW-2 | 14 | — | 8 | 15 | 25 | 43 | 58 | 96 | 168 | 254 | 332 | 424 | — | — |
| | 12 | — | 6 | 11 | 19 | 33 | 45 | 74 | 129 | 195 | 255 | 326 | — | — |
| | 10 | — | 5 | 8 | 14 | 24 | 33 | 55 | 96 | 145 | 190 | 243 | — | — |
| | 8 | — | 2 | 5 | 8 | 13 | 18 | 30 | 53 | 81 | 105 | 135 | — | — |
| RHH*, RHW*, RHW-2* | 14 | — | 6 | 10 | 16 | 28 | 39 | 64 | 112 | 169 | 221 | 282 | — | — |
| | 12 | — | 4 | 8 | 13 | 23 | 31 | 51 | 90 | 136 | 177 | 227 | — | — |
| | 10 | — | 3 | 6 | 10 | 18 | 24 | 40 | 70 | 106 | 138 | 177 | — | — |
| | 8 | — | 1 | 4 | 6 | 10 | 14 | 24 | 42 | 63 | 83 | 106 | — | — |
| TW, THW, THHW, THW-2, RHH*, RHW*, RHW-2* | 6 | — | 1 | 3 | 4 | 8 | 11 | 18 | 32 | 48 | 63 | 81 | — | — |
| | 4 | — | 1 | 1 | 3 | 6 | 8 | 13 | 24 | 36 | 47 | 60 | — | — |
| | 3 | — | 1 | 1 | 3 | 5 | 7 | 12 | 20 | 31 | 40 | 52 | — | — |
| | 2 | — | 1 | 1 | 2 | 4 | 6 | 10 | 17 | 26 | 34 | 44 | — | — |
| | 1 | — | 1 | 1 | 1 | 3 | 4 | 7 | 12 | 18 | 24 | 31 | — | — |
| | 1/0 | — | 0 | 1 | 1 | 2 | 3 | 6 | 10 | 16 | 20 | 26 | — | — |
| | 2/0 | — | 0 | 1 | 1 | 1 | 3 | 5 | 9 | 13 | 17 | 22 | — | — |
| | 3/0 | — | 0 | 1 | 1 | 1 | 2 | 4 | 7 | 11 | 15 | 19 | — | — |
| | 4/0 | — | 0 | 0 | 1 | 1 | 1 | 3 | 6 | 9 | 12 | 16 | — | — |
| | 250 | — | 0 | 0 | 1 | 1 | 1 | 3 | 5 | 7 | 10 | 13 | — | — |
| | 300 | — | 0 | 0 | 1 | 1 | 1 | 2 | 4 | 6 | 8 | 11 | — | — |
| | 350 | — | 0 | 0 | 0 | 1 | 1 | 1 | 4 | 6 | 7 | 10 | — | — |
| | 400 | — | 0 | 0 | 0 | 1 | 1 | 1 | 3 | 5 | 7 | 9 | — | — |
| | 500 | — | 0 | 0 | 0 | 1 | 1 | 1 | 3 | 4 | 6 | 7 | — | — |
| | 600 | — | 0 | 0 | 0 | 1 | 1 | 1 | 2 | 3 | 4 | 6 | — | — |
| | 700 | — | 0 | 0 | 0 | 0 | 1 | 1 | 1 | 3 | 4 | 5 | — | — |
| | 750 | — | 0 | 0 | 0 | 0 | 1 | 1 | 1 | 3 | 4 | 5 | — | — |
| | 800 | — | 0 | 0 | 0 | 0 | 1 | 1 | 1 | 3 | 3 | 5 | — | — |
| | 900 | — | 0 | 0 | 0 | 0 | 0 | 1 | 1 | 2 | 3 | 4 | — | — |

***TABLE C.1*** *Continued*

| Type | Conductor Size (AWG/kcmil) | Trade Size (Metric Designator) | | | | | | | | | | | | |
|---|---|---|---|---|---|---|---|---|---|---|---|---|---|---|
| | | ⅜ (12) | ½ (16) | ¾ (21) | 1 (27) | 1¼ (35) | 1½ (41) | 2 (53) | 2½ (63) | 3 (78) | 3½ (91) | 4 (103) | 5 (129) | 6 (155) |
| | 1000 | — | 0 | 0 | 0 | 0 | 0 | 1 | 1 | 2 | 3 | 4 | — | — |
| | 1250 | — | 0 | 0 | 0 | 0 | 0 | 1 | 1 | 1 | 2 | 3 | — | — |
| | 1500 | — | 0 | 0 | 0 | 0 | 0 | 1 | 1 | 1 | 1 | 2 | — | — |
| | 1750 | — | 0 | 0 | 0 | 0 | 0 | 0 | 1 | 1 | 1 | 2 | — | — |
| | 2000 | — | 0 | 0 | 0 | 0 | 0 | 0 | 1 | 1 | 1 | 1 | — | — |
| THHN, THWN, THWN-2 | 14 | — | 12 | 22 | 35 | 61 | 84 | 138 | 241 | 364 | 476 | 608 | — | — |
| | 12 | — | 9 | 16 | 26 | 45 | 61 | 101 | 176 | 266 | 347 | 443 | — | — |
| | 10 | — | 5 | 10 | 16 | 28 | 38 | 63 | 111 | 167 | 219 | 279 | — | — |
| | 8 | — | 3 | 6 | 9 | 16 | 22 | 36 | 64 | 96 | 126 | 161 | — | — |
| | 6 | — | 2 | 4 | 7 | 12 | 16 | 26 | 46 | 69 | 91 | 116 | — | — |
| | 4 | — | 1 | 2 | 4 | 7 | 10 | 16 | 28 | 43 | 56 | 71 | — | — |
| | 3 | — | 1 | 1 | 3 | 6 | 8 | 13 | 24 | 36 | 47 | 60 | — | — |
| | 2 | — | 1 | 1 | 3 | 5 | 7 | 11 | 20 | 30 | 40 | 51 | — | — |
| | 1 | — | 1 | 1 | 1 | 4 | 5 | 8 | 15 | 22 | 29 | 37 | — | — |
| | 1/0 | — | 1 | 1 | 1 | 3 | 4 | 7 | 12 | 19 | 25 | 32 | — | — |
| | 2/0 | — | 0 | 1 | 1 | 2 | 3 | 6 | 10 | 16 | 20 | 26 | — | — |
| | 3/0 | — | 0 | 1 | 1 | 1 | 3 | 5 | 8 | 13 | 17 | 22 | — | — |
| | 4/0 | — | 0 | 1 | 1 | 1 | 2 | 4 | 7 | 11 | 14 | 18 | — | — |
| | 250 | — | 0 | 0 | 1 | 1 | 1 | 3 | 6 | 9 | 11 | 15 | — | — |
| | 300 | — | 0 | 0 | 1 | 1 | 1 | 3 | 5 | 7 | 10 | 13 | — | — |
| | 350 | — | 0 | 0 | 1 | 1 | 1 | 2 | 4 | 6 | 9 | 11 | — | — |
| | 400 | — | 0 | 0 | 0 | 1 | 1 | 1 | 4 | 6 | 8 | 10 | — | — |
| | 500 | — | 0 | 0 | 0 | 1 | 1 | 1 | 3 | 5 | 6 | 8 | — | — |
| | 600 | — | 0 | 0 | 0 | 1 | 1 | 1 | 2 | 4 | 5 | 7 | — | — |
| | 700 | — | 0 | 0 | 0 | 1 | 1 | 1 | 2 | 3 | 4 | 6 | — | — |
| | 750 | — | 0 | 0 | 0 | 0 | 1 | 1 | 1 | 3 | 4 | 5 | — | — |
| | 800 | — | 0 | 0 | 0 | 0 | 1 | 1 | 1 | 3 | 4 | 5 | — | — |
| | 900 | — | 0 | 0 | 0 | 0 | 1 | 1 | 1 | 3 | 3 | 4 | — | — |
| | 1000 | — | 0 | 0 | 0 | 0 | 1 | 1 | 1 | 2 | 3 | 4 | — | — |
| FEP, FEPB, PFA, PFAH, TFE | 14 | — | 12 | 21 | 34 | 60 | 81 | 134 | 234 | 354 | 462 | 590 | — | — |
| | 12 | — | 9 | 15 | 25 | 43 | 59 | 98 | 171 | 258 | 337 | 430 | — | — |
| | 10 | — | 6 | 11 | 18 | 31 | 42 | 70 | 122 | 185 | 241 | 309 | — | — |
| | 8 | — | 3 | 6 | 10 | 18 | 24 | 40 | 70 | 106 | 138 | 177 | — | — |
| | 6 | — | 2 | 4 | 7 | 12 | 17 | 28 | 50 | 75 | 98 | 126 | — | — |
| | 4 | — | 1 | 3 | 5 | 9 | 12 | 20 | 35 | 53 | 69 | 88 | — | — |
| | 3 | — | 1 | 2 | 4 | 7 | 10 | 16 | 29 | 44 | 57 | 73 | — | — |
| | 2 | — | 1 | 1 | 3 | 6 | 8 | 13 | 24 | 36 | 47 | 60 | — | — |
| PFA, PFAH, TFE | 1 | — | 1 | 1 | 2 | 4 | 6 | 9 | 16 | 25 | 33 | 42 | — | — |
| PFA, PFAH, TFE, Z | 1/0 | — | 1 | 1 | 1 | 3 | 5 | 8 | 14 | 21 | 27 | 35 | — | — |
| | 2/0 | — | 0 | 1 | 1 | 3 | 4 | 6 | 11 | 17 | 22 | 29 | — | — |
| | 3/0 | — | 0 | 1 | 1 | 2 | 3 | 5 | 9 | 14 | 18 | 24 | — | — |
| | 4/0 | — | 0 | 1 | 1 | 1 | 2 | 4 | 8 | 11 | 15 | 19 | — | — |
| Z | 14 | — | 14 | 25 | 41 | 72 | 98 | 161 | 282 | 426 | 556 | 711 | — | — |
| | 12 | — | 10 | 18 | 29 | 51 | 69 | 114 | 200 | 302 | 394 | 504 | — | — |
| | 10 | — | 6 | 11 | 18 | 31 | 42 | 70 | 122 | 185 | 241 | 309 | — | — |
| | 8 | — | 4 | 7 | 11 | 20 | 27 | 44 | 77 | 117 | 153 | 195 | — | — |
| | 6 | — | 3 | 5 | 8 | 14 | 19 | 31 | 54 | 82 | 107 | 137 | — | — |

*(continues)*

***TABLE C.1** Continued*

| Type | Conductor Size (AWG/ kcmil) | Trade Size (Metric Designator) 3/8 (12) | 1/2 (16) | 3/4 (21) | 1 (27) | 1¼ (35) | 1½ (41) | 2 (53) | 2½ (63) | 3 (78) | 3½ (91) | 4 (103) | 5 (129) | 6 (155) |
|---|---|---|---|---|---|---|---|---|---|---|---|---|---|---|
| | 4 | — | 1 | 3 | 5 | 9 | 13 | 21 | 37 | 56 | 74 | 94 | — | — |
| | 3 | — | 1 | 2 | 4 | 7 | 9 | 15 | 27 | 41 | 54 | 69 | — | — |
| | 2 | — | 1 | 1 | 3 | 6 | 8 | 13 | 22 | 34 | 45 | 57 | — | — |
| | 1 | — | 1 | 1 | 2 | 4 | 6 | 10 | 18 | 28 | 36 | 46 | — | — |
| XHHW, ZW, XHHW-2, XHH | 14 | — | 8 | 15 | 25 | 43 | 58 | 96 | 168 | 254 | 332 | 424 | — | — |
| | 12 | — | 6 | 11 | 19 | 33 | 45 | 74 | 129 | 195 | 255 | 326 | — | — |
| | 10 | — | 5 | 8 | 14 | 24 | 33 | 55 | 96 | 145 | 190 | 243 | — | — |
| | 8 | — | 2 | 5 | 8 | 13 | 18 | 30 | 53 | 81 | 105 | 135 | — | — |
| | 6 | — | 1 | 3 | 6 | 10 | 14 | 22 | 39 | 60 | 78 | 100 | — | — |
| | 4 | — | 1 | 2 | 4 | 7 | 10 | 16 | 28 | 43 | 56 | 72 | — | — |
| | 3 | — | 1 | 1 | 3 | 6 | 8 | 14 | 24 | 36 | 48 | 61 | — | — |
| | 2 | — | 1 | 1 | 3 | 5 | 7 | 11 | 20 | 31 | 40 | 51 | — | — |
| XHHW, XHHW-2, XHH | 1 | — | 1 | 1 | 1 | 4 | 5 | 8 | 15 | 23 | 30 | 38 | — | — |
| | 1/0 | — | 1 | 1 | 1 | 3 | 4 | 7 | 13 | 19 | 25 | 32 | — | — |
| | 2/0 | — | 0 | 1 | 1 | 2 | 3 | 6 | 10 | 16 | 21 | 27 | — | — |
| | 3/0 | — | 0 | 1 | 1 | 1 | 3 | 5 | 9 | 13 | 17 | 22 | — | — |
| | 4/0 | — | 0 | 1 | 1 | 1 | 2 | 4 | 7 | 11 | 14 | 18 | — | — |
| | 250 | — | 0 | 0 | 1 | 1 | 1 | 3 | 6 | 9 | 12 | 15 | — | — |
| | 300 | — | 0 | 0 | 1 | 1 | 1 | 3 | 5 | 8 | 10 | 13 | — | — |
| | 350 | — | 0 | 0 | 1 | 1 | 1 | 2 | 4 | 7 | 9 | 11 | — | — |
| | 400 | — | 0 | 0 | 0 | 1 | 1 | 1 | 4 | 6 | 8 | 10 | — | — |
| | 500 | — | 0 | 0 | 0 | 1 | 1 | 1 | 3 | 5 | 6 | 8 | — | — |
| | 600 | — | 0 | 0 | 0 | 1 | 1 | 1 | 2 | 4 | 5 | 6 | — | — |
| | 700 | — | 0 | 0 | 0 | 0 | 1 | 1 | 2 | 3 | 4 | 6 | — | — |
| | 750 | — | 0 | 0 | 0 | 0 | 1 | 1 | 1 | 3 | 4 | 5 | — | — |
| | 800 | — | 0 | 0 | 0 | 0 | 1 | 1 | 1 | 3 | 4 | 5 | — | — |
| | 900 | — | 0 | 0 | 0 | 0 | 1 | 1 | 1 | 3 | 3 | 4 | — | — |
| | 1000 | — | 0 | 0 | 0 | 0 | 0 | 1 | 1 | 2 | 3 | 4 | — | — |
| | 1250 | — | 0 | 0 | 0 | 0 | 0 | 1 | 1 | 1 | 2 | 3 | — | — |
| | 1500 | — | 0 | 0 | 0 | 0 | 0 | 1 | 1 | 1 | 1 | 3 | — | — |
| | 1750 | — | 0 | 0 | 0 | 0 | 0 | 0 | 1 | 1 | 1 | 2 | — | — |
| | 2000 | — | 0 | 0 | 0 | 0 | 0 | 0 | 1 | 1 | 1 | 1 | — | — |
| **FIXTURE WIRES** | | | | | | | | | | | | | | |
| RFH-2, FFH-2, RFHH-2 | 18 | — | 8 | 14 | 24 | 41 | 56 | 92 | 161 | 244 | 318 | 407 | — | — |
| | 16 | — | 7 | 12 | 20 | 34 | 47 | 78 | 136 | 205 | 268 | 343 | — | — |
| SF-2, SFF-2 | 18 | — | 10 | 18 | 30 | 52 | 71 | 116 | 203 | 307 | 401 | 513 | — | — |
| | 16 | — | 8 | 15 | 25 | 43 | 58 | 96 | 168 | 254 | 332 | 424 | — | — |
| | 14 | — | 7 | 12 | 20 | 34 | 47 | 78 | 136 | 205 | 268 | 343 | — | — |
| SF-1, SFF-1 | 18 | — | 18 | 33 | 53 | 92 | 125 | 206 | 360 | 544 | 710 | 908 | — | — |
| RFH-1, TF, TFF, XF, XFF | 18 | — | 14 | 24 | 39 | 68 | 92 | 152 | 266 | 402 | 524 | 670 | — | — |
| | 16 | — | 11 | 19 | 31 | 55 | 74 | 123 | 215 | 324 | 423 | 541 | — | — |
| XF, XFF | 14 | — | 8 | 15 | 25 | 43 | 58 | 96 | 168 | 254 | 332 | 424 | — | — |
| TFN, TFFN | 18 | — | 22 | 38 | 63 | 109 | 148 | 244 | 426 | 643 | 839 | 1073 | — | — |
| | 16 | — | 17 | 29 | 48 | 83 | 113 | 186 | 325 | 491 | 641 | 819 | — | — |

***TABLE C.1*** *Continued*

| Type | Conductor Size (AWG/ kcmil) | Trade Size (Metric Designator) | | | | | | | | | | | | |
|---|---|---|---|---|---|---|---|---|---|---|---|---|---|---|
| | | ⅜ (12) | ½ (16) | ¾ (21) | 1 (27) | 1¼ (35) | 1½ (41) | 2 (53) | 2½ (63) | 3 (78) | 3½ (91) | 4 (103) | 5 (129) | 6 (155) |
| PF, PFF, PGF, PGFF, PAF, PTF, PTFF, PAFF | 18 | — | 21 | 36 | 59 | 103 | 140 | 231 | 404 | 610 | 796 | 1017 | — | — |
| | 16 | — | 16 | 28 | 46 | 79 | 108 | 179 | 312 | 471 | 615 | 787 | — | — |
| | 14 | — | 12 | 21 | 34 | 60 | 81 | 134 | 234 | 354 | 462 | 590 | — | — |
| ZF, ZFF, ZHF | 18 | — | 27 | 47 | 77 | 133 | 181 | 298 | 520 | 786 | 1026 | 1311 | — | — |
| | 16 | — | 20 | 35 | 56 | 98 | 133 | 220 | 384 | 580 | 757 | 967 | — | — |
| | 14 | — | 14 | 25 | 41 | 72 | 98 | 161 | 282 | 426 | 556 | 711 | — | — |
| KF-2, KFF-2 | 18 | — | 40 | 71 | 115 | 199 | 271 | 447 | 781 | 1179 | 1539 | 1967 | — | — |
| | 16 | — | 28 | 49 | 80 | 139 | 189 | 312 | 545 | 823 | 1074 | 1372 | — | — |
| | 14 | — | 19 | 33 | 54 | 93 | 127 | 209 | 366 | 553 | 721 | 922 | — | — |
| | 12 | — | 13 | 23 | 37 | 65 | 88 | 146 | 254 | 384 | 502 | 641 | — | — |
| | 10 | — | 8 | 15 | 25 | 43 | 58 | 96 | 168 | 254 | 332 | 424 | — | — |
| KF-1, KFF-1 | 18 | — | 46 | 82 | 133 | 230 | 313 | 516 | 901 | 1361 | 1776 | 2269 | — | — |
| | 16 | — | 33 | 57 | 93 | 161 | 220 | 363 | 633 | 956 | 1248 | 1595 | — | — |
| | 14 | — | 22 | 38 | 63 | 109 | 148 | 244 | 426 | 643 | 839 | 1073 | — | — |
| | 12 | — | 14 | 25 | 41 | 72 | 98 | 161 | 282 | 426 | 556 | 711 | — | — |
| | 10 | — | 9 | 16 | 27 | 47 | 64 | 105 | 184 | 278 | 363 | 464 | — | — |
| XF, XFF | 12 | — | 4 | 8 | 13 | 23 | 31 | 51 | 90 | 136 | 177 | 227 | — | — |
| | 10 | — | 3 | 6 | 10 | 18 | 24 | 40 | 70 | 106 | 138 | 177 | — | — |

Notes:

1. This table is for concentric stranded conductors only. For compact stranded conductors, Table C.1(A) should be used.

2. Two-hour fire-rated RHH cable has ceramifiable insulation, which has much larger diameters than other RHH wires. Consult manufacturer's conduit fill tables.

*Types RHH, RHW, and RHW-2 without outer covering.

***TABLE C.1(A)*** *Maximum Number of Conductors or Fixture Wires in Electrical Metallic Tubing (EMT)*
(Based on Chapter 9: Table 1, Table 4, and Table 5A)

| Type | Conductor Size (AWG/ kcmil) | Trade Size (Metric Designator) | | | | | | | | | | | | |
|---|---|---|---|---|---|---|---|---|---|---|---|---|---|---|
| | | ⅜ (12) | ½ (16) | ¾ (21) | 1 (27) | 1¼ (35) | 1½ (41) | 2 (53) | 2½ (63) | 3 (78) | 3½ (91) | 4 (103) | 5 (129) | 6 (155) |
| **COMPACT CONDUCTORS** | | | | | | | | | | | | | | |
| THW, THW-2, THHW | 8 | — | 2 | 4 | 6 | 11 | 16 | 26 | 46 | 69 | 90 | 115 | — | — |
| | 6 | — | 1 | 3 | 5 | 9 | 12 | 20 | 35 | 53 | 70 | 89 | — | — |
| | 4 | — | 1 | 2 | 4 | 6 | 9 | 15 | 26 | 40 | 52 | 67 | — | — |
| | 2 | — | 1 | 1 | 3 | 5 | 7 | 11 | 19 | 29 | 38 | 49 | — | — |
| | 1 | — | 1 | 1 | 1 | 3 | 4 | 8 | 13 | 21 | 27 | 34 | — | — |
| | 1/0 | — | 1 | 1 | 1 | 3 | 4 | 7 | 12 | 18 | 23 | 30 | — | — |
| | 2/0 | — | 0 | 1 | 1 | 2 | 3 | 5 | 10 | 15 | 20 | 25 | — | — |
| | 3/0 | — | 0 | 1 | 1 | 1 | 3 | 5 | 8 | 13 | 17 | 21 | — | — |
| | 4/0 | — | 0 | 1 | 1 | 1 | 2 | 4 | 7 | 11 | 14 | 18 | — | — |
| | 250 | — | 0 | 0 | 1 | 1 | 1 | 3 | 5 | 8 | 11 | 14 | — | — |
| | 300 | — | 0 | 0 | 1 | 1 | 1 | 3 | 5 | 7 | 9 | 12 | — | — |
| | 350 | — | 0 | 0 | 1 | 1 | 1 | 2 | 4 | 6 | 8 | 11 | — | — |
| | 400 | — | 0 | 0 | 0 | 1 | 1 | 1 | 4 | 6 | 8 | 10 | — | — |
| | 500 | — | 0 | 0 | 0 | 1 | 1 | 1 | 3 | 5 | 6 | 8 | — | — |
| | 600 | — | 0 | 0 | 0 | 1 | 1 | 1 | 2 | 4 | 5 | 7 | — | — |
| | 700 | — | 0 | 0 | 0 | 1 | 1 | 1 | 2 | 3 | 4 | 6 | — | — |
| | 750 | — | 0 | 0 | 0 | 0 | 1 | 1 | 1 | 3 | 4 | 5 | — | — |
| | 900 | — | 0 | 0 | 0 | 0 | 1 | 1 | 1 | 3 | 4 | 5 | — | — |
| | 1000 | — | 0 | 0 | 0 | 0 | 1 | 1 | 1 | 2 | 3 | 4 | — | — |
| THHN, THWN, THWN-2 | 8 | — | — | — | — | — | — | — | — | — | — | — | — | — |
| | 6 | — | 2 | 4 | 7 | 13 | 18 | 29 | 52 | 78 | 102 | 130 | — | — |
| | 4 | — | 1 | 3 | 4 | 8 | 11 | 18 | 32 | 48 | 63 | 81 | — | — |
| | 2 | — | 1 | 1 | 3 | 6 | 8 | 13 | 23 | 34 | 45 | 58 | — | — |
| | 1 | — | 1 | 1 | 2 | 4 | 6 | 10 | 17 | 26 | 34 | 43 | — | — |
| | 1/0 | — | 1 | 1 | 1 | 3 | 5 | 8 | 14 | 22 | 29 | 37 | — | — |
| | 2/0 | — | 1 | 1 | 1 | 3 | 4 | 7 | 12 | 18 | 24 | 30 | — | — |
| | 3/0 | — | 0 | 1 | 1 | 2 | 3 | 6 | 10 | 15 | 20 | 25 | — | — |
| | 4/0 | — | 0 | 1 | 1 | 1 | 3 | 5 | 8 | 12 | 16 | 21 | — | — |
| | 250 | — | 0 | 1 | 1 | 1 | 1 | 4 | 6 | 10 | 13 | 16 | — | — |
| | 300 | — | 0 | 0 | 1 | 1 | 1 | 3 | 5 | 8 | 11 | 14 | — | — |
| | 350 | — | 0 | 0 | 1 | 1 | 1 | 3 | 5 | 7 | 10 | 12 | — | — |
| | 400 | — | 0 | 0 | 1 | 1 | 1 | 2 | 4 | 6 | 9 | 11 | — | — |
| | 500 | — | 0 | 0 | 0 | 1 | 1 | 1 | 4 | 5 | 7 | 9 | — | — |
| | 600 | — | 0 | 0 | 0 | 1 | 1 | 1 | 3 | 4 | 6 | 7 | — | — |
| | 700 | — | 0 | 0 | 0 | 1 | 1 | 1 | 2 | 4 | 5 | 7 | — | — |
| | 750 | — | 0 | 0 | 0 | 1 | 1 | 1 | 2 | 4 | 5 | 6 | — | — |
| | 900 | — | 0 | 0 | 0 | 0 | 1 | 1 | 1 | 3 | 4 | 5 | — | — |
| | 1000 | — | 0 | 0 | 0 | 0 | 1 | 1 | 1 | 3 | 3 | 4 | — | — |

***TABLE C.1(A)*** *Continued*

| Type | Conductor Size (AWG/ kcmil) | Trade Size (Metric Designator) | | | | | | | | | | | | |
|---|---|---|---|---|---|---|---|---|---|---|---|---|---|---|
| | | **⅜ (12)** | **½ (16)** | **¾ (21)** | **1 (27)** | **1¼ (35)** | **1½ (41)** | **2 (53)** | **2½ (63)** | **3 (78)** | **3½ (91)** | **4 (103)** | **5 (129)** | **6 (155)** |
| XHHW, XHHW-2 | 8 | — | 3 | 5 | 8 | 15 | 20 | 34 | 59 | 90 | 117 | 149 | — | — |
| | 6 | — | 1 | 4 | 6 | 11 | 15 | 25 | 44 | 66 | 87 | 111 | — | — |
| | 4 | — | 1 | 3 | 4 | 8 | 11 | 18 | 32 | 48 | 63 | 81 | — | — |
| | 2 | — | 1 | 1 | 3 | 6 | 8 | 13 | 23 | 34 | 45 | 58 | — | — |
| | 1 | — | 1 | 1 | 2 | 4 | 6 | 10 | 17 | 26 | 34 | 43 | — | — |
| | 1/0 | — | 1 | 1 | 1 | 3 | 5 | 8 | 14 | 22 | 29 | 37 | — | — |
| | 2/0 | — | 1 | 1 | 1 | 3 | 4 | 7 | 12 | 18 | 24 | 31 | — | — |
| | 3/0 | — | 0 | 1 | 1 | 2 | 3 | 6 | 10 | 15 | 20 | 25 | — | — |
| | 4/0 | — | 0 | 1 | 1 | 1 | 3 | 5 | 8 | 13 | 17 | 21 | — | — |
| | 250 | — | 0 | 1 | 1 | 1 | 2 | 4 | 7 | 10 | 13 | 17 | — | — |
| | 300 | — | 0 | 0 | 1 | 1 | 1 | 3 | 6 | 9 | 11 | 14 | — | — |
| | 350 | — | 0 | 0 | 1 | 1 | 1 | 3 | 5 | 8 | 10 | 13 | — | — |
| | 400 | — | 0 | 0 | 1 | 1 | 1 | 2 | 4 | 7 | 9 | 11 | — | — |
| | 500 | — | 0 | 0 | 0 | 1 | 1 | 1 | 4 | 6 | 7 | 9 | — | — |
| | 600 | — | 0 | 0 | 0 | 1 | 1 | 1 | 3 | 4 | 6 | 8 | — | — |
| | 700 | — | 0 | 0 | 0 | 1 | 1 | 1 | 2 | 4 | 5 | 7 | — | — |
| | 750 | — | 0 | 0 | 0 | 1 | 1 | 1 | 2 | 3 | 5 | 6 | — | — |
| | 900 | — | 0 | 0 | 0 | 0 | 1 | 1 | 1 | 3 | 4 | 5 | — | — |
| | 1000 | — | 0 | 0 | 0 | 0 | 1 | 1 | 1 | 3 | 4 | 5 | — | — |

Definition: *Compact stranding* is the result of a manufacturing process where the stranded conductor is compressed to the extent that the interstices (voids between strand wires) are virtually eliminated.

***TABLE C.2*** *Maximum Number of Conductors or Fixture Wires in Electrical Nonmetallic Tubing (ENT)*
(Based on Chapter 9: Table 1, Table 4, and Table 5)

| Type | Conductor Size (AWG/ kcmil) | Trade Size (Metric Designator) | | | | | | | | | | | | |
|---|---|---|---|---|---|---|---|---|---|---|---|---|---|---|
| | | ⅜ (12) | ½ (16) | ¾ (21) | 1 (27) | 1¼ (35) | 1½ (41) | 2 (53) | 2½ (63) | 3 (78) | 3½ (91) | 4 (103) | 5 (129) | 6 (155) |
| **CONDUCTORS** | | | | | | | | | | | | | | |
| RHH, RHW, RHW-2 | 14 | — | 4 | 7 | 11 | 20 | 27 | 45 | — | — | — | — | — | — |
| | 12 | — | 3 | 5 | 9 | 16 | 22 | 37 | — | — | — | — | — | — |
| | 10 | — | 2 | 4 | 7 | 13 | 18 | 30 | — | — | — | — | — | — |
| | 8 | — | 1 | 2 | 4 | 7 | 9 | 15 | — | — | — | — | — | — |
| | 6 | — | 1 | 1 | 3 | 5 | 7 | 12 | — | — | — | — | — | — |
| | 4 | — | 1 | 1 | 2 | 4 | 6 | 10 | — | — | — | — | — | — |
| | 3 | — | 1 | 1 | 1 | 4 | 5 | 8 | — | — | — | — | — | — |
| | 2 | — | 1 | 1 | 1 | 3 | 4 | 7 | — | — | — | — | — | — |
| | 1 | — | 0 | 1 | 1 | 1 | 3 | 5 | — | — | — | — | — | — |
| | 1/0 | — | 0 | 1 | 1 | 1 | 2 | 4 | — | — | — | — | — | — |
| | 2/0 | — | 0 | 0 | 1 | 1 | 1 | 3 | — | — | — | — | — | — |
| | 3/0 | — | 0 | 0 | 1 | 1 | 1 | 3 | — | — | — | — | — | — |
| | 4/0 | — | 0 | 0 | 1 | 1 | 1 | 2 | — | — | — | — | — | — |
| | 250 | — | 0 | 0 | 0 | 1 | 1 | 1 | — | — | — | — | — | — |
| | 300 | — | 0 | 0 | 0 | 1 | 1 | 1 | — | — | — | — | — | — |
| | 350 | — | 0 | 0 | 0 | 1 | 1 | 1 | — | — | — | — | — | — |
| | 400 | — | 0 | 0 | 0 | 1 | 1 | 1 | — | — | — | — | — | — |
| | 500 | — | 0 | 0 | 0 | 0 | 1 | 1 | — | — | — | — | — | — |
| | 600 | — | 0 | 0 | 0 | 0 | 1 | 1 | — | — | — | — | — | — |
| | 700 | — | 0 | 0 | 0 | 0 | 0 | 1 | — | — | — | — | — | — |
| | 750 | — | 0 | 0 | 0 | 0 | 0 | 1 | — | — | — | — | — | — |
| | 800 | — | 0 | 0 | 0 | 0 | 0 | 1 | — | — | — | — | — | — |
| | 900 | — | 0 | 0 | 0 | 0 | 0 | 1 | — | — | — | — | — | — |
| | 1000 | — | 0 | 0 | 0 | 0 | 0 | 1 | — | — | — | — | — | — |
| | 1250 | — | 0 | 0 | 0 | 0 | 0 | 0 | — | — | — | — | — | — |
| | 1500 | — | 0 | 0 | 0 | 0 | 0 | 0 | — | — | — | — | — | — |
| | 1750 | — | 0 | 0 | 0 | 0 | 0 | 0 | — | — | — | — | — | — |
| | 2000 | — | 0 | 0 | 0 | 0 | 0 | 0 | — | — | — | — | — | — |
| TW, THHW, THW, THW-2 | 14 | — | 8 | 14 | 24 | 42 | 57 | 94 | — | — | — | — | — | — |
| | 12 | — | 6 | 11 | 18 | 32 | 44 | 72 | — | — | — | — | — | — |
| | 10 | — | 4 | 8 | 13 | 24 | 32 | 54 | — | — | — | — | — | — |
| | 8 | — | 2 | 4 | 7 | 13 | 18 | 30 | — | — | — | — | — | — |
| RHH*, RHW*, RHW-2* | 14 | — | 5 | 9 | 16 | 28 | 38 | 63 | — | — | — | — | — | — |
| | 12 | — | 4 | 8 | 13 | 22 | 30 | 50 | — | — | — | — | — | — |
| | 10 | — | 3 | 6 | 10 | 17 | 24 | 39 | — | — | — | — | — | — |
| | 8 | — | 1 | 3 | 6 | 10 | 14 | 23 | — | — | — | — | — | — |
| TW, THW, THHW, THW-2, RHH*, RHW*, RHW-2* | 6 | — | 1 | 2 | 4 | 8 | 11 | 18 | — | — | — | — | — | — |
| | 4 | — | 1 | 1 | 3 | 6 | 8 | 13 | — | — | — | — | — | — |
| | 3 | — | 1 | 1 | 3 | 5 | 7 | 11 | — | — | — | — | — | — |
| | 2 | — | 1 | 1 | 2 | 4 | 6 | 10 | — | — | — | — | — | — |
| | 1 | — | 0 | 1 | 1 | 3 | 4 | 7 | — | — | — | — | — | — |
| | 1/0 | — | 0 | 1 | 1 | 2 | 3 | 6 | — | — | — | — | — | — |
| | 2/0 | — | 0 | 1 | 1 | 1 | 3 | 5 | — | — | — | — | — | — |
| | 3/0 | — | 0 | 1 | 1 | 1 | 2 | 4 | — | — | — | — | — | — |
| | 4/0 | — | 0 | 0 | 1 | 1 | 1 | 3 | — | — | — | — | — | — |

***TABLE C.2*** *Continued*

| Type | Conductor Size (AWG/ kcmil) | Trade Size (Metric Designator) | | | | | | | | | | | | |
|---|---|---|---|---|---|---|---|---|---|---|---|---|---|---|
| | | **⅜ (12)** | **½ (16)** | **¾ (21)** | **1 (27)** | **1¼ (35)** | **1½ (41)** | **2 (53)** | **2½ (63)** | **3 (78)** | **3½ (91)** | **4 (103)** | **5 (129)** | **6 (155)** |
| | 250 | — | 0 | 0 | 1 | 1 | 1 | 3 | — | — | — | — | — | — |
| | 300 | — | 0 | 0 | 1 | 1 | 1 | 2 | — | — | — | — | — | — |
| | 350 | — | 0 | 0 | 0 | 1 | 1 | 1 | — | — | — | — | — | — |
| | 400 | — | 0 | 0 | 0 | 1 | 1 | 1 | — | — | — | — | — | — |
| | 500 | — | 0 | 0 | 0 | 1 | 1 | 1 | — | — | — | — | — | — |
| | 600 | — | 0 | 0 | 0 | 0 | 1 | 1 | — | — | — | — | — | — |
| | 700 | — | 0 | 0 | 0 | 0 | 1 | 1 | — | — | — | — | — | — |
| | 750 | — | 0 | 0 | 0 | 0 | 1 | 1 | — | — | — | — | — | — |
| | 800 | — | 0 | 0 | 0 | 0 | 1 | 1 | — | — | — | — | — | — |
| | 900 | — | 0 | 0 | 0 | 0 | 0 | 1 | — | — | — | — | — | — |
| | 1000 | — | 0 | 0 | 0 | 0 | 0 | 1 | — | — | — | — | — | — |
| | 1250 | — | 0 | 0 | 0 | 0 | 0 | 1 | — | — | — | — | — | — |
| | 1500 | — | 0 | 0 | 0 | 0 | 0 | 1 | — | — | — | — | — | — |
| | 1750 | — | 0 | 0 | 0 | 0 | 0 | 0 | — | — | — | — | — | — |
| | 2000 | — | 0 | 0 | 0 | 0 | 0 | 0 | — | — | — | — | — | — |
| THHN, THWN, THWN-2 | 14 | — | 11 | 21 | 34 | 60 | 82 | 135 | — | — | — | — | — | — |
| | 12 | — | 8 | 15 | 25 | 43 | 59 | 99 | — | — | — | — | — | — |
| | 10 | — | 5 | 9 | 15 | 27 | 37 | 62 | — | — | — | — | — | — |
| | 8 | — | 3 | 5 | 9 | 16 | 21 | 36 | — | — | — | — | — | — |
| | 6 | — | 1 | 4 | 6 | 11 | 15 | 26 | — | — | — | — | — | — |
| | 4 | — | 1 | 2 | 4 | 7 | 9 | 16 | — | — | — | — | — | — |
| | 3 | — | 1 | 1 | 3 | 6 | 8 | 13 | — | — | — | — | — | — |
| | 2 | — | 1 | 1 | 3 | 5 | 7 | 11 | — | — | — | — | — | — |
| | 1 | — | 1 | 1 | 1 | 3 | 5 | 8 | — | — | — | — | — | — |
| | 1/0 | — | 1 | 1 | 1 | 3 | 4 | 7 | — | — | — | — | — | — |
| | 2/0 | — | 0 | 1 | 1 | 2 | 3 | 6 | — | — | — | — | — | — |
| | 3/0 | — | 0 | 1 | 1 | 1 | 3 | 5 | — | — | — | — | — | — |
| | 4/0 | — | 0 | 1 | 1 | 1 | 2 | 4 | — | — | — | — | — | — |
| | 250 | — | 0 | 0 | 1 | 1 | 1 | 3 | — | — | — | — | — | — |
| | 300 | — | 0 | 0 | 1 | 1 | 1 | 3 | — | — | — | — | — | — |
| | 350 | — | 0 | 0 | 1 | 1 | 1 | 2 | — | — | — | — | — | — |
| | 400 | — | 0 | 0 | 0 | 1 | 1 | 1 | — | — | — | — | — | — |
| | 500 | — | 0 | 0 | 0 | 1 | 1 | 1 | — | — | — | — | — | — |
| | 600 | — | 0 | 0 | 0 | 1 | 1 | 1 | — | — | — | — | — | — |
| | 700 | — | 0 | 0 | 0 | 0 | 1 | 1 | — | — | — | — | — | — |
| | 750 | — | 0 | 0 | 0 | 0 | 1 | 1 | — | — | — | — | — | — |
| | 800 | — | 0 | 0 | 0 | 0 | 1 | 1 | — | — | — | — | — | — |
| | 900 | — | 0 | 0 | 0 | 0 | 1 | 1 | — | — | — | — | — | — |
| | 1000 | — | 0 | 0 | 0 | 0 | 0 | 1 | — | — | — | — | — | — |
| FEP, FEPB, PFA, PFAH, TFE | 14 | — | 11 | 20 | 33 | 58 | 79 | 131 | — | — | — | — | — | — |
| | 12 | — | 8 | 15 | 24 | 42 | 58 | 96 | — | — | — | — | — | — |
| | 10 | — | 6 | 10 | 17 | 30 | 41 | 69 | — | — | — | — | — | — |
| | 8 | — | 3 | 6 | 10 | 17 | 24 | 39 | — | — | — | — | — | — |
| | 6 | — | 2 | 4 | 7 | 12 | 17 | 28 | — | — | — | — | — | — |
| | 4 | — | 1 | 3 | 5 | 8 | 12 | 19 | — | — | — | — | — | — |
| | 3 | — | 1 | 2 | 4 | 7 | 10 | 16 | — | — | — | — | — | — |
| | 2 | — | 1 | 1 | 3 | 6 | 8 | 13 | — | — | — | — | — | — |

*(continues)*

*TABLE C.2 Continued*

| Type | Conductor Size (AWG/ kcmil) | Trade Size (Metric Designator) | | | | | | | | | | | | |
|---|---|---|---|---|---|---|---|---|---|---|---|---|---|---|
| | | ⅜ (12) | ½ (16) | ¾ (21) | 1 (27) | 1¼ (35) | 1½ (41) | 2 (53) | 2½ (63) | 3 (78) | 3½ (91) | 4 (103) | 5 (129) | 6 (155) |
| PFA, PFAH, TFE | 1 | — | 1 | 1 | 2 | 4 | 5 | 9 | — | — | — | — | — | — |
| PFA, PFAH, TFE, Z | 1/0 | — | 1 | 1 | 1 | 3 | 4 | 8 | — | — | — | — | — | — |
| | 2/0 | — | 0 | 1 | 1 | 3 | 4 | 6 | — | — | — | — | — | — |
| | 3/0 | — | 0 | 1 | 1 | 2 | 3 | 5 | — | — | — | — | — | — |
| | 4/0 | — | 0 | 1 | 1 | 1 | 2 | 4 | — | — | — | — | — | — |
| Z | 14 | — | 13 | 24 | 40 | 70 | 95 | 158 | — | — | — | — | — | — |
| | 12 | — | 9 | 17 | 28 | 49 | 68 | 112 | — | — | — | — | — | — |
| | 10 | — | 6 | 10 | 17 | 30 | 41 | 69 | — | — | — | — | — | — |
| | 8 | — | 3 | 6 | 11 | 19 | 26 | 43 | — | — | — | — | — | — |
| | 6 | — | 2 | 4 | 7 | 13 | 18 | 30 | — | — | — | — | — | — |
| | 4 | — | 1 | 3 | 5 | 9 | 12 | 21 | — | — | — | — | — | — |
| | 3 | — | 1 | 2 | 4 | 6 | 9 | 15 | — | — | — | — | — | — |
| | 2 | — | 1 | 1 | 3 | 5 | 7 | 12 | — | — | — | — | — | — |
| | 1 | — | 1 | 1 | 2 | 4 | 6 | 10 | — | — | — | — | — | — |
| XHHW, ZW, XHHW-2, XHH | 14 | — | 8 | 14 | 24 | 42 | 57 | 94 | — | — | — | — | — | — |
| | 12 | — | 6 | 11 | 18 | 32 | 44 | 72 | — | — | — | — | — | — |
| | 10 | — | 4 | 8 | 13 | 24 | 32 | 54 | — | — | — | — | — | — |
| | 8 | — | 2 | 4 | 7 | 13 | 18 | 30 | — | — | — | — | — | — |
| | 6 | — | 1 | 3 | 5 | 10 | 13 | 22 | — | — | — | — | — | — |
| | 4 | — | 1 | 2 | 4 | 7 | 9 | 16 | — | — | — | — | — | — |
| | 3 | — | 1 | 1 | 3 | 6 | 8 | 13 | — | — | — | — | — | — |
| | 2 | — | 1 | 1 | 3 | 5 | 7 | 11 | — | — | — | — | — | — |
| XHHW, XHHW-2, XHH | 1 | — | 1 | 1 | 1 | 3 | 5 | 8 | — | — | — | — | — | — |
| | 1/0 | — | 0 | 1 | 1 | 3 | 4 | 7 | — | — | — | — | — | — |
| | 2/0 | — | 0 | 1 | 1 | 2 | 3 | 6 | — | — | — | — | — | — |
| | 3/0 | — | 0 | 1 | 1 | 1 | 3 | 5 | — | — | — | — | — | — |
| | 4/0 | — | 0 | 1 | 1 | 1 | 2 | 4 | — | — | — | — | — | — |
| | 250 | — | 0 | 0 | 1 | 1 | 1 | 3 | — | — | — | — | — | — |
| | 300 | — | 0 | 0 | 1 | 1 | 1 | 3 | — | — | — | — | — | — |
| | 350 | — | 0 | 0 | 1 | 1 | 1 | 2 | — | — | — | — | — | — |
| | 400 | — | 0 | 0 | 0 | 1 | 1 | 1 | — | — | — | — | — | — |
| | 500 | — | 0 | 0 | 0 | 1 | 1 | 1 | — | — | — | — | — | — |
| | 600 | — | 0 | 0 | 0 | 1 | 1 | 1 | — | — | — | — | — | — |
| | 700 | — | 0 | 0 | 0 | 0 | 1 | 1 | — | — | — | — | — | — |
| | 750 | — | 0 | 0 | 0 | 0 | 1 | 1 | — | — | — | — | — | — |
| | 800 | — | 0 | 0 | 0 | 0 | 1 | 1 | — | — | — | — | — | — |
| | 900 | — | 0 | 0 | 0 | 0 | 1 | 1 | — | — | — | — | — | — |
| | 1000 | — | 0 | 0 | 0 | 0 | 0 | 1 | — | — | — | — | — | — |
| | 1250 | — | 0 | 0 | 0 | 0 | 0 | 1 | — | — | — | — | — | — |
| | 1500 | — | 0 | 0 | 0 | 0 | 0 | 1 | — | — | — | — | — | — |
| | 1750 | — | 0 | 0 | 0 | 0 | 0 | 0 | — | — | — | — | — | — |
| | 2000 | — | 0 | 0 | 0 | 0 | 0 | 0 | — | — | — | — | — | — |

***TABLE C.2** Continued*

| Type | Conductor Size (AWG/ kcmil) | Trade Size (Metric Designator) ⅜ (12) | ½ (16) | ¾ (21) | 1 (27) | 1¼ (35) | 1½ (41) | 2 (53) | 2½ (63) | 3 (78) | 3½ (91) | 4 (103) | 5 (129) | 6 (155) |
|---|---|---|---|---|---|---|---|---|---|---|---|---|---|---|
| | | | | | | | | FIXTURE WIRES | | | | | | |
| RFH-2, FFH-2, RFHH-2 | 18 | — | 8 | 14 | 23 | 40 | 54 | 90 | — | — | — | — | — | — |
| | 16 | — | 6 | 12 | 19 | 33 | 46 | 76 | — | — | — | — | — | — |
| SF-2, SFF-2 | 18 | — | 10 | 17 | 29 | 50 | 69 | 114 | — | — | — | — | — | — |
| | 16 | — | 8 | 14 | 24 | 42 | 57 | 94 | — | — | — | — | — | — |
| | 14 | — | 6 | 12 | 19 | 33 | 46 | 76 | — | — | — | — | — | — |
| SF-1, SFF-1 | 18 | — | 17 | 31 | 51 | 89 | 122 | 202 | — | — | — | — | — | — |
| RFH-1, TF, TFF, XF, XFF | 18 | — | 13 | 23 | 38 | 66 | 90 | 149 | — | — | — | — | — | — |
| | 16 | — | 10 | 18 | 30 | 53 | 73 | 120 | — | — | — | — | — | — |
| • XF, XFF | 14 | — | 8 | 14 | 24 | 42 | 57 | 94 | — | — | — | — | — | — |
| TFN, TFFN | 18 | — | 20 | 37 | 60 | 105 | 144 | 239 | — | — | — | — | — | — |
| | 16 | — | 16 | 28 | 46 | 80 | 110 | 183 | — | — | — | — | — | — |
| PF, PFF, PGF, PGFF, PAF, PTF, PTFF, PAFF | 18 | — | 19 | 35 | 57 | 100 | 137 | 227 | — | — | — | — | — | — |
| | 16 | — | 15 | 27 | 44 | 77 | 106 | 175 | — | — | — | — | — | — |
| | 14 | — | 11 | 20 | 33 | 58 | 79 | 131 | — | — | — | — | — | — |
| ZF, ZFF, ZHF | 18 | — | 25 | 45 | 74 | 129 | 176 | 292 | — | — | — | — | — | — |
| | 16 | — | 18 | 33 | 54 | 95 | 130 | 216 | — | — | — | — | — | — |
| | 14 | — | 13 | 24 | 40 | 70 | 95 | 158 | — | — | — | — | — | — |
| KF-2, KFF-2 | 18 | — | 38 | 67 | 111 | 193 | 265 | 439 | — | — | — | — | — | — |
| | 16 | — | 26 | 47 | 77 | 135 | 184 | 306 | — | — | — | — | — | — |
| | 14 | — | 18 | 31 | 52 | 91 | 124 | 205 | — | — | — | — | — | — |
| | 12 | — | 12 | 22 | 36 | 63 | 86 | 143 | — | — | — | — | — | — |
| | 10 | — | 8 | 14 | 24 | 42 | 57 | 94 | — | — | — | — | — | — |
| KF-1, KFF-1 | 18 | — | 44 | 78 | 128 | 223 | 305 | 506 | — | — | — | — | — | — |
| | 16 | — | 31 | 55 | 90 | 157 | 214 | 355 | — | — | — | — | — | — |
| | 14 | — | 20 | 37 | 60 | 105 | 144 | 239 | — | — | — | — | — | — |
| | 12 | — | 13 | 24 | 40 | 70 | 95 | 158 | — | — | — | — | — | — |
| | 10 | — | 9 | 16 | 26 | 45 | 62 | 103 | — | — | — | — | — | — |
| XF, XFF | 12 | — | 4 | 8 | 13 | 22 | 30 | 50 | — | — | — | — | — | — |
| | 10 | — | 3 | 6 | 10 | 17 | 24 | 39 | — | — | — | — | — | — |

Notes:

1. This table is for concentric stranded conductors only. For compact stranded conductors, Table C.2(A) should be used.

2. Two-hour fire-rated RHH cable has ceramifiable insulation, which has much larger diameters than other RHH wires. Consult manufacturer's conduit fill tables.

*Types RHH, RHW, and RHW-2 without outer covering.

***TABLE C.2(A)*** *Maximum Number of Conductors or Fixture Wires in Electrical Nonmetallic Tubing (ENT)* (Based on Chapter 9: Table 1, Table 4, and Table 5A)

| Type | Conductor Size (AWG/ kcmil) | Trade Size (Metric Designator) | | | | | | | | | | | | |
|---|---|---|---|---|---|---|---|---|---|---|---|---|---|---|
| | | ⅜ (12) | ½ (16) | ¾ (21) | 1 (27) | 1¼ (35) | 1½ (41) | 2 (53) | 2½ (63) | 3 (78) | 3½ (91) | 4 (103) | 5 (129) | 6 (155) |
| **COMPACT CONDUCTORS** | | | | | | | | | | | | | | |
| THW, THW-2, THHW | 8 | — | 1 | 4 | 6 | 11 | 15 | 26 | — | — | — | — | — | — |
| | 6 | — | 1 | 3 | 5 | 9 | 12 | 20 | — | — | — | — | — | — |
| | 4 | — | 1 | 1 | 3 | 6 | 9 | 15 | — | — | — | — | — | — |
| | 2 | — | 1 | 1 | 2 | 5 | 6 | 11 | — | — | — | — | — | — |
| | 1 | — | 1 | 1 | 1 | 3 | 4 | 7 | — | — | — | — | — | — |
| | 1/0 | — | 0 | 1 | 1 | 3 | 4 | 6 | — | — | — | — | — | — |
| | 2/0 | — | 0 | 1 | 1 | 2 | 3 | 5 | — | — | — | — | — | — |
| | 3/0 | — | 0 | 1 | 1 | 1 | 3 | 5 | — | — | — | — | — | — |
| | 4/0 | — | 0 | 1 | 1 | 1 | 2 | 4 | — | — | — | — | — | — |
| | 250 | — | 0 | 0 | 1 | 1 | 1 | 3 | — | — | — | — | — | — |
| | 300 | — | 0 | 0 | 1 | 1 | 1 | 2 | — | — | — | — | — | — |
| | 350 | — | 0 | 0 | 1 | 1 | 1 | 2 | — | — | — | — | — | — |
| | 400 | — | 0 | 0 | 0 | 1 | 1 | 1 | — | — | — | — | — | — |
| | 500 | — | 0 | 0 | 0 | 1 | 1 | 1 | — | — | — | — | — | — |
| | 600 | — | 0 | 0 | 0 | 1 | 1 | 1 | — | — | — | — | — | — |
| | 700 | — | 0 | 0 | 0 | 0 | 1 | 1 | — | — | — | — | — | — |
| | 750 | — | 0 | 0 | 0 | 0 | 1 | 1 | — | — | — | — | — | — |
| | 900 | — | 0 | 0 | 0 | 0 | 1 | 1 | — | — | — | — | — | — |
| | 1000 | — | 0 | 0 | 0 | 0 | 1 | 1 | — | — | — | — | — | — |
| THHN, THWN, THWN-2 | 8 | — | — | — | — | — | — | — | — | — | — | — | — | — |
| | 6 | — | 2 | 4 | 7 | 13 | 17 | 29 | — | — | — | — | — | — |
| | 4 | — | 1 | 2 | 4 | 8 | 11 | 18 | — | — | — | — | — | — |
| | 2 | — | 1 | 1 | 3 | 5 | 8 | 13 | — | — | — | — | — | — |
| | 1 | — | 1 | 1 | 2 | 4 | 6 | 9 | — | — | — | — | — | — |
| | 1/0 | — | 1 | 1 | 1 | 3 | 5 | 8 | — | — | — | — | — | — |
| | 2/0 | — | 0 | 1 | 1 | 3 | 4 | 7 | — | — | — | — | — | — |
| | 3/0 | — | 0 | 1 | 1 | 2 | 3 | 5 | — | — | — | — | — | — |
| | 4/0 | — | 0 | 1 | 1 | 1 | 3 | 4 | — | — | — | — | — | — |
| | 250 | — | 0 | 0 | 1 | 1 | 1 | 3 | — | — | — | — | — | — |
| | 300 | — | 0 | 0 | 1 | 1 | 1 | 3 | — | — | — | — | — | — |
| | 350 | — | 0 | 0 | 1 | 1 | 1 | 3 | — | — | — | — | — | — |
| | 400 | — | 0 | 0 | 1 | 1 | 1 | 2 | — | — | — | — | — | — |
| | 500 | — | 0 | 0 | 0 | 1 | 1 | 1 | — | — | — | — | — | — |
| | 600 | — | 0 | 0 | 0 | 1 | 1 | 1 | — | — | — | — | — | — |
| | 700 | — | 0 | 0 | 0 | 1 | 1 | 1 | — | — | — | — | — | — |
| | 750 | — | 0 | 0 | 0 | 1 | 1 | 1 | — | — | — | — | — | — |
| | 900 | — | 0 | 0 | 0 | 0 | 1 | 1 | — | — | — | — | — | — |
| | 1000 | — | 0 | 0 | 0 | 0 | 1 | 1 | — | — | — | — | — | — |

*TABLE C.2(A) Continued*

| Type | Conductor Size (AWG/ kcmil) | Trade Size (Metric Designator) | | | | | | | | | | | | |
|---|---|---|---|---|---|---|---|---|---|---|---|---|---|---|
| | | 3/8 (12) | 1/2 (16) | 3/4 (21) | 1 (27) | 1 1/4 (35) | 1 1/2 (41) | 2 (53) | 2 1/2 (63) | 3 (78) | 3 1/2 (91) | 4 (103) | 5 (129) | 6 (155) |
| XHHW, XHHW-2 | 8 | — | 3 | 5 | 8 | 14 | 20 | 33 | — | — | — | — | — | — |
| | 6 | — | 1 | 4 | 6 | 11 | 15 | 25 | — | — | — | — | — | — |
| | 4 | — | 1 | 2 | 4 | 8 | 11 | 18 | — | — | — | — | — | — |
| | 2 | — | 1 | 1 | 3 | 5 | 8 | 13 | — | — | — | — | — | — |
| | 1 | — | 1 | 1 | 2 | 4 | 6 | 9 | — | — | — | — | — | — |
| | 1/0 | — | 1 | 1 | 1 | 3 | 5 | 8 | — | — | — | — | — | — |
| | 2/0 | — | 1 | 1 | 1 | 3 | 4 | 7 | — | — | — | — | — | — |
| | 3/0 | — | 0 | 1 | 1 | 2 | 3 | 5 | — | — | — | — | — | — |
| | 4/0 | — | 0 | 1 | 1 | 1 | 3 | 5 | — | — | — | — | — | — |
| | 250 | — | 0 | 0 | 1 | 1 | 1 | 4 | — | — | — | — | — | — |
| | 300 | — | 0 | 0 | 1 | 1 | 1 | 3 | — | — | — | — | — | — |
| | 350 | — | 0 | 0 | 1 | 1 | 1 | 3 | — | — | — | — | — | — |
| | 400 | — | 0 | 0 | 1 | 1 | 1 | 2 | — | — | — | — | — | — |
| | 500 | — | 0 | 0 | 0 | 1 | 1 | 1 | — | — | — | — | — | — |
| | 600 | — | 0 | 0 | 0 | 1 | 1 | 1 | — | — | — | — | — | — |
| | 700 | — | 0 | 0 | 0 | 1 | 1 | 1 | — | — | — | — | — | — |
| | 750 | — | 0 | 0 | 0 | 1 | 1 | 1 | — | — | — | — | — | — |
| | 900 | — | 0 | 0 | 0 | 0 | 1 | 1 | — | — | — | — | — | — |
| | 1000 | — | 0 | 0 | 0 | 0 | 1 | 1 | — | — | — | — | — | — |

Definition: *Compact stranding* is the result of a manufacturing process where the stranded conductor is compressed to the extent that the interstices (voids between strand wires) are virtually eliminated.

***TABLE C.3*** *Maximum Number of Conductors or Fixture Wires in Flexible Metal Conduit (FMC)* (Based on Chapter 9: Table 1, Table 4, and Table 5)

| Type | Conductor Size (AWG/ kcmil) | Trade Size (Metric Designator) | | | | | | | | | | | | |
|---|---|---|---|---|---|---|---|---|---|---|---|---|---|---|
| | | ⅜ (12) | ½ (16) | ¾ (21) | 1 (27) | 1¼ (35) | 1½ (41) | 2 (53) | 2½ (63) | 3 (78) | 3½ (91) | 4 (103) | 5 (129) | 6 (155) |
| **CONDUCTORS** | | | | | | | | | | | | | | |
| RHH, RHW, RHW-2 | 14 | 1 | 4 | 7 | 11 | 17 | 25 | 44 | 67 | 96 | 131 | 171 | — | — |
| | 12 | 1 | 3 | 6 | 9 | 14 | 21 | 37 | 55 | 80 | 109 | 142 | — | — |
| | 10 | 1 | 3 | 5 | 7 | 11 | 17 | 30 | 45 | 64 | 88 | 115 | — | — |
| | 8 | 0 | 1 | 2 | 4 | 6 | 9 | 15 | 23 | 34 | 46 | 60 | — | — |
| | 6 | 0 | 1 | 1 | 3 | 5 | 7 | 12 | 19 | 27 | 37 | 48 | — | — |
| | 4 | 0 | 1 | 1 | 2 | 4 | 5 | 10 | 14 | 21 | 29 | 37 | — | — |
| | 3 | 0 | 1 | 1 | 1 | 3 | 5 | 8 | 13 | 18 | 25 | 33 | — | — |
| | 2 | 0 | 1 | 1 | 1 | 3 | 4 | 7 | 11 | 16 | 22 | 28 | — | — |
| | 1 | 0 | 0 | 1 | 1 | 1 | 2 | 5 | 7 | 10 | 14 | 19 | — | — |
| | 1/0 | 0 | 0 | 1 | 1 | 1 | 2 | 4 | 6 | 9 | 12 | 16 | — | — |
| | 2/0 | 0 | 0 | 1 | 1 | 1 | 1 | 3 | 5 | 8 | 11 | 14 | — | — |
| | 3/0 | 0 | 0 | 0 | 1 | 1 | 1 | 3 | 5 | 7 | 9 | 12 | — | — |
| | 4/0 | 0 | 0 | 0 | 1 | 1 | 1 | 2 | 4 | 6 | 8 | 10 | — | — |
| | 250 | 0 | 0 | 0 | 0 | 1 | 1 | 1 | 3 | 4 | 6 | 8 | — | — |
| | 300 | 0 | 0 | 0 | 0 | 1 | 1 | 1 | 2 | 4 | 5 | 7 | — | — |
| | 350 | 0 | 0 | 0 | 0 | 1 | 1 | 1 | 2 | 3 | 5 | 6 | — | — |
| | 400 | 0 | 0 | 0 | 0 | 0 | 1 | 1 | 1 | 3 | 4 | 6 | — | — |
| | 500 | 0 | 0 | 0 | 0 | 0 | 1 | 1 | 1 | 3 | 4 | 5 | — | — |
| | 600 | 0 | 0 | 0 | 0 | 0 | 1 | 1 | 1 | 2 | 3 | 4 | — | — |
| | 700 | 0 | 0 | 0 | 0 | 0 | 0 | 1 | 1 | 1 | 3 | 3 | — | — |
| | 750 | 0 | 0 | 0 | 0 | 0 | 0 | 1 | 1 | 1 | 2 | 3 | — | — |
| | 800 | 0 | 0 | 0 | 0 | 0 | 0 | 1 | 1 | 1 | 2 | 3 | — | — |
| | 900 | 0 | 0 | 0 | 0 | 0 | 0 | 1 | 1 | 1 | 2 | 3 | — | — |
| | 1000 | 0 | 0 | 0 | 0 | 0 | 0 | 1 | 1 | 1 | 1 | 3 | — | — |
| | 1250 | 0 | 0 | 0 | 0 | 0 | 0 | 0 | 1 | 1 | 1 | 1 | — | — |
| | 1500 | 0 | 0 | 0 | 0 | 0 | 0 | 0 | 1 | 1 | 1 | 1 | — | — |
| | 1750 | 0 | 0 | 0 | 0 | 0 | 0 | 0 | 1 | 1 | 1 | 1 | — | — |
| | 2000 | 0 | 0 | 0 | 0 | 0 | 0 | 0 | 0 | 1 | 1 | 1 | — | — |
| TW, THHW, THW, THW-2 | 14 | 3 | 9 | 15 | 23 | 36 | 53 | 94 | 141 | 203 | 277 | 361 | — | — |
| | 12 | 2 | 7 | 11 | 18 | 28 | 41 | 72 | 108 | 156 | 212 | 277 | — | — |
| | 10 | 1 | 5 | 8 | 13 | 21 | 30 | 54 | 81 | 116 | 158 | 207 | — | — |
| | 8 | 1 | 3 | 5 | 7 | 11 | 17 | 30 | 45 | 64 | 88 | 115 | — | — |
| RHH*, RHW*, RHW-2* | 14 | 1 | 6 | 10 | 15 | 24 | 35 | 62 | 94 | 135 | 184 | 240 | — | — |
| | 12 | 1 | 5 | 8 | 12 | 19 | 28 | 50 | 75 | 108 | 148 | 193 | — | — |
| | 10 | 1 | 4 | 6 | 10 | 15 | 22 | 39 | 59 | 85 | 115 | 151 | — | — |
| | 8 | 1 | 1 | 4 | 6 | 9 | 13 | 23 | 35 | 51 | 69 | 90 | — | — |
| TW, THW, THHW, THW-2, RHH*, RHW*, RHW-2* | 6 | 1 | 1 | 3 | 4 | 7 | 10 | 18 | 27 | 39 | 53 | 69 | — | — |
| | 4 | 0 | 1 | 1 | 3 | 5 | 7 | 13 | 20 | 29 | 39 | 51 | — | — |
| | 3 | 0 | 1 | 1 | 3 | 4 | 6 | 11 | 17 | 25 | 34 | 44 | — | — |
| | 2 | 0 | 1 | 1 | 2 | 4 | 5 | 10 | 14 | 21 | 29 | 37 | — | — |
| | 1 | 0 | 1 | 1 | 1 | 2 | 4 | 7 | 10 | 15 | 20 | 26 | — | — |
| | 1/0 | 0 | 0 | 1 | 1 | 1 | 3 | 6 | 9 | 12 | 17 | 22 | — | — |
| | 2/0 | 0 | 0 | 1 | 1 | 1 | 3 | 5 | 7 | 10 | 14 | 19 | — | — |
| | 3/0 | 0 | 0 | 1 | 1 | 1 | 2 | 4 | 6 | 9 | 12 | 16 | — | — |
| | 4/0 | 0 | 0 | 0 | 1 | 1 | 1 | 3 | 5 | 7 | 10 | 13 | — | — |

***TABLE C.3*** *Continued*

| Type | Conductor Size (AWG/ kcmil) | Trade Size (Metric Designator) | | | | | | | | | | | | |
|---|---|---|---|---|---|---|---|---|---|---|---|---|---|---|
| | | ⅜ (12) | ½ (16) | ¾ (21) | 1 (27) | 1¼ (35) | 1½ (41) | 2 (53) | 2½ (63) | 3 (78) | 3½ (91) | 4 (103) | 5 (129) | 6 (155) |
| | 250 | 0 | 0 | 0 | 1 | 1 | 1 | 3 | 4 | 6 | 8 | 11 | — | — |
| | 300 | 0 | 0 | 0 | 1 | 1 | 1 | 2 | 3 | 5 | 7 | 9 | — | — |
| | 350 | 0 | 0 | 0 | 0 | 1 | 1 | 1 | 3 | 4 | 6 | 8 | — | — |
| | 400 | 0 | 0 | 0 | 0 | 1 | 1 | 1 | 3 | 4 | 6 | 7 | — | — |
| | 500 | 0 | 0 | 0 | 0 | 1 | 1 | 1 | 2 | 3 | 5 | 6 | — | — |
| | 600 | 0 | 0 | 0 | 0 | 0 | 1 | 1 | 1 | 3 | 4 | 5 | — | — |
| | 700 | 0 | 0 | 0 | 0 | 0 | 1 | 1 | 1 | 2 | 3 | 4 | — | — |
| | 750 | 0 | 0 | 0 | 0 | 0 | 1 | 1 | 1 | 2 | 3 | 4 | — | — |
| | 800 | 0 | 0 | 0 | 0 | 0 | 1 | 1 | 1 | 1 | 3 | 4 | — | — |
| | 900 | 0 | 0 | 0 | 0 | 0 | 0 | 1 | 1 | 1 | 3 | 3 | — | — |
| | 1000 | 0 | 0 | 0 | 0 | 0 | 0 | 1 | 1 | 1 | 2 | 3 | — | — |
| | 1250 | 0 | 0 | 0 | 0 | 0 | 0 | 1 | 1 | 1 | 1 | 2 | — | — |
| | 1500 | 0 | 0 | 0 | 0 | 0 | 0 | 0 | 1 | 1 | 1 | 1 | — | — |
| | 1750 | 0 | 0 | 0 | 0 | 0 | 0 | 0 | 1 | 1 | 1 | 1 | — | — |
| | 2000 | 0 | 0 | 0 | 0 | 0 | 0 | 0 | 1 | 1 | 1 | 1 | — | — |
| THHN, THWN, THWN-2 | 14 | 4 | 13 | 22 | 33 | 52 | 76 | 135 | 202 | 291 | 396 | 518 | — | — |
| | 12 | 3 | 9 | 16 | 24 | 38 | 56 | 98 | 147 | 212 | 289 | 378 | — | — |
| | 10 | 1 | 6 | 10 | 15 | 24 | 35 | 62 | 93 | 134 | 182 | 238 | — | — |
| | 8 | 1 | 3 | 6 | 9 | 14 | 20 | 35 | 53 | 77 | 105 | 137 | — | — |
| | 6 | 1 | 2 | 4 | 6 | 10 | 14 | 25 | 38 | 55 | 76 | 99 | — | — |
| | 4 | 0 | 1 | 2 | 4 | 6 | 9 | 16 | 24 | 34 | 46 | 61 | — | — |
| | 3 | 0 | 1 | 1 | 3 | 5 | 7 | 13 | 20 | 29 | 39 | 51 | — | — |
| | 2 | 0 | 1 | 1 | 3 | 4 | 6 | 11 | 17 | 24 | 33 | 43 | — | — |
| | 1 | 0 | 1 | 1 | 1 | 3 | 4 | 8 | 12 | 18 | 24 | 32 | — | — |
| | 1/0 | 0 | 1 | 1 | 1 | 2 | 4 | 7 | 10 | 15 | 20 | 27 | — | — |
| | 2/0 | 0 | 0 | 1 | 1 | 1 | 3 | 6 | 9 | 12 | 17 | 22 | — | — |
| | 3/0 | 0 | 0 | 1 | 1 | 1 | 2 | 5 | 7 | 10 | 14 | 18 | — | — |
| | 4/0 | 0 | 0 | 1 | 1 | 1 | 1 | 4 | 6 | 8 | 12 | 15 | — | — |
| | 250 | 0 | 0 | 0 | 1 | 1 | 1 | 3 | 5 | 7 | 9 | 12 | — | — |
| | 300 | 0 | 0 | 0 | 1 | 1 | 1 | 3 | 4 | 6 | 8 | 11 | — | — |
| | 350 | 0 | 0 | 0 | 1 | 1 | 1 | 2 | 3 | 5 | 7 | 9 | — | — |
| | 400 | 0 | 0 | 0 | 0 | 1 | 1 | 1 | 3 | 5 | 6 | 8 | — | — |
| | 500 | 0 | 0 | 0 | 0 | 1 | 1 | 1 | 2 | 4 | 5 | 7 | — | — |
| | 600 | 0 | 0 | 0 | 0 | 0 | 1 | 1 | 1 | 3 | 4 | 5 | — | — |
| | 700 | 0 | 0 | 0 | 0 | 0 | 1 | 1 | 1 | 3 | 4 | 5 | — | — |
| | 750 | 0 | 0 | 0 | 0 | 0 | 1 | 1 | 1 | 2 | 3 | 4 | — | — |
| | 800 | 0 | 0 | 0 | 0 | 0 | 1 | 1 | 1 | 2 | 3 | 4 | — | — |
| | 900 | 0 | 0 | 0 | 0 | 0 | 0 | 1 | 1 | 1 | 3 | 4 | — | — |
| | 1000 | 0 | 0 | 0 | 0 | 0 | 0 | 1 | 1 | 1 | 3 | 3 | — | — |
| FEP, FEPB, PFA, PFAH, TFE | 14 | 4 | 12 | 21 | 32 | 51 | 74 | 130 | 196 | 282 | 385 | 502 | — | — |
| | 12 | 3 | 9 | 15 | 24 | 37 | 54 | 95 | 143 | 206 | 281 | 367 | — | — |
| | 10 | 2 | 6 | 11 | 17 | 26 | 39 | 68 | 103 | 148 | 201 | 263 | — | — |
| | 8 | 1 | 4 | 6 | 10 | 15 | 22 | 39 | 59 | 85 | 115 | 151 | — | — |
| | 6 | 1 | 2 | 4 | 7 | 11 | 16 | 28 | 42 | 60 | 82 | 107 | — | — |
| | 4 | 1 | 1 | 3 | 5 | 7 | 11 | 19 | 29 | 42 | 57 | 75 | — | — |
| | 3 | 0 | 1 | 2 | 4 | 6 | 9 | 16 | 24 | 35 | 48 | 62 | — | — |
| | 2 | 0 | 1 | 1 | 3 | 5 | 7 | 13 | 20 | 29 | 39 | 51 | — | — |

*(continues)*

***TABLE C.3*** *Continued*

| Type | Conductor Size (AWG/ kcmil) | Trade Size (Metric Designator) | | | | | | | | | | | | |
|---|---|---|---|---|---|---|---|---|---|---|---|---|---|---|
| | | 3/8 (12) | 1/2 (16) | 3/4 (21) | 1 (27) | 1¼ (35) | 1½ (41) | 2 (53) | 2½ (63) | 3 (78) | 3½ (91) | 4 (103) | 5 (129) | 6 (155) |
| PFA, PFAH, TFE | 1 | 0 | 1 | 1 | 2 | 3 | 5 | 9 | 14 | 20 | 27 | 36 | — | — |
| PFA, PFAH, TFE, Z | 1/0 | 0 | 1 | 1 | 1 | 3 | 4 | 8 | 11 | 17 | 23 | 30 | — | — |
| | 2/0 | 0 | 1 | 1 | 1 | 2 | 3 | 6 | 9 | 14 | 19 | 24 | — | — |
| | 3/0 | 0 | 0 | 1 | 1 | 1 | 3 | 5 | 8 | 11 | 15 | 20 | — | — |
| | 4/0 | 0 | 0 | 1 | 1 | 1 | 2 | 4 | 6 | 9 | 13 | 16 | — | — |
| Z | 14 | 5 | 15 | 25 | 39 | 61 | 89 | 157 | 236 | 340 | 463 | 605 | — | — |
| | 12 | 4 | 11 | 18 | 28 | 43 | 63 | 111 | 168 | 241 | 329 | 429 | — | — |
| | 10 | 2 | 6 | 11 | 17 | 26 | 39 | 68 | 103 | 148 | 201 | 263 | — | — |
| | 8 | 1 | 4 | 7 | 11 | 17 | 24 | 43 | 65 | 93 | 127 | 166 | — | — |
| | 6 | 1 | 3 | 5 | 7 | 12 | 17 | 30 | 45 | 65 | 89 | 117 | — | — |
| | 4 | 1 | 1 | 3 | 5 | 8 | 12 | 21 | 31 | 45 | 61 | 80 | — | — |
| | 3 | 0 | 1 | 2 | 4 | 6 | 8 | 15 | 23 | 33 | 45 | 58 | — | — |
| | 2 | 0 | 1 | 1 | 3 | 5 | 7 | 12 | 19 | 27 | 37 | 49 | — | — |
| | 1 | 0 | 1 | 1 | 2 | 4 | 6 | 10 | 15 | 22 | 30 | 39 | — | — |
| XHHW, ZW, XHHW-2, XHH | 14 | 3 | 9 | 15 | 23 | 36 | 53 | 94 | 141 | 203 | 277 | 361 | — | — |
| | 12 | 2 | 7 | 11 | 18 | 28 | 41 | 72 | 108 | 156 | 212 | 277 | — | — |
| | 10 | 1 | 5 | 8 | 13 | 21 | 30 | 54 | 81 | 116 | 158 | 207 | — | — |
| | 8 | 1 | 3 | 5 | 7 | 11 | 17 | 30 | 45 | 64 | 88 | 115 | — | — |
| | 6 | 1 | 1 | 3 | 5 | 8 | 12 | 22 | 33 | 48 | 65 | 85 | — | — |
| | 4 | 0 | 1 | 2 | 4 | 6 | 9 | 16 | 24 | 34 | 47 | 61 | — | — |
| | 3 | 0 | 1 | 1 | 3 | 5 | 7 | 13 | 20 | 29 | 40 | 52 | — | — |
| | 2 | 0 | 1 | 1 | 3 | 4 | 6 | 11 | 17 | 24 | 33 | 44 | — | — |
| XHH, XHHW, XHHW-2 | 1 | 0 | 1 | 1 | 1 | 3 | 5 | 8 | 13 | 18 | 25 | 32 | — | — |
| | 1/0 | 0 | 1 | 1 | 1 | 2 | 4 | 7 | 10 | 15 | 21 | 27 | — | — |
| | 2/0 | 0 | 0 | 1 | 1 | 2 | 3 | 6 | 9 | 13 | 17 | 23 | — | — |
| | 3/0 | 0 | 0 | 1 | 1 | 1 | 3 | 5 | 7 | 10 | 14 | 19 | — | — |
| | 4/0 | 0 | 0 | 1 | 1 | 1 | 2 | 4 | 6 | 9 | 12 | 15 | — | — |
| | 250 | 0 | 0 | 0 | 1 | 1 | 1 | 3 | 5 | 7 | 10 | 13 | — | — |
| | 300 | 0 | 0 | 0 | 1 | 1 | 1 | 3 | 4 | 6 | 8 | 11 | — | — |
| | 350 | 0 | 0 | 0 | 1 | 1 | 1 | 2 | 4 | 5 | 7 | 9 | — | — |
| | 400 | 0 | 0 | 0 | 0 | 1 | 1 | 1 | 3 | 5 | 6 | 8 | — | — |
| | 500 | 0 | 0 | 0 | 0 | 1 | 1 | 1 | 3 | 4 | 5 | 7 | — | — |
| | 600 | 0 | 0 | 0 | 0 | 0 | 1 | 1 | 1 | 3 | 4 | 5 | — | — |
| | 700 | 0 | 0 | 0 | 0 | 0 | 1 | 1 | 1 | 3 | 4 | 5 | — | — |
| | 750 | 0 | 0 | 0 | 0 | 0 | 1 | 1 | 1 | 2 | 3 | 4 | — | — |
| | 800 | 0 | 0 | 0 | 0 | 0 | 1 | 1 | 1 | 2 | 3 | 4 | — | — |
| | 900 | 0 | 0 | 0 | 0 | 0 | 0 | 1 | 1 | 1 | 3 | 4 | — | — |
| | 1000 | 0 | 0 | 0 | 0 | 0 | 0 | 1 | 1 | 1 | 3 | 3 | — | — |
| | 1250 | 0 | 0 | 0 | 0 | 0 | 0 | 1 | 1 | 1 | 1 | 3 | — | — |
| | 1500 | 0 | 0 | 0 | 0 | 0 | 0 | 1 | 1 | 1 | 1 | 2 | — | — |
| | 1750 | 0 | 0 | 0 | 0 | 0 | 0 | 0 | 1 | 1 | 1 | 1 | — | — |
| | 2000 | 0 | 0 | 0 | 0 | 0 | 0 | 0 | 1 | 1 | 1 | 1 | — | — |

***TABLE C.3*** *Continued*

| Type | Conductor Size (AWG/ kcmil) | Trade Size (Metric Designator) ⅜ (12) | ½ (16) | ¾ (21) | 1 (27) | 1¼ (35) | 1½ (41) | 2 (53) | 2½ (63) | 3 (78) | 3½ (91) | 4 (103) | 5 (129) | 6 (155) |
|---|---|---|---|---|---|---|---|---|---|---|---|---|---|---|
| **FIXTURE WIRES** | | | | | | | | | | | | | | |
| RFH-2, FFH-2, RFHH-2 | 18 | 3 | 8 | 14 | 22 | 35 | 51 | 90 | 135 | 195 | 265 | 346 | — | — |
| | 16 | 2 | 7 | 12 | 19 | 29 | 43 | 76 | 114 | 164 | 223 | 292 | — | — |
| SF-2, SFF-2 | 18 | 4 | 11 | 18 | 28 | 44 | 64 | 113 | 170 | 246 | 334 | 437 | — | — |
| | 16 | 3 | 9 | 15 | 23 | 36 | 53 | 94 | 141 | 203 | 277 | 361 | — | — |
| | 14 | 2 | 7 | 12 | 19 | 29 | 43 | 76 | 114 | 164 | 223 | 292 | — | — |
| SF-1, SFF-1 | 18 | 7 | 19 | 33 | 50 | 78 | 114 | 201 | 302 | 435 | 592 | 773 | — | — |
| RFH-1, TF, TFF, XF, XFF | 18 | 5 | 14 | 24 | 37 | 58 | 84 | 148 | 223 | 321 | 437 | 571 | — | — |
| | 16 | 4 | 11 | 19 | 30 | 47 | 68 | 120 | 180 | 259 | 353 | 461 | — | — |
| • XF, XFF | 14 | 3 | 9 | 15 | 23 | 36 | 53 | 94 | 141 | 203 | 277 | 361 | — | — |
| TFN, TFFN | 18 | 8 | 23 | 38 | 59 | 93 | 135 | 237 | 357 | 514 | 699 | 914 | — | — |
| | 16 | 6 | 17 | 29 | 45 | 71 | 103 | 181 | 272 | 392 | 534 | 698 | — | — |
| PF, PFF, PGF, PGFF, PAF, PTF, PTFF, PAFF | 18 | 8 | 22 | 36 | 56 | 88 | 128 | 225 | 338 | 487 | 663 | 866 | — | — |
| | 16 | 6 | 17 | 28 | 43 | 68 | 99 | 174 | 262 | 377 | 513 | 670 | — | — |
| | 14 | 4 | 12 | 21 | 32 | 51 | 74 | 130 | 196 | 282 | 385 | 502 | — | — |
| ZF, ZFF, ZHF | 18 | 10 | 28 | 47 | 72 | 113 | 165 | 290 | 436 | 628 | 855 | 1117 | — | — |
| | 16 | 7 | 20 | 35 | 53 | 83 | 122 | 214 | 322 | 463 | 631 | 824 | — | — |
| | 14 | 5 | 15 | 25 | 39 | 61 | 89 | 157 | 236 | 340 | 463 | 605 | — | — |
| KF-2, KFF-2 | 18 | 15 | 42 | 71 | 109 | 170 | 247 | 436 | 654 | 942 | 1282 | 1675 | — | — |
| | 16 | 10 | 29 | 49 | 76 | 118 | 173 | 304 | 456 | 657 | 895 | 1169 | — | — |
| | 14 | 7 | 20 | 33 | 51 | 80 | 116 | 204 | 307 | 442 | 601 | 785 | — | — |
| | 12 | 5 | 13 | 23 | 35 | 55 | 80 | 142 | 213 | 307 | 418 | 546 | — | — |
| | 10 | 3 | 9 | 15 | 23 | 36 | 53 | 94 | 141 | 203 | 277 | 361 | — | — |
| KF-1, KFF-1 | 18 | 18 | 48 | 82 | 125 | 196 | 286 | 503 | 755 | 1087 | 1480 | 1933 | — | — |
| | 16 | 12 | 34 | 57 | 88 | 138 | 201 | 353 | 530 | 764 | 1040 | 1358 | — | — |
| | 14 | 8 | 23 | 38 | 59 | 93 | 135 | 237 | 357 | 514 | 699 | 914 | — | — |
| | 12 | 5 | 15 | 25 | 39 | 61 | 89 | 157 | 236 | 340 | 463 | 605 | — | — |
| | 10 | 3 | 10 | 16 | 25 | 40 | 58 | 103 | 154 | 222 | 303 | 395 | — | — |
| XF, XFF | 12 | 1 | 5 | 8 | 12 | 19 | 28 | 50 | 75 | 108 | 148 | 193 | — | — |
| | 10 | 1 | 4 | 6 | 10 | 15 | 22 | 39 | 59 | 85 | 115 | 151 | — | — |

Notes:

1. This table is for concentric stranded conductors only. For compact stranded conductors, Table C.3(A) should be used.
2. Two-hour fire-rated RHH cable has ceramifiable insulation, which has much larger diameters than other RHH wires. Consult manufacturer's conduit fill tables.

*Types RHH, RHW, and RHW-2 without outer covering.

***TABLE C.3(A)*** *Maximum Number of Conductors or Fixture Wires in Flexible Metal Conduit (FMC)*
(Based on Chapter 9: Table 1, Table 4, and Table 5A)

| Type | Conductor Size (AWG/ kcmil) | Trade Size (Metric Designator) | | | | | | | | | | | | |
|---|---|---|---|---|---|---|---|---|---|---|---|---|---|---|
| | | 3⁄8 (12) | 1⁄2 (16) | 3⁄4 (21) | 1 (27) | 1¼ (35) | 1½ (41) | 2 (53) | 2½ (63) | 3 (78) | 3½ (91) | 4 (103) | 5 (129) | 6 (155) |
| **COMPACT CONDUCTORS** | | | | | | | | | | | | | | |
| THW, THW-2, THHW | 8 | 1 | 2 | 4 | 6 | 10 | 14 | 25 | 38 | 55 | 75 | 98 | — | — |
| | 6 | 1 | 1 | 3 | 5 | 7 | 11 | 20 | 29 | 43 | 58 | 76 | — | — |
| | 4 | 0 | 1 | 2 | 3 | 5 | 8 | 15 | 22 | 32 | 43 | 57 | — | — |
| | 2 | 0 | 1 | 1 | 2 | 4 | 6 | 11 | 16 | 23 | 32 | 42 | — | — |
| | 1 | 0 | 1 | 1 | 1 | 3 | 4 | 7 | 11 | 16 | 22 | 29 | — | — |
| | 1/0 | 0 | 1 | 1 | 1 | 2 | 3 | 6 | 10 | 14 | 19 | 25 | — | — |
| | 2/0 | 0 | 0 | 1 | 1 | 1 | 3 | 5 | 8 | 12 | 16 | 21 | — | — |
| | 3/0 | 0 | 0 | 1 | 1 | 1 | 2 | 4 | 7 | 10 | 14 | 18 | — | — |
| | 4/0 | 0 | 0 | 1 | 1 | 1 | 1 | 4 | 6 | 8 | 11 | 15 | — | — |
| | 250 | 0 | 0 | 0 | 1 | 1 | 1 | 3 | 4 | 7 | 9 | 12 | — | — |
| | 300 | 0 | 0 | 0 | 1 | 1 | 1 | 2 | 4 | 6 | 8 | 10 | — | — |
| | 350 | 0 | 0 | 0 | 1 | 1 | 1 | 2 | 3 | 5 | 7 | 9 | — | — |
| | 400 | 0 | 0 | 0 | 0 | 1 | 1 | 1 | 3 | 5 | 6 | 8 | — | — |
| | 500 | 0 | 0 | 0 | 0 | 1 | 1 | 1 | 3 | 4 | 5 | 7 | — | — |
| | 600 | 0 | 0 | 0 | 0 | 0 | 1 | 1 | 1 | 3 | 4 | 6 | — | — |
| | 700 | 0 | 0 | 0 | 0 | 0 | 1 | 1 | 1 | 3 | 4 | 5 | — | — |
| | 750 | 0 | 0 | 0 | 0 | 0 | 1 | 1 | 1 | 2 | 3 | 5 | — | — |
| | 900 | 0 | 0 | 0 | 0 | 0 | 1 | 1 | 1 | 2 | 3 | 4 | — | — |
| | 1000 | 0 | 0 | 0 | 0 | 0 | 0 | 1 | 1 | 1 | 3 | 4 | — | — |
| THHN, THWN, THWN-2 | 8 | — | — | — | — | — | — | — | — | — | — | — | — | — |
| | 6 | 1 | 3 | 4 | 7 | 11 | 16 | 29 | 43 | 62 | 85 | 111 | — | — |
| | 4 | 1 | 1 | 3 | 4 | 7 | 10 | 18 | 27 | 38 | 52 | 69 | — | — |
| | 2 | 0 | 1 | 1 | 3 | 5 | 7 | 13 | 19 | 28 | 38 | 49 | — | — |
| | 1 | 0 | 1 | 1 | 2 | 3 | 5 | 9 | 14 | 21 | 28 | 37 | — | — |
| | 1/0 | 0 | 1 | 1 | 1 | 3 | 4 | 8 | 12 | 17 | 24 | 31 | — | — |
| | 2/0 | 0 | 1 | 1 | 1 | 2 | 4 | 6 | 10 | 14 | 20 | 26 | — | — |
| | 3/0 | 0 | 0 | 1 | 1 | 1 | 3 | 5 | 8 | 12 | 17 | 22 | — | — |
| | 4/0 | 0 | 0 | 1 | 1 | 1 | 2 | 4 | 7 | 10 | 14 | 18 | — | — |
| | 250 | 0 | 0 | 1 | 1 | 1 | 1 | 3 | 5 | 8 | 11 | 14 | — | — |
| | 300 | 0 | 0 | 0 | 1 | 1 | 1 | 3 | 5 | 7 | 9 | 12 | — | — |
| | 350 | 0 | 0 | 0 | 1 | 1 | 1 | 3 | 4 | 6 | 8 | 10 | — | — |
| | 400 | 0 | 0 | 0 | 1 | 1 | 1 | 2 | 3 | 5 | 7 | 9 | — | — |
| | 500 | 0 | 0 | 0 | 0 | 1 | 1 | 1 | 3 | 4 | 6 | 8 | — | — |
| | 600 | 0 | 0 | 0 | 0 | 1 | 1 | 1 | 2 | 3 | 5 | 6 | — | — |
| | 700 | 0 | 0 | 0 | 0 | 0 | 1 | 1 | 1 | 3 | 4 | 6 | — | — |
| | 750 | 0 | 0 | 0 | 0 | 0 | 1 | 1 | 1 | 3 | 4 | 5 | — | — |
| | 900 | 0 | 0 | 0 | 0 | 0 | 1 | 1 | 1 | 2 | 3 | 4 | — | — |
| | 1000 | 0 | 0 | 0 | 0 | 0 | 0 | 1 | 1 | 1 | 3 | 4 | — | — |

*TABLE C.3(A) Continued*

| Type | Conductor Size (AWG/ kcmil) | Trade Size (Metric Designator) | | | | | | | | | | | | |
|---|---|---|---|---|---|---|---|---|---|---|---|---|---|---|
| | | ⅜ (12) | ½ (16) | ¾ (21) | 1 (27) | 1¼ (35) | 1½ (41) | 2 (53) | 2½ (63) | 3 (78) | 3½ (91) | 4 (103) | 5 (129) | 6 (155) |
| XHHW, XHHW-2 | 8 | 1 | 3 | 5 | 8 | 13 | 19 | 33 | 50 | 71 | 97 | 127 | — | — |
| | 6 | 1 | 2 | 4 | 6 | 9 | 14 | 24 | 37 | 53 | 72 | 95 | — | — |
| | 4 | 1 | 1 | 3 | 4 | 7 | 10 | 18 | 27 | 38 | 52 | 69 | — | — |
| | 2 | 0 | 1 | 1 | 3 | 5 | 7 | 13 | 19 | 28 | 38 | 49 | — | — |
| | 1 | 0 | 1 | 1 | 2 | 3 | 5 | 9 | 14 | 21 | 28 | 37 | — | — |
| | 1/0 | 0 | 1 | 1 | 1 | 3 | 4 | 8 | 12 | 17 | 24 | 31 | — | — |
| | 2/0 | 0 | 1 | 1 | 1 | 2 | 4 | 7 | 10 | 15 | 20 | 26 | — | — |
| | 3/0 | 0 | 0 | 1 | 1 | 1 | 3 | 5 | 8 | 12 | 17 | 22 | — | — |
| | 4/0 | 0 | 0 | 1 | 1 | 1 | 2 | 4 | 7 | 10 | 14 | 18 | — | — |
| | 250 | 0 | 0 | 1 | 1 | 1 | 1 | 4 | 5 | 8 | 11 | 14 | — | — |
| | 300 | 0 | 0 | 0 | 1 | 1 | 1 | 3 | 5 | 7 | 9 | 12 | — | — |
| | 350 | 0 | 0 | 0 | 1 | 1 | 1 | 3 | 4 | 6 | 8 | 11 | — | — |
| | 400 | 0 | 0 | 0 | 1 | 1 | 1 | 2 | 4 | 5 | 7 | 10 | — | — |
| | 500 | 0 | 0 | 0 | 0 | 1 | 1 | 1 | 3 | 4 | 6 | 8 | — | — |
| | 600 | 0 | 0 | 0 | 0 | 1 | 1 | 1 | 2 | 3 | 5 | 6 | — | — |
| | 700 | 0 | 0 | 0 | 0 | 0 | 1 | 1 | 1 | 3 | 4 | 6 | — | — |
| | 750 | 0 | 0 | 0 | 0 | 0 | 1 | 1 | 1 | 3 | 4 | 5 | — | — |
| | 900 | 0 | 0 | 0 | 0 | 0 | 1 | 1 | 1 | 2 | 3 | 4 | — | — |
| | 1000 | 0 | 0 | 0 | 0 | 0 | 1 | 1 | 1 | 2 | 3 | 4 | — | — |

Definition: *Compact stranding* is the result of a manufacturing process where the stranded conductor is compressed to the extent that the interstices (voids between strand wires) are virtually eliminated.

***TABLE C.4*** *Maximum Number of Conductors or Fixture Wires in Intermediate Metal Conduit (IMC)*
(Based on Chapter 9: Table 1, Table 4, and Table 5)

| Type | Conductor Size (AWG/ kcmil) | Trade Size (Metric Designator) ⅜ (12) | ½ (16) | ¾ (21) | 1 (27) | 1¼ (35) | 1½ (41) | 2 (53) | 2½ (63) | 3 (78) | 3½ (91) | 4 (103) | 5 (129) | 6 (155) |
|---|---|---|---|---|---|---|---|---|---|---|---|---|---|---|
| | | | | | | **CONDUCTORS** | | | | | | | | |
| RHH, RHW, RHW-2 | 14 | — | 4 | 8 | 13 | 22 | 30 | 49 | 70 | 108 | 144 | 186 | — | — |
| | 12 | — | 4 | 6 | 11 | 18 | 25 | 41 | 58 | 89 | 120 | 154 | — | — |
| | 10 | — | 3 | 5 | 8 | 15 | 20 | 33 | 47 | 72 | 97 | 124 | — | — |
| | 8 | — | 1 | 3 | 4 | 8 | 10 | 17 | 24 | 38 | 50 | 65 | — | — |
| | 6 | — | 1 | 1 | 3 | 6 | 8 | 14 | 19 | 30 | 40 | 52 | — | — |
| | 4 | — | 1 | 1 | 3 | 5 | 6 | 11 | 15 | 23 | 31 | 41 | — | — |
| | 3 | — | 1 | 1 | 2 | 4 | 6 | 9 | 13 | 21 | 28 | 36 | — | — |
| | 2 | — | 1 | 1 | 1 | 3 | 5 | 8 | 11 | 18 | 24 | 31 | — | — |
| | 1 | — | 0 | 1 | 1 | 2 | 3 | 5 | 7 | 12 | 16 | 20 | — | — |
| | 1/0 | — | 0 | 1 | 1 | 1 | 3 | 4 | 6 | 10 | 14 | 18 | — | — |
| | 2/0 | — | 0 | 1 | 1 | 1 | 2 | 4 | 6 | 9 | 12 | 15 | — | — |
| | 3/0 | — | 0 | 0 | 1 | 1 | 1 | 3 | 5 | 7 | 10 | 13 | — | — |
| | 4/0 | — | 0 | 0 | 1 | 1 | 1 | 3 | 4 | 6 | 9 | 11 | — | — |
| | 250 | — | 0 | 0 | 1 | 1 | 1 | 1 | 3 | 5 | 6 | 8 | — | — |
| | 300 | — | 0 | 0 | 0 | 1 | 1 | 1 | 3 | 4 | 6 | 7 | — | — |
| | 350 | — | 0 | 0 | 0 | 1 | 1 | 1 | 2 | 4 | 5 | 7 | — | — |
| | 400 | — | 0 | 0 | 0 | 1 | 1 | 1 | 2 | 3 | 5 | 6 | — | — |
| | 500 | — | 0 | 0 | 0 | 1 | 1 | 1 | 1 | 3 | 4 | 5 | — | — |
| | 600 | — | 0 | 0 | 0 | 0 | 1 | 1 | 1 | 2 | 3 | 4 | — | — |
| | 700 | — | 0 | 0 | 0 | 0 | 1 | 1 | 1 | 2 | 3 | 4 | — | — |
| | 750 | — | 0 | 0 | 0 | 0 | 1 | 1 | 1 | 1 | 3 | 4 | — | — |
| | 800 | — | 0 | 0 | 0 | 0 | 0 | 1 | 1 | 1 | 3 | 3 | — | — |
| | 900 | — | 0 | 0 | 0 | 0 | 0 | 1 | 1 | 1 | 2 | 3 | — | — |
| | 1000 | — | 0 | 0 | 0 | 0 | 0 | 1 | 1 | 1 | 2 | 3 | — | — |
| | 1250 | — | 0 | 0 | 0 | 0 | 0 | 1 | 1 | 1 | 1 | 1 | — | — |
| | 1500 | — | 0 | 0 | 0 | 0 | 0 | 0 | 1 | 1 | 1 | 1 | — | — |
| | 1750 | — | 0 | 0 | 0 | 0 | 0 | 0 | 1 | 1 | 1 | 1 | — | — |
| | 2000 | — | 0 | 0 | 0 | 0 | 0 | 0 | 1 | 1 | 1 | 1 | — | — |
| TW, THHW, THW, THW-2 | 14 | — | 10 | 17 | 27 | 47 | 64 | 104 | 147 | 228 | 304 | 392 | — | — |
| | 12 | — | 7 | 13 | 21 | 36 | 49 | 80 | 113 | 175 | 234 | 301 | — | — |
| | 10 | — | 5 | 9 | 15 | 27 | 36 | 59 | 84 | 130 | 174 | 224 | — | — |
| | 8 | — | 3 | 5 | 8 | 15 | 20 | 33 | 47 | 72 | 97 | 124 | — | — |
| RHH*, RHW*, RHW-2* | 14 | — | 6 | 11 | 18 | 31 | 42 | 69 | 98 | 151 | 202 | 261 | — | — |
| | 12 | — | 5 | 9 | 14 | 25 | 34 | 56 | 79 | 122 | 163 | 209 | — | — |
| | 10 | — | 4 | 7 | 11 | 19 | 26 | 43 | 61 | 95 | 127 | 163 | — | — |
| | 8 | — | 2 | 4 | 7 | 12 | 16 | 26 | 37 | 57 | 76 | 98 | — | — |
| • TW, THW, THHW, THW-2, RHH*, RHW*, RHW-2* | 6 | — | 1 | 3 | 5 | 9 | 12 | 20 | 28 | 43 | 58 | 75 | — | — |
| | 4 | — | 1 | 2 | 4 | 6 | 9 | 15 | 21 | 32 | 43 | 56 | — | — |
| | 3 | — | 1 | 1 | 3 | 6 | 8 | 13 | 18 | 28 | 37 | 48 | — | — |
| | 2 | — | 1 | 1 | 3 | 5 | 6 | 11 | 15 | 23 | 31 | 41 | — | — |
| | 1 | — | 1 | 1 | 1 | 3 | 4 | 7 | 11 | 16 | 22 | 28 | — | — |
| | 1/0 | — | 1 | 1 | 1 | 3 | 4 | 6 | 9 | 14 | 19 | 24 | — | — |
| | 2/0 | — | 0 | 1 | 1 | 2 | 3 | 5 | 8 | 12 | 16 | 20 | — | — |
| | 3/0 | — | 0 | 1 | 1 | 1 | 3 | 4 | 6 | 10 | 13 | 17 | — | — |
| | 4/0 | — | 0 | 1 | 1 | 1 | 2 | 4 | 5 | 8 | 11 | 14 | — | — |

***TABLE C.4*** *Continued*

| Type | Conductor Size (AWG/ kcmil) | Trade Size (Metric Designator) ⅜ (12) | ½ (16) | ¾ (21) | 1 (27) | 1¼ (35) | 1½ (41) | 2 (53) | 2½ (63) | 3 (78) | 3½ (91) | 4 (103) | 5 (129) | 6 (155) |
|---|---|---|---|---|---|---|---|---|---|---|---|---|---|---|
| | 250 | — | 0 | 0 | 1 | 1 | 1 | 3 | 4 | 7 | 9 | 12 | — | — |
| | 300 | — | 0 | 0 | 1 | 1 | 1 | 2 | 4 | 6 | 8 | 10 | — | — |
| | 350 | — | 0 | 0 | 1 | 1 | 1 | 2 | 3 | 5 | 7 | 9 | — | — |
| | 400 | — | 0 | 0 | 0 | 1 | 1 | 1 | 3 | 4 | 6 | 8 | — | — |
| | 500 | — | 0 | 0 | 0 | 1 | 1 | 1 | 2 | 4 | 5 | 7 | — | — |
| | 600 | — | 0 | 0 | 0 | 1 | 1 | 1 | 1 | 3 | 4 | 5 | — | — |
| | 700 | — | 0 | 0 | 0 | 0 | 1 | 1 | 1 | 3 | 4 | 5 | — | — |
| | 750 | — | 0 | 0 | 0 | 0 | 1 | 1 | 1 | 2 | 3 | 4 | — | — |
| | 800 | — | 0 | 0 | 0 | 0 | 1 | 1 | 1 | 2 | 3 | 4 | — | — |
| | 900 | — | 0 | 0 | 0 | 0 | 1 | 1 | 1 | 2 | 3 | 4 | — | — |
| | 1000 | — | 0 | 0 | 0 | 0 | 0 | 1 | 1 | 1 | 3 | 3 | — | — |
| | 1250 | — | 0 | 0 | 0 | 0 | 0 | 1 | 1 | 1 | 1 | 3 | — | — |
| | 1500 | — | 0 | 0 | 0 | 0 | 0 | 1 | 1 | 1 | 1 | 2 | — | — |
| | 1750 | — | 0 | 0 | 0 | 0 | 0 | 0 | 1 | 1 | 1 | 1 | — | — |
| | 2000 | — | 0 | 0 | 0 | 0 | 0 | 0 | 1 | 1 | 1 | 1 | — | — |
| THHN, THWN, THWN-2 | 14 | — | 14 | 24 | 39 | 68 | 91 | 149 | 211 | 326 | 436 | 562 | — | — |
| | 12 | — | 10 | 17 | 29 | 49 | 67 | 109 | 154 | 238 | 318 | 410 | — | — |
| | 10 | — | 6 | 11 | 18 | 31 | 42 | 69 | 97 | 150 | 200 | 258 | — | — |
| | 8 | — | 3 | 6 | 10 | 18 | 24 | 39 | 56 | 86 | 115 | 149 | — | — |
| | 6 | — | 2 | 4 | 7 | 13 | 17 | 28 | 40 | 62 | 83 | 107 | — | — |
| | 4 | — | 1 | 3 | 4 | 8 | 11 | 17 | 25 | 38 | 51 | 66 | — | — |
| | 3 | — | 1 | 2 | 4 | 6 | 9 | 15 | 21 | 32 | 43 | 56 | — | — |
| | 2 | — | 1 | 1 | 3 | 5 | 7 | 12 | 17 | 27 | 36 | 47 | — | — |
| | 1 | — | 1 | 1 | 2 | 4 | 5 | 9 | 13 | 20 | 27 | 35 | — | — |
| | 1/0 | — | 1 | 1 | 1 | 3 | 4 | 8 | 11 | 17 | 23 | 29 | — | — |
| | 2/0 | — | 1 | 1 | 1 | 3 | 4 | 6 | 9 | 14 | 19 | 24 | — | — |
| | 3/0 | — | 0 | 1 | 1 | 2 | 3 | 5 | 7 | 12 | 16 | 20 | — | — |
| | 4/0 | — | 0 | 1 | 1 | 1 | 2 | 4 | 6 | 9 | 13 | 17 | — | — |
| | 250 | — | 0 | 0 | 1 | 1 | 1 | 3 | 5 | 8 | 10 | 13 | — | — |
| | 300 | — | 0 | 0 | 1 | 1 | 1 | 3 | 4 | 7 | 9 | 12 | — | — |
| | 350 | — | 0 | 0 | 1 | 1 | 1 | 2 | 4 | 6 | 8 | 10 | — | — |
| | 400 | — | 0 | 0 | 1 | 1 | 1 | 2 | 3 | 5 | 7 | 9 | — | — |
| | 500 | — | 0 | 0 | 0 | 1 | 1 | 1 | 3 | 4 | 6 | 7 | — | — |
| | 600 | — | 0 | 0 | 0 | 1 | 1 | 1 | 2 | 3 | 5 | 6 | — | — |
| | 700 | — | 0 | 0 | 0 | 1 | 1 | 1 | 1 | 3 | 4 | 5 | — | — |
| | 750 | — | 0 | 0 | 0 | 1 | 1 | 1 | 1 | 3 | 4 | 5 | — | — |
| | 800 | — | 0 | 0 | 0 | 0 | 1 | 1 | 1 | 3 | 4 | 5 | — | — |
| | 900 | — | 0 | 0 | 0 | 0 | 1 | 1 | 1 | 2 | 3 | 4 | — | — |
| | 1000 | — | 0 | 0 | 0 | 0 | 1 | 1 | 1 | 2 | 3 | 4 | — | — |
| FEP, FEPB, PFA, PFAH, TFE | 14 | — | 13 | 23 | 38 | 66 | 89 | 145 | 205 | 317 | 423 | 545 | — | — |
| | 12 | — | 10 | 17 | 28 | 48 | 65 | 106 | 150 | 231 | 309 | 398 | — | — |
| | 10 | — | 7 | 12 | 20 | 34 | 46 | 76 | 107 | 166 | 221 | 285 | — | — |
| | 8 | — | 4 | 7 | 11 | 19 | 26 | 43 | 61 | 95 | 127 | 163 | — | — |
| | 6 | — | 3 | 5 | 8 | 14 | 19 | 31 | 44 | 67 | 90 | 116 | — | — |
| | 4 | — | 1 | 3 | 5 | 10 | 13 | 21 | 30 | 47 | 63 | 81 | — | — |
| | 3 | — | 1 | 3 | 4 | 8 | 11 | 18 | 25 | 39 | 52 | 68 | — | — |
| | 2 | — | 1 | 2 | 4 | 6 | 9 | 15 | 21 | 32 | 43 | 56 | — | — |

*(continues)*

***TABLE C.4*** *Continued*

| Type | Conductor Size (AWG/ kcmil) | Trade Size (Metric Designator) | | | | | | | | | | | | |
|---|---|---|---|---|---|---|---|---|---|---|---|---|---|---|
| | | 3/8 (12) | 1/2 (16) | 3/4 (21) | 1 (27) | 1 1/4 (35) | 1 1/2 (41) | 2 (53) | 2 1/2 (63) | 3 (78) | 3 1/2 (91) | 4 (103) | 5 (129) | 6 (155) |
| PFA, PFAH, TFE | 1 | — | 1 | 1 | 2 | 4 | 6 | 10 | 14 | 22 | 30 | 39 | — | — |
| PFA, PFAH, TFE, Z | 1/0 | — | 1 | 1 | 1 | 4 | 5 | 8 | 12 | 19 | 25 | 32 | — | — |
| | 2/0 | — | 1 | 1 | 1 | 3 | 4 | 7 | 10 | 15 | 21 | 27 | — | — |
| | 3/0 | — | 0 | 1 | 1 | 2 | 3 | 6 | 8 | 13 | 17 | 22 | — | — |
| | 4/0 | — | 0 | 1 | 1 | 1 | 3 | 5 | 7 | 10 | 14 | 18 | — | — |
| Z | 14 | — | 16 | 28 | 46 | 79 | 107 | 175 | 247 | 381 | 510 | 657 | — | — |
| | 12 | — | 11 | 20 | 32 | 56 | 76 | 124 | 175 | 271 | 362 | 466 | — | — |
| | 10 | — | 7 | 12 | 20 | 34 | 46 | 76 | 107 | 166 | 221 | 285 | — | — |
| | 8 | — | 4 | 7 | 12 | 22 | 29 | 48 | 68 | 105 | 140 | 180 | — | — |
| | 6 | — | 3 | 5 | 9 | 15 | 20 | 33 | 47 | 73 | 98 | 127 | — | — |
| | 4 | — | 1 | 3 | 6 | 10 | 14 | 23 | 33 | 50 | 67 | 87 | — | — |
| | 3 | — | 1 | 2 | 4 | 7 | 10 | 17 | 24 | 37 | 49 | 63 | — | — |
| | 2 | — | 1 | 1 | 3 | 6 | 8 | 14 | 20 | 30 | 41 | 53 | — | — |
| | 1 | — | 1 | 1 | 3 | 5 | 7 | 11 | 16 | 25 | 33 | 43 | — | — |
| XHHW, ZW, XHHW-2, XHH | 14 | — | 10 | 17 | 27 | 47 | 64 | 104 | 147 | 228 | 304 | 392 | — | — |
| | 12 | — | 7 | 13 | 21 | 36 | 49 | 80 | 113 | 175 | 234 | 301 | — | — |
| | 10 | — | 5 | 9 | 15 | 27 | 36 | 59 | 84 | 130 | 174 | 224 | — | — |
| | 8 | — | 3 | 5 | 8 | 15 | 20 | 33 | 47 | 72 | 97 | 124 | — | — |
| | 6 | — | 1 | 4 | 6 | 11 | 15 | 24 | 35 | 53 | 71 | 92 | — | — |
| | 4 | — | 1 | 3 | 4 | 8 | 11 | 18 | 25 | 39 | 52 | 67 | — | — |
| | 3 | — | 1 | 2 | 4 | 7 | 9 | 15 | 21 | 33 | 44 | 56 | — | — |
| | 2 | — | 1 | 1 | 3 | 5 | 7 | 12 | 18 | 27 | 37 | 47 | — | — |
| XHHW, XHHW-2, XHH | 1 | — | 1 | 1 | 2 | 4 | 6 | 9 | 13 | 20 | 27 | 35 | — | — |
| | 1/0 | — | 1 | 1 | 1 | 3 | 5 | 8 | 11 | 17 | 23 | 30 | — | — |
| | 2/0 | — | 1 | 1 | 1 | 3 | 4 | 6 | 9 | 14 | 19 | 25 | — | — |
| | 3/0 | — | 0 | 1 | 1 | 2 | 3 | 5 | 7 | 12 | 16 | 20 | — | — |
| | 4/0 | — | 0 | 1 | 1 | 1 | 2 | 4 | 6 | 10 | 13 | 17 | — | — |
| | 250 | — | 0 | 0 | 1 | 1 | 1 | 3 | 5 | 8 | 11 | 14 | — | — |
| | 300 | — | 0 | 0 | 1 | 1 | 1 | 3 | 4 | 7 | 9 | 12 | — | — |
| | 350 | — | 0 | 0 | 1 | 1 | 1 | 3 | 4 | 6 | 8 | 10 | — | — |
| | 400 | — | 0 | 0 | 1 | 1 | 1 | 2 | 3 | 5 | 7 | 9 | — | — |
| | 500 | — | 0 | 0 | 0 | 1 | 1 | 1 | 3 | 4 | 6 | 8 | — | — |
| | 600 | — | 0 | 0 | 0 | 1 | 1 | 1 | 2 | 3 | 5 | 6 | — | — |
| | 700 | — | 0 | 0 | 0 | 1 | 1 | 1 | 1 | 3 | 4 | 5 | — | — |
| | 750 | — | 0 | 0 | 0 | 1 | 1 | 1 | 1 | 3 | 4 | 5 | — | — |
| | 800 | — | 0 | 0 | 0 | 0 | 1 | 1 | 1 | 3 | 4 | 5 | — | — |
| | 900 | — | 0 | 0 | 0 | 0 | 1 | 1 | 1 | 2 | 3 | 4 | — | — |
| | 1000 | — | 0 | 0 | 0 | 0 | 1 | 1 | 1 | 2 | 3 | 4 | — | — |
| | 1250 | — | 0 | 0 | 0 | 0 | 0 | 1 | 1 | 1 | 2 | 3 | — | — |
| | 1500 | — | 0 | 0 | 0 | 0 | 0 | 1 | 1 | 1 | 1 | 2 | — | — |
| | 1750 | — | 0 | 0 | 0 | 0 | 0 | 1 | 1 | 1 | 1 | 2 | — | — |
| | 2000 | — | 0 | 0 | 0 | 0 | 0 | 0 | 1 | 1 | 1 | 1 | — | — |

***TABLE C.4*** *Continued*

| Type | Conductor Size (AWG/ kcmil) | Trade Size (Metric Designator) | | | | | | | | | | | | |
|---|---|---|---|---|---|---|---|---|---|---|---|---|---|---|
| | | ⅜ (12) | ½ (16) | ¾ (21) | 1 (27) | 1¼ (35) | 1½ (41) | 2 (53) | 2½ (63) | 3 (78) | 3½ (91) | 4 (103) | 5 (129) | 6 (155) |
| **FIXTURE WIRES** | | | | | | | | | | | | | | |
| RFH-2, FFH-2, RFHH-2 | 18 | — | 9 | 16 | 26 | 45 | 61 | 100 | 141 | 218 | 292 | 376 | — | — |
| | 16 | — | 8 | 13 | 22 | 38 | 51 | 84 | 119 | 184 | 246 | 317 | — | — |
| SF-2, SFF-2 | 18 | — | 12 | 20 | 33 | 57 | 77 | 126 | 178 | 275 | 368 | 474 | — | — |
| | 16 | — | 10 | 17 | 27 | 47 | 64 | 104 | 147 | 228 | 304 | 392 | — | — |
| | 14 | — | 8 | 13 | 22 | 38 | 51 | 84 | 119 | 184 | 246 | 317 | — | — |
| SF-1, SFF-1 | 18 | — | 21 | 36 | 59 | 101 | 137 | 223 | 316 | 487 | 651 | 839 | — | — |
| RFH-1, TF, TFF, XF, XFF | 18 | — | 15 | 26 | 43 | 75 | 101 | 165 | 233 | 360 | 481 | 619 | — | — |
| | 16 | — | 12 | 21 | 35 | 60 | 81 | 133 | 188 | 290 | 388 | 500 | — | — |
| • XF, XFF | 14 | — | 10 | 17 | 27 | 47 | 64 | 104 | 147 | 228 | 304 | 392 | — | — |
| TFN, TFFN | 18 | — | 25 | 42 | 69 | 119 | 162 | 264 | 373 | 576 | 769 | 991 | — | — |
| | 16 | — | 19 | 32 | 53 | 91 | 123 | 201 | 285 | 440 | 588 | 757 | — | — |
| PF, PFF, PGF, PGFF, PAF, PTF, PTFF, PAFF | 18 | — | 23 | 40 | 66 | 113 | 153 | 250 | 354 | 546 | 730 | 940 | — | — |
| | 16 | — | 18 | 31 | 51 | 88 | 118 | 193 | 274 | 422 | 564 | 727 | — | — |
| | 14 | — | 13 | 23 | 38 | 66 | 89 | 145 | 205 | 317 | 423 | 545 | — | — |
| ZF, ZFF, ZHF | 18 | — | 30 | 52 | 85 | 146 | 197 | 322 | 456 | 704 | 941 | 1211 | — | — |
| | 16 | — | 22 | 38 | 63 | 108 | 146 | 238 | 336 | 519 | 694 | 894 | — | — |
| | 14 | — | 16 | 28 | 46 | 79 | 107 | 175 | 247 | 381 | 510 | 657 | — | — |
| KF-2, KFF-2 | 18 | — | 45 | 78 | 128 | 219 | 296 | 484 | 684 | 1056 | 1411 | 1817 | — | — |
| | 16 | — | 32 | 54 | 89 | 153 | 207 | 337 | 477 | 737 | 984 | 1268 | — | — |
| | 14 | — | 21 | 36 | 60 | 103 | 139 | 227 | 321 | 495 | 661 | 852 | — | — |
| | 12 | — | 15 | 25 | 41 | 71 | 96 | 158 | 223 | 344 | 460 | 592 | — | — |
| | 10 | — | 10 | 17 | 27 | 47 | 64 | 104 | 147 | 228 | 304 | 392 | — | — |
| KF-1, KFF-1 | 18 | — | 52 | 90 | 147 | 253 | 342 | 558 | 790 | 1218 | 1628 | 2097 | — | — |
| | 16 | — | 37 | 63 | 103 | 178 | 240 | 392 | 555 | 856 | 1144 | 1473 | — | — |
| | 14 | — | 25 | 42 | 69 | 119 | 162 | 264 | 373 | 576 | 769 | 991 | — | — |
| | 12 | — | 16 | 28 | 46 | 79 | 107 | 175 | 247 | 381 | 510 | 657 | — | — |
| | 10 | — | 10 | 18 | 30 | 52 | 70 | 114 | 161 | 249 | 333 | 429 | — | — |
| XF, XFF | 12 | — | 5 | 9 | 14 | 25 | 34 | 56 | 79 | 122 | 163 | 209 | — | — |
| | 10 | — | 4 | 7 | 11 | 19 | 26 | 43 | 61 | 95 | 127 | 163 | — | — |

Notes:

1. This table is for concentric stranded conductors only. For compact stranded conductors, Table C.4(A) should be used.
2. Two-hour fire-rated RHH cable has ceramifiable insulation, which has much larger diameters than other RHH wires. Consult manufacturer's conduit fill tables.

*Types RHH, RHW, and RHW-2 without outer covering.

***TABLE C.4(A)*** *Maximum Number of Conductors or Fixture Wires in Intermediate Metal Conduit (IMC)* (Based on Chapter 9: Table 1, Table 4, and Table 5A)

| Type | Conductor Size (AWG/kcmil) | ⅜ (12) | ½ (16) | ¾ (21) | 1 (27) | 1¼ (35) | 1½ (41) | 2 (53) | 2½ (63) | 3 (78) | 3½ (91) | 4 (103) | 5 (129) | 6 (155) |
|---|---|---|---|---|---|---|---|---|---|---|---|---|---|---|
| | | Trade Size (Metric Designator) | | | | | | | | | | | | |
| **COMPACT CONDUCTORS** | | | | | | | | | | | | | | |
| THW, THW-2, THHW | 8 | — | 2 | 4 | 7 | 13 | 17 | 28 | 40 | 62 | 83 | 107 | — | — |
| | 6 | — | 1 | 3 | 6 | 10 | 13 | 22 | 31 | 48 | 64 | 82 | — | — |
| | 4 | — | 1 | 2 | 4 | 7 | 10 | 16 | 23 | 36 | 48 | 62 | — | — |
| | 2 | — | 1 | 1 | 3 | 5 | 7 | 12 | 17 | 26 | 35 | 45 | — | — |
| | 1 | — | 1 | 1 | 1 | 4 | 5 | 8 | 12 | 18 | 25 | 32 | — | — |
| | 1/0 | — | 1 | 1 | 1 | 3 | 4 | 7 | 10 | 16 | 21 | 27 | — | — |
| | 2/0 | — | 0 | 1 | 1 | 3 | 4 | 6 | 9 | 13 | 18 | 23 | — | — |
| | 3/0 | — | 0 | 1 | 1 | 2 | 3 | 5 | 7 | 11 | 15 | 20 | — | — |
| | 4/0 | — | 0 | 1 | 1 | 1 | 2 | 4 | 6 | 9 | 13 | 16 | — | — |
| | 250 | — | 0 | 0 | 1 | 1 | 1 | 3 | 5 | 7 | 10 | 13 | — | — |
| | 300 | — | 0 | 0 | 1 | 1 | 1 | 3 | 4 | 6 | 9 | 11 | — | — |
| | 350 | — | 0 | 0 | 1 | 1 | 1 | 2 | 4 | 6 | 8 | 10 | — | — |
| | 400 | — | 0 | 0 | 1 | 1 | 1 | 2 | 3 | 5 | 7 | 9 | — | — |
| | 500 | — | 0 | 0 | 0 | 1 | 1 | 1 | 3 | 4 | 6 | 8 | — | — |
| | 600 | — | 0 | 0 | 0 | 1 | 1 | 1 | 2 | 3 | 5 | 6 | — | — |
| | 700 | — | 0 | 0 | 0 | 1 | 1 | 1 | 1 | 3 | 4 | 5 | — | — |
| | 750 | — | 0 | 0 | 0 | 1 | 1 | 1 | 1 | 3 | 4 | 5 | — | — |
| | 900 | — | 0 | 0 | 0 | 0 | 1 | 1 | 1 | 2 | 3 | 4 | — | — |
| | 1000 | — | 0 | 0 | 0 | 0 | 1 | 1 | 1 | 2 | 3 | 4 | — | — |
| THHN, THWN, THWN-2 | 8 | — | — | — | — | — | — | — | — | — | — | — | — | — |
| | 6 | — | 3 | 5 | 8 | 14 | 19 | 32 | 45 | 70 | 93 | 120 | — | — |
| | 4 | — | 1 | 3 | 5 | 9 | 12 | 20 | 28 | 43 | 58 | 74 | — | — |
| | 2 | — | 1 | 1 | 3 | 6 | 8 | 14 | 20 | 31 | 41 | 53 | — | — |
| | 1 | — | 1 | 1 | 3 | 5 | 6 | 10 | 15 | 23 | 31 | 40 | — | — |
| | 1/0 | — | 1 | 1 | 2 | 4 | 5 | 9 | 13 | 20 | 26 | 34 | — | — |
| | 2/0 | — | 1 | 1 | 1 | 3 | 4 | 7 | 10 | 16 | 22 | 28 | — | — |
| | 3/0 | — | 0 | 1 | 1 | 3 | 4 | 6 | 9 | 14 | 18 | 24 | — | — |
| | 4/0 | — | 0 | 1 | 1 | 2 | 3 | 5 | 7 | 11 | 15 | 19 | — | — |
| | 250 | — | 0 | 1 | 1 | 1 | 2 | 4 | 6 | 9 | 12 | 15 | — | — |
| | 300 | — | 0 | 0 | 1 | 1 | 1 | 3 | 5 | 7 | 10 | 13 | — | — |
| | 350 | — | 0 | 0 | 1 | 1 | 1 | 3 | 4 | 7 | 9 | 11 | — | — |
| | 400 | — | 0 | 0 | 1 | 1 | 1 | 2 | 4 | 6 | 8 | 10 | — | — |
| | 500 | — | 0 | 0 | 1 | 1 | 1 | 2 | 3 | 5 | 7 | 9 | — | — |
| | 600 | — | 0 | 0 | 0 | 1 | 1 | 1 | 2 | 4 | 5 | 7 | — | — |
| | 700 | — | 0 | 0 | 0 | 1 | 1 | 1 | 2 | 3 | 5 | 6 | — | — |
| | 750 | — | 0 | 0 | 0 | 1 | 1 | 1 | 1 | 3 | 4 | 6 | — | — |
| | 900 | — | 0 | 0 | 0 | 0 | 1 | 1 | 1 | 3 | 3 | 5 | — | — |
| | 1000 | — | 0 | 0 | 0 | 0 | 1 | 1 | 1 | 2 | 3 | 4 | — | — |

***TABLE C.4(A)*** *Continued*

| Type | Conductor Size (AWG/ kcmil) | Trade Size (Metric Designator) | | | | | | | | | | | | |
|---|---|---|---|---|---|---|---|---|---|---|---|---|---|---|
| | | 3⁄8 (12) | 1⁄2 (16) | 3⁄4 (21) | 1 (27) | 1¼ (35) | 1½ (41) | 2 (53) | 2½ (63) | 3 (78) | 3½ (91) | 4 (103) | 5 (129) | 6 (155) |
| XHHW, XHHW-2 | 8 | — | 3 | 6 | 9 | 16 | 22 | 37 | 52 | 80 | 107 | 138 | — | — |
| | 6 | — | 2 | 4 | 7 | 12 | 16 | 27 | 38 | 59 | 80 | 103 | — | — |
| | 4 | — | 1 | 3 | 5 | 9 | 12 | 20 | 28 | 43 | 58 | 74 | — | — |
| | 2 | — | 1 | 1 | 3 | 6 | 8 | 14 | 20 | 31 | 41 | 53 | — | — |
| | 1 | — | 1 | 1 | 3 | 5 | 6 | 10 | 15 | 23 | 31 | 40 | — | — |
| | 1/0 | — | 1 | 1 | 2 | 4 | 5 | 9 | 13 | 20 | 26 | 34 | — | — |
| | 2/0 | — | 1 | 1 | 1 | 3 | 4 | 7 | 11 | 17 | 22 | 29 | — | — |
| | 3/0 | — | 0 | 1 | 1 | 3 | 4 | 6 | 9 | 14 | 18 | 24 | — | — |
| | 4/0 | — | 0 | 1 | 1 | 2 | 3 | 5 | 7 | 11 | 15 | 20 | — | — |
| | 250 | — | 0 | 1 | 1 | 1 | 2 | 4 | 6 | 9 | 12 | 16 | — | — |
| | 300 | — | 0 | 0 | 1 | 1 | 1 | 3 | 5 | 8 | 10 | 13 | — | — |
| | 350 | — | 0 | 0 | 1 | 1 | 1 | 3 | 4 | 7 | 9 | 12 | — | — |
| | 400 | — | 0 | 0 | 1 | 1 | 1 | 3 | 4 | 6 | 8 | 11 | — | — |
| | 500 | — | 0 | 0 | 1 | 1 | 1 | 2 | 3 | 5 | 7 | 9 | — | — |
| | 600 | — | 0 | 0 | 0 | 1 | 1 | 1 | 2 | 4 | 5 | 7 | — | — |
| | 700 | — | 0 | 0 | 0 | 1 | 1 | 1 | 2 | 3 | 5 | 6 | — | — |
| | 750 | — | 0 | 0 | 0 | 1 | 1 | 1 | 1 | 3 | 4 | 6 | — | — |
| | 900 | — | 0 | 0 | 0 | 1 | 1 | 1 | 1 | 3 | 4 | 5 | — | — |
| | 1000 | — | 0 | 0 | 0 | 0 | 1 | 1 | 1 | 2 | 3 | 4 | — | — |

Definition: *Compact stranding* is the result of a manufacturing process where the stranded conductor is compressed to the extent that interstices (voids between strand wires) are virtually eliminated.

***TABLE C.5*** *Maximum Number of Conductors or Fixture Wires in Liquidtight Flexible Nonmetallic Conduit (Type LFNC-B*)* (Based on Chapter 9: Table 1, Table 4, and Table 5)

| Type | Conductor Size (AWG/ kcmil) | Trade Size (Metric Designator) ⅜ (12) | ½ (16) | ¾ (21) | 1 (27) | 1¼ (35) | 1½ (41) | 2 (53) | 2½ (63) | 3 (78) | 3½ (91) | 4 (103) | 5 (129) | 6 (155) |
|---|---|---|---|---|---|---|---|---|---|---|---|---|---|---|
| | | **CONDUCTORS** | | | | | | | | | | | | |
| RHH, RHW, RHW-2 | 14 | 2 | 4 | 7 | 12 | 21 | 27 | 44 | — | — | — | — | — | — |
| | 12 | 1 | 3 | 6 | 10 | 17 | 22 | 36 | — | — | — | — | — | — |
| | 10 | 1 | 3 | 5 | 8 | 14 | 18 | 29 | — | — | — | — | — | — |
| | 8 | 1 | 1 | 2 | 4 | 7 | 9 | 15 | — | — | — | — | — | — |
| | 6 | 1 | 1 | 1 | 3 | 6 | 7 | 12 | — | — | — | — | — | — |
| | 4 | 0 | 1 | 1 | 2 | 4 | 6 | 9 | — | — | — | — | — | — |
| | 3 | 0 | 1 | 1 | 1 | 4 | 5 | 8 | — | — | — | — | — | — |
| | 2 | 0 | 1 | 1 | 1 | 3 | 4 | 7 | — | — | — | — | — | — |
| | 1 | 0 | 0 | 1 | 1 | 1 | 3 | 5 | — | — | — | — | — | — |
| | 1/0 | 0 | 0 | 1 | 1 | 1 | 2 | 4 | — | — | — | — | — | — |
| | 2/0 | 0 | 0 | 1 | 1 | 1 | 1 | 3 | — | — | — | — | — | — |
| | 3/0 | 0 | 0 | 0 | 1 | 1 | 1 | 3 | — | — | — | — | — | — |
| | 4/0 | 0 | 0 | 0 | 1 | 1 | 1 | 2 | — | — | — | — | — | — |
| | 250 | 0 | 0 | 0 | 0 | 1 | 1 | 1 | — | — | — | — | — | — |
| | 300 | 0 | 0 | 0 | 0 | 1 | 1 | 1 | — | — | — | — | — | — |
| | 350 | 0 | 0 | 0 | 0 | 1 | 1 | 1 | — | — | — | — | — | — |
| | 400 | 0 | 0 | 0 | 0 | 1 | 1 | 1 | — | — | — | — | — | — |
| | 500 | 0 | 0 | 0 | 0 | 1 | 1 | 1 | — | — | — | — | — | — |
| | 600 | 0 | 0 | 0 | 0 | 0 | 1 | 1 | — | — | — | — | — | — |
| | 700 | 0 | 0 | 0 | 0 | 0 | 0 | 1 | — | — | — | — | — | — |
| | 750 | 0 | 0 | 0 | 0 | 0 | 0 | 1 | — | — | — | — | — | — |
| | 800 | 0 | 0 | 0 | 0 | 0 | 0 | 1 | — | — | — | — | — | — |
| | 900 | 0 | 0 | 0 | 0 | 0 | 0 | 1 | — | — | — | — | — | — |
| | 1000 | 0 | 0 | 0 | 0 | 0 | 0 | 1 | — | — | — | — | — | — |
| | 1250 | 0 | 0 | 0 | 0 | 0 | 0 | 0 | — | — | — | — | — | — |
| | 1500 | 0 | 0 | 0 | 0 | 0 | 0 | 0 | — | — | — | — | — | — |
| | 1750 | 0 | 0 | 0 | 0 | 0 | 0 | 0 | — | — | — | — | — | — |
| | 2000 | 0 | 0 | 0 | 0 | 0 | 0 | 0 | — | — | — | — | — | — |
| TW, THHW, THW, THW-2 | 14 | 5 | 9 | 15 | 25 | 44 | 57 | 93 | — | — | — | — | — | — |
| | 12 | 4 | 7 | 12 | 19 | 33 | 43 | 71 | — | — | — | — | — | — |
| | 10 | 3 | 5 | 9 | 14 | 25 | 32 | 53 | — | — | — | — | — | — |
| | 8 | 1 | 3 | 5 | 8 | 14 | 18 | 29 | — | — | — | — | — | — |
| RHH*, RHW*, RHW-2* | 14 | 3 | 6 | 10 | 16 | 29 | 38 | 62 | — | — | — | — | — | — |
| | 12 | 3 | 5 | 8 | 13 | 23 | 30 | 50 | — | — | — | — | — | — |
| | 10 | 1 | 3 | 6 | 10 | 18 | 23 | 39 | — | — | — | — | — | — |
| | 8 | 1 | 1 | 4 | 6 | 11 | 14 | 23 | — | — | — | — | — | — |
| TW, THW, THHW, THW-2, RHH*, RHW*, RHW-2* | 6 | 1 | 1 | 3 | 5 | 8 | 11 | 18 | — | — | — | — | — | — |
| | 4 | 1 | 1 | 1 | 3 | 6 | 8 | 13 | — | — | — | — | — | — |
| | 3 | 1 | 1 | 1 | 3 | 5 | 7 | 11 | — | — | — | — | — | — |
| | 2 | 0 | 1 | 1 | 2 | 4 | 6 | 9 | — | — | — | — | — | — |
| | 1 | 0 | 1 | 1 | 1 | 3 | 4 | 7 | — | — | — | — | — | — |
| | 1/0 | 0 | 0 | 1 | 1 | 2 | 3 | 6 | — | — | — | — | — | — |
| | 2/0 | 0 | 0 | 1 | 1 | 2 | 3 | 5 | — | — | — | — | — | — |
| | 3/0 | 0 | 0 | 1 | 1 | 1 | 2 | 4 | — | — | — | — | — | — |
| | 4/0 | 0 | 0 | 0 | 1 | 1 | 1 | 3 | — | — | — | — | — | — |

*TABLE C.5 Continued*

| Type | Conductor Size (AWG/ kcmil) | Trade Size (Metric Designator) | | | | | | | | | | | | |
|---|---|---|---|---|---|---|---|---|---|---|---|---|---|---|
| | | ⅜ (12) | ½ (16) | ¾ (21) | 1 (27) | 1¼ (35) | 1½ (41) | 2 (53) | 2½ (63) | 3 (78) | 3½ (91) | 4 (103) | 5 (129) | 6 (155) |
| | 250 | 0 | 0 | 0 | 1 | 1 | 1 | 3 | — | — | — | — | — | — |
| | 300 | 0 | 0 | 0 | 1 | 1 | 1 | 2 | — | — | — | — | — | — |
| | 350 | 0 | 0 | 0 | 0 | 1 | 1 | 1 | — | — | — | — | — | — |
| | 400 | 0 | 0 | 0 | 0 | 1 | 1 | 1 | — | — | — | — | — | — |
| | 500 | 0 | 0 | 0 | 0 | 1 | 1 | 1 | — | — | — | — | — | — |
| | 600 | 0 | 0 | 0 | 0 | 1 | 1 | 1 | — | — | — | — | — | — |
| | 700 | 0 | 0 | 0 | 0 | 0 | 1 | 1 | — | — | — | — | — | — |
| | 750 | 0 | 0 | 0 | 0 | 0 | 1 | 1 | — | — | — | — | — | — |
| | 800 | 0 | 0 | 0 | 0 | 0 | 1 | 1 | — | — | — | — | — | — |
| | 900 | 0 | 0 | 0 | 0 | 0 | 0 | 1 | — | — | — | — | — | — |
| | 1000 | 0 | 0 | 0 | 0 | 0 | 0 | 1 | — | — | — | — | — | — |
| | 1250 | 0 | 0 | 0 | 0 | 0 | 0 | 1 | — | — | — | — | — | — |
| | 1500 | 0 | 0 | 0 | 0 | 0 | 0 | 0 | — | — | — | — | — | — |
| | 1750 | 0 | 0 | 0 | 0 | 0 | 0 | 0 | — | — | — | — | — | — |
| | 2000 | 0 | 0 | 0 | 0 | 0 | 0 | 0 | — | — | — | — | — | — |
| THHN, THWN, THWN-2 | 14 | 8 | 13 | 22 | 36 | 63 | 81 | 134 | — | — | — | — | — | — |
| | 12 | 5 | 9 | 16 | 26 | 46 | 59 | 97 | — | — | — | — | — | — |
| | 10 | 3 | 6 | 10 | 16 | 29 | 37 | 61 | — | — | — | — | — | — |
| | 8 | 1 | 3 | 6 | 9 | 16 | 21 | 35 | — | — | — | — | — | — |
| | 6 | 1 | 2 | 4 | 7 | 12 | 15 | 25 | — | — | — | — | — | — |
| | 4 | 1 | 1 | 2 | 4 | 7 | 9 | 15 | — | — | — | — | — | — |
| | 3 | 1 | 1 | 1 | 3 | 6 | 8 | 13 | — | — | — | — | — | — |
| | 2 | 1 | 1 | 1 | 3 | 5 | 7 | 11 | — | — | — | — | — | — |
| | 1 | 0 | 1 | 1 | 1 | 4 | 5 | 8 | — | — | — | — | — | — |
| | 1/0 | 0 | 1 | 1 | 1 | 3 | 4 | 7 | — | — | — | — | — | — |
| | 2/0 | 0 | 0 | 1 | 1 | 2 | 3 | 6 | — | — | — | — | — | — |
| | 3/0 | 0 | 0 | 1 | 1 | 1 | 3 | 5 | — | — | — | — | — | — |
| | 4/0 | 0 | 0 | 1 | 1 | 1 | 2 | 4 | — | — | — | — | — | — |
| | 250 | 0 | 0 | 0 | 1 | 1 | 1 | 3 | — | — | — | — | — | — |
| | 300 | 0 | 0 | 0 | 1 | 1 | 1 | 3 | — | — | — | — | — | — |
| | 350 | 0 | 0 | 0 | 1 | 1 | 1 | 2 | — | — | — | — | — | — |
| | 400 | 0 | 0 | 0 | 0 | 1 | 1 | 1 | — | — | — | — | — | — |
| | 500 | 0 | 0 | 0 | 0 | 1 | 1 | 1 | — | — | — | — | — | — |
| | 600 | 0 | 0 | 0 | 0 | 1 | 1 | 1 | — | — | — | — | — | — |
| | 700 | 0 | 0 | 0 | 0 | 1 | 1 | 1 | — | — | — | — | — | — |
| | 750 | 0 | 0 | 0 | 0 | 0 | 1 | 1 | — | — | — | — | — | — |
| | 800 | 0 | 0 | 0 | 0 | 0 | 1 | 1 | — | — | — | — | — | — |
| | 900 | 0 | 0 | 0 | 0 | 0 | 1 | 1 | — | — | — | — | — | — |
| | 1000 | 0 | 0 | 0 | 0 | 0 | 0 | 1 | — | — | — | — | — | — |
| FEP, FEPB, PFA, PFAH, TFE | 14 | 7 | 12 | 21 | 35 | 61 | 79 | 130 | — | — | — | — | — | — |
| | 12 | 5 | 9 | 15 | 25 | 44 | 58 | 94 | — | — | — | — | — | — |
| | 10 | 4 | 6 | 11 | 18 | 32 | 41 | 68 | — | — | — | — | — | — |
| | 8 | 1 | 3 | 6 | 10 | 18 | 23 | 39 | — | — | — | — | — | — |
| | 6 | 1 | 2 | 4 | 7 | 13 | 17 | 27 | — | — | — | — | — | — |
| | 4 | 1 | 1 | 3 | 5 | 9 | 12 | 19 | — | — | — | — | — | — |
| | 3 | 1 | 1 | 2 | 4 | 7 | 10 | 16 | — | — | — | — | — | — |
| | 2 | 1 | 1 | 1 | 3 | 6 | 8 | 13 | — | — | — | — | — | — |

*(continues)*

*TABLE C.5 Continued*

| Type | Conductor Size (AWG/ kcmil) | Trade Size (Metric Designator) | | | | | | | | | | | | |
|---|---|---|---|---|---|---|---|---|---|---|---|---|---|---|
| | | ⅜ (12) | ½ (16) | ¾ (21) | 1 (27) | 1¼ (35) | 1½ (41) | 2 (53) | 2½ (63) | 3 (78) | 3½ (91) | 4 (103) | 5 (129) | 6 (155) |
| PFA, PFAH, TFE | 1 | 0 | 1 | 1 | 2 | 4 | 5 | 9 | — | — | — | — | — | — |
| PFA, PFAH, TFE, Z | 1/0 | 0 | 1 | 1 | 1 | 3 | 4 | 7 | — | — | — | — | — | — |
| | 2/0 | 0 | 1 | 1 | 1 | 3 | 4 | 6 | — | — | — | — | — | — |
| | 3/0 | 0 | 0 | 1 | 1 | 2 | 3 | 5 | — | — | — | — | — | — |
| | 4/0 | 0 | 0 | 1 | 1 | 1 | 2 | 4 | — | — | — | — | — | — |
| Z | 14 | 9 | 15 | 26 | 42 | 73 | 95 | 156 | — | — | — | — | — | — |
| | 12 | 6 | 10 | 18 | 30 | 52 | 67 | 111 | — | — | — | — | — | — |
| | 10 | 4 | 6 | 11 | 18 | 32 | 41 | 68 | — | — | — | — | — | — |
| | 8 | 2 | 4 | 7 | 11 | 20 | 26 | 43 | — | — | — | — | — | — |
| | 6 | 1 | 3 | 5 | 8 | 14 | 18 | 30 | — | — | — | — | — | — |
| | 4 | 1 | 1 | 3 | 5 | 9 | 12 | 20 | — | — | — | — | — | — |
| | 3 | 1 | 1 | 2 | 4 | 7 | 9 | 15 | — | — | — | — | — | — |
| | 2 | 1 | 1 | 1 | 3 | 6 | 7 | 12 | — | — | — | — | — | — |
| | 1 | 1 | 1 | 1 | 2 | 5 | 6 | 10 | — | — | — | — | — | — |
| XHHW, ZW, XHHW-2, XHH | 14 | 5 | 9 | 15 | 25 | 44 | 57 | 93 | — | — | — | — | — | — |
| | 12 | 4 | 7 | 12 | 19 | 33 | 43 | 71 | — | — | — | — | — | — |
| | 10 | 3 | 5 | 9 | 14 | 25 | 32 | 53 | — | — | — | — | — | — |
| | 8 | 1 | 3 | 5 | 8 | 14 | 18 | 29 | — | — | — | — | — | — |
| | 6 | 1 | 1 | 3 | 6 | 10 | 13 | 22 | — | — | — | — | — | — |
| | 4 | 1 | 1 | 2 | 4 | 7 | 9 | 16 | — | — | — | — | — | — |
| | 3 | 1 | 1 | 1 | 3 | 6 | 8 | 13 | — | — | — | — | — | — |
| | 2 | 1 | 1 | 1 | 3 | 5 | 7 | 11 | — | — | — | — | — | — |
| XHHW, XHHW-2, XHH | 1 | 0 | 1 | 1 | 1 | 4 | 5 | 8 | — | — | — | — | — | — |
| | 1/0 | 0 | 1 | 1 | 1 | 3 | 4 | 7 | — | — | — | — | — | — |
| | 2/0 | 0 | 0 | 1 | 1 | 2 | 3 | 6 | — | — | — | — | — | — |
| | 3/0 | 0 | 0 | 1 | 1 | 1 | 3 | 5 | — | — | — | — | — | — |
| | 4/0 | 0 | 0 | 1 | 1 | 1 | 2 | 4 | — | — | — | — | — | — |
| | 250 | 0 | 0 | 0 | 1 | 1 | 1 | 3 | — | — | — | — | — | — |
| | 300 | 0 | 0 | 0 | 1 | 1 | 1 | 3 | — | — | — | — | — | — |
| | 350 | 0 | 0 | 0 | 1 | 1 | 1 | 2 | — | — | — | — | — | — |
| | 400 | 0 | 0 | 0 | 1 | 1 | 1 | 1 | — | — | — | — | — | — |
| | 500 | 0 | 0 | 0 | 0 | 1 | 1 | 1 | — | — | — | — | — | — |
| | 600 | 0 | 0 | 0 | 0 | 1 | 1 | 1 | — | — | — | — | — | — |
| | 700 | 0 | 0 | 0 | 0 | 1 | 1 | 1 | — | — | — | — | — | — |
| | 750 | 0 | 0 | 0 | 0 | 0 | 1 | 1 | — | — | — | — | — | — |
| | 800 | 0 | 0 | 0 | 0 | 0 | 1 | 1 | — | — | — | — | — | — |
| | 900 | 0 | 0 | 0 | 0 | 0 | 1 | 1 | — | — | — | — | — | — |
| | 1000 | 0 | 0 | 0 | 0 | 0 | 0 | 1 | — | — | — | — | — | — |
| | 1250 | 0 | 0 | 0 | 0 | 0 | 0 | 1 | — | — | — | — | — | — |
| | 1500 | 0 | 0 | 0 | 0 | 0 | 0 | 1 | — | — | — | — | — | — |
| | 1750 | 0 | 0 | 0 | 0 | 0 | 0 | 0 | — | — | — | — | — | — |
| | 2000 | 0 | 0 | 0 | 0 | 0 | 0 | 0 | — | — | — | — | — | — |

***TABLE C.5*** *Continued*

| Type | Conductor Size (AWG/ kcmil) | Trade Size (Metric Designator) | | | | | | | | | | | | |
|---|---|---|---|---|---|---|---|---|---|---|---|---|---|---|
| | | 3/8 (12) | 1/2 (16) | 3/4 (21) | 1 (27) | 1 1/4 (35) | 1 1/2 (41) | 2 (53) | 2 1/2 (63) | 3 (78) | 3 1/2 (91) | 4 (103) | 5 (129) | 6 (155) |
| **FIXTURE WIRES** | | | | | | | | | | | | | | |
| RFH-2, FFH-2, RFHH-2 | 18 | 5 | 8 | 15 | 24 | 42 | 54 | 89 | — | — | — | — | — | — |
| | 16 | 4 | 7 | 12 | 20 | 35 | 46 | 75 | — | — | — | — | — | — |
| SF-2, SFF-2 | 18 | 6 | 11 | 19 | 30 | 53 | 69 | 113 | — | — | — | — | — | — |
| | 16 | 5 | 9 | 15 | 25 | 44 | 57 | 93 | — | — | — | — | — | — |
| | 14 | 4 | 7 | 12 | 20 | 35 | 46 | 75 | — | — | — | — | — | — |
| SF-1, SFF-1 | 18 | 12 | 19 | 33 | 53 | 94 | 122 | 199 | — | — | — | — | — | — |
| RFH-1, TF, TFF, XF, XFF | 18 | 8 | 14 | 24 | 39 | 69 | 90 | 147 | — | — | — | — | — | — |
| | 16 | 7 | 11 | 20 | 32 | 56 | 72 | 119 | — | — | — | — | — | — |
| • XF, XFF | 14 | 5 | 9 | 15 | 25 | 44 | 57 | 93 | — | — | — | — | — | — |
| TFN, TFFN | 18 | 14 | 23 | 39 | 63 | 111 | 144 | 236 | — | — | — | — | — | — |
| | 16 | 10 | 17 | 30 | 48 | 85 | 110 | 180 | — | — | — | — | — | — |
| PF, PFF, PGF, PGFF, PAF, PTF, PTFF, PAFF | 18 | 13 | 21 | 37 | 60 | 105 | 136 | 224 | — | — | — | — | — | — |
| | 16 | 10 | 16 | 29 | 46 | 81 | 105 | 173 | — | — | — | — | — | — |
| | 14 | 7 | 12 | 21 | 35 | 61 | 79 | 130 | — | — | — | — | — | — |
| ZF, ZFF, ZHF | 18 | 17 | 28 | 48 | 77 | 136 | 176 | 288 | — | — | — | — | — | — |
| | 16 | 12 | 20 | 35 | 57 | 100 | 130 | 213 | — | — | — | — | — | — |
| | 14 | 9 | 15 | 26 | 42 | 73 | 95 | 156 | — | — | — | — | — | — |
| KF-2, KFF-2 | 18 | 25 | 42 | 72 | 116 | 203 | 264 | 433 | — | — | — | — | — | — |
| | 16 | 18 | 29 | 50 | 81 | 142 | 184 | 302 | — | — | — | — | — | — |
| | 14 | 12 | 19 | 34 | 54 | 95 | 124 | 203 | — | — | — | — | — | — |
| | 12 | 8 | 13 | 23 | 38 | 66 | 86 | 141 | — | — | — | — | — | — |
| | 10 | 5 | 9 | 15 | 25 | 44 | 57 | 93 | — | — | — | — | — | — |
| KF-1, KFF-1 | 18 | 29 | 48 | 83 | 134 | 235 | 304 | 499 | — | — | — | — | — | — |
| | 16 | 20 | 34 | 58 | 94 | 165 | 214 | 351 | — | — | — | — | — | — |
| | 14 | 14 | 23 | 39 | 63 | 111 | 144 | 236 | — | — | — | — | — | — |
| | 12 | 9 | 15 | 26 | 42 | 73 | 95 | 156 | — | — | — | — | — | — |
| | 10 | 6 | 10 | 17 | 27 | 48 | 62 | 102 | — | — | — | — | — | — |
| XF, XFF | 12 | 3 | 5 | 8 | 13 | 23 | 30 | 50 | — | — | — | — | — | — |
| | 10 | 1 | 3 | 6 | 10 | 18 | 23 | 39 | — | — | — | — | — | — |

Notes:

1. This table is for concentric stranded conductors only. For compact stranded conductors, Table C.5(A) should be used.

2. Two-hour fire-rated RHH cable has ceramifiable insulation, which has much larger diameters than other RHH wires. Consult manufacturer's conduit fill tables.

*Types RHH, RHW, and RHW-2 without outer covering.

***TABLE C.5(A)*** *Maximum Number of Conductors or Fixture Wires in Liquidtight Flexible Nonmetallic Conduit (Type LFNC-B)* (Based on Chapter 9: Table 1, Table 4, and Table 5A)

| **Type** | **Conductor Size (AWG/ kcmil)** | **Trade Size (Metric Designator)** | | | | | | | | | | | | |
|---|---|---|---|---|---|---|---|---|---|---|---|---|---|---|
| | | **⅜ (12)** | **½ (16)** | **¾ (21)** | **1 (27)** | **1¼ (35)** | **1½ (41)** | **2 (53)** | **2½ (63)** | **3 (78)** | **3½ (91)** | **4 (103)** | **5 (129)** | **6 (155)** |
| | | | | | | **COMPACT CONDUCTORS** | | | | | | | | |
| THW, THW-2, THHW | 8 | 1 | 2 | 4 | 7 | 12 | 15 | 25 | — | — | — | — | — | — |
| | 6 | 1 | 1 | 3 | 5 | 9 | 12 | 19 | — | — | — | — | — | — |
| | 4 | 1 | 1 | 2 | 4 | 7 | 9 | 14 | — | — | — | — | — | — |
| | 2 | 1 | 1 | 1 | 3 | 5 | 6 | 11 | — | — | — | — | — | — |
| | 1 | 0 | 1 | 1 | 1 | 3 | 4 | 7 | — | — | — | — | — | — |
| | 1/0 | 0 | 1 | 1 | 1 | 3 | 4 | 6 | — | — | — | — | — | — |
| | 2/0 | 0 | 0 | 1 | 1 | 2 | 3 | 5 | — | — | — | — | — | — |
| | 3/0 | 0 | 0 | 1 | 1 | 1 | 3 | 4 | — | — | — | — | — | — |
| | 4/0 | 0 | 0 | 1 | 1 | 1 | 2 | 4 | — | — | — | — | — | — |
| | 250 | 0 | 0 | 0 | 1 | 1 | 1 | 3 | — | — | — | — | — | — |
| | 300 | 0 | 0 | 0 | 1 | 1 | 1 | 2 | — | — | — | — | — | — |
| | 350 | 0 | 0 | 0 | 1 | 1 | 1 | 2 | — | — | — | — | — | — |
| | 400 | 0 | 0 | 0 | 0 | 1 | 1 | 1 | — | — | — | — | — | — |
| | 500 | 0 | 0 | 0 | 0 | 1 | 1 | 1 | — | — | — | — | — | — |
| | 600 | 0 | 0 | 0 | 0 | 1 | 1 | 1 | — | — | — | — | — | — |
| | 700 | 0 | 0 | 0 | 0 | 1 | 1 | 1 | — | — | — | — | — | — |
| | 750 | 0 | 0 | 0 | 0 | 0 | 1 | 1 | — | — | — | — | — | — |
| | 900 | 0 | 0 | 0 | 0 | 0 | 1 | 1 | — | — | — | — | — | — |
| | 1000 | 0 | 0 | 0 | 0 | 0 | 1 | 1 | — | — | — | — | — | — |
| THHN, THWN, THWN-2 | 8 | — | — | — | — | — | — | — | — | — | — | — | — | — |
| | 6 | 1 | 2 | 4 | 7 | 13 | 17 | 28 | — | — | — | — | — | — |
| | 4 | 1 | 1 | 3 | 4 | 8 | 11 | 17 | — | — | — | — | — | — |
| | 2 | 1 | 1 | 1 | 3 | 6 | 7 | 12 | — | — | — | — | — | — |
| | 1 | 0 | 1 | 1 | 2 | 4 | 6 | 9 | — | — | — | — | — | — |
| | 1/0 | 0 | 1 | 1 | 1 | 4 | 5 | 8 | — | — | — | — | — | — |
| | 2/0 | 0 | 1 | 1 | 1 | 3 | 4 | 6 | — | — | — | — | — | — |
| | 3/0 | 0 | 0 | 1 | 1 | 2 | 3 | 5 | — | — | — | — | — | — |
| | 4/0 | 0 | 0 | 1 | 1 | 1 | 3 | 4 | — | — | — | — | — | — |
| | 250 | 0 | 0 | 1 | 1 | 1 | 1 | 3 | — | — | — | — | — | — |
| | 300 | 0 | 0 | 0 | 1 | 1 | 1 | 3 | — | — | — | — | — | — |
| | 350 | 0 | 0 | 0 | 1 | 1 | 1 | 2 | — | — | — | — | — | — |
| | 400 | 0 | 0 | 0 | 1 | 1 | 1 | 2 | — | — | — | — | — | — |
| | 500 | 0 | 0 | 0 | 0 | 1 | 1 | 1 | — | — | — | — | — | — |
| | 600 | 0 | 0 | 0 | 0 | 1 | 1 | 1 | — | — | — | — | — | — |
| | 700 | 0 | 0 | 0 | 0 | 1 | 1 | 1 | — | — | — | — | — | — |
| | 750 | 0 | 0 | 0 | 0 | 1 | 1 | 1 | — | — | — | — | — | — |
| | 900 | 0 | 0 | 0 | 0 | 0 | 1 | 1 | — | — | — | — | — | — |
| | 1000 | 0 | 0 | 0 | 0 | 0 | 1 | 1 | — | — | — | — | — | — |

***TABLE C.5(A)*** *Continued*

| Type | Conductor Size (AWG/ kcmil) | Trade Size (Metric Designator) | | | | | | | | | | | | |
|---|---|---|---|---|---|---|---|---|---|---|---|---|---|---|
| | | 3⁄8 (12) | 1⁄2 (16) | 3⁄4 (21) | 1 (27) | 1¼ (35) | 1½ (41) | 2 (53) | 2½ (63) | 3 (78) | 3½ (91) | 4 (103) | 5 (129) | 6 (155) |
| XHHW, XHHW-2 | 8 | 1 | 3 | 5 | 9 | 15 | 20 | 33 | — | — | — | — | — | — |
| | 6 | 1 | 2 | 4 | 6 | 11 | 15 | 24 | — | — | — | — | — | — |
| | 4 | 1 | 1 | 3 | 4 | 8 | 11 | 17 | — | — | — | — | — | — |
| | 2 | 1 | 1 | 1 | 3 | 6 | 7 | 12 | — | — | — | — | — | — |
| | 1 | 0 | 1 | 1 | 2 | 4 | 6 | 9 | — | — | — | — | — | — |
| | 1/0 | 0 | 1 | 1 | 1 | 4 | 5 | 8 | — | — | — | — | — | — |
| | 2/0 | 0 | 1 | 1 | 1 | 3 | 4 | 7 | — | — | — | — | — | — |
| | 3/0 | 0 | 0 | 1 | 1 | 2 | 3 | 5 | — | — | — | — | — | — |
| | 4/0 | 0 | 0 | 1 | 1 | 1 | 3 | 4 | — | — | — | — | — | — |
| | 250 | 0 | 0 | 1 | 1 | 1 | 1 | 3 | — | — | — | — | — | — |
| | 300 | 0 | 0 | 0 | 1 | 1 | 1 | 3 | — | — | — | — | — | — |
| | 350 | 0 | 0 | 0 | 1 | 1 | 1 | 3 | — | — | — | — | — | — |
| | 400 | 0 | 0 | 0 | 1 | 1 | 1 | 2 | — | — | — | — | — | — |
| | 500 | 0 | 0 | 0 | 0 | 1 | 1 | 1 | — | — | — | — | — | — |
| | 600 | 0 | 0 | 0 | 0 | 1 | 1 | 1 | — | — | — | — | — | — |
| | 700 | 0 | 0 | 0 | 0 | 1 | 1 | 1 | — | — | — | — | — | — |
| | 750 | 0 | 0 | 0 | 0 | 1 | 1 | 1 | — | — | — | — | — | — |
| | 900 | 0 | 0 | 0 | 0 | 0 | 1 | 1 | — | — | — | — | — | — |
| | 1000 | 0 | 0 | 0 | 0 | 0 | 1 | 1 | — | — | — | — | — | — |

Definition: *Compact stranding* is the result of a manufacturing process where the stranded conductor is compressed to the extent that the interstices (voids between strand wires) are virtually eliminated.

***TABLE C.6*** *Maximum Number of Conductors or Fixture Wires in Liquidtight Flexible Nonmetallic Conduit (Type LFNC-A)* (Based on Chapter 9: Table 1, Table 4, and Table 5)

| | | **Trade Size (Metric Designator)** | | | | | | | | | | | | |
|---|---|---|---|---|---|---|---|---|---|---|---|---|---|---|
| **Type** | **Conductor Size (AWG/ kcmil)** | **⅜ (12)** | **½ (16)** | **¾ (21)** | **1 (27)** | **1¼ (35)** | **1½ (41)** | **2 (53)** | **2½ (63)** | **3 (78)** | **3½ (91)** | **4 (103)** | **5 (129)** | **6 (155)** |
| | | | | | | **CONDUCTORS** | | | | | | | | |
| RHH, RHW, RHW-2 | 14 | 2 | 4 | 7 | 11 | 20 | 27 | 45 | — | — | — | — | — | — |
| | 12 | 1 | 3 | 6 | 9 | 17 | 23 | 38 | — | — | — | — | — | — |
| | 10 | 1 | 3 | 5 | 8 | 13 | 18 | 30 | — | — | — | — | — | — |
| | 8 | 1 | 1 | 2 | 4 | 7 | 9 | 16 | — | — | — | — | — | — |
| | 6 | 1 | 1 | 1 | 3 | 5 | 7 | 13 | — | — | — | — | — | — |
| | 4 | 0 | 1 | 1 | 2 | 4 | 6 | 10 | — | — | — | — | — | — |
| | 3 | 0 | 1 | 1 | 1 | 4 | 5 | 8 | — | — | — | — | — | — |
| | 2 | 0 | 1 | 1 | 1 | 3 | 4 | 7 | — | — | — | — | — | — |
| | 1 | 0 | 0 | 1 | 1 | 1 | 3 | 5 | — | — | — | — | — | — |
| | 1/0 | 0 | 0 | 1 | 1 | 1 | 2 | 4 | — | — | — | — | — | — |
| | 2/0 | 0 | 0 | 1 | 1 | 1 | 1 | 4 | — | — | — | — | — | — |
| | 3/0 | 0 | 0 | 0 | 1 | 1 | 1 | 3 | — | — | — | — | — | — |
| | 4/0 | 0 | 0 | 0 | 1 | 1 | 1 | 3 | — | — | — | — | — | — |
| | 250 | 0 | 0 | 0 | 0 | 1 | 1 | 1 | — | — | — | — | — | — |
| | 300 | 0 | 0 | 0 | 0 | 1 | 1 | 1 | — | — | — | — | — | — |
| | 350 | 0 | 0 | 0 | 0 | 1 | 1 | 1 | — | — | — | — | — | — |
| | 400 | 0 | 0 | 0 | 0 | 1 | 1 | 1 | — | — | — | — | — | — |
| | 500 | 0 | 0 | 0 | 0 | 0 | 1 | 1 | — | — | — | — | — | — |
| | 600 | 0 | 0 | 0 | 0 | 0 | 1 | 1 | — | — | — | — | — | — |
| | 700 | 0 | 0 | 0 | 0 | 0 | 0 | 1 | — | — | — | — | — | — |
| | 750 | 0 | 0 | 0 | 0 | 0 | 0 | 1 | — | — | — | — | — | — |
| | 800 | 0 | 0 | 0 | 0 | 0 | 0 | 1 | — | — | — | — | — | — |
| | 900 | 0 | 0 | 0 | 0 | 0 | 0 | 1 | — | — | — | — | — | — |
| | 1000 | 0 | 0 | 0 | 0 | 0 | 0 | 1 | — | — | — | — | — | — |
| | 1250 | 0 | 0 | 0 | 0 | 0 | 0 | 0 | — | — | — | — | — | — |
| | 1500 | 0 | 0 | 0 | 0 | 0 | 0 | 0 | — | — | — | — | — | — |
| | 1750 | 0 | 0 | 0 | 0 | 0 | 0 | 0 | — | — | — | — | — | — |
| | 2000 | 0 | 0 | 0 | 0 | 0 | 0 | 0 | — | — | — | — | — | — |
| TW, THHW, THW, THW-2 | 14 | 5 | 9 | 15 | 24 | 43 | 58 | 96 | — | — | — | — | — | — |
| | 12 | 4 | 7 | 12 | 19 | 33 | 44 | 74 | — | — | — | — | — | — |
| | 10 | 3 | 5 | 9 | 14 | 24 | 33 | 55 | — | — | — | — | — | — |
| | 8 | 1 | 3 | 5 | 8 | 13 | 18 | 30 | — | — | — | — | — | — |
| RHH*, RHW*, RHW-2* | 14 | 3 | 6 | 10 | 16 | 28 | 38 | 64 | — | — | — | — | — | — |
| | 12 | 3 | 5 | 8 | 13 | 23 | 31 | 51 | — | — | — | — | — | — |
| | 10 | 1 | 3 | 6 | 10 | 18 | 24 | 40 | — | — | — | — | — | — |
| | 8 | 1 | 1 | 4 | 6 | 11 | 14 | 24 | — | — | — | — | — | — |
| • TW, THW, THHW, THW-2, RHH*, RHW*, RHW-2* | 6 | 1 | 1 | 3 | 4 | 8 | 11 | 18 | — | — | — | — | — | — |
| | 4 | 1 | 1 | 1 | 3 | 6 | 8 | 13 | — | — | — | — | — | — |
| | 3 | 1 | 1 | 1 | 3 | 5 | 7 | 11 | — | — | — | — | — | — |
| | 2 | 0 | 1 | 1 | 2 | 4 | 6 | 10 | — | — | — | — | — | — |
| | 1 | 0 | 1 | 1 | 1 | 3 | 4 | 7 | — | — | — | — | — | — |
| | 1/0 | 0 | 0 | 1 | 1 | 2 | 3 | 6 | — | — | — | — | — | — |
| | 2/0 | 0 | 0 | 1 | 1 | 1 | 3 | 5 | — | — | — | — | — | — |
| | 3/0 | 0 | 0 | 1 | 1 | 1 | 2 | 4 | — | — | — | — | — | — |
| | 4/0 | 0 | 0 | 0 | 1 | 1 | 1 | 3 | — | — | — | — | — | — |

***TABLE C.6*** *Continued*

| Type | Conductor Size (AWG/ kcmil) | Trade Size (Metric Designator) | | | | | | | | | | | | |
|---|---|---|---|---|---|---|---|---|---|---|---|---|---|---|
| | | **3/8 (12)** | **1/2 (16)** | **3/4 (21)** | **1 (27)** | **1 1/4 (35)** | **1 1/2 (41)** | **2 (53)** | **2 1/2 (63)** | **3 (78)** | **3 1/2 (91)** | **4 (103)** | **5 (129)** | **6 (155)** |
| | 250 | 0 | 0 | 0 | 1 | 1 | 1 | 3 | — | — | — | — | — | — |
| | 300 | 0 | 0 | 0 | 1 | 1 | 1 | 2 | — | — | — | — | — | — |
| | 350 | 0 | 0 | 0 | 0 | 1 | 1 | 1 | — | — | — | — | — | — |
| | 400 | 0 | 0 | 0 | 0 | 1 | 1 | 1 | — | — | — | — | — | — |
| | 500 | 0 | 0 | 0 | 0 | 1 | 1 | 1 | — | — | — | — | — | — |
| | 600 | 0 | 0 | 0 | 0 | 1 | 1 | 1 | — | — | — | — | — | — |
| | 700 | 0 | 0 | 0 | 0 | 0 | 1 | 1 | — | — | — | — | — | — |
| | 750 | 0 | 0 | 0 | 0 | 0 | 1 | 1 | — | — | — | — | — | — |
| | 800 | 0 | 0 | 0 | 0 | 0 | 1 | 1 | — | — | — | — | — | — |
| | 900 | 0 | 0 | 0 | 0 | 0 | 0 | 1 | — | — | — | — | — | — |
| | 1000 | 0 | 0 | 0 | 0 | 0 | 0 | 1 | — | — | — | — | — | — |
| | 1250 | 0 | 0 | 0 | 0 | 0 | 0 | 1 | — | — | — | — | — | — |
| | 1500 | 0 | 0 | 0 | 0 | 0 | 0 | 1 | — | — | — | — | — | — |
| | 1750 | 0 | 0 | 0 | 0 | 0 | 0 | 0 | — | — | — | — | — | — |
| | 2000 | 0 | 0 | 0 | 0 | 0 | 0 | 0 | — | — | — | — | — | — |
| THHN, THWN, THWN-2 | 14 | 8 | 13 | 22 | 35 | 62 | 83 | 138 | — | — | — | — | — | — |
| | 12 | 5 | 9 | 16 | 25 | 45 | 60 | 100 | — | — | — | — | — | — |
| | 10 | 3 | 6 | 10 | 16 | 28 | 38 | 63 | — | — | — | — | — | — |
| | 8 | 1 | 3 | 6 | 9 | 16 | 22 | 36 | — | — | — | — | — | — |
| | 6 | 1 | 2 | 4 | 6 | 12 | 16 | 26 | — | — | — | — | — | — |
| | 4 | 1 | 1 | 2 | 4 | 7 | 9 | 16 | — | — | — | — | — | — |
| | 3 | 1 | 1 | 1 | 3 | 6 | 8 | 13 | — | — | — | — | — | — |
| | 2 | 1 | 1 | 1 | 3 | 5 | 7 | 11 | — | — | — | — | — | — |
| | 1 | 0 | 1 | 1 | 1 | 4 | 5 | 8 | — | — | — | — | — | — |
| | 1/0 | 0 | 1 | 1 | 1 | 3 | 4 | 7 | — | — | — | — | — | — |
| | 2/0 | 0 | 0 | 1 | 1 | 2 | 3 | 6 | — | — | — | — | — | — |
| | 3/0 | 0 | 0 | 1 | 1 | 1 | 3 | 5 | — | — | — | — | — | — |
| | 4/0 | 0 | 0 | 1 | 1 | 1 | 2 | 4 | — | — | — | — | — | — |
| | 250 | 0 | 0 | 0 | 1 | 1 | 1 | 3 | — | — | — | — | — | — |
| | 300 | 0 | 0 | 0 | 1 | 1 | 1 | 3 | — | — | — | — | — | — |
| | 350 | 0 | 0 | 0 | 1 | 1 | 1 | 2 | — | — | — | — | — | — |
| | 400 | 0 | 0 | 0 | 0 | 1 | 1 | 1 | — | — | — | — | — | — |
| | 500 | 0 | 0 | 0 | 0 | 1 | 1 | 1 | — | — | — | — | — | — |
| | 600 | 0 | 0 | 0 | 0 | 1 | 1 | 1 | — | — | — | — | — | — |
| | 700 | 0 | 0 | 0 | 0 | 1 | 1 | 1 | — | — | — | — | — | — |
| | 750 | 0 | 0 | 0 | 0 | 0 | 1 | 1 | — | — | — | — | — | — |
| | 800 | 0 | 0 | 0 | 0 | 0 | 1 | 1 | — | — | — | — | — | — |
| | 900 | 0 | 0 | 0 | 0 | 0 | 1 | 1 | — | — | — | — | — | — |
| | 1000 | 0 | 0 | 0 | 0 | 0 | 0 | 1 | — | — | — | — | — | — |
| FEP, FEPB, PFA, PFAH, TFE | 14 | 7 | 12 | 21 | 34 | 60 | 80 | 133 | — | — | — | — | — | — |
| | 12 | 5 | 9 | 15 | 25 | 44 | 59 | 97 | — | — | — | — | — | — |
| | 10 | 4 | 6 | 11 | 18 | 31 | 42 | 70 | — | — | — | — | — | — |
| | 8 | 1 | 3 | 6 | 10 | 18 | 24 | 40 | — | — | — | — | — | — |
| | 6 | 1 | 2 | 4 | 7 | 13 | 17 | 28 | — | — | — | — | — | — |
| | 4 | 1 | 1 | 3 | 5 | 9 | 12 | 20 | — | — | — | — | — | — |
| | 3 | 1 | 1 | 2 | 4 | 7 | 10 | 16 | — | — | — | — | — | — |
| | 2 | 1 | 1 | 1 | 3 | 6 | 8 | 13 | — | — | — | — | — | — |

*(continues)*

***TABLE C.6*** *Continued*

| Type | Conductor Size (AWG/ kcmil) | Trade Size (Metric Designator) | | | | | | | | | | | | |
|---|---|---|---|---|---|---|---|---|---|---|---|---|---|---|
| | | **⅜ (12)** | **½ (16)** | **¾ (21)** | **1 (27)** | **1¼ (35)** | **1½ (41)** | **2 (53)** | **2½ (63)** | **3 (78)** | **3½ (91)** | **4 (103)** | **5 (129)** | **6 (155)** |
| PFA, PFAH, TFE | 1 | 0 | 1 | 1 | 2 | 4 | 5 | 9 | — | — | — | — | — | — |
| PFA, PFAH, TFE, Z | 1/0 | 0 | 1 | 1 | 1 | 3 | 5 | 8 | — | — | — | — | — | — |
| | 2/0 | 0 | 1 | 1 | 1 | 3 | 4 | 6 | — | — | — | — | — | — |
| | 3/0 | 0 | 0 | 1 | 1 | 2 | 3 | 5 | — | — | — | — | — | — |
| | 4/0 | 0 | 0 | 1 | 1 | 1 | 2 | 4 | — | — | — | — | — | — |
| Z | 14 | 9 | 15 | 25 | 41 | 72 | 97 | 161 | — | — | — | — | — | — |
| | 12 | 6 | 10 | 18 | 29 | 51 | 69 | 114 | — | — | — | — | — | — |
| | 10 | 4 | 6 | 11 | 18 | 31 | 42 | 70 | — | — | — | — | — | — |
| | 8 | 2 | 4 | 7 | 11 | 20 | 26 | 44 | — | — | — | — | — | — |
| | 6 | 1 | 3 | 5 | 8 | 14 | 18 | 31 | — | — | — | — | — | — |
| | 4 | 1 | 1 | 3 | 5 | 9 | 13 | 21 | — | — | — | — | — | — |
| | 3 | 1 | 1 | 2 | 4 | 7 | 9 | 15 | — | — | — | — | — | — |
| | 2 | 1 | 1 | 1 | 3 | 6 | 8 | 13 | — | — | — | — | — | — |
| | 1 | 1 | 1 | 1 | 2 | 4 | 6 | 10 | — | — | — | — | — | — |
| XHHW, ZW, XHHW-2, XHH | 14 | 5 | 9 | 15 | 24 | 43 | 58 | 96 | — | — | — | — | — | — |
| | 12 | 4 | 7 | 12 | 19 | 33 | 44 | 74 | — | — | — | — | — | — |
| | 10 | 3 | 5 | 9 | 14 | 24 | 33 | 55 | — | — | — | — | — | — |
| | 8 | 1 | 3 | 5 | 8 | 13 | 18 | 30 | — | — | — | — | — | — |
| | 6 | 1 | 1 | 3 | 5 | 10 | 13 | 22 | — | — | — | — | — | — |
| | 4 | 1 | 1 | 2 | 4 | 7 | 10 | 16 | — | — | — | — | — | — |
| | 3 | 1 | 1 | 1 | 3 | 6 | 8 | 14 | — | — | — | — | — | — |
| | 2 | 1 | 1 | 1 | 3 | 5 | 7 | 11 | — | — | — | — | — | — |
| XHHW, XHHW-2, XHH | 1 | 0 | 1 | 1 | 1 | 4 | 5 | 8 | — | — | — | — | — | — |
| | 1/0 | 0 | 1 | 1 | 1 | 3 | 4 | 7 | — | — | — | — | — | — |
| | 2/0 | 0 | 0 | 1 | 1 | 2 | 3 | 6 | — | — | — | — | — | — |
| | 3/0 | 0 | 0 | 1 | 1 | 1 | 3 | 5 | — | — | — | — | — | — |
| | 4/0 | 0 | 0 | 1 | 1 | 1 | 2 | 4 | — | — | — | — | — | — |
| | 250 | 0 | 0 | 0 | 1 | 1 | 1 | 3 | — | — | — | — | — | — |
| | 300 | 0 | 0 | 0 | 1 | 1 | 1 | 3 | — | — | — | — | — | — |
| | 350 | 0 | 0 | 0 | 1 | 1 | 1 | 2 | — | — | — | — | — | — |
| | 400 | 0 | 0 | 0 | 0 | 1 | 1 | 1 | — | — | — | — | — | — |
| | 500 | 0 | 0 | 0 | 0 | 1 | 1 | 1 | — | — | — | — | — | — |
| | 600 | 0 | 0 | 0 | 0 | 1 | 1 | 1 | — | — | — | — | — | — |
| | 700 | 0 | 0 | 0 | 0 | 1 | 1 | 1 | — | — | — | — | — | — |
| | 750 | 0 | 0 | 0 | 0 | 0 | 1 | 1 | — | — | — | — | — | — |
| | 800 | 0 | 0 | 0 | 0 | 0 | 1 | 1 | — | — | — | — | — | — |
| | 900 | 0 | 0 | 0 | 0 | 0 | 1 | 1 | — | — | — | — | — | — |
| | 1000 | 0 | 0 | 0 | 0 | 0 | 0 | 1 | — | — | — | — | — | — |
| | 1250 | 0 | 0 | 0 | 0 | 0 | 0 | 1 | — | — | — | — | — | — |
| | 1500 | 0 | 0 | 0 | 0 | 0 | 0 | 1 | — | — | — | — | — | — |
| | 1750 | 0 | 0 | 0 | 0 | 0 | 0 | 0 | — | — | — | — | — | — |
| | 2000 | 0 | 0 | 0 | 0 | 0 | 0 | 0 | — | — | — | — | — | — |

***TABLE C.6*** *Continued*

| Type | Conductor Size (AWG/ kcmil) | Trade Size (Metric Designator) | | | | | | | | | | | | |
|---|---|---|---|---|---|---|---|---|---|---|---|---|---|---|
| | | 3/8 (12) | 1/2 (16) | 3/4 (21) | 1 (27) | 1¼ (35) | 1½ (41) | 2 (53) | 2½ (63) | 3 (78) | 3½ (91) | 4 (103) | 5 (129) | 6 (155) |
| **FIXTURE WIRES** | | | | | | | | | | | | | | |
| RFH-2, FFH-2, RFHH-2 | 18 | 5 | 8 | 14 | 23 | 41 | 55 | 92 | — | — | — | — | — | — |
| | 16 | 4 | 7 | 12 | 20 | 35 | 47 | 77 | — | — | — | — | — | — |
| SF-2, SFF-2 | 18 | 6 | 11 | 18 | 29 | 52 | 70 | 116 | — | — | — | — | — | — |
| | 16 | 5 | 9 | 15 | 24 | 43 | 58 | 96 | — | — | — | — | — | — |
| | 14 | 4 | 7 | 12 | 20 | 35 | 47 | 77 | — | — | — | — | — | — |
| SF-1, SFF-1 | 18 | 12 | 19 | 33 | 52 | 92 | 124 | 205 | — | — | — | — | — | — |
| RFH-1, TF, TFF, XF, XFF | 18 | 8 | 14 | 24 | 39 | 68 | 91 | 152 | — | — | — | — | — | — |
| | 16 | 7 | 11 | 19 | 31 | 55 | 74 | 122 | — | — | — | — | — | — |
| XF, XFF | 14 | 5 | 9 | 15 | 24 | 43 | 58 | 96 | — | — | — | — | — | — |
| • TFN, TFFN | 18 | 14 | 22 | 39 | 62 | 109 | 146 | 243 | — | — | — | — | — | — |
| | 16 | 10 | 17 | 29 | 47 | 83 | 112 | 185 | — | — | — | — | — | — |
| PF, PFF, PGF, PGFF, PAF, PTF, PTFF, PAFF | 18 | 13 | 21 | 37 | 59 | 103 | 139 | 230 | — | — | — | — | — | — |
| | 16 | 10 | 16 | 28 | 45 | 80 | 107 | 178 | — | — | — | — | — | — |
| | 14 | 7 | 12 | 21 | 34 | 60 | 80 | 133 | — | — | — | — | — | — |
| ZF, ZFF, ZHF | 18 | 17 | 27 | 47 | 76 | 133 | 179 | 297 | — | — | — | — | — | — |
| | 16 | 12 | 20 | 35 | 56 | 98 | 132 | 219 | — | — | — | — | — | — |
| | 14 | 9 | 15 | 25 | 41 | 72 | 97 | 161 | — | — | — | — | — | — |
| KF-2, KFF-2 | 18 | 25 | 41 | 71 | 114 | 200 | 269 | 445 | — | — | — | — | — | — |
| | 16 | 18 | 29 | 49 | 79 | 139 | 187 | 311 | — | — | — | — | — | — |
| | 14 | 12 | 19 | 33 | 53 | 94 | 126 | 209 | — | — | — | — | — | — |
| | 12 | 8 | 13 | 23 | 37 | 65 | 87 | 145 | — | — | — | — | — | — |
| | 10 | 5 | 9 | 15 | 24 | 43 | 58 | 96 | — | — | — | — | — | — |
| KF-1, KFF-1 | 18 | 29 | 48 | 82 | 131 | 231 | 310 | 514 | — | — | — | — | — | — |
| | 16 | 20 | 33 | 58 | 92 | 162 | 218 | 361 | — | — | — | — | — | — |
| | 14 | 14 | 22 | 39 | 62 | 109 | 146 | 243 | — | — | — | — | — | — |
| | 12 | 9 | 15 | 25 | 41 | 72 | 97 | 161 | — | — | — | — | — | — |
| | 10 | 6 | 10 | 17 | 27 | 47 | 63 | 105 | — | — | — | — | — | — |
| XF, XFF | 12 | 3 | 5 | 8 | 13 | 23 | 31 | 51 | — | — | — | — | — | — |
| | 10 | 1 | 3 | 6 | 10 | 18 | 24 | 40 | — | — | — | — | — | — |

Notes:

1. This table is for concentric stranded conductors only. For compact stranded conductors, Table C.6(A) should be used.

2. Two-hour fire-rated RHH cable has ceramifiable insulation, which has much larger diameters than other RHH wires. Consult manufacturer's conduit fill tables.

•

*Types RHH, RHW, and RHW-2 without outer covering.

***TABLE C.6(A)*** *Maximum Number of Conductors or Fixture Wires in Liquidtight Flexible Nonmetallic Conduit (Type LFNC-A)* (Based on Chapter 9: Table 1, Table 4, and Table 5A)

| Type | Conductor Size (AWG/ kcmil) | Trade Size (Metric Designator) | | | | | | | | | | | | |
|---|---|---|---|---|---|---|---|---|---|---|---|---|---|---|
| | | 3⁄8 (12) | 1⁄2 (16) | 3⁄4 (21) | 1 (27) | 1¼ (35) | 1½ (41) | 2 (53) | 2½ (63) | 3 (78) | 3½ (91) | 4 (103) | 5 (129) | 6 (155) |
| | | **COMPACT CONDUCTORS** | | | | | | | | | | | | |
| THW, THW-2, THHW | 8 | 1 | 2 | 4 | 6 | 11 | 16 | 26 | — | — | — | — | — | — |
| | 6 | 1 | 1 | 3 | 5 | 9 | 12 | 20 | — | — | — | — | — | — |
| | 4 | 1 | 1 | 2 | 4 | 7 | 9 | 15 | — | — | — | — | — | — |
| | 2 | 1 | 1 | 1 | 3 | 5 | 6 | 11 | — | — | — | — | — | — |
| | 1 | 0 | 1 | 1 | 1 | 3 | 4 | 8 | — | — | — | — | — | — |
| | 1/0 | 0 | 1 | 1 | 1 | 3 | 4 | 7 | — | — | — | — | — | — |
| | 2/0 | 0 | 0 | 1 | 1 | 2 | 3 | 5 | — | — | — | — | — | — |
| | 3/0 | 0 | 0 | 1 | 1 | 1 | 3 | 5 | — | — | — | — | — | — |
| | 4/0 | 0 | 0 | 1 | 1 | 1 | 2 | 4 | — | — | — | — | — | — |
| | 250 | 0 | 0 | 0 | 1 | 1 | 1 | 3 | — | — | — | — | — | — |
| | 300 | 0 | 0 | 0 | 1 | 1 | 1 | 3 | — | — | — | — | — | — |
| | 350 | 0 | 0 | 0 | 1 | 1 | 1 | 2 | — | — | — | — | — | — |
| | 400 | 0 | 0 | 0 | 0 | 1 | 1 | 1 | — | — | — | — | — | — |
| | 500 | 0 | 0 | 0 | 0 | 1 | 1 | 1 | — | — | — | — | — | — |
| | 600 | 0 | 0 | 0 | 0 | 1 | 1 | 1 | — | — | — | — | — | — |
| | 700 | 0 | 0 | 0 | 0 | 1 | 1 | 1 | — | — | — | — | — | — |
| | 750 | 0 | 0 | 0 | 0 | 0 | 1 | 1 | — | — | — | — | — | — |
| | 900 | 0 | 0 | 0 | 0 | 0 | 1 | 1 | — | — | — | — | — | — |
| | 1000 | 0 | 0 | 0 | 0 | 0 | 1 | 1 | — | — | — | — | — | — |
| THHN, THWN, THWN-2 | 8 | — | — | — | — | — | — | — | — | — | — | — | — | — |
| | 6 | 1 | 2 | 4 | 7 | 13 | 18 | 29 | — | — | — | — | — | — |
| | 4 | 1 | 1 | 3 | 4 | 8 | 11 | 18 | — | — | — | — | — | — |
| | 2 | 1 | 1 | 1 | 3 | 6 | 8 | 13 | — | — | — | — | — | — |
| | 1 | 0 | 1 | 1 | 2 | 4 | 6 | 10 | — | — | — | — | — | — |
| | 1/0 | 0 | 1 | 1 | 1 | 3 | 5 | 8 | — | — | — | — | — | — |
| | 2/0 | 0 | 1 | 1 | 1 | 3 | 4 | 7 | — | — | — | — | — | — |
| | 3/0 | 0 | 0 | 1 | 1 | 2 | 3 | 6 | — | — | — | — | — | — |
| | 4/0 | 0 | 0 | 1 | 1 | 1 | 3 | 5 | — | — | — | — | — | — |
| | 250 | 0 | 0 | 1 | 1 | 1 | 1 | 3 | — | — | — | — | — | — |
| | 300 | 0 | 0 | 0 | 1 | 1 | 1 | 3 | — | — | — | — | — | — |
| | 350 | 0 | 0 | 0 | 1 | 1 | 1 | 3 | — | — | — | — | — | — |
| | 400 | 0 | 0 | 0 | 1 | 1 | 1 | 2 | — | — | — | — | — | — |
| | 500 | 0 | 0 | 0 | 0 | 1 | 1 | 1 | — | — | — | — | — | — |
| | 600 | 0 | 0 | 0 | 0 | 1 | 1 | 1 | — | — | — | — | — | — |
| | 700 | 0 | 0 | 0 | 0 | 1 | 1 | 1 | — | — | — | — | — | — |
| | 750 | 0 | 0 | 0 | 0 | 1 | 1 | 1 | — | — | — | — | — | — |
| | 900 | 0 | 0 | 0 | 0 | 0 | 1 | 1 | — | — | — | — | — | — |
| | 1000 | 0 | 0 | 0 | 0 | 0 | 1 | 1 | — | — | — | — | — | — |

***TABLE C.6(A)*** *Continued*

| Type | Conductor Size (AWG/kcmil) | Trade Size (Metric Designator) | | | | | | | | | | | | |
|---|---|---|---|---|---|---|---|---|---|---|---|---|---|---|
| | | ⅜ (12) | ½ (16) | ¾ (21) | 1 (27) | 1¼ (35) | 1½ (41) | 2 (53) | 2½ (63) | 3 (78) | 3½ (91) | 4 (103) | 5 (129) | 6 (155) |
| XHHW, XHHW-2 | 8 | 1 | 3 | 5 | 8 | 15 | 20 | 34 | — | — | — | — | — | — |
| | 6 | 1 | 2 | 4 | 6 | 11 | 15 | 25 | — | — | — | — | — | — |
| | 4 | 1 | 1 | 3 | 4 | 8 | 11 | 18 | — | — | — | — | — | — |
| | 2 | 1 | 1 | 1 | 3 | 6 | 8 | 13 | — | — | — | — | — | — |
| | 1 | 0 | 1 | 1 | 2 | 4 | 6 | 10 | — | — | — | — | — | — |
| | 1/0 | 0 | 1 | 1 | 1 | 3 | 5 | 8 | — | — | — | — | — | — |
| | 2/0 | 0 | 1 | 1 | 1 | 3 | 4 | 7 | — | — | — | — | — | — |
| | 3/0 | 0 | 0 | 1 | 1 | 2 | 3 | 6 | — | — | — | — | — | — |
| | 4/0 | 0 | 0 | 1 | 1 | 1 | 3 | 5 | — | — | — | — | — | — |
| | 250 | 0 | 0 | 1 | 1 | 1 | 2 | 4 | — | — | — | — | — | — |
| | 300 | 0 | 0 | 0 | 1 | 1 | 1 | 3 | — | — | — | — | — | — |
| | 350 | 0 | 0 | 0 | 1 | 1 | 1 | 3 | — | — | — | — | — | — |
| | 400 | 0 | 0 | 0 | 1 | 1 | 1 | 2 | — | — | — | — | — | — |
| | 500 | 0 | 0 | 0 | 0 | 1 | 1 | 1 | — | — | — | — | — | — |
| | 600 | 0 | 0 | 0 | 0 | 1 | 1 | 1 | — | — | — | — | — | — |
| | 700 | 0 | 0 | 0 | 0 | 1 | 1 | 1 | — | — | — | — | — | — |
| | 750 | 0 | 0 | 0 | 0 | 1 | 1 | 1 | — | — | — | — | — | — |
| | 900 | 0 | 0 | 0 | 0 | 0 | 1 | 1 | — | — | — | — | — | — |
| | 1000 | 0 | 0 | 0 | 0 | 0 | 1 | 1 | — | — | — | — | — | — |

•

Definition: *Compact stranding* is the result of a manufacturing process where the stranded conductor is compressed to the extent that the interstices (voids between strand wires) are virtually eliminated.

***TABLE C.7*** *Maximum Number of Conductors or Fixture Wires in Liquidtight Flexible Metal Conduit (LFMC)* (Based on Chapter 9: Table 1, Table 4, and Table 5)

| Type | Conductor Size (AWG/ kcmil) | Trade Size (Metric Designator) ⅜ (12) | ½ (16) | ¾ (21) | 1 (27) | 1¼ (35) | 1½ (41) | 2 (53) | 2½ (63) | 3 (78) | 3½ (91) | 4 (103) | 5 (129) | 6 (155) |
|---|---|---|---|---|---|---|---|---|---|---|---|---|---|---|
| **CONDUCTORS** | | | | | | | | | | | | | | |
| RHH, RHW, RHW-2 | 14 | 2 | 4 | 7 | 12 | 21 | 27 | 44 | 66 | 102 | 133 | 173 | — | — |
| | 12 | 1 | 3 | 6 | 10 | 17 | 22 | 36 | 55 | 84 | 110 | 144 | — | — |
| | 10 | 1 | 3 | 5 | 8 | 14 | 18 | 29 | 44 | 68 | 89 | 116 | — | — |
| | 8 | 1 | 1 | 2 | 4 | 7 | 9 | 15 | 23 | 36 | 46 | 61 | — | — |
| | 6 | 1 | 1 | 1 | 3 | 6 | 7 | 12 | 18 | 28 | 37 | 48 | — | — |
| | 4 | 0 | 1 | 1 | 2 | 4 | 6 | 9 | 14 | 22 | 29 | 38 | — | — |
| | 3 | 0 | 1 | 1 | 1 | 4 | 5 | 8 | 13 | 19 | 25 | 33 | — | — |
| | 2 | 0 | 1 | 1 | 1 | 3 | 4 | 7 | 11 | 17 | 22 | 29 | — | — |
| | 1 | 0 | 0 | 1 | 1 | 1 | 3 | 5 | 7 | 11 | 14 | 19 | — | — |
| | 1/0 | 0 | 0 | 1 | 1 | 1 | 2 | 4 | 6 | 10 | 13 | 16 | — | — |
| | 2/0 | 0 | 0 | 1 | 1 | 1 | 1 | 3 | 5 | 8 | 11 | 14 | — | — |
| | 3/0 | 0 | 0 | 0 | 1 | 1 | 1 | 3 | 4 | 7 | 9 | 12 | — | — |
| | 4/0 | 0 | 0 | 0 | 1 | 1 | 1 | 2 | 4 | 6 | 8 | 10 | — | — |
| | 250 | 0 | 0 | 0 | 0 | 1 | 1 | 1 | 3 | 4 | 6 | 8 | — | — |
| | 300 | 0 | 0 | 0 | 0 | 1 | 1 | 1 | 2 | 4 | 5 | 7 | — | — |
| | 350 | 0 | 0 | 0 | 0 | 1 | 1 | 1 | 2 | 3 | 5 | 6 | — | — |
| | 400 | 0 | 0 | 0 | 0 | 1 | 1 | 1 | 1 | 3 | 4 | 6 | — | — |
| | 500 | 0 | 0 | 0 | 0 | 1 | 1 | 1 | 1 | 3 | 4 | 5 | — | — |
| | 600 | 0 | 0 | 0 | 0 | 0 | 1 | 1 | 1 | 2 | 3 | 4 | — | — |
| | 700 | 0 | 0 | 0 | 0 | 0 | 0 | 1 | 1 | 1 | 3 | 3 | — | — |
| | 750 | 0 | 0 | 0 | 0 | 0 | 0 | 1 | 1 | 1 | 2 | 3 | — | — |
| | 800 | 0 | 0 | 0 | 0 | 0 | 0 | 1 | 1 | 1 | 2 | 3 | — | — |
| | 900 | 0 | 0 | 0 | 0 | 0 | 0 | 1 | 1 | 1 | 2 | 3 | — | — |
| | 1000 | 0 | 0 | 0 | 0 | 0 | 0 | 1 | 1 | 1 | 1 | 3 | — | — |
| | 1250 | 0 | 0 | 0 | 0 | 0 | 0 | 0 | 1 | 1 | 1 | 1 | — | — |
| | 1500 | 0 | 0 | 0 | 0 | 0 | 0 | 0 | 1 | 1 | 1 | 1 | — | — |
| | 1750 | 0 | 0 | 0 | 0 | 0 | 0 | 0 | 1 | 1 | 1 | 1 | — | — |
| | 2000 | 0 | 0 | 0 | 0 | 0 | 0 | 0 | 0 | 1 | 1 | 1 | — | — |
| TW, THHW, THW, THW-2 | 14 | 5 | 9 | 15 | 25 | 44 | 57 | 93 | 140 | 215 | 280 | 365 | — | — |
| | 12 | 4 | 7 | 12 | 19 | 33 | 43 | 71 | 108 | 165 | 215 | 280 | — | — |
| | 10 | 3 | 5 | 9 | 14 | 25 | 32 | 53 | 80 | 123 | 160 | 209 | — | — |
| | 8 | 1 | 3 | 5 | 8 | 14 | 18 | 29 | 44 | 68 | 89 | 116 | — | — |
| RHH*, RHW*, RHW-2* | 14 | 3 | 6 | 10 | 16 | 29 | 38 | 62 | 93 | 143 | 186 | 243 | — | — |
| | 12 | 3 | 5 | 8 | 13 | 23 | 30 | 50 | 75 | 115 | 149 | 195 | — | — |
| | 10 | 1 | 3 | 6 | 10 | 18 | 23 | 39 | 58 | 89 | 117 | 152 | — | — |
| | 8 | 1 | 1 | 4 | 6 | 11 | 14 | 23 | 35 | 53 | 70 | 91 | — | — |
| • TW, THW, THHW, THW-2, RHH*, RHW*, RHW-2* | 6 | 1 | 1 | 3 | 5 | 8 | 11 | 18 | 27 | 41 | 53 | 70 | — | — |
| | 4 | 1 | 1 | 1 | 3 | 6 | 8 | 13 | 20 | 30 | 40 | 52 | — | — |
| | 3 | 1 | 1 | 1 | 3 | 5 | 7 | 11 | 17 | 26 | 34 | 44 | — | — |
| | 2 | 0 | 1 | 1 | 2 | 4 | 6 | 9 | 14 | 22 | 29 | 38 | — | — |
| | 1 | 0 | 1 | 1 | 1 | 3 | 4 | 7 | 10 | 15 | 20 | 26 | — | — |
| | 1/0 | 0 | 0 | 1 | 1 | 2 | 3 | 6 | 8 | 13 | 17 | 23 | — | — |
| | 2/0 | 0 | 0 | 1 | 1 | 2 | 3 | 5 | 7 | 11 | 15 | 19 | — | — |
| | 3/0 | 0 | 0 | 1 | 1 | 1 | 2 | 4 | 6 | 9 | 12 | 16 | — | — |
| | 4/0 | 0 | 0 | 0 | 1 | 1 | 1 | 3 | 5 | 8 | 10 | 13 | — | — |

***TABLE C.7*** *Continued*

| Type | Conductor Size (AWG/ kcmil) | Trade Size (Metric Designator) | | | | | | | | | | | | |
|---|---|---|---|---|---|---|---|---|---|---|---|---|---|---|
| | | 3⁄8 (12) | 1⁄2 (16) | 3⁄4 (21) | 1 (27) | 1¼ (35) | 1½ (41) | 2 (53) | 2½ (63) | 3 (78) | 3½ (91) | 4 (103) | 5 (129) | 6 (155) |
| | 250 | 0 | 0 | 0 | 1 | 1 | 1 | 3 | 4 | 6 | 8 | 11 | — | — |
| | 300 | 0 | 0 | 0 | 1 | 1 | 1 | 2 | 3 | 5 | 7 | 9 | — | — |
| | 350 | 0 | 0 | 0 | 0 | 1 | 1 | 1 | 3 | 5 | 6 | 8 | — | — |
| | 400 | 0 | 0 | 0 | 0 | 1 | 1 | 1 | 3 | 4 | 6 | 7 | — | — |
| | 500 | 0 | 0 | 0 | 0 | 1 | 1 | 1 | 2 | 3 | 5 | 6 | — | — |
| | 600 | 0 | 0 | 0 | 0 | 1 | 1 | 1 | 1 | 3 | 4 | 5 | — | — |
| | 700 | 0 | 0 | 0 | 0 | 0 | 1 | 1 | 1 | 2 | 3 | 4 | — | — |
| | 750 | 0 | 0 | 0 | 0 | 0 | 1 | 1 | 1 | 2 | 3 | 4 | — | — |
| | 800 | 0 | 0 | 0 | 0 | 0 | 1 | 1 | 1 | 2 | 3 | 4 | — | — |
| | 900 | 0 | 0 | 0 | 0 | 0 | 0 | 1 | 1 | 1 | 3 | 3 | — | — |
| | 1000 | 0 | 0 | 0 | 0 | 0 | 0 | 1 | 1 | 1 | 2 | 3 | — | — |
| | 1250 | 0 | 0 | 0 | 0 | 0 | 0 | 1 | 1 | 1 | 1 | 2 | — | — |
| | 1500 | 0 | 0 | 0 | 0 | 0 | 0 | 0 | 1 | 1 | 1 | 2 | — | — |
| | 1750 | 0 | 0 | 0 | 0 | 0 | 0 | 0 | 1 | 1 | 1 | 1 | — | — |
| | 2000 | 0 | 0 | 0 | 0 | 0 | 0 | 0 | 1 | 1 | 1 | 1 | — | — |
| THHN, THWN, THWN-2 | 14 | 8 | 13 | 22 | 36 | 63 | 81 | 134 | 201 | 308 | 401 | 523 | — | — |
| | 12 | 5 | 9 | 16 | 26 | 46 | 59 | 97 | 146 | 225 | 292 | 381 | — | — |
| | 10 | 3 | 6 | 10 | 16 | 29 | 37 | 61 | 92 | 141 | 184 | 240 | — | — |
| | 8 | 1 | 3 | 6 | 9 | 16 | 21 | 35 | 53 | 81 | 106 | 138 | — | — |
| | 6 | 1 | 2 | 4 | 7 | 12 | 15 | 25 | 38 | 59 | 76 | 100 | — | — |
| | 4 | 1 | 1 | 2 | 4 | 7 | 9 | 15 | 23 | 36 | 47 | 61 | — | — |
| | 3 | 1 | 1 | 1 | 3 | 6 | 8 | 13 | 20 | 30 | 40 | 52 | — | — |
| | 2 | 1 | 1 | 1 | 3 | 5 | 7 | 11 | 17 | 26 | 33 | 44 | — | — |
| | 1 | 0 | 1 | 1 | 1 | 4 | 5 | 8 | 12 | 19 | 25 | 32 | — | — |
| | 1/0 | 0 | 1 | 1 | 1 | 3 | 4 | 7 | 10 | 16 | 21 | 27 | — | — |
| | 2/0 | 0 | 0 | 1 | 1 | 2 | 3 | 6 | 8 | 13 | 17 | 23 | — | — |
| | 3/0 | 0 | 0 | 1 | 1 | 1 | 3 | 5 | 7 | 11 | 14 | 19 | — | — |
| | 4/0 | 0 | 0 | 1 | 1 | 1 | 2 | 4 | 6 | 9 | 12 | 15 | — | — |
| | 250 | 0 | 0 | 0 | 1 | 1 | 1 | 3 | 5 | 7 | 10 | 12 | — | — |
| | 300 | 0 | 0 | 0 | 1 | 1 | 1 | 3 | 4 | 6 | 8 | 11 | — | — |
| | 350 | 0 | 0 | 0 | 1 | 1 | 1 | 2 | 3 | 5 | 7 | 9 | — | — |
| | 400 | 0 | 0 | 0 | 0 | 1 | 1 | 1 | 3 | 5 | 6 | 8 | — | — |
| | 500 | 0 | 0 | 0 | 0 | 1 | 1 | 1 | 2 | 4 | 5 | 7 | — | — |
| | 600 | 0 | 0 | 0 | 0 | 1 | 1 | 1 | 1 | 3 | 4 | 6 | — | — |
| | 700 | 0 | 0 | 0 | 0 | 1 | 1 | 1 | 1 | 3 | 4 | 5 | — | — |
| | 750 | 0 | 0 | 0 | 0 | 0 | 1 | 1 | 1 | 3 | 3 | 5 | — | — |
| | 800 | 0 | 0 | 0 | 0 | 0 | 1 | 1 | 1 | 2 | 3 | 4 | — | — |
| | 900 | 0 | 0 | 0 | 0 | 0 | 1 | 1 | 1 | 2 | 3 | 4 | — | — |
| | 1000 | 0 | 0 | 0 | 0 | 0 | 0 | 1 | 1 | 1 | 3 | 3 | — | — |
| FEP, FEPB, PFA, PFAH, TFE | 14 | 7 | 12 | 21 | 35 | 61 | 79 | 130 | 195 | 299 | 389 | 507 | — | — |
| | 12 | 5 | 9 | 15 | 25 | 44 | 58 | 94 | 142 | 218 | 284 | 370 | — | — |
| | 10 | 4 | 6 | 11 | 18 | 32 | 41 | 68 | 102 | 156 | 203 | 266 | — | — |
| | 8 | 1 | 3 | 6 | 10 | 18 | 23 | 39 | 58 | 89 | 117 | 152 | — | — |
| | 6 | 1 | 2 | 4 | 7 | 13 | 17 | 27 | 41 | 64 | 83 | 108 | — | — |
| | 4 | 1 | 1 | 3 | 5 | 9 | 12 | 19 | 29 | 44 | 58 | 75 | — | — |
| | 3 | 1 | 1 | 2 | 4 | 7 | 10 | 16 | 24 | 37 | 48 | 63 | — | — |
| | 2 | 1 | 1 | 1 | 3 | 6 | 8 | 13 | 20 | 30 | 40 | 52 | — | — |

*(continues)*

***TABLE C.7*** *Continued*

| Type | Conductor Size (AWG/ kcmil) | Trade Size (Metric Designator) | | | | | | | | | | | | |
|---|---|---|---|---|---|---|---|---|---|---|---|---|---|---|
| | | ⅜ (12) | ½ (16) | ¾ (21) | 1 (27) | 1¼ (35) | 1½ (41) | 2 (53) | 2½ (63) | 3 (78) | 3½ (91) | 4 (103) | 5 (129) | 6 (155) |
| PFA, PFAH, TFE | 1 | 0 | 1 | 1 | 2 | 4 | 5 | 9 | 14 | 21 | 28 | 36 | — | — |
| PFA, PFAH, TFE, Z | 1/0 | 0 | 1 | 1 | 1 | 3 | 4 | 7 | 11 | 18 | 23 | 30 | — | — |
| | 2/0 | 0 | 1 | 1 | 1 | 3 | 4 | 6 | 9 | 14 | 19 | 25 | — | — |
| | 3/0 | 0 | 0 | 1 | 1 | 2 | 3 | 5 | 8 | 12 | 16 | 20 | — | — |
| | 4/0 | 0 | 0 | 1 | 1 | 1 | 2 | 4 | 6 | 10 | 13 | 17 | — | — |
| Z | 14 | 9 | 15 | 26 | 42 | 73 | 95 | 156 | 235 | 360 | 469 | 611 | — | — |
| | 12 | 6 | 10 | 18 | 30 | 52 | 67 | 111 | 167 | 255 | 332 | 434 | — | — |
| | 10 | 4 | 6 | 11 | 18 | 32 | 41 | 68 | 102 | 156 | 203 | 266 | — | — |
| | 8 | 2 | 4 | 7 | 11 | 20 | 26 | 43 | 64 | 99 | 129 | 168 | — | — |
| | 6 | 1 | 3 | 5 | 8 | 14 | 18 | 30 | 45 | 69 | 90 | 118 | — | — |
| | 4 | 1 | 1 | 3 | 5 | 9 | 12 | 20 | 31 | 48 | 62 | 81 | — | — |
| | 3 | 1 | 1 | 2 | 4 | 7 | 9 | 15 | 23 | 35 | 45 | 59 | — | — |
| | 2 | 1 | 1 | 1 | 3 | 6 | 7 | 12 | 19 | 29 | 38 | 49 | — | — |
| | 1 | 1 | 1 | 1 | 2 | 5 | 6 | 10 | 15 | 23 | 30 | 40 | — | — |
| XHHW, ZW, XHHW-2, XHH | 14 | 5 | 9 | 15 | 25 | 44 | 57 | 93 | 140 | 215 | 280 | 365 | — | — |
| | 12 | 4 | 7 | 12 | 19 | 33 | 43 | 71 | 108 | 165 | 215 | 280 | — | — |
| | 10 | 3 | 5 | 9 | 14 | 25 | 32 | 53 | 80 | 123 | 160 | 209 | — | — |
| | 8 | 1 | 3 | 5 | 8 | 14 | 18 | 29 | 44 | 68 | 89 | 116 | — | — |
| | 6 | 1 | 1 | 3 | 6 | 10 | 13 | 22 | 33 | 50 | 66 | 86 | — | — |
| | 4 | 1 | 1 | 2 | 4 | 7 | 9 | 16 | 24 | 36 | 48 | 62 | — | — |
| | 3 | 1 | 1 | 1 | 3 | 6 | 8 | 13 | 20 | 31 | 40 | 52 | — | — |
| | 2 | 1 | 1 | 1 | 3 | 5 | 7 | 11 | 17 | 26 | 34 | 44 | — | — |
| XHHW, XHHW-2, XHH | 1 | 0 | 1 | 1 | 1 | 4 | 5 | 8 | 12 | 19 | 25 | 33 | — | — |
| | 1/0 | 0 | 1 | 1 | 1 | 3 | 4 | 7 | 10 | 16 | 21 | 28 | — | — |
| | 2/0 | 0 | 0 | 1 | 1 | 2 | 3 | 6 | 9 | 13 | 17 | 23 | — | — |
| | 3/0 | 0 | 0 | 1 | 1 | 1 | 3 | 5 | 7 | 11 | 14 | 19 | — | — |
| | 4/0 | 0 | 0 | 1 | 1 | 1 | 2 | 4 | 6 | 9 | 12 | 16 | — | — |
| | 250 | 0 | 0 | 0 | 1 | 1 | 1 | 3 | 5 | 7 | 10 | 13 | — | — |
| | 300 | 0 | 0 | 0 | 1 | 1 | 1 | 3 | 4 | 6 | 8 | 11 | — | — |
| | 350 | 0 | 0 | 0 | 1 | 1 | 1 | 2 | 3 | 5 | 7 | 10 | — | — |
| | 400 | 0 | 0 | 0 | 1 | 1 | 1 | 1 | 3 | 5 | 6 | 8 | — | — |
| | 500 | 0 | 0 | 0 | 0 | 1 | 1 | 1 | 2 | 4 | 5 | 7 | — | — |
| | 600 | 0 | 0 | 0 | 0 | 1 | 1 | 1 | 1 | 3 | 4 | 6 | — | — |
| | 700 | 0 | 0 | 0 | 0 | 1 | 1 | 1 | 1 | 3 | 4 | 5 | — | — |
| | 750 | 0 | 0 | 0 | 0 | 0 | 1 | 1 | 1 | 3 | 3 | 5 | — | — |
| | 800 | 0 | 0 | 0 | 0 | 0 | 1 | 1 | 1 | 2 | 3 | 4 | — | — |
| | 900 | 0 | 0 | 0 | 0 | 0 | 1 | 1 | 1 | 2 | 3 | 4 | — | — |
| | 1000 | 0 | 0 | 0 | 0 | 0 | 0 | 1 | 1 | 1 | 3 | 3 | — | — |
| | 1250 | 0 | 0 | 0 | 0 | 0 | 0 | 1 | 1 | 1 | 1 | 3 | — | — |
| | 1500 | 0 | 0 | 0 | 0 | 0 | 0 | 1 | 1 | 1 | 1 | 2 | — | — |
| | 1750 | 0 | 0 | 0 | 0 | 0 | 0 | 0 | 1 | 1 | 1 | 1 | — | — |
| | 2000 | 0 | 0 | 0 | 0 | 0 | 0 | 0 | 1 | 1 | 1 | 1 | — | — |

***TABLE C.7*** *Continued*

| Type | Conductor Size (AWG/kcmil) | Trade Size (Metric Designator) | | | | | | | | | | | | |
|---|---|---|---|---|---|---|---|---|---|---|---|---|---|---|
| | | ⅜ (12) | ½ (16) | ¾ (21) | 1 (27) | 1¼ (35) | 1½ (41) | 2 (53) | 2½ (63) | 3 (78) | 3½ (91) | 4 (103) | 5 (129) | 6 (155) |
| **FIXTURE WIRES** | | | | | | | | | | | | | | |
| RFH-2, FFH-2, RFHH-2 | 18 | 5 | 8 | 15 | 24 | 42 | 54 | 89 | 134 | 206 | 268 | 350 | — | — |
| | 16 | 4 | 7 | 12 | 20 | 35 | 46 | 75 | 113 | 174 | 226 | 295 | — | — |
| SF-2, SFF-2 | 18 | 6 | 11 | 19 | 30 | 53 | 69 | 113 | 169 | 260 | 338 | 441 | — | — |
| | 16 | 5 | 9 | 15 | 25 | 44 | 57 | 93 | 140 | 215 | 280 | 365 | — | — |
| | 14 | 4 | 7 | 12 | 20 | 35 | 46 | 75 | 113 | 174 | 226 | 295 | — | — |
| SF-1, SFF-1 | 18 | 12 | 19 | 33 | 53 | 94 | 122 | 199 | 300 | 460 | 599 | 781 | — | — |
| RFH-1, TF, TFF, XF, XFF | 18 | 8 | 14 | 24 | 39 | 69 | 90 | 147 | 222 | 339 | 442 | 577 | — | — |
| | 16 | 7 | 11 | 20 | 32 | 56 | 72 | 119 | 179 | 274 | 357 | 465 | — | — |
| XF, XFF | 14 | 5 | 9 | 15 | 25 | 44 | 57 | 93 | 140 | 215 | 280 | 365 | — | — |
| TFN, TFFN | 18 | 14 | 23 | 39 | 63 | 111 | 144 | 236 | 355 | 543 | 707 | 923 | — | — |
| | 16 | 10 | 17 | 30 | 48 | 85 | 110 | 180 | 271 | 415 | 540 | 705 | — | — |
| PF, PFF, PGF, PGFF, PAF, PTF, PTFF, PAFF | 18 | 13 | 21 | 37 | 60 | 105 | 136 | 224 | 336 | 515 | 671 | 875 | — | — |
| | 16 | 10 | 16 | 29 | 46 | 81 | 105 | 173 | 260 | 398 | 519 | 677 | — | — |
| | 14 | 7 | 12 | 21 | 35 | 61 | 79 | 130 | 195 | 299 | 389 | 507 | — | — |
| ZF, ZFF, ZHF | 18 | 17 | 28 | 48 | 77 | 136 | 176 | 288 | 434 | 664 | 865 | 1128 | — | — |
| | 16 | 12 | 20 | 35 | 57 | 100 | 130 | 213 | 320 | 490 | 638 | 832 | — | — |
| | 14 | 9 | 15 | 26 | 42 | 73 | 95 | 156 | 235 | 360 | 469 | 611 | — | — |
| KF-2, KFF-2 | 18 | 25 | 42 | 72 | 116 | 203 | 264 | 433 | 651 | 996 | 1297 | 1692 | — | — |
| | 16 | 18 | 29 | 50 | 81 | 142 | 184 | 302 | 454 | 695 | 905 | 1180 | — | — |
| | 14 | 12 | 19 | 34 | 54 | 95 | 124 | 203 | 305 | 467 | 608 | 793 | — | — |
| | 12 | 8 | 13 | 23 | 38 | 66 | 86 | 141 | 212 | 325 | 423 | 552 | — | — |
| | 10 | 5 | 9 | 15 | 25 | 44 | 57 | 93 | 140 | 215 | 280 | 365 | — | — |
| KF-1, KFF-1 | 18 | 29 | 48 | 83 | 134 | 235 | 304 | 499 | 751 | 1150 | 1497 | 1952 | — | — |
| | 16 | 20 | 34 | 58 | 94 | 165 | 214 | 351 | 527 | 808 | 1052 | 1372 | — | — |
| | 14 | 14 | 23 | 39 | 63 | 111 | 144 | 236 | 355 | 543 | 707 | 923 | — | — |
| | 12 | 9 | 15 | 26 | 42 | 73 | 95 | 156 | 235 | 360 | 469 | 611 | — | — |
| | 10 | 6 | 10 | 17 | 27 | 48 | 62 | 102 | 153 | 235 | 306 | 399 | — | — |
| XF, XFF | 12 | 3 | 5 | 8 | 13 | 23 | 30 | 50 | 75 | 115 | 149 | 195 | — | — |
| | 10 | 1 | 3 | 6 | 10 | 18 | 23 | 39 | 58 | 89 | 117 | 152 | — | — |

Notes:

1. This table is for concentric stranded conductors only. For compact stranded conductors, Table C.7(A) should be used.

2. Two-hour fire-rated RHH cable has ceramifiable insulation, which has much larger diameters than other RHH wires. Consult manufacturer's conduit fill tables.

*Types RHH, RHW, and RHW-2 without outer covering.

***TABLE C.7(A)*** *Maximum Number of Conductors or Fixture Wires in Liquidtight Flexible Metal Conduit (LFMC)* (Based on Chapter 9: Table 1, Table 4, and Table 5A)

| Type | Conductor Size (AWG/ kcmil) | Trade Size (Metric Designator) ⅜ (12) | ½ (16) | ¾ (21) | 1 (27) | 1¼ (35) | 1½ (41) | 2 (53) | 2½ (63) | 3 (78) | 3½ (91) | 4 (103) | 5 (129) | 6 (155) |
|---|---|---|---|---|---|---|---|---|---|---|---|---|---|---|
| **COMPACT CONDUCTORS** | | | | | | | | | | | | | | |
| THW, THW-2, THHW | 8 | 1 | 2 | 4 | 7 | 12 | 15 | 25 | 38 | 58 | 76 | 99 | — | — |
| | 6 | 1 | 1 | 3 | 5 | 9 | 12 | 19 | 29 | 45 | 59 | 77 | — | — |
| | 4 | 1 | 1 | 2 | 4 | 7 | 9 | 14 | 22 | 34 | 44 | 57 | — | — |
| | 2 | 1 | 1 | 1 | 3 | 5 | 6 | 11 | 16 | 25 | 32 | 42 | — | — |
| | 1 | 0 | 1 | 1 | 1 | 3 | 4 | 7 | 11 | 17 | 23 | 30 | — | — |
| | 1/0 | 0 | 1 | 1 | 1 | 3 | 4 | 6 | 10 | 15 | 20 | 26 | — | — |
| | 2/0 | 0 | 0 | 1 | 1 | 2 | 3 | 5 | 8 | 13 | 16 | 21 | — | — |
| | 3/0 | 0 | 0 | 1 | 1 | 1 | 3 | 4 | 7 | 11 | 14 | 18 | — | — |
| | 4/0 | 0 | 0 | 1 | 1 | 1 | 2 | 4 | 6 | 9 | 12 | 15 | — | — |
| | 250 | 0 | 0 | 0 | 1 | 1 | 1 | 3 | 4 | 7 | 9 | 12 | — | — |
| | 300 | 0 | 0 | 0 | 1 | 1 | 1 | 2 | 4 | 6 | 8 | 10 | — | — |
| | 350 | 0 | 0 | 0 | 1 | 1 | 1 | 2 | 3 | 5 | 7 | 9 | — | — |
| | 400 | 0 | 0 | 0 | 0 | 1 | 1 | 1 | 3 | 5 | 6 | 8 | — | — |
| | 500 | 0 | 0 | 0 | 0 | 1 | 1 | 1 | 3 | 4 | 5 | 7 | — | — |
| | 600 | 0 | 0 | 0 | 0 | 1 | 1 | 1 | 1 | 3 | 4 | 6 | — | — |
| | 700 | 0 | 0 | 0 | 0 | 1 | 1 | 1 | 1 | 3 | 4 | 5 | — | — |
| | 750 | 0 | 0 | 0 | 0 | 0 | 1 | 1 | 1 | 3 | 3 | 5 | — | — |
| | 900 | 0 | 0 | 0 | 0 | 0 | 1 | 1 | 1 | 2 | 3 | 4 | — | — |
| | 1000 | 0 | 0 | 0 | 0 | 0 | 1 | 1 | 1 | 1 | 3 | 4 | — | — |
| THHN, THWN, THWN-2 | 8 | — | — | — | — | — | — | — | — | — | — | — | — | — |
| | 6 | 1 | 2 | 4 | 7 | 13 | 17 | 28 | 43 | 66 | 86 | 112 | — | — |
| | 4 | 1 | 1 | 3 | 4 | 8 | 11 | 17 | 26 | 41 | 53 | 69 | — | — |
| | 2 | 1 | 1 | 1 | 3 | 6 | 7 | 12 | 19 | 29 | 38 | 50 | — | — |
| | 1 | 0 | 1 | 1 | 2 | 4 | 6 | 9 | 14 | 22 | 28 | 37 | — | — |
| | 1/0 | 0 | 1 | 1 | 1 | 4 | 5 | 8 | 12 | 19 | 24 | 32 | — | — |
| | 2/0 | 0 | 1 | 1 | 1 | 3 | 4 | 6 | 10 | 15 | 20 | 26 | — | — |
| | 3/0 | 0 | 0 | 1 | 1 | 2 | 3 | 5 | 8 | 13 | 17 | 22 | — | — |
| | 4/0 | 0 | 0 | 1 | 1 | 1 | 3 | 4 | 7 | 10 | 14 | 18 | — | — |
| | 250 | 0 | 0 | 1 | 1 | 1 | 1 | 3 | 5 | 8 | 11 | 14 | — | — |
| | 300 | 0 | 0 | 0 | 1 | 1 | 1 | 3 | 4 | 7 | 9 | 12 | — | — |
| | 350 | 0 | 0 | 0 | 1 | 1 | 1 | 2 | 4 | 6 | 8 | 11 | — | — |
| | 400 | 0 | 0 | 0 | 1 | 1 | 1 | 2 | 3 | 5 | 7 | 9 | — | — |
| | 500 | 0 | 0 | 0 | 0 | 1 | 1 | 1 | 3 | 5 | 6 | 8 | — | — |
| | 600 | 0 | 0 | 0 | 0 | 1 | 1 | 1 | 2 | 4 | 5 | 6 | — | — |
| | 700 | 0 | 0 | 0 | 0 | 1 | 1 | 1 | 1 | 3 | 4 | 6 | — | — |
| | 750 | 0 | 0 | 0 | 0 | 1 | 1 | 1 | 1 | 3 | 4 | 5 | — | — |
| | 900 | 0 | 0 | 0 | 0 | 0 | 1 | 1 | 1 | 2 | 3 | 4 | — | — |
| | 1000 | 0 | 0 | 0 | 0 | 0 | 1 | 1 | 1 | 2 | 3 | 4 | — | — |

***TABLE C.7(A)*** *Continued*

| Type | Conductor Size (AWG/ kcmil) | Trade Size (Metric Designator) | | | | | | | | | | | | |
|---|---|---|---|---|---|---|---|---|---|---|---|---|---|---|
| | | 3/8 (12) | 1/2 (16) | 3/4 (21) | 1 (27) | 1¼ (35) | 1½ (41) | 2 (53) | 2½ (63) | 3 (78) | 3½ (91) | 4 (103) | 5 (129) | 6 (155) |
| XHHW, XHHW-2 | 8 | 1 | 3 | 5 | 9 | 15 | 20 | 33 | 49 | 76 | 98 | 129 | — | — |
| | 6 | 1 | 2 | 4 | 6 | 11 | 15 | 24 | 37 | 56 | 73 | 95 | — | — |
| | 4 | 1 | 1 | 3 | 4 | 8 | 11 | 17 | 26 | 41 | 53 | 69 | — | — |
| | 2 | 1 | 1 | 1 | 3 | 6 | 7 | 12 | 19 | 29 | 38 | 50 | — | — |
| | 1 | 0 | 1 | 1 | 2 | 4 | 6 | 9 | 14 | 22 | 28 | 37 | — | — |
| | 1/0 | 0 | 1 | 1 | 1 | 4 | 5 | 8 | 12 | 19 | 24 | 32 | — | — |
| | 2/0 | 0 | 1 | 1 | 1 | 3 | 4 | 7 | 10 | 16 | 20 | 27 | — | — |
| | 3/0 | 0 | 0 | 1 | 1 | 2 | 3 | 5 | 8 | 13 | 17 | 22 | — | — |
| | 4/0 | 0 | 0 | 1 | 1 | 1 | 3 | 4 | 7 | 11 | 14 | 18 | — | — |
| | 250 | 0 | 0 | 1 | 1 | 1 | 1 | 3 | 5 | 8 | 11 | 15 | — | — |
| | 300 | 0 | 0 | 0 | 1 | 1 | 1 | 3 | 5 | 7 | 9 | 12 | — | — |
| | 350 | 0 | 0 | 0 | 1 | 1 | 1 | 3 | 4 | 6 | 8 | 11 | — | — |
| | 400 | 0 | 0 | 0 | 1 | 1 | 1 | 2 | 4 | 6 | 7 | 10 | — | — |
| | 500 | 0 | 0 | 0 | 0 | 1 | 1 | 1 | 3 | 5 | 6 | 8 | — | — |
| | 600 | 0 | 0 | 0 | 0 | 1 | 1 | 1 | 2 | 4 | 5 | 6 | — | — |
| | 700 | 0 | 0 | 0 | 0 | 1 | 1 | 1 | 1 | 3 | 4 | 6 | — | — |
| | 750 | 0 | 0 | 0 | 0 | 1 | 1 | 1 | 1 | 3 | 4 | 5 | — | — |
| | 900 | 0 | 0 | 0 | 0 | 0 | 1 | 1 | 1 | 2 | 3 | 4 | — | — |
| | 1000 | 0 | 0 | 0 | 0 | 0 | 1 | 1 | 1 | 2 | 3 | 4 | — | — |

Definition: *Compact stranding* is the result of a manufacturing process where the stranded conductor is compressed to the extent that the interstices (voids between strand wires) are virtually eliminated.

***TABLE C.8*** *Maximum Number of Conductors or Fixture Wires in Rigid Metal Conduit (RMC)* (Based on Chapter 9: Table 1, Table 4, and Table 5)

| Type | Conductor Size (AWG/kcmil) | Trade Size (Metric Designator) ⅜ (12) | ½ (16) | ¾ (21) | 1 (27) | 1¼ (35) | 1½ (41) | 2 (53) | 2½ (63) | 3 (78) | 3½ (91) | 4 (103) | 5 (129) | 6 (155) |
|---|---|---|---|---|---|---|---|---|---|---|---|---|---|---|
| | | | | | | | CONDUCTORS | | | | | | | |
| RHH, RHW, RHW-2 | 14 | — | 4 | 7 | 12 | 21 | 28 | 46 | 66 | 102 | 136 | 176 | 276 | 398 |
| | 12 | — | 3 | 6 | 10 | 17 | 23 | 38 | 55 | 85 | 113 | 146 | 229 | 330 |
| | 10 | — | 3 | 5 | 8 | 14 | 19 | 31 | 44 | 68 | 91 | 118 | 185 | 267 |
| | 8 | — | 1 | 2 | 4 | 7 | 10 | 16 | 23 | 36 | 48 | 61 | 97 | 139 |
| | 6 | — | 1 | 1 | 3 | 6 | 8 | 13 | 18 | 29 | 38 | 49 | 77 | 112 |
| | 4 | — | 1 | 1 | 2 | 4 | 6 | 10 | 14 | 22 | 30 | 38 | 60 | 87 |
| | 3 | — | 1 | 1 | 2 | 4 | 5 | 9 | 12 | 19 | 26 | 34 | 53 | 76 |
| | 2 | — | 1 | 1 | 1 | 3 | 4 | 7 | 11 | 17 | 23 | 29 | 46 | 66 |
| | 1 | — | 0 | 1 | 1 | 1 | 3 | 5 | 7 | 11 | 15 | 19 | 30 | 44 |
| | 1/0 | — | 0 | 1 | 1 | 1 | 2 | 4 | 6 | 10 | 13 | 17 | 26 | 38 |
| | 2/0 | — | 0 | 1 | 1 | 1 | 2 | 4 | 5 | 8 | 11 | 14 | 23 | 33 |
| | 3/0 | — | 0 | 0 | 1 | 1 | 1 | 3 | 4 | 7 | 10 | 12 | 20 | 28 |
| | 4/0 | — | 0 | 0 | 1 | 1 | 1 | 3 | 4 | 6 | 8 | 11 | 17 | 24 |
| | 250 | — | 0 | 0 | 0 | 1 | 1 | 1 | 3 | 4 | 6 | 8 | 13 | 18 |
| | 300 | — | 0 | 0 | 0 | 1 | 1 | 1 | 2 | 4 | 5 | 7 | 11 | 16 |
| | 350 | — | 0 | 0 | 0 | 1 | 1 | 1 | 2 | 4 | 5 | 6 | 10 | 15 |
| | 400 | — | 0 | 0 | 0 | 1 | 1 | 1 | 1 | 3 | 4 | 6 | 9 | 13 |
| | 500 | — | 0 | 0 | 0 | 1 | 1 | 1 | 1 | 3 | 4 | 5 | 8 | 11 |
| | 600 | — | 0 | 0 | 0 | 0 | 1 | 1 | 1 | 2 | 3 | 4 | 6 | 9 |
| | 700 | — | 0 | 0 | 0 | 0 | 1 | 1 | 1 | 1 | 3 | 3 | 6 | 8 |
| | 750 | — | 0 | 0 | 0 | 0 | 0 | 1 | 1 | 1 | 3 | 3 | 5 | 8 |
| | 800 | — | 0 | 0 | 0 | 0 | 0 | 1 | 1 | 1 | 2 | 3 | 5 | 7 |
| | 900 | — | 0 | 0 | 0 | 0 | 0 | 1 | 1 | 1 | 2 | 3 | 5 | 7 |
| | 1000 | — | 0 | 0 | 0 | 0 | 0 | 1 | 1 | 1 | 1 | 3 | 4 | 6 |
| | 1250 | — | 0 | 0 | 0 | 0 | 0 | 0 | 1 | 1 | 1 | 1 | 3 | 5 |
| | 1500 | — | 0 | 0 | 0 | 0 | 0 | 0 | 1 | 1 | 1 | 1 | 3 | 4 |
| | 1750 | — | 0 | 0 | 0 | 0 | 0 | 0 | 1 | 1 | 1 | 1 | 2 | 4 |
| | 2000 | — | 0 | 0 | 0 | 0 | 0 | 0 | 0 | 1 | 1 | 1 | 2 | 3 |
| TW, THHW, THW, THW-2 | 14 | — | 9 | 15 | 25 | 44 | 59 | 98 | 140 | 215 | 288 | 370 | 581 | 839 |
| | 12 | — | 7 | 12 | 19 | 33 | 45 | 75 | 107 | 165 | 221 | 284 | 446 | 644 |
| | 10 | — | 5 | 9 | 14 | 25 | 34 | 56 | 80 | 123 | 164 | 212 | 332 | 480 |
| | 8 | — | 3 | 5 | 8 | 14 | 19 | 31 | 44 | 68 | 91 | 118 | 185 | 267 |
| RHH*, RHW*, RHW-2* | 14 | — | 6 | 10 | 17 | 29 | 39 | 65 | 93 | 143 | 191 | 246 | 387 | 558 |
| | 12 | — | 5 | 8 | 13 | 23 | 32 | 52 | 75 | 115 | 154 | 198 | 311 | 448 |
| | 10 | — | 3 | 6 | 10 | 18 | 25 | 41 | 58 | 90 | 120 | 154 | 242 | 350 |
| | 8 | — | 1 | 4 | 6 | 11 | 15 | 24 | 35 | 54 | 72 | 92 | 145 | 209 |
| TW, THW, THHW, THW-2, RHH*, RHW*, RHW-2* | 6 | — | 1 | 3 | 5 | 8 | 11 | 18 | 27 | 41 | 55 | 71 | 111 | 160 |
| | 4 | — | 1 | 1 | 3 | 6 | 8 | 14 | 20 | 31 | 41 | 53 | 83 | 120 |
| | 3 | — | 1 | 1 | 3 | 5 | 7 | 12 | 17 | 26 | 35 | 45 | 71 | 103 |
| | 2 | — | 1 | 1 | 2 | 4 | 6 | 10 | 14 | 22 | 30 | 38 | 60 | 87 |
| | 1 | — | 1 | 1 | 1 | 3 | 4 | 7 | 10 | 15 | 21 | 27 | 42 | 61 |
| | 1/0 | — | 0 | 1 | 1 | 2 | 3 | 6 | 8 | 13 | 18 | 23 | 36 | 52 |
| | 2/0 | — | 0 | 1 | 1 | 2 | 3 | 5 | 7 | 11 | 15 | 19 | 31 | 44 |
| | 3/0 | — | 0 | 1 | 1 | 1 | 2 | 4 | 6 | 9 | 13 | 16 | 26 | 37 |
| | 4/0 | — | 0 | 0 | 1 | 1 | 1 | 3 | 5 | 8 | 10 | 14 | 21 | 31 |

***TABLE C.8*** *Continued*

| Type | Conductor Size (AWG/ kcmil) | Trade Size (Metric Designator) | | | | | | | | | | | | |
|---|---|---|---|---|---|---|---|---|---|---|---|---|---|---|
| | | ⅜ (12) | ½ (16) | ¾ (21) | 1 (27) | 1¼ (35) | 1½ (41) | 2 (53) | 2½ (63) | 3 (78) | 3½ (91) | 4 (103) | 5 (129) | 6 (155) |
| | 250 | — | 0 | 0 | 1 | 1 | 1 | 3 | 4 | 6 | 8 | 11 | 17 | 25 |
| | 300 | — | 0 | 0 | 1 | 1 | 1 | 2 | 3 | 5 | 7 | 9 | 15 | 22 |
| | 350 | — | 0 | 0 | 0 | 1 | 1 | 1 | 3 | 5 | 6 | 8 | 13 | 19 |
| | 400 | — | 0 | 0 | 0 | 1 | 1 | 1 | 3 | 4 | 6 | 7 | 12 | 17 |
| | 500 | — | 0 | 0 | 0 | 1 | 1 | 1 | 2 | 3 | 5 | 6 | 10 | 14 |
| | 600 | — | 0 | 0 | 0 | 1 | 1 | 1 | 1 | 3 | 4 | 5 | 8 | 12 |
| | 700 | — | 0 | 0 | 0 | 0 | 1 | 1 | 1 | 2 | 3 | 4 | 7 | 10 |
| | 750 | — | 0 | 0 | 0 | 0 | 1 | 1 | 1 | 2 | 3 | 4 | 7 | 10 |
| | 800 | — | 0 | 0 | 0 | 0 | 1 | 1 | 1 | 2 | 3 | 4 | 6 | 9 |
| | 900 | — | 0 | 0 | 0 | 0 | 1 | 1 | 1 | 1 | 3 | 3 | 6 | 8 |
| | 1000 | — | 0 | 0 | 0 | 0 | 0 | 1 | 1 | 1 | 2 | 3 | 5 | 8 |
| | 1250 | — | 0 | 0 | 0 | 0 | 0 | 1 | 1 | 1 | 1 | 2 | 4 | 6 |
| | 1500 | — | 0 | 0 | 0 | 0 | 0 | 1 | 1 | 1 | 1 | 2 | 3 | 5 |
| | 1750 | — | 0 | 0 | 0 | 0 | 0 | 0 | 1 | 1 | 1 | 1 | 3 | 4 |
| | 2000 | — | 0 | 0 | 0 | 0 | 0 | 0 | 1 | 1 | 1 | 1 | 3 | 4 |
| THHN, THWN, THWN-2 | 14 | — | 13 | 22 | 36 | 63 | 85 | 140 | 200 | 309 | 412 | 531 | 833 | 1202 |
| | 12 | — | 9 | 16 | 26 | 46 | 62 | 102 | 146 | 225 | 301 | 387 | 608 | 877 |
| | 10 | — | 6 | 10 | 17 | 29 | 39 | 64 | 92 | 142 | 189 | 244 | 383 | 552 |
| | 8 | — | 3 | 6 | 9 | 16 | 22 | 37 | 53 | 82 | 109 | 140 | 221 | 318 |
| | 6 | — | 2 | 4 | 7 | 12 | 16 | 27 | 38 | 59 | 79 | 101 | 159 | 230 |
| | 4 | — | 1 | 2 | 4 | 7 | 10 | 16 | 23 | 36 | 48 | 62 | 98 | 141 |
| | 3 | — | 1 | 1 | 3 | 6 | 8 | 14 | 20 | 31 | 41 | 53 | 83 | 120 |
| | 2 | — | 1 | 1 | 3 | 5 | 7 | 11 | 17 | 26 | 34 | 44 | 70 | 100 |
| | 1 | — | 1 | 1 | 1 | 4 | 5 | 8 | 12 | 19 | 25 | 33 | 51 | 74 |
| | 1/0 | — | 1 | 1 | 1 | 3 | 4 | 7 | 10 | 16 | 21 | 27 | 43 | 63 |
| | 2/0 | — | 0 | 1 | 1 | 2 | 3 | 6 | 8 | 13 | 18 | 23 | 36 | 52 |
| | 3/0 | — | 0 | 1 | 1 | 1 | 3 | 5 | 7 | 11 | 15 | 19 | 30 | 43 |
| | 4/0 | — | 0 | 1 | 1 | 1 | 2 | 4 | 6 | 9 | 12 | 16 | 25 | 36 |
| | 250 | — | 0 | 0 | 1 | 1 | 1 | 3 | 5 | 7 | 10 | 13 | 20 | 29 |
| | 300 | — | 0 | 0 | 1 | 1 | 1 | 3 | 4 | 6 | 8 | 11 | 17 | 25 |
| | 350 | — | 0 | 0 | 1 | 1 | 1 | 2 | 3 | 5 | 7 | 10 | 15 | 22 |
| | 400 | — | 0 | 0 | 1 | 1 | 1 | 2 | 3 | 5 | 7 | 8 | 13 | 20 |
| | 500 | | 0 | 0 | 0 | 1 | 1 | 1 | 2 | 4 | 5 | 7 | 11 | 16 |
| | 600 | — | 0 | 0 | 0 | 1 | 1 | 1 | 1 | 3 | 4 | 6 | 9 | 13 |
| | 700 | — | 0 | 0 | 0 | 1 | 1 | 1 | 1 | 3 | 4 | 5 | 8 | 11 |
| | 750 | — | 0 | 0 | 0 | 0 | 1 | 1 | 1 | 3 | 4 | 5 | 7 | 11 |
| | 800 | — | 0 | 0 | 0 | 0 | 1 | 1 | 1 | 2 | 3 | 4 | 7 | 10 |
| | 900 | — | 0 | 0 | 0 | 0 | 1 | 1 | 1 | 2 | 3 | 4 | 6 | 9 |
| | 1000 | — | 0 | 0 | 0 | 0 | 1 | 1 | 1 | 1 | 3 | 4 | 6 | 8 |
| FEP, FEPB, PFA, PFAH, TFE | 14 | — | 12 | 22 | 35 | 61 | 83 | 136 | 194 | 300 | 400 | 515 | 808 | 1166 |
| | 12 | — | 9 | 16 | 26 | 44 | 60 | 99 | 142 | 219 | 292 | 376 | 590 | 851 |
| | 10 | — | 6 | 11 | 18 | 32 | 43 | 71 | 102 | 157 | 209 | 269 | 423 | 610 |
| | 8 | — | 3 | 6 | 10 | 18 | 25 | 41 | 58 | 90 | 120 | 154 | 242 | 350 |
| | 6 | — | 2 | 4 | 7 | 13 | 17 | 29 | 41 | 64 | 85 | 110 | 172 | 249 |
| | 4 | — | 1 | 3 | 5 | 9 | 12 | 20 | 29 | 44 | 59 | 77 | 120 | 174 |
| | 3 | — | 1 | 2 | 4 | 7 | 10 | 17 | 24 | 37 | 50 | 64 | 100 | 145 |
| | 2 | | 1 | 1 | 3 | 6 | 8 | 14 | 20 | 31 | 41 | 53 | 83 | 120 |

*(continues)*

*TABLE C.8 Continued*

| | | Trade Size (Metric Designator) | | | | | | | | | | | | |
|---|---|---|---|---|---|---|---|---|---|---|---|---|---|---|
| **Type** | **Conductor Size (AWG/ kcmil)** | **⅜ (12)** | **½ (16)** | **¾ (21)** | **1 (27)** | **1¼ (35)** | **1½ (41)** | **2 (53)** | **2½ (63)** | **3 (78)** | **3½ (91)** | **4 (103)** | **5 (129)** | **6 (155)** |
| PFA, PFAH, TFE | 1 | — | 1 | 1 | 2 | 4 | 6 | 9 | 14 | 21 | 28 | 37 | 57 | 83 |
| PFA, PFAH, TFE, Z | 1/0 | — | 1 | 1 | 1 | 3 | 5 | 8 | 11 | 18 | 24 | 30 | 48 | 69 |
| | 2/0 | — | 1 | 1 | 1 | 3 | 4 | 6 | 9 | 14 | 19 | 25 | 40 | 57 |
| | 3/0 | — | 0 | 1 | 1 | 2 | 3 | 5 | 8 | 12 | 16 | 21 | 33 | 47 |
| | 4/0 | — | 0 | 1 | 1 | 1 | 2 | 4 | 6 | 10 | 13 | 17 | 27 | 39 |
| Z | 14 | — | 15 | 26 | 42 | 73 | 100 | 164 | 234 | 361 | 482 | 621 | 974 | 1405 |
| | 12 | — | 10 | 18 | 30 | 52 | 71 | 116 | 166 | 256 | 342 | 440 | 691 | 997 |
| | 10 | — | 6 | 11 | 18 | 32 | 43 | 71 | 102 | 157 | 209 | 269 | 423 | 610 |
| | 8 | — | 4 | 7 | 11 | 20 | 27 | 45 | 64 | 99 | 132 | 170 | 267 | 386 |
| | 6 | — | 3 | 5 | 8 | 14 | 19 | 31 | 45 | 69 | 93 | 120 | 188 | 271 |
| | 4 | — | 1 | 3 | 5 | 9 | 13 | 22 | 31 | 48 | 64 | 82 | 129 | 186 |
| | 3 | — | 1 | 2 | 4 | 7 | 9 | 16 | 22 | 35 | 47 | 60 | 94 | 136 |
| | 2 | — | 1 | 1 | 3 | 6 | 8 | 13 | 19 | 29 | 39 | 50 | 78 | 113 |
| | 1 | — | 1 | 1 | 2 | 5 | 6 | 10 | 15 | 23 | 31 | 40 | 63 | 92 |
| XHHW, ZW. XHHW-2, XHH | 14 | — | 9 | 15 | 25 | 44 | 59 | 98 | 140 | 215 | 288 | 370 | 581 | 839 |
| | 12 | — | 7 | 12 | 19 | 33 | 45 | 75 | 107 | 165 | 221 | 284 | 446 | 644 |
| | 10 | — | 5 | 9 | 14 | 25 | 34 | 56 | 80 | 123 | 164 | 212 | 332 | 480 |
| | 8 | — | 3 | 5 | 8 | 14 | 19 | 31 | 44 | 68 | 91 | 118 | 185 | 267 |
| | 6 | — | 1 | 3 | 6 | 10 | 14 | 23 | 33 | 51 | 68 | 87 | 137 | 197 |
| | 4 | — | 1 | 2 | 4 | 7 | 10 | 16 | 24 | 37 | 49 | 63 | 99 | 143 |
| | 3 | — | 1 | 1 | 3 | 6 | 8 | 14 | 20 | 31 | 41 | 53 | 84 | 121 |
| | 2 | — | 1 | 1 | 3 | 5 | 7 | 12 | 17 | 26 | 35 | 45 | 70 | 101 |
| XHHW, XHHW-2, XHH | 1 | — | 1 | 1 | 1 | 4 | 5 | 9 | 12 | 19 | 26 | 33 | 52 | 76 |
| | 1/0 | — | 1 | 1 | 1 | 3 | 4 | 7 | 10 | 16 | 22 | 28 | 44 | 64 |
| | 2/0 | — | 0 | 1 | 1 | 2 | 3 | 6 | 9 | 13 | 18 | 23 | 37 | 53 |
| | 3/0 | — | 0 | 1 | 1 | 1 | 3 | 5 | 7 | 11 | 15 | 19 | 30 | 44 |
| | 4/0 | — | 0 | 1 | 1 | 1 | 2 | 4 | 6 | 9 | 12 | 16 | 25 | 36 |
| | 250 | — | 0 | 0 | 1 | 1 | 1 | 3 | 5 | 7 | 10 | 13 | 20 | 30 |
| | 300 | — | 0 | 0 | 1 | 1 | 1 | 3 | 4 | 6 | 9 | 11 | 18 | 25 |
| | 350 | — | 0 | 0 | 1 | 1 | 1 | 2 | 3 | 6 | 7 | 10 | 15 | 22 |
| | 400 | — | 0 | 0 | 1 | 1 | 1 | 2 | 3 | 5 | 7 | 9 | 14 | 20 |
| | 500 | — | 0 | 0 | 0 | 1 | 1 | 1 | 2 | 4 | 5 | 7 | 11 | 16 |
| | 600 | — | 0 | 0 | 0 | 1 | 1 | 1 | 1 | 3 | 4 | 6 | 9 | 13 |
| | 700 | — | 0 | 0 | 0 | 1 | 1 | 1 | 1 | 3 | 4 | 5 | 8 | 11 |
| | 750 | — | 0 | 0 | 0 | 0 | 1 | 1 | 1 | 3 | 4 | 5 | 7 | 11 |
| | 800 | — | 0 | 0 | 0 | 0 | 1 | 1 | 1 | 2 | 3 | 4 | 7 | 10 |
| | 900 | — | 0 | 0 | 0 | 0 | 1 | 1 | 1 | 2 | 3 | 4 | 6 | 9 |
| | 1000 | — | 0 | 0 | 0 | 0 | 1 | 1 | 1 | 1 | 3 | 4 | 6 | 8 |
| | 1250 | — | 0 | 0 | 0 | 0 | 0 | 1 | 1 | 1 | 2 | 3 | 4 | 6 |
| | 1500 | — | 0 | 0 | 0 | 0 | 0 | 1 | 1 | 1 | 1 | 2 | 4 | 5 |
| | 1750 | — | 0 | 0 | 0 | 0 | 0 | 0 | 1 | 1 | 1 | 1 | 3 | 5 |
| | 2000 | — | 0 | 0 | 0 | 0 | 0 | 0 | 1 | 1 | 1 | 1 | 3 | 4 |

***TABLE C.8** Continued*

| Type | Conductor Size (AWG/ kcmil) | Trade Size (Metric Designator) | | | | | | | | | | | | |
|---|---|---|---|---|---|---|---|---|---|---|---|---|---|---|
| | | 3/8 (12) | 1/2 (16) | 3/4 (21) | 1 (27) | 1 1/4 (35) | 1 1/2 (41) | 2 (53) | 2 1/2 (63) | 3 (78) | 3 1/2 (91) | 4 (103) | 5 (129) | 6 (155) |
| **FIXTURE WIRES** | | | | | | | | | | | | | | |
| RFH-2, FFH-2, RFHH-2 | 18 | — | 8 | 15 | 24 | 42 | 57 | 94 | 134 | 207 | 276 | 355 | 557 | 804 |
| | 16 | — | 7 | 12 | 20 | 35 | 48 | 79 | 113 | 174 | 232 | 299 | 470 | 678 |
| SF-2, SFF-2 | 18 | — | 11 | 19 | 31 | 53 | 72 | 118 | 169 | 261 | 348 | 448 | 703 | 1014 |
| | 16 | — | 9 | 15 | 25 | 44 | 59 | 98 | 140 | 215 | 288 | 370 | 581 | 839 |
| | 14 | — | 7 | 12 | 20 | 35 | 48 | 79 | 113 | 174 | 232 | 299 | 470 | 678 |
| SF-1, SFF-1 | 18 | — | 19 | 33 | 54 | 94 | 127 | 209 | 299 | 461 | 616 | 792 | 1244 | 1794 |
| RFH-1, TF, TFF, XF, XFF | 18 | — | 14 | 25 | 40 | 69 | 94 | 155 | 221 | 341 | 455 | 585 | 918 | 1325 |
| | 16 | — | 11 | 20 | 32 | 56 | 76 | 125 | 178 | 275 | 367 | 472 | 741 | 1070 |
| • XF, XFF | 14 | — | 9 | 15 | 25 | 44 | 59 | 98 | 140 | 215 | 288 | 370 | 581 | 839 |
| TFN, TFFN | 18 | — | 23 | 40 | 64 | 111 | 150 | 248 | 354 | 545 | 728 | 937 | 1470 | 2120 |
| | 16 | — | 17 | 30 | 49 | 84 | 115 | 189 | 270 | 416 | 556 | 715 | 1123 | 1620 |
| PF, PFF, PGF, PGFF, PAF, PTF, PTFF, PAFF | 18 | — | 21 | 38 | 61 | 105 | 143 | 235 | 335 | 517 | 690 | 888 | 1394 | 2011 |
| | 16 | — | 16 | 29 | 47 | 81 | 110 | 181 | 259 | 400 | 534 | 687 | 1078 | 1555 |
| | 14 | — | 12 | 22 | 35 | 61 | 83 | 136 | 194 | 300 | 400 | 515 | 808 | 1166 |
| ZF, ZFF, ZHF | 18 | — | 28 | 49 | 79 | 135 | 184 | 303 | 432 | 666 | 889 | 1145 | 1796 | 2592 |
| | 16 | — | 20 | 36 | 58 | 100 | 136 | 223 | 319 | 491 | 656 | 844 | 1325 | 1912 |
| | 14 | — | 15 | 26 | 42 | 73 | 100 | 164 | 234 | 361 | 482 | 621 | 974 | 1405 |
| KF-2, KFF-2 | 18 | — | 42 | 73 | 118 | 203 | 276 | 454 | 648 | 1000 | 1334 | 1717 | 2695 | 3887 |
| | 16 | — | 29 | 51 | 82 | 142 | 192 | 317 | 452 | 697 | 931 | 1198 | 1880 | 2712 |
| | 14 | — | 19 | 34 | 55 | 95 | 129 | 213 | 304 | 468 | 625 | 805 | 1263 | 1822 |
| | 12 | — | 13 | 24 | 38 | 66 | 90 | 148 | 211 | 326 | 435 | 560 | 878 | 1267 |
| | 10 | — | 9 | 15 | 25 | 44 | 59 | 98 | 140 | 215 | 288 | 370 | 581 | 839 |
| KF-1, KFF-1 | 18 | — | 48 | 84 | 136 | 234 | 318 | 524 | 748 | 1153 | 1540 | 1982 | 3109 | 4486 |
| | 16 | — | 34 | 59 | 96 | 165 | 224 | 368 | 526 | 810 | 1082 | 1392 | 2185 | 3152 |
| | 14 | — | 23 | 40 | 64 | 111 | 150 | 248 | 354 | 545 | 728 | 937 | 1470 | 2120 |
| | 12 | — | 15 | 26 | 42 | 73 | 100 | 164 | 234 | 361 | 482 | 621 | 974 | 1405 |
| | 10 | — | 10 | 17 | 28 | 48 | 65 | 107 | 153 | 236 | 315 | 405 | 636 | 918 |
| XF, XFF | 12 | — | 5 | 8 | 13 | 23 | 32 | 52 | 75 | 115 | 154 | 198 | 311 | 448 |
| | 10 | — | 3 | 6 | 10 | 18 | 25 | 41 | 58 | 90 | 120 | 154 | 242 | 350 |

Notes:

1. This table is for concentric stranded conductors only. For compact stranded conductors, Table C.8(A) should be used.

2. Two-hour fire-rated RHH cable has ceramifiable insulation, which has much larger diameters than other RHH wires. Consult manufacturer's conduit fill tables.

*Types RHH, RHW, and RHW-2 without outer covering.

***TABLE C.8(A)*** *Maximum Number of Conductors or Fixture Wires in Rigid Metal Conduit (RMC)*
(Based on Chapter 9: Table 1, Table 4, and Table 5A)

| Type | Conductor Size (AWG/kcmil) | Trade Size (Metric Designator) ⅜ (12) | ½ (16) | ¾ (21) | 1 (27) | 1¼ (35) | 1½ (41) | 2 (53) | 2½ (63) | 3 (78) | 3½ (91) | 4 (103) | 5 (129) | 6 (155) |
|---|---|---|---|---|---|---|---|---|---|---|---|---|---|---|
| **COMPACT CONDUCTORS** | | | | | | | | | | | | | | |
| THW, THW-2, THHW | 8 | — | 2 | 4 | 7 | 12 | 16 | 26 | 38 | 59 | 78 | 101 | 158 | 228 |
| | 6 | — | 1 | 3 | 5 | 9 | 12 | 20 | 29 | 45 | 60 | 78 | 122 | 176 |
| | 4 | — | 1 | 2 | 4 | 7 | 9 | 15 | 22 | 34 | 45 | 58 | 91 | 132 |
| | 2 | — | 1 | 1 | 3 | 5 | 7 | 11 | 16 | 25 | 33 | 43 | 67 | 97 |
| | 1 | — | 1 | 1 | 1 | 3 | 5 | 8 | 11 | 17 | 23 | 30 | 47 | 68 |
| | 1/0 | — | 1 | 1 | 1 | 3 | 4 | 7 | 10 | 15 | 20 | 26 | 41 | 59 |
| | 2/0 | — | 0 | 1 | 1 | 2 | 3 | 6 | 8 | 13 | 17 | 22 | 34 | 50 |
| | 3/0 | — | 0 | 1 | 1 | 1 | 3 | 5 | 7 | 11 | 14 | 19 | 29 | 42 |
| | 4/0 | — | 0 | 1 | 1 | 1 | 2 | 4 | 6 | 9 | 12 | 15 | 24 | 35 |
| | 250 | — | 0 | 0 | 1 | 1 | 1 | 3 | 4 | 7 | 9 | 12 | 19 | 28 |
| | 300 | — | 0 | 0 | 1 | 1 | 1 | 3 | 4 | 6 | 8 | 11 | 17 | 24 |
| | 350 | — | 0 | 0 | 1 | 1 | 1 | 2 | 3 | 5 | 7 | 9 | 15 | 22 |
| | 400 | — | 0 | 0 | 1 | 1 | 1 | 1 | 3 | 5 | 7 | 8 | 13 | 20 |
| | 500 | — | 0 | 0 | 0 | 1 | 1 | 1 | 3 | 4 | 5 | 7 | 11 | 17 |
| | 600 | — | 0 | 0 | 0 | 1 | 1 | 1 | 1 | 3 | 4 | 6 | 9 | 13 |
| | 700 | — | 0 | 0 | 0 | 1 | 1 | 1 | 1 | 3 | 4 | 5 | 8 | 12 |
| | 750 | — | 0 | 0 | 0 | 0 | 1 | 1 | 1 | 3 | 4 | 5 | 7 | 11 |
| | 900 | — | 0 | 0 | 0 | 0 | 1 | 1 | 1 | 2 | 3 | 4 | 7 | 10 |
| | 1000 | — | 0 | 0 | 0 | 0 | 1 | 1 | 1 | 1 | 3 | 4 | 6 | 9 |
| THHN, THWN, THWN-2 | 8 | — | — | — | — | — | — | — | — | — | — | — | — | — |
| | 6 | — | 2 | 5 | 8 | 13 | 18 | 30 | 43 | 66 | 88 | 114 | 179 | 258 |
| | 4 | — | 1 | 3 | 5 | 8 | 11 | 18 | 26 | 41 | 55 | 70 | 110 | 159 |
| | 2 | — | 1 | 1 | 3 | 6 | 8 | 13 | 19 | 29 | 39 | 50 | 79 | 114 |
| | 1 | — | 1 | 1 | 2 | 4 | 6 | 10 | 14 | 22 | 29 | 38 | 59 | 86 |
| | 1/0 | — | 1 | 1 | 1 | 4 | 5 | 8 | 12 | 19 | 25 | 32 | 51 | 73 |
| | 2/0 | — | 1 | 1 | 1 | 3 | 4 | 7 | 10 | 15 | 21 | 26 | 42 | 60 |
| | 3/0 | — | 0 | 1 | 1 | 2 | 3 | 6 | 8 | 13 | 17 | 22 | 35 | 51 |
| | 4/0 | — | 0 | 1 | 1 | 1 | 3 | 5 | 7 | 10 | 14 | 18 | 29 | 42 |
| | 250 | — | 0 | 1 | 1 | 1 | 2 | 4 | 5 | 8 | 11 | 14 | 23 | 33 |
| | 300 | — | 0 | 0 | 1 | 1 | 1 | 3 | 4 | 7 | 10 | 12 | 20 | 28 |
| | 350 | — | 0 | 0 | 1 | 1 | 1 | 3 | 4 | 6 | 8 | 11 | 17 | 25 |
| | 400 | — | 0 | 0 | 1 | 1 | 1 | 2 | 3 | 5 | 7 | 10 | 15 | 22 |
| | 500 | — | 0 | 0 | 0 | 1 | 1 | 1 | 3 | 5 | 6 | 8 | 13 | 19 |
| | 600 | — | 0 | 0 | 0 | 1 | 1 | 1 | 2 | 4 | 5 | 6 | 10 | 15 |
| | 700 | — | 0 | 0 | 0 | 1 | 1 | 1 | 1 | 3 | 4 | 6 | 9 | 13 |
| | 750 | — | 0 | 0 | 0 | 1 | 1 | 1 | 1 | 3 | 4 | 5 | 9 | 13 |
| | 900 | — | 0 | 0 | 0 | 0 | 1 | 1 | 1 | 2 | 3 | 4 | 7 | 10 |
| | 1000 | — | 0 | 0 | 0 | 0 | 1 | 1 | 1 | 2 | 3 | 4 | 6 | 9 |

***TABLE C.8(A)*** *Continued*

| Type | Conductor Size (AWG/ kcmil) | Trade Size (Metric Designator) | | | | | | | | | | | | |
|---|---|---|---|---|---|---|---|---|---|---|---|---|---|---|
| | | ⅜ (12) | ½ (16) | ¾ (21) | 1 (27) | 1¼ (35) | 1½ (41) | 2 (53) | 2½ (63) | 3 (78) | 3½ (91) | 4 (103) | 5 (129) | 6 (155) |
| XHHW, XHHW-2 | 8 | — | 3 | 5 | 9 | 15 | 21 | 34 | 49 | 76 | 101 | 130 | 205 | 296 |
| | 6 | — | 2 | 4 | 6 | 11 | 15 | 25 | 36 | 56 | 75 | 97 | 152 | 220 |
| | 4 | — | 1 | 3 | 5 | 8 | 11 | 18 | 26 | 41 | 55 | 70 | 110 | 159 |
| | 2 | — | 1 | 1 | 3 | 6 | 8 | 13 | 19 | 29 | 39 | 50 | 79 | 114 |
| | 1 | — | 1 | 1 | 2 | 4 | 6 | 10 | 14 | 22 | 29 | 38 | 59 | 86 |
| | 1/0 | — | 1 | 1 | 1 | 4 | 5 | 8 | 12 | 19 | 25 | 32 | 51 | 73 |
| | 2/0 | — | 1 | 1 | 1 | 3 | 4 | 7 | 10 | 16 | 21 | 27 | 43 | 62 |
| | 3/0 | — | 0 | 1 | 1 | 2 | 3 | 6 | 8 | 13 | 17 | 22 | 35 | 51 |
| | 4/0 | — | 0 | 1 | 1 | 1 | 3 | 5 | 7 | 11 | 14 | 19 | 29 | 42 |
| | 250 | — | 0 | 1 | 1 | 1 | 2 | 4 | 5 | 8 | 11 | 15 | 23 | 34 |
| | 300 | — | 0 | 0 | 1 | 1 | 1 | 3 | 5 | 7 | 10 | 13 | 20 | 29 |
| | 350 | — | 0 | 0 | 1 | 1 | 1 | 3 | 4 | 6 | 9 | 11 | 18 | 25 |
| | 400 | — | 0 | 0 | 1 | 1 | 1 | 2 | 4 | 6 | 8 | 10 | 16 | 23 |
| | 500 | — | 0 | 0 | 0 | 1 | 1 | 1 | 3 | 5 | 6 | 8 | 13 | 19 |
| | 600 | — | 0 | 0 | 0 | 1 | 1 | 1 | 2 | 4 | 5 | 7 | 10 | 15 |
| | 700 | — | 0 | 0 | 0 | 1 | 1 | 1 | 1 | 3 | 4 | 6 | 9 | 13 |
| | 750 | — | 0 | 0 | 0 | 1 | 1 | 1 | 1 | 3 | 4 | 5 | 8 | 12 |
| | 900 | — | 0 | 0 | 0 | 0 | 1 | 1 | 1 | 2 | 3 | 5 | 7 | 11 |
| | 1000 | — | 0 | 0 | 0 | 0 | 1 | 1 | 1 | 2 | 3 | 4 | 7 | 10 |

Definition: *Compact stranding* is the result of a manufacturing process where the stranded conductor is compressed to the extent that the interstices (voids between strand wires) are virtually eliminated.

***TABLE C.9*** *Maximum Number of Conductors or Fixture Wires in Rigid PVC Conduit, Schedule 80*
(Based on Chapter 9: Table 1, Table 4, and Table 5)

| Type | Conductor Size (AWG/ kcmil) | Trade Size (Metric Designator) | | | | | | | | | | | | |
|---|---|---|---|---|---|---|---|---|---|---|---|---|---|---|
| | | ⅜ (12) | ½ (16) | ¾ (21) | 1 (27) | 1¼ (35) | 1½ (41) | 2 (53) | 2½ (63) | 3 (78) | 3½ (91) | 4 (103) | 5 (129) | 6 (155) |
| | | | | | | CONDUCTORS | | | | | | | | |
| RHH, RHW, RHW-2 | 14 | — | 3 | 5 | 9 | 17 | 23 | 39 | 56 | 88 | 118 | 153 | 243 | 349 |
| | 12 | — | 2 | 4 | 7 | 14 | 19 | 32 | 46 | 73 | 98 | 127 | 202 | 290 |
| | 10 | — | 1 | 3 | 6 | 11 | 15 | 26 | 37 | 59 | 79 | 103 | 163 | 234 |
| | 8 | — | 1 | 1 | 3 | 6 | 8 | 13 | 19 | 31 | 41 | 54 | 85 | 122 |
| | 6 | — | 1 | 1 | 2 | 4 | 6 | 11 | 16 | 24 | 33 | 43 | 68 | 98 |
| | 4 | — | 1 | 1 | 1 | 3 | 5 | 8 | 12 | 19 | 26 | 33 | 53 | 77 |
| | 3 | — | 0 | 1 | 1 | 3 | 4 | 7 | 11 | 17 | 23 | 29 | 47 | 67 |
| | 2 | — | 0 | 1 | 1 | 3 | 4 | 6 | 9 | 14 | 20 | 25 | 41 | 58 |
| | 1 | — | 0 | 1 | 1 | 1 | 2 | 4 | 6 | 9 | 13 | 17 | 27 | 38 |
| | 1/0 | — | 0 | 0 | 1 | 1 | 1 | 3 | 5 | 8 | 11 | 15 | 23 | 33 |
| | 2/0 | — | 0 | 0 | 1 | 1 | 1 | 3 | 4 | 7 | 10 | 13 | 20 | 29 |
| | 3/0 | — | 0 | 0 | 1 | 1 | 1 | 3 | 4 | 6 | 8 | 11 | 17 | 25 |
| | 4/0 | — | 0 | 0 | 0 | 1 | 1 | 2 | 3 | 5 | 7 | 9 | 15 | 21 |
| | 250 | — | 0 | 0 | 0 | 1 | 1 | 1 | 2 | 4 | 5 | 7 | 11 | 16 |
| | 300 | — | 0 | 0 | 0 | 1 | 1 | 1 | 2 | 3 | 5 | 6 | 10 | 14 |
| | 350 | — | 0 | 0 | 0 | 1 | 1 | 1 | 1 | 3 | 4 | 5 | 9 | 13 |
| | 400 | — | 0 | 0 | 0 | 0 | 1 | 1 | 1 | 3 | 4 | 5 | 8 | 12 |
| | 500 | — | 0 | 0 | 0 | 0 | 1 | 1 | 1 | 2 | 3 | 4 | 7 | 10 |
| | 600 | — | 0 | 0 | 0 | 0 | 0 | 1 | 1 | 1 | 3 | 3 | 6 | 8 |
| | 700 | — | 0 | 0 | 0 | 0 | 0 | 1 | 1 | 1 | 2 | 3 | 5 | 7 |
| | 750 | — | 0 | 0 | 0 | 0 | 0 | 1 | 1 | 1 | 2 | 3 | 5 | 7 |
| | 800 | — | 0 | 0 | 0 | 0 | 0 | 1 | 1 | 1 | 2 | 3 | 4 | 7 |
| | 900 | — | 0 | 0 | 0 | 0 | 0 | 1 | 1 | 1 | 1 | 2 | 4 | 6 |
| | 1000 | — | 0 | 0 | 0 | 0 | 0 | 1 | 1 | 1 | 1 | 2 | 4 | 5 |
| | 1250 | — | 0 | 0 | 0 | 0 | 0 | 0 | 1 | 1 | 1 | 1 | 3 | 4 |
| | 1500 | — | 0 | 0 | 0 | 0 | 0 | 0 | 1 | 1 | 1 | 1 | 2 | 4 |
| | 1750 | — | 0 | 0 | 0 | 0 | 0 | 0 | 0 | 1 | 1 | 1 | 2 | 3 |
| | 2000 | — | 0 | 0 | 0 | 0 | 0 | 0 | 0 | 1 | 1 | 1 | 1 | 3 |
| TW, THHW, THW, THW-2 | 14 | — | 6 | 11 | 19 | 35 | 49 | 82 | 118 | 185 | 250 | 324 | 514 | 736 |
| | 12 | — | 4 | 9 | 15 | 27 | 38 | 63 | 91 | 142 | 192 | 248 | 394 | 565 |
| | 10 | — | 3 | 6 | 11 | 20 | 28 | 47 | 68 | 106 | 143 | 185 | 294 | 421 |
| | 8 | — | 1 | 3 | 6 | 11 | 15 | 26 | 37 | 59 | 79 | 103 | 163 | 234 |
| RHH*, RHW*, RHW-2* | 14 | — | 4 | 8 | 13 | 23 | 32 | 55 | 79 | 123 | 166 | 215 | 341 | 490 |
| | 12 | — | 3 | 6 | 10 | 19 | 26 | 44 | 63 | 99 | 133 | 173 | 274 | 394 |
| | 10 | — | 2 | 5 | 8 | 15 | 20 | 34 | 49 | 77 | 104 | 135 | 214 | 307 |
| | 8 | — | 1 | 3 | 5 | 9 | 12 | 20 | 29 | 46 | 62 | 81 | 128 | 184 |
| • TW, THW, THHW, THW-2, RHH*, RHW*, RHW-2* | 6 | — | 1 | 1 | 3 | 7 | 9 | 16 | 22 | 35 | 48 | 62 | 98 | 141 |
| | 4 | — | 1 | 1 | 3 | 5 | 7 | 12 | 17 | 26 | 35 | 46 | 73 | 105 |
| | 3 | — | 1 | 1 | 2 | 4 | 6 | 10 | 14 | 22 | 30 | 39 | 63 | 90 |
| | 2 | — | 1 | 1 | 1 | 3 | 5 | 8 | 12 | 19 | 26 | 33 | 53 | 77 |
| | 1 | — | 0 | 1 | 1 | 2 | 3 | 6 | 8 | 13 | 18 | 23 | 37 | 54 |
| | 1/0 | — | 0 | 1 | 1 | 1 | 3 | 5 | 7 | 11 | 15 | 20 | 32 | 46 |
| | 2/0 | — | 0 | 1 | 1 | 1 | 2 | 4 | 6 | 10 | 13 | 17 | 27 | 39 |
| | 3/0 | — | 0 | 0 | 1 | 1 | 1 | 3 | 5 | 8 | 11 | 14 | 23 | 33 |
| | 4/0 | — | 0 | 0 | 1 | 1 | 1 | 3 | 4 | 7 | 9 | 12 | 19 | 27 |

***TABLE C.9*** *Continued*

| Type | Conductor Size (AWG/ kcmil) | Trade Size (Metric Designator) | | | | | | | | | | | | |
|---|---|---|---|---|---|---|---|---|---|---|---|---|---|---|
| | | ⅜ (12) | ½ (16) | ¾ (21) | 1 (27) | 1¼ (35) | 1½ (41) | 2 (53) | 2½ (63) | 3 (78) | 3½ (91) | 4 (103) | 5 (129) | 6 (155) |
| | 250 | — | 0 | 0 | 0 | 1 | 1 | 2 | 3 | 5 | 7 | 9 | 15 | 22 |
| | 300 | — | 0 | 0 | 0 | 1 | 1 | 1 | 3 | 5 | 6 | 8 | 13 | 19 |
| | 350 | — | 0 | 0 | 0 | 1 | 1 | 1 | 2 | 4 | 6 | 7 | 12 | 17 |
| | 400 | — | 0 | 0 | 0 | 1 | 1 | 1 | 2 | 4 | 5 | 7 | 10 | 15 |
| | 500 | — | 0 | 0 | 0 | 1 | 1 | 1 | 1 | 3 | 4 | 5 | 9 | 13 |
| | 600 | — | 0 | 0 | 0 | 0 | 1 | 1 | 1 | 2 | 3 | 4 | 7 | 10 |
| | 700 | — | 0 | 0 | 0 | 0 | 1 | 1 | 1 | 2 | 3 | 4 | 6 | 9 |
| | 750 | — | 0 | 0 | 0 | 0 | 0 | 1 | 1 | 1 | 3 | 4 | 6 | 8 |
| | 800 | — | 0 | 0 | 0 | 0 | 0 | 1 | 1 | 1 | 3 | 3 | 6 | 8 |
| | 900 | — | 0 | 0 | 0 | 0 | 0 | 1 | 1 | 1 | 2 | 3 | 5 | 7 |
| | 1000 | — | 0 | 0 | 0 | 0 | 0 | 1 | 1 | 1 | 2 | 3 | 5 | 7 |
| | 1250 | — | 0 | 0 | 0 | 0 | 0 | 1 | 1 | 1 | 1 | 2 | 4 | 5 |
| | 1500 | — | 0 | 0 | 0 | 0 | 0 | 0 | 1 | 1 | 1 | 1 | 3 | 4 |
| | 1750 | — | 0 | 0 | 0 | 0 | 0 | 0 | 1 | 1 | 1 | 1 | 3 | 4 |
| | 2000 | — | 0 | 0 | 0 | 0 | 0 | 0 | 0 | 1 | 1 | 1 | 2 | 3 |
| THHN, THWN, THWN-2 | 14 | — | 9 | 17 | 28 | 51 | 70 | 118 | 170 | 265 | 358 | 464 | 736 | 1055 |
| | 12 | — | 6 | 12 | 20 | 37 | 51 | 86 | 124 | 193 | 261 | 338 | 537 | 770 |
| | 10 | — | 4 | 7 | 13 | 23 | 32 | 54 | 78 | 122 | 164 | 213 | 338 | 485 |
| | 8 | — | 2 | 4 | 7 | 13 | 18 | 31 | 45 | 70 | 95 | 123 | 195 | 279 |
| | 6 | — | 1 | 3 | 5 | 9 | 13 | 22 | 32 | 51 | 68 | 89 | 141 | 202 |
| | 4 | — | 1 | 1 | 3 | 6 | 8 | 14 | 20 | 31 | 42 | 54 | 86 | 124 |
| | 3 | — | 1 | 1 | 3 | 5 | 7 | 12 | 17 | 26 | 35 | 46 | 73 | 105 |
| | 2 | — | 1 | 1 | 2 | 4 | 6 | 10 | 14 | 22 | 30 | 39 | 61 | 88 |
| | 1 | — | 0 | 1 | 1 | 3 | 4 | 7 | 10 | 16 | 22 | 29 | 45 | 65 |
| | 1/0 | — | 0 | 1 | 1 | 2 | 3 | 6 | 9 | 14 | 18 | 24 | 38 | 55 |
| | 2/0 | — | 0 | 1 | 1 | 1 | 3 | 5 | 7 | 11 | 15 | 20 | 32 | 46 |
| | 3/0 | — | 0 | 1 | 1 | 1 | 2 | 4 | 6 | 9 | 13 | 17 | 26 | 38 |
| | 4/0 | — | 0 | 0 | 1 | 1 | 1 | 3 | 5 | 8 | 10 | 14 | 22 | 31 |
| | 250 | — | 0 | 0 | 1 | 1 | 1 | 3 | 4 | 6 | 8 | 11 | 18 | 25 |
| | 300 | — | 0 | 0 | 0 | 1 | 1 | 2 | 3 | 5 | 7 | 9 | 15 | 22 |
| | 350 | — | 0 | 0 | 0 | 1 | 1 | 1 | 3 | 5 | 6 | 8 | 13 | 19 |
| | 400 | — | 0 | 0 | 0 | 1 | 1 | 1 | 3 | 4 | 6 | 7 | 12 | 17 |
| | 500 | — | 0 | 0 | 0 | 1 | 1 | 1 | 2 | 3 | 5 | 6 | 10 | 14 |
| | 600 | — | 0 | 0 | 0 | 0 | 1 | 1 | 1 | 3 | 4 | 5 | 8 | 12 |
| | 700 | — | 0 | 0 | 0 | 0 | 1 | 1 | 1 | 2 | 3 | 4 | 7 | 10 |
| | 750 | — | 0 | 0 | 0 | 0 | 1 | 1 | 1 | 2 | 3 | 4 | 7 | 9 |
| | 800 | — | 0 | 0 | 0 | 0 | 1 | 1 | 1 | 2 | 3 | 4 | 6 | 9 |
| | 900 | — | 0 | 0 | 0 | 0 | 0 | 1 | 1 | 1 | 3 | 3 | 6 | 8 |
| | 1000 | — | 0 | 0 | 0 | 0 | 0 | 1 | 1 | 1 | 2 | 3 | 5 | 7 |
| FEP, FEPB, PFA, PFAH, TFE | 14 | — | 8 | 16 | 27 | 49 | 68 | 115 | 164 | 257 | 347 | 450 | 714 | 1024 |
| | 12 | — | 6 | 12 | 20 | 36 | 50 | 84 | 120 | 188 | 253 | 328 | 521 | 747 |
| | 10 | — | 4 | 8 | 14 | 26 | 36 | 60 | 86 | 135 | 182 | 235 | 374 | 536 |
| | 8 | — | 2 | 5 | 8 | 15 | 20 | 34 | 49 | 77 | 104 | 135 | 214 | 307 |
| | 6 | — | 1 | 3 | 6 | 10 | 14 | 24 | 35 | 55 | 74 | 96 | 152 | 218 |
| | 4 | — | 1 | 2 | 4 | 7 | 10 | 17 | 24 | 38 | 52 | 67 | 106 | 153 |
| | 3 | — | 1 | 1 | 3 | 6 | 8 | 14 | 20 | 32 | 43 | 56 | 89 | 127 |
| | 2 | — | 1 | 1 | 3 | 5 | 7 | 12 | 17 | 26 | 35 | 46 | 73 | 105 |

*(continues)*

***TABLE C.9*** *Continued*

| Type | Conductor Size (AWG/ kcmil) | Trade Size (Metric Designator) | | | | | | | | | | | | |
|---|---|---|---|---|---|---|---|---|---|---|---|---|---|---|
| | | ⅜ (12) | ½ (16) | ¾ (21) | 1 (27) | 1¼ (35) | 1½ (41) | 2 (53) | 2½ (63) | 3 (78) | 3½ (91) | 4 (103) | 5 (129) | 6 (155) |
| PFA, PFAH, TFE | 1 | — | 1 | 1 | 1 | 3 | 5 | 8 | 11 | 18 | 25 | 32 | 51 | 73 |
| PFA, PFAH, TFE, Z | 1/0 | — | 0 | 1 | 1 | 3 | 4 | 7 | 10 | 15 | 20 | 27 | 42 | 61 |
| | 2/0 | — | 0 | 1 | 1 | 2 | 3 | 5 | 8 | 12 | 17 | 22 | 35 | 50 |
| | 3/0 | — | 0 | 1 | 1 | 1 | 2 | 4 | 6 | 10 | 14 | 18 | 29 | 41 |
| | 4/0 | — | 0 | 0 | 1 | 1 | 1 | 4 | 5 | 8 | 11 | 15 | 24 | 34 |
| Z | 14 | — | 10 | 19 | 33 | 59 | 82 | 138 | 198 | 310 | 418 | 542 | 860 | 1233 |
| | 12 | — | 7 | 14 | 23 | 42 | 58 | 98 | 141 | 220 | 297 | 385 | 610 | 875 |
| | 10 | — | 4 | 8 | 14 | 26 | 36 | 60 | 86 | 135 | 182 | 235 | 374 | 536 |
| | 8 | — | 3 | 5 | 9 | 16 | 22 | 38 | 54 | 85 | 115 | 149 | 236 | 339 |
| | 6 | — | 1 | 4 | 6 | 11 | 16 | 26 | 38 | 60 | 81 | 104 | 166 | 238 |
| | 4 | — | 1 | 2 | 4 | 8 | 11 | 18 | 26 | 41 | 55 | 72 | 114 | 164 |
| | 3 | — | 1 | 1 | 3 | 5 | 8 | 13 | 19 | 30 | 40 | 52 | 83 | 119 |
| | 2 | — | 1 | 1 | 2 | 5 | 6 | 11 | 16 | 25 | 33 | 43 | 69 | 99 |
| | 1 | — | 1 | 1 | 1 | 4 | 5 | 9 | 13 | 20 | 27 | 35 | 56 | 80 |
| XHHW, ZW, XHHW-2, XHH | 14 | — | 6 | 11 | 19 | 35 | 49 | 82 | 118 | 185 | 250 | 324 | 514 | 736 |
| | 12 | — | 4 | 9 | 15 | 27 | 38 | 63 | 91 | 142 | 192 | 248 | 394 | 565 |
| | 10 | — | 3 | 6 | 11 | 20 | 28 | 47 | 68 | 106 | 143 | 185 | 294 | 421 |
| | 8 | — | 1 | 3 | 6 | 11 | 15 | 26 | 37 | 59 | 79 | 103 | 163 | 234 |
| | 6 | — | 1 | 2 | 4 | 8 | 11 | 19 | 28 | 43 | 59 | 76 | 121 | 173 |
| | 4 | — | 1 | 1 | 3 | 6 | 8 | 14 | 20 | 31 | 42 | 55 | 87 | 125 |
| | 3 | — | 1 | 1 | 3 | 5 | 7 | 12 | 17 | 26 | 36 | 47 | 74 | 106 |
| | 2 | — | 1 | 1 | 2 | 4 | 6 | 10 | 14 | 22 | 30 | 39 | 62 | 89 |
| XHHW, XHHW-2, XHH | 1 | — | 0 | 1 | 1 | 3 | 4 | 7 | 10 | 16 | 22 | 29 | 46 | 66 |
| | 1/0 | — | 0 | 1 | 1 | 2 | 3 | 6 | 9 | 14 | 19 | 24 | 39 | 56 |
| | 2/0 | — | 0 | 1 | 1 | 1 | 3 | 5 | 7 | 11 | 16 | 20 | 32 | 46 |
| | 3/0 | — | 0 | 1 | 1 | 1 | 2 | 4 | 6 | 9 | 13 | 17 | 27 | 38 |
| | 4/0 | — | 0 | 0 | 1 | 1 | 1 | 3 | 5 | 8 | 11 | 14 | 22 | 32 |
| | 250 | — | 0 | 0 | 1 | 1 | 1 | 3 | 4 | 6 | 9 | 11 | 18 | 26 |
| | 300 | — | 0 | 0 | 1 | 1 | 1 | 2 | 3 | 5 | 7 | 10 | 15 | 22 |
| | 350 | — | 0 | 0 | 0 | 1 | 1 | 1 | 3 | 5 | 6 | 8 | 14 | 20 |
| | 400 | — | 0 | 0 | 0 | 1 | 1 | 1 | 3 | 4 | 6 | 7 | 12 | 17 |
| | 500 | — | 0 | 0 | 0 | 1 | 1 | 1 | 2 | 3 | 5 | 6 | 10 | 14 |
| | 600 | — | 0 | 0 | 0 | 0 | 1 | 1 | 1 | 3 | 4 | 5 | 8 | 11 |
| | 700 | — | 0 | 0 | 0 | 0 | 1 | 1 | 1 | 2 | 3 | 4 | 7 | 10 |
| | 750 | — | 0 | 0 | 0 | 0 | 1 | 1 | 1 | 2 | 3 | 4 | 6 | 9 |
| | 800 | — | 0 | 0 | 0 | 0 | 1 | 1 | 1 | 1 | 3 | 4 | 6 | 9 |
| | 900 | — | 0 | 0 | 0 | 0 | 0 | 1 | 1 | 1 | 3 | 3 | 5 | 8 |
| | 1000 | — | 0 | 0 | 0 | 0 | 0 | 1 | 1 | 1 | 2 | 3 | 5 | 7 |
| | 1250 | — | 0 | 0 | 0 | 0 | 0 | 1 | 1 | 1 | 1 | 2 | 4 | 6 |
| | 1500 | — | 0 | 0 | 0 | 0 | 0 | 0 | 1 | 1 | 1 | 1 | 3 | 5 |
| | 1750 | — | 0 | 0 | 0 | 0 | 0 | 0 | 1 | 1 | 1 | 1 | 3 | 4 |
| | 2000 | — | 0 | 0 | 0 | 0 | 0 | 0 | 1 | 1 | 1 | 1 | 2 | 4 |

***TABLE C.9*** *Continued*

| Type | Conductor Size (AWG/ kcmil) | Trade Size (Metric Designator) | | | | | | | | | | | | |
|---|---|---|---|---|---|---|---|---|---|---|---|---|---|---|
| | | ⅜ (12) | ½ (16) | ¾ (21) | 1 (27) | 1¼ (35) | 1½ (41) | 2 (53) | 2½ (63) | 3 (78) | 3½ (91) | 4 (103) | 5 (129) | 6 (155) |
| **FIXTURE WIRES** | | | | | | | | | | | | | | |
| RFH-2, FFH-2, RFHH-2 | 18 | — | 6 | 11 | 19 | 34 | 47 | 79 | 113 | 177 | 239 | 310 | 492 | 706 |
| | 16 | — | 5 | 9 | 16 | 28 | 39 | 67 | 95 | 150 | 202 | 262 | 415 | 595 |
| SF-2, SFF-2 | 18 | — | 7 | 14 | 24 | 43 | 59 | 100 | 143 | 224 | 302 | 391 | 621 | 890 |
| | 16 | — | 6 | 11 | 19 | 35 | 49 | 82 | 118 | 185 | 250 | 324 | 514 | 736 |
| | 14 | — | 5 | 9 | 16 | 28 | 39 | 67 | 95 | 150 | 202 | 262 | 415 | 595 |
| SF-1, SFF-1 | 18 | — | 13 | 25 | 42 | 76 | 105 | 177 | 253 | 396 | 534 | 692 | 1098 | 1575 |
| RFH-1, TF, TFF, XF, XFF | 18 | — | 10 | 18 | 31 | 56 | 77 | 130 | 187 | 293 | 395 | 511 | 811 | 1163 |
| | 16 | — | 8 | 15 | 25 | 45 | 62 | 105 | 151 | 236 | 319 | 413 | 655 | 939 |
| • XF, XFF | 14 | — | 6 | 11 | 19 | 35 | 49 | 82 | 118 | 185 | 250 | 324 | 514 | 736 |
| TFN, TFFN | 18 | — | 15 | 29 | 50 | 90 | 124 | 209 | 299 | 468 | 632 | 818 | 1298 | 1861 |
| | 16 | — | 12 | 22 | 38 | 68 | 95 | 159 | 229 | 358 | 482 | 625 | 992 | 1422 |
| PF, PFF, PGF, PGFF, PAF, PTF, PTFF, PAFF | 18 | — | 15 | 28 | 47 | 85 | 118 | 198 | 284 | 444 | 599 | 776 | 1231 | 1765 |
| | 16 | — | 11 | 22 | 36 | 66 | 91 | 153 | 219 | 343 | 463 | 600 | 952 | 1365 |
| | 14 | — | 8 | 16 | 27 | 49 | 68 | 115 | 164 | 257 | 347 | 450 | 714 | 1024 |
| ZF, ZFF, ZHF | 18 | — | 19 | 36 | 61 | 110 | 152 | 255 | 366 | 572 | 772 | 1000 | 1587 | 2275 |
| | 16 | — | 14 | 27 | 45 | 81 | 112 | 188 | 270 | 422 | 569 | 738 | 1171 | 1678 |
| | 14 | — | 10 | 19 | 33 | 59 | 82 | 138 | 198 | 310 | 418 | 542 | 860 | 1233 |
| KF-2, KFF-2 | 18 | — | 29 | 54 | 91 | 165 | 228 | 383 | 549 | 859 | 1158 | 1501 | 2380 | 3413 |
| | 16 | — | 20 | 38 | 64 | 115 | 159 | 267 | 383 | 599 | 808 | 1047 | 1661 | 2381 |
| | 14 | — | 13 | 25 | 43 | 77 | 107 | 179 | 257 | 402 | 543 | 703 | 1116 | 1600 |
| | 12 | — | 9 | 17 | 30 | 53 | 74 | 125 | 179 | 280 | 377 | 489 | 776 | 1113 |
| | 10 | — | 6 | 11 | 19 | 35 | 49 | 82 | 118 | 185 | 250 | 324 | 514 | 736 |
| KF-1, KFF-1 | 18 | — | 33 | 63 | 106 | 190 | 263 | 442 | 633 | 991 | 1336 | 1732 | 2747 | 3938 |
| | 16 | — | 23 | 44 | 74 | 133 | 185 | 310 | 445 | 696 | 939 | 1217 | 1930 | 2767 |
| | 14 | — | 15 | 29 | 50 | 90 | 124 | 209 | 299 | 468 | 632 | 818 | 1298 | 1861 |
| | 12 | — | 10 | 19 | 33 | 59 | 82 | 138 | 198 | 310 | 418 | 542 | 860 | 1233 |
| | 10 | — | 7 | 13 | 21 | 39 | 54 | 90 | 129 | 203 | 273 | 354 | 562 | 806 |
| XF, XFF | 12 | — | 3 | 6 | 10 | 19 | 26 | 44 | 63 | 99 | 133 | 173 | 274 | 394 |
| | 10 | — | 2 | 5 | 8 | 15 | 20 | 34 | 49 | 77 | 104 | 135 | 214 | 307 |

Notes:

1. This table is for concentric stranded conductors only. For compact stranded conductors, Table C.9(A) should be used.

2. Two-hour fire-rated RHH cable has ceramifiable insulation, which has much larger diameters than other RHH wires. Consult manufacturer's conduit fill tables.

*Types RHH, RHW, and RHW-2 without outer covering.

***TABLE C.9(A)*** *Maximum Number of Conductors or Fixture Wires in Rigid PVC Conduit, Schedule 80*
(Based on Chapter 9: Table 1, Table 4, and Table 5A)

| Type | Conductor Size (AWG/kcmil) | Trade Size (Metric Designator) | | | | | | | | | | | | |
|---|---|---|---|---|---|---|---|---|---|---|---|---|---|---|
| | | ⅜ (12) | ½ (16) | ¾ (21) | 1 (27) | 1¼ (35) | 1½ (41) | 2 (53) | 2½ (63) | 3 (78) | 3½ (91) | 4 (103) | 5 (129) | 6 (155) |
| **COMPACT CONDUCTORS** | | | | | | | | | | | | | | |
| THW, THW-2, THHW | 8 | — | 1 | 3 | 5 | 9 | 13 | 22 | 32 | 50 | 68 | 88 | 140 | 200 |
| | 6 | — | 1 | 2 | 4 | 7 | 10 | 17 | 25 | 39 | 52 | 68 | 108 | 155 |
| | 4 | — | 1 | 1 | 3 | 5 | 7 | 13 | 18 | 29 | 39 | 51 | 81 | 116 |
| | 2 | — | 1 | 1 | 1 | 4 | 5 | 9 | 13 | 21 | 29 | 37 | 60 | 85 |
| | 1 | — | 0 | 1 | 1 | 3 | 4 | 6 | 9 | 15 | 20 | 26 | 42 | 60 |
| | 1/0 | — | 0 | 1 | 1 | 2 | 3 | 6 | 8 | 13 | 17 | 23 | 36 | 52 |
| | 2/0 | — | 0 | 1 | 1 | 1 | 3 | 5 | 7 | 11 | 15 | 19 | 30 | 44 |
| | 3/0 | — | 0 | 0 | 1 | 1 | 2 | 4 | 6 | 9 | 12 | 16 | 26 | 37 |
| | 4/0 | — | 0 | 0 | 1 | 1 | 1 | 3 | 5 | 8 | 10 | 13 | 22 | 31 |
| | 250 | — | 0 | 0 | 1 | 1 | 1 | 2 | 4 | 6 | 8 | 11 | 17 | 25 |
| | 300 | — | 0 | 0 | 0 | 1 | 1 | 2 | 3 | 5 | 7 | 9 | 15 | 21 |
| | 350 | — | 0 | 0 | 0 | 1 | 1 | 1 | 3 | 5 | 6 | 8 | 13 | 19 |
| | 400 | — | 0 | 0 | 0 | 1 | 1 | 1 | 3 | 4 | 6 | 7 | 12 | 17 |
| | 500 | — | 0 | 0 | 0 | 1 | 1 | 1 | 2 | 3 | 5 | 6 | 10 | 14 |
| | 600 | — | 0 | 0 | 0 | 0 | 1 | 1 | 1 | 3 | 4 | 5 | 8 | 12 |
| | 700 | — | 0 | 0 | 0 | 0 | 1 | 1 | 1 | 2 | 3 | 4 | 7 | 10 |
| | 750 | — | 0 | 0 | 0 | 0 | 1 | 1 | 1 | 2 | 3 | 4 | 7 | 10 |
| | 900 | — | 0 | 0 | 0 | 0 | 0 | 1 | 1 | 1 | 3 | 4 | 6 | 8 |
| | 1000 | — | 0 | 0 | 0 | 0 | 0 | 1 | 1 | 1 | 2 | 3 | 5 | 8 |
| THHN, THWN, THWN-2 | 8 | — | — | — | — | — | — | — | — | — | — | — | — | — |
| | 6 | — | 1 | 3 | 6 | 11 | 15 | 25 | 36 | 57 | 77 | 99 | 158 | 226 |
| | 4 | — | 1 | 1 | 3 | 6 | 9 | 15 | 22 | 35 | 47 | 61 | 98 | 140 |
| | 2 | — | 1 | 1 | 2 | 5 | 6 | 11 | 16 | 25 | 34 | 44 | 70 | 100 |
| | 1 | — | 1 | 1 | 1 | 3 | 5 | 8 | 12 | 19 | 25 | 33 | 53 | 75 |
| | 1/0 | — | 0 | 1 | 1 | 3 | 4 | 7 | 10 | 16 | 22 | 28 | 45 | 64 |
| | 2/0 | — | 0 | 1 | 1 | 2 | 3 | 6 | 8 | 13 | 18 | 23 | 37 | 53 |
| | 3/0 | — | 0 | 1 | 1 | 1 | 3 | 5 | 7 | 11 | 15 | 19 | 31 | 44 |
| | 4/0 | — | 0 | 0 | 1 | 1 | 2 | 4 | 6 | 9 | 12 | 16 | 25 | 37 |
| | 250 | — | 0 | 0 | 1 | 1 | 1 | 3 | 4 | 7 | 10 | 12 | 20 | 29 |
| | 300 | — | 0 | 0 | 1 | 1 | 1 | 3 | 4 | 6 | 8 | 11 | 17 | 25 |
| | 350 | — | 0 | 0 | 0 | 1 | 1 | 2 | 3 | 5 | 7 | 9 | 15 | 22 |
| | 400 | — | 0 | 0 | 0 | 1 | 1 | 1 | 3 | 5 | 6 | 8 | 13 | 19 |
| | 500 | — | 0 | 0 | 0 | 1 | 1 | 1 | 2 | 4 | 5 | 7 | 11 | 16 |
| | 600 | — | 0 | 0 | 0 | 1 | 1 | 1 | 1 | 3 | 4 | 6 | 9 | 13 |
| | 700 | — | 0 | 0 | 0 | 0 | 1 | 1 | 1 | 3 | 4 | 5 | 8 | 12 |
| | 750 | — | 0 | 0 | 0 | 0 | 1 | 1 | 1 | 3 | 4 | 5 | 8 | 11 |
| | 900 | — | 0 | 0 | 0 | 0 | 1 | 1 | 1 | 1 | 3 | 4 | 6 | 9 |
| | 1000 | — | 0 | 0 | 0 | 0 | 0 | 1 | 1 | 1 | 3 | 3 | 5 | 8 |

***TABLE C.9(A)*** *Continued*

| Type | Conductor Size (AWG/ kcmil) | Trade Size (Metric Designator) | | | | | | | | | | | | |
|---|---|---|---|---|---|---|---|---|---|---|---|---|---|---|
| | | ⅜ (12) | ½ (16) | ¾ (21) | 1 (27) | 1¼ (35) | 1½ (41) | 2 (53) | 2½ (63) | 3 (78) | 3½ (91) | 4 (103) | 5 (129) | 6 (155) |
| XHHW, XHHW-2 | 8 | — | 1 | 4 | 7 | 12 | 17 | 29 | 42 | 65 | 88 | 114 | 181 | 260 |
| | 6 | — | 1 | 3 | 5 | 9 | 13 | 21 | 31 | 48 | 65 | 85 | 134 | 193 |
| | 4 | — | 1 | 1 | 3 | 6 | 9 | 15 | 22 | 35 | 47 | 61 | 98 | 140 |
| | 2 | — | 1 | 1 | 2 | 5 | 6 | 11 | 16 | 25 | 34 | 44 | 70 | 100 |
| | 1 | — | 1 | 1 | 1 | 3 | 5 | 8 | 12 | 19 | 25 | 33 | 53 | 75 |
| | 1/0 | — | 0 | 1 | 1 | 3 | 4 | 7 | 10 | 16 | 22 | 28 | 45 | 64 |
| | 2/0 | — | 0 | 1 | 1 | 2 | 3 | 6 | 8 | 13 | 18 | 24 | 38 | 54 |
| | 3/0 | — | 0 | 1 | 1 | 1 | 3 | 5 | 7 | 11 | 15 | 19 | 31 | 44 |
| | 4/0 | — | 0 | 0 | 1 | 1 | 2 | 4 | 6 | 9 | 12 | 16 | 26 | 37 |
| | 250 | — | 0 | 0 | 1 | 1 | 1 | 3 | 5 | 7 | 10 | 13 | 21 | 30 |
| | 300 | — | 0 | 0 | 1 | 1 | 1 | 3 | 4 | 6 | 8 | 11 | 17 | 25 |
| | 350 | — | 0 | 0 | 1 | 1 | 1 | 2 | 3 | 5 | 7 | 10 | 15 | 22 |
| | 400 | — | 0 | 0 | 0 | 1 | 1 | 1 | 3 | 5 | 7 | 9 | 14 | 20 |
| | 500 | — | 0 | 0 | 0 | 1 | 1 | 1 | 2 | 4 | 5 | 7 | 11 | 17 |
| | 600 | — | 0 | 0 | 0 | 1 | 1 | 1 | 1 | 3 | 4 | 6 | 9 | 13 |
| | 700 | — | 0 | 0 | 0 | 0 | 1 | 1 | 1 | 3 | 4 | 5 | 8 | 12 |
| | 750 | — | 0 | 0 | 0 | 0 | 1 | 1 | 1 | 2 | 3 | 5 | 7 | 11 |
| | 900 | — | 0 | 0 | 0 | 0 | 1 | 1 | 1 | 2 | 3 | 4 | 6 | 9 |
| | 1000 | — | 0 | 0 | 0 | 0 | 0 | 1 | 1 | 1 | 3 | 3 | 6 | 8 |

Definition: *Compact stranding* is the result of a manufacturing process where the stranded conductor is compressed to the extent that the interstices (voids between strand wires) are virtually eliminated.

***TABLE C.10*** *Maximum Number of Conductors or Fixture Wires in Rigid PVC Conduit, Schedule 40 and HDPE Conduit* (Based on Chapter 9: Table 1, Table 4, and Table 5)

| Type | Conductor Size (AWG/kcmil) | Trade Size (Metric Designator) | | | | | | | | | | | | |
|---|---|---|---|---|---|---|---|---|---|---|---|---|---|---|
| | | ⅜ (12) | ½ (16) | ¾ (21) | 1 (27) | 1¼ (35) | 1½ (41) | 2 (53) | 2½ (63) | 3 (78) | 3½ (91) | 4 (103) | 5 (129) | 6 (155) |
| **CONDUCTORS** | | | | | | | | | | | | | | |
| RHH, RHW, RHW-2 | 14 | — | 4 | 7 | 11 | 20 | 27 | 45 | 64 | 99 | 133 | 171 | 269 | 390 |
| | 12 | — | 3 | 5 | 9 | 16 | 22 | 37 | 53 | 82 | 110 | 142 | 224 | 323 |
| | 10 | — | 2 | 4 | 7 | 13 | 18 | 30 | 43 | 66 | 89 | 115 | 181 | 261 |
| | 8 | — | 1 | 2 | 4 | 7 | 9 | 15 | 22 | 35 | 46 | 60 | 94 | 137 |
| | 6 | — | 1 | 1 | 3 | 5 | 7 | 12 | 18 | 28 | 37 | 48 | 76 | 109 |
| | 4 | — | 1 | 1 | 2 | 4 | 6 | 10 | 14 | 22 | 29 | 37 | 59 | 85 |
| | 3 | — | 1 | 1 | 1 | 4 | 5 | 8 | 12 | 19 | 25 | 33 | 52 | 75 |
| | 2 | — | 1 | 1 | 1 | 3 | 4 | 7 | 10 | 16 | 22 | 28 | 45 | 65 |
| | 1 | — | 0 | 1 | 1 | 1 | 3 | 5 | 7 | 11 | 14 | 19 | 29 | 43 |
| | 1/0 | — | 0 | 1 | 1 | 1 | 2 | 4 | 6 | 9 | 13 | 16 | 26 | 37 |
| | 2/0 | — | 0 | 0 | 1 | 1 | 1 | 3 | 5 | 8 | 11 | 14 | 22 | 32 |
| | 3/0 | — | 0 | 0 | 1 | 1 | 1 | 3 | 4 | 7 | 9 | 12 | 19 | 28 |
| | 4/0 | — | 0 | 0 | 1 | 1 | 1 | 2 | 4 | 6 | 8 | 10 | 16 | 24 |
| | 250 | — | 0 | 0 | 0 | 1 | 1 | 1 | 3 | 4 | 6 | 8 | 12 | 18 |
| | 300 | — | 0 | 0 | 0 | 1 | 1 | 1 | 2 | 4 | 5 | 7 | 11 | 16 |
| | 350 | — | 0 | 0 | 0 | 1 | 1 | 1 | 2 | 3 | 5 | 6 | 10 | 14 |
| | 400 | — | 0 | 0 | 0 | 1 | 1 | 1 | 1 | 3 | 4 | 6 | 9 | 13 |
| | 500 | — | 0 | 0 | 0 | 0 | 1 | 1 | 1 | 3 | 4 | 5 | 8 | 11 |
| | 600 | — | 0 | 0 | 0 | 0 | 1 | 1 | 1 | 2 | 3 | 4 | 6 | 9 |
| | 700 | — | 0 | 0 | 0 | 0 | 0 | 1 | 1 | 1 | 3 | 3 | 6 | 8 |
| | 750 | — | 0 | 0 | 0 | 0 | 0 | 1 | 1 | 1 | 2 | 3 | 5 | 8 |
| | 800 | — | 0 | 0 | 0 | 0 | 0 | 1 | 1 | 1 | 2 | 3 | 5 | 7 |
| | 900 | — | 0 | 0 | 0 | 0 | 0 | 1 | 1 | 1 | 2 | 3 | 5 | 7 |
| | 1000 | — | 0 | 0 | 0 | 0 | 0 | 1 | 1 | 1 | 1 | 3 | 4 | 6 |
| | 1250 | — | 0 | 0 | 0 | 0 | 0 | 0 | 1 | 1 | 1 | 1 | 3 | 5 |
| | 1500 | — | 0 | 0 | 0 | 0 | 0 | 0 | 1 | 1 | 1 | 1 | 3 | 4 |
| | 1750 | — | 0 | 0 | 0 | 0 | 0 | 0 | 1 | 1 | 1 | 1 | 2 | 3 |
| | 2000 | — | 0 | 0 | 0 | 0 | 0 | 0 | 0 | 1 | 1 | 1 | 2 | 3 |
| TW, THHW, THW, THW-2 | 14 | — | 8 | 14 | 24 | 42 | 57 | 94 | 135 | 209 | 280 | 361 | 568 | 822 |
| | 12 | — | 6 | 11 | 18 | 32 | 44 | 72 | 103 | 160 | 215 | 277 | 436 | 631 |
| | 10 | — | 4 | 8 | 13 | 24 | 32 | 54 | 77 | 119 | 160 | 206 | 325 | 470 |
| | 8 | — | 2 | 4 | 7 | 13 | 18 | 30 | 43 | 66 | 89 | 115 | 181 | 261 |
| RHH*, RHW*, RHW-2* | 14 | — | 5 | 9 | 16 | 28 | 38 | 63 | 90 | 139 | 186 | 240 | 378 | 546 |
| | 12 | — | 4 | 8 | 13 | 22 | 30 | 50 | 72 | 112 | 150 | 193 | 304 | 439 |
| | 10 | — | 3 | 6 | 10 | 17 | 24 | 39 | 56 | 87 | 117 | 150 | 237 | 343 |
| | 8 | — | 1 | 3 | 6 | 10 | 14 | 23 | 33 | 52 | 70 | 90 | 142 | 205 |
| • TW, THW, THHW, THW-2, RHH*, RHW*, RHW-2* | 6 | — | 1 | 2 | 4 | 8 | 11 | 18 | 26 | 40 | 53 | 69 | 109 | 157 |
| | 4 | — | 1 | 1 | 3 | 6 | 8 | 13 | 19 | 30 | 40 | 51 | 81 | 117 |
| | 3 | — | 1 | 1 | 3 | 5 | 7 | 11 | 16 | 25 | 34 | 44 | 69 | 100 |
| | 2 | — | 1 | 1 | 2 | 4 | 6 | 10 | 14 | 22 | 29 | 37 | 59 | 85 |
| | 1 | — | 0 | 1 | 1 | 3 | 4 | 7 | 10 | 15 | 20 | 26 | 41 | 60 |
| | 1/0 | — | 0 | 1 | 1 | 2 | 3 | 6 | 8 | 13 | 17 | 22 | 35 | 51 |
| | 2/0 | — | 0 | 1 | 1 | 1 | 3 | 5 | 7 | 11 | 15 | 19 | 30 | 43 |
| | 3/0 | — | 0 | 1 | 1 | 1 | 2 | 4 | 6 | 9 | 12 | 16 | 25 | 36 |
| | 4/0 | — | 0 | 0 | 1 | 1 | 1 | 3 | 5 | 8 | 10 | 13 | 21 | 30 |

***TABLE C.10*** *Continued*

| Type | Conductor Size (AWG/ kcmil) | Trade Size (Metric Designator) ⅜ (12) | ½ (16) | ¾ (21) | 1 (27) | 1¼ (35) | 1½ (41) | 2 (53) | 2½ (63) | 3 (78) | 3½ (91) | 4 (103) | 5 (129) | 6 (155) |
|---|---|---|---|---|---|---|---|---|---|---|---|---|---|---|
| | 250 | — | 0 | 0 | 1 | 1 | 1 | 3 | 4 | 6 | 8 | 11 | 17 | 25 |
| | 300 | — | 0 | 0 | 1 | 1 | 1 | 2 | 3 | 5 | 7 | 9 | 15 | 21 |
| | 350 | — | 0 | 0 | 0 | 1 | 1 | 1 | 3 | 5 | 6 | 8 | 13 | 19 |
| | 400 | — | 0 | 0 | 0 | 1 | 1 | 1 | 3 | 4 | 6 | 7 | 12 | 17 |
| | 500 | — | 0 | 0 | 0 | 1 | 1 | 1 | 2 | 3 | 5 | 6 | 10 | 14 |
| | 600 | — | 0 | 0 | 0 | 0 | 1 | 1 | 1 | 3 | 4 | 5 | 8 | 11 |
| | 700 | — | 0 | 0 | 0 | 0 | 1 | 1 | 1 | 2 | 3 | 4 | 7 | 10 |
| | 750 | — | 0 | 0 | 0 | 0 | 1 | 1 | 1 | 2 | 3 | 4 | 6 | 10 |
| | 800 | — | 0 | 0 | 0 | 0 | 1 | 1 | 1 | 2 | 3 | 4 | 6 | 9 |
| | 900 | — | 0 | 0 | 0 | 0 | 0 | 1 | 1 | 1 | 3 | 3 | 6 | 8 |
| | 1000 | — | 0 | 0 | 0 | 0 | 0 | 1 | 1 | 1 | 2 | 3 | 5 | 7 |
| | 1250 | — | 0 | 0 | 0 | 0 | 0 | 1 | 1 | 1 | 1 | 2 | 4 | 6 |
| | 1500 | — | 0 | 0 | 0 | 0 | 0 | 1 | 1 | 1 | 1 | 1 | 3 | 5 |
| | 1750 | — | 0 | 0 | 0 | 0 | 0 | 0 | 1 | 1 | 1 | 1 | 3 | 4 |
| | 2000 | — | 0 | 0 | 0 | 0 | 0 | 0 | 1 | 1 | 1 | 1 | 3 | 4 |
| THHN, THWN, THWN-2 | 14 | — | 11 | 21 | 34 | 60 | 82 | 135 | 193 | 299 | 401 | 517 | 815 | 1178 |
| | 12 | — | 8 | 15 | 25 | 43 | 59 | 99 | 141 | 218 | 293 | 377 | 594 | 859 |
| | 10 | — | 5 | 9 | 15 | 27 | 37 | 62 | 89 | 137 | 184 | 238 | 374 | 541 |
| | 8 | — | 3 | 5 | 9 | 16 | 21 | 36 | 51 | 79 | 106 | 137 | 216 | 312 |
| | 6 | — | 1 | 4 | 6 | 11 | 15 | 26 | 37 | 57 | 77 | 99 | 156 | 225 |
| | 4 | — | 1 | 2 | 4 | 7 | 9 | 16 | 22 | 35 | 47 | 61 | 96 | 138 |
| | 3 | — | 1 | 1 | 3 | 6 | 8 | 13 | 19 | 30 | 40 | 51 | 81 | 117 |
| | 2 | — | 1 | 1 | 3 | 5 | 7 | 11 | 16 | 25 | 33 | 43 | 68 | 98 |
| | 1 | — | 1 | 1 | 1 | 3 | 5 | 8 | 12 | 18 | 25 | 32 | 50 | 73 |
| | 1/0 | — | 1 | 1 | 1 | 3 | 4 | 7 | 10 | 15 | 21 | 27 | 42 | 61 |
| | 2/0 | — | 0 | 1 | 1 | 2 | 3 | 6 | 8 | 13 | 17 | 22 | 35 | 51 |
| | 3/0 | — | 0 | 1 | 1 | 1 | 3 | 5 | 7 | 11 | 14 | 18 | 29 | 42 |
| | 4/0 | — | 0 | 1 | 1 | 1 | 2 | 4 | 6 | 9 | 12 | 15 | 24 | 35 |
| | 250 | — | 0 | 0 | 1 | 1 | 1 | 3 | 4 | 7 | 10 | 12 | 20 | 28 |
| | 300 | — | 0 | 0 | 1 | 1 | 1 | 3 | 4 | 6 | 8 | 11 | 17 | 24 |
| | 350 | — | 0 | 0 | 1 | 1 | 1 | 2 | 3 | 5 | 7 | 9 | 15 | 21 |
| | 400 | — | 0 | 0 | 0 | 1 | 1 | 1 | 3 | 5 | 6 | 8 | 13 | 19 |
| | 500 | — | 0 | 0 | 0 | 1 | 1 | 1 | 2 | 4 | 5 | 7 | 11 | 16 |
| | 600 | — | 0 | 0 | 0 | 1 | 1 | 1 | 1 | 3 | 4 | 5 | 9 | 13 |
| | 700 | — | 0 | 0 | 0 | 0 | 1 | 1 | 1 | 3 | 4 | 5 | 8 | 11 |
| | 750 | — | 0 | 0 | 0 | 0 | 1 | 1 | 1 | 2 | 3 | 4 | 7 | 11 |
| | 800 | — | 0 | 0 | 0 | 0 | 1 | 1 | 1 | 2 | 3 | 4 | 7 | 10 |
| | 900 | — | 0 | 0 | 0 | 0 | 1 | 1 | 1 | 2 | 3 | 4 | 6 | 9 |
| | 1000 | — | 0 | 0 | 0 | 0 | 0 | 1 | 1 | 1 | 3 | 3 | 6 | 8 |
| FEP, FEPB, PFA, PFAH, TFE | 14 | — | 11 | 20 | 33 | 58 | 79 | 131 | 188 | 290 | 389 | 502 | 790 | 1142 |
| | 12 | — | 8 | 15 | 24 | 42 | 58 | 96 | 137 | 212 | 284 | 366 | 577 | 834 |
| | 10 | — | 6 | 10 | 17 | 30 | 41 | 69 | 98 | 152 | 204 | 263 | 414 | 598 |
| | 8 | — | 3 | 6 | 10 | 17 | 24 | 39 | 56 | 87 | 117 | 150 | 237 | 343 |
| | 6 | — | 2 | 4 | 7 | 12 | 17 | 28 | 40 | 62 | 83 | 107 | 169 | 244 |
| | 4 | — | 1 | 3 | 5 | 8 | 12 | 19 | 28 | 43 | 58 | 75 | 118 | 170 |
| | 3 | — | 1 | 2 | 4 | 7 | 10 | 16 | 23 | 36 | 48 | 62 | 98 | 142 |
| | 2 | — | 1 | 1 | 3 | 6 | 8 | 13 | 19 | 30 | 40 | 51 | 81 | 117 |

*(continues)*

***TABLE C.10*** *Continued*

| Type | Conductor Size (AWG/ kcmil) | Trade Size (Metric Designator) ⅜ (12) | ½ (16) | ¾ (21) | 1 (27) | 1¼ (35) | 1½ (41) | 2 (53) | 2½ (63) | 3 (78) | 3½ (91) | 4 (103) | 5 (129) | 6 (155) |
|---|---|---|---|---|---|---|---|---|---|---|---|---|---|---|
| PFA, PFAH, TFE | 1 | — | 1 | 1 | 2 | 4 | 5 | 9 | 13 | 20 | 28 | 36 | 56 | 81 |
| PFA, PFAH, TFE, Z | 1/0 | — | 1 | 1 | 1 | 3 | 4 | 8 | 11 | 17 | 23 | 30 | 47 | 68 |
| | 2/0 | — | 0 | 1 | 1 | 3 | 4 | 6 | 9 | 14 | 19 | 24 | 39 | 56 |
| | 3/0 | — | 0 | 1 | 1 | 2 | 3 | 5 | 7 | 12 | 16 | 20 | 32 | 46 |
| | 4/0 | — | 0 | 1 | 1 | 1 | 2 | 4 | 6 | 9 | 13 | 16 | 26 | 38 |
| Z | 14 | — | 13 | 24 | 40 | 70 | 95 | 158 | 226 | 350 | 469 | 605 | 952 | 1376 |
| | 12 | — | 9 | 17 | 28 | 49 | 68 | 112 | 160 | 248 | 333 | 429 | 675 | 976 |
| | 10 | — | 6 | 10 | 17 | 30 | 41 | 69 | 98 | 152 | 204 | 263 | 414 | 598 |
| | 8 | — | 3 | 6 | 11 | 19 | 26 | 43 | 62 | 96 | 129 | 166 | 261 | 378 |
| | 6 | — | 2 | 4 | 7 | 13 | 18 | 30 | 43 | 67 | 90 | 116 | 184 | 265 |
| | 4 | — | 1 | 3 | 5 | 9 | 12 | 21 | 30 | 46 | 62 | 80 | 126 | 183 |
| | 3 | — | 1 | 2 | 4 | 6 | 9 | 15 | 22 | 34 | 45 | 58 | 92 | 133 |
| | 2 | — | 1 | 1 | 3 | 5 | 7 | 12 | 18 | 28 | 38 | 49 | 77 | 111 |
| | 1 | — | 1 | 1 | 2 | 4 | 6 | 10 | 14 | 23 | 30 | 39 | 62 | 90 |
| XHHW, ZW, XHHW-2, XHH | 14 | — | 8 | 14 | 24 | 42 | 57 | 94 | 135 | 209 | 280 | 361 | 568 | 822 |
| | 12 | — | 6 | 11 | 18 | 32 | 44 | 72 | 103 | 160 | 215 | 277 | 436 | 631 |
| | 10 | — | 4 | 8 | 13 | 24 | 32 | 54 | 77 | 119 | 160 | 206 | 325 | 470 |
| | 8 | — | 2 | 4 | 7 | 13 | 18 | 30 | 43 | 66 | 89 | 115 | 181 | 261 |
| | 6 | — | 1 | 3 | 5 | 10 | 13 | 22 | 32 | 49 | 66 | 85 | 134 | 193 |
| | 4 | — | 1 | 2 | 4 | 7 | 9 | 16 | 23 | 35 | 48 | 61 | 97 | 140 |
| | 3 | — | 1 | 1 | 3 | 6 | 8 | 13 | 19 | 30 | 40 | 52 | 82 | 118 |
| | 2 | — | 1 | 1 | 3 | 5 | 7 | 11 | 16 | 25 | 34 | 44 | 69 | 99 |
| XHHW, XHHW-2, XHH | 1 | — | 1 | 1 | 1 | 3 | 5 | 8 | 12 | 19 | 25 | 32 | 51 | 74 |
| | 1/0 | — | 1 | 1 | 1 | 3 | 4 | 7 | 10 | 16 | 21 | 27 | 43 | 62 |
| | 2/0 | — | 0 | 1 | 1 | 2 | 3 | 6 | 8 | 13 | 17 | 23 | 36 | 52 |
| | 3/0 | — | 0 | 1 | 1 | 1 | 3 | 5 | 7 | 11 | 14 | 19 | 30 | 43 |
| | 4/0 | — | 0 | 1 | 1 | 1 | 2 | 4 | 6 | 9 | 12 | 15 | 24 | 35 |
| | 250 | — | 0 | 0 | 1 | 1 | 1 | 3 | 5 | 7 | 10 | 13 | 20 | 29 |
| | 300 | — | 0 | 0 | 1 | 1 | 1 | 3 | 4 | 6 | 8 | 11 | 17 | 25 |
| | 350 | — | 0 | 0 | 1 | 1 | 1 | 2 | 3 | 5 | 7 | 9 | 15 | 22 |
| | 400 | — | 0 | 0 | 0 | 1 | 1 | 1 | 3 | 5 | 6 | 8 | 13 | 19 |
| | 500 | — | 0 | 0 | 0 | 1 | 1 | 1 | 2 | 4 | 5 | 7 | 11 | 16 |
| | 600 | — | 0 | 0 | 0 | 1 | 1 | 1 | 1 | 3 | 4 | 5 | 9 | 13 |
| | 700 | — | 0 | 0 | 0 | 0 | 1 | 1 | 1 | 3 | 4 | 5 | 8 | 11 |
| | 750 | — | 0 | 0 | 0 | 0 | 1 | 1 | 1 | 2 | 3 | 4 | 7 | 11 |
| | 800 | — | 0 | 0 | 0 | 0 | 1 | 1 | 1 | 2 | 3 | 4 | 7 | 10 |
| | 900 | — | 0 | 0 | 0 | 0 | 1 | 1 | 1 | 2 | 3 | 4 | 6 | 9 |
| | 1000 | — | 0 | 0 | 0 | 0 | 0 | 1 | 1 | 1 | 3 | 3 | 6 | 8 |
| | 1250 | — | 0 | 0 | 0 | 0 | 0 | 1 | 1 | 1 | 1 | 3 | 4 | 6 |
| | 1500 | — | 0 | 0 | 0 | 0 | 0 | 1 | 1 | 1 | 1 | 2 | 4 | 5 |
| | 1750 | — | 0 | 0 | 0 | 0 | 0 | 0 | 1 | 1 | 1 | 1 | 3 | 5 |
| | 2000 | — | 0 | 0 | 0 | 0 | 0 | 0 | 1 | 1 | 1 | 1 | 3 | 4 |

***TABLE C.10*** *Continued*

| Type | Conductor Size (AWG/ kcmil) | Trade Size (Metric Designator) | | | | | | | | | | | | |
|---|---|---|---|---|---|---|---|---|---|---|---|---|---|---|
| | | ⅜ (12) | ½ (16) | ¾ (21) | 1 (27) | 1¼ (35) | 1½ (41) | 2 (53) | 2½ (63) | 3 (78) | 3½ (91) | 4 (103) | 5 (129) | 6 (155) |
| **FIXTURE WIRES** | | | | | | | | | | | | | | |
| RFH-2, FFH-2, RFHH-2 | 18 | — | 8 | 14 | 23 | 40 | 54 | 90 | 129 | 200 | 268 | 346 | 545 | 788 |
| | 16 | — | 6 | 12 | 19 | 33 | 46 | 76 | 109 | 169 | 226 | 292 | 459 | 664 |
| SF-2, SFF-2 | 18 | — | 10 | 17 | 29 | 50 | 69 | 114 | 163 | 253 | 338 | 436 | 687 | 993 |
| | 16 | — | 8 | 14 | 24 | 42 | 57 | 94 | 135 | 209 | 280 | 361 | 568 | 822 |
| | 14 | — | 6 | 12 | 19 | 33 | 46 | 76 | 109 | 169 | 226 | 292 | 459 | 664 |
| SF-1, SFF-1 | 18 | — | 17 | 31 | 51 | 89 | 122 | 202 | 289 | 447 | 599 | 772 | 1216 | 1758 |
| RFH-1, TF, TFF, XF, XFF | 18 | — | 13 | 23 | 38 | 66 | 90 | 149 | 213 | 330 | 442 | 570 | 898 | 1298 |
| | 16 | — | 10 | 18 | 30 | 53 | 73 | 120 | 172 | 266 | 357 | 460 | 725 | 1048 |
| • XF, XFF | 14 | — | 8 | 14 | 24 | 42 | 57 | 94 | 135 | 209 | 280 | 361 | 568 | 822 |
| TFN, TFFN | 18 | — | 20 | 37 | 60 | 105 | 144 | 239 | 341 | 528 | 708 | 913 | 1437 | 2077 |
| | 16 | — | 16 | 28 | 46 | 80 | 110 | 183 | 261 | 403 | 541 | 697 | 1098 | 1587 |
| PF, PFF, PGF, PGFF, PAF, PTF, PTFF, PAFF | 18 | — | 19 | 35 | 57 | 100 | 137 | 227 | 323 | 501 | 671 | 865 | 1363 | 1970 |
| | 16 | — | 15 | 27 | 44 | 77 | 106 | 175 | 250 | 387 | 519 | 669 | 1054 | 1523 |
| | 14 | — | 11 | 20 | 33 | 58 | 79 | 131 | 188 | 290 | 389 | 502 | 790 | 1142 |
| ZF, ZFF, ZHF | 18 | — | 25 | 45 | 74 | 129 | 176 | 292 | 417 | 646 | 865 | 1116 | 1756 | 2539 |
| | 16 | — | 18 | 33 | 54 | 95 | 130 | 216 | 308 | 476 | 638 | 823 | 1296 | 1873 |
| | 14 | — | 13 | 24 | 40 | 70 | 95 | 158 | 226 | 350 | 469 | 605 | 952 | 1376 |
| KF-2, KFF-2 | 18 | — | 38 | 67 | 111 | 193 | 265 | 439 | 626 | 969 | 1298 | 1674 | 2634 | 3809 |
| | 16 | — | 26 | 47 | 77 | 135 | 184 | 306 | 436 | 676 | 905 | 1168 | 1838 | 2657 |
| | 14 | — | 18 | 31 | 52 | 91 | 124 | 205 | 293 | 454 | 608 | 784 | 1235 | 1785 |
| | 12 | — | 12 | 22 | 36 | 63 | 86 | 143 | 204 | 316 | 423 | 546 | 859 | 1242 |
| | 10 | — | 8 | 14 | 24 | 42 | 57 | 94 | 135 | 209 | 280 | 361 | 568 | 822 |
| KF-1, KFF-1 | 18 | — | 44 | 78 | 128 | 223 | 305 | 506 | 722 | 1118 | 1498 | 1931 | 3040 | 4395 |
| | 16 | — | 31 | 55 | 90 | 157 | 214 | 355 | 507 | 785 | 1052 | 1357 | 2136 | 3088 |
| | 14 | — | 20 | 37 | 60 | 105 | 144 | 239 | 341 | 528 | 708 | 913 | 1437 | 2077 |
| | 12 | — | 13 | 24 | 40 | 70 | 95 | 158 | 226 | 350 | 469 | 605 | 952 | 1376 |
| | 10 | — | 9 | 16 | 26 | 45 | 62 | 103 | 148 | 229 | 306 | 395 | 622 | 899 |
| XF, XFF | 12 | — | 4 | 8 | 13 | 22 | 30 | 50 | 72 | 112 | 150 | 193 | 304 | 439 |
| | 10 | — | 3 | 6 | 10 | 17 | 24 | 39 | 56 | 87 | 117 | 150 | 237 | 343 |

Notes:

1. This table is for concentric stranded conductors only. For compact stranded conductors, Table C.10(A) should be used.

2. Two-hour fire-rated RHH cable has ceramifiable insulation, which has much larger diameters than other RHH wires. Consult manufacturer's conduit fill tables.

*Types RHH, RHW, and RHW-2 without outer covering.

***TABLE C.10(A)*** *Maximum Number of Conductors or Fixture Wires in Rigid PVC Conduit, Schedule 40 and HDPE Conduit* (Based on Chapter 9: Table 1, Table 4, and Table 5A)

| Type | Conductor Size (AWG/ kcmil) | Trade Size (Metric Designator) ⅜ (12) | ½ (16) | ¾ (21) | 1 (27) | 1¼ (35) | 1½ (41) | 2 (53) | 2½ (63) | 3 (78) | 3½ (91) | 4 (103) | 5 (129) | 6 (155) |
|---|---|---|---|---|---|---|---|---|---|---|---|---|---|---|
| **COMPACT CONDUCTORS** | | | | | | | | | | | | | | |
| THW, THW-2, THHW | 8 | — | 1 | 4 | 6 | 11 | 15 | 26 | 37 | 57 | 76 | 98 | 155 | 224 |
| | 6 | — | 1 | 3 | 5 | 9 | 12 | 20 | 28 | 44 | 59 | 76 | 119 | 173 |
| | 4 | — | 1 | 1 | 3 | 6 | 9 | 15 | 21 | 33 | 44 | 57 | 89 | 129 |
| | 2 | — | 1 | 1 | 2 | 5 | 6 | 11 | 15 | 24 | 32 | 42 | 66 | 95 |
| | 1 | — | 1 | 1 | 1 | 3 | 4 | 7 | 11 | 17 | 23 | 29 | 46 | 67 |
| | 1/0 | — | 0 | 1 | 1 | 3 | 4 | 6 | 9 | 15 | 20 | 25 | 40 | 58 |
| | 2/0 | — | 0 | 1 | 1 | 2 | 3 | 5 | 8 | 12 | 16 | 21 | 34 | 49 |
| | 3/0 | — | 0 | 1 | 1 | 1 | 3 | 5 | 7 | 10 | 14 | 18 | 29 | 42 |
| | 4/0 | — | 0 | 1 | 1 | 1 | 2 | 4 | 5 | 9 | 12 | 15 | 24 | 35 |
| | 250 | — | 0 | 0 | 1 | 1 | 1 | 3 | 4 | 7 | 9 | 12 | 19 | 27 |
| | 300 | — | 0 | 0 | 1 | 1 | 1 | 2 | 4 | 6 | 8 | 10 | 16 | 24 |
| | 350 | — | 0 | 0 | 1 | 1 | 1 | 2 | 3 | 5 | 7 | 9 | 15 | 21 |
| | 400 | — | 0 | 0 | 0 | 1 | 1 | 1 | 3 | 5 | 6 | 8 | 13 | 19 |
| | 500 | — | 0 | 0 | 0 | 1 | 1 | 1 | 2 | 4 | 5 | 7 | 11 | 16 |
| | 600 | — | 0 | 0 | 0 | 1 | 1 | 1 | 1 | 3 | 4 | 5 | 9 | 13 |
| | 700 | — | 0 | 0 | 0 | 0 | 1 | 1 | 1 | 3 | 4 | 5 | 8 | 12 |
| | 750 | — | 0 | 0 | 0 | 0 | 1 | 1 | 1 | 2 | 3 | 5 | 7 | 11 |
| | 900 | — | 0 | 0 | 0 | 0 | 1 | 1 | 1 | 2 | 3 | 4 | 6 | 9 |
| | 1000 | — | 0 | 0 | 0 | 0 | 1 | 1 | 1 | 1 | 3 | 4 | 6 | 9 |
| THHN, THWN, THWN-2 | 8 | — | — | — | — | — | — | — | — | — | — | — | — | — |
| | 6 | — | 2 | 4 | 7 | 13 | 17 | 29 | 41 | 64 | 86 | 111 | 175 | 253 |
| | 4 | — | 1 | 2 | 4 | 8 | 11 | 18 | 25 | 40 | 53 | 68 | 108 | 156 |
| | 2 | — | 1 | 1 | 3 | 5 | 8 | 13 | 18 | 28 | 38 | 49 | 77 | 112 |
| | 1 | — | 1 | 1 | 2 | 4 | 6 | 9 | 14 | 21 | 29 | 37 | 58 | 84 |
| | 1/0 | — | 1 | 1 | 1 | 3 | 5 | 8 | 12 | 18 | 24 | 31 | 49 | 72 |
| | 2/0 | — | 0 | 1 | 1 | 3 | 4 | 7 | 9 | 15 | 20 | 26 | 41 | 59 |
| | 3/0 | — | 0 | 1 | 1 | 2 | 3 | 5 | 8 | 12 | 17 | 22 | 34 | 50 |
| | 4/0 | — | 0 | 1 | 1 | 1 | 3 | 4 | 6 | 10 | 14 | 18 | 28 | 41 |
| | 250 | — | 0 | 0 | 1 | 1 | 1 | 3 | 5 | 8 | 11 | 14 | 22 | 32 |
| | 300 | — | 0 | 0 | 1 | 1 | 1 | 3 | 4 | 7 | 9 | 12 | 19 | 28 |
| | 350 | — | 0 | 0 | 1 | 1 | 1 | 3 | 4 | 6 | 8 | 10 | 17 | 24 |
| | 400 | — | 0 | 0 | 1 | 1 | 1 | 2 | 3 | 5 | 7 | 9 | 15 | 22 |
| | 500 | — | 0 | 0 | 0 | 1 | 1 | 1 | 3 | 4 | 6 | 8 | 13 | 18 |
| | 600 | — | 0 | 0 | 0 | 1 | 1 | 1 | 2 | 4 | 5 | 6 | 10 | 15 |
| | 700 | — | 0 | 0 | 0 | 1 | 1 | 1 | 1 | 3 | 4 | 5 | 9 | 13 |
| | 750 | — | 0 | 0 | 0 | 1 | 1 | 1 | 1 | 3 | 4 | 5 | 8 | 12 |
| | 900 | — | 0 | 0 | 0 | 0 | 1 | 1 | 1 | 2 | 3 | 4 | 7 | 10 |
| | 1000 | — | 0 | 0 | 0 | 0 | 1 | 1 | 1 | 2 | 3 | 4 | 6 | 9 |

***TABLE C.10(A)*** *Continued*

| Type | Conductor Size (AWG/ kcmil) | Trade Size (Metric Designator) | | | | | | | | | | | | |
|---|---|---|---|---|---|---|---|---|---|---|---|---|---|---|
| | | ⅜ (12) | ½ (16) | ¾ (21) | 1 (27) | 1¼ (35) | 1½ (41) | 2 (53) | 2½ (63) | 3 (78) | 3½ (91) | 4 (103) | 5 (129) | 6 (155) |
| XHHW, XHHW-2 | 8 | — | 3 | 5 | 8 | 14 | 20 | 33 | 47 | 73 | 99 | 127 | 200 | 290 |
| | 6 | — | 1 | 4 | 6 | 11 | 15 | 25 | 35 | 55 | 73 | 94 | 149 | 215 |
| | 4 | — | 1 | 2 | 4 | 8 | 11 | 18 | 25 | 40 | 53 | 68 | 108 | 156 |
| | 2 | — | 1 | 1 | 3 | 5 | 8 | 13 | 18 | 28 | 38 | 49 | 77 | 112 |
| | 1 | — | 1 | 1 | 2 | 4 | 6 | 9 | 14 | 21 | 29 | 37 | 58 | 84 |
| | 1/0 | — | 1 | 1 | 1 | 3 | 5 | 8 | 12 | 18 | 24 | 31 | 49 | 72 |
| | 2/0 | — | 1 | 1 | 1 | 3 | 4 | 7 | 10 | 15 | 20 | 26 | 42 | 60 |
| | 3/0 | — | 0 | 1 | 1 | 2 | 3 | 5 | 8 | 12 | 17 | 22 | 34 | 50 |
| | 4/0 | — | 0 | 1 | 1 | 1 | 3 | 5 | 7 | 10 | 14 | 18 | 29 | 42 |
| | 250 | — | 0 | 0 | 1 | 1 | 1 | 4 | 5 | 8 | 11 | 14 | 23 | 33 |
| | 300 | — | 0 | 0 | 1 | 1 | 1 | 3 | 4 | 7 | 9 | 12 | 19 | 28 |
| | 350 | — | 0 | 0 | 1 | 1 | 1 | 3 | 4 | 6 | 8 | 11 | 17 | 25 |
| | 400 | — | 0 | 0 | 1 | 1 | 1 | 2 | 3 | 5 | 7 | 10 | 15 | 22 |
| | 500 | — | 0 | 0 | 0 | 1 | 1 | 1 | 3 | 4 | 6 | 8 | 13 | 18 |
| | 600 | — | 0 | 0 | 0 | 1 | 1 | 1 | 2 | 4 | 5 | 6 | 10 | 15 |
| | 700 | — | 0 | 0 | 0 | 1 | 1 | 1 | 1 | 3 | 4 | 5 | 9 | 13 |
| | 750 | — | 0 | 0 | 0 | 1 | 1 | 1 | 1 | 3 | 4 | 5 | 8 | 12 |
| | 900 | — | 0 | 0 | 0 | 0 | 1 | 1 | 1 | 2 | 3 | 4 | 7 | 10 |
| | 1000 | — | 0 | 0 | 0 | 0 | 1 | 1 | 1 | 2 | 3 | 4 | 6 | 9 |

Definition: *Compact stranding* is the result of a manufacturing process where the stranded conductor is compressed to the extent that the interstices (voids between strand wires) are virtually eliminated.

***TABLE C.11*** *Maximum Number of Conductors or Fixture Wires in Type A, Rigid PVC Conduit* (Based on Chapter 9: Table 1, Table 4, and Table 5)

| Type | Conductor Size (AWG/ kcmil) | Trade Size (Metric Designator) ⅜ (12) | ½ (16) | ¾ (21) | 1 (27) | 1¼ (35) | 1½ (41) | 2 (53) | 2½ (63) | 3 (78) | 3½ (91) | 4 (103) | 5 (129) | 6 (155) |
|---|---|---|---|---|---|---|---|---|---|---|---|---|---|---|
| **CONDUCTORS** | | | | | | | | | | | | | | |
| RHH, RHW, RHW-2 | 14 | — | 5 | 9 | 14 | 24 | 31 | 49 | 74 | 112 | 146 | 187 | — | — |
| | 12 | — | 4 | 7 | 12 | 20 | 26 | 41 | 61 | 93 | 121 | 155 | — | — |
| | 10 | — | 3 | 6 | 10 | 16 | 21 | 33 | 50 | 75 | 98 | 125 | — | — |
| | 8 | — | 1 | 3 | 5 | 8 | 11 | 17 | 26 | 39 | 51 | 65 | — | — |
| | 6 | — | 1 | 2 | 4 | 6 | 9 | 14 | 21 | 31 | 41 | 52 | — | — |
| | 4 | — | 1 | 1 | 3 | 5 | 7 | 11 | 16 | 24 | 32 | 41 | — | — |
| | 3 | — | 1 | 1 | 3 | 4 | 6 | 9 | 14 | 21 | 28 | 36 | — | — |
| | 2 | — | 1 | 1 | 2 | 4 | 5 | 8 | 12 | 18 | 24 | 31 | — | — |
| | 1 | — | 0 | 1 | 1 | 2 | 3 | 5 | 8 | 12 | 16 | 20 | — | — |
| | 1/0 | — | 0 | 1 | 1 | 2 | 3 | 5 | 7 | 10 | 14 | 18 | — | — |
| | 2/0 | — | 0 | 1 | 1 | 1 | 2 | 4 | 6 | 9 | 12 | 15 | — | — |
| | 3/0 | — | 0 | 1 | 1 | 1 | 1 | 3 | 5 | 8 | 10 | 13 | — | — |
| | 4/0 | — | 0 | 0 | 1 | 1 | 1 | 3 | 4 | 7 | 9 | 11 | — | — |
| | 250 | — | 0 | 0 | 1 | 1 | 1 | 1 | 3 | 5 | 6 | 8 | — | — |
| | 300 | — | 0 | 0 | 1 | 1 | 1 | 1 | 3 | 4 | 6 | 7 | — | — |
| | 350 | — | 0 | 0 | 0 | 1 | 1 | 1 | 2 | 4 | 5 | 7 | — | — |
| | 400 | — | 0 | 0 | 0 | 1 | 1 | 1 | 2 | 3 | 5 | 6 | — | — |
| | 500 | — | 0 | 0 | 0 | 1 | 1 | 1 | 1 | 3 | 4 | 5 | — | — |
| | 600 | — | 0 | 0 | 0 | 0 | 1 | 1 | 1 | 2 | 3 | 4 | — | — |
| | 700 | — | 0 | 0 | 0 | 0 | 1 | 1 | 1 | 2 | 3 | 4 | — | — |
| | 750 | — | 0 | 0 | 0 | 0 | 1 | 1 | 1 | 1 | 3 | 4 | — | — |
| | 800 | — | 0 | 0 | 0 | 0 | 1 | 1 | 1 | 1 | 3 | 3 | — | — |
| | 900 | — | 0 | 0 | 0 | 0 | 0 | 1 | 1 | 1 | 2 | 3 | — | — |
| | 1000 | — | 0 | 0 | 0 | 0 | 0 | 1 | 1 | 1 | 2 | 3 | — | — |
| | 1250 | — | 0 | 0 | 0 | 0 | 0 | 1 | 1 | 1 | 1 | 2 | — | — |
| | 1500 | — | 0 | 0 | 0 | 0 | 0 | 0 | 1 | 1 | 1 | 1 | — | — |
| | 1750 | — | 0 | 0 | 0 | 0 | 0 | 0 | 1 | 1 | 1 | 1 | — | — |
| | 2000 | — | 0 | 0 | 0 | 0 | 0 | 0 | 1 | 1 | 1 | 1 | — | — |
| TW, THHW, THW, THW-2 | 14 | — | 11 | 18 | 31 | 51 | 67 | 105 | 157 | 235 | 307 | 395 | — | — |
| | 12 | — | 8 | 14 | 24 | 39 | 51 | 80 | 120 | 181 | 236 | 303 | — | — |
| | 10 | — | 6 | 10 | 18 | 29 | 38 | 60 | 89 | 135 | 176 | 226 | — | — |
| | 8 | — | 3 | 6 | 10 | 16 | 21 | 33 | 50 | 75 | 98 | 125 | — | — |
| RHH*, RHW*, RHW-2* | 14 | — | 7 | 12 | 20 | 34 | 44 | 69 | 104 | 157 | 204 | 262 | — | — |
| | 12 | — | 6 | 10 | 16 | 27 | 35 | 56 | 84 | 126 | 164 | 211 | — | — |
| | 10 | — | 4 | 8 | 13 | 21 | 28 | 44 | 65 | 98 | 128 | 165 | — | — |
| | 8 | — | 2 | 4 | 7 | 12 | 16 | 26 | 39 | 59 | 77 | 98 | — | — |
| • TW, THW, THHW, THW-2, RHH*, RHW*, RHW-2* | 6 | — | 1 | 3 | 6 | 9 | 13 | 20 | 30 | 45 | 59 | 75 | — | — |
| | 4 | — | 1 | 2 | 4 | 7 | 9 | 15 | 22 | 33 | 44 | 56 | — | — |
| | 3 | — | 1 | 1 | 4 | 6 | 8 | 13 | 19 | 29 | 37 | 48 | — | — |
| | 2 | — | 1 | 1 | 3 | 5 | 7 | 11 | 16 | 24 | 32 | 41 | — | — |
| | 1 | — | 1 | 1 | 1 | 3 | 5 | 7 | 11 | 17 | 22 | 29 | — | — |
| | 1/0 | — | 1 | 1 | 1 | 3 | 4 | 6 | 10 | 14 | 19 | 24 | — | — |
| | 2/0 | — | 0 | 1 | 1 | 2 | 3 | 5 | 8 | 12 | 16 | 21 | — | — |
| | 3/0 | — | 0 | 1 | 1 | 1 | 3 | 4 | 7 | 10 | 13 | 17 | — | — |
| | 4/0 | — | 0 | 1 | 1 | 1 | 2 | 4 | 6 | 9 | 11 | 14 | — | — |

***TABLE C.11*** *Continued*

| Type | Conductor Size (AWG/ kcmil) | Trade Size (Metric Designator) ⅜ (12) | ½ (16) | ¾ (21) | 1 (27) | 1¼ (35) | 1½ (41) | 2 (53) | 2½ (63) | 3 (78) | 3½ (91) | 4 (103) | 5 (129) | 6 (155) |
|---|---|---|---|---|---|---|---|---|---|---|---|---|---|---|
| | 250 | — | 0 | 0 | 1 | 1 | 1 | 3 | 4 | 7 | 9 | 12 | — | — |
| | 300 | — | 0 | 0 | 1 | 1 | 1 | 2 | 4 | 6 | 8 | 10 | — | — |
| | 350 | — | 0 | 0 | 1 | 1 | 1 | 2 | 3 | 5 | 7 | 9 | — | — |
| | 400 | — | 0 | 0 | 1 | 1 | 1 | 1 | 3 | 5 | 6 | 8 | — | — |
| | 500 | — | 0 | 0 | 0 | 1 | 1 | 1 | 2 | 4 | 5 | 7 | — | — |
| | 600 | — | 0 | 0 | 0 | 1 | 1 | 1 | 1 | 3 | 4 | 5 | — | — |
| | 700 | — | 0 | 0 | 0 | 1 | 1 | 1 | 1 | 3 | 4 | 5 | — | — |
| | 750 | — | 0 | 0 | 0 | 1 | 1 | 1 | 1 | 3 | 3 | 4 | — | — |
| | 800 | — | 0 | 0 | 0 | 0 | 1 | 1 | 1 | 2 | 3 | 4 | — | — |
| | 900 | — | 0 | 0 | 0 | 0 | 1 | 1 | 1 | 2 | 3 | 4 | — | — |
| | 1000 | — | 0 | 0 | 0 | 0 | 1 | 1 | 1 | 1 | 3 | 3 | — | — |
| | 1250 | — | 0 | 0 | 0 | 0 | 0 | 1 | 1 | 1 | 1 | 3 | — | — |
| | 1500 | — | 0 | 0 | 0 | 0 | 0 | 1 | 1 | 1 | 1 | 2 | — | — |
| | 1750 | — | 0 | 0 | 0 | 0 | 0 | 0 | 1 | 1 | 1 | 1 | — | — |
| | 2000 | — | 0 | 0 | 0 | 0 | 0 | 0 | 1 | 1 | 1 | 1 | — | — |
| THHN, THWN, THWN-2 | 14 | — | 16 | 27 | 44 | 73 | 96 | 150 | 225 | 338 | 441 | 566 | — | — |
| | 12 | — | 11 | 19 | 32 | 53 | 70 | 109 | 164 | 246 | 321 | 412 | — | — |
| | 10 | — | 7 | 12 | 20 | 33 | 44 | 69 | 103 | 155 | 202 | 260 | — | — |
| | 8 | — | 4 | 7 | 12 | 19 | 25 | 40 | 59 | 89 | 117 | 150 | — | — |
| | 6 | — | 3 | 5 | 8 | 14 | 18 | 28 | 43 | 64 | 84 | 108 | — | — |
| | 4 | — | 1 | 3 | 5 | 8 | 11 | 17 | 26 | 39 | 52 | 66 | — | — |
| | 3 | — | 1 | 2 | 4 | 7 | 9 | 15 | 22 | 33 | 44 | 56 | — | — |
| | 2 | — | 1 | 1 | 3 | 6 | 8 | 12 | 19 | 28 | 37 | 47 | — | — |
| | 1 | — | 1 | 1 | 2 | 4 | 6 | 9 | 14 | 21 | 27 | 35 | — | — |
| | 1/0 | — | 1 | 1 | 2 | 4 | 5 | 8 | 11 | 17 | 23 | 29 | — | — |
| | 2/0 | — | 1 | 1 | 1 | 3 | 4 | 6 | 10 | 14 | 19 | 24 | — | — |
| | 3/0 | — | 0 | 1 | 1 | 2 | 3 | 5 | 8 | 12 | 16 | 20 | — | — |
| | 4/0 | — | 0 | 1 | 1 | 1 | 3 | 4 | 6 | 10 | 13 | 17 | — | — |
| | 250 | — | 0 | 1 | 1 | 1 | 2 | 3 | 5 | 8 | 10 | 14 | — | — |
| | 300 | — | 0 | 0 | 1 | 1 | 1 | 3 | 4 | 7 | 9 | 12 | — | — |
| | 350 | — | 0 | 0 | 1 | 1 | 1 | 2 | 4 | 6 | 8 | 10 | — | — |
| | 400 | — | 0 | 0 | 1 | 1 | 1 | 2 | 3 | 5 | 7 | 9 | — | — |
| | 500 | — | 0 | 0 | 1 | 1 | 1 | 1 | 3 | 4 | 6 | 7 | — | — |
| | 600 | — | 0 | 0 | 0 | 1 | 1 | 1 | 2 | 3 | 5 | 6 | — | — |
| | 700 | — | 0 | 0 | 0 | 1 | 1 | 1 | 1 | 3 | 4 | 5 | — | — |
| | 750 | — | 0 | 0 | 0 | 1 | 1 | 1 | 1 | 3 | 4 | 5 | — | — |
| | 800 | — | 0 | 0 | 0 | 1 | 1 | 1 | 1 | 3 | 4 | 5 | — | — |
| | 900 | — | 0 | 0 | 0 | 0 | 1 | 1 | 1 | 2 | 3 | 4 | — | — |
| | 1000 | — | 0 | 0 | 0 | 0 | 1 | 1 | 1 | 2 | 3 | 4 | — | — |
| FEP, FEPB, PFA, PFAH, TFE | 14 | — | 15 | 26 | 43 | 70 | 93 | 146 | 218 | 327 | 427 | 549 | — | — |
| | 12 | — | 11 | 19 | 31 | 51 | 68 | 106 | 159 | 239 | 312 | 400 | — | — |
| | 10 | — | 8 | 13 | 22 | 37 | 48 | 76 | 114 | 171 | 224 | 287 | — | — |
| | 8 | — | 4 | 8 | 13 | 21 | 28 | 44 | 65 | 98 | 128 | 165 | — | — |
| | 6 | — | 3 | 5 | 9 | 15 | 20 | 31 | 46 | 70 | 91 | 117 | — | — |
| | 4 | — | 1 | 4 | 6 | 10 | 14 | 21 | 32 | 49 | 64 | 82 | — | — |
| | 3 | — | 1 | 3 | 5 | 8 | 11 | 18 | 27 | 40 | 53 | 68 | — | — |
| | 2 | — | 1 | 2 | 4 | 7 | 9 | 15 | 22 | 33 | 44 | 56 | — | — |

*(continues)*

***TABLE C.11*** *Continued*

| Type | Conductor Size (AWG/ kcmil) | Trade Size (Metric Designator) | | | | | | | | | | | | |
|---|---|---|---|---|---|---|---|---|---|---|---|---|---|---|
| | | 3/8 (12) | 1/2 (16) | 3/4 (21) | 1 (27) | 1 1/4 (35) | 1 1/2 (41) | 2 (53) | 2 1/2 (63) | 3 (78) | 3 1/2 (91) | 4 (103) | 5 (129) | 6 (155) |
| PFA, PFAH, TFE | 1 | — | 1 | 1 | 3 | 5 | 6 | 10 | 15 | 23 | 30 | 39 | — | — |
| PFA, PFAH, TFE, Z | 1/0 | — | 1 | 1 | 2 | 4 | 5 | 8 | 13 | 19 | 25 | 32 | — | — |
| | 2/0 | — | 1 | 1 | 1 | 3 | 4 | 7 | 10 | 16 | 21 | 27 | — | — |
| | 3/0 | — | 1 | 1 | 1 | 3 | 3 | 6 | 9 | 13 | 17 | 22 | — | — |
| | 4/0 | — | 0 | 1 | 1 | 2 | 3 | 5 | 7 | 11 | 14 | 18 | — | — |
| Z | 14 | — | 18 | 31 | 52 | 85 | 112 | 175 | 262 | 395 | 515 | 661 | — | — |
| | 12 | — | 13 | 22 | 37 | 60 | 79 | 124 | 186 | 280 | 365 | 469 | — | — |
| | 10 | — | 8 | 13 | 22 | 37 | 48 | 76 | 114 | 171 | 224 | 287 | — | — |
| | 8 | — | 5 | 8 | 14 | 23 | 30 | 48 | 72 | 108 | 141 | 181 | — | — |
| | 6 | — | 3 | 6 | 10 | 16 | 21 | 34 | 50 | 76 | 99 | 127 | — | — |
| | 4 | — | 2 | 4 | 7 | 11 | 15 | 23 | 35 | 52 | 68 | 88 | — | — |
| | 3 | — | 1 | 3 | 5 | 8 | 11 | 17 | 25 | 38 | 50 | 64 | — | — |
| | 2 | — | 1 | 2 | 4 | 7 | 9 | 14 | 21 | 32 | 41 | 53 | — | — |
| | 1 | — | 1 | 1 | 3 | 5 | 7 | 11 | 17 | 26 | 33 | 43 | — | — |
| XHHW, ZW, XHHW-2, XHH | 14 | — | 11 | 18 | 31 | 51 | 67 | 105 | 157 | 235 | 307 | 395 | — | — |
| | 12 | — | 8 | 14 | 24 | 39 | 51 | 80 | 120 | 181 | 236 | 303 | — | — |
| | 10 | — | 6 | 10 | 18 | 29 | 38 | 60 | 89 | 135 | 176 | 226 | — | — |
| | 8 | — | 3 | 6 | 10 | 16 | 21 | 33 | 50 | 75 | 98 | 125 | — | — |
| | 6 | — | 2 | 4 | 7 | 12 | 15 | 24 | 37 | 55 | 72 | 93 | — | — |
| | 4 | — | 1 | 3 | 5 | 8 | 11 | 18 | 26 | 40 | 52 | 67 | — | — |
| | 3 | — | 1 | 2 | 4 | 7 | 9 | 15 | 22 | 34 | 44 | 57 | — | — |
| | 2 | — | 1 | 1 | 3 | 6 | 8 | 12 | 19 | 28 | 37 | 48 | — | — |
| XHHW, XHHW-2, XHH | 1 | — | 1 | 1 | 3 | 4 | 6 | 9 | 14 | 21 | 28 | 35 | — | — |
| | 1/0 | — | 1 | 1 | 2 | 4 | 5 | 8 | 12 | 18 | 23 | 30 | — | — |
| | 2/0 | — | 1 | 1 | 1 | 3 | 4 | 6 | 10 | 15 | 19 | 25 | — | — |
| | 3/0 | — | 0 | 1 | 1 | 2 | 3 | 5 | 8 | 12 | 16 | 20 | — | — |
| | 4/0 | — | 0 | 1 | 1 | 1 | 3 | 4 | 7 | 10 | 13 | 17 | — | — |
| | 250 | — | 0 | 1 | 1 | 1 | 2 | 3 | 5 | 8 | 11 | 14 | — | — |
| | 300 | — | 0 | 0 | 1 | 1 | 1 | 3 | 5 | 7 | 9 | 12 | — | — |
| | 350 | — | 0 | 0 | 1 | 1 | 1 | 3 | 4 | 6 | 8 | 10 | — | — |
| | 400 | — | 0 | 0 | 1 | 1 | 1 | 2 | 3 | 5 | 7 | 9 | — | — |
| | 500 | — | 0 | 0 | 1 | 1 | 1 | 1 | 3 | 4 | 6 | 8 | — | — |
| | 600 | — | 0 | 0 | 0 | 1 | 1 | 1 | 2 | 3 | 5 | 6 | — | — |
| | 700 | — | 0 | 0 | 0 | 1 | 1 | 1 | 1 | 3 | 4 | 5 | — | — |
| | 750 | — | 0 | 0 | 0 | 1 | 1 | 1 | 1 | 3 | 4 | 5 | — | — |
| | 800 | — | 0 | 0 | 0 | 1 | 1 | 1 | 1 | 3 | 4 | 5 | — | — |
| | 900 | — | 0 | 0 | 0 | 0 | 1 | 1 | 1 | 2 | 3 | 4 | — | — |
| | 1000 | — | 0 | 0 | 0 | 0 | 1 | 1 | 1 | 2 | 3 | 4 | — | — |
| | 1250 | — | 0 | 0 | 0 | 0 | 0 | 1 | 1 | 1 | 2 | 3 | — | — |
| | 1500 | — | 0 | 0 | 0 | 0 | 0 | 1 | 1 | 1 | 1 | 2 | — | — |
| | 1750 | — | 0 | 0 | 0 | 0 | 0 | 1 | 1 | 1 | 1 | 2 | — | — |
| | 2000 | — | 0 | 0 | 0 | 0 | 0 | 0 | 1 | 1 | 1 | 1 | — | — |

***TABLE C.11*** *Continued*

| Type | Conductor Size (AWG/ kcmil) | ⅜ (12) | ½ (16) | ¾ (21) | 1 (27) | 1¼ (35) | 1½ (41) | 2 (53) | 2½ (63) | 3 (78) | 3½ (91) | 4 (103) | 5 (129) | 6 (155) |
|---|---|---|---|---|---|---|---|---|---|---|---|---|---|---|
| | | | | | | | **FIXTURE WIRES** | | | | | | | |
| RFH-2, FFH-2, RFHH-2 | 18 | — | 10 | 18 | 30 | 48 | 64 | 100 | 150 | 226 | 295 | 378 | — | — |
| | 16 | — | 9 | 15 | 25 | 41 | 54 | 85 | 127 | 190 | 248 | 319 | — | — |
| SF-2, SFF-2 | 18 | — | 13 | 22 | 37 | 61 | 81 | 127 | 189 | 285 | 372 | 477 | — | — |
| | 16 | — | 11 | 18 | 31 | 51 | 67 | 105 | 157 | 235 | 307 | 395 | — | — |
| | 14 | — | 9 | 15 | 25 | 41 | 54 | 85 | 127 | 190 | 248 | 319 | — | — |
| SF-1, SFF-1 | 18 | — | 23 | 40 | 66 | 108 | 143 | 224 | 335 | 504 | 658 | 844 | — | — |
| RFH-1, TF, TFF, XF, XFF | 18 | — | 17 | 29 | 49 | 80 | 105 | 165 | 248 | 372 | 486 | 623 | — | — |
| | 16 | — | 14 | 24 | 39 | 65 | 85 | 134 | 200 | 300 | 392 | 503 | — | — |
| • XF, XFF | 14 | — | 11 | 18 | 31 | 51 | 67 | 105 | 157 | 235 | 307 | 395 | — | — |
| TFN, TFFN | 18 | — | 28 | 47 | 79 | 128 | 169 | 265 | 396 | 596 | 777 | 998 | — | — |
| | 16 | — | 21 | 36 | 60 | 98 | 129 | 202 | 303 | 455 | 594 | 762 | — | — |
| PF, PFF, PGF, PGFF, PAF, PTF, PTFF, PAFF | 18 | — | 26 | 45 | 74 | 122 | 160 | 251 | 376 | 565 | 737 | 946 | — | — |
| | 16 | — | 20 | 34 | 58 | 94 | 124 | 194 | 291 | 437 | 570 | 732 | — | — |
| | 14 | — | 15 | 26 | 43 | 70 | 93 | 146 | 218 | 327 | 427 | 549 | — | — |
| ZF, ZFF, ZHF | 18 | — | 34 | 57 | 96 | 157 | 206 | 324 | 484 | 728 | 950 | 1220 | — | — |
| | 16 | — | 25 | 42 | 71 | 116 | 152 | 239 | 357 | 537 | 701 | 900 | — | — |
| | 14 | — | 18 | 31 | 52 | 85 | 112 | 175 | 262 | 395 | 515 | 661 | — | — |
| KF-2, KFF-2 | 18 | — | 51 | 86 | 144 | 235 | 310 | 486 | 727 | 1092 | 1426 | 1829 | — | — |
| | 16 | — | 36 | 60 | 101 | 164 | 216 | 339 | 507 | 762 | 994 | 1276 | — | — |
| | 14 | — | 24 | 40 | 67 | 110 | 145 | 228 | 341 | 512 | 668 | 857 | — | — |
| | 12 | — | 16 | 28 | 47 | 77 | 101 | 158 | 237 | 356 | 465 | 596 | — | — |
| | 10 | — | 11 | 18 | 31 | 51 | 67 | 105 | 157 | 235 | 307 | 395 | — | — |
| KF-1, KFF-1 | 18 | — | 59 | 100 | 166 | 272 | 357 | 561 | 839 | 1260 | 1645 | 2111 | — | — |
| | 16 | — | 41 | 70 | 117 | 191 | 251 | 394 | 589 | 886 | 1156 | 1483 | — | — |
| | 14 | — | 28 | 47 | 79 | 128 | 169 | 265 | 396 | 596 | 777 | 998 | — | — |
| | 12 | — | 18 | 31 | 52 | 85 | 112 | 175 | 262 | 395 | 515 | 661 | — | — |
| | 10 | — | 12 | 20 | 34 | 55 | 73 | 115 | 171 | 258 | 337 | 432 | — | — |
| XF, XFF | 12 | — | 6 | 10 | 16 | 27 | 35 | 56 | 84 | 126 | 164 | 211 | — | — |
| | 10 | — | 4 | 8 | 13 | 21 | 28 | 44 | 65 | 98 | 128 | 165 | — | — |

Notes:

1. This table is for concentric stranded conductors only. For compact stranded conductors, Table C.11(A) should be used.
2. Two-hour fire-rated RHH cable has ceramifiable insulation, which has much larger diameters than other RHH wires. Consult manufacturer's conduit fill tables.

*Types RHH, RHW, and RHW-2 without outer covering.

***TABLE C.11(A)*** *Maximum Number of Conductors or Fixture Wires in Type A, Rigid PVC Conduit*
(Based on Chapter 9: Table 1, Table 4, and Table 5A)

| Type | Conductor Size (AWG/ kcmil) | Trade Size (Metric Designator) ⅜ (12) | ½ (16) | ¾ (21) | 1 (27) | 1¼ (35) | 1½ (41) | 2 (53) | 2½ (63) | 3 (78) | 3½ (91) | 4 (103) | 5 (129) | 6 (155) |
|---|---|---|---|---|---|---|---|---|---|---|---|---|---|---|
| **COMPACT CONDUCTORS** | | | | | | | | | | | | | | |
| THW, THW-2, THHW | 8 | — | 3 | 5 | 8 | 14 | 18 | 28 | 42 | 64 | 84 | 107 | — | — |
| | 6 | — | 2 | 4 | 6 | 10 | 14 | 22 | 33 | 49 | 65 | 83 | — | — |
| | 4 | — | 1 | 3 | 5 | 8 | 10 | 16 | 24 | 37 | 48 | 62 | — | — |
| | 2 | — | 1 | 1 | 3 | 6 | 7 | 12 | 18 | 27 | 36 | 46 | — | — |
| | 1 | — | 1 | 1 | 2 | 4 | 5 | 8 | 13 | 19 | 25 | 32 | — | — |
| | 1/0 | — | 1 | 1 | 1 | 3 | 4 | 7 | 11 | 16 | 21 | 28 | — | — |
| | 2/0 | — | 1 | 1 | 1 | 3 | 4 | 6 | 9 | 14 | 18 | 23 | — | — |
| | 3/0 | — | 0 | 1 | 1 | 2 | 3 | 5 | 8 | 12 | 15 | 20 | — | — |
| | 4/0 | — | 0 | 1 | 1 | 1 | 3 | 4 | 6 | 10 | 13 | 17 | — | — |
| | 250 | — | 0 | 1 | 1 | 1 | 1 | 3 | 5 | 8 | 10 | 13 | — | — |
| | 300 | — | 0 | 0 | 1 | 1 | 1 | 3 | 4 | 7 | 9 | 11 | — | — |
| | 350 | — | 0 | 0 | 1 | 1 | 1 | 2 | 4 | 6 | 8 | 10 | — | — |
| | 400 | — | 0 | 0 | 1 | 1 | 1 | 2 | 3 | 5 | 7 | 9 | — | — |
| | 500 | — | 0 | 0 | 1 | 1 | 1 | 1 | 3 | 4 | 6 | 8 | — | — |
| | 600 | — | 0 | 0 | 0 | 1 | 1 | 1 | 2 | 3 | 5 | 6 | — | — |
| | 700 | — | 0 | 0 | 0 | 1 | 1 | 1 | 1 | 3 | 4 | 5 | — | — |
| | 750 | — | 0 | 0 | 0 | 1 | 1 | 1 | 1 | 3 | 4 | 5 | — | — |
| | 900 | — | 0 | 0 | 0 | 0 | 1 | 1 | 1 | 2 | 3 | 4 | — | — |
| | 1000 | — | 0 | 0 | 0 | 0 | 1 | 1 | 1 | 2 | 3 | 4 | — | — |
| THHN, THWN, THWN-2 | 8 | — | — | — | — | — | — | — | — | — | — | — | — | — |
| | 6 | — | 3 | 5 | 9 | 15 | 20 | 32 | 48 | 72 | 94 | 121 | — | — |
| | 4 | — | 1 | 3 | 6 | 9 | 12 | 20 | 30 | 45 | 58 | 75 | — | — |
| | 2 | — | 1 | 2 | 4 | 7 | 9 | 14 | 21 | 32 | 42 | 54 | — | — |
| | 1 | — | 1 | 1 | 3 | 5 | 7 | 10 | 16 | 24 | 31 | 40 | — | — |
| | 1/0 | — | 1 | 1 | 2 | 4 | 6 | 9 | 13 | 20 | 27 | 34 | — | — |
| | 2/0 | — | 1 | 1 | 1 | 3 | 5 | 7 | 11 | 17 | 22 | 28 | — | — |
| | 3/0 | — | 1 | 1 | 1 | 3 | 4 | 6 | 9 | 14 | 18 | 24 | — | — |
| | 4/0 | — | 0 | 1 | 1 | 2 | 3 | 5 | 8 | 11 | 15 | 19 | — | — |
| | 250 | — | 0 | 1 | 1 | 1 | 2 | 4 | 6 | 9 | 12 | 15 | — | — |
| | 300 | — | 0 | 1 | 1 | 1 | 1 | 3 | 5 | 8 | 10 | 13 | — | — |
| | 350 | — | 0 | 0 | 1 | 1 | 1 | 3 | 4 | 7 | 9 | 11 | — | — |
| | 400 | — | 0 | 0 | 1 | 1 | 1 | 2 | 4 | 6 | 8 | 10 | — | — |
| | 500 | — | 0 | 0 | 1 | 1 | 1 | 2 | 3 | 5 | 7 | 9 | — | — |
| | 600 | — | 0 | 0 | 0 | 1 | 1 | 1 | 3 | 4 | 5 | 7 | — | — |
| | 700 | — | 0 | 0 | 0 | 1 | 1 | 1 | 2 | 3 | 5 | 6 | — | — |
| | 750 | — | 0 | 0 | 0 | 1 | 1 | 1 | 2 | 3 | 4 | 6 | — | — |
| | 900 | — | 0 | 0 | 0 | 1 | 1 | 1 | 1 | 3 | 4 | 5 | — | — |
| | 1000 | — | 0 | 0 | 0 | 0 | 1 | 1 | 1 | 2 | 3 | 4 | — | — |

***TABLE C.11(A)*** *Continued*

| Type | Conductor Size (AWG/ kcmil) | Trade Size (Metric Designator) | | | | | | | | | | | | |
|---|---|---|---|---|---|---|---|---|---|---|---|---|---|---|
| | | 3/8 (12) | 1/2 (16) | 3/4 (21) | 1 (27) | 1¼ (35) | 1½ (41) | 2 (53) | 2½ (63) | 3 (78) | 3½ (91) | 4 (103) | 5 (129) | 6 (155) |
| XHHW, XHHW-2 | 8 | — | 4 | 6 | 11 | 18 | 23 | 37 | 55 | 83 | 108 | 139 | — | — |
| | 6 | — | 3 | 5 | 8 | 13 | 17 | 27 | 41 | 62 | 80 | 103 | — | — |
| | 4 | — | 1 | 3 | 6 | 9 | 12 | 20 | 30 | 45 | 58 | 75 | — | — |
| | 2 | — | 1 | 2 | 4 | 7 | 9 | 14 | 21 | 32 | 42 | 54 | — | — |
| | 1 | — | 1 | 1 | 3 | 5 | 7 | 10 | 16 | 24 | 31 | 40 | — | — |
| | 1/0 | — | 1 | 1 | 2 | 4 | 6 | 9 | 13 | 20 | 27 | 34 | — | — |
| | 2/0 | — | 1 | 1 | 1 | 3 | 5 | 7 | 11 | 17 | 22 | 29 | — | — |
| | 3/0 | — | 1 | 1 | 1 | 3 | 4 | 6 | 9 | 14 | 18 | 24 | — | — |
| | 4/0 | — | 0 | 1 | 1 | 2 | 3 | 5 | 8 | 12 | 15 | 20 | — | — |
| | 250 | — | 0 | 1 | 1 | 1 | 2 | 4 | 6 | 9 | 12 | 16 | — | — |
| | 300 | — | 0 | 1 | 1 | 1 | 1 | 3 | 5 | 8 | 10 | 13 | — | — |
| | 350 | — | 0 | 0 | 1 | 1 | 1 | 3 | 5 | 7 | 9 | 12 | — | — |
| | 400 | — | 0 | 0 | 1 | 1 | 1 | 3 | 4 | 6 | 8 | 11 | — | — |
| | 500 | — | 0 | 0 | 1 | 1 | 1 | 2 | 3 | 5 | 7 | 9 | — | — |
| | 600 | — | 0 | 0 | 0 | 1 | 1 | 1 | 3 | 4 | 5 | 7 | — | — |
| | 700 | — | 0 | 0 | 0 | 1 | 1 | 1 | 2 | 3 | 5 | 6 | — | — |
| | 750 | — | 0 | 0 | 0 | 1 | 1 | 1 | 2 | 3 | 4 | 6 | — | — |
| | 900 | — | 0 | 0 | 0 | 1 | 1 | 1 | 1 | 3 | 4 | 5 | — | — |
| | 1000 | — | 0 | 0 | 0 | 0 | 1 | 1 | 1 | 2 | 3 | 4 | — | — |

Definition: *Compact stranding* is the result of a manufacturing process where the stranded conductor is compressed to the extent that the interstices (voids between strand wires) are virtually eliminated.

***TABLE C.12*** *Maximum Number of Conductors or Fixture Wires in Type EB, PVC Conduit* (Based on Chapter 9: Table 1, Table 4, and Table 5)

| Type | Conductor Size (AWG/ kcmil) | Trade Size (Metric Designator) ⅜ (12) | ½ (16) | ¾ (21) | 1 (27) | 1¼ (35) | 1½ (41) | 2 (53) | 2½ (63) | 3 (78) | 3½ (91) | 4 (103) | 5 (129) | 6 (155) |
|---|---|---|---|---|---|---|---|---|---|---|---|---|---|---|
| | | | | | | | CONDUCTORS | | | | | | | |
| RHH, RHW, RHW-2 | 14 | — | — | — | — | — | — | 53 | — | 119 | 155 | 197 | 303 | 430 |
| | 12 | — | — | — | — | — | — | 44 | — | 98 | 128 | 163 | 251 | 357 |
| | 10 | — | — | — | — | — | — | 35 | — | 79 | 104 | 132 | 203 | 288 |
| | 8 | — | — | — | — | — | — | 18 | — | 41 | 54 | 69 | 106 | 151 |
| | 6 | — | — | — | — | — | — | 15 | — | 33 | 43 | 55 | 85 | 121 |
| | 4 | — | — | — | — | — | — | 11 | — | 26 | 34 | 43 | 66 | 94 |
| | 3 | — | — | — | — | — | — | 10 | — | 23 | 30 | 38 | 58 | 83 |
| | 2 | — | — | — | — | — | — | 9 | — | 20 | 26 | 33 | 50 | 72 |
| | 1 | — | — | — | — | — | — | 6 | — | 13 | 17 | 21 | 33 | 47 |
| | 1/0 | — | — | — | — | — | — | 5 | — | 11 | 15 | 19 | 29 | 41 |
| | 2/0 | — | — | — | — | — | — | 4 | — | 10 | 13 | 16 | 25 | 36 |
| | 3/0 | — | — | — | — | — | — | 4 | — | 8 | 11 | 14 | 22 | 31 |
| | 4/0 | — | — | — | — | — | — | 3 | — | 7 | 9 | 12 | 18 | 26 |
| | 250 | — | — | — | — | — | — | 2 | — | 5 | 7 | 9 | 14 | 20 |
| | 300 | — | — | — | — | — | — | 1 | — | 5 | 6 | 8 | 12 | 17 |
| | 350 | — | — | — | — | — | — | 1 | — | 4 | 5 | 7 | 11 | 16 |
| | 400 | — | — | — | — | — | — | 1 | — | 4 | 5 | 6 | 10 | 14 |
| | 500 | — | — | — | — | — | — | 1 | — | 3 | 4 | 5 | 9 | 12 |
| | 600 | — | — | — | — | — | — | 1 | — | 3 | 3 | 4 | 7 | 10 |
| | 700 | — | — | — | — | — | — | 1 | — | 2 | 3 | 4 | 6 | 9 |
| | 750 | — | — | — | — | — | — | 1 | — | 2 | 3 | 4 | 6 | 9 |
| | 800 | — | — | — | — | — | — | 1 | — | 2 | 3 | 4 | 6 | 8 |
| | 900 | — | — | — | — | — | — | 1 | — | 1 | 2 | 3 | 5 | 7 |
| | 1000 | — | — | — | — | — | — | 1 | — | 1 | 2 | 3 | 5 | 7 |
| | 1250 | — | — | — | — | — | — | 1 | — | 1 | 1 | 2 | 3 | 5 |
| | 1500 | — | — | — | — | — | — | 0 | — | 1 | 1 | 1 | 3 | 4 |
| | 1750 | — | — | — | — | — | — | 0 | — | 1 | 1 | 1 | 3 | 4 |
| | 2000 | — | — | — | — | — | — | 0 | — | 1 | 1 | 1 | 2 | 3 |
| TW, THHW, THW, THW-2 | 14 | — | — | — | — | — | — | 111 | — | 250 | 327 | 415 | 638 | 907 |
| | 12 | — | — | — | — | — | — | 85 | — | 192 | 251 | 319 | 490 | 696 |
| | 10 | — | — | — | — | — | — | 63 | — | 143 | 187 | 238 | 365 | 519 |
| | 8 | — | — | — | — | — | — | 35 | — | 79 | 104 | 132 | 203 | 288 |
| RHH*, RHW*, RHW-2* | 14 | — | — | — | — | — | — | 74 | — | 166 | 217 | 276 | 424 | 603 |
| | 12 | — | — | — | — | — | — | 59 | — | 134 | 175 | 222 | 341 | 485 |
| | 10 | — | — | — | — | — | — | 46 | — | 104 | 136 | 173 | 266 | 378 |
| | 8 | — | — | — | — | — | — | 28 | — | 62 | 81 | 104 | 159 | 227 |
| • TW, THW, THHW, THW-2, RHH*, RHW*, RHW-2* | 6 | — | — | — | — | — | — | 21 | — | 48 | 62 | 79 | 122 | 173 |
| | 4 | — | — | — | — | — | — | 16 | — | 36 | 46 | 59 | 91 | 129 |
| | 3 | — | — | — | — | — | — | 13 | — | 30 | 40 | 51 | 78 | 111 |
| | 2 | — | — | — | — | — | — | 11 | — | 26 | 34 | 43 | 66 | 94 |
| | 1 | — | — | — | — | — | — | 8 | — | 18 | 24 | 30 | 46 | 66 |
| | 1/0 | — | — | — | — | — | — | 7 | — | 15 | 20 | 26 | 40 | 56 |
| | 2/0 | — | — | — | — | — | — | 6 | — | 13 | 17 | 22 | 34 | 48 |
| | 3/0 | — | — | — | — | — | — | 5 | — | 11 | 14 | 18 | 28 | 40 |
| | 4/0 | — | — | — | — | — | — | 4 | — | 9 | 12 | 15 | 24 | 34 |

***TABLE C.12** Continued*

| Type | Conductor Size (AWG/ kcmil) | Trade Size (Metric Designator) | | | | | | | | | | | | |
|---|---|---|---|---|---|---|---|---|---|---|---|---|---|---|
| | | ⅜ (12) | ½ (16) | ¾ (21) | 1 (27) | 1¼ (35) | 1½ (41) | 2 (53) | 2½ (63) | 3 (78) | 3½ (91) | 4 (103) | 5 (129) | 6 (155) |
| | 250 | — | — | — | — | — | — | 3 | — | 7 | 10 | 12 | 19 | 27 |
| | 300 | — | — | — | — | — | — | 3 | — | 6 | 8 | 11 | 17 | 24 |
| | 350 | — | — | — | — | — | — | 2 | — | 6 | 7 | 9 | 15 | 21 |
| | 400 | — | — | — | — | — | — | 2 | — | 5 | 7 | 8 | 13 | 19 |
| | 500 | — | — | — | — | — | — | 1 | — | 4 | 5 | 7 | 11 | 16 |
| | 600 | — | — | — | — | — | — | 1 | — | 3 | 4 | 6 | 9 | 13 |
| | 700 | — | — | — | — | — | — | 1 | — | 3 | 4 | 5 | 8 | 11 |
| | 750 | — | — | — | — | — | — | 1 | — | 3 | 4 | 5 | 7 | 11 |
| | 800 | — | — | — | — | — | — | 1 | — | 3 | 3 | 4 | 7 | 10 |
| | 900 | — | — | — | — | — | — | 1 | — | 2 | 3 | 4 | 6 | 9 |
| | 1000 | — | — | — | — | — | — | 1 | — | 2 | 3 | 4 | 6 | 8 |
| | 1250 | — | — | — | — | — | — | 1 | — | 1 | 2 | 3 | 4 | 6 |
| | 1500 | — | — | — | — | — | — | 1 | — | 1 | 1 | 2 | 4 | 6 |
| | 1750 | — | — | — | — | — | — | 1 | — | 1 | 1 | 2 | 3 | 5 |
| | 2000 | — | — | — | — | — | — | 0 | — | 1 | 1 | 1 | 3 | 4 |
| THHN, THWN, THWN-2 | 14 | — | — | — | — | — | — | 159 | — | 359 | 468 | 595 | 915 | 1300 |
| | 12 | — | — | — | — | — | — | 116 | — | 262 | 342 | 434 | 667 | 948 |
| | 10 | — | — | — | — | — | — | 73 | — | 165 | 215 | 274 | 420 | 597 |
| | 8 | — | — | — | — | — | — | 42 | — | 95 | 124 | 158 | 242 | 344 |
| | 6 | — | — | — | — | — | — | 30 | — | 68 | 89 | 114 | 175 | 248 |
| | 4 | — | — | — | — | — | — | 19 | — | 42 | 55 | 70 | 107 | 153 |
| | 3 | — | — | — | — | — | — | 16 | — | 36 | 46 | 59 | 91 | 129 |
| | 2 | — | — | — | — | — | — | 13 | — | 30 | 39 | 50 | 76 | 109 |
| | 1 | — | — | — | — | — | — | 10 | — | 22 | 29 | 37 | 57 | 80 |
| | 1/0 | — | — | — | — | — | — | 8 | — | 18 | 24 | 31 | 48 | 68 |
| | 2/0 | — | — | — | — | — | — | 7 | — | 15 | 20 | 26 | 40 | 56 |
| | 3/0 | — | — | — | — | — | — | 5 | — | 13 | 17 | 21 | 33 | 47 |
| | 4/0 | — | — | — | — | — | — | 4 | — | 10 | 14 | 18 | 27 | 39 |
| | 250 | — | — | — | — | — | — | 4 | — | 8 | 11 | 14 | 22 | 31 |
| | 300 | — | — | — | — | — | — | 3 | — | 7 | 10 | 12 | 19 | 27 |
| | 350 | — | — | — | — | — | — | 3 | — | 6 | 8 | 11 | 17 | 24 |
| | 400 | — | — | — | — | — | — | 2 | — | 6 | 7 | 10 | 15 | 21 |
| | 500 | — | — | — | — | — | — | 1 | — | 5 | 6 | 8 | 12 | 18 |
| | 600 | — | — | — | — | — | — | 1 | — | 4 | 5 | 6 | 10 | 14 |
| | 700 | — | — | — | — | — | — | 1 | — | 3 | 4 | 6 | 9 | 12 |
| | 750 | — | — | — | — | — | — | 1 | — | 3 | 4 | 5 | 8 | 12 |
| | 800 | — | — | — | — | — | — | 1 | — | 3 | 4 | 5 | 8 | 11 |
| | 900 | — | — | — | — | — | — | 1 | — | 3 | 3 | 4 | 7 | 10 |
| | 1000 | — | — | — | — | — | — | 1 | — | 2 | 3 | 4 | 6 | 9 |
| FEP, FEPB, PFA, PFAH, TFE | 14 | — | — | — | — | — | — | 155 | — | 348 | 454 | 578 | 887 | 1261 |
| | 12 | — | — | — | — | — | — | 113 | — | 254 | 332 | 422 | 648 | 920 |
| | 10 | — | — | — | — | — | — | 81 | — | 182 | 238 | 302 | 465 | 660 |
| | 8 | — | — | — | — | — | — | 46 | — | 104 | 136 | 173 | 266 | 378 |
| | 6 | — | — | — | — | — | — | 33 | — | 74 | 97 | 123 | 189 | 269 |
| | 4 | — | — | — | — | — | — | 23 | — | 52 | 68 | 86 | 132 | 188 |
| | 3 | — | — | — | — | — | — | 19 | — | 43 | 56 | 72 | 110 | 157 |
| | 2 | — | — | — | — | — | — | 16 | — | 36 | 46 | 59 | 91 | 129 |

*(continues)*

***TABLE C.12** Continued*

| Type | Conductor Size (AWG/ kcmil) | Trade Size (Metric Designator) | | | | | | | | | | | | |
|---|---|---|---|---|---|---|---|---|---|---|---|---|---|---|
| | | ⅜ (12) | ½ (16) | ¾ (21) | 1 (27) | 1¼ (35) | 1½ (41) | 2 (53) | 2½ (63) | 3 (78) | 3½ (91) | 4 (103) | 5 (129) | 6 (155) |
| PFA, PFAH, TFE | 1 | — | — | — | — | — | — | 11 | — | 25 | 32 | 41 | 63 | 90 |
| PFA, PFAH, TFE, Z | 1/0 | — | — | — | — | — | — | 9 | — | 20 | 27 | 34 | 53 | 75 |
| | 2/0 | — | — | — | — | — | — | 7 | — | 17 | 22 | 28 | 43 | 62 |
| | 3/0 | — | — | — | — | — | — | 6 | — | 14 | 18 | 23 | 36 | 51 |
| | 4/0 | — | — | — | — | — | — | 5 | — | 11 | 15 | 19 | 29 | 42 |
| Z | 14 | — | — | — | — | — | — | 186 | — | 419 | 547 | 696 | 1069 | 1519 |
| | 12 | — | — | — | — | — | — | 132 | — | 297 | 388 | 494 | 759 | 1078 |
| | 10 | — | — | — | — | — | — | 81 | — | 182 | 238 | 302 | 465 | 660 |
| | 8 | — | — | — | — | — | — | 51 | — | 115 | 150 | 191 | 294 | 417 |
| | 6 | — | — | — | — | — | — | 36 | — | 81 | 105 | 134 | 206 | 293 |
| | 4 | — | — | — | — | — | — | 24 | — | 55 | 72 | 92 | 142 | 201 |
| | 3 | — | — | — | — | — | — | 18 | — | 40 | 53 | 67 | 104 | 147 |
| | 2 | — | — | — | — | — | — | 15 | — | 34 | 44 | 56 | 86 | 122 |
| | 1 | — | — | — | — | — | — | 12 | — | 27 | 36 | 45 | 70 | 99 |
| XHHW, ZW, XHHW-2, XHH | 14 | — | — | — | — | — | — | 111 | — | 250 | 327 | 415 | 638 | 907 |
| | 12 | — | — | — | — | — | — | 85 | — | 192 | 251 | 319 | 490 | 696 |
| | 10 | — | — | — | — | — | — | 63 | — | 143 | 187 | 238 | 365 | 519 |
| | 8 | — | — | — | — | — | — | 35 | — | 79 | 104 | 132 | 203 | 288 |
| | 6 | — | — | — | — | — | — | 26 | — | 59 | 77 | 98 | 150 | 213 |
| | 4 | — | — | — | — | — | — | 19 | — | 42 | 56 | 71 | 109 | 155 |
| | 3 | — | — | — | — | — | — | 16 | — | 36 | 47 | 60 | 92 | 131 |
| | 2 | — | — | — | — | — | — | 13 | — | 30 | 39 | 50 | 77 | 110 |
| XHHW, XHHW-2, XHH | 1 | — | — | — | — | — | — | 10 | — | 22 | 29 | 37 | 58 | 82 |
| | 1/0 | — | — | — | — | — | — | 8 | — | 19 | 25 | 31 | 48 | 69 |
| | 2/0 | — | — | — | — | — | — | 7 | — | 16 | 20 | 26 | 40 | 57 |
| | 3/0 | — | — | — | — | — | — | 6 | — | 13 | 17 | 22 | 33 | 47 |
| | 4/0 | — | — | — | — | — | — | 5 | — | 11 | 14 | 18 | 27 | 39 |
| | 250 | — | — | — | — | — | — | 4 | — | 9 | 11 | 15 | 22 | 32 |
| | 300 | — | — | — | — | — | — | 3 | — | 7 | 10 | 12 | 19 | 28 |
| | 350 | — | — | — | — | — | — | 3 | — | 6 | 8 | 11 | 17 | 24 |
| | 400 | — | — | — | — | — | — | 2 | — | 6 | 8 | 10 | 15 | 22 |
| | 500 | — | — | — | — | — | — | 1 | — | 5 | 6 | 8 | 12 | 18 |
| | 600 | — | — | — | — | — | — | 1 | — | 4 | 5 | 6 | 10 | 14 |
| | 700 | — | — | — | — | — | — | 1 | — | 3 | 4 | 6 | 9 | 12 |
| | 750 | — | — | — | — | — | — | 1 | — | 3 | 4 | 5 | 8 | 12 |
| | 800 | — | — | — | — | — | — | 1 | — | 3 | 4 | 5 | 8 | 11 |
| | 900 | — | — | — | — | — | — | 1 | — | 3 | 3 | 4 | 7 | 10 |
| | 1000 | — | — | — | — | — | — | 1 | — | 2 | 3 | 4 | 6 | 9 |
| | 1250 | — | — | — | — | — | — | 1 | — | 1 | 2 | 3 | 5 | 7 |
| | 1500 | — | — | — | — | — | — | 1 | — | 1 | 1 | 3 | 4 | 6 |
| | 1750 | — | — | — | — | — | — | 1 | — | 1 | 1 | 2 | 4 | 5 |
| | 2000 | — | — | — | — | — | — | 0 | — | 1 | 1 | 1 | 3 | 5 |

***TABLE C.12*** *Continued*

| Type | Conductor Size (AWG/kcmil) | Trade Size (Metric Designator) | | | | | | | | | | | | |
|---|---|---|---|---|---|---|---|---|---|---|---|---|---|---|
| | | ⅜ (12) | ½ (16) | ¾ (21) | 1 (27) | 1¼ (35) | 1½ (41) | 2 (53) | 2½ (63) | 3 (78) | 3½ (91) | 4 (103) | 5 (129) | 6 (155) |
| **FIXTURE WIRES** | | | | | | | | | | | | | | |
| RFH-2, FFH-2, RFHH-2 | 18 | — | — | — | — | — | — | 107 | — | 240 | 313 | 398 | 612 | 869 |
| | 16 | — | — | — | — | — | — | 90 | — | 202 | 264 | 336 | 516 | 733 |
| SF-2, SFF-2 | 18 | — | — | — | — | — | — | 134 | — | 303 | 395 | 502 | 772 | 1096 |
| | 16 | — | — | — | — | — | — | 111 | — | 250 | 327 | 415 | 638 | 907 |
| | 14 | — | — | — | — | — | — | 90 | — | 202 | 264 | 336 | 516 | 733 |
| SF-1, SFF-1 | 18 | — | — | — | — | — | — | 238 | — | 536 | 699 | 889 | 1366 | 1940 |
| RFH-1, TF, TFF, XF, XFF | 18 | — | — | — | — | — | — | 176 | — | 396 | 516 | 656 | 1009 | 1433 |
| | 16 | — | — | — | — | — | — | 142 | — | 319 | 417 | 530 | 814 | 1157 |
| XF, XFF | 14 | — | — | — | — | — | — | 111 | — | 250 | 327 | 415 | 638 | 907 |
| TFN, TFFN | 18 | — | — | — | — | — | — | 281 | — | 633 | 826 | 1050 | 1614 | 2293 |
| | 16 | — | — | — | — | — | — | 215 | — | 484 | 631 | 802 | 1233 | 1751 |
| PF, PFF, PGF, PGFF, PAF, PTF, PTFF, PAFF | 18 | — | — | — | — | — | — | 267 | — | 600 | 783 | 996 | 1530 | 2174 |
| | 16 | — | — | — | — | — | — | 206 | — | 464 | 606 | 770 | 1183 | 1681 |
| | 14 | — | — | — | — | — | — | 155 | — | 348 | 454 | 578 | 887 | 1261 |
| ZF, ZFF, ZHF | 18 | — | — | — | — | — | — | 344 | — | 774 | 1010 | 1284 | 1973 | 2802 |
| | 16 | — | — | — | — | — | — | 254 | — | 571 | 745 | 947 | 1455 | 2067 |
| | 14 | — | — | — | — | — | — | 186 | — | 419 | 547 | 696 | 1069 | 1519 |
| KF-2, KFF-2 | 18 | — | — | — | — | — | — | 516 | — | 1161 | 1515 | 1926 | 2959 | 4204 |
| | 16 | — | — | — | — | — | — | 360 | — | 810 | 1057 | 1344 | 2064 | 2933 |
| | 14 | — | — | — | — | — | — | 242 | — | 544 | 710 | 903 | 1387 | 1970 |
| | 12 | — | — | — | — | — | — | 168 | — | 378 | 494 | 628 | 965 | 1371 |
| | 10 | — | — | — | — | — | — | 111 | — | 250 | 327 | 415 | 638 | 907 |
| KF-1, KFF-1 | 18 | — | — | — | — | — | — | 596 | — | 1340 | 1748 | 2222 | 3414 | 4850 |
| | 16 | — | — | — | — | — | — | 419 | — | 941 | 1228 | 1562 | 2399 | 3408 |
| | 14 | — | — | — | — | — | — | 281 | — | 633 | 826 | 1050 | 1614 | 2293 |
| | 12 | — | — | — | — | — | — | 186 | — | 419 | 547 | 696 | 1069 | 1519 |
| | 10 | — | — | — | — | — | — | 122 | — | 274 | 358 | 455 | 699 | 993 |
| XF, XFF | 12 | — | — | — | — | — | — | 59 | — | 134 | 175 | 222 | 341 | 485 |
| | 10 | — | — | — | — | — | — | 46 | — | 104 | 136 | 173 | 266 | 378 |

Notes:

1. This table is for concentric stranded conductors only. For compact stranded conductors, Table C.12(A) should be used.

2. Two-hour fire-rated RHH cable has ceramifiable insulation, which has much larger diameters than other RHH wires. Consult manufacturer's conduit fill tables.

*Types RHH, RHW, and RHW-2 without outer covering.

***TABLE C.12(A)*** *Maximum Number of Conductors or Fixture Wires in Type EB, PVC Conduit* (Based on Chapter 9: Table 1, Table 4, and Table 5A)

| Type | Conductor Size (AWG/ kcmil) | Trade Size (Metric Designator) | | | | | | | | | | | | |
|---|---|---|---|---|---|---|---|---|---|---|---|---|---|---|
| | | **⅜ (12)** | **½ (16)** | **¾ (21)** | **1 (27)** | **1¼ (35)** | **1½ (41)** | **2 (53)** | **2½ (63)** | **3 (78)** | **3½ (91)** | **4 (103)** | **5 (129)** | **6 (155)** |
| | | | | | | | **COMPACT CONDUCTORS** | | | | | | | |
| THW, THW-2, THHW | 8 | — | — | — | — | — | — | 30 | — | 68 | 89 | 113 | 174 | 247 |
| | 6 | — | — | — | — | — | — | 23 | — | 52 | 69 | 87 | 134 | 191 |
| | 4 | — | — | — | — | — | — | 17 | — | 39 | 51 | 65 | 100 | 143 |
| | 2 | — | — | — | — | — | — | 13 | — | 29 | 38 | 48 | 74 | 105 |
| | 1 | — | — | — | — | — | — | 9 | — | 20 | 26 | 34 | 52 | 74 |
| | 1/0 | — | — | — | — | — | — | 8 | — | 17 | 23 | 29 | 45 | 64 |
| | 2/0 | — | — | — | — | — | — | 6 | — | 15 | 19 | 24 | 38 | 54 |
| | 3/0 | — | — | — | — | — | — | 5 | — | 12 | 16 | 21 | 32 | 46 |
| | 4/0 | — | — | — | — | — | — | 4 | — | 10 | 14 | 17 | 27 | 38 |
| | 250 | — | — | — | — | — | — | 3 | — | 8 | 11 | 14 | 21 | 30 |
| | 300 | — | — | — | — | — | — | 3 | — | 7 | 9 | 12 | 19 | 26 |
| | 350 | — | — | — | — | — | — | 3 | — | 6 | 8 | 11 | 17 | 24 |
| | 400 | — | — | — | — | — | — | 2 | — | 6 | 7 | 10 | 15 | 21 |
| | 500 | — | — | — | — | — | — | 1 | — | 5 | 6 | 8 | 12 | 18 |
| | 600 | — | — | — | — | — | — | 1 | — | 4 | 5 | 6 | 10 | 14 |
| | 700 | — | — | — | — | — | — | 1 | — | 3 | 4 | 6 | 9 | 13 |
| | 750 | — | — | — | — | — | — | 1 | — | 3 | 4 | 5 | 8 | 12 |
| | 900 | — | — | — | — | — | — | 1 | — | 3 | 4 | 5 | 7 | 10 |
| | 1000 | — | — | — | — | — | — | 1 | — | 2 | 3 | 4 | 7 | 9 |
| THHN, THWN, THWN-2 | 8 | — | — | — | — | — | — | — | — | — | — | — | — | — |
| | 6 | — | — | — | — | — | — | 34 | — | 77 | 100 | 128 | 196 | 279 |
| | 4 | — | — | — | — | — | — | 21 | — | 47 | 62 | 79 | 121 | 172 |
| | 2 | — | — | — | — | — | — | 15 | — | 34 | 44 | 57 | 87 | 124 |
| | 1 | — | — | — | — | — | — | 11 | — | 25 | 33 | 42 | 65 | 93 |
| | 1/0 | — | — | — | — | — | — | 9 | — | 22 | 28 | 36 | 56 | 79 |
| | 2/0 | — | — | — | — | — | — | 8 | — | 18 | 23 | 30 | 46 | 65 |
| | 3/0 | — | — | — | — | — | — | 6 | — | 15 | 20 | 25 | 38 | 55 |
| | 4/0 | — | — | — | — | — | — | 5 | — | 12 | 16 | 20 | 32 | 45 |
| | 250 | — | — | — | — | — | — | 4 | — | 10 | 13 | 16 | 25 | 35 |
| | 300 | — | — | — | — | — | — | 4 | — | 8 | 11 | 14 | 22 | 31 |
| | 350 | — | — | — | — | — | — | 3 | — | 7 | 9 | 12 | 19 | 27 |
| | 400 | — | — | — | — | — | — | 3 | — | 6 | 8 | 11 | 17 | 24 |
| | 500 | — | — | — | — | — | — | 2 | — | 5 | 7 | 9 | 14 | 20 |
| | 600 | — | — | — | — | — | — | 1 | — | 4 | 6 | 7 | 11 | 16 |
| | 700 | — | — | — | — | — | — | 1 | — | 4 | 5 | 6 | 10 | 14 |
| | 750 | — | — | — | — | — | — | 1 | — | 4 | 5 | 6 | 9 | 14 |
| | 900 | — | — | — | — | — | — | 1 | — | 3 | 4 | 5 | 8 | 11 |
| | 1000 | — | — | — | — | — | — | 1 | — | 3 | 3 | 4 | 7 | 10 |

***TABLE C.12(A)*** *Continued*

| Type | Conductor Size (AWG/ kcmil) | Trade Size (Metric Designator) | | | | | | | | | | | | |
|---|---|---|---|---|---|---|---|---|---|---|---|---|---|---|
| | | ⅜ (12) | ½ (16) | ¾ (21) | 1 (27) | 1¼ (35) | 1½ (41) | 2 (53) | 2½ (63) | 3 (78) | 3½ (91) | 4 (103) | 5 (129) | 6 (155) |
| XHHW, XHHW-2 | 8 | — | — | — | — | — | — | 39 | — | 88 | 115 | 146 | 225 | 320 |
| | 6 | — | — | — | — | — | — | 29 | — | 65 | 85 | 109 | 167 | 238 |
| | 4 | — | — | — | — | — | — | 21 | — | 47 | 62 | 79 | 121 | 172 |
| | 2 | — | — | — | — | — | — | 15 | — | 34 | 44 | 57 | 87 | 124 |
| | 1 | — | — | — | — | — | — | 11 | — | 25 | 33 | 42 | 65 | 93 |
| | 1/0 | — | — | — | — | — | — | 9 | — | 22 | 28 | 36 | 56 | 79 |
| | 2/0 | — | — | — | — | — | — | 8 | — | 18 | 24 | 30 | 47 | 67 |
| | 3/0 | — | — | — | — | — | — | 6 | — | 15 | 20 | 25 | 38 | 55 |
| | 4/0 | — | — | — | — | — | — | 5 | — | 12 | 16 | 21 | 32 | 46 |
| | 250 | — | — | — | — | — | — | 4 | — | 10 | 13 | 17 | 26 | 37 |
| | 300 | — | — | — | — | — | — | 4 | — | 8 | 11 | 14 | 22 | 31 |
| | 350 | — | — | — | — | — | — | 3 | — | 7 | 10 | 12 | 19 | 28 |
| | 400 | — | — | — | — | — | — | 3 | — | 7 | 9 | 11 | 17 | 25 |
| | 500 | — | — | — | — | — | — | 2 | — | 5 | 7 | 9 | 14 | 20 |
| | 600 | — | — | — | — | — | — | 1 | — | 4 | 6 | 7 | 11 | 16 |
| | 700 | — | — | — | — | — | — | 1 | — | 4 | 5 | 6 | 10 | 14 |
| | 750 | — | — | — | — | — | — | 1 | — | 3 | 5 | 6 | 9 | 13 |
| | 900 | — | — | — | — | — | — | 1 | — | 3 | 4 | 5 | 8 | 11 |
| | 1000 | — | — | — | — | — | — | 1 | — | 3 | 4 | 5 | 7 | 10 |

Definition: *Compact stranding* is the result of a manufacturing process where the stranded conductor is compressed to the extent that the interstices (voids between strand wires) are virtually eliminated.

The Informative Annex C conduit and tubing conductor fill tables are provided only as an informational tool and are not part of the mandatory requirements of the *Code*. These conductor fill values have been calculated based on the conductor fill requirements, conduit and tubing dimensions, and conductor dimensions from Chapter 9, Tables 1, 4, 5, and 5A. Where all of the conductors in conduit or tubing are all of the same physical size and insulation characteristic, the use of the Informative Annex C tables to determine the maximum number of conductors ensures compliance with the *Code* requirements for raceway fill found in 300.17 and the respective conduit and tubing articles. Although the values for maximum conduit or tubing fill do not exceed those permitted by Chapter 9, Table 1, the advice provided in Informational Note No. 1 to Table 1 on considering a lesser conductor fill or a larger size conduit to that table should always be considered where the wire-pulling conditions are not optimum.

Chapter 9, Table 1, sets forth the percentage fill required, and Tables 4, 5, and 5A list the accurate conduit, tubing, and wire dimensions. Users can calculate the percent fill or use the tables in Informative Annex C, all of which were generated using Chapter 9, Tables 1, 4, 5, and 5A.

The 12 sets of tables in Informative Annex C correspond to the 12 conduit and tubing wiring methods in Chapter 3 of the *Code*. Each set of tables is subdivided into three conductor categories: (1) conductors for general wiring (Article 310), (2) fixture wires (Article 402), and (3) compact stranded conductors (Article 310). In Informative Annex C, tables that use compact stranding are listed as "A" tables.

To select the correct metric designator or trade size conduit or tubing, proceed as follows:

STEP 1. Select the appropriate table using the lists of metal and nonmetallic wiring methods.

STEP 2. Choose the appropriate conductors (general wiring conductors, fixture wires, or compact conductors).

STEP 3. Choose the appropriate insulation.

STEP 4. Select the correct trade size conduit or tubing for the given quantity and size of conductors required.

The following examples show how to determine the correct trade size conduit or tubing.

### Application Example 1

An installation requires ten 10 AWG THWN-2 copper conductors in an underground conduit across a parking lot for exterior lighting. What size PVC conduit will be required?

*Solution*

Commentary Table C.1 lists minimum sizes of PVC conduit for this application.

**COMMENTARY TABLE C.1** Types of PVC Conduit

| Wiring Method | Table | Minimum Trade Size PVC Conduit |
|---|---|---|
| Rigid PVC Conduit Schedule 40 and HDPE | Table C.10 | 1 |
| Type A, Rigid PVC Conduit | Table C.11 | ¾ |
| Rigid PVC Conduit Schedule 80 | Table C.9 | 1 |
| Type EB, PVC Conduit Type EB | Table C.12 | Available only in trade sizes 2 and larger |

### Application Example 2

An underground service lateral requires four 600 kcmil XHHW compact-stranded aluminum conductors. What trade size conduit will be required for RMC, IMC, PVC Schedule 40, PVC Schedule 80, and PVC Type EB?

*Solution*

See Commentary Table C.2 for the minimum trade size for the respective conduit types needed for the four 600 kcmil aluminum conductors.

**COMMENTARY TABLE C.2** Minimum Trade Size Conduit

| Wiring Method | Table | Minimum Trade Size Conduit |
|---|---|---|
| RMC | Table C.8A | 3 |
| IMC | Table C.4A | 3 |
| Rigid PVC Conduit Schedule 40 | Table C.10A | 3 |
| Rigid PVC Conduit Schedule 80 | Table C.9A | 3½ |
| Type EB PVC Conduit | Table C.12A | 3 |

Most aluminum building wire in Types THW, THWN/THHN, and XHHW is compact stranded.

### Application Example 3

A motor will be supplied by three 4 AWG THW conductors. What size metal conduit or tubing will be required? What size metal flex will be required at the motor termination?

*Solution*

See Commentary Table C.3 for the minimum size metal conduit or metal tubing for this application.

**COMMENTARY TABLE C.3** Types of Metal Conduit and Metal Flex

| Wiring Method | Table | Trade Size Conduit or Tubing |
|---|---|---|
| EMT | Table C.1 | 1 |
| IMC | Table C.4 | 1 |
| RMC | Table C.8 | 1 |
| FMC | Table C.3 | 1 |
| Liquidtight FMC | Table C.7 | 1 |

If a wire-type equipment grounding conductor (EGC) (likely to be smaller than 4 AWG based on rating or size of motor circuit short-circuit, ground-fault protective device, and Table 250.122) is installed in any one of these raceway types, there is a mixture of conductor sizes. For that reason, the minimum size conduit or tubing has to be calculated based on Chapter 9, Tables 4, 5, and 8 if a wire-type EGC is installed in the conduit or tubing. Of course, based on the Informative Annex C tables, the conduit size could be increased to trade size 1¼. Because six 4 AWG THW conductors can be installed in trade size 1¼, it would be of sufficient size for the three circuit conductors and the wire-type EGC. According to 250.122(A), a wire-type EGC is not required to be larger than the circuit conductors. This approach may not yield the minimum conduit or tubing size but will result in a compliant installation and may facilitate an easier installation if the conduit run is particularly long or has a substantial number of bends. See the commentary following Informational Note No. 1 to Chapter 9, Table 1.

### Application Example 4

A fire alarm system installation requires the riser to contain twenty-one 16 AWG TFF conductors. If the riser conductors are installed in electrical metallic tubing, what minimum size tubing is required?

*Solution*

According to Table C.1, trade size 1 EMT is required.

# D Informative Annex
## Examples

Annex D provides the following sample calculations:



*This informative annex is not a part of the requirements of this NFPA document but is included for informational purposes only.*

**Selection of Conductors.** In the following examples, the results are generally expressed in amperes (A). To select conductor sizes, refer to the 0 through 2000 volt (V) ampacity tables of Article 310 and the rules of 310.15 that pertain to these tables.

**Voltage.** For uniform application of Articles 210, 215, and 220, a nominal voltage of 120, 120/240, 240, and 208Y/120 V is used in calculating the ampere load on the conductor.

**Fractions of an Ampere.** Except where the calculations result in a major fraction of an ampere (0.5 or larger), such fractions are permitted to be dropped.

**Power Factor.** Calculations in the following examples are based, for convenience, on the assumption that all loads have the same power factor (PF).

**Ranges.** For the calculation of the range loads in these examples, Column C of Table 220.55 has been used. For optional methods, see Columns A and B of Table 220.55. Except where the calculations result in a major fraction of a kilowatt (0.5 or larger), such fractions are permitted to be dropped.

**SI Units.** For metric conversions, 0.093 $m^2$ = 1 $ft^2$ and 0.3048 m = 1 ft.

In the informative annex examples, loads are assumed to be properly balanced on the system. If loads are not properly balanced, additional feeder capacity may be required. The calculations are based on the standard method using Parts I, II, and III of Article 220.

### Example D1(a) One-Family Dwelling

The dwelling has a floor area of 1500 $ft^2$, exclusive of an unfinished cellar not adaptable for future use, unfinished attic, and open porches. Appliances are a 12-kW range and a 5.5-kW, 240-V dryer. Assume range and dryer kW ratings equivalent to kVA ratings in accordance with 220.54 and 220.55.

The general lighting and general-use receptacle load is computed from the outside dimensions of the building, apartment, or other area involved. For a dwelling unit, the computed floor area does not include open porches, garages, or, as stated in the opening paragraph, "an unfinished cellar not adaptable for future use." A point to consider regarding this statement is that many of today's homes with basements or cellars do have space that is suitable for conversion to family rooms, bedrooms, home offices, or other habitable areas. In such instances, the basement or cellar space suitable for conversion needs to be included in the general lighting load calculation. See 220.12 and 220.14(J) for requirements on how to calculate the lighting and general-use receptacle load for this occupancy.

The two-story dwelling measures 30 ft × 30 ft for the first floor and 30 ft × 20 ft for the second floor.

First-floor area: 30 ft × 30 ft = 900 ft²
Second-floor area: 30 ft × 20 ft = 600 ft²
Total area = 1500 ft²

**Calculated Load** *[see 220.40]*

**General Lighting Load** 1500 ft² at 3 VA/ft² = 4500 VA

**Minimum Number of Branch Circuits Required** *[see 210.11(A)]*

**General Lighting Load:** 4500 VA ÷ 120 V = 38 A

This requires three 15-A, 2-wire or two 20-A, 2-wire circuits.
Small-Appliance Load: Two 2-wire, 20-A circuits *[see 210.11(C)(1)]*
Laundry Load: One 2-wire, 20-A circuit *[see 210.11(C)(2)]*
Bathroom Branch Circuit: One 2-wire, 20-A circuit (no additional load calculation is required for this circuit) *[see 210.11(C)(3)]*

**Minimum Size Feeder Required** *[see 220.40]*

| | | |
|---|---|---|
| General Lighting | | 4,500 VA |
| Small Appliance | | 3,000 VA |
| Laundry | | 1,500 VA |
| | Total | 9,000 VA |
| 3000 VA at 100% | | 3,000 VA |
| 9000 VA – 3000 VA = 6000 VA at 35% | | 2,100 VA |
| | Net Load | 5,100 VA |
| Range *(see Table 220.55)* | | 8,000 VA |
| Dryer Load *(see Table 220.54)* | | 5,500 VA |
| Net Calculated Load | | 18,600 VA |

**Net Calculated Load for 120/240-V, 3-wire, single-phase service or feeder**

18,600 VA ÷ 240 V = 78 A

Sections 230.42(B) and 230.79 require service conductors and disconnecting means rated not less than 100 amperes.

**Calculation for Neutral for Feeder and Service**

| | | |
|---|---|---|
| Lighting and Small-Appliance Load | | 5,100 VA |
| Range: 8000 VA at 70% *(see 220.61)* | | 5,600 VA |
| Dryer: 5500 VA at 70% *(see 220.61)* | | 3,850 VA |
| | Total | 14,550 VA |

**Calculated Load for Neutral**

14,550 VA ÷ 240 V = 61 A

### Example D1(b) One-Family Dwelling

Assume same conditions as Example No. D1(a), plus addition of one 6-A, 230-V, room air-conditioning unit and one 12-A, 115-V, room air-conditioning unit,* one 8-A, 115-V, rated waste disposer, and one 10-A, 120-V, rated dishwasher. See Article 430 for general motors and Article 440, Part VII, for air-conditioning equipment. Motors have nameplate ratings of 115 V and 230 V for use on 120-V and 240-V nominal voltage systems.

*(For feeder neutral, use larger of the two appliances for unbalance.)

From Example D1(a), feeder current is 78 A (3-wire, 240 V)

| | Line A | Neutral | Line B |
|---|---|---|---|
| Amperes from Example D1(a) | 78 | 61 | 78 |
| One 230-V air conditioner | 6 | — | 6 |
| One 115-V air conditioner and 120-V dishwasher | 12 | 12 | 10 |
| One 115-V disposer | — | 8 | 8 |
| 25% of largest motor *(see 430.24)* | 3 | 3 | 2 |
| Total amperes per conductor | 99 | 84 | 104 |

Therefore, the service would be rated 110 A.

The air-conditioning load is calculated at 100 percent and is calculated separately to comply with the requirements of 220.82(C)(1).

### Example D2(a) Optional Calculation for One-Family Dwelling, Heating Larger Than Air Conditioning *[see 220.82]*

The dwelling has a floor area of 1500 ft², exclusive of an unfinished cellar not adaptable for future use, unfinished attic, and open porches. It has a 12-kW range, a 2.5-kW water heater, a 1.2-kW dishwasher, 9 kW of electric space heating installed in five rooms, a 5-kW clothes dryer, and a 6-A, 230-V, room air-conditioning unit. Assume range, water heater, dishwasher, space heating, and clothes dryer kW ratings equivalent to kVA.

**Air Conditioner kVA Calculation**

6 A × 230 V ÷ 1000 = 1.38 kVA

This 1.38 kVA [item 1 from 220.82(C)] is less than 40% of 9 kVA of separately controlled electric heat [item 6 from 220.82(C)], so the 1.38 kVA need not be included in the service calculation.

**General Load**

| | | |
|---|---|---|
| 1500 ft² at 3 VA | | 4,500 VA |
| Two 20-A appliance outlet circuits at 1500 VA each | | 3,000 VA |
| Laundry circuit | | 1,500 VA |
| Range (at nameplate rating) | | 12,000 VA |
| Water heater | | 2,500 VA |
| Dishwasher | | 1,200 VA |
| Clothes dryer | | 5,000 VA |
| | Total | 29,700 VA |

**Application of Demand Factor** *[see 220.82(B)]*

| | | |
|---|---|---|
| First 10 kVA of general load at 100% | | 10,000 VA |
| Remainder of general load at 40% (19.7 kVA × 0.4) | | 7,880 VA |
| | Total of general load | 17,880 VA |
| 9 kVA of heat at 40% (9000 VA × 0.4) = | | 3,600 VA |
| | Total | 21,480 VA |

**Calculated Load for Service Size**

21.48 kVA = 21,480 VA
21,480 VA ÷ 240 V = 90 A

Therefore, the minimum service rating would be 100 A in accordance with 230.42 and 230.79.

**Feeder Neutral Load in Accordance with 220.61**

| | | |
|---|---|---|
| 1500 ft² at 3 VA | | 4,500 VA |
| Three 20-A circuits at 1500 VA | | 4,500 VA |
| | Total | 9,000 VA |

| | |
|---|---|
| 3000 VA at 100% | 3,000 VA |
| 9000 VA – 3000 VA = 6000 VA at 35% | 2,100 VA |
| Subtotal | 5,100 VA |
| Range: 8 kVA at 70% | 5,600 VA |
| Clothes dryer: 5 kVA at 70% | 3,500 VA |
| Dishwasher | 1,200 VA |
| Total | 15,400 VA |

**Calculated Load for Neutral**

15,400 VA ÷ 240 V= 64 A

**Example D2(b) Optional Calculation for One-Family Dwelling, Air Conditioning Larger Than Heating**

*[see 220.82(A) and 220.82(C)]*

The dwelling has a floor area of 1500 ft$^2$, exclusive of an unfinished cellar not adaptable for future use, unfinished attic, and open porches. It has two 20-A small appliance circuits, one 20-A laundry circuit, two 4-kW wall-mounted ovens, one 5.1-kW counter-mounted cooking unit, a 4.5-kW water heater, a 1.2-kW dishwasher, a 5-kW combination clothes washer and dryer, six 7-A, 230-V room air-conditioning units, and a 1.5-kW permanently installed bathroom space heater. Assume wall-mounted ovens, counter-mounted cooking unit, water heater, dishwasher, and combination clothes washer and dryer kW ratings equivalent to kVA.

**Air Conditioning kVA Calculation**

Total amperes = 6 units × 7 A = 42 A

42 A × 240 V ÷ 1000 = 10.08 kVA (assume PF = 1.0)

**Load Included at 100%**

**Air Conditioning:** Included below *[see item 1 in* 220.82(C)*]*

**Space Heater:** Omit *[see item 5 in 220.82(C)]*

**General Load**

| | |
|---|---|
| 1500 ft$^2$ at 3 VA | 4,500 VA |
| Two 20-A small-appliance circuits at 1500 VA each | 3,000 VA |
| Laundry circuit | 1,500 VA |
| Two ovens | 8,000 VA |
| One cooking unit | 5,100 VA |
| Water heater | 4,500 VA |
| Dishwasher | 1,200 VA |
| Washer/dryer | 5,000 VA |
| Total general load | 32,800 VA |
| First 10 kVA at 100% | 10,000 VA |
| Remainder at 40% (22.8 kVA × 0.4 × 1000) | 9,120 VA |
| Subtotal general load | 19,120 VA |
| Air conditioning | 10,080 VA |
| Total | 29,200 VA |

**Calculated Load for Service**

29,200 VA ÷ 240 V = 122 A (service rating)

**Feeder Neutral Load, in accordance with 220.61**

Assume that the two 4-kVA wall-mounted ovens are supplied by one branch circuit, the 5.1-kVA counter-mounted cooking unit by a separate circuit.

| | |
|---|---|
| 1500 ft$^2$ at 3 VA | 4,500 VA |
| Three 20-A circuits at 1500 VA | 4,500 VA |
| Subtotal | 9,000 VA |
| 3000 VA at 100% | 3,000 VA |
| 9000 VA – 3000 VA = 6000 VA at 35% | 2,100 VA |
| Subtotal | 5,100 VA |

Two 4-kVA ovens plus one 5.1-kVA cooking unit = 13.1 kVA. Table 220.55 permits 55% demand factor or 13.1 kVA × 0.55 = 7.2 kVA feeder capacity.

| | |
|---|---|
| Subtotal from above | 5,100 VA |
| Ovens and cooking unit: 7200 VA × 70% for neutral load | 5,040 VA |
| Clothes washer/dryer: 5 kVA × 70% for neutral load | 3,500 VA |
| Dishwasher | 1,200 VA |
| Total | 14,840 VA |

**Calculated Load for Neutral**

14,840 VA ÷ 240 V = 62

**Example D2(c) Optional Calculation for One-Family Dwelling with Heat Pump (Single-Phase, 240/120-Volt Service)** *(see 220.82)*

The dwelling has a floor area of 2000 ft$^2$, exclusive of an unfinished cellar not adaptable for future use, unfinished attic, and open porches. It has a 12-kW range, a 4.5-kW water heater, a 1.2-kW dishwasher, a 5-kW clothes dryer, and a 2½-ton (24-A) heat pump with 15 kW of backup heat.

**Heat Pump kVA Calculation**

24 A × 240 V ÷ 1000 = 5.76 kVA

This 5.76 kVA is less than 15 kVA of the backup heat; therefore, the heat pump load need not be included in the service calculation *[see 220.82(C)]*.

**General Load**

| | |
|---|---|
| 2000 ft$^2$ at 3 VA | 6,000 VA |
| Two 20-A appliance outlet circuits at 1500 VA each | 3,000 VA |
| Laundry circuit | 1,500 VA |
| Range (at nameplate rating) | 12,000 VA |
| Water heater | 4,500 VA |
| Dishwasher | 1,200 VA |
| Clothes dryer | 5,000 VA |
| Subtotal general load | 33,200 VA |
| First 10 kVA at 100% | 10,000 VA |
| Remainder of general load at 40% (23,200 VA× 0.4) | 9,280 VA |
| Total net general load | 19,280 VA |

**Heat Pump and Supplementary Heat***

240 V × 24 A = 5760 VA

**15 kW Electric Heat:**

5760 VA + (15,000 VA × 65%) = 5.76 kVA + 9.75 kVA = 15.51 kVA

***If supplementary heat is not on at same time as heat pump, heat pump kVA need not be added to total.**

**Totals**

| | |
|---|---|
| Net general load | 19,280 VA |
| Heat pump and supplementary heat | 15,510 VA |
| Total | 34,790 VA |

**Calculated Load for Service**

34.79 kVA × 1000 ÷ 240 V= 145 A

Therefore, this dwelling unit would be permitted to be served by a 150-A service.

### Example D3 Store Building

A store 50 ft by 60 ft, or 3000 ft², has 30 ft of show window. There are a total of 80 duplex receptacles. The service is 120/240 V, single phase 3-wire service. Actual connected lighting load is 8500 VA.

**Calculated Load** *(see 220.40)*

**Noncontinuous Loads**

| | |
|---|---|
| Receptacle Load *(see 220.44)* | |
| 80 receptacles at 180 VA | 14,400 VA |
| 10,000 VA at 100% | 10,000 VA |
| 14,400 VA − 10,000 VA = 4400 at 50% | 2,200 VA |
| Subtotal | 12,200 VA |

**Continuous Loads**

| | |
|---|---|
| General Lighting* | |
| 3000 ft² at 3 VA/ft² | 9,000 VA |
| Show Window Lighting Load | |
| 30 ft at 200 VA/ft *[see 220.14(G)]* | 6,000 VA |
| Outside Sign Circuit *[see 220.14(F)]* | 1,200 VA |
| Subtotal | 16,200 VA |
| Subtotal from noncontinuous | 12,200 VA |
| Total noncontinuous loads + continuous loads = | 28,400 VA |

*In the example, 125% of the actual connected lighting load (8500 VA × 1.25 = 10,625 VA) is less than 125% of the load from Table 220.12, so the minimum lighting load from Table 220.12 is used in the calculation. Had the actual lighting load been greater than the value calculated from Table 220.12, 125% of the actual connected lighting load would have been used.

**Minimum Number of Branch Circuits Required**

General Lighting: Branch circuits need only be installed to supply the actual connected load *[see 210.11(B)]*.

8500 VA × 1.25 = 10,625 VA

10,625 VA ÷ 240 V = 44 A for 3-wire, 120/240 V

The lighting load would be permitted to be served by 2-wire or 3-wire, 15- or 20-A circuits with combined capacity equal to 44 A or greater for 3-wire circuits or 88 A or greater for 2-wire circuits. The feeder capacity as well as the number of branch-circuit positions available for lighting circuits in the panelboard must reflect the full calculated load of 9000 VA × 1.25 = 11,250 VA.

**Show Window**

6000 VA × 1.25 = 7500 VA

7500 VA ÷ 240 V = 31 A for 3-wire, 120/240 V

The show window lighting is permitted to be served by 2-wire or 3-wire circuits with a capacity equal to 31 A or greater for 3-wire circuits or 62 A or greater for 2-wire circuits.

Receptacles required by 210.62 are assumed to be included in the receptacle load above if these receptacles do not supply the show window lighting load.

**Receptacles**

Receptacle Load: 14,400 VA ÷ 240 V = 60 A for 3-wire, 120/240 V

The receptacle load would be permitted to be served by 2-wire or 3-wire circuits with a capacity equal to 60 A or greater for 3-wire circuits or 120 A or greater for 2-wire circuits.

**Minimum Size Feeder (or Service) Overcurrent Protection** *[see 215.3 or 230.90]*

| | |
|---|---|
| Subtotal noncontinuous loads | 12,200 VA |
| Subtotal continuous load at 125% (16,200 VA × 1.25) | 20,250 VA |
| Total | 32,450 VA |

32,450 VA ÷ 240 V = 135 A

The next higher standard size is 150 A *(see 240.6)*.

**Minimum Size Feeders (or Service Conductors) Required** *[see 215.2, 230.42(A)]*

For 120/240 V, 3-wire system,

32,450 VA ÷ 240 V = 135 A

Service or feeder conductor is 1/0 Cu in accordance with 215.3 and Table 310.15(B)(16) (with 75°C terminations).

### Example D3(a) Industrial Feeders in a Common Raceway

An industrial multi-building facility has its service at the rear of its main building, and then provides 480Y/277-volt feeders to additional buildings behind the main building in order to segregate certain processes. The facility supplies its remote buildings through a partially enclosed access corridor that extends from the main switchboard rearward along a path that provides convenient access to services within 15 m (50 ft) of each additional building supplied. Two building feeders share a common raceway for approximately 45 m (150 ft) and run in the access corridor along with process steam and control and communications cabling. The steam raises the ambient temperature around the power raceway to as much as 35°C. At a tee fitting, the individual building feeders then run to each of the two buildings involved. The feeder neutrals are not connected to the equipment grounding conductors in the remote buildings. All distribution equipment terminations are listed as being suitable for 75°C connections.

Each of the two buildings has the following loads:

Lighting, 11,600 VA, comprised of electric-discharge luminaires connected at 277 V

Receptacles, 22 125-volt, 20-ampere receptacles on general-purpose branch circuits, supplied by separately derived systems in each of the buildings

1 Air compressor, 460 volt, three phase, 5 hp

1 Grinder, 460 volt, three phase, 1.5 hp

3 Welders, AC transformer type (nameplate: 23 amperes, 480 volts, 60 percent duty cycle)

3 Industrial Process Dryers, 480 volt, three phase,15 kW each (assume continuous use throughout certain shifts)

Determine the overcurrent protection and conductor size for the feeders in the common raceway, assuming the use of XHHW-2 insulation (90°C):

**Calculated Load** {Note: For reasonable precision, volt-ampere calculations are carried to three significant figures only; where loads are converted to amperes, the results are rounded to the nearest ampere *[see 220.5(B)]*.

**Noncontinuous Loads**

| | |
|---|---|
| Receptacle Load (see 220.44) | |
| 22 receptacles at 180 VA | 3,960 VA |
| Welder Load *[see 630.11(A), Table 630.11(A)]* | |
| Each welder: 480V × 23A × 0.78 = 8,610 VA | |
| All 3 welders *[see 630.11(B)]* (demand factors 100%, 100%, 85% respectively) | |
| 8,610 VA + 8,610 VA + 7,320 VA = | 24,500 VA |
| **Subtotal, Noncontinuous Loads** | **28,500 VA** |

**Motor Loads (see 430.24, Table 430.250)**

| | | |
|---|---|---|
| Air compressor: 7.6 A × 480 V × $\sqrt{3}$ = | | 6,310 VA |
| Grinder: 3 A × 480 V × $\sqrt{3}$ = | | 2,490 VA |
| Largest motor, additional 25%: | | 1,580 VA |
| **Subtotal, Motor Loads** | | **10,400 VA** |
| By using 430.24, the motor loads and the noncontinuous loads can be combined for the remaining calculation. | | |
| **Subtotal for load calculations, Noncontinuous Loads** | | **38,900 VA** |
| **Continuous Loads** | | |
| General Lighting | | 11,600 VA |
| 3 Industrial Process Dryers | 15 kW each | 45,000 VA |
| **Subtotal, Continuous Loads:** | | **56,600 VA** |

**Overcurrent protection** *(see 215.3)*

The overcurrent protective device must accommodate 125% of the continuous load, plus the noncontinuous load:

| | |
|---|---|
| Continuous load | 56,600 VA |
| Noncontinuous load | 38,900 VA |
| **Subtotal, actual load [actual load in amperes]** [99,000 VA ÷ (480V × $\sqrt{3}$) = 119 A] | **95,500 VA** |
| (25% of 56,600 VA) *(See 215.3)* | **14,200 VA** |
| **Total VA** | **109,700 VA** |

Conversion to amperes using three significant figures:
109,700 VA / (480V × $\sqrt{3}$) = 132 A

Minimum size overcurrent protective device: 132 A

Minimum standard size overcurrent protective device *(see 240.6)*: 150 amperes

Where the overcurrent protective device and its assembly are listed for operation at 100 percent of its rating, a 125 ampere overcurrent protective device would be permitted. However, overcurrent protective device assemblies listed for 100 percent of their rating are typically not available at the 125-ampere rating. *(See 215.3 Exception.)*

**Ungrounded Feeder Conductors**

The conductors must independently meet requirements for (1) terminations, and (2) conditions of use throughout the raceway run.

Minimum size conductor at the overcurrrent device termination *[see 110.14(C) and 215.2(A)(1), using 75°C ampacity column in Table 310.15(B)(16)]*: 1/0 AWG.

Minimum size conductors in the raceway based on actual load *[see Article 100, Ampacity, and 310.15(B)(3)(a) and correction factors to Table 310.15(B)(16)]*:

95,500 VA / 0.7 / 0.96 = 142,000 VA

[70% = 310.15(B)(3)(a)] & [0.96 = Correction factors to Table 310.15(B)(16)]

Conversion to amperes:

142,000 VA / (480V × $\sqrt{3}$) = 171 A

Note that the neutral conductors are counted as current-carrying conductors *[see 310.15(B)(5)(c)]* in this example because the discharge lighting has substantial nonlinear content. This requires a 2/0 AWG conductor based on the 90°C column of Table 310.15(B)(16). Therefore, the worst case is given by the raceway conditions, and 2/0 AWG conductors must be used. If the utility corridor was at normal temperatures [(30°C (86°F)], and if the lighting at each building were supplied from the local separately derived system (thus requiring no neutrals in the supply feeders) the raceway result (95,500 VA / 0.8 = 119,000 VA; 119,000 VA / (480V × $\sqrt{3}$) = 143 A, or a 1 AWG conductor @90°C) could not be used because the termination result (1/0 AWG based on the 75°C column of Table 310.15(B)(16) would become the worst case, requiring the larger conductor.

In every case, the overcurrent protective device shall provide overcurrent protection for the feeder conductors in accordance with their ampacity as provided by this *Code (see 240.4)*. A 90°C 2/0 AWG conductor has a Table 310.15(B)(16) ampacity of 195 amperes. Adjusting for the conditions of use (35°C ambient temperature, 8 current-carrying conductors in the common raceway),

195 amperes × 0.96 × 0.7 = 131 A

The 150-ampere circuit breaker protects the 2/0 AWG feeder conductors, because 240.4(B) permits the use of the next higher standard size overcurrent protective device. Note that the feeder layout precludes the application of 310.15(A)(2) Exception.

**Feeder Neutral Conductor** *(see 220.61)*

Because 210.11(B) does not apply to these buildings, the load cannot be assumed to be evenly distributed across phases. Therefore the maximum imbalance must be assumed to be the full lighting load in this case, or 11,600 VA. (11,600 VA / 277V = 42 amperes.) The ability of the neutral to return fault current *[see 250.32(B)Exception(2)]* is not a factor in this calculation.

Because the neutral runs between the main switchboard and the building panelboard, likely terminating on a busbar at both locations, and not on overcurrent devices, the effects of continuous loading can be disregarded in evaluating its terminations *[see 215.2(A)(1) Exception No. 2]*. That calculation is (11,600 VA ÷ 277V) = 42 amperes, to be evaluated under the 75°C column of Table 310.15(B)(16). The minimum size of the neutral might seem to be 8 AWG, but that size would not be sufficient to be depended upon in the event of a line-to-neutral short circuit *[see 215.2(A)(1), second paragraph]*. Therefore, since the minimum size equipment grounding conductor for a 150 ampere circuit, as covered in Table 250.122, is 6 AWG, that is the minimum neutral size required for this feeder.

### Example D4(a) Multifamily Dwelling

A multifamily dwelling has 40 dwelling units.

Meters are in two banks of 20 each with individual feeders to each dwelling unit.

One-half of the dwelling units are equipped with electric ranges not exceeding 12 kW each. Assume range kW rating equivalent to kVA rating in accordance with 220.55. Other half of ranges are gas ranges.

Area of each dwelling unit is 840 ft$^2$.

Laundry facilities on premises are available to all tenants. Add no circuit to individual dwelling unit.

**Calculated Load for Each Dwelling Unit** *(see Article 220)*

General Lighting: 840 ft$^2$ at 3 VA/ft$^2$ = 2520 VA

Special Appliance: Electric range *(see 220.55)* = 8000 VA

**Minimum Number of Branch Circuits Required for Each Dwelling Unit** *[see 210.11(A)]*

General Lighting Load: 2520 VA ÷ 120 V = 21 A or two 15-A, 2-wire circuits; or two 20-A, 2-wire circuits

Small-Appliance Load: Two 2-wire circuits of 12 AWG wire *[see 210.11(C)(1)]*

Range Circuit: 8000 VA ÷ 240 V = 33 A or a circuit of two 8 AWG conductors and one 10 AWG conductor in accordance with 210.19(A)(3)

**Minimum Size Feeder Required for Each Dwelling Unit** *(see 215.2)*

Calculated Load *(see Article 220)*:

| | |
|---|---|
| General Lighting | 2,520 VA |
| Small Appliance (two 20-ampere circuits) | 3,000 VA |
| Subtotal Calculated Load (without ranges) | 5,520 VA |

**Application of Demand Factor** *(see Table 220.42)*

| | |
|---|---|
| First 3000 VA at 100% | 3,000 VA |
| 5520 VA – 3000 VA = 2520 VA at 35% | 882 VA |
| Net Calculated Load (without ranges) | 3,882 VA |
| Range Load | 8,000 VA |
| Net Calculated Load (with ranges) | 11,882 VA |

**Size of Each Feeder** *(see 215.2)*

For 120/240-V, 3-wire system (without ranges)
Net calculated load of 3882 VA ÷ 240 V = 16 A
For 120/240-V, 3-wire system (with ranges)
Net calculated load, 11,882 VA ÷ 240 V = 50 A

**Feeder Neutral**

| | |
|---|---|
| Lighting and Small-Appliance Load | 3,882 VA |
| Range Load: 8000 VA at 70% *(see 220.61)* | 5,600 VA |
| (only for apartments with electric range) | 5,600 VA |
| Net Calculated Load (neutral) | 9,482 VA |

**Calculated Load for Neutral**

9482 VA ÷ 240 V= 39.5 A

**Minimum Size Feeders Required from Service Equipment to Meter Bank (For 20 Dwelling Units — 10 with Ranges)**

Total Calculated Load:

| | |
|---|---|
| Lighting and Small Appliance | |
| 20 units × 5520 VA | 110,400 VA |
| Application of Demand Factor | |
| First 3000 VA at 100% | 3,000 VA |
| 110,400 VA – 3000 VA = 107,400 VA at 35% | 37,590 VA |
| Net Calculated Load | 40,590 VA |
| Range Load: 10 ranges (not over 12 kVA) *(see Col. C, Table 220.55, 25 kW)* | 25,000 VA |
| Net Calculated Load (with ranges) | 65,590 VA |

Net calculated load for 120/240-V, 3-wire system,

65,590 VA ÷ 240 V = 273 A

**Feeder Neutral**

| | |
|---|---|
| Lighting and Small-Appliance Load | 40,590 VA |
| Range Load: 25,000 VA at 70% *[see 220.61(B)]* | 17,500 VA |
| Calculated Load (neutral) | 58,090 VA |

**Calculated Load for Neutral**

58,090 VA ÷ 240 V = 242 A

**Further Demand Factor** *[220.61(B)]*

| | |
|---|---|
| 200 A at 100% | 200 A |
| 242 A – 200 A = 42 A at 70% | 29 A |
| Net Calculated Load (neutral) | 229 A |

**Minimum Size Main Feeders (or Service Conductors) Required (Less House Load) (For 40 Dwelling Units — 20 with Ranges)**

Total Calculated Load:

| | |
|---|---|
| Lighting and Small-Appliance Load 40 units × 5520 VA | 220,800 VA |

**Application of Demand Factor** *(from Table 220.42)*

| | |
|---|---|
| First 3000 VA at 100% | 3,000 VA |
| Next 120,000 VA – 3000 VA = 117,000 VA at 35% | 40,950 VA |
| Remainder 220,800 VA – 120,000 VA = 100,800 VA at 25% | 25,200 VA |
| Net Calculated Load | 69,150 VA |
| Range Load: 20 ranges (less than 12 kVA) *(see Col. C, Table 220.55)* | 35,000 VA |
| Net Calculated Load | 104,150 VA |

For 120/240-V, 3-wire system
Net calculated load of 104,150 VA ÷ 240 V = 434 A

**Feeder Neutral**

| | |
|---|---|
| Lighting and Small-Appliance Load | 69,150 VA |
| Range: 35,000 VA at 70% *[see 220.61(B)]* | 24,500 VA |
| Calculated Load (neutral) | 93,650 VA |

93,650 VA ÷ 240 V = 390 A

**Further Demand Factor** *[see 220.61(B)]*

| | |
|---|---|
| 200 A at 100% | 200 A |
| 390 A – 200 A = 190 A at 70% | 133 A |
| Net Calculated Load (neutral) | 333 A |

*[See Table 310.15(B)(16) through Table 310.15(B)(21), and 310.15(B)(2), (B)(3), and (B)(5).]*

### Example D4(b) Optional Calculation for Multifamily Dwelling

A multifamily dwelling equipped with electric cooking and space heating or air conditioning has 40 dwelling units.

Meters are in two banks of 20 each plus house metering and individual feeders to each dwelling unit.

Each dwelling unit is equipped with an electric range of 8-kW nameplate rating, four 1.5-kW separately controlled 240-V electric space heaters, and a 2.5-kW, 240-V electric water heater. Assume range, space heater, and water heater kW ratings equivalent to kVA. Calculate the load for the individual dwelling unit by the standard calculation (Part III of Article 220).

A common laundry facility is available to all tenants *[see 210.52(F), Exception No. 1]*.

Area of each dwelling unit is 840 ft².

**Calculated Load for Each Dwelling Unit** *(see Part II and Part III of Article 220)*

General Lighting Load:

| | |
|---|---|
| 840 ft² at 3 VA/ft² | 2,520 VA |
| Electric range | 8,000 VA |
| Electric heat: 6 kVA (or air conditioning if larger) | 6,000 VA |
| Electric water heater | 2,500 VA |

**Minimum Number of Branch Circuits Required for Each Dwelling Unit**

General Lighting Load: 2520 VA ÷ 120 V = 21 A or two 15-A, 2-wire circuits, or two 20-A, 2-wire circuits

Small-Appliance Load: Two 2-wire circuits of 12 AWG *[see 210.11(C)(1)]*

Range Circuit *(See Table 220.55, Column B)*:

8000 VA × 80% ÷ 240 V = 27 A on a circuit of three 10 AWG conductors in accordance with 210.19(A)(3)

Space Heating: 6000 VA ÷ 240 V = 25 A

Number of circuits *(see 210.11)*

**Minimum Size Feeder Required for Each Dwelling Unit** *(see 215.2)*

Calculated Load *(see Article 220)*:

| | |
|---|---|
| General Lighting | 2,520 VA |
| Small Appliance (two 20-A circuits) | 3,000 VA |
| Subtotal Calculated Load (without range and space heating) | 5,520 VA |

**Application of Demand Factor**

| | |
|---|---|
| First 3000 VA at 100% | 3,000 VA |
| 5520 VA – 3000 VA = 2520 VA at 35% | 882 VA |
| Net Calculated Load (without range and space heating) | 3,882 VA |
| Range | 6,400 VA |
| Space Heating *(see 220.51)* | 6,000 VA |
| Water Heater | 2,500 VA |
| Net Calculated Load (for individual dwelling unit) | 18,782 VA |

**Size of Each Feeder**

For 120/240-V, 3-wire system,

Net calculated load of 18,782 VA ÷ 240 V = 78 A

**Feeder Neutral** *(see 220.61)*

| | |
|---|---|
| Lighting and Small Appliance | 3,882 VA |
| Range Load: 6400 VA at 70% *[see 220.61(B)]* | 4,480 VA |
| Space and Water Heating (no neutral): 240 V | 0 VA |
| Net Calculated Load (neutral) | 8,362 VA |

**Calculated Load for Neutral**

8362 VA ÷ 240 V = 35 A

**Minimum Size Feeder Required from Service Equipment to Meter Bank (For 20 Dwelling Units**

Total Calculated Load:

| | |
|---|---|
| Lighting and Small-Appliance Load | |
| 20 units × 5520 VA | 110,400 VA |
| Water and Space Heating Load | |
| 20 units × 8500 VA | 170,000 VA |
| Range Load: 20 × 8000 VA | 160,000 VA |
| Net Calculated Load (20 dwelling units) | 440,400 VA |
| Net Calculated Load Using Optional Calculation *(see Table 220.84)* | |
| 440,400 VA × 0.38 | 167,352 VA |

167,352 VA ÷ 240 V = 697 A

**Minimum Size Main Feeder Required (Less House Load) (For 40 Dwelling Units)**

Calculated Load:

| | |
|---|---|
| Lighting and Small-Appliance Load | |
| 40 units × 5520 VA | 220,800 VA |
| Water and Space Heating Load | |
| 40 units × 8500 VA | 340,000 VA |
| Range: 40 ranges × 8000 VA | 320,000 VA |
| Net Calculated Load (40 dwelling units) | 880,800 VA |

Net Calculated Load Using Optional Calculation *(see Table 220.84)*

880,800 VA × 0.28 = 246,624 VA

246,624 VA ÷ 240 V = 1028 A

**Feeder Neutral Load for Feeder from Service Equipment to Meter Bank (For 20 Dwelling Units)**

| | |
|---|---|
| Lighting and Small-Appliance Load | |
| 20 units × 5520 VA | 110,400 VA |
| First 3000 VA at 100% | 3,000 VA |
| 110,400 VA – 3000 VA = 107,400 VA at 35% | 37,590 VA |
| Net Calculated Load | 40,590 VA |
| 20 ranges: 35,000 VA at 70% *[see Table 220.55 and 220.61(B)]* | 24,500 VA |
| Total | 65,090 VA |

65,090 VA ÷ 240 V = 271 A

**Further Demand Factor** *[see 220.61(B)]*

| | |
|---|---|
| First 200 A at 100% | 200 A |
| Balance: 271 A – 200 A = 71 A at 70% | 50 A |
| Total | 250 A |

**Feeder Neutral Load of Main Feeder (Less House Load) (For 40 Dwelling Units)**

| | |
|---|---|
| Lighting and Small-Appliance Load | |
| 40 units × 5520 VA | 220,800 VA |
| First 3000 VA at 100% | 3,000 VA |
| Next 120,000 VA – 3000 VA = 117,000 VA at 35% | 40,950 VA |
| Remainder 220,800 VA – 120,000 VA = 100,800 VA at 25% | 25,200 VA |
| Net Calculated Load | 69,150 VA |
| 40 ranges: 55,000 VA at 70% *[see Table 220.55 and 220.61(B)]* | 38,500 VA |
| Total | 107,650 VA |

107,650 VA ÷ 240 V = 449 A

**Further Demand Factor** *[see 220.61(B)]*

| | |
|---|---|
| First 200 A at 100% | 200 A |
| Balance: 449 – 200 A = 249 A at 70% | 174 A |
| Total | 374 A |

### Example D5(a) Multifamily Dwelling Served at 208Y/120 Volts, Three Phase

All conditions and calculations are the same as for the multifamily dwelling [Example D4(a)] served at 120/240 V, single phase except as follows: Service to each dwelling unit would be two phase legs and neutral.

**Minimum Number of Branch Circuits Required for Each Dwelling Unit** *(see 210.11)*

Range Circuit: 8000 VA ÷ 208 V = 38 A or a circuit of two 8 AWG conductors and one 10 AWG conductor in accordance with 210.19(A)(3)

**Minimum Size Feeder Required for Each Dwelling Unit** *(see 215.2)*

For 120/208-V, 3-wire system (without ranges),

Net calculated load of 3882 VA ÷ 2 legs ÷ 120 V/leg = 16 A

For 120/208-V, 3-wire system (with ranges),

Net calculated load (range) of 8000 VA ÷ 208 V = 39 A

Total load (range + lighting) = 39 A + 16 A = 55 A

•

Reducing the neutral load on the feeder to each dwelling unit is not permitted *[see 220.61(C)(1)]*.

**Minimum Size Feeders Required from Service Equipment to Meter Bank (For 20 Dwelling Units — 10 with Ranges)**

For 208Y/120-V, 3-phase, 4-wire system,

Ranges: Maximum number between any two phase legs = 4
2 × 4 = 8.

Table 220.55 demand = 23,000 VA

Per phase demand = 23,000 VA ÷ 2 = 11,500 VA

Equivalent 3-phase load = 34,500 VA

Net Calculated Load (total):

40,590 VA + 34,500 VA = 75,090 VA
75,090 VA ÷ (208 V)(1.732) = 208 A

**Feeder Neutral Size**
Net Calculated Lighting and Appliance Load & Equivalent Range Load:

40,590 VA + (34,500 VA at 70%) = 64,700 VA

Net Calculated Neutral Load:

64,700 VA ÷ (208 V)(1.732) = 180 A

**Minimum Size Main Feeder (Less House Load) (For 40 Dwelling Units — 20 with Ranges)**

For 208Y/120-V, 3-phase, 4-wire system,

Ranges:

Maximum number between any two phase legs = 7
2 × 7 = 14.

Table 220.55 demand = 29,000 VA

Per phase demand = 29,000 VA ÷ 2 = 14,500 VA

Equivalent 3-phase load = 43,500 VA

Net Calculated Load (total):

69,150 VA + 43,500 VA = 112,650 VA
112,650 VA ÷ (208 V)(1.732) = 313 A

Main Feeder Neutral Size:

69,150 VA + (43,500 VA at 70%) = 99,600 VA
99,600 VA ÷ (208 V)(1.732) = 277 A

**Further Demand Factor** *(see 220.61)*

| | |
|---|---|
| 200 A at 100% | 200.0 A |
| 277 A – 200 A = 77 A at 70% | 54 A |
| Net Calculated Load (neutral) | 254 A |

**Example D5(b) Optional Calculation for Multifamily Dwelling Served at 208Y/120 Volts, Three Phase**

All conditions and calculations are the same as for Optional Calculation for the Multifamily Dwelling [Example D4(b)] served at 120/240 V, single phase except as follows:

Service to each dwelling unit would be two phase legs and neutral.

**Minimum Number of Branch Circuits Required for Each Dwelling Unit** *(see 210.11)*

Range Circuit *(see Table 220.55, Column B)*: 8000 VA at 80% ÷ 208 V = 31 A or a circuit of two 8 AWG conductors and one 10 AWG conductor in accordance with 210.19(A)(3)

Space Heating: 6000 VA ÷ 208 V = 29 A

Two 20-ampere, 2-pole circuits required, 12 AWG conductors

**Minimum Size Feeder Required for Each Dwelling Unit**
120/208-V, 3-wire circuit

Net calculated load of 18,782 VA ÷ 208 V = 90 A

Net calculated load (lighting line to neutral):

3882 VA ÷ 2 legs ÷ 120 V per leg = 16 amperes

Line to line = 14,900 VA ÷ 208 V = 72 A

Total load = 16.2 A + 71.6 A = 88 A

**Minimum Size Feeder Required for Service Equipment to Meter Bank (For 20 Dwelling Units)**

**Net Calculated Load**

167,352 VA ÷ (208 V)(1.732) = 465 A

**Feeder Neutral Load**

65,080 VA ÷ (208 V)(1.732) = 181 A

**Minimum Size Main Feeder Required (Less House Load) (For 40 Dwelling Units)**

**Net Calculated Load**

246,624 VA ÷ (208 V)(1.732) = 685 A

**Main Feeder Neutral Load**

107,650 VA ÷ (208 V)(1.732) = 299 A

**Further Demand Factor** *[see 220.61(B)]*

| | |
|---|---|
| 200 A at 100% | 200.0 A |
| 299 A – 200 A = 99 A at 70% | 69 A |
| Net Calculated Load (neutral) | 269 A |

**Example D6 Maximum Demand for Range Loads**

Table 220.55, Column C, applies to ranges not over 12 kW. The application of Note 1 to ranges over 12 kW (and not over 27 kW) and Note 2 to ranges over 8¾ kW (and not over 27 kW) is illustrated in the following two examples.

**A. Ranges All the Same Rating** *(see Table 220.55, Note 1)*
Assume 24 ranges, each rated 16 kW.
From Table 220.55, Column C, the maximum demand for 24 ranges of 12-kW rating is 39 kW. 16 kW exceeds 12 kW by 4.
5% × 4 = 20% (5% increase for each kW in excess of 12)
39 kW × 20% = 7.8 kW increase
39 + 7.8 = 46.8 kW (value to be used in selection of feeders)

**B. Ranges of Unequal Rating** *(see Table 220.55, Note 2)*
Assume 5 ranges, each rated 11 kW; 2 ranges, each rated 12 kW; 20 ranges, each rated 13.5 kW; 3 ranges, each rated 18 kW.

| | | |
|---|---|---|
| 5 ranges | × 12 kW = | 60 kW (use 12 kW for range rated less than 12) |
| 2 ranges | × 12 kW = | 24 kW |
| 20 ranges | × 13.5 kW = | 270 kW |
| 3 ranges | × 18 kW = | 54 kW |
| 30 ranges, Total kW = | | 408 kW |

408 ÷ 30 ranges = 13.6 kW (average to be used for calculation)

From Table 220.55, Column C, the demand for 30 ranges of 12-kW rating is 15 kW + 30 (1 kW × 30 ranges) = 45 kW. 13.6 kW exceeds 12 kW by 1.6 kW (use 2 kW).

5% × 2 = 10% (5% increase for each kW in excess of 12 kW)

45 kW × 10% = 4.5 kW increase

45 kW + 4.5 kW = 49.5 kW (value to be used in selection of feeders)

### Example D7 Sizing of Service Conductors for Dwelling(s) *[see 310.15(B)(7)]*

Service conductors and feeders for certain dwellings are permitted to be sized in accordance with 310.15(B)(7).

If a 175-ampere service rating is selected, a service conductor is then sized as follows:

175 amperes × 0.83 = 145.25 amperes per 310.15(B)(7).

If no other adjustments or corrections are required for the installation, then, in accordance with Table 310.15(B)(16), a 1/0 AWG Cu or a 3/0 AWG Al meets this rating at 75°C (167°F).

### Example D8 Motor Circuit Conductors, Overload Protection, and Short-Circuit and Ground-Fault Protection

*(see 240.6, 430.6, 430.22, 430.23, 430.24, 430.32, 430.52, and 430.62, Table 430.52, and Table 430.250)*

Determine the minimum required conductor ampacity, the motor overload protection, the branch-circuit short-circuit and ground-fault protection, and the feeder protection, for three induction-type motors on a 480-V, 3-phase feeder, as follows:

(a) One 25-hp, 460-V, 3-phase, squirrel-cage motor, nameplate full-load current 32 A, Design B, Service Factor 1.15

(b) Two 30-hp, 460-V, 3-phase, wound-rotor motors, nameplate primary full-load current 38 A, nameplate secondary full-load current 65 A, 40°C rise.

**Conductor Ampacity**

The full-load current value used to determine the minimum required conductor ampacity is obtained from Table 430.250 *[see 430.6(A)]* for the squirrel-cage motor and the primary of the wound-rotor motors. To obtain the minimum required conductor ampacity, the full-load current is multiplied by 1.25 *[see 430.22 and 430.23(A)]*.

For the 25-hp motor,

34 A × 1.25 = 43 A

For the 30-horsepower motors,

40 A × 1.25 = 50 A

65 A × 1.25 = 81 A

**Motor Overload Protection**

Where protected by a separate overload device, the motors are required to have overload protection rated or set to trip at not more than 125% of the nameplate full-load current *[see 430.6(A) and 430.32(A)(1)]*.

For the 25-hp motor,

32 A × 1.25 = 40.0 A

For the 30-hp motors,

38 A × 1.25 = 48 A

Where the separate overload device is an overload relay (not a fuse or circuit breaker), and the overload device selected at 125% is not sufficient to start the motor or carry the load, the trip setting is permitted to be increased in accordance with 430.32(C).

**Branch-Circuit Short-Circuit and Ground-Fault Protection**

The selection of the rating of the protective device depends on the type of protective device selected, in accordance with 430.52 and Table 430.52. The following is for the 25-hp motor.

(a) Nontime-Delay Fuse: The fuse rating is 300% × 34 A = 102 A. The next larger standard fuse is 110 A *[see 240.6 and 430.52(C)(1), Exception No. 1]*. If the motor will not start with a 110-A nontime-delay fuse, the fuse rating is permitted to be increased to 125 A because this rating does not exceed 400% *[see 430.52(C)(1), Exception No. 2(a)]*.

(b) Time-Delay Fuse: The fuse rating is 175% × 34 A = 59.5 A. The next larger standard fuse is 60 A *[see 240.6 and 430.52(C)(1), Exception No. 1]*. If the motor will not start with a 60-A time-delay fuse, the fuse rating is permitted to be increased to 70 A because this rating does not exceed 225% *[see 430.52(C)(1), Exception No. 2(b)]*.

**Feeder Short-Circuit and Ground-Fault Protection**

The rating of the feeder protective device is based on the sum of the largest branch-circuit protective device (example is 110 A) plus the sum of the full-load currents of the other motors, or 110 A + 40 A + 40 A = 190 A. The nearest standard fuse that does not exceed this value is 175 A *[see 240.6 and 430.62(A)]*.

### Example D9 Feeder Ampacity Determination for Generator Field Control *[see 215.2, 430.24, 430.24 Exception No. 1, 620.13, 620.14, 620.61, and Table 430.22(E) and 620.14]*

Determine the conductor ampacity for a 460-V 3-phase, 60-Hz ac feeder supplying a group of six elevators. The 460-V ac drive motor nameplate rating of the largest MG set for one elevator is 40 hp and 52 A, and the remaining elevators each have a 30-hp, 40-A, ac drive motor rating for their MG sets. In addition to a motor controller, each elevator has a separate motion/operation controller rated 10 A continuous to operate microprocessors, relays, power supplies, and the elevator car door operator. The MG sets are rated continuous.

**Conductor Ampacity. Conductor ampacity is determined as follows:**

(a) In accordance with 620.13(D) and 620.61(B)(1), use Table 430.22(E), for intermittent duty (elevators). For intermittent duty using a continuous rated motor, the percentage of nameplate current rating to be used is 140%.

(b) For the 30-hp ac drive motor,

140% × 40 A = 56 A.

(c) For the 40-hp ac drive motor,

140% × 52 A = 73 A.

(d) The total conductor ampacity is the sum of all the motor currents: (1 motor × 73 A) + (5 motors × 56 A) = 353 A

(e) In accordance with 620.14 and Table 620.14, the conductor (feeder) ampacity would be permitted to be reduced by the use of a demand factor. Constant loads are not included *(see 620.14, Informational Note)*. For six elevators, the demand factor is 0.79. The feeder diverse ampacity is, therefore, 0.79 × 353 A = 279 A.

(f) In accordance with 430.24 and 215.3, the controller continuous current is 125% × 10 A = 13 A

(g) The total feeder ampacity is the sum of the diverse current and all the controller continuous current.

$$I_{total} = 279 \text{ A} + (6 \text{ elevators} \times 12.5 \text{ A}) = 354 \text{ A}$$

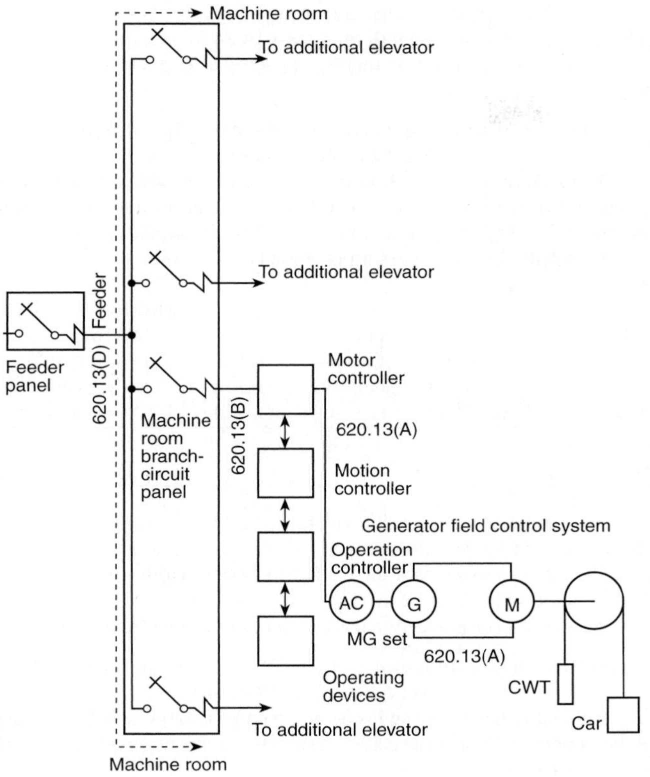


*FIGURE D9 Generator Field Control.*

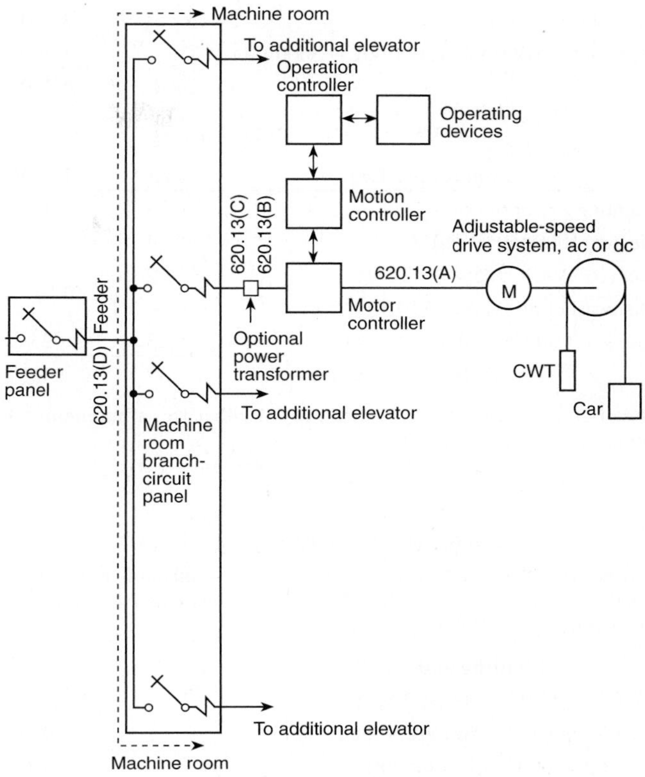


*FIGURE D10 Adjustable Speed Drive Control.*

(h) This ampacity would be permitted to be used to select the wire size.

See Figure D9.

### Example D10 Feeder Ampacity Determination for Adjustable Speed Drive Control *[see 215.2, 430.24, 620.13, 620.14, 620.61, and Table 430.22(E)]*

Determine the conductor ampacity for a 460-V, 3-phase, 60-Hz ac feeder supplying a group of six identical elevators. The system is adjustable-speed SCR dc drive. The power transformers are external to the drive (motor controller) cabinet. Each elevator has a separate motion/operation controller connected to the load side of the main line disconnect switch rated 10 A continuous to operate microprocessors, relays, power supplies, and the elevator car door operator. Each transformer is rated 95 kVA with an efficiency of 90%.

**Conductor Ampacity**

Conductor ampacity is determined as follows:

(a) Calculate the nameplate rating of the transformer:

$$I = \frac{95 \text{ kVA} \times 1000}{\sqrt{3} \times 460 \text{ V} \times 0.90_{\text{eff.}}} = 133 \text{ A}$$

(b) In accordance with 620.13(D), for six elevators, the total conductor ampacity is the sum of all the currents. 6 elevators × 133 A = 798 A

(c) In accordance with 620.14 and Table 620.14, the conductor (feeder) ampacity would be permitted to be reduced by the use of a demand factor. Constant loads are not included *(see 620.13, Informational Note No. 2)*. For six elevators, the demand factor is 0.79. The feeder diverse ampacity is, therefore, 0.79 × 798 A = 630 A.

(d) In accordance with 430.24 and 215.3, the controller continuous current is 125% × 10 A = 13 A.

(e) The total feeder ampacity is the sum of the diverse current and all the controller constant current.

$$I_{\text{total}} = 630 \text{ A} + (6 \text{ elevators} \times 12.5 \text{ A}) = 705 \text{ A}$$

(f) This ampacity would be permitted to be used to select the wire size.

See Figure D10.

### Example D11 Mobile Home *(see 550.18)*

A mobile home floor is 70 ft by 10 ft and has two small appliance circuits; a 1000-VA, 240-V heater; a 200-VA, 120-V exhaust fan; a 400-VA, 120-V dishwasher; and a 7000-VA electric range.

**Lighting and Small-Appliance Load**

| | | |
|---|---|---|
| Lighting (70 ft × 10 ft × 3 VA per ft$^2$) | | 2,100 VA |
| Small-appliance (1500 VA× 2 circuits) | | 3,000 VA |
| Laundry (1500 VA × 1 circuit) | | 1,500 VA |
| | Subtotal | 6,600 VA |

| | |
|---|---|
| First 3000 VA at 100% | 3,000 VA |
| Remainder (6600 VA – 3000 VA = 3600 VA) × 35% | 1,260 VA |
| Total | 4,260 VA |

4260 VA ÷ 240 V = 17.75 A per leg

| Amperes per Leg | Leg A | Leg B |
|---|---|---|
| Lighting and appliances | 18 | 18 |
| Heater (1000 VA ÷ 240 V) | 4 | 4 |
| Fan (200 VA × 125% ÷ 120 V) | 2 | — |
| Dishwasher (400 VA ÷ 120 V) | — | 3 |
| Range (7000 VA × 0.8 ÷ 240 V) | 23 | 23 |
| Total amperes per leg | 47 | 48 |

Based on the higher current calculated for either leg, a minimum 50-A supply cord would be required.

For SI units, 0.093 m² = 1 ft² and 0.3048 m = 1 ft.

### Example D12 Park Trailer *(see 552.47)*

A park trailer floor is 40 ft by 10 ft and has two small appliance circuits, a 1000-VA, 240-V heater, a 200-VA, 120-V exhaust fan, a 400-VA, 120-V dishwasher, and a 7000-VA electric range.

**Lighting and Small-Appliance Load**

| | |
|---|---|
| Lighting (40 ft × 10 ft × 3 VA per ft²) | 1,200 VA |
| Small-appliance (1500 VA × 2 circuits) | 3,000 VA |
| Laundry (1500 VA × 1 circuit) | 1,500 VA |
| Subtotal | 5,700 VA |
| First 3000 VA at 100% | 3,000 VA |
| Remainder (5700 VA – 3000 VA = 2700 VA) × 35% | 945 VA |
| Total | 3,945 VA |

3945 VA÷ 240 V = 16.44 A per leg

| Amperes per Leg | Leg A | Leg B |
|---|---|---|
| Lighting and appliances | 16 | 16 |
| Heater (1000 VA ÷ 240 V) | 4 | 4 |
| Fan (200 VA × 125% ÷ 120 V) | 2 | — |
| Dishwasher (400 VA ÷ 120 V) | — | 3 |
| Range (7000 VA × 0.8 ÷ 240 V) | 23 | 23 |
| Totals | 45 | 46 |

Based on the higher current calculated for either leg, a minimum 50-A supply cord would be required.

For SI units, 0.093 m² = 1 ft² and 0.3048 m = 1 ft.

### Example D13 Cable Tray Calculations *(See Article 392)*

#### D13(a) Multiconductor Cables 4/0 AWG and Larger

**Use:** *NEC* 392.22(A)(1)(a)

Cable tray must have an inside width equal to or greater than the sum of the diameters (Sd) of the cables, which must be installed in a single layer.

**Example:** Cable tray width is obtained as follows:

| Cable Size Being Used | (OD) Cable Outside Diameters (in.) | (N) Number of Cables | SD = (OD) × (N) (Sum of the Cable Diameters) (in.) |
|---|---|---|---|
| 3–conductor Type MC cable — 4/0 AWG | 1.57 | 12 | 18.84 |

The sum of the diameters (Sd) of all cables = 18.84 in., therefore a cable tray with an inside width of at least 18.84 in. is required.

Note: Cable outside diameter is a nominal diameter from catalog data.

#### D13(b) Multiconductor Cables Smaller Than 4/0 AWG

**Use:** *NEC* 392.22(A)(1)(b)

The sum of the cross-sectional areas of all the cables to be installed in the cable tray must be equal to or less than the allowable cable area for the tray width, as indicated in Table 392.22(A), Column 1.

**Example:** Cable tray width is obtained as follows:

| Cable Size Being Used | (A) Cable Cross-Sectional Area (in.²) | (N) Number of Cables | Multiply (A) × (N) (Which Is a Total Cable Cross-Sectional Area in in.²) |
|---|---|---|---|
| 4-conductor Type MC cable — 1 AWG | 1.1350 | 9 | 12.15 |

The total cable cross-sectional area is 12.15 in.². Using Table D13(b) above, the next higher allowable cable area must be used, which is 14.0 in.². The table specifies that the cable tray inside width for an allowable cable area of 14.0 in.² is 12 in.

Note: Cable cross-sectional area is a nominal area from catalog data.

#### D13(c) Single Conductor Cables 1/0 AWG through 4/0 AWG

**Use:** *NEC* 392.22(B)(1)(d)

Cable tray must have an inside width equal to or greater than the sum of the diameters (Sd) of the cables. The cables must be evenly distributed across the cable tray.

**Example:** Cable tray width is obtained as follows:

| Single Conductor Cable Size Being Used | (OD) Cable Outside Diameters (in.) | (N) Number of Cables | Sd = (OD) × (N) (Sum of the Cable Diameters) (in.) |
|---|---|---|---|
| THHN — 4/0 AWG | 0.642 | 18 | 11.556 |

The sum of the diameters (Sd) of all cables = 11.56 in., therefore, a cable tray with an inside width of at least 11.56 in. is required.

Note: Cable outside diameter from Chapter 9, Table 5.

#### D13(d) Single Conductor Cables 250 kcmil through 900 kcmil

**Use:** *NEC* 392.22(B)(1)(b)

The sum of the cross-sectional areas of all the cables to be installed in the cable tray must be equal to or less than the allowable cable area for the tray width, as indicated in Table 392.22(B)(1), Column 1.

**Example:** Cable tray width is obtained as follows:

| Cable Size Being Used | (A) Cable Cross-Sectional Area (in.²) | (N) Number of Cables | Multiply (A) × (N) (Which Is a Total Cable Cross-Sectional Area in in.²) |
|---|---|---|---|
| THHN — 500 kcmil | 0.707 | 9 | 6.36 |

The total cable cross-sectional area is 6.36 in.². Using Table D13(d), the next higher allowable cable area must be used, which is 6.5 in.². The table specifies that the cable tray inside width for an allowable cable area of 6.5 in.² is 6 in.

Note: Single-conductor cable cross-sectional area from Chapter 9, Table 5.

***TABLE D13(b)*** *from Table 392.22(A), Column 1*

| Inside Width of Cable Tray (in.) | Allowable Cable Area ($in.^2$) |
|---|---|
| 6 | 7.0 |
| 9 | 10.5 |
| 12 | 14.0 |
| 18 | 21.0 |
| 24 | 28.0 |
| 30 | 35.0 |
| 36 | 42.0 |

***TABLE D13(d)*** *from Table 392.22(B)(1), Column 1*

| Inside Width of Cable Tray (in.) | Allowable Cable Area ($in.^2$) |
|---|---|
| 6 | 6.5 |
| 9 | 9.5 |
| 12 | 13.0 |
| 18 | 19.5 |
| 24 | 26.0 |
| 30 | 32.5 |
| 36 | 39.0 |

# E Informative Annex
## Types of Construction

*This informative annex is not a part of the requirements of this NFPA document but is included for informational purposes only.*

Table E.1 contains the fire resistance rating, in hours, for Types I through V construction. The five different types of construction can be summarized briefly as follows (see also Table E.2):

***TABLE E.1*** *Fire Resistance Ratings for Type I Through Type V Construction (hr)*

| | Type I | | Type II | | | Type III | | Type IV | Type V | |
|---|---|---|---|---|---|---|---|---|---|---|
| | 442 | 332 | 222 | 111 | 000 | 211 | 200 | 2HH | 111 | 000 |
| **Exterior Bearing Walls**[a] | | | | | | | | | | |
| Supporting more than one floor, columns, or other bearing walls | 4 | 3 | 2 | 1 | 0[b] | 2 | 2 | 2 | 1 | 0[b] |
| Supporting one floor only | 4 | 3 | 2 | 1 | 0[b] | 2 | 2 | 2 | 1 | 0[b] |
| Supporting a roof only | 4 | 3 | 1 | 1 | 0[b] | 2 | 2 | 2 | 1 | 0[b] |
| **Interior Bearing Walls** | | | | | | | | | | |
| Supporting more than one floor, columns, or other bearing walls | 4 | 3 | 2 | 1 | 0 | 1 | 0 | 2 | 1 | 0 |
| Supporting one floor only | 3 | 2 | 2 | 1 | 0 | 1 | 0 | 1 | 1 | 0 |
| Supporting roofs only | 3 | 2 | 1 | 1 | 0 | 1 | 0 | 1 | 1 | 0 |
| **Columns** | | | | | | | | | | |
| Supporting more than one floor, columns, or other bearing walls | 4 | 3 | 2 | 1 | 0 | 1 | 0 | H | 1 | 0 |
| Supporting one floor only | 3 | 2 | 2 | 1 | 0 | 1 | 0 | H | 1 | 0 |
| Supporting roofs only | 3 | 2 | 1 | 1 | 0 | 1 | 0 | H | 1 | 0 |
| **Beams, Girders, Trusses, and Arches** | | | | | | | | | | |
| Supporting more than one floor, columns, or other bearing walls | 4 | 3 | 2 | 1 | 0 | 1 | 0 | H | 1 | 0 |
| Supporting one floor only | 2 | 2 | 2 | 1 | 0 | 1 | 0 | H | 1 | 0 |
| Supporting roofs only | 2 | 2 | 1 | 1 | 0 | 1 | 0 | H | 1 | 0 |
| **Floor/Ceiling Assemblies** | 2 | 2 | 2 | 1 | 0 | 1 | 0 | H | 1 | 0 |
| **Roof/Ceiling Assemblies** | 2 | 1½ | 1 | 1 | 0 | 1 | 0 | H | 1 | 0 |
| **Interior Nonbearing Walls** | 0 | 0 | 0 | 0 | 0 | 0 | 0 | 0 | 0 | 0 |
| **Exterior Nonbearing Walls**[c] | 0[b] | 0[b] | 0[b] | 0[b] | 0[b] | 0[b] | 0[b] | 0[b] | 0[b] | 0[b] |

Source: Table 7.2.1.1 from *NFPA 5000, Building Construction and Safety Code*, 2012 edition.
H: Heavy timber members.
[a]See 7.3.2.1 in *NFPA 5000*.
[b]See Section 7.3 in *NFPA 5000*.
[c]See 7.2.3.2.12, 7.2.4.2.3, and 7.2.5.6.8 in *NFPA 5000*.

***TABLE E.2*** *Maximum Number of Stories for Types V, IV, and III Construction*

| Construction Type | Maximum Number of Stories Permitted |
|---|---|
| V Rated | 2 |
| V Rated, Sprinklered | 3 |
| V One-Hour Rated | 3 |
| V One-Hour Rated, Sprinklered | 4 |
| IV Heavy Timber | 4 |
| IV Heavy Timber, Sprinklered | 5 |
| III Rated | 2 |
| III Rated, Sprinklered | 3 |
| III One-Hour Rated | 4 |
| III One-Hour Rated, Sprinklered | 5 |

Type I is a Fire-Resistive construction type. All structural elements and most interior elements are required to be noncombustible. Interior, nonbearing partitions are permitted to be 1 or 2 hour rated. For nearly all occupancy types, Type I construction can be of unlimited height.

Type II construction has 3 categories: Fire-Resistive, One-Hour Rated, and Rated. The number of stories permitted for multifamily dwellings varies from two for Rated and four for One-Hour Rated to 12 for Fire-Resistive construction.

Type III construction has two categories: One-Hour Rated and Rated. Both categories require the structural framework and exterior walls to be of noncombustible material. One-Hour Rated construction requires all interior partitions to be one-hour rated. Rated construction allows nonbearing interior partitions to be of non-rated construction. The maximum permitted number of stories for multifamily dwellings and other structures is two for Rated and four for One-Hour Rated.

Type IV is a single construction category that provides for heavy timber construction. Both the structural framework and the exterior walls are required to be noncombustible except that wood members of certain minimum sizes are allowed. This construction type is seldom used for multifamily dwellings but, if used, would be permitted to be four stories high.

Type V construction has two categories: One-Hour Rated and Rated. One-Hour Rated construction requires a minimum of one-hour rated construction throughout the building. Rated construction allows non-rated interior partitions with certain restrictions. The maximum permitted number of stories for multifamily dwellings and other structures is 2 for Rated and 3 for One-Hour Rated.

In Table E.1 the system of designating types of construction also includes a specific breakdown of the types of construction through the use of arabic numbers. These arabic numbers follow the roman numeral notation where identifying a type of construction [for example, Type I(442), Type II(111), Type III(200)] and indicate the fire resistance rating requirements for certain structural elements as follows:

(1) First arabic number — exterior bearing walls
(2) Second arabic number — columns, beams, girders, trusses and arches, supporting bearing walls, columns, or loads from more than one floor
(3) Third arabic number — floor construction

Table E.3 provides a comparison of the types of construction for various model building codes. [***5000:*** A.7.2.1.1]

Table E.3 is reproduced from *NFPA 5000®, Building Construction and Safety Code®*. This table cross-references the building construction types described in NFPA 220, *Standard on Types of Building Construction*, to the construction types described in four other model building codes. For AHJs in a municipality where one of these model building codes is used, this table provides helpful information to assist in the proper application of 334.10(2), 334.10(3), and 334.10(5) covering the permitted use of Type NM cable based on building construction type. The types of construction are based on NFPA 220, and Table E.3 facilitates assimilation of the 334.10

***TABLE E.3*** *Cross-Reference of Building Construction Types*

| *NFPA 5000* | I(442) | I(332) | II(222) | II(111) | II(000) | III(211) | III(200) | IV(2HH) | V(111) | V(000) |
|---|---|---|---|---|---|---|---|---|---|---|
| UBC | — | I FR | II FR | II 1 hr | II N | III 1 hr | III N | IV HT | V 1 hr | V N |
| B/NBC | 1A | 1B | 2A | 2B | 2C | 3A | 3B | 4 | 5A | 5B |
| SBC | I | II | — | IV 1 hr | IV UNP | V 1 hr | V UNP | III | VI 1 hr | VI UNP |
| IBC | — | IA | IB | IIA | IIB | IIIA | IIIB | IV | VA | VB |

Source: Table A.7.2.1.1 from *NFPA 5000, Building Construction and Safety Code*, 2012 edition.
UBC: *Uniform Building Code.*
FR: Fire rated.
N: Nonsprinklered.
HT: Heavy timber.
B/NBC: *National Building Code.*
SBC: *Standard Building Code.*
UNP: Unprotected.
IBC: *International Building Code.*

and 334.12 requirements with the construction types contained within the building code that is adopted by a jurisdiction.

The following is a description of the model building code acronyms contained in Table E.3:

UBC – *Uniform Building Code,* International Conference of Building Officials, Whittier, CA

B/NBC – *BOCA National Building Code,* Building Officials and Code Administrators International Inc., Country Club Hills, IL

SBC – *Standard Building Code,* Southern Building Code Congress International, Inc., Birmingham, AL

IBC – *International Building Code,* International Code Council, Inc., Falls Church, VA

The information for the UBC, B/NBC, and SBC is included as reference for jurisdictions using these building codes. The publishers of these three codes founded the International Code Council in 1994 and combined the regional codes into a single document, the *International Building Code (IBC).*

# F Informative Annex

# Availability and Reliability for Critical Operations Power Systems; and Development and Implementation of Functional Performance Tests (FPTs) for Critical Operations Power Systems

*This informative annex is not a part of the requirements of this NFPA document but is included for informational purposes only.*

**I. Availability and Reliability for Critical Operations Power Systems.** Critical operations power systems may support facilities with a variety of objectives that are vital to public safety. Often these objectives are of such critical importance that system downtime is costly in terms of economic losses, loss of security, or loss of mission. For those reasons, the availability of the critical operations power system, the percentage of time that the system is in service, is important to those facilities. Given a specified level of availability, the reliability and maintainability requirements are then derived based on that availability requirement.

*Availability.* Availability is defined as the percentage of time that a system is available to perform its function(s). Availability is measured in a variety of ways, including the following:

$$\text{Availability} = \frac{MTBF}{MTBF + MTTR}$$

where:

$MTBF$ = mean time between failures
$MTTF$ = mean time to failure
$MTTR$ = mean time to repair

See the following table for an example of how to establish required availability for critical operation power systems:

| Availability | Hours of Downtime* |
|---|---|
| 0.9 | 876 |
| 0.99 | 87.6 |
| 0.999 | 8.76 |
| 0.9999 | 0.876 |
| 0.99999 | 0.0876 |
| 0.999999 | 0.00876 |
| 0.9999999 | 0.000876 |

*Based on a year of 8760 hours.

Availability of a system in actual operations is determined by the following:

(1) The frequency of occurrence of failures. Failures may prevent the system from performing its function or may cause a degraded effect on system operation. Frequency of failures is directly related to the system's level of reliability.
(2) The time required to restore operations following a system failure or the time required to perform maintenance to prevent a failure. These times are determined in part by the system's level of maintainability.
(3) The logistics provided to support maintenance of the system. The number and availability of spares, maintenance personnel, and other logistics resources (refueling, etc.) combined with the system's level of maintainability determine the total downtime following a system failure.

*Reliability.* Reliability is concerned with the probability and frequency of failures (or lack of failures). A commonly used measure of reliability for repairable systems is *MTBF*. The equivalent measure for nonrepairable items is *MTTF*. Reliability is more accurately expressed as a probability over a given duration of time, cycles, or other parameter. For example, the reliability of a power plant might be stated as 95 percent probability of no failure over a 1000-hour operating period while generating a certain level of power. Reliability is usually defined in two ways (the electrical power industry has historically not used these definitions):

(1) The duration or probability of failure-free performance under stated conditions
(2) The probability that an item can perform its intended function for a specified interval under stated conditions [For nonredundant items, this is equivalent to the preceding definition (1). For redundant items this is equivalent to the definition of mission reliability.]

*Maintainability.* Maintainability is a measure of how quickly and economically failures can be prevented through preventive maintenance, or system operation can be restored following failure through corrective maintenance. A commonly used measure of maintainability in terms of corrective maintenance is the mean time to repair (*MTTR*). Maintainability is not the same thing as maintenance. It is a design parameter, while maintenance consists of actions to correct or prevent a failure event.

*Improving Availability.* The appropriate methods to use for improving availability depend on whether the facility is being designed or is already in use. For both cases, a reliability/availability analysis should be performed to determine the availability of the old system or proposed new system in order to ascertain the hours of downtime (see the preceding table). The AHJ or government agency should dictate how much downtime is acceptable.

Existing facilities: For a facility that is being operated, two basic methods are available for improving availability when the current level of availability is unacceptable: (1) Selectively adding redundant units (e.g., generators, chillers, fuel supply to eliminate sources of single-point failure, and (2) optimizing maintenance using a reliability-centered maintenance (RCM) approach to minimize downtime. [Refer to NFPA 70B-2010, *Recommended Practice for Electrical Equipment Maintenance*.] A combination of the previous two methods can also be implemented. A third very expensive method is to redesign subsystems or to replace components and subsystems with higher reliability items. [Refer to NFPA 70B.]

New facilities: The opportunity for high availability and reliability is greatest when designing a new facility. By applying an effective reliability strategy, designing for maintainability, and ensuring that manufacturing and commissioning do not negatively affect the inherent levels of reliability and maintainability, a highly available facility will result. The approach should be as follows:

(1) *Develop and determine a reliability strategy* (establish goals, develop a system model, design for reliability, conduct reliability development testing, conduct reliability acceptance testing, design system delivery, maintain design reliability, maintain design reliability in operation).
(2) *Develop a reliability program.* This is the application of the reliability strategy to a specific system, process, or function. Each step in the preceding strategy requires the selection and use of specific methods and tools. For example, various tools can be used to develop requirements or evaluate potential failures. To derive requirements, analytical models can be used, for example, quality function development (a technique for deriving more detailed, lower-level requirements from one level to another, beginning with mission requirements, i.e., customer needs). This model was developed as part of the total quality management movement. Parametric models can also be used to derive design values of reliability from operational values and vice versa. Analytical methods include but are not limited to things such as thermal analysis, durability analysis, and predictions. Finally, one should evaluate possible failures. A failure modes and effects criticality analysis (FMECA) and fault tree analysis (FTA) are two methods for evaluating possible failures. The mission facility engineer should determine which method to use or whether to use both.
(3) *Identify Reliability Requirements.* The entire effort for designing for reliability begins with identifying the mission critical facility's reliability requirements. These requirements are stated in a variety of ways, depending on the customer and the specific system. For a mission critical facility, it would be the mission success probability.

## II. Development and Implementation of Functional Performance Tests (FPTs) for Critical Operations Power Systems Development of FPT

**(1) Submit Functional Performance Tests (FPTs).** System/component tests or FPTs are developed from submitted drawings, systems operating documents (SODs), and systems operation and maintenance manuals (SOMMs), including large component testing (i.e., transformers, cable, generators, UPS), and how components operate as part of the total system. The commissioning authority develops the test and cannot be the installation contractor (or subcontractor).

As the equipment/components/systems are installed, quality assurance procedures are administered to verify that components are installed in accordance with minimum manufacturers' recommendations, safety codes, and acceptable installation practices. Quality assurance discrepancies are then identified and added to a "commissioning action list" that must be rectified as part of the commissioning program. These items would usually be discussed during commissioning meetings. Discrepancies are usually identified initially by visual inspection.

**(2) Review FPTs.** The tests must be reviewed by the customer, electrical contractors, quality assurance personnel, maintenance personnel, and other key personnel (the commissioning team). Areas of concern include, among others, all functions of the system being tested, all major components included, whether the tests reflect the system operating documents, and verification that the tests make sense.

**(3) Make Changes to FPTs as Required.** The commissioning authority then implements the corrections, questions answered, and additions.

**(4) FPTs Approval.** After the changes are made to the FPTs, they are submitted to the commissioning team. When it is acceptable, the customer or the designated approval authority approves the FPTs. It should be noted that even though the FPT is approved, problems that arise during the test (or areas not covered) must be addressed.

**Testing Implementation for FPTs.** The final step in the successful commissioning plan is testing and proper execution of system-integrated tests.

**(1) Systems Ready to Operate.** The FPTs can be implemented as various systems become operative (i.e., test for the generator system) or when the entire system is installed. However, the final "pull the plug" test is performed only after all systems are completely installed. If the electrical contractor (or subcontractor) implements the FPTs, a witness must initial each step of the test. The electrical contractor cannot employ the witness directly or indirectly.

**(2) Perform Tests (FPTs).** If the system fails the test, the problem must be resolved and the equipment or system retested or the testing requirements re-analyzed until successful tests are witnessed. Once the system or equipment passes testing, it is verified by designated commissioning official.

**(3) Customer Receives System.** After all tests are completed (including the "pull the plug" test), the system is turned over to the customer.

# G Informative Annex
## Supervisory Control and Data Acquisition (SCADA)

*This informative annex is not a part of the requirements of this NFPA document, but is included for informational purposes only.*

**(A) General.** Where provided, the general requirements in (A)(1) through (A)(11) shall apply to SCADA systems. The SCADA system for the COPS loads shall be separate from the building management SCADA system. No single point failure shall be able to disable the SCADA system.

(1) The SCADA system for the COPS loads shall be separate from the building management SCADA system.
(2) No single point failure shall be able to disable the SCADA system.
(3) The SCADA system shall be permitted to provide control and monitor electrical and mechanical utility systems related to mission critical loads, including, but not limited to the following:
   a. The fire alarm system
   b. The security system
   c. Power distribution
   d. Power generation
   e. HVAC and ventilation (damper position, airflow speed and direction)
   f. Load shedding
   g. Fuel levels or hours of operation
(4) Before installing or employing a SCADA system, an operations and maintenance analysis and risk assessment shall be performed to provide the maintenance parameter data.
(5) A redundant system shall be provided in either warm or hot standby.
(6) The controller shall be a programmable logic controller (PLC).
(7) The SCADA system shall utilize open, not proprietary, protocols.
(8) The SCADA system shall be able to assess the damage and determine system integrity after the "event."
(9) The monitor display shall provide graphical user interface for all major components monitored and controlled by the SCADA system, with color schemes readily recognized by the typical user.
(10) The SCADA system shall have the capability to provide storage of critical system parameters at a 15-minute rate or more often when out-of-limit conditions exist.
(11) The SCADA system shall have a separate data storage facility not located in the same vicinity.

**(B) Power Supply.** The SCADA system power supply shall comply with (B)(1) through (B)(3):

(1) The power supply shall be provided with a direct-current station battery system, rated between 24 and 125 volts dc, with a 72-hour capacity.
(2) The batteries of the SCADA system shall be separate from the batteries for other electrical systems.
(3) The power supply shall be provided with a properly installed surge-protective device (TVSS) at its terminals with a direct low-impedance path to ground. Protected and unprotected circuits shall be physically separated to prevent coupling.

**(C) Security Against Hazards.** Security against hazards shall be provided in accordance with (C)(1) through (C)(6):

(1) Controlled physical access by authorized personnel to only the system operational controls and software shall be provided.
(2) The SCADA system shall be protected against dust, dirt, water, and other contaminants by specifying enclosures appropriate for the environment.
(3) Conduit and tubing shall not violate the integrity of the SCADA system enclosure.
(4) The SCADA system shall be located in the same secure locations as the secured systems that they monitor and control.
(5) The SCADA system shall be provided with dry agent fire protection systems or double interlocked preaction sprinkler systems using cross-zoned detection, to minimize the threat of accidental water discharge into unprotected equipment. The fire protection systems shall be monitored by the fire alarm system in accordance with *NFPA 72*-2013, *National Fire Alarm and Signaling Code*.

(6) The SCADA system shall not be connected to other network communications outside the secure locations without encryption or use of fiber optics.

**(D) Maintenance and Testing.** SCADA systems shall be maintained and tested in accordance with (D)(1) and (D)(2).

**(1) Maintenance.** The maintenance program for SCADA systems shall consist of the following components:

(1) A documented preventive maintenance program
(2) Concurrent maintenance capabilities, to allow the testing, troubleshooting, repair, and/or replacement of a component or subsystem while redundant component(s) or subsystem(s) are serving the load
(3) Retention of operational data — the deleted material goes well beyond requirements to ensure proper maintenance and operation

**(2) Testing.** SCADA systems shall be tested periodically under actual or simulated contingency conditions.

Informational Note No. 1: Periodic system testing procedures can duplicate or be derived from the recommended functional performance testing procedures of individual components, as provided by the manufacturers.

Informational Note No. 2: For more information on maintenance and testing of SCADA, see NFPA 70B-2013, *Recommended Practice for Electrical Equipment Maintenance.*

# H Informative Annex
## Administration and Enforcement

Informative Annex H is a model set of rules for adoption by jurisdictions, in whole or in part, to administer an electrical inspection program. Unless specifically adopted by a governmental or other entity charged with enforcing electrical installation requirements, the material is advisory only. This annex can serve as a template that can be used verbatim or with amendments tailored to fit that jurisdiction's needs. It provides an inspection process covering plan review, issuance of permits, number and types of inspections, and connection to or disconnection from the electrical supply system. Processes to file and adjudicate appeals and to adopt the current edition of the *Code* are also provided. In addition, requirements for qualified electrical inspection personnel and for establishing an electrical board are provided.

Informative Annex H, Article 80, is based on and replaces NFPA 70L, *Model State Law, Inspection of Electrical Installations*, which was adopted by NFPA on May 15, 1973, and a second edition was approved on March 27, 1987.

For most political subdivisions, adoption of the *NEC* can occur in two ways. It can be incorporated in a law, or a law can be enacted authorizing a governmental agency or board to adopt it. Rule-making and legislative processes in a particular jurisdiction determine which alternative is more appropriate.

*Informative Annex H is not a part of the requirements of this NFPA document and is included for informational purposes only. This informative annex is informative unless specifically adopted by the local jurisdiction adopting the National Electrical Code®.*

### 80.1 Scope

The following functions are covered:

(1) The inspection of electrical installations as covered by 90.2
(2) The investigation of fires caused by electrical installations
(3) The review of construction plans, drawings, and specifications for electrical systems
(4) The design, alteration, modification, construction, maintenance, and testing of electrical systems and equipment
(5) The regulation and control of electrical installations at special events including but not limited to exhibits, trade shows, amusement parks, and other similar special occupancies

### 80.2 Definitions

**Authority Having Jurisdiction.** The organization, office, or individual responsible for approving equipment, materials, an installation, or a procedure.

**Chief Electrical Inspector.** An electrical inspector who either is the authority having jurisdiction or is designated by the authority having jurisdiction and is responsible for administering the requirements of this *Code*.

**Electrical Inspector.** An individual meeting the requirements of 80.27 and authorized to perform electrical inspections.

### 80.3 Purpose

The purpose of this article shall be to provide requirements for administration and enforcement of the *National Electrical Code*.

### 80.5 Adoption

Article 80 shall not apply unless specifically adopted by the local jurisdiction adopting the *National Electrical Code*.

### 80.7 Title

The title of this *Code* shall be NFPA 70, *National Electrical Code*®, of the National Fire Protection Association. The short title of this *Code* shall be the *NEC*®.

### 80.9 Application

**(A) New Installations.** This *Code* applies to new installations. Buildings with construction permits dated after adoption of this *Code* shall comply with its requirements.

**(B) Existing Installations.** Existing electrical installations that do not comply with the provisions of this *Code* shall be permitted to be continued in use unless the authority having jurisdiction determines that the lack of conformity with this *Code* presents an imminent danger to occupants. Where changes are required for correction of hazards, a reasonable amount of time shall be given for compliance, depending on the degree of the hazard.

**(C) Additions, Alterations, or Repairs.** Additions, alterations, or repairs to any building, structure, or premises shall conform to that required of a new building without requiring the existing building to comply with all the requirements of this *Code*. Additions, alterations, installations, or repairs shall not cause an existing building to become unsafe or to adversely affect the performance of the building as determined by the authority having jurisdiction. Electrical wiring added to an existing service, feeder, or branch circuit shall not result in an installation that violates the provisions of the *Code* in force at the time the additions are made.

## 80.11 Occupancy of Building or Structure

**(A) New Construction.** No newly constructed building shall be occupied in whole or in part in violation of the provisions of this *Code*.

**(B) Existing Buildings.** Existing buildings that are occupied at the time of adoption of this *Code* shall be permitted to remain in use provided the following conditions apply:

(1) The occupancy classification remains unchanged
(2) There exists no condition deemed hazardous to life or property that would constitute an imminent danger

## 80.13 Authority

Where used in this article, the term *authority having jurisdiction* shall include the chief electrical inspector or other individuals designated by the governing body. This *Code* shall be administered and enforced by the authority having jurisdiction designated by the governing authority as follows.

(1) The authority having jurisdiction shall be permitted to render interpretations of this *Code* in order to provide clarification to its requirements, as permitted by 90.4.
(2) When the use of any electrical equipment or its installations is found to be dangerous to human life or property, the authority having jurisdiction shall be empowered to have the premises disconnected from its source of electric supply, as established by the Board. When such equipment or installation has been so condemned or disconnected, a notice shall be placed thereon listing the causes for the condemnation, the disconnection, or both, and the penalty under 80.23 for the unlawful use thereof. Written notice of such condemnation or disconnection and the causes therefor shall be given within 24 hours to the owners, the occupant, or both, of such building, structure, or premises. It shall be unlawful for any person to remove said notice, to reconnect the electrical equipment to its source of electric supply, or to use or permit to be used electric power in any such electrical equipment until such causes for the condemnation or disconnection have been remedied to the satisfaction of the inspection authorities.
(3) The authority having jurisdiction shall be permitted to delegate to other qualified individuals such powers as necessary for the proper administration and enforcement of this *Code*.
(4) Police, fire, and other enforcement agencies shall have authority to render necessary assistance in the enforcement of this *Code* when requested to do so by the authority having jurisdiction.
(5) The authority having jurisdiction shall be authorized to inspect, at all reasonable times, any building or premises for dangerous or hazardous conditions or equipment as set forth in this *Code*. The authority having jurisdiction shall be permitted to order any person(s) to remove or remedy such dangerous or hazardous condition or equipment. Any person(s) failing to comply with such order shall be in violation of this *Code*.
(6) Where the authority having jurisdiction deems that conditions hazardous to life and property exist, he or she shall be permitted to require that such hazardous conditions in violation of this *Code* be corrected.
(7) To the full extent permitted by law, any authority having jurisdiction engaged in inspection work shall be authorized at all reasonable times to enter and examine any building, structure, or premises for the purpose of making electrical inspections. Before entering a premises, the authority having jurisdiction shall obtain the consent of the occupant thereof or obtain a court warrant authorizing entry for the purpose of inspection except in those instances where an emergency exists. As used in this section, *emergency* means circumstances that the authority having jurisdiction knows, or has reason to believe, exist and that reasonably can constitute immediate danger
(8) Persons authorized to enter and inspect buildings, structures, and premises as herein set forth shall be identified by proper credentials issued by this governing authority.
(9) Persons shall not interfere with an authority having jurisdiction carrying out any duties or functions prescribed by this *Code*.
(10) Persons shall not use a badge, uniform, or other credentials to impersonate the authority having jurisdiction.
(11) The authority having jurisdiction shall be permitted to investigate the cause, origin, and circumstances of any fire, explosion, or other hazardous condition.
(12) The authority having jurisdiction shall be permitted to require plans and specifications to ensure compliance with this *Code*.
(13) Whenever any installation subject to inspection prior to use is covered or concealed without having first been inspected, the authority having jurisdiction shall be permitted to require that such work be exposed for inspection. The authority having jurisdiction shall be notified when the installation is ready for inspection and shall conduct the inspection within ___ days.

(14) The authority having jurisdiction shall be permitted to order the immediate evacuation of any occupied building deemed unsafe when such building has hazardous conditions that present imminent danger to building occupants.
(15) The authority having jurisdiction shall be permitted to waive specific requirements in this *Code* or permit alternative methods where it is assured that equivalent objectives can be achieved by establishing and maintaining effective safety. Technical documentation shall be submitted to the authority having jurisdiction to demonstrate equivalency and that the system, method, or device is approved for the intended purpose.
(16) Each application for a waiver of a specific electrical requirement shall be filed with the authority having jurisdiction and shall be accompanied by such evidence, letters, statements, results of tests, or other supporting information as required to justify the request. The authority having jurisdiction shall keep a record of actions on such applications, and a signed copy of the authority having jurisdiction's decision shall be provided for the applicant.

## 80.15 Electrical Board

**(A) Creation of the Electrical Board.** There is hereby created the Electrical Board of the _________ of _________, hereinafter designated as the Board.

**(B) Appointments.** Board members shall be appointed by the Governor with the advice and consent of the Senate (or by the Mayor with the advice and consent of the Council, or the equivalent).

(1) Members of the Board shall be chosen in a manner to reflect a balanced representation of individuals or organizations. The Chair of the Board shall be elected by the Board membership.
(2) The Chief Electrical Inspector in the jurisdiction adopting this Article authorized in (B)(3)(a) shall be the nonvoting secretary of the Board. Where the Chief Electrical Inspector of a local municipality serves a Board at a state level, he or she shall be permitted to serve as a voting member of the Board.
(3) The board shall consist of not fewer than five voting members. Board members shall be selected from the following:
   a. Chief Electrical Inspector from a local government (for State Board only)
   b. An electrical contractor operating in the jurisdiction
   c. A licensed professional engineer engaged primarily in the design or maintenance of electrical installations
   d. A journeyman electrician
(4) Additional membership shall be selected from the following:
   a. A master (supervising) electrician
   b. The Fire Marshal (or Fire Chief)
   c. A representative of the property/casualty insurance industry
   d. A representative of an electric power utility operating in the jurisdiction
   e. A representative of electrical manufacturers primarily and actively engaged in producing materials, fittings, devices, appliances, luminaires, or apparatus used as part of or in connection with electrical installations
   f. A member of the labor organization that represents the primary electrical workforce
   g. A member from the public who is not affiliated with any other designated group
   h. A representative of a telecommunications utility operating in the jurisdiction

**(C) Terms.** Of the members first appointed, _____ shall be appointed for a term of 1 year, _____ for a term of 2 years, _____ for a term of 3 years, and _____ for a term of 4 years, and thereafter each appointment shall be for a term of 4 years or until a successor is appointed. The Chair of the Board shall be appointed for a term not to exceed ____ years.

**(D) Compensation.** Each appointed member shall receive the sum of ______dollars ($______) for each day during which the member attends a meeting of the Board and, in addition thereto, shall be reimbursed for direct lodging, travel, and meal expenses as covered by policies and procedures established by the jurisdiction.

**(E) Quorum.** A quorum as established by the Board operating procedures shall be required to conduct Board business. The Board shall hold such meetings as necessary to carry out the purposes of Article 80. The Chair or a majority of the members of the Board shall have the authority to call meetings of the Board.

**(F) Duties.** It shall be the duty of the Board to perform the following:

(1) Adopt the necessary rules and regulations to administer and enforce Article 80.
(2) Establish qualifications of electrical inspectors.
(3) Revoke or suspend the recognition of any inspector's certificate for the jurisdiction.
(4) After advance notice of the public hearings and the execution of such hearings, as established by law, the Board is authorized to establish and update the provisions for the safety of electrical installations to conform to the current edition of the *National Electrical Code* (NFPA 70) and other nationally recognized safety standards for electrical installations.
(5) Establish procedures for recognition of electrical safety standards and acceptance of equipment conforming to these standards.

**(G) Appeals.**

(1) *Review of Decisions.* Any person, firm, or corporation may register an appeal with the Board for a review of any decision of the Chief Electrical Inspector or of any Electrical Inspector, provided that such appeal is made in writing

within fifteen (15) days after such person, firm, or corporation shall have been notified. Upon receipt of such appeal, said Board shall, if requested by the person making the appeal, hold a public hearing and proceed to determine whether the action of the Board, or of the Chief Electrical Inspector, or of the Electrical Inspector complies with this law and, within fifteen (15) days after receipt of the appeal or after holding the hearing, shall make a decision in accordance with its findings.

(2) *Conditions.* Any person shall be permitted to appeal a decision of the authority having jurisdiction to the Board when it is claimed that any one or more of the following conditions exist:

a. The true intent of the codes or ordinances described in this *Code* has been incorrectly interpreted.
b. The provisions of the codes or ordinances do not fully apply.
c. A decision is unreasonable or arbitrary as it applies to alternatives or new materials.

(3) *Submission of Appeals.* A written appeal, outlining the *Code* provision from which relief is sought and the remedy proposed, shall be submitted to the authority having jurisdiction within 15 calendar days of notification of violation.

**(H) Meetings and Records.** Meetings and records of the Board shall conform to the following:

(1) Meetings of the Board shall be open to the public as required by law.
(2) Records of meetings of the Board shall be available for review during normal business hours, as required by law.

## 80.17 Records and Reports

The authority having jurisdiction shall retain records in accordance with (A) and (B).

**(A) Retention.** The authority having jurisdiction shall keep a record of all electrical inspections, including the date of such inspections and a summary of any violations found to exist, the date of the services of notices, and a record of the final disposition of all violations. All required records shall be maintained until their usefulness has been served or as otherwise required by law.

**(B) Availability.** A record of examinations, approvals, and variances granted shall be maintained by the authority having jurisdiction and shall be available for public review as prescribed by law during normal business hours.

## 80.19 Permits and Approvals

Permits and approvals shall conform to (A) through (H).

**(A) Application.**

(1) Activity authorized by a permit issued under this *Code* shall be conducted by the permittee or the permittee's agents or employees in compliance with all requirements of this *Code* applicable thereto and in accordance with the approved plans and specifications. No permit issued under this *Code* shall be interpreted to justify a violation of any provision of this *Code* or any other applicable law or regulation. Any addition or alteration of approved plans or specifications shall be approved in advance by the authority having jurisdiction, as evidenced by the issuance of a new or amended permit.
(2) A copy of the permit shall be posted or otherwise readily accessible at each work site or carried by the permit holder as specified by the authority having jurisdiction.

**(B) Content.** Permits shall be issued by the authority having jurisdiction and shall bear the name and signature of the authority having jurisdiction or that of the authority having jurisdiction's designated representative. In addition, the permit shall indicate the following:

(1) Operation or activities for which the permit is issued
(2) Address or location where the operation or activity is to be conducted
(3) Name and address of the permittee
(4) Permit number and date of issuance
(5) Period of validity of the permit
(6) Inspection requirements

**(C) Issuance of Permits.** The authority having jurisdiction shall be authorized to establish and issue permits, certificates, notices, and approvals, or orders pertaining to electrical safety hazards pursuant to 80.23, except that no permit shall be required to execute any of the classes of electrical work specified in the following:

(1) Installation or replacement of equipment such as lamps and of electric utilization equipment approved for connection to suitable permanently installed receptacles. Replacement of flush or snap switches, fuses, lamp sockets, and receptacles, and other minor maintenance and repair work, such as replacing worn cords and tightening connections on a wiring device
(2) The process of manufacturing, testing, servicing, or repairing electrical equipment or apparatus

**(D) Annual Permits.** In lieu of an individual permit for each installation or alteration, an annual permit shall, upon application, be issued to any person, firm, or corporation regularly employing one or more employees for the installation, alteration, and maintenance of electrical equipment in or on buildings or premises owned or occupied by the applicant for the permit. Upon application, an electrical contractor as agent for the owner or tenant shall be issued an annual permit. The applicant shall keep records of all work done, and the records shall be transmitted periodically to the electrical inspector.

**(E) Fees.** Any political subdivision that has been provided for electrical inspection in accordance with the provisions of Article 80 may establish fees that shall be paid by the applicant for a permit before the permit is issued.

**(F) Inspection and Approvals.**

(1) Upon the completion of any installation of electrical equipment that has been made under a permit other than an annual permit, it shall be the duty of the person, firm, or corporation making the installation to notify the Electrical Inspector having jurisdiction, who shall inspect the work within a reasonable time.

(2) Where the Inspector finds the installation to be in conformity with the statutes of all applicable local ordinances and all rules and regulations, the Inspector shall issue to the person, firm, or corporation making the installation a certificate of approval, with duplicate copy for delivery to the owner, authorizing the connection to the supply of electricity and shall send written notice of such authorization to the supplier of electric service. When a certificate of temporary approval is issued authorizing the connection of an installation, such certificates shall be issued to expire at a time to be stated therein and shall be revocable by the Electrical Inspector for cause.

(3) When any portion of the electrical installation within the jurisdiction of an Electrical Inspector is to be hidden from view by the permanent placement of parts of the building, the person, firm, or corporation installing the equipment shall notify the Electrical Inspector, and the equipment shall not be concealed until it has been approved by the Electrical Inspector or until _____ days have elapsed from the time of such notification, provided that on large installations, where the concealment of equipment proceeds continuously, the person, firm, or corporation installing the equipment shall give the Electrical Inspector due notice in advance, and inspections shall be made periodically during the progress of the work.

(4) At regular intervals, the Electrical Inspector having jurisdiction shall visit all buildings and premises where work may be done under annual permits and shall inspect all electrical equipment installed under such permits since the date of the previous inspection. The Electrical Inspector shall issue a certificate of approval for such work as is found to be in conformity with the provisions of Article 80 and all applicable ordinances, orders, rules, and regulations, after payments of all required fees.

(5) If, upon inspection, any installation is found not to be fully in conformity with the provisions of Article 80, and all applicable ordinances, rules, and regulations, the Inspector making the inspection shall at once forward to the person, firm, or corporation making the installation a written notice stating the defects that have been found to exist.

**(G) Revocation of Permits.** Revocation of permits shall conform to the following:

(1) The authority having jurisdiction shall be permitted to revoke a permit or approval issued if any violation of this *Code* is found upon inspection or in case there have been any false statements or misrepresentations submitted in the application or plans on which the permit or approval was based.

(2) Any attempt to defraud or otherwise deliberately or knowingly design, install, service, maintain, operate, sell, represent for sale, falsify records, reports, or applications, or other related activity in violation of the requirements prescribed by this *Code* shall be a violation of this *Code*. Such violations shall be cause for immediate suspension or revocation of any related licenses, certificates, or permits issued by this jurisdiction. In addition, any such violation shall be subject to any other criminal or civil penalties as available by the laws of this jurisdiction.

(3) Revocation shall be constituted when the permittee is duly notified by the authority having jurisdiction.

(4) Any person who engages in any business, operation, or occupation, or uses any premises, after the permit issued therefor has been suspended or revoked pursuant to the provisions of this *Code*, and before such suspended permit has been reinstated or a new permit issued, shall be in violation of this *Code*.

(5) A permit shall be predicated upon compliance with the requirements of this *Code* and shall constitute written authority issued by the authority having jurisdiction to install electrical equipment. Any permit issued under this *Code* shall not take the place of any other license or permit required by other regulations or laws of this jurisdiction.

(6) The authority having jurisdiction shall be permitted to require an inspection prior to the issuance of a permit.

(7) A permit issued under this *Code* shall continue until revoked or for the period of time designated on the permit. The permit shall be issued to one person or business only and for the location or purpose described in the permit. Any change that affects any of the conditions of the permit shall require a new or amended permit.

**(H) Applications and Extensions.** Applications and extensions of permits shall conform to the following:

(1) The authority having jurisdiction shall be permitted to grant an extension of the permit time period upon presentation by the permittee of a satisfactory reason for failure to start or complete the work or activity authorized by the permit.

(2) Applications for permits shall be made to the authority having jurisdiction on forms provided by the jurisdiction and shall include the applicant's answers in full to inquiries set forth on such forms. Applications for permits shall be accompanied by such data as required by the authority having jurisdiction, such as plans and specifications, location, and so forth. Fees shall be determined as required by local laws.

(3) The authority having jurisdiction shall review all applications submitted and issue permits as required. If an application for a permit is rejected by the authority having

jurisdiction, the applicant shall be advised of the reasons for such rejection. Permits for activities requiring evidence of financial responsibility by the jurisdiction shall not be issued unless proof of required financial responsibility is furnished.

## 80.21 Plans Review

Review of plans and specifications shall conform to (A) through (C).

**(A) Authority.** For new construction, modification, or rehabilitation, the authority having jurisdiction shall be permitted to review construction documents and drawings.

**(B) Responsibility of the Applicant.** It shall be the responsibility of the applicant to ensure the following:

(1) The construction documents include all of the electrical requirements.
(2) The construction documents and drawings are correct and in compliance with the applicable codes and standards.

**(C) Responsibility of the Authority Having Jurisdiction.** It shall be the responsibility of the authority having jurisdiction to promulgate rules that cover the following:

(1) Review of construction documents and drawings shall be completed within established time frames for the purpose of acceptance or to provide reasons for nonacceptance.
(2) Review and approval by the authority having jurisdiction shall not relieve the applicant of the responsibility of compliance with this *Code*.
(3) Where field conditions necessitate any substantial change from the approved plan, the authority having jurisdiction shall be permitted to require that the corrected plans be submitted for approval.

## 80.23 Notice of Violations, Penalties

Notice of violations and penalties shall conform to (A) and (B).

**(A) Violations.**

(1) Whenever the authority having jurisdiction determines that there are violations of this *Code*, a written notice shall be issued to confirm such findings.
(2) Any order or notice issued pursuant to this *Code* shall be served upon the owner, operator, occupant, or other person responsible for the condition or violation, either by personal service or mail or by delivering the same to, and leaving it with, some person of responsibility upon the premises. For unattended or abandoned locations, a copy of such order or notice shall be posted on the premises in a conspicuous place at or near the entrance to such premises and the order or notice shall be mailed by registered or certified mail, with return receipt requested, to the last known address of the owner, occupant, or both.

**(B) Penalties.**

(1) Any person who fails to comply with the provisions of this *Code* or who fails to carry out an order made pursuant to this *Code* or violates any condition attached to a permit, approval, or certificate shall be subject to the penalties established by this jurisdiction.
(2) Failure to comply with the time limits of an abatement notice or other corrective notice issued by the authority having jurisdiction shall result in each day that such violation continues being regarded as a new and separate offense.
(3) Any person, firm, or corporation who shall willfully violate any of the applicable provisions of this article shall be guilty of a misdemeanor and, upon conviction thereof, shall be punished by a fine of not less than _____dollars ($_____) or more than _______dollars ($_____) for each offense, together with the costs of prosecution, imprisonment, or both, for not less than ___________ (_________) days or more than _______ (______) days.

## 80.25 Connection to Electricity Supply

Connections to the electric supply shall conform to (A) through (E).

**(A) Authorization.** Except where work is done under an annual permit and except as otherwise provided in 80.25, it shall be unlawful for any person, firm, or corporation to make connection to a supply of electricity or to supply electricity to any electrical equipment installation for which a permit is required or that has been disconnected or ordered to be disconnected.

**(B) Special Consideration.** By special permission of the authority having jurisdiction, temporary power shall be permitted to be supplied to the premises for specific needs of the construction project. The Board shall determine what needs are permitted under this provision.

**(C) Notification.** If, within _____ business days after the Electrical Inspector is notified of the completion of an installation of electric equipment, other than a temporary approval installation, the Electrical Inspector has neither authorized connection nor disapproved the installation, the supplier of electricity is authorized to make connections and supply electricity to such installation.

**(D) Other Territories.** If an installation or electric equipment is located in any territory where an Electrical Inspector has not been authorized or is not required to make inspections, the supplier of electricity is authorized to make connections and supply electricity to such installations.

**(E) Disconnection.** Where a connection is made to an installation that has not been inspected, as outlined in the preceding paragraphs of this section, the supplier of electricity shall immediately report such connection to the Chief Electrical Inspector. If, upon subsequent inspection, it is found that the installation is not in conformity with the provisions of Article 80, the Chief

Electrical Inspector shall notify the person, firm, or corporation making the installation to rectify the defects and, if such work is not completed within fifteen (15) business days or a longer period as may be specified by the Board, the Board shall have the authority to cause the disconnection of that portion of the installation that is not in conformity.

## 80.27 Inspector's Qualifications

**(A) Certificate.** All electrical inspectors shall be certified by a nationally recognized inspector certification program accepted by the Board. The certification program shall specifically qualify the inspector in electrical inspections. No person shall be employed as an Electrical Inspector unless that person is the holder of an Electrical Inspector's certificate of qualification issued by the Board, except that any person who on the date on which this law went into effect was serving as a legally appointed Electrical Inspector of _________ shall, upon application and payment of the prescribed fee and without examination, be issued a special certificate permitting him or her to continue to serve as an Electrical Inspector in the same territory.

**(B) Experience.** Electrical inspector applicants shall demonstrate the following:

(1) Have a demonstrated knowledge of the standard materials and methods used in the installation of electric equipment
(2) Be well versed in the approved methods of construction for safety to persons and property
(3) Be well versed in the statutes of ________ relating to electrical work and the *National Electrical Code*, as approved by the American National Standards Institute
(4) Have had at least ____ years' experience as an Electrical Inspector or ____ years in the installation of electrical equipment. In lieu of such experience, the applicant shall be a graduate in electrical engineering or of a similar curriculum of a college or university considered by the Board as having suitable requirements for graduation and shall have had two years' practical electrical experience.

**(C) Recertification.** Electrical inspectors shall be recertified as established by provisions of the applicable certification program.

**(D) Revocation and Suspension of Authority.** The Board shall have the authority to revoke an inspector's authority to conduct inspections within a jurisdiction.

## 80.29 Liability for Damages

Article 80 shall not be construed to affect the responsibility or liability of any party owning, designing, operating, controlling, or installing any electrical equipment for damages to persons or property caused by a defect therein, nor shall the _____ or any of its employees be held as assuming any such liability by reason of the inspection, reinspection, or other examination authorized.

## 80.31 Validity

If any section, subsection, sentence, clause, or phrase of Article 80 is for any reason held to be unconstitutional, such decision shall not affect the validity of the remaining portions of Article 80.

## 80.33 Repeal of Conflicting Acts

All acts or parts of acts in conflict with the provisions of Article 80 are hereby repealed.

## 80.35 Effective Date

Article 80 shall take effect _________ (_________) days after its passage and publication.

# Informative Annex I
# Recommended Tightening Torque Tables from UL Standard 486A-B

*This informative annex is not a part of the requirements of this NFPA document, but is included for informational purposes only.*

In the absence of connector or equipment manufacturer's recommended torque values, Table I.1, Table I.2, and Table I.3 may be used to correctly tighten screw-type connections for power and lighting circuits*. Control and signal circuits may require different torque values, and the manufacturer should be contacted for guidance.

*For proper termination of conductors, it is very important that field connections be properly tightened. In the absence of manufacturer's instructions on the equipment, the torque values given in these tables are recommended. Because it is normal for some relaxation to occur in service, checking torque values sometime after installation is not a reliable means of determining the values of torque applied at installation.

***TABLE I.1*** *Tightening Torque for Screws*

| Test Conductor Installed in Connector | | Tightening Torque, N-m (lbf-in.) | | | | | | | |
|---|---|---|---|---|---|---|---|---|---|
| | | Slotted head No. 10 and larger* | | | | | | | |
| AWG or kcmil | mm² | Slot width 1.2 mm (0.047 in.) or less and slot length 6.4 mm (¼ in.) or less | | Slot width over 1.2 mm (0.047 in.) or slot length over 8.4 mm (1.4 in.) | | Split-bolt connectors | | Other connectors | |
| 30–10 | 0.05–5.3 | 2.3 | (20) | 4.0 | (35) | 9.0 | (80) | 8.5 | (75) |
| 8 | 8.4 | 2.8 | (25) | 4.5 | (40) | 9.0 | (80) | 8.5 | (75) |
| 6–4 | 13.2–21.2 | 4.0 | (35) | 5.1 | (45) | 18.5 | (165) | 12.4 | (110) |
| 3 | 26.7 | 4.0 | (35) | 5.6 | (50) | 31.1 | (275) | 16.9 | (150) |
| 2 | 33.6 | 4.5 | (40) | 5.6 | (50) | 31.1 | (275) | 16.9 | (150) |
| 1 | 42.4 | — | | 5.6 | (50) | 31.1 | (275) | 16.9 | (150) |
| 1/0–2/0 | 53.5–67.4 | — | | 5.6 | (50) | 43.5 | (385) | 20.3 | (180) |
| 3/0–4/0 | 85.0–107.2 | — | | 5.6 | (50) | 56.5 | (500) | 28.2 | (250) |
| 250–350 | 127–177 | — | | 5.6 | (50) | 73.4 | (650) | 36.7 | (325) |
| 400 | 203 | — | | 5.6 | (50) | 93.2 | (825) | 36.7 | (325) |
| 500 | 253 | — | | 5.6 | (50) | 93.2 | (825) | 42.4 | (375) |
| 600–750 | 304–380 | — | | 5.6 | (50) | 113.0 | (1000) | 42.4 | (375) |
| 800–1000 | 405–508 | — | | 5.6 | (50) | 124.3 | (1100) | 56.5 | (500) |
| 1250–2000 | 635–1010 | — | | — | | 124.3 | (1100) | 67.8 | (600) |

*For values of slot width or length not corresponding to those specified, select the largest torque value associated with the conductor size. Slot width is the nominal design value. Slot length shall be measured at the bottom of the slot.

***TABLE I.2*** *Tightening Torque for Slotted Head Screws Smaller Than No. 10 Intended for Use with 8 AWG (8.4 mm²) or Smaller Conductors*

| Slot Length of Screw[a] | | Tightening Torque, N-m (lbf-in.) | |
|---|---|---|---|
| **mm** | **in.** | **Slot width of screw smaller than 1.2 mm (0.047 in.)[b]** | **Slot width of screw 1.2 mm (0.047 in.) and larger[b]** |
| Less than 4 | Less than 5⁄32 | 0.79 (7) | 1.0 (9) |
| 4 | 5⁄32 | 0.79 (7) | 1.4 (12) |
| 4.8 | 3⁄16 | 0.79 (7) | 1.4 (12) |
| 5.5 | 7⁄32 | 0.79 (7) | 1.4 (12) |
| 6.4 | 1⁄4 | 1.0 (9) | 1.4 (12) |
| 7.1 | 9⁄32 | | 1.7 (15) |
| Above 7.1 | Above 9⁄32 | | 2.3 (20) |

[a]For slot lengths of intermediate values, select torques pertaining to next shorter slot lengths. Also, see 9.1.9.6 of UL 486A-2003, *Wire Connectors and Soldering Lugs for Use with Copper Conductors*, for screws with multiple tightening means. Slot length shall be measured at the bottom of the slot.

[b]Slot width is the nominal design value.

***TABLE I.3*** *Tightening Torque for Screws with Recessed Allen or Square Drives*

| Socket Width Across Flats[a] | | Tightening Torque, N-m (lbf-in.) | |
|---|---|---|---|
| **mm** | **in.** | | |
| 3.2 | 1⁄8 | 5.1 | (45) |
| 4.0 | 5⁄32 | 11.3 | (100) |
| 4.8 | 3⁄16 | 13.5 | (120) |
| 5.5 | 7⁄32 | 16.9 | (150) |
| 6.4 | 1⁄4 | 22.5 | (200) |
| 7.9 | 5⁄16 | 31.1 | (275) |
| 9.5 | 3⁄8 | 42.4 | (375) |
| 12.7 | 1⁄2 | 56.5 | (500) |
| 14.3 | 9⁄16 | 67.8 | (600) |

[a]See 9.1.9.6 of UL 486A-2003, *Wire Connectors and Soldering Lugs for Use with Copper Conductors*, for screws with multiple tightening means.

# Informative Annex J ADA Standards for Accessible Design

*This informative annex is not a part of the requirements of this NFPA document, but is included for informational purposes only.*

The provisions cited in Informative Annex J are intended to assist the users of the *Code* in properly considering the various electrical design constraints of other building systems and are part of the 2010 ADA Standards for Accessible Design. They are the same provisions as those found in ANSI/ICC A117.1-2009, *Accessible and Usable Buildings and Facilities*.

**J.1 Protruding Objects.** Protruding objects shall comply with Section J.2.

**J.2 Protrusion Limits.** Objects with leading edges more than 685 mm (27 in.) and not more than 2030 mm (80 in.) above the finish floor or ground shall protrude a maximum of 100 mm (4 in.) horizontally into the circulation path. *(See Figure J.2.)*

*Exception: Handrails shall be permitted to protrude 115 mm (4½ in.) maximum.*

**J.3 Post-Mounted Objects.** Freestanding objects mounted on posts or pylons shall overhang circulation paths 305 mm (12 in.) maximum where located 685 mm (27 in.) minimum and 2030 mm (80 in.) maximum above the finish floor or ground. Where a sign or other obstruction is mounted between posts or pylons, and the clear distance between the posts or pylons is greater than 305 mm (12 in.), the lowest edge of such sign or obstruction shall be 685 mm (27 in.) maximum or 2030 mm (80 in.) minimum above the finish floor or ground. *(See Figure J.3.)*

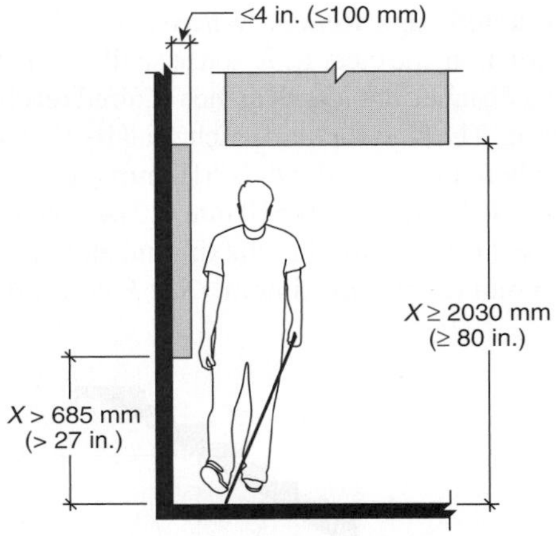


***FIGURE J.2*** *Limits of Protruding Objects.*

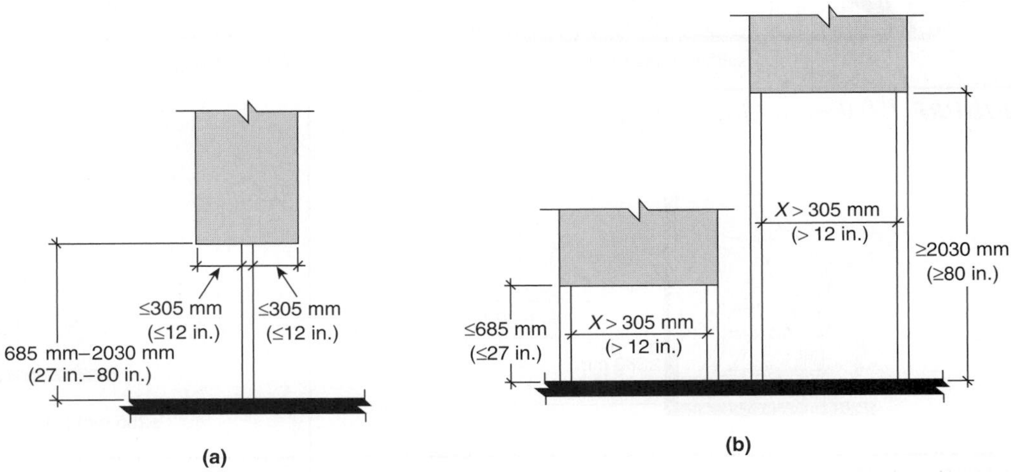


***FIGURE J.3*** *Post-Mounted Protruding Objects.*

*Exception: The sloping portions of handrails serving stairs and ramps shall not be required to comply with Section J.3.*

**J.4 Vertical Clearance.** Vertical clearance shall be 2030 mm (80 in.) high minimum. Guardrails or other barriers shall be provided where the vertical clearance is less than 2030 mm (80 in.) high. The leading edge of such guardrail or barrier shall be located 685 mm (27 in.) maximum above the finish floor or ground. *(See Figure J.4.)*

*Exception: Door closers and door stops shall be permitted to be 1980 mm (78 in.) minimum above the finish floor or ground.*

**J.5 Required Clear Width.** Protruding objects shall not reduce the clear width required for accessible routes.

**J.6 Forward Reach.**

**J.6.1 Unobstructed.** Where a forward reach is unobstructed, the high forward reach shall be 1220 mm (48 in.) maximum and the low forward reach shall be 380 mm (15 in.) minimum above the finish floor or ground. *(See Figure J.6.1.)*

**J.6.2 Obstructed High Reach.** Where a high forward reach is over an obstruction, the clear floor space shall extend beneath the element for a distance not less than the required reach depth over the obstruction. The high forward reach shall be 1220 mm (48 in.) maximum where the reach depth is 510 mm (20 in.) maximum. Where the reach depth exceeds 510 mm (20 in.), the high forward reach shall be 1120 mm (44 in.) maximum, and the reach depth shall be 635 mm (25 in.) maximum. *(See Figure J.6.2.)*

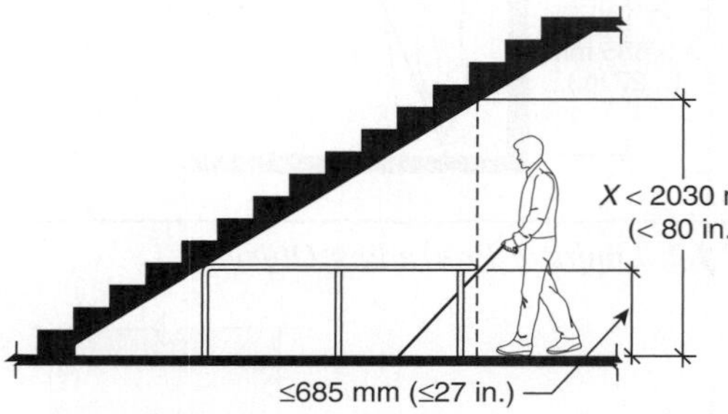


***FIGURE J.4*** *Vertical Clearance.*

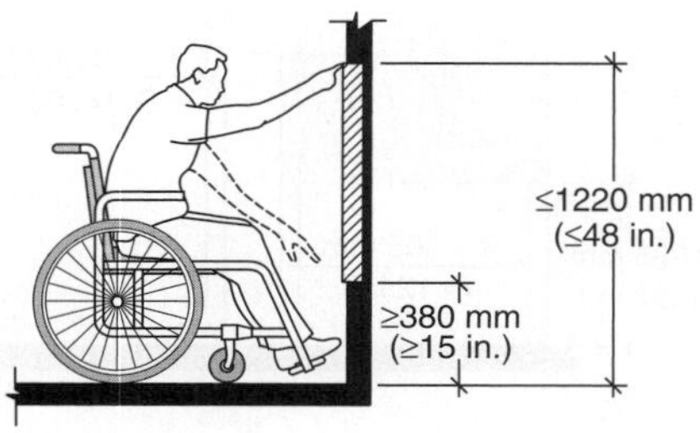


***FIGURE J.6.1*** *Unobstructed Forward Reach.*

**J.7 Side Reach.**

**J.7.1 Unobstructed.** Where a clear floor or ground space allows a parallel approach to an element, and the side reach is unobstructed, the high side reach shall be 1220 mm (48 in.) maximum, and the low side reach shall be 380 mm (15 in.) minimum above the finish floor or ground. *(See Figure J.7.1.)*

*Exception No. 1: An obstruction shall be permitted between the clear floor or ground space and the element where the depth of the obstruction is 255 mm (10 in.) maximum.*

*Exception No. 2: Operable parts of fuel dispensers shall be permitted to be 1370 mm (54 in.) maximum, measured from the surface of the vehicular way where fuel dispensers are installed on existing curbs.*

**J.7.2 Obstructed High Reach.** Where a clear floor or ground space allows a parallel approach to an element and the high side reach is over an obstruction, the height of the obstruction shall be 865 mm (34 in.) maximum, and the depth of the obstruction shall be 610 mm (24 in.) maximum. The high side reach shall be 1220 mm (48 in.) maximum for a reach depth of 255 mm (10 in.) maximum. Where the reach depth exceeds 255 mm (10 in.), the high side reach shall be 1170 mm (46 in.) maximum for a reach depth of 610 mm (24 in.) maximum. *(See Figure J.7.2.)*

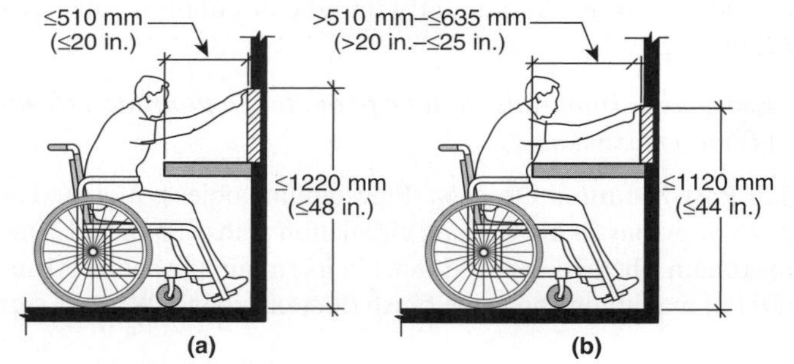


***FIGURE J.6.2*** *Obstructed High Forward Reach.*

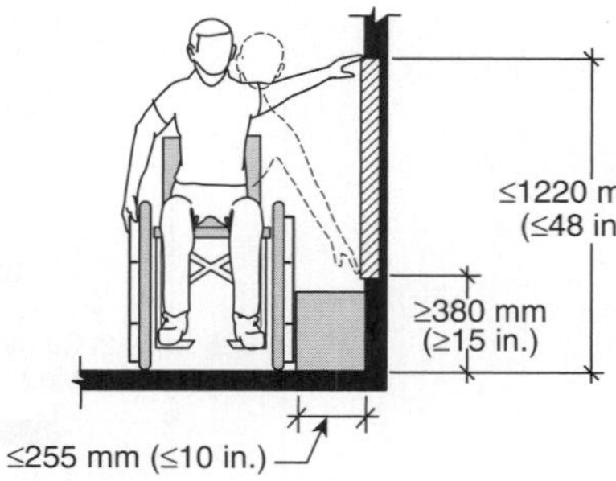


***FIGURE J.7.1*** *Unobstructed Side Reach.*

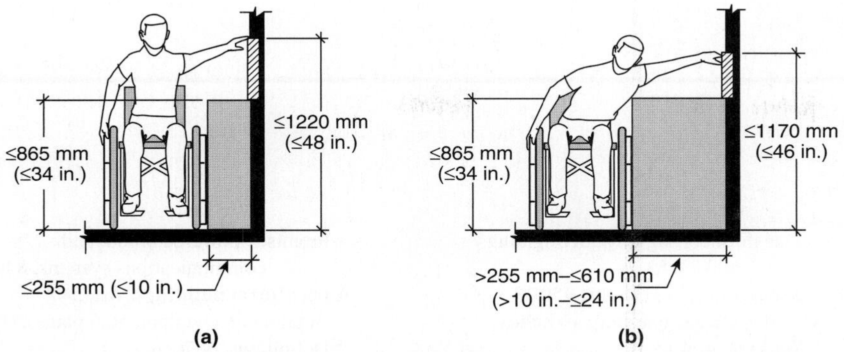


***FIGURE J.7.2*** *Obstructed High Side Reach.*

*Exception No. 1: The top of washing machines and clothes dryers shall be permitted to be 915 mm (36 in.) maximum above the finish floor.*

*Exception No. 2: Operable parts of fuel dispensers shall be permitted to be 1370 mm (54 in.) maximum, measured from the surface of the vehicular way where fuel dispensers are installed on existing curbs.*

# Index

**B**

## C

## D

## F

## G

## H

## J



## K

## M

## O

## P

## S

## U

## X



## Z

# Nonlinear Loads

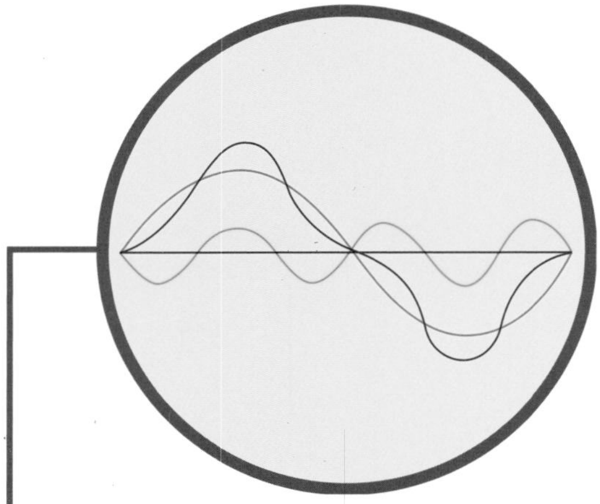

## The Subcommittee on Nonlinear Loads

During the 1996 *NEC*® development cycle, a task group was formed to study the impact of nonlinear loads. This group was designated the *NEC Correlating Committee Ad Hoc Subcommittee on Nonlinear Loads*. The subcommittee reviewed technical literature and electrical theory on the fundamental nature of harmonic distortion, as well as the requirements in and proposals for the 1993 *NEC* regarding nonlinear loads. Their research and study yielded some changes to the *Code* for 1996 and beyond.

The *Ad Hoc Subcommittee on Nonlinear Loads*, was asked to do the following:

1. Study the effects of electrical loads producing substantial current distortion upon electrical system distribution components including but not limited to
   a. Distribution transformers, current transformers, and other transformers
   b. Switchboards and panelboards
   c. Phase and neutral feeder conductors
   d. Phase and neutral branch-circuit conductors
   e. Proximate data and communications conductors
2. Study harmful effects, if any, to the system components from overheating resulting from nonlinear load characteristics
3. Make recommendations for methods to minimize the harmful effects of nonlinear loads considering all means, including compensating methods at load sources
4. Prepare proposals, if necessary, to amend the 1996 *National Electrical Code*®, to achieve greater safety

After much study of both theoretical and practical data concerning nonlinear loads—obtained from consultants, equipment manufacturers, and testing laboratories—the subcommittee concluded that, while nonlinear loads can cause undesirable operational effects—including additional heating—no significant threat to persons and property had been substantiated.

Although they agreed with the existing *Code* text regarding nonlinear loads, the subcommittee submitted several proposals, including a definition of *nonlinear load*, revised text reflecting that definition, informational notes calling attention to the effects of nonlinear loads, and proposals permitting the paralleling of neutral conductors in existing installations under engineering supervision.